WITHDRAWN

WITHDRAWN

GC-MS Guide to Ignitable Liquids

Reta Newman
Michael Gilbert
Kevin Lothridge

CRC Press
Boca Raton London New York Washington, D.C.

Library of Congress Cataloging-in-Publication Data

Newman, Reta.
GC-MS Guide to Ignitable Liquids / by Reta Newman, Michael Gilbert, Kevin Lothridge.
p. cm.
Includes bibliographical references and index.
ISBN 0-8493-3107-2 (alk. paper)
1. Chemistry, Forensic. 2. Inflammable liquids. 3. Fire Investigation.
I. Gilbert, Michael. II. Lothridge, Kevin. III. Title.
HV8073.N47 1998
363.252—dc21 97-24293
CIP

Direct all inquiries to CRC Press LLC, 2000 N.W. Corporate Blvd., Boca Raton, Florida 33431.

Visit the CRC Press Web site at www.crcpress.com

International Standard Book Number 0-8493-3107-2
Library of Congress Card Number 97-24293
Printed in the United States of America 5 6 7 8 9 0
Printed on acid-free paper

PREFACE

There exist, in most areas of forensic chemistry, compilations of standards and data for sample comparison. There is a notable lack of such a work in the area of ignitable liquid data for fire debris analysis. In the past such a compilation was not warranted, as the vast majority of ignitable liquids of interest to the fire debris analyst were comprised of simple distillates and gasoline. Recently, however, advances in the petroleum industry have led to the production of a variety of new formulations which have inundated the consumer markets.

This text is comprised of over 100 different ignitable liquid formulations representing tens of thousands of commercial products. Both total ion chromatograms and extracted ion chromatograms (mass chromatograms) are included for each formulation.

Analysts faced with unusual chromatographic patterns will find this text helpful in locating possible ignitable liquid standards for comparison. This text is not intended to supplant the laboratory's own standard collection. Due to variations in chromatographic conditions, it is imperative that standard to sample comparisons are conducted on data obtained on the *same* instrument and under the *same* conditions. The intent of the authors is merely to provide an easy reference to aid in ignitable liquid identification and classification.

Reta Newman
Michael Gilbert
Kevin Lothridge

ACKNOWLEDGMENTS

The authors greatly appreciate the assistance and encouragement of Dr. Joan E. Wood, Executive Director, Pinellas County Forensic Laboratory; Dr. Mary Lou Fultz, Bureau of Alcohol, Tobacco and Firearms; William R. Dietz, Bureau of Alcohol, Tobacco and Firearms; Jamie Crippin, Colorado Bureau of Investigation; Jose Amirall, Metro-Dade Crime Laboratory; Restek Corporation; and Louise Bell, Ric Nagle, Patricia Pattee, and Joan Ring, Pinellas County Forensic Laboratory.

We also wish to express our graditude to colleagues and family members across the country who graciously provided us with numerous ignitable liquid products.

CONTENTS

INTRODUCTION

In the 1990s, the petroleum industry began mass-marketing products which were more than simple distillates and gasoline. By making more efficient use of waste streams, they began marketing petroleum products made of various combinations of naphthenic, paraffinic, isoparaffinic, and aromatic compounds. These exotic formulation have resulted in the "odorless", "superclean", and "extra strength" products which have inundated the consumer markets. Their eventual presence in fire debris samples, whether intentional or incidental, is inevitable.

This book was created to provide the fire debris analyst with a source of ignitable liquid data for substances which they may encounter. Gas chromatograms and mass chromatograms are presented for over 100 unique ignitable liquid formulations representing tens of thousand of commercial products. The data are by no means all inclusive. Every effort was made to locate and include as many ignitable liquid formulations as possible. However, the sheer number of products and brands produced makes a comprehensive listing impossible.

Comparison of sample and text data will aid the analyst in locating suitable ignitable liquid standards for further comparison and confirmation. Sample to standard comparison/identification must be made on data obtained on the same instruments with the same chromatographic conditions.

Total ion chromatograms, expanded total ion chromatograms, and extracted ion chromatograms (mass chromatograms) for specific ions are included for each formulation. Other specific information is also provided, including ASTM classification, major peak identification, comparable commercial products, and potential product uses. A brief introduction to extracted ion chromatography, ASTM classification, and product identification precedes the ignitable liquid data.

Extracted Ion Chromatography

One of the most significant advances in fire debris analysis is the widespread availability and use of benchtop mass spectrometers. These instruments, coupled with user-friendly software, allow for routine use of mass spectrometry. The most common mass spectrometry application in fire debris is extracted ion chromatography or mass chromatography.

Extracted ion chromatography is based on the premise that specific ions are indicators of certain classes of compounds. For example, the ions 43, 57, 71, and 85 are indicative of aliphatic compounds (Figure 1); 91, 105, and 119 are indicative of alkylbenzene (aromatic) compounds (Figure 2); etc. The extracted ion profiles in this text (Table 1) were created with ions recommended by the American Society for Testing and Materials (ASTM) 1618-94: Standard Guide for Ignitable Liquid Residues in Extracts from Fire Debris Samples by Gas Chromatography-Mass Spectrometry.[1]

Alkane	**43, 57, 71, 85**
Cycloalkane/alkene	**55, 69, 82, 83**
Aromatic (alkylbenzene)	**91, 105, 119**
Naphthalene	**128, 142, 156**

Table 1: Selected ions for extracted ion chromatograms.

There are other ions inherent in compounds common to ignitable liquids.[2,3] However, the ion combinations listed in Table 1 and used in this text provide the most chromatographic information for the majority of ignitable liquid formulations.

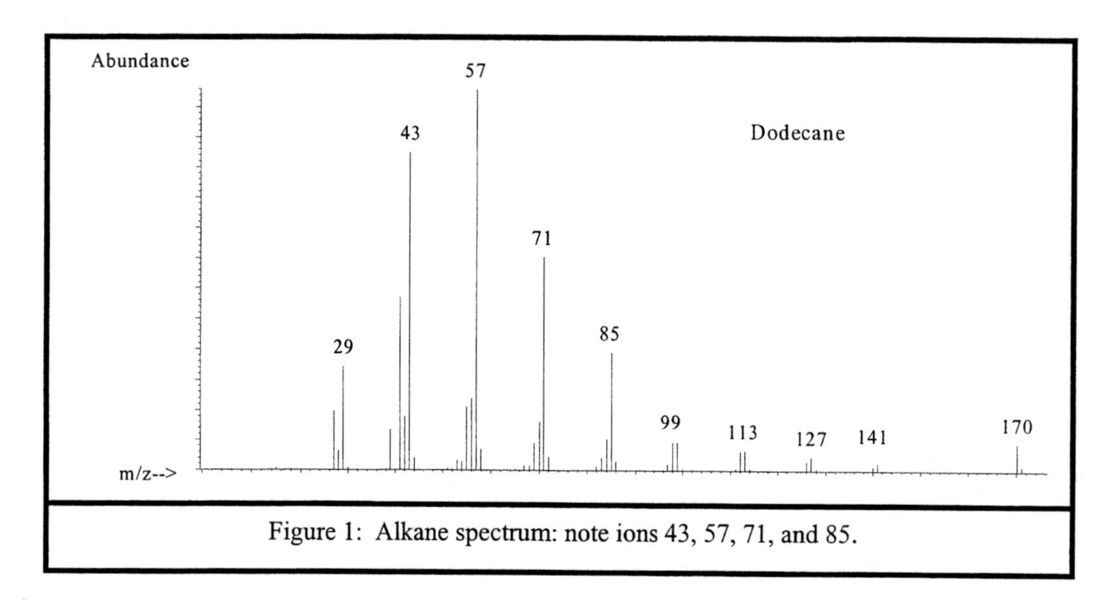

Figure 1: Alkane spectrum: note ions 43, 57, 71, and 85.

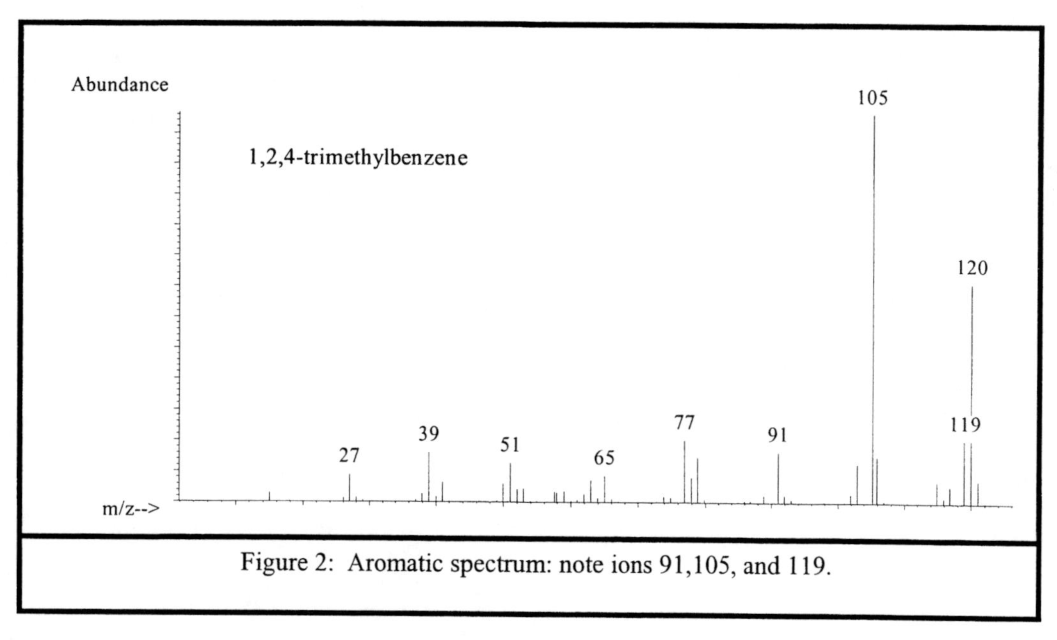

Figure 2: Aromatic spectrum: note ions 91,105, and 119.

It should be noted that ions may not be unique to specific compounds or classes of compounds. For example, while ions 55 and 69 are indicative of cycloaliphatic compounds, they also occur in alkenes (Figure 3). For this reason, the profile headings "alkane", "aromatic", "cycloparaffin," and "naphthalene" should not be interpreted as exclusive. The analyst using mass chromatography must be aware that the profile peaks may not uniquely represent a specific class of compounds. Peaks may also be due to minor components of other compounds.

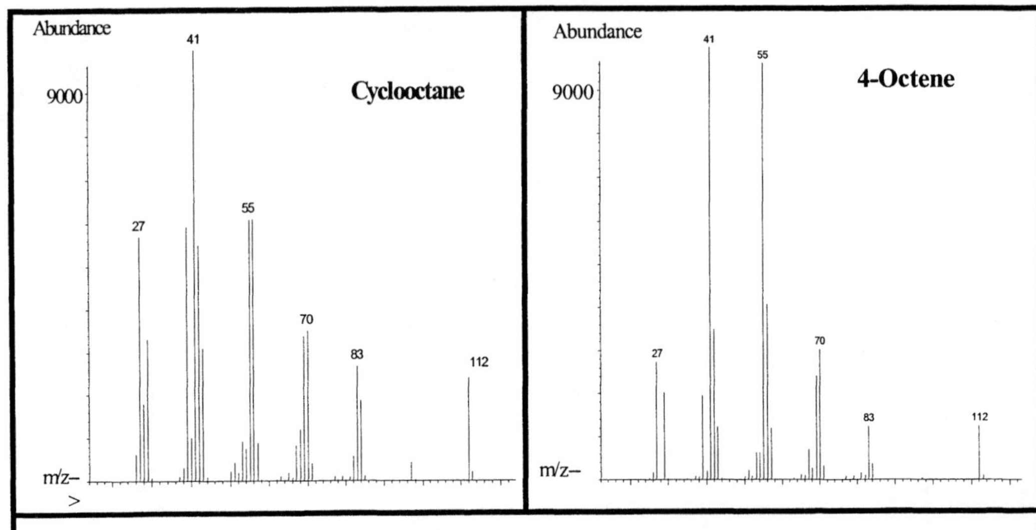

Figure 3: Mass spectra of cycloalkane and alkene. Note the presence of ions 55, 69, and 83 in both compounds.

The extracted ion profiles are presented in two forms: individual ion chromatograms and summed ion chromatograms. Both individual and summed profiles are used by analysts for standard to sample comparisons, based primarily on personal preference. A computer program (macros) for producing comparable data for both individual and summed profiles from sample data is included in Appendix A.

Summed ion chromatograms are created by extracting ions common for a specific class of compounds and adding the mass chromatograms together. For example, the ions 91, 105, and 119 are extracted individually and the resultant chromatograms are summed together to produce an aromatic profile (Figure 4). With summed ions, the peaks of a class of compounds are viewed proportionately. Individual ion profiles are created by extracting specific ions from the total ion chromatogram and displaying each mass chromatogram produced (Figure 5). Individual profiles allow the analyst to see less abundant ions indicative of certain classes of compounds in greater detail than summed profiles.

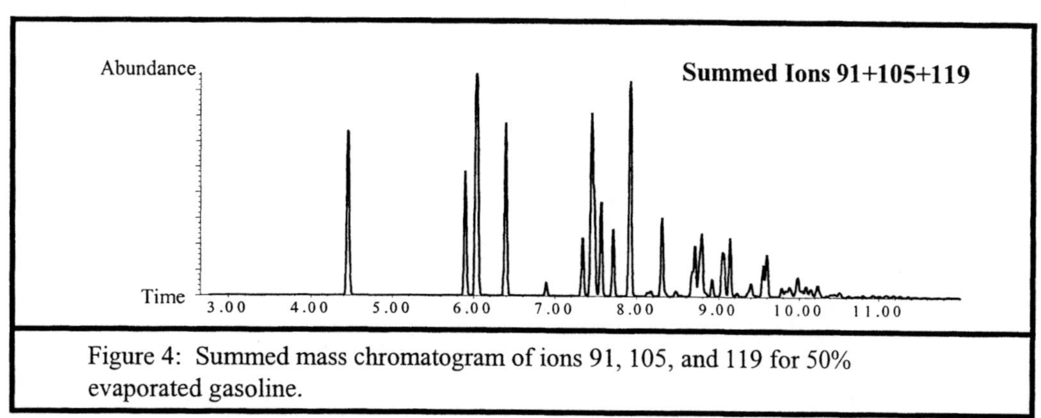

Figure 4: Summed mass chromatogram of ions 91, 105, and 119 for 50% evaporated gasoline.

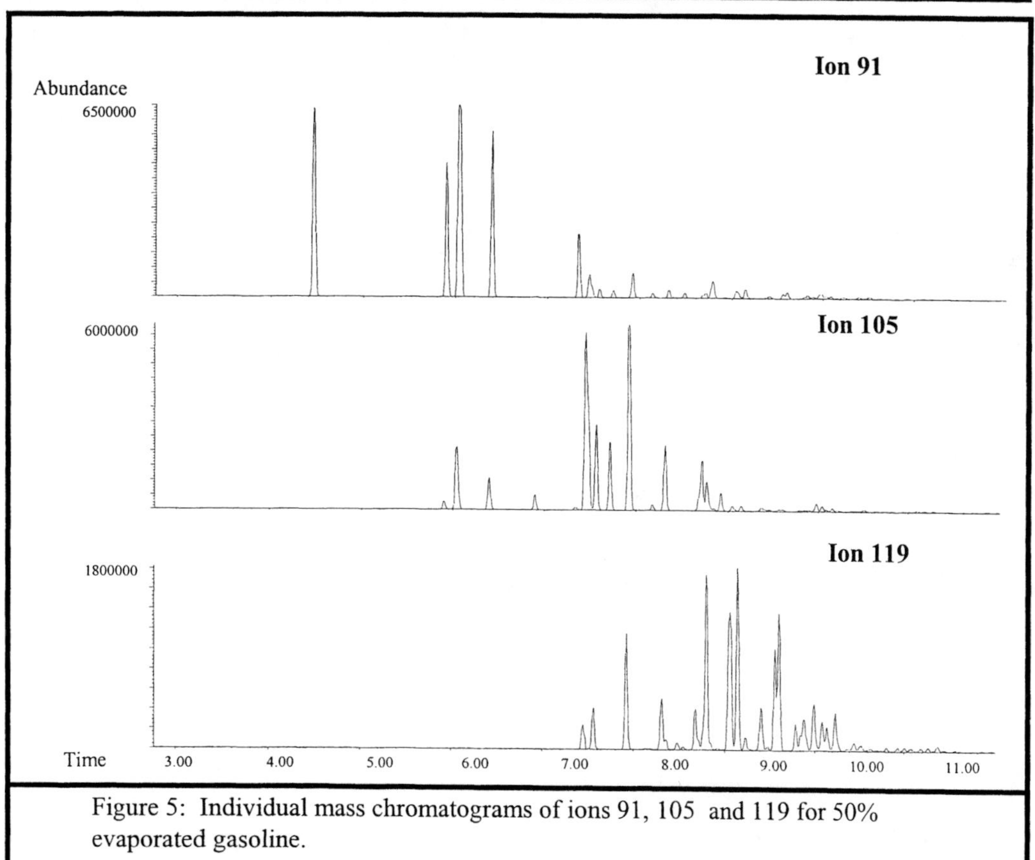

Figure 5: Individual mass chromatograms of ions 91, 105 and 119 for 50% evaporated gasoline.

IGNITABLE LIQUID CLASSIFICATION

ASTM 1387-95: Standard Method for Ignitable Liquid Residues in Extracts from Fire Debris Samples by Gas Chromatography[4] and ASTM 1618-94: Standard Guide for Ignitable Liquid Residues in Extracts from Fire Debris Samples by Gas Chromatography-Mass Spectrometry[5] include classification schemes for ignitable liquids (see Table 2). Originally, the classification scheme emphasized distillates and gasoline; however, with the vast number of new formulations, the classification scheme has been updated to include as many of these products as possible. The data in this text are divided into 8 sections based primarily on ASTM classification.

Class 1:	Light petroleum distillates
Class 2:	Gasoline
Class 3:	Medium petroleum distillates
Class 4:	Kerosene (Kersosine)
Class 5:	Heavy petroleum distillates
Class 0.1:	Oxygenated solvents
Class 0.2:	Isoparaffinic products
Class 0.3:	Normal alkane product
Class 0.4:	Aromatic solvents
Class 0.5:	Naphthenic-paraffinic products
Class 0:	Miscellaneous

Table 2: ASTM 1387-95 Classification

DISTILLATES

Distillates are easily distinguished by a predominate homologous normal alkane series in a bell shaped pattern (Figure 6). Cycloparaffinic and isoparaffinic compounds appear among the normal alkanes and occur in characteristic patterns. Aromatic compounds are present and may mimic the aromatic profile of gasoline. A common error made by analysts new to extracted ion chromatography is to mistake a simple distillate as a distillate/gasoline mixture. Differentiation between distillates and distillate/gasoline mixtures can easily be achieved by comparing the abundance of n-alkanes to aromatics in both the total ion and extracted ion chromatograms. Distillates are divided into four ASTM classes based on volatility.

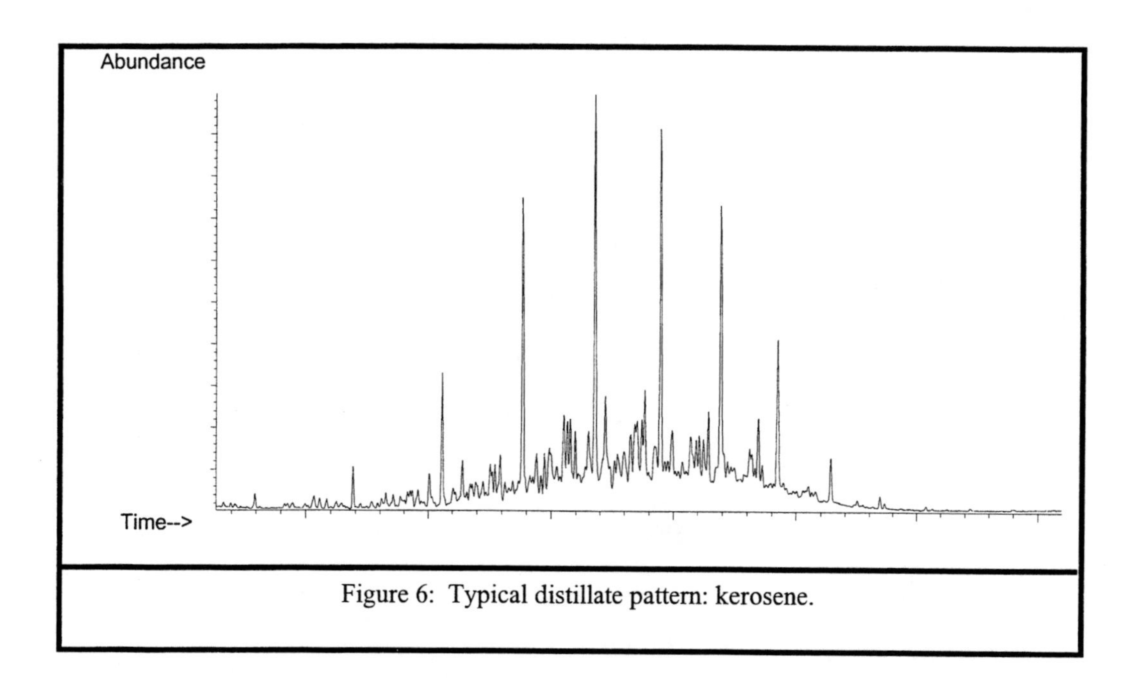

Figure 6: Typical distillate pattern: kerosene.

Light petroleum distillates (ASTM Class 1) consist of distillate components with the majority of the pattern between butane and undecane (C4-C11) with the most abundant n-alkane preceeding nonane (C9). Data before hexane (C6) are not included in this text. Extracted ion data are typified by an abundant alkane pattern with the predominate n-alkane being C6, C7, or C8. Cycloparaffin and aromatic patterns are generally present but are significantly less abundant than the alkanes.

Medium petroleum distillates (ASTM Class 3) consist of distillate components, with the majority of the components in the range of octane to dodecane (C8-C12). Medium petroleum distillates generally have a peak spread of 3 to 4 n-alkanes. Cycloparaffins and aromatics are present but are significantly less abundant than the alkanes. Small amounts of naphthalenes may be present depending on the alkane range.

Kerosene (ASTM Class 4) is a broader distillate class than the medium distillate class. Kerosene products consist of distillate components between nonane and hexadecane (C9-C16). At least 5 consecutive n-alkanes in this range must be present. Cycloparaffins, aromatics, and naphthalenes are present, but are significantly less abundant than aliphatics. Small amounts of heptadecane (C17) and octadecane (C18) may be present. If C17 and C18 are present, pristane and phytane must also be present.

Heavy petroleum distillates (ASTM Class 5) consist of distillate components, with the bulk of the pattern between decane and tricosane (C10-C23). The most abundant n-alkane should be dodecane (C12) or higher, and pristane and phytane must be present. Lower levels of cycloparaffins and aromatics are generally present but may be absent or diminished in more evaporated samples. Naphthalenes are present in a characteristic pattern.

Dearomatized distillates (ASTM Class 0) are not considered distillates by the ASTM classification system; however, due to the similarity of the data, they are included in the distillate section of this book. These products contain abundant normal alkanes with the characteristic cycloparaffin and isoparaffin components typical of distillates but with the aromatic compounds removed. These products have some similarity to naphthenic paraffinic products (ASTM Class 0.5); however, naphthenic paraffinic products have both aromatics and normal alkanes removed.

GASOLINES

Gasoline (ASTM Class 2) consists of all automotive gasolines, including reformulated gasoline. The peak spread is commonly butane to dodecane (C4-C12), although it may vary depending on geographic region. Gasoline has a characteristic, but by no means unique, aromatic profile (Figure 7). The aromatic content of gasoline is similar to that of petroleum distillates; however, in gasoline, the aromatics are significantly more abundant than the aliphatic content. The aromatic profile of aromatic solvents may also mimic gasoline, although aromatic solvents (Class 0.4) usually are of a narrower range with aliphatic compounds notably absent. The aliphatic profile of gasoline will vary by brand, grade, and lot, but is always present. Most gasolines contain characteristic naphthalenes; however, they may be absent in some northern winter markets. Several different gasolines in various stages of evaporation, including examples of California reformulated gasolines, are included in this collection. Brand and grade information are specified solely to illustrate variation which may exist in the alkane and cycloparaffin content of different gasolines. Variations from station to station, pump to pump, and lot to lot make brand identification from this data impossible.

OXYGENATED SOLVENTS

Oxygenated solvents (ASTM Class 0.1) are generally very volatile with major components before octane (C8). Major peaks must include an oxygenated compound: i.e., alcohol, ketone, ester. Other compounds may be present, including toluene, xylenes, petroleum distillates, and aromatic solvents. Extracted ion chromatograms may be helpful for comparing some of the additional constituents of Class 0.1 products; however, mass spectral peak identification of the oxygenated compounds is strongly recommended.

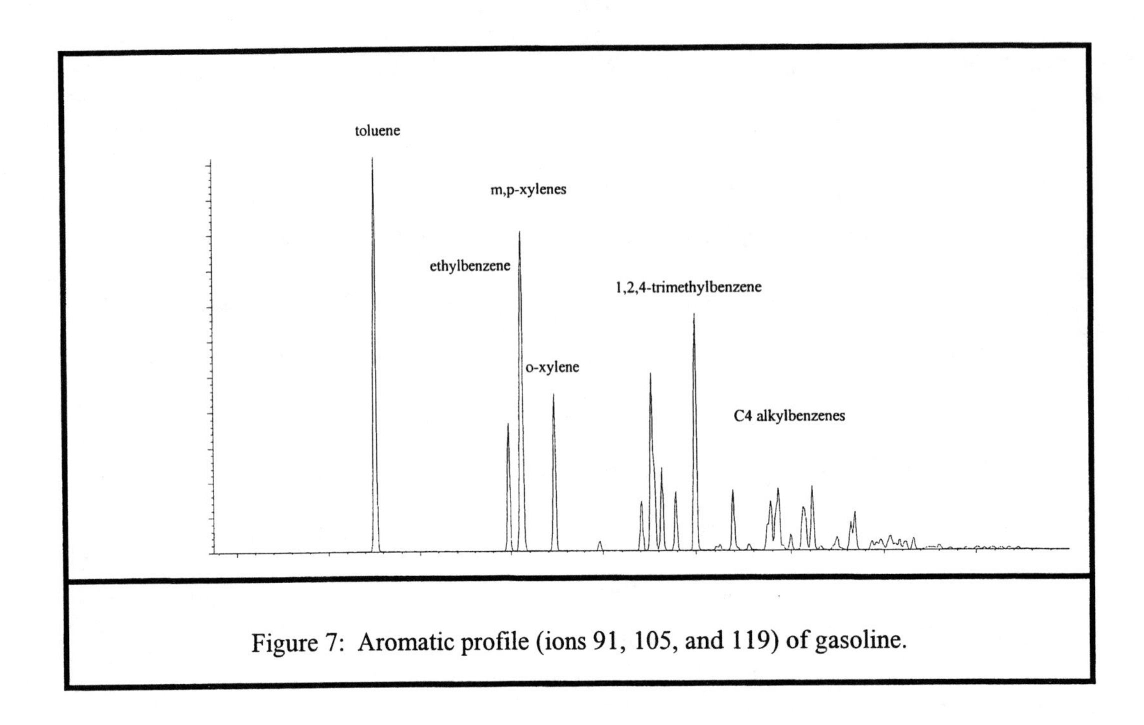

Figure 7: Aromatic profile (ions 91, 105, and 119) of gasoline.

ISOPARAFFINIC PRODUCTS

Isoparaffins (ASTM Class 0.2) are branched chained alkanes. Isoparaffinic products consist solely of isoparaffinic compounds. Normal (straight chained) alkanes, cycloparaffins, and naphthalenes are not present. The hydrocarbon range for isoparaffinic products varies. Chromatographic data are distinguished by similar total ion chromatograms, alkane, and cycloparaffin profiles (Figure 8). The alkane profile is most abundant (normal alkanes are absent); "cycloparaffin" profiles are much less abundant than the alkanes but with a similar pattern. The compounds represented in the "cycloparaffin" profile are actually the same compounds represented in the alkane profile. They do not represent cycloparaffins, but are produced by the minor ions of the substituted alkanes represented by the alkane profile (Figure 9).

NORMAL ALKANE PRODUCTS

Normal alkane (ASTM Class 0.3) products consist exclusively of straight chain alkanes. Isoparaffins, cycloparaffins, aromatics, and naphthalenes are notably absent. Most products contain five or fewer peaks in the range of decane to hexadecane (C10-C16) (Figure 10). Identification is made by traditional mass spectrometry with retention time and standard spectra comparison.

AROMATIC PRODUCTS

Aromatic products (ASTM Class 0.4) are comprised, almost exclusively, of aromatic compounds in the range of C6-C14. Aliphatic and cycloaliphatic compounds are not present in significant amounts. The aromatic profile often looks similar to that of gasoline. Differentiation is achieved by the absence of aliphatics in the aromatic products. The aliphatic content in gasoline varies in both composition and proportion, but is notably present (Figure 11).

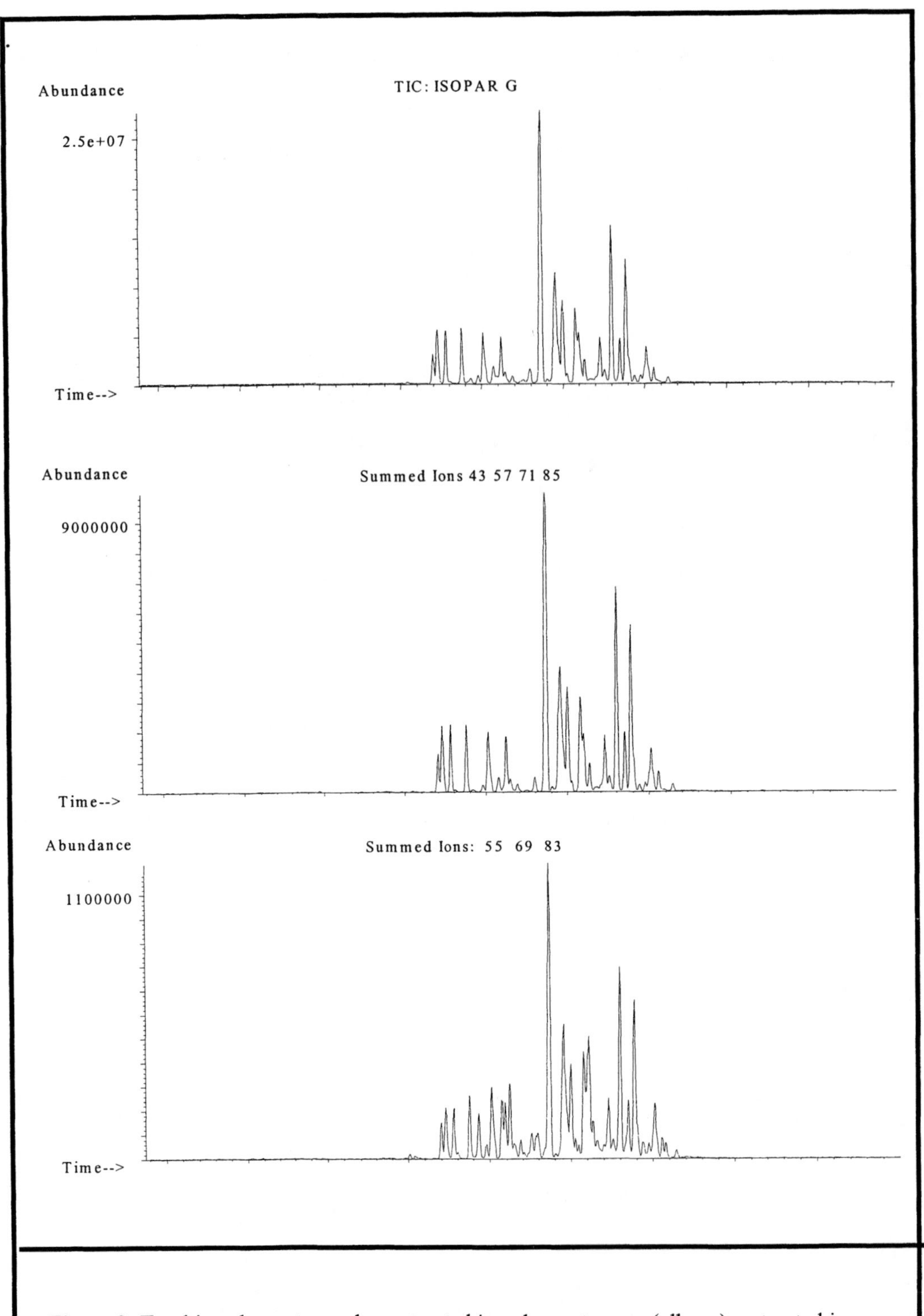

Figure 8: Total ion chromatography, extracted ion chromatogram (alkane), extracted ion chromatogram ("cycloparaffin") for Isopar G. Note similarities of the three patterns in isoparaffinic products.

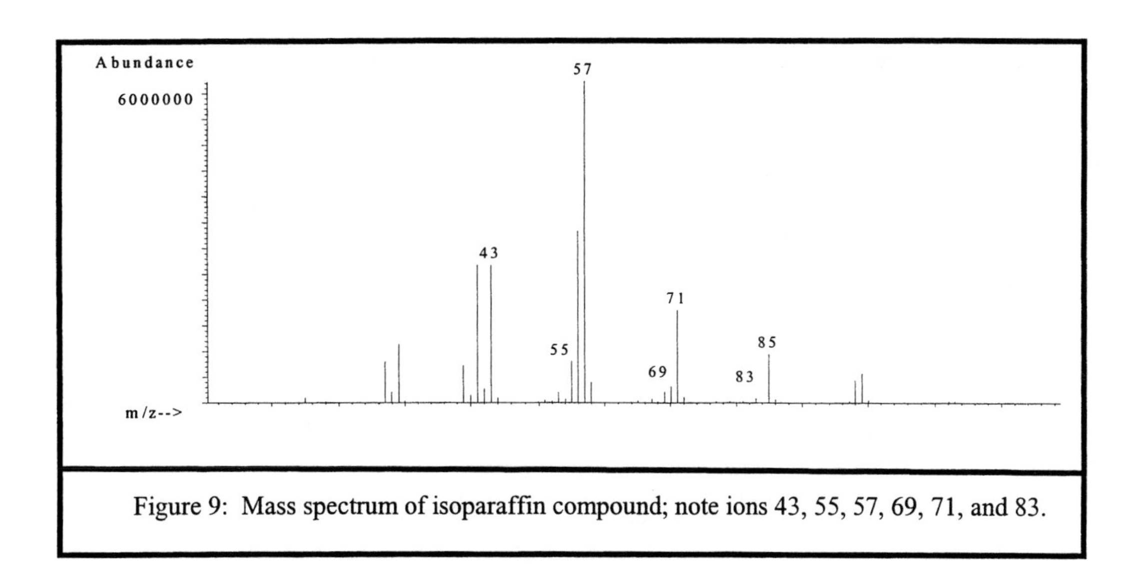

Figure 9: Mass spectrum of isoparaffin compound; note ions 43, 55, 57, 69, 71, and 83.

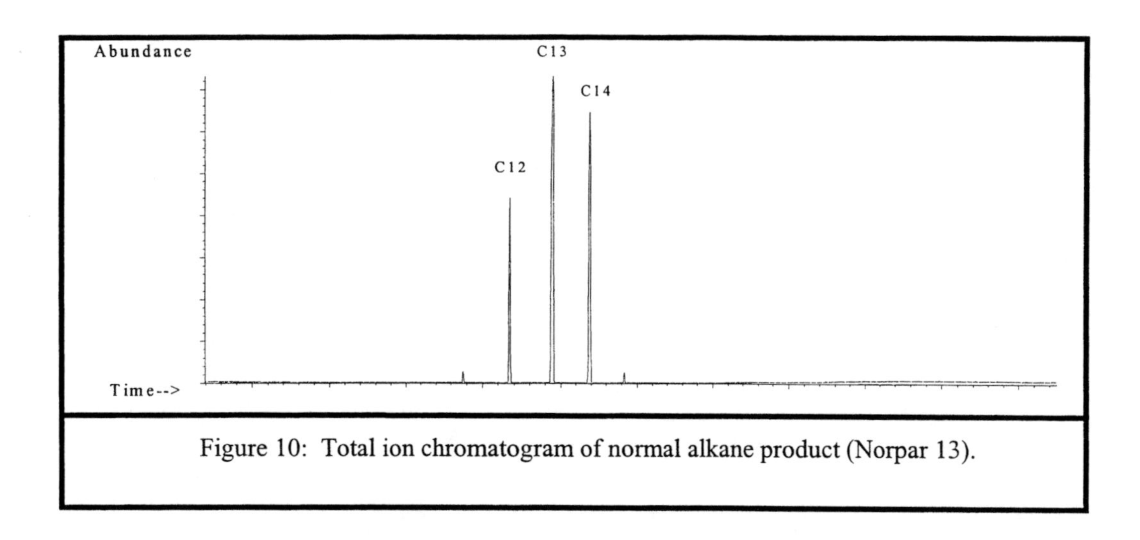

Figure 10: Total ion chromatogram of normal alkane product (Norpar 13).

NAPHTHENIC PARAFFINIC PRODUCTS

Naphthenic-Paraffinic (ASTM Class 0.5) products consist of isoparaffinic (branched chained alkane) compounds and cycloparaffinic (cycloalkane) compounds. Naphthenic-paraffinic products are produced by removing normal

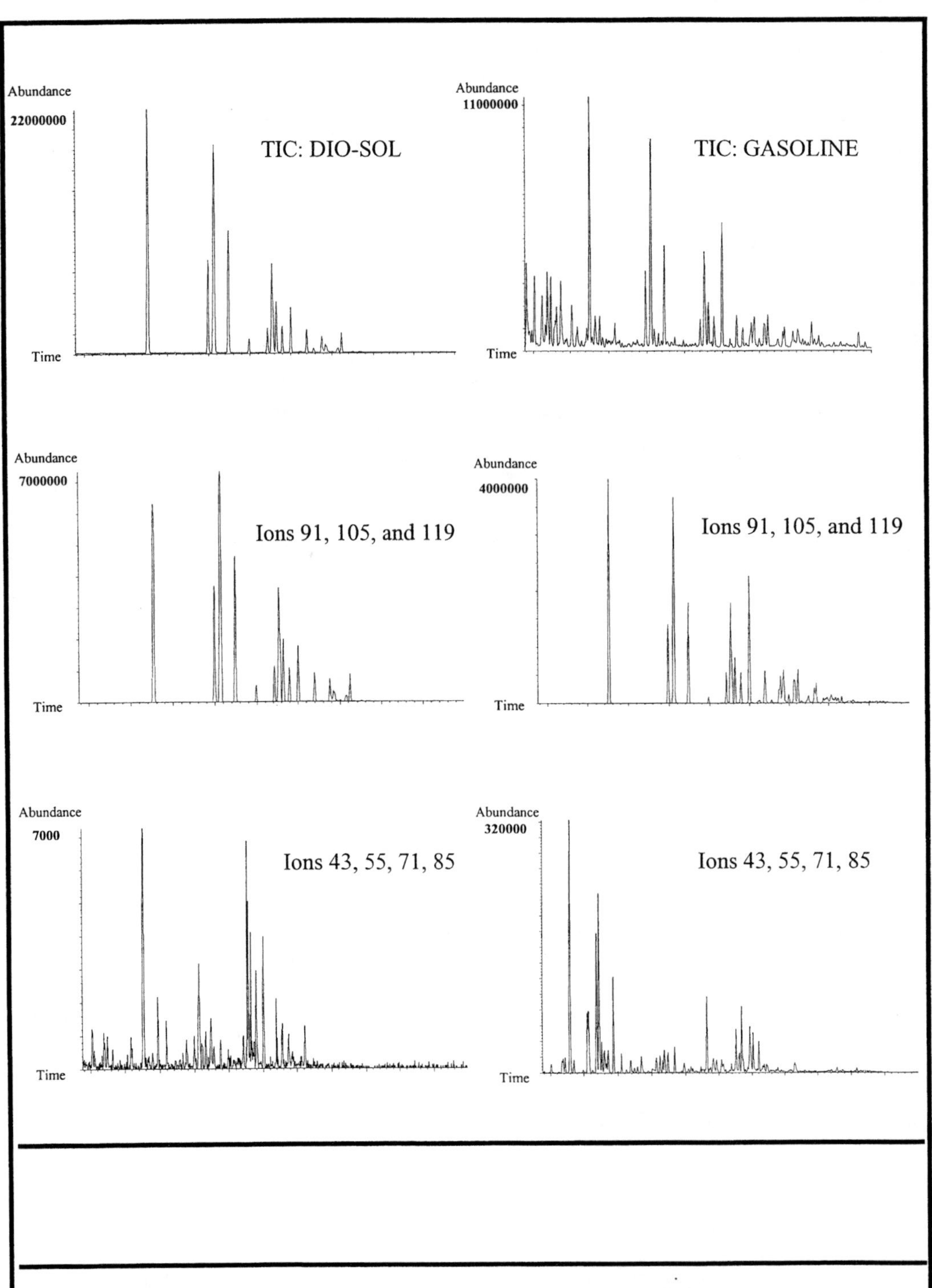

Figure 11: Total ion chromatograms, aromatic and alkane profiles for an aromatic product (Dio-Sol) and gasoline. Note the similarity of the aromatic profiles and the abundance of the "alkane" profile in gasoline compared to the aromatic solvent.

alkanes and aromatics from petroleum distillates. Most products in this class are in the same range as Class 3 (C8-C12) and Class 4 (C9-C16) distillates. Total ion chromatograms are characterized by an unresolved envelope of compounds (Figure 12). Alkane and cycloparaffin profiles are abundant; however, unlike the isoparaffinic products, the two patterns are significantly different. Normal alkanes, aromatics and naphthalenes are not present.

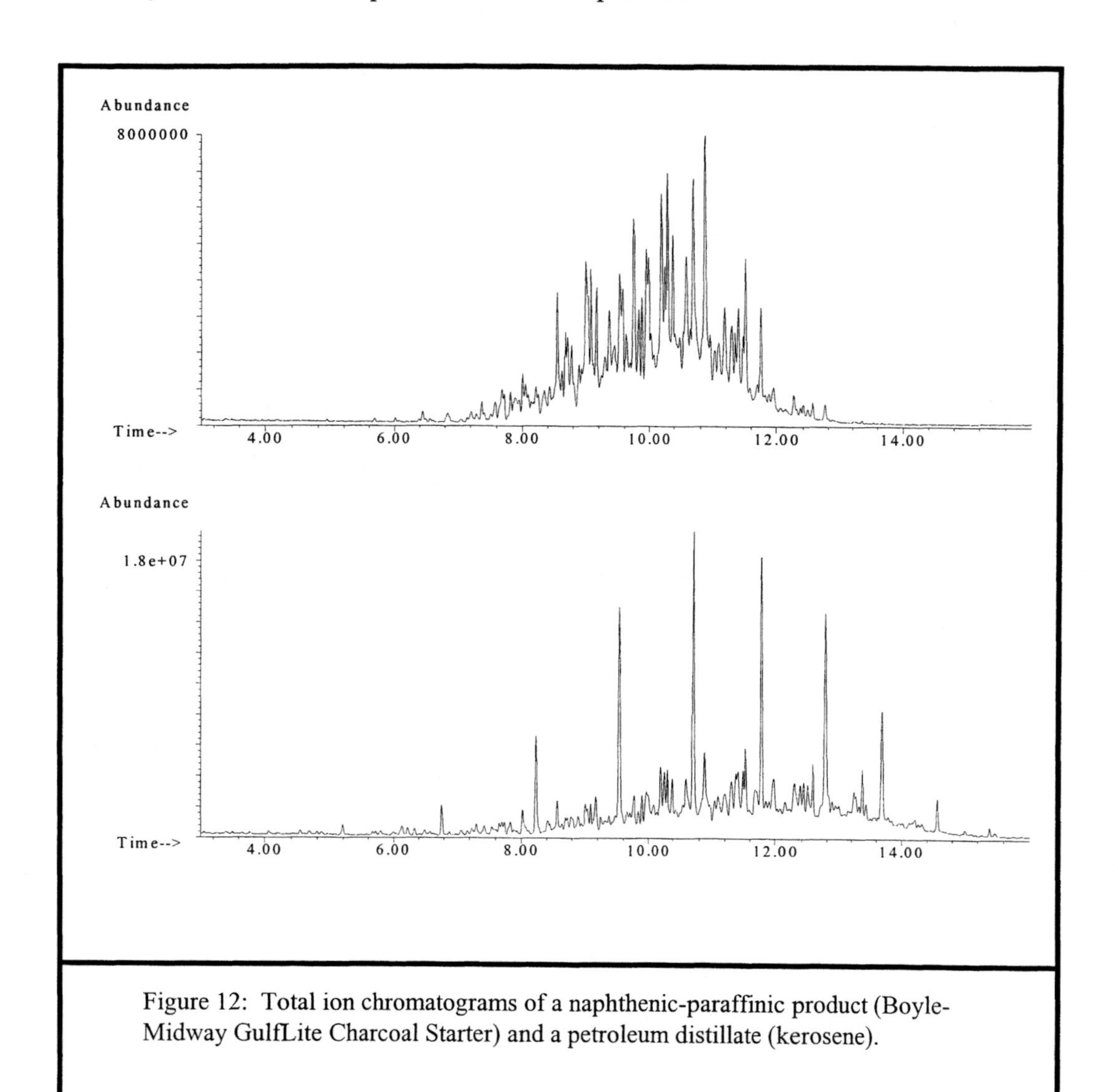

Figure 12: Total ion chromatograms of a naphthenic-paraffinic product (Boyle-Midway GulfLite Charcoal Starter) and a petroleum distillate (kerosene).

PRODUCT IDENTIFICATION/BRAND IDENTIFICATION

Brand names and product uses are included in this text to provide possible sources of ignitable liquid standards for sample comparison. In many cases, several brand named products are listed for the sample formulation. Under no circumstances is brand identification advocated from the data presented. Numerous products have similar if not identical compositions. Furthermore, many manufacturers change product formulations on a regular basis. For example, four distinctly different GulfLite Charcoal Starters and two different Zippo Lighter Fluids are included in this text. Manufacturers often produce different formulations with the same product name for different geographic markets.

The data in this text are by no means all inclusive. Tens, possibly hundreds, of thousands of ignitable liquid products are available. It is not possible or practical to obtain and analyze all these products. Every effort was made to include as many unique formulations as possible.

DATA COLLECTION

The data were collected on a Hewlett Packard 5890 Gas Chromatograph with a 5970 Mass Selective Detector, a 7673A Autosampler, and Chemstation software.

The instrumental and column conditions were as follows:

GC Conditions

Column	Rtx-1 (polymethylsiloxane) Restek Corporation, Bellefonte, PA
Length	30 meters
Inner diameter	0.25 millimeters
Film thickness	0.25 microns
Carrier gas	Helium
Flow rate	0.68 ml/min @ 60°C
Linear velocity	23 cm/sec
Split ratio	20:1
Injection volume	1 microliter
Injection temperature	250°C
Initial temperature	50°C
Initial time	2.5 minutes
Rate	15° C/min
Final temperature	300°C
Final time	5.83 minutes
Total run time	25.00 minutes

MS Conditions

Transfer line temperature	280°C
Electron multiplier voltage	Relative to tune + 200
Acquisition mode	Scan
Scan parameters	10 - 400 amu
Solvent delay	2.75 minutes

SAMPLE PREPARATION

Samples were obtained through numerous sources. Most were received in the form of neat liquids. These samples were diluted from 20 microliters to 1 milliliter with pentane. Other samples were received on activated charcoal strips. These were eluted with 1 milliliter of carbon disulfide.

Evaporated (weathered) standards are also included in this text. They were prepared by evaporating (by volume) neat liquids on a hot plate at low setting.

Most of the liquids were evaporated by 25, 50, and 75%. Gasolines were further evaporated to 90 and 98%. Evaporated liquids were diluted from 20 microliters to 1 milliliter with pentane.

DATA PRESENTATION

Total ion chromatograms, single ion mass chromatograms, and summed ion mass chromatograms are presented on consecutive pages for each formulation. The ions used for mass chromatograms are listed in Table 1. Total ion chromatograms are presented in two scales on the first page. The first presents the data in its entirety. The expanded total ion chromatogram is presented in one of three hydrocarbon ranges to provide data in greater detail. The range is based loosely on the volatility of the product:

Light products & gasoline	C6-C13	2.75 - 12.00 minutes
Medium products	C8-C16	5.00 - 15.00 minutes
Heavy products	C8-C25	5.00 - 21.00 minutes

A hydrocarbon scale is presented with each total ion chromatogram.

Summed and individual ion mass chromatograms are presented for alkanes, aromatics, cycloparaffins, and naphthalenes. Data are presented in scales designed to provide the most possible detail:

Aromatics	C6 - C14	2.75 - 14.00 minutes	
Naphthalenes	C12-C15	10.00 - 15.00 minutes	
Alkanes/ cycloparaffins	Light products	C6-C11	2.75-10.00 minutes
	Medium products	C7-C14	4.00-13.00 minutes
	Heavy products	C6-C26	2.75-22.00 minutes
	Gasoline	C6-C14	2.75-14.00 minute

ASTM Classifications are based on the specific wording and authors' interpretation of ASTM 1387-95 and 1618-94.

Major peak identification is provided for most products. Identifications were made by computerized mass spectral library searches, retention times, and literature data. Normal text indicates peak identification with high certainty based on retention time data and mass spectral identification. Italicized text indicates that the identification is more tentative, either due to library matches of less than 95% or the number of compounds (generally isomers) with similar mass spectral data.

References

(1) ASTM Method E 1618-94 Standard Guide for Ignitable Liquid Residue in Extracts from Fire Debris Samples by Gas Chromatography-Mass Spectrometry, *1996 Annual Book of ASTM Standards*, Volume 14.02, 1995, pp. 1023-1028.

(2) Smith, Martin R. Analytical Chemistry, Volume 54, Number 13, November 1982, pp. 1399A-1409A.

(3) ASTM Method E 1618-94.

(4) ASTM Method E 1387-95 Standard Test Method for Ignitable Liquid Residues in Extracts from Fire Debris Samples by Gas Chromatography, *1996 Annual Book of ASTM Standards*, Volume 4.02, 1995, pp. 871-879.

(5) ASTM Method E 1618-94.

HOW TO USE THIS BOOK

Ignitable liquid identification of fire debris extracts is a systematic endeavor. Ideally, comparison should be made with sample and standard data produced and displayed under the same conditions. The steps below outline the ignitable liquid process utilizing the data presented in this text.

Step 1: Identify predominent n-alkane range.

Step 2: Produce sample chromatograms with extraction and output parameters suitable for comparison. In order to provide the most detail, the data are presented based on hydrocarbon (n-alkane) range (see Data Presentation, page 18).

Light products:	C6-C12
Medium products:	C8-C16
Heavy products:	C8-C25

A code is provided in the upper right hand corner of the total ion chromatogram pages that indicates the range for the data displayed. Appendix A contains a computer program designed to allow the analyst to produce comparable data. Of course, comparable data may also be produced manually.

Step 3: Identify potential ignitable liquid classification. This can be done by using ion profiles to notate the abundance of different types of compounds (Table 3). The flowchart presented on pages 24-25 is designed to aid the analyst in ASTM classification of sample chromatograms. The data in this text are separated primarily by classification.

Section 1:	Hydrocarbon and Aromatic Standard Solutions
Section 2:	Distillates
Section 3:	Gasolines
Section 4:	Oxygenated Solvents
Section 5:	Isoparaffinic products
Section 6:	Normal alkane products

Section 7: Aromatic solvents
Section 8: Naphthenic-paraffinic products
Section 9: Miscellaneous and mixture

Data in each section are organized by volatility.

Step 4: Compare total ion and extracted ion chromatograms of sample and text data.

Step 5: Compare major peak identification where applicable.

Step 6: Locate and analyze comparable ignitable liquid standards.

Compare sample to standard data to confirm identification.

Class	Range	Alkanes	Aromatics	Cycloparaffins	Naphthalenes
1 LPD	C4-C11	Yes*	Yes	Yes	No
2 Gasoline	C4-C12	Yes	Yes*	Yes	Yes
3 MPD	C8-C12	Yes*	Yes	Yes	Varies
4 Kerosene	C9-C16	Yes*	Yes	Yes	Yes
5 HPD	C10-C23	Yes*	Yes	Yes	Yes
0.1 Oxygenated Solvents	Varies	Varies	Varies	Varies	Varies
0.2 Isoparaffinic Products	Varies	Yes*	No	No†	No
0.3 N-alkane Products	C8-C18	Yes*	No	No	No
0.4 Aromatic Solvents	Varies	No	Yes*	No	Varies
0.5 Naphthenic-Paraffinics	C8-C16	Yes*	No	Yes*	No

Table 3: Types of compounds found in various classes of ignitable liquids.

*denotes most abundant compounds, † denotes "cycloparaffin" profile may be present in ion chromatograms but does not actually represent cycloaliphatic compounds.

Classification Flow Chart

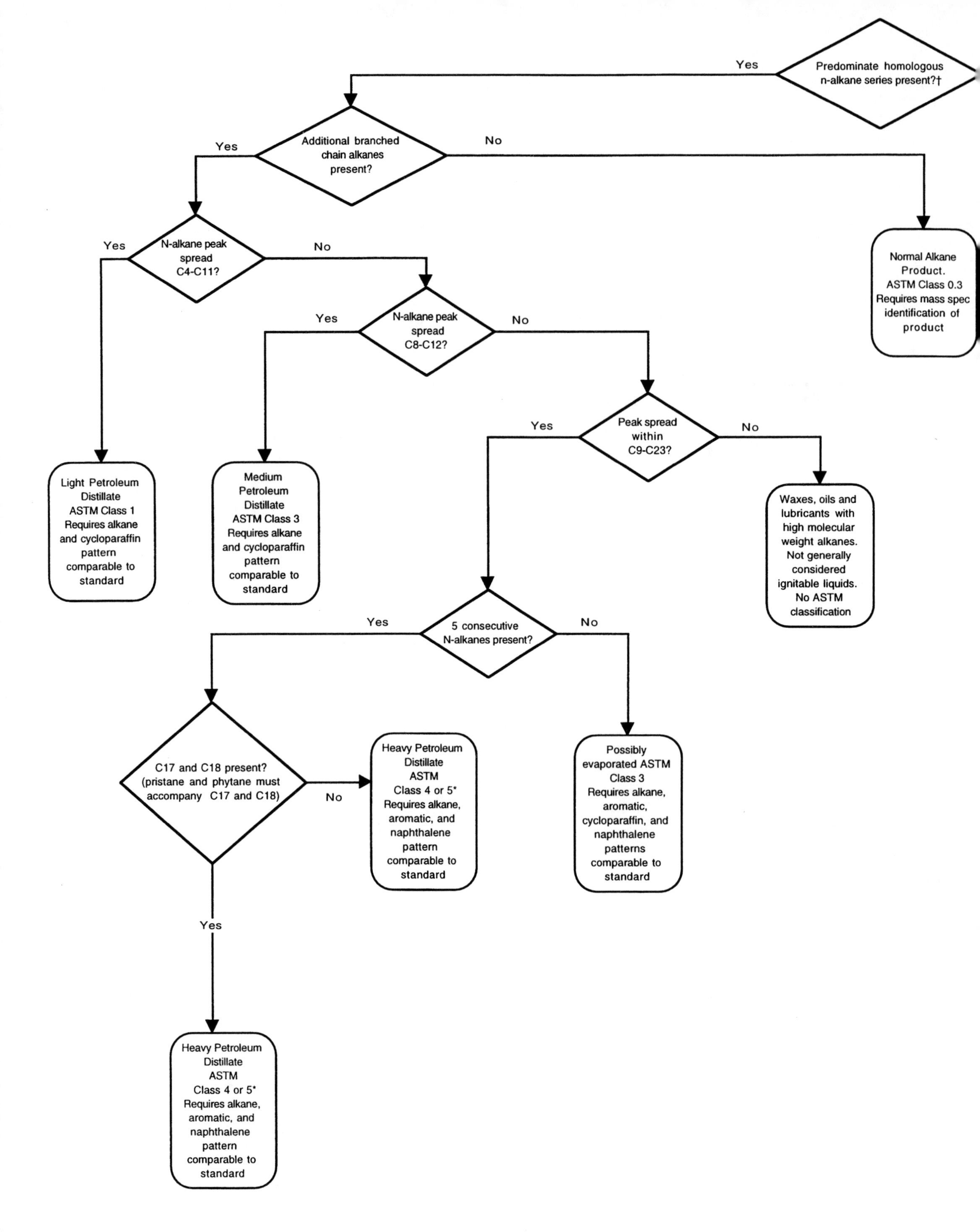

* Differentiation of Class 4 and 5 Dependent on Recovery technique

† If homologous series includes triplets and doublets with alkenes present and no pristane or phytane following C17 and C18 indicates presence of plastic; alkenes with pristane and phytane possible asphalt

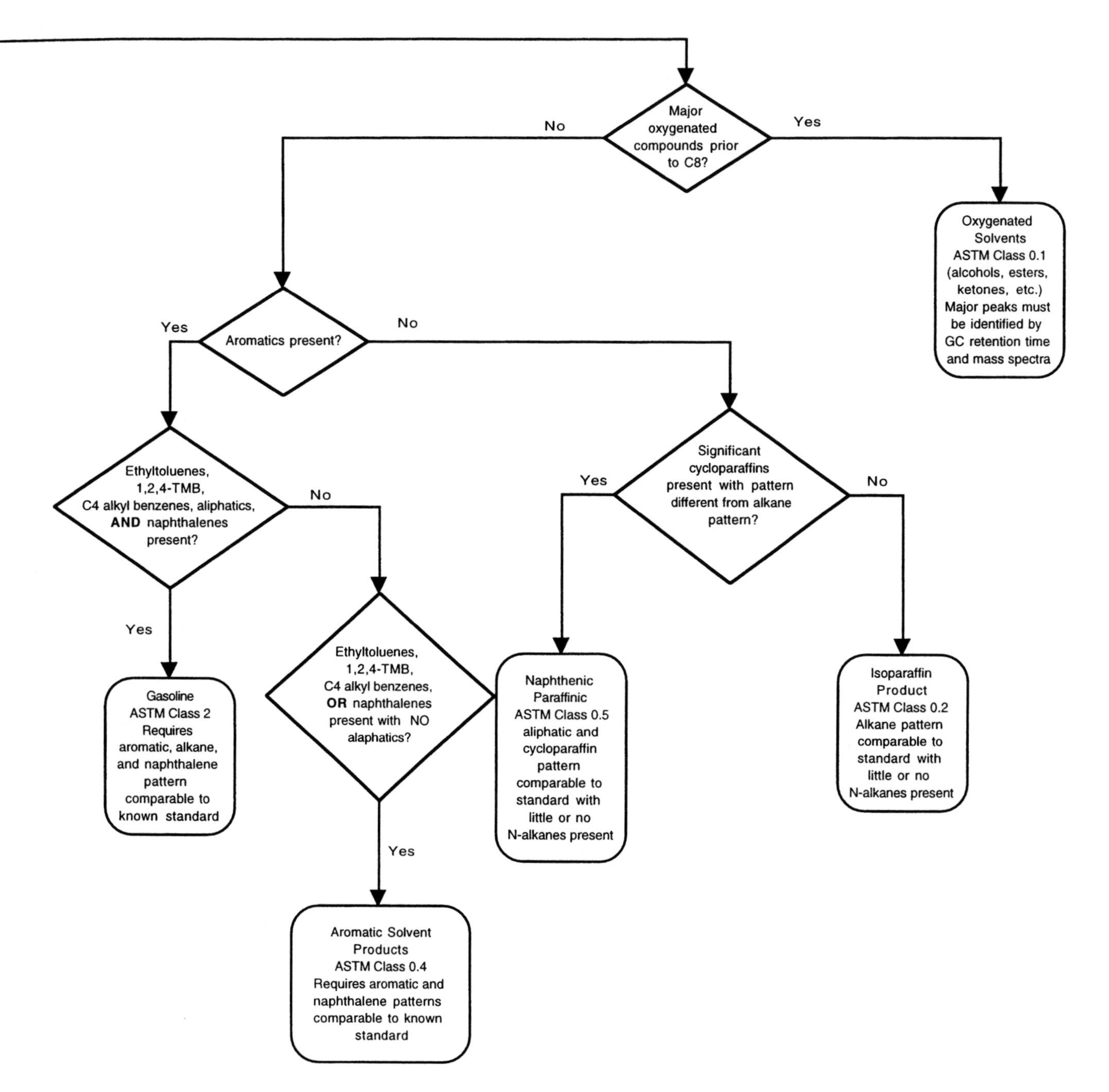

Classification Flow Chart for Fire Debris

Hydrocarbons and Aromatics Standards

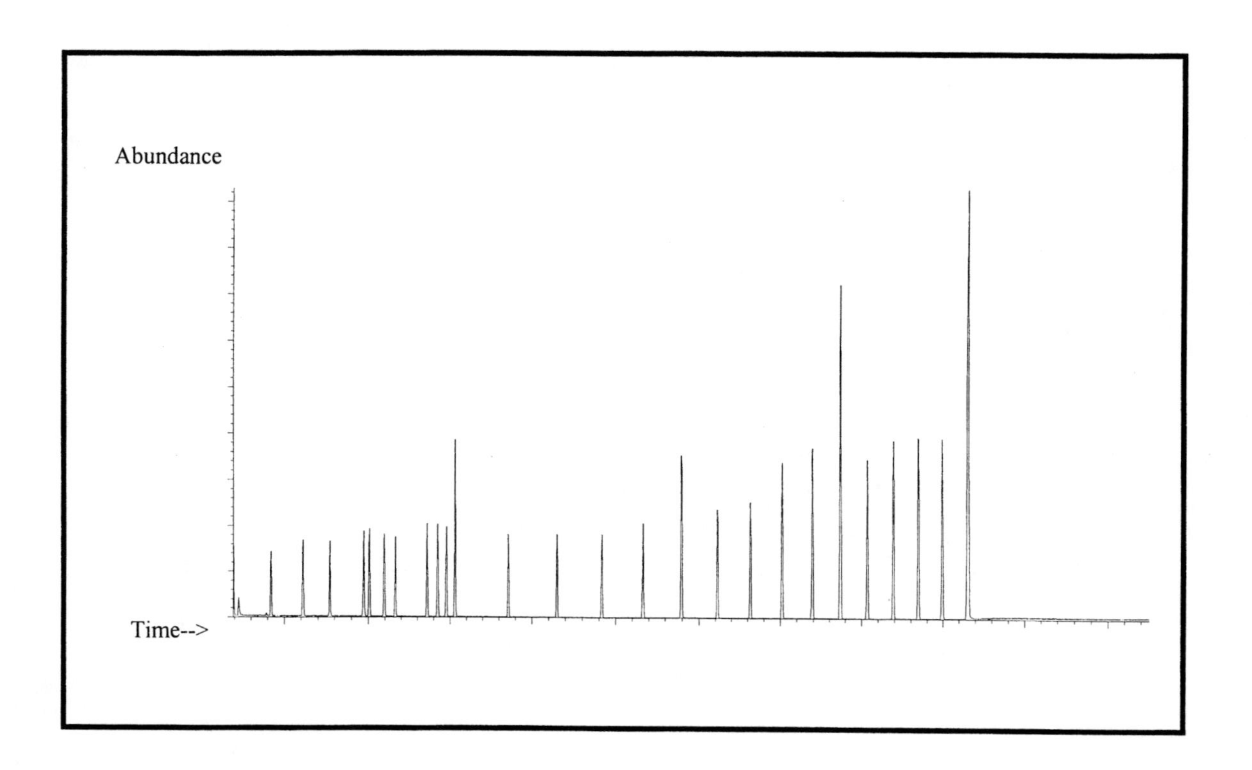

Hydrocarbons and Aromatics Standard

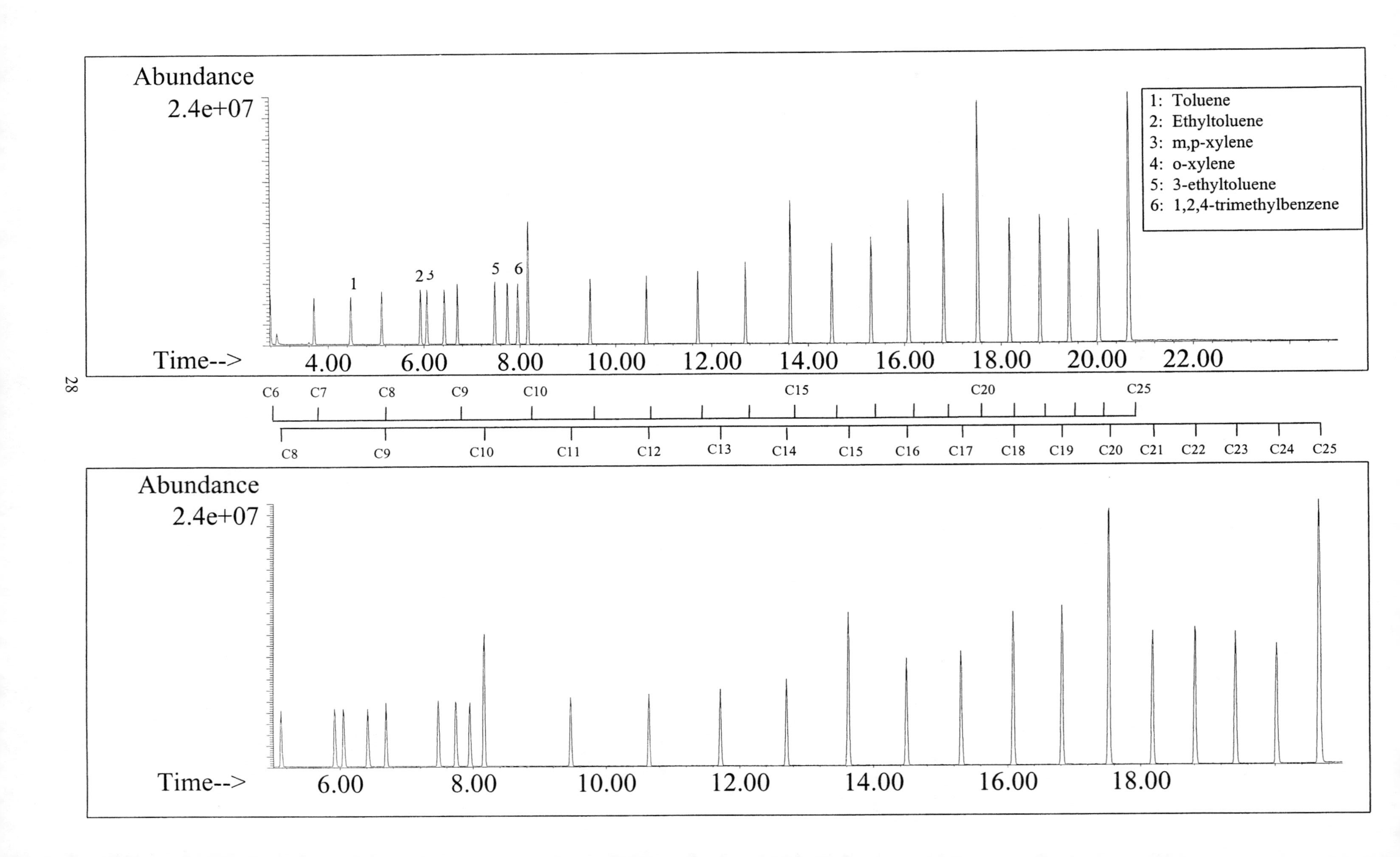

ALKANES

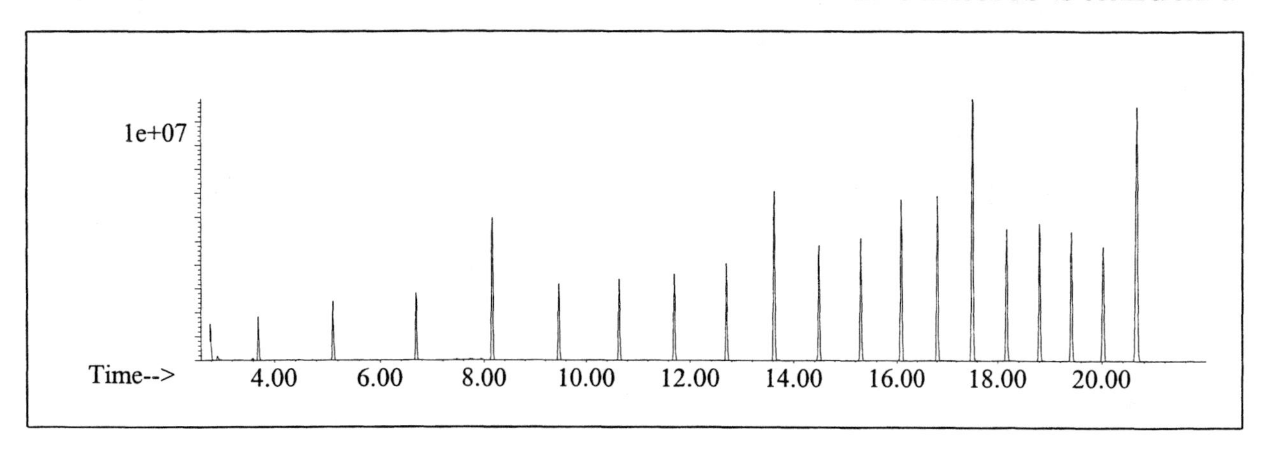

AROMATICS

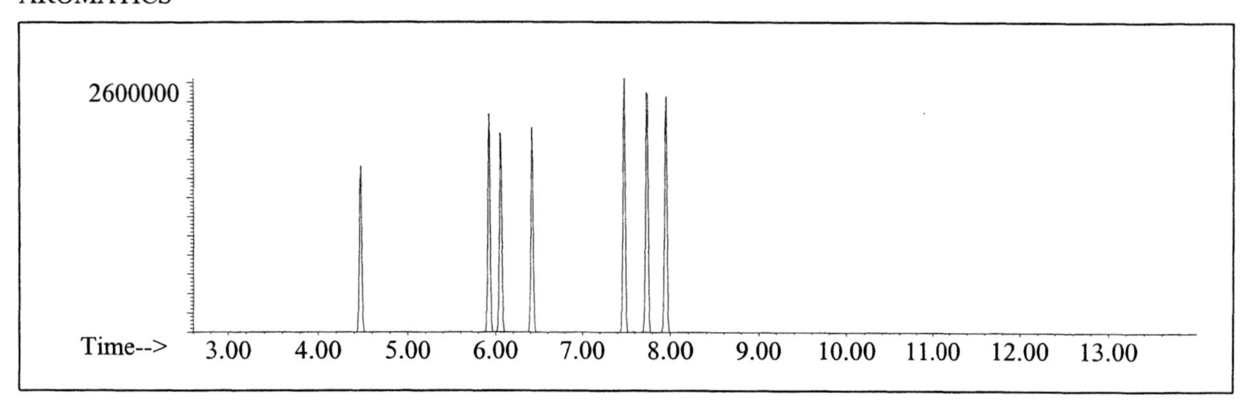

CYCLOPARAFFINS AND ALKENES

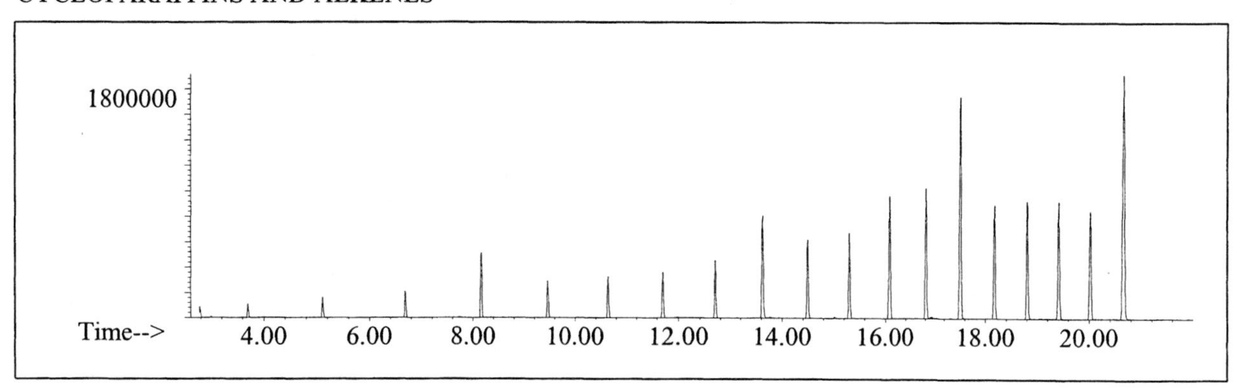

NAPHTHALENES

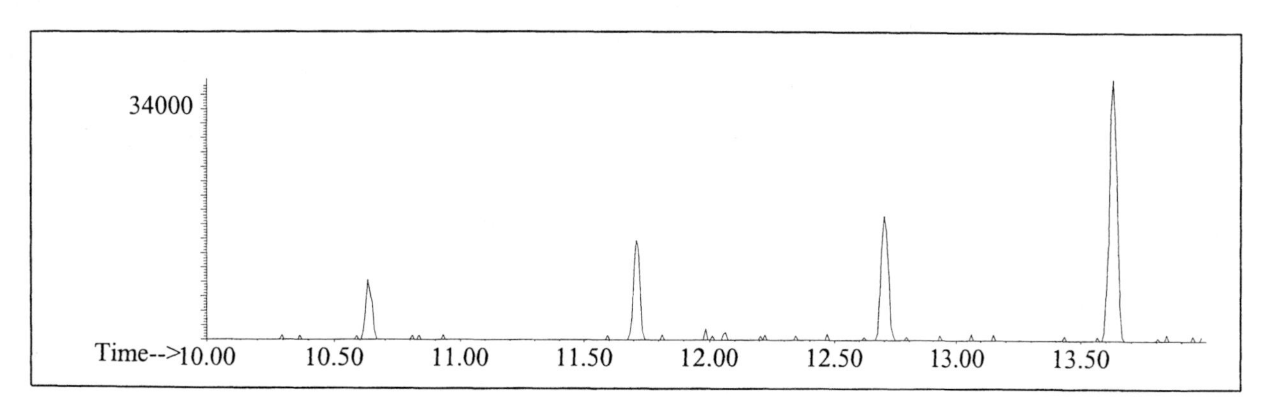

Hydrocarbons and Aromatics Standard (Restek)

20uL/mL Pentane

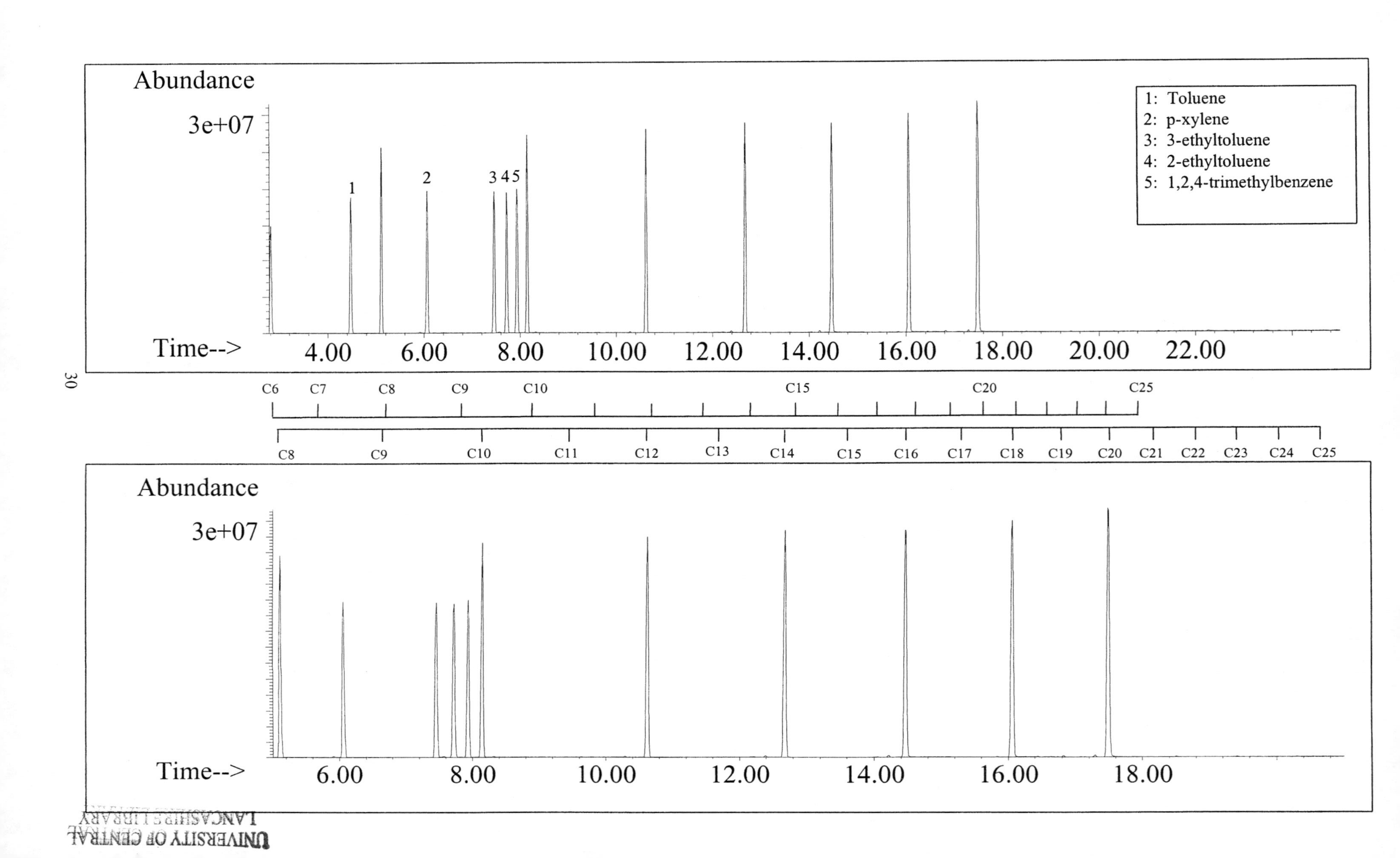

SUMMED PROFILES

Hydrocarbons and Aromatics Standard (Restek)

ALKANES

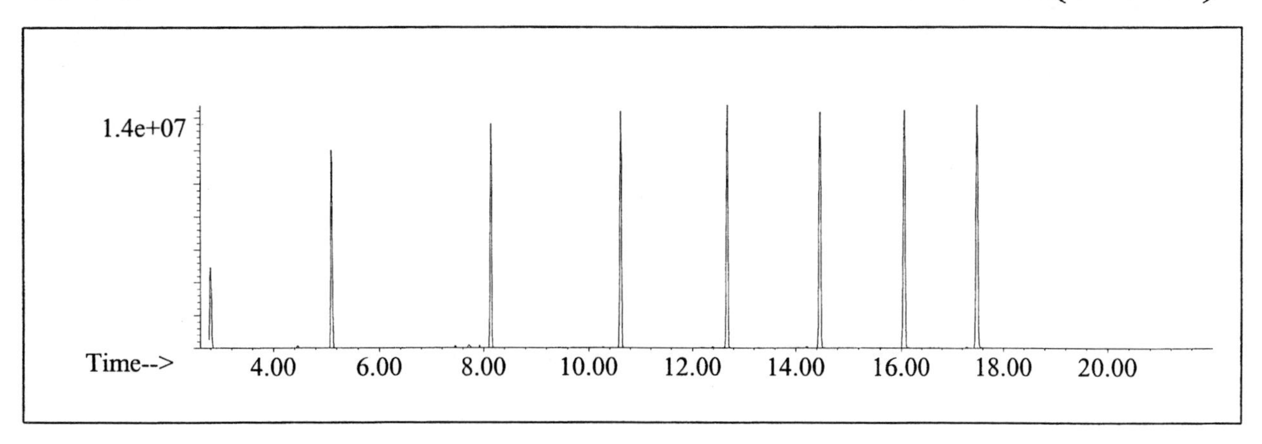

AROMATICS

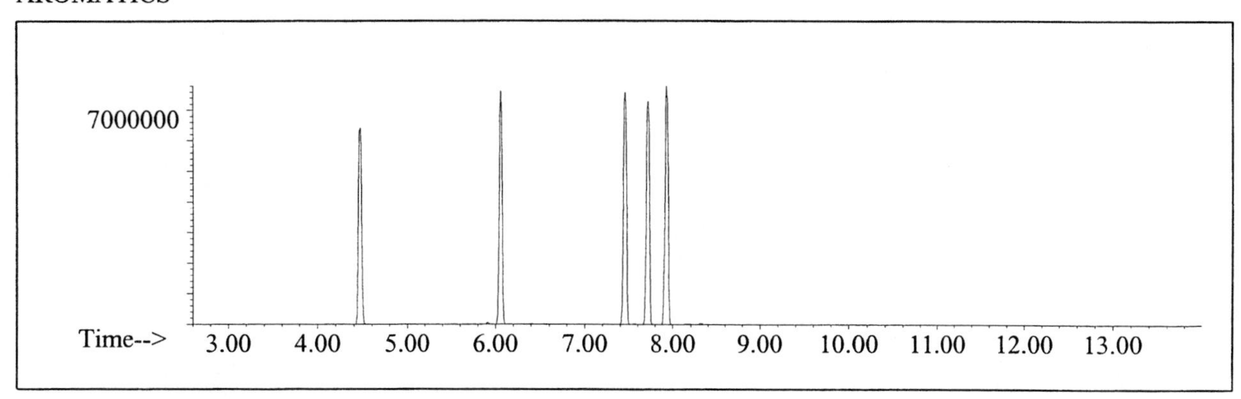

CYCLOPARAFFINS AND ALKENES

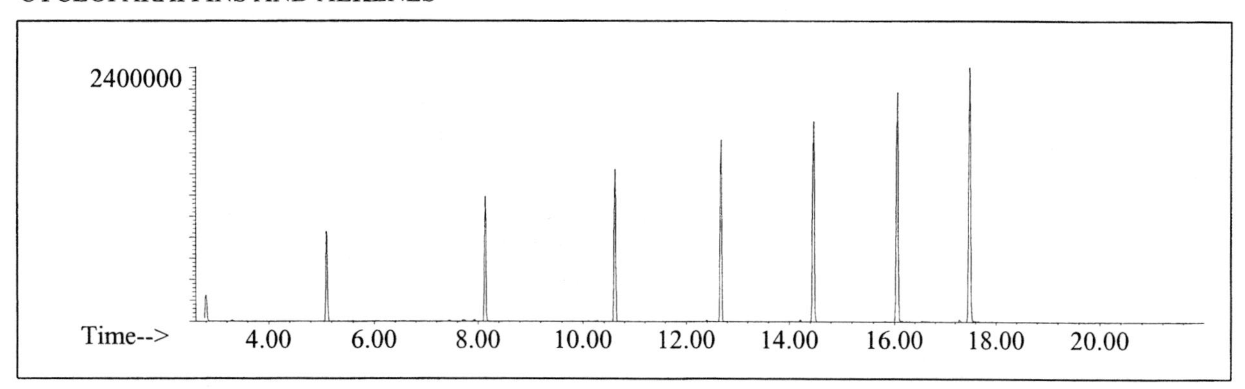

NAPHTHALENES

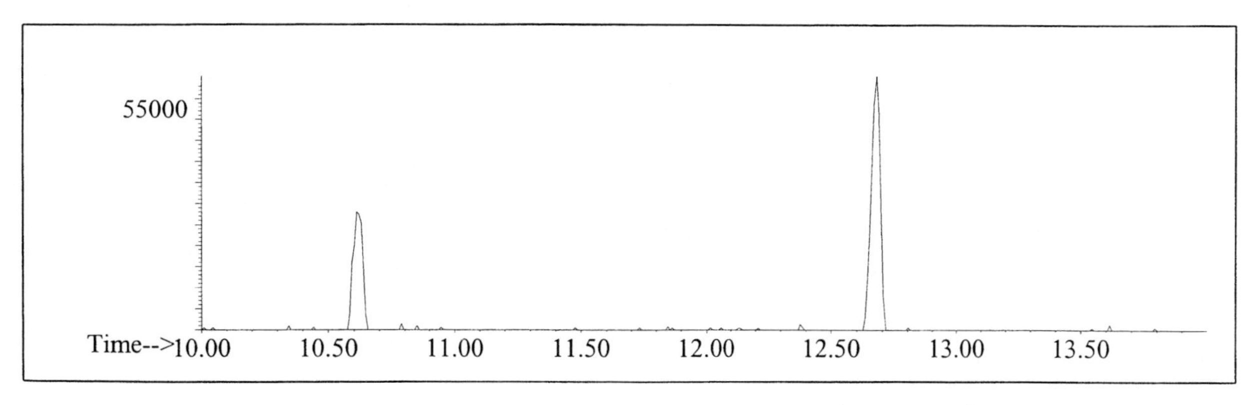

Distillates

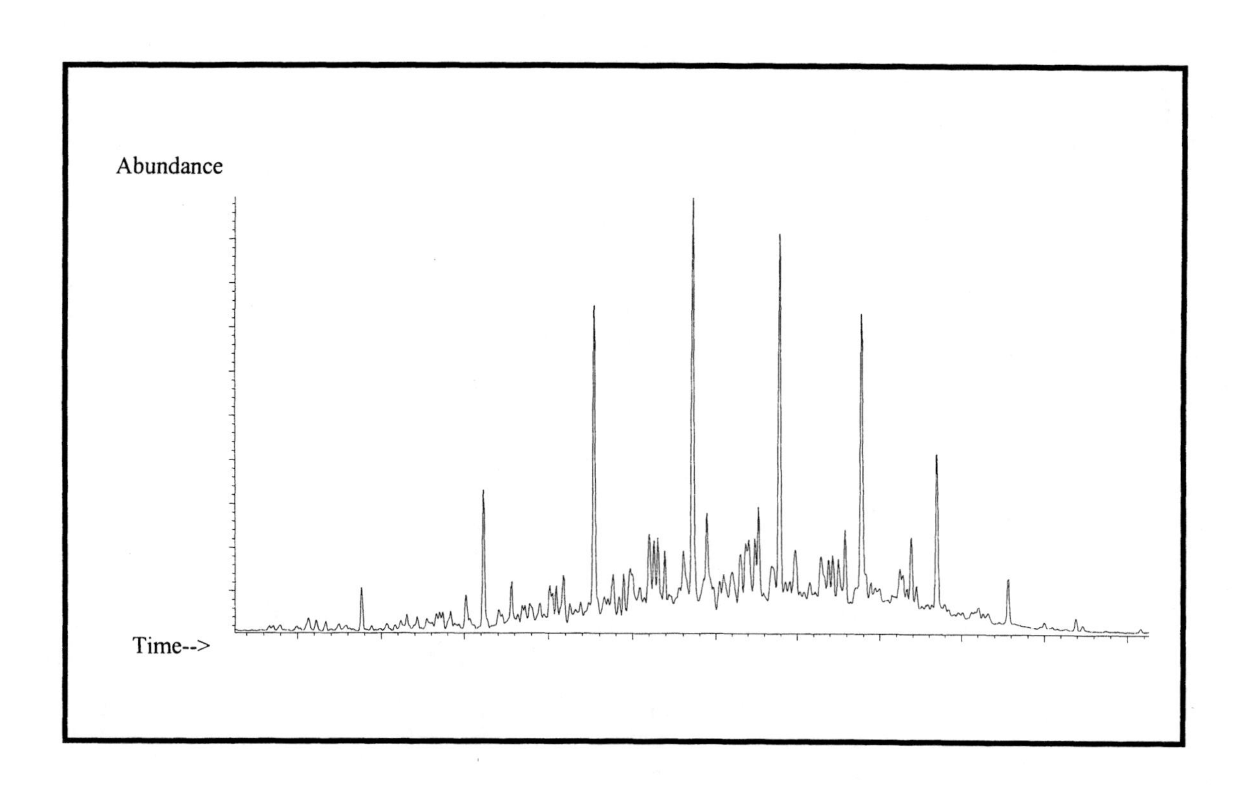

Distillate #1

20uL/mL Pentane

Product Displayed: Snap Instant Starting Fluid

Other Similar Products: 500XL Engine Starter Fluid

ASTM: Class 1 (Light Petroleum Distillate)

Product Uses: Engine starter fluid

Macro Code: L

Abundance

1.4e+07

1 2 3 4 5

Time-->

4.00 6.00 8.00 10.00 12.00 14.00 16.00 18.00 20.00 22.00

1: Hexane
2: Methylcyclopentane
3: 2-methylheptane
4: 3-methylheptane
5: Heptane

C6 C7 C8 C9 C10 C15 C20 C25

C6 C7 C8 C9 C10 C11 C12 C13

Abundance

1.4e+07

Time-->

3.00 4.00 5.00 6.00 7.00 8.00 9.00 10.00 11.00

SUMMED PROFILES

Distillate #1

ALKANES

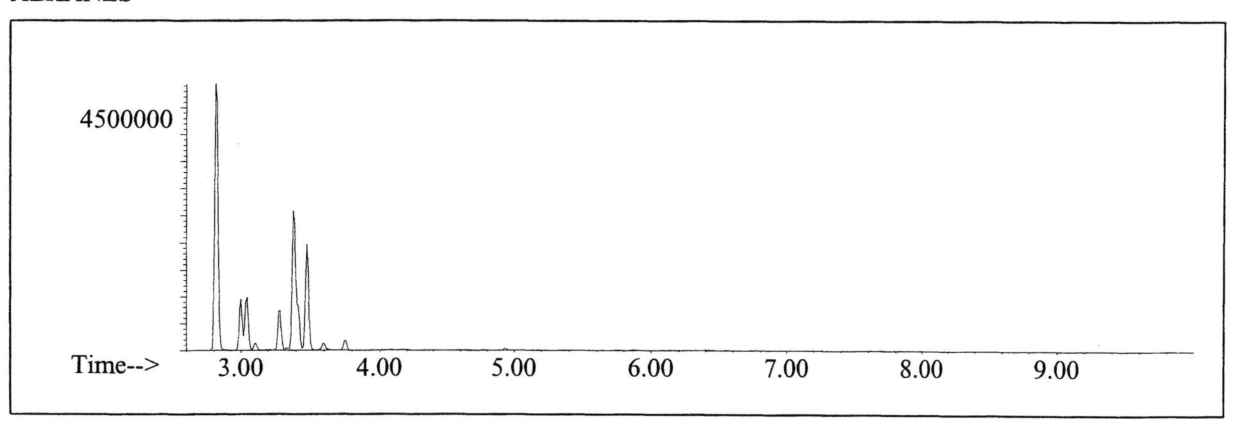

AROMATICS

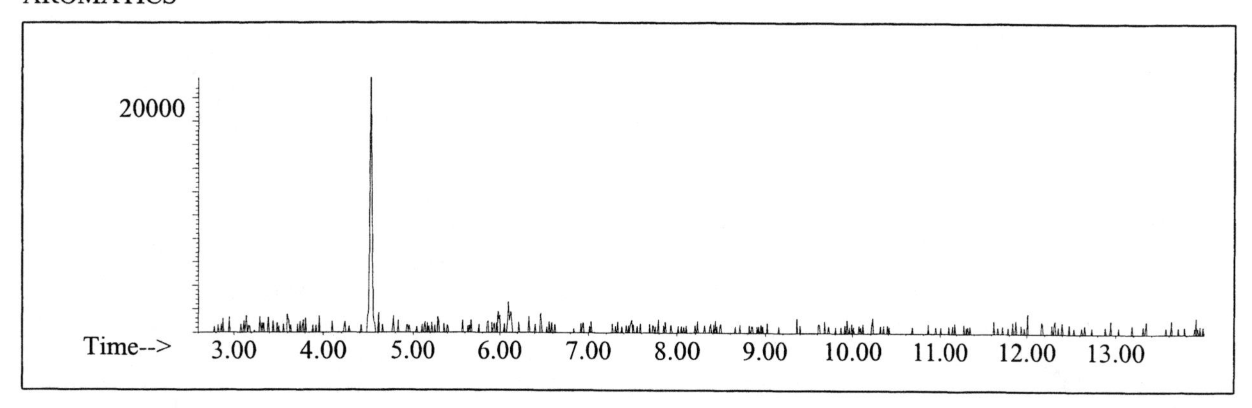

CYCLOPARAFFINS AND ALKENES

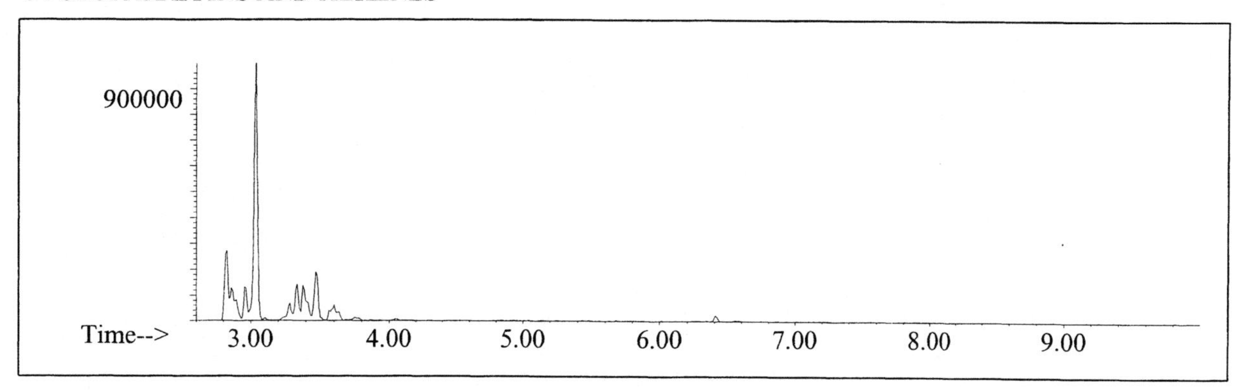

NAPHTHALENES

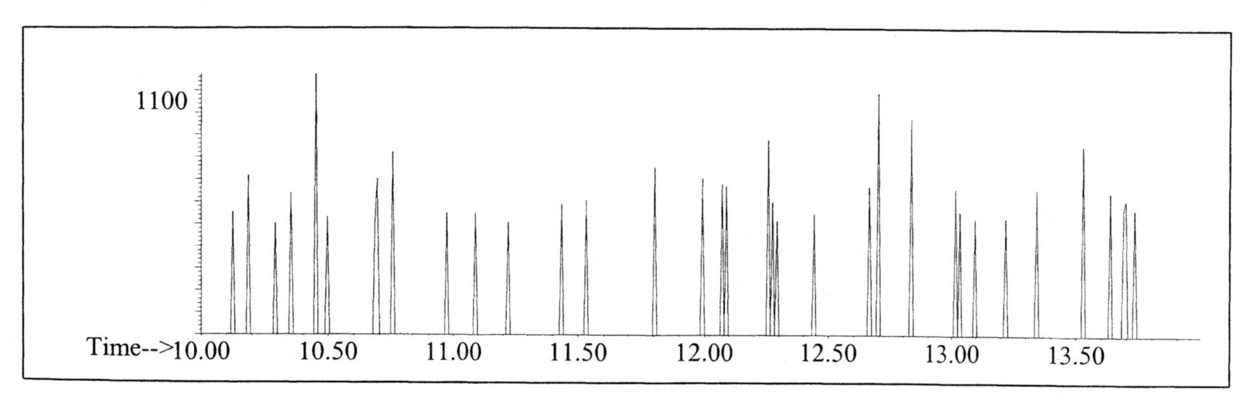

INDIVIDUAL PROFILES Distillate #1

Alkane

Ion 43

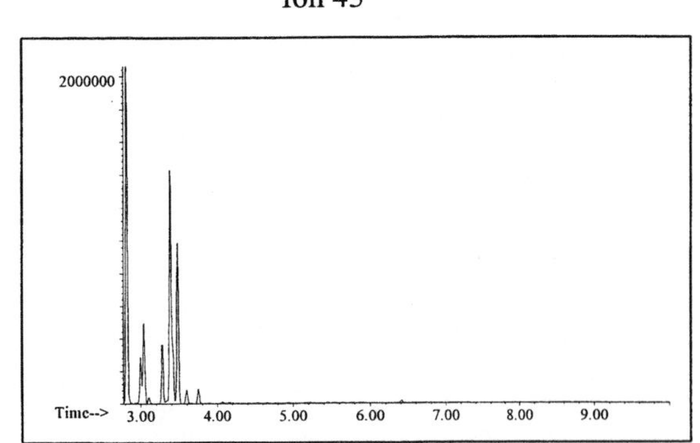

Ion 57

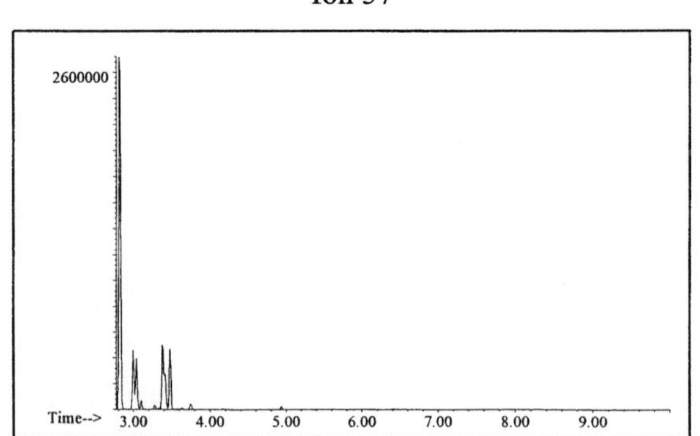

Ion 71

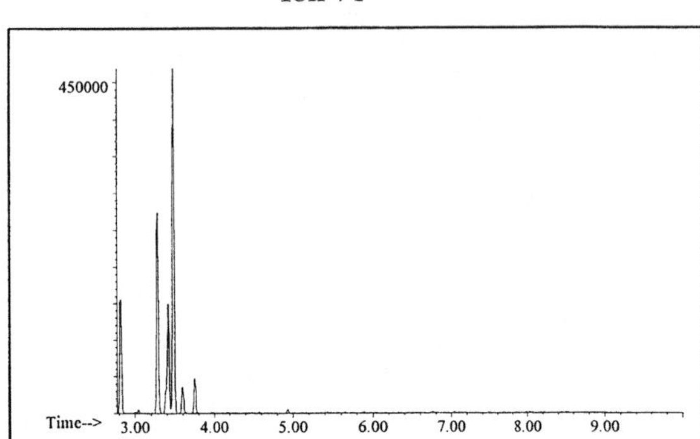

Ion 85

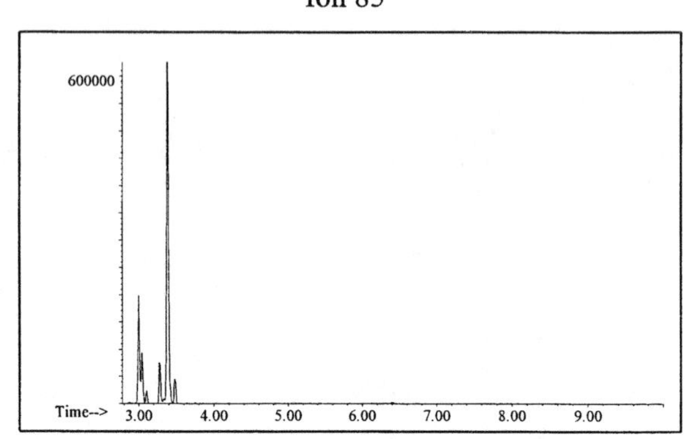

Aromatic

Ion 91

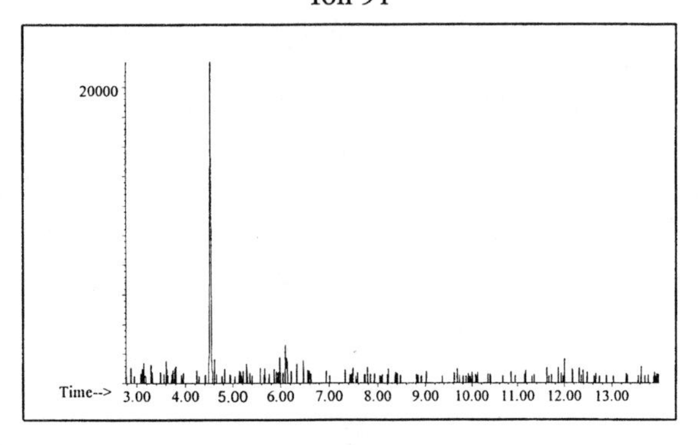

Ion 105

No Useful Data Obtained

Ion 119

No Useful Data Obtained

INDIVIDUAL PROFILES

Distillate #1

Cycloparaffin

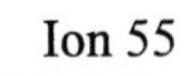
Ion 55

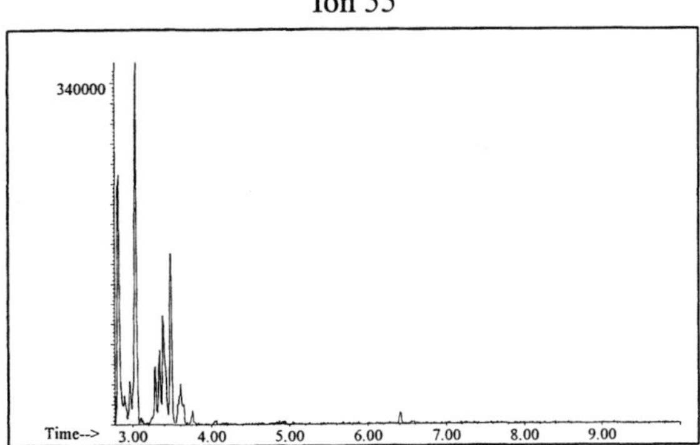

Ion 69

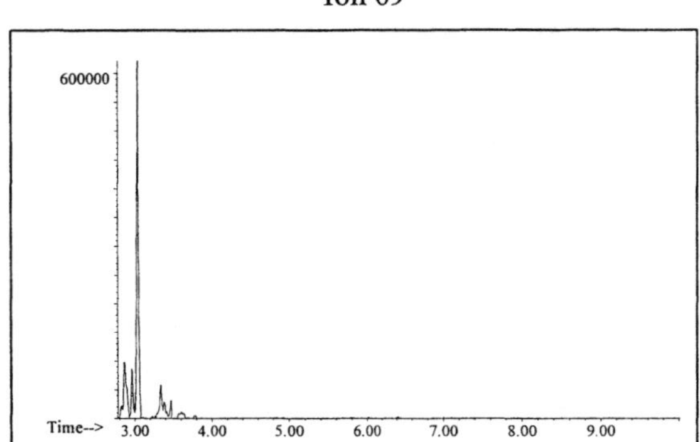

Ion 83

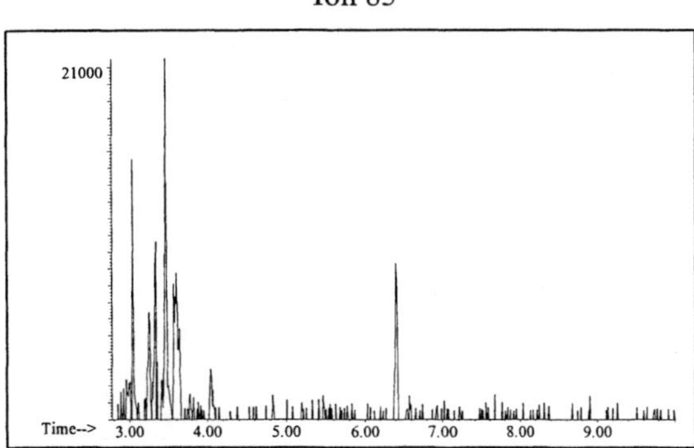

Naphthalene

Ion 128

No Useful Data Obtained

Ion 142

No Useful Data Obtained

Ion 156

No Useful Data Obtained

Distillate #2

20uL/mL Pentane
Product Displayed: Patch Rubber Company Cleaner
Other Similar Products:

ASTM: Class 1 (Light Petroleum Distillate)
Product Uses: Cleaning solvent

Macro Code: L

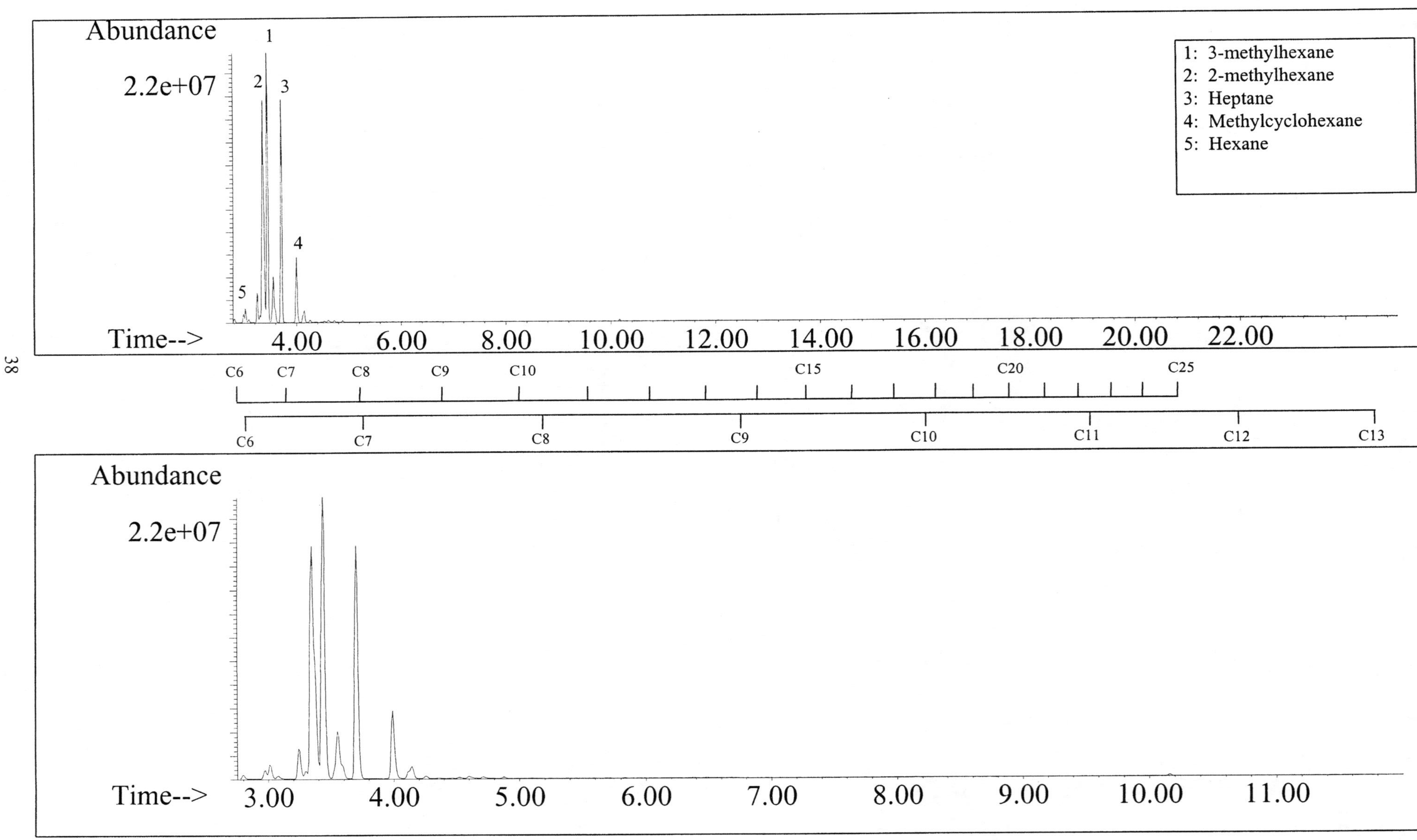

SUMMED PROFILES — Distillate #2

ALKANES

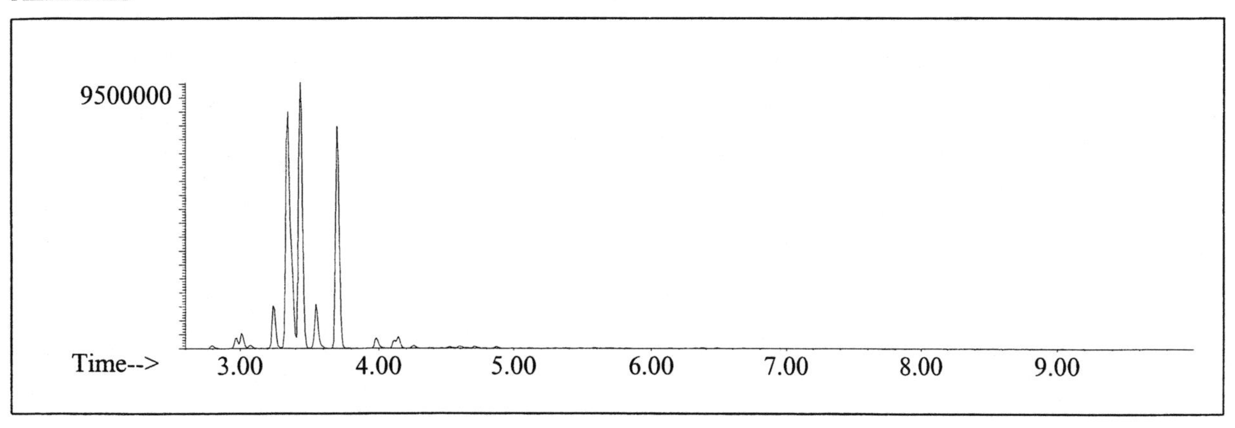

AROMATICS

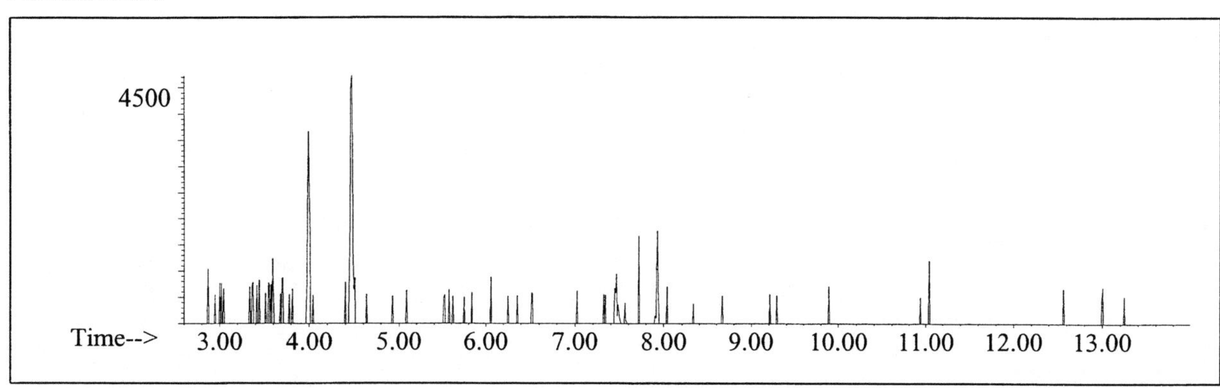

CYCLOPARAFFINS AND ALKENES

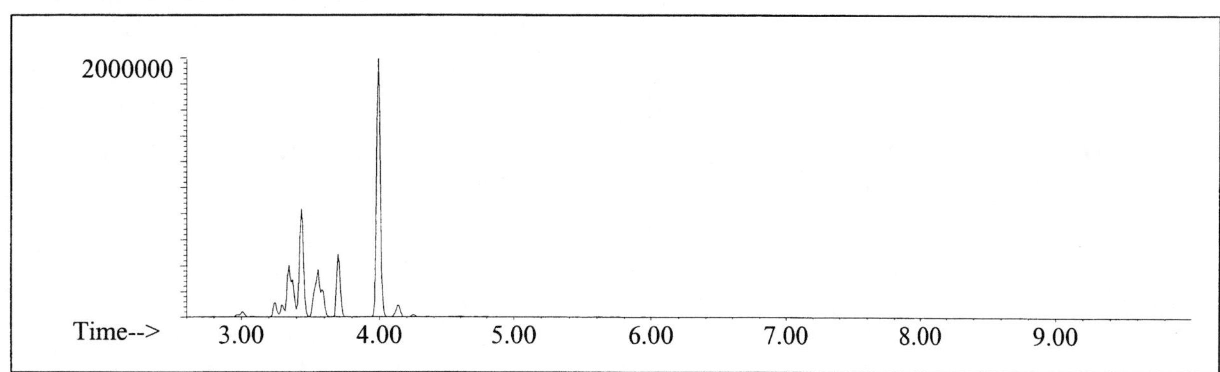

NAPHTHALENES

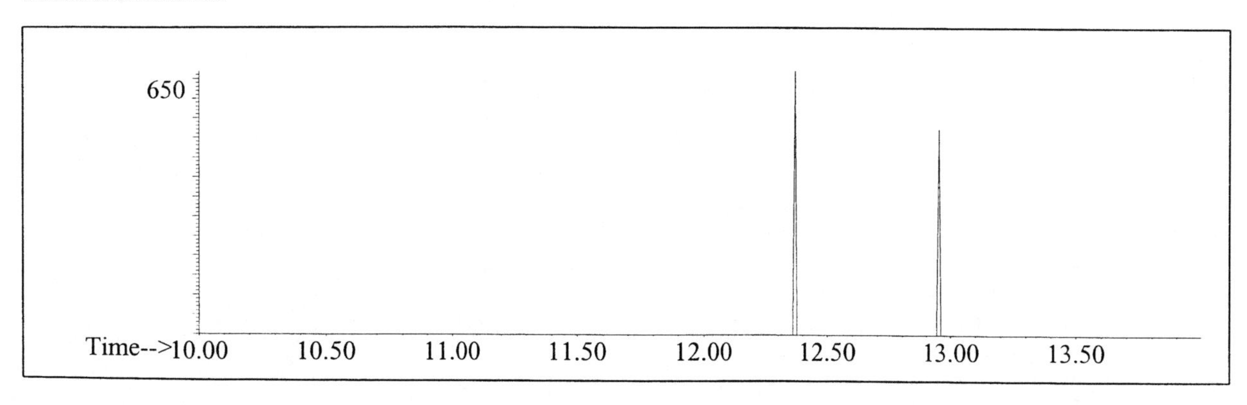

INDIVIDUAL PROFILES

Distillate #2

Alkane

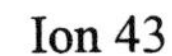

Ion 43

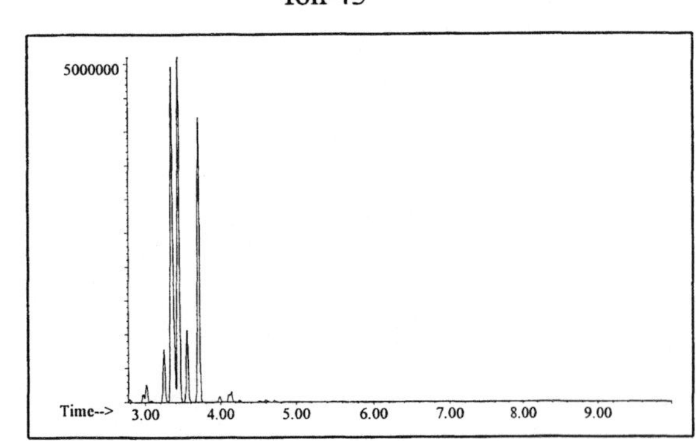

Ion 57

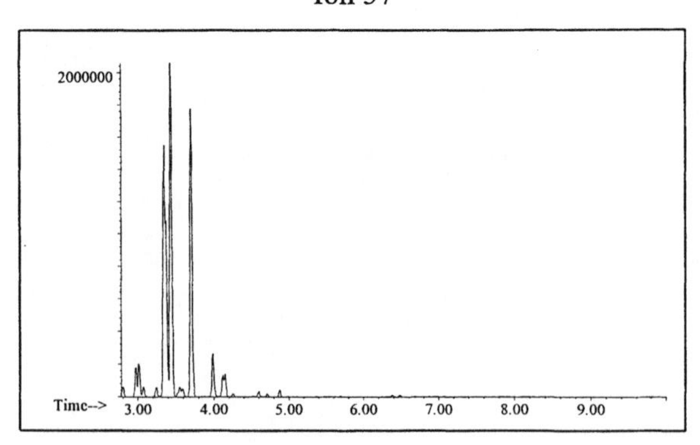

Ion 71

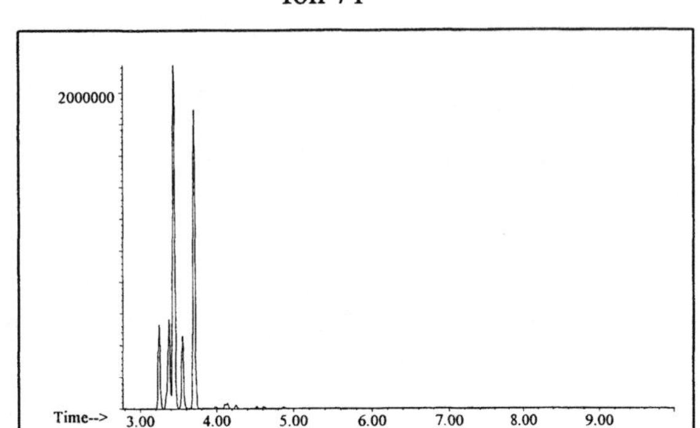

Ion 85

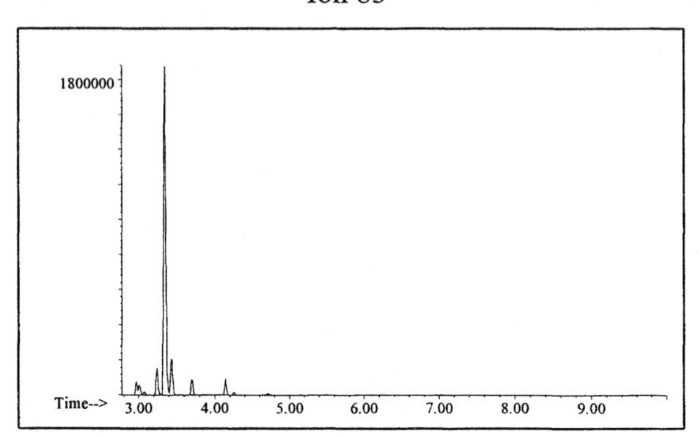

Aromatic

Ion 91

No Useful Data Obtained

Ion 105

No Useful Data Obtained

Ion 119

No Useful Data Obtained

INDIVIDUAL PROFILES Distillate #2

Cycloparaffin

Ion 55

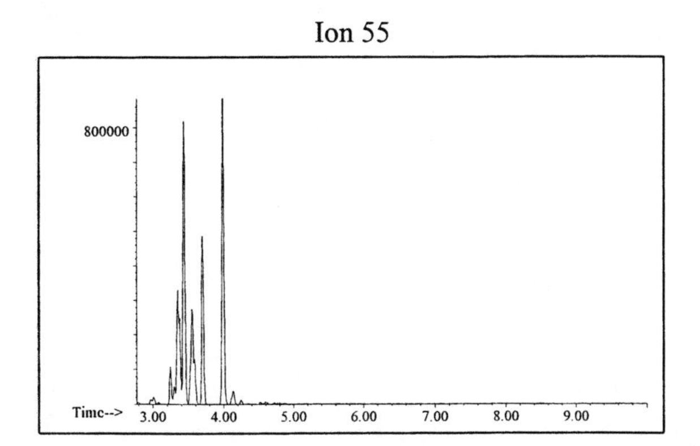

Ion 69

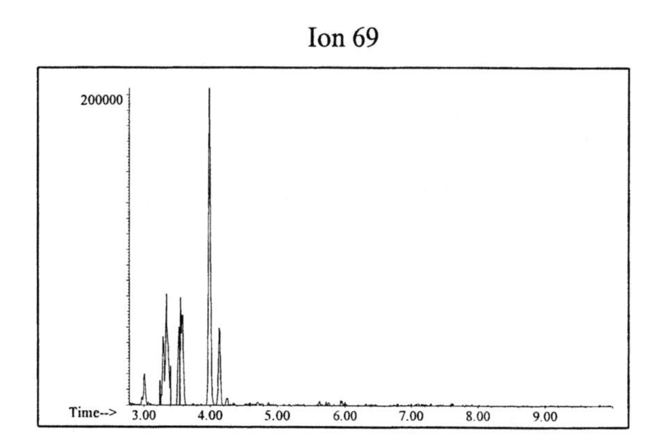

Ion 83

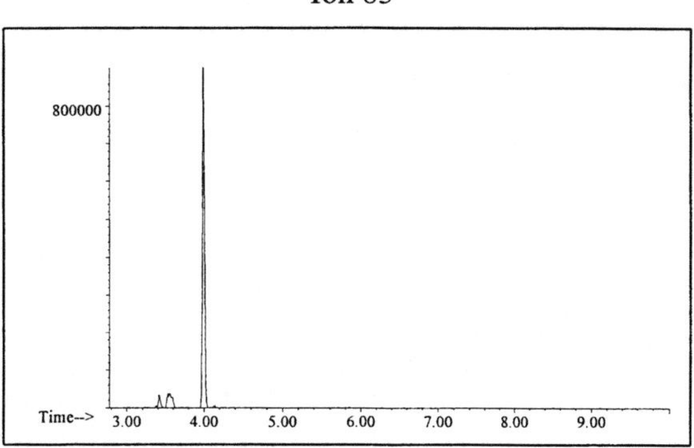

Naphthalene

Ion 128

No Useful Data Obtained

Ion 142

No Useful Data Obtained

Ion 156

No Useful Data Obtained

Distillate #3

20uL/mL Pentane

Product Displayed: Chevron Blazo Camp Fuel

Other Similar Products:

ASTM: Class 1 (Light Petroleum Distillate)

Product Uses: Camping Fuel

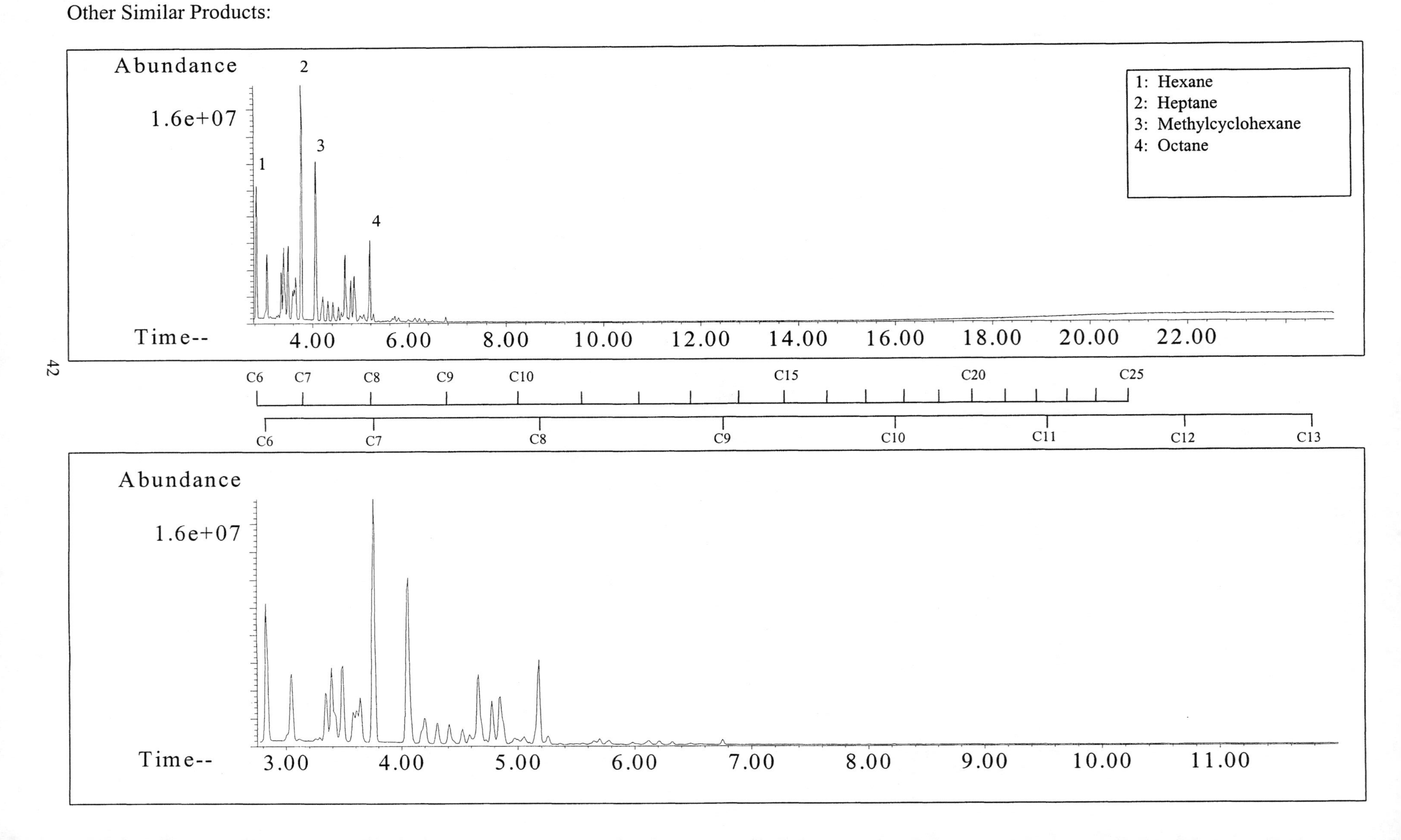

SUMMED PROFILES　　Distillate #3

ALKANES

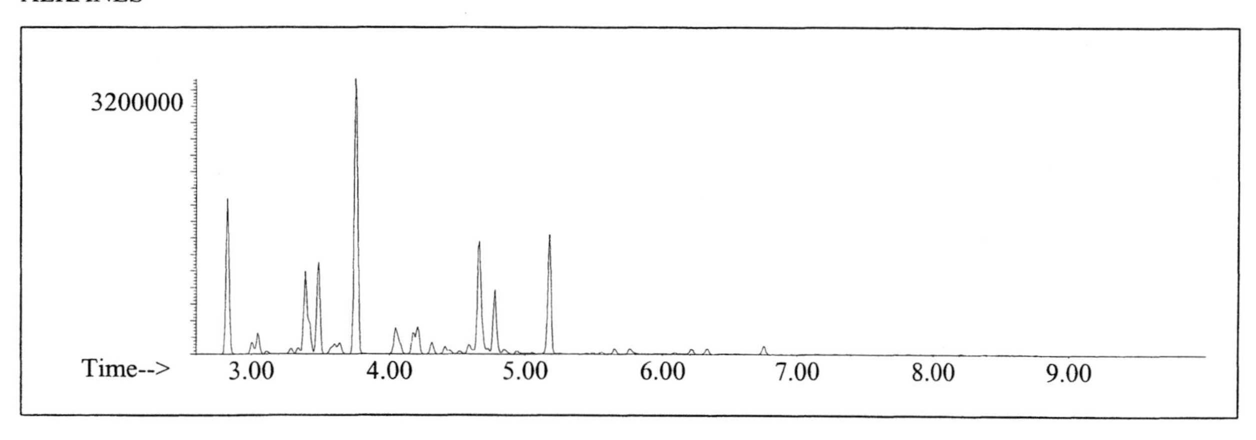

AROMATICS

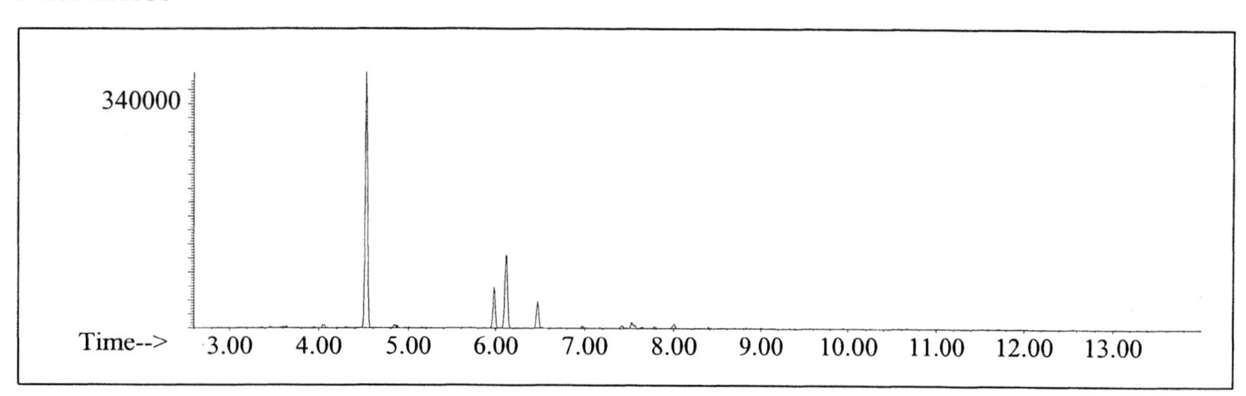

CYCLOPARAFFINS AND ALKENES

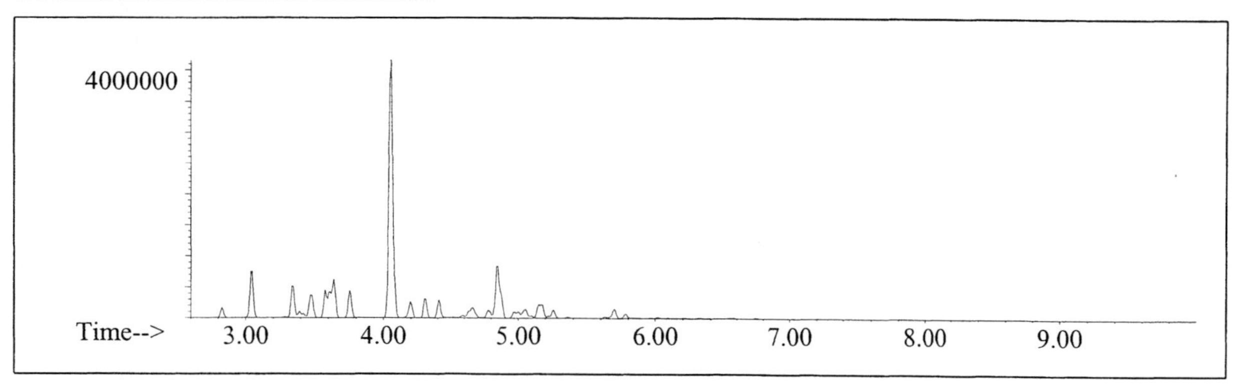

NAPHTHALENES

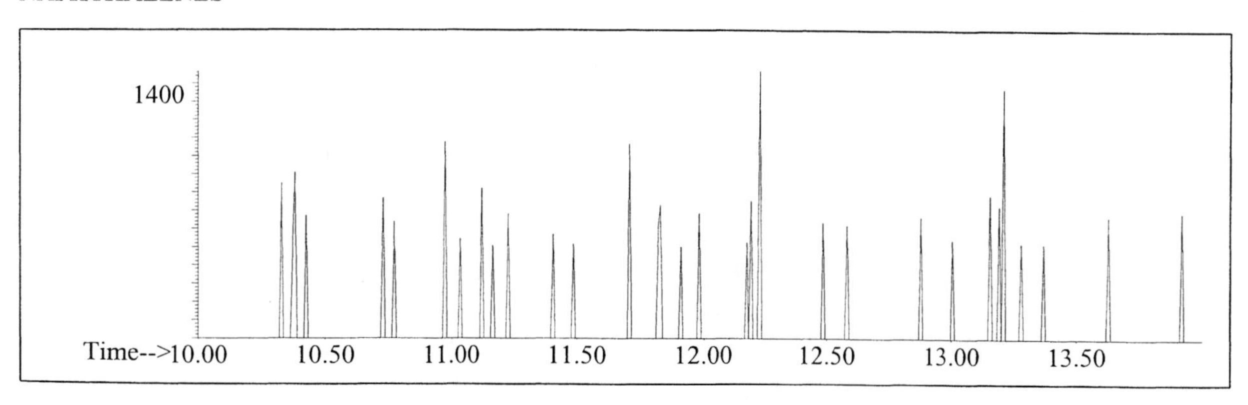

INDIVIDUAL PROFILES

Distillate #3

Alkane

Ion 43

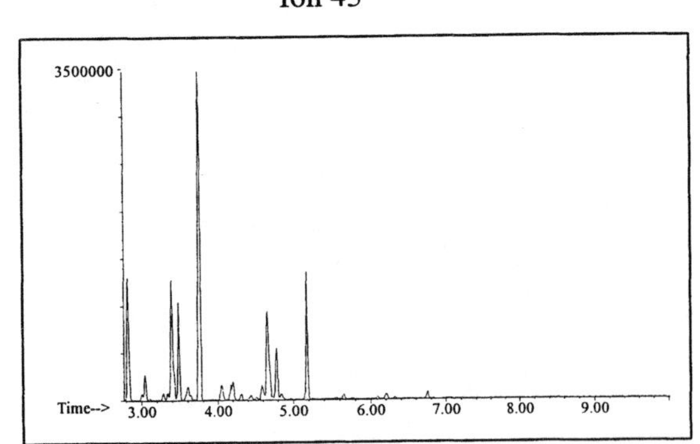

Ion 57

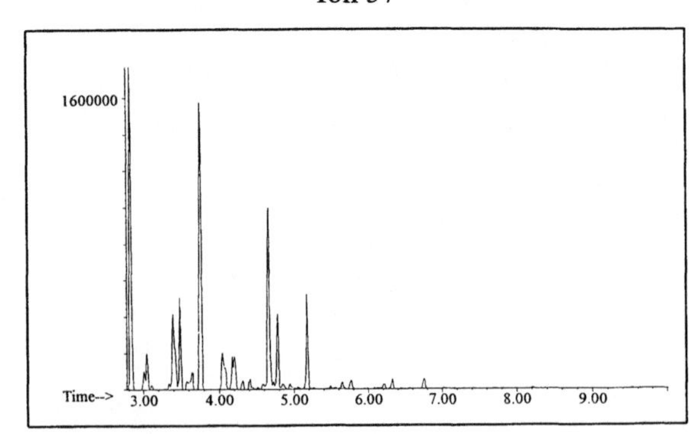

Ion 71

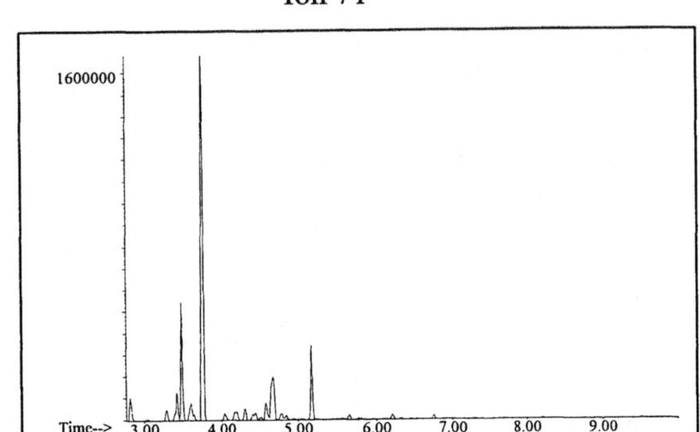

Ion 85

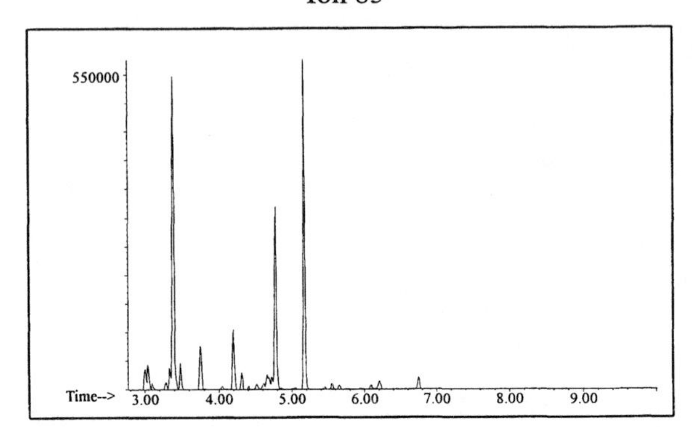

Aromatic

Ion 91

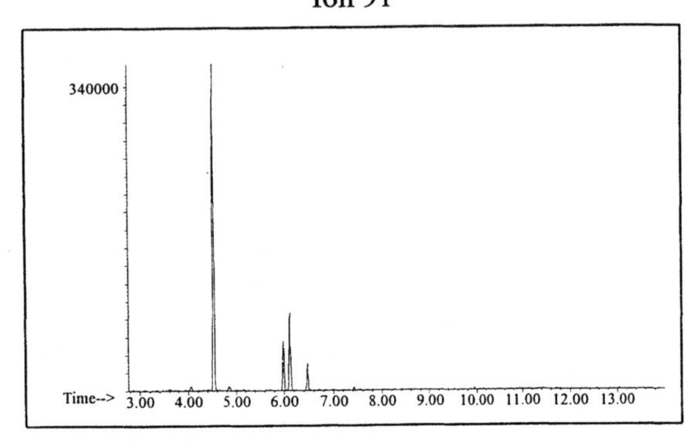

Ion 105

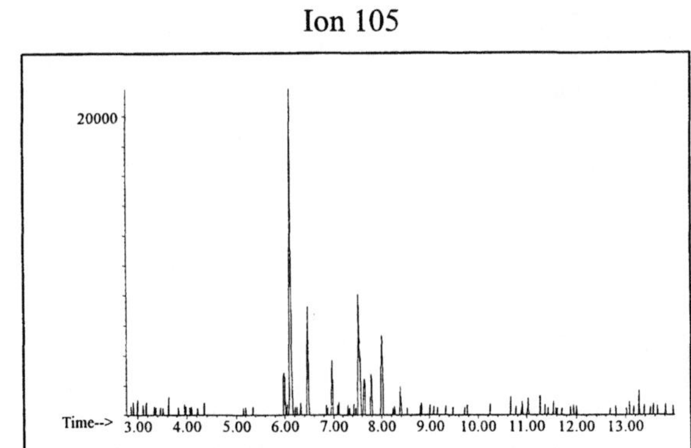

Ion 119

No Useful Data Obtained

INDIVIDUAL PROFILES

Distillate #3

Cycloparaffin

Ion 55

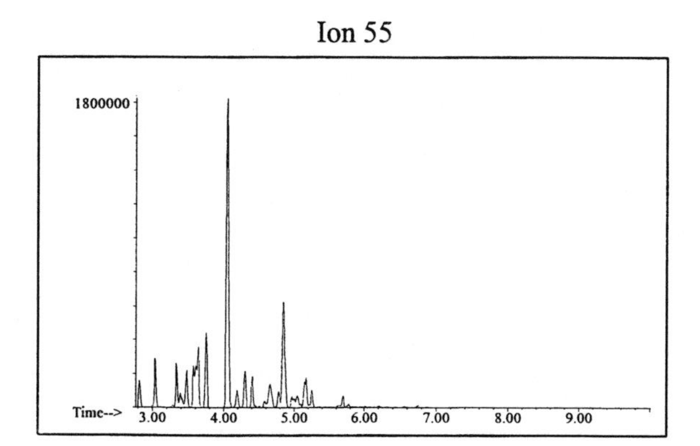

Ion 69

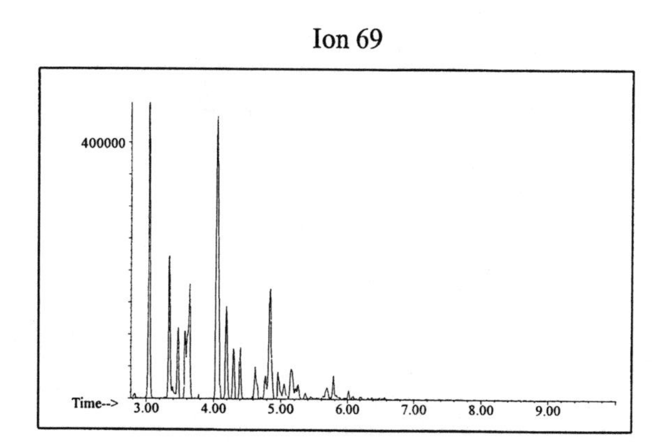

Ion 83

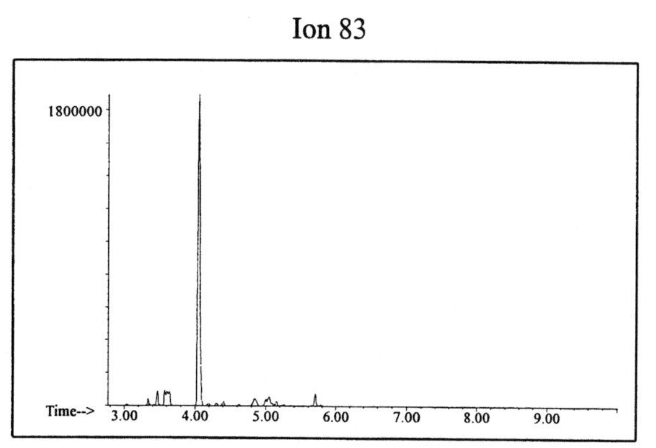

Naphthalene

Ion 128

No Useful Data Obtained

Ion 142

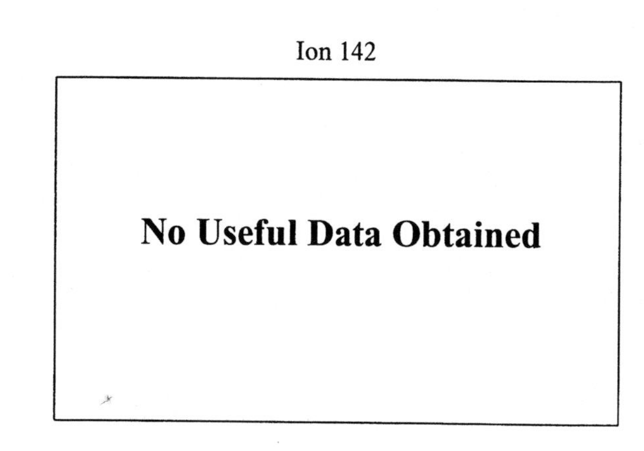

Ion 156

No Useful Data Obtained

Distillate #4

20uL/mL Pentane

Product Displayed: Energine Spot Remover

Other Similar Products:

ASTM: Class 1 (Light Petroleum Distillate)

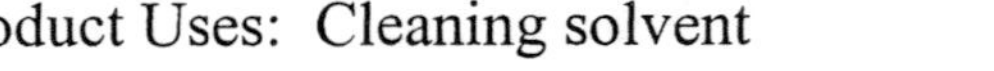

Product Uses: Cleaning solvent

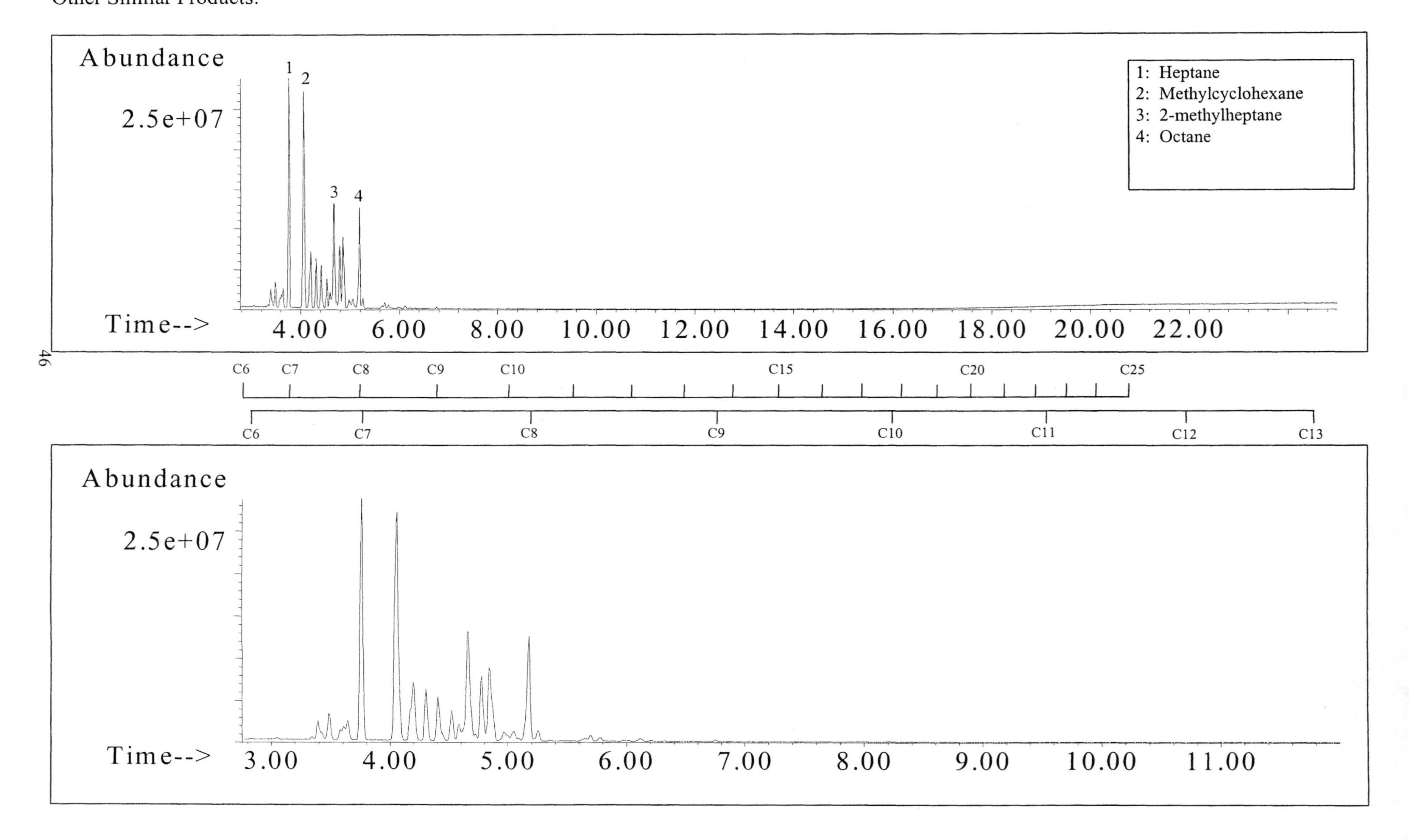

SUMMED PROFILES

Distillate #4

ALKANES

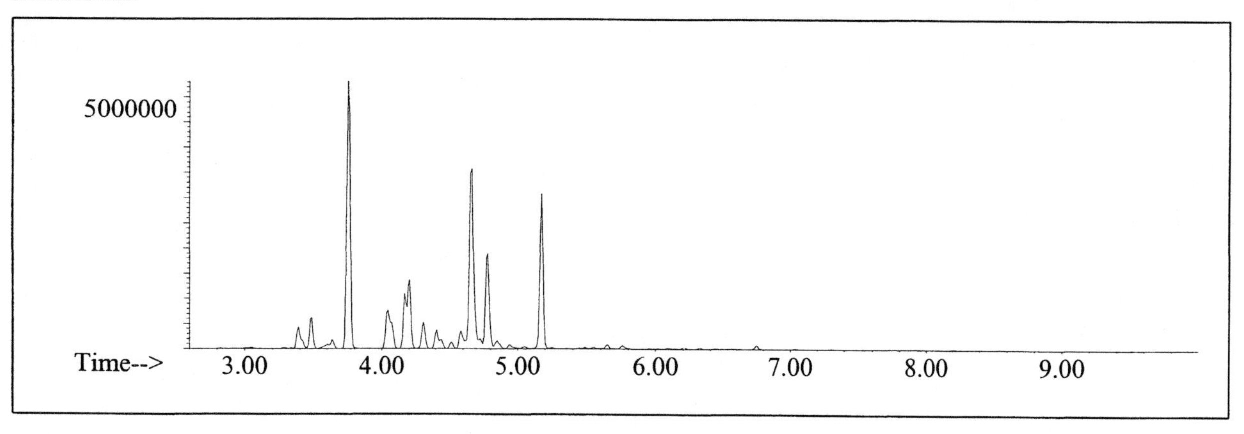

AROMATICS

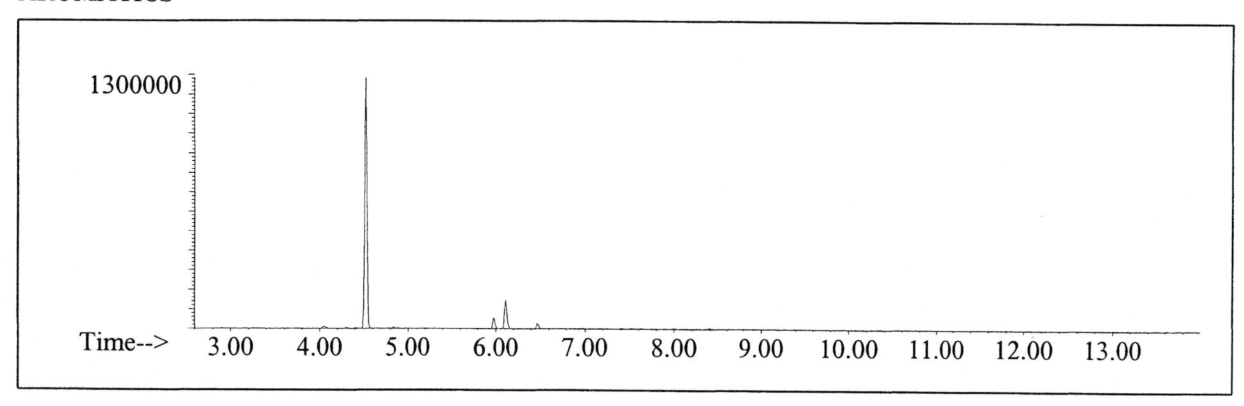

CYCLOPARAFFINS AND ALKENES

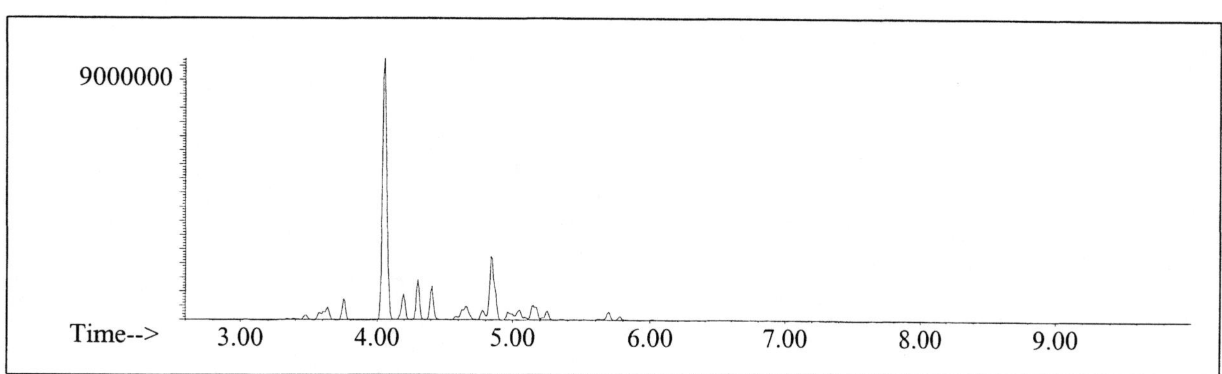

NAPHTHALENES

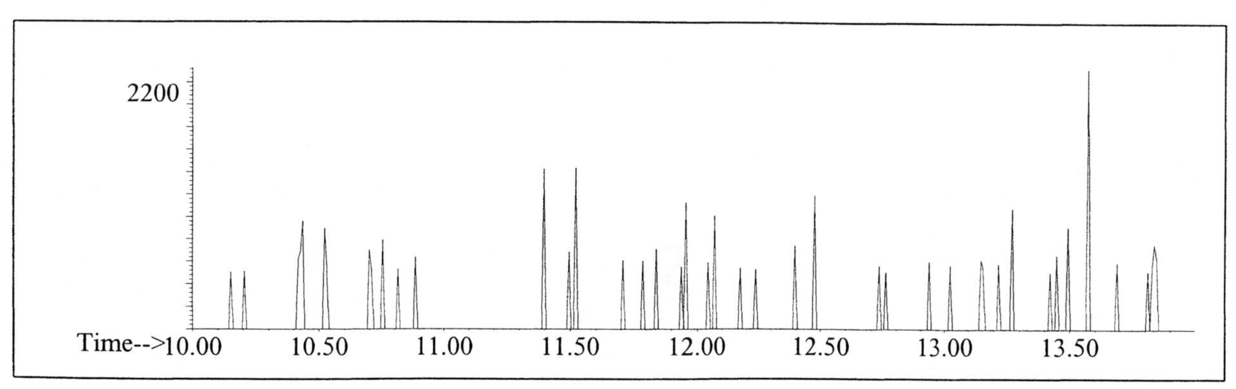

INDIVIDUAL PROFILES

Distillate #4

Alkane

Ion 43

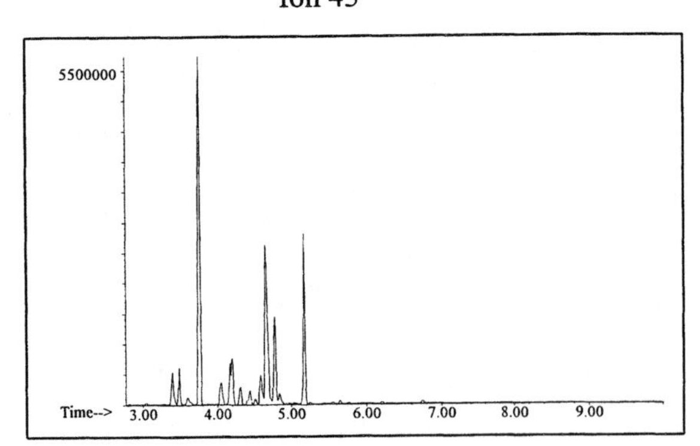

Ion 57

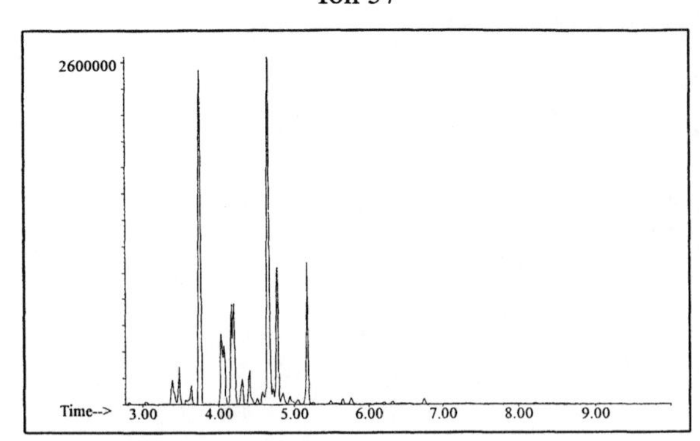

Ion 71

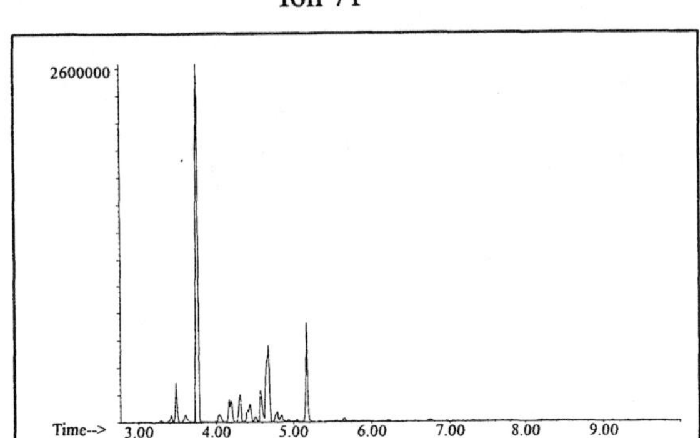

Ion 85

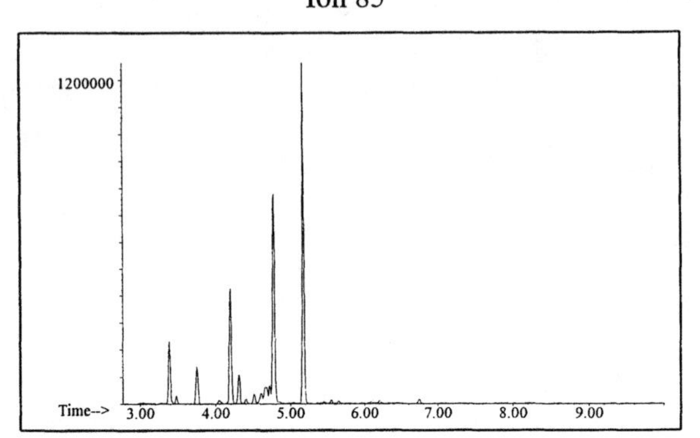

Aromatic

Ion 91

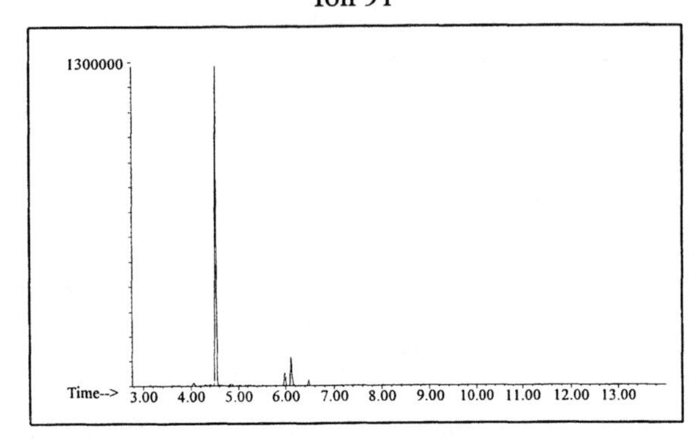

Ion 105

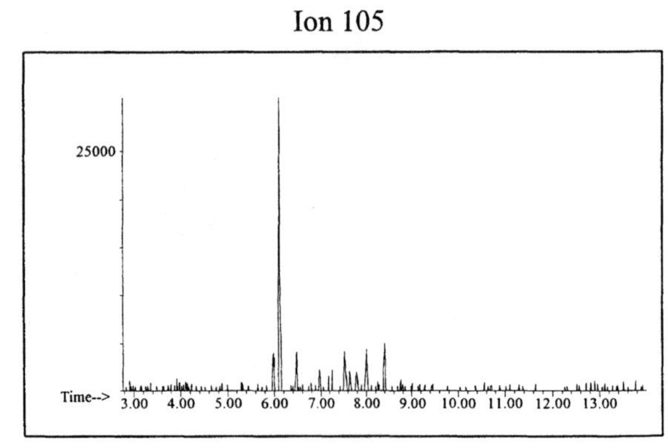

Ion 119

No Useful Data Obtained

INDIVIDUAL PROFILES

Distillate #4

Cycloparaffin

Ion 55

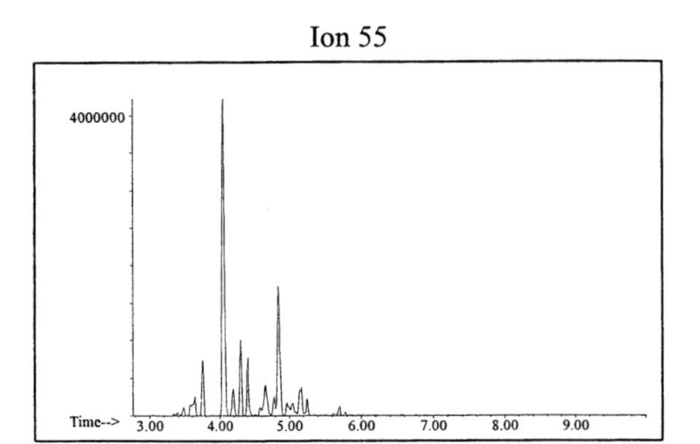

Ion 69

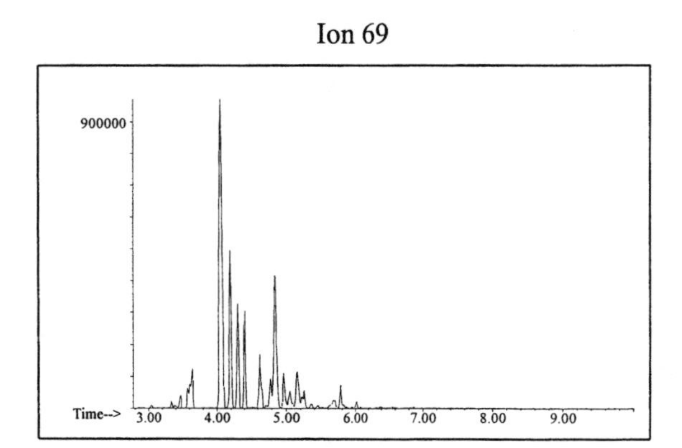

Ion 83

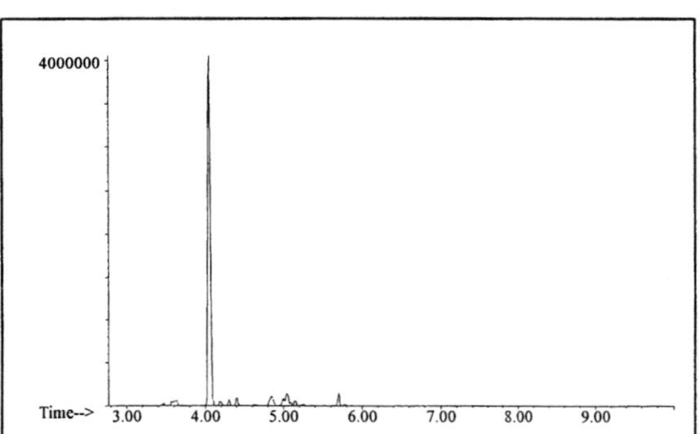

Naphthalene

Ion 128

No Useful Data Obtained

Ion 142

No Useful Data Obtained

Ion 156

No Useful Data Obtained

Distillate #5

20uL/mL Pentane

Product Displayed: Fred Meyer Camp Fuel

Other Similar Products:

ASTM: Class 1 (Light Petroleum Distillate)

Product Uses: Camping fuel

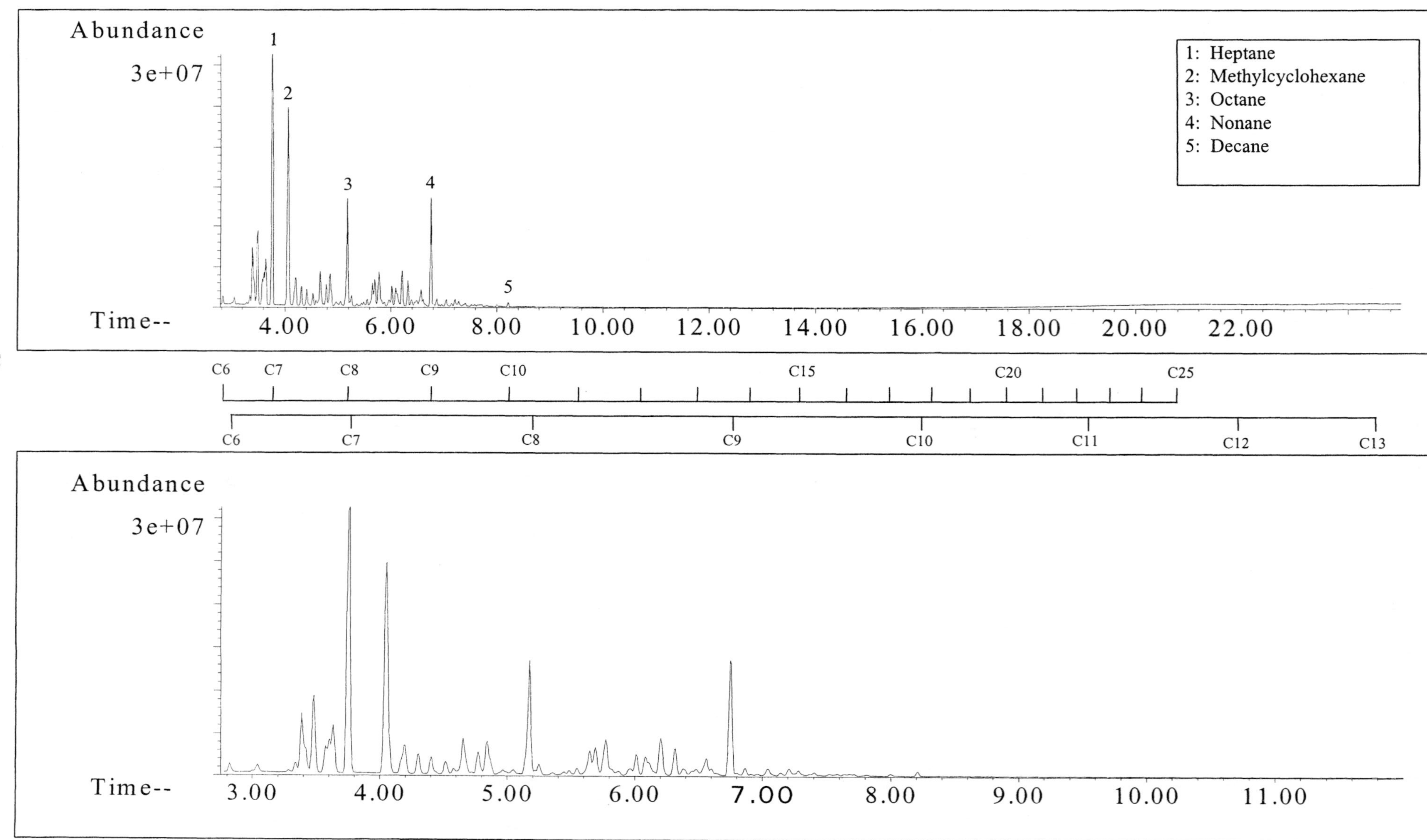

SUMMED PROFILES Distillate #5

ALKANES

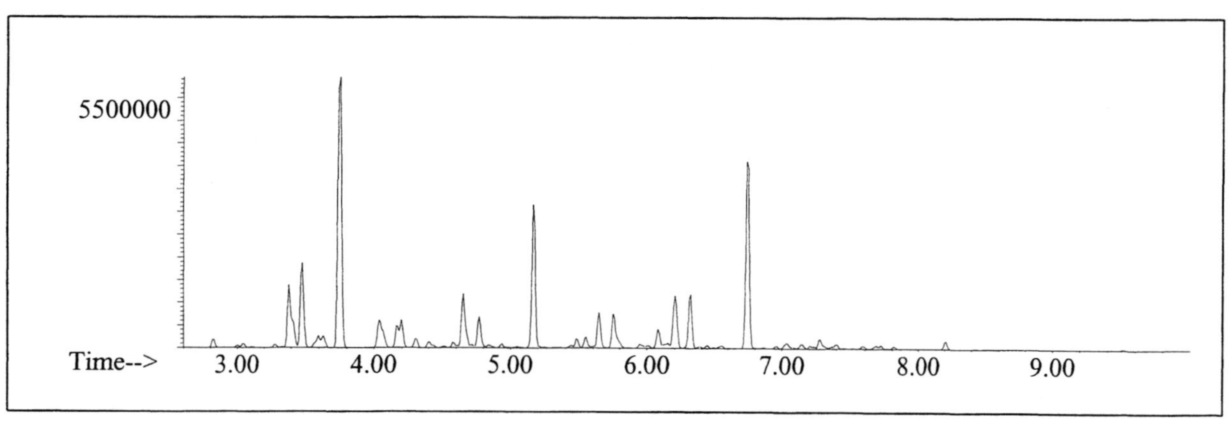

AROMATICS

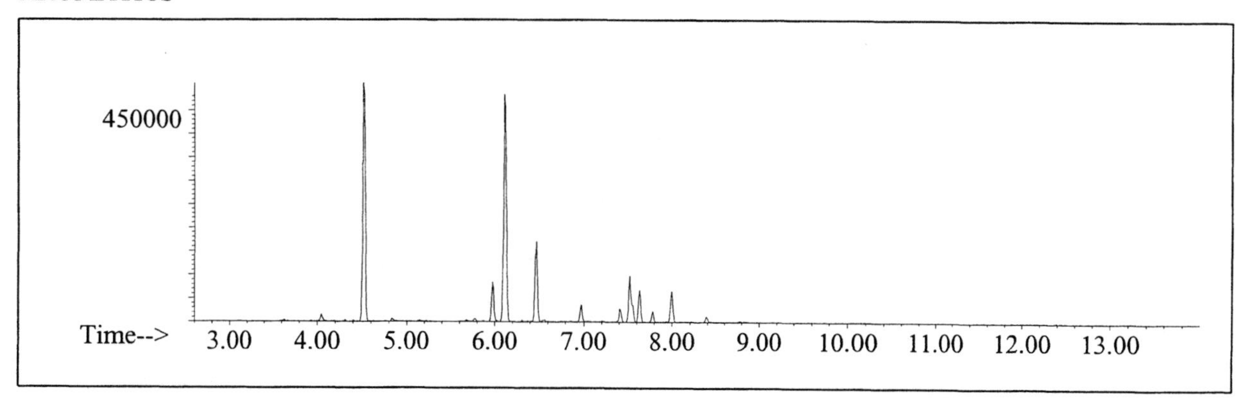

CYCLOPARAFFINS AND ALKENES

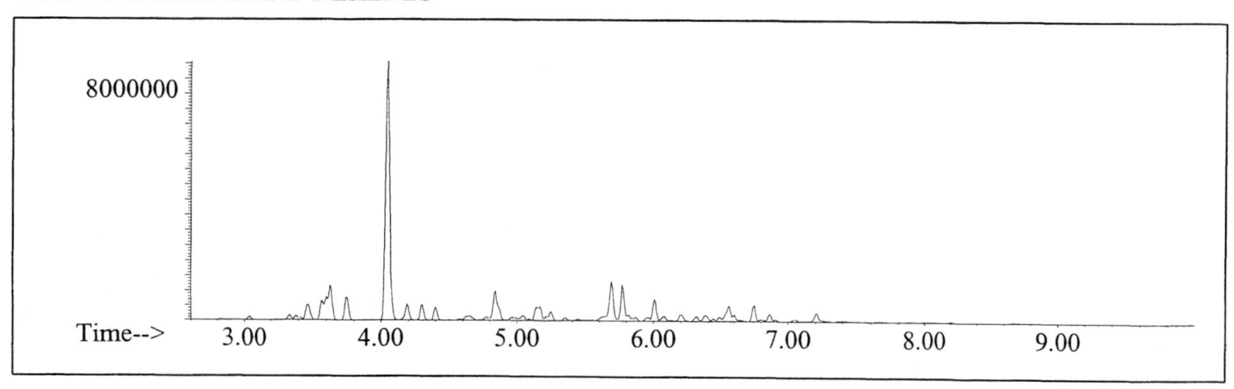

NAPHTHALENES

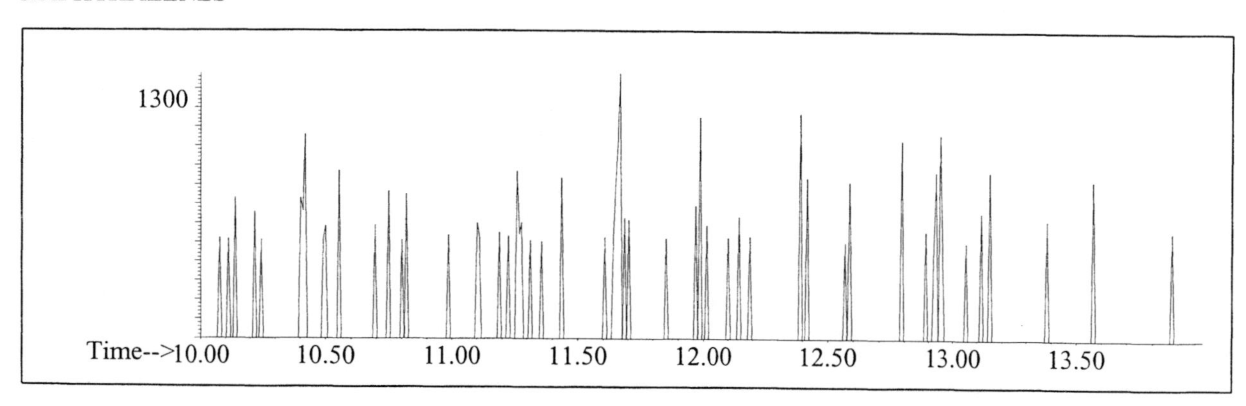

INDIVIDUAL PROFILES

Distillate #5

Alkane

Ion 43

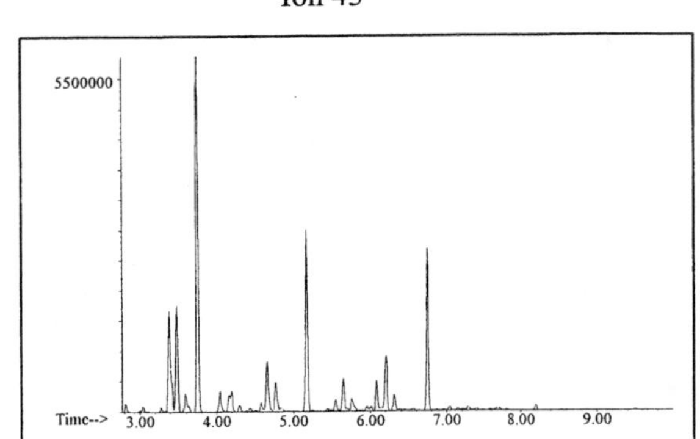

Ion 57

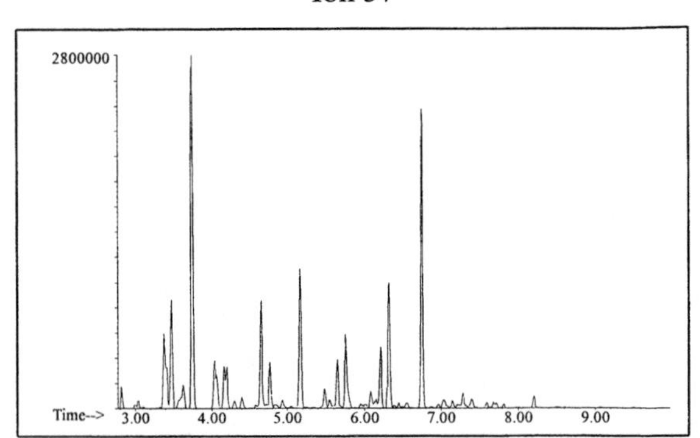

Ion 71

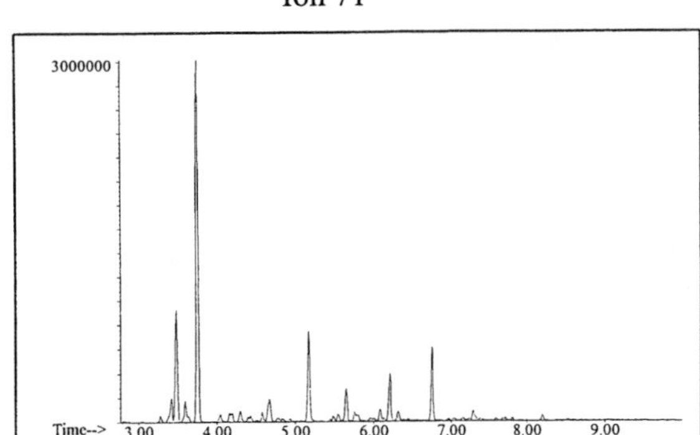

Ion 85

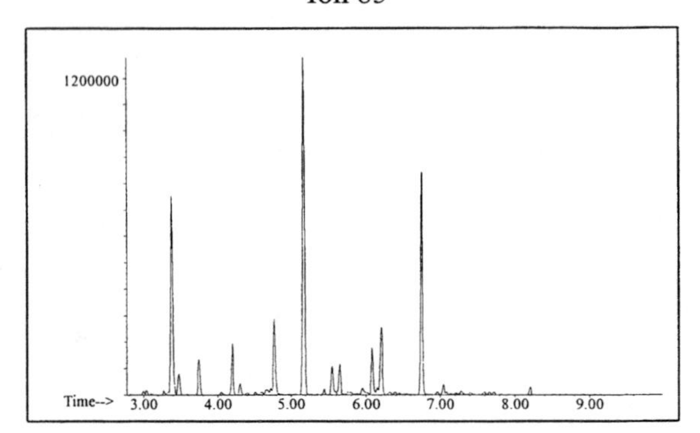

Aromatic

Ion 91

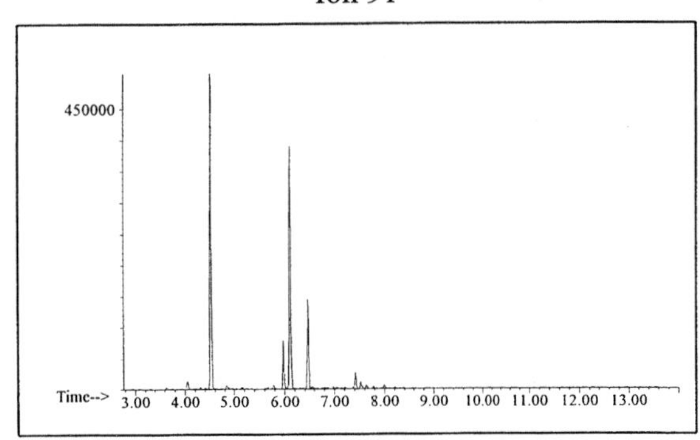

Ion 105

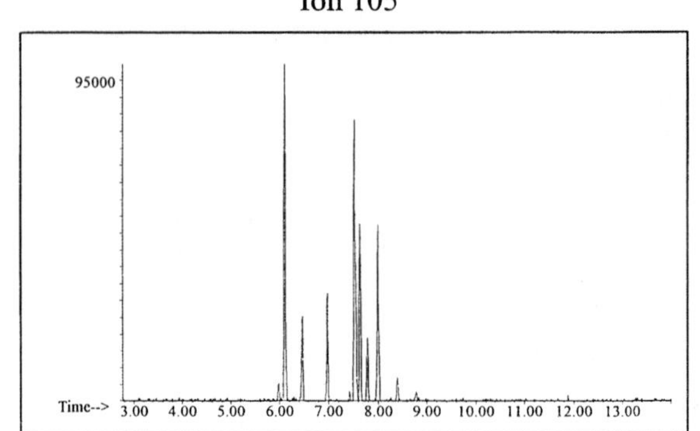

Ion 119

No Useful Data Obtained

INDIVIDUAL PROFILES

Distillate #5

Cycloparaffin

Ion 55

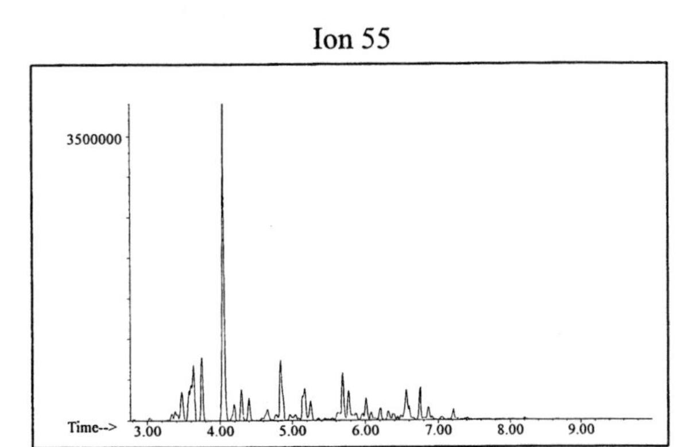

Ion 69

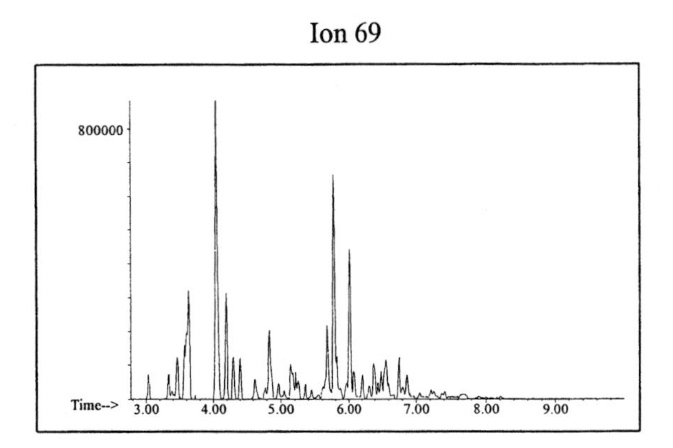

Ion 83

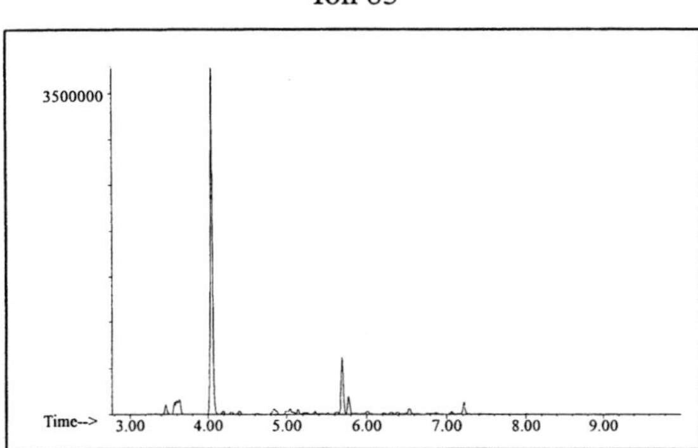

Naphthalene

Ion 128

No Useful Data Obtained

Ion 142

No Useful Data Obtained

Ion 156

No Useful Data Obtained

Distillate #6

20uL/mL Pentane

ASTM: Class 1 (Light Petroleum Distillate)

Macro Code: L

Product Displayed: Zippo Lighter Fluid #1

Product Uses: Pocket lighter fluid

Other Similar Products:

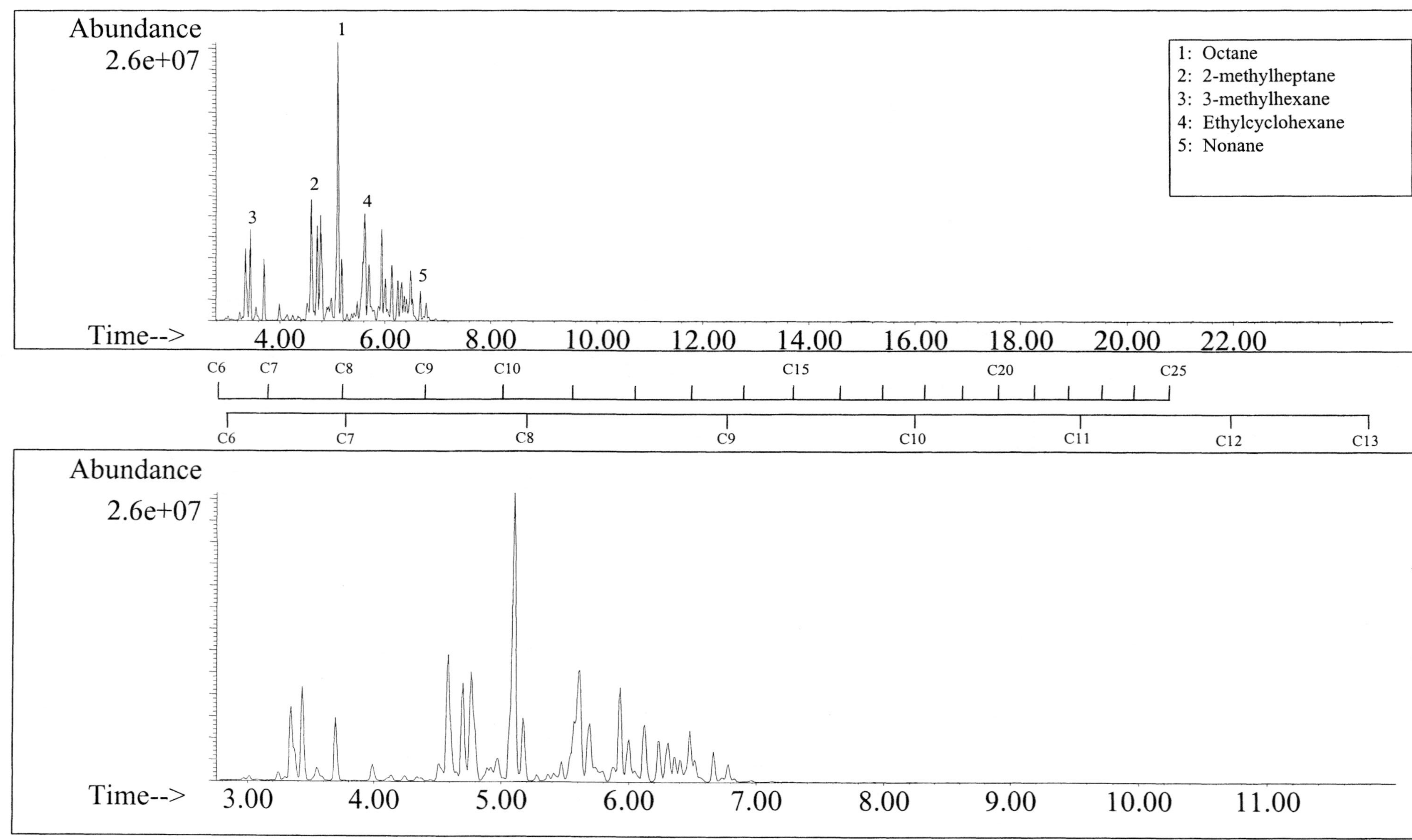

SUMMED PROFILES

Distillate #6

ALKANES

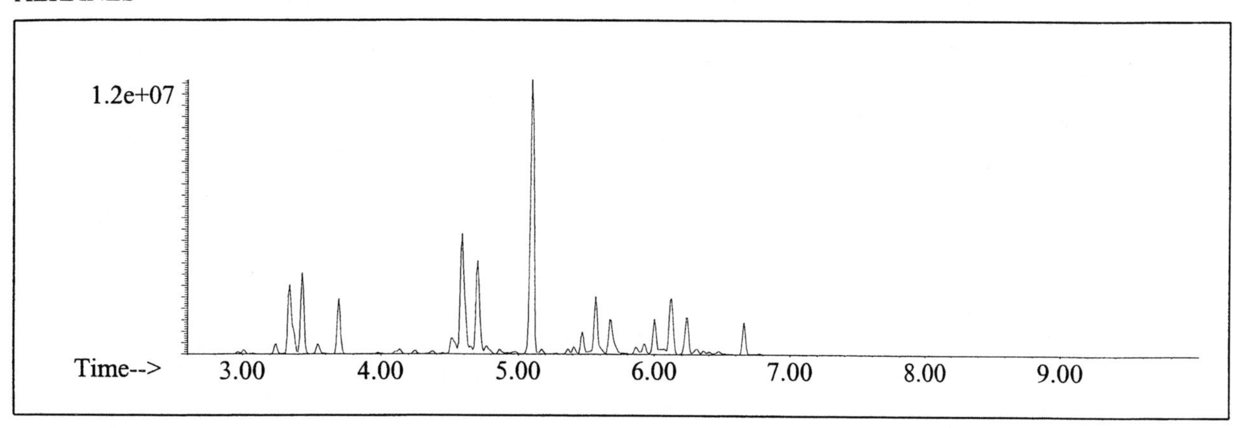

AROMATICS

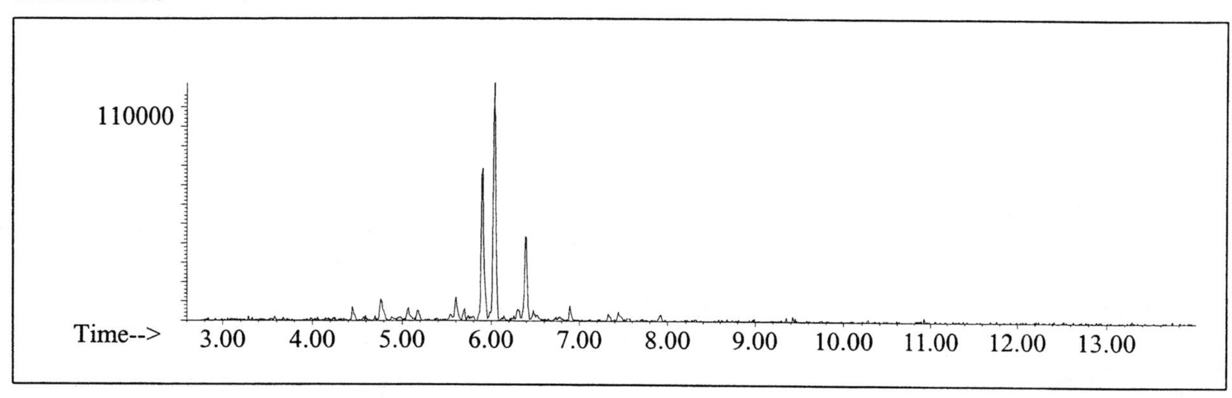

CYCLOPARAFFINS AND ALKENES

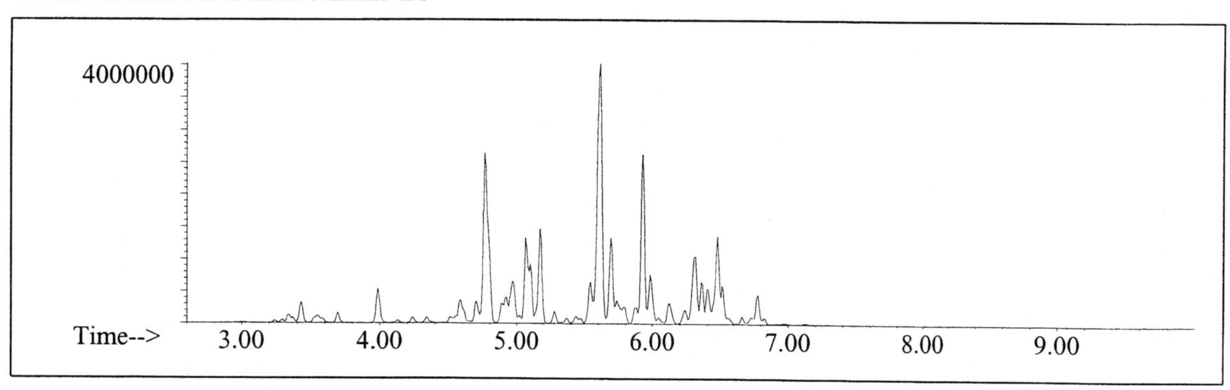

NAPHTHALENES

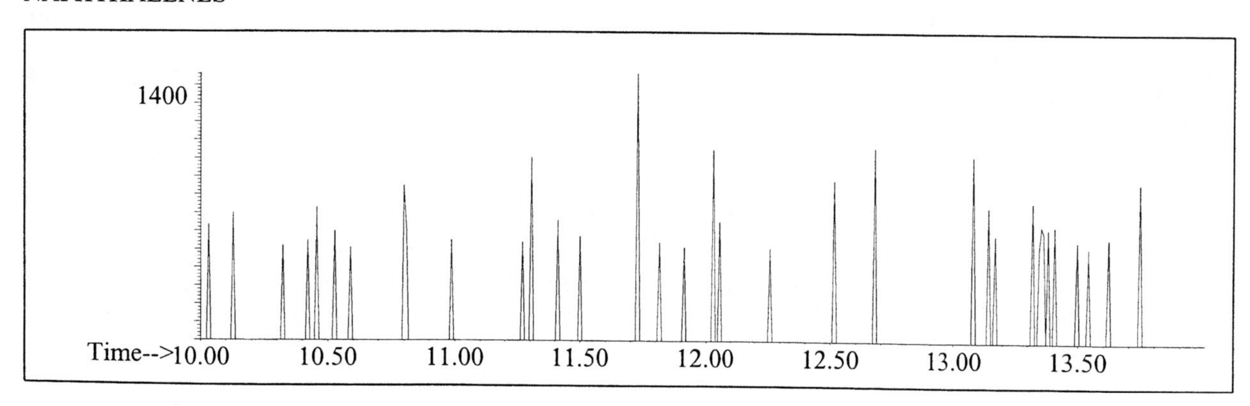

INDIVIDUAL PROFILES

Distillate #6

Alkane

Ion 43

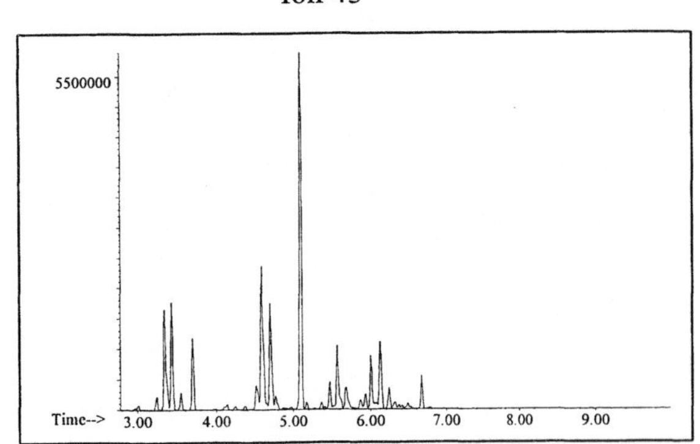

Ion 57

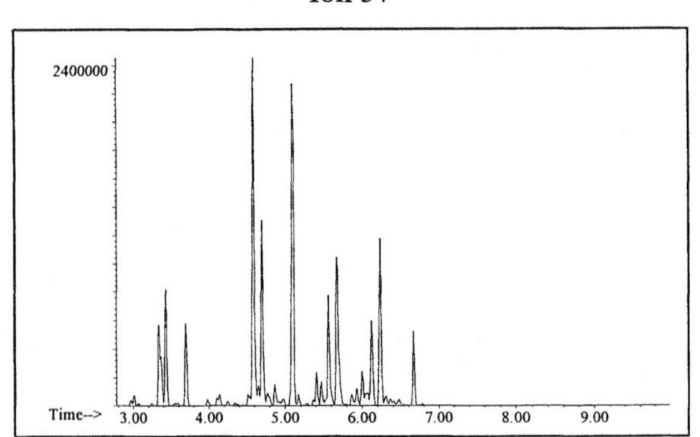

Ion 71

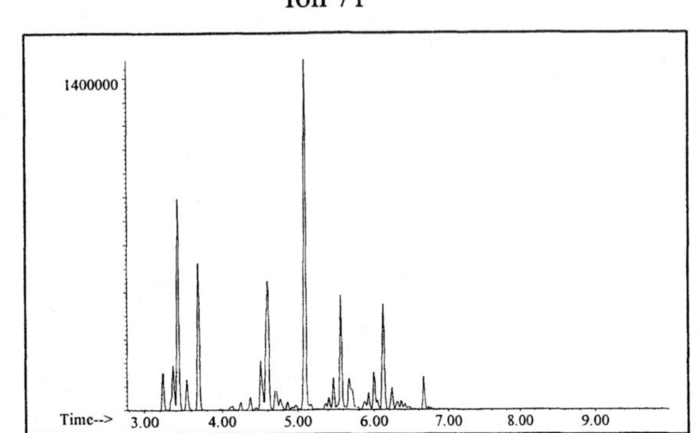

Ion 85

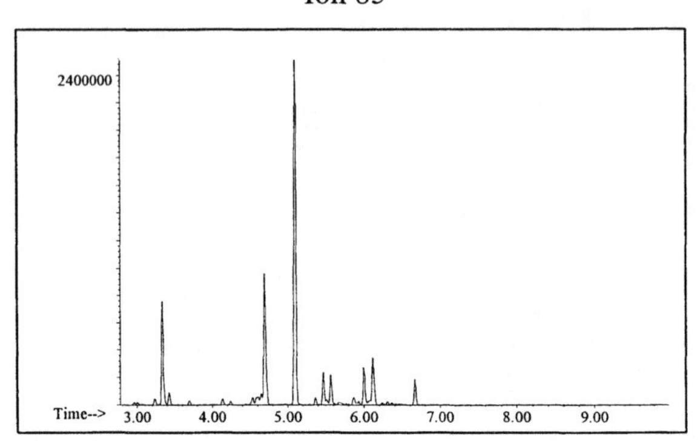

Aromatic

Ion 91

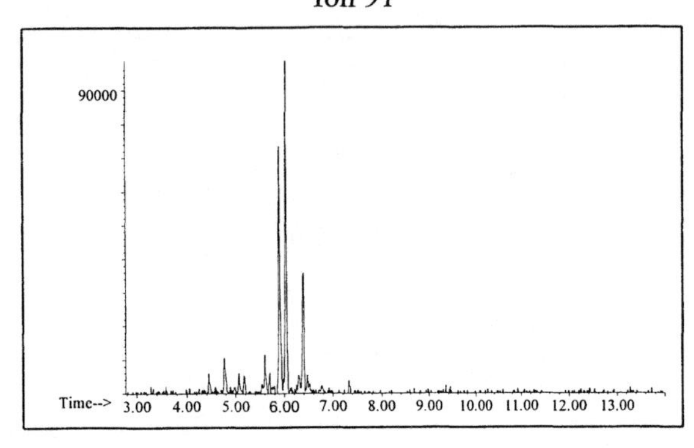

Ion 105

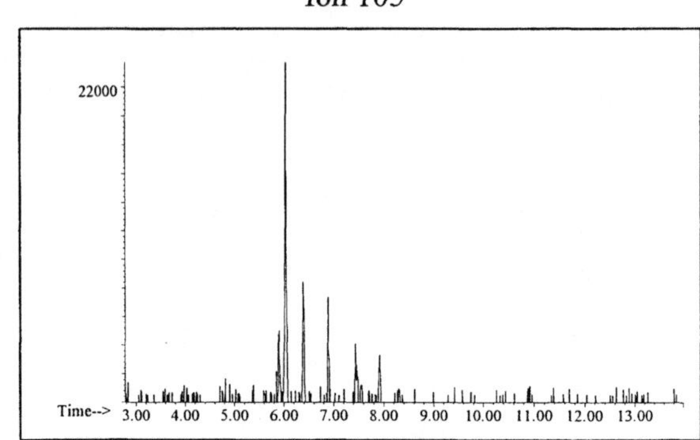

Ion 119

No Useful Data Obtained

INDIVIDUAL PROFILES

Distillate #6

Cycloparaffin

Ion 55

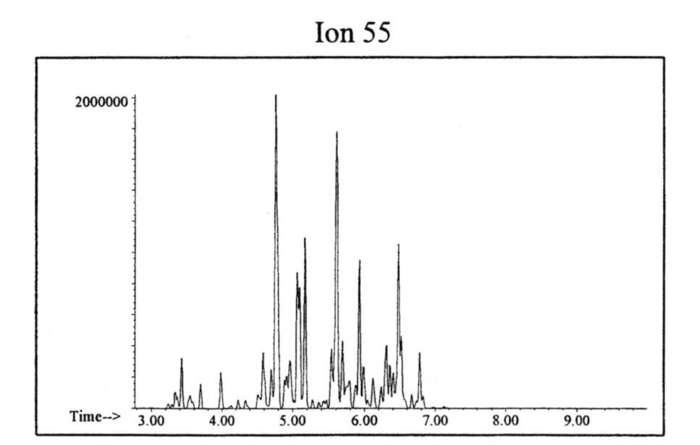

Ion 69

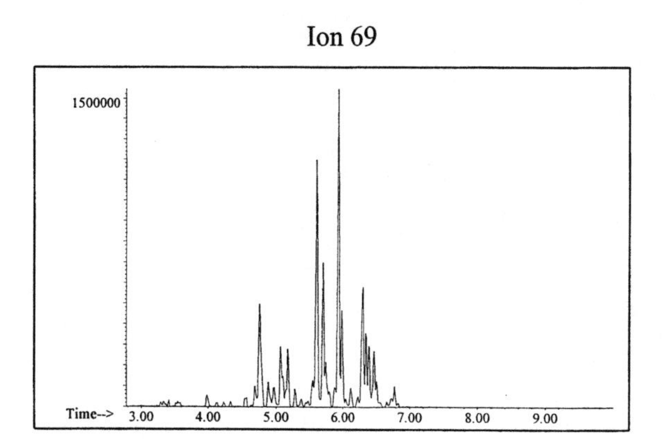

Ion 83

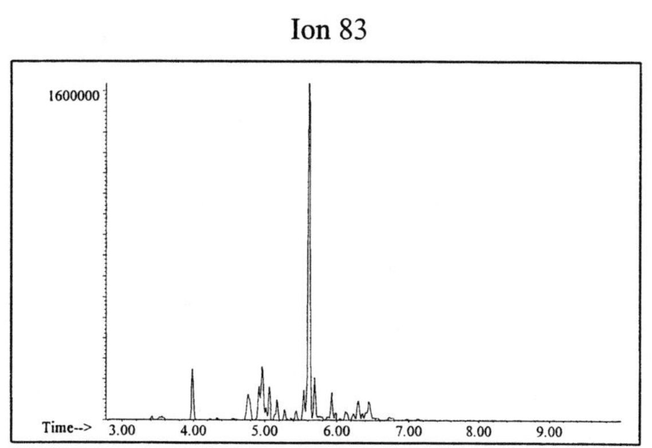

Naphthalene

Ion 128

No Useful Data Obtained

Ion 142

No Useful Data Obtained

Ion 156

No Useful Data Obtained

Distillate #7

20uL/mL Pentane

Product Displayed: 25% Evaporated Zippo Lighter Fluid #1

Other Similar Products:

ASTM: Class 1 (Light Petroleum Distillate)

Product Uses: Pocket lighter fluid

Macro Code: L

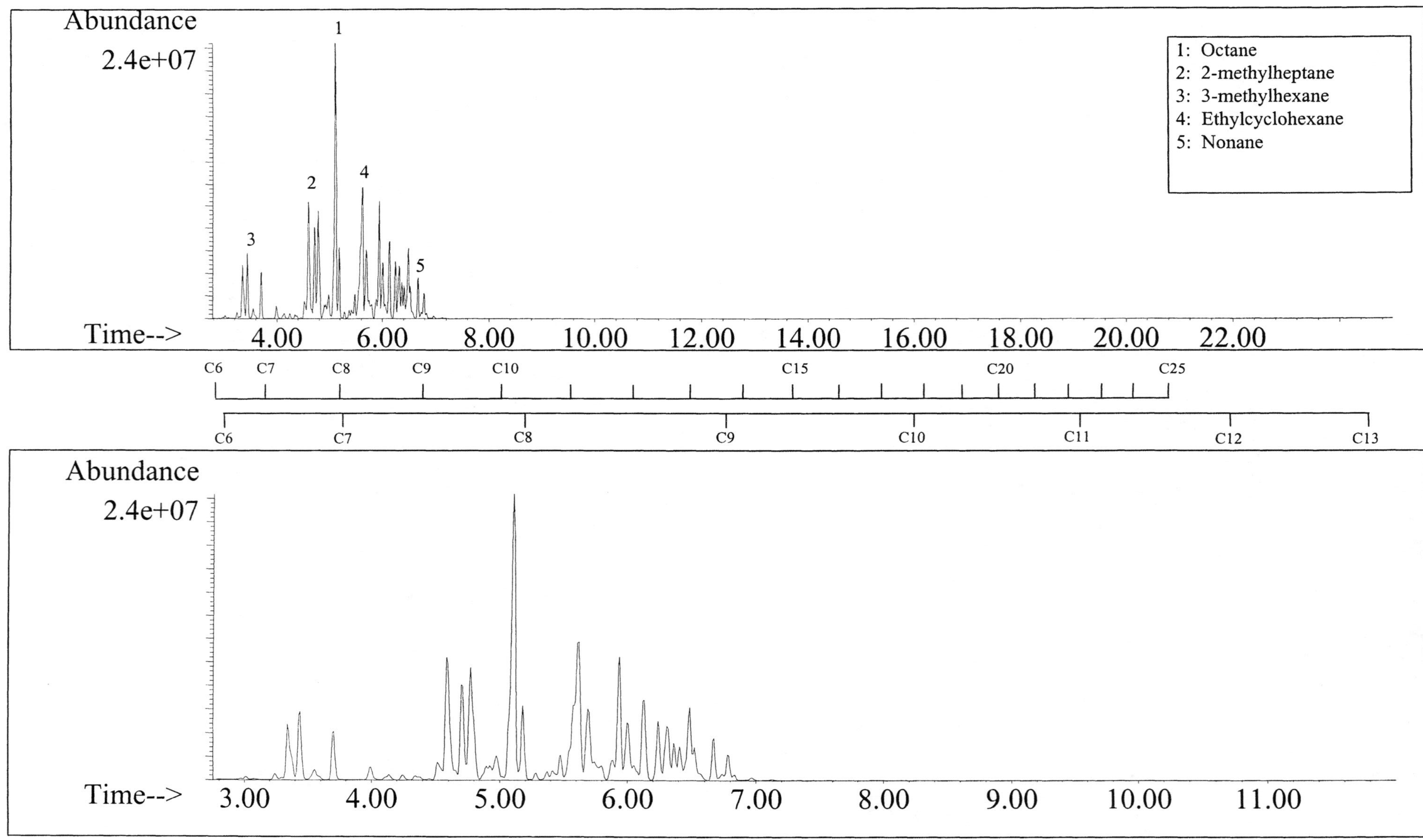

SUMMED PROFILES Distillate #7

ALKANES

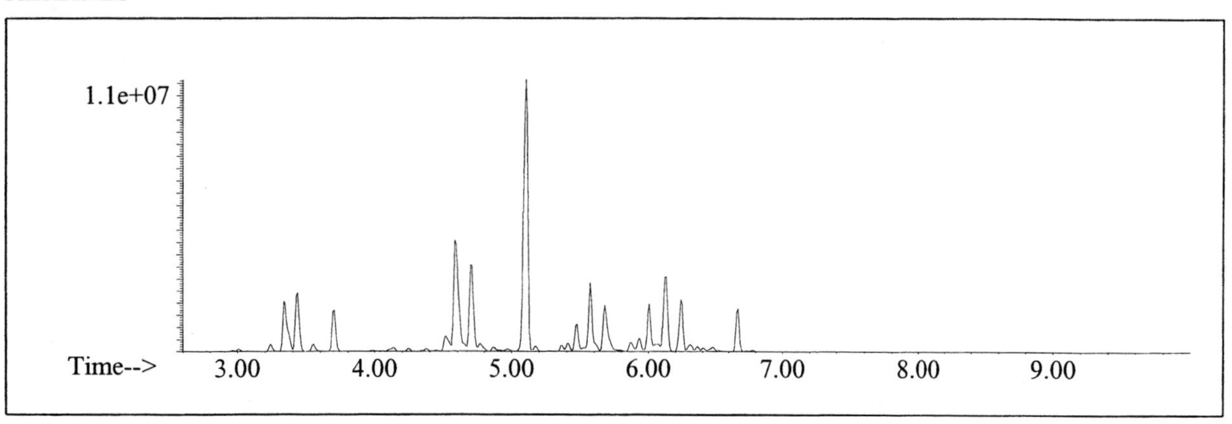

AROMATICS

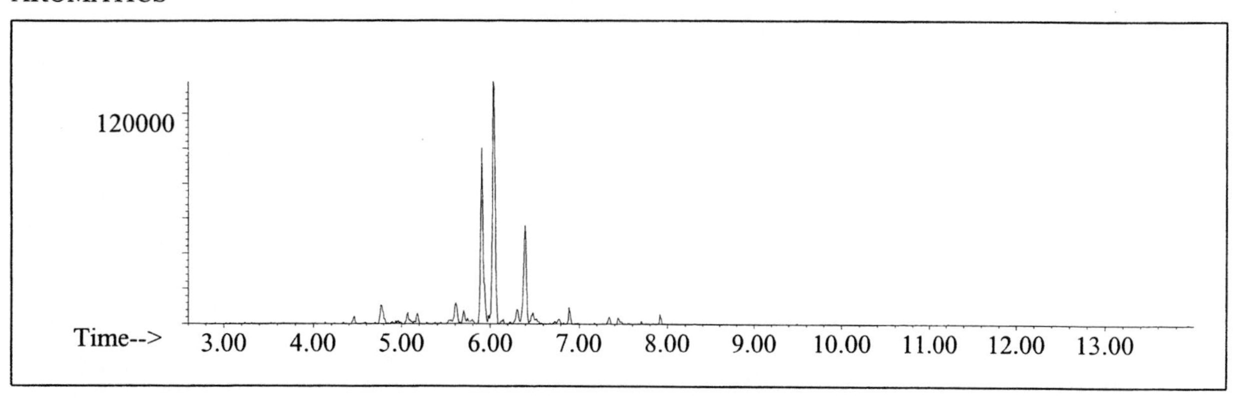

CYCLOPARAFFINS AND ALKENES

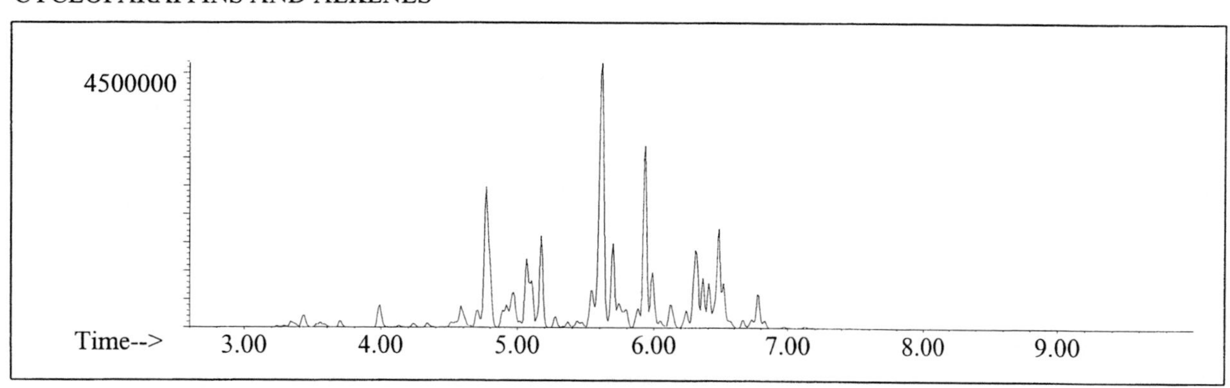

NAPHTHALENES

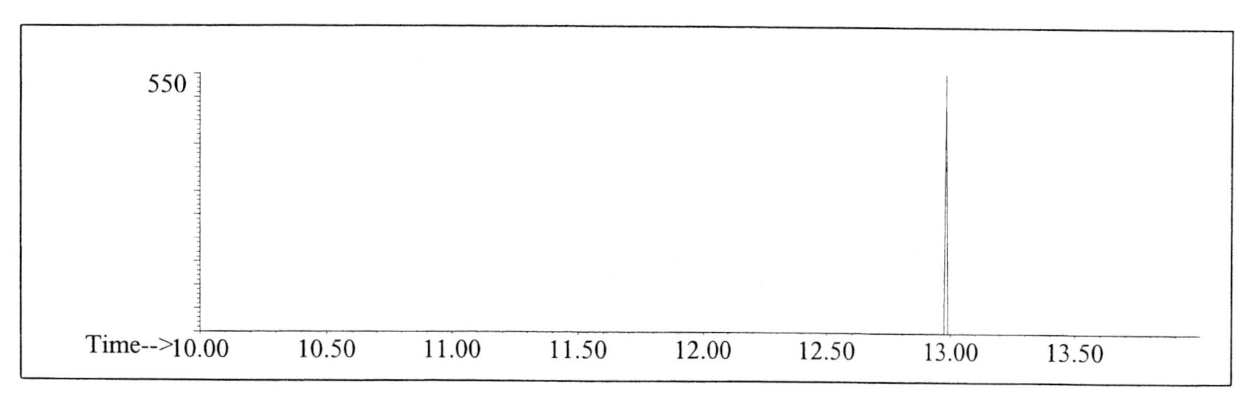

INDIVIDUAL PROFILES Distillate #7

Alkane

Ion 43

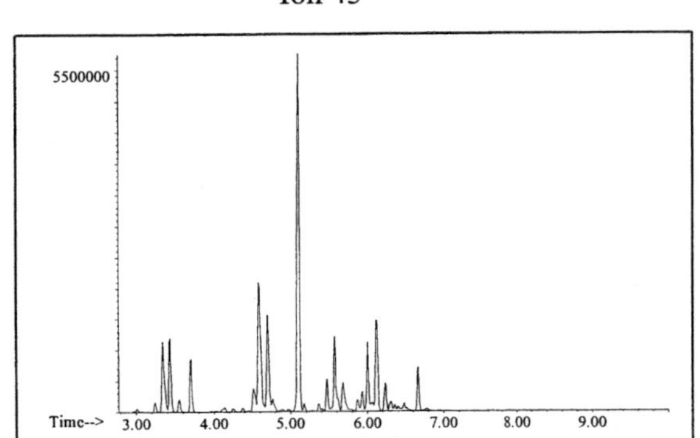

Ion 57

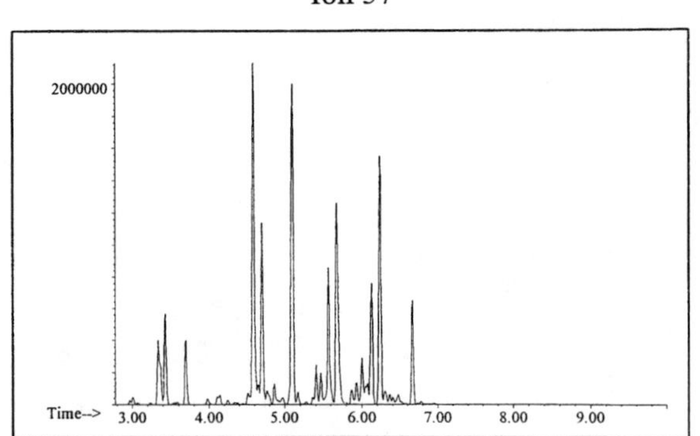

Ion 71

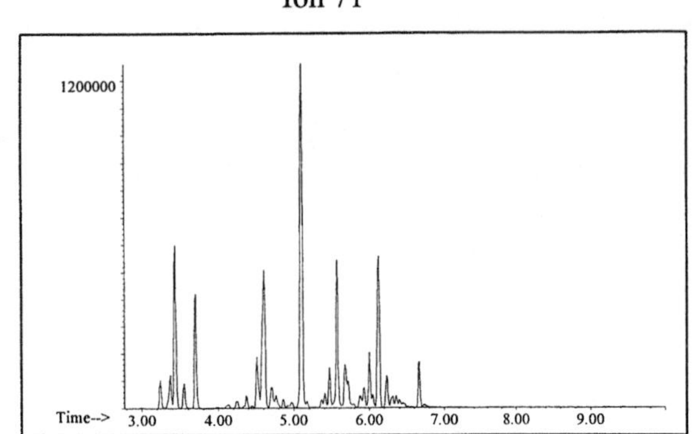

Ion 85

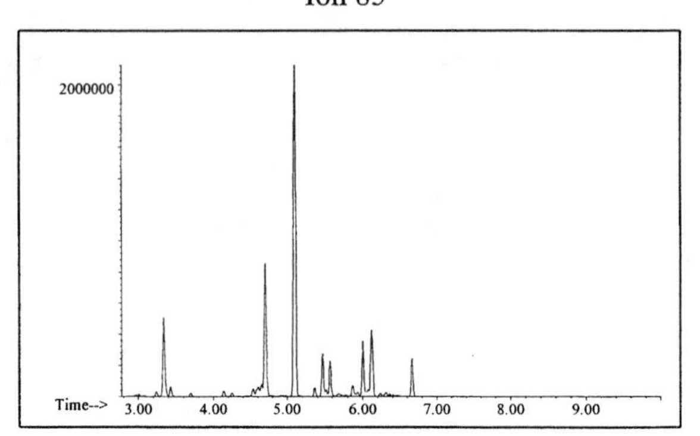

Aromatic

Ion 91

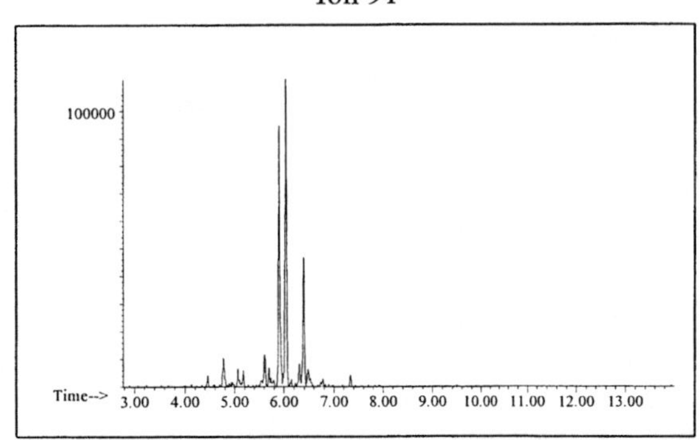

Ion 105

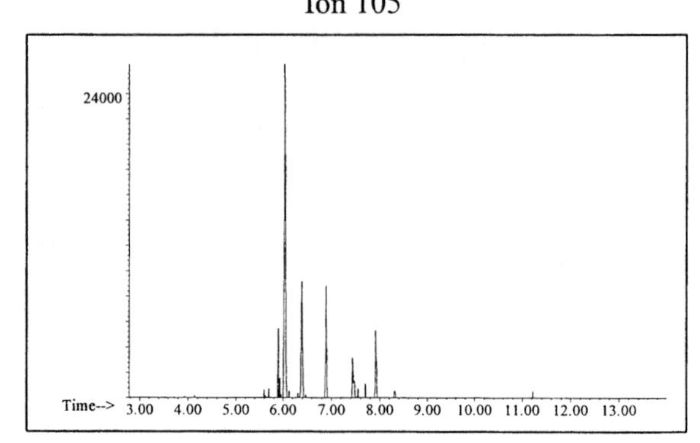

Ion 119

No Useful Data Obtained

INDIVIDUAL PROFILES Distillate #7

Cycloparaffin

Ion 55

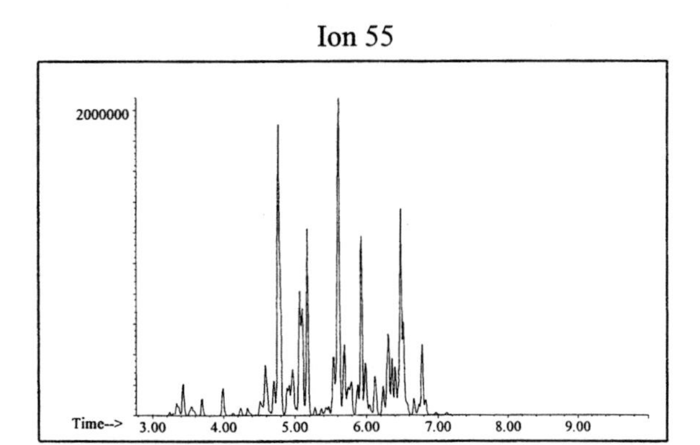

Ion 69

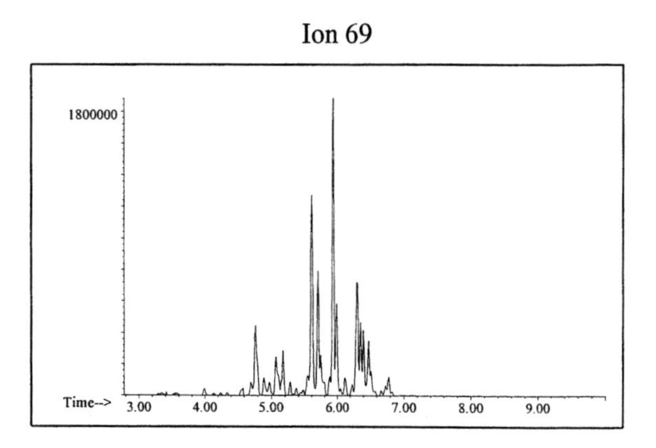

Ion 83

1800000
Time--> 3.00 4.00 5.00 6.00 7.00 8.00 9.00

Naphthalene

Ion 128

No Useful Data Obtained

Ion 142

No Useful Data Obtained

Ion 156

No Useful Data Obtained

Distillate #8

20uL/mL Pentane

Product Displayed: 50% Evaporated Zippo Lighter Fluid #1

Other Similar Products:

ASTM: Class 1 (Light Petroleum Distillate)

Product Uses: Pocket lighter fluid

Macro Code: L

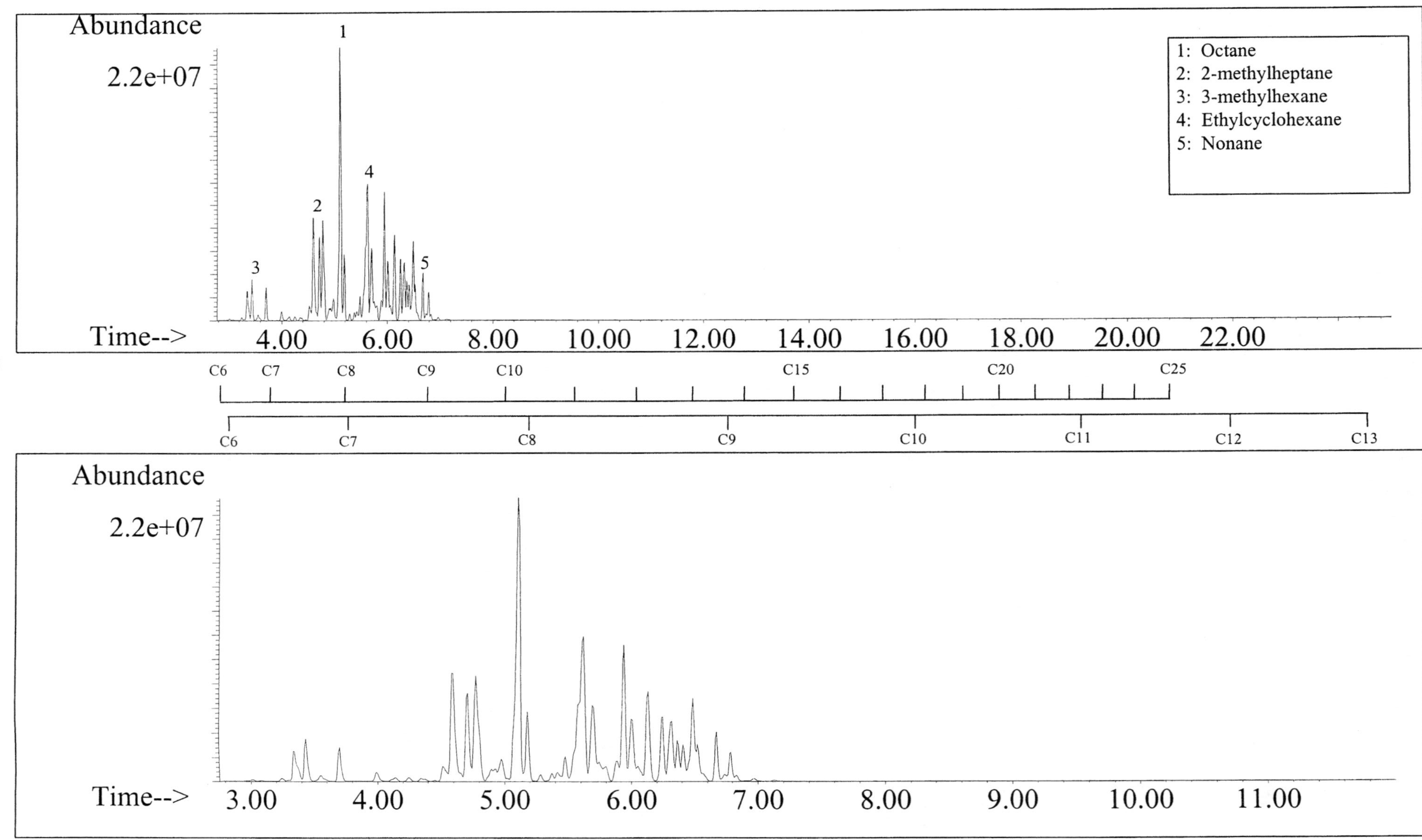

SUMMED PROFILES

Distillate #8

ALKANES

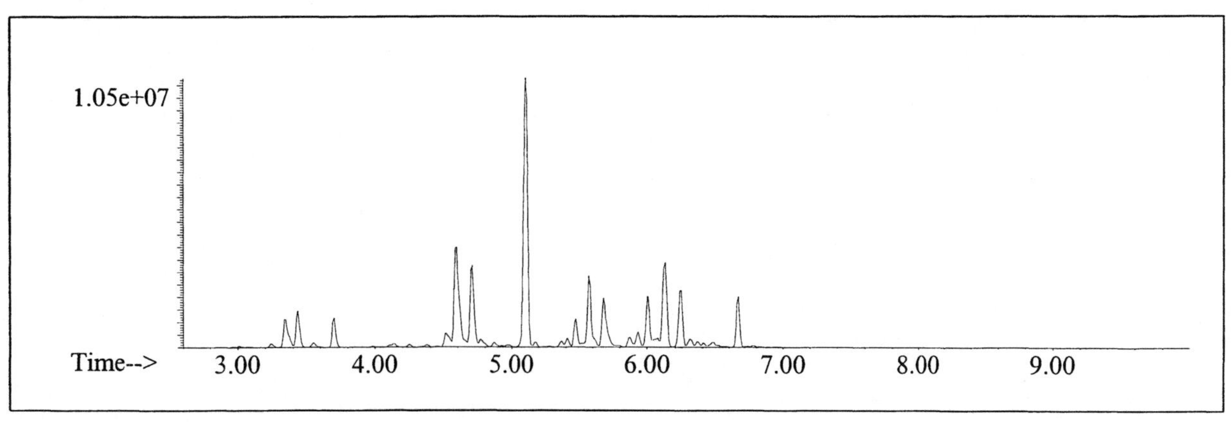

AROMATICS

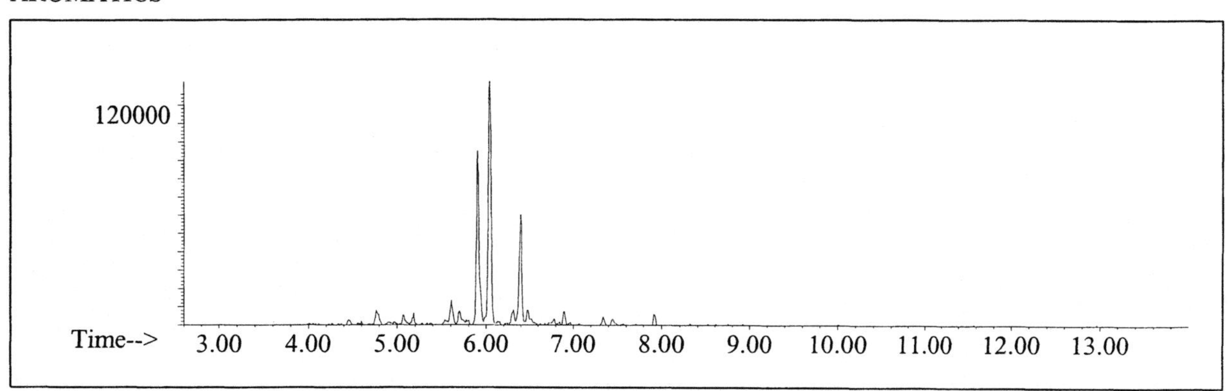

CYCLOPARAFFINS AND ALKENES

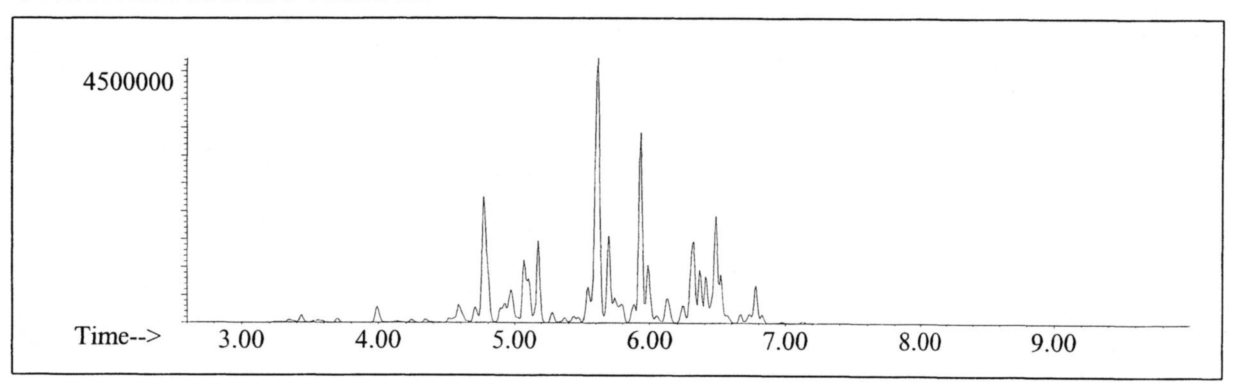

NAPHTHALENES

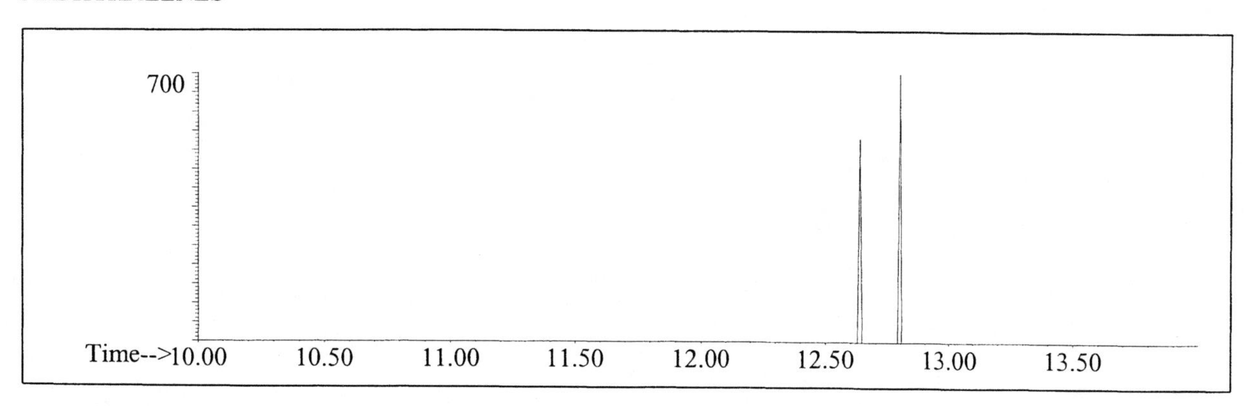

INDIVIDUAL PROFILES

Distillate #8

Alkane

Ion 43

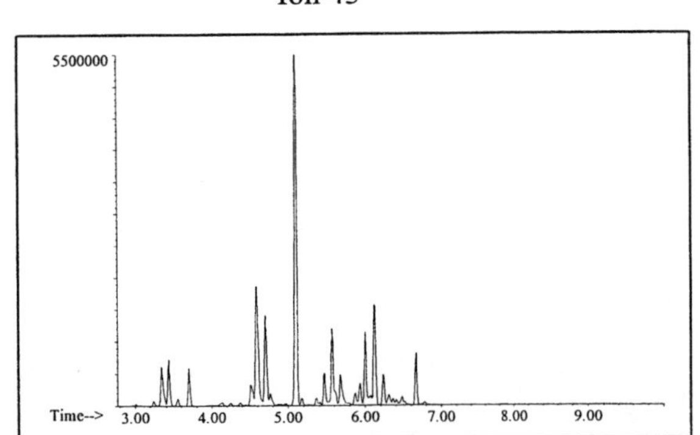

Ion 57

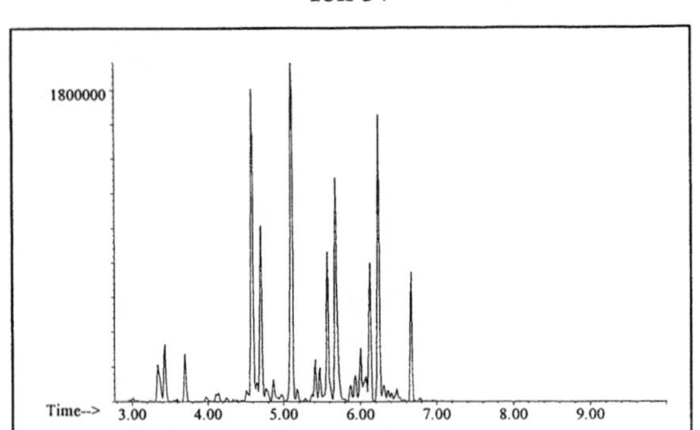

Ion 71

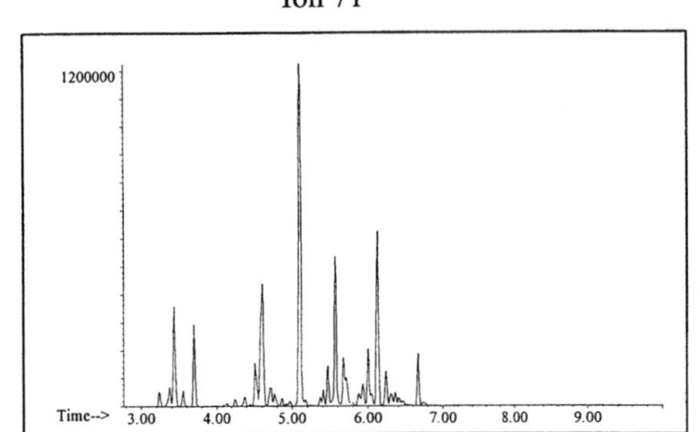

Ion 85

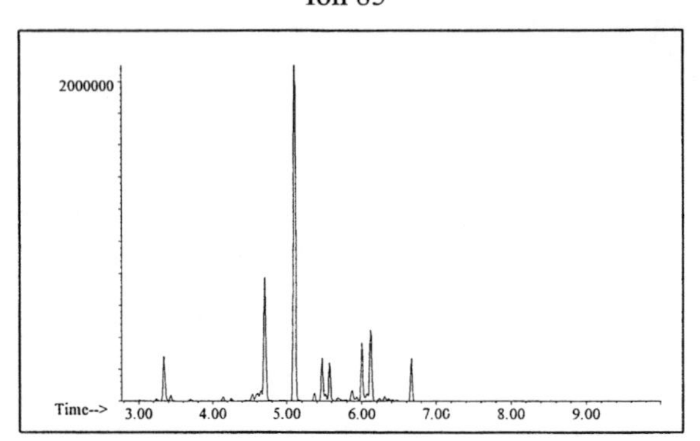

Aromatic

Ion 91

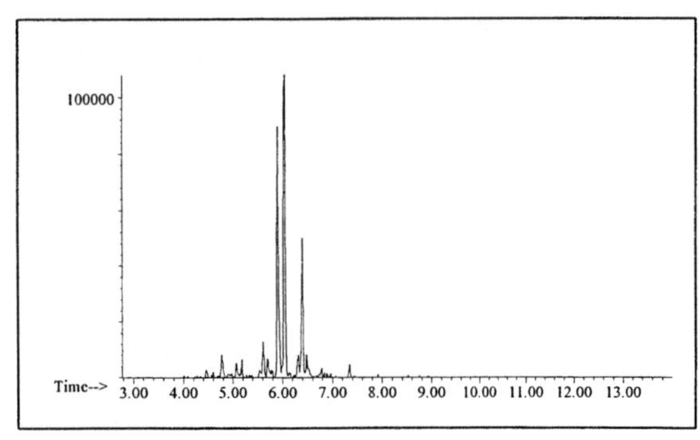

Ion 105

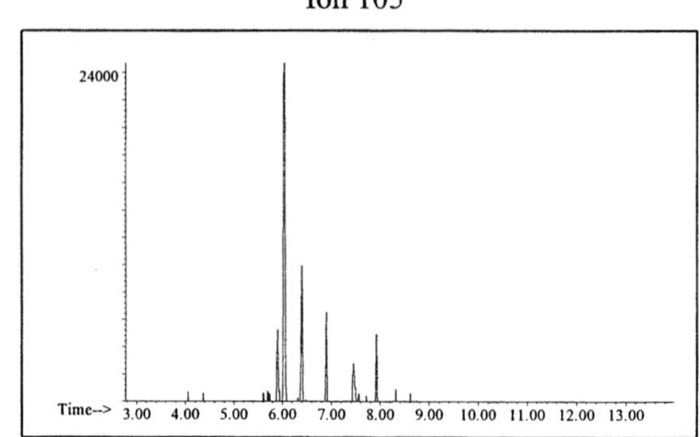

Ion 119

No Useful Data Obtained

INDIVIDUAL PROFILES Distillate #8

Cycloparaffin

Ion 55

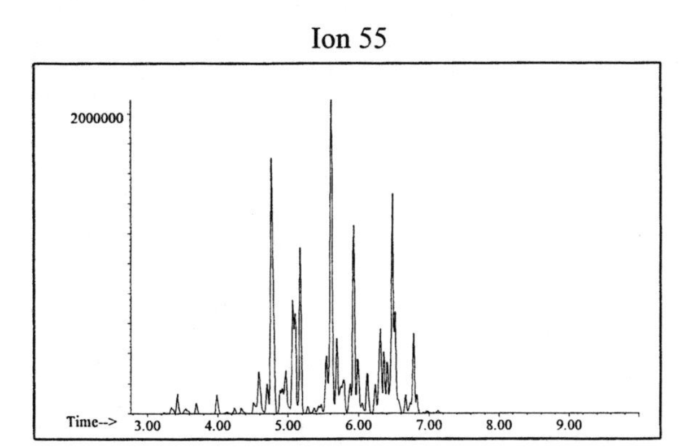

Ion 69

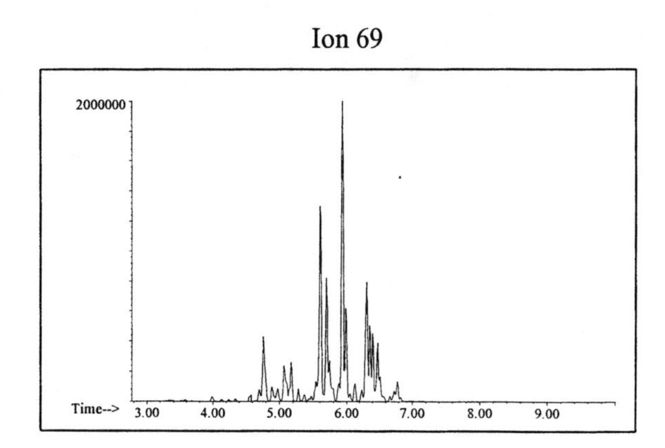

Ion 83

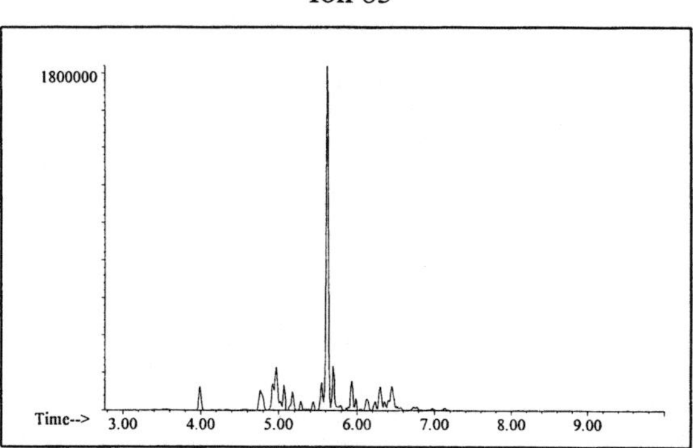

Naphthalene

Ion 128

No Useful Data Obtained

Ion 142

No Useful Data Obtained

Ion 156

No Useful Data Obtained

Distillate #9

20uL/mL Pentane

Product Displayed: 75% Evaporated Zippo Lighter Fluid #1

Other Similar Products:

ASTM: Class 1 (Light Petroleum Distillate)

Product Uses: Pocket lighter fluid

Macro Code: L

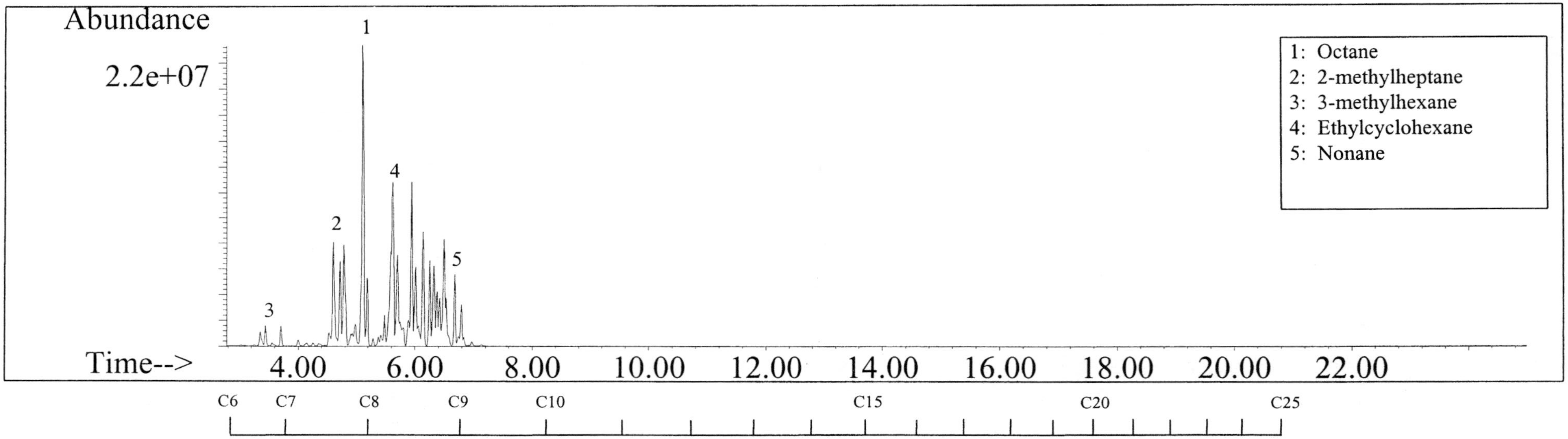

Abundance

2.2e+07

1: Octane
2: 2-methylheptane
3: 3-methylhexane
4: Ethylcyclohexane
5: Nonane

Time--> 4.00 6.00 8.00 10.00 12.00 14.00 16.00 18.00 20.00 22.00

C6 C7 C8 C9 C10 C15 C20 C25

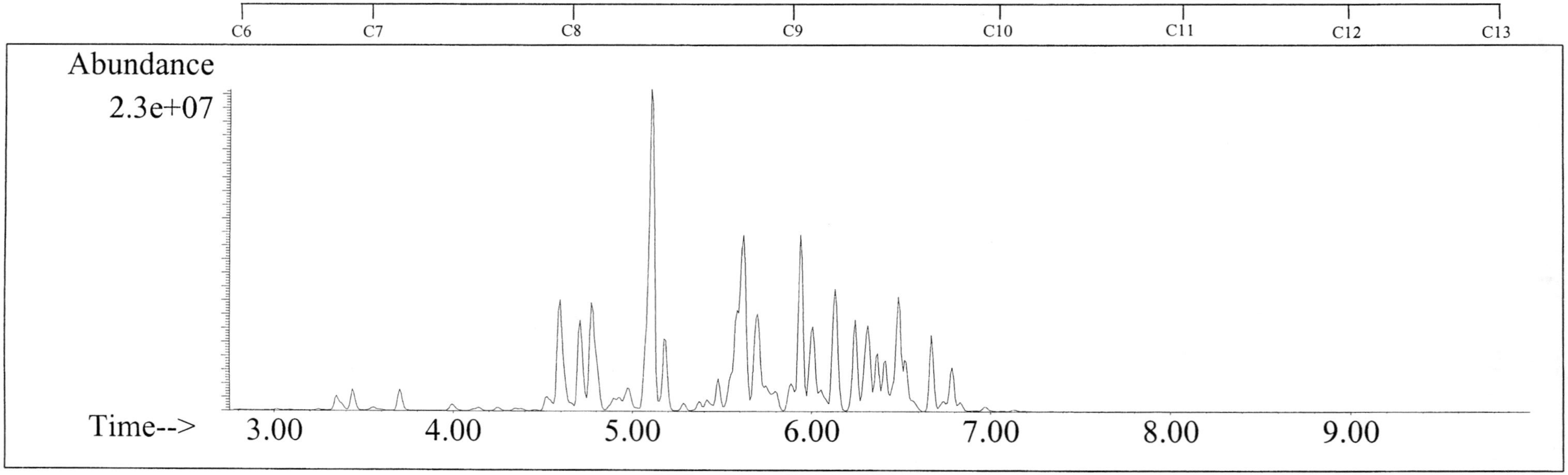

C6 C7 C8 C9 C10 C11 C12 C13

Abundance

2.3e+07

Time--> 3.00 4.00 5.00 6.00 7.00 8.00 9.00

SUMMED PROFILES

Distillate #9

ALKANES

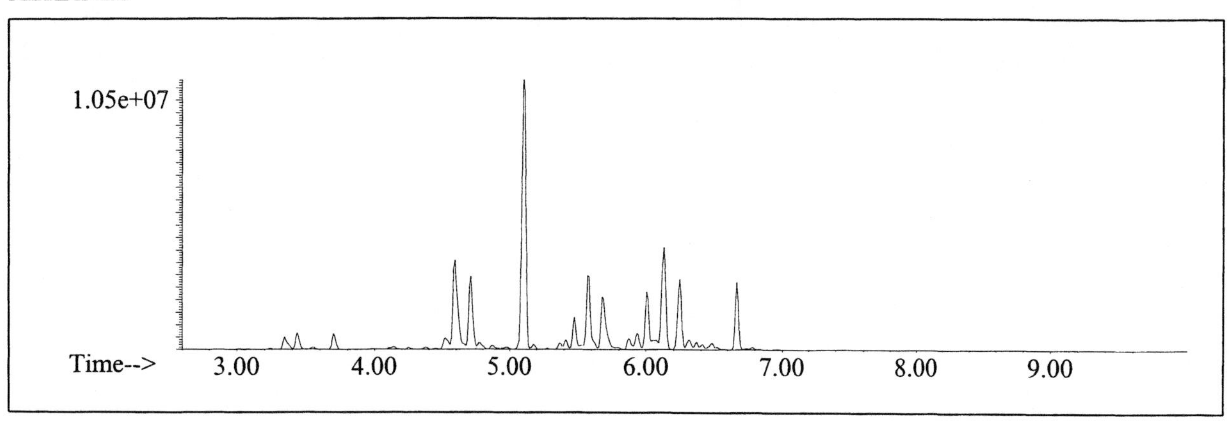

AROMATICS

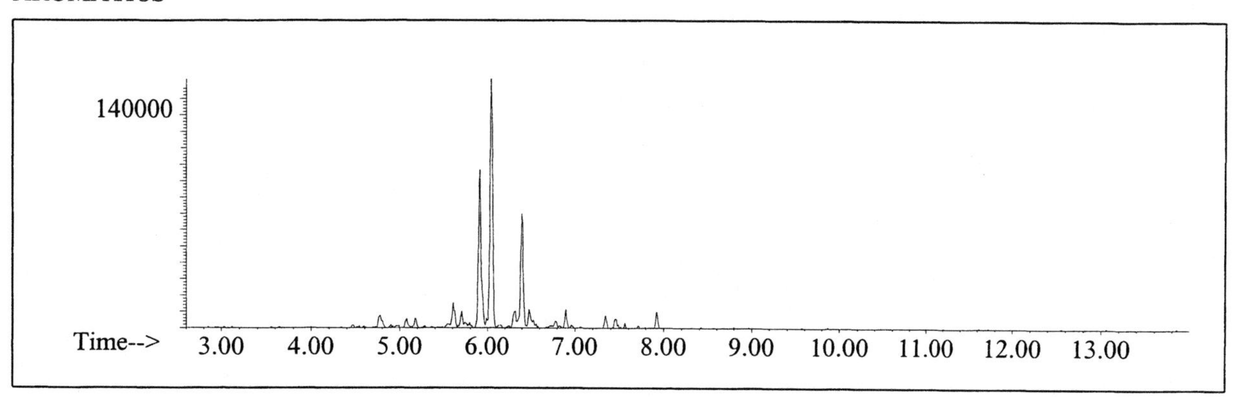

CYCLOPARAFFINS AND ALKENES

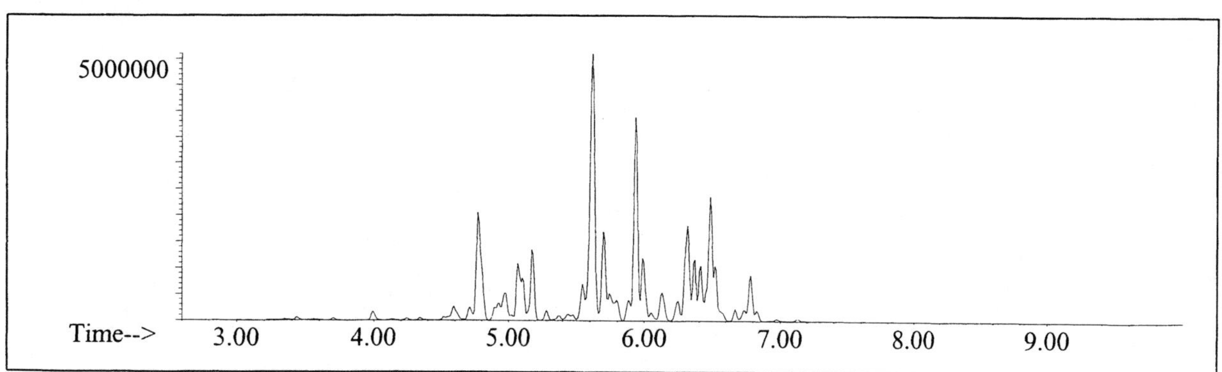

NAPHTHALENES

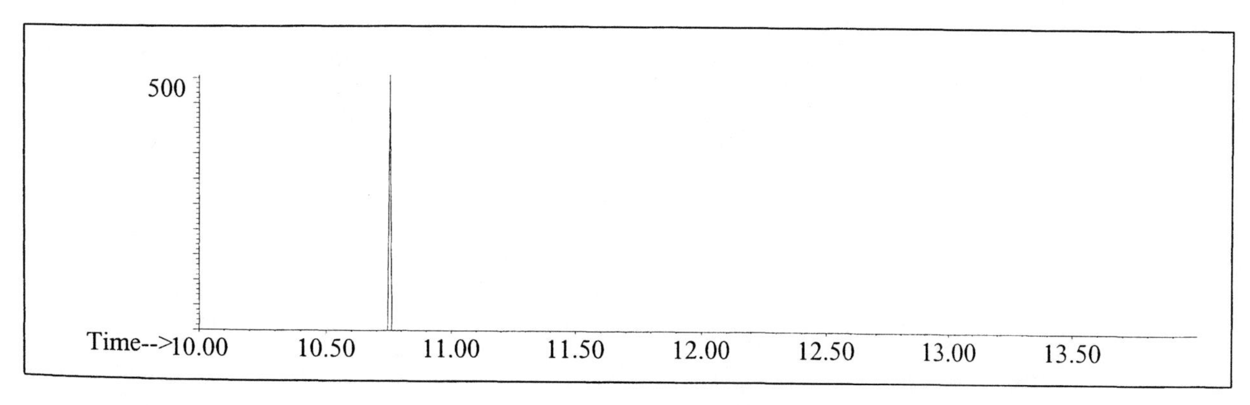

INDIVIDUAL PROFILES

Distillate #9

Alkane

Ion 43

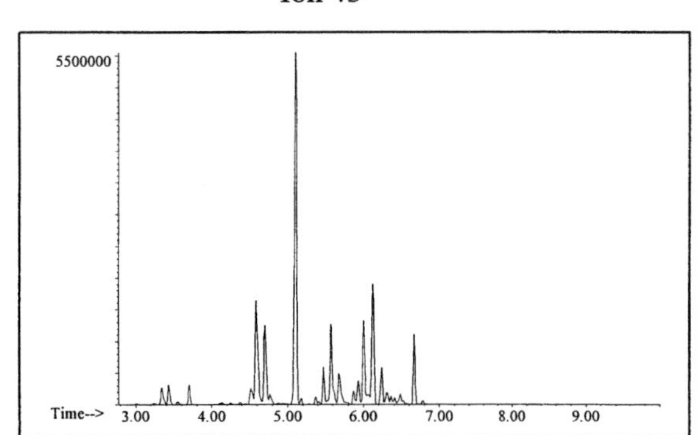

Ion 57

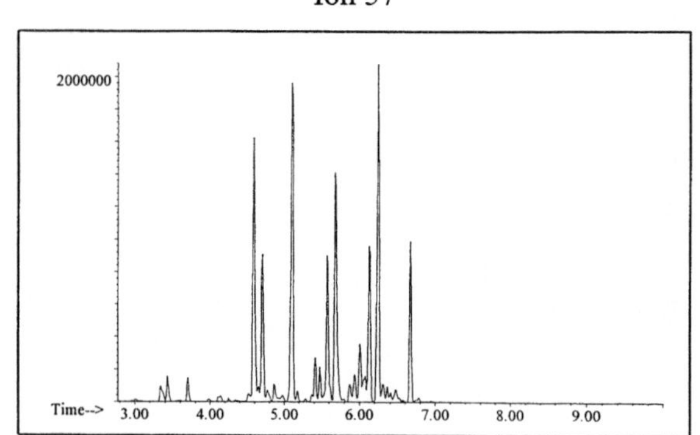

Ion 71

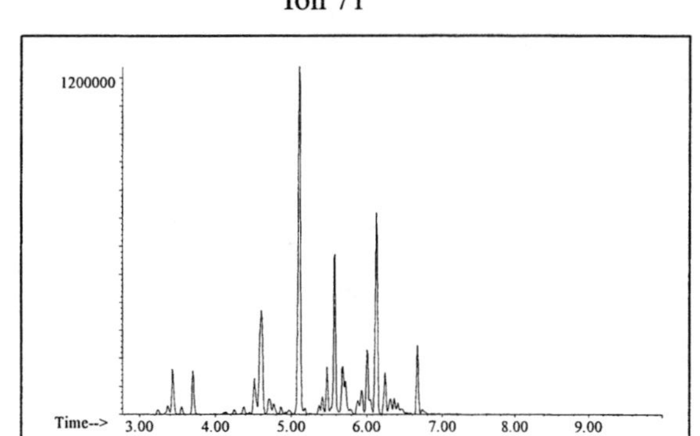

Ion 85

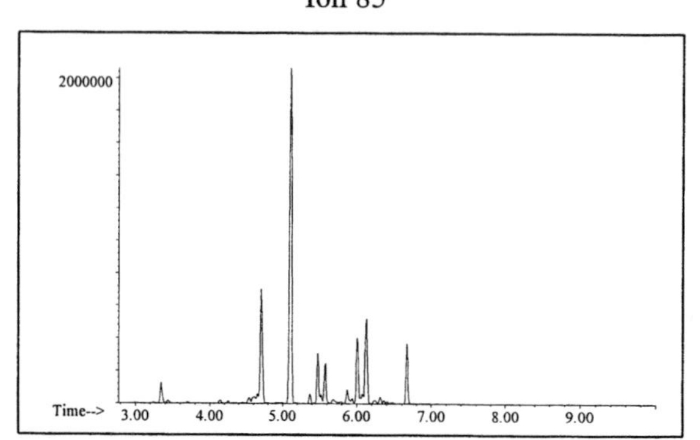

Aromatic

Ion 91

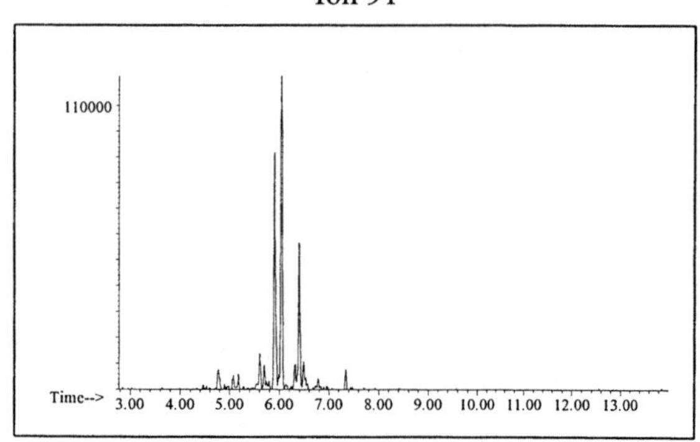

Ion 105

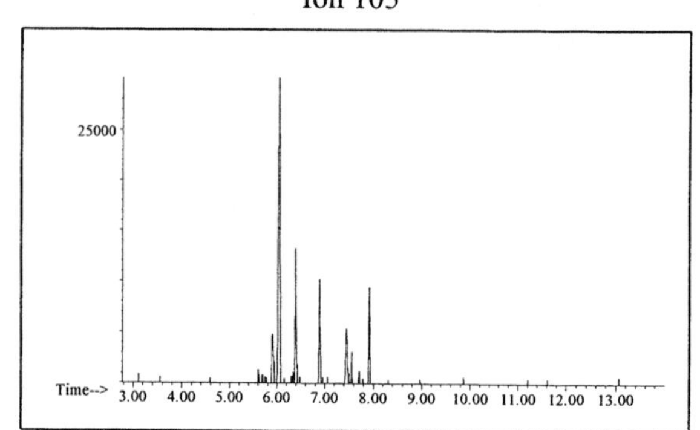

Ion 119

No Useful Data Obtained

INDIVIDUAL PROFILES Distillate #9

Cycloparaffin

Ion 55

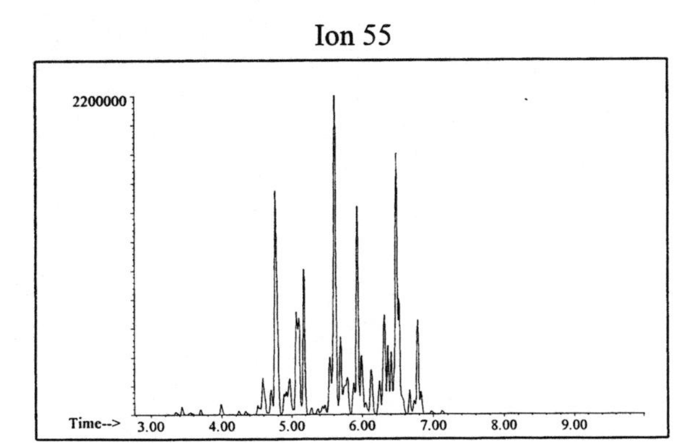

Ion 69

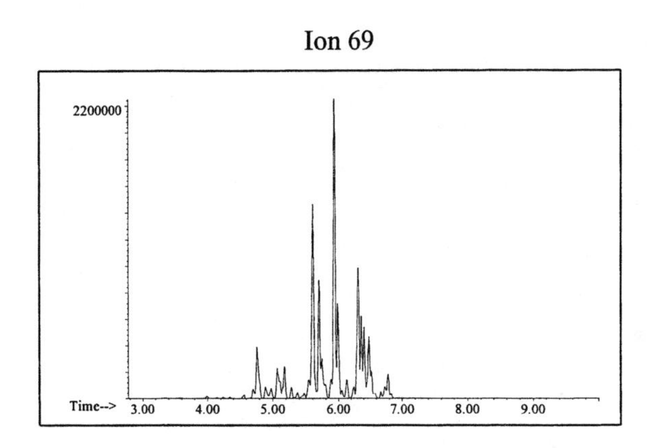

Ion 83

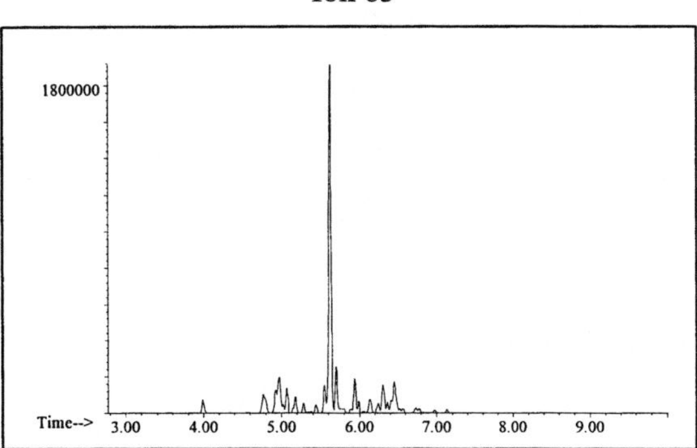

Naphthalene

Ion 128

No Useful Data Obtained

Ion 142

No Useful Data Obtained

Ion 156

No Useful Data Obtained

Distillate #10

20uL/mL Pentane

Product Displayed: 90% Evaporated Zippo Lighter Fluid #1

Other Similar Products:

ASTM: Class 1 (Light Petroleum Distillate)

Product Uses: Pocket lighter fluid

Macro Code: L

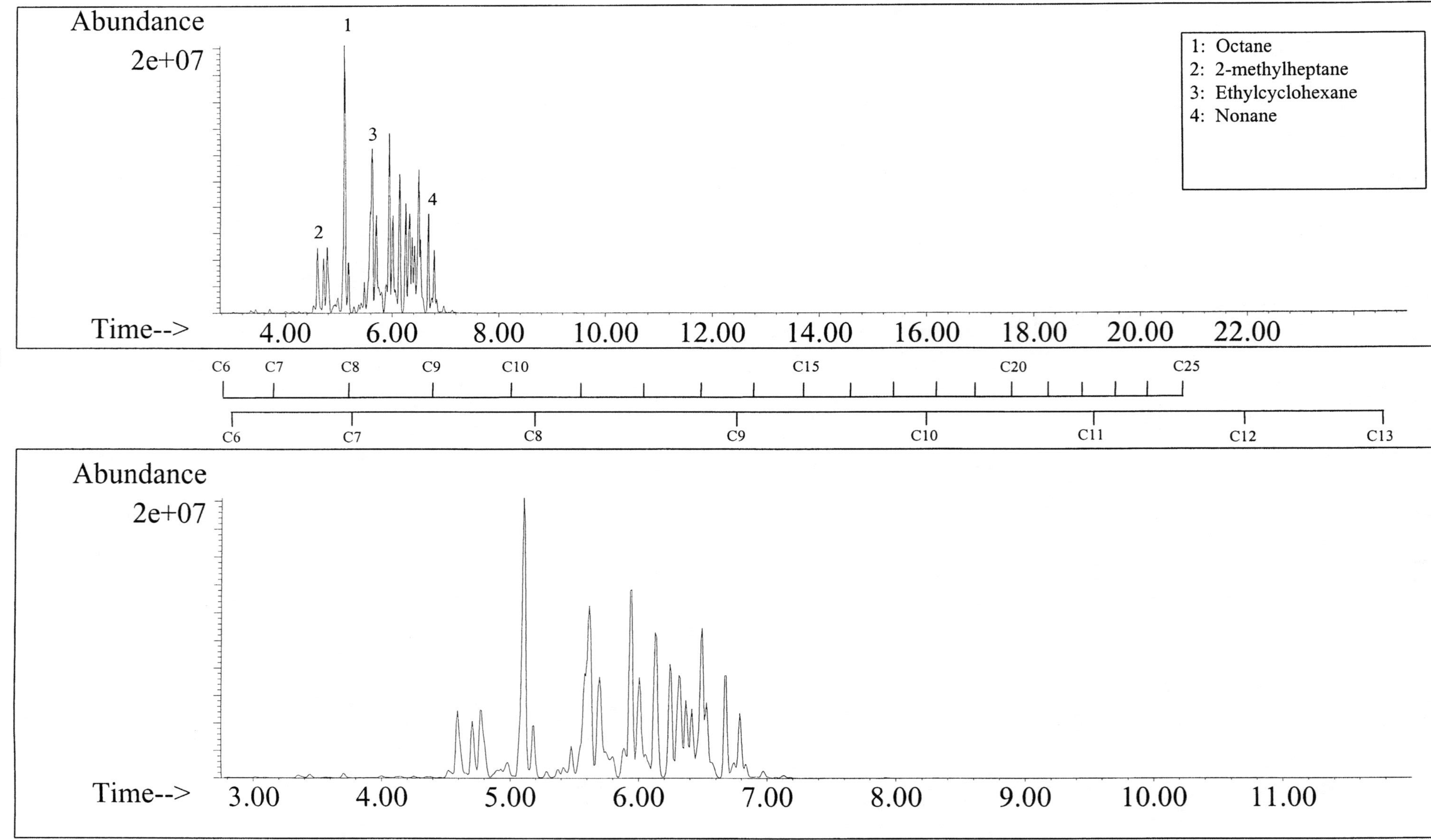

SUMMED PROFILES Distillate #10

ALKANES

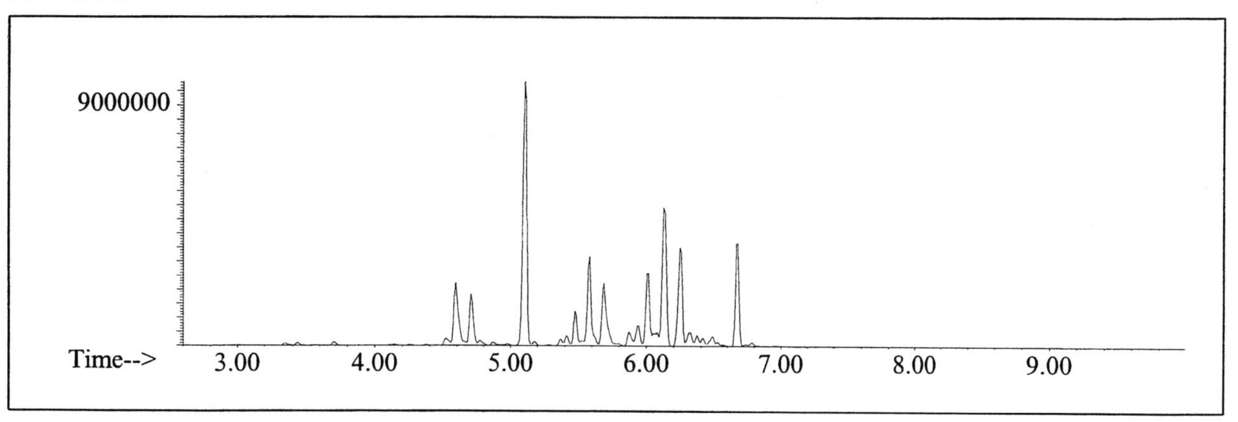

AROMATICS

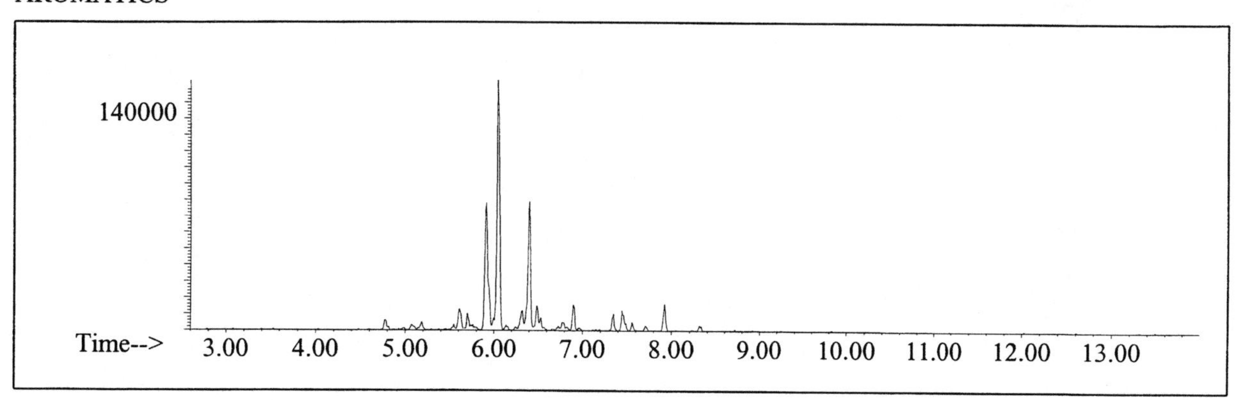

CYCLOPARAFFINS AND ALKENES

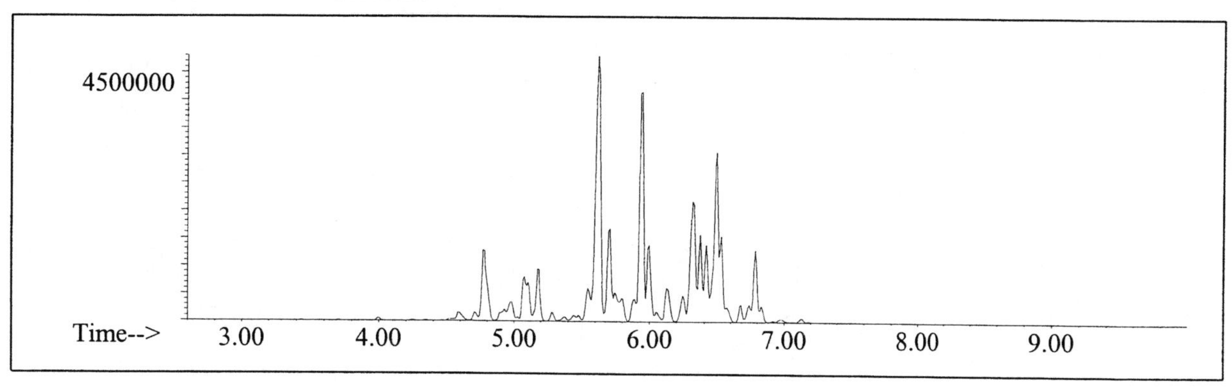

NAPHTHALENES

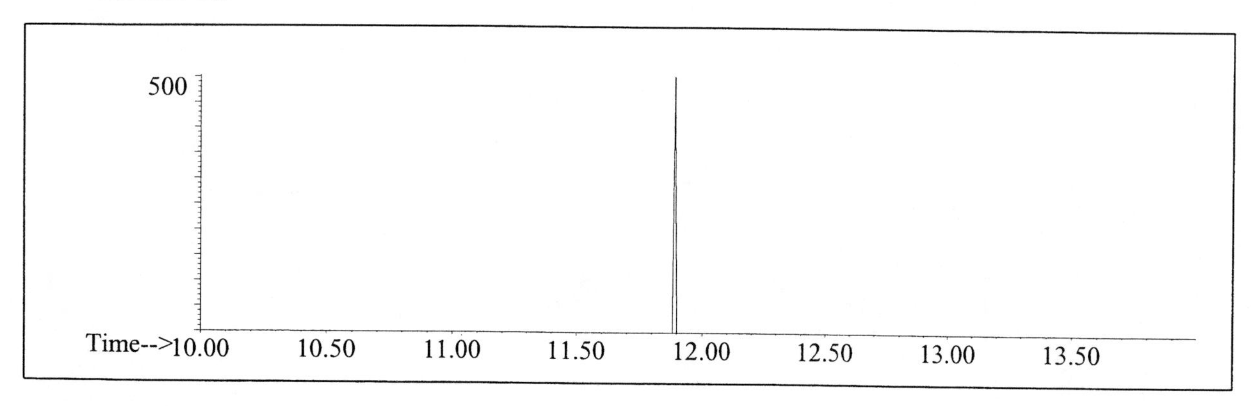

INDIVIDUAL PROFILES

Distillate #10

Alkane

Ion 43

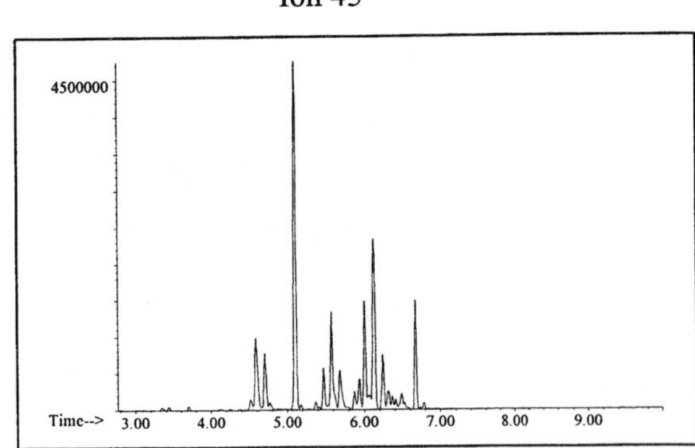

Ion 57

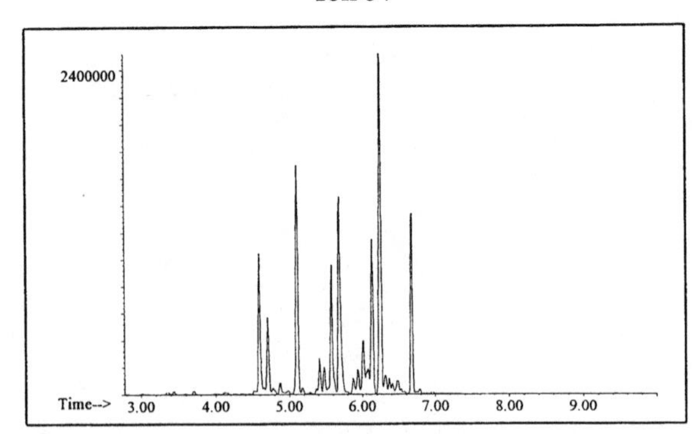

Ion 71

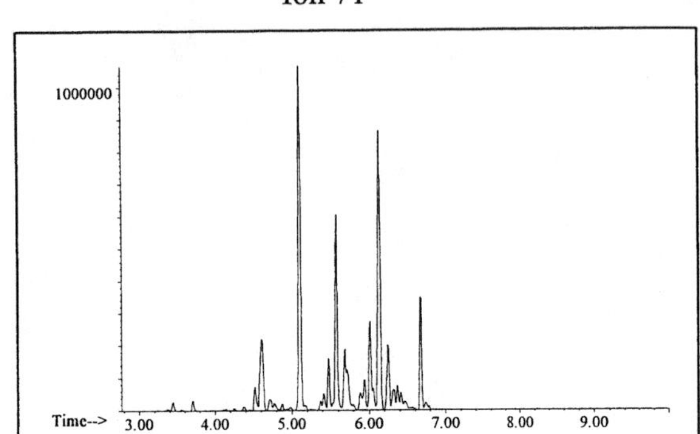

Ion 85

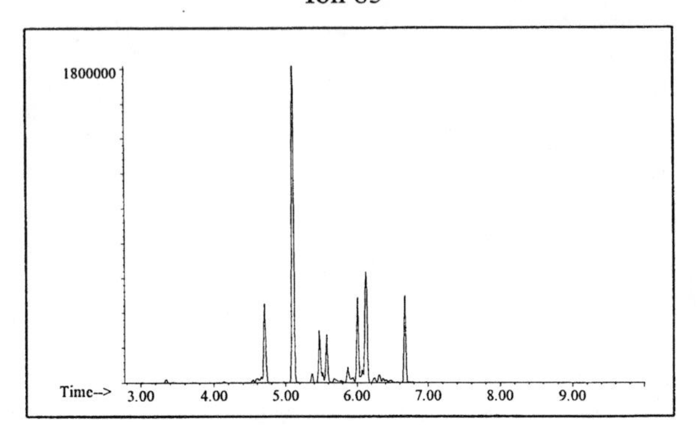

Aromatic

Ion 91

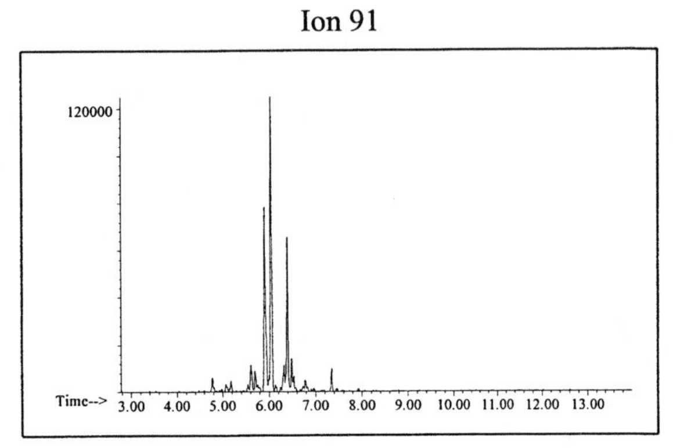

Ion 105

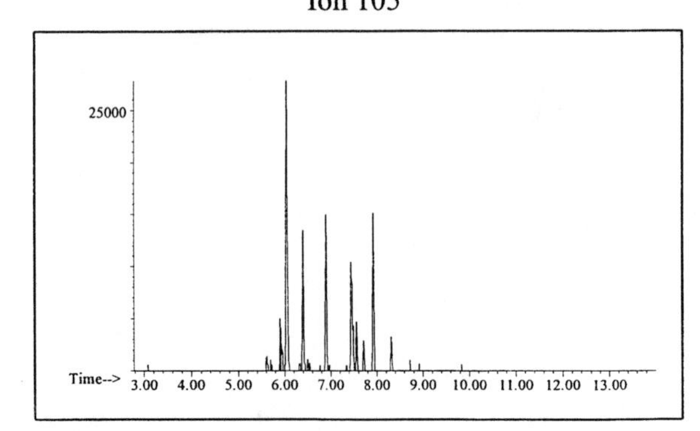

Ion 119

No Useful Data Obtained

INDIVIDUAL PROFILES

Distillate #10

Cycloparaffin

Ion 55

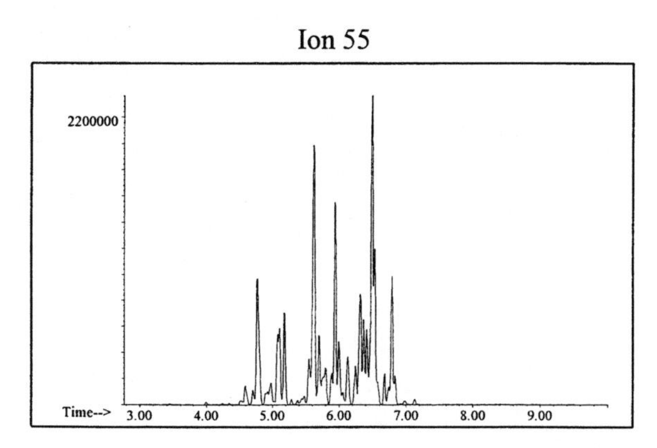

Ion 69

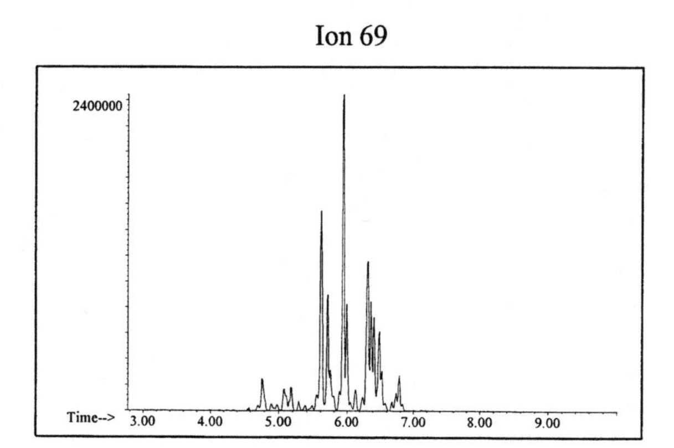

Ion 83

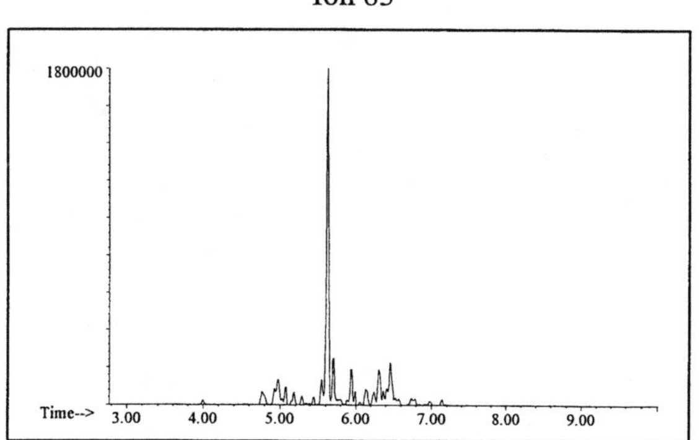

Naphthalene

Ion 128

No Useful Data Obtained

Ion 142

No Useful Data Obtained

Ion 156

No Useful Data Obtained

Distillate #11

20uL/mL Pentane

Product Displayed: USA VM&P Naphtha

Other Similar Products:

ASTM: Class 1 (Light Petroleum Distillate)

Product Uses: Paint thinner, solvent

Macro Code: L

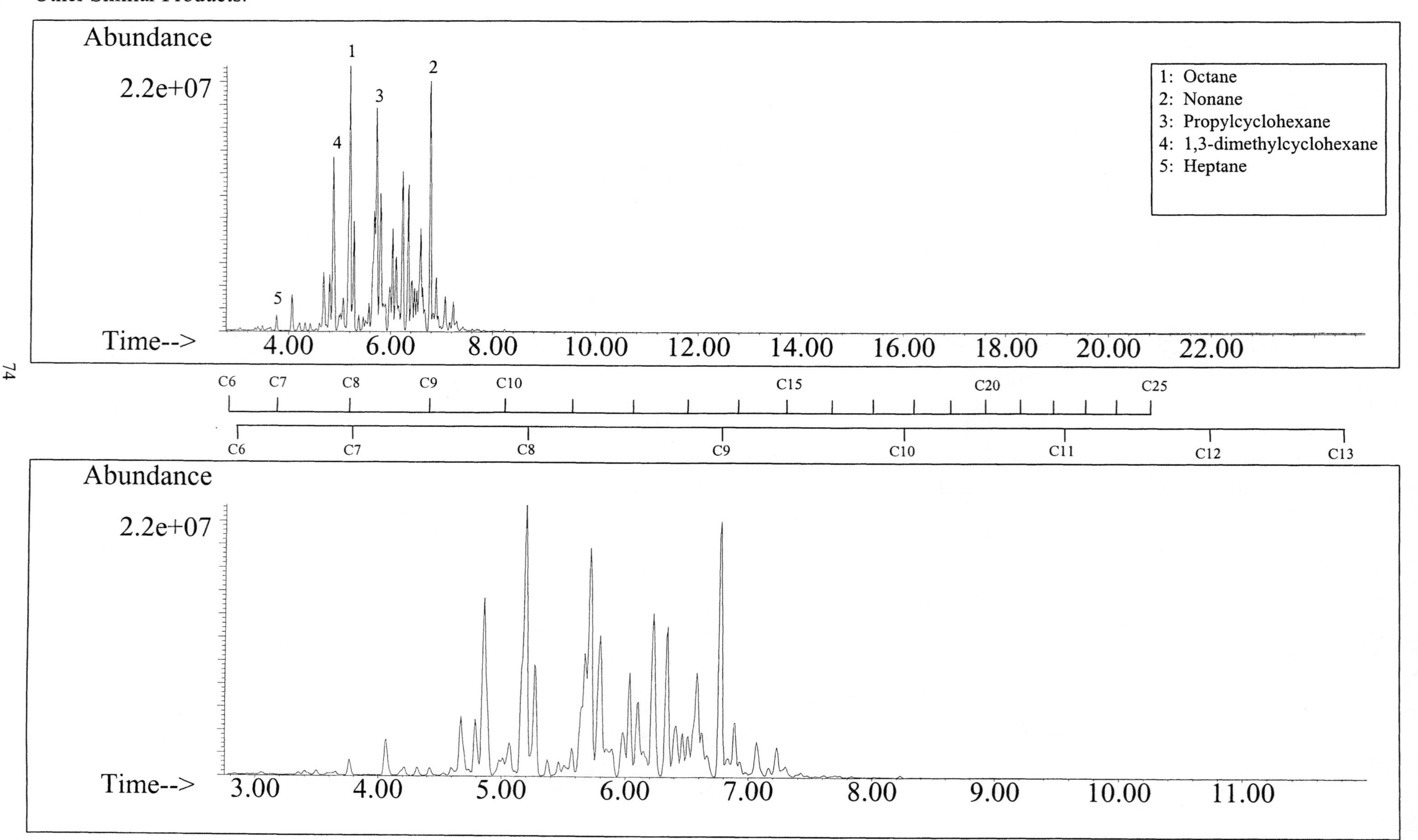

SUMMED PROFILES

Distillate #11

ALKANES

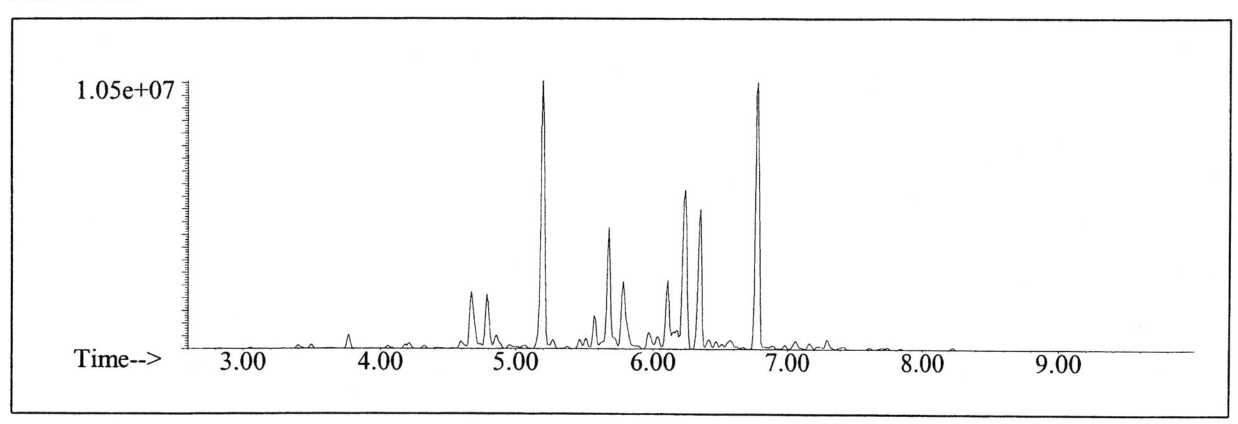

AROMATICS

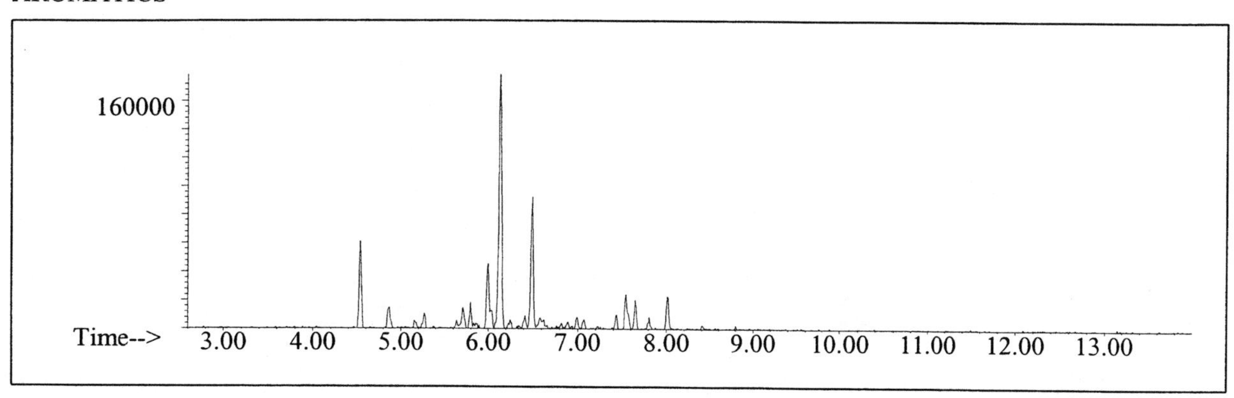

CYCLOPARAFFINS AND ALKENES

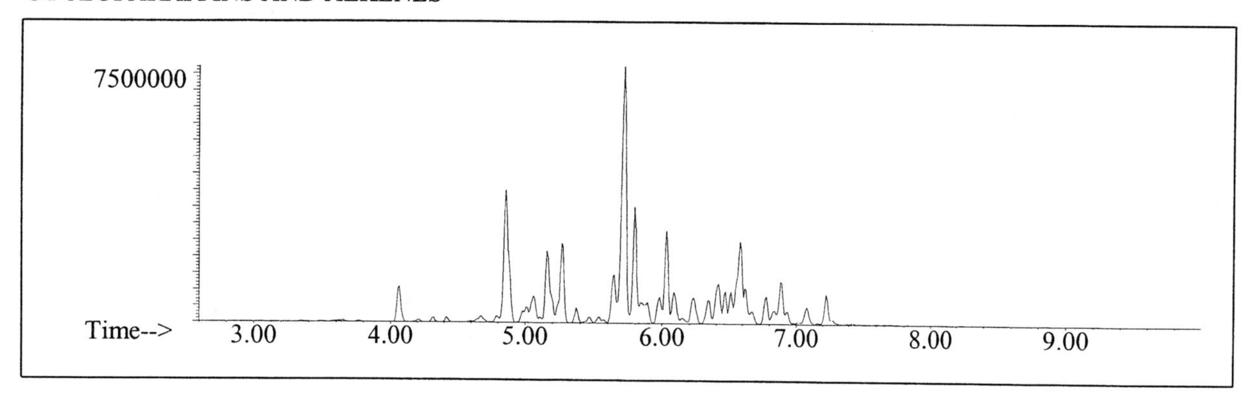

NAPHTHALENES

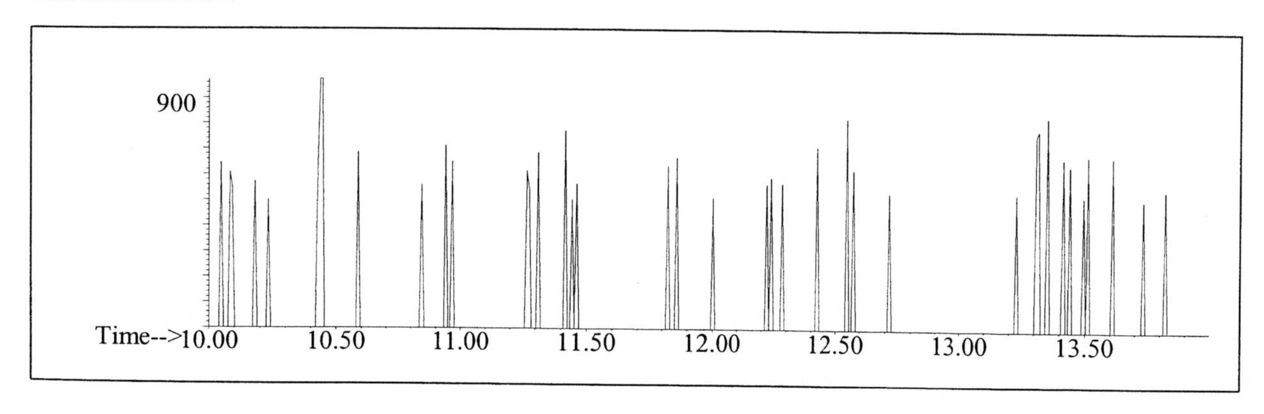

INDIVIDUAL PROFILES

Distillate #11

Alkane

Ion 43

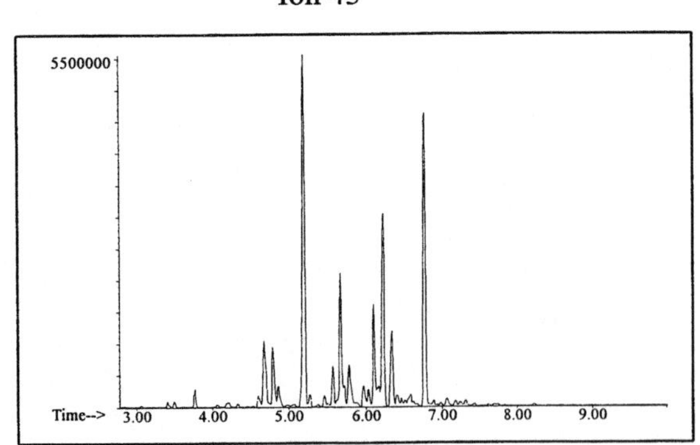

Ion 57

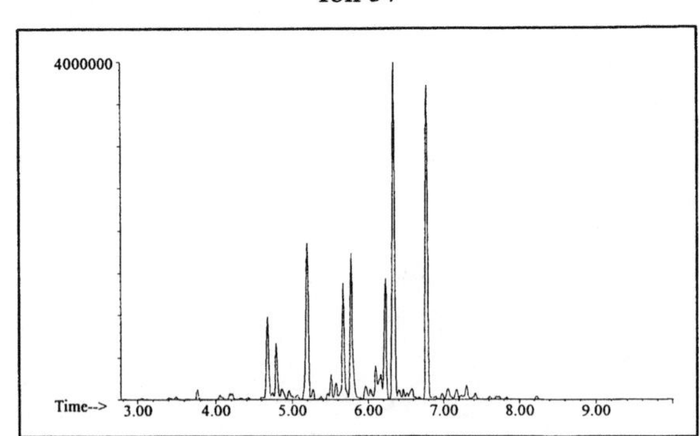

Ion 71

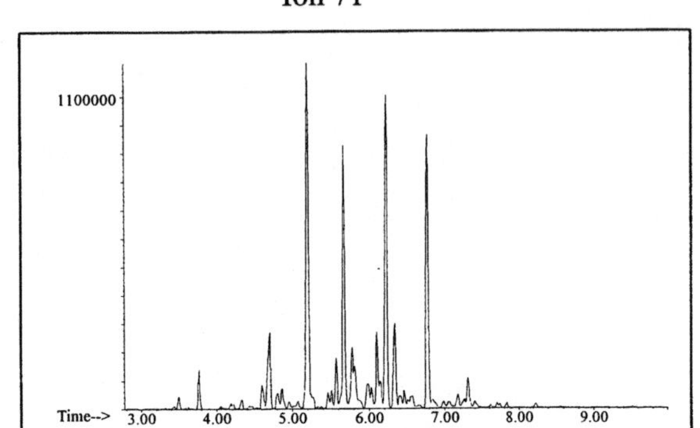

Ion 85

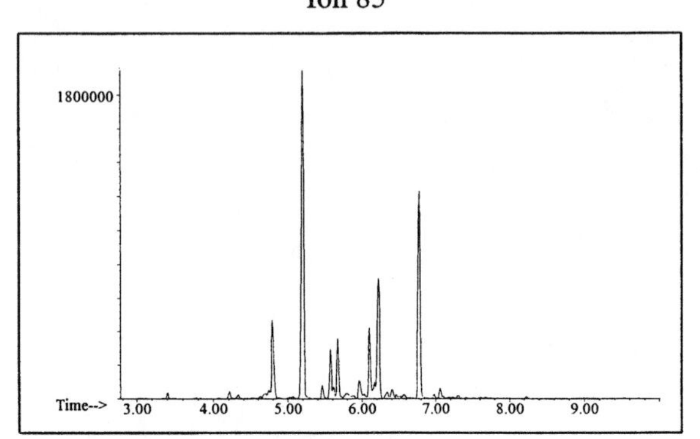

Aromatic

Ion 91

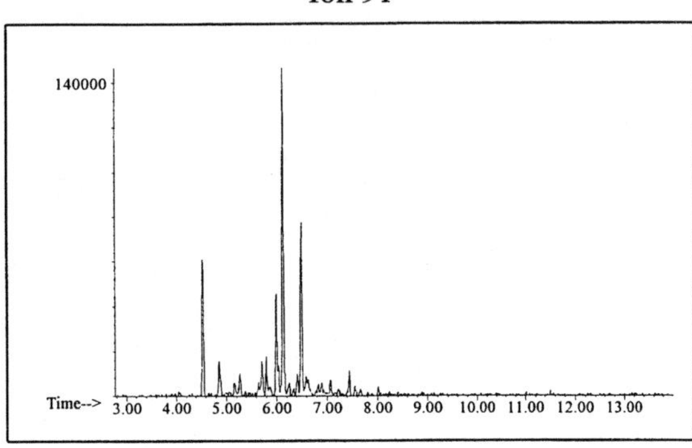

Ion 105

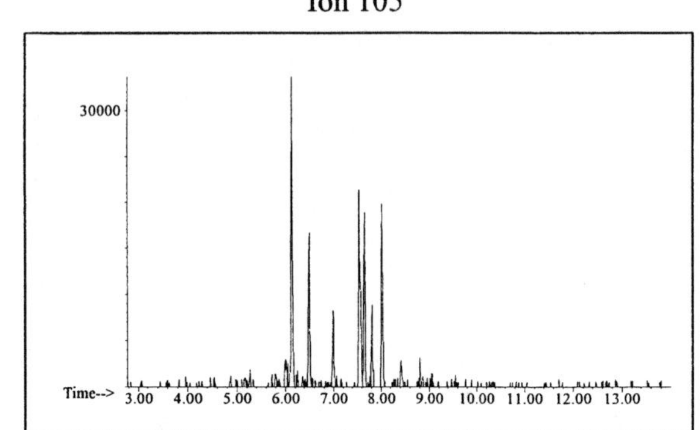

Ion 119

No Useful Data Obtained

INDIVIDUAL PROFILES

Distillate #11

Cycloparaffin

Ion 55

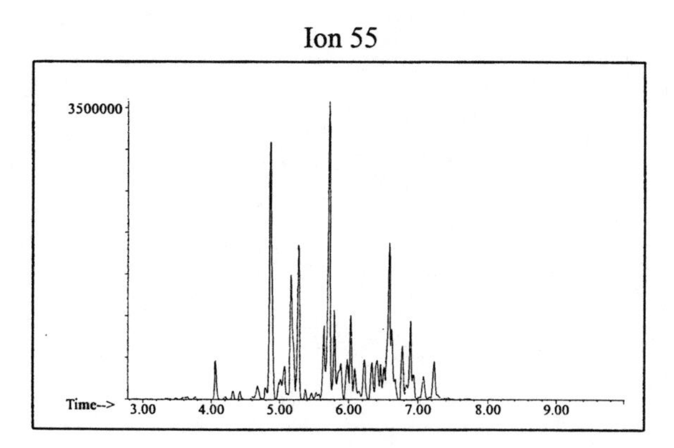

Ion 69

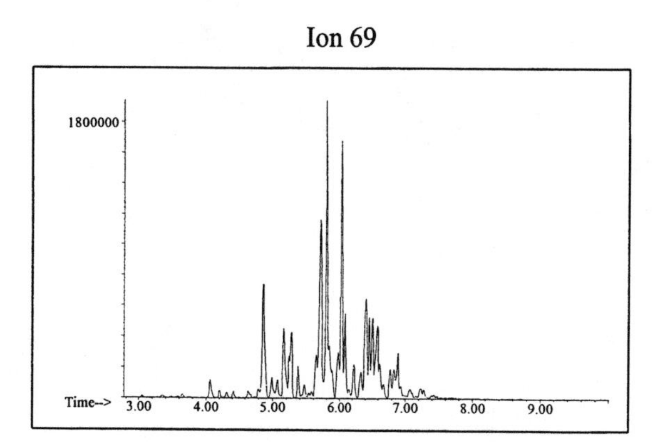

Ion 83

3200000
Time--> 3.00 4.00 5.00 6.00 7.00 8.00 9.00

Naphthalene

Ion 128

No Useful Data Obtained

Ion 142

No Useful Data Obtained

Ion 156

No Useful Data Obtained

Distillate #12

20uL/mL Pentane

Product Displayed: Ronsonol Pocket Lighter Fluid

Other Similar Products:

ASTM: Class 1 (Light Petroleum Distillate)

Product Uses: Pocket lighter fluid

Macro Code: L

Abundance

1.6e+07

1: Nonane
2: Octane
3: 1.3-dimethylcyclohexane
4: Heptane
5: 3-methylheptane

Time--> 4.00 6.00 8.00 10.00 12.00 14.00 16.00 18.00 20.00 22.00

C6 C7 C8 C9 C10 C15 C20 C25

C6 C7 C8 C9 C10 C11 C12 C13

Abundance

1.6e+07

Time--> 3.00 4.00 5.00 6.00 7.00 8.00 9.00 10.00 11.00

SUMMED PROFILES

ALKANES

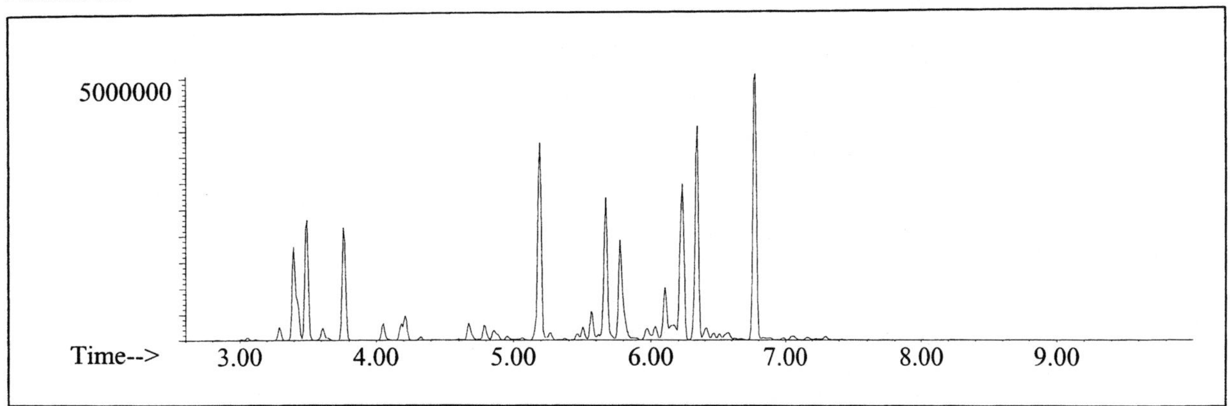

AROMATICS

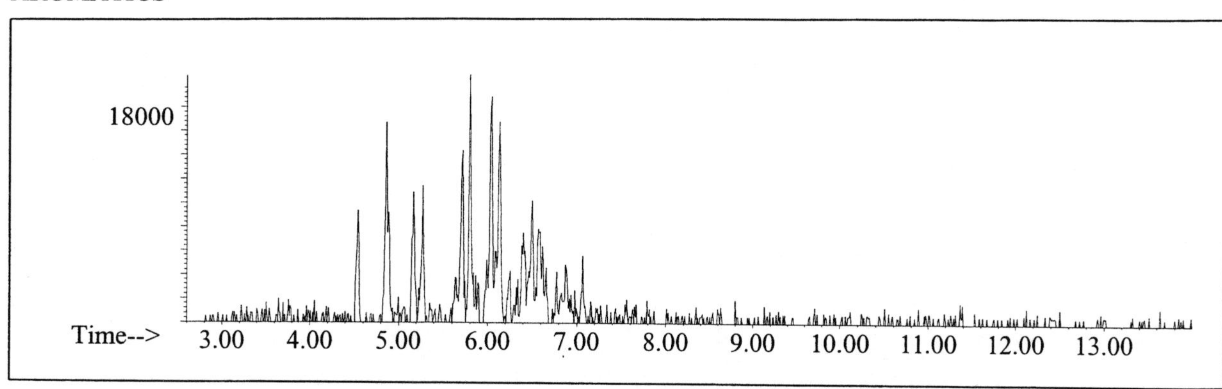

CYCLOPARAFFINS AND ALKENES

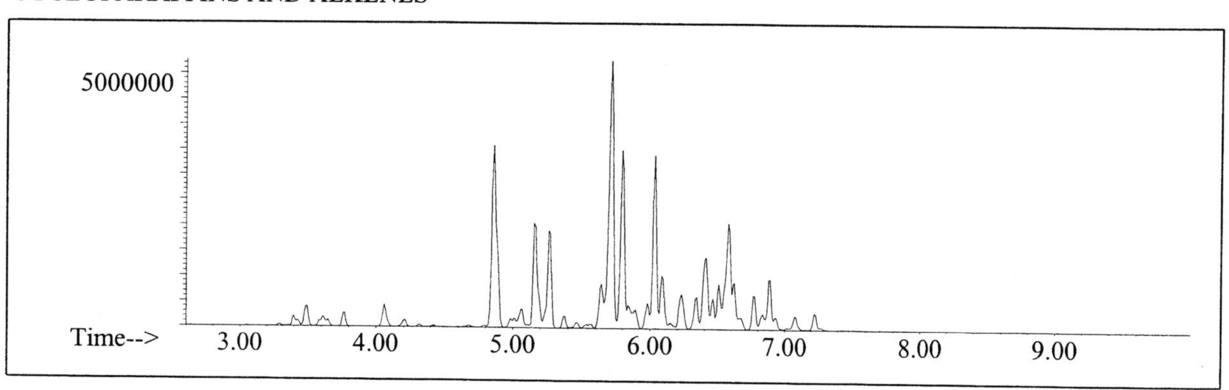

NAPHTHALENES

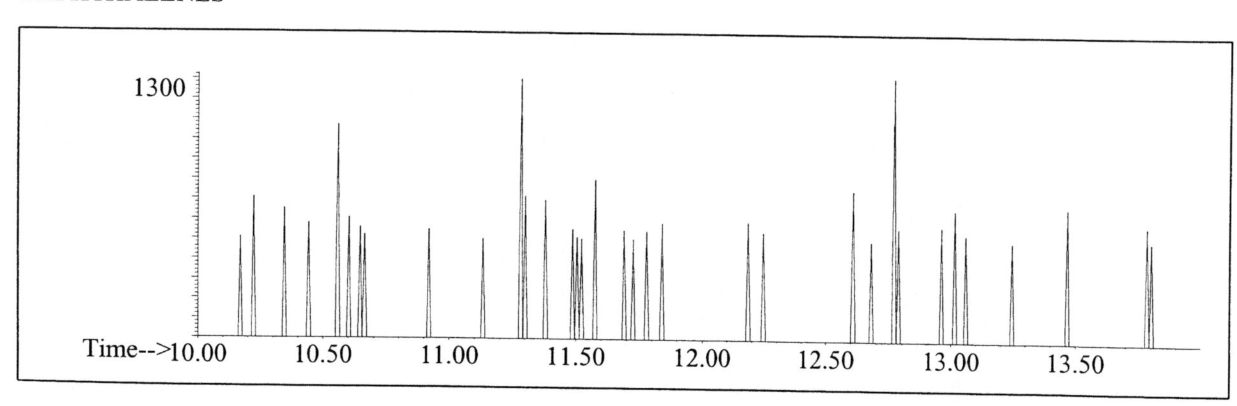

INDIVIDUAL PROFILES

Distillate #12

Alkane

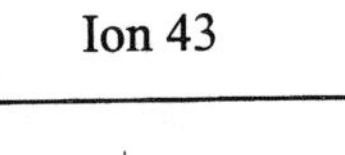

Ion 43

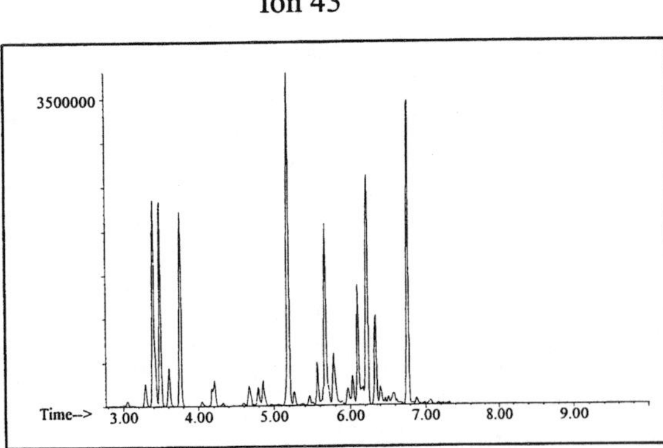

Ion 57

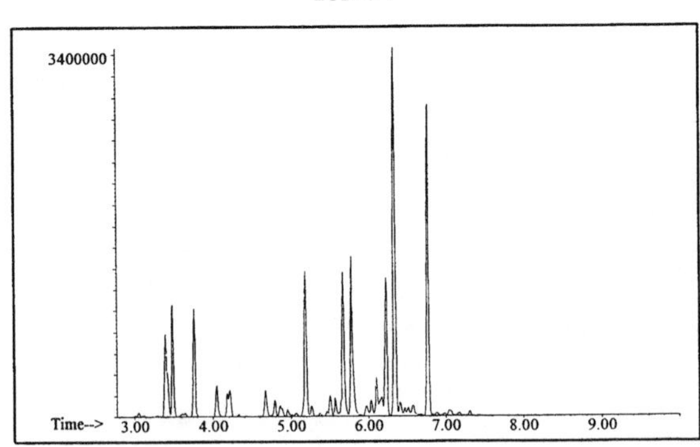

Ion 71

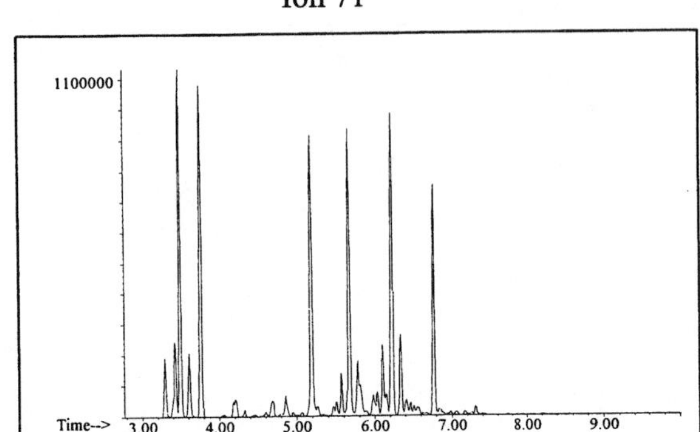

Ion 85

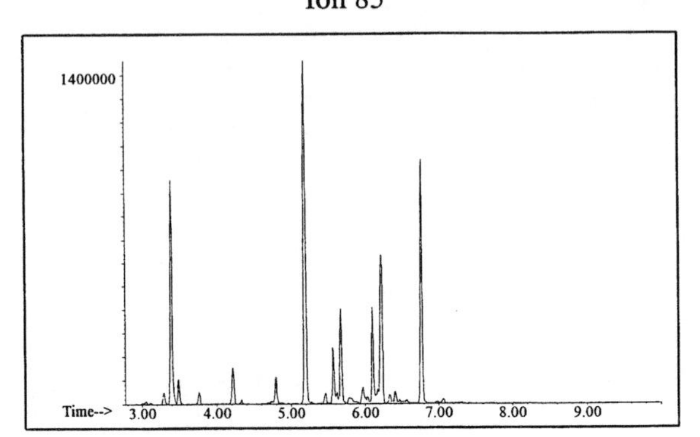

Aromatic

Ion 91

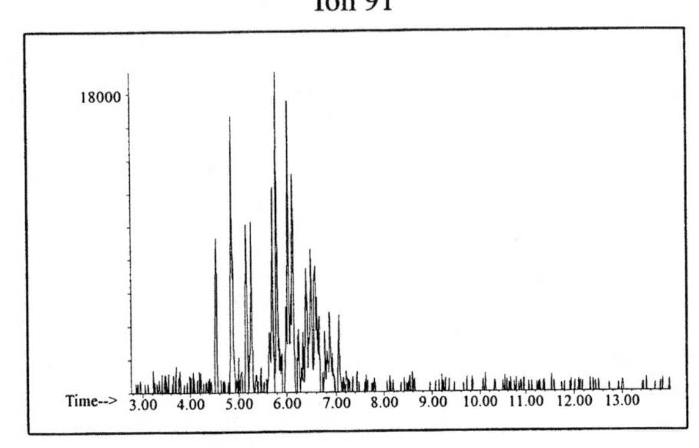

Ion 105

No Useful Data Obtained

Ion 119

No Useful Data Obtained

INDIVIDUAL PROFILES Distillate #12

Cycloparaffin

Ion 55

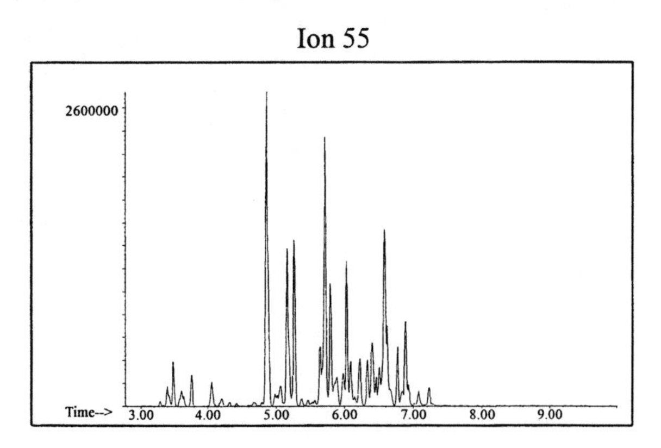

Ion 69

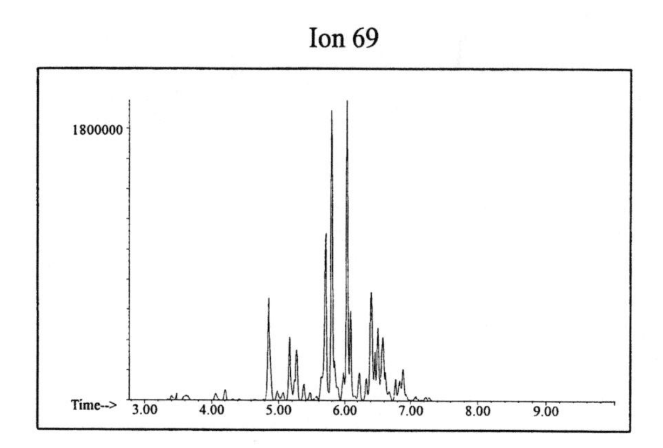

Ion 83

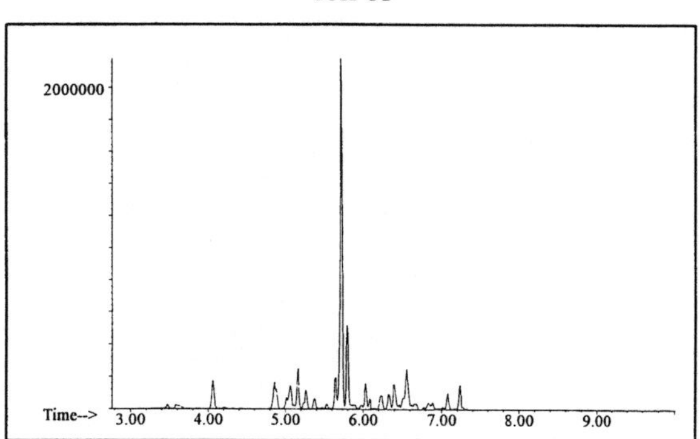

Naphthalene

Ion 128

No Useful Data Obtained

Ion 142

No Useful Data Obtained

Ion 156

No Useful Data Obtained

Distillate #13

20uL/mL Pentane

Product Displayed: 25% Evaporated Ronsonol Lighter Fluid

Other Similar Products:

ASTM: Class 1 (Light Petroleum Distillate)

Product Uses: Pocket lighter fluid

Macro Code: L

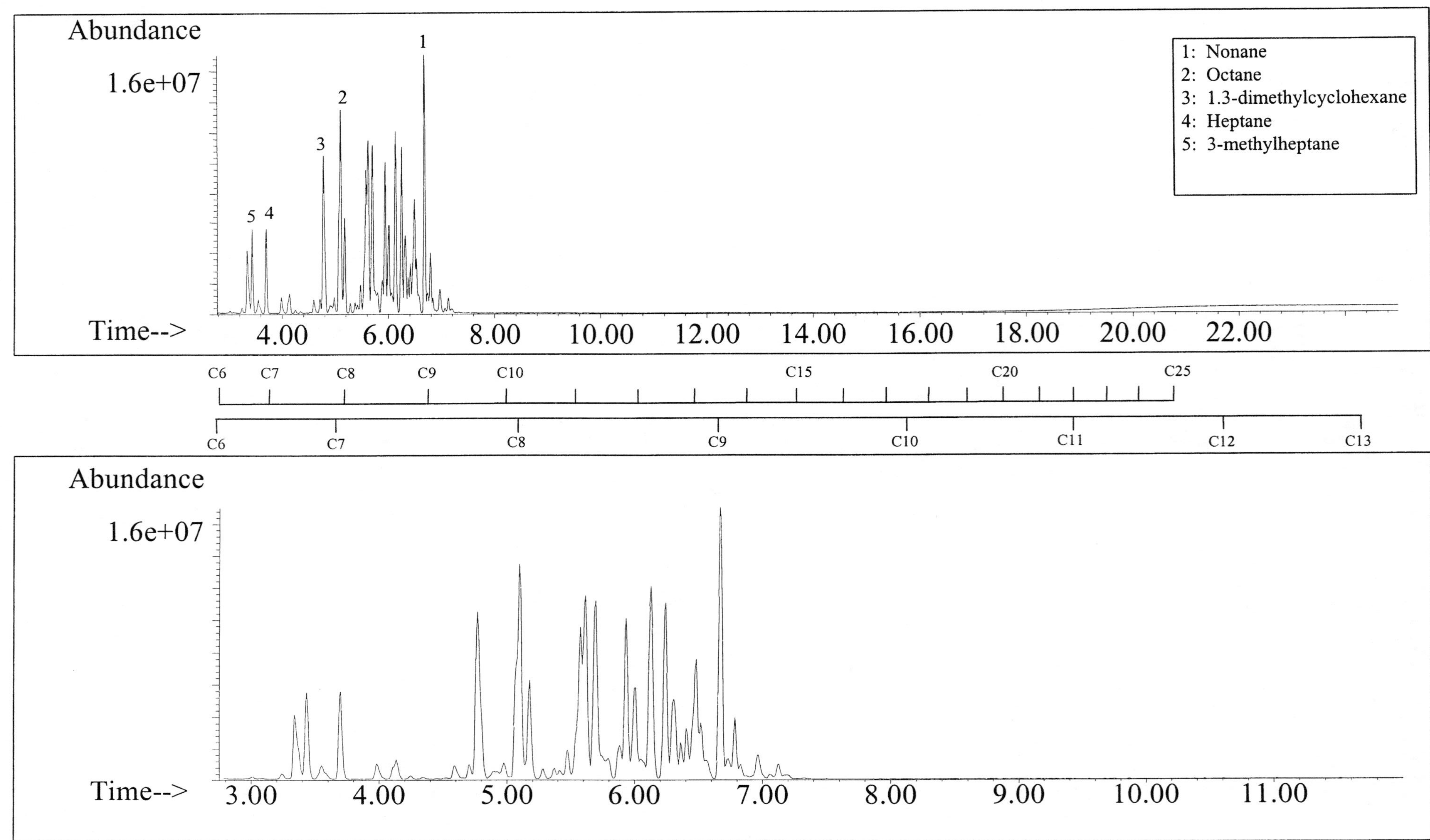

SUMMED PROFILES

Distillate #13

ALKANES

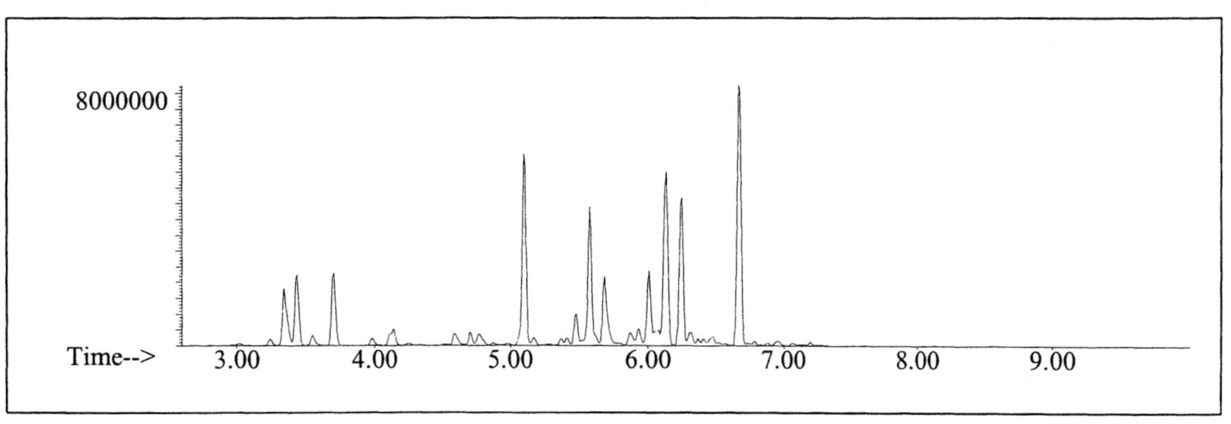

AROMATICS

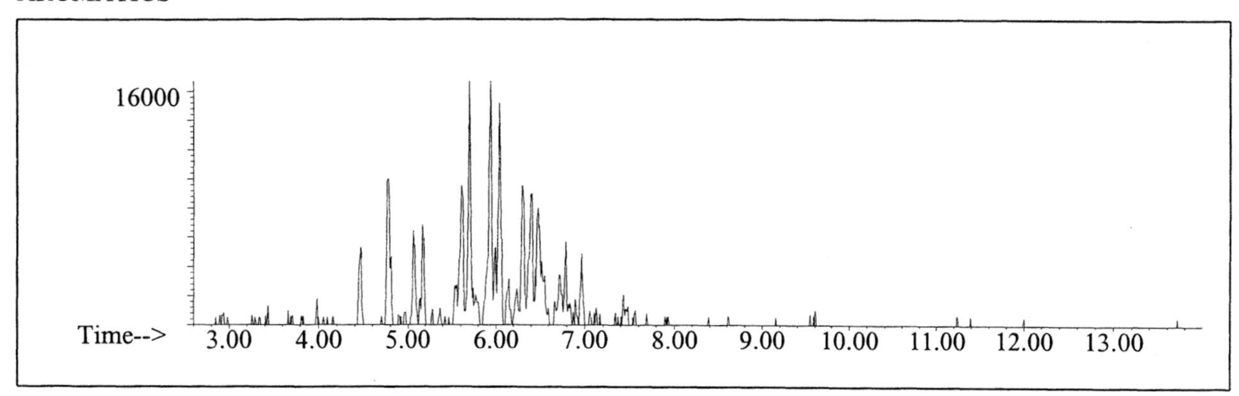

CYCLOPARAFFINS AND ALKENES

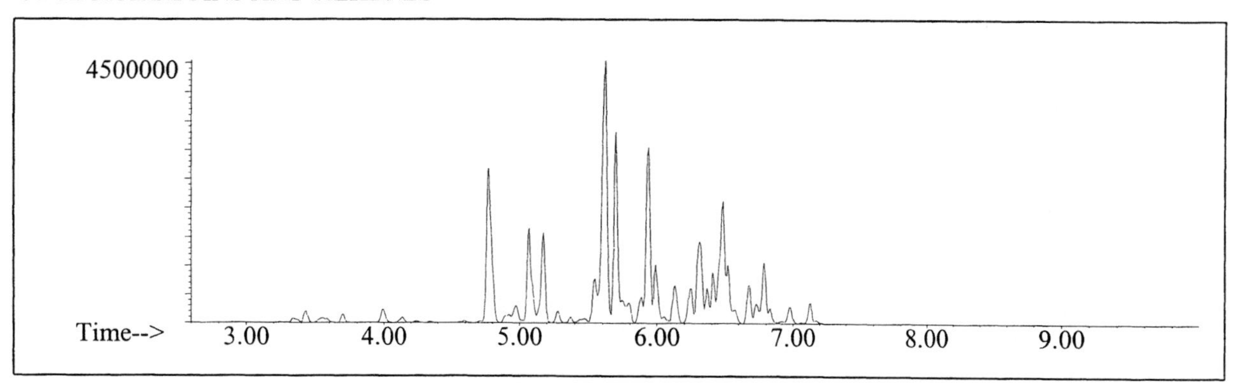

NAPHTHALENES

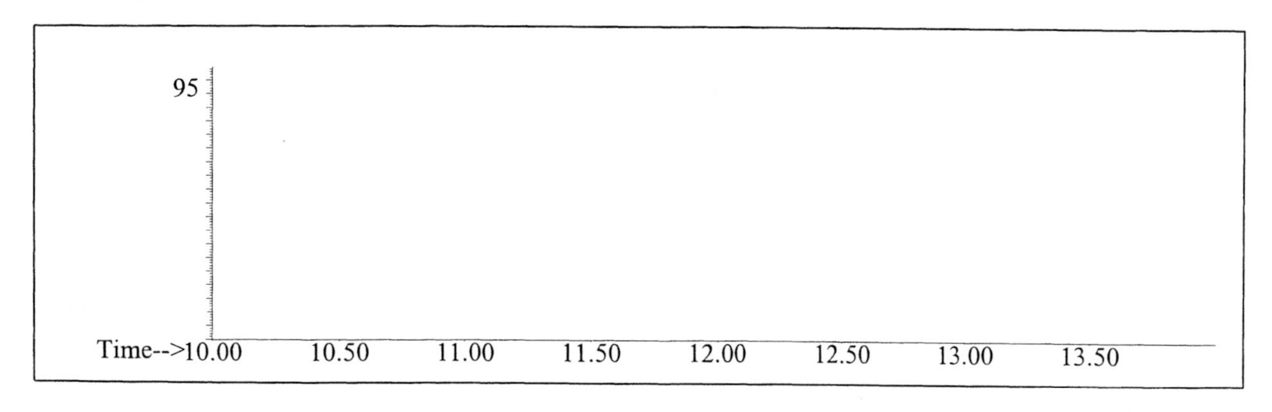

INDIVIDUAL PROFILES

Distillate #13

Alkane

Ion 43

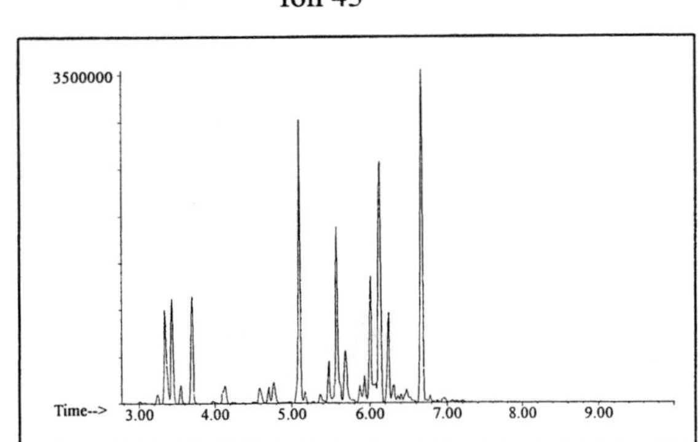

Ion 57

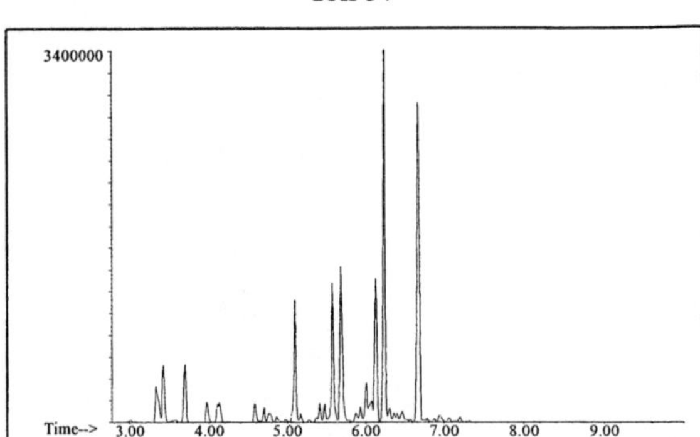

Ion 71

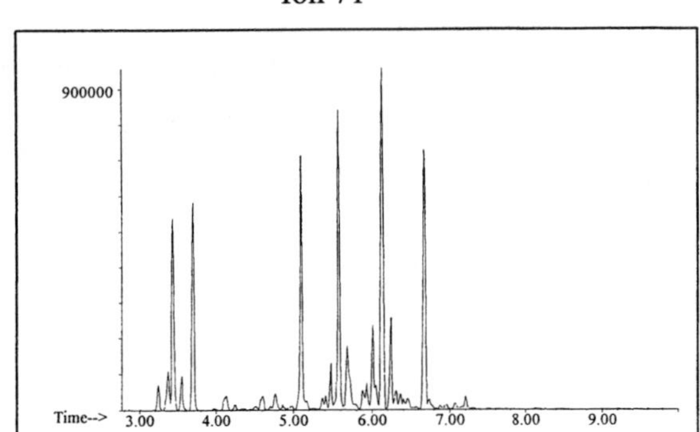

Ion 85

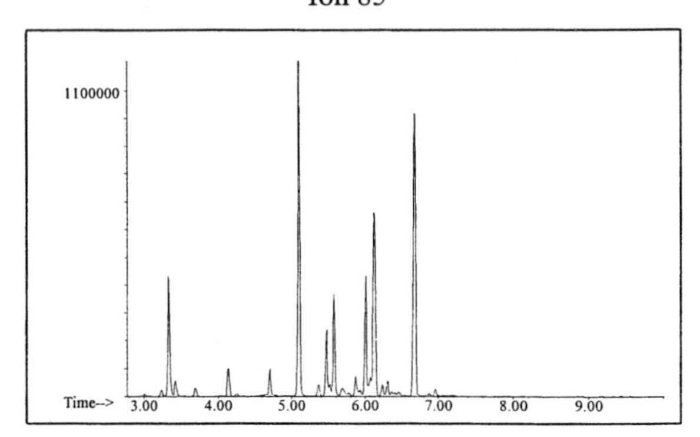

Aromatic

Ion 91

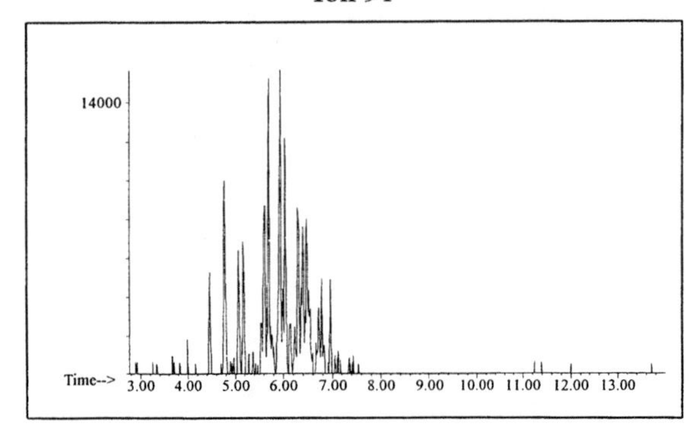

Ion 105

No Useful Data Obtained

Ion 119

No Useful Data Obtained

INDIVIDUAL PROFILES Distillate #13

Cycloparaffin

Ion 55

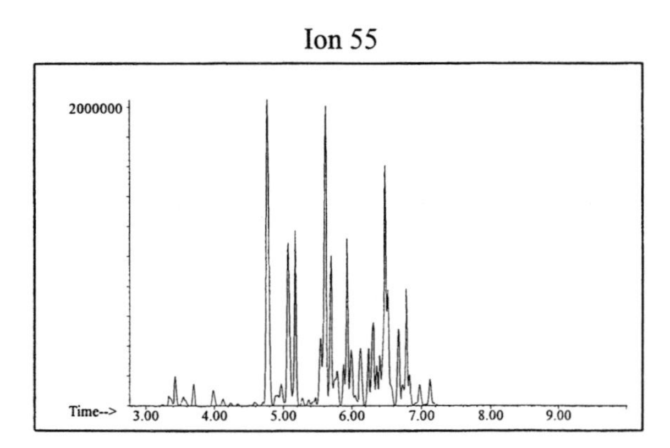

Ion 69

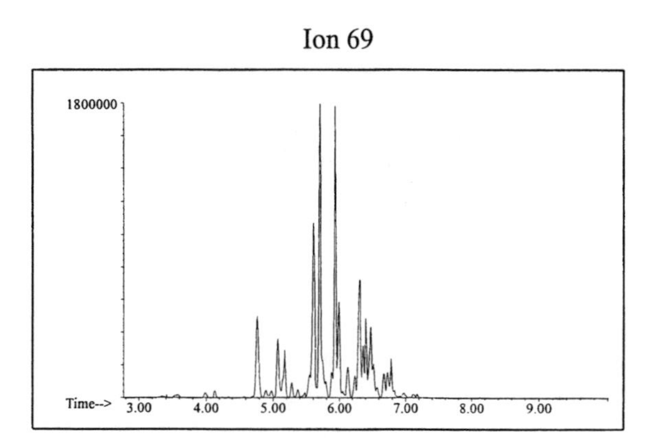

Ion 83

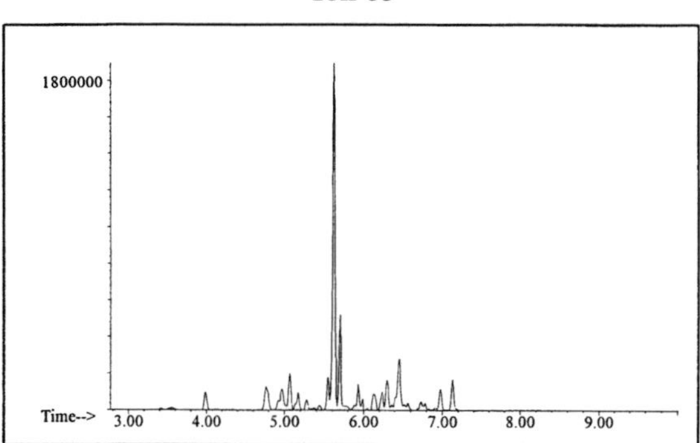

Naphthalene

Ion 128

No Useful Data Obtained

Ion 142

No Useful Data Obtained

Ion 156

No Useful Data Obtained

Distillate #14

20uL/mL Pentane

Product Displayed: 50% Evaporated Ronsonol Lighter Fluid

Other Similar Products:

ASTM: Class 1 (Light Petroleum Distillate)

Product Uses: Pocket lighter fluid

Macro Code: L

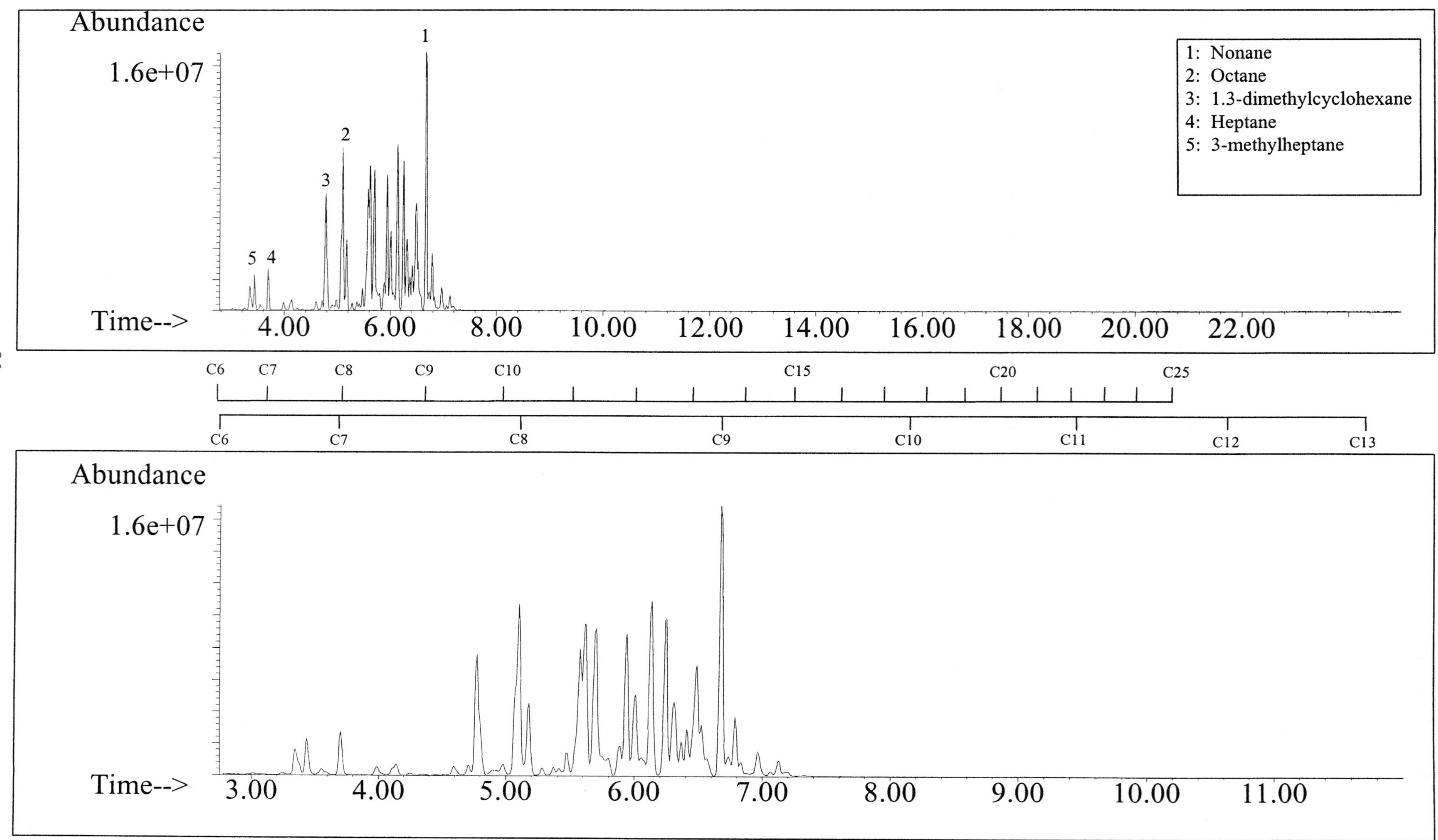

SUMMED PROFILES

Distillate #14

ALKANES

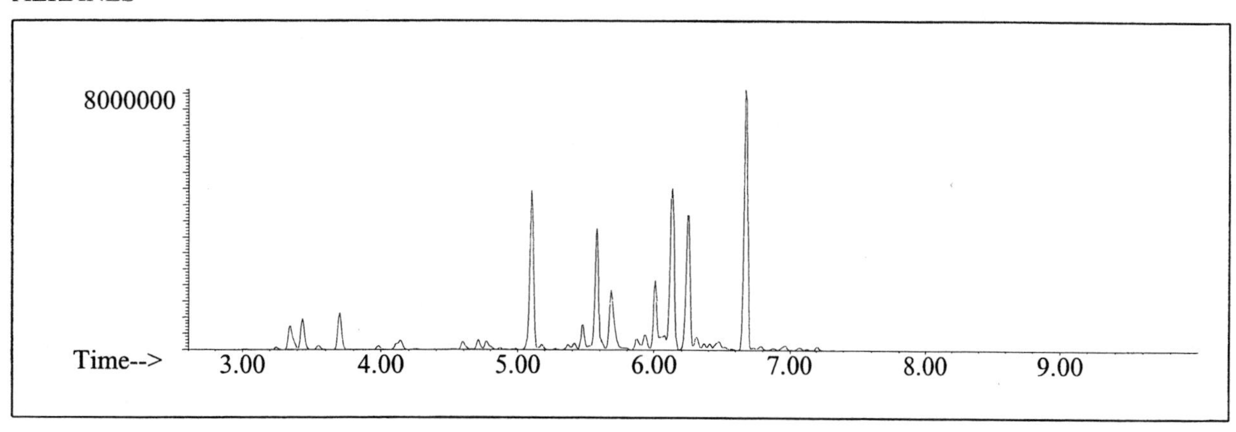

AROMATICS

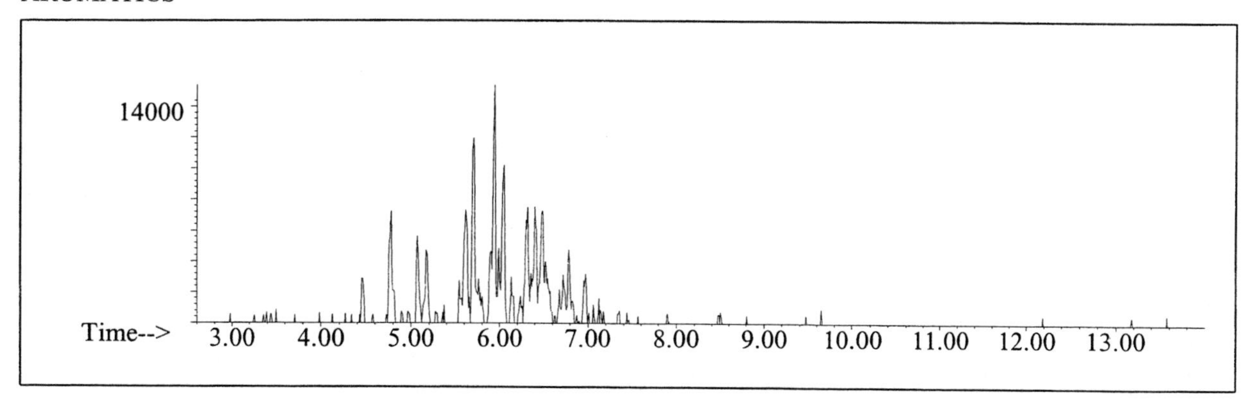

CYCLOPARAFFINS AND ALKENES

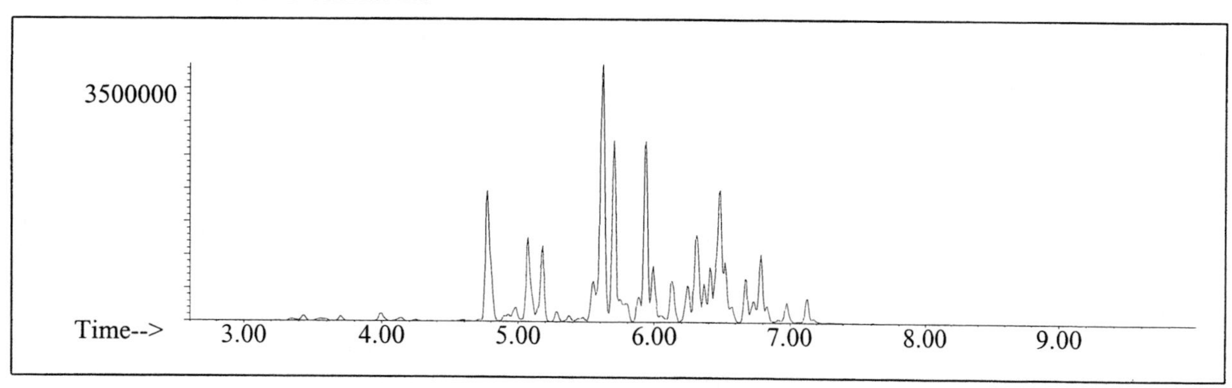

NAPHTHALENES

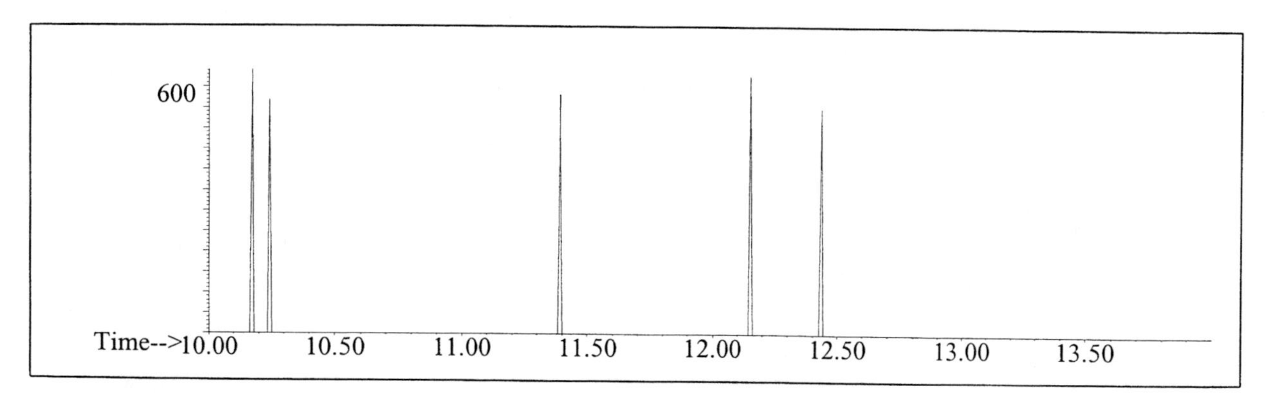

INDIVIDUAL PROFILES

Distillate #14

Alkane

Ion 43

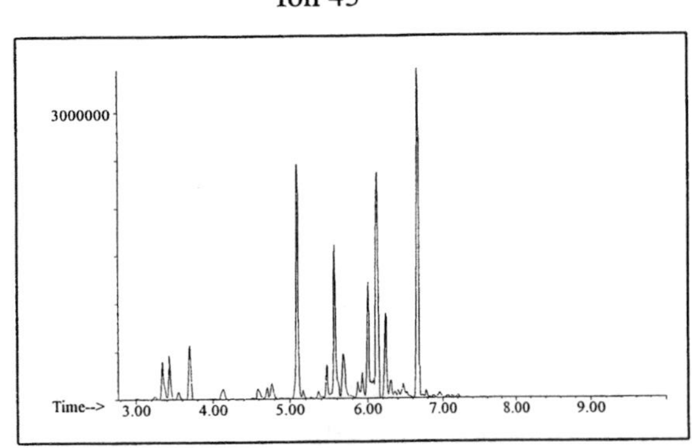

Ion 57

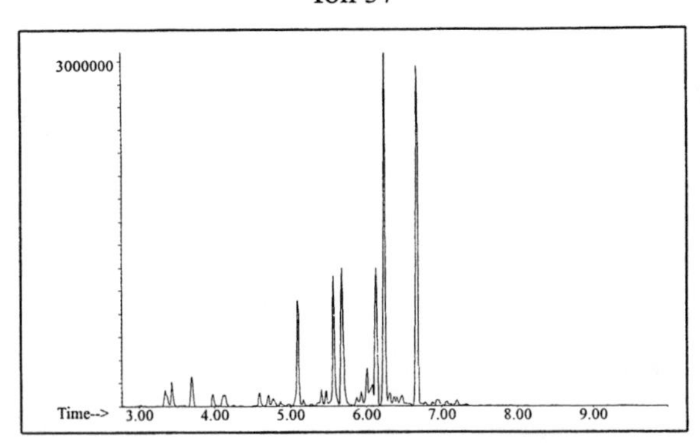

Ion 71

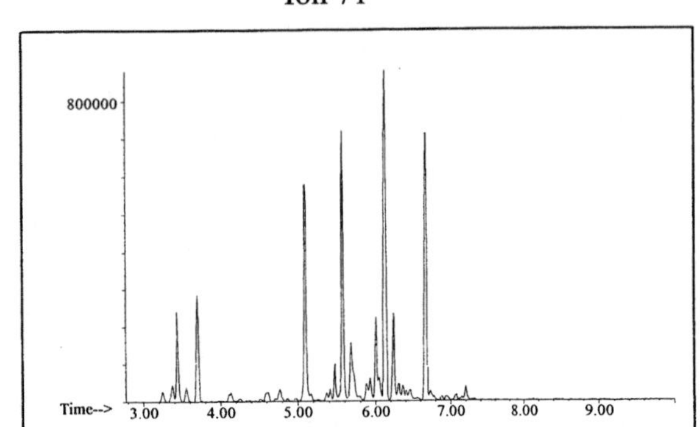

Ion 85

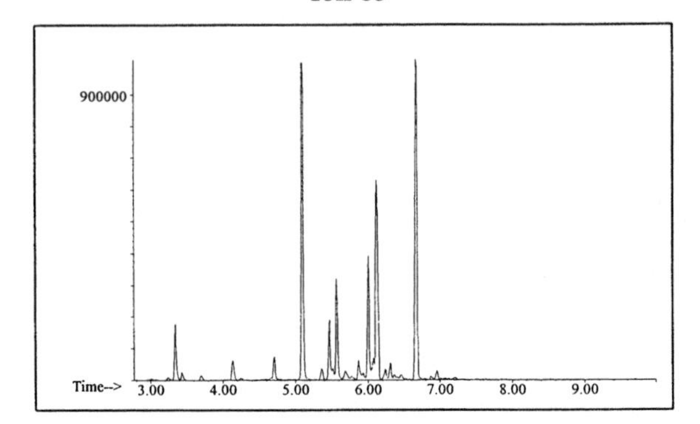

Aromatic

Ion 91

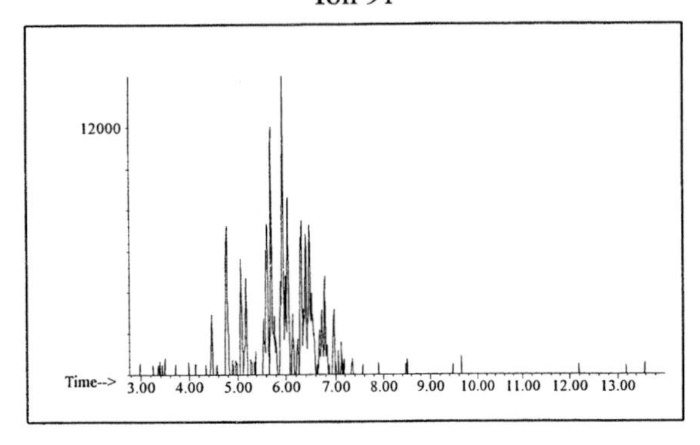

Ion 105

No Useful Data Obtained

Ion 119

No Useful Data Obtained

INDIVIDUAL PROFILES

Distillate #14

Cycloparaffin

Ion 55

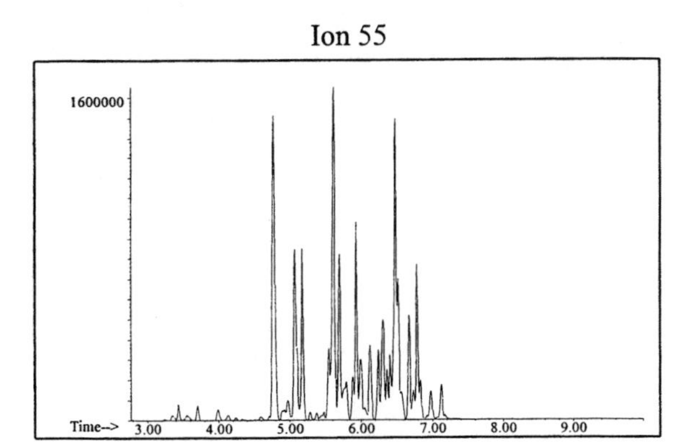

Ion 69

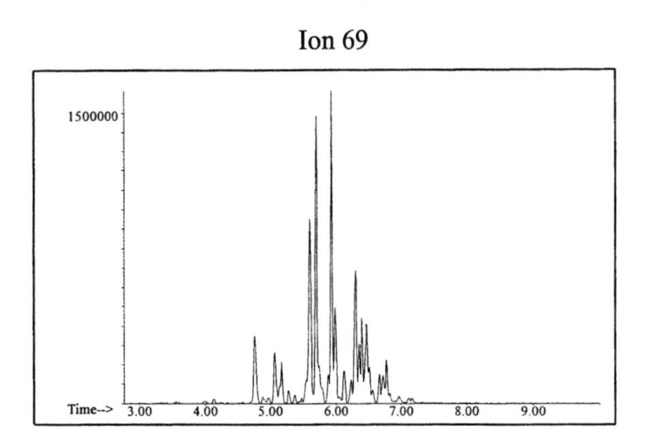

Ion 83

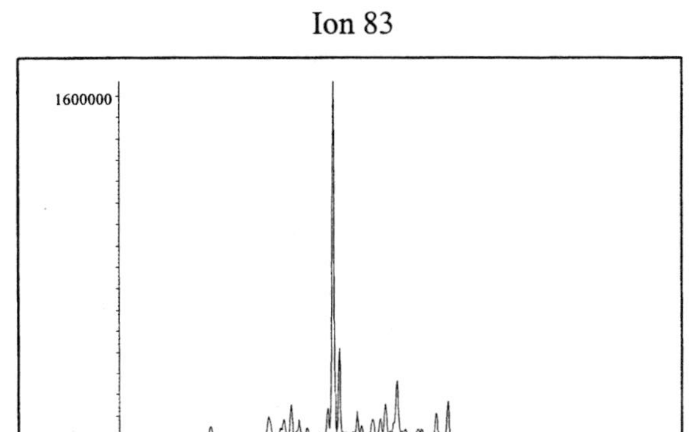

Naphthalene

Ion 128

No Useful Data Obtained

Ion 142

No Useful Data Obtained

Ion 156

No Useful Data Obtained

Distillate #15

20uL/mL Pentane

ASTM: Class 1 (Light Petroleum Distillate)

Macro Code: L

Product Displayed: 75% Evaporated Ronsonol Lighter Fluid

Product Uses: Pocket lighter fluid

Other Similar Products:

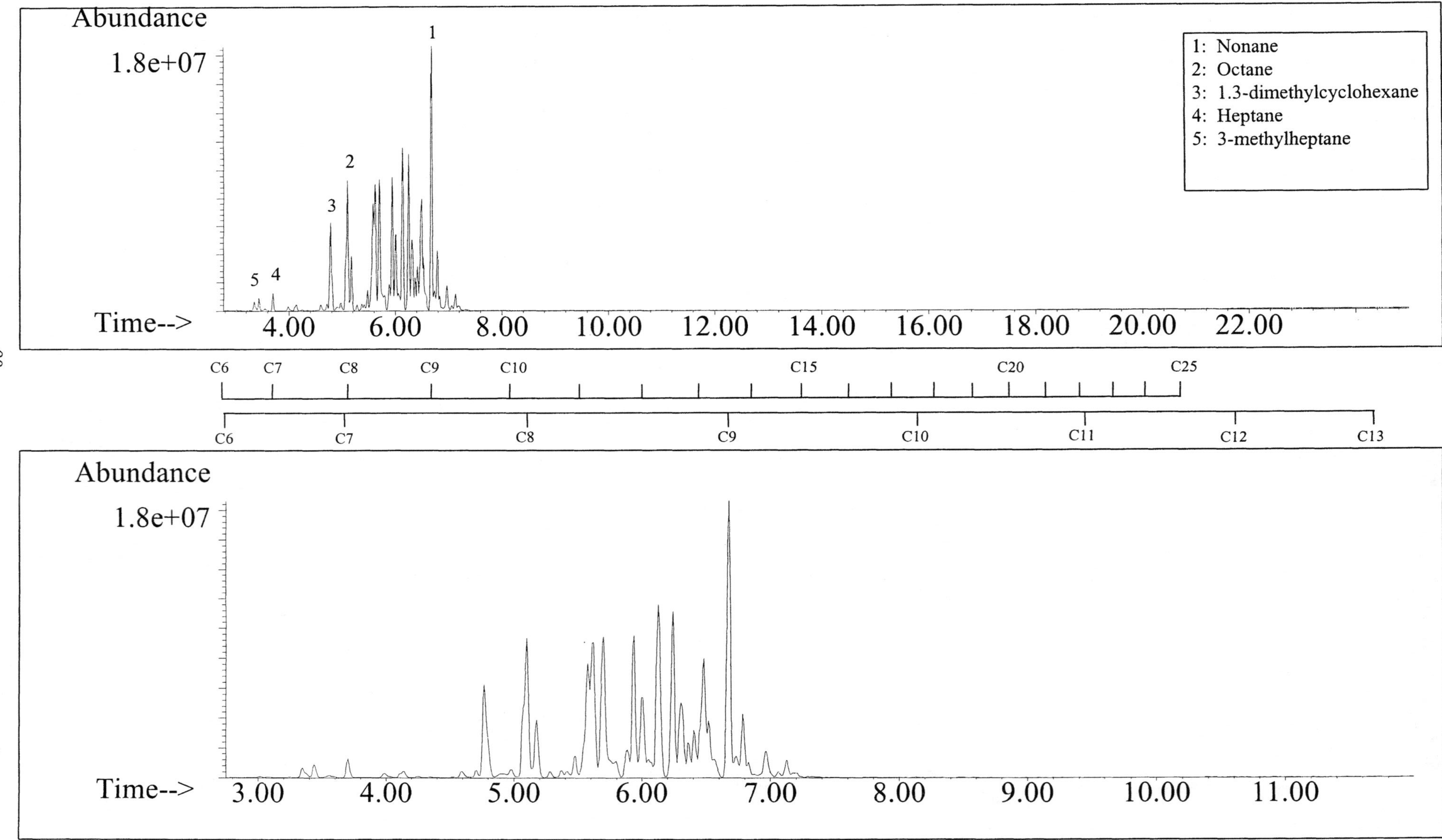

SUMMED PROFILES Distillate #15

ALKANES

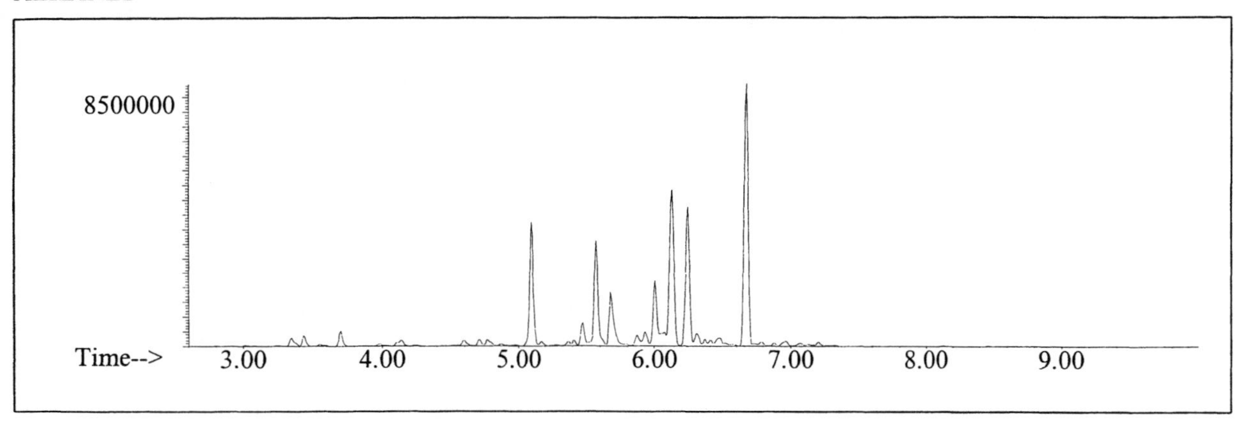

AROMATICS

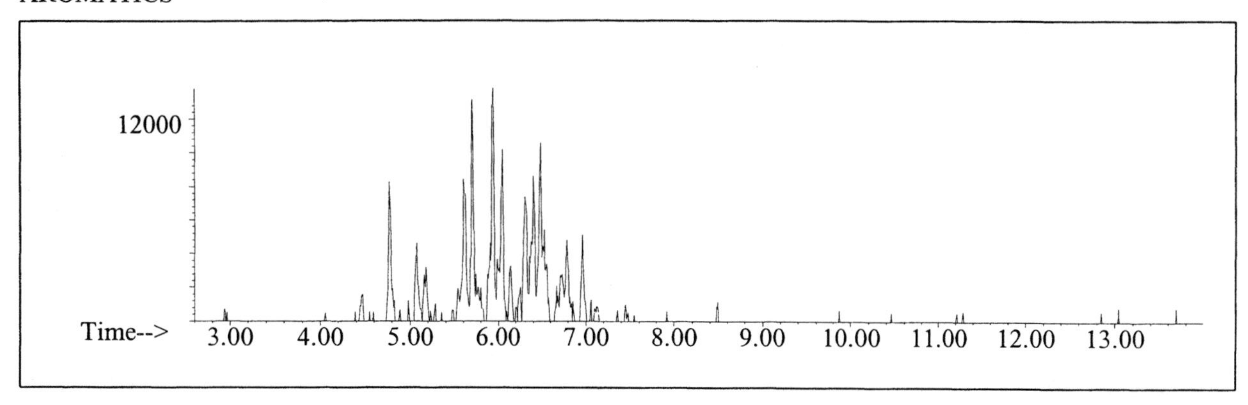

CYCLOPARAFFINS AND ALKENES

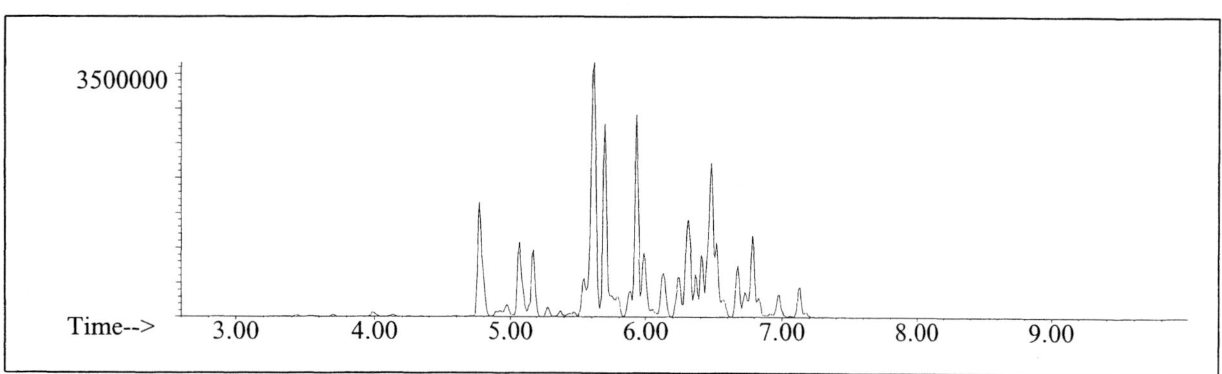

NAPHTHALENES

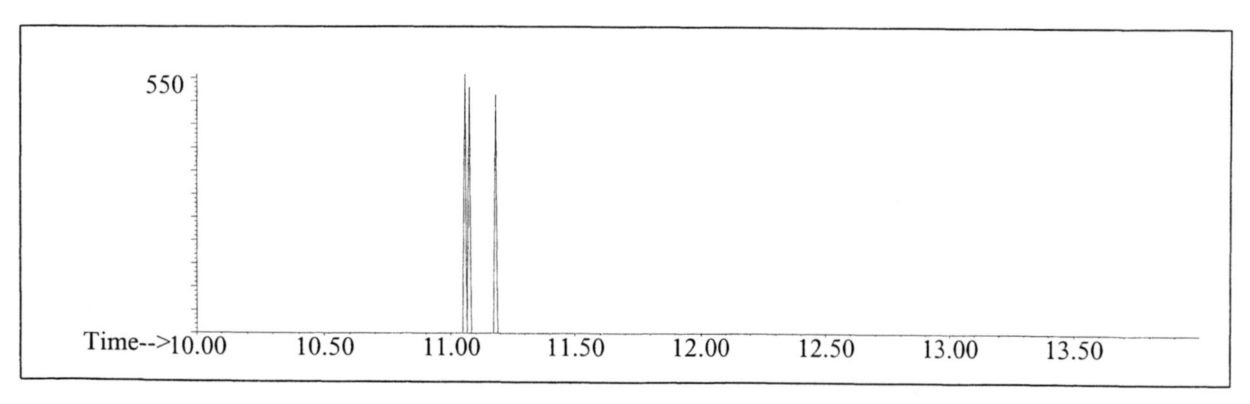

INDIVIDUAL PROFILES

Distillate #15

Alkane

Ion 43

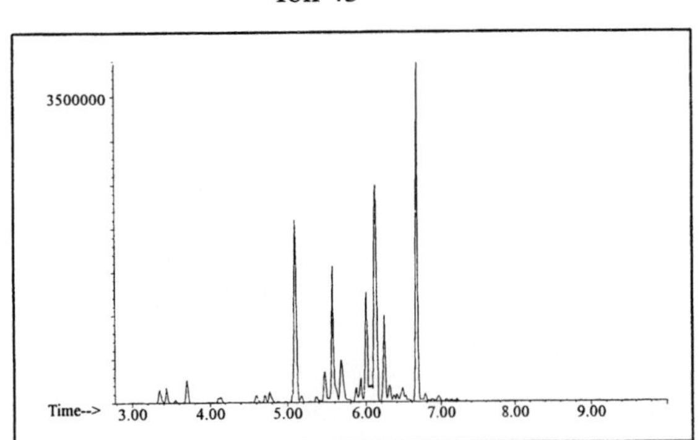

Ion 57

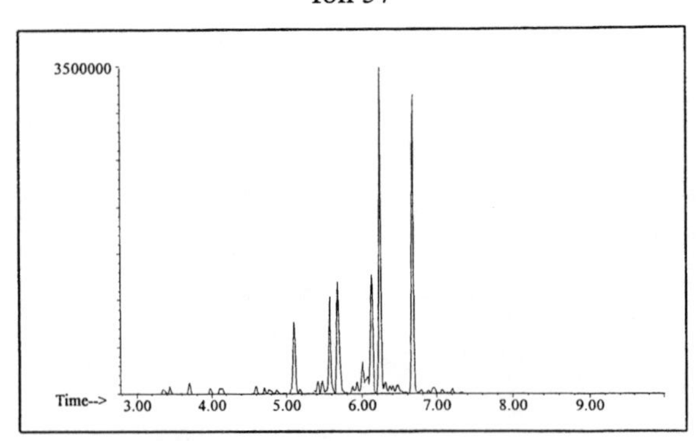

Ion 71

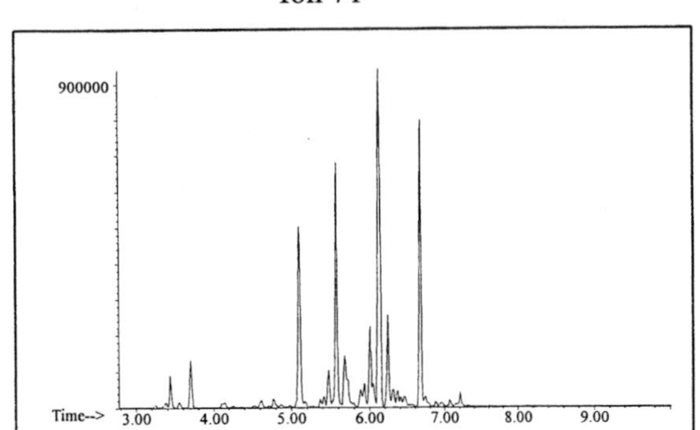

Ion 85

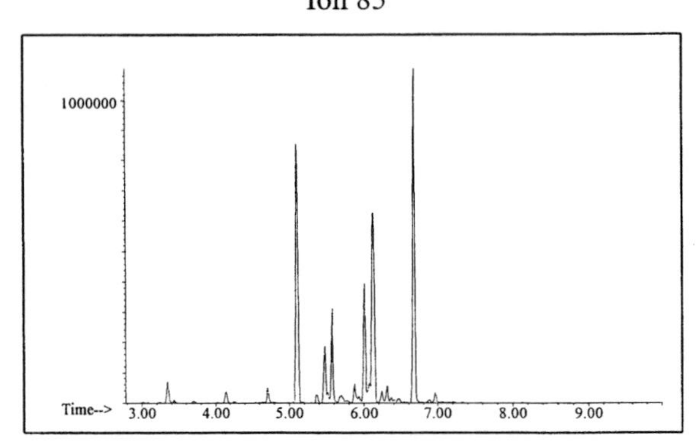

Aromatic

Ion 91

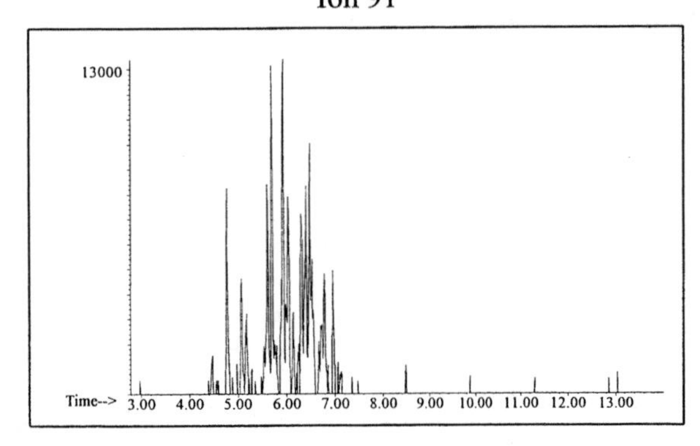

Ion 105

No Useful Data Obtained

Ion 119

No Useful Data Obtained

INDIVIDUAL PROFILES

Distillate #15

Cycloparaffin

Ion 55

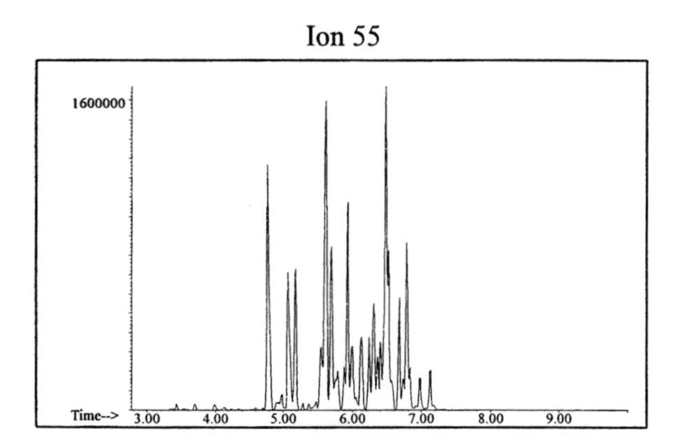

Ion 69

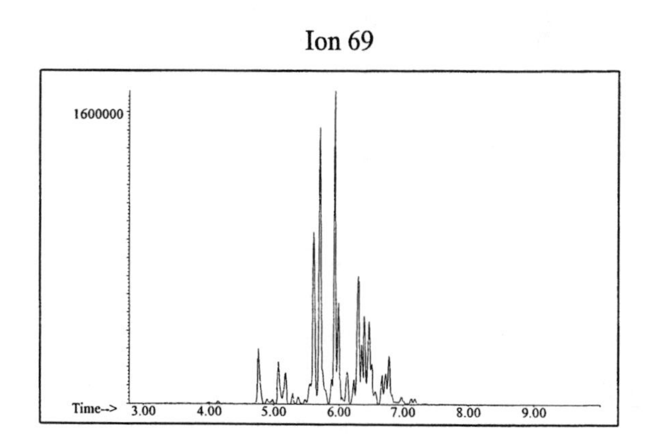

Ion 83

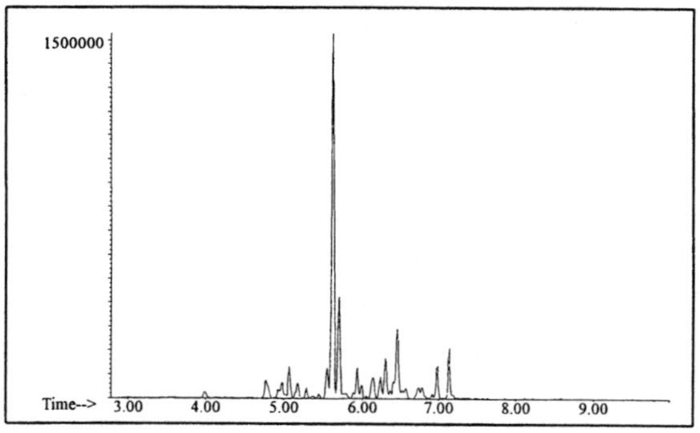

1500000

Time--> 3.00 4.00 5.00 6.00 7.00 8.00 9.00

Naphthalene

Ion 128

No Useful Data Obtained

Ion 142

No Useful Data Obtained

Ion 156

No Useful Data Obtained

Distillate #16

20uL/mL Pentane

Product Displayed: 90% Evaporated Ronsonol Lighter Fluid

Other Similar Products:

ASTM: Class 1 (Light Petroleum Distillate)

Product Uses: Pocket lighter fluid

Macro Code: L

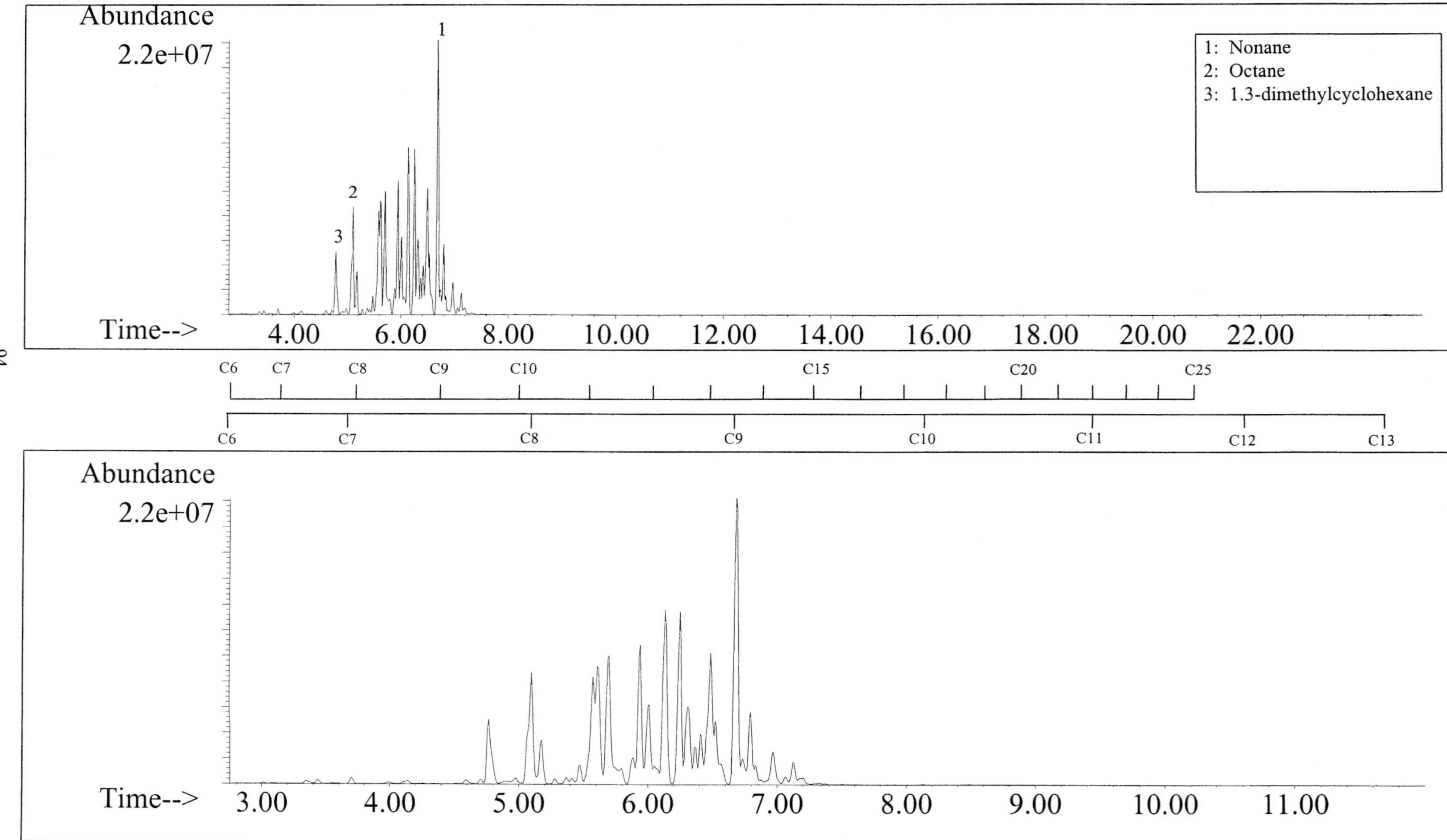

SUMMED PROFILES

Distillate #16

ALKANES

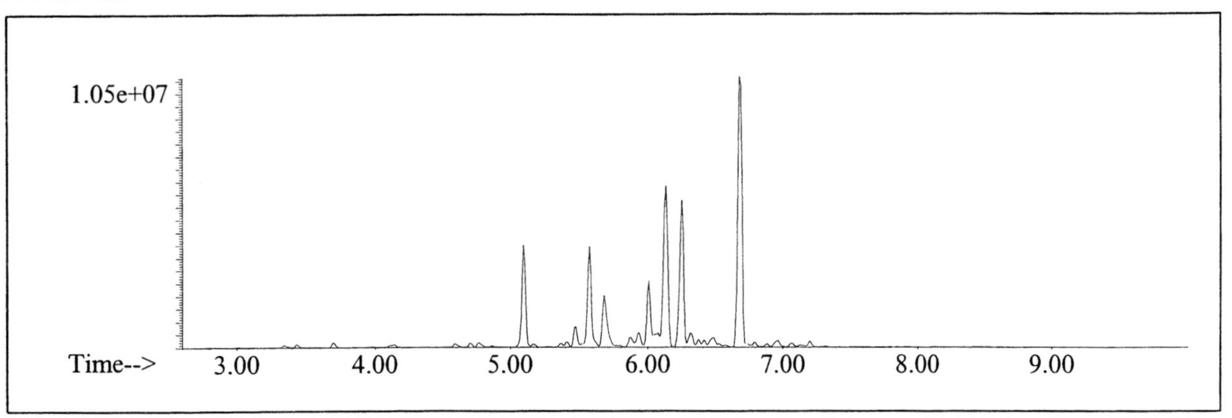

AROMATICS

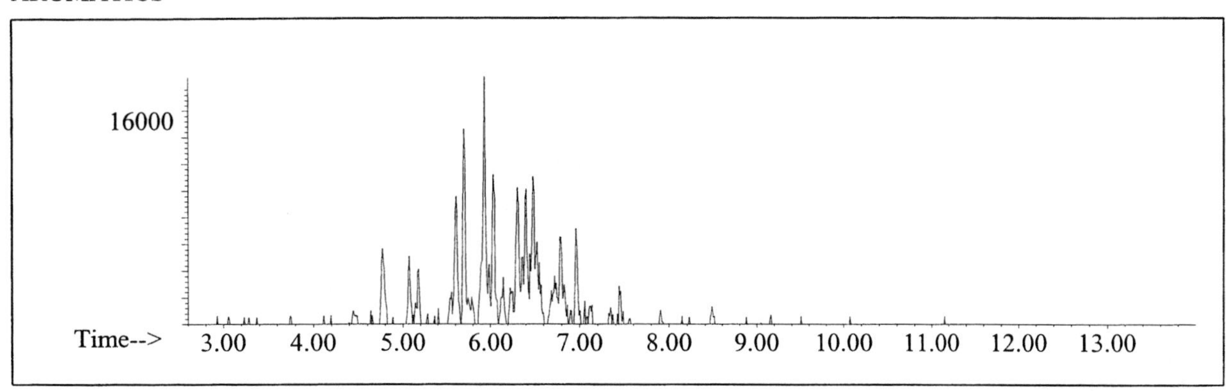

CYCLOPARAFFINS AND ALKENES

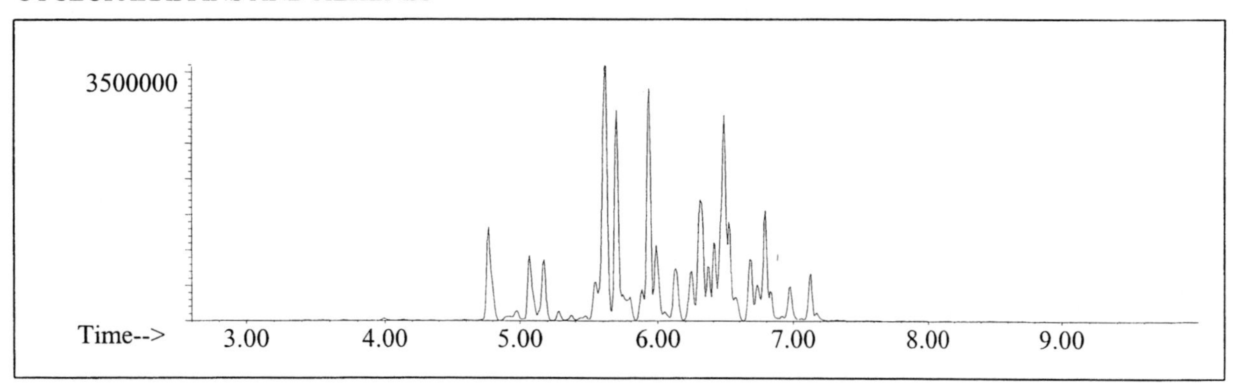

NAPHTHALENES

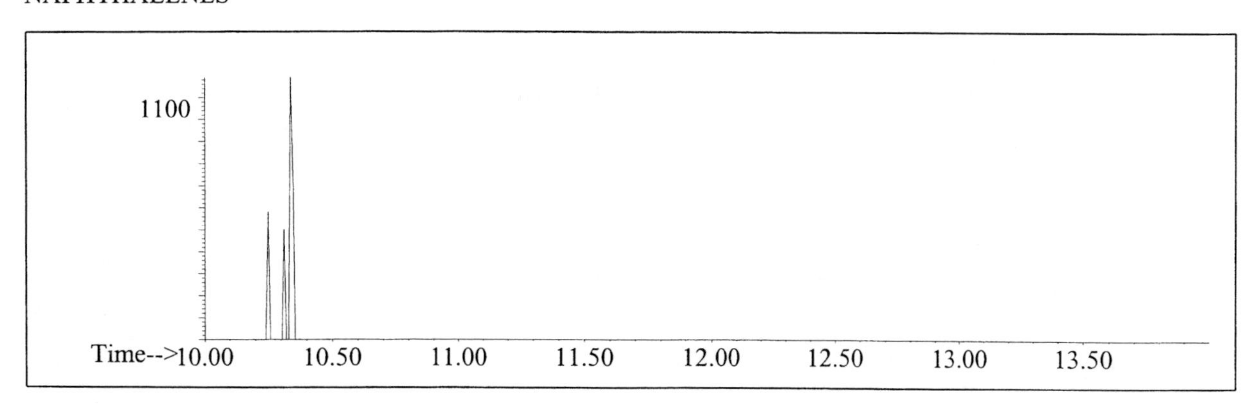

INDIVIDUAL PROFILES

Distillate #16

Alkane

Ion 43

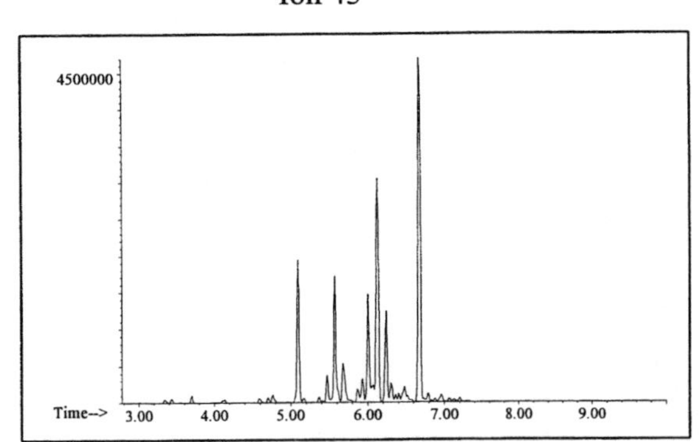

Ion 57

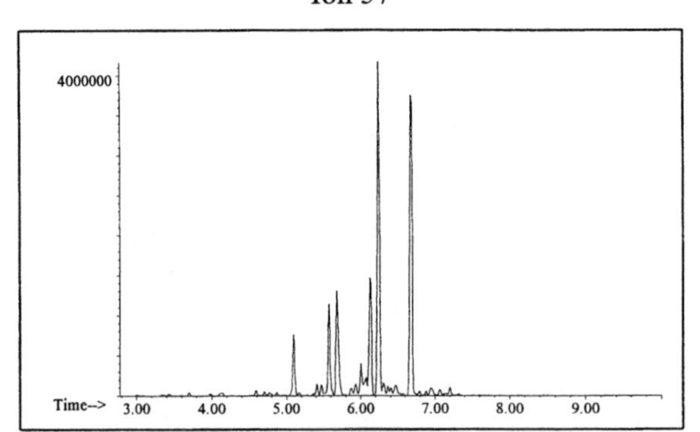

Ion 71

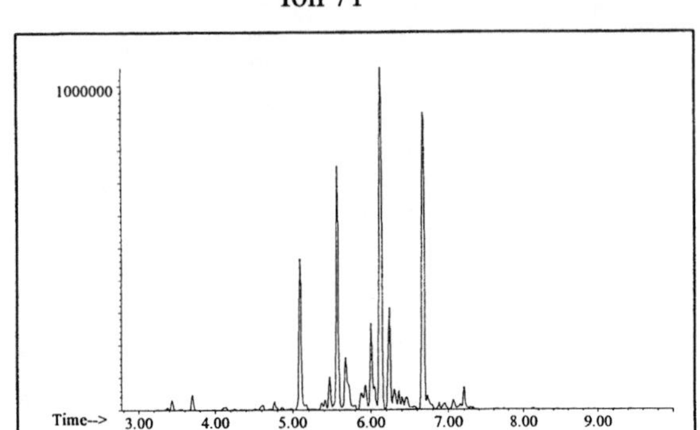

Ion 85

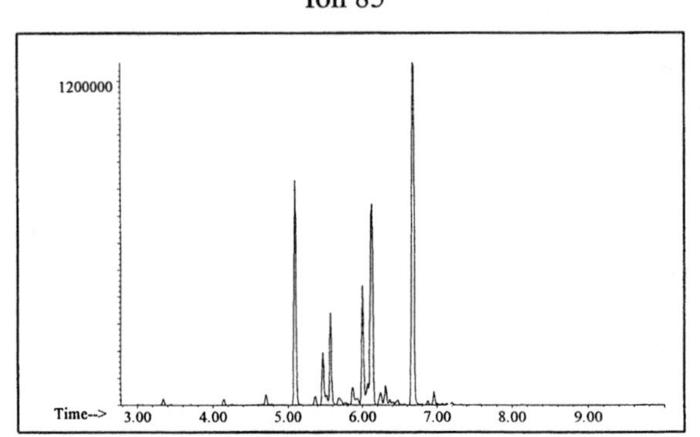

Aromatic

Ion 91

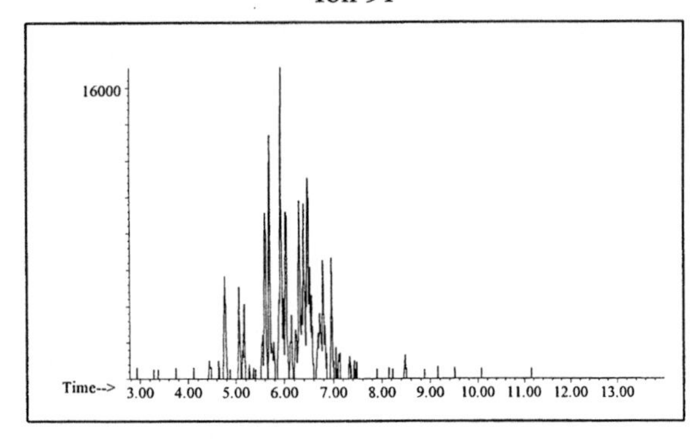

Ion 105

No Useful Data Obtained

Ion 119

No Useful Data Obtained

INDIVIDUAL PROFILES

Distillate #16

Cycloparaffin

Ion 55

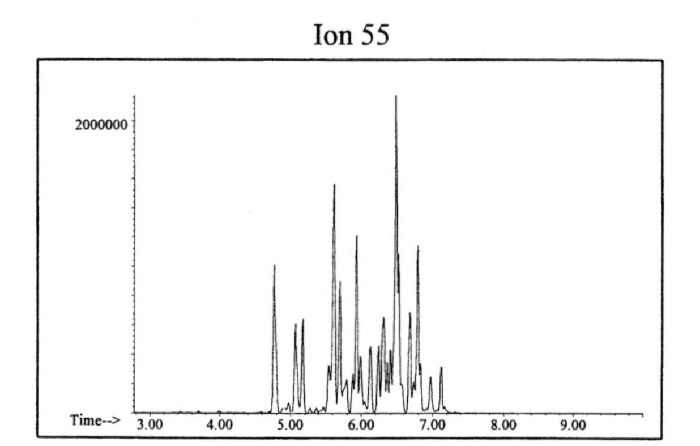

Ion 69

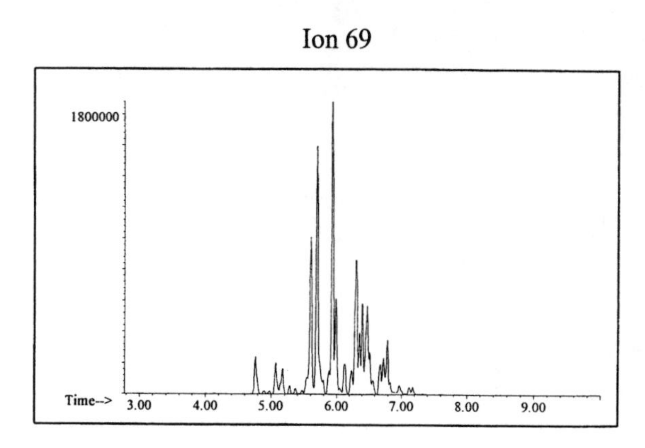

Ion 83

1400000

Time--> 3.00 4.00 5.00 6.00 7.00 8.00 9.00

Naphthalene

Ion 128

No Useful Data Obtained

Ion 142

No Useful Data Obtained

Ion 156

No Useful Data Obtained

Distillate #17

20uL/mL Pentane

Product Displayed: Camplite Camp Fuel

Other Similar Products:

ASTM: Class 1 (Light Petroleum Distillate)

Product Uses: Camping fuel

Macro Code: L

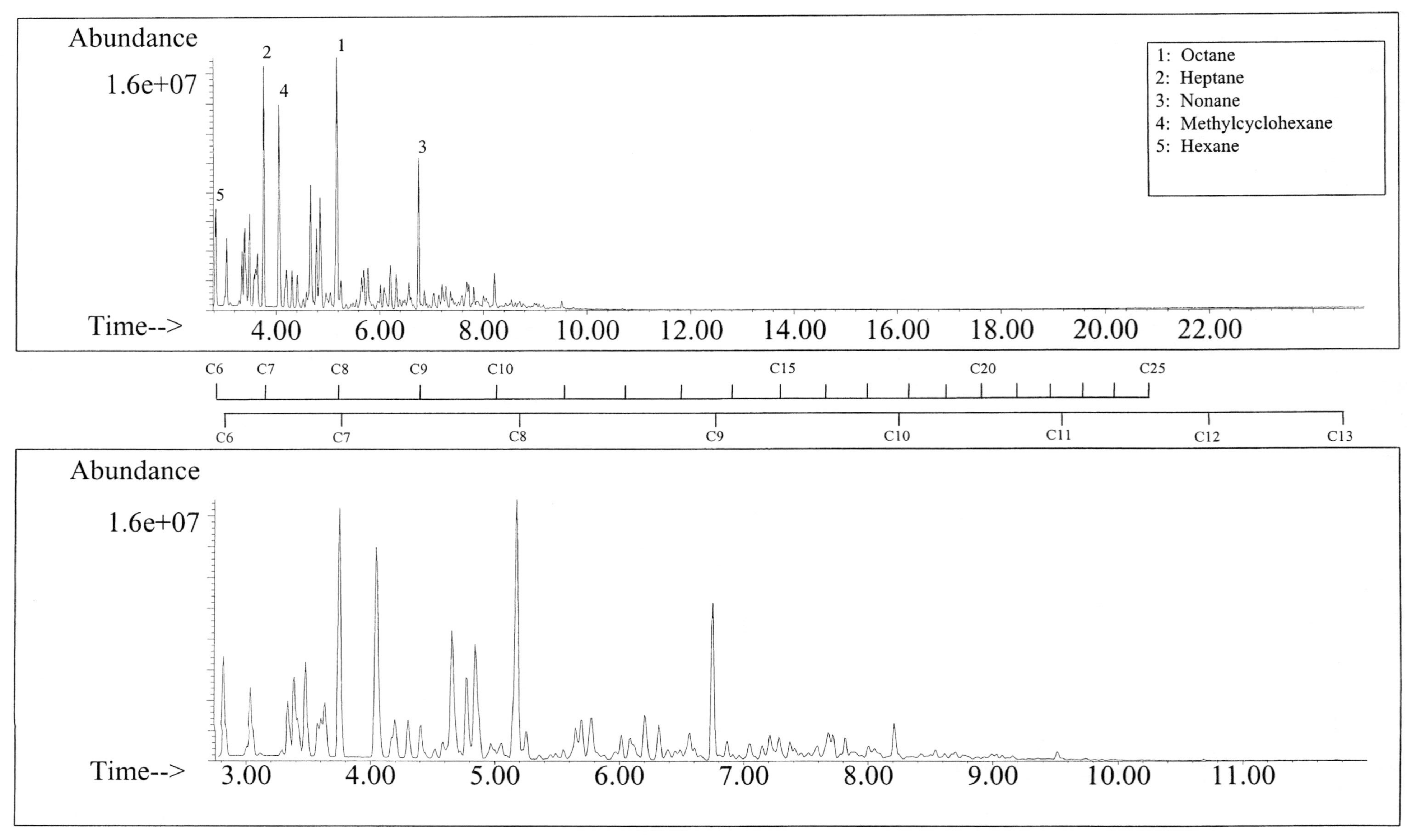

SUMMED PROFILES

Distillate #17

ALKANES

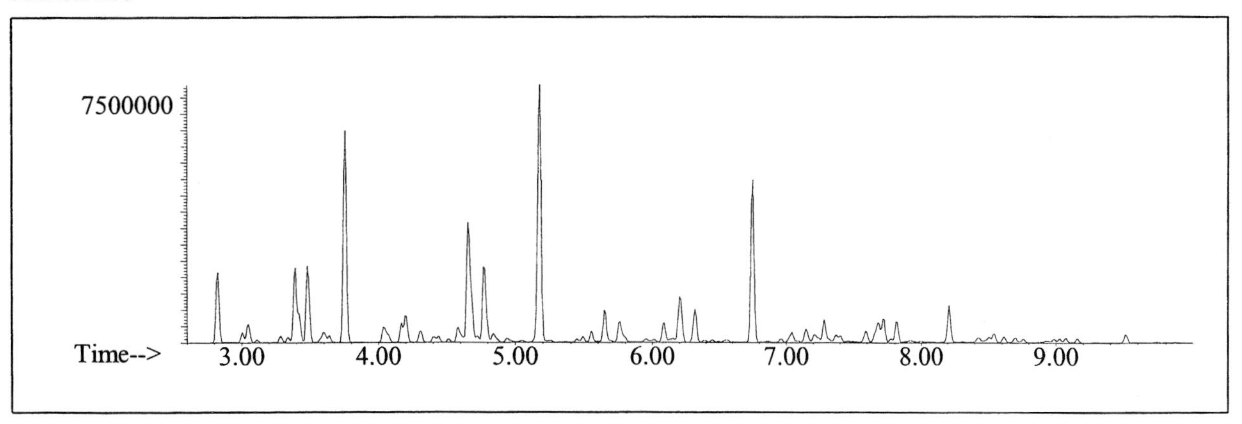

AROMATICS

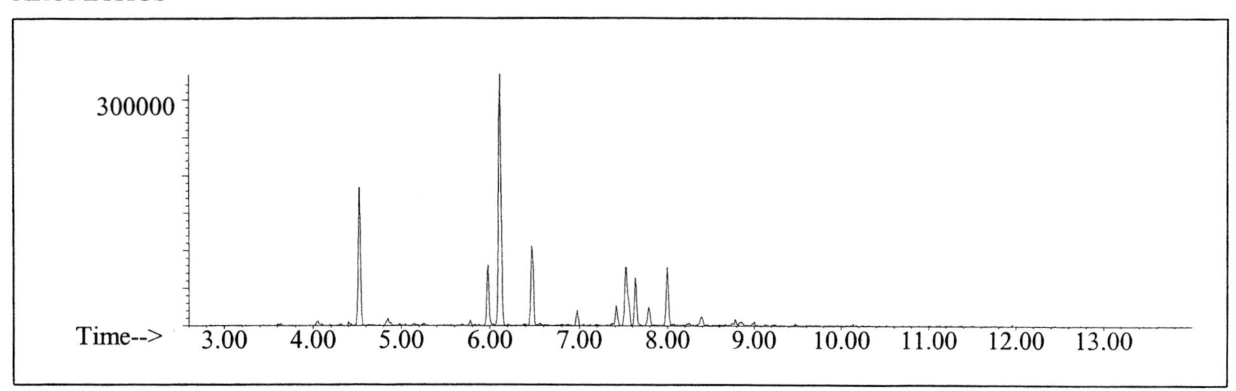

CYCLOPARAFFINS AND ALKENES

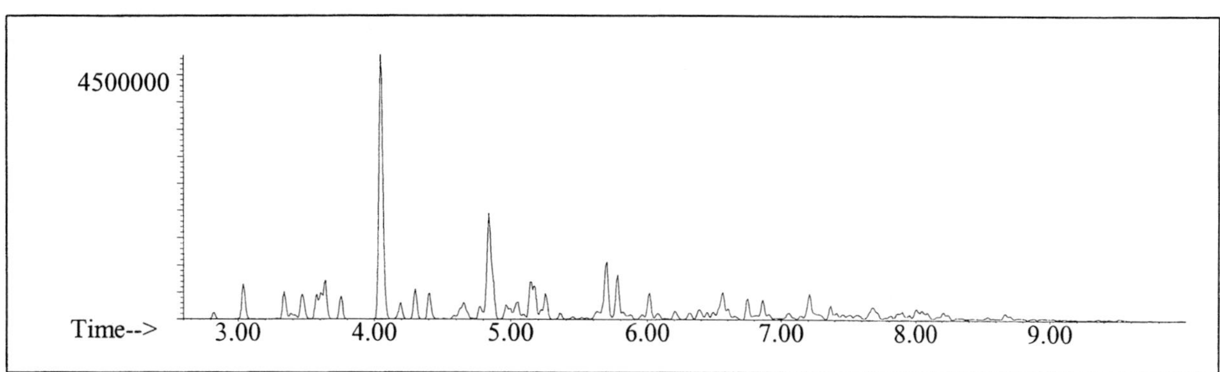

NAPHTHALENES

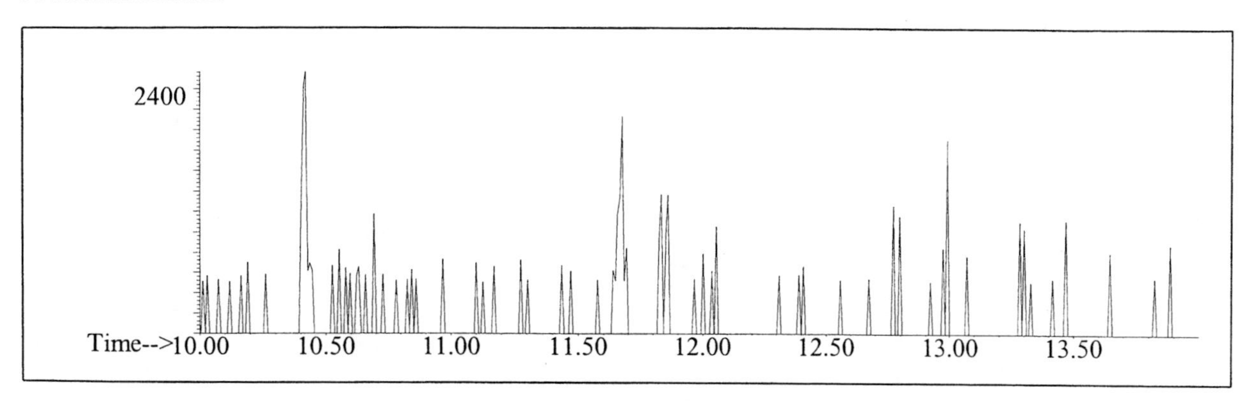

INDIVIDUAL PROFILES Distillate #17

Alkane

Ion 43

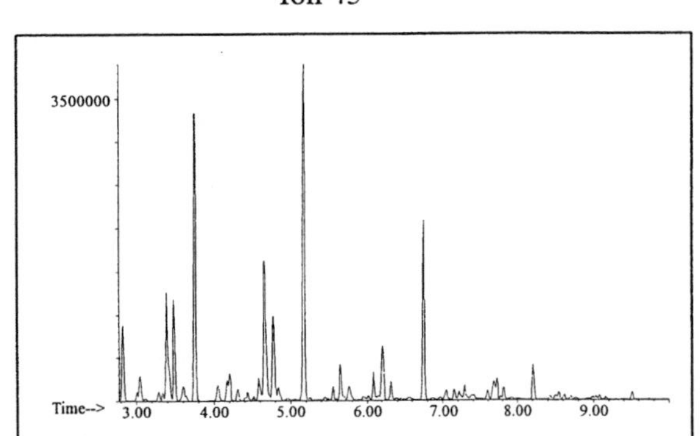

Ion 57

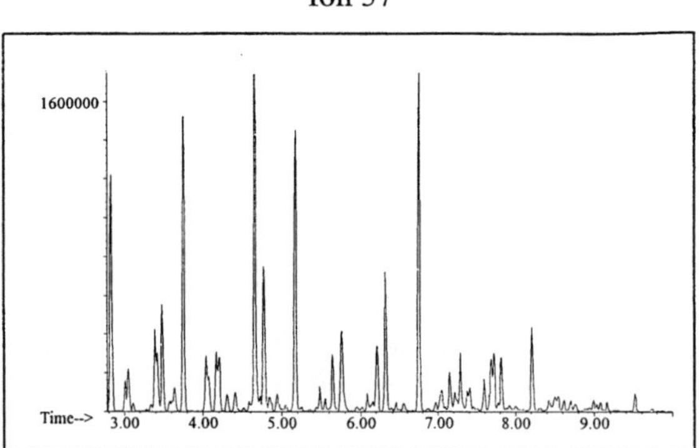

Ion 71

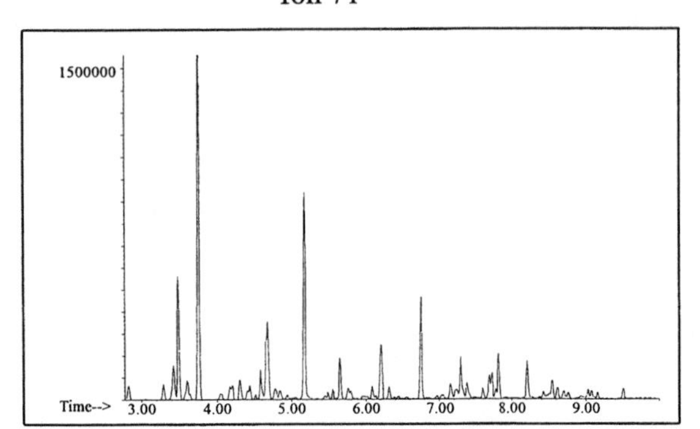

Ion 85

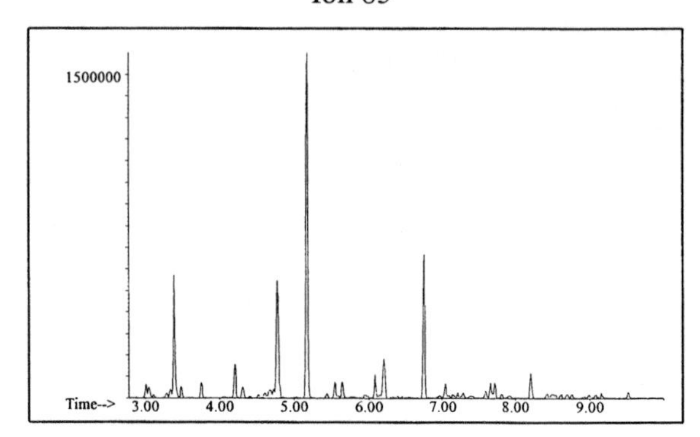

Aromatic

Ion 91

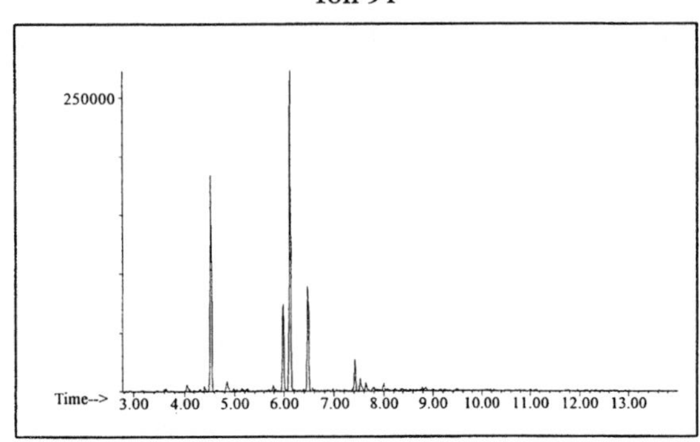

Ion 105

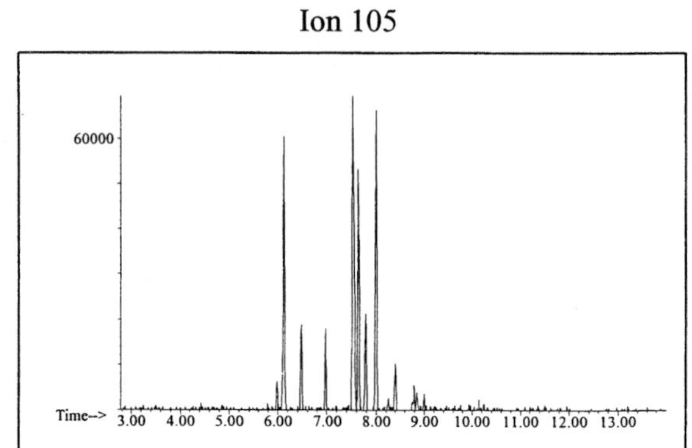

Ion 119

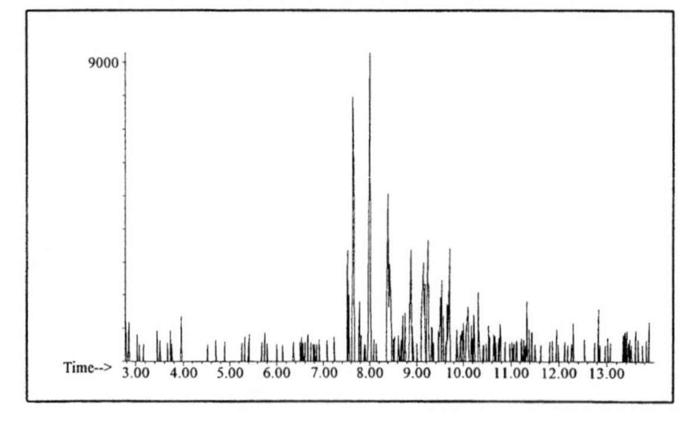

INDIVIDUAL PROFILES

Distillate #17

Cycloparaffin

Ion 55

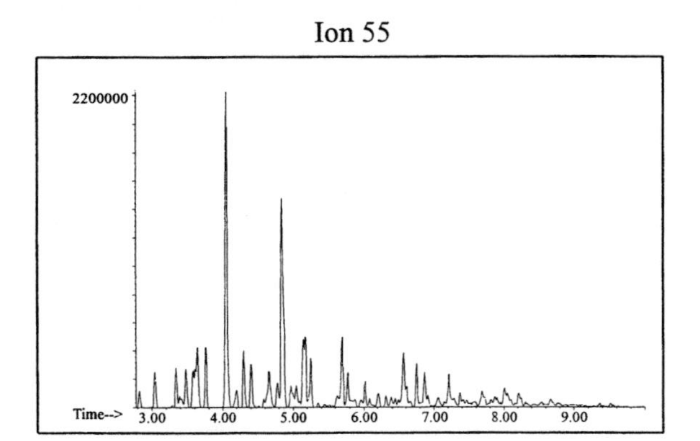

Ion 69

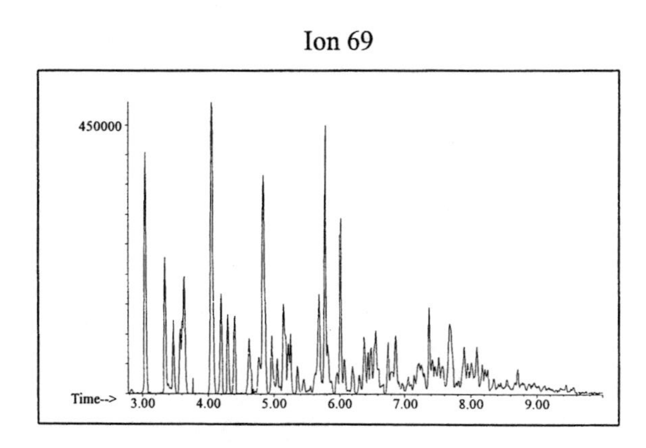

Ion 83

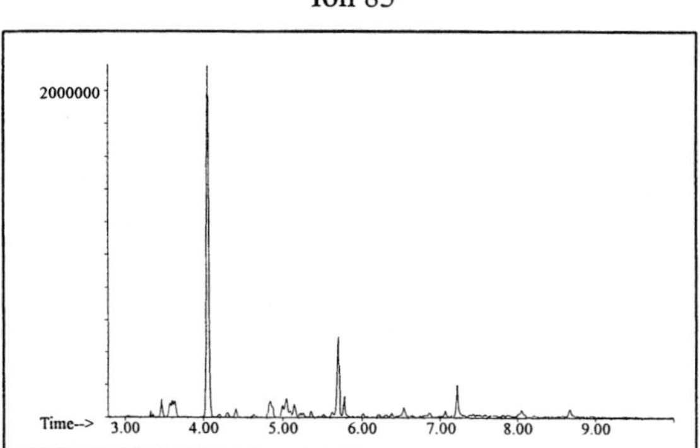

Naphthalene

Ion 128

No Useful Data Obtained

Ion 142

No Useful Data Obtained

Ion 156

No Useful Data Obtained

Distillate #18

20uL/mL Pentane

Product Displayed: Chemical Corp Mineral Spirits

Other Similar Products:

ASTM: Class 3 (Medium Petroleum Distillate) Macro Code: M

Product Uses: Mineral spirits

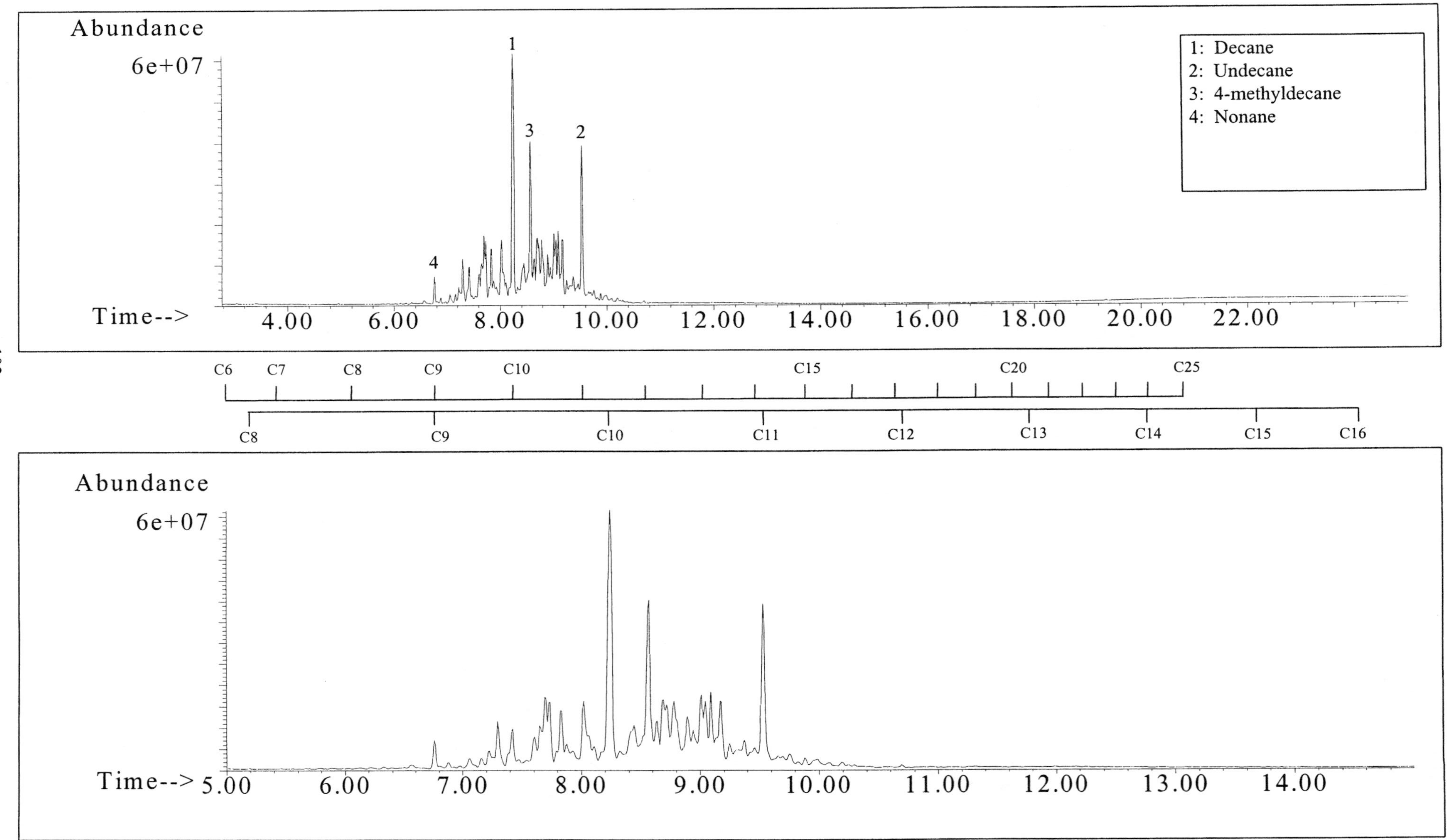

SUMMED PROFILES

Distillate #18

ALKANES

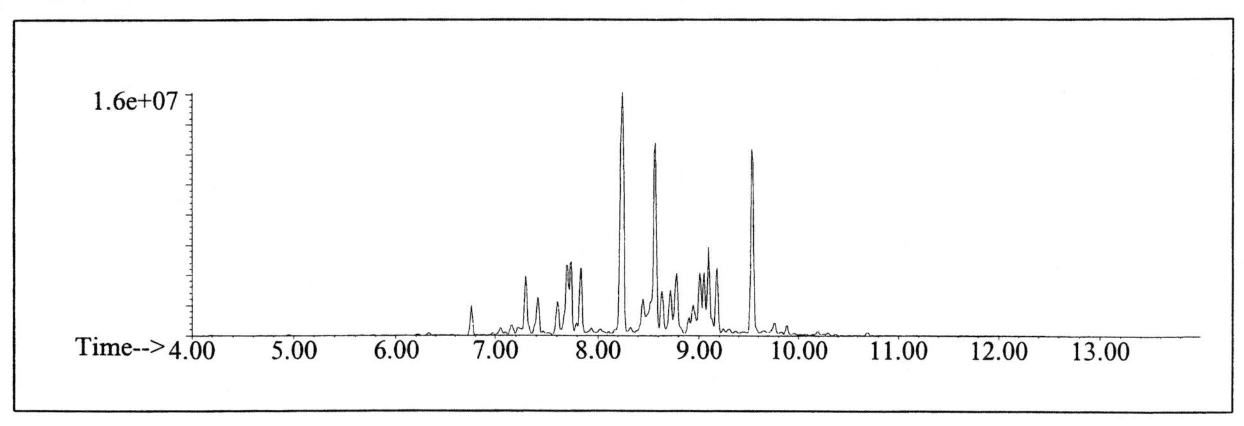

AROMATICS

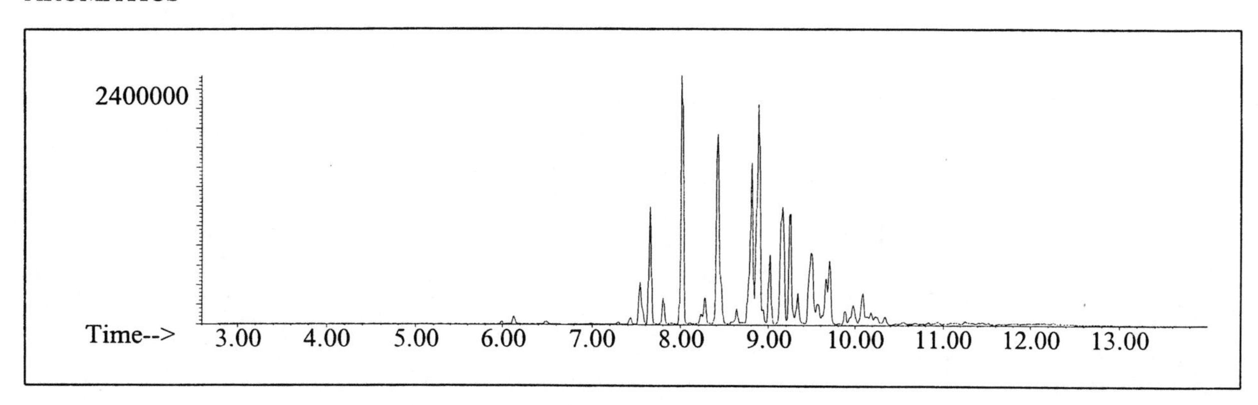

CYCLOPARAFFINS AND ALKENES

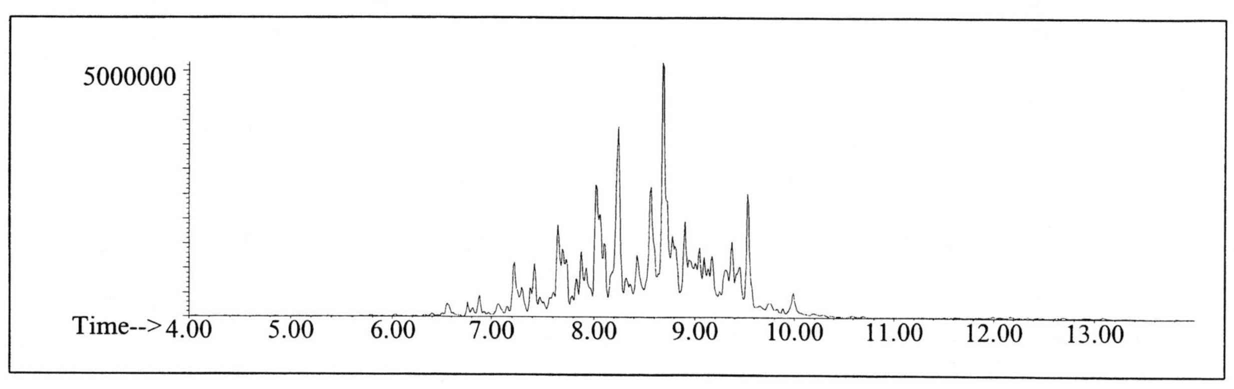

NAPHTHALENES

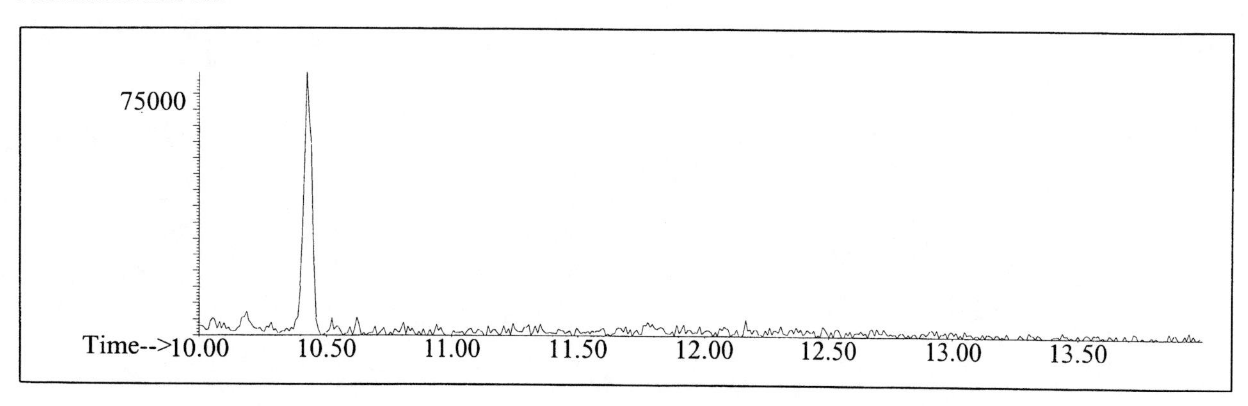

INDIVIDUAL PROFILES

Distillate #18

Alkane

Ion 43

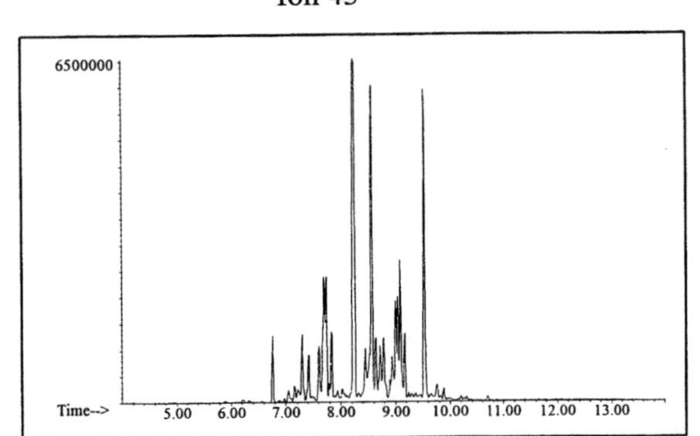

Ion 57

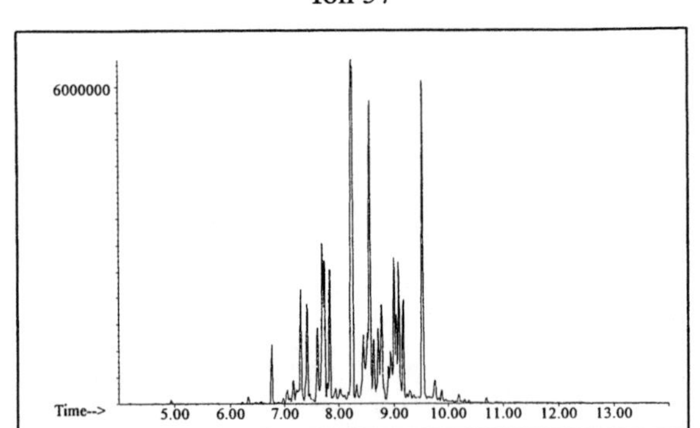

Ion 71

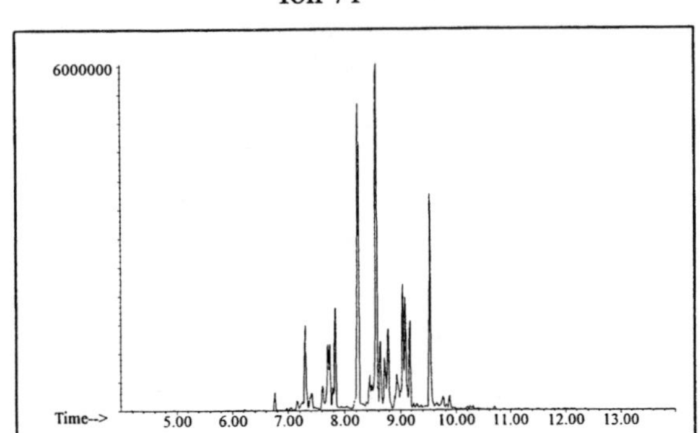

Ion 85

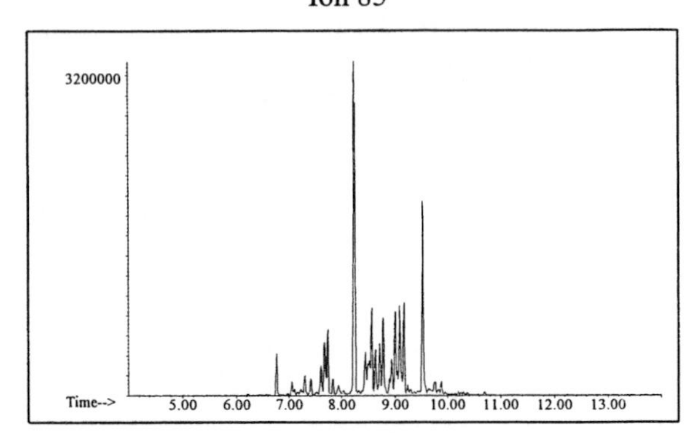

Aromatic

Ion 91

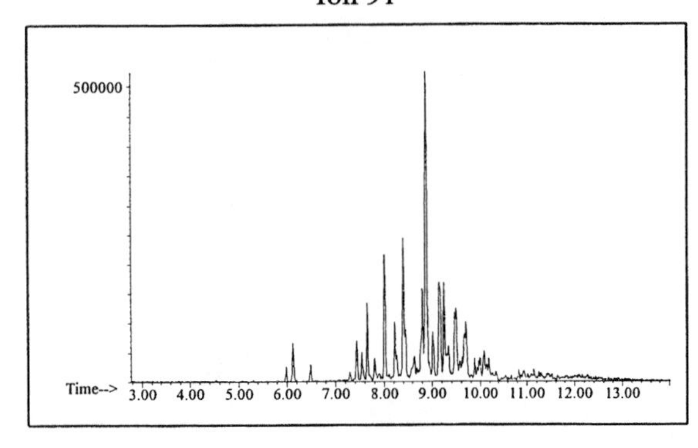

Ion 105

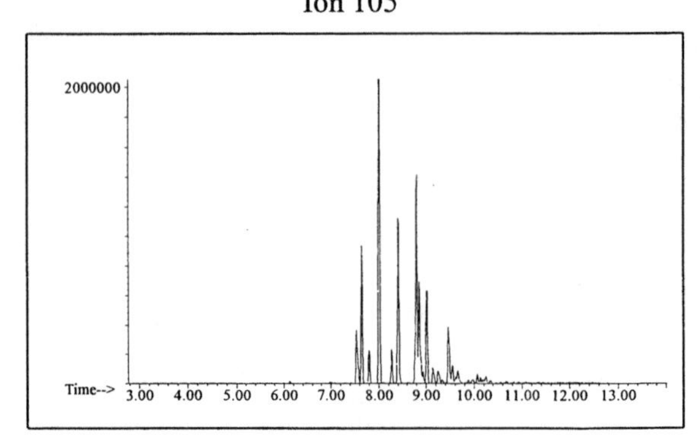

Ion 119

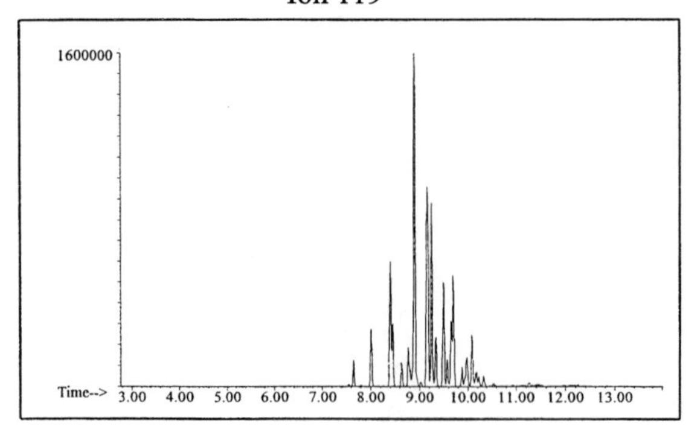

INDIVIDUAL PROFILES

Distillate #18

Cycloparaffin

Ion 55

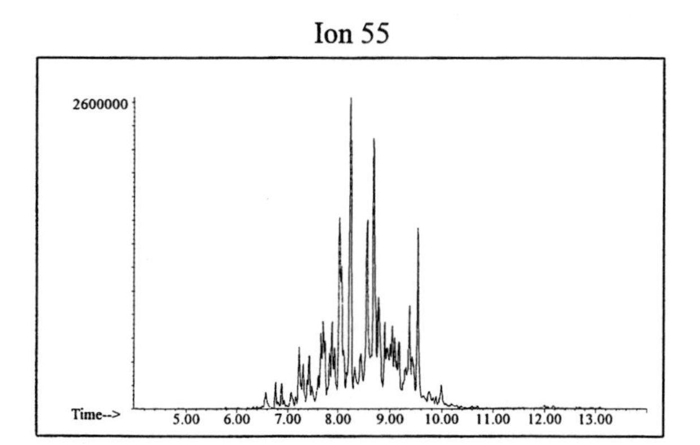

Ion 69

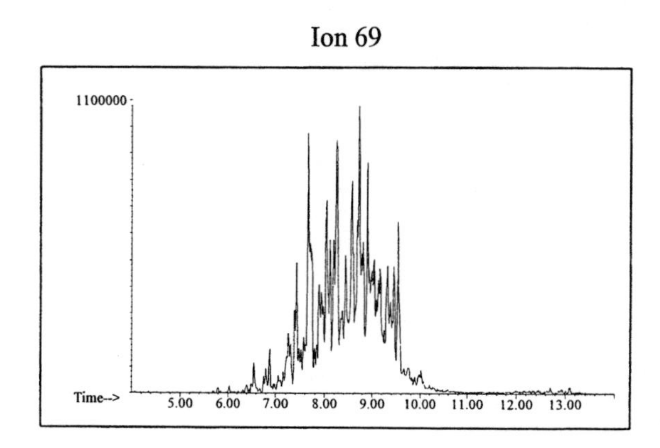

Ion 83

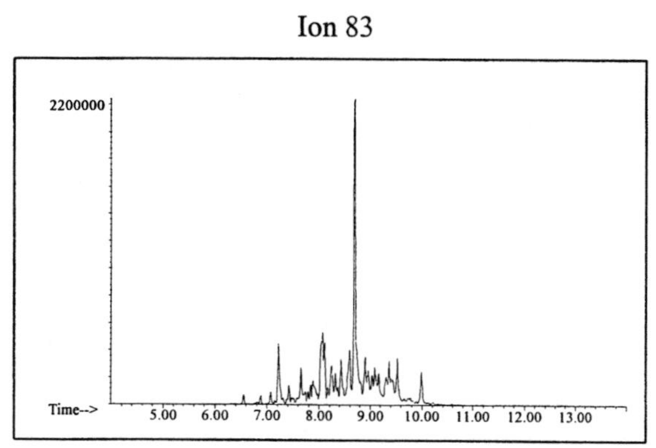

Naphthalene

Ion 128

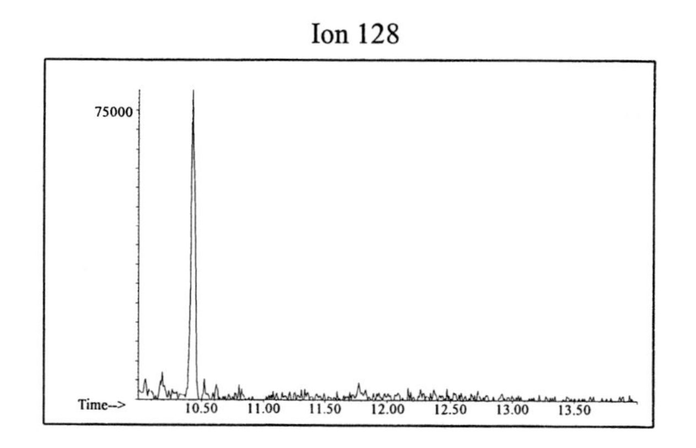

Ion 142

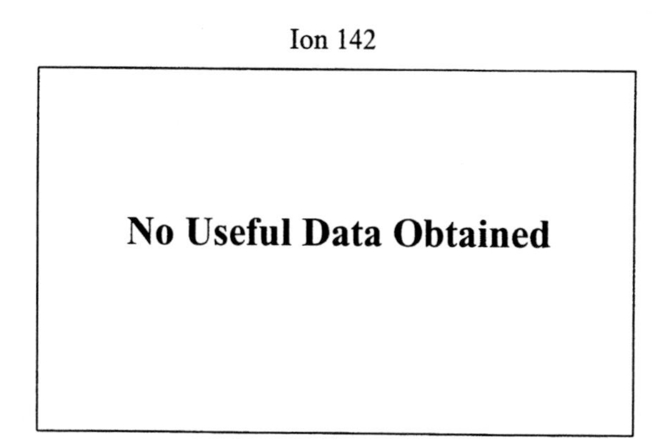

Ion 156

No Useful Data Obtained

Distillate #19

20uL/mL Pentane

Product Displayed: Kleen Strip Mineral Spirits

Other Similar Products: DuPont Prep Sol

ASTM: Class 3 (Medium Petroleum Distillate) Macro Code: M

Product Uses: Mineral Spirits

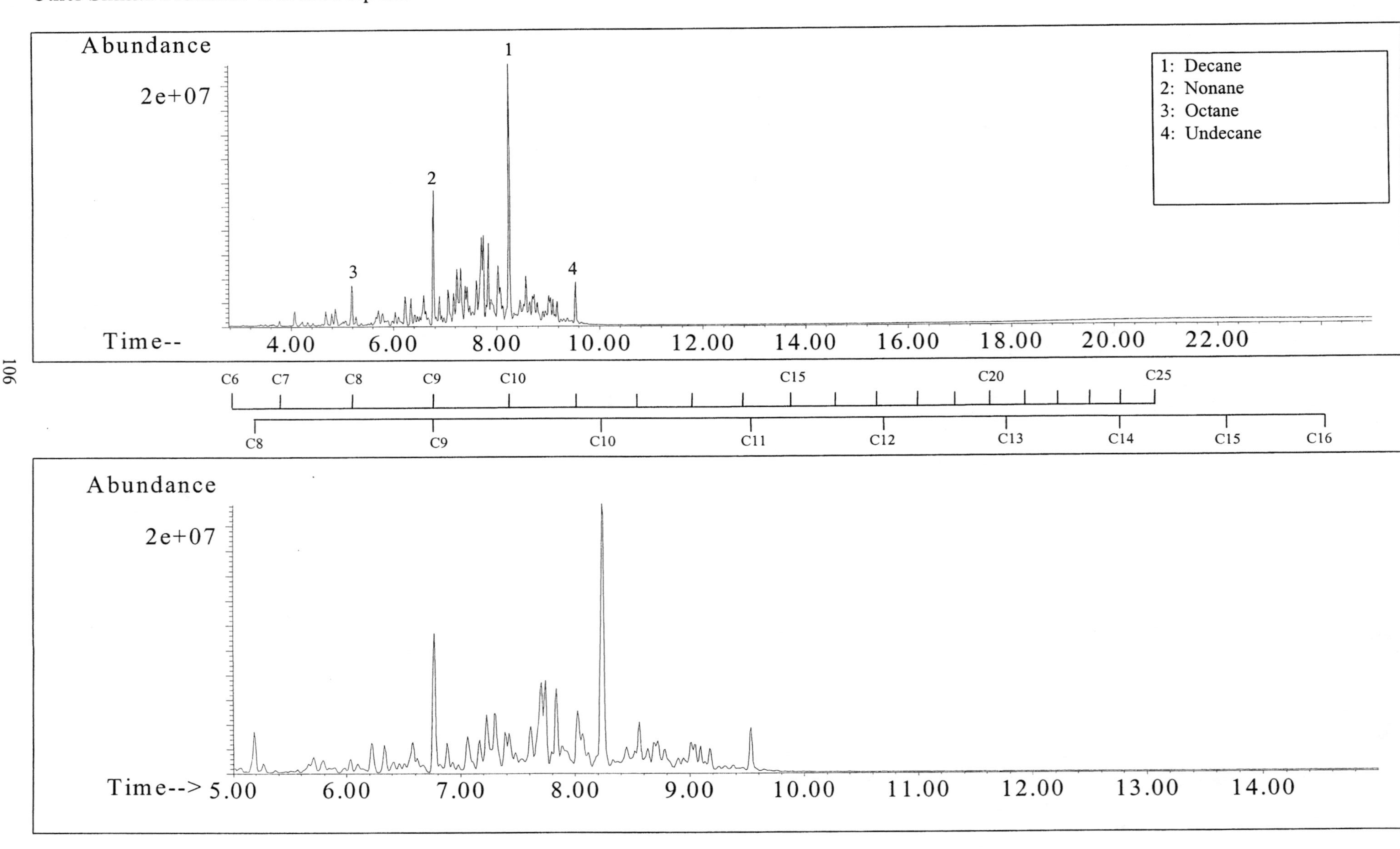

SUMMED PROFILES

Distillate #19

ALKANES

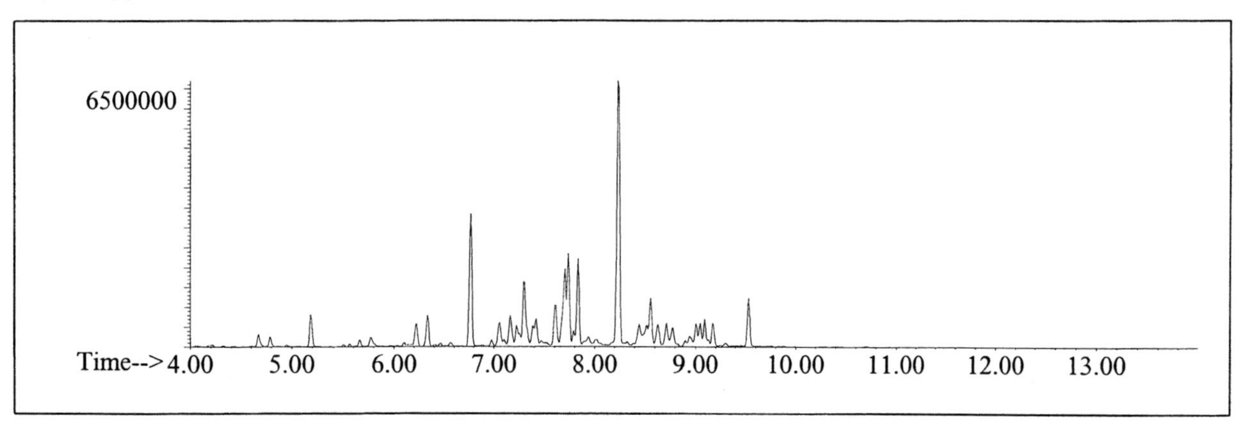

AROMATICS

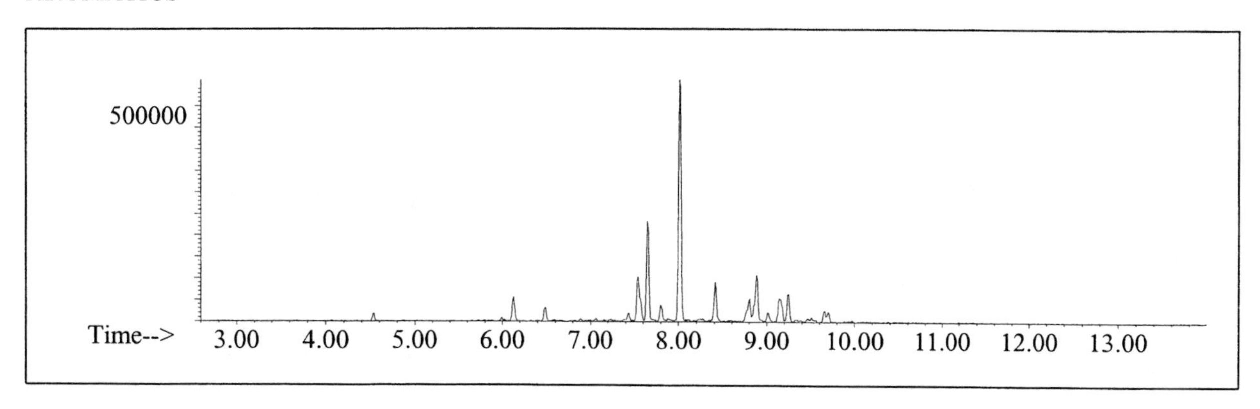

CYCLOPARAFFINS AND ALKENES

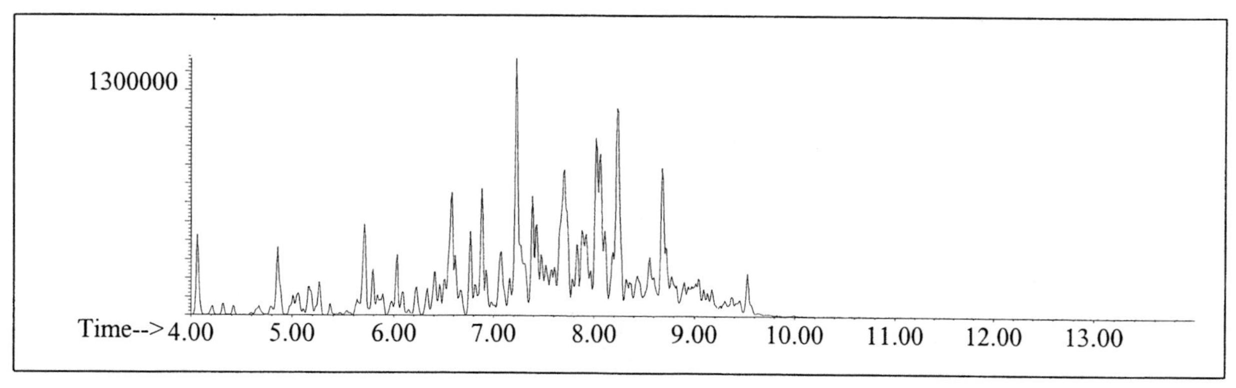

NAPHTHALENES

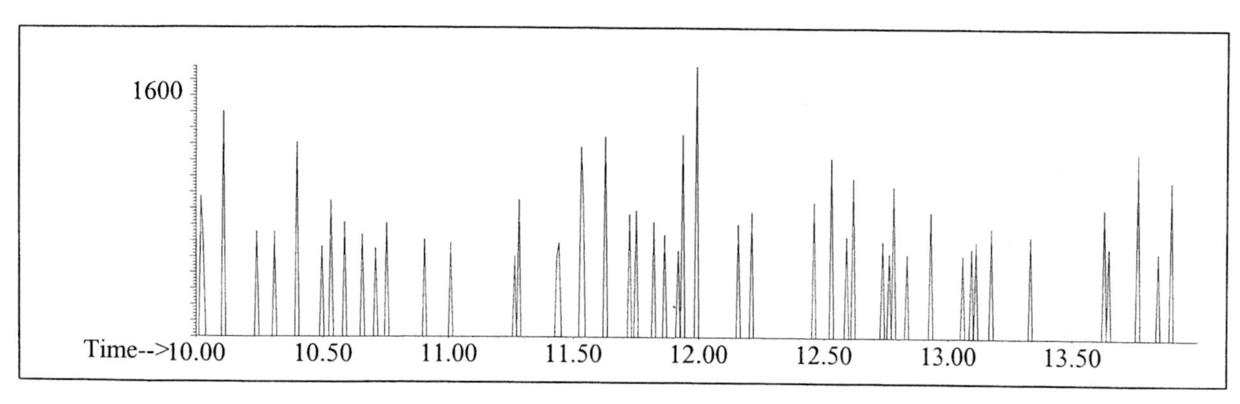

INDIVIDUAL PROFILES

Distillate #19

Alkane

Ion 43

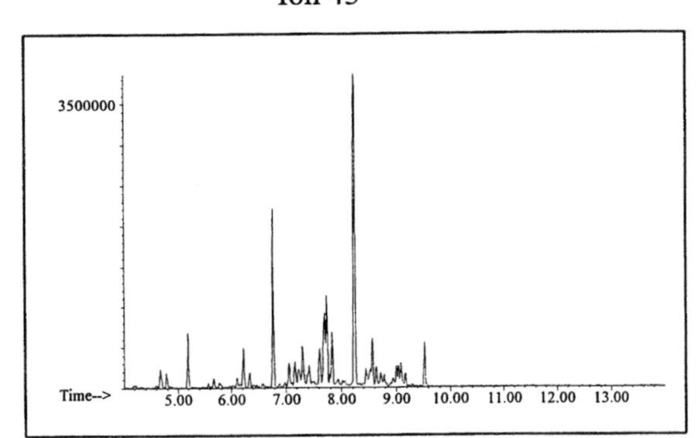

Ion 57

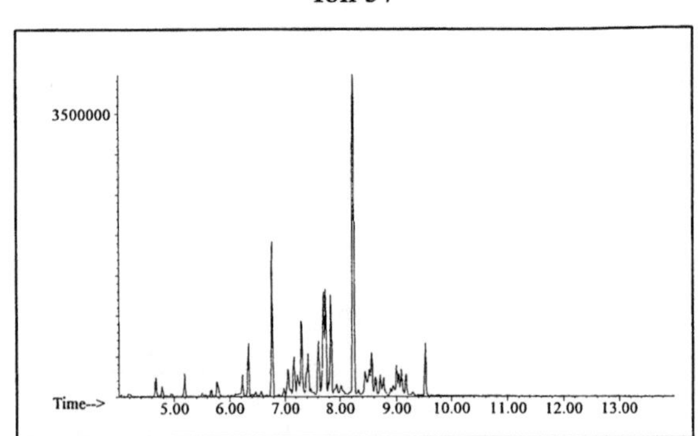

Ion 71

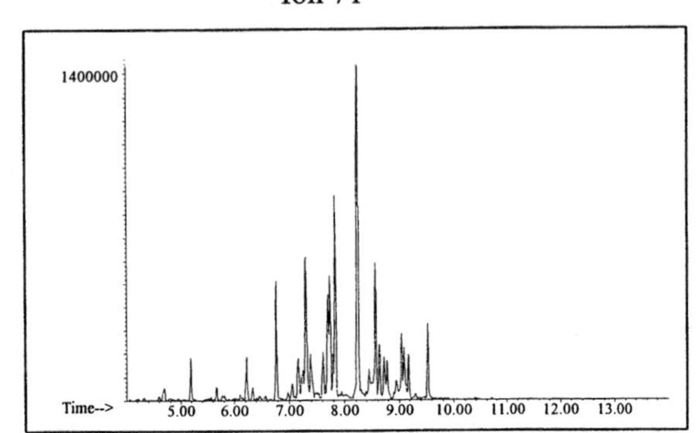

Ion 85

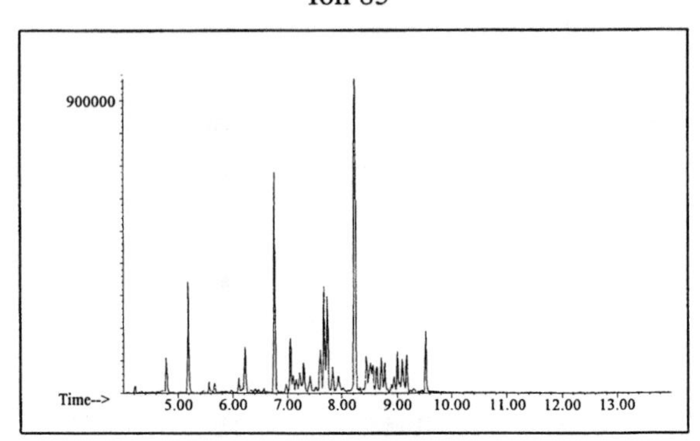

Aromatic

Ion 91

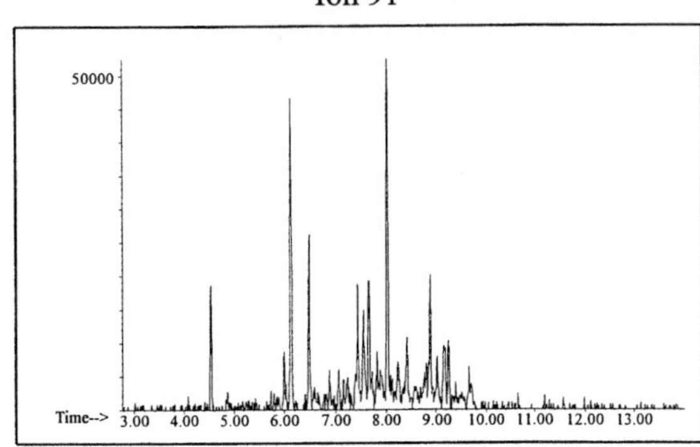

Ion 105

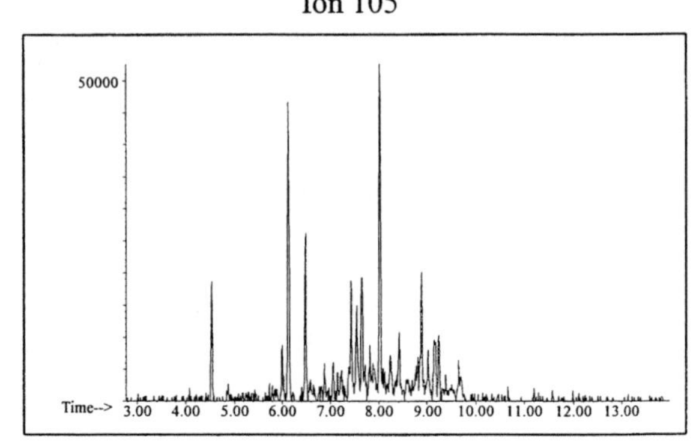

Ion 119

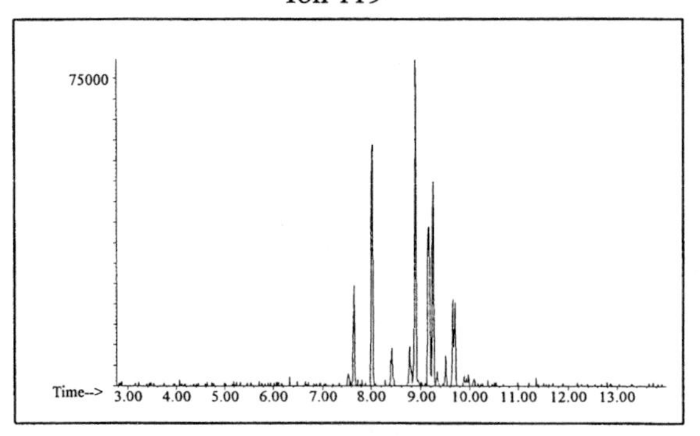

INDIVIDUAL PROFILES

Distillate #19

Cycloparaffin

Ion 55

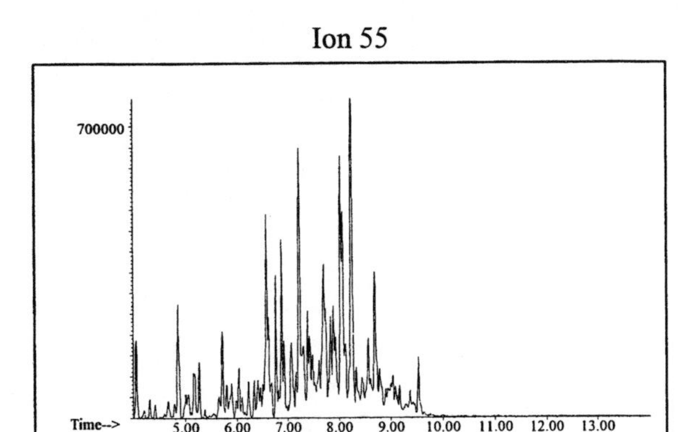

Ion 69

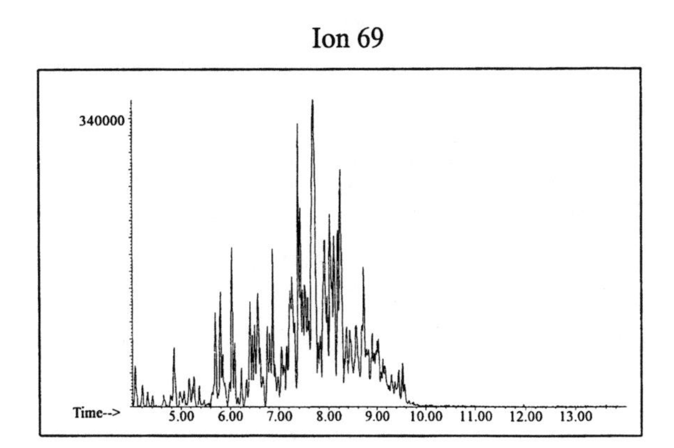

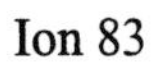

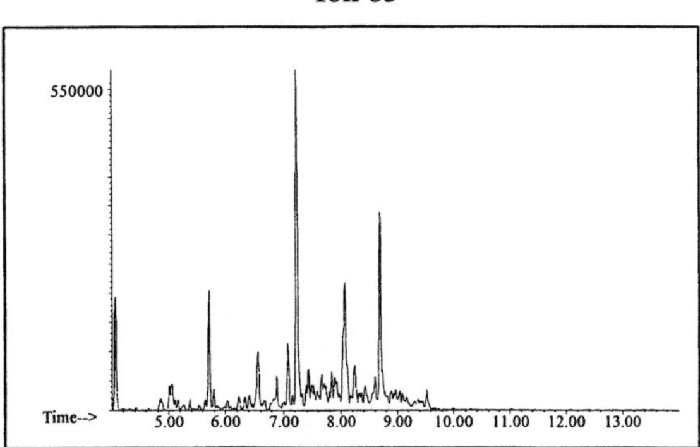

Naphthalene

Ion 128

No Useful Data Obtained

Ion 142

No Useful Data Obtained

Ion 156

No Useful Data Obtained

Distillate #20

20uL/mL Pentane

ASTM: Class 3 (Medium Petroleum Distillate) Macro Code: M

Product Displayed: Gum Out Xtra Fuel Injector Cleaner

Product Uses: Cleaner, charcoal lighter, mineral spirits, gas treatment

Other Similar Products: Bruce Wood Floor Cleaner, Sparky Charcoal Lighter, K-Mart Gas Treatment, Porter Mineral Spirits, Marvel Mystery Oil, Marks Charcoal Lighter, Parks Odorless Mineral Spirits

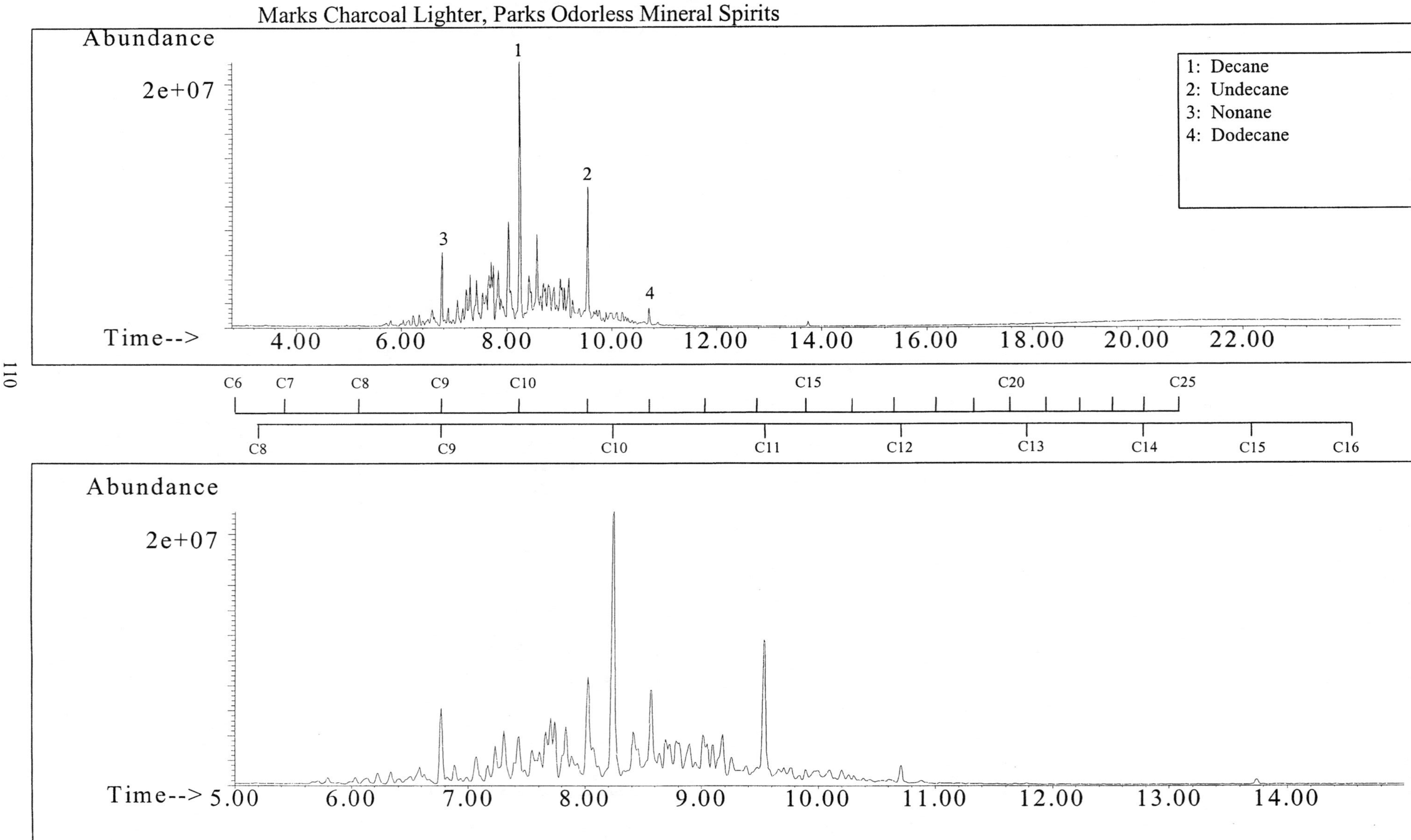

SUMMED PROFILES

Distillate #20

ALKANES

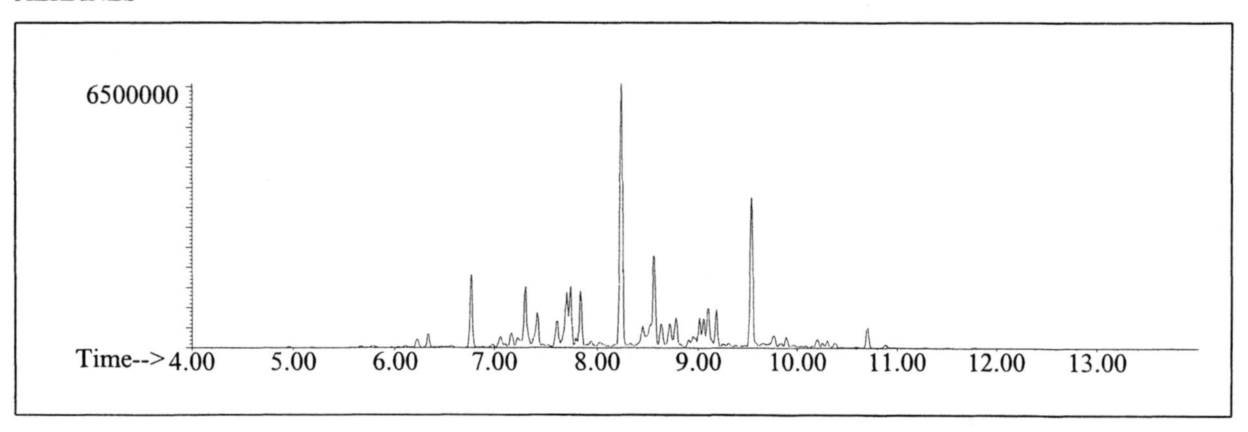

AROMATICS

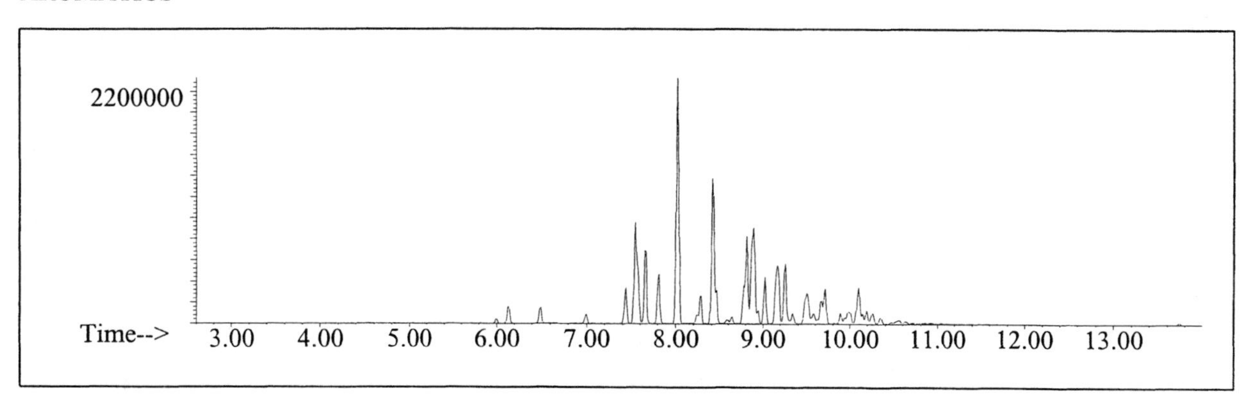

CYCLOPARAFFINS AND ALKENES

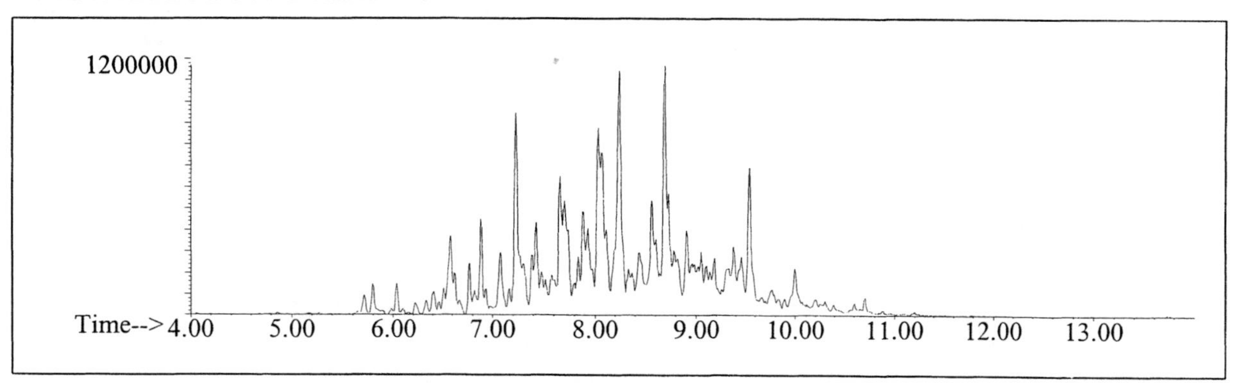

NAPHTHALENES

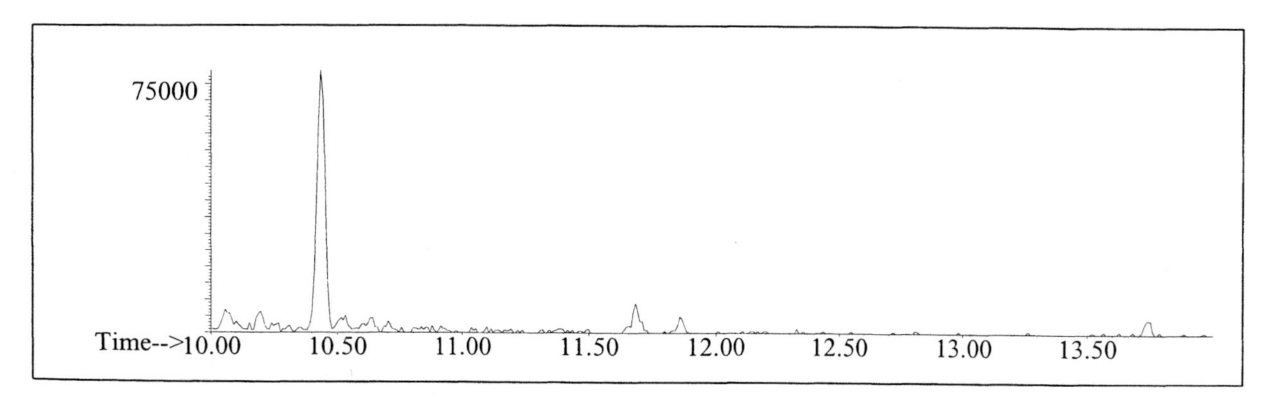

INDIVIDUAL PROFILES Distillate #20

Alkane

Ion 43

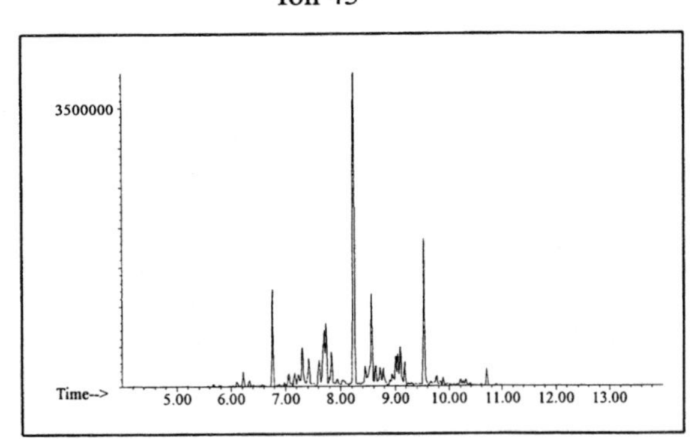

Ion 57

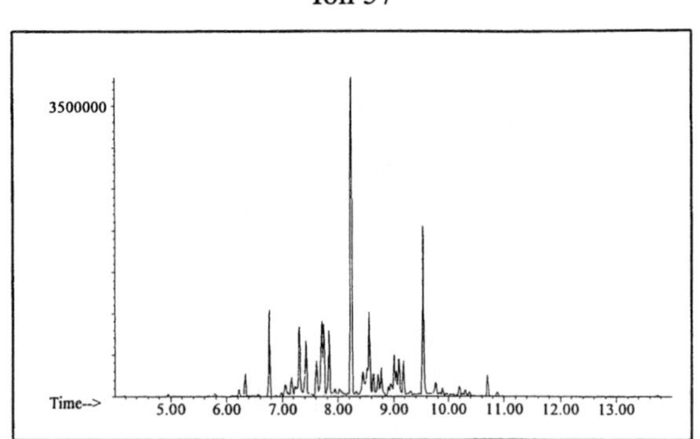

Ion 71

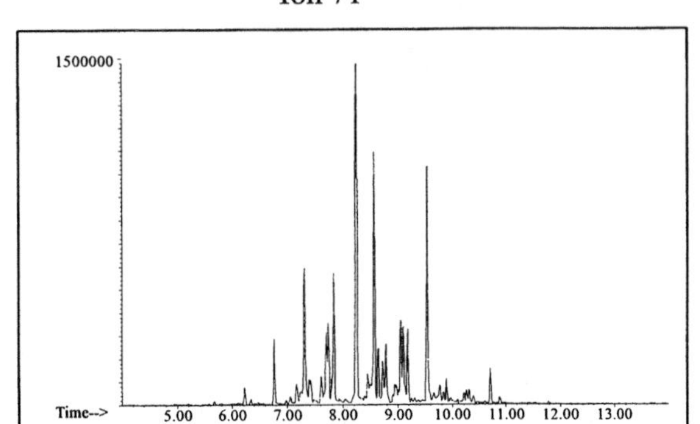

Ion 85

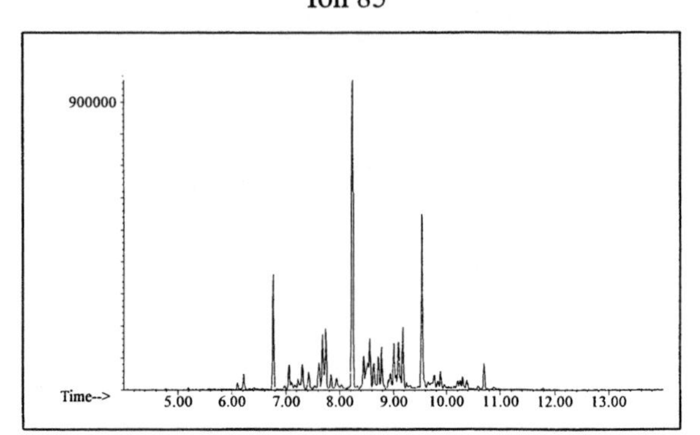

Aromatic

Ion 91

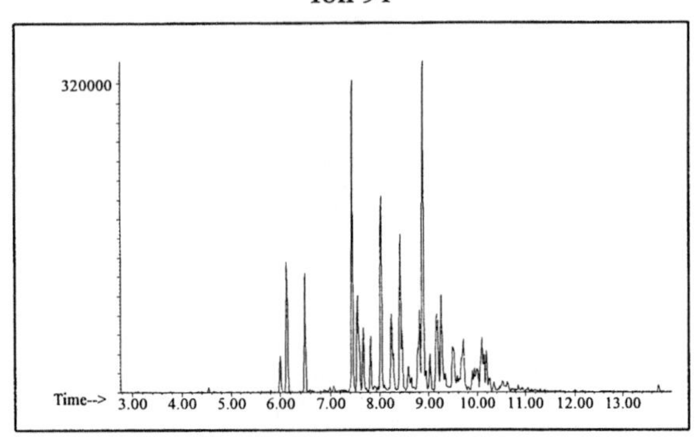

Ion 105

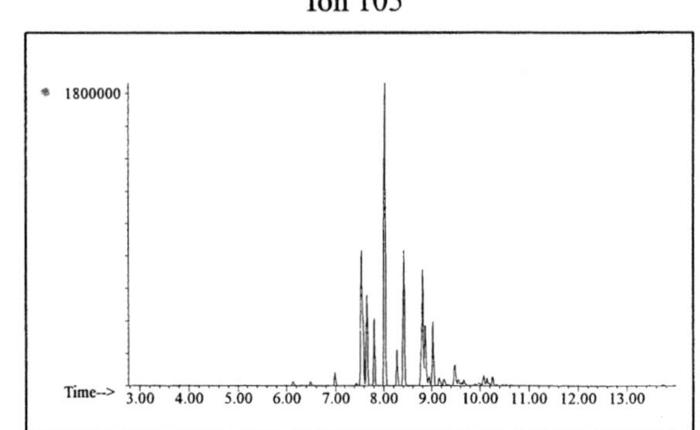

Ion 119

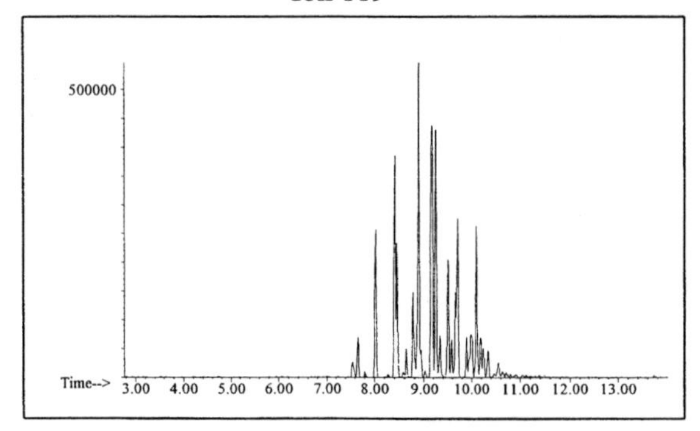

INDIVIDUAL PROFILES

Distillate #20

Cycloparaffin

Ion 55

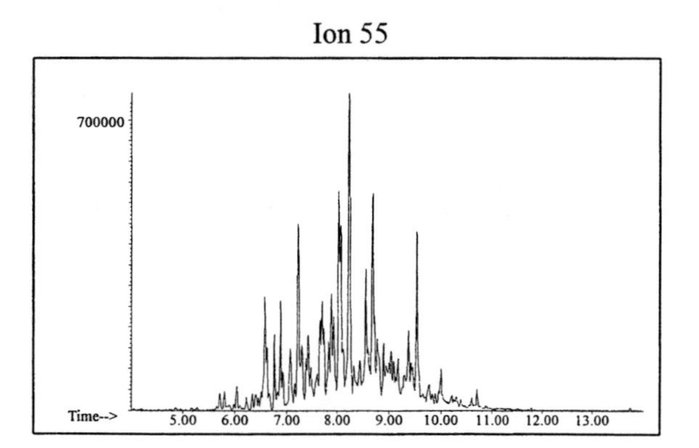

Ion 69

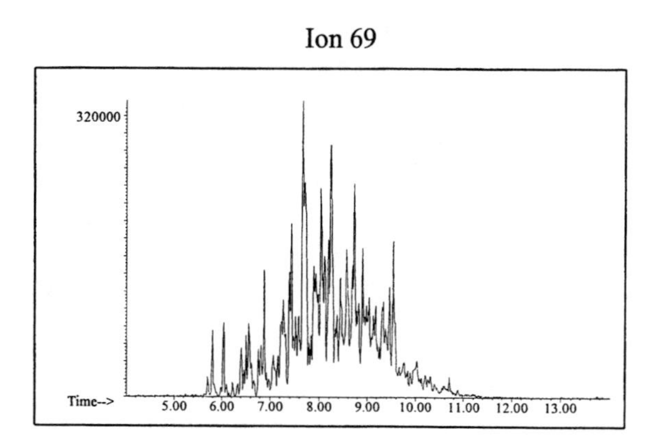

Ion 83

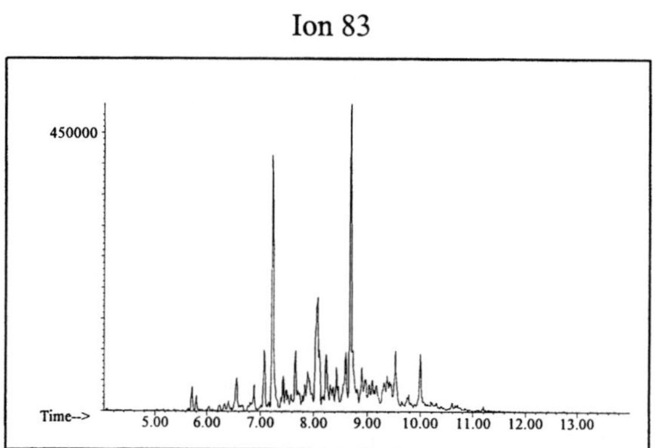

Naphthalene

Ion 128

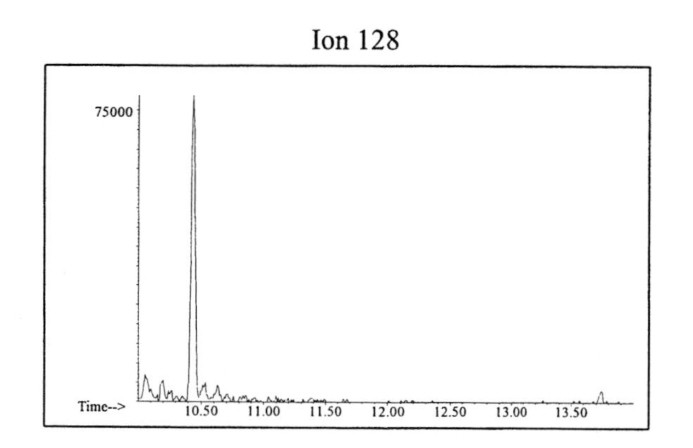

Ion 142

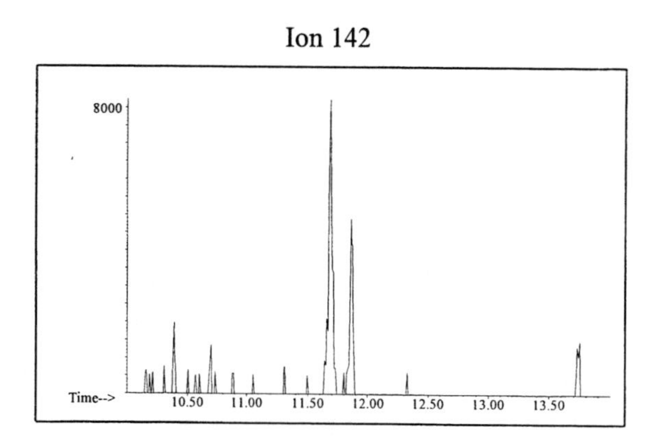

Ion 156

No Useful Data Obtained

Distillate #21

20uL/mL Pentane

ASTM: Class 3 (Medium Petroleum Distillate) Macro Code: M

Product Displayed: Publix Charcoal Starter

Product Uses: Charcoal lighter

Other Similar Products:

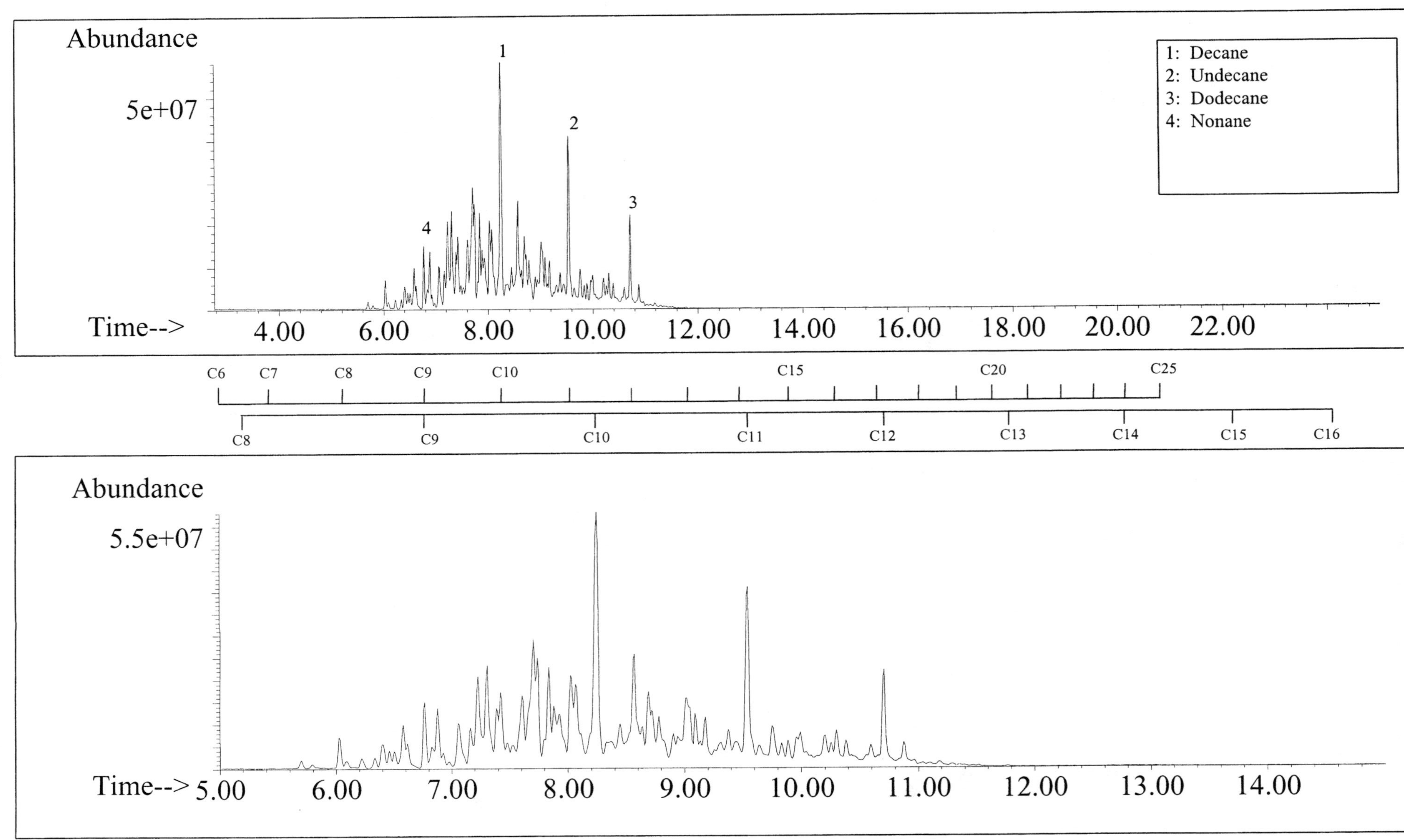

SUMMED PROFILES Distillate #21

ALKANES

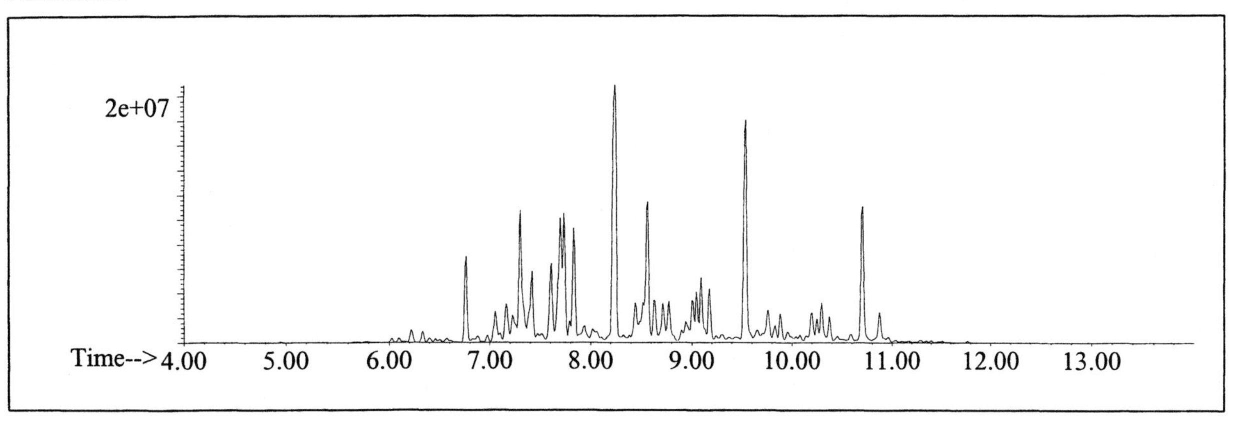

AROMATICS

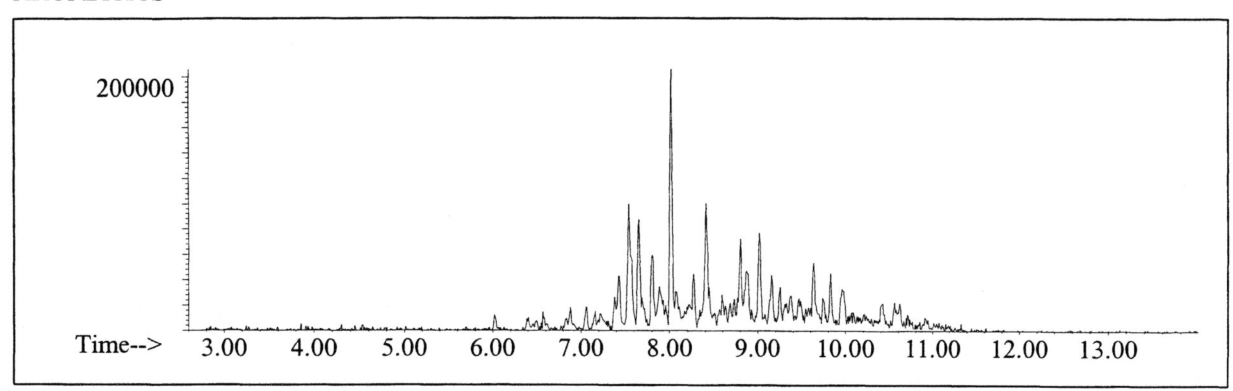

CYCLOPARAFFINS AND ALKENES

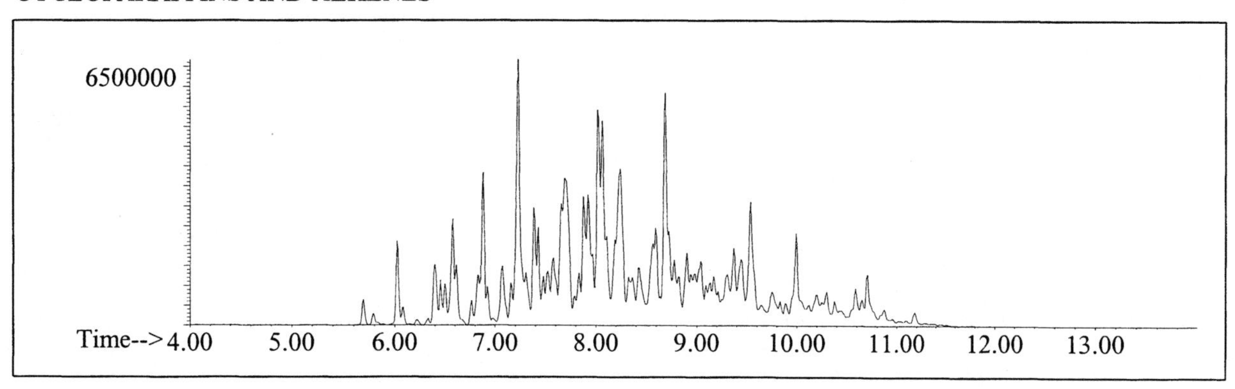

NAPHTHALENES

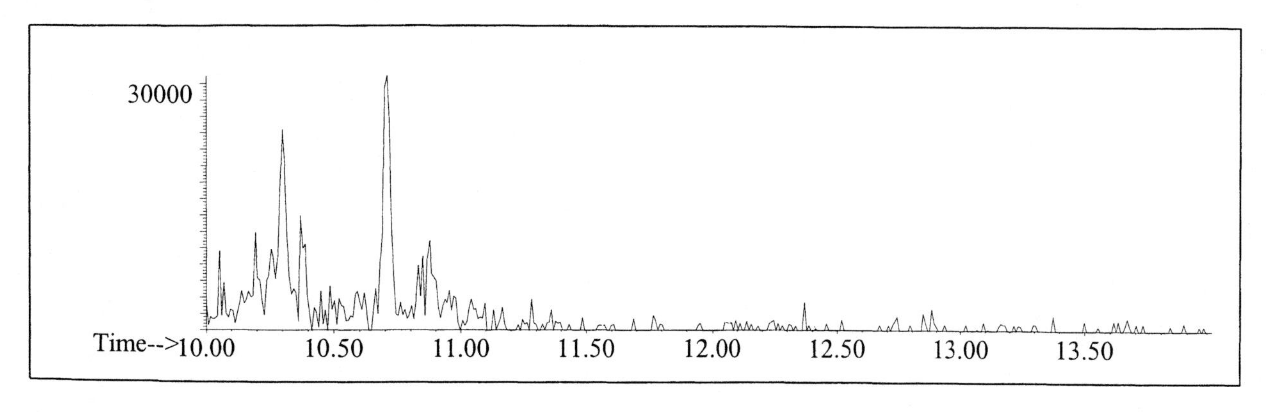

INDIVIDUAL PROFILES Distillate #21

Alkane

Ion 43

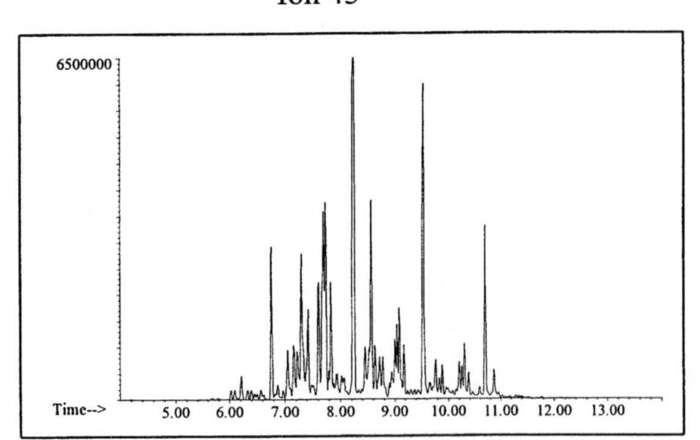

Ion 57

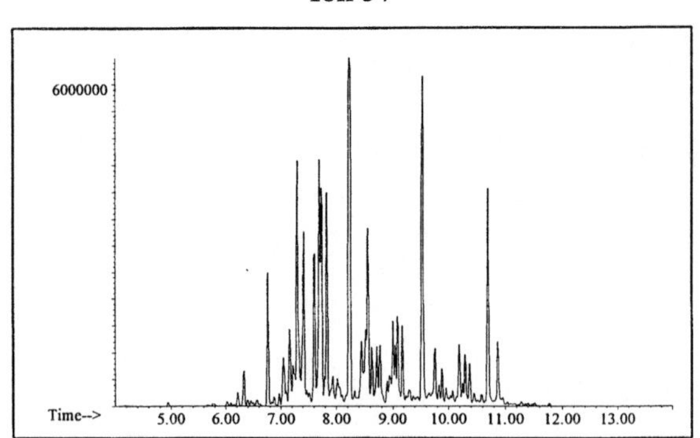

Ion 71

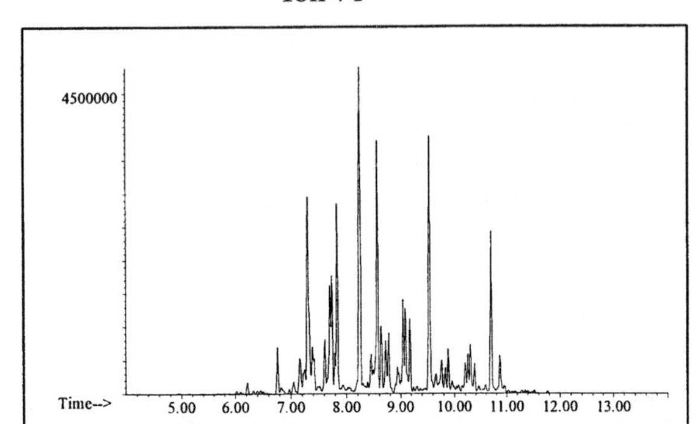

Ion 85

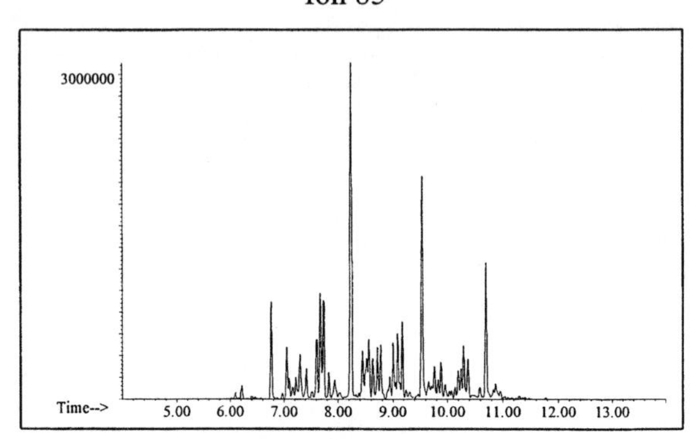

Aromatic

Ion 91

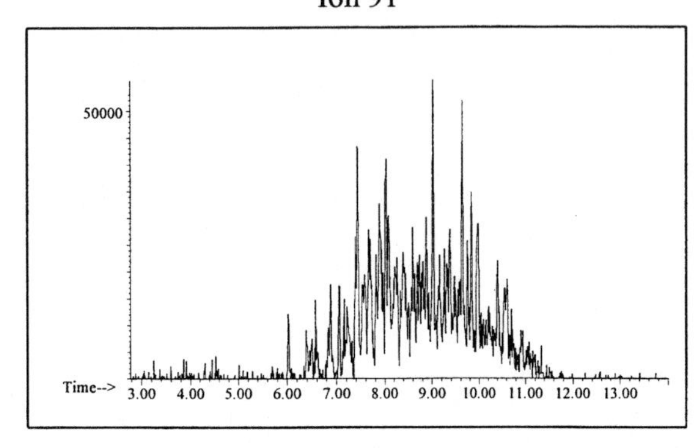

Ion 105

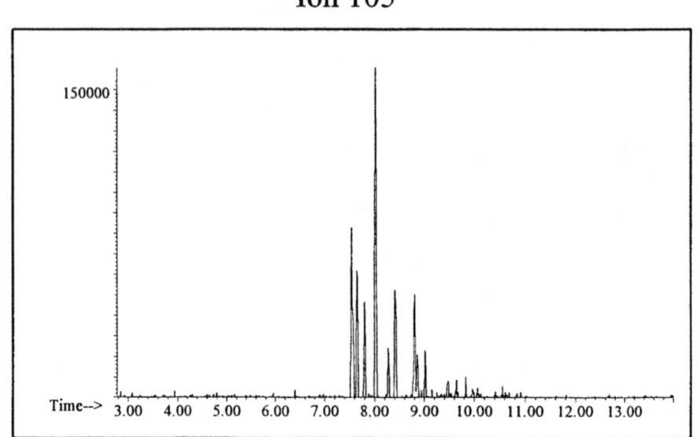

Ion 119

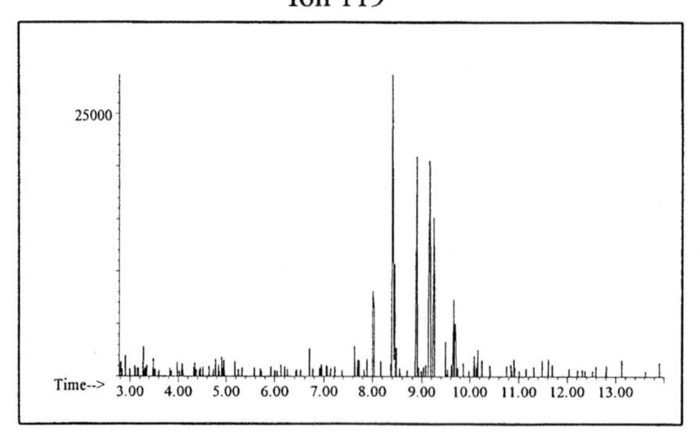

INDIVIDUAL PROFILES

Distillate #21

Cycloparaffin

Ion 55

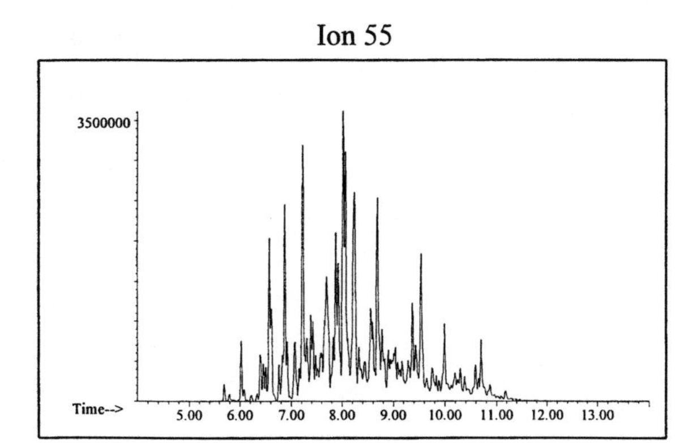

Ion 69

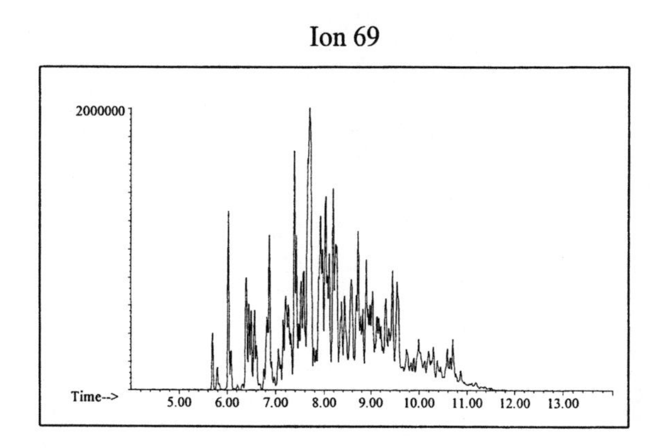

Ion 83

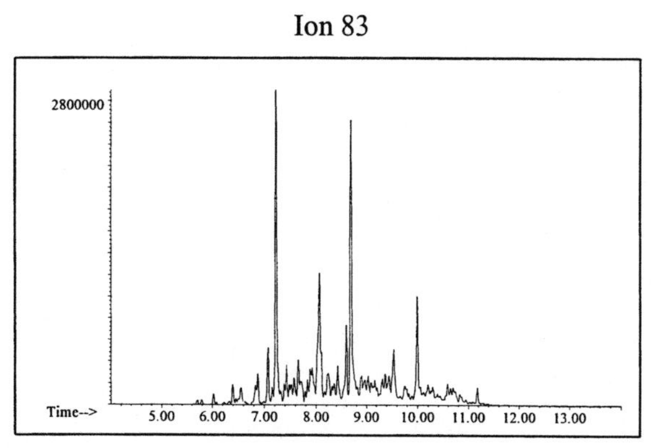

Naphthalene

Ion 128

No Useful Data Obtained

Ion 142

No Useful Data Obtained

Ion 156

No Useful Data Obtained

Distillate #22

20uL/mL Pentane

Product Displayed: Walmart Charcoal Lighter Fluid

Other Similar Products:

ASTM: Class 3 (Medium Petroleum Distillate)

Macro Code: M

Product Uses: Charcoal lighter

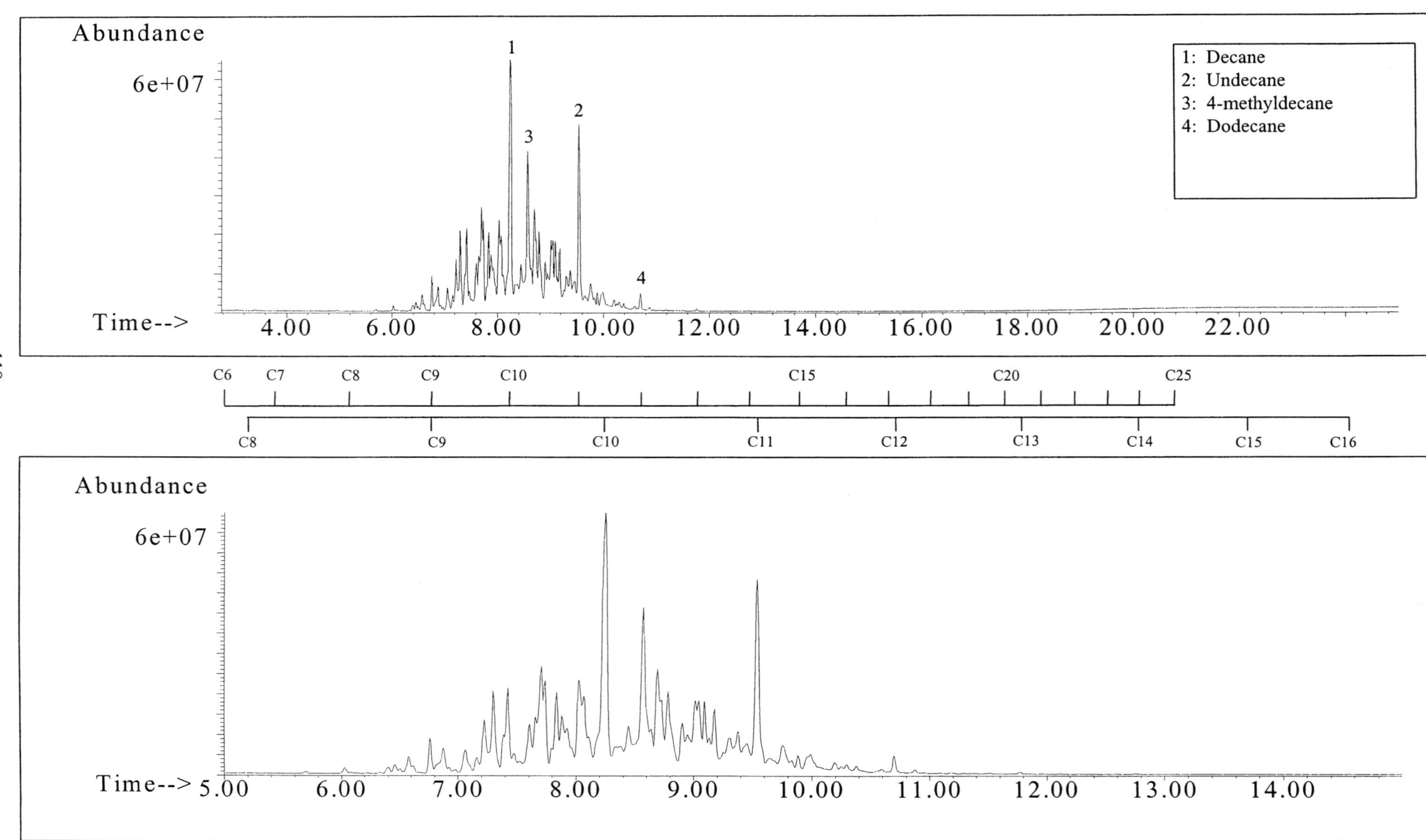

SUMMED PROFILES Distillate #22

ALKANES

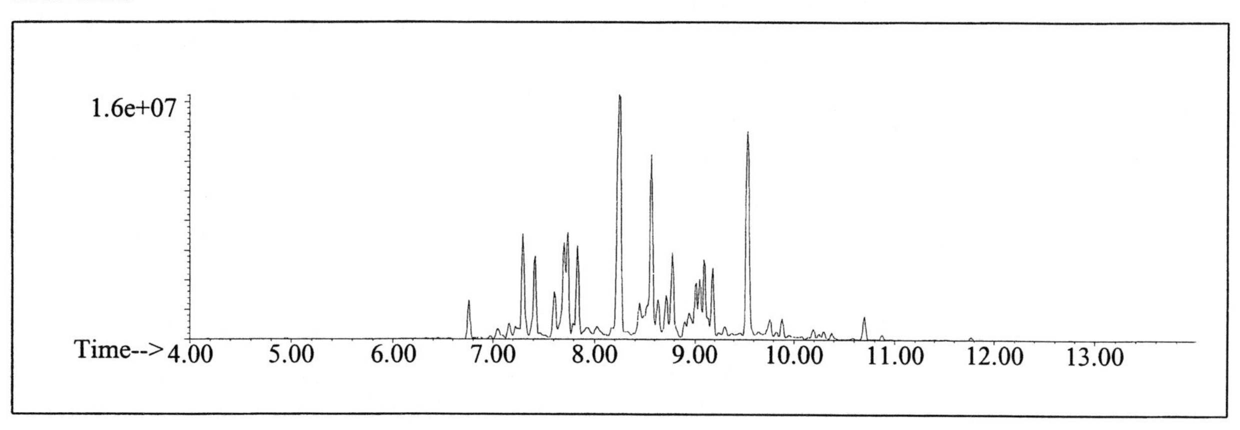

AROMATICS

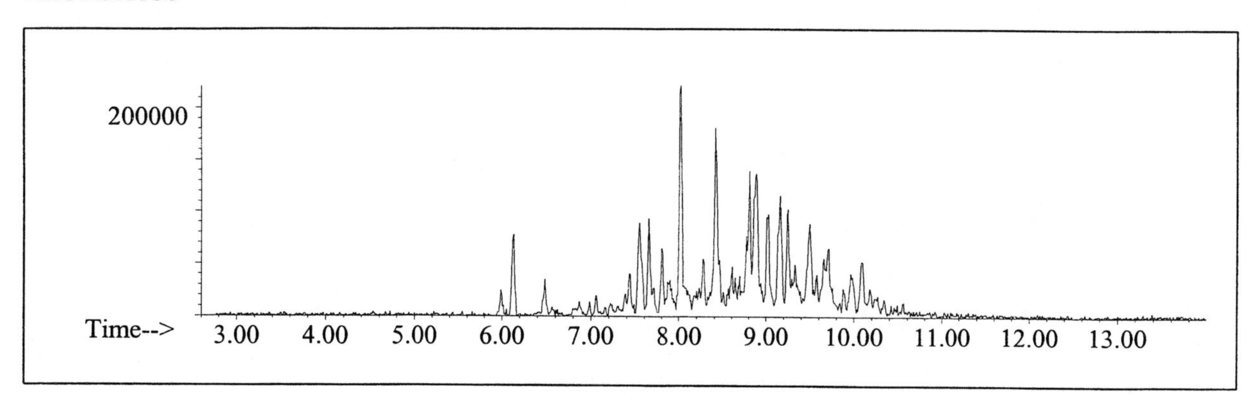

CYCLOPARAFFINS AND ALKENES

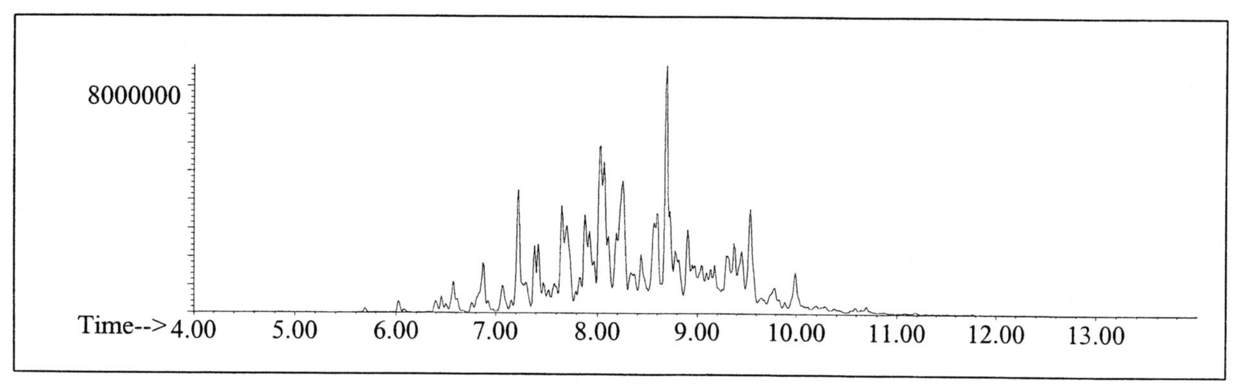

NAPHTHALENES

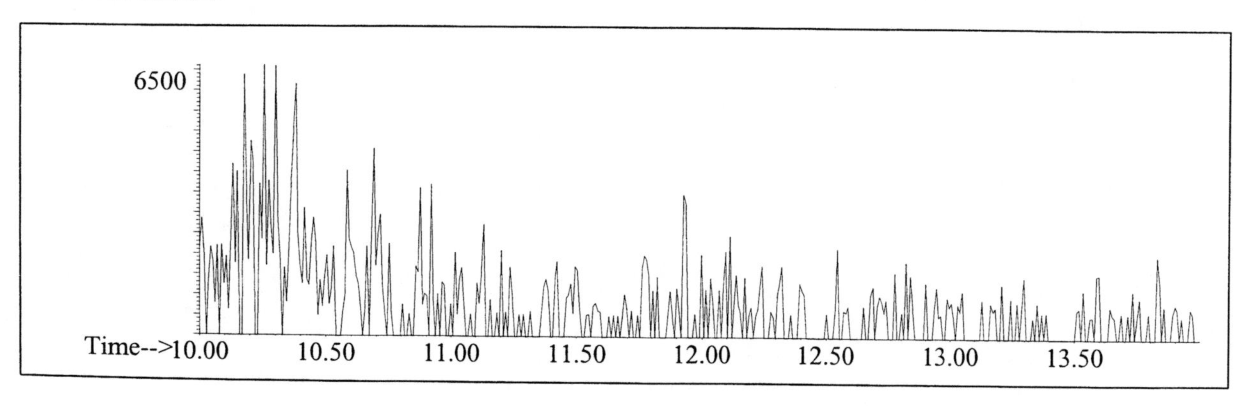

INDIVIDUAL PROFILES

Distillate #22

Alkane

Ion 43

Ion 57

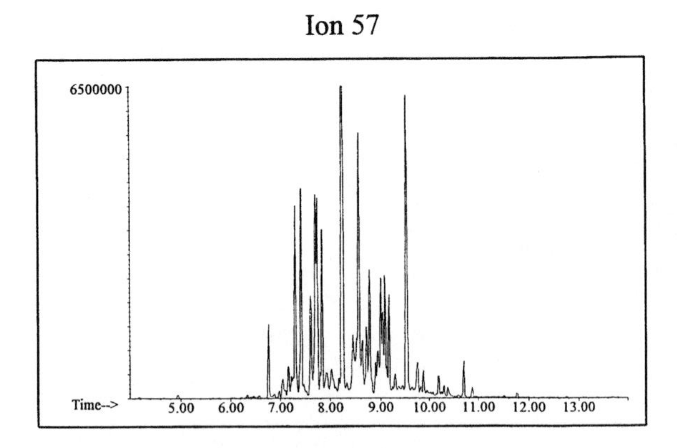

Ion 71

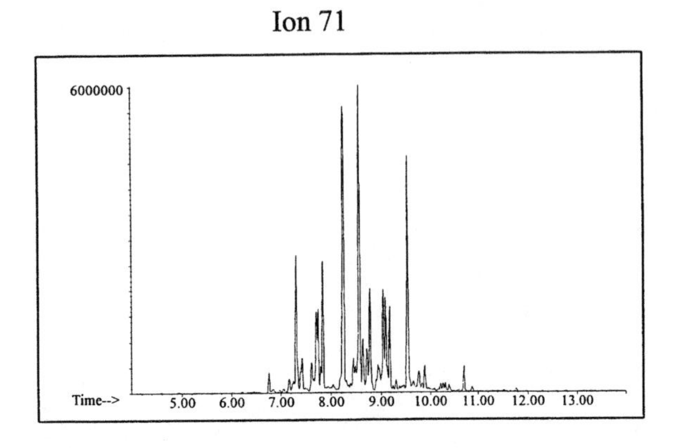

Ion 85

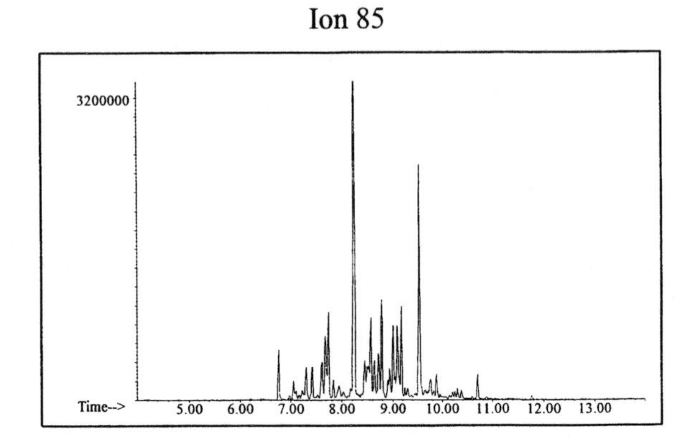

Aromatic

Ion 91

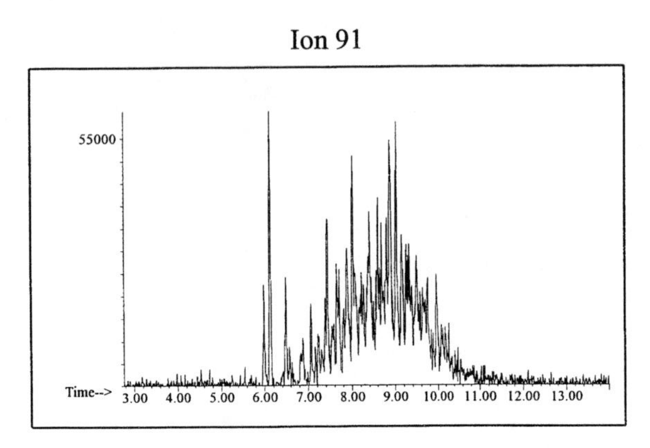

Ion 105

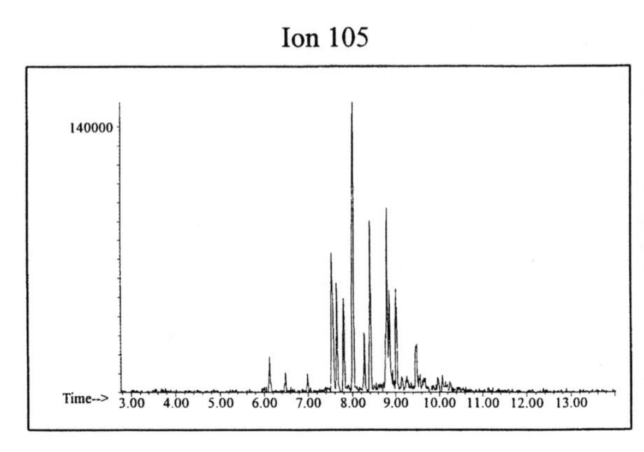

Ion 119

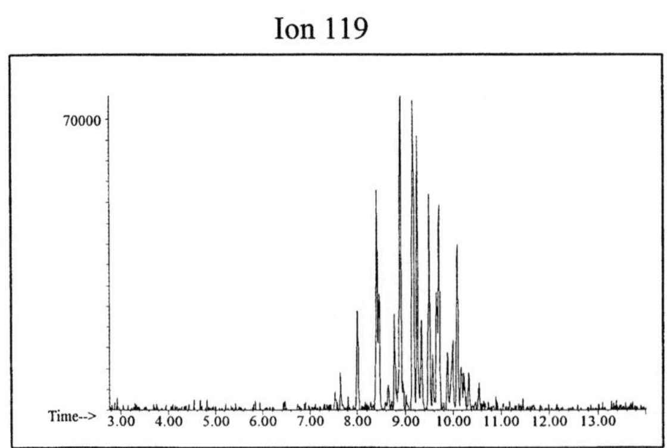

INDIVIDUAL PROFILES Distillate #22

Cycloparaffin

Ion 55

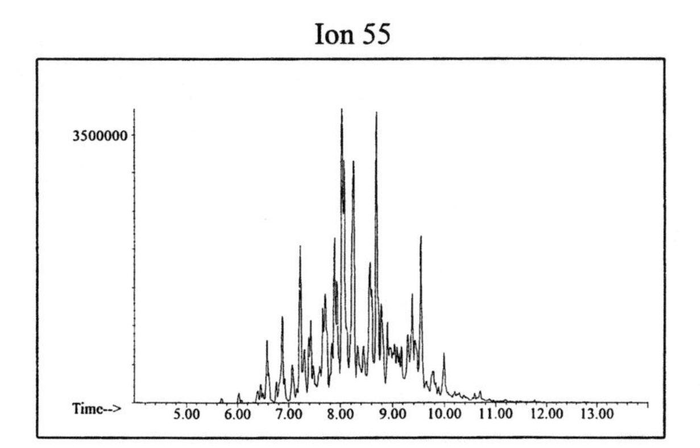

Ion 69

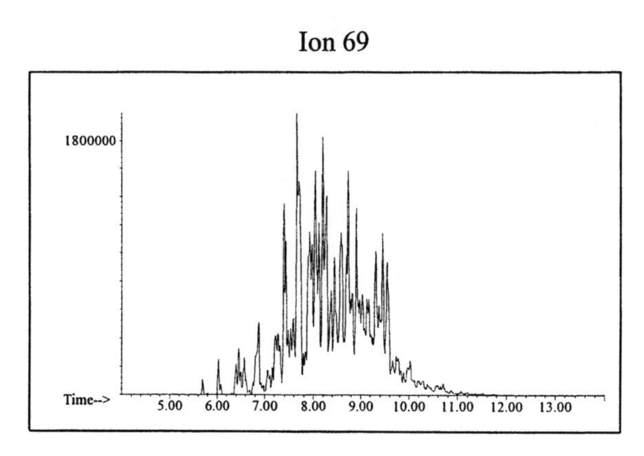

Ion 83

Naphthalene

Ion 128

No Useful Data Obtained

Ion 142

No Useful Data Obtained

Ion 156

No Useful Data Obtained

Distillate #23

20uL/mL Pentane

Product Displayed: Sunnyside Paint Thinner #2

Other Similar Products: Kingsford Odorless Charcoal Starter

ASTM: Class 0 (Miscellaneous)

Product Uses: Paint thinner, charcoal lighter

Macro Code: M

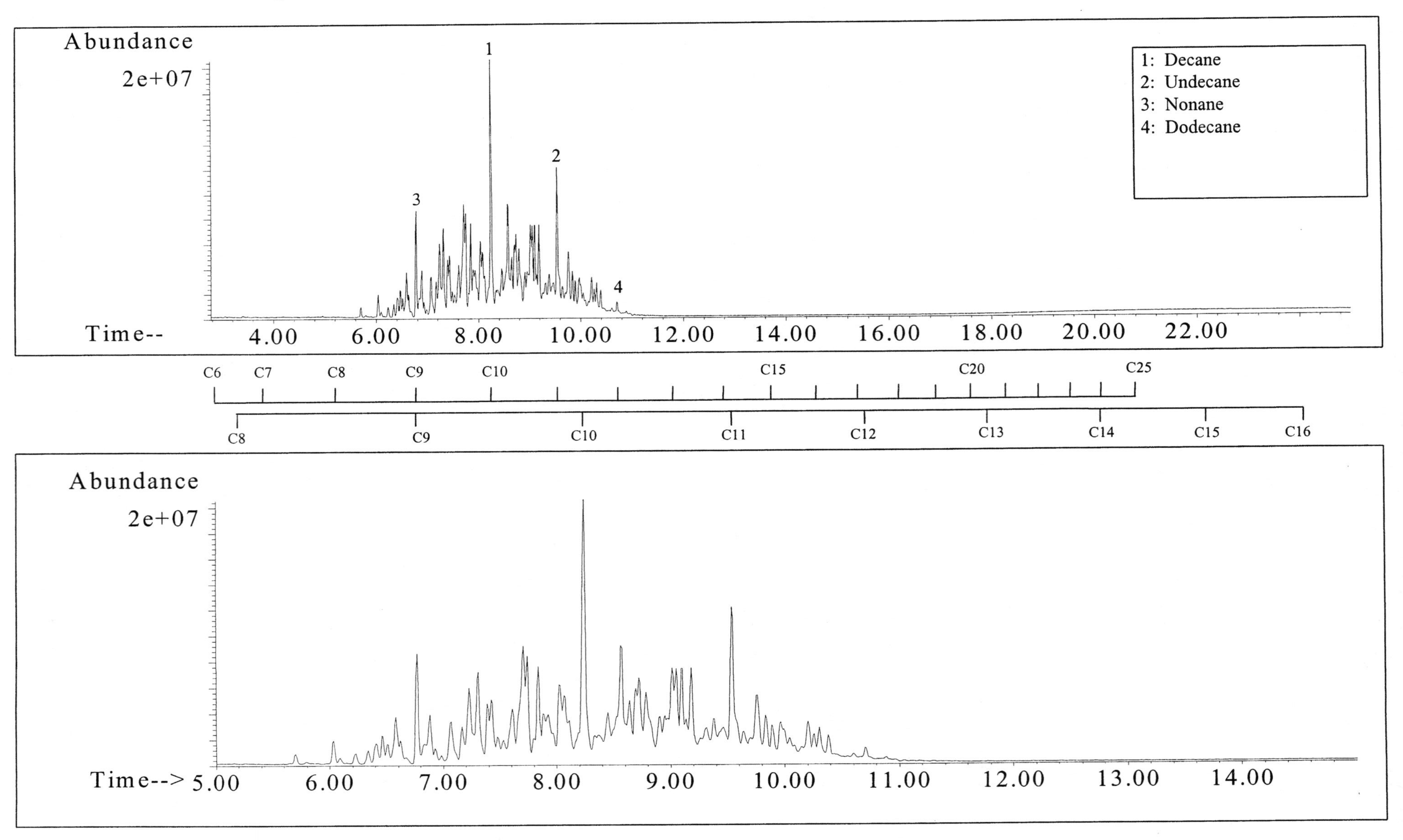

SUMMED PROFILES Distillate #23

ALKANES

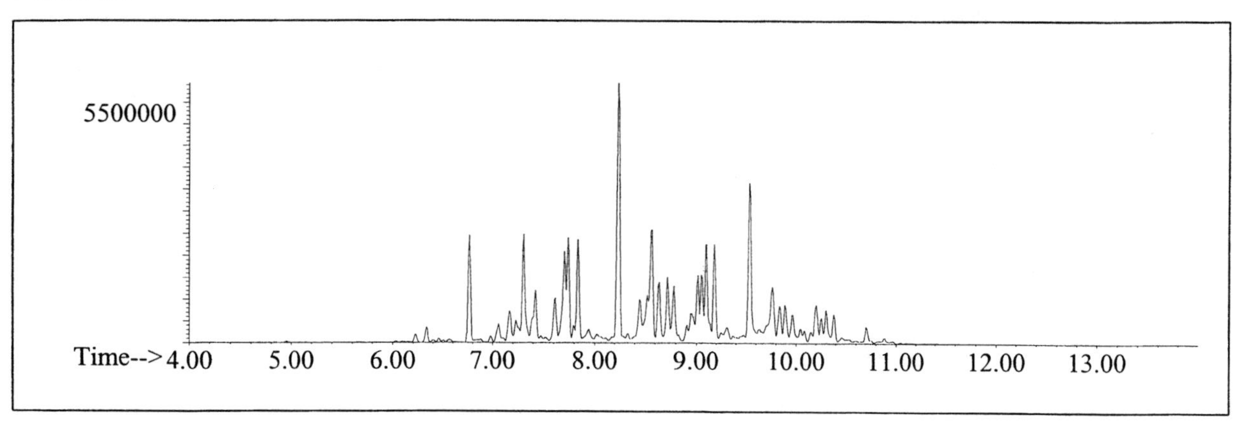

AROMATICS

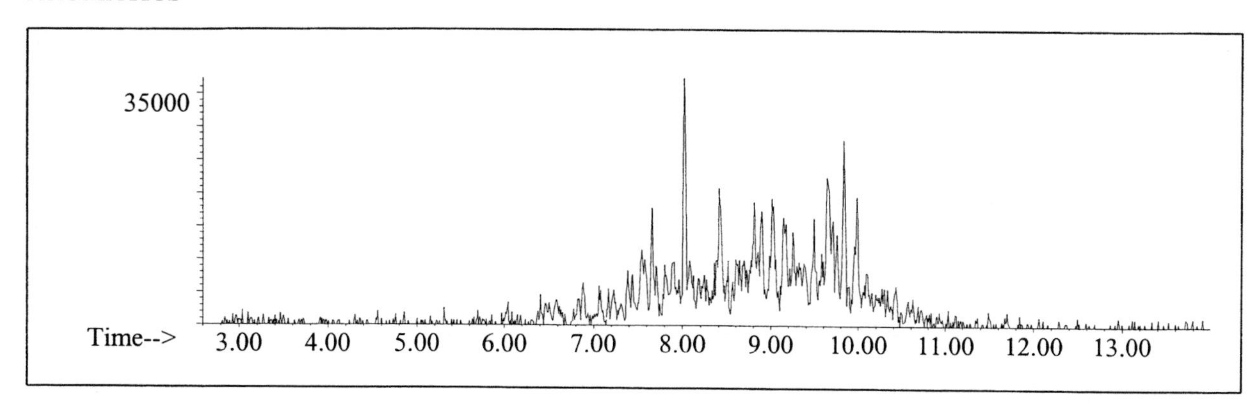

CYCLOPARAFFINS AND ALKENES

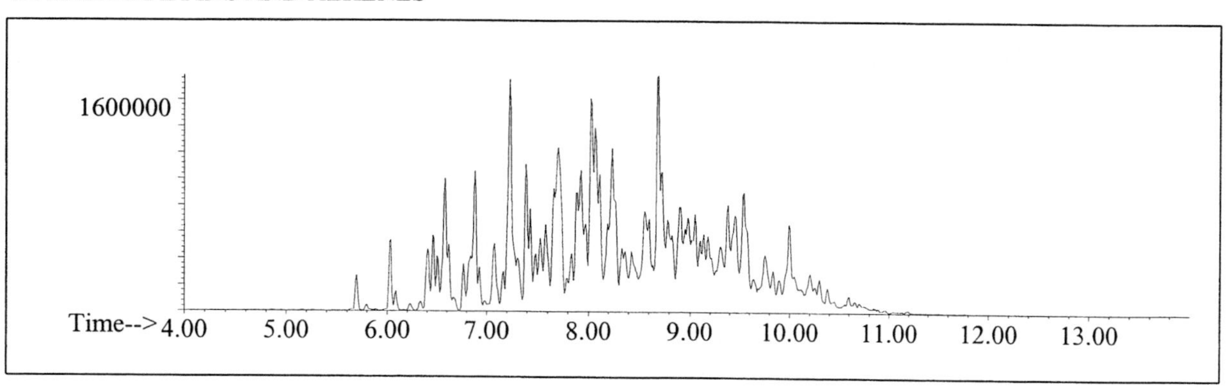

NAPHTHALENES

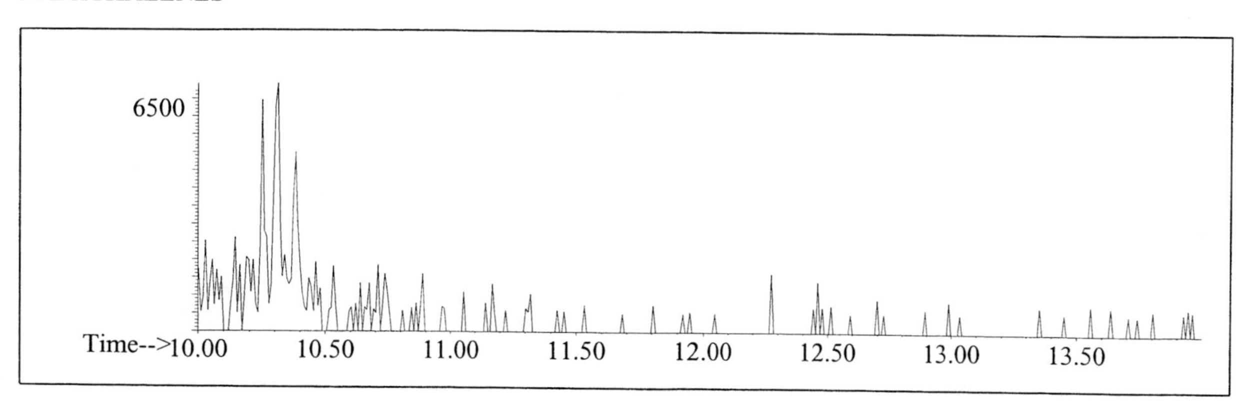

INDIVIDUAL PROFILES

Distillate #23

Alkane

Ion 43

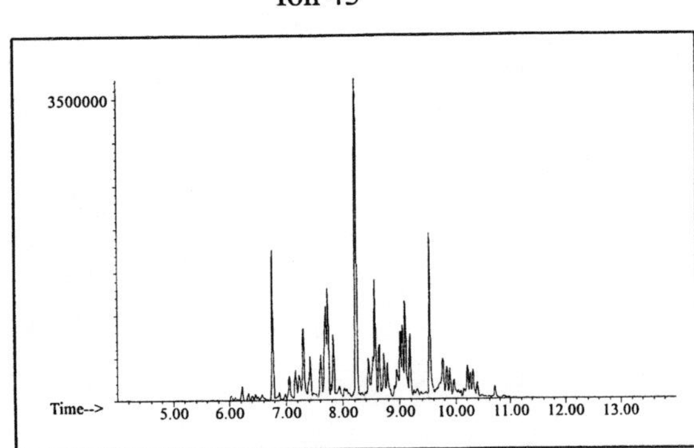

Ion 57

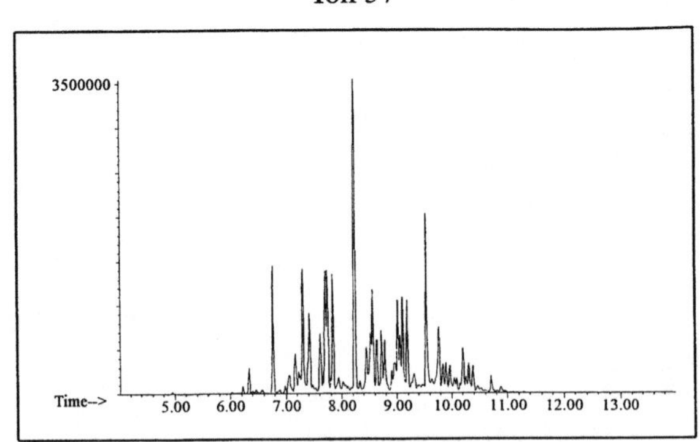

Ion 71

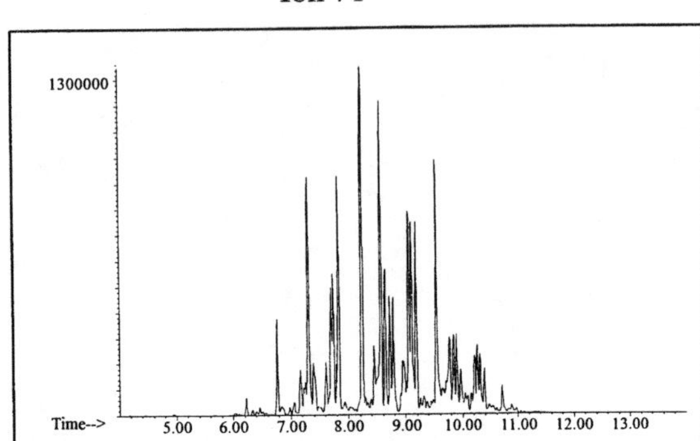

Ion 85

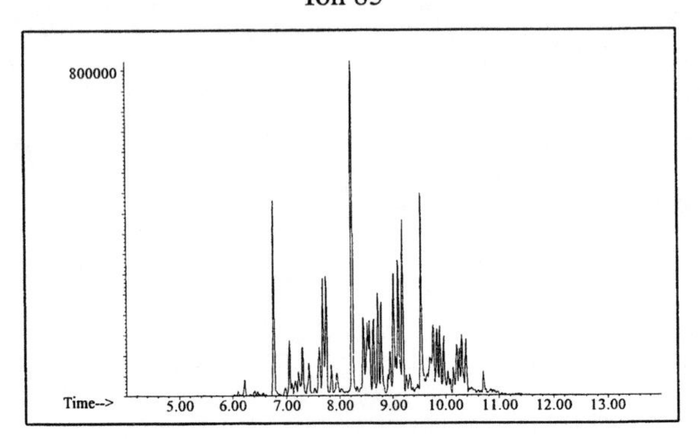

Aromatic

Ion 91

No Useful Data Obtained

Ion 105

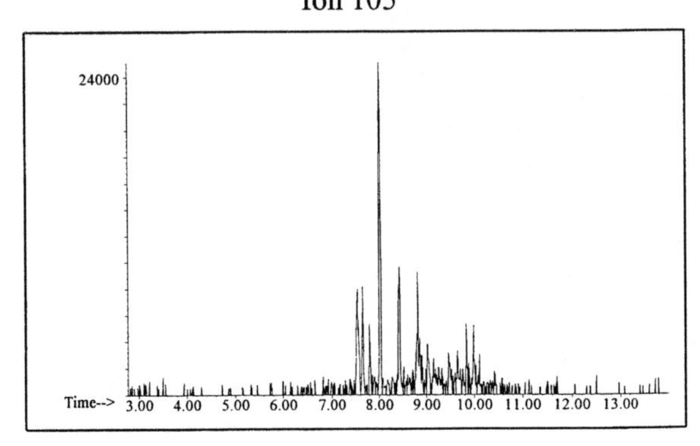

Ion 119

No Useful Data Obtained

INDIVIDUAL PROFILES

Distillate #23

Cycloparaffin

Ion 55

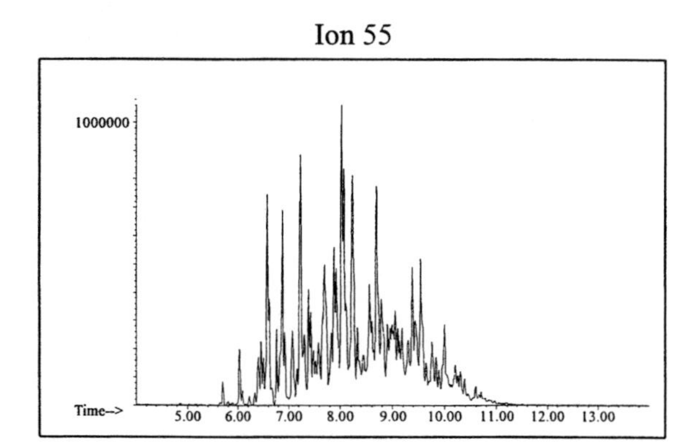

Ion 69

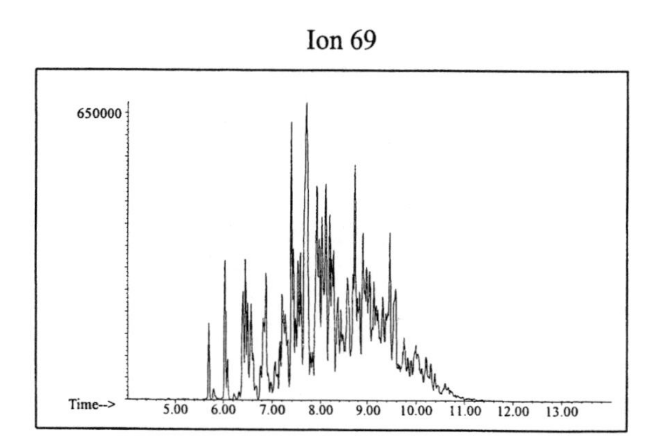

Ion 83

700000

Time-->

5.00 6.00 7.00 8.00 9.00 10.00 11.00 12.00 13.00

Naphthalene

Ion 128

No Useful Data Obtained

Ion 142

No Useful Data Obtained

Ion 156

No Useful Data Obtained

Distillate #24

20uL/mL Pentane

Product Displayed: Varsol 18

Other Similar Products:

ASTM: Class 3 (Medium Petroleum Distillate) Macro Code: M

Product Uses: Solvent, feedstock

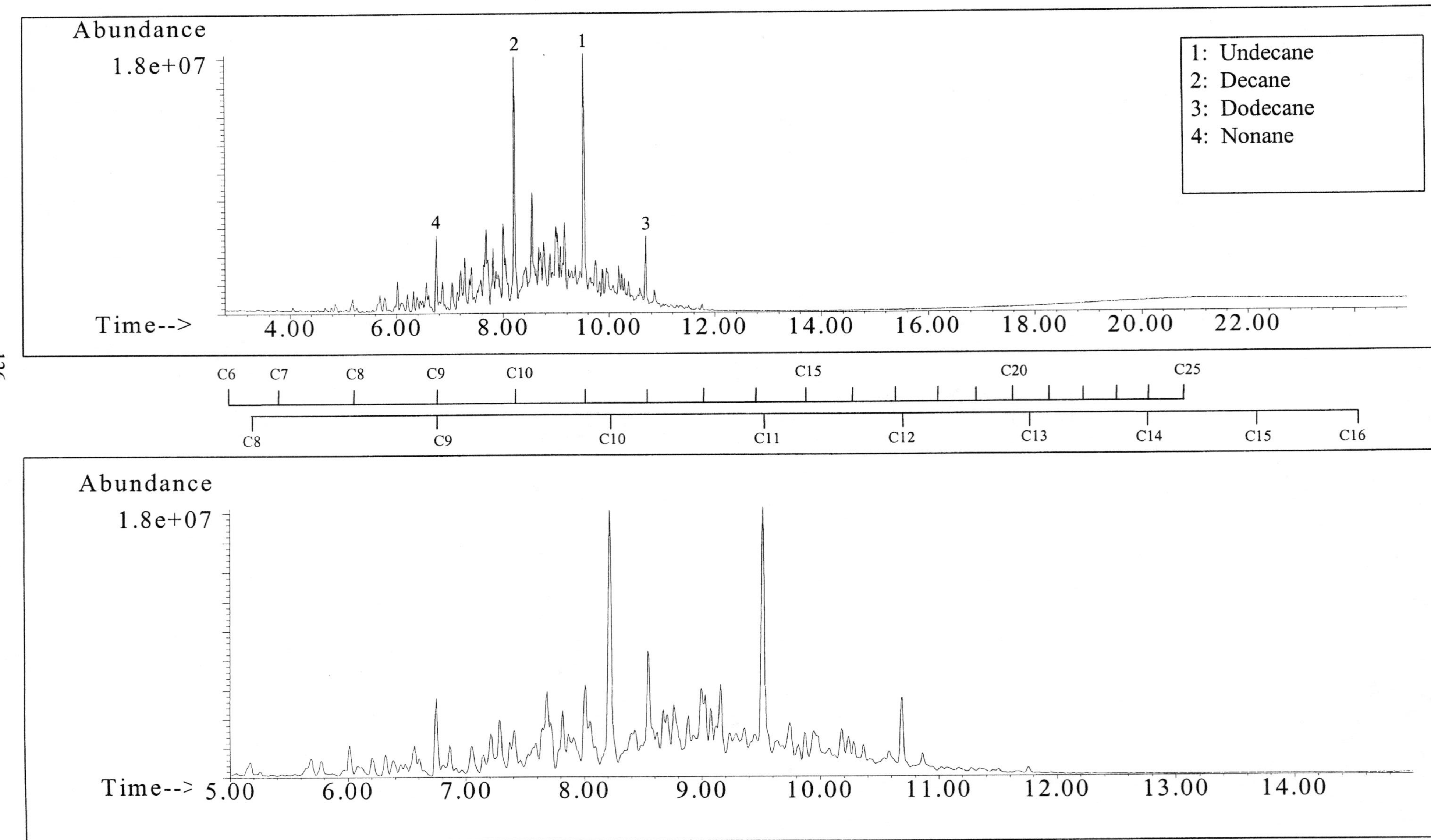

SUMMED PROFILES Distillate #24

ALKANES

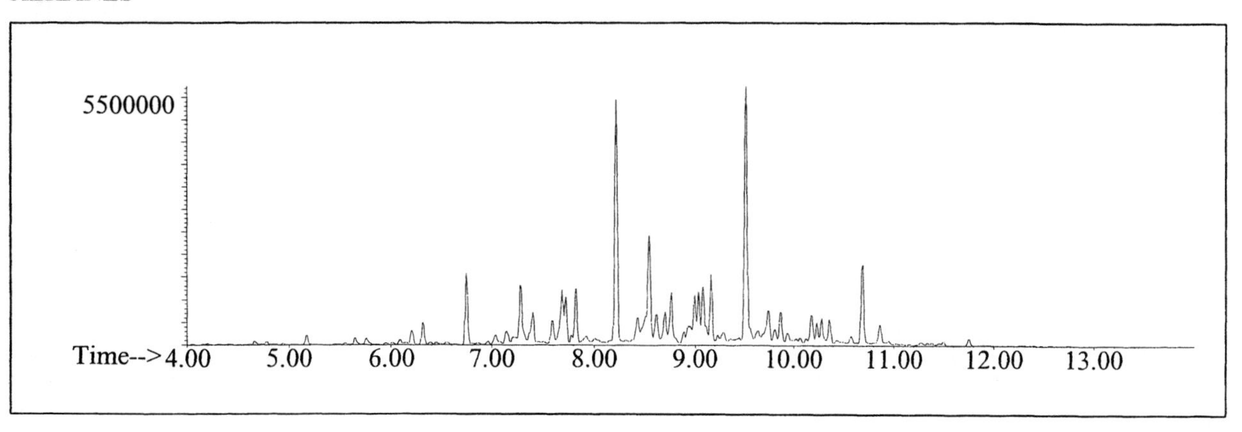

AROMATICS

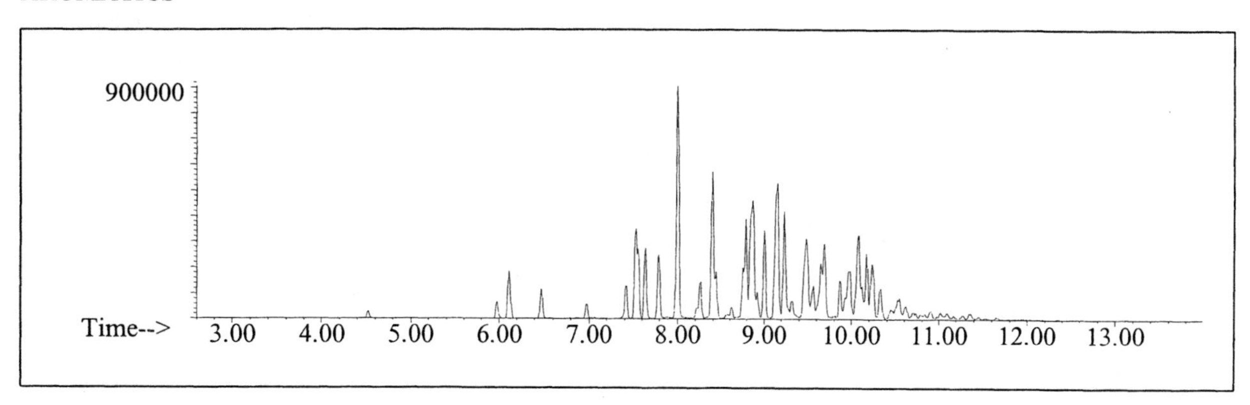

CYCLOPARAFFINS AND ALKENES

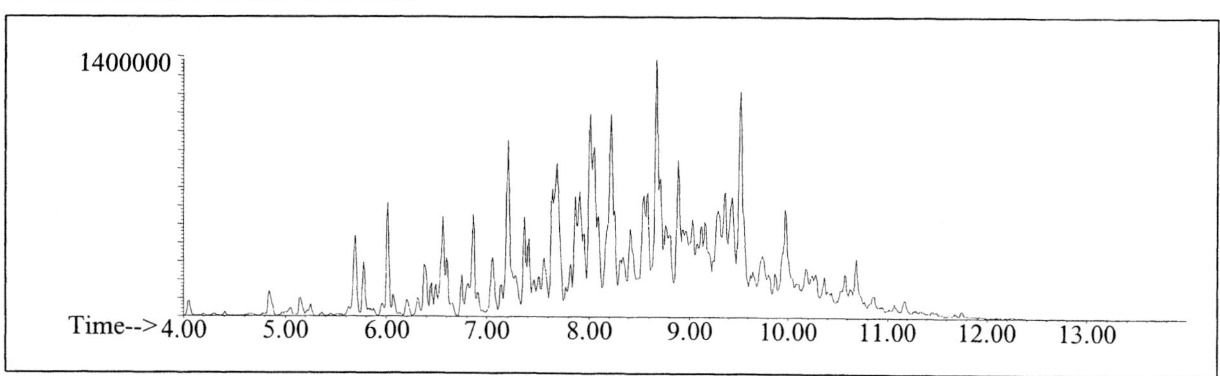

NAPHTHALENES

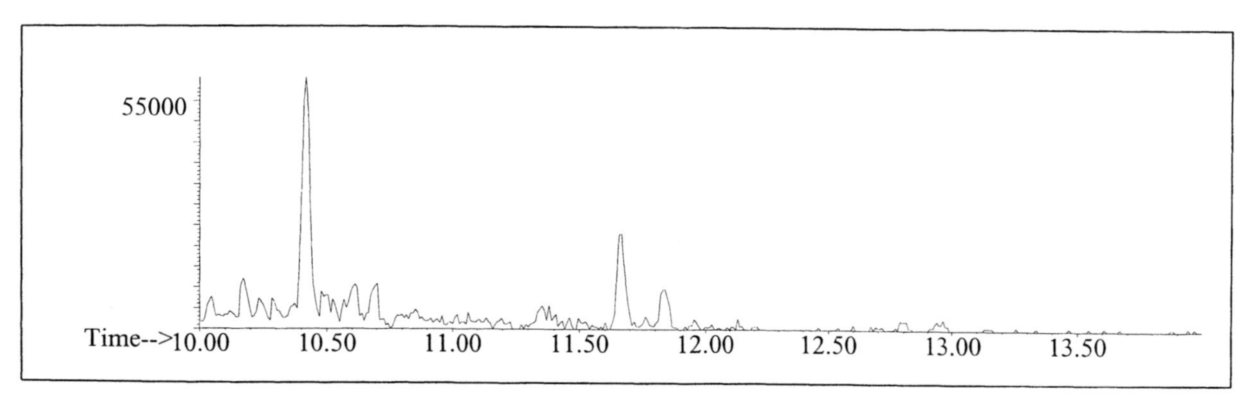

INDIVIDUAL PROFILES

Distillate #24

Alkane

Ion 43

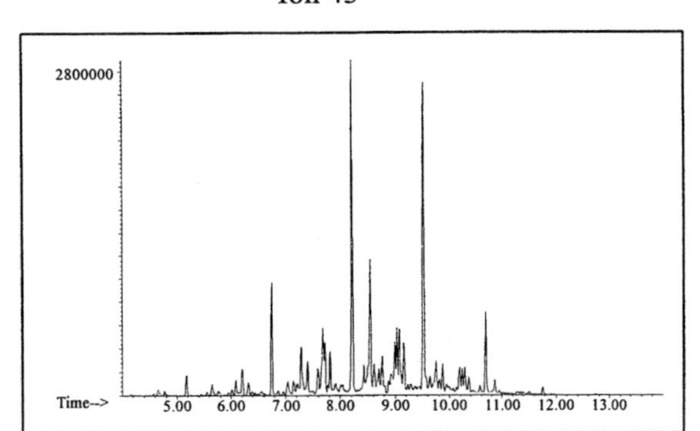

Ion 57

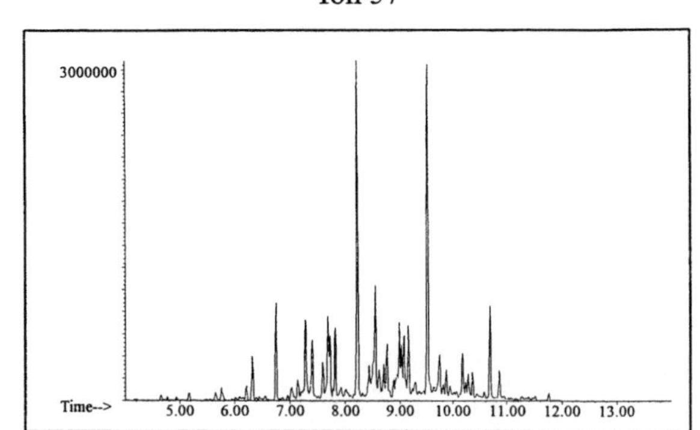

Ion 71

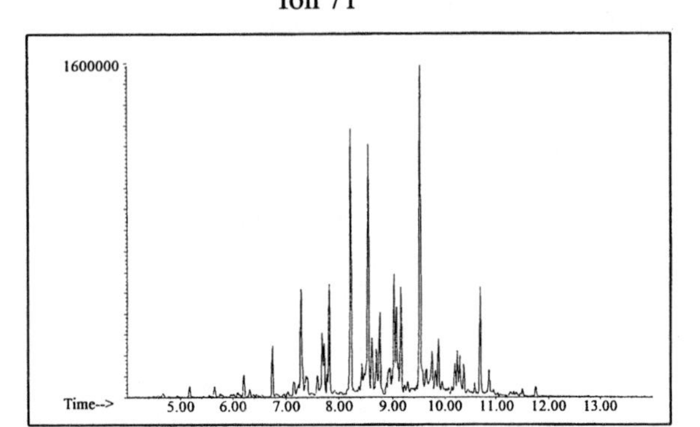

Ion 85

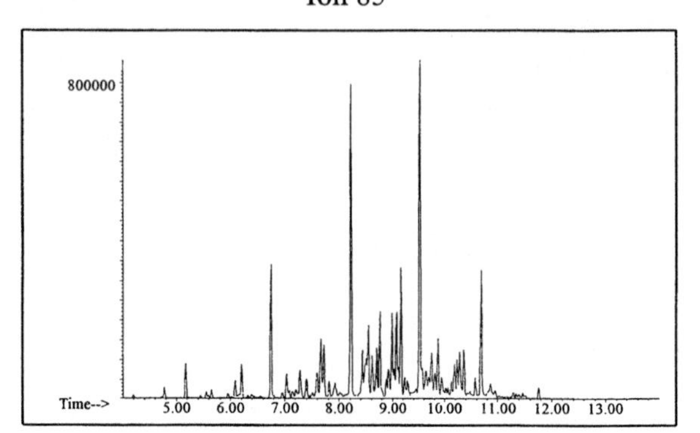

Aromatic

Ion 91

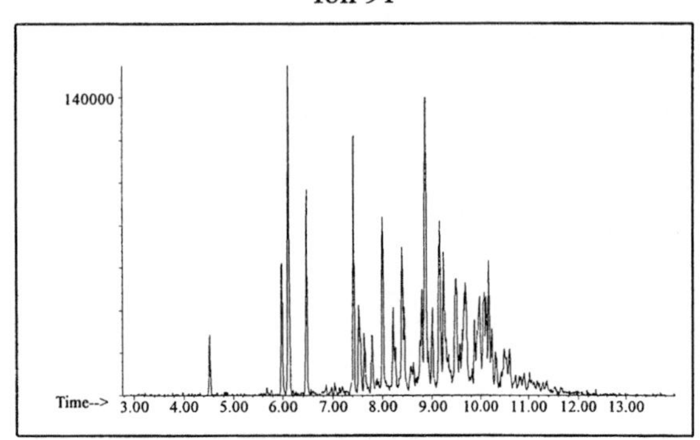

Ion 105

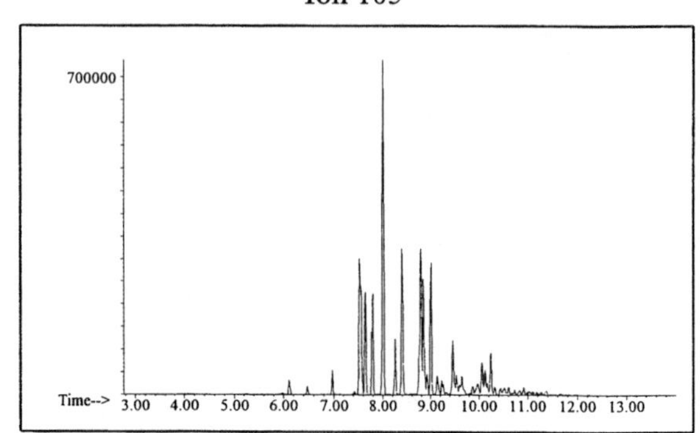

Ion 119

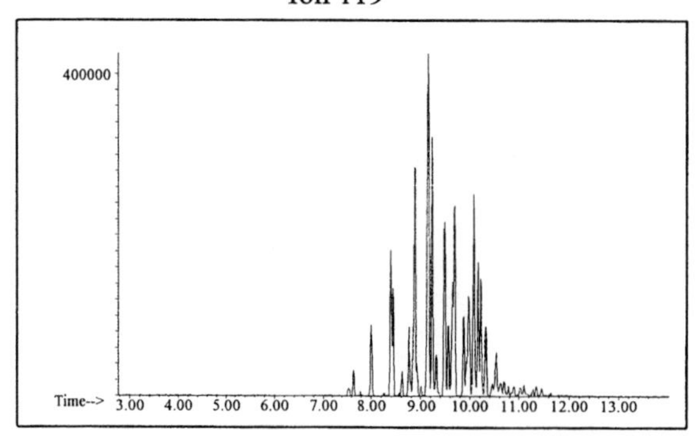

Cycloparaffin

Ion 55

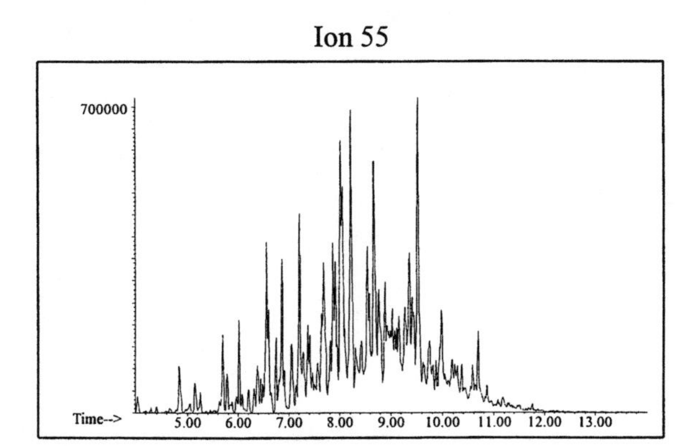

Ion 69

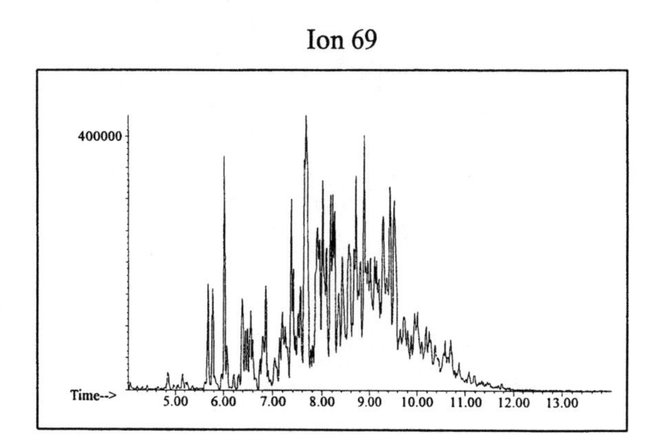

Ion 83

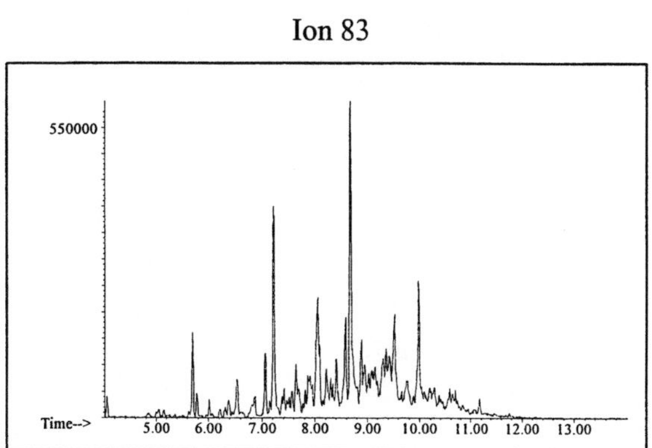

Naphthalene

Ion 128

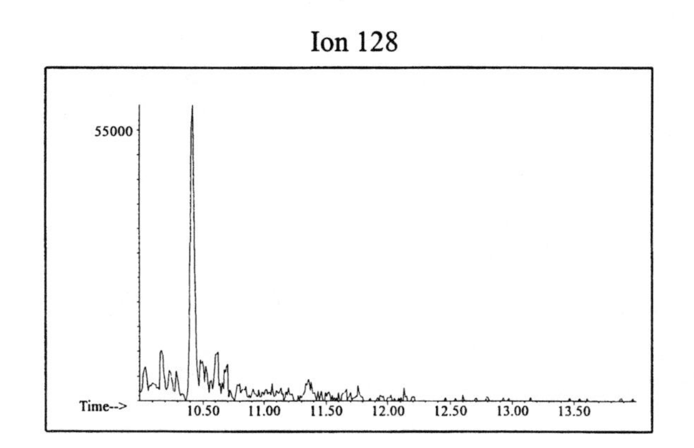

Ion 142

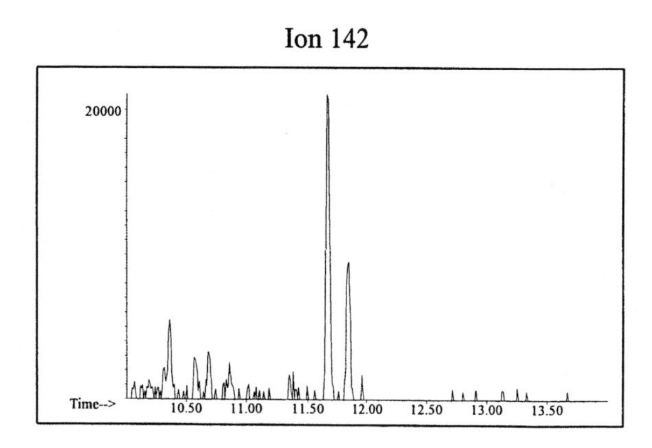

Ion 156

No Useful Data Obtained

Distillate #25

20uL/mL Pentane

ASTM: Class 3 (Medium Petroleum Distillate) Macro Code: M

Product Displayed: Dyco Mineral Spirits

Product Uses: Mineral spirits, charcoal lighter, heating fuel

Other Similar Products: Varsol 1, Lonestar Mineral Spirits, Kardol Mineral Spirits, K-1 Kerosene

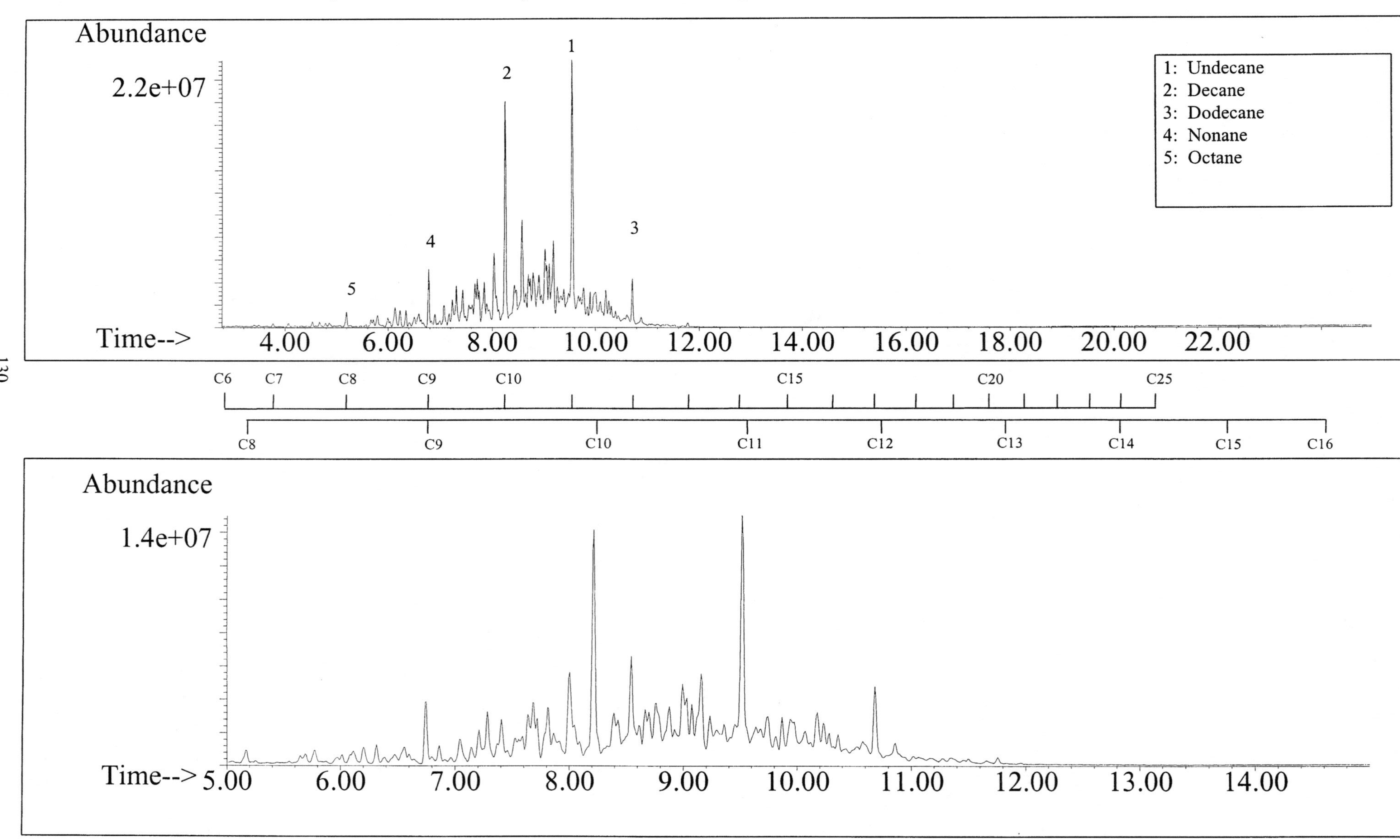

SUMMED PROFILES Distillate #25

ALKANES

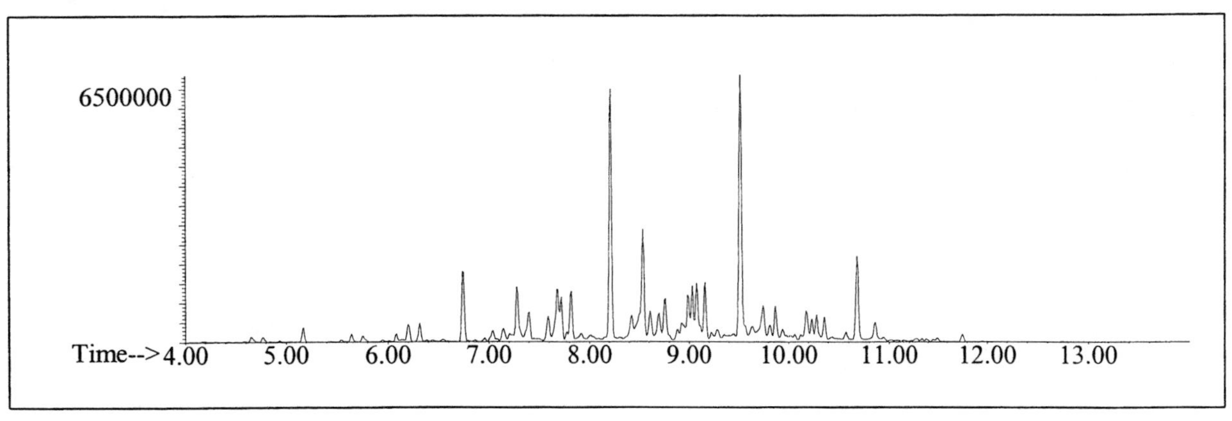

AROMATICS

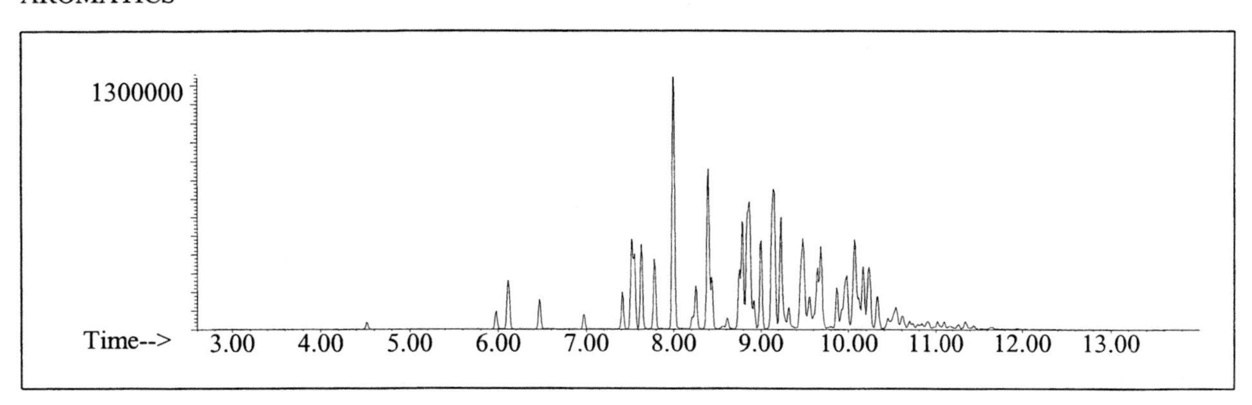

CYCLOPARAFFINS AND ALKENES

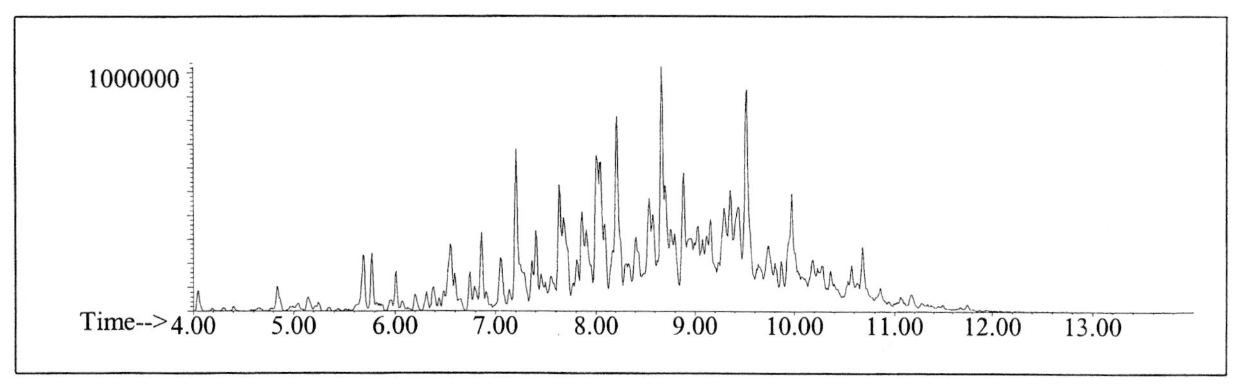

NAPHTHALENES

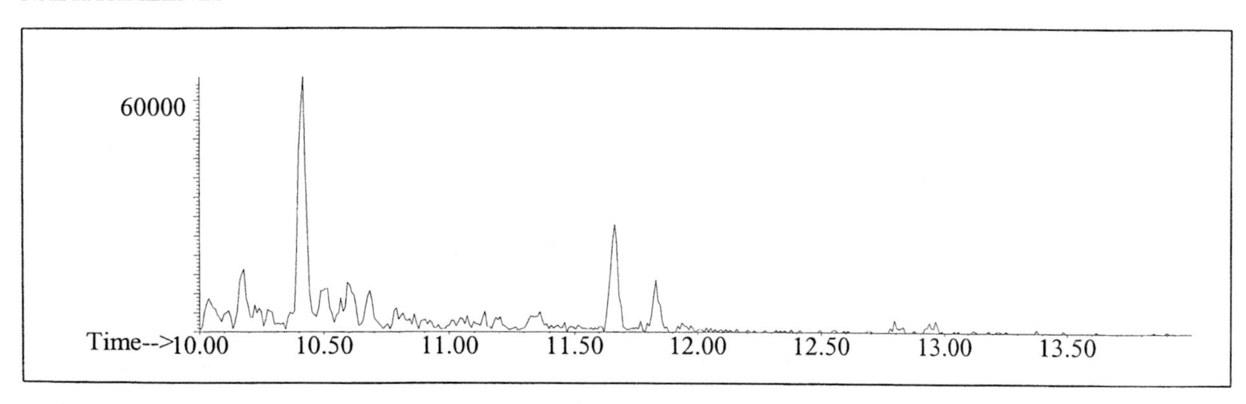

INDIVIDUAL PROFILES Distillate #25

Alkane

Ion 43

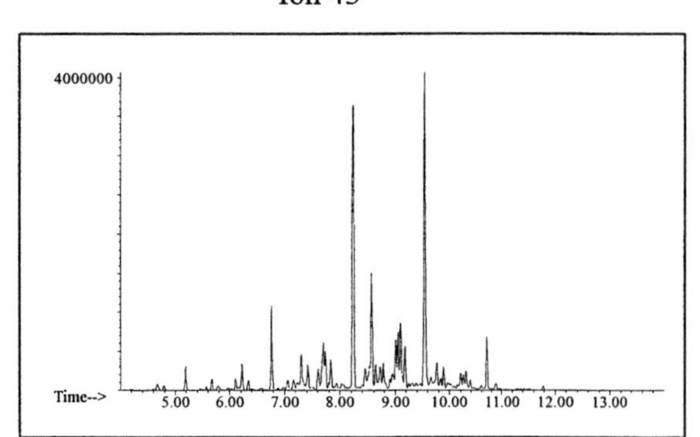

Ion 57

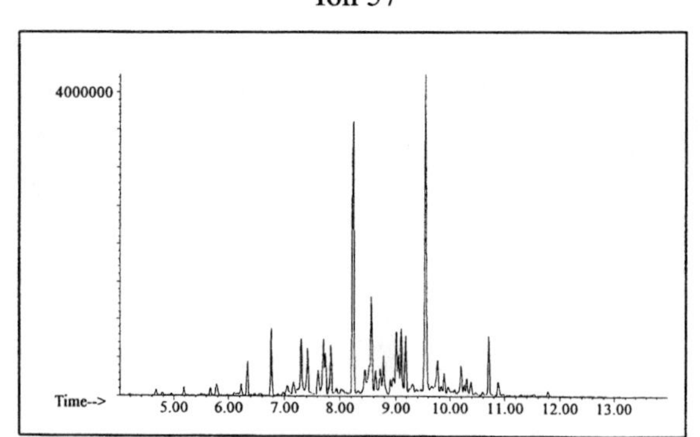

Ion 71

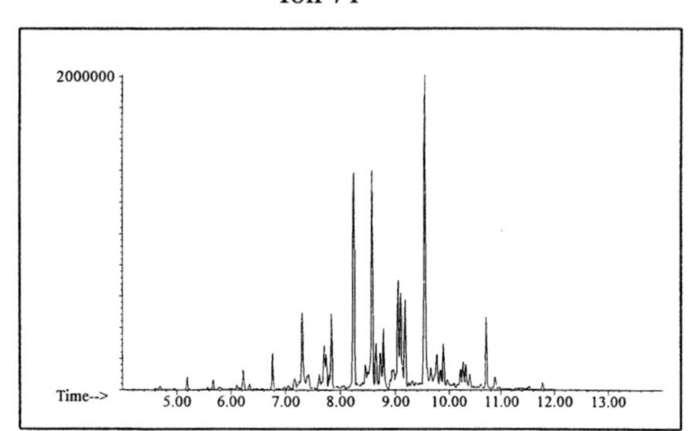

Ion 85

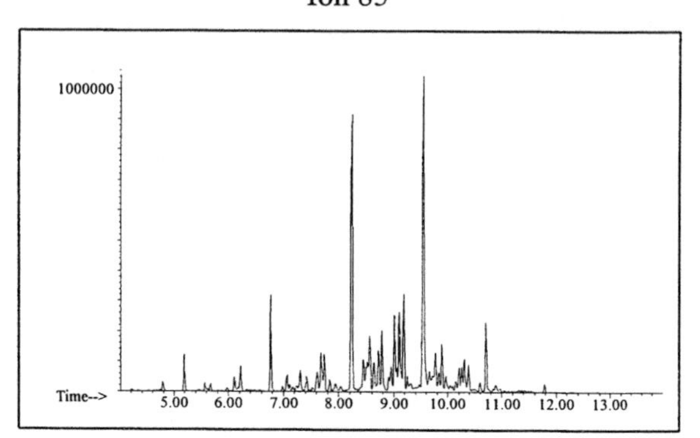

Aromatic

Ion 91

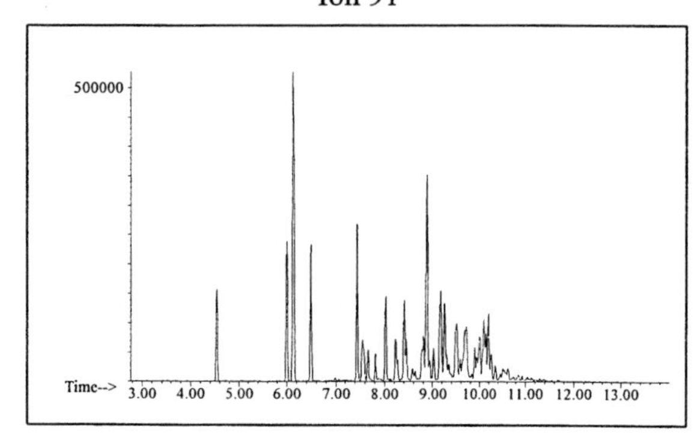

Ion 105

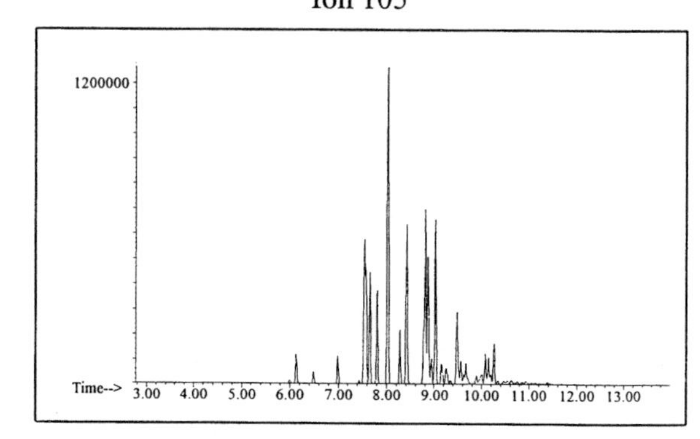

Ion 119

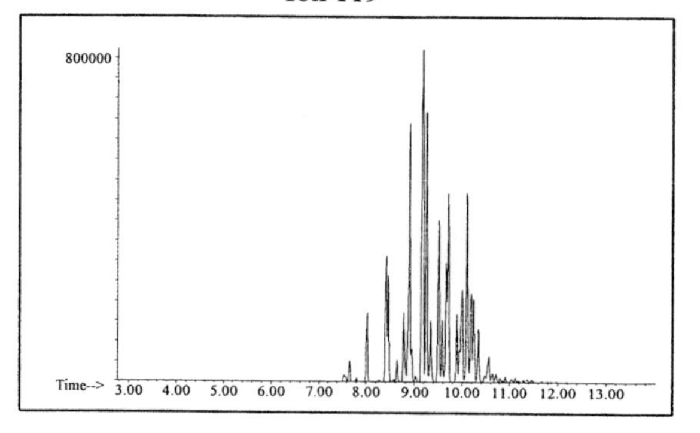

INDIVIDUAL PROFILES

Distillate #25

Cycloparaffin

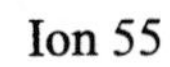

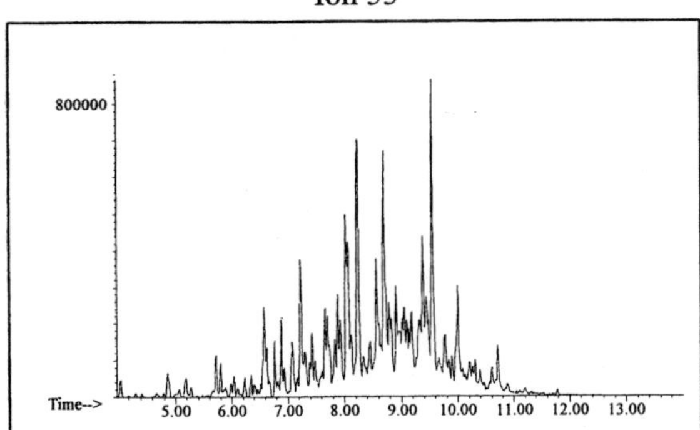

Ion 69

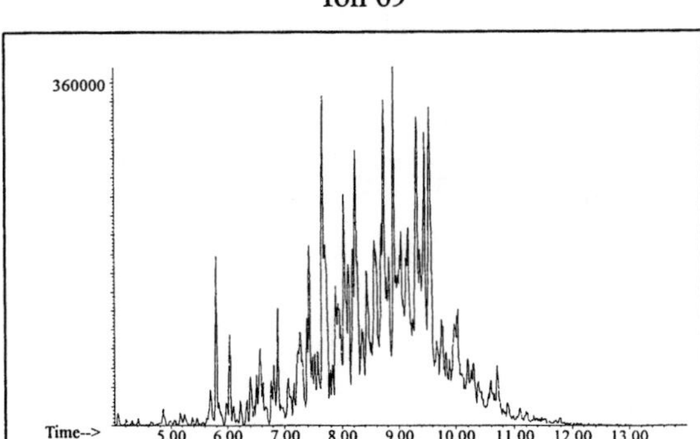

Ion 83

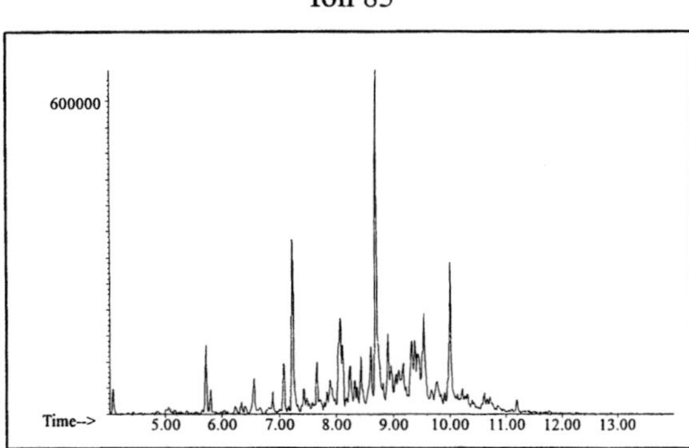

Naphthalene

Ion 128

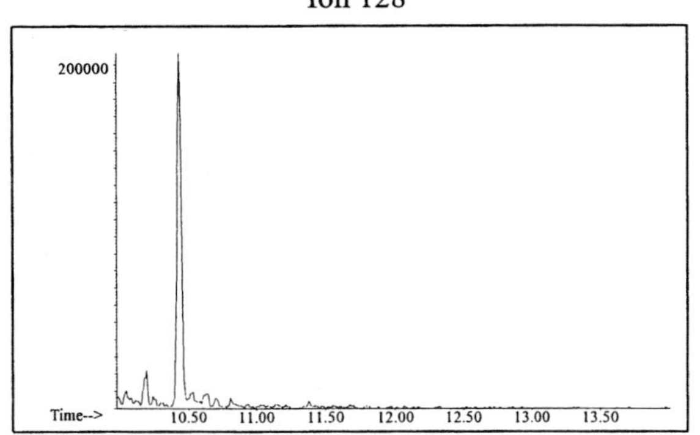

Ion 142

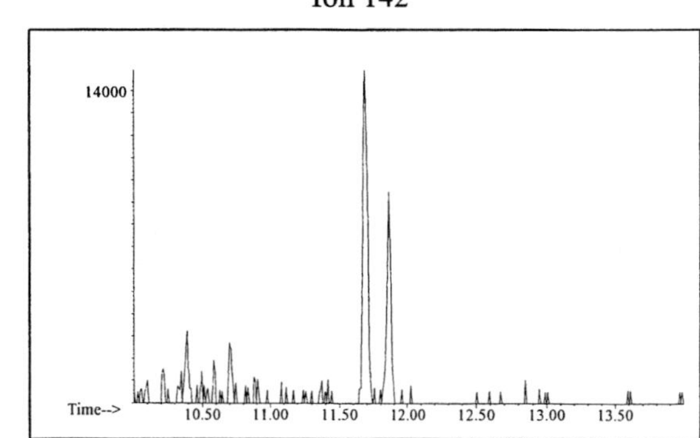

Ion 156

No Useful Data Obtained

Distillate #26

20uL/mL Pentane

ASTM: Class 3 (Medium Petroleum Distillate) Macro Code: M

Product Displayed: 25% Evaporated Dyco Mineral Spirits

Product Uses: Mineral spirits, charcoal lighter, heating fuel

Other Similar Products: Varsol 1, Lonestar Mineral Spirits, Kardol Mineral Spirits, K-1 Kerosene

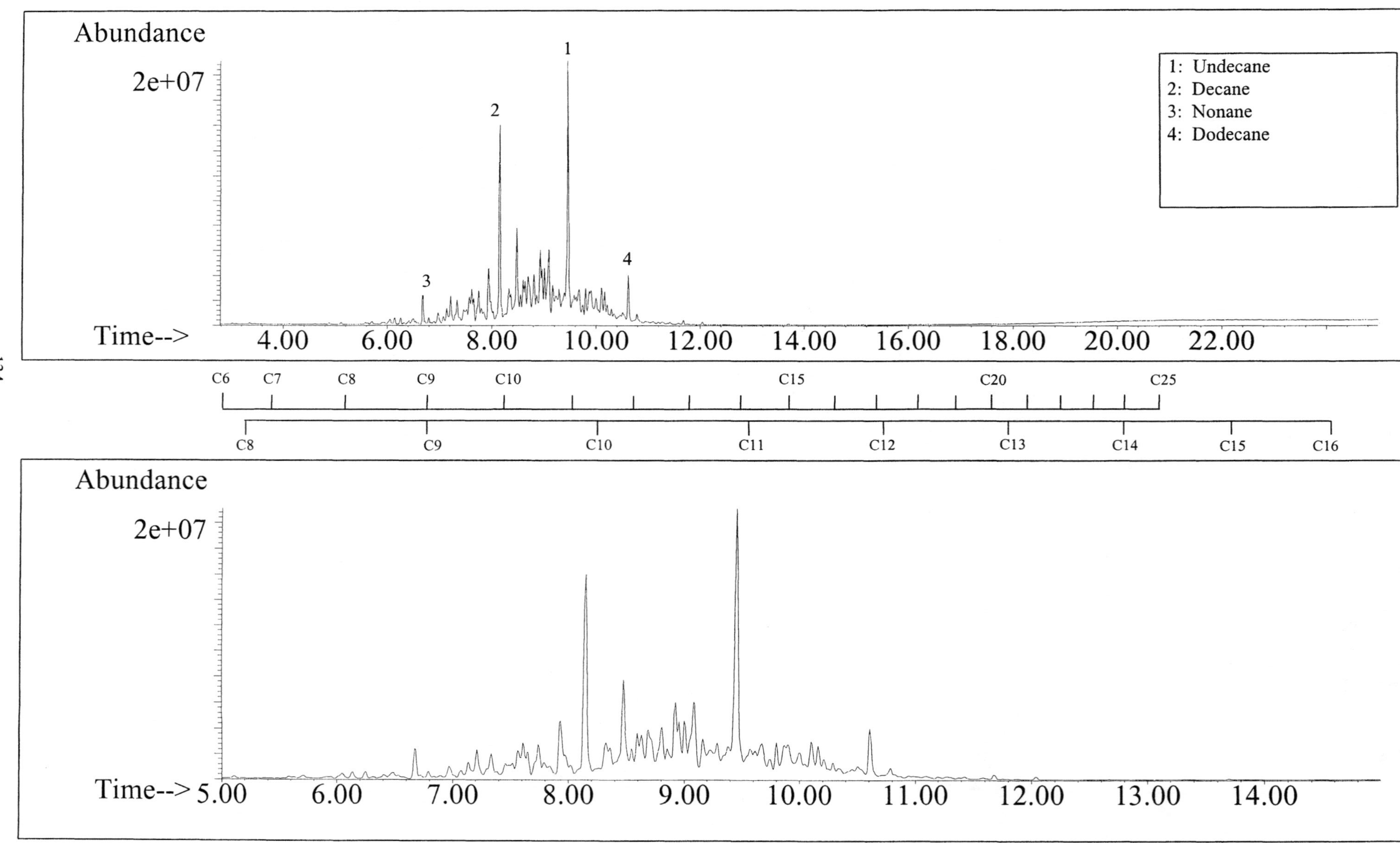

SUMMED PROFILES Distillate #26

ALKANES

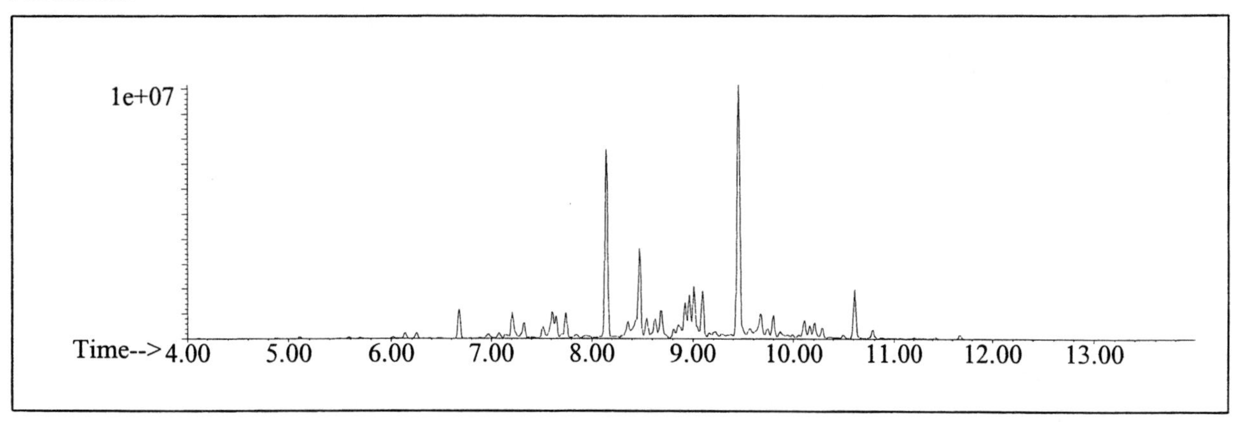

AROMATICS

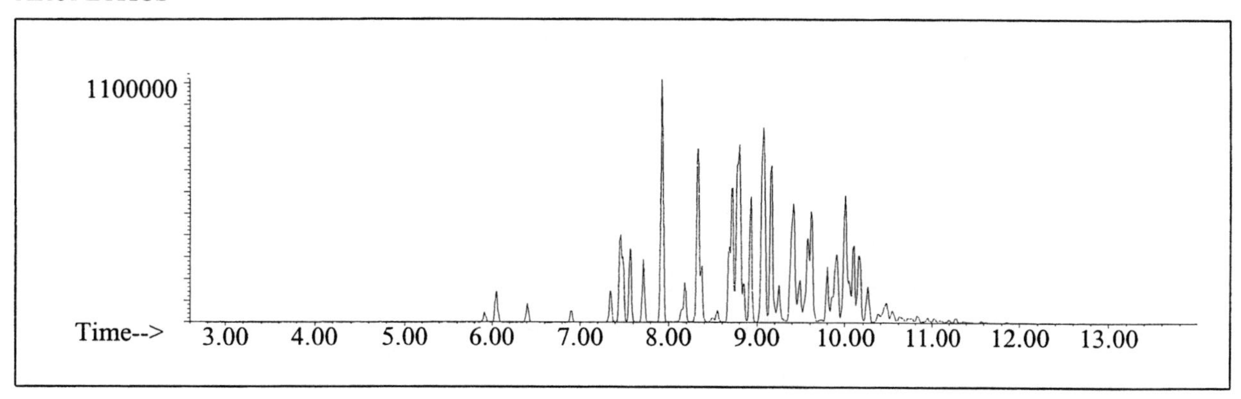

CYCLOPARAFFINS AND ALKENES

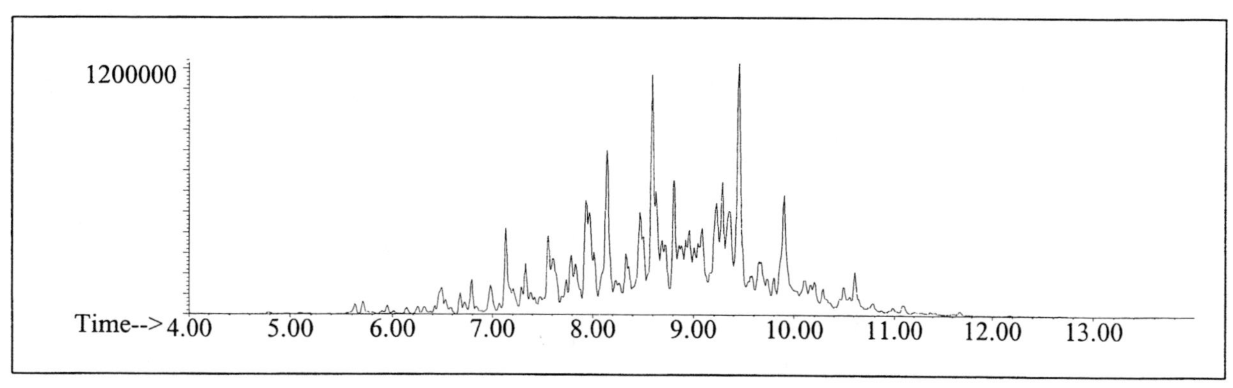

NAPHTHALENES

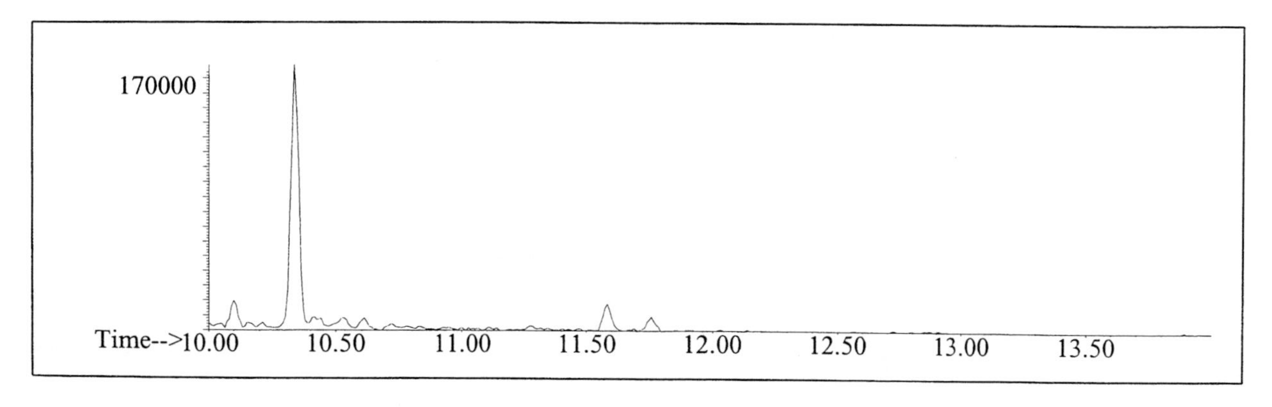

INDIVIDUAL PROFILES

Distillate #26

Alkane

Ion 43

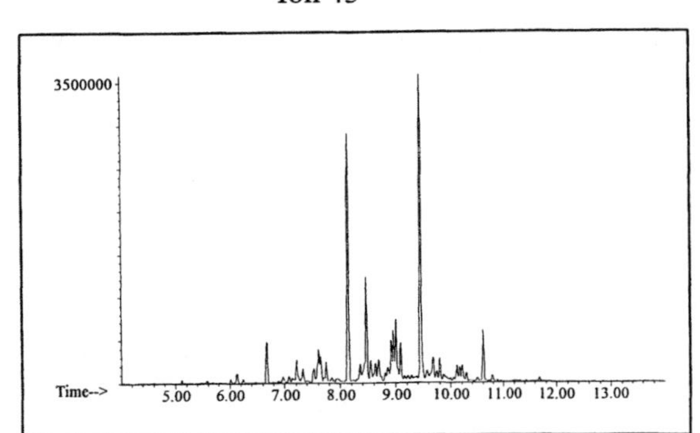

Ion 57

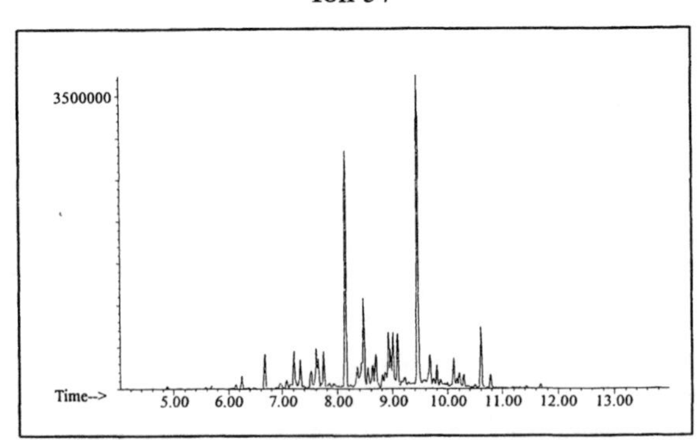

Ion 71

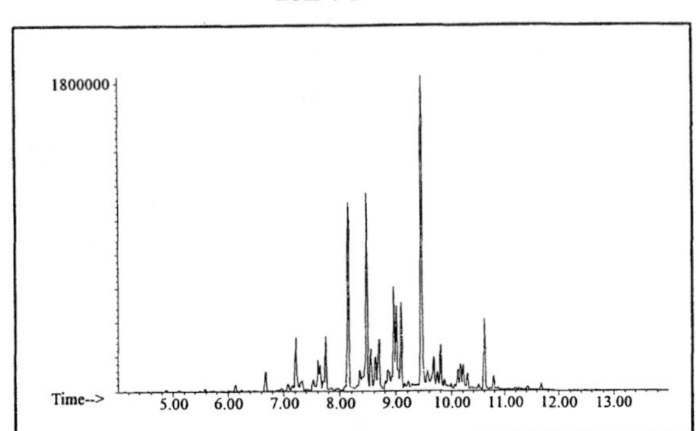

Ion 85

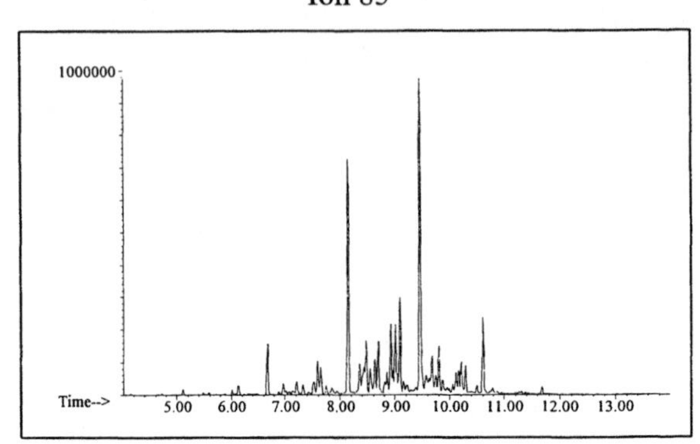

Aromatic

Ion 91

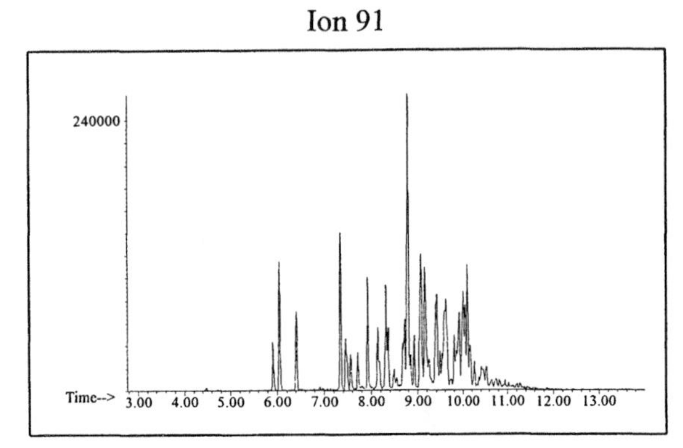

Ion 105

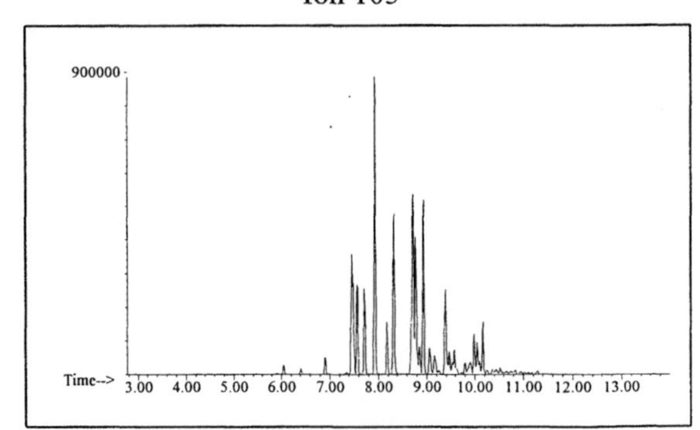

Ion 119

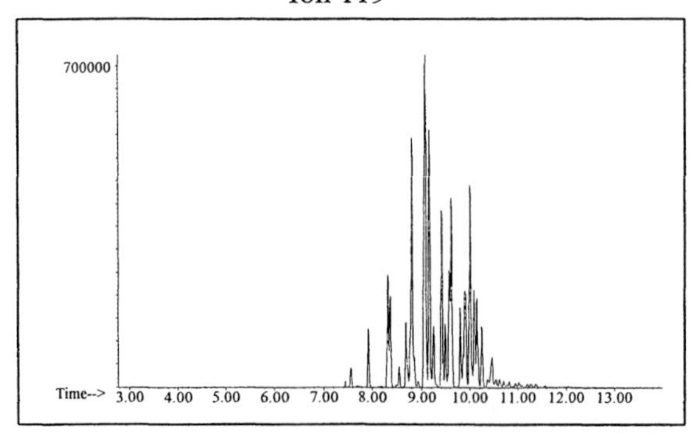

INDIVIDUAL PROFILES

Distillate #26

Cycloparaffin

Ion 55

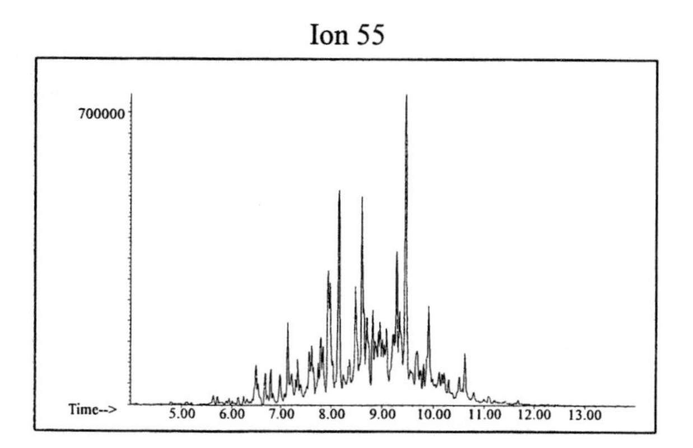

Ion 69

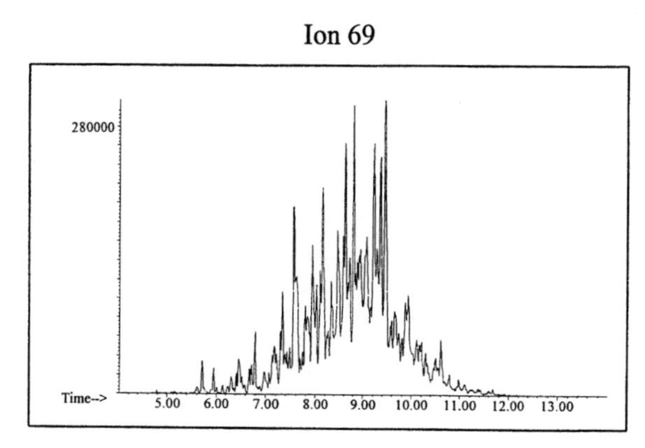

Ion 83

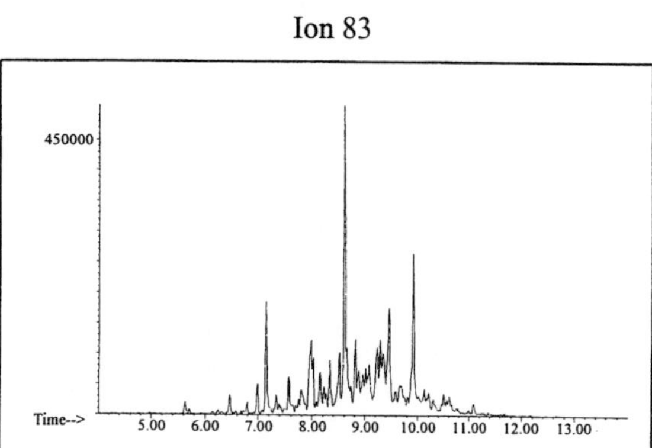

Naphthalene

Ion 128

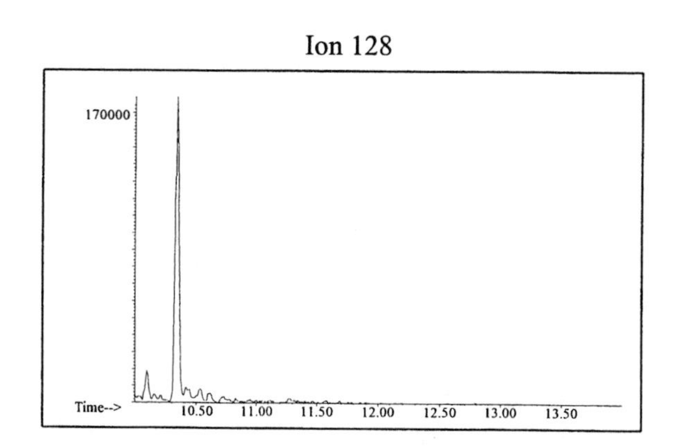

Ion 142

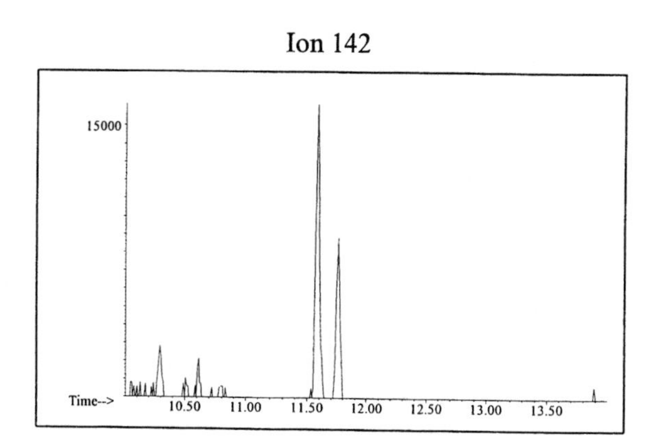

Ion 156

No Useful Data Obtained

Distillate #27

20uL/mL Pentane

ASTM: Class 3 (Medium Petroleum Distillate) Macro Code: M

Product Displayed: 50% Evaporated Dyco Mineral Spirits

Product Uses: Mineral spirits, charcoal lighter, heating fuel

Other Similar Products: Varsol 1, Lonestar Mineral Spirits, Kardol Mineral Spirits, K-1 Kerosene

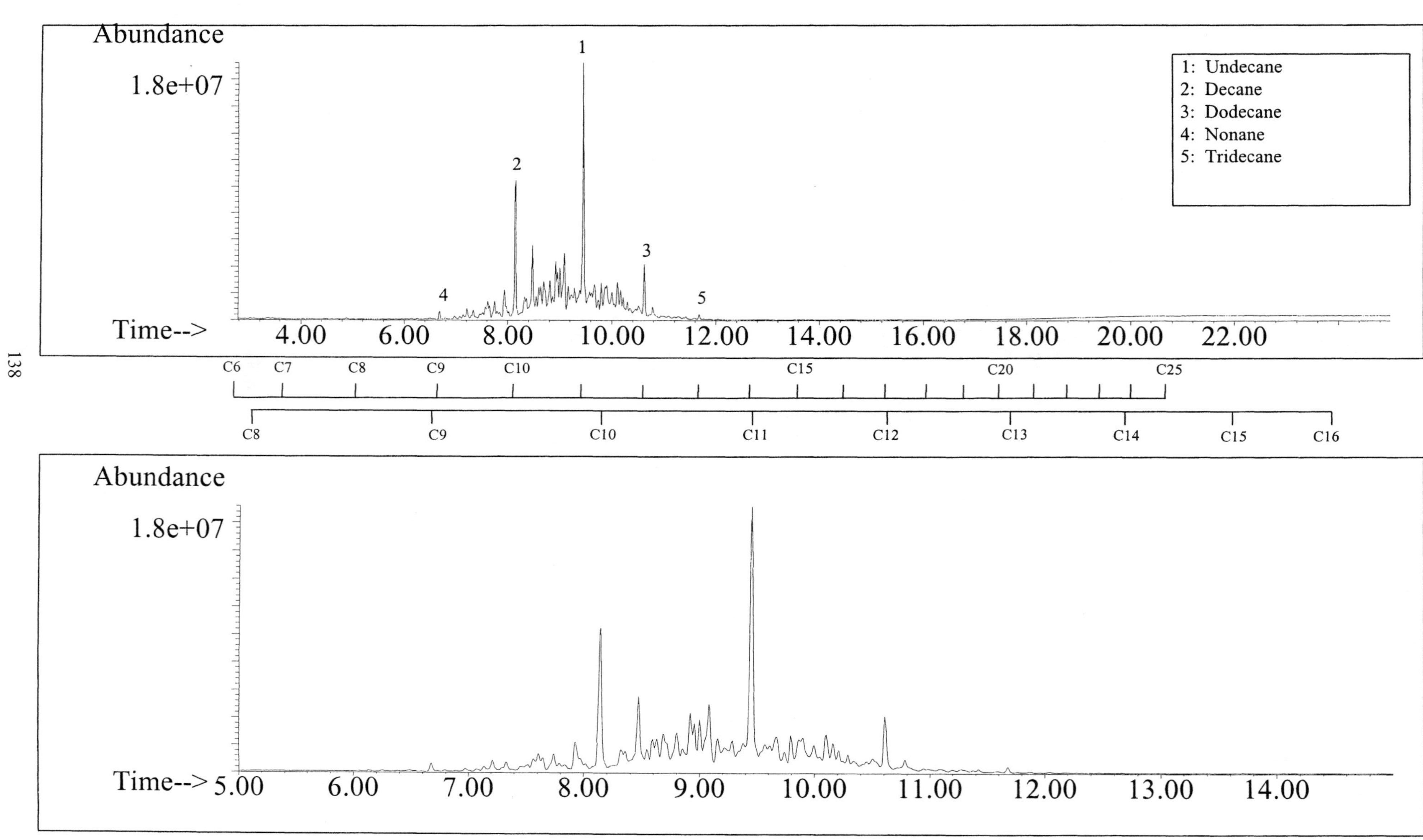

SUMMED PROFILES Distillate #27

ALKANES

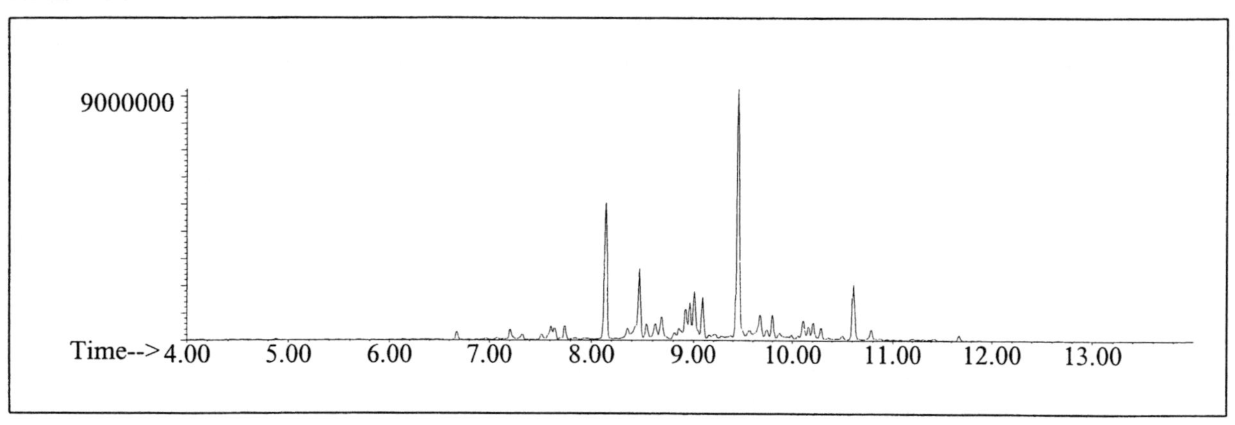

AROMATICS

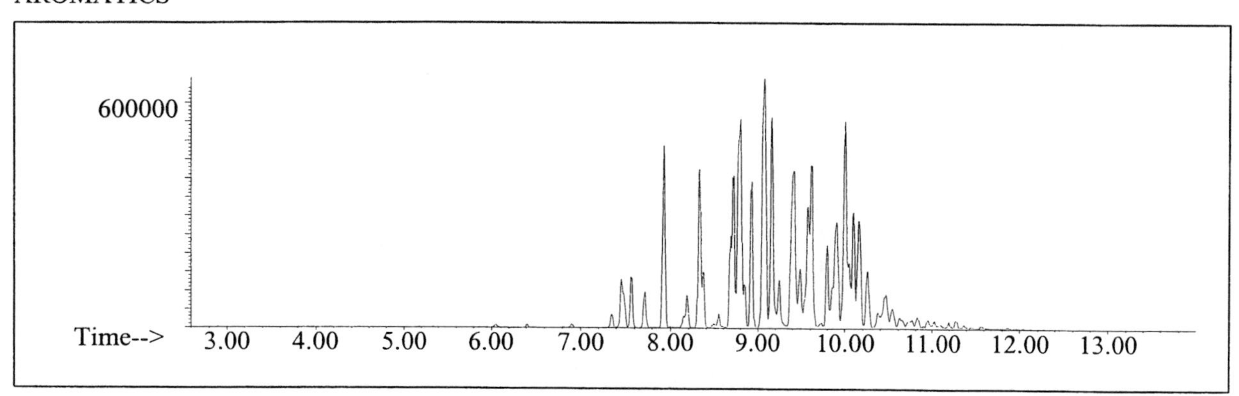

CYCLOPARAFFINS AND ALKENES

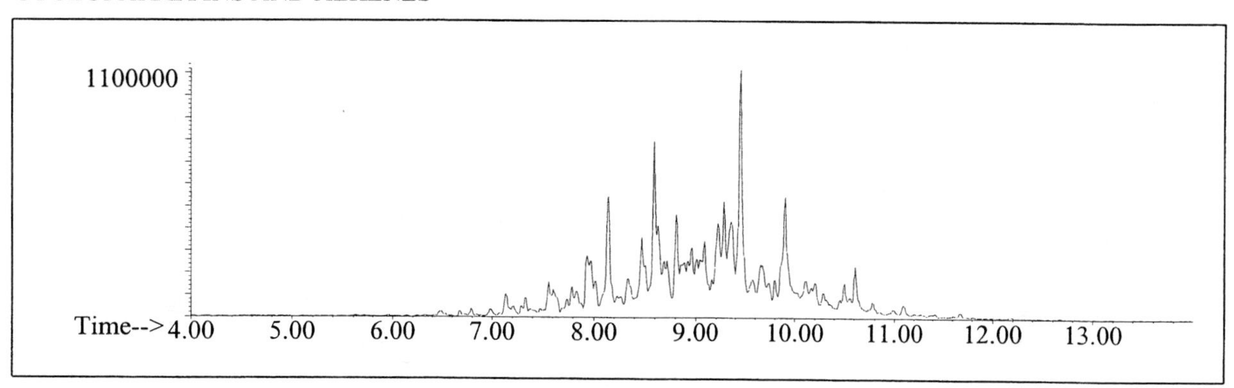

NAPHTHALENES

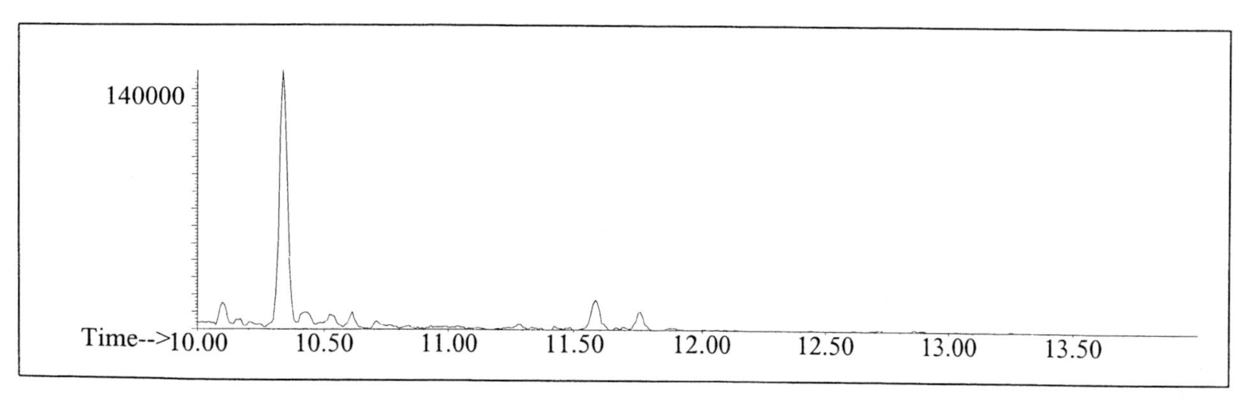

INDIVIDUAL PROFILES

Distillate #27

Alkane

Ion 43

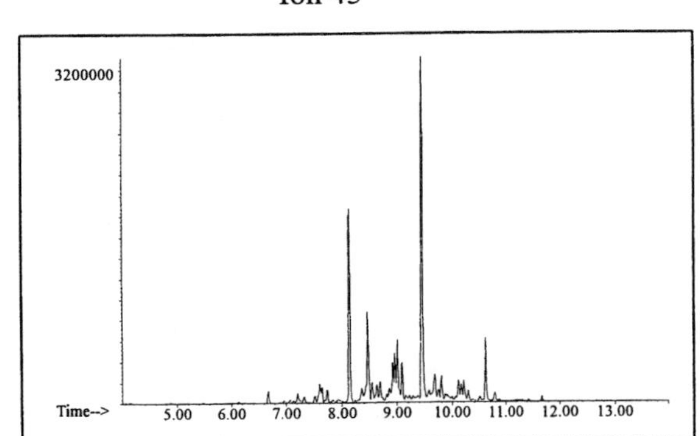

Ion 57

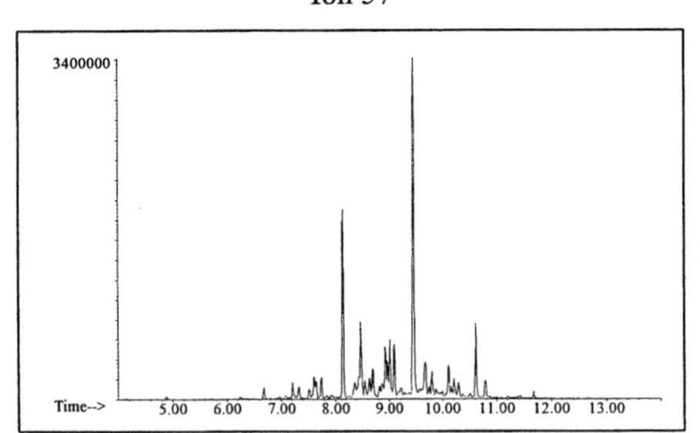

Ion 71

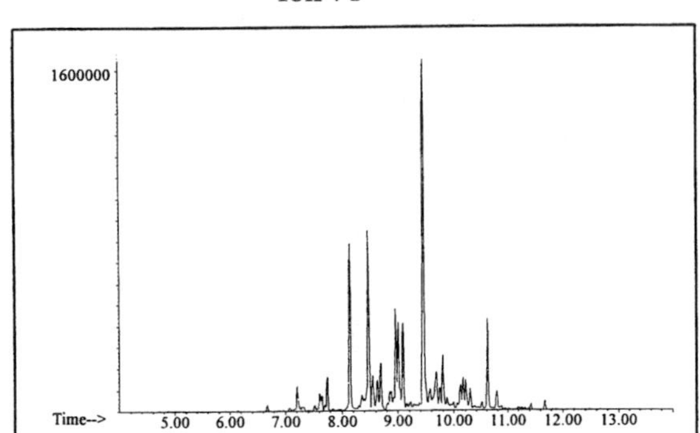

Ion 85

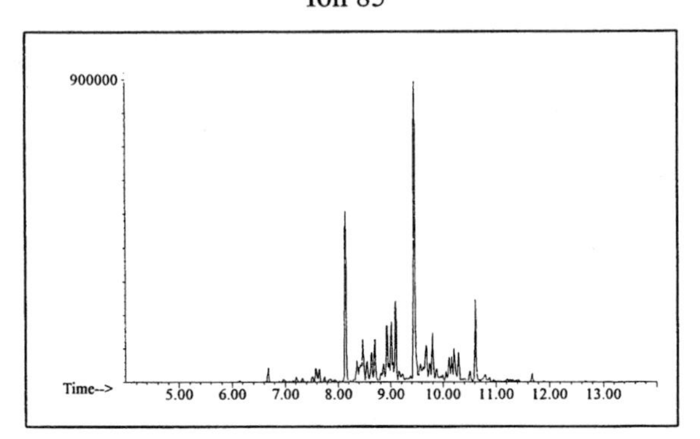

Aromatic

Ion 91

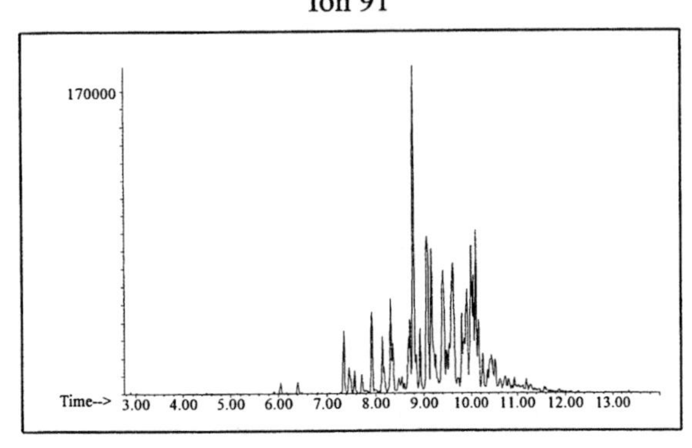

Ion 105

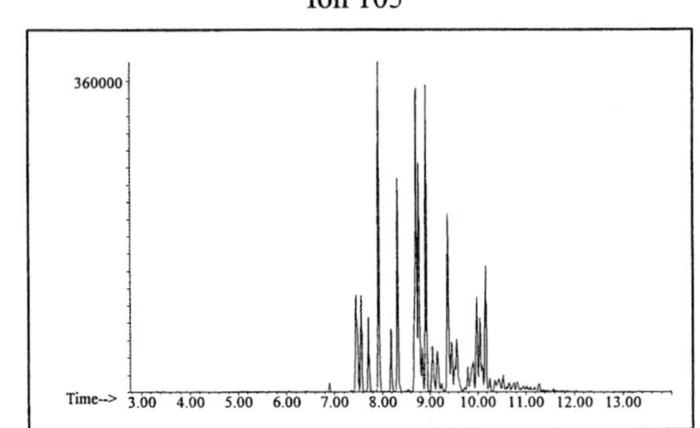

Ion 119

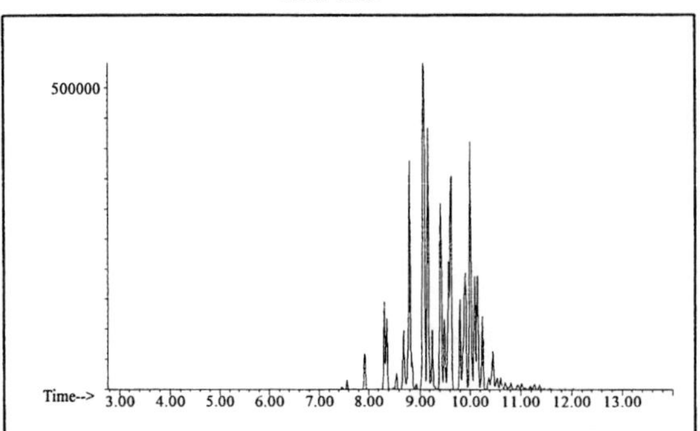

INDIVIDUAL PROFILES Distillate #27

Cycloparaffin

Ion 55

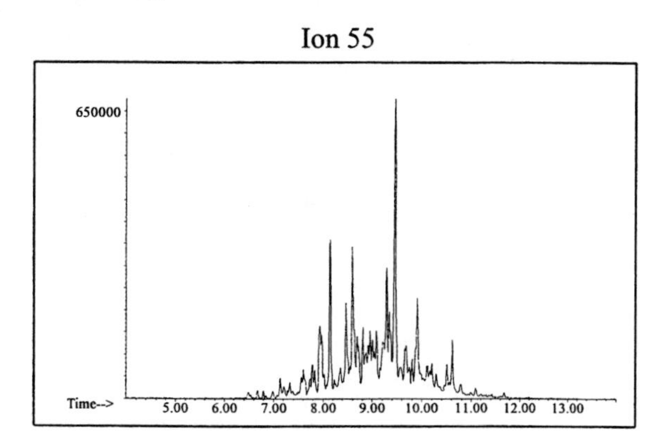

Ion 69

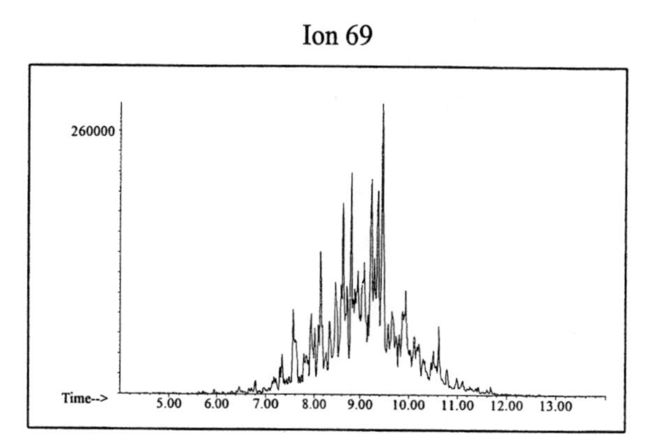

Ion 83

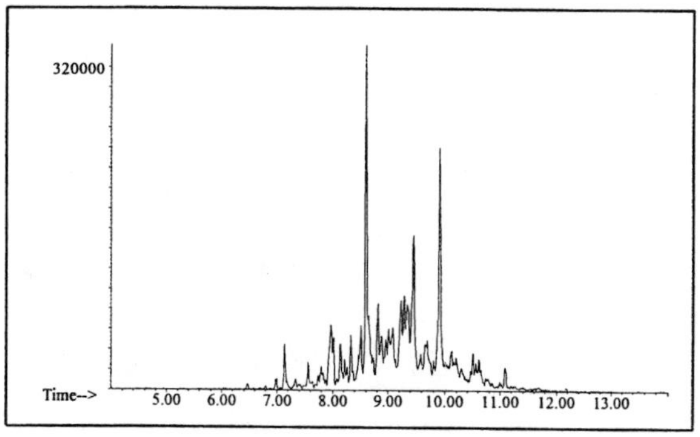

Naphthalene

Ion 128

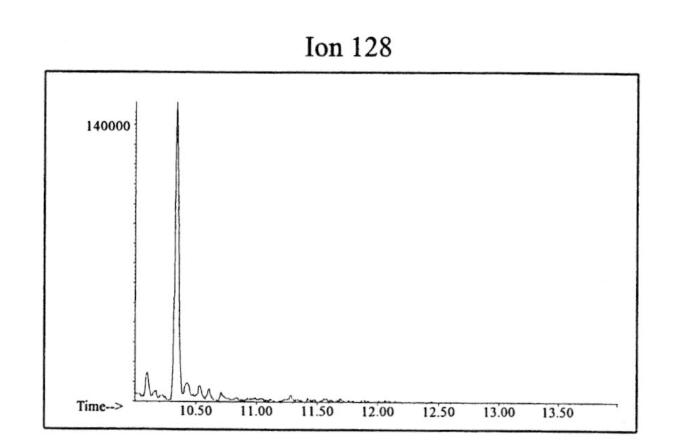

Ion 142

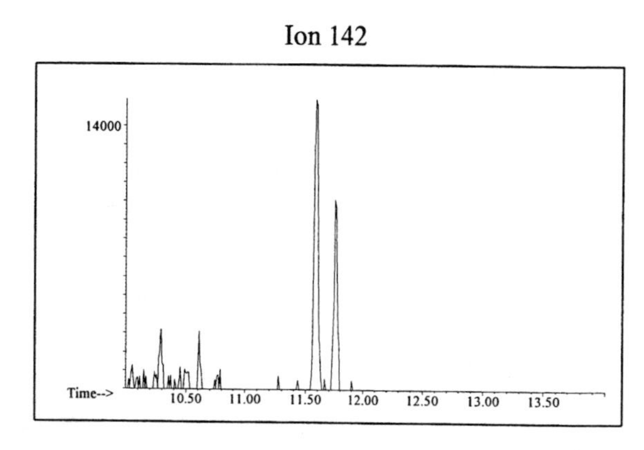

Ion 156

No Useful Data Obtained

Distillate #28

20uL/mL Pentane

ASTM: Class 3 (Medium Petroleum Distillate) Macro Code: M

Product Displayed: 75% Evaporated Dyco Mineral Spirits

Product Uses: Mineral spirits, charcoal lighter, heating fuel

Other Similar Products: Varsol 1, Lonestar Mineral Spirits, Kardol Mineral Spirits, K-1 Kerosene

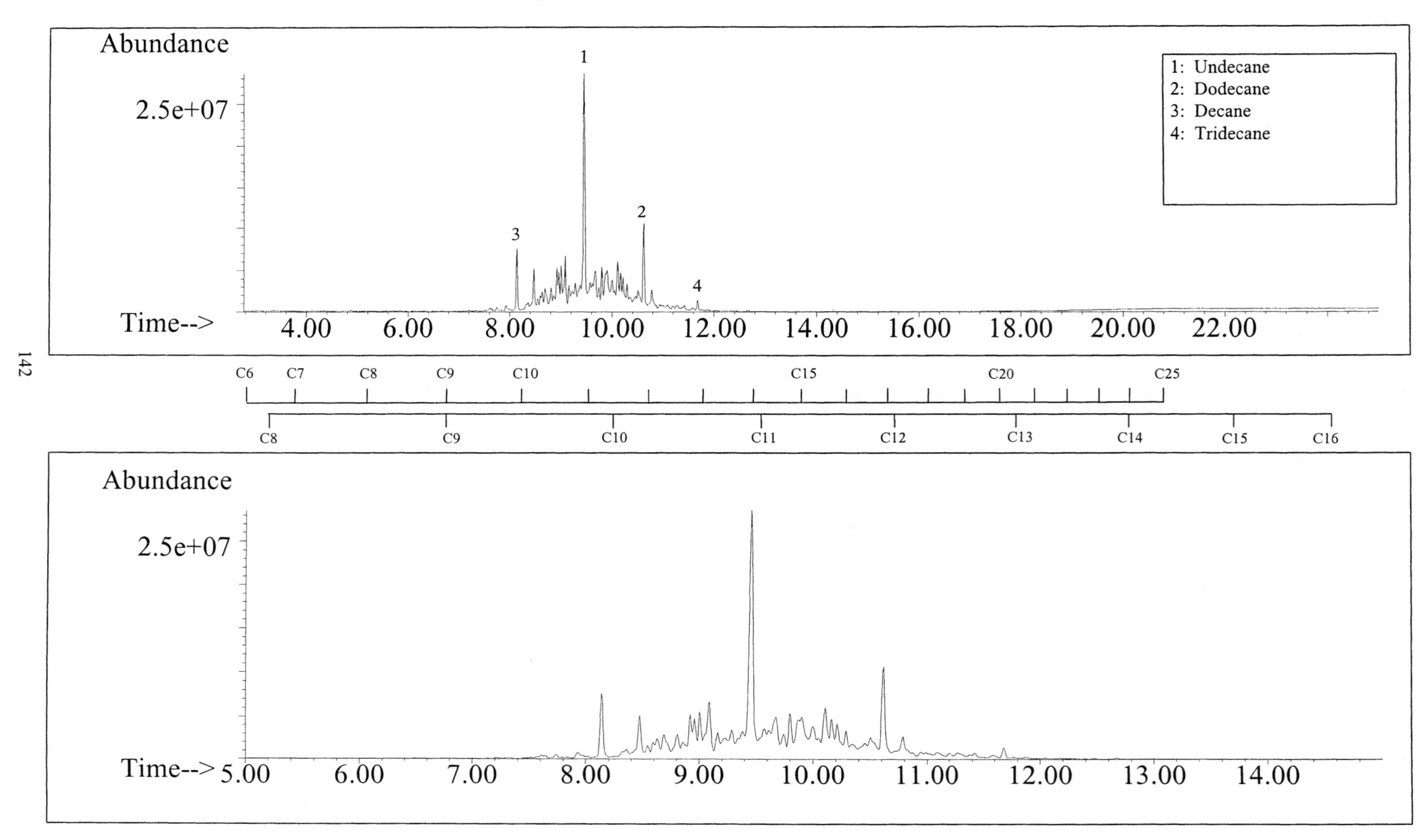

SUMMED PROFILES Distillate #28

ALKANES

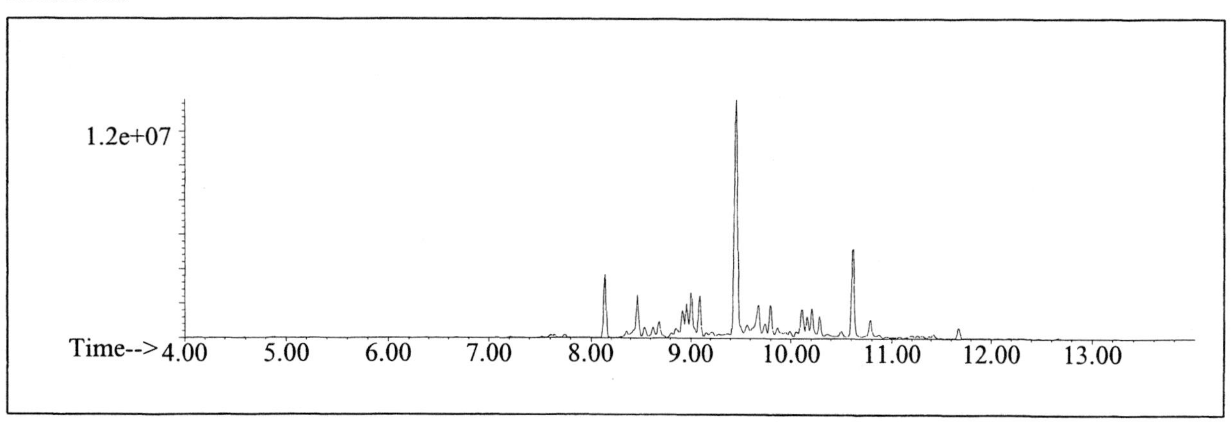

AROMATICS

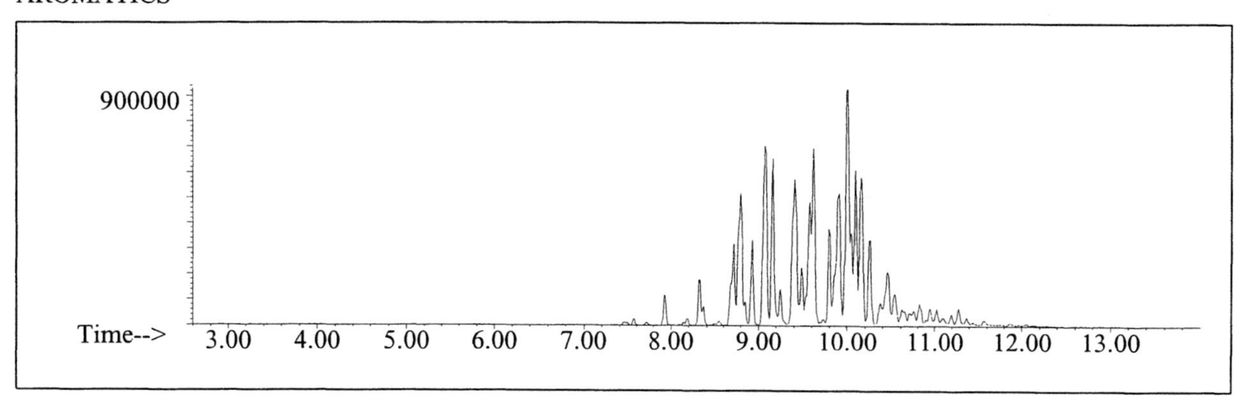

CYCLOPARAFFINS AND ALKENES

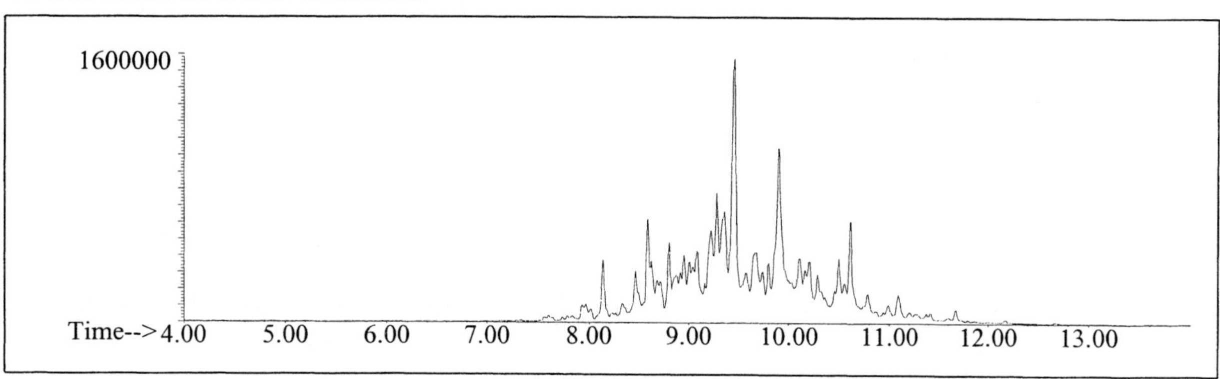

NAPHTHALENES

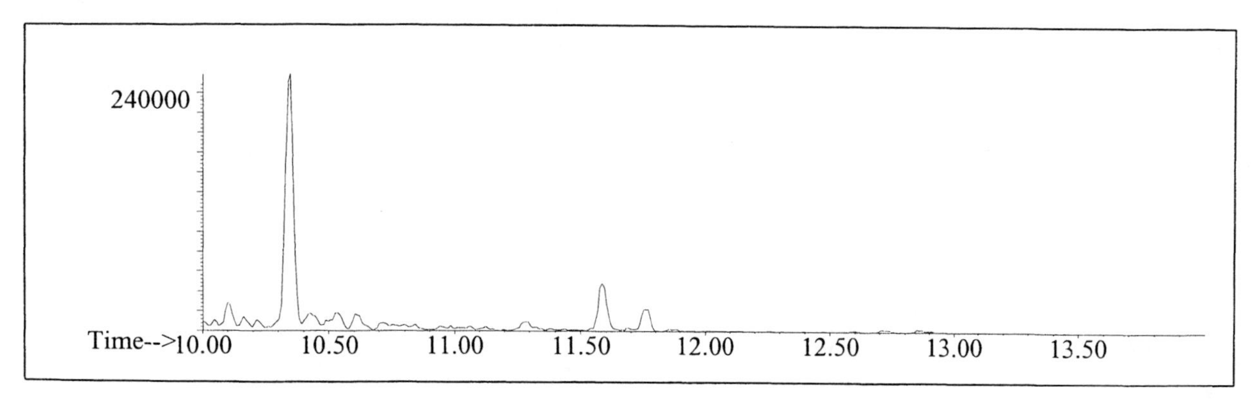

INDIVIDUAL PROFILES

Distillate #28

Alkane

Ion 43

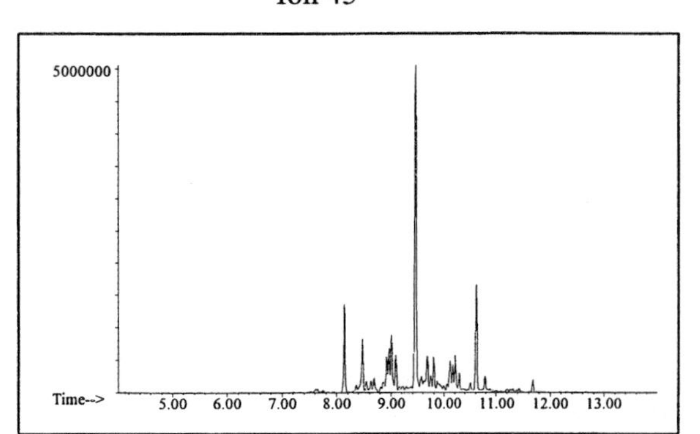

Ion 57

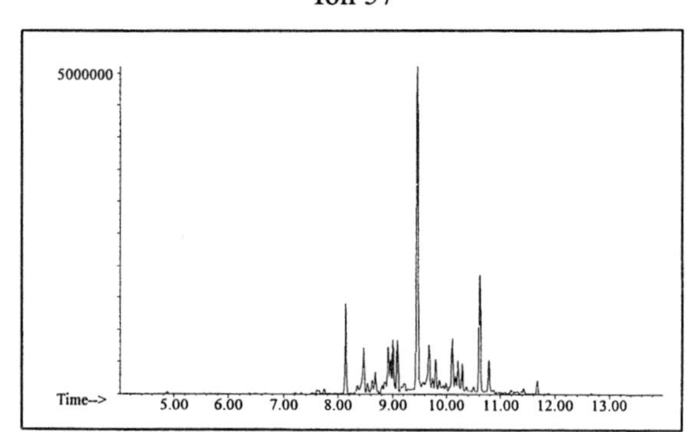

Ion 71

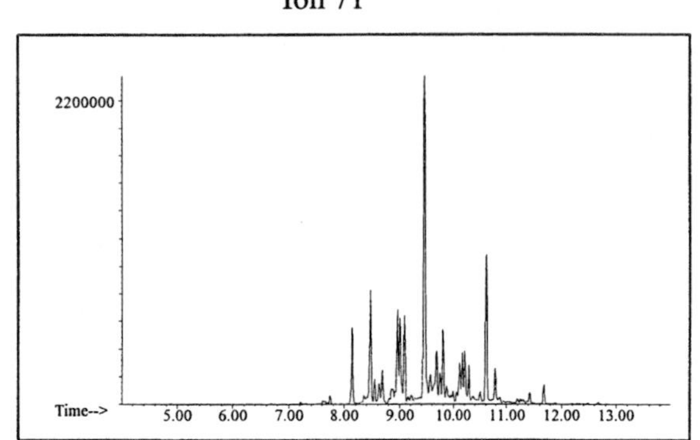

Ion 85

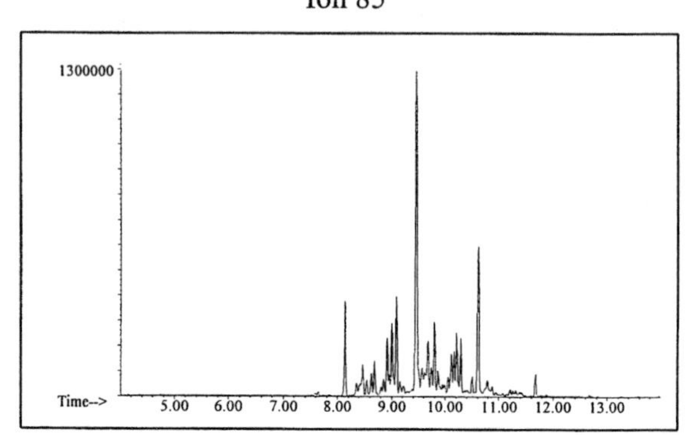

Aromatic

Ion 91

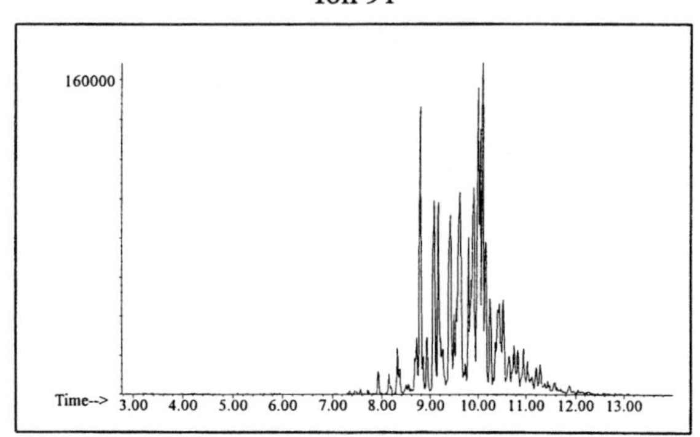

Ion 105

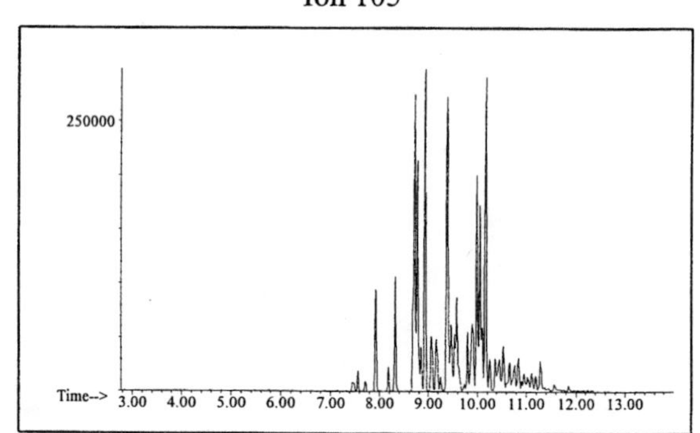

Ion 119

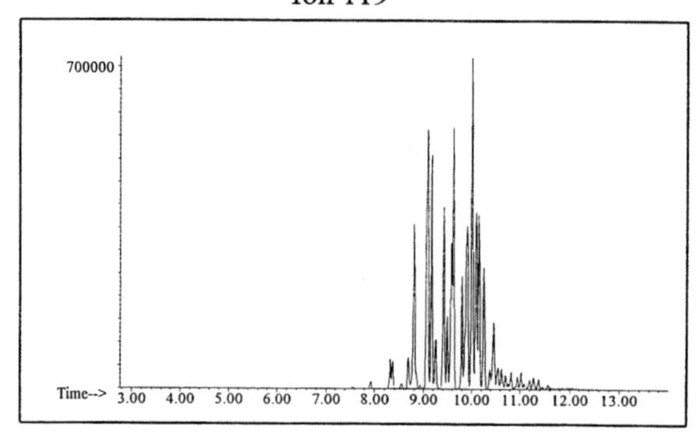

INDIVIDUAL PROFILES

Distillate #28

Cycloparaffin

Ion 55

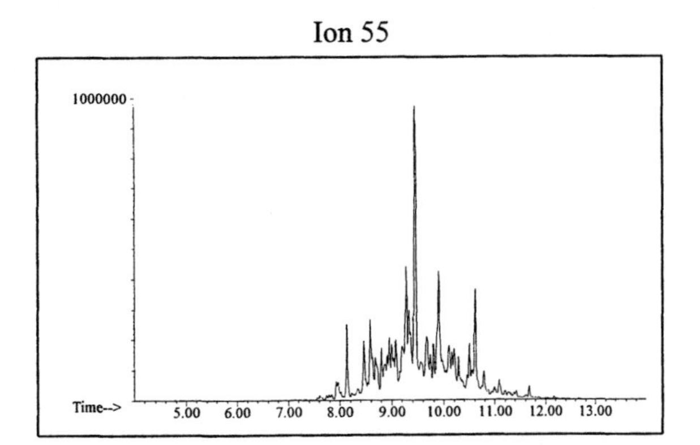

Ion 69

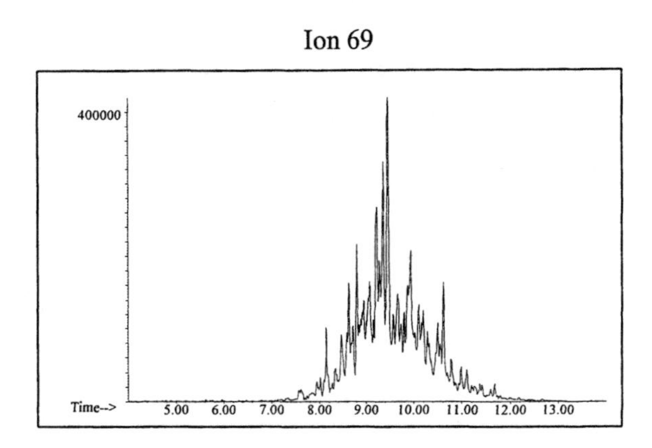

Ion 83

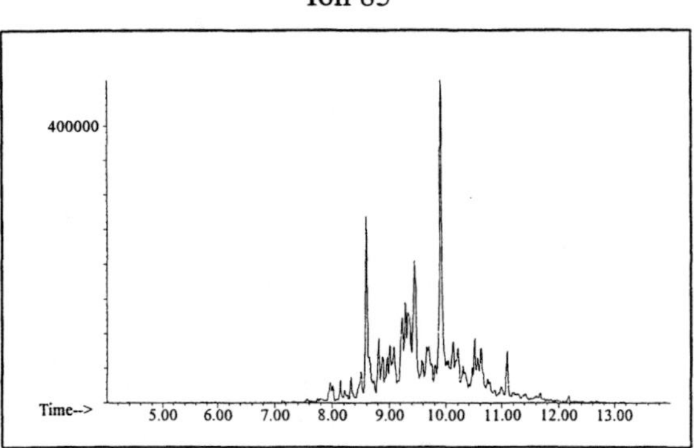

Naphthalene

Ion 128

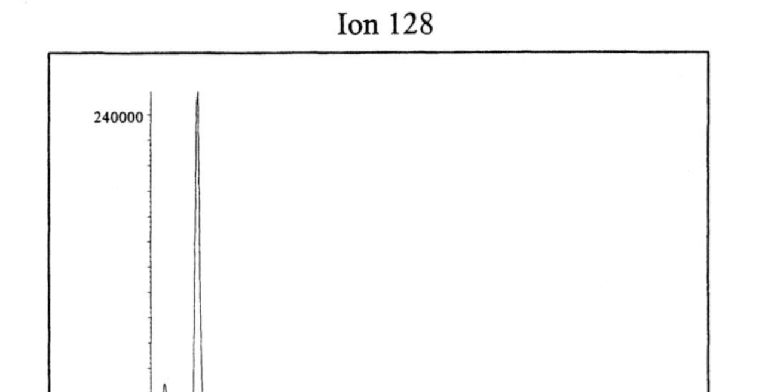

Ion 142

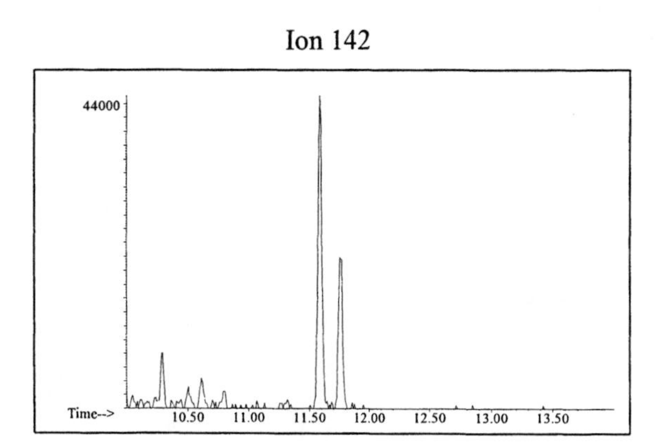

Ion 156

No Useful Data Obtained

Distillate #29

20uL/mL Pentane

ASTM: Class 3 (Medium Petroleum Distillate) Macro Code: M

Product Displayed: Stoddard Solvent

Product Uses: Charcoal lighter, paint thinner

Other Similar Products: Magic Chef Charcoal Starter, Piggly Wiggly Charcoal Lighter

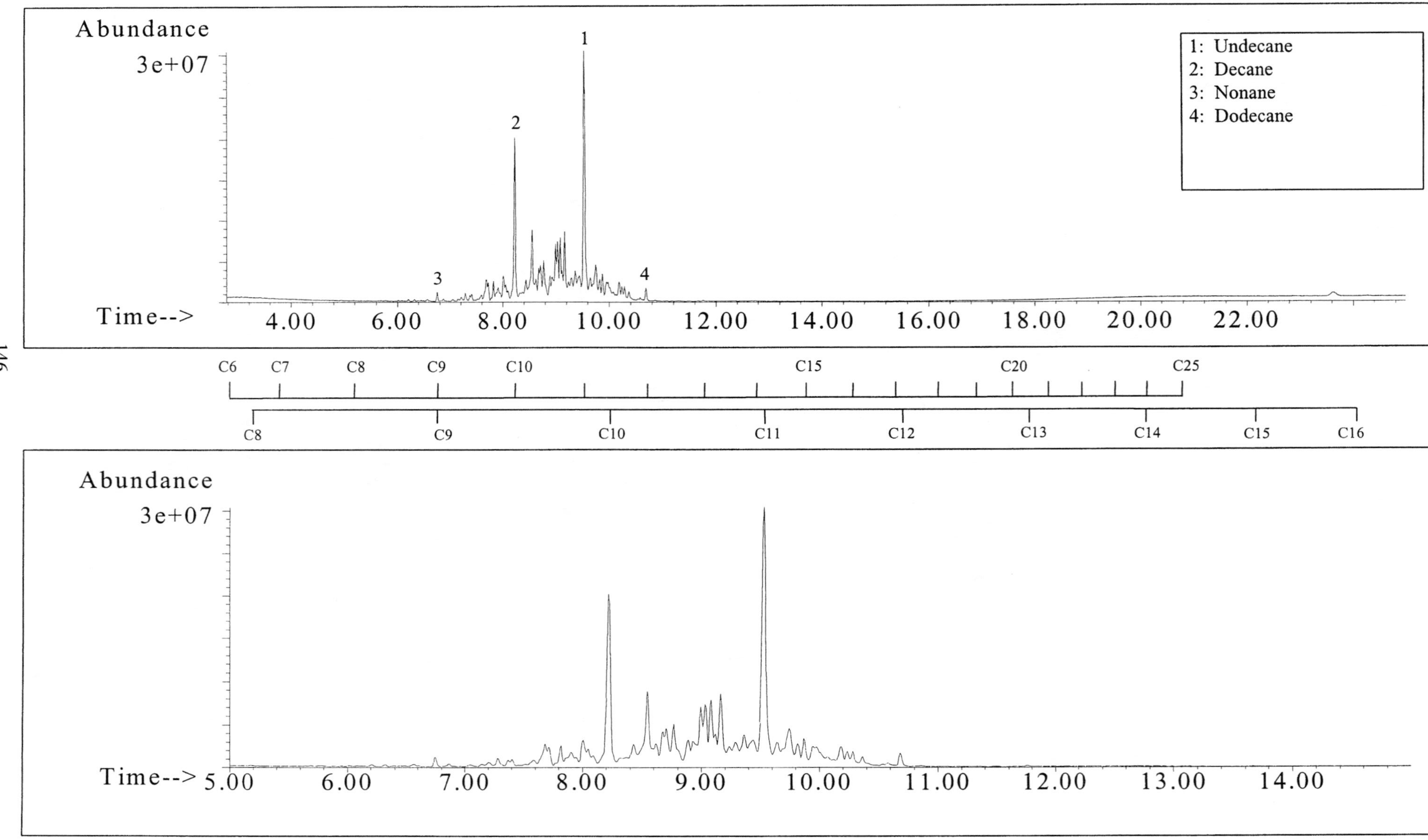

SUMMED PROFILES **Distillate #29**

ALKANES

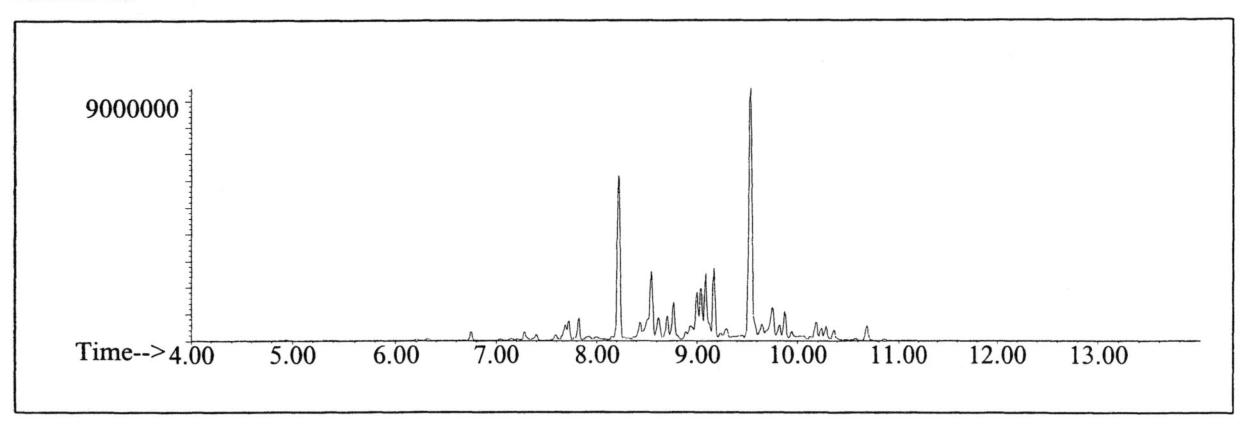

AROMATICS

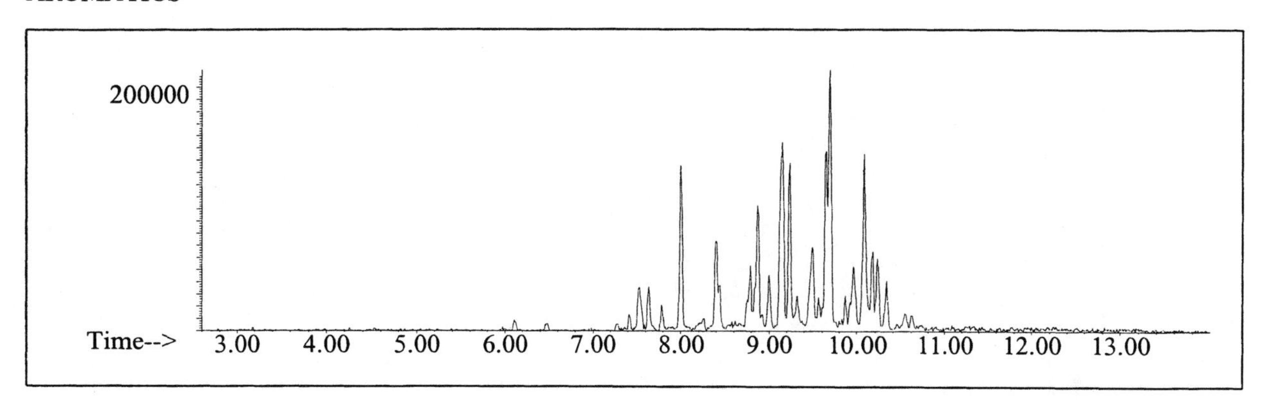

CYCLOPARAFFINS AND ALKENES

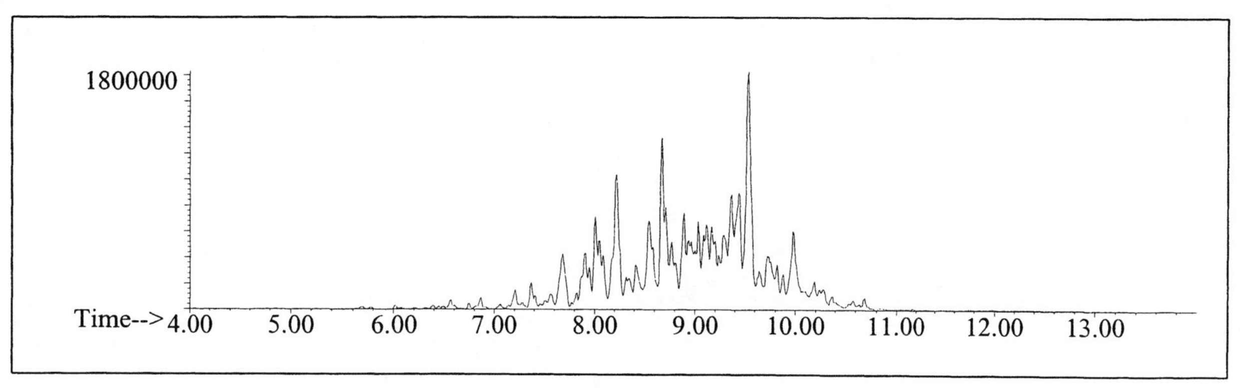

NAPHTHALENES

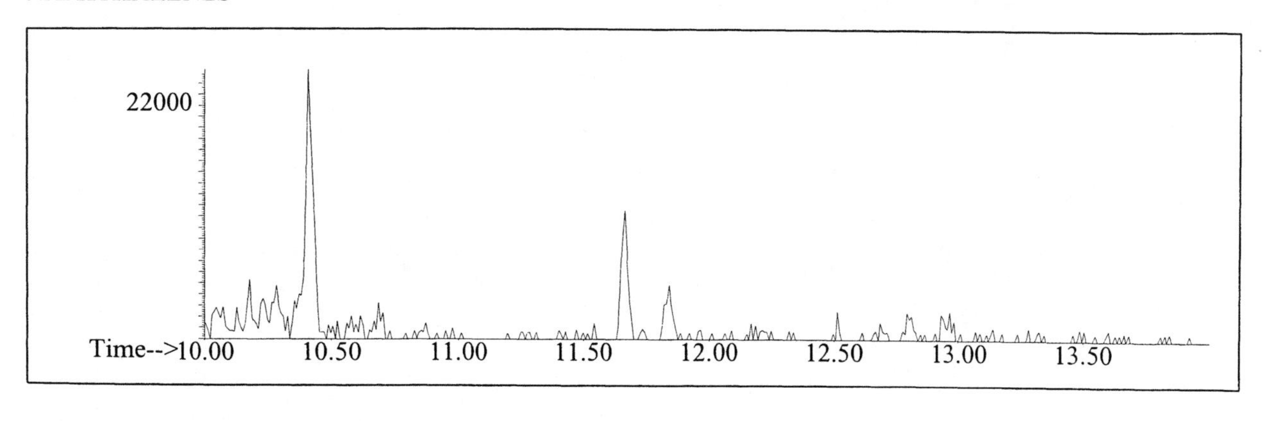

INDIVIDUAL PROFILES Distillate #29

Alkane

Ion 43

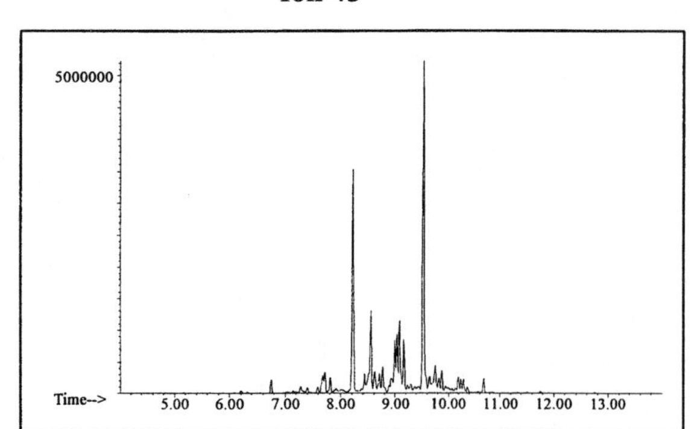

Ion 57

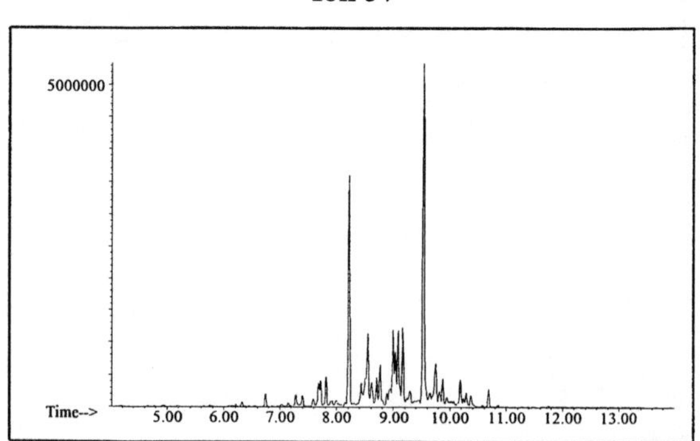

Ion 71

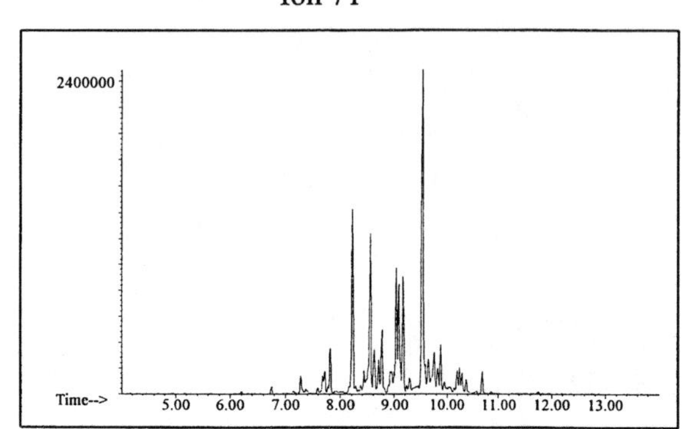

Ion 85

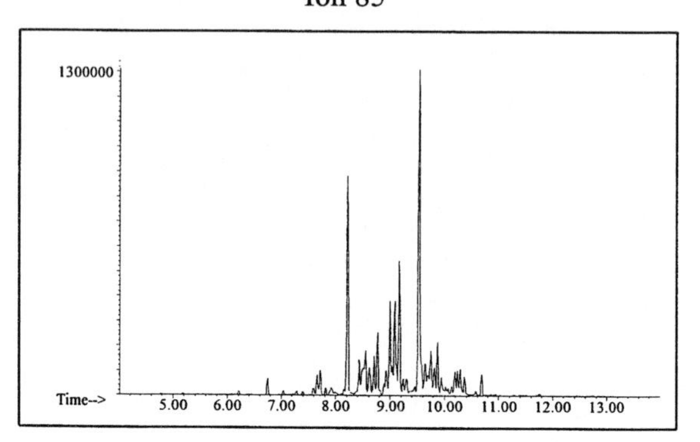

Aromatic

Ion 91

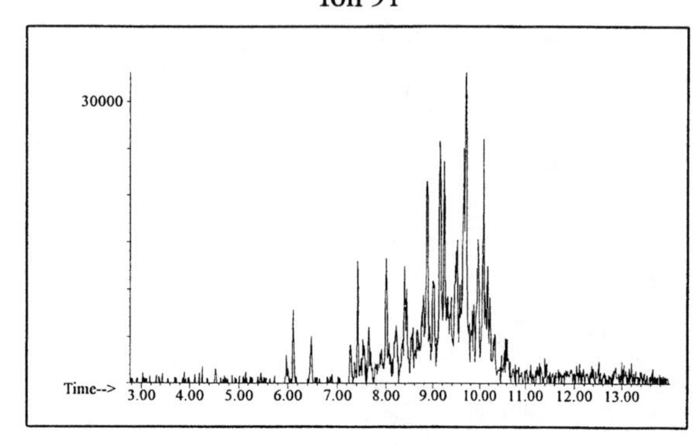

Ion 105

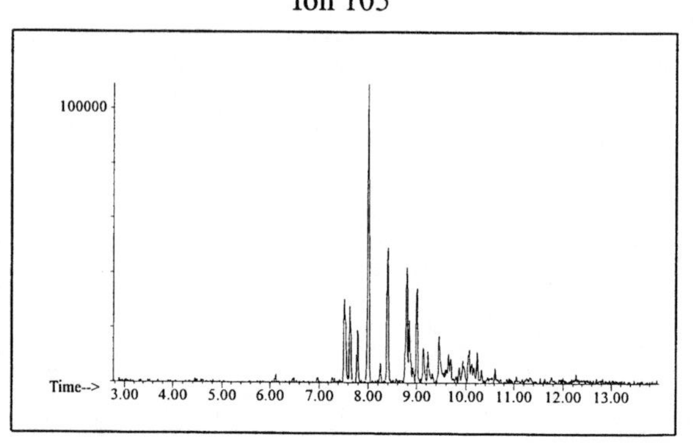

Ion 119

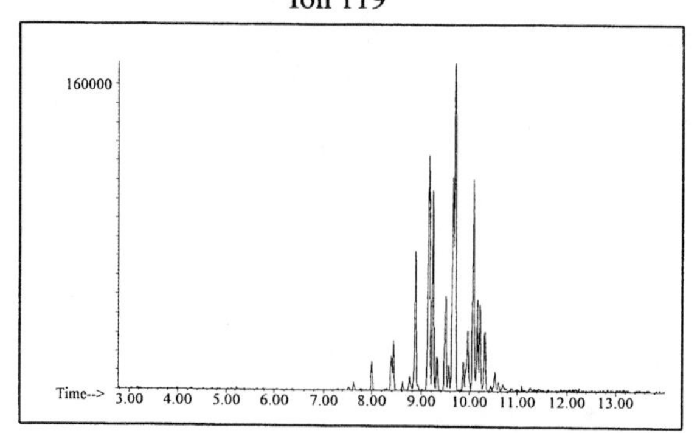

INDIVIDUAL PROFILES Distillate #29

Cycloparaffin

Ion 55

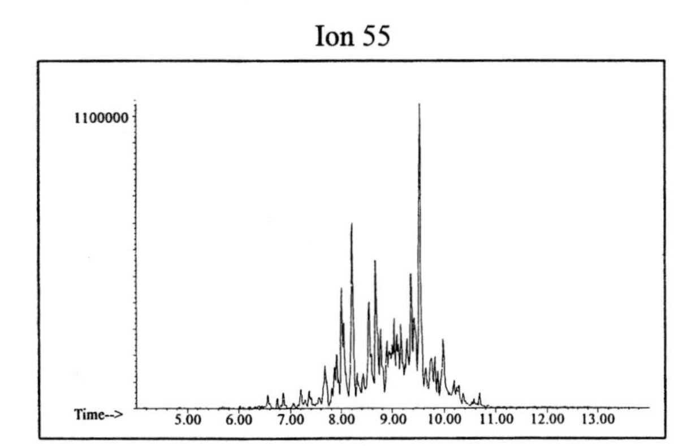

Ion 69

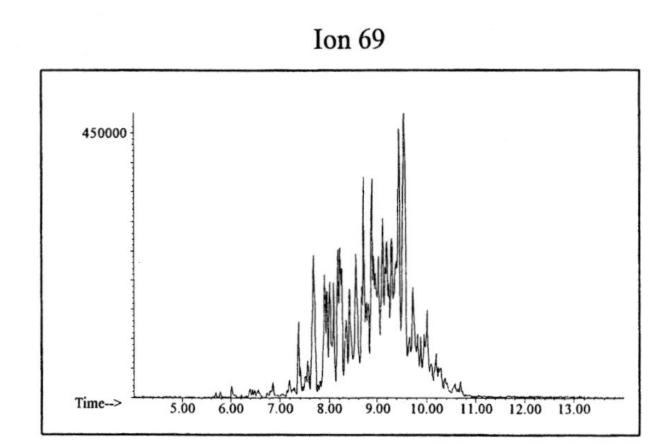

Ion 83

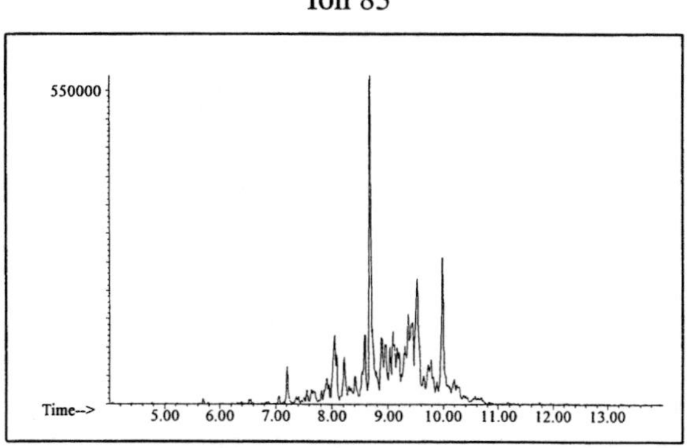

Naphthalene

Ion 128

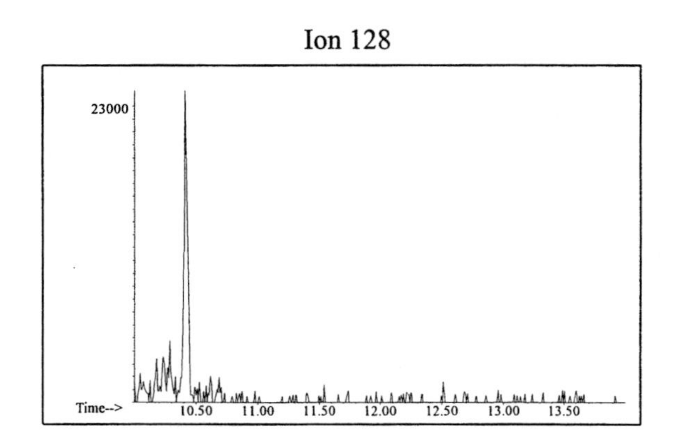

Ion 142

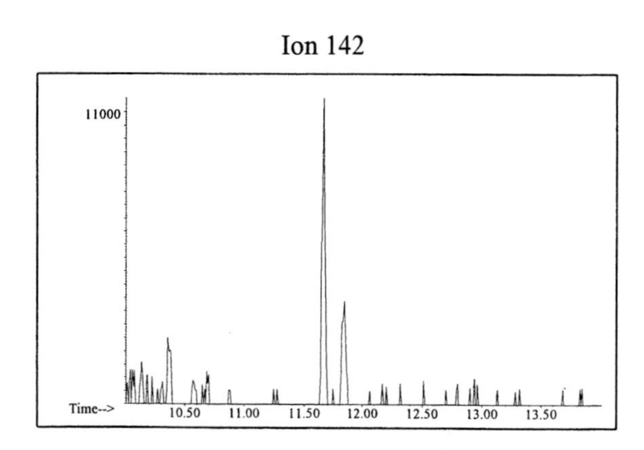

Ion 156

No Useful Data Obtained

Distillate #30

20uL/mL Pentane

Product Displayed: K-Mart TruBurn Charcoal Starter

Other Similar Products:

ASTM: Class 3 (Medium Petroleum Distillate)

Macro Code: M

Product Uses: Charcoal lighter

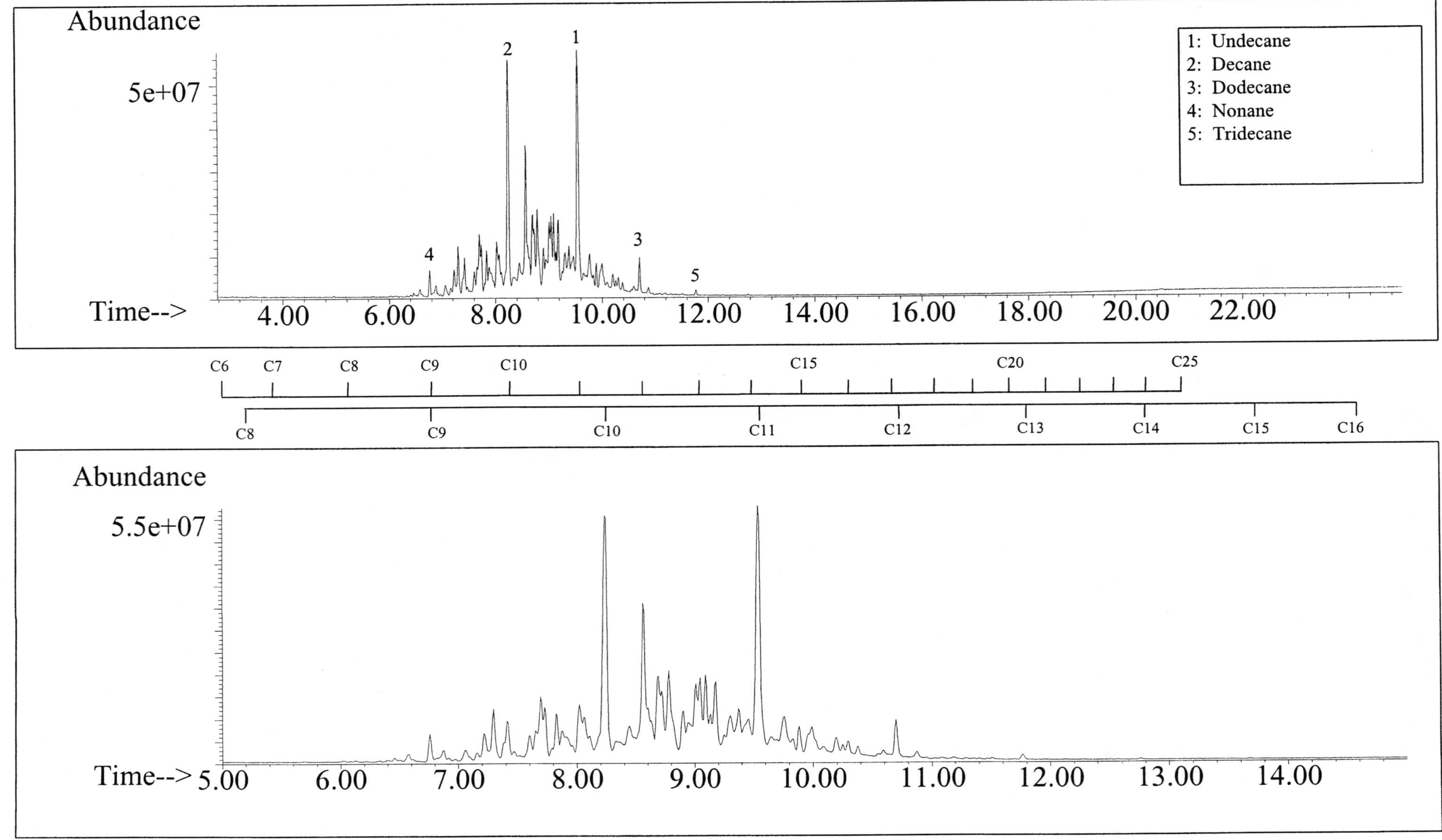

SUMMED PROFILES Distillate #30

ALKANES

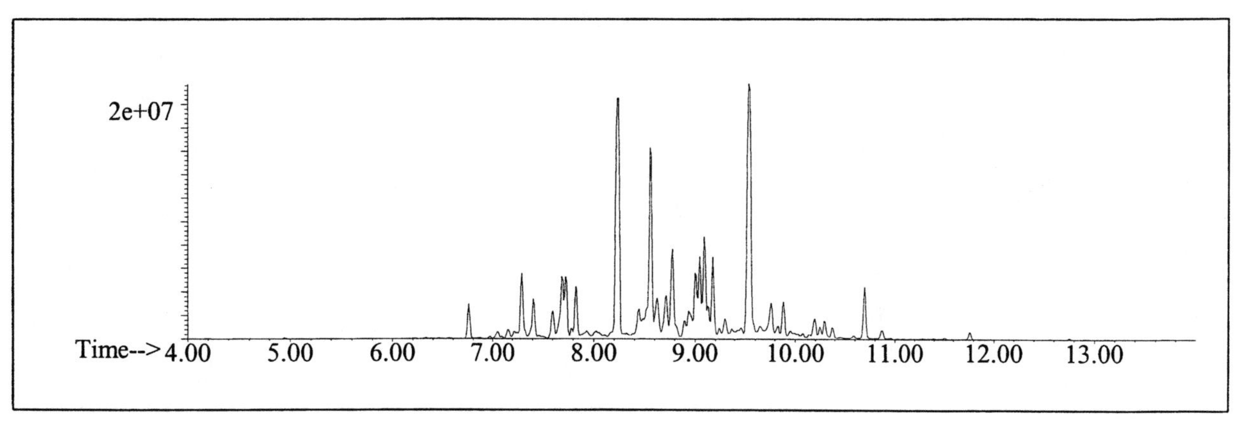

AROMATICS

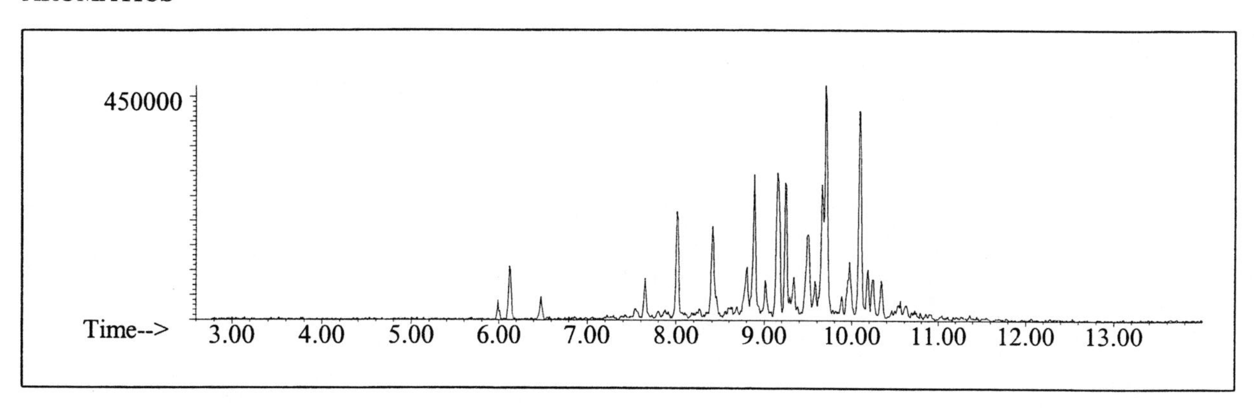

CYCLOPARAFFINS AND ALKENES

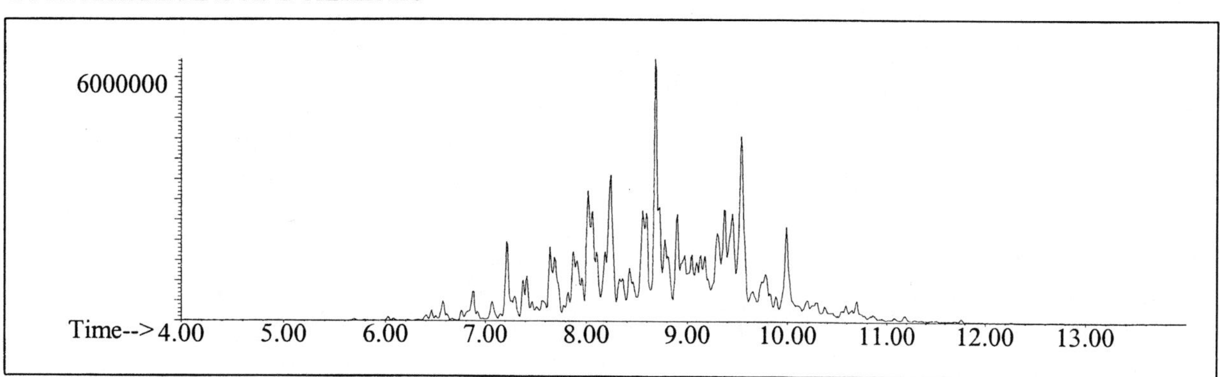

NAPHTHALENES

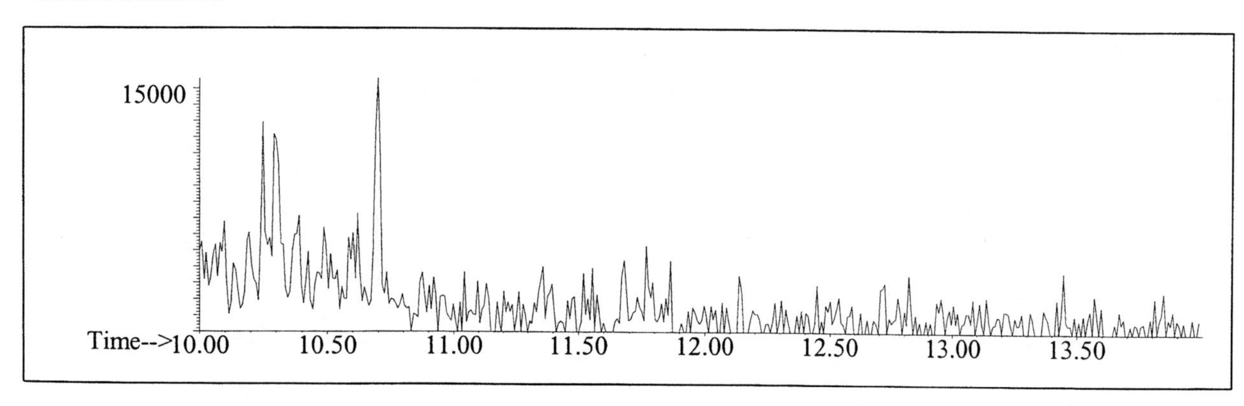

INDIVIDUAL PROFILES

Distillate #30

Alkane

Ion 43

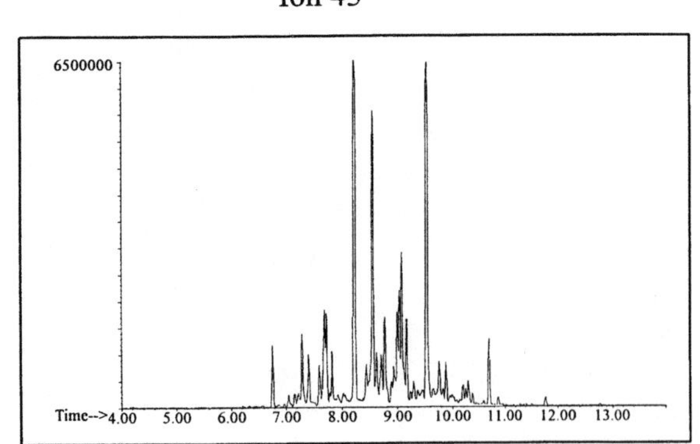

Ion 57

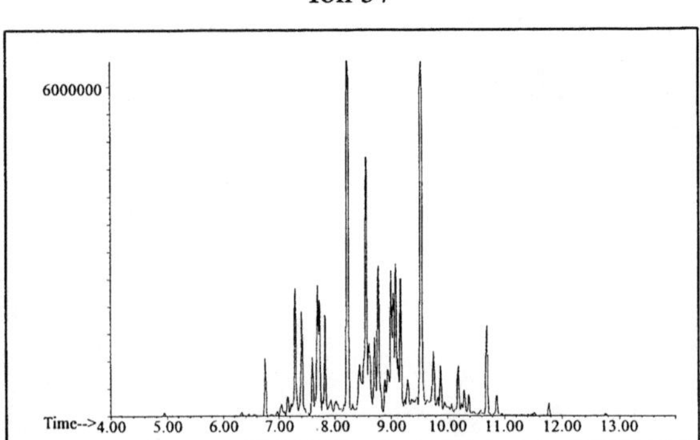

Ion 71

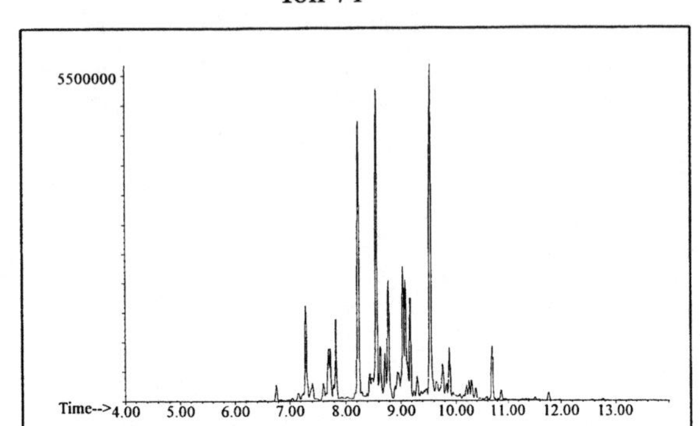

Ion 85

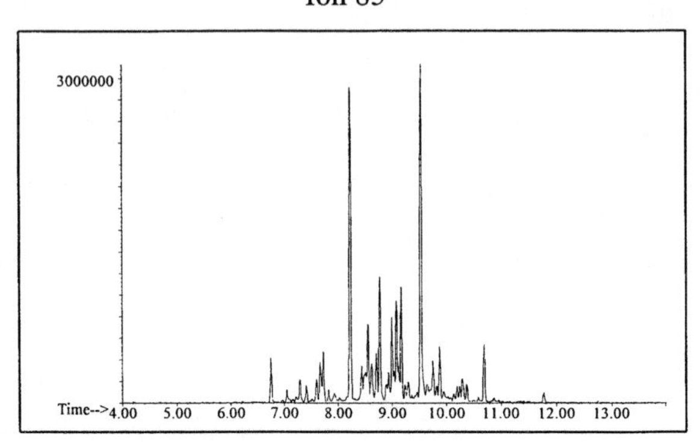

Aromatic

Ion 91

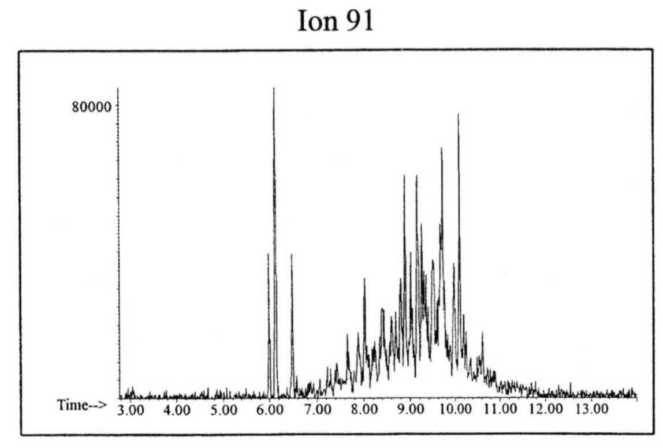

Ion 105

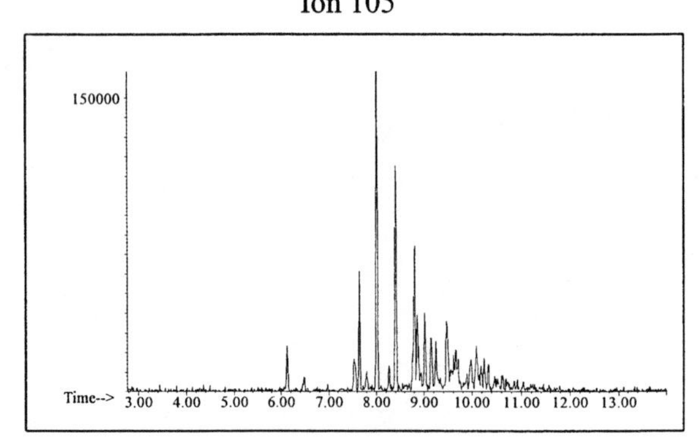

Ion 119

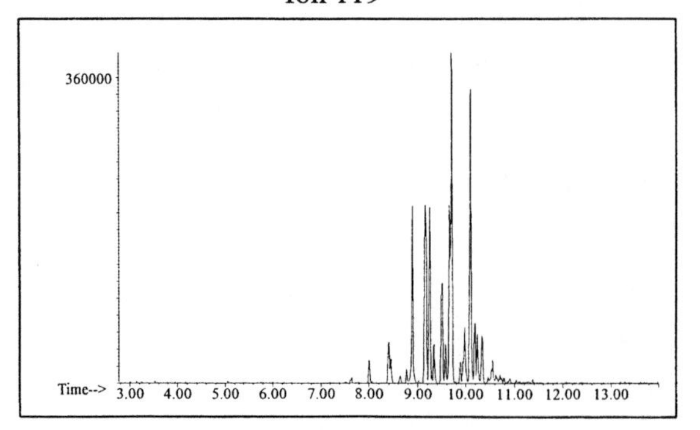

INDIVIDUAL PROFILES

Distillate #30

Cycloparaffin

Ion 55

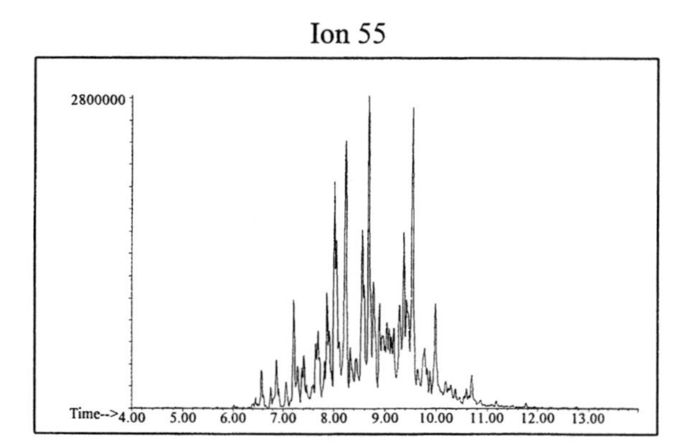

Ion 69

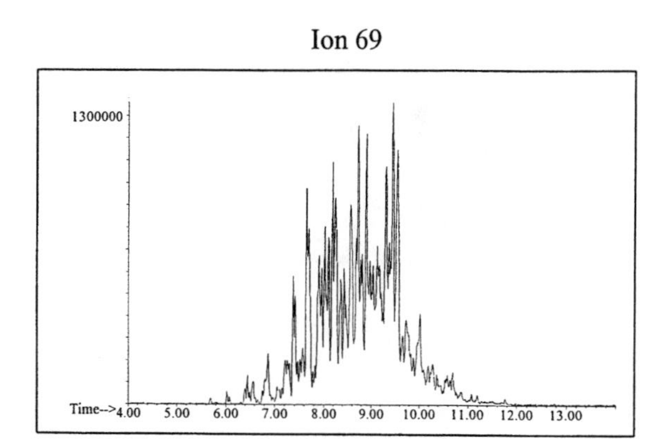

Ion 83

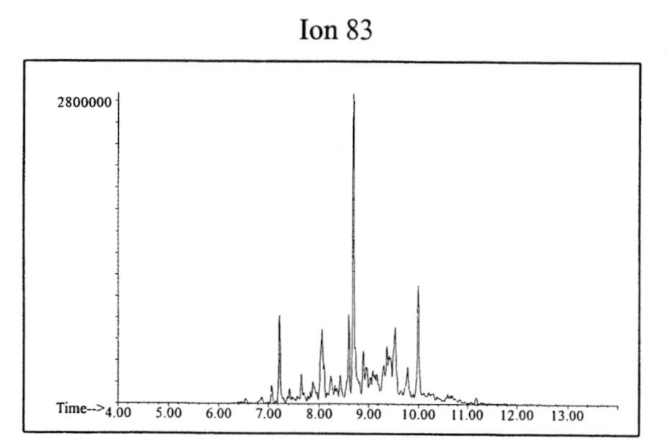

Naphthalene

Ion 128

No Useful Data Obtained

Ion 142

No Useful Data Obtained

Ion 156

No Useful Data Obtained

Distillate #31

20uL/mL Pentane

ASTM: Class 3 (Medium Petroleum Distillate)

Macro Code: M

Product Displayed: Kountry Cookin' Charcoal Starter

Product Uses: Charcoal lighter

Other Similar Products: Chef's Choice Charcoal Lighter, Better Value Charcoal Starter

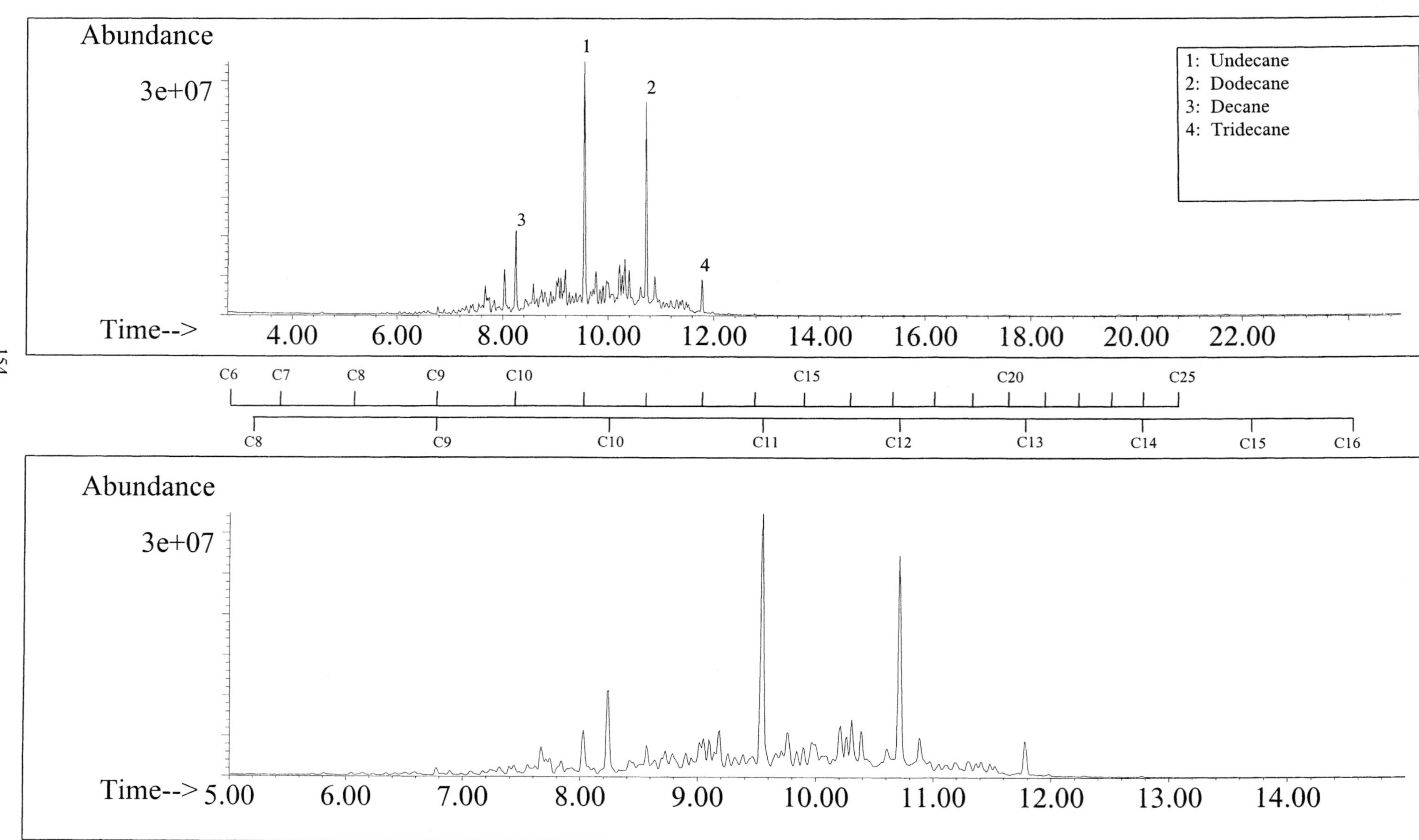

SUMMED PROFILES

Distillate #31

ALKANES

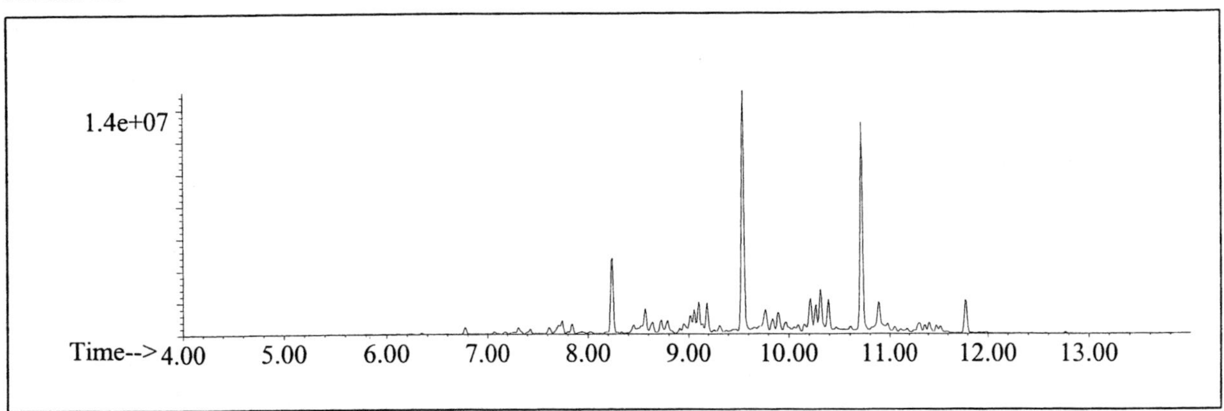

AROMATICS

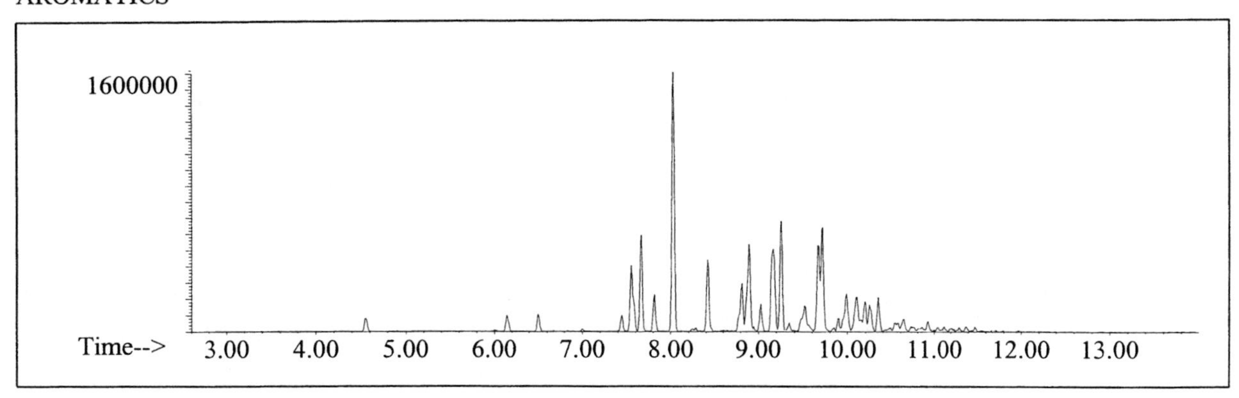

CYCLOPARAFFINS AND ALKENES

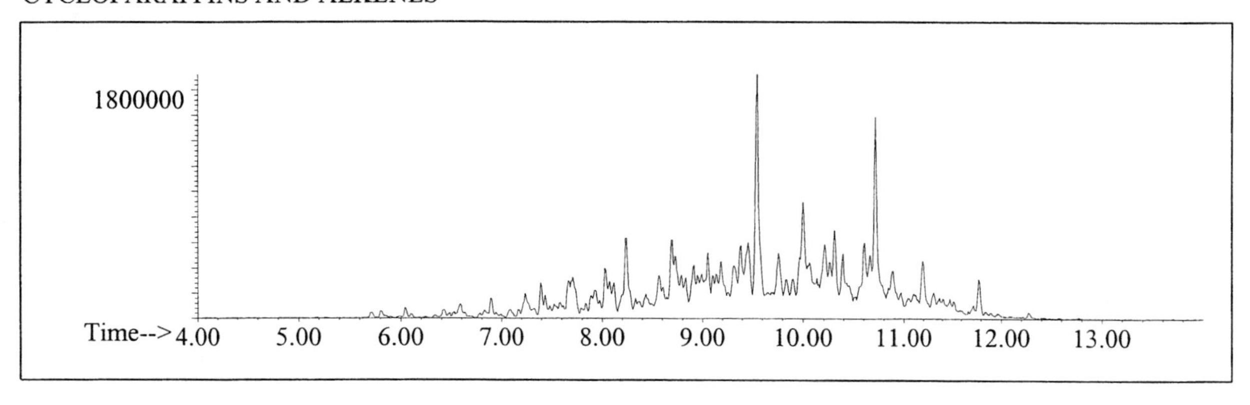

NAPHTHALENES

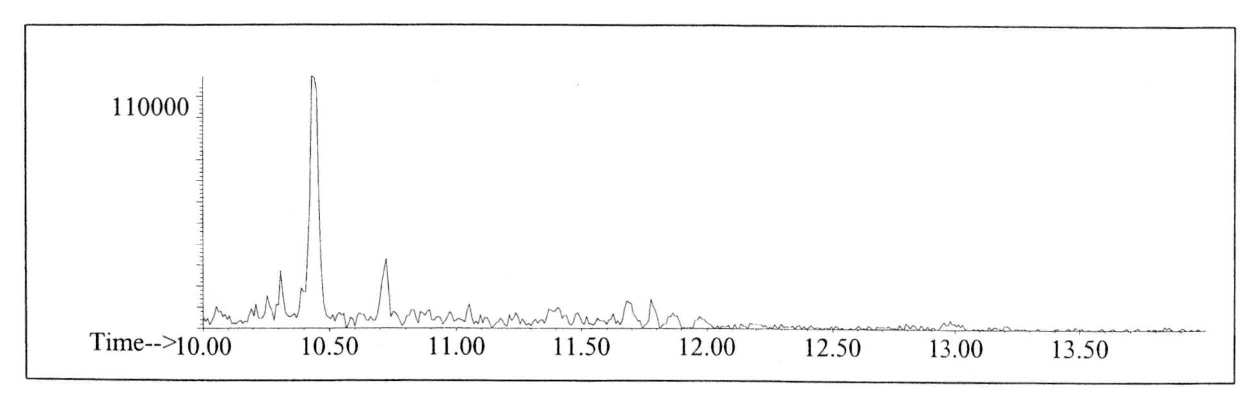

INDIVIDUAL PROFILES

Distillate #31

Alkane

Ion 43

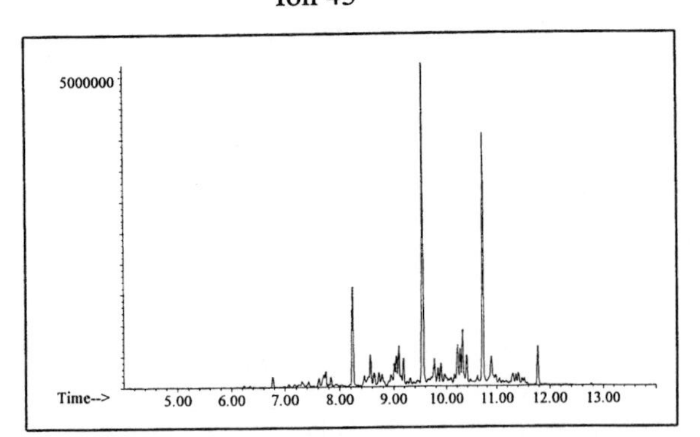

Ion 57

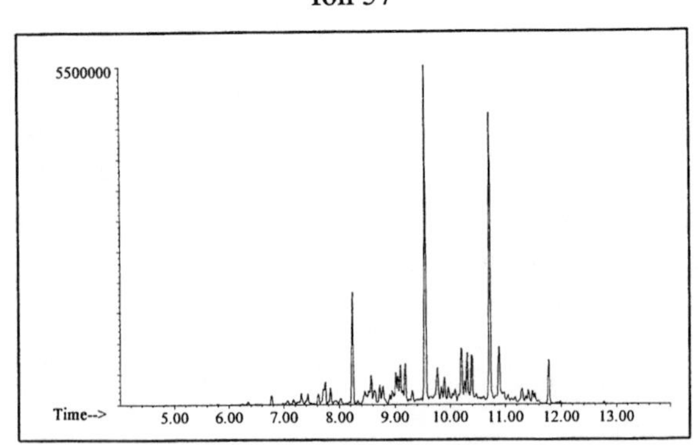

Ion 71

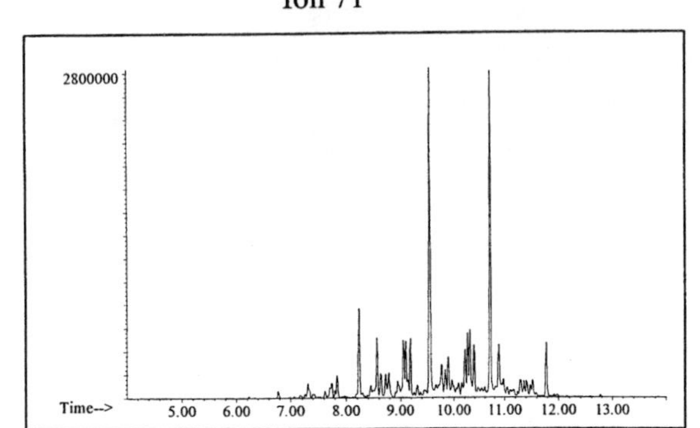

Ion 85

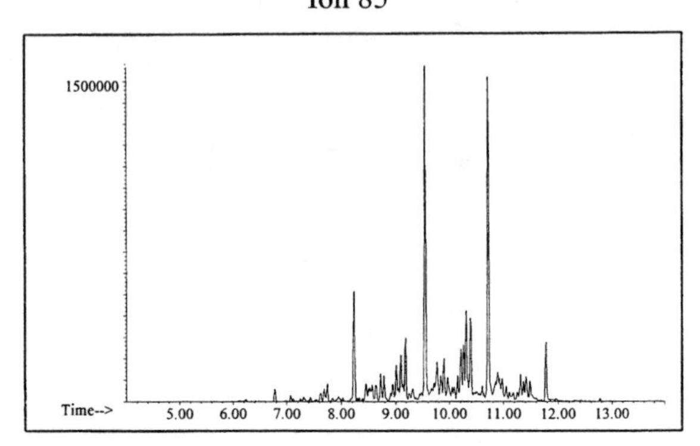

Aromatic

Ion 91

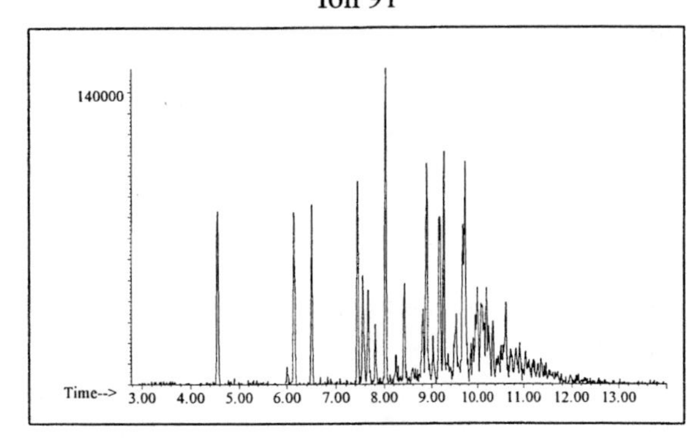

Ion 105

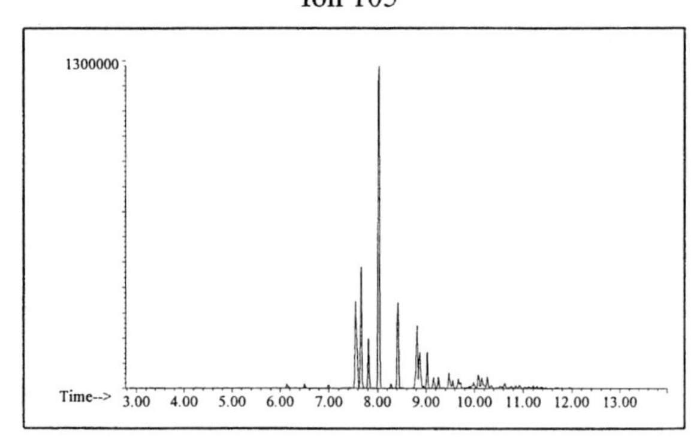

Ion 119

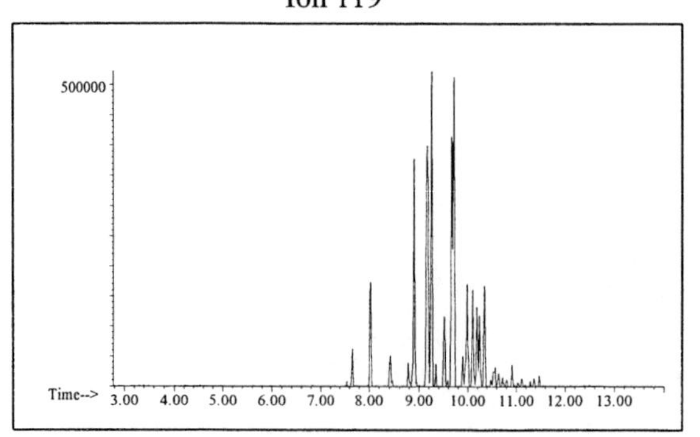

INDIVIDUAL PROFILES

Distillate #31

Cycloparaffin

Ion 55

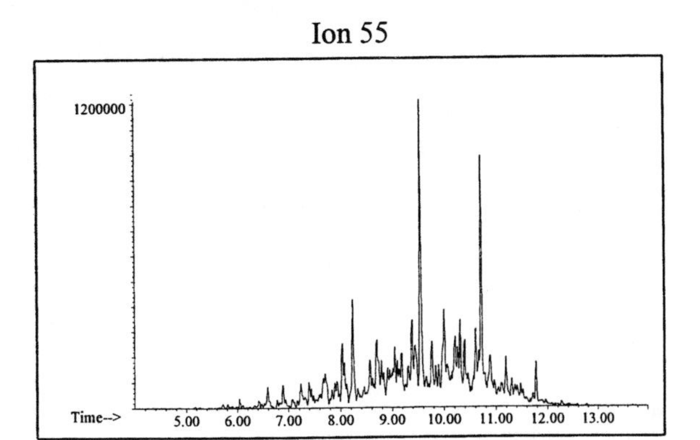

Ion 69

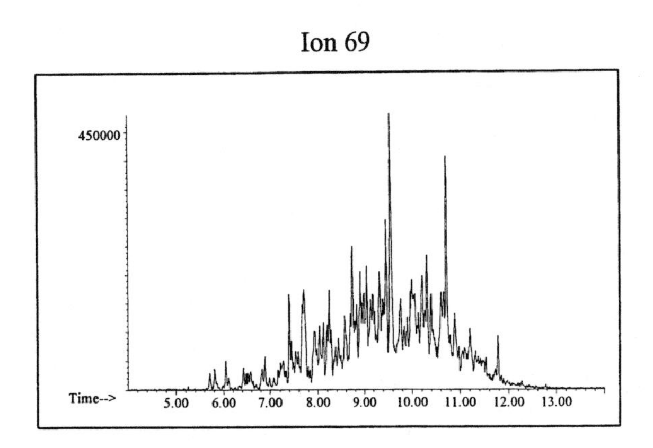

Ion 83

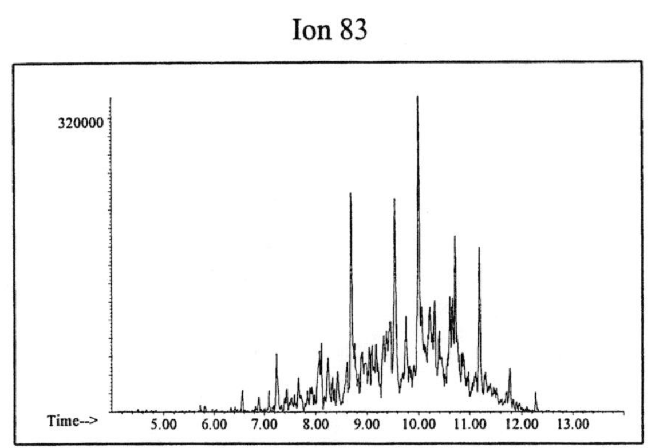

Naphthalene

Ion 128

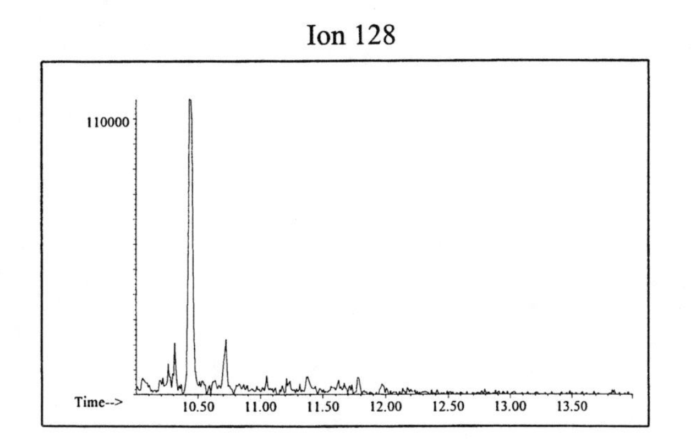

Ion 142

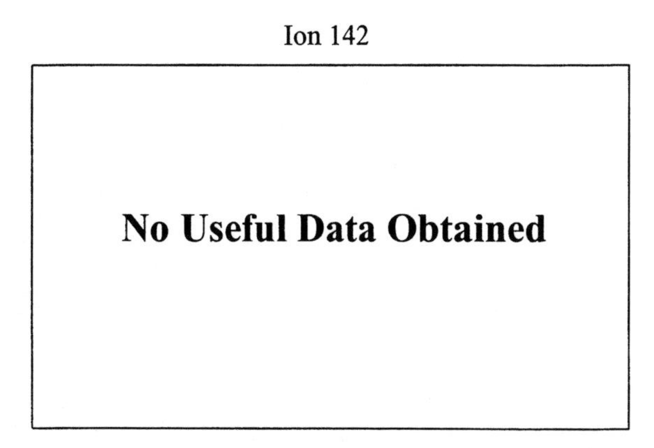

Ion 156

No Useful Data Obtained

Distillate #32

20uL/mL Pentane

Product Displayed: Kleenstrip Paint Thinner

Other Similar Products: Kwal Howell Mineral Spirits

ASTM: Class 3 (Medium Petroleum Distillate)

Macro Code: M

Product Uses: Paint thinner, mineral spirits

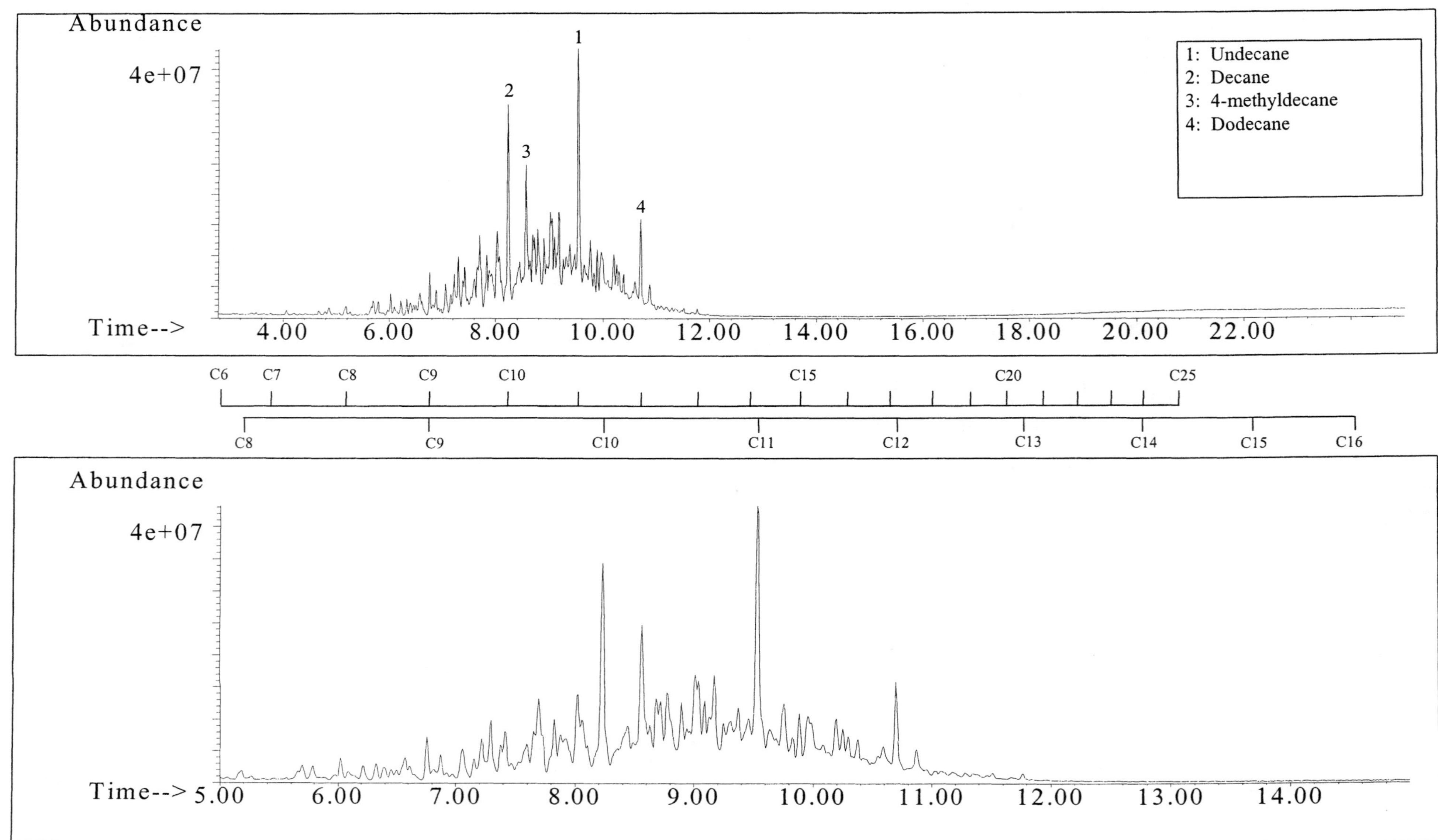

SUMMED PROFILES

Distillate #32

ALKANES

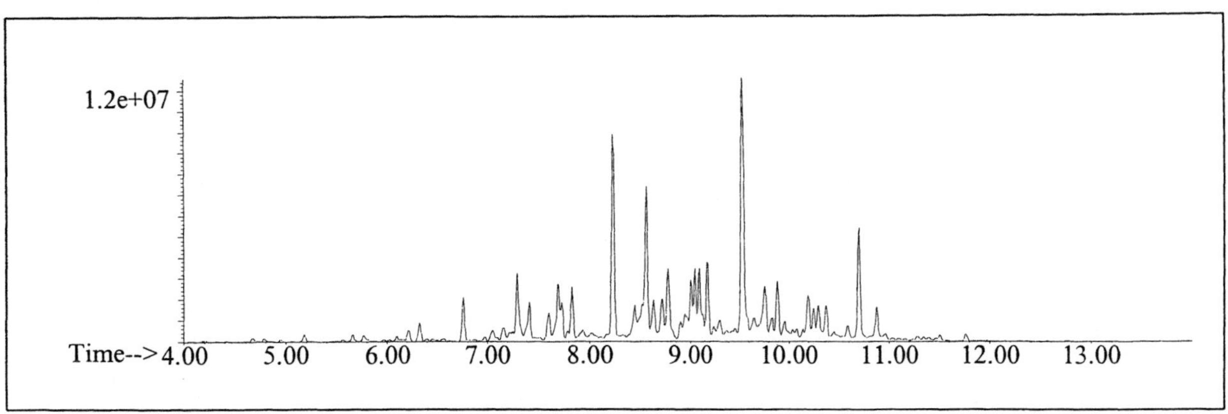

AROMATICS

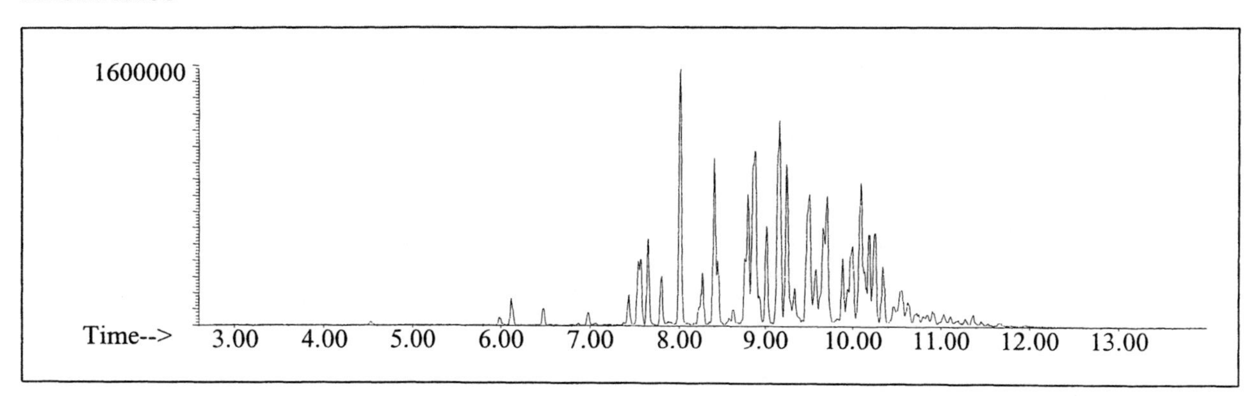

CYCLOPARAFFINS AND ALKENES

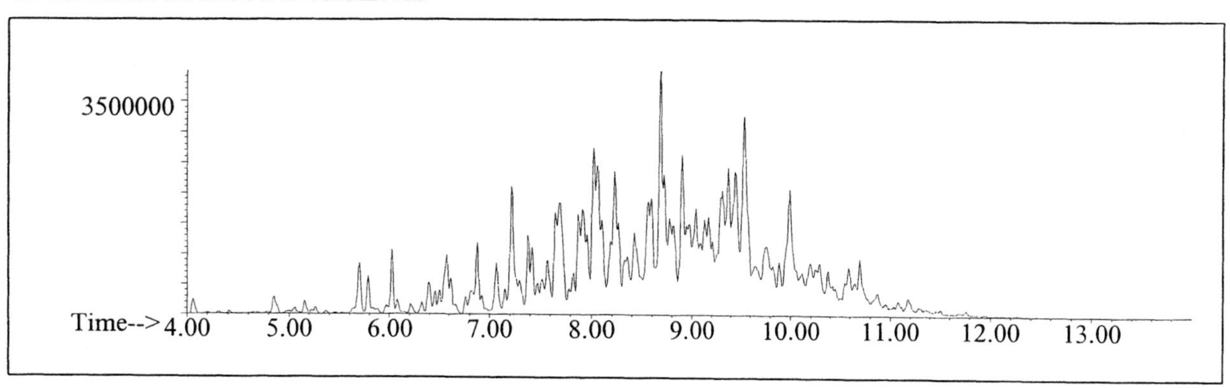

NAPHTHALENES

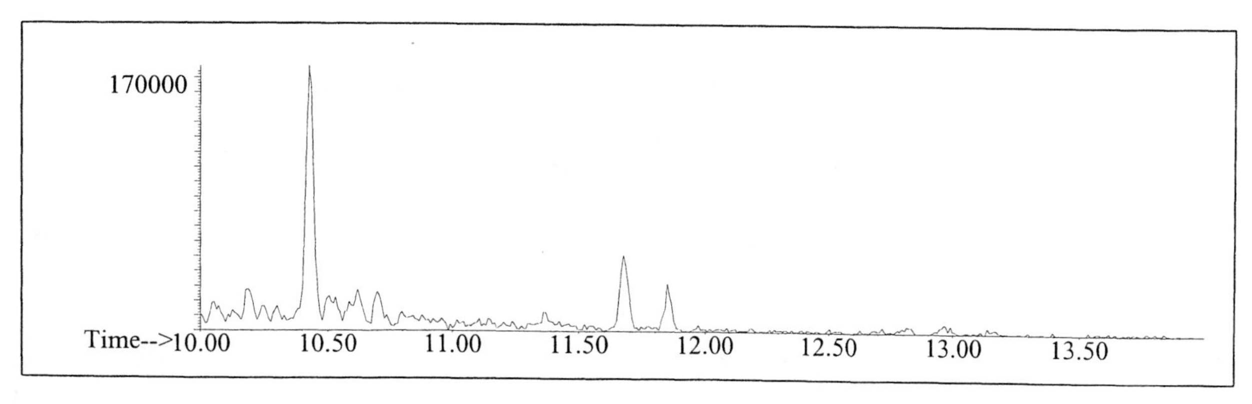

INDIVIDUAL PROFILES

Distillate #32

Alkane

Ion 43

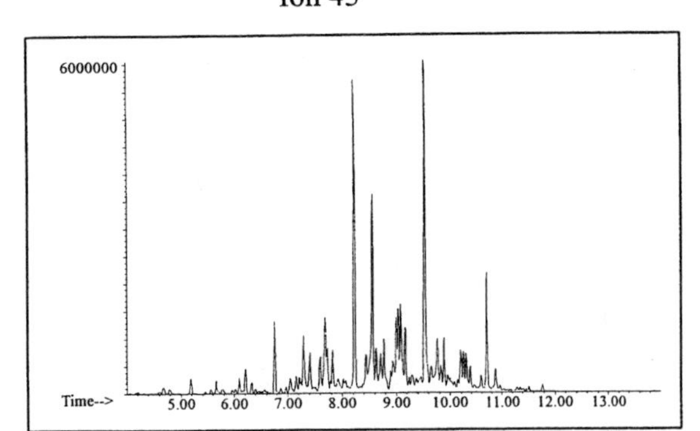

Ion 57

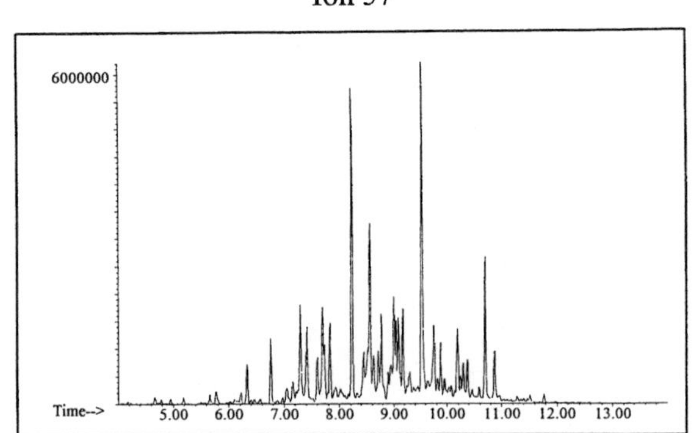

Ion 71

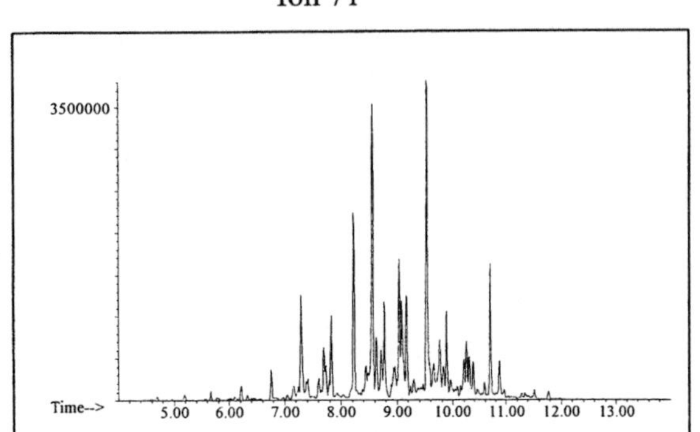

Ion 85

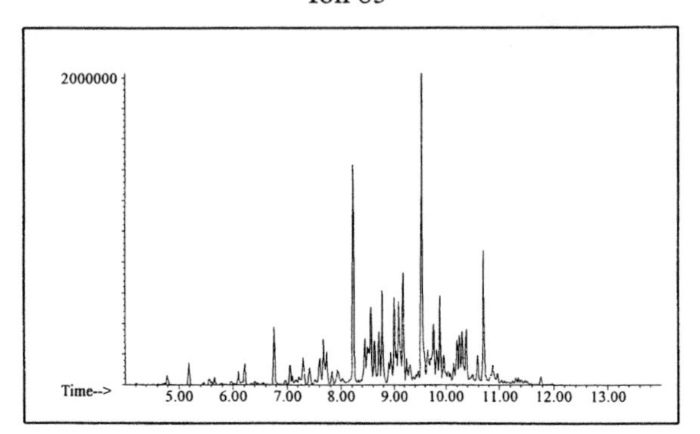

Aromatic

Ion 91

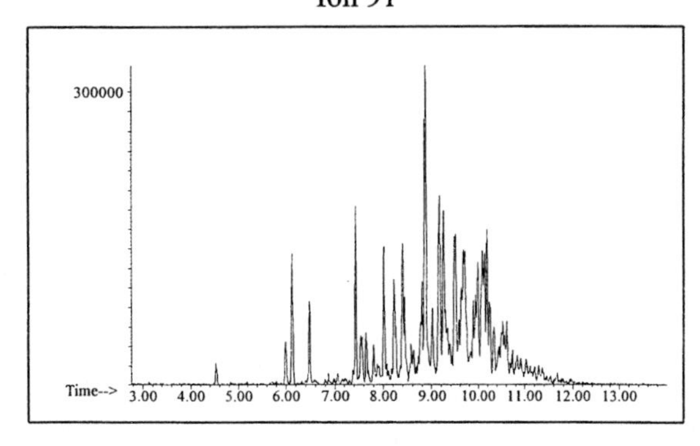

Ion 105

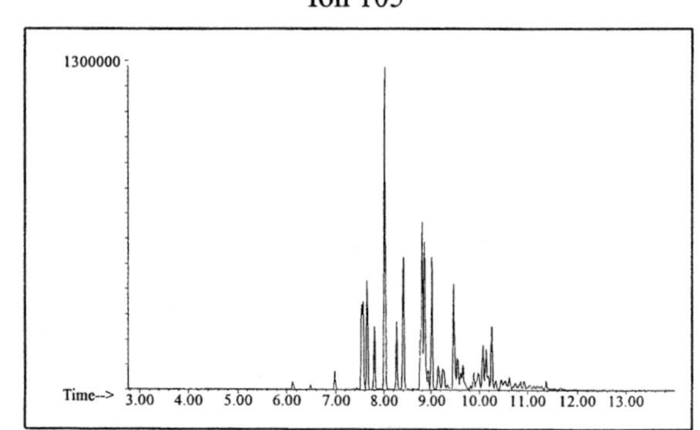

Ion 119

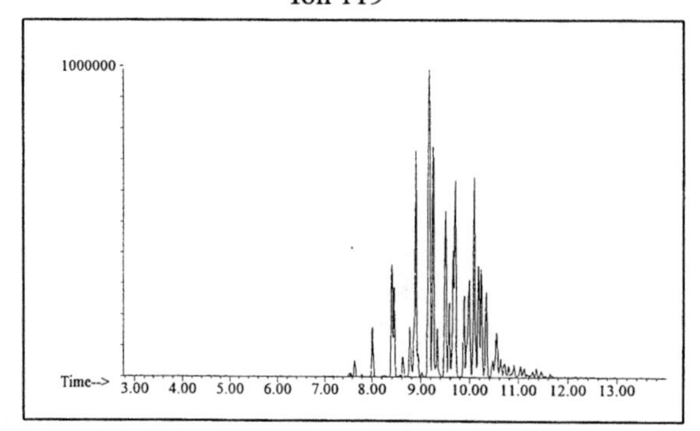

INDIVIDUAL PROFILES

Distillate #32

Cycloparaffin

Ion 55

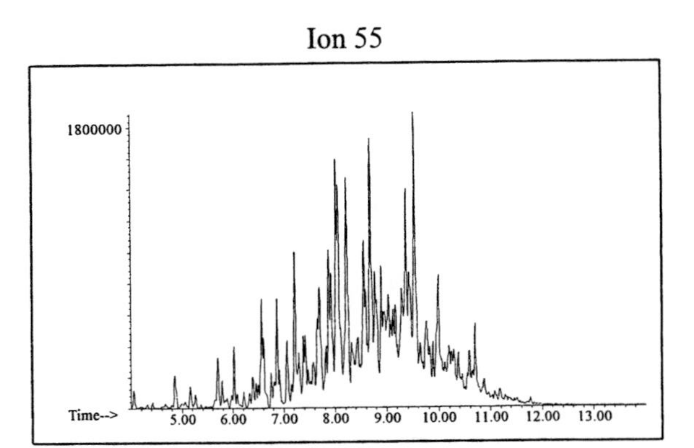

Ion 69

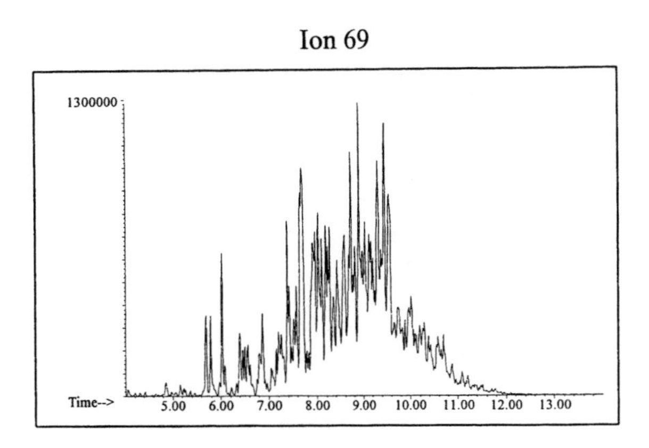

Ion 83

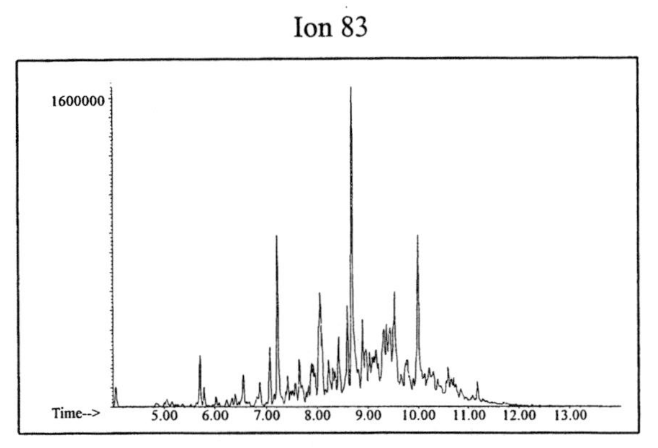

Naphthalene

Ion 128

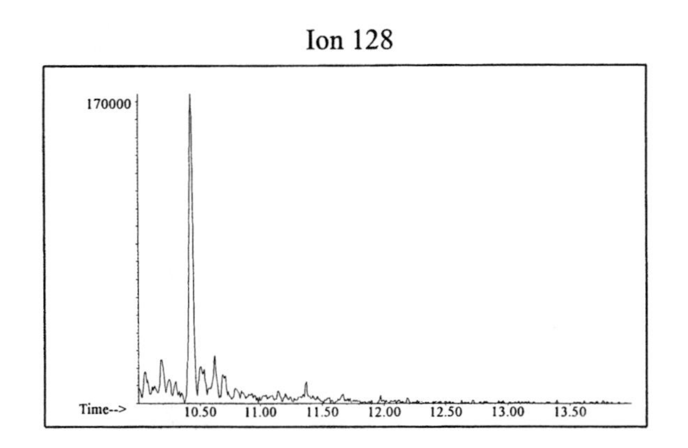

Ion 142

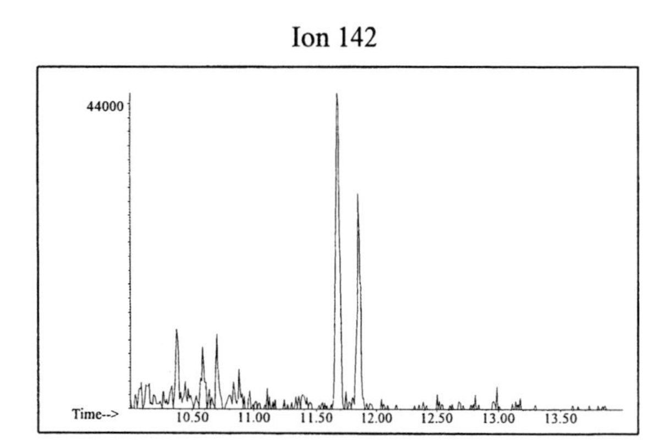

Ion 156

No Useful Data Obtained

Distillate #33

20uL/mL Pentane

ASTM: Class 0 (Miscellaneous)

Macro Code: M

Product Displayed: Jasco Deodorized Kerosene

Product Uses: Heating fuel, charcoal lighter

Other Similar Products: Ralph's Hickory Scent Charcoal Lighter

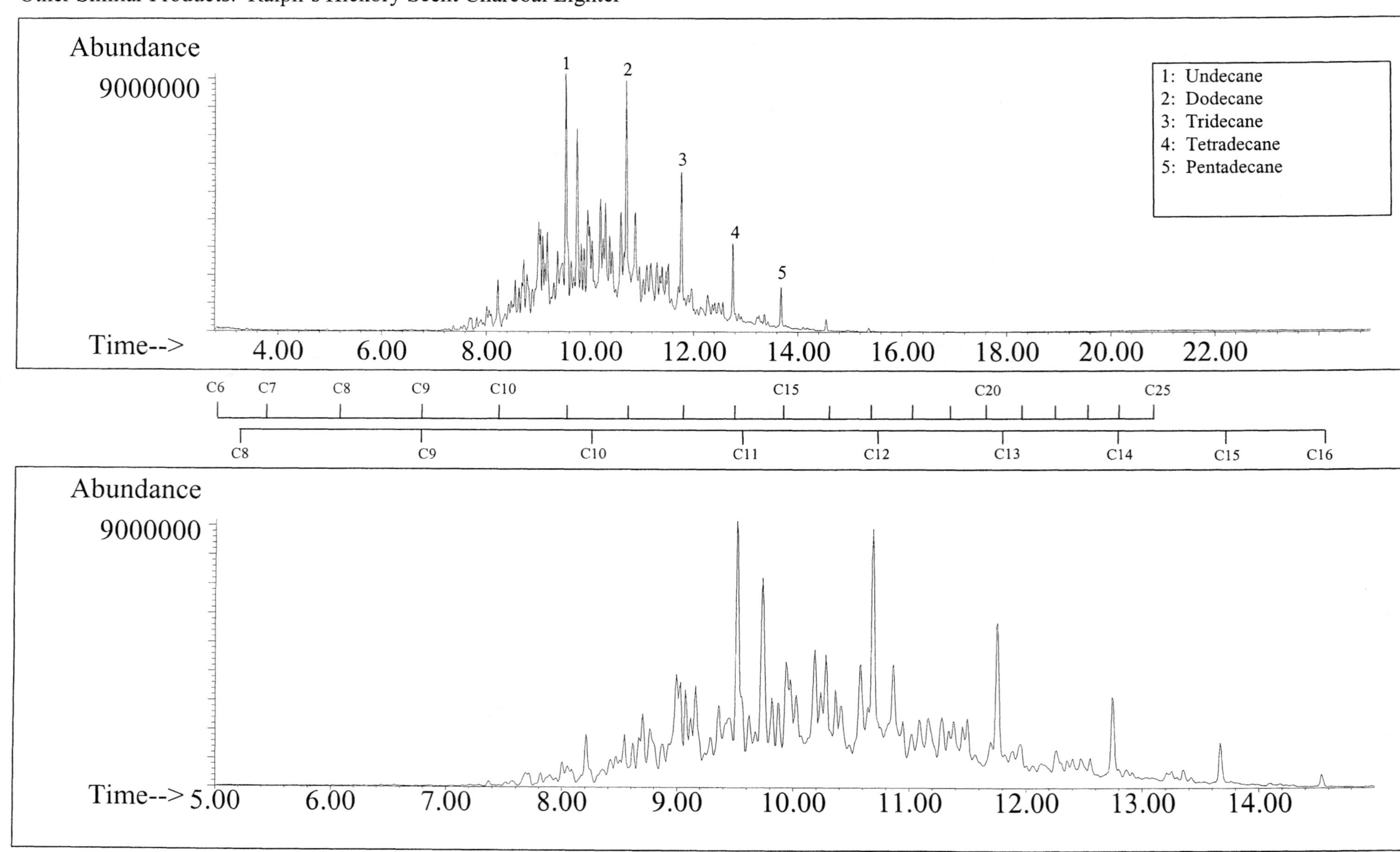

SUMMED PROFILES

Distillate #33

ALKANES

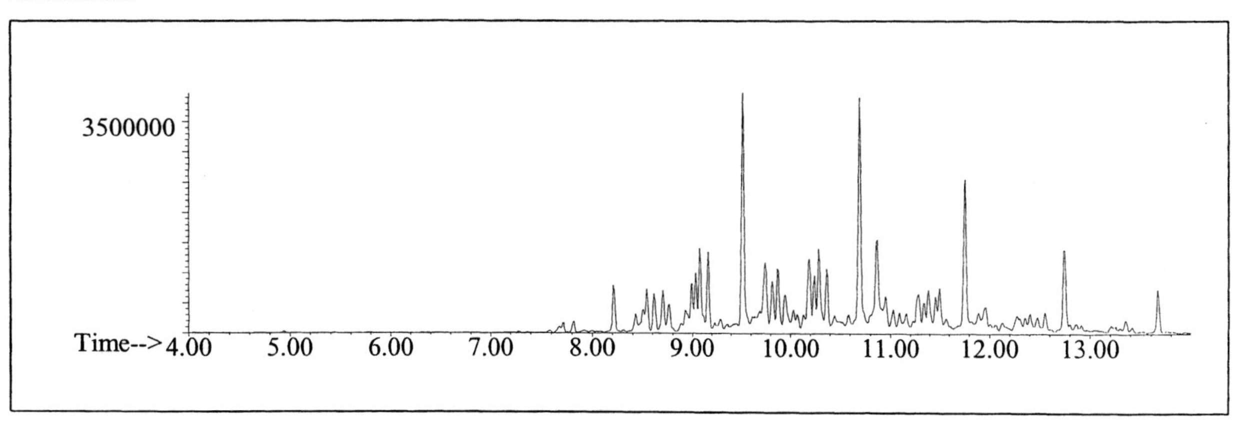

AROMATICS

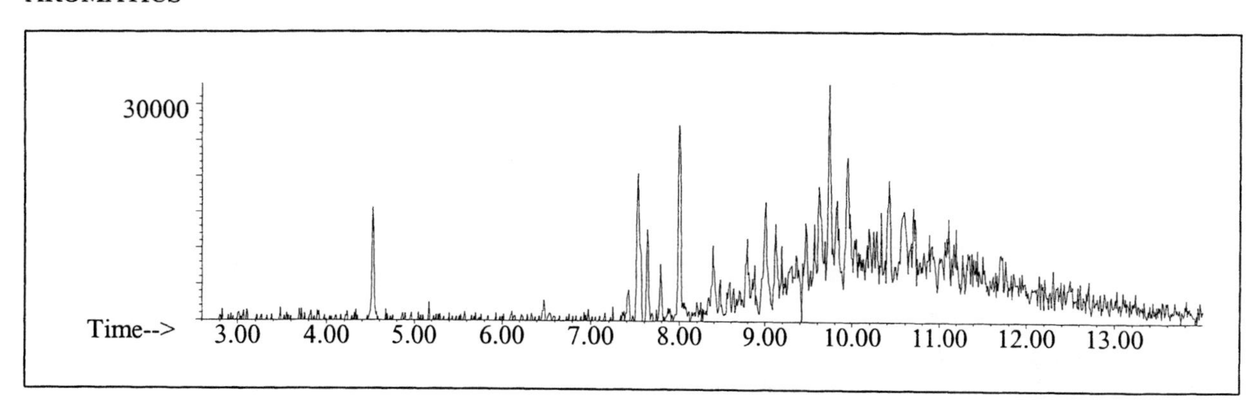

CYCLOPARAFFINS AND ALKENES

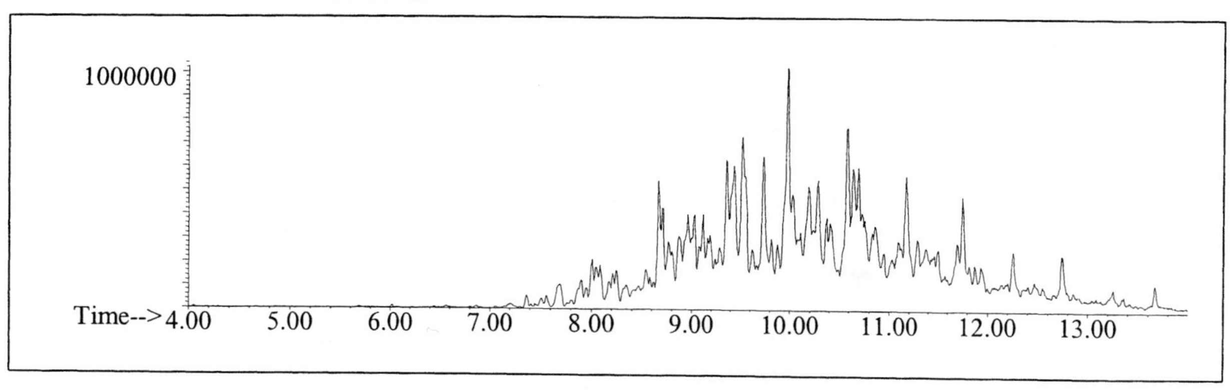

NAPHTHALENES

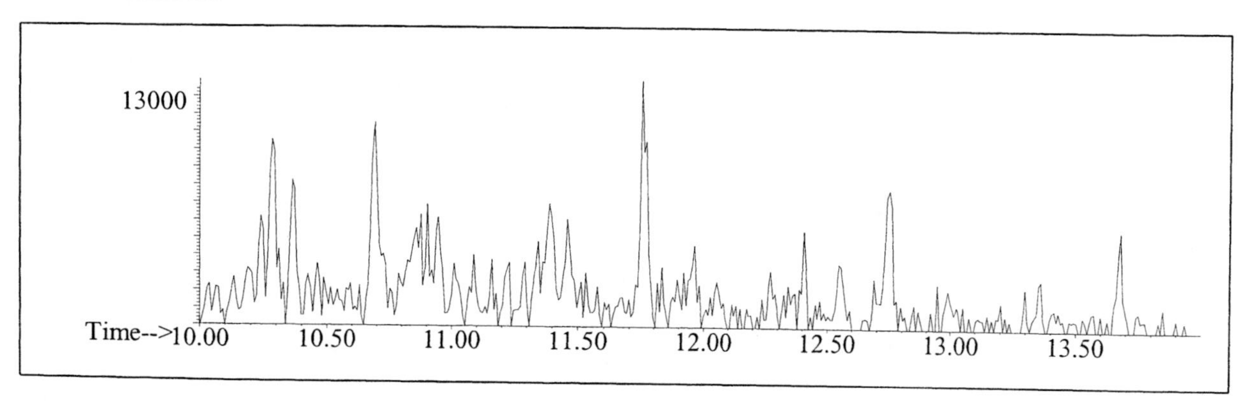

INDIVIDUAL PROFILES

Distillate #33

Alkane

Ion 43

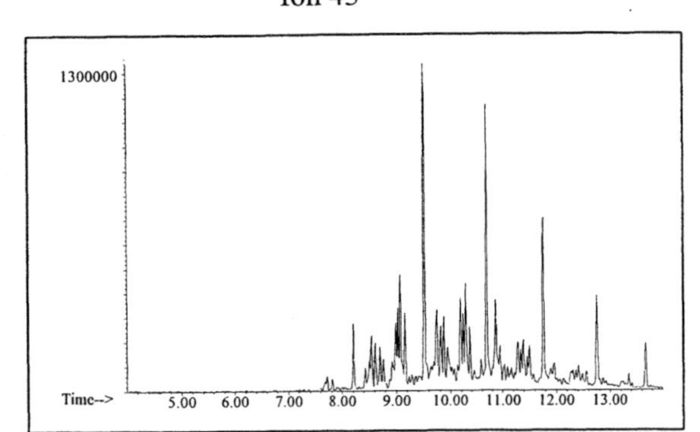

Ion 57

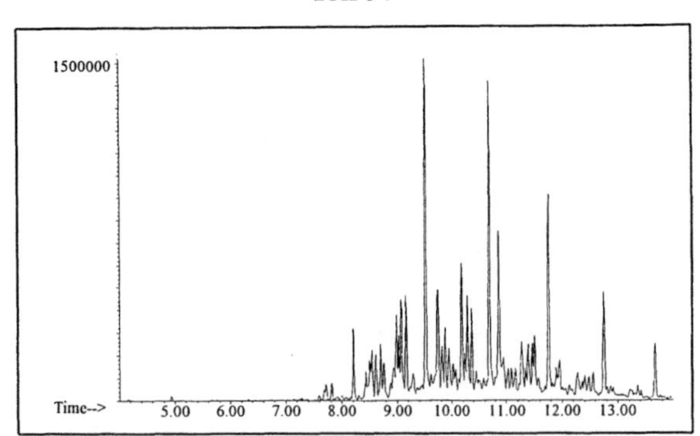

Ion 71

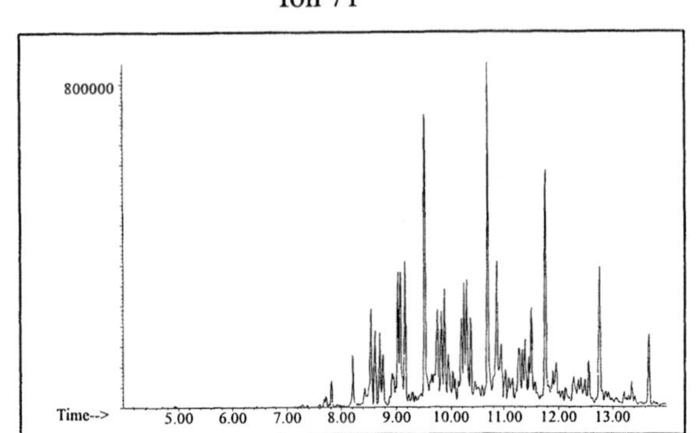

Ion 85

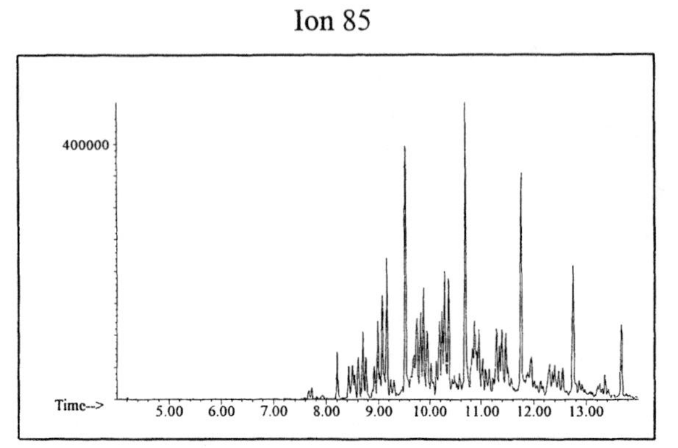

Aromatic

Ion 91

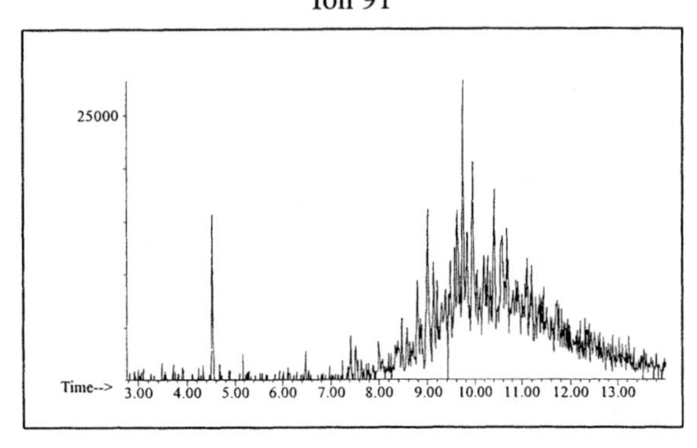

Ion 105

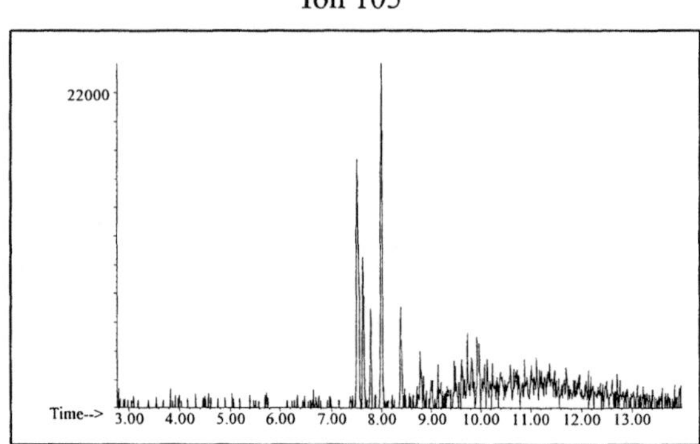

Ion 119

No Useful Data Obtained

Cycloparaffin

Ion 55

400000
Time-->
5.00 6.00 7.00 8.00 9.00 10.00 11.00 12.00 13.00

Ion 69

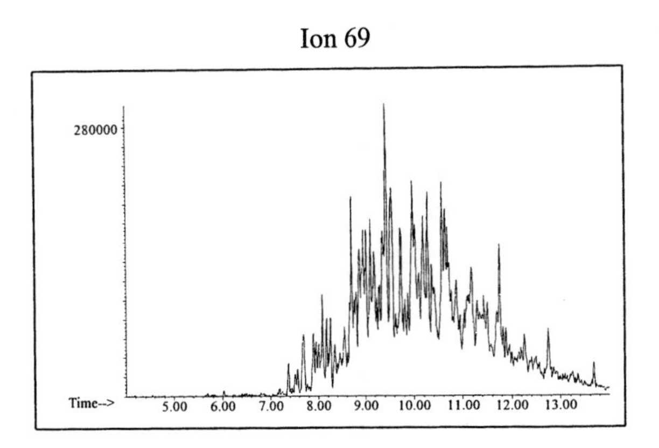

Ion 83

380000
Time-->
5.00 6.00 7.00 8.00 9.00 10.00 11.00 12.00 13.00

Naphthalene

Ion 128

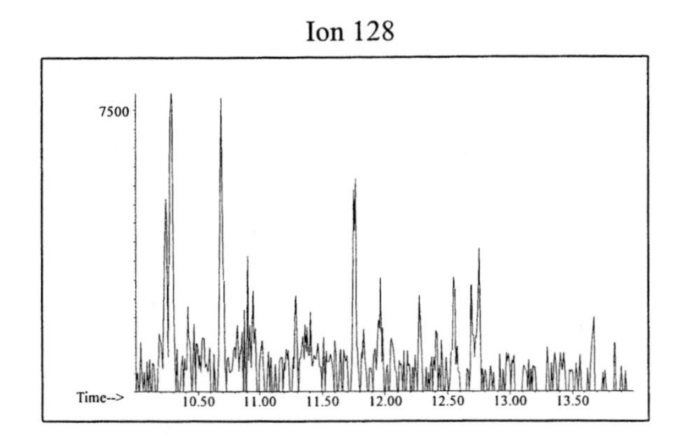

Ion 142

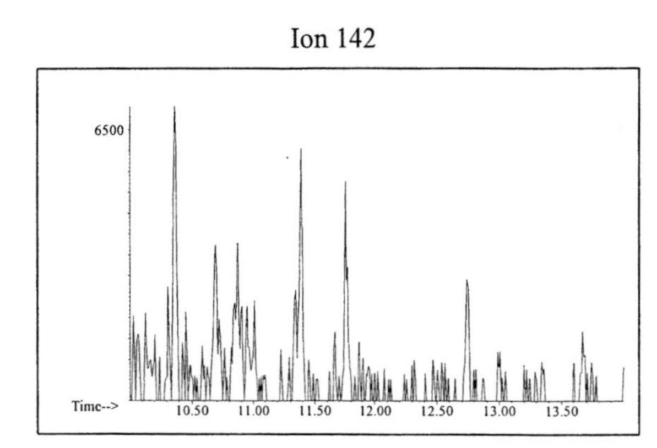

Ion 156

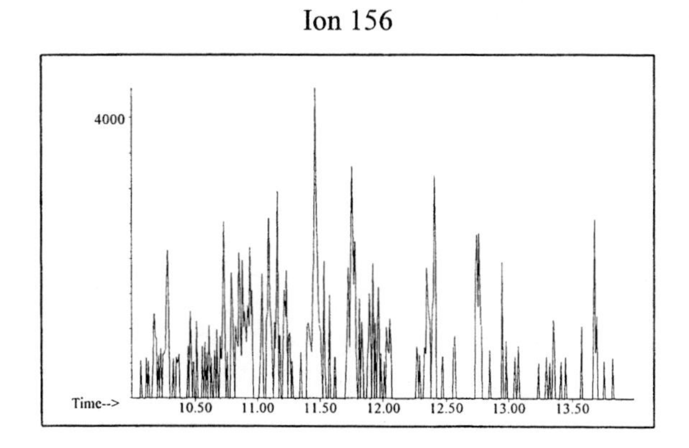

Distillate #34

20uL/mL Pentane

ASTM: Class 3 (Medium Petroleum Distillate)

Macro Code: M

Product Displayed: STP Fuel Injector and Carburetor Cleaner

Product Uses: Cleaning solvent

Other Similar Products:

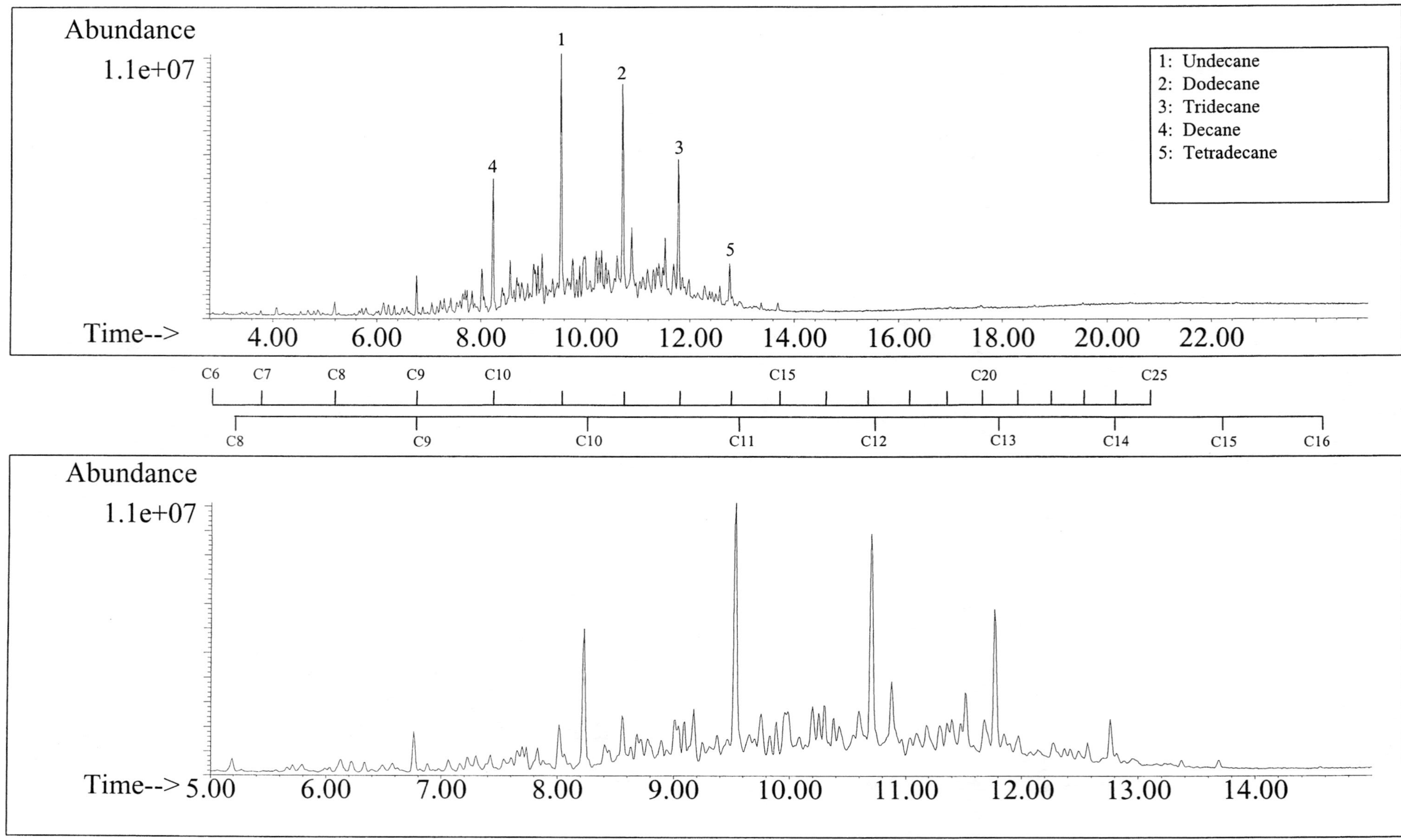

SUMMED PROFILES Distillate #34

ALKANES

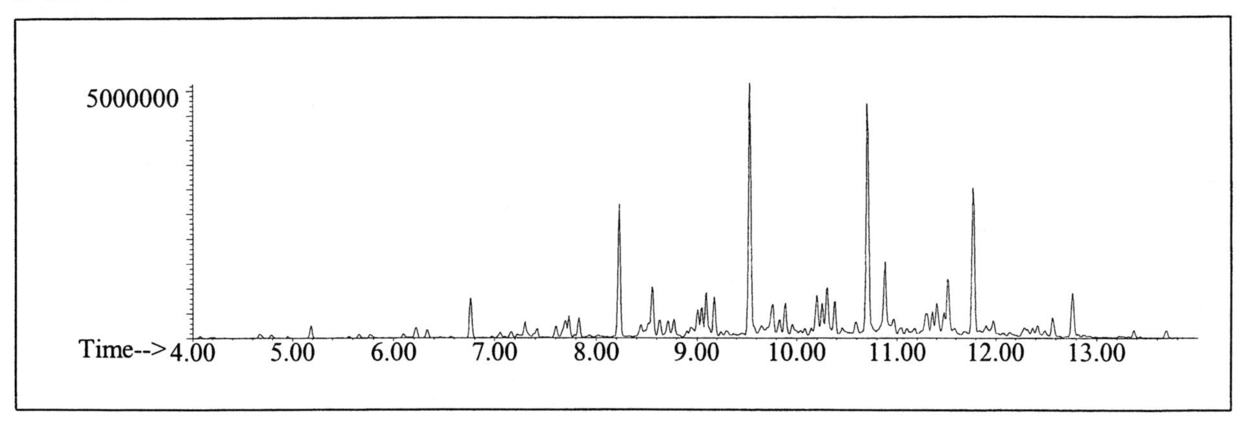

AROMATICS

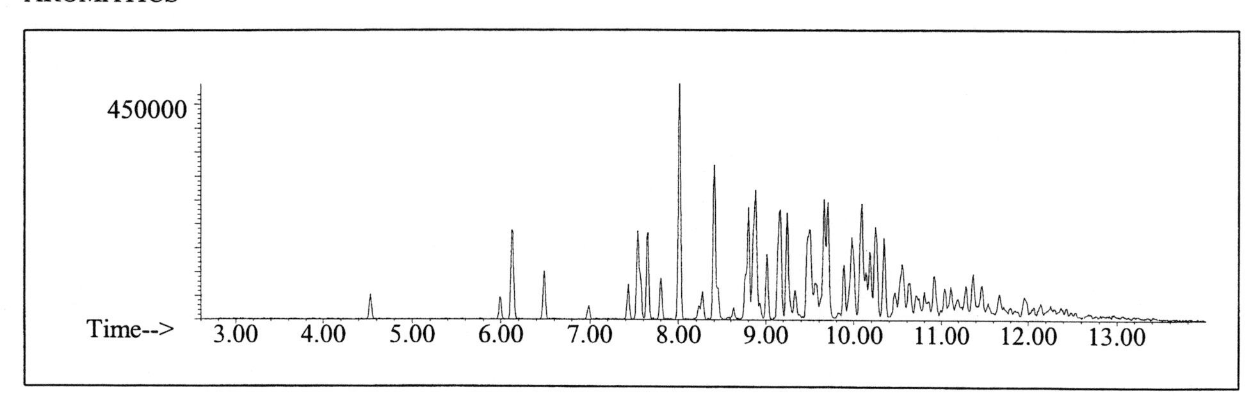

CYCLOPARAFFINS AND ALKENES

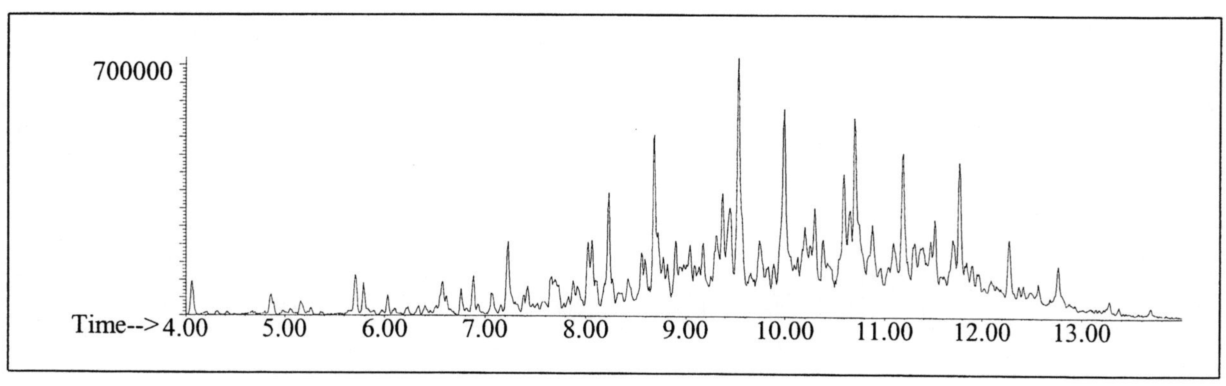

NAPHTHALENES

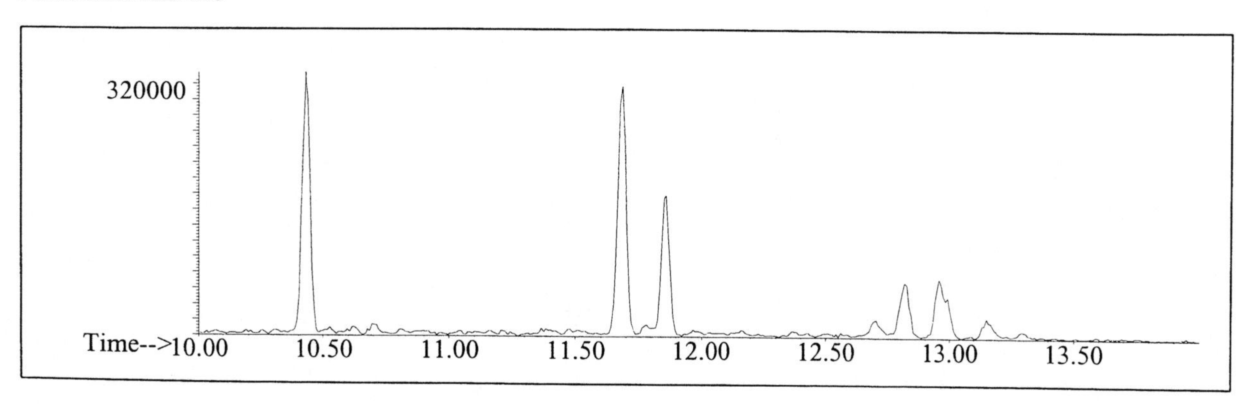

INDIVIDUAL PROFILES

Distillate #34

Alkane

Ion 43

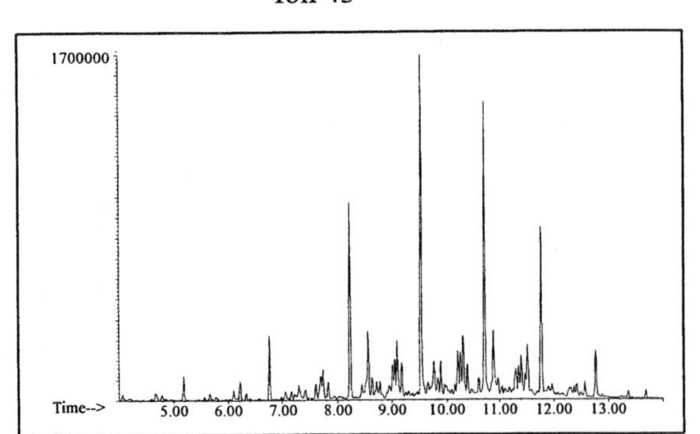

Ion 57

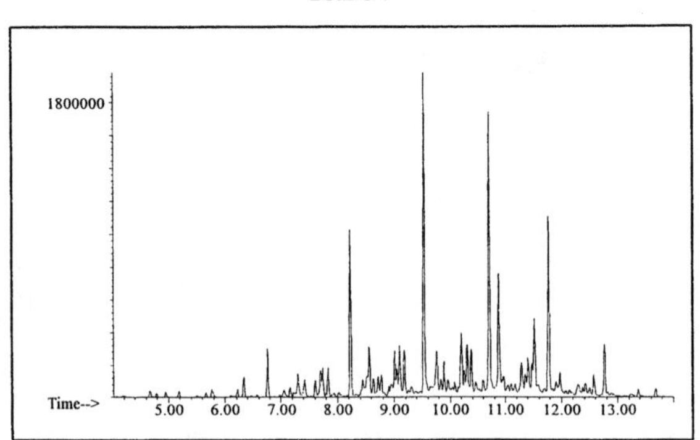

Ion 71

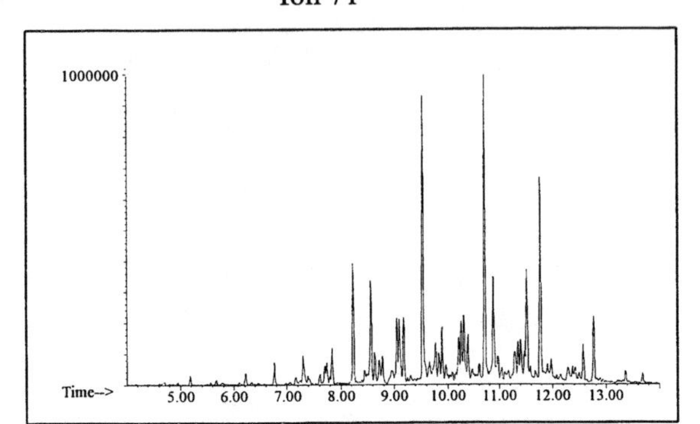

Ion 85

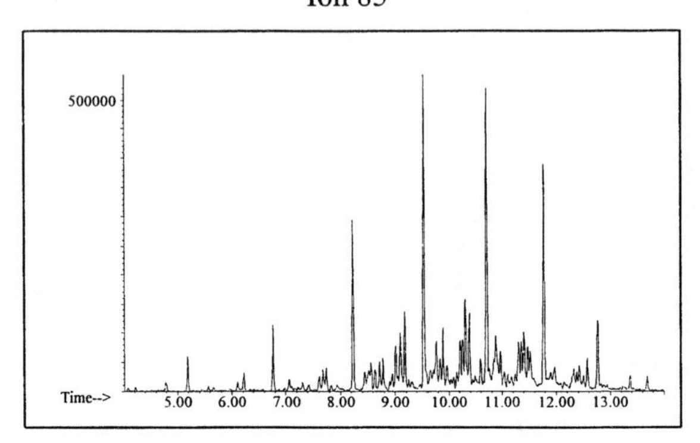

Aromatic

Ion 91

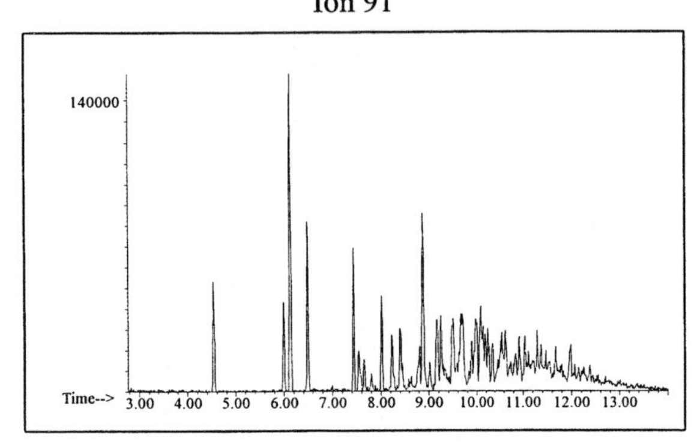

Ion 105

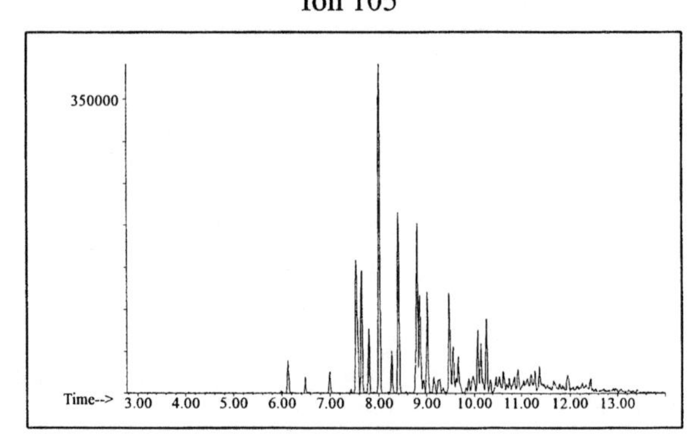

Ion 119

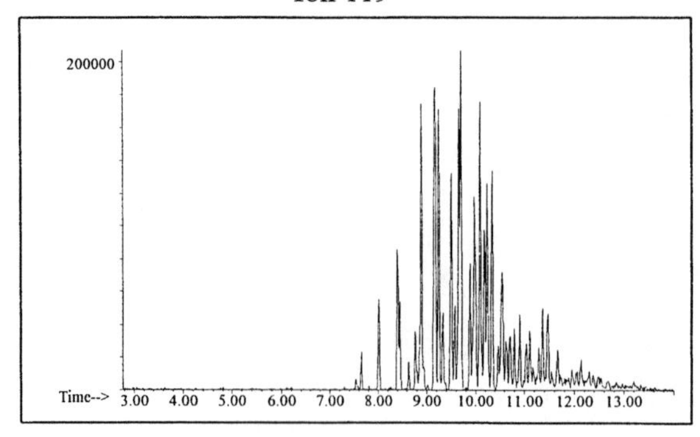

INDIVIDUAL PROFILES

Distillate #34

Cycloparaffin

Ion 55

Ion 69

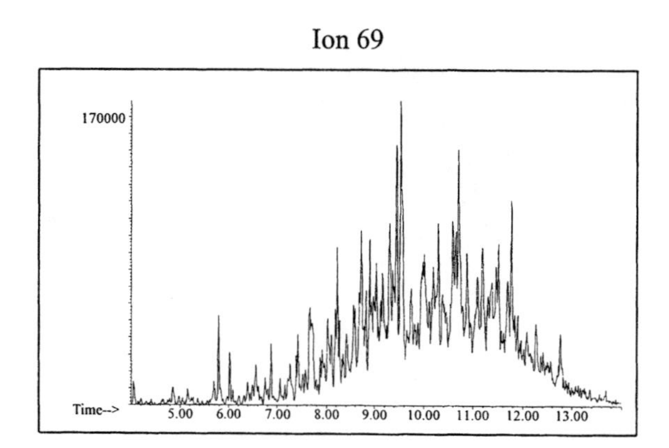

Ion 83

Naphthalene

Ion 128

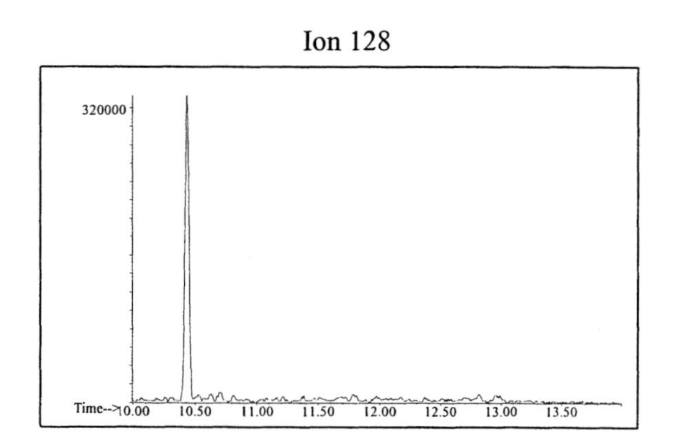

Ion 142

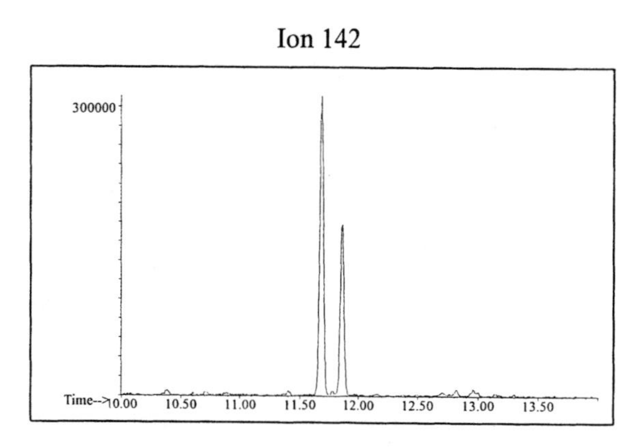

Ion 156

Distillate #35

20uL/mL Pentane

Product Displayed: Grill King Charcoal Lighter

Other Similar Products:

ASTM: Class 4 (Kerosene)

Product Uses: Charcoal lighter

Macro Code: H

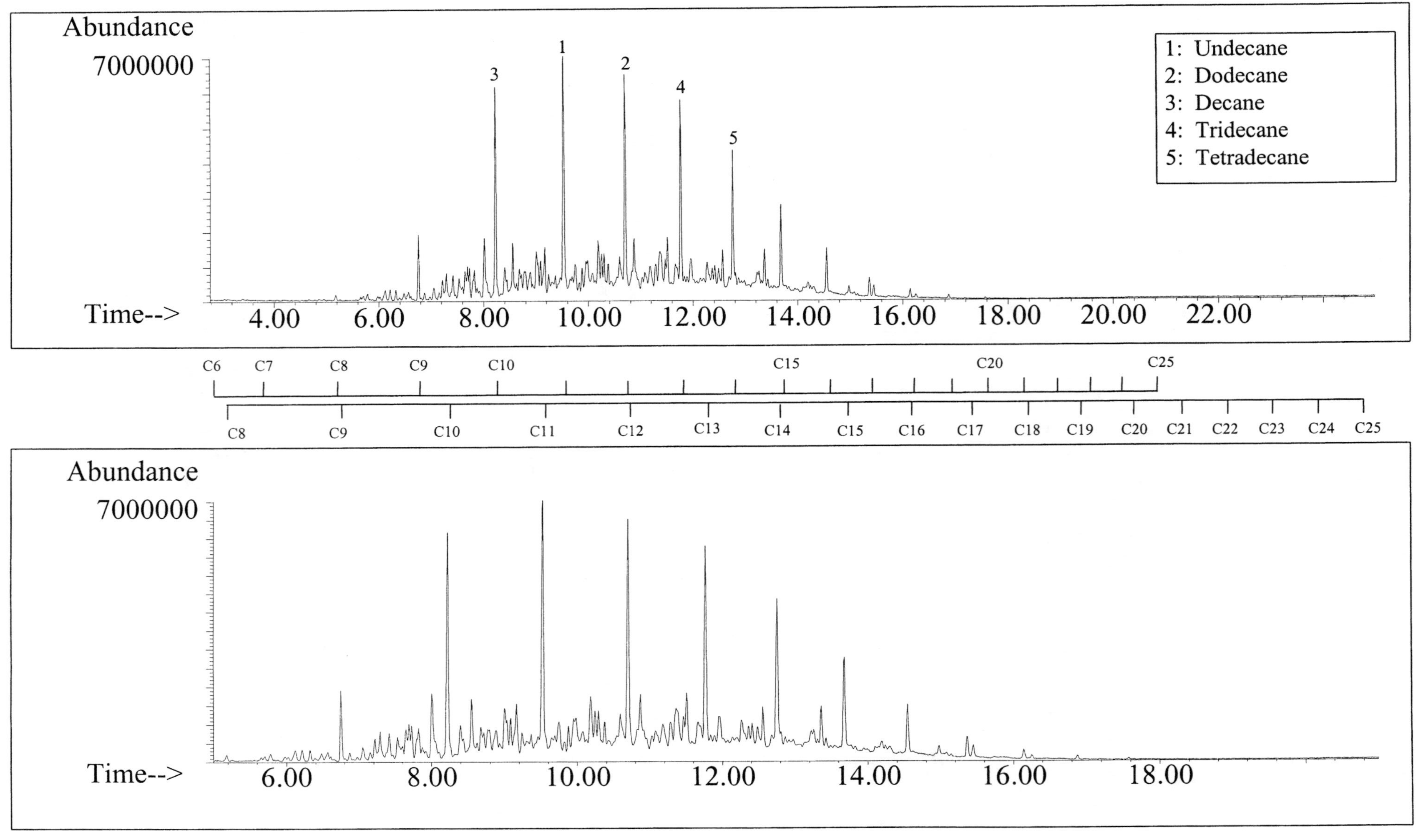

SUMMED PROFILES

Distillate #35

ALKANES

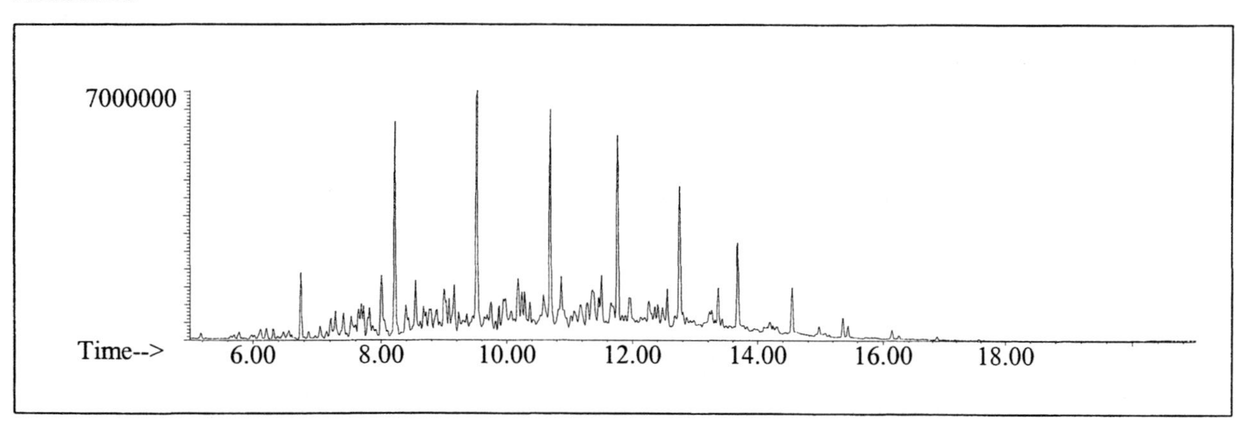

AROMATICS

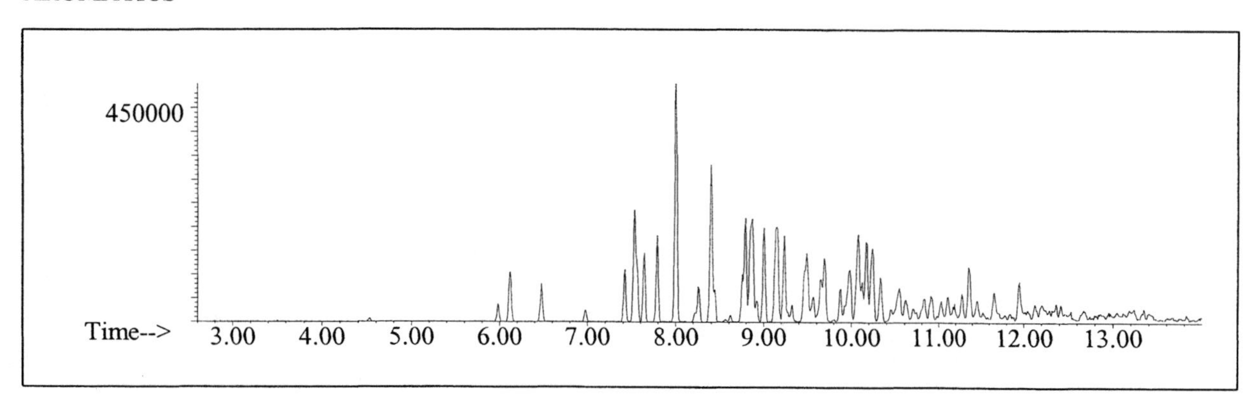

CYCLOPARAFFINS AND ALKENES

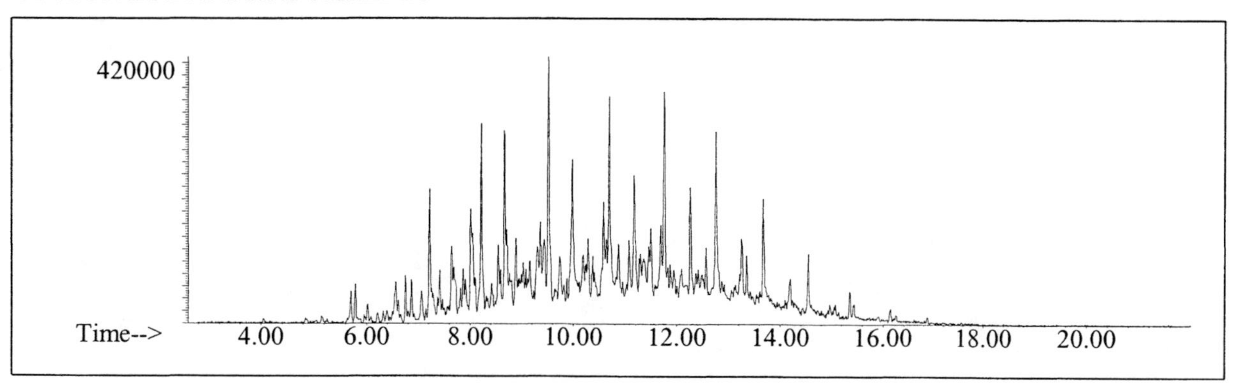

NAPHTHALENES

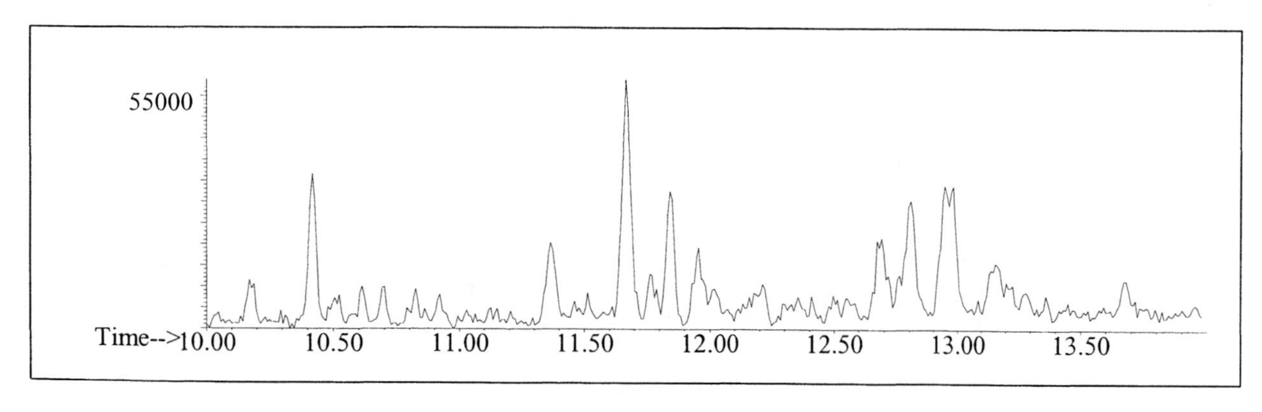

INDIVIDUAL PROFILES Distillate #35

Alkane

Ion 43

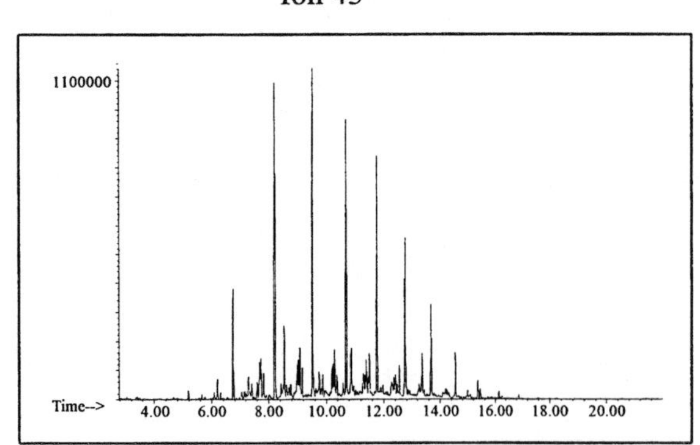

Ion 57

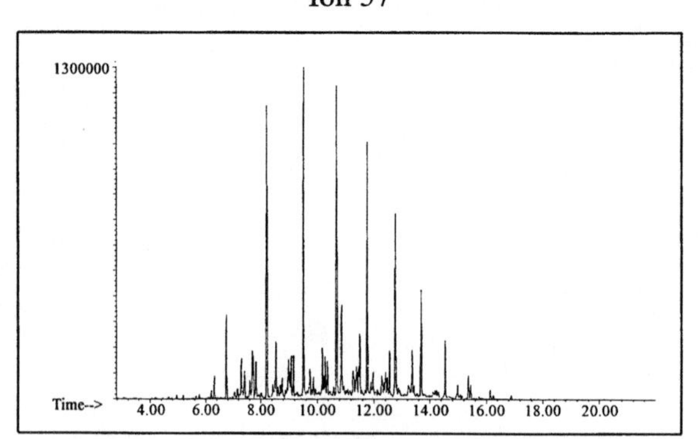

Ion 71

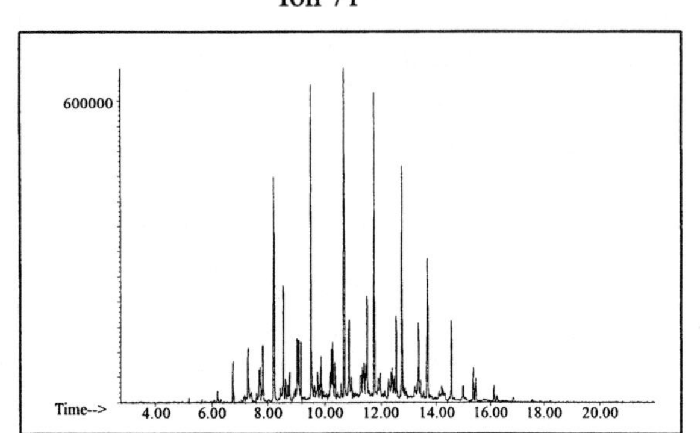

Ion 85

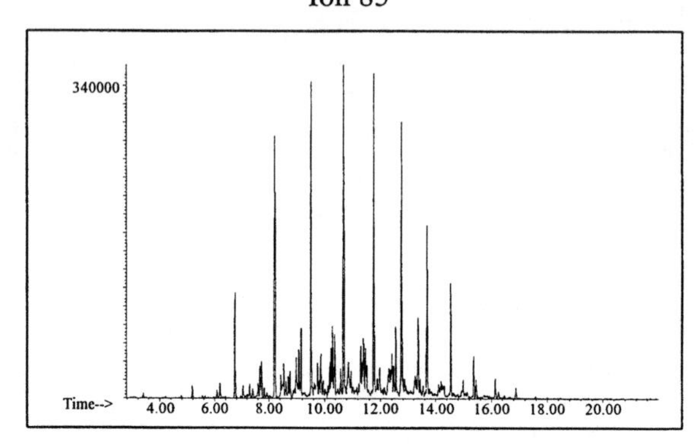

Aromatic

Ion 91

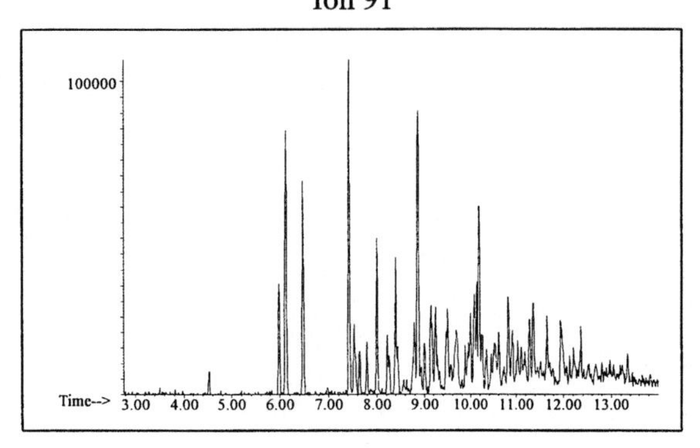

Ion 105

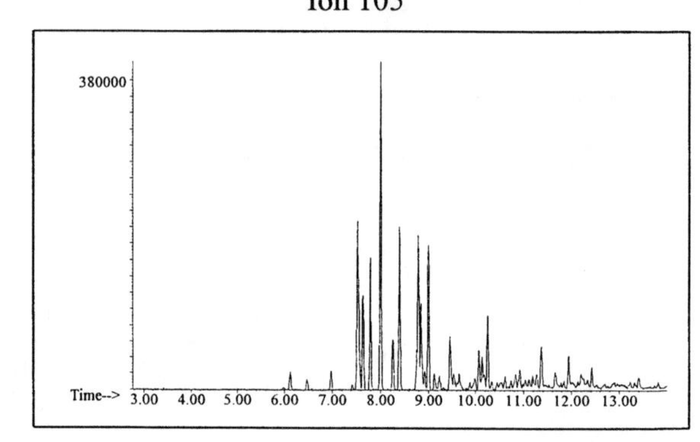

Ion 119

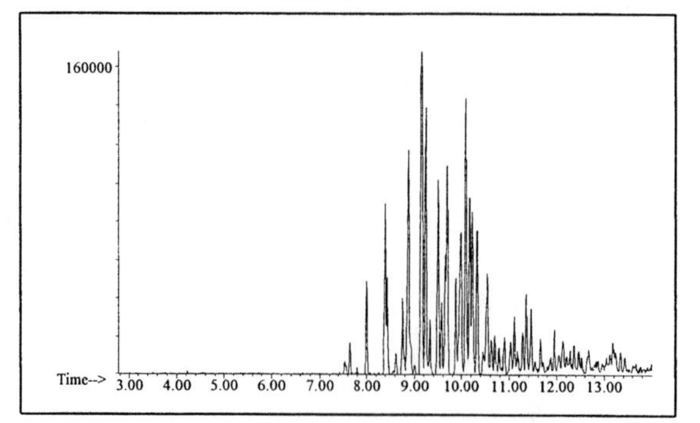

INDIVIDUAL PROFILES

Distillate #35

Cycloparaffin

Ion 55

Ion 69

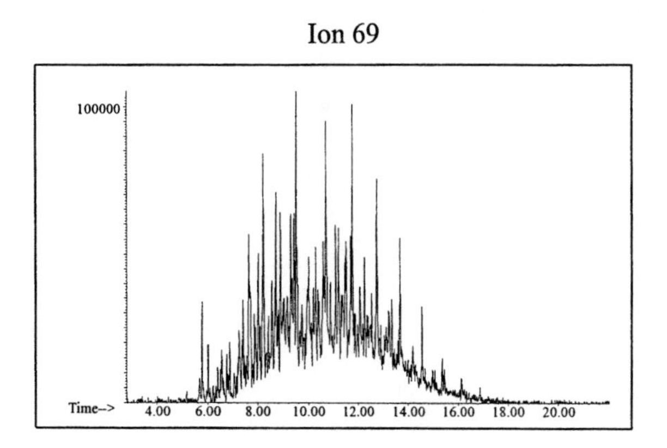

Ion 83

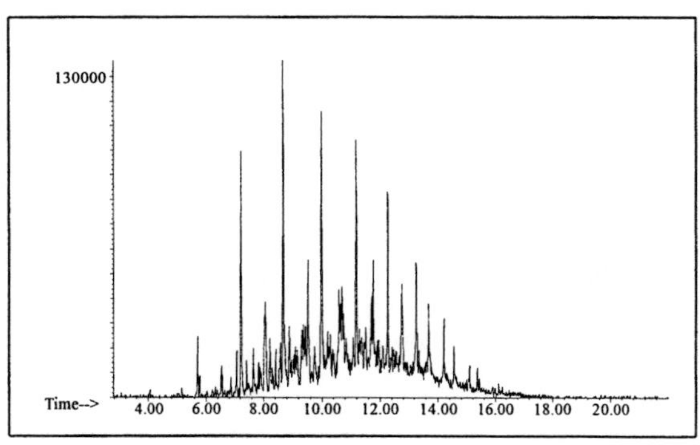

Naphthalene

Ion 128

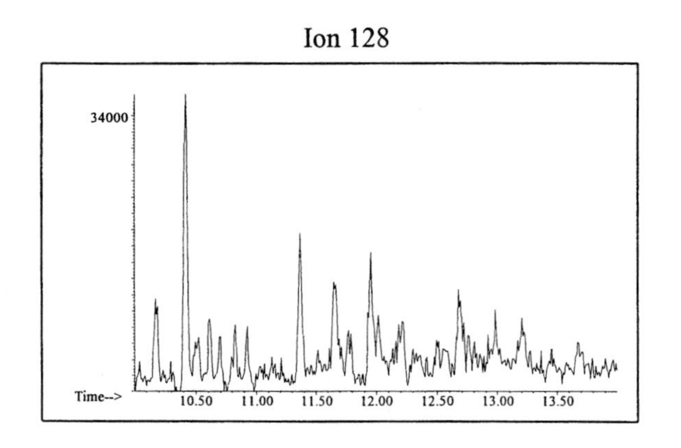

Ion 142

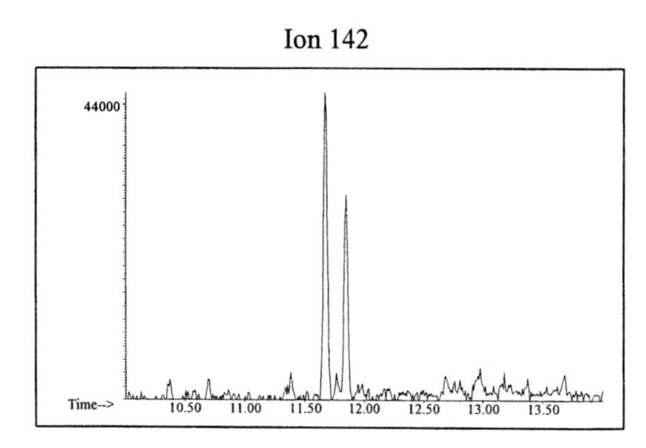

Ion 156

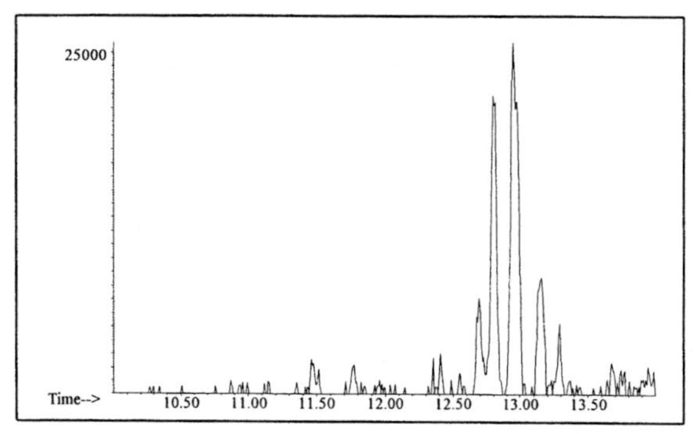

Distillate #36

20uL/mL Pentane

ASTM: Class 4 (Kerosene)

Macro Code: H

Product Displayed: Kerosene

Product Uses: Heating fuel, jet fuel, lamp oil

Other Similar Products: Olde Village Lamp Oil, Sears Penetrating Oil, JP-5

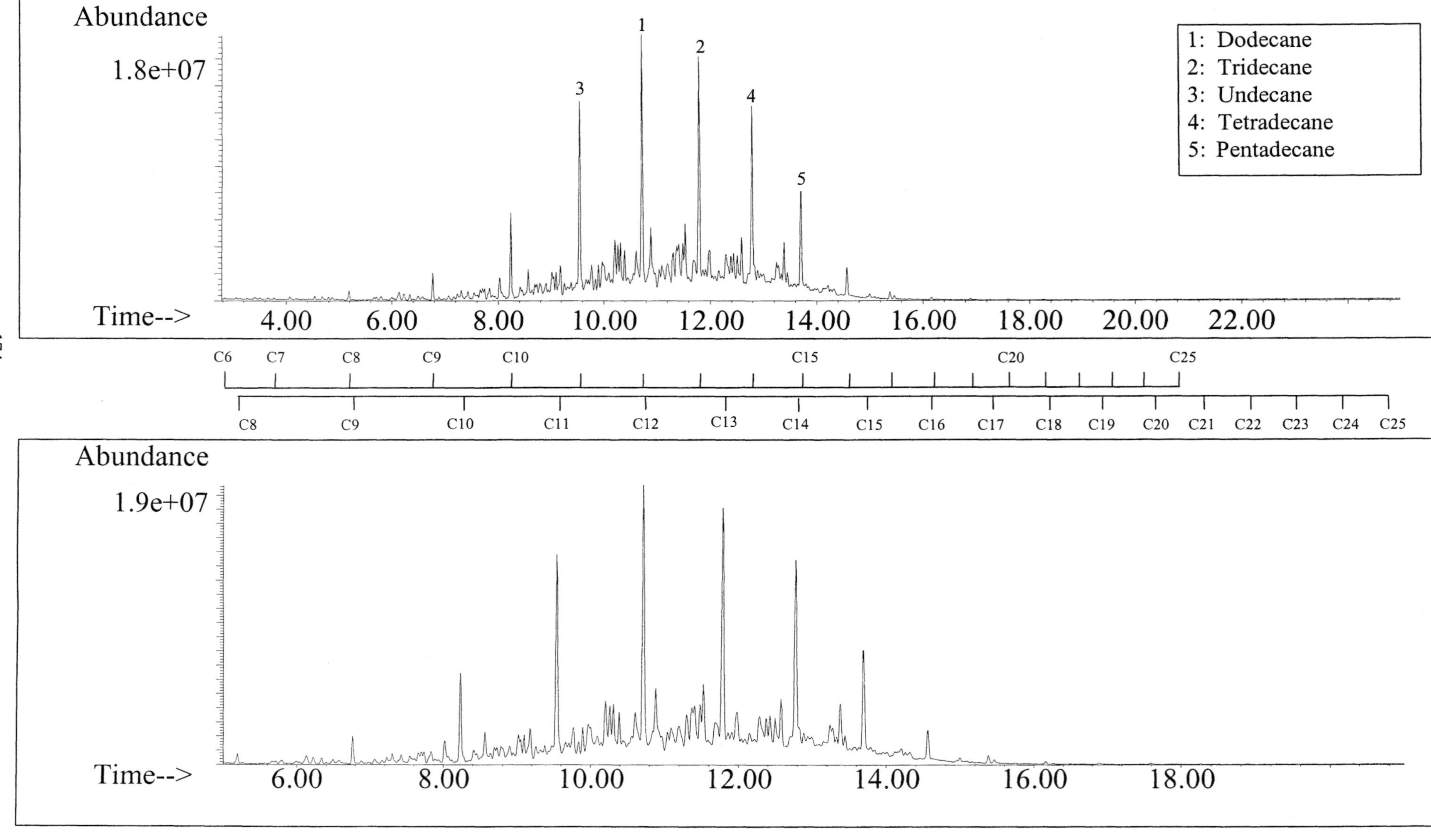

SUMMED PROFILES Distillate #36

ALKANES

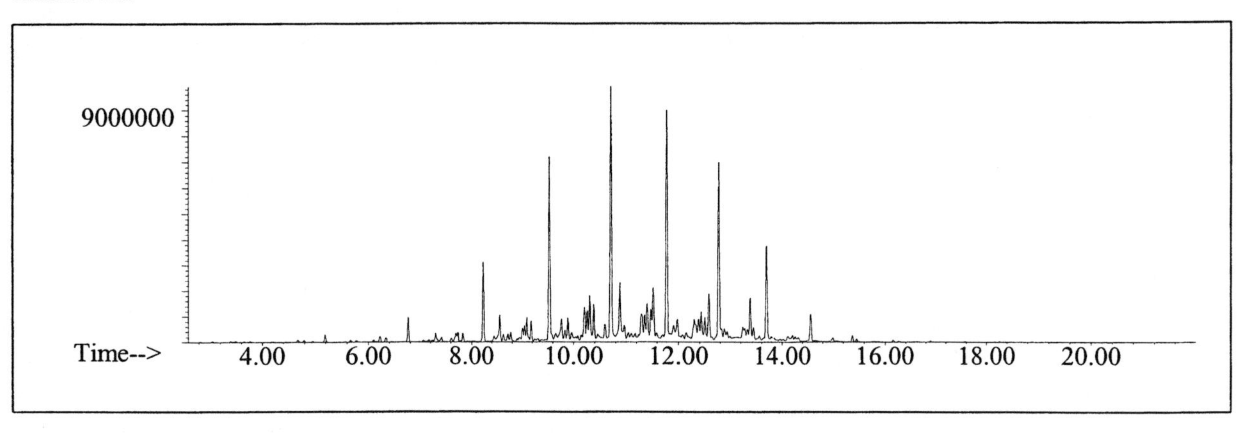

AROMATICS

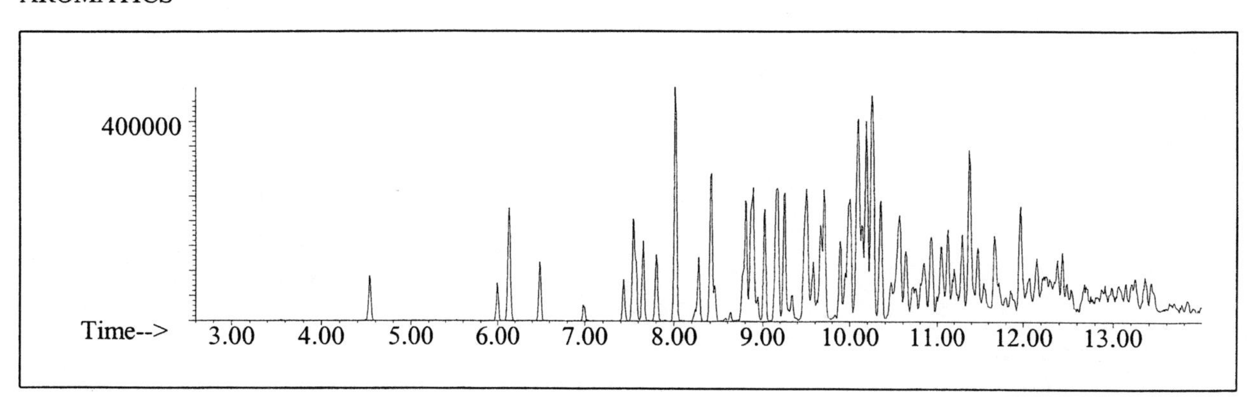

CYCLOPARAFFINS AND ALKENES

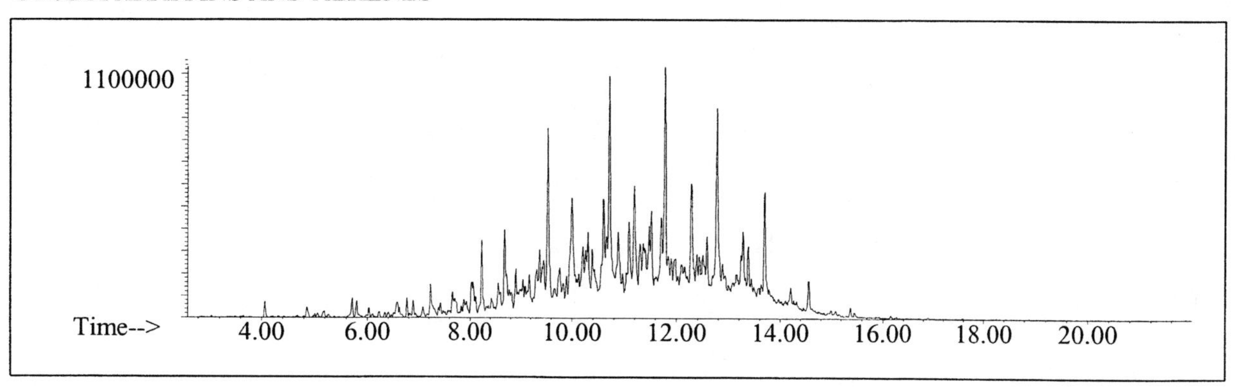

NAPHTHALENES

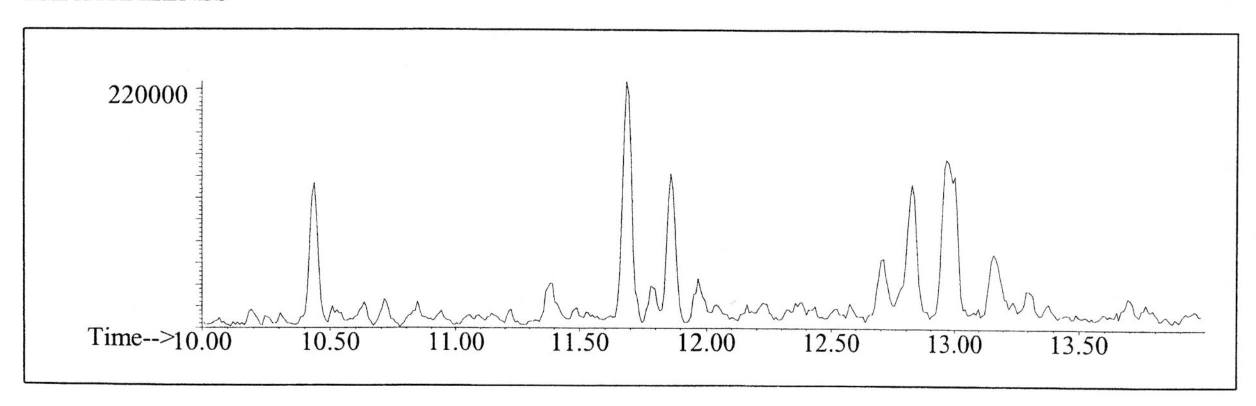

INDIVIDUAL PROFILES

Distillate #36

Alkane

Ion 43

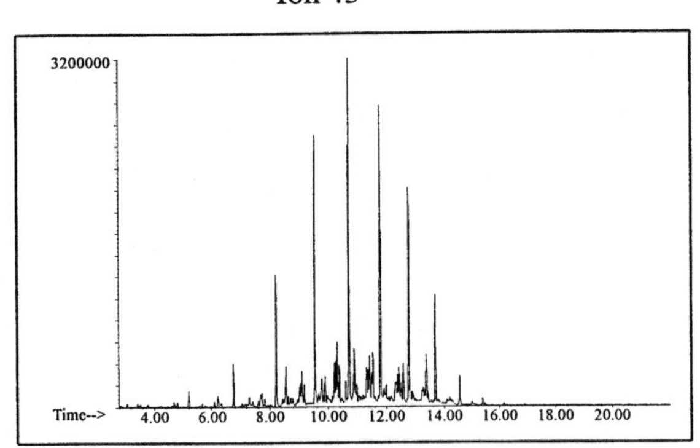

Ion 57

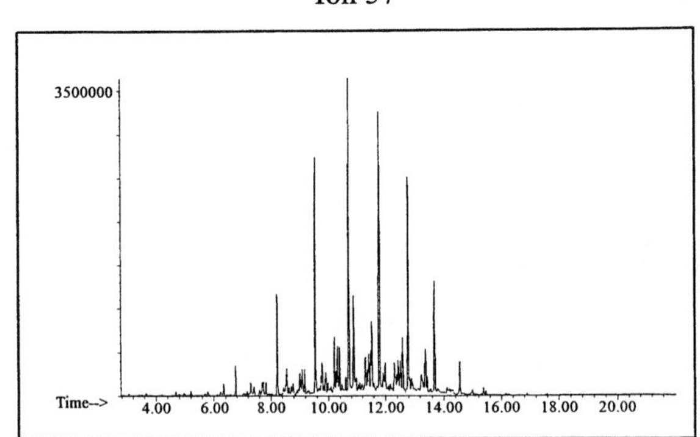

Ion 71

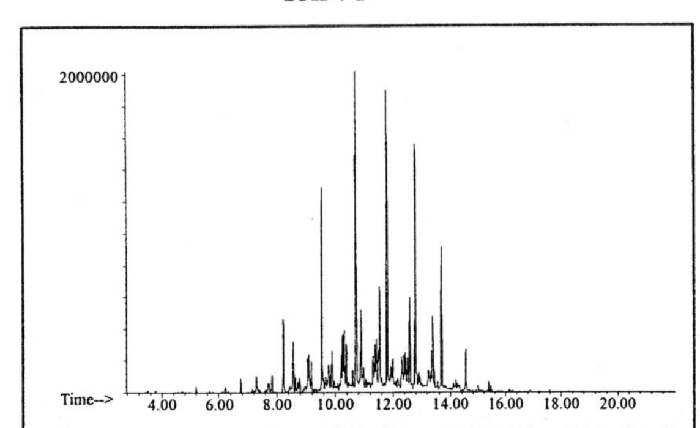

Ion 85

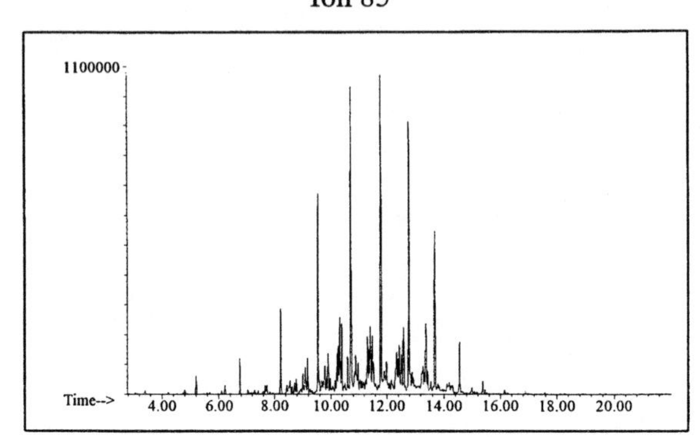

Aromatic

Ion 91

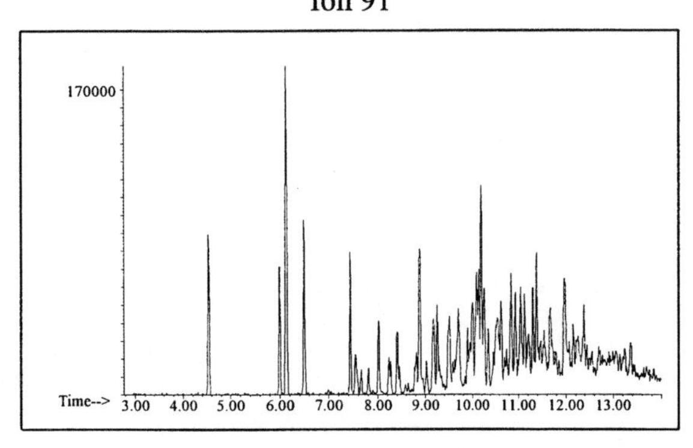

Ion 105

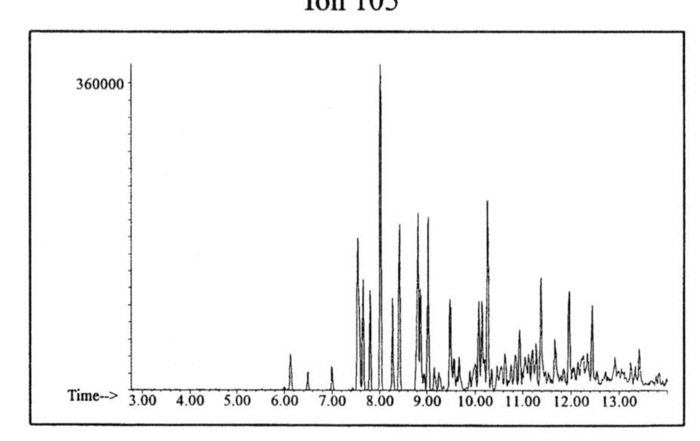

Ion 119

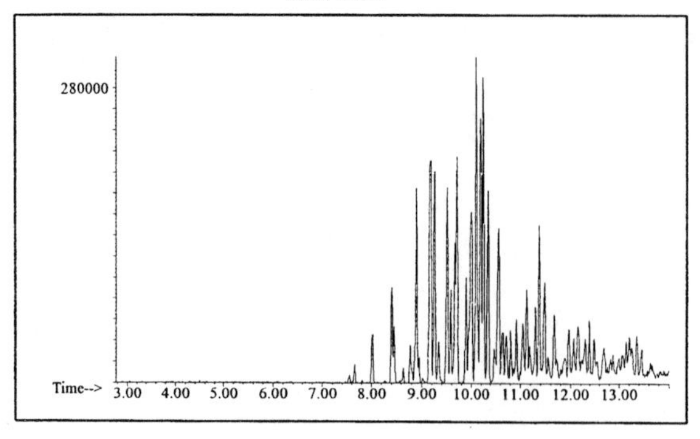

INDIVIDUAL PROFILES

Cycloparaffin

Ion 55

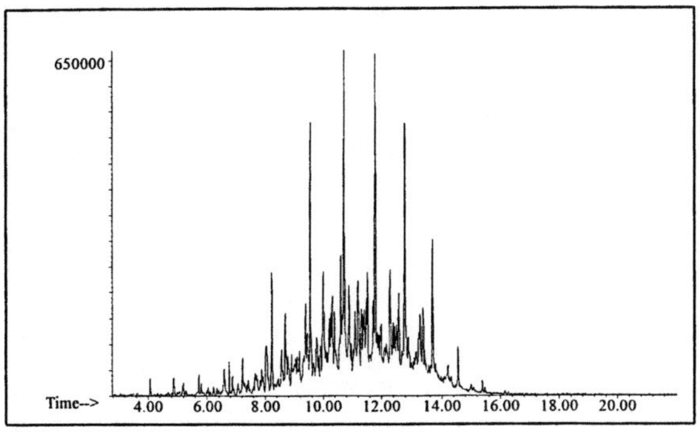

Ion 69

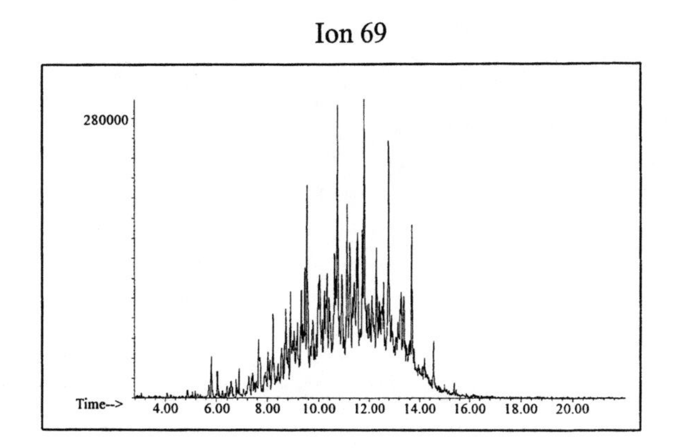

Ion 83

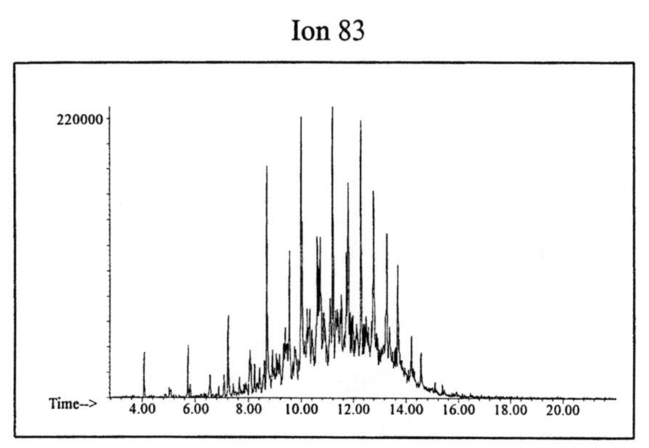

Naphthalene

Ion 128

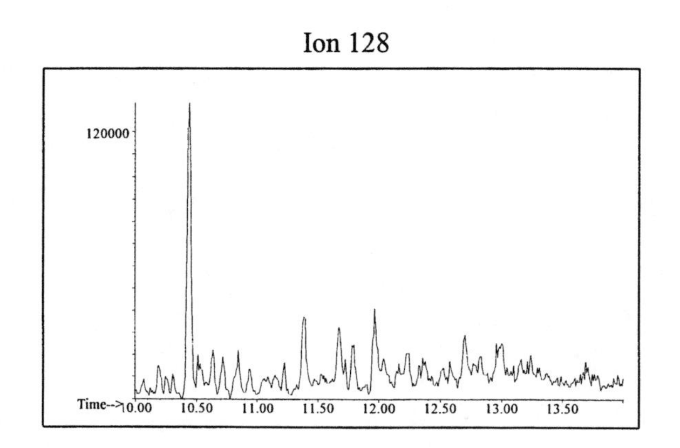

Ion 142

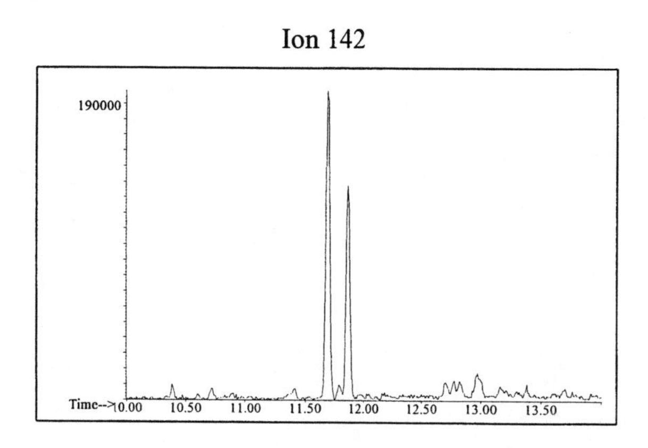

Ion 156

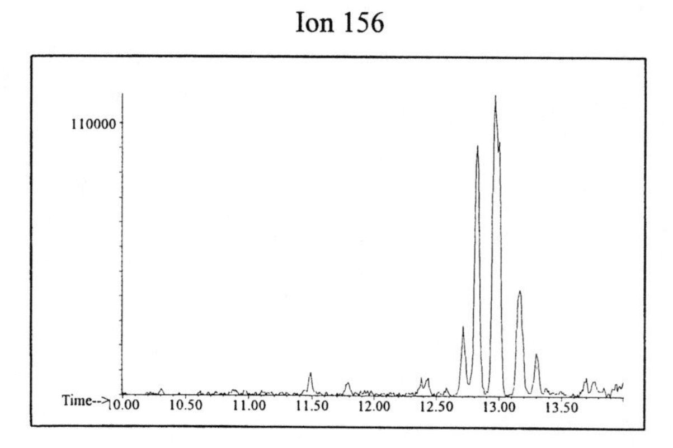

Distillate #37

20uL/mL Pentane

Product Displayed: 25% Evaporated Kerosene

Other Similar Products: Olde Village Lamp Oil, Sears Penetrating Oil, JP-5

ASTM: Class 4 (Kerosene)

Product Uses: Heating fuel, jet fuel, lamp oil

Macro Code: H

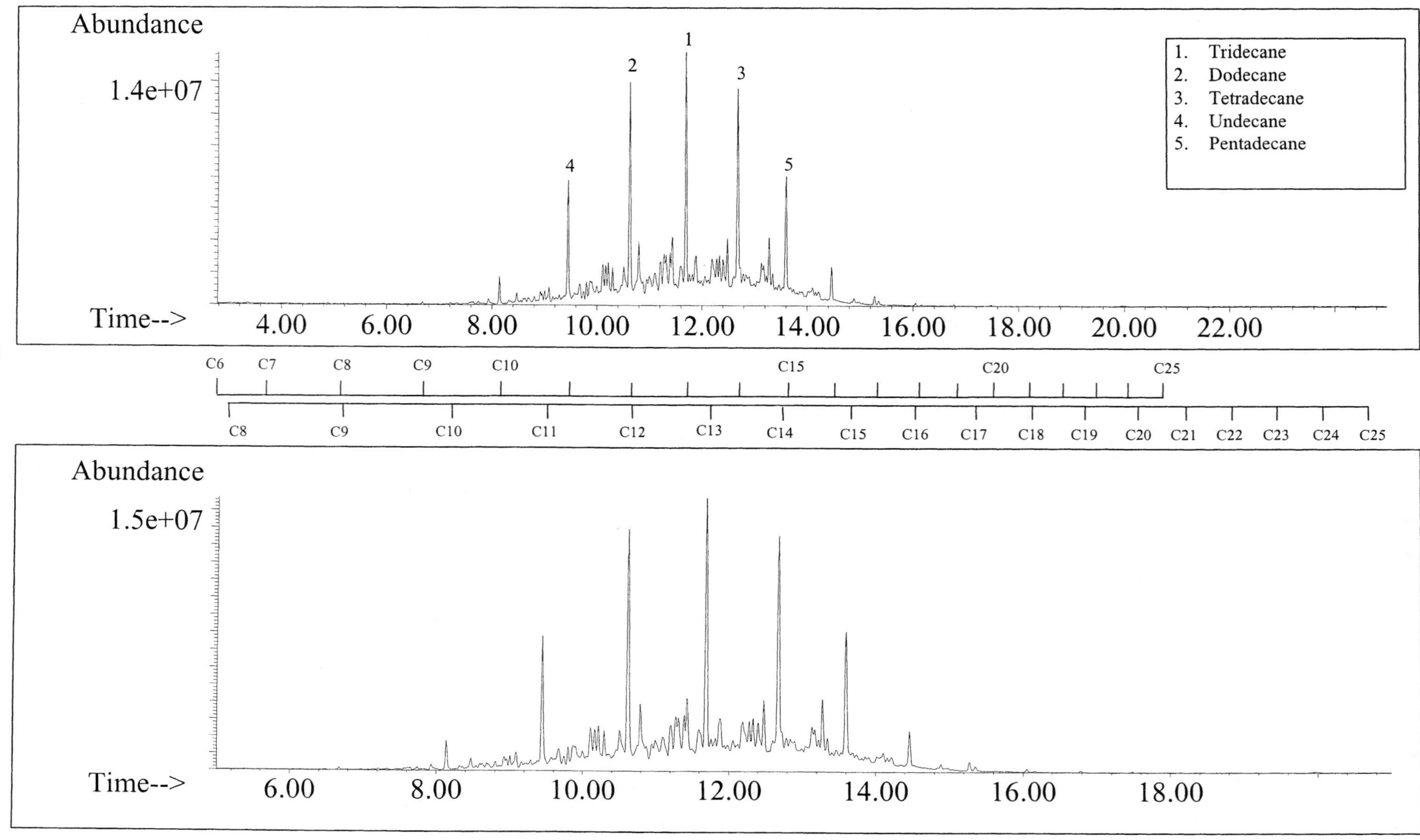

SUMMED PROFILES

Distillate #37

ALKANES

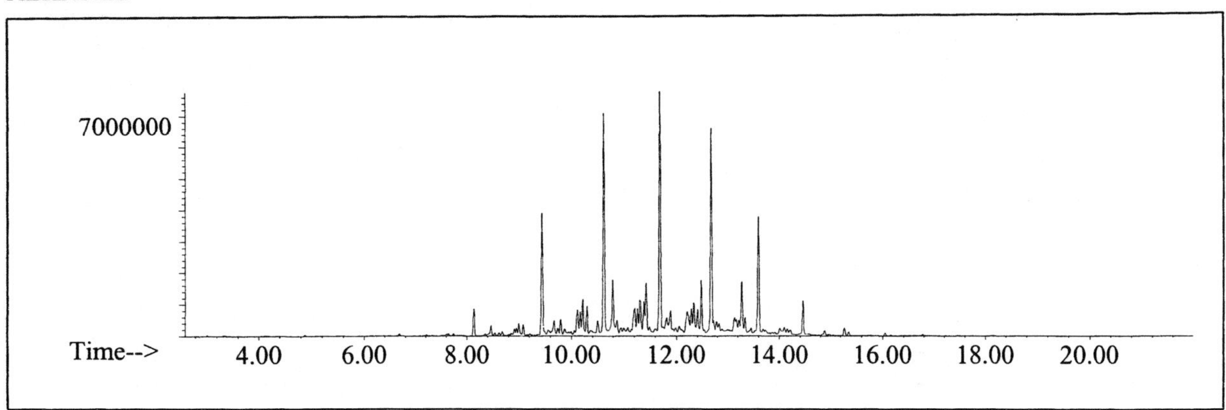

AROMATICS

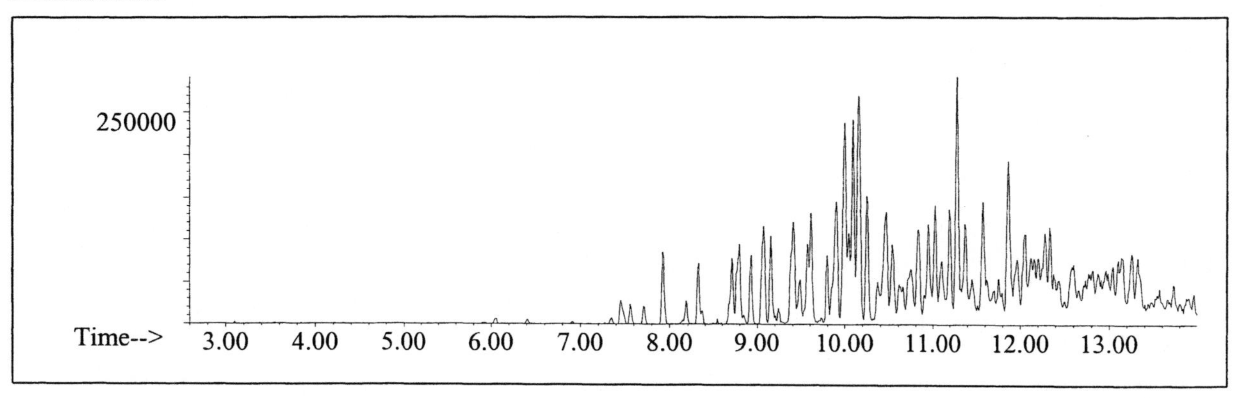

CYCLOPARAFFINS AND ALKENES

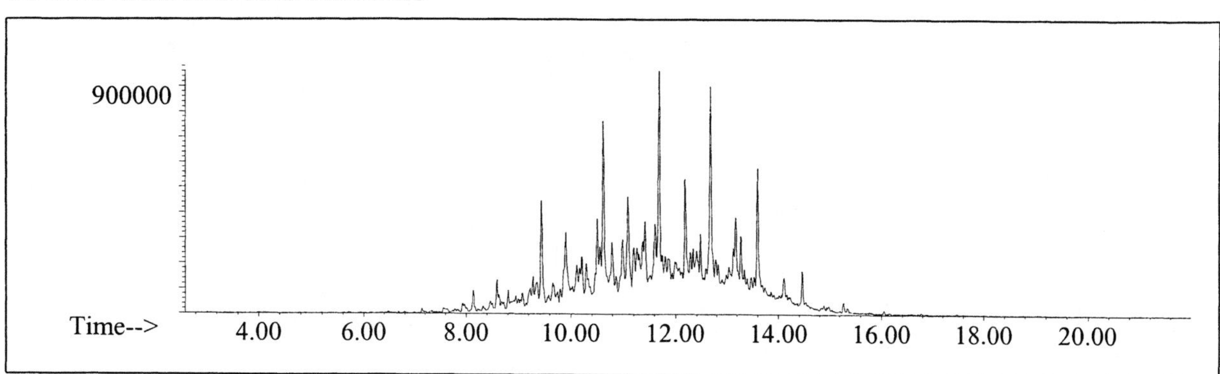

NAPHTHALENES

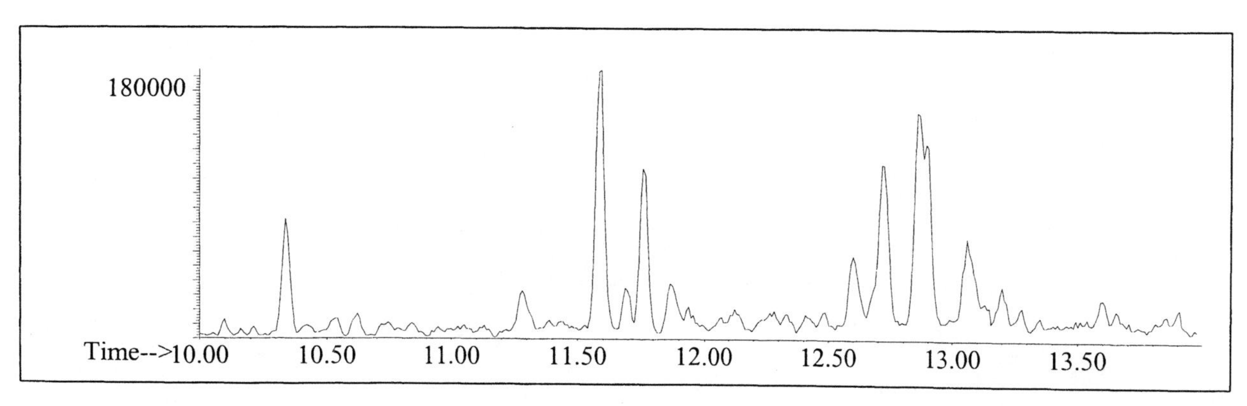

INDIVIDUAL PROFILES

Distillate #37

Alkane

Ion 43

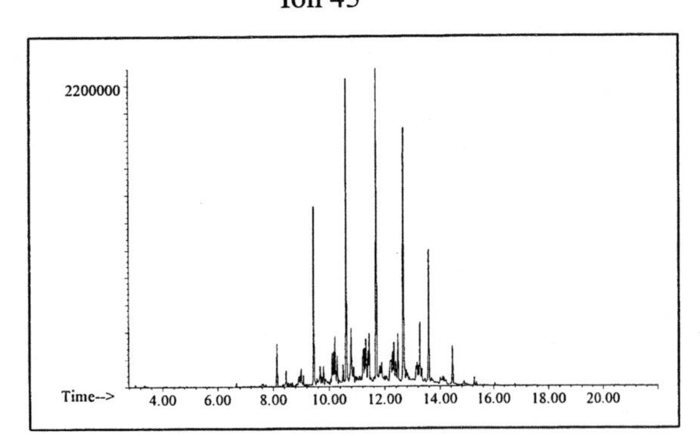

Ion 57

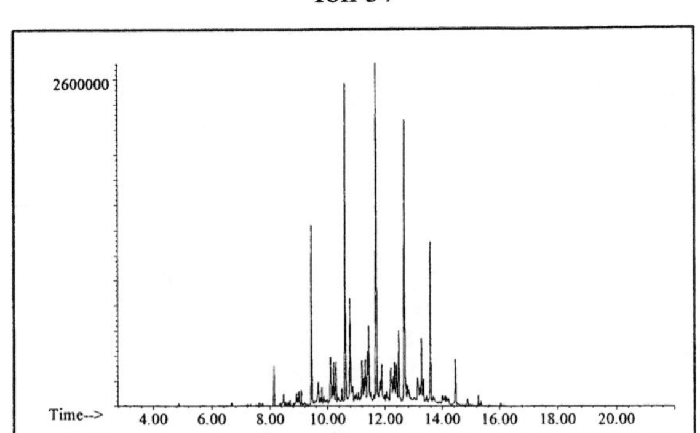

Ion 71

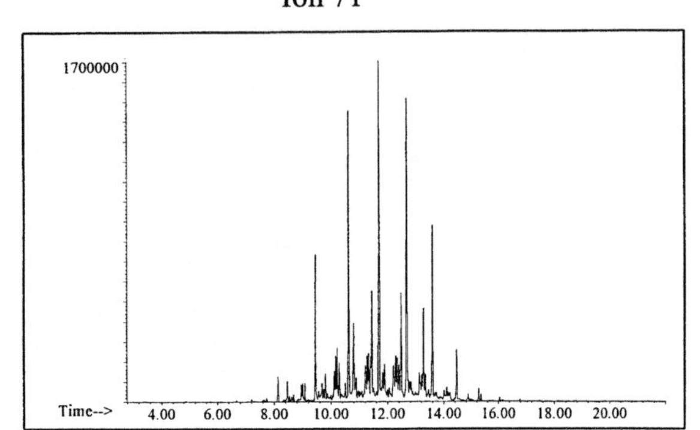

Ion 85

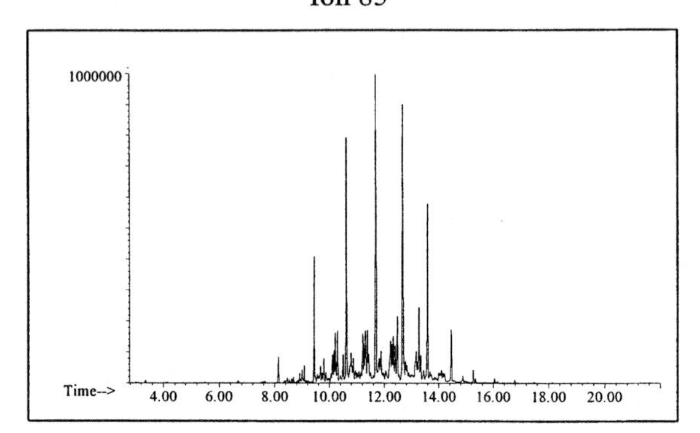

Aromatic

Ion 91

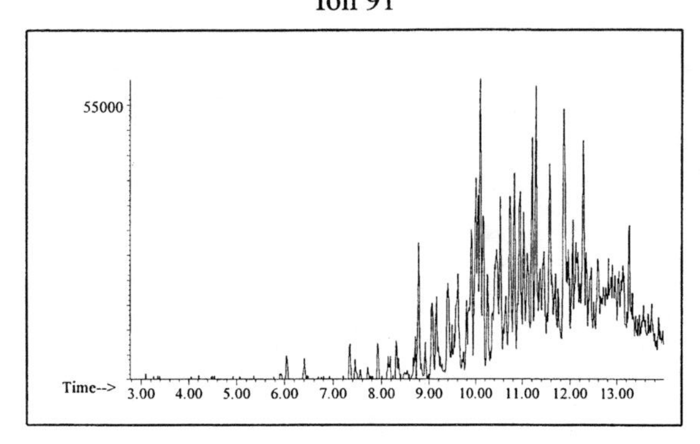

Ion 105

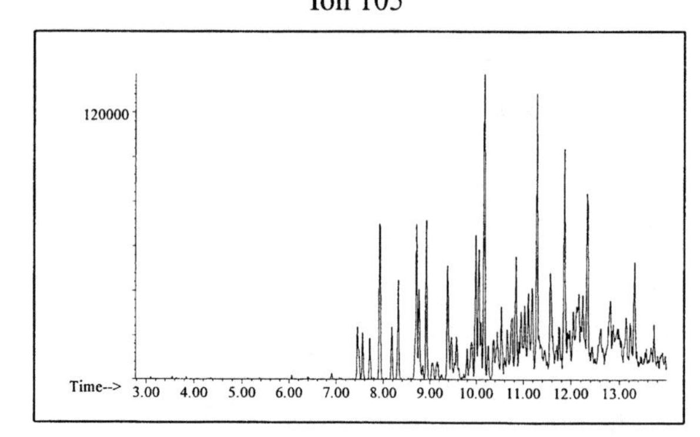

Ion 119

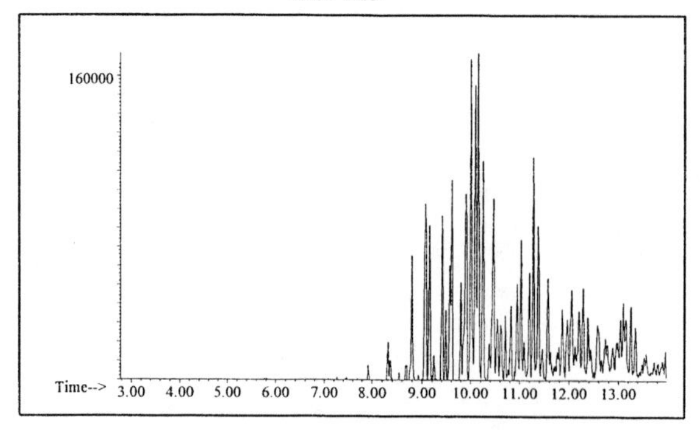

INDIVIDUAL PROFILES

Distillate #37

Cycloparaffin

Ion 55

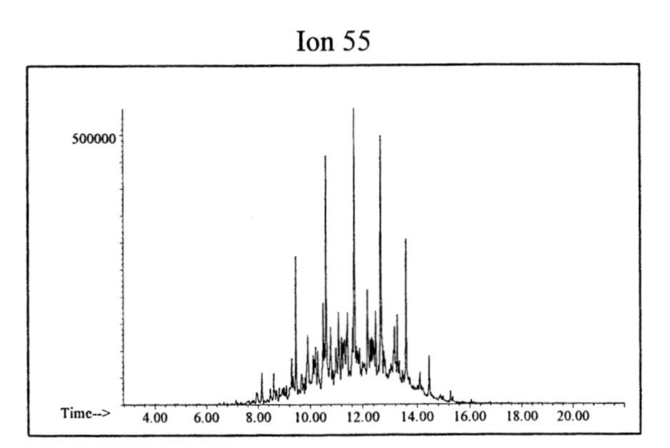

Ion 69

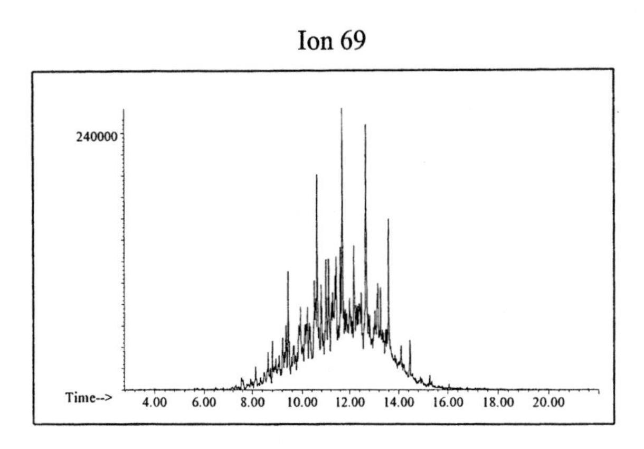

Ion 83

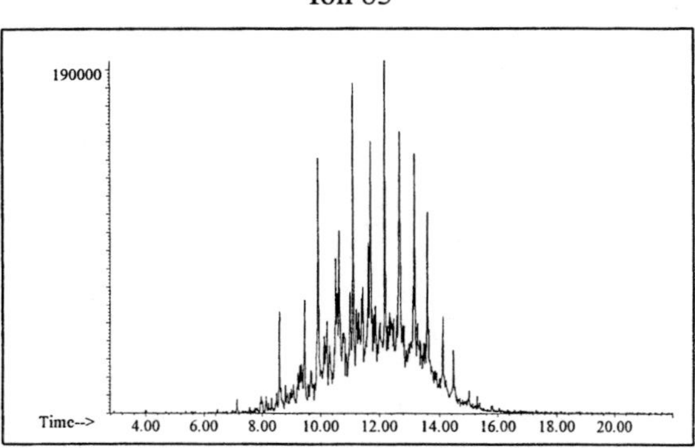

Naphthalene

Ion 128

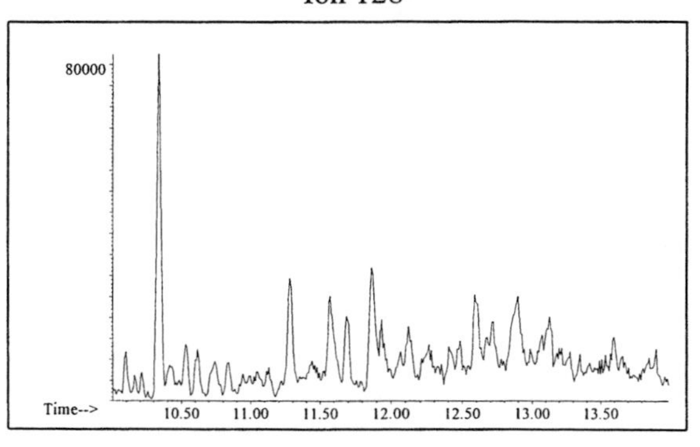

Ion 142

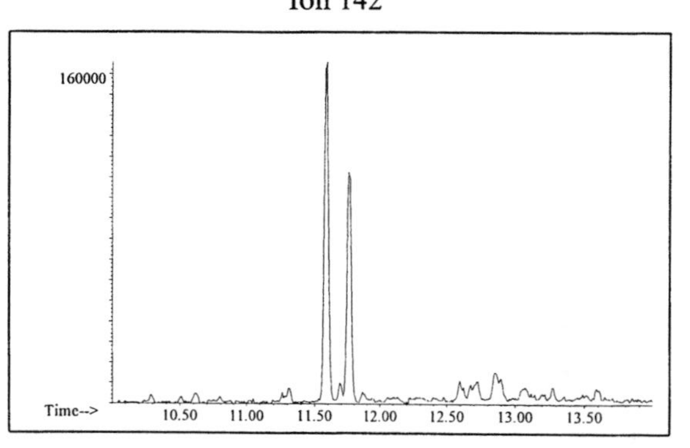

Ion 156

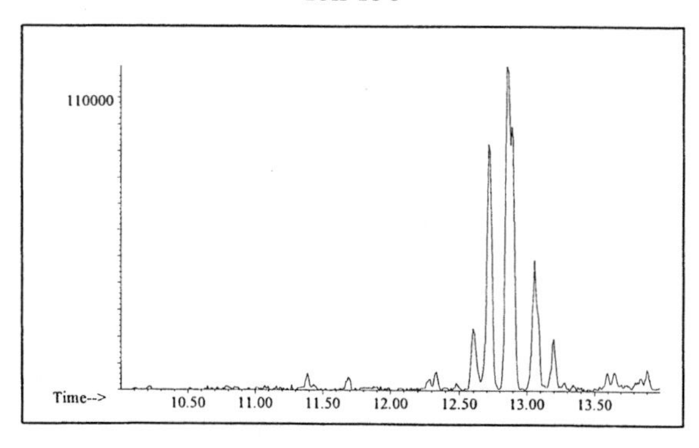

Distillate #38

20uL/mL Pentane

ASTM: Class 4 (Kerosene)

Macro Code: H

Product Displayed: 50% Evaporated Kerosene

Product Uses: Heating fuel, jet fuel, lamp oil

Other Similar Products: Olde Village Lamp Oil, Sears Penetrating Oil, JP-5

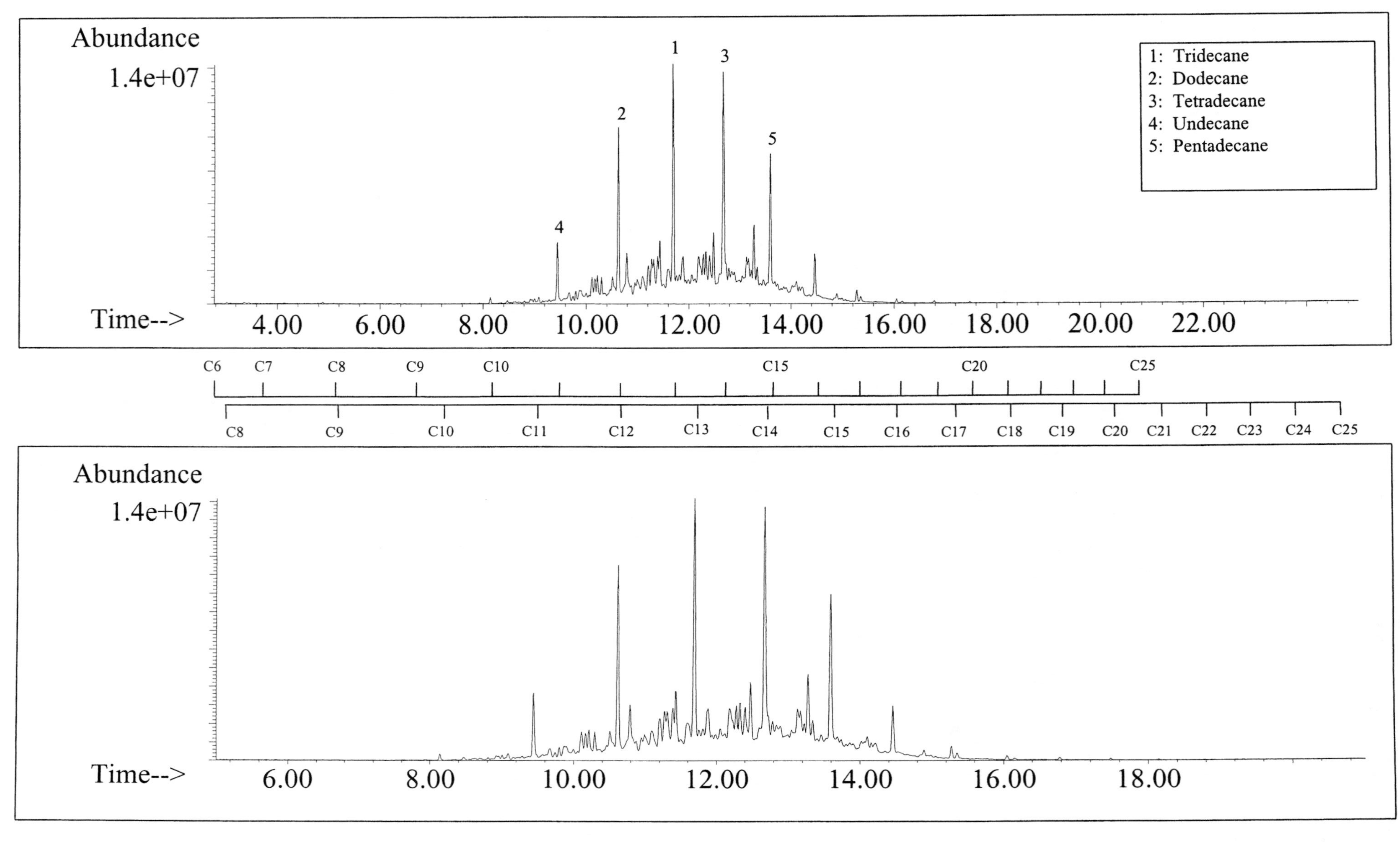

SUMMED PROFILES — Distillate #38

ALKANES

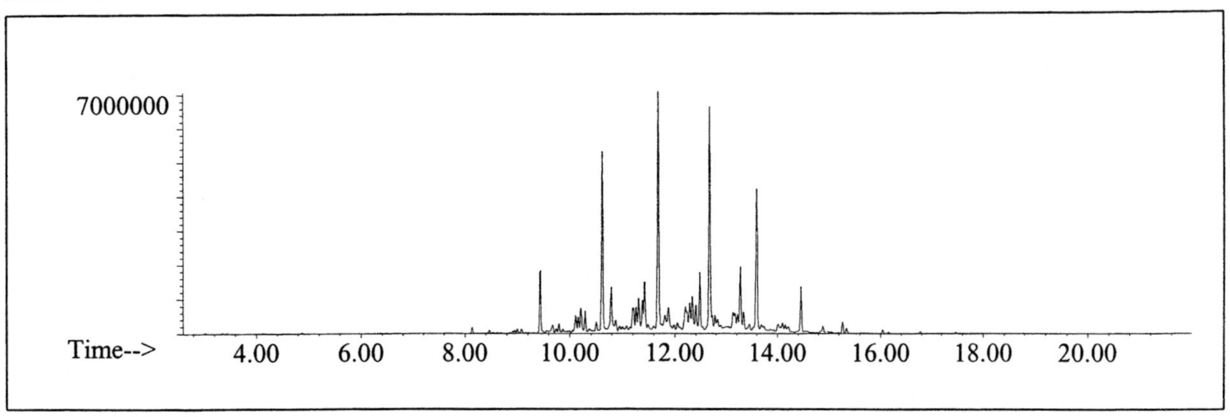

AROMATICS

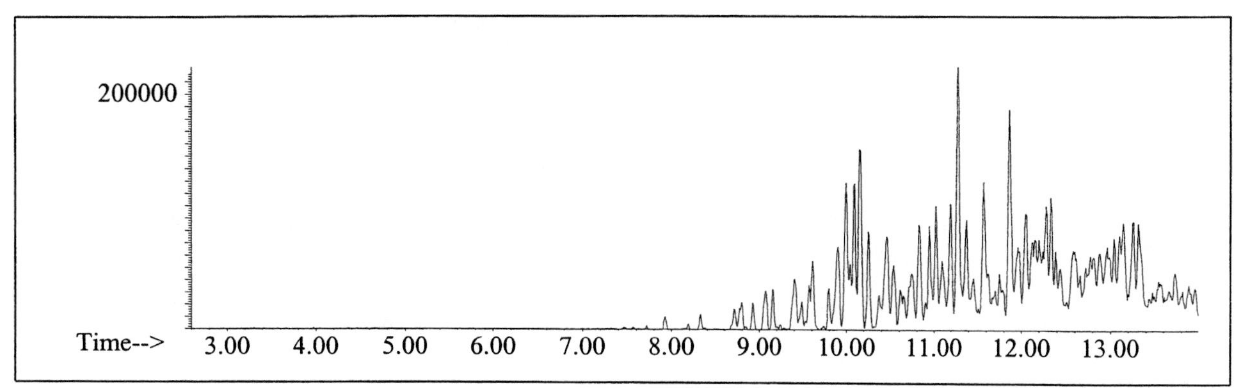

CYCLOPARAFFINS AND ALKENES

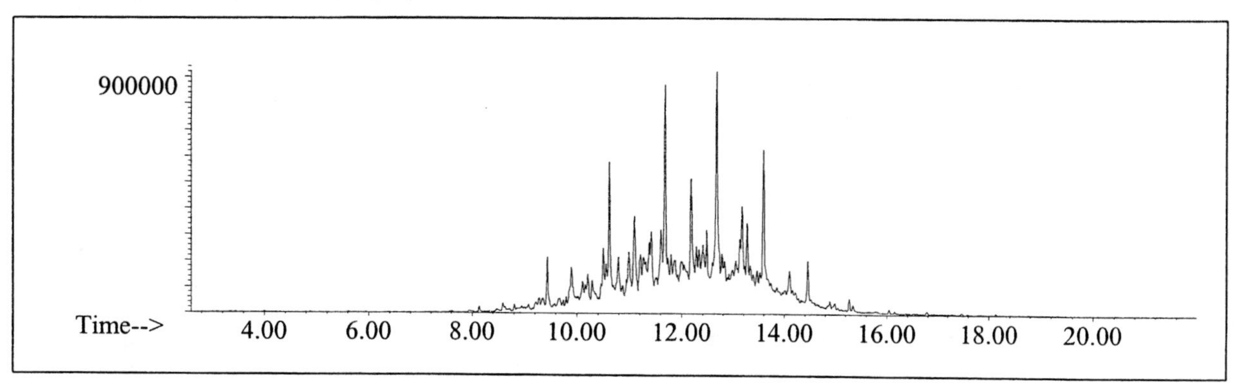

NAPHTHALENES

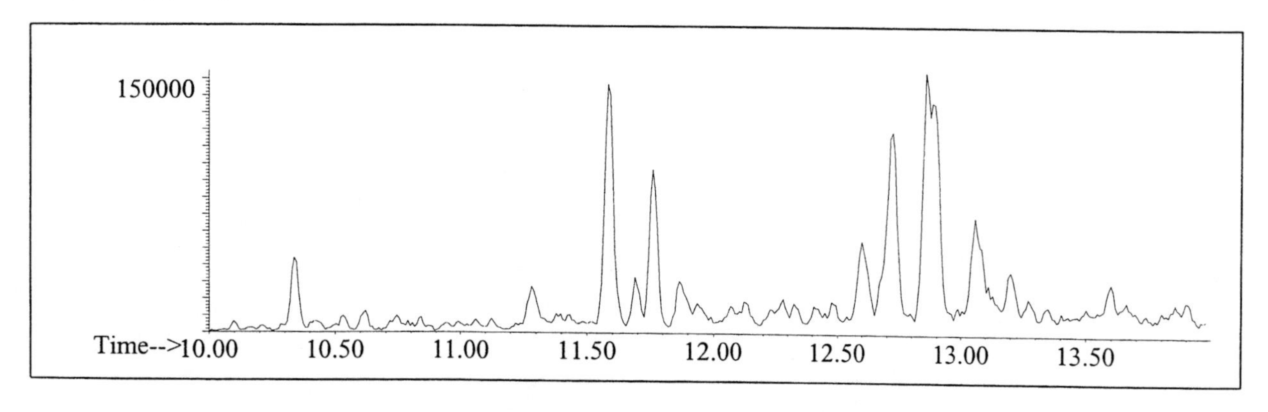

INDIVIDUAL PROFILES

Distillate #38

Alkane

Ion 43

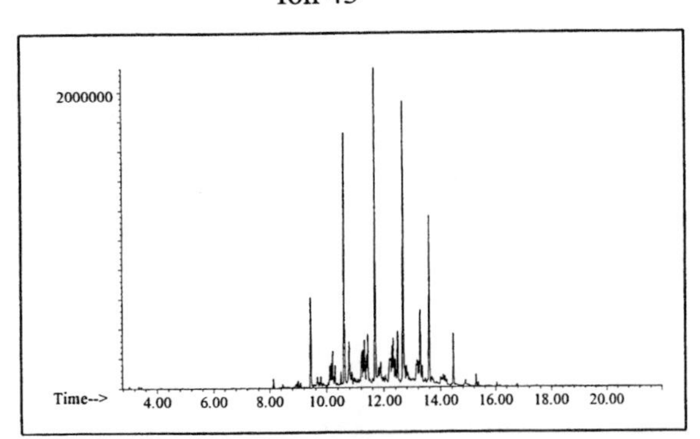

Ion 57

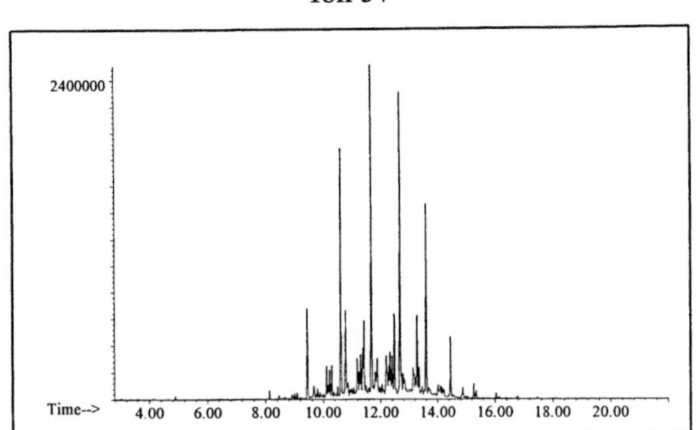

Ion 71

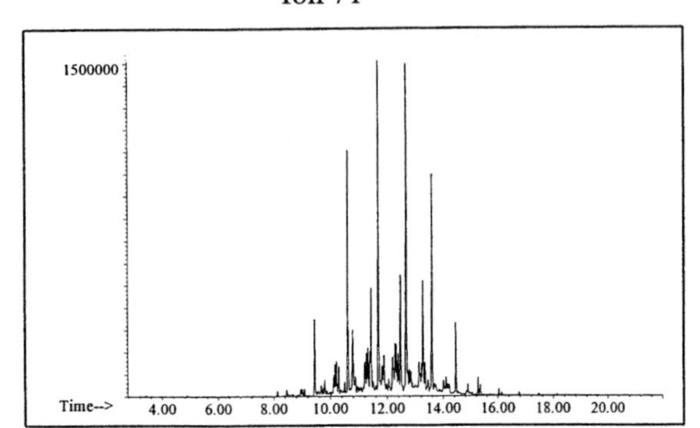

Ion 85

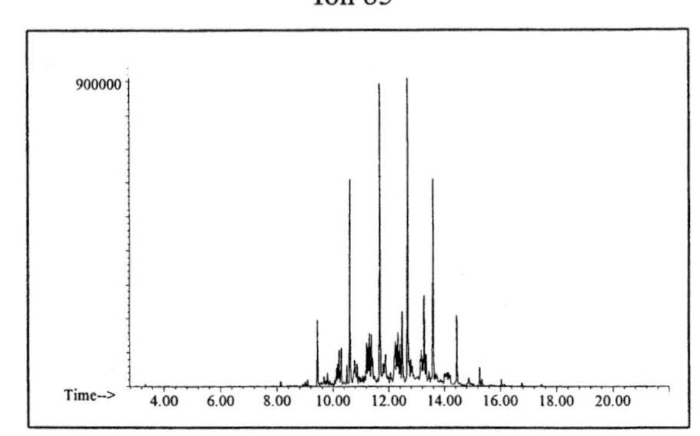

Aromatic

Ion 91

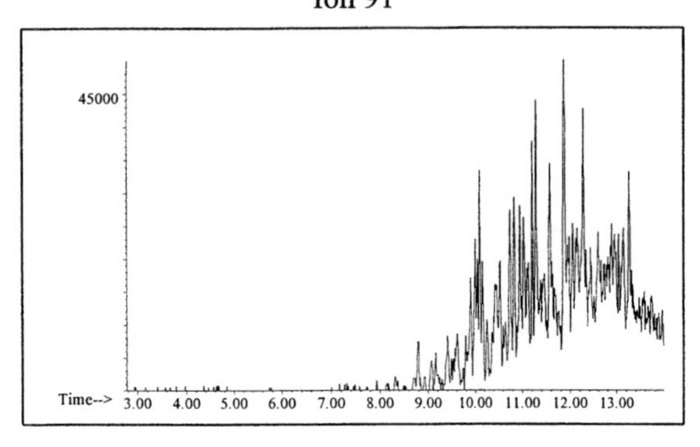

Ion 105

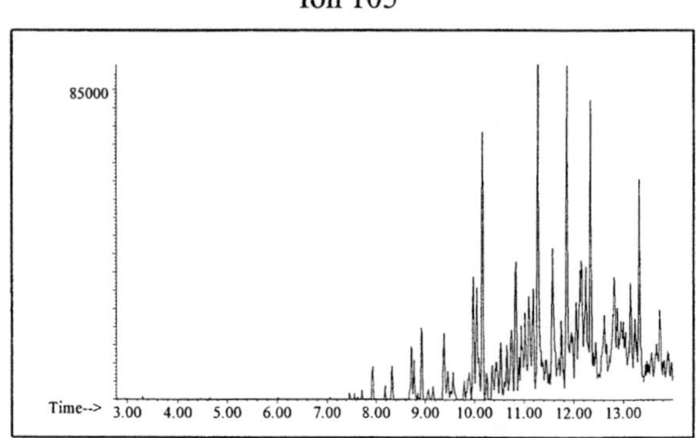

Ion 119

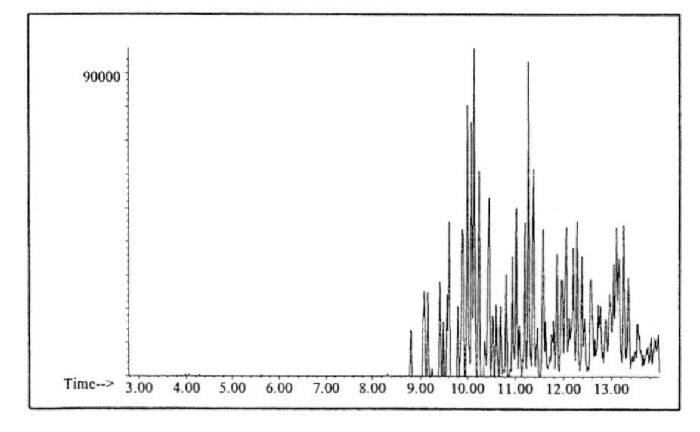

INDIVIDUAL PROFILES

Distillate #38

Cycloparaffin

Ion 55

Ion 69

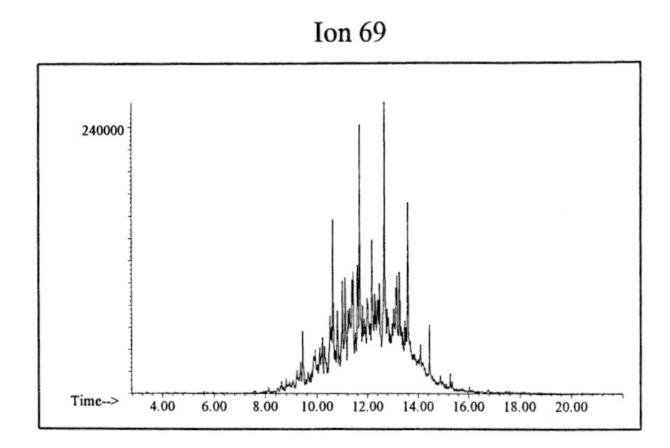

Ion 83

Naphthalene

Ion 128

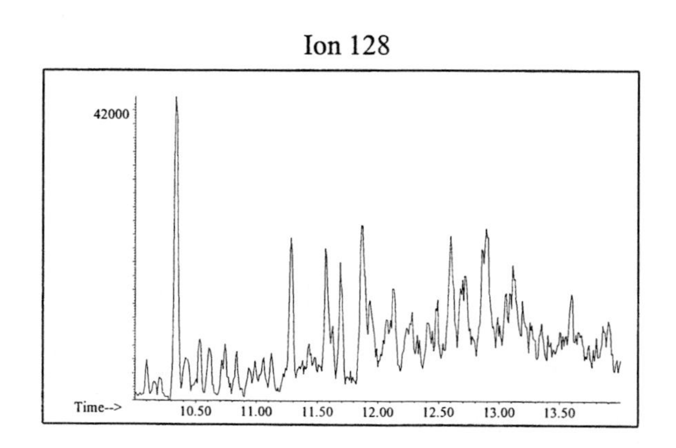

Ion 142

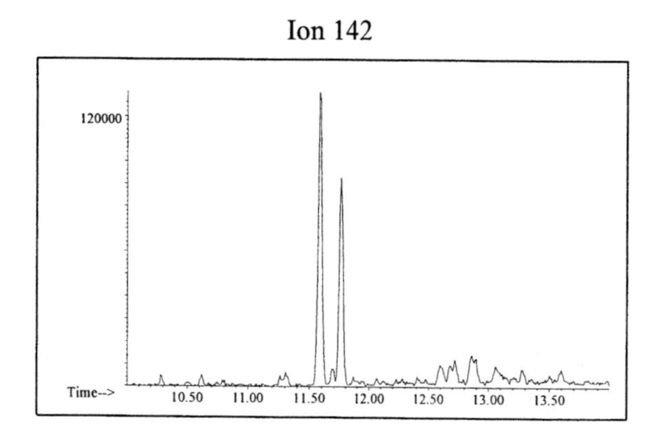

Ion 156

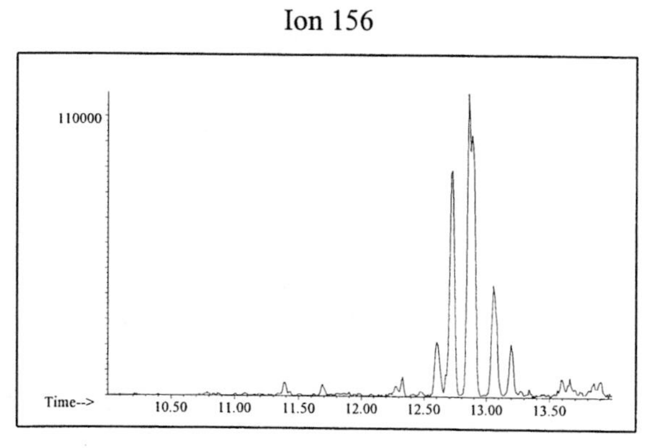

Distillate #39

20uL/mL Pentane

ASTM: Class 4 (Kerosene)

Macro Code: H

Product Displayed: 75% Evaporated Kerosene

Product Uses: Heating fuel, jet fuel, lamp oil

Other Similar Products: Olde Village Lamp Oil, Sears Penetrating Oil, JP-5

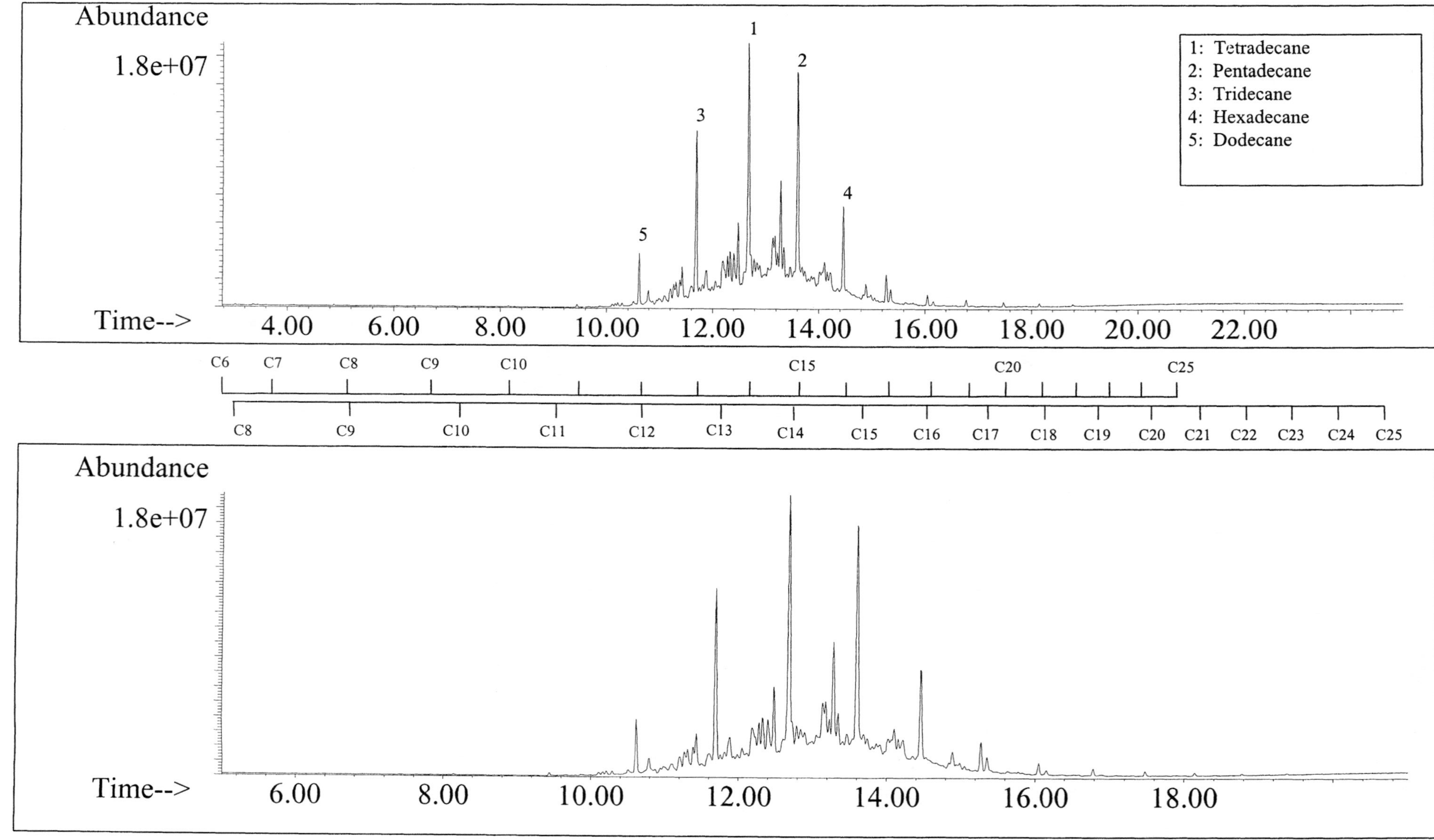

SUMMED PROFILES

Distillate #39

ALKANES

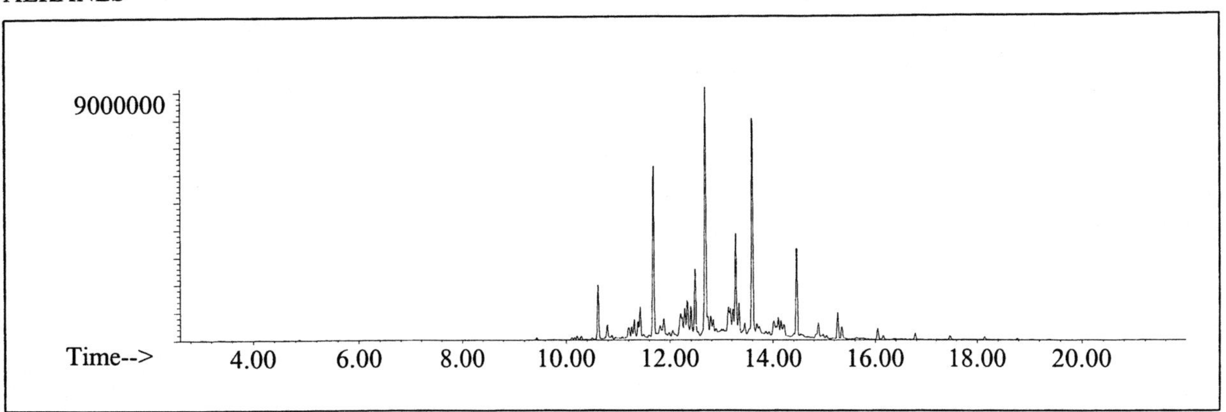

AROMATICS

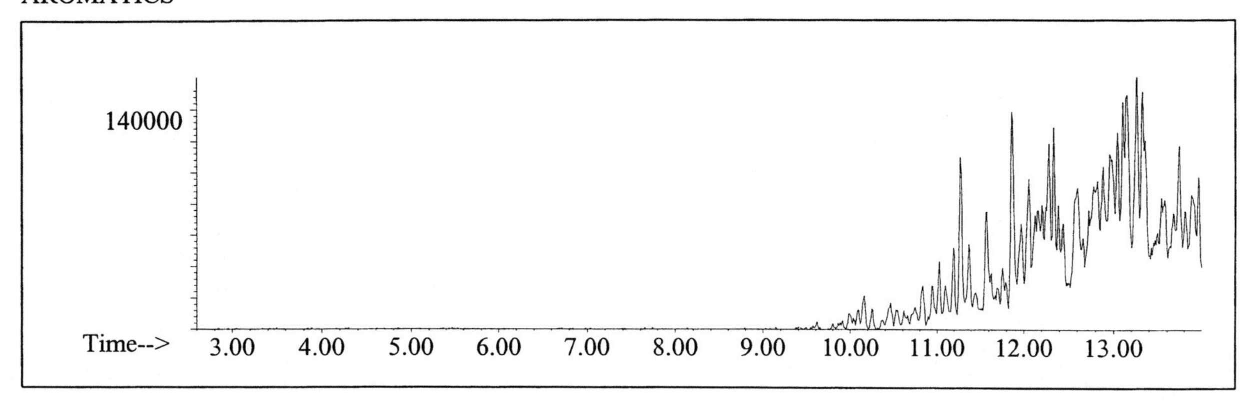

CYCLOPARAFFINS AND ALKENES

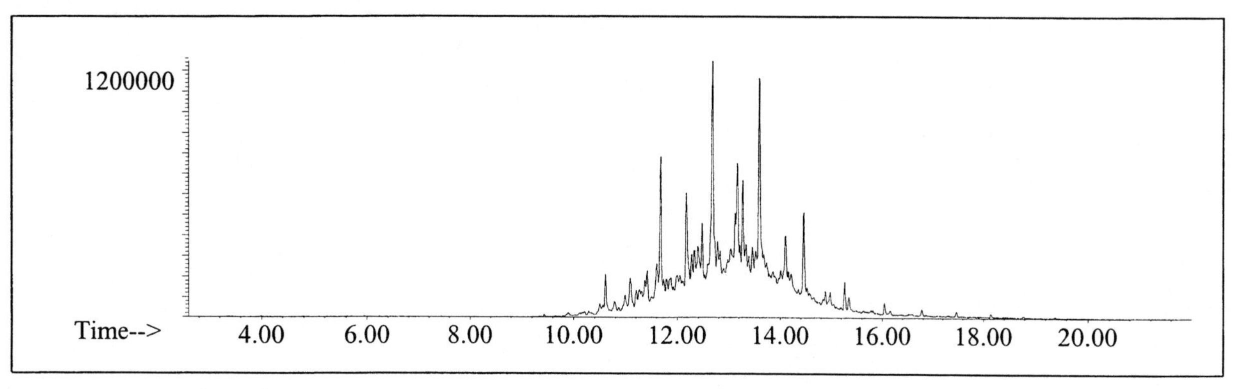

NAPHTHALENES

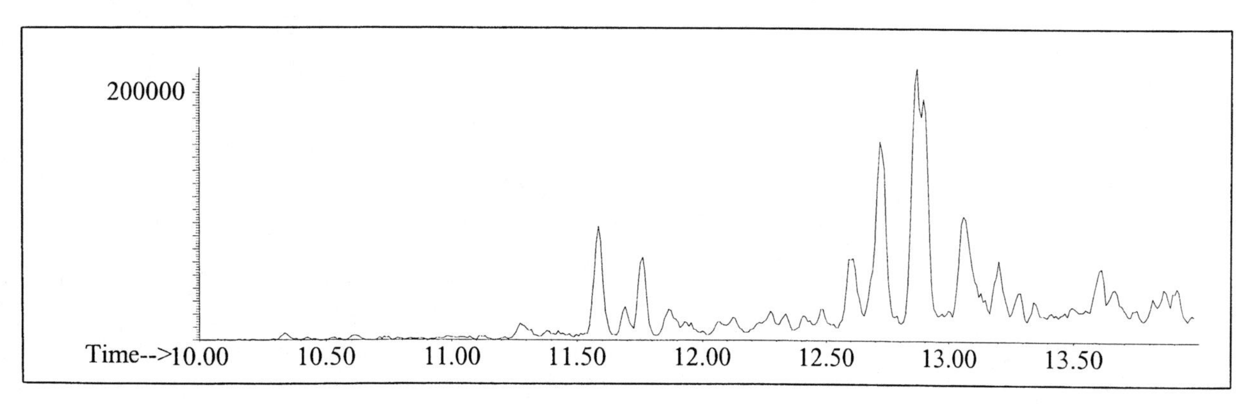

INDIVIDUAL PROFILES

Distillate #39

Alkane

Ion 43

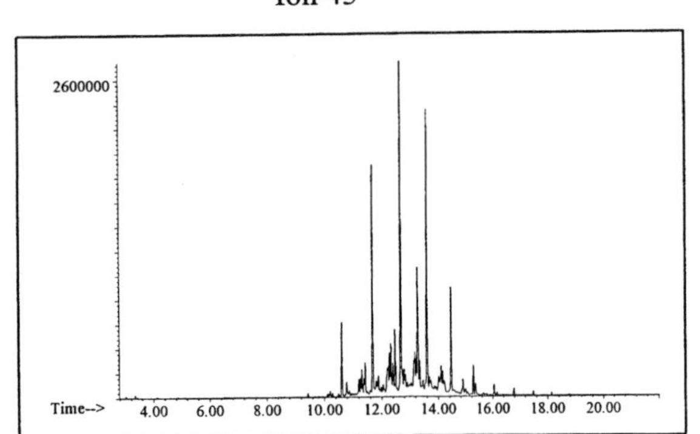

Ion 57

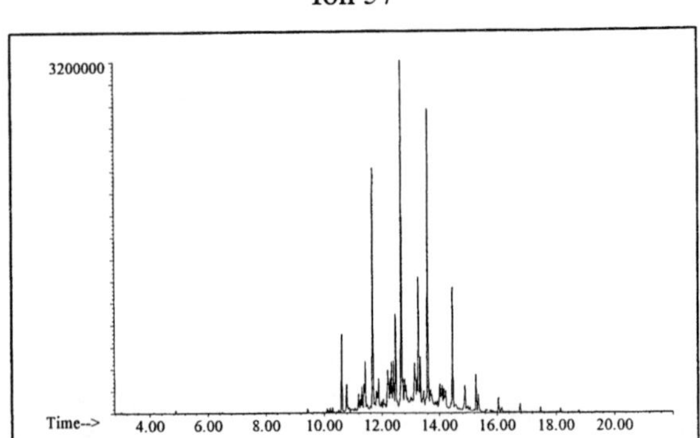

Ion 71

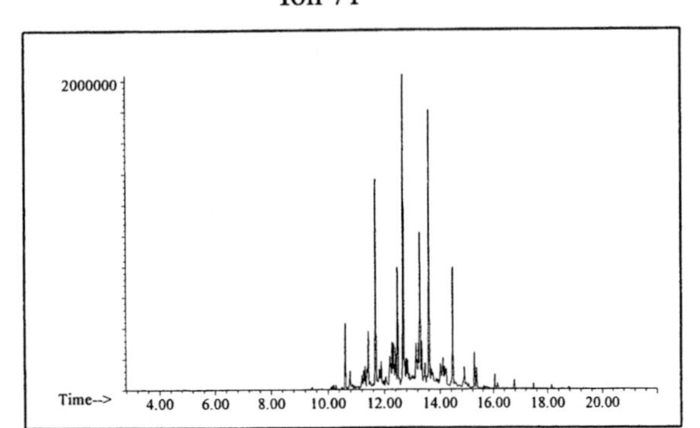

Ion 85

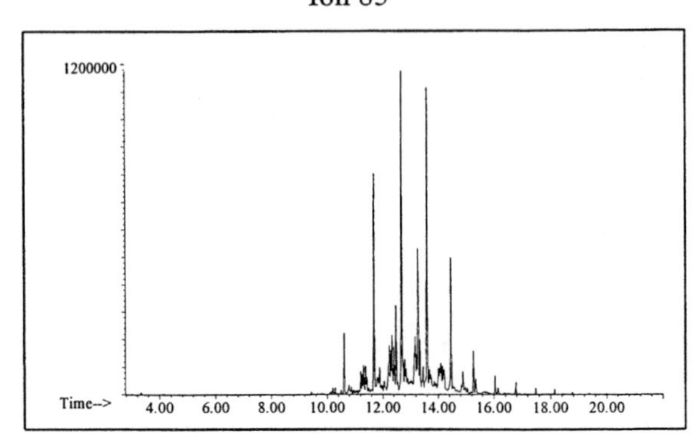

Aromatic

Ion 91

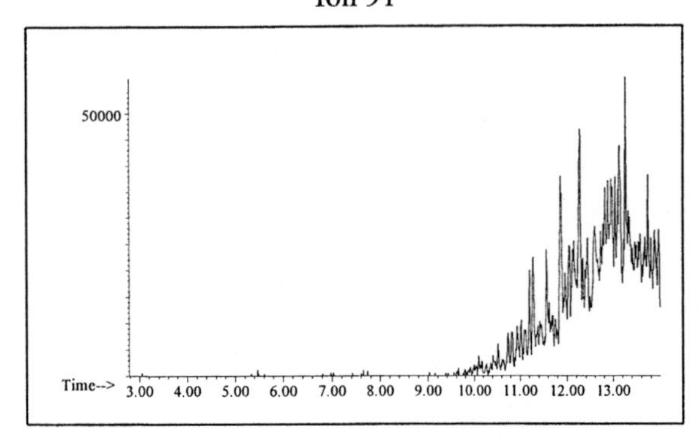

Ion 105

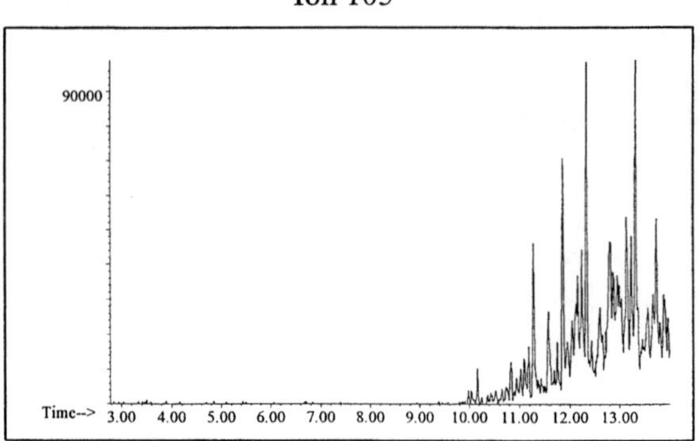

Ion 119

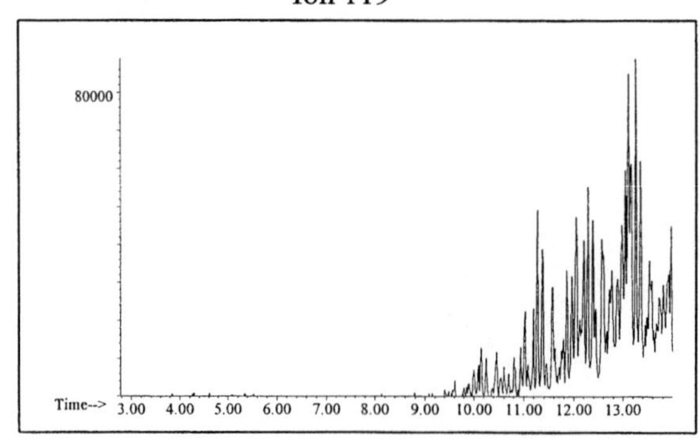

INDIVIDUAL PROFILES

Distillate #39

Cycloparaffin

Ion 55

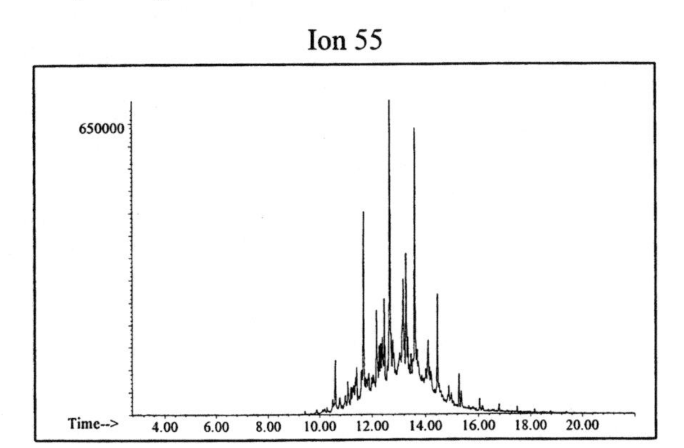

Ion 69

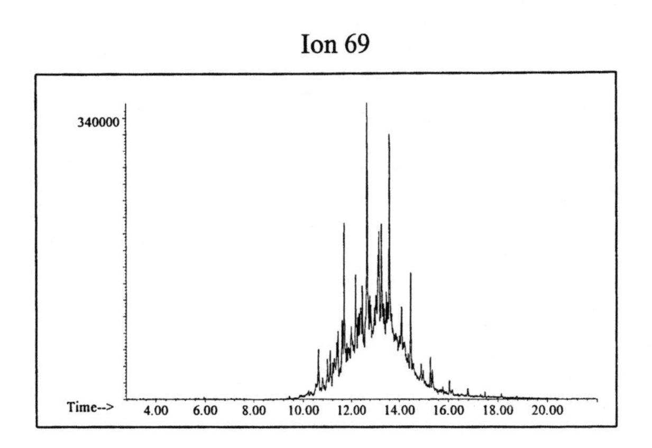

Ion 83

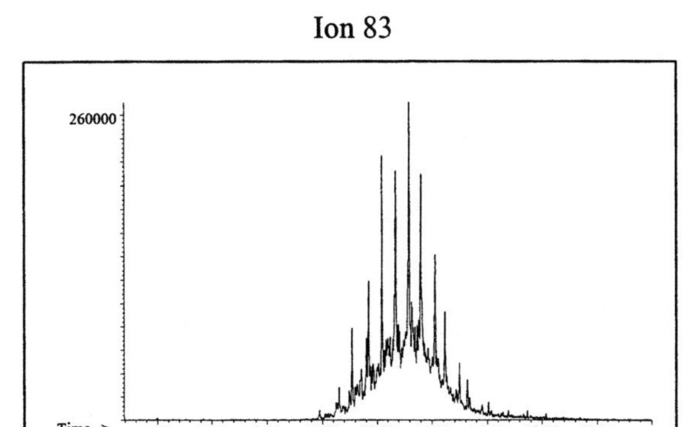

Naphthalene

Ion 128

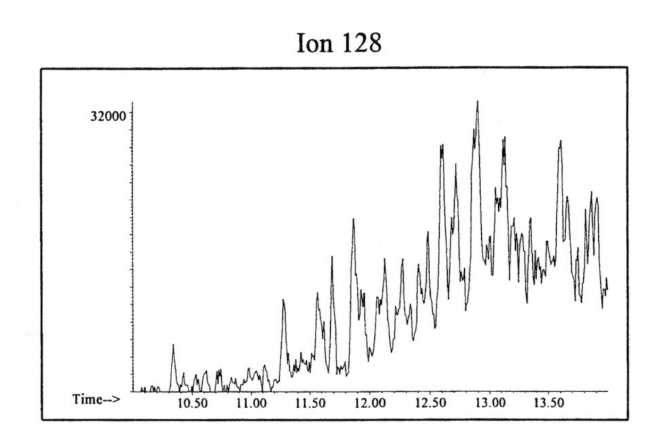

Ion 142

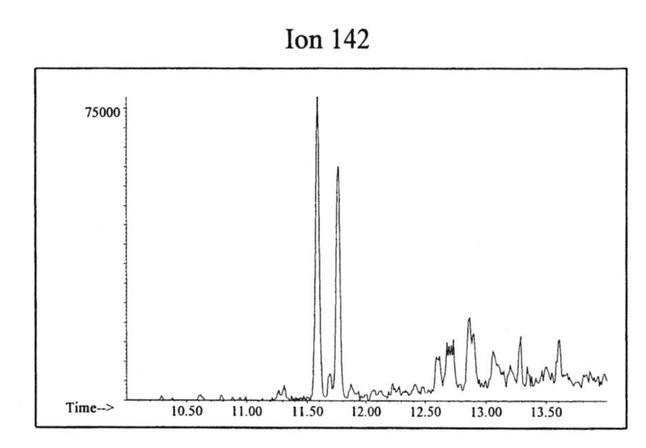

Ion 156

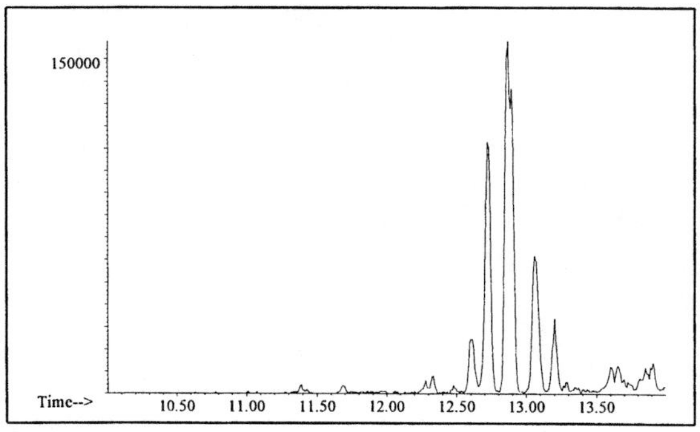

150000

Time-->

10.50 11.00 11.50 12.00 12.50 13.00 13.50

Distillate #40

20uL/mL Pentane

Product Displayed: Ozark Lighter Fluid

Other Similar Products:

ASTM: Class 4 (Kerosene)

Product Uses: Charcoal lighter

Macro Code: H

SUMMED PROFILES

Distillate #40

ALKANES

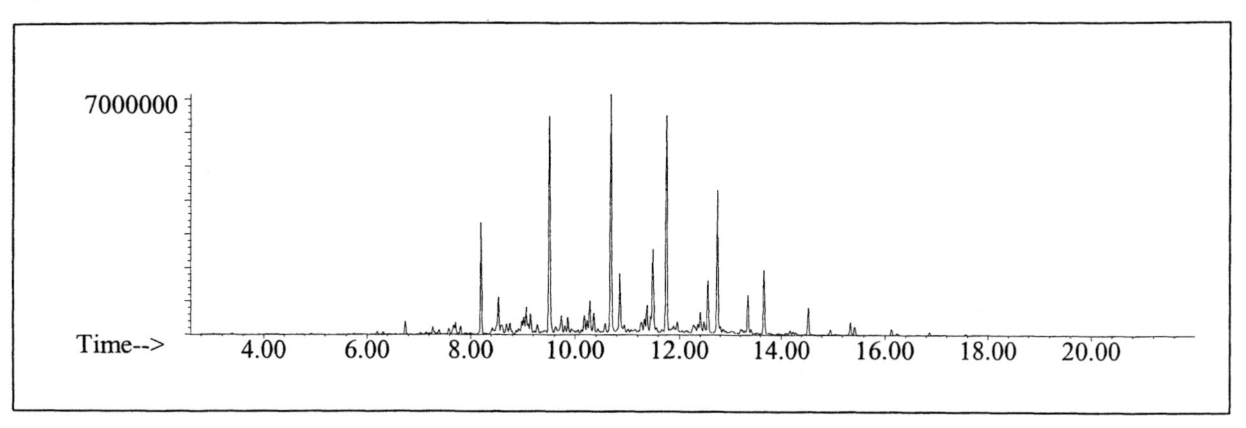

AROMATICS

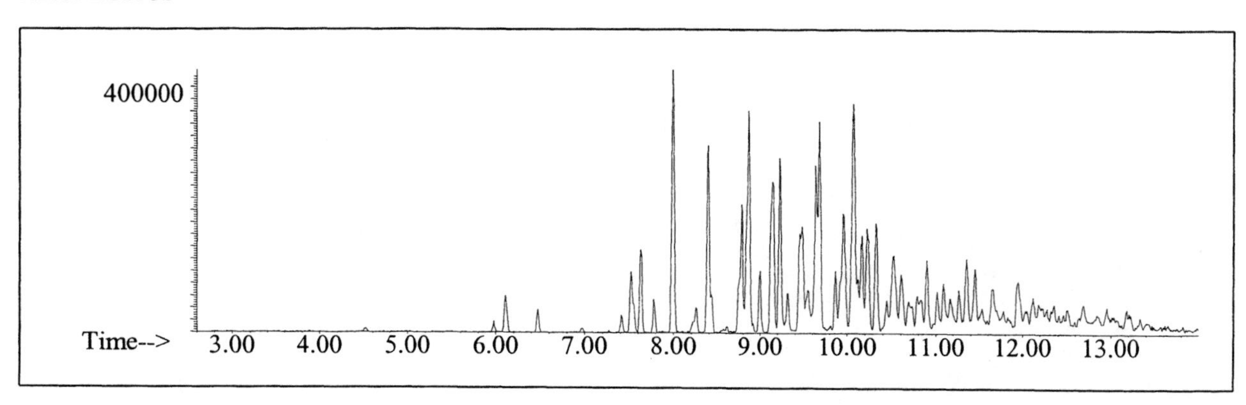

CYCLOPARAFFINS AND ALKENES

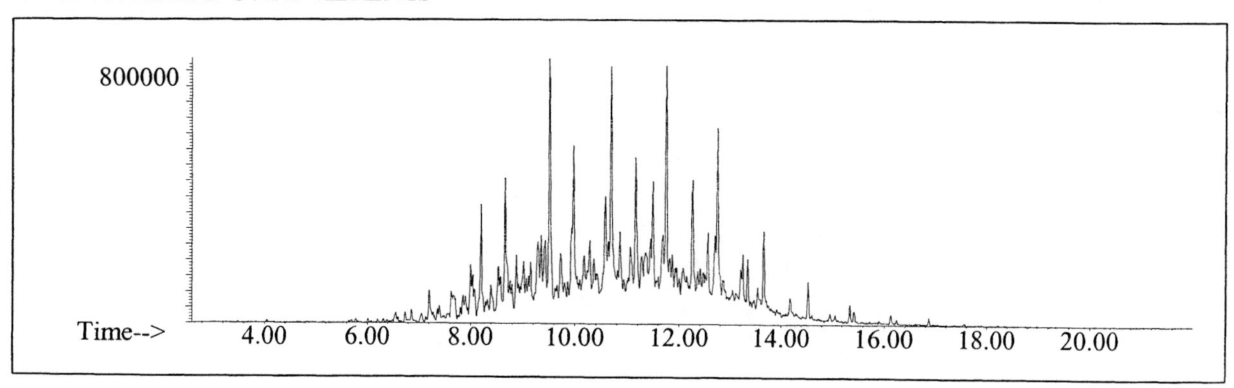

NAPHTHALENES

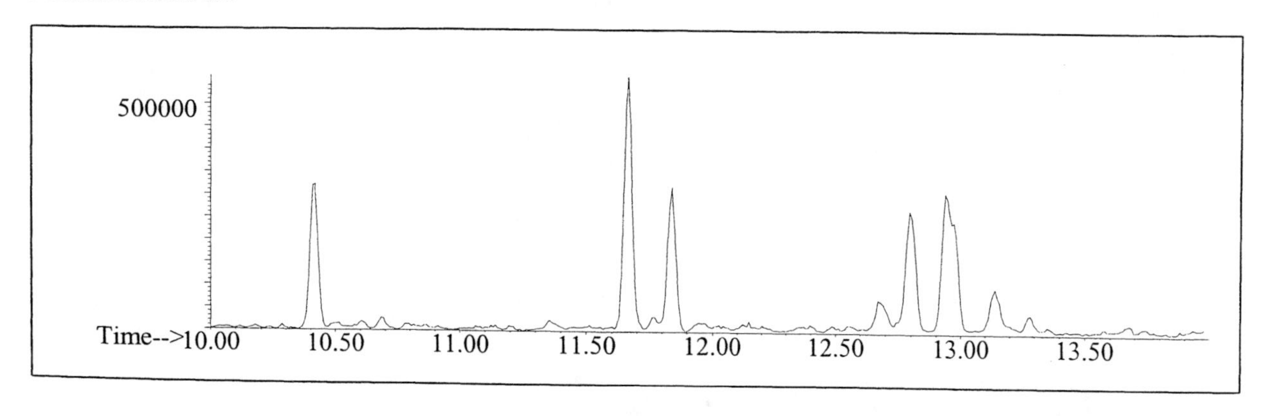

INDIVIDUAL PROFILES

Distillate #40

Alkane

Ion 43

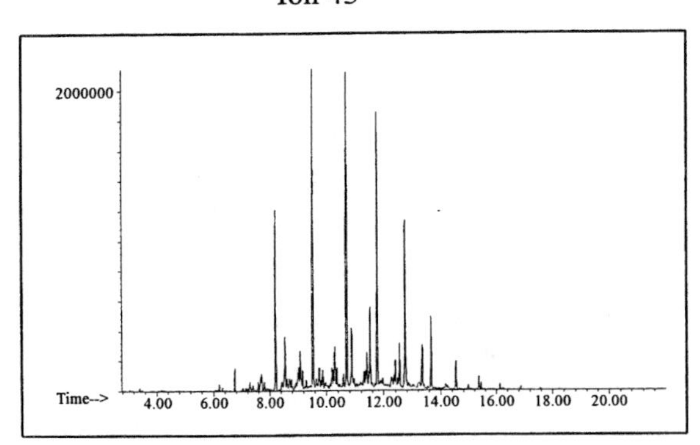

Ion 57

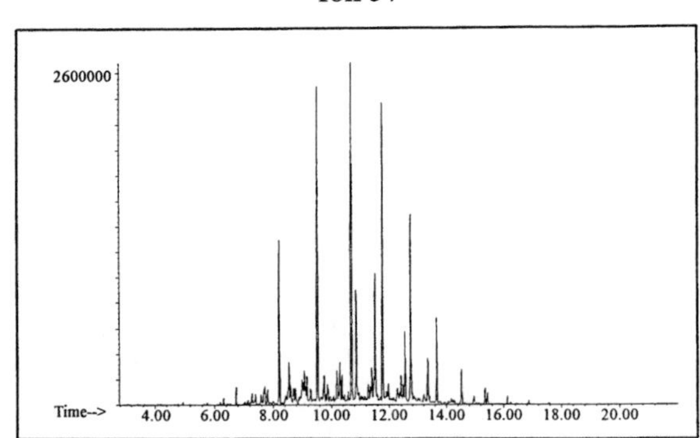

Ion 71

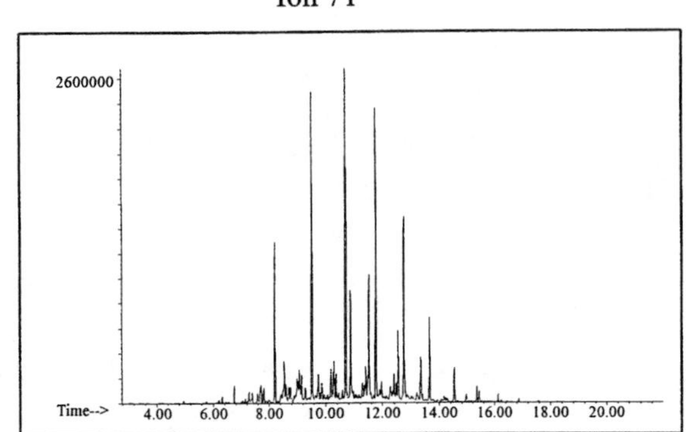

Ion 85

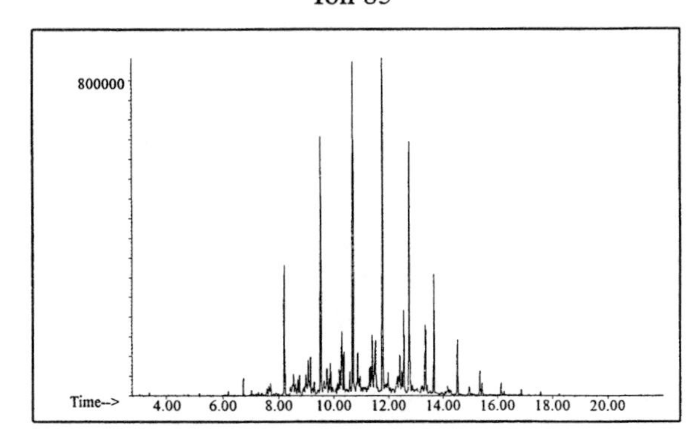

Aromatic

Ion 91

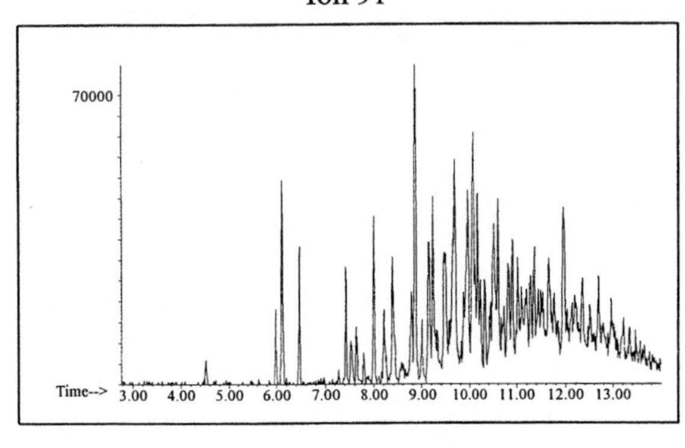

Ion 105

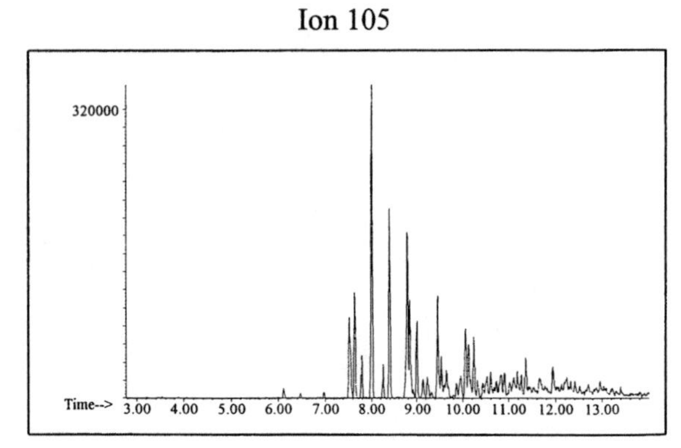

Ion 119

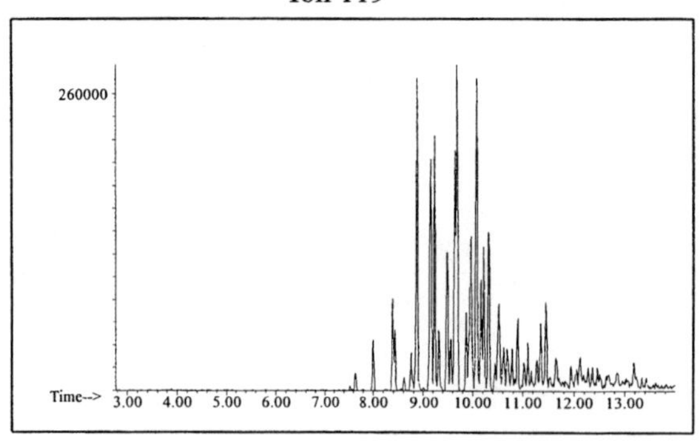

INDIVIDUAL PROFILES

Distillate #40

Cycloparaffin

Ion 55

Ion 69

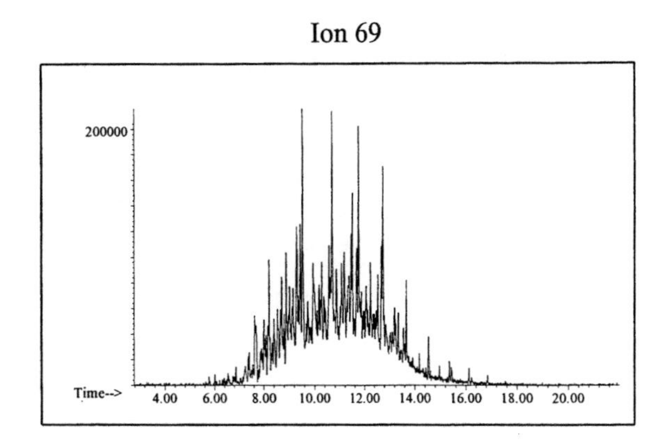

Ion 83

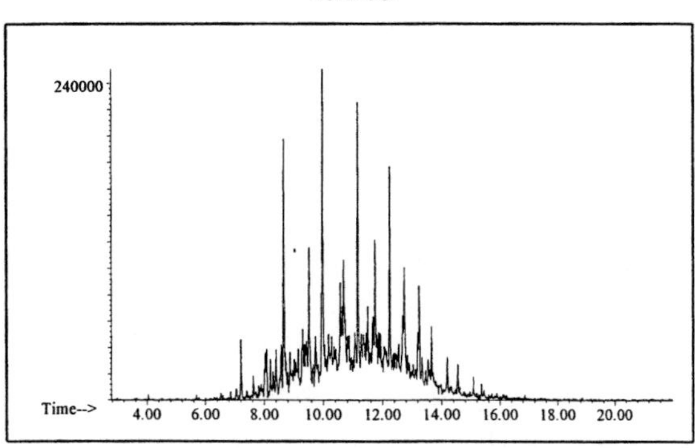

Naphthalene

Ion 128

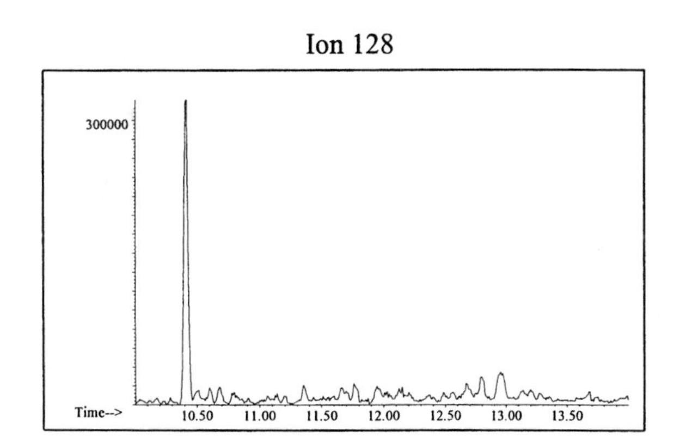

Ion 142

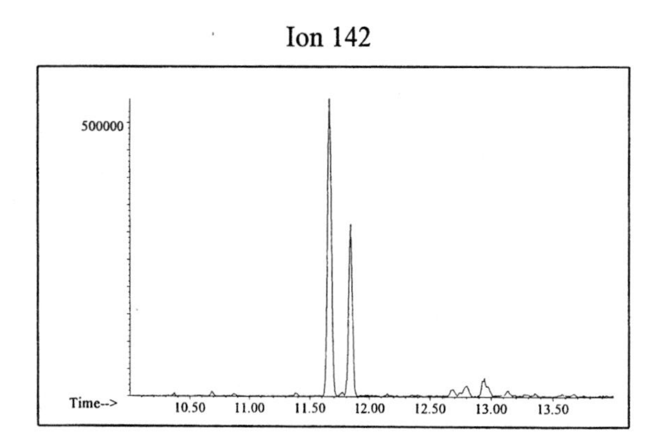

Ion 156

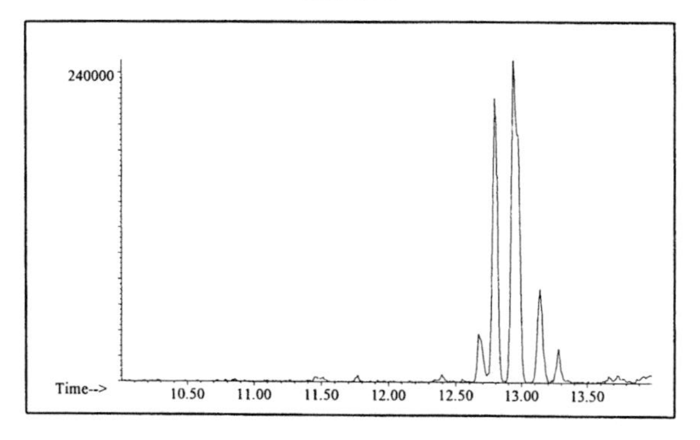

Distillate #41

20uL/mL Pentane
Product Displayed: JP-7
Other Similar Products: Xerox Film Remover

ASTM: Class 4 (Kerosene)
Product Uses: Jet fuel, solvent

Macro Code: M

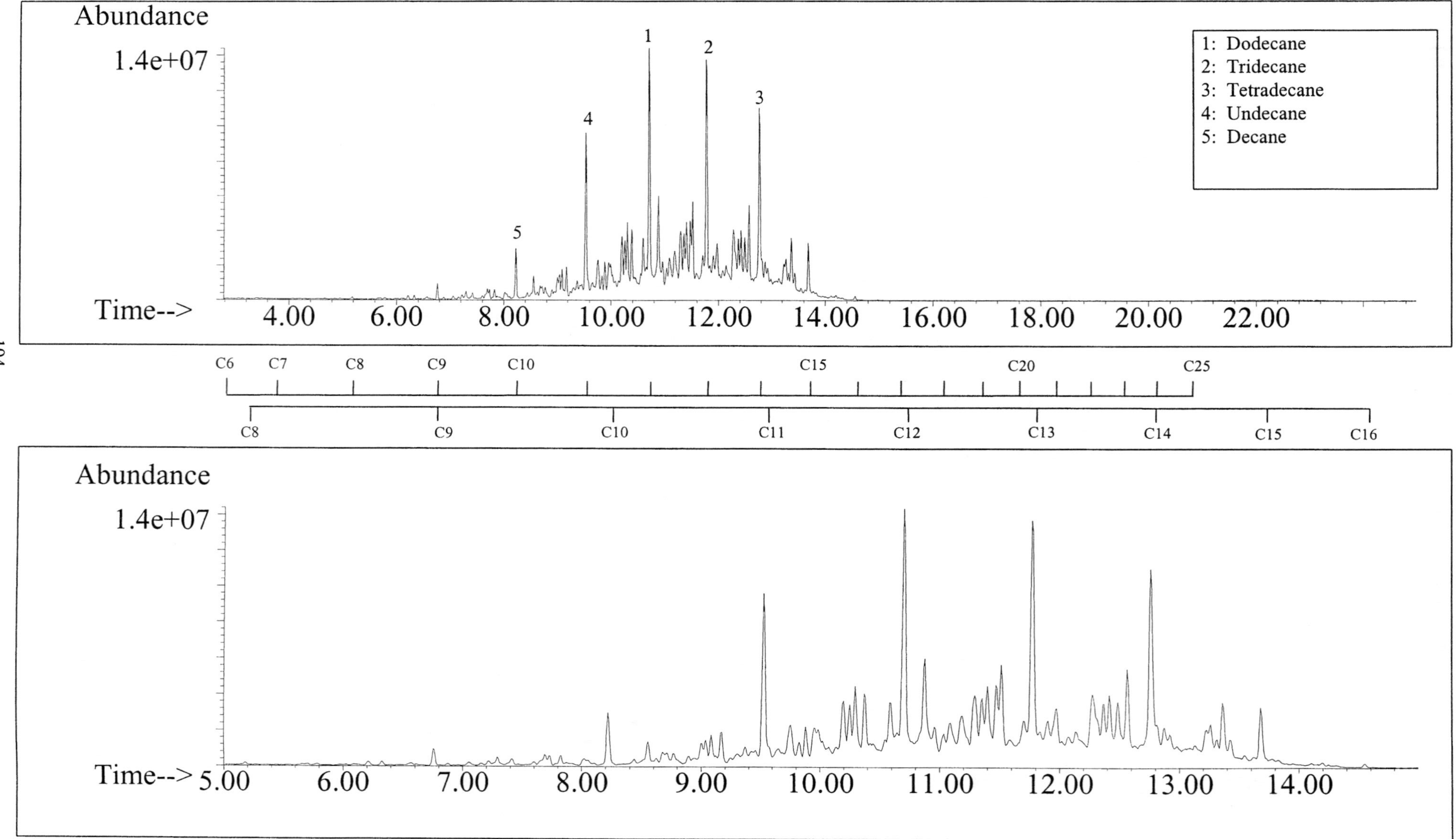

SUMMED PROFILES

Distillate #41

ALKANES

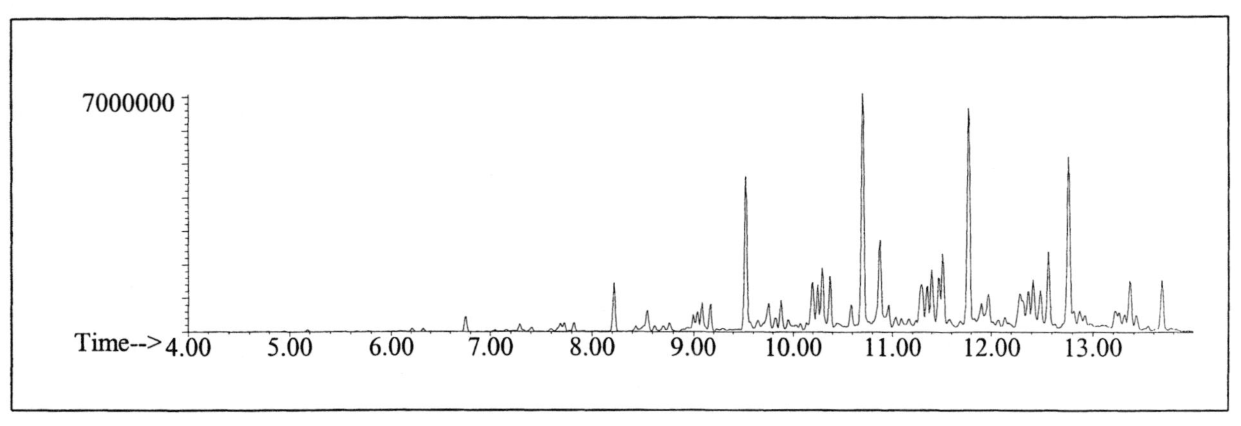

AROMATICS

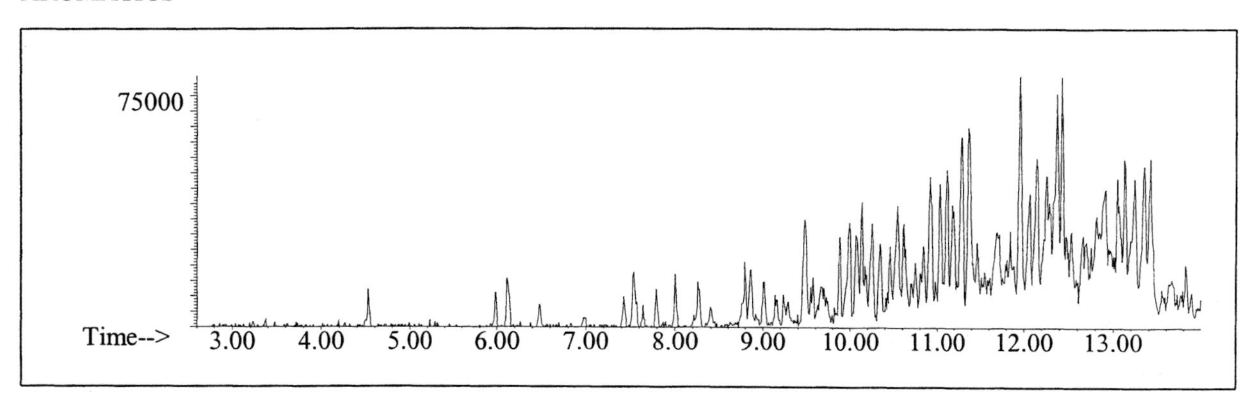

CYCLOPARAFFINS AND ALKENES

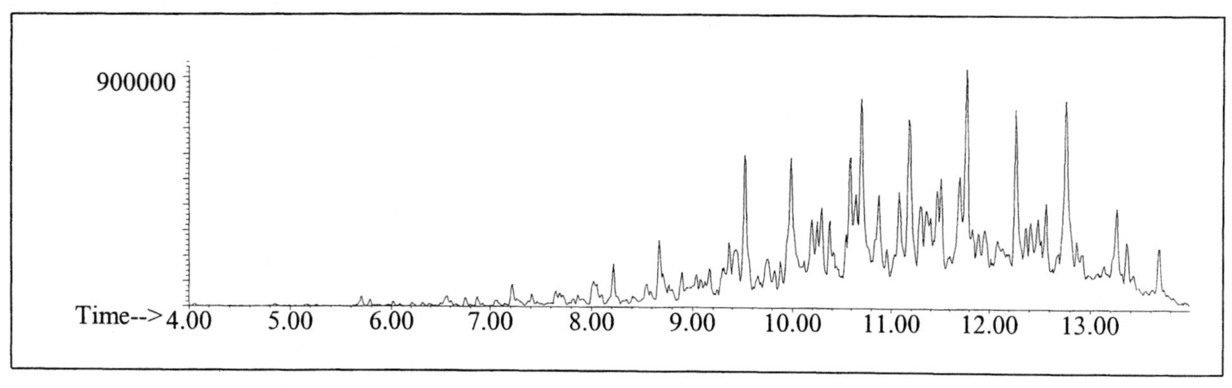

NAPHTHALENES

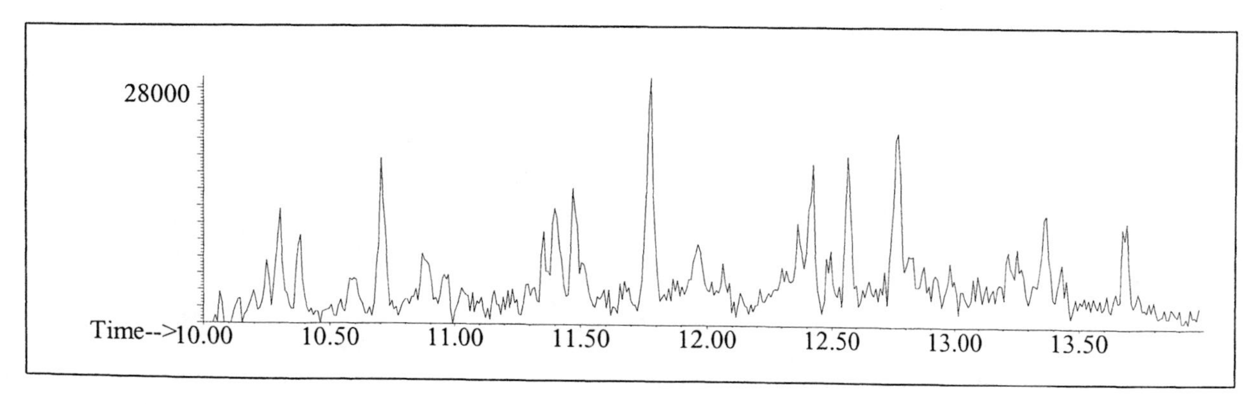

INDIVIDUAL PROFILES

Distillate #41

Alkane

Ion 43

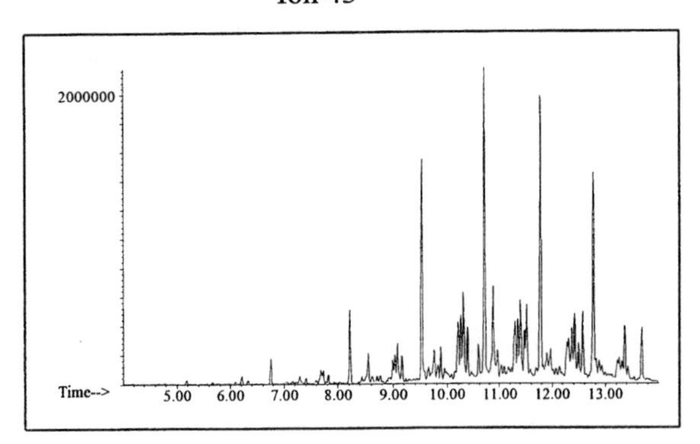

Ion 57

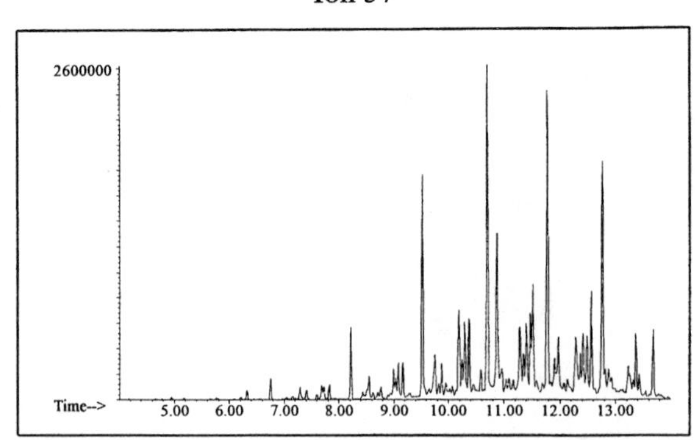

Ion 71

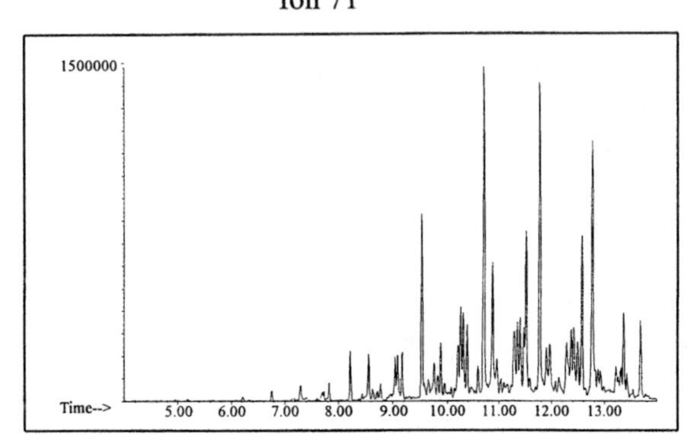

Ion 85

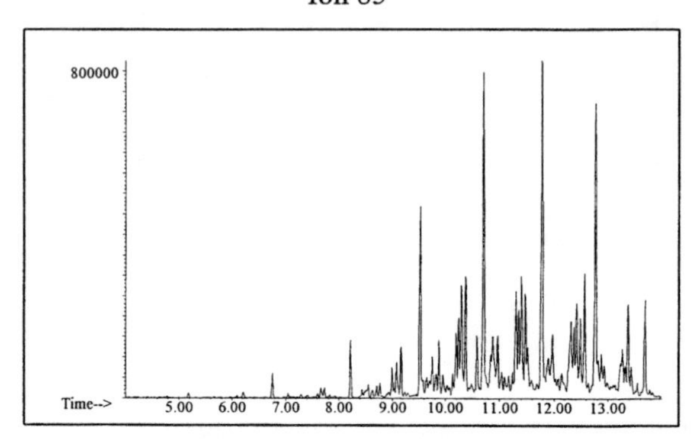

Aromatic

Ion 91

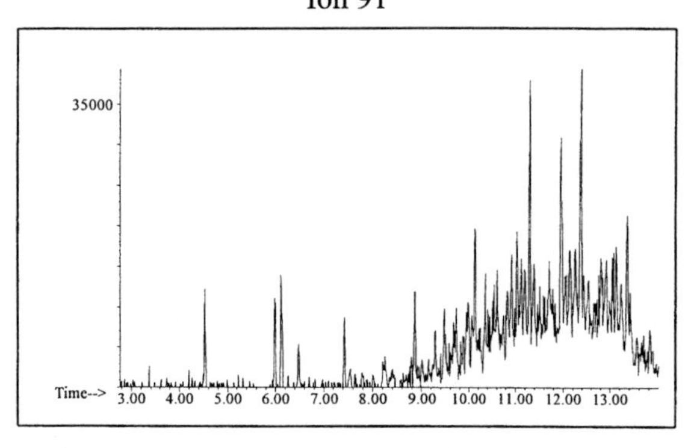

Ion 105

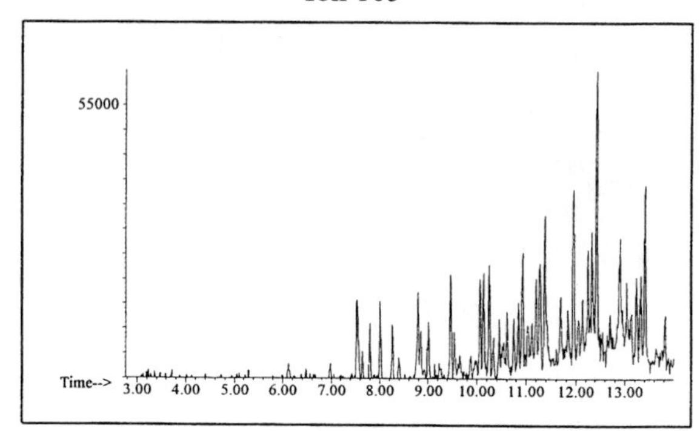

Ion 119

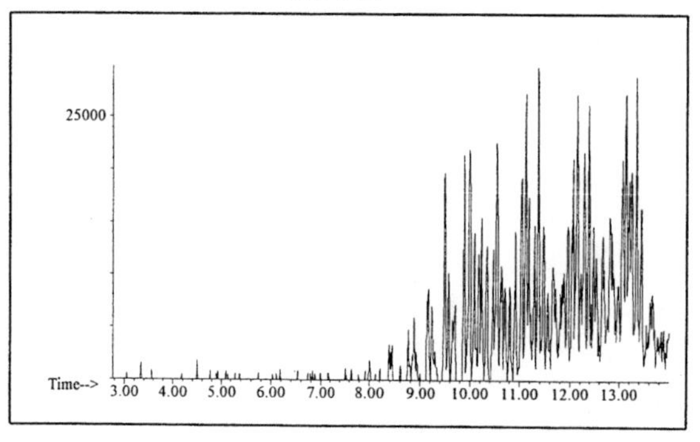

INDIVIDUAL PROFILES

Distillate #41

Cycloparaffin

Ion 55

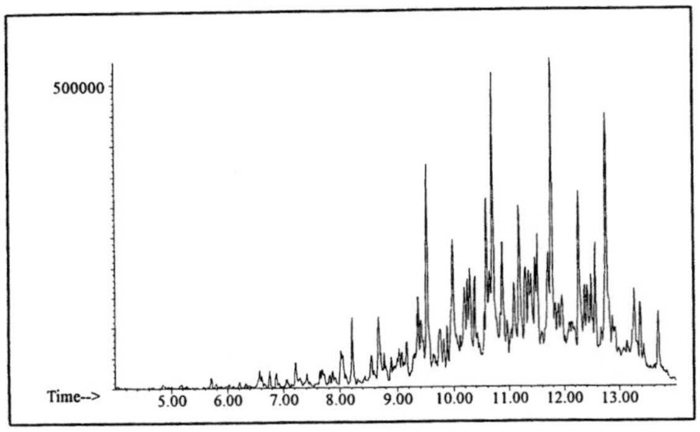

Ion 69

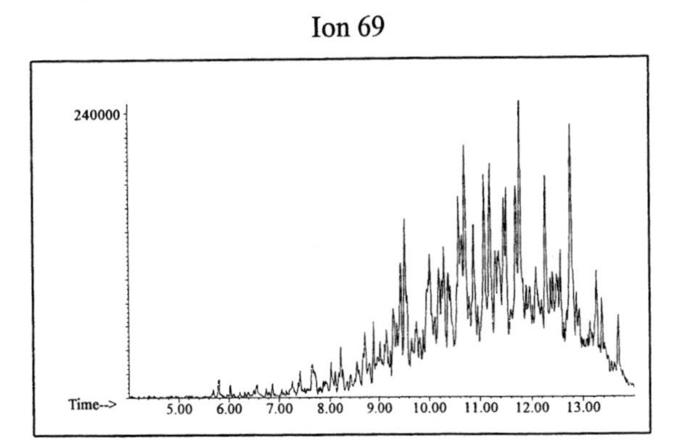

Ion 83

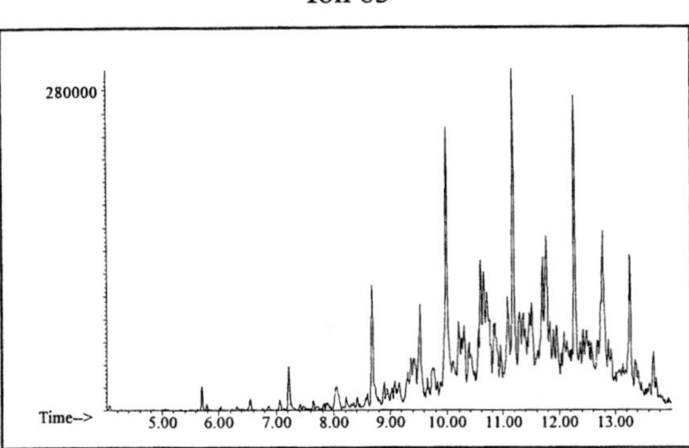

Naphthalene

Ion 128

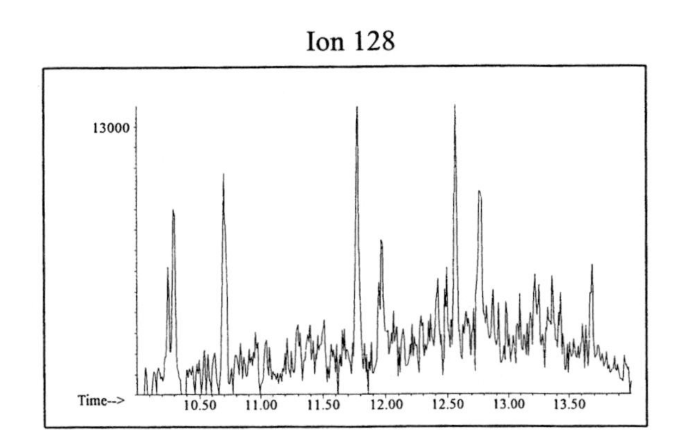

Ion 142

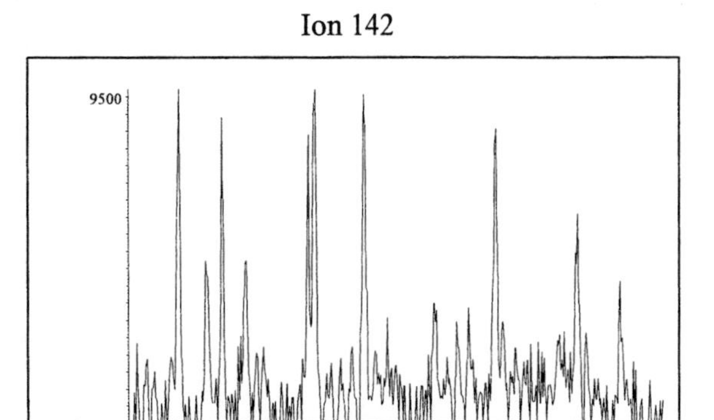

Ion 156

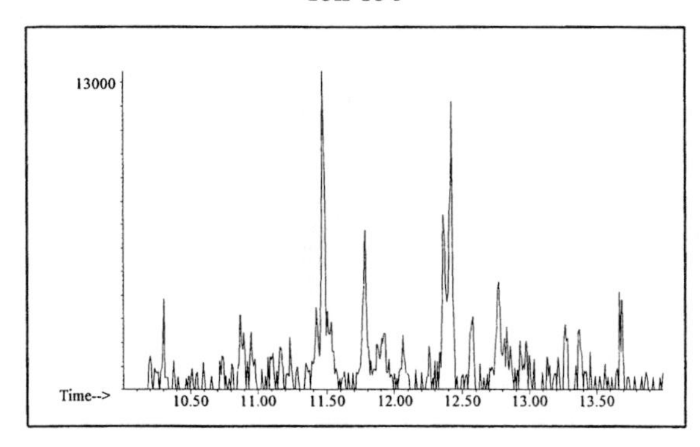

Distillate #42

20uL/mL Pentane

Product Displayed: Exxsol D80

Other Similar Products:

ASTM: Class 0 (Miscellaneous)

Product Uses: Solvent, feedstock

Macro Code: M

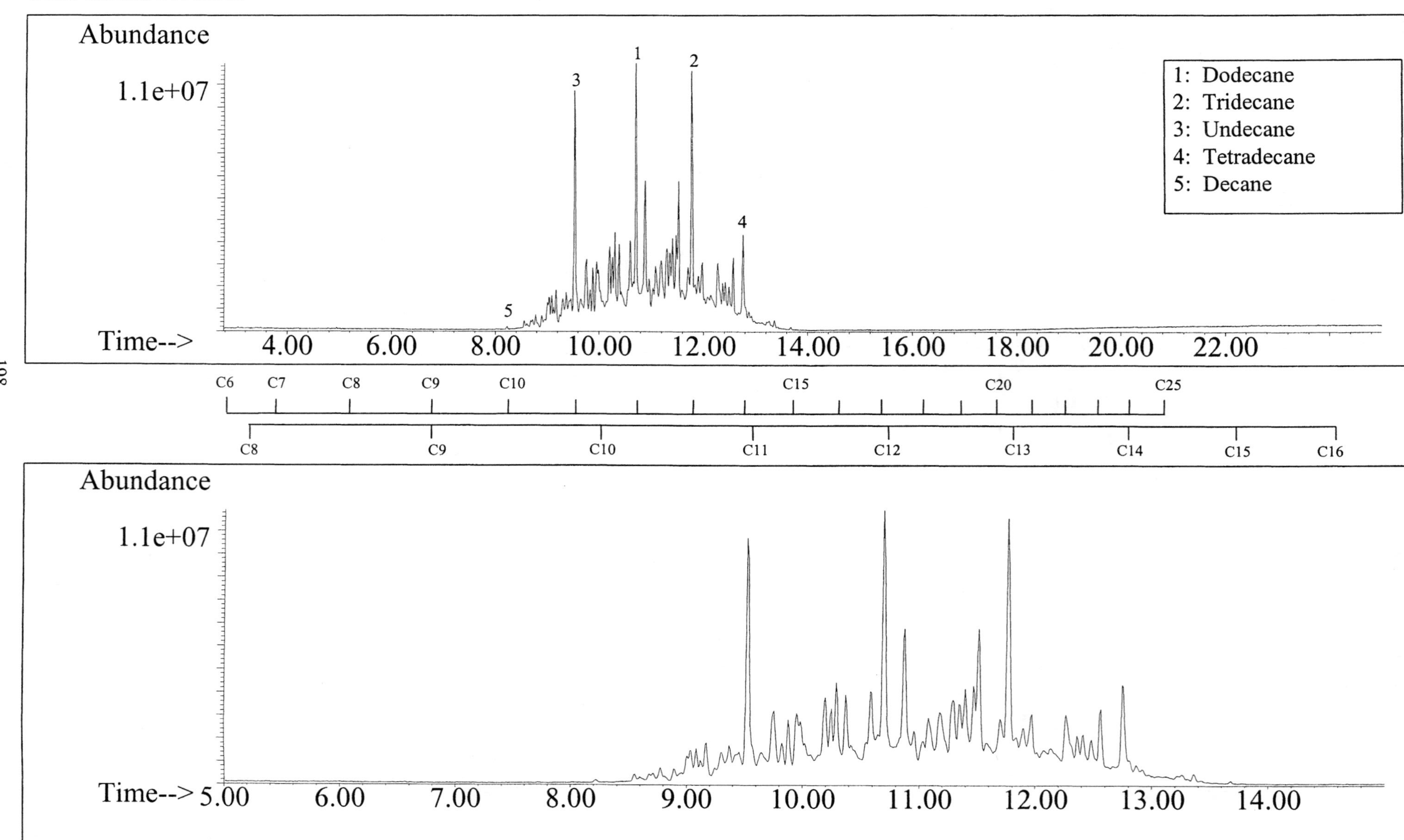

SUMMED PROFILES Distillate #42

ALKANES

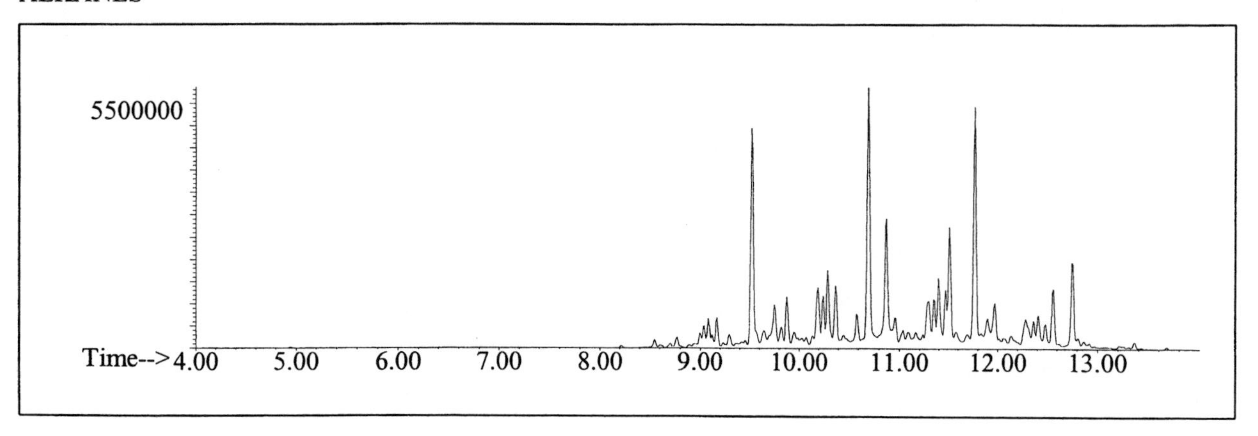

AROMATICS

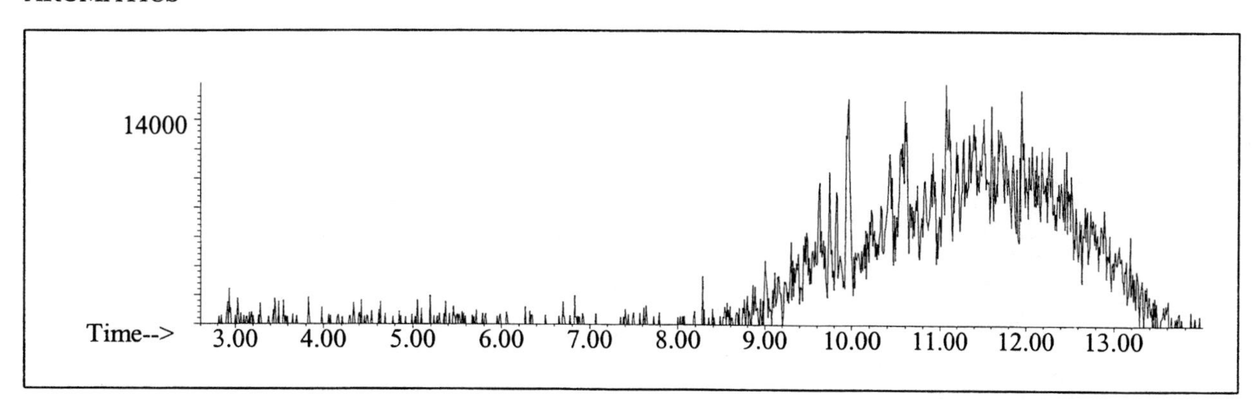

CYCLOPARAFFINS AND ALKENES

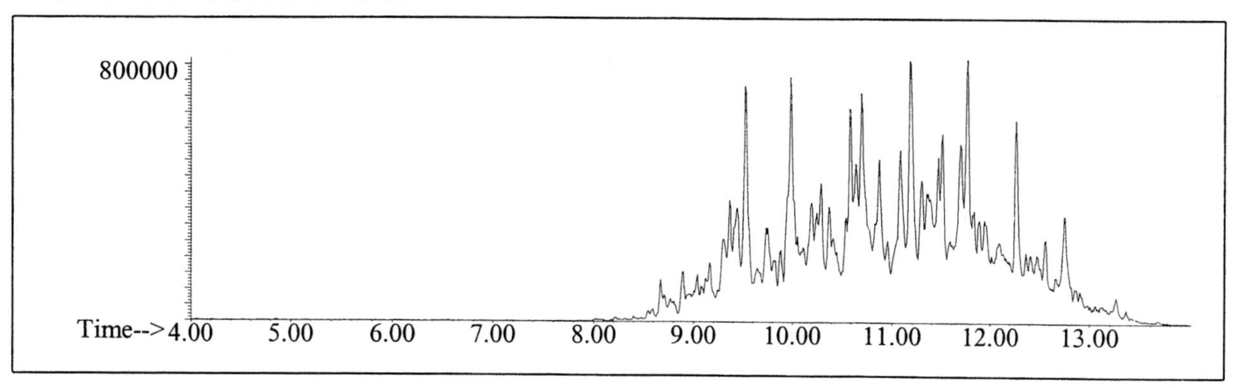

NAPHTHALENES

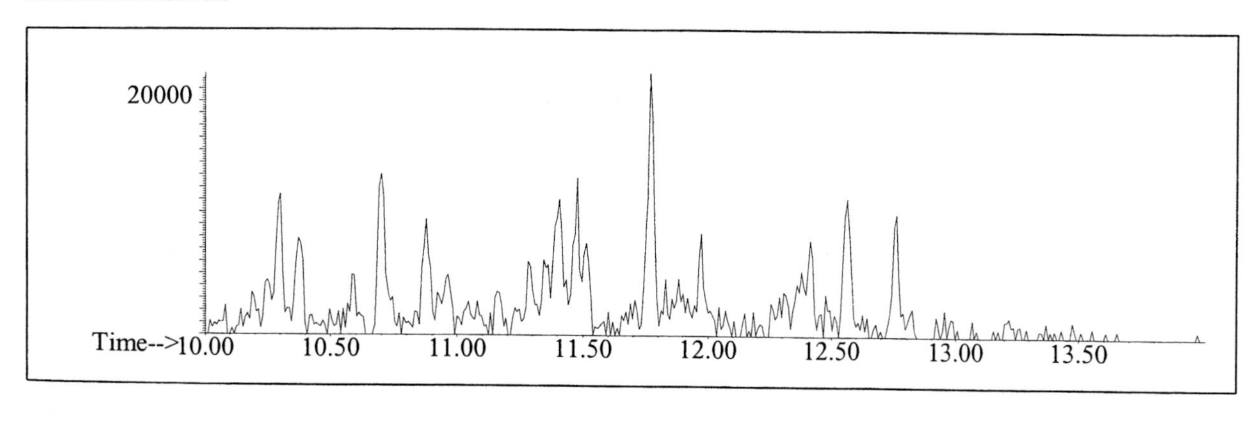

INDIVIDUAL PROFILES

Distillate #42

Alkane

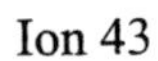
Ion 43

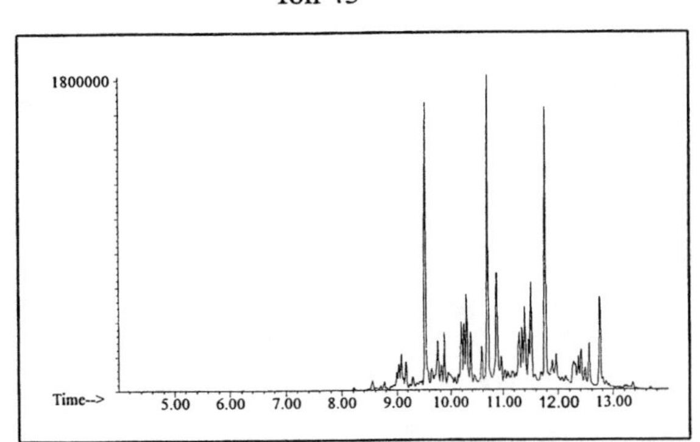

Ion 57

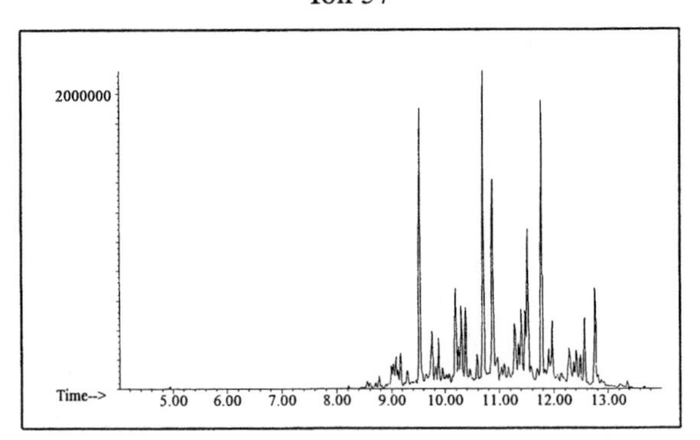

Ion 71

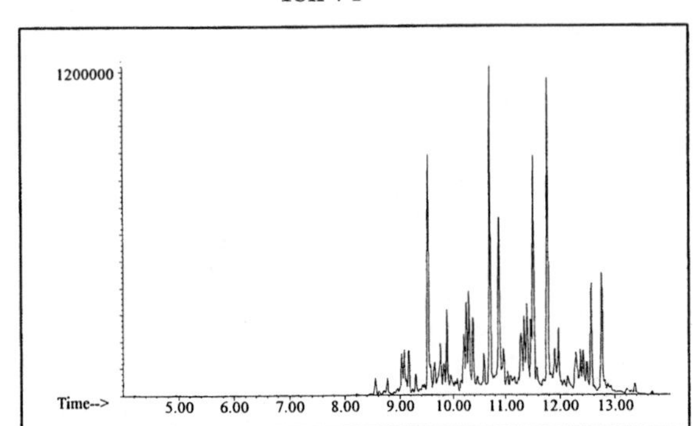

Ion 85

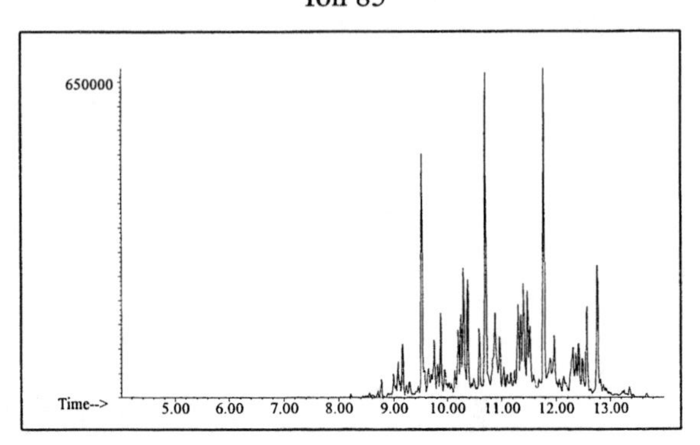

Aromatic

Ion 91

No Useful Data Obtained

Ion 105

No Useful Data Obtained

Ion 119

No Useful Data Obtained

INDIVIDUAL PROFILES

Cycloparaffin

Ion 55

Ion 69

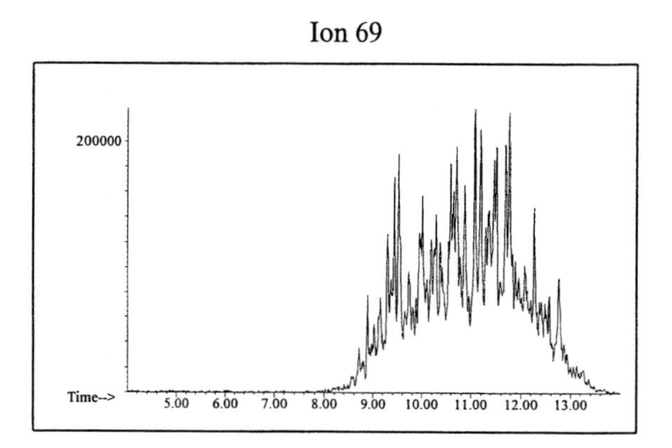

Ion 83

Naphthalene

Ion 128

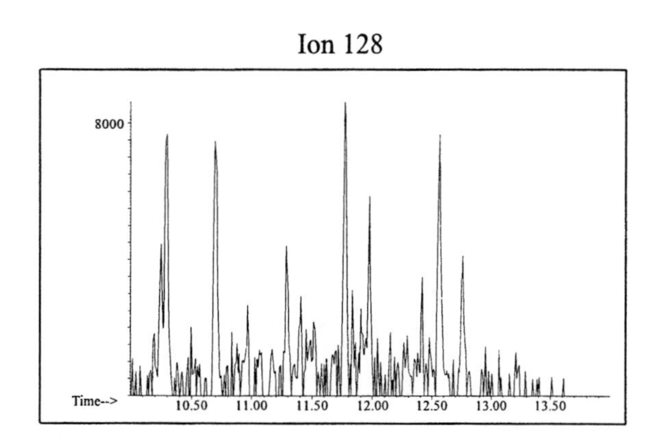

Ion 142

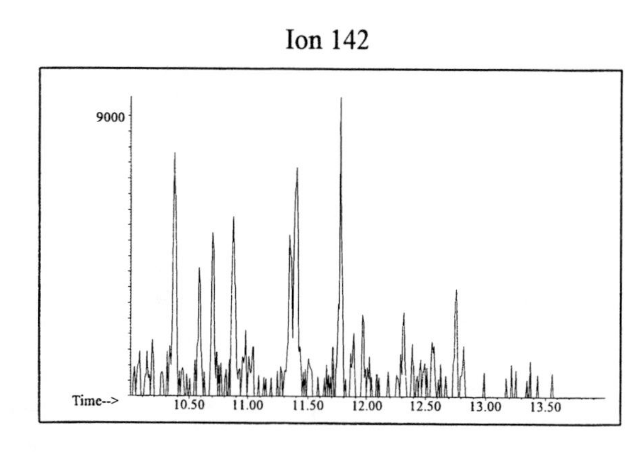

Ion 156

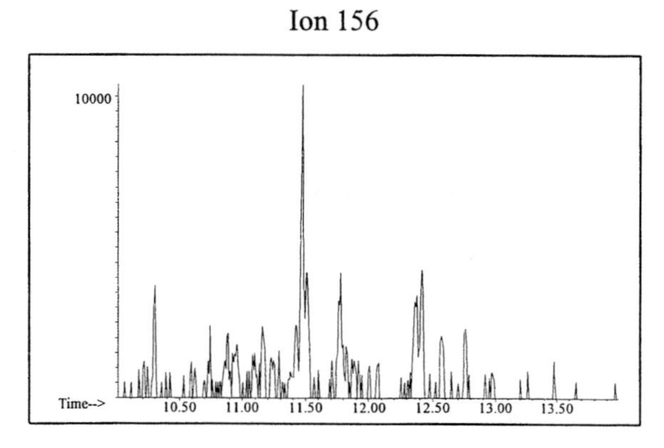

Distillate #43

20uL/mL Pentane

ASTM: Class 4 (Kerosene)

Macro Code: H

Product Displayed: Classic Glo-Light Torch Fuel

Product Uses: Torch fuel, lamp oil

Other Similar Products:

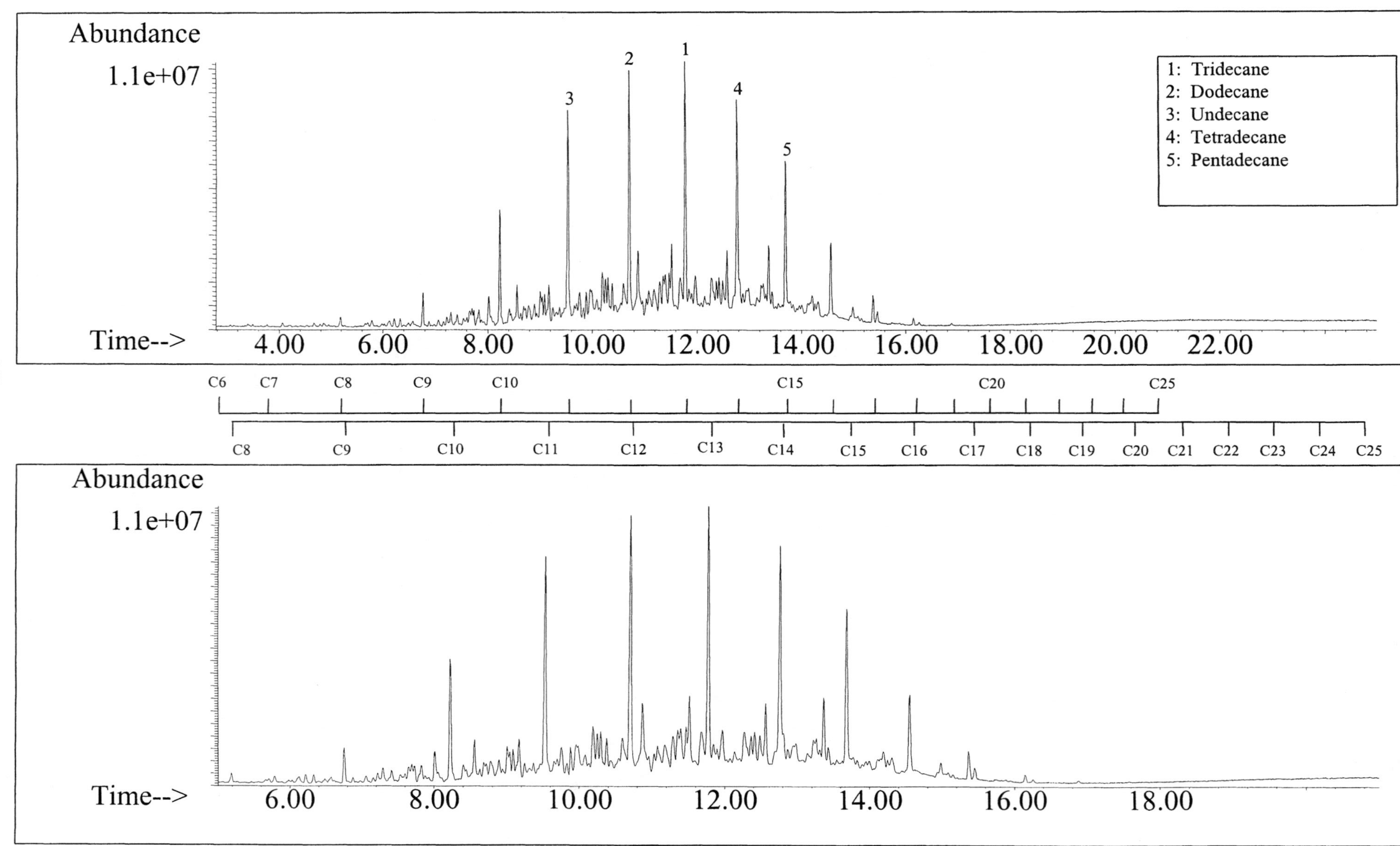

SUMMED PROFILES Distillate #43

ALKANES

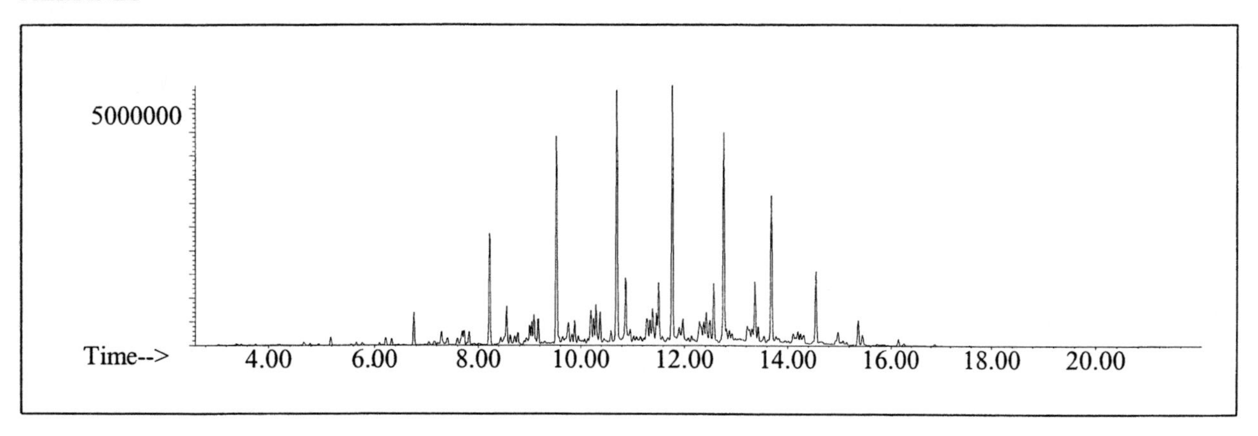

AROMATICS

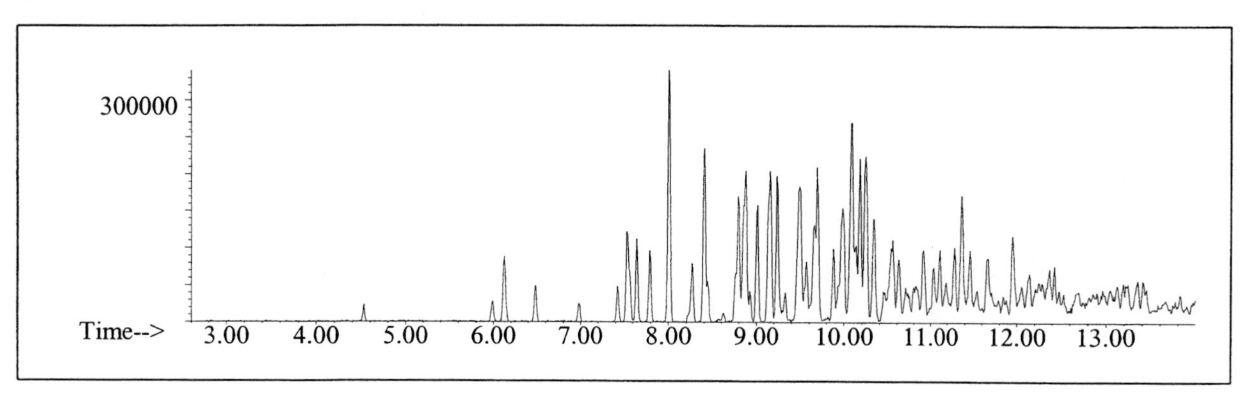

CYCLOPARAFFINS AND ALKENES

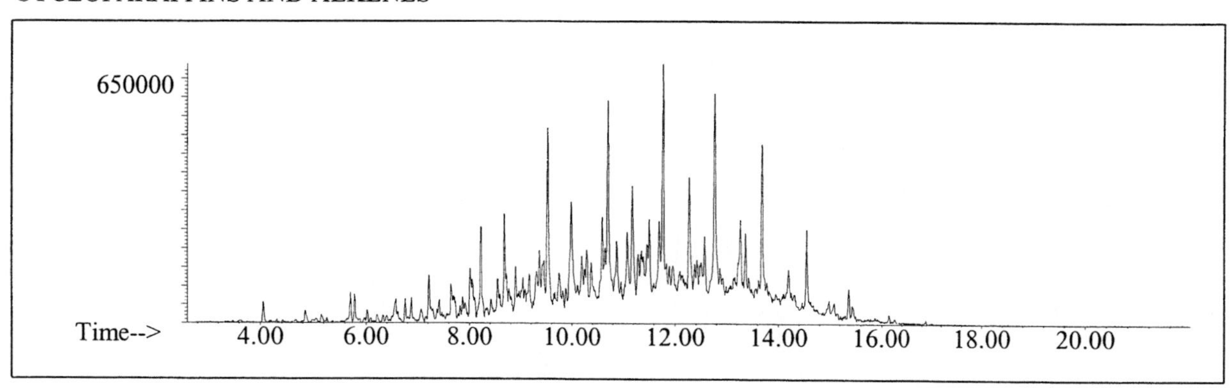

NAPHTHALENES

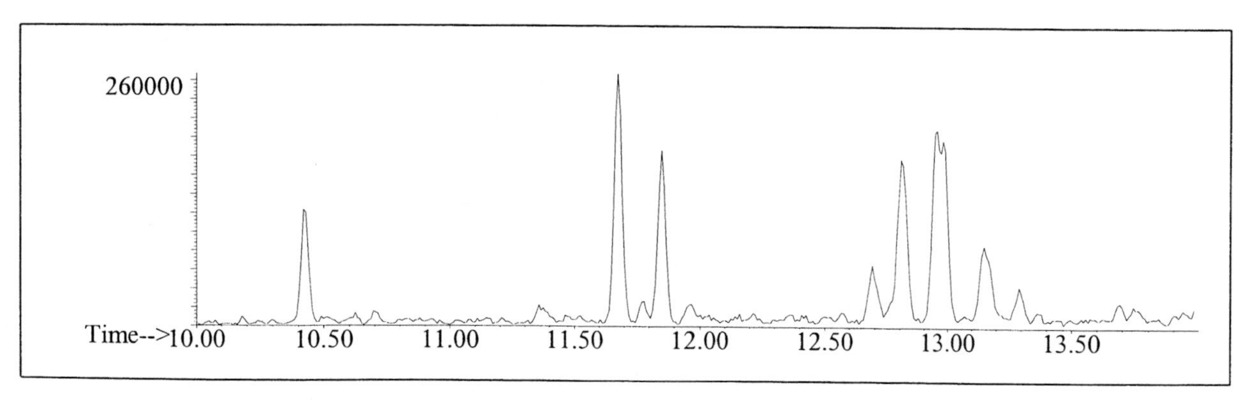

INDIVIDUAL PROFILES

Distillate #43

Alkane

Ion 43

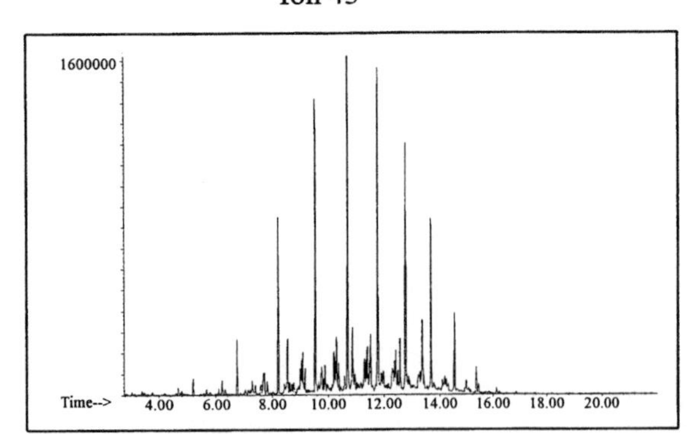

Ion 57

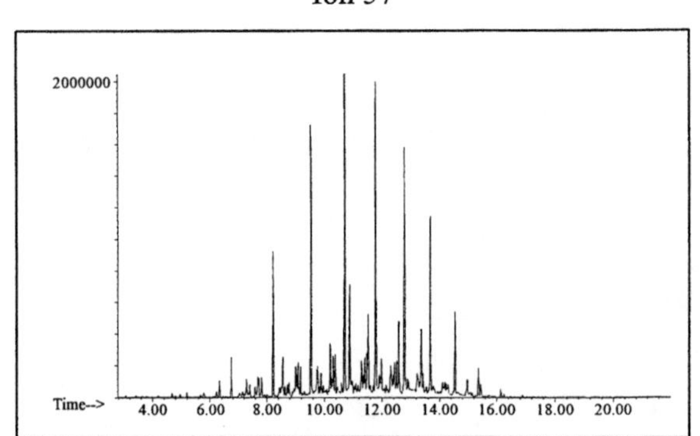

Ion 71

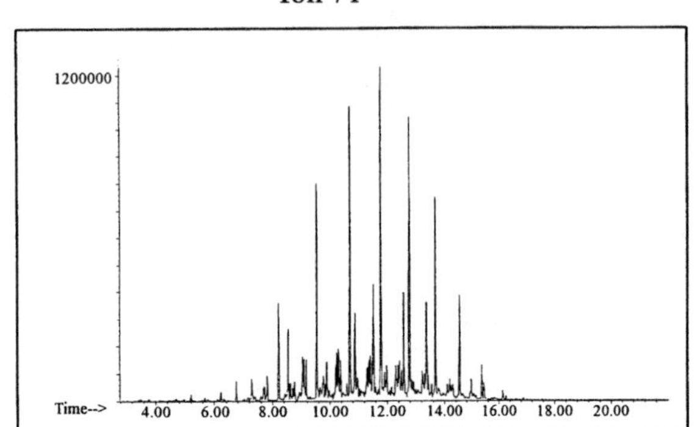

Ion 85

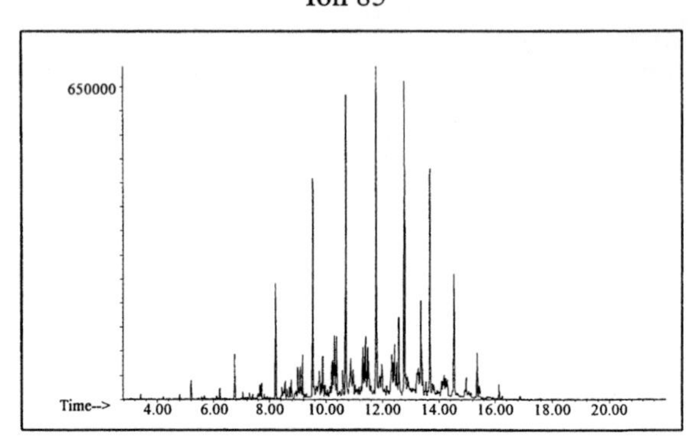

Aromatic

Ion 91

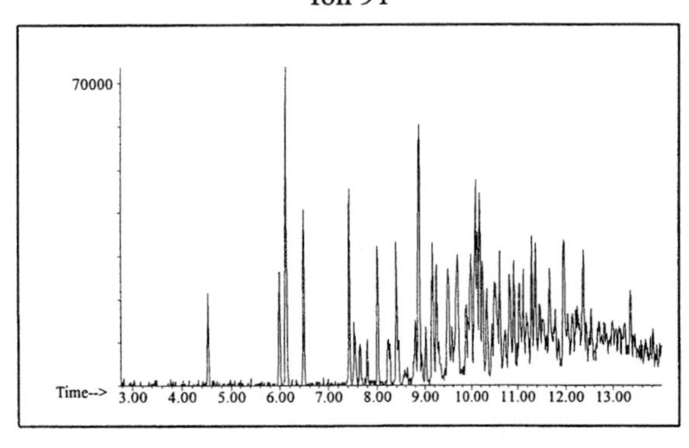

Ion 105

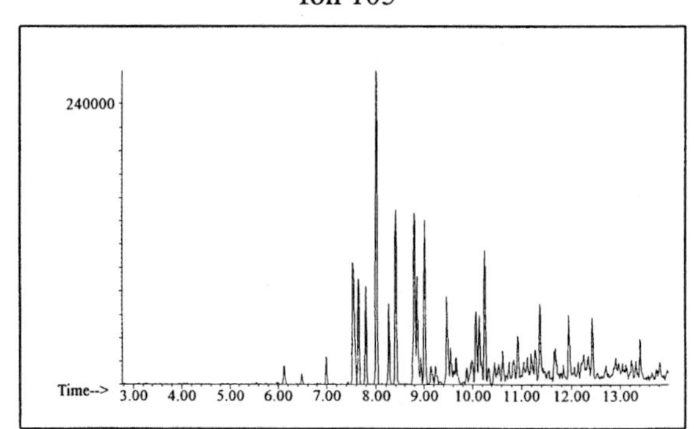

Ion 119

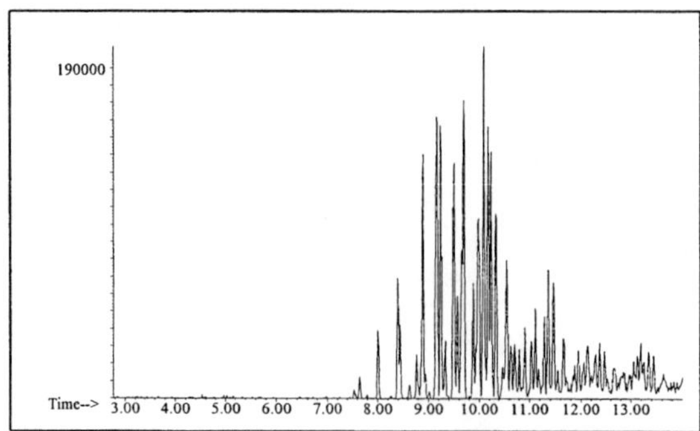

INDIVIDUAL PROFILES

Distillate #43

Cycloparaffin

Ion 55

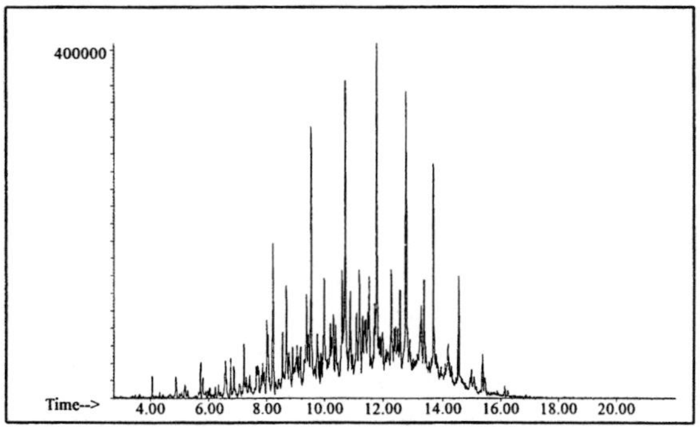

Ion 69

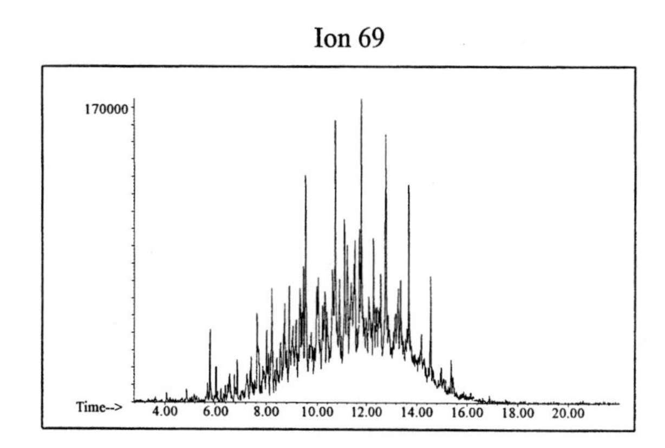

Ion 83

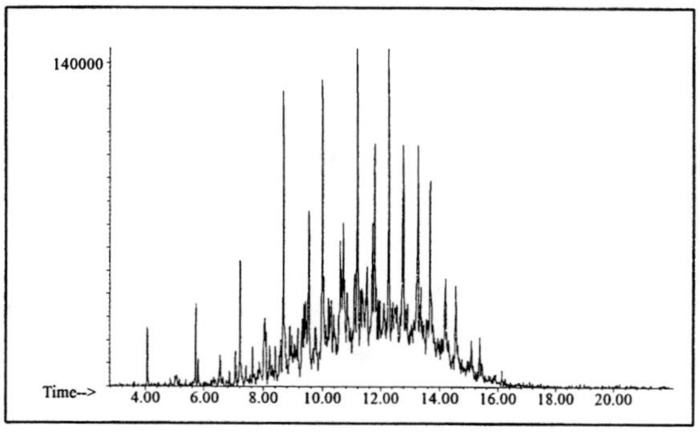

Naphthalene

Ion 128

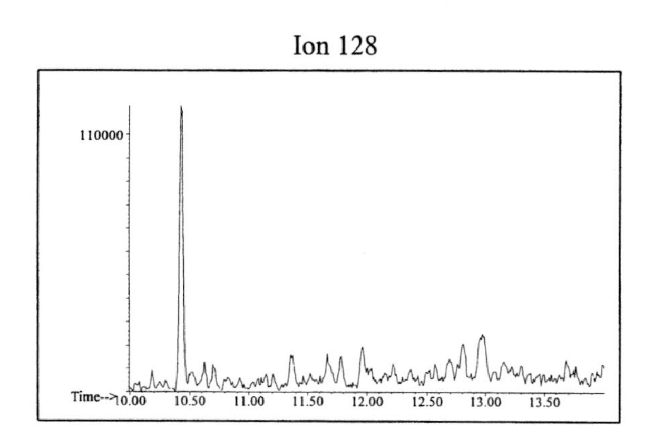

Ion 142

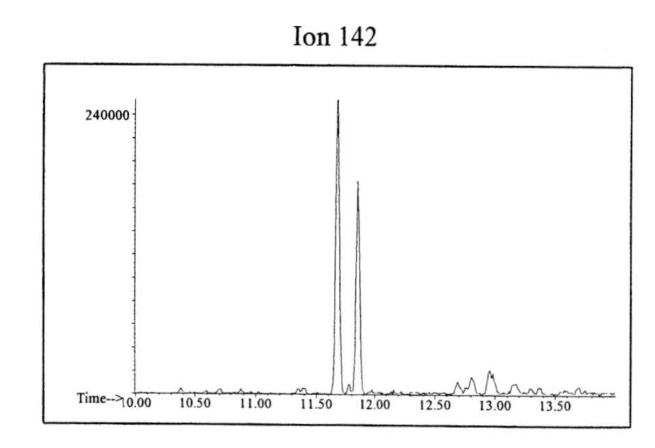

Ion 156

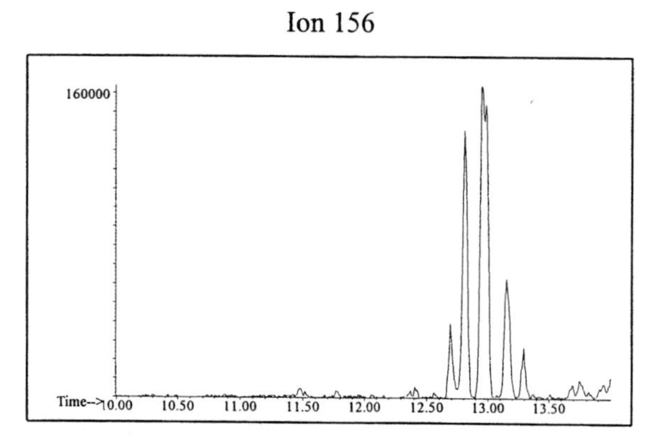

Distillate #44

20uL/mL Pentane

Product Displayed: Ozark Charcoal Lighter #2

Other Similar Products:

ASTM: Class 5 (Heavy Petroleum Distillate)

Product Uses: Charcoal lighter

Macro Code: H

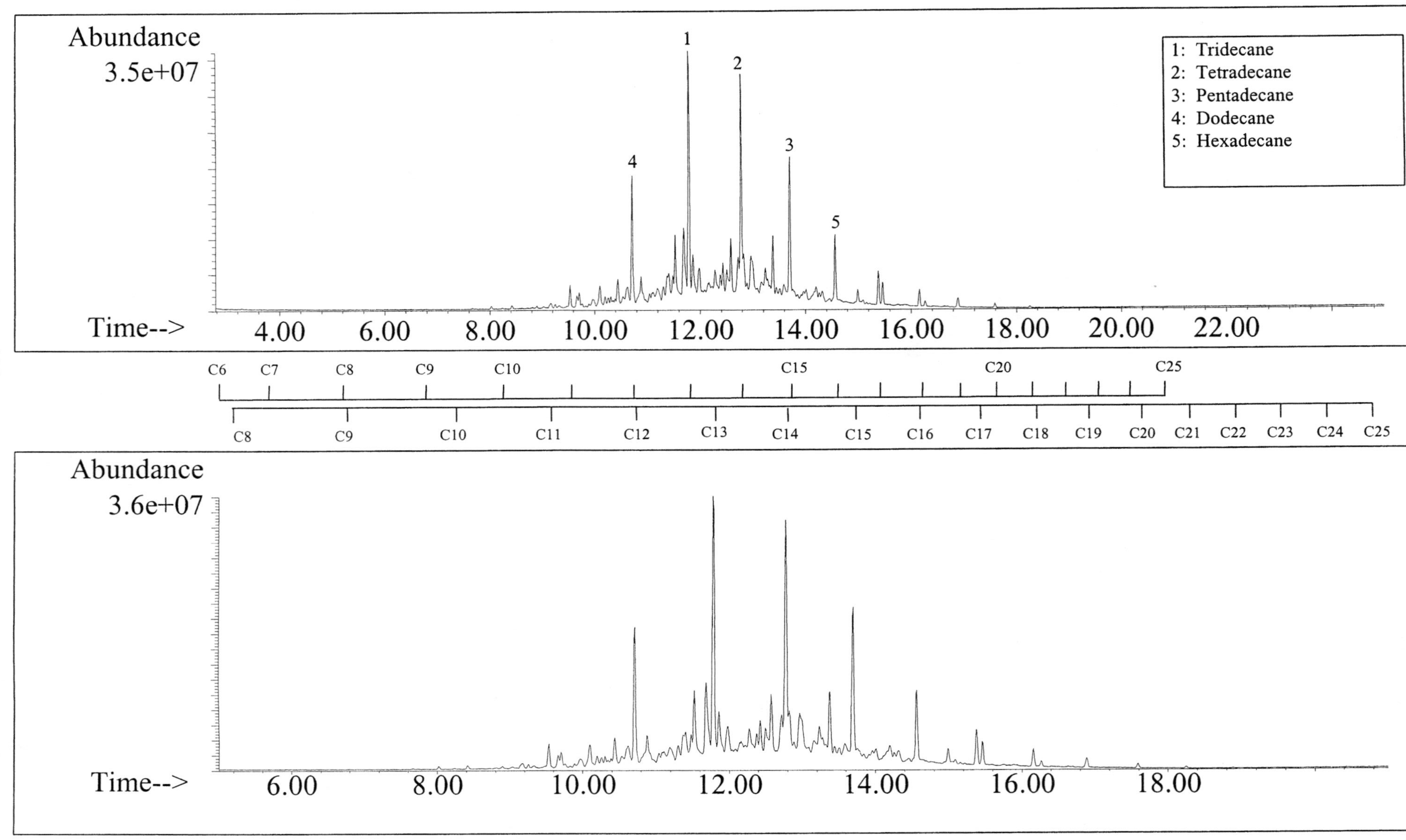

SUMMED PROFILES

Distillate #44

ALKANES

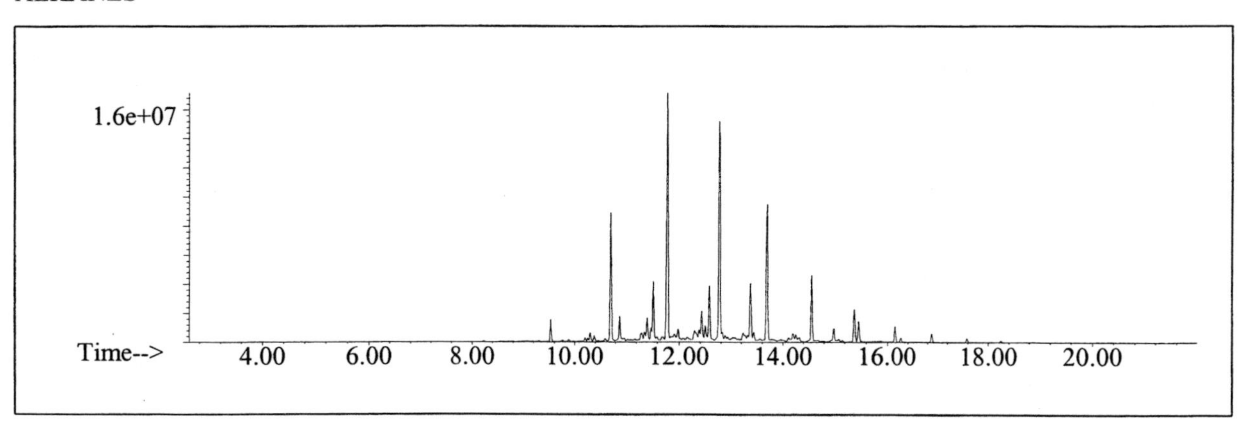

AROMATICS

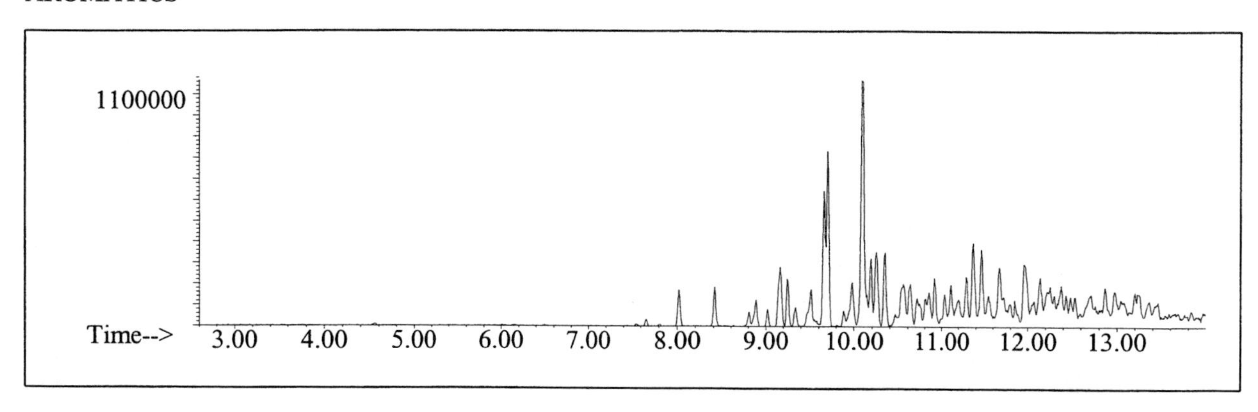

CYCLOPARAFFINS AND ALKENES

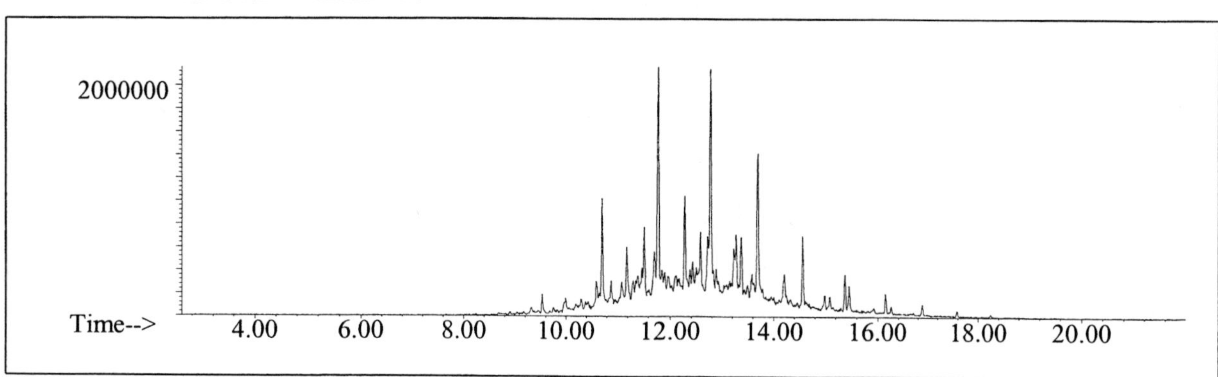

NAPHTHALENES

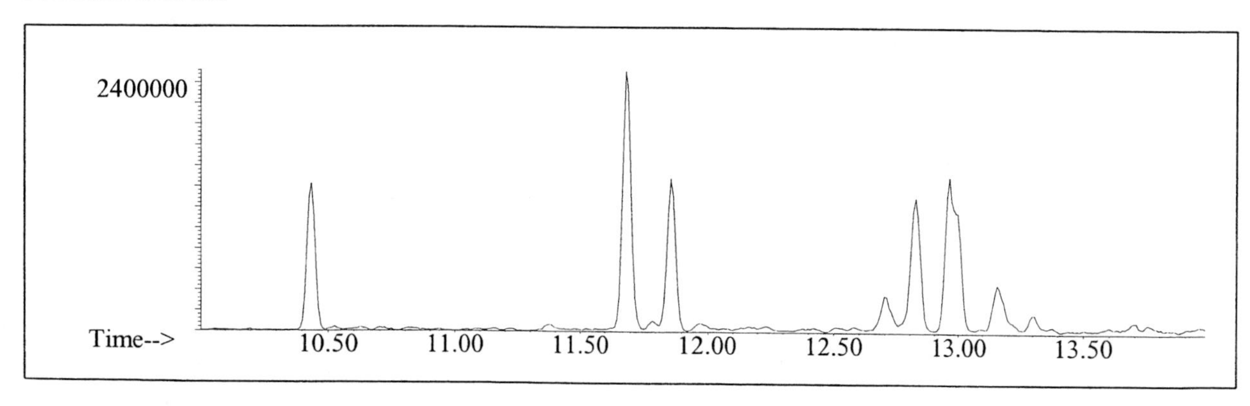

INDIVIDUAL PROFILES

Distillate #44

Alkane

Ion 43

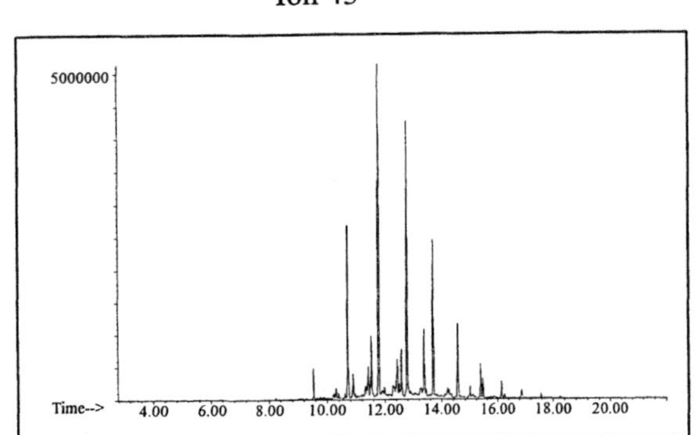

Ion 57

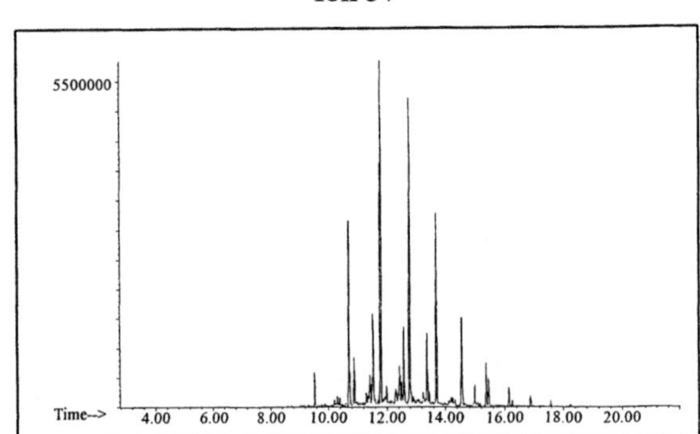

Ion 71

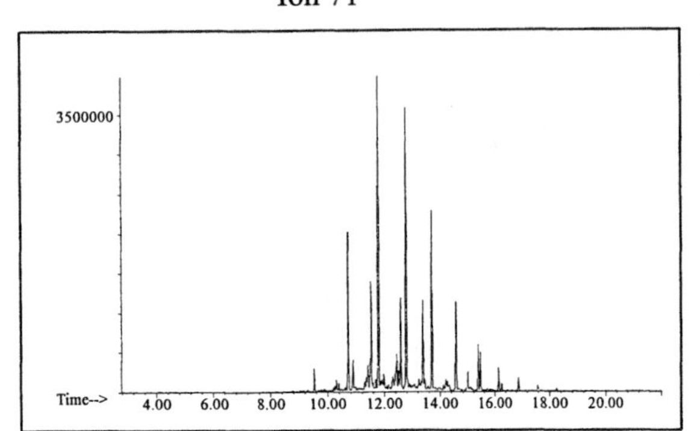

Ion 85

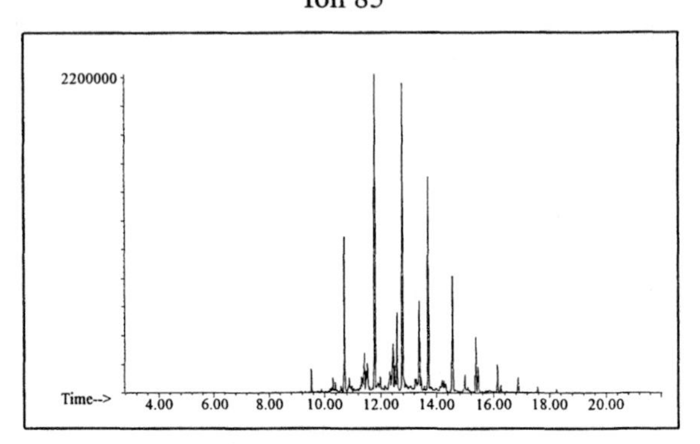

Aromatic

Ion 91

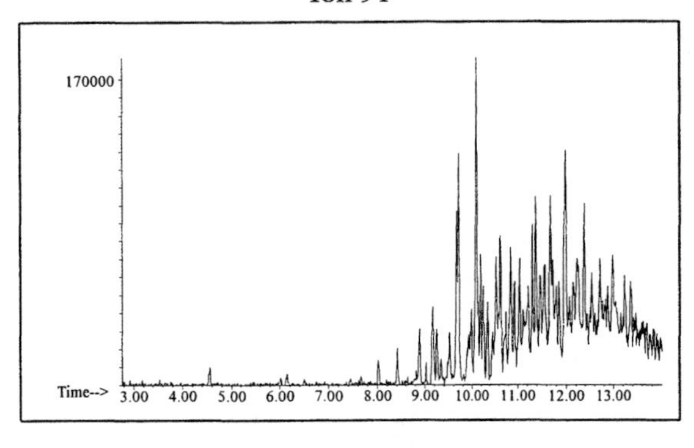

Ion 105

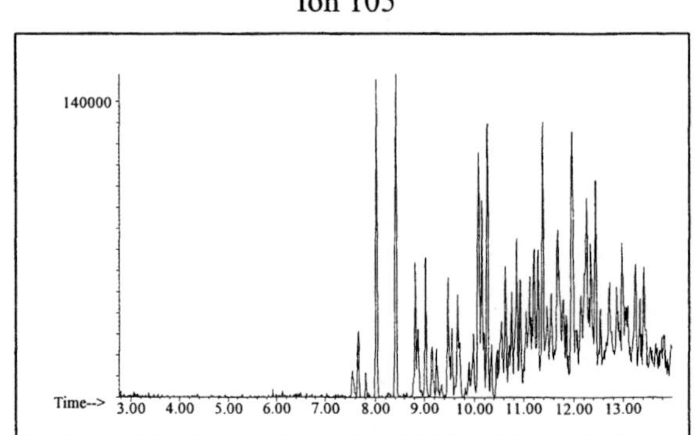

Ion 119

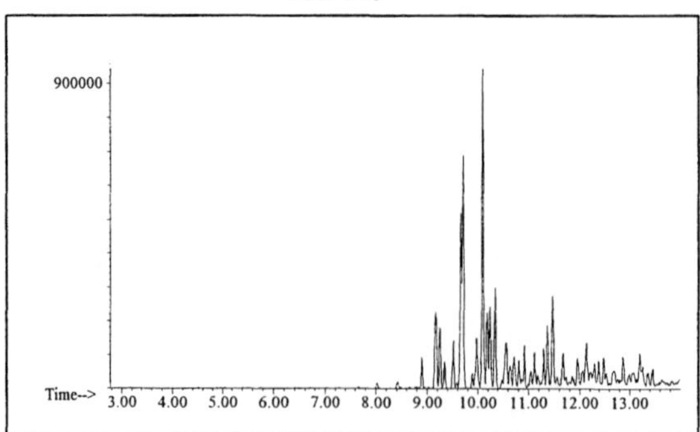

INDIVIDUAL PROFILES

Distillate #44

Cycloparaffin

Ion 55

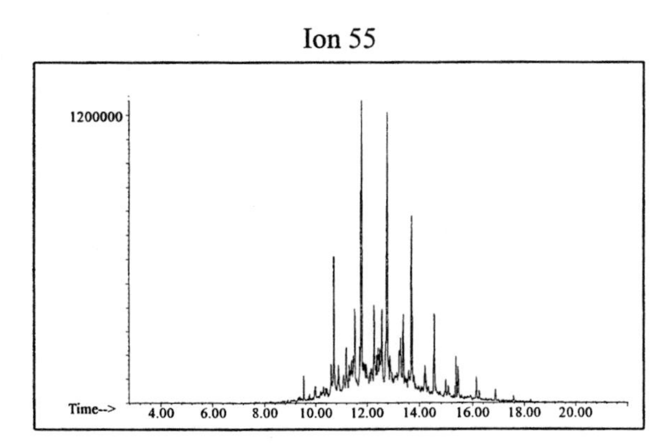

Ion 69

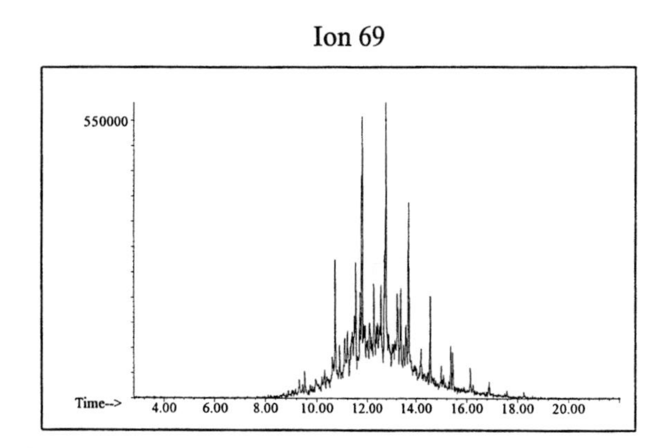

Ion 83

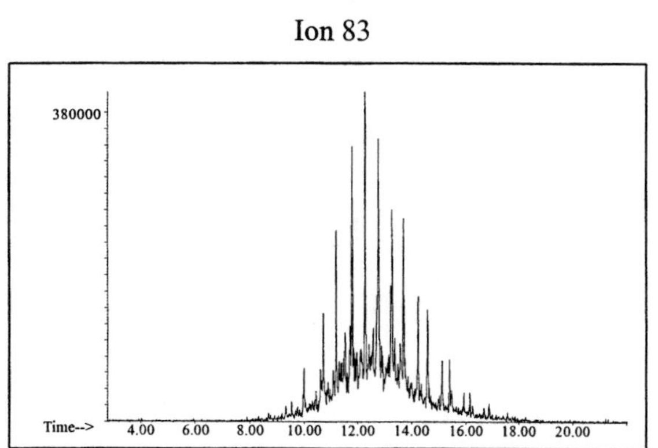

Naphthalene

Ion 128

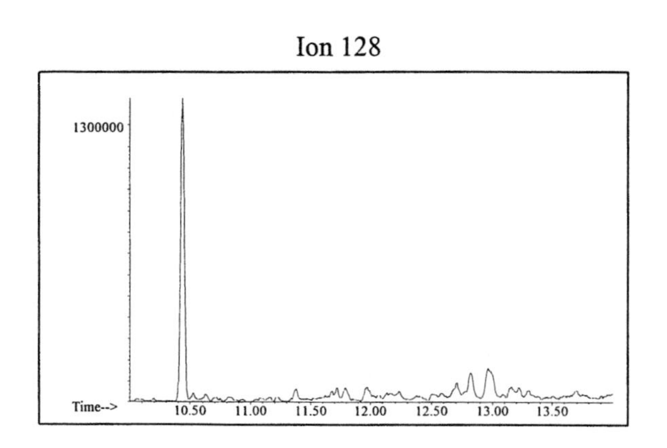

Ion 142

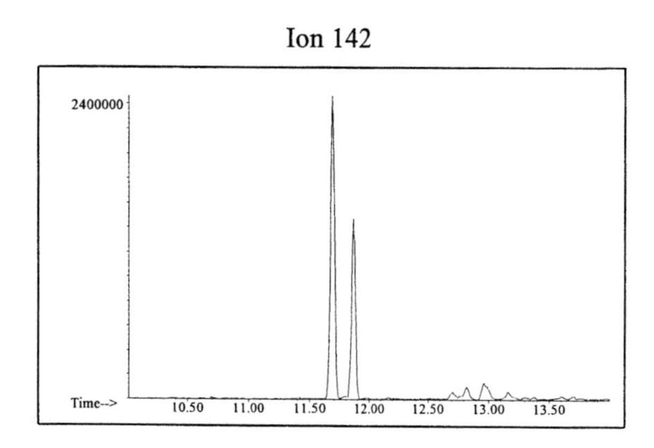

Ion 156

Distillate #45

20uL/mL Pentane

Product Displayed: Getty Fuel Oil #2

Other Similar Products:

ASTM: Class 5 (Heavy Petroleum Distillate)

Product Uses: Heating oil

Macro Code: H

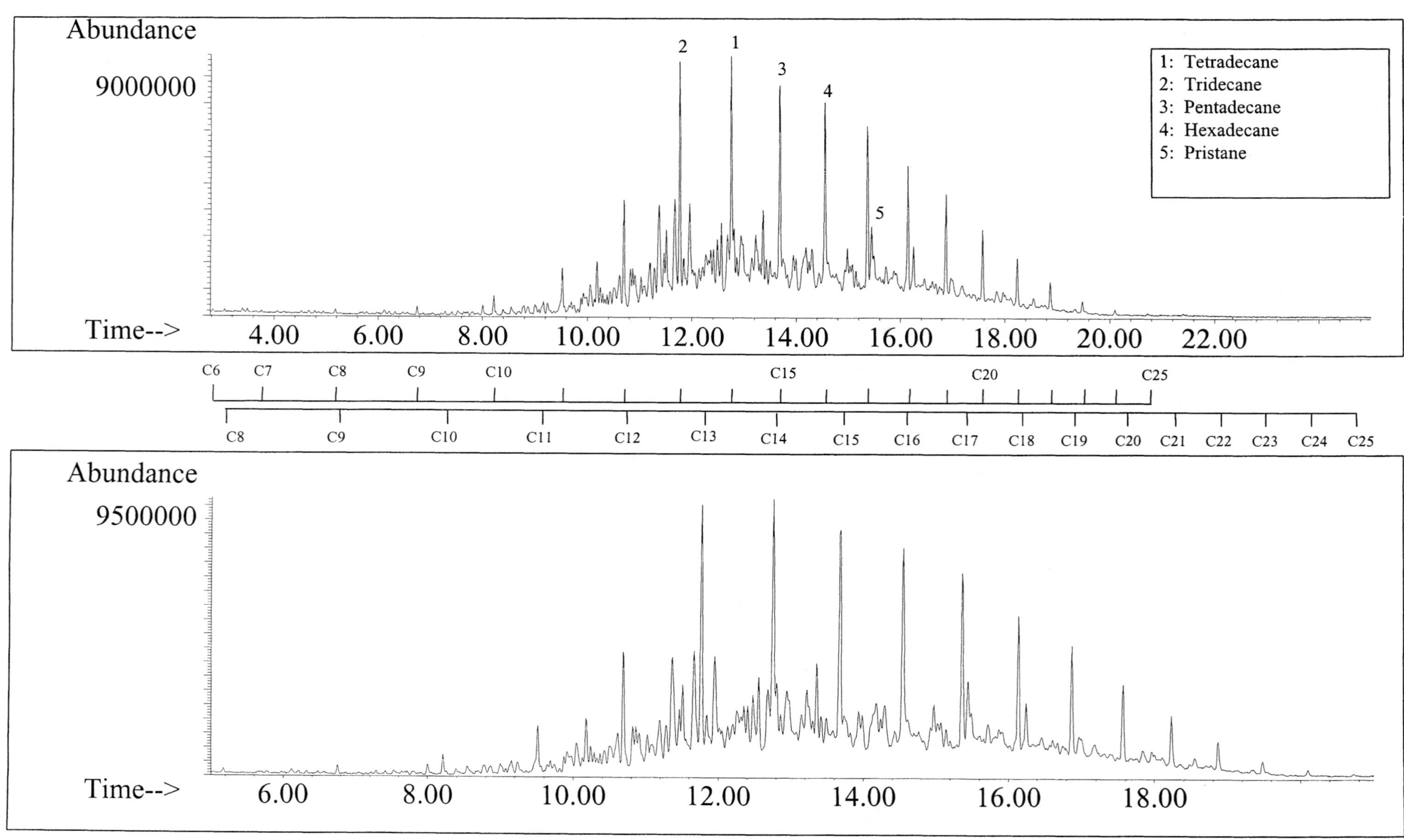

SUMMED PROFILES

Distillate #45

ALKANES

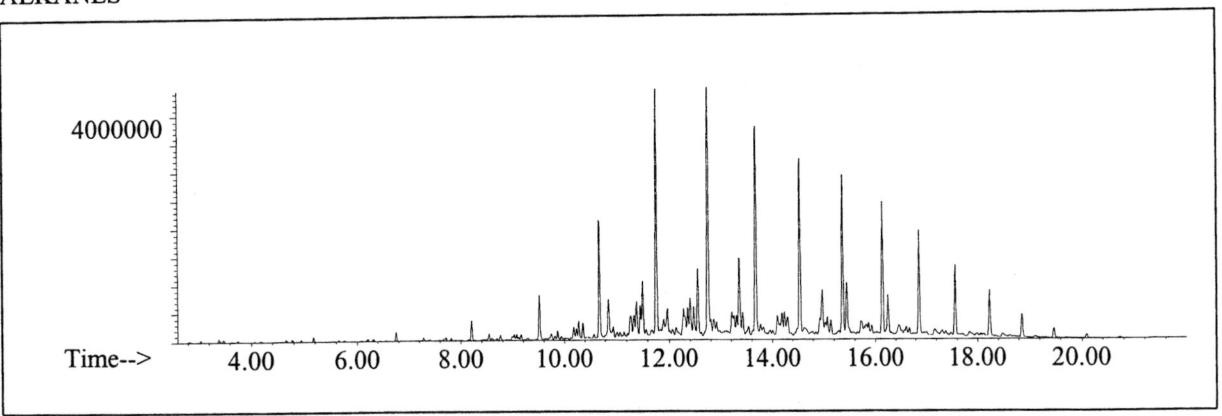

AROMATICS

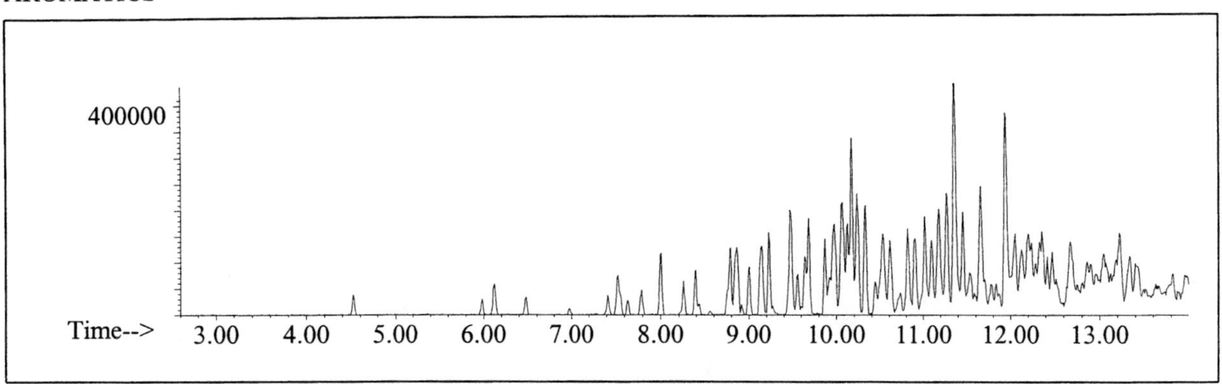

CYCLOPARAFFINS AND ALKENES

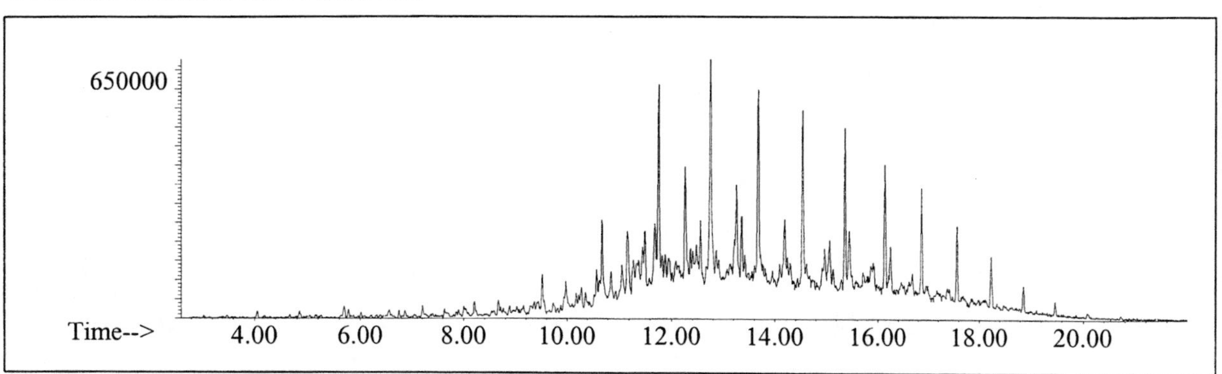

NAPHTHALENES

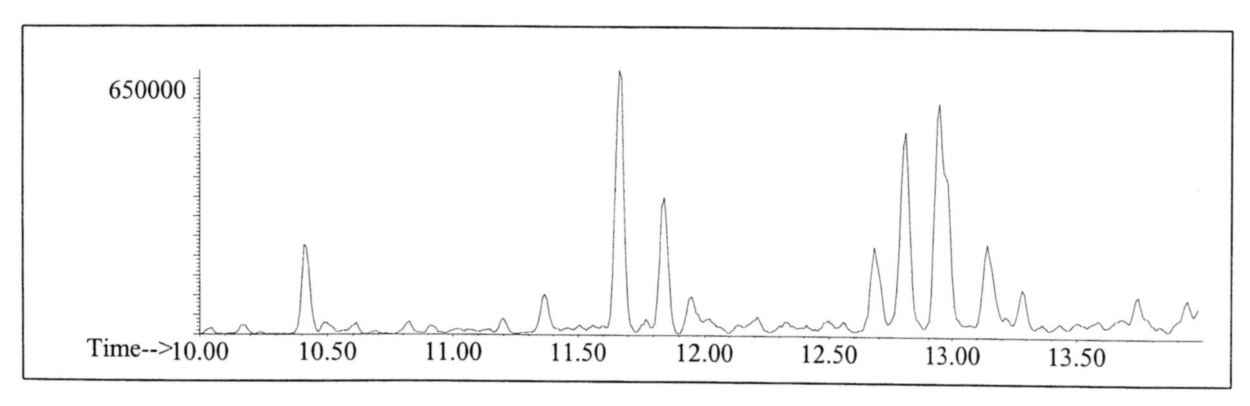

INDIVIDUAL PROFILES

Distillate #45

Alkane

Ion 43

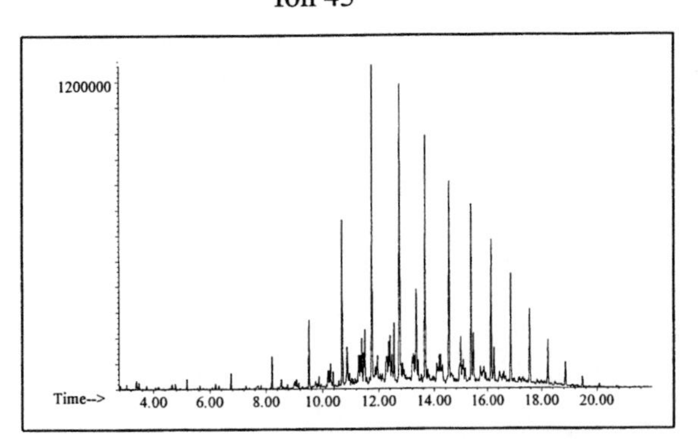

Ion 57

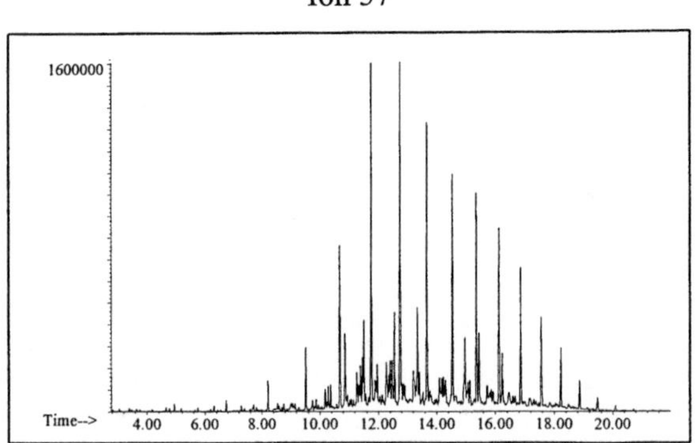

Ion 71

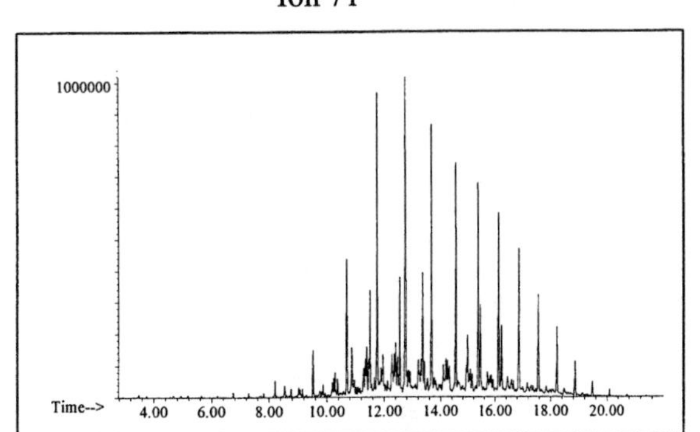

Ion 85

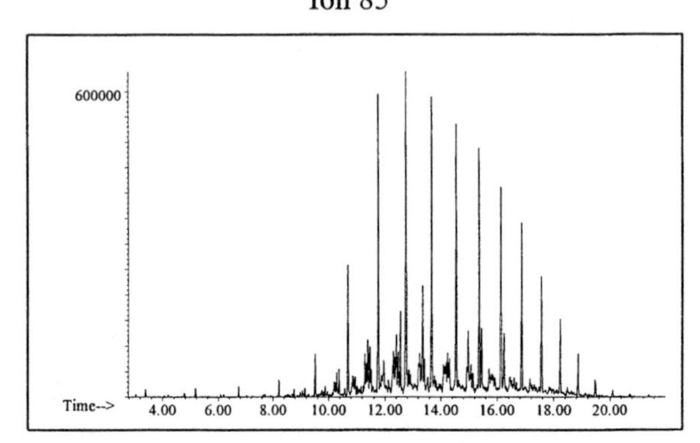

Aromatic

Ion 91

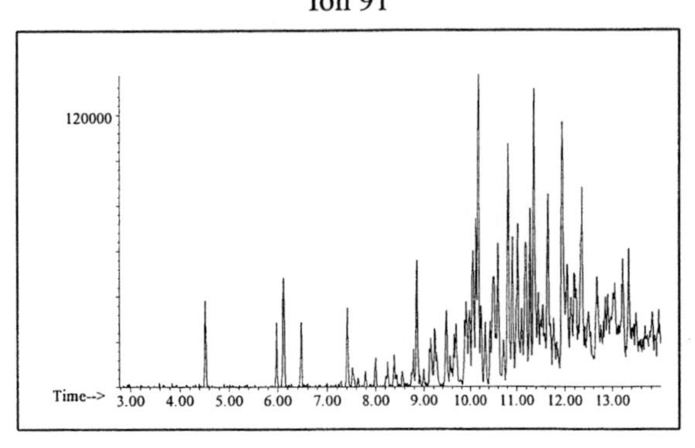

Ion 105

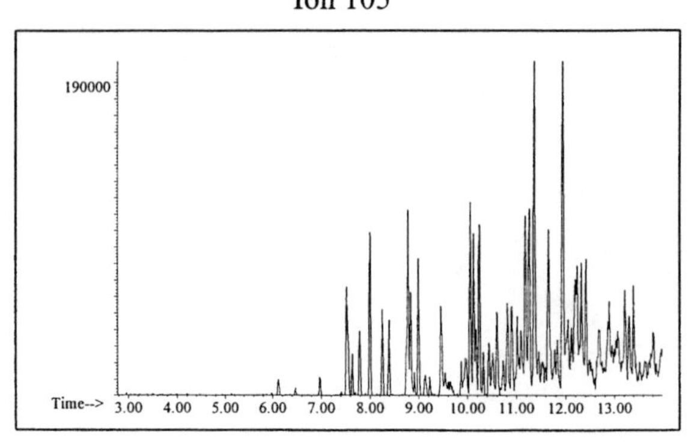

Ion 119

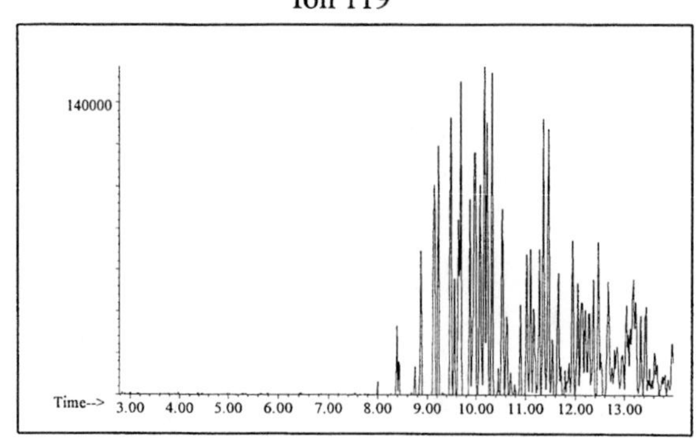

INDIVIDUAL PROFILES

Distillate #45

Cycloparaffin

Ion 55

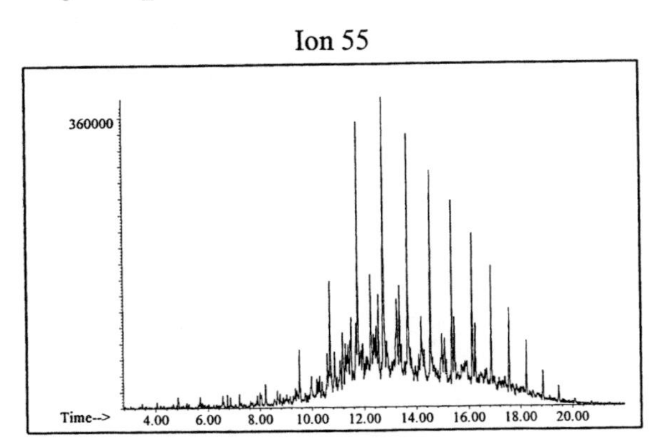

Ion 69

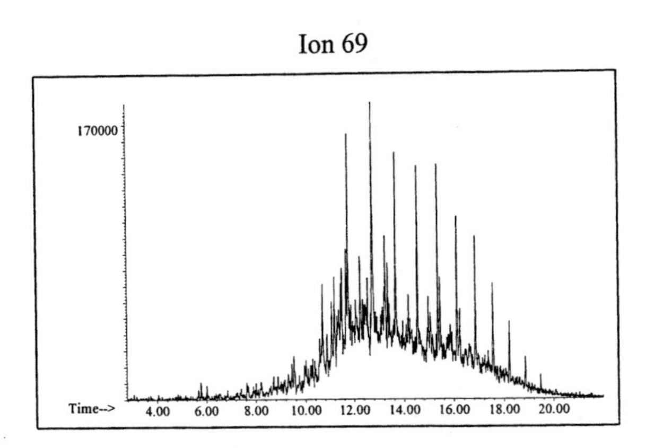

Ion 83

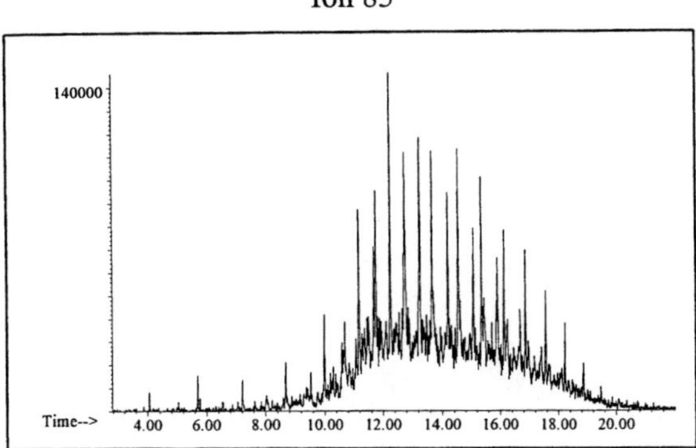

Naphthalene

Ion 128

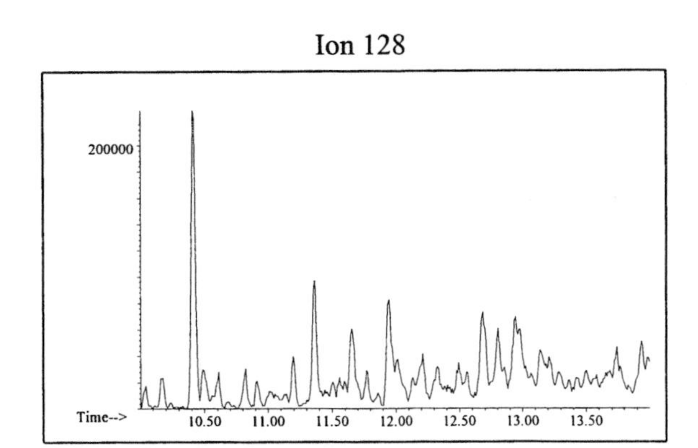

Ion 142

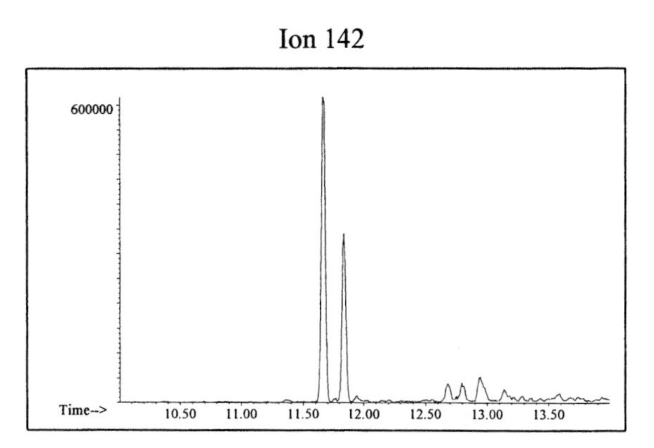

Ion 156

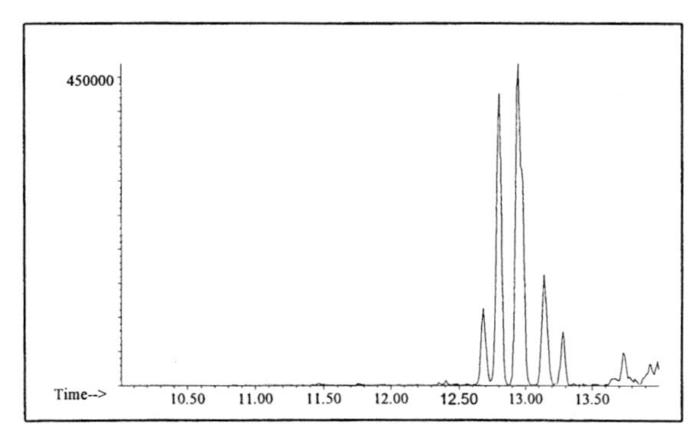

Distillate #46

20uL/mL Pentane

ASTM: Class 4 (Kerosene)

Macro Code: H

Product Displayed: Classic Bar-O-Lite Charcoal Starter

Product Uses: charcoal lighter fluids, heating fuels

Other Similar Products:

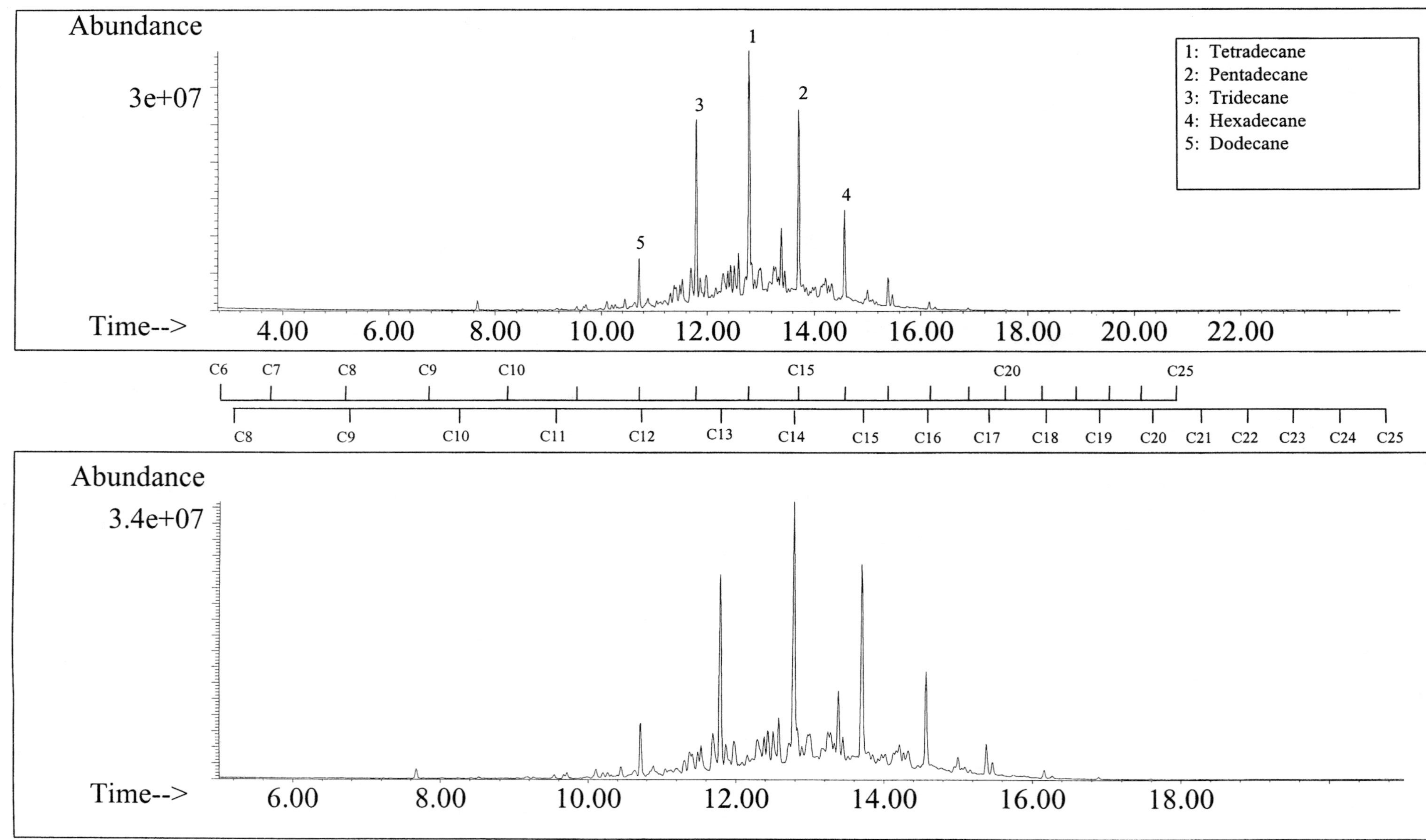

ALKANES

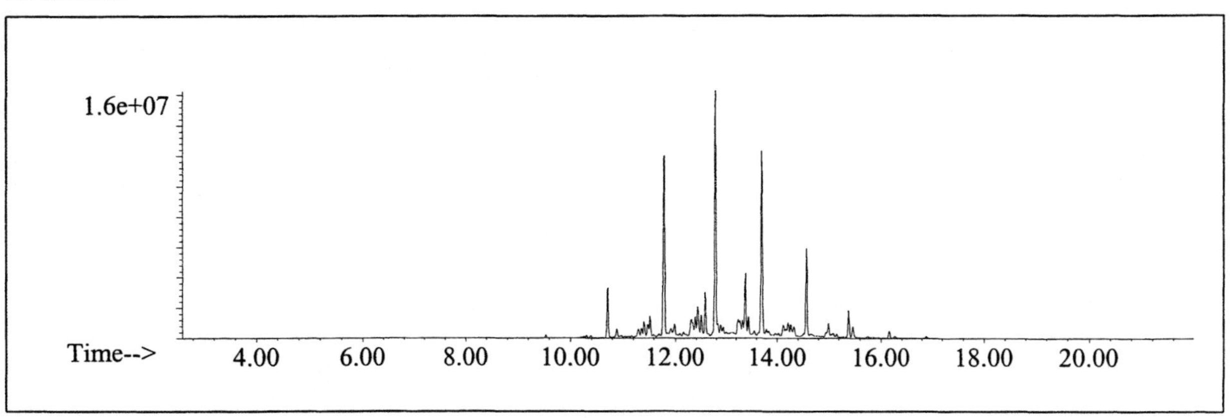

AROMATICS

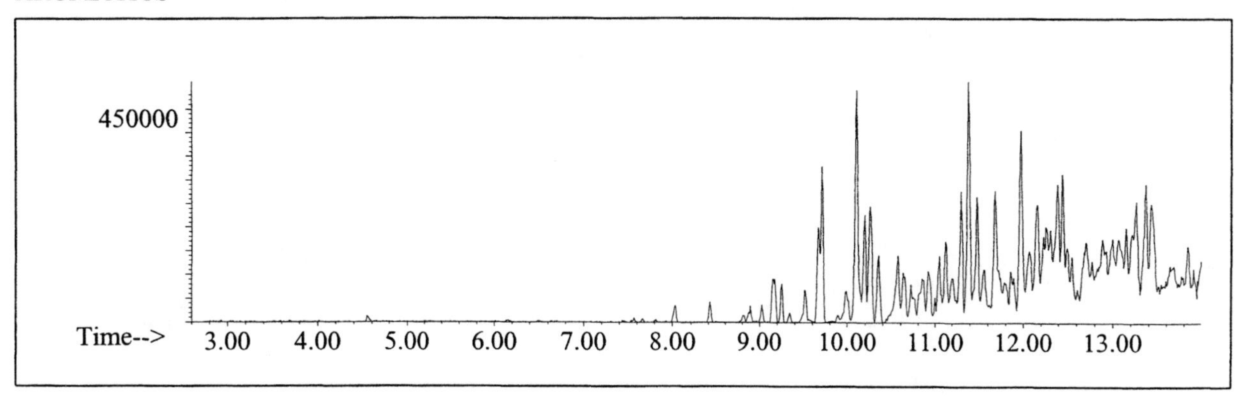

CYCLOPARAFFINS AND ALKENES

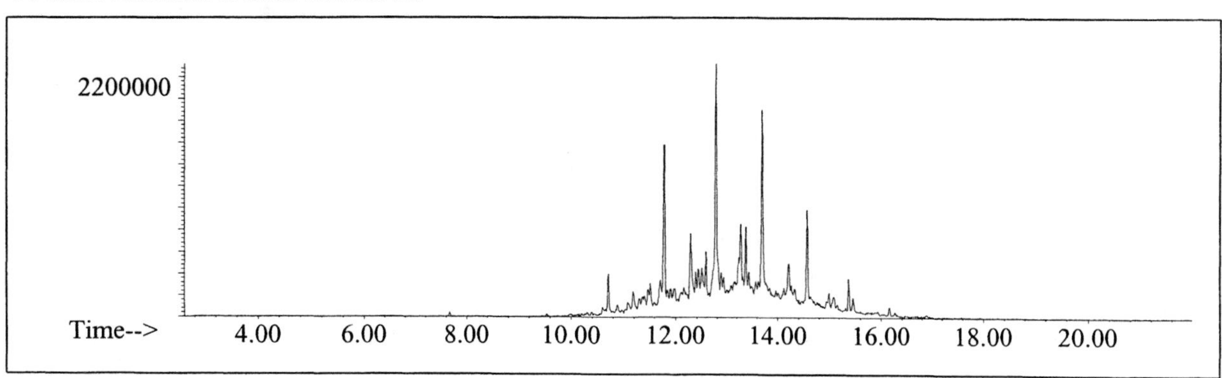

NAPHTHALENES

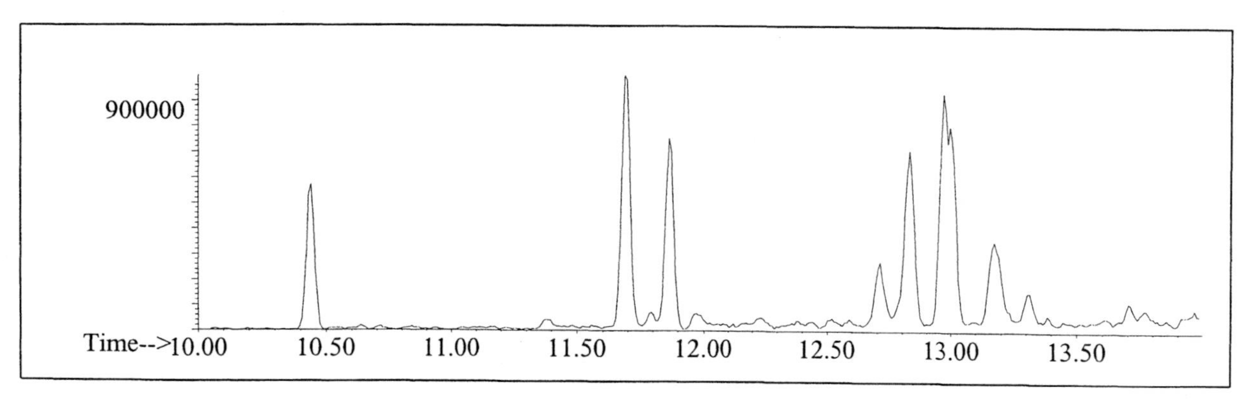

INDIVIDUAL PROFILES

Distillate #46

Alkane

Ion 43

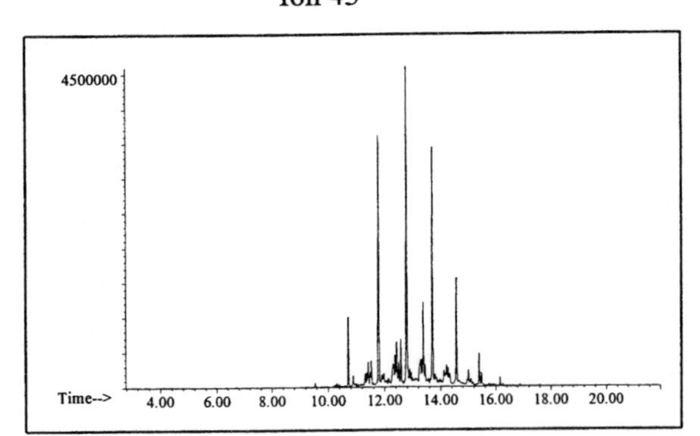

Ion 57

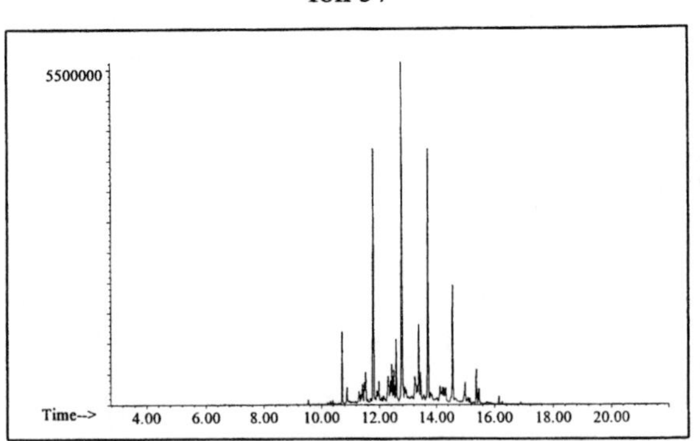

Ion 71

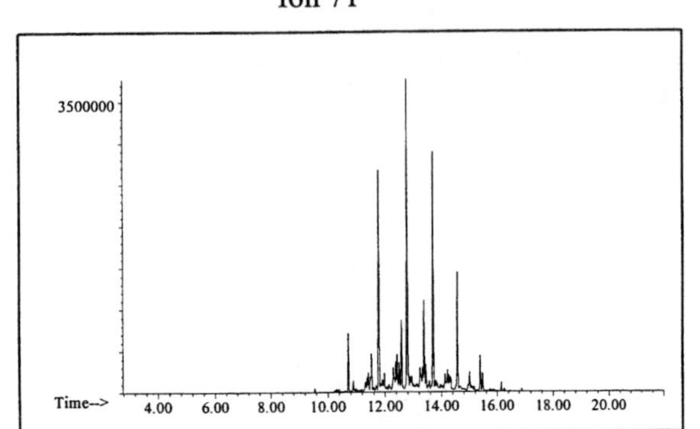

Ion 85

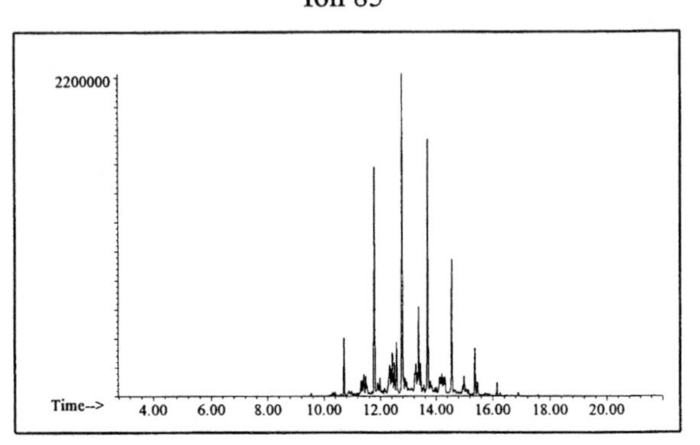

Aromatic

Ion 91

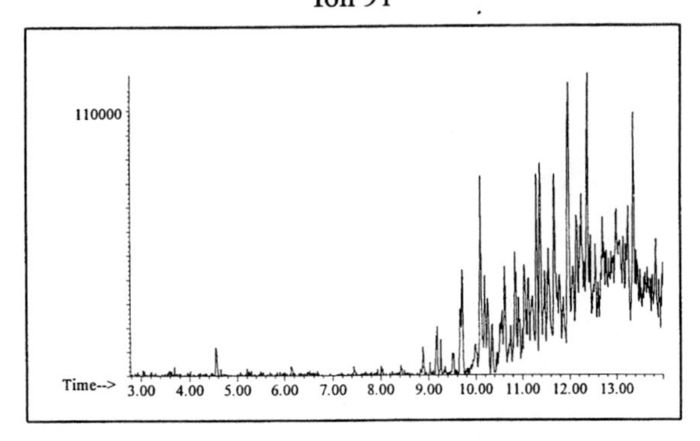

Ion 105

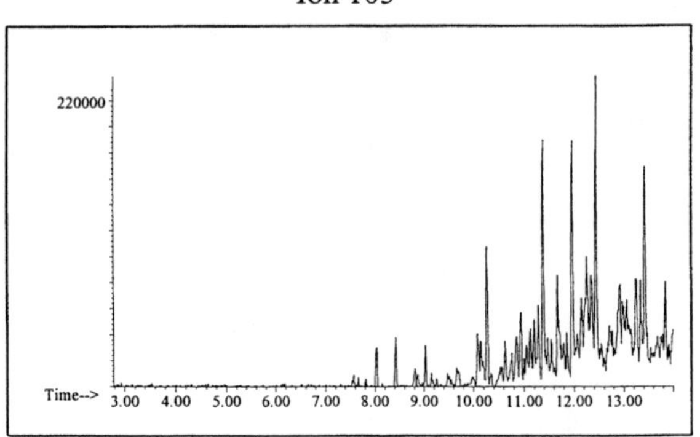

Ion 119

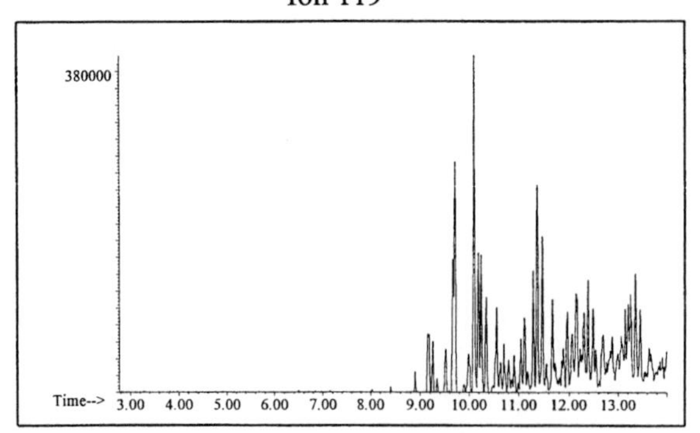

INDIVIDUAL PROFILES

Distillate #46

Cycloparaffin

Ion 55

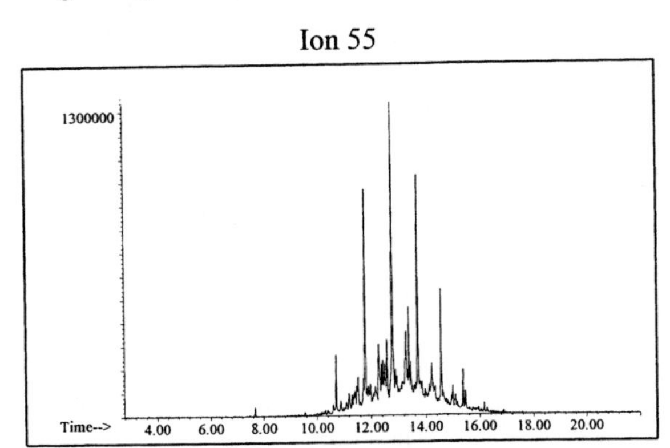

Ion 69

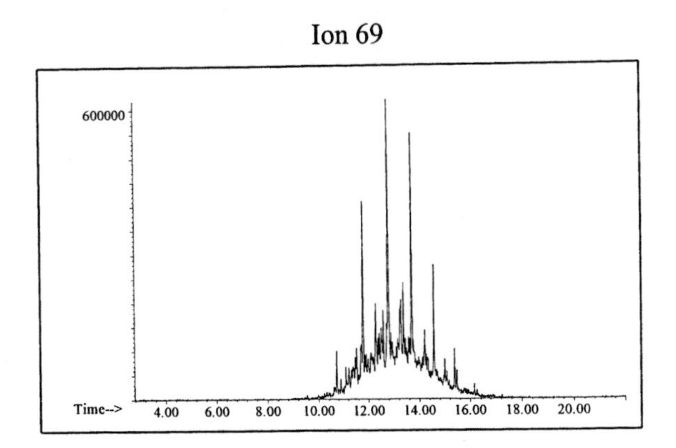

Ion 83

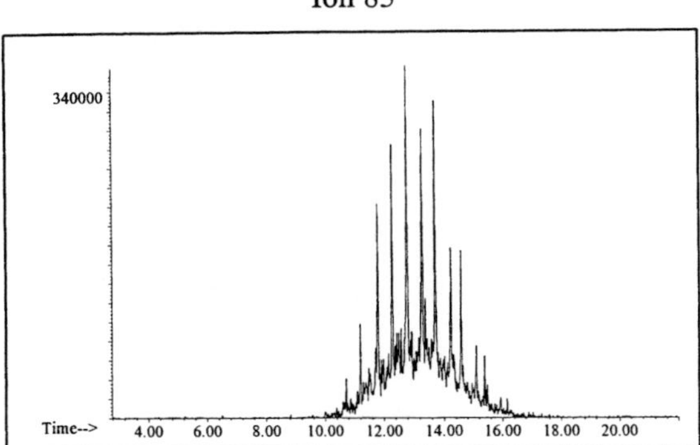

Naphthalene

Ion 128

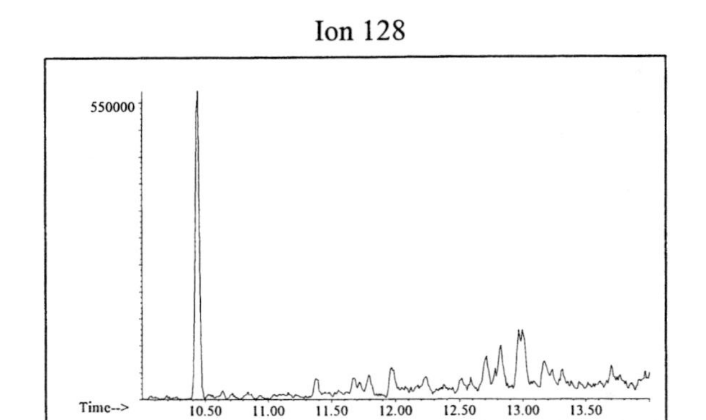

Ion 142

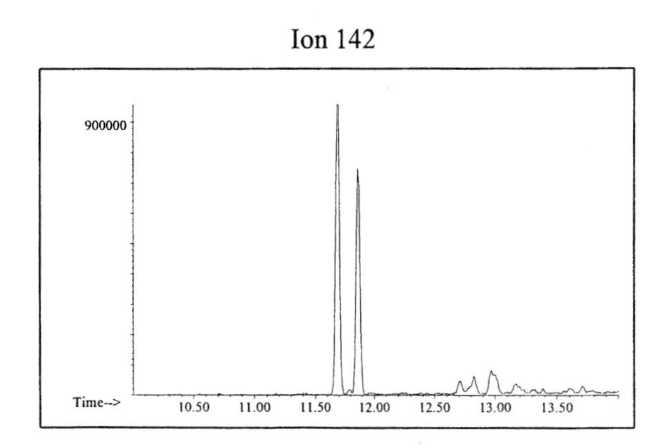

Ion 156

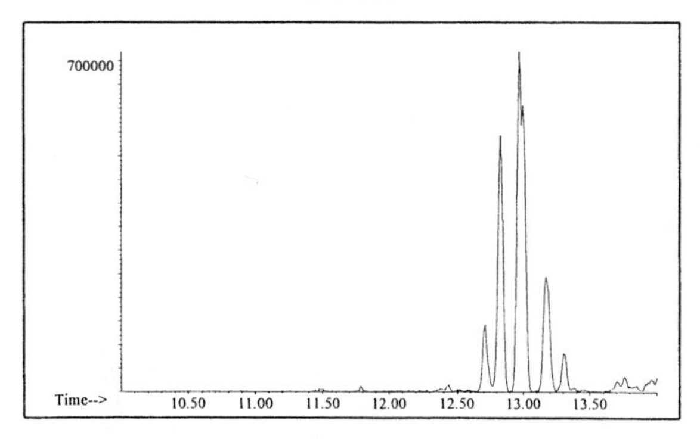

Distillate #47

20uL/mL Pentane
Product Displayed: Exxsol D110
Other Similar Products:

ASTM: Class 0 (Miscellaneous)
Product Uses: Solvent, feedstock

Macro Code: H

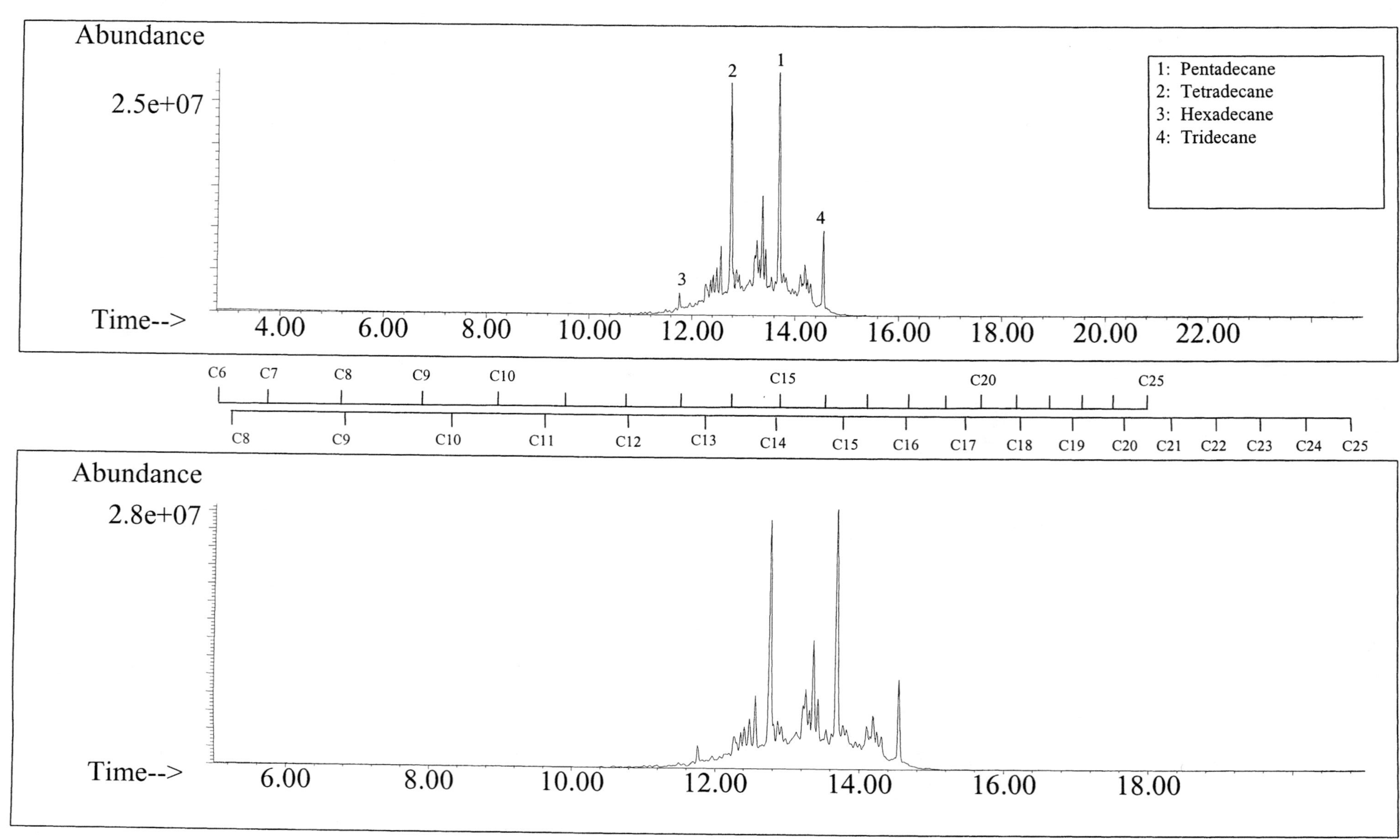

SUMMED PROFILES

Distillate #47

ALKANES

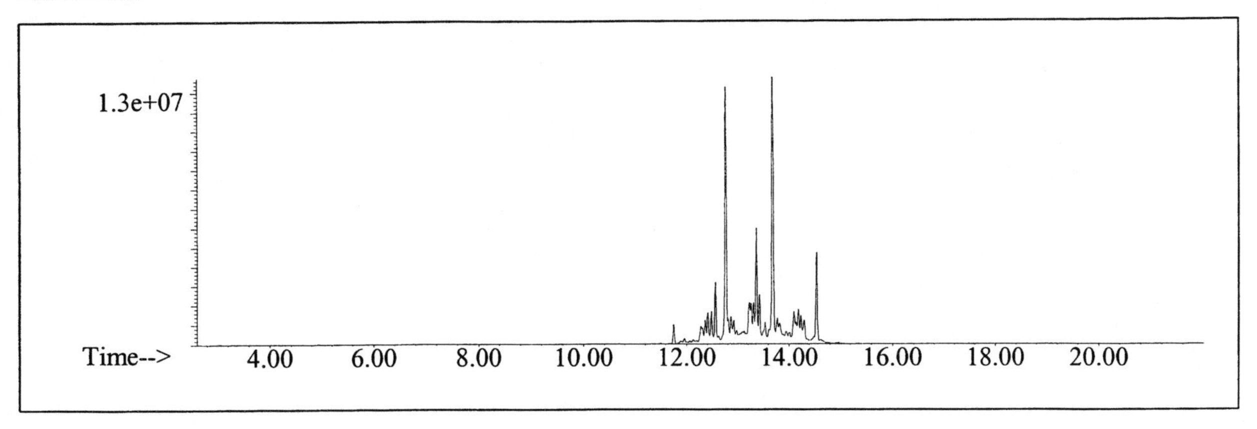

AROMATICS

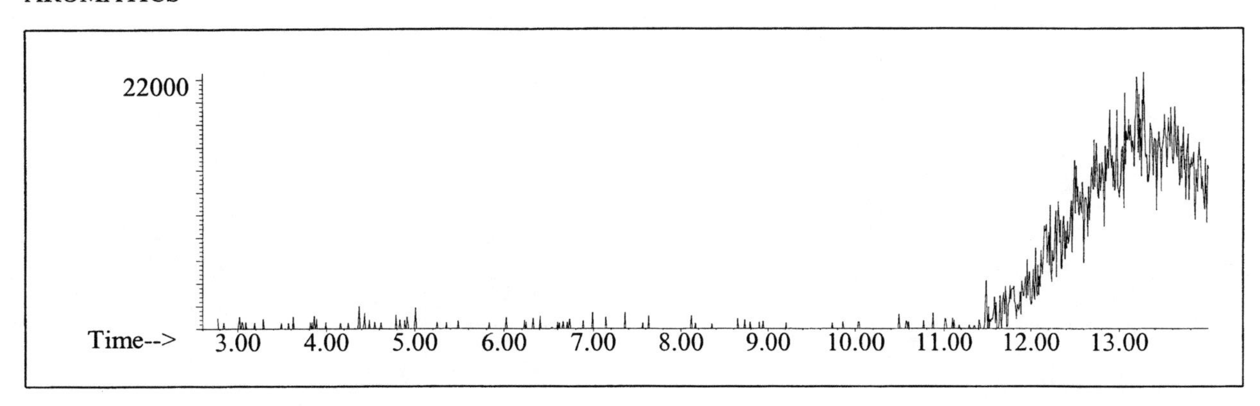

CYCLOPARAFFINS AND ALKENES

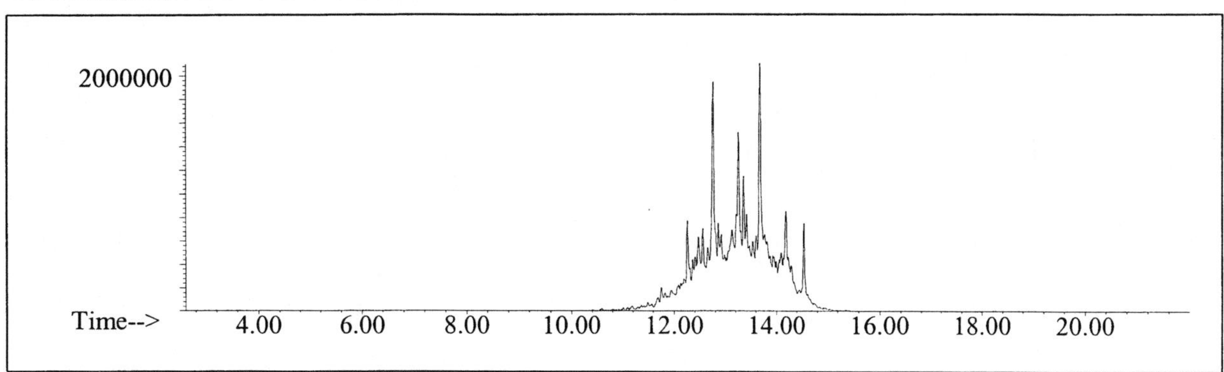

NAPHTHALENES

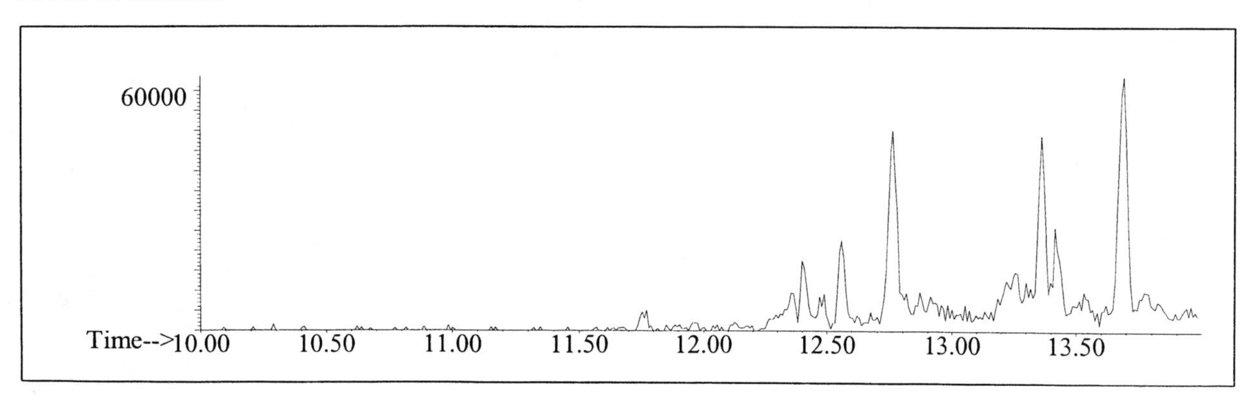

INDIVIDUAL PROFILES

Distillate #47

Alkane

Ion 43

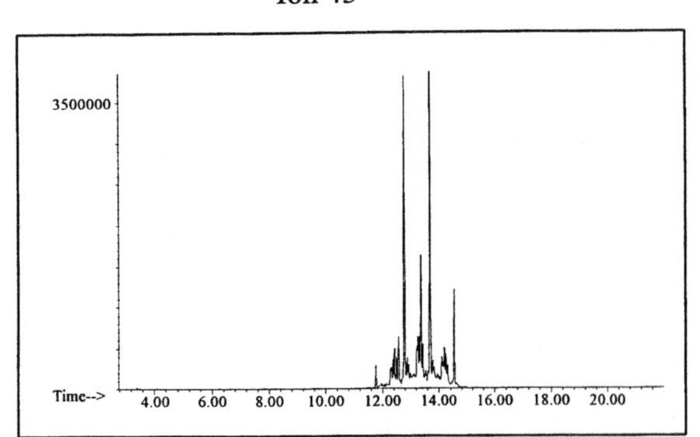

Ion 57

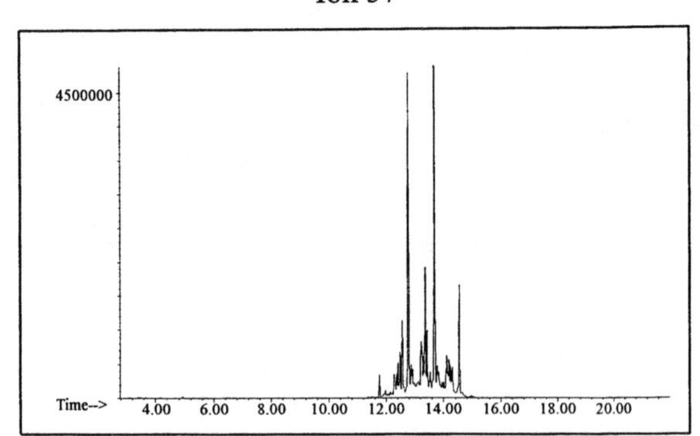

Ion 71

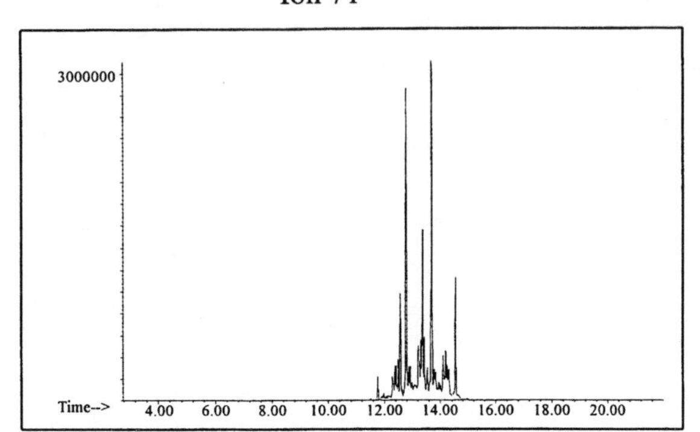

Ion 85

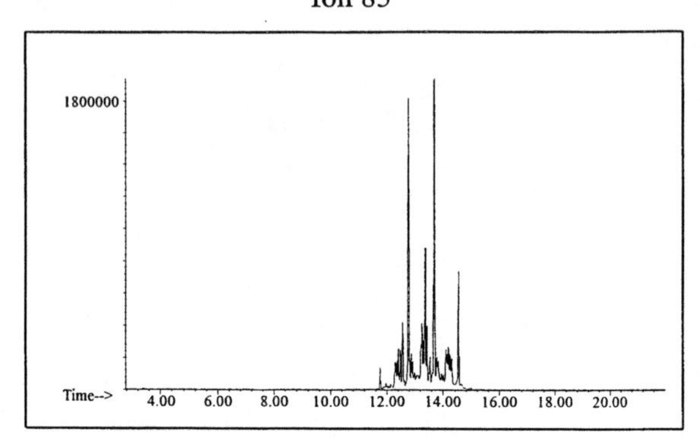

Aromatic

Ion 91

No Useful Data Obtained

Ion 105

No Useful Data Obtained

Ion 119

No Useful Data Obtained

INDIVIDUAL PROFILES

Distillate #47

Cycloparaffin

Ion 55

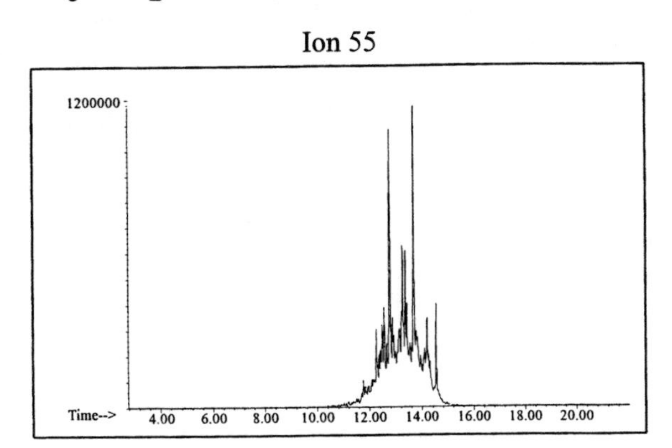

Ion 69

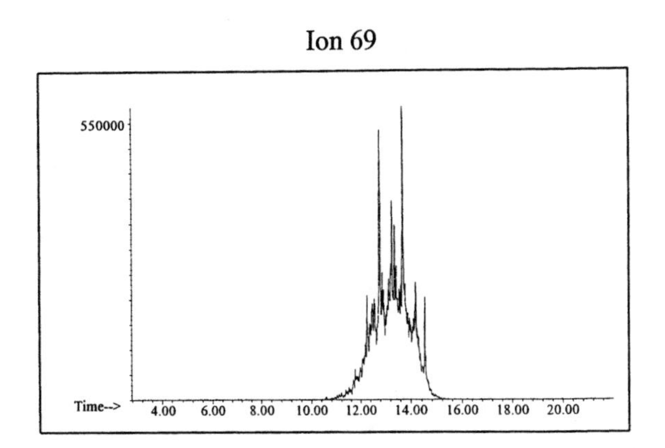

Ion 83

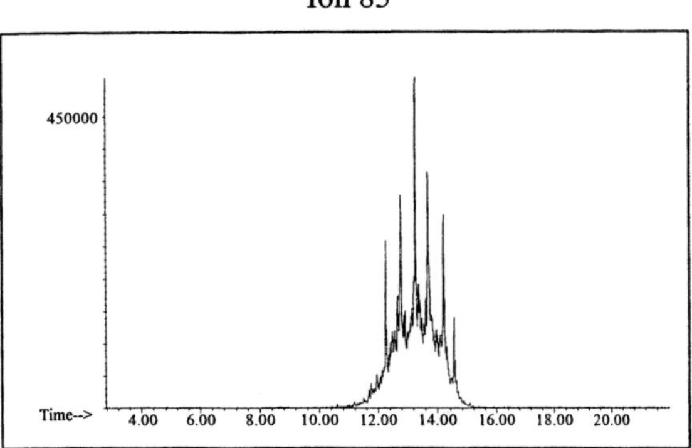

Naphthalene

Ion 128

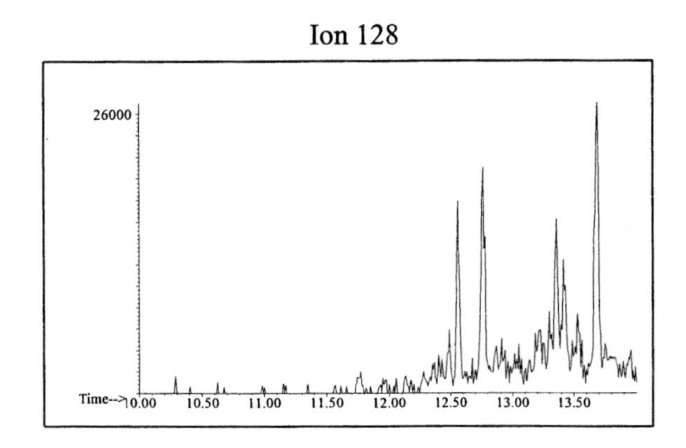

Ion 142

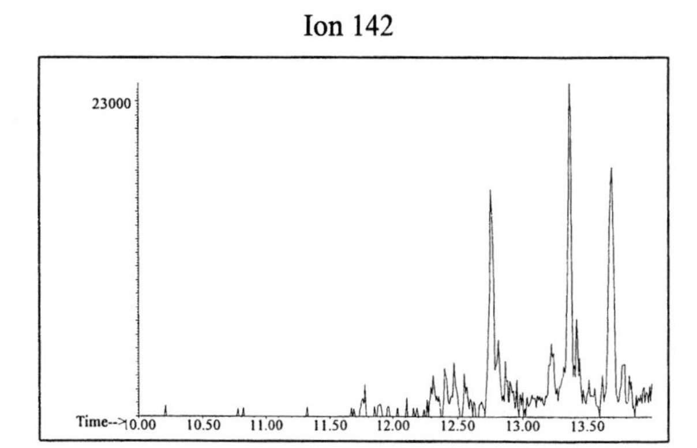

Ion 156

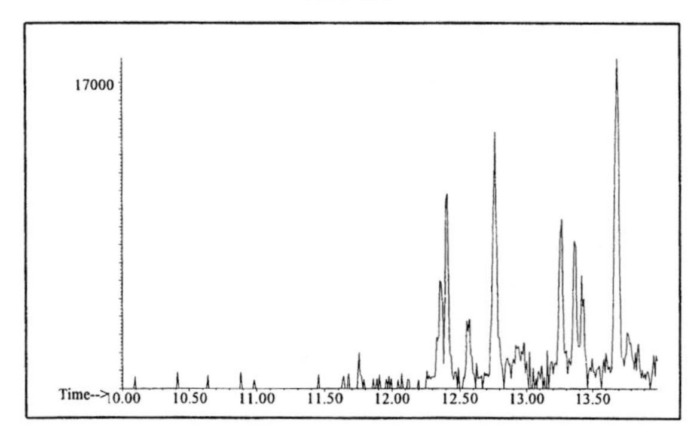

Distillate #48

20uL/mL Pentane

ASTM: Class 5 (Heavy Petroleum Distillate)

Macro Code: H

Product Displayed: Boron Gardens Torches Lamp Oil

Product Uses: Lamp oil

Other Similar Products:

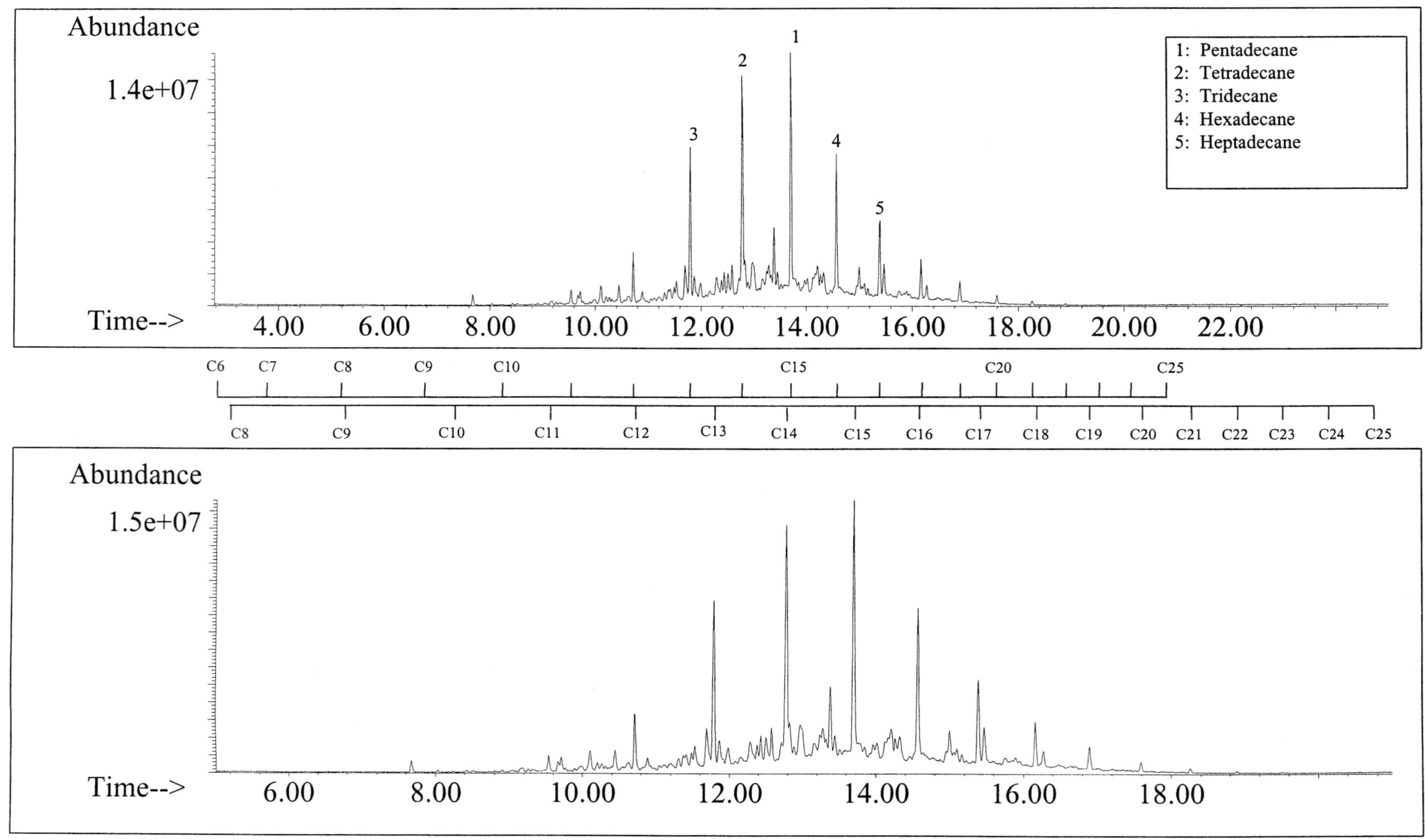

SUMMED PROFILES

Distillate #48

ALKANES

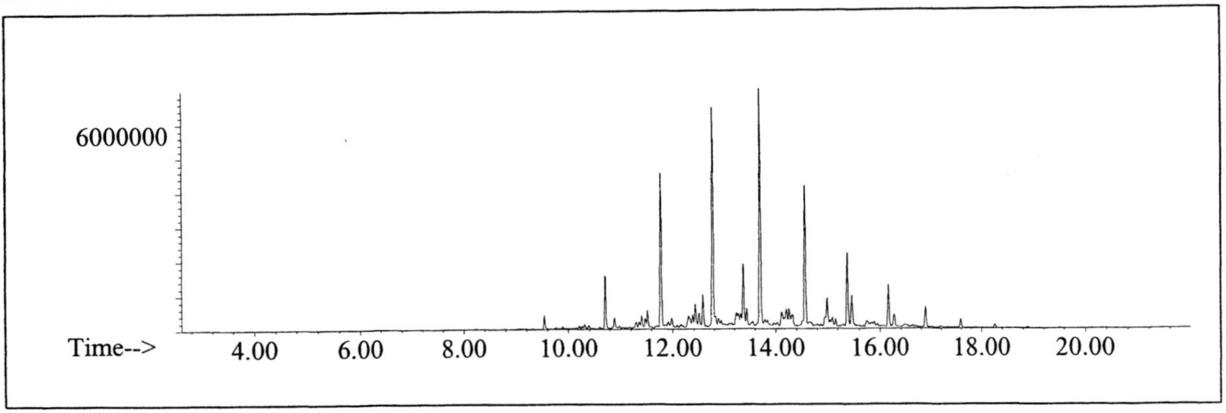

AROMATICS

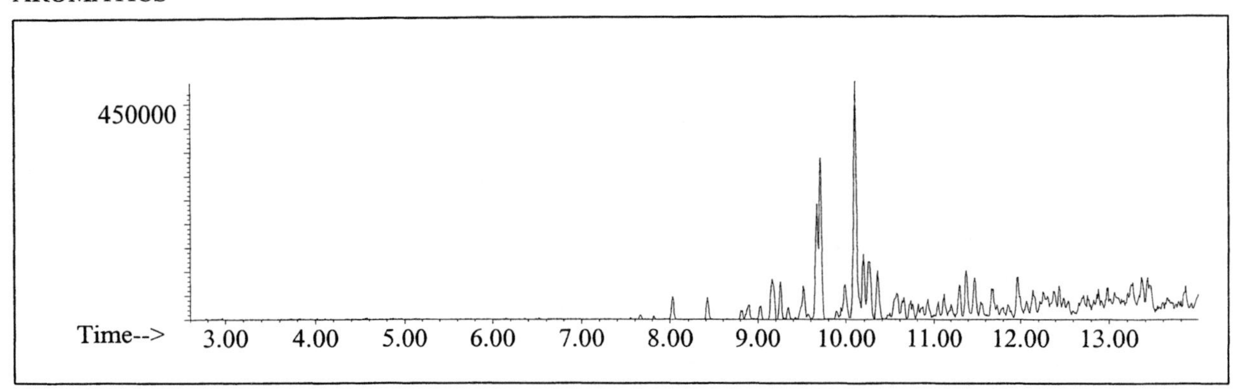

CYCLOPARAFFINS AND ALKENES

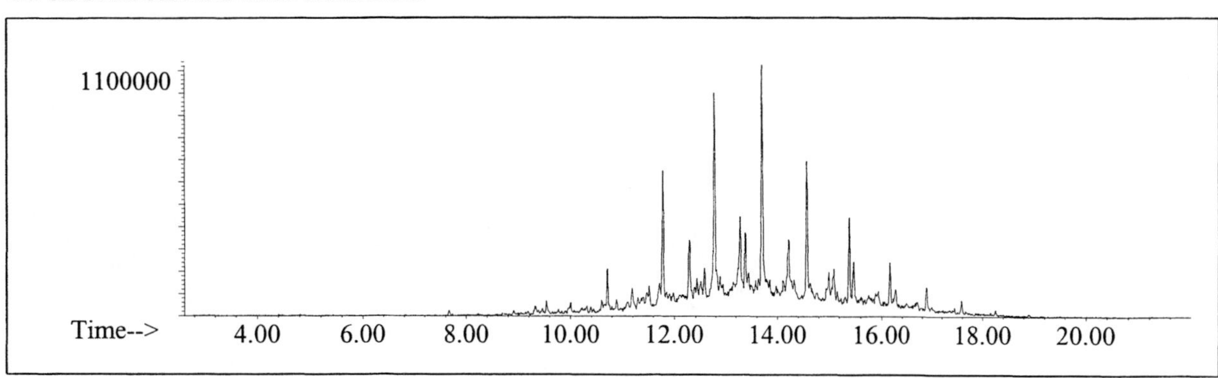

NAPHTHALENES

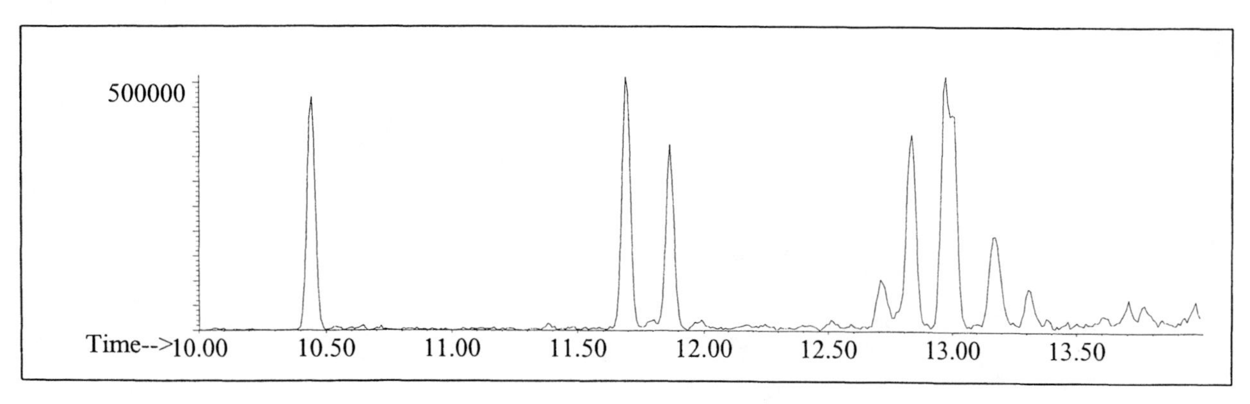

INDIVIDUAL PROFILES Distillate #48

Alkane

Ion 43

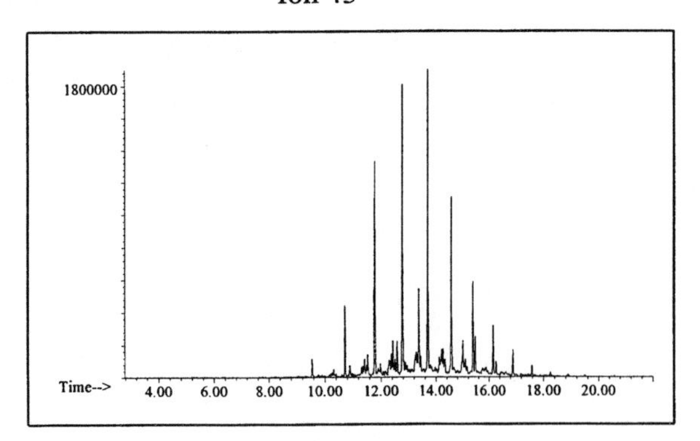

Ion 57

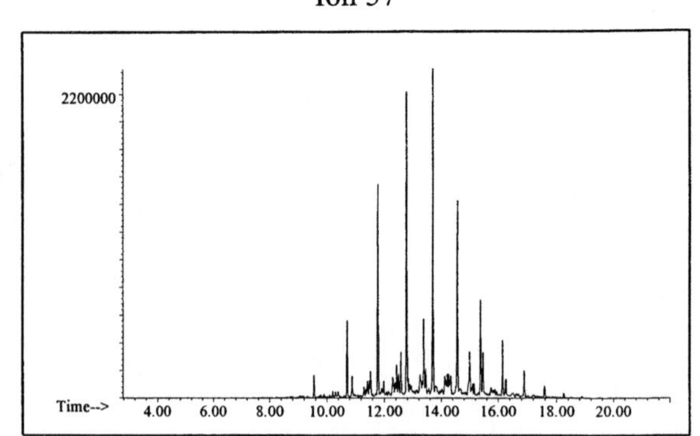

Ion 71

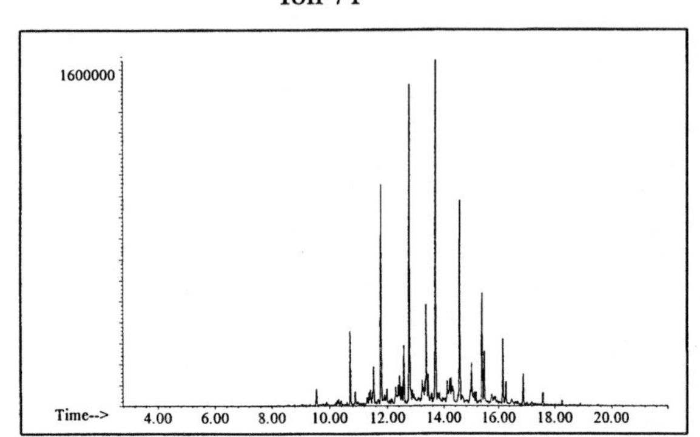

Ion 85

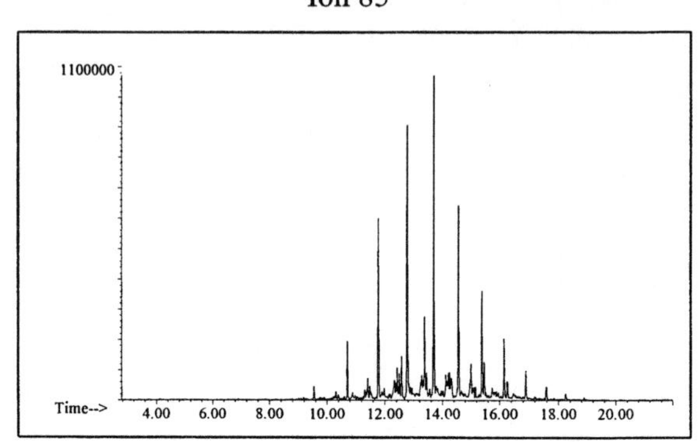

Aromatic

Ion 91

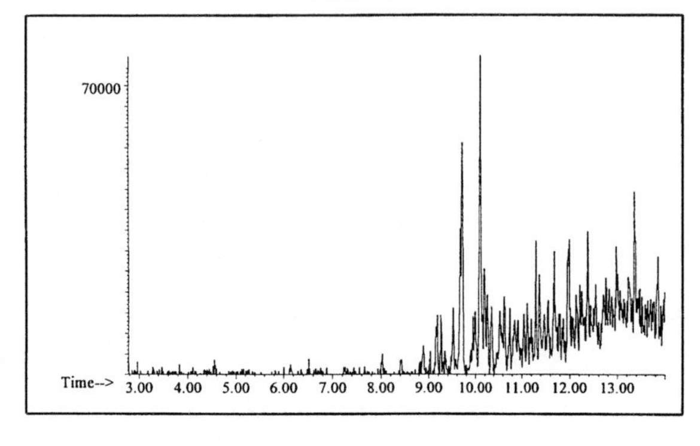

Ion 105

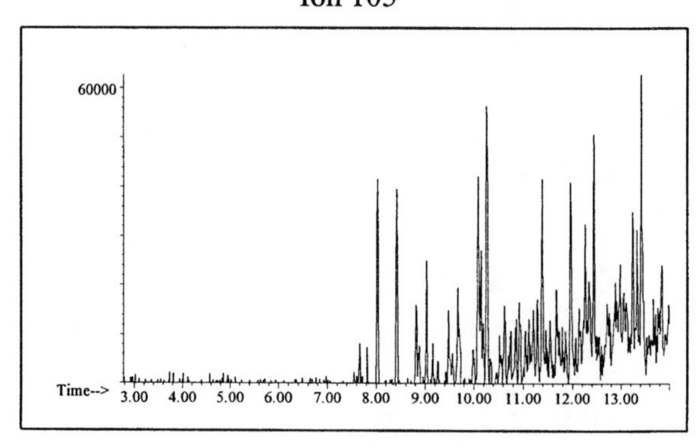

Ion 119

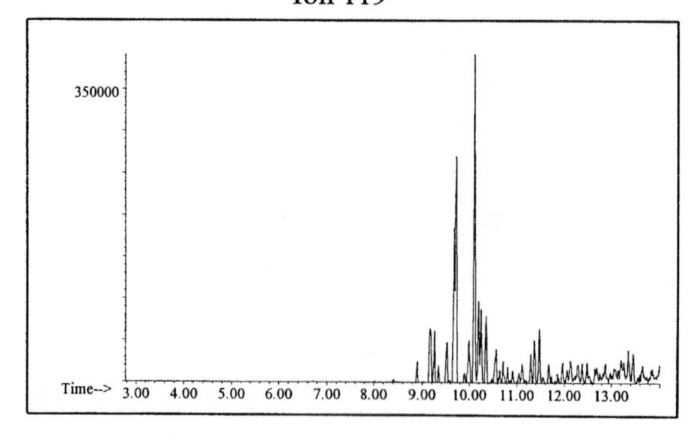

INDIVIDUAL PROFILES

Distillate #48

Cycloparaffin

Ion 55

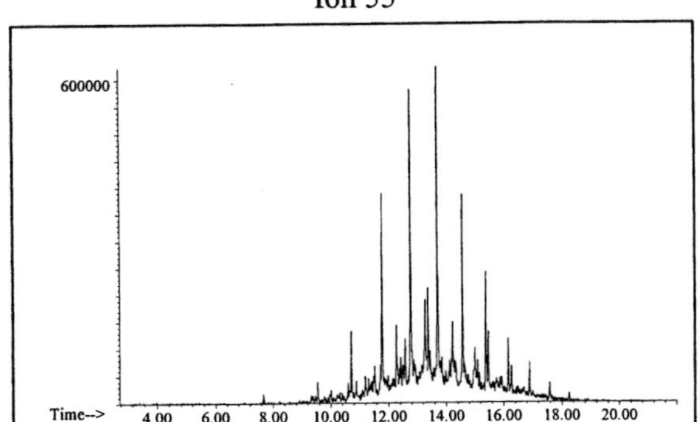

Ion 69

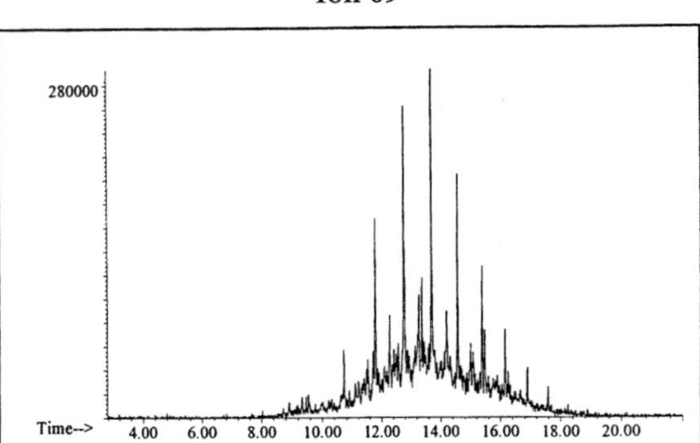

Ion 83

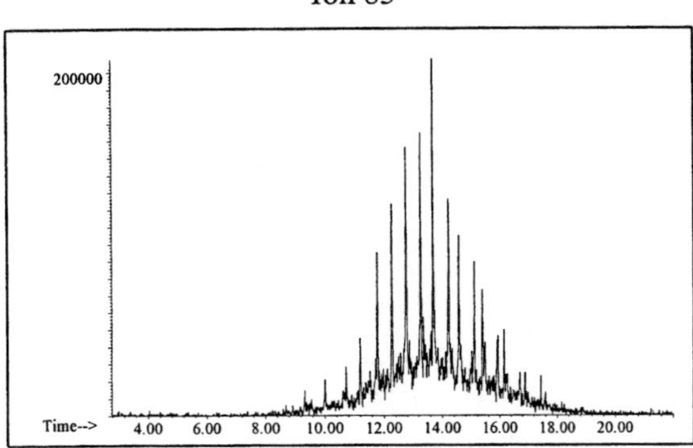

Naphthalene

Ion 128

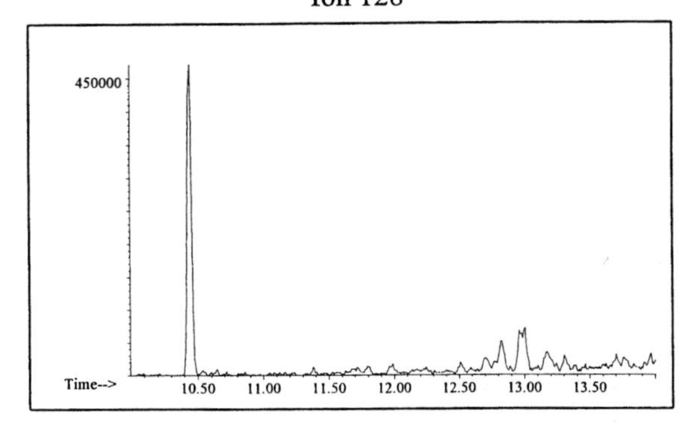

Ion 142

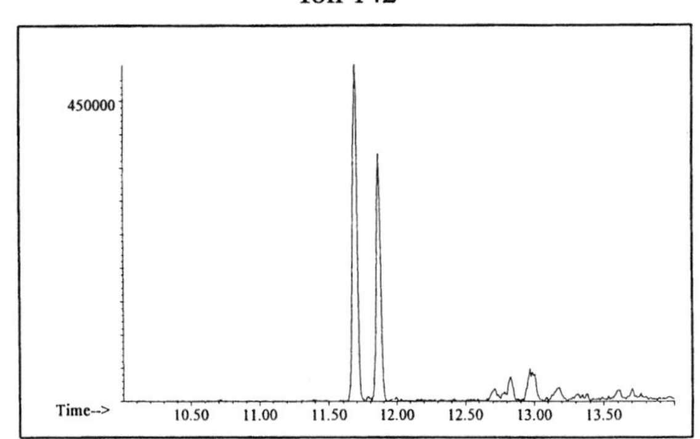

Ion 156

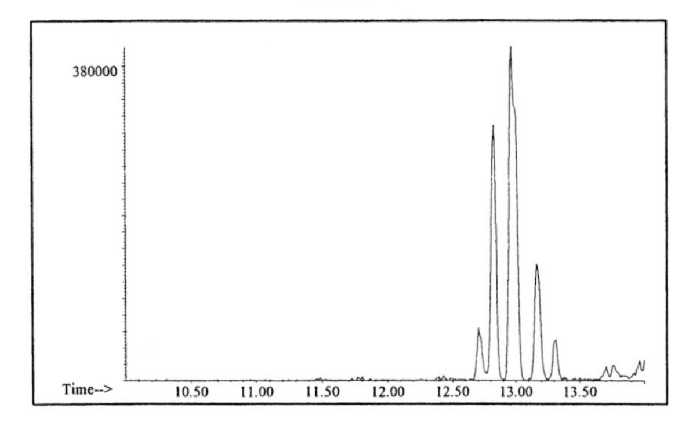

Distillate #49

20uL/mL Pentane

Product Displayed: Diesel Fuel

Other Similar Products:

ASTM: Class 5 (Heavy Petroleum Distillate)

Product Uses: Fuel

Macro Code: H

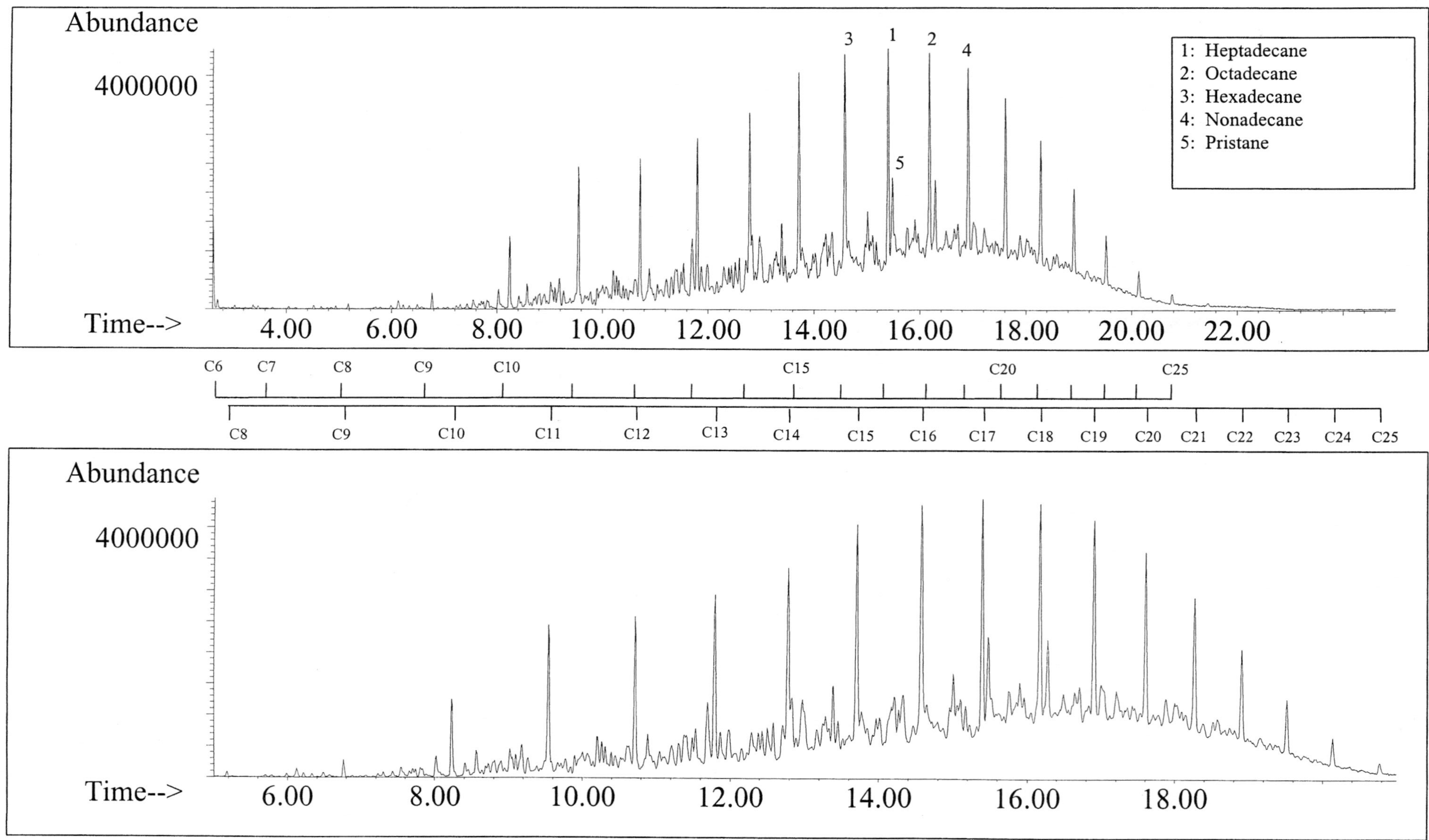

SUMMED PROFILES

Distillate #49

ALKANES

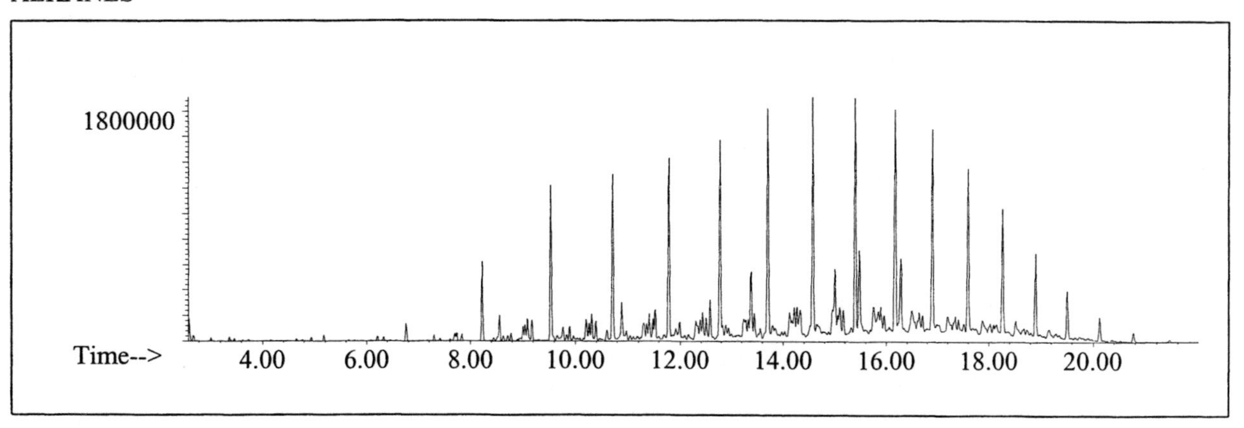

AROMATICS

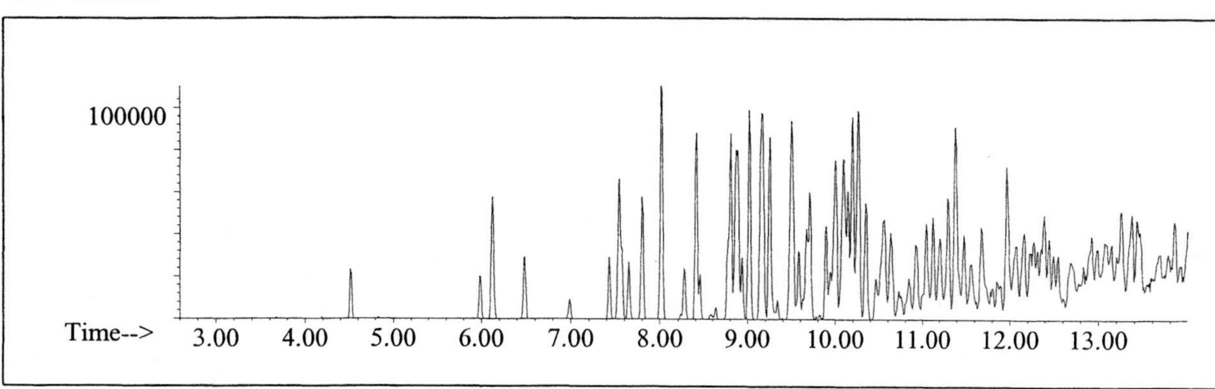

CYCLOPARAFFINS AND ALKENES

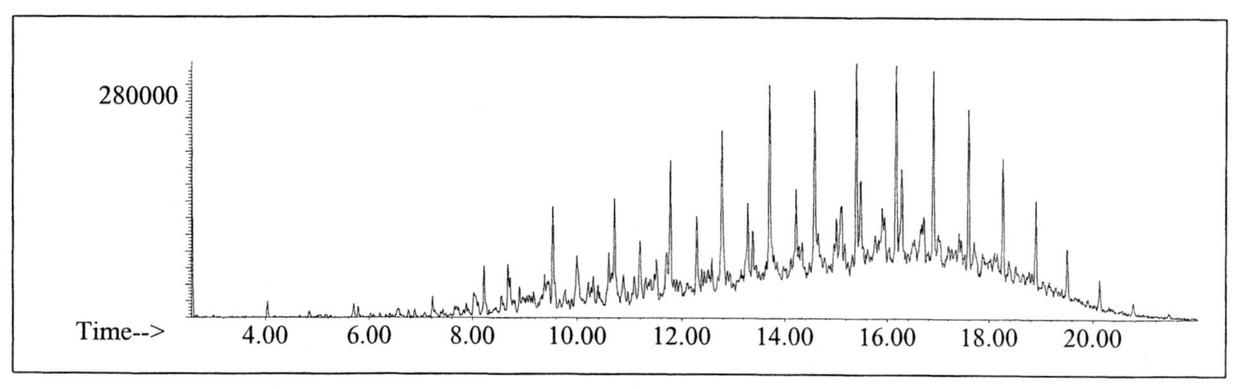

NAPHTHALENES

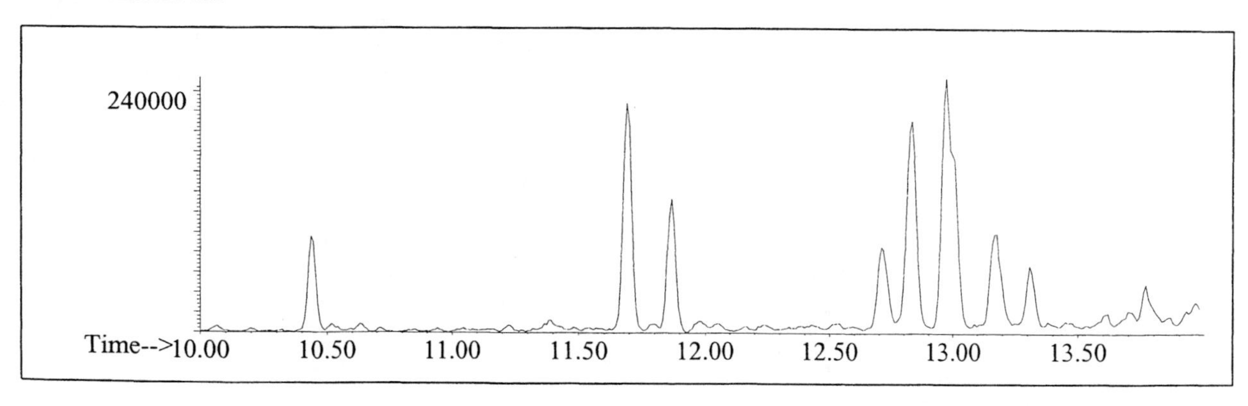

INDIVIDUAL PROFILES Distillate #49

Alkane

Ion 43

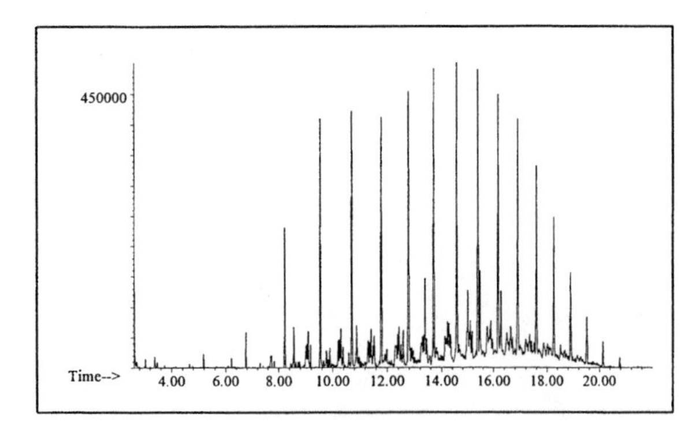

Ion 57

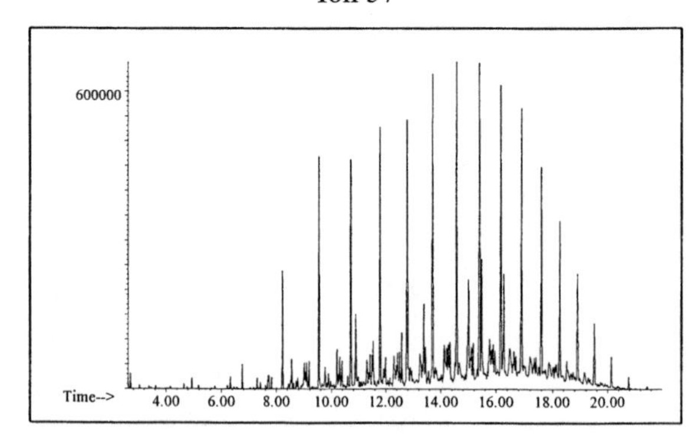

Ion 71

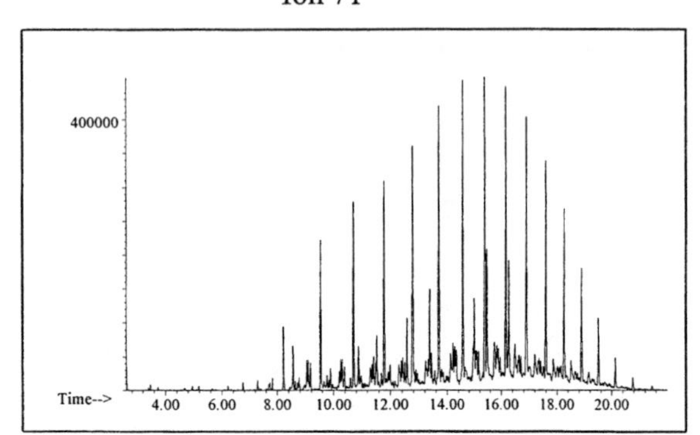

Ion 85

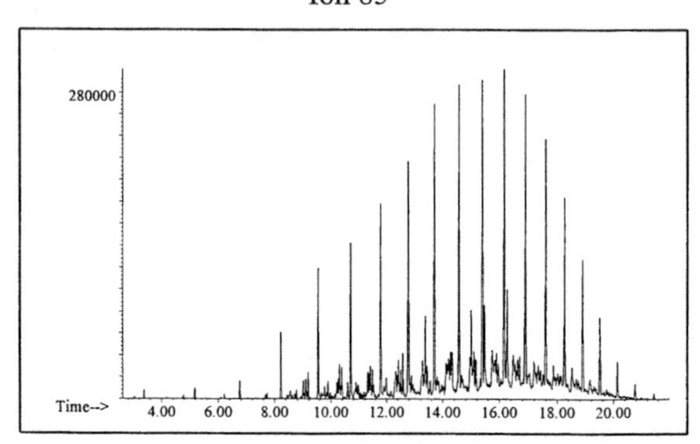

Aromatic

Ion 91

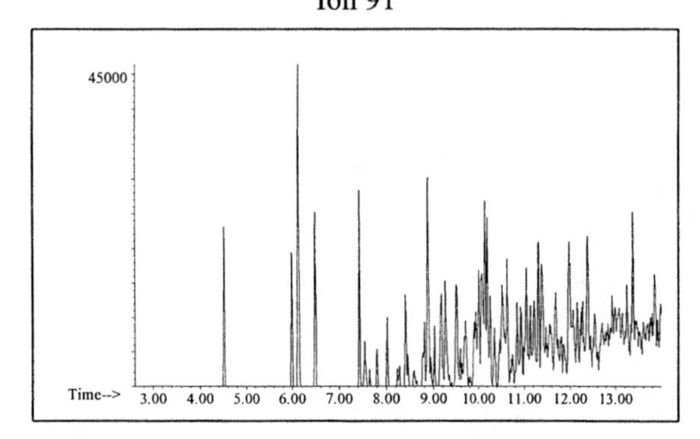

Ion 105

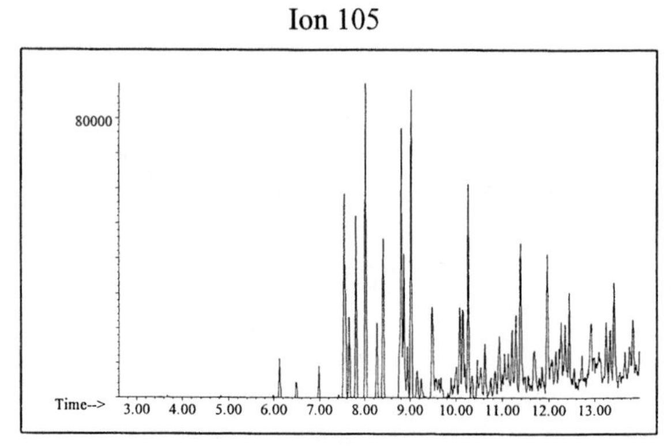

Ion 119

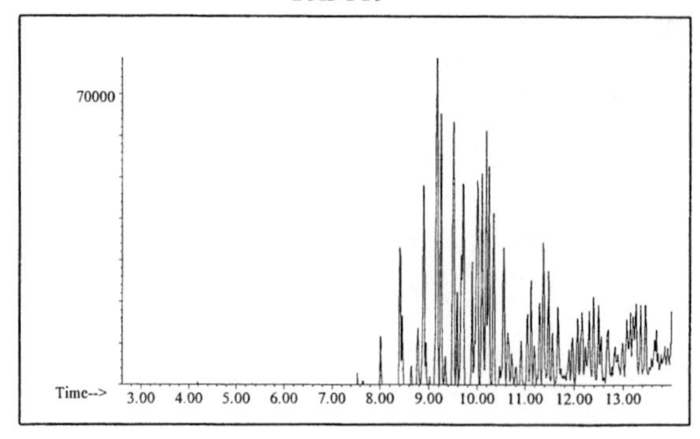

INDIVIDUAL PROFILES Distillate #49

Cycloparaffin

Ion 55

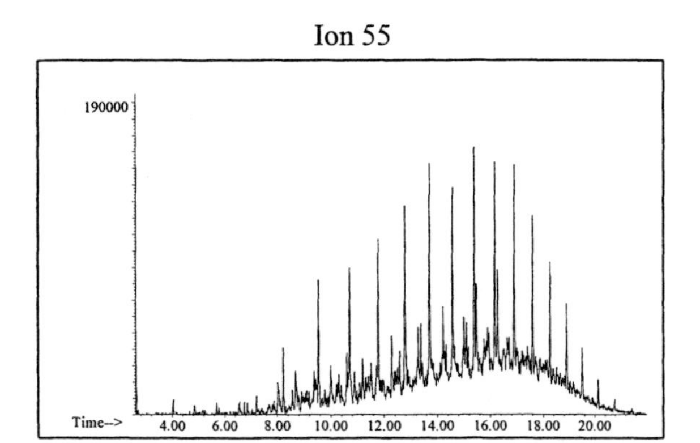

Ion 69

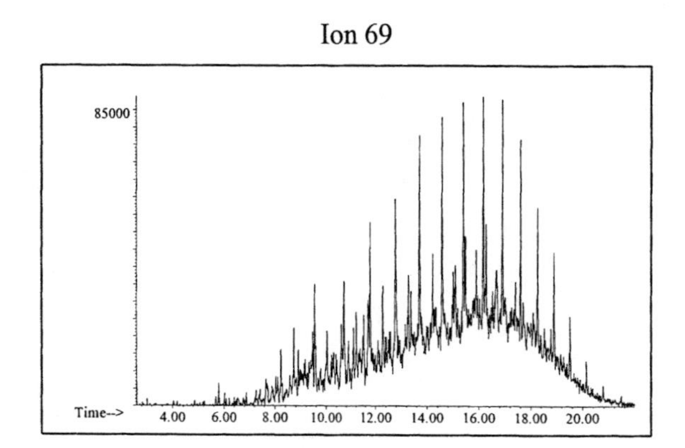

Ion 83

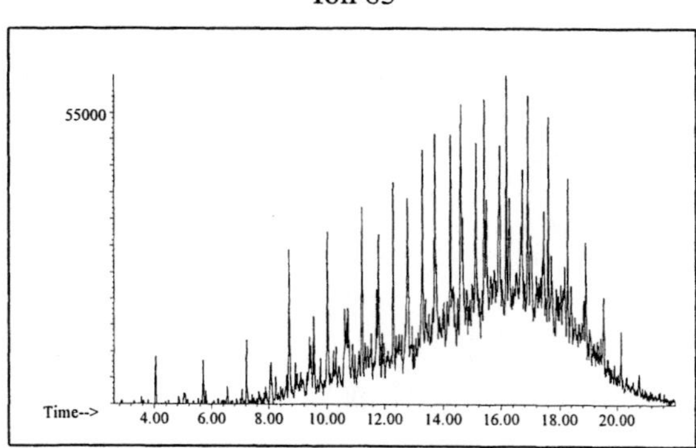

Naphthalene

Ion 128

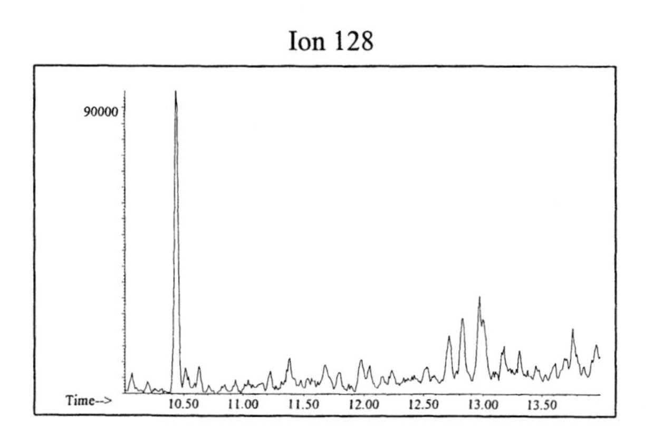

Ion 142

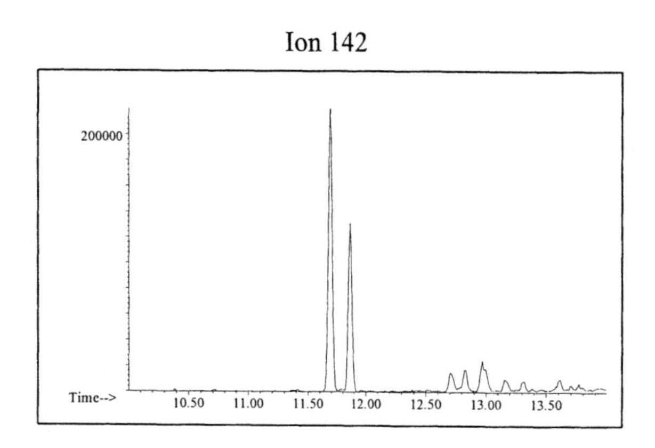

Ion 156

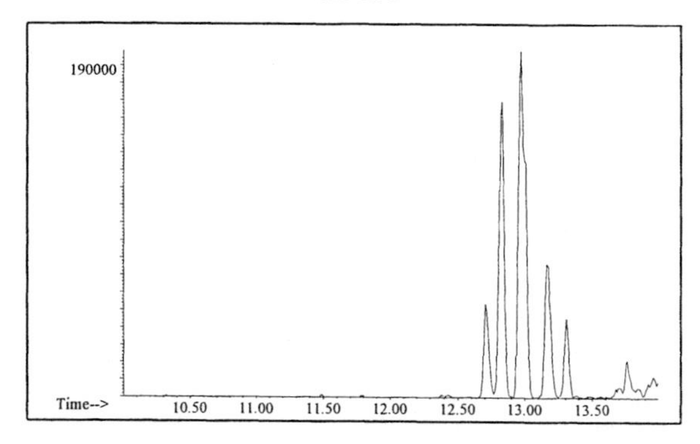

Distillate #50

20uL/mL Pentane

Product Displayed: 25% Evaporated Diesel Fuel

Other Similar Products:

ASTM: Class 5 (Heavy Petroleum Distillate)

Product Uses: Fuel

Macro Code: H

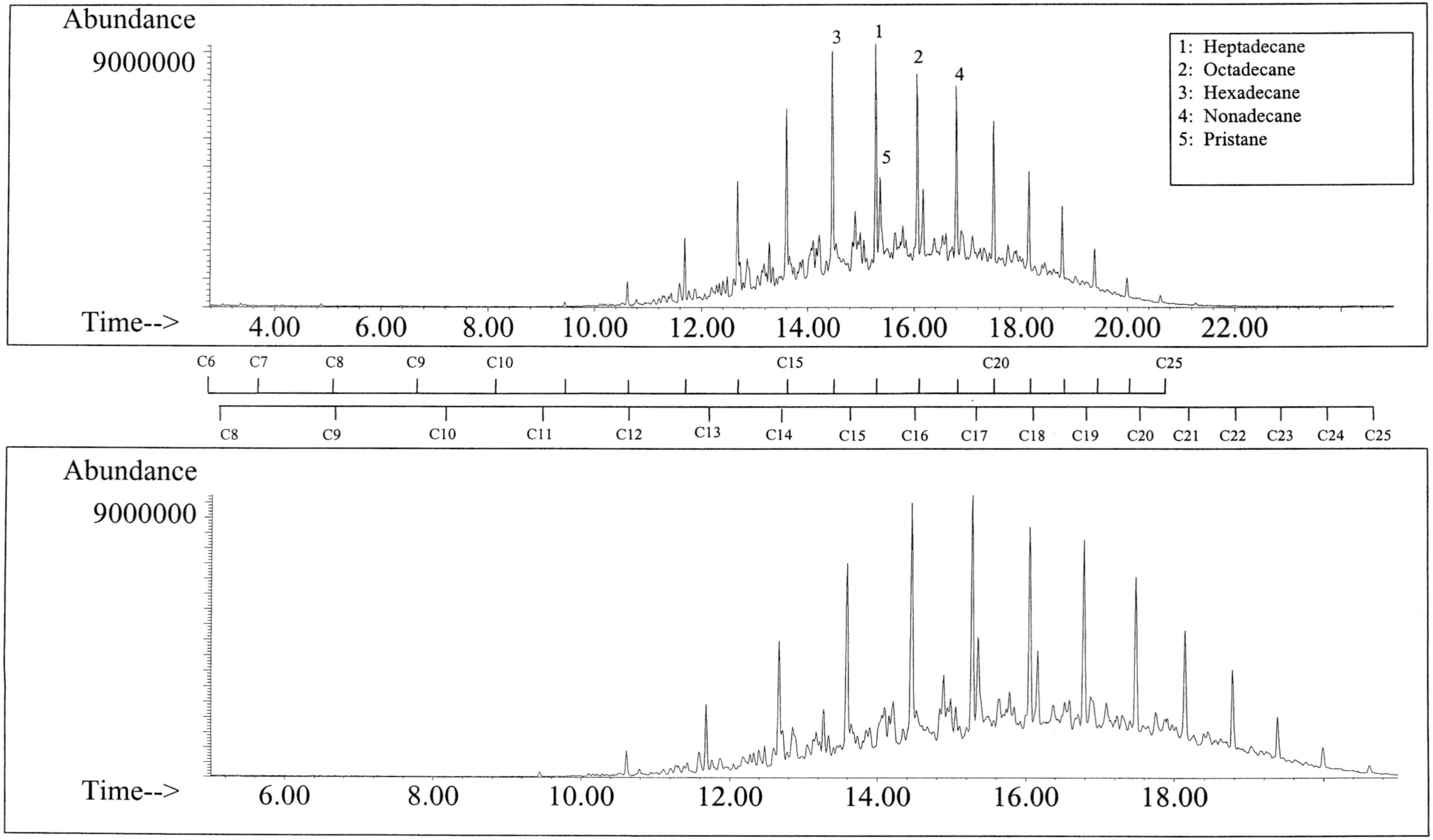

SUMMED PROFILES

Distillate #50

ALKANES

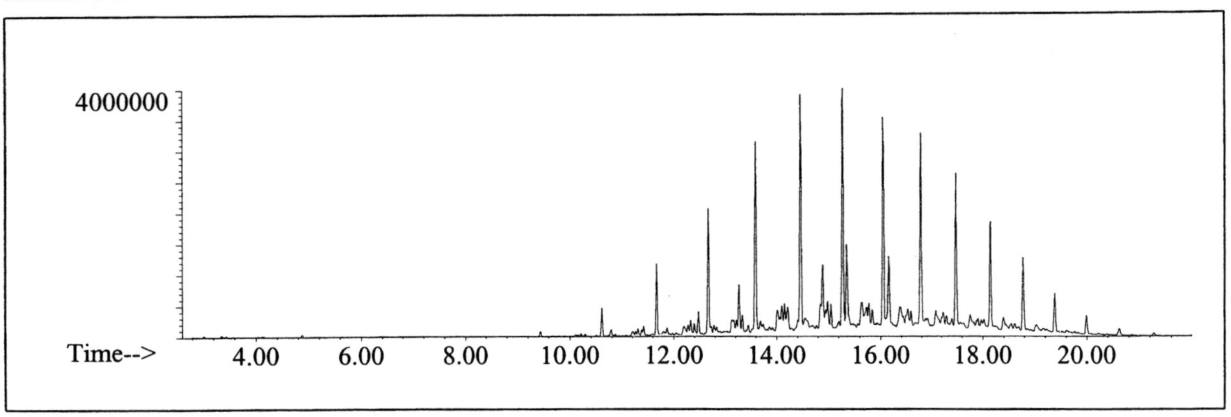

AROMATICS

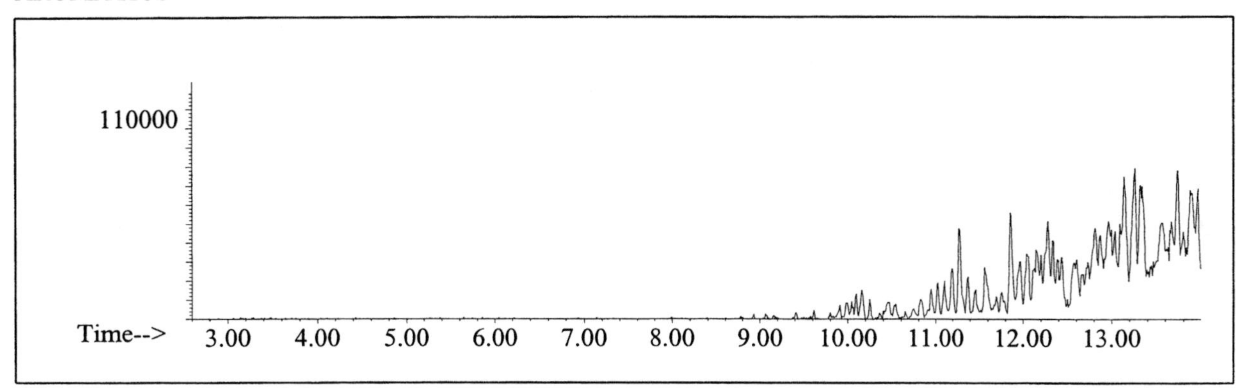

CYCLOPARAFFINS AND ALKENES

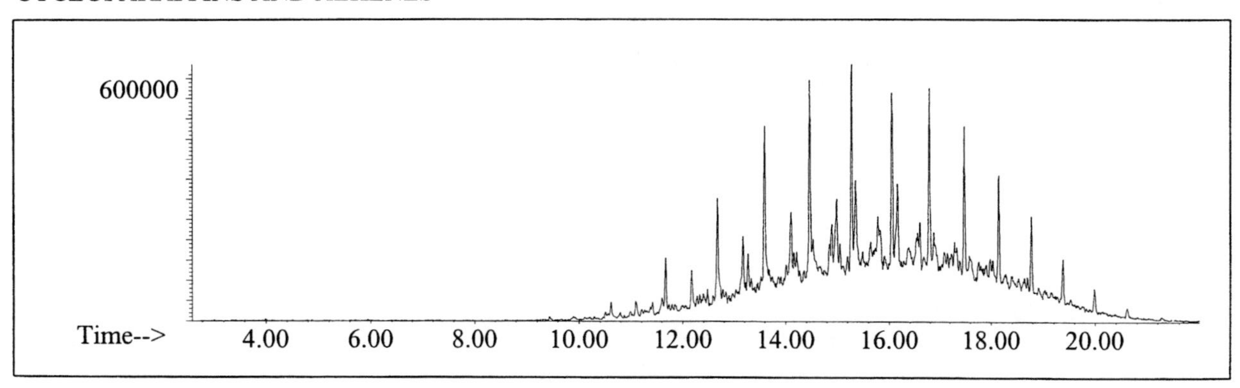

NAPHTHALENES

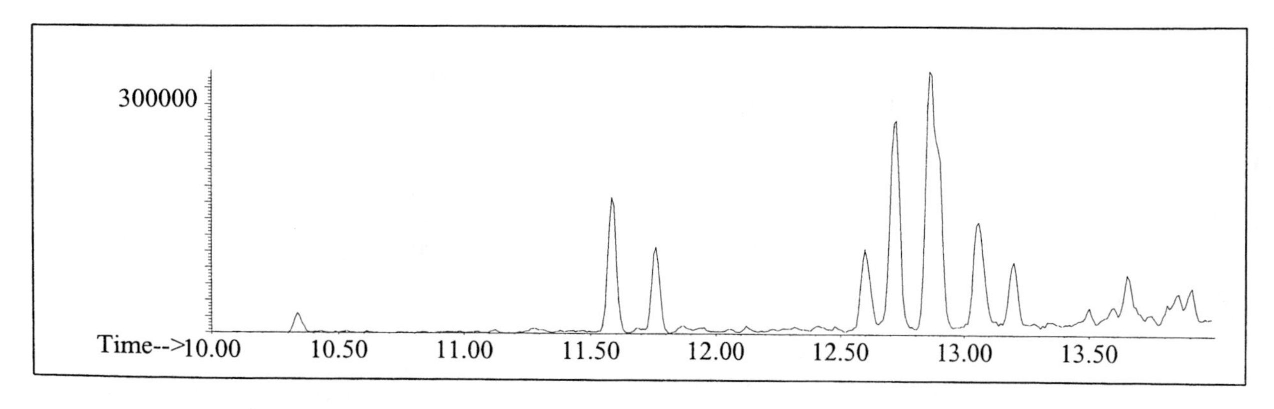

INDIVIDUAL PROFILES

Distillate #50

Alkane

Ion 43

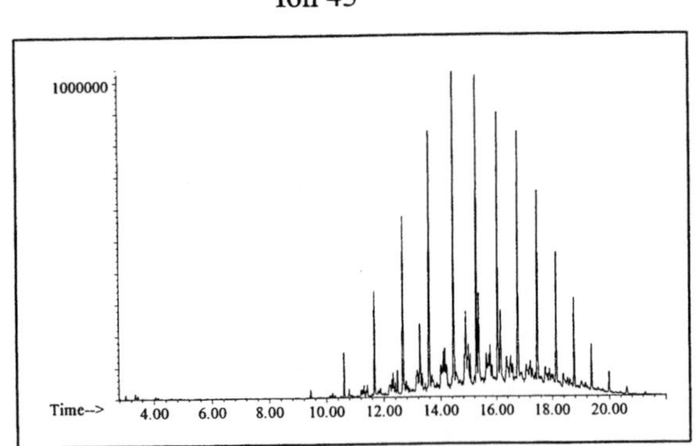

Ion 57

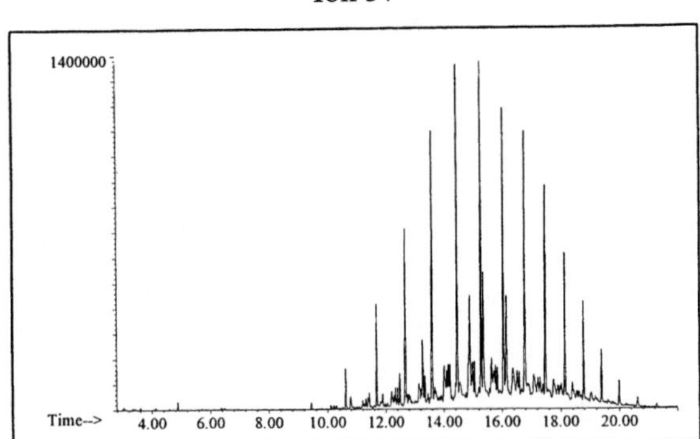

Ion 71

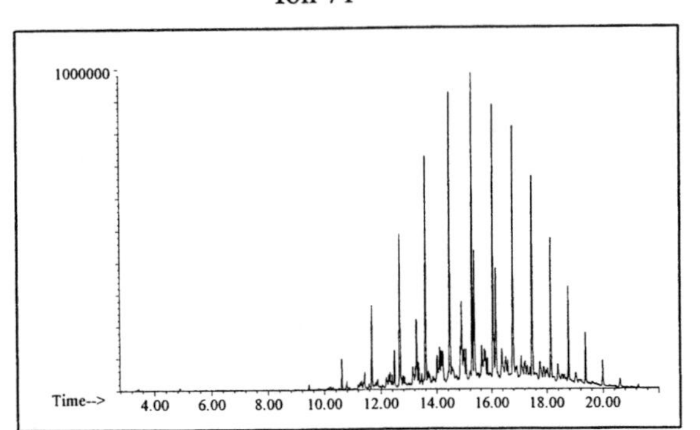

Ion 85

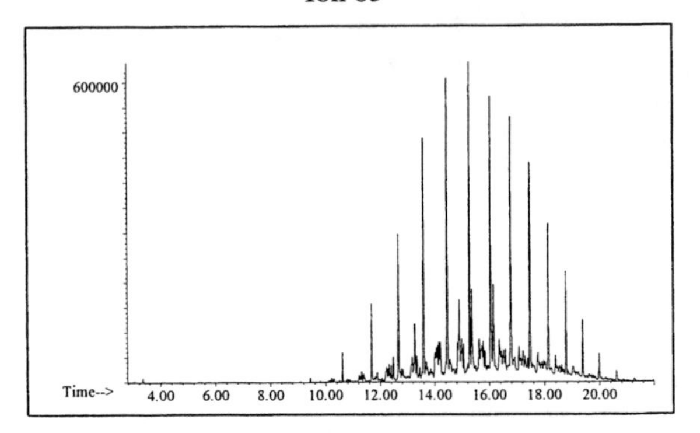

Aromatic

Ion 91

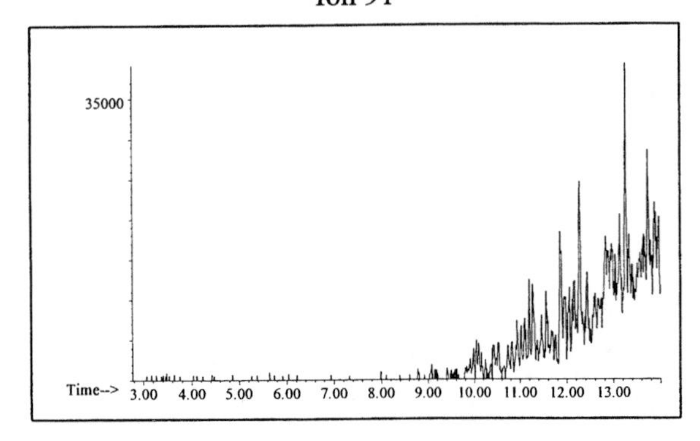

Ion 105

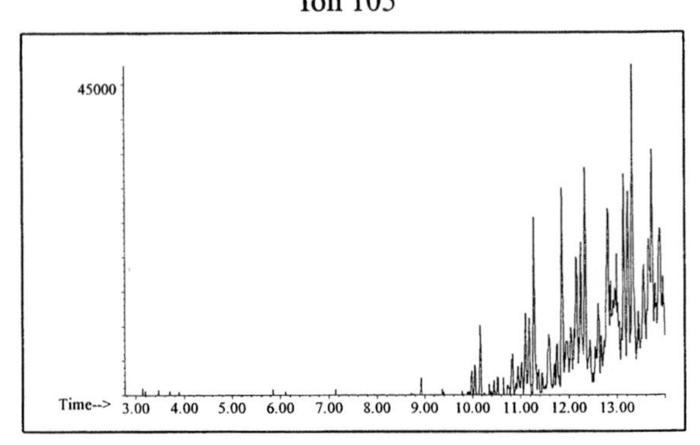

Ion 119

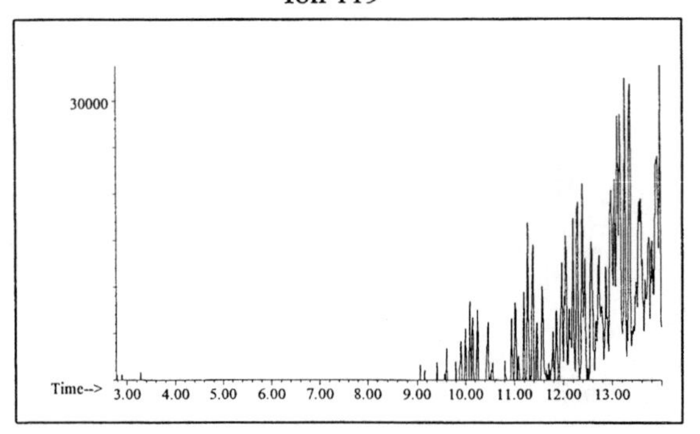

INDIVIDUAL PROFILES

Distillate #50

Cycloparaffin

Ion 55

Ion 69

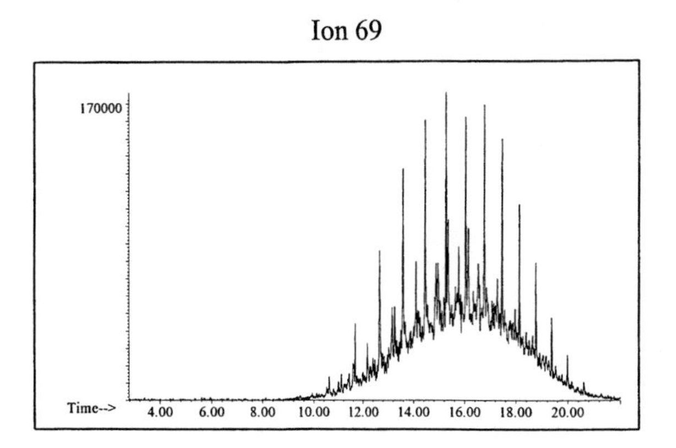

Ion 83

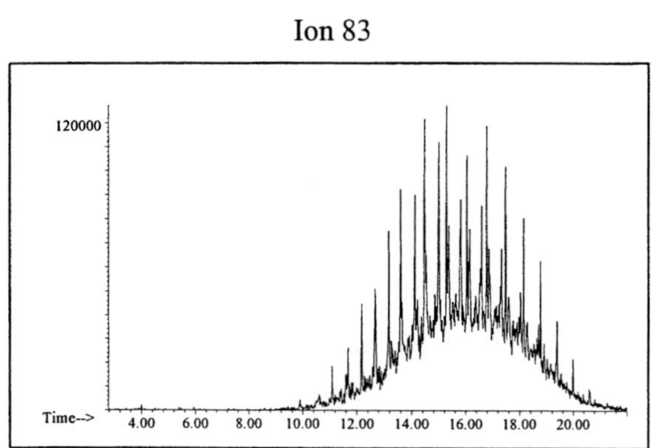

Naphthalene

Ion 128

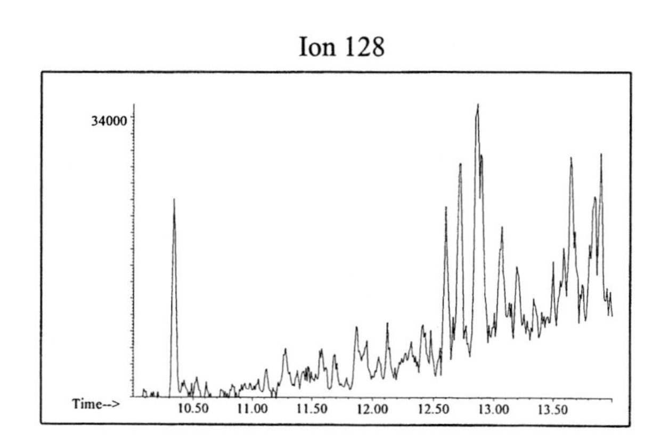

Ion 142

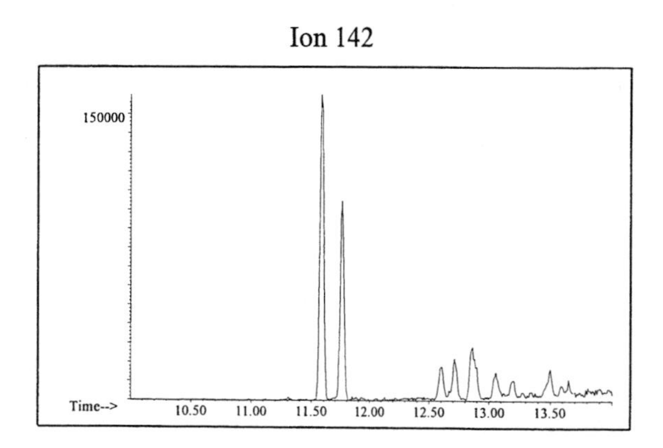

Ion 156

Distillate #51

20uL/mL Pentane

Product Displayed: 50% Evaporated Diesel Fuel

Other Similar Products:

ASTM: Class 5 (Heavy Petroleum Distillate)

Product Uses: Automotive fuels, marine fuels

Macro Code: H

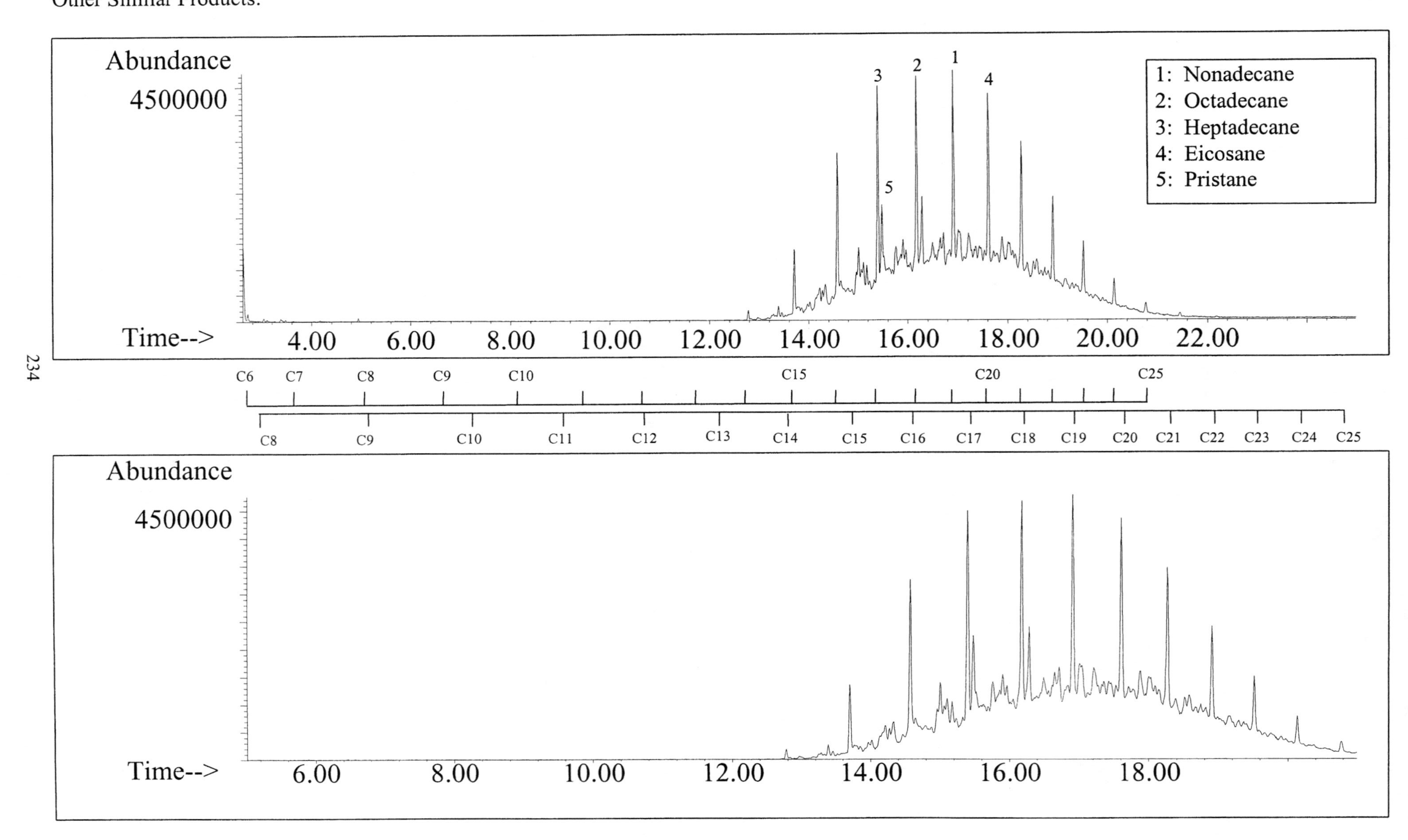

SUMMED PROFILES

Distillate #51

ALKANES

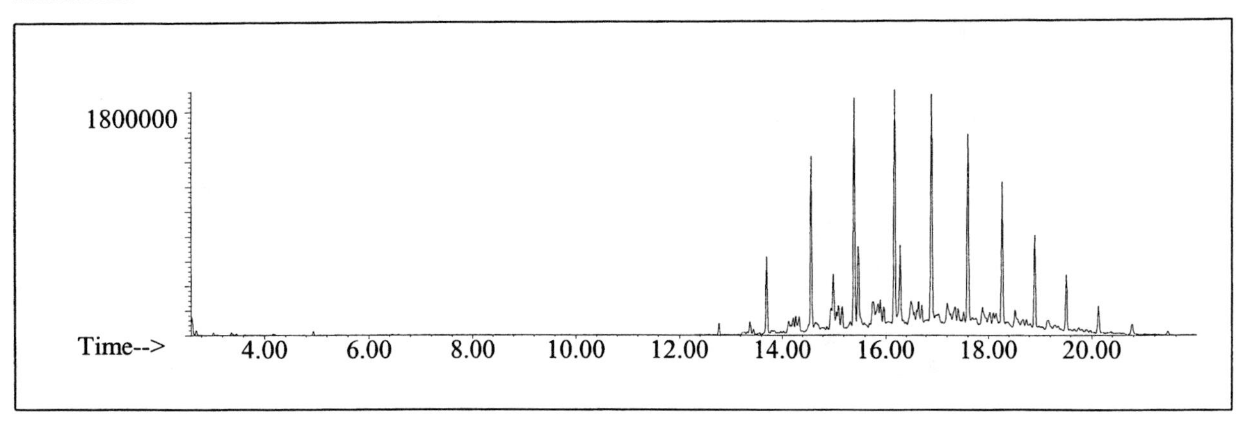

AROMATICS

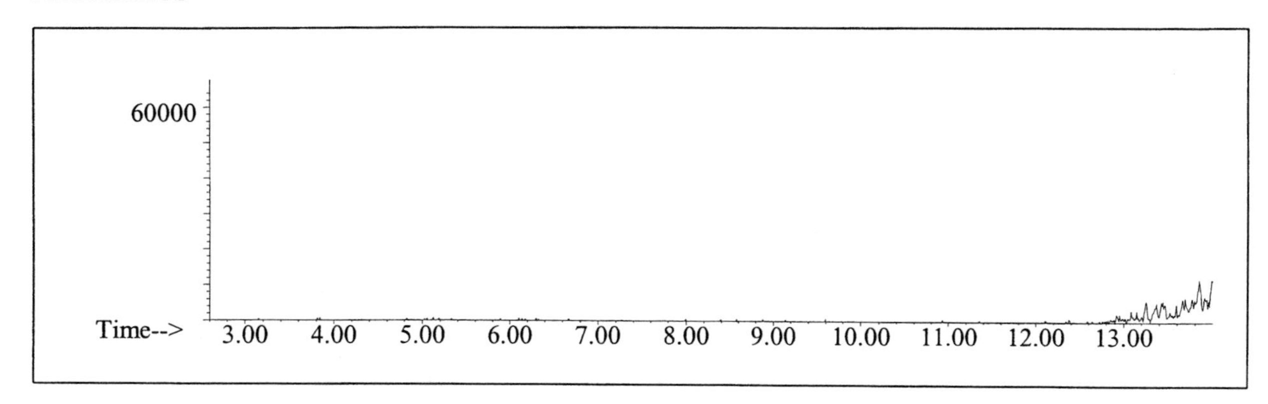

CYCLOPARAFFINS AND ALKENES

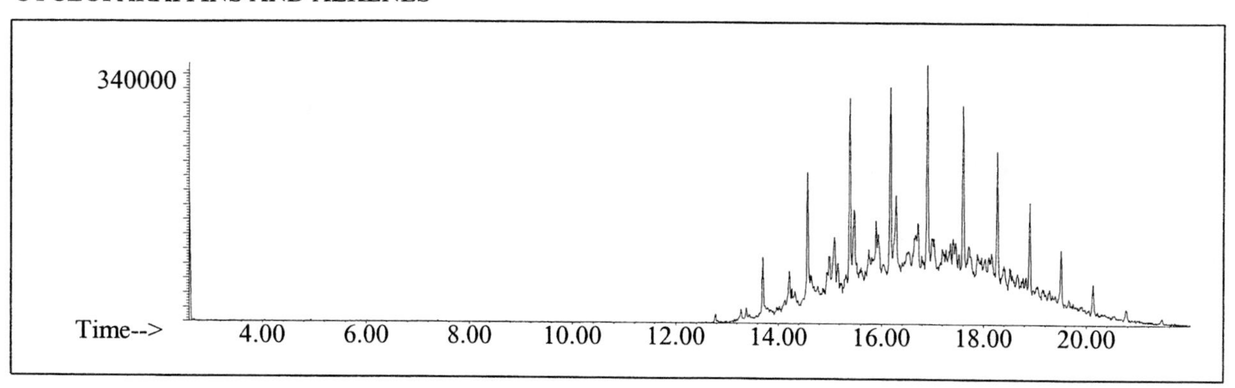

NAPHTHALENES

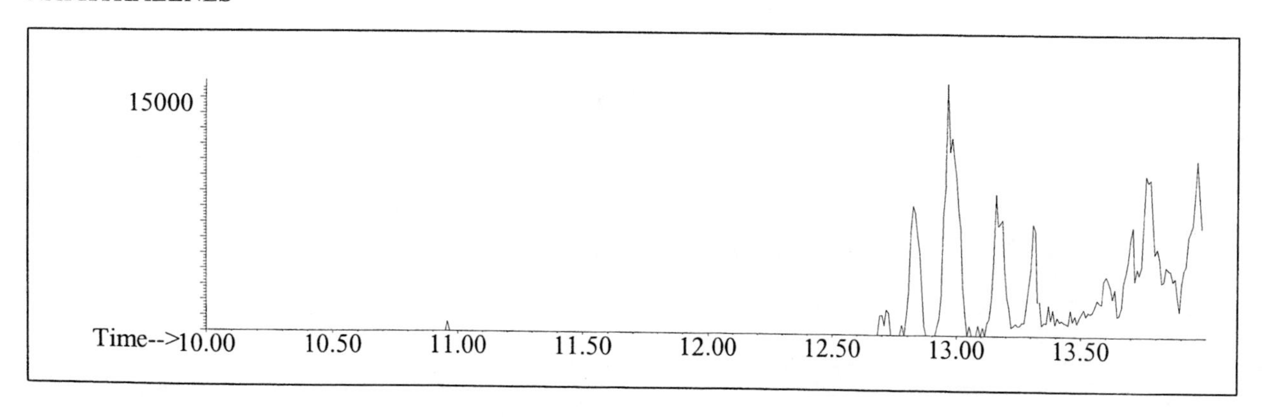

INDIVIDUAL PROFILES

Distillate #51

Alkane

Ion 43

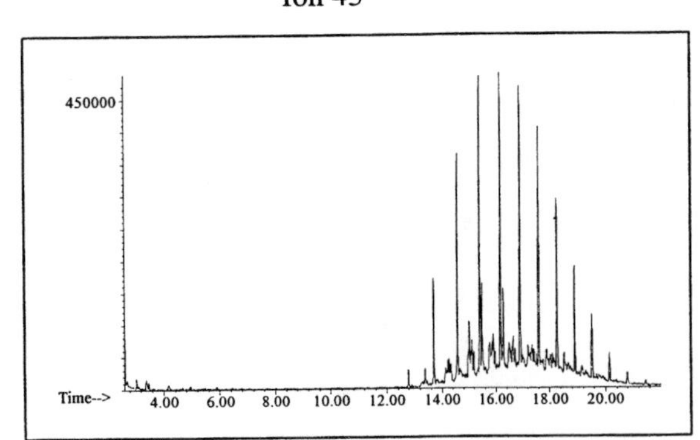

Ion 57

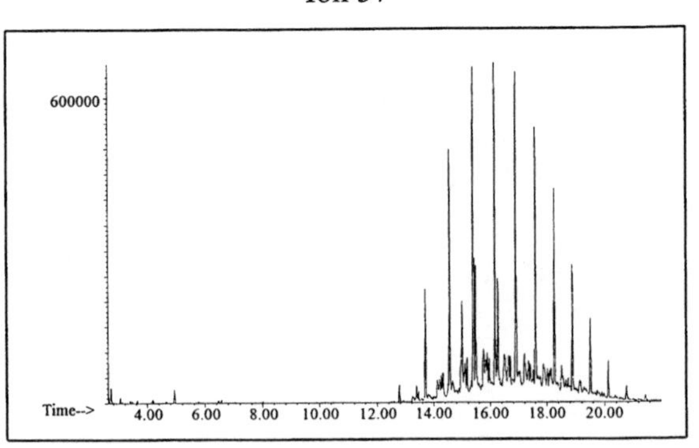

Ion 77

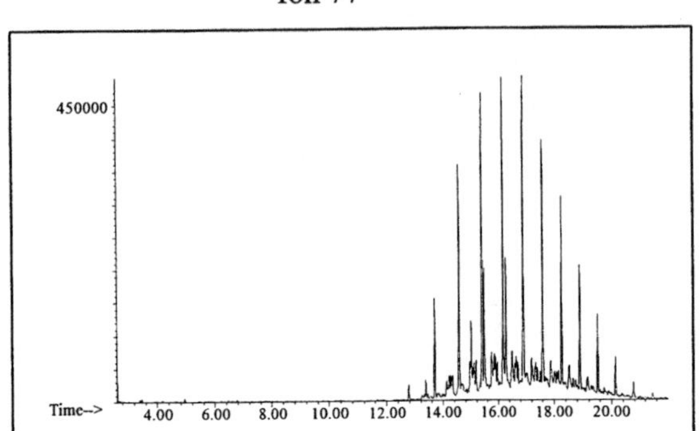

Ion 85

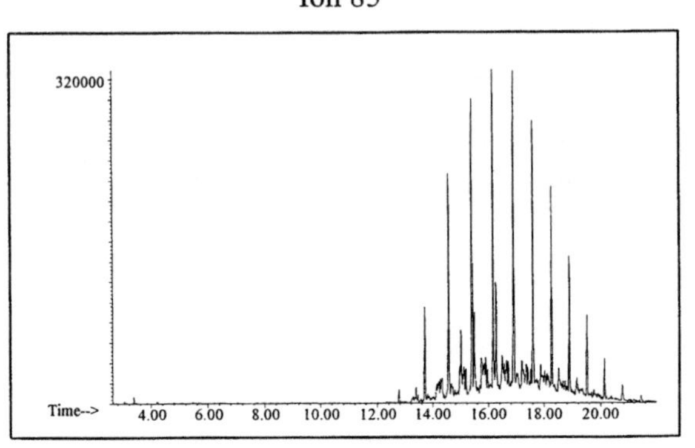

Aromatic

Ion 91

No Useful Data Obtained

Ion 105

No Useful Data Obtained

Ion 119

No Useful Data Obtained

INDIVIDUAL PROFILES

Distillate #51

Cycloparaffin

Ion 55

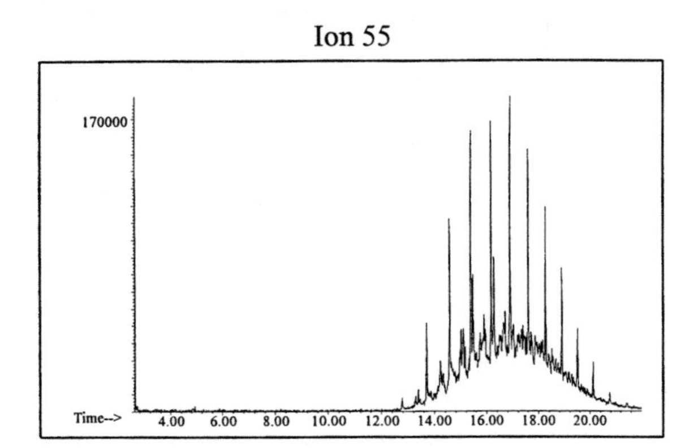

Ion 69

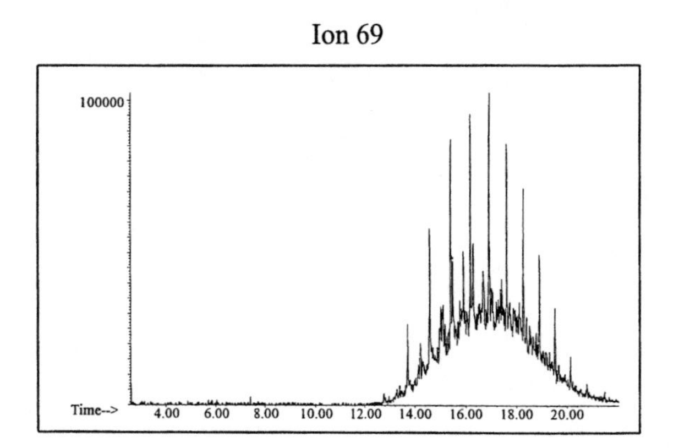

Ion 83

100000

Time-->

4.00 6.00 8.00 10.00 12.00 14.00 16.00 18.00 20.00

Naphthalene

Ion 128

No Useful Data Obtained

Ion 142

No Useful Data Obtained

Ion 156

12500

Time-->

10.50 11.00 11.50 12.00 12.50 13.00 13.50

Distillate #52

20uL/mL Pentane

Product Displayed: 75% Evaporated Diesel Fuel

Other Similar Products:

ASTM: Class 5 (Heavy Petroleum Distillate)

Product Uses: Fuel

Macro Code: H

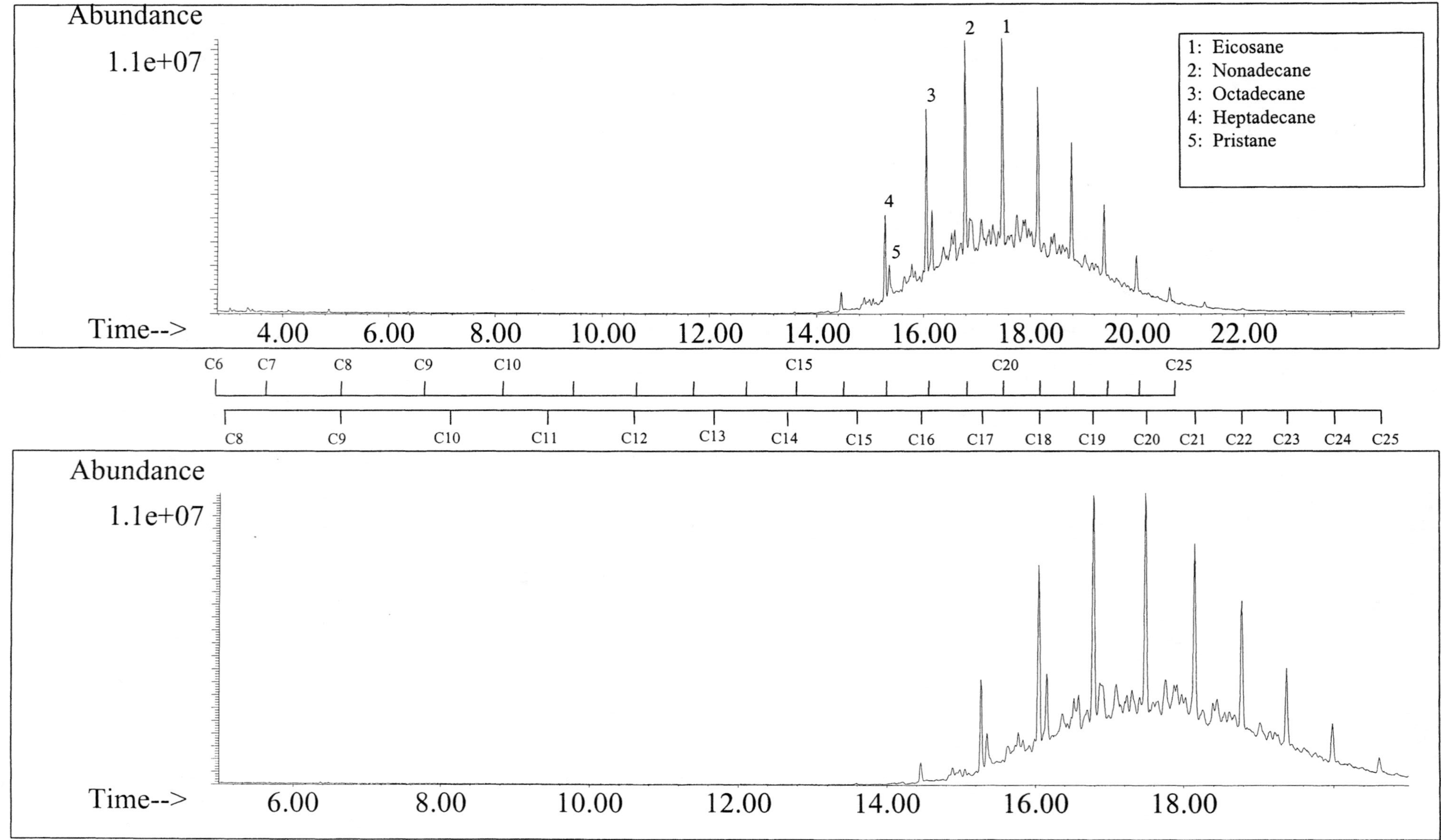

SUMMED PROFILES

Distillate #52

ALKANES

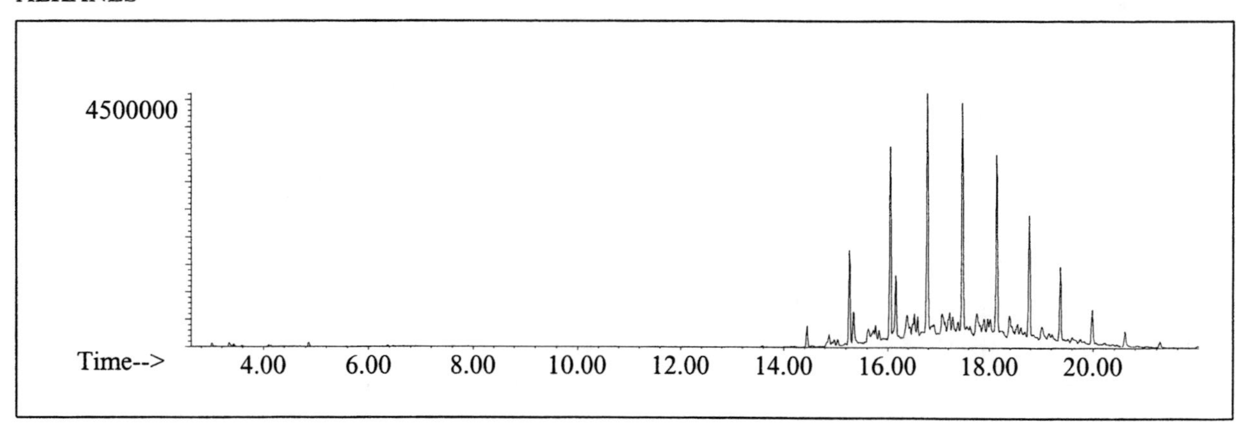

AROMATICS

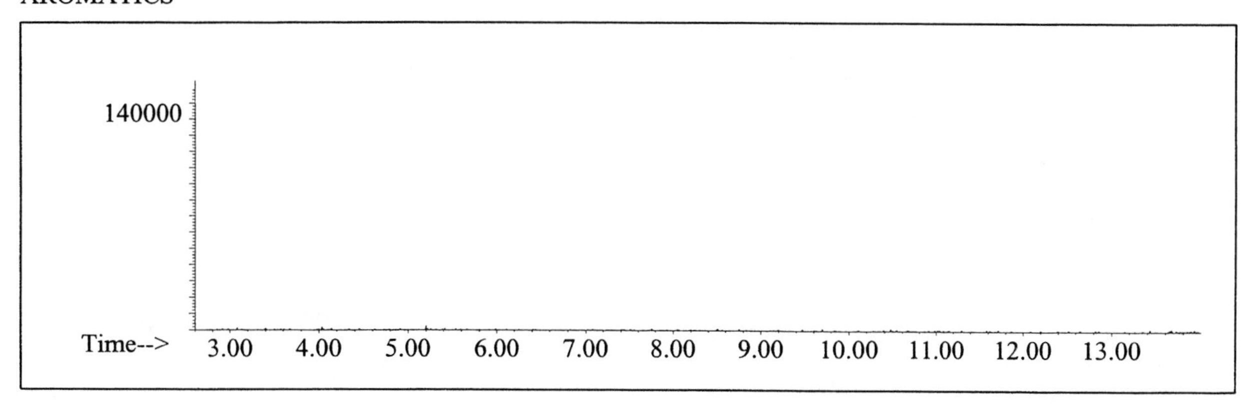

CYCLOPARAFFINS AND ALKENES

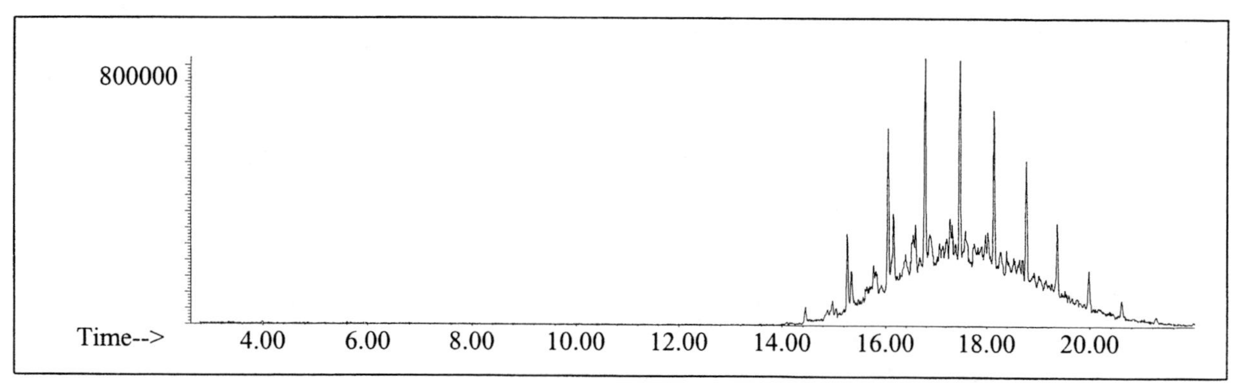

NAPHTHALENES

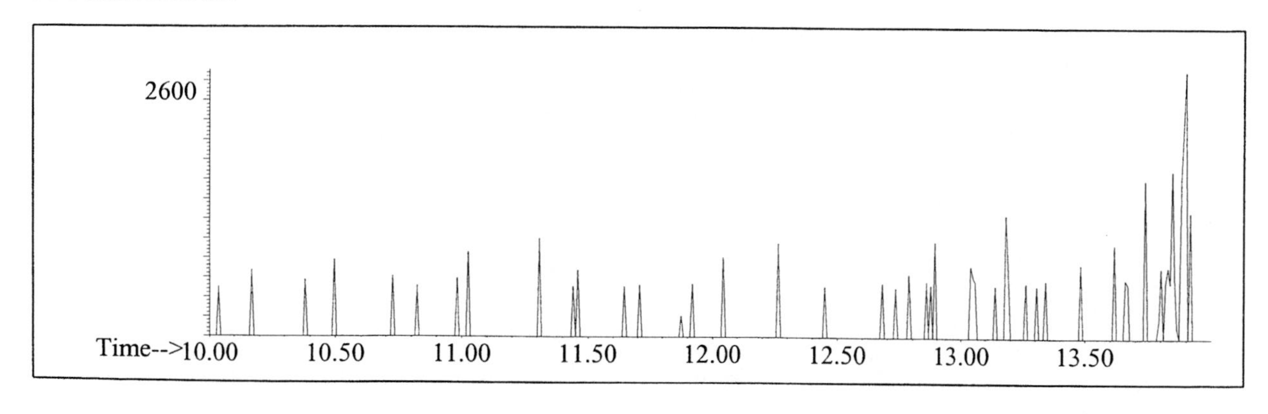

INDIVIDUAL PROFILES

Distillate #52

Alkane

Ion 43

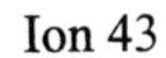

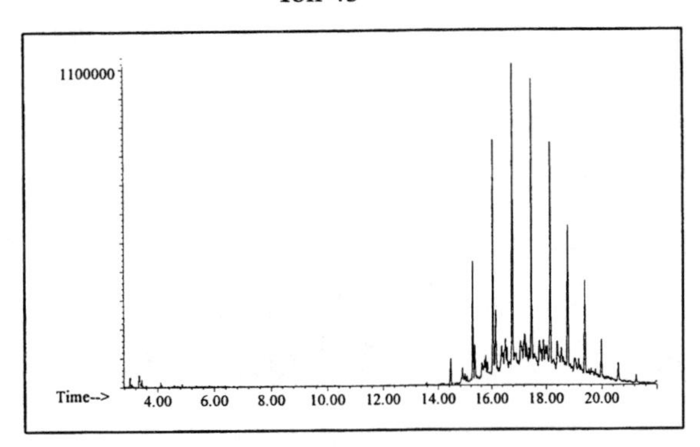

Ion 57

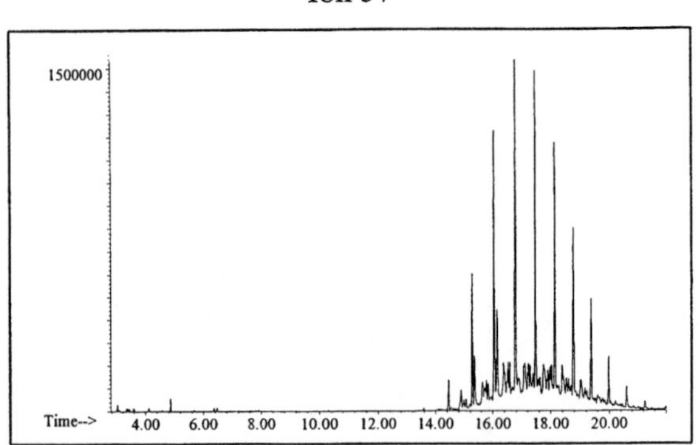

Ion 71

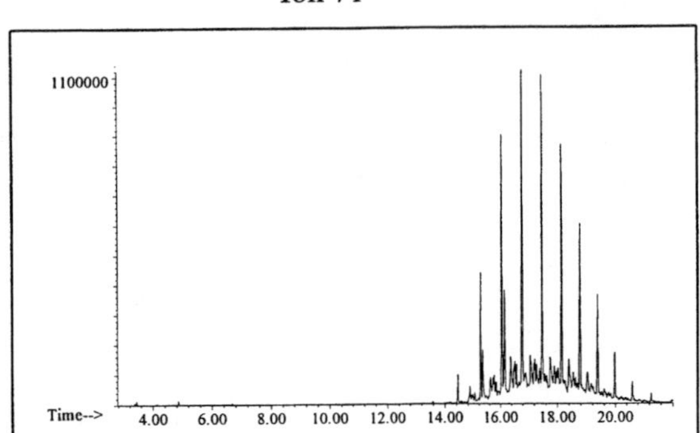

Ion 85

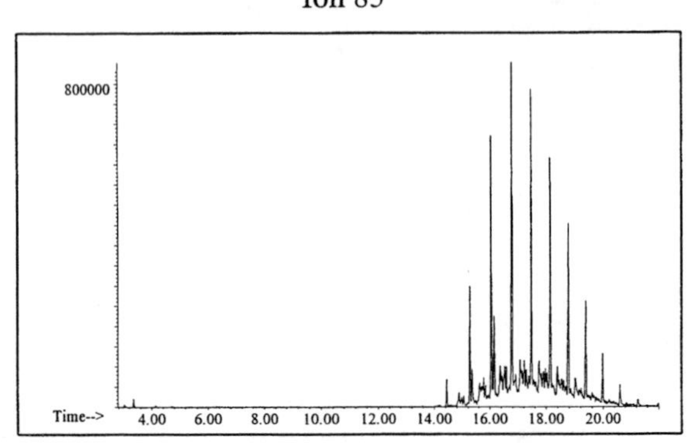

Aromatic

Ion 91

No Useful Data Obtained

Ion 105

No Useful Data Obtained

Ion 119

No Useful Data Obtained

INDIVIDUAL PROFILES

Distillate #52

Cycloparaffin

Ion 55

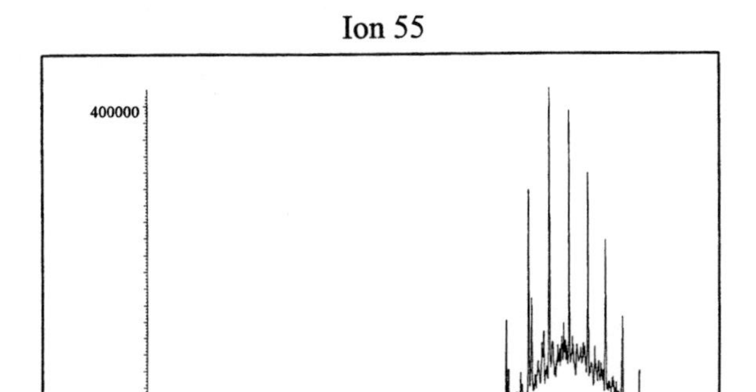

Ion 69

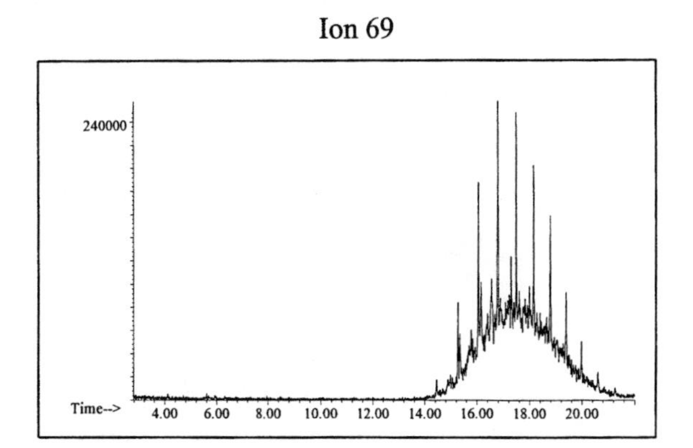

Ion 83

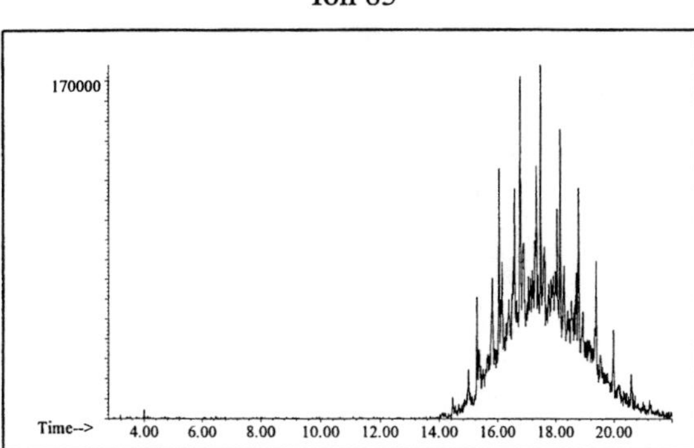

Naphthalene

Ion 128

No Useful Data Obtained

Ion 142

No Useful Data Obtained

Ion 156

No Useful Data Obtained

Distillate #53

20uL/mL Pentane

Product Displayed: Snap Octane Treatment

Other Similar Products:

ASTM: Class 5 (Heavy Petroleum Distillate)

Macro Code: H

Product Uses: Engine treatment

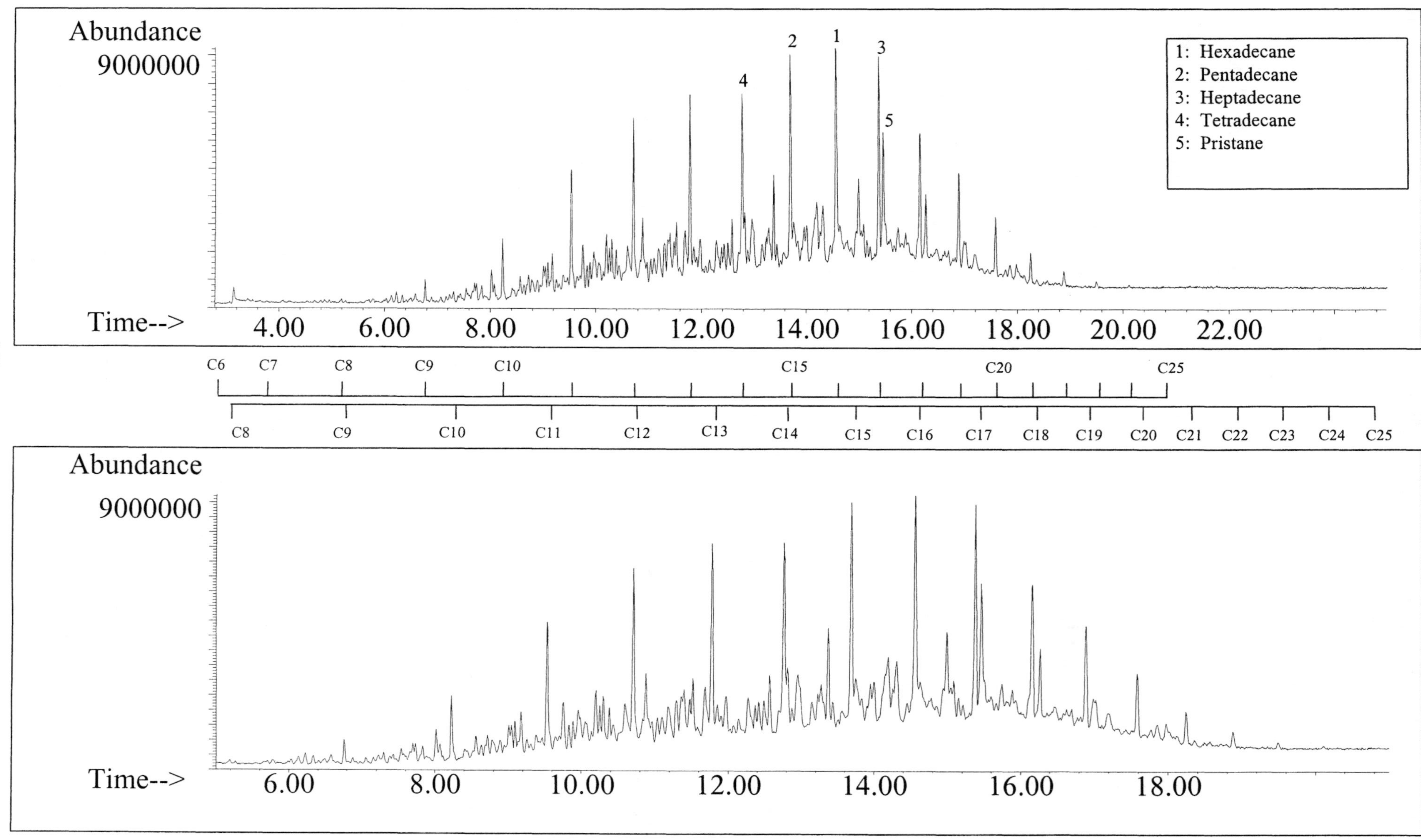

SUMMED PROFILES Distillate #53

ALKANES

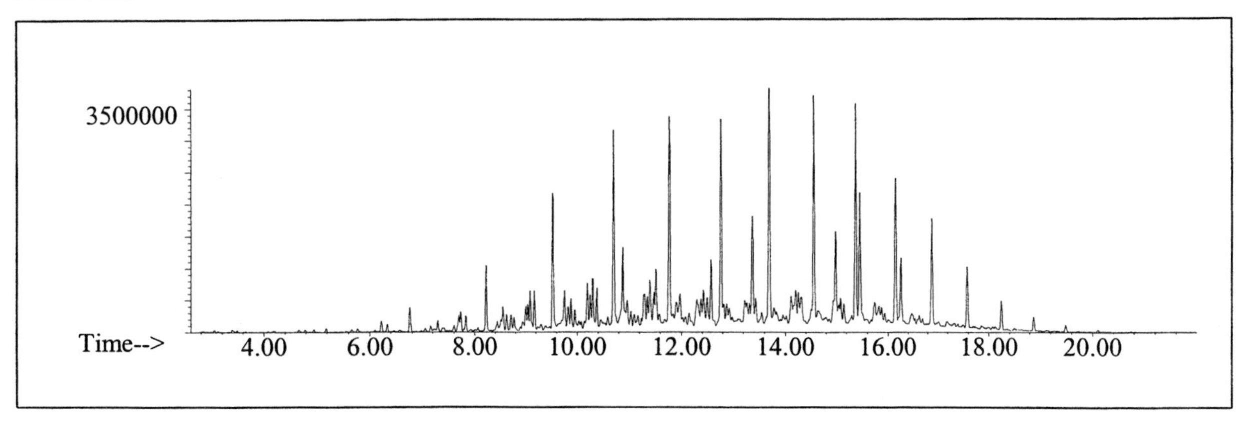

AROMATICS

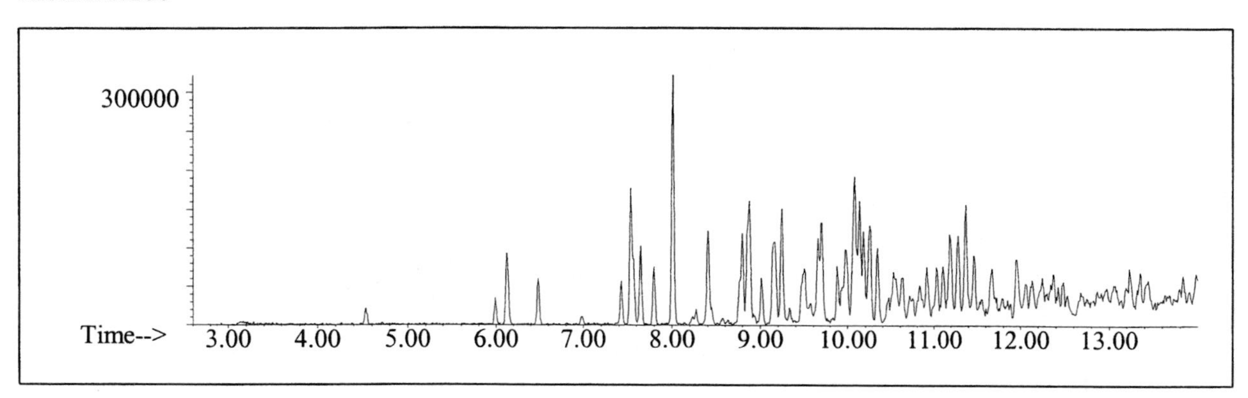

CYCLOPARAFFINS AND ALKENES

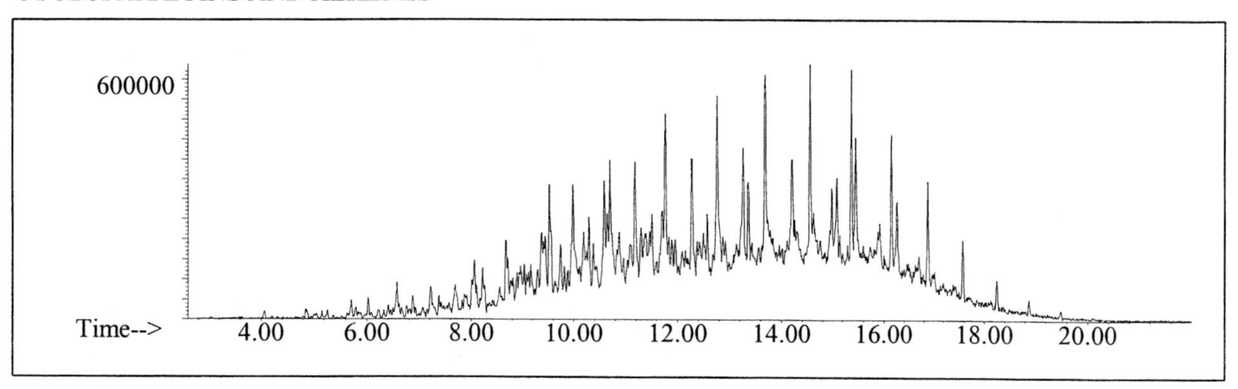

NAPHTHALENES

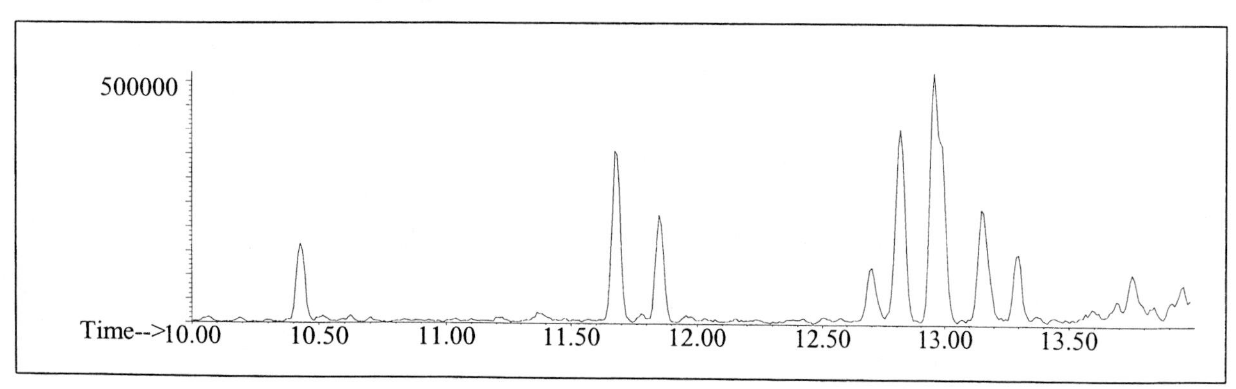

INDIVIDUAL PROFILES

Distillate #53

Alkane

Ion 43

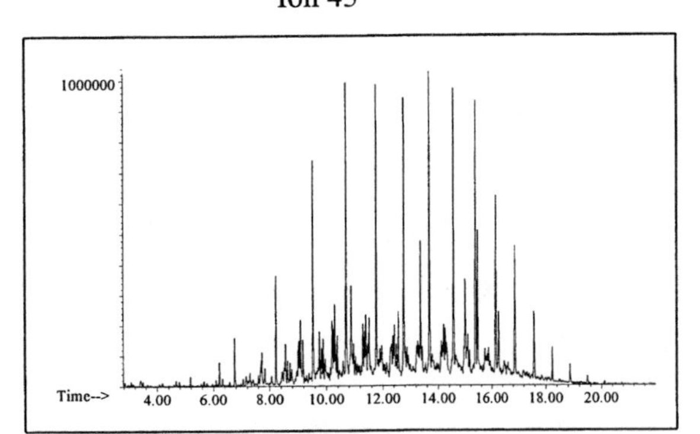

Ion 57

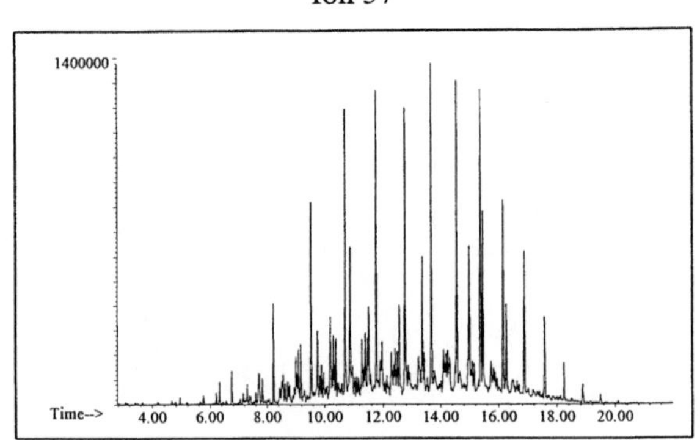

Ion 71

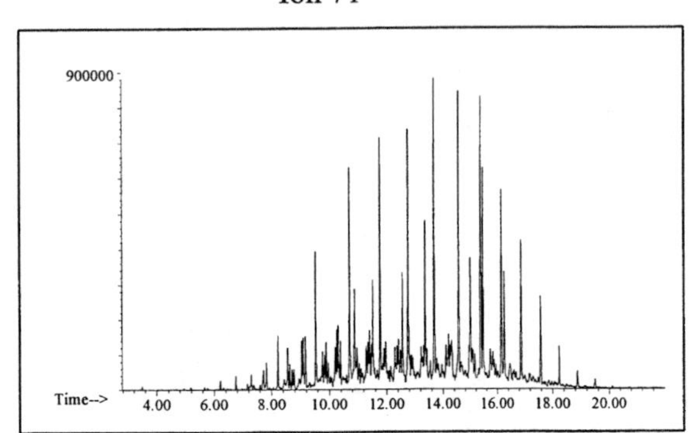

Ion 85

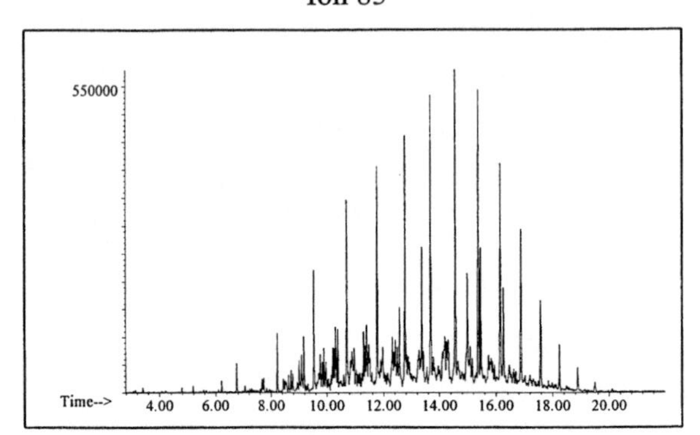

Aromatic

Ion 91

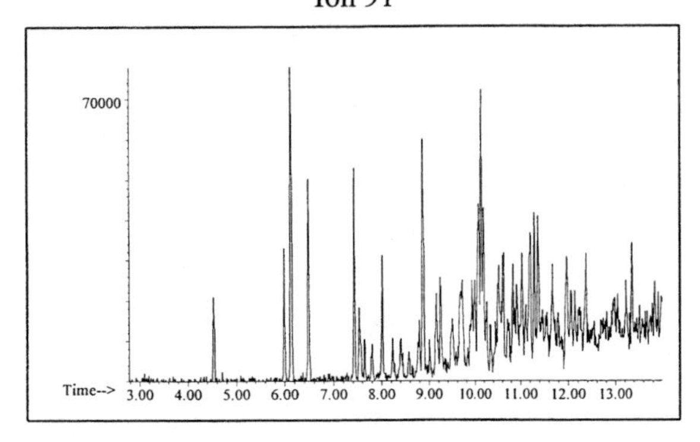

Ion 105

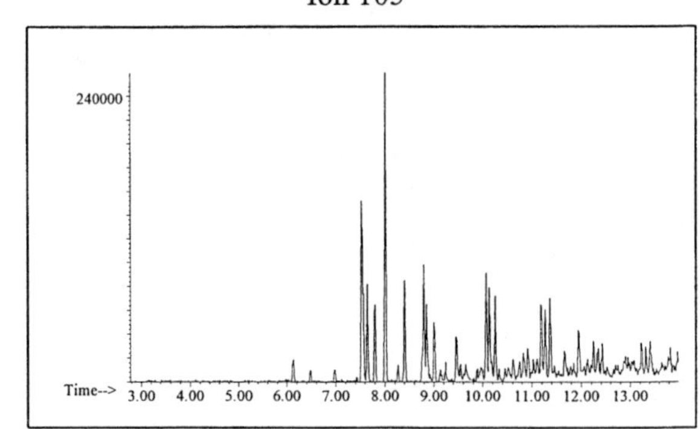

Ion 119

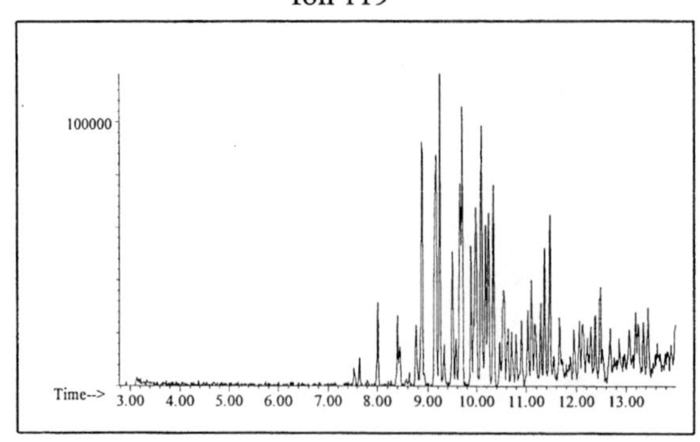

INDIVIDUAL PROFILES

Distillate #53

Cycloparaffin

Ion 55

320000
Time-->
4.00 6.00 8.00 10.00 12.00 14.00 16.00 18.00 20.00

Ion 69

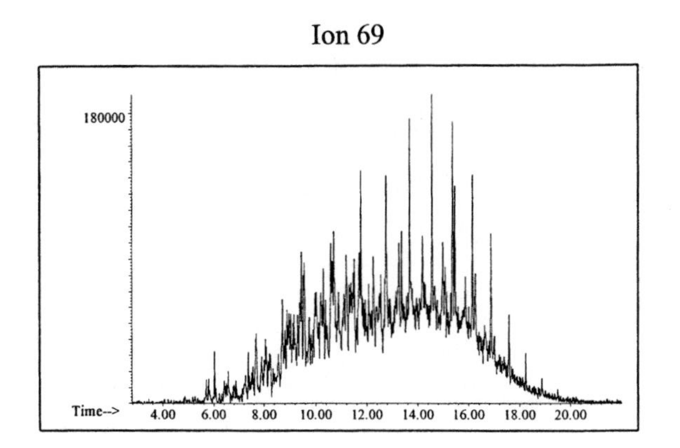

Ion 83

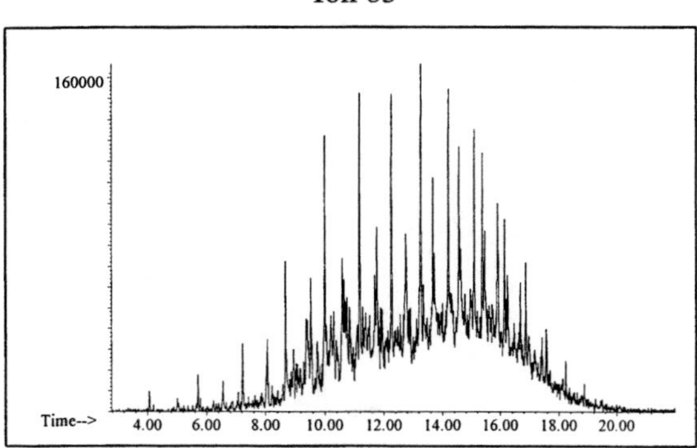

Naphthalene

Ion 128

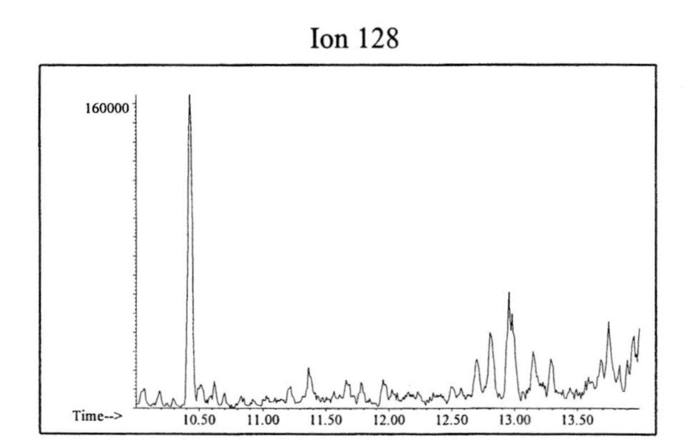

Ion 142

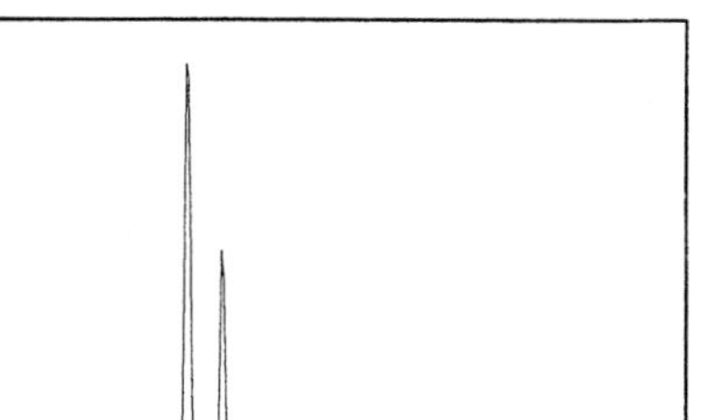

Ion 156

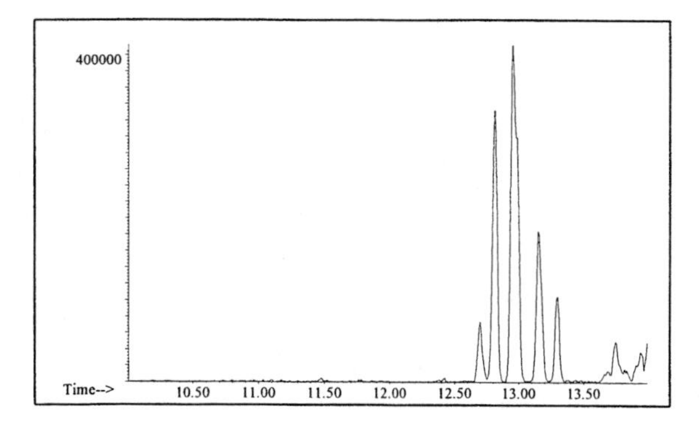

Distillate #54

20uL/mL Pentane

Product Displayed: Old Harbor Lamp Oil

Other Similar Products:

ASTM: Class 5 (Heavy Petroleum Distillate)

Product Uses: Lamp oil

Macro Code: H

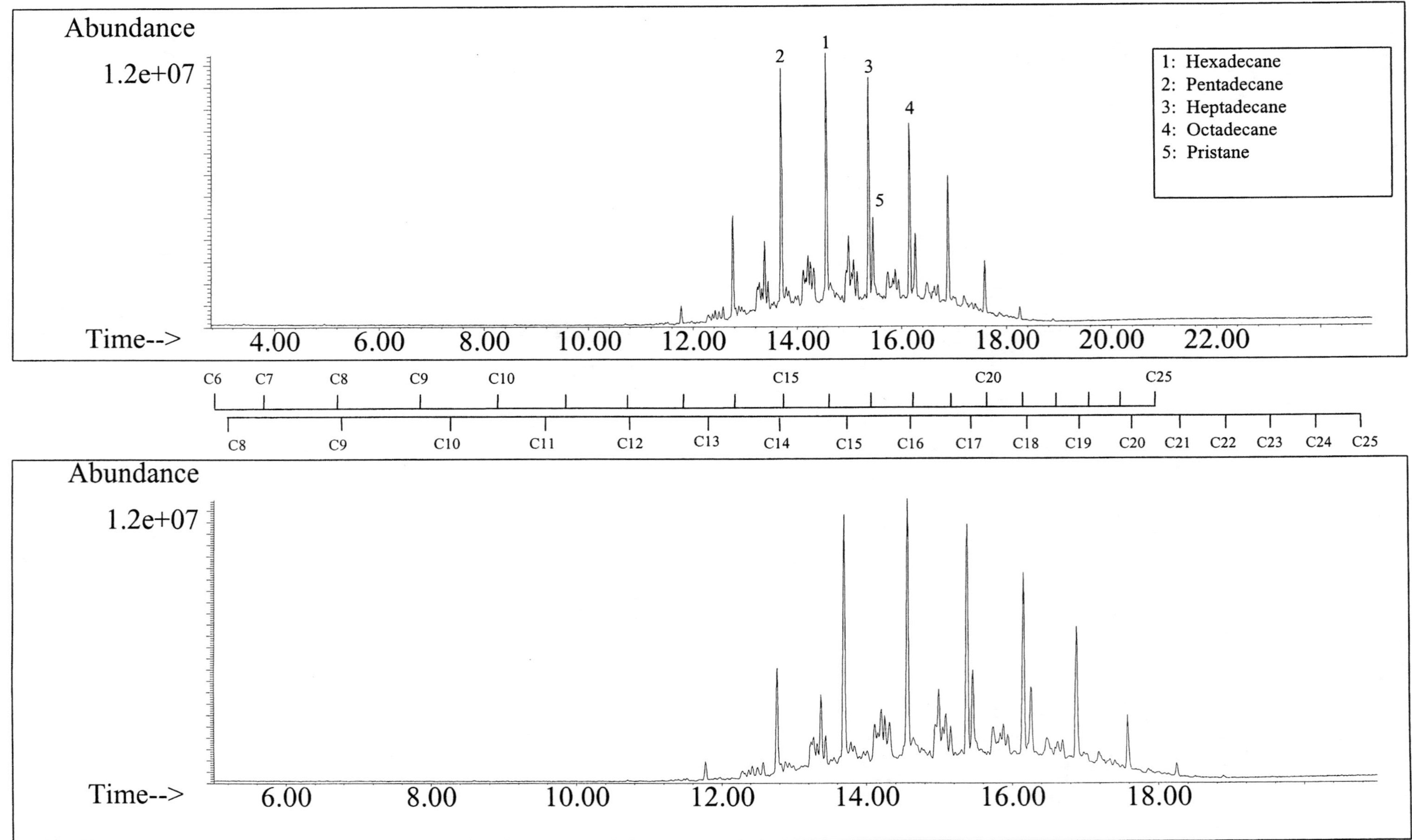

SUMMED PROFILES

Distillate #54

ALKANES

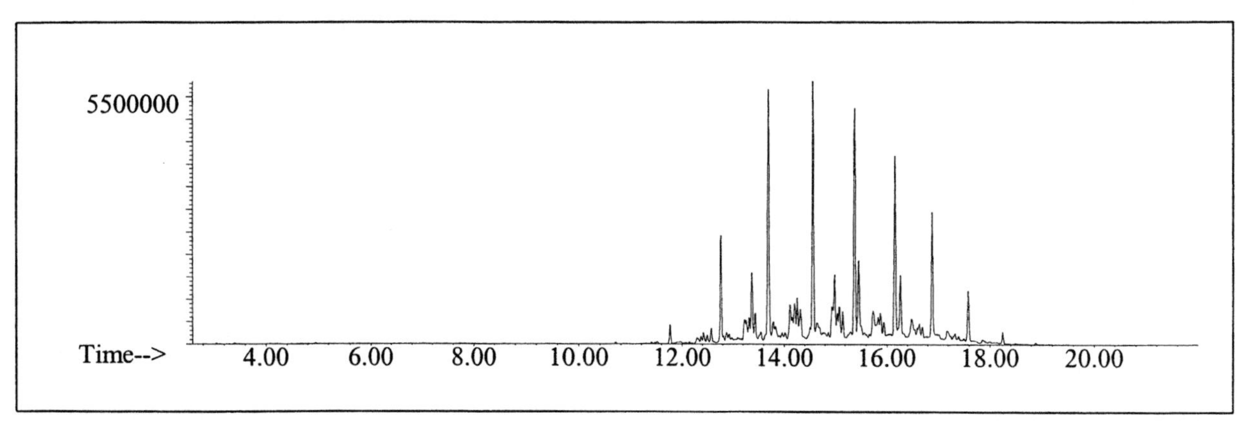

AROMATICS

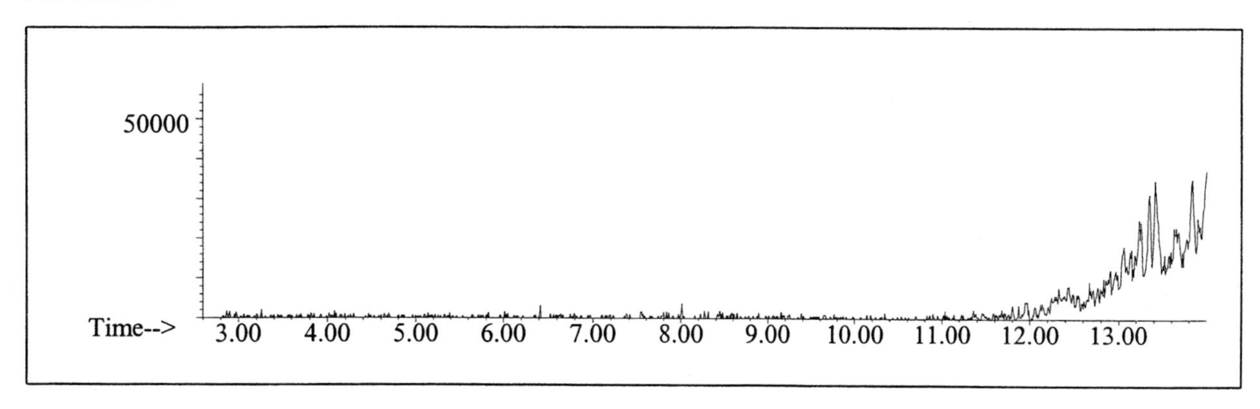

CYCLOPARAFFINS AND ALKENES

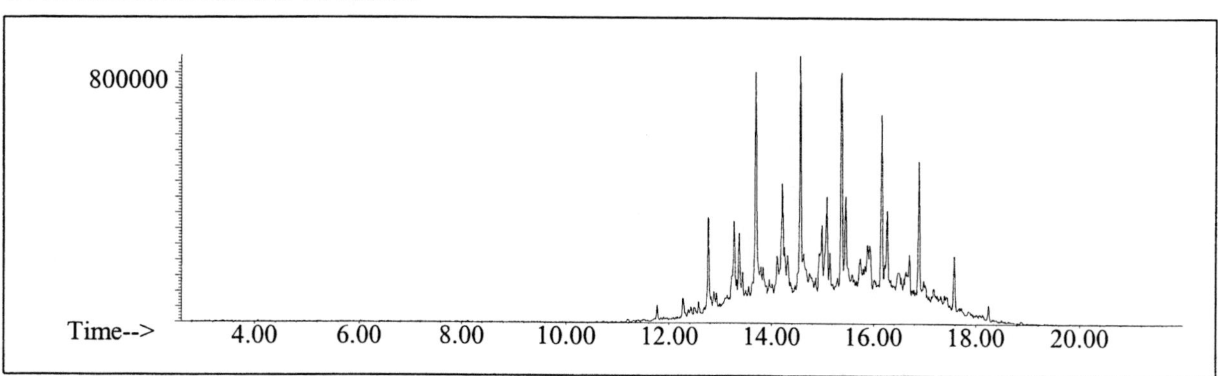

NAPHTHALENES

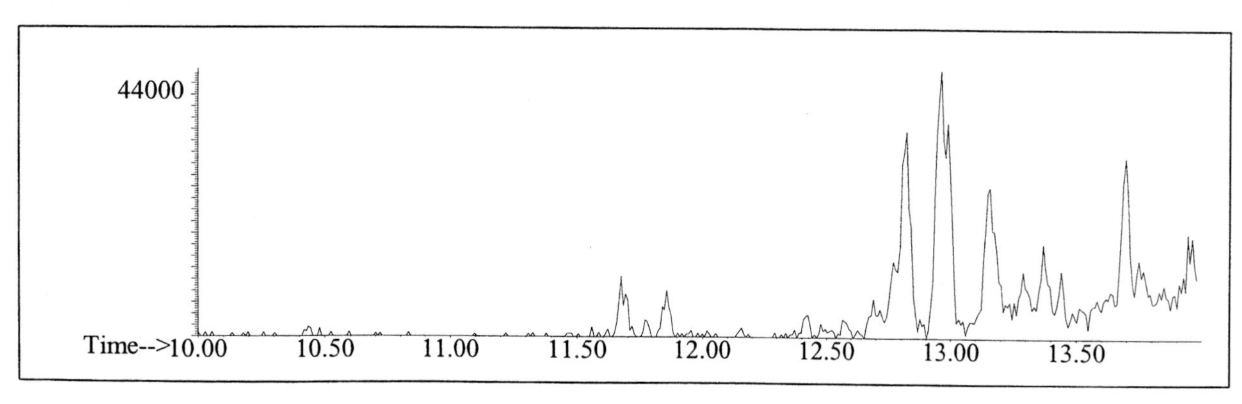

INDIVIDUAL PROFILES

Distillate #54

Alkane

Ion 43

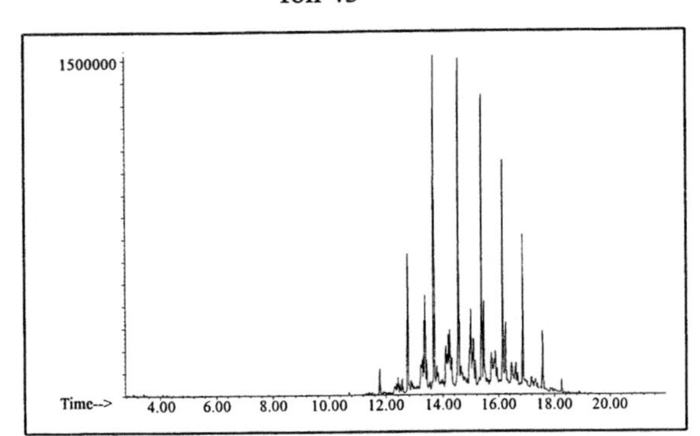

Ion 57

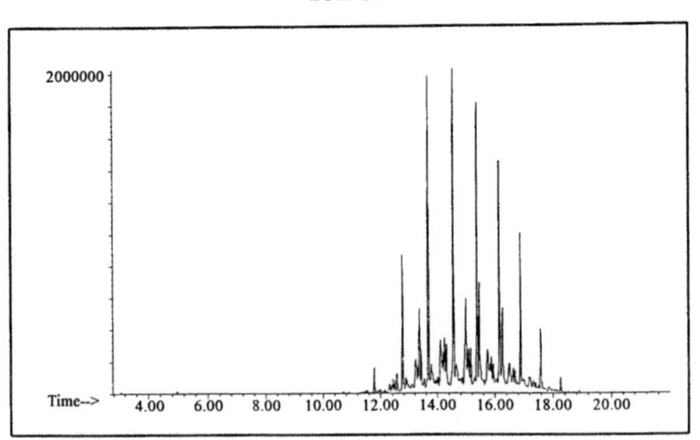

Ion 71

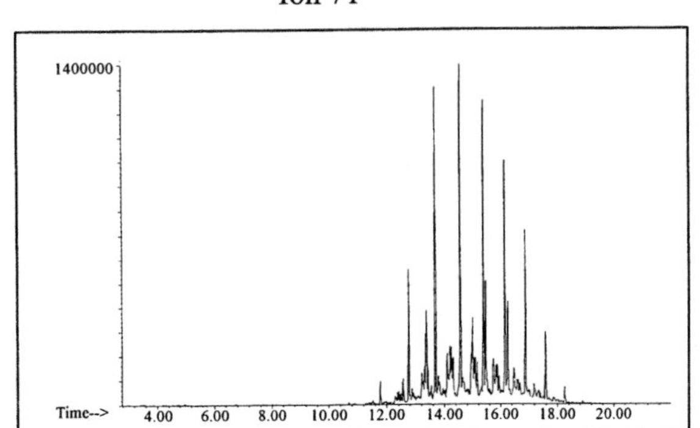

Ion 85

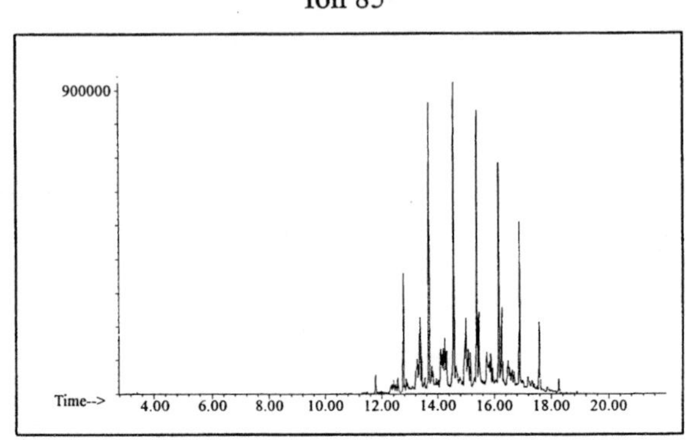

Aromatic

Ion 91

No Useful Data Obtained

Ion 105

No Useful Data Obtained

Ion 119

No Useful Data Obtained

INDIVIDUAL PROFILES

Distillate #54

Cycloparaffin

Ion 55

Ion 69

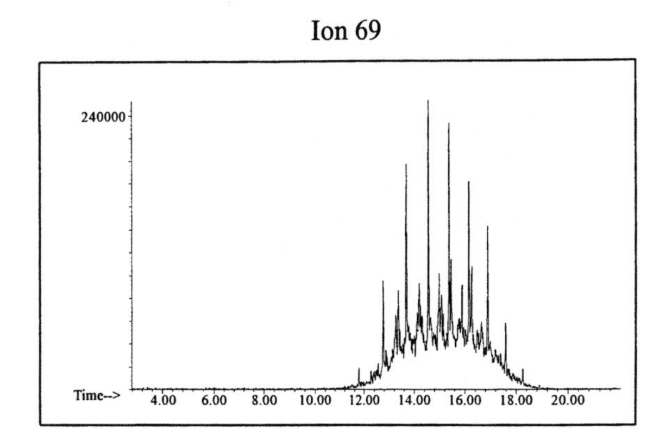

Ion 83

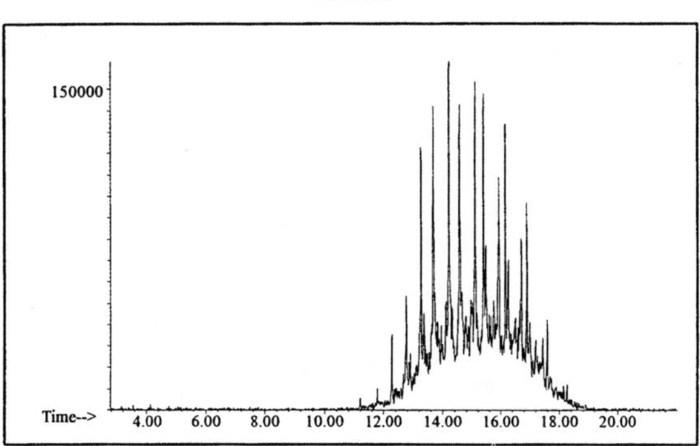

Naphthalene

Ion 128

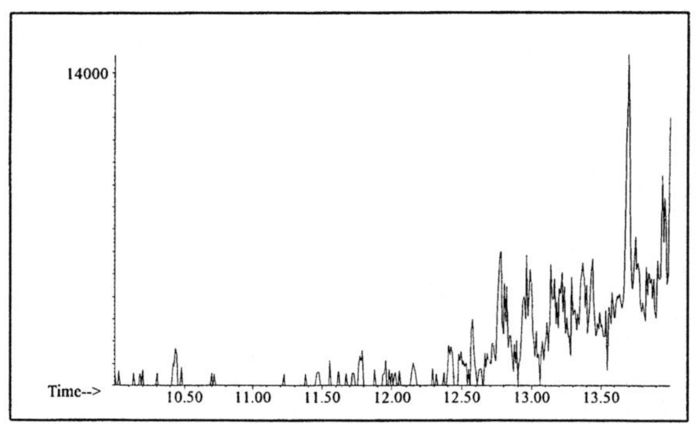

Ion 142

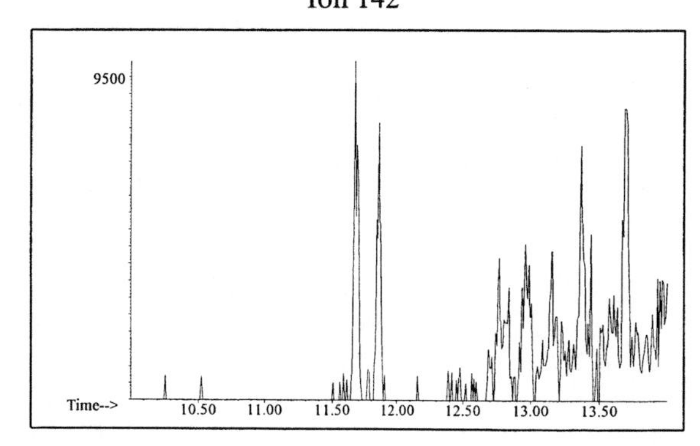

Ion 156

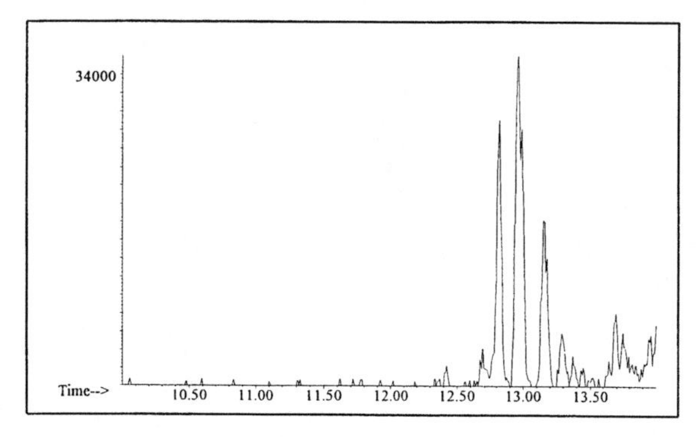

Distillate #55

20uL/mL Pentane

Product Displayed: Exxsol D130

Other Similar Products:

ASTM: Class 0 (Miscellaneous)

Product Uses: Solvent, feedstock

Macro Code: H

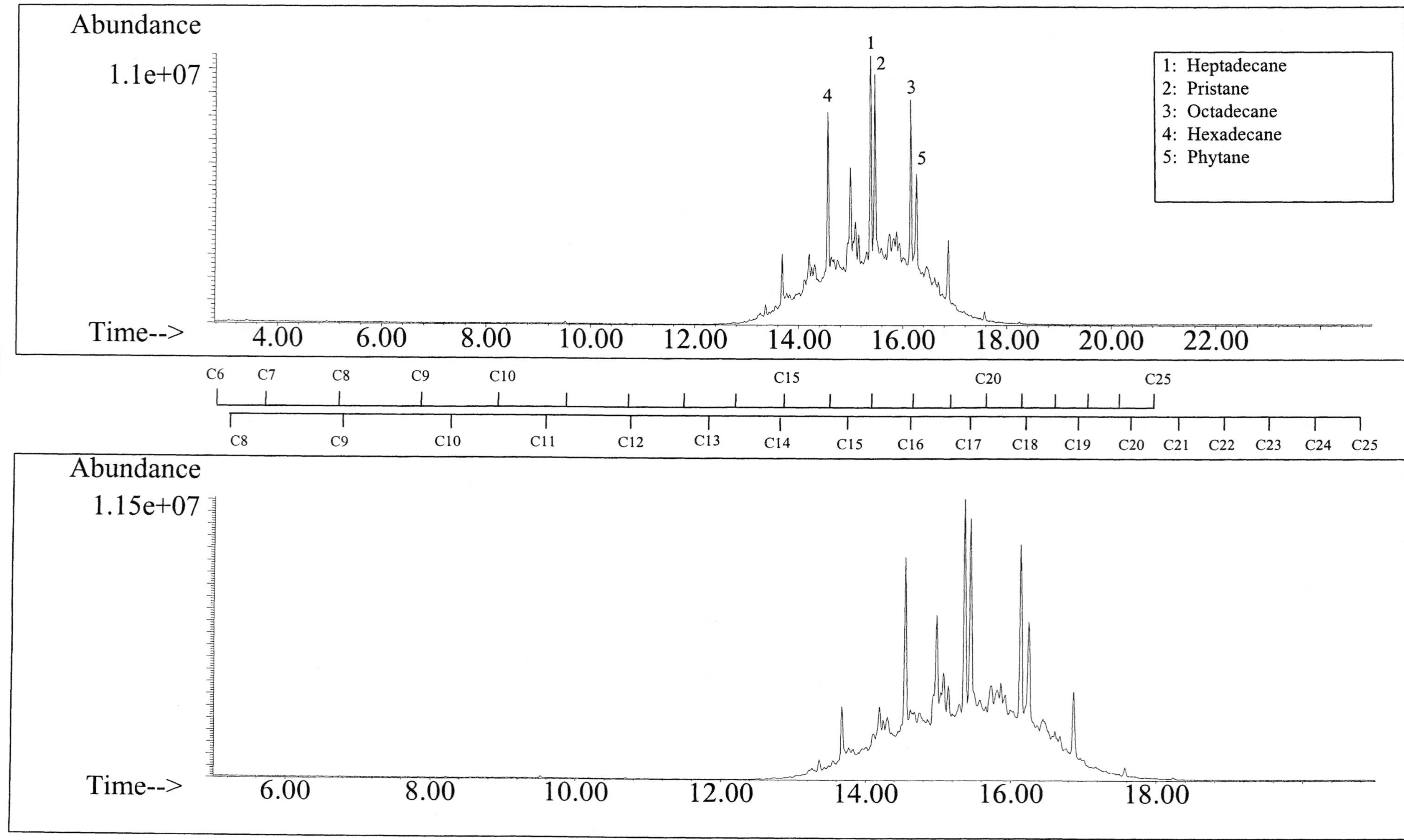

SUMMED PROFILES

Distillate #55

ALKANES

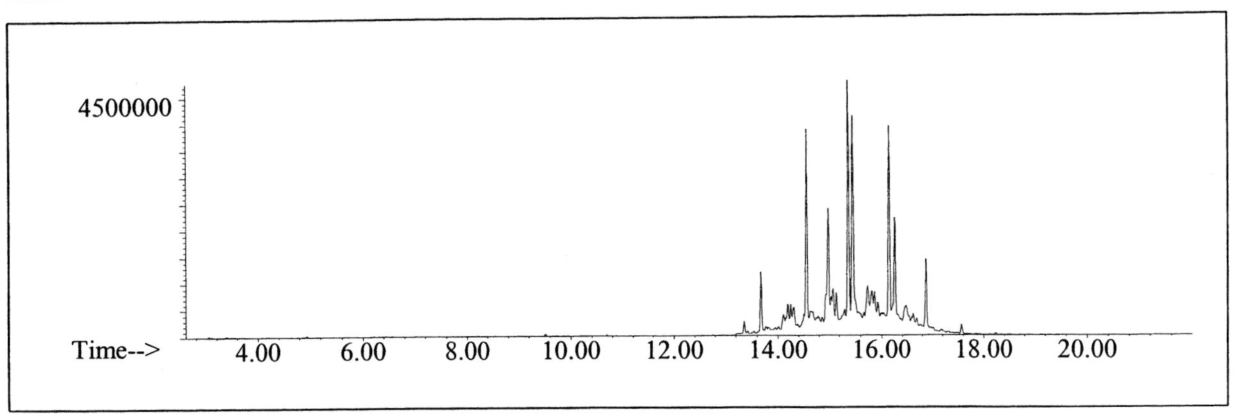

AROMATICS

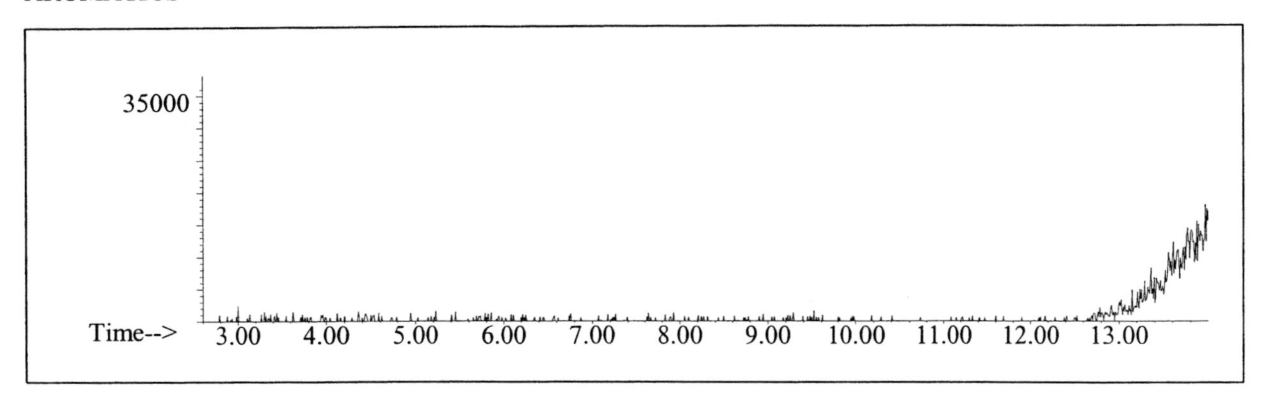

CYCLOPARAFFINS AND ALKENES

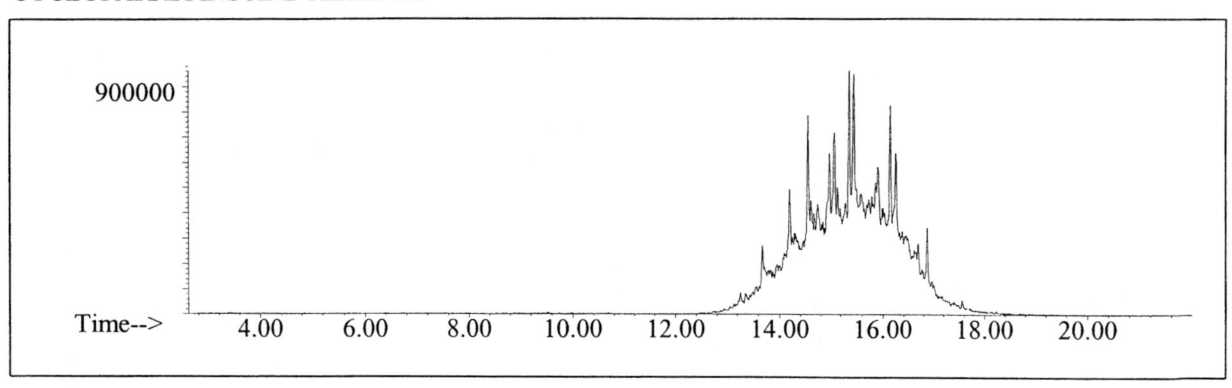

NAPHTHALENES

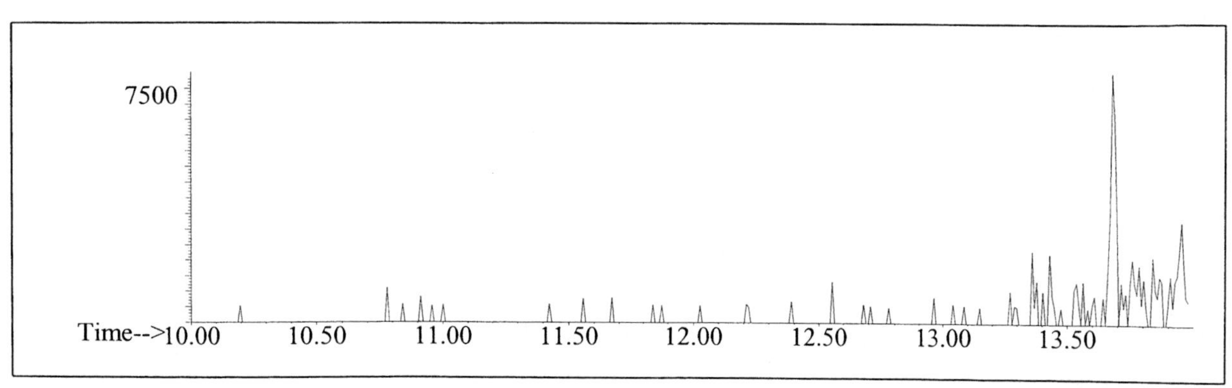

INDIVIDUAL PROFILES

Distillate #55

Alkane

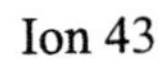

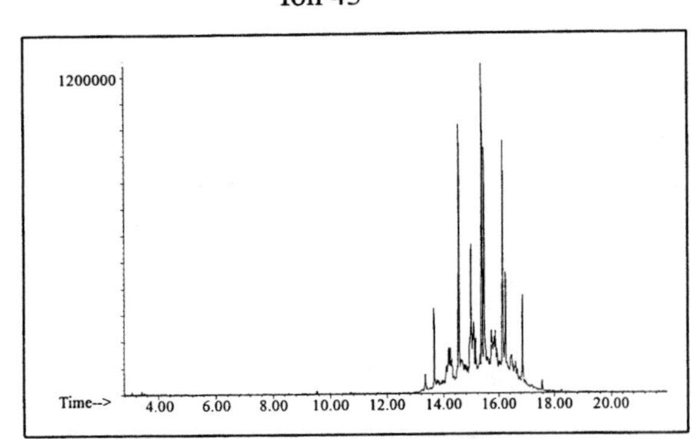

Ion 57

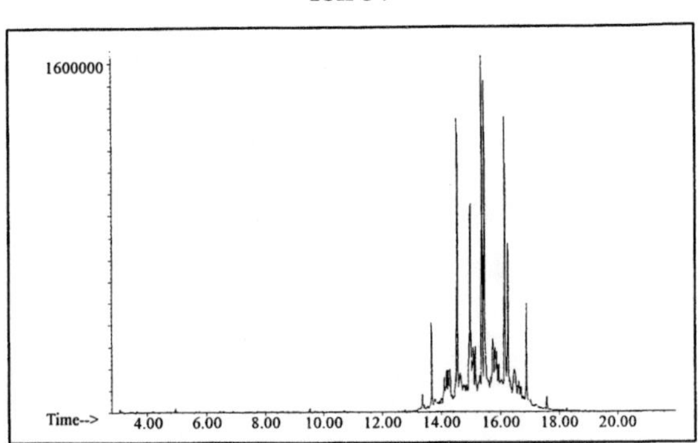

Ion 71

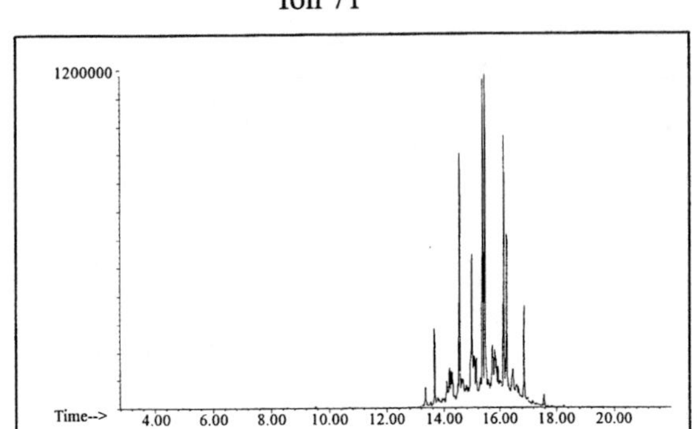

Ion 85

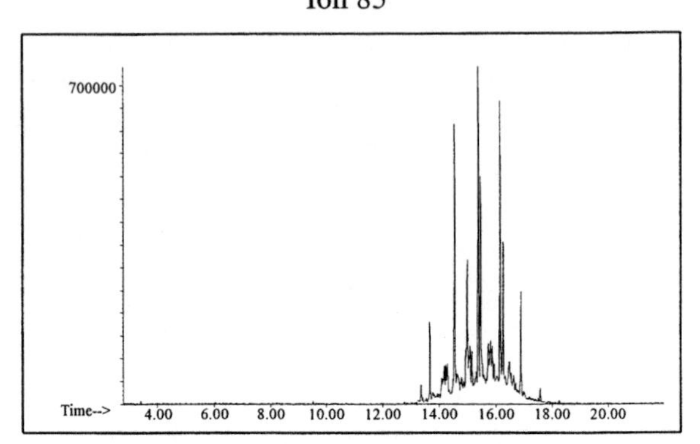

Aromatic

Ion 91

No Useful Data Obtained

Ion 105

No Useful Data Obtained

Ion 119

No Useful Data Obtained

INDIVIDUAL PROFILES

Distillate #55

Cycloparaffin

Ion 55

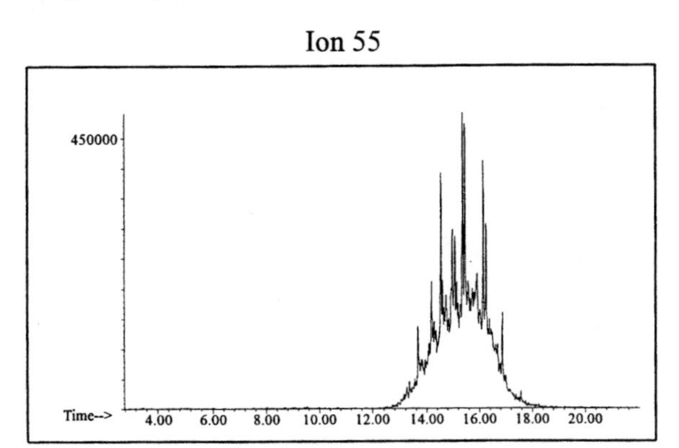

Ion 69

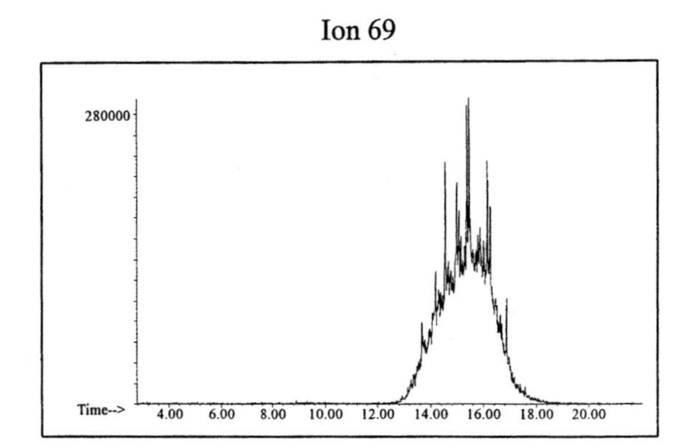

Ion 83

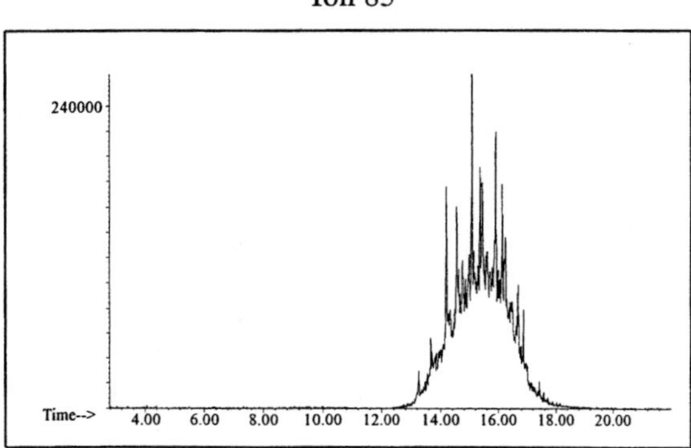

Naphthalene

Ion 128

No Useful Data Obtained

Ion 142

No Useful Data Obtained

Ion 156

No Useful Data Obtained

Distillate #56

20uL/mL Pentane

Product Displayed: Marine Diesel Fuel

Other Similar Products:

ASTM: Class 5 (Heavy Petroleum Distillate)

Product Uses: Fuel

Macro Code: H

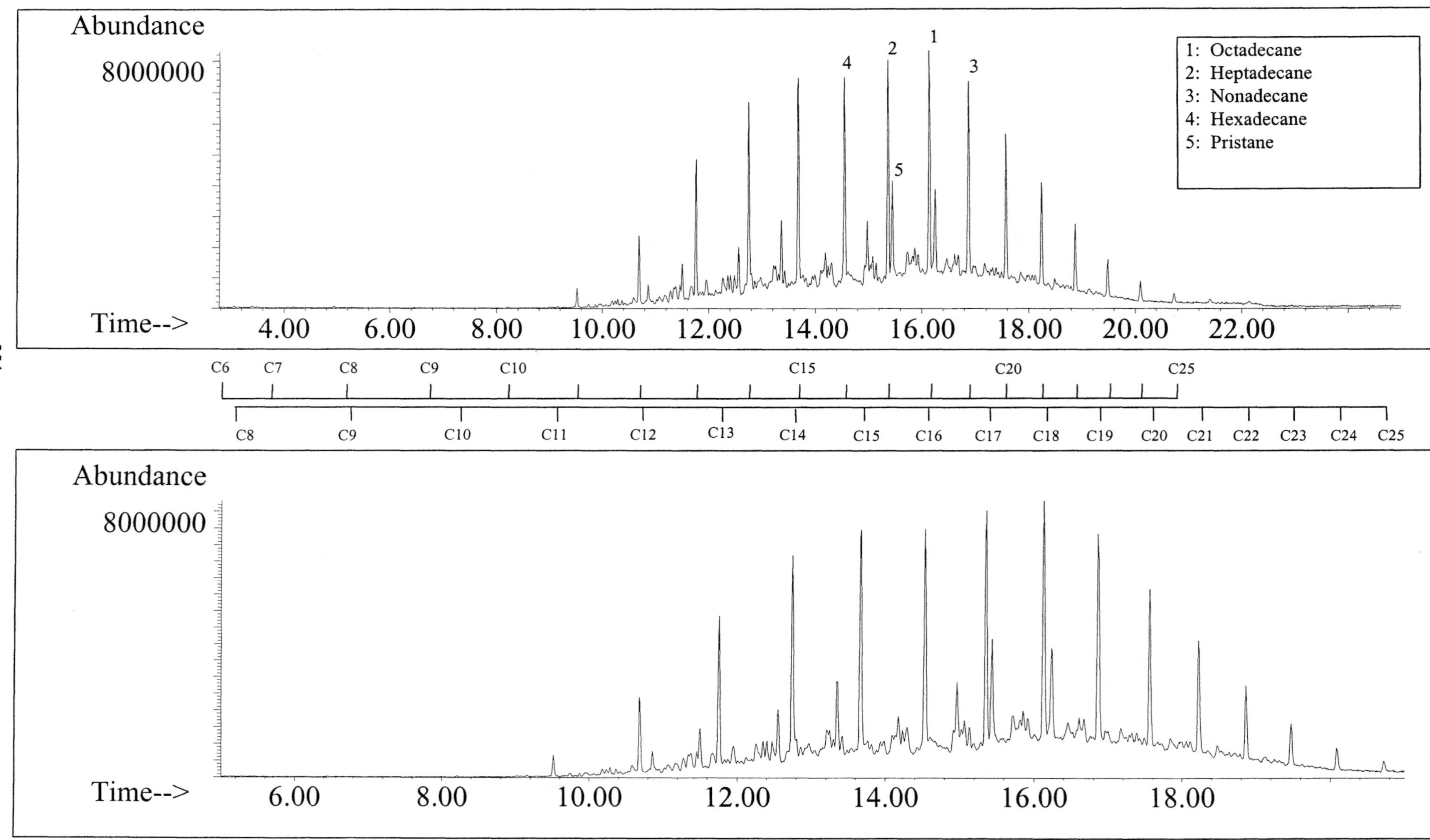

SUMMED PROFILES

Distillate #56

ALKANES

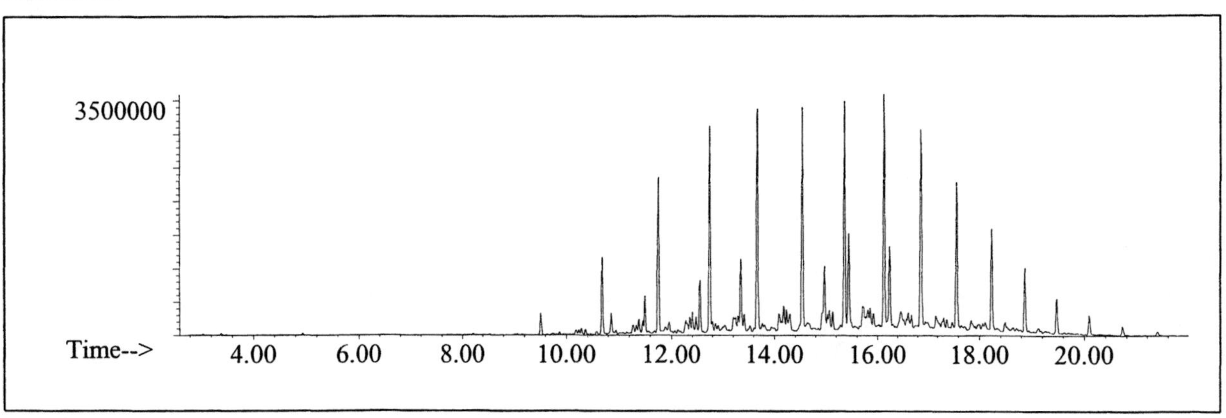

AROMATICS

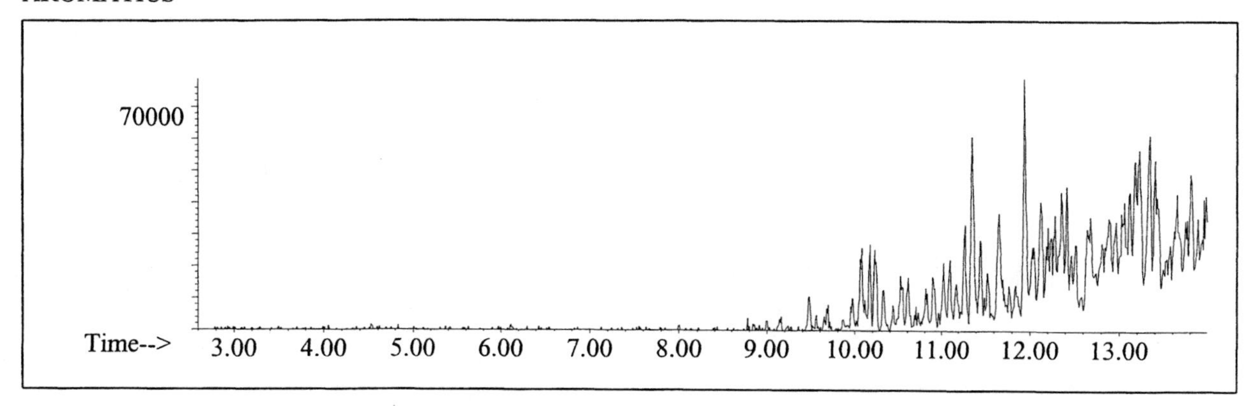

CYCLOPARAFFINS AND ALKENES

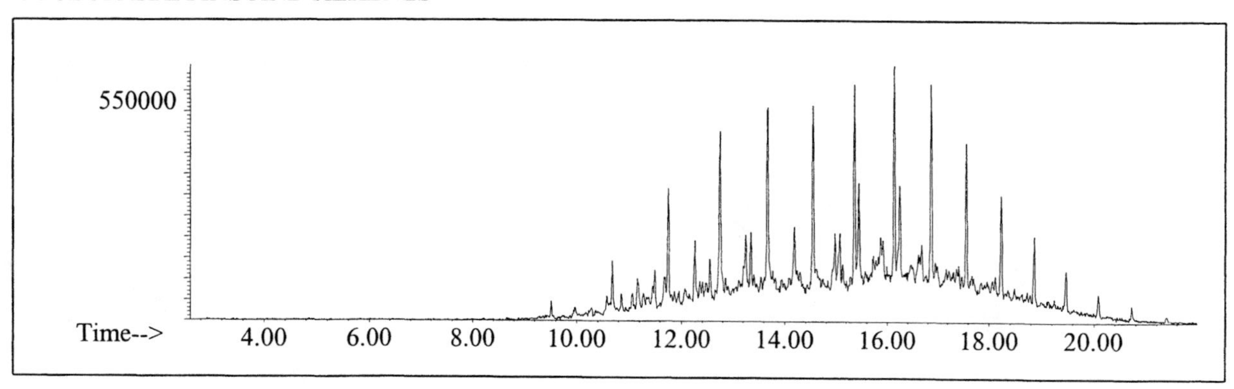

NAPHTHALENES

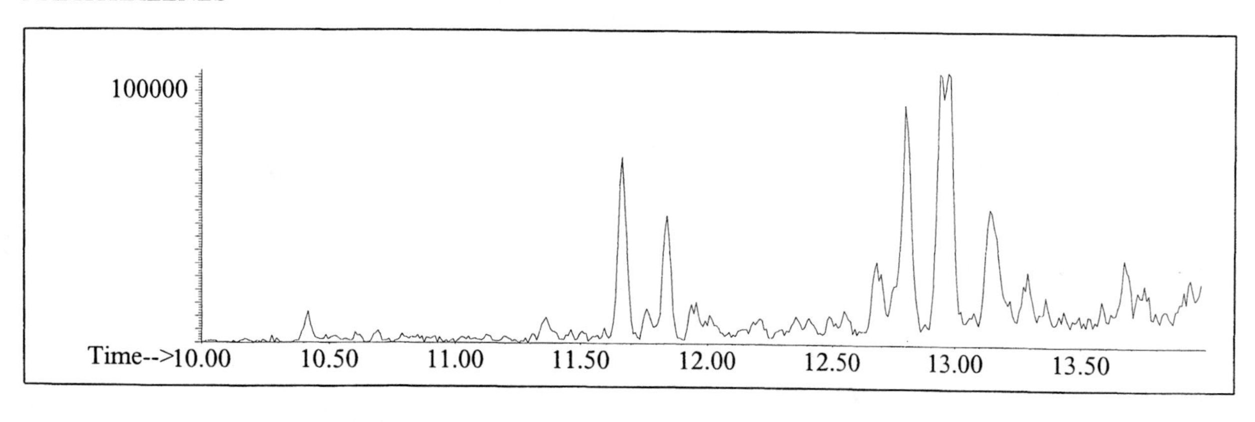

INDIVIDUAL PROFILES

Distillate #56

Alkane

Ion 43

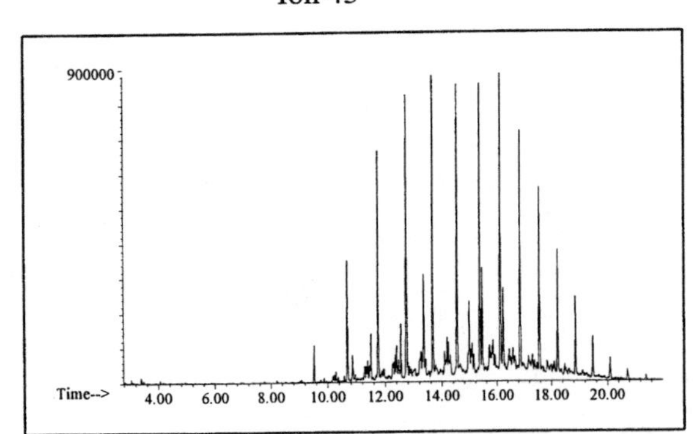

Ion 57

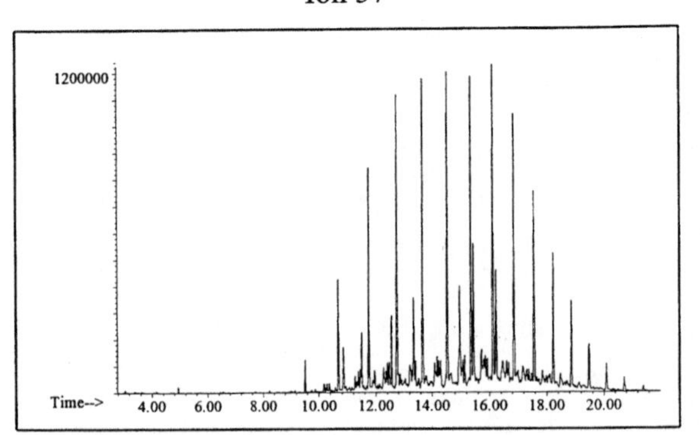

Ion 71

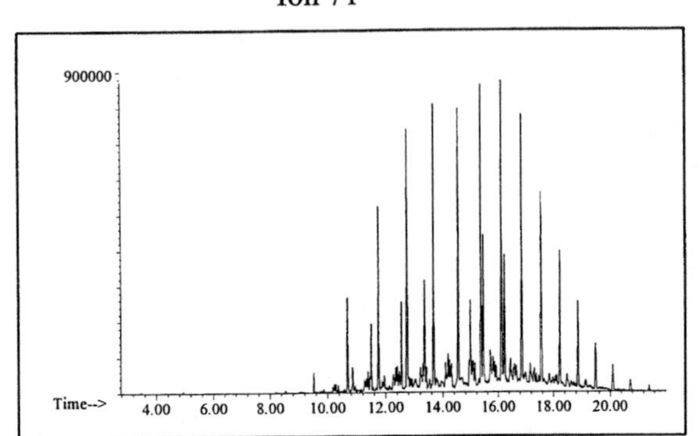

Ion 85

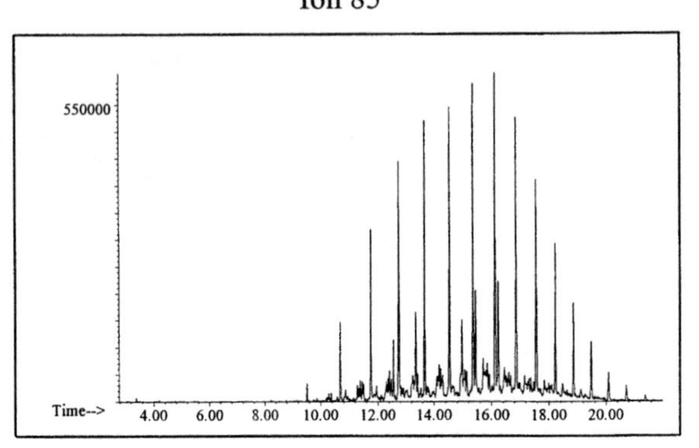

Aromatic

Ion 91

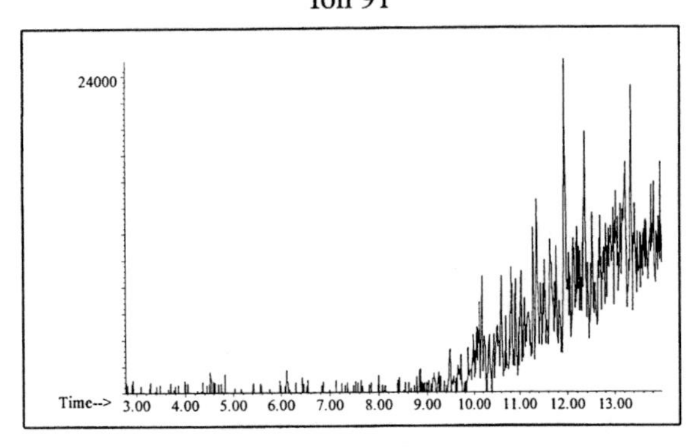

Ion 105

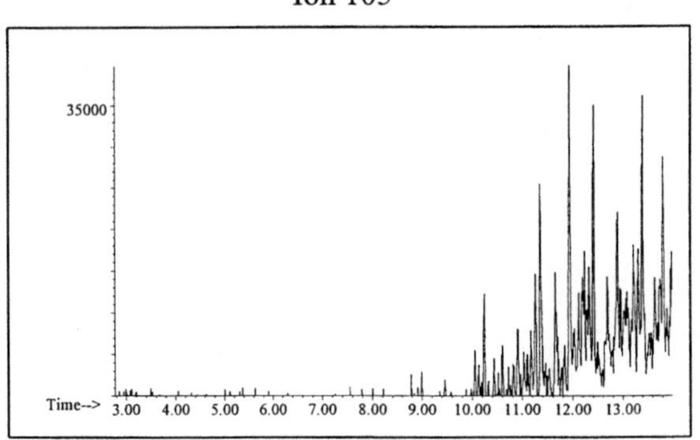

Ion 119

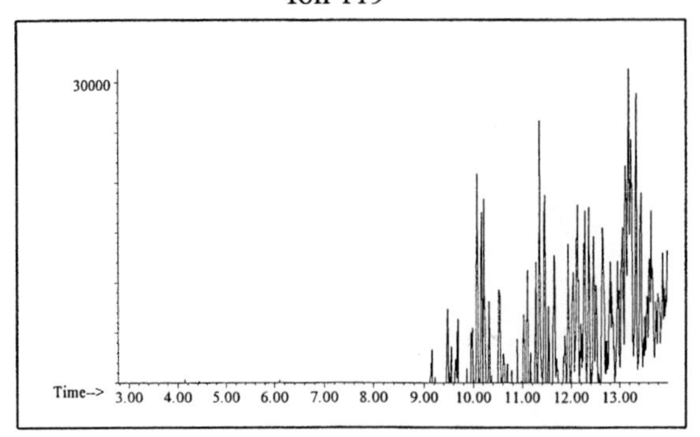

INDIVIDUAL PROFILES

Distillate #56

Cycloparaffin

Ion 55

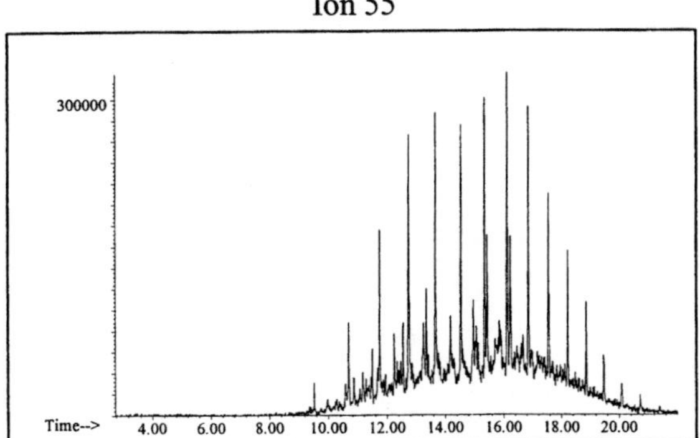

Ion 69

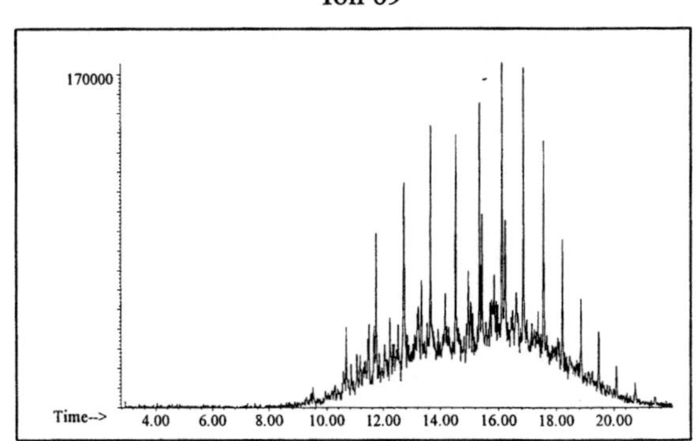

Ion 83

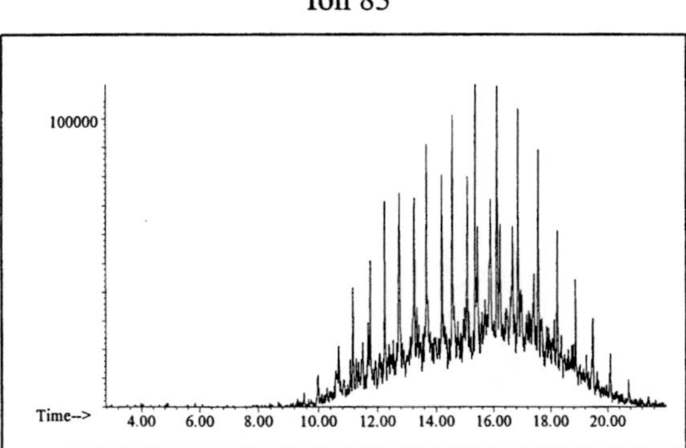

Naphthalene

Ion 128

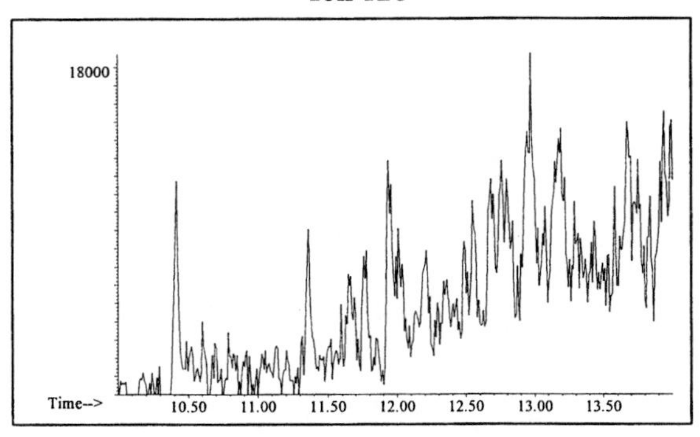

Ion 142

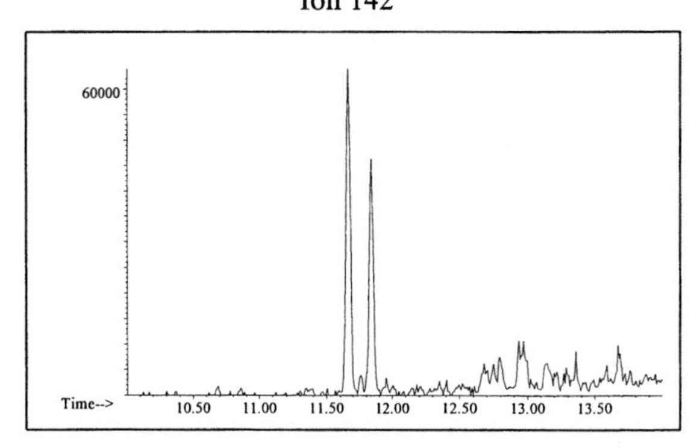

Ion 156

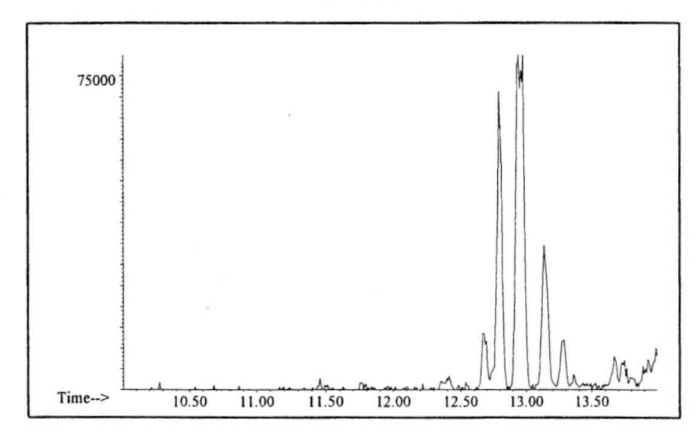

Gasolines

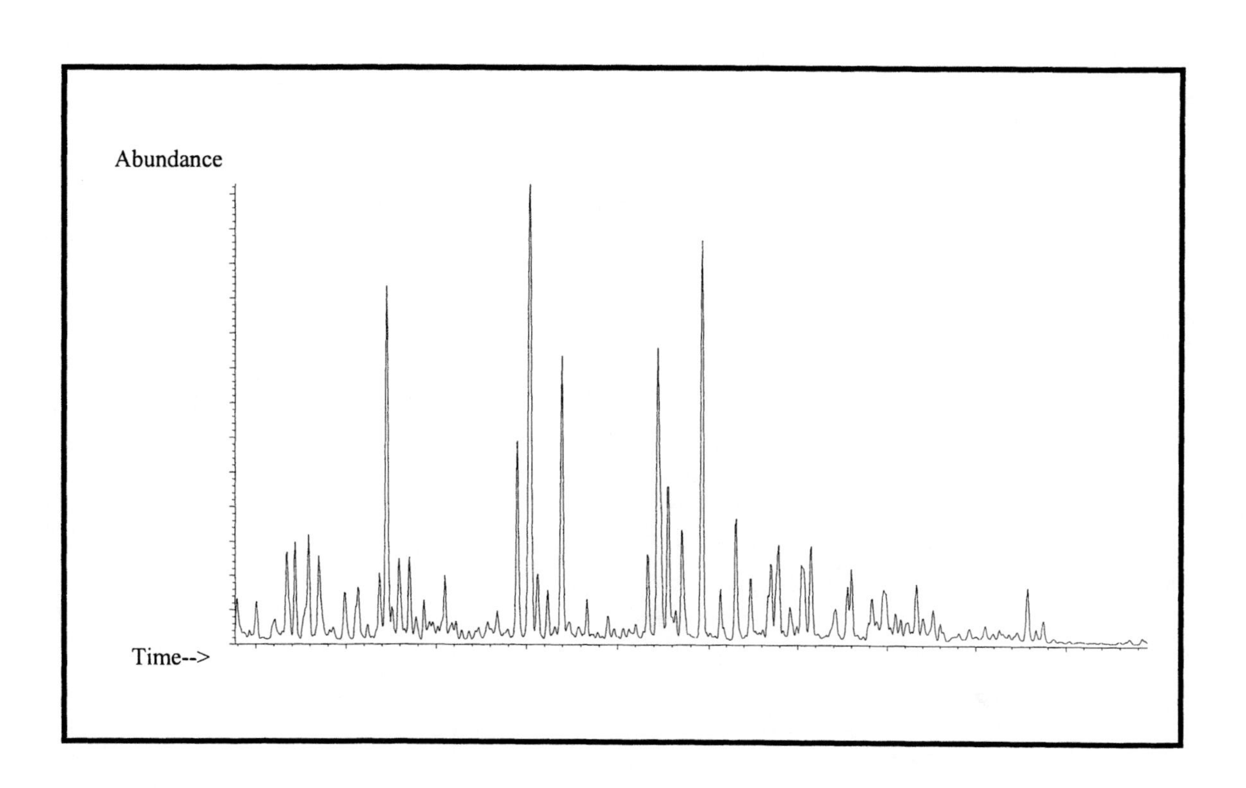

Gasoline #1

20uL/mL Pentane

ASTM: Class 2 (Gasoline)

Macro Code: L

Product Displayed: Mobil Regular Unleaded Gasoline

Product Uses: Automotive fuel

Other Similar Products: Various brands and grades of gasoline

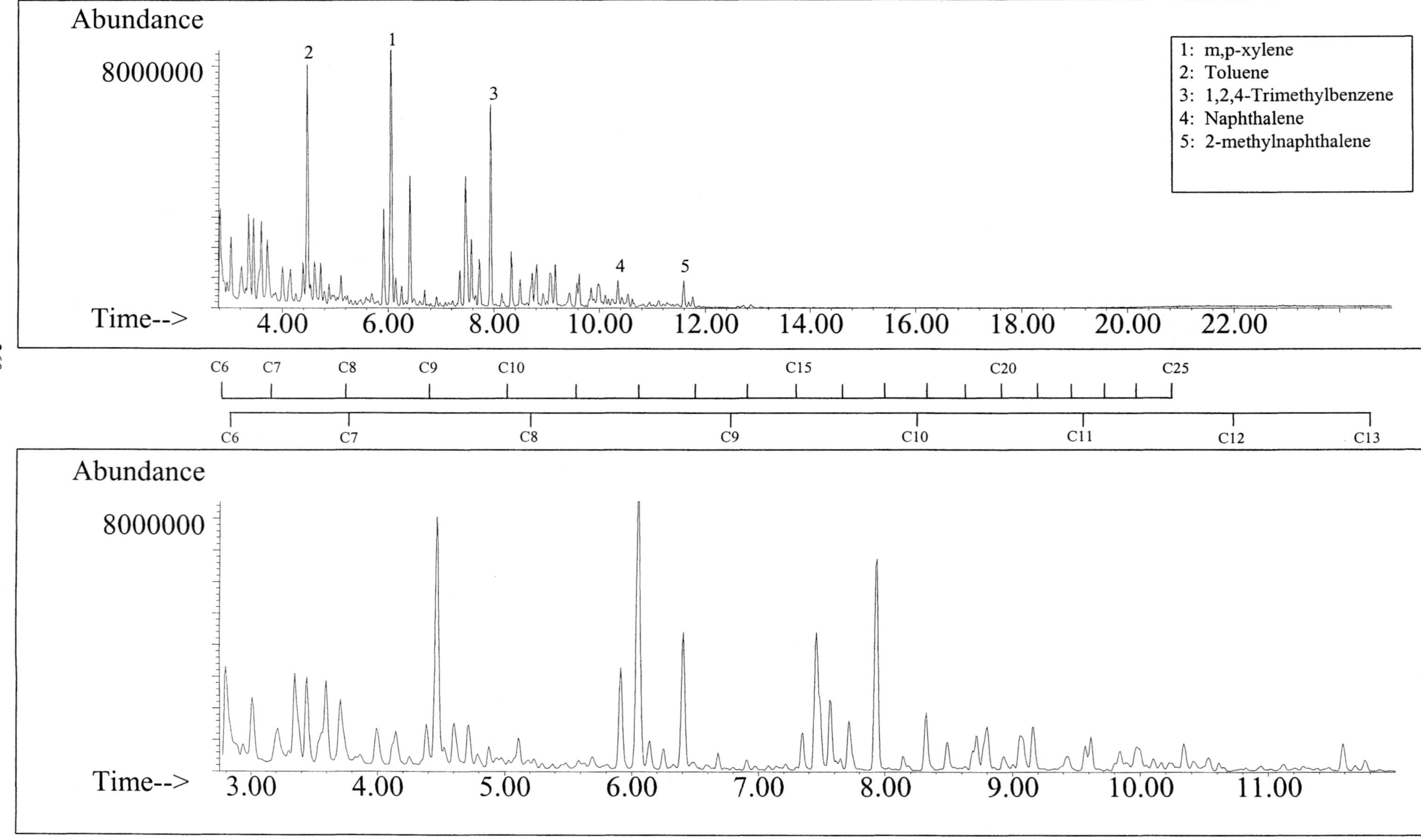

SUMMED PROFILES

Gasoline #1

ALKANES

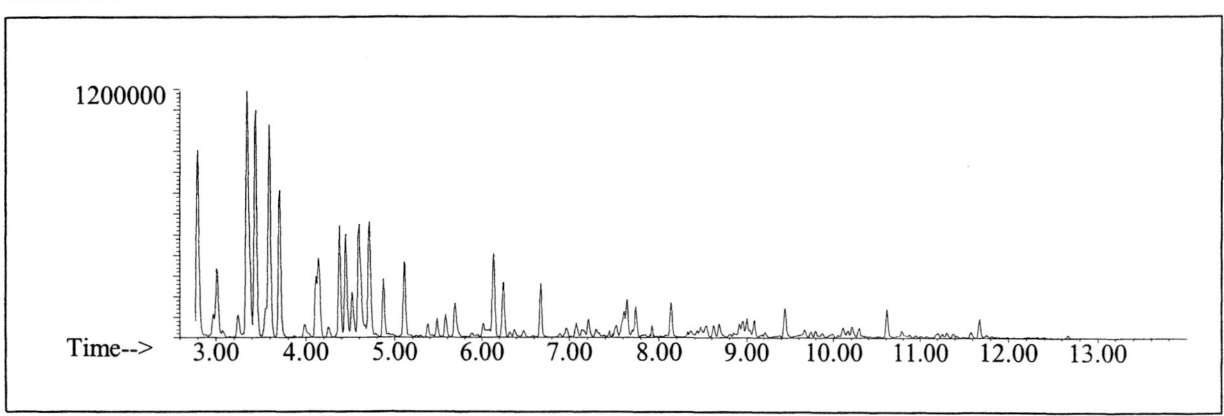

AROMATICS

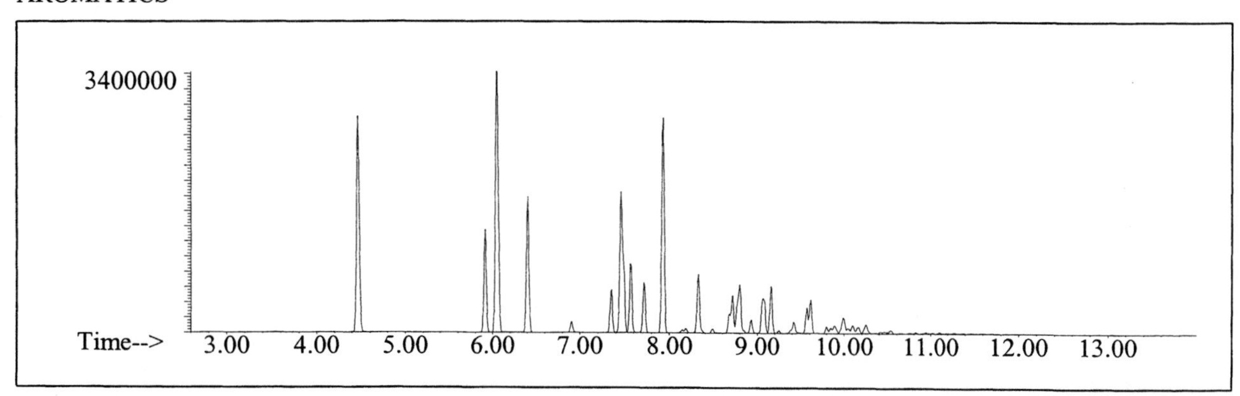

CYCLOPARAFFINS AND ALKENES

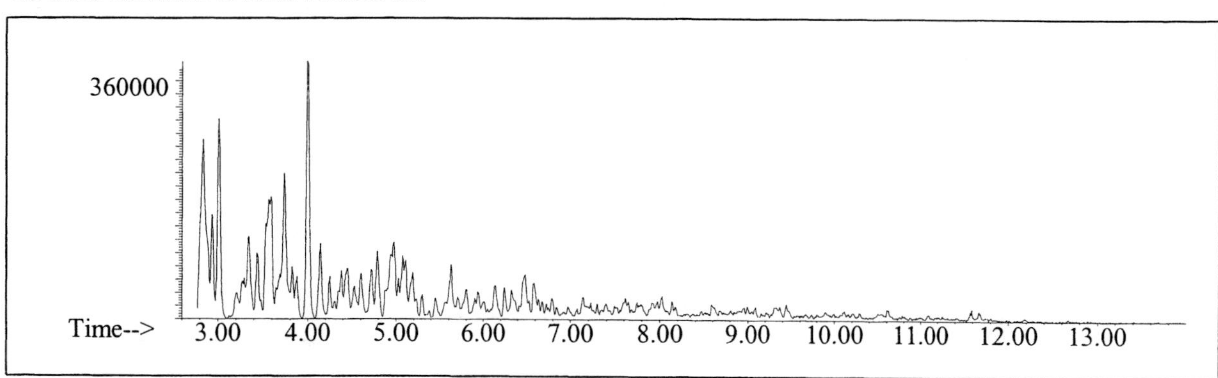

NAPHTHALENES

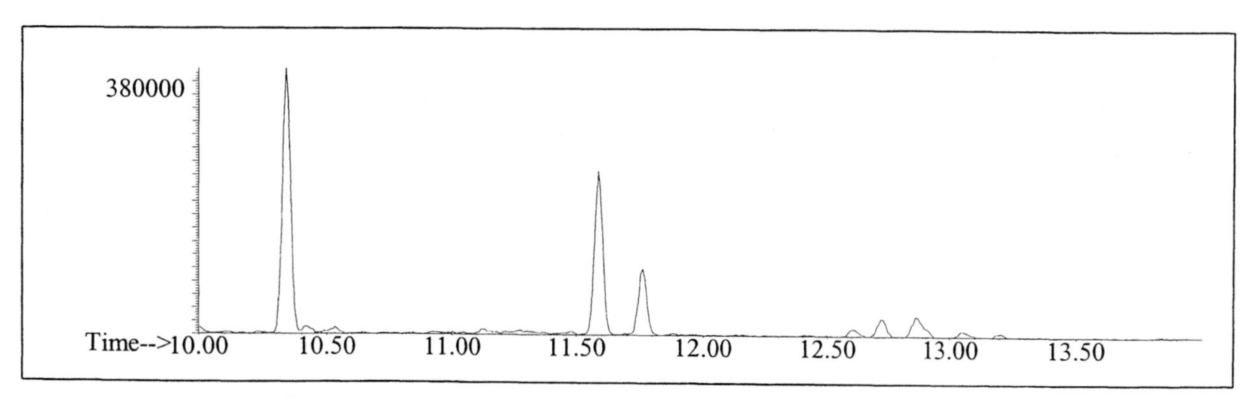

INDIVIDUAL PROFILES — Gasoline #1

Alkane

Ion 43

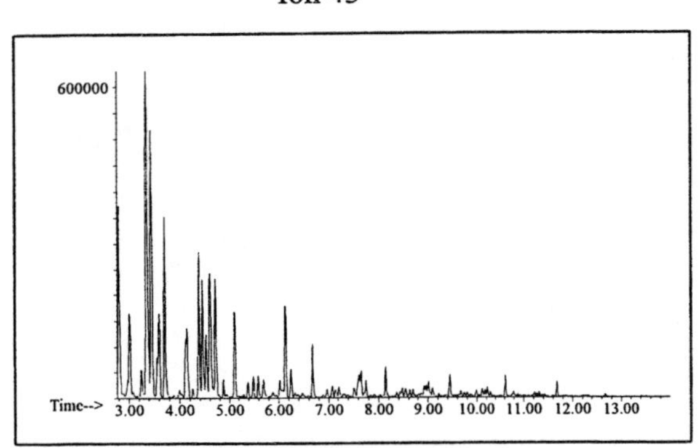

Ion 57

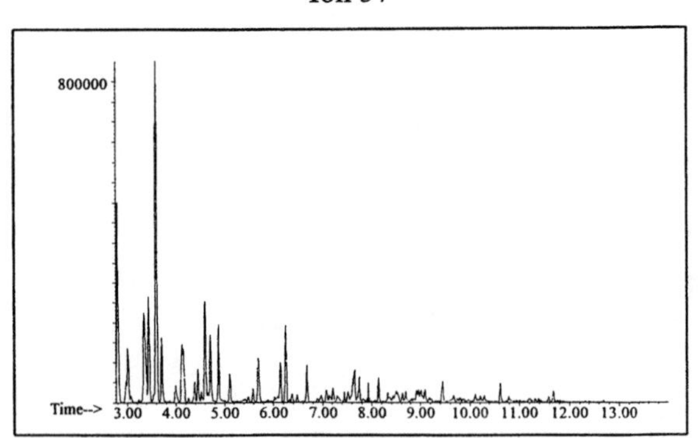

Ion 71

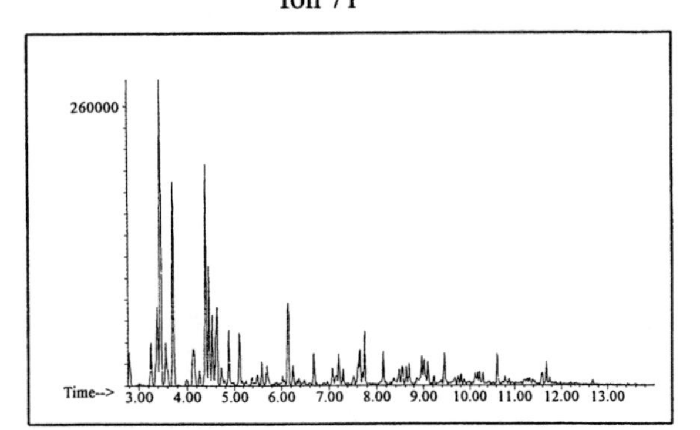

Ion 85

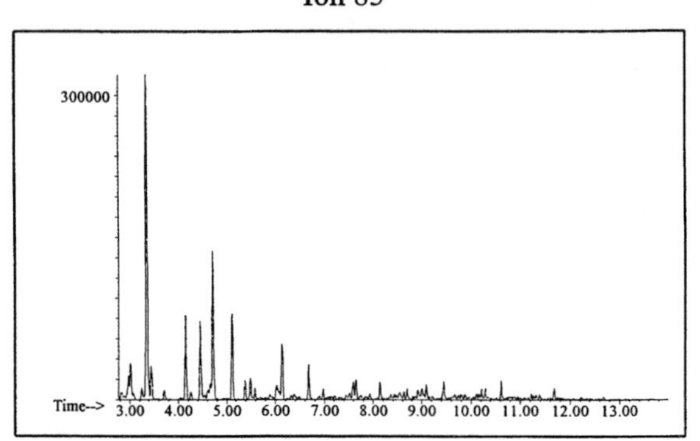

Aromatic

Ion 91

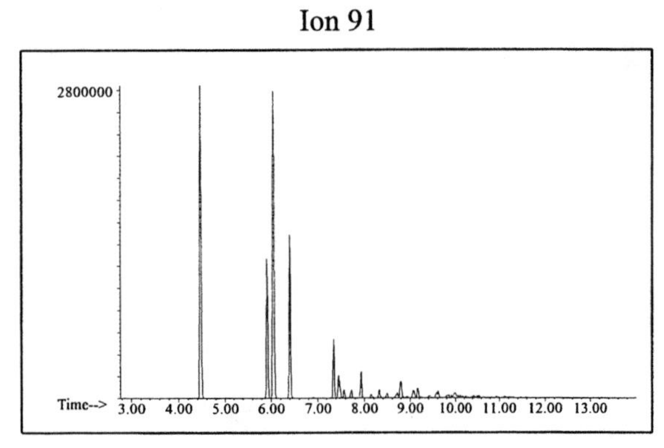

Ion 105

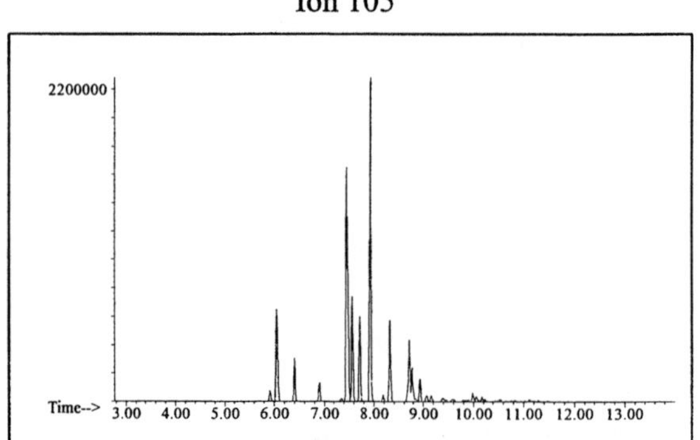

Ion 119

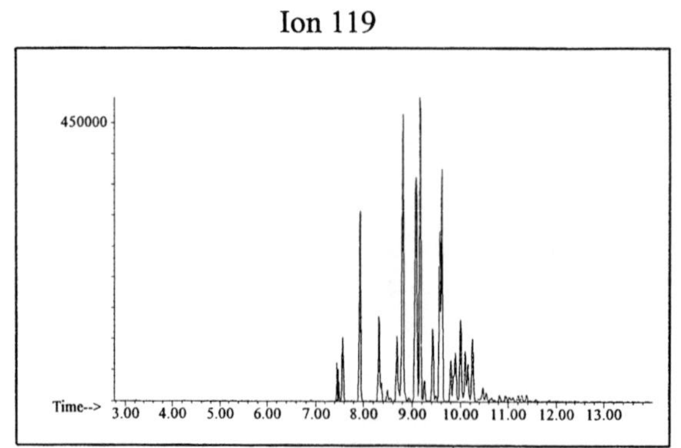

INDIVIDUAL PROFILES

Gasoline #1

Cycloparaffin

Ion 55

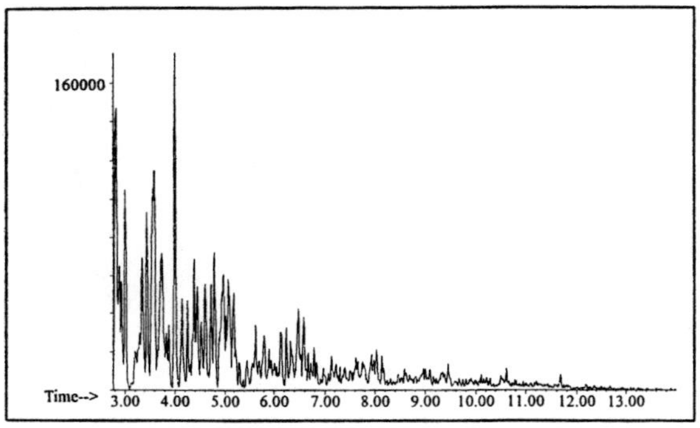

Ion 69

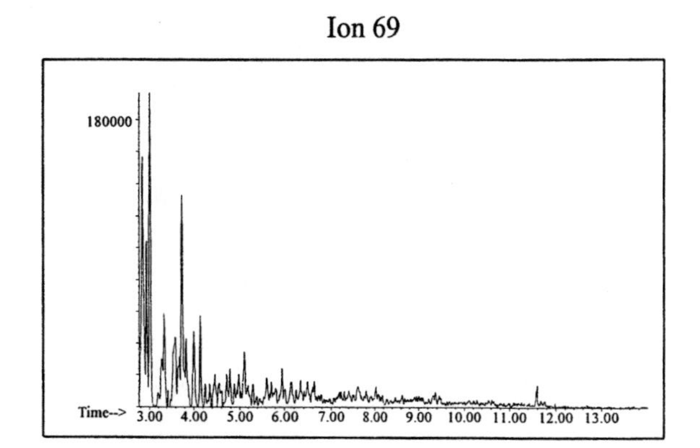

Ion 83

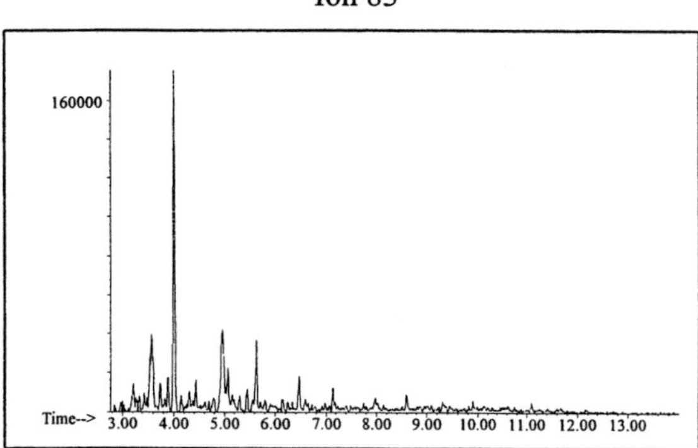

Naphthalene

Ion 128

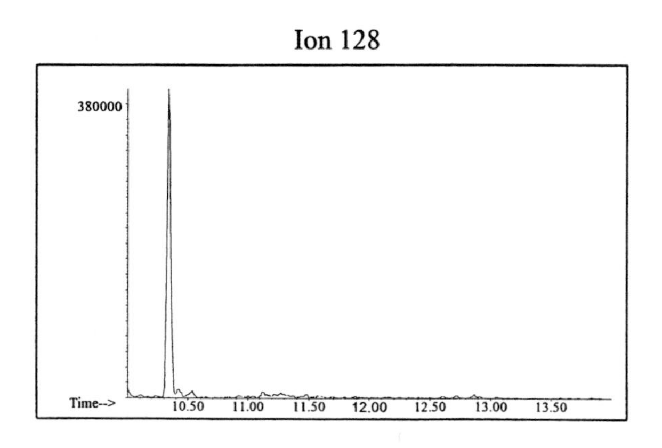

Ion 142

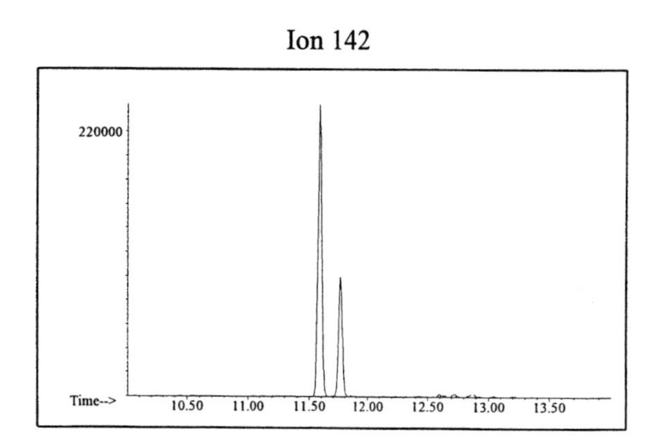

Ion 156

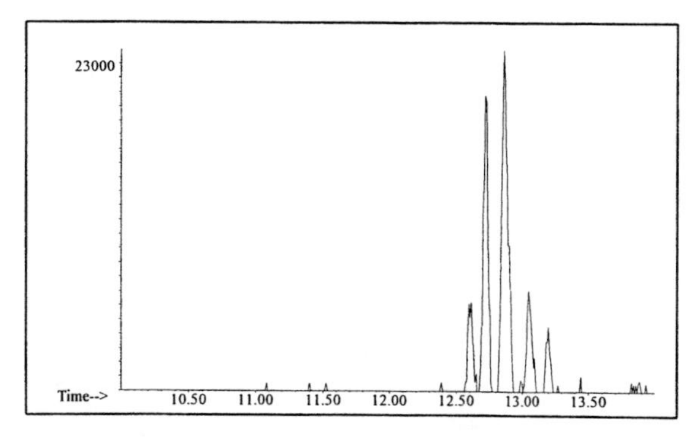

Gasoline #2

20uL/mL Pentane

Product Displayed: 25% Evaporated Mobil Regular Gasoline

Other Similar Products: Other brands and grades of gasoline

ASTM: Class 2 (Gasoline)

Product Uses: Automotive fuel

Macro Code: L

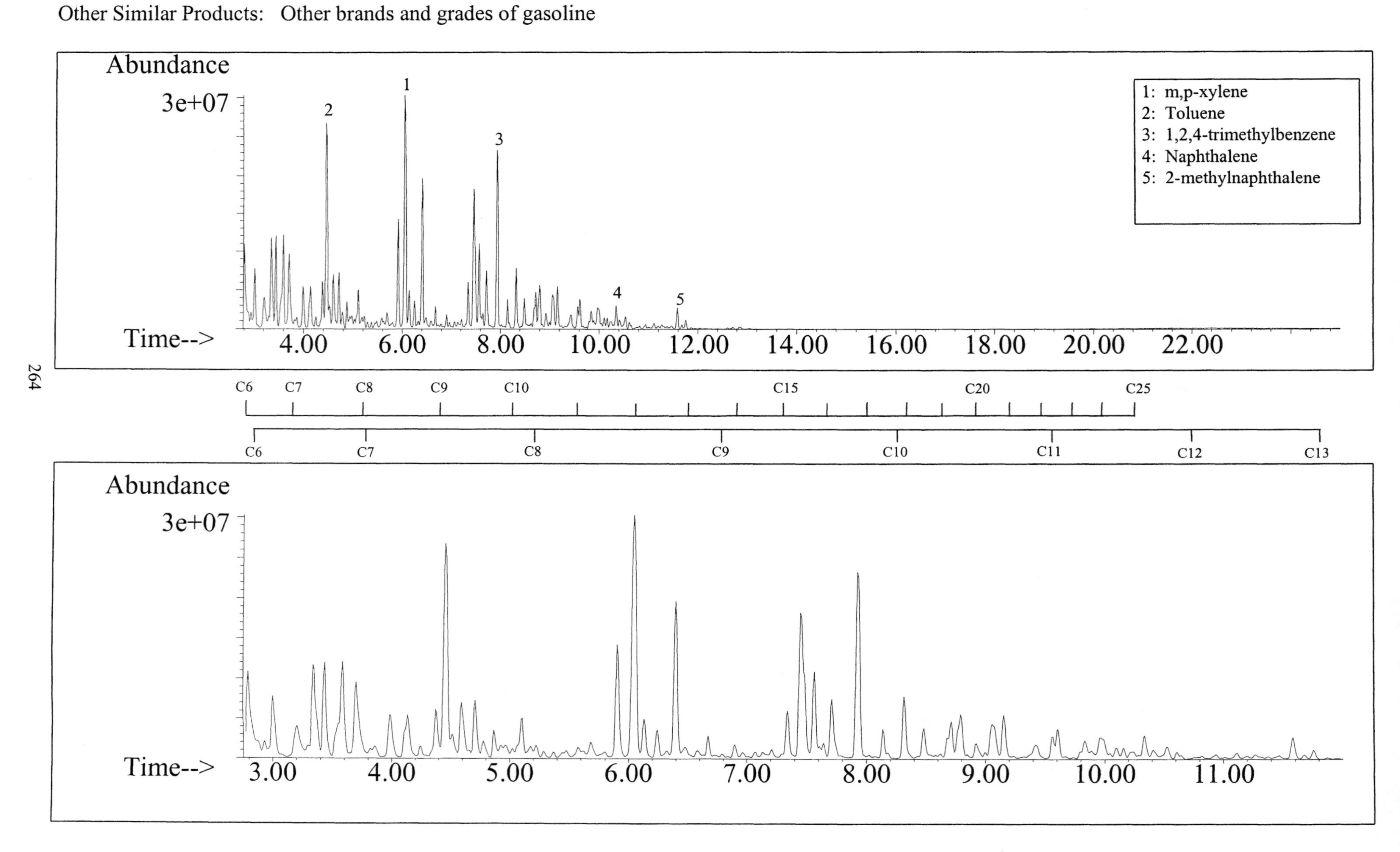

SUMMED PROFILES -Gasoline #2

ALKANES

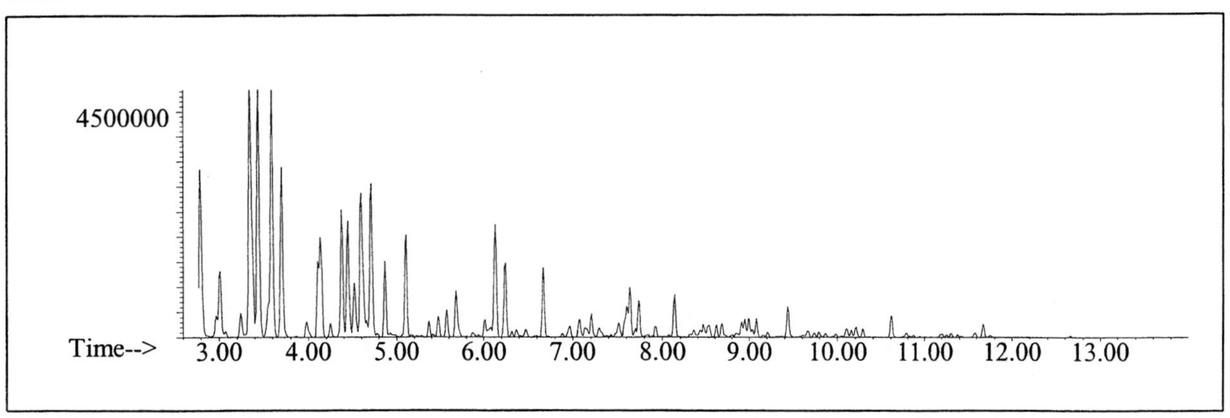

AROMATICS

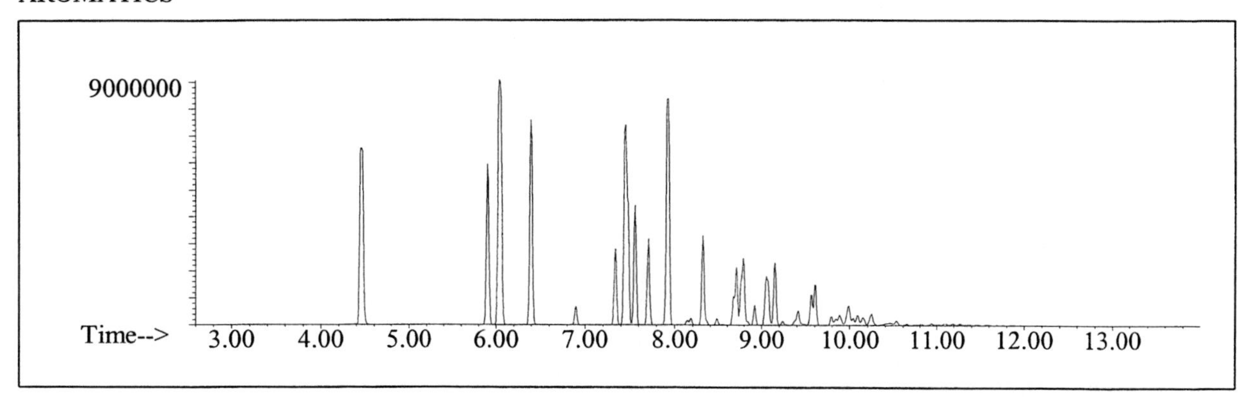

CYCLOPARAFFINS AND ALKENES

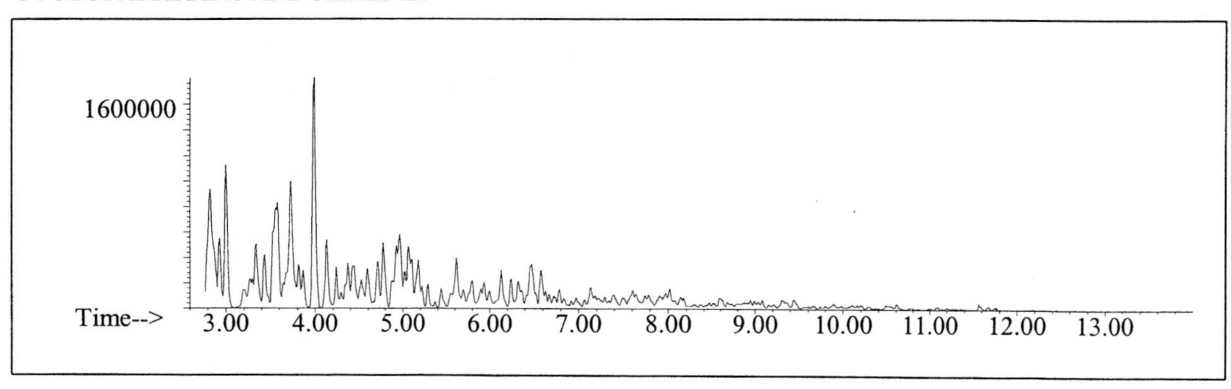

NAPHTHALENES

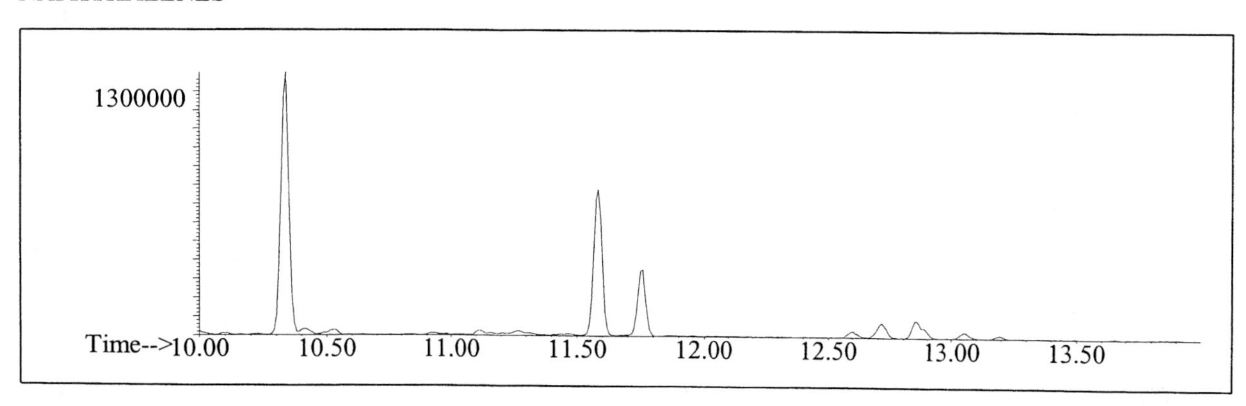

INDIVIDUAL PROFILES

Gasoline #2

Alkane

Ion 43

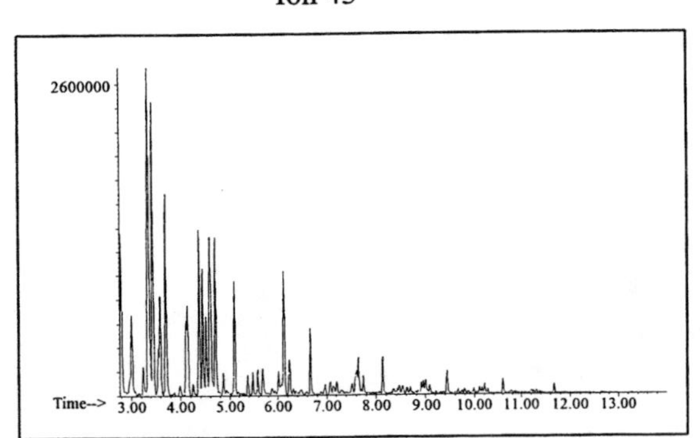

Ion 57

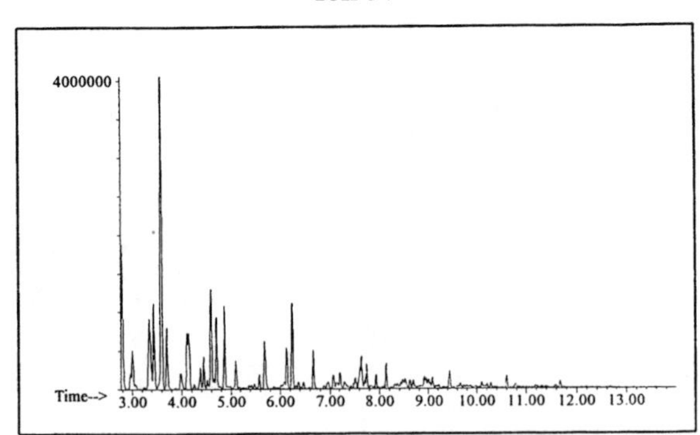

Ion 71

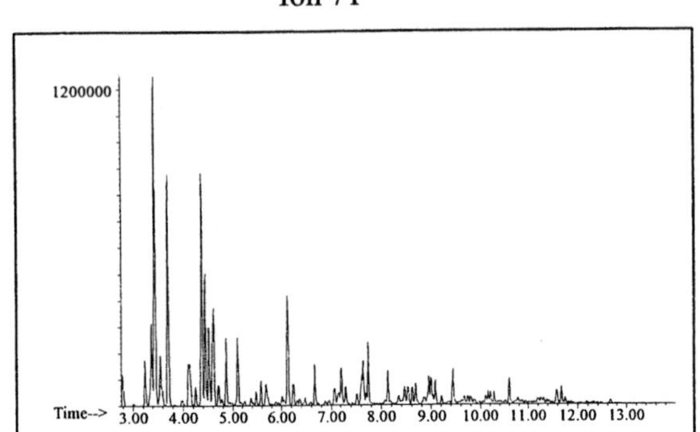

Ion 85

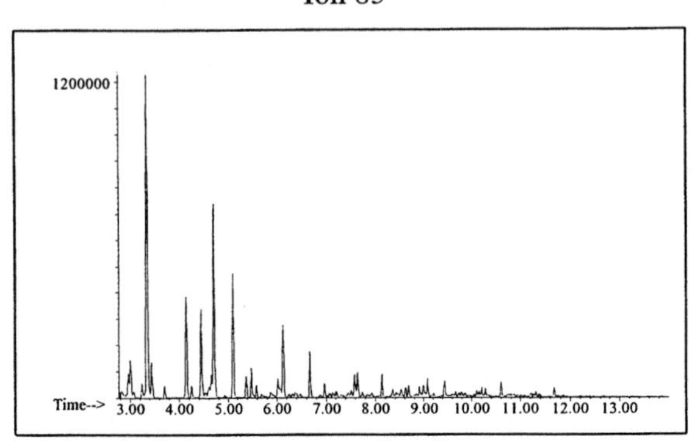

Aromatic

Ion 91

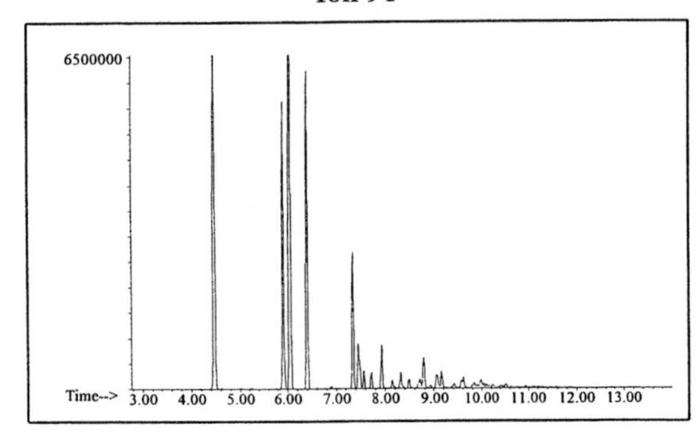

Ion 105

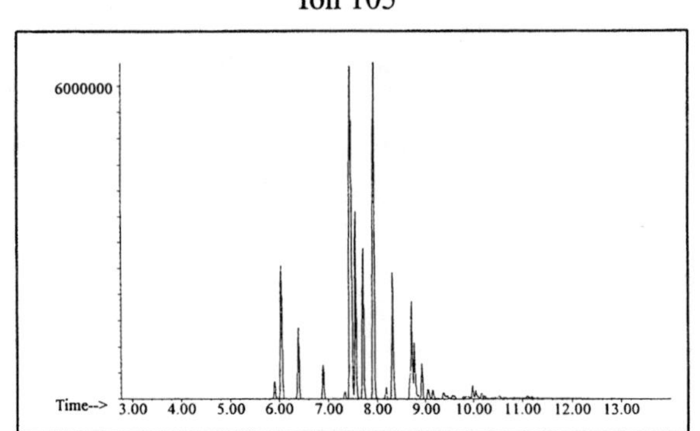

Ion 119

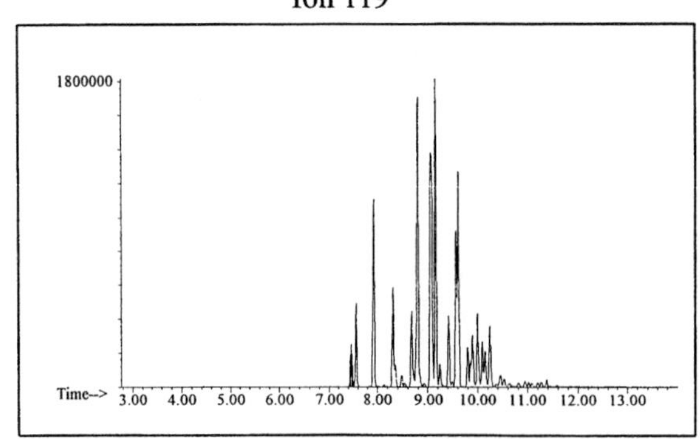

INDIVIDUAL PROFILES Gasoline #2

Cycloparaffin

Ion 55

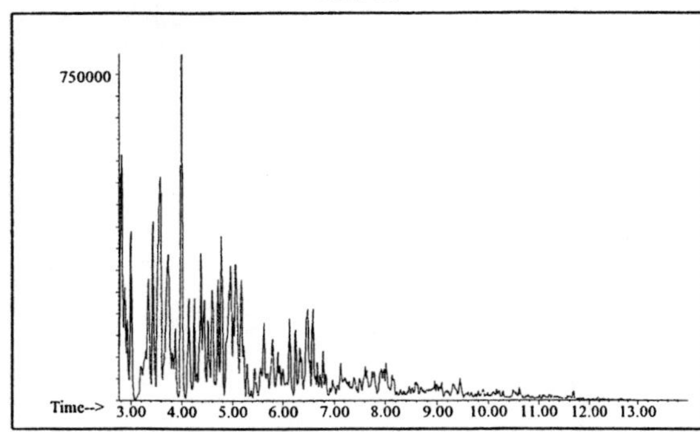

Ion 69

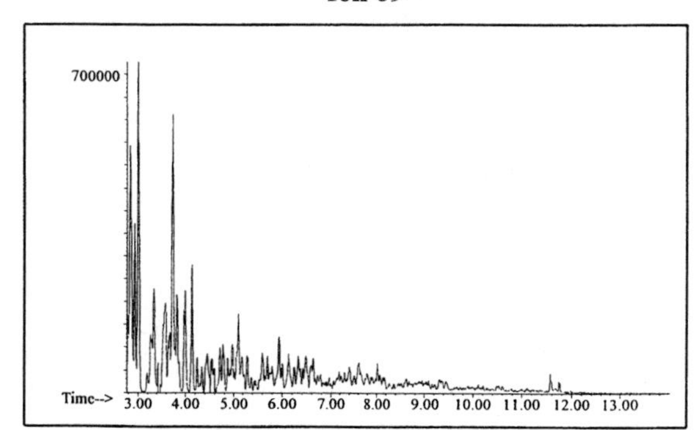

Ion 83

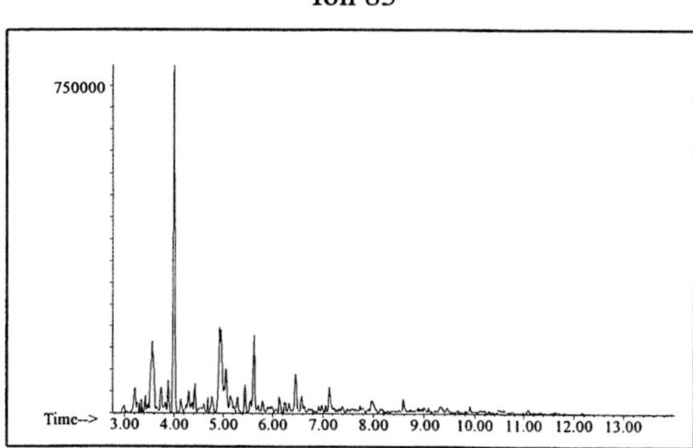

Naphthalene

Ion 128

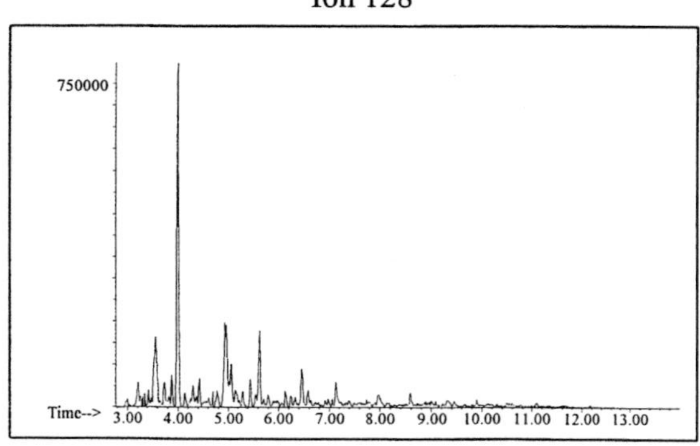

Ion 142

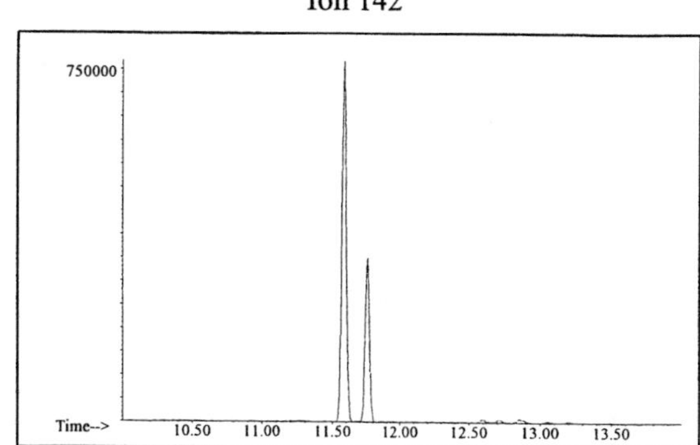

Ion 156

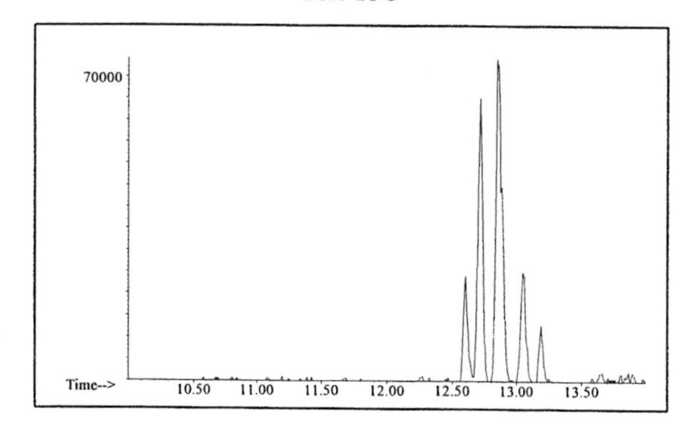

Gasoline #3

20uL/mL Pentane

Product Displayed: 50% Evaporated Mobil Regular Gasoline

Other Similar Products: Other brands and grades of gasoline

ASTM: Class 2 (Gasoline)

Product Uses: Automotive fuel

Macro Code: L

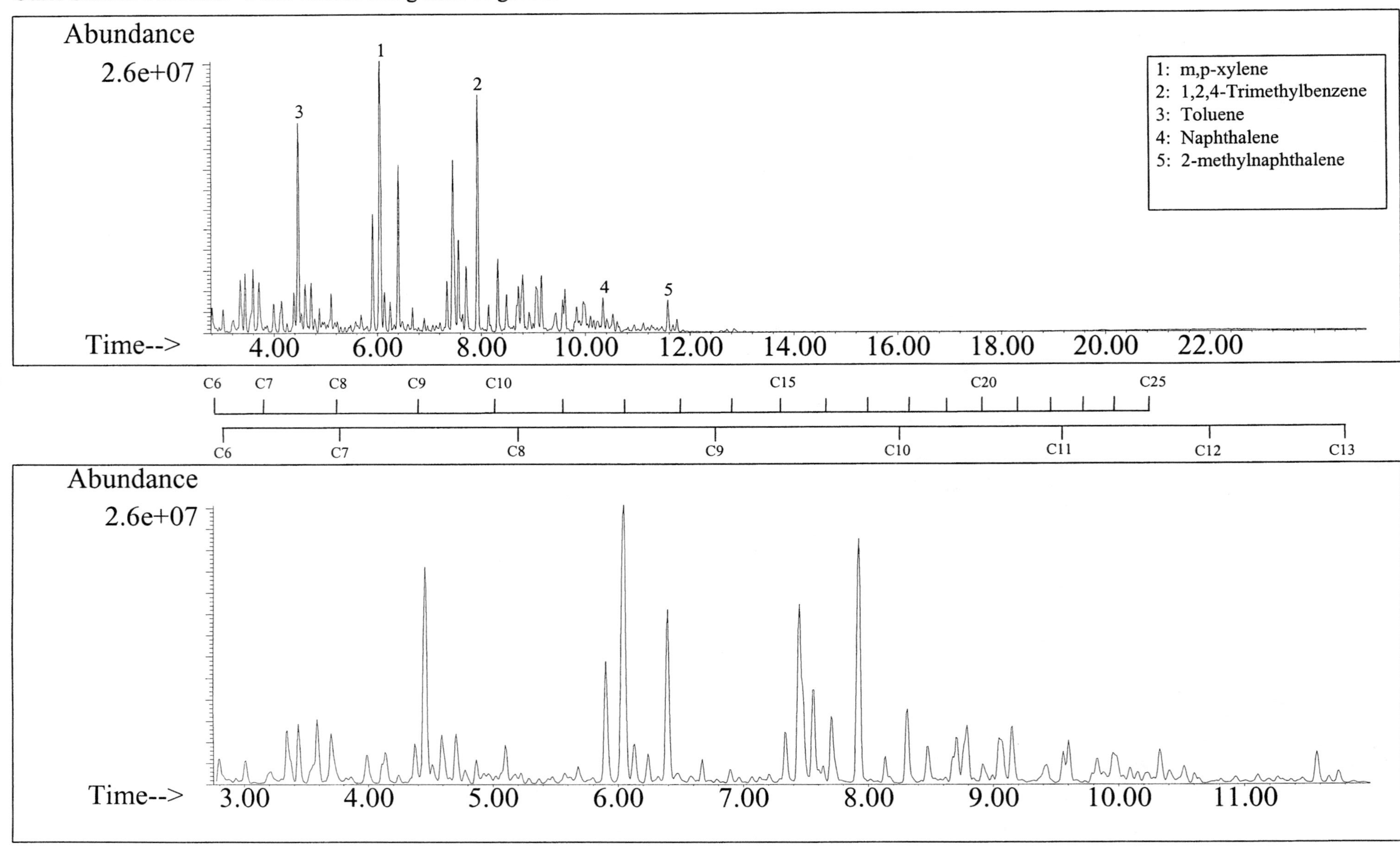

SUMMED PROFILES

Gasoline #3

ALKANES

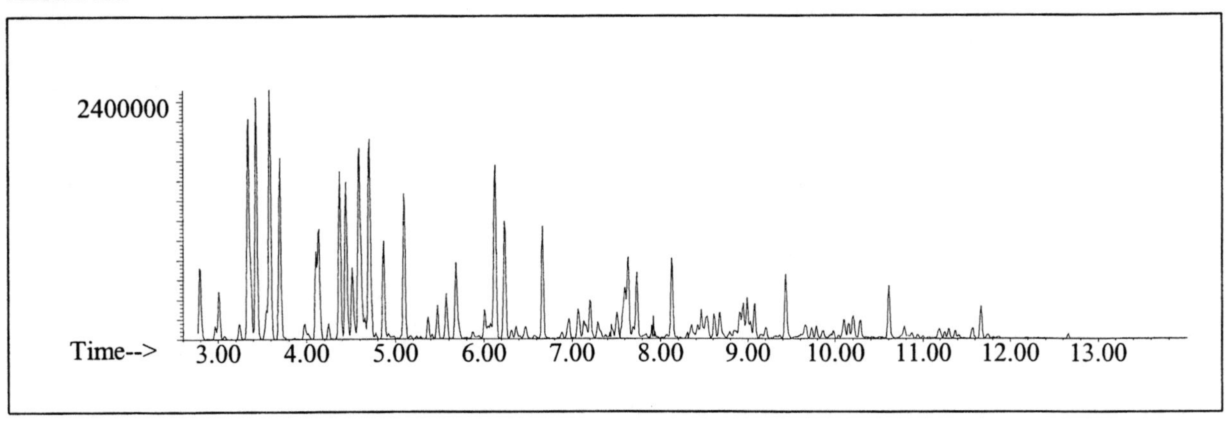

AROMATICS

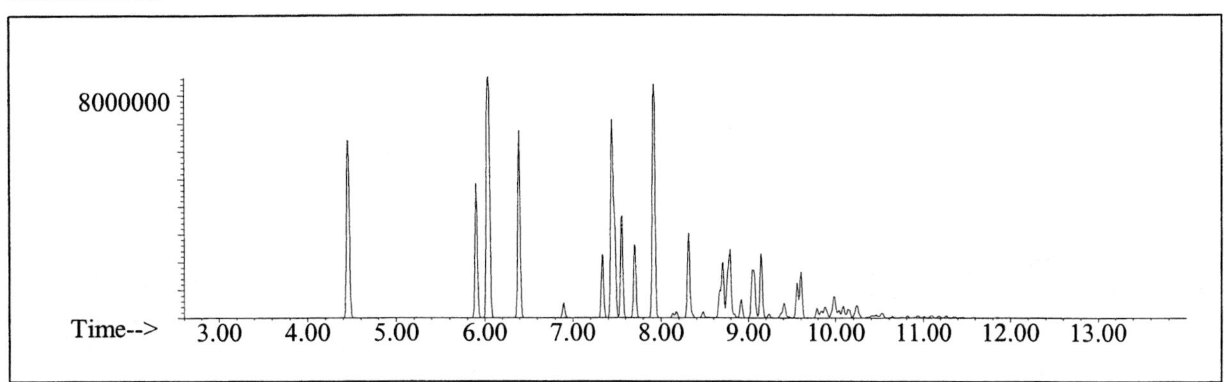

CYCLOPARAFFINS AND ALKENES

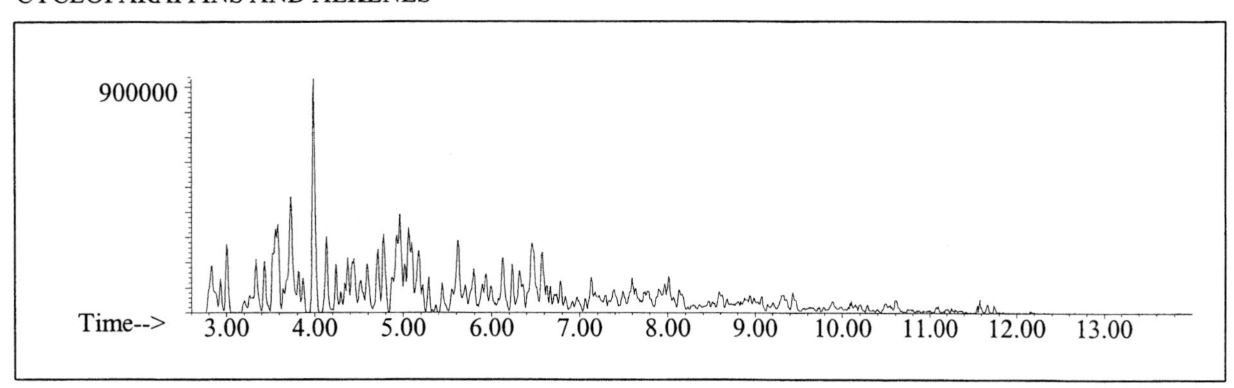

NAPHTHALENES

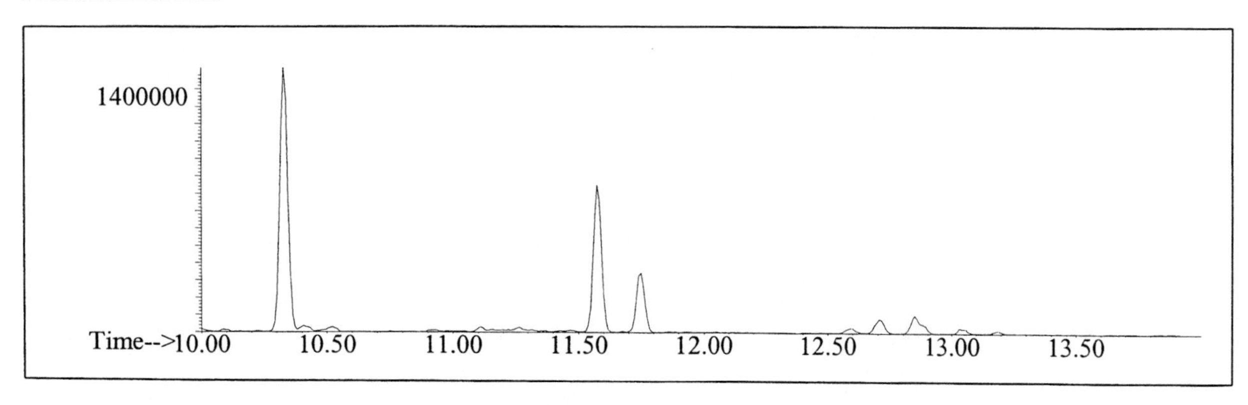

INDIVIDUAL PROFILES Gasoline #3

Alkane

Ion 43

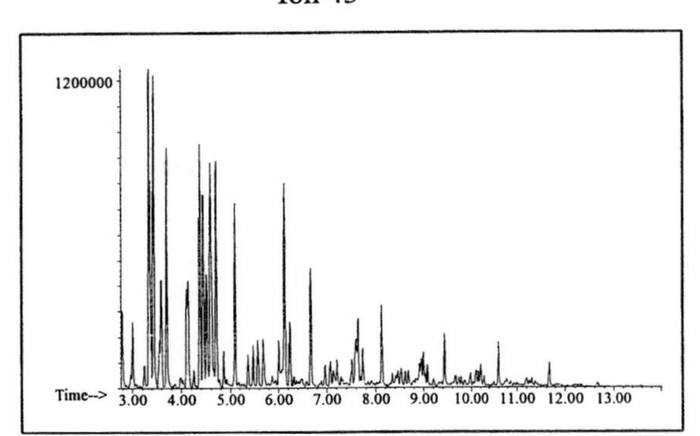

Ion 57

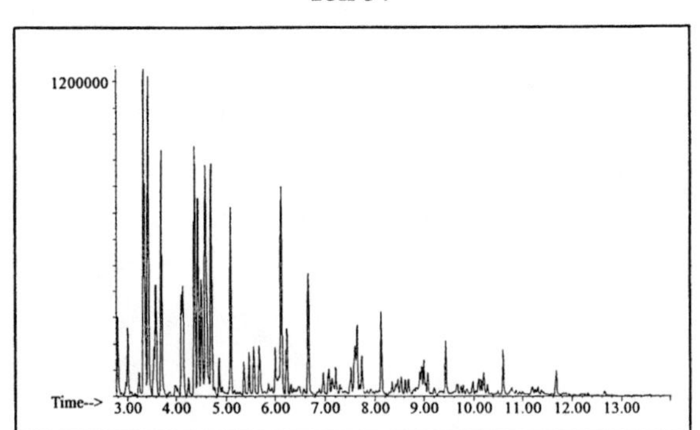

Ion 71

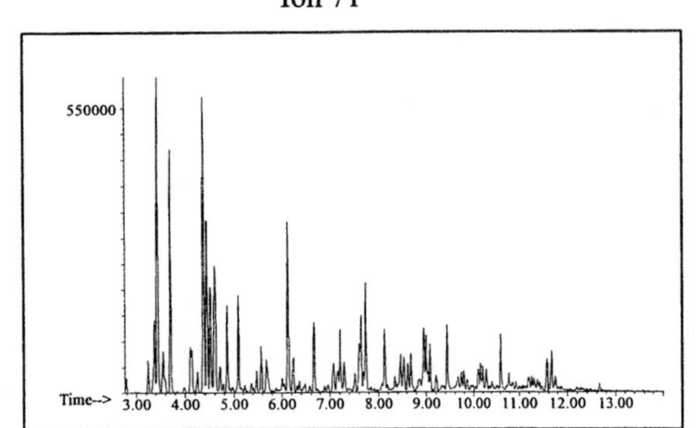

Ion 85

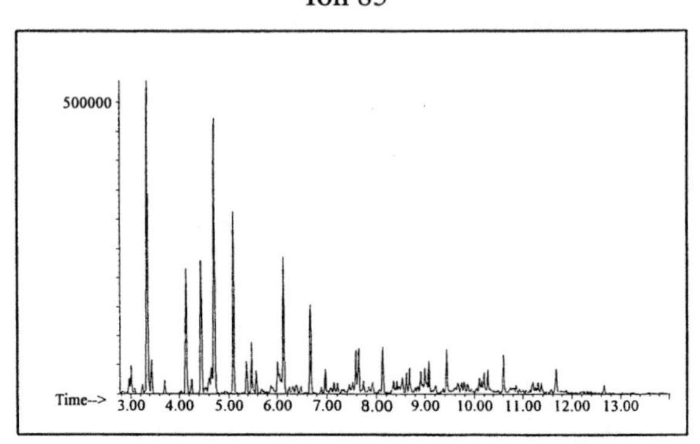

Aromatic

Ion 91

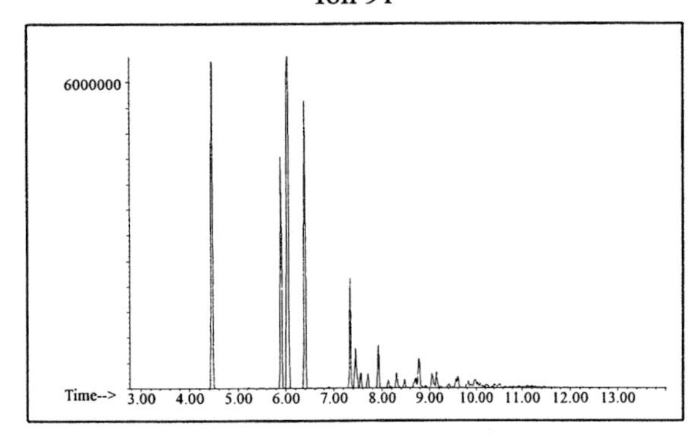

Ion 105

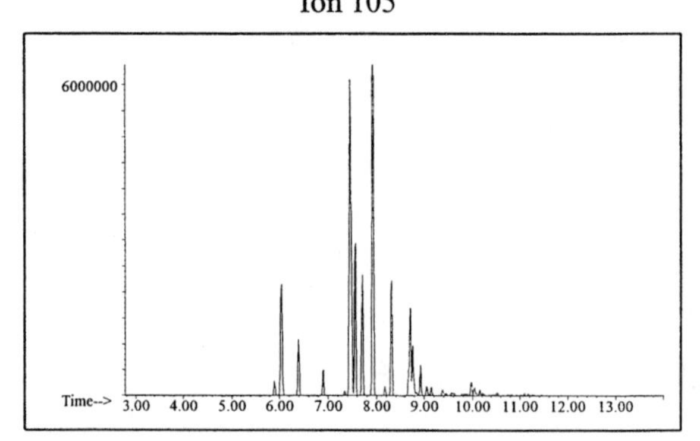

Ion 119

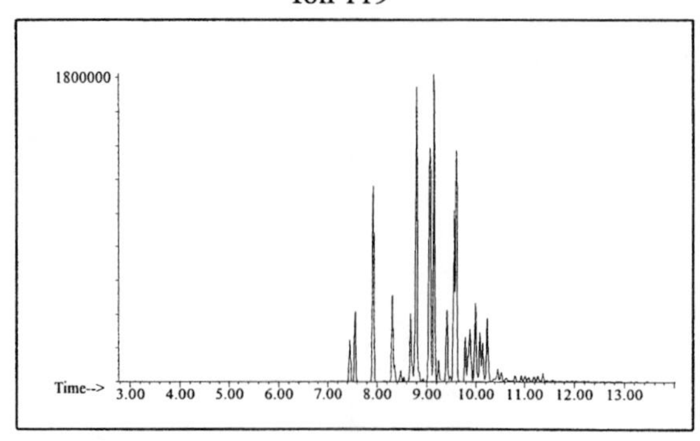

INDIVIDUAL PROFILES Gasoline #3

Cycloparaffin

Ion 55

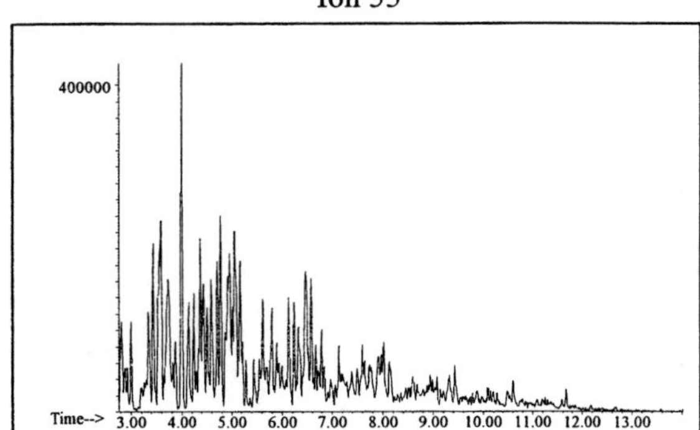

Ion 69

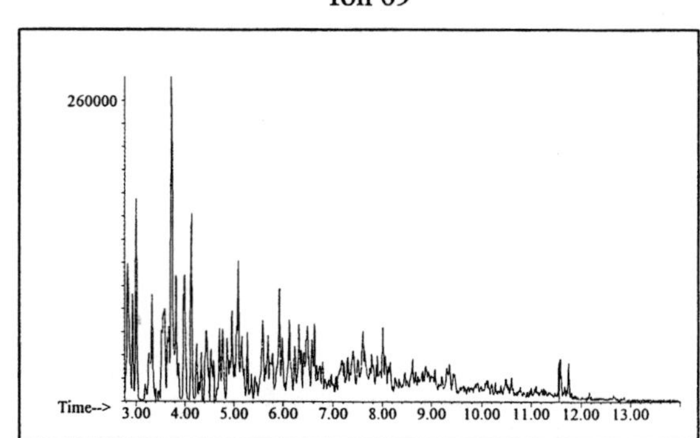

Ion 83

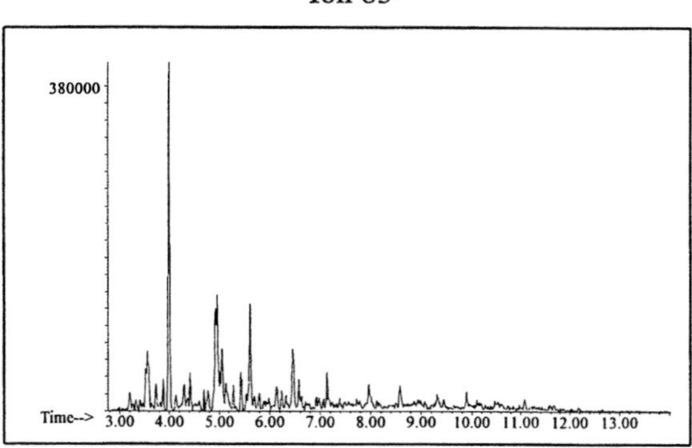

Naphthalene

Ion 128

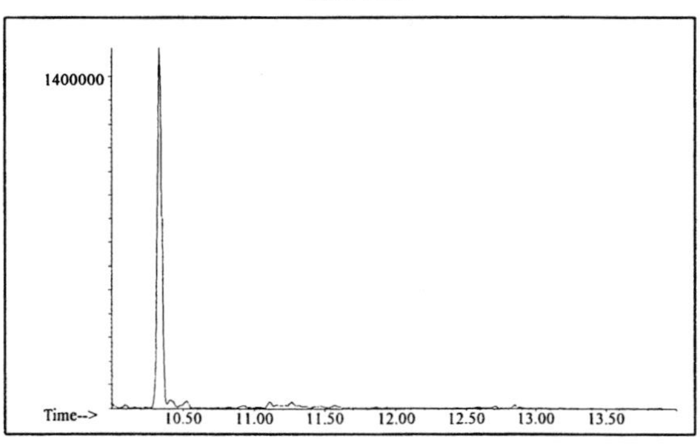

Ion 142

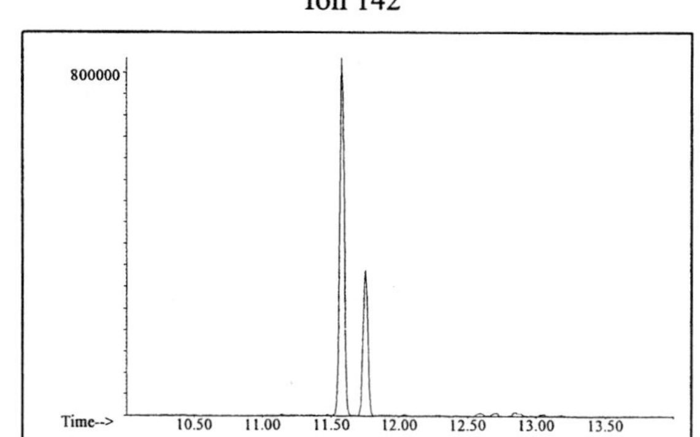

Ion 156

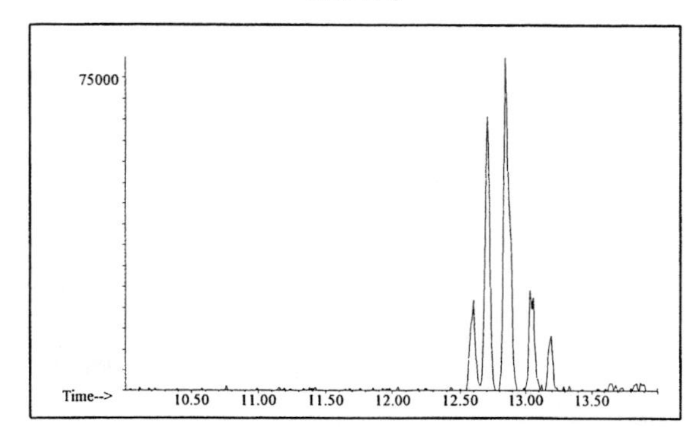

Gasoline #4

20uL/mL Pentane

ASTM: Class 2 (Gasoline)

Macro Code: L

Product Displayed: 75% Evaporated Mobil Regular Gasoline

Product Uses: Automotive fuel

Other Similar Products: Other brands and grades of gasoline

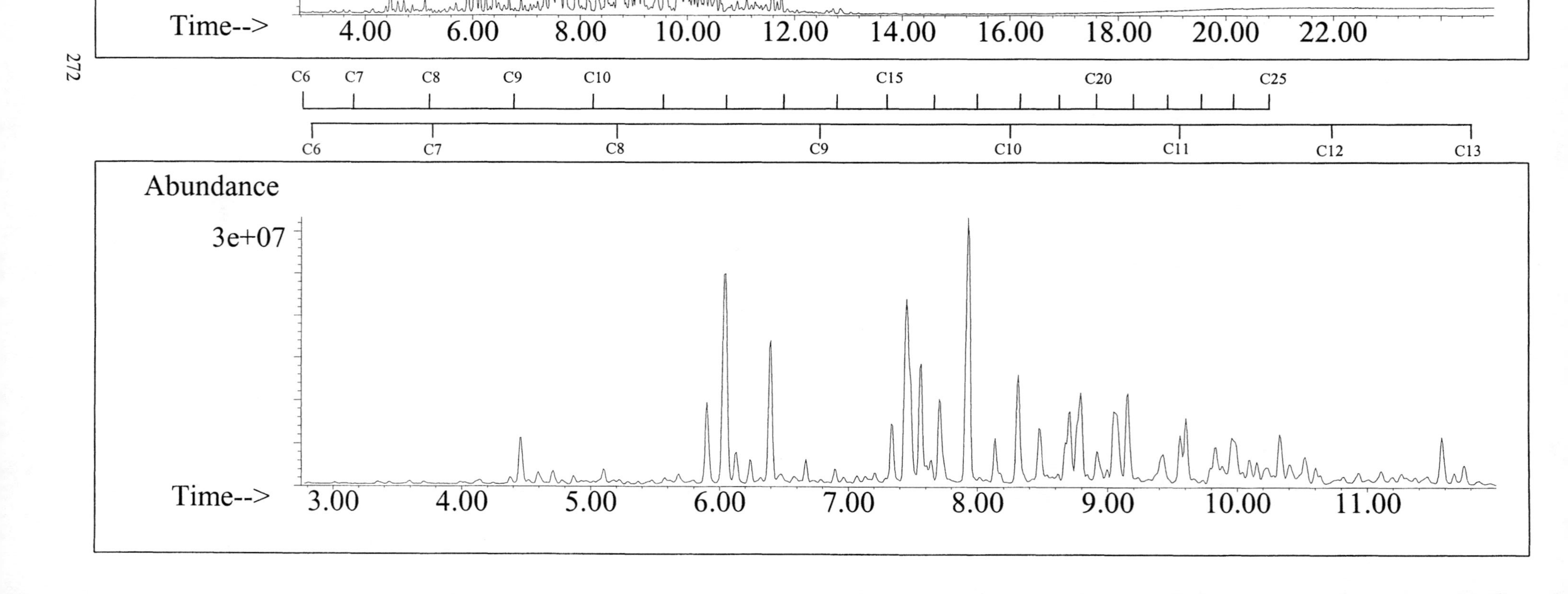

ALKANES

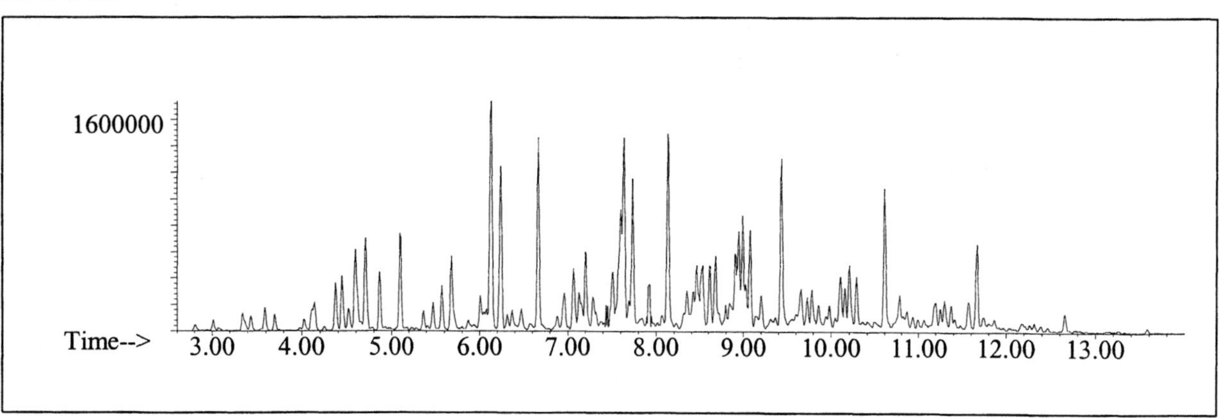

AROMATICS

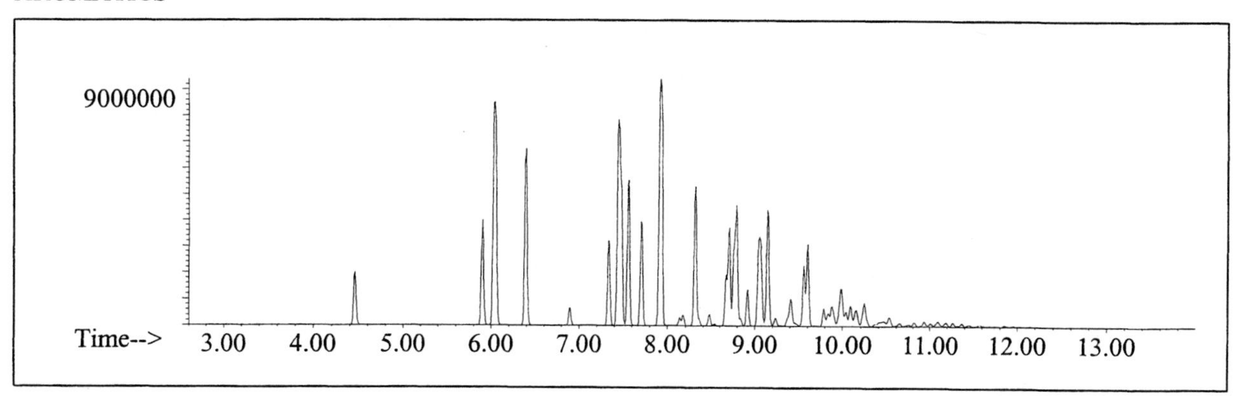

CYCLOPARAFFINS AND ALKENES

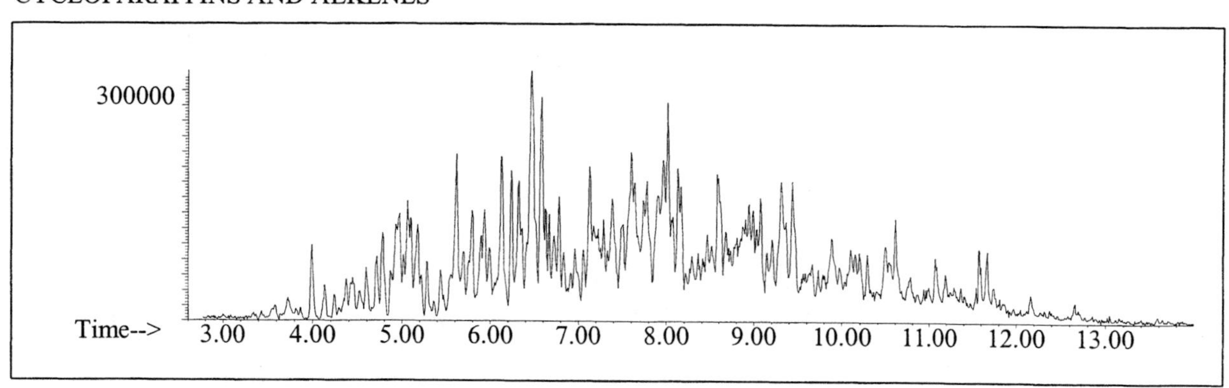

NAPHTHALENES

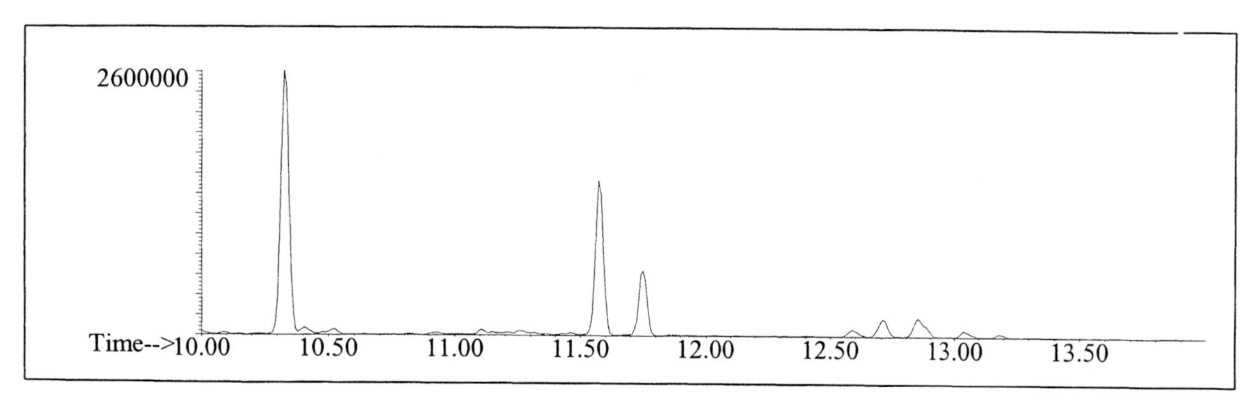

INDIVIDUAL PROFILES

Gasoline #4

Alkane

Ion 43

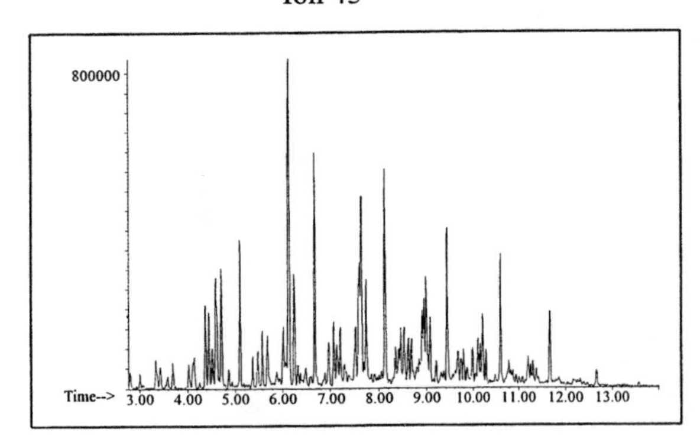

Ion 57

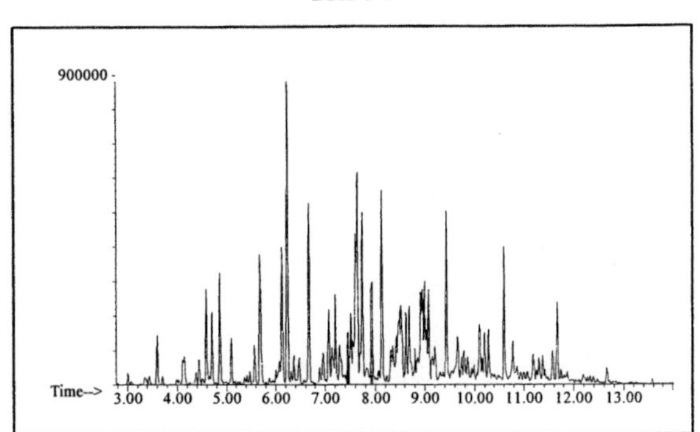

Ion 71

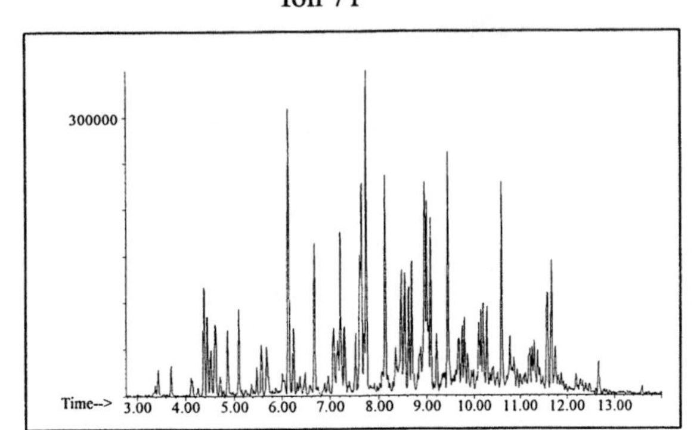

Ion 85

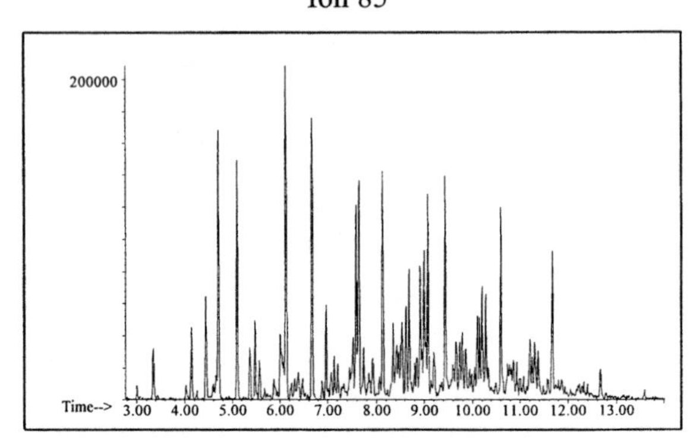

Aromatic

Ion 91

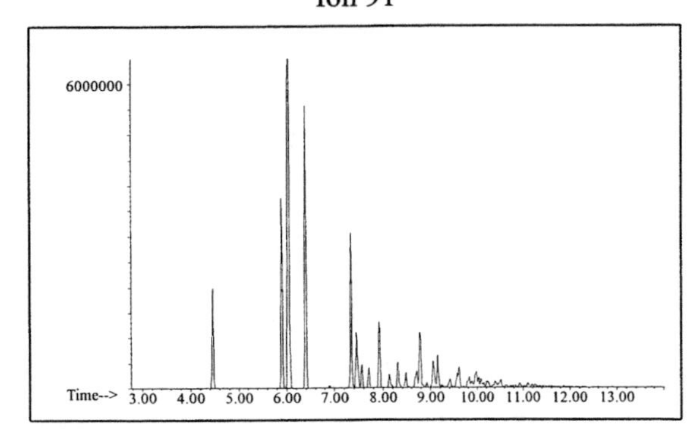

Ion 105

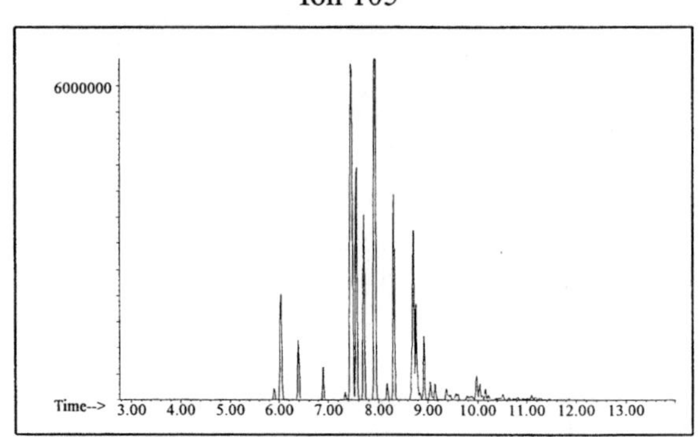

Ion 119

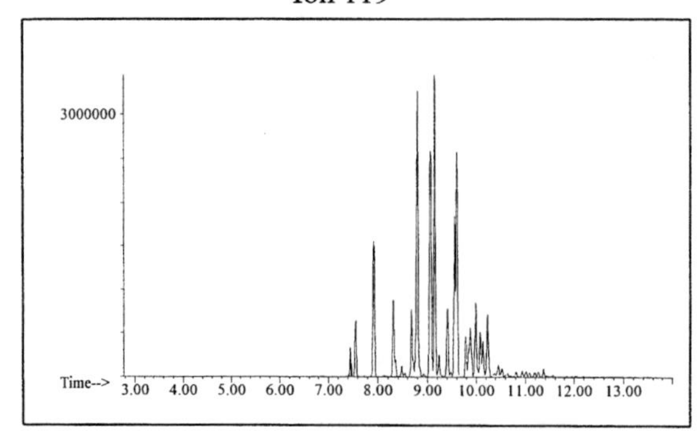

INDIVIDUAL PROFILES

Gasoline #4

Cycloparaffin

Ion 55

190000

Time--> 3.00 4.00 5.00 6.00 7.00 8.00 9.00 10.00 11.00 12.00 13.00

Ion 69

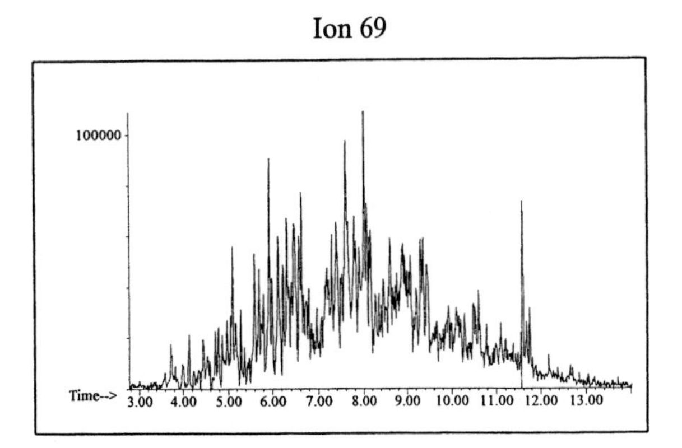

Ion 83

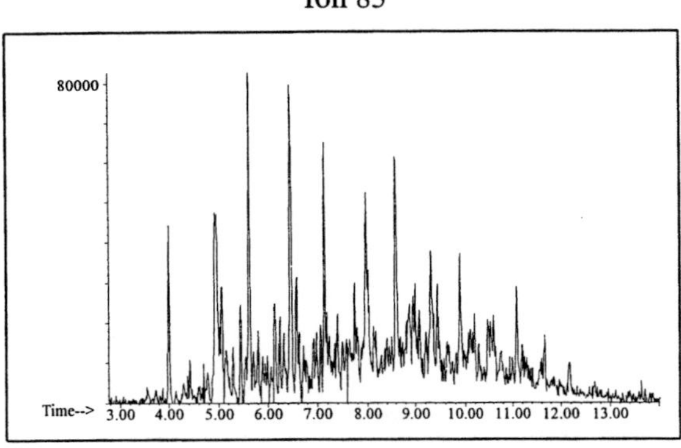

Naphthalene

Ion 128

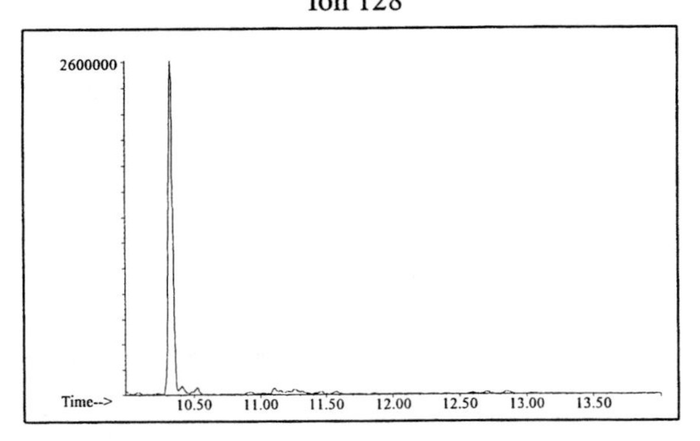

Ion 142

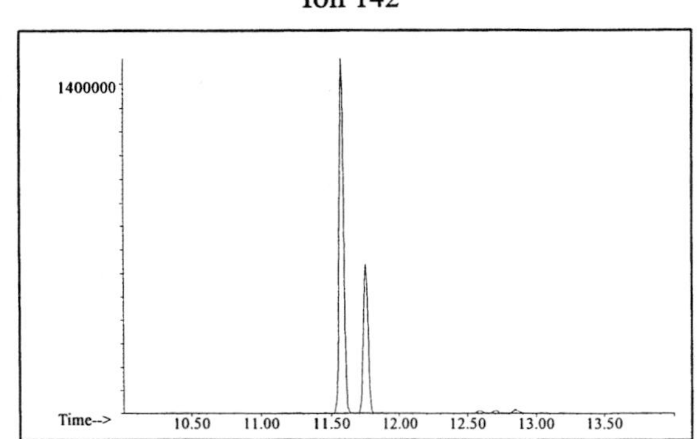

Ion 156

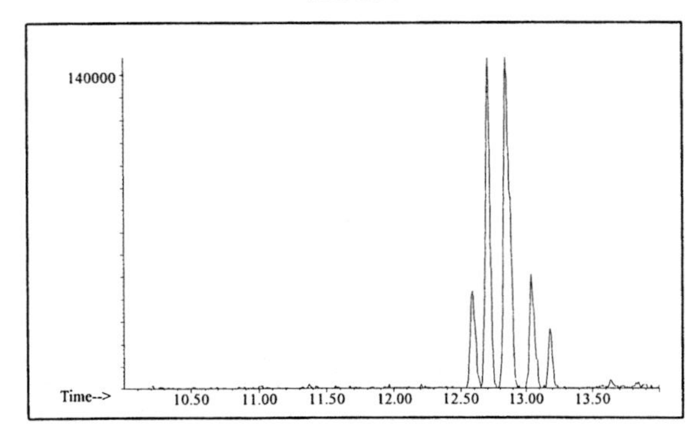

Gasoline #5

20uL/mL Pentane

Product Displayed: 90% Evaporated Mobil Regular Gasoline

Other Similar Products: Other grades and brands of gasoline

ASTM: Class 2 (Gasoline)

Product Uses: Automotive fuel

Macro Code: L

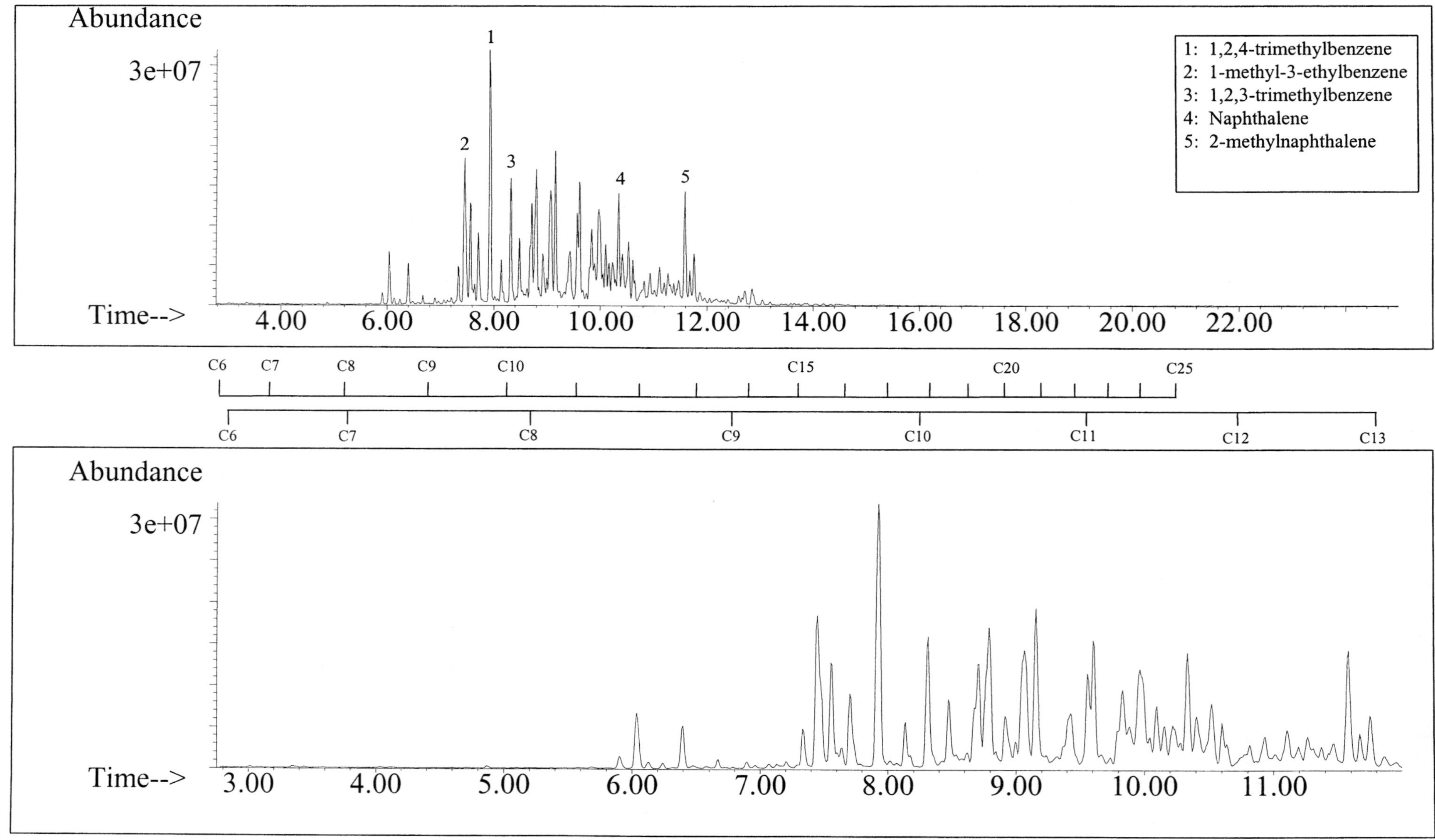

SUMMED PROFILES

Gasoline #5

ALKANES

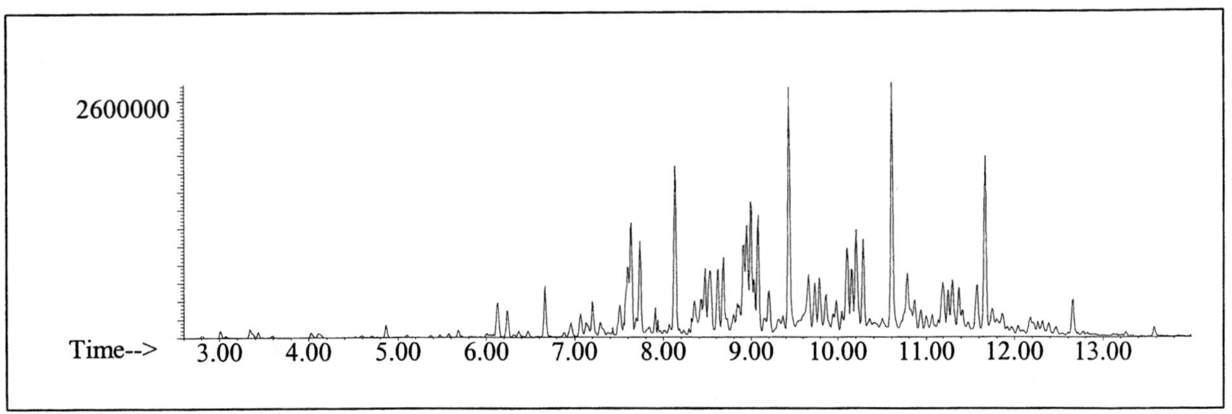

AROMATICS

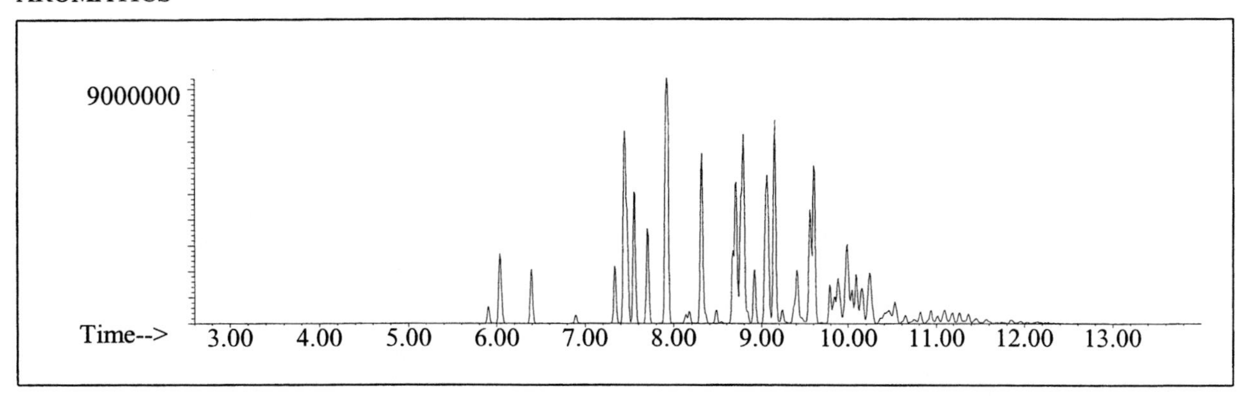

CYCLOPARAFFINS AND ALKENES

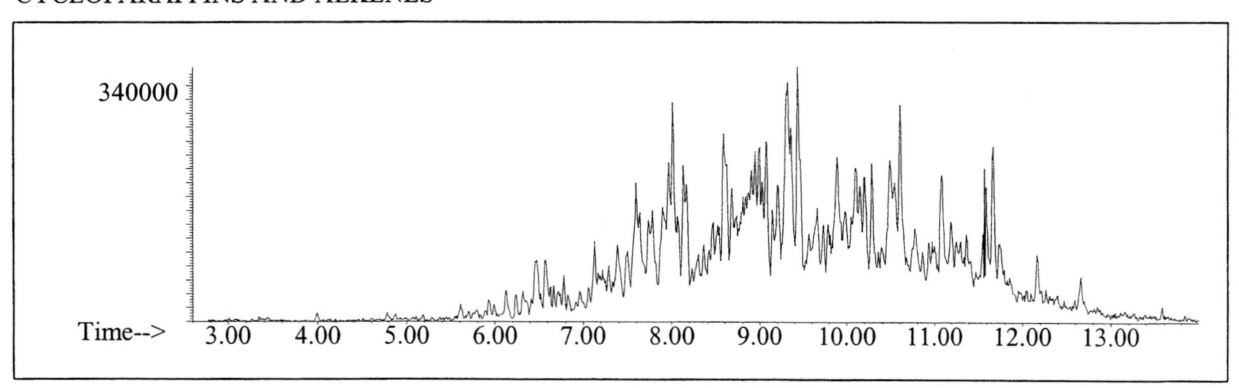

NAPHTHALENES

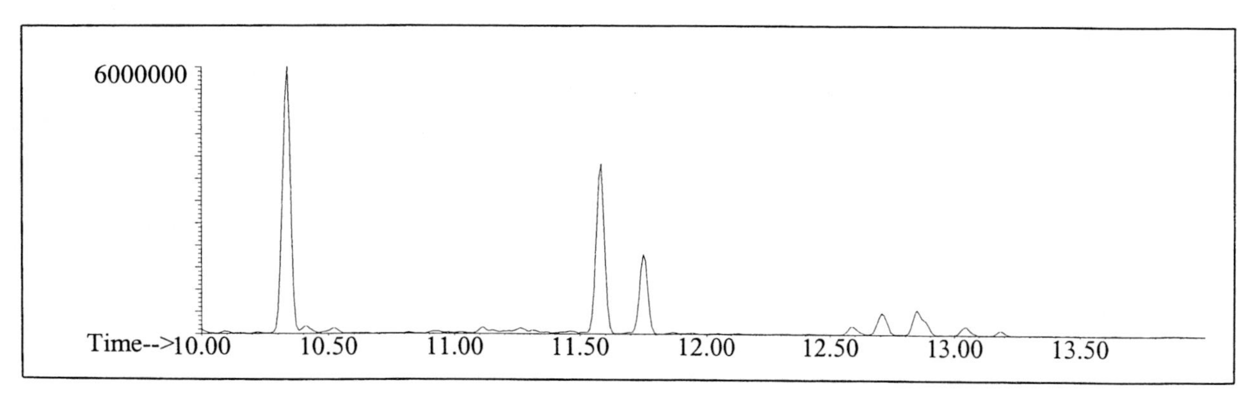

Alkane

Ion 43

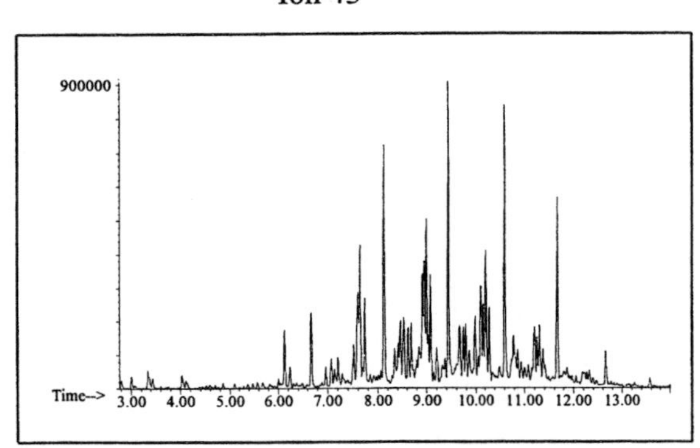

Ion 57

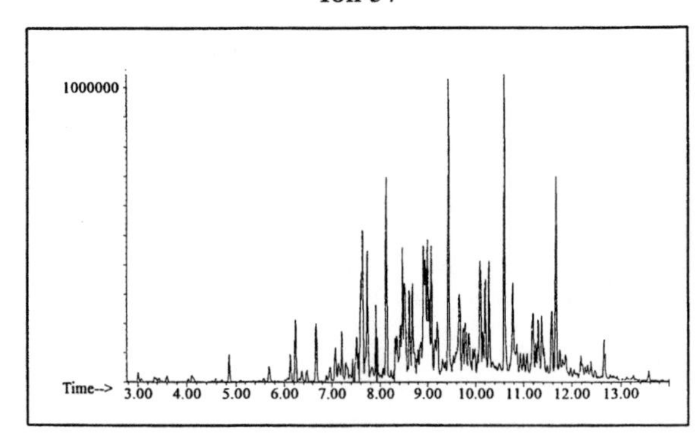

Ion 71

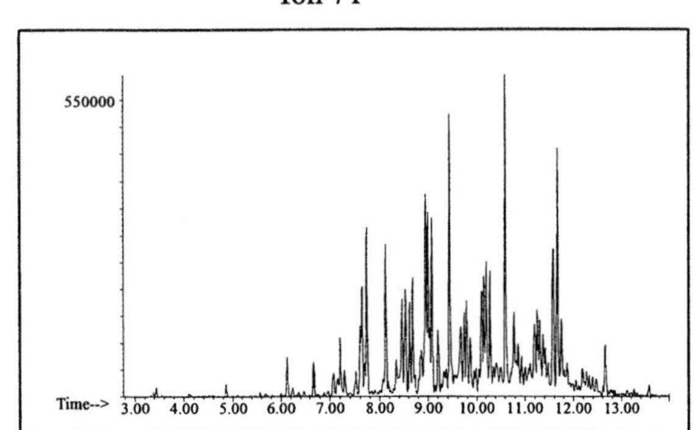

Ion 85

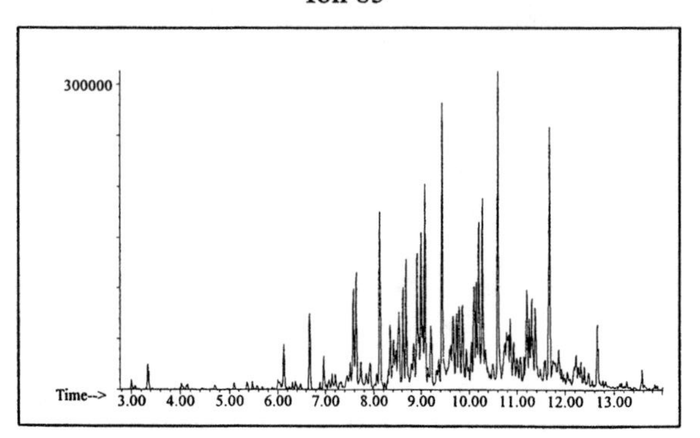

Aromatic

Ion 91

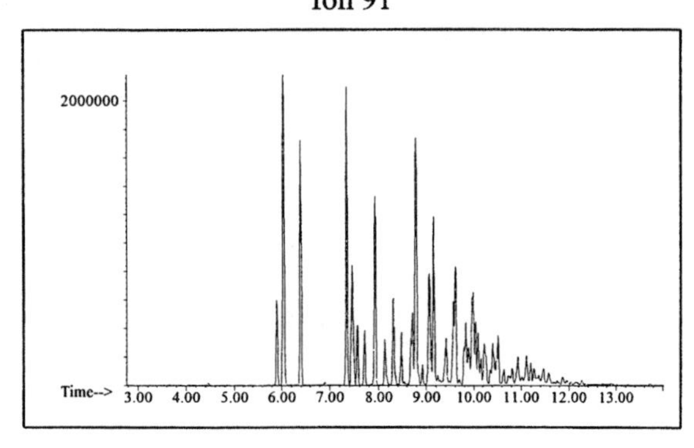

Ion 105

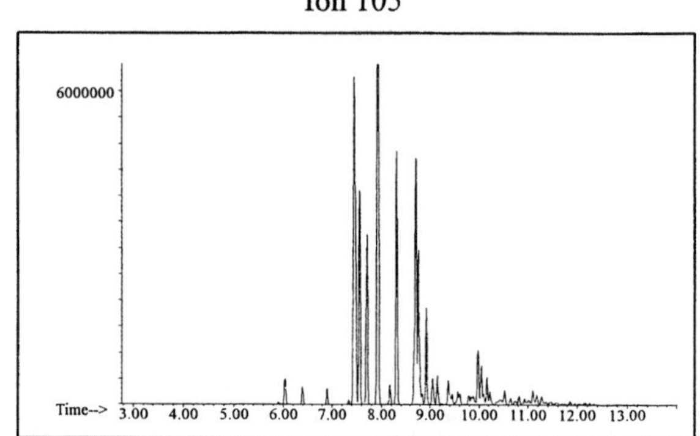

Ion 119

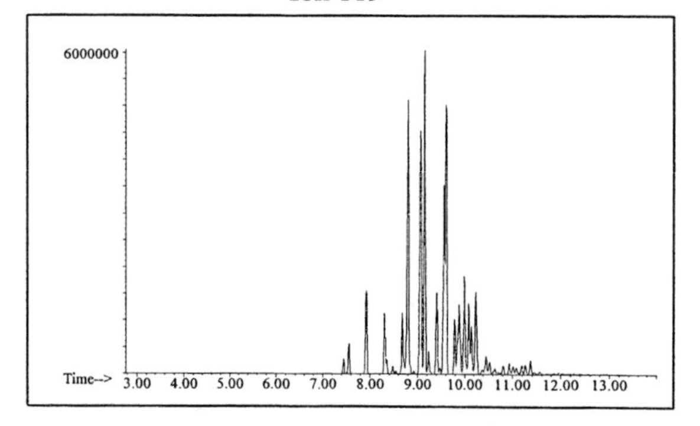

INDIVIDUAL PROFILES Gasoline #5

Cycloparaffin

Ion 55

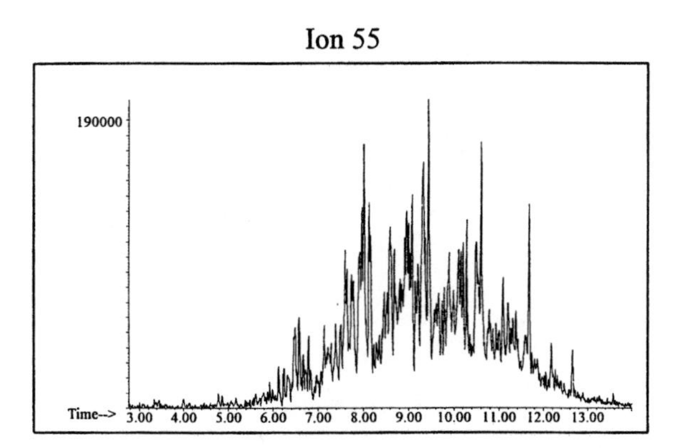

Ion 69

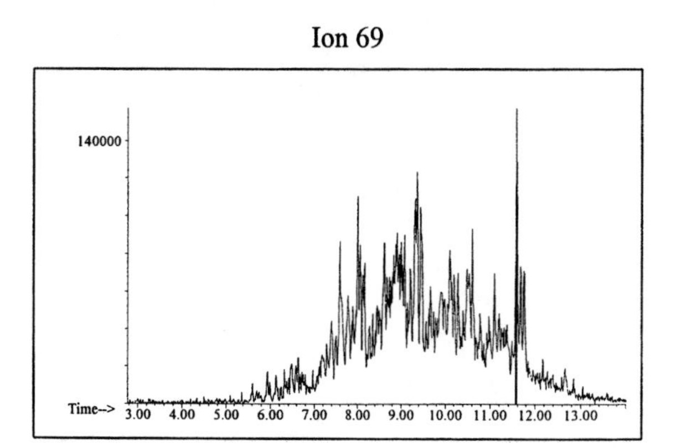

Ion 83

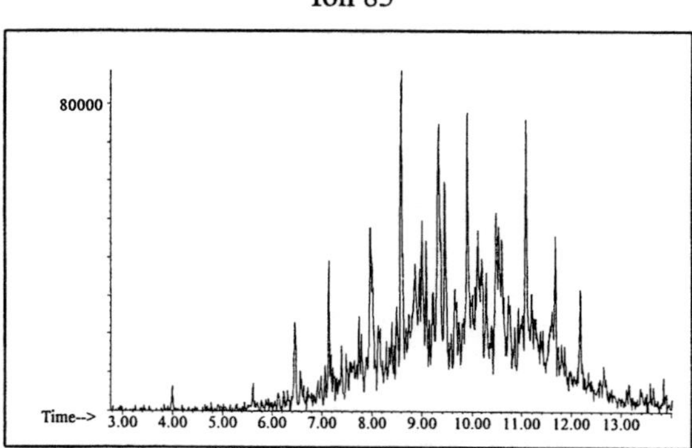

Naphthalene

Ion 128

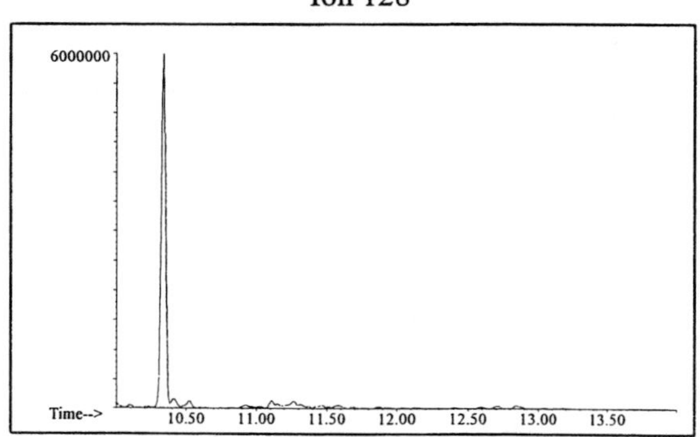

Ion 142

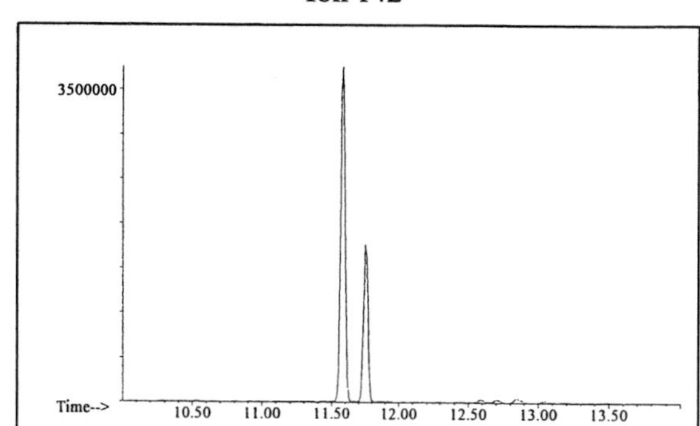

Ion 156

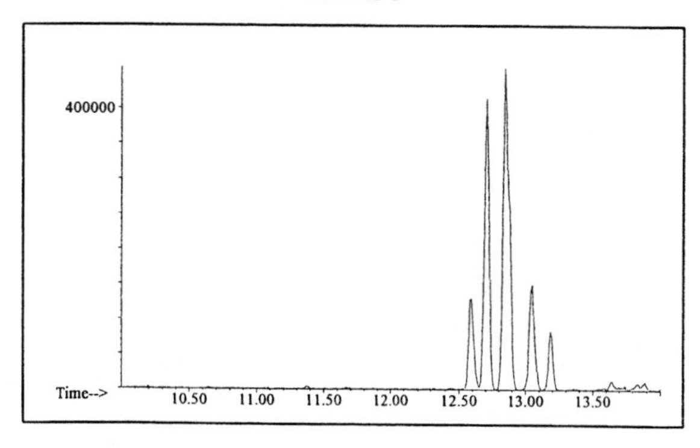

Gasoline #6

20uL/mL Pentane

Product Displayed: 98% Evaporated Mobil Regular Gasoline

Other Similar Products: Other brands and grades of gasoline

ASTM: Class 2 (Gasoline)

Product Uses: Automotive fuel

Macro Code: M

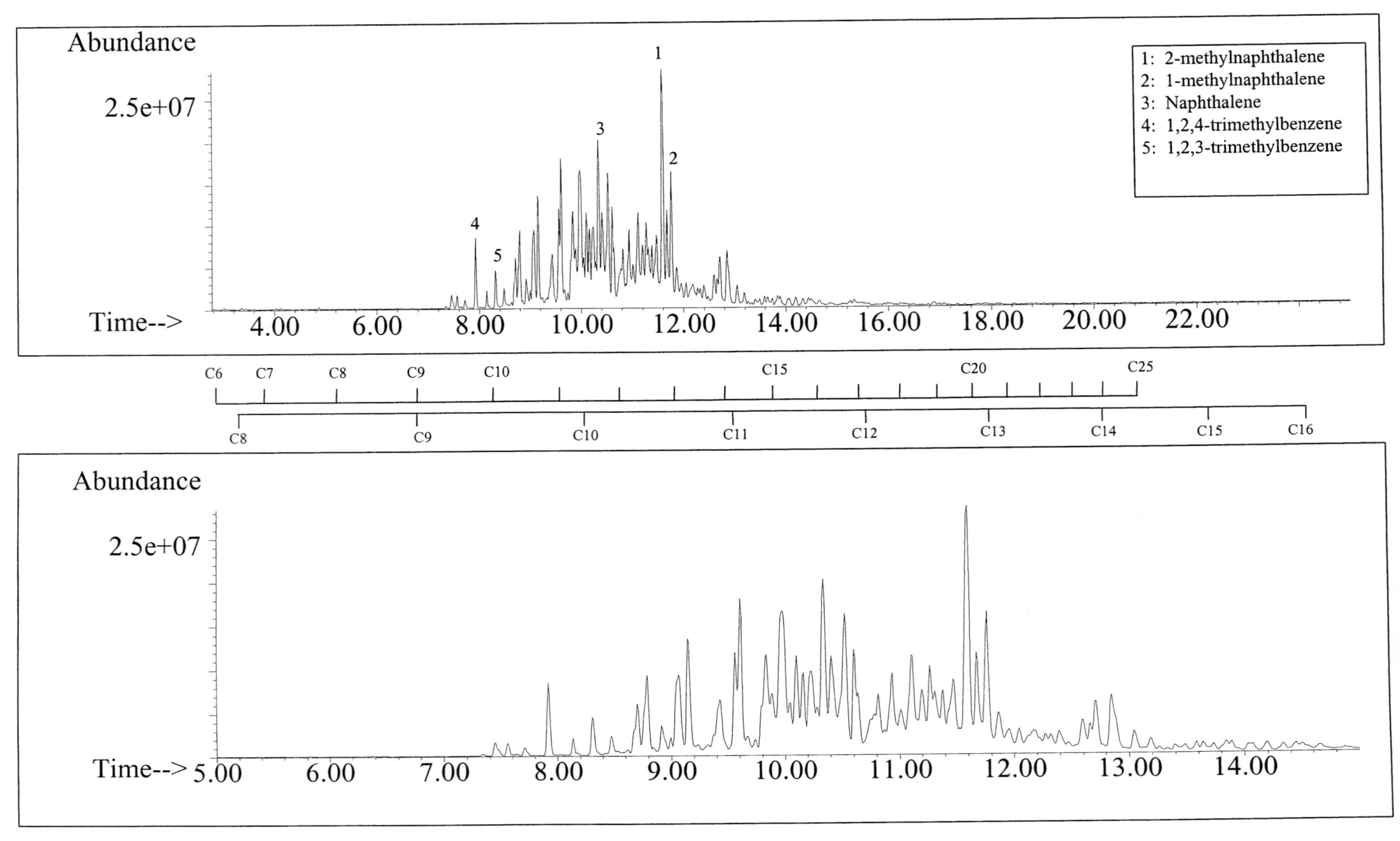

SUMMED PROFILES

Gasoline #6

ALKANES

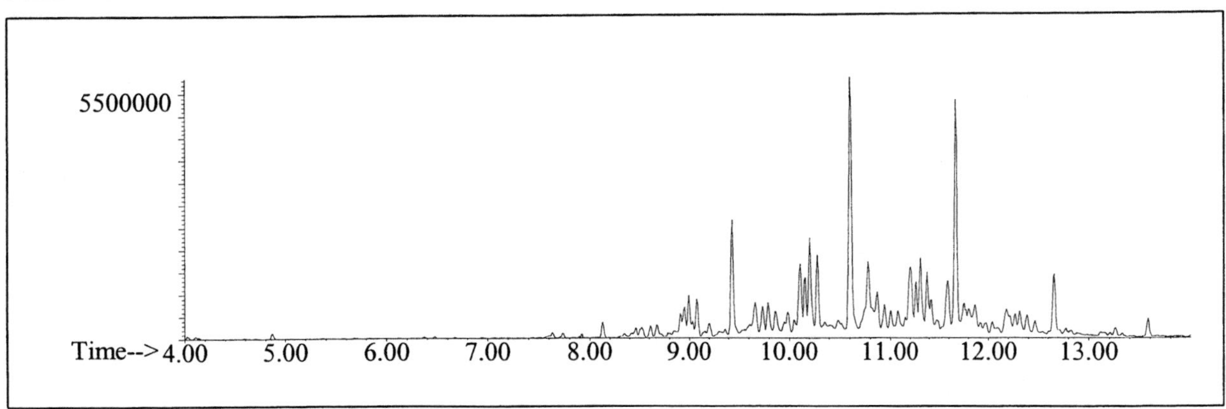

AROMATICS

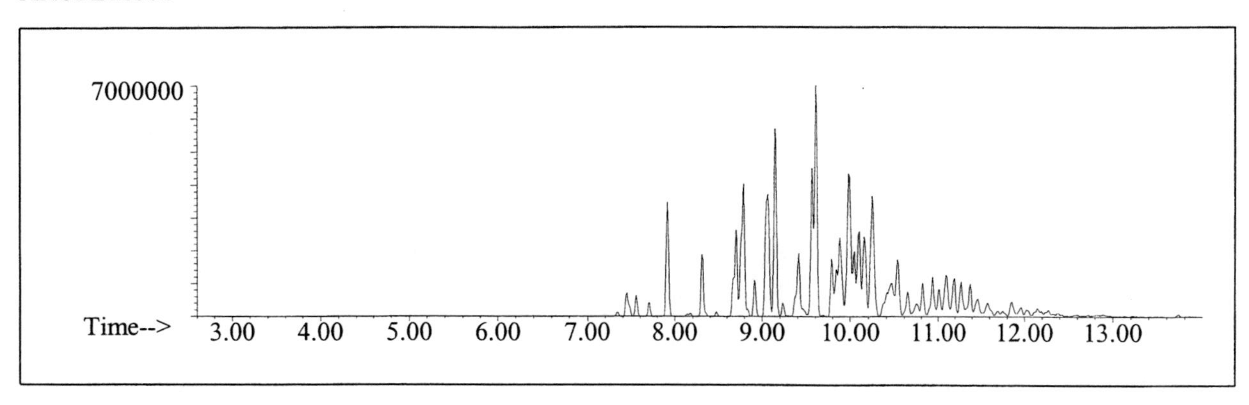

CYCLOPARAFFINS AND ALKENES

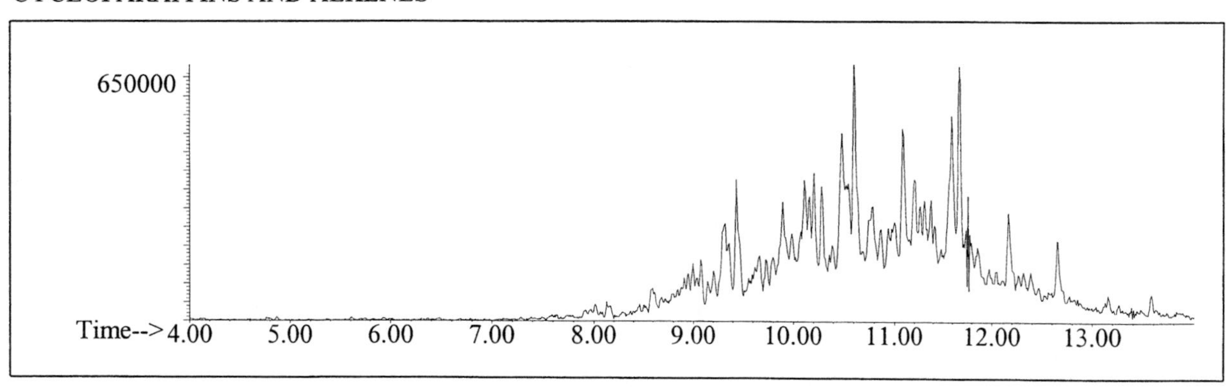

NAPHTHALENES

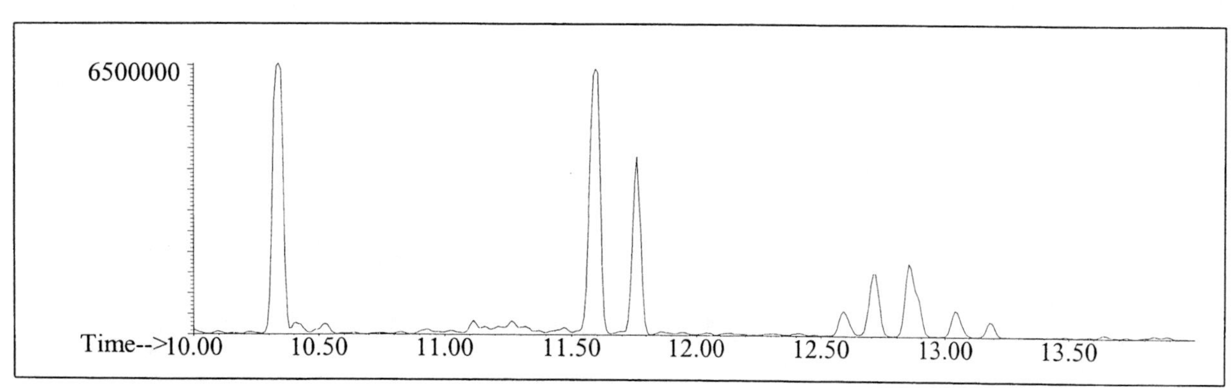

INDIVIDUAL PROFILES

Gasoline #6

Alkane

Ion 43

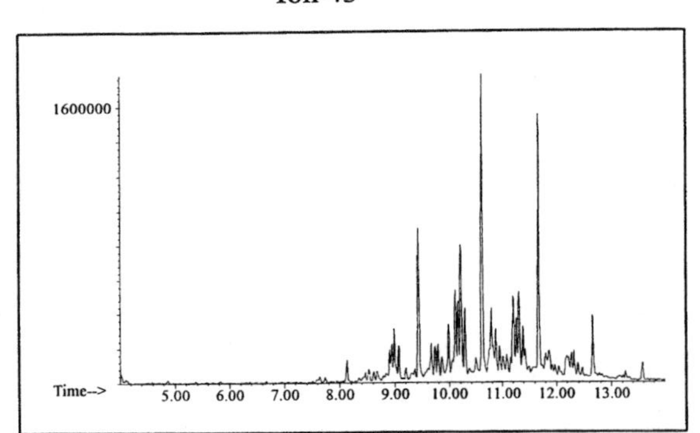

Ion 57

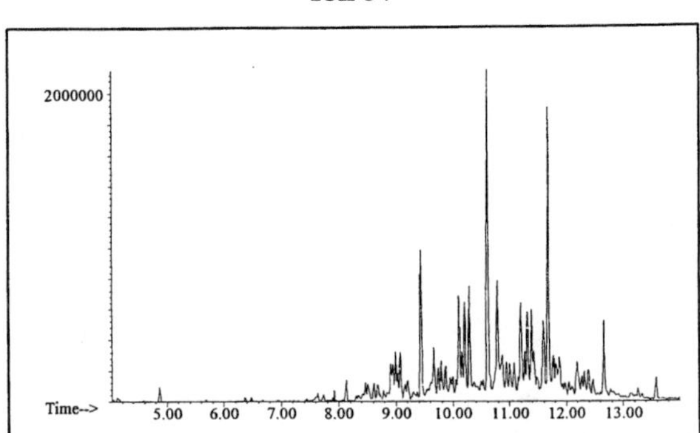

Ion 71

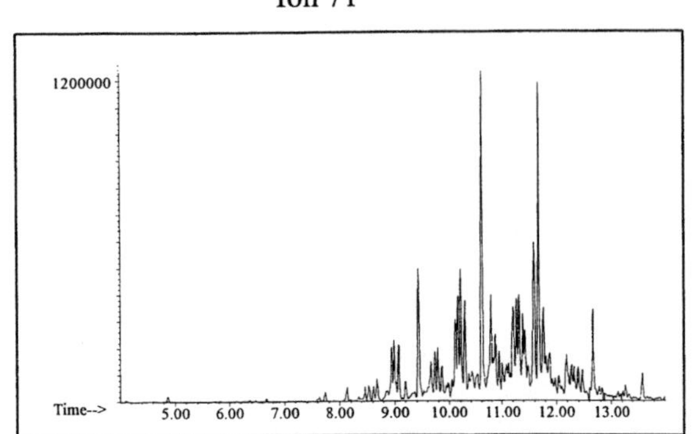

Ion 85

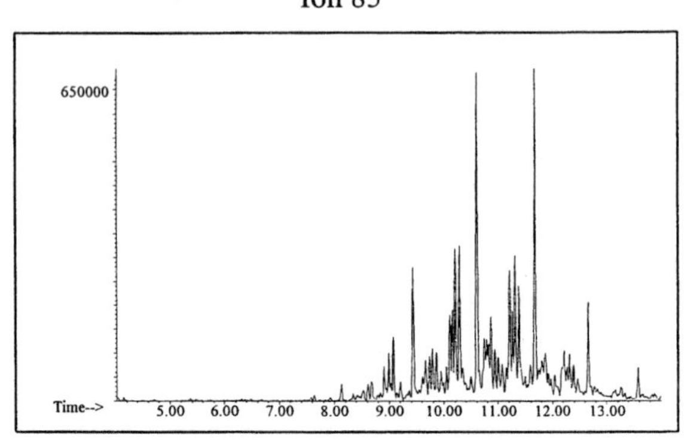

Aromatic

Ion 91

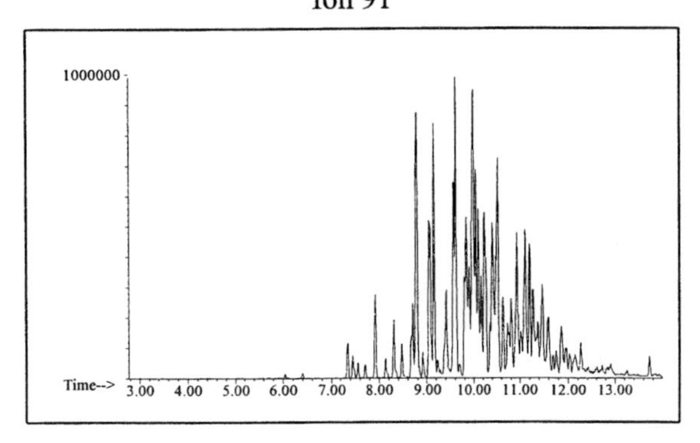

Ion 105

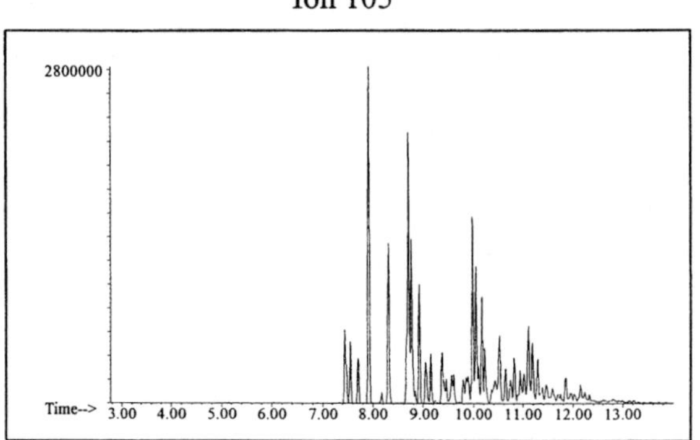

Ion 119

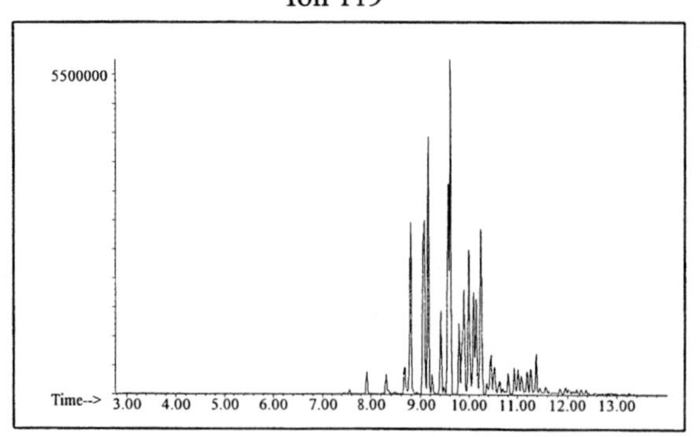

INDIVIDUAL PROFILES Gasoline #6

Cycloparaffin

Ion 55

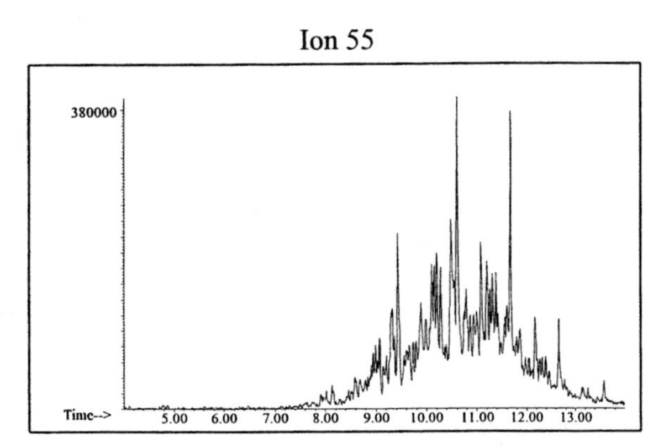

Ion 69

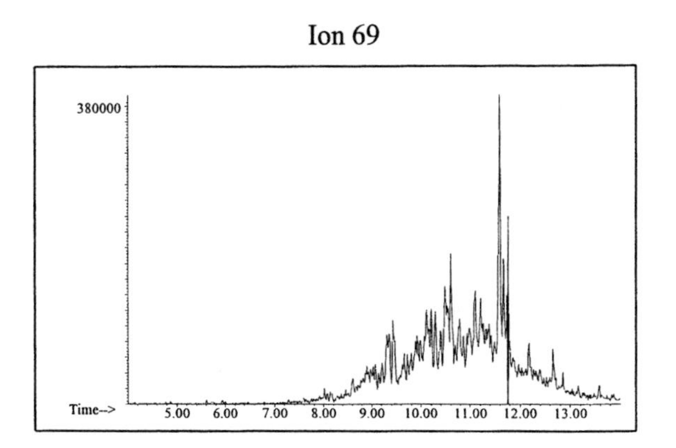

Ion 83

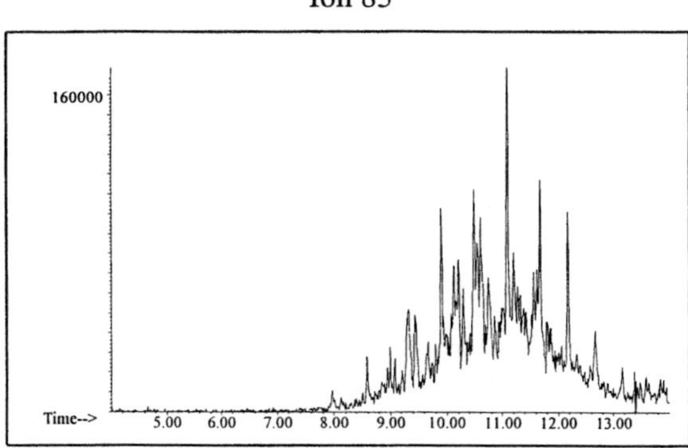

Naphthalene

Ion 128

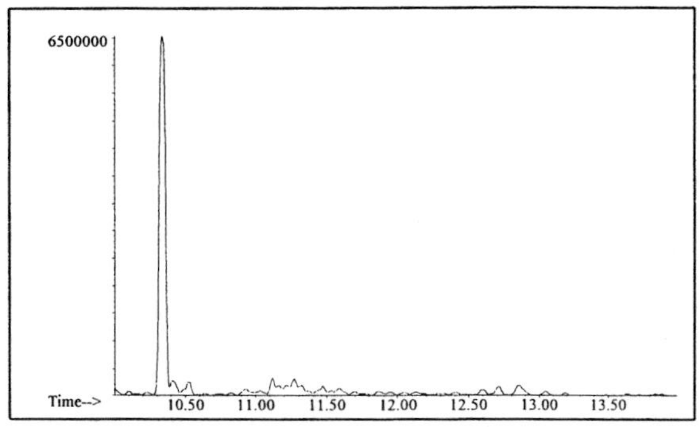

Ion 142

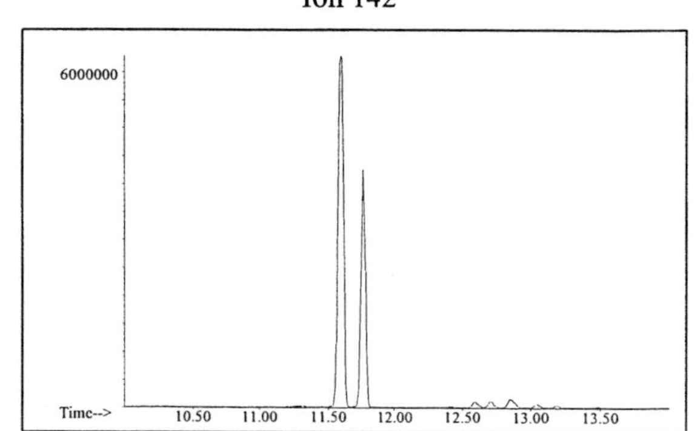

Ion 156

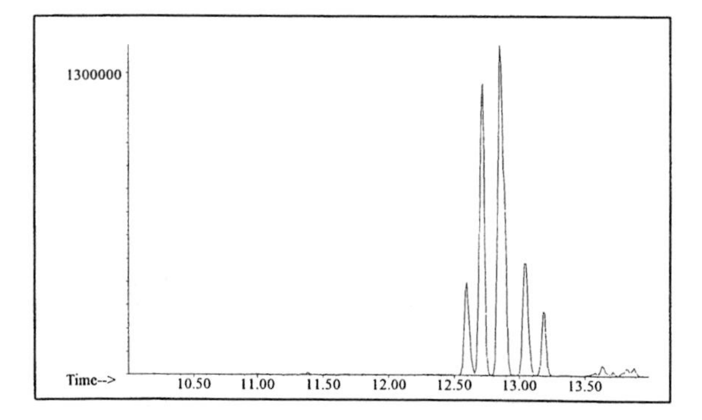

Gasoline #7

20uL/mL Pentane

Product Displayed: Mobil Super Premium Gasoline

Other Similar Products: Other grades and brands of gasoline

ASTM: Class 2 (Gasoline)

Product Uses: Automotive fuel

Macro Code: L

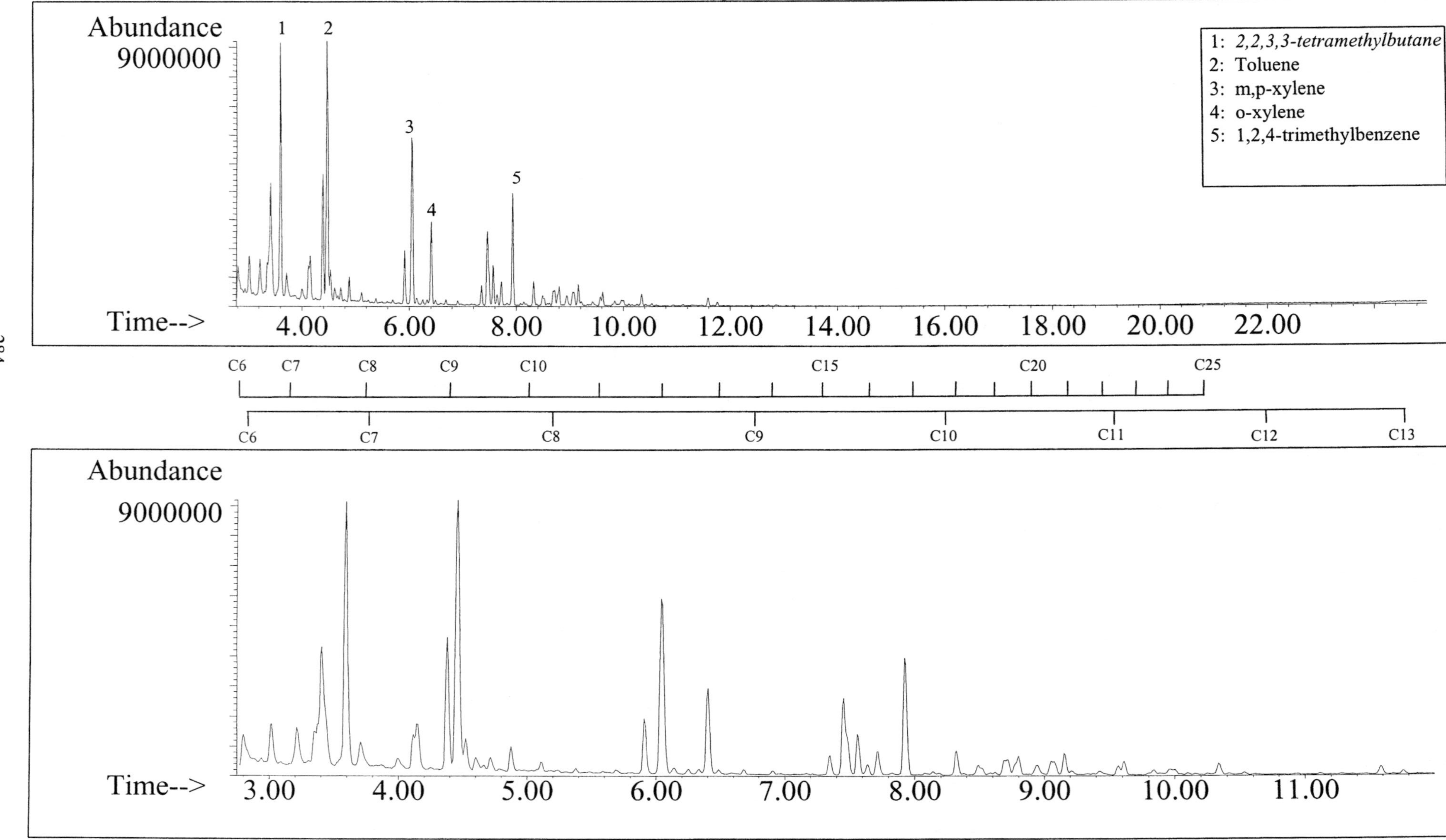

SUMMED PROFILES **Gasoline #7**

ALKANES

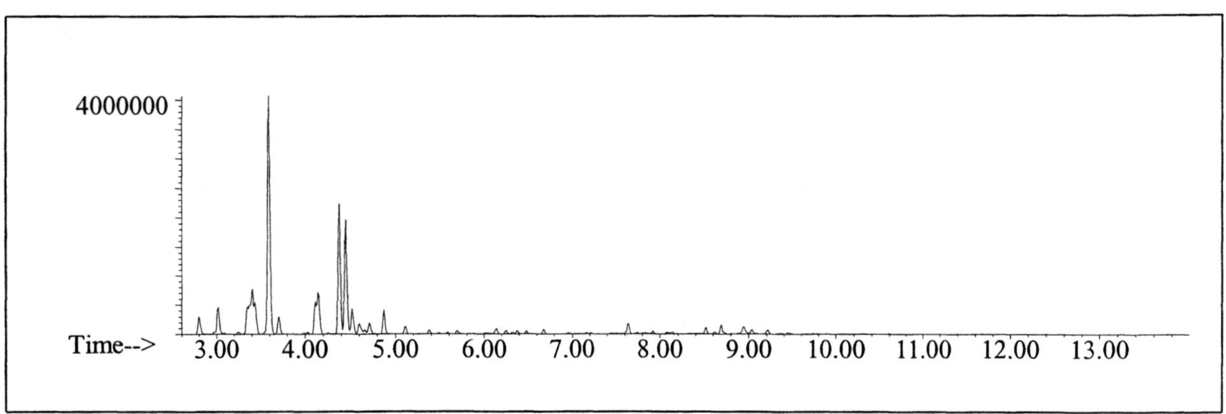

AROMATICS

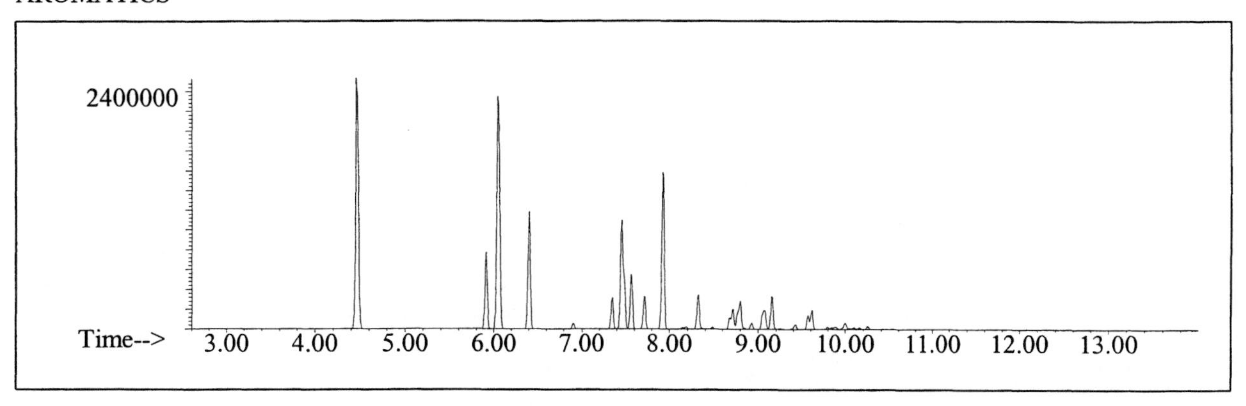

CYCLOPARAFFINS AND ALKENES

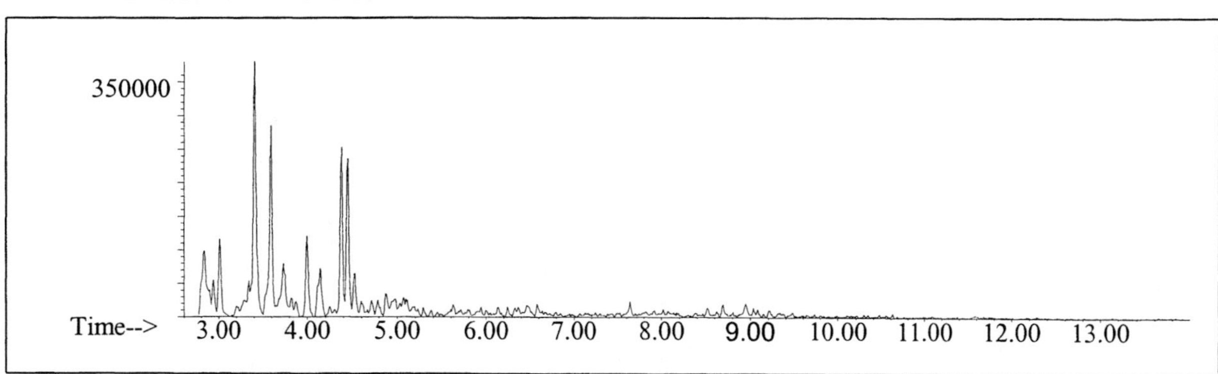

NAPHTHALENES

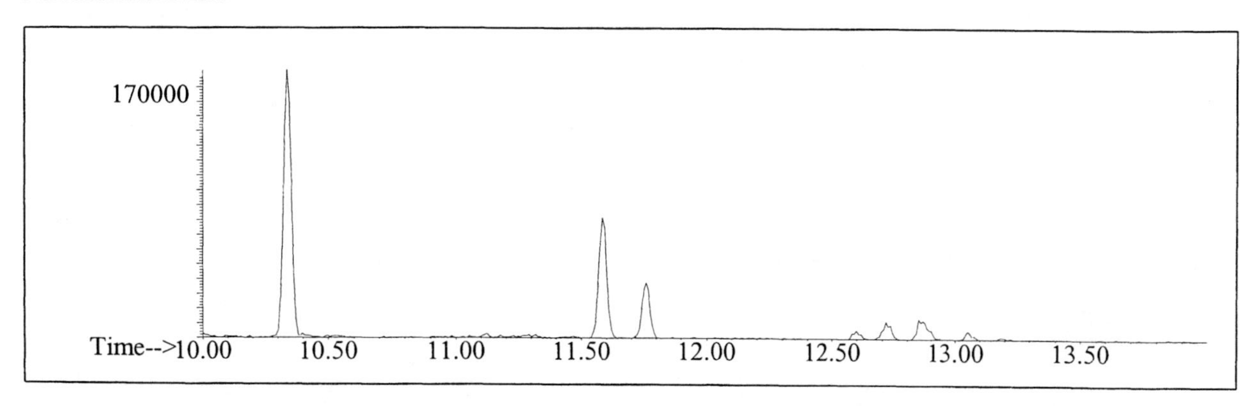

INDIVIDUAL PROFILES

Gasoline #7

Alkane

Ion 43

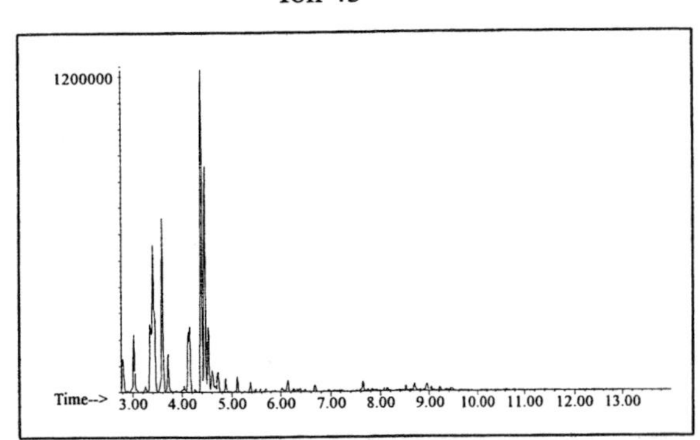

Ion 57

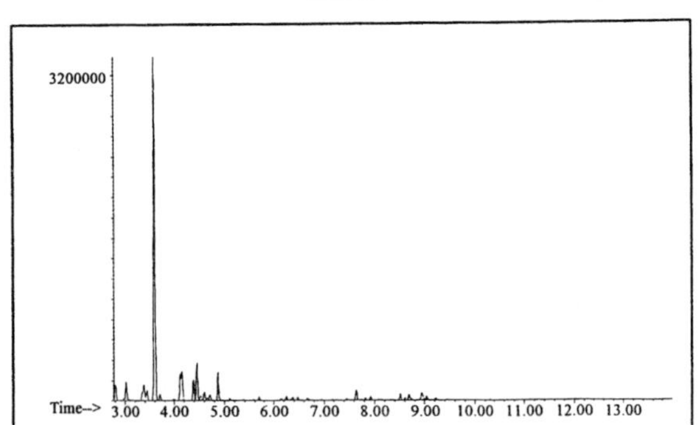

Ion 71

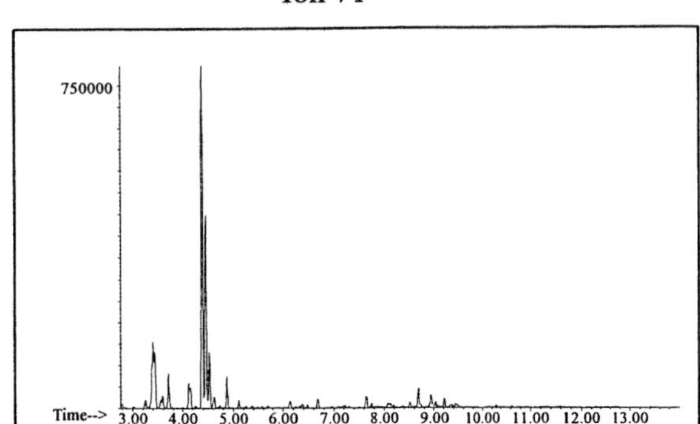

Ion 85

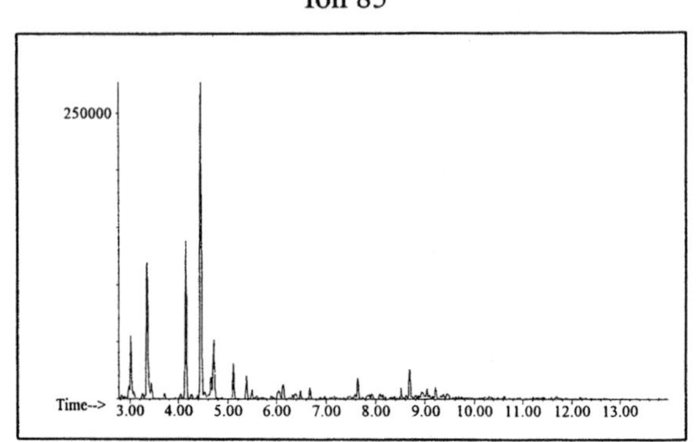

Aromatic

Ion 91

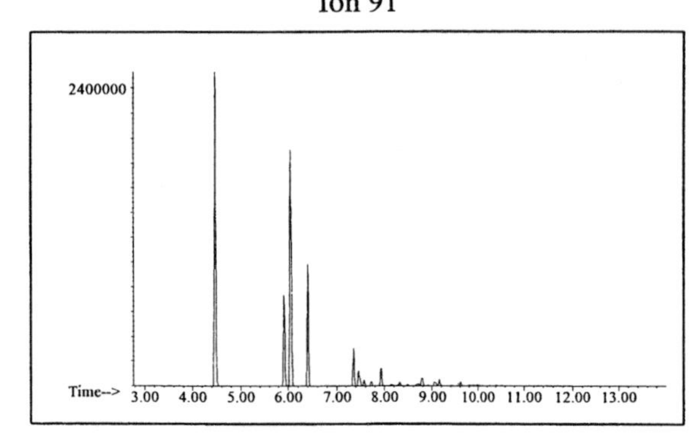

Ion 105

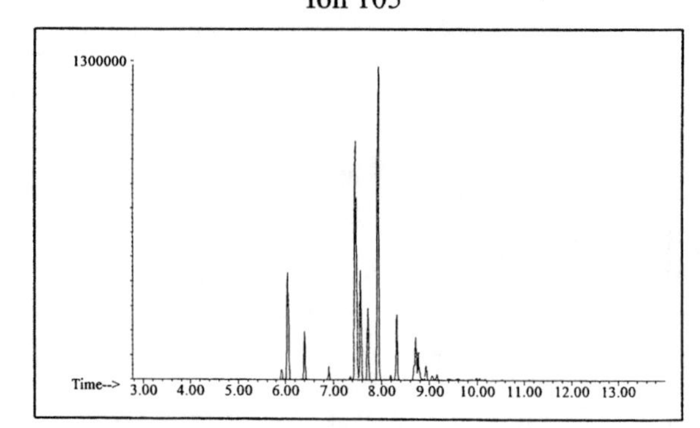

Ion 119

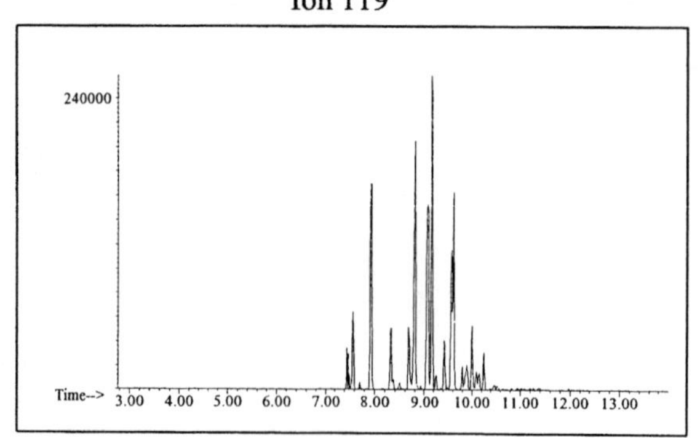

INDIVIDUAL PROFILES

Gasoline #7

Cycloparaffin

Ion 55

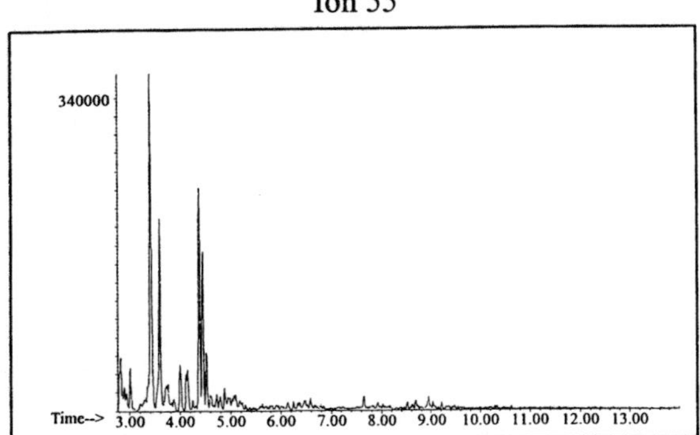

Ion 69

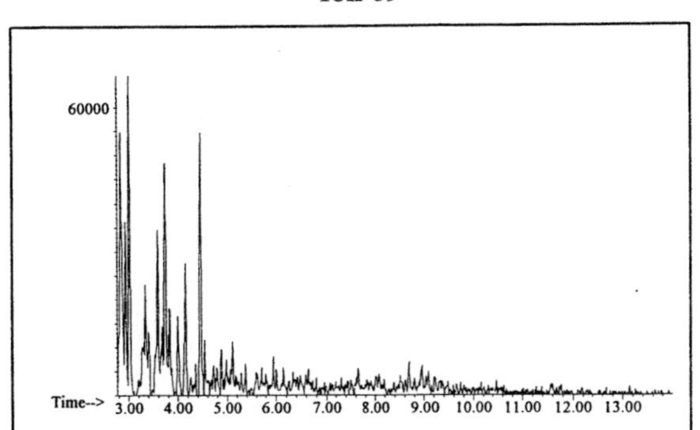

Ion 83

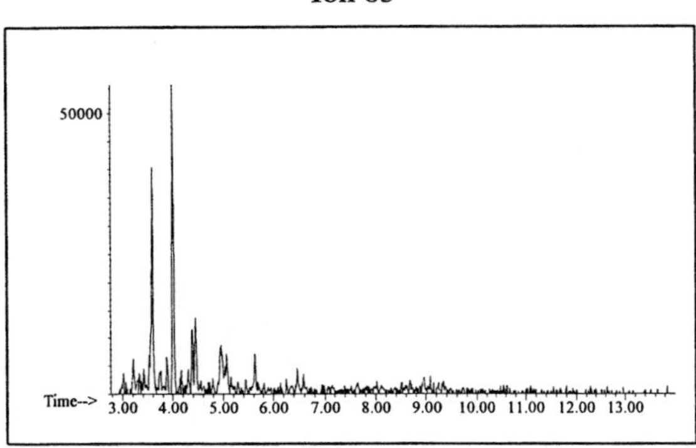

Naphthalene

Ion 128

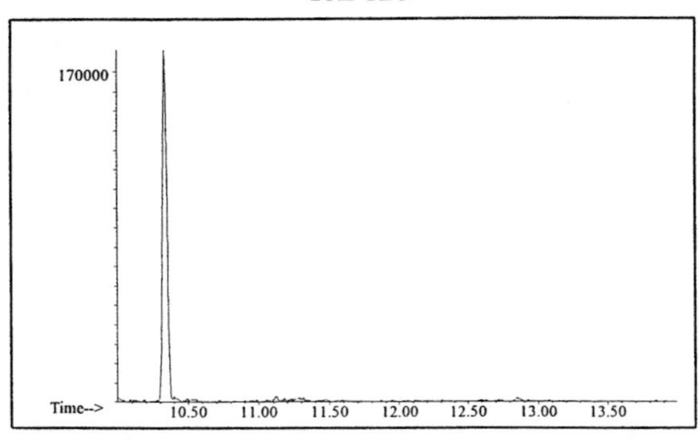

Ion 142

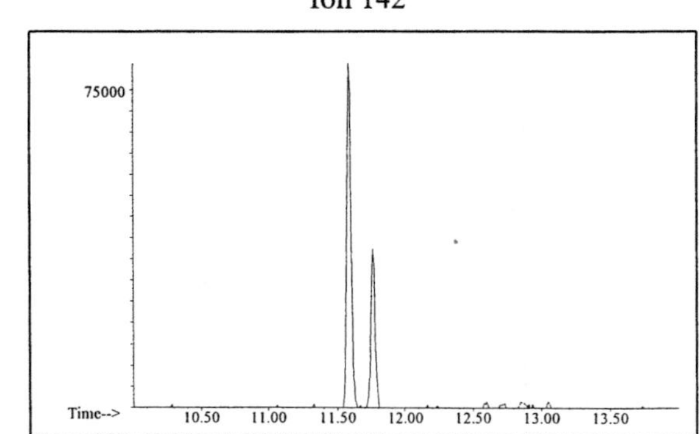

Ion 156

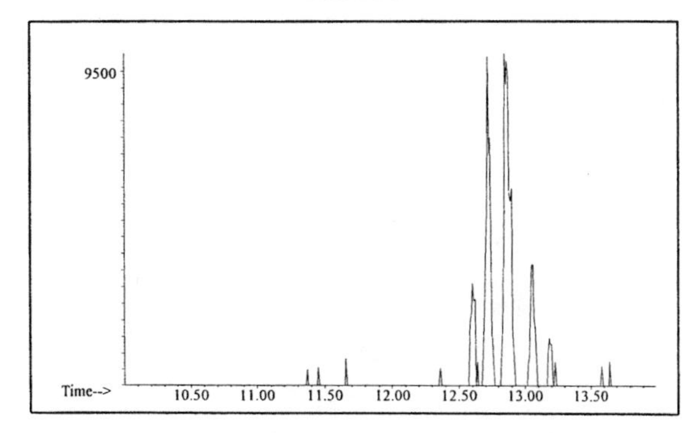

Gasoline #8

20uL/mL Pentane

Product Displayed: 25% Evaporated Mobil Super Premium Gasoline

Other Similar Products: Other grades and brands of gasoline

ASTM: Class 2 (Gasoline)

Product Uses: Automotive fuel

Macro Code: L

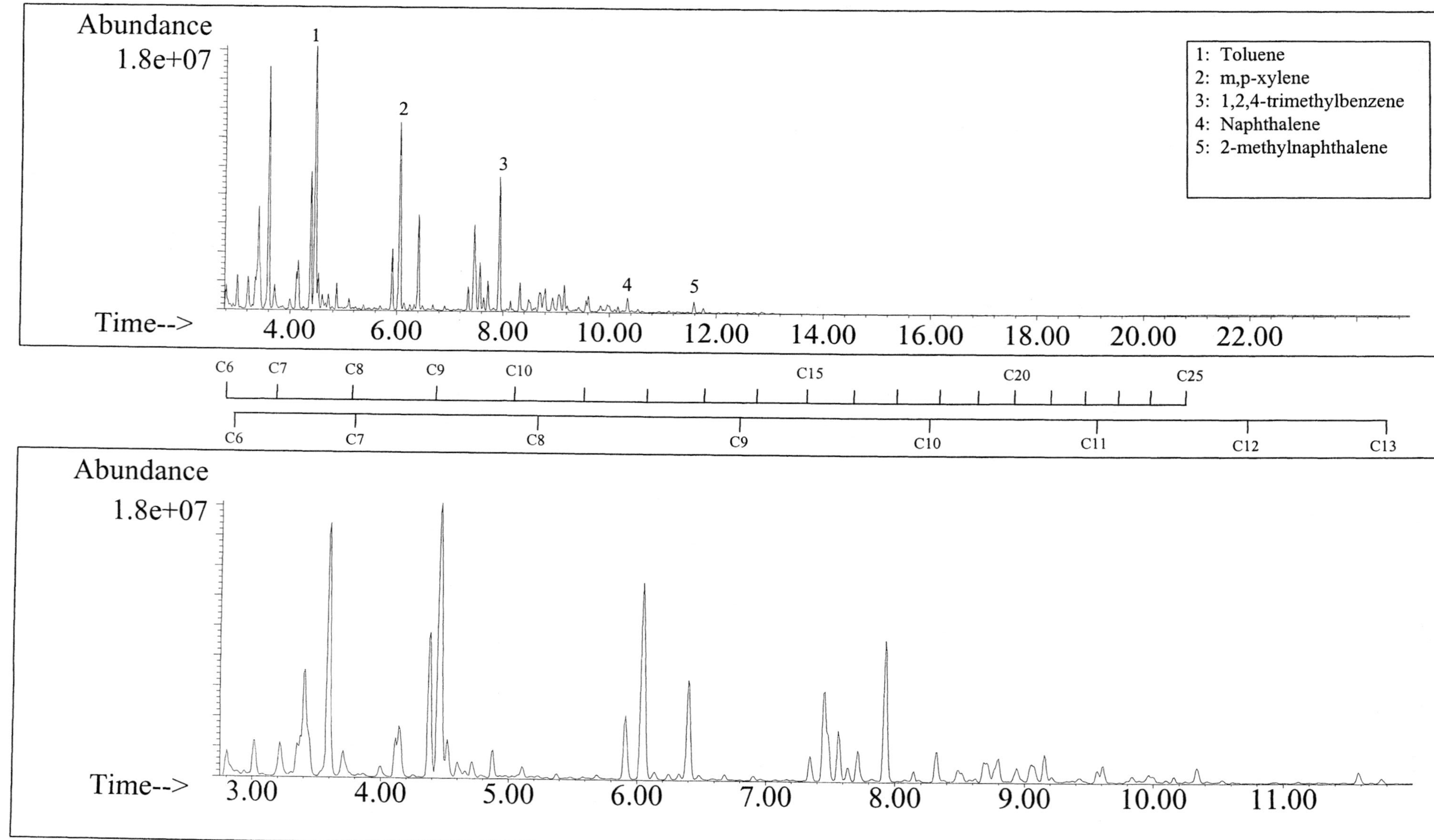

SUMMED PROFILES Gasoline #8

ALKANES

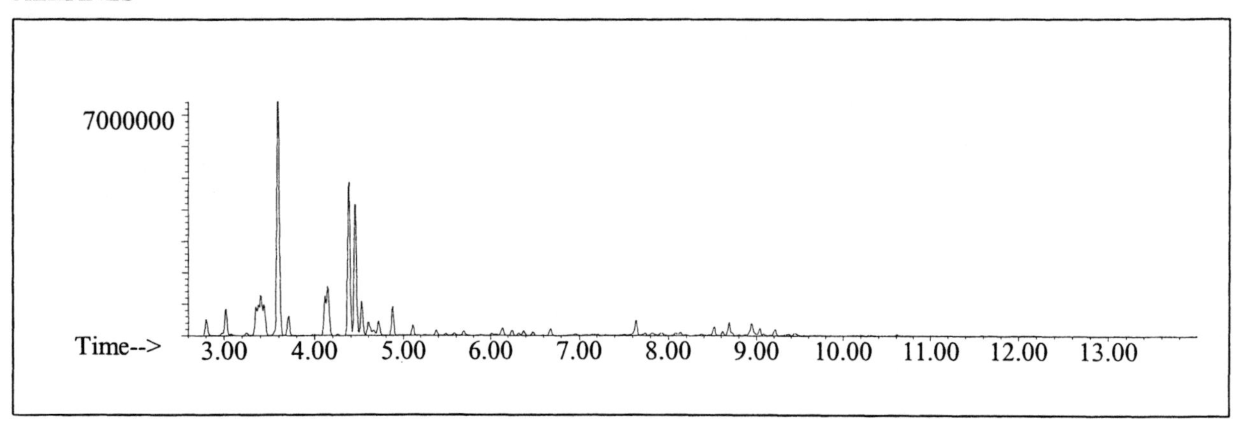

AROMATICS

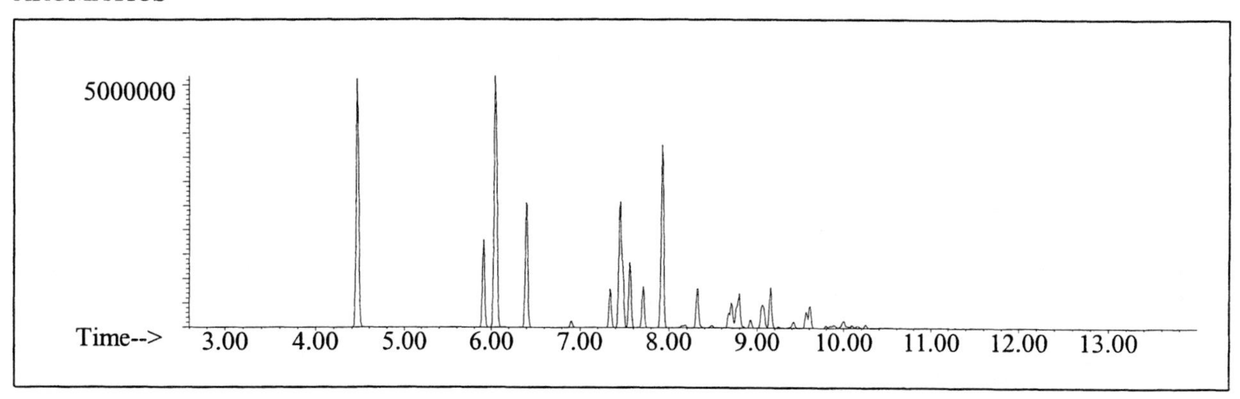

CYCLOPARAFFINS AND ALKENES

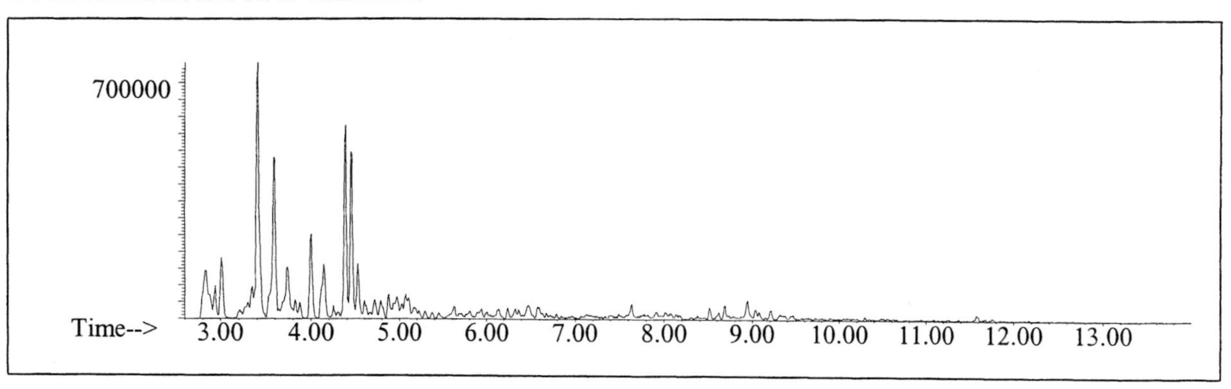

NAPHTHALENES

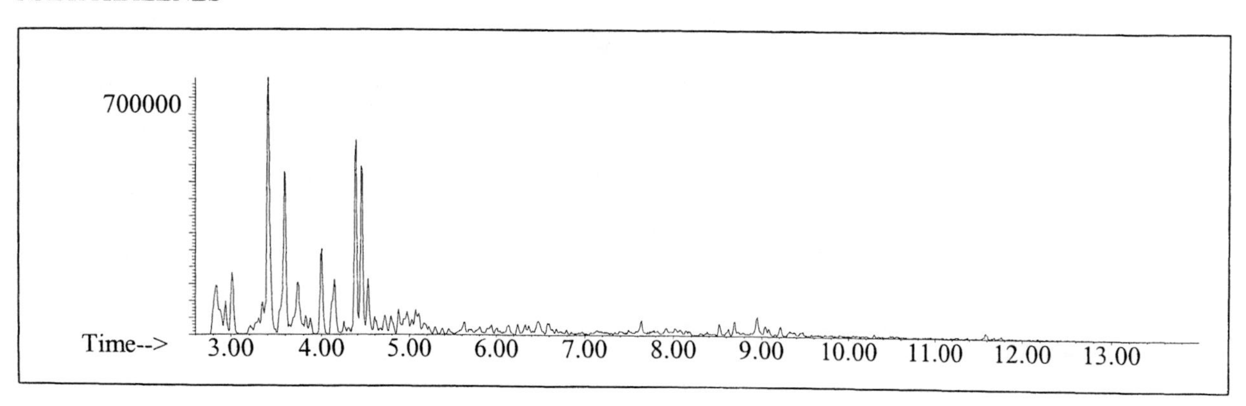

INDIVIDUAL PROFILES

Gasoline #8

Alkane

Ion 43

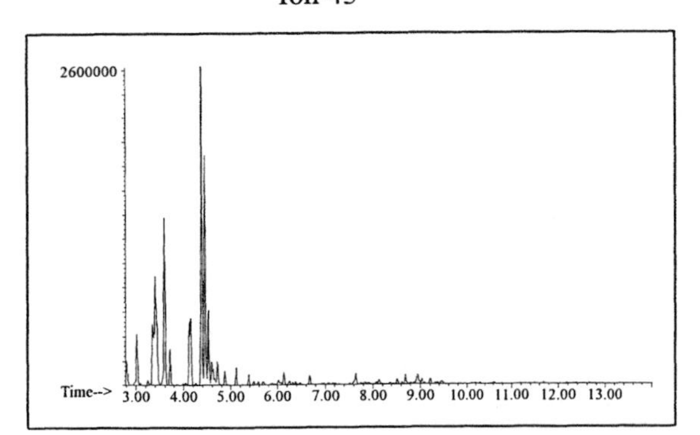

Ion 57

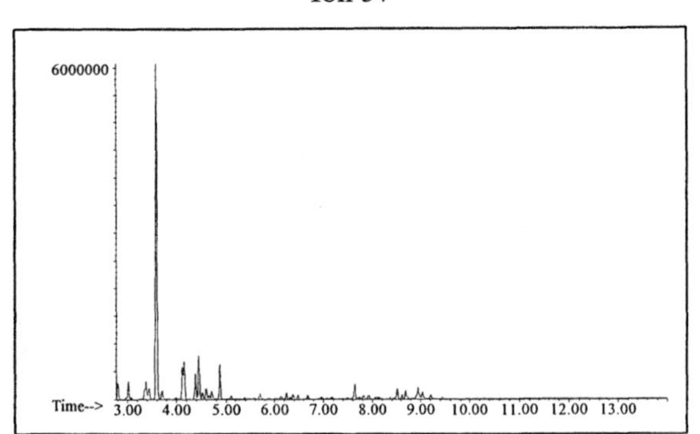

Ion 71

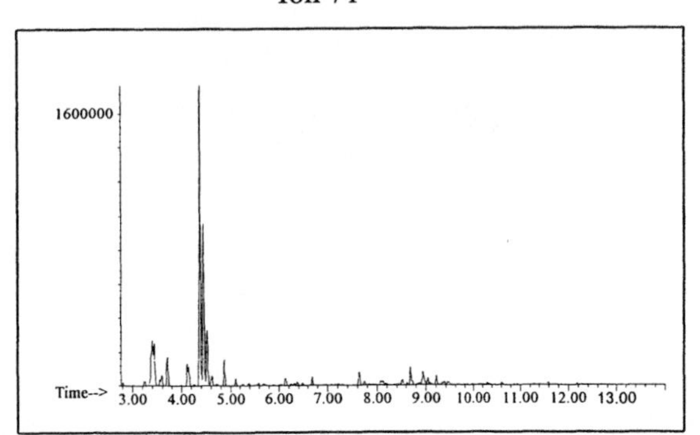

Ion 85

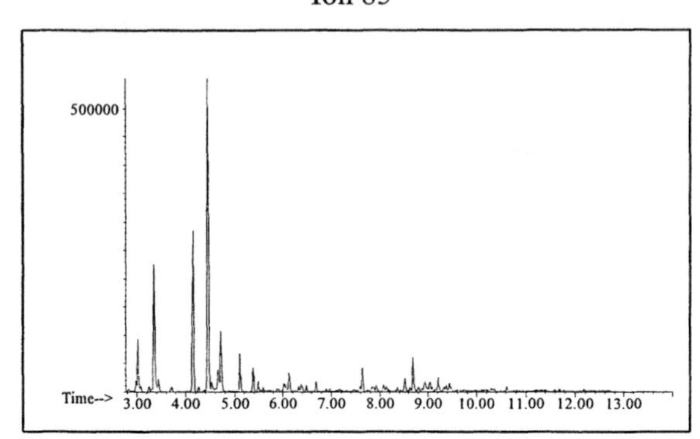

Aromatic

Ion 91

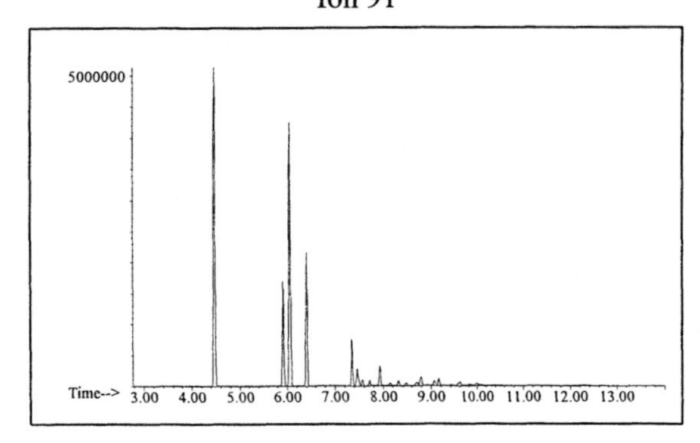

Ion 105

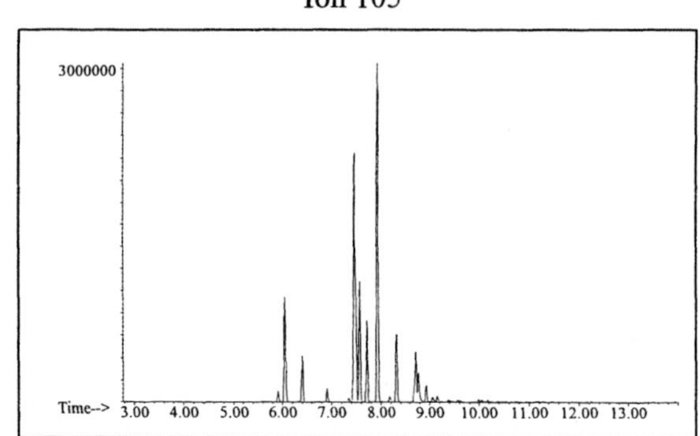

Ion 119

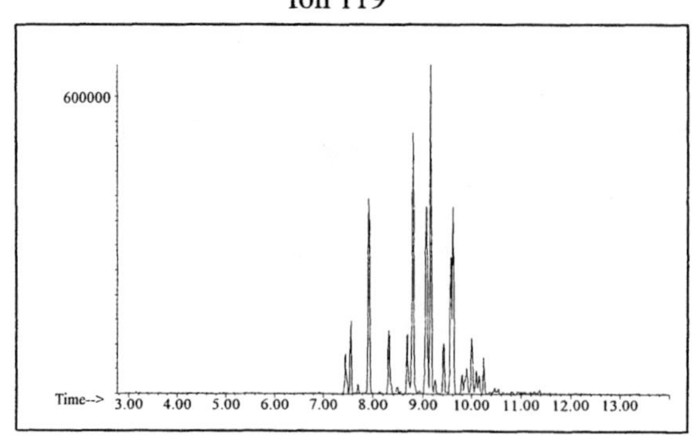

INDIVIDUAL PROFILES

Gasoline #8

Cycloparaffin

Ion 55

Ion 69

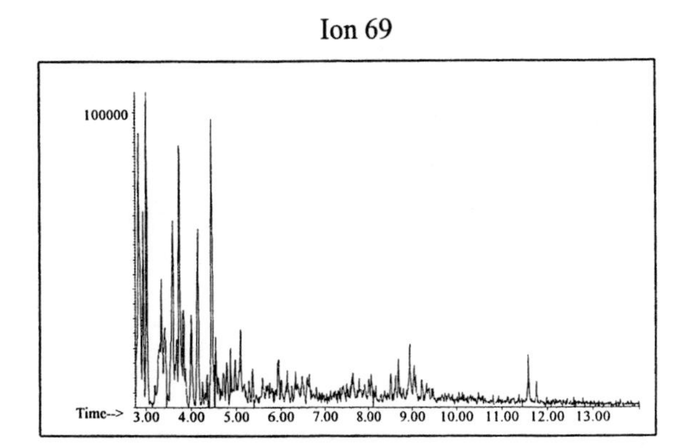

Ion 83

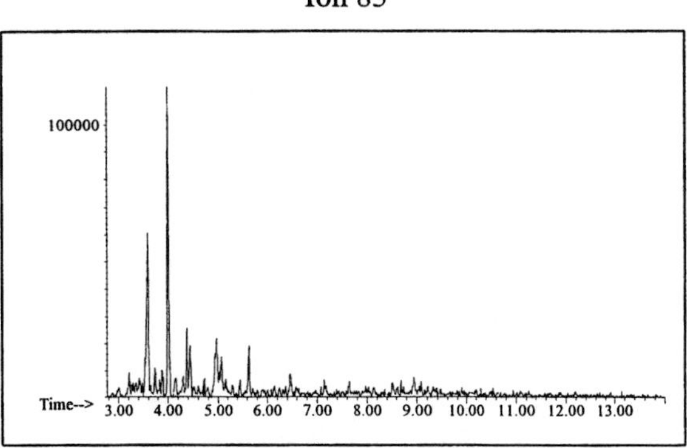

Naphthalene

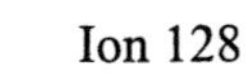

Ion 128

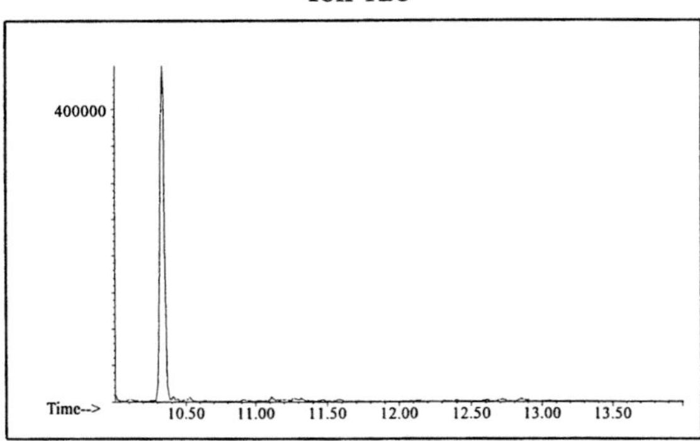

Ion 142

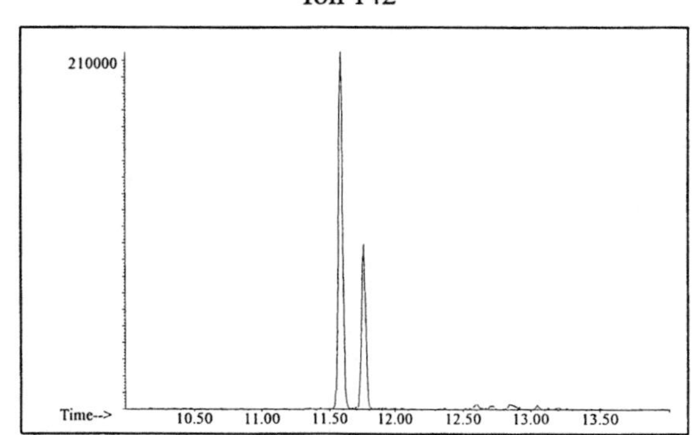

Ion 156

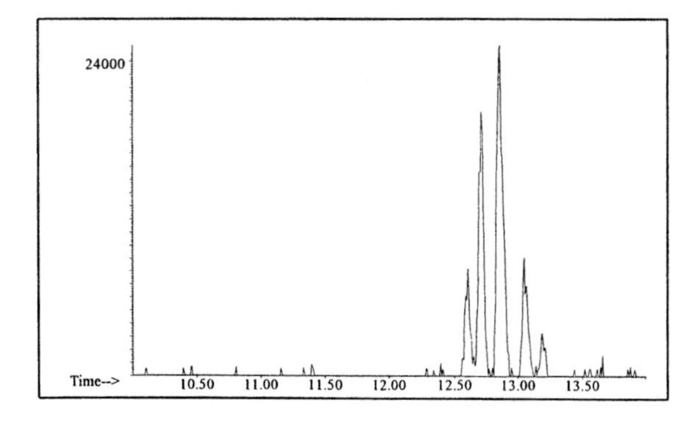

Gasoline #9

20uL/mL Pentane

Product Displayed: 50% Evaporated Mobil Super Premium Gasoline

Other Similar Products: Other grades and brands of gasoline

ASTM: Class 2 (Gasoline)

Product Uses: Automotive fuel

Macro Code: L

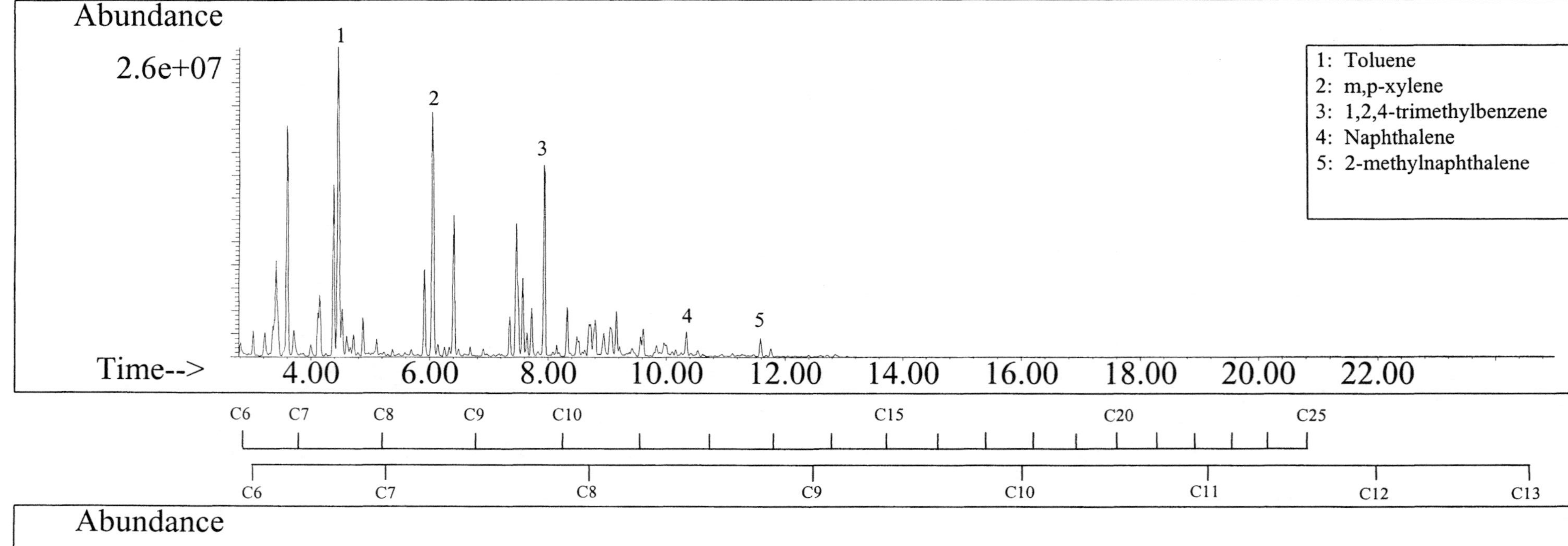

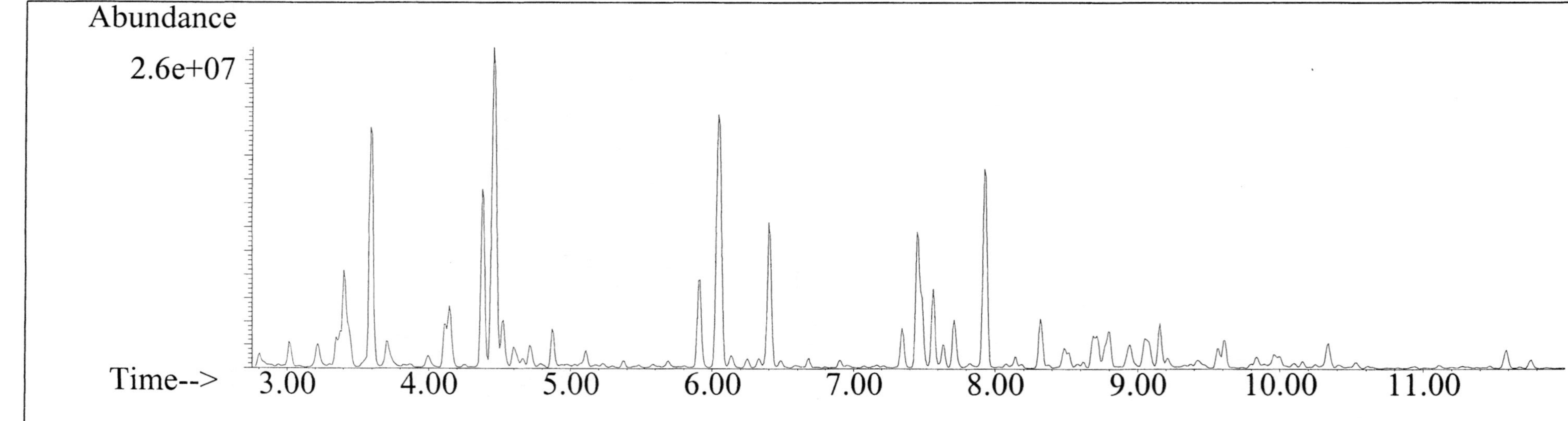

SUMMED PROFILES — Gasoline #9

ALKANES

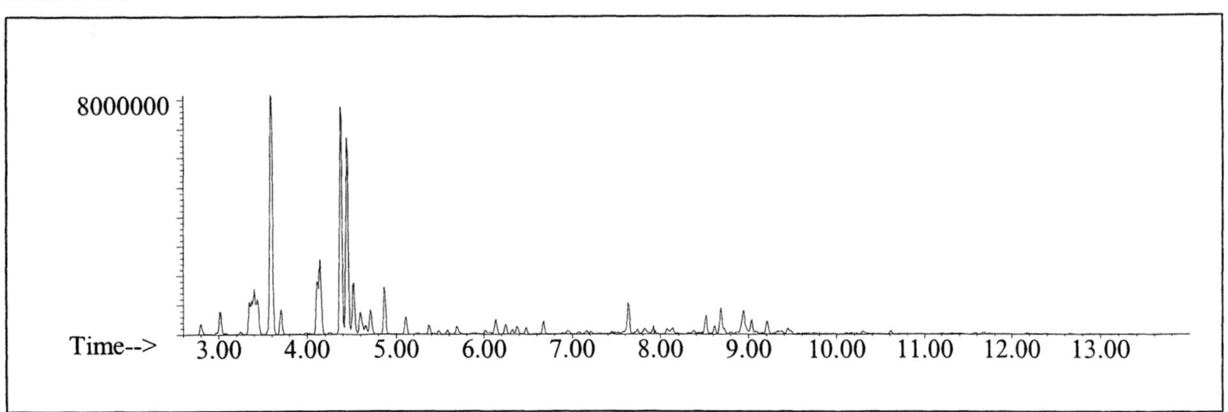

AROMATICS

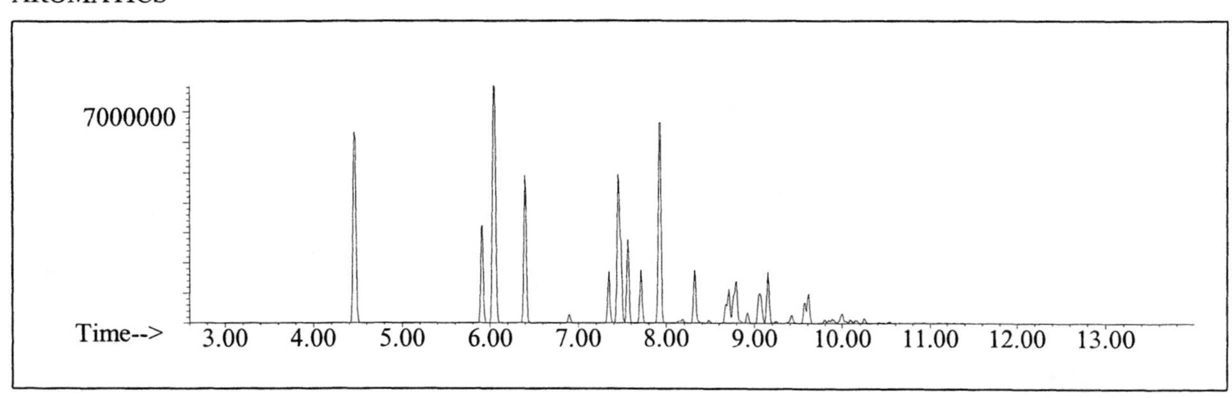

CYCLOPARAFFINS AND ALKENES

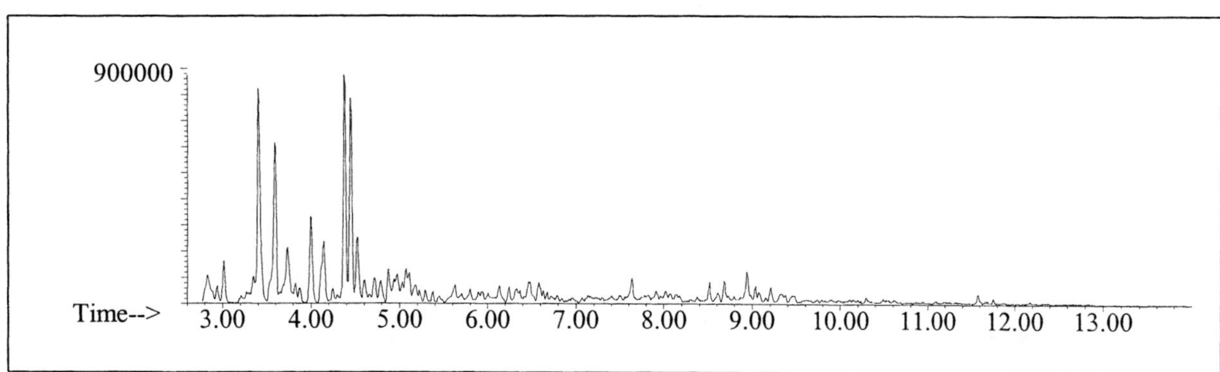

NAPHTHALENES

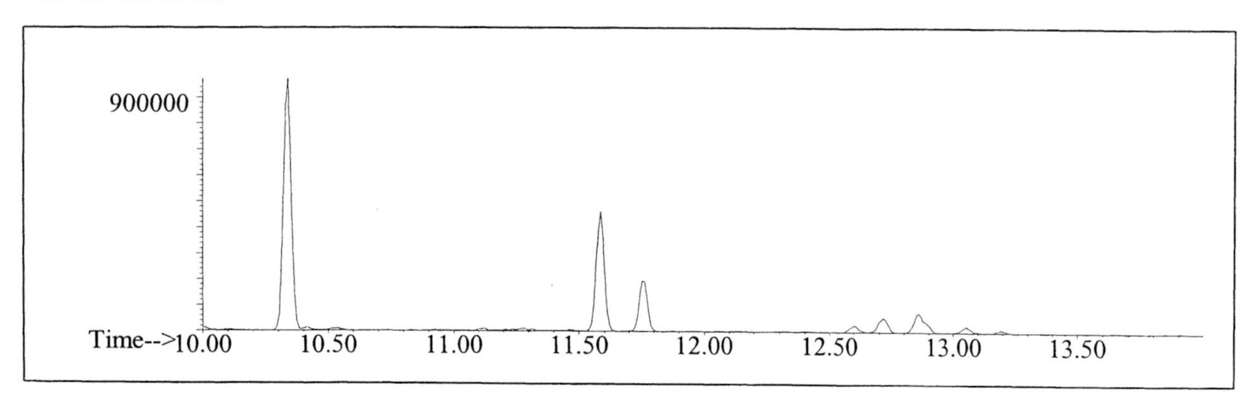

INDIVIDUAL PROFILES

Gasoline #9

Alkane

Ion 43

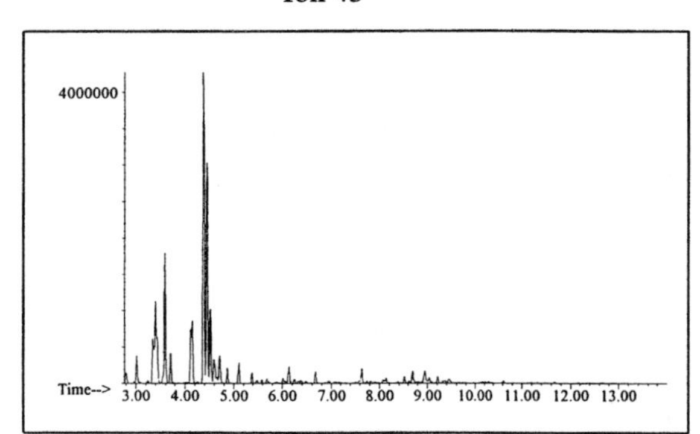

Ion 57

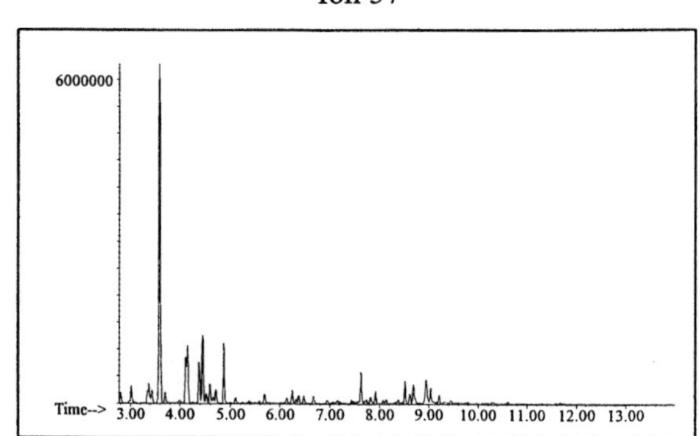

Ion 71

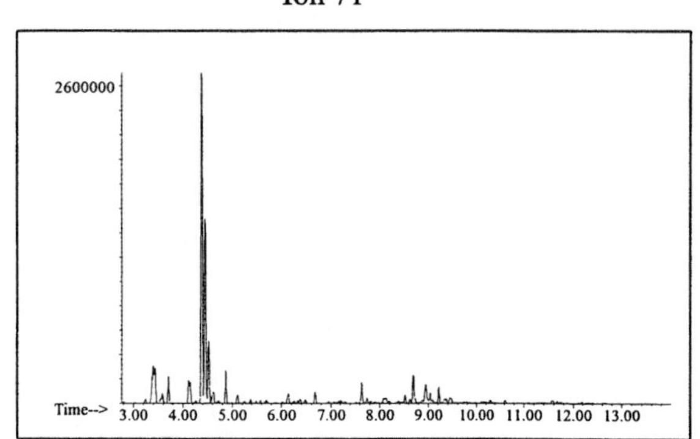

Ion 85

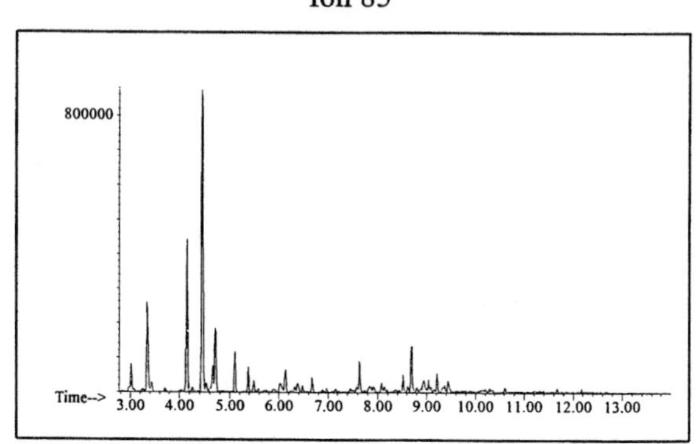

Aromatic

Ion 91

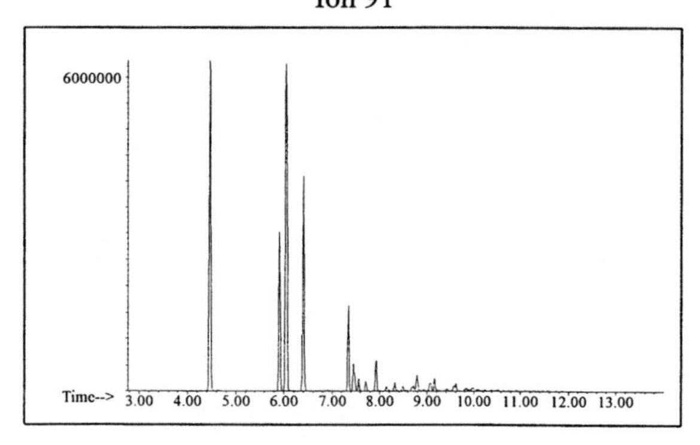

Ion 105

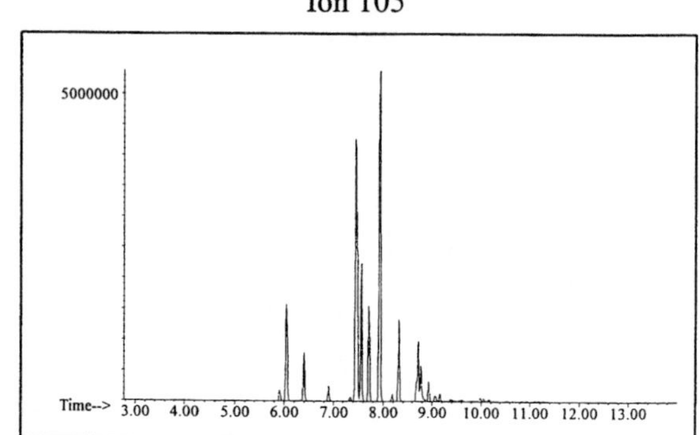

Ion 119

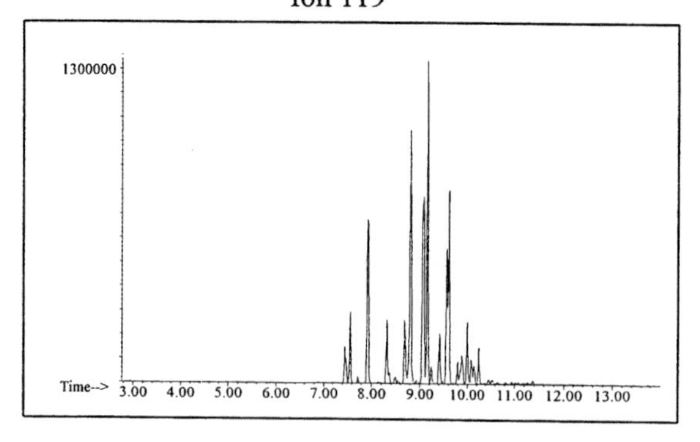

INDIVIDUAL PROFILES

Cycloparaffin

Ion 55

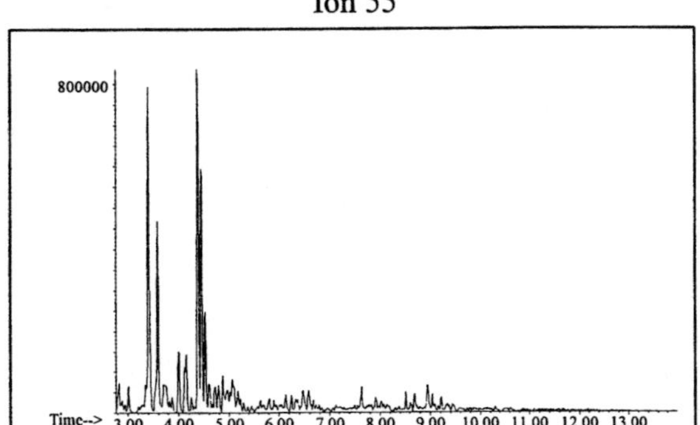

Ion 69

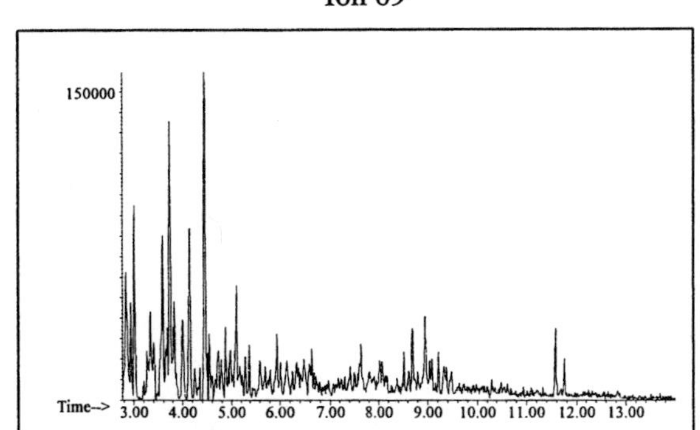

Ion 83

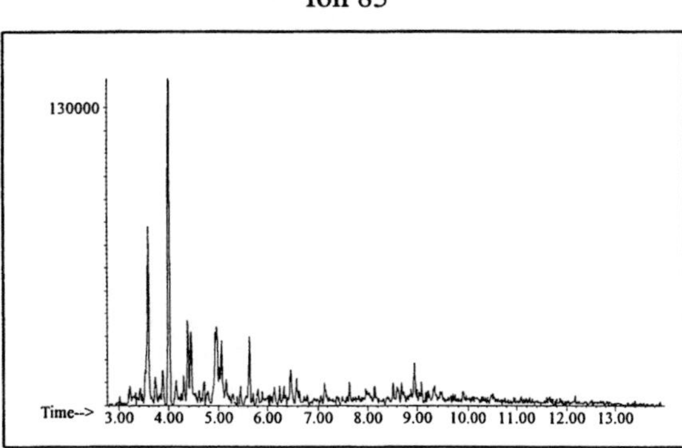

Naphthalene

Ion 128

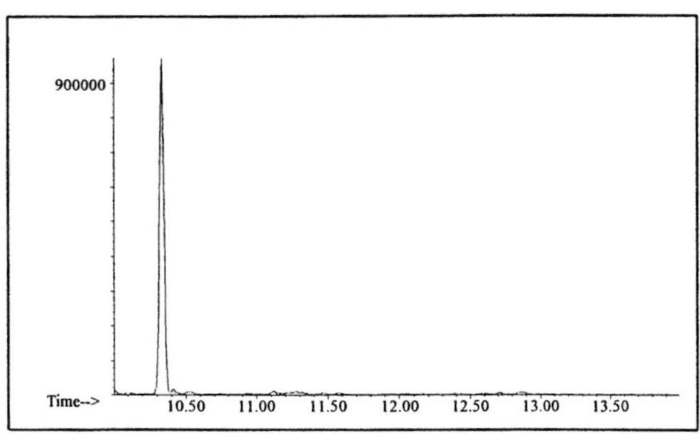

Ion 142

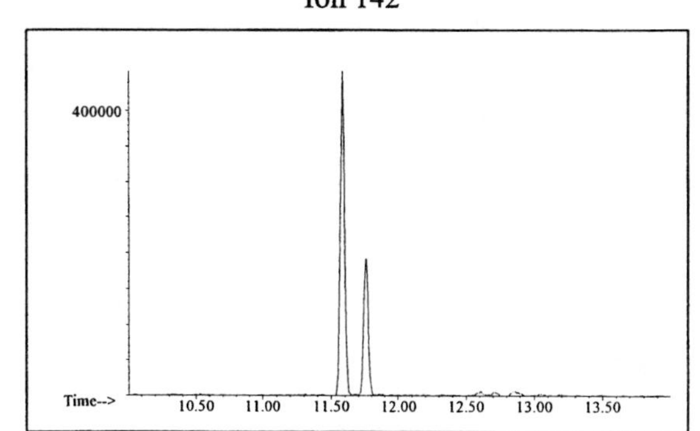

Ion 156

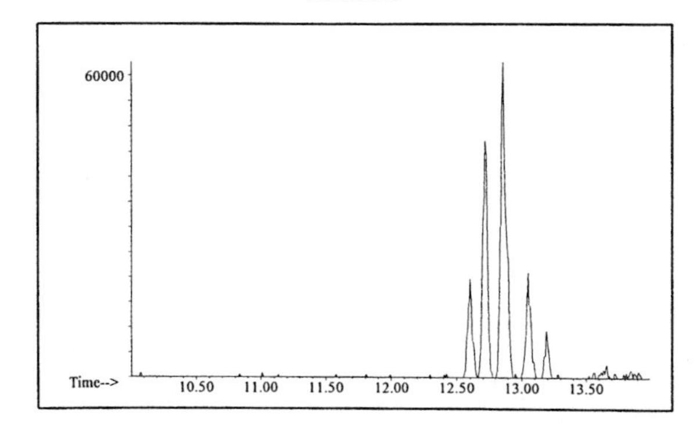

Gasoline #10

20uL/mL Pentane

ASTM: Class 2 (Gasoline)

Macro Code: L

Product Displayed: 75% Evaporated Mobil Super Premium Gasoline

Product Uses: Automotive fuel

Other Similar Products: Other grades and brands of gasoline

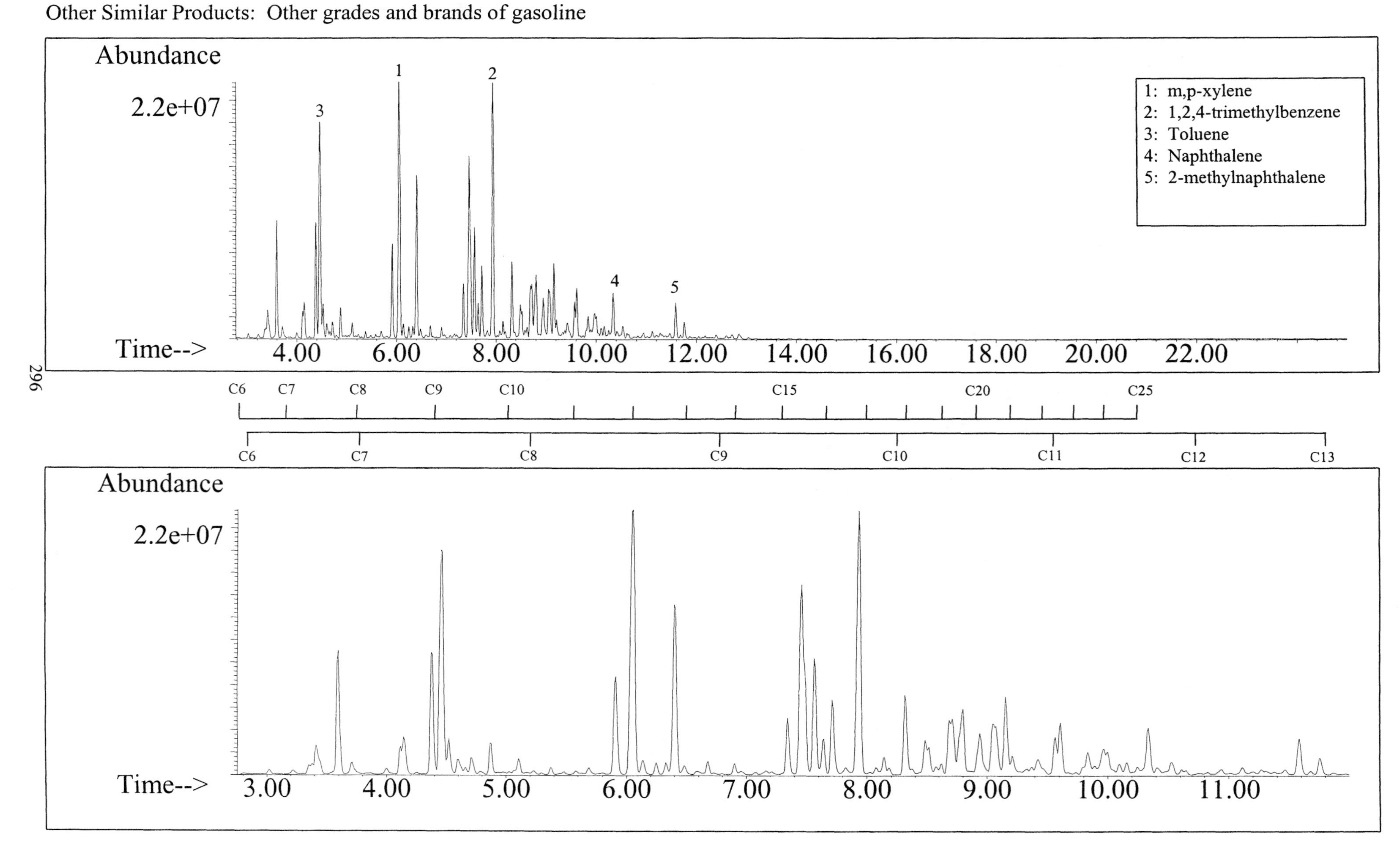

SUMMED PROFILES

Gasoline #10

ALKANES

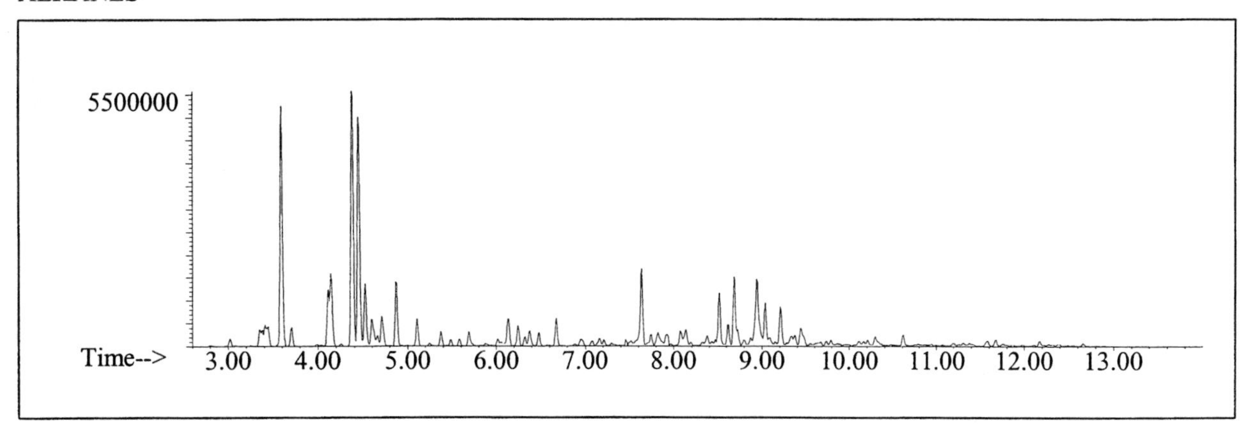

AROMATICS

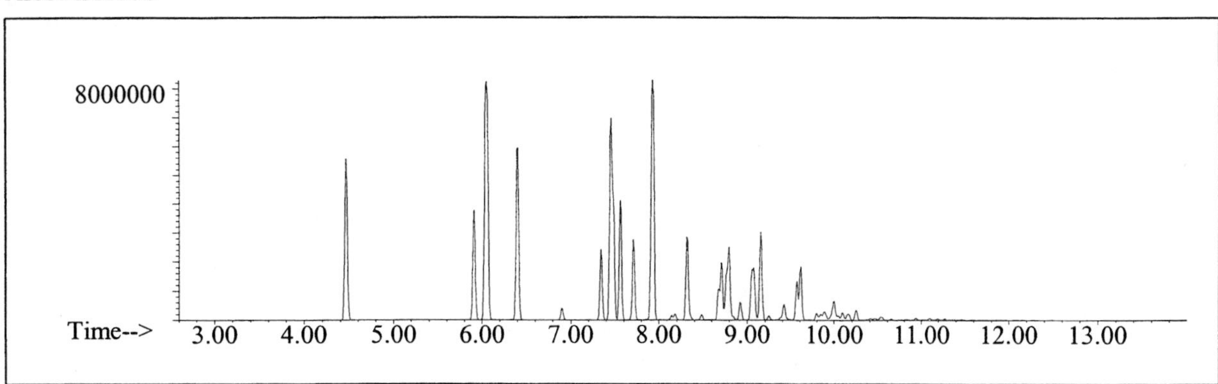

CYCLOPARAFFINS AND ALKENES

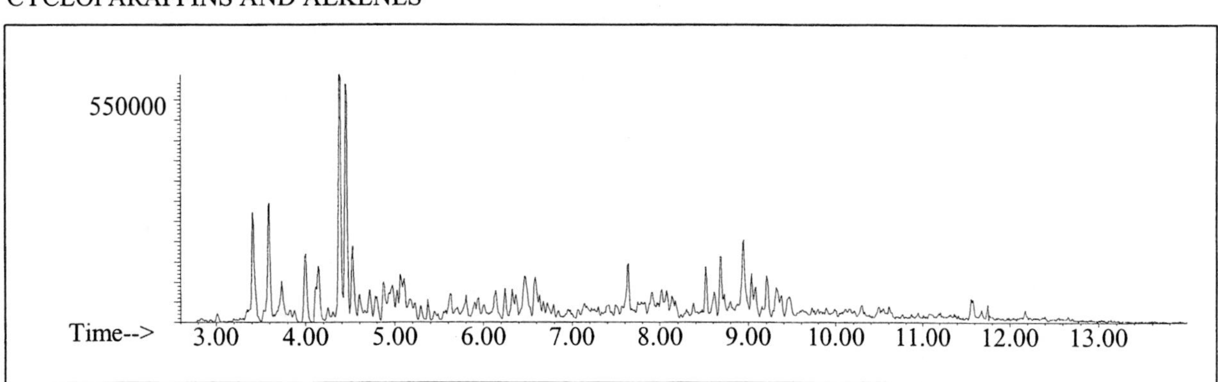

NAPHTHALENES

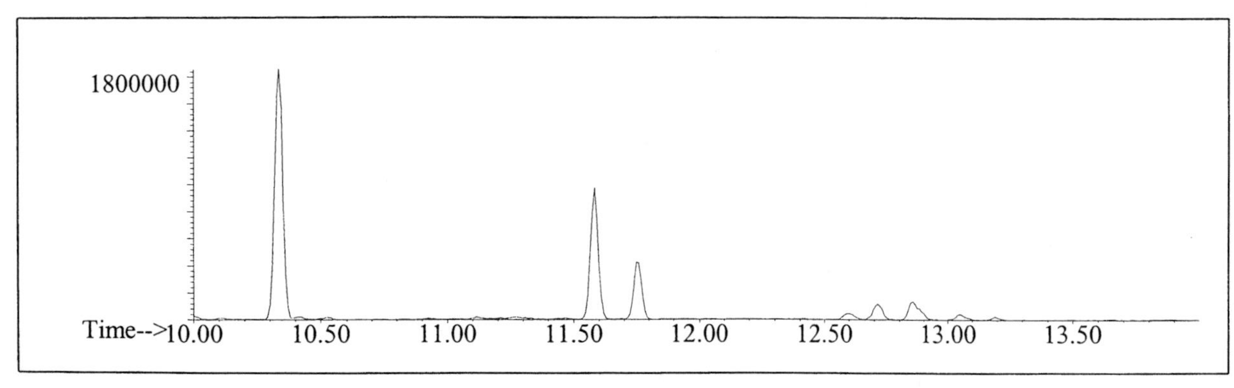

INDIVIDUAL PROFILES

Gasoline #10

Alkane

Ion 43

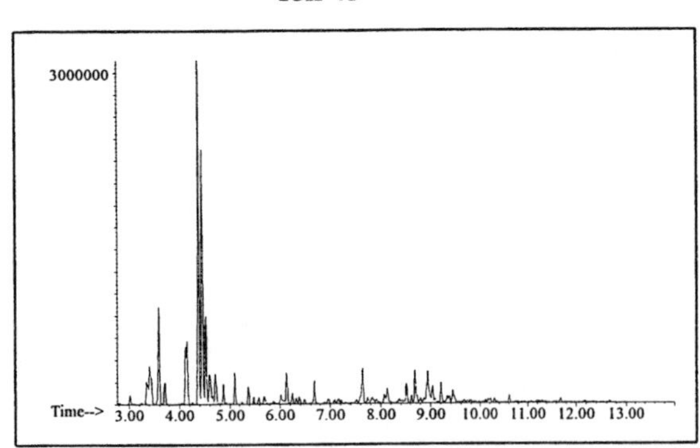

Ion 57

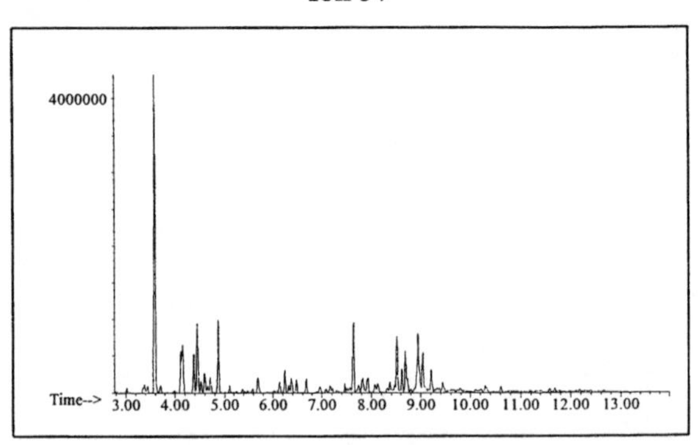

Ion 71

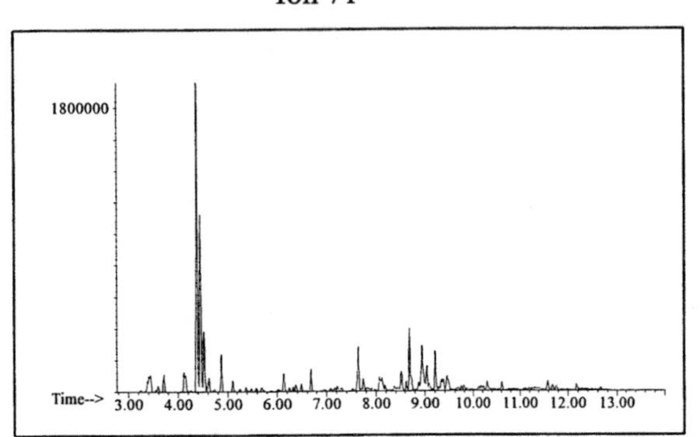

Ion 85

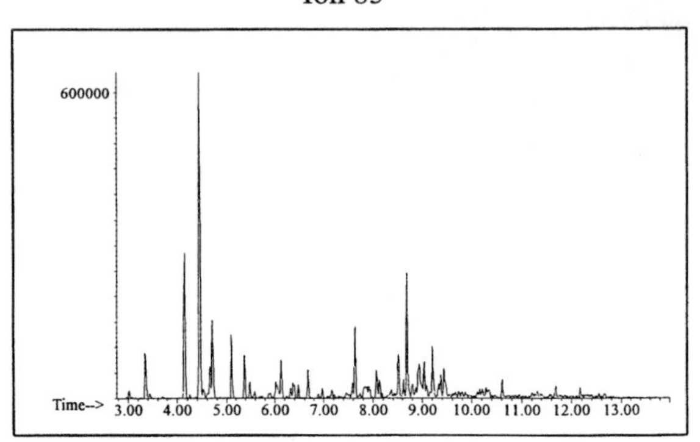

Aromatic

Ion 91

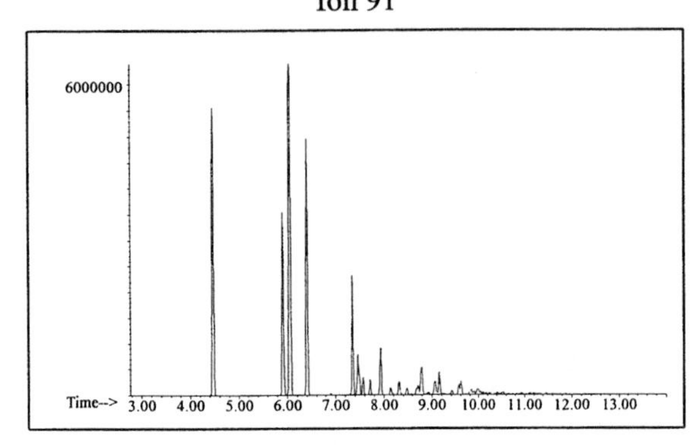

Ion 105

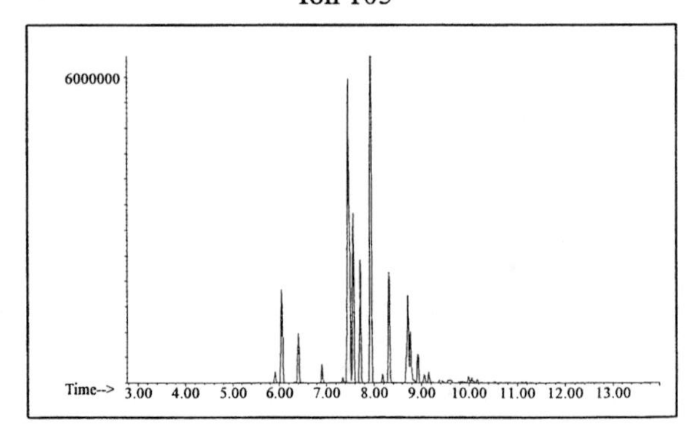

Ion 119

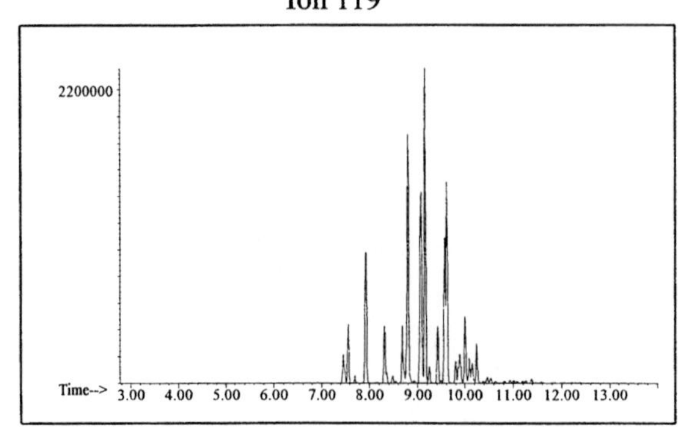

Cycloparaffin

Ion 55

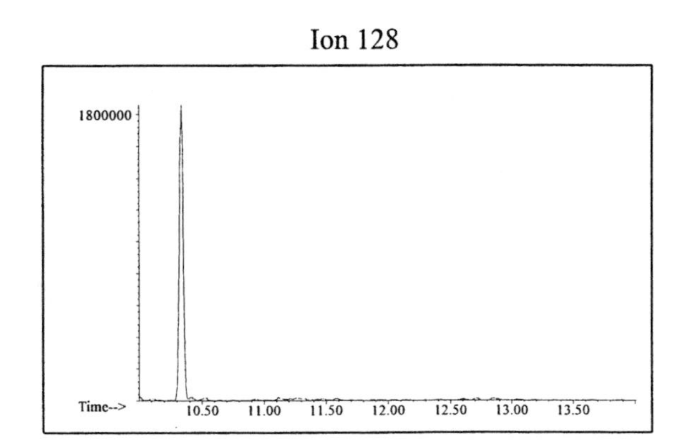

Ion 69

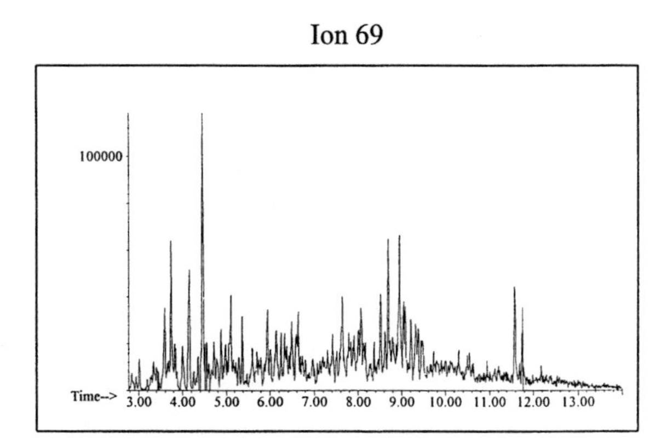

Ion 83

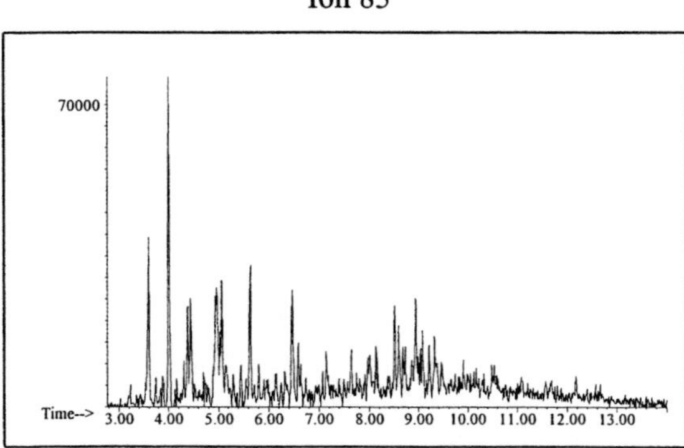

Naphthalene

Ion 128

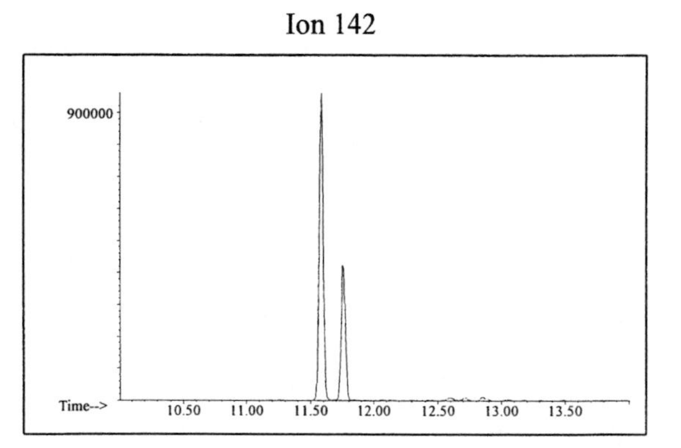

Ion 142

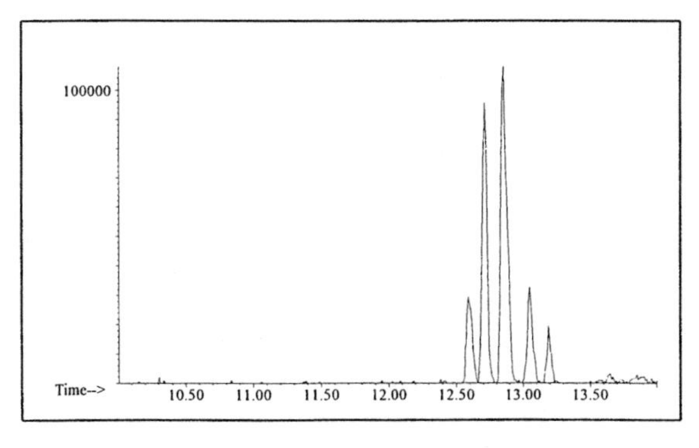

Ion 156

Gasoline #11

20uL/mL Pentane

Product Displayed: 90% Evaporated Mobil Super Premium Gasoline

Other Similar Products: Other grades and brands of gasoline

ASTM: Class 2 (Gasoline)

Product Uses: Automotive fuel

Macro Code: L

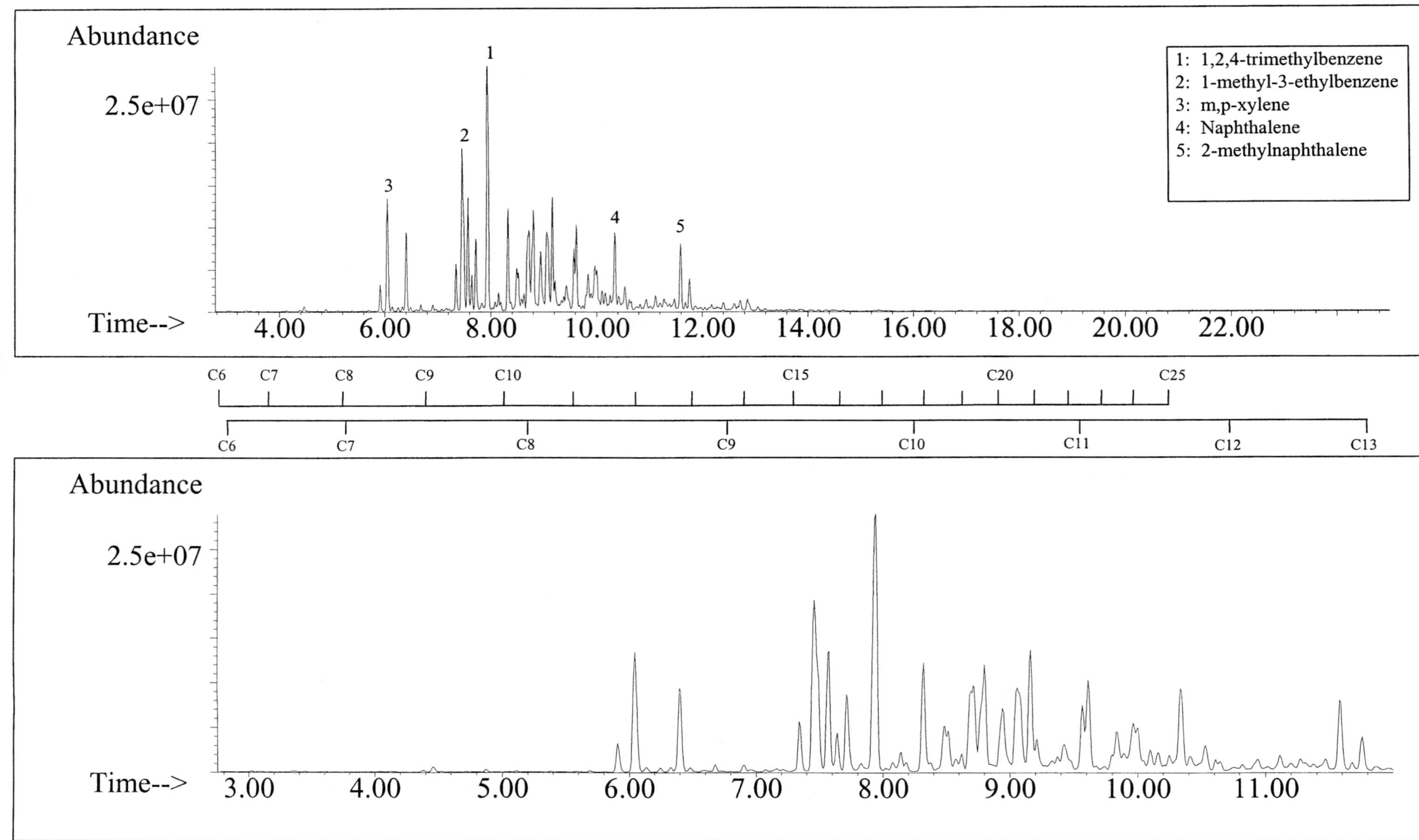

SUMMED PROFILES

Gasoline #11

ALKANES

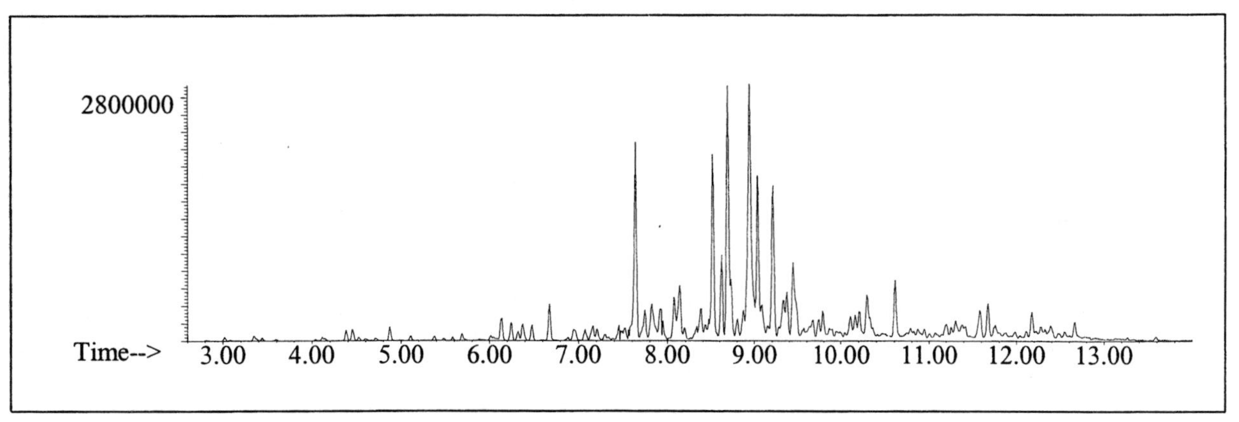

AROMATICS

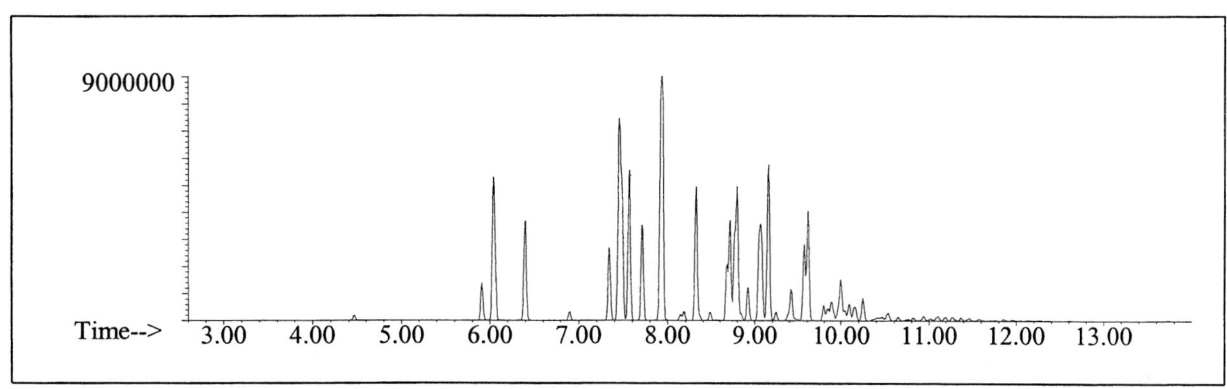

CYCLOPARAFFINS AND ALKENES

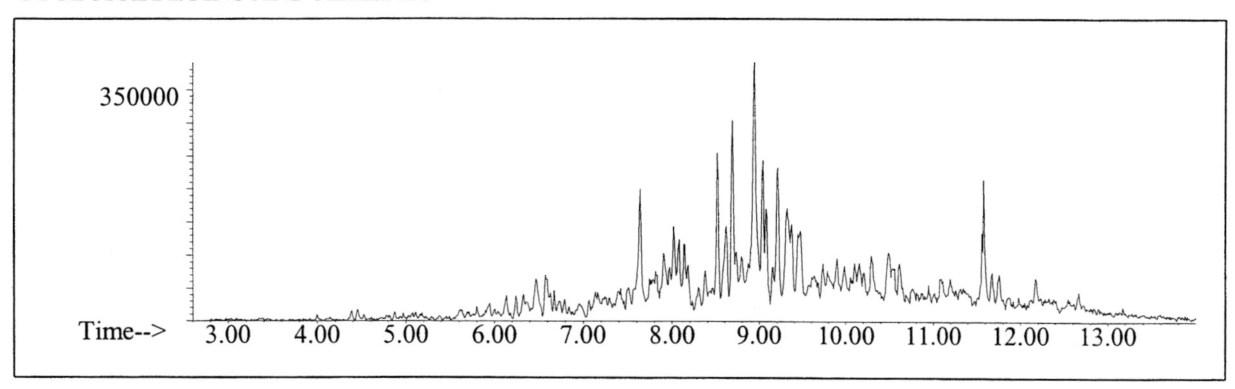

NAPHTHALENES

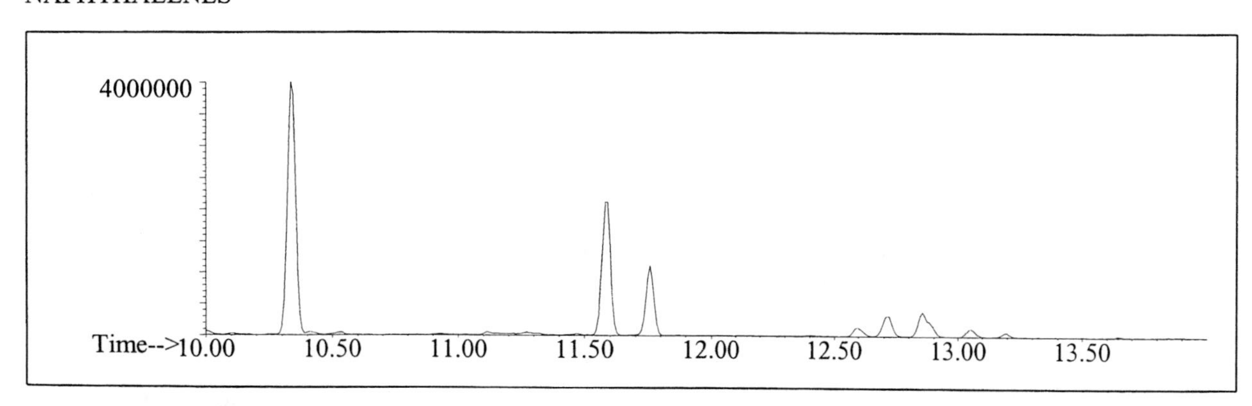

INDIVIDUAL PROFILES

Alkane

Ion 43

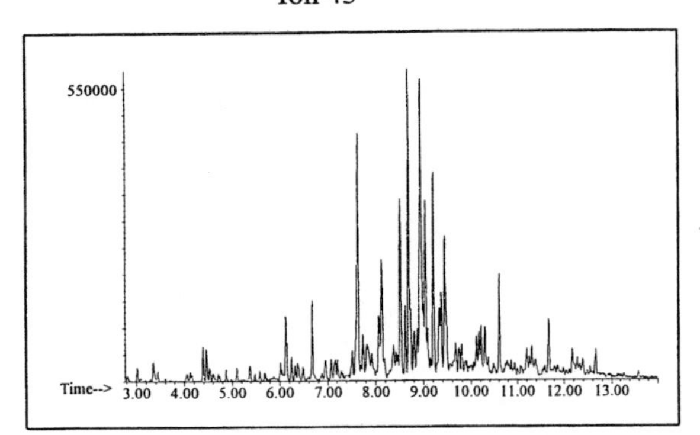

Ion 57

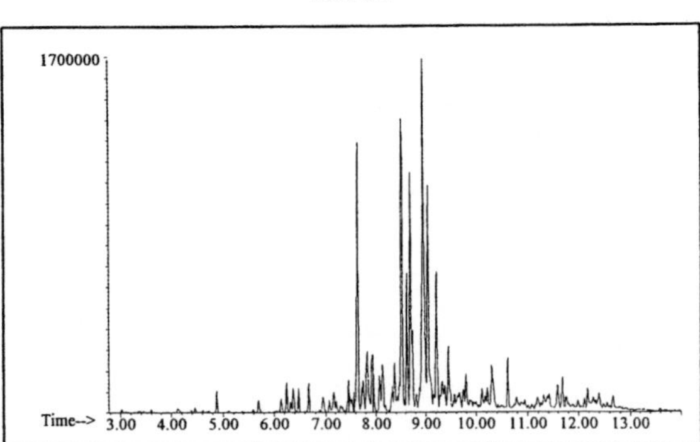

Ion 71

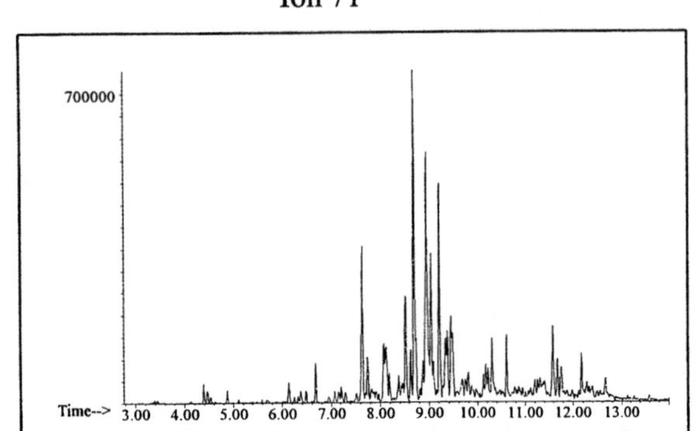

Ion 85

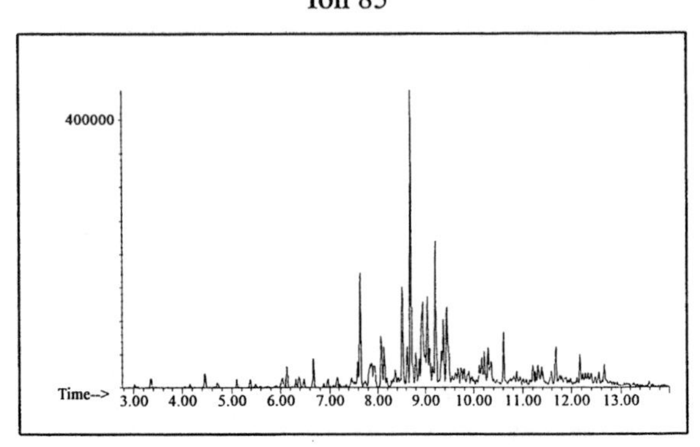

Aromatic

Ion 91

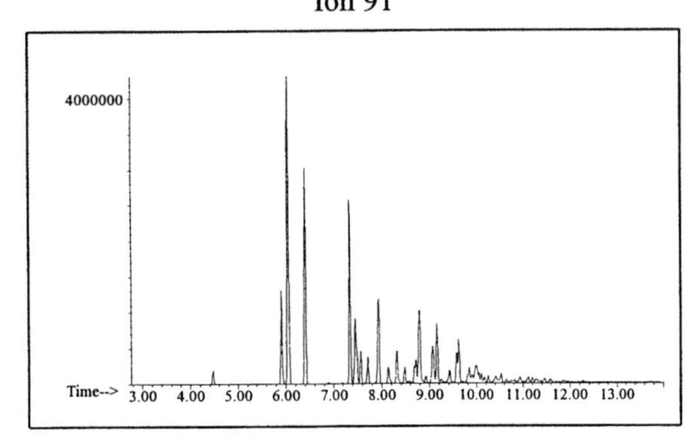

Ion 105

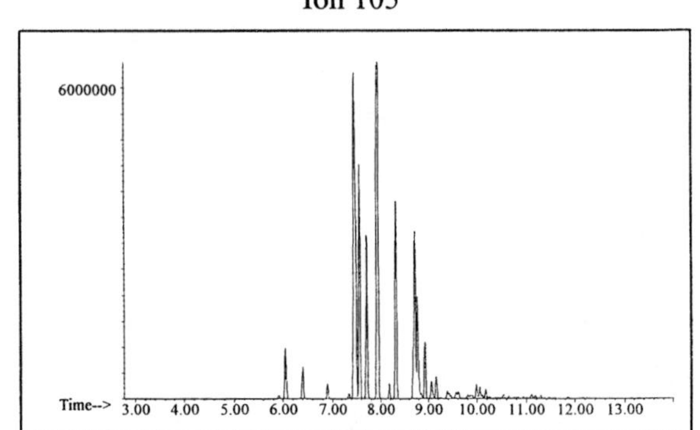

Ion 119

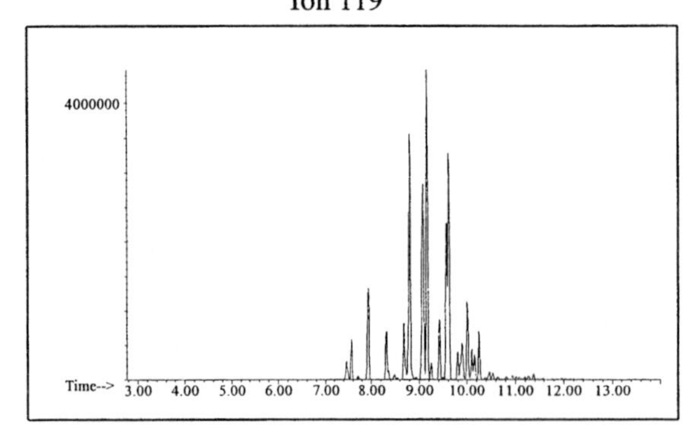

Cycloparaffin

Ion 55

200000

Time--> 3.00 4.00 5.00 6.00 7.00 8.00 9.00 10.00 11.00 12.00 13.00

Ion 69

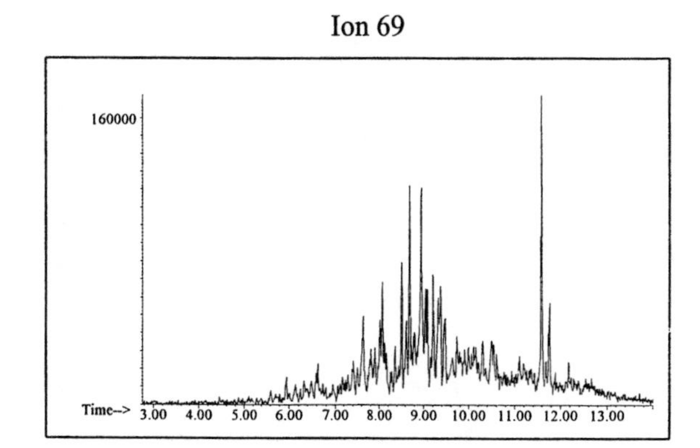

Ion 83

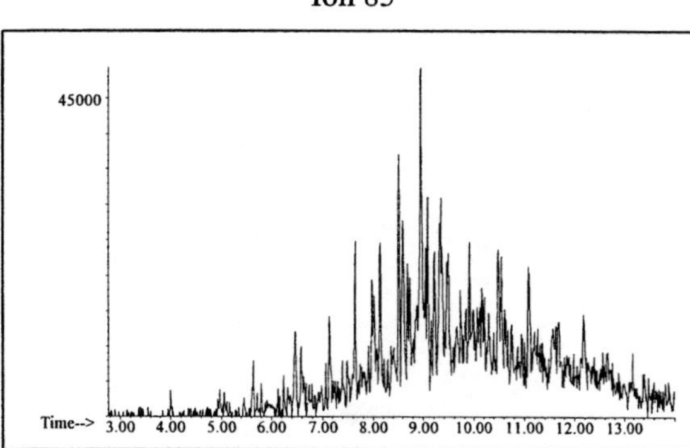

Naphthalene

Ion 128

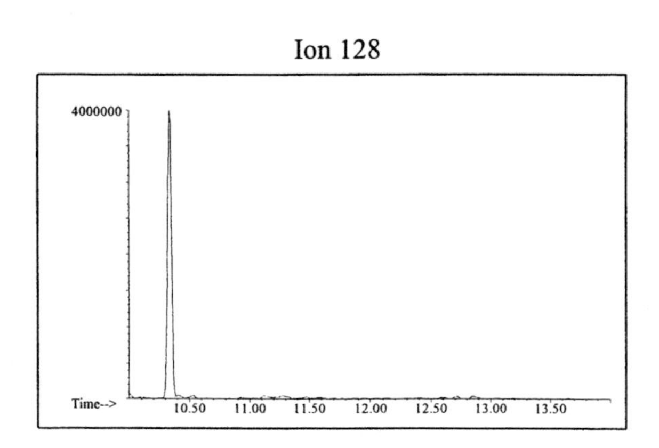

Ion 142

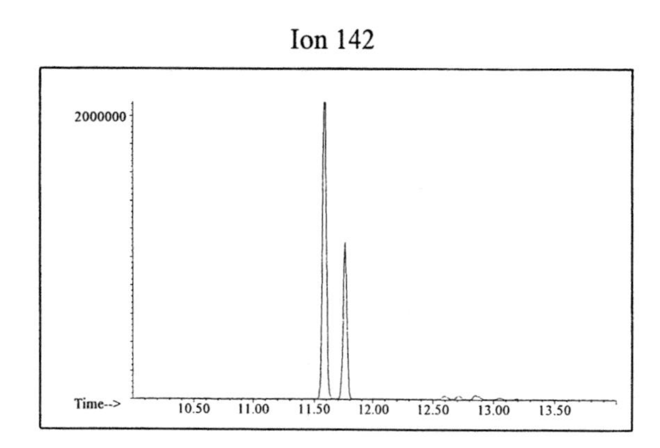

Ion 156

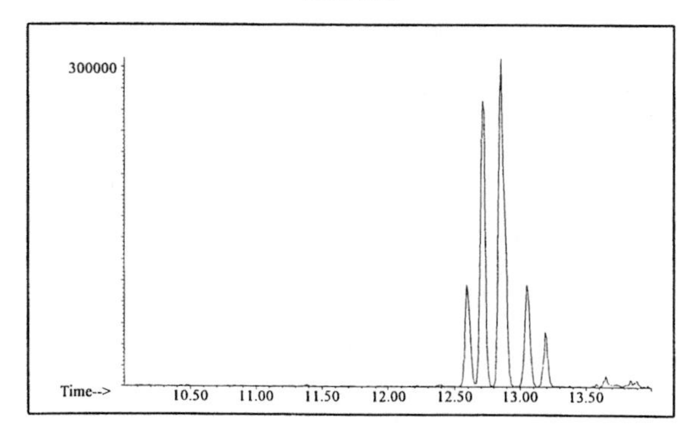

Gasoline #12

20uL/mL Pentane

ASTM: Class 2 (Gasoline)

Macro Code: M

Product Displayed: 98% Evaporated Mobil Super Premium Gasoline

Product Uses: Automotive fuel

Other Similar Products: Other brands and grades of gasoline

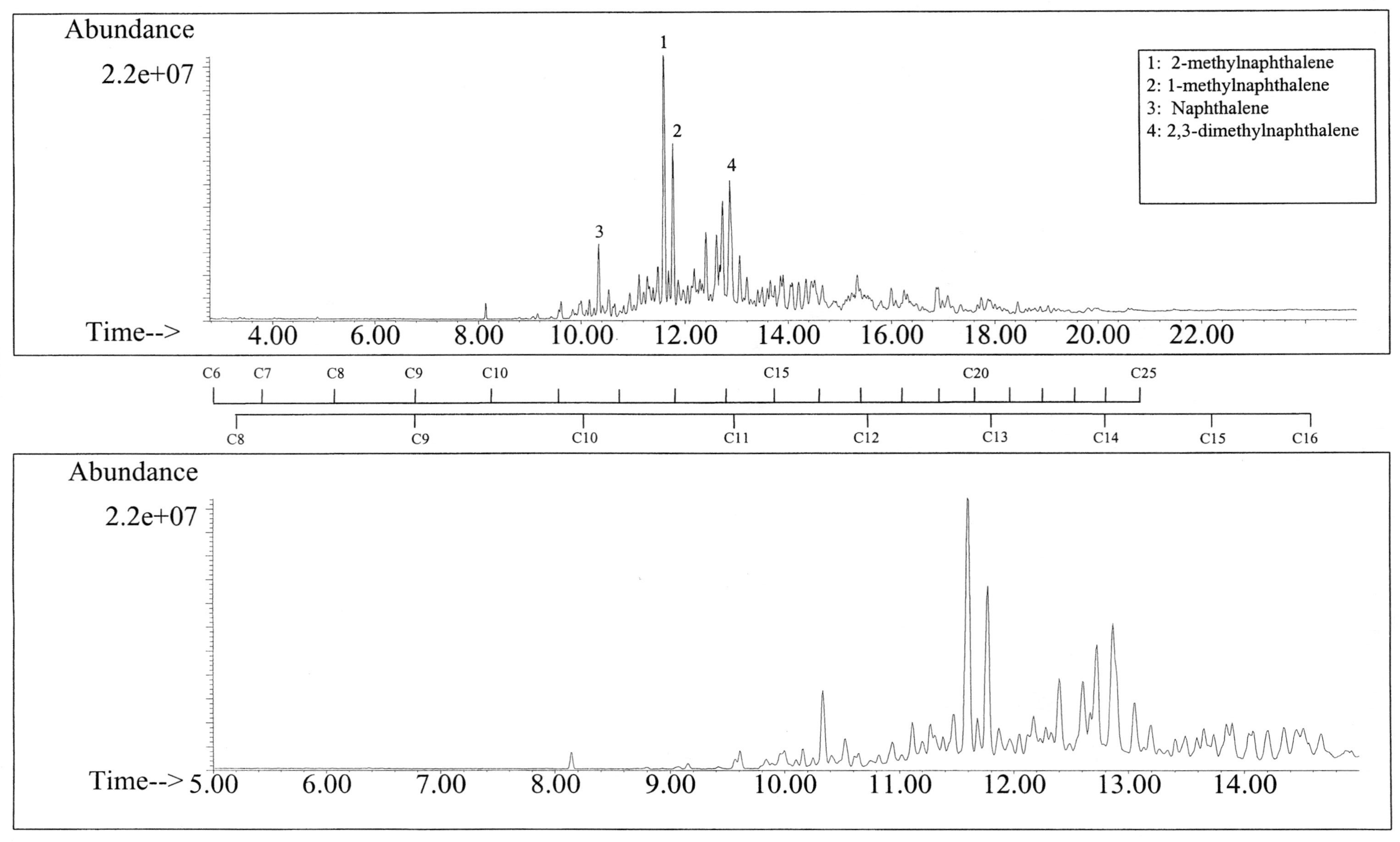

SUMMED PROFILES Gasoline #12

ALKANES

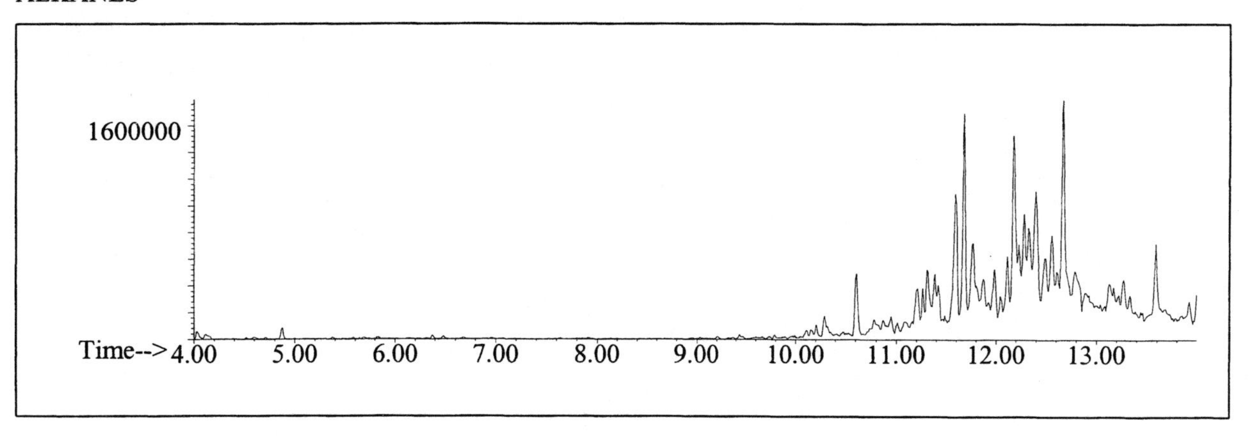

AROMATICS

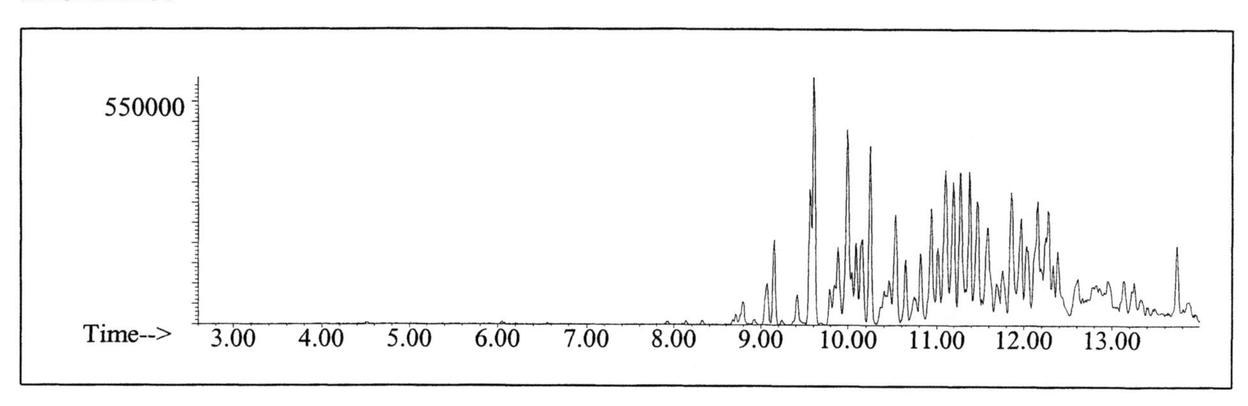

CYCLOPARAFFINS AND ALKENES

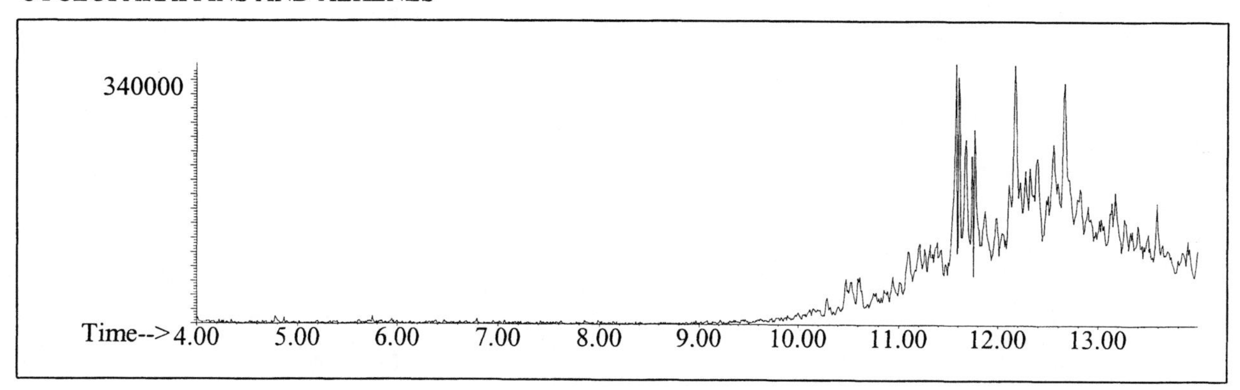

NAPHTHALENES

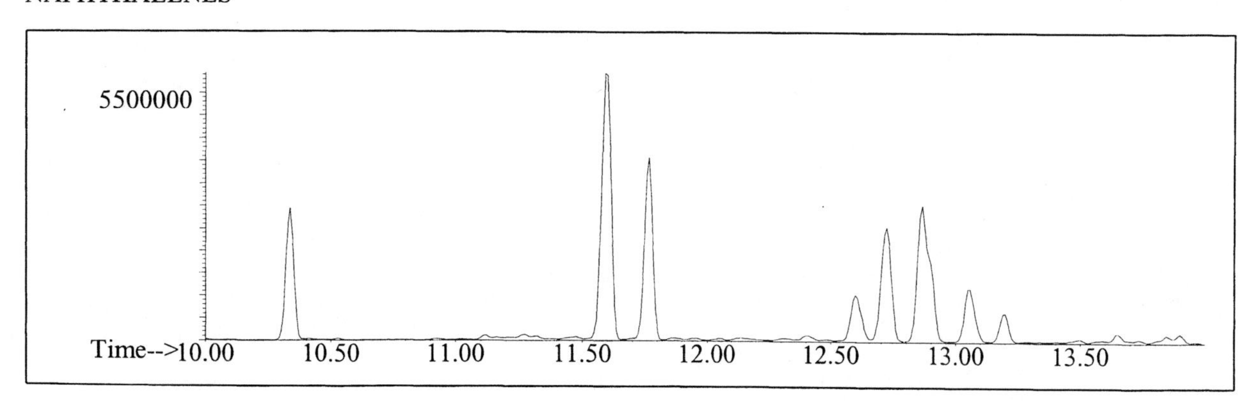

INDIVIDUAL PROFILES Gasoline #12

Alkane

Ion 43

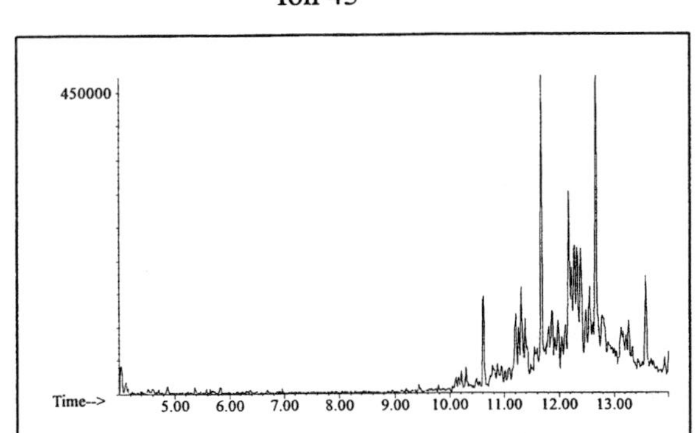

Ion 57

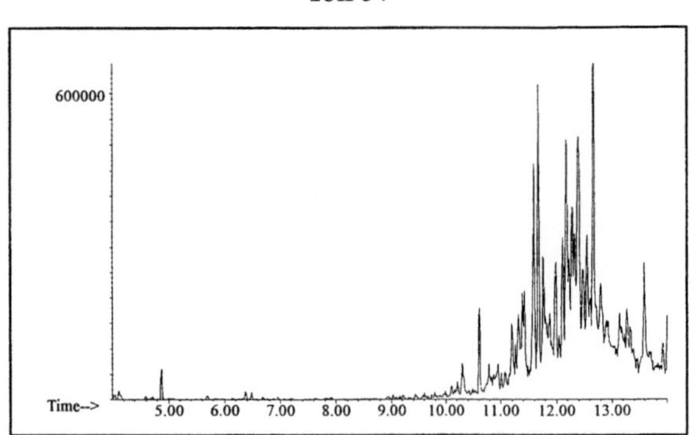

Ion 71

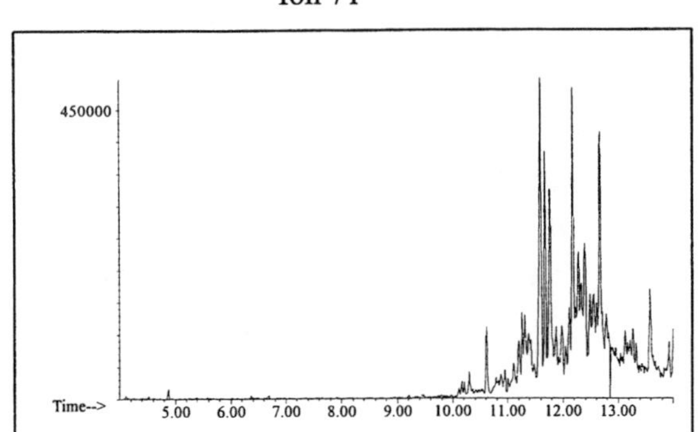

Ion 85

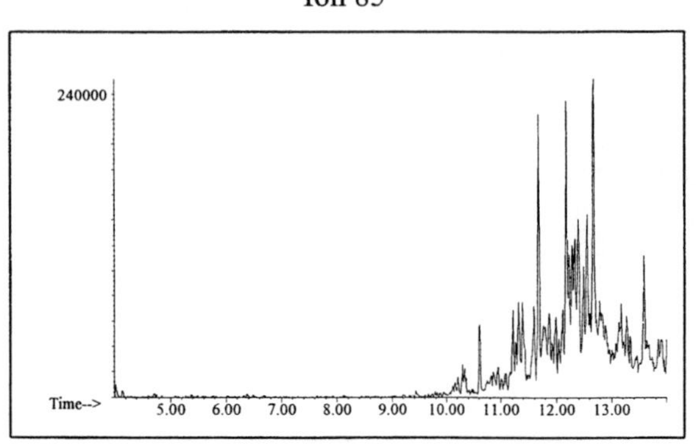

Aromatic

Ion 91

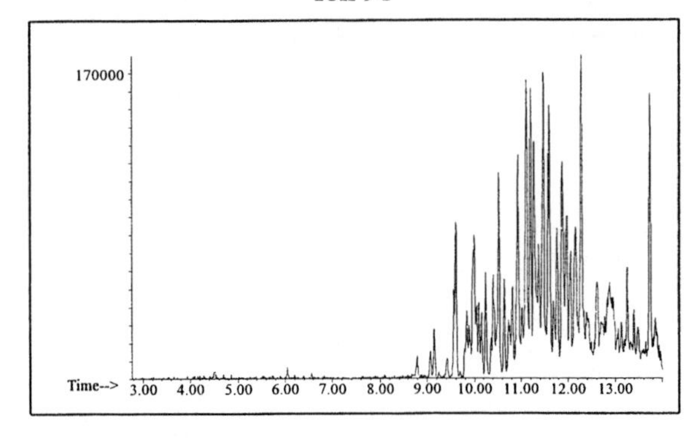

Ion 105

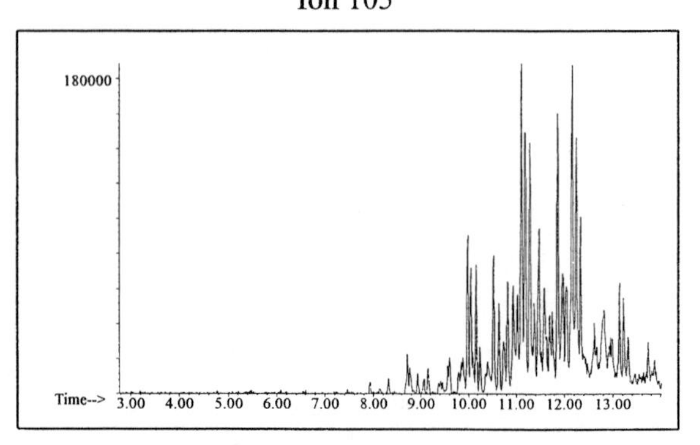

Ion 119

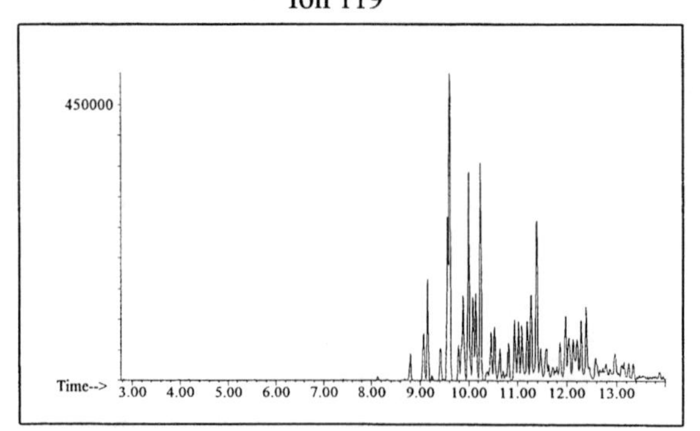

INDIVIDUAL PROFILES

Gasoline #12

Cycloparaffin

Ion 55

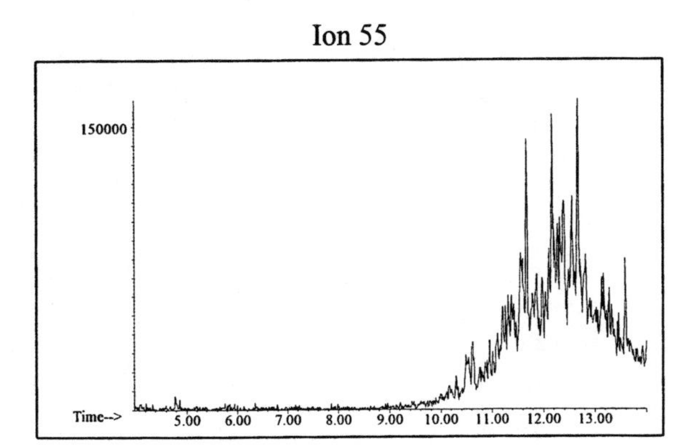

Ion 69

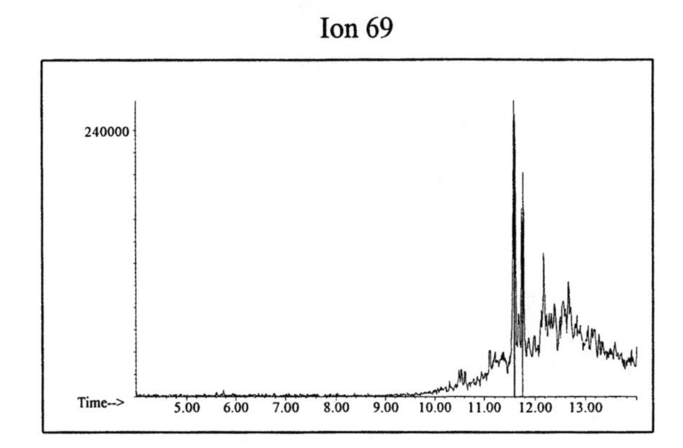

Ion 83

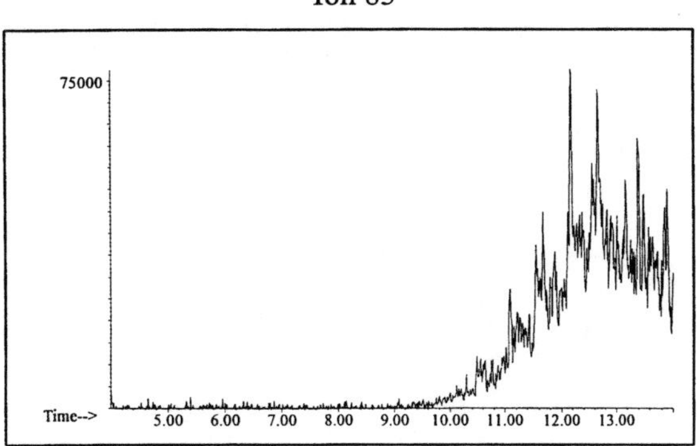

Naphthalene

Ion 128

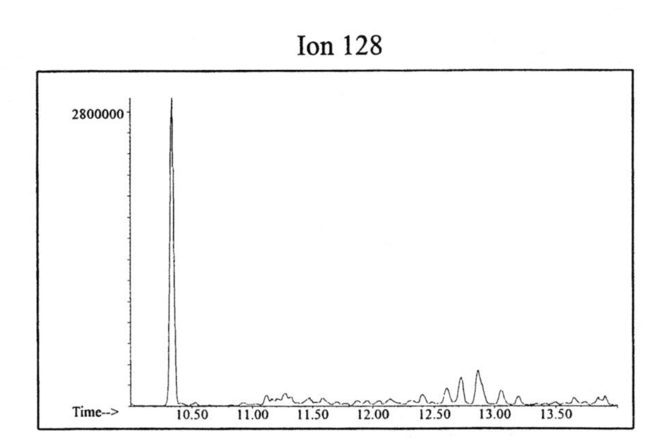

Ion 142

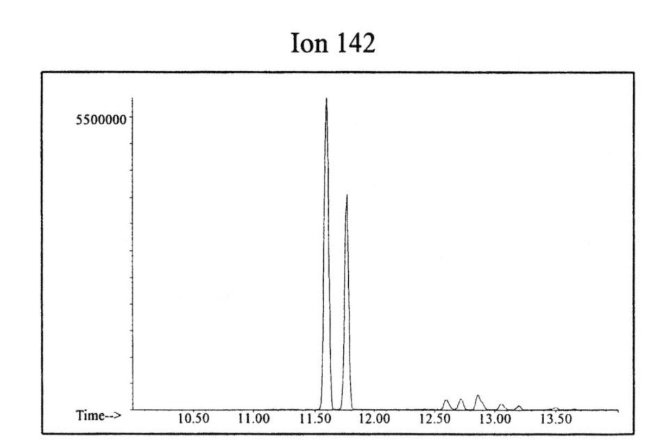

Ion 156

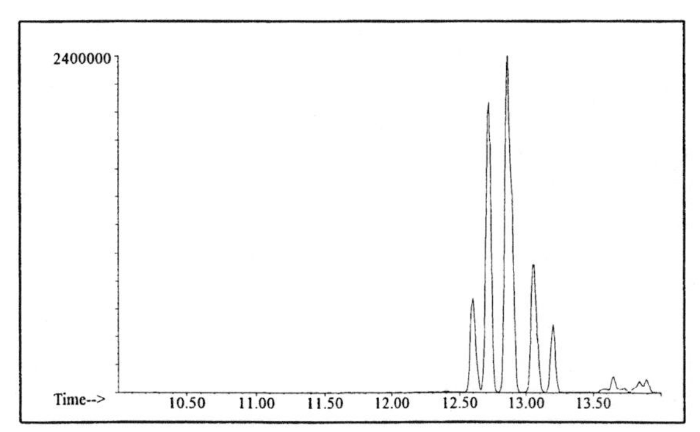

Gasoline #13

20uL/mL Pentane

ASTM: Class 2 (Gasoline)

Macro Code: M

Product Displayed: Shell Premium Unleaded Gasoline

Product Uses: Automotive fuel

Other Similar Products: Other grades and brands of gasoline

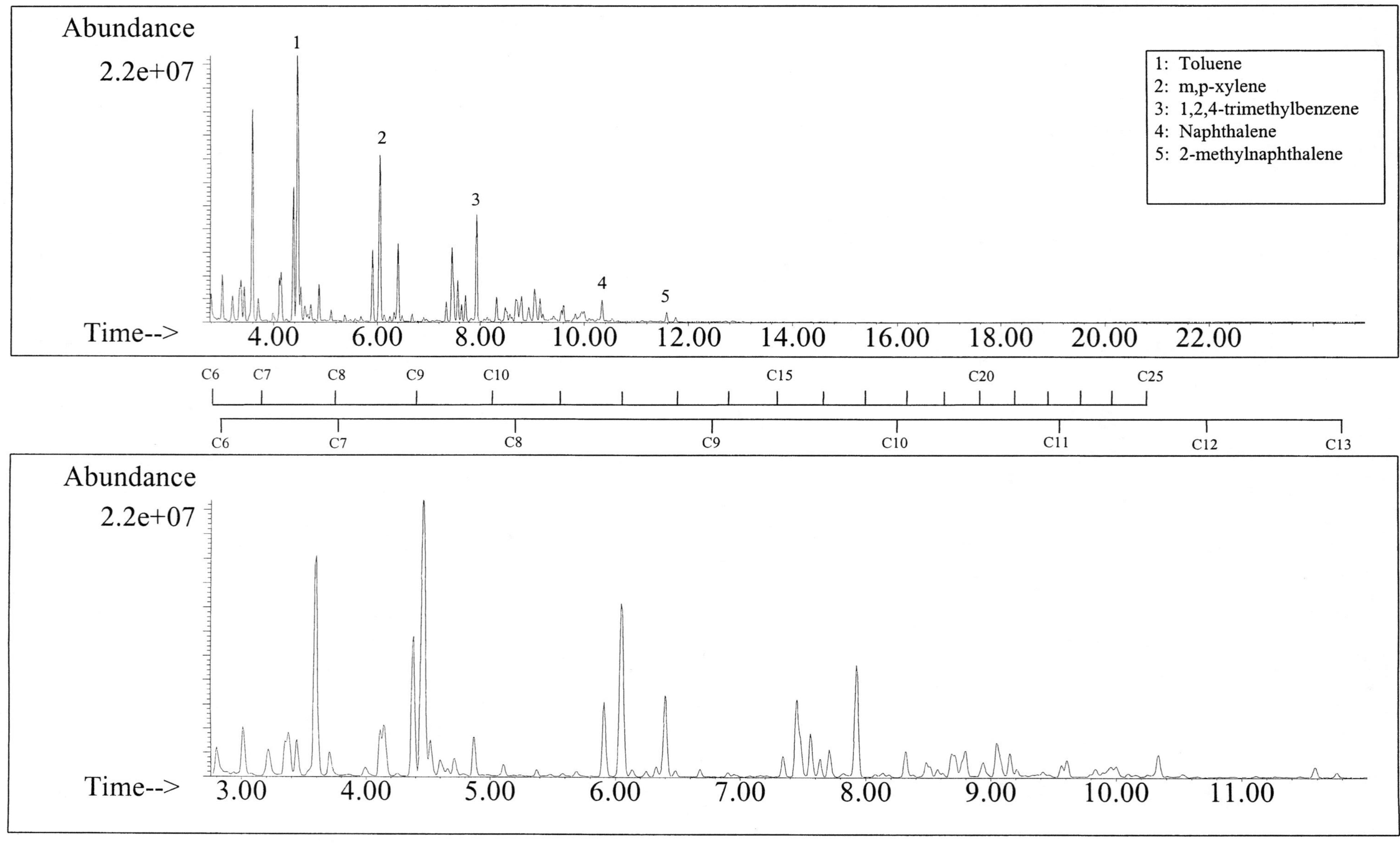

SUMMED PROFILES

Gasoline #13

ALKANES

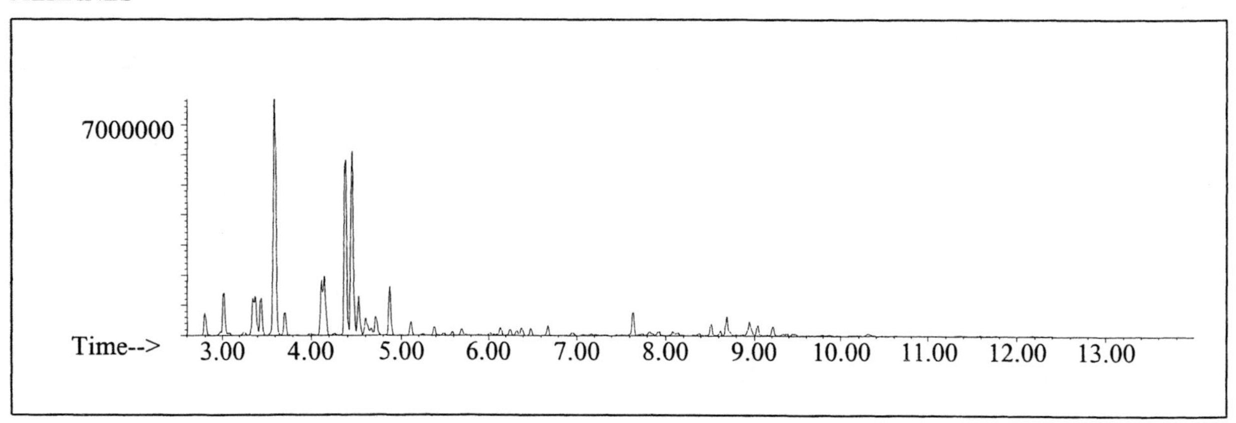

AROMATICS

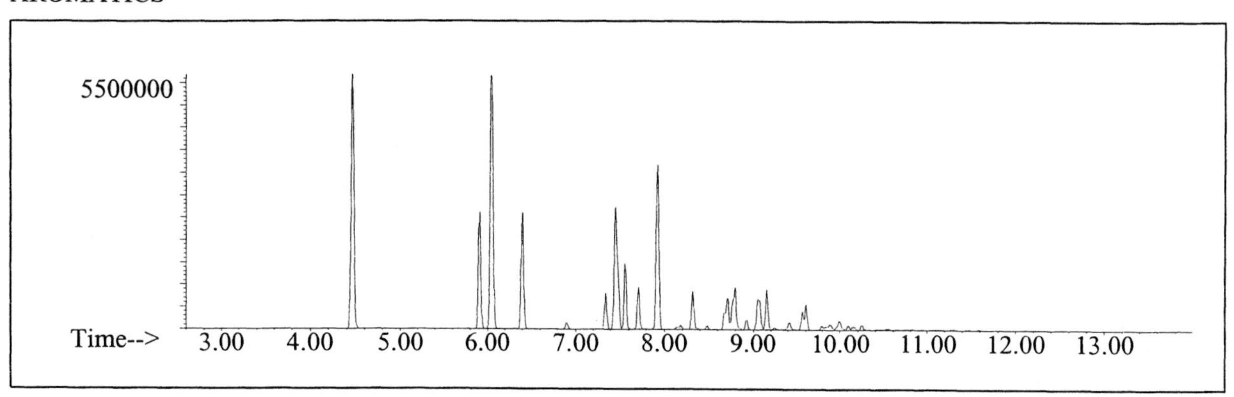

CYCLOPARAFFINS AND ALKENES

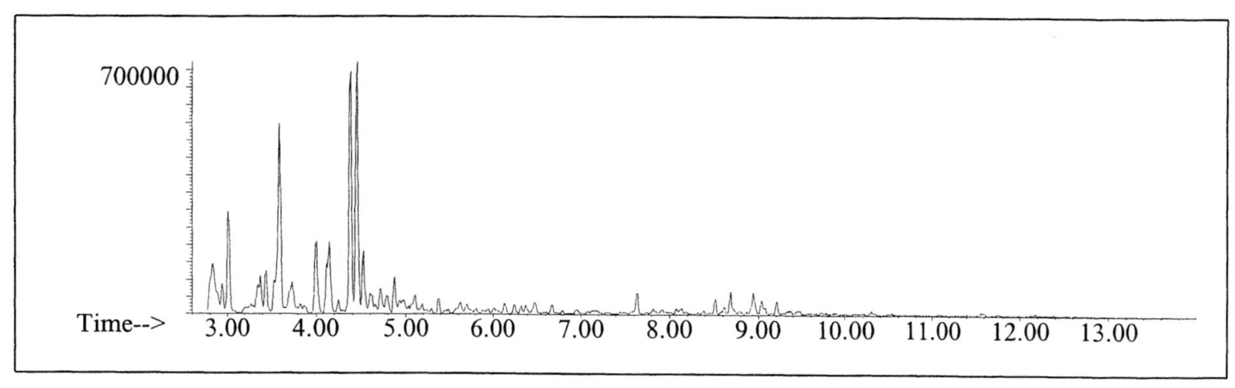

NAPHTHALENES

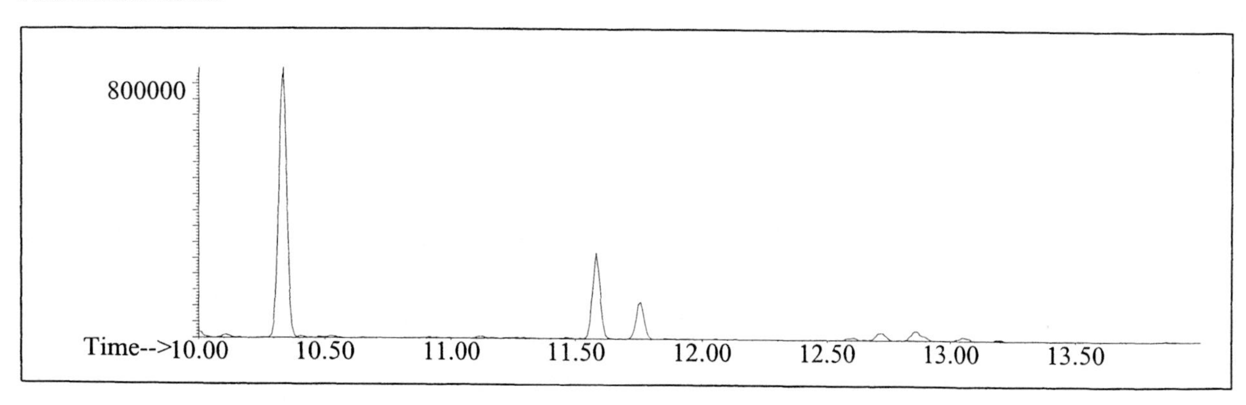

INDIVIDUAL PROFILES

Alkane

Ion 43

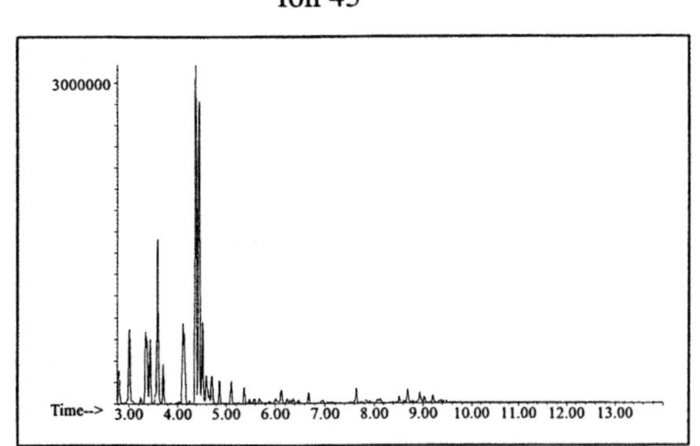

Ion 57

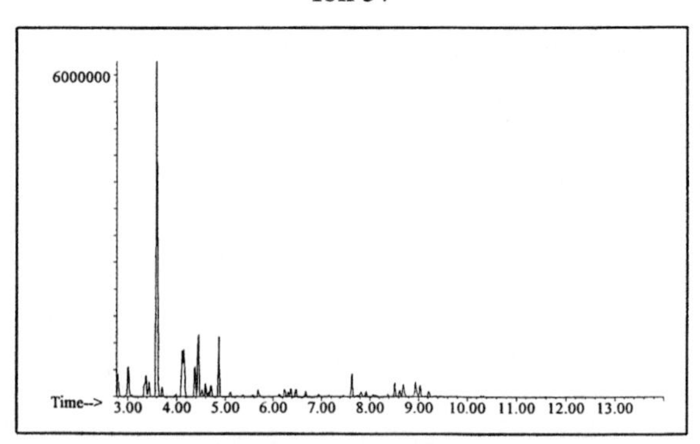

Ion 71

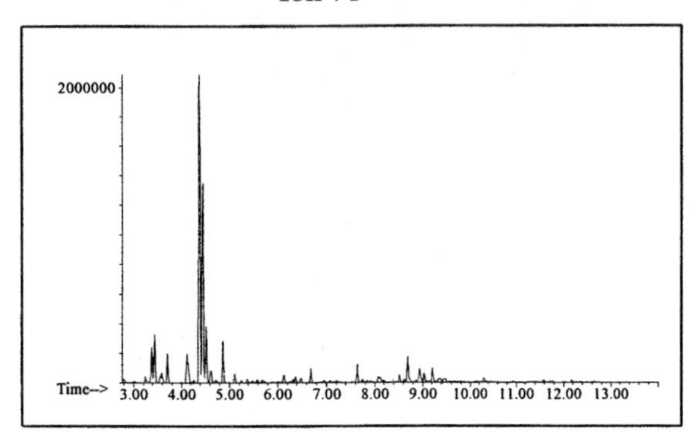

Ion 85

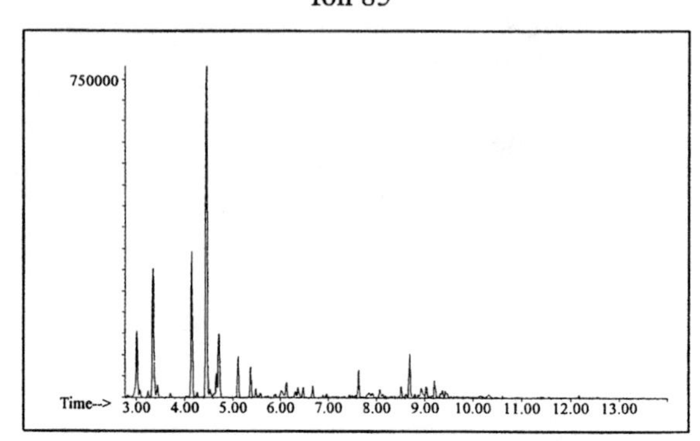

Aromatic

Ion 91

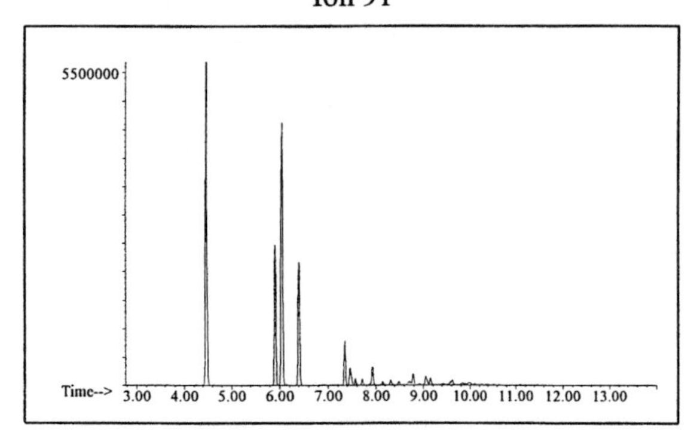

Ion 105

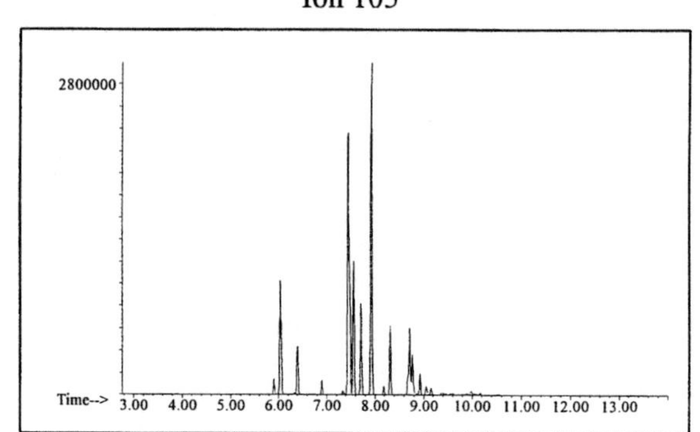

Ion 119

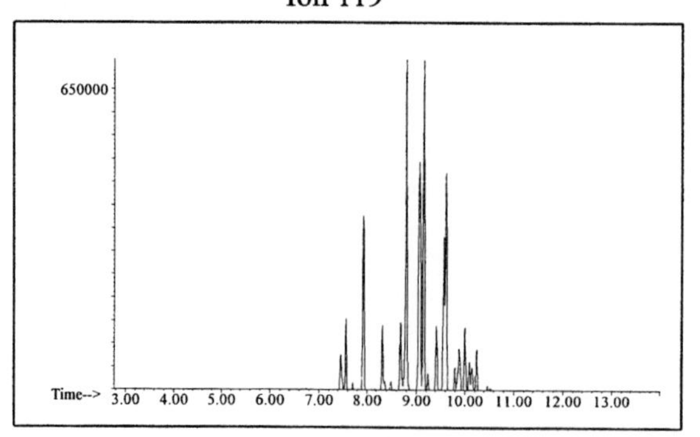

Cycloparaffin

Ion 55

Ion 69

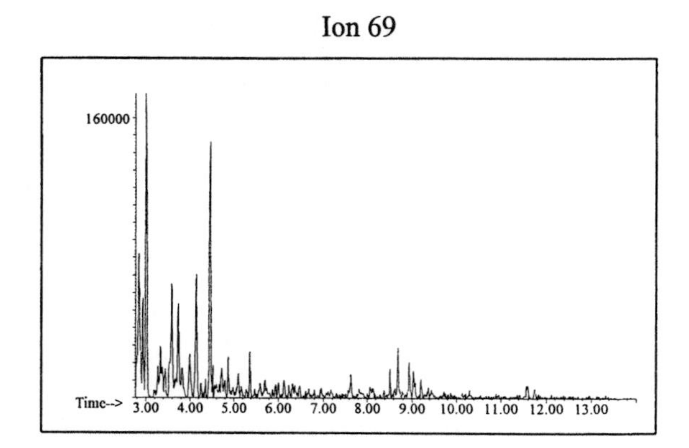

Ion 83

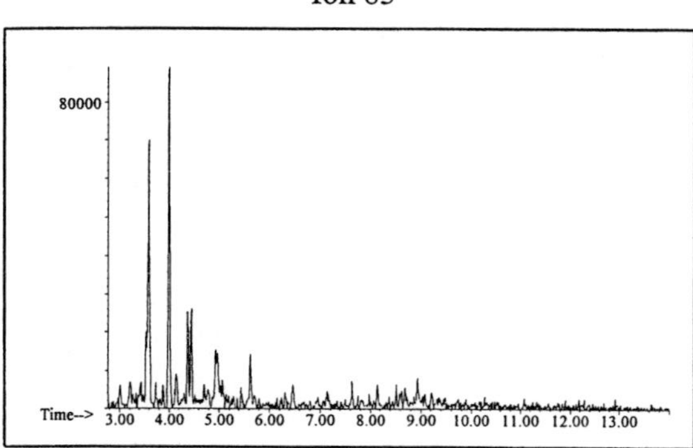

Naphthalene

Ion 128

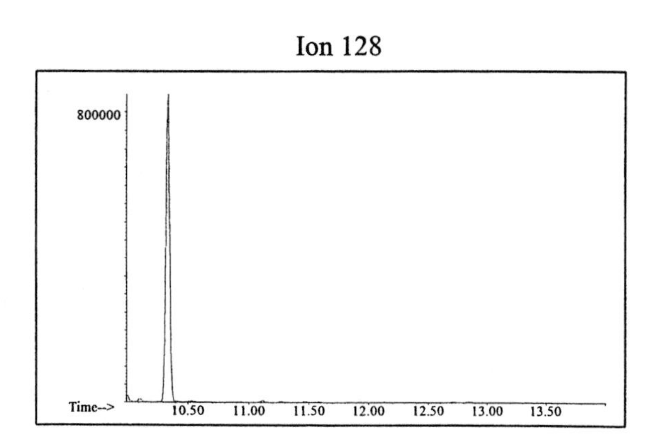

Ion 142

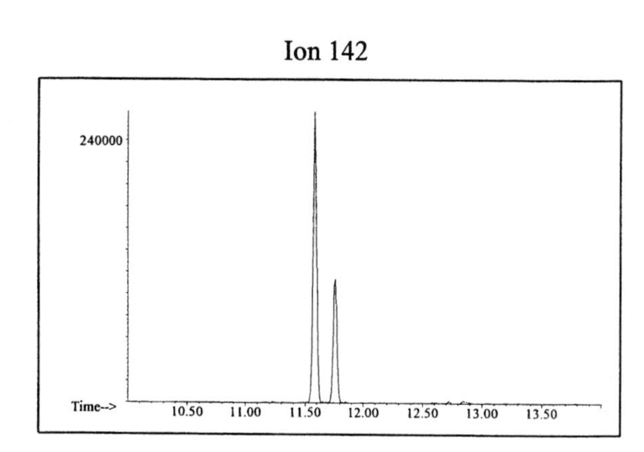

Ion 156

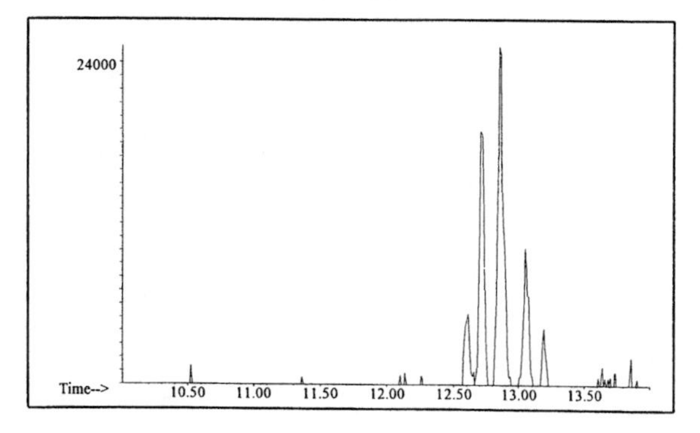

Gasoline #14

20uL/mL Pentane

ASTM: Class 2 (Gasoline)

Macro Code: L

Product Displayed: 25% Evaporated Shell Premium Gasoline

Product Uses: Automotive fuel

Other Similar Products: Other grades and brands of gasoline

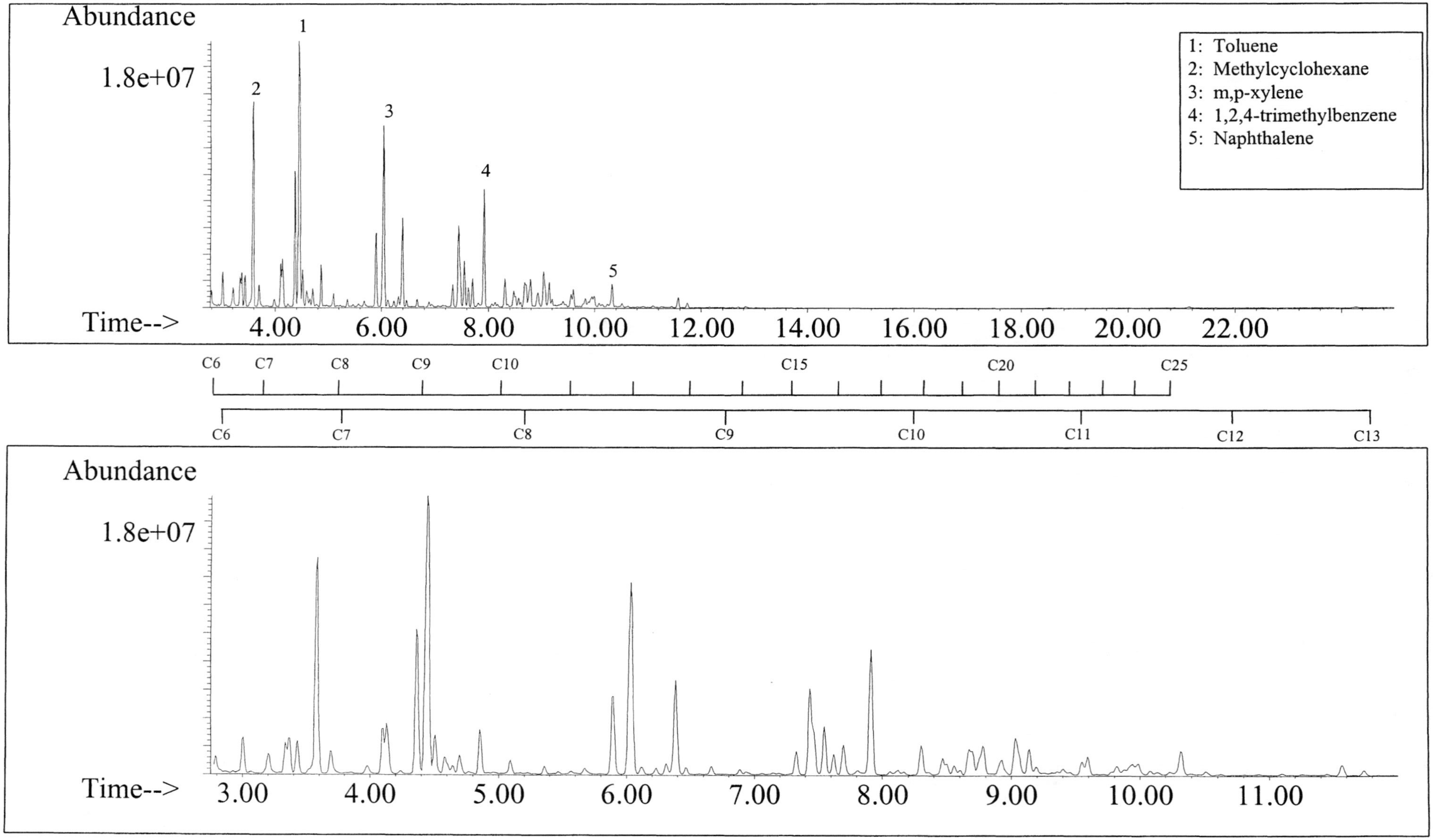

SUMMED PROFILES

Gasoline #14

ALKANES

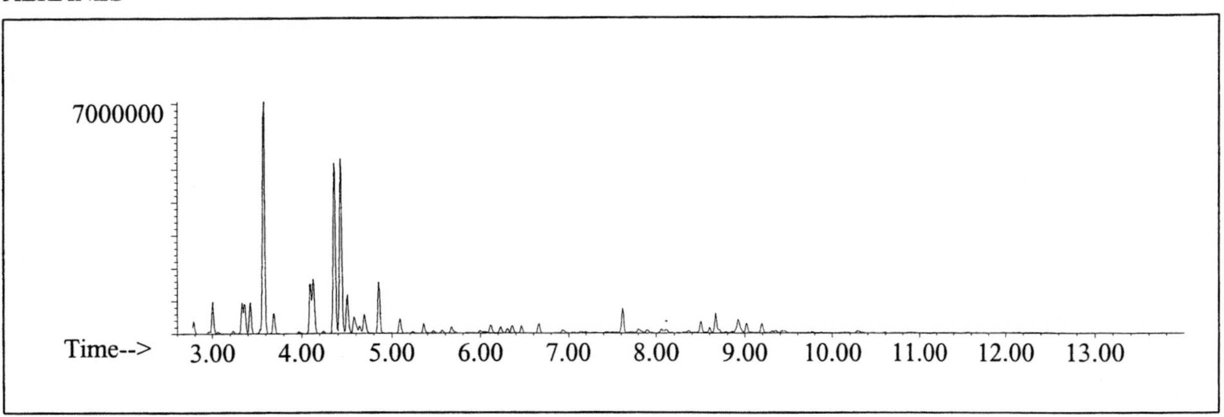

AROMATICS

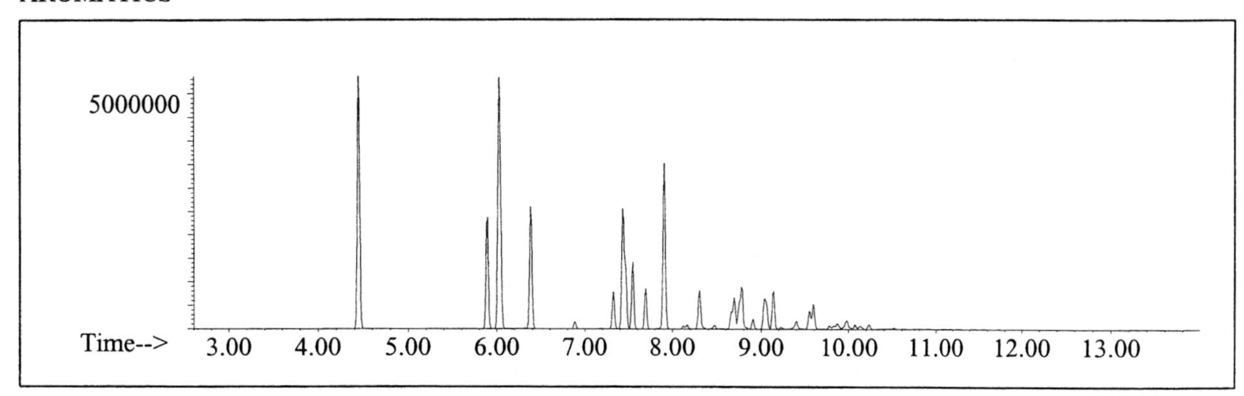

CYCLOPARAFFINS AND ALKENES

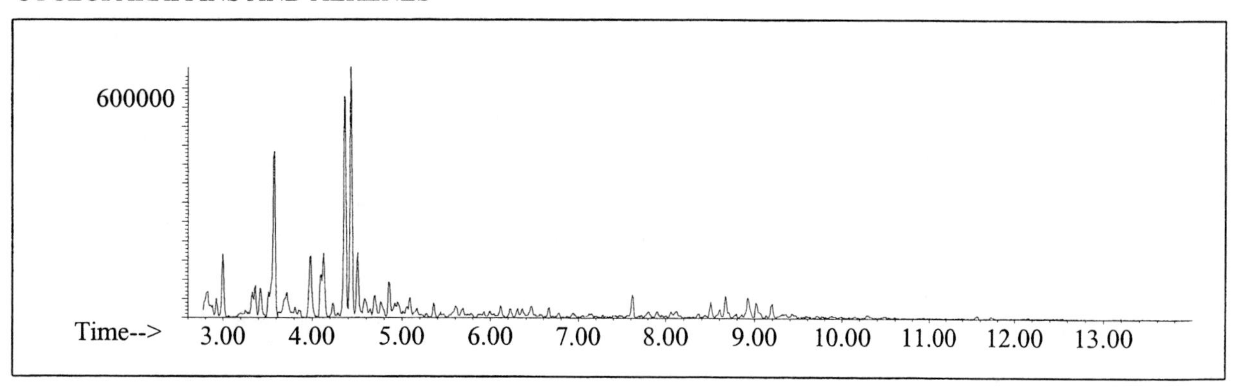

NAPHTHALENES

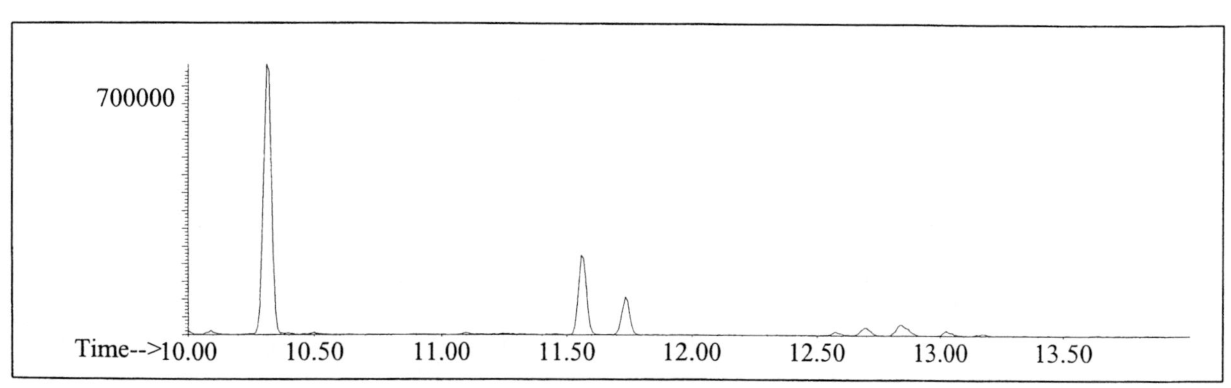

INDIVIDUAL PROFILES

Gasoline #14

Alkane

Ion 43

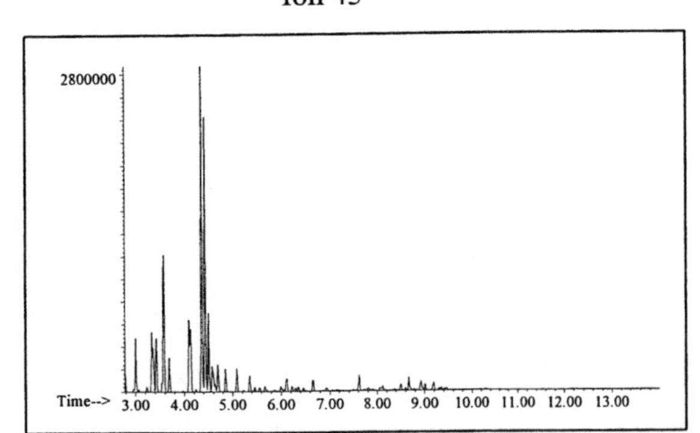

Ion 57

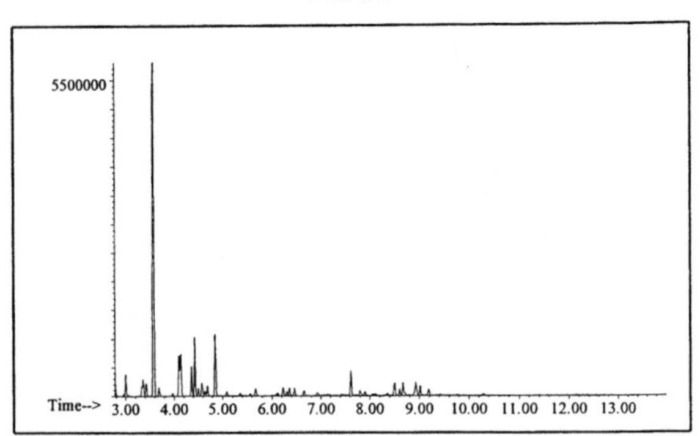

Ion 71

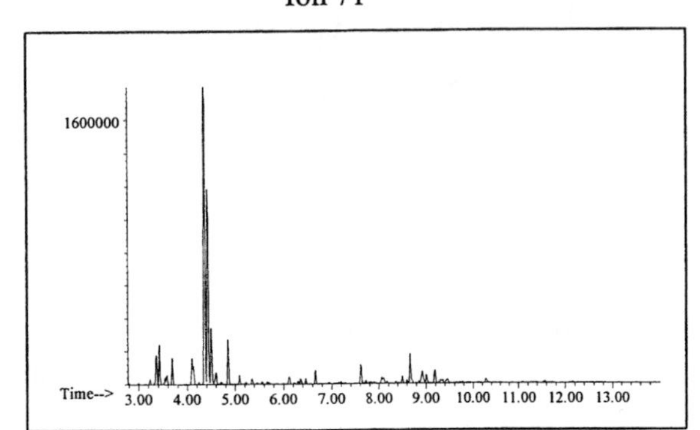

Ion 85

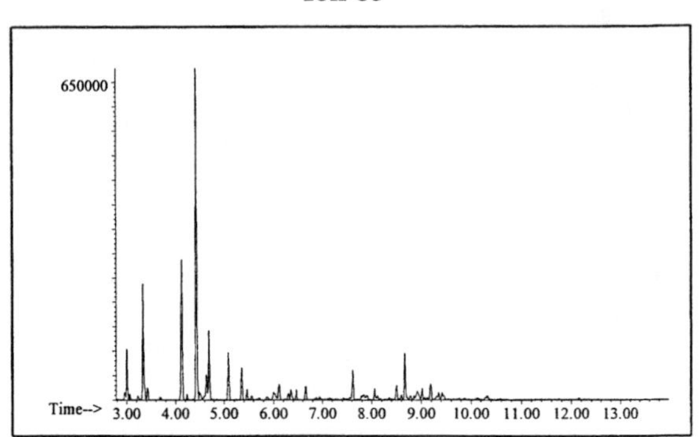

Aromatic

Ion 91

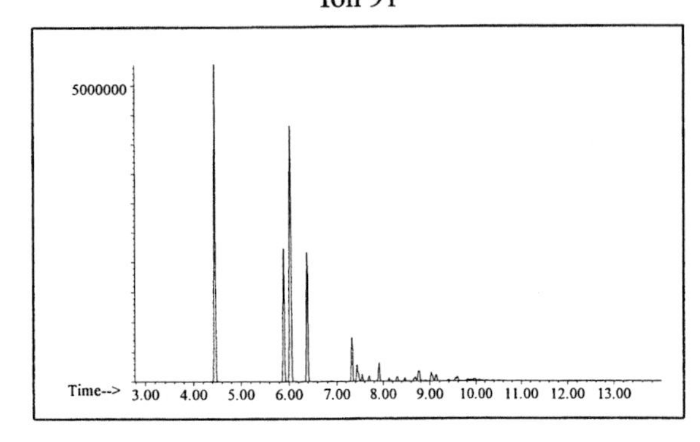

Ion 105

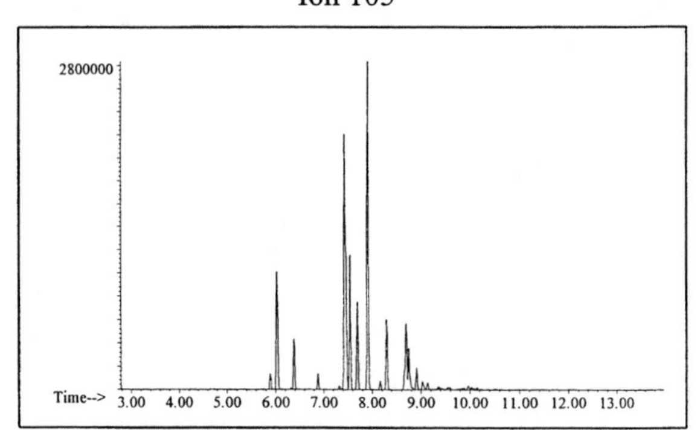

Ion 119

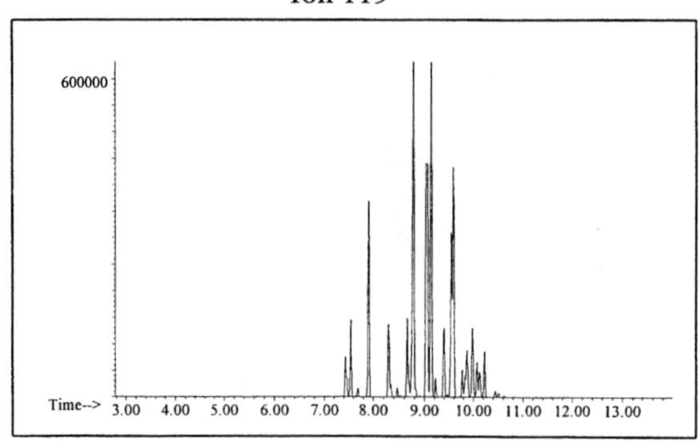

INDIVIDUAL PROFILES

Gasoline #14

Cycloparaffin

Ion 55

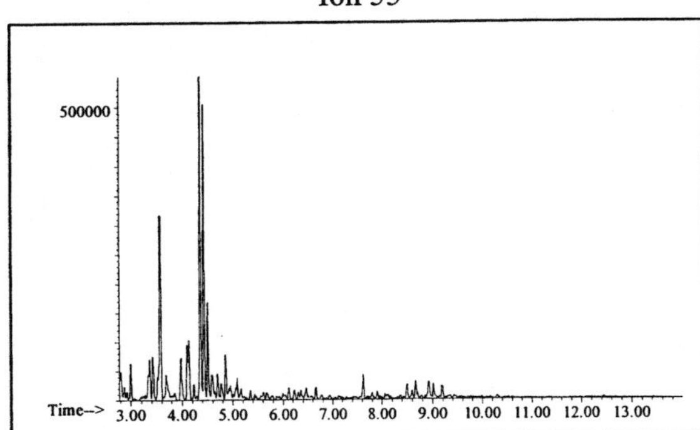

Ion 69

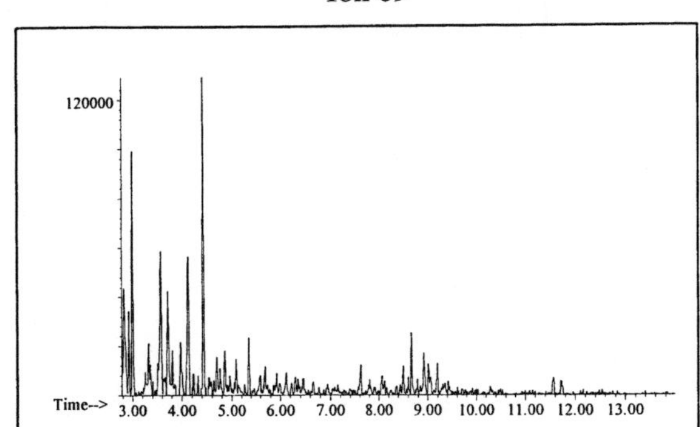

Ion 83

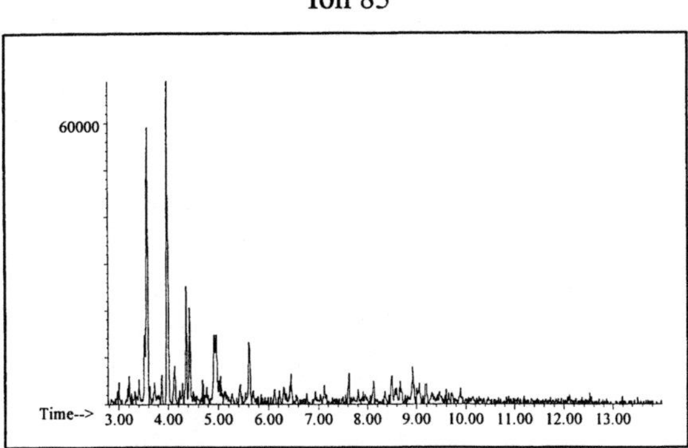

Naphthalene

Ion 128

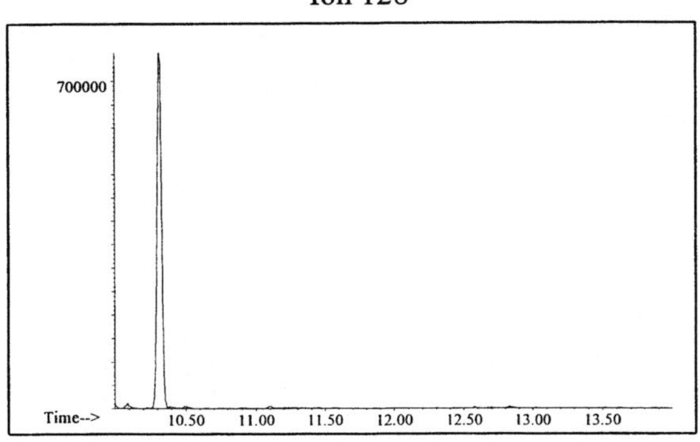

Ion 142

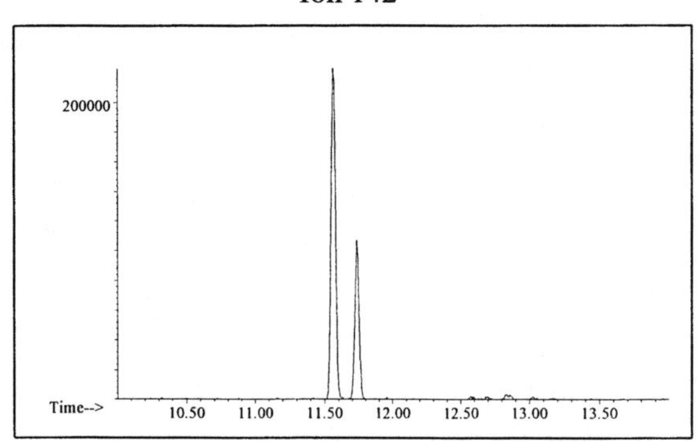

Ion 156

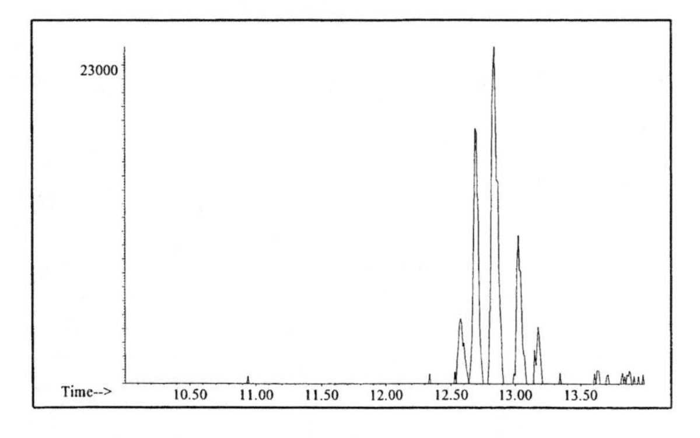

Gasoline #15

20uL/mL Pentane

Product Displayed: 50% Evaporated Shell Premium Gasoline

Other Similar Products: Other grades and brands of gasoline

ASTM: Class 2 (Gasoline)

Product Uses: Automotive fuel

Macro Code: M

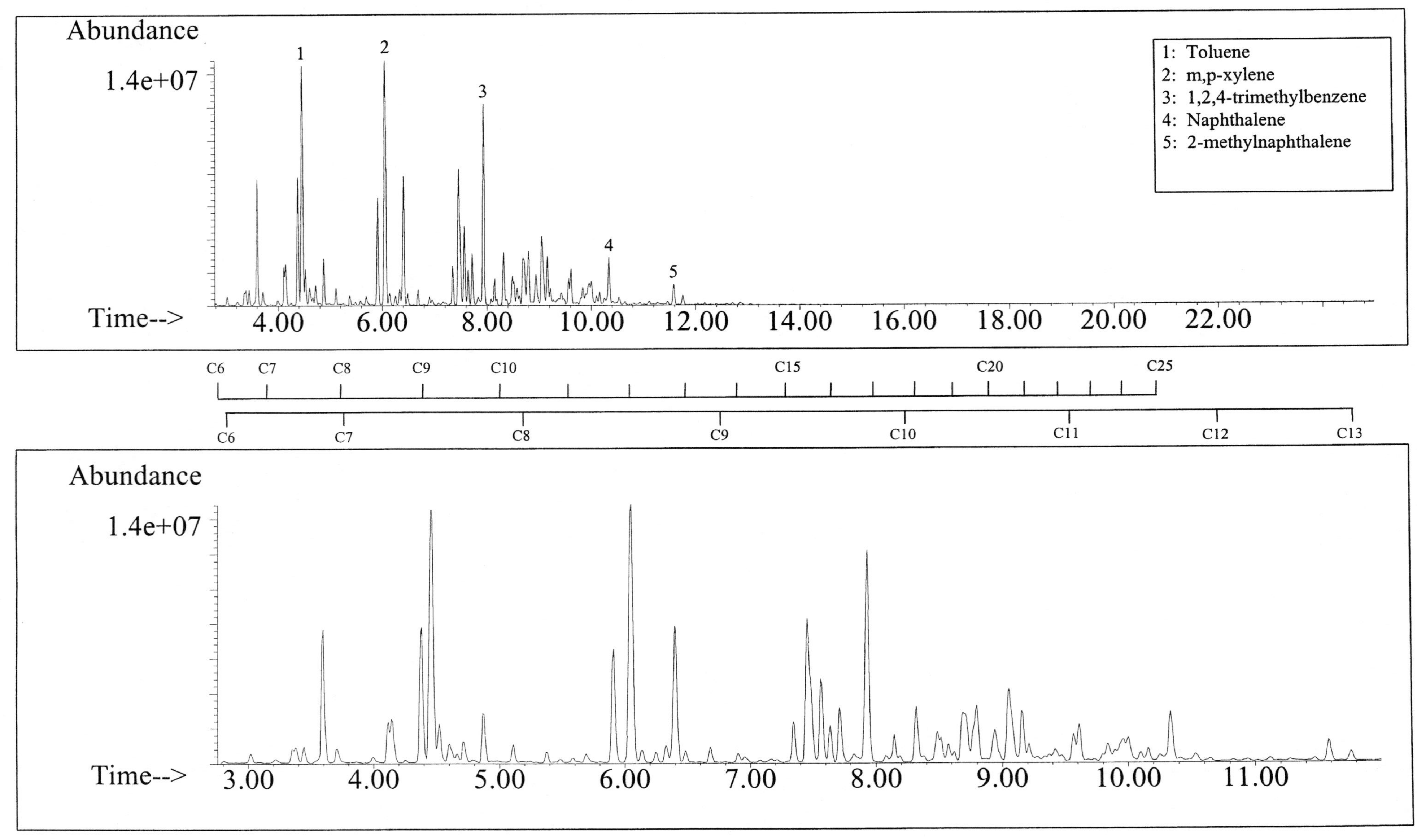

SUMMED PROFILES Gasoline #15

ALKANES

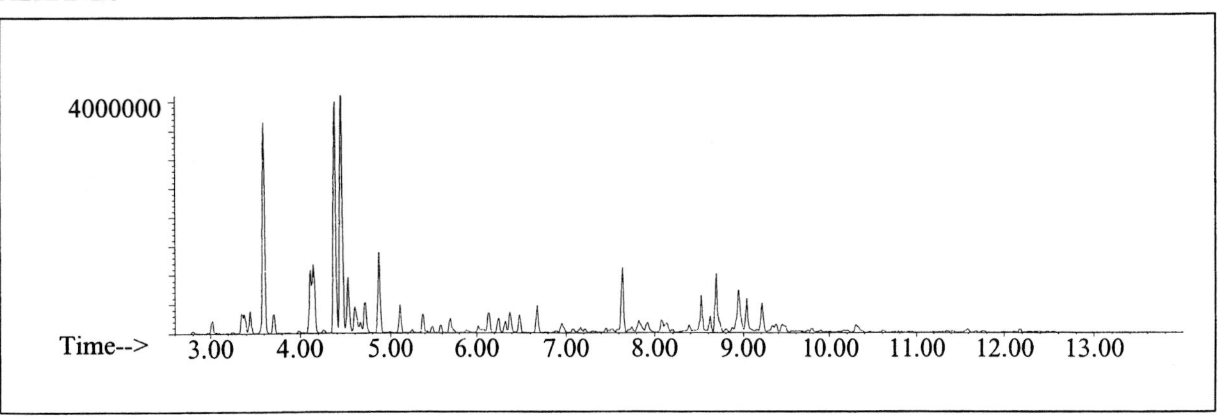

AROMATICS

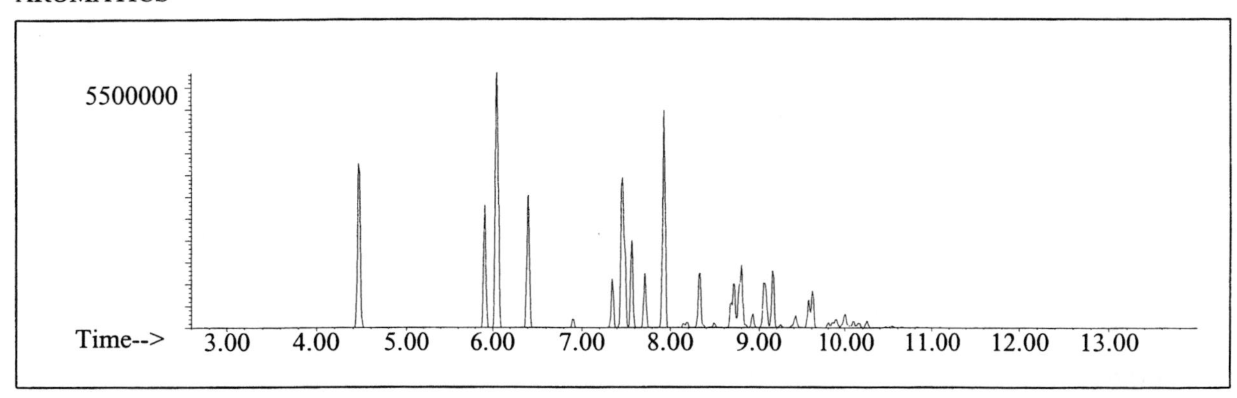

CYCLOPARAFFINS AND ALKENES

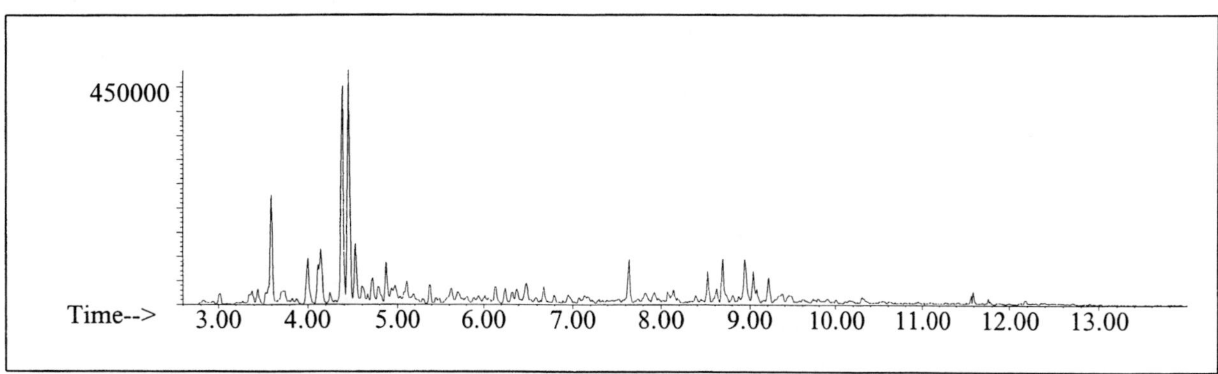

NAPHTHALENES

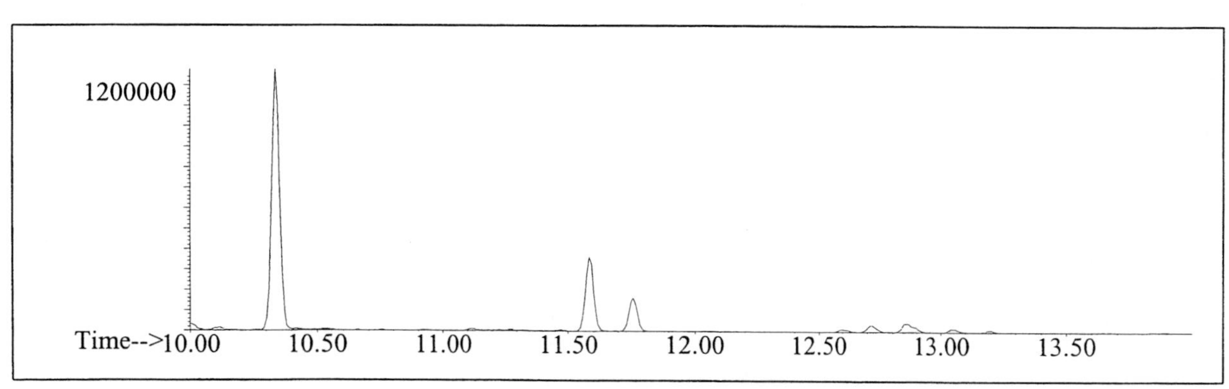

INDIVIDUAL PROFILES

Gasoline #15

Alkane

Ion 43

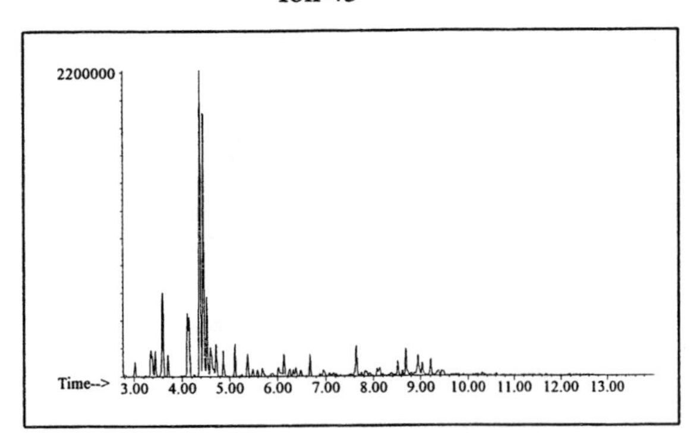

Ion 57

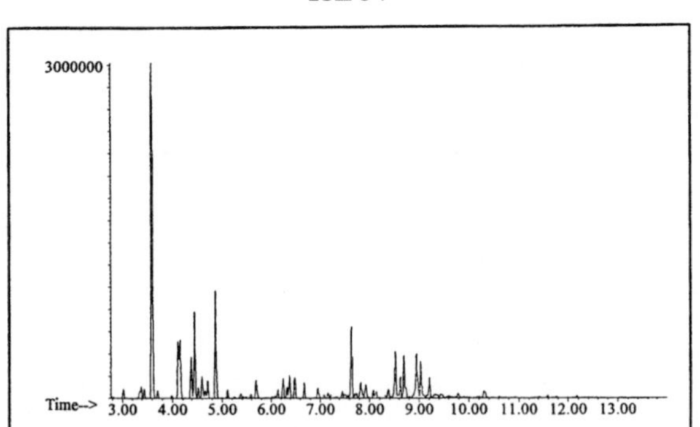

Ion 71

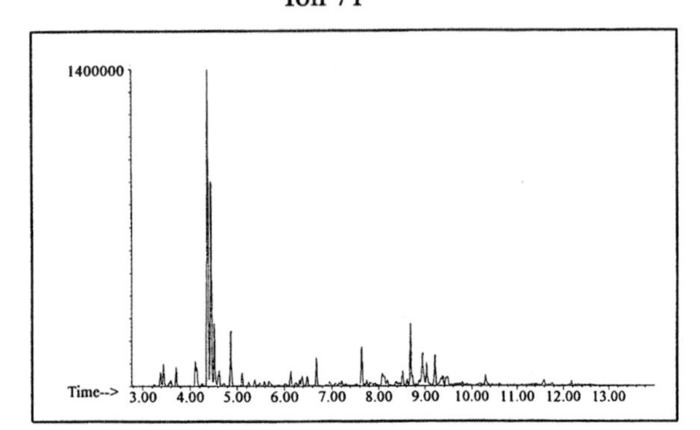

Ion 85

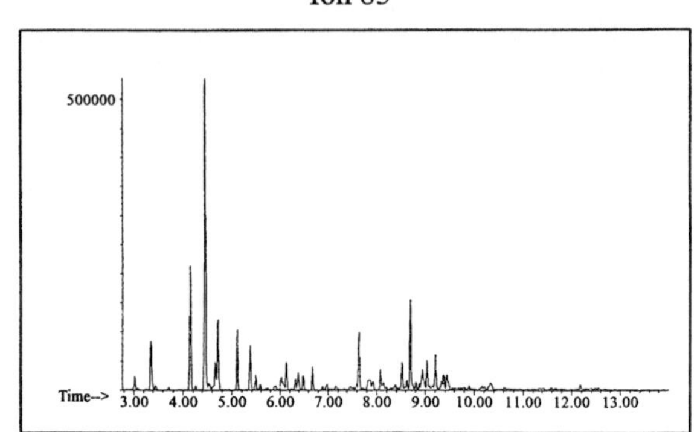

Aromatic

Ion 91

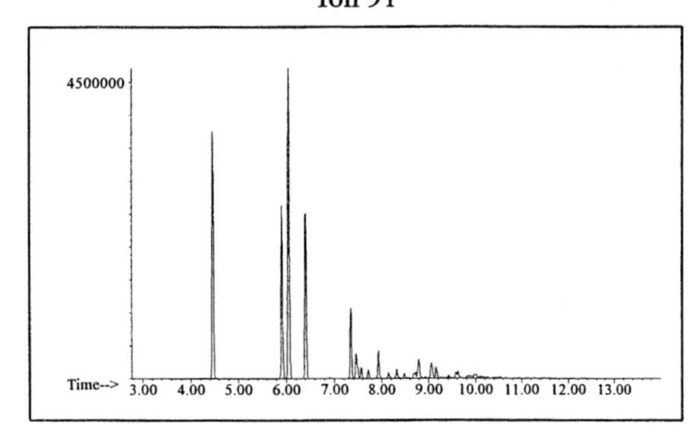

Ion 105

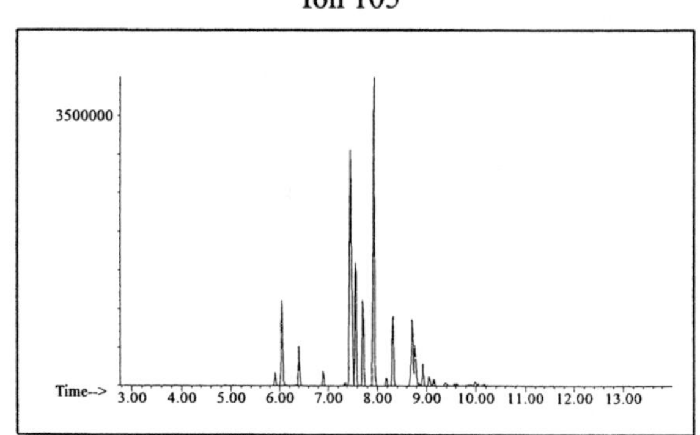

Ion 119

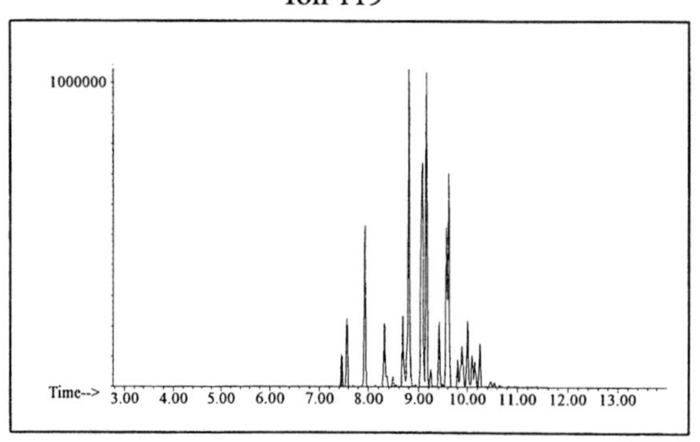

INDIVIDUAL PROFILES

Gasoline #15

Cycloparaffin

Ion 55

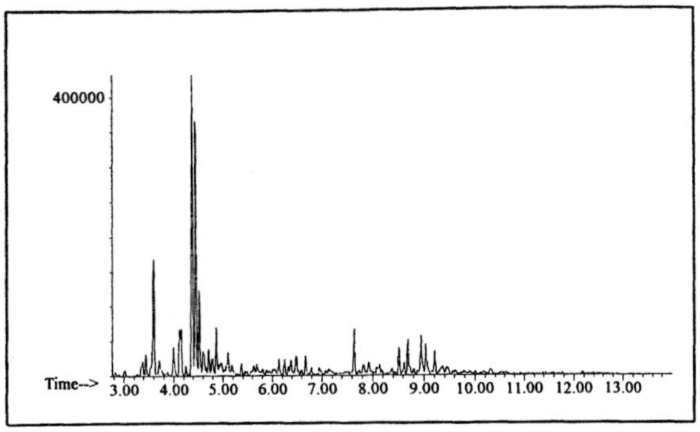

Ion 69

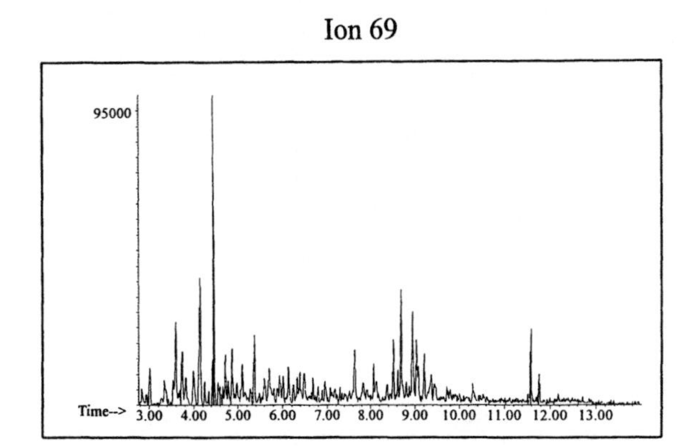

Ion 83

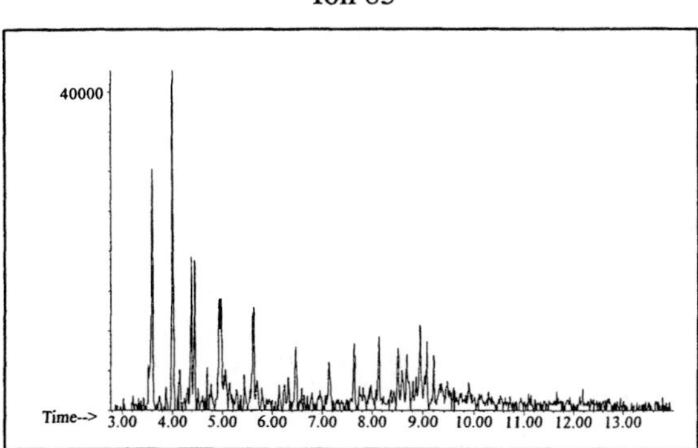

Naphthalene

Ion 128

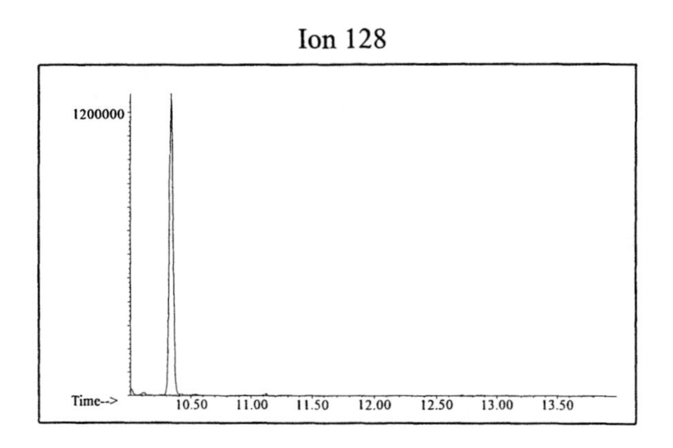

Ion 142

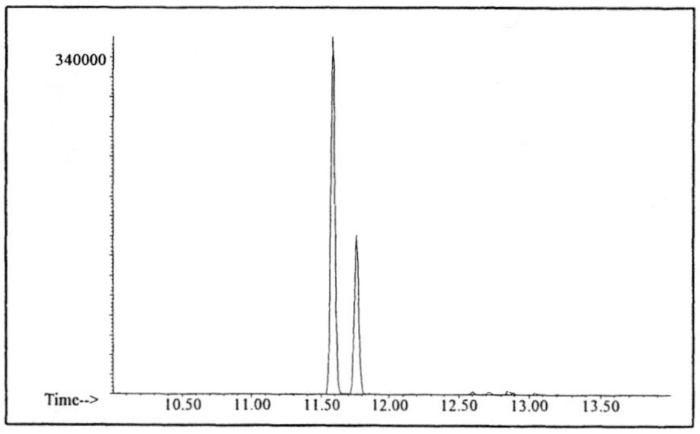

Ion 156

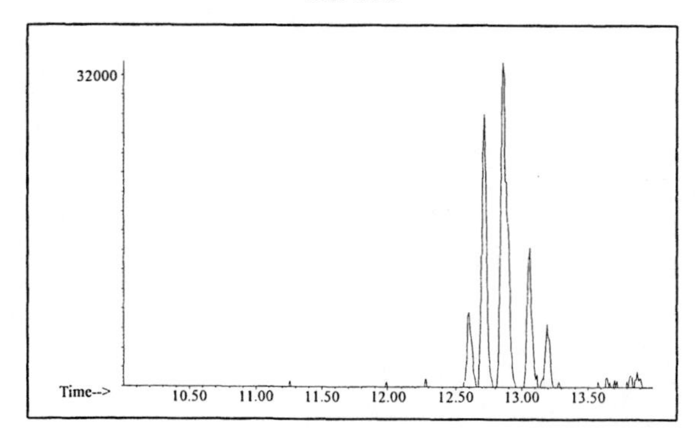

Gasoline #16

20uL/mL Pentane

Product Displayed: 75% Evaporated Shell Premium Gasoline

Other Similar Products: Other grades and brands of gasoline

ASTM: Class 2 (Gasoline)

Product Uses: Automotive fuel

Macro Code: M

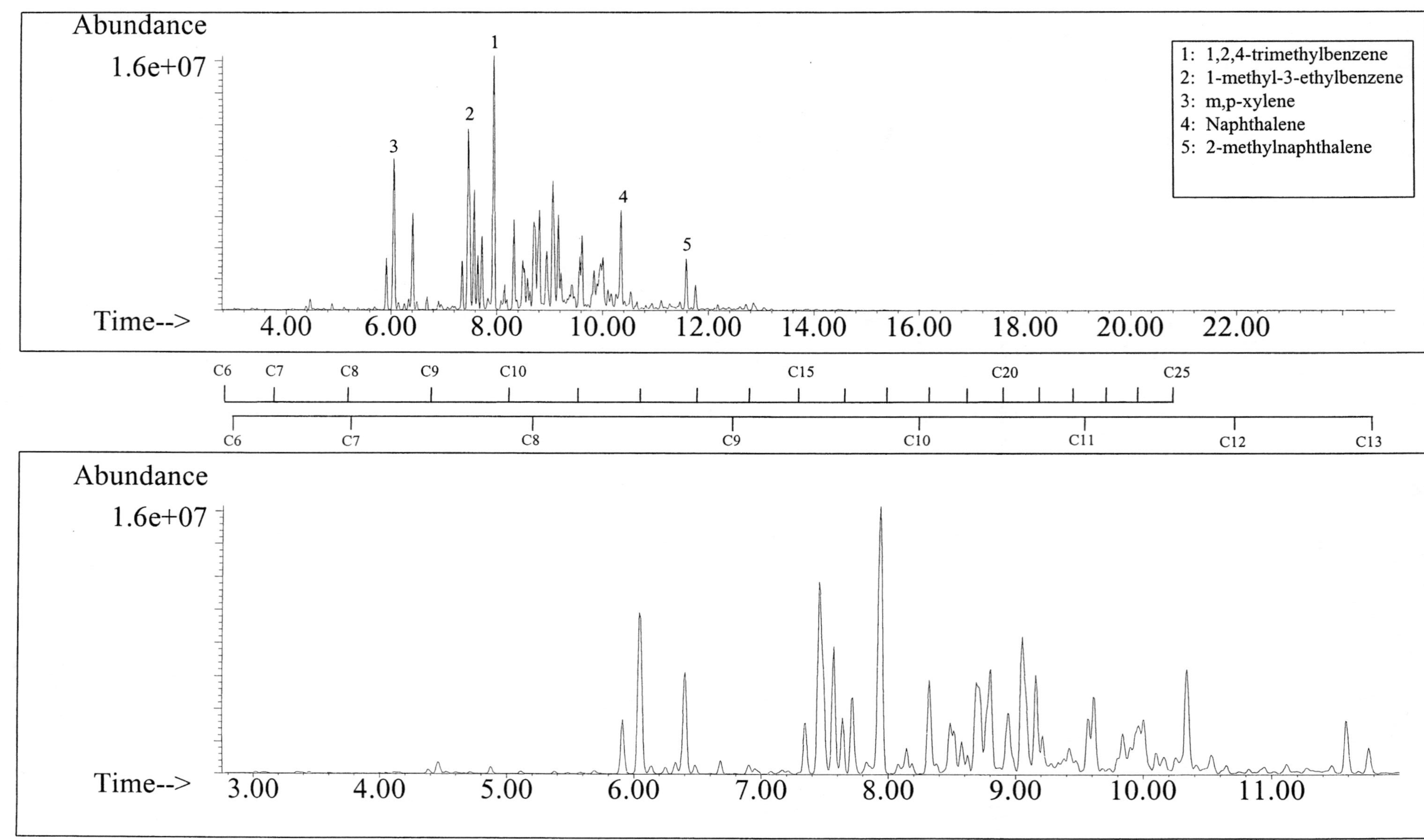

SUMMED PROFILES Gasoline #16

ALKANES

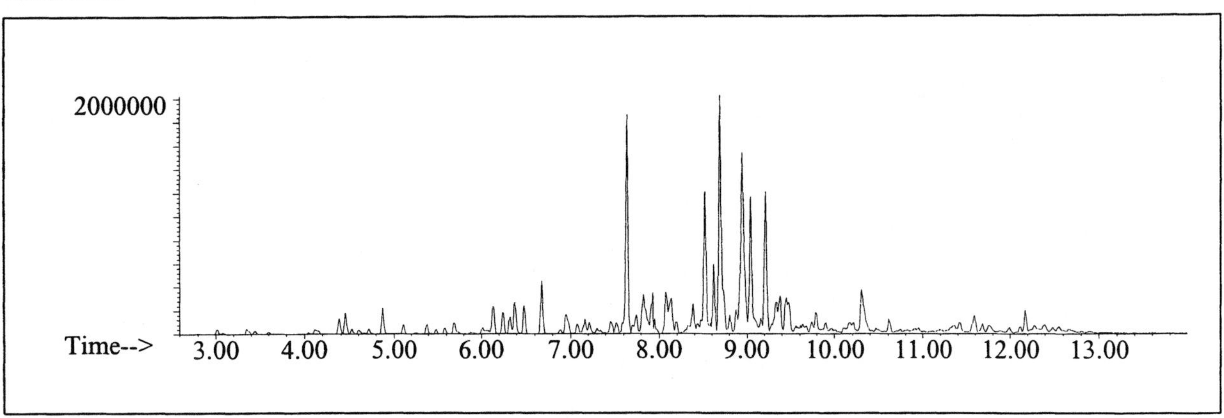

AROMATICS

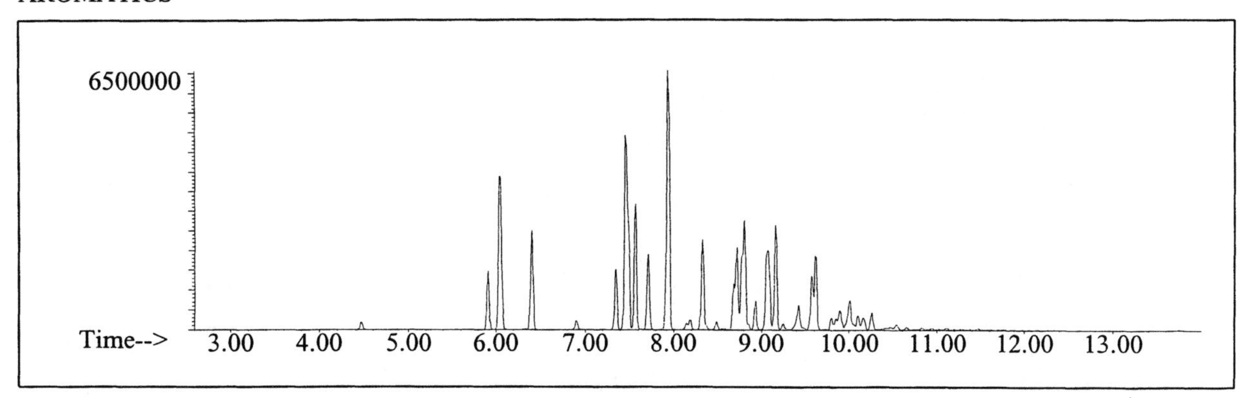

CYCLOPARAFFINS AND ALKENES

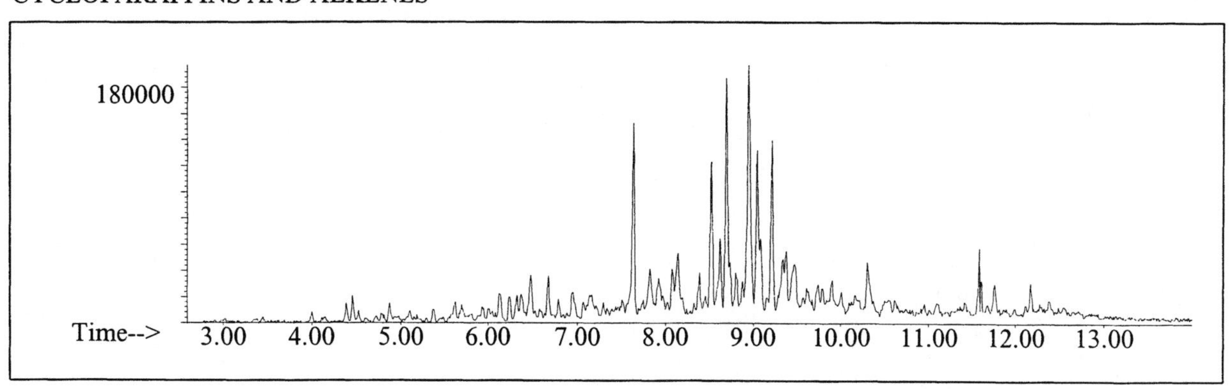

NAPHTHALENES

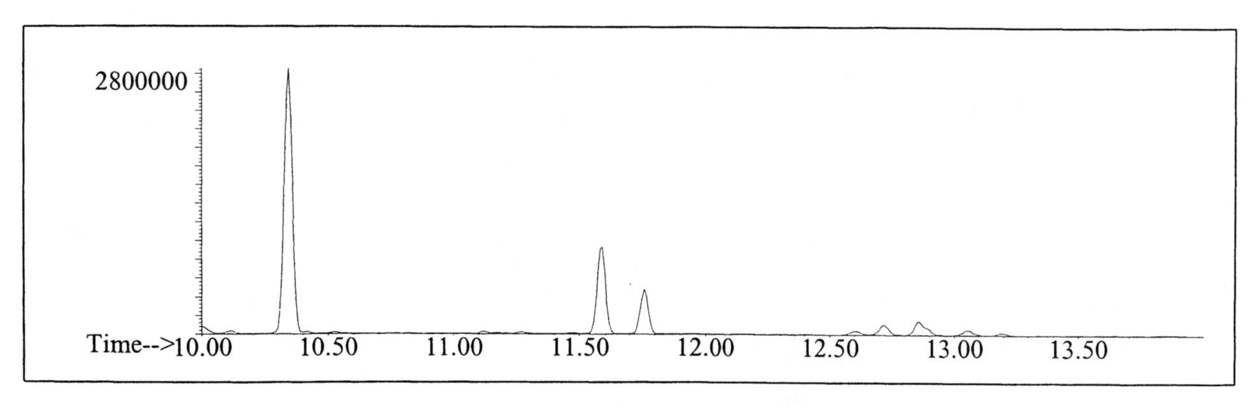

INDIVIDUAL PROFILES

Gasoline #16

Alkane

Ion 43

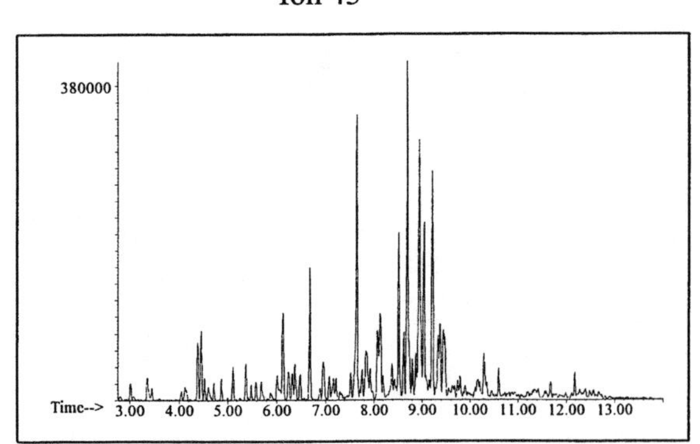

Ion 57

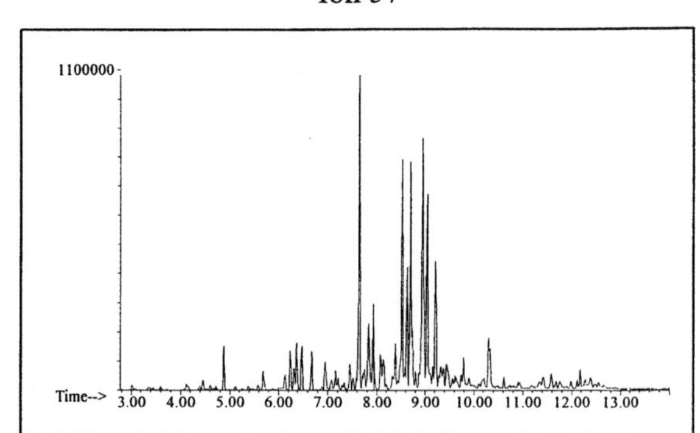

Ion 71

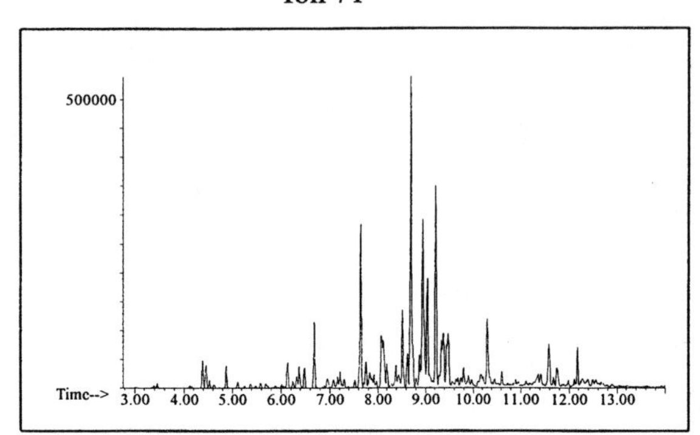

Ion 85

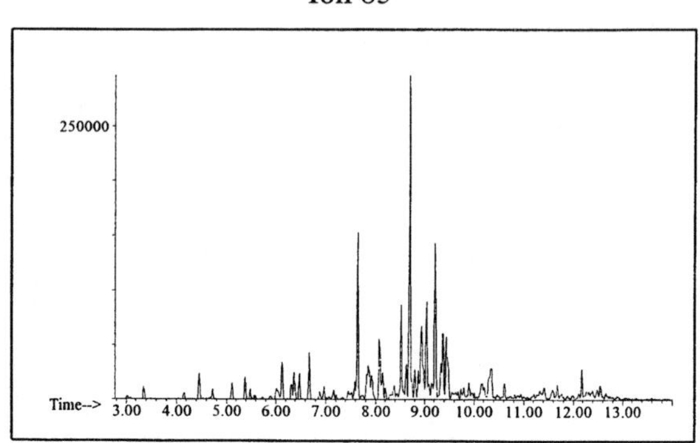

Aromatic

Ion 91

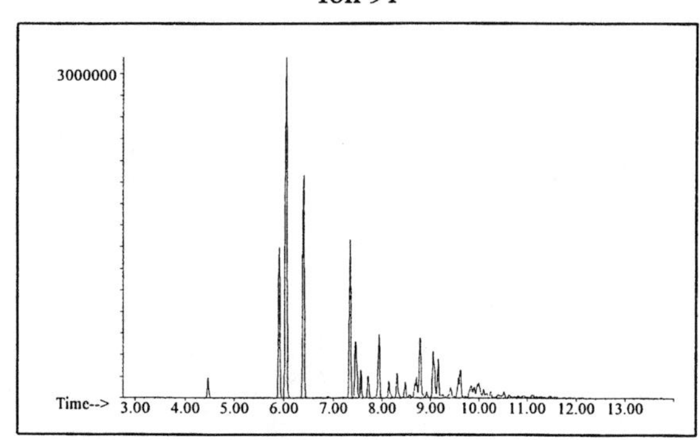

Ion 105

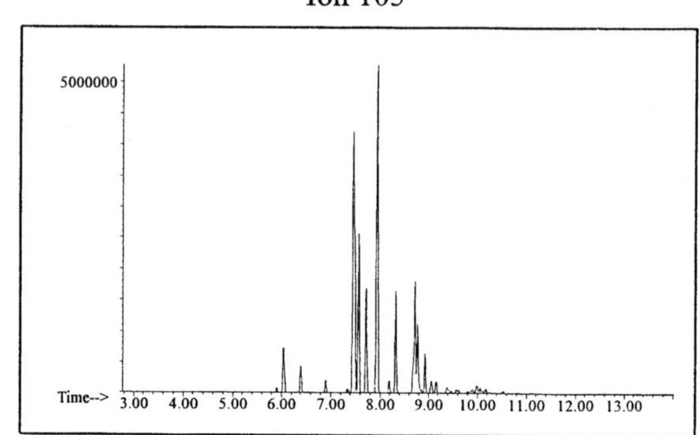

Ion 119

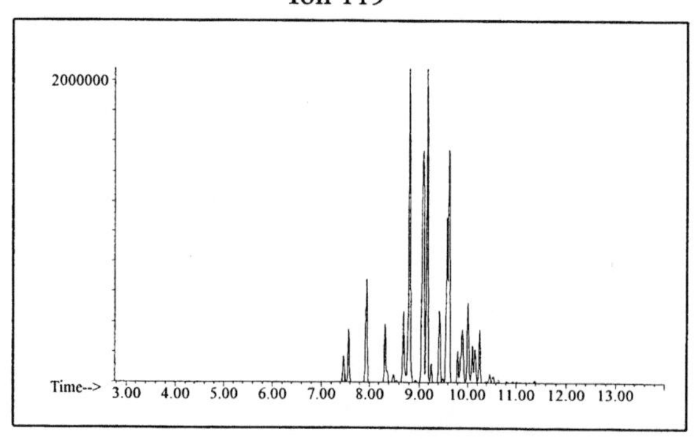

INDIVIDUAL PROFILES

Gasoline #16

Cycloparaffin

Ion 55

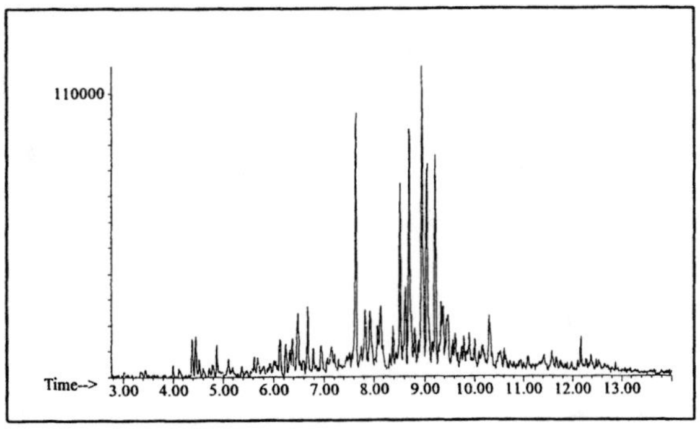

110000

Time--> 3.00 4.00 5.00 6.00 7.00 8.00 9.00 10.00 11.00 12.00 13.00

Ion 69

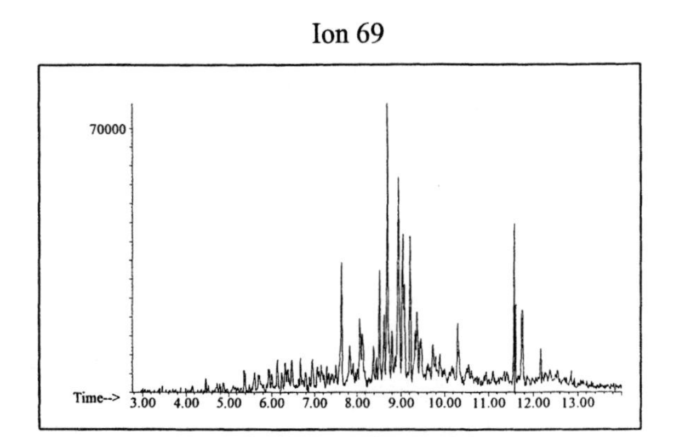

Ion 83

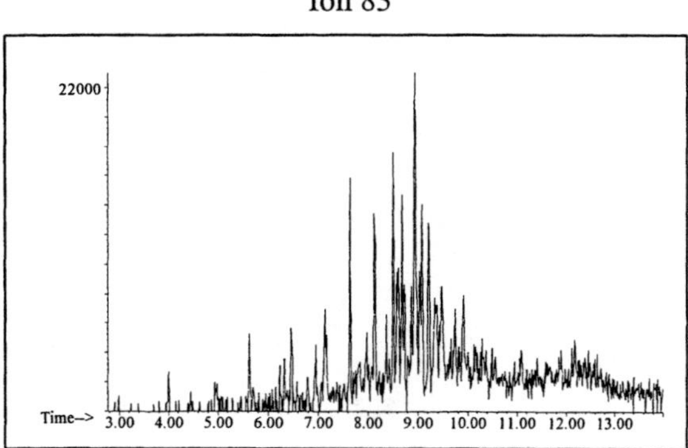

Naphthalene

Ion 128

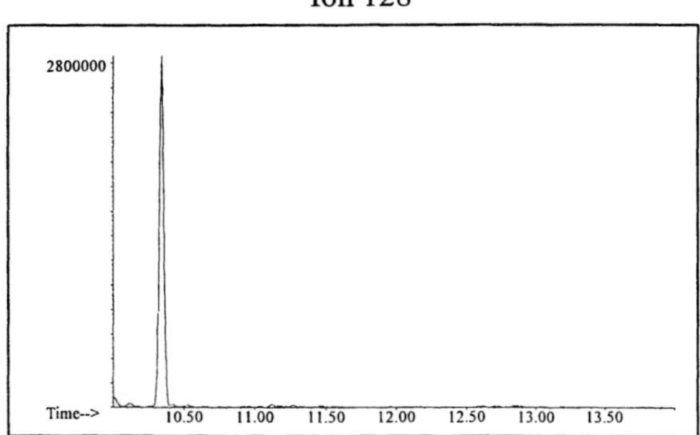

Ion 142

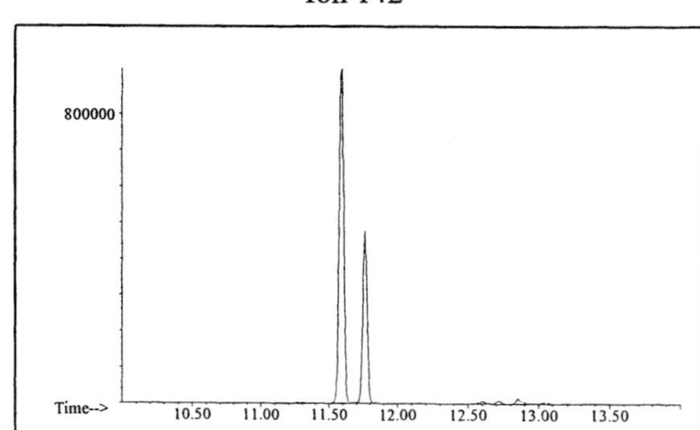

Ion 156

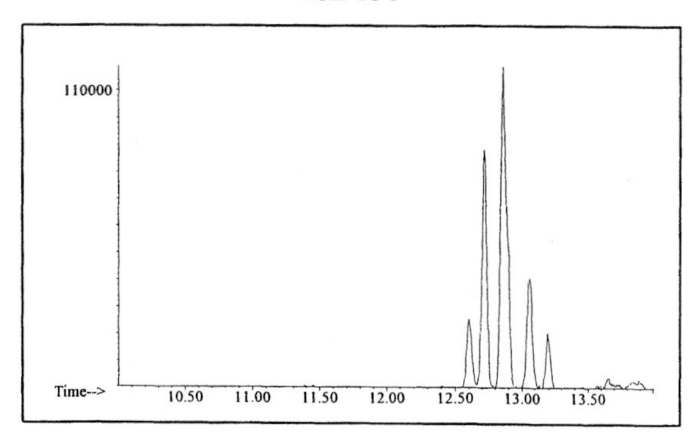

Gasoline #17

20uL/mL Pentane

Product Displayed: 90% Evaporated Shell Premium Gasoline

Other Similar Products: Other grades and brands of gasoline

ASTM: Class 2 (Gasoline)

Product Uses: Automotive fuel

Macro Code: M

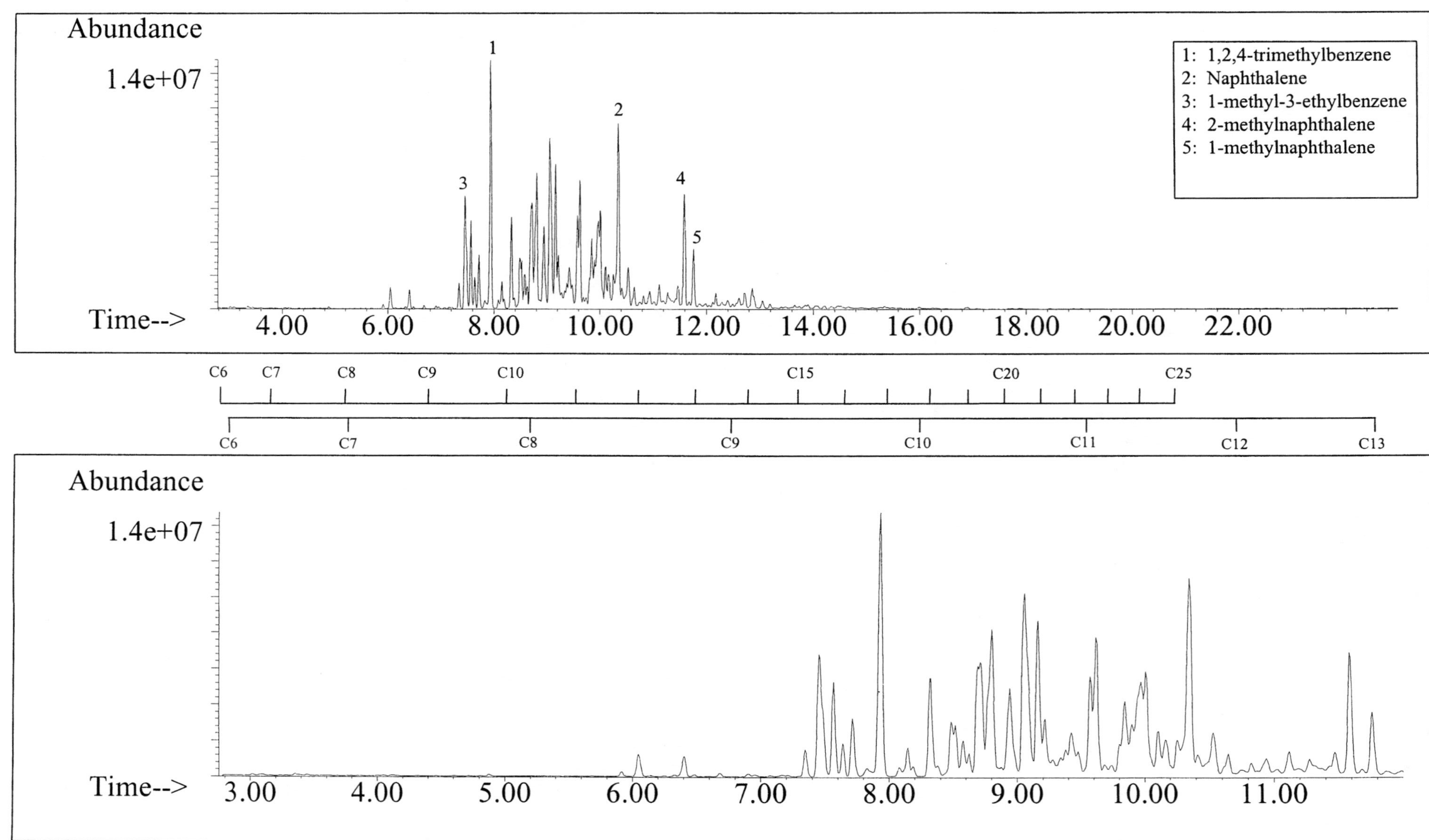

SUMMED PROFILES

Gasoline #17

ALKANES

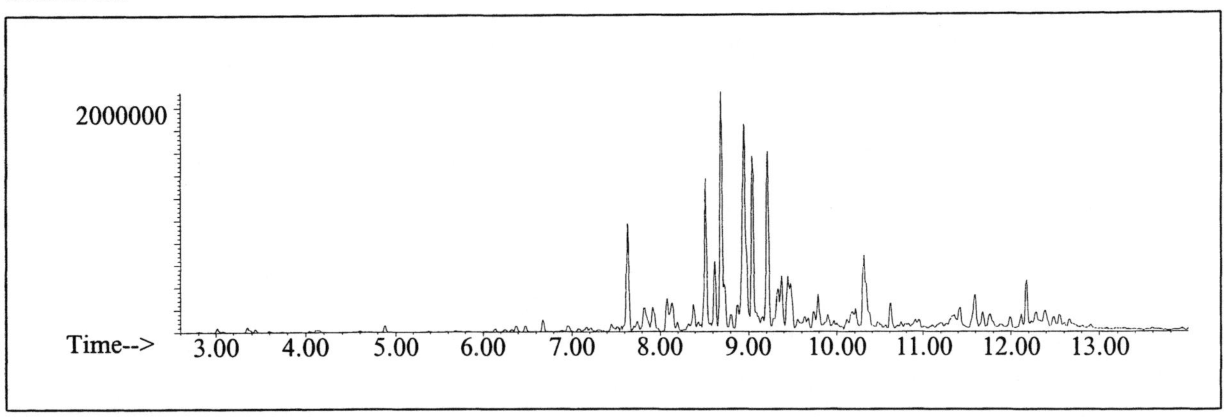

AROMATICS

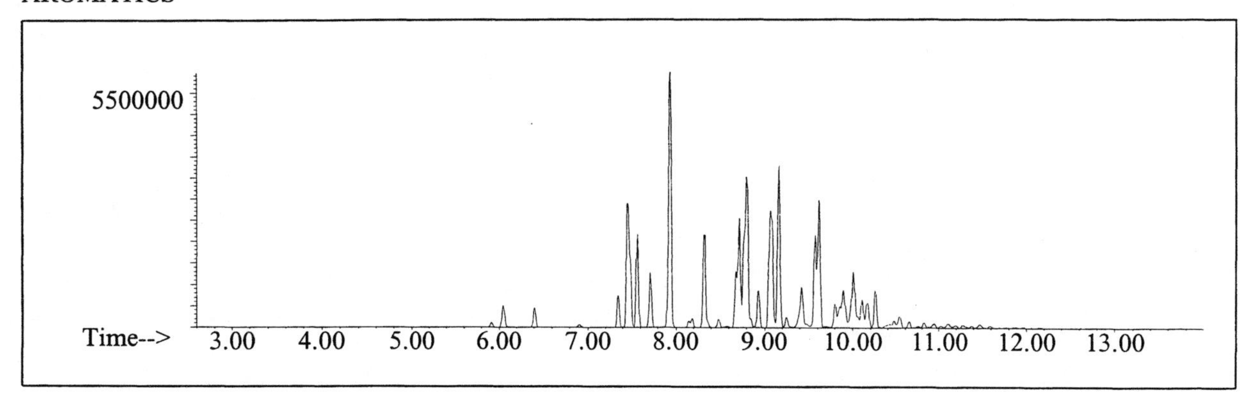

CYCLOPARAFFINS AND ALKENES

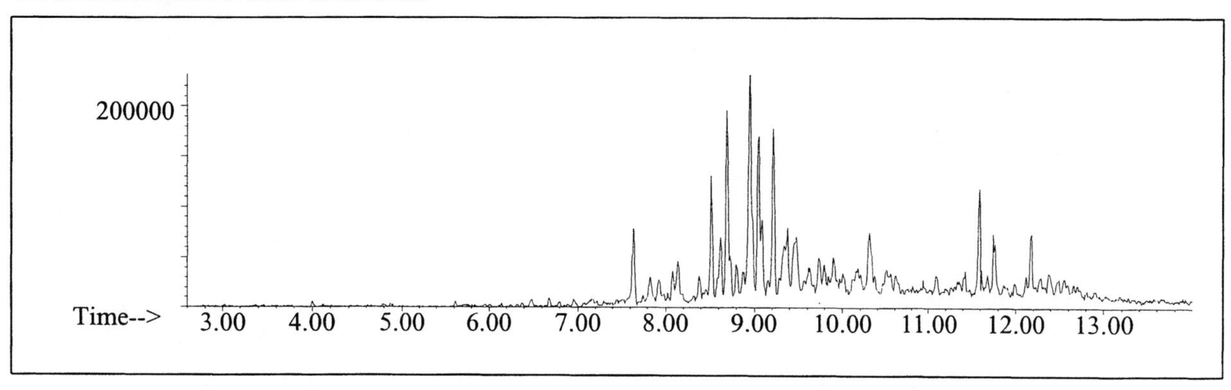

NAPHTHALENES

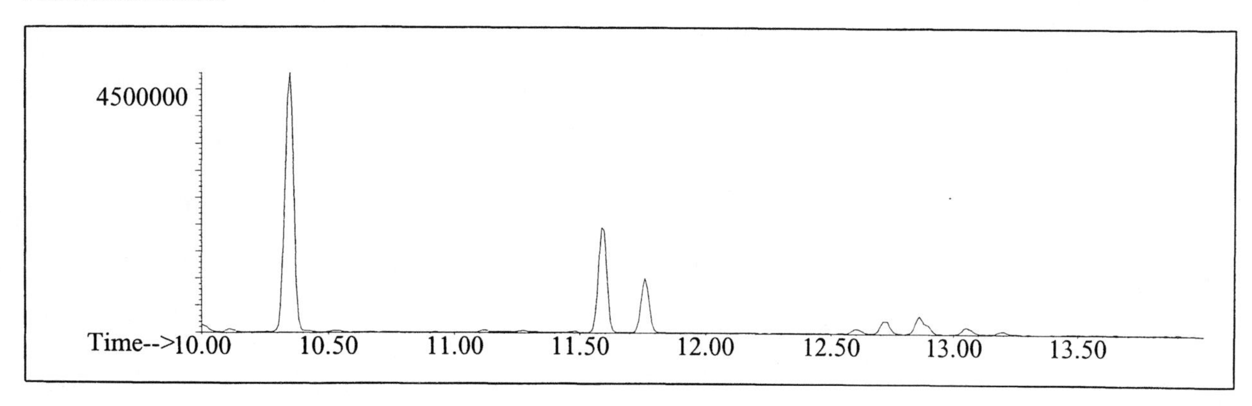

INDIVIDUAL PROFILES

Gasoline #17

Alkane

Ion 43

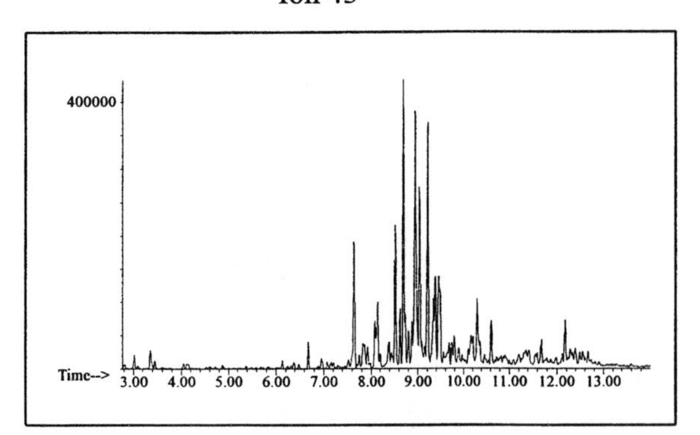

Ion 57

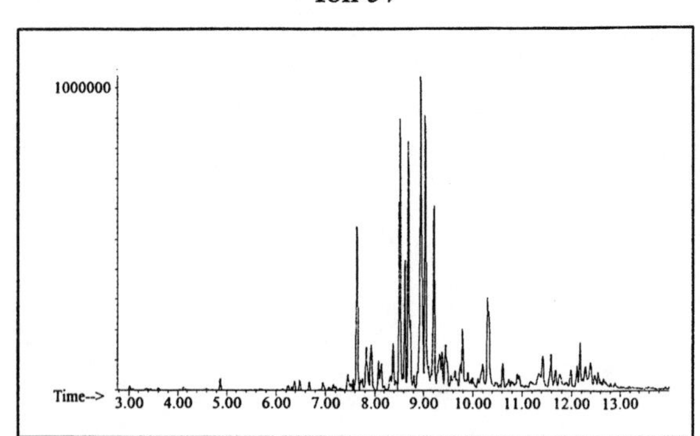

Ion 71

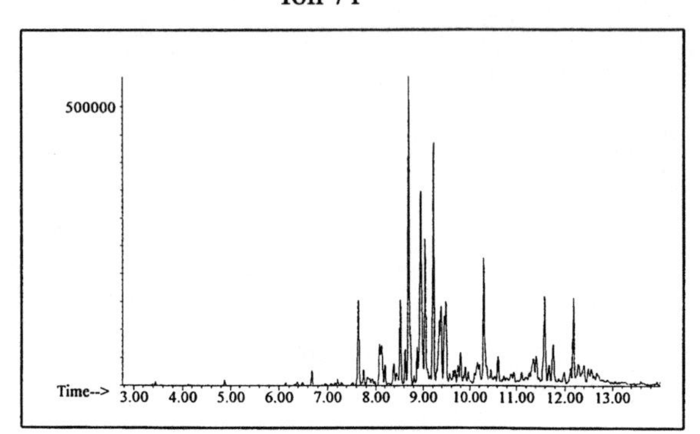

Ion 85

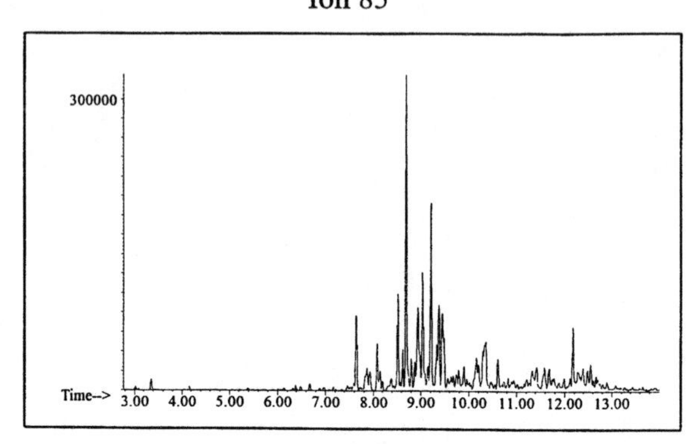

Aromatic

Ion 91

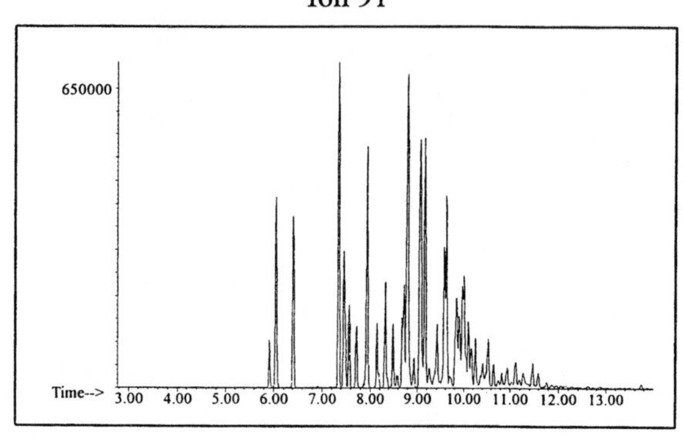

Ion 105

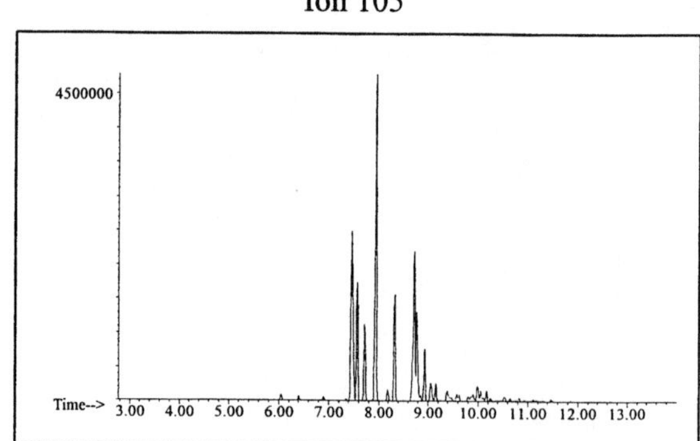

Ion 119

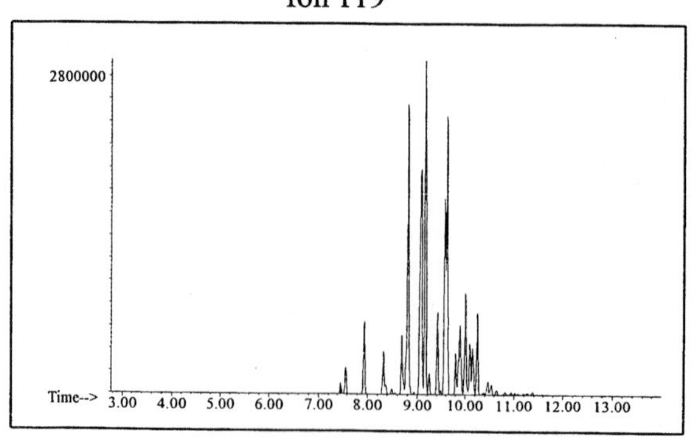

Cycloparaffin

Ion 55

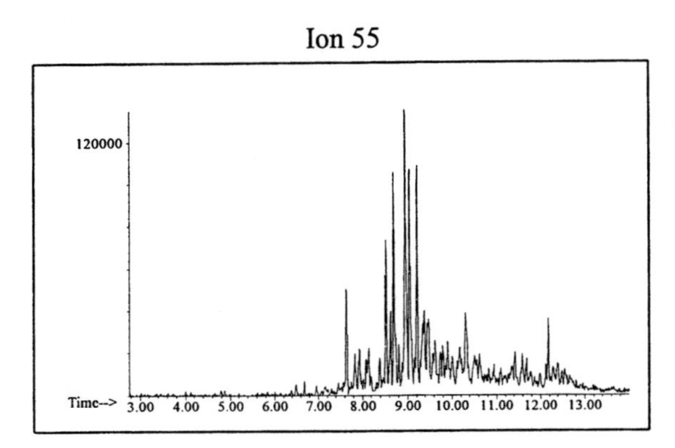

Ion 69

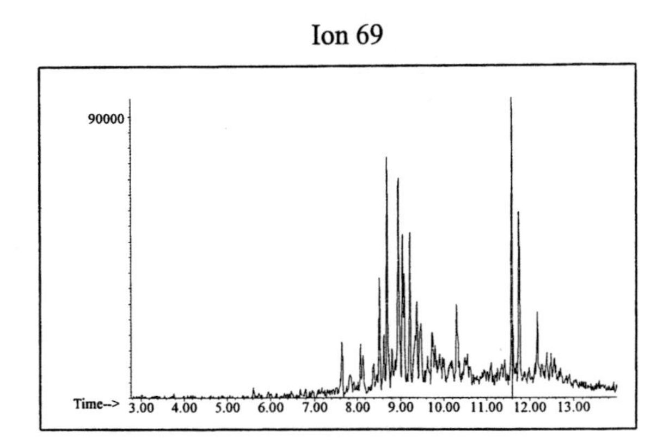

Ion 83

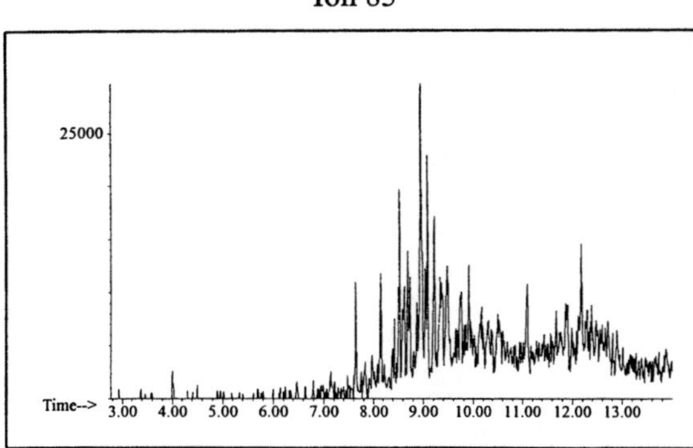

Naphthalene

Ion 128

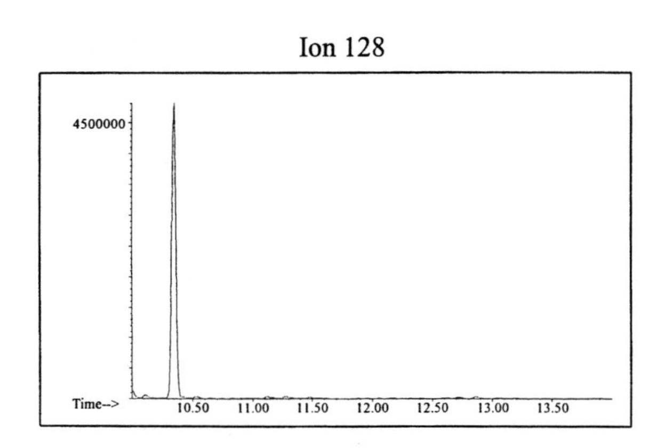

Ion 142

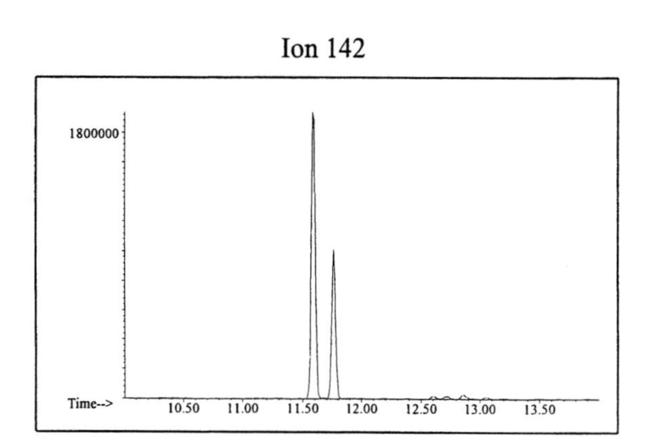

Ion 156

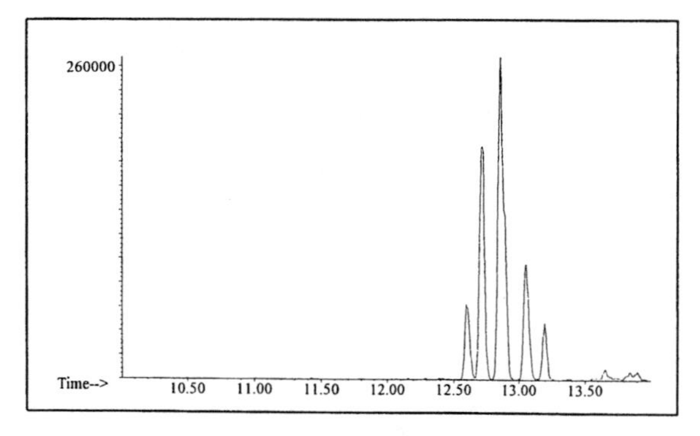

Gasoline #18

20uL/mL Pentane

Product Displayed: 98% Evaporated Shell Premium Gasoline

Other Similar Products: Other brands and grades of gasoline

ASTM: Class 2 (Gasoline)

Product Uses: Automotive fuel

Macro Code: M

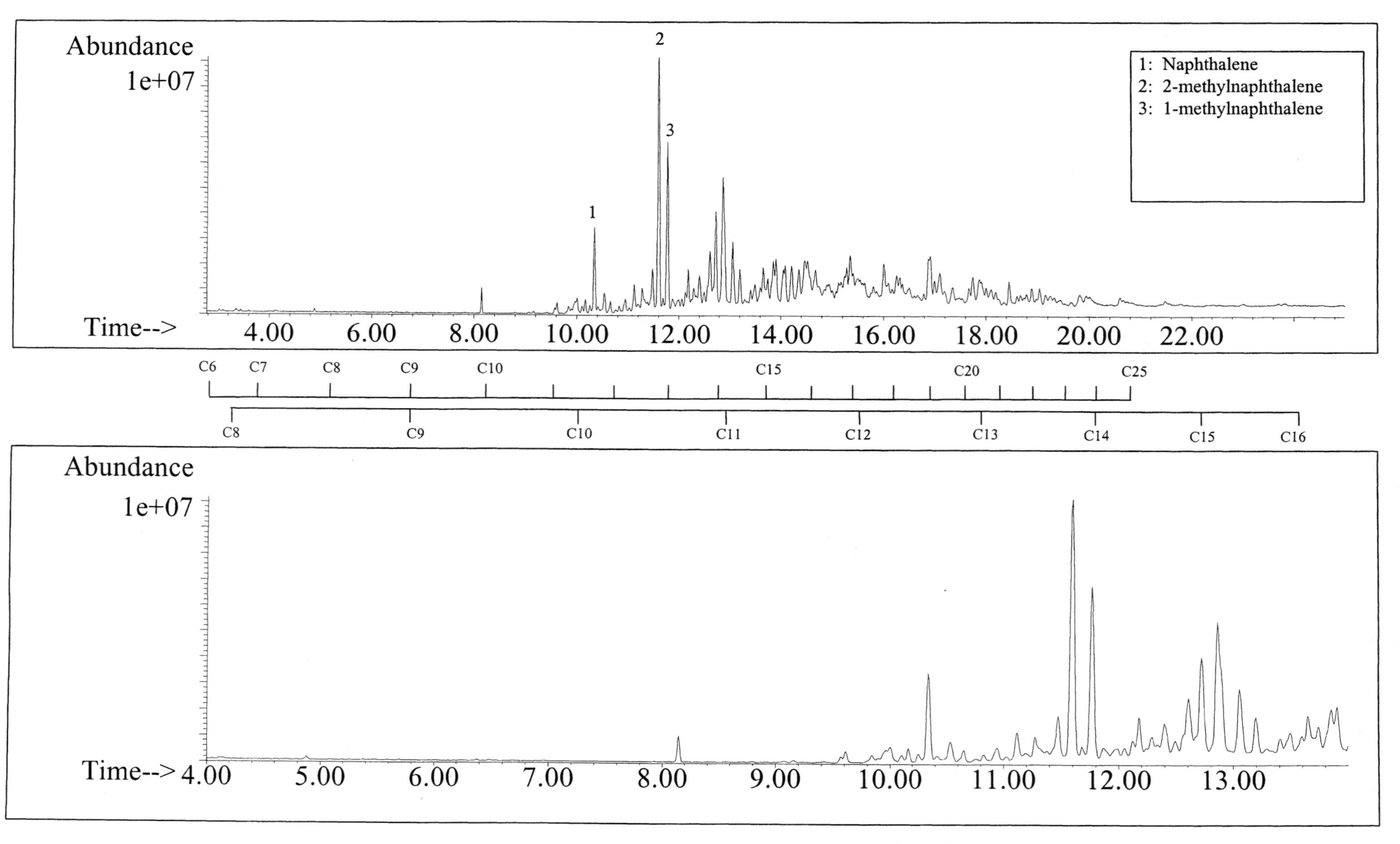

SUMMED PROFILES Gasoline #18

ALKANES

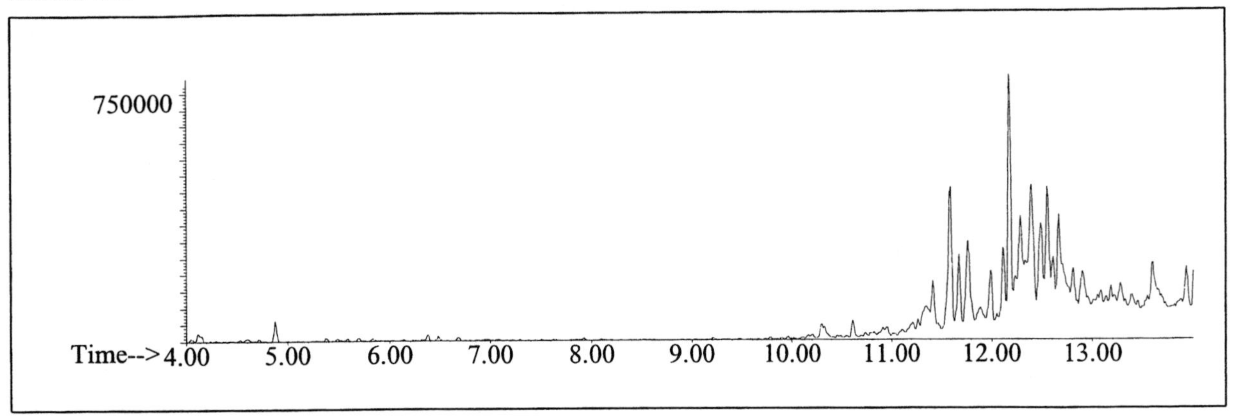

AROMATICS

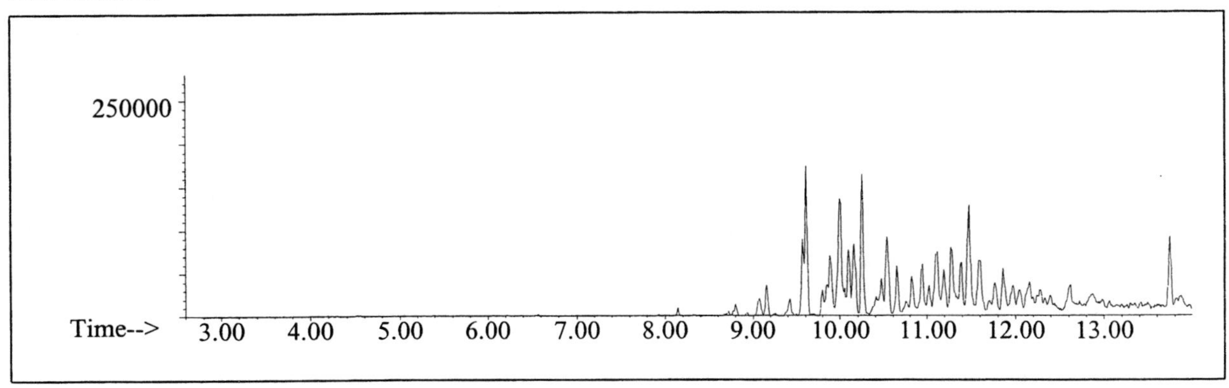

CYCLOPARAFFINS AND ALKENES

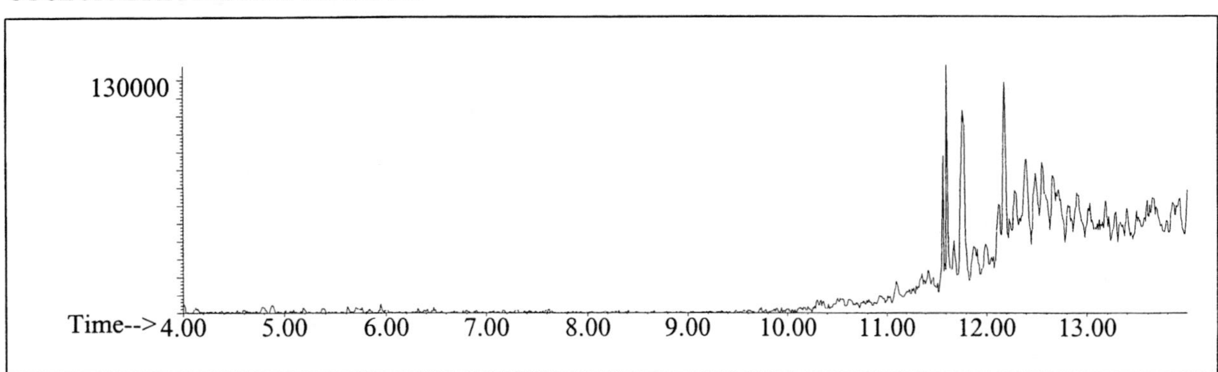

NAPHTHALENES

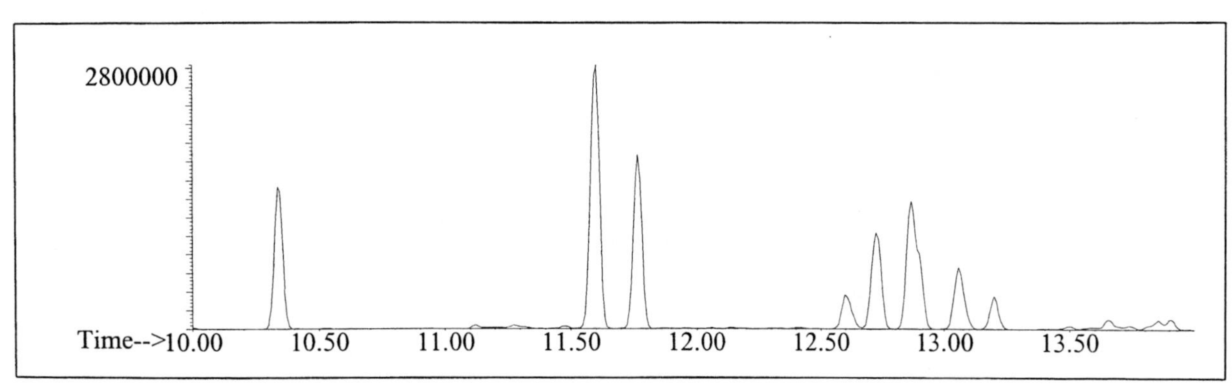

INDIVIDUAL PROFILES

Gasoline #18

Alkane

Ion 43

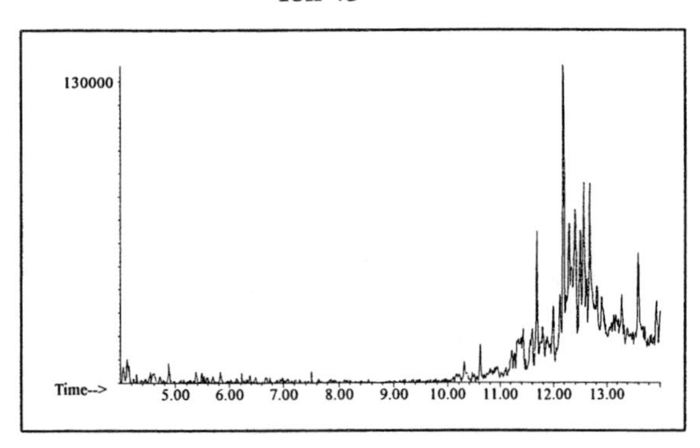

Ion 57

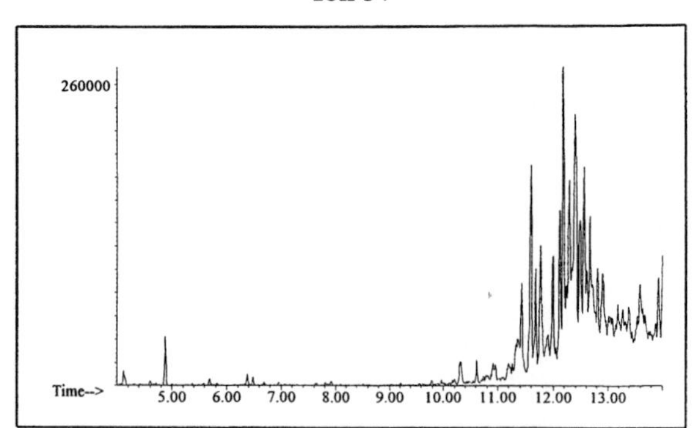

Ion 71

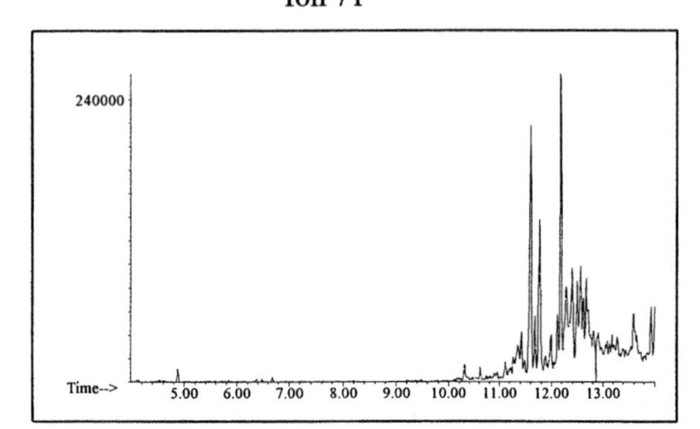

Ion 85

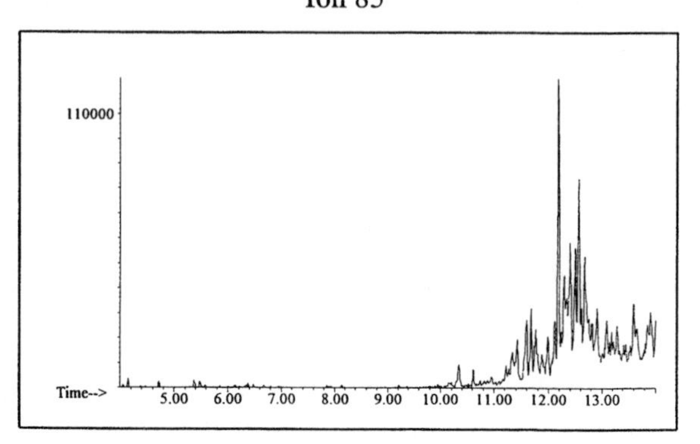

Aromatic

Ion 91

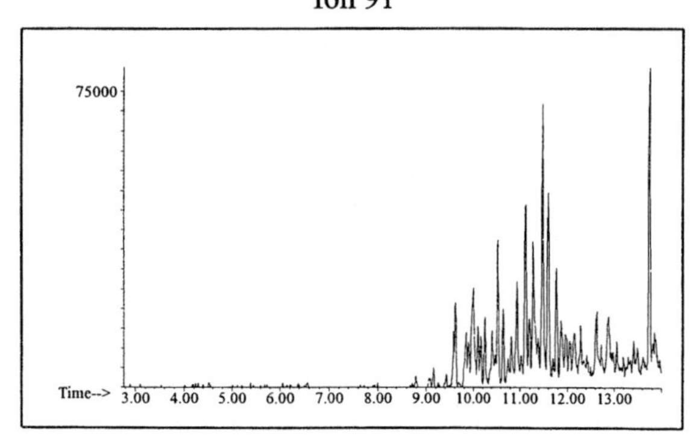

Ion 105

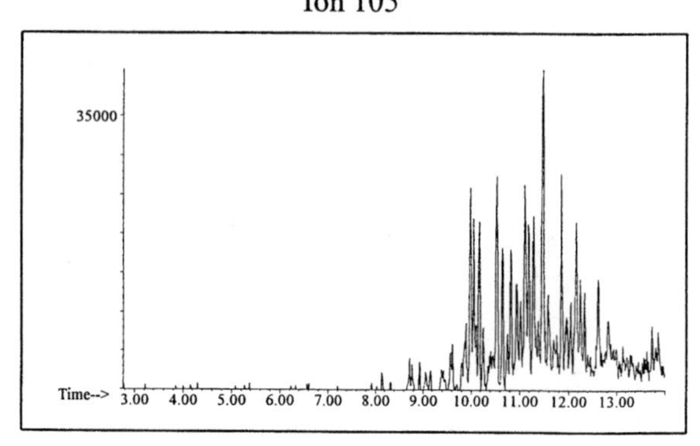

Ion 119

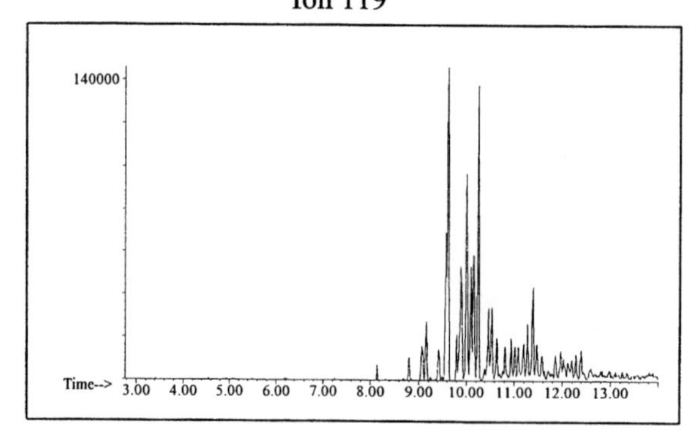

INDIVIDUAL PROFILES

Gasoline #18

Cycloparaffin

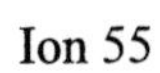

Ion 55

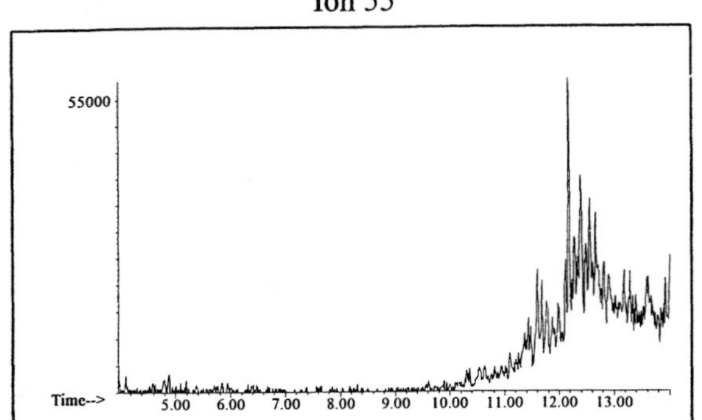

Ion 69

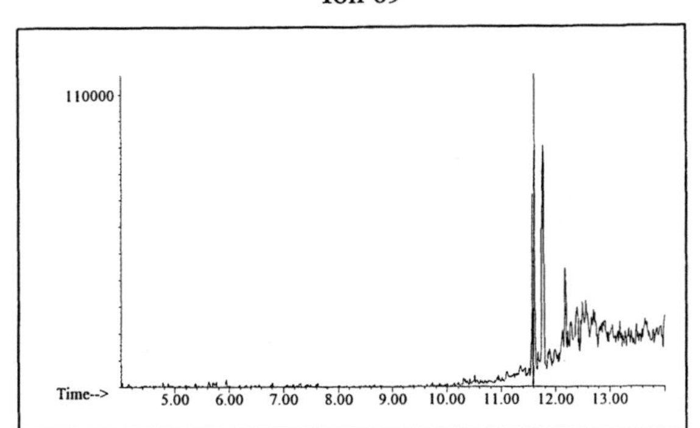

Ion 83

No Useful Data Obtained

Naphthalene

Ion 128

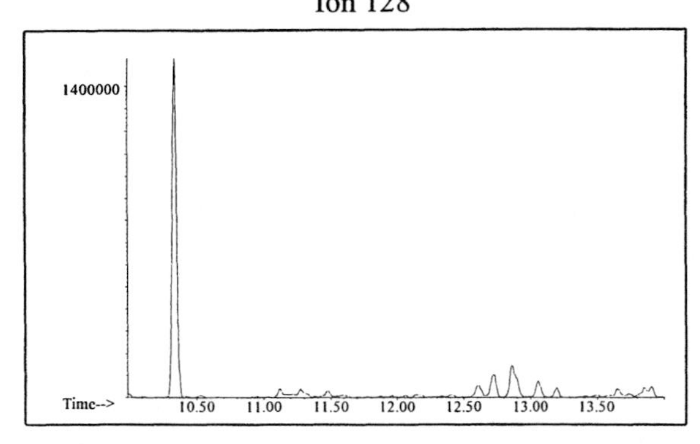

Ion 142

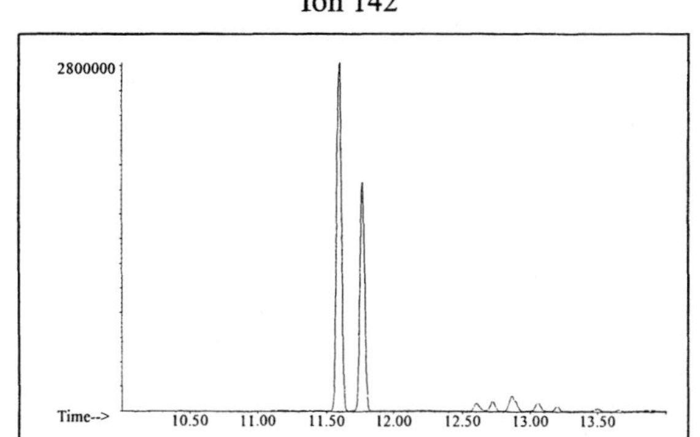

Ion 156

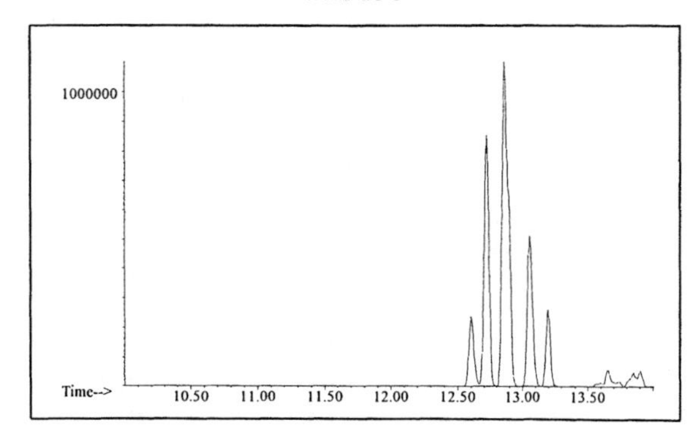

Gasoline #19

20uL/mL Pentane

Product Displayed: Amoco Ultimate Unleaded Gasoline

Other Similar Products: Other grades and brands of gasoline

ASTM: Class 2 (Gasoline)

Product Uses: Automotive fuel

Macro Code: M

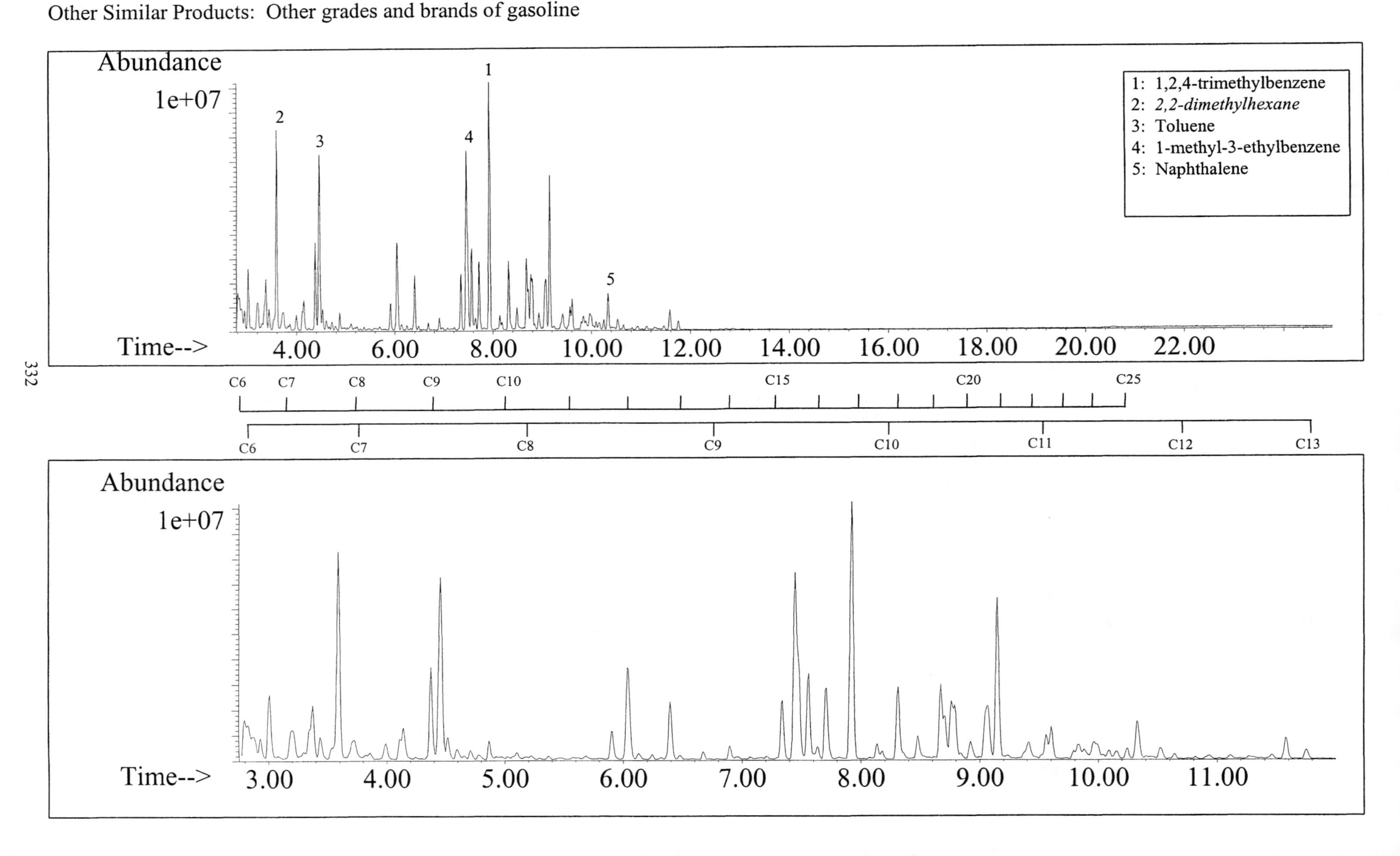

SUMMED PROFILES

Gasoline #19

ALKANES

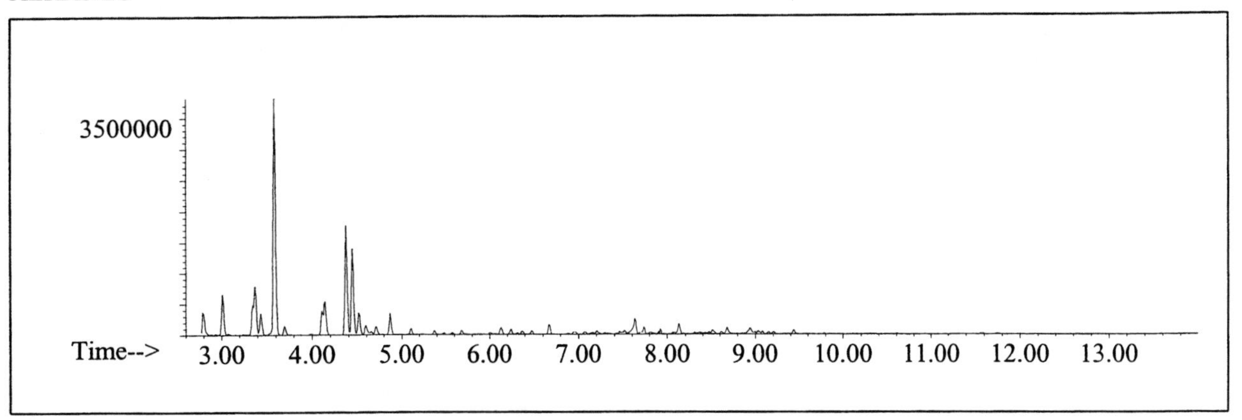

AROMATICS

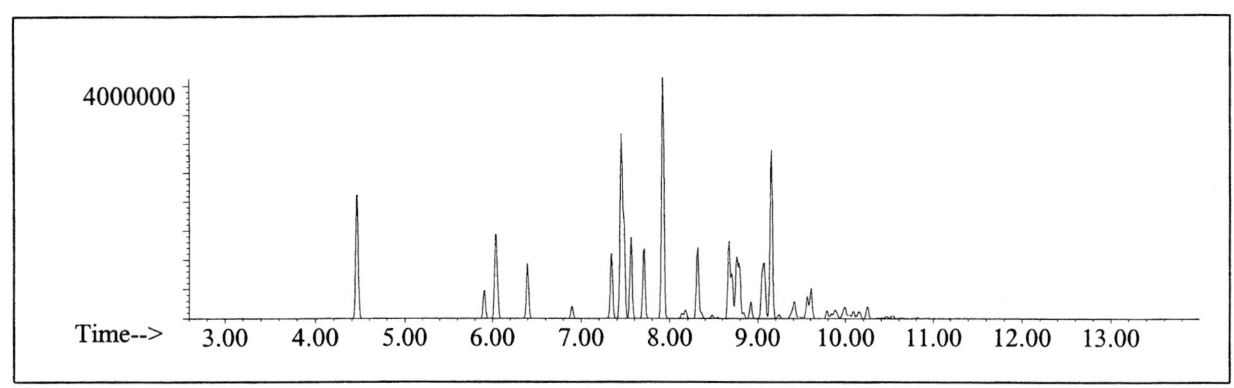

CYCLOPARAFFINS AND ALKENES

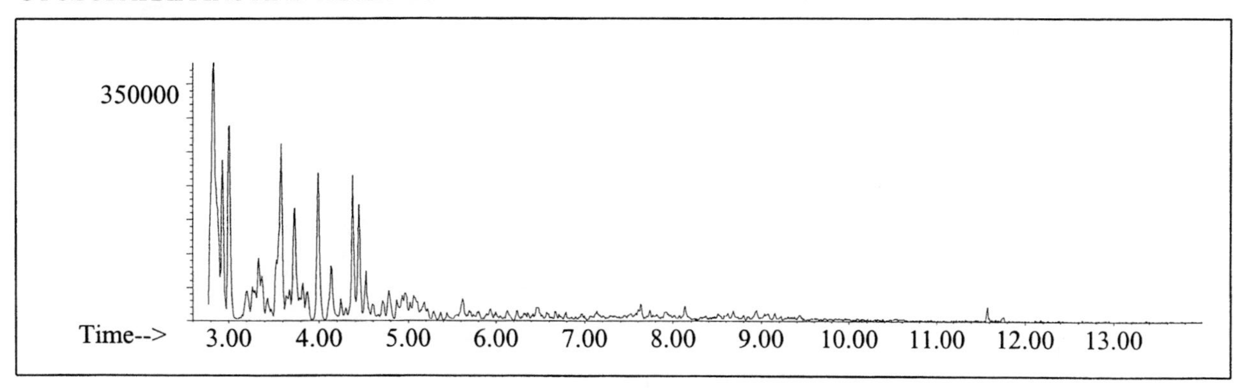

NAPHTHALENES

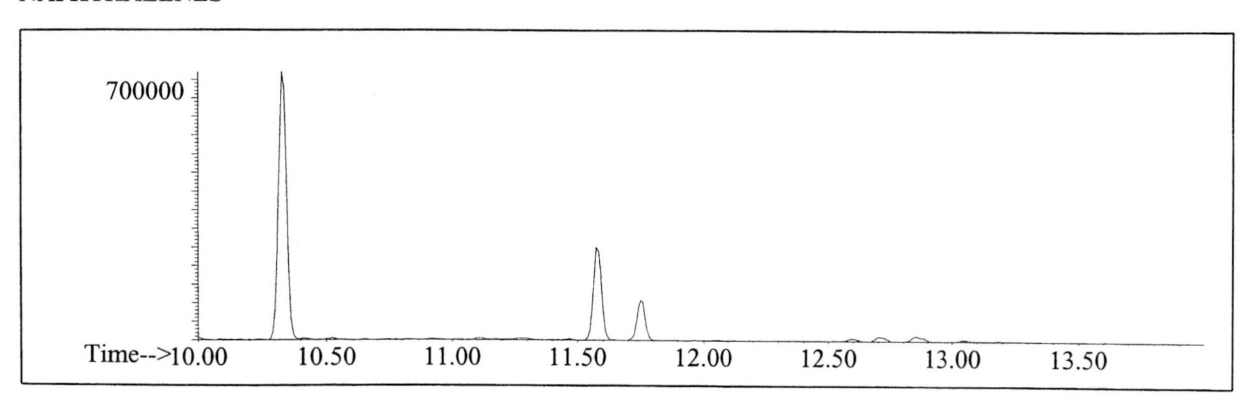

INDIVIDUAL PROFILES

Gasoline #19

Alkane

Ion 43

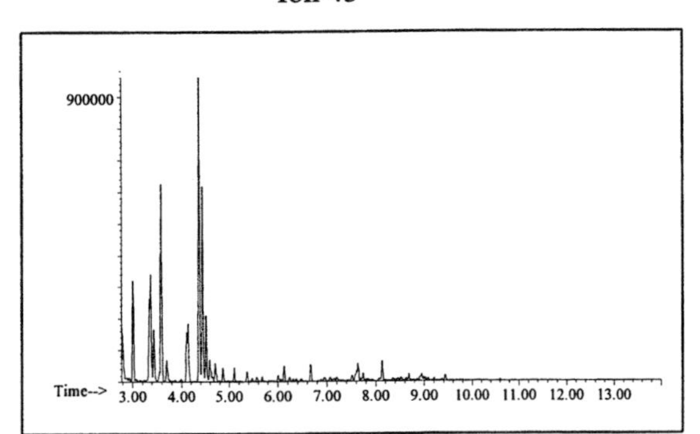

Ion 57

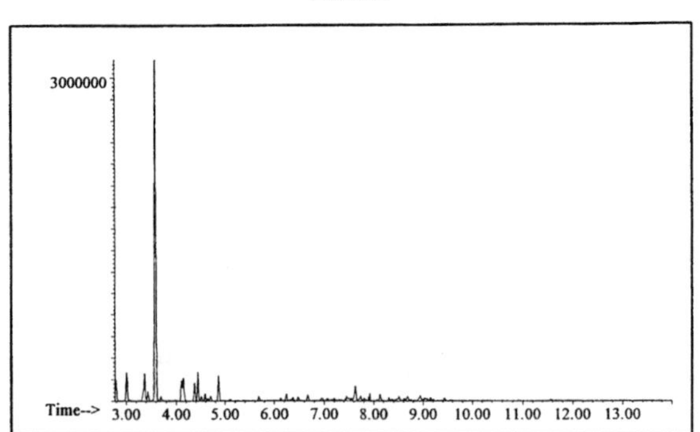

Ion 71

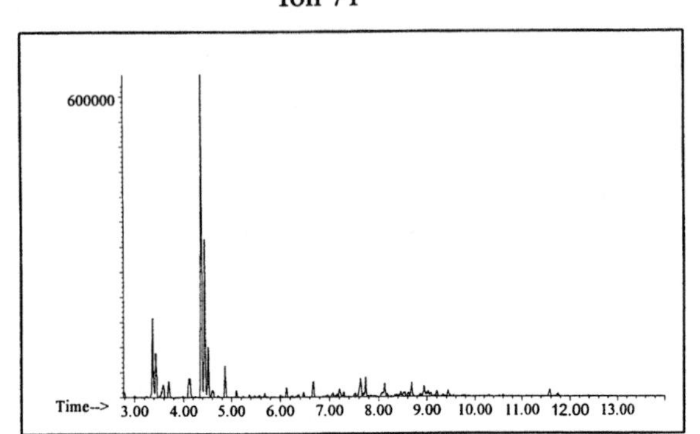

Ion 85

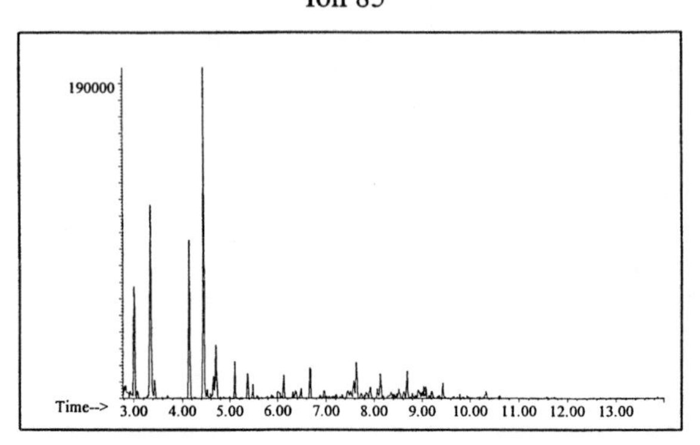

Aromatic

Ion 91

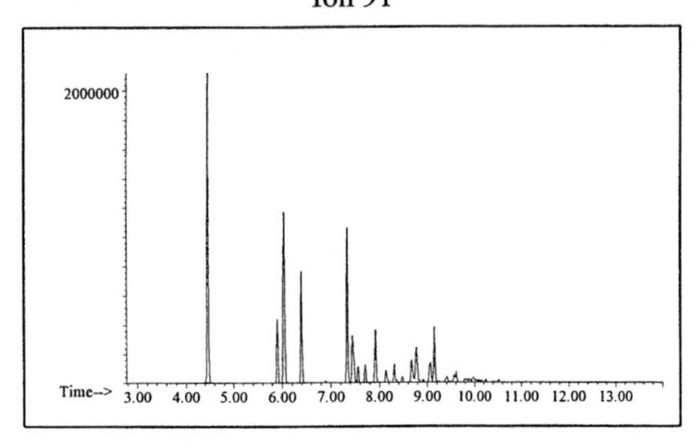

Ion 105

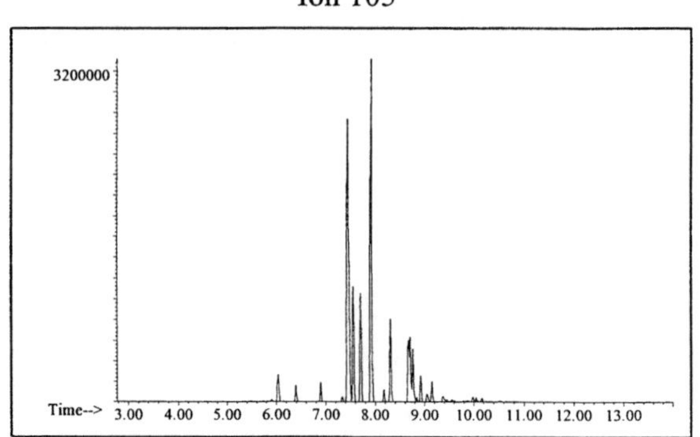

Ion 119

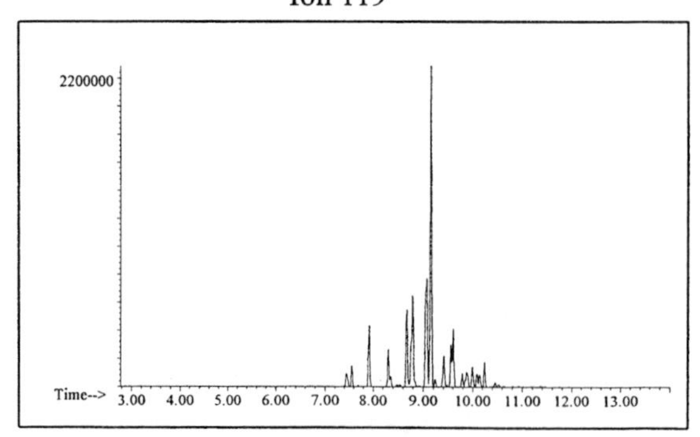

Cycloparaffin

Ion 55

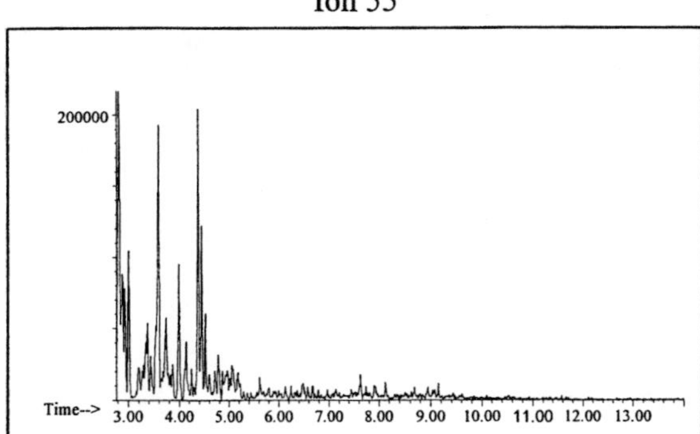

Ion 69

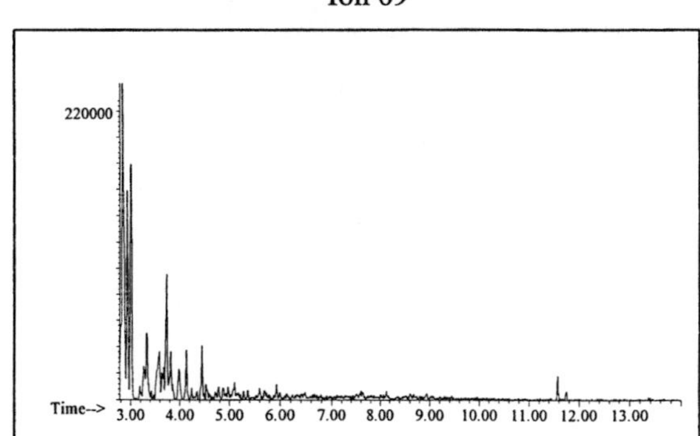

Ion 83

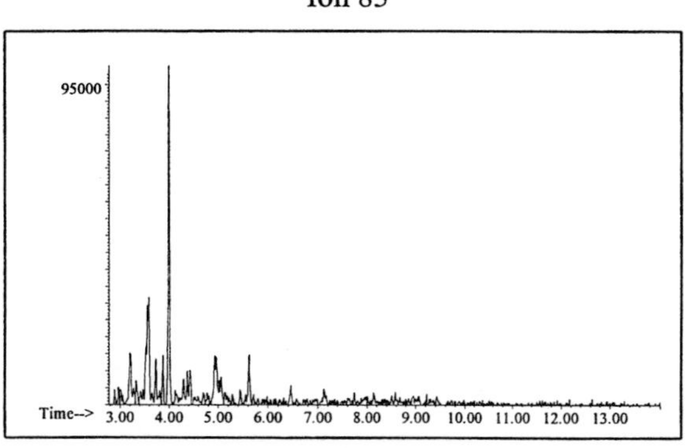

Naphthalene

Ion 128

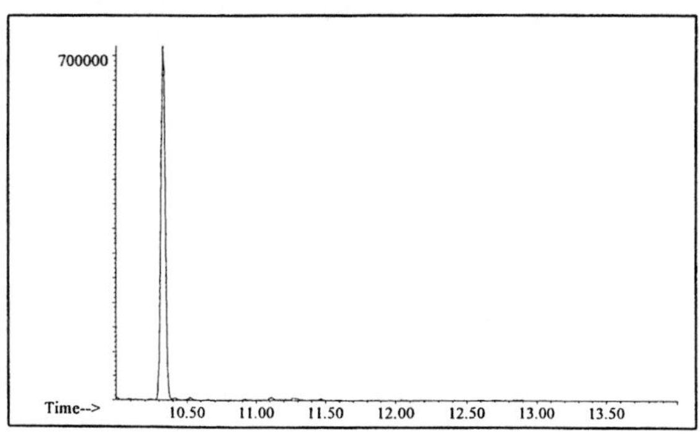

Ion 142

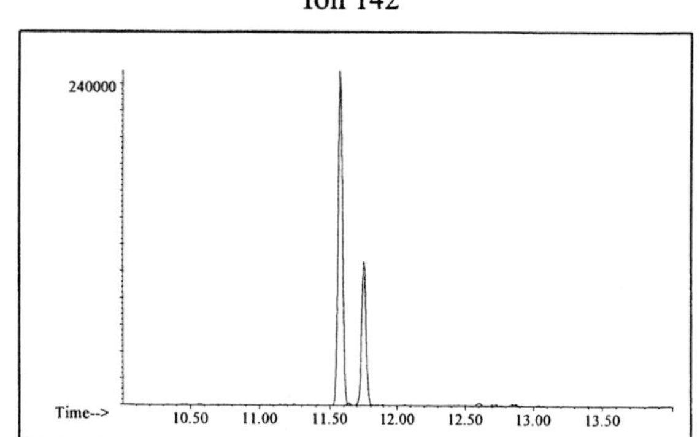

Ion 156

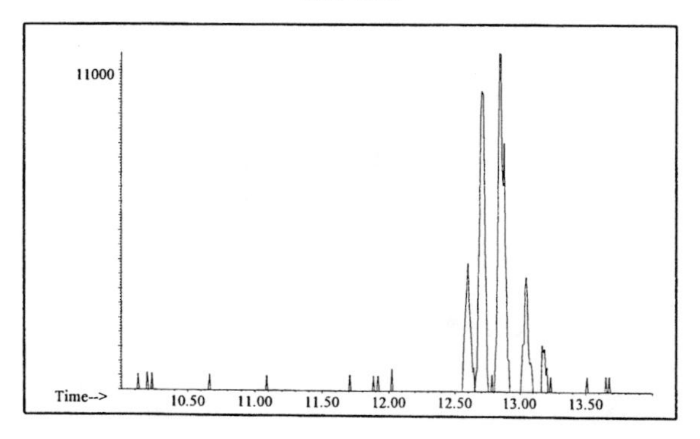

Gasoline #20

20uL/mL Pentane

ASTM: Class 2 (Gasoline)

Macro Code: M

Product Displayed: 25% Evaporated Amoco Ultimate Gasoline

Product Uses: Automotive fuel

Other Similar Products: Other grades and brands of gasoline

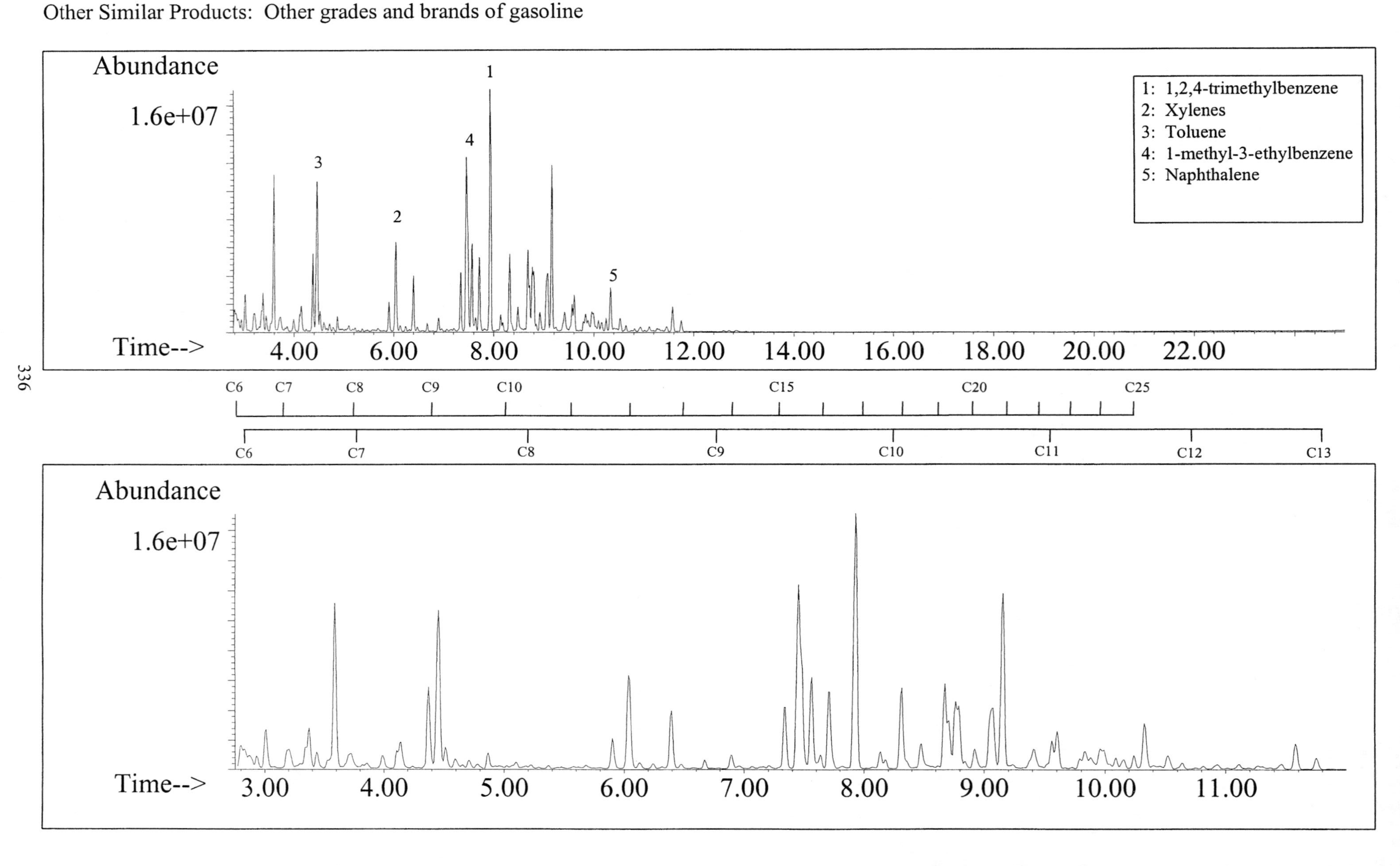

SUMMED PROFILES Gasoline #20

ALKANES

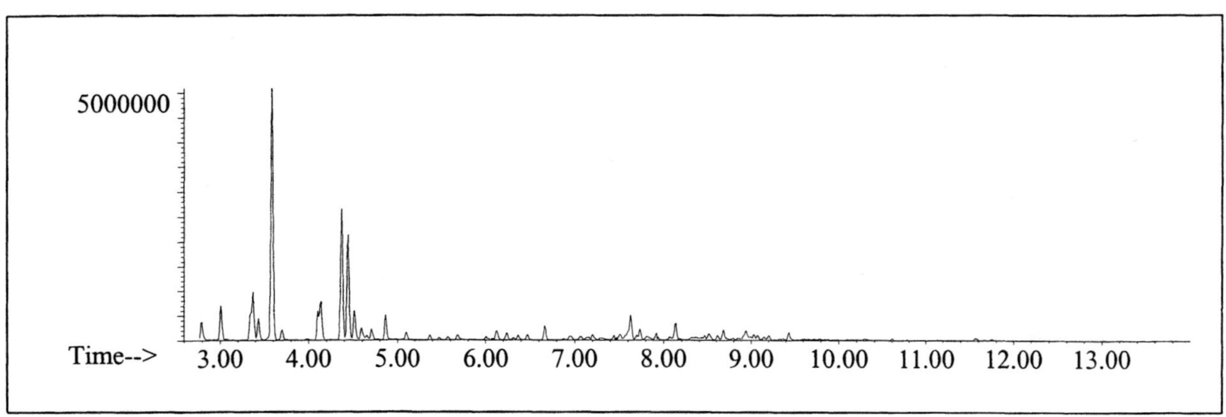

AROMATICS

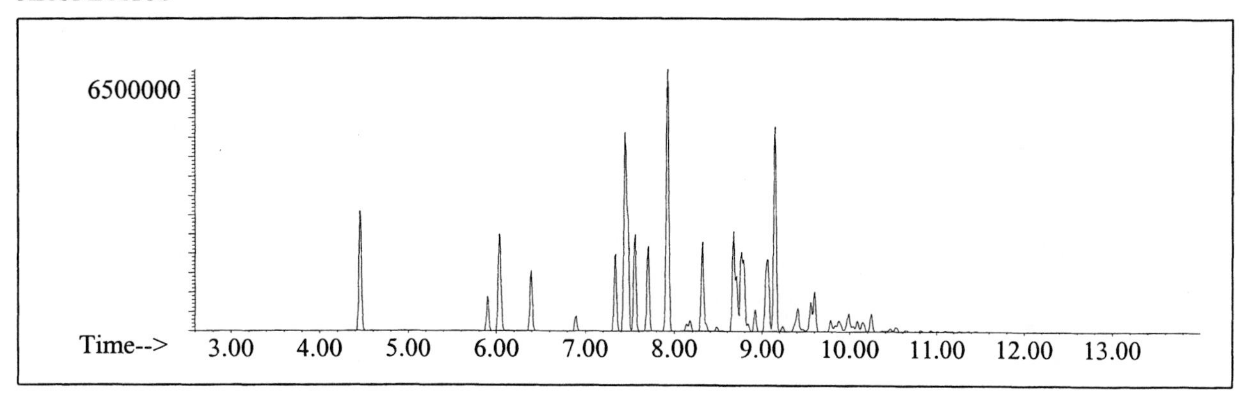

CYCLOPARAFFINS AND ALKENES

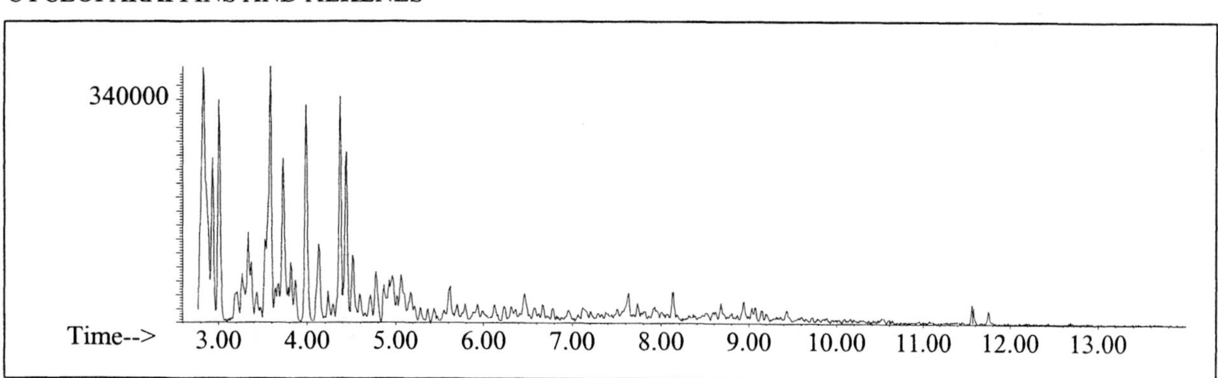

NAPHTHALENES

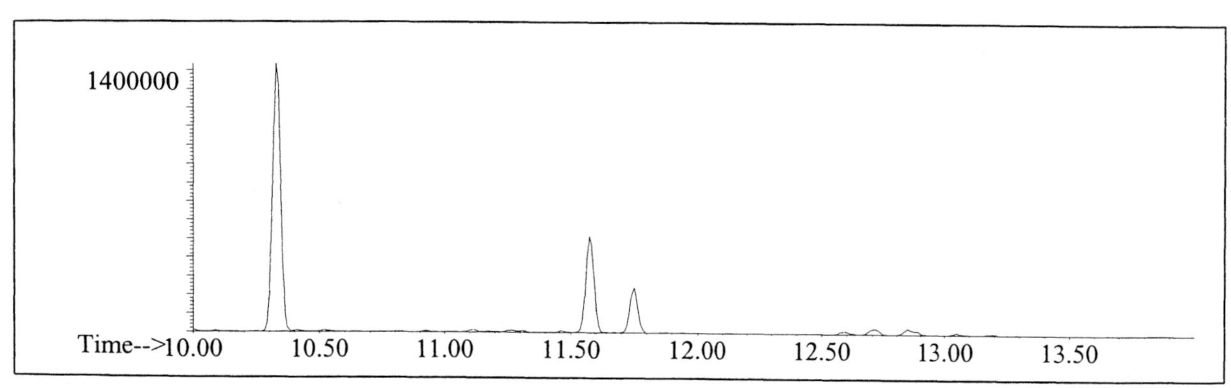

INDIVIDUAL PROFILES

Gasoline #20

Alkane

Ion 43

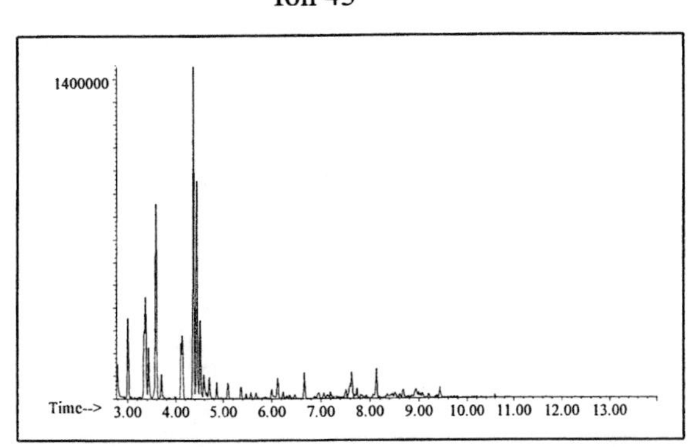

Ion 57

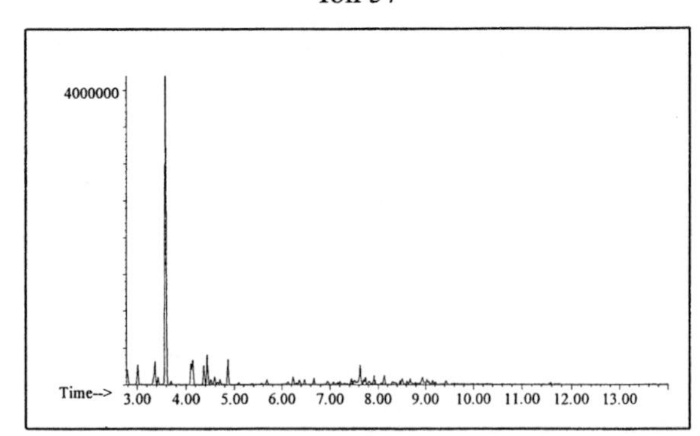

Ion 71

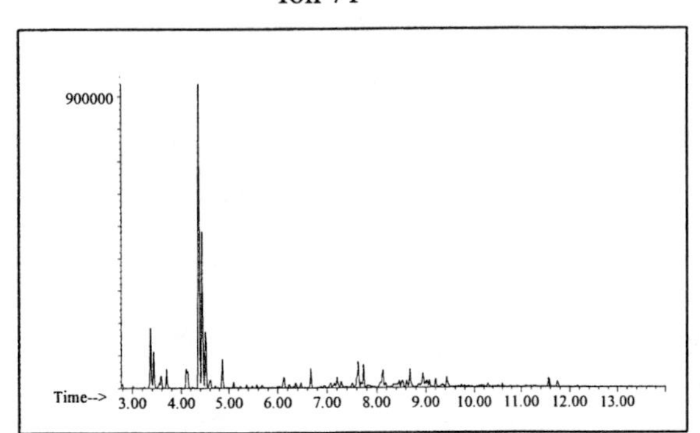

Ion 85

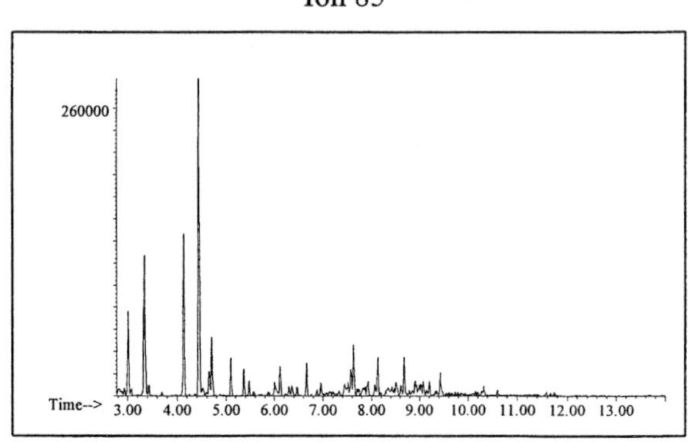

Aromatic

Ion 91

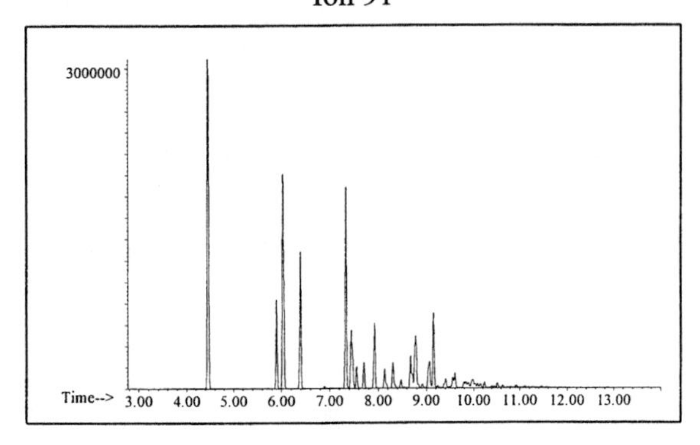

Ion 105

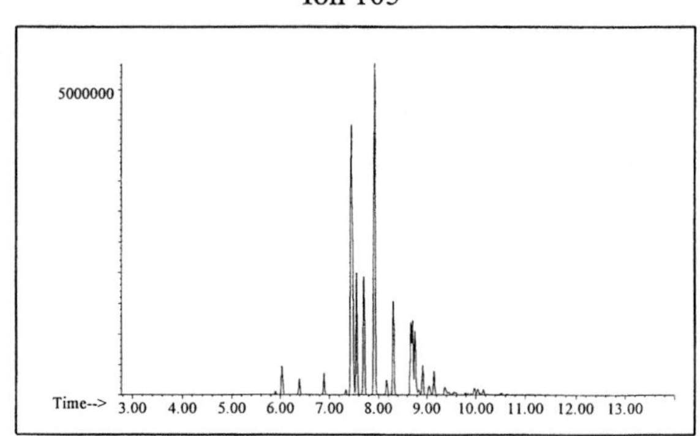

Ion 119

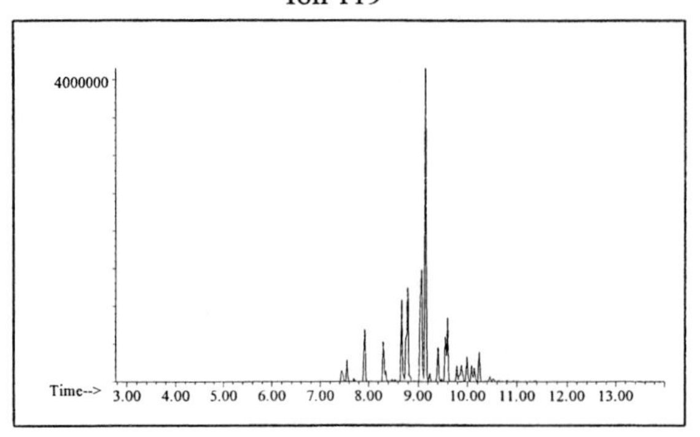

Cycloparaffin

Ion 55

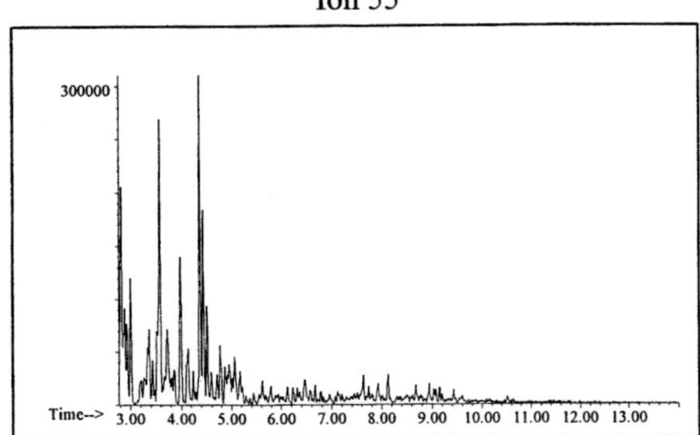

Ion 69

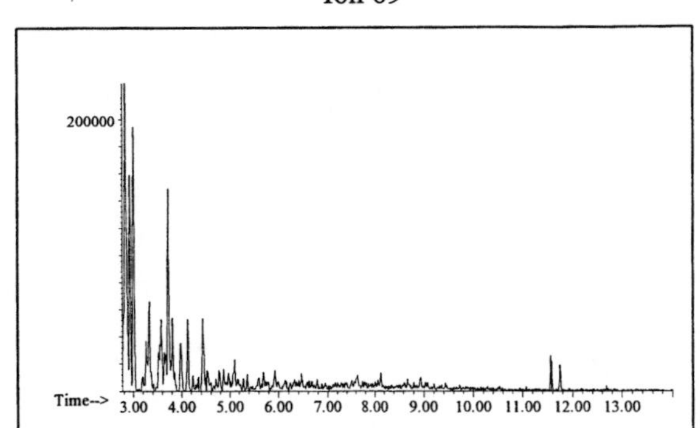

Ion 83

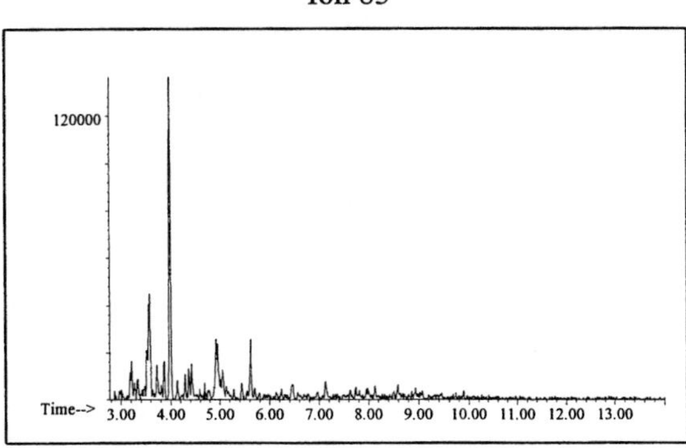

Naphthalene

Ion 128

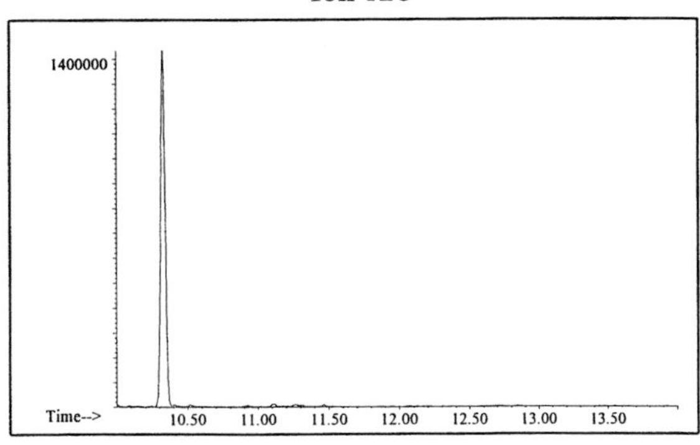

Ion 142

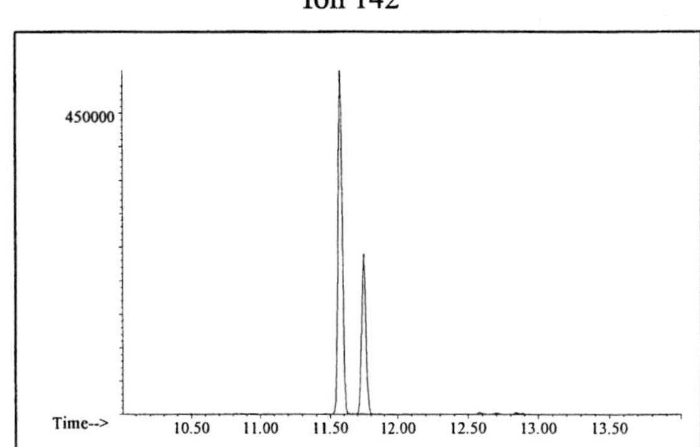

Ion 156

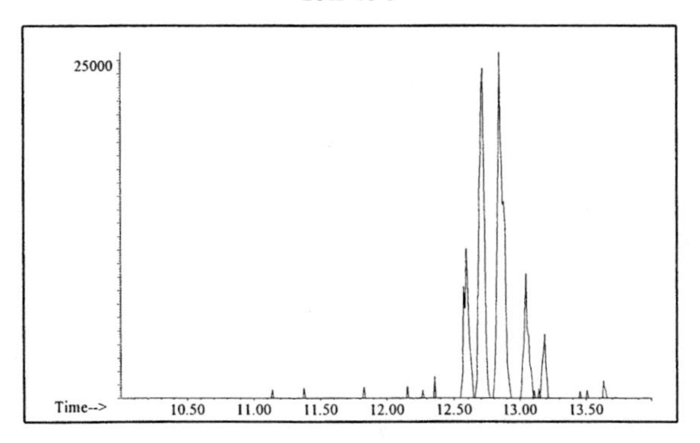

Gasoline #21

20uL/mL Pentane

Product Displayed: 50% Evaporated Amoco Ultimate Gasoline

Other Similar Products: Other grades and brands of gasoline

ASTM: Class 2 (Gasoline)

Product Uses: Automotive fuel

Macro Code: M

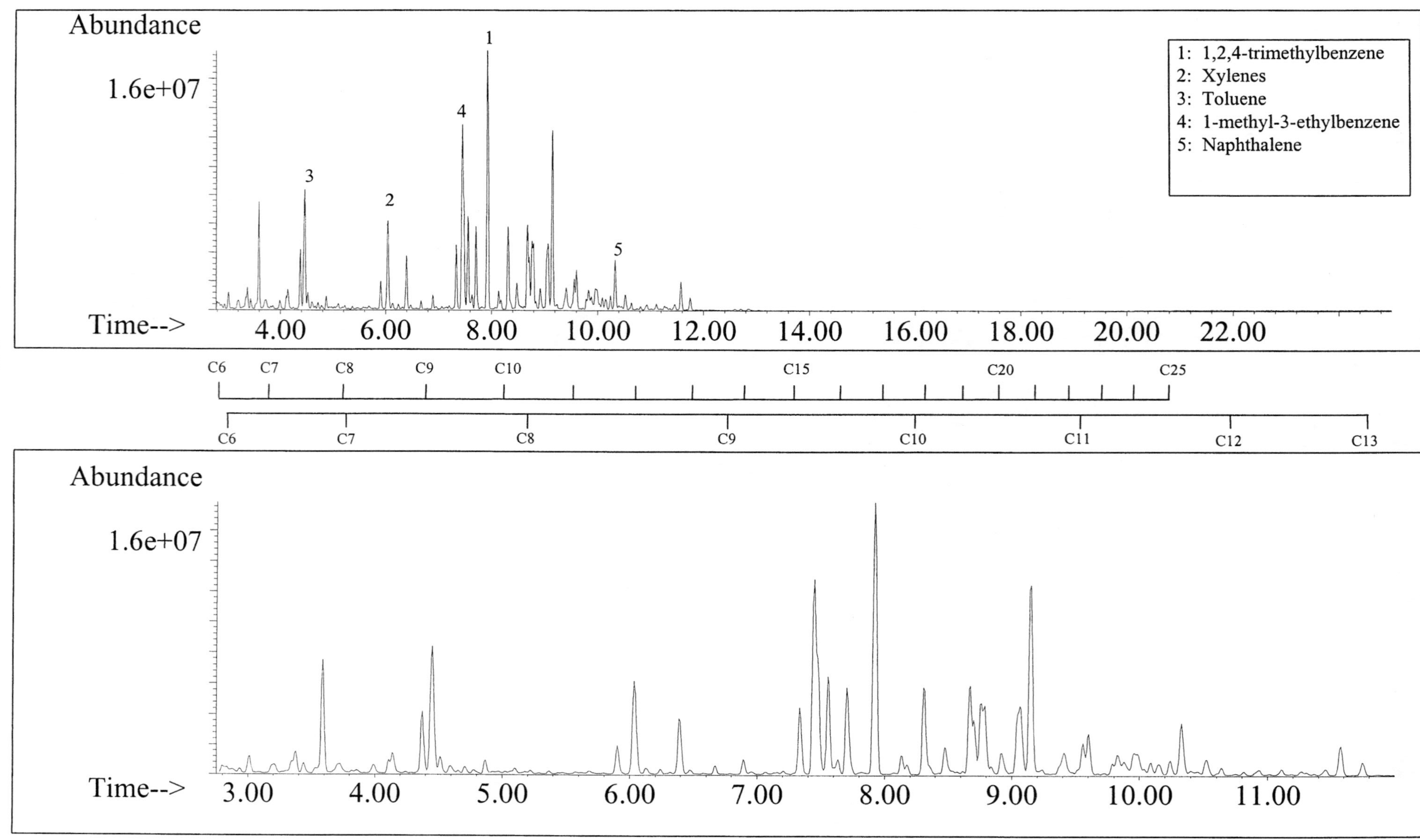

SUMMED PROFILES Gasoline #21

ALKANES

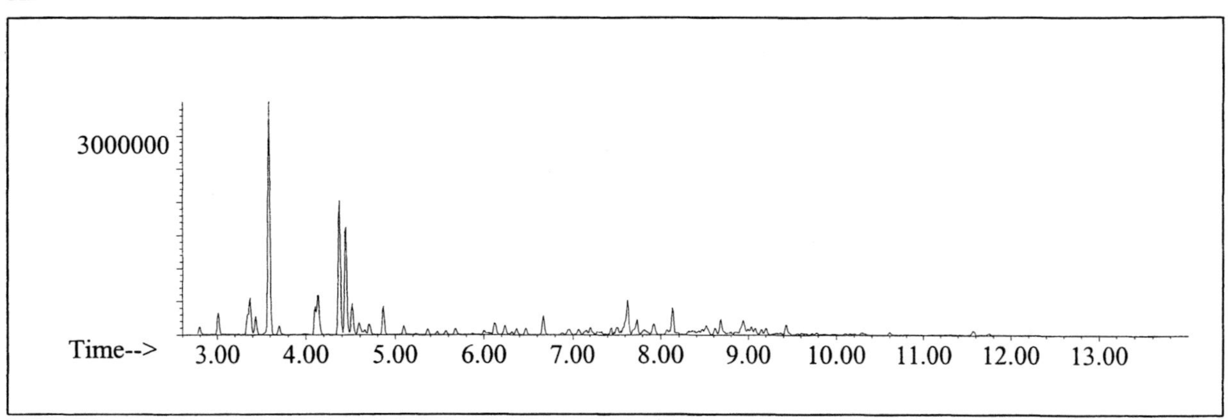

AROMATICS

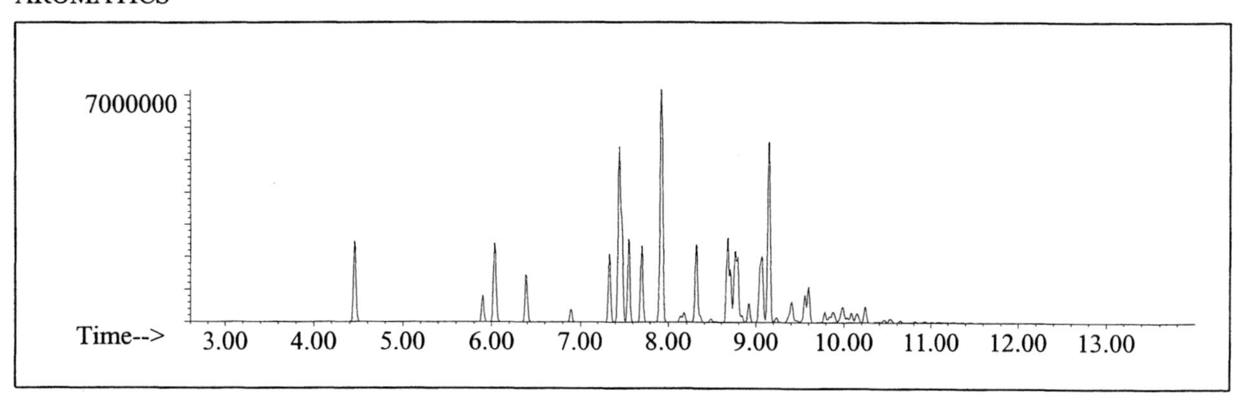

CYCLOPARAFFINS AND ALKENES

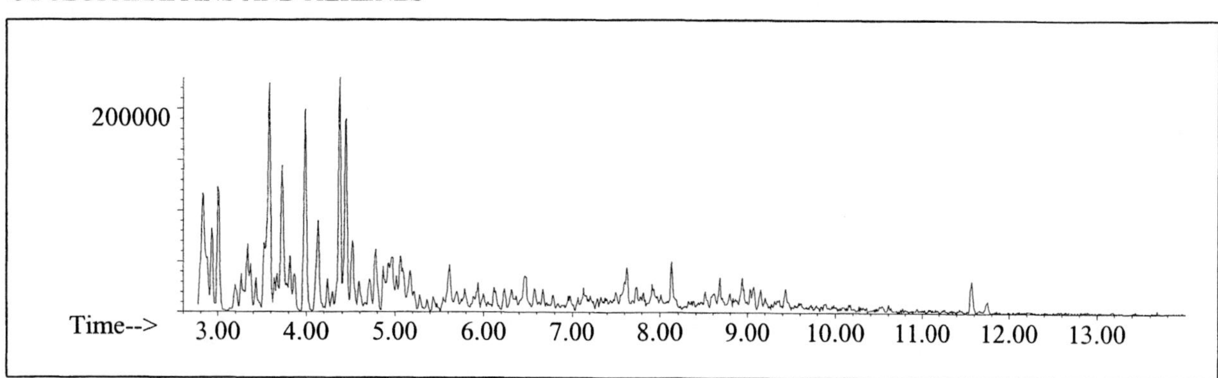

NAPHTHALENES

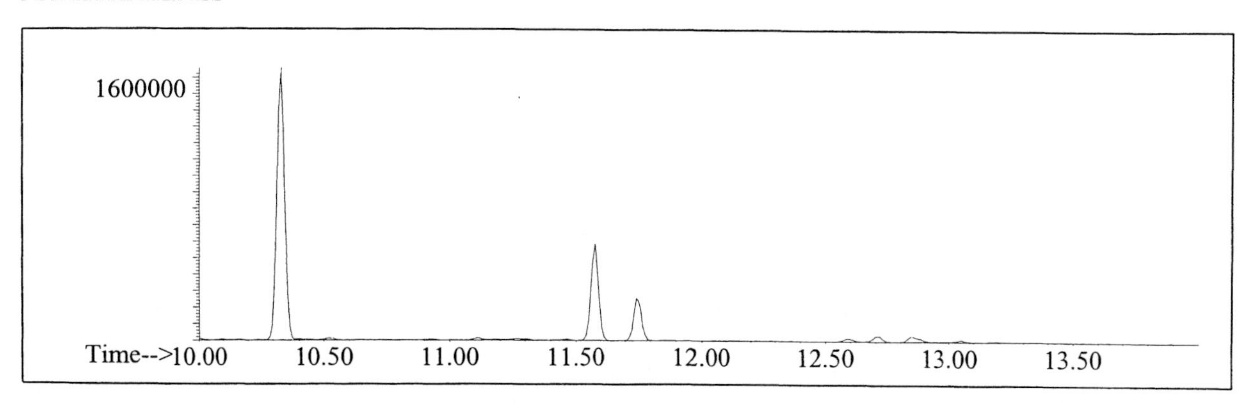

Alkane

Ion 43

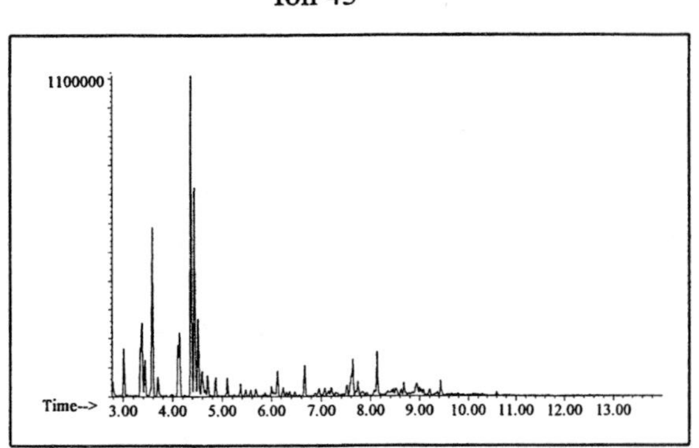

Ion 57

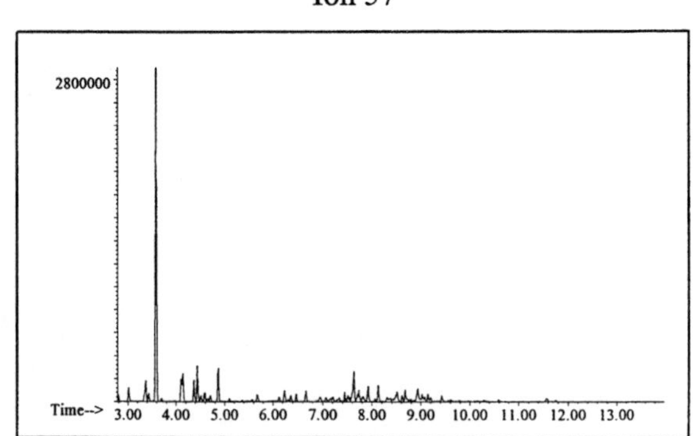

Ion 71

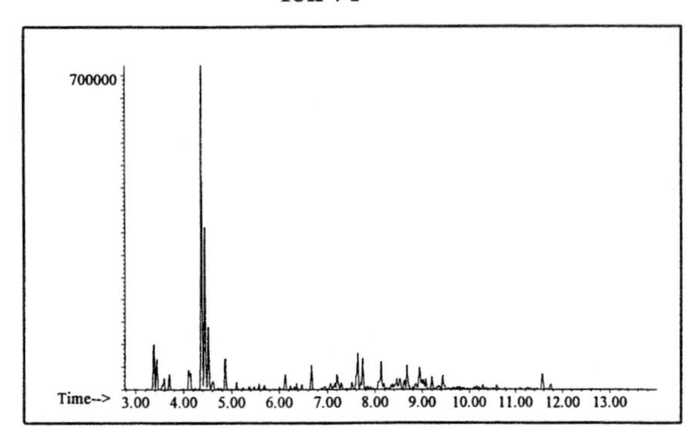

Ion 85

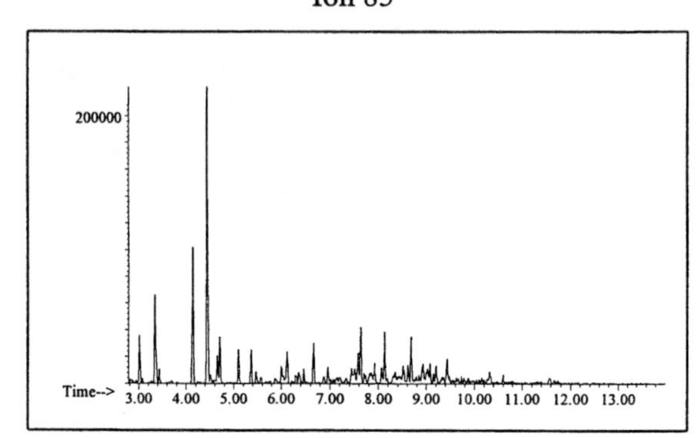

Aromatic

Ion 91

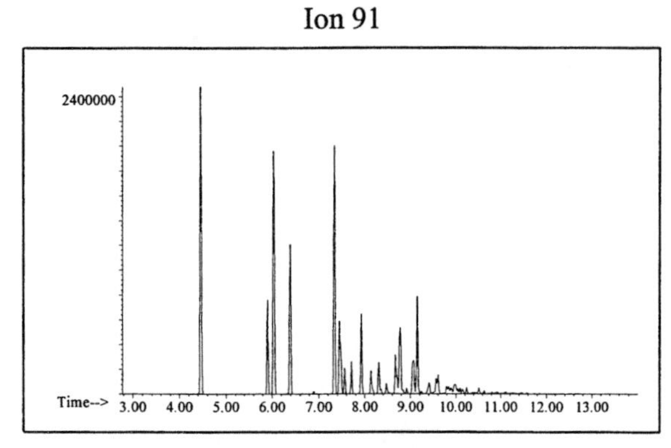

Ion 105

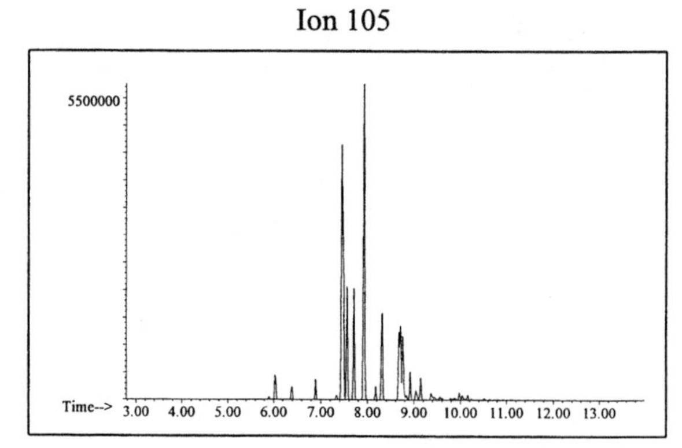

Ion 119

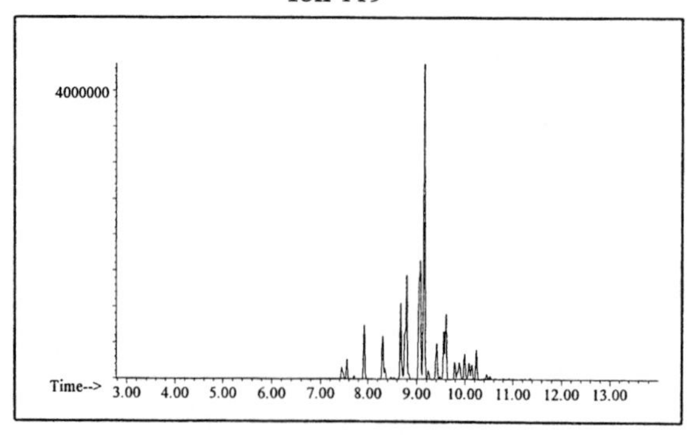

Cycloparaffin

Ion 55

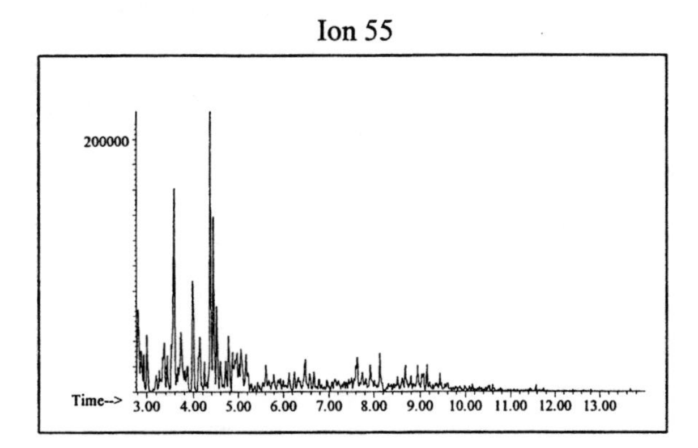

Ion 69

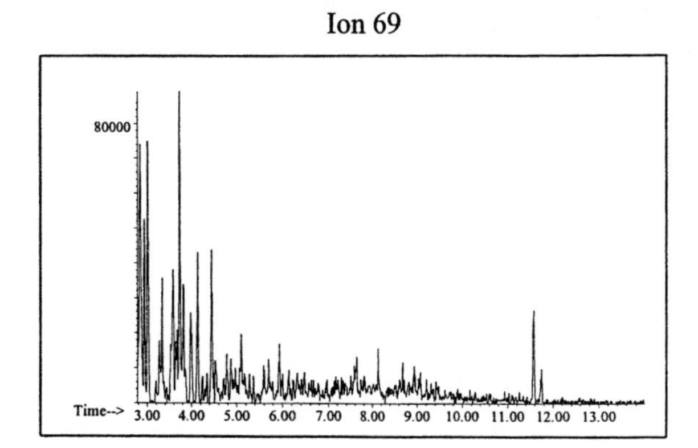

Ion 83

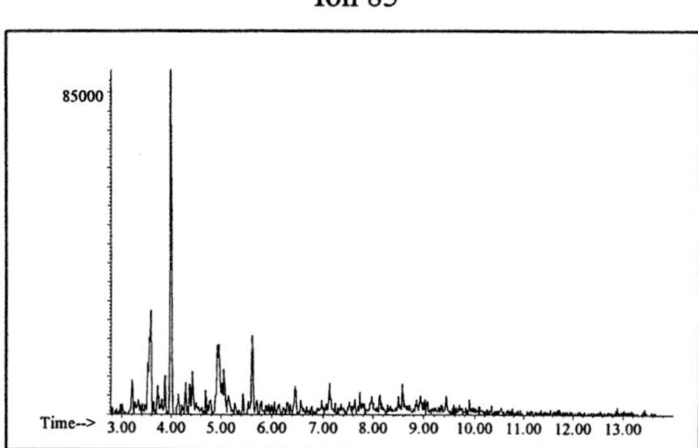

Naphthalene

Ion 128

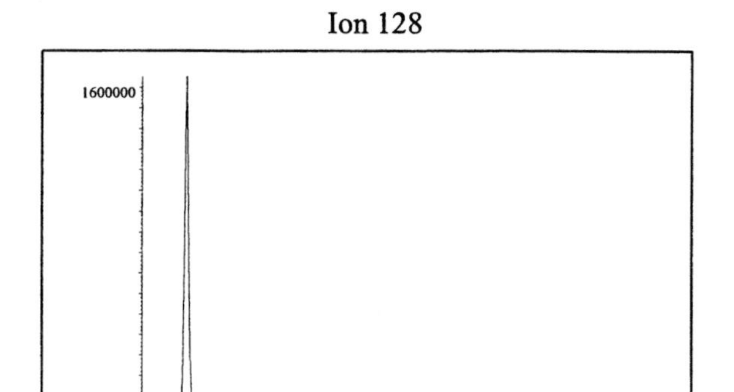

Ion 142

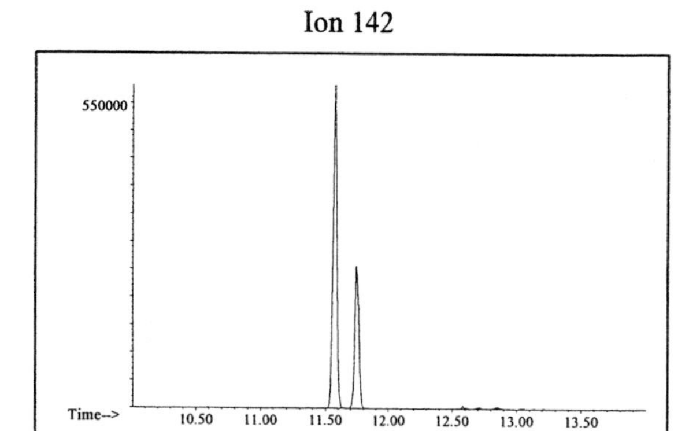

Ion 156

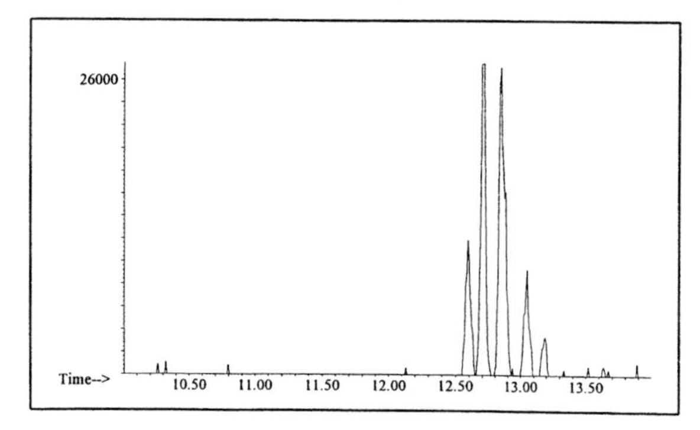

Gasoline #22

20uL/mL Pentane

Product Displayed: 75% Evaporated Amoco Ultimate Gasoline

Other Similar Products: Other grades and brands of gasoline

ASTM: Class 2 (Gasoline)

Product Uses: Automotive fuel

Macro Code: M

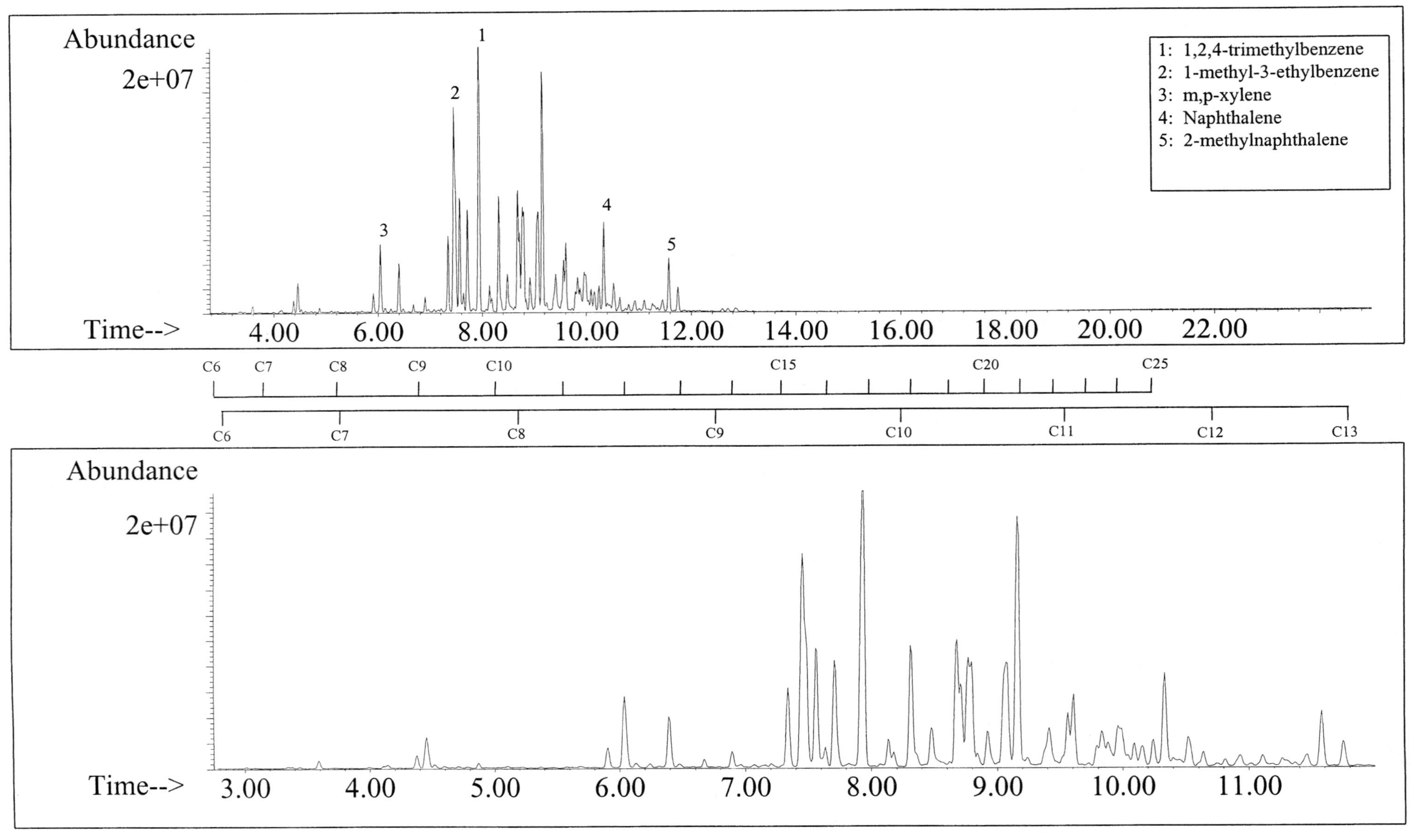

SUMMED PROFILES

Gasoline #22

ALKANES

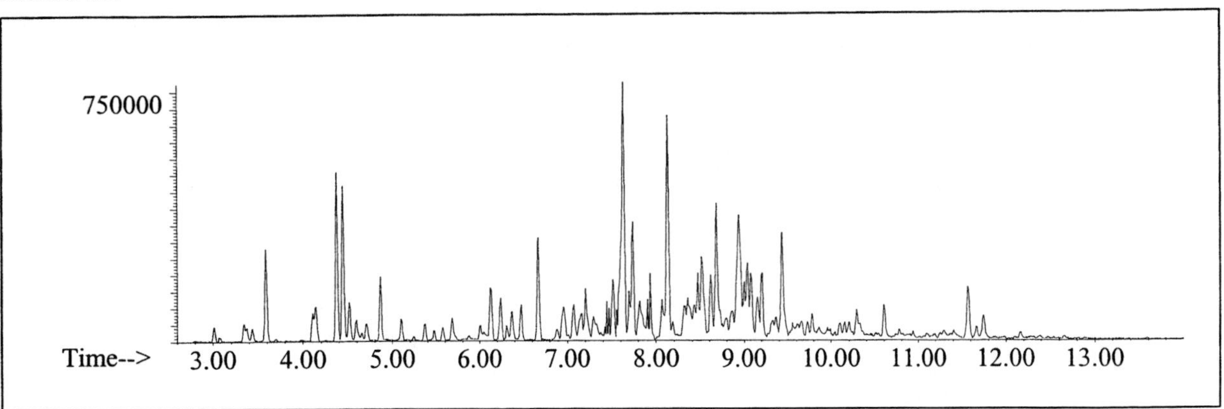

AROMATICS

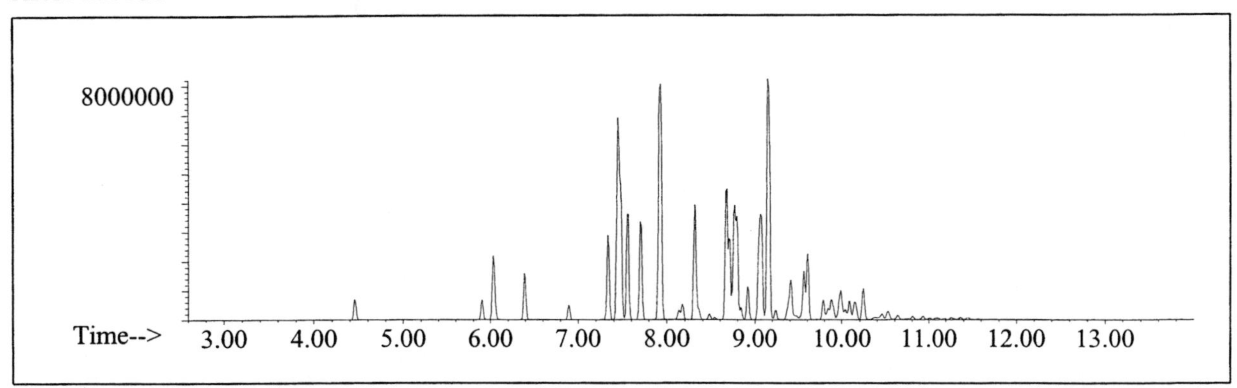

CYCLOPARAFFINS AND ALKENES

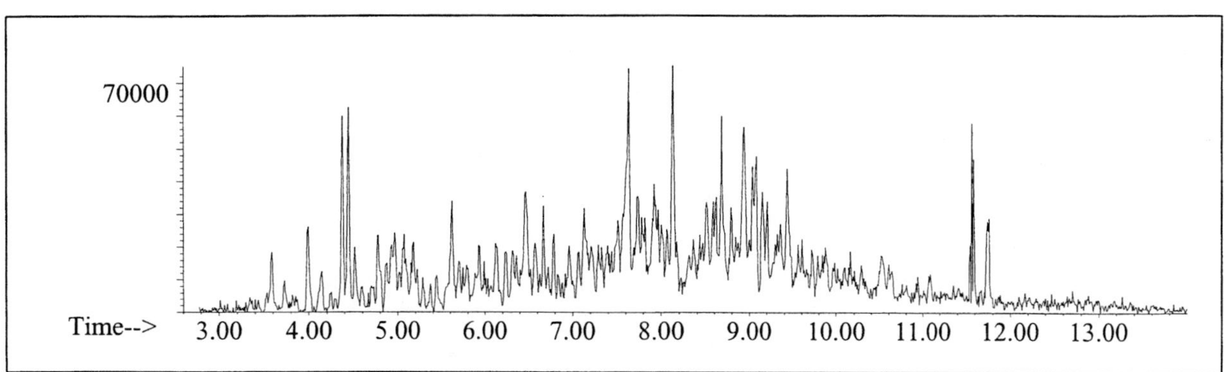

NAPHTHALENES

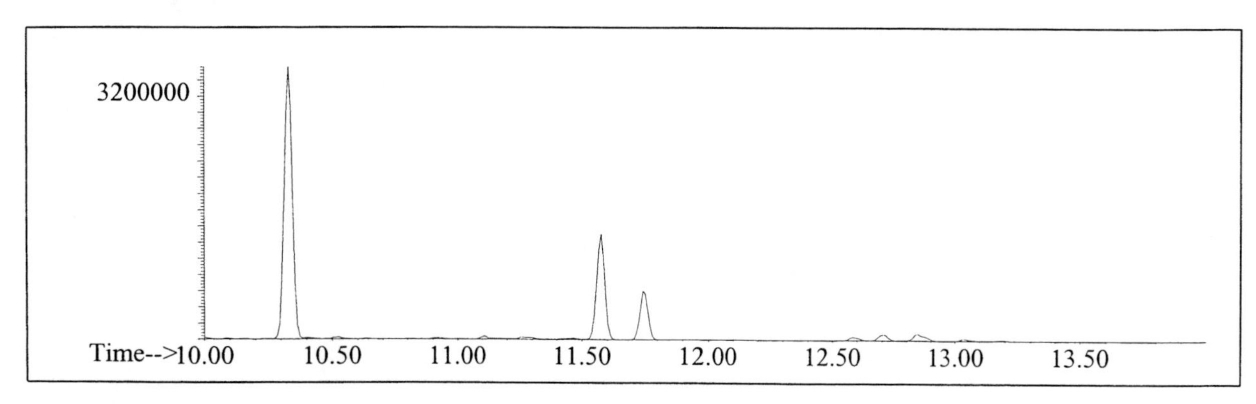

INDIVIDUAL PROFILES

Gasoline #22

Alkane

Ion 43

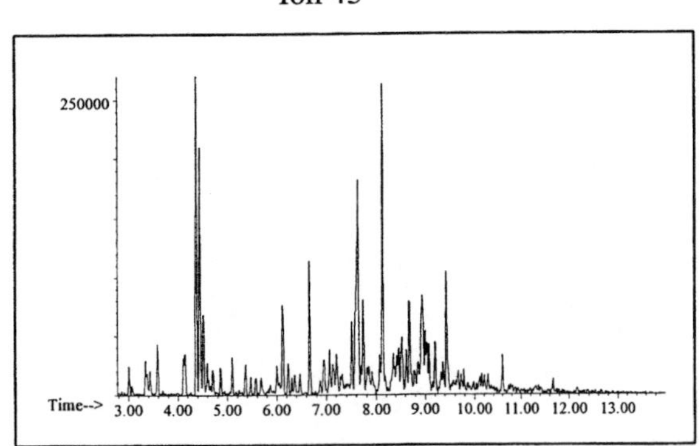

Ion 57

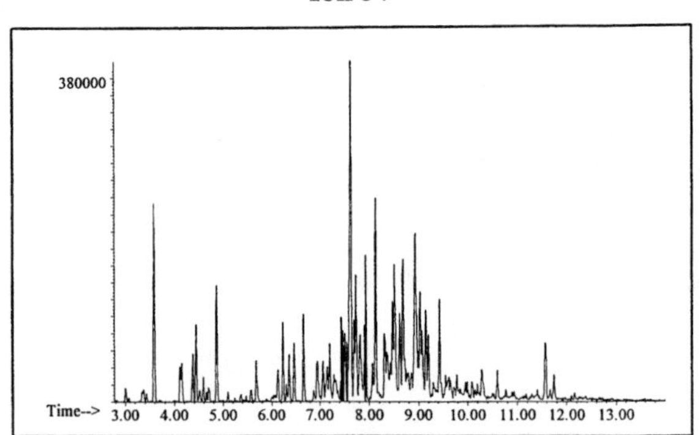

Ion 71

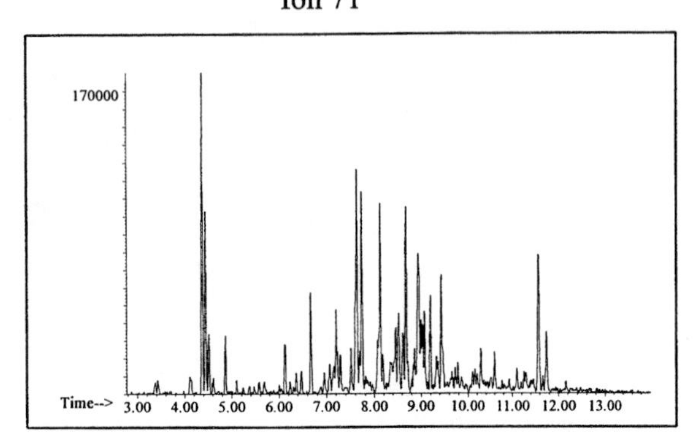

Ion 85

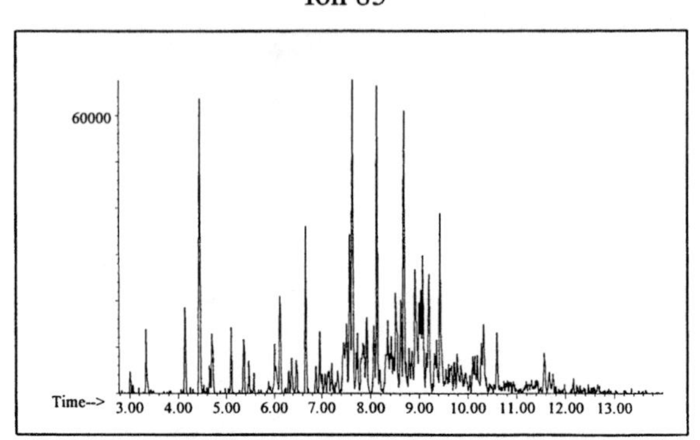

Aromatic

Ion 91

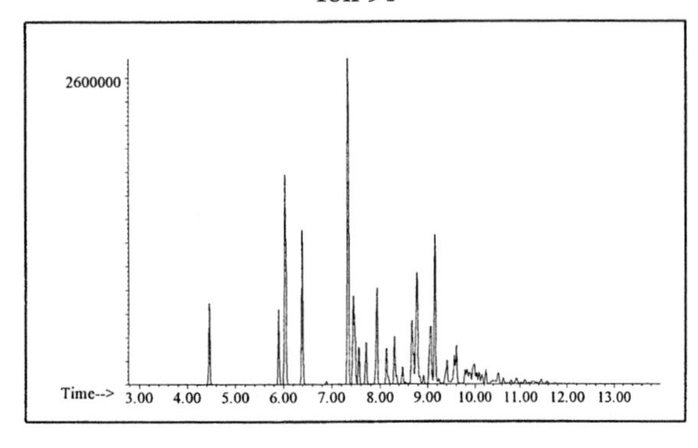

Ion 105

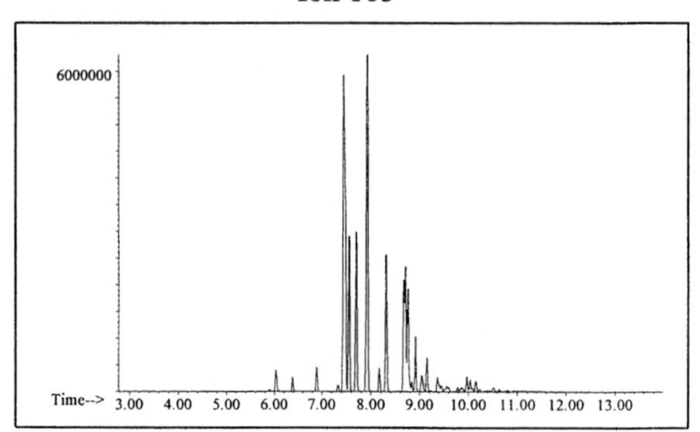

Ion 119

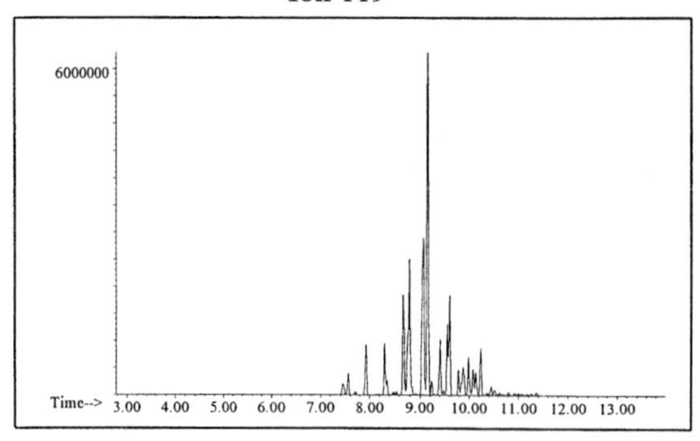

INDIVIDUAL PROFILES

Gasoline #22

Cycloparaffin

Ion 55

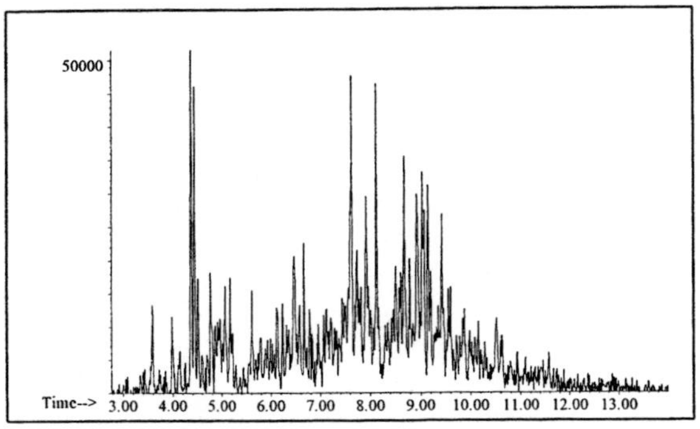

Ion 69

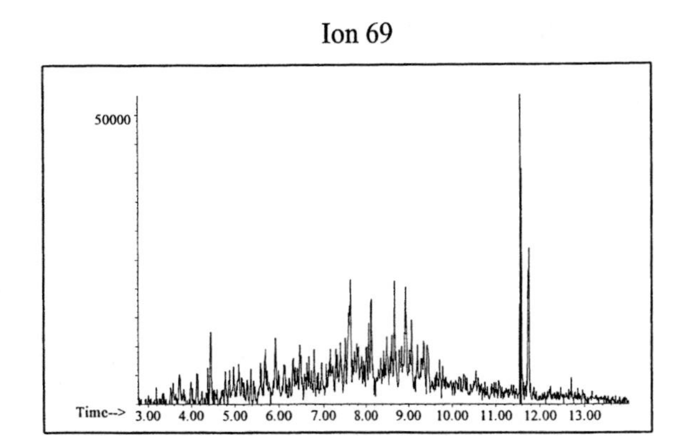

Ion 83

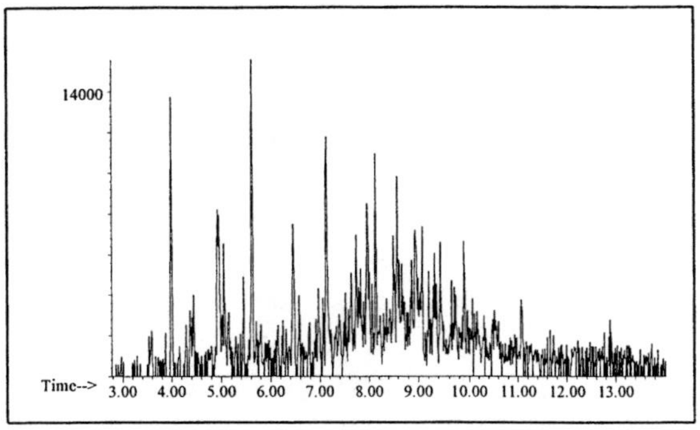

Naphthalene

Ion 128

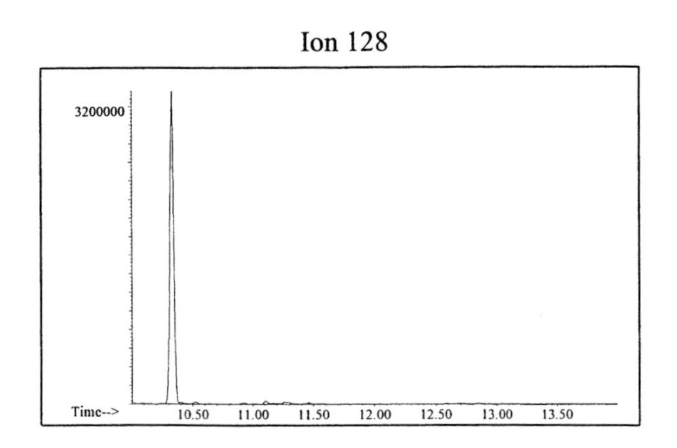

Ion 142

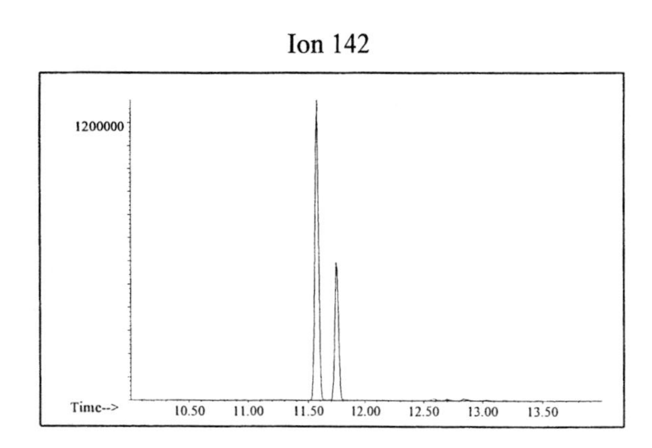

Ion 156

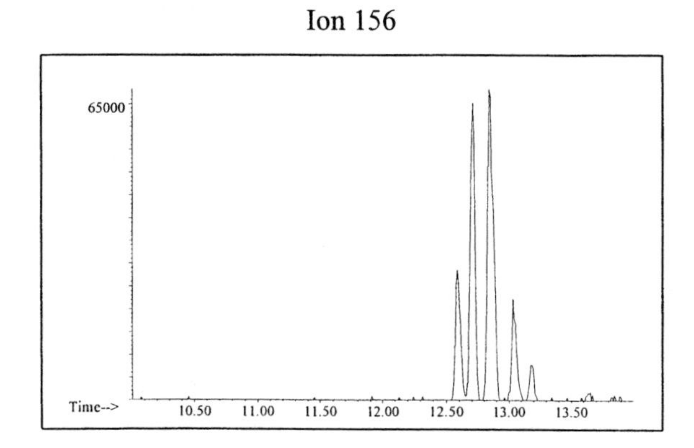

Gasoline #23

20uL/mL Pentane

Product Displayed: 90% Evaporated Amoco Ultimate Gasoline

Other Similar Products: Other grades and brands of gasoline

ASTM: Class 2 (Gasoline)

Product Uses: Automotive fuel

Macro Code: L

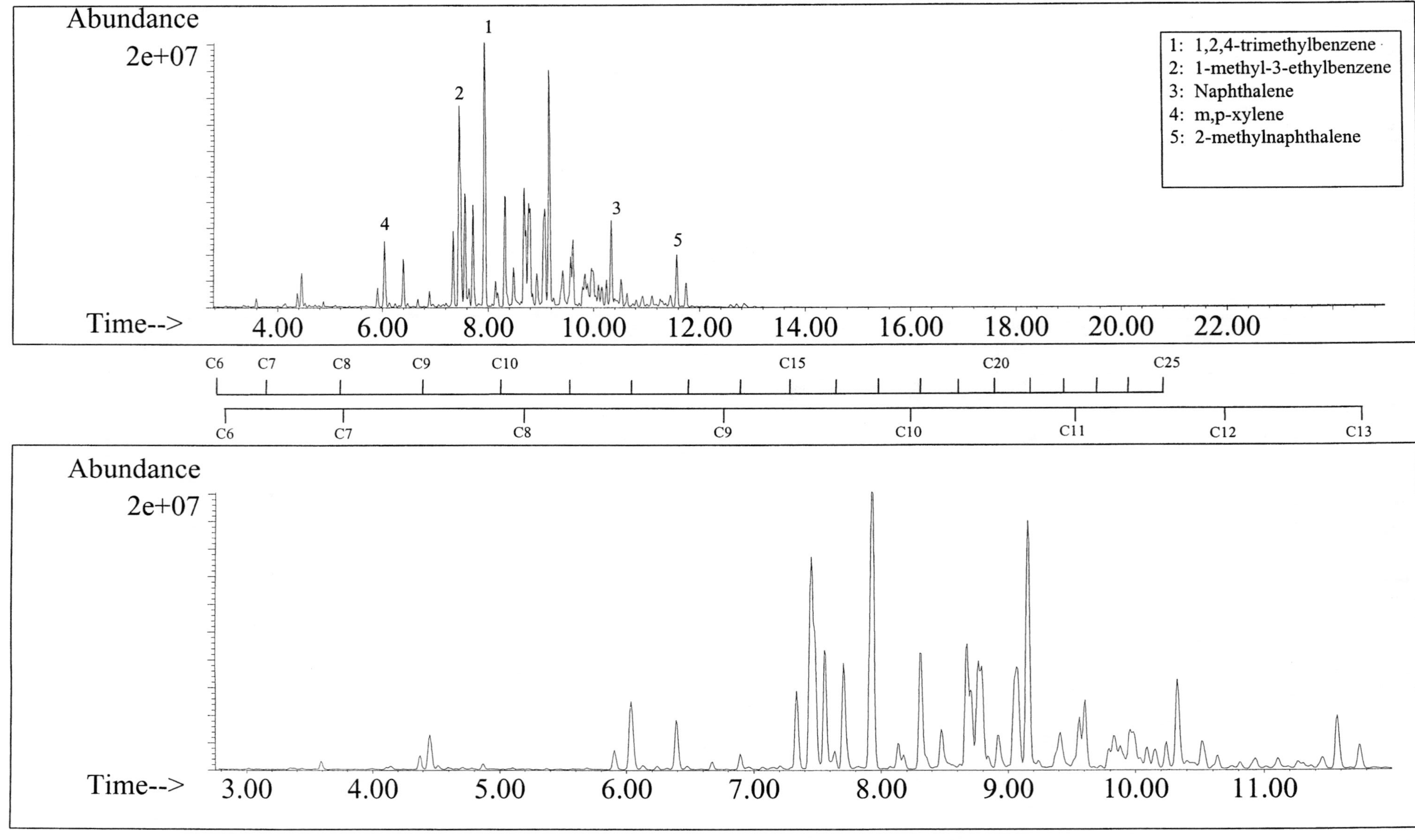

SUMMED PROFILES Gasoline #23

ALKANES

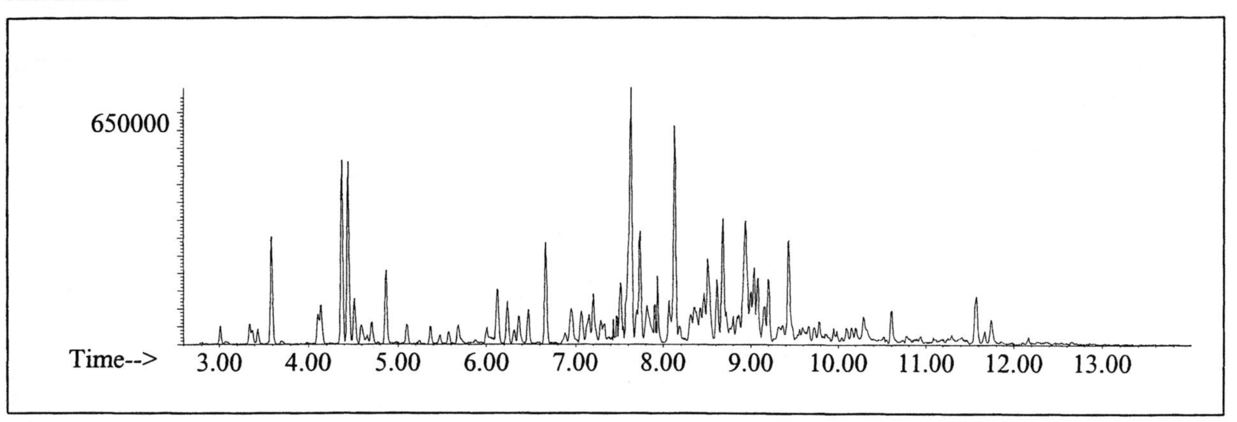

AROMATICS

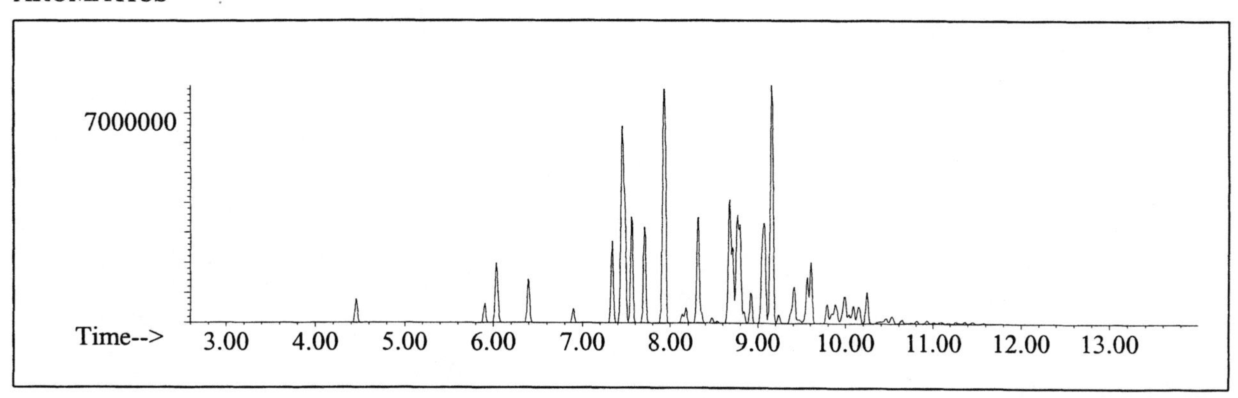

CYCLOPARAFFINS AND ALKENES

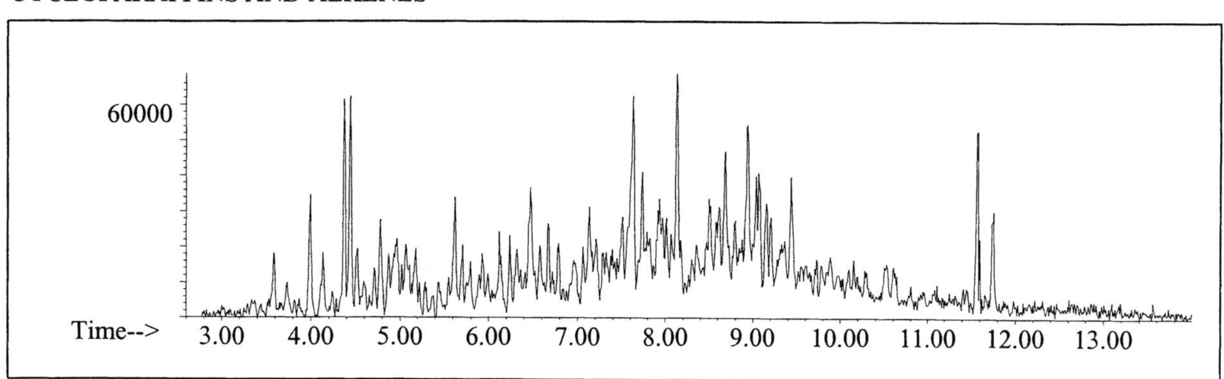

NAPHTHALENES

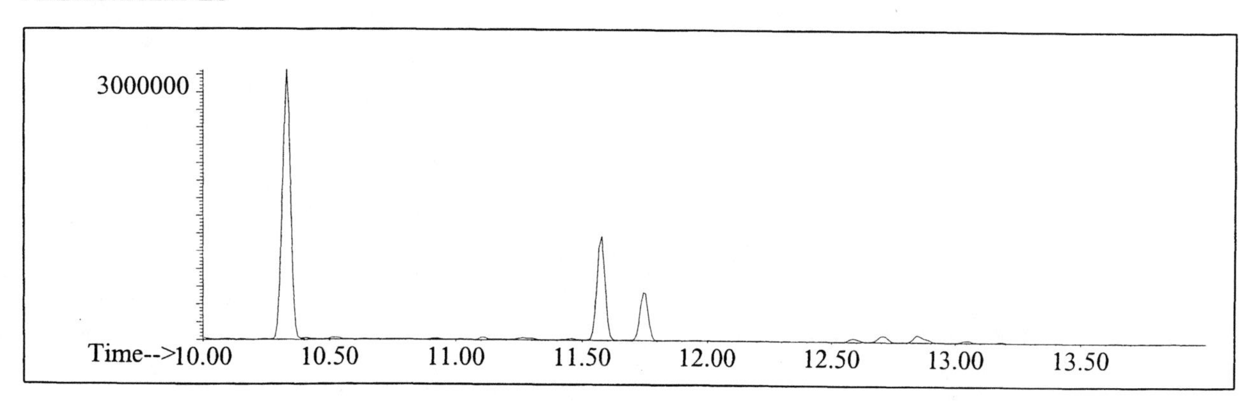

INDIVIDUAL PROFILES

Gasoline #23

Alkane

Ion 43

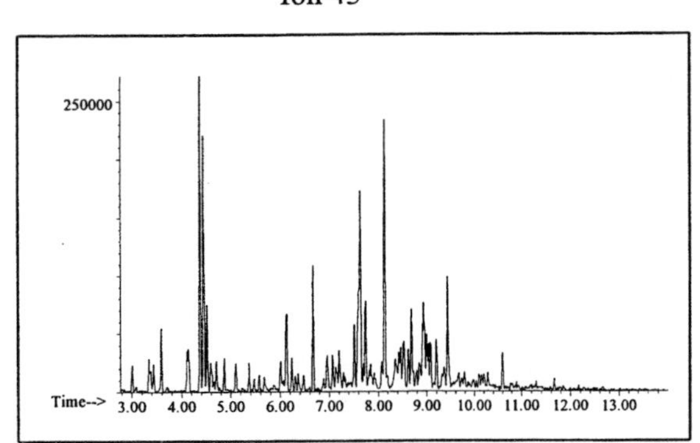

Ion 57

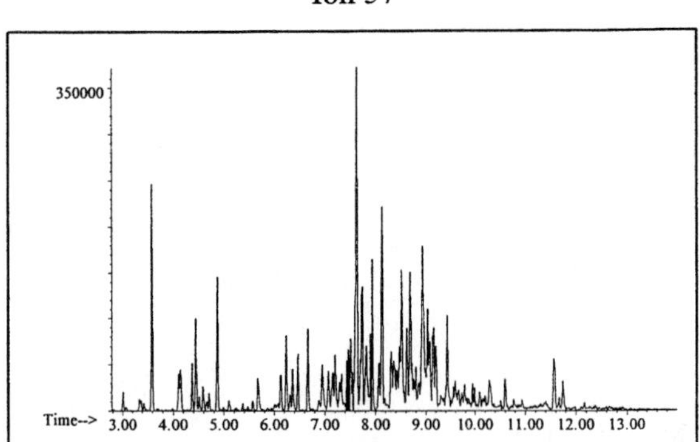

Ion 71

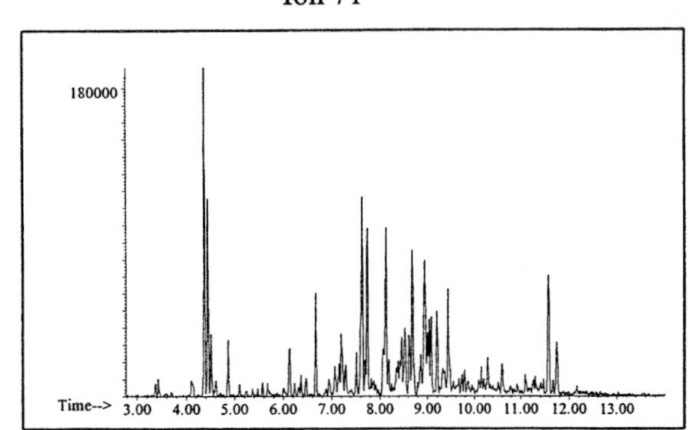

Ion 85

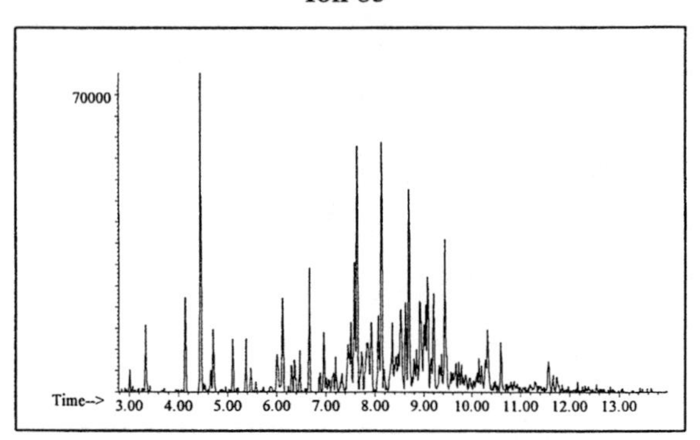

Aromatic

Ion 91

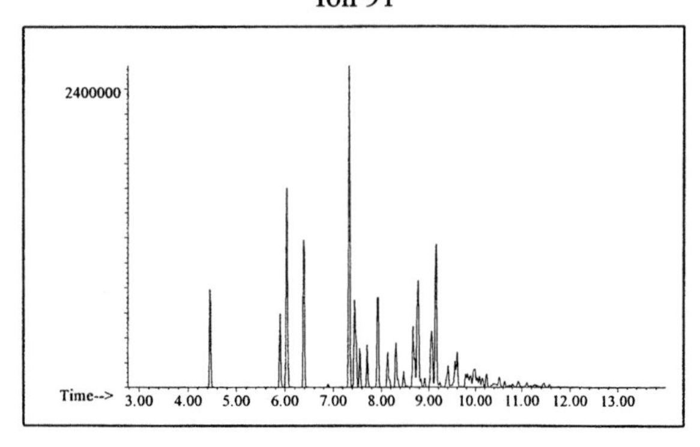

Ion 105

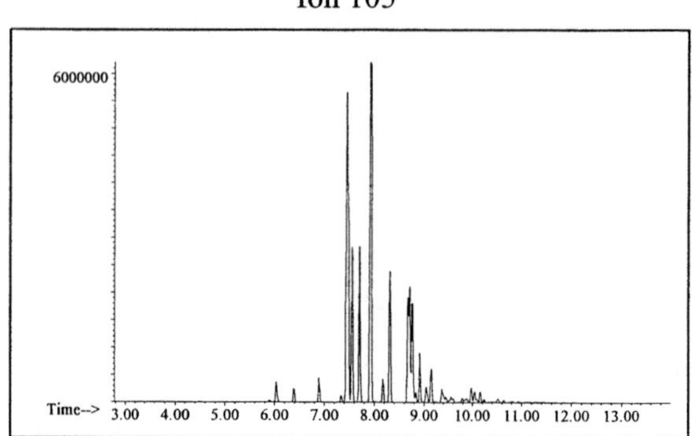

Ion 119

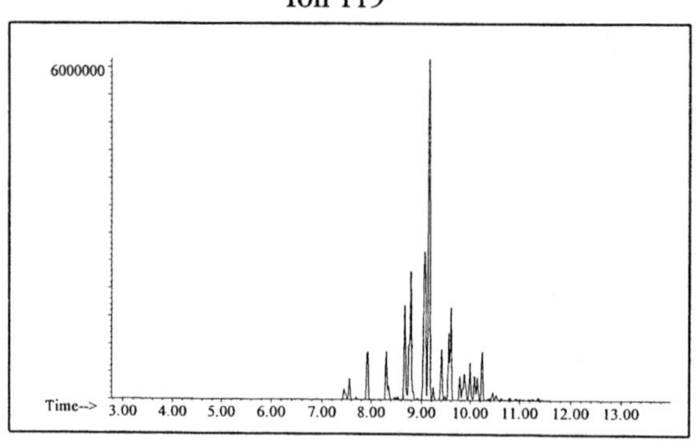

INDIVIDUAL PROFILES

Gasoline #23

Cycloparaffin

Ion 55

Ion 69

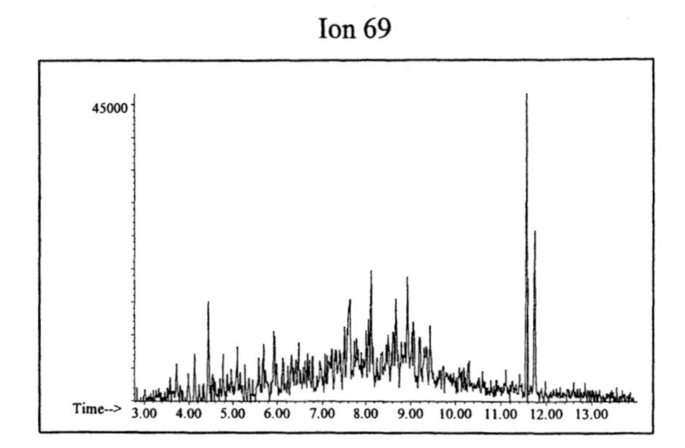

Ion 83

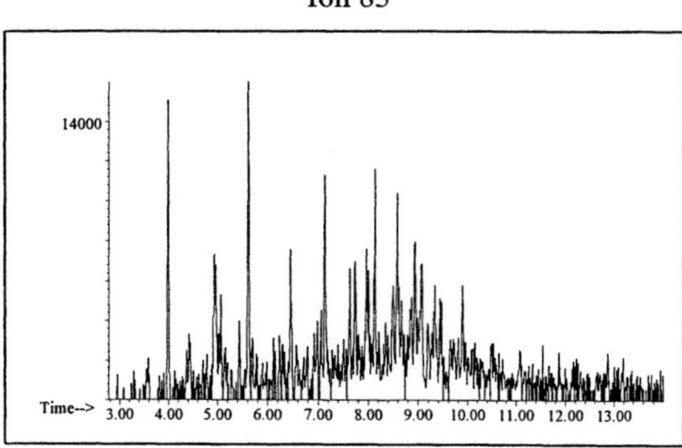

Naphthalene

Ion 128

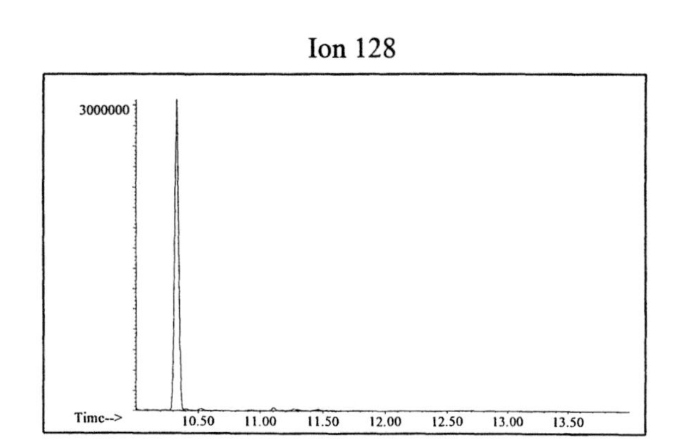

Ion 142

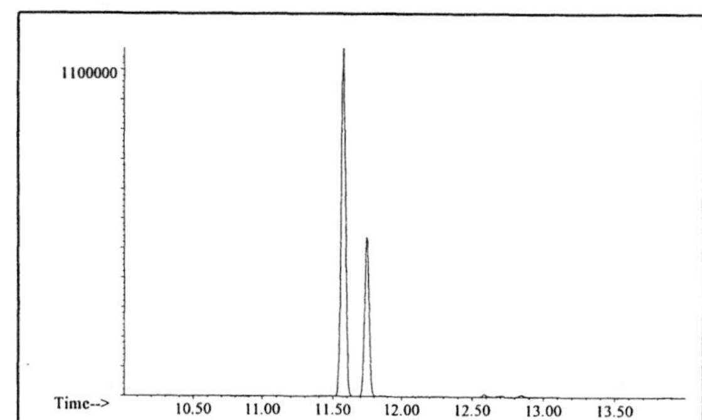

Ion 156

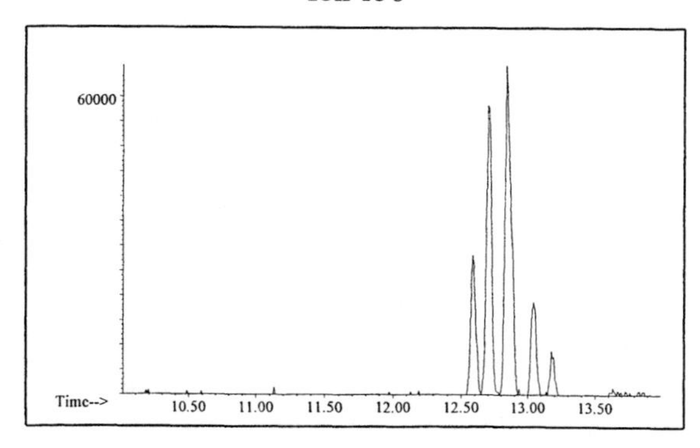

Gasoline #24

20uL/mL Pentane

Product Displayed: 98% Evaporated Amoco Ultimate Gasoline

Other Similar Products: Other grades and brands of gasoline

ASTM: Class 2 (Gasoline)

Product Uses: Automotive fuel

Macro Code: M

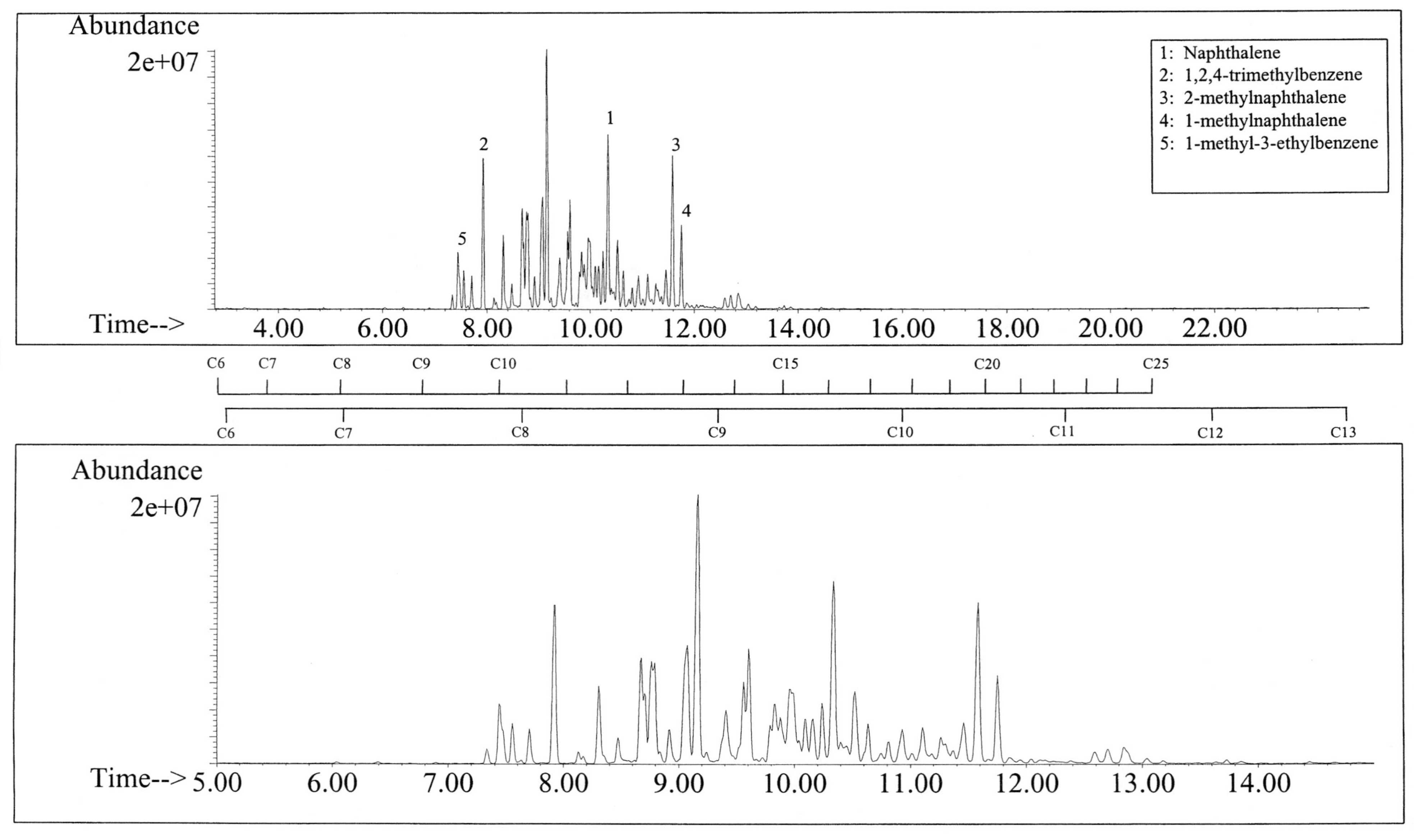

SUMMED PROFILES

Gasoline #24

ALKANES

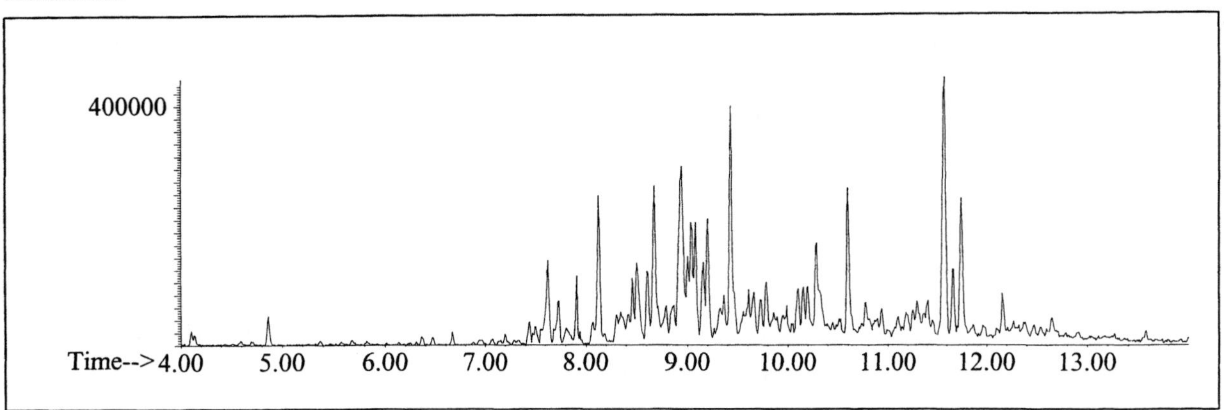

AROMATICS

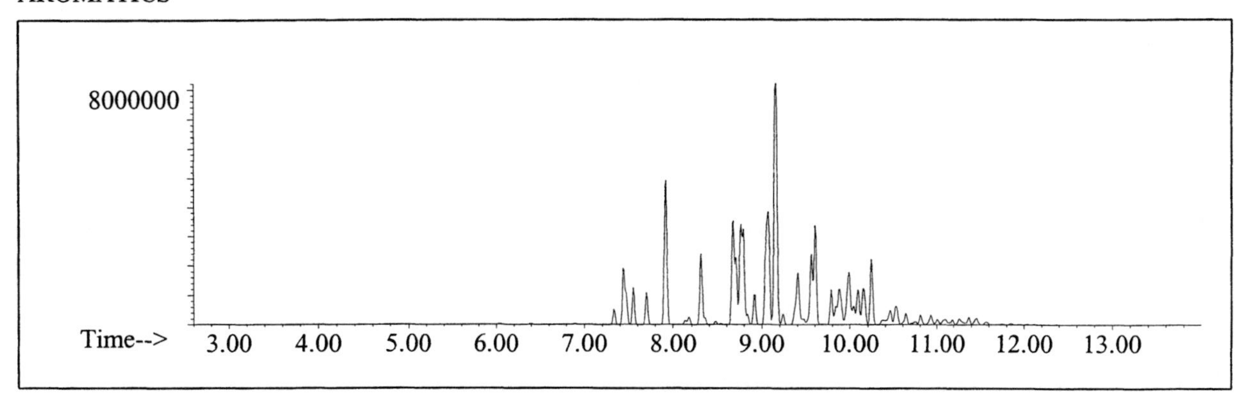

CYCLOPARAFFINS AND ALKENES

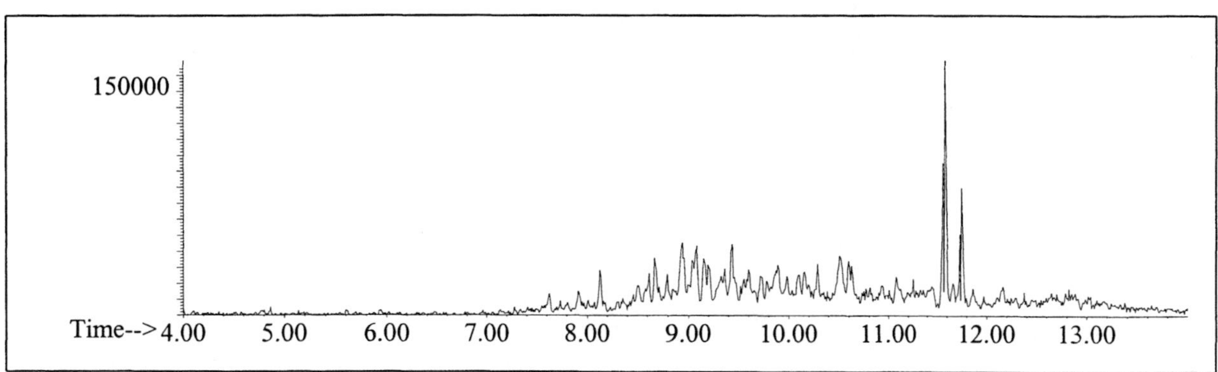

NAPHTHALENES

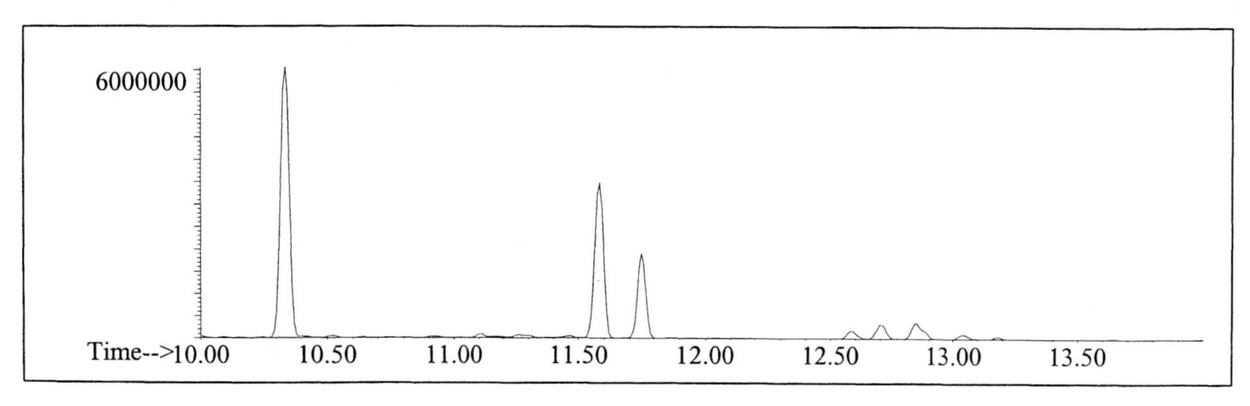

INDIVIDUAL PROFILES Gasoline #24

Alkane

Ion 43

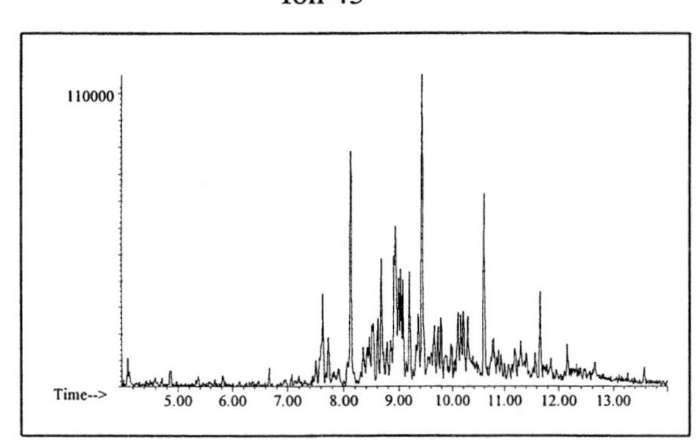

Ion 57

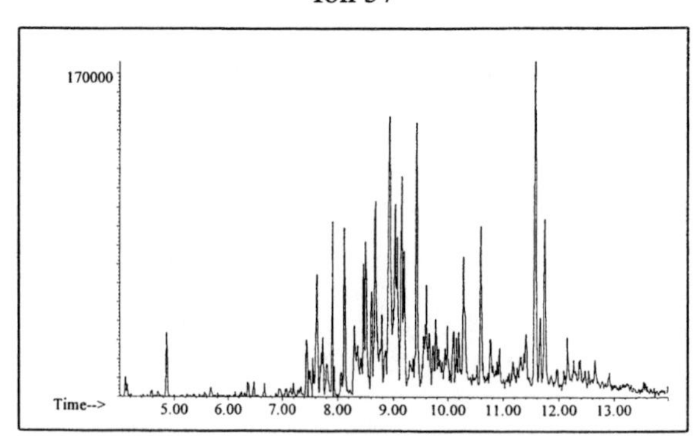

Ion 71

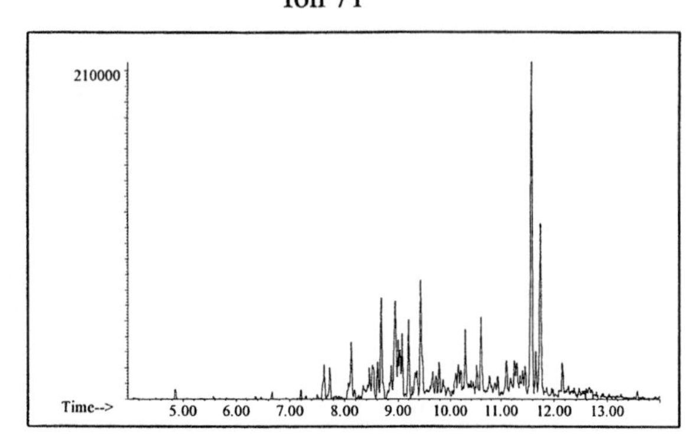

Ion 85

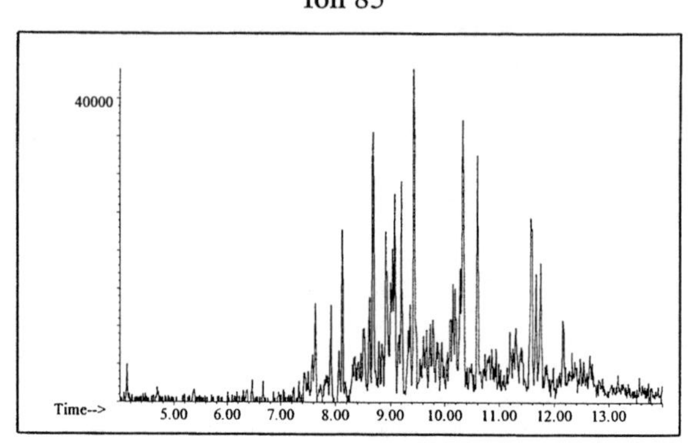

Aromatic

Ion 91

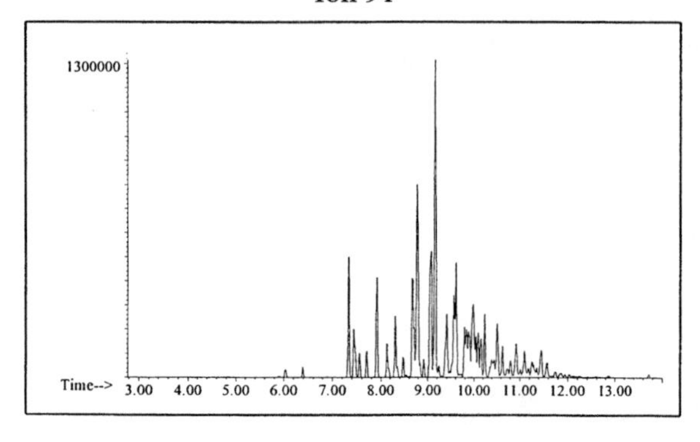

Ion 105

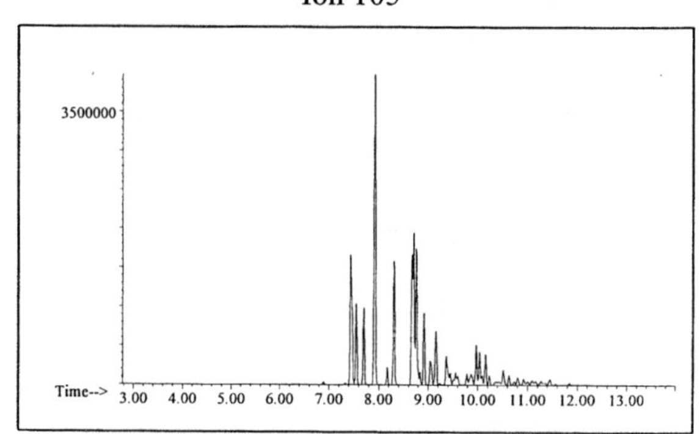

Ion 119

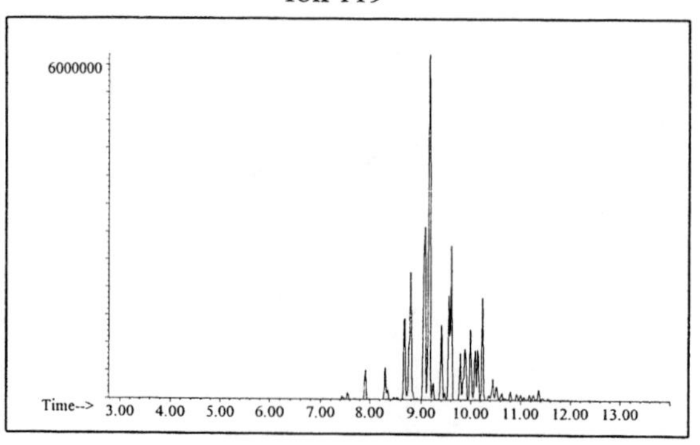

Cycloparaffin

Ion 55

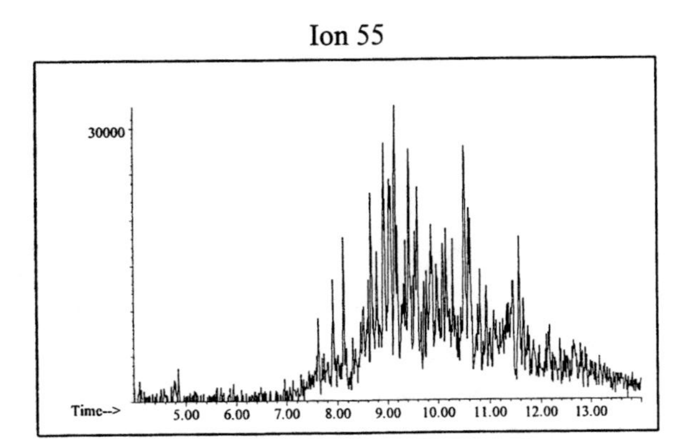

Ion 69

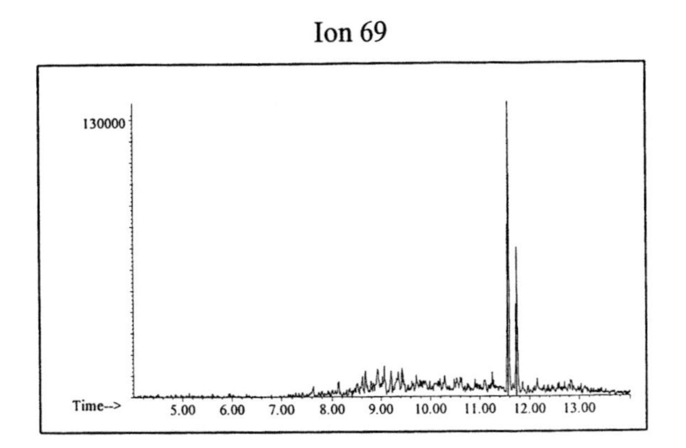

Ion 83

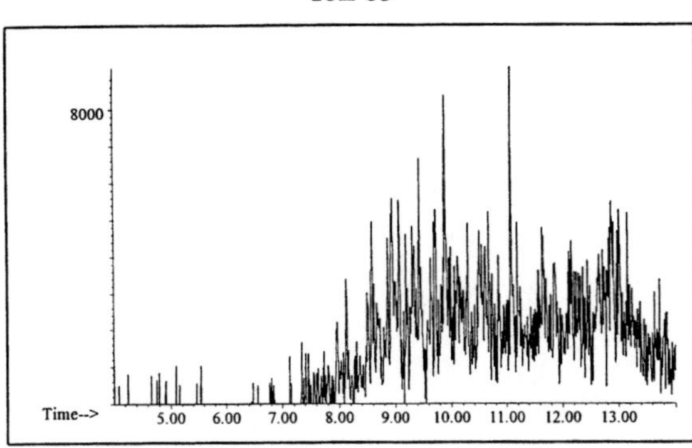

Naphthalene

Ion 128

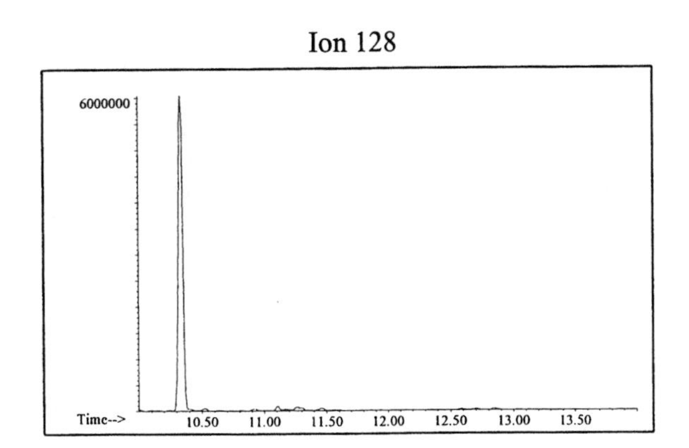

Ion 142

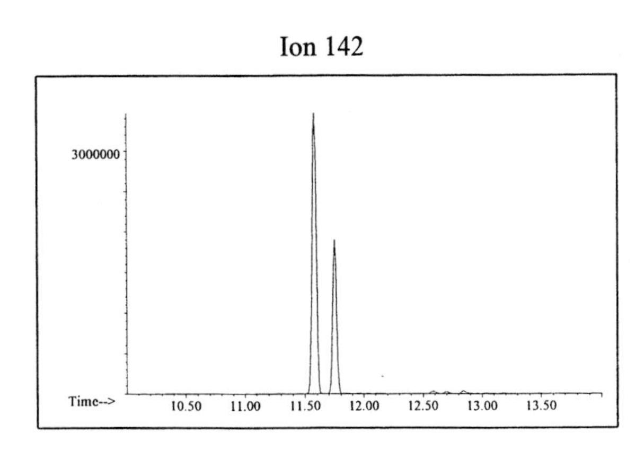

Ion 156

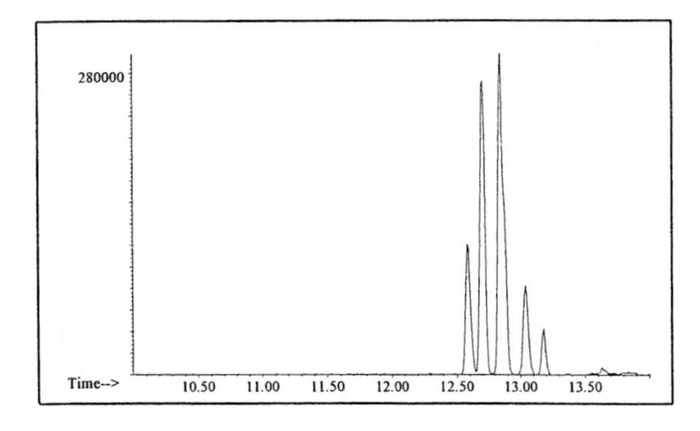

Gasoline #25

20uL/mL Pentane

ASTM: Class 2 (Gasoline)

Macro Code: M

Product Displayed: Chevron 87 Reformulated Gasoline

Product Uses: Automotive fuel

Other Similar Products: Other grades and brands of gasoline

Abundance

9000000

Time-->

4.00 6.00 8.00 10.00 12.00 14.00 16.00 18.00 20.00 22.00

1: Toluene
2: m,p-xylene
3: 1,2,4-trimethylbenzene
4: o-xylene
5: Naphthalene

C6 C7 C8 C9 C10 C15 C20 C25

C6 C7 C8 C9 C10 C11 C12 C13

Abundance

9000000

Time-->

3.00 4.00 5.00 6.00 7.00 8.00 9.00 10.00 11.00

SUMMED PROFILES

Gasoline #25

ALKANES

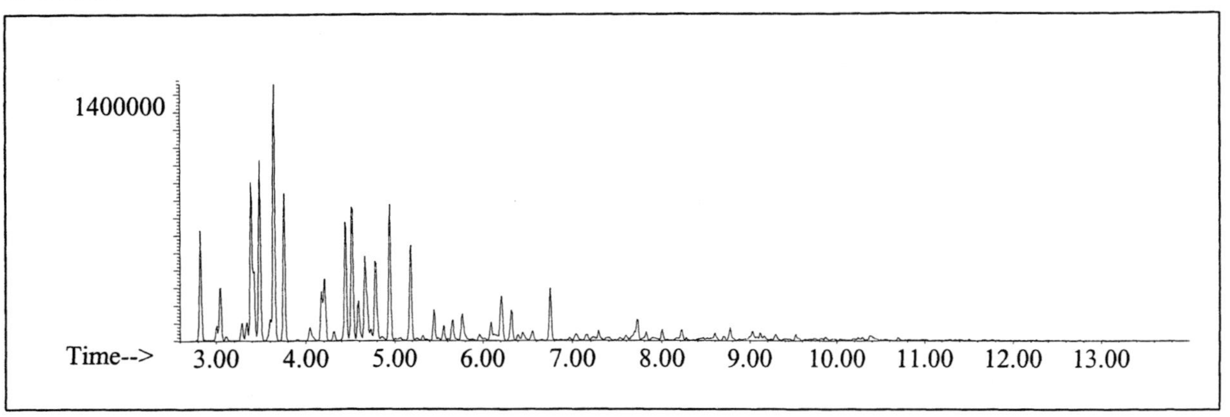

AROMATICS

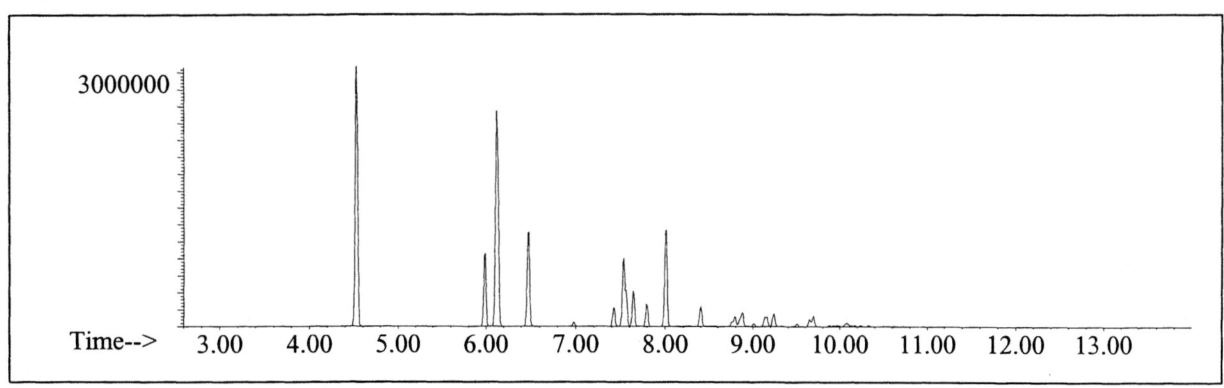

CYCLOPARAFFINS AND ALKENES

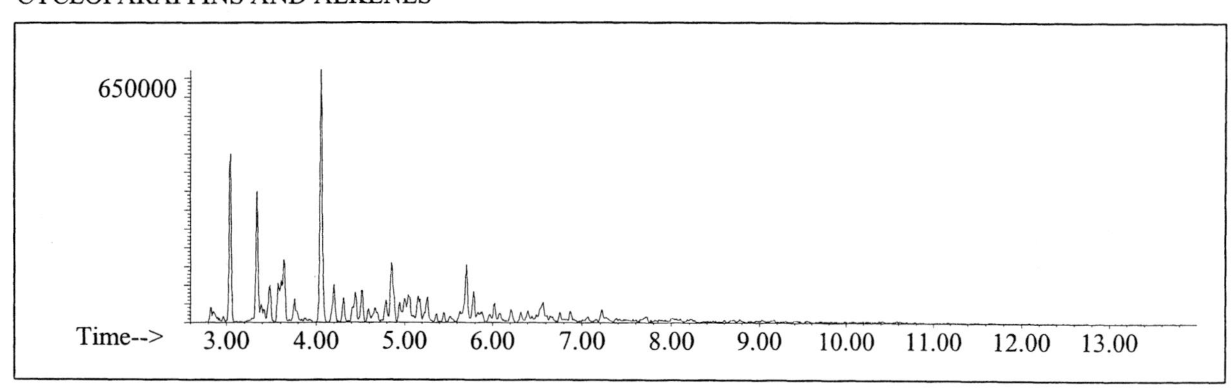

NAPHTHALENES

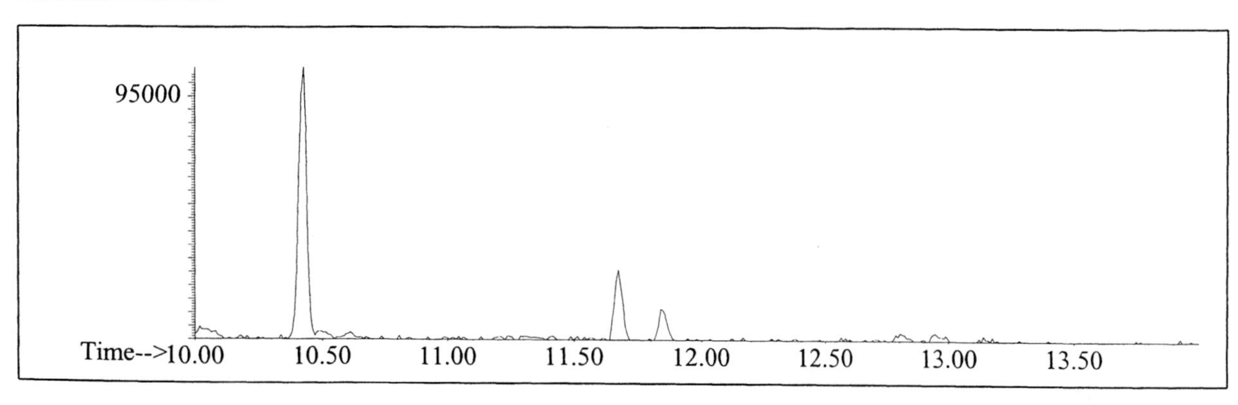

INDIVIDUAL PROFILES

Gasoline #25

Alkane

Ion 43

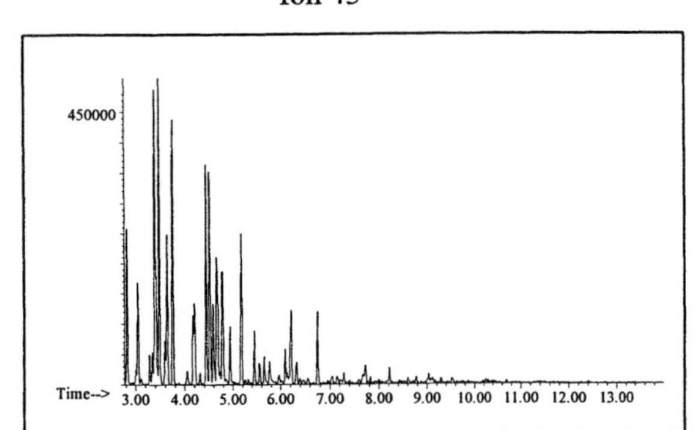

Ion 57

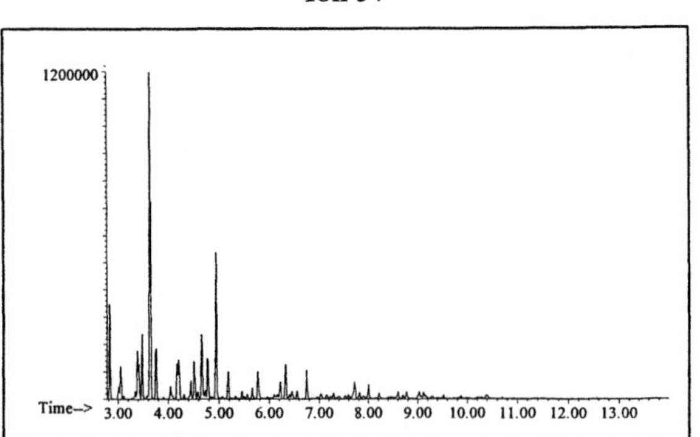

Ion 71

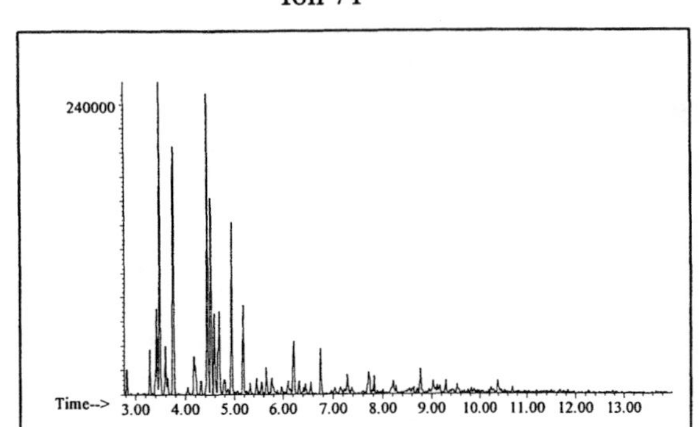

Ion 85

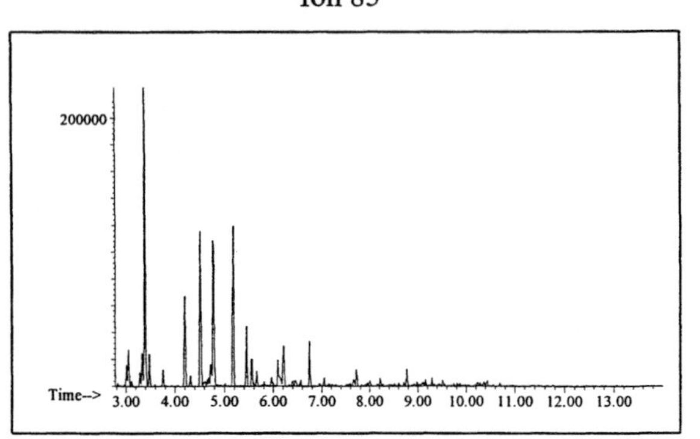

Aromatic

Ion 91

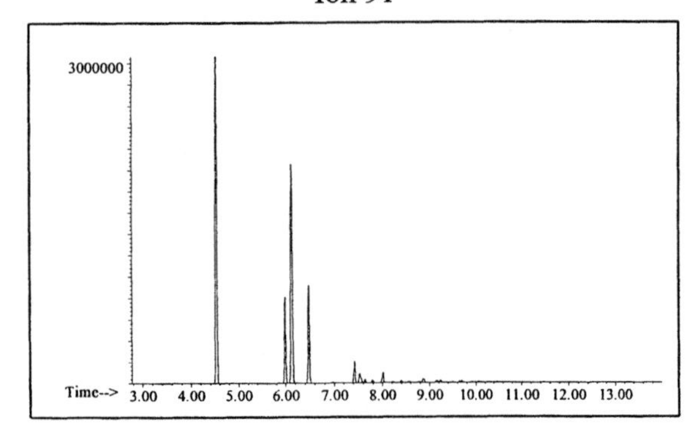

Ion 105

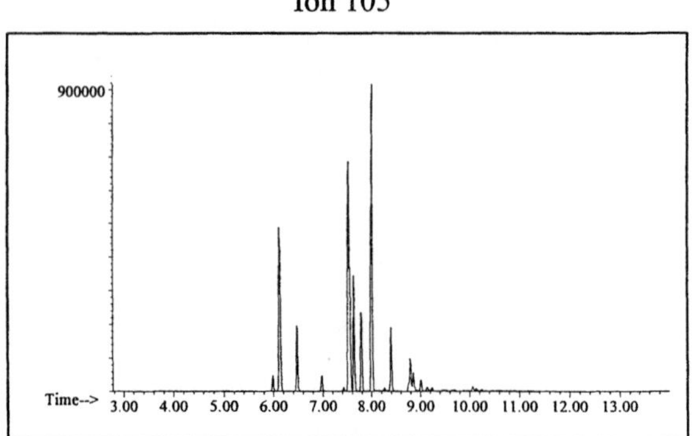

Ion 119

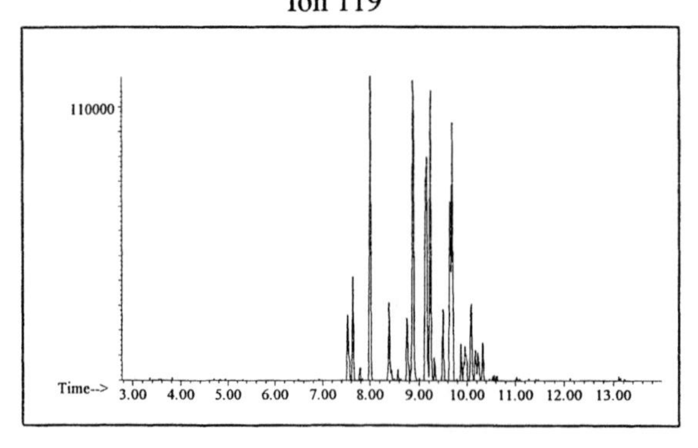

INDIVIDUAL PROFILES

Gasoline #25

Cycloparaffin

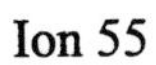
Ion 55

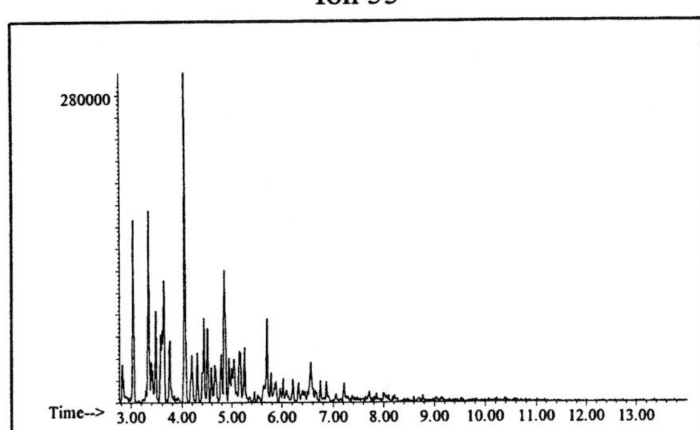

Ion 69

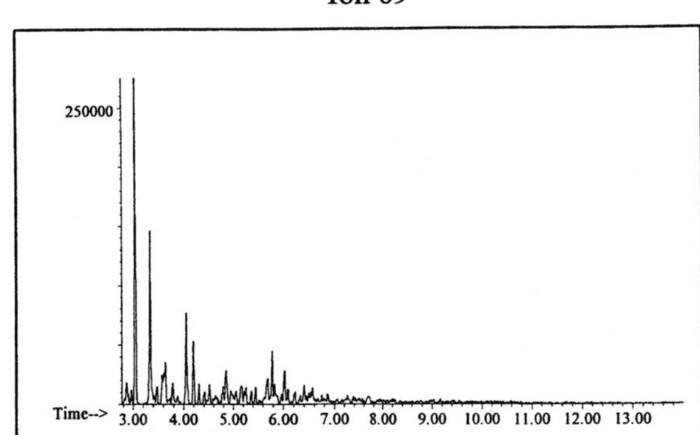

Ion 83

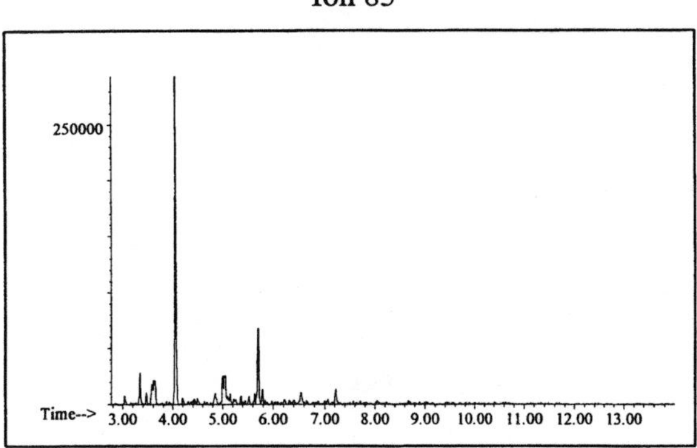

Naphthalene

Ion 128

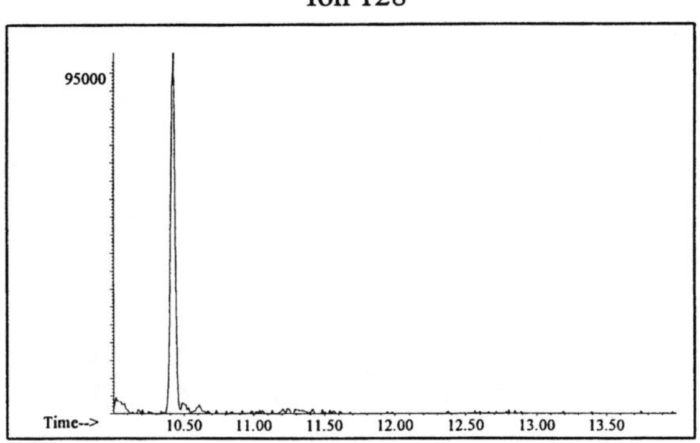

Ion 142

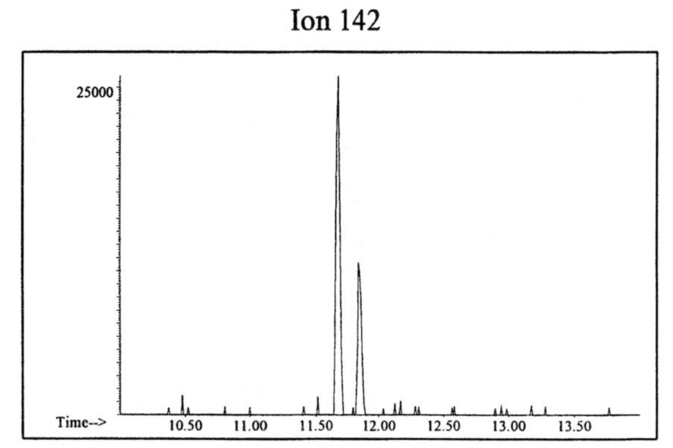

Ion 156

No Useful Data Obtained

Gasoline #26

20uL/mL Pentane

Product Displayed: Chevron Regular Reformulated Gasoline

Other Similar Products: Other grades and brands of gasoline

ASTM: Class 2 (Gasoline)

Product Uses: Automotive fuel

Macro Code: M

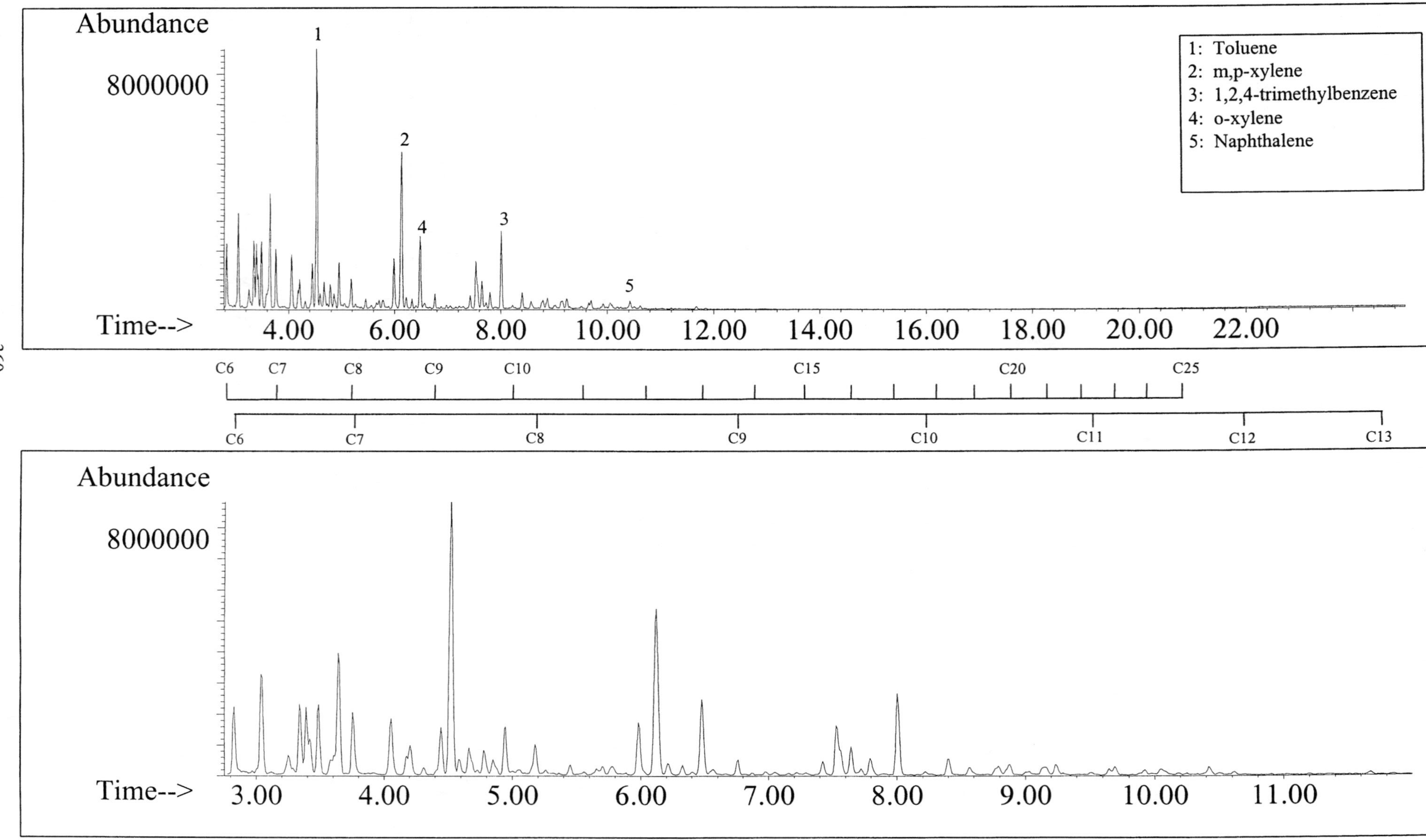

SUMMED PROFILES

Gasoline #26

ALKANES

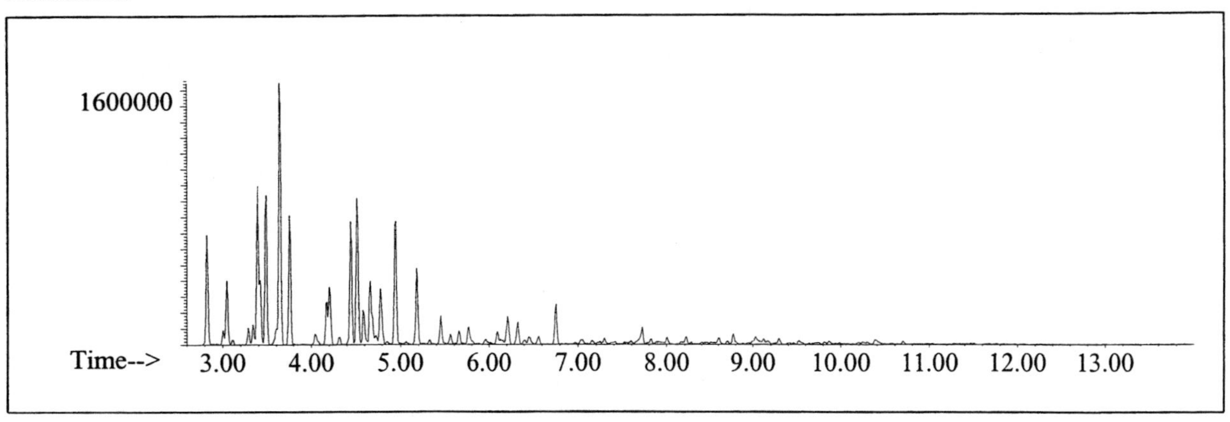

AROMATICS

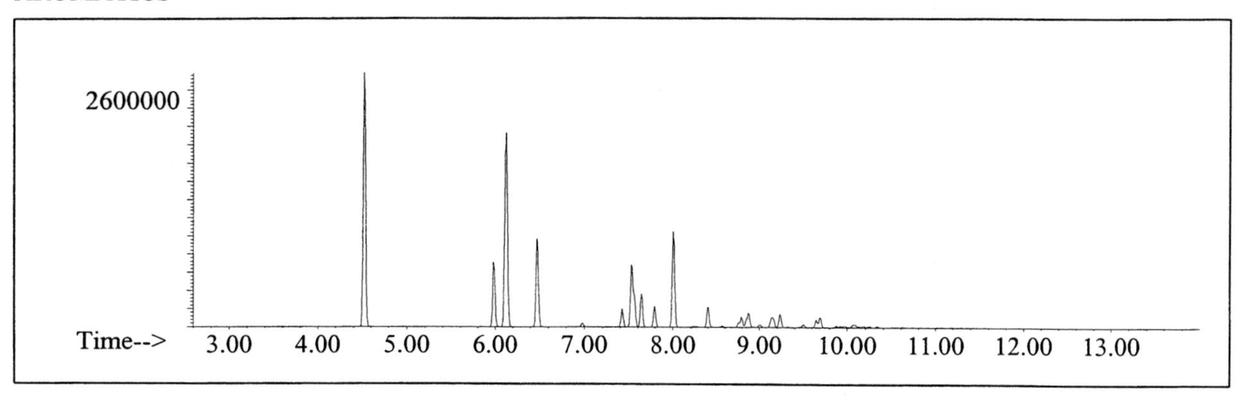

CYCLOPARAFFINS AND ALKENES

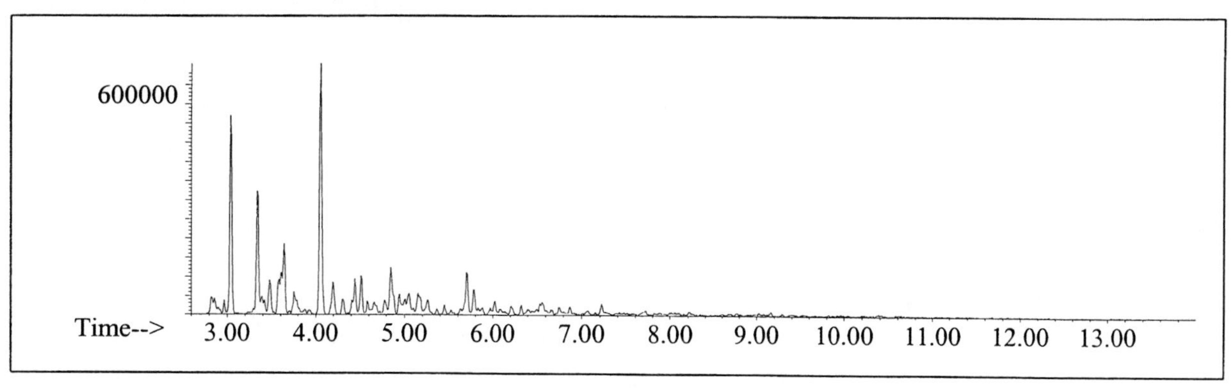

NAPHTHALENES

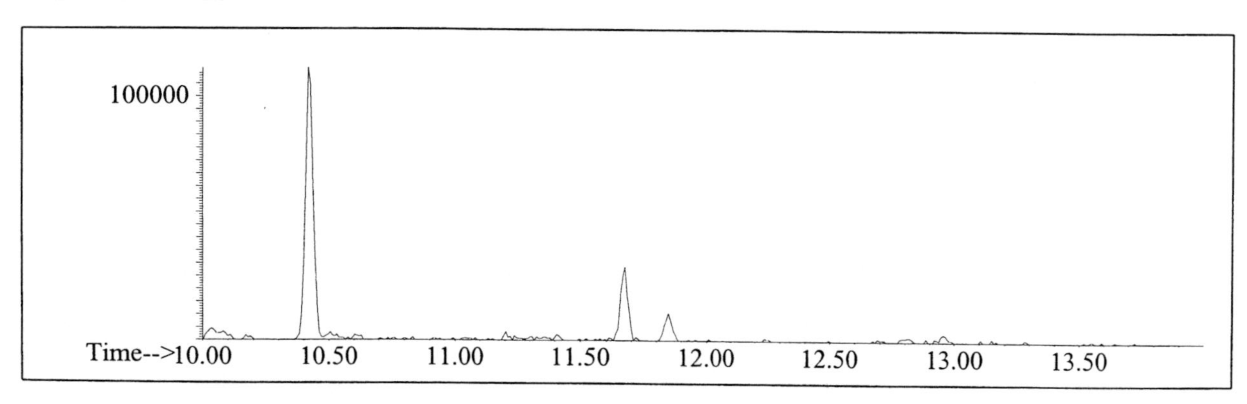

INDIVIDUAL PROFILES

Gasoline #26

Alkane

Ion 43

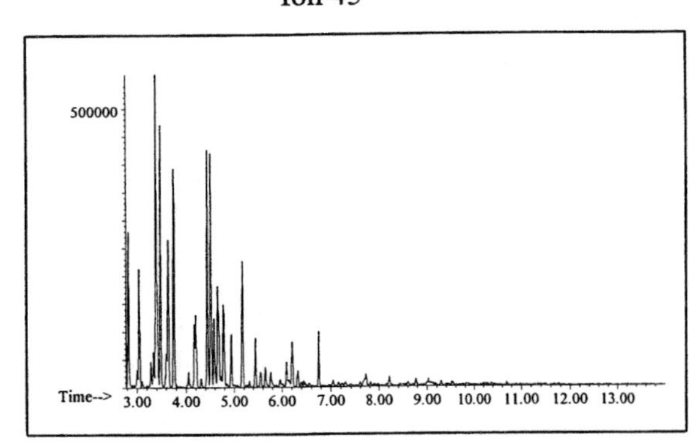

Ion 57

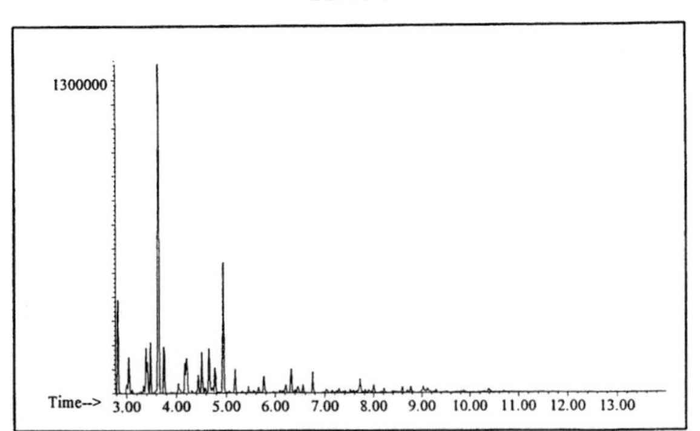

Ion 71

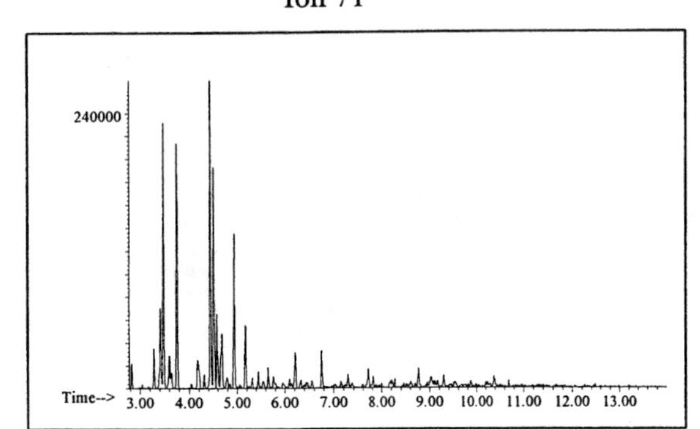

Ion 85

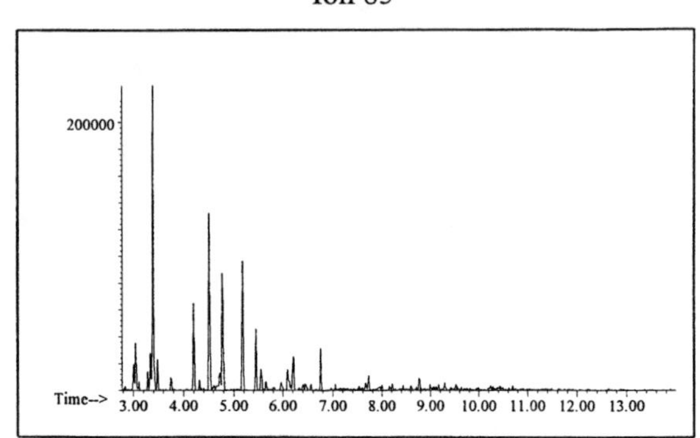

Aromatic

Ion 91

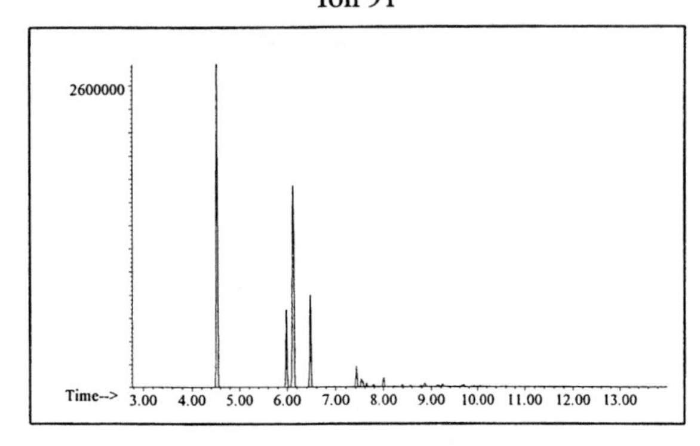

Ion 105

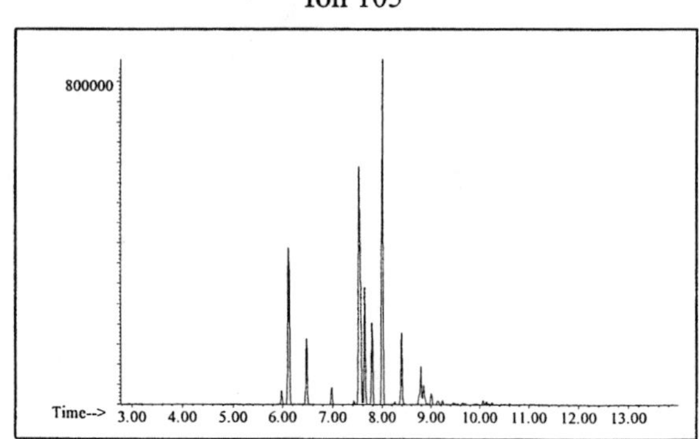

Ion 119

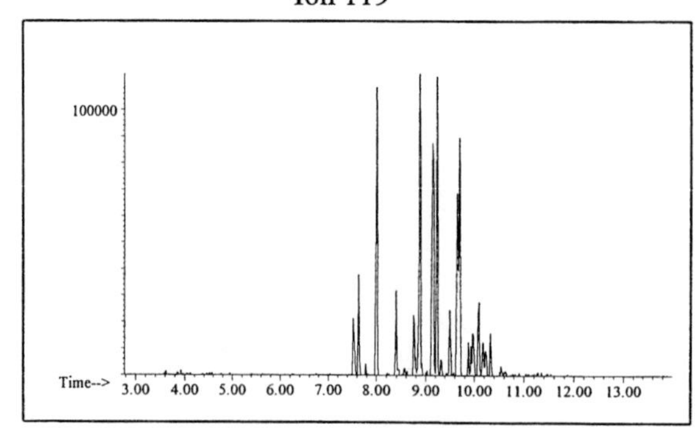

INDIVIDUAL PROFILES

Gasoline #26

Cycloparaffin

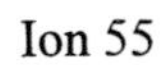

Ion 55

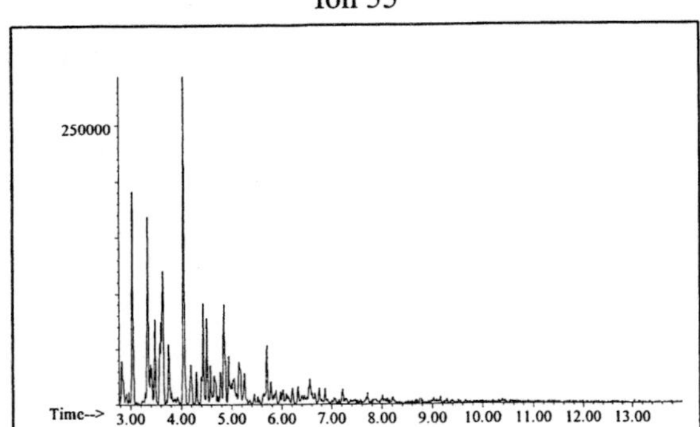

Ion 69

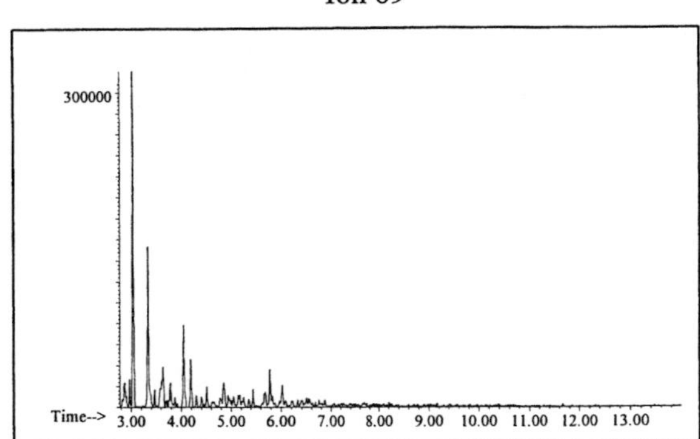

Ion 83

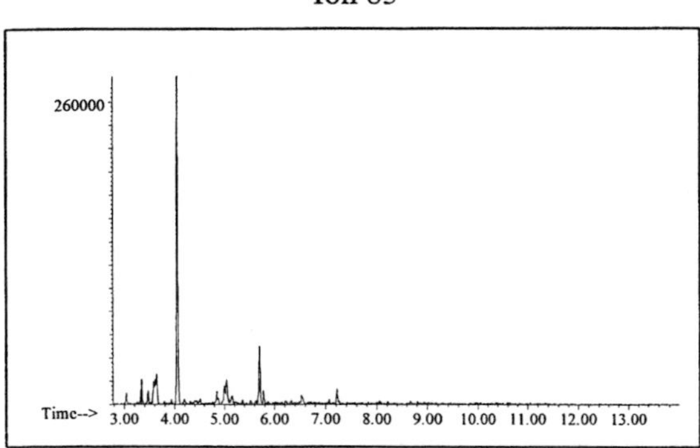

Naphthalene

Ion 128

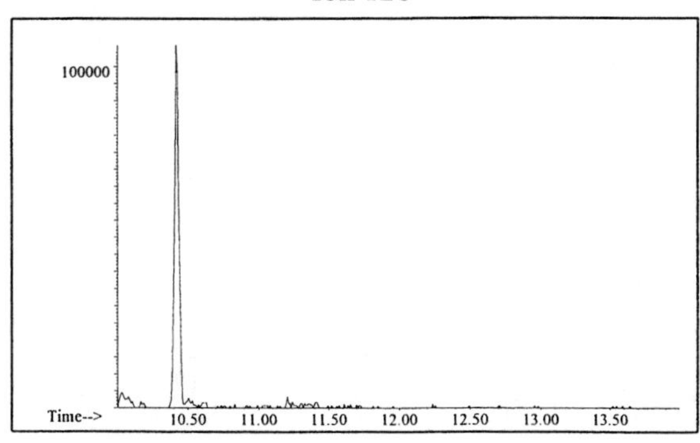

Ion 142

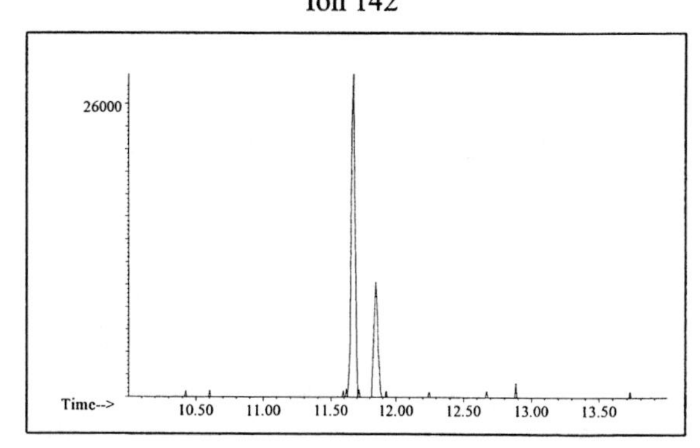

Ion 156

No Useful Data Obtained

Gasoline #27

20uL/mL Pentane

Product Displayed: Shell Premium Reformulated Gasoline

Other Similar Products: Other grades and brands of gasoline

ASTM: Class 2 (Gasoline)

Product Uses: Automotive fuel

Macro Code: M

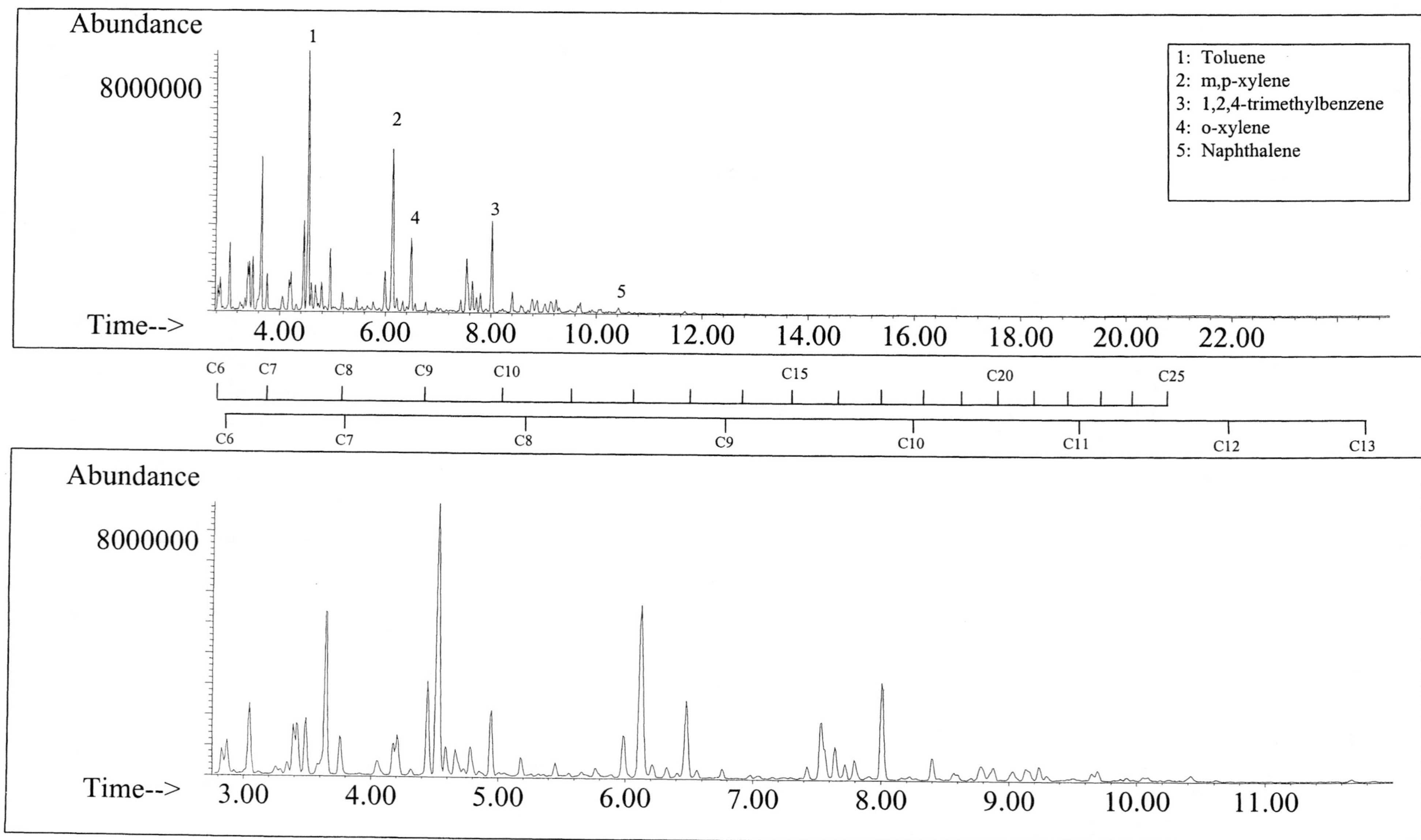

SUMMED PROFILES   Gasoline #27

ALKANES

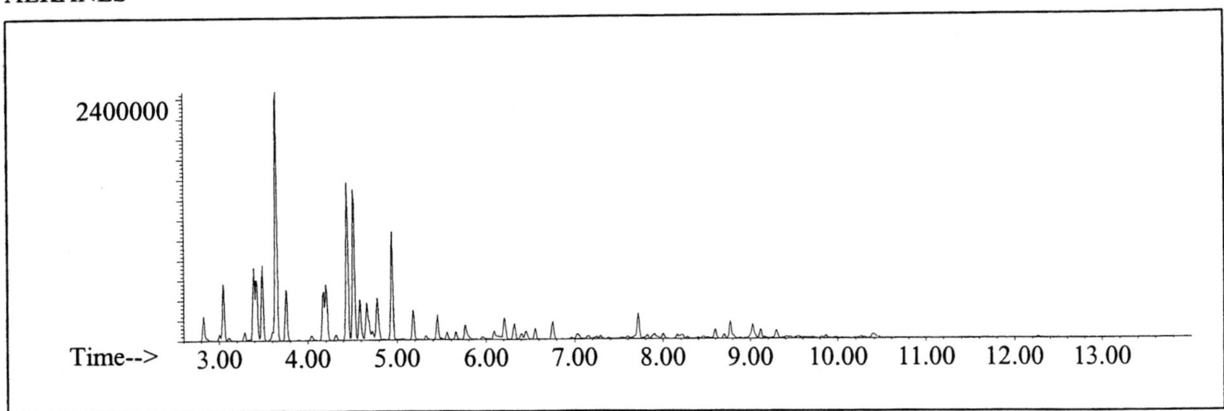

AROMATICS

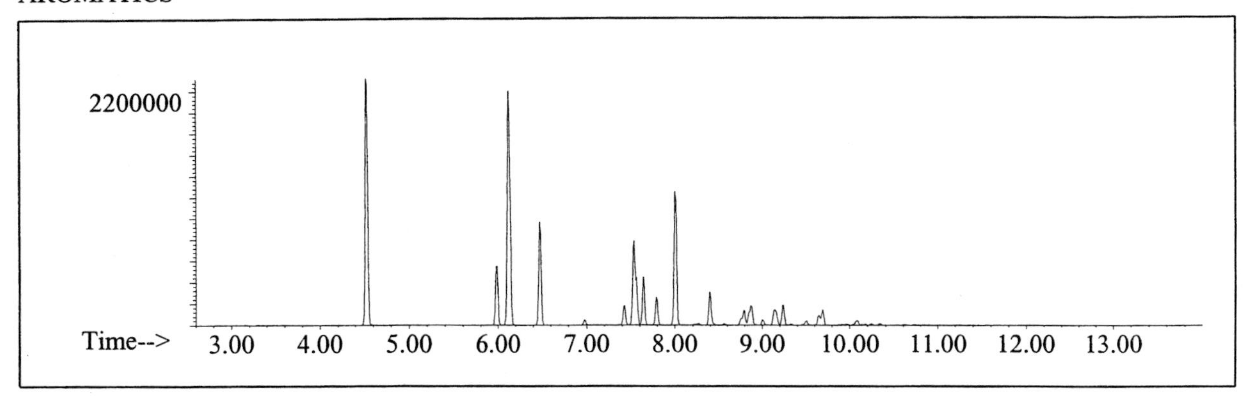

CYCLOPARAFFINS AND ALKENES

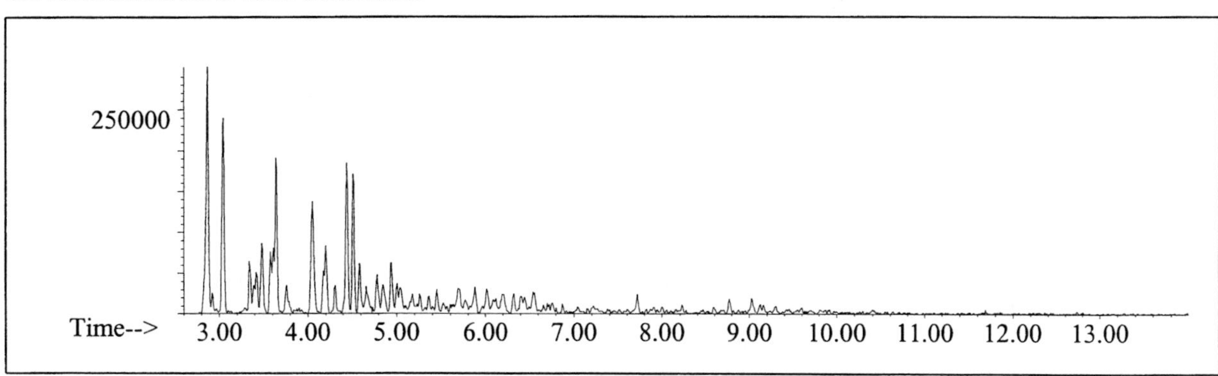

NAPHTHALENES

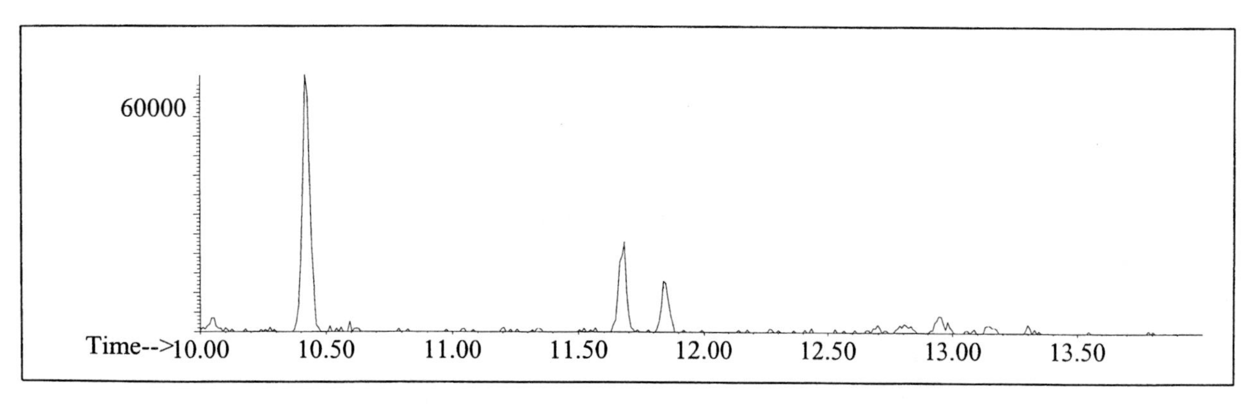

INDIVIDUAL PROFILES

Gasoline #27

Alkane

Ion 43

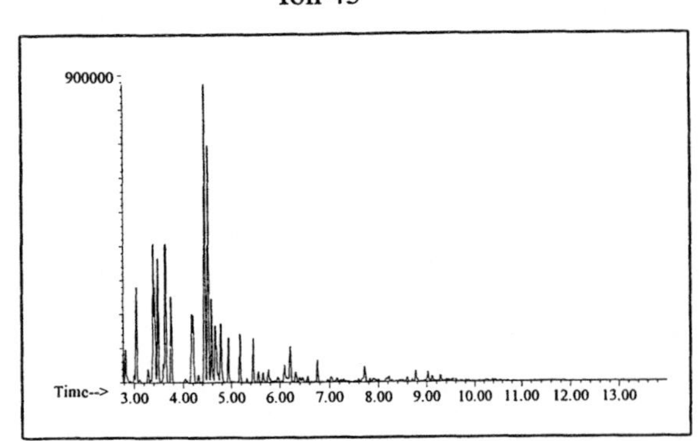

Ion 57

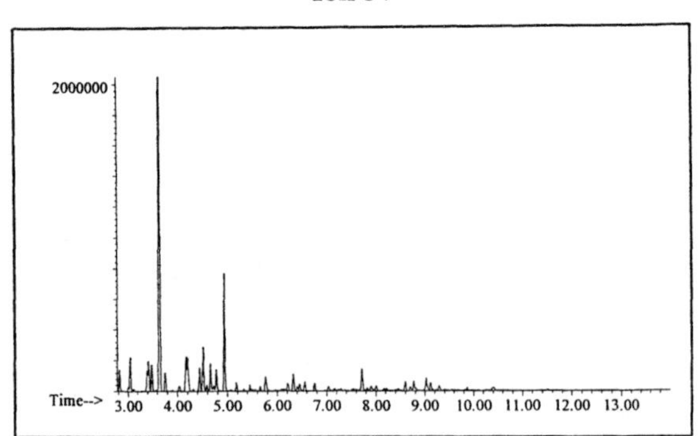

Ion 71

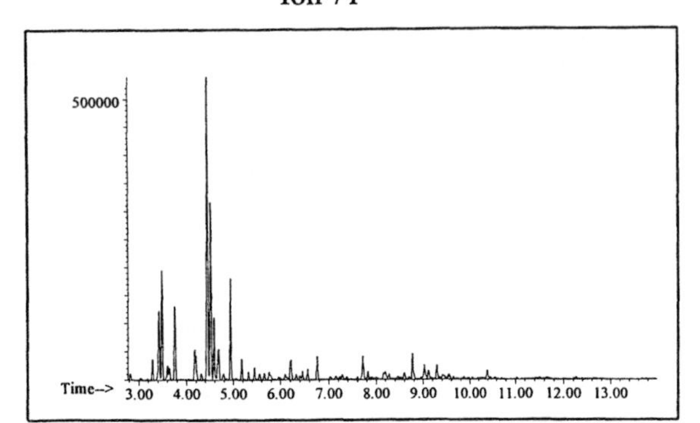

Ion 85

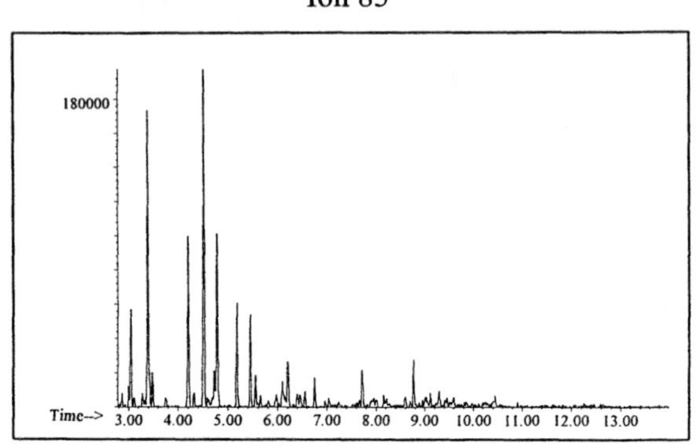

Aromatic

Ion 91

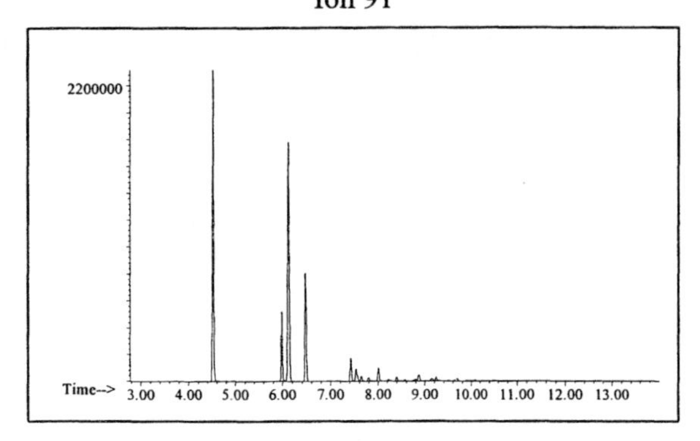

Ion 105

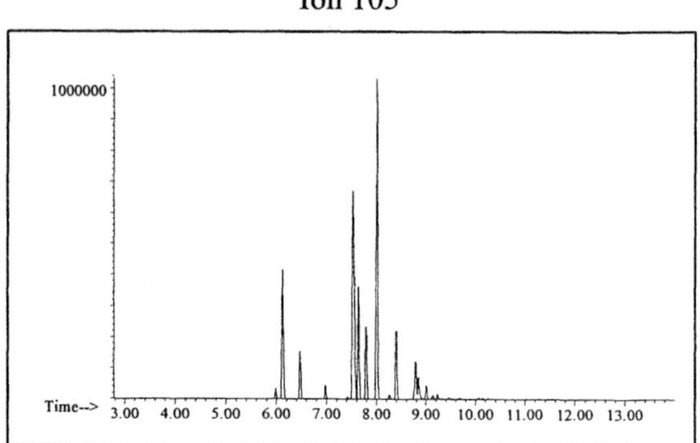

Ion 119

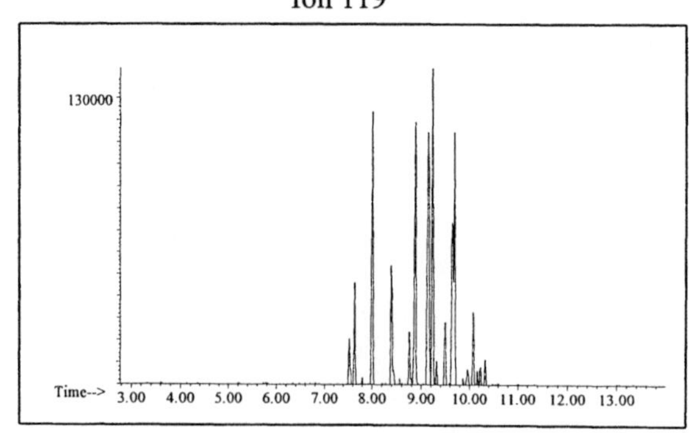

INDIVIDUAL PROFILES

Gasoline #27

Cycloparaffin

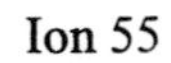

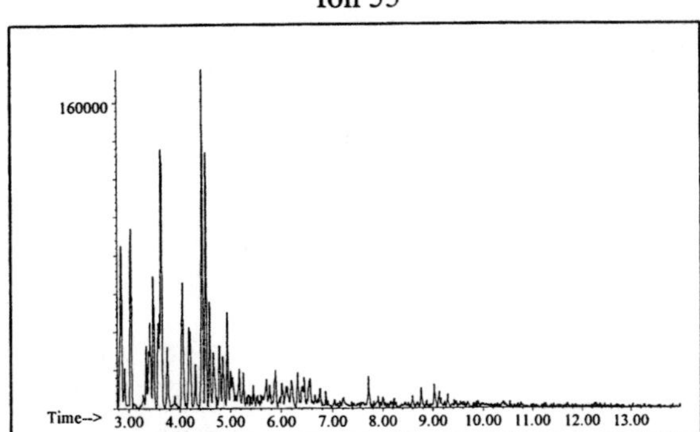

Ion 69

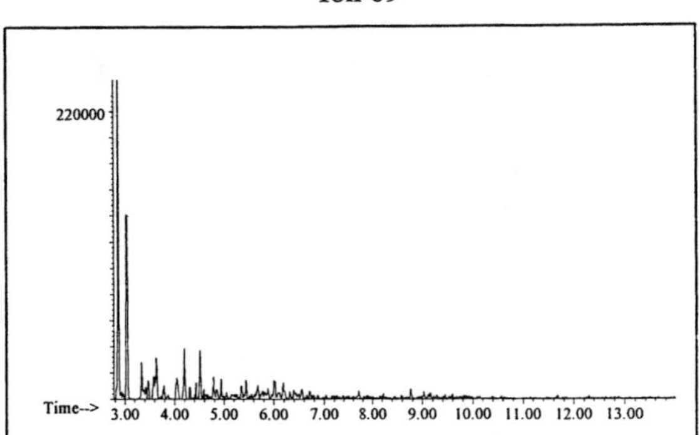

Ion 83

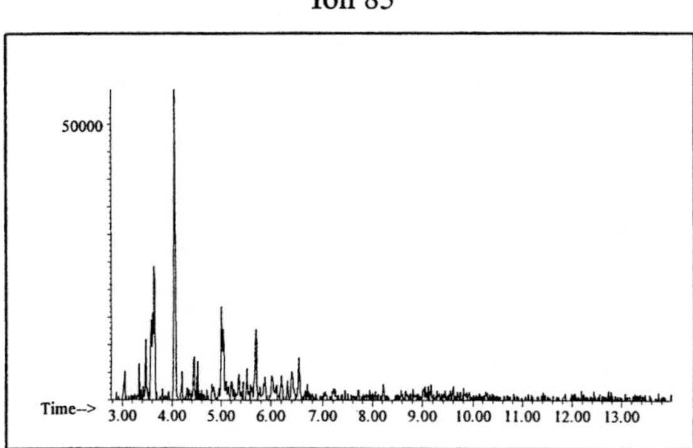

Naphthalene

Ion 128

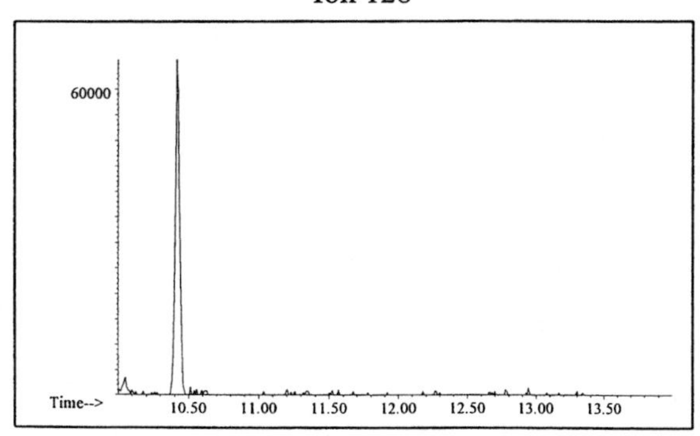

Ion 142

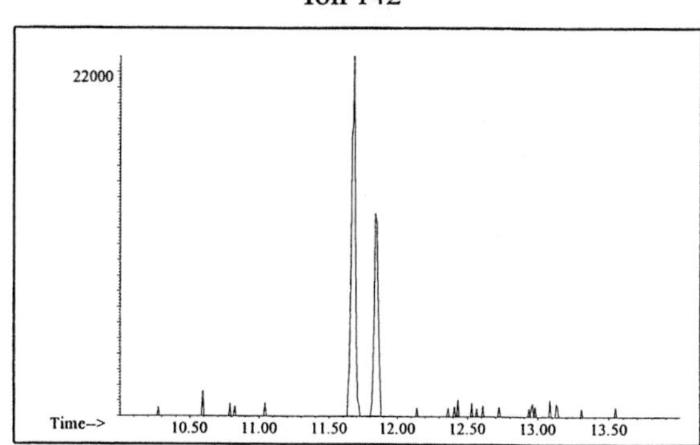

Ion 156

No Useful Data Obtained

Gasoline #28

20uL/mL Pentane

Product Displayed: Shell Regular Reformulated Gasoline

Other Similar Products: Other grades and brands of gasoline

ASTM: Class 2 (Gasoline)

Product Uses: Automotive fuel

Macro Code: M

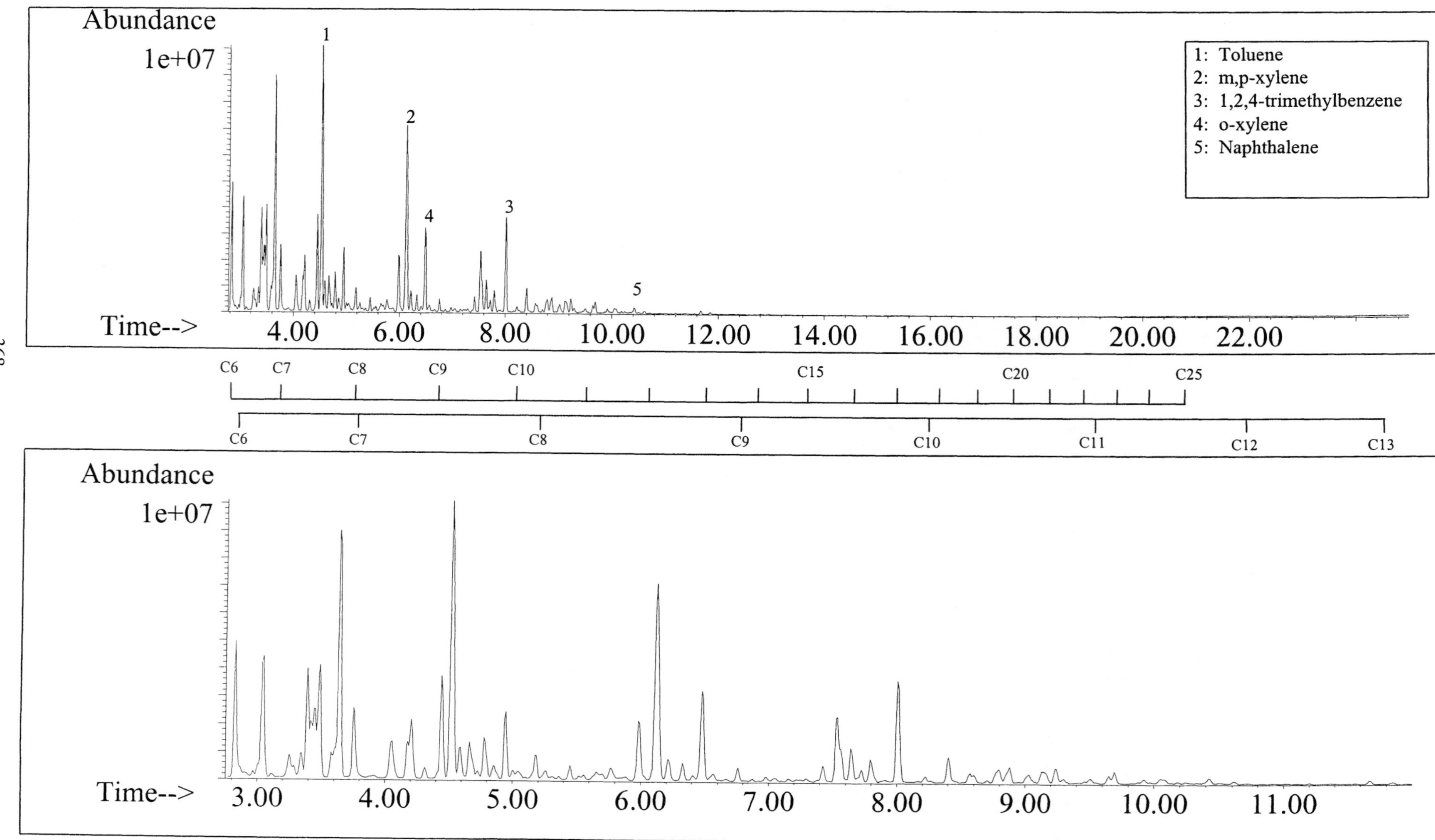

SUMMED PROFILES

Gasoline #28

ALKANES

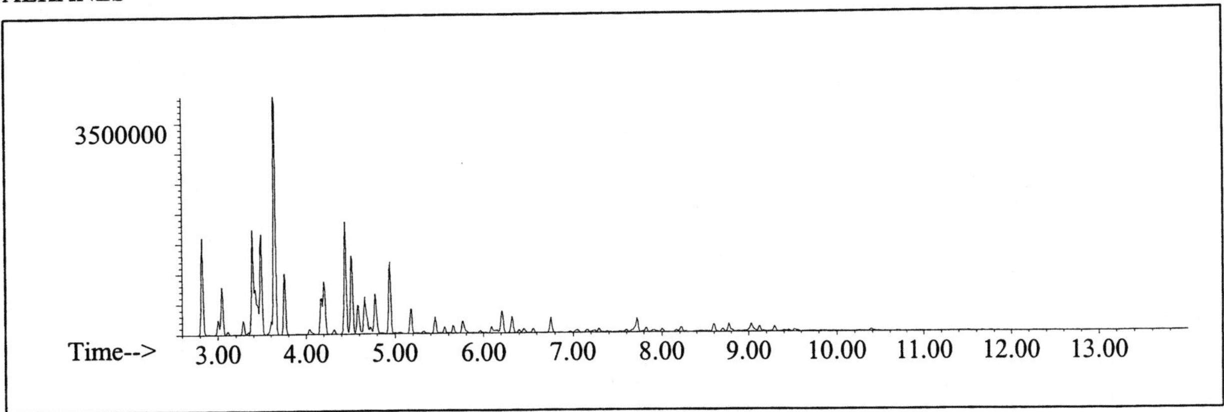

AROMATICS

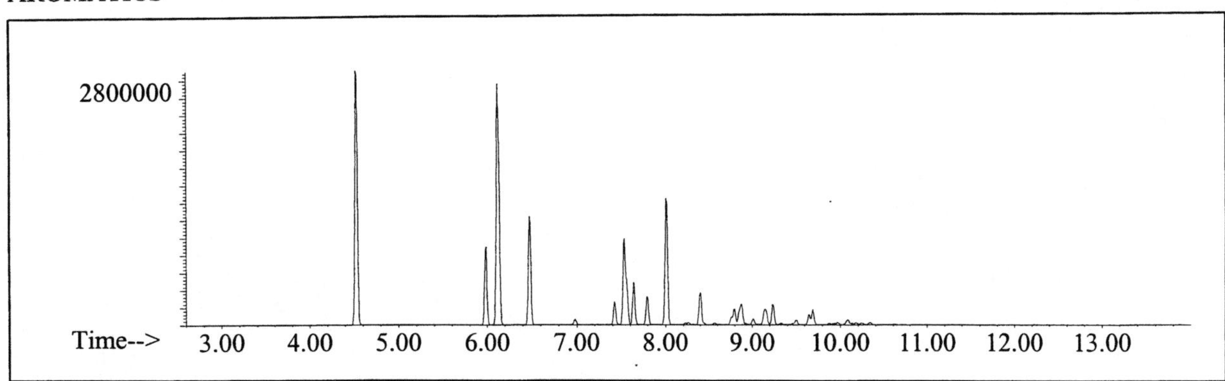

CYCLOPARAFFINS AND ALKENES

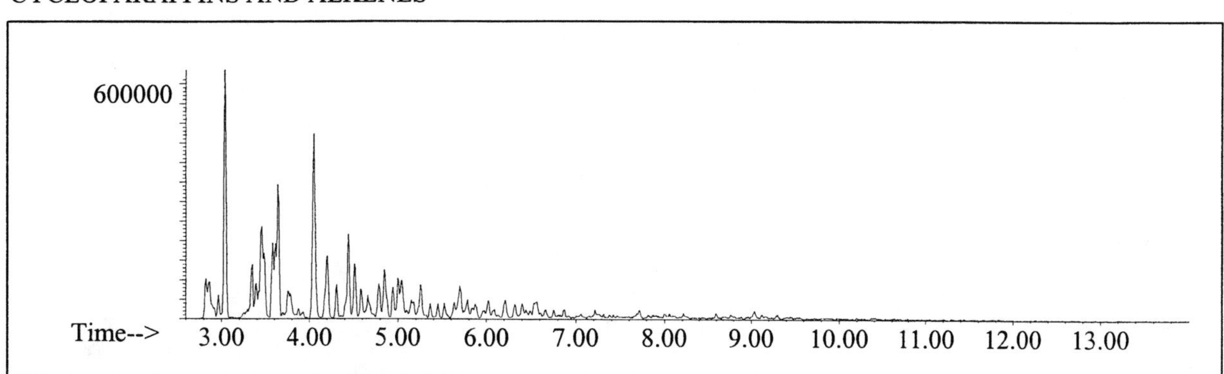

NAPHTHALENES

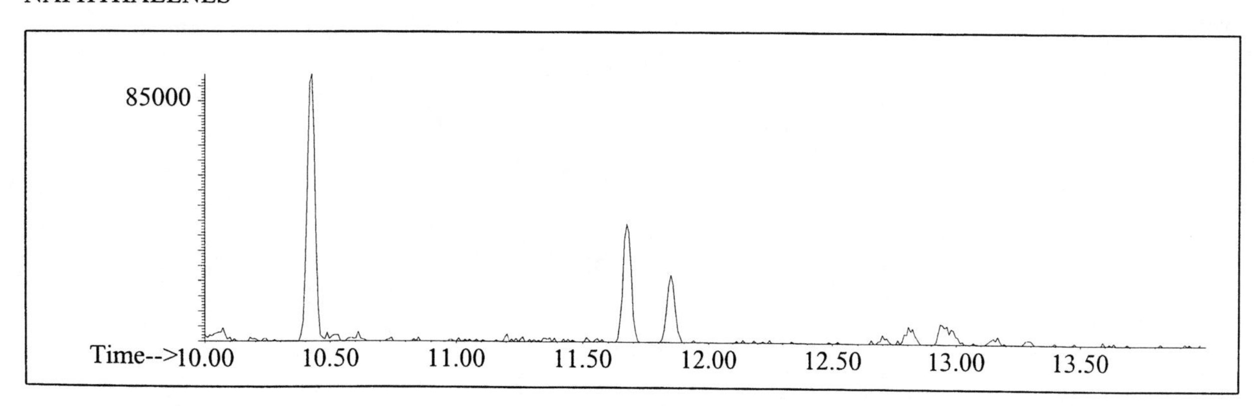

INDIVIDUAL PROFILES

Gasoline #28

Alkane

Ion 43

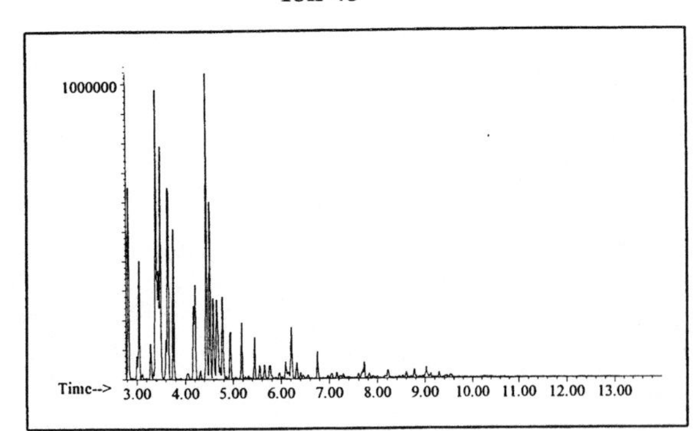

Ion 57

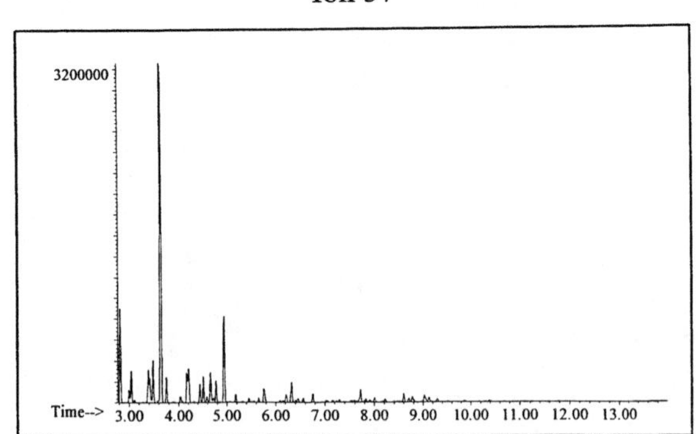

Ion 71

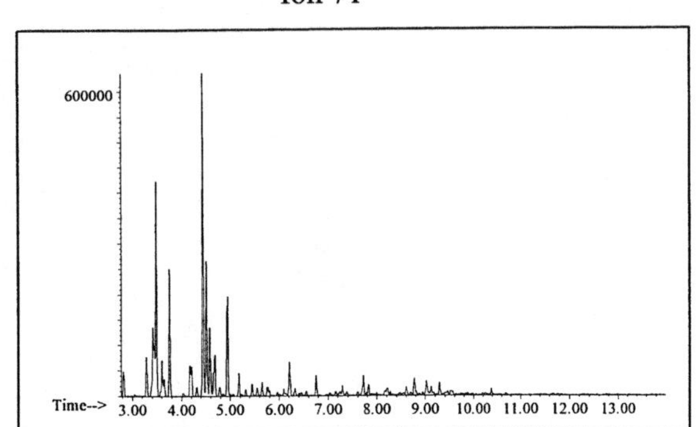

Ion 85

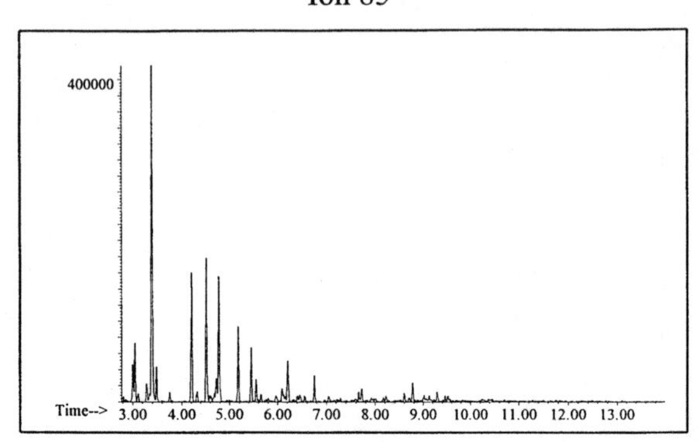

Aromatic

Ion 91

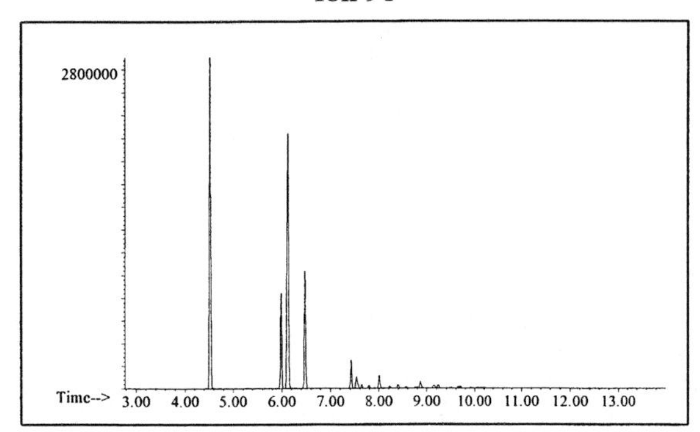

Ion 105

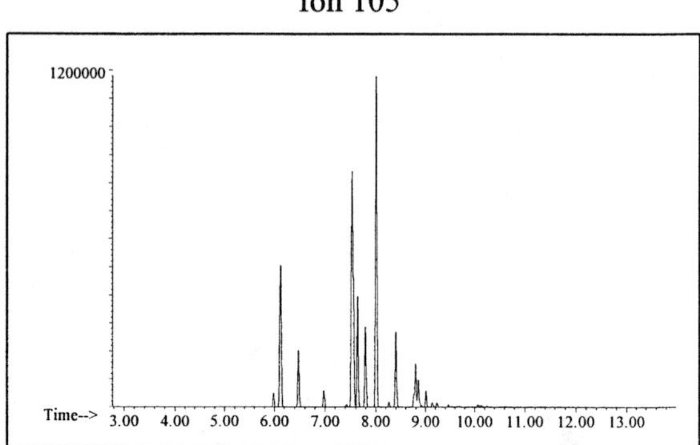

Ion 119

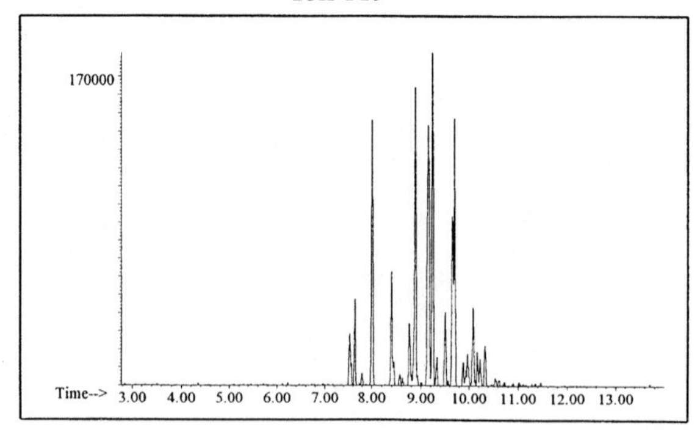

INDIVIDUAL PROFILES

Gasoline #28

Cycloparaffin

Ion 55

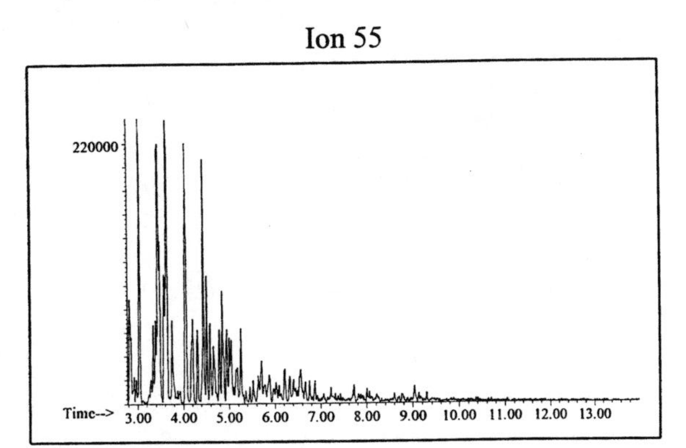

Ion 69

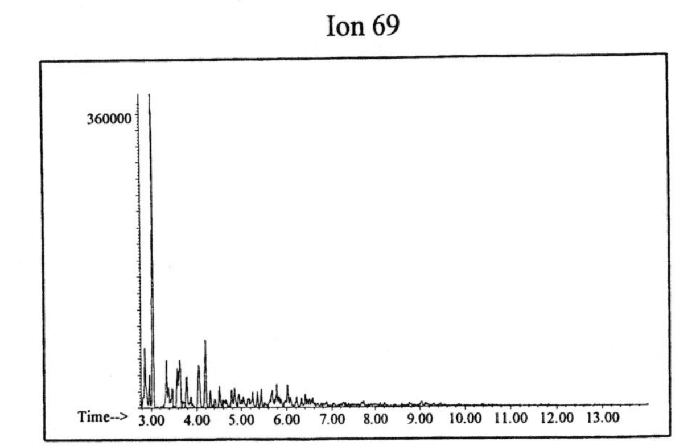

Ion 83

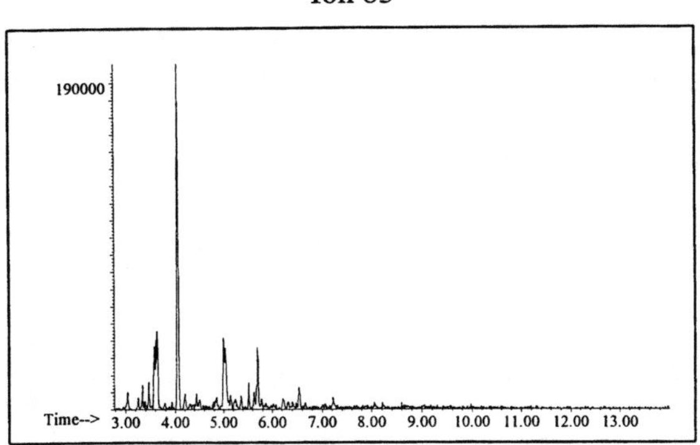

Naphthalene

Ion 128

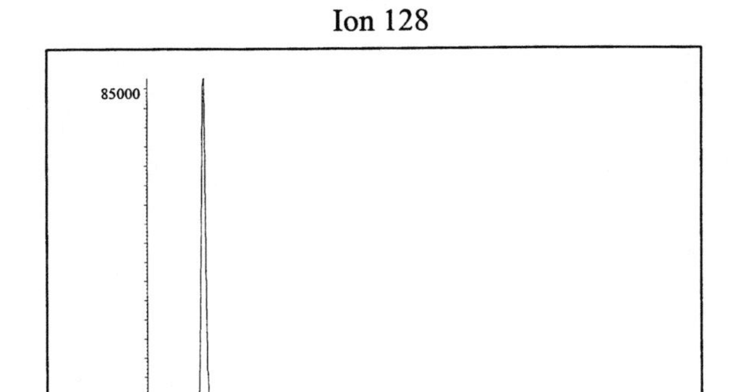

Ion 142

38000
Time--> 10.50 11.00 11.50 12.00 12.50 13.00 13.50

Ion 156

No Useful Data Obtained

Oxygenated Solvents

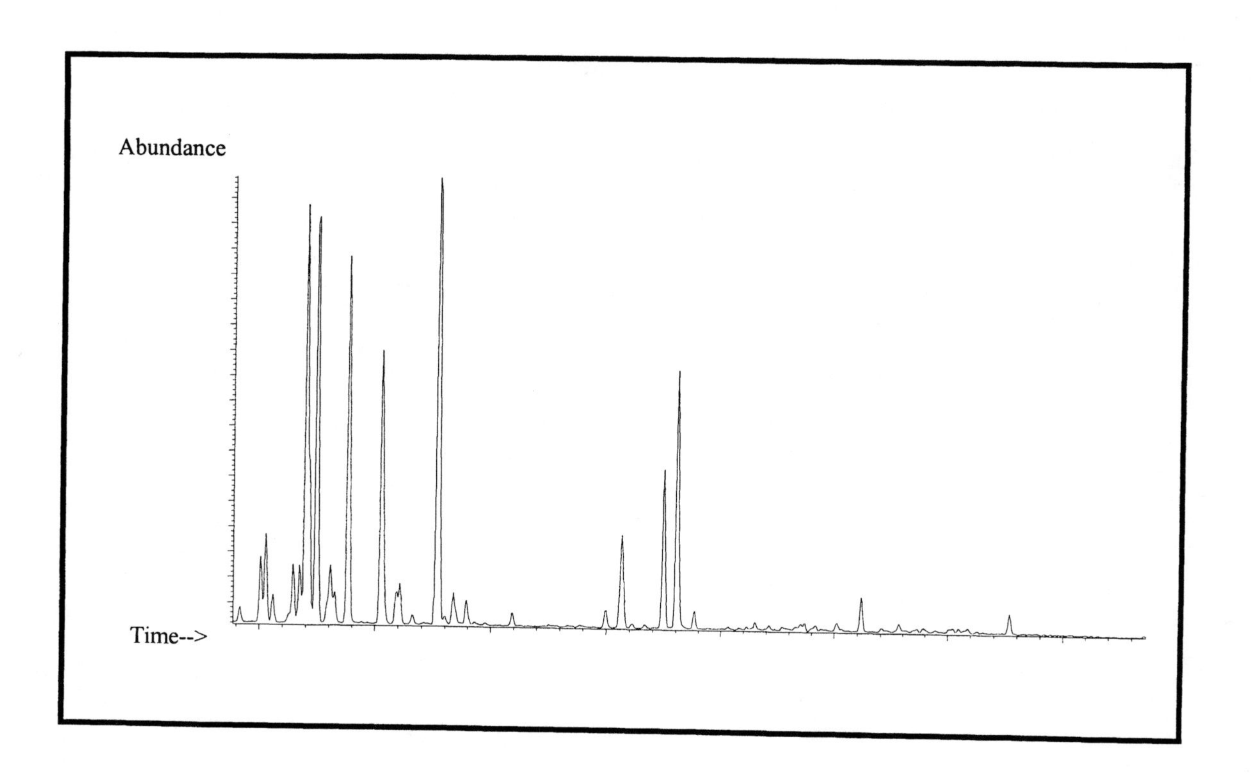

Oxygenate #1

20uL/mL Pentane

ASTM: Class 0.1 (Oxygenate)

Macro Code: L

Product Displayed: Prist Fuel Additive

Product Uses: Fuel treatment

Other Similar Products:

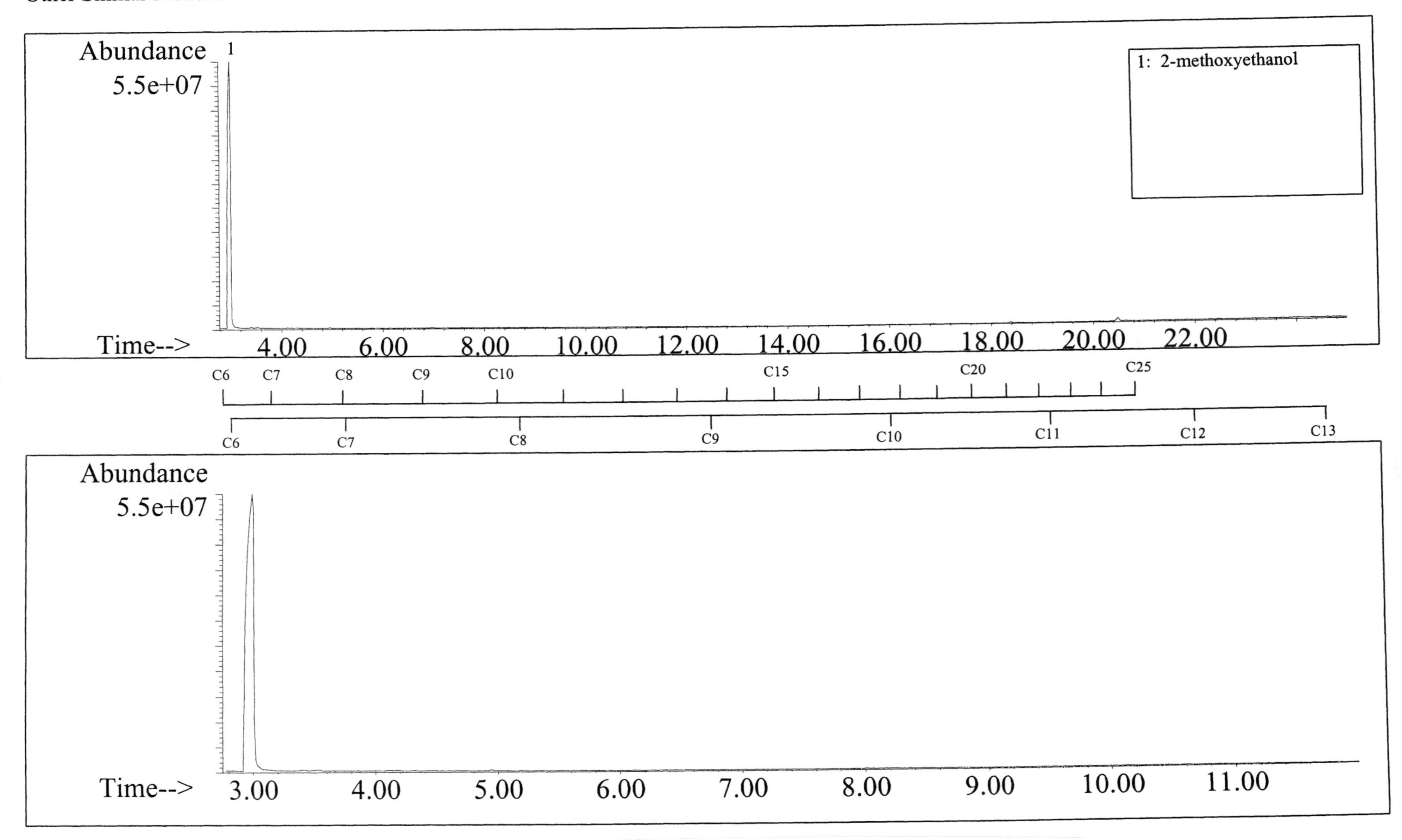

SUMMED PROFILES

Oxygenate #1

ALKANES

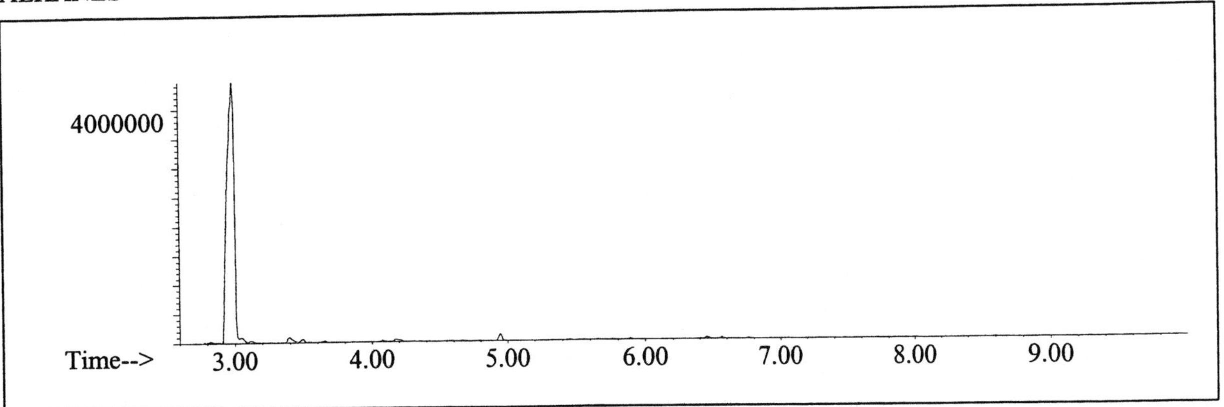

AROMATICS

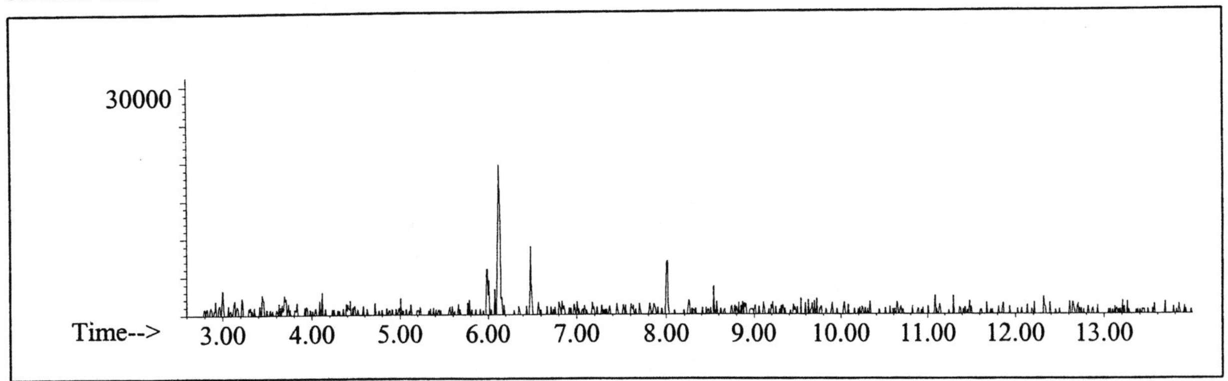

CYCLOPARAFFINS AND ALKENES

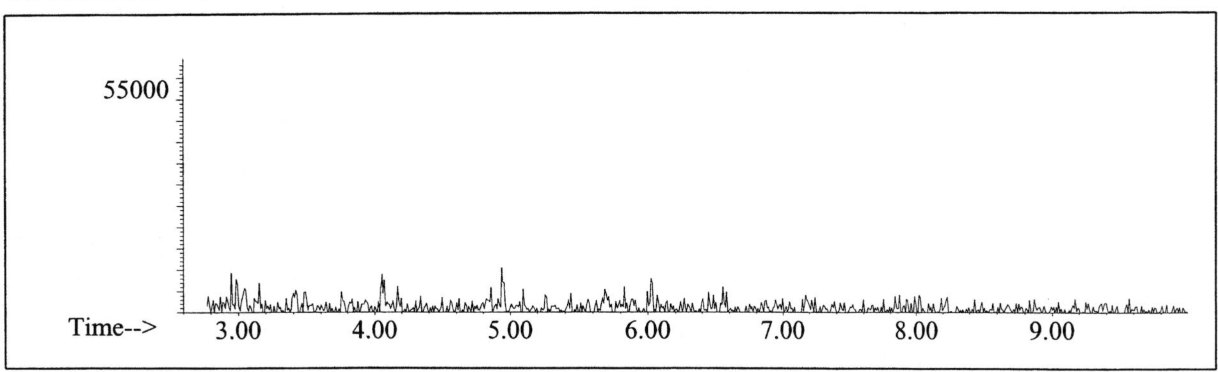

NAPHTHALENES

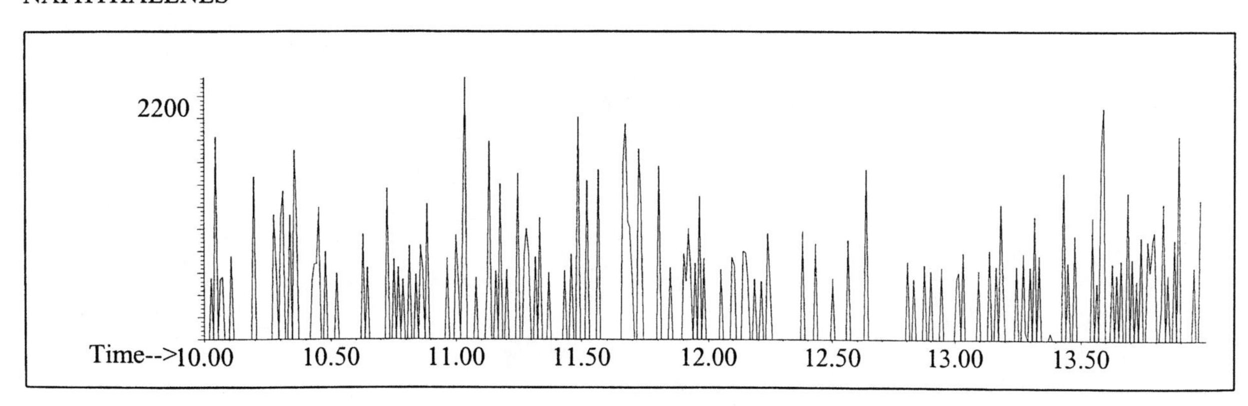

INDIVIDUAL PROFILES

Oxygenate #1

Alkane

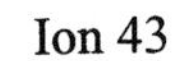

Ion 43

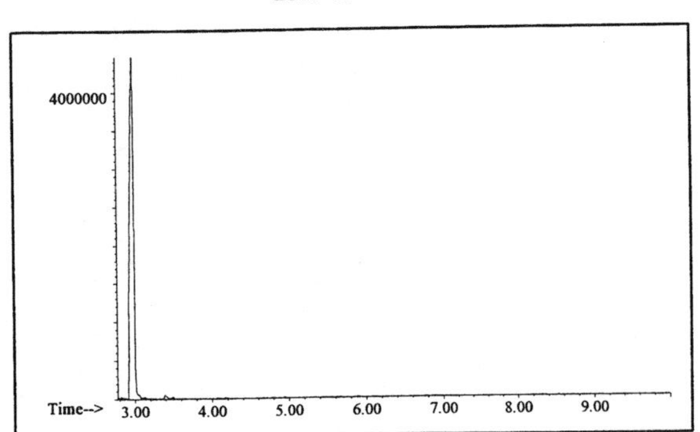

Ion 57

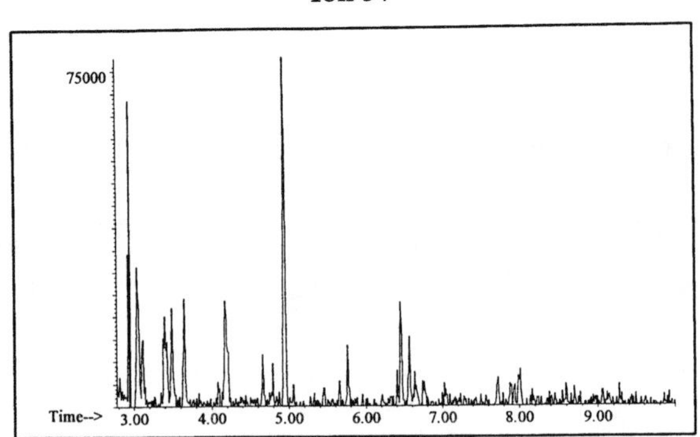

Ion 71

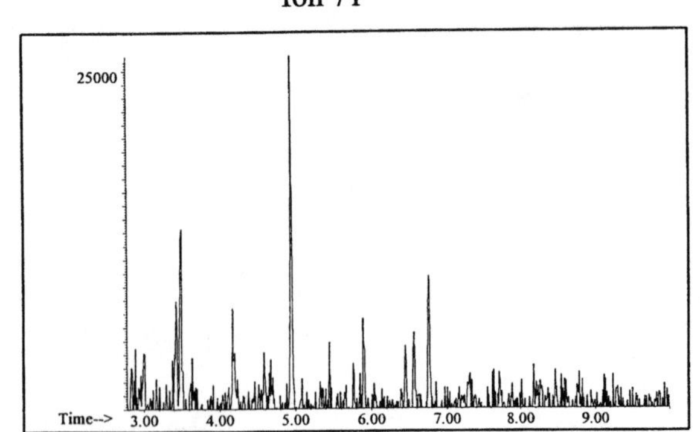

Ion 85

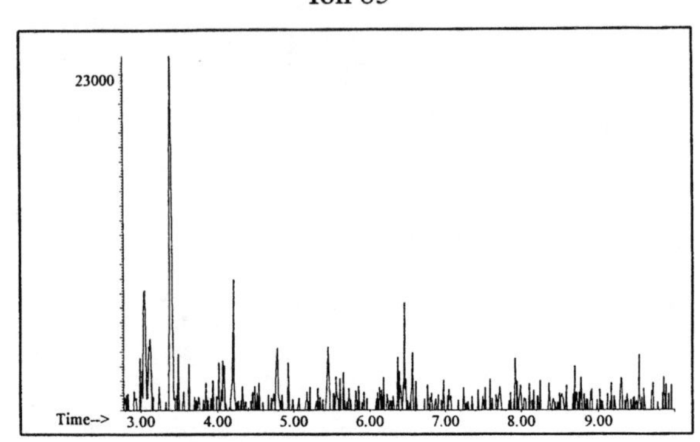

Aromatic

Ion 91

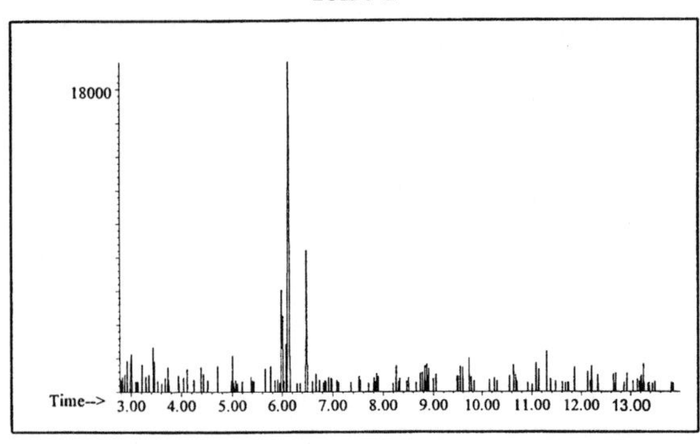

Ion 105

No Useful Data Obtained

Ion 119

No Useful Data Obtained

INDIVIDUAL PROFILES

Oxygenate #1

Cycloparaffin

Ion 55

No Useful Data Obtained

Ion 69

No Useful Data Obtained

Ion 83

No Useful Data Obtained

Naphthalene

Ion 128

No Useful Data Obtained

Ion 142

No Useful Data Obtained

Ion 156

No Useful Data Obtained

Oxygenate #2

20uL/mL Pentane

ASTM: Class 0.1 (Oxygenate)

Macro Code: L

Product Displayed: Wilbond Surface Prep

Product Uses: Cleaning solvent

Other Similar Products:

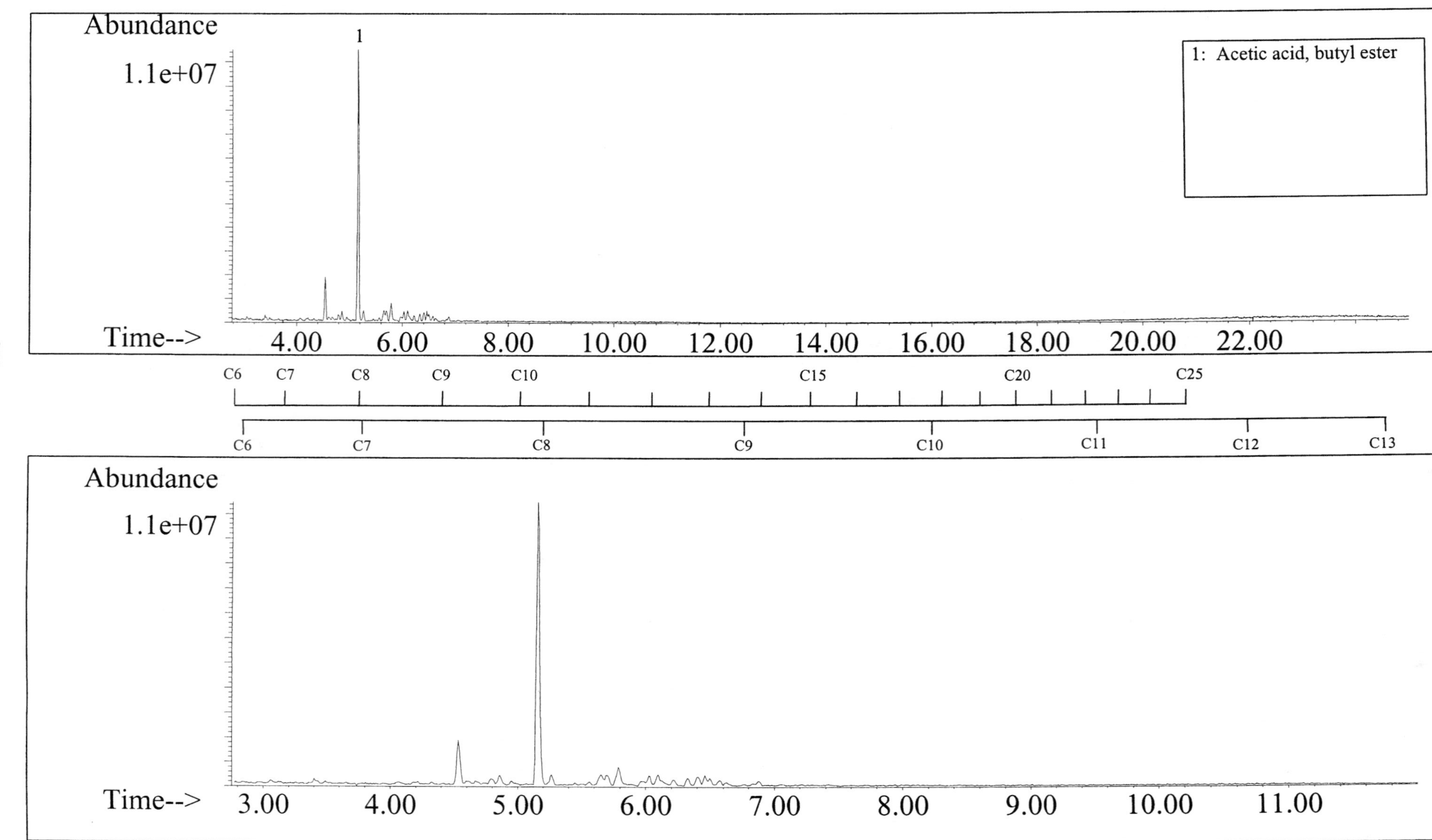

SUMMED PROFILES Oxygenate #2

ALKANES

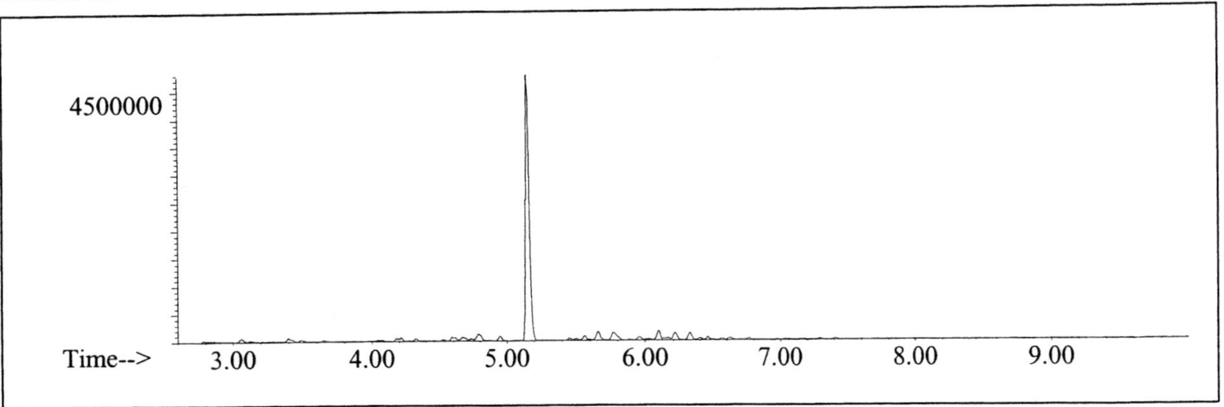

AROMATICS

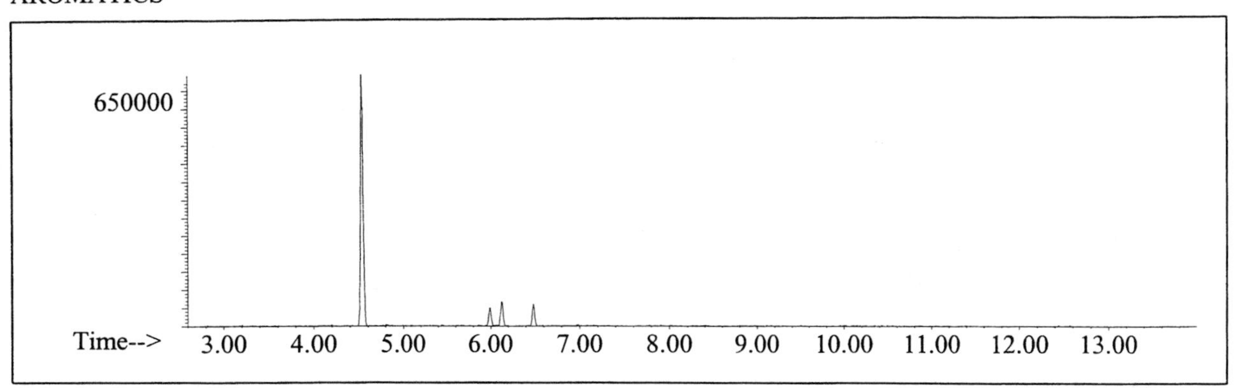

CYCLOPARAFFINS AND ALKENES

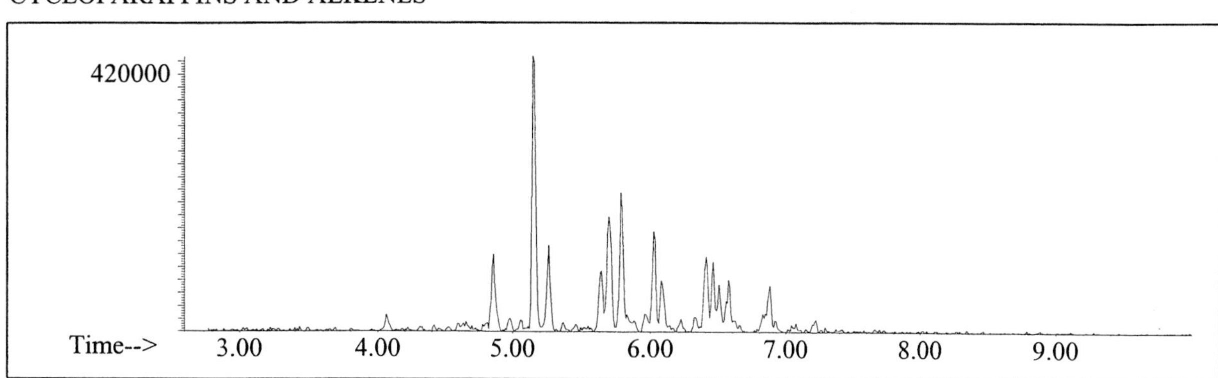

NAPHTHALENES

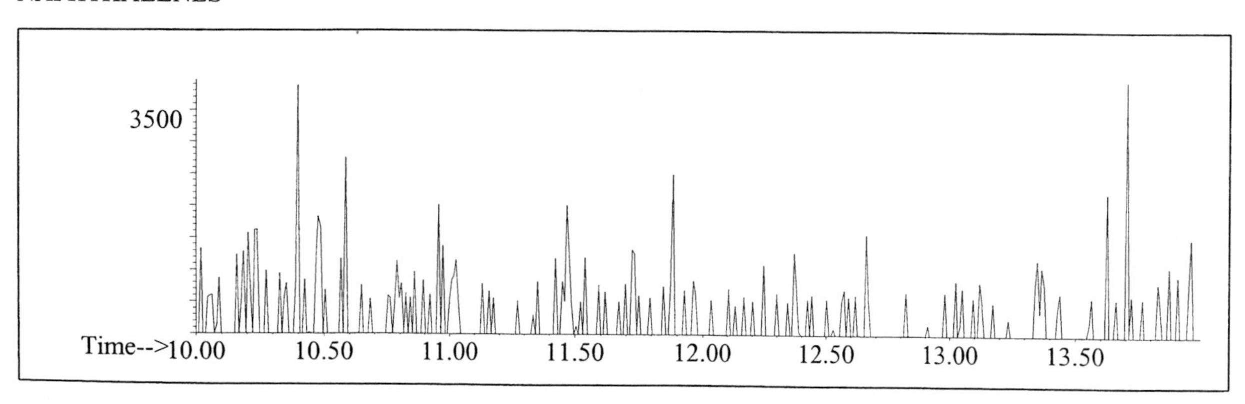

INDIVIDUAL PROFILES

Oxygenate #2

Alkane

Ion 43

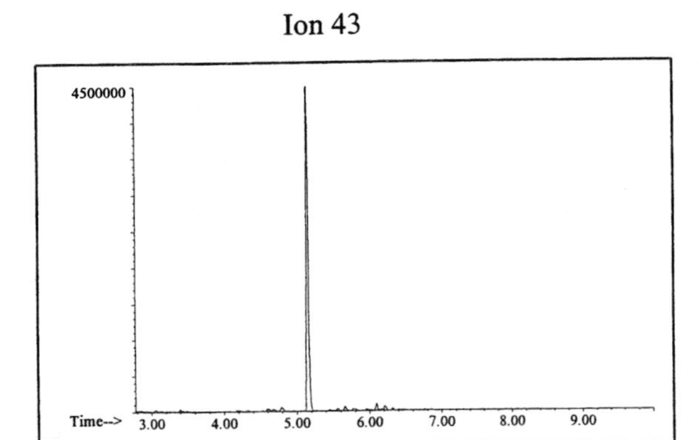

Ion 57

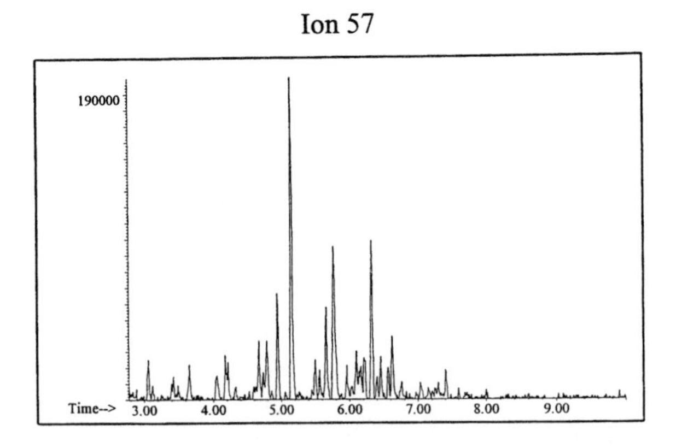

Ion 71

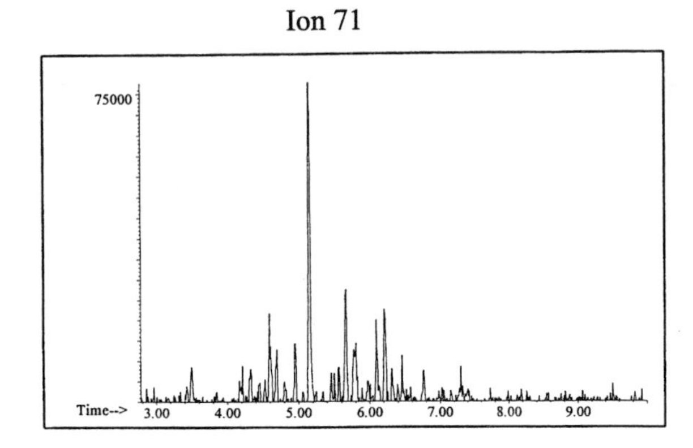

Ion 85

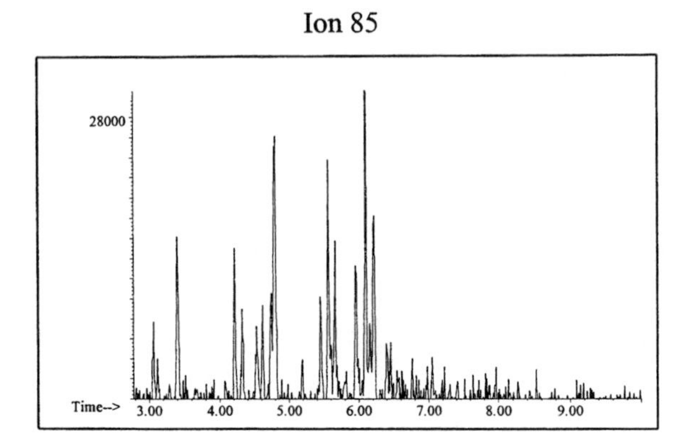

Aromatic

Ion 91

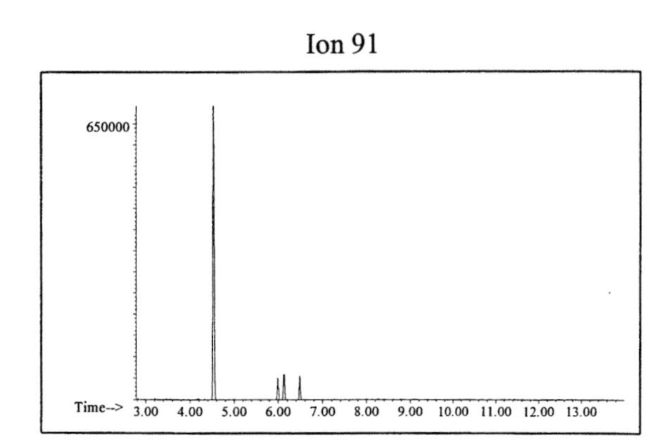

Ion 105

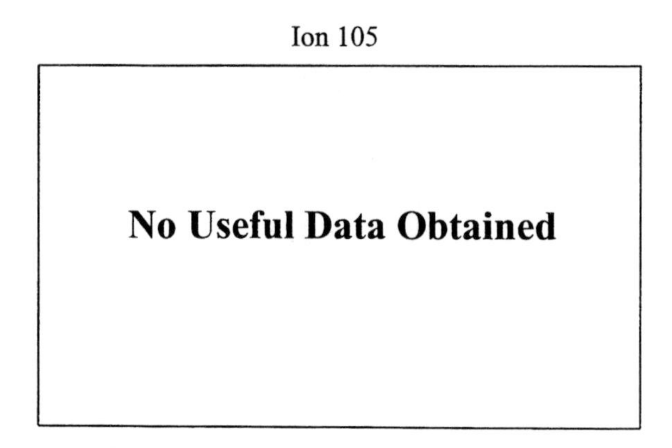

Ion 119

No Useful Data Obtained

INDIVIDUAL PROFILES

Cycloparaffin

Ion 55

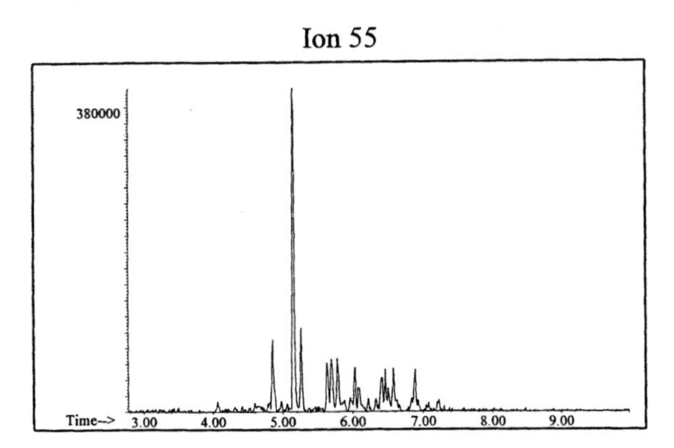

Ion 69

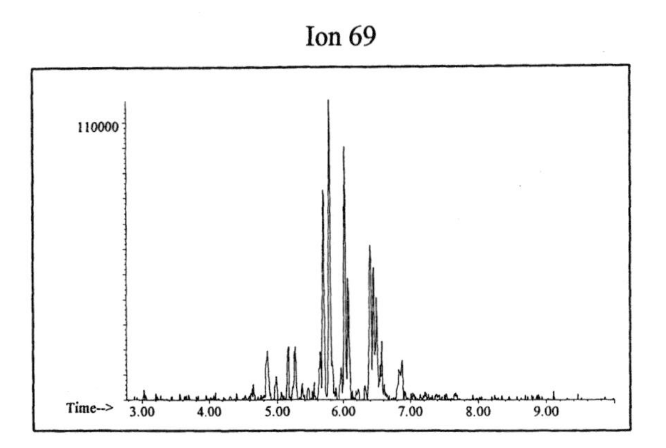

Ion 83

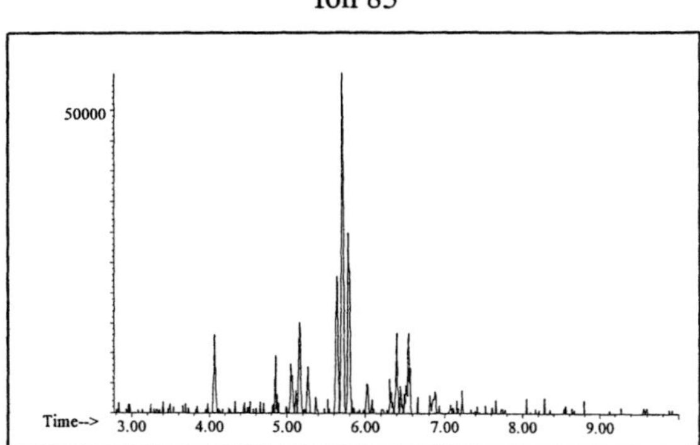

Naphthalene

Ion 128

No Useful Data Obtained

Ion 142

No Useful Data Obtained

Ion 156

No Useful Data Obtained

Oxygenate #3

20uL/mL Pentane

Product Displayed: L&H Lacquer Thinner

Other Similar Products:

ASTM: Class 0.1 (Oxygenate)

Product Uses: Lacquer thinner

Macro Code: L

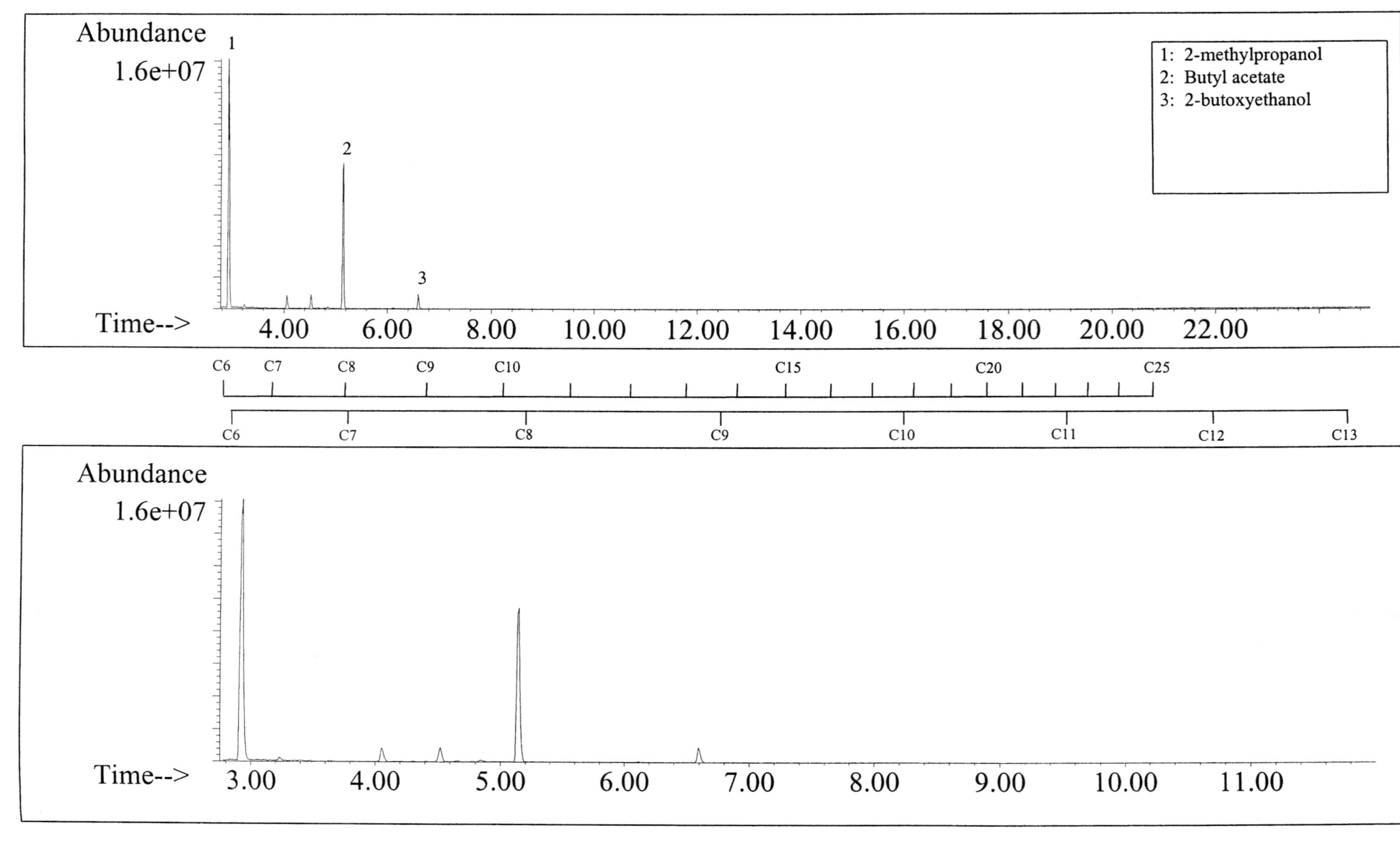

SUMMED PROFILES

Oxygenate #3

ALKANES

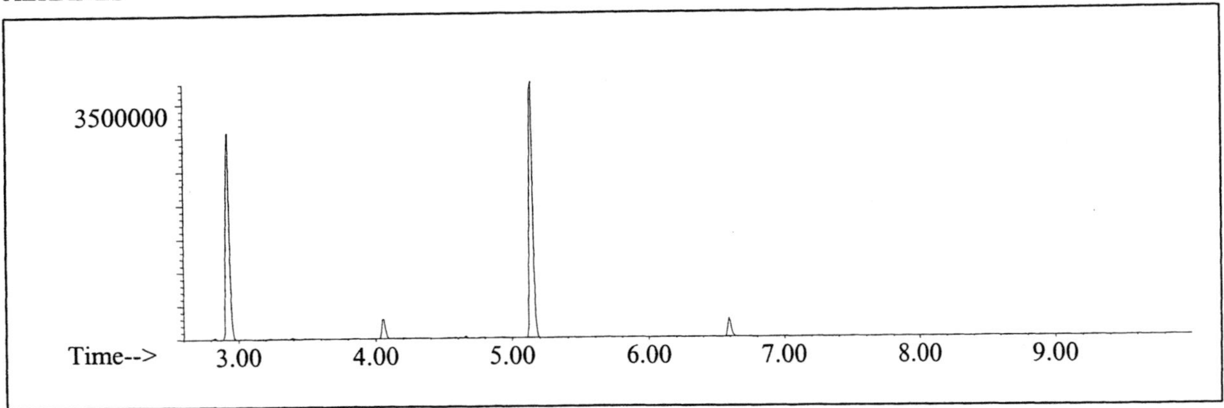

AROMATICS

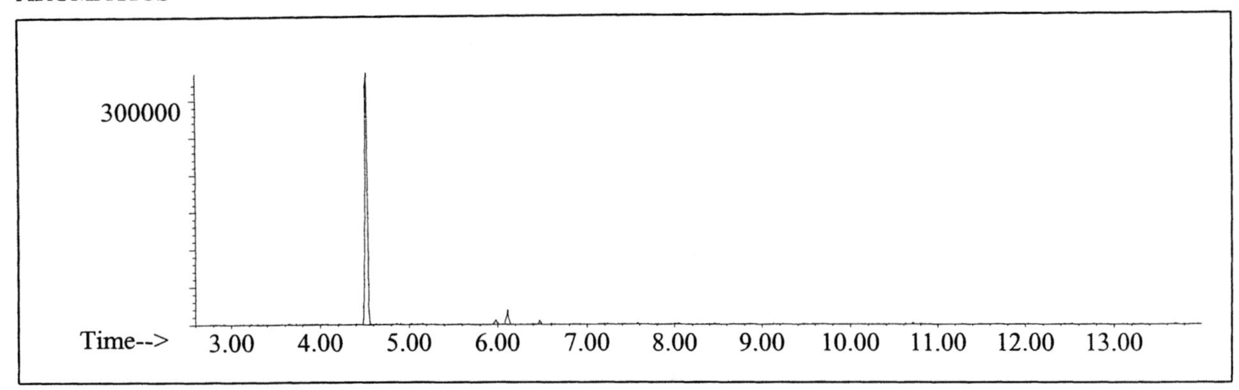

CYCLOPARAFFINS AND ALKENES

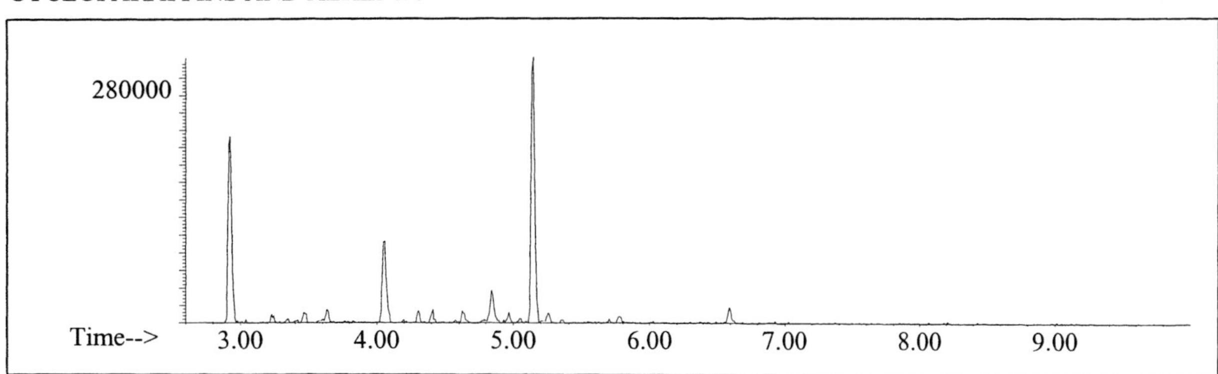

NAPHTHALENES

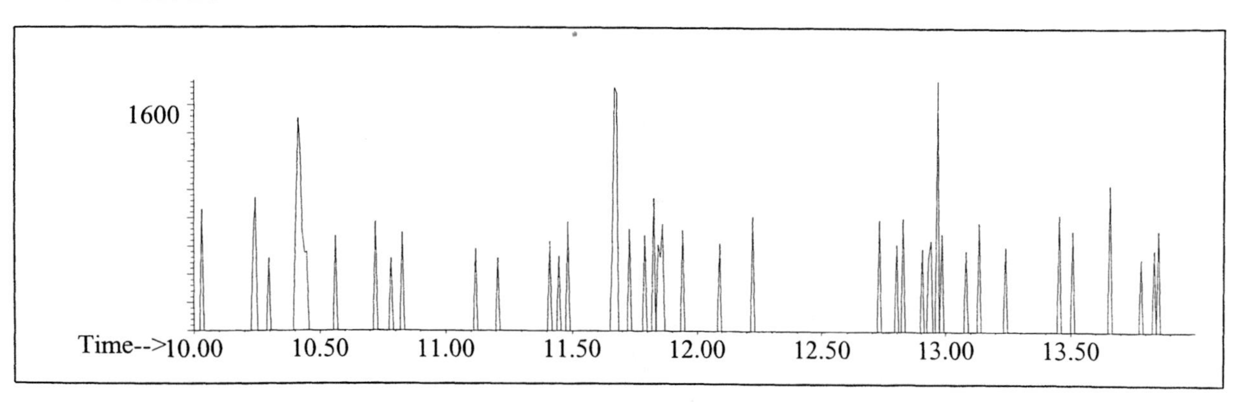

INDIVIDUAL PROFILES

Oxygenate #3

Alkane

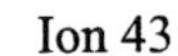

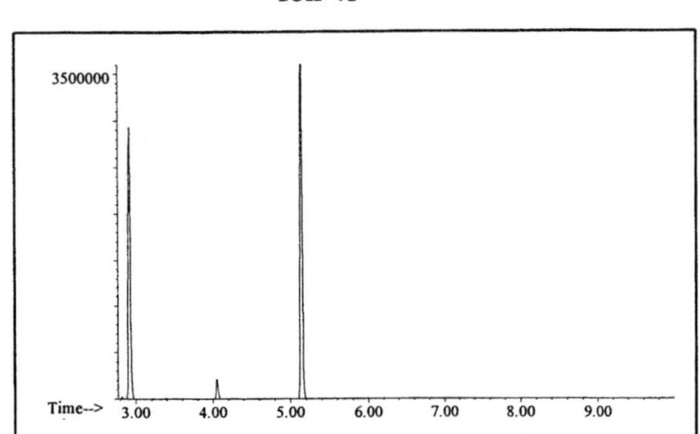

Ion 57

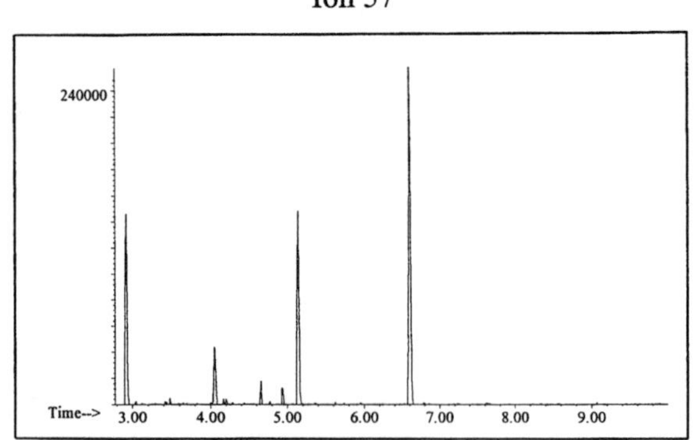

Ion 71

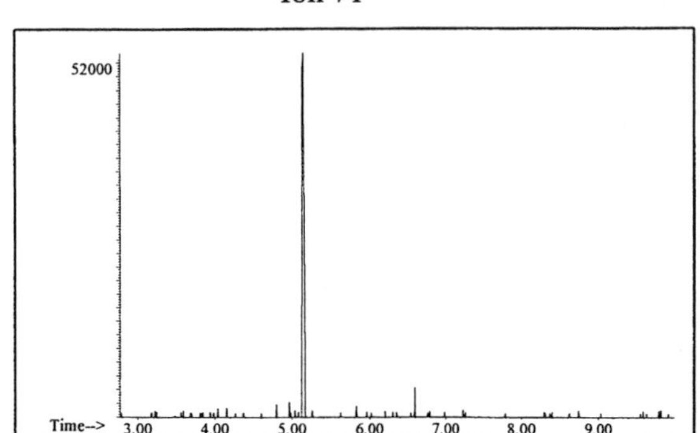

Ion 85

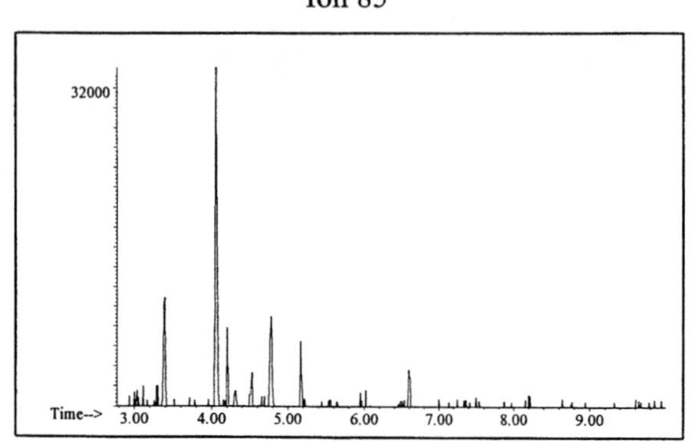

Aromatic

Ion 91

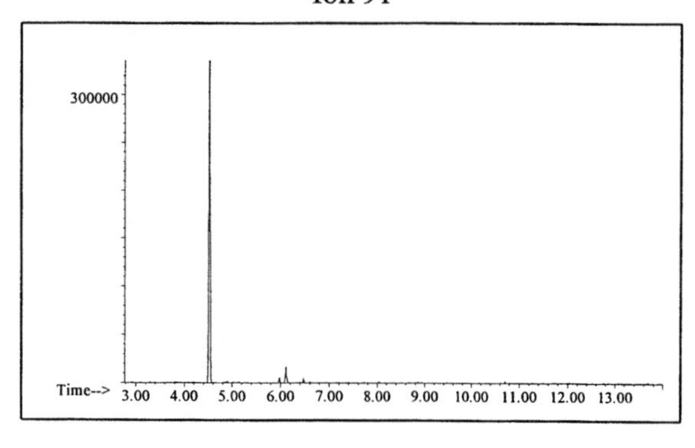

Ion 105

No Useful Data Obtained

Ion 119

No Useful Data Obtained

INDIVIDUAL PROFILES

Oxygenate #3

Cycloparaffin

Ion 55

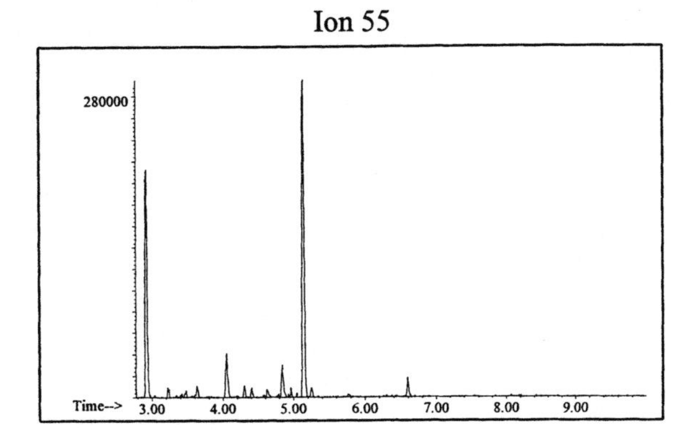

Ion 69

No Useful Data Obtained

Ion 83

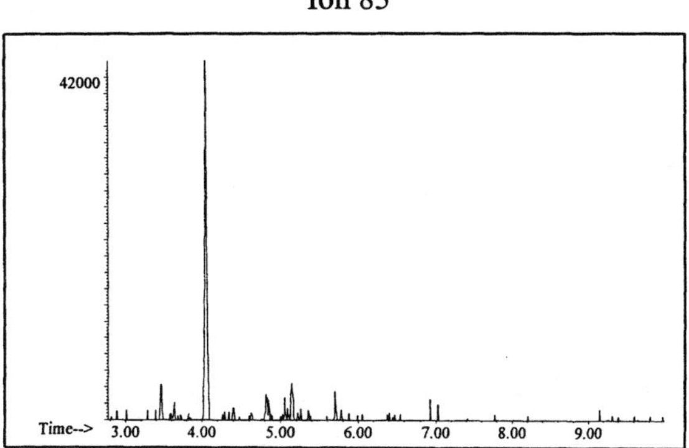

Naphthalene

Ion 128

No Useful Data Obtained

Ion 142

No Useful Data Obtained

Ion 156

No Useful Data Obtained

Oxygenate #4

20uL/mL Pentane

ASTM: Class 0.1 (Oxygenate)

Macro Code: L

Product Displayed: IPS Weld-on Primer P70 for PVC

Product Uses: Surface preparation solvent

Other Similar Products:

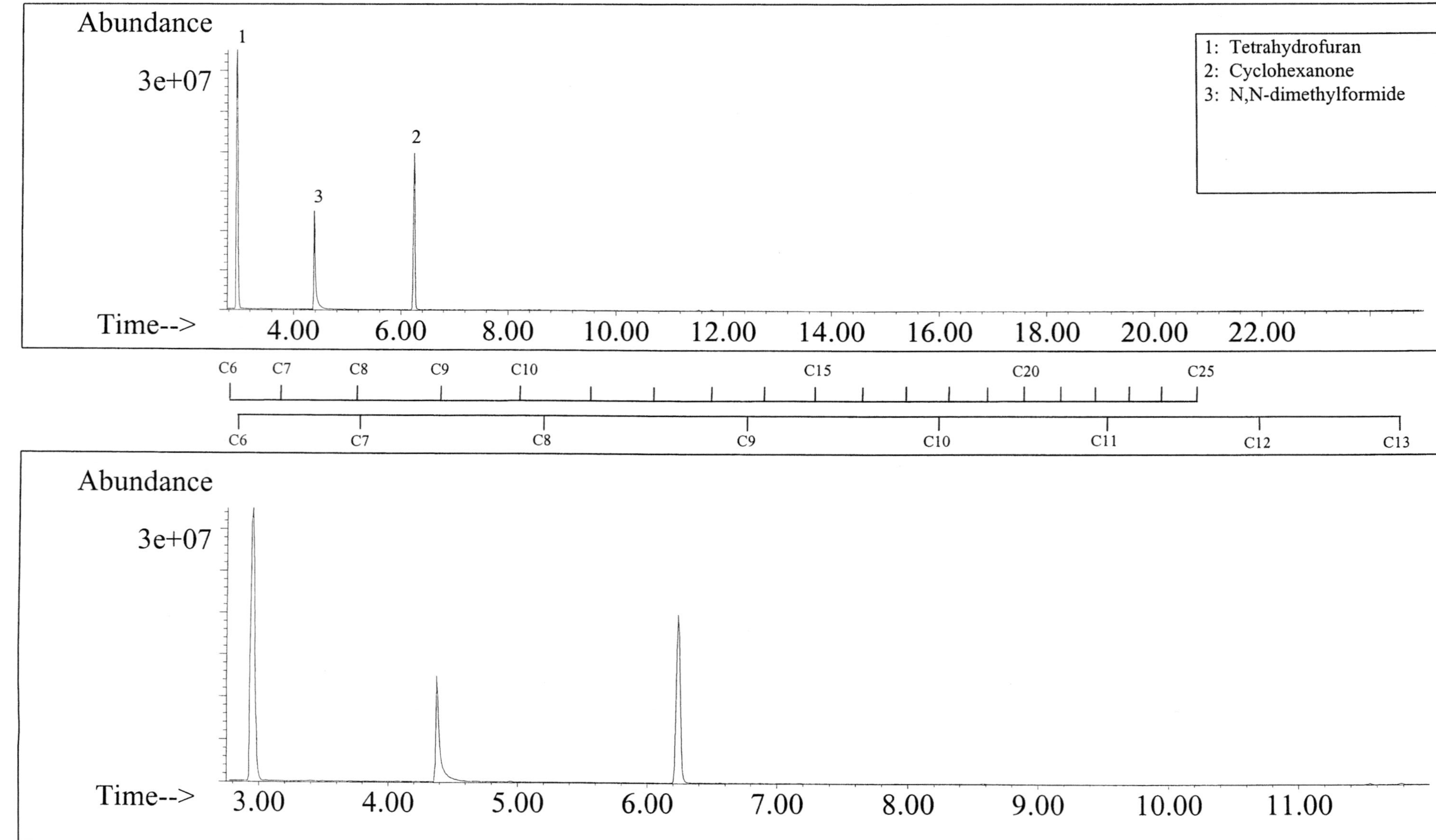

SUMMED PROFILES

Oxygenate #4

ALKANES

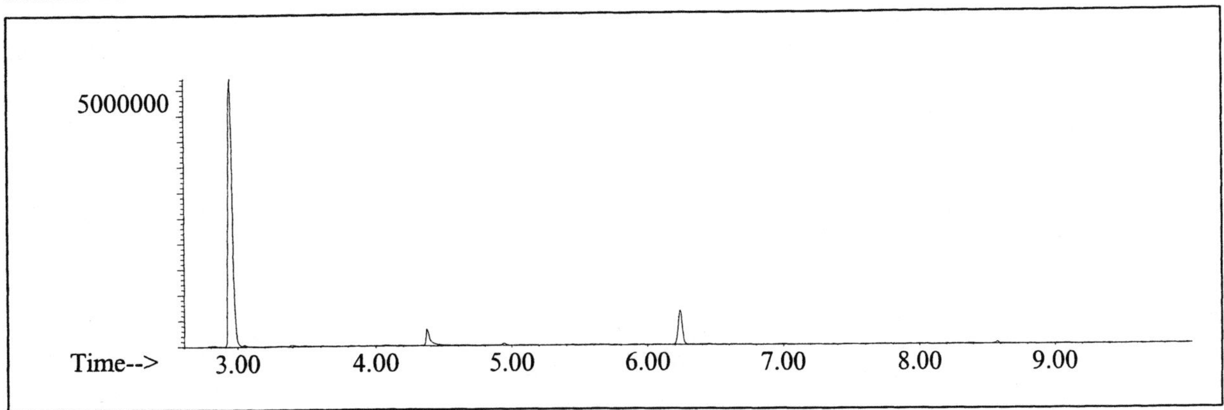

AROMATICS

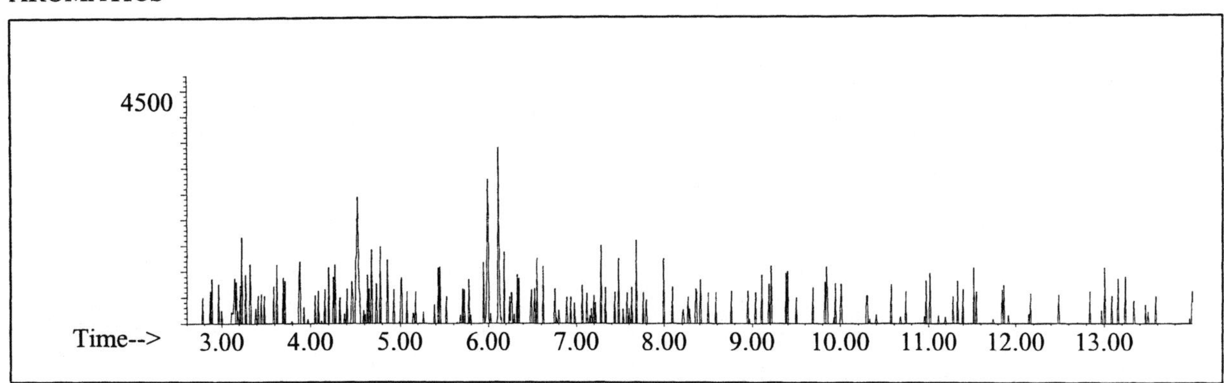

CYCLOPARAFFINS AND ALKENES

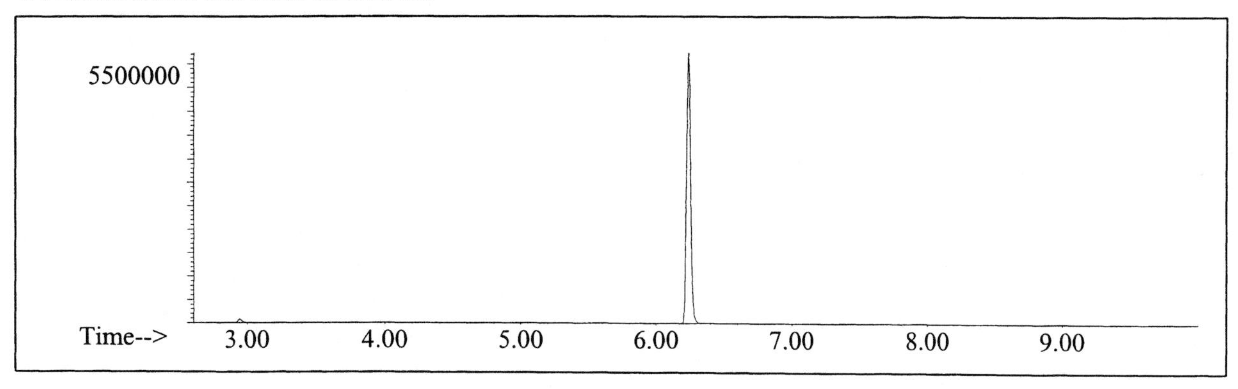

NAPHTHALENES

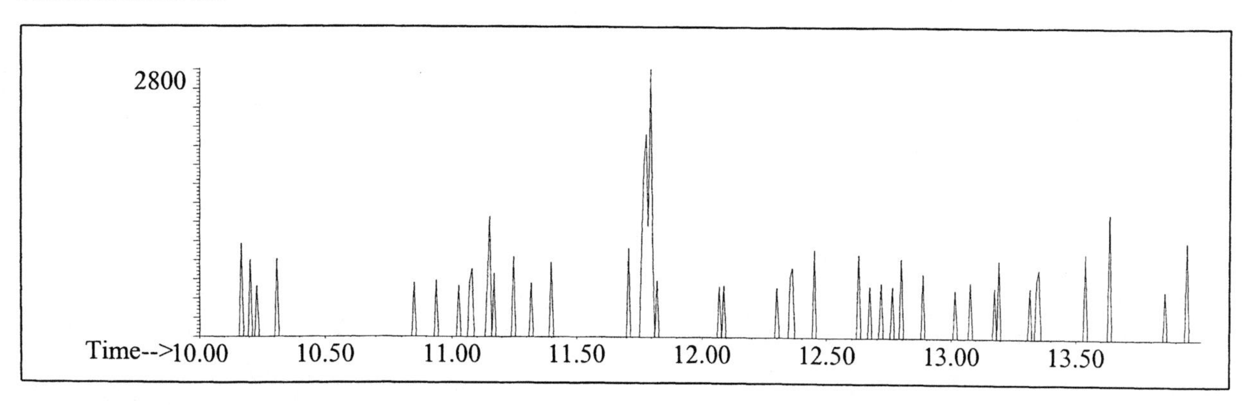

INDIVIDUAL PROFILES

Oxygenate #4

Alkane

Ion 43

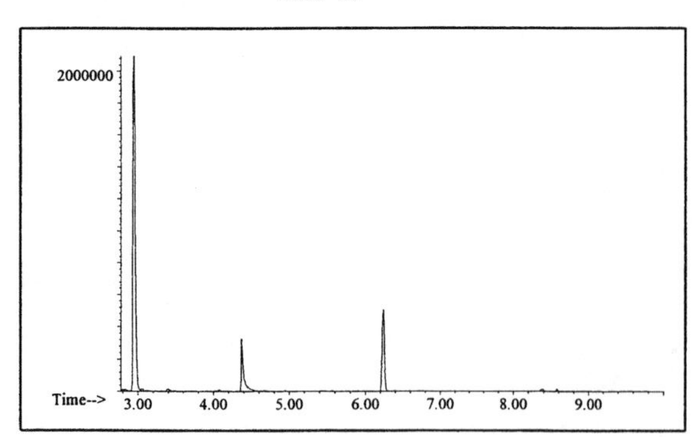

Ion 57

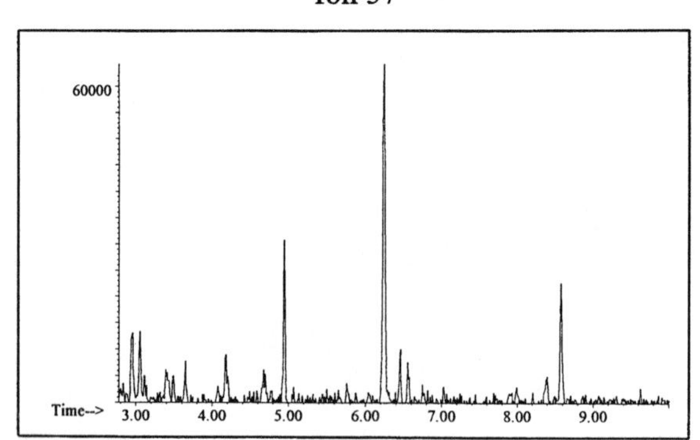

Ion 71

Ion 85

No Useful Data Obtained

Aromatic

Ion 91

No Useful Data Obtained

Ion 105

No Useful Data Obtained

Ion 119

No Useful Data Obtained

INDIVIDUAL PROFILES

Oxygenate #4

Cycloparaffin

Ion 55

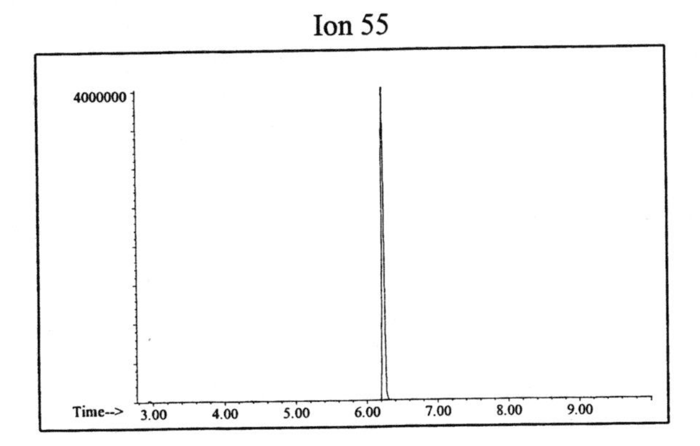

Ion 69

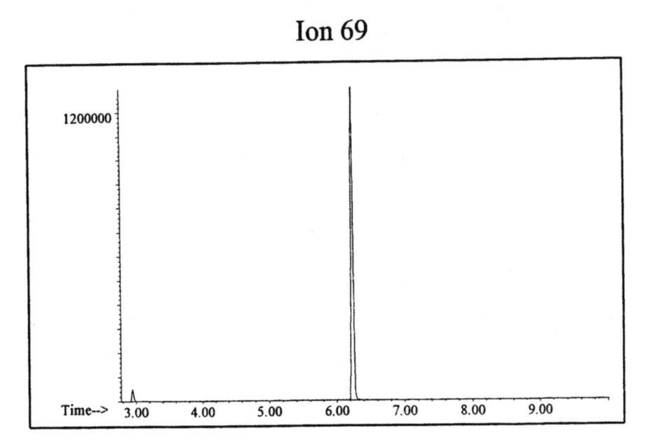

Ion 83

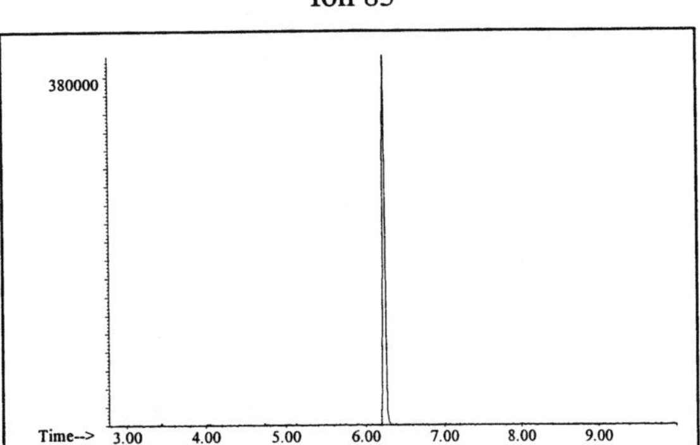

Naphthalene

Ion 128

No Useful Data Obtained

Ion 142

No Useful Data Obtained

Ion 156

No Useful Data Obtained

Oxygenate #5

20uL/mL Pentane

ASTM: Class 0.1 (Oxygenate)

Macro Code: L

Product Displayed: Kleenstrip K6 Lacquer Thinner

Product Uses: Lacquer thinner

Other Similar Products:

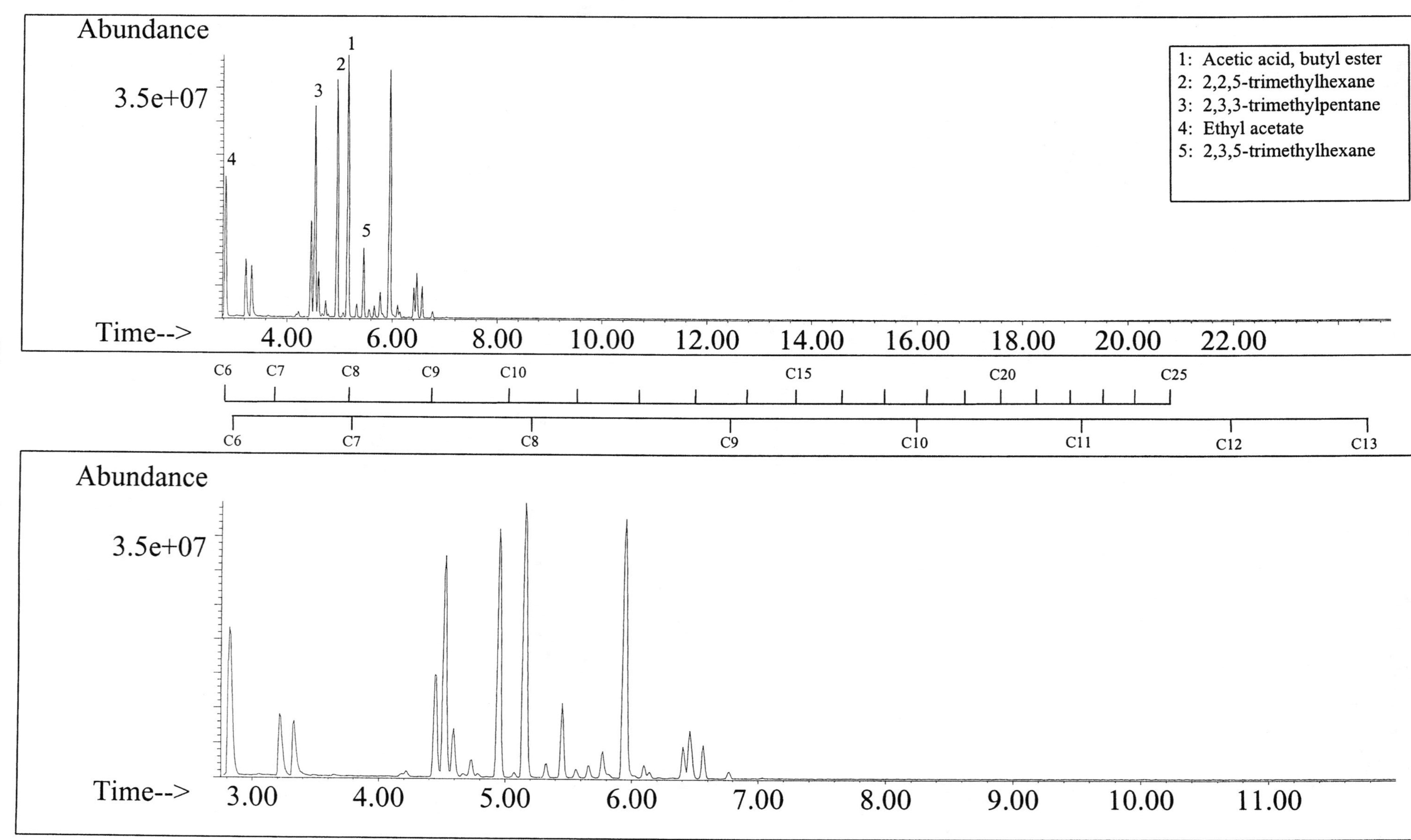

SUMMED PROFILES

Oxygenate #5

ALKANES

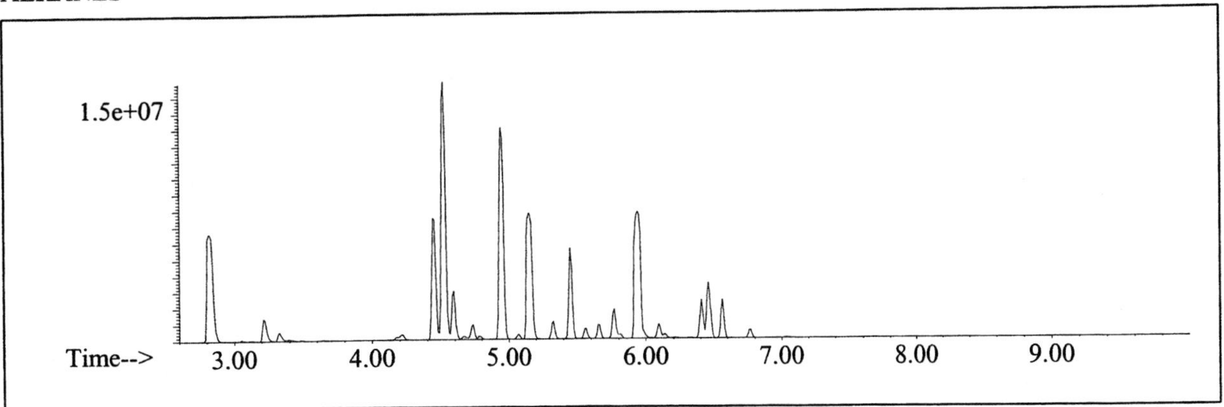

AROMATICS

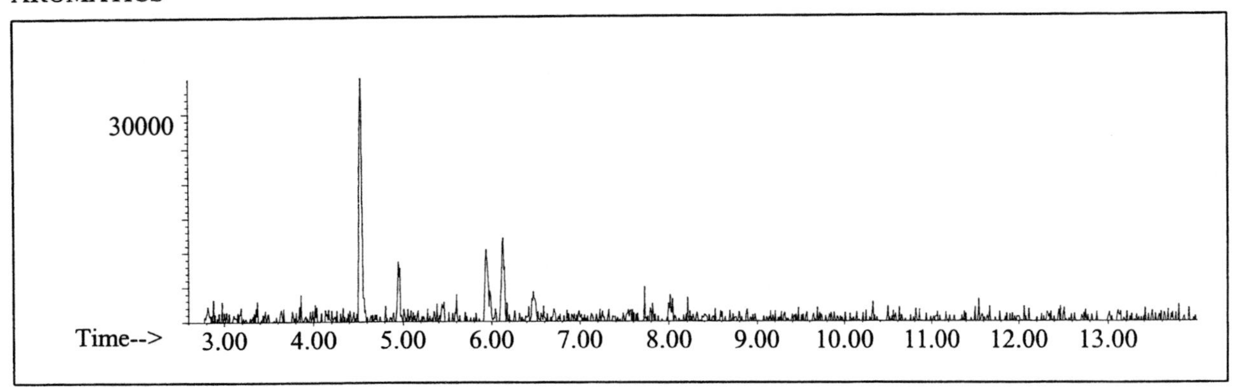

CYCLOPARAFFINS AND ALKENES

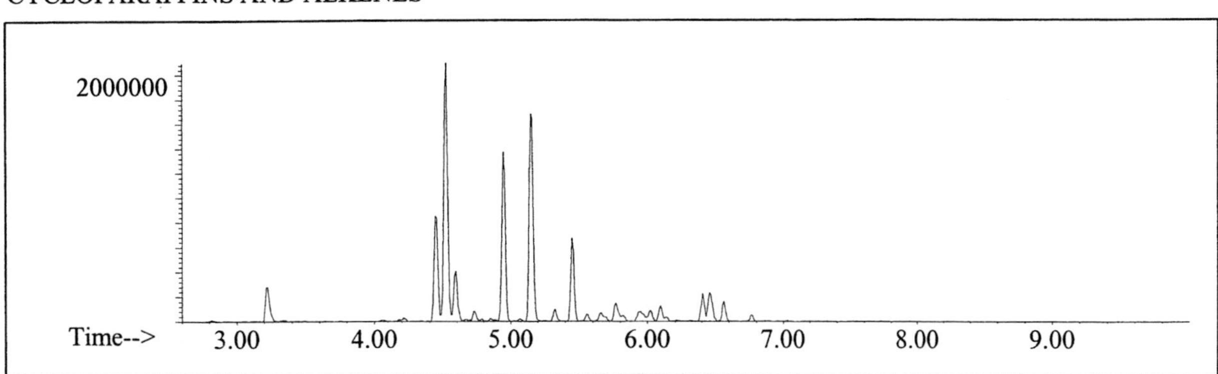

NAPHTHALENES

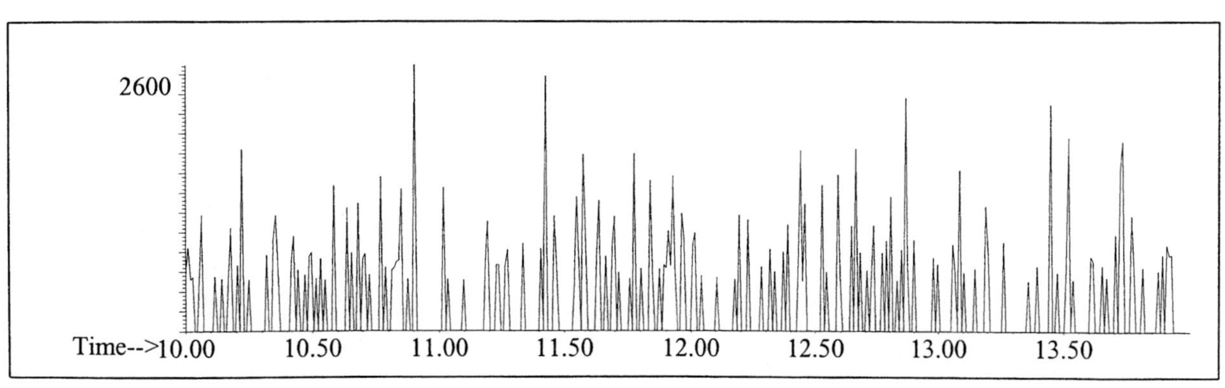

INDIVIDUAL PROFILES

Oxygenate #5

Alkane

Ion 43

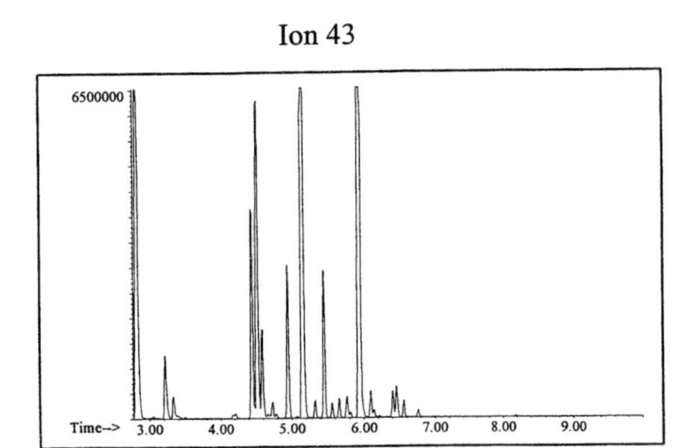

Ion 57

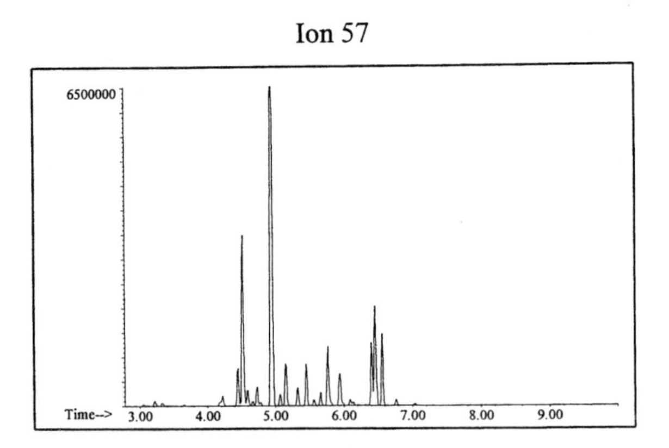

Ion 71

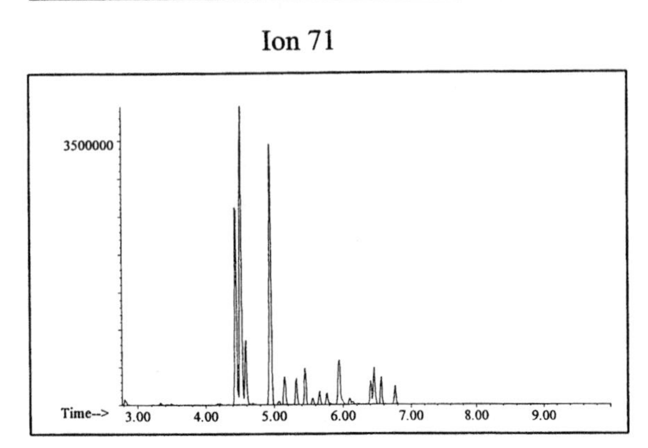

Ion 85

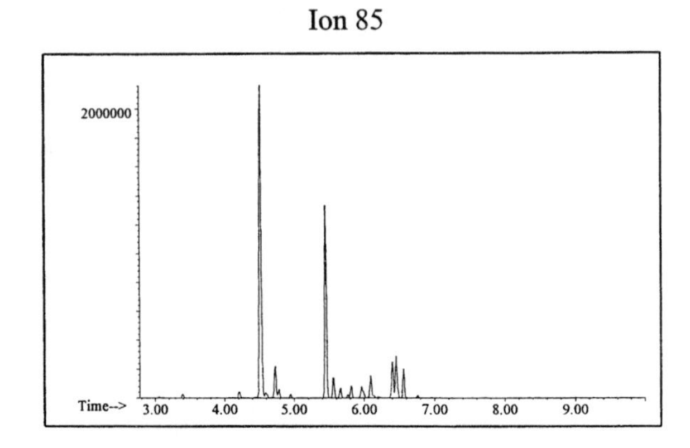

Aromatic

Ion 91

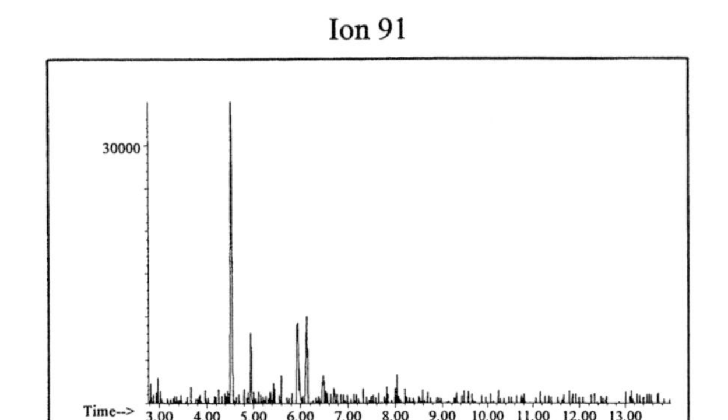

Ion 105

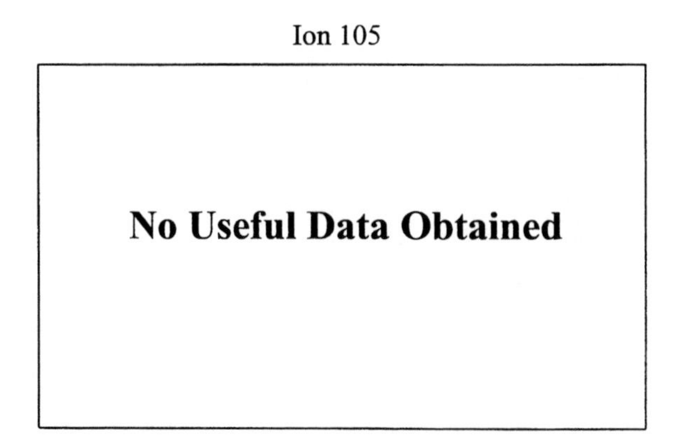

Ion 119

No Useful Data Obtained

INDIVIDUAL PROFILES

Oxygenate #5

Cycloparaffin

Ion 55

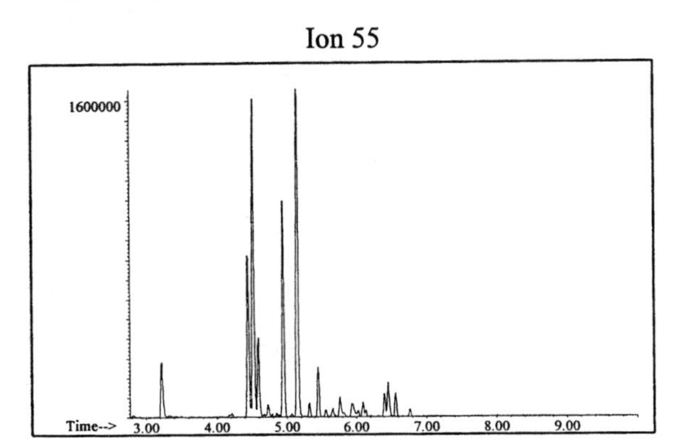

Ion 69

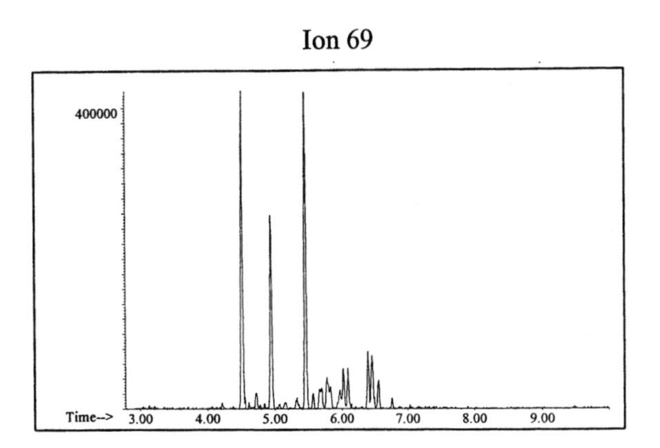

Ion 83

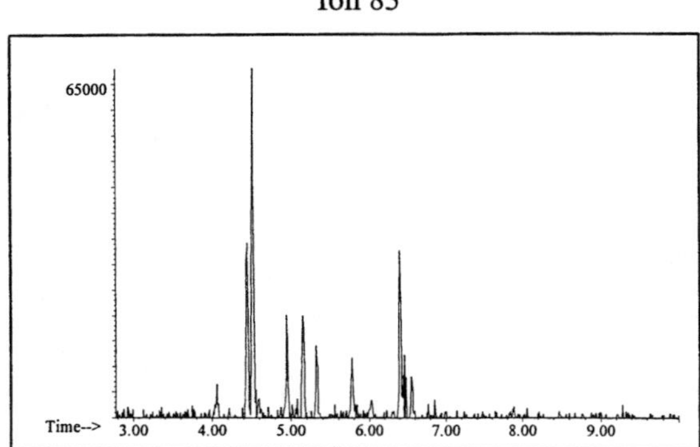

Naphthalene

Ion 128

No Useful Data Obtained

Ion 142

No Useful Data Obtained

Ion 156

No Useful Data Obtained

Oxygenate #6

20uL/mL Pentane

ASTM: Class 0.1 (Oxygenate)

Macro Code: L

Product Displayed: USA Lacquer Thinner

Product Uses: Lacquer thinner

Other Similar Products:

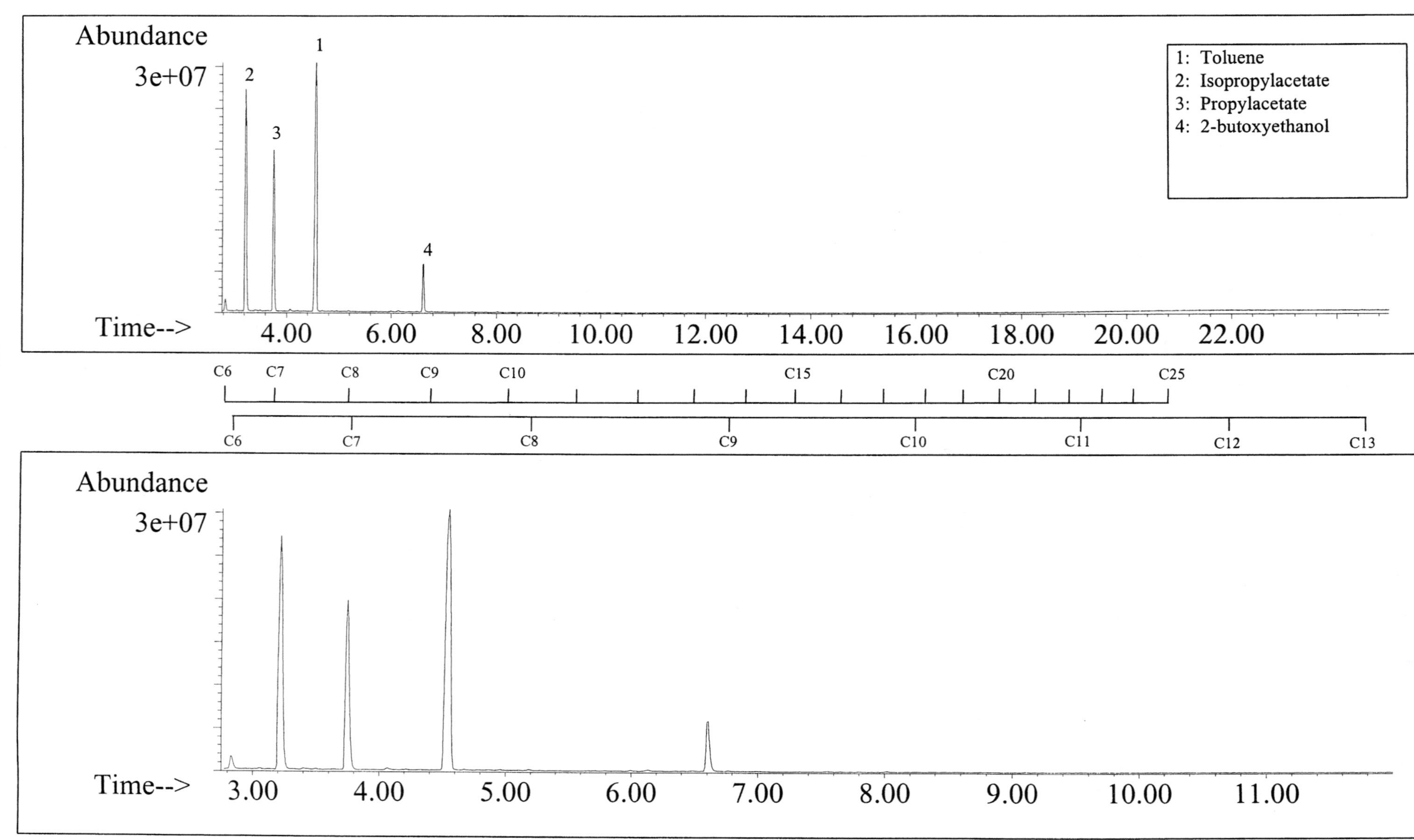

SUMMED PROFILES Oxygenate #6

ALKANES

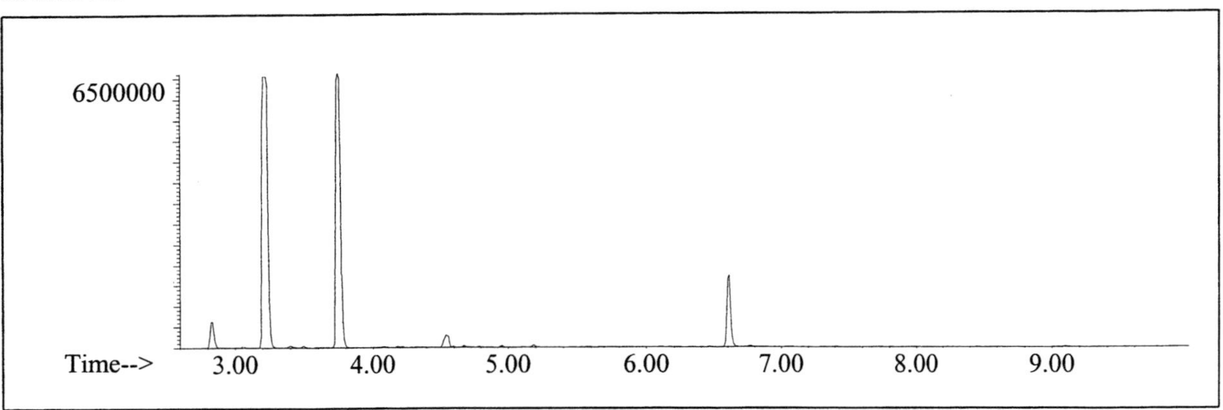

AROMATICS

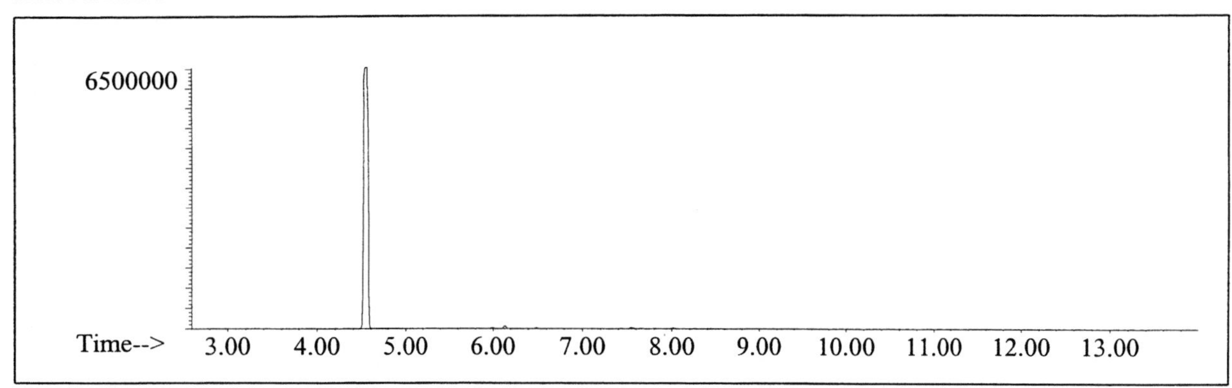

CYCLOPARAFFINS AND ALKENES

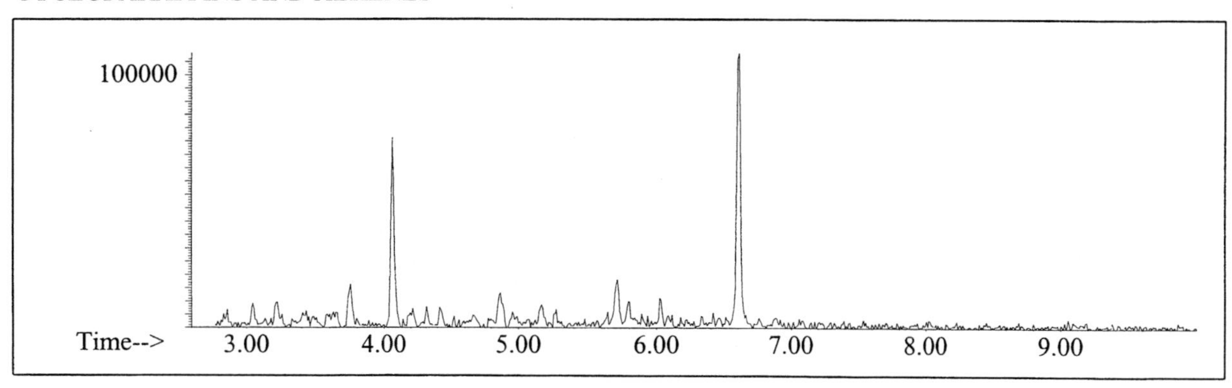

NAPHTHALENES

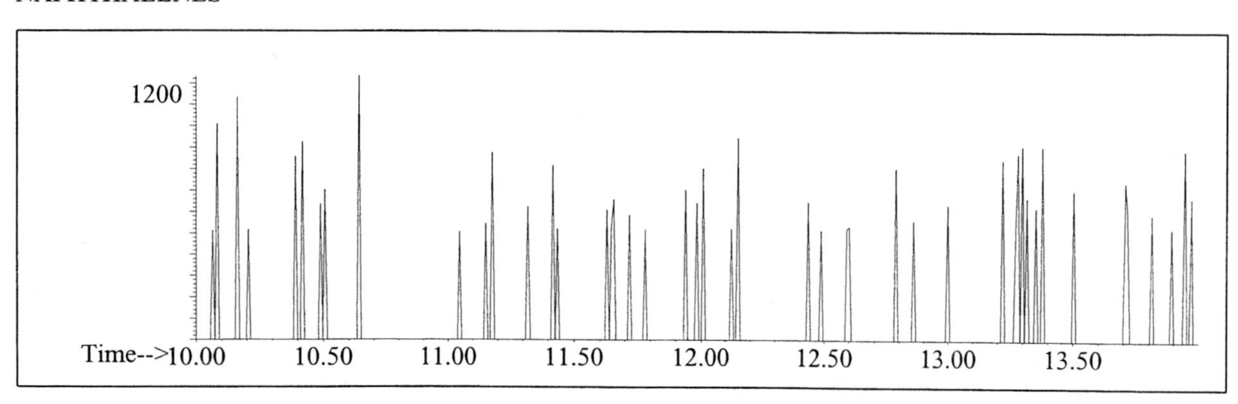

INDIVIDUAL PROFILES

Oxygenate #6

Alkane

Ion 43

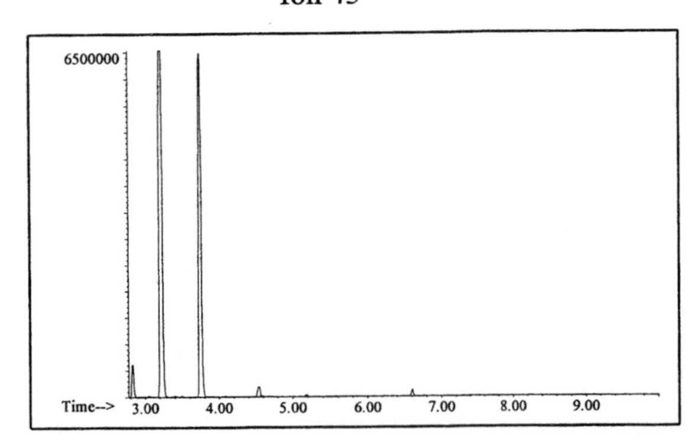

Ion 57

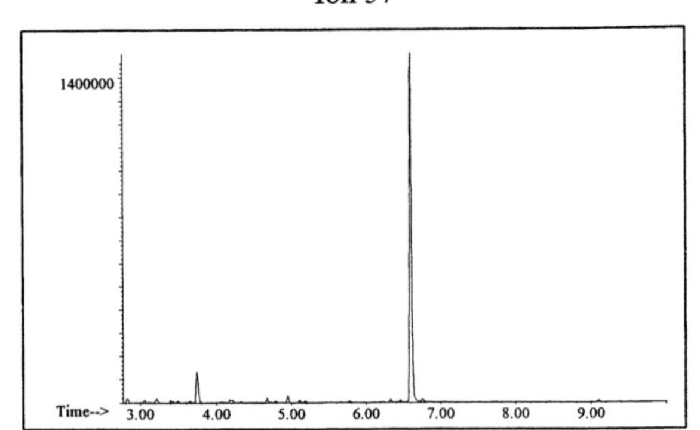

Ion 71

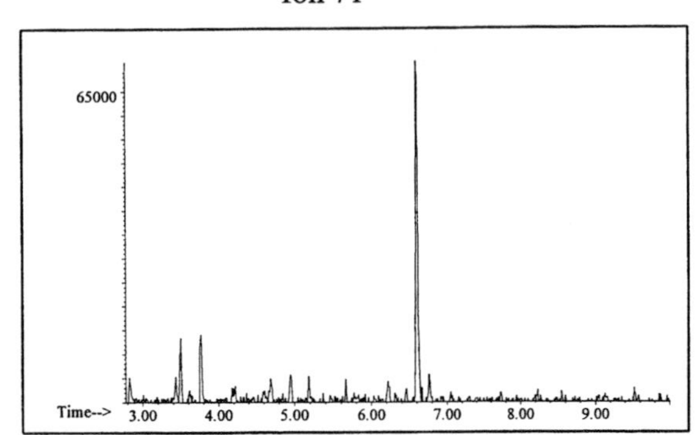

Ion 85

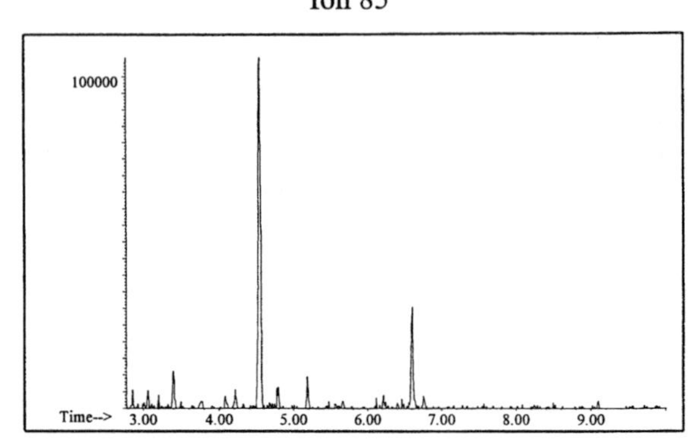

Aromatic

Ion 91

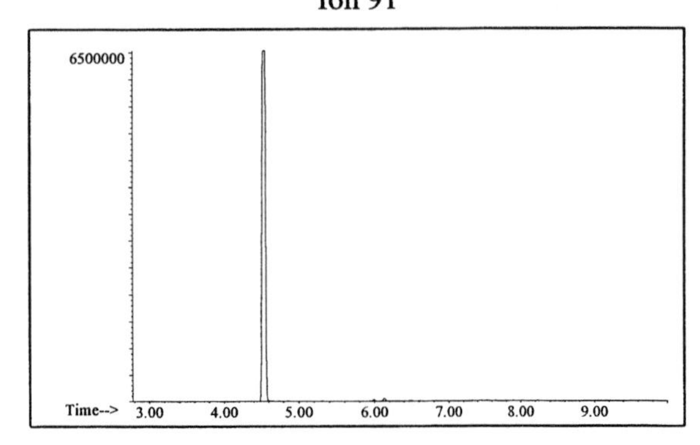

Ion 105

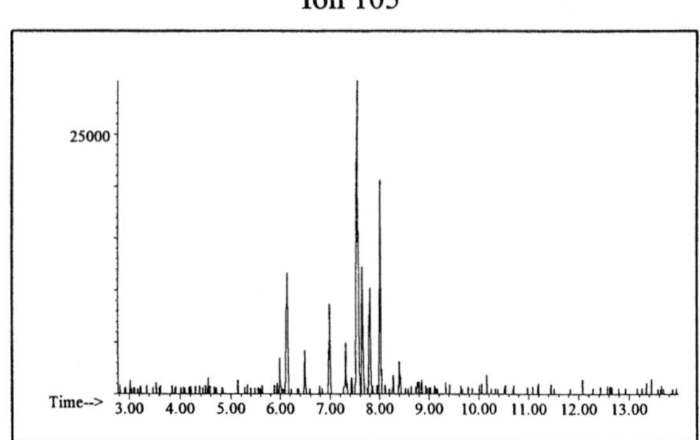

Ion 119

No Useful Data Obtained

INDIVIDUAL PROFILES

Oxygenate #6

Cycloparaffin

Ion 55

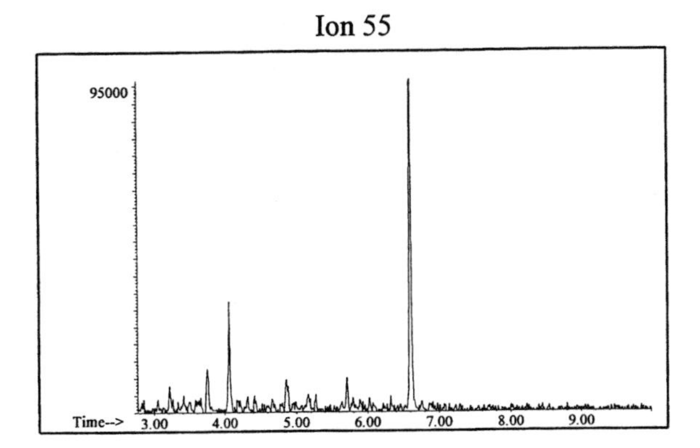

Ion 69

No Useful Data Obtained

Ion 83

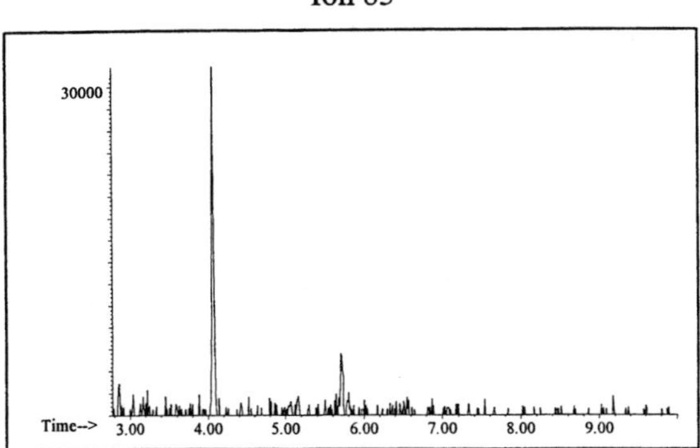

Naphthalene

Ion 128

No Useful Data Obtained

Ion 142

No Useful Data Obtained

Ion 156

No Useful Data Obtained

Oxygenate #7

20uL/mL Pentane

Product Displayed: Tru-Test Lacquer Thinner

Other Similar Products:

ASTM: Class 0.1 (Oxygenate)

Product Uses: Lacquer thinner

Macro Code: L

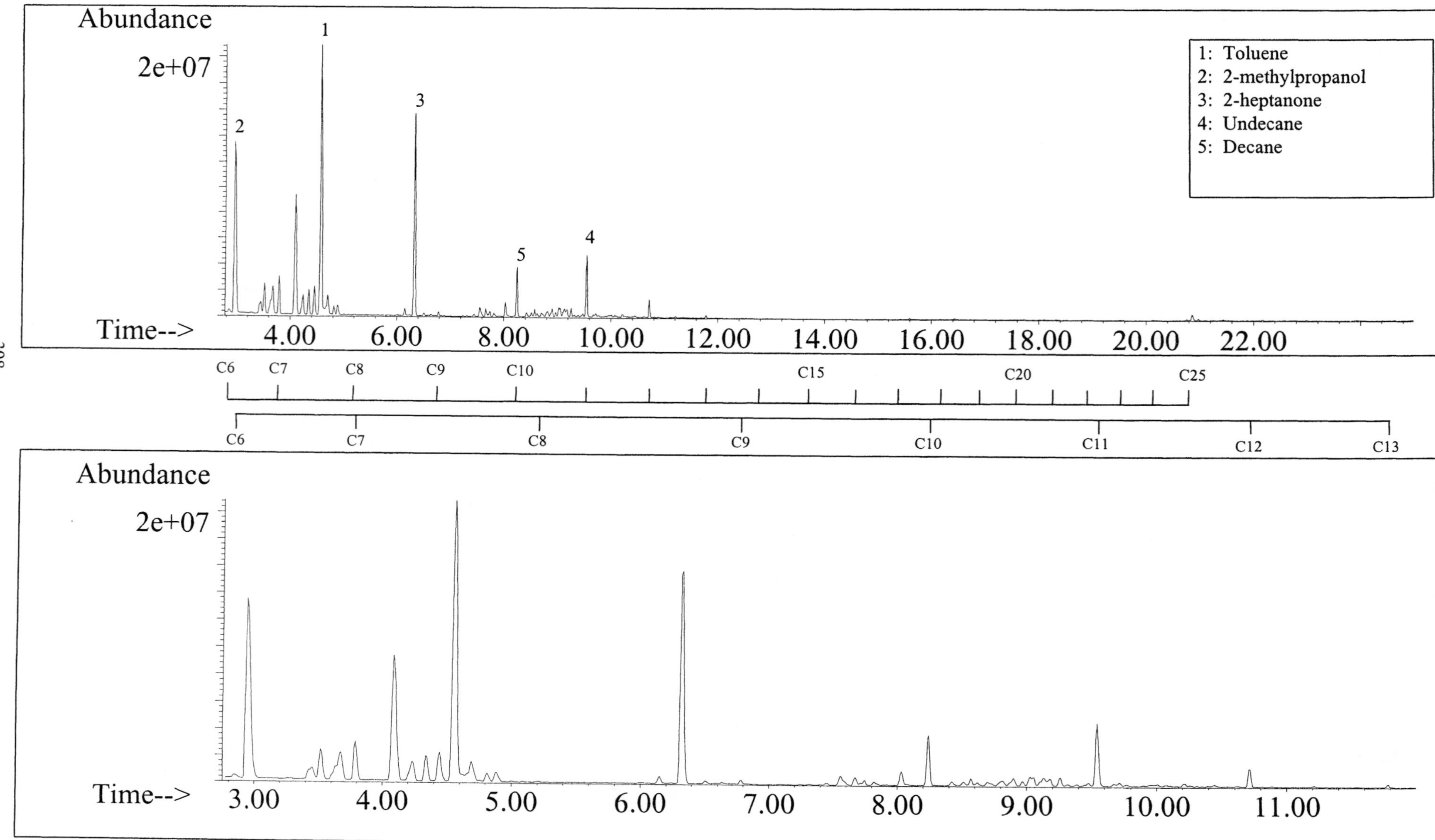

SUMMED PROFILES Oxygenate #7

ALKANES

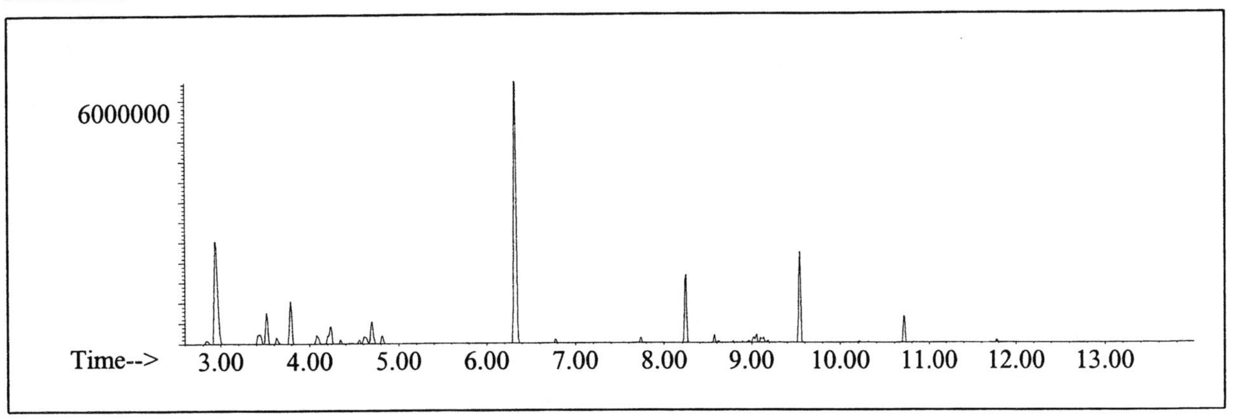

AROMATICS

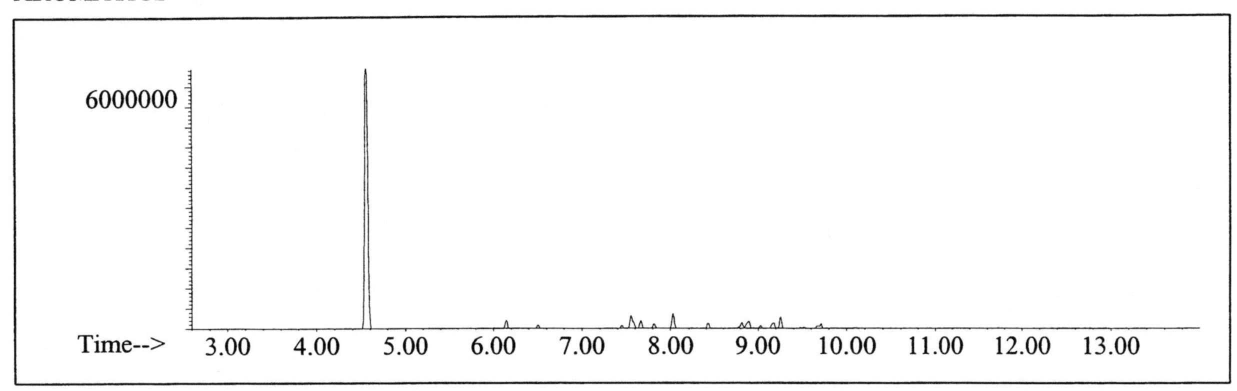

CYCLOPARAFFINS AND ALKENES

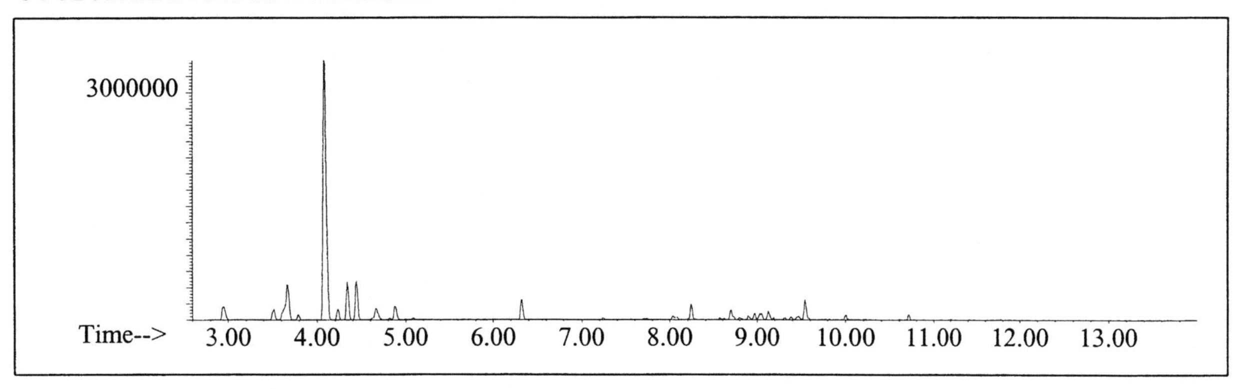

NAPHTHALENES

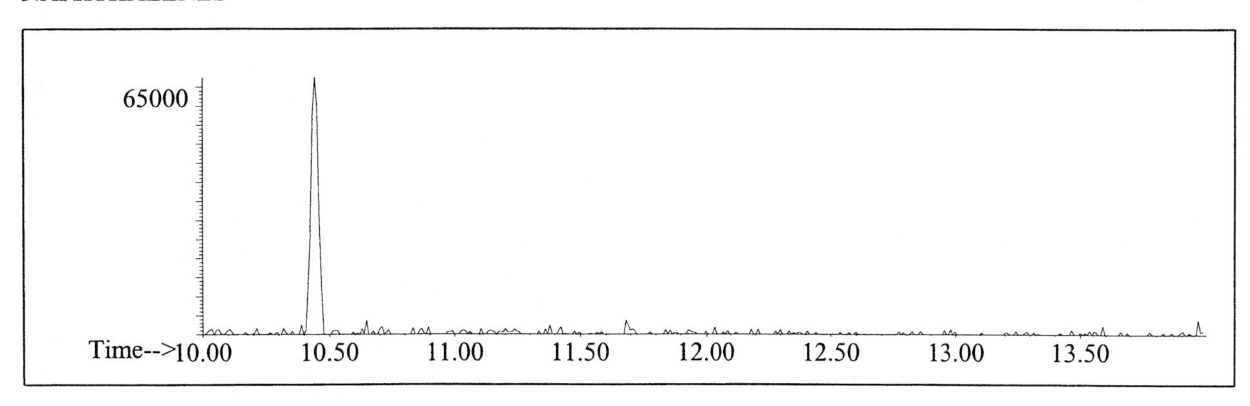

INDIVIDUAL PROFILES

Oxygenate #7

Alkane

Ion 43

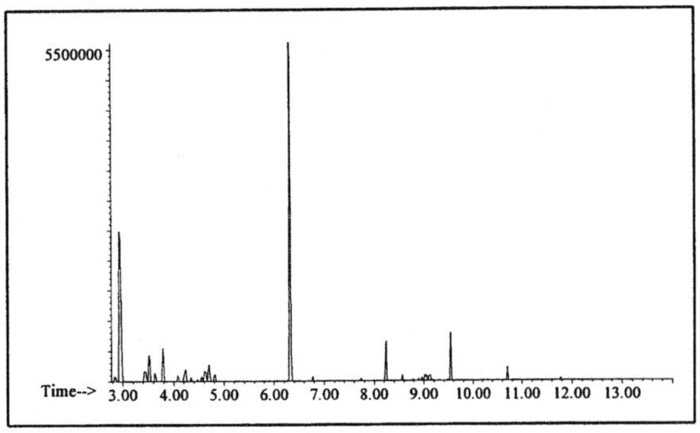

5500000

Time--> 3.00 4.00 5.00 6.00 7.00 8.00 9.00 10.00 11.00 12.00 13.00

Ion 57

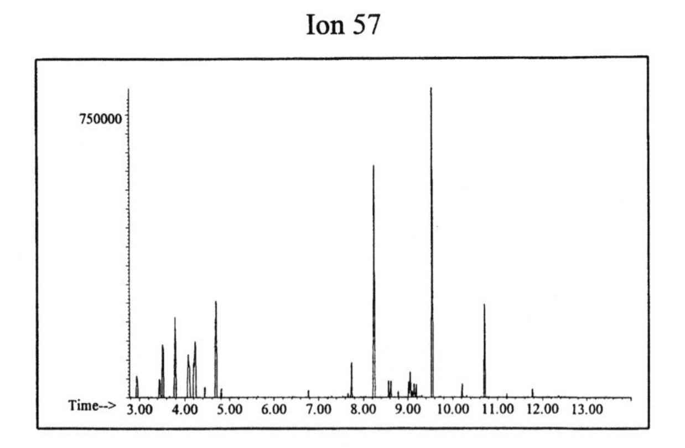

Ion 71

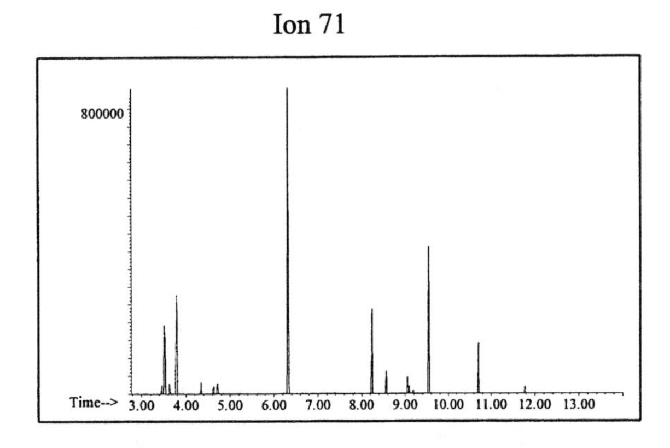

Ion 85

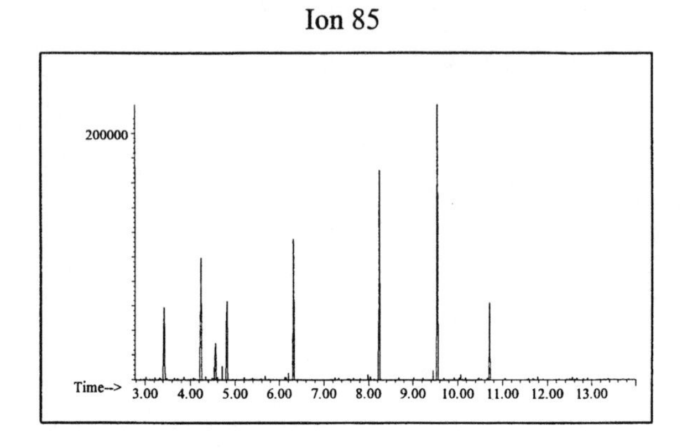

Aromatic

Ion 91

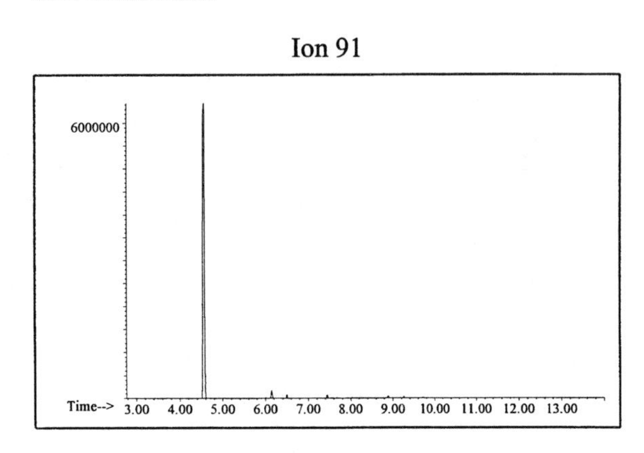

Ion 105

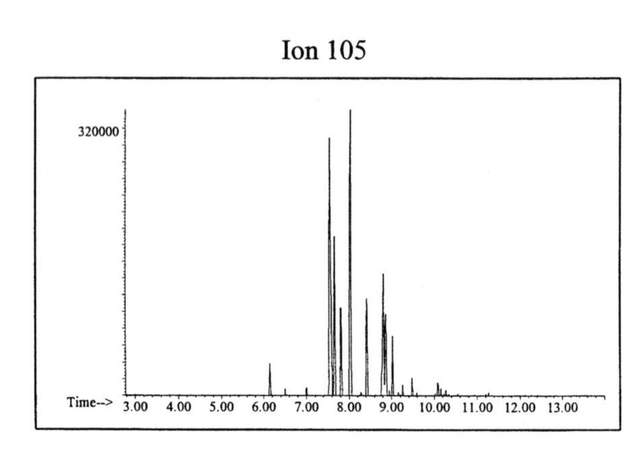

Ion 119

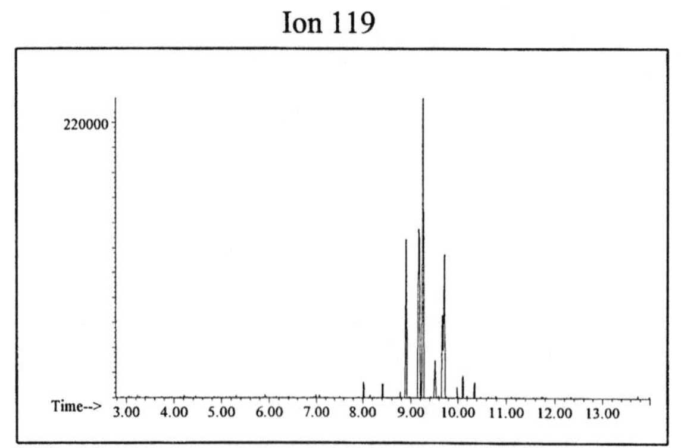

INDIVIDUAL PROFILES

Oxygenate #7

Cycloparaffin

Ion 55

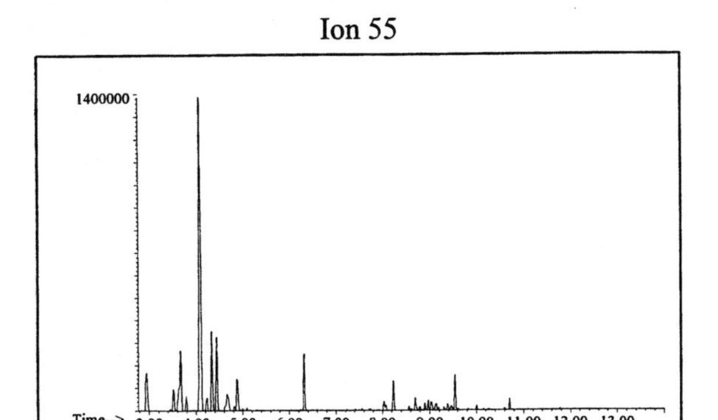

Ion 69

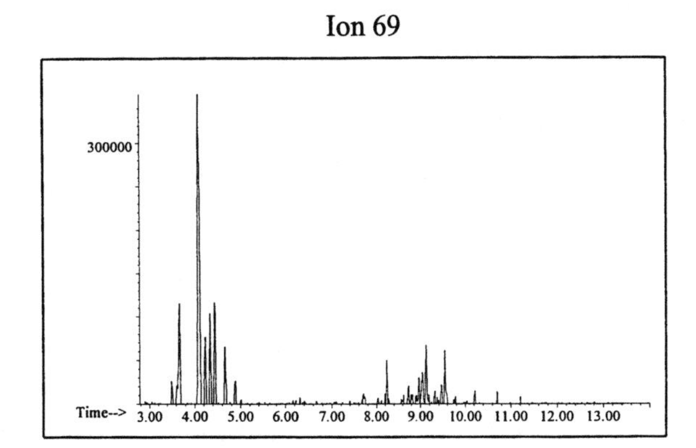

Ion 83

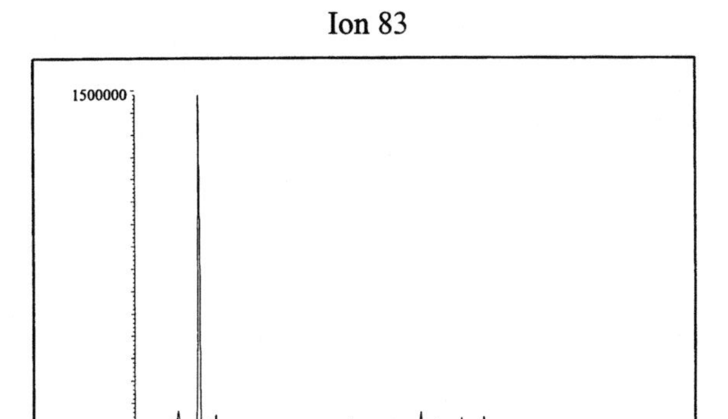

Naphthalene

Ion 128

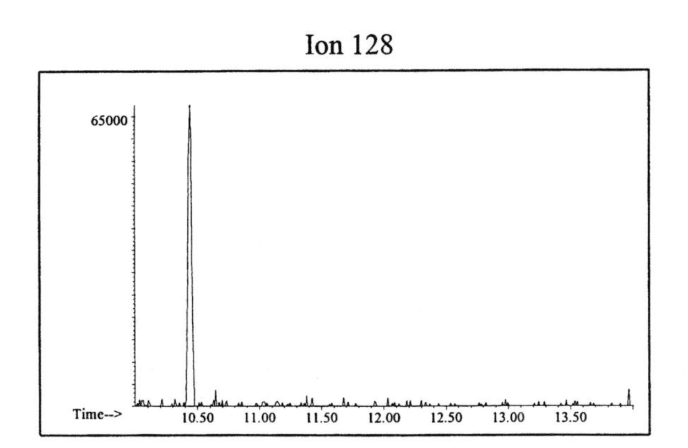

Ion 142

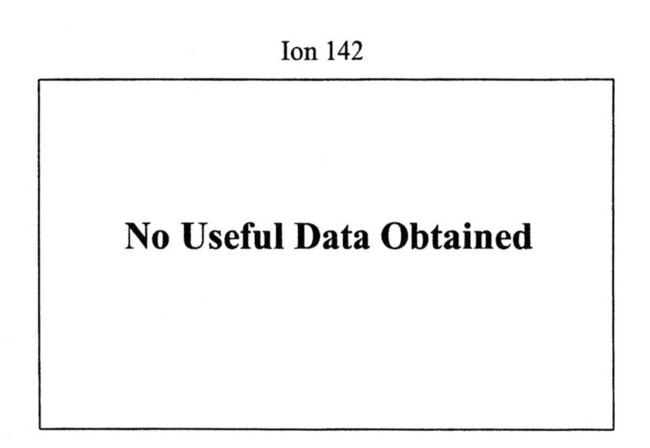

Ion 156

No Useful Data Obtained

Oxygenate #8

20uL/mL Pentane

Product Displayed: Jasco Lacquer Thinner

Other Similar Products:

ASTM: Class 0.1 (Oxygenate)

Product Uses: Lacquer thinner

Macro Code: L

Abundance 2e+07

Time-->

1: Toluene
2: Heptane
3: Methylcyclohexane
4: 2-butoxyethanol
5: Ethyl acetate

SUMMED PROFILES Oxygenate #8

ALKANES

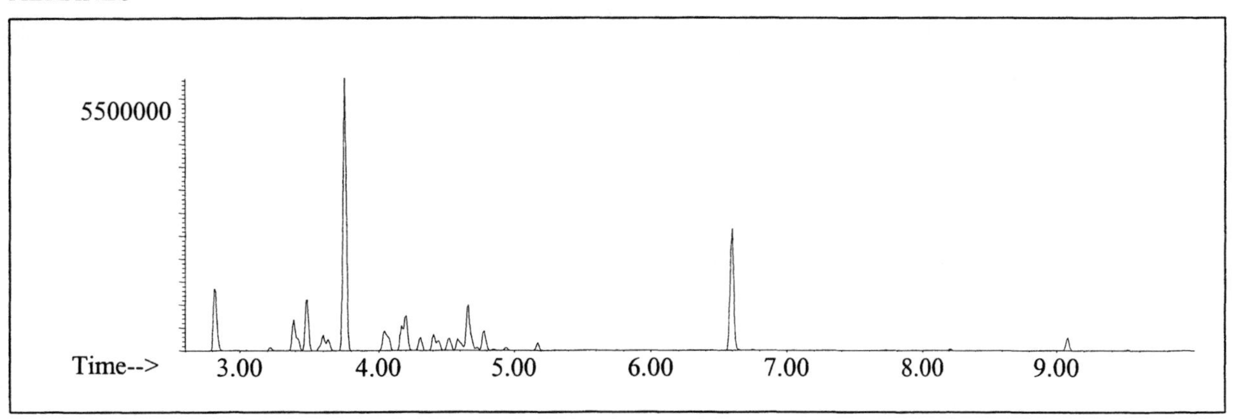

AROMATICS

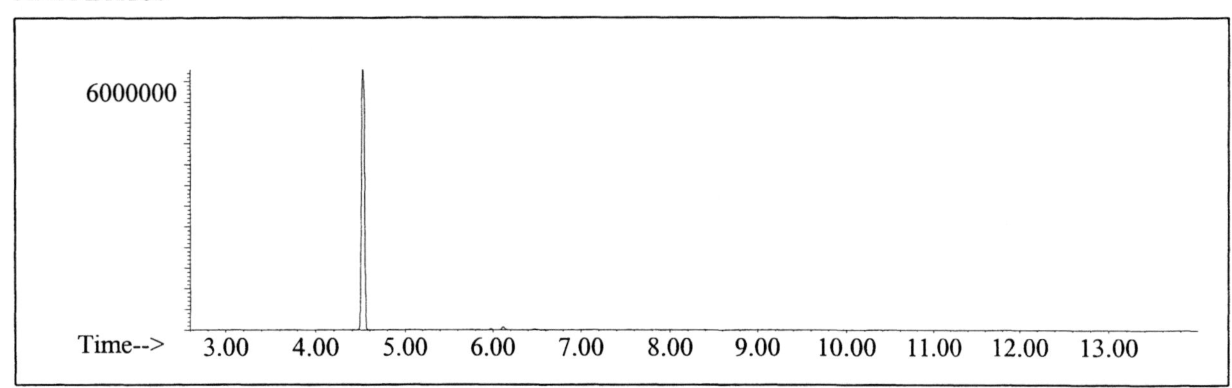

CYCLOPARAFFINS AND ALKENES

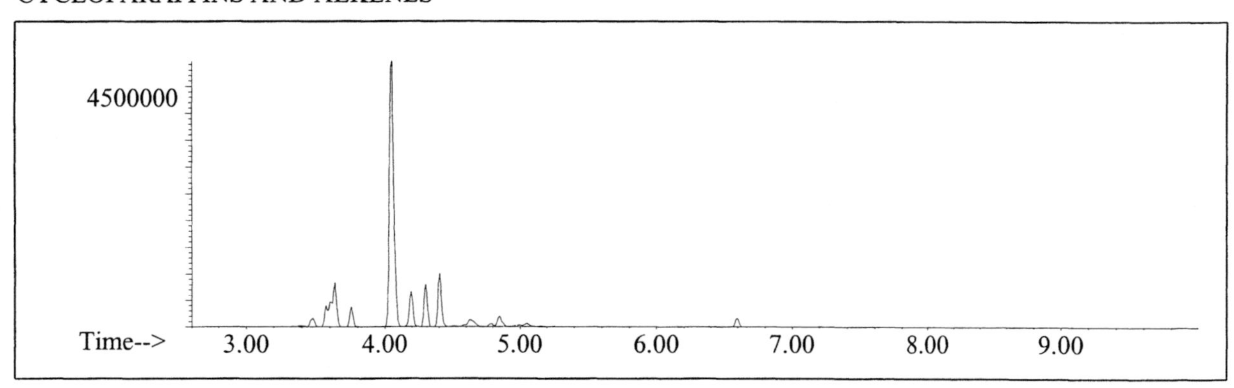

NAPHTHALENES

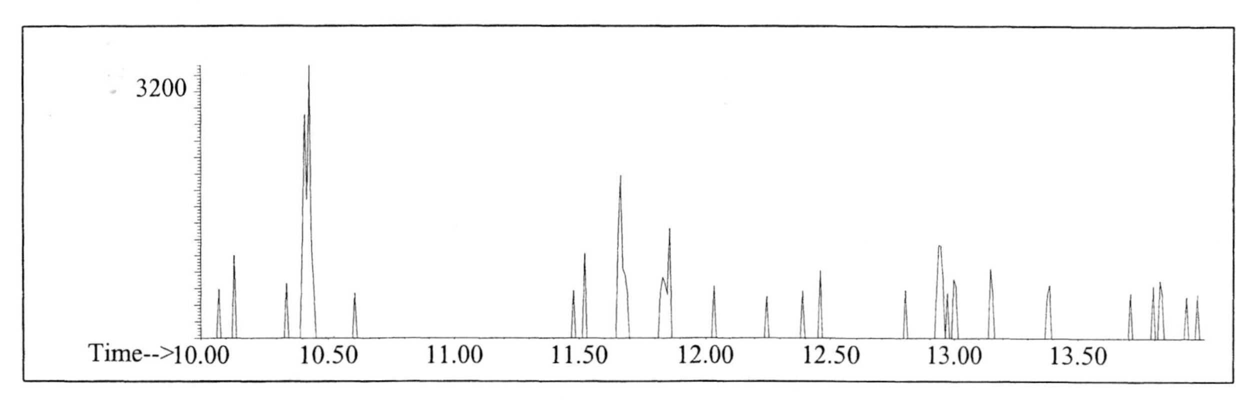

INDIVIDUAL PROFILES

Oxygenate #8

Alkane

Ion 43

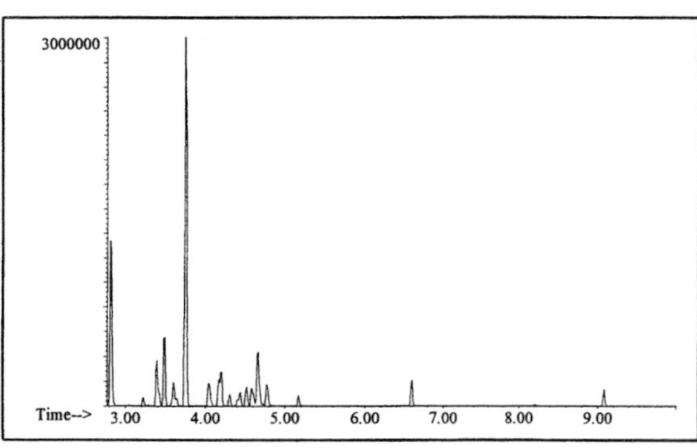

Ion 57

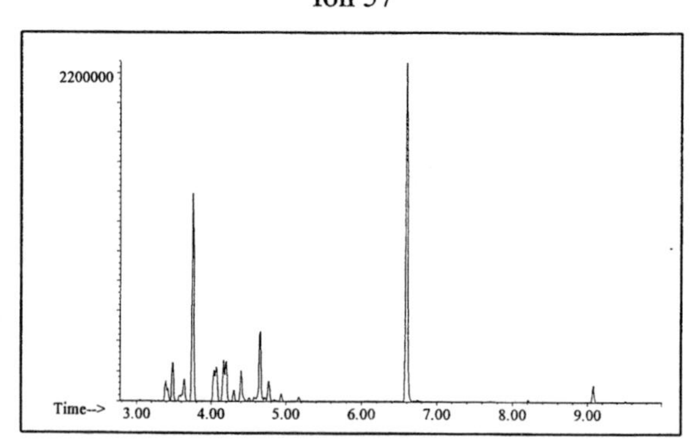

Ion 71

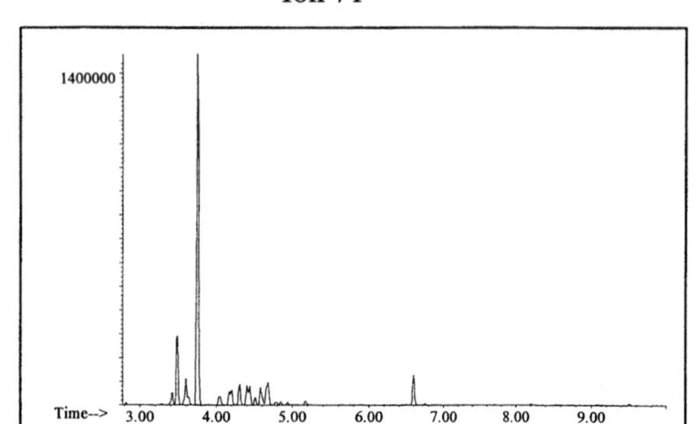

Ion 85

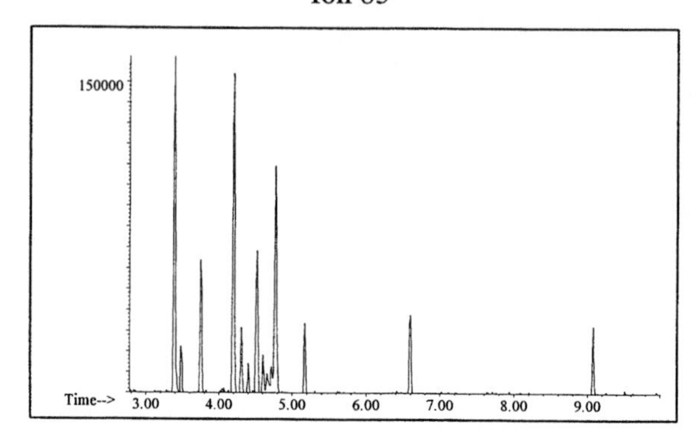

Aromatic

Ion 91

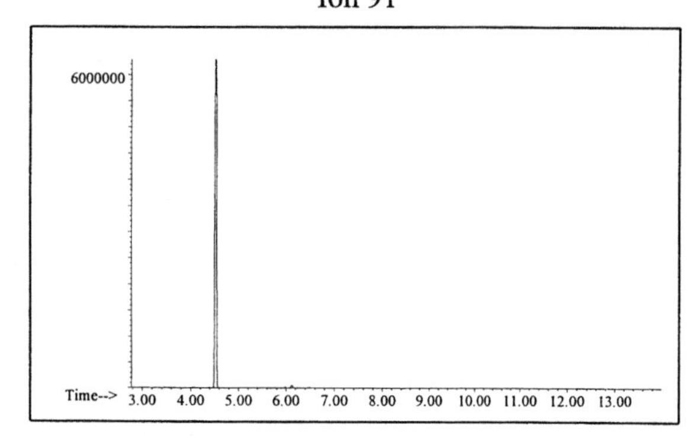

Ion 105

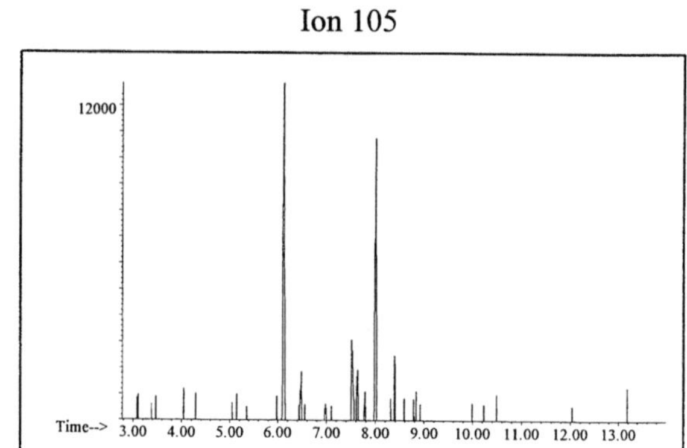

Ion 119

No Useful Data Obtained

INDIVIDUAL PROFILES

Oxygenate #8

Cycloparaffin

Ion 55

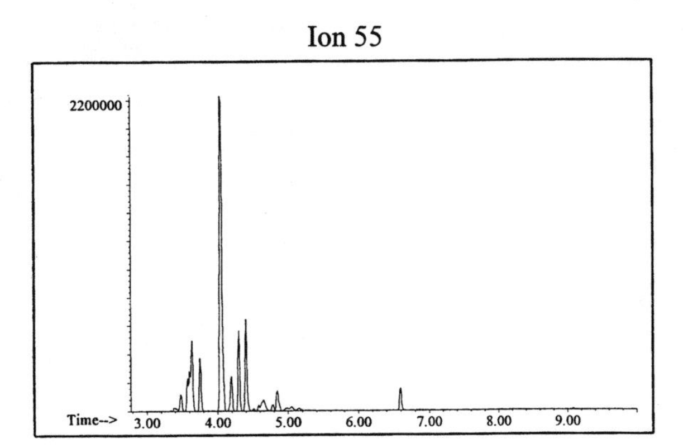

Ion 69

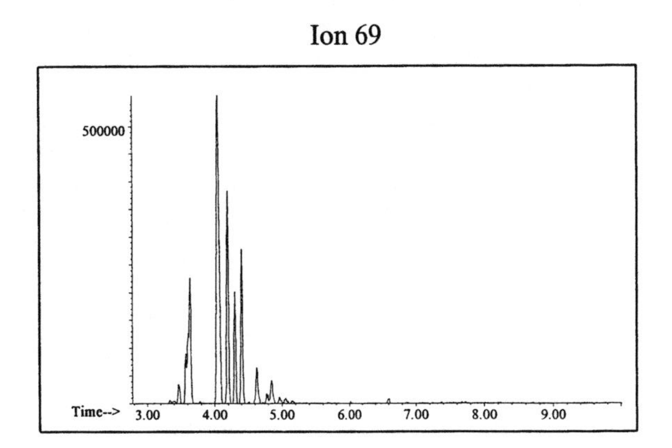

Ion 83

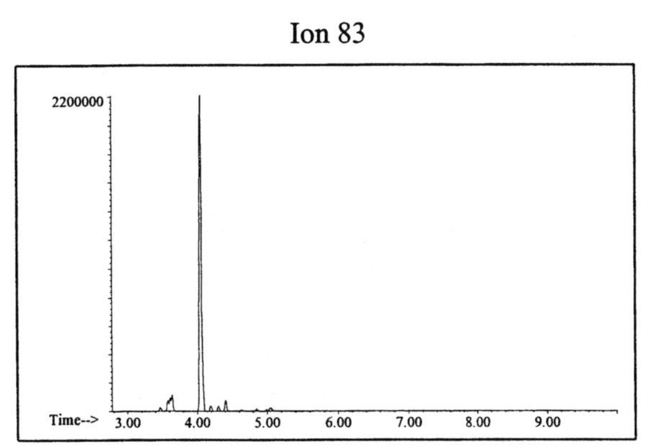

Naphthalene

Ion 128

No Useful Data Obtained

Ion 142

No Useful Data Obtained

Ion 156

No Useful Data Obtained

Oxygenate #9

20uL/mL Pentane

ASTM: Class 0.1 (Oxygenate)

Macro Code: L

Product Displayed: Willard Lacquer Thinner

Product Uses: Lacquer thinner

Other Similar Products:

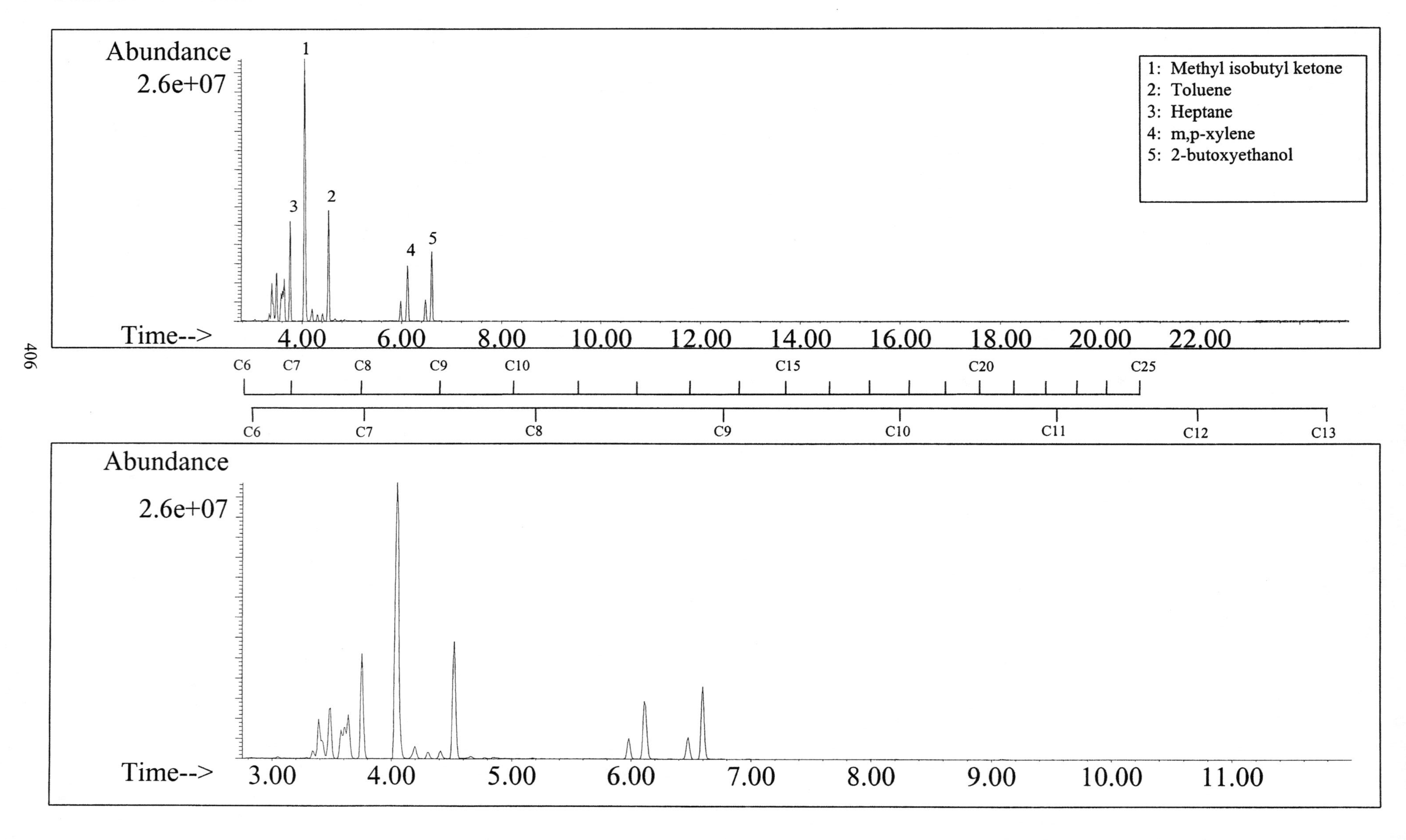

SUMMED PROFILES Oxygenate #9

ALKANES

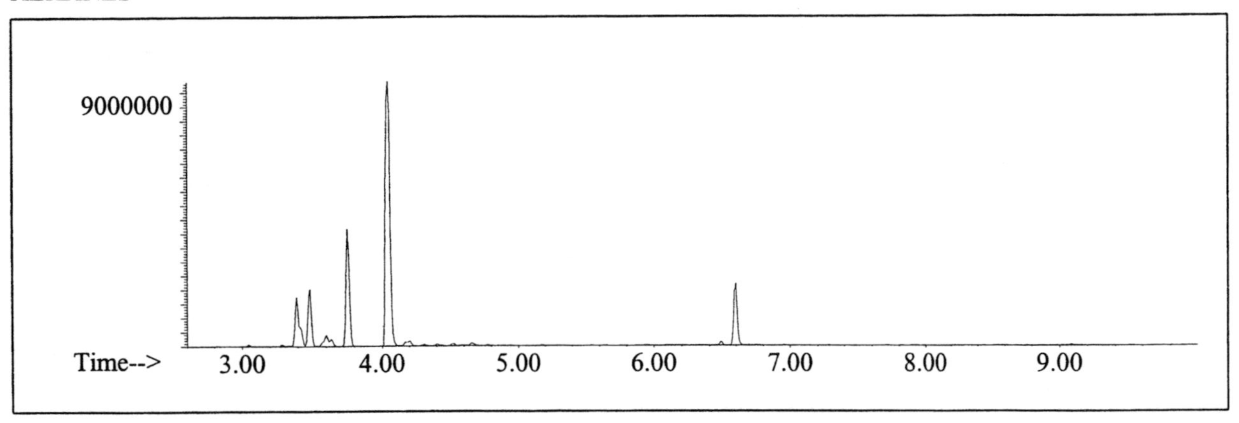

AROMATICS

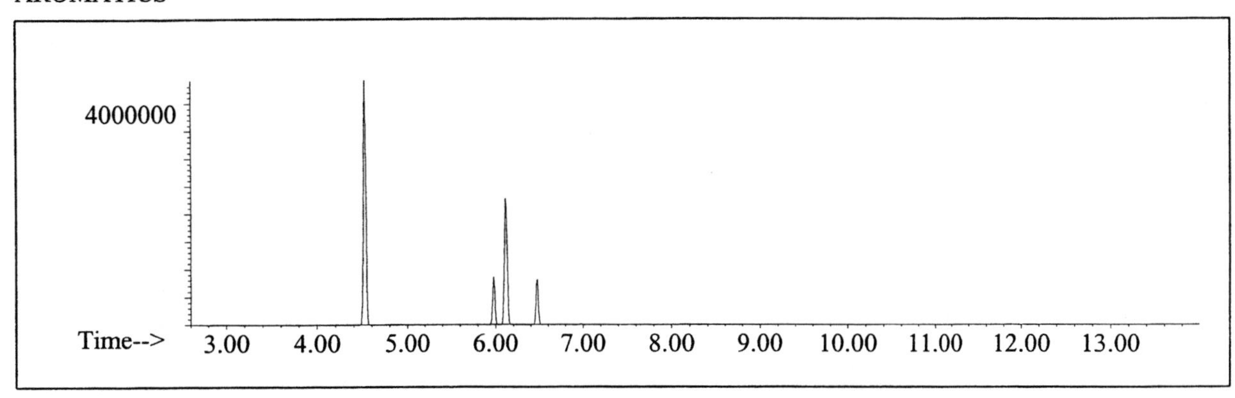

CYCLOPARAFFINS AND ALKENES

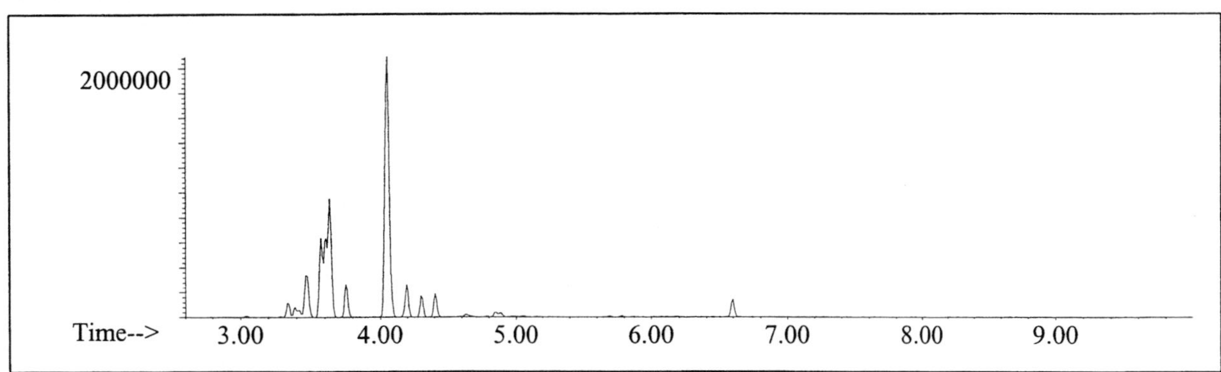

NAPHTHALENES

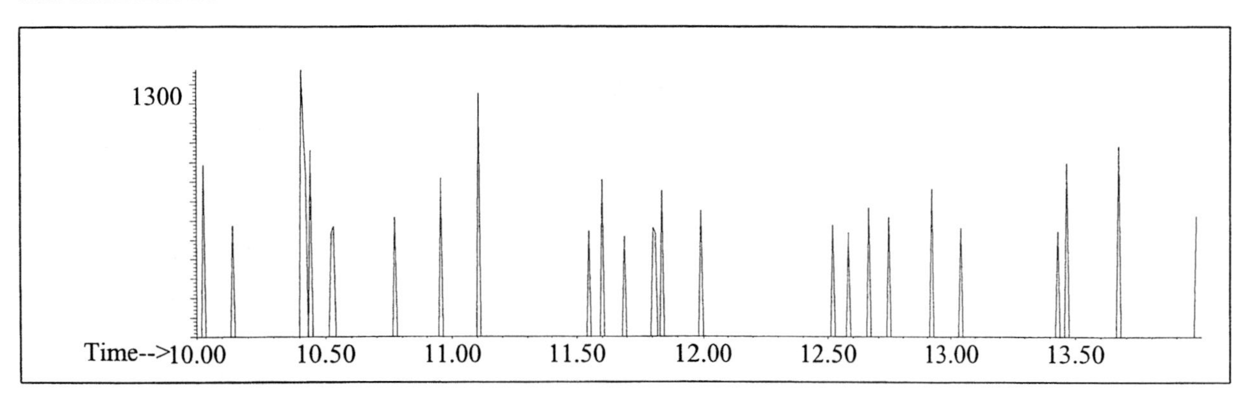

INDIVIDUAL PROFILES Oxygenate #9

Alkane

Ion 43

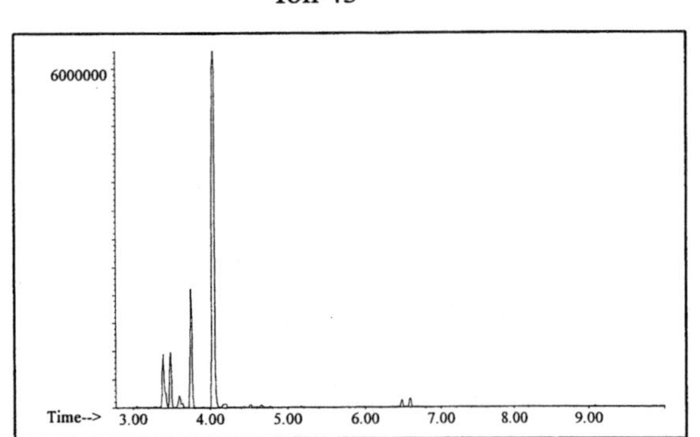

Ion 57

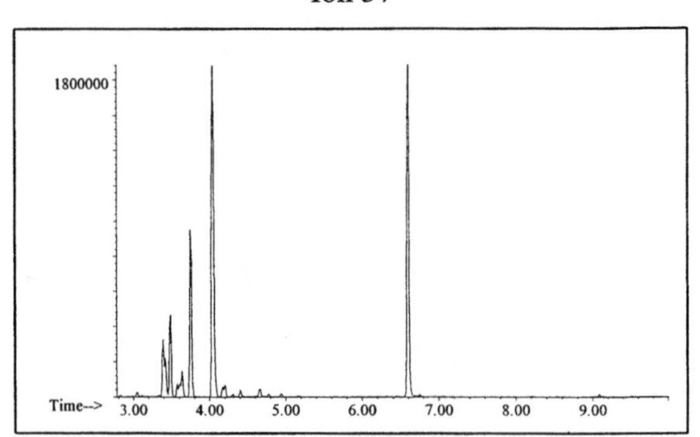

Ion 71

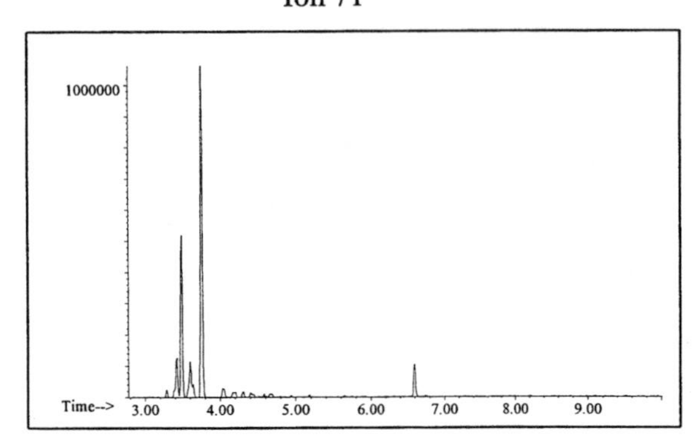

Ion 85

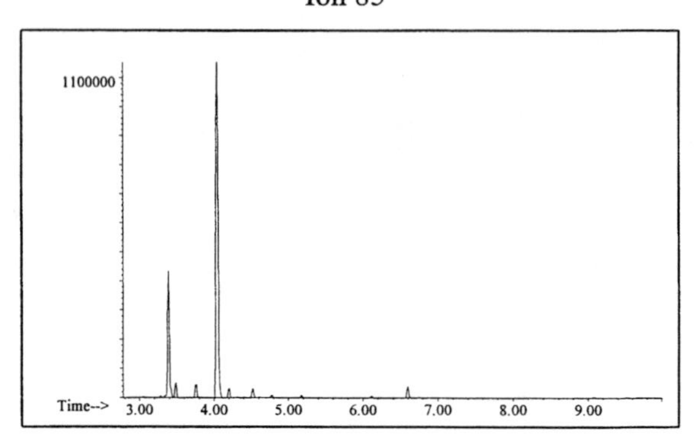

Aromatic

Ion 91

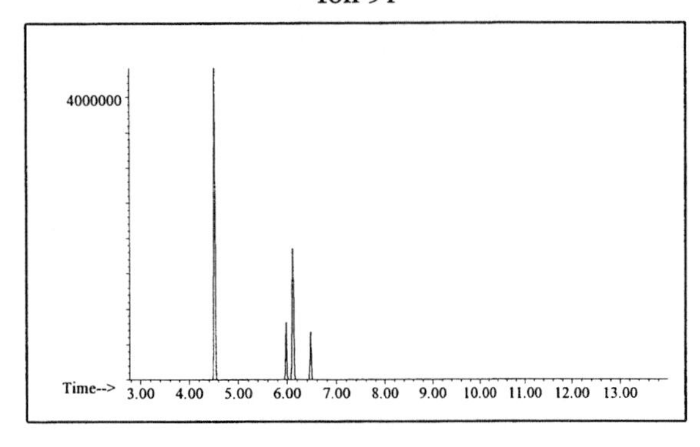

Ion 105

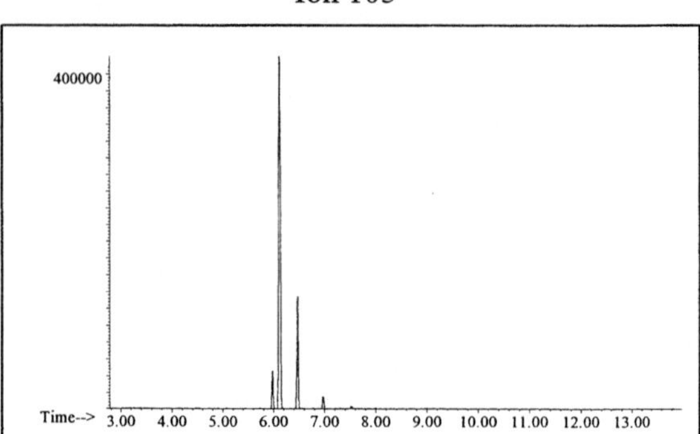

Ion 119

No Useful Data Obtained

INDIVIDUAL PROFILES

Oxygenate #9

Cycloparaffin

Ion 55

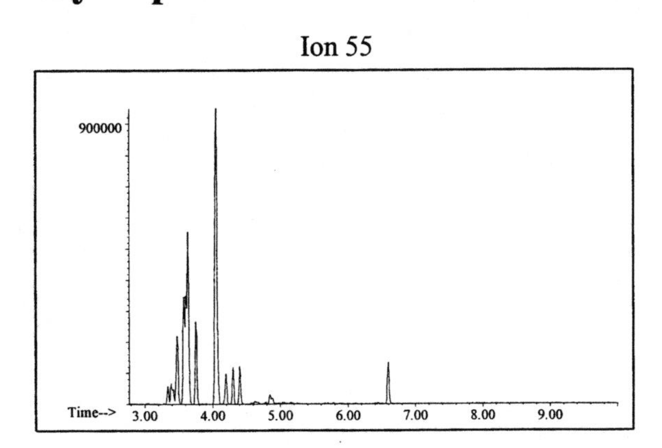

Ion 69

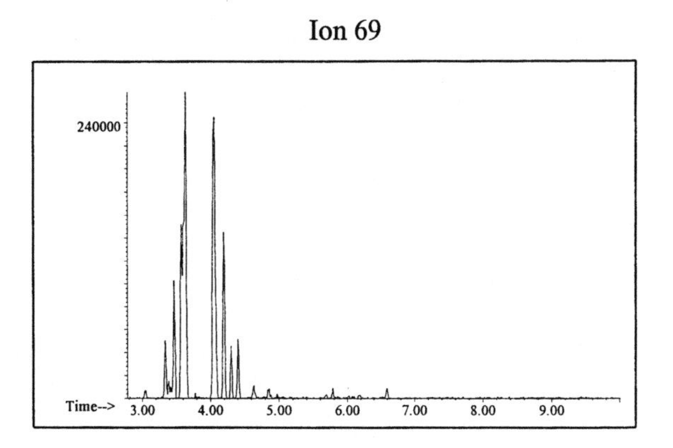

Ion 83

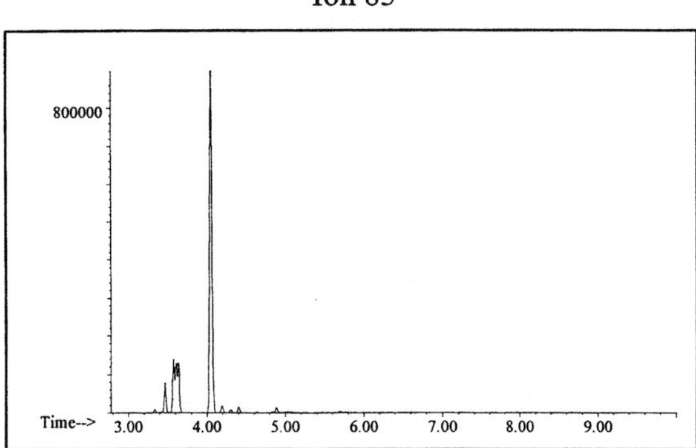

Naphthalene

Ion 128

No Useful Data Obtained

Ion 142

No Useful Data Obtained

Ion 156

No Useful Data Obtained

Oxygenate #10

20uL/mL Pentane

Product Displayed: Blake Lacquer Thinner

Other Similar Products:

ASTM: Class 0.1 (Oxygenate)

Product Uses: Lacquer thinner

Macro Code: L

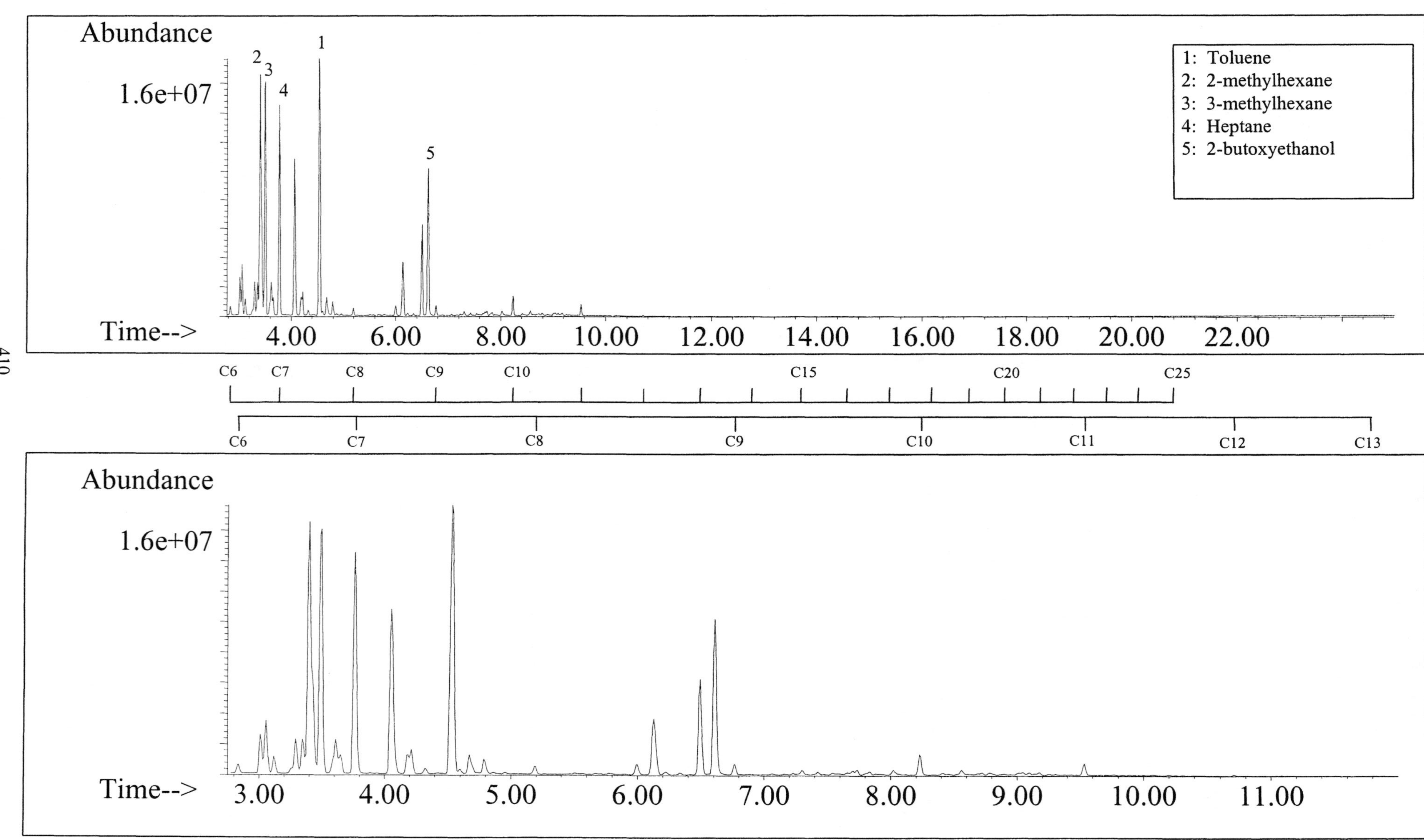

SUMMED PROFILES Oxygenate #10

ALKANES

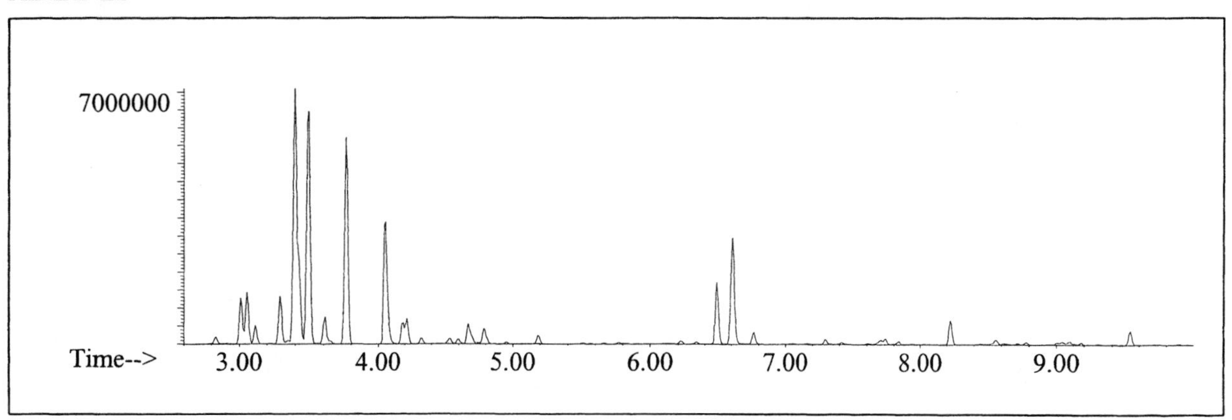

AROMATICS

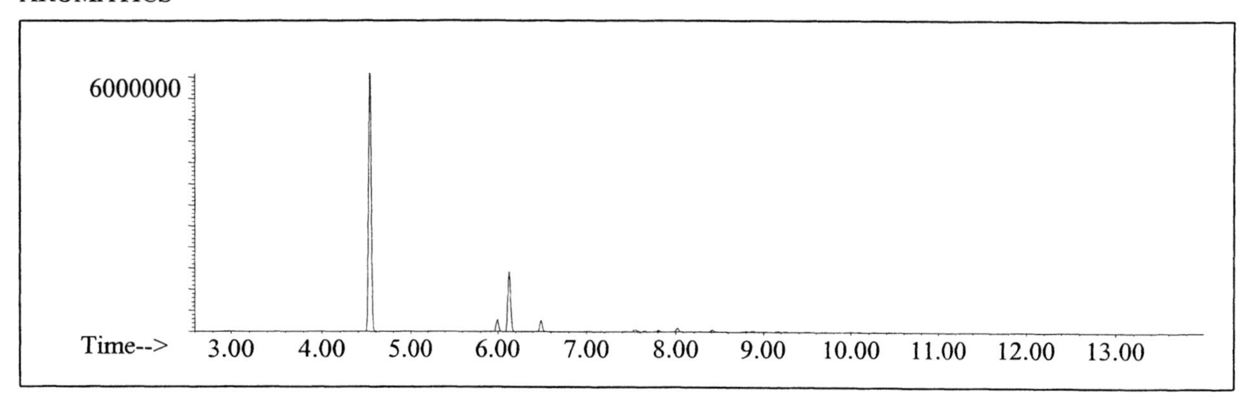

CYCLOPARAFFINS AND ALKENES

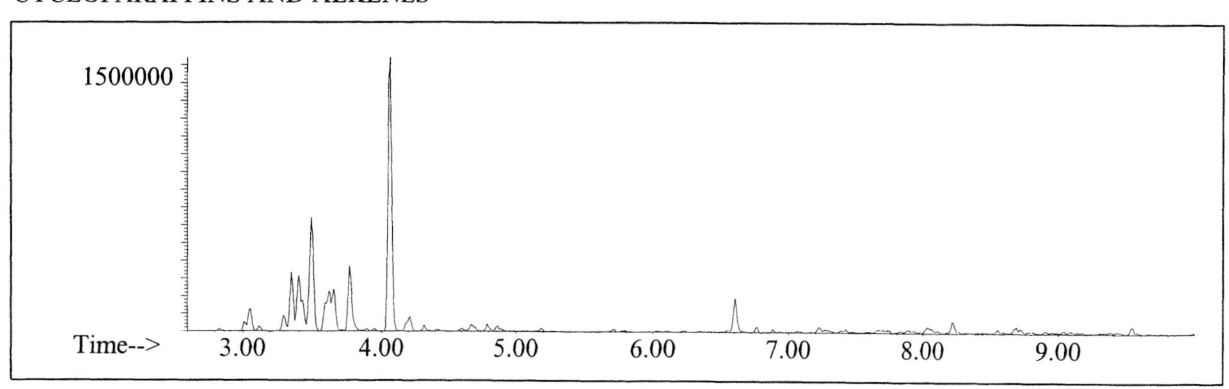

NAPHTHALENES

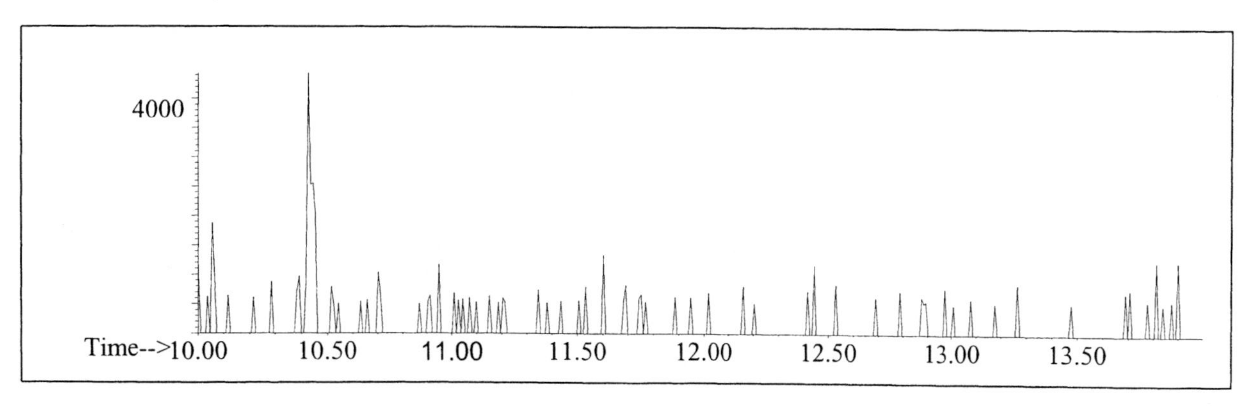

INDIVIDUAL PROFILES

Oxygenate #10

Alkane

Ion 43

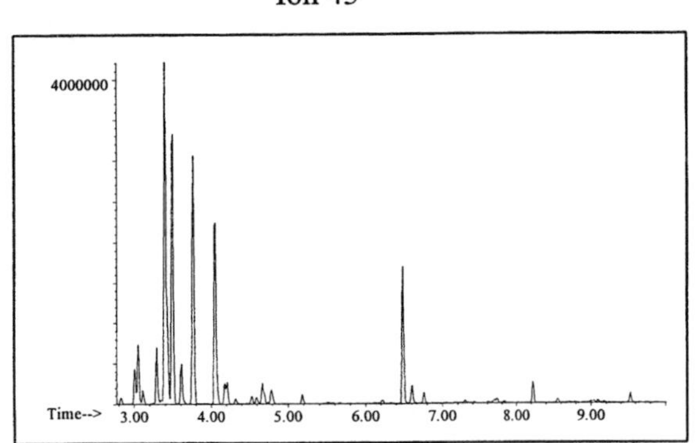

Ion 57

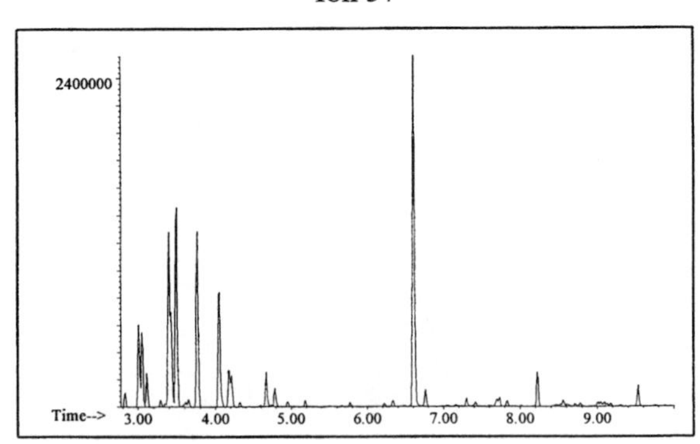

Ion 71

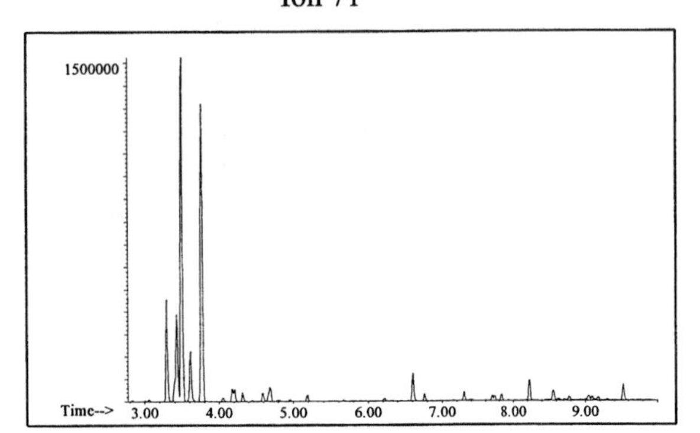

Ion 85

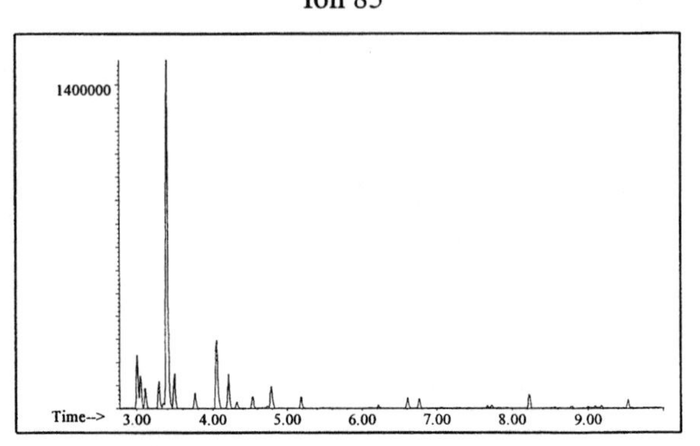

Aromatic

Ion 91

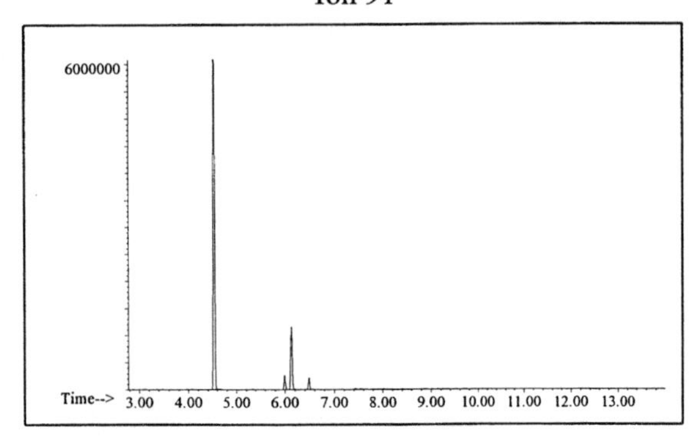

Ion 105

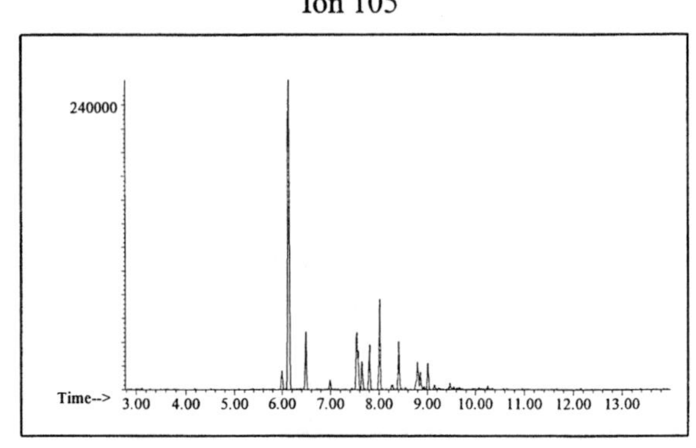

Ion 119

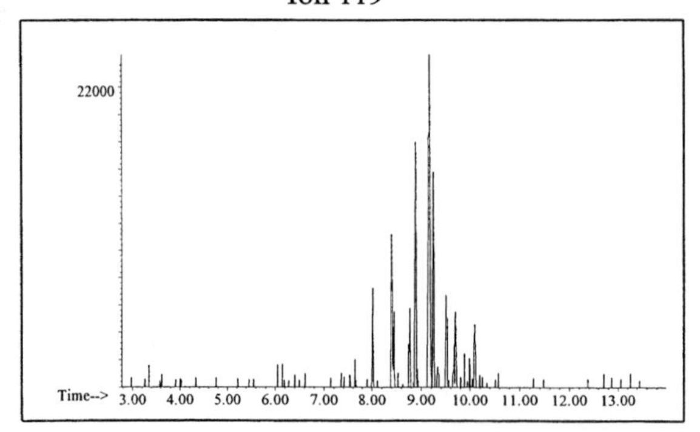

Cycloparaffin

Ion 55

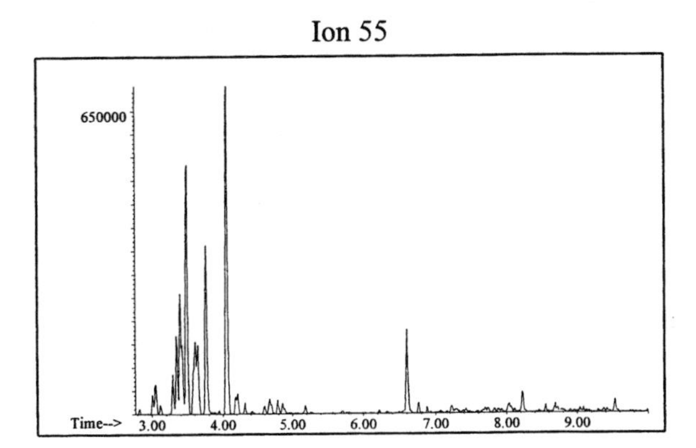

Ion 69

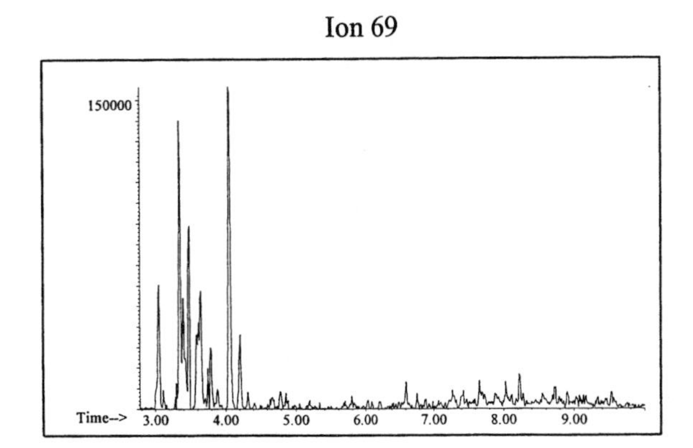

Ion 83

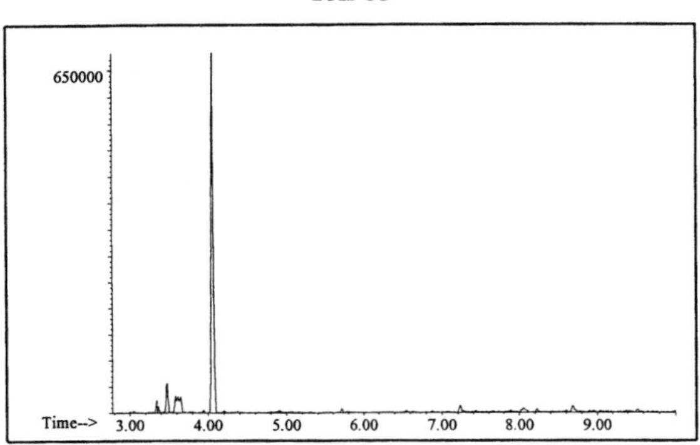

Naphthalene

Ion 128

No Useful Data Obtained

Ion 142

No Useful Data Obtained

Ion 156

No Useful Data Obtained

Oxygenate #11

20uL/mL Pentane

ASTM: Class 0.1 (Oxygenate)

Macro Code: L

Product Displayed: JC Penney Gloss Remover

Product Uses: Surface preparation, solvent

Other Similar Products:

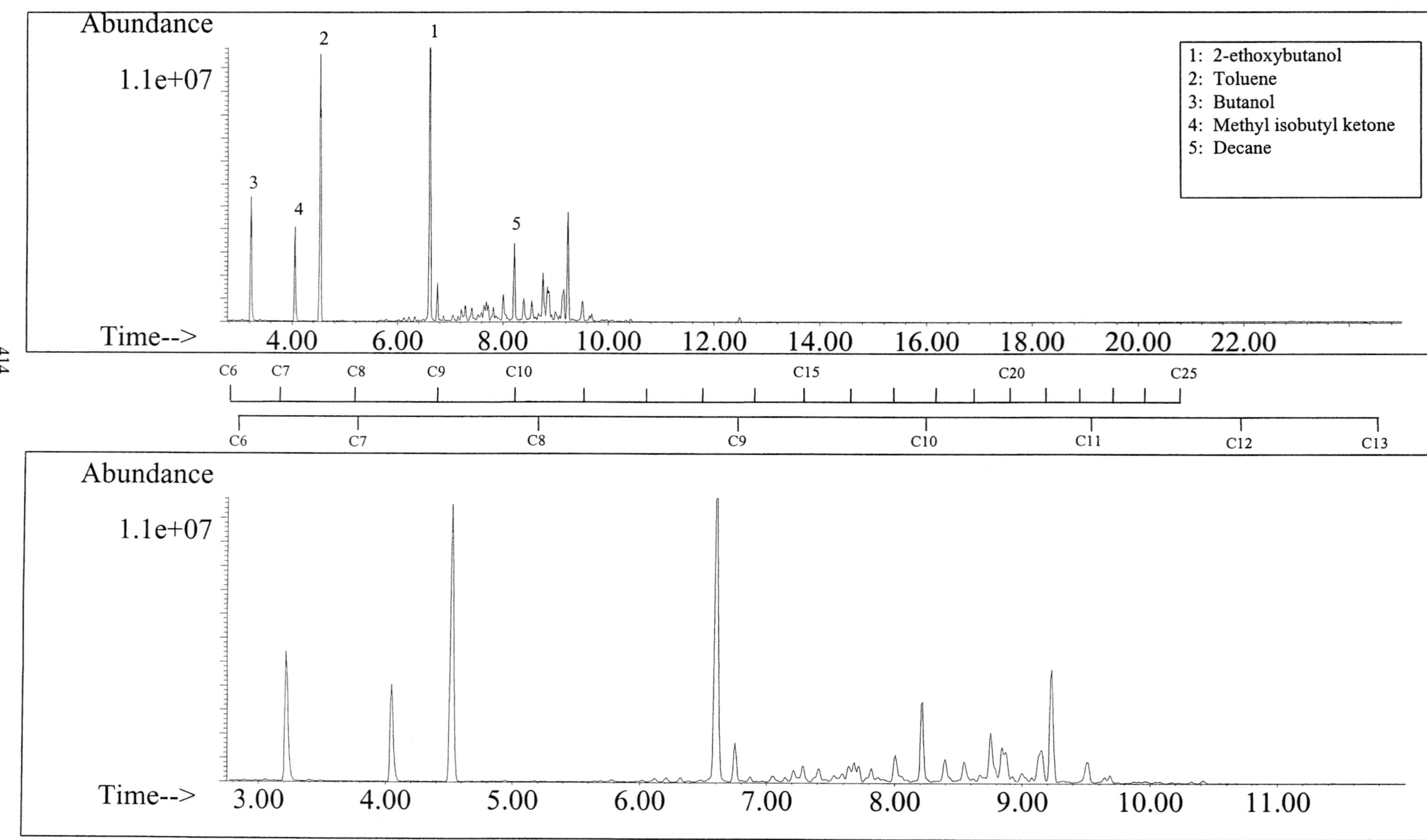

SUMMED PROFILES

Oxygenate #11

ALKANES

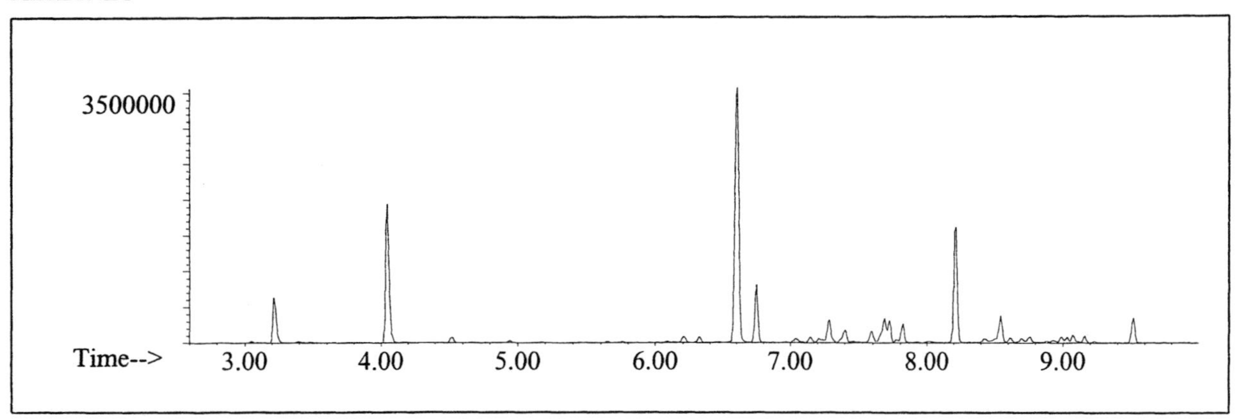

AROMATICS

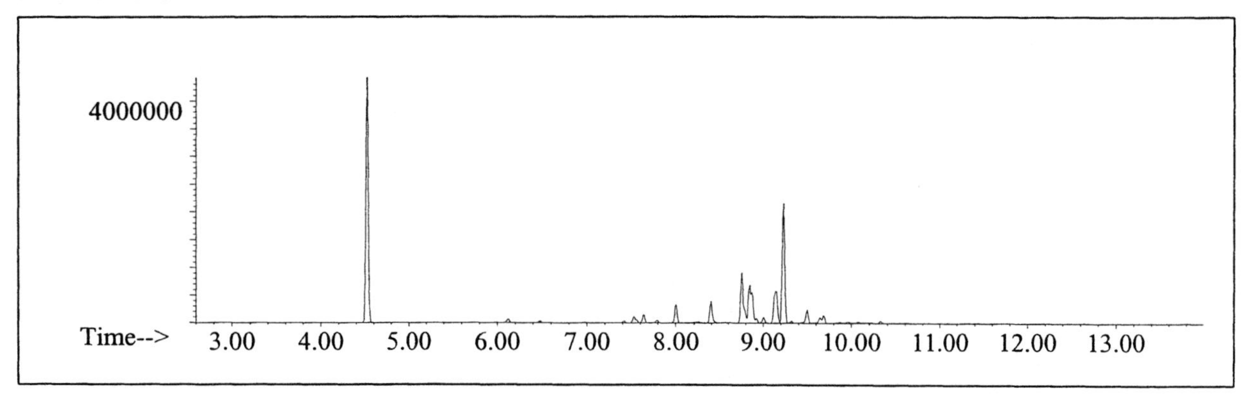

CYCLOPARAFFINS AND ALKENES

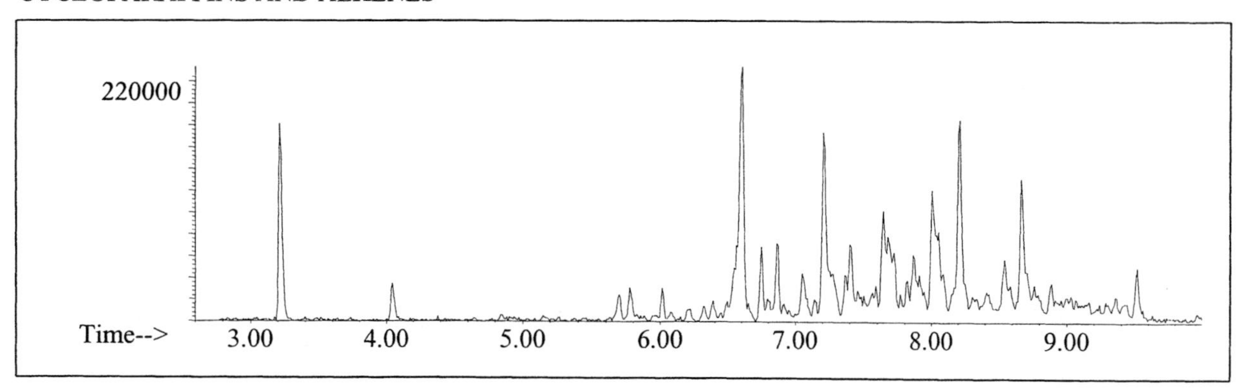

NAPHTHALENES

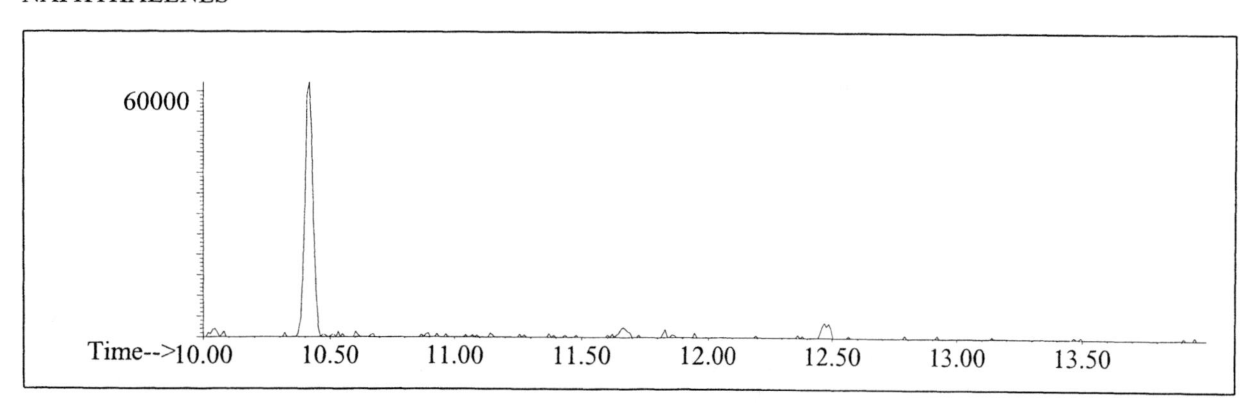

INDIVIDUAL PROFILES

Oxygenate #11

Alkane

Ion 43

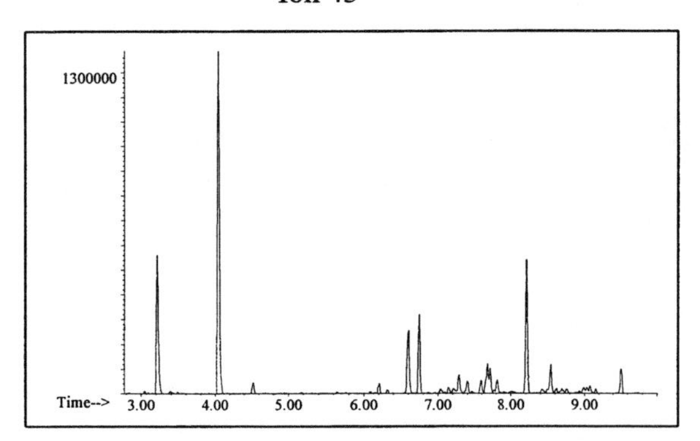

Ion 57

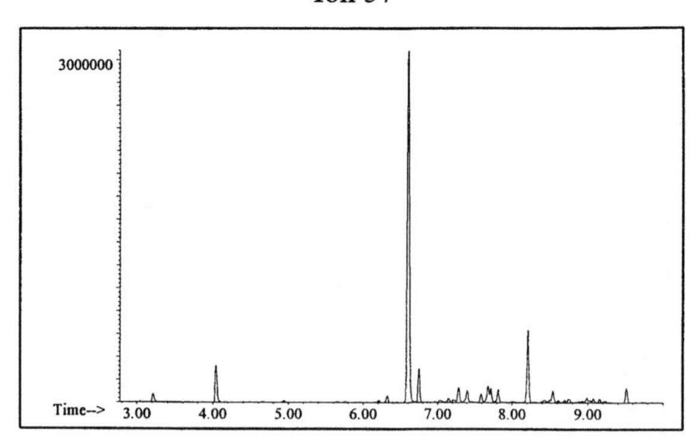

Ion 71

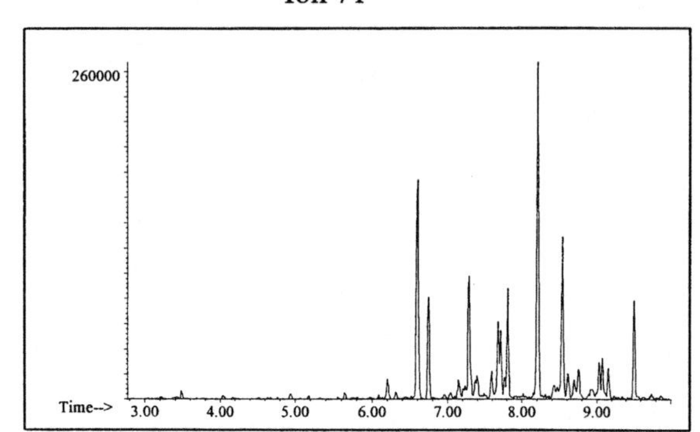

Ion 85

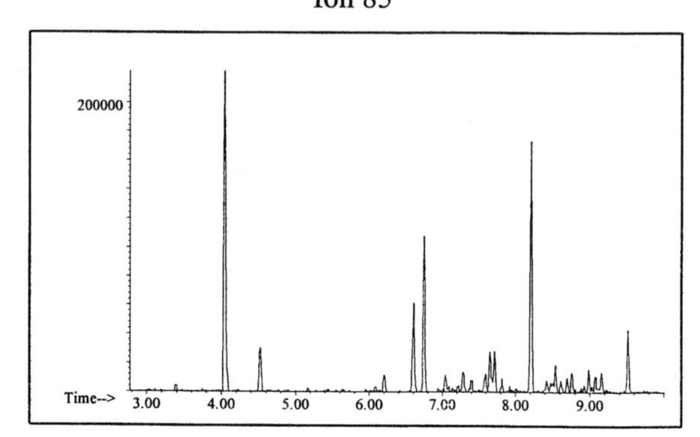

Aromatic

Ion 91

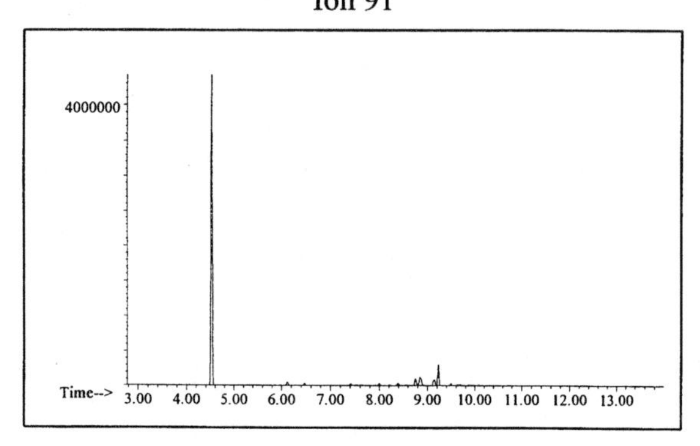

Ion 105

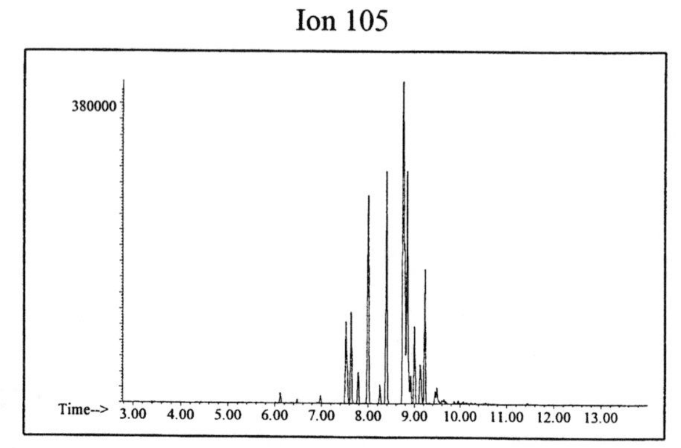

Ion 119

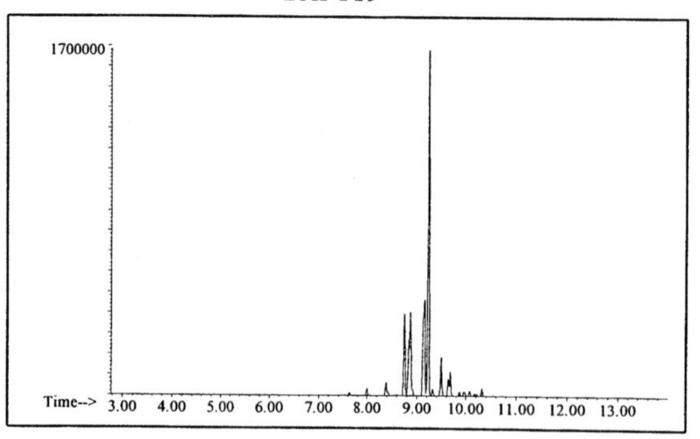

INDIVIDUAL PROFILES

Oxygenate #11

Cycloparaffin

Ion 55

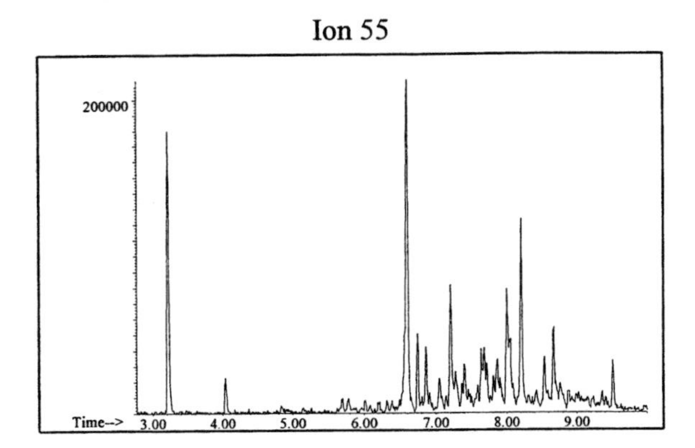

Ion 69

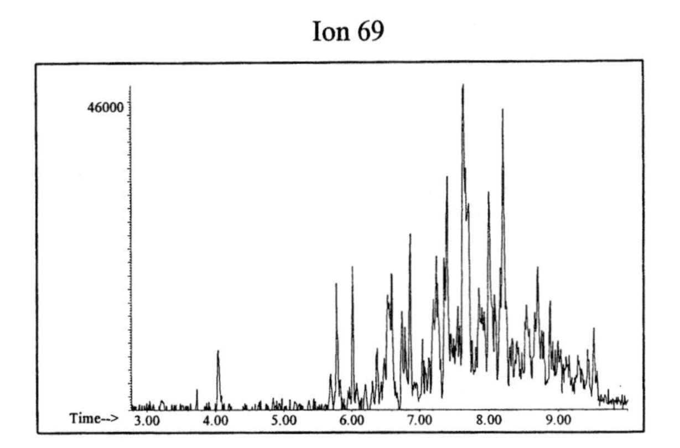

Ion 83

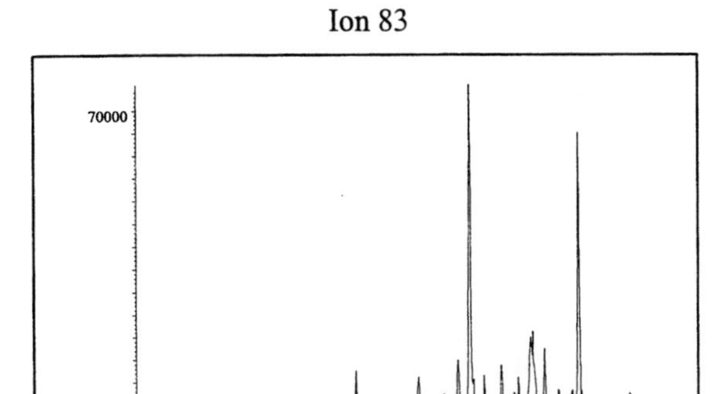

Naphthalene

Ion 128

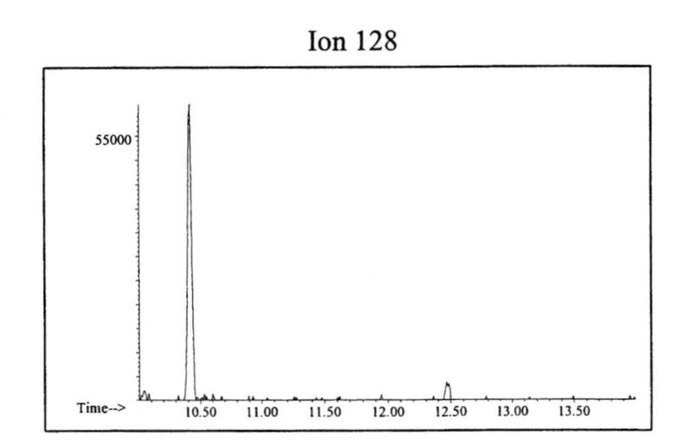

Ion 142

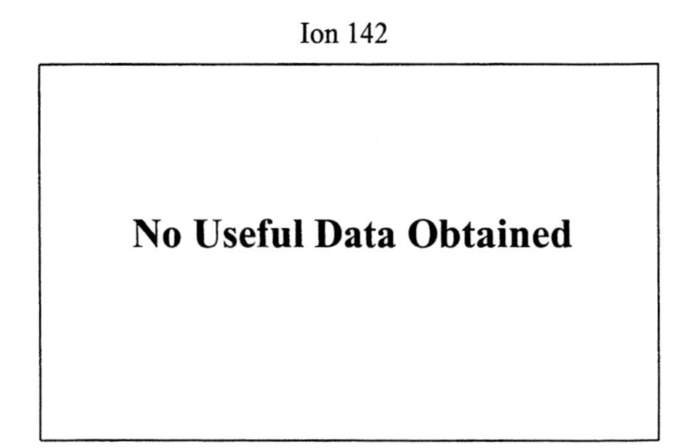

Ion 156

No Useful Data Obtained

Oxygenate #12

20uL/mL Pentane

ASTM: Class 0.1 (Oxygenate)

Macro Code: L

Product Displayed: Kleenstrip M170 Lacquer Thinner

Product Uses: Lacquer thinner

Other Similar Products:

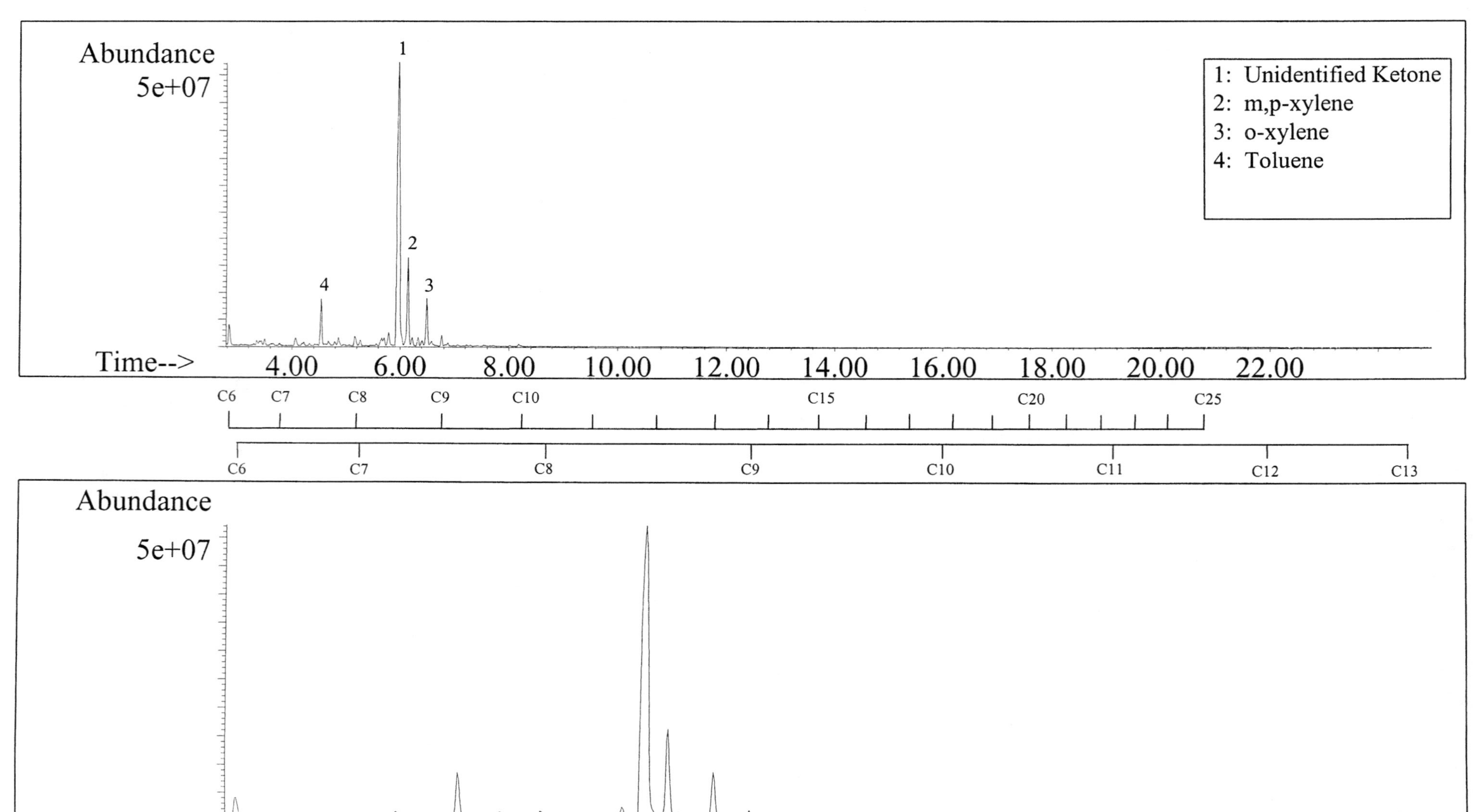

SUMMED PROFILES Oxygenate #12

ALKANES

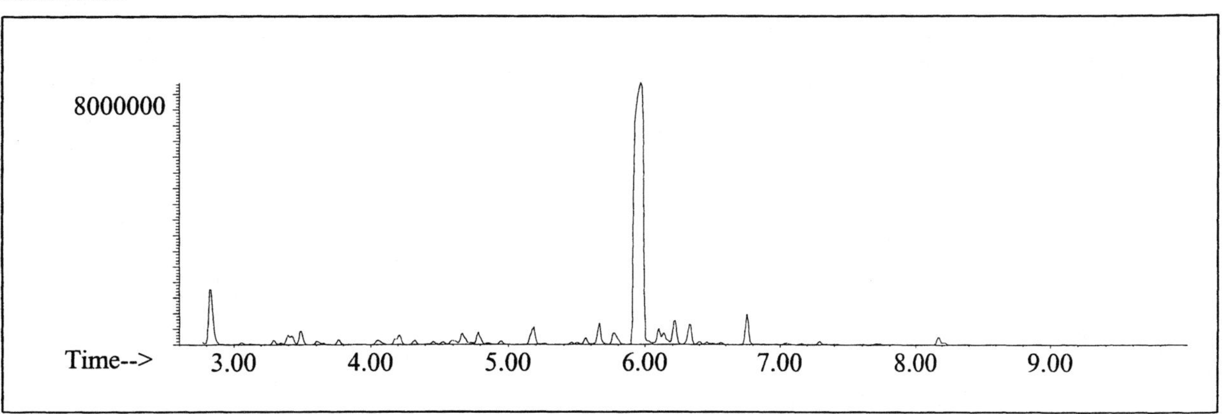

AROMATICS

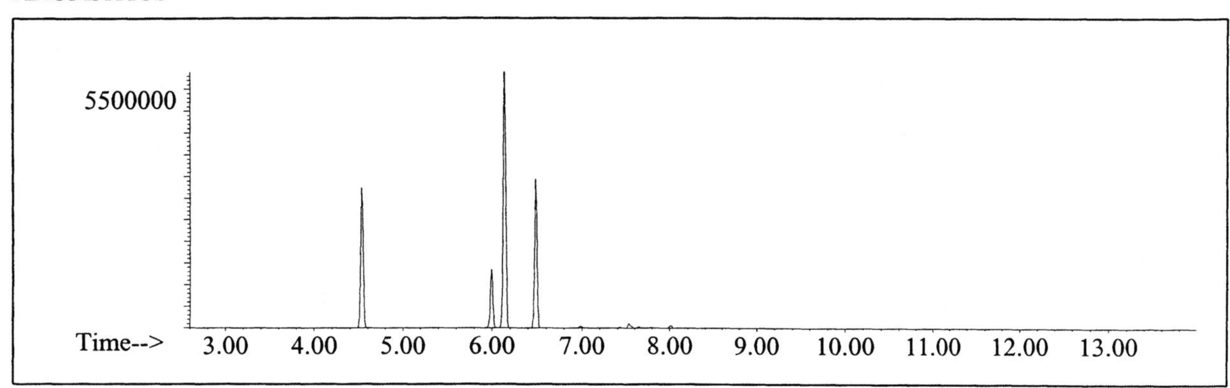

CYCLOPARAFFINS AND ALKENES

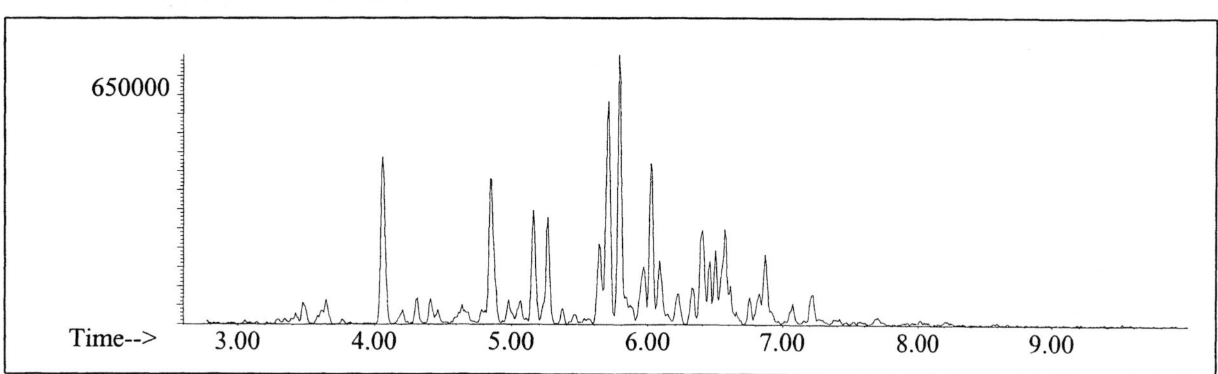

NAPHTHALENES

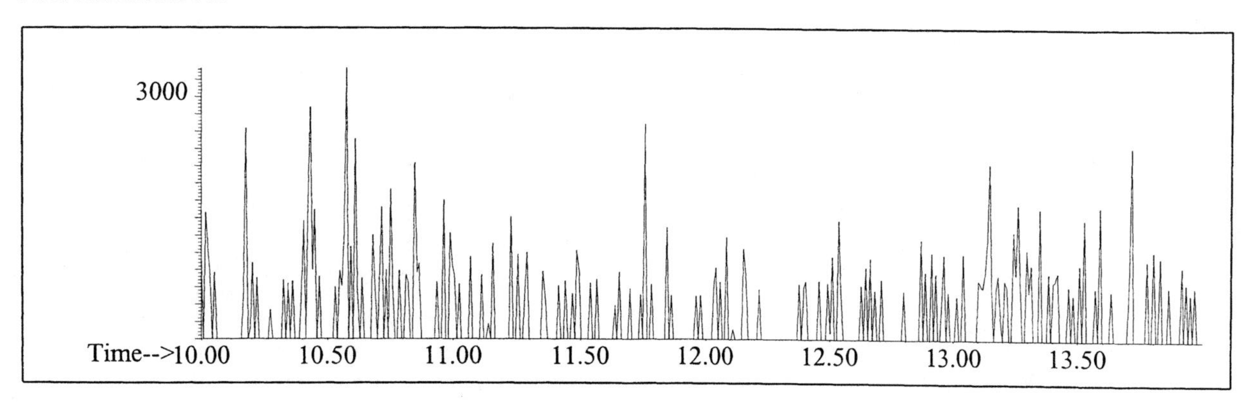

INDIVIDUAL PROFILES

Oxygenate #12

Alkane

Ion 43

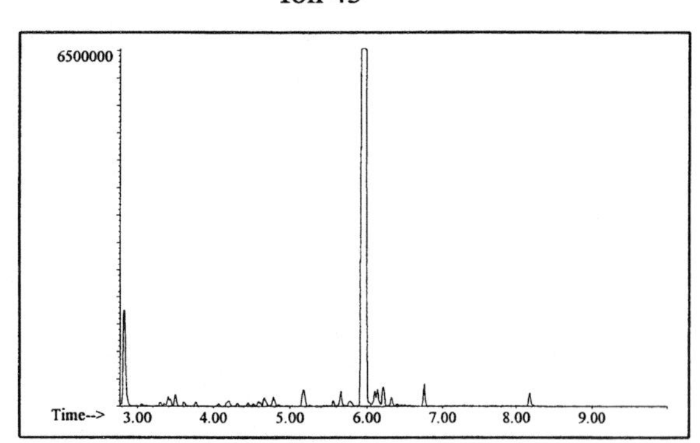

Ion 57

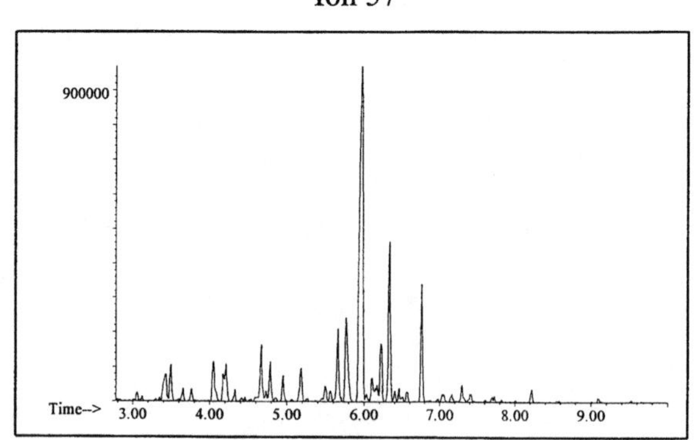

Ion 71

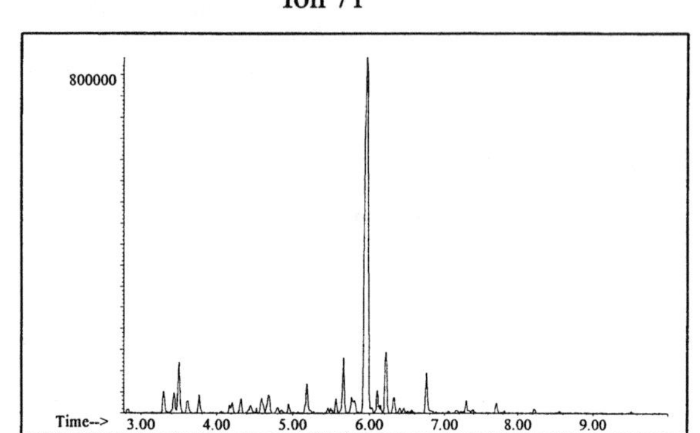

Ion 85

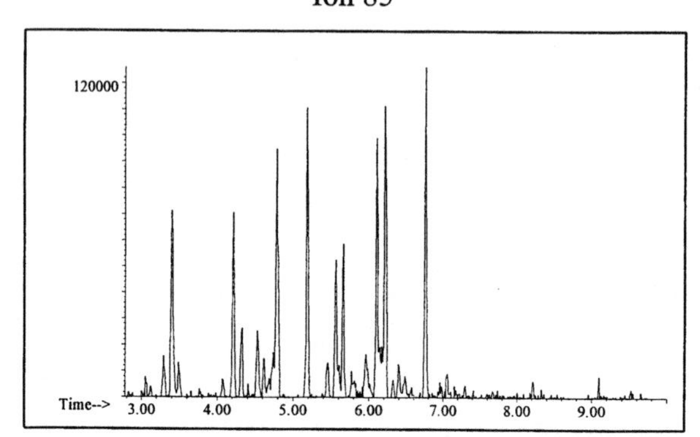

Aromatic

Ion 91

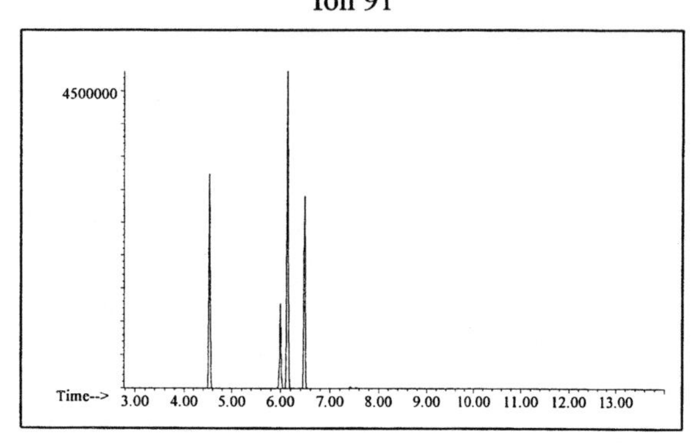

Ion 105

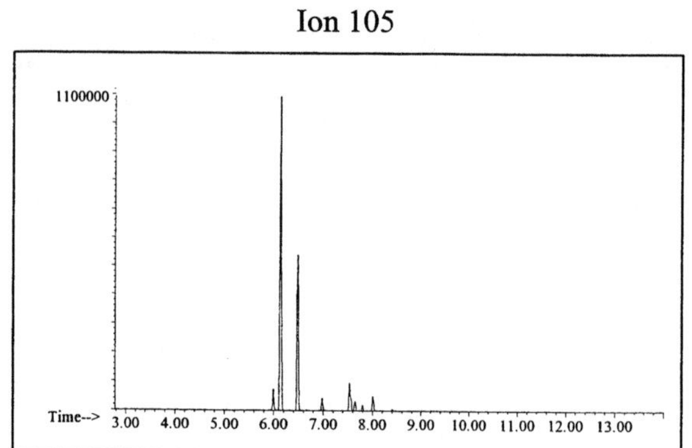

Ion 119

No Useful Data Obtained

Cycloparaffin

Ion 55

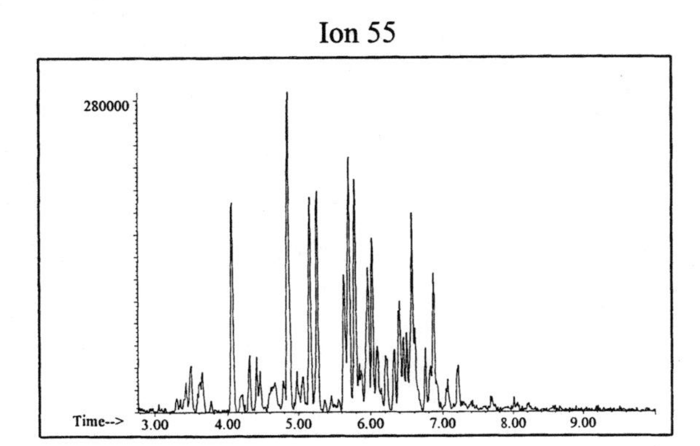

Ion 69

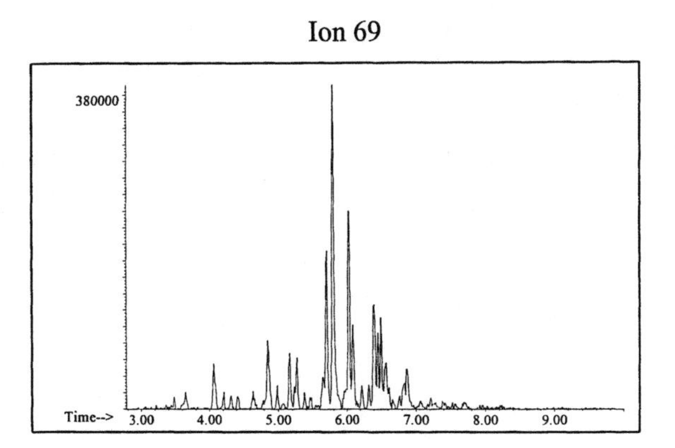

Ion 83

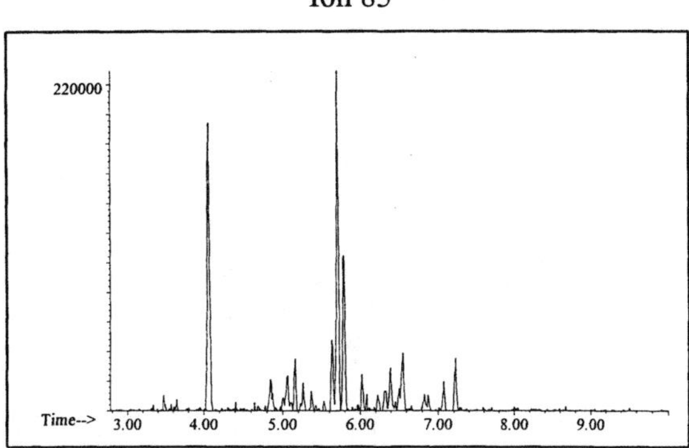

Naphthalene

Ion 128

No Useful Data Obtained

Ion 142

No Useful Data Obtained

Ion 156

No Useful Data Obtained

Oxygenate #13

20uL/mL Pentane

Product Displayed: Horizon Lacquer Thinner

Other Similar Products:

ASTM: Class 0.1 (Oxygenate)

Product Uses: Lacquer thinner

Macro Code: L

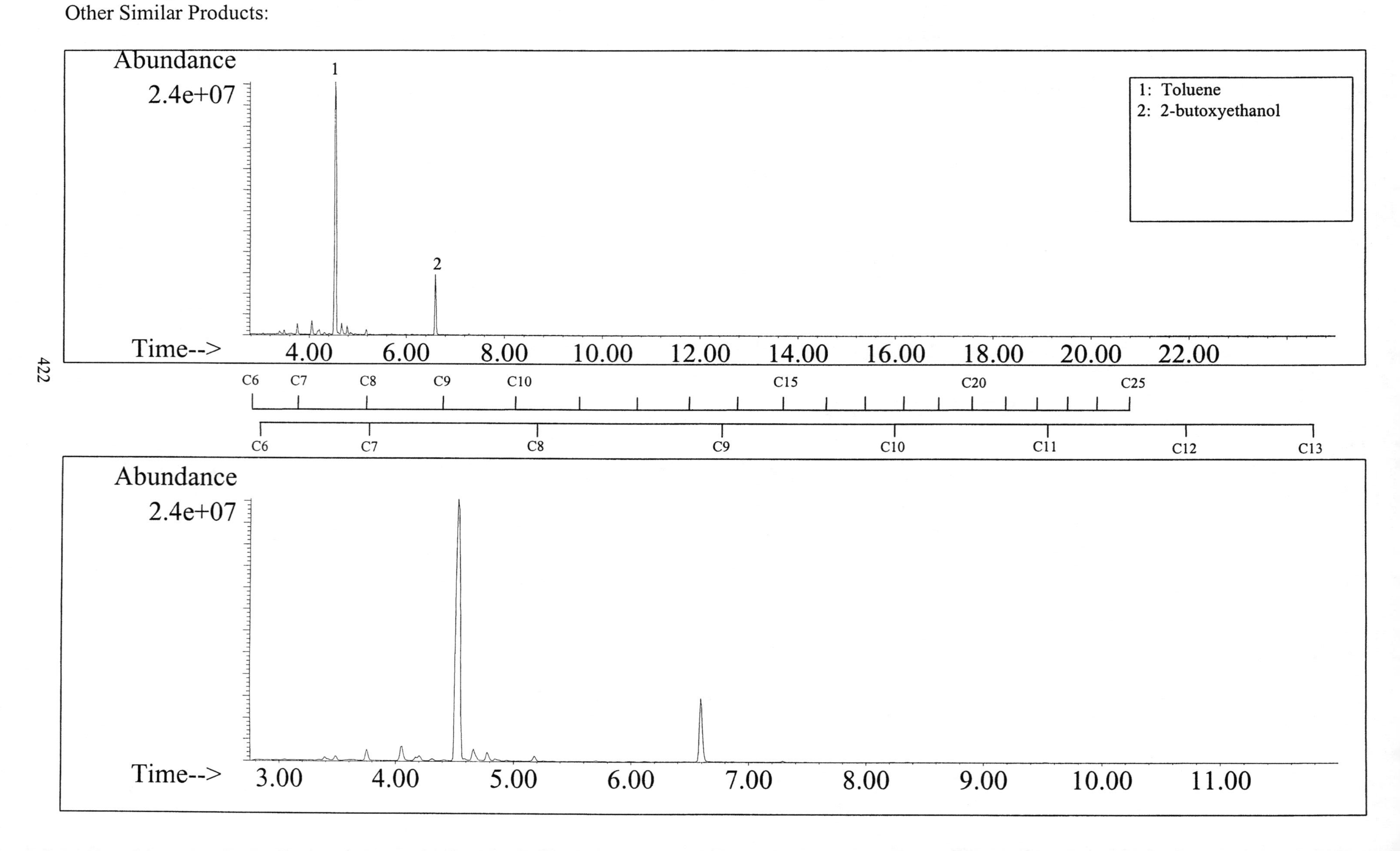

SUMMED PROFILES

Oxygenate #13

ALKANES

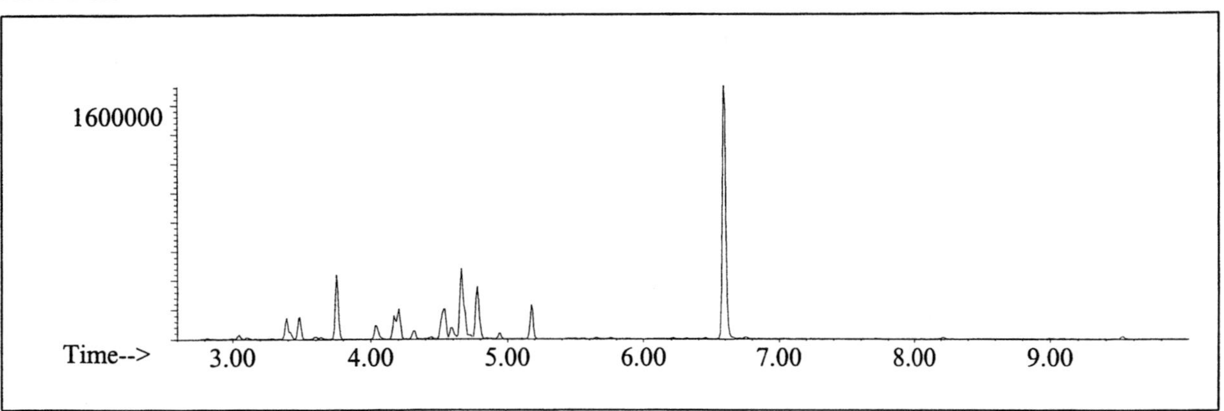

AROMATICS

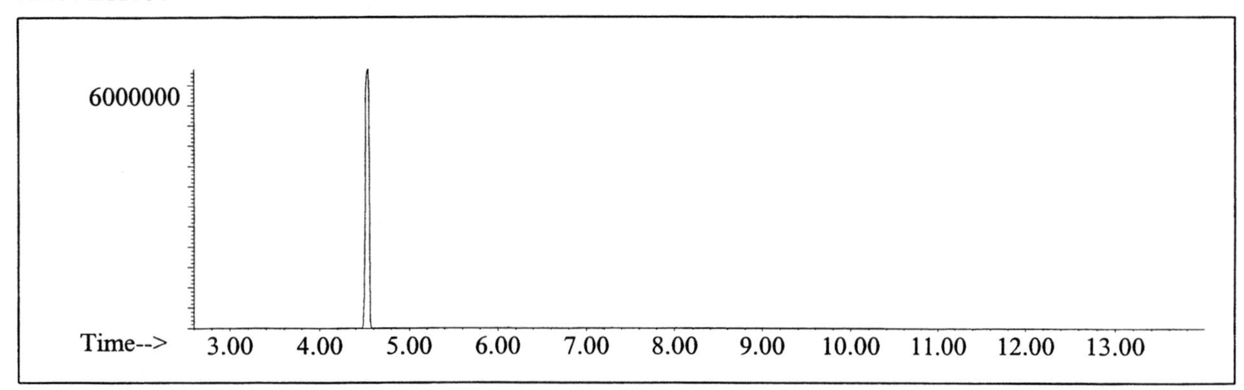

CYCLOPARAFFINS AND ALKENES

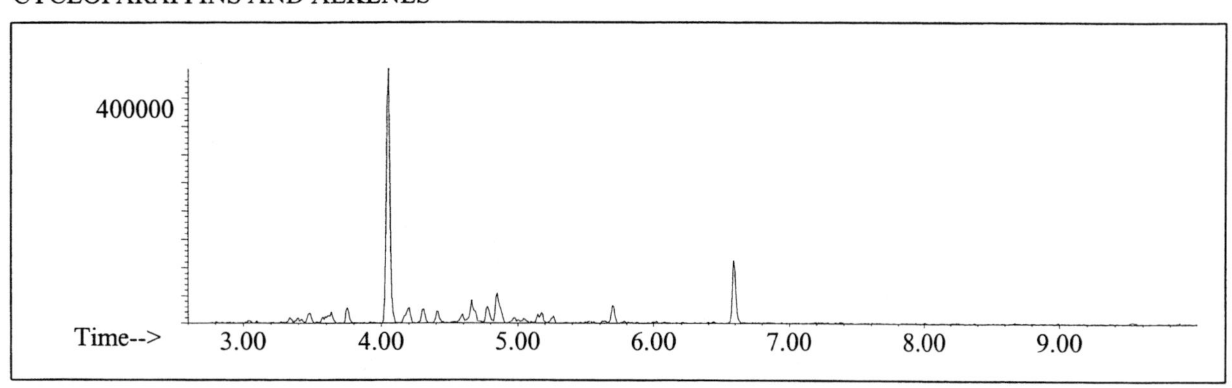

NAPHTHALENES

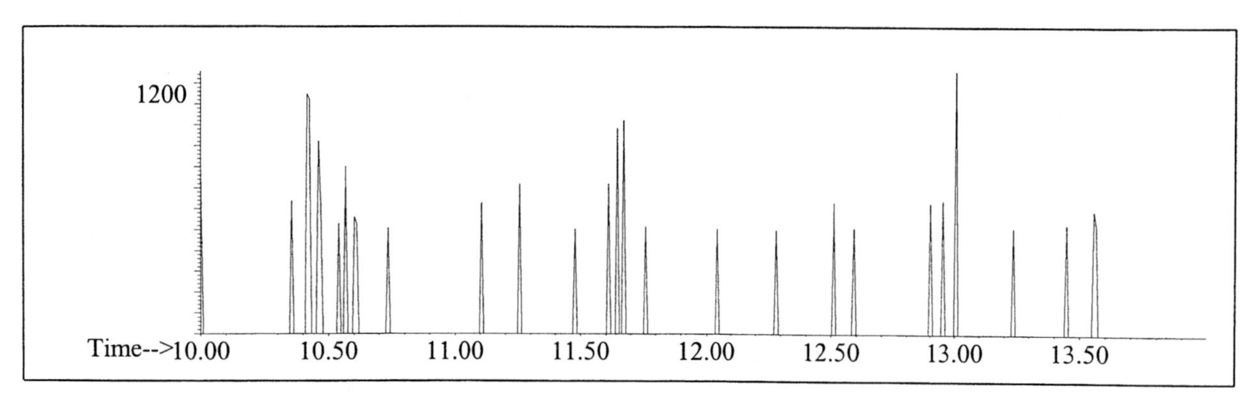

INDIVIDUAL PROFILES

Oxygenate #13

Alkane

Ion 43

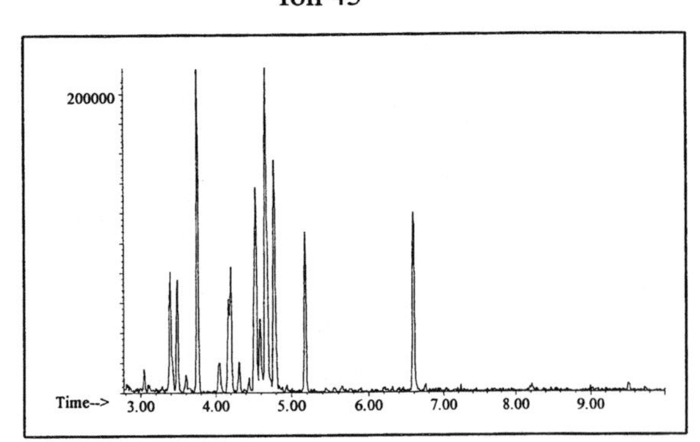

Ion 57

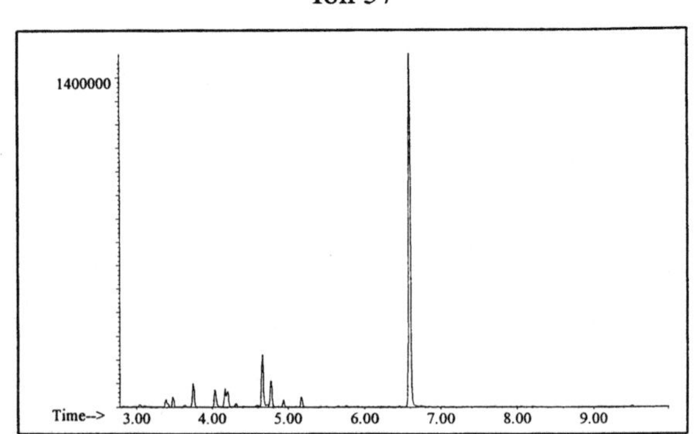

Ion 71

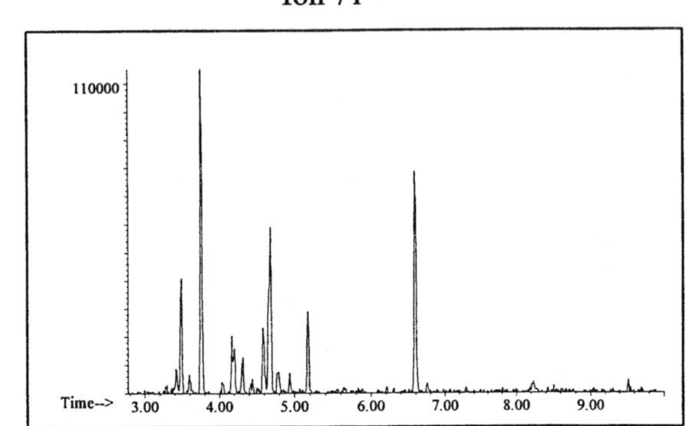

Ion 85

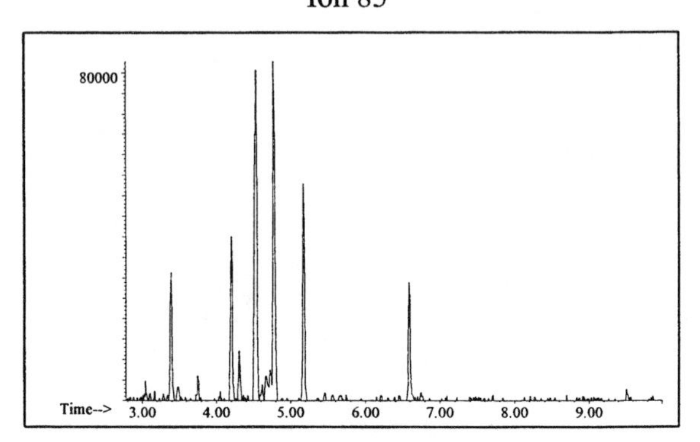

Aromatic

Ion 91

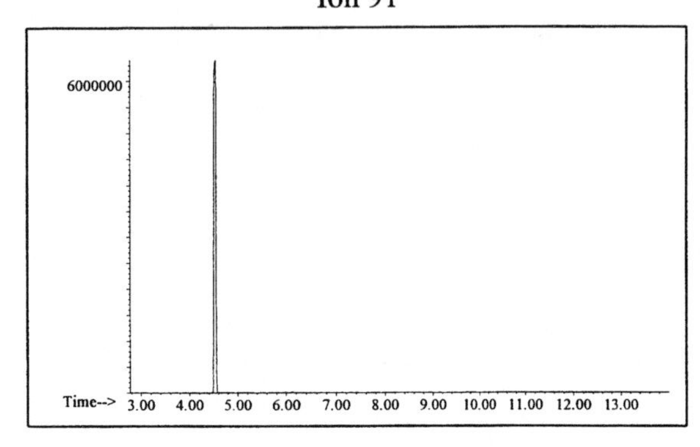

Ion 105

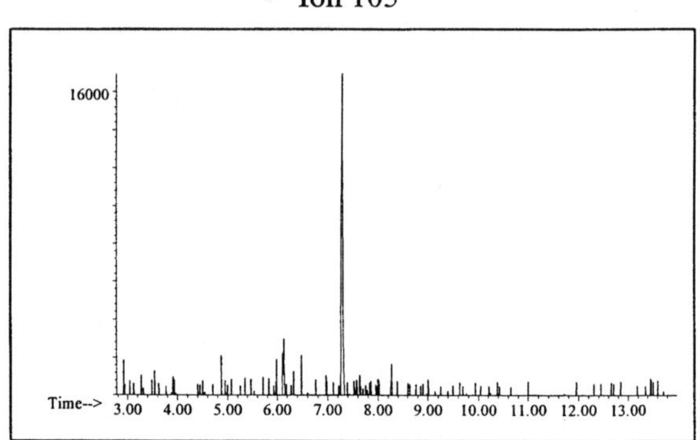

Ion 119

No Useful Data Obtained

Cycloparaffin

Ion 55

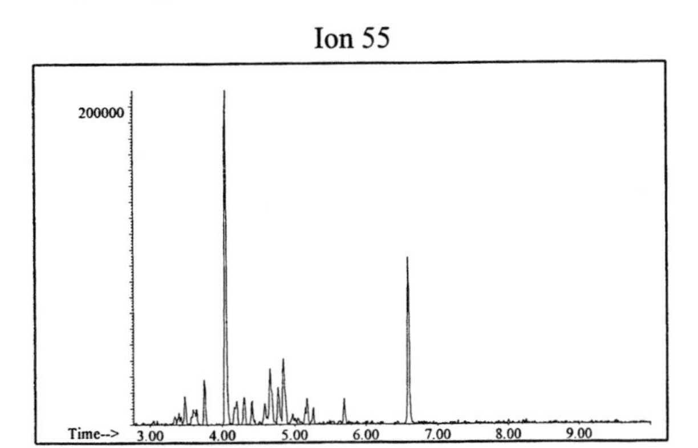

Ion 69

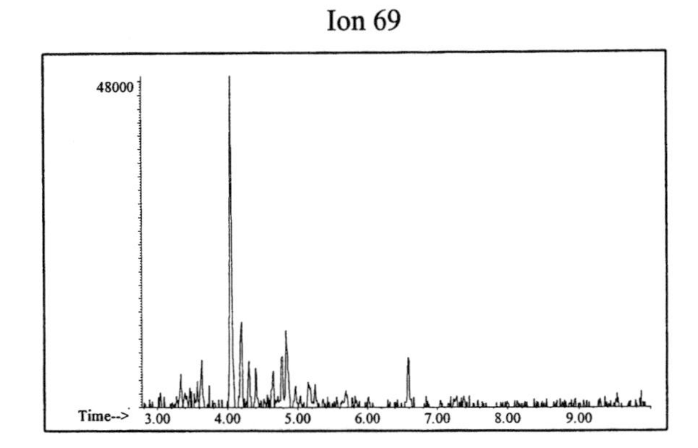

Ion 83

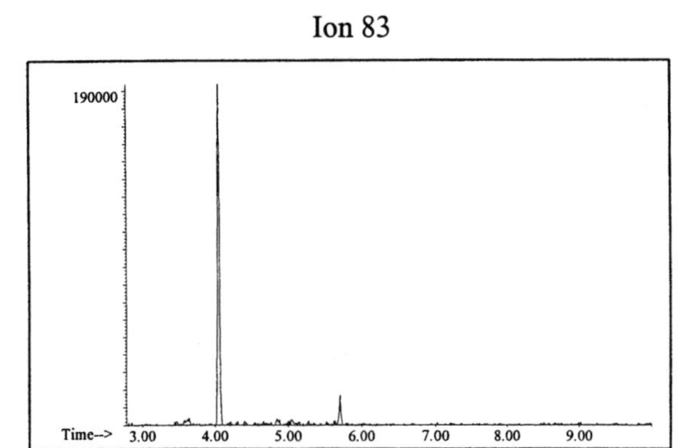

Naphthalene

Ion 128

No Useful Data Obtained

Ion 142

No Useful Data Obtained

Ion 156

No Useful Data Obtained

Oxygenate #14

20uL/mL Pentane

Product Displayed: Sears Epoxy and Lacquer Thinner

Other Similar Products:

ASTM: Class 0.1 (Oxygenate)

Product Uses: Lacquer thinner

Macro Code: L

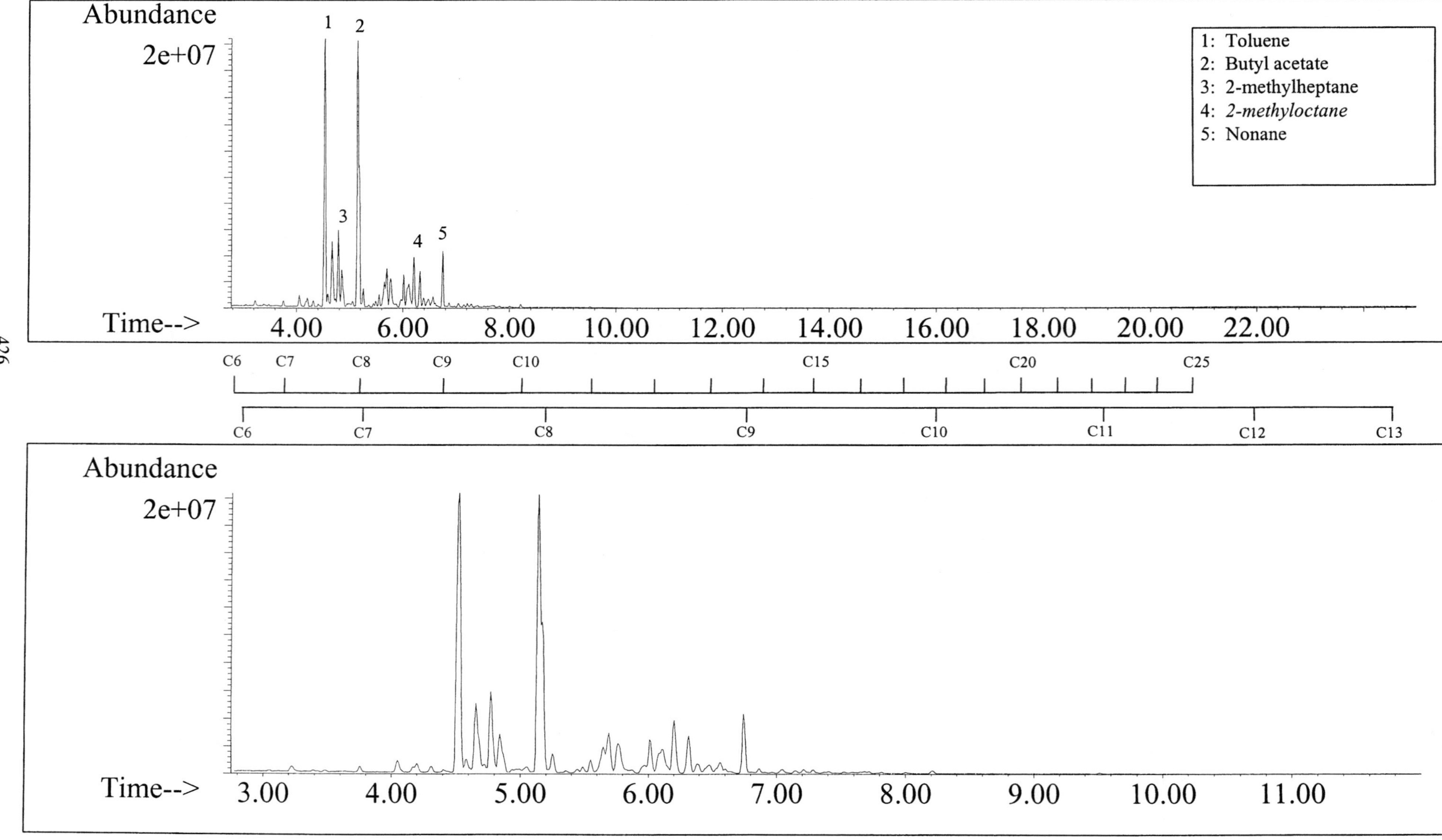

SUMMED PROFILES

Oxygenate #14

ALKANES

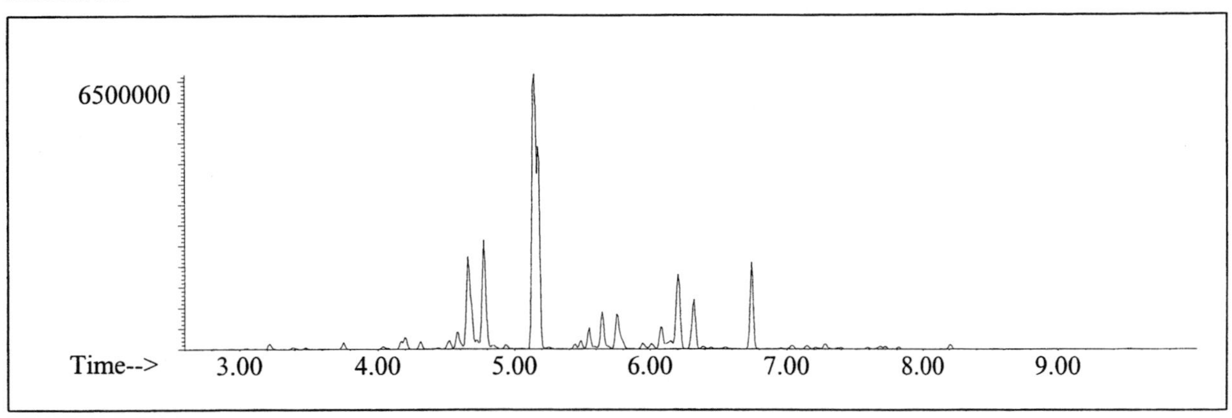

AROMATICS

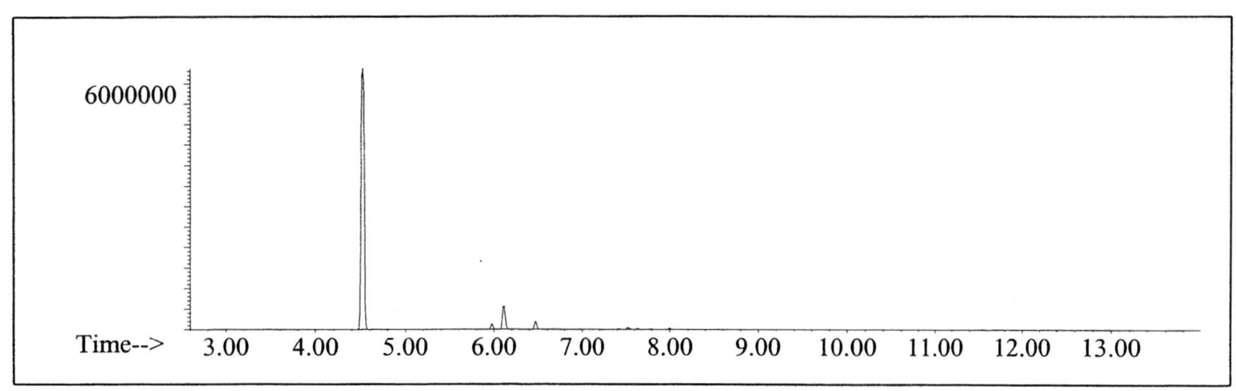

CYCLOPARAFFINS AND ALKENES

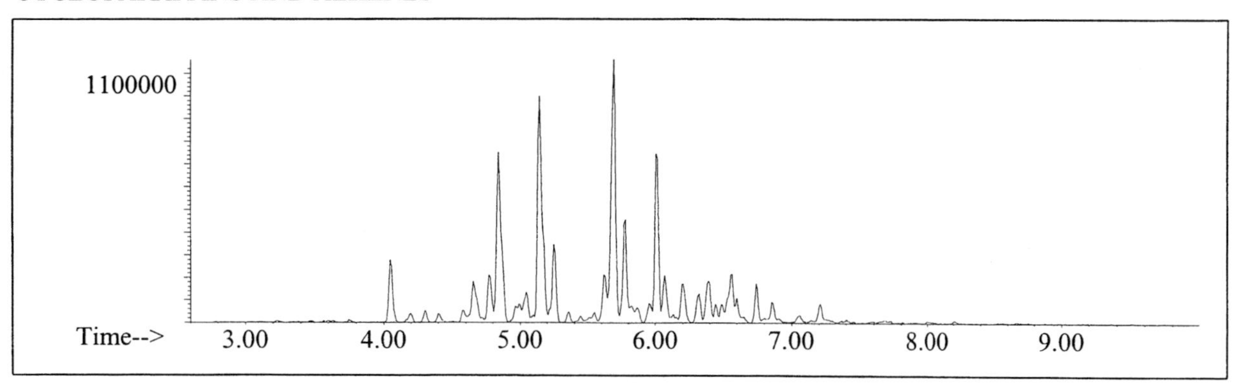

NAPHTHALENES

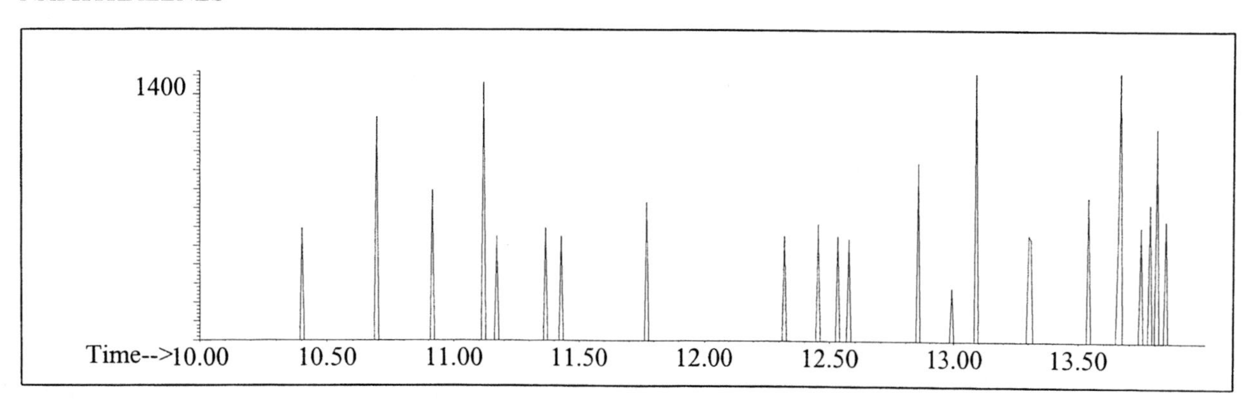

INDIVIDUAL PROFILES

Oxygenate #14

Alkane

Ion 43

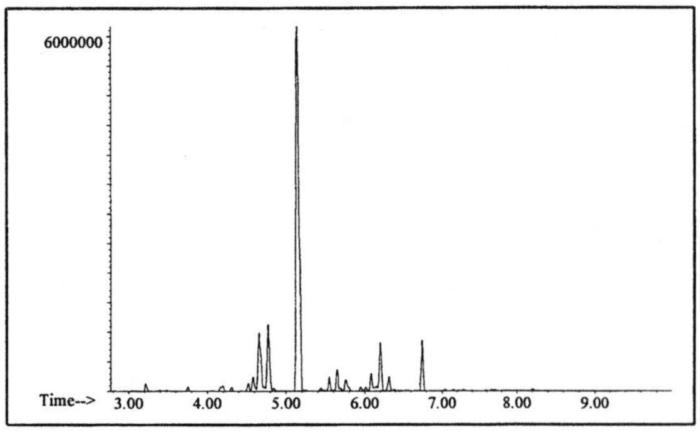

Ion 57

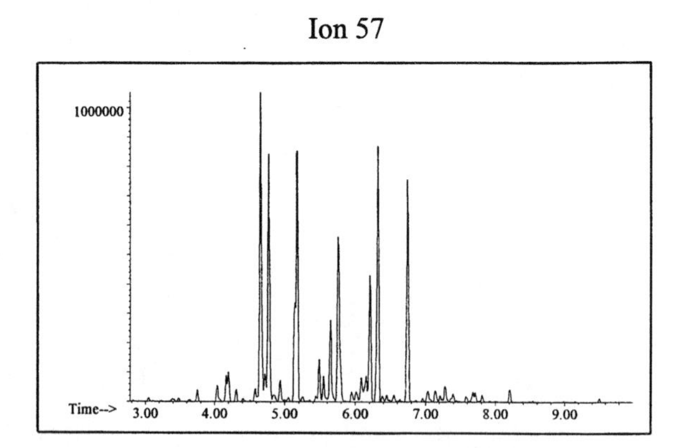

Ion 71

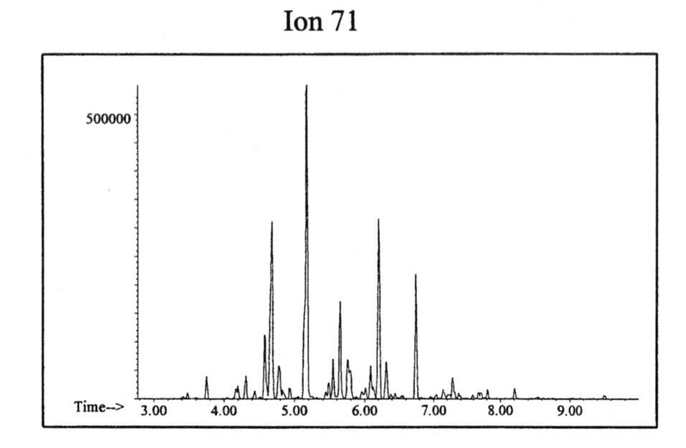

Ion 85

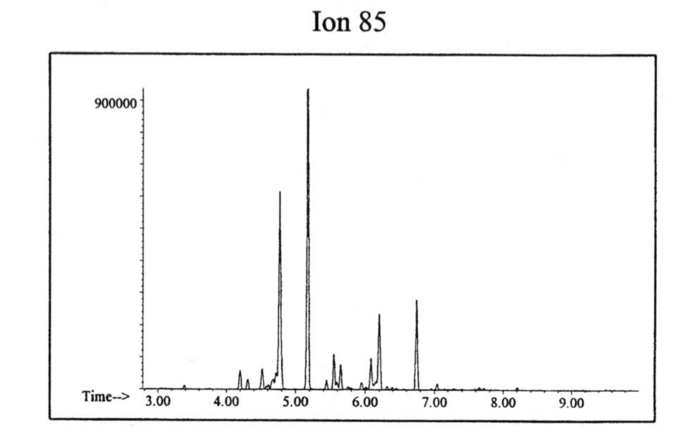

Aromatic

Ion 91

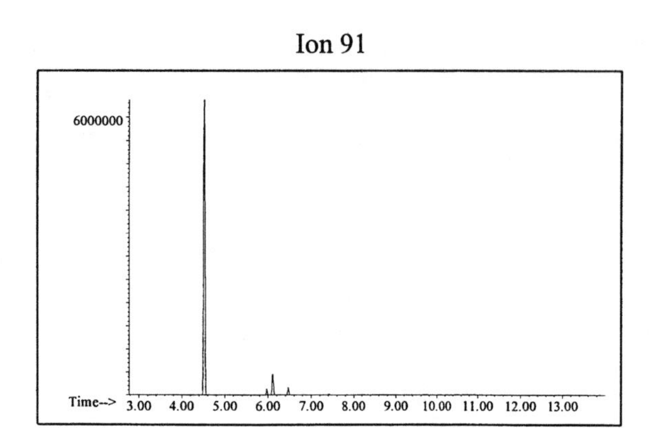

Ion 105

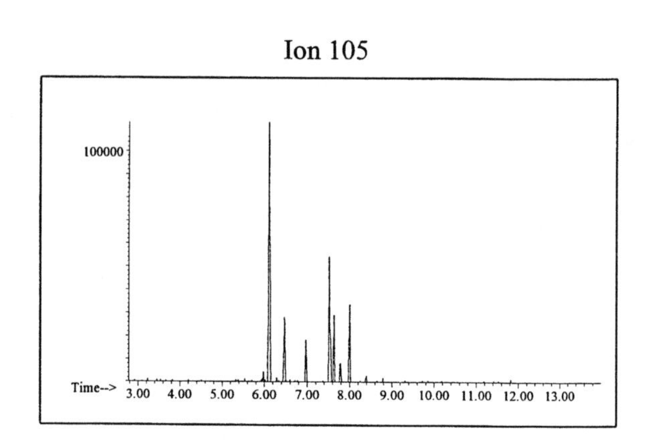

Ion 119

No Useful Data Obtained

INDIVIDUAL PROFILES

Cycloparaffin

Ion 55

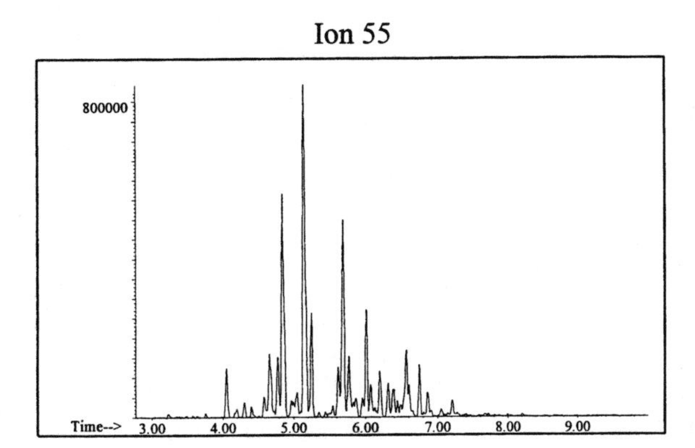

Ion 69

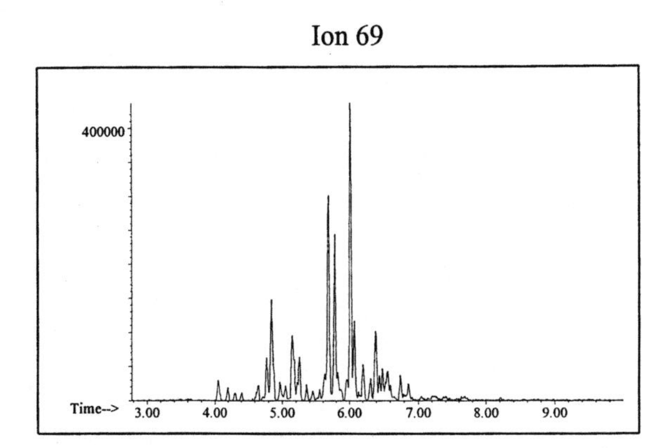

Ion 83

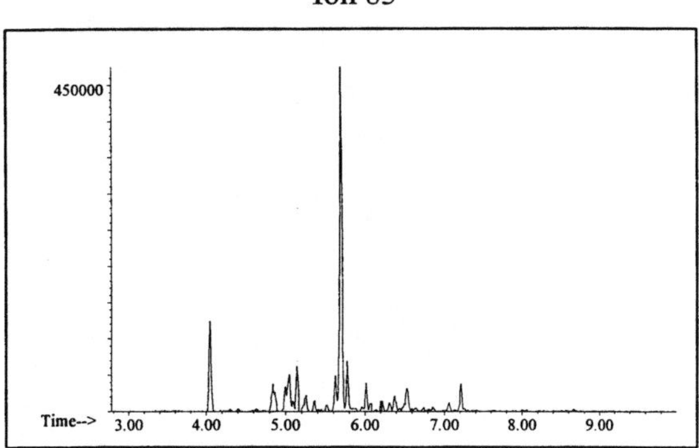

Naphthalene

Ion 128

No Useful Data Obtained

Ion 142

No Useful Data Obtained

Ion 156

No Useful Data Obtained

Oxygenate #15

20uL/mL Pentane
Product Displayed: Exxon TEI 600
Other Similar Products:

ASTM: Class 0.1 (Oxygenate)
Product Uses: Feedstock

Macro Code: L

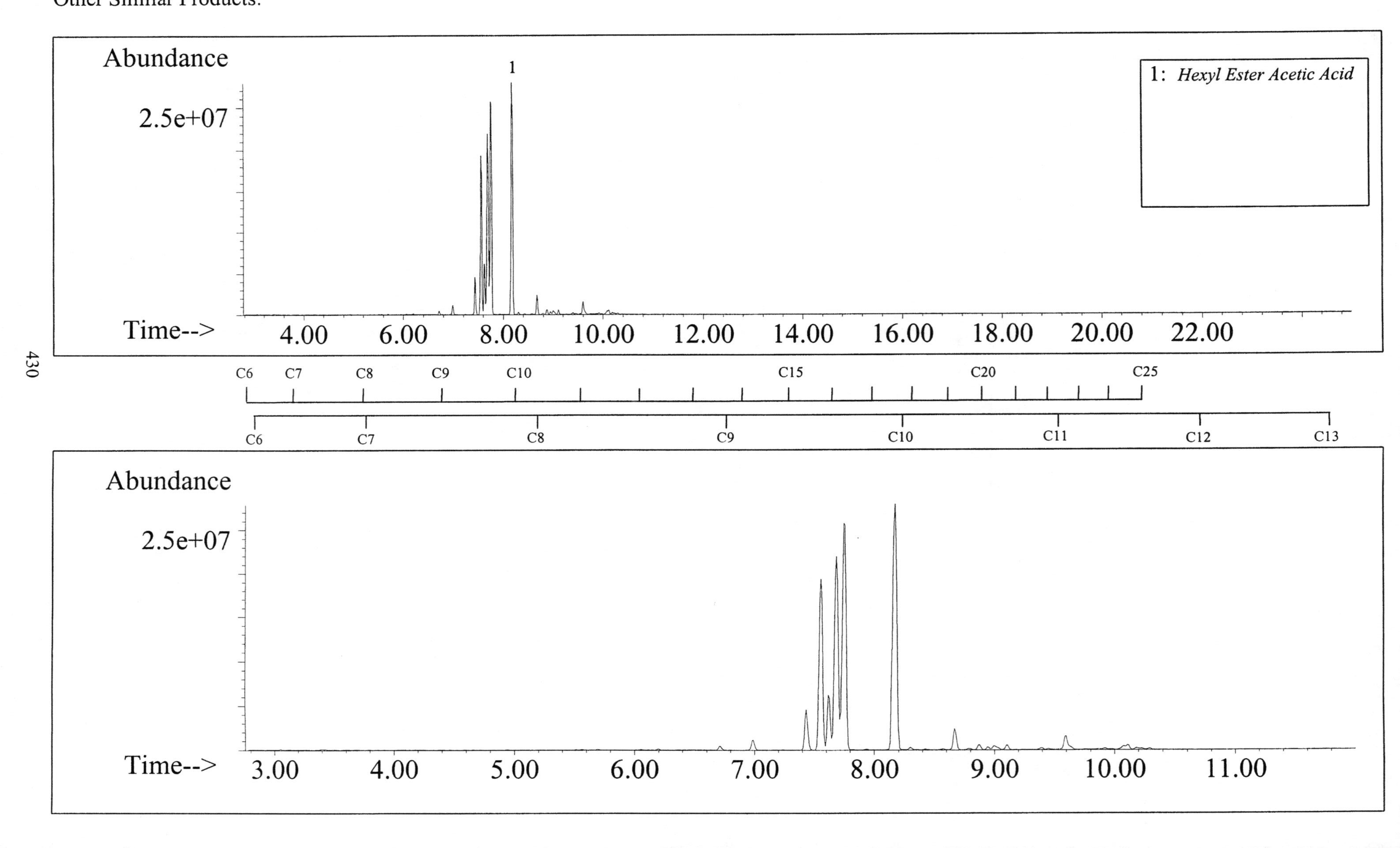

SUMMED PROFILES Oxygenate #15

ALKANES

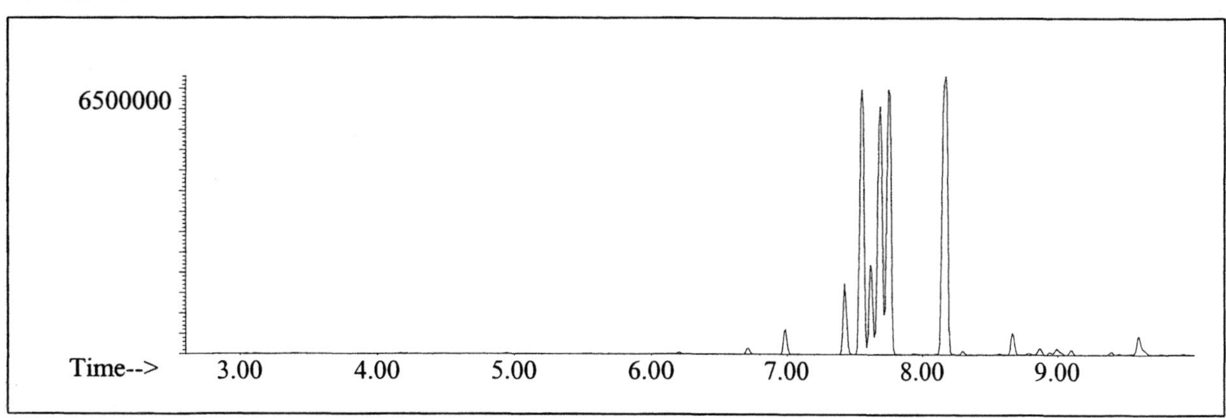

AROMATICS

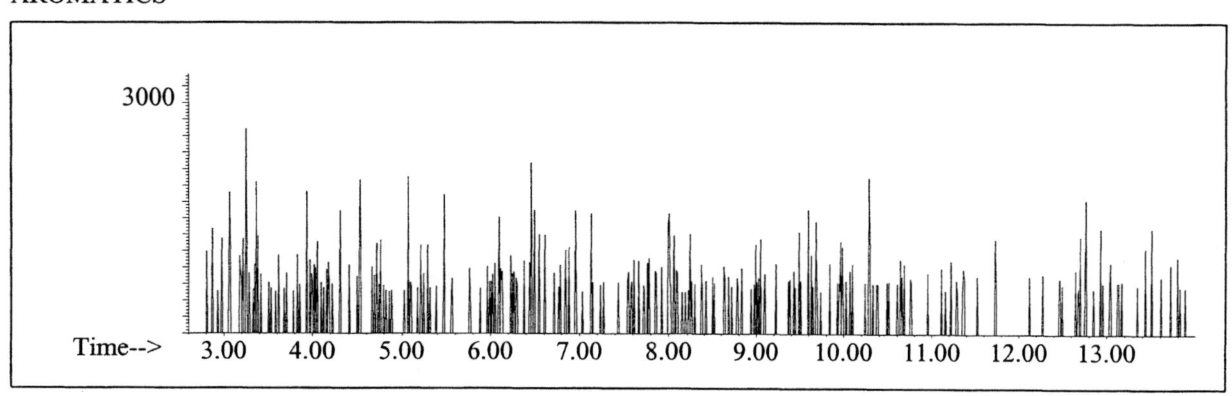

CYCLOPARAFFINS AND ALKENES

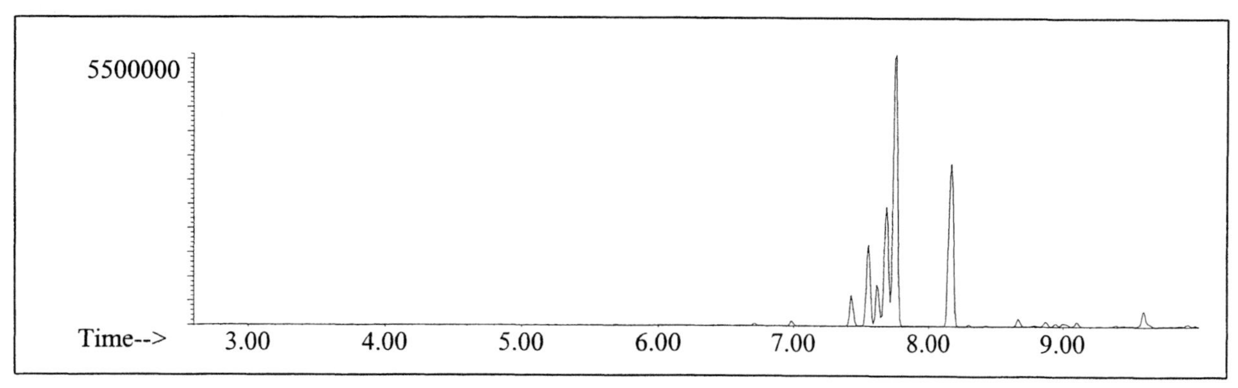

NAPHTHALENES

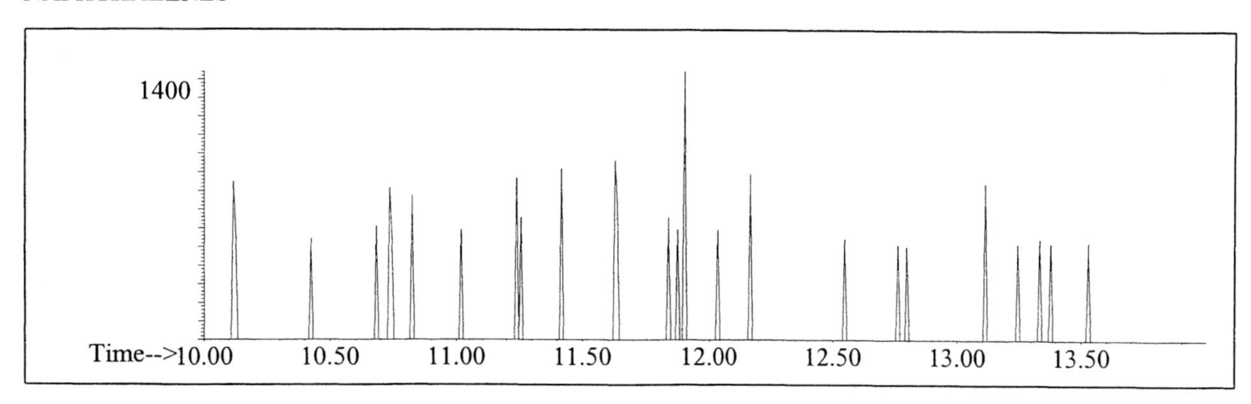

INDIVIDUAL PROFILES

Oxygenate #15

Alkane

Ion 43

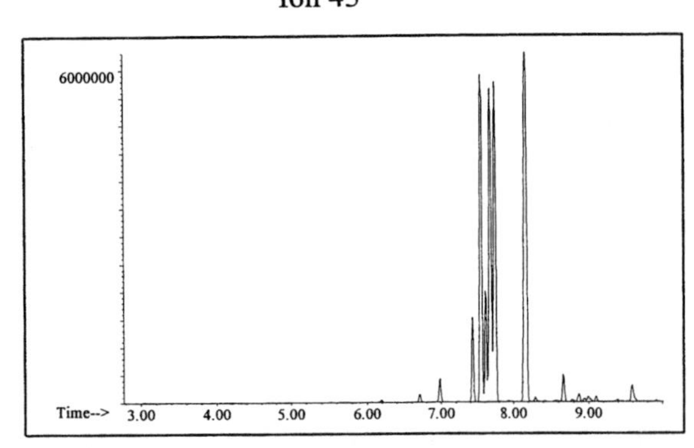

Ion 57

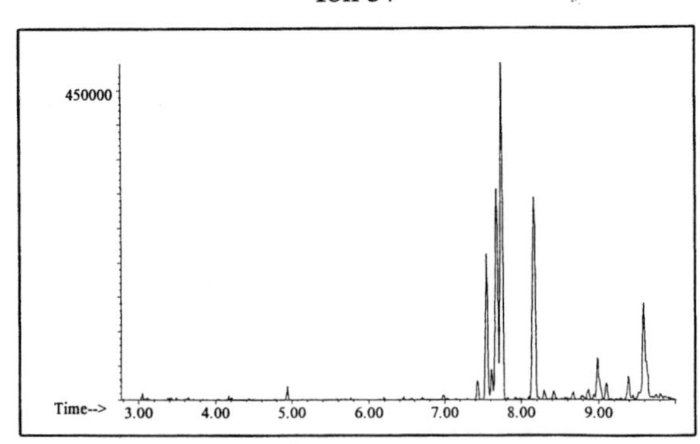

Ion 71

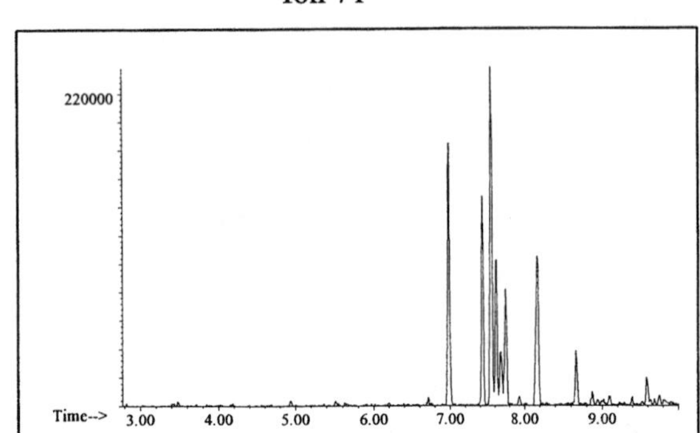

Ion 85

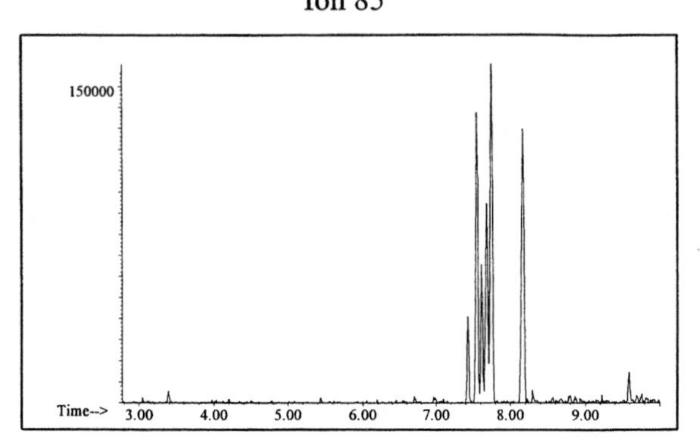

Aromatic

Ion 91

No Useful Data Obtained

Ion 105

No Useful Data Obtained

Ion 119

No Useful Data Obtained

Cycloparaffin

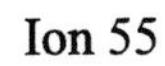
Ion 55

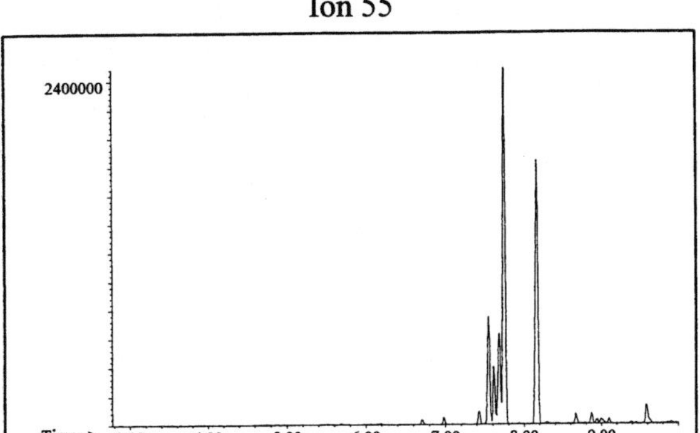

Ion 69

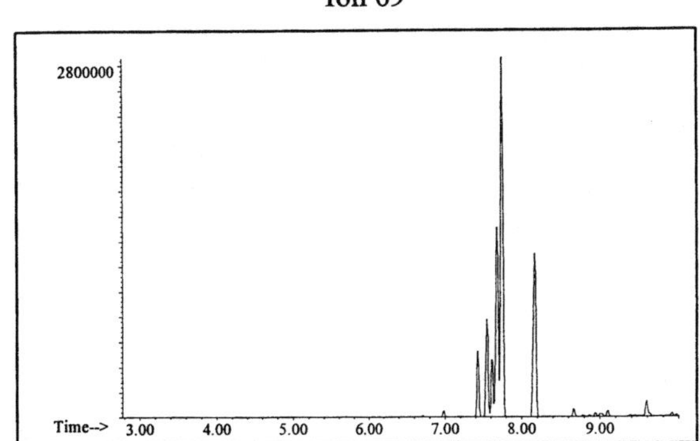

Ion 83

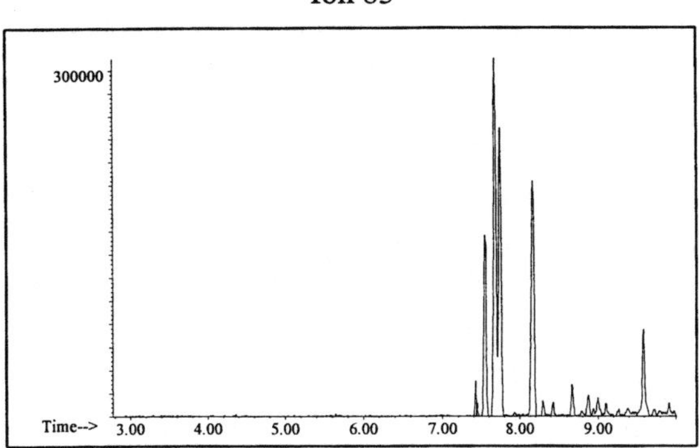

Naphthalene

Ion 128

No Useful Data Obtained

Ion 142

No Useful Data Obtained

Ion 156

No Useful Data Obtained

Isoparaffinic Products

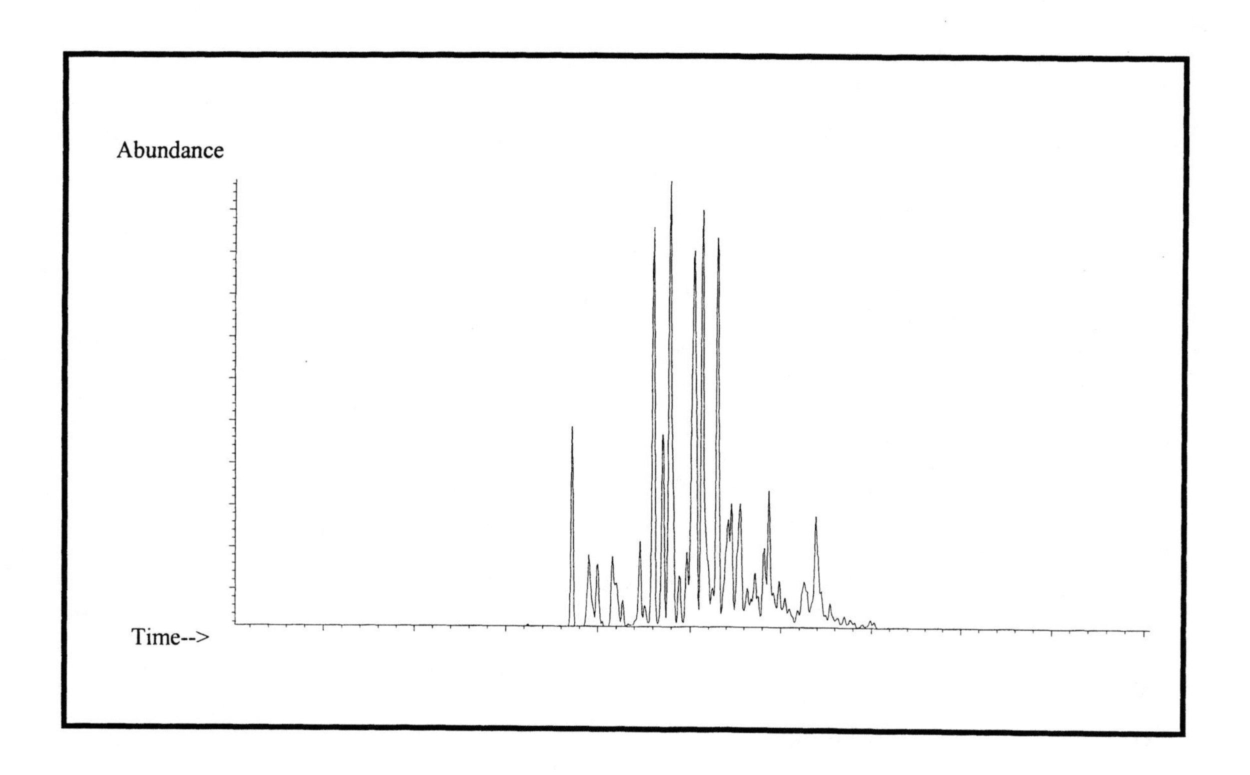

Isoparaffin #1

20uL/mL Pentane

Product Displayed: Phillips 66 Soltrol 10

Other Similar Products:

ASTM: Class 0.2 (Isoparaffin)

Product Uses: Solvent, feedstock

Macro Code: L

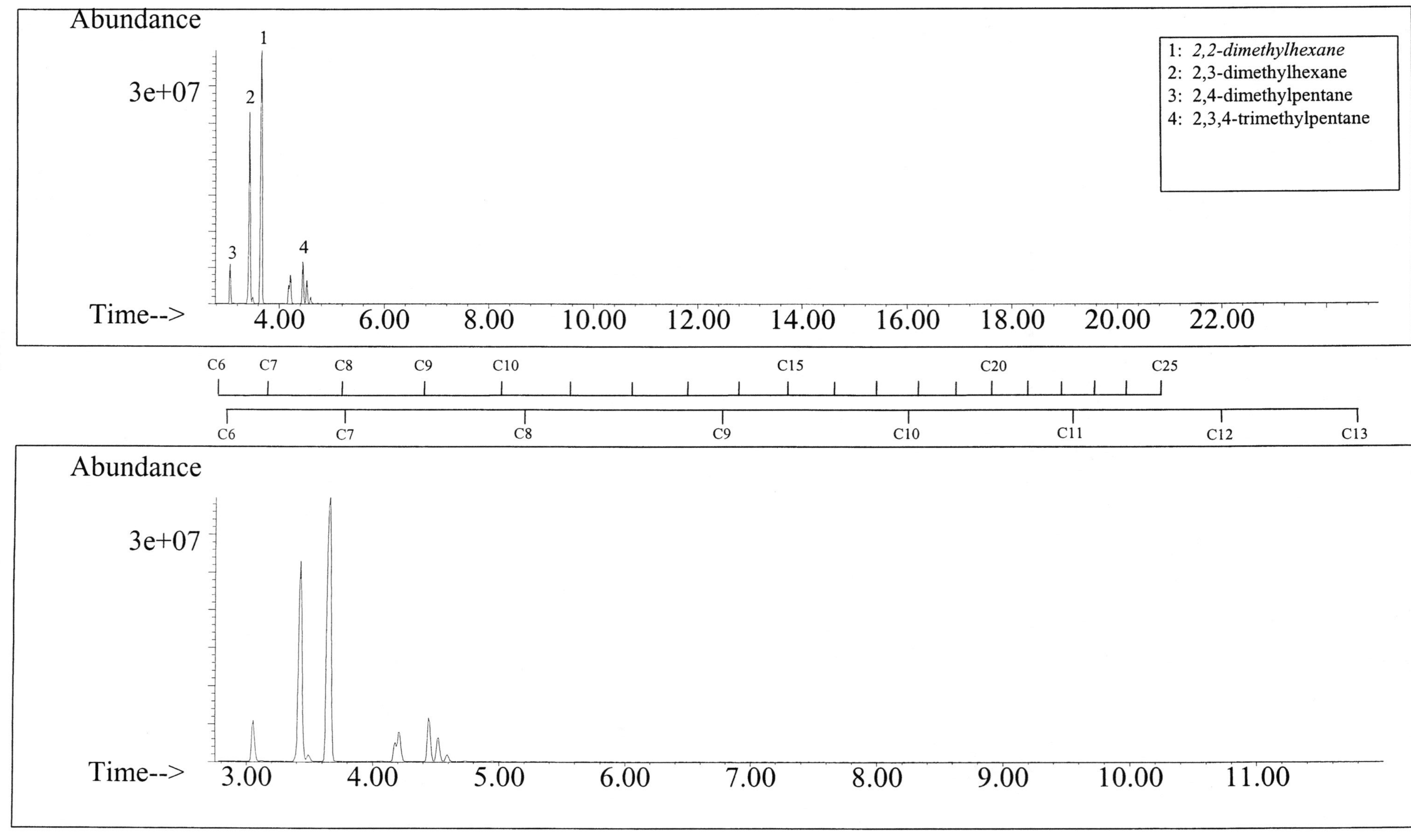

ALKANES

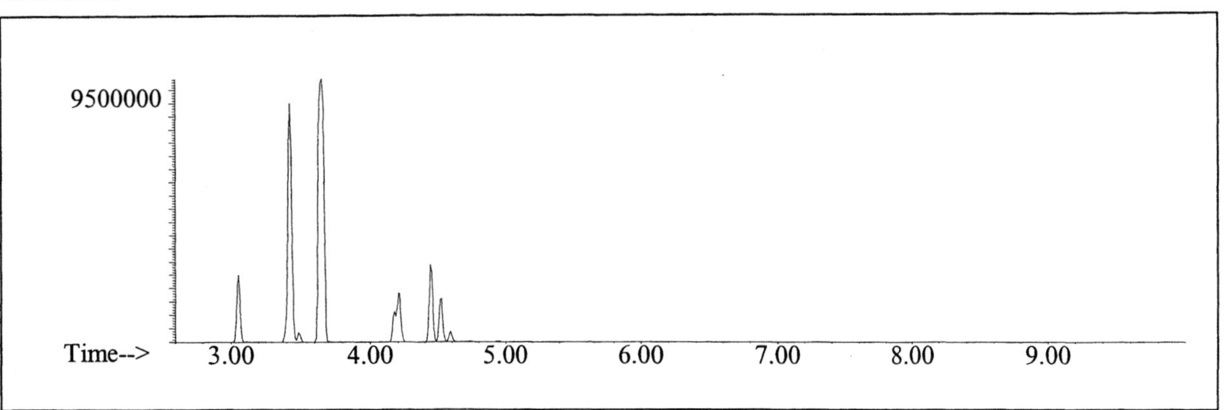

AROMATICS

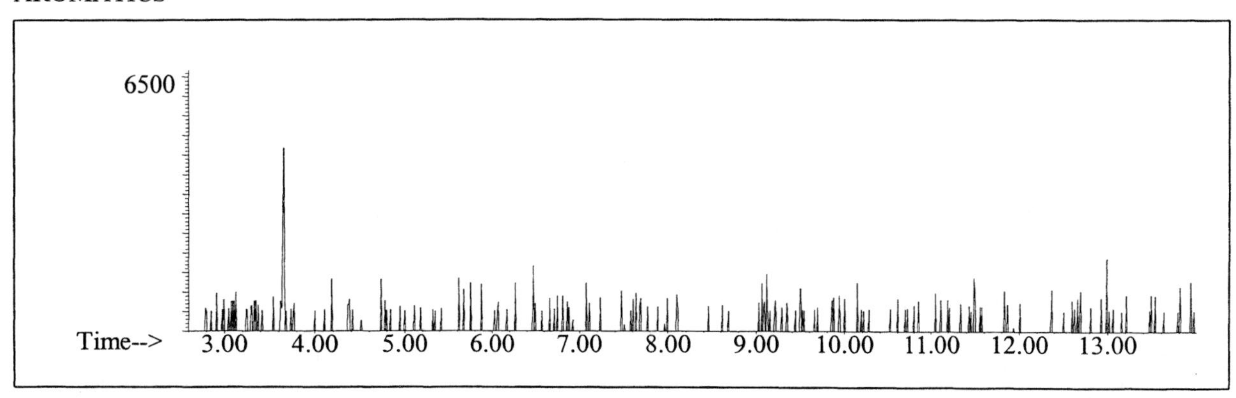

CYCLOPARAFFINS AND ALKENES

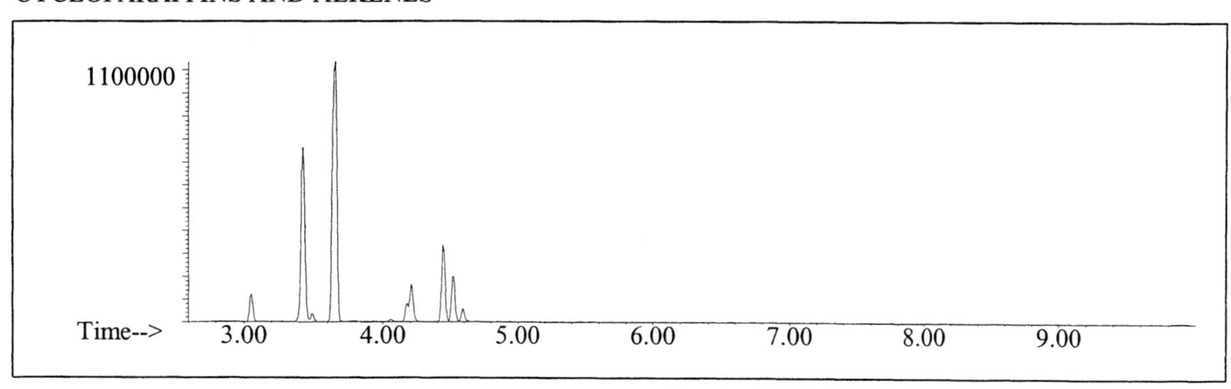

NAPHTHALENES

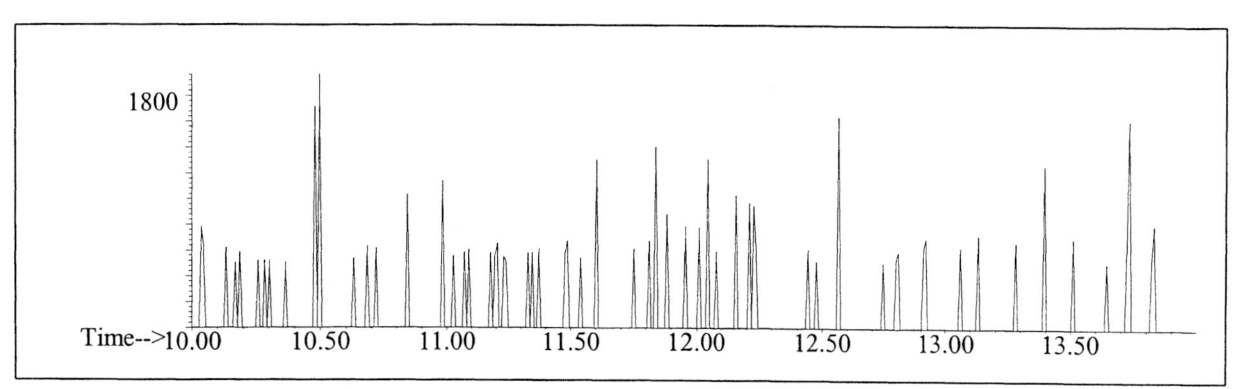

INDIVIDUAL PROFILES

Isoparaffin #1

Alkane

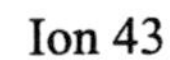

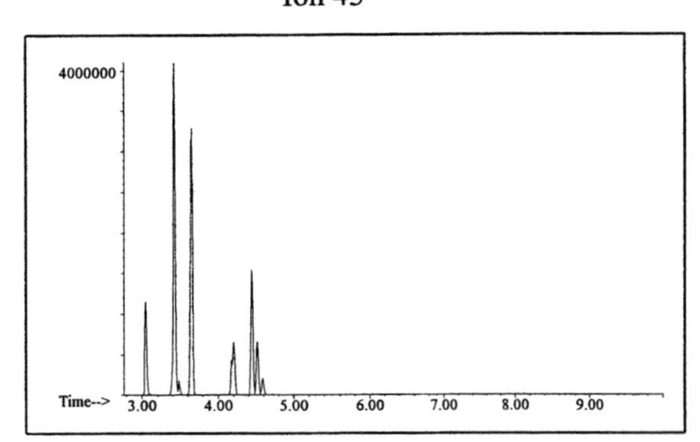

Ion 57

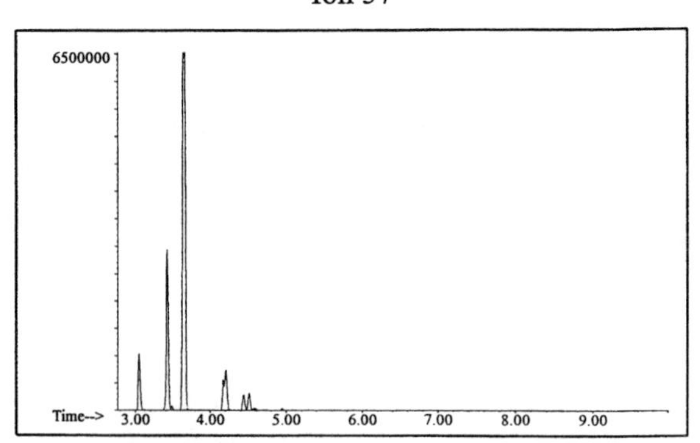

Ion 71

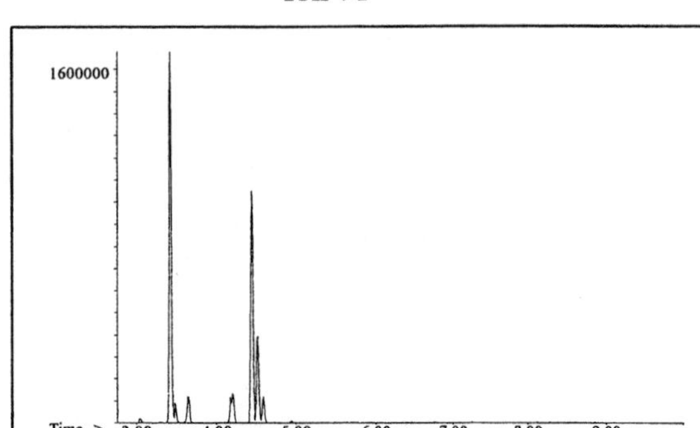

Ion 85

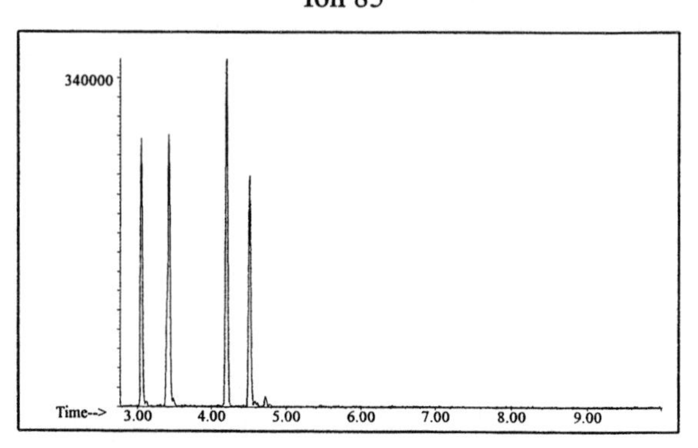

Aromatic

Ion 91

No Useful Data Obtained

Ion 105

No Useful Data Obtained

Ion 119

No Useful Data Obtained

INDIVIDUAL PROFILES

Isoparaffin #1

Cycloparaffin

Ion 55

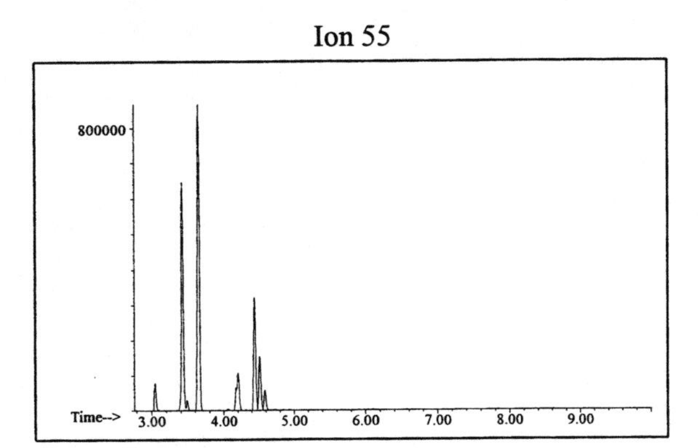

Ion 69

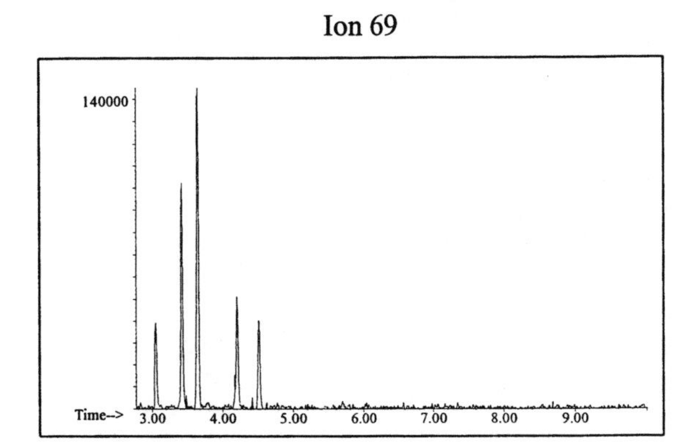

Ion 83

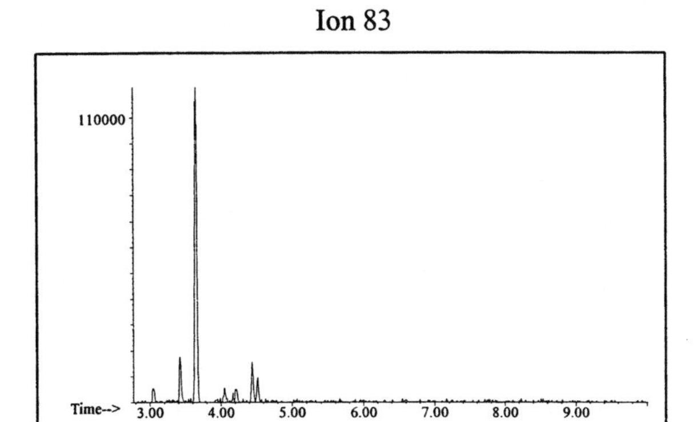

Naphthalene

Ion 128

No Useful Data Obtained

Ion 142

No Useful Data Obtained

Ion 156

No Useful Data Obtained

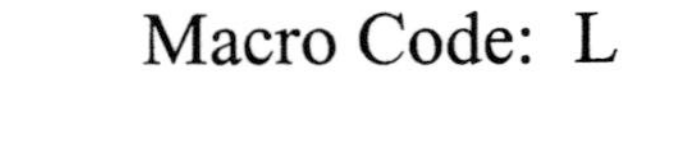

Isoparaffin #2

20uL/mL Pentane

Product Displayed: Isopar C

Other Similar Products:

ASTM: Class 0.2 (Isoparaffin)

Product Uses: Solvent, feedstock

Macro Code: L

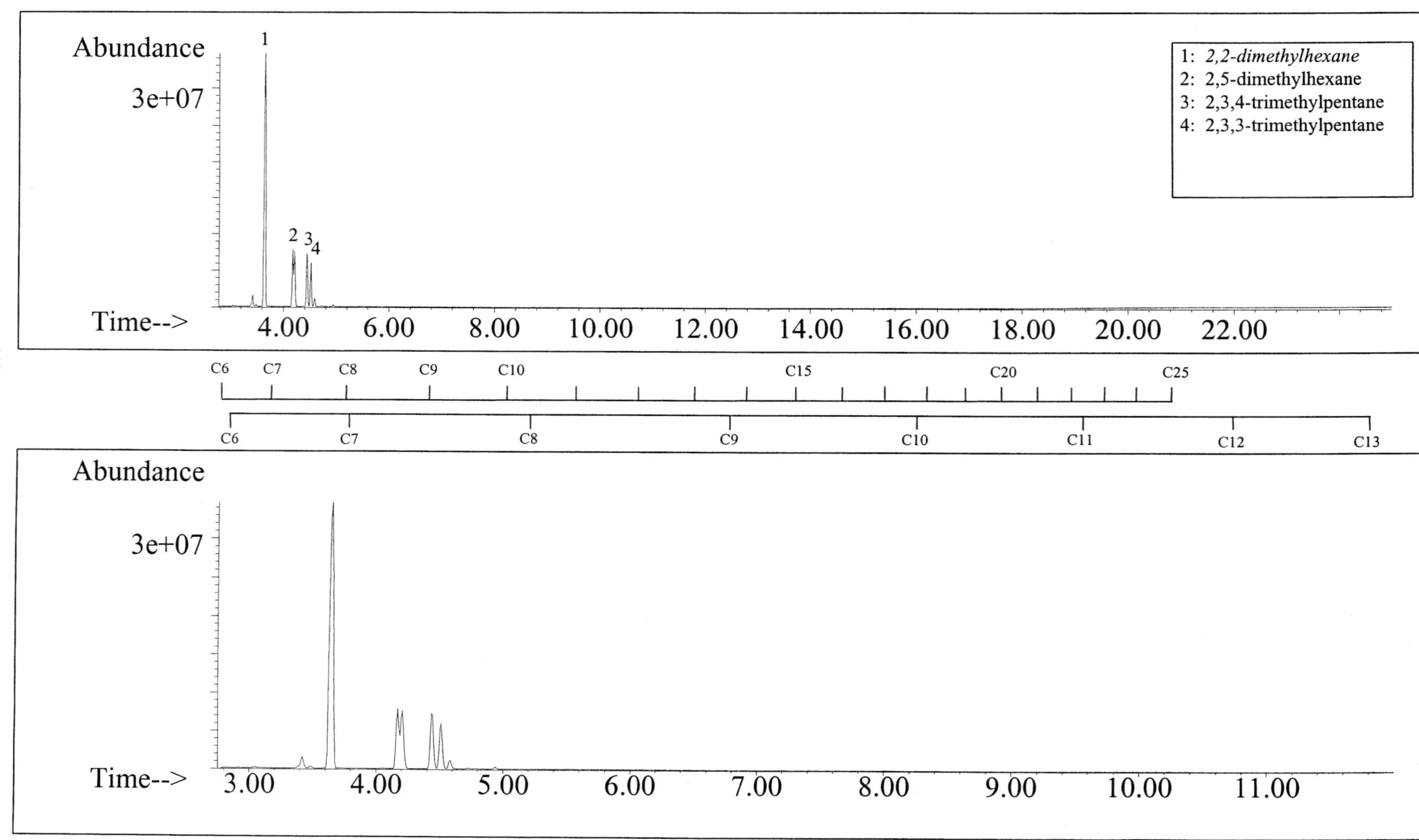

SUMMED PROFILES Isoparaffin #2

ALKANES

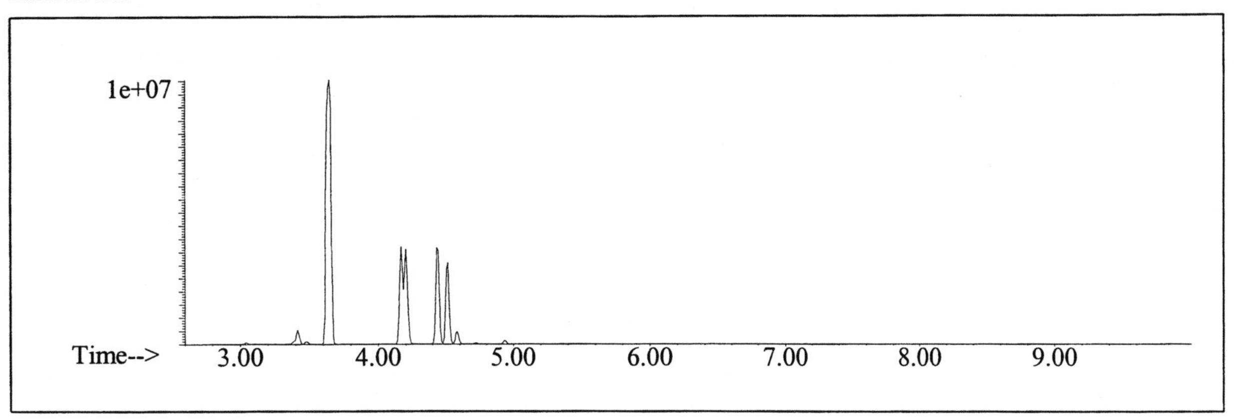

AROMATICS

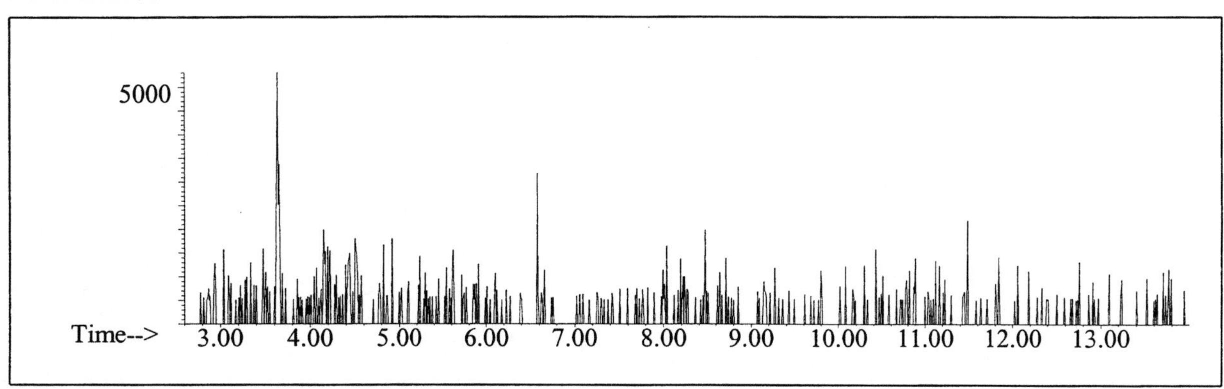

CYCLOPARAFFINS AND ALKENES

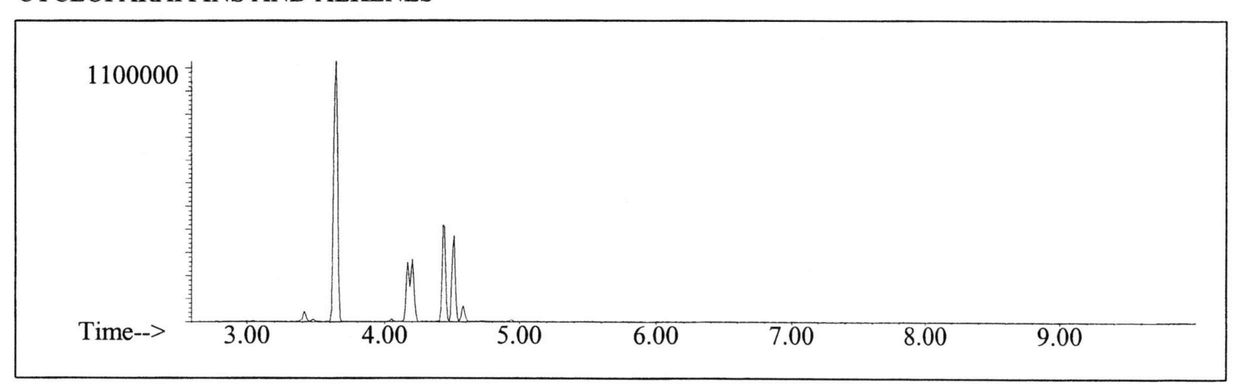

NAPHTHALENES

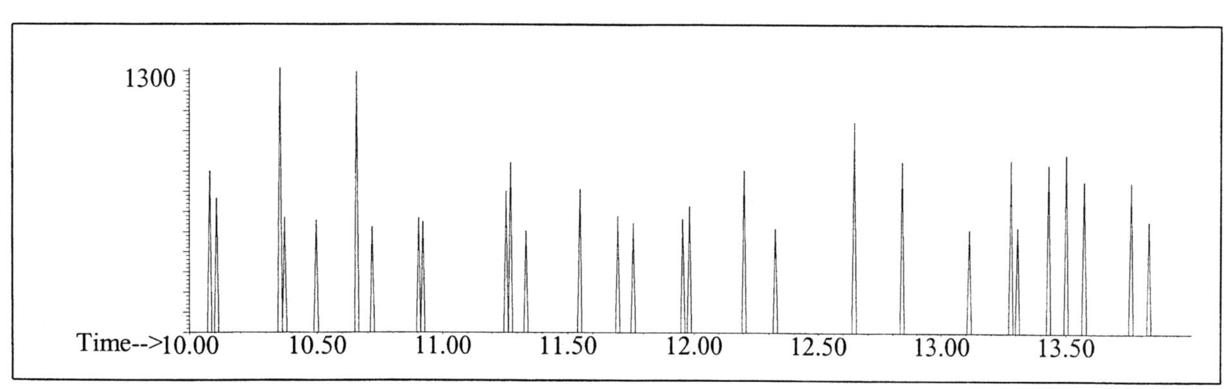

INDIVIDUAL PROFILES

Isoparaffin #2

Alkane

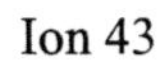

Ion 43

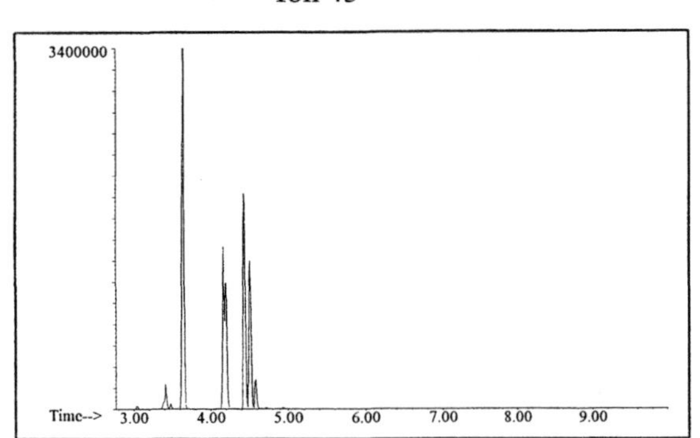

Ion 57

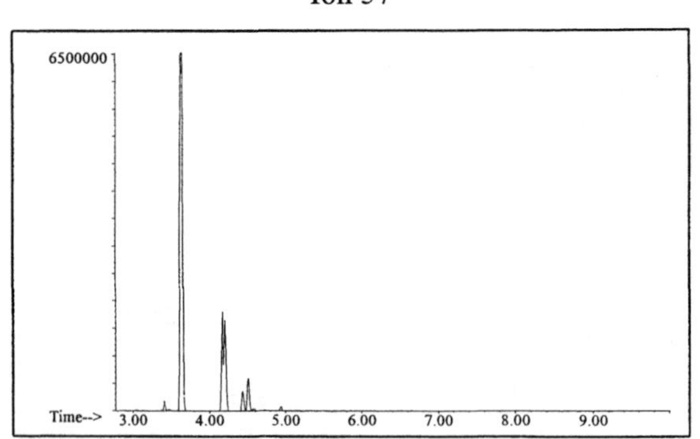

Ion 71

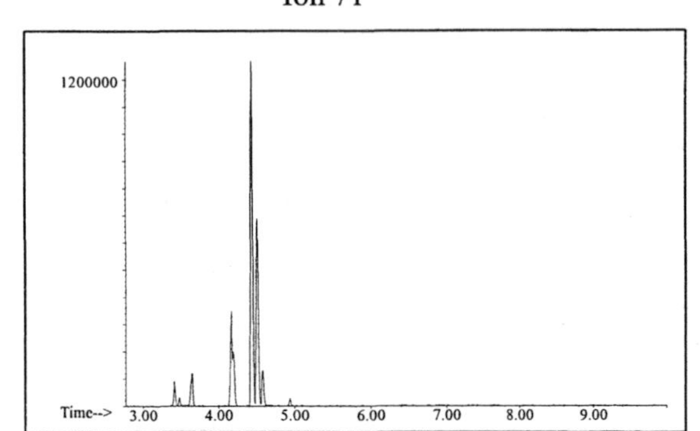

Ion 85

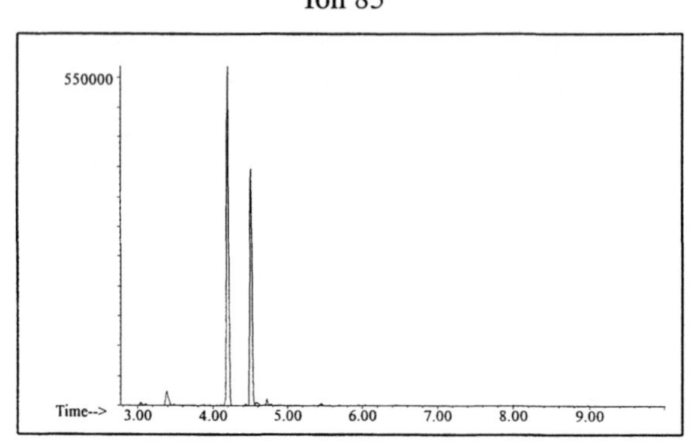

Aromatic

Ion 91

No Useful Data Obtained

Ion 105

No Useful Data Obtained

Ion 119

No Useful Data Obtained

INDIVIDUAL PROFILES

Cycloparaffin

Ion 55

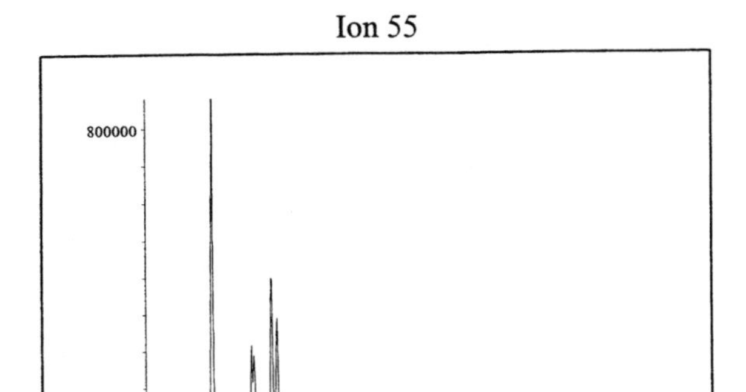

Ion 69

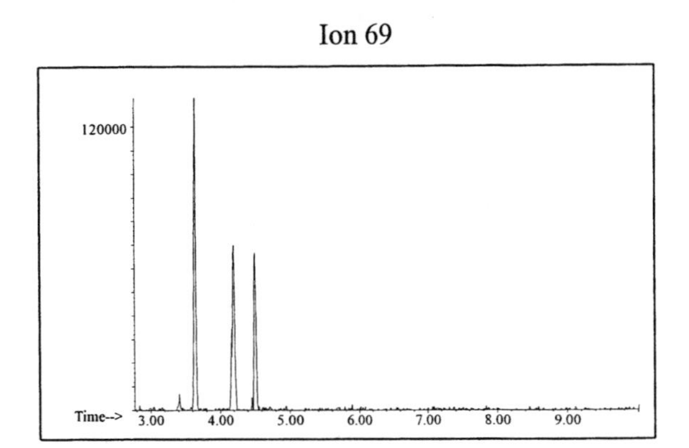

Ion 83

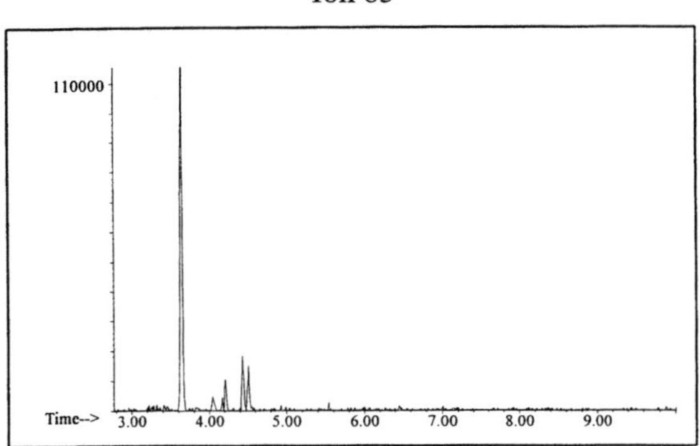

Naphthalene

Ion 128

No Useful Data Obtained

Ion 142

No Useful Data Obtained

Ion 156

No Useful Data Obtained

Isoparaffin #3

20uL/mL Pentane

Product Displayed: Aviation Gas

ASTM: Class 0.2 (Isoparaffin)

Product Uses: Aviation fuel

Macro Code: L

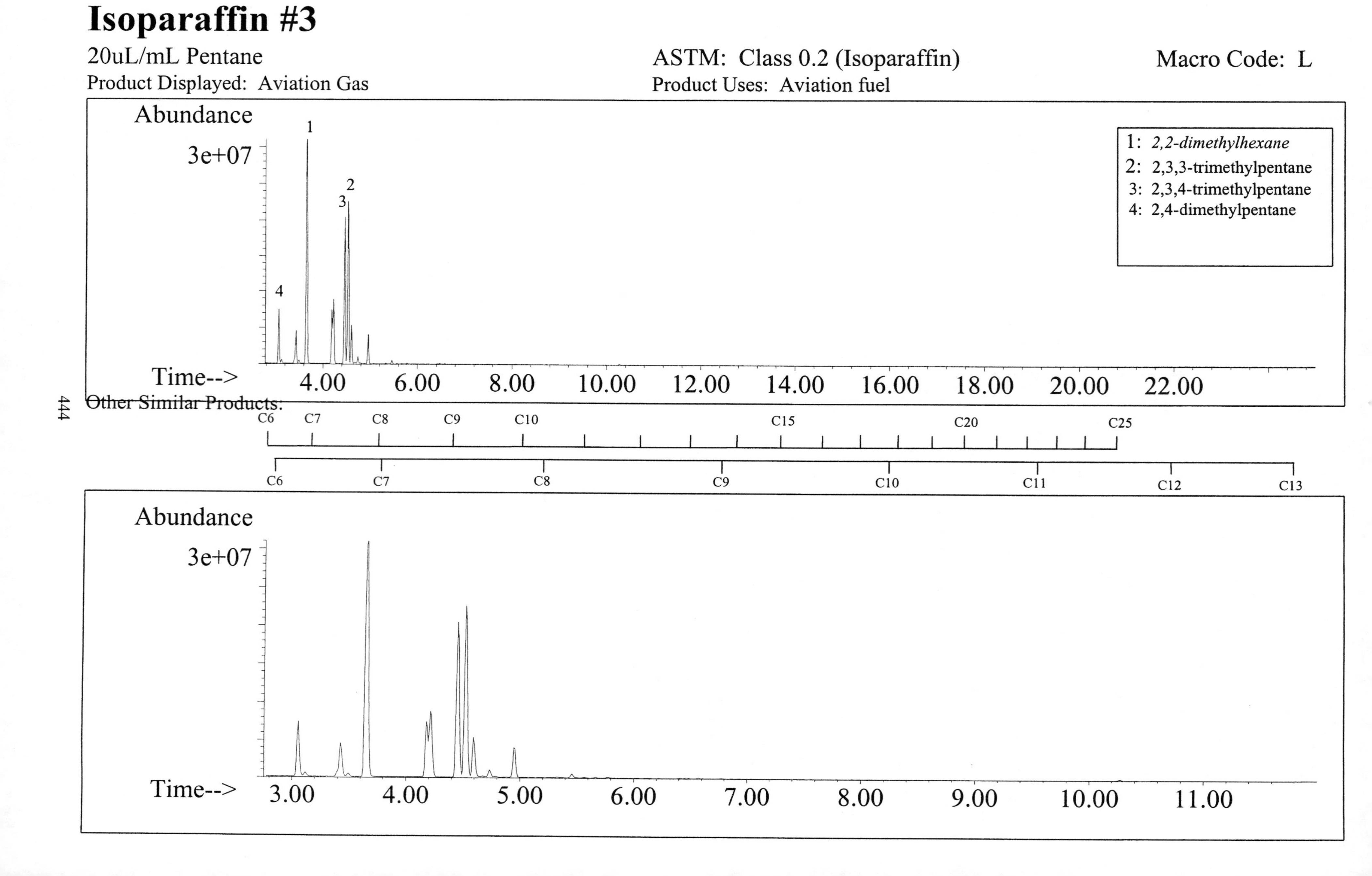

Other Similar Products:

SUMMED PROFILES   Isoparaffin #3

ALKANES

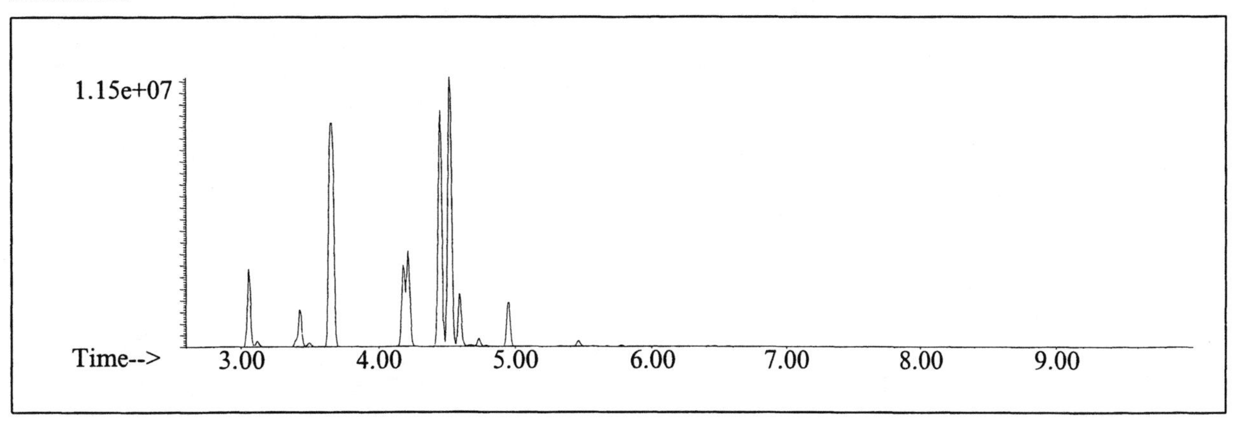

AROMATICS

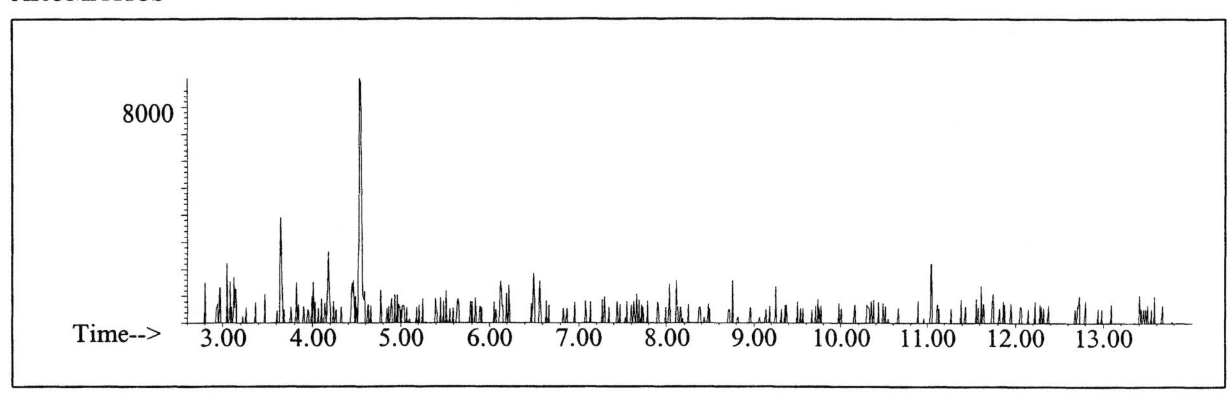

CYCLOPARAFFINS AND ALKENES

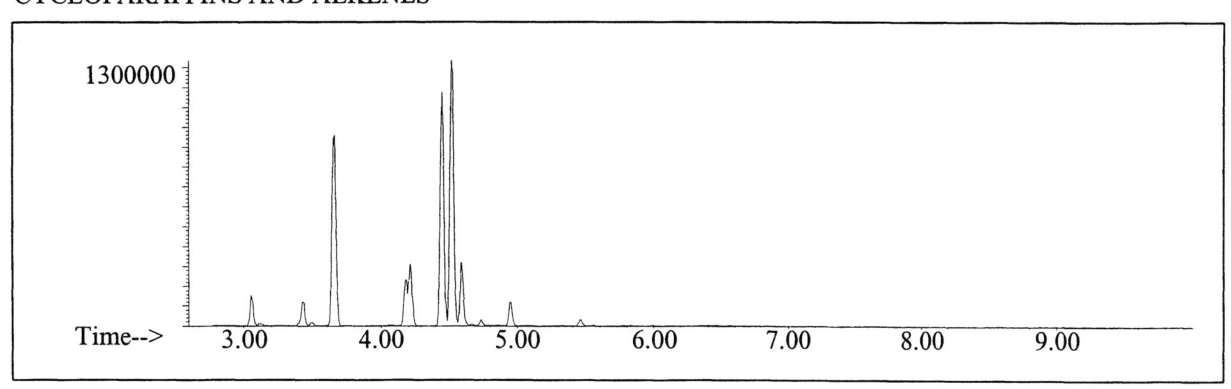

NAPHTHALENES

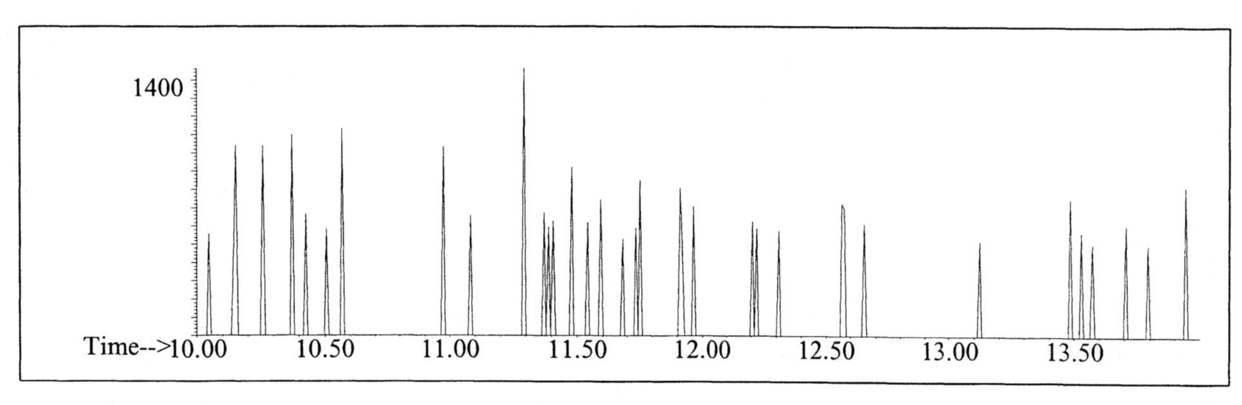

INDIVIDUAL PROFILES Isoparaffin #3

Alkane

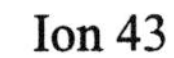

Ion 43

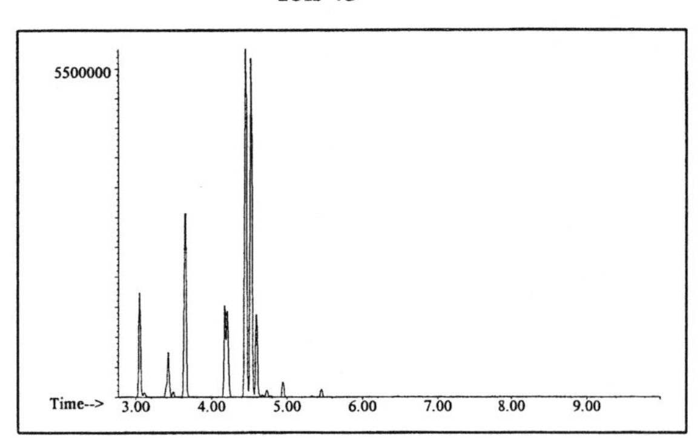

Ion 57

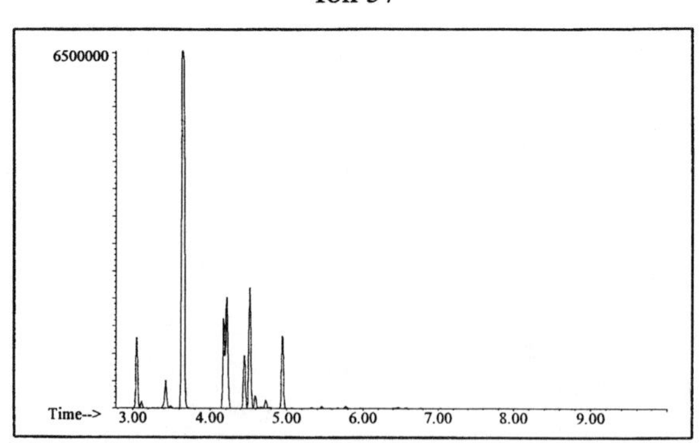

Ion 71

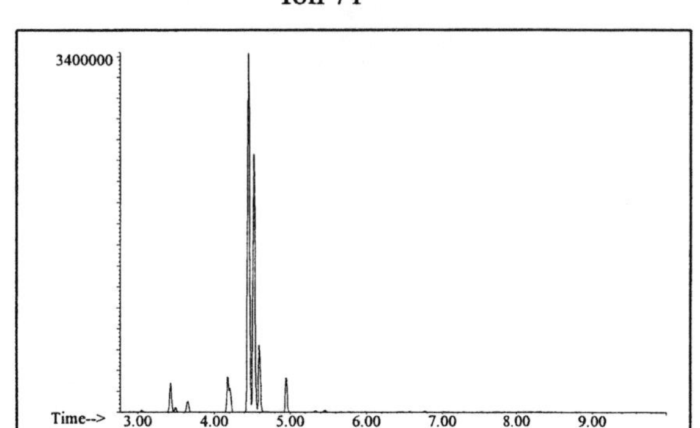

Ion 85

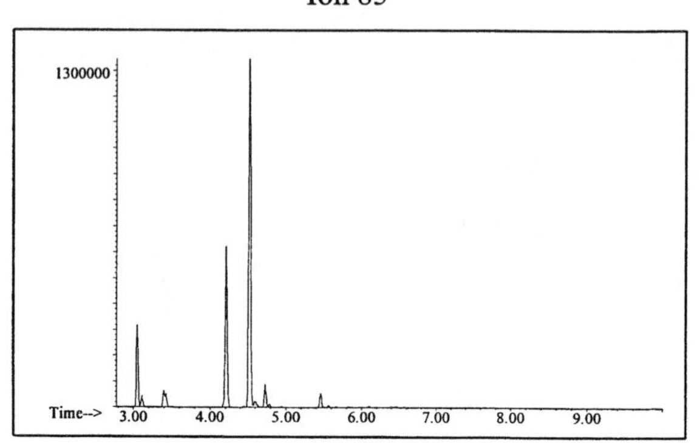

Aromatic

Ion 91

No Useful Data Obtained

Ion 105

No Useful Data Obtained

Ion 119

No Useful Data Obtained

Cycloparaffin

Ion 55

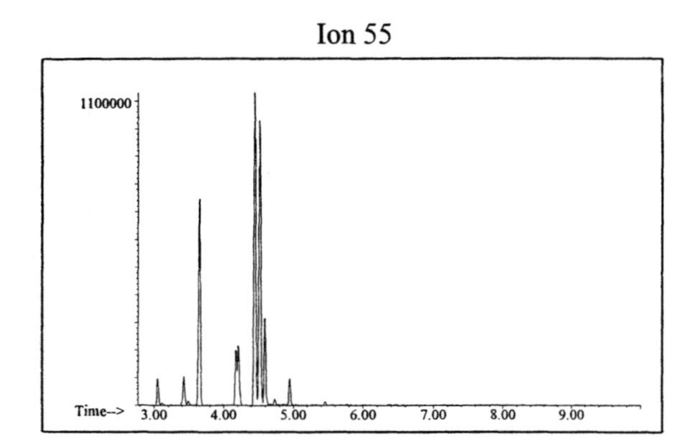

Ion 69

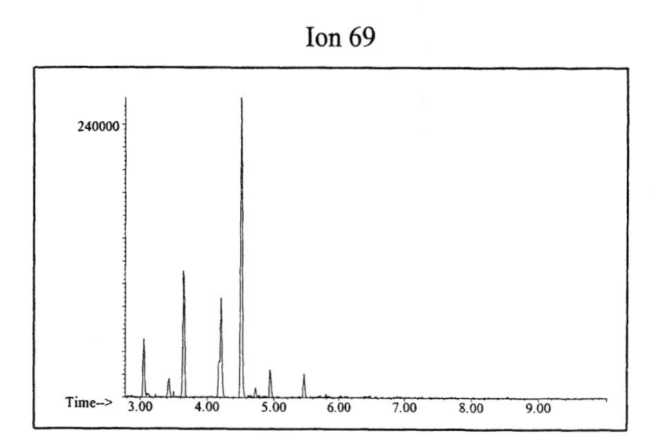

Ion 83

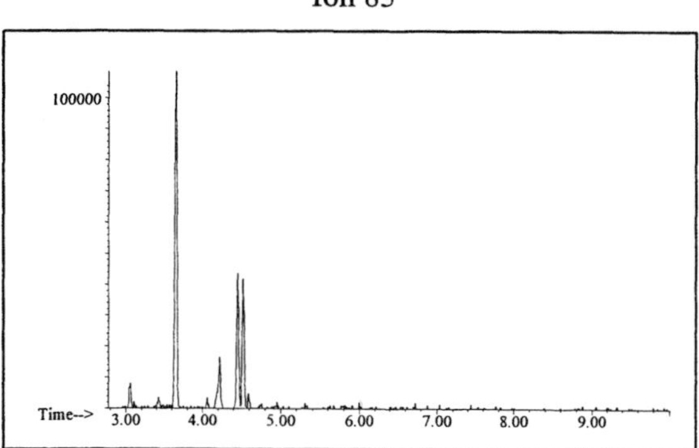

Naphthalene

Ion 128

No Useful Data Obtained

Ion 142

No Useful Data Obtained

Ion 156

No Useful Data Obtained

Isoparaffin #4

20uL/mL Pentane
Product Displayed: Isopar E
Other Similar Products:

ASTM: Class 0.2 (Isoparaffin)
Product Uses: Solvent, feestock

Macro Code: L

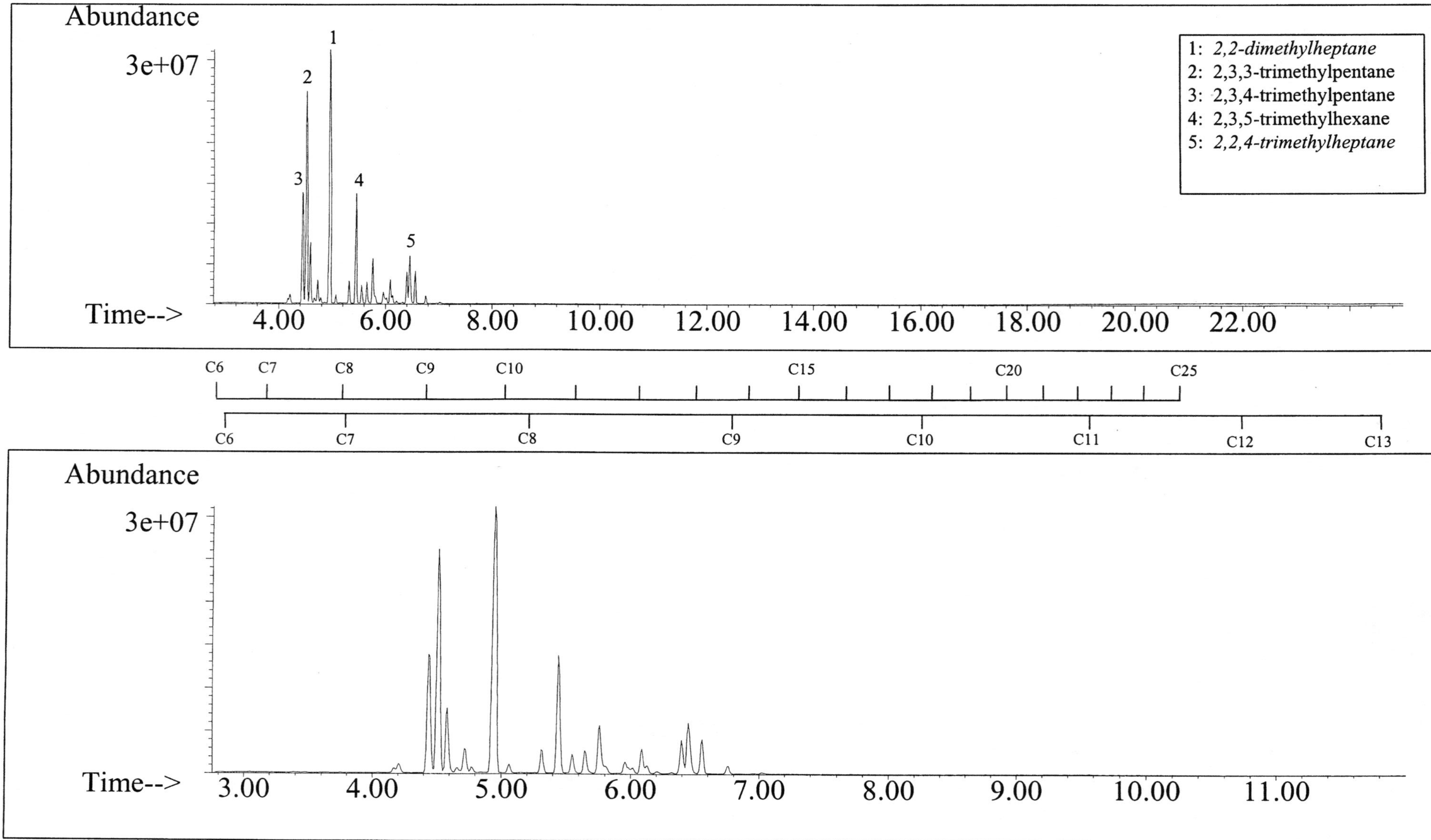

ALKANES

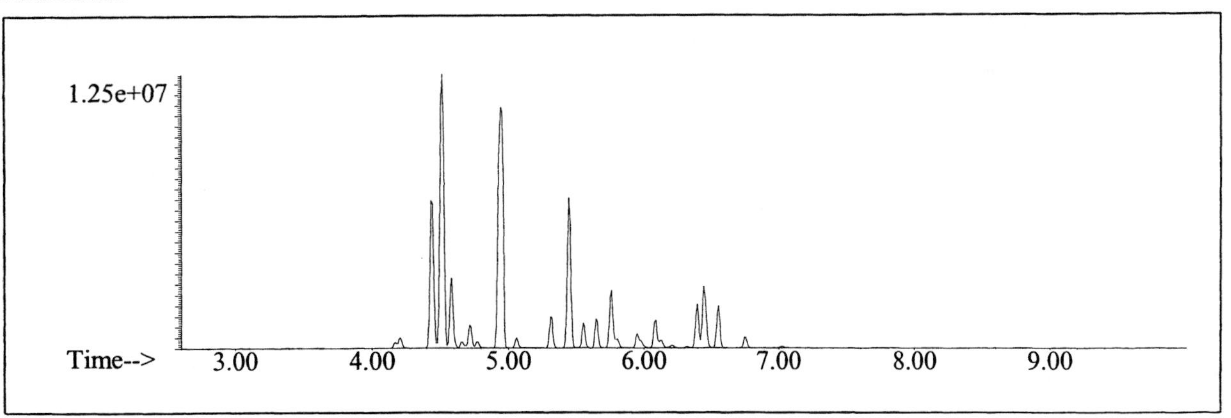

AROMATICS

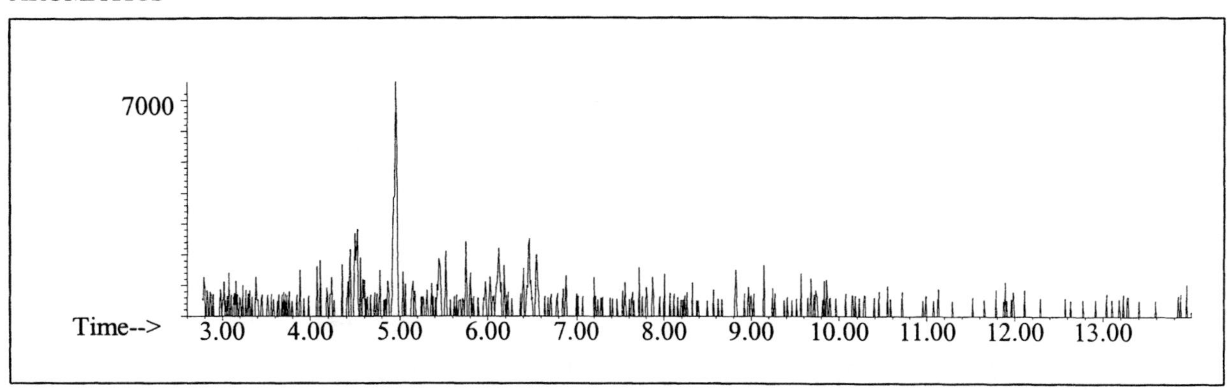

CYCLOPARAFFINS AND ALKENES

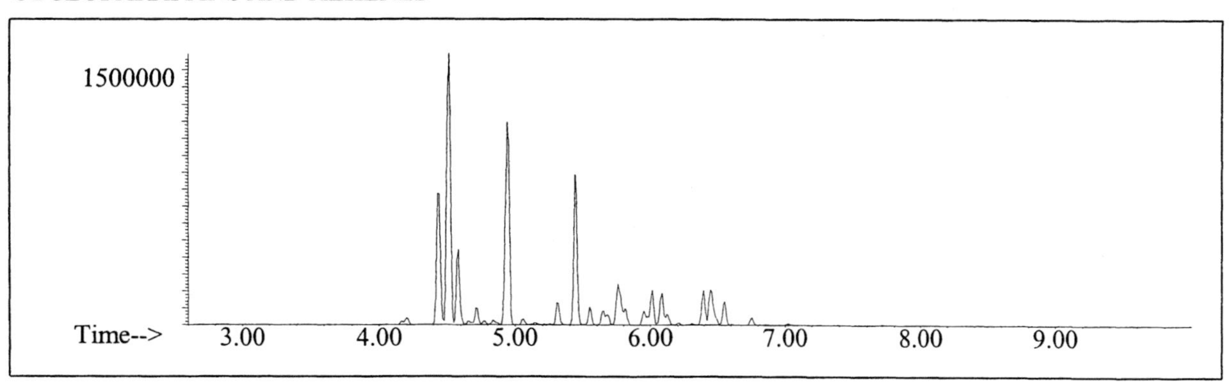

NAPHTHALENES

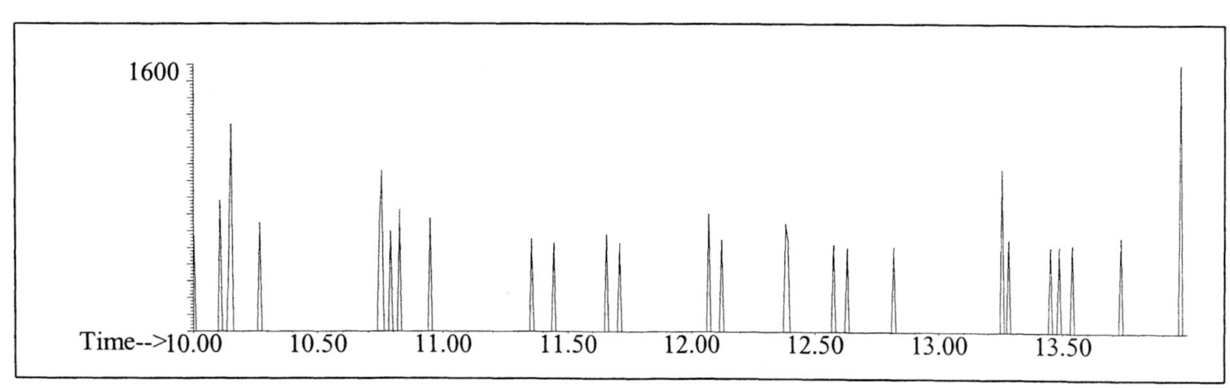

INDIVIDUAL PROFILES

Isoparaffin #4

Alkane

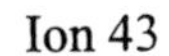

Ion 43

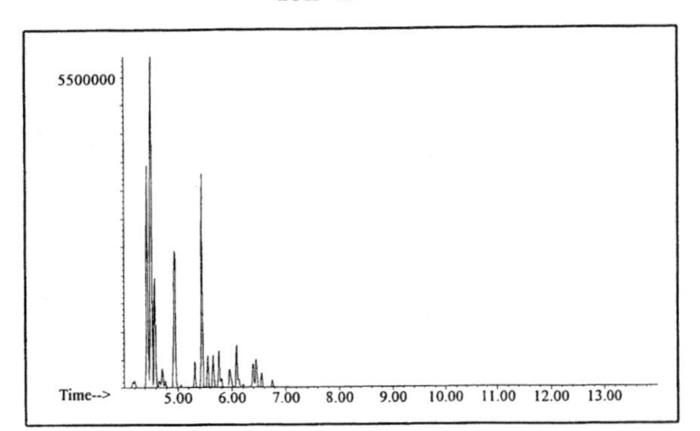

Ion 57

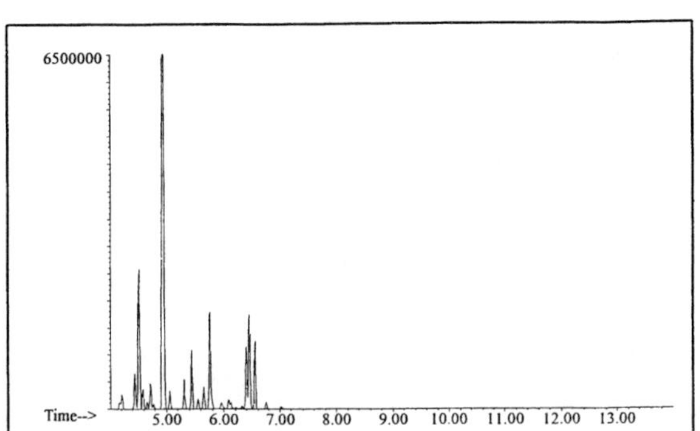

Ion 71

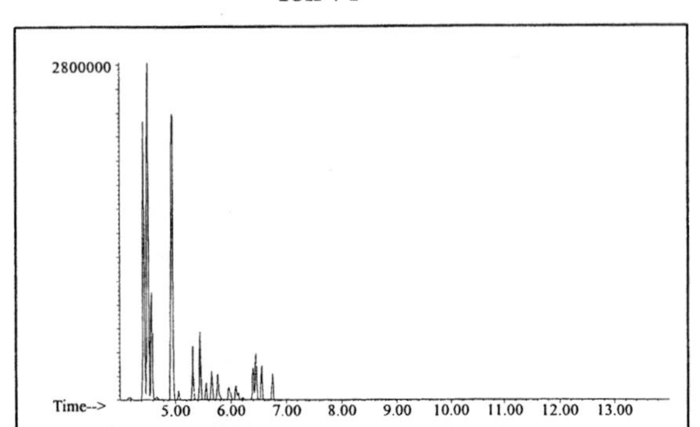

Ion 85

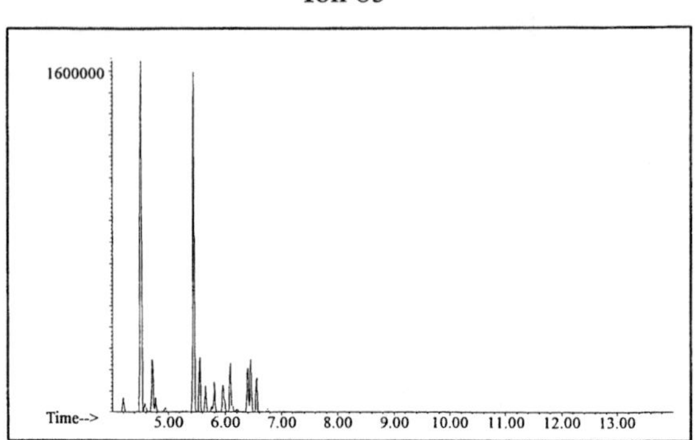

Aromatic

Ion 91

No Useful Data Obtained

Ion 105

No Useful Data Obtained

Ion 119

No Useful Data Obtained

Cycloparaffin

Ion 55

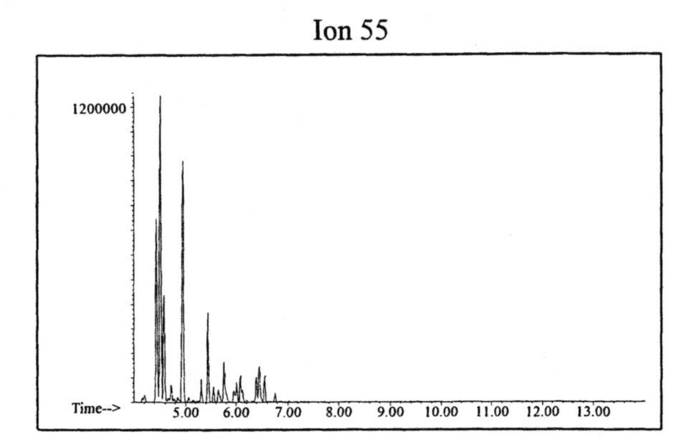

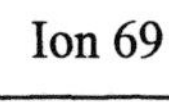

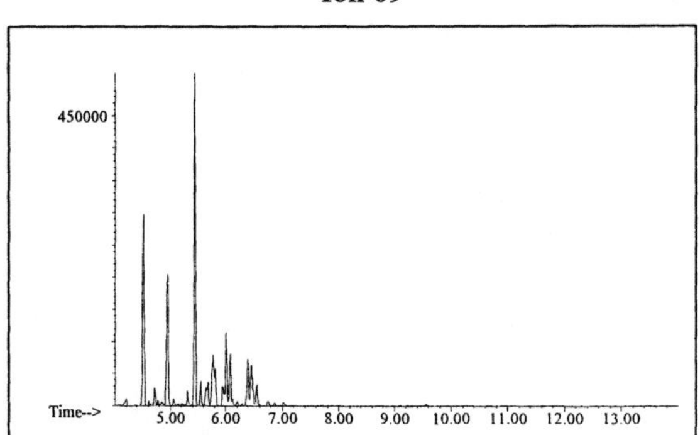

Ion 83

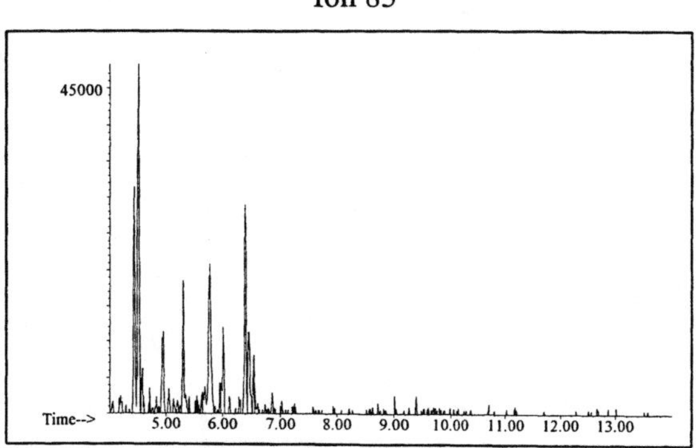

Naphthalene

Ion 128

No Useful Data Obtained

Ion 142

No Useful Data Obtained

Ion 156

No Useful Data Obtained

Isoparaffin #5

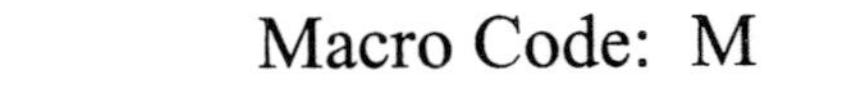

20uL/mL Pentane
Product Displayed: Phillips 66 Soltrol 100
Other Similar Products: Isopar G

ASTM: Class 0.2 (Isoparaffin)
Product Uses: Solvent, feedstock

Macro Code: M

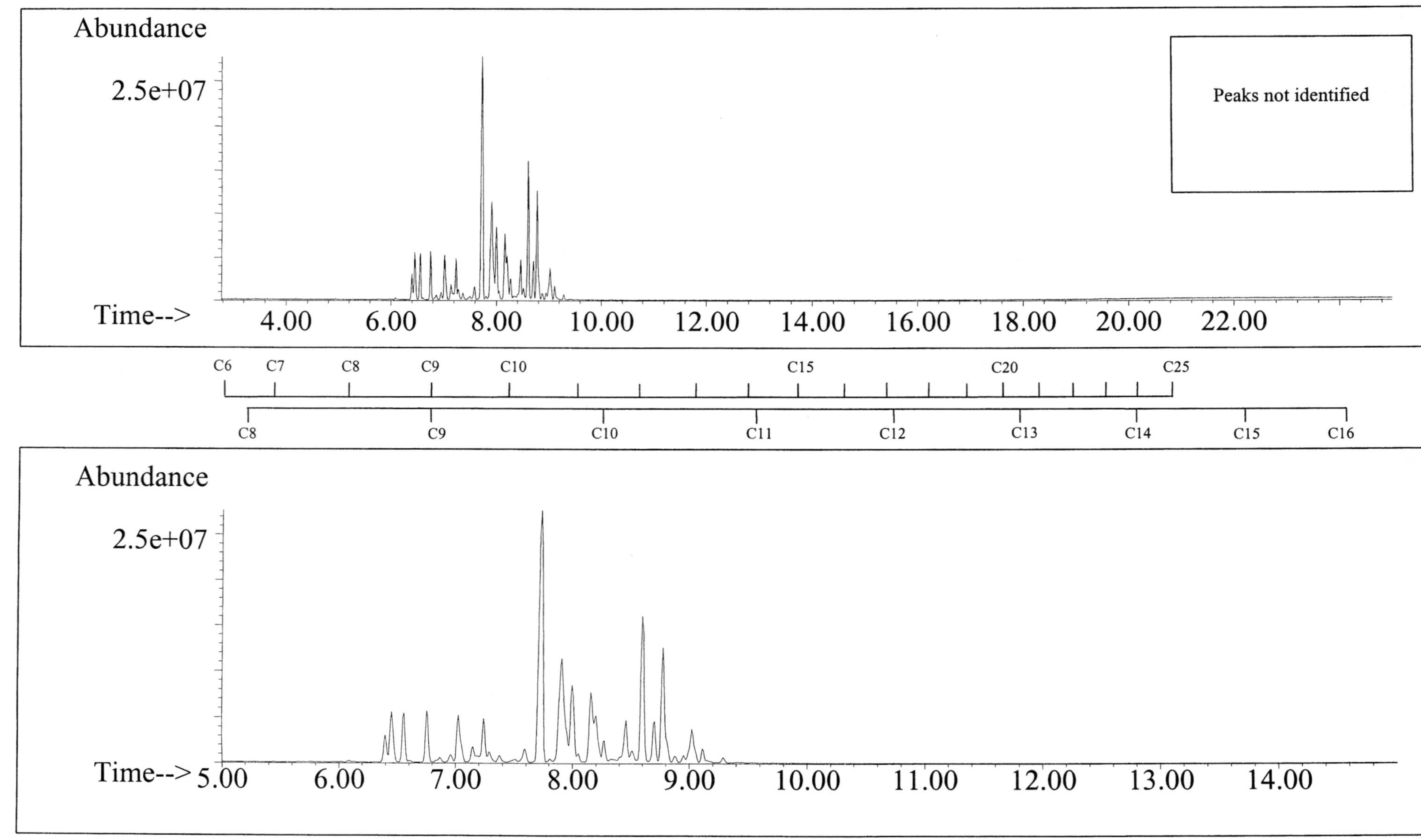

SUMMED PROFILES

Isoparaffin #5

ALKANES

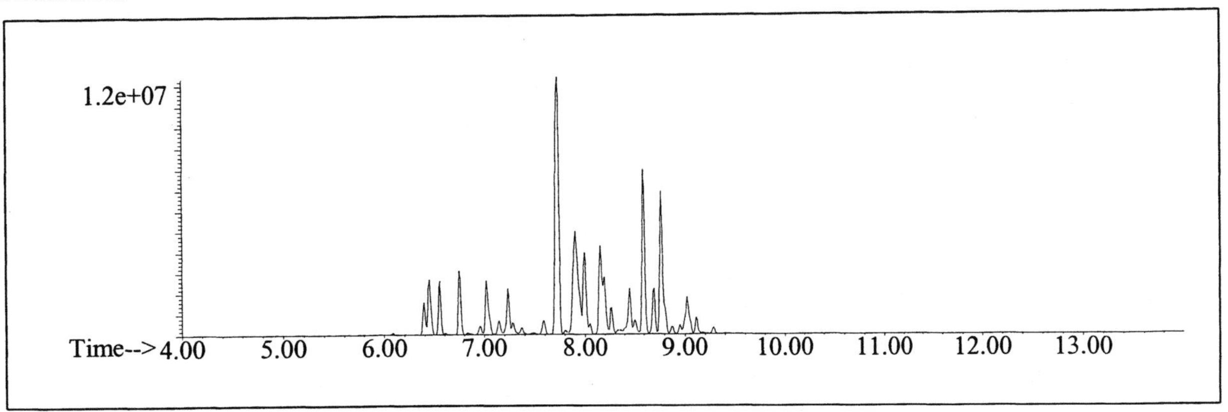

AROMATICS

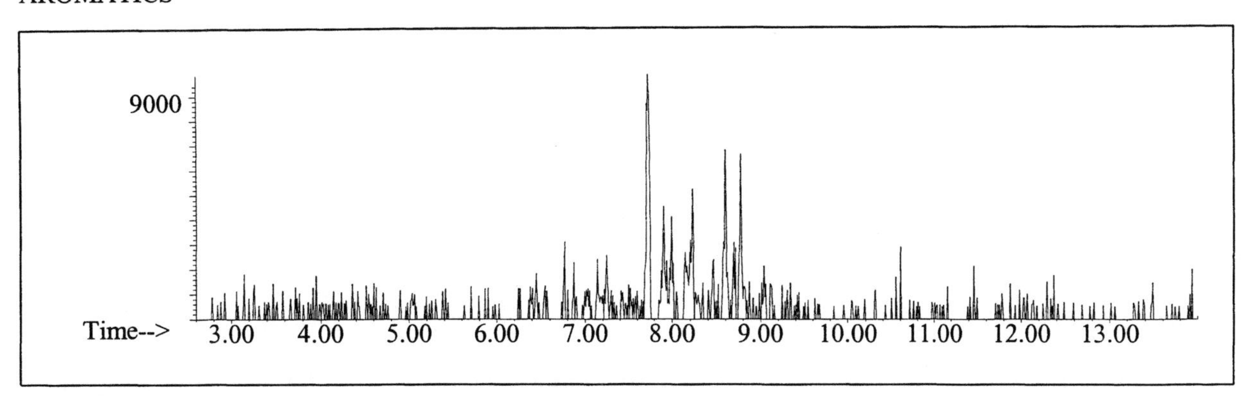

CYCLOPARAFFINS AND ALKENES

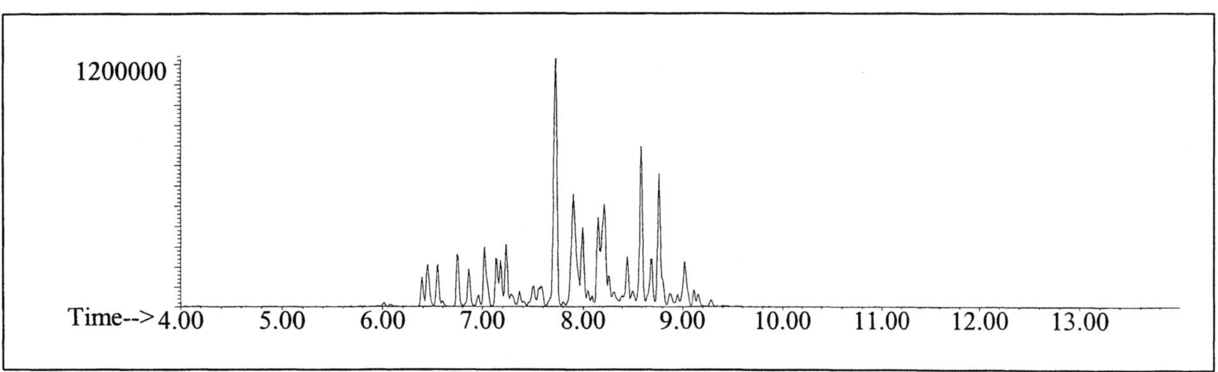

NAPHTHALENES

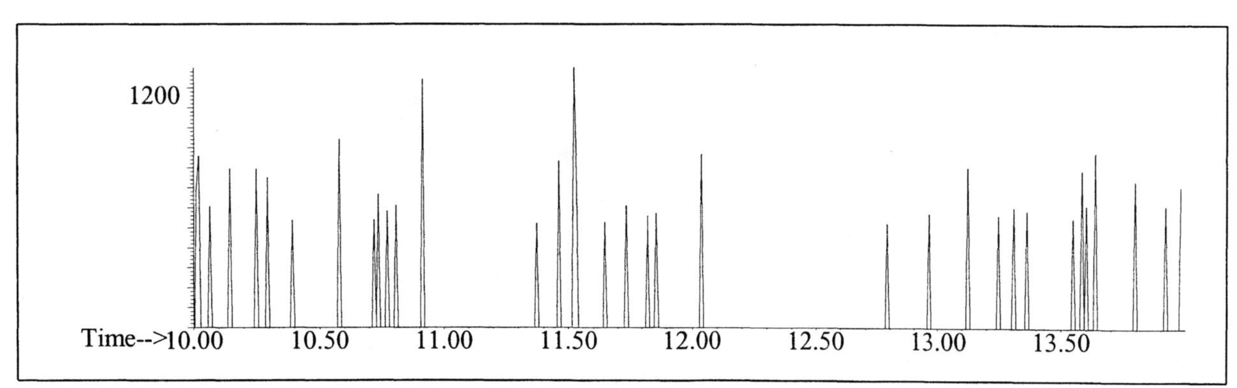

INDIVIDUAL PROFILES

Isoparaffin #5

Alkane

Ion 43

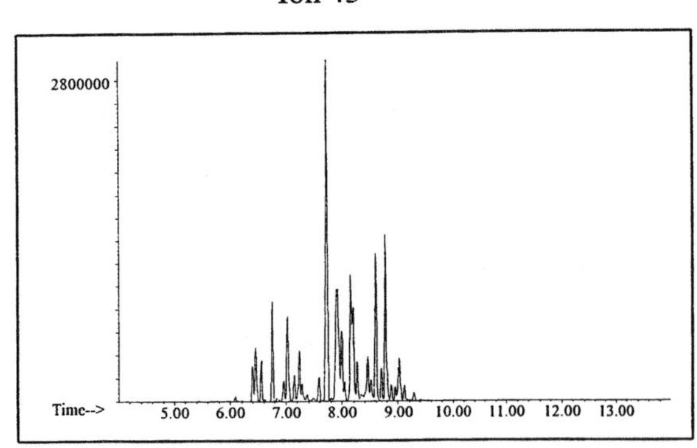

Ion 57

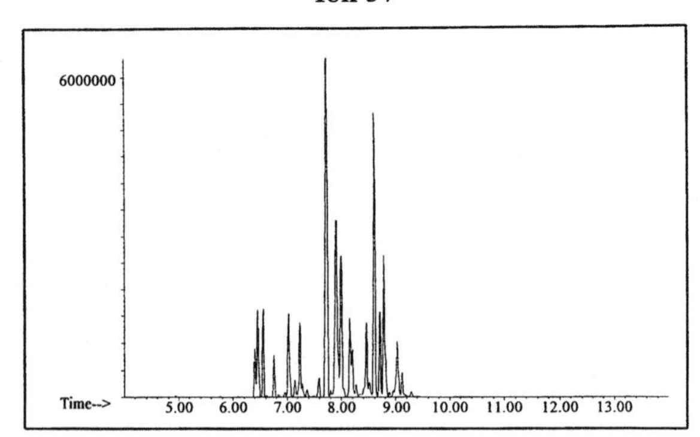

Ion 71

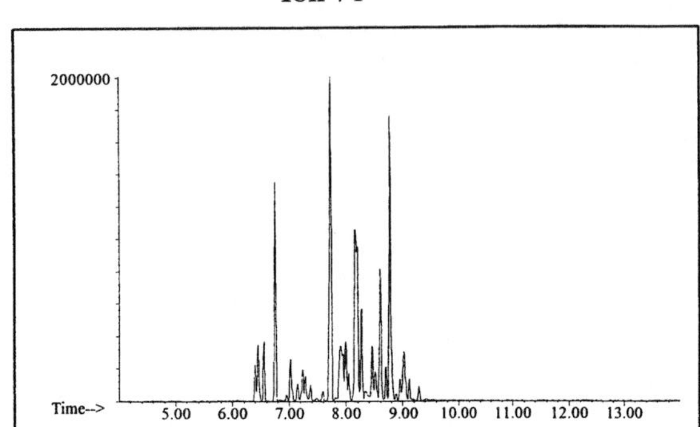

Ion 85

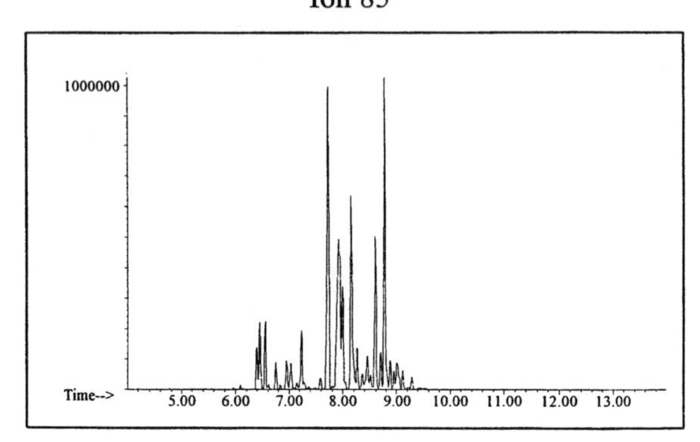

Aromatic

Ion 91

No Useful Data Obtained

Ion 105

No Useful Data Obtained

Ion 119

No Useful Data Obtained

INDIVIDUAL PROFILES

Cycloparaffin

Ion 55

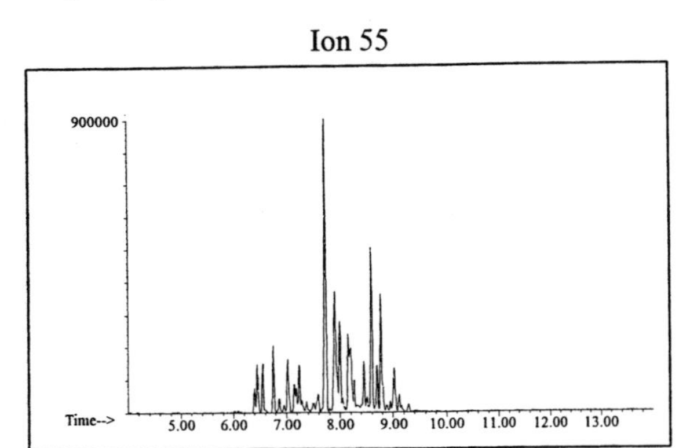

Ion 69

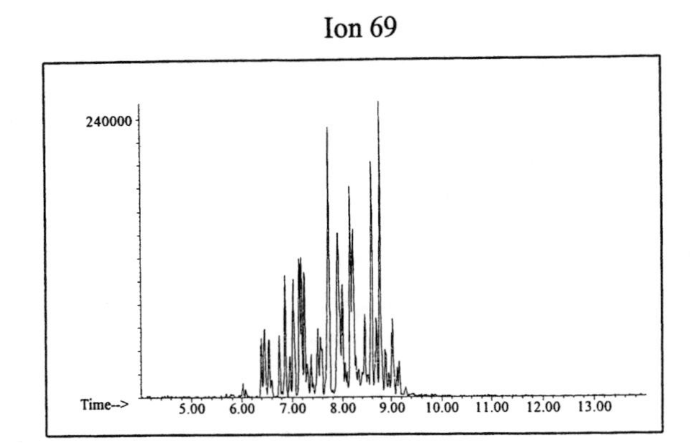

Ion 83

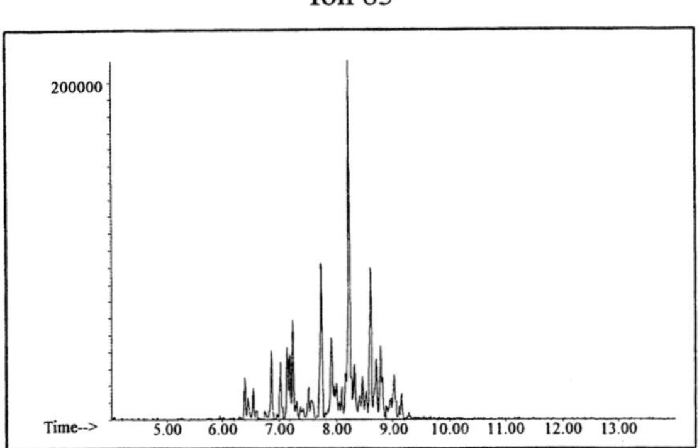

Naphthalene

Ion 128

No Useful Data Obtained

Ion 142

No Useful Data Obtained

Ion 156

No Useful Data Obtained

Isoparaffin #6

20uL/mL Pentane

Product Displayed: Wizard Charcoal Lighter

Other Similar Products:

ASTM: Class 0.2 (Isoparaffin)

Product Uses: Charcoal lighter

Macro Code: M

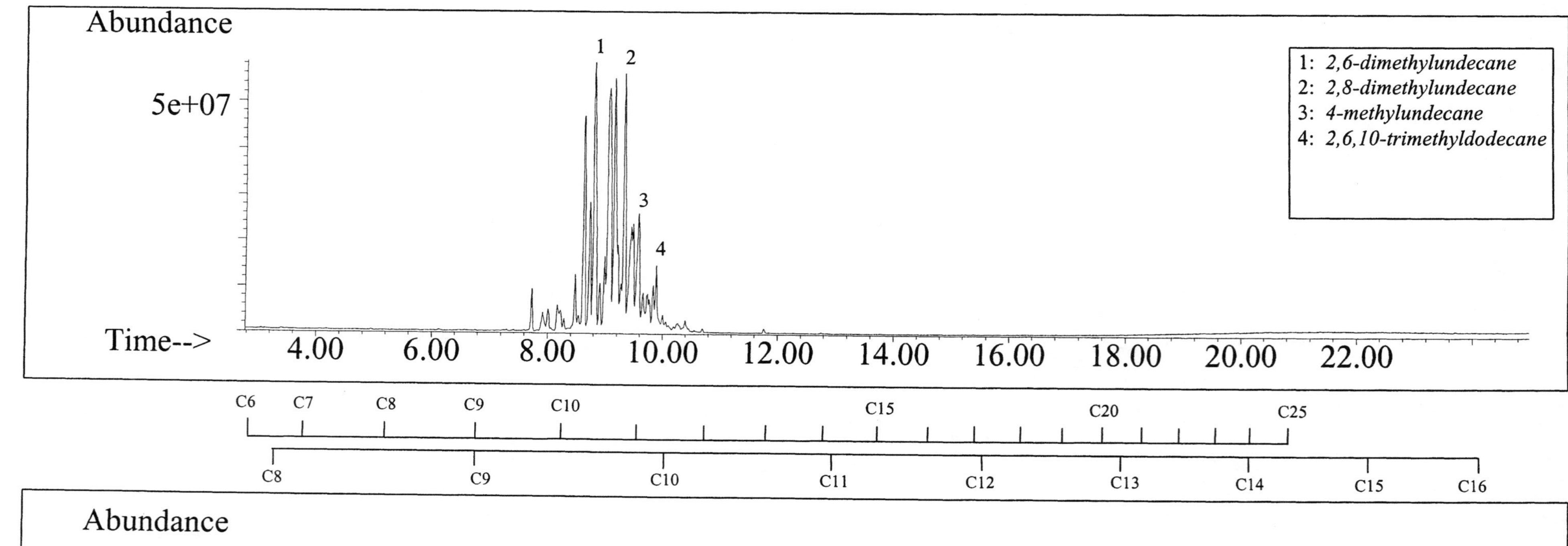

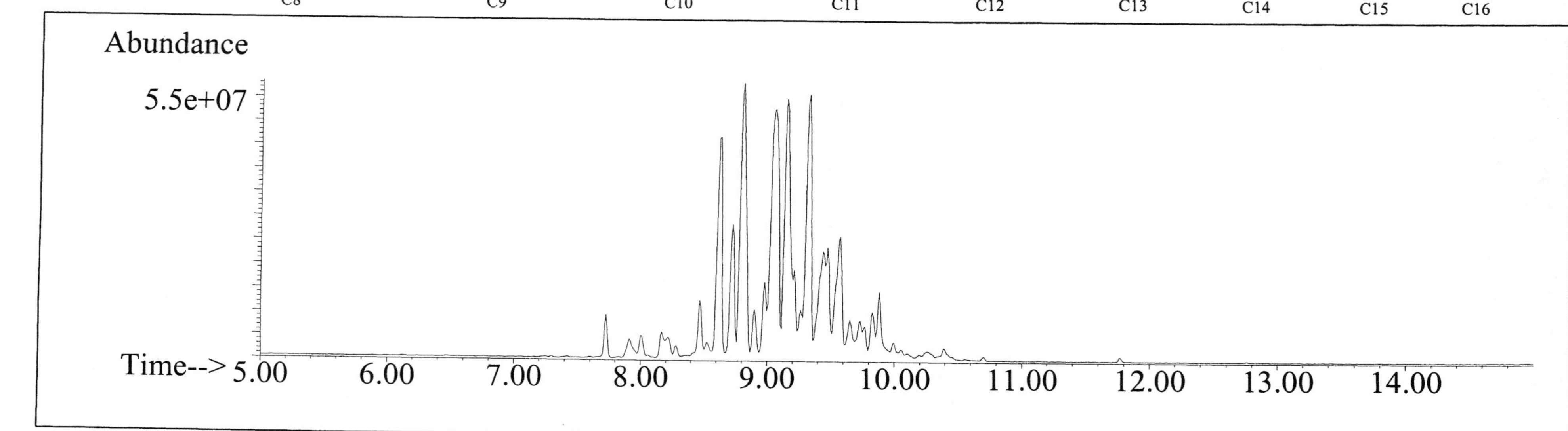

SUMMED PROFILES

Isoparaffin #6

ALKANES

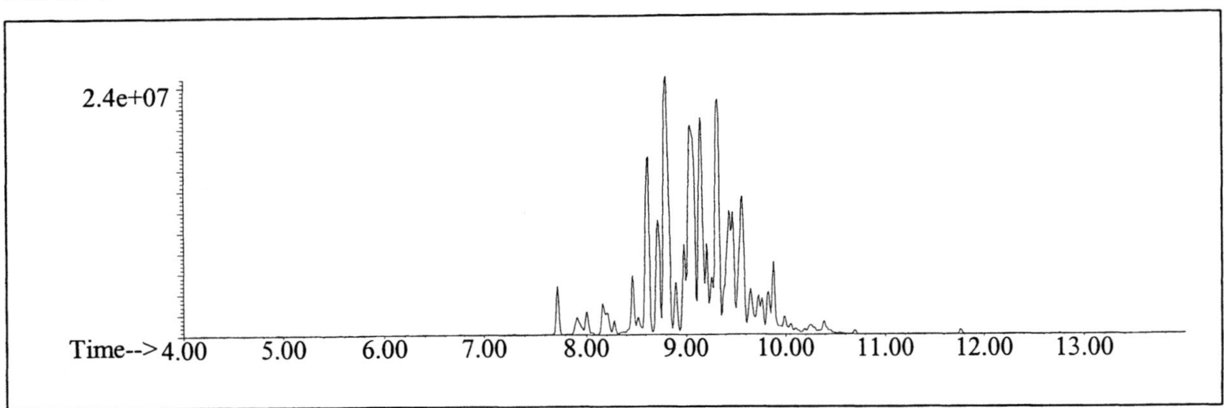

AROMATICS

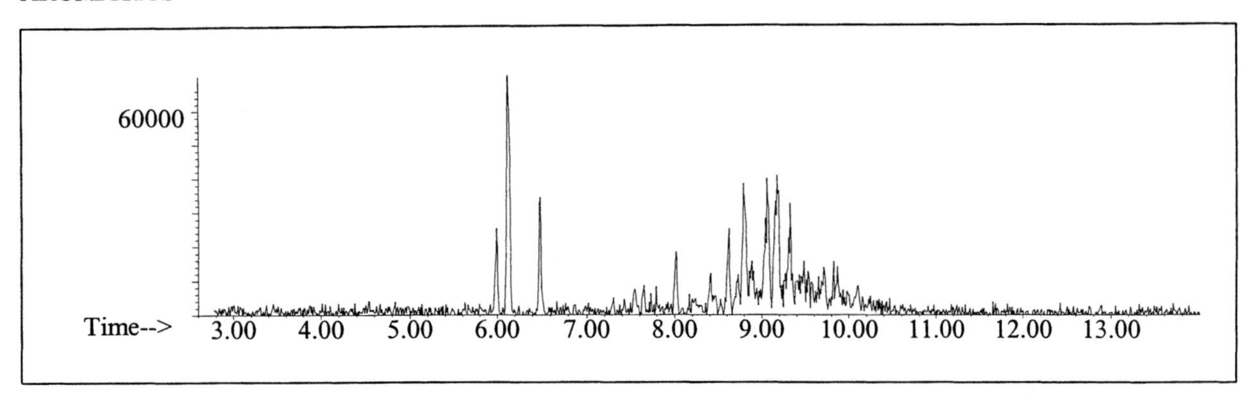

CYCLOPARAFFINS AND ALKENES

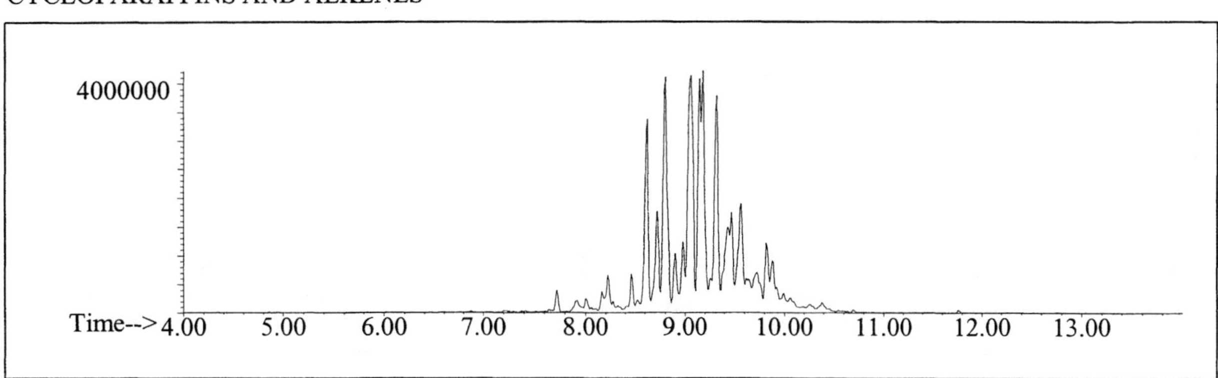

NAPHTHALENES

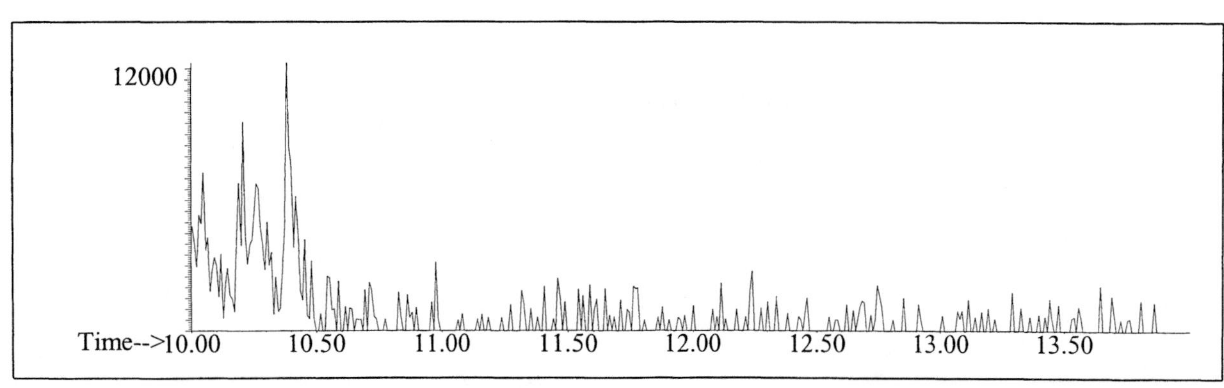

INDIVIDUAL PROFILES

Isoparaffin #6

Alkane

Ion 43

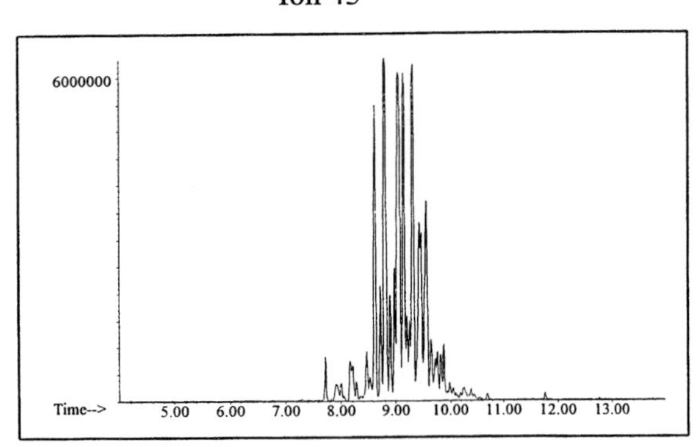

Ion 57

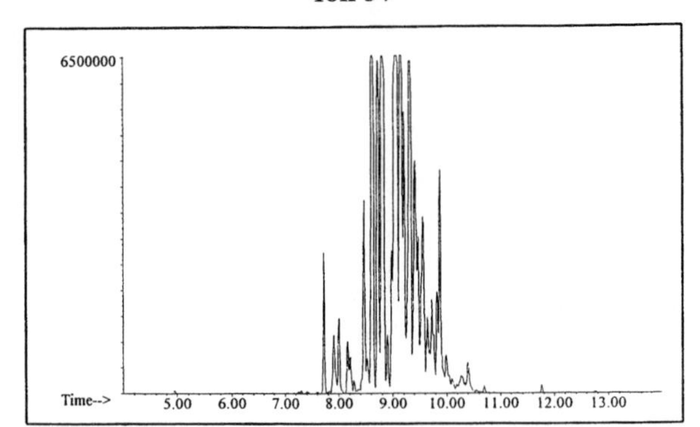

Ion 71

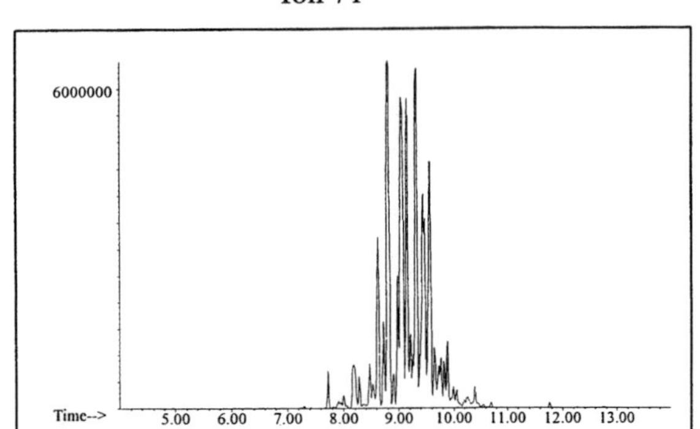

Ion 85

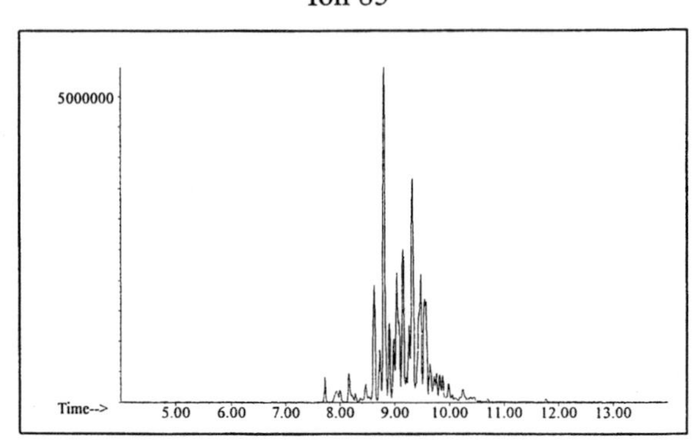

Aromatic

Ion 91

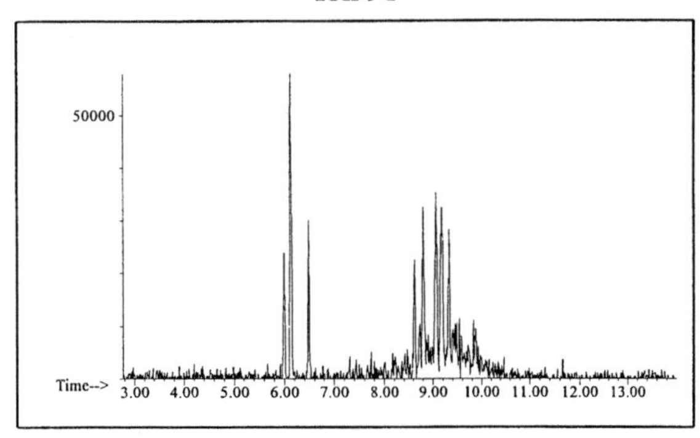

Ion 105

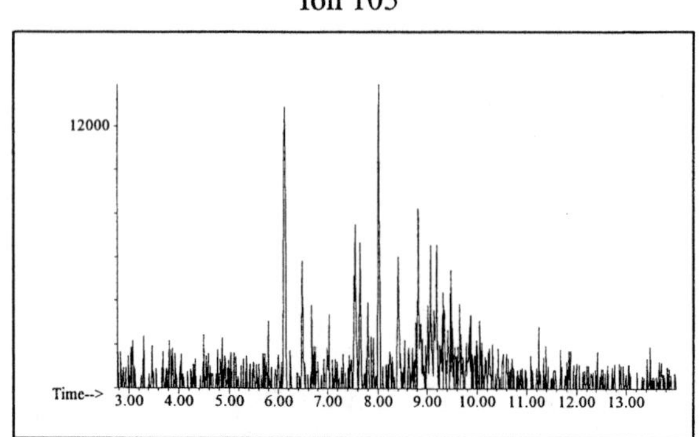

Ion 119

No Useful Data Obtained

Cycloparaffin

Ion 55

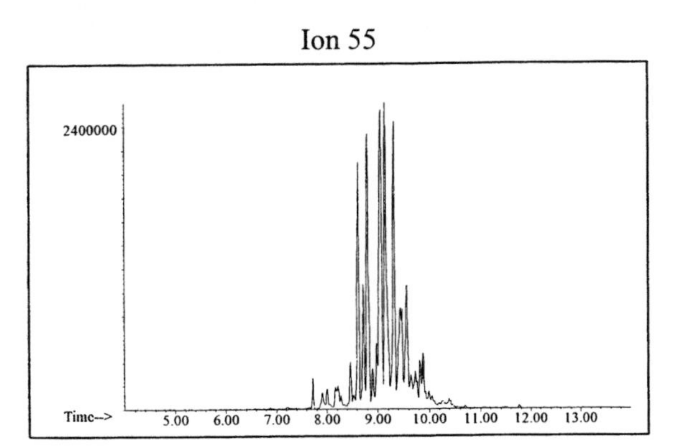

Ion 69

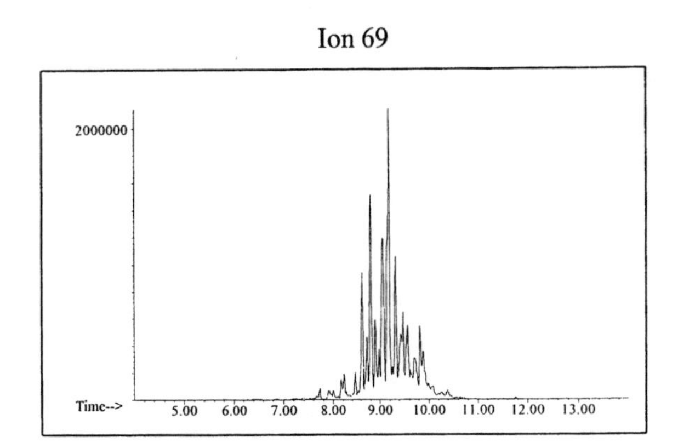

Ion 83

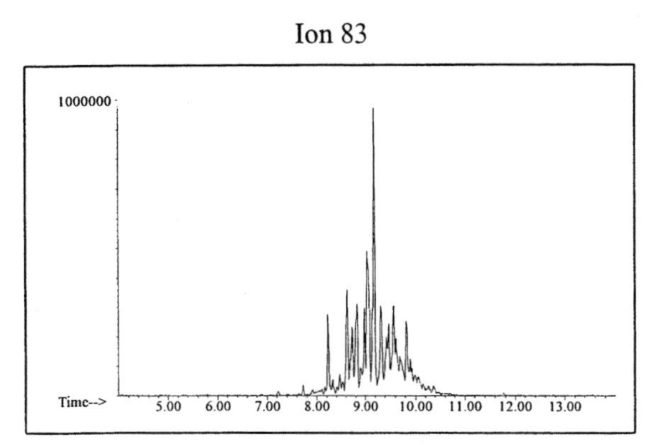

Naphthalene

Ion 128

No Useful Data Obtained

Ion 142

No Useful Data Obtained

Ion 156

No Useful Data Obtained

Isoparaffin #7

20uL/mL Pentane

ASTM: Class 0.2 (Isoparaffin)

Macro Code: M

Product Displayed: Isopar H

Product Uses: Copier toner, paint thinnners, solvent

Other Similar Products: Saxon 3 Plain Paper Toner, Liquidex Permitine Turpentine

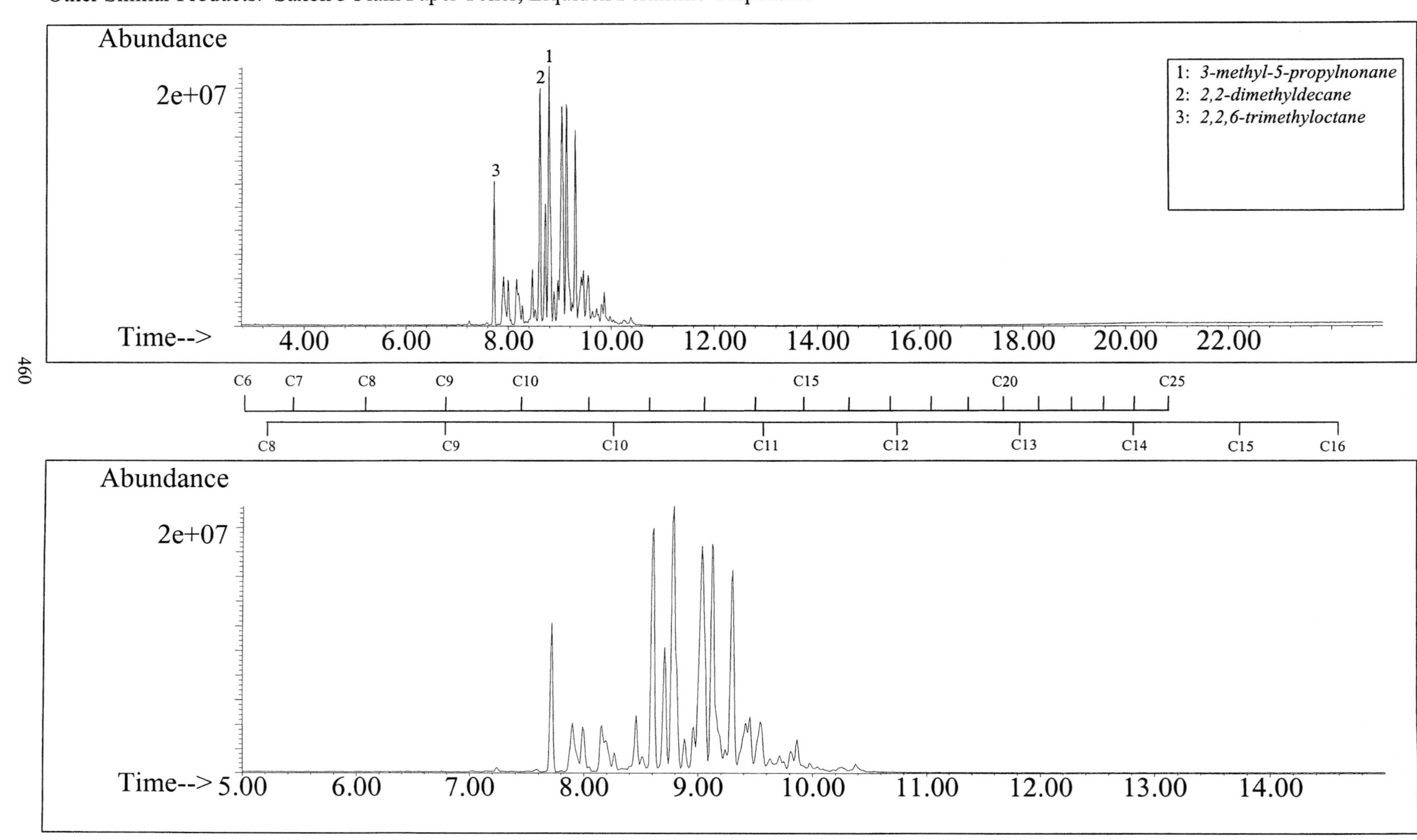

SUMMED PROFILES

Isoparaffin #7

ALKANES

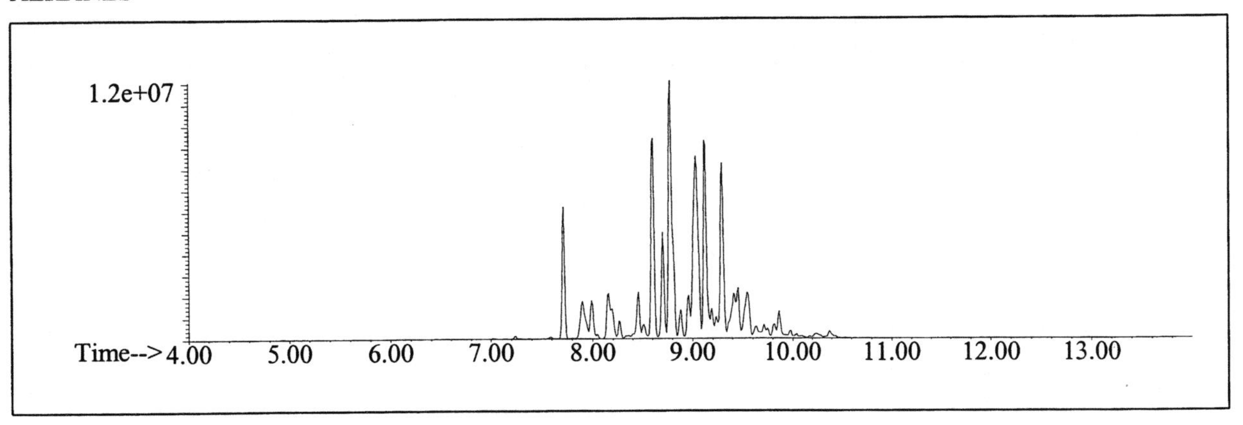

AROMATICS

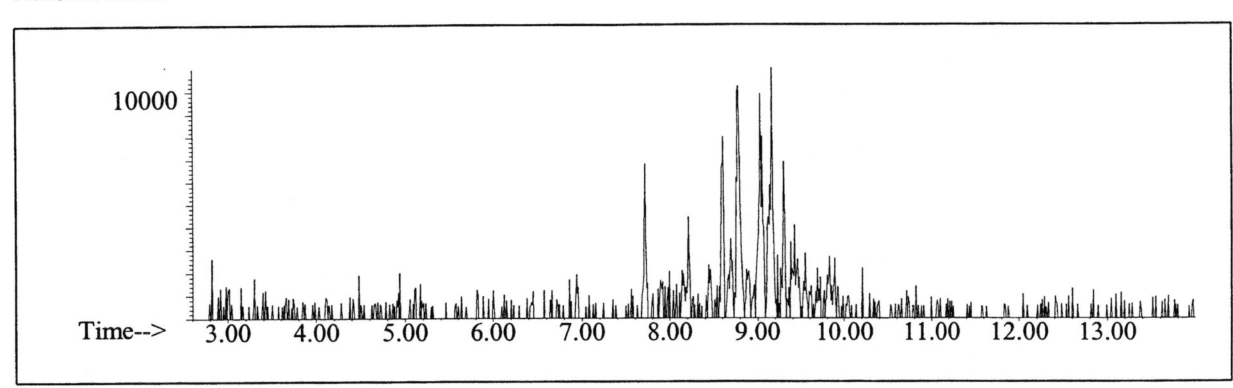

CYCLOPARAFFINS AND ALKENES

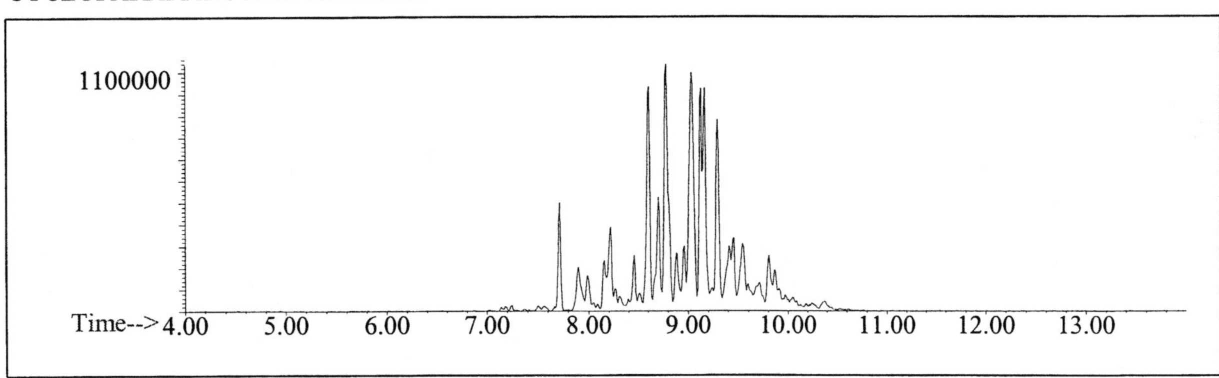

NAPHTHALENES

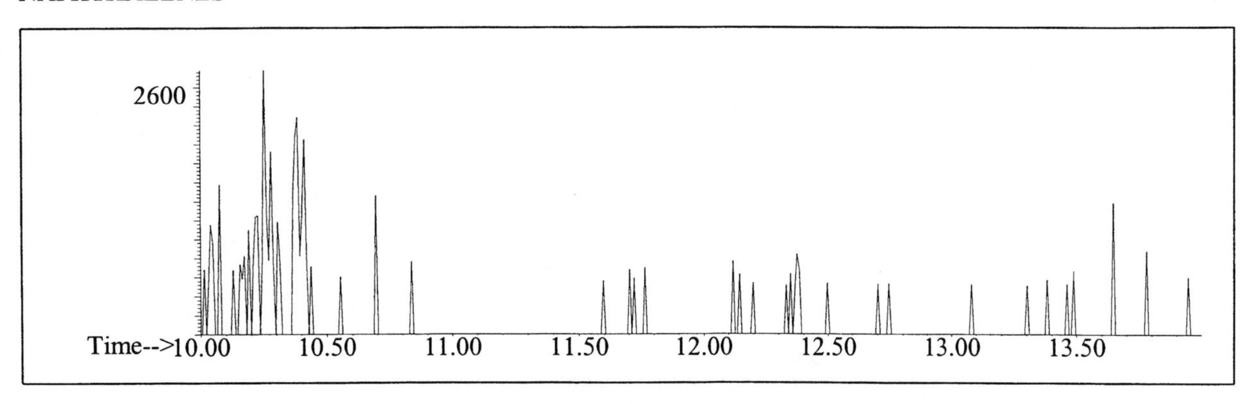

INDIVIDUAL PROFILES

Isoparaffin #7

Alkane

Ion 43

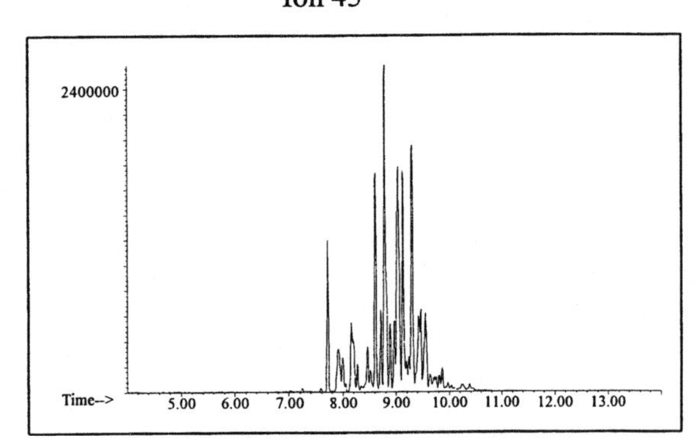

Ion 57

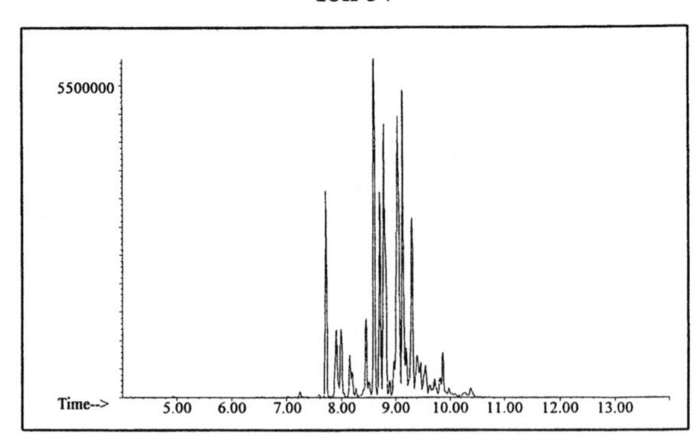

Ion 71

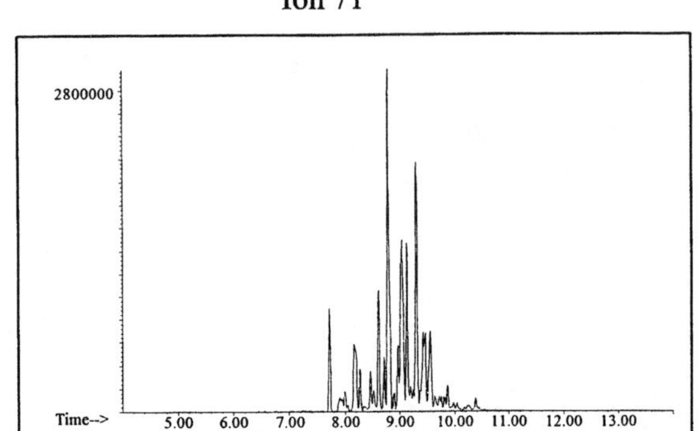

Ion 85

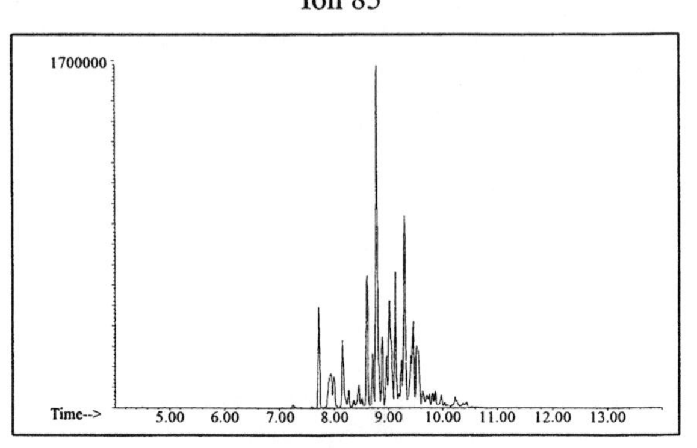

Aromatic

Ion 91

No Useful Data Obtained

Ion 105

No Useful Data Obtained

Ion 119

No Useful Data Obtained

INDIVIDUAL PROFILES

Isoparaffin #7

Cycloparaffin

Ion 55

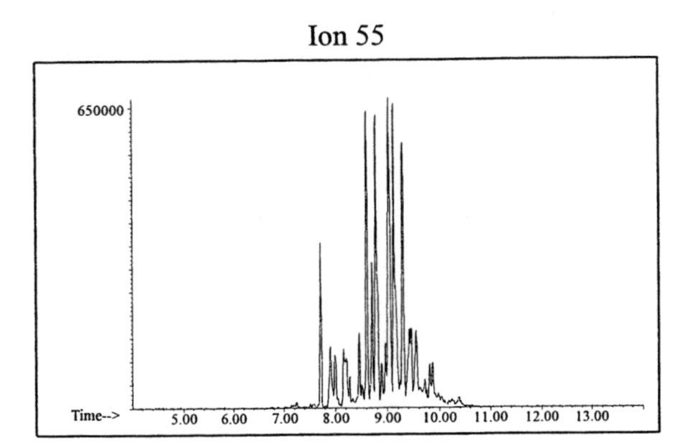

Ion 69

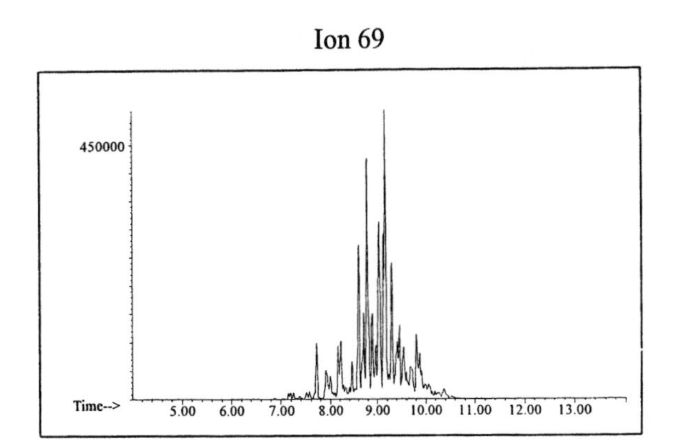

Ion 83

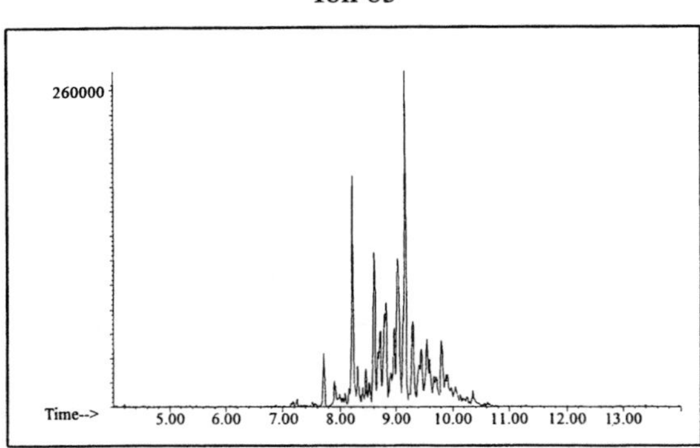

Naphthalene

Ion 128

No Useful Data Obtained

Ion 142

No Useful Data Obtained

Ion 156

No Useful Data Obtained

Isoparaffin #8

20uL/mL Pentane

Product Displayed: Isopar L

Other Similar Products:

ASTM: Class 0.2 (Isoparaffin)

Product Uses: Solvent, feedstock

Macro Code: M

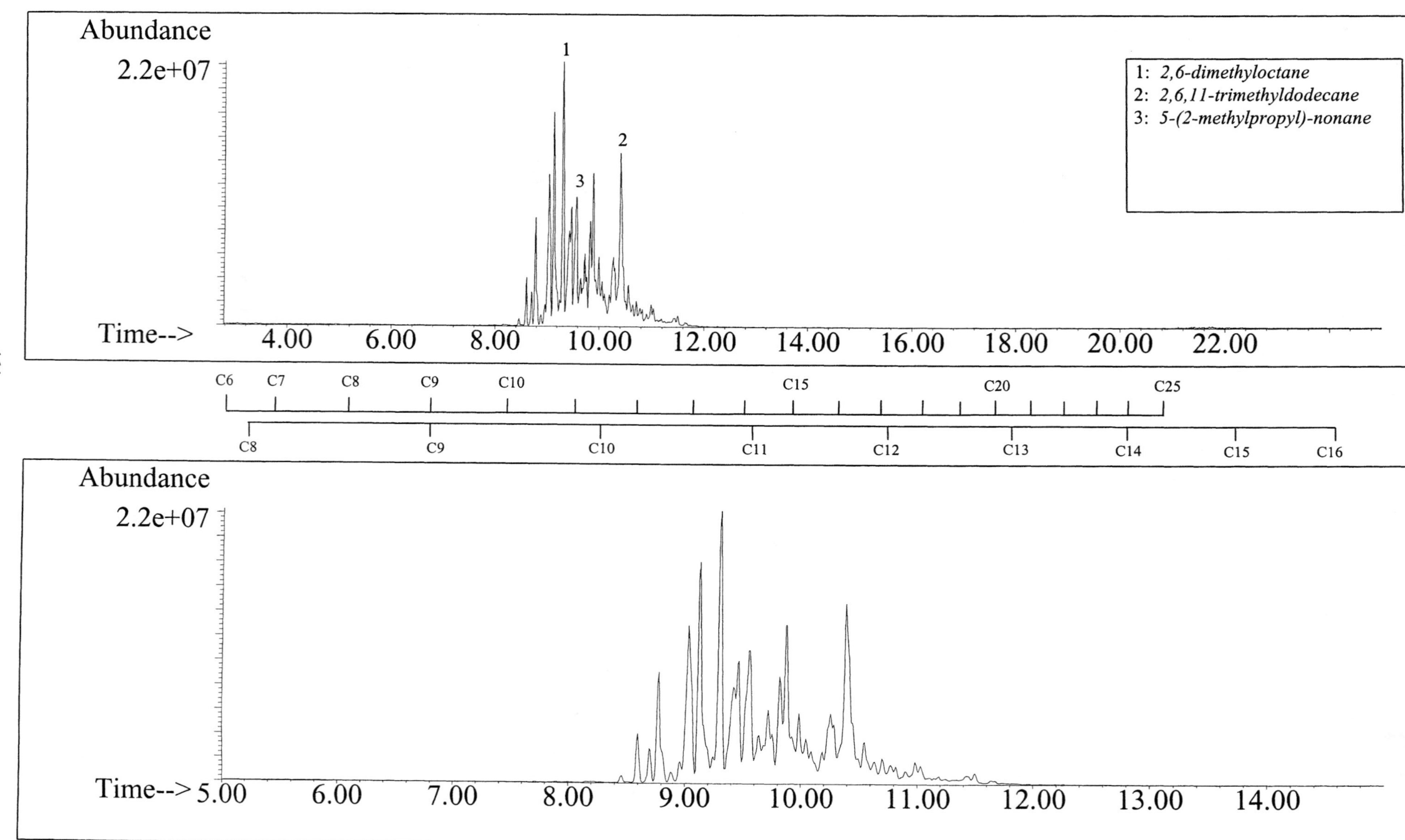

SUMMED PROFILES

Isoparaffin #8

ALKANES

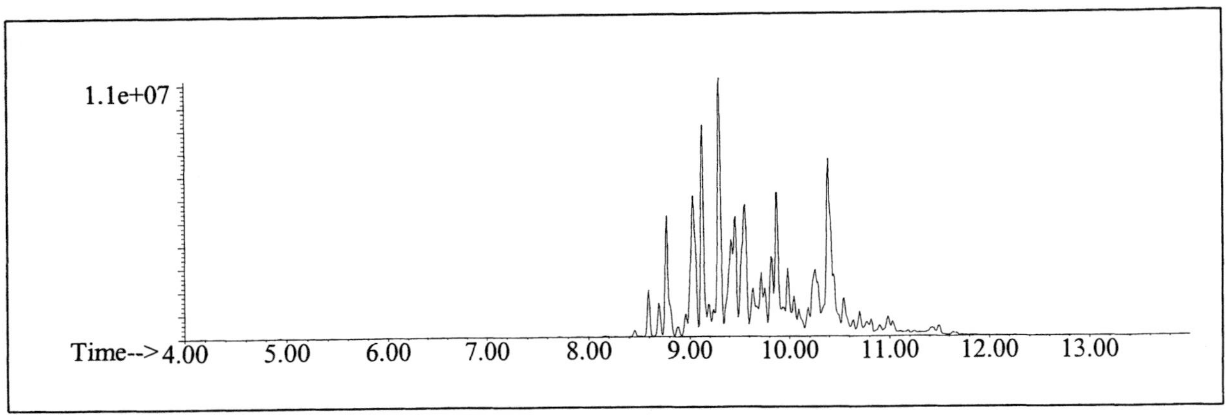

AROMATICS

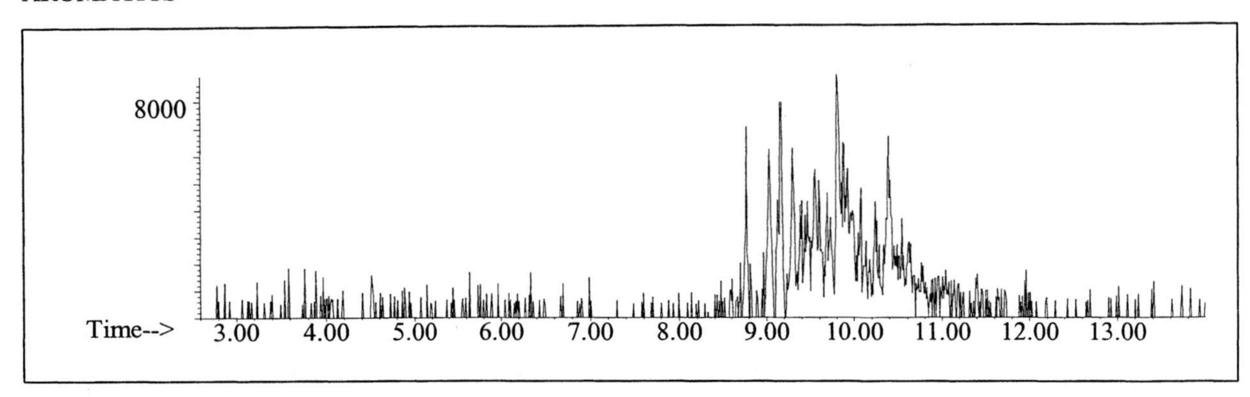

CYCLOPARAFFINS AND ALKENES

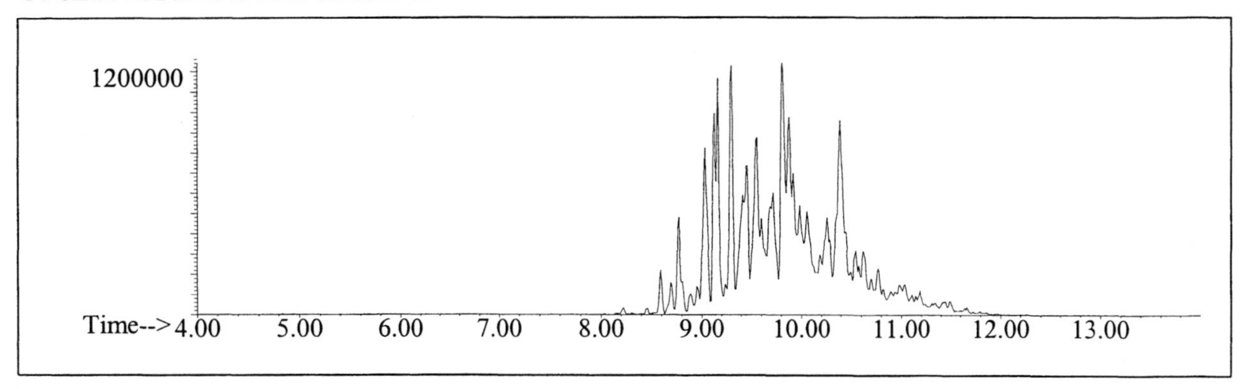

NAPHTHALENES

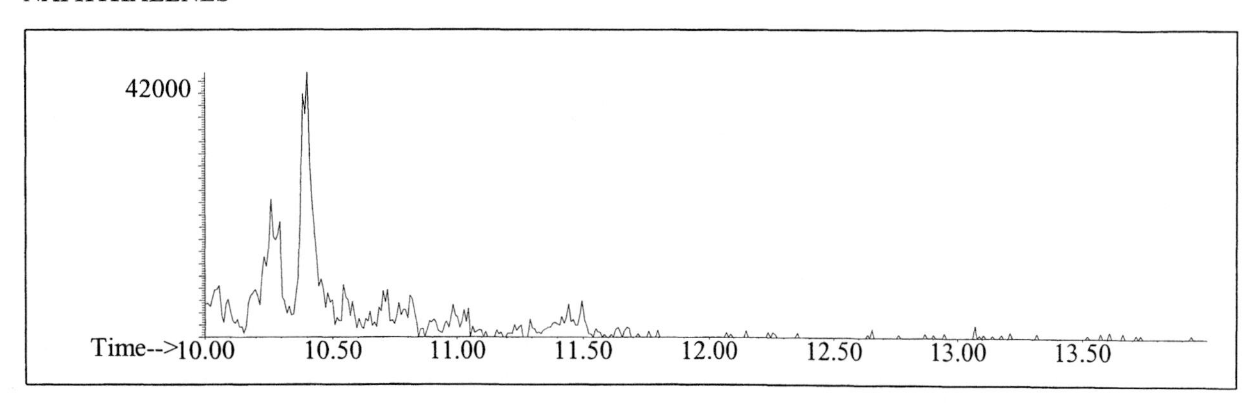

INDIVIDUAL PROFILES

Isoparaffin #8

Alkane

Ion 43

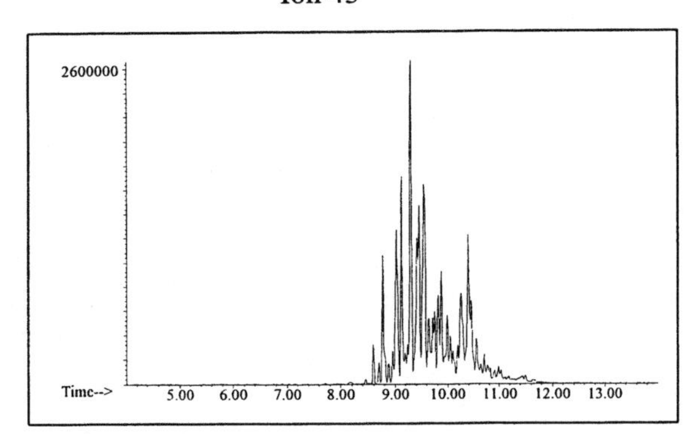

Ion 57

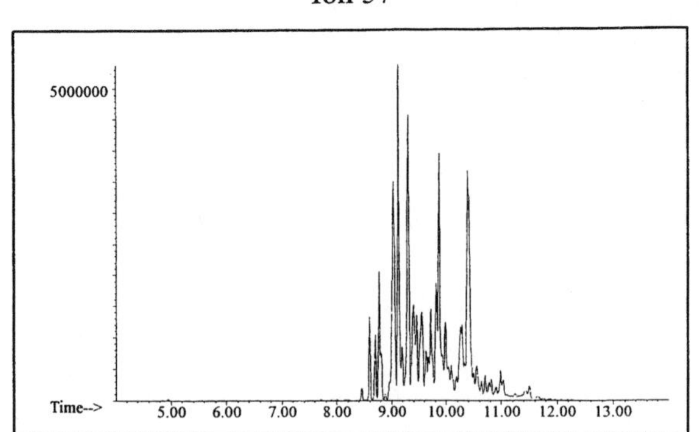

Ion 71

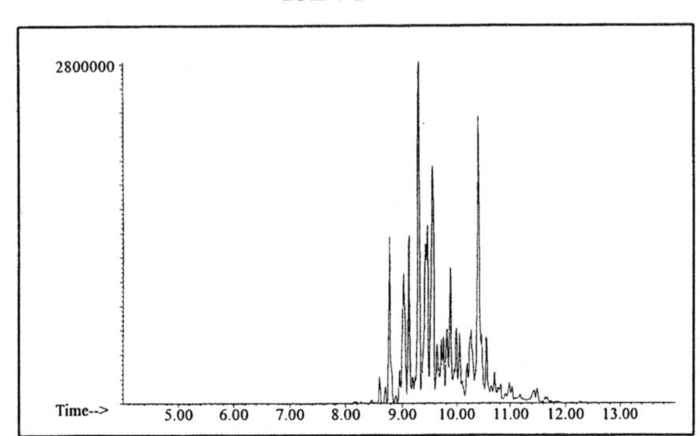

Ion 85

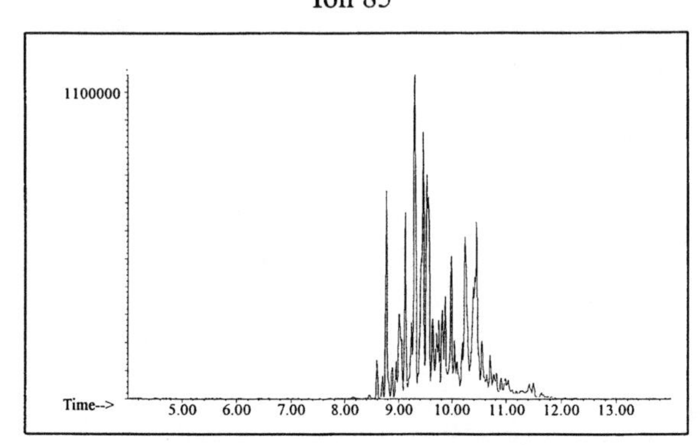

Aromatic

Ion 91

No Useful Data Obtained

Ion 105

No Useful Data Obtained

Ion 119

No Useful Data Obtained

INDIVIDUAL PROFILES

Isoparaffin #8

Cycloparaffin

Ion 55

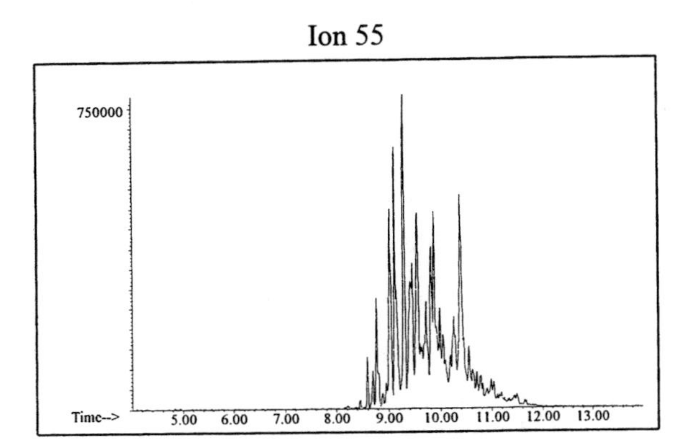

Ion 69

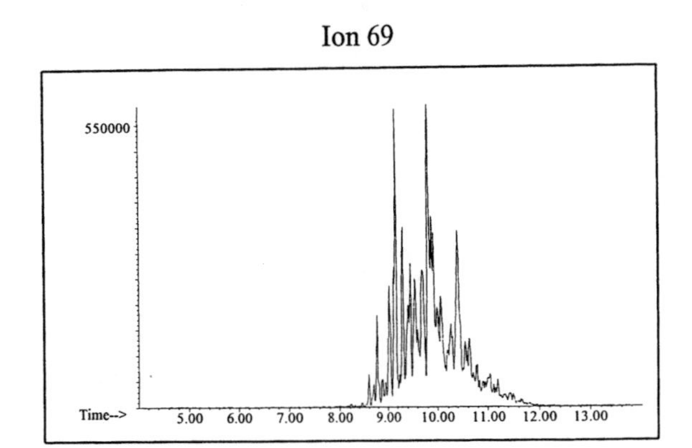

Ion 83

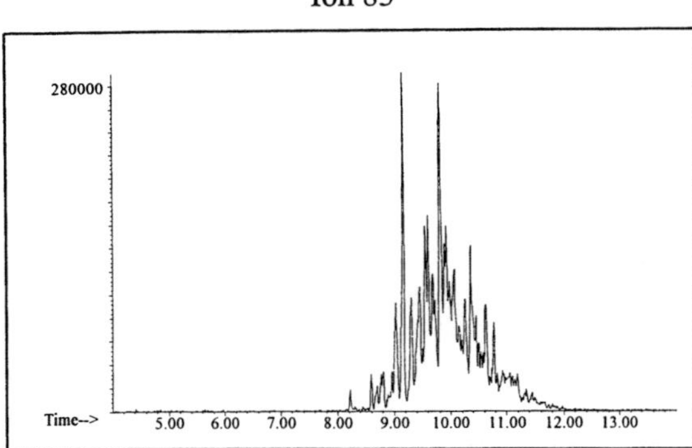

Naphthalene

Ion 128

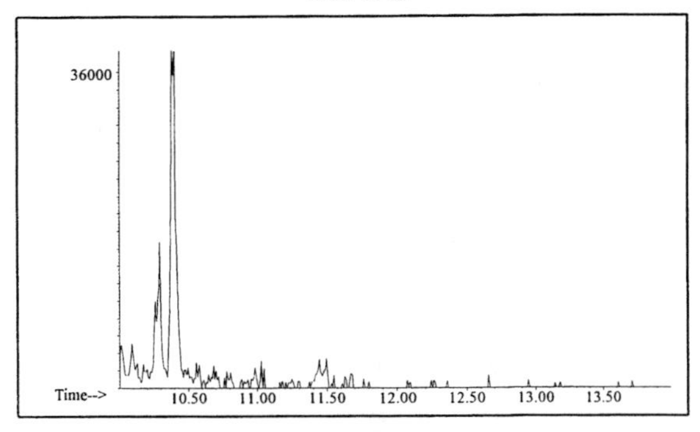

Ion 142

No Useful Data Obtained

Ion 156

No Useful Data Obtained

Isoparaffin #9

20uL/mL Pentane

Product Displayed: Isopar K

Other Similar Products: Shell SO 71 Isoparaffin Solvent, Gulf Lite Charcoal Lighter (Gulf)

ASTM: Class 0.2 (Isoparaffin)

Product Uses: Charcoal lighter, solvent, feedstock

Macro Code: M

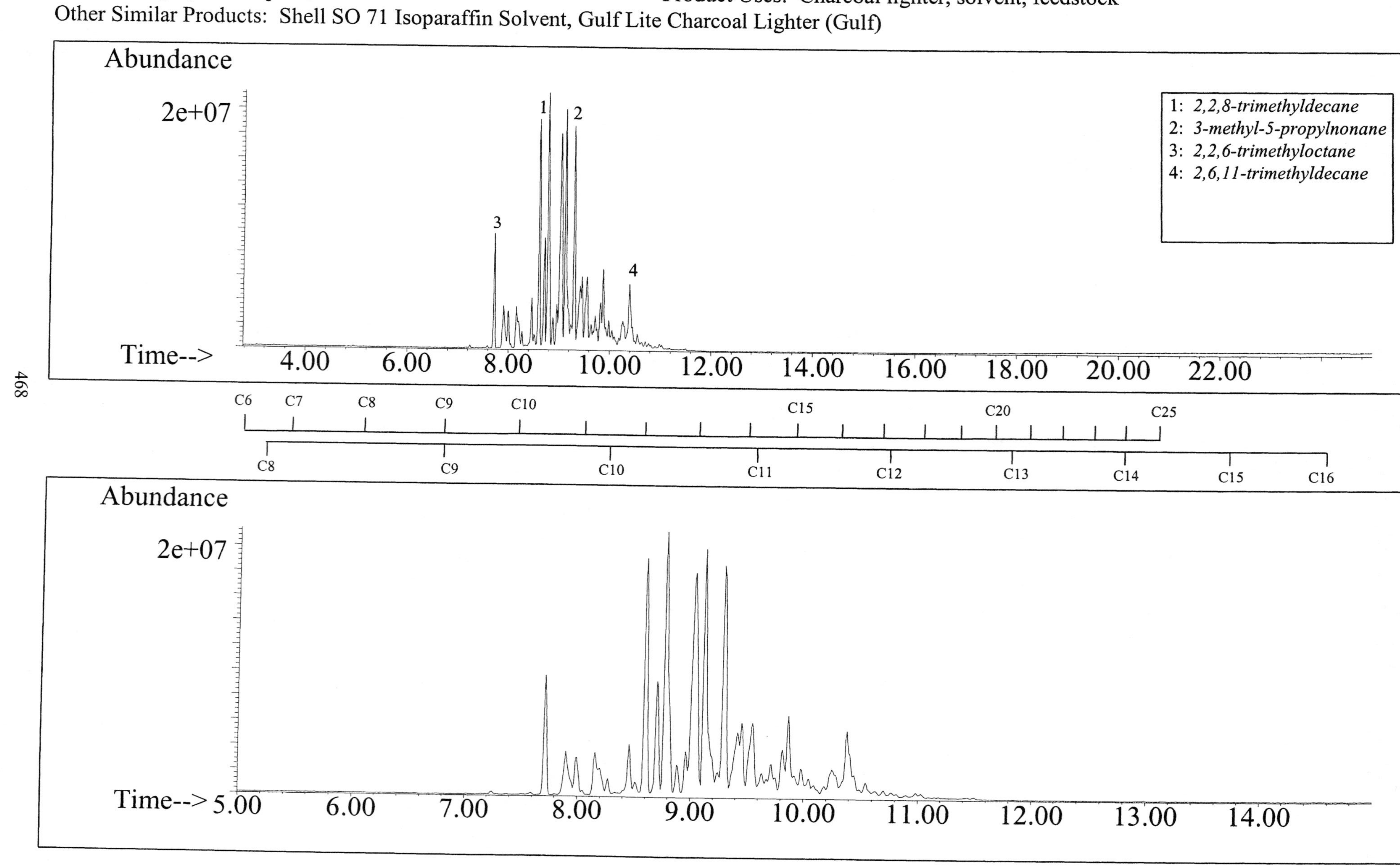

SUMMED PROFILES

Isoparaffin #9

ALKANES

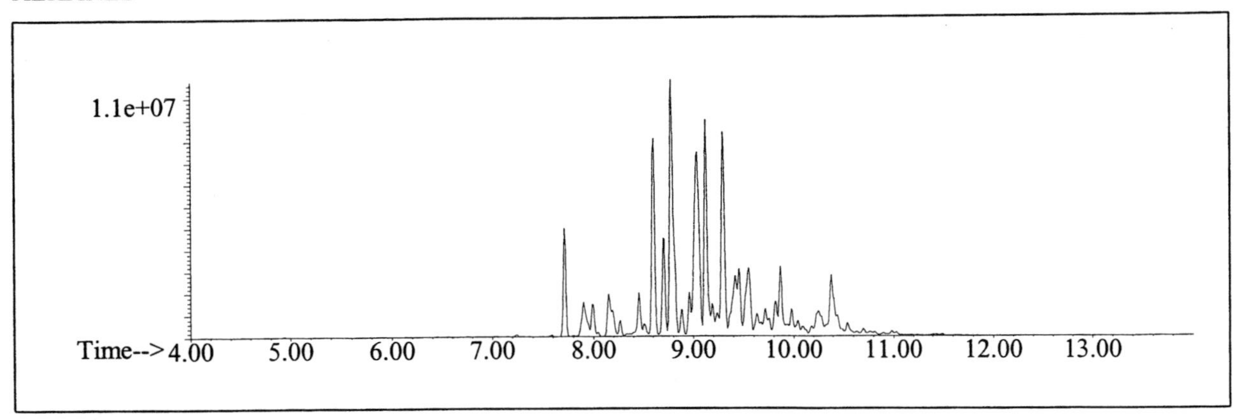

AROMATICS

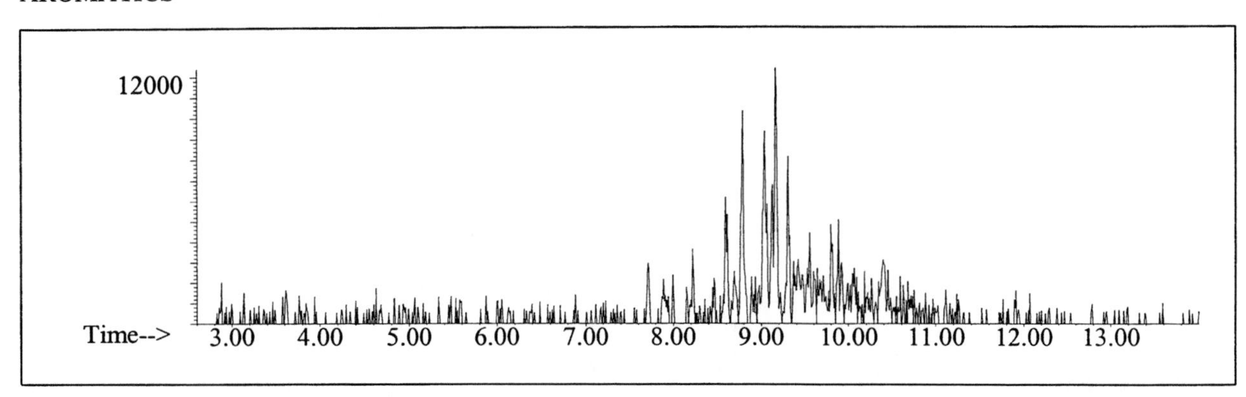

CYCLOPARAFFINS AND ALKENES

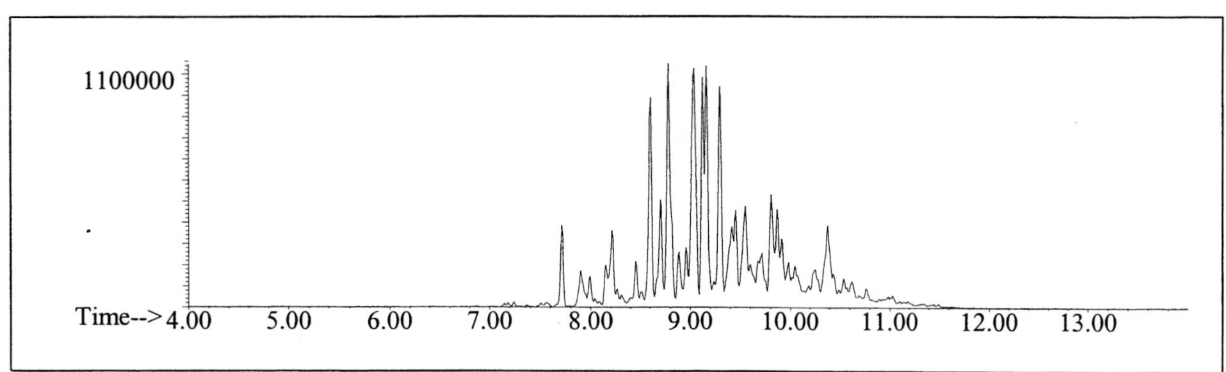

NAPHTHALENES

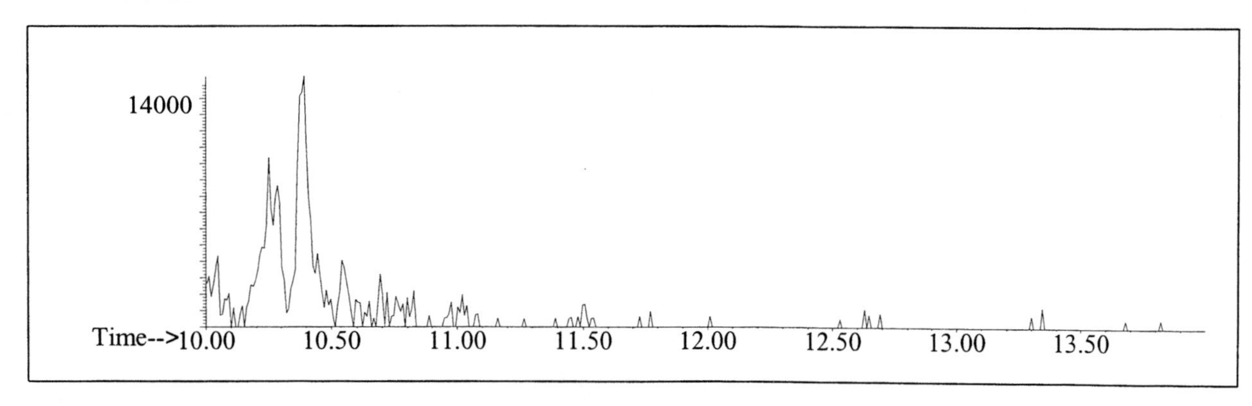

INDIVIDUAL PROFILES

Isoparaffin #9

Alkane

Ion 43

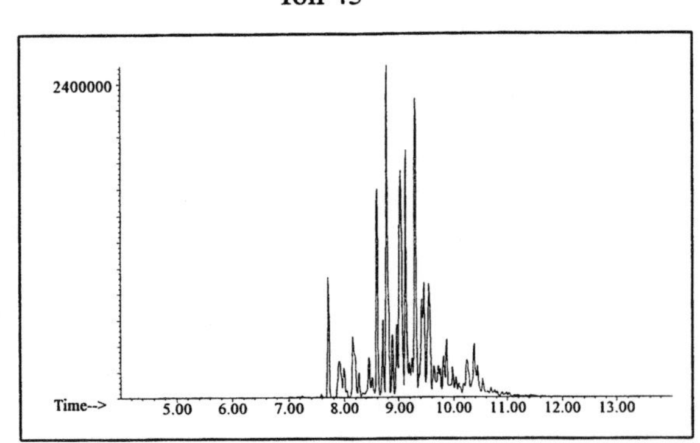

Ion 57

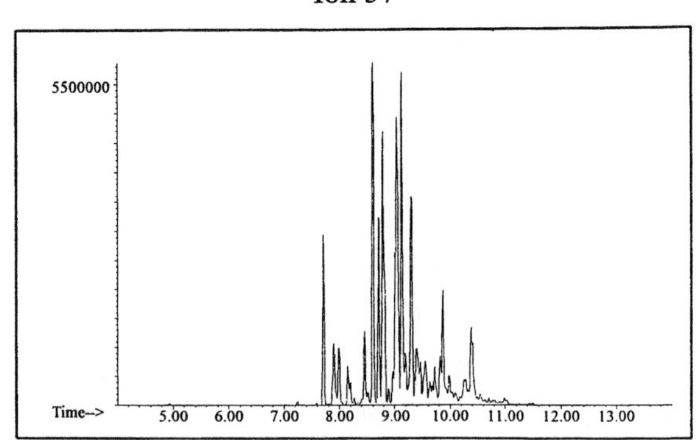

Ion 71

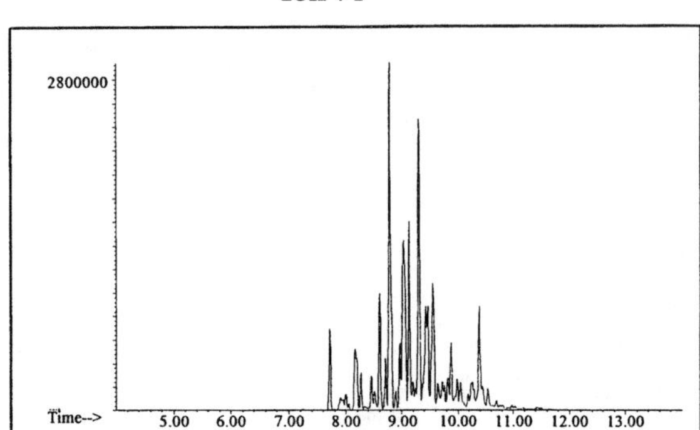

Ion 85

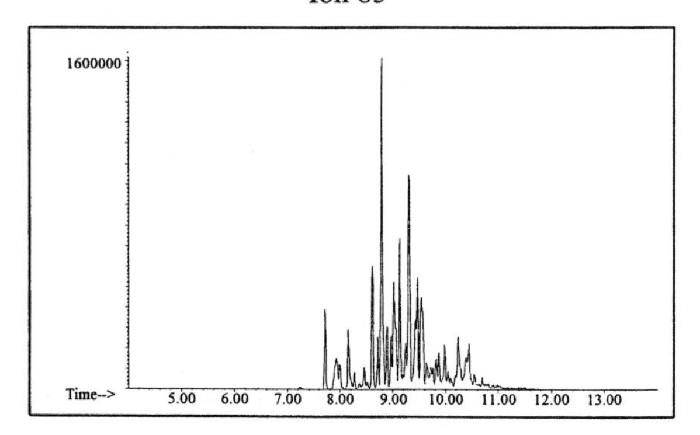

Aromatic

Ion 91

No Useful Data Obtained

Ion 105

No Useful Data Obtained

Ion 119

No Useful Data Obtained

Cycloparaffin

Ion 55

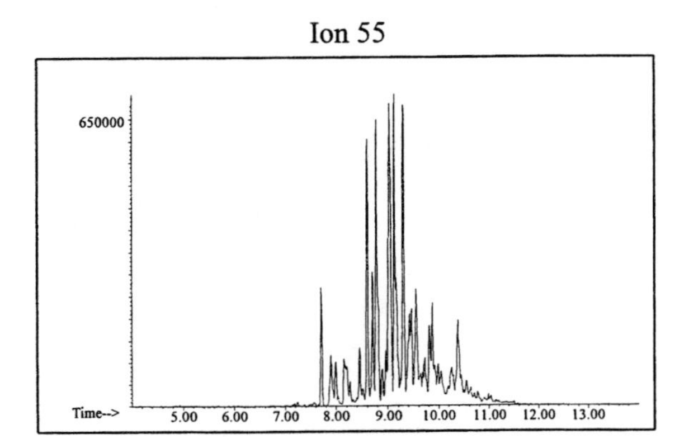

Ion 69

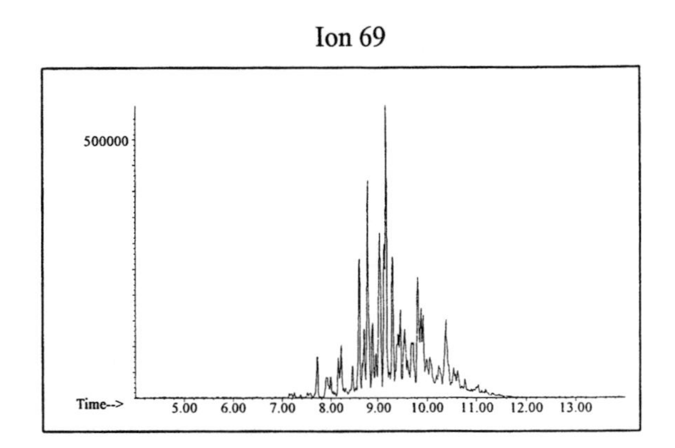

Ion 83

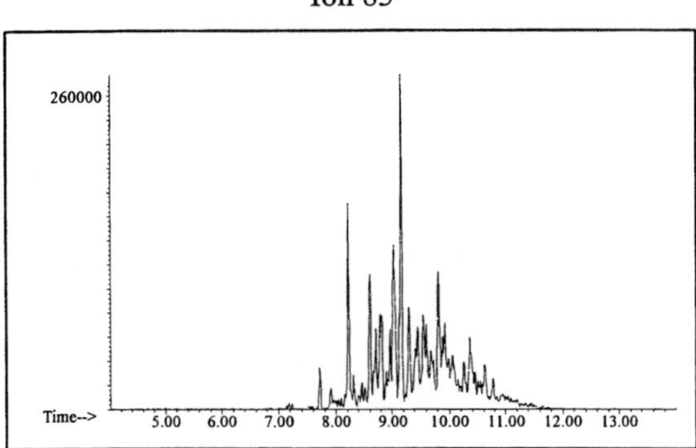

Naphthalene

Ion 128

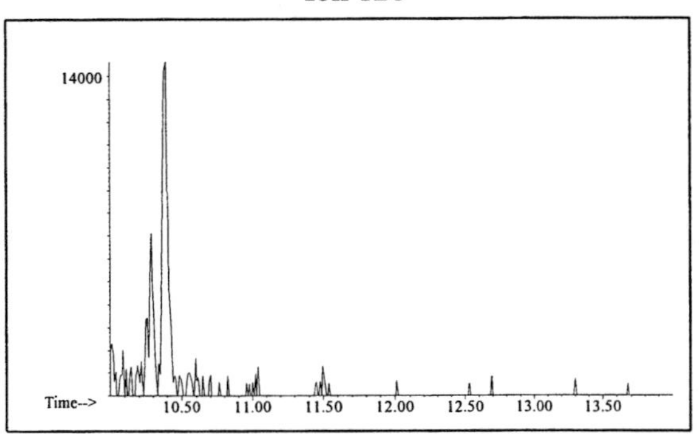

Ion 142

No Useful Data Obtained

Ion 156

No Useful Data Obtained

Isoparaffin #10

20uL/mL Pentane

ASTM: Class 0.2 (Isoparaffin)

Macro Code: M

Product Displayed: USA Odorless Mineral Spirits

Product Uses: Mineral spirits

Other Similar Products: Phillips 66 Soltrol 130

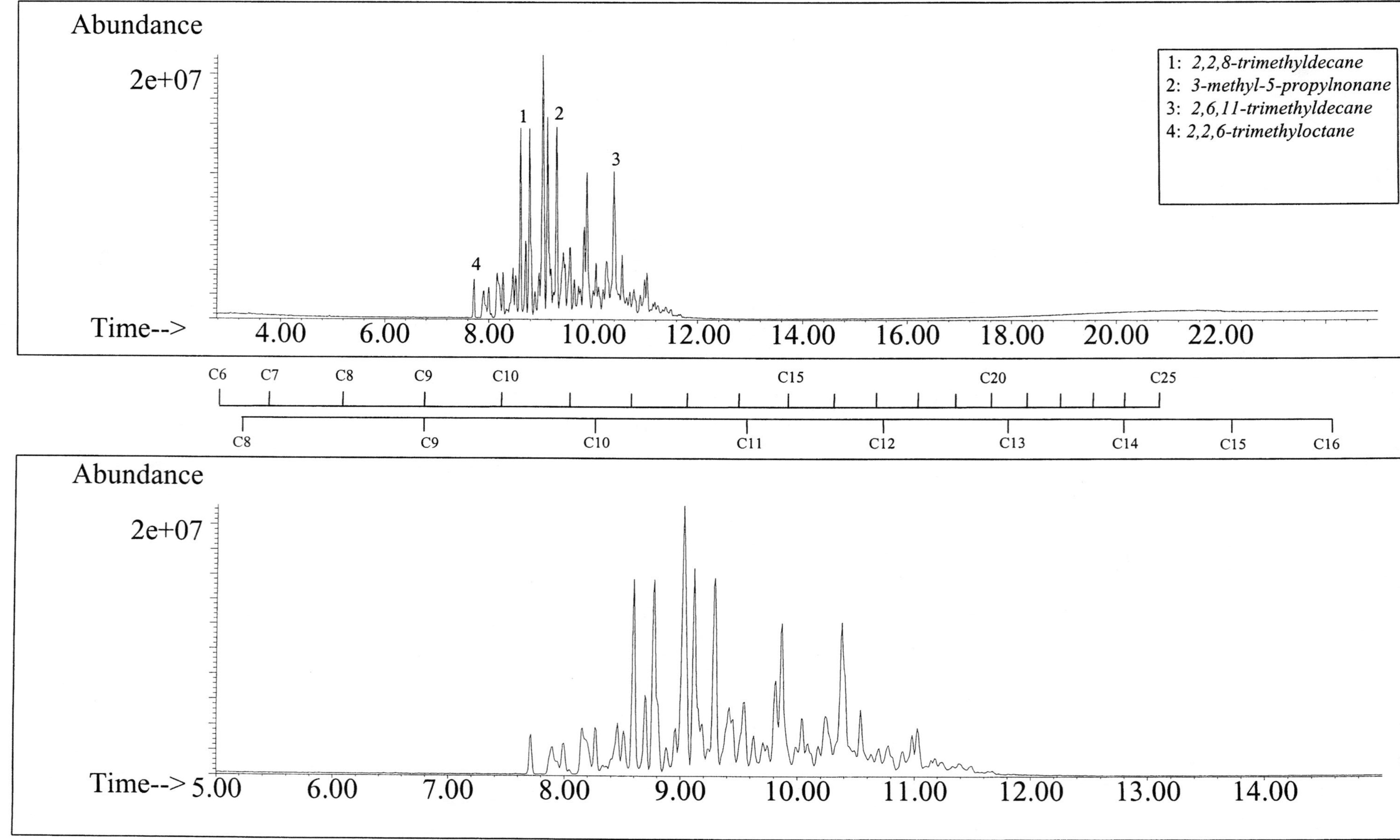

SUMMED PROFILES

Isoparaffin #10

ALKANES

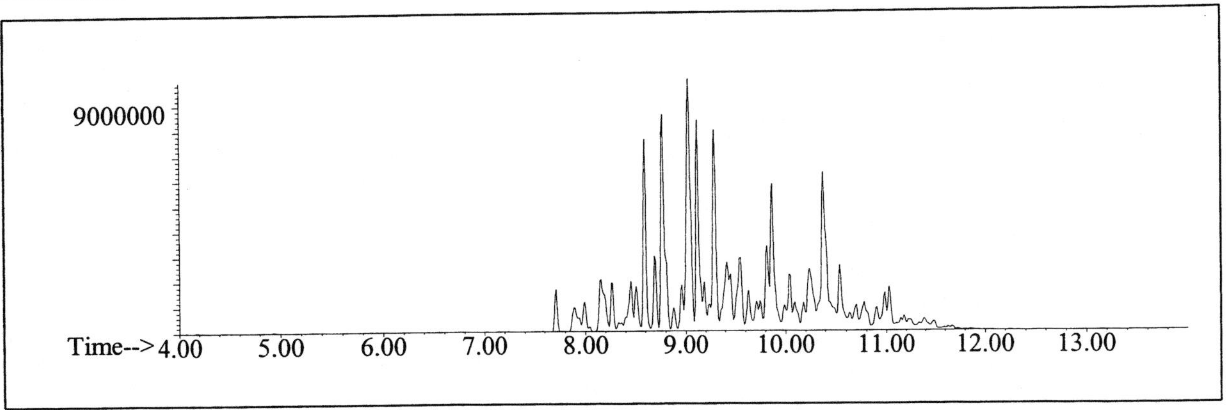

AROMATICS

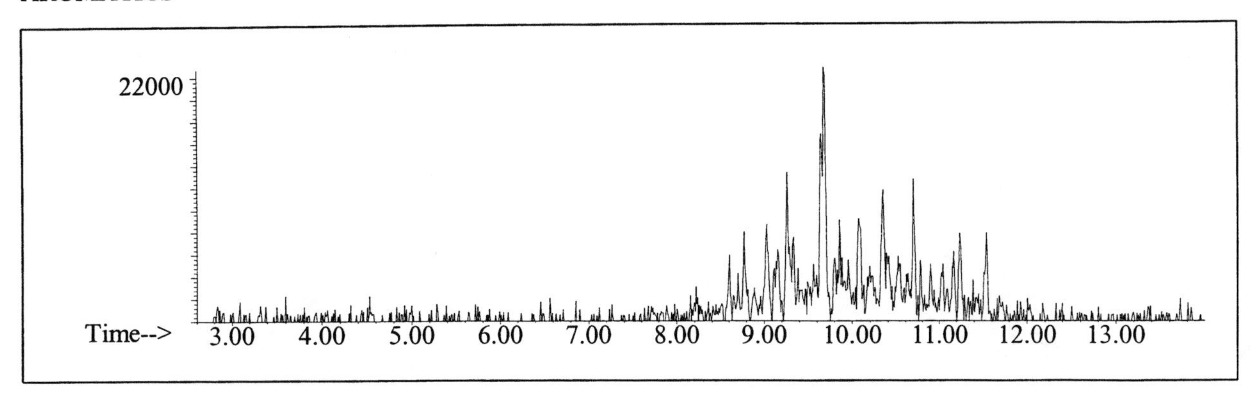

CYCLOPARAFFINS AND ALKENES

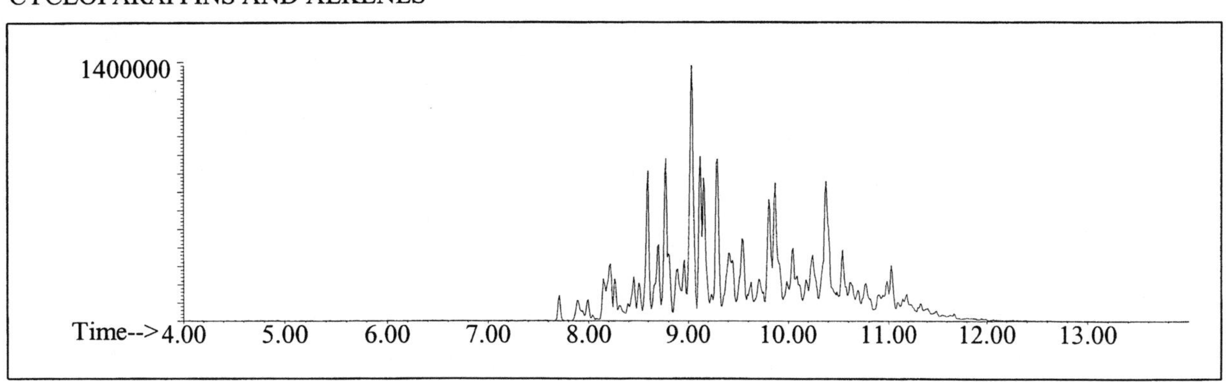

NAPHTHALENES

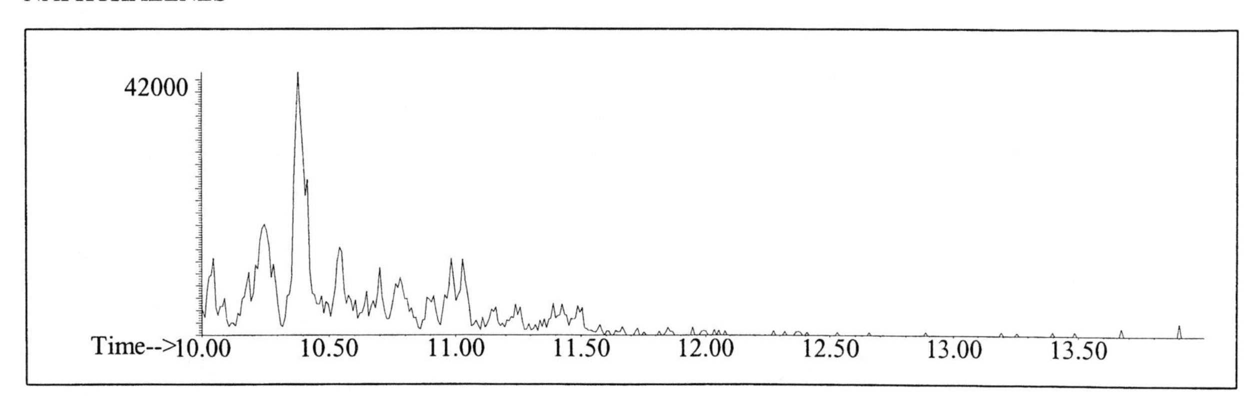

INDIVIDUAL PROFILES

Isoparaffin #10

Alkane

Ion 43

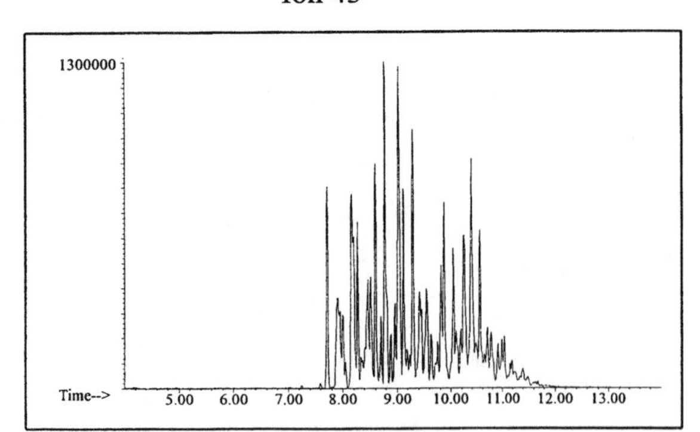

Ion 57

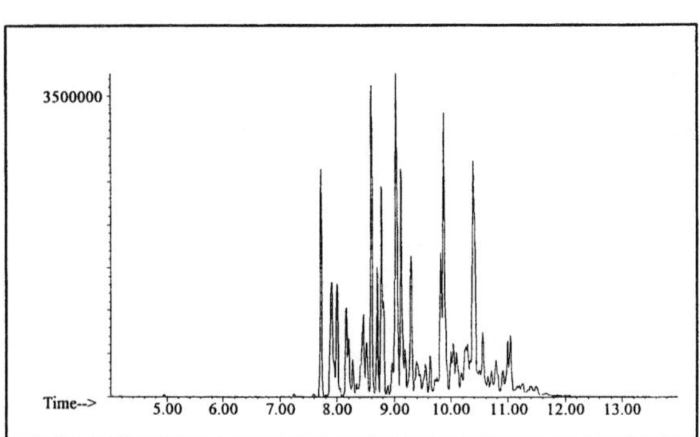

Ion 71

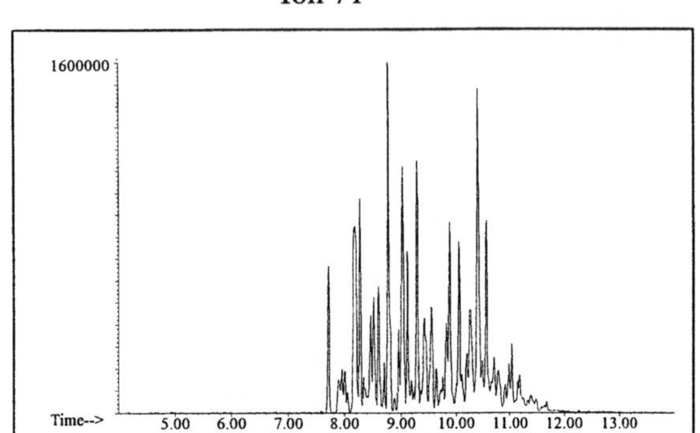

Ion 85

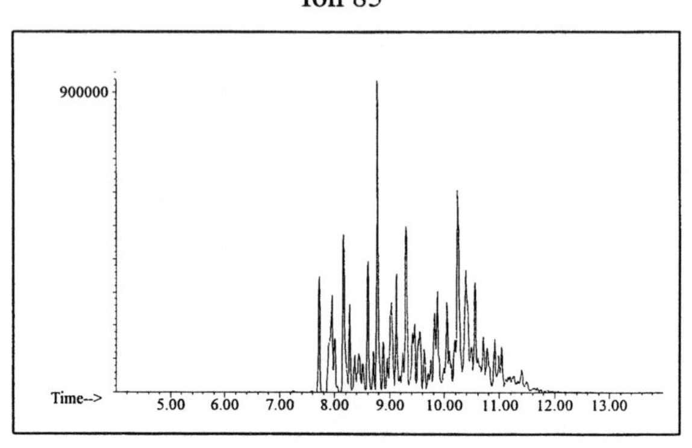

Aromatic

Ion 91

No Useful Data Obtained

Ion 105

No Useful Data Obtained

Ion 119

No Useful Data Obtained

Cycloparaffin

Ion 55

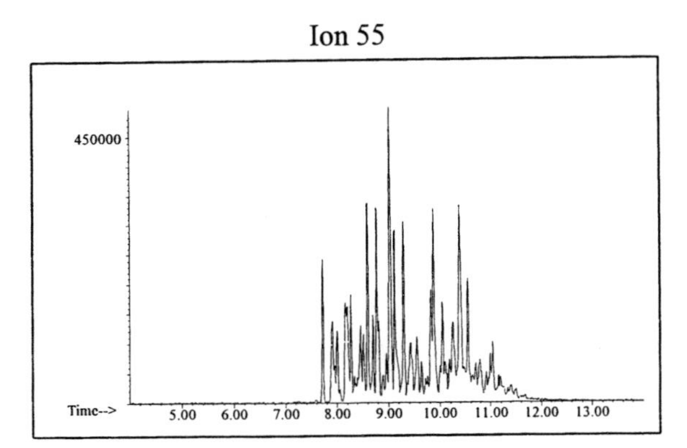

Ion 69

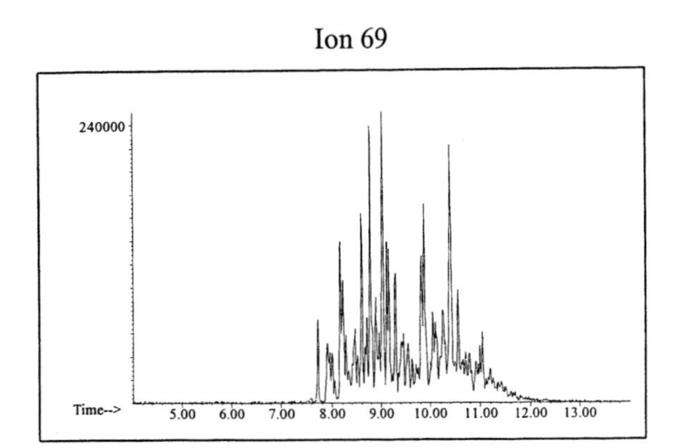

Ion 83

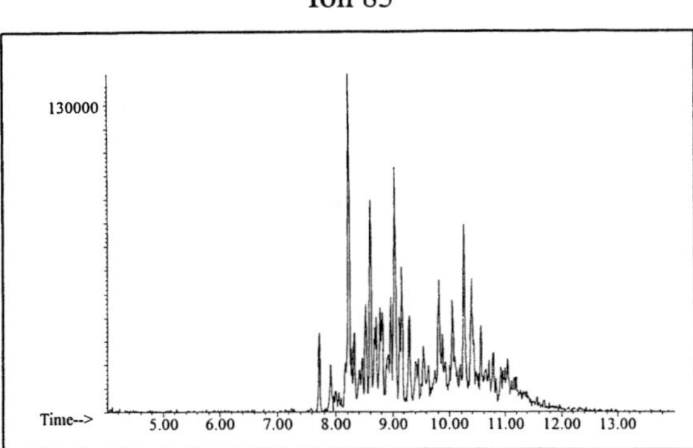

Naphthalene

Ion 128

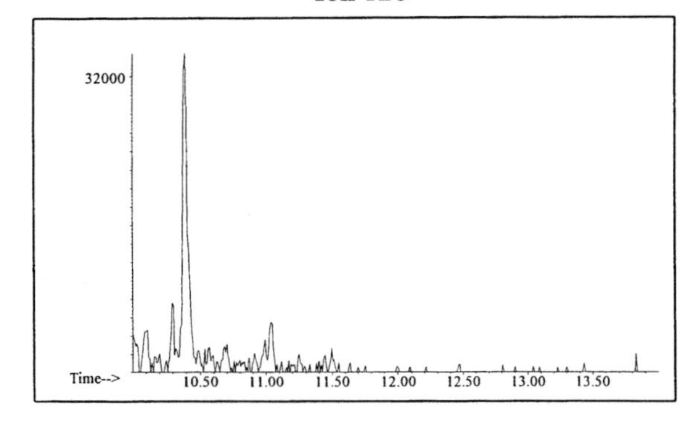

Ion 142

No Useful Data Obtained

Ion 156

No Useful Data Obtained

Isoparaffin #11

20uL/mL Pentane

Product Displayed: Isopar M

Other Similar Products:

ASTM: Class 0.2 (Isoparaffin)

Product Uses: Solvent, feedstock

Macro Code: M

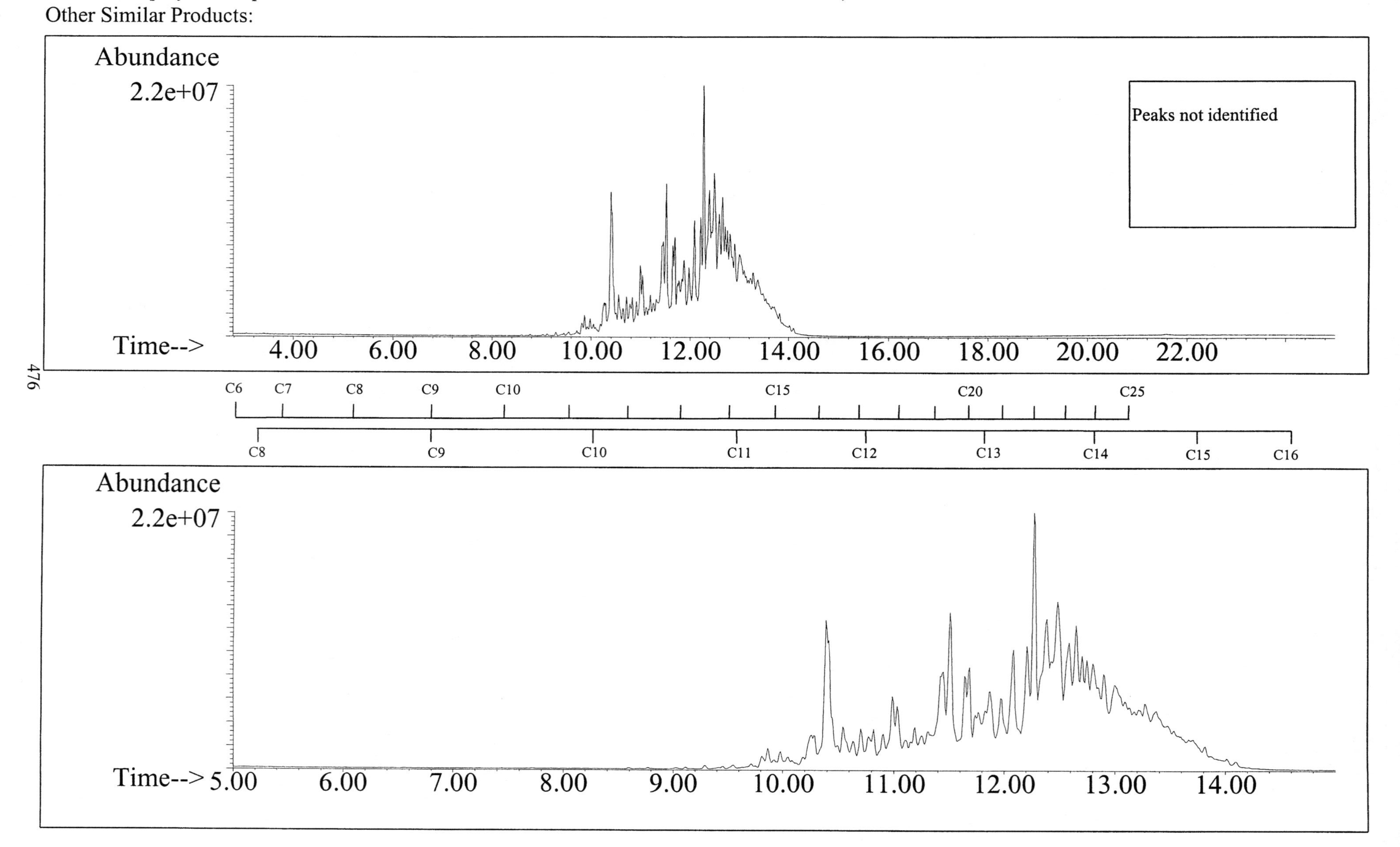

SUMMED PROFILES

Isoparaffin #11

ALKANES

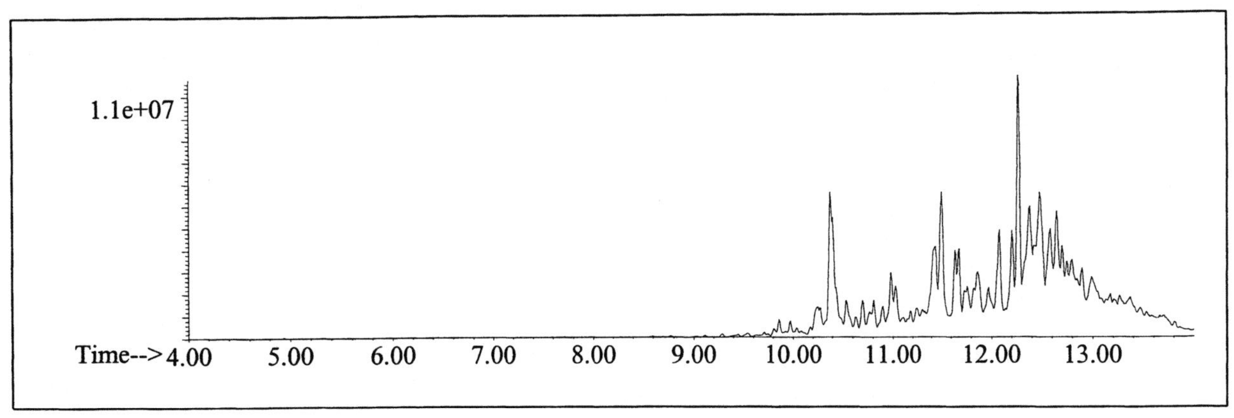

AROMATICS

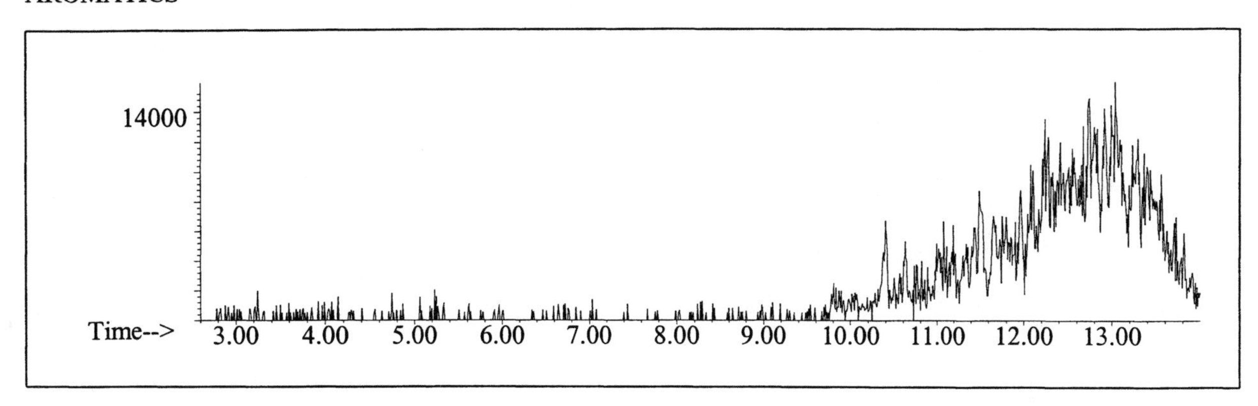

CYCLOPARAFFINS AND ALKENES

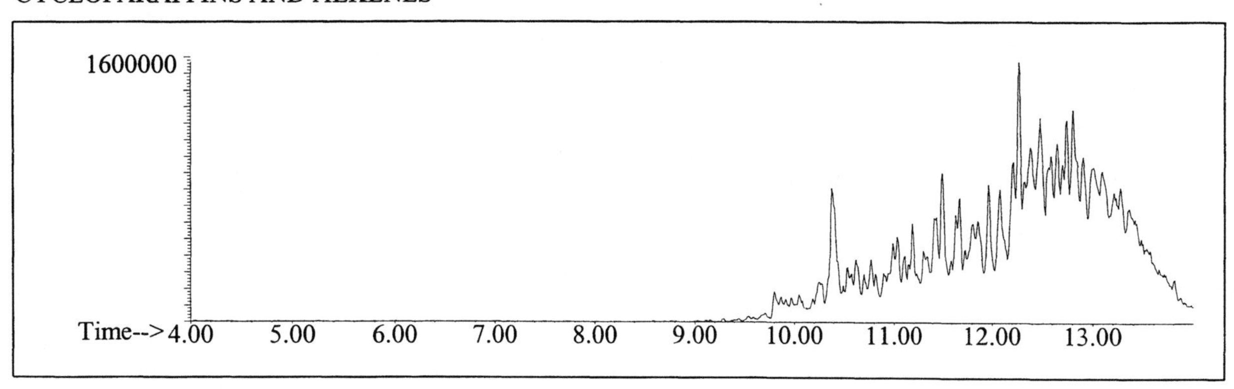

NAPHTHALENES

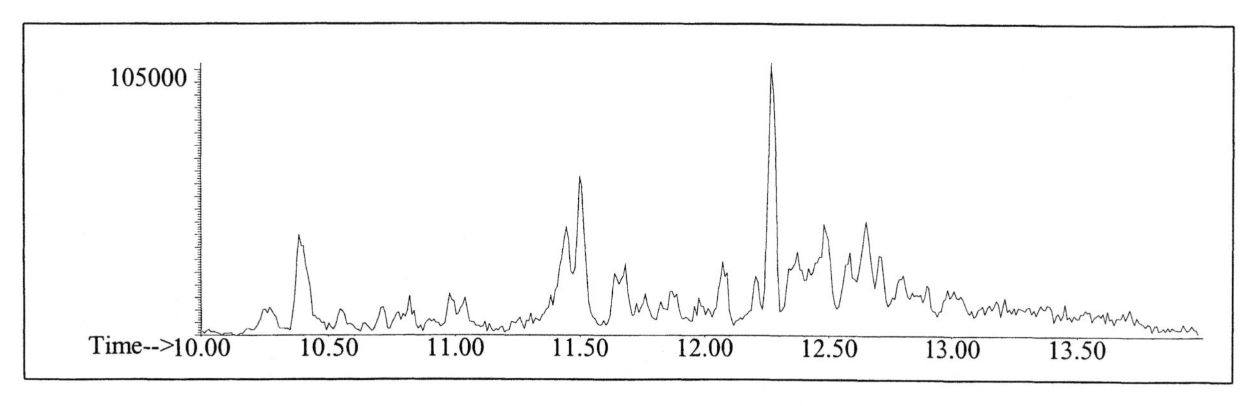

INDIVIDUAL PROFILES

Isoparaffin #11

Alkane

Ion 43

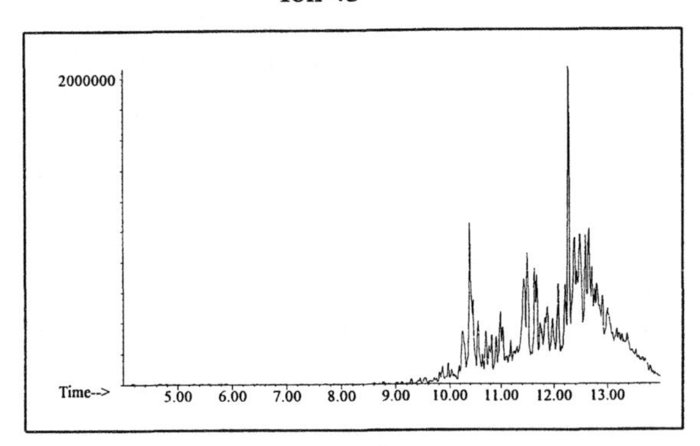

Ion 57

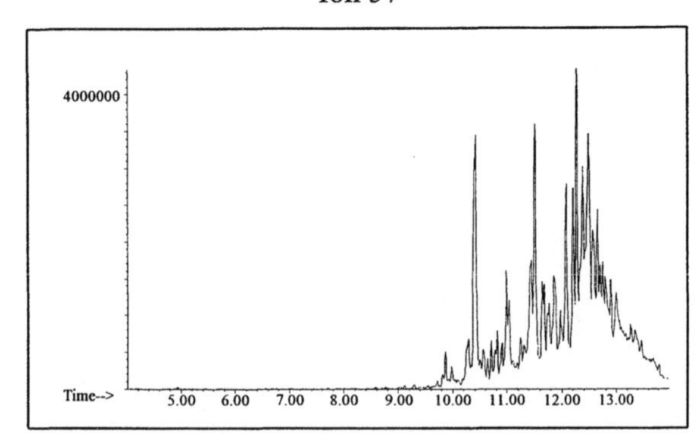

Ion 71

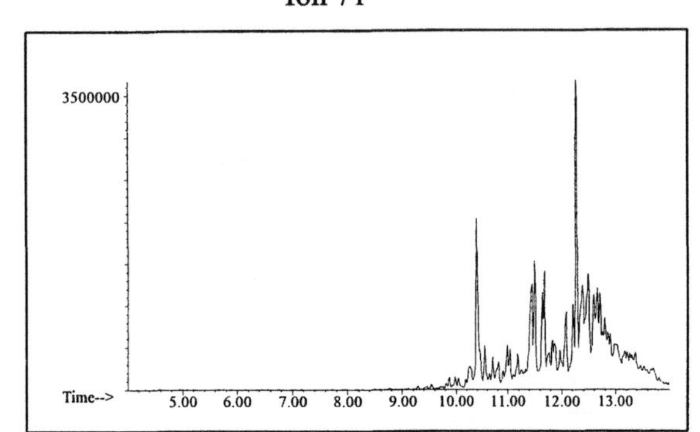

Ion 85

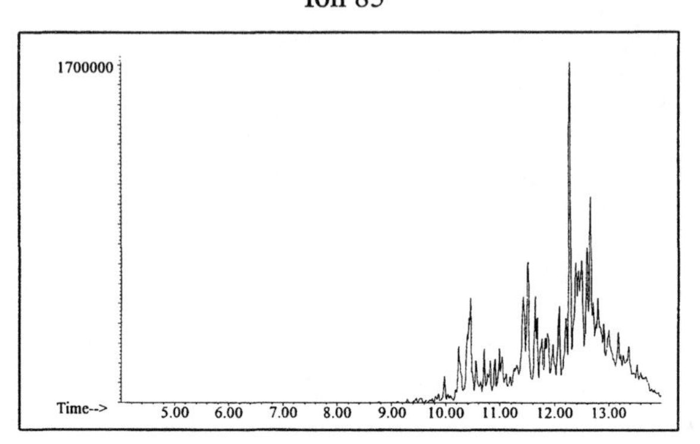

Aromatic

Ion 91

No Useful Data Obtained

Ion 105

No Useful Data Obtained

Ion 119

No Useful Data Obtained

Cycloparaffin

Ion 55

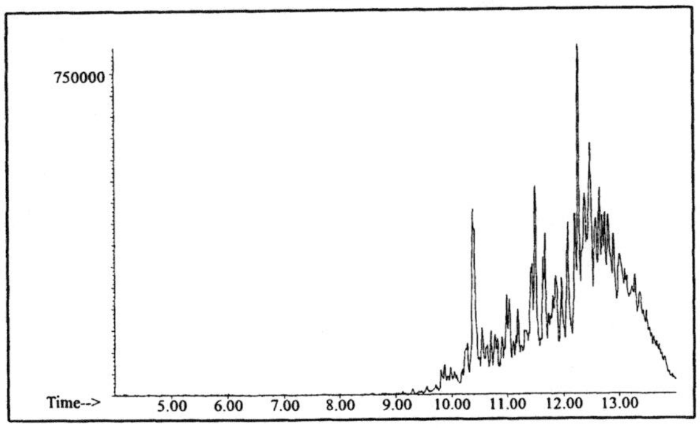

Ion 69

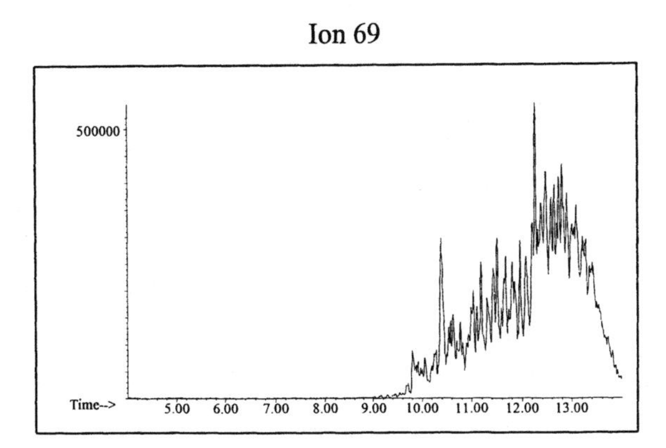

Ion 83

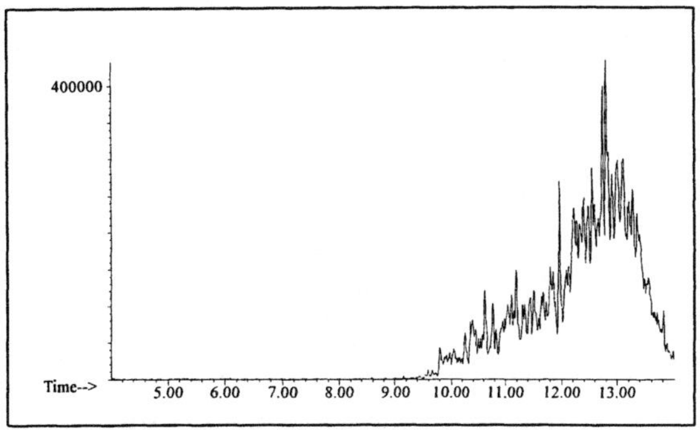

Naphthalene

Ion 128

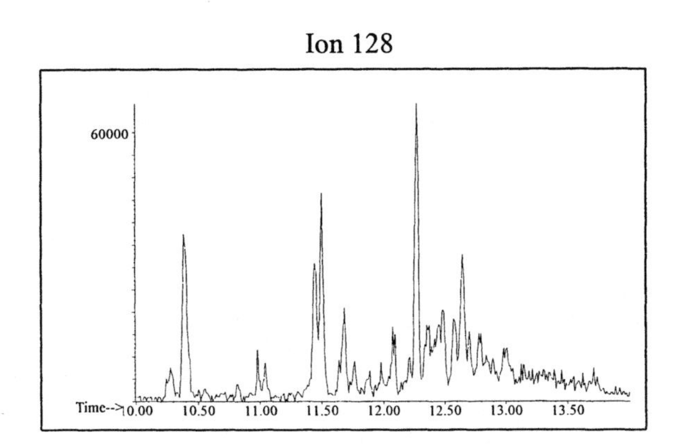

Ion 142

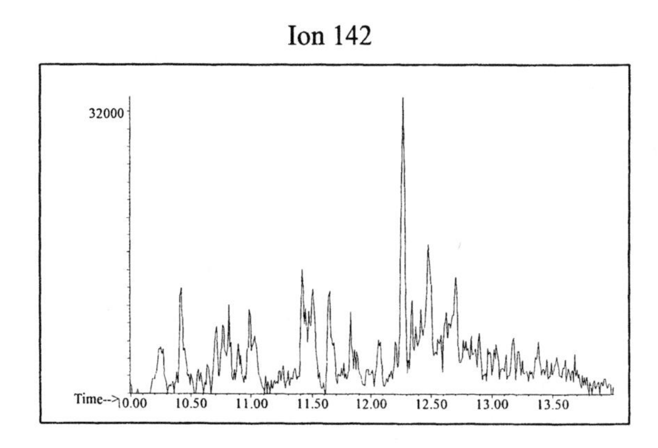

Ion 156

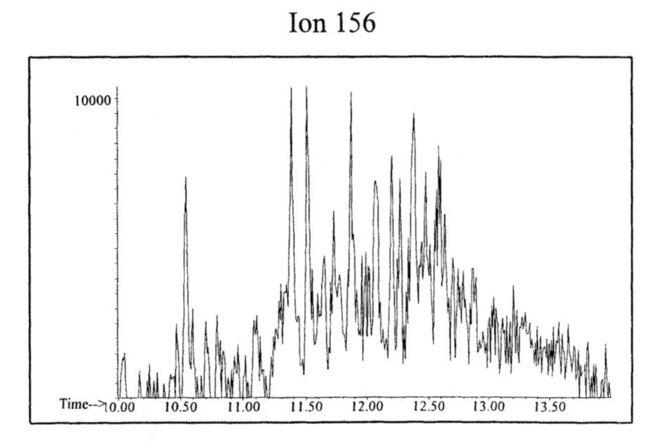

Isoparaffin #12

20uL/mL Pentane

Product Displayed: Phillips 66 Soltrol 170

Other Similar Products:

ASTM: Class 0.2 (Isoparaffin)

Product Uses: Solvent, feedstock

Macro Code: M

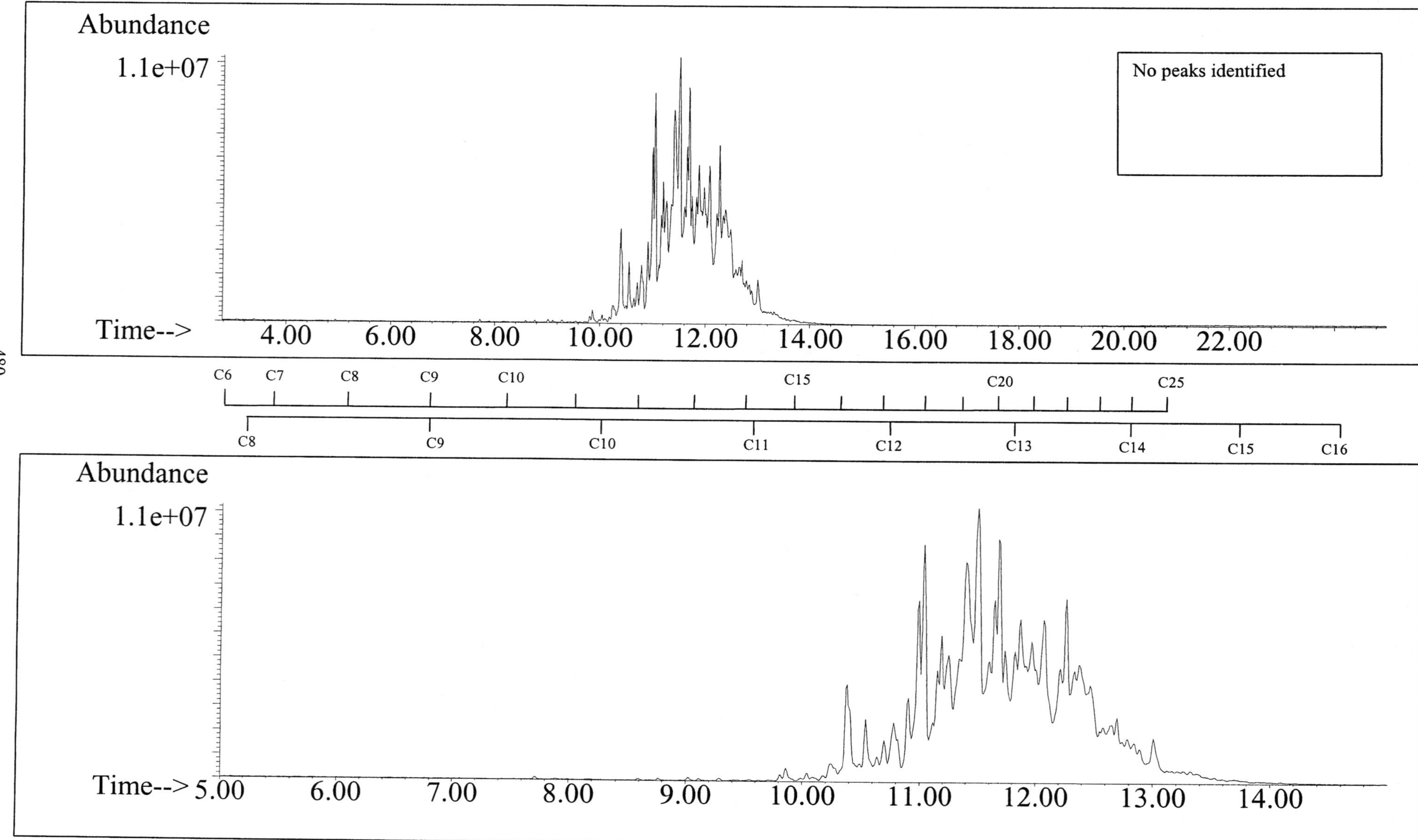

SUMMED PROFILES

Isoparaffin #12

ALKANES

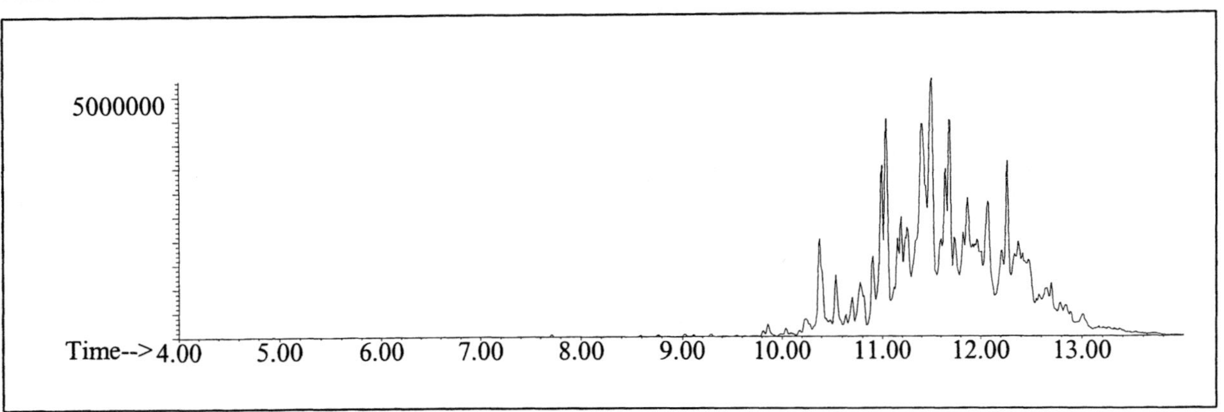

AROMATICS

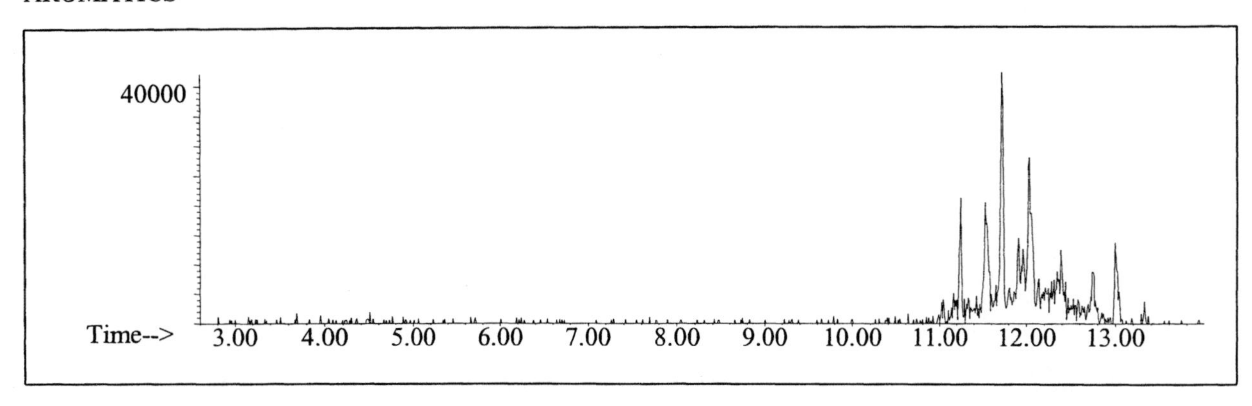

CYCLOPARAFFINS AND ALKENES

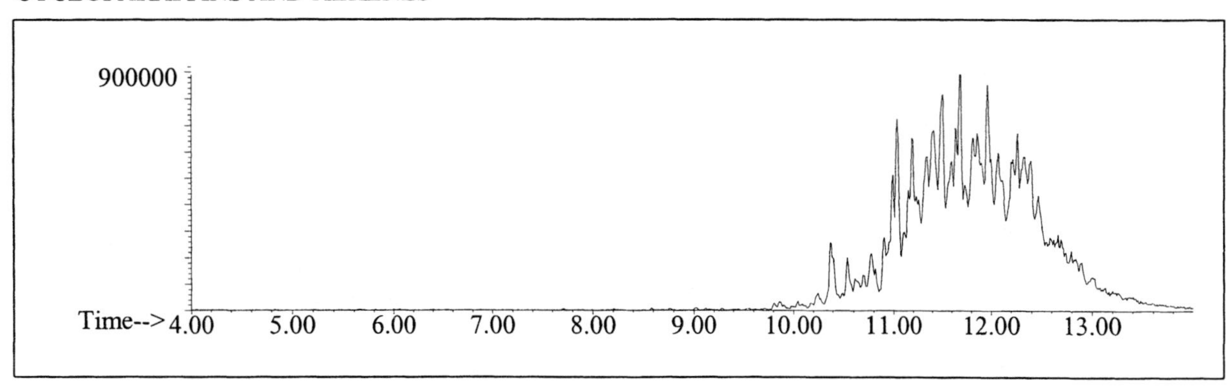

NAPHTHALENES

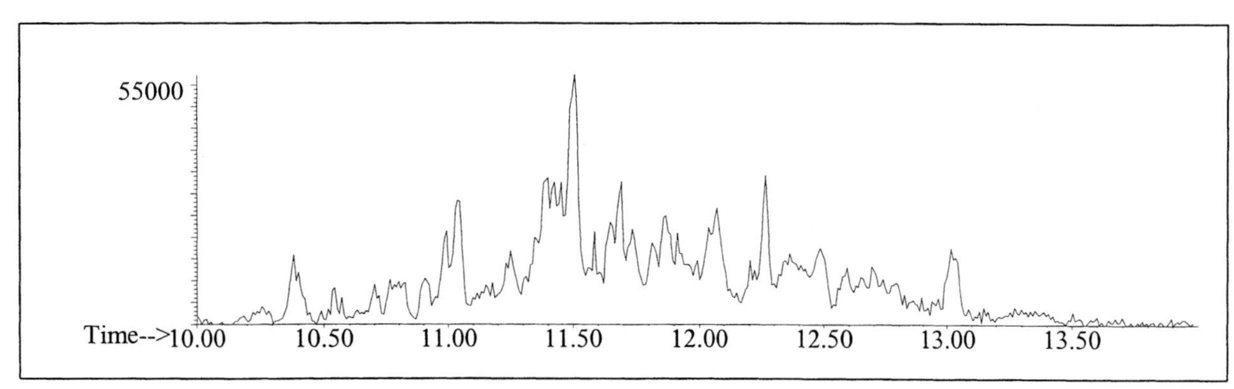

Alkane

Ion 43

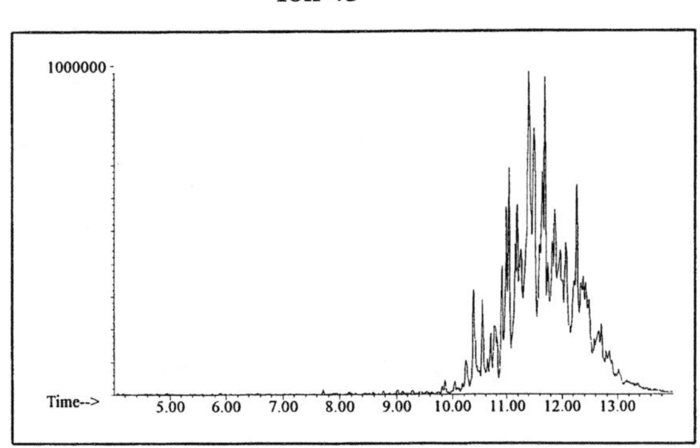

Ion 57

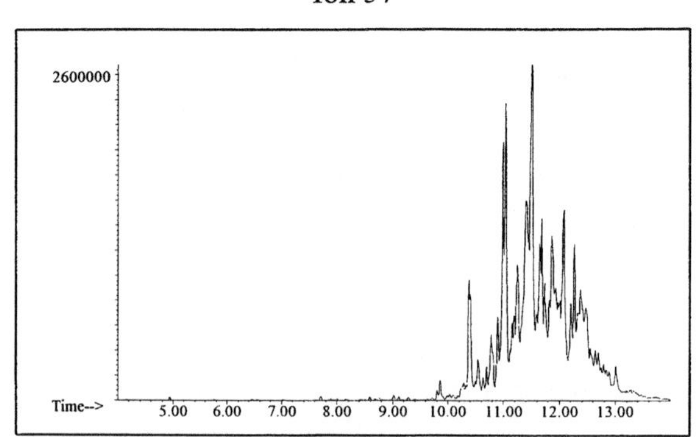

Ion 71

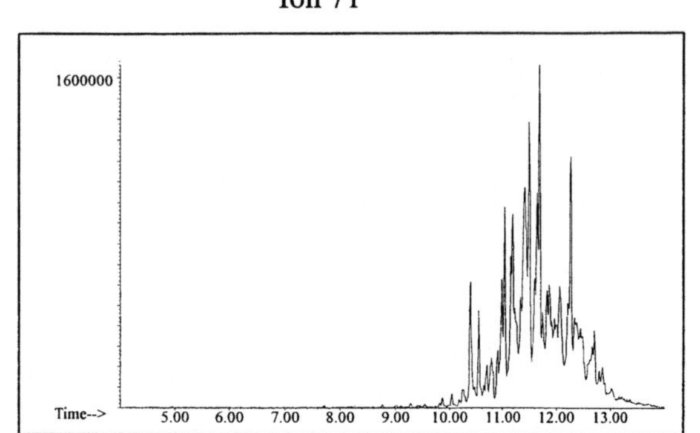

Ion 85

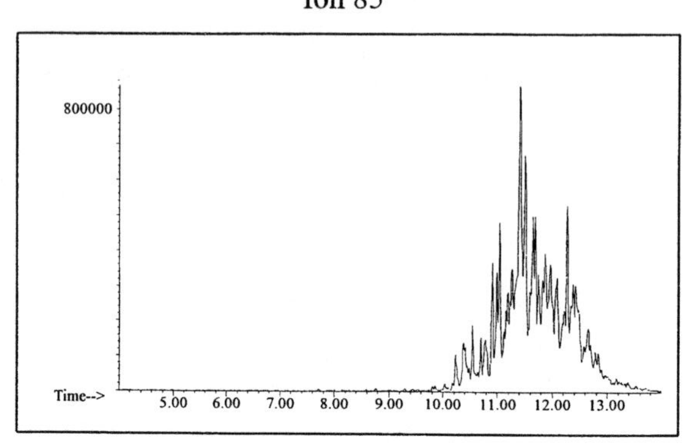

Aromatic

Ion 91

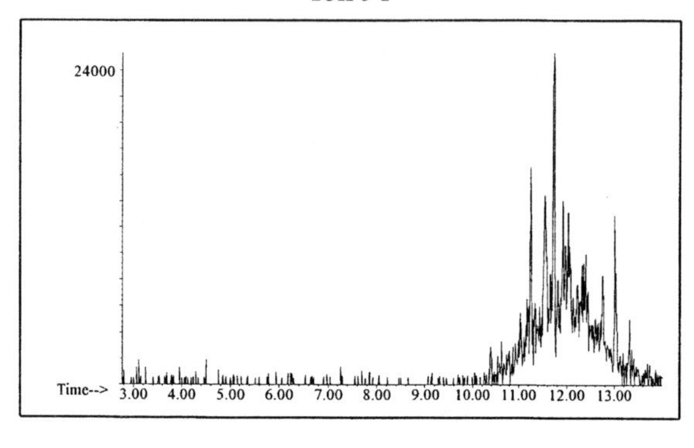

Ion 105

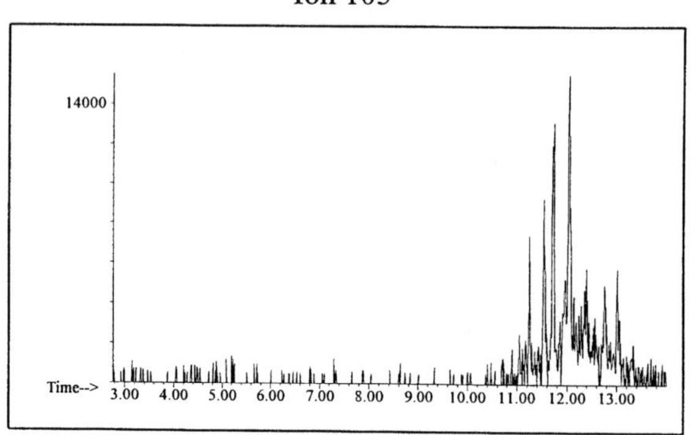

Ion 119

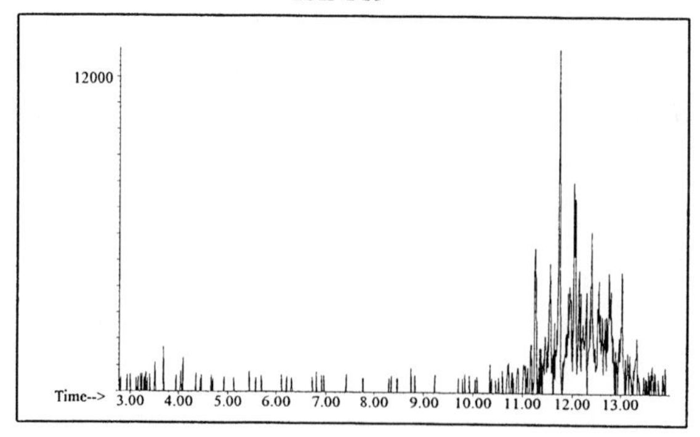

INDIVIDUAL PROFILES

Isoparaffin #12

Cycloparaffin

Ion 55

Ion 69

Ion 83

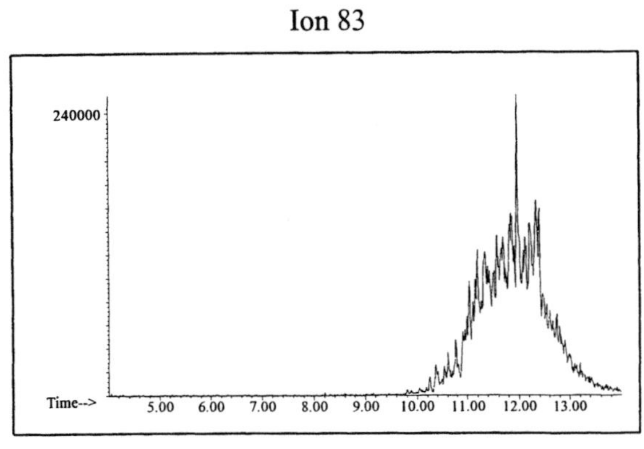

Naphthalene

Ion 128

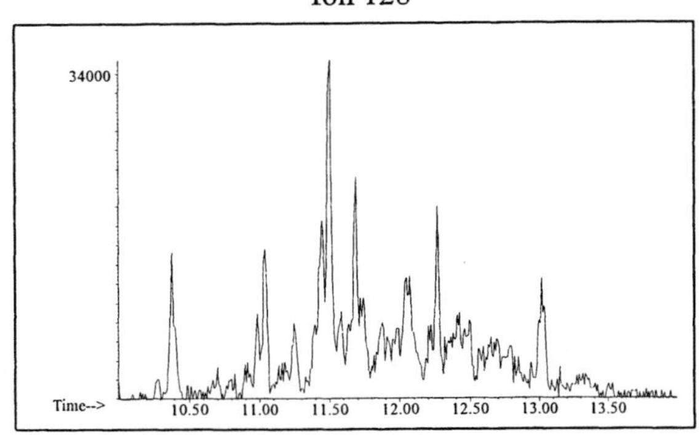

Ion 142

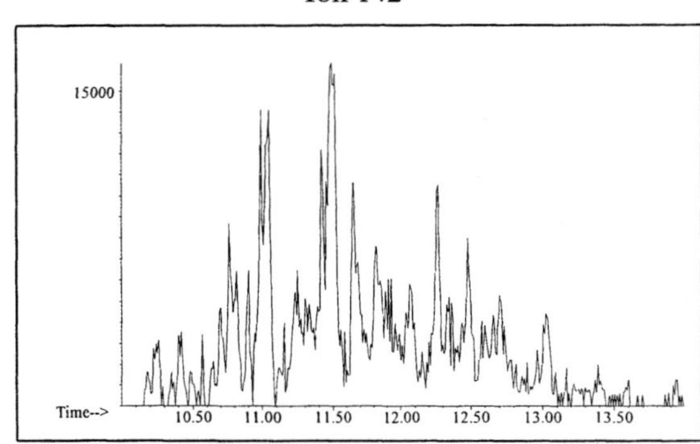

Ion 156

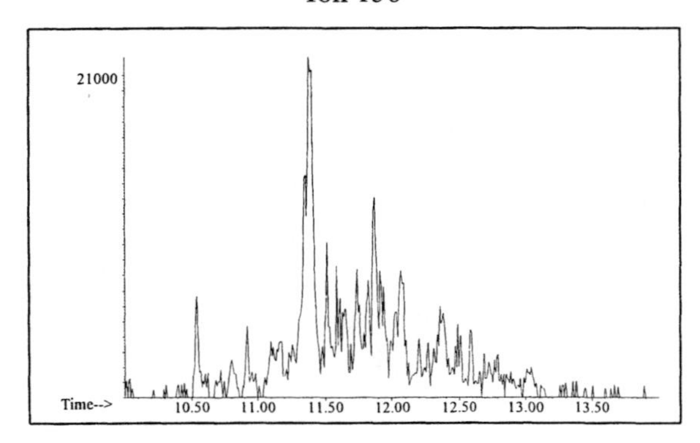

Isoparaffin #13

20uL/mL Pentane

Product Displayed: Phillips 66 Soltrol 220

Other Similar Products:

ASTM: Class 0.2 (Isoparaffin)

Product Uses: Solvent, feedstock

Macro Code: H

Abundance

5000000

No peaks identified

Time-->

4.00 6.00 8.00 10.00 12.00 14.00 16.00 18.00 20.00 22.00

C6 C7 C8 C9 C10 C15 C20 C25

C8 C9 C10 C11 C12 C13 C14 C15 C16 C17 C18 C19 C20 C21 C22 C23 C24 C25

Abundance

5000000

Time-->

6.00 8.00 10.00 12.00 14.00 16.00 18.00

SUMMED PROFILES

Isoparaffin #13

ALKANES

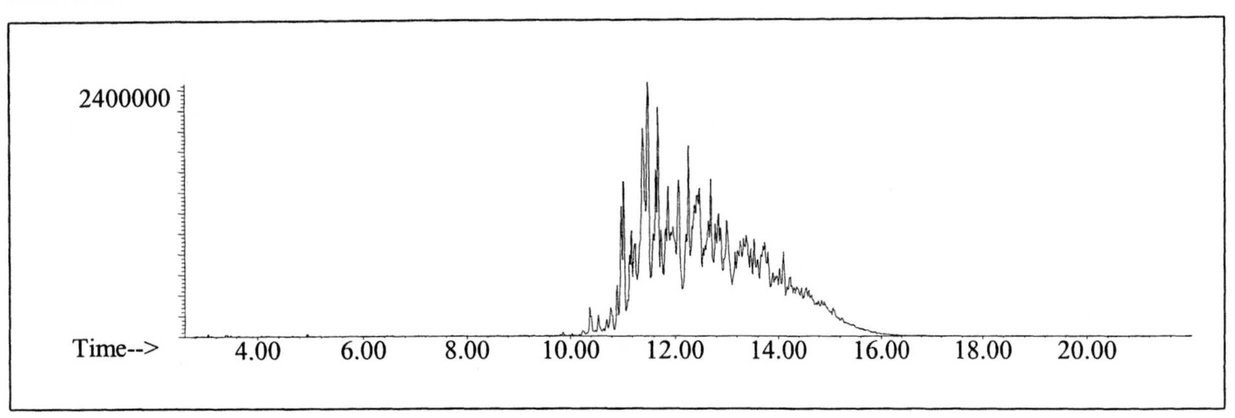

AROMATICS

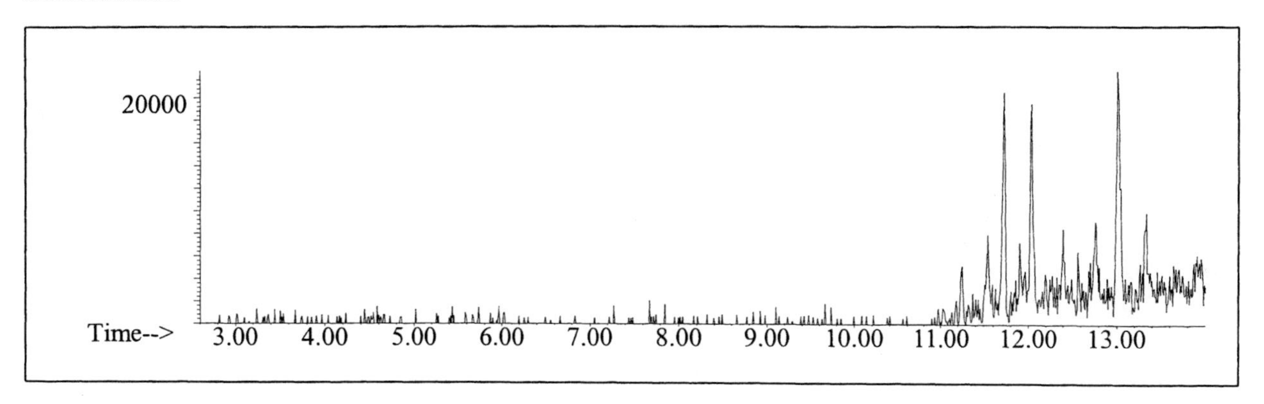

CYCLOPARAFFINS AND ALKENES

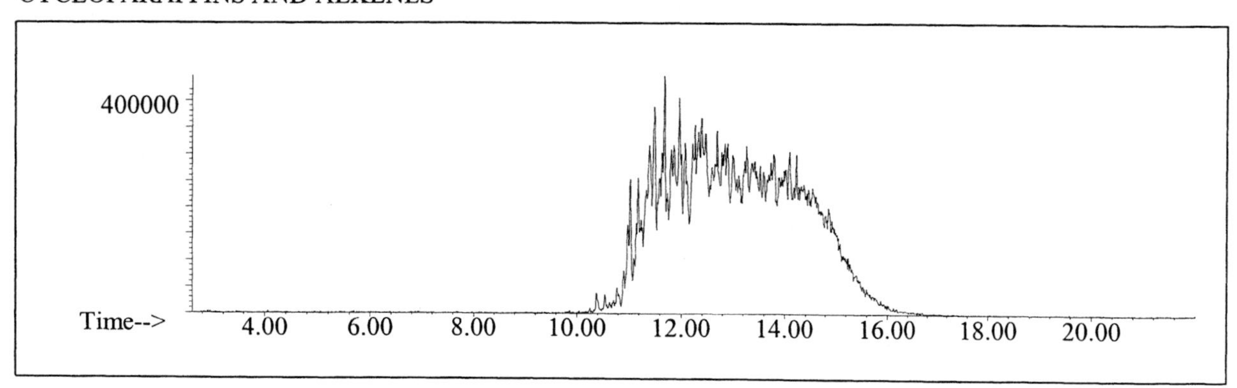

NAPHTHALENES

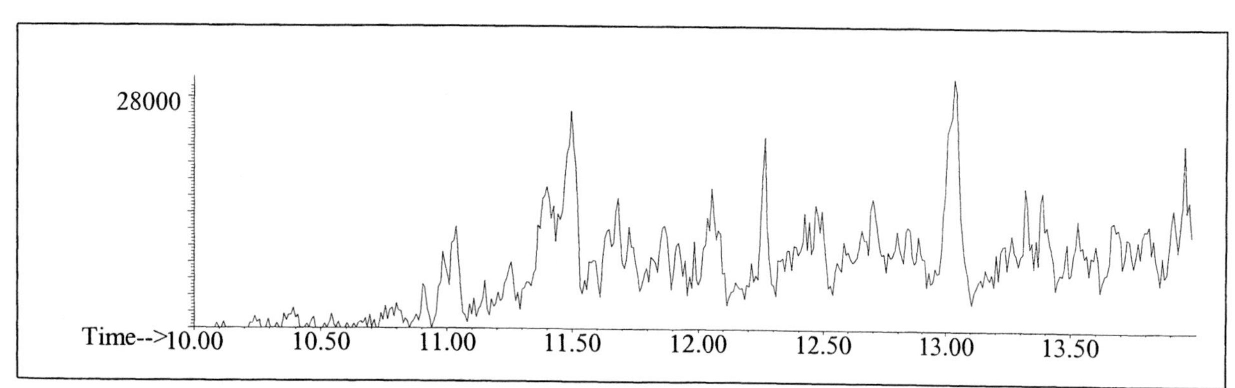

INDIVIDUAL PROFILES

Isoparaffin #13

Alkane

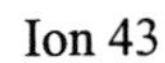

Ion 43

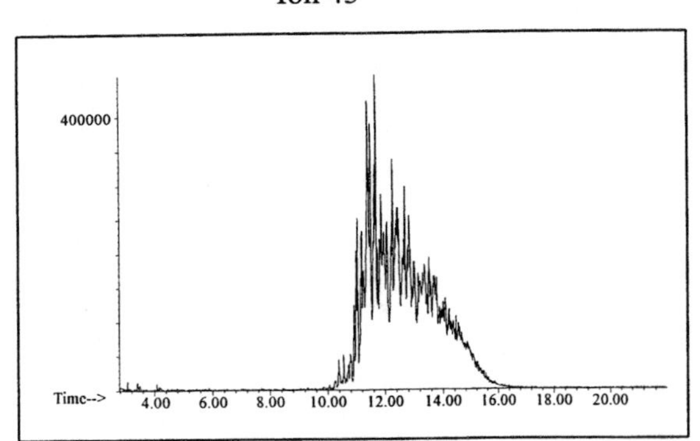

Ion 57

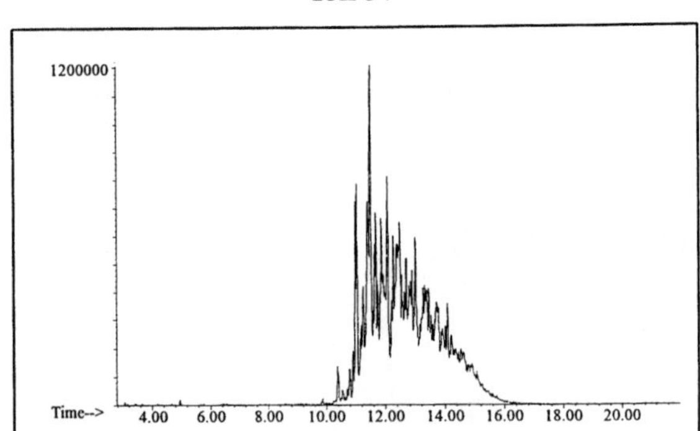

Ion 71

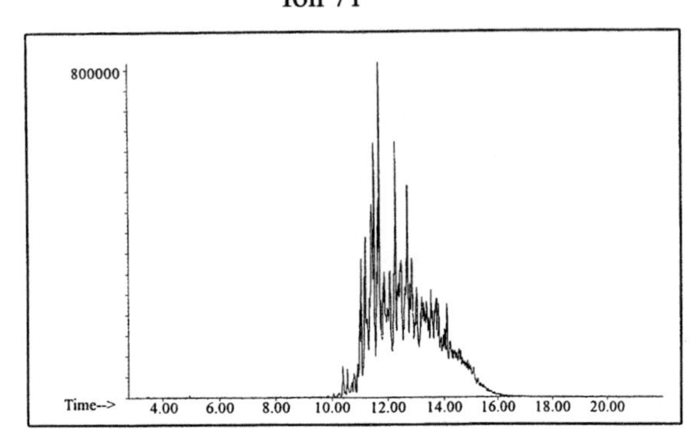

Ion 85

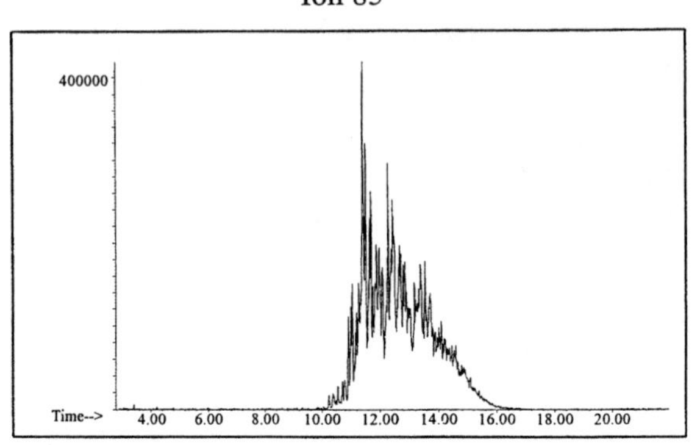

Aromatic

Ion 91

No Useful Data Obtained

Ion 105

No Useful Data Obtained

Ion 119

No Useful Data Obtained

INDIVIDUAL PROFILES

Isoparaffin #13

Cycloparaffin

Ion 55

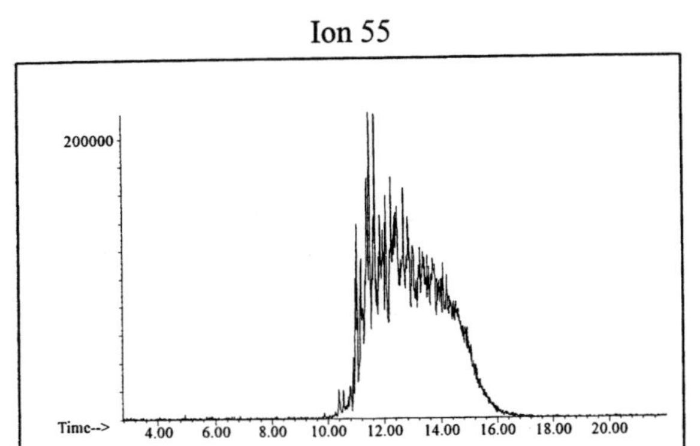

Ion 69

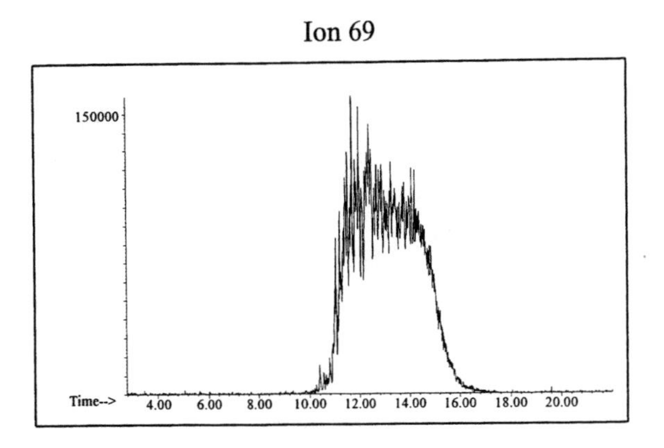

Ion 83

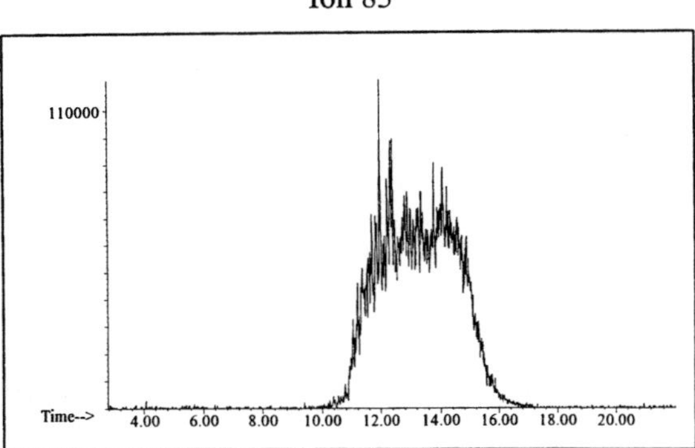

Naphthalene

Ion 128

No Useful Data Obtained

Ion 142

No Useful Data Obtained

Ion 156

No Useful Data Obtained

Normal Alkane Products

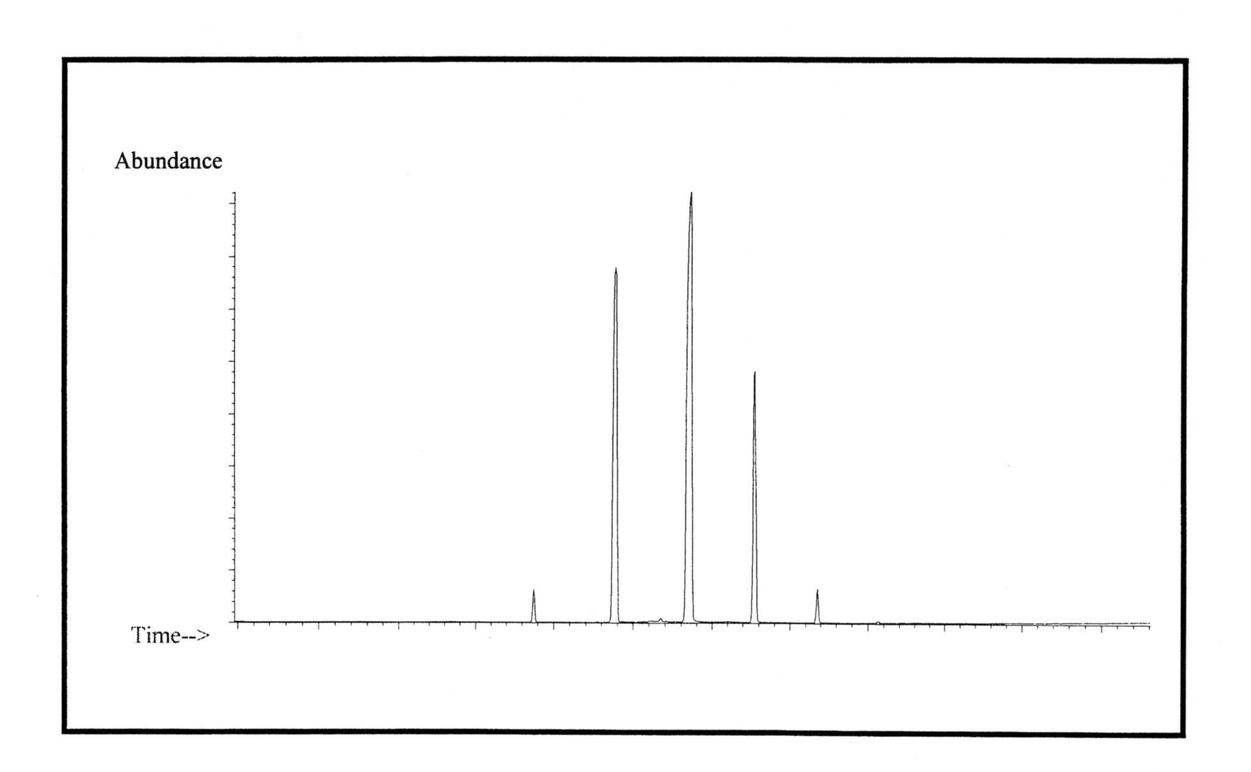

Alkane #1

20uL/mL Pentane

Product Displayed: Norpar 12

Other Similar Products: Northern Lights Lamp Oil, Candleman Premium Lamp Fuel

ASTM: Class 0.3 (Alkane)

Product Uses: Lamp oil

Macro Code: M

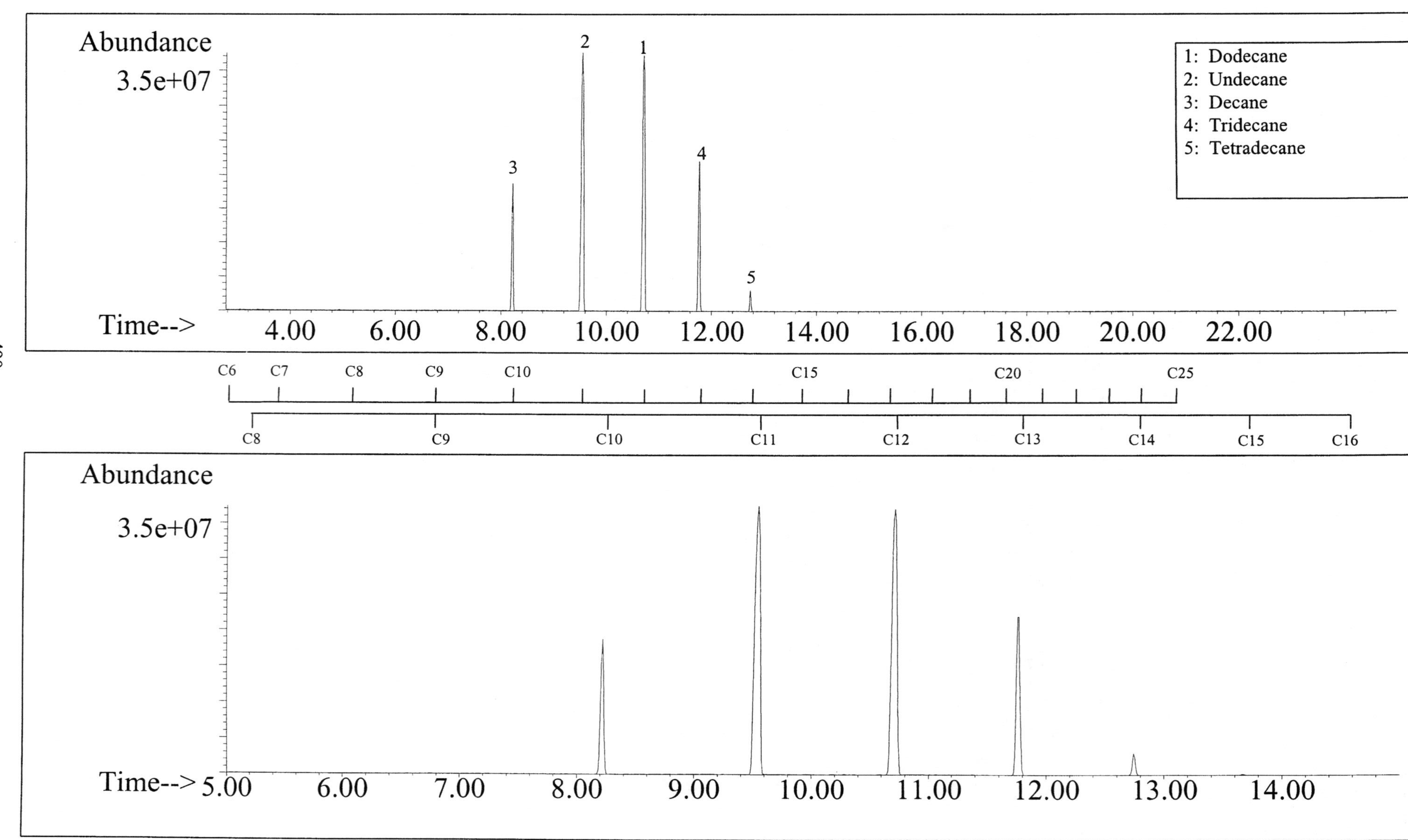

SUMMED PROFILES

Alkane #1

ALKANES

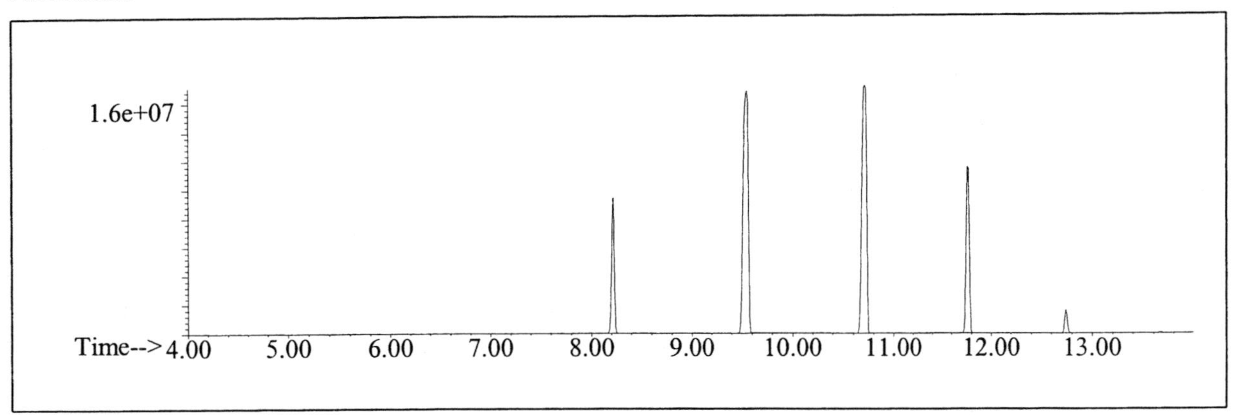

AROMATICS

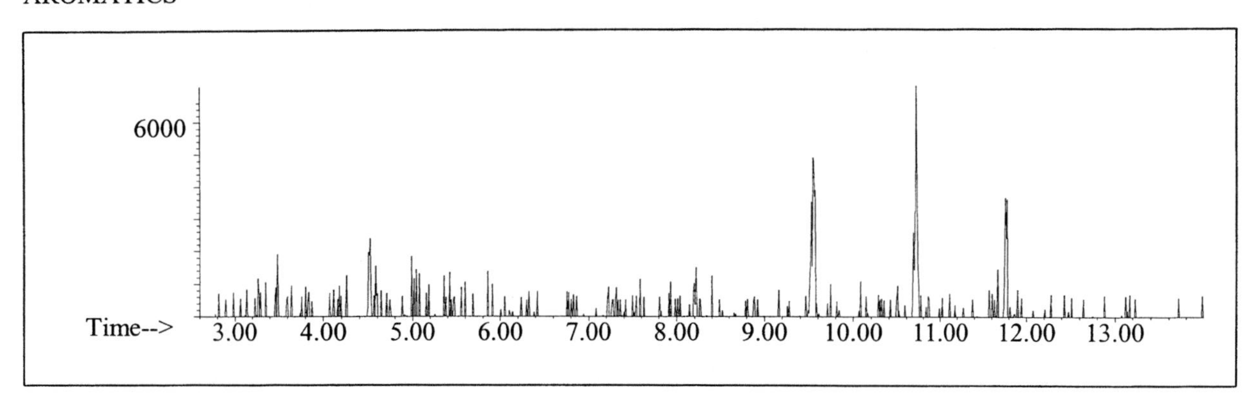

CYCLOPARAFFINS AND ALKENES

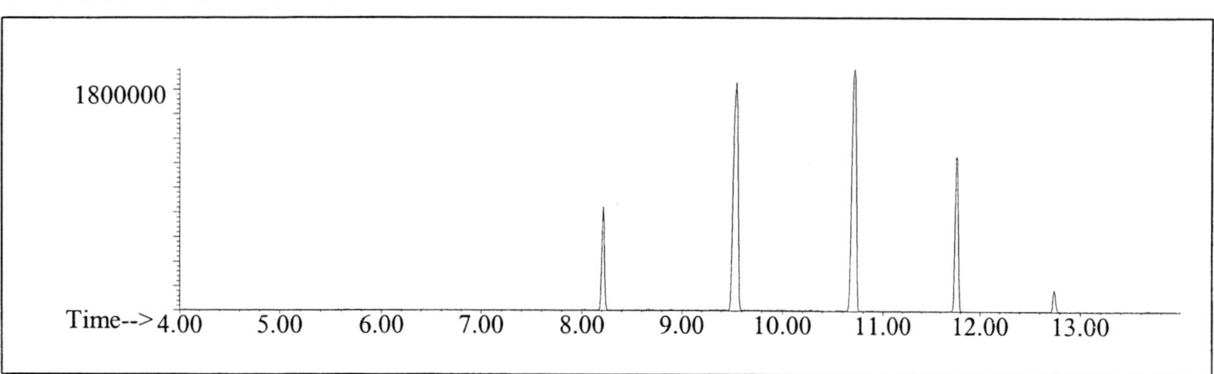

NAPHTHALENES

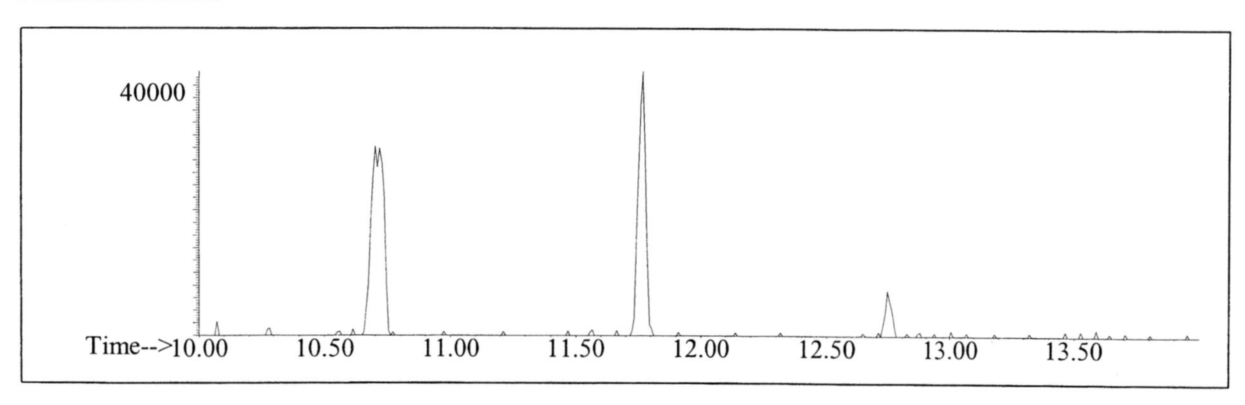

INDIVIDUAL PROFILES

Alkane #1

Alkane

Ion 43

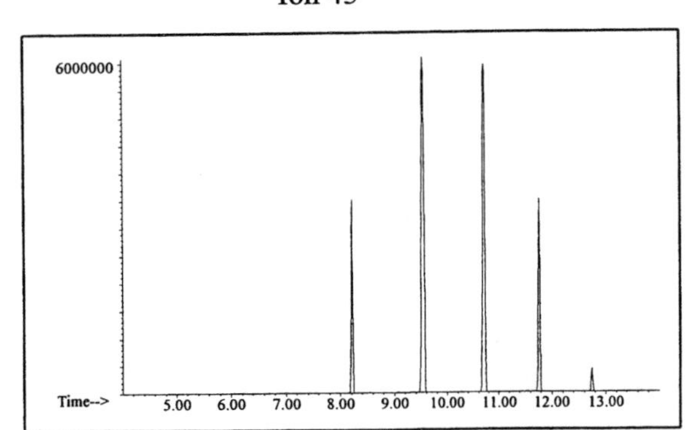

Ion 57

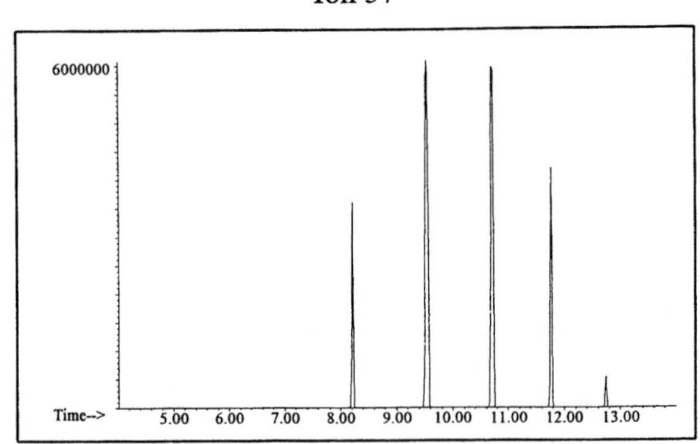

Ion 71

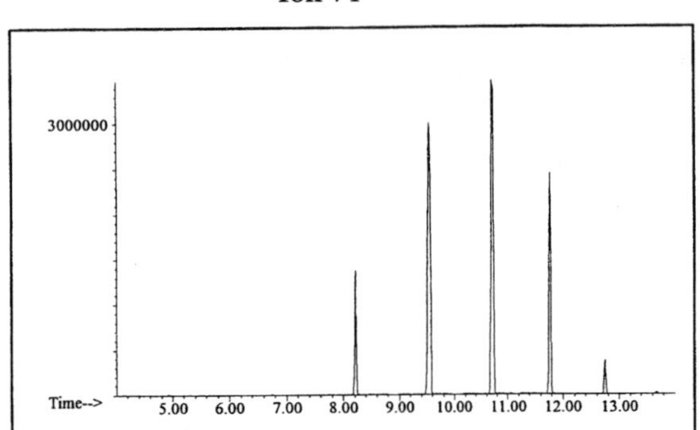

Ion 85

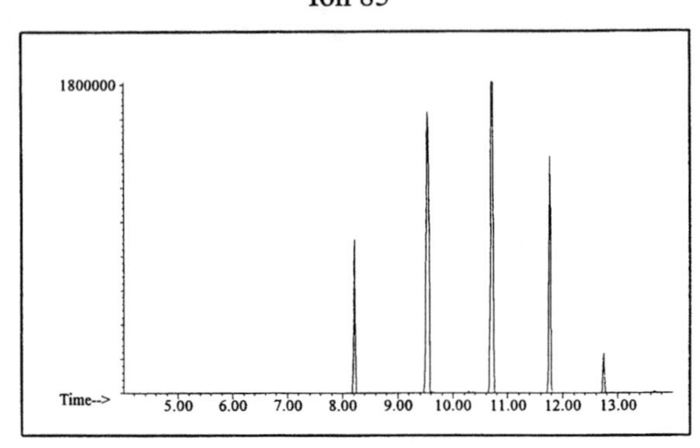

Aromatic

Ion 91

No Useful Data Obtained

Ion 105

No Useful Data Obtained

Ion 119

No Useful Data Obtained

INDIVIDUAL PROFILES

Alkane #1

Cycloparaffin

Ion 55

Ion 69

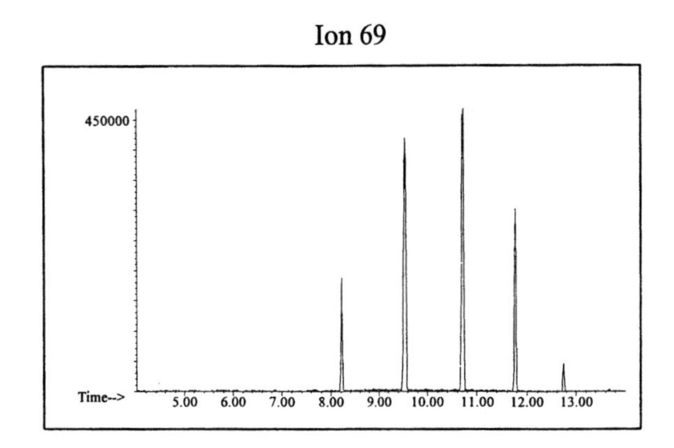

Ion 83

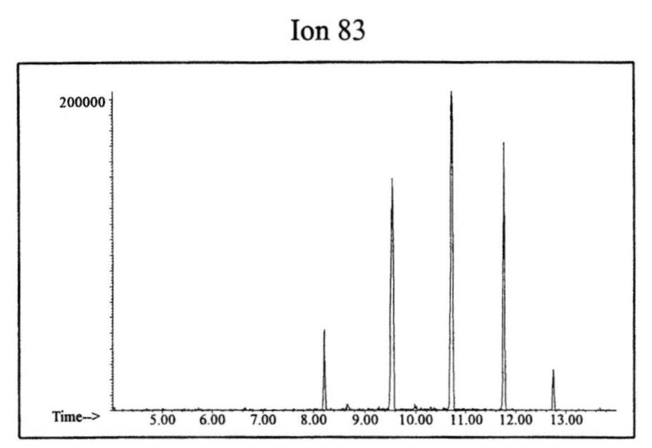

Naphthalene

Ion 128

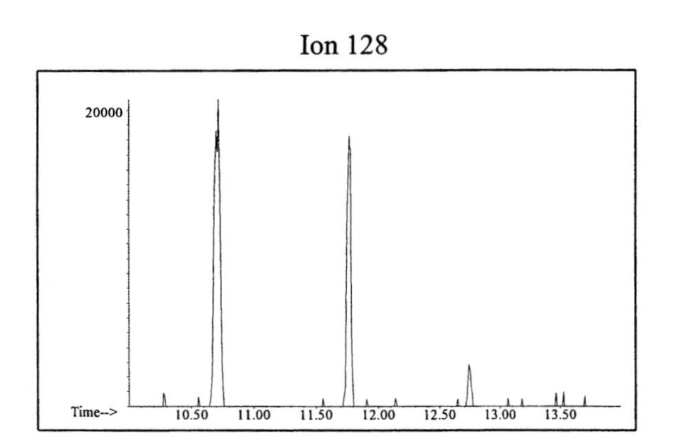

Ion 142

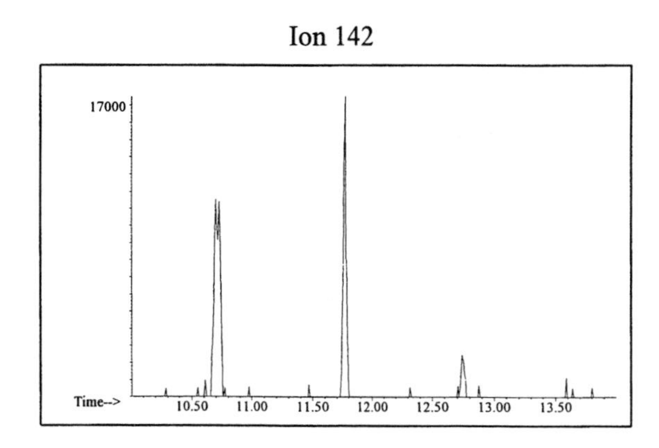

Ion 156

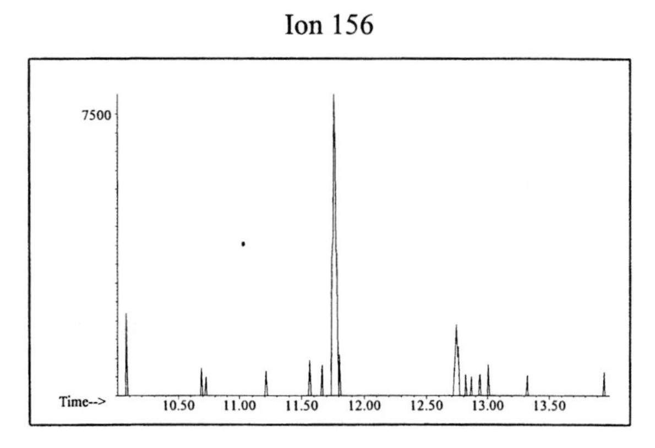

Alkane #2

20uL/mL Pentane

Product Displayed: Norpar 13

Other Similar Products:

ASTM: Class 0.3 (Alkane)

Product Uses: Solvent, feedstock, lamp oil

Macro Code: M

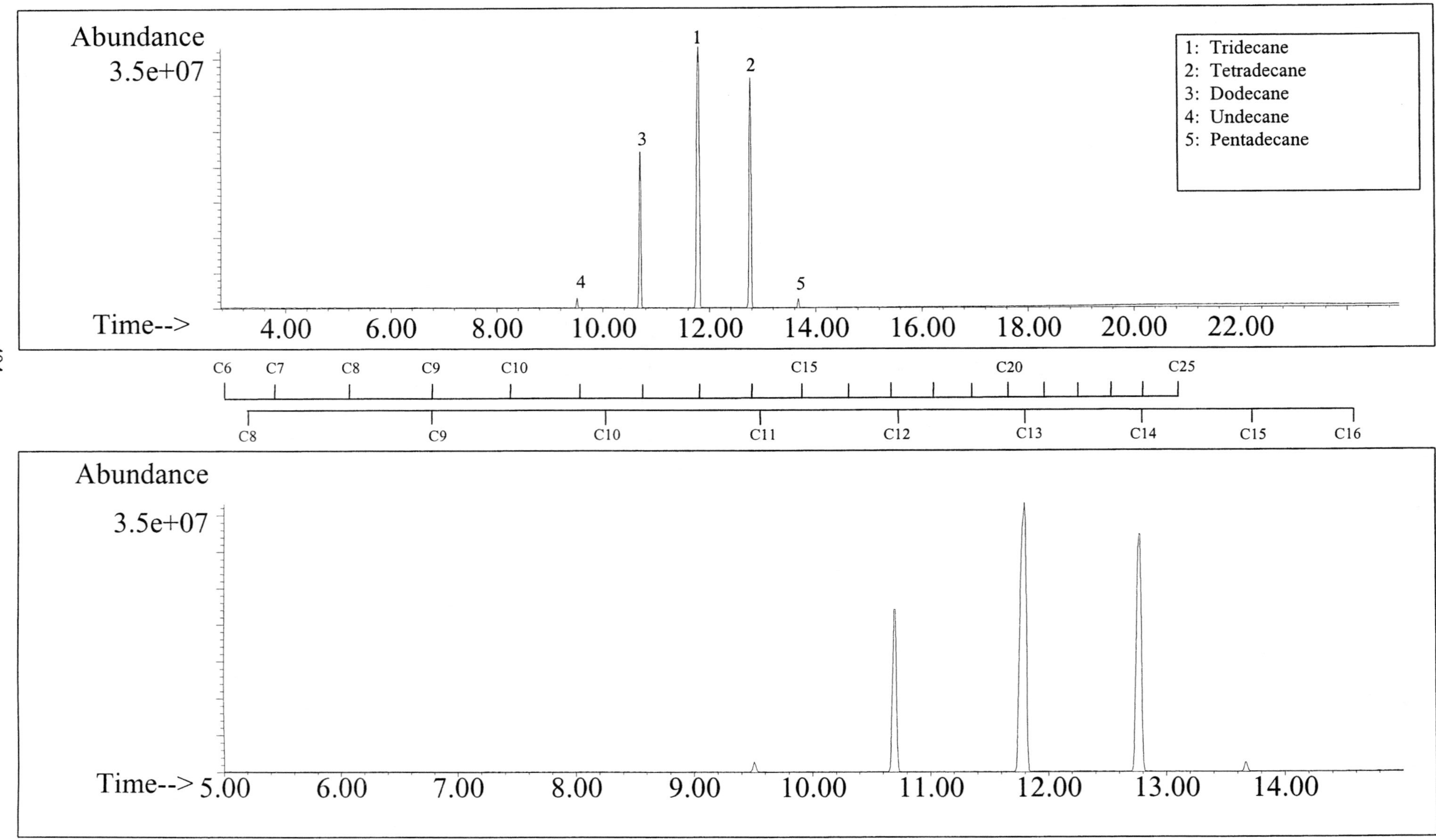

SUMMED PROFILES

Alkane #2

ALKANES

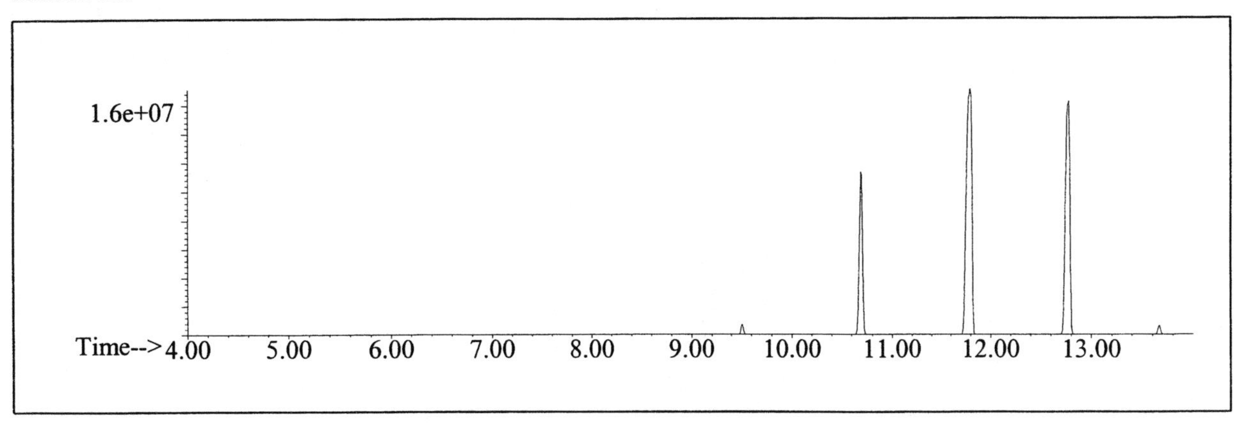

AROMATICS

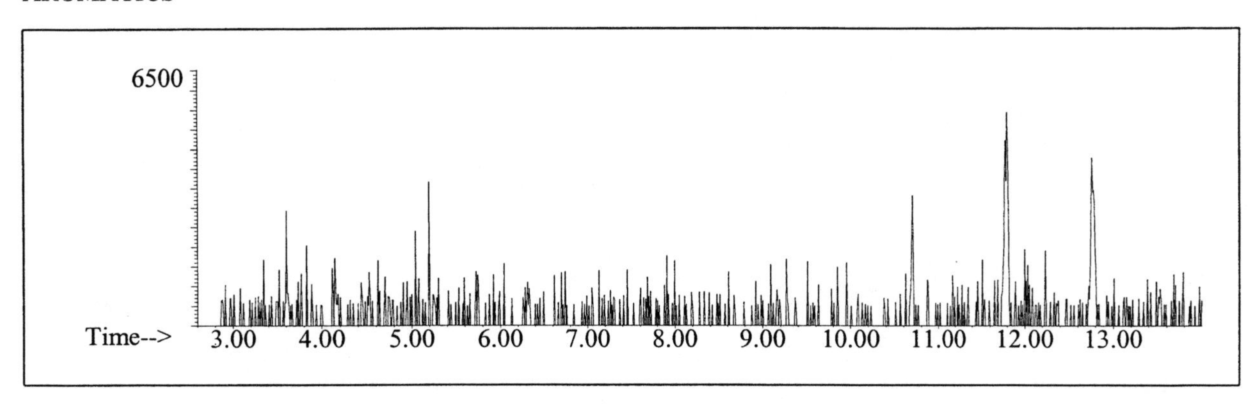

CYCLOPARAFFINS AND ALKENES

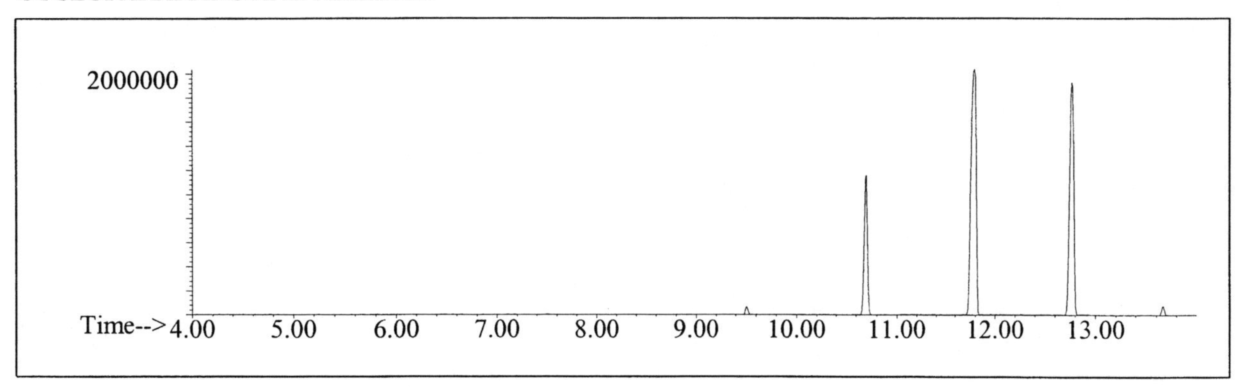

NAPHTHALENES

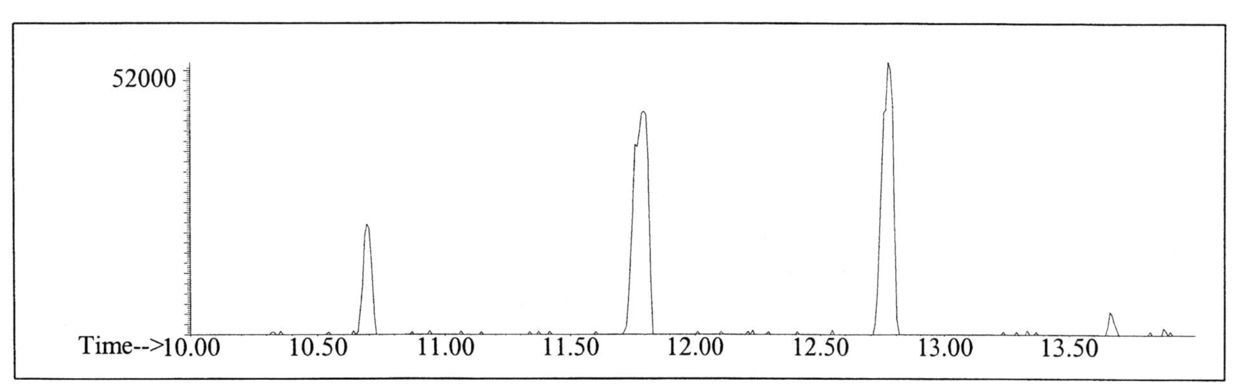

INDIVIDUAL PROFILES

Alkane #2

Alkane

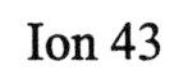

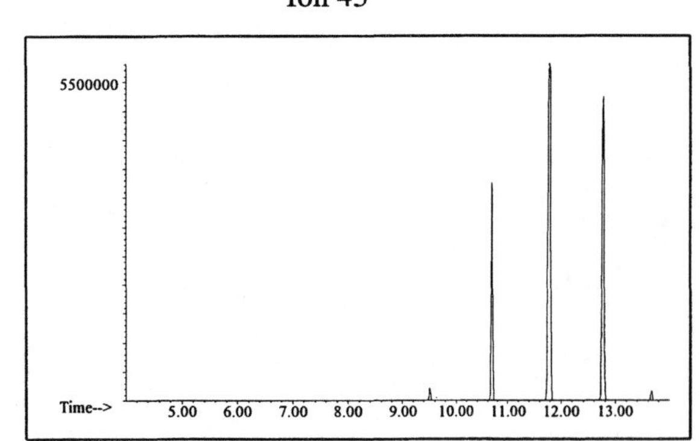

Ion 57

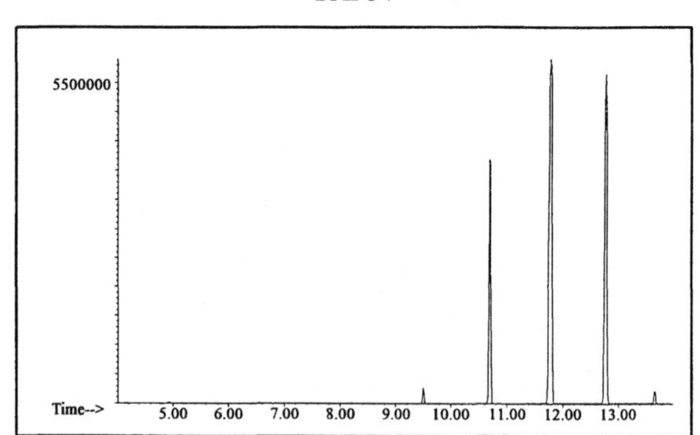

Ion 71

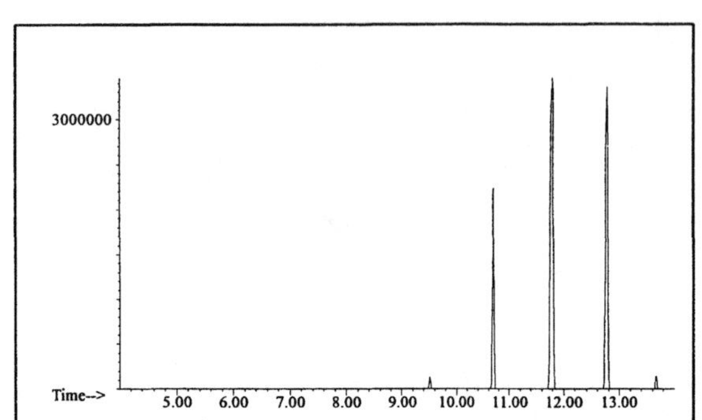

Ion 85

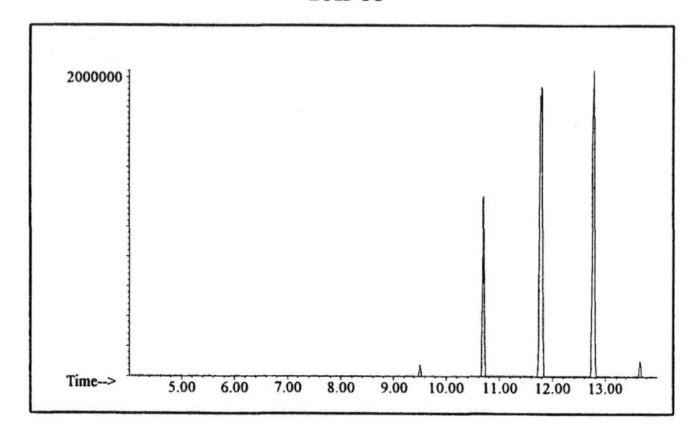

Aromatic

Ion 91

No Useful Data Obtained

Ion 105

No Useful Data Obtained

Ion 119

No Useful Data Obtained

INDIVIDUAL PROFILES

Alkane #2

Cycloparaffin

Ion 55

Ion 69

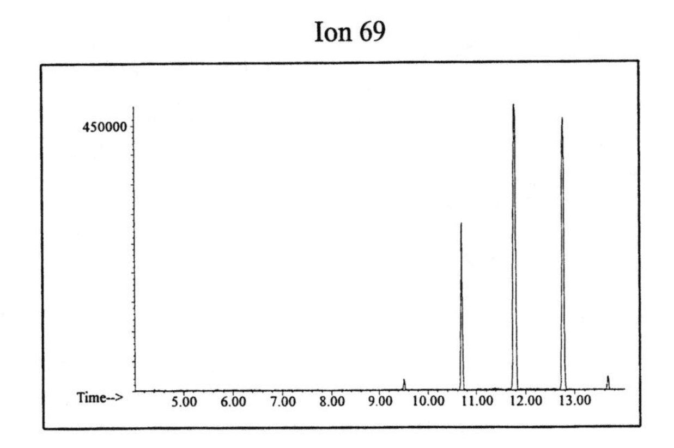

Ion 83

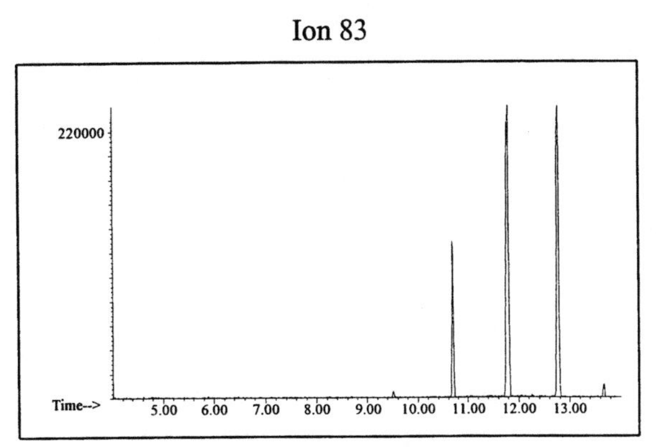

Naphthalene

Ion 128

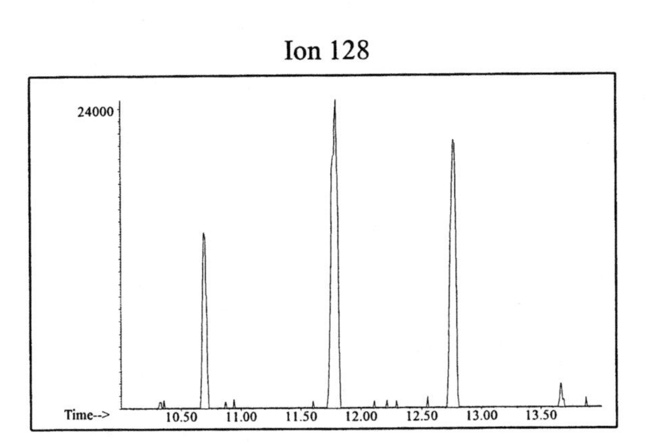

Ion 142

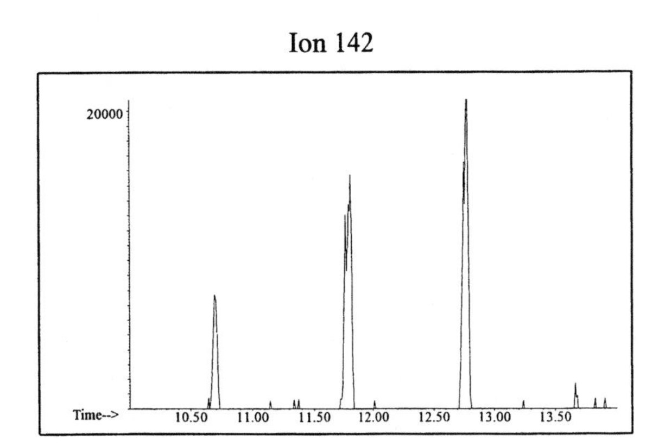

Ion 156

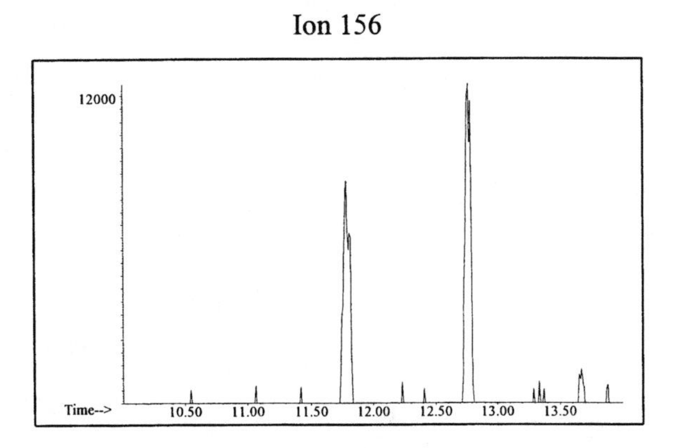

Alkane #3

20uL/mL Pentane

Product Displayed: Norpar 15

Other Similar Products:

ASTM: Class 0.3 (Alkane)

Product Uses: Solvent, feedstock, lamp oil

Macro Code: H

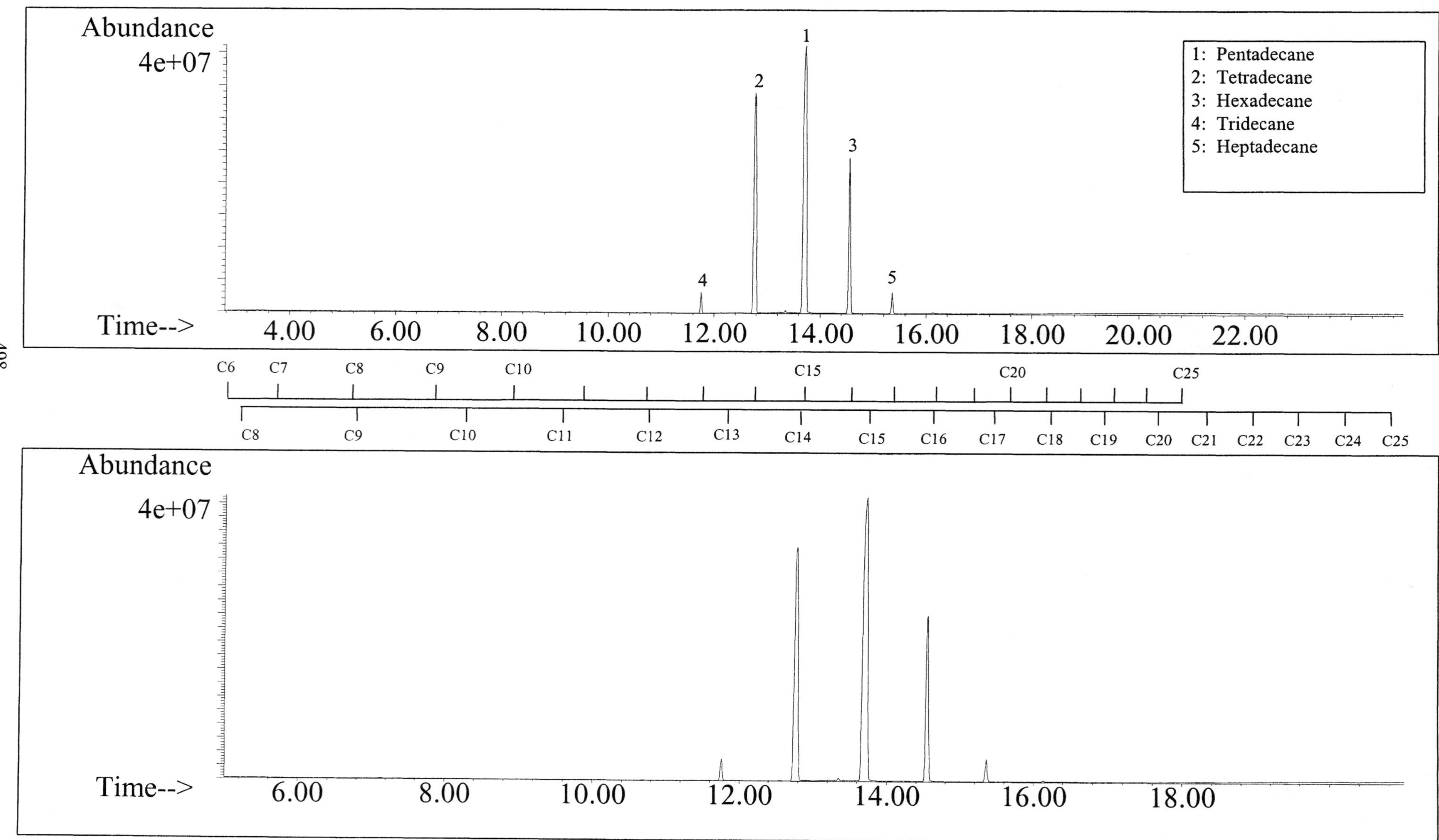

SUMMED PROFILES

Alkane #3

ALKANES

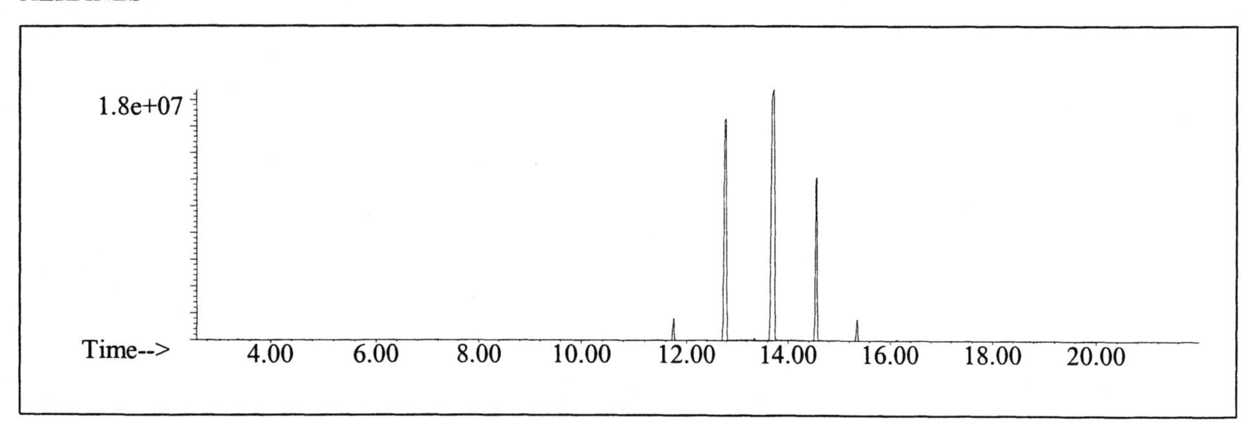

AROMATICS

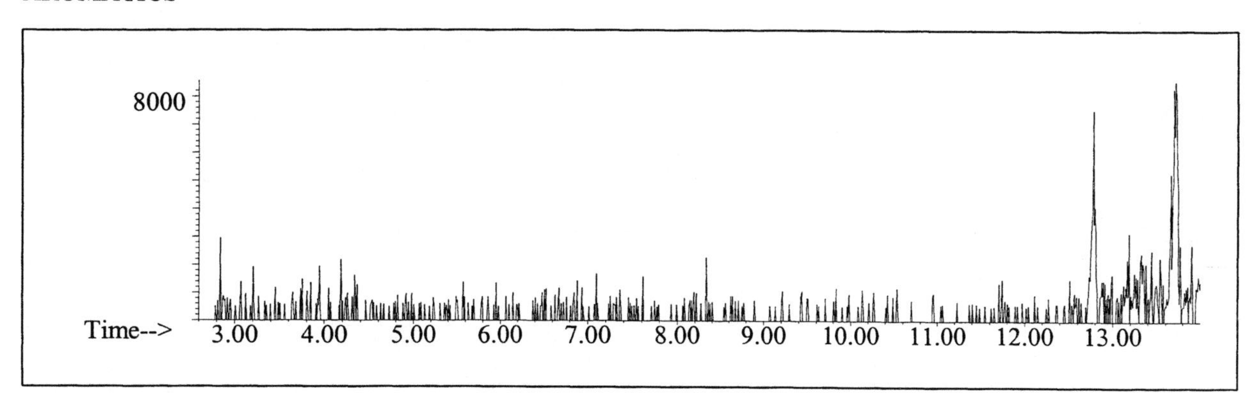

CYCLOPARAFFINS AND ALKENES

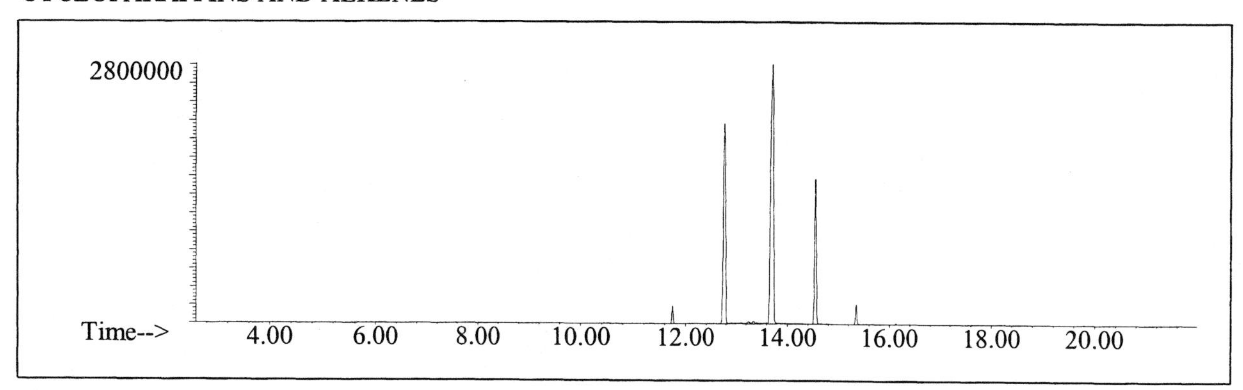

NAPHTHALENES

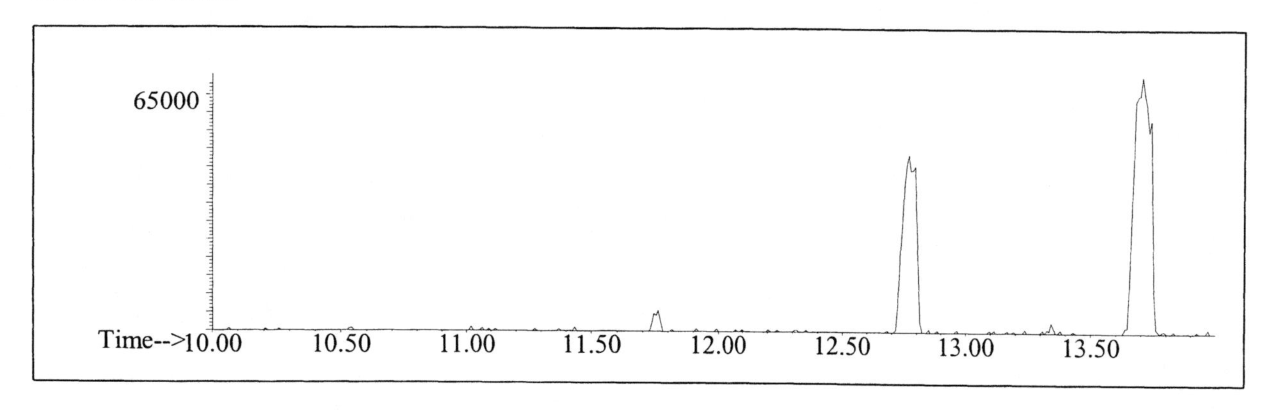

INDIVIDUAL PROFILES Alkane #3

Alkane

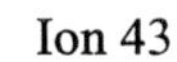

Ion 43

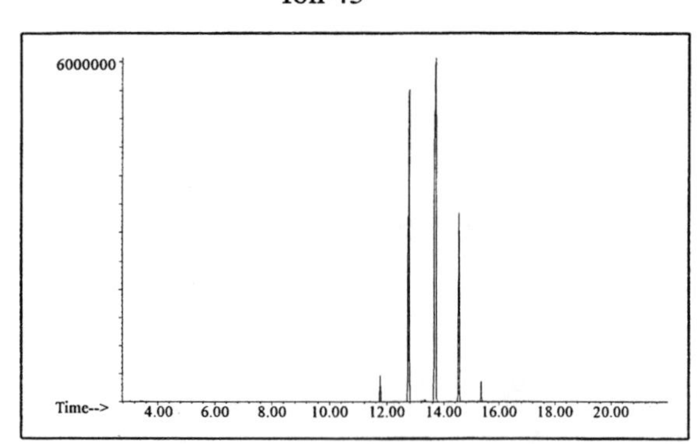

Ion 57

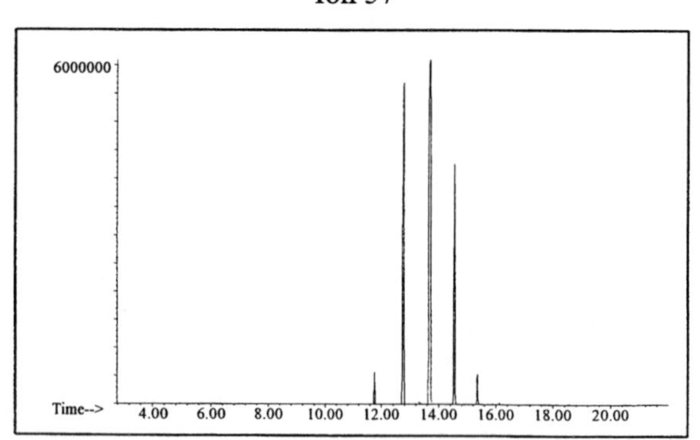

Ion 71

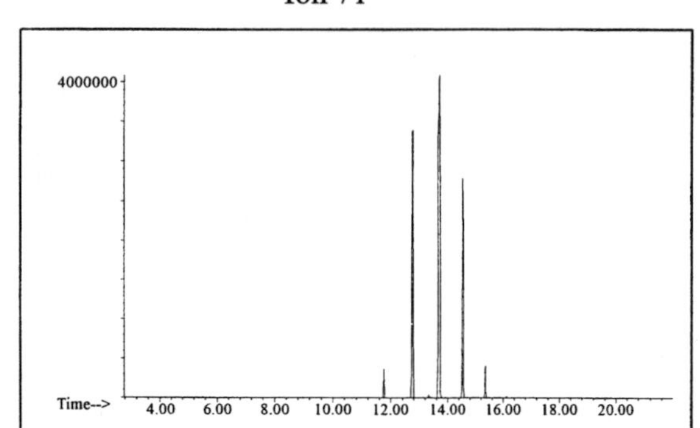

Ion 85

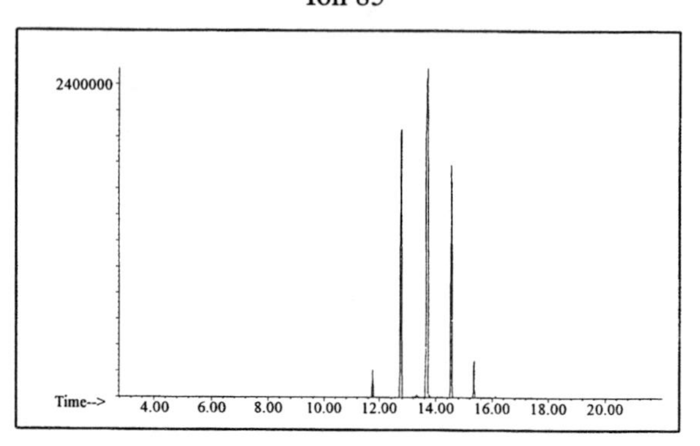

Aromatic

Ion 91

No Useful Data Obtained

Ion 105

No Useful Data Obtained

Ion 119

No Useful Data Obtained

INDIVIDUAL PROFILES

Cycloparaffin

Ion 55

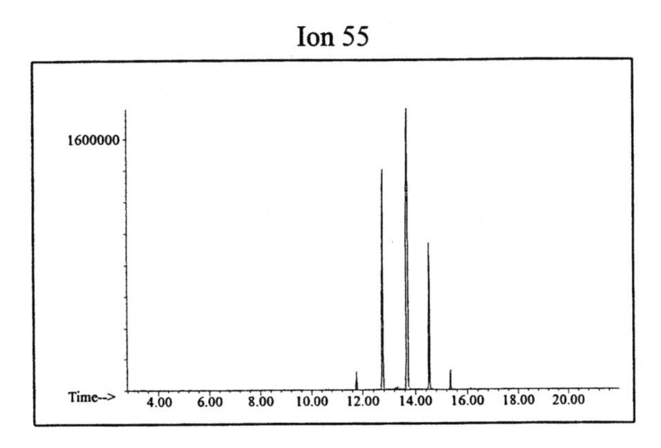

Ion 69

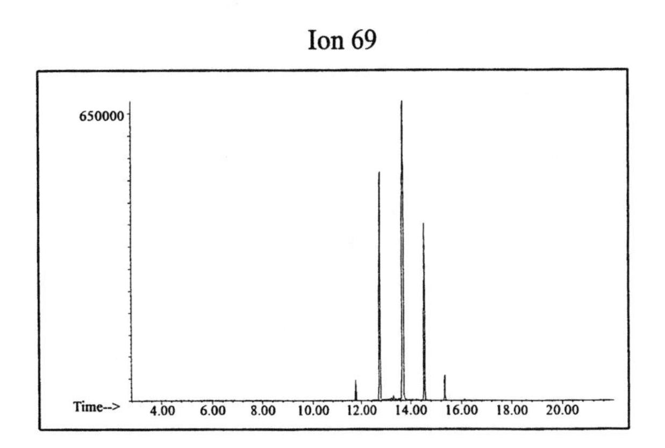

Ion 83

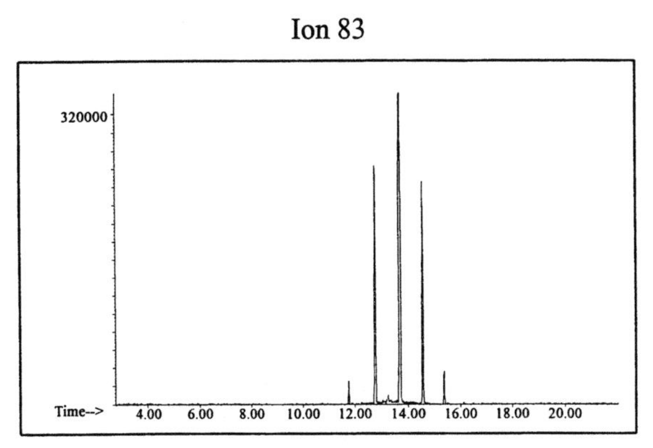

Naphthalene

Ion 128

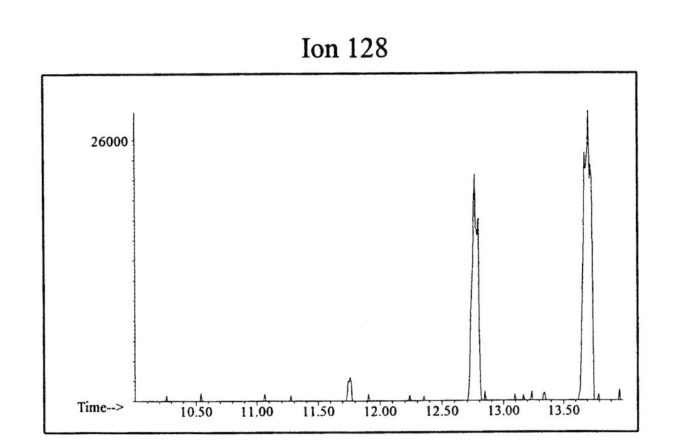

Ion 142

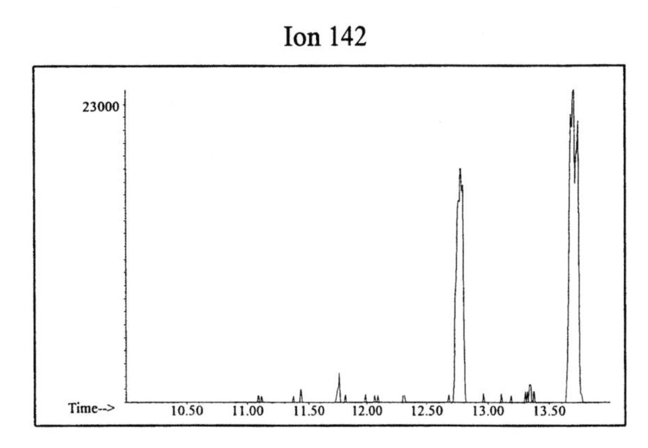

Ion 156

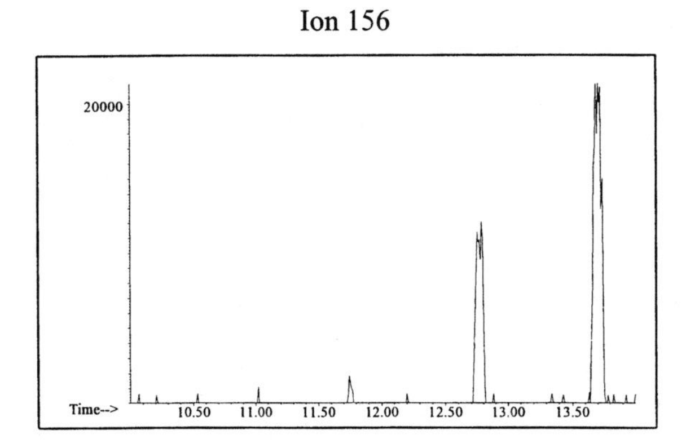

Aromatic Solvents

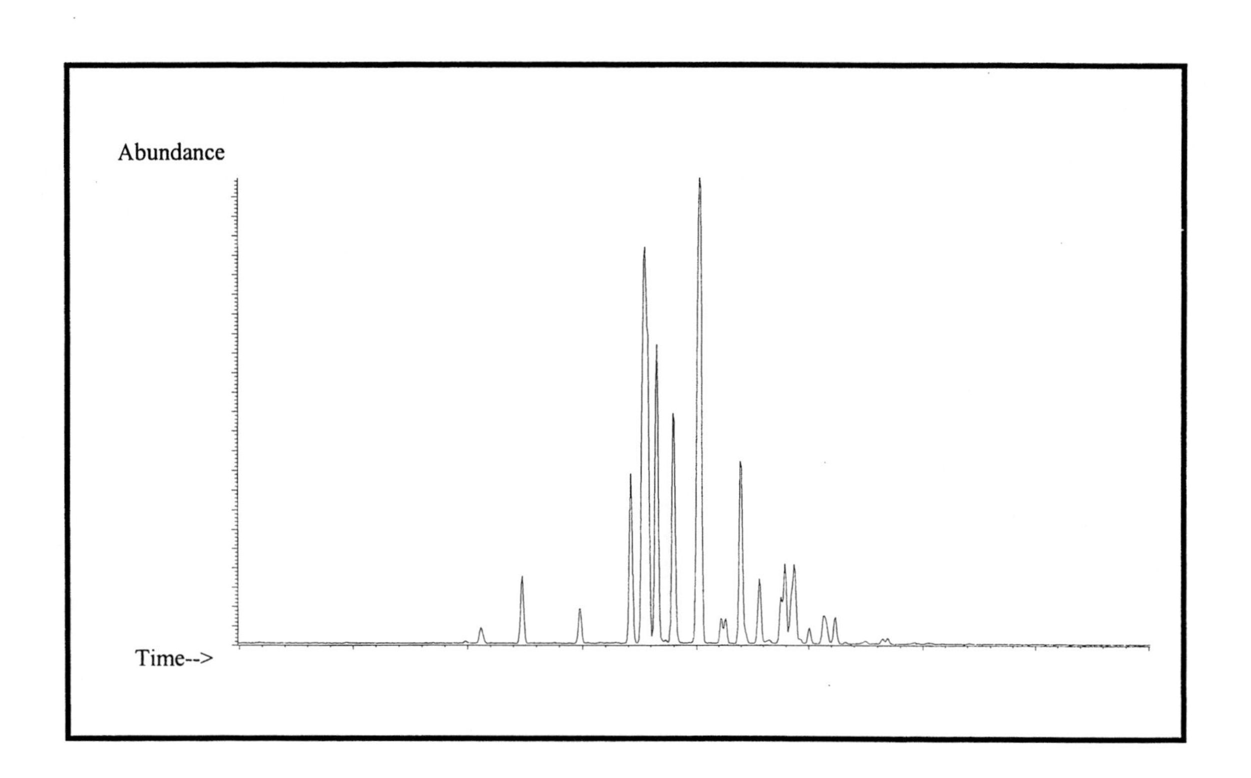

Aromatic #1

20uL/mL Pentane

Product Displayed: Glaze'n Seal Thinner

Other Similar Products:

ASTM: Class 0.4 (Aromatic)

Product Uses: Suface preparation

Macro Code: L

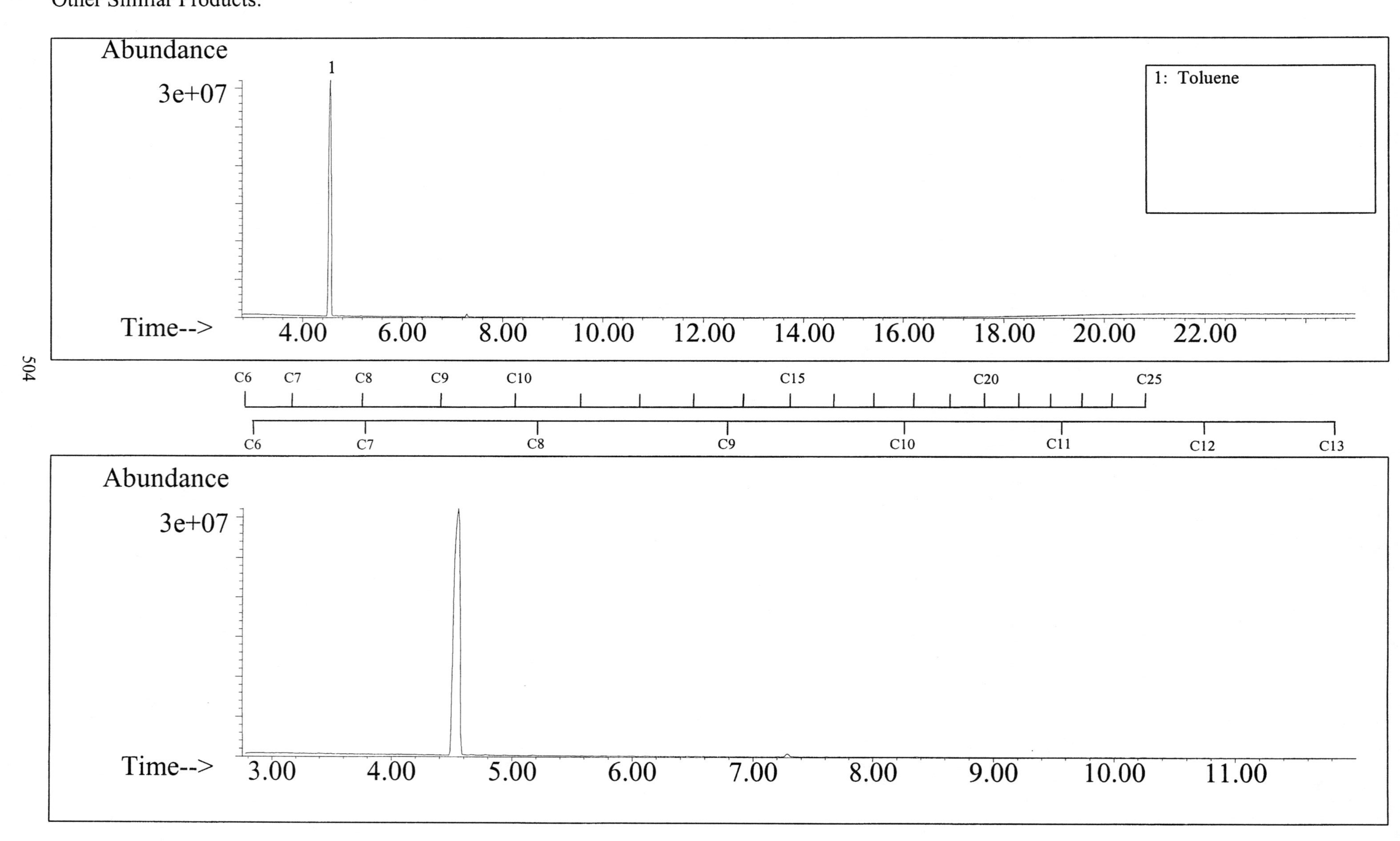

SUMMED PROFILES

Aromatic #1

ALKANES

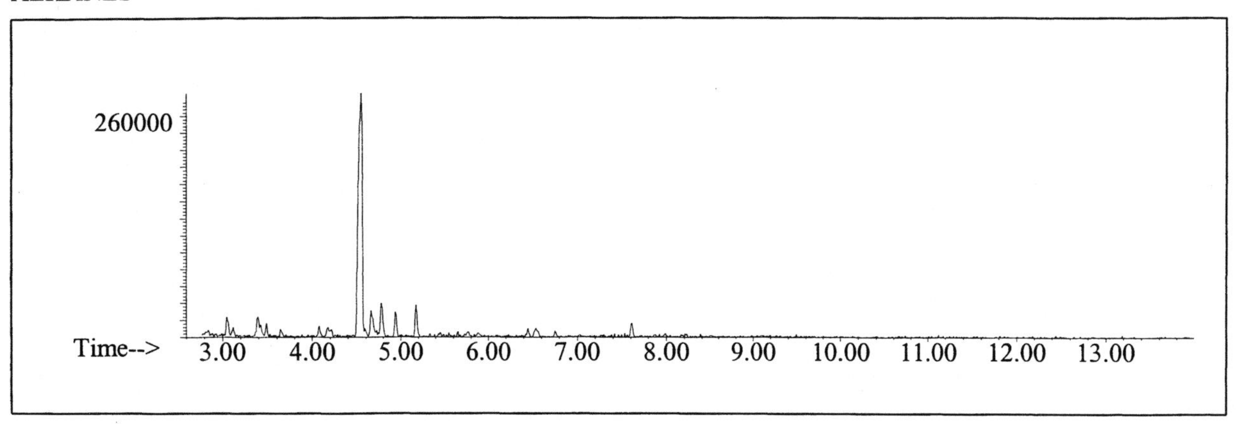

AROMATICS

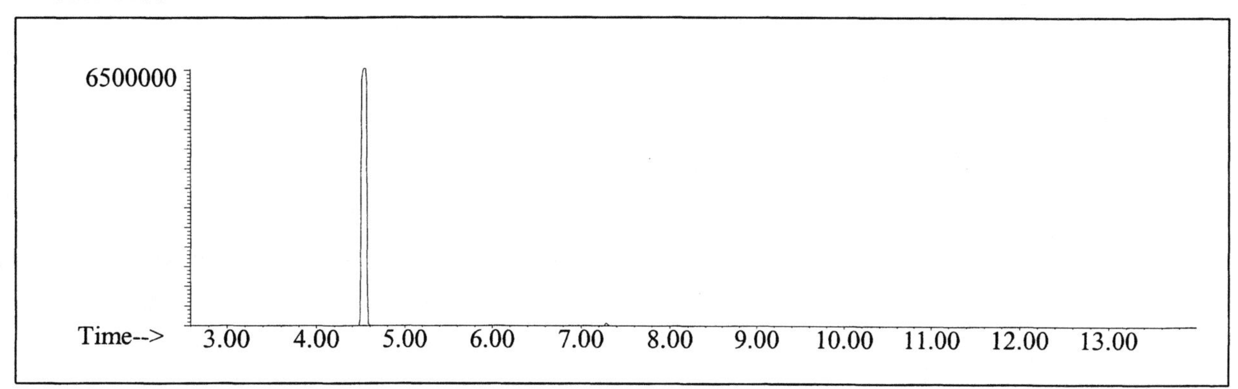

CYCLOPARAFFINS AND ALKENES

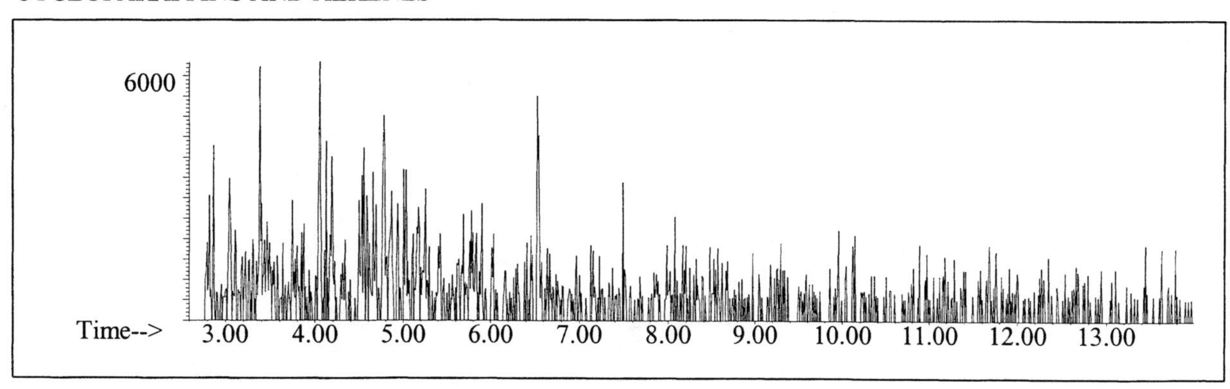

NAPHTHALENES

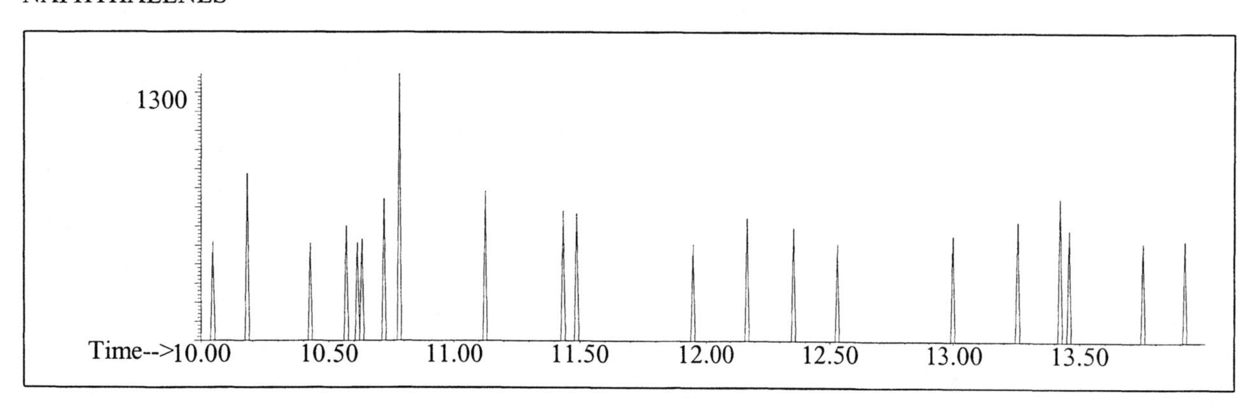

INDIVIDUAL PROFILES

Aromatic #1

Alkane

Ion 43

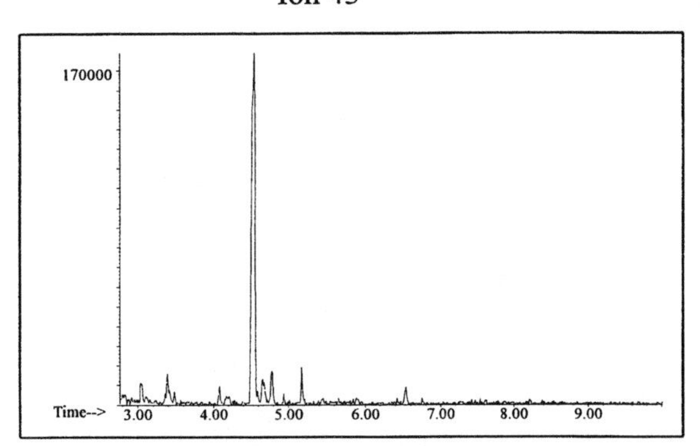

Ion 57

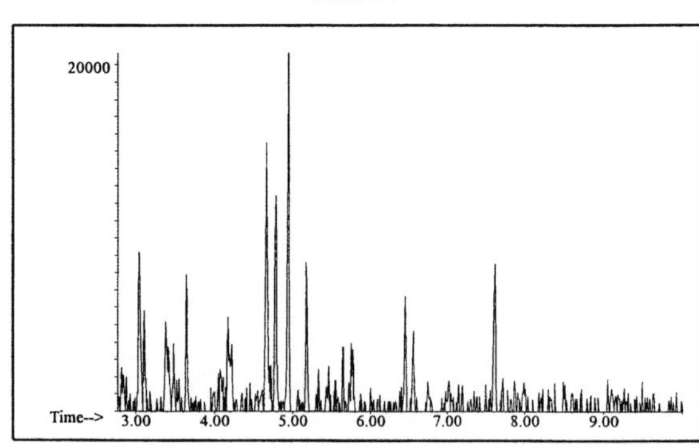

Ion 71

No Useful Data Obtained

Ion 85

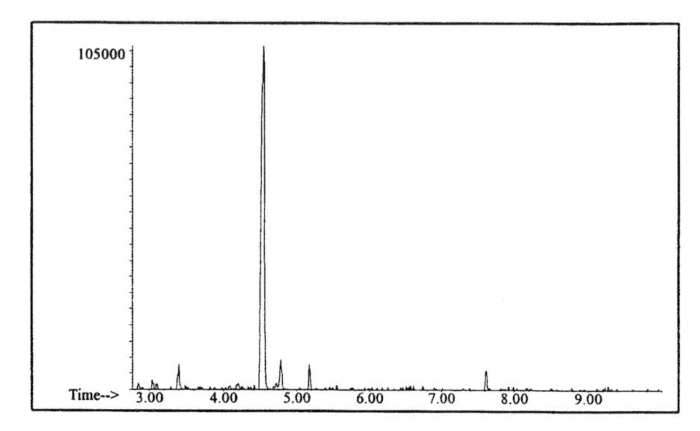

Aromatic

Ion 91

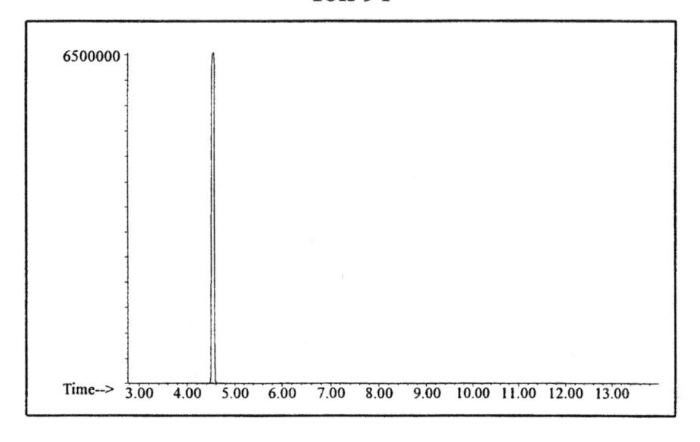

Ion 105

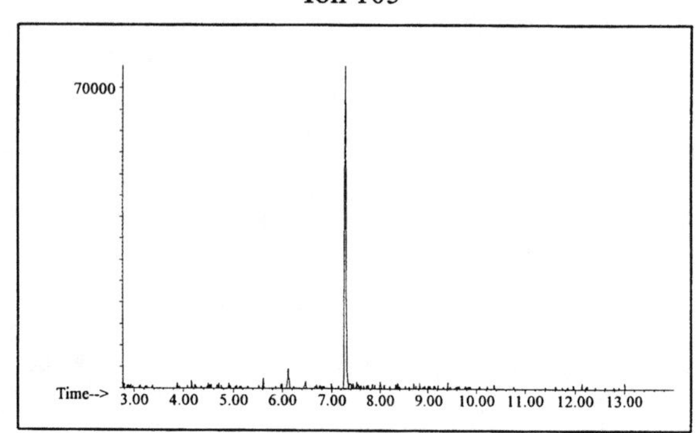

Ion 119

No Useful Data Obtained

Cycloparaffin

Ion 55

No Useful Data Obtained

Ion 69

No Useful Data Obtained

Ion 83

No Useful Data Obtained

Naphthalene

Ion 128

No Useful Data Obtained

Ion 142

No Useful Data Obtained

Ion 156

No Useful Data Obtained

Aromatic #2

20uL/mL Pentane

Product Displayed: Red Devil Paint/Varnish Remover

Other Similar Products:

ASTM: Class 0.4 (Aromatic)

Product Uses: Paint thinner, paint remover, solvent

Macro Code: L

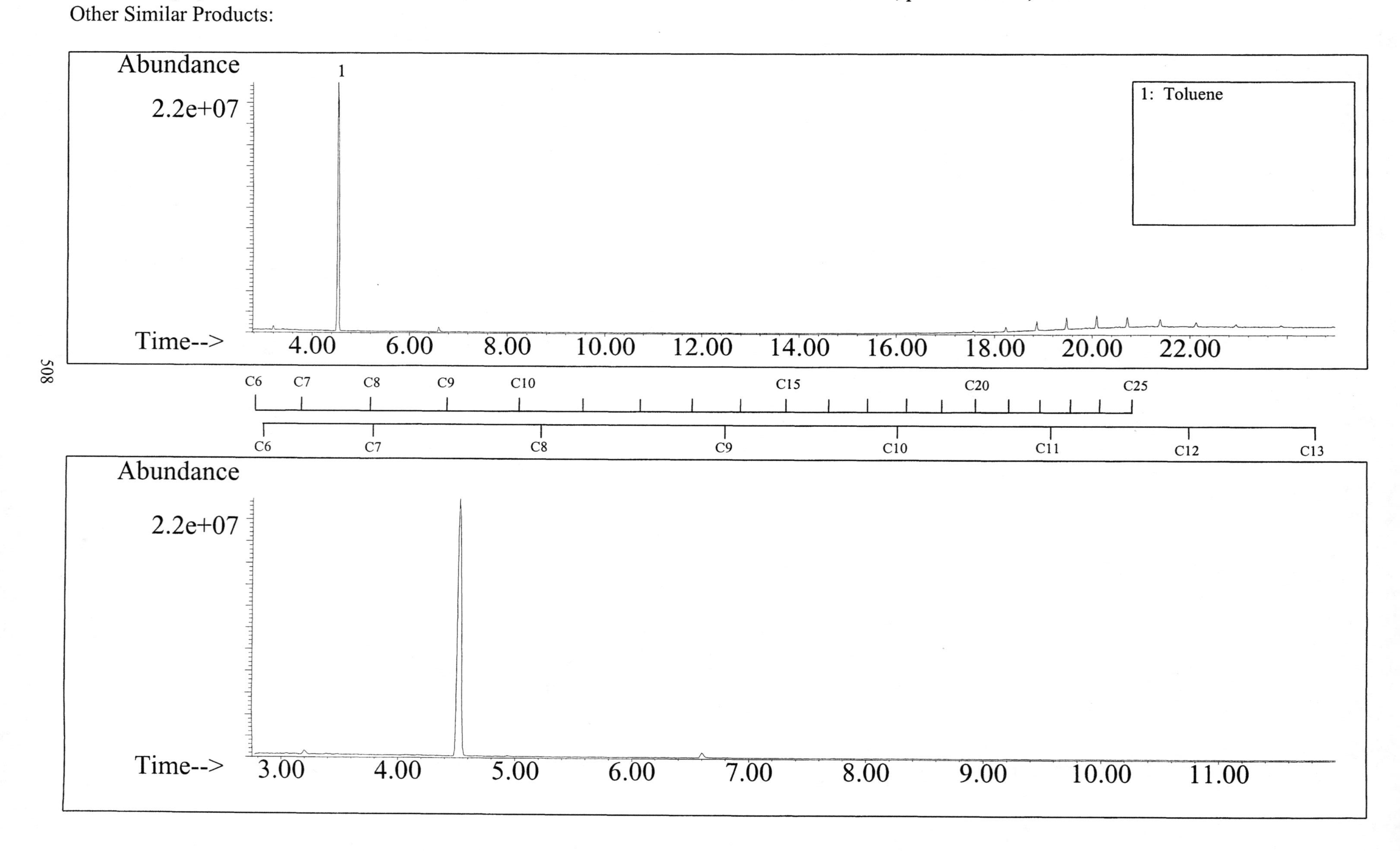

ALKANES

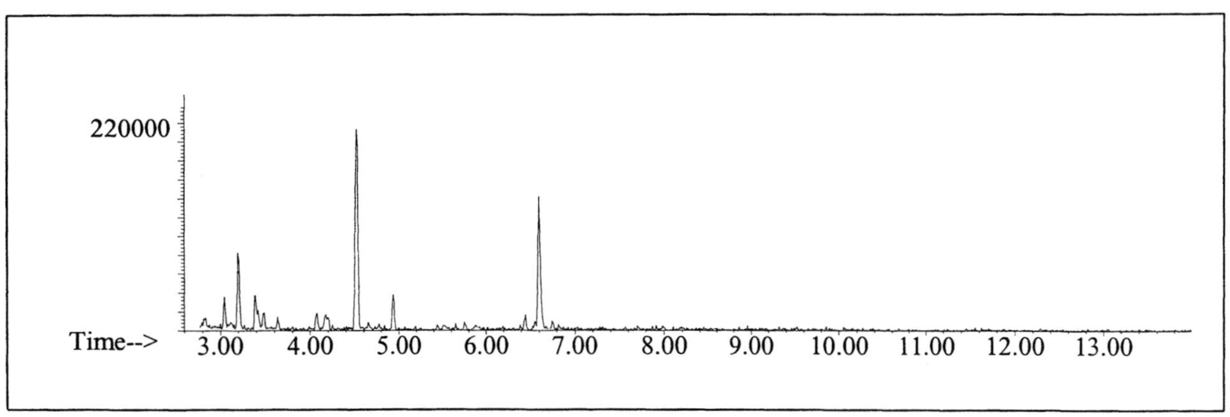

AROMATICS

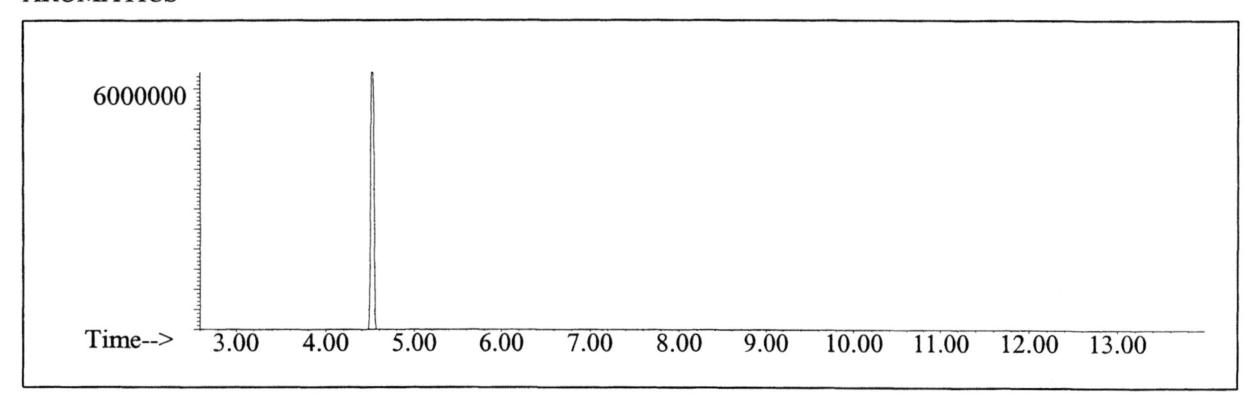

CYCLOPARAFFINS AND ALKENES

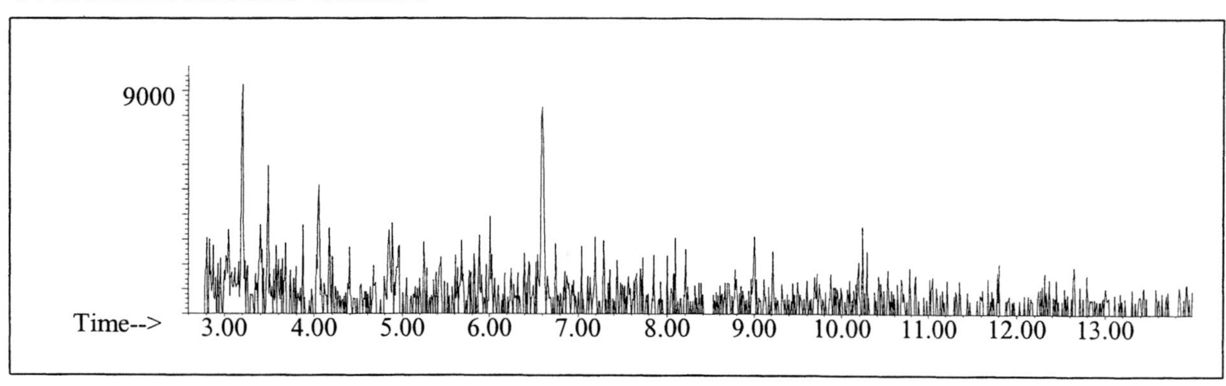

NAPHTHALENES

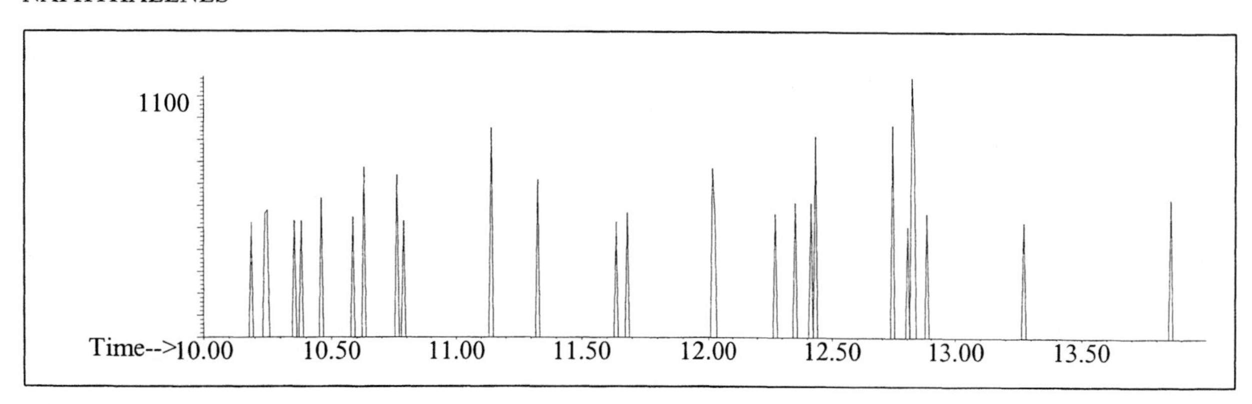

INDIVIDUAL PROFILES

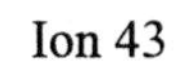

Aromatic #2

Alkane

Ion 43

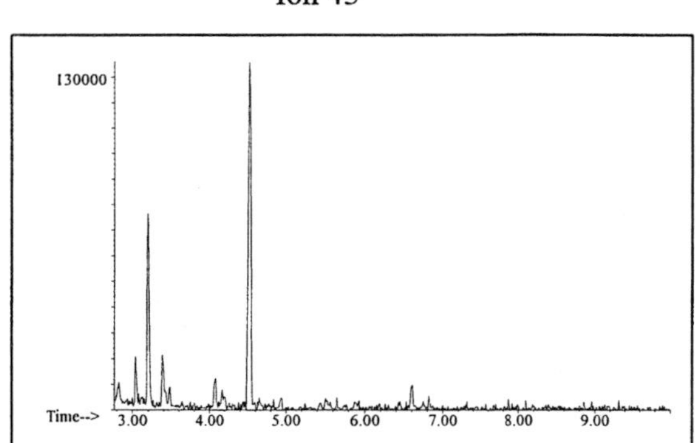

Ion 57

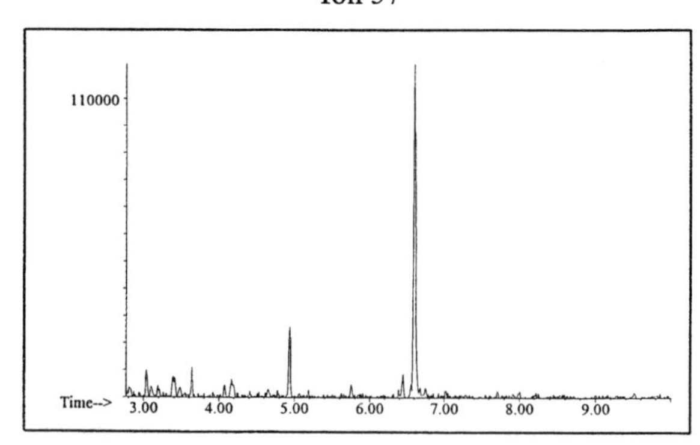

Ion 71

No Useful Data Obtained

Ion 85

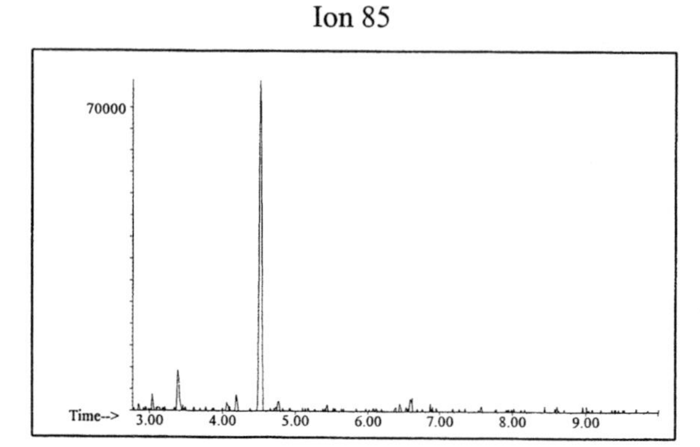

Aromatic

Ion 91

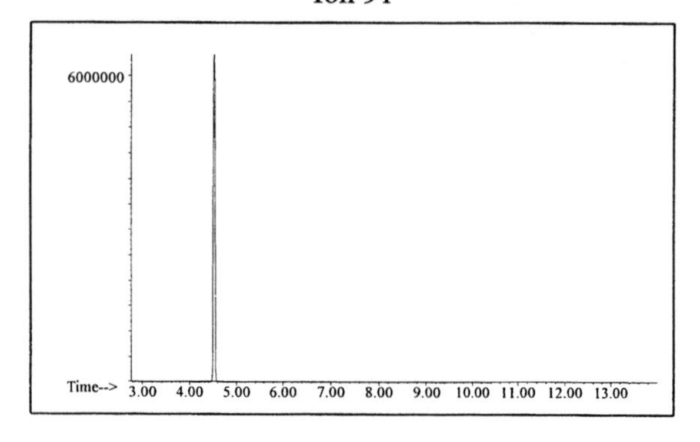

Ion 105

No Useful Data Obtained

Ion 119

No Useful Data Obtained

INDIVIDUAL PROFILES

Cycloparaffin

Ion 55

No Useful Data Obtained

Ion 69

No Useful Data Obtained

Ion 83

No Useful Data Obtained

Naphthalene

Ion 128

No Useful Data Obtained

Ion 142

No Useful Data Obtained

Ion 156

No Useful Data Obtained

Aromatic #3

20uL/mL Pentane
Product Displayed: Dio-Sol Solvent
Other Similar Products:

ASTM: Class 0.4 (Aromatic)
Product Uses: Solvent

Macro Code: L

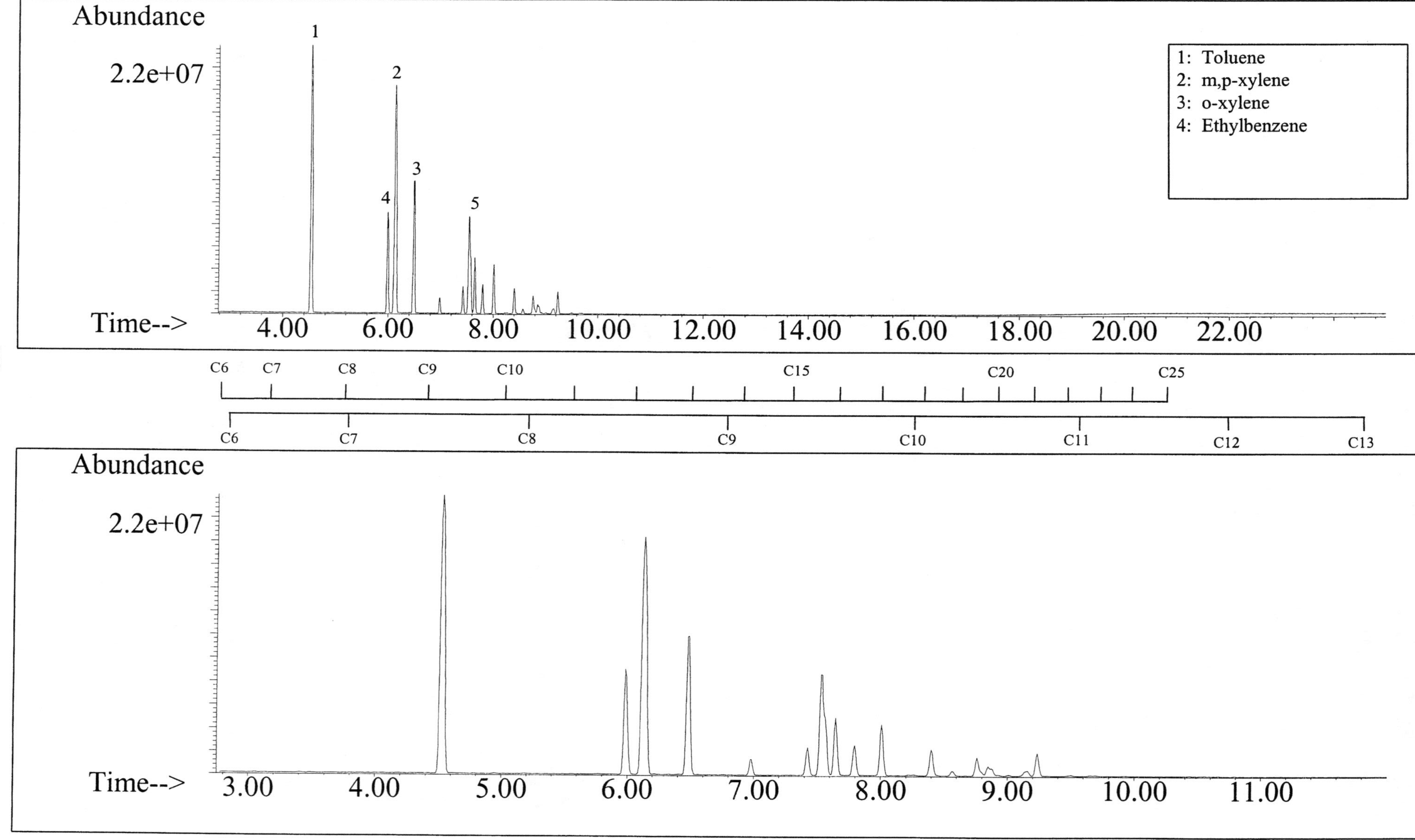

SUMMED PROFILES

Aromatic #3

ALKANES

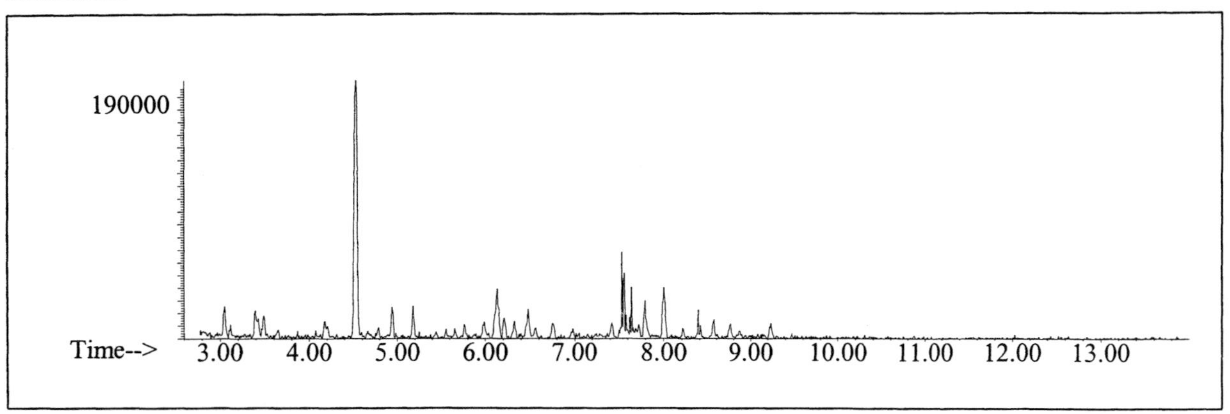

AROMATICS

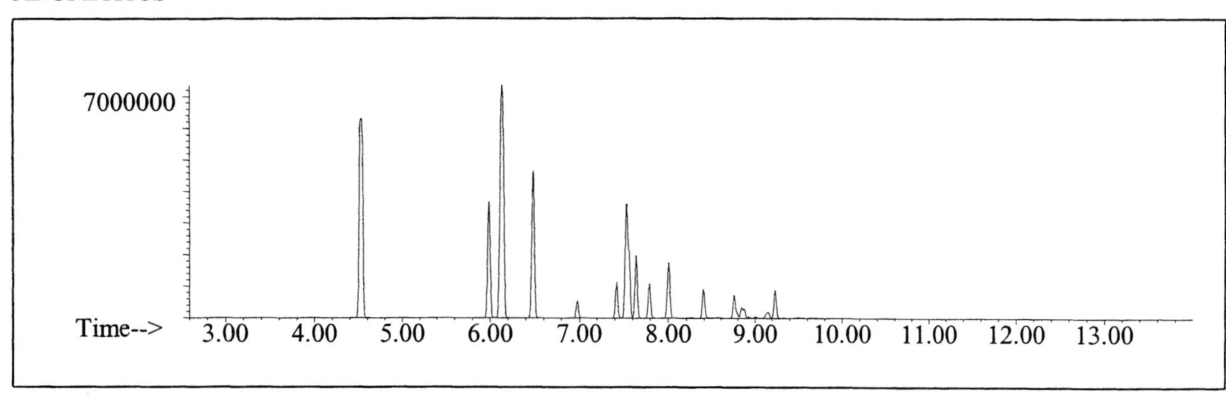

CYCLOPARAFFINS AND ALKENES

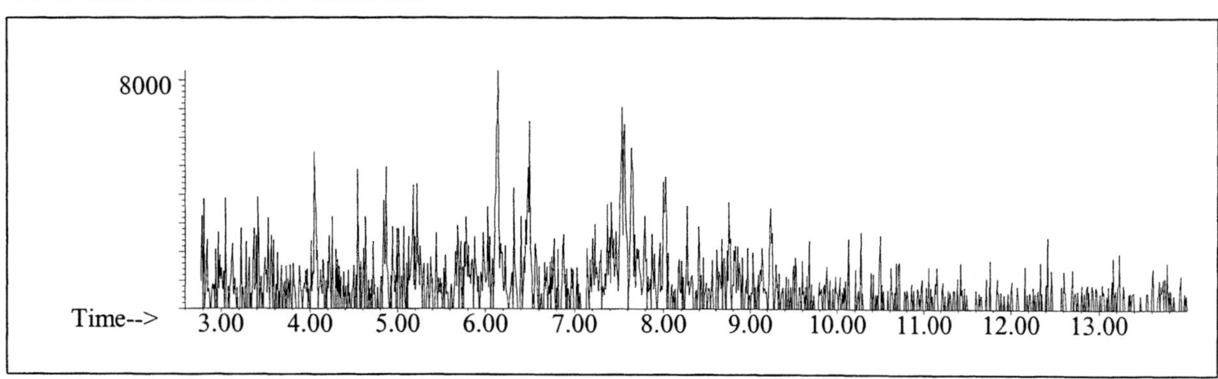

NAPHTHALENES

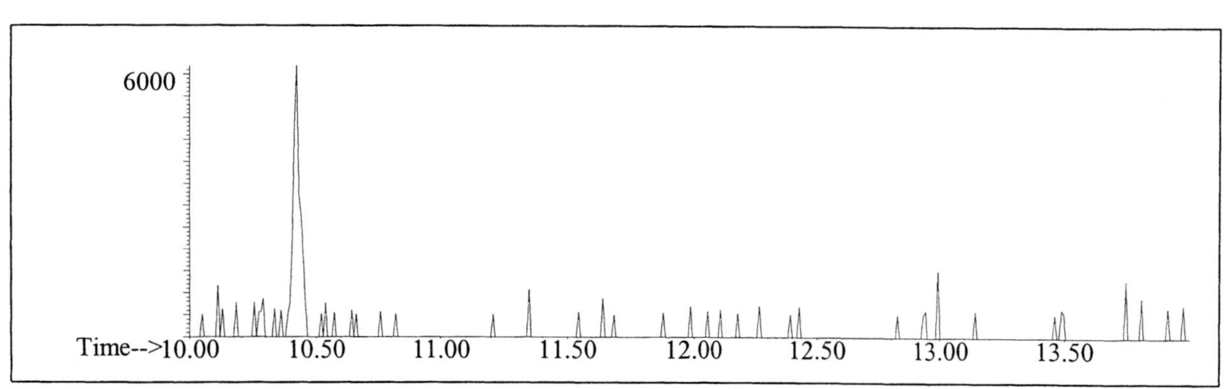

INDIVIDUAL PROFILES

Aromatic #3

Alkane

Ion 43

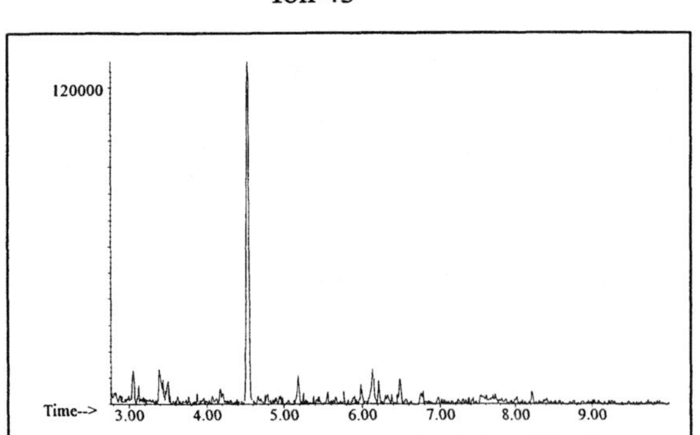

Ion 57

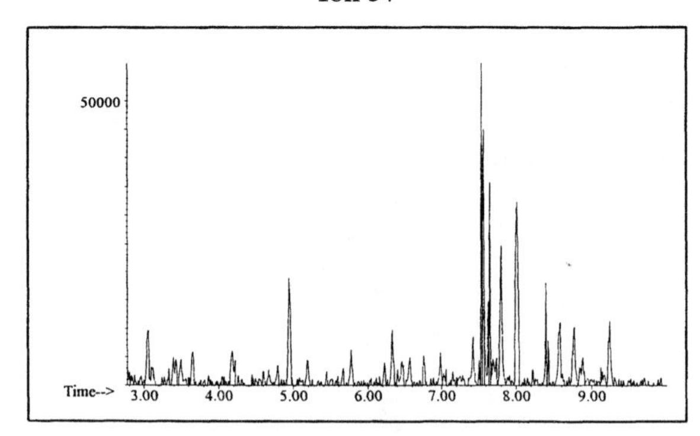

Ion 71

No Useful Data Obtained

Ion 85

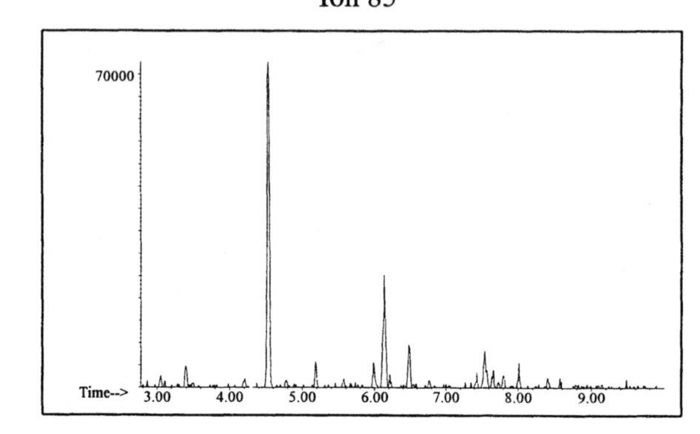

Aromatic

Ion 91

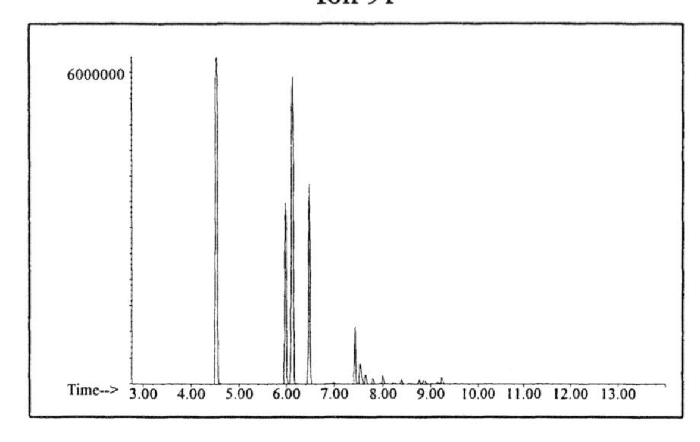

Ion 105

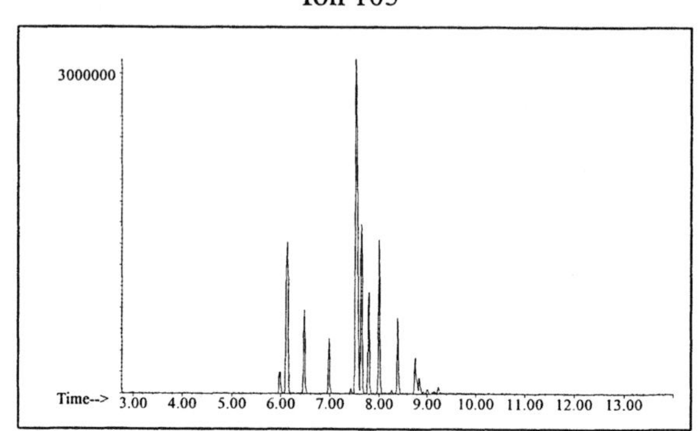

Ion 119

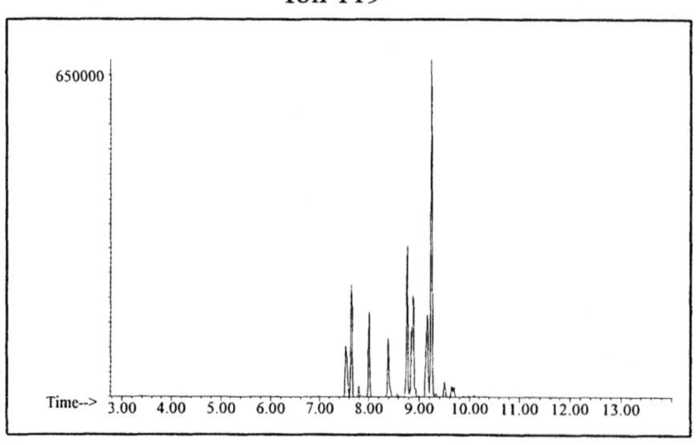

INDIVIDUAL PROFILES

Aromatic #3

Cycloparaffin

Ion 55

No Useful Data Obtained

Ion 69

No Useful Data Obtained

Ion 83

No Useful Data Obtained

Naphthalene

Ion 128

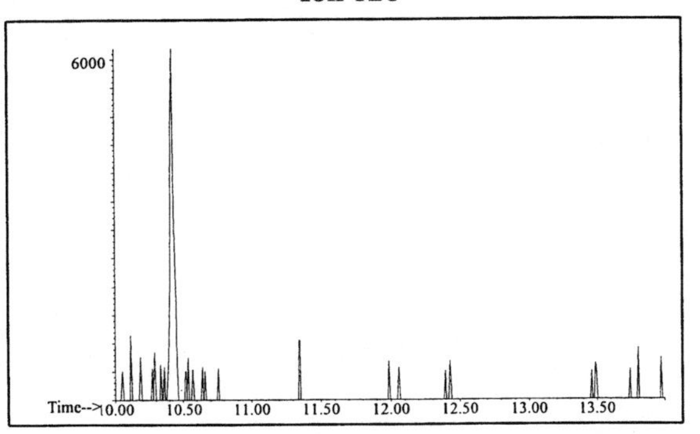

Ion 142

No Useful Data Obtained

Ion 156

No Useful Data Obtained

Aromatic #4

20uL/mL Pentane

ASTM: Class 0.4 (Aromatic)

Macro Code: L

Product Displayed: Gumout Choke and Carburetor Cleaner

Product Uses: Automotive parts cleaner

Other Similar Products:

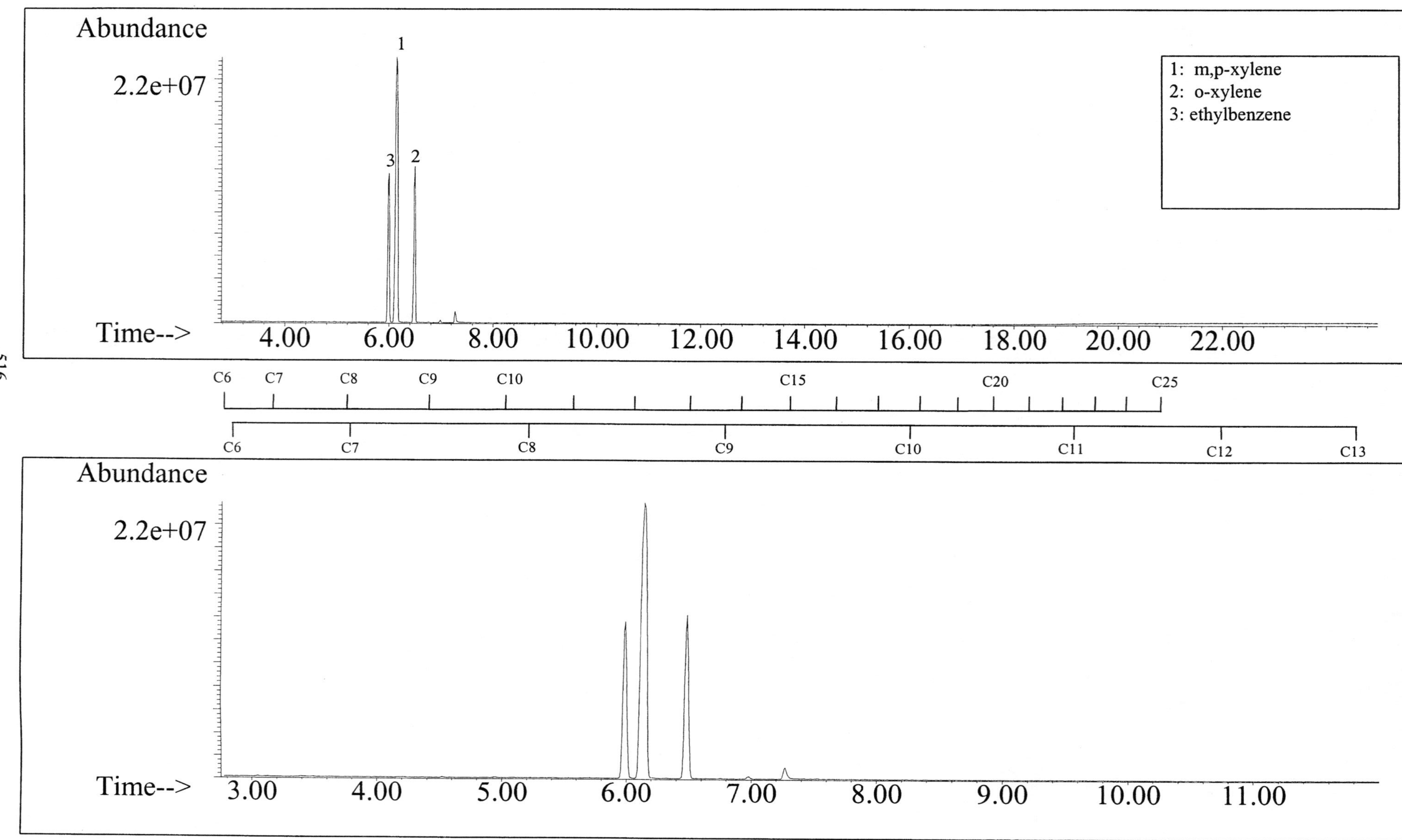

SUMMED PROFILES

Aromatic #4

ALKANES

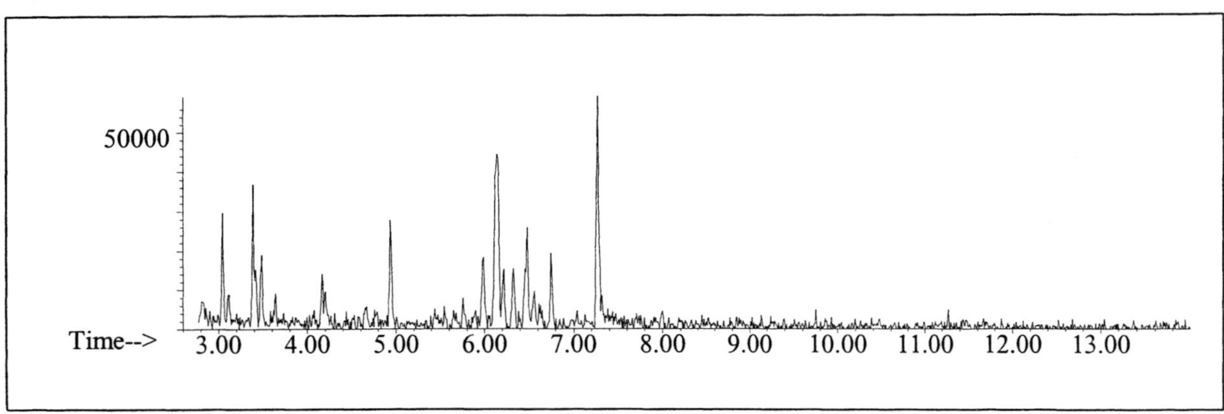

AROMATICS

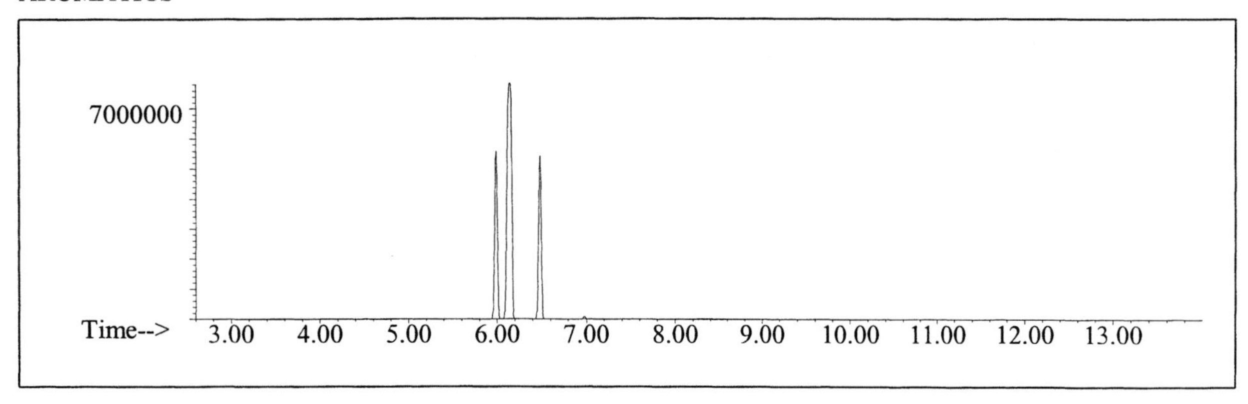

CYCLOPARAFFINS AND ALKENES

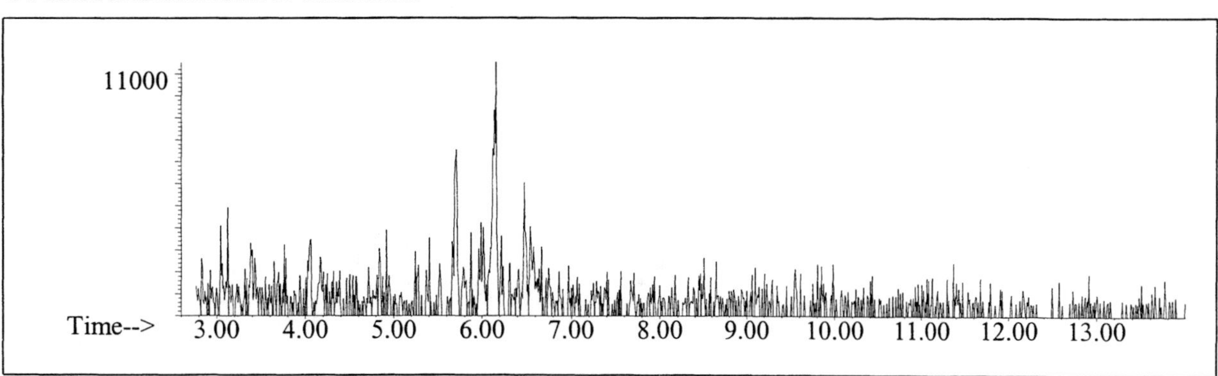

NAPHTHALENES

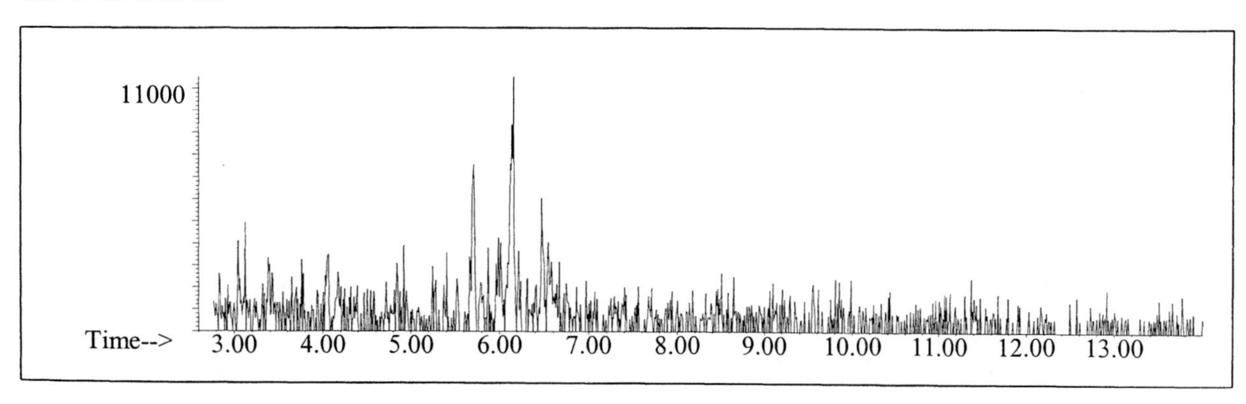

INDIVIDUAL PROFILES

Aromatic #4

Alkane

Ion 43

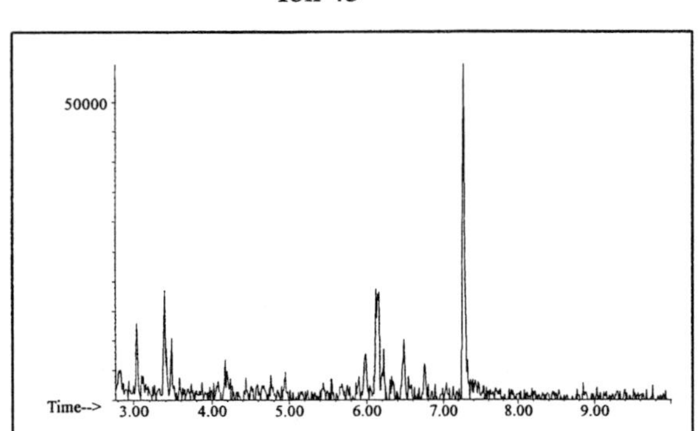

Ion 57

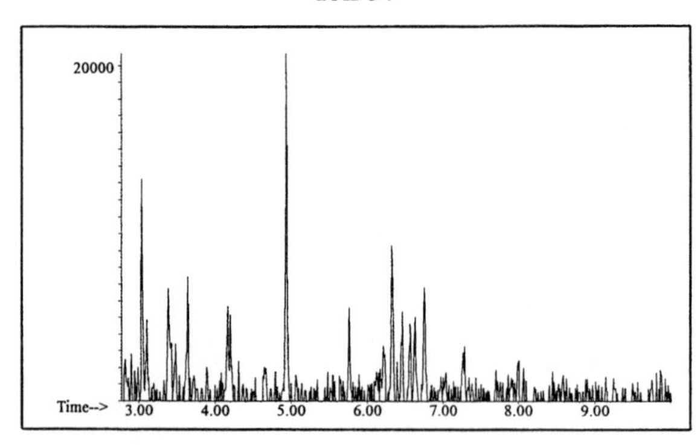

Ion 71

No Useful Data Obtained

Ion 85

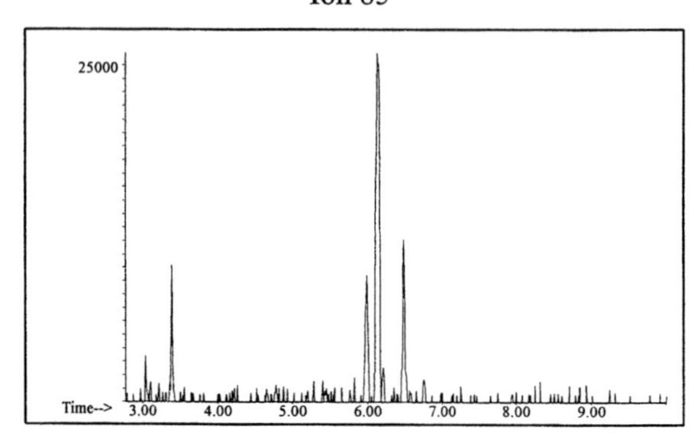

Aromatic

Ion 91

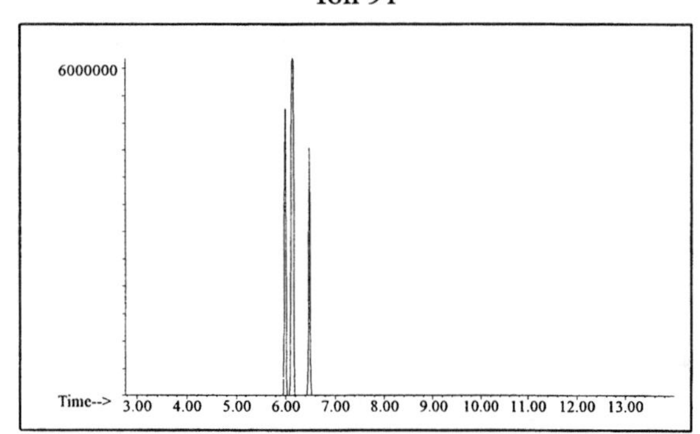

Ion 105

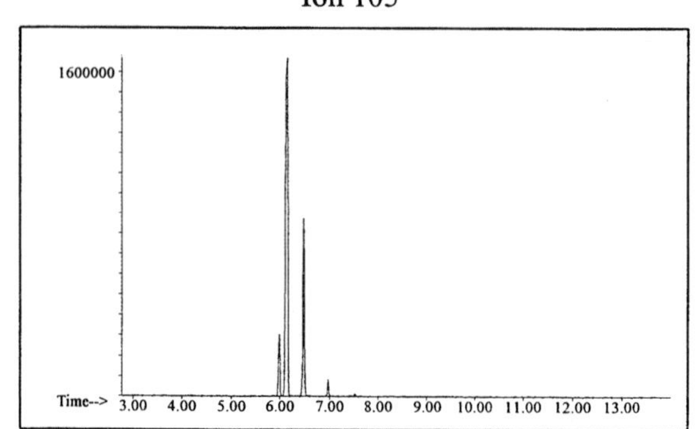

Ion 119

No Useful Data Obtained

Cycloparaffin

Ion 55

No Useful Data Obtained

Ion 69

No Useful Data Obtained

Ion 83

No Useful Data Obtained

Naphthalene

Ion 128

No Useful Data Obtained

Ion 142

No Useful Data Obtained

Ion 156

No Useful Data Obtained

Aromatic #5

20uL/mL Pentane

Product Displayed: Diazonon 500

Other Similar Products:

ASTM: Class 0.4 (Aromatic)

Product Uses: Insecticide

Macro Code: L

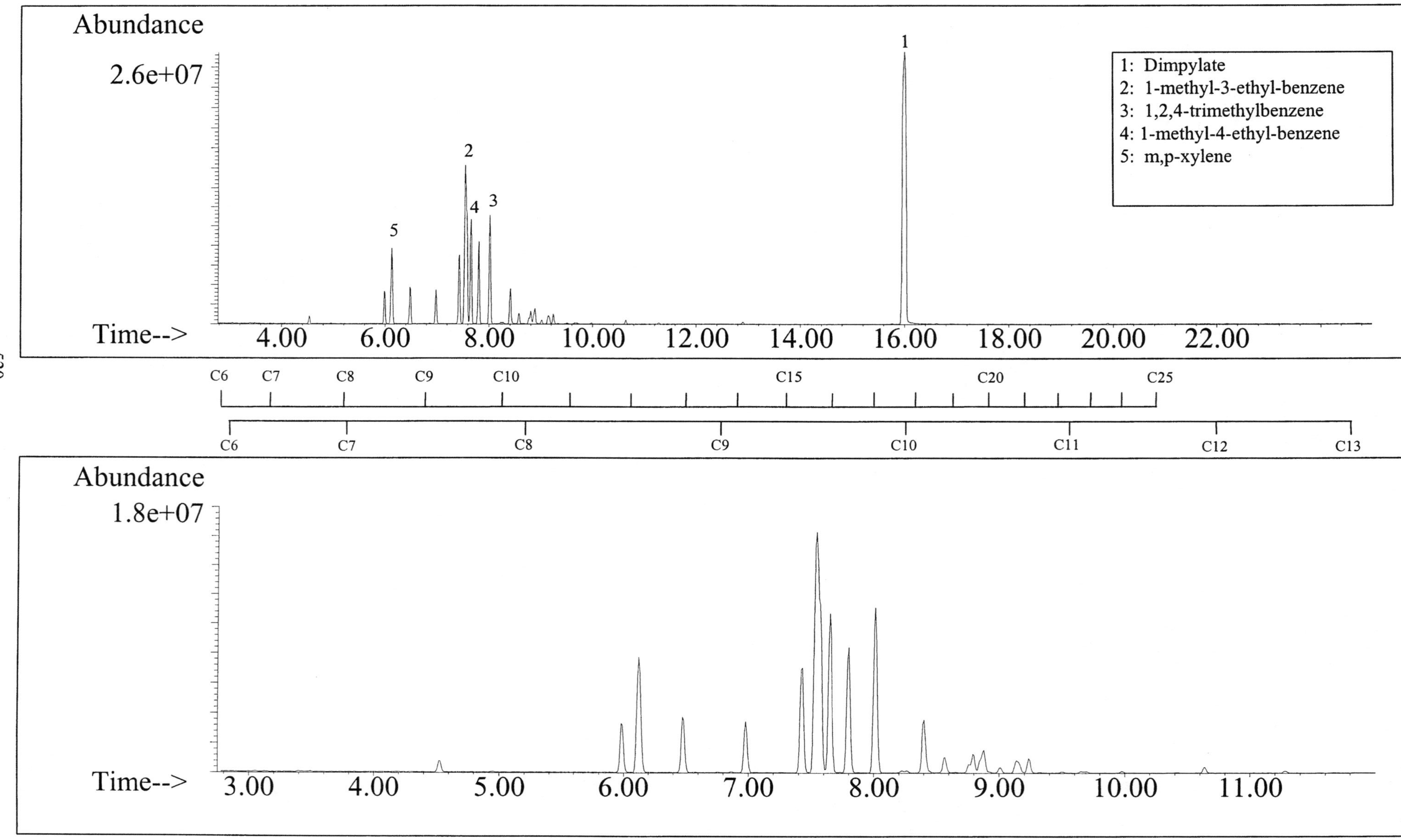

SUMMED PROFILES

Aromatic #5

ALKANES

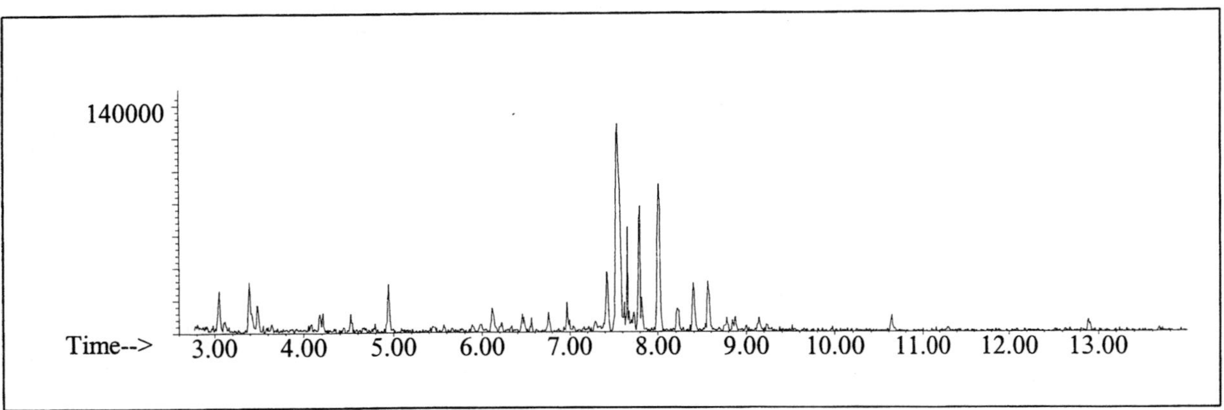

AROMATICS

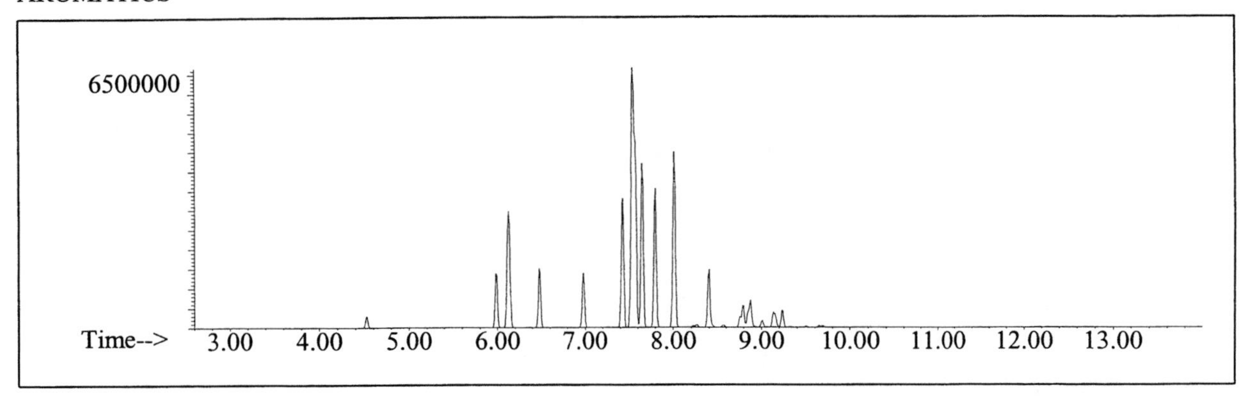

CYCLOPARAFFINS AND ALKENES

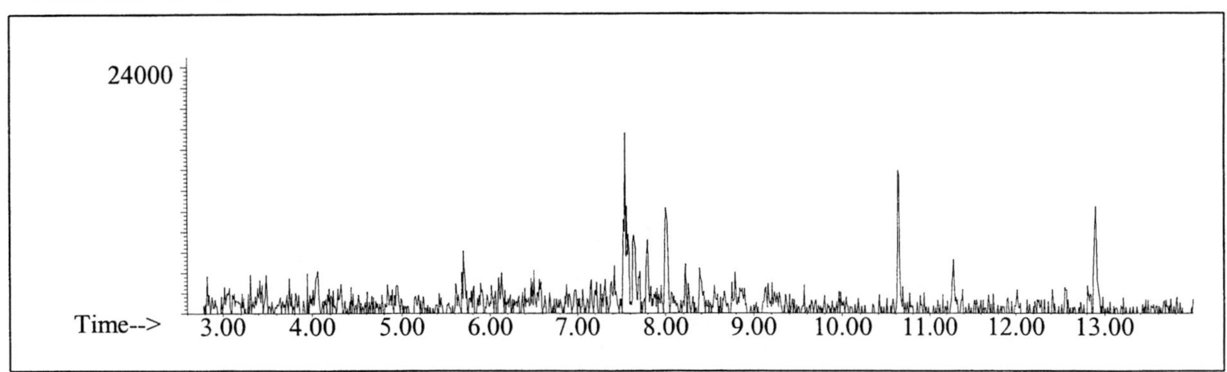

NAPHTHALENES

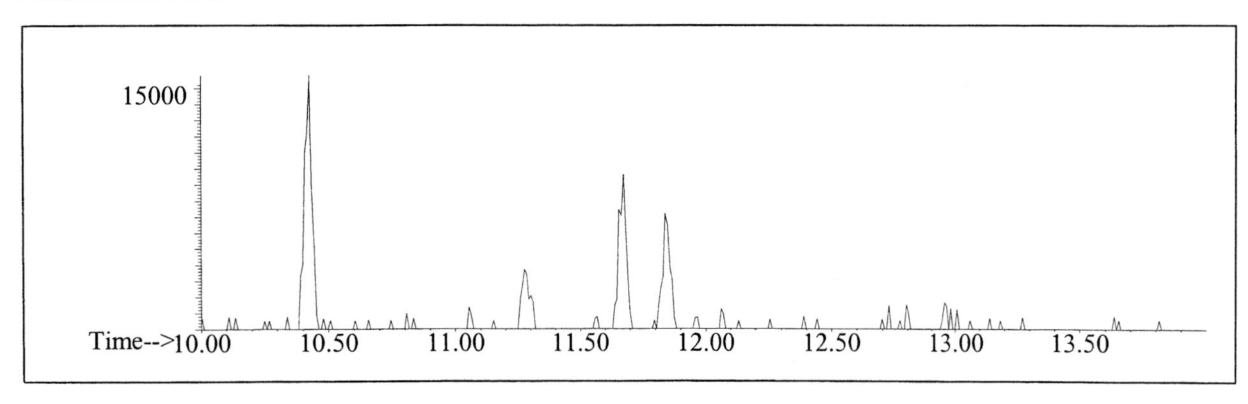

INDIVIDUAL PROFILES

Aromatic #5

Alkane

Ion 43

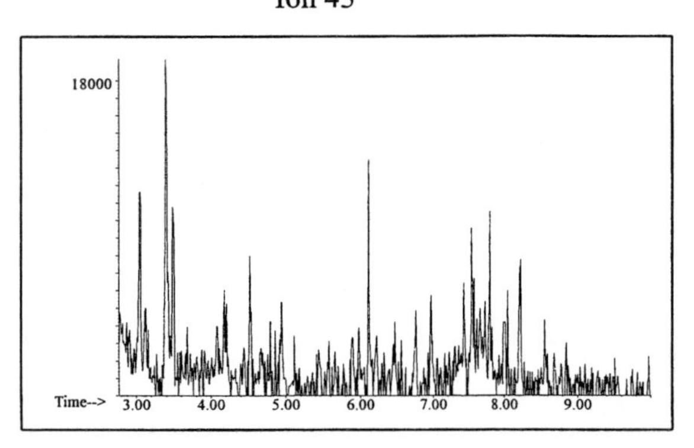

Ion 57

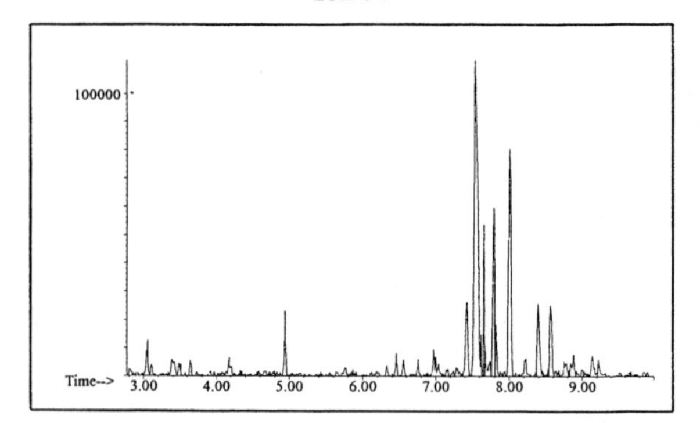

Ion 71

No Useful Data Obtained

Ion 85

No Useful Data Obtained

Aromatic

Ion 91

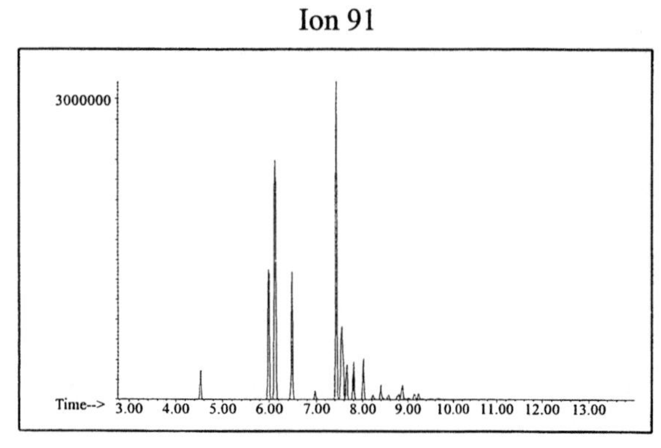

Ion 105

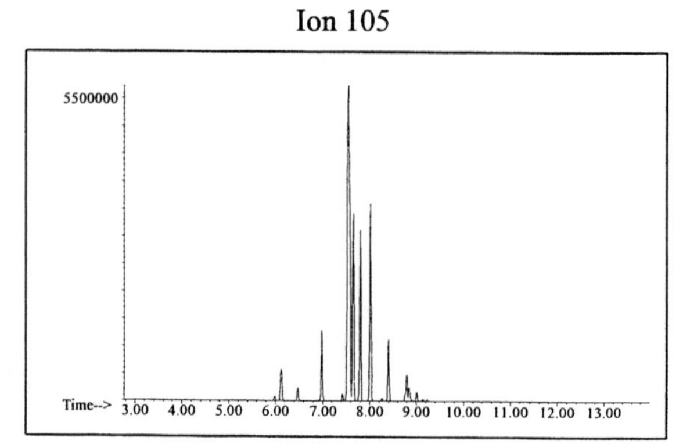

Ion 119

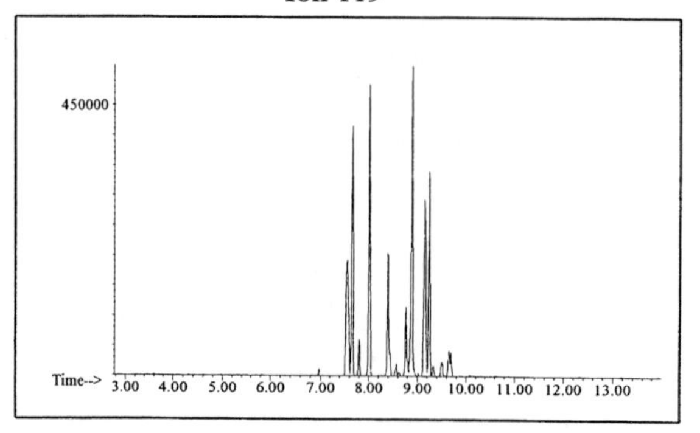

INDIVIDUAL PROFILES

Aromatic #5

Cycloparaffin

Ion 55

No Useful Data Obtained

Ion 69

No Useful Data Obtained

Ion 83

No Useful Data Obtained

Naphthalene

Ion 128

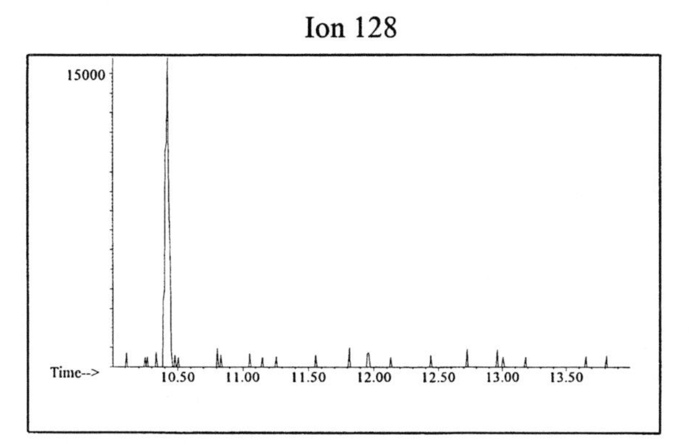

Ion 142

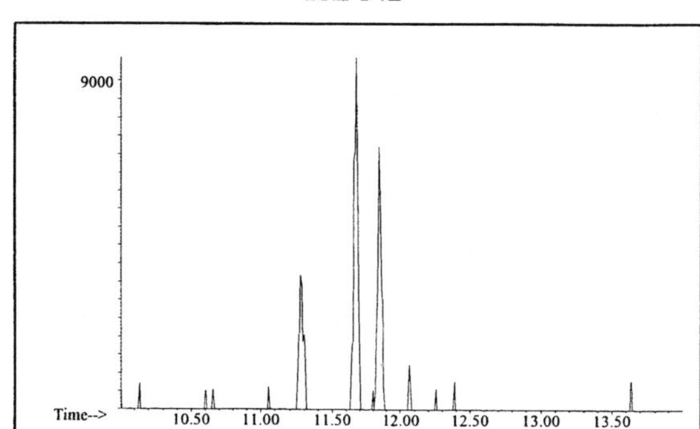

Ion 156

No Useful Data Obtained

Aromatic #6

20uL/mL Pentane

Product Displayed: Steel Seal Power Steering Sealant

Other Similar Products:

ASTM: Class 0.4 (Aromatic)

Product Uses: Automotive sealant

Macro Code: L

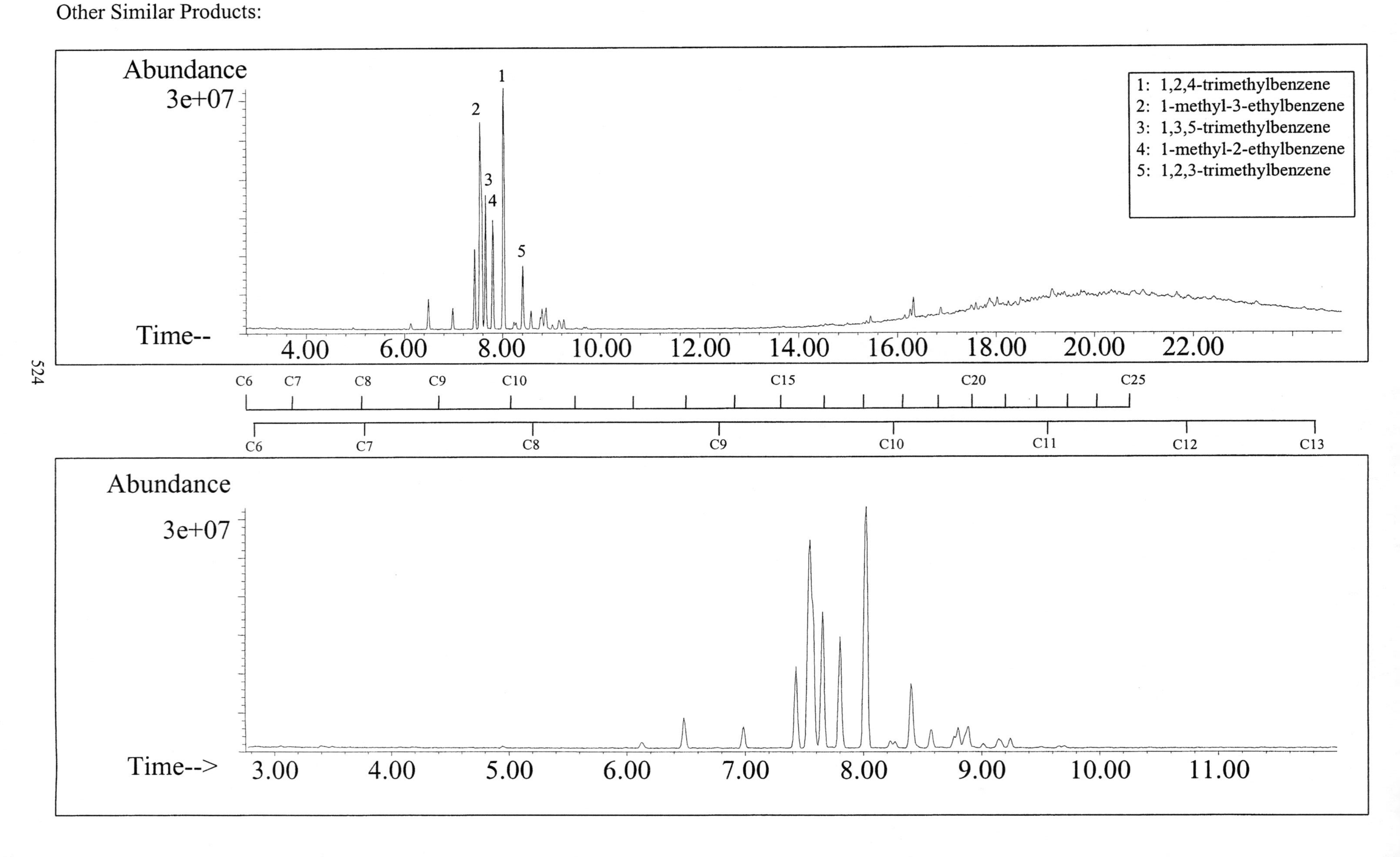

SUMMED PROFILES

Aromatic #6

ALKANES

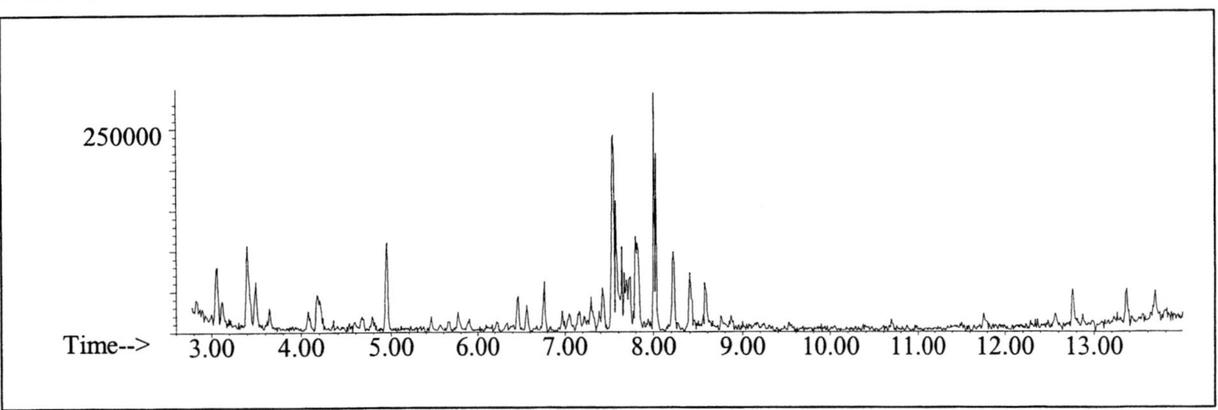

AROMATICS

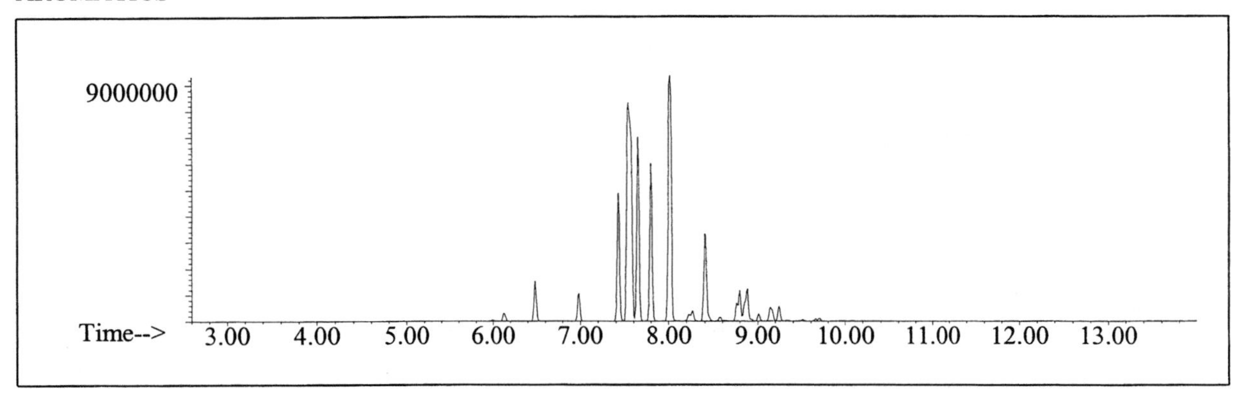

CYCLOPARAFFINS AND ALKENES

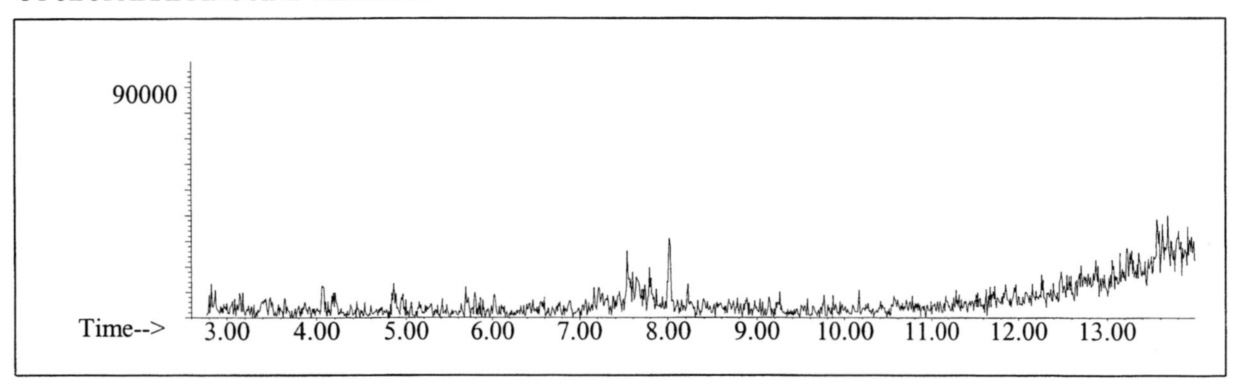

NAPHTHALENES

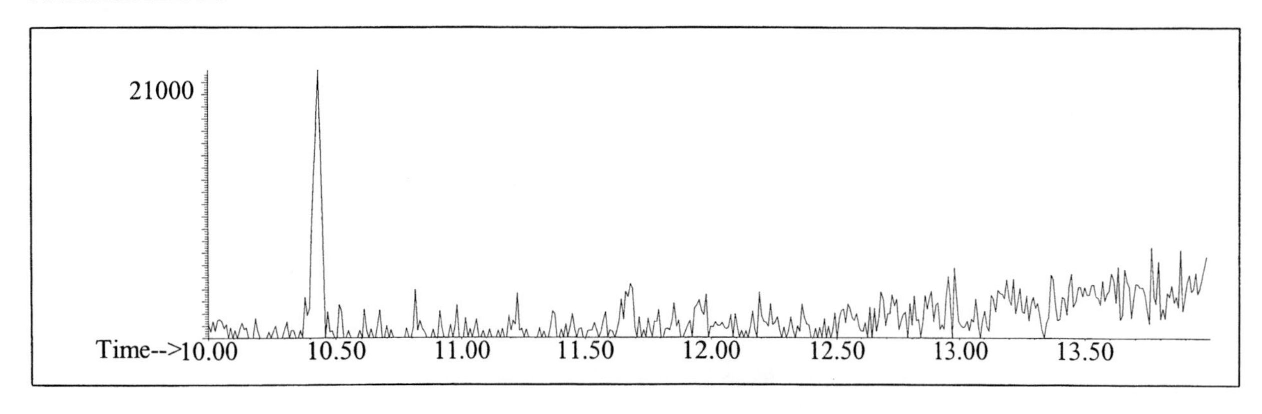

INDIVIDUAL PROFILES

Aromatic #6

Alkane

Ion 43

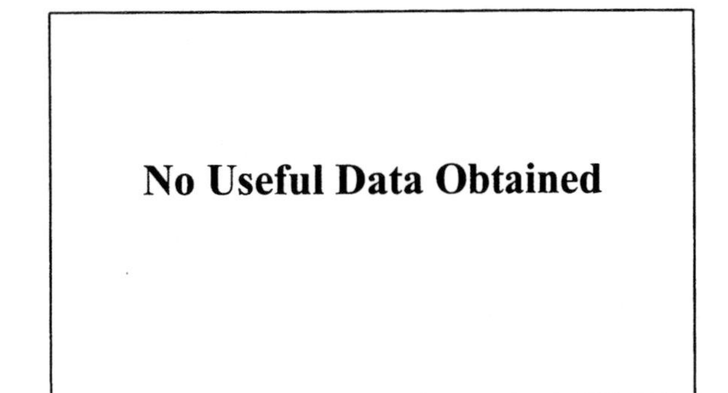

Ion 57

240000

Time--> 3.00 4.00 5.00 6.00 7.00 8.00 9.00

Ion 71

No Useful Data Obtained

Ion 85

No Useful Data Obtained

Aromatic

Ion 91

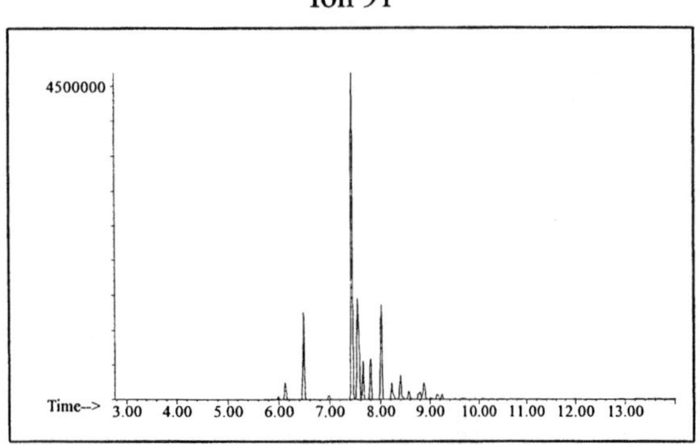

Ion 105

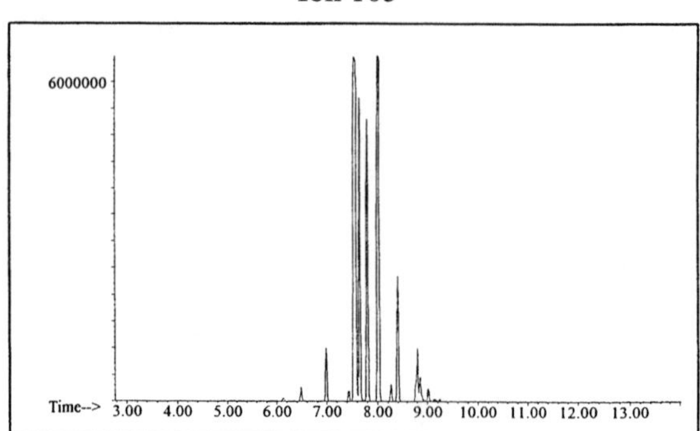

Ion 119

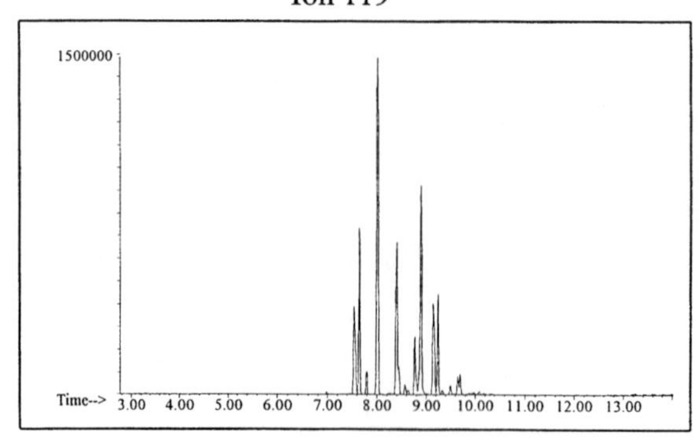

INDIVIDUAL PROFILES

Aromatic #6

Cycloparaffin

Ion 55

No Useful Data Obtained

Ion 69

No Useful Data Obtained

Ion 83

No Useful Data Obtained

Naphthalene

Ion 128

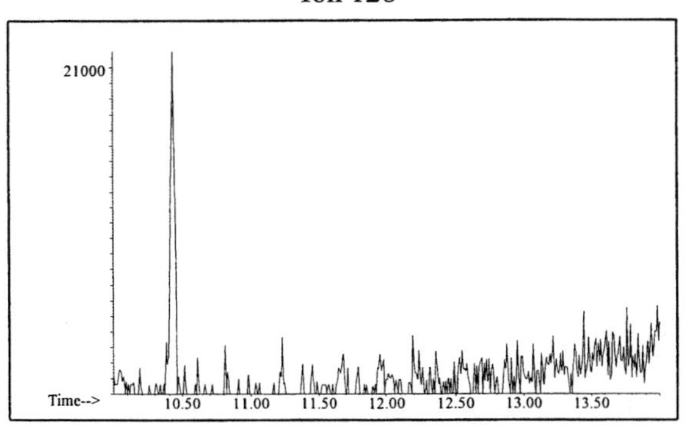

Ion 142

No Useful Data Obtained

Ion 156

No Useful Data Obtained

Aromatic #7

20uL/mL Pentane

Product Displayed: Exxon 100

Other Similar Products:

ASTM: Class 0.4 (Aromatic)

Product Uses: Solvent, feedstock

Macro Code: L

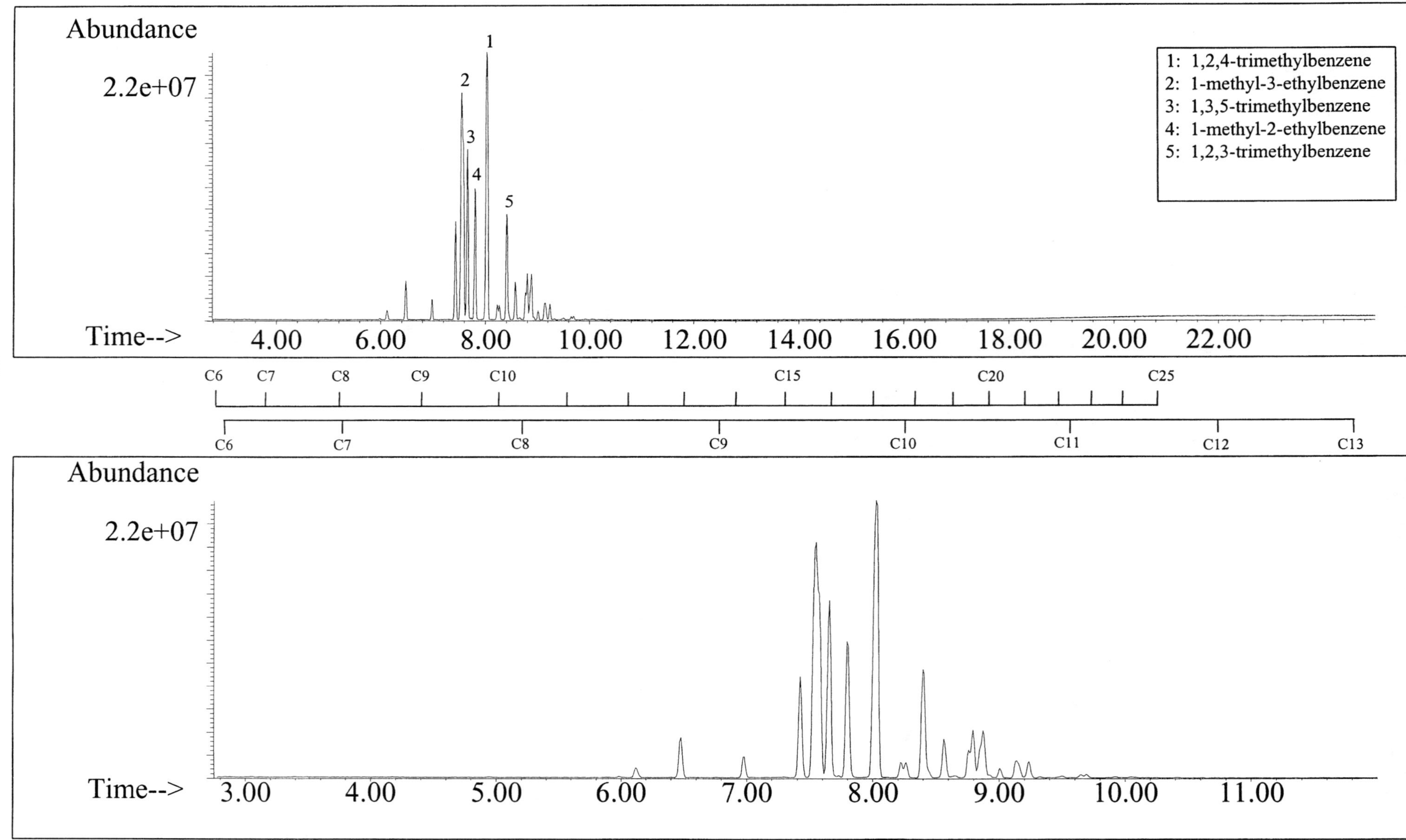

SUMMED PROFILES

Aromatic #7

ALKANES

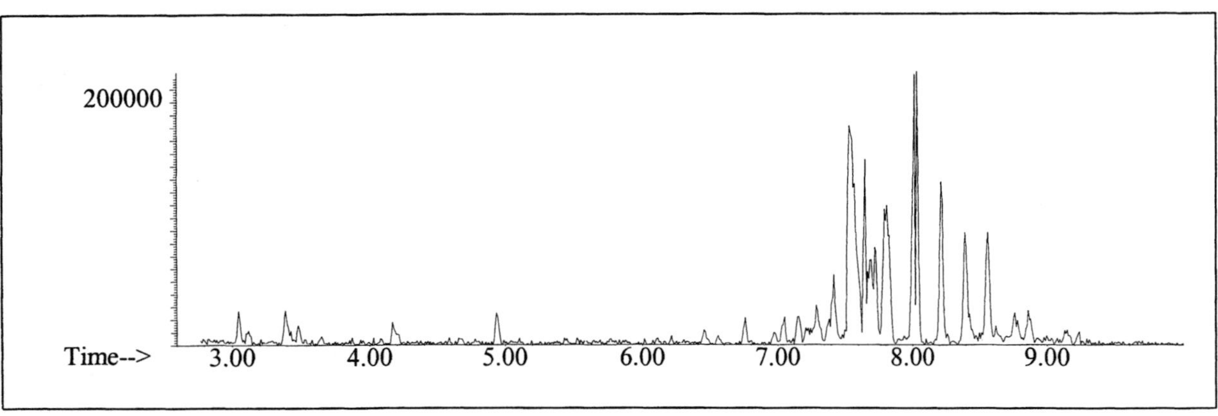

AROMATICS

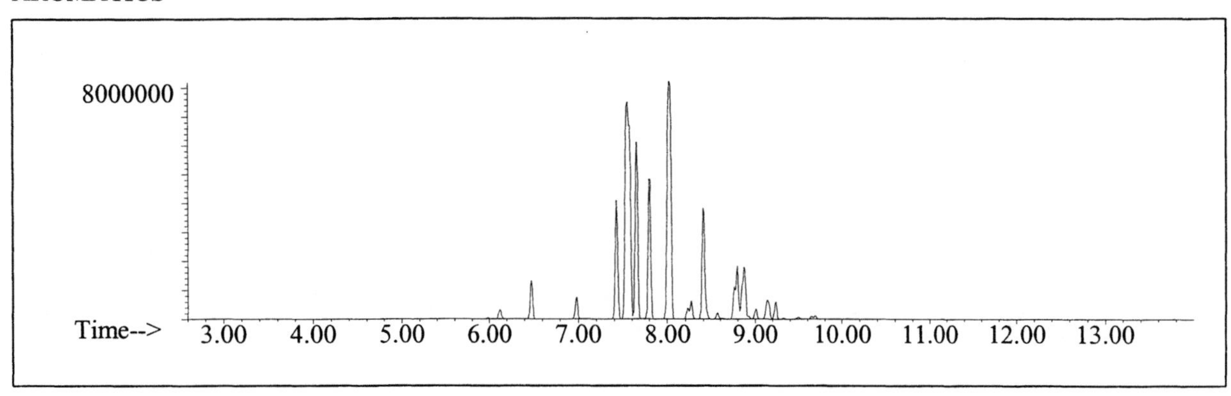

CYCLOPARAFFINS AND ALKENES

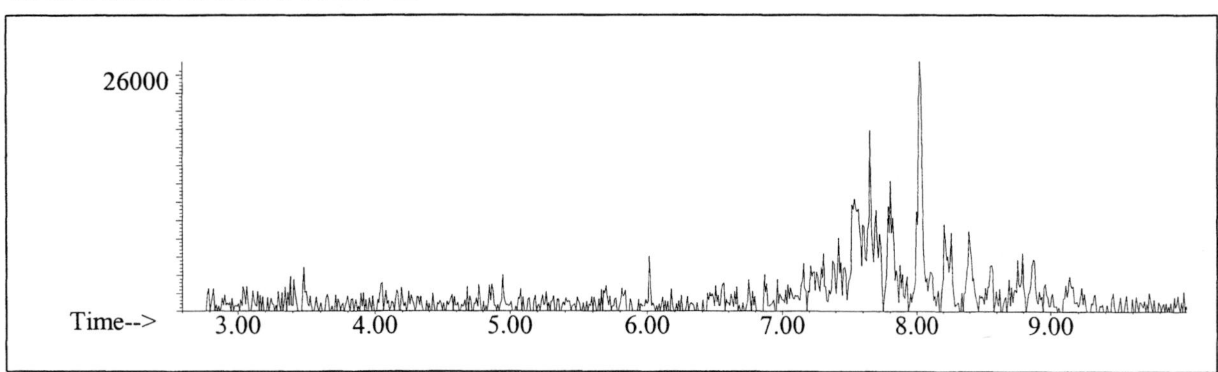

NAPHTHALENES

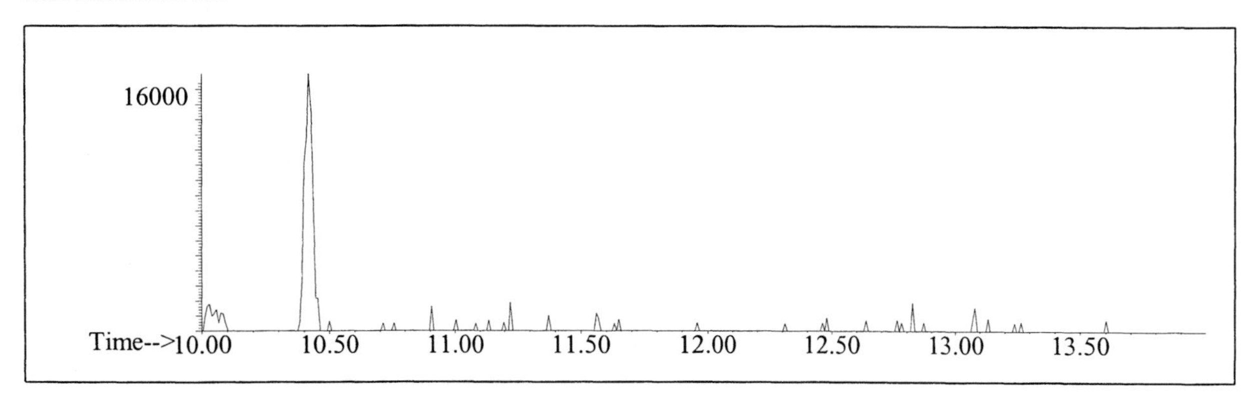

INDIVIDUAL PROFILES

Aromatic #7

Alkane

Ion 43

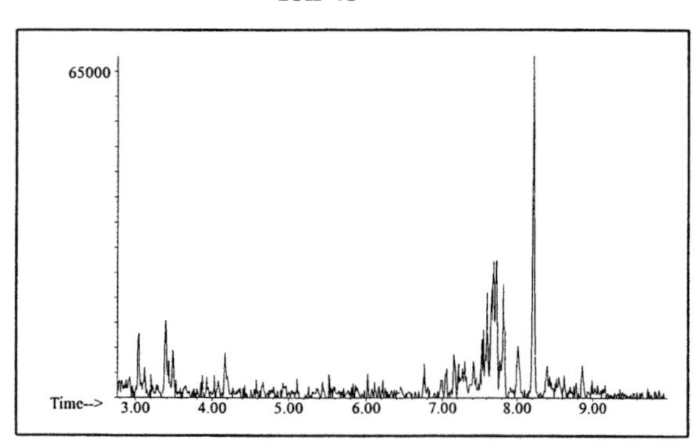

Ion 57

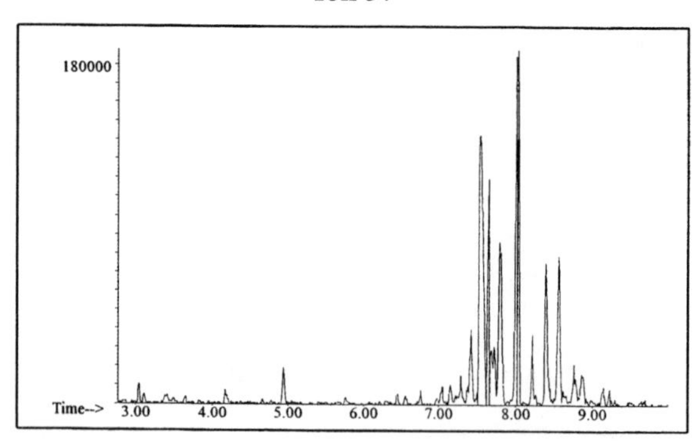

Ion 71

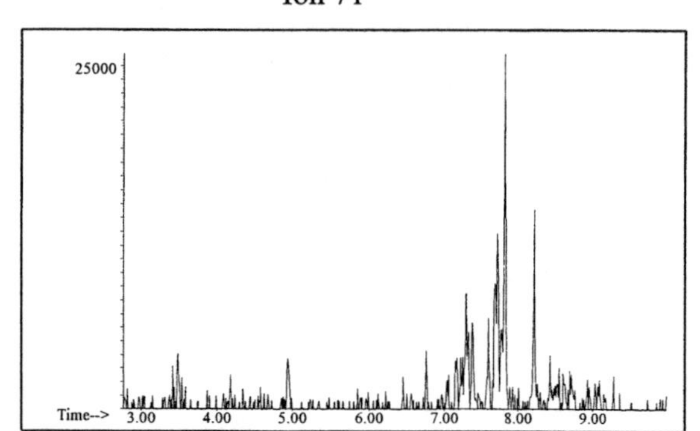

Ion 85

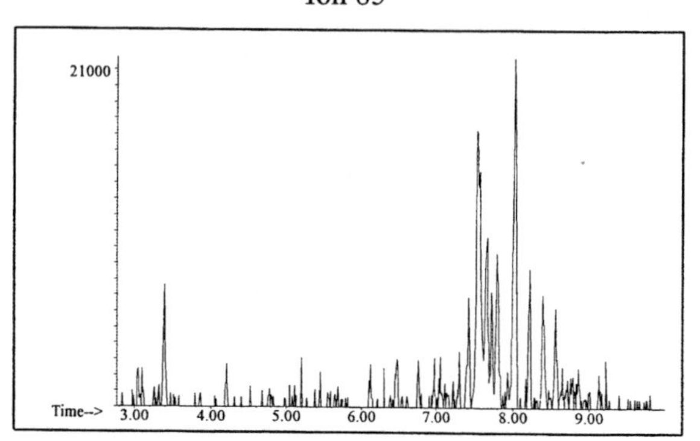

Aromatic

Ion 91

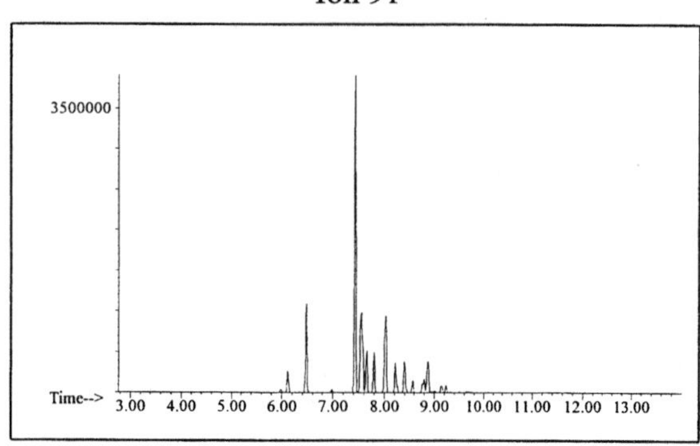

Ion 105

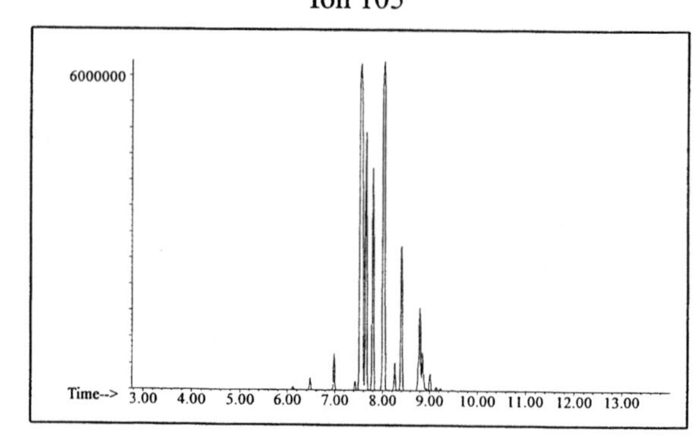

Ion 119

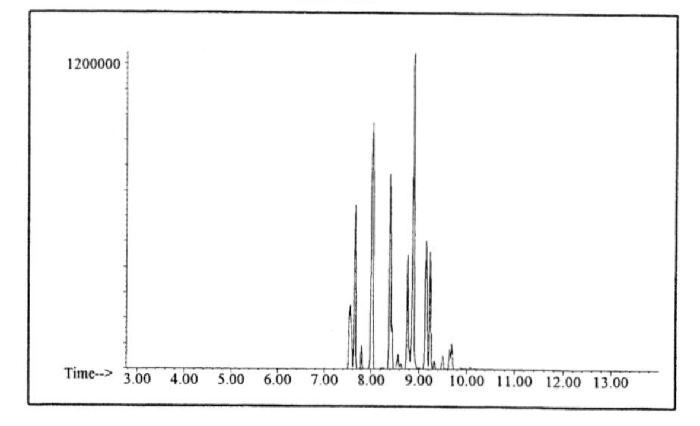

INDIVIDUAL PROFILES

Cycloparaffin

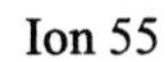
Ion 55

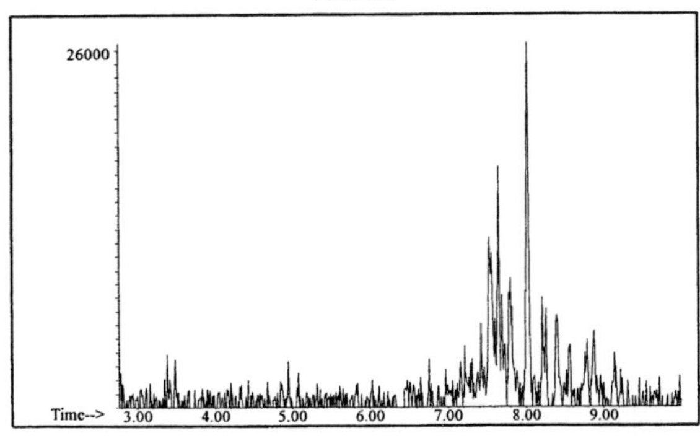

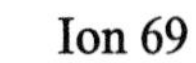
Ion 69

No Useful Data Obtained

Ion 83

No Useful Data Obtained

Naphthalene

Ion 128

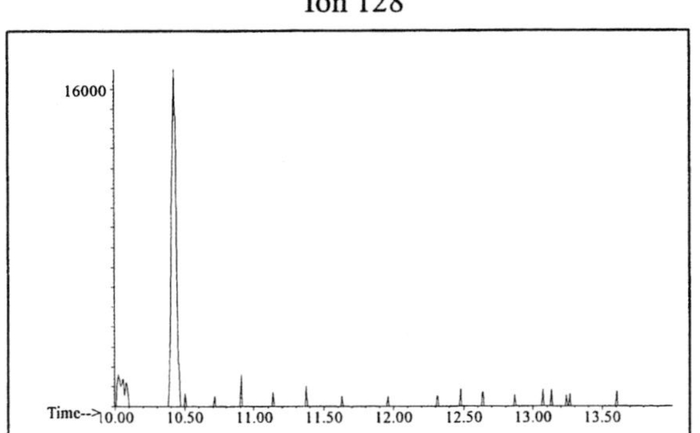

Ion 142

No Useful Data Obtained

Ion 156

No Useful Data Obtained

Aromatic #8

20uL/mL Pentane

ASTM: Class 0.4 (Aromatic)

Macro Code: M

Product Displayed: Gold Eagle Stabil Fuel Stabilizer

Product Uses: Automotive fuel additive

Other Similar Products:

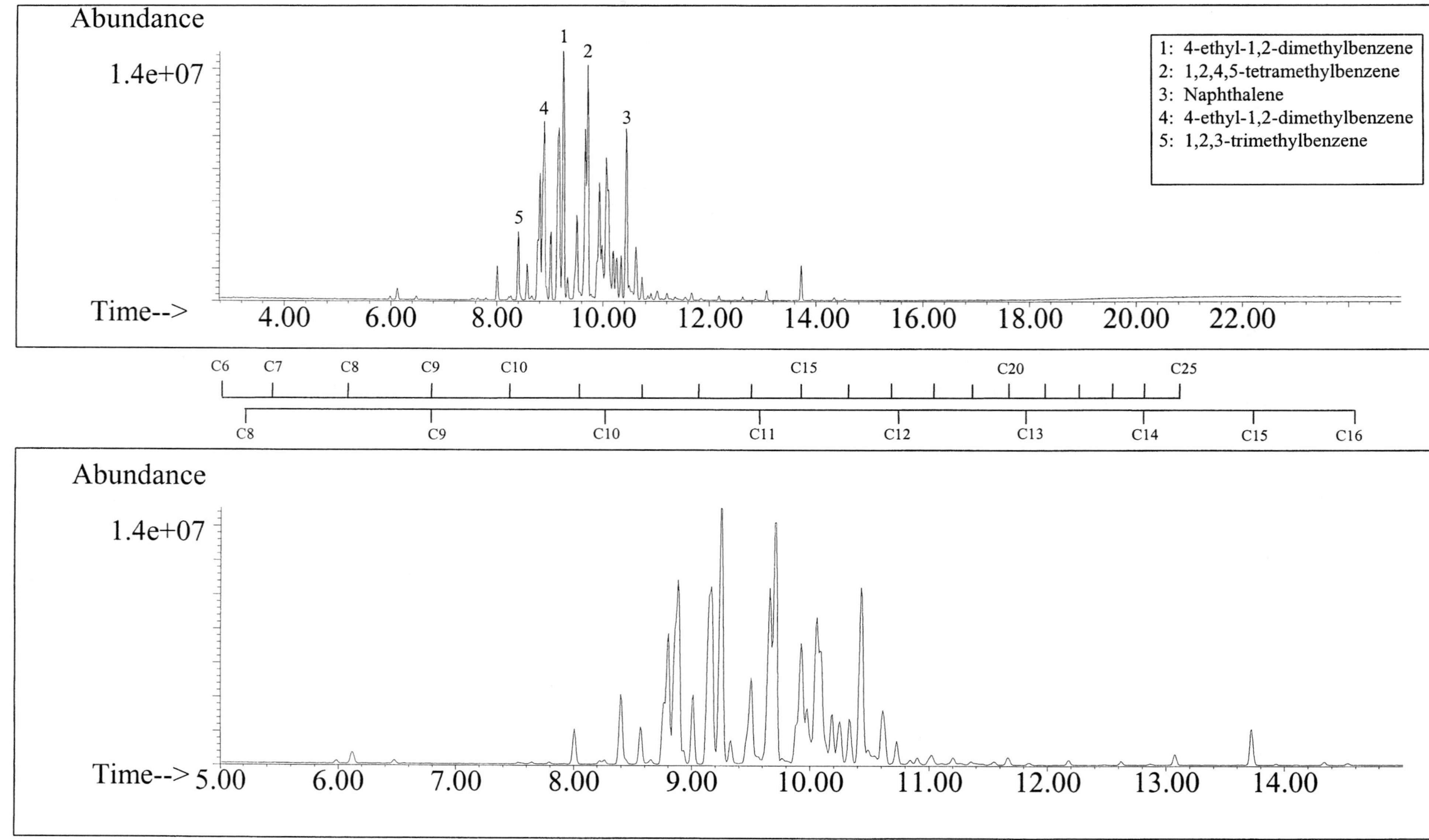

SUMMED PROFILES Aromatic #8

ALKANES

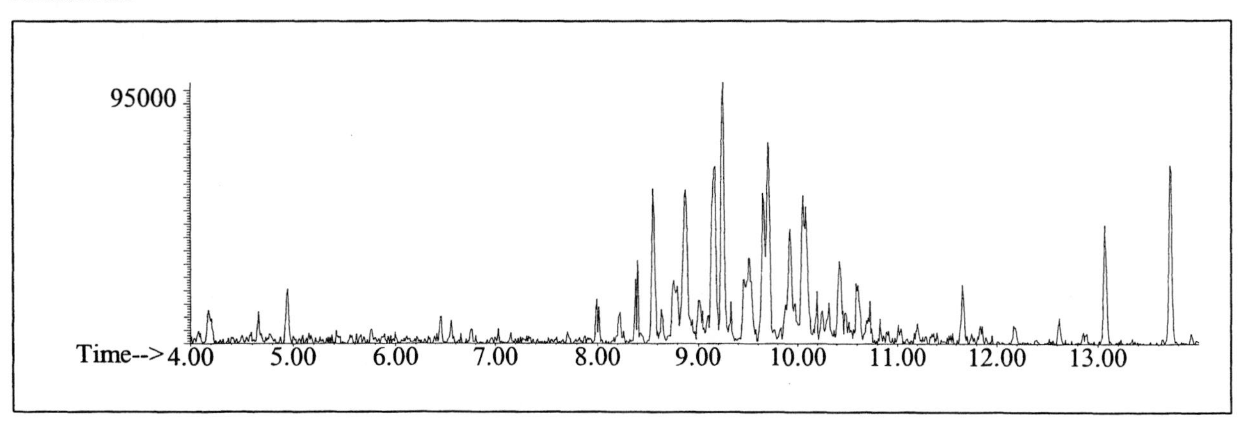

AROMATICS

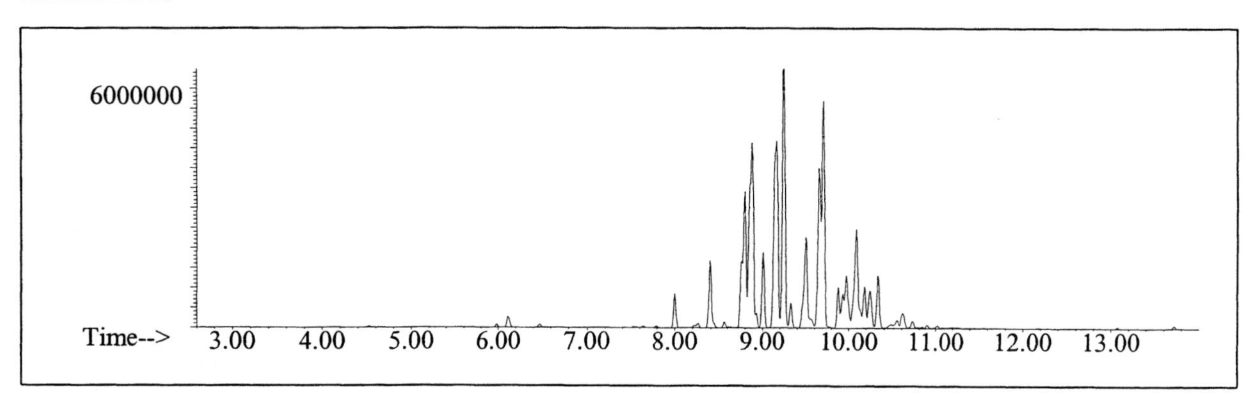

CYCLOPARAFFINS AND ALKENES

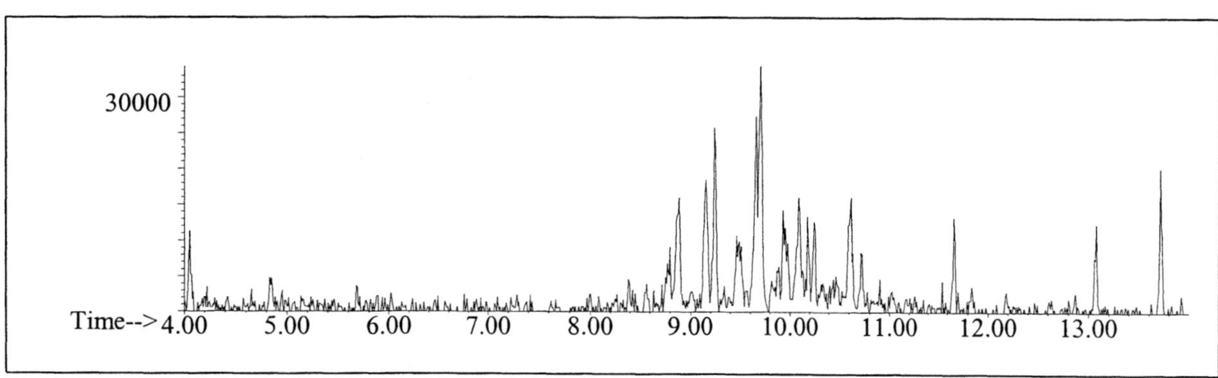

NAPHTHALENES

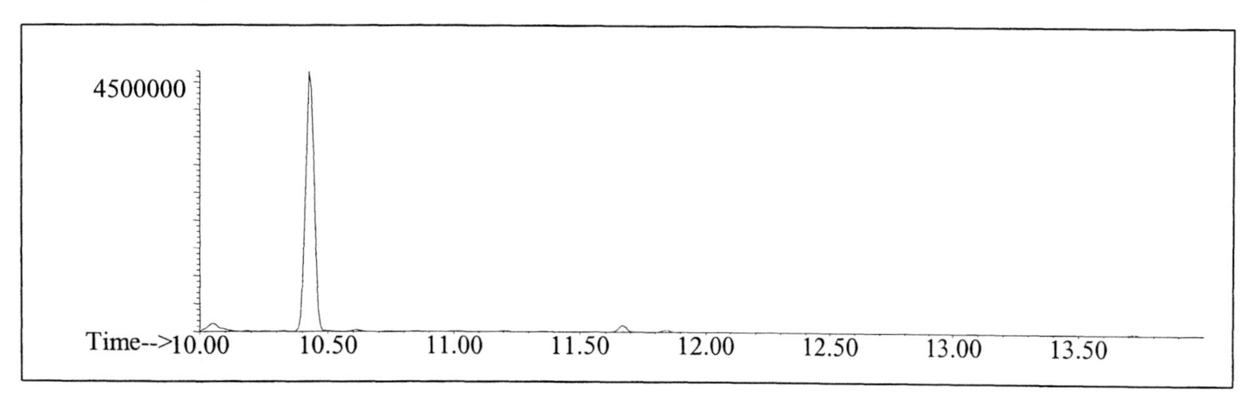

INDIVIDUAL PROFILES

Aromatic #8

Alkane

Ion 43

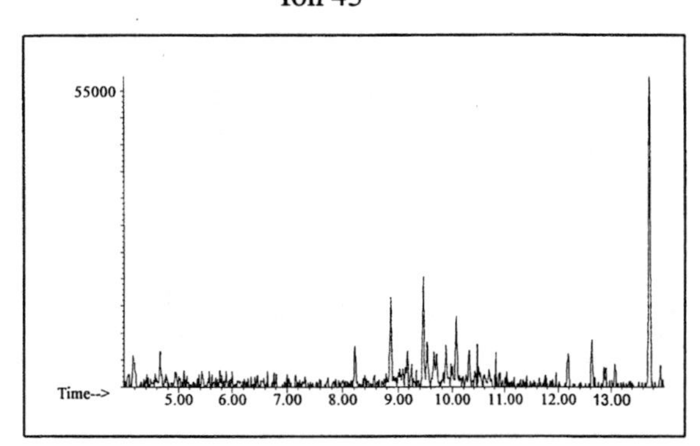

Ion 57

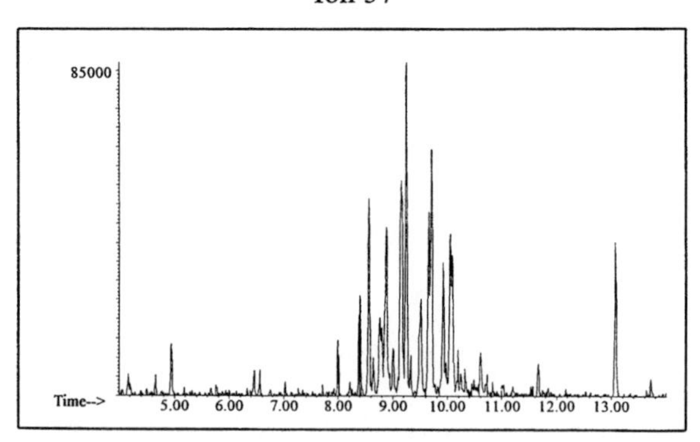

Ion 71

No Useful Data Obtained

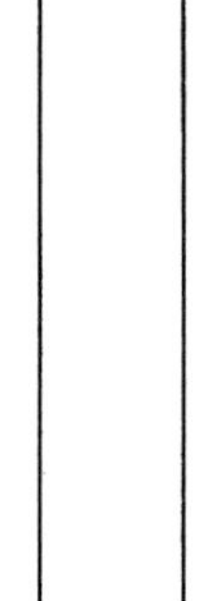

Ion 85

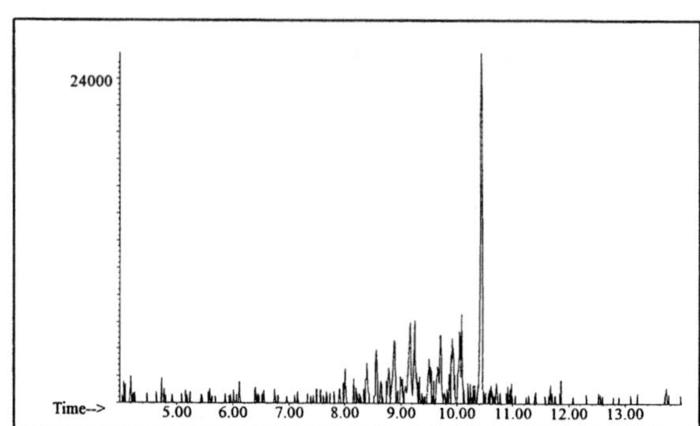

Aromatic

Ion 91

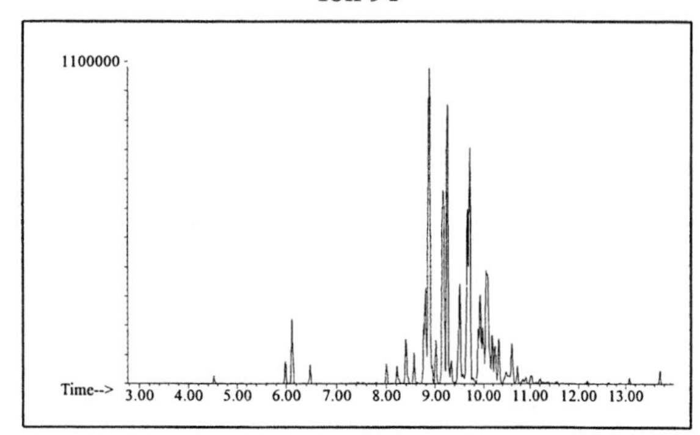

Ion 105

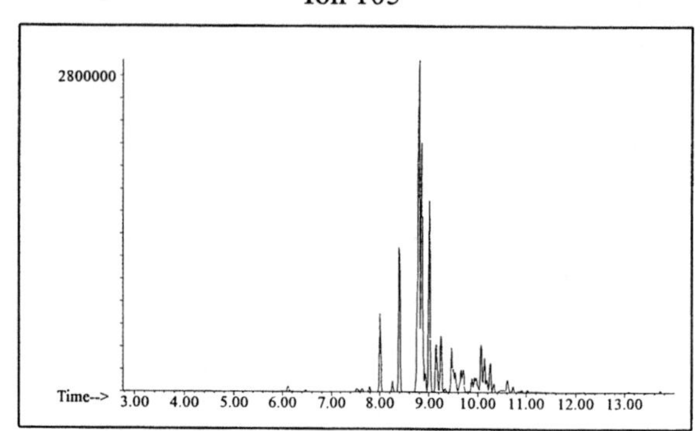

Ion 119

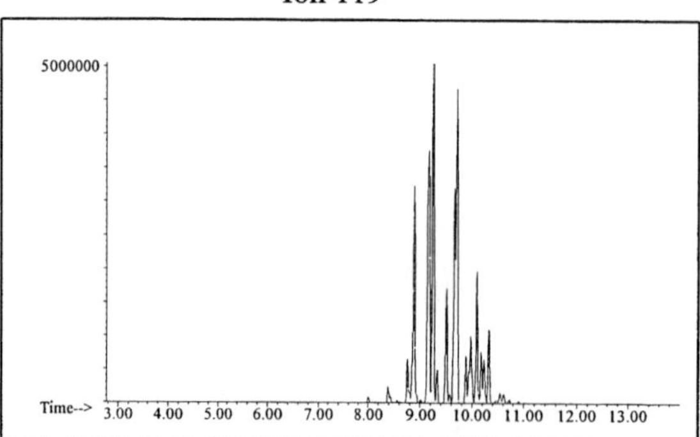

INDIVIDUAL PROFILES

Aromatic #8

Cycloparaffin

Ion 55

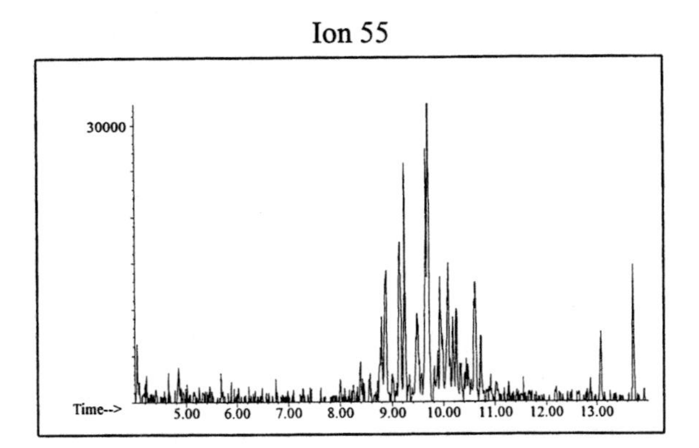

Ion 69

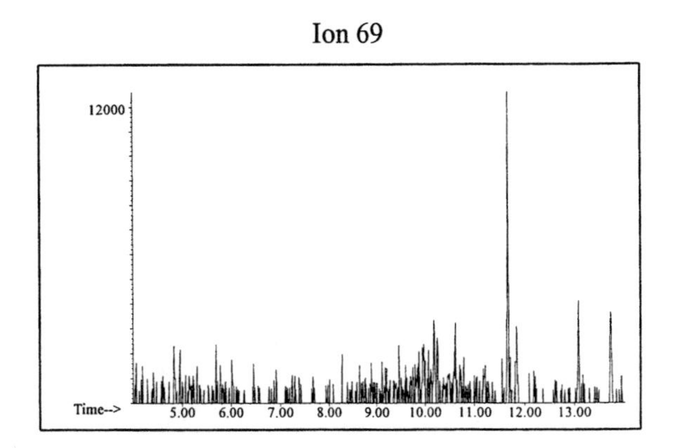

Ion 83

No Useful Data Obtained

Naphthalene

Ion 128

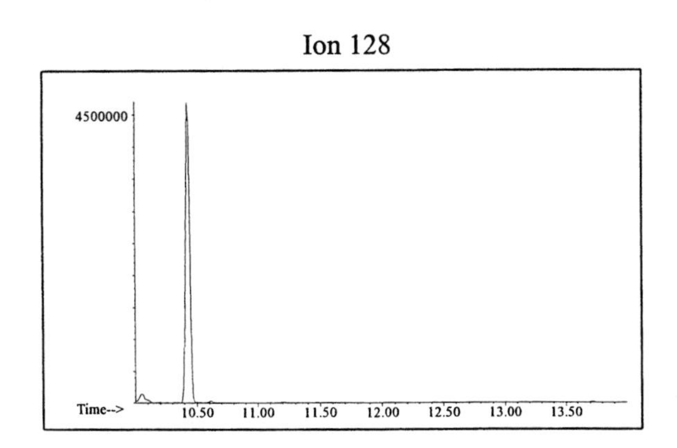

Ion 142

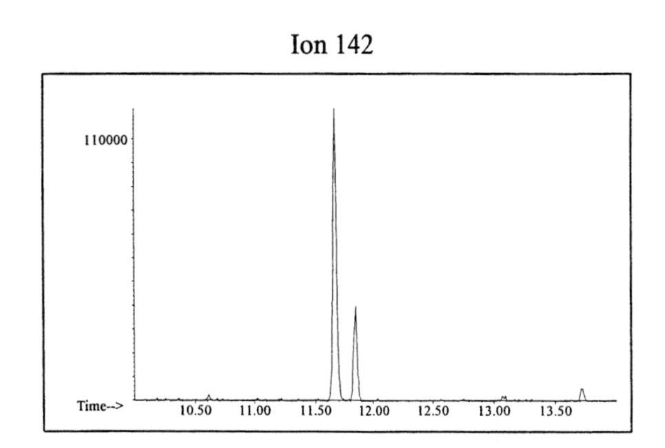

Ion 156

No Useful Data Obtained

Aromatic #9

20uL/mL Pentane

Product Displayed: Rousel Uclaf Cyberactive Insecticide

Other Similar Products:

ASTM: Class 0.4 (Aromatic)

Product Uses: Insecticides

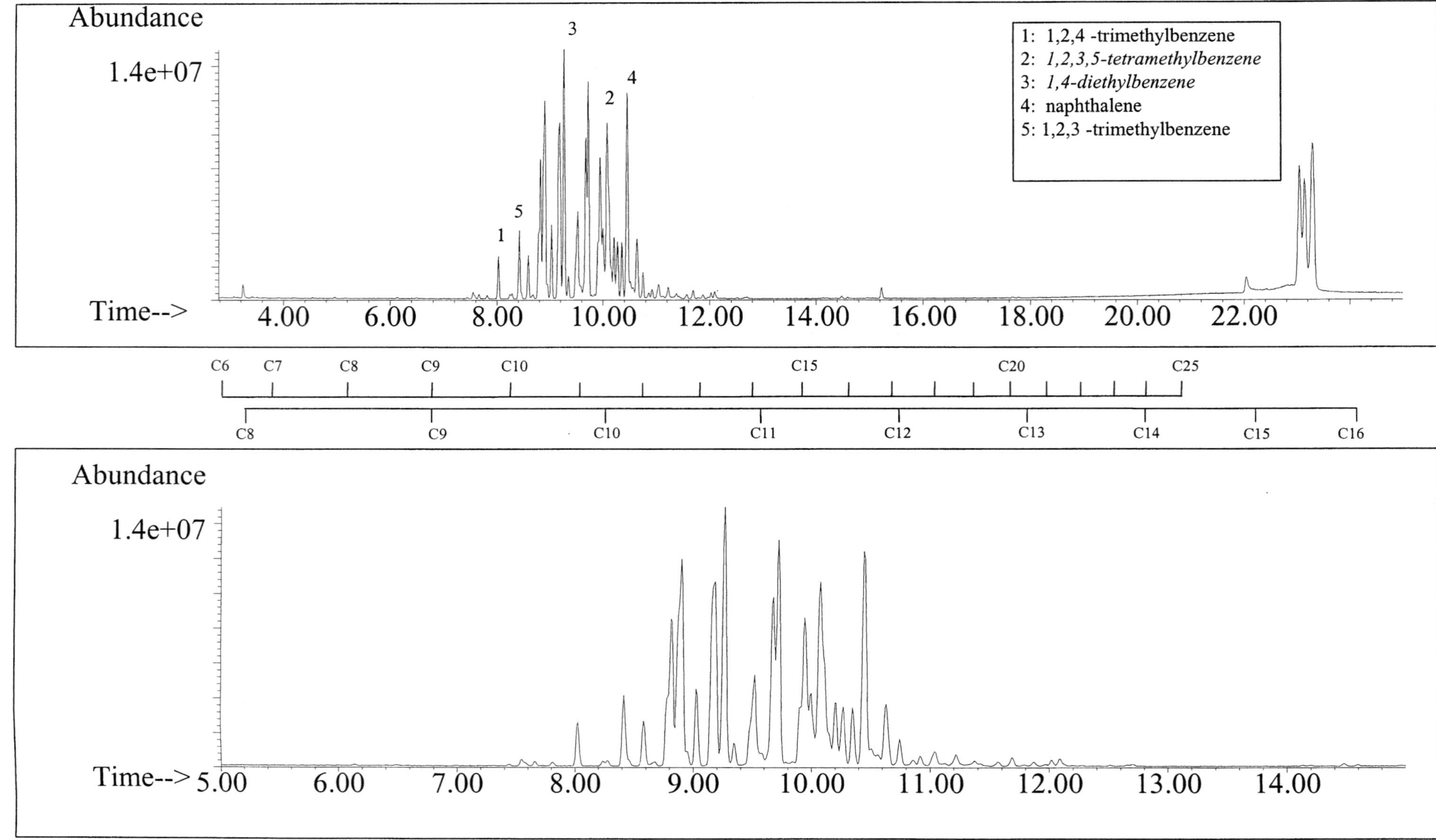

SUMMED PROFILES

Aromatic #9

ALKANES

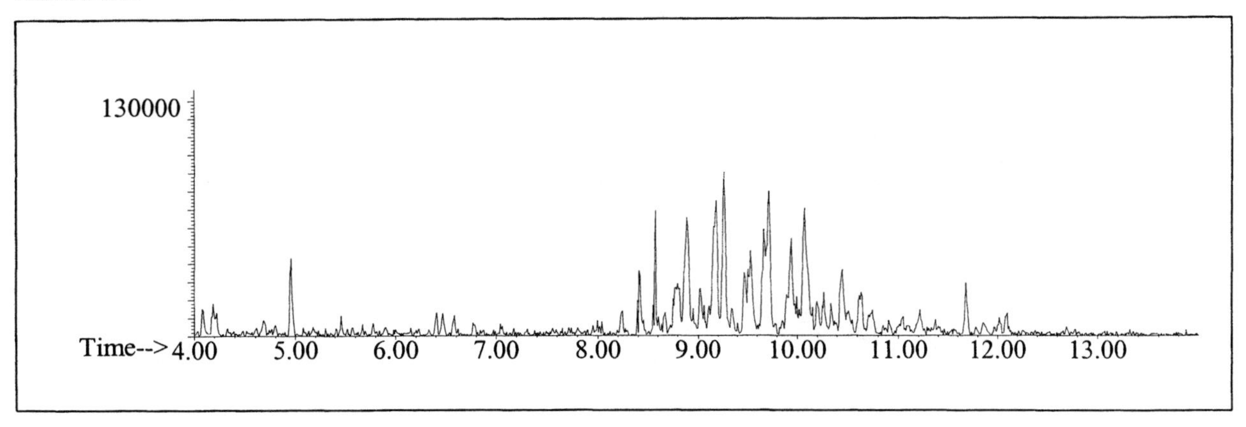

AROMATICS

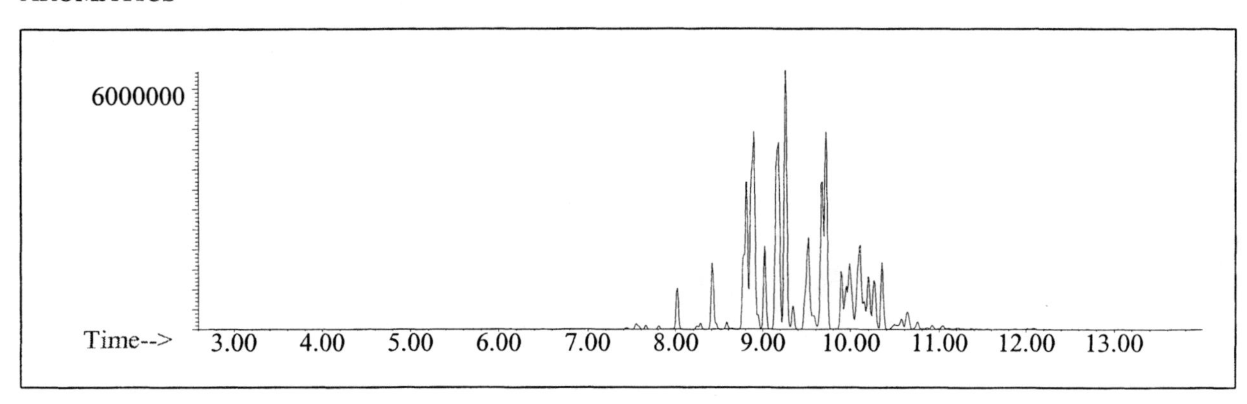

CYCLOPARAFFINS AND ALKENES

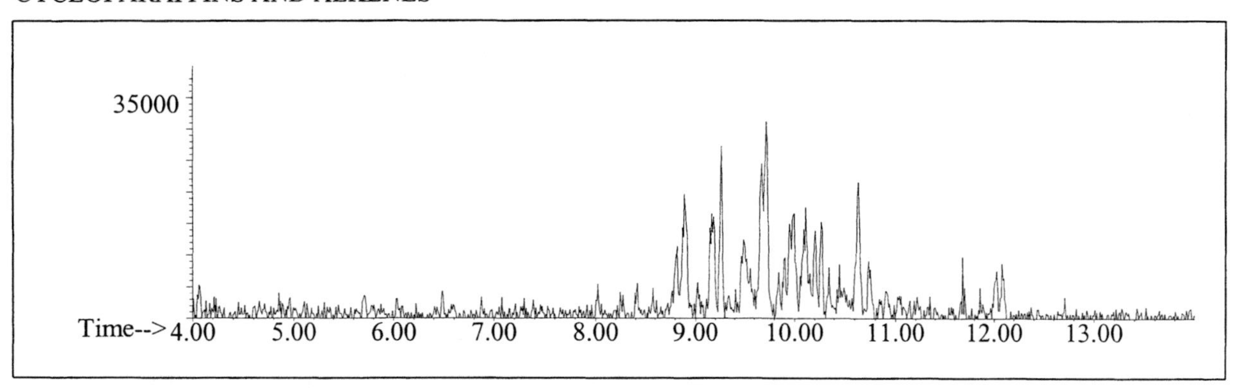

NAPHTHALENES

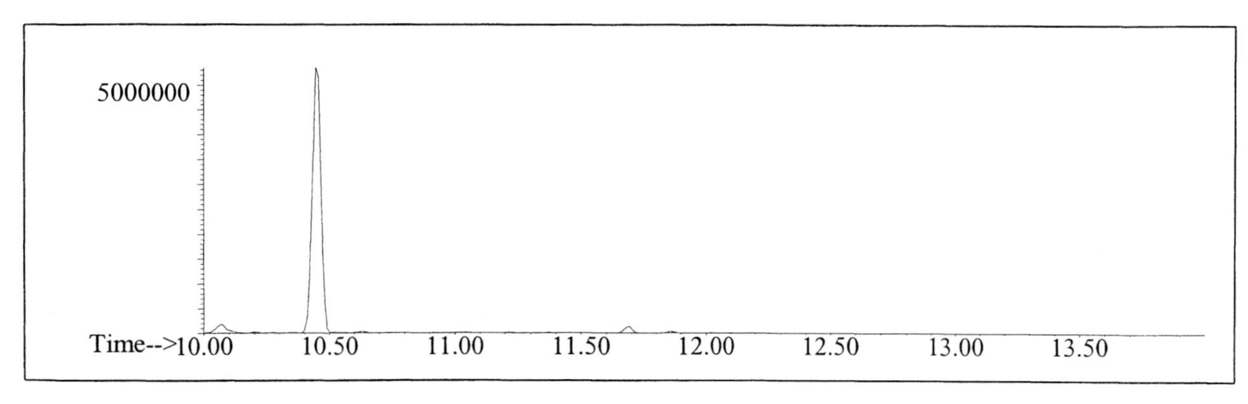

INDIVIDUAL PROFILES

Aromatic #9

Alkane

Ion 43

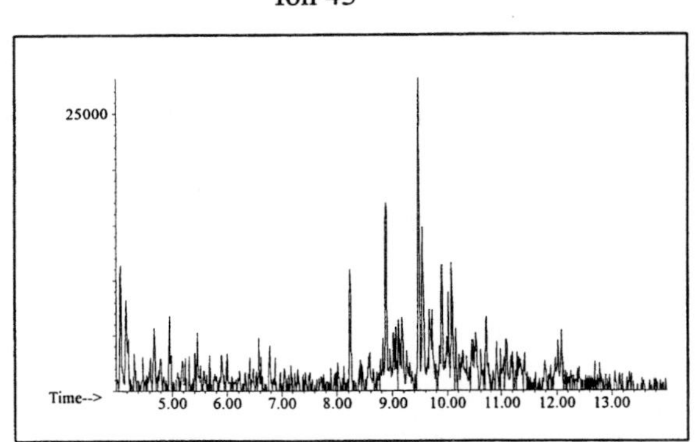

Ion 57

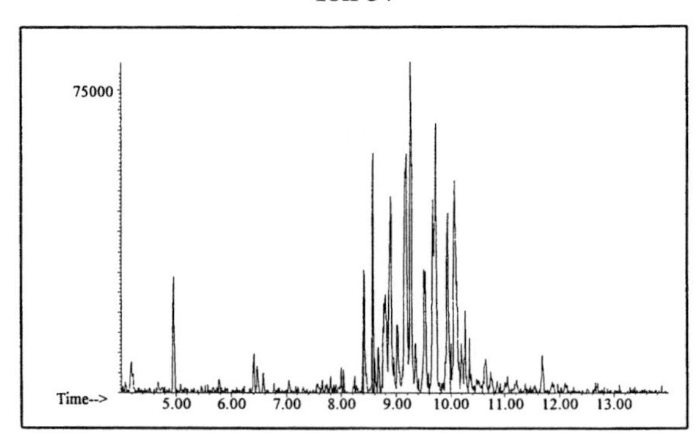

Ion 71

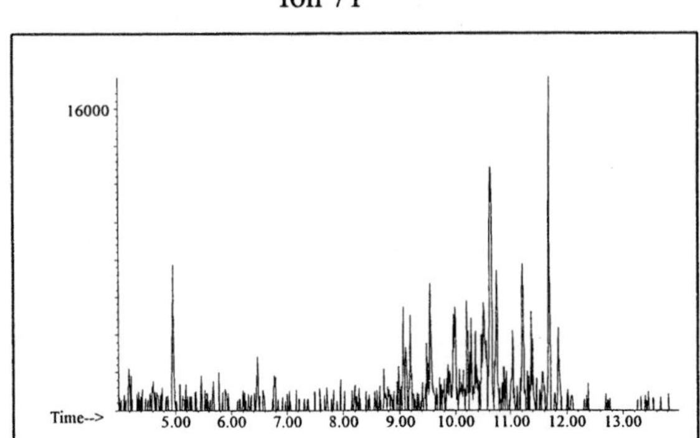

Ion 85

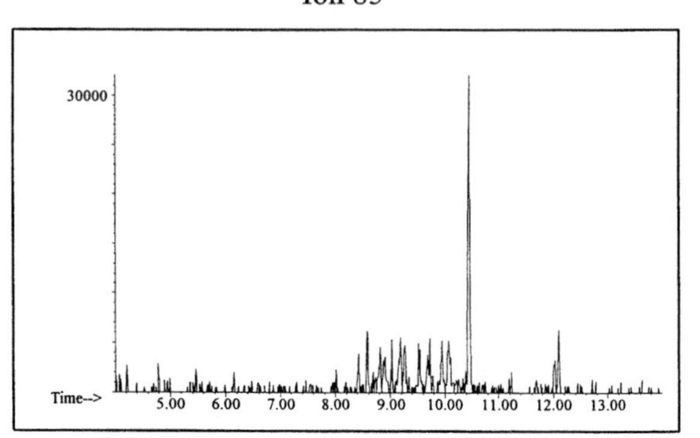

Aromatic

Ion 91

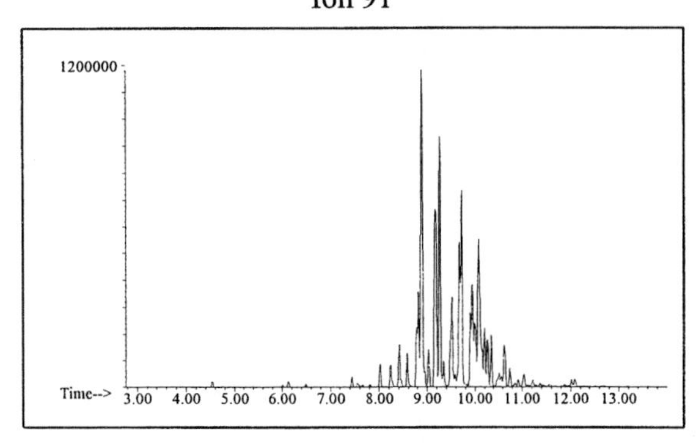

Ion 105

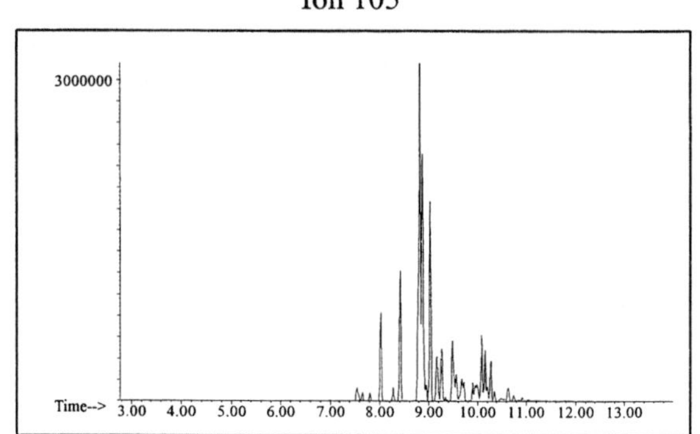

Ion 119

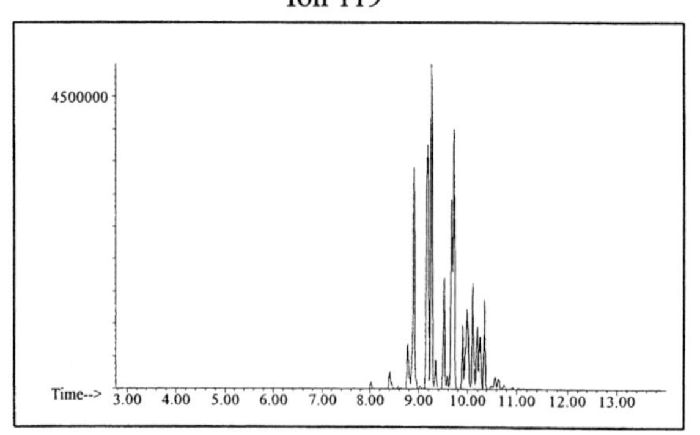

INDIVIDUAL PROFILES

Cycloparaffin

Ion 55

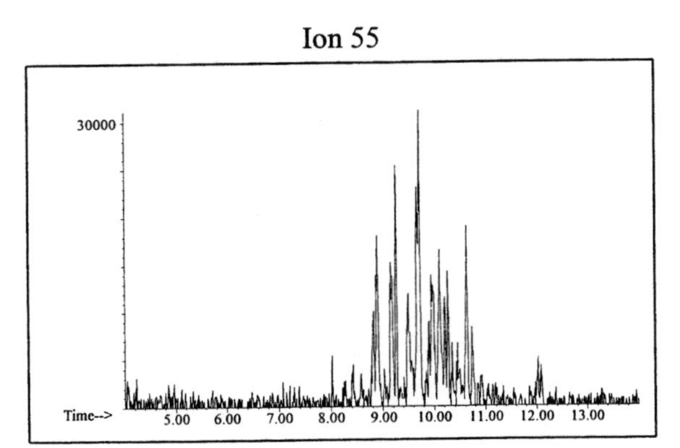

Ion 69

No Useful Data Obtained

Ion 83

No Useful Data Obtained

Naphthalene

Ion 128

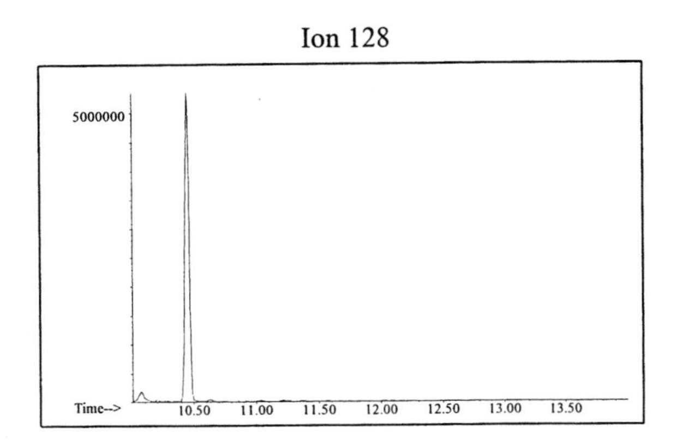

Ion 142

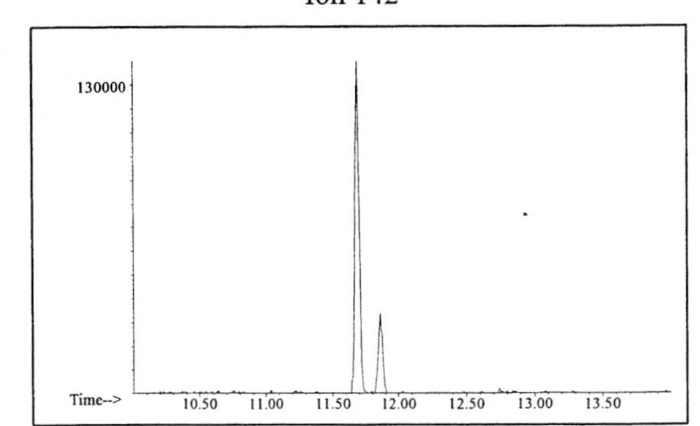

Ion 156

No Useful Data Obtained

Aromatic #10

20uL/mL Pentane

Product Displayed: Exxon 150

Other Similar Products:

ASTM: Class 0.4 (Aromatic)

Product Uses: Solvent, feedstock

Macro Code: M

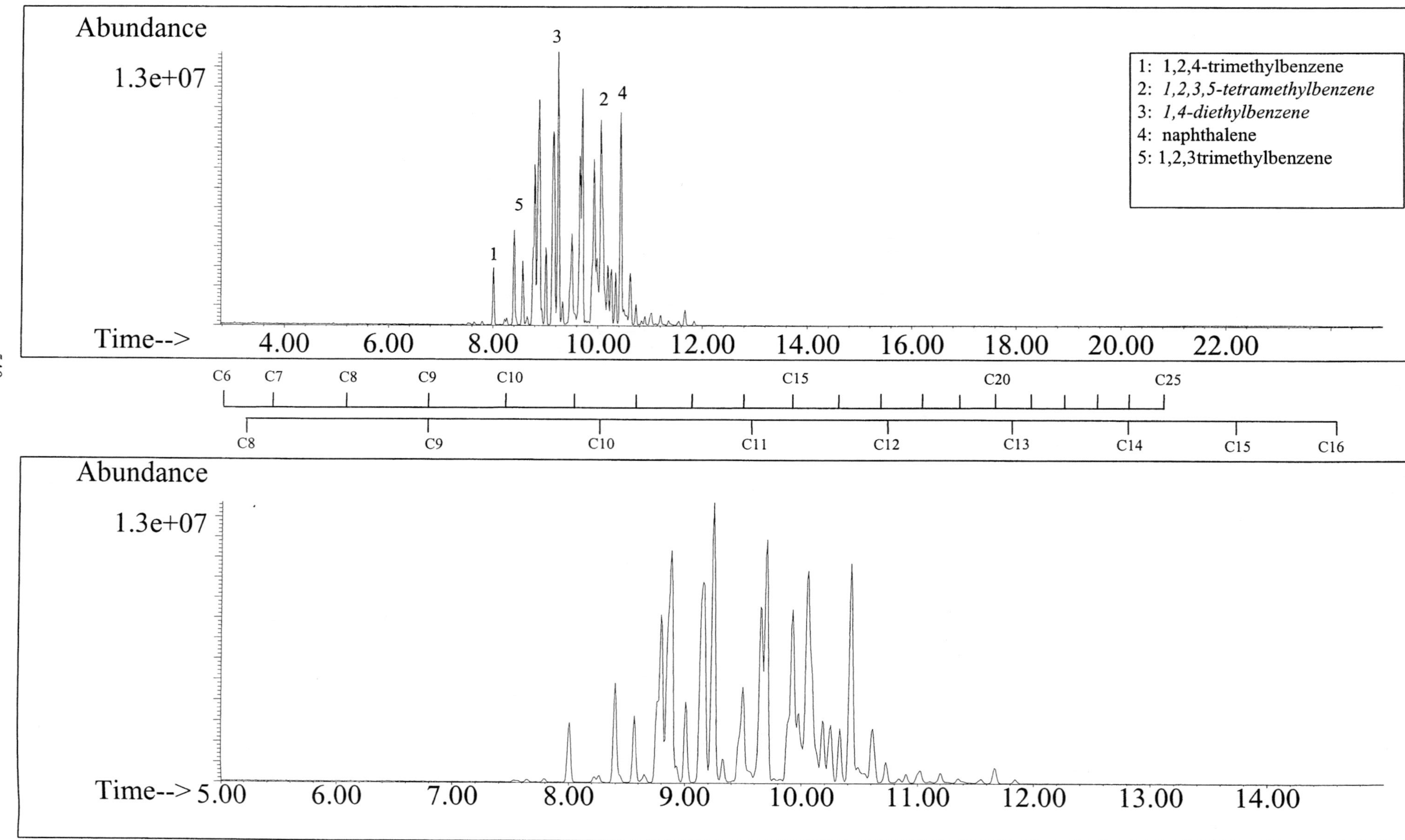

SUMMED PROFILES

ALKANES

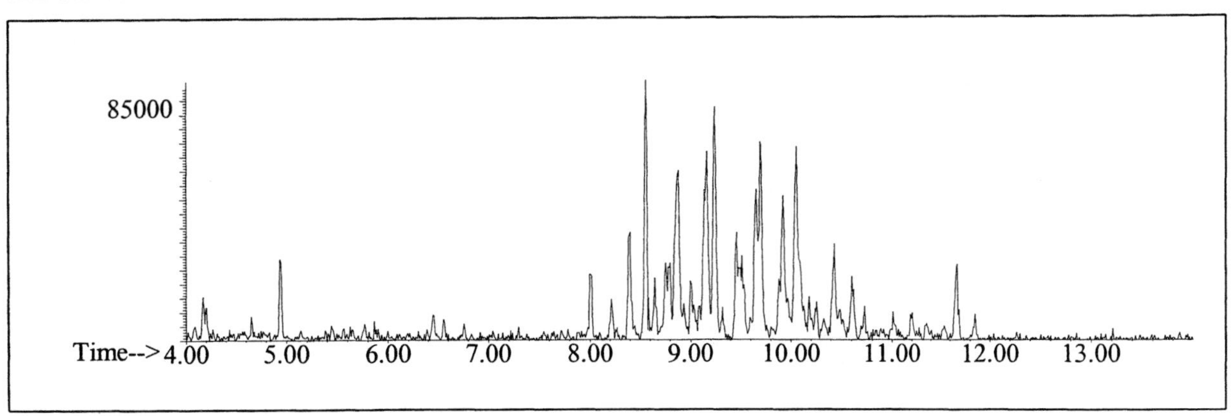

AROMATICS

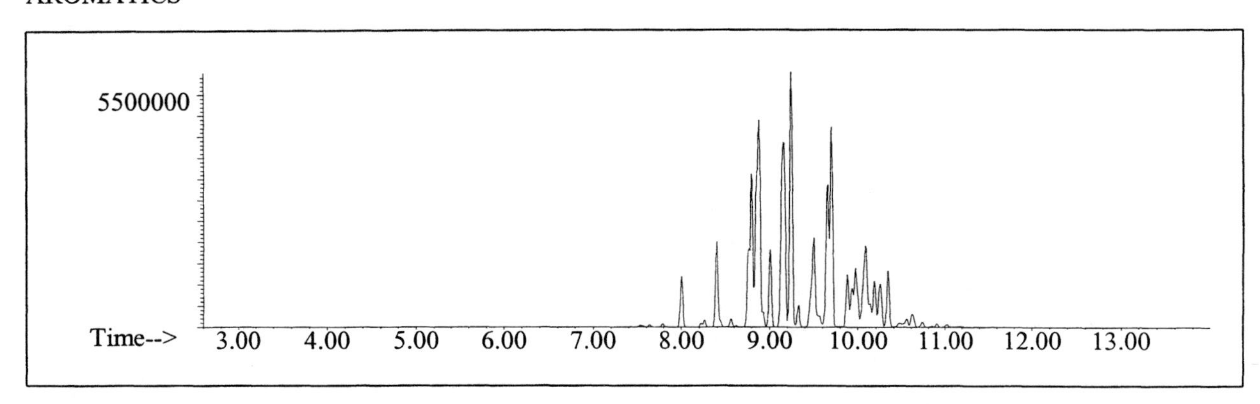

CYCLOPARAFFINS AND ALKENES

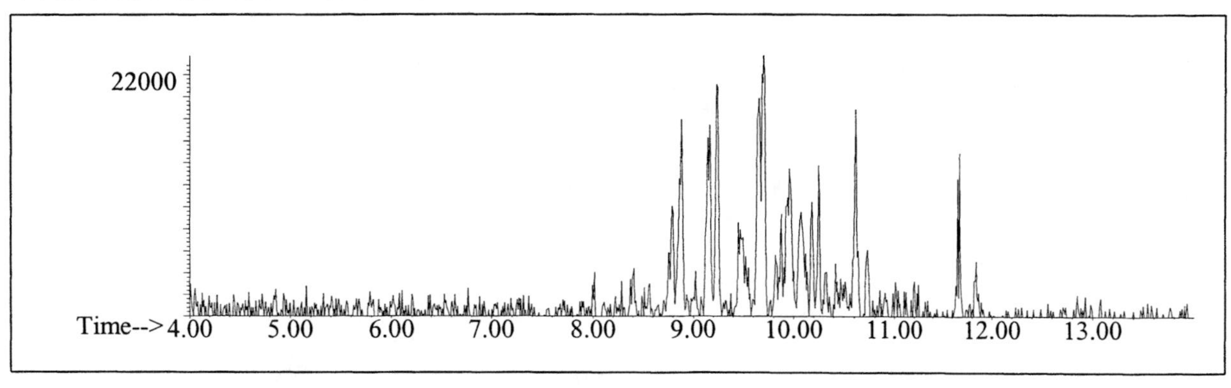

NAPHTHALENES

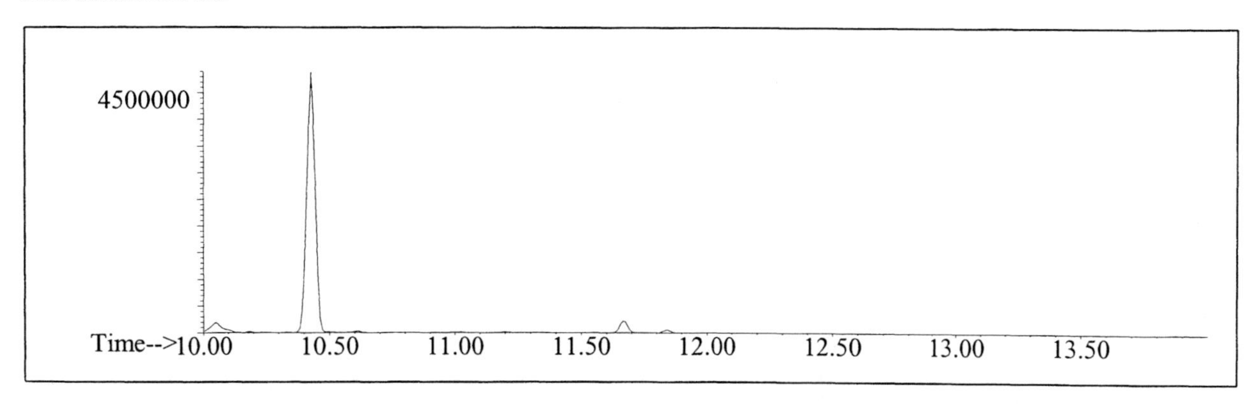

INDIVIDUAL PROFILES

Aromatic #10

Alkane

Ion 43

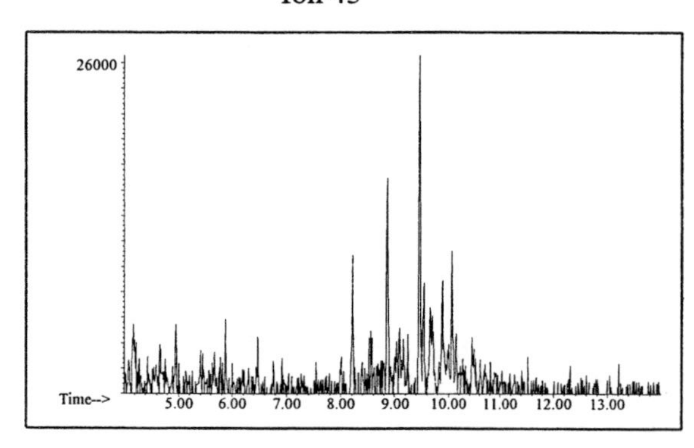

Ion 57

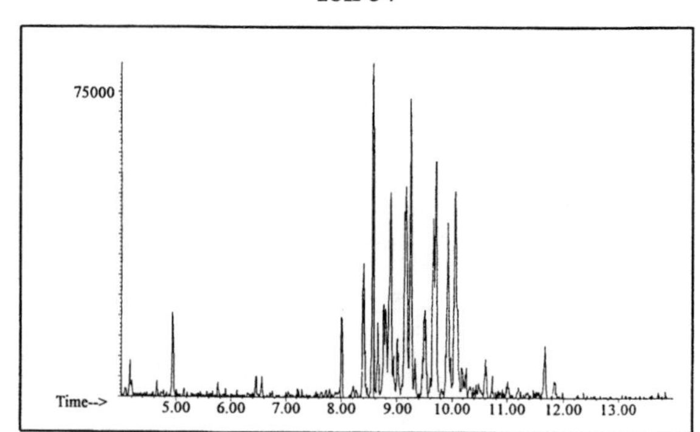

Ion 71

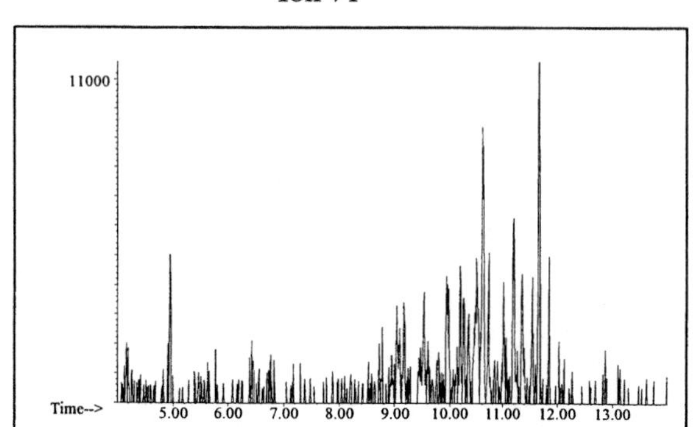

Ion 85

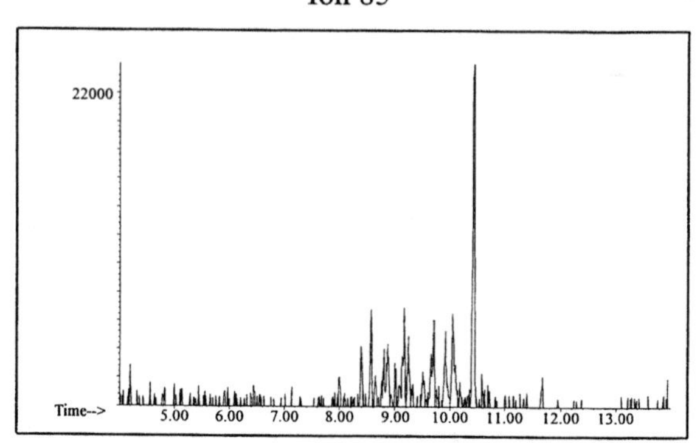

Aromatic

Ion 91

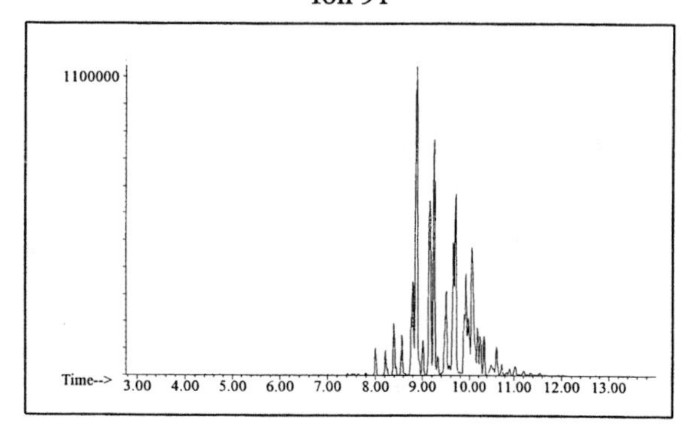

Ion 105

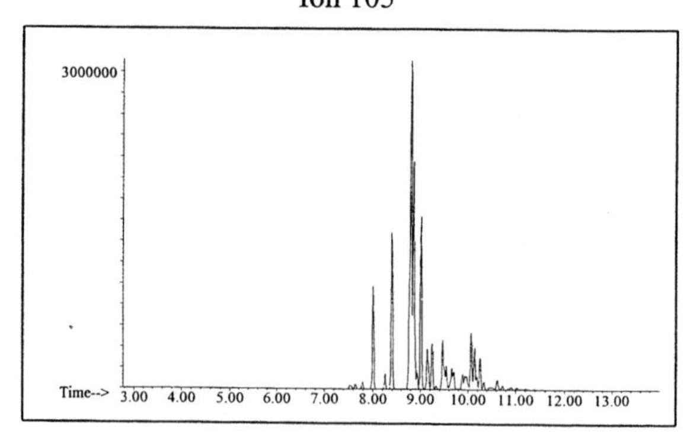

Ion 119

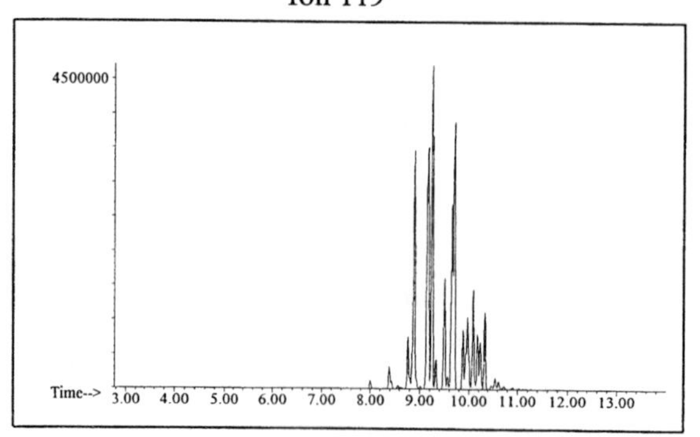

INDIVIDUAL PROFILES

Cycloparaffin

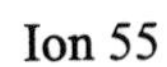

Ion 55

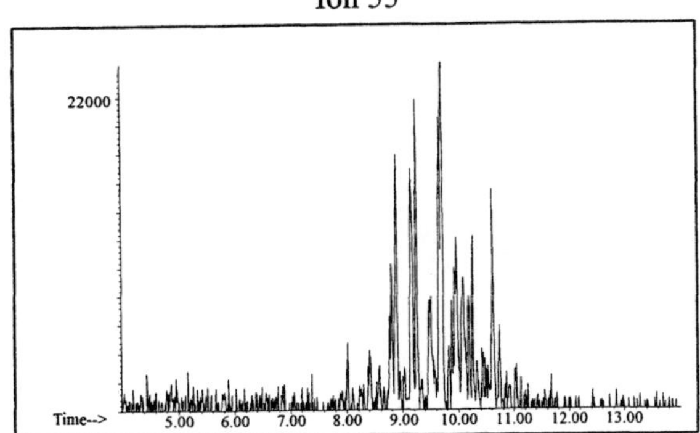

Ion 69

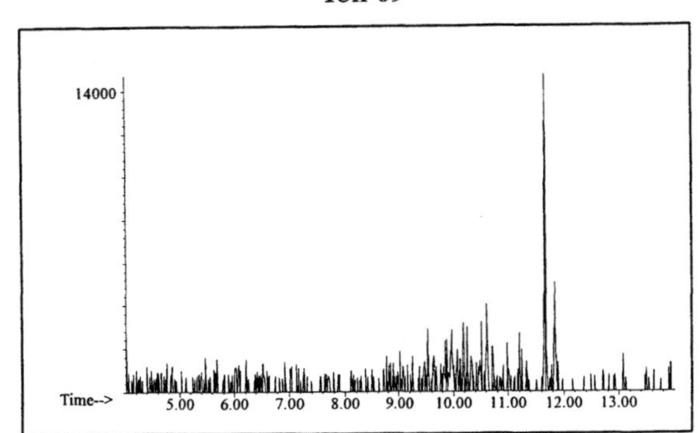

Ion 83

No Useful Data Obtained

Naphthalene

Ion 128

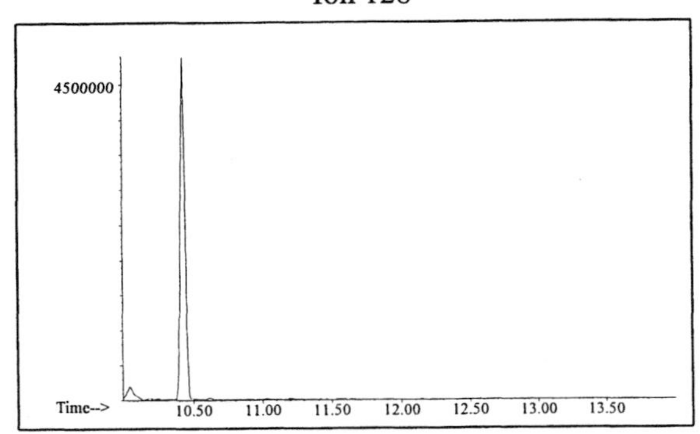

Ion 142

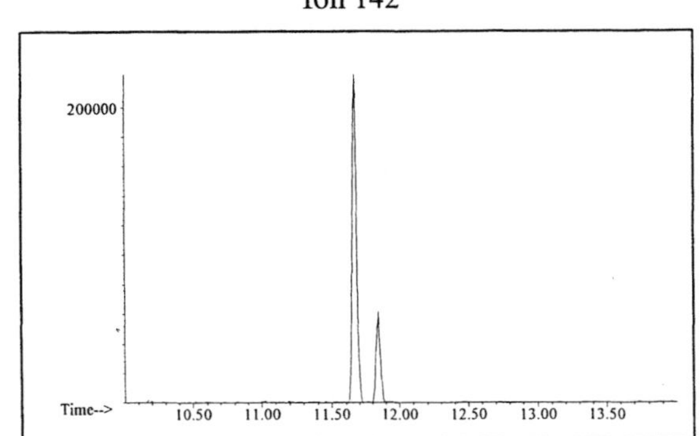

Ion 156

No Useful Data Obtained

Aromatic #11

20uL/mL Pentane

Product Displayed: Kleanstrip Concrete and Drive Cleaner

Other Similar Products:

ASTM: Class 0.4 (Aromatic)

Product Uses: Cleaning solvent

Macro Code: M

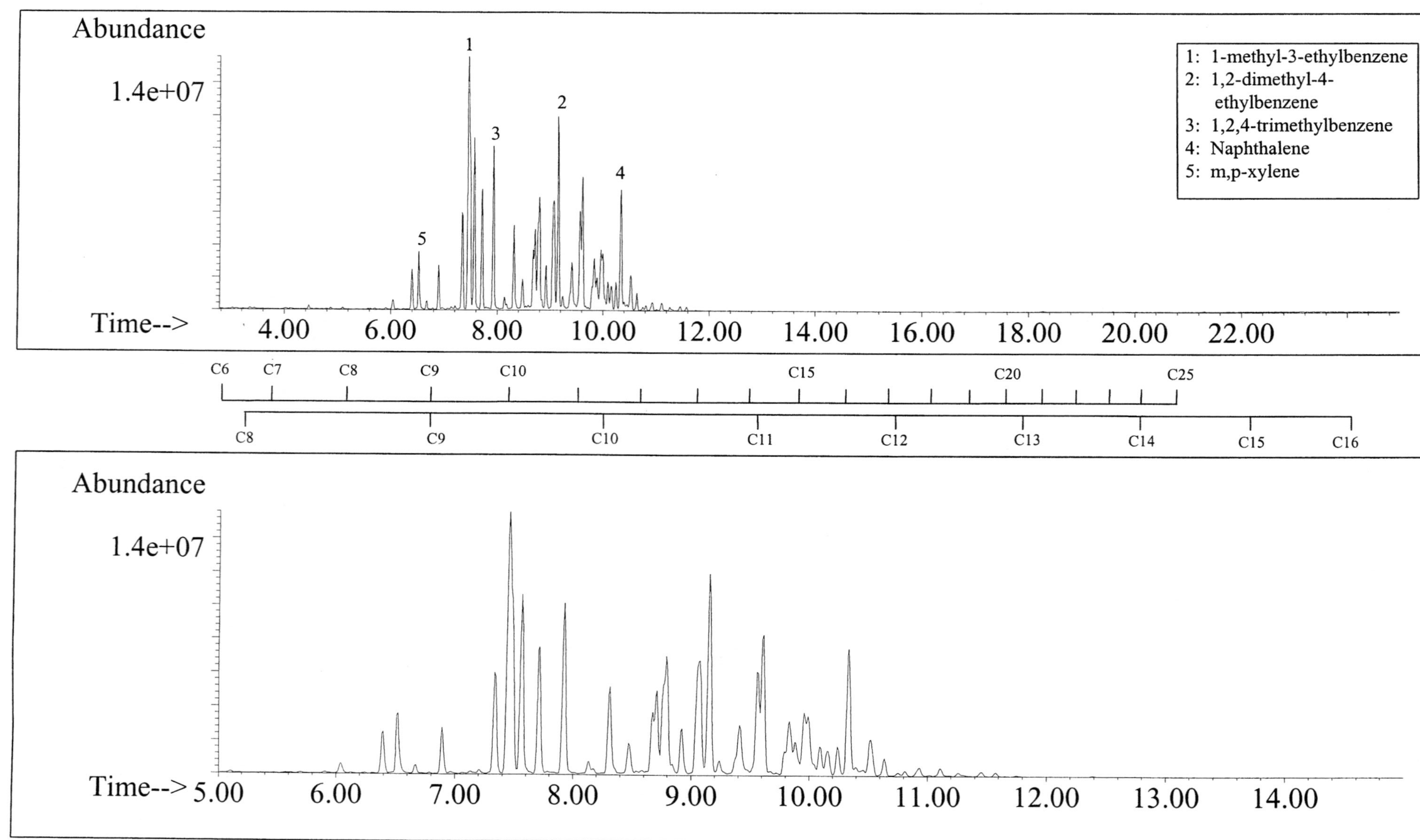

SUMMED PROFILES

Aromatic #11

ALKANES

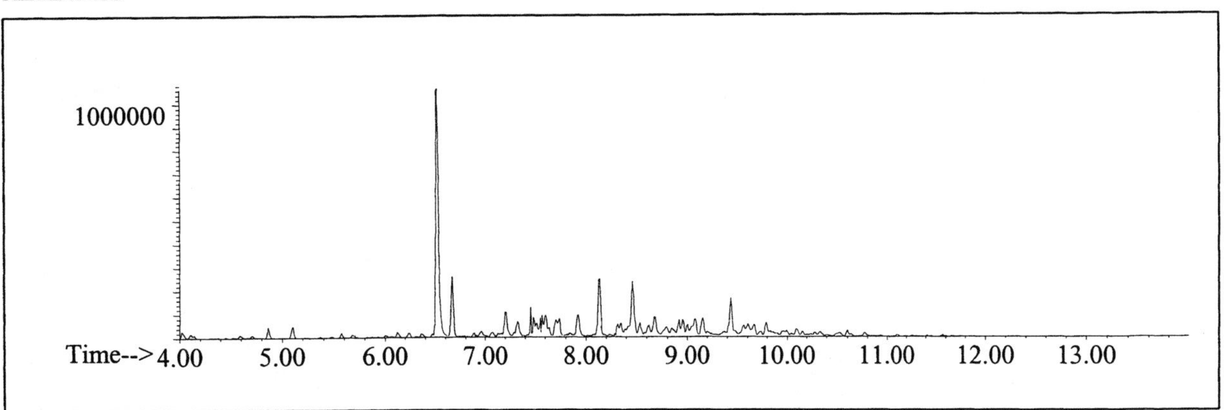

AROMATICS

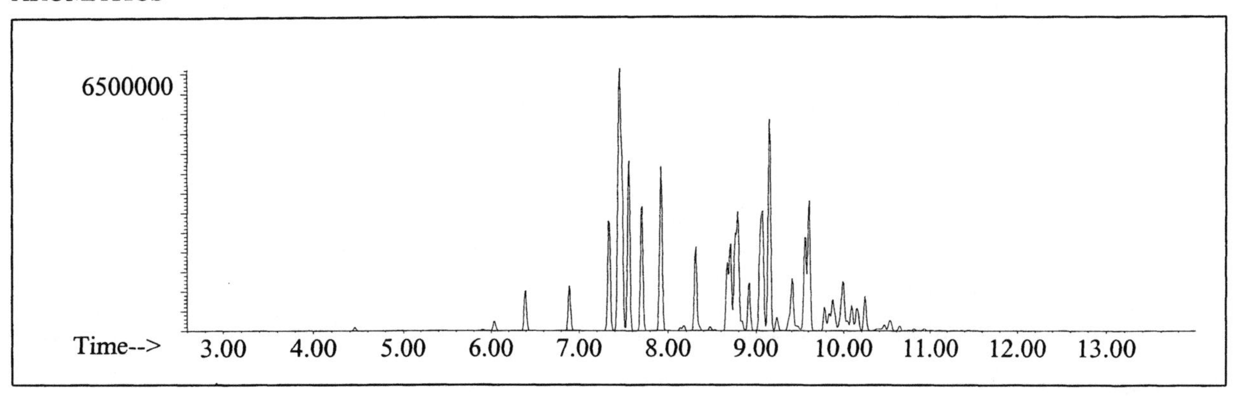

CYCLOPARAFFINS AND ALKENES

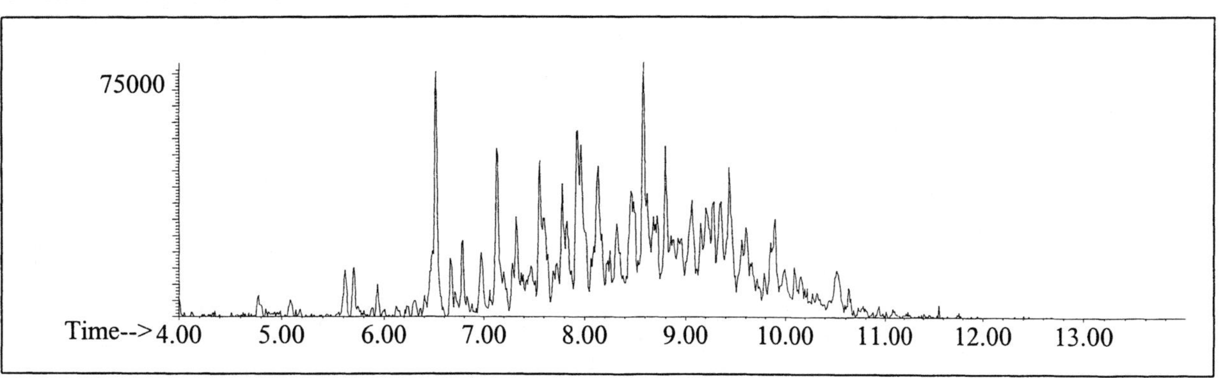

NAPHTHALENES

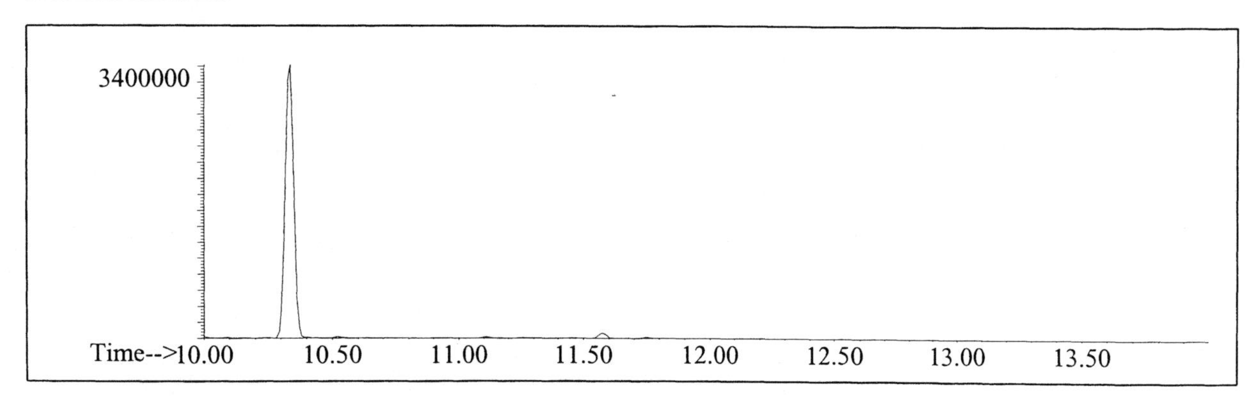

INDIVIDUAL PROFILES

Aromatic #11

Alkane

Ion 43

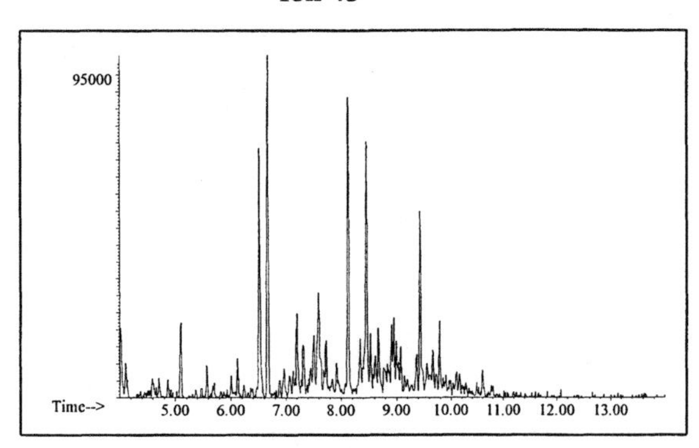

Ion 57

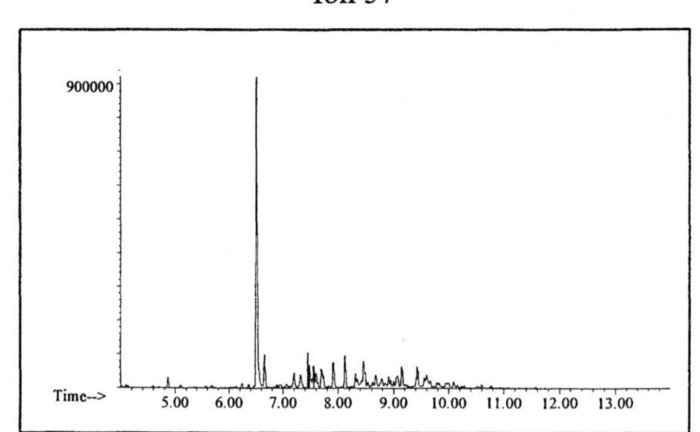

Ion 71

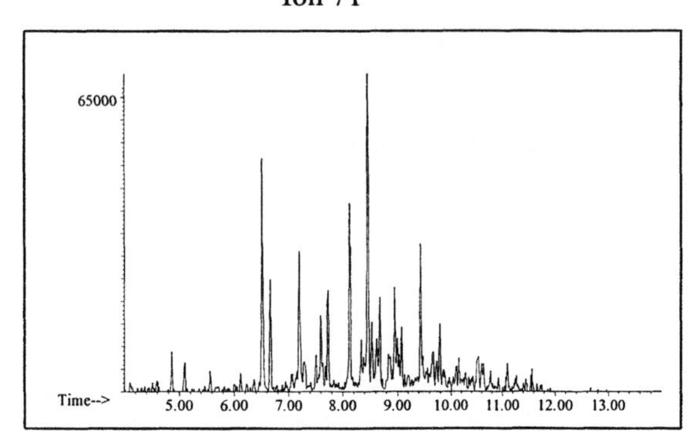

Ion 85

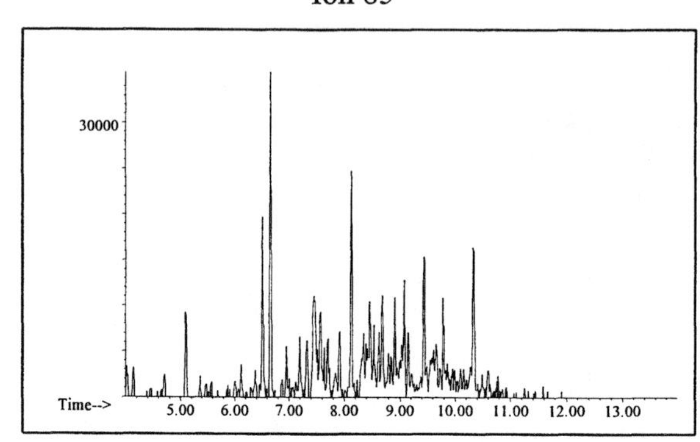

Aromatic

Ion 91

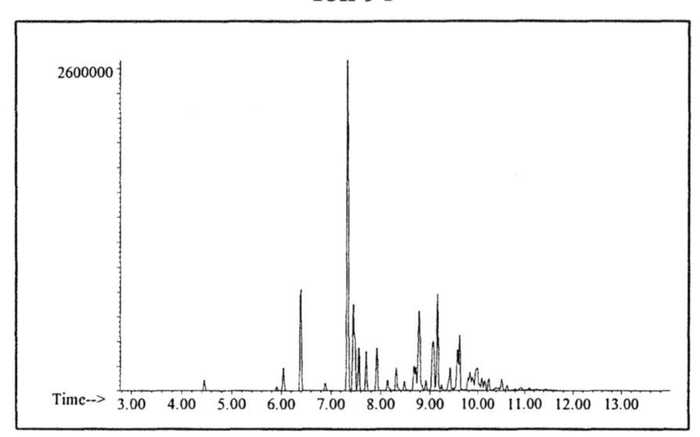

Ion 105

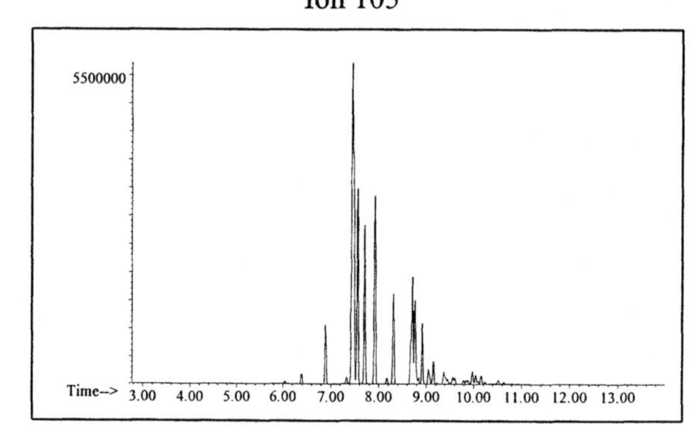

Ion 119

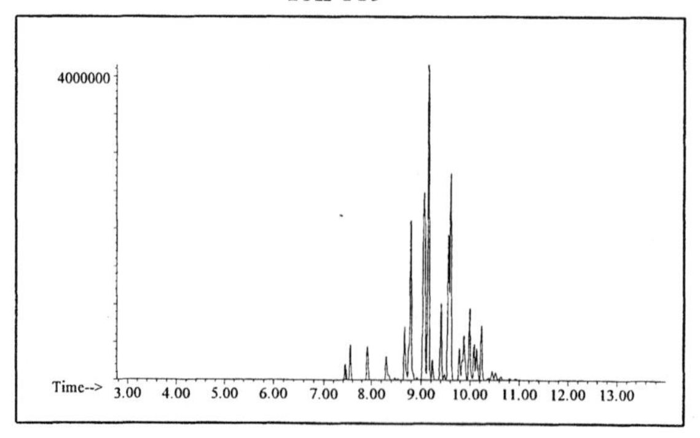

INDIVIDUAL PROFILES

Aromatic #11

Cycloparaffin

Ion 55

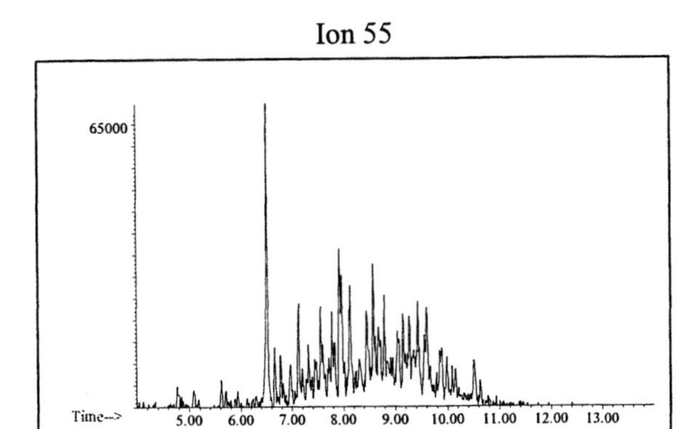

Ion 69

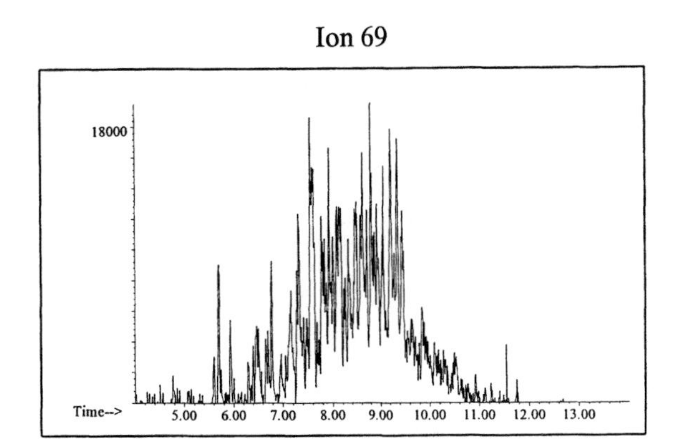

Ion 83

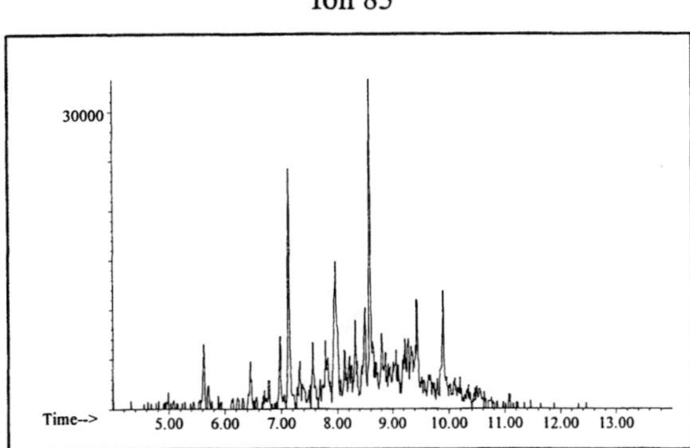

Naphthalene

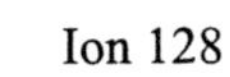

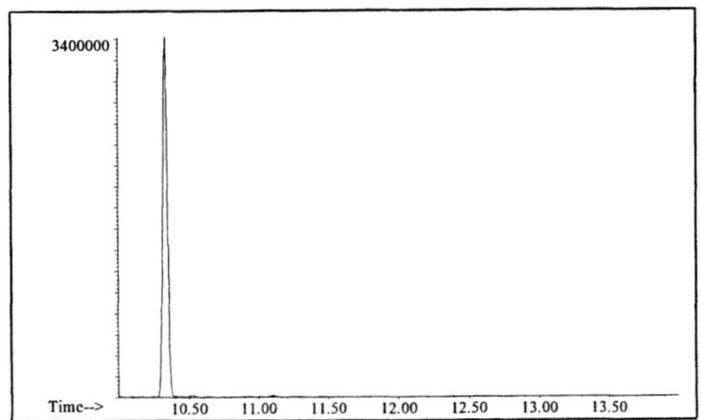

Ion 142

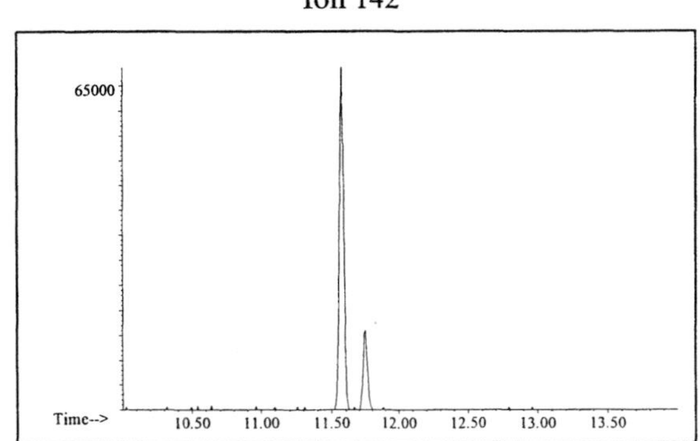

Ion 156

No Useful Data Obtained

Aromatic #12

20uL/mL Pentane

Product Displayed: Exxon 200

Other Similar Products:

ASTM: Class 0.4 (Aromatic)

Product Uses: Solvent, feedstock

Macro Code: M

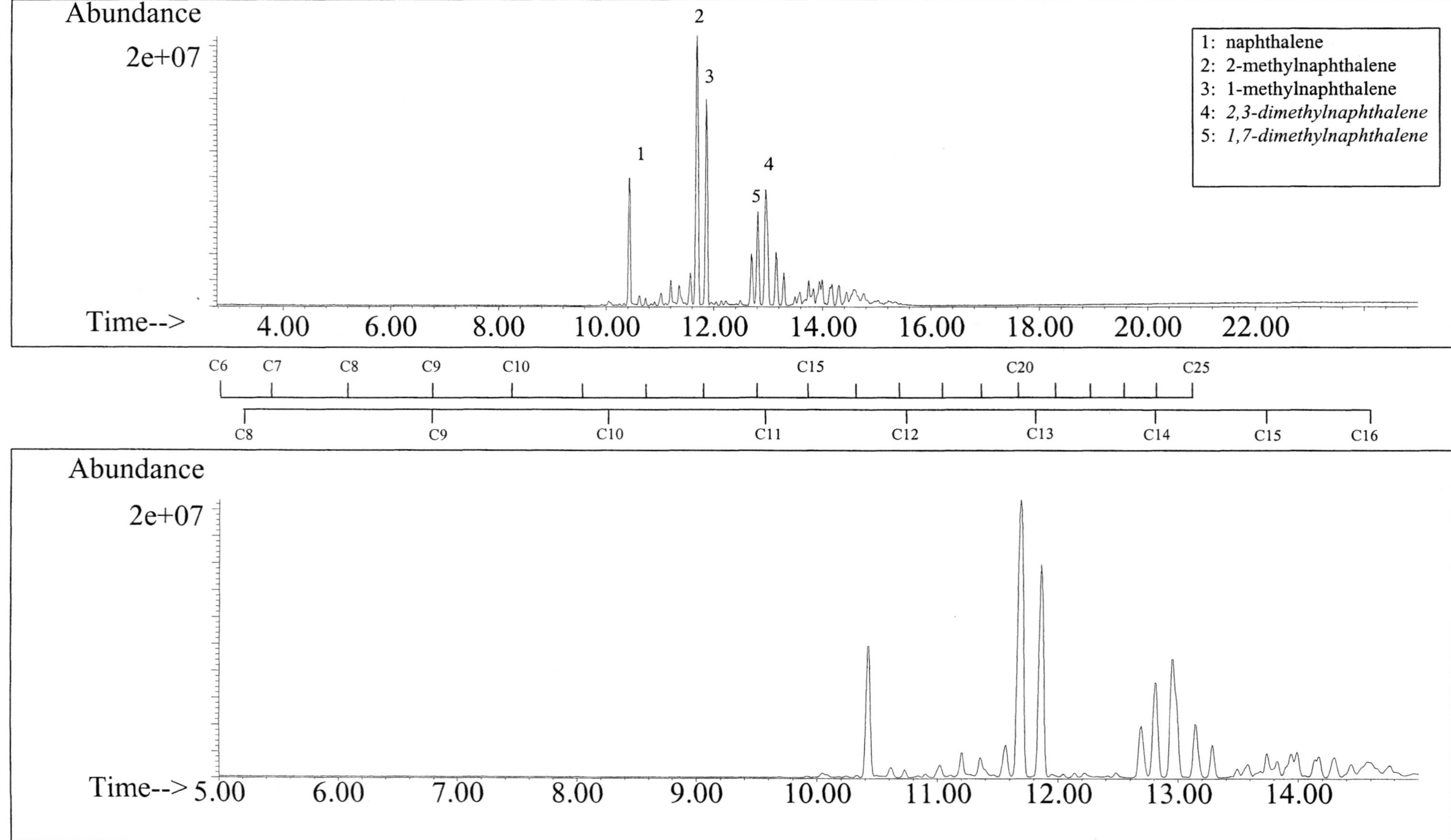

ALKANES

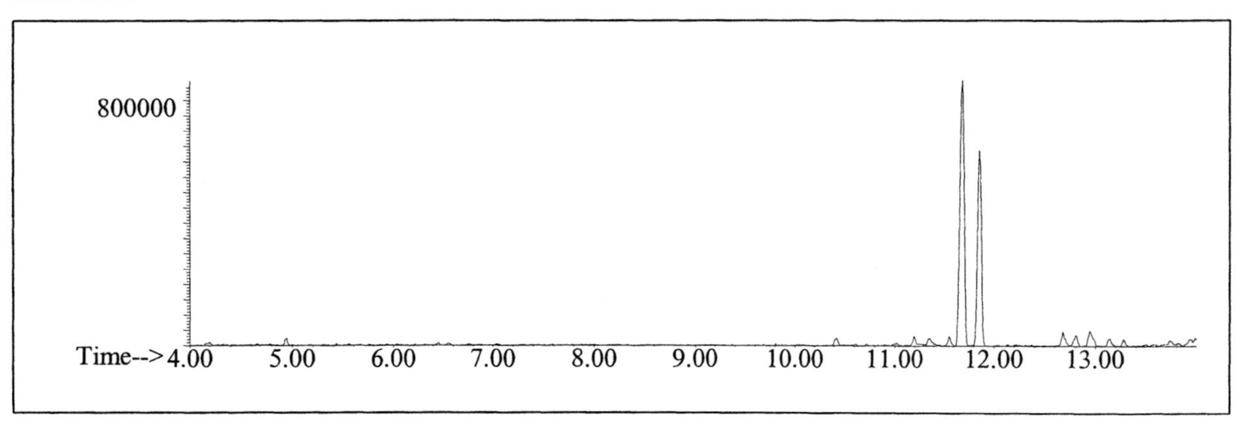

AROMATICS

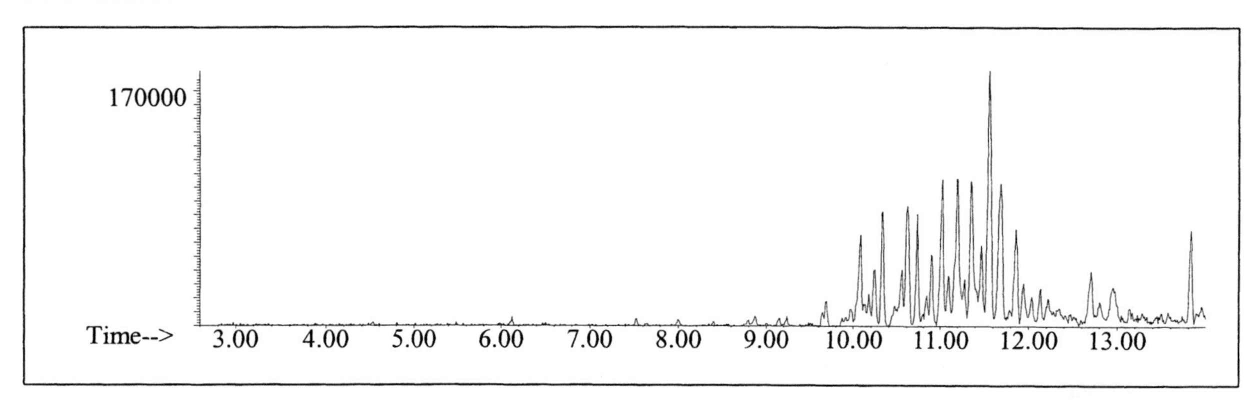

CYCLOPARAFFINS AND ALKENES

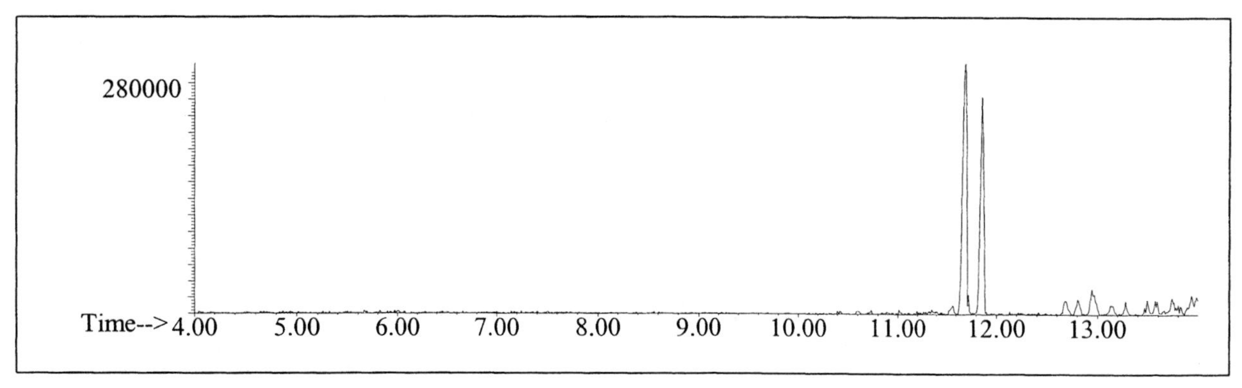

NAPHTHALENES

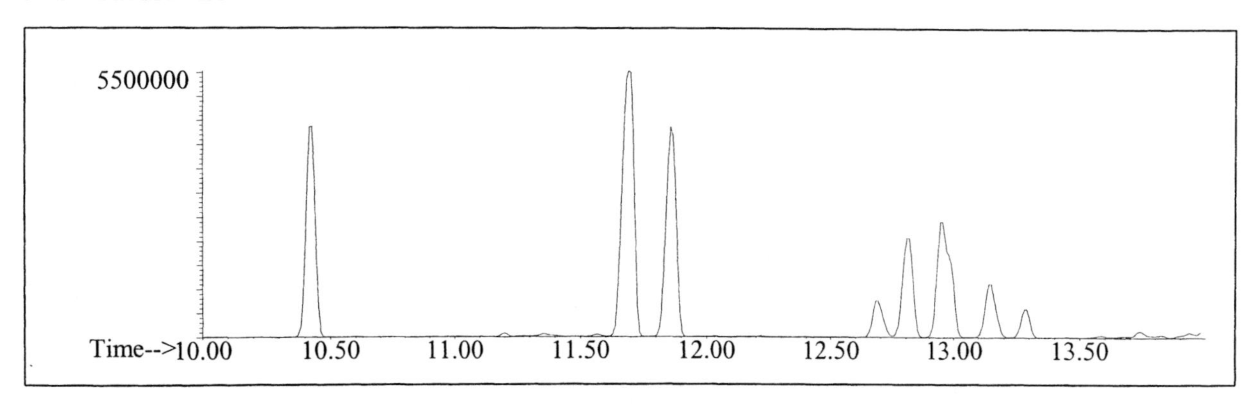

INDIVIDUAL PROFILES

Aromatic #12

Alkane

Ion 43

No Useful Data Obtained

Ion 57

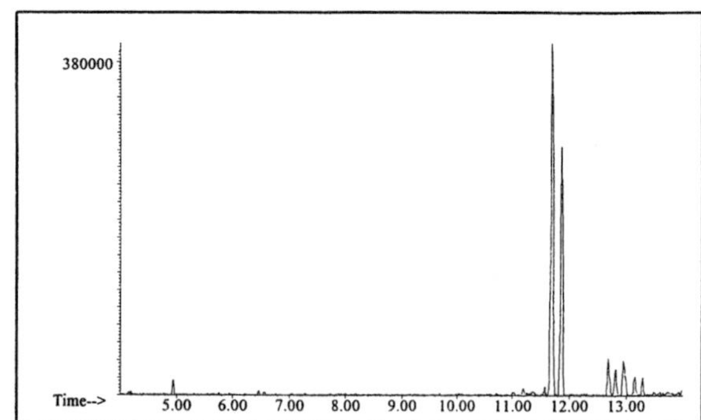

Ion 71

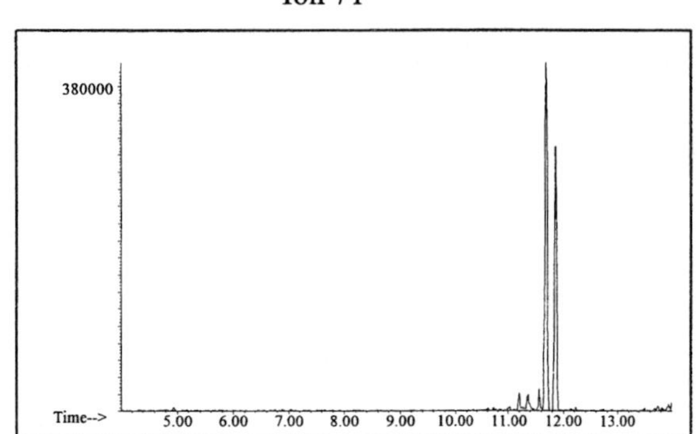

Ion 85

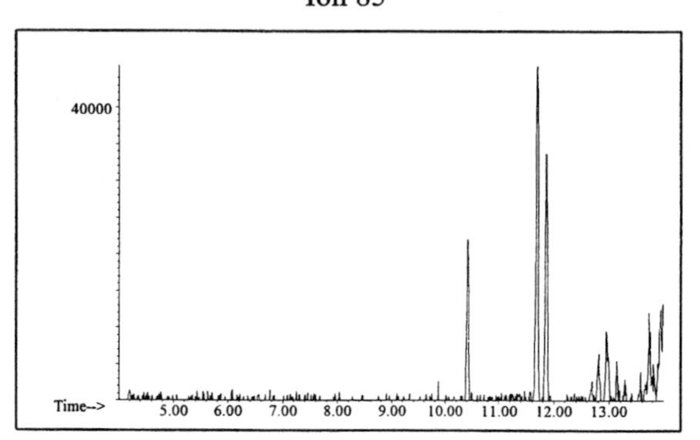

Aromatic

Ion 91

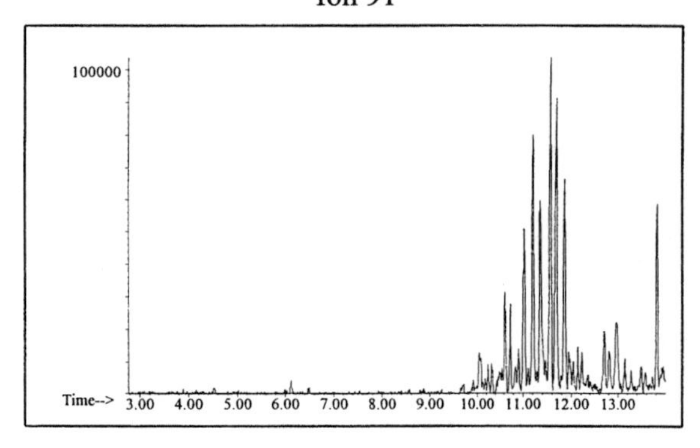

Ion 105

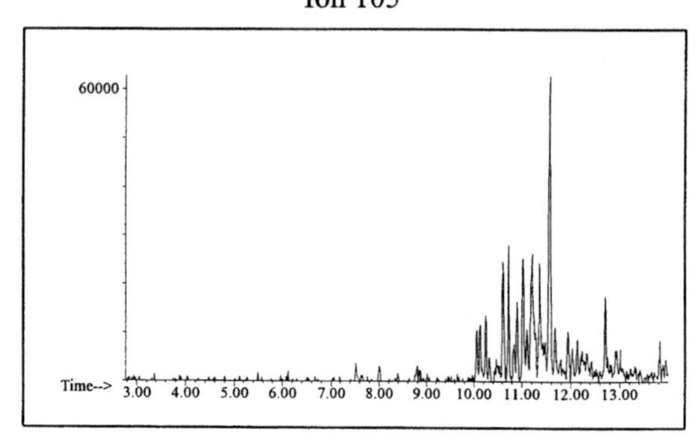

Ion 119

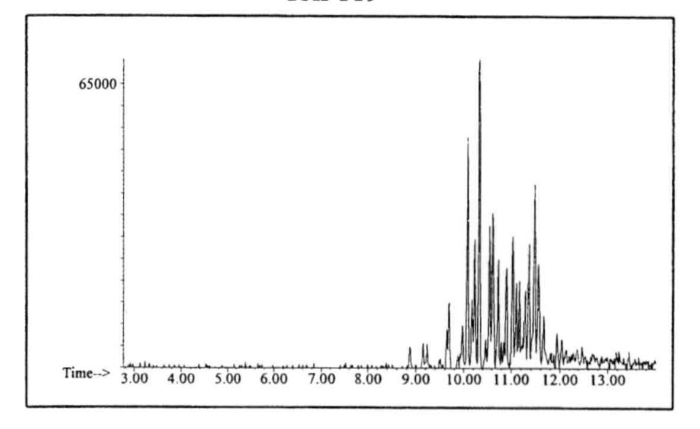

Cycloparaffin

Ion 55

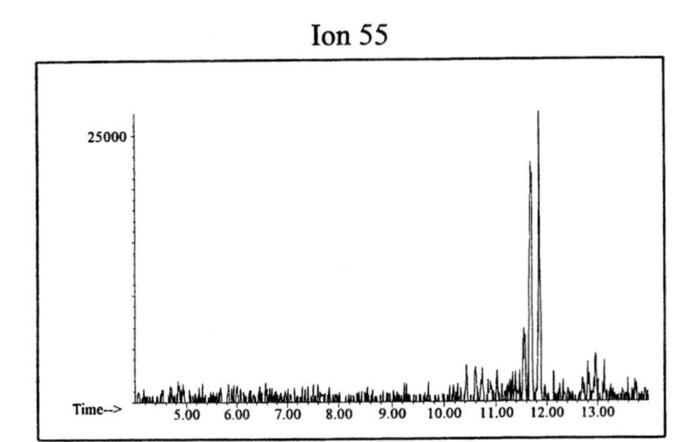

Ion 69

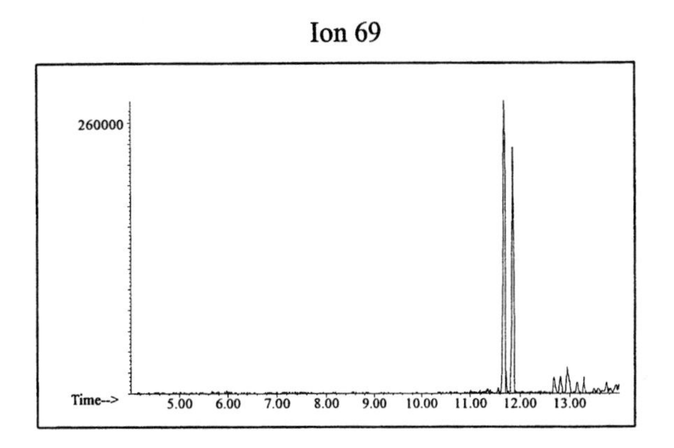

Ion 83

No Useful Data Obtained

Naphthalene

Ion 128

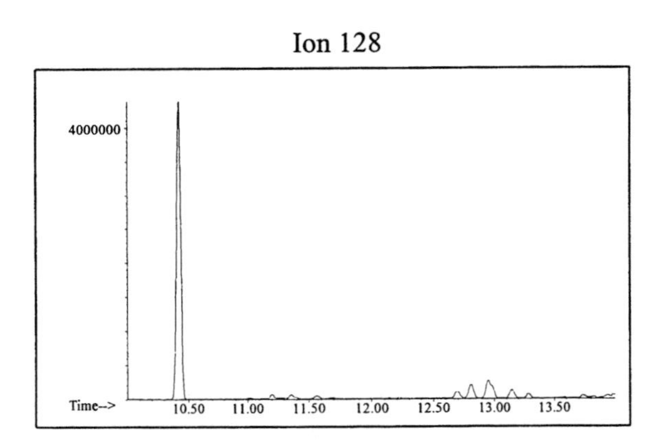

Ion 142

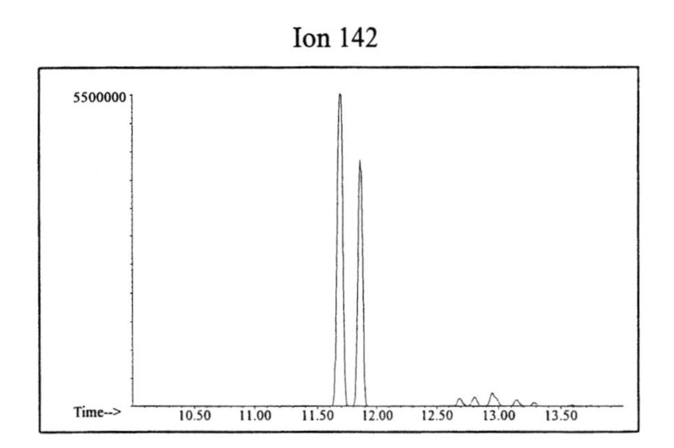

Ion 156

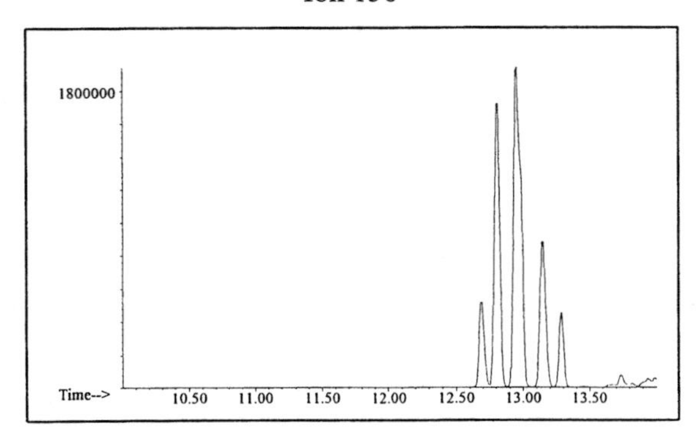

Naphthenic-Paraffinic Products

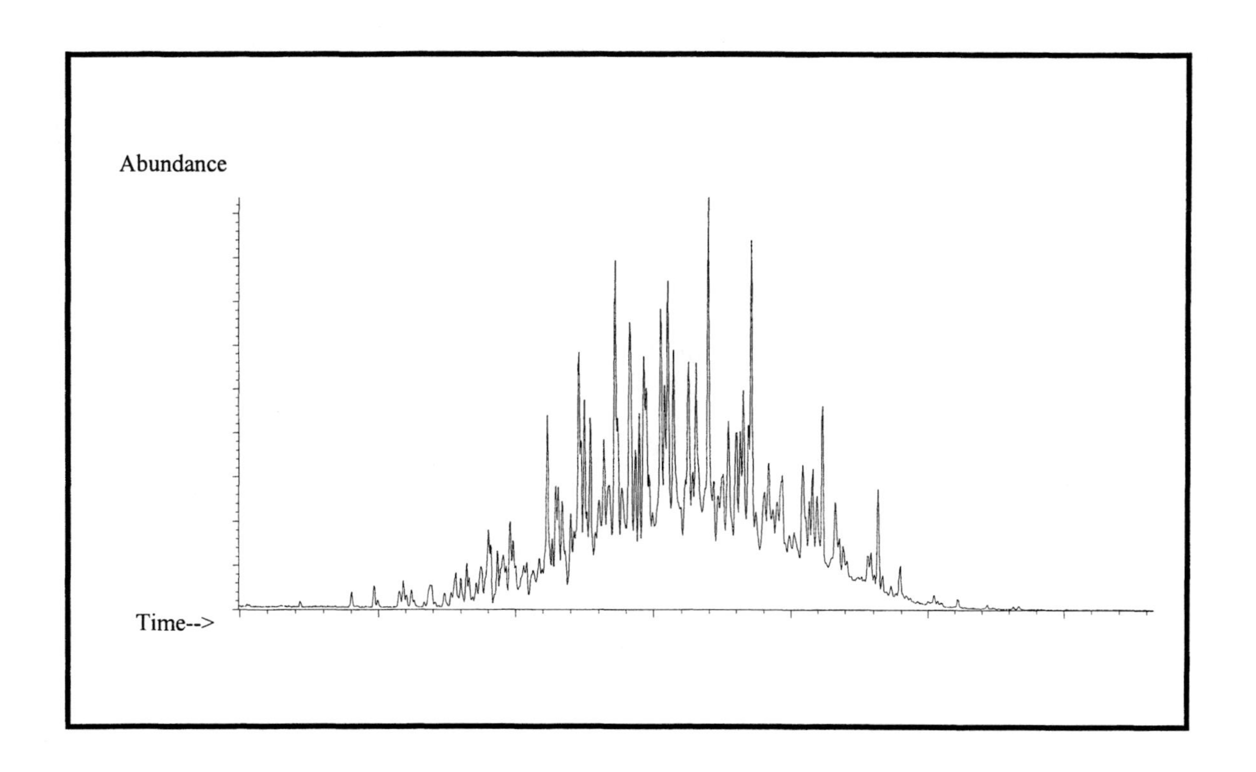

Naphthenic-Paraffinic #1

20uL/mL Pentane

Product Displayed: Gulflite One Match Charcoal Starter

Other Similar Products: Boyle Midway Gulflite

ASTM: Class 0.5 (Naphthenic-paraffinic)

Product Uses: Charcoal lighter

Macro Code: M

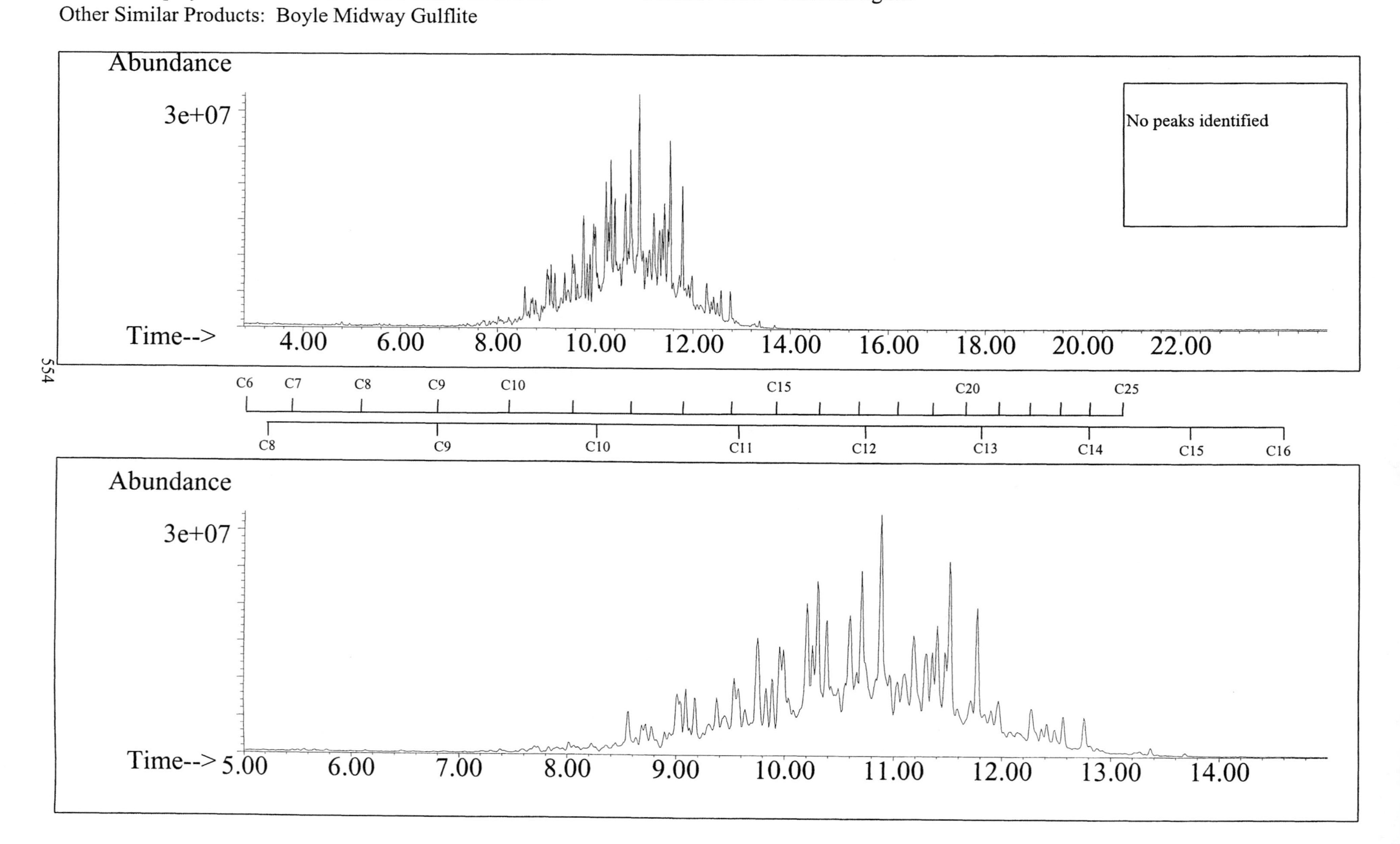

ALKANES

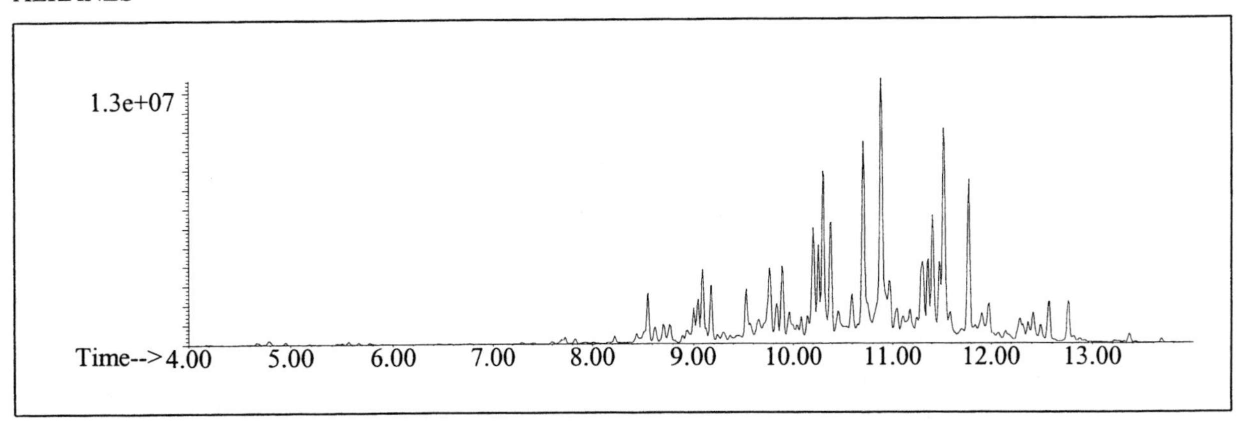

AROMATICS

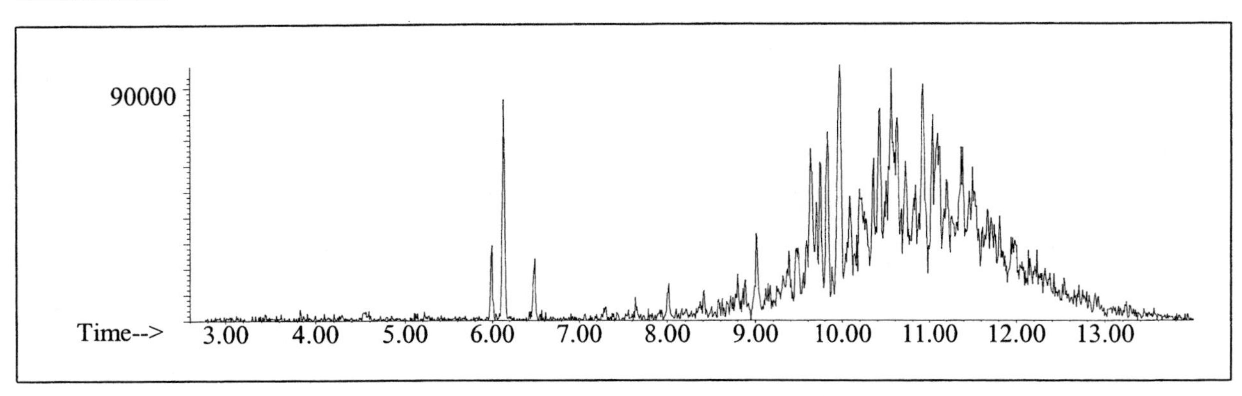

CYCLOPARAFFINS AND ALKENES

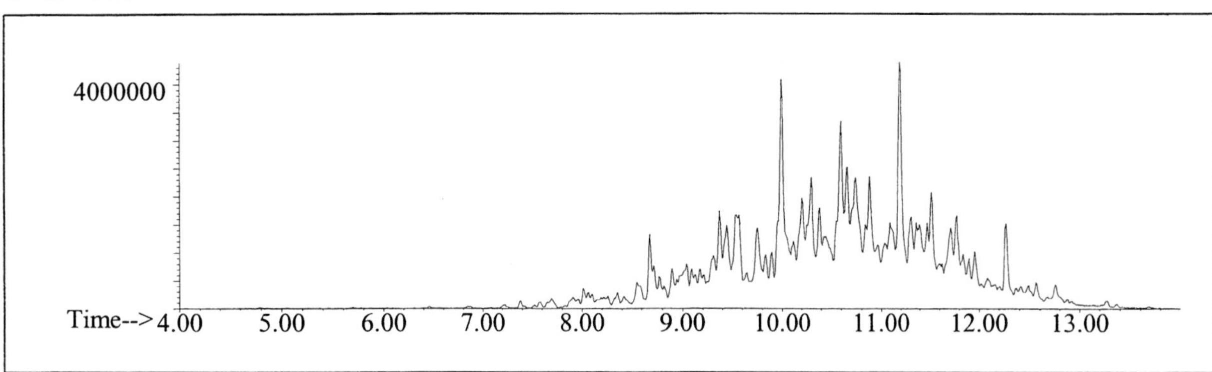

NAPHTHALENES

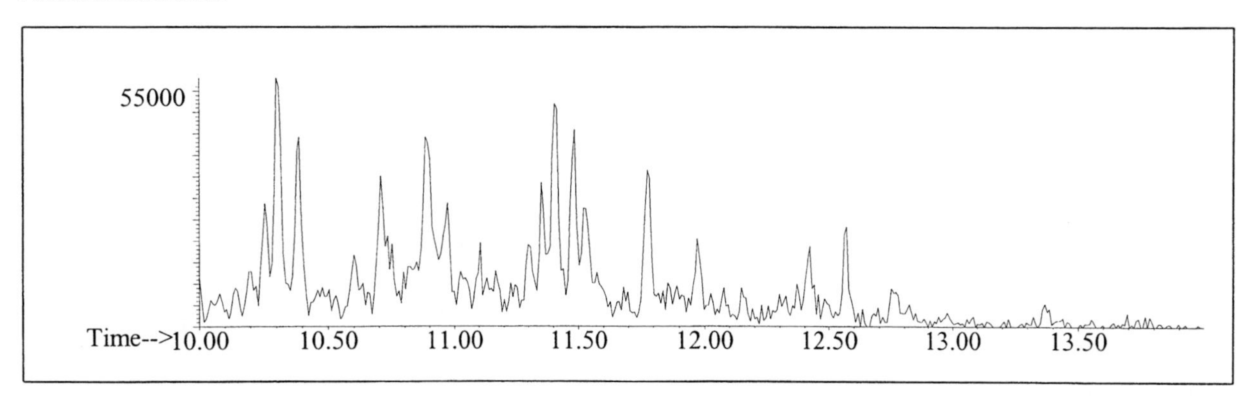

INDIVIDUAL PROFILES Naphthenic-Paraffinic #1

Alkane

Ion 43

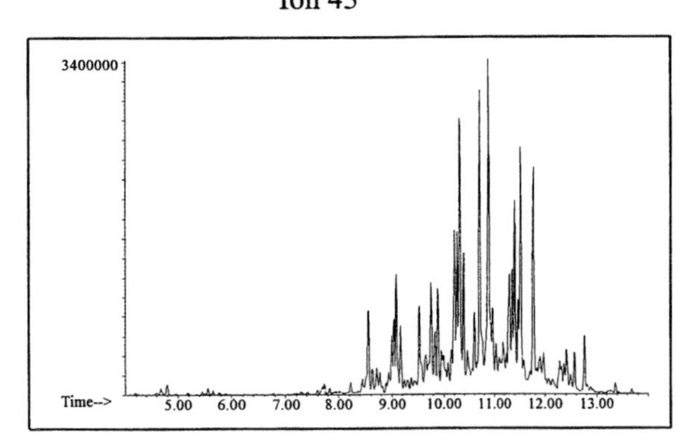

Ion 57

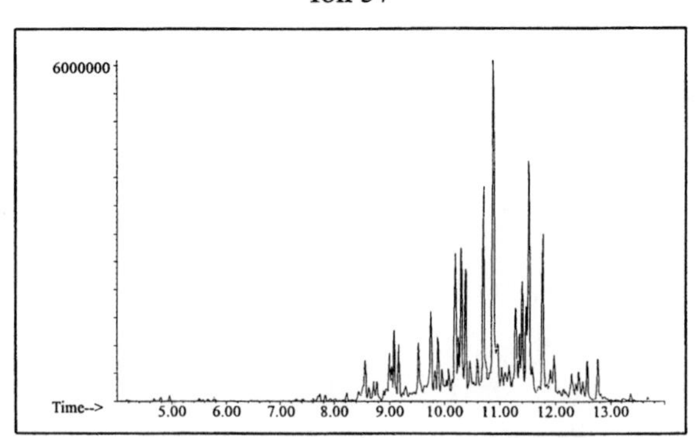

Ion 71

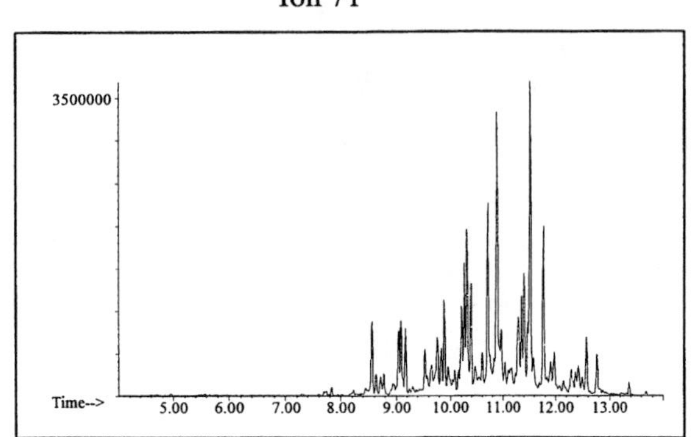

Ion 85

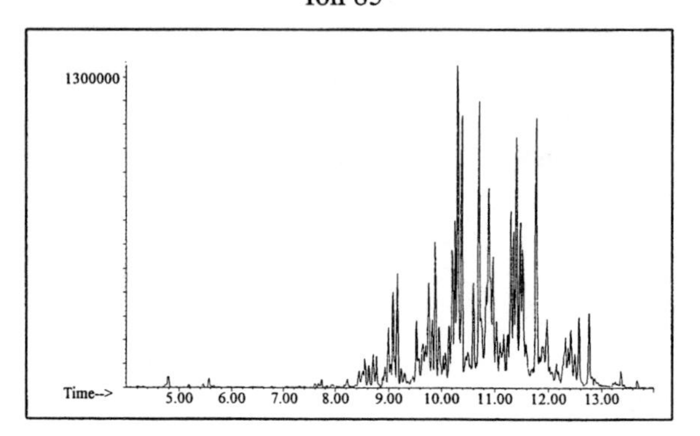

Aromatic

Ion 91

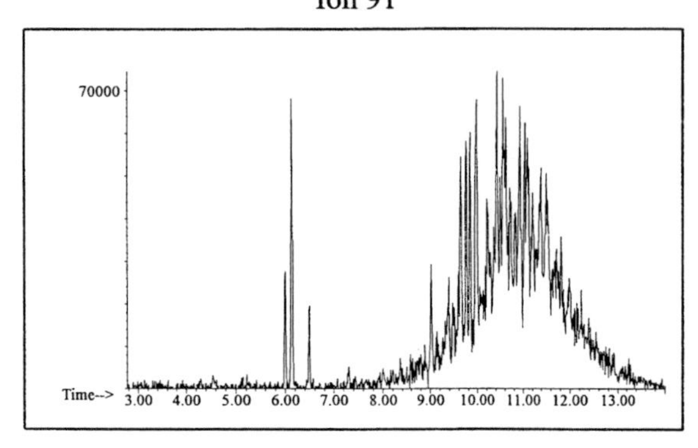

Ion 105

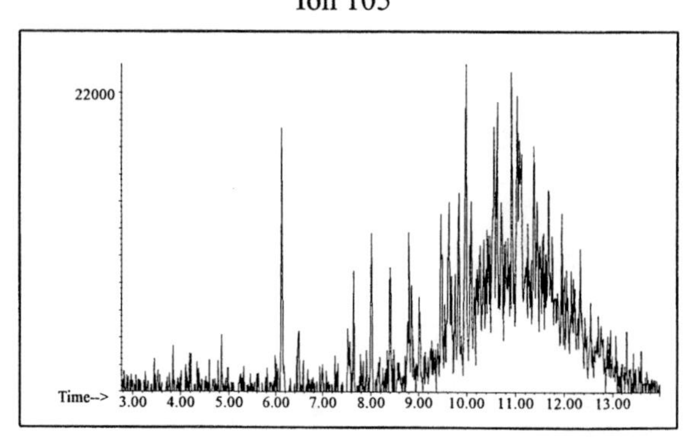

Ion 119

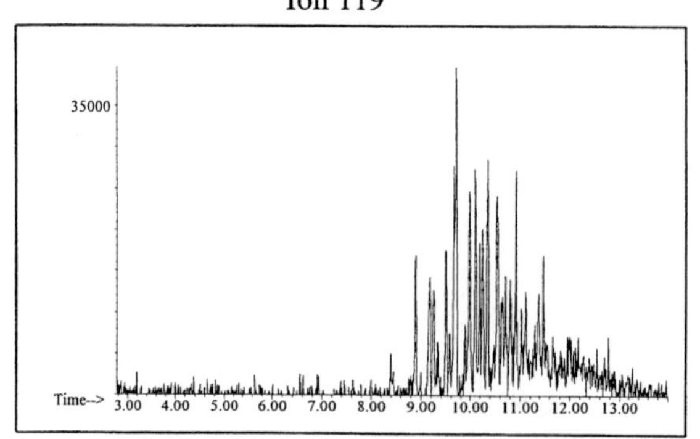

INDIVIDUAL PROFILES Naphthenic-Paraffinic #1

Cycloparaffin

Ion 55

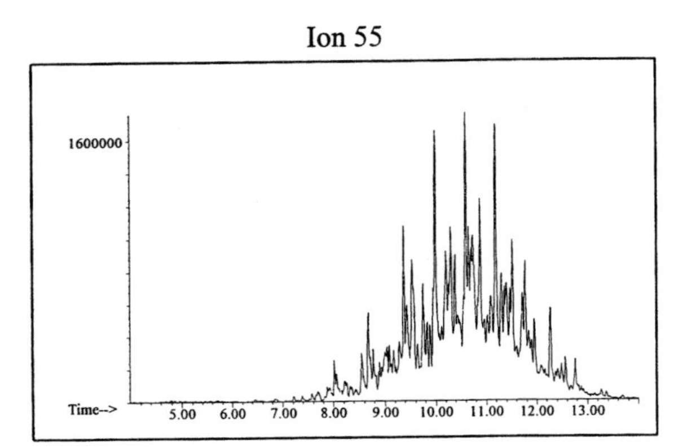

Ion 69

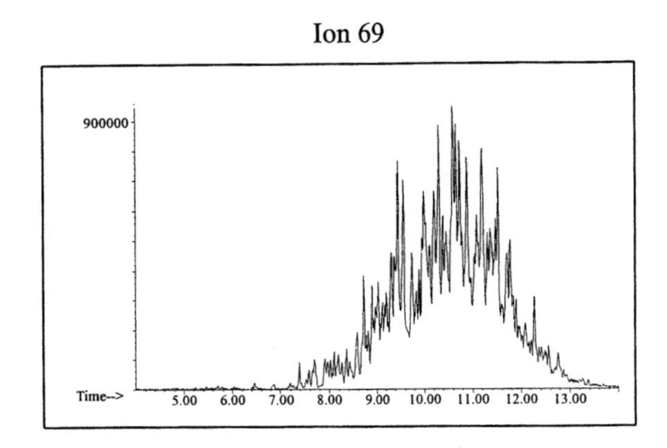

Ion 83

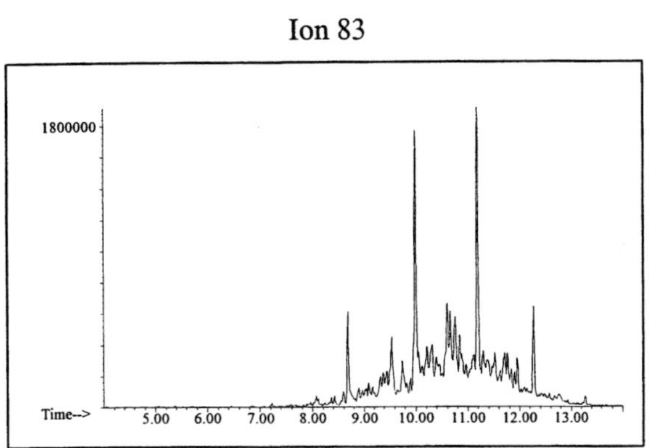

Naphthalene

Ion 128

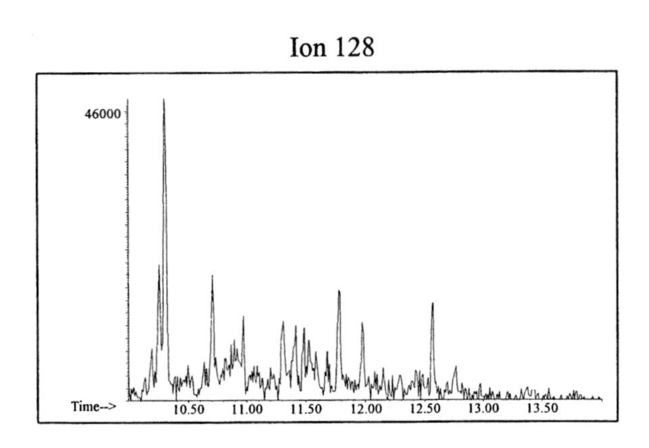

Ion 142

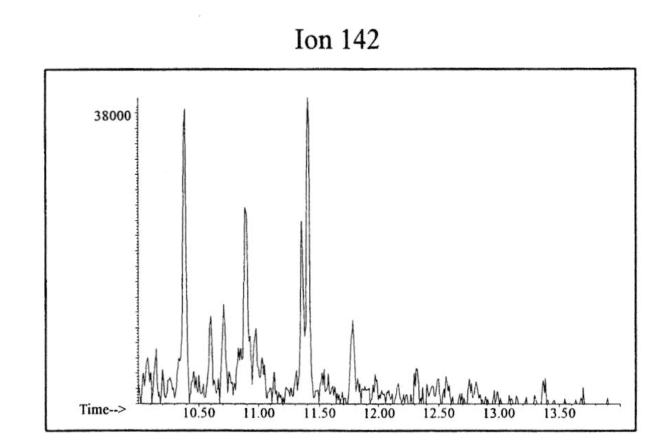

Ion 156

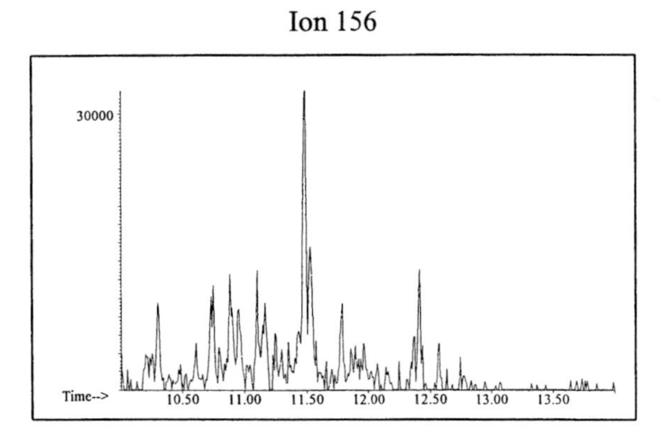

Naphthenic-Paraffinic #2

20uL/mL Pentane

Product Displayed: Enforcer Wasp and Hornet Spray

Other Similar Products:

ASTM: Class 0.5 (Naphthenic-paraffinic)

Product Uses: Insecticide vehicle

Macro Code: M

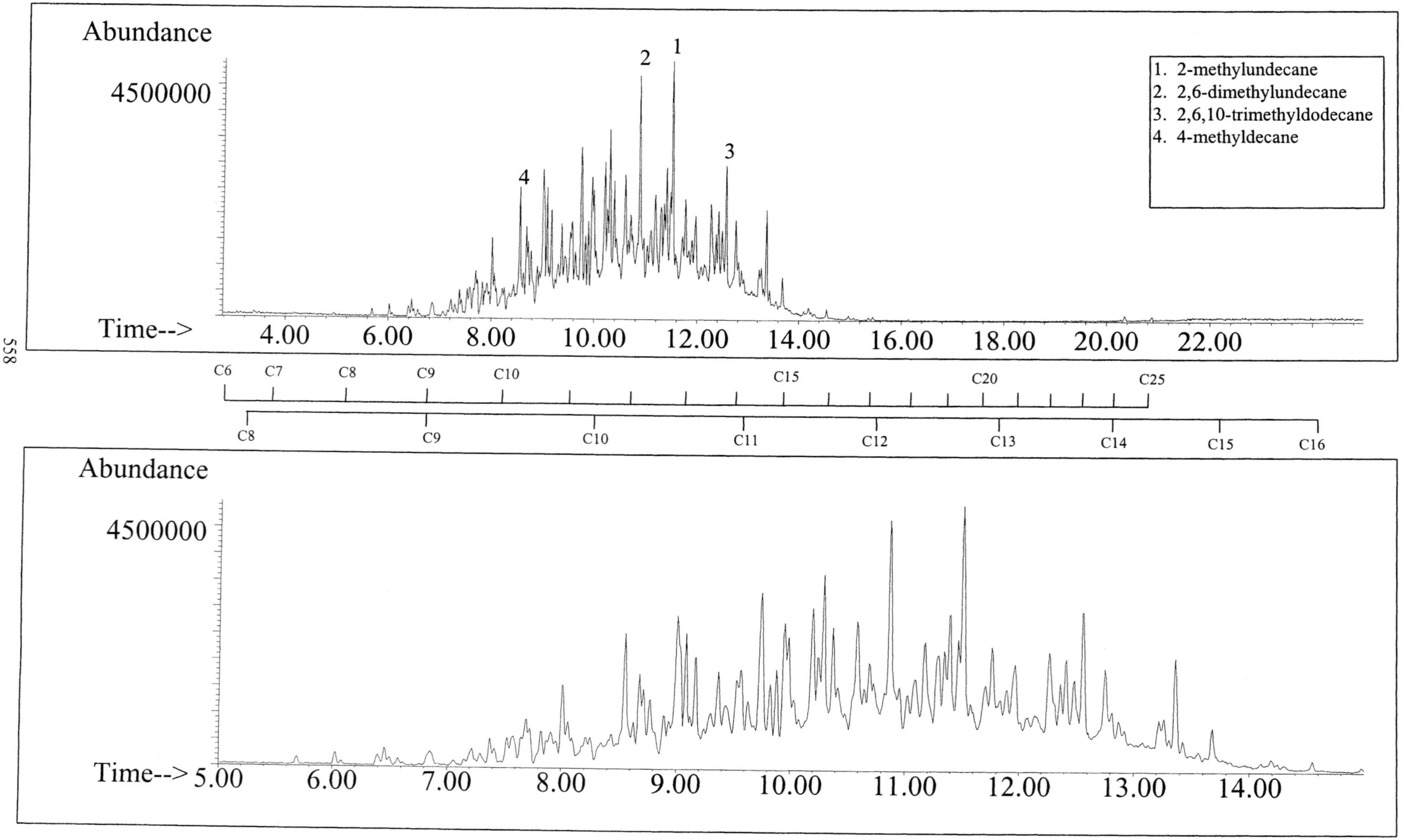

SUMMED PROFILES Naphthenic-Paraffinic #2

ALKANES

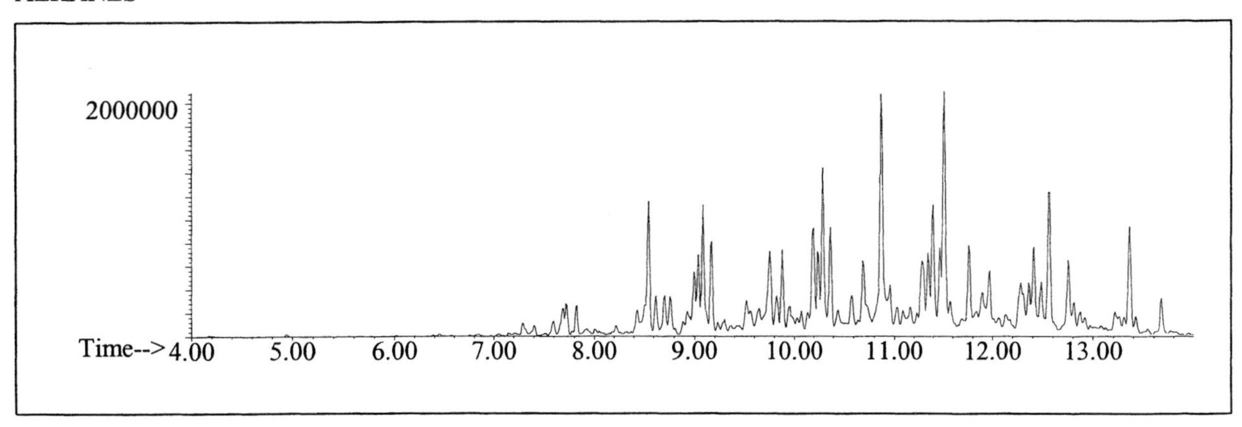

AROMATICS

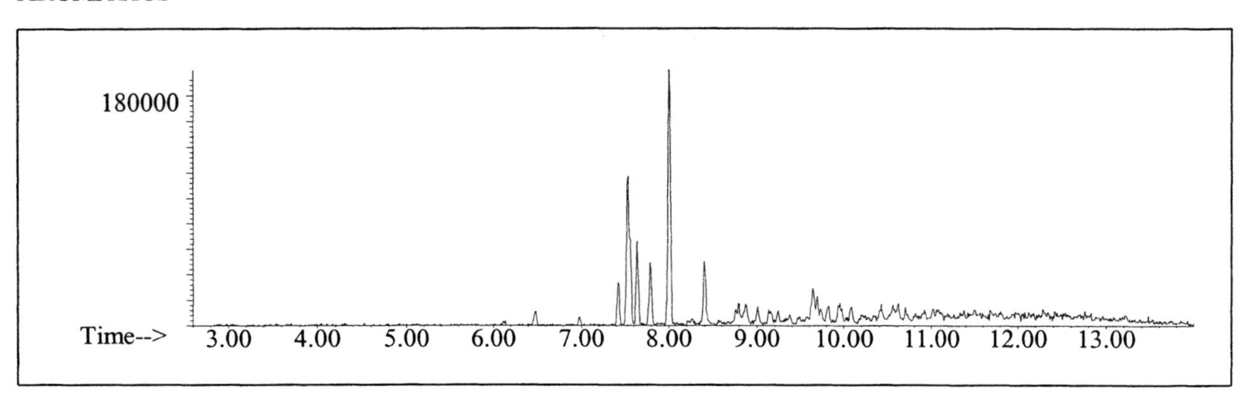

CYCLOPARAFFINS AND ALKENES

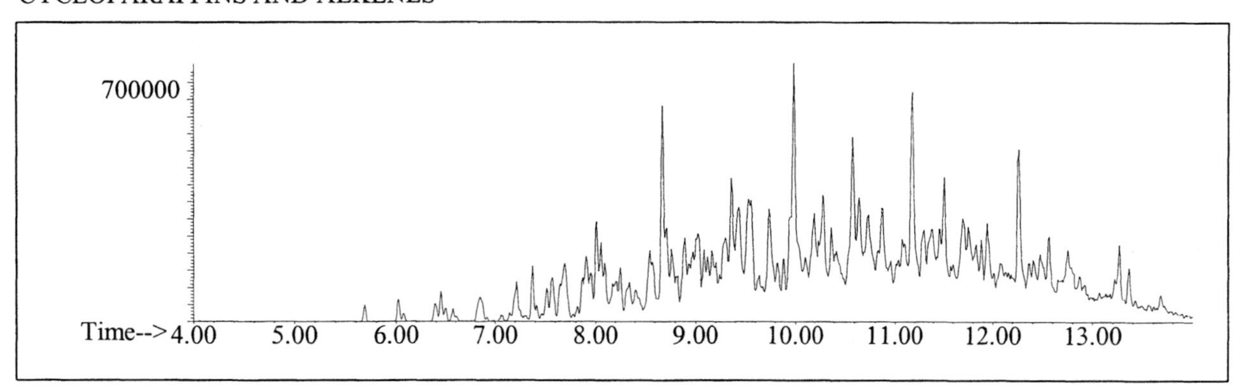

NAPHTHALENES

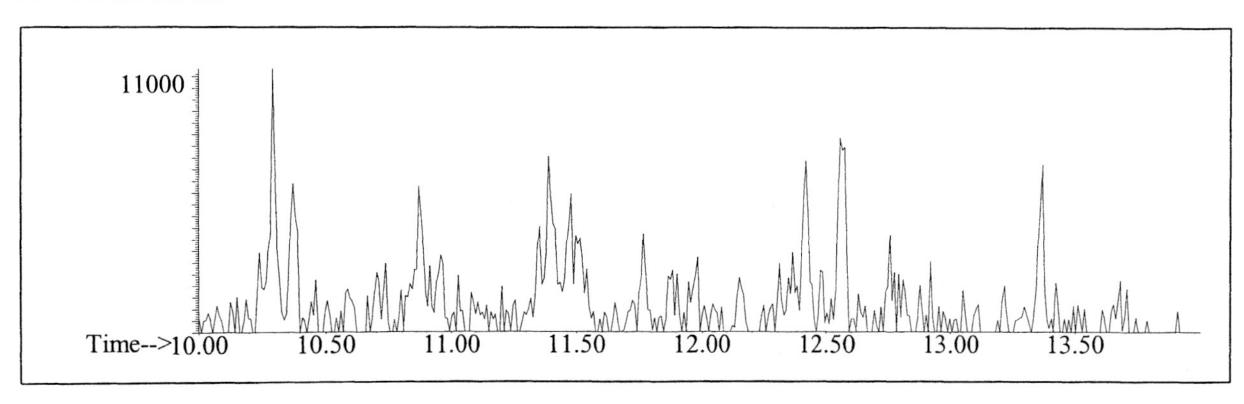

INDIVIDUAL PROFILES Naphthenic-Paraffinic #2

Alkane

Ion 43

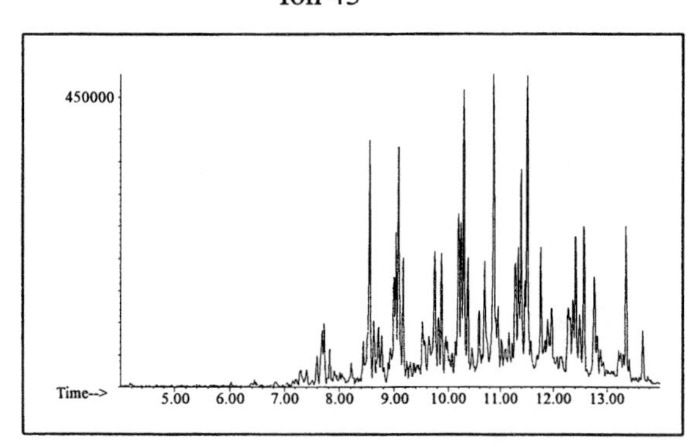

Ion 57

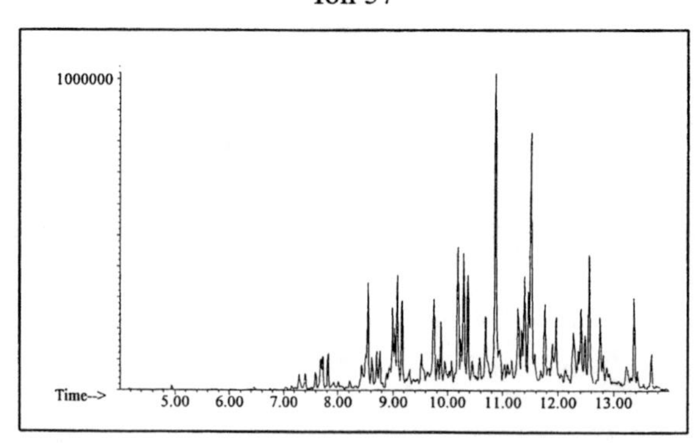

Ion 71

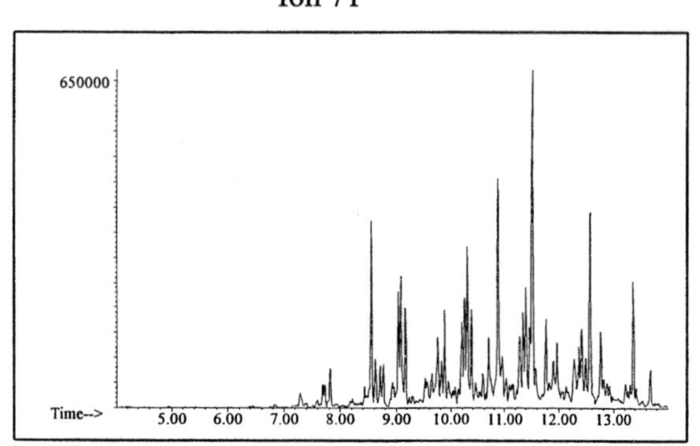

Ion 85

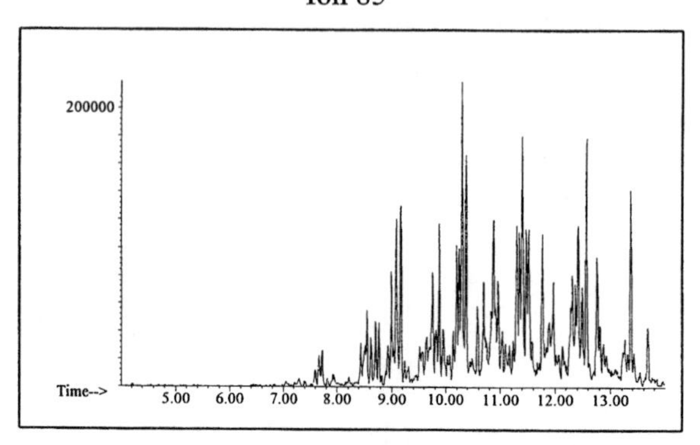

Aromatic

Ion 91

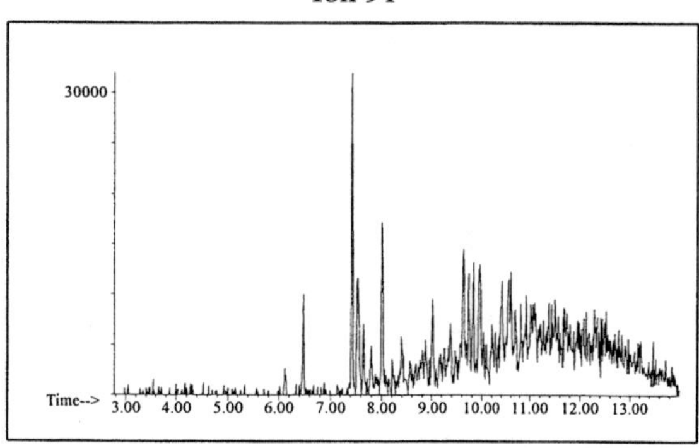

Ion 105

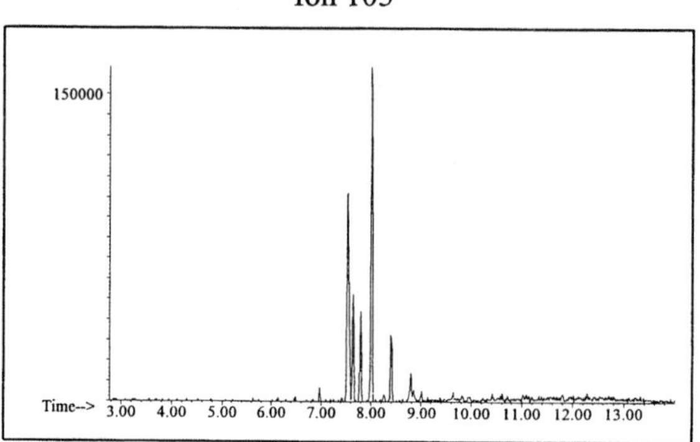

Ion 119

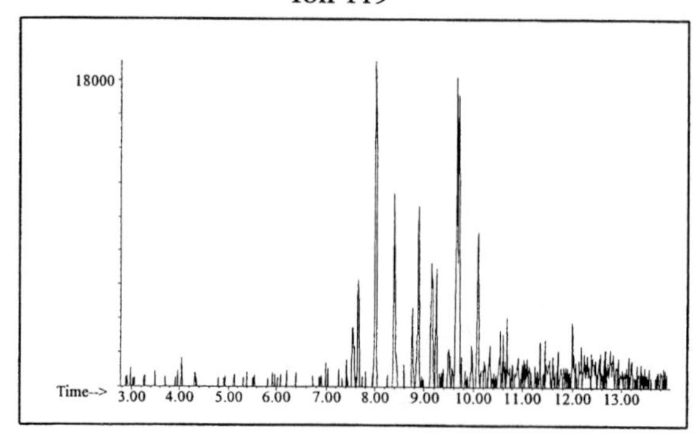

INDIVIDUAL PROFILES Naphthenic-Paraffinic #2

Cycloparaffin

Ion 55

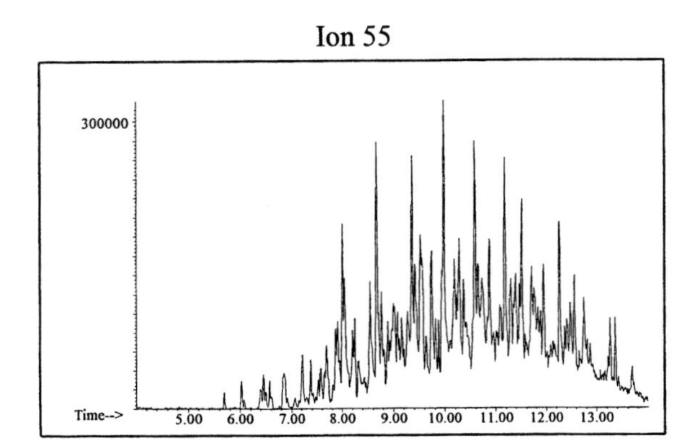

Ion 69

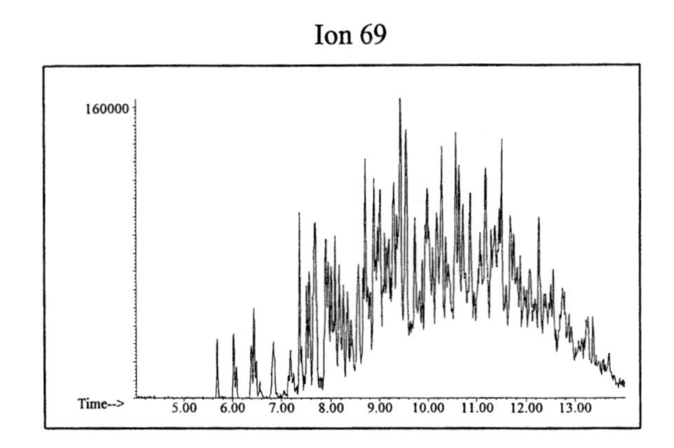

Ion 83

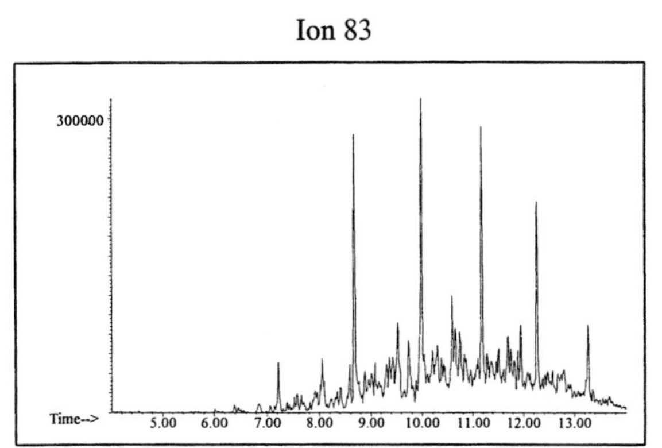

Naphthalene

Ion 128

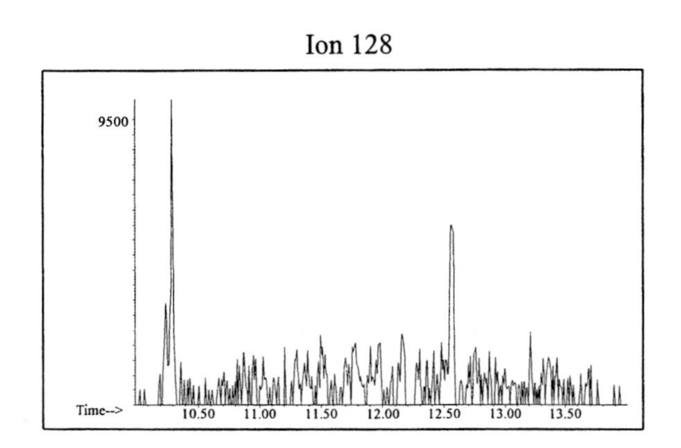

Ion 142

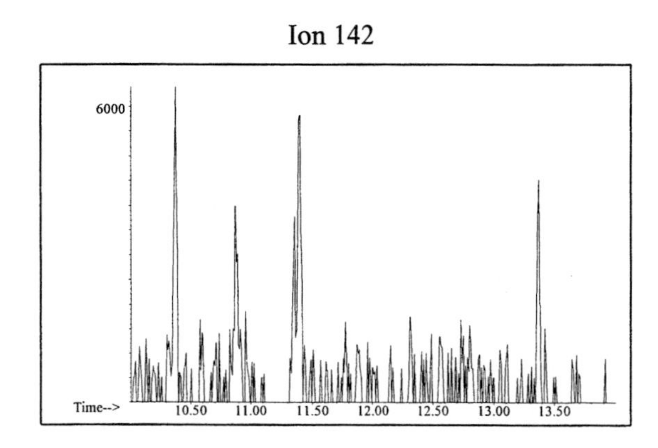

Ion 156

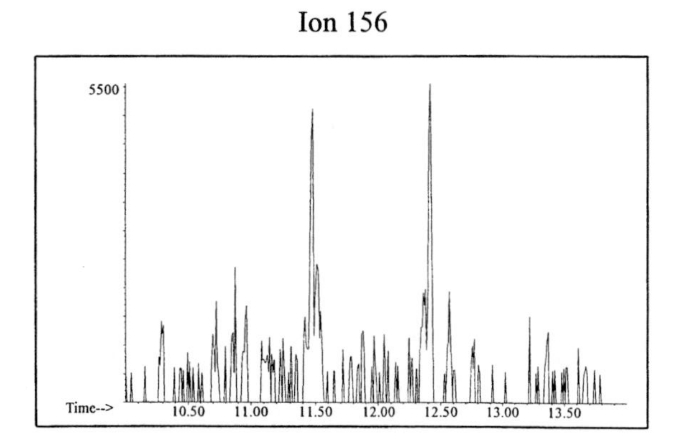

Naphthenic-Paraffinic #3

20uL/mL Pentane

Product Displayed: Beachcomber Citronella Lamp Fuel

Other Similar Products:

ASTM: Class 0.5 (Naphthenic-Paraffinic)

Product Uses: Lamp oil

Macro Code: M

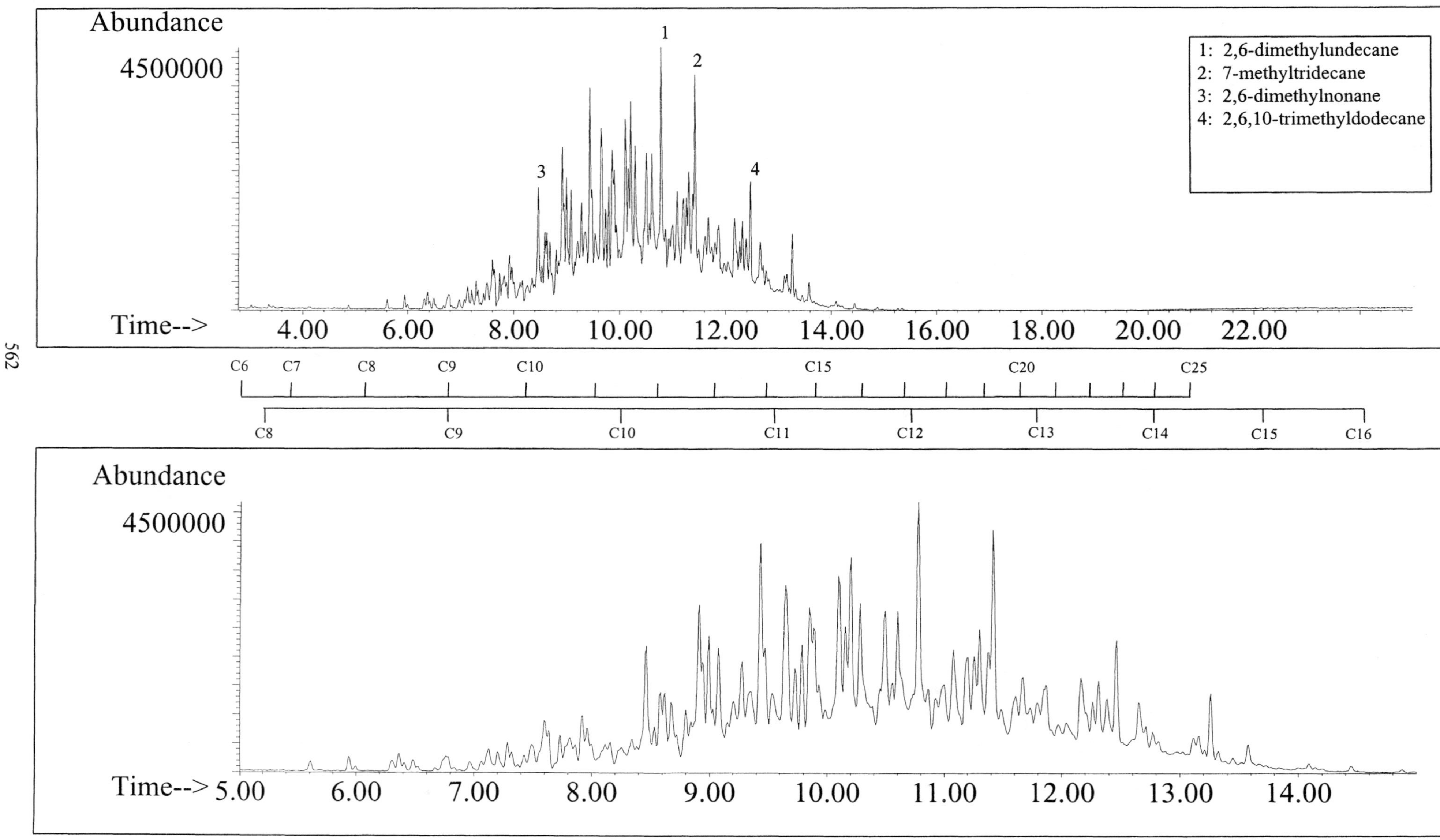

SUMMED PROFILES Naphthenic-Paraffinic #3

ALKANES

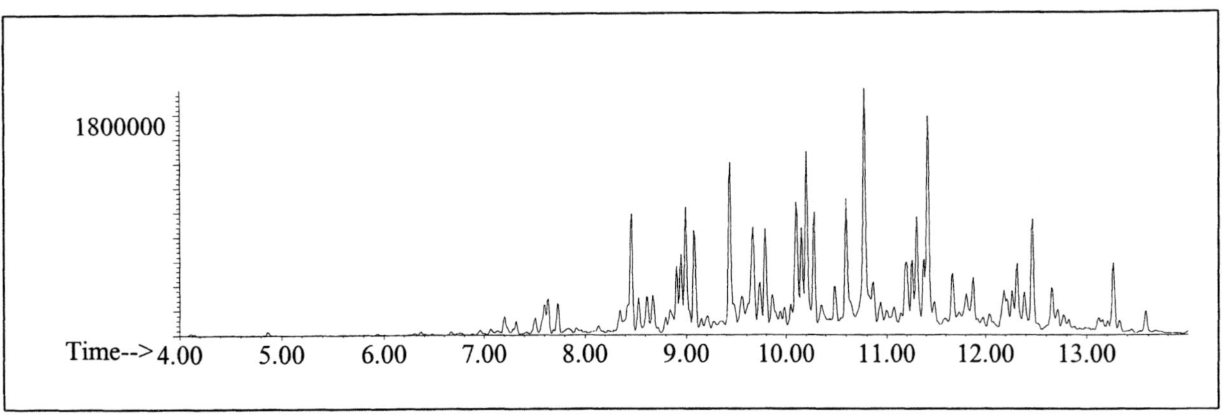

AROMATICS

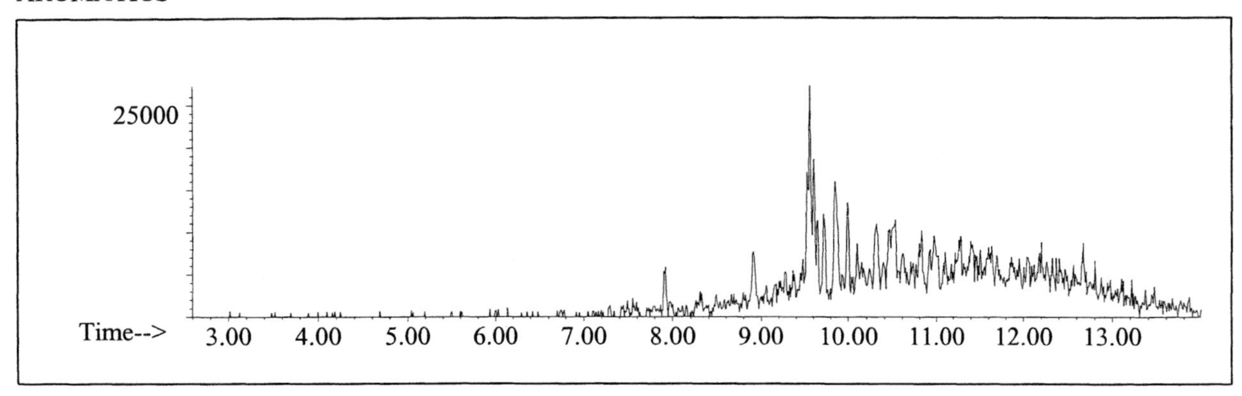

CYCLOPARAFFINS AND ALKENES

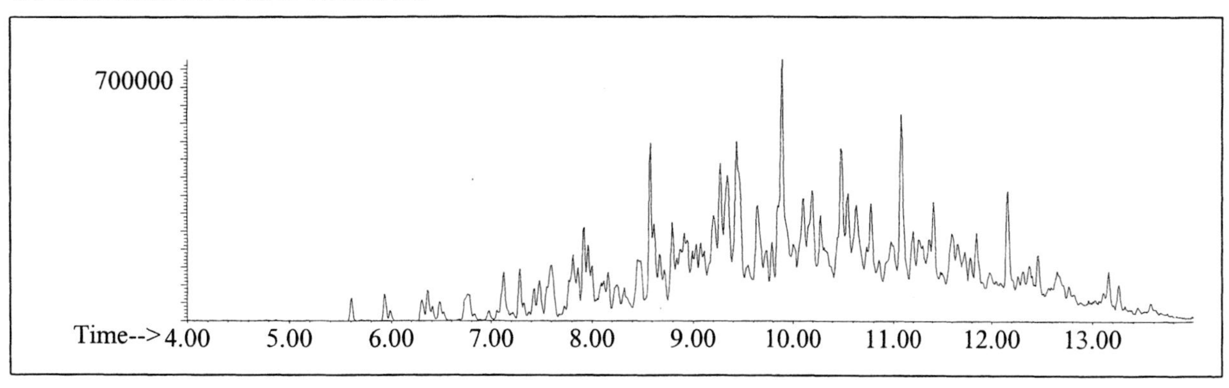

NAPHTHALENES

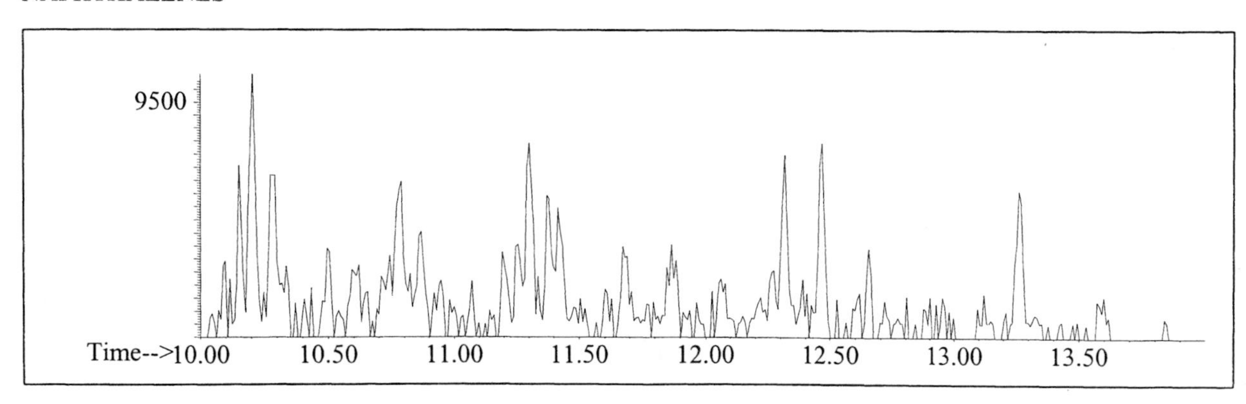

INDIVIDUAL PROFILES Naphthenic-Paraffinic #3

Alkane

Ion 43

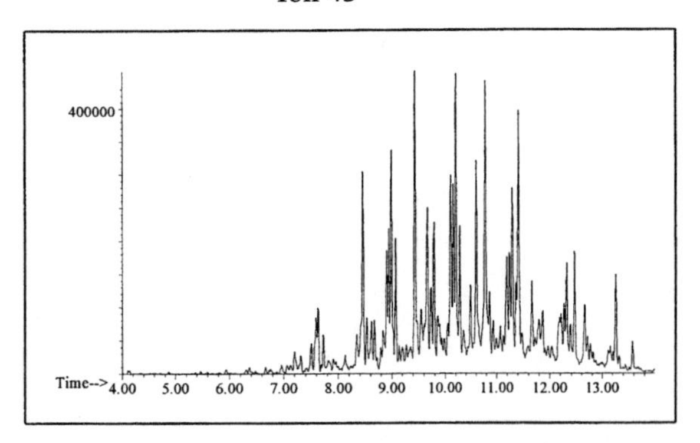

Ion 57

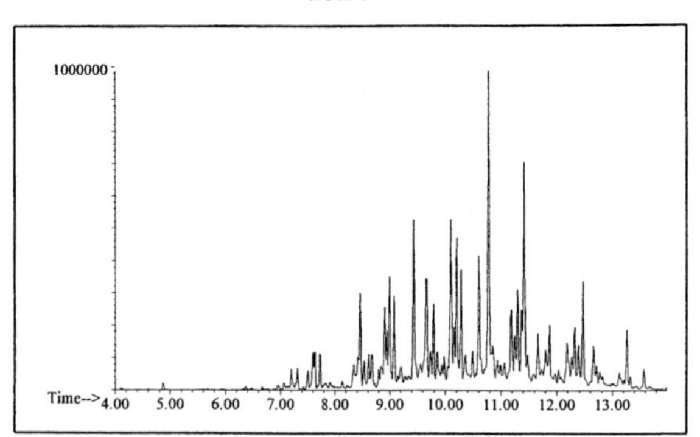

Ion 71

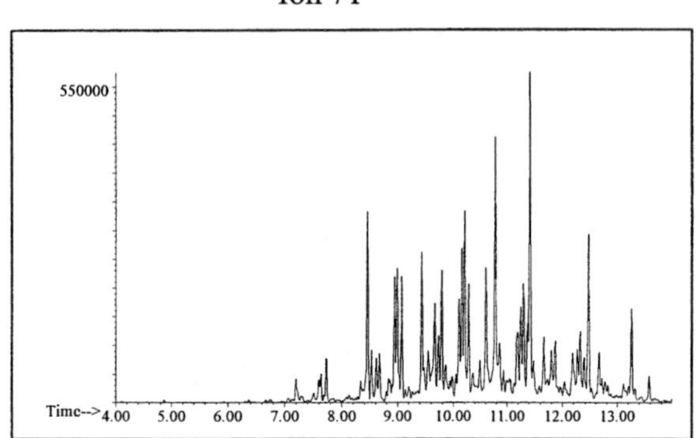

Ion 85

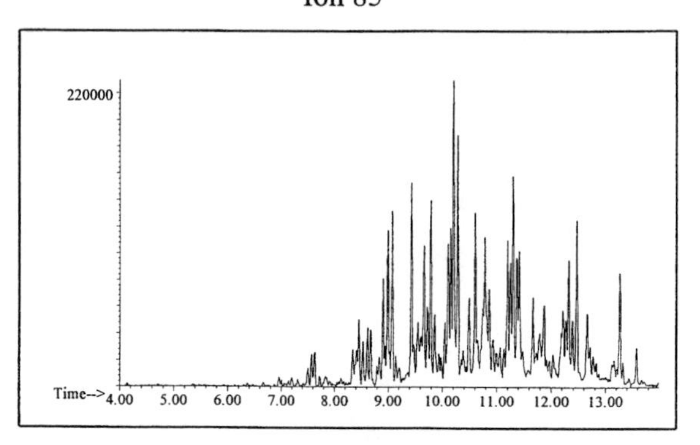

Aromatic

Ion 91

No Useful Data Obtained

Ion 105

No Useful Data Obtained

Ion 119

No Useful Data Obtained

INDIVIDUAL PROFILES Naphthenic-Paraffinic #3

Cycloparaffin

Ion 55

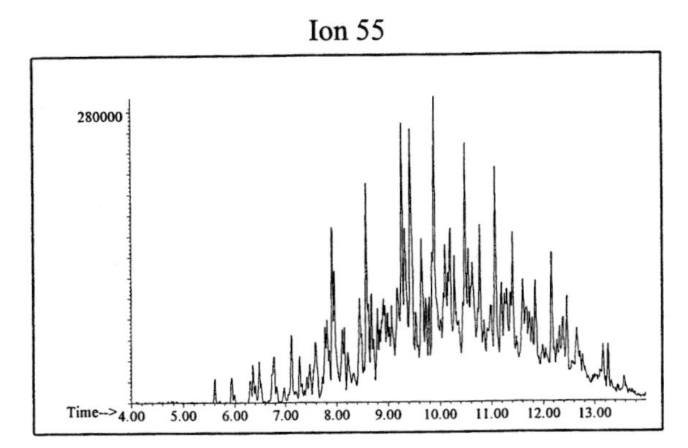

Ion 69

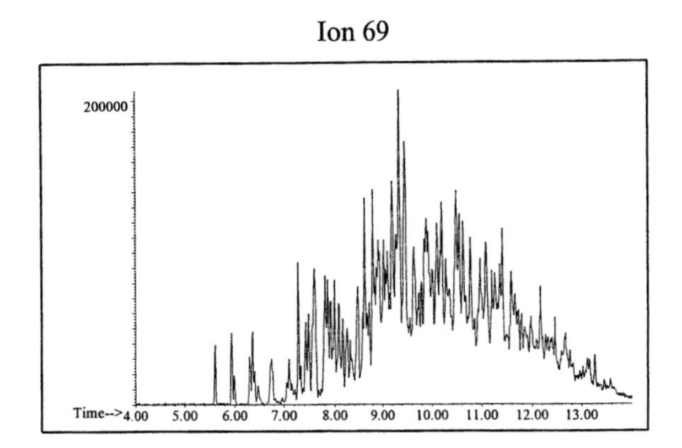

Ion 83

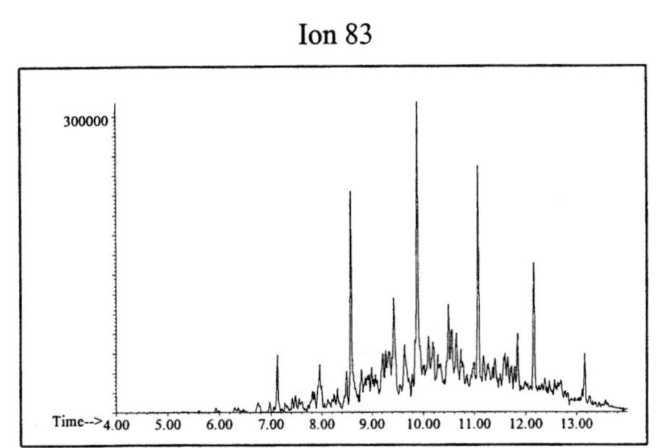

Naphthalene

Ion 128

No Useful Data Obtained

Ion 142

No Useful Data Obtained

Ion 156

No Useful Data Obtained

Naphthenic-Paraffinic #4

20uL/mL Pentane

ASTM: Class 0.5 (Naphthenic-paraffinic)

Macro Code: H

Product Displayed: Lamplight Farms Lamp Oil

Product Uses: Lamp oil

Other Similar Products:

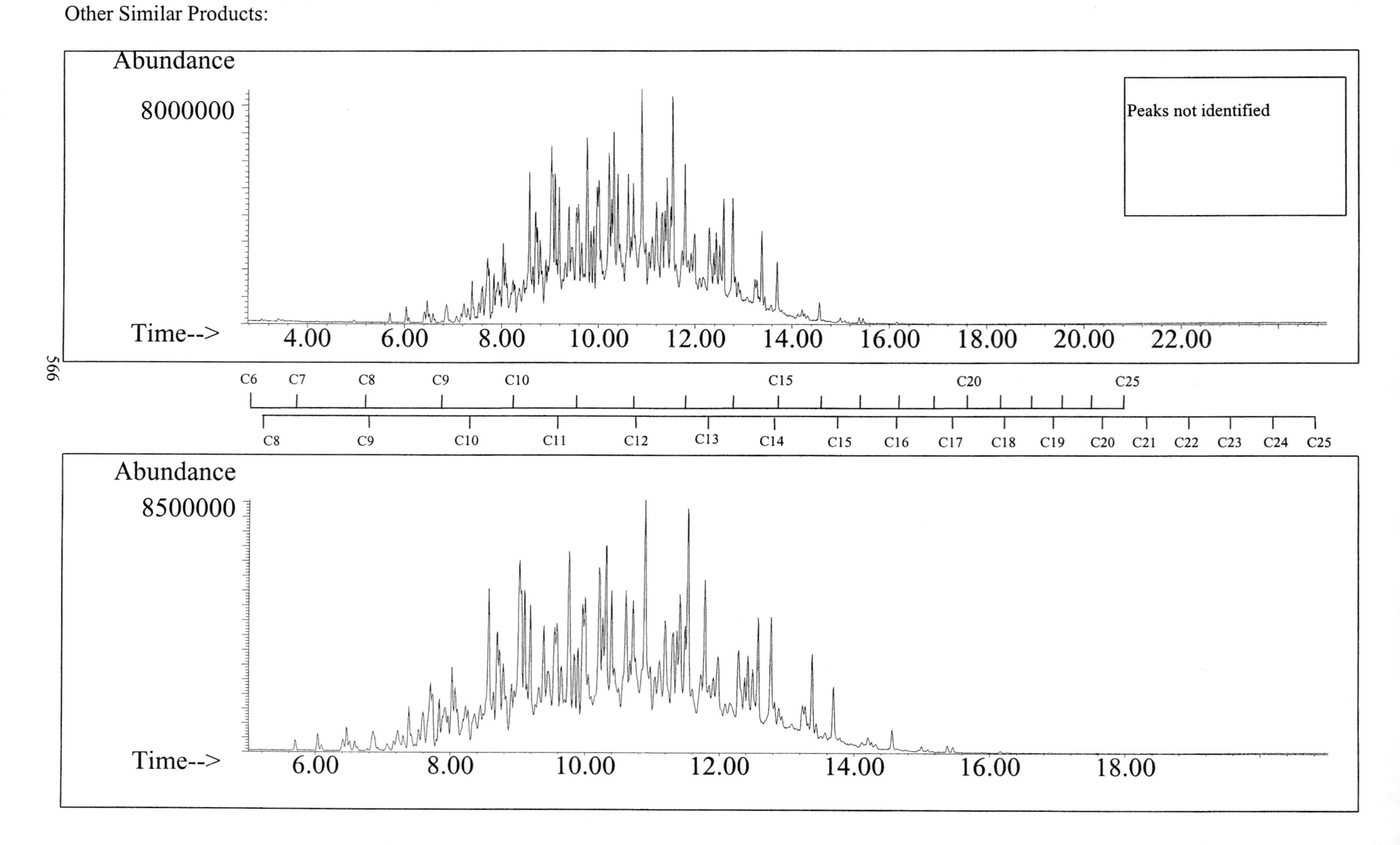

ALKANES

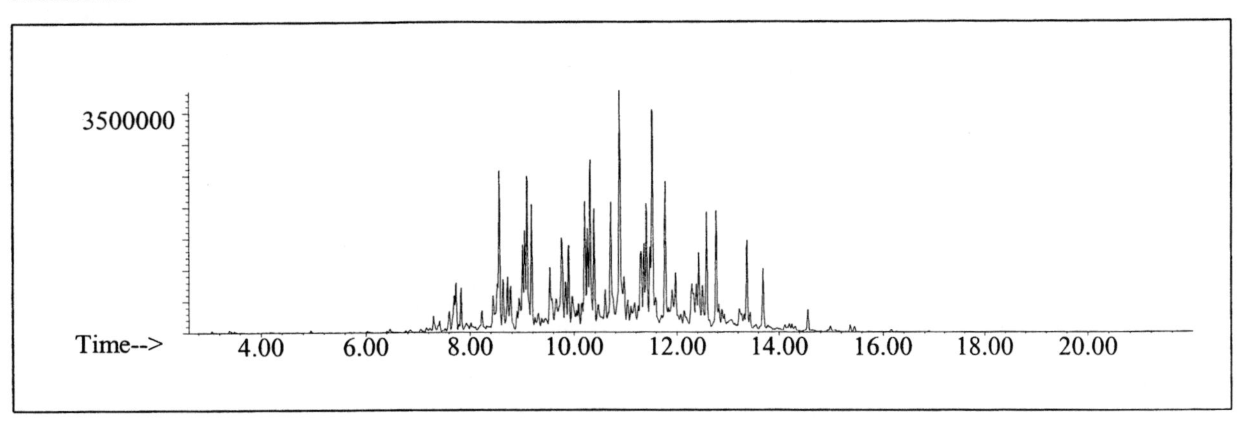

AROMATICS

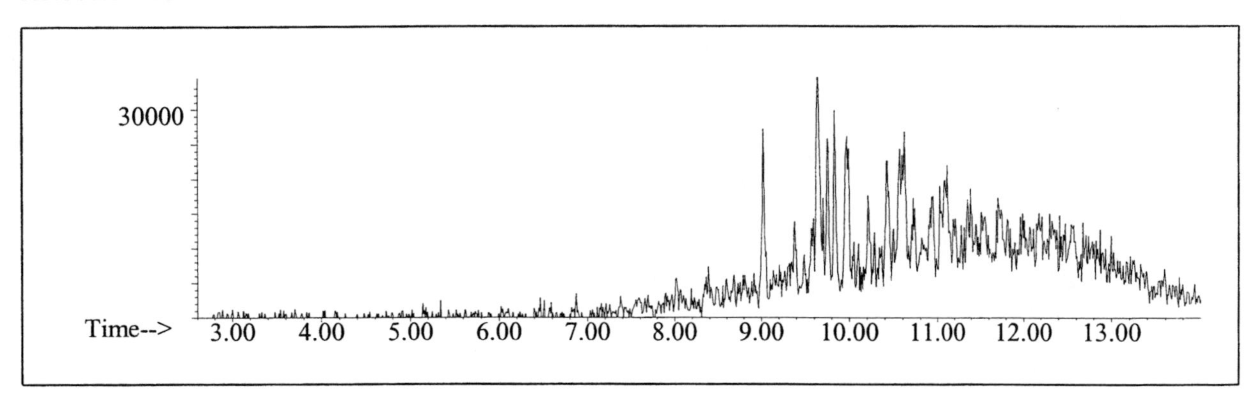

CYCLOPARAFFINS AND ALKENES

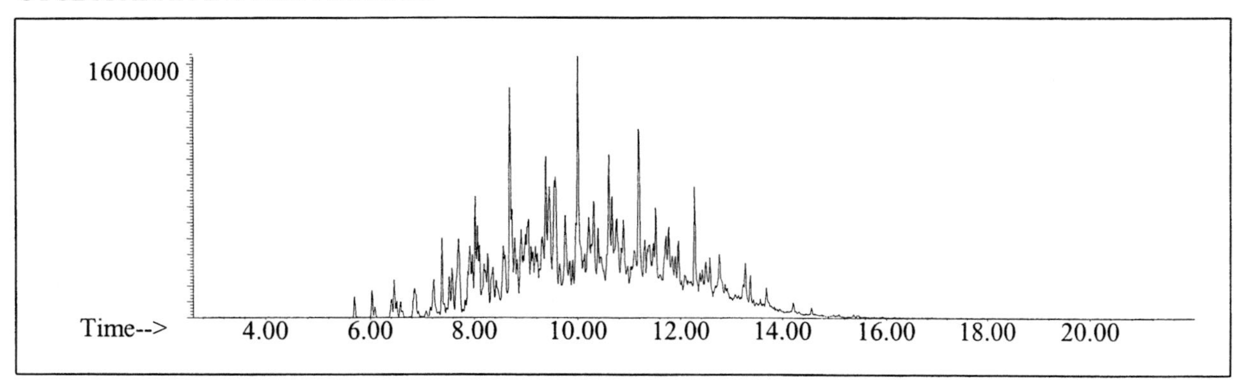

NAPHTHALENES

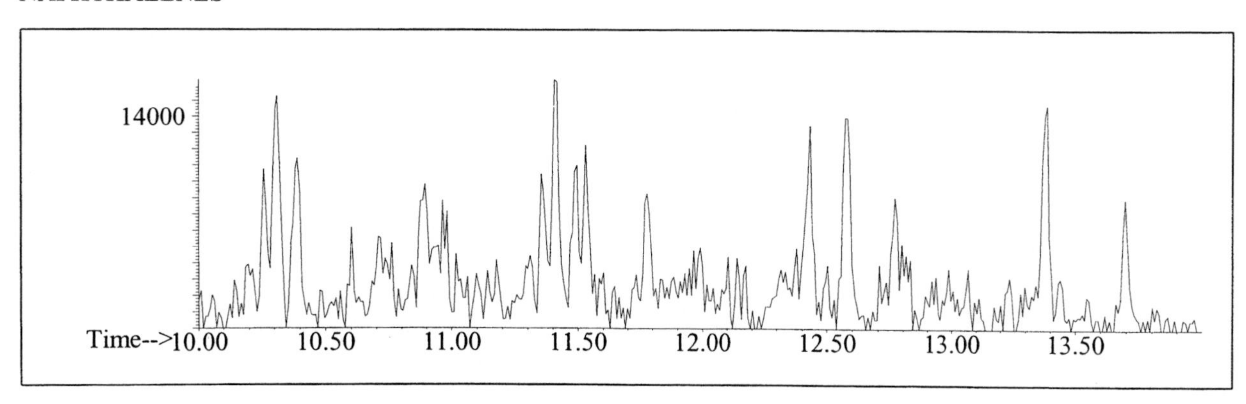

INDIVIDUAL PROFILES

Naphthenic-Paraffinic #4

Alkane

Ion 43

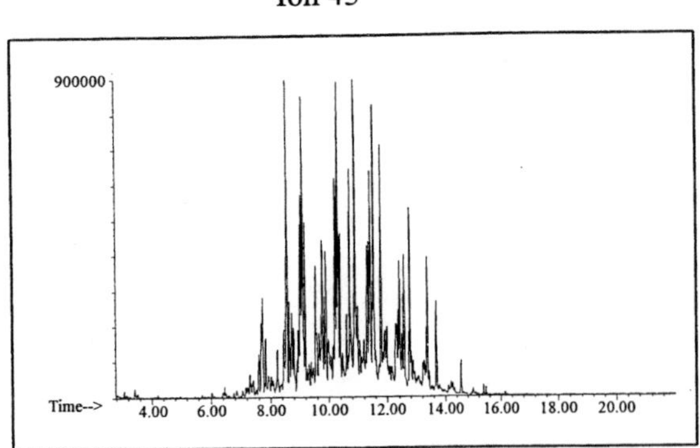

Ion 57

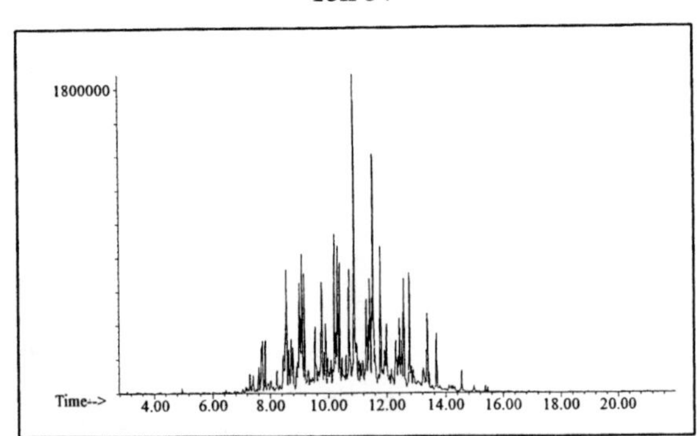

Ion 71

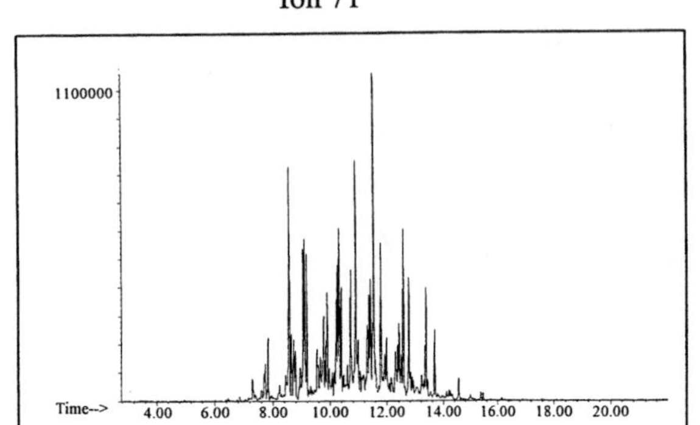

Ion 85

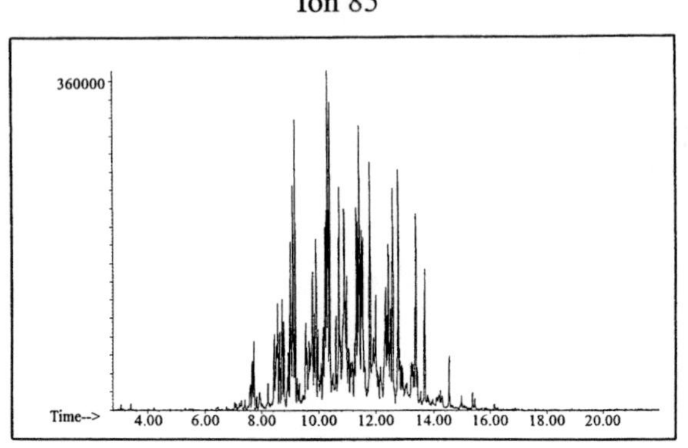

Aromatic

Ion 91

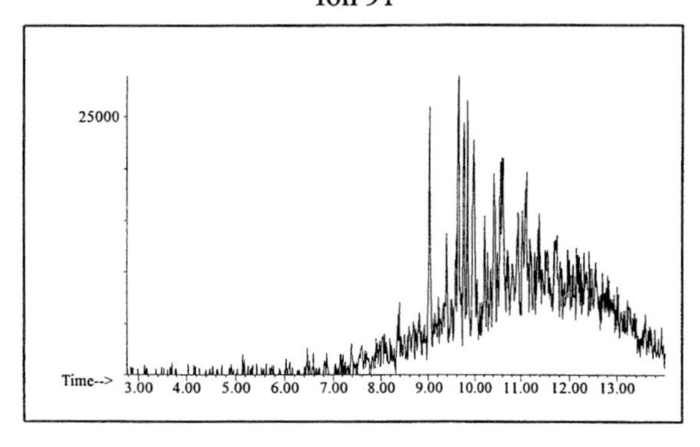

Ion 105

No Useful Data Obtained

Ion 119

No Useful Data Obtained

INDIVIDUAL PROFILES Naphthenic-Paraffinic #4

Cycloparaffin

Ion 55

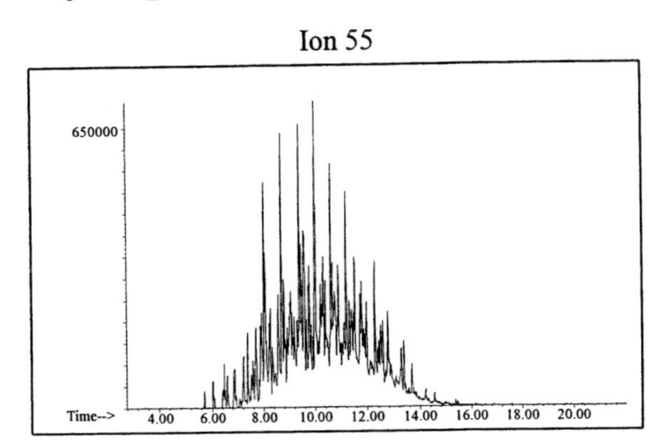

Ion 69

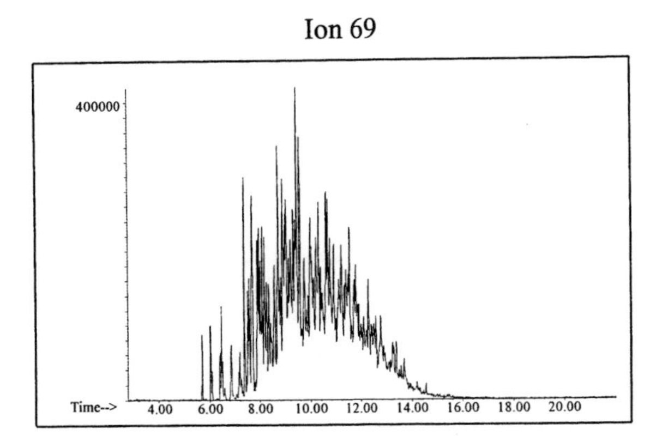

Ion 83

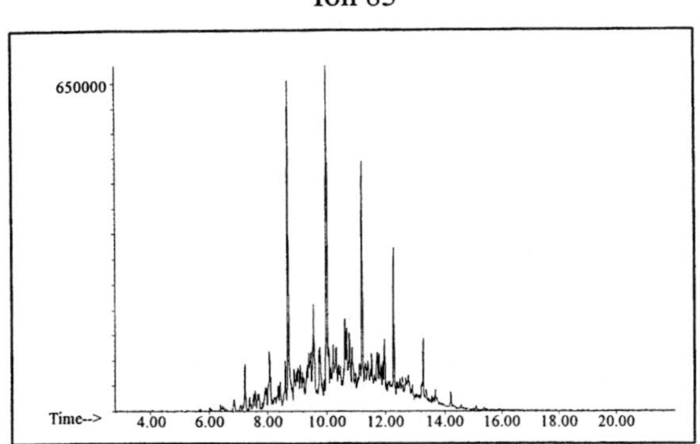

Naphthalene

Ion 128

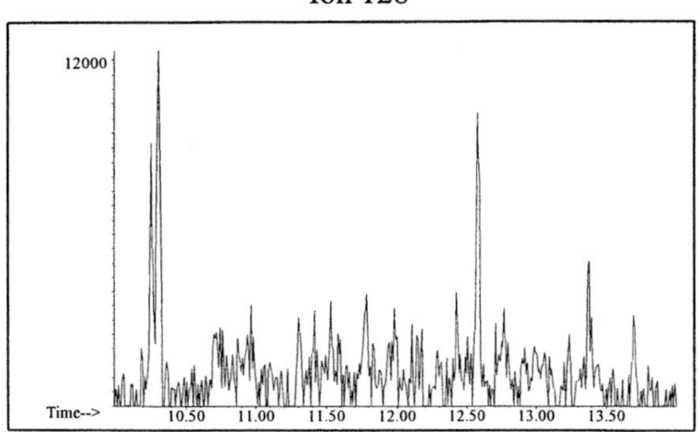

Ion 142

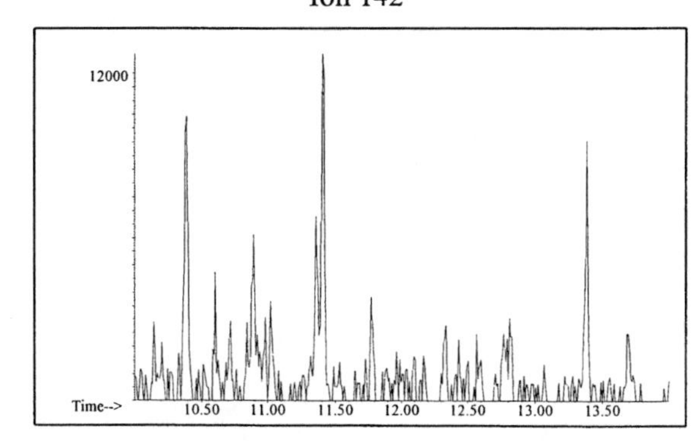

Ion 156

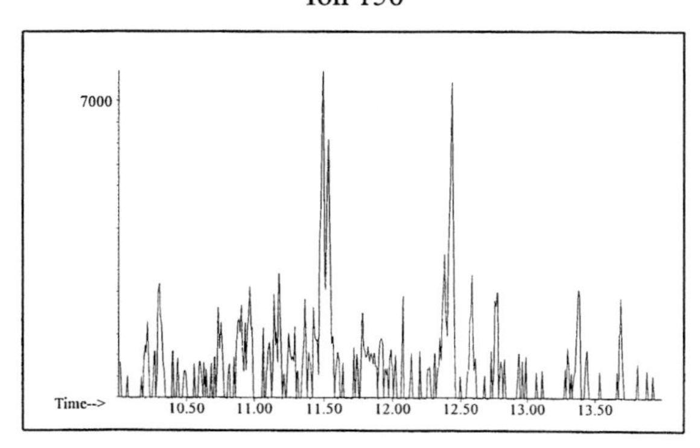

Naphthenic-Paraffinic #5

20uL/mL Pentane

Product Displayed: Vista LPA 120

Other Similar Products:

ASTM: Class 0.5 (Naphthenic-paraffinic)

Product Uses: Solvent, feedstock

Macro Code: H

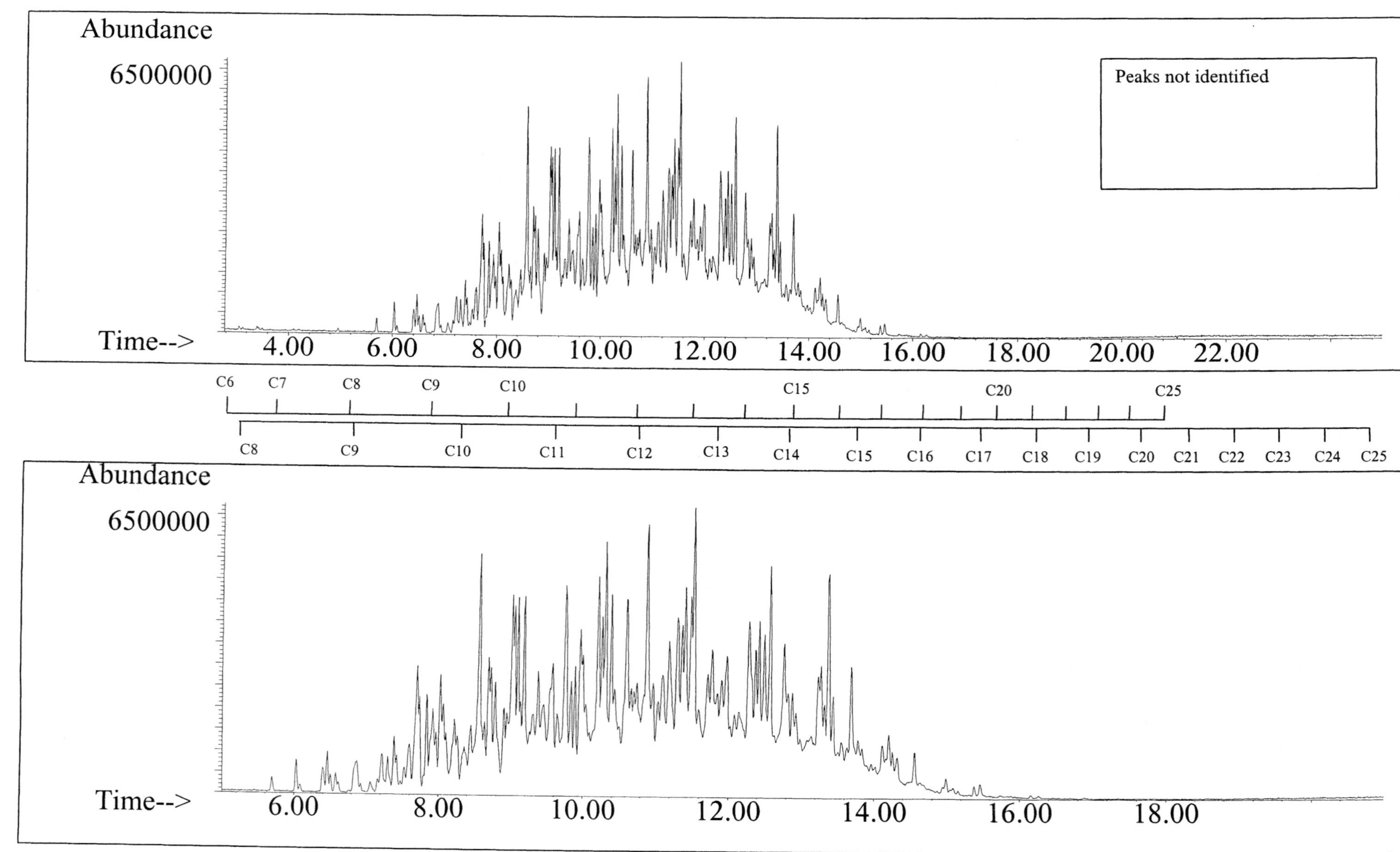

SUMMED PROFILES

Naphthenic-Paraffinic #5

ALKANES

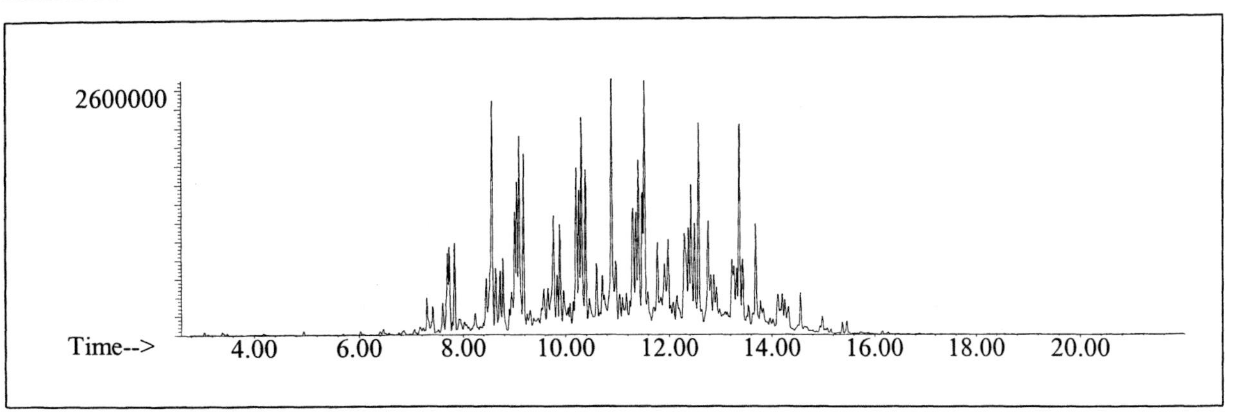

AROMATICS

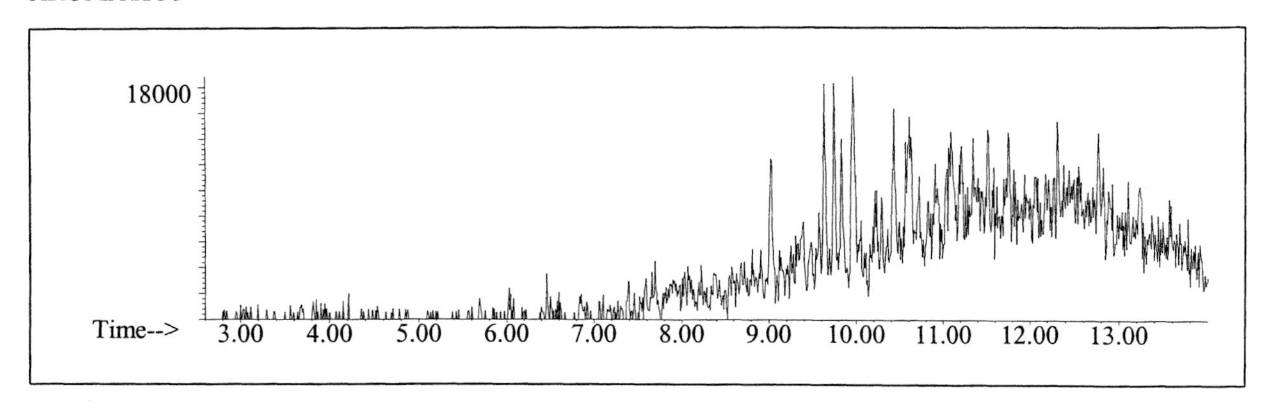

CYCLOPARAFFINS AND ALKENES

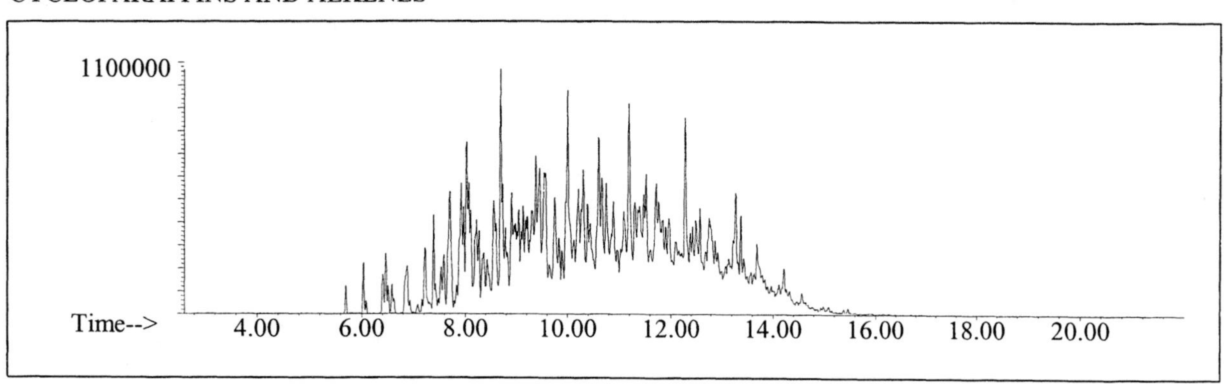

NAPHTHALENES

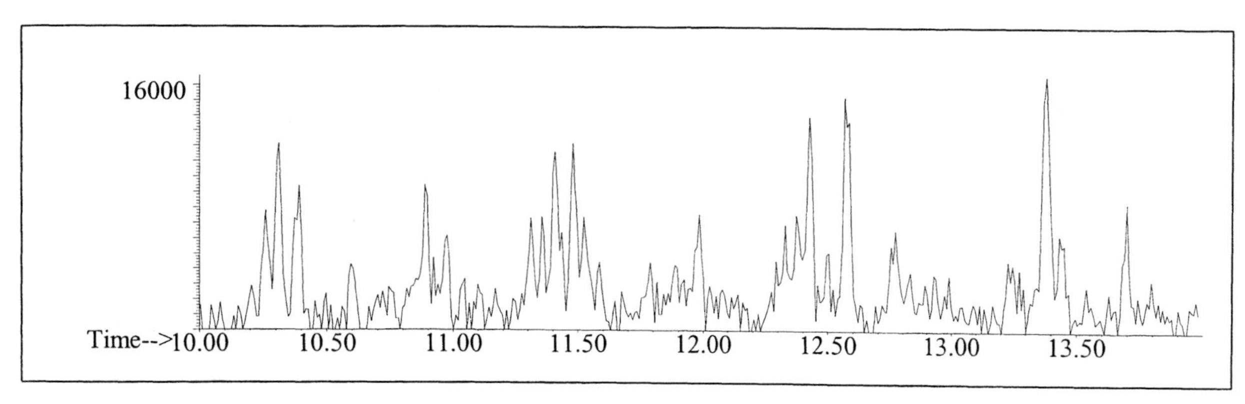

INDIVIDUAL PROFILES Naphthenic-Paraffinic #5

Ion 43

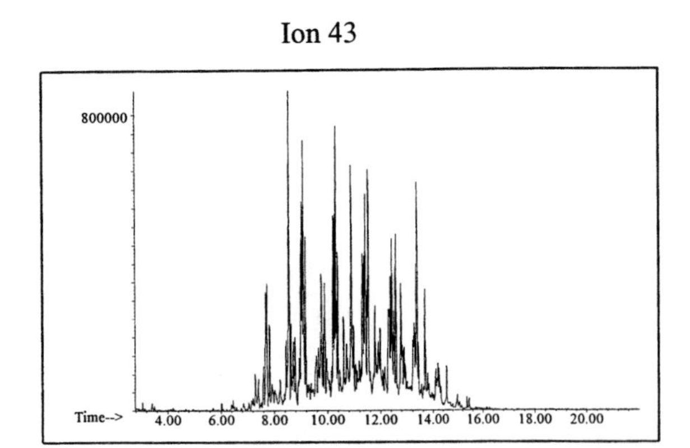

Ion 57

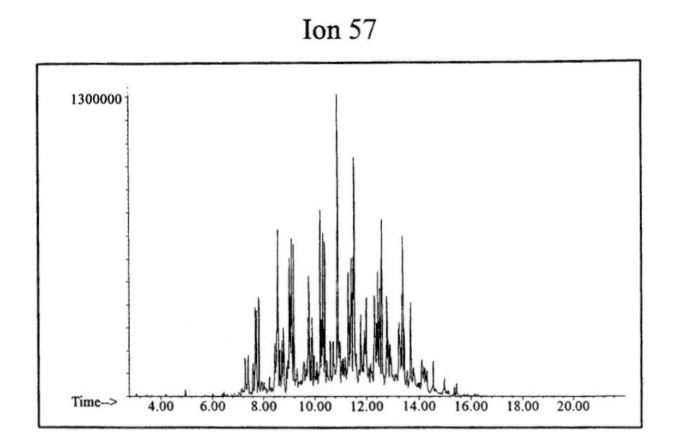

Ion 71

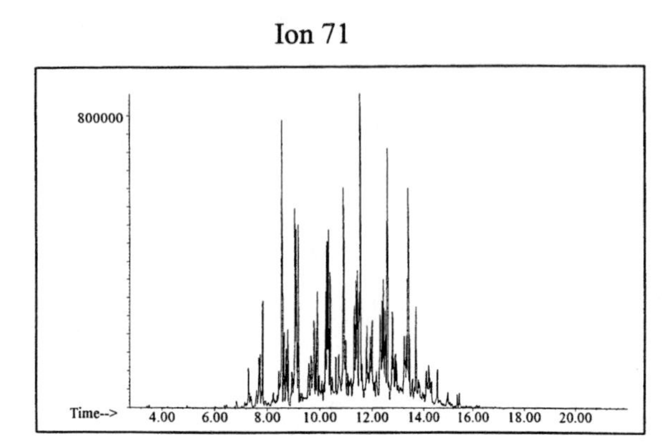

Ion 85

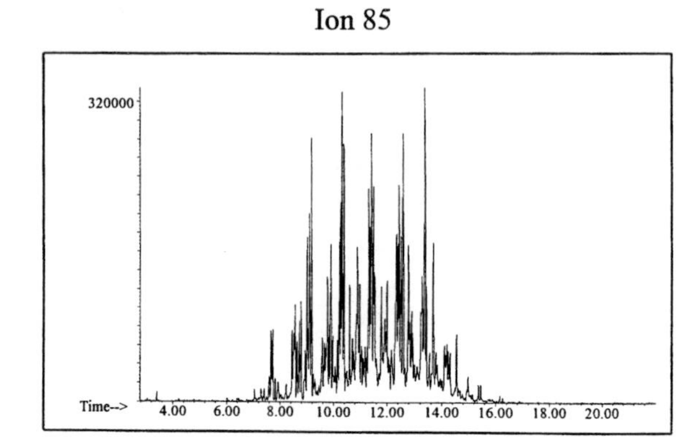

Aromatic

Ion 91

No Useful Data Obtained

Ion 105

No Useful Data Obtained

Ion 119

No Useful Data Obtained

INDIVIDUAL PROFILES Naphthenic-Paraffinic #5

Cycloparaffin

Ion 55

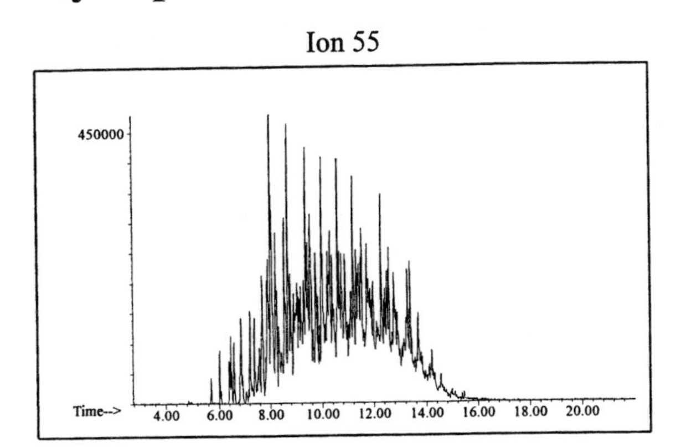

Ion 69

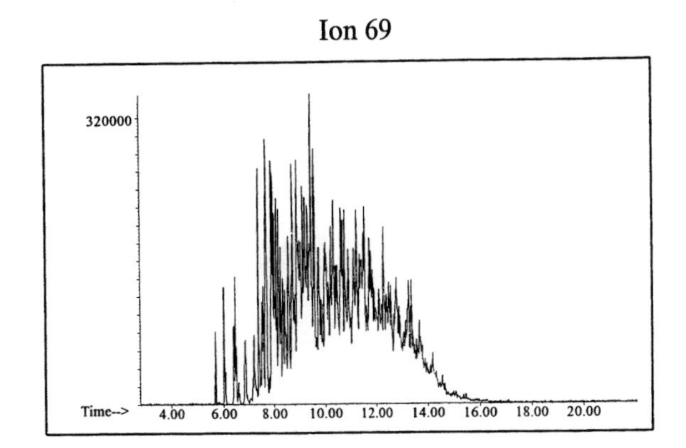

Ion 83

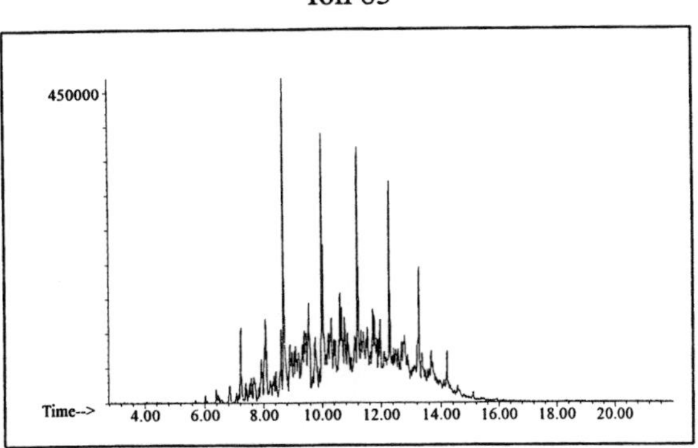

Naphthalene

Ion 128

No Useful Data Obtained

Ion 142

No Useful Data Obtained

Ion 156

No Useful Data Obtained

Naphthenic-Paraffinic #6

20uL/mL Pentane
Product Displayed: Vista LPA 142
Other Similar Products:

ASTM: Class 0.5 (Naphthenic-paraffinic)
Product Uses: Solvent, feedstock

Macro Code: H

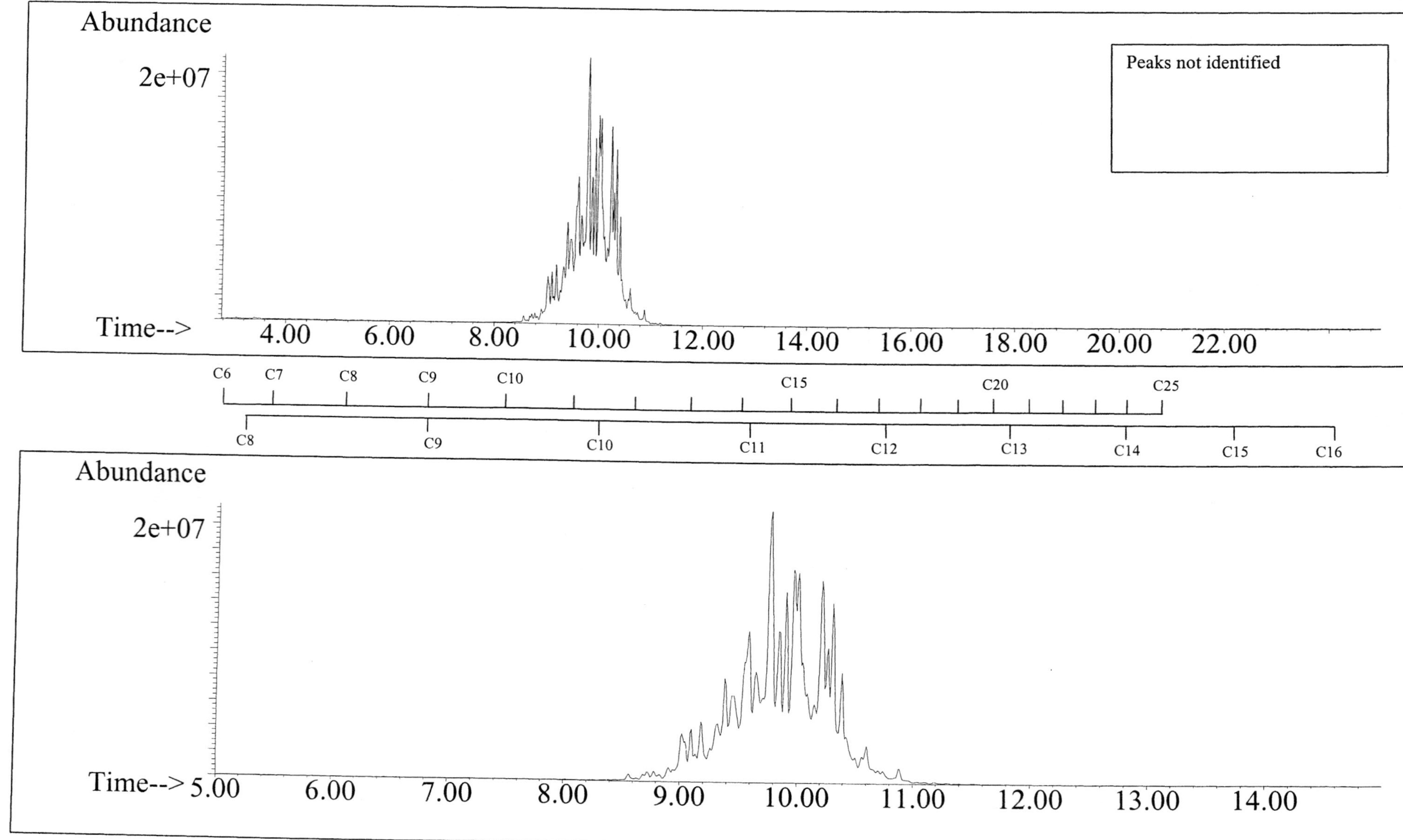

ALKANES

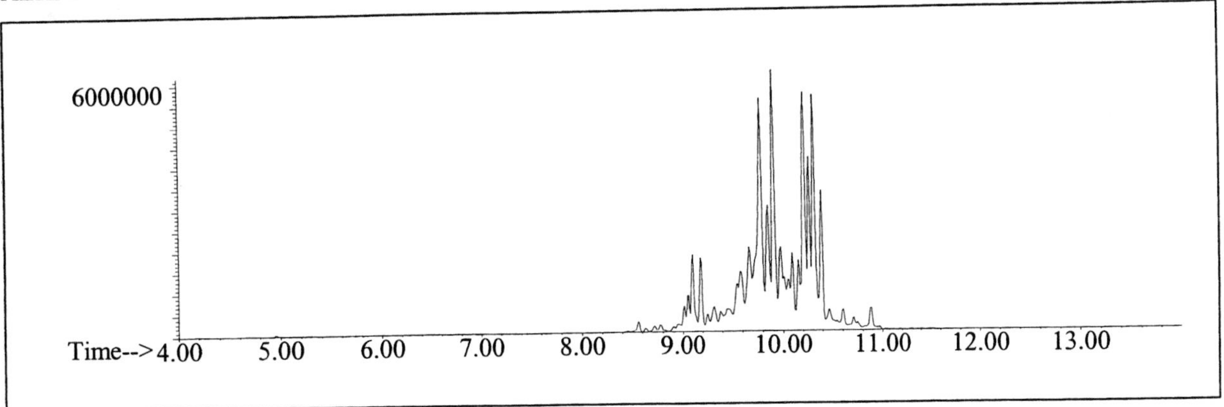

AROMATICS

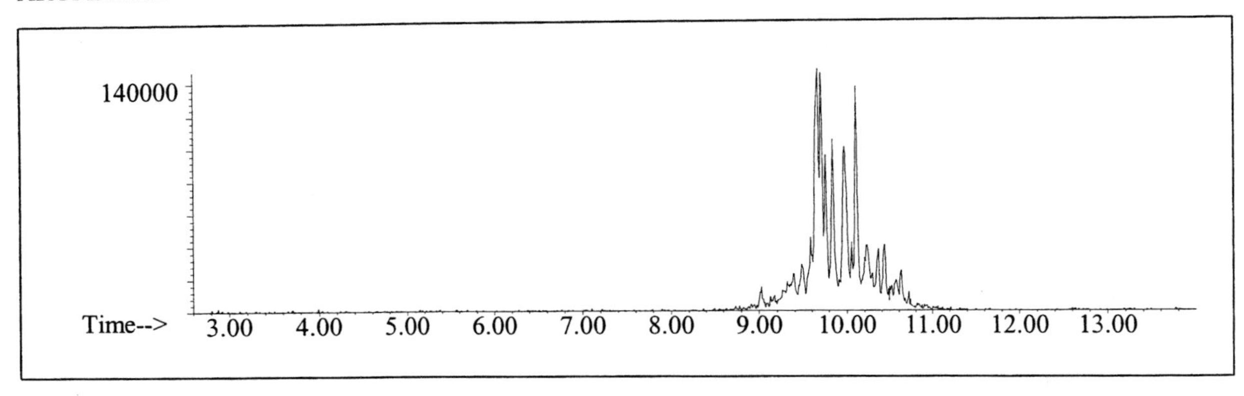

CYCLOPARAFFINS AND ALKENES

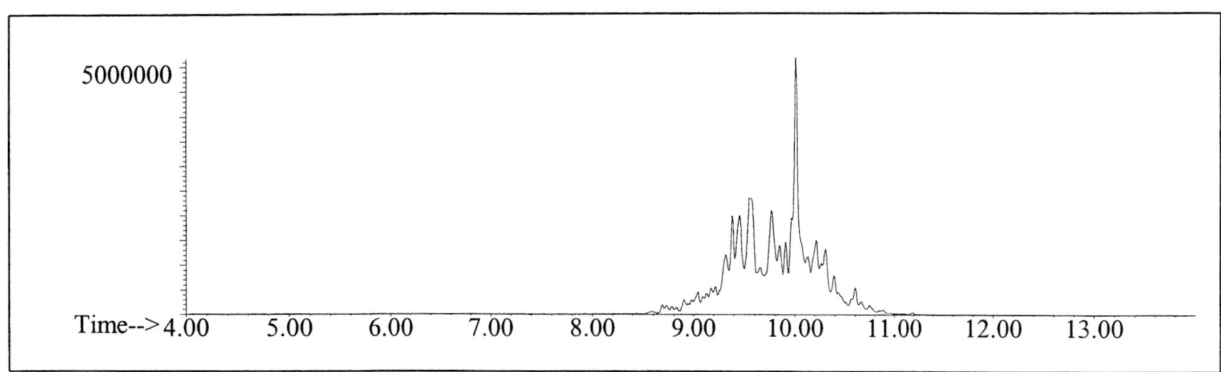

NAPHTHALENES

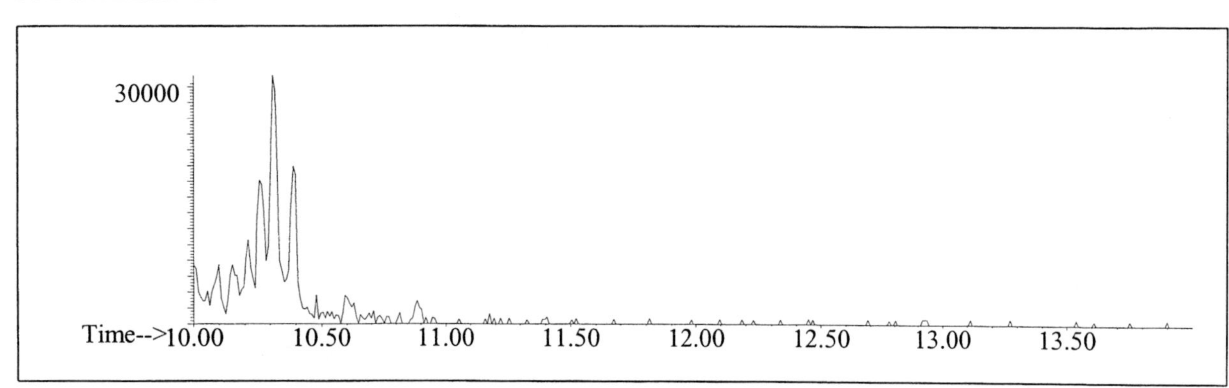

INDIVIDUAL PROFILES Naphthenic-Paraffinic #6

Alkane

Ion 43

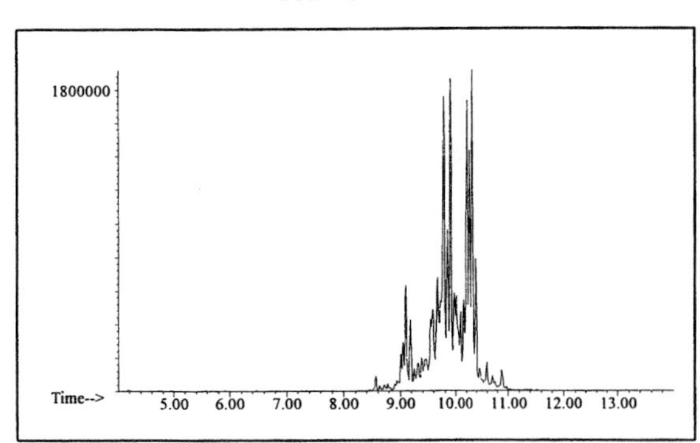

Ion 57

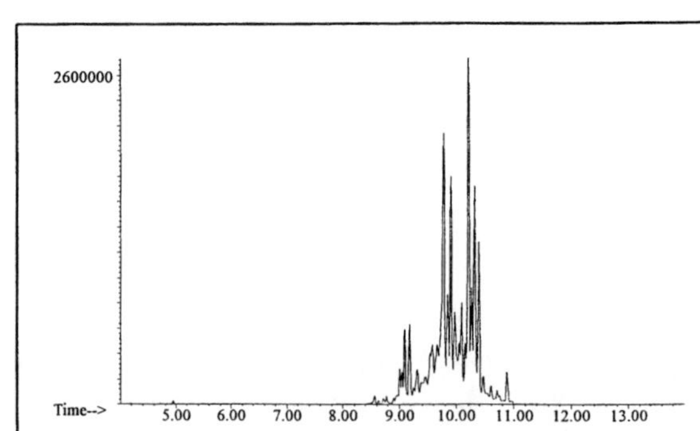

Ion 71

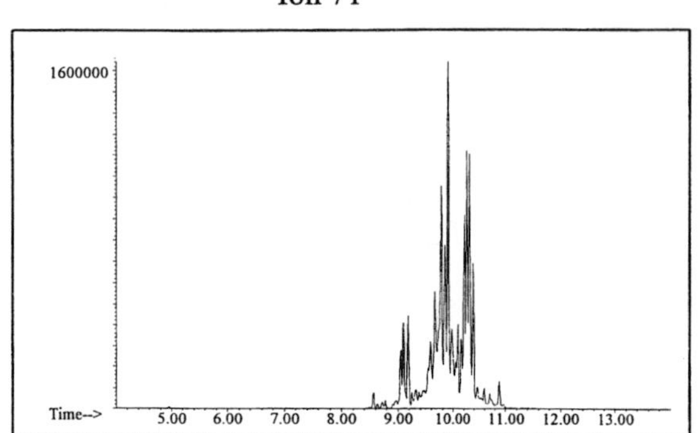

Ion 85

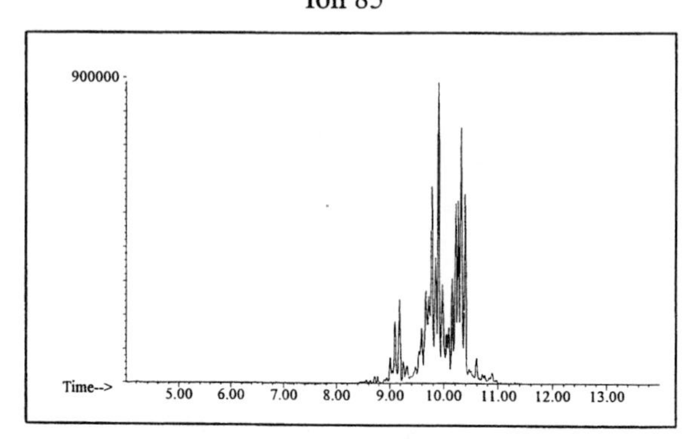

Aromatic

Ion 91

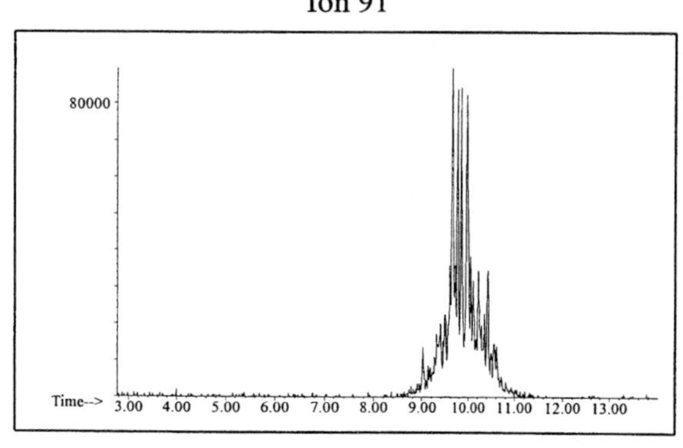

Ion 105

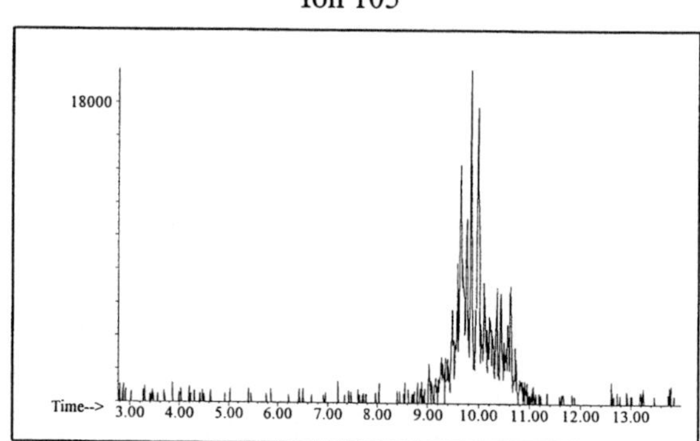

Ion 119

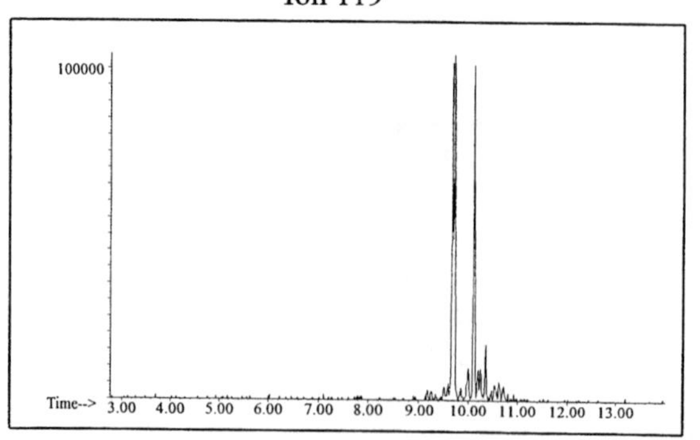

INDIVIDUAL PROFILES Naphthenic-Paraffinic #6

Cycloparaffin

Ion 55

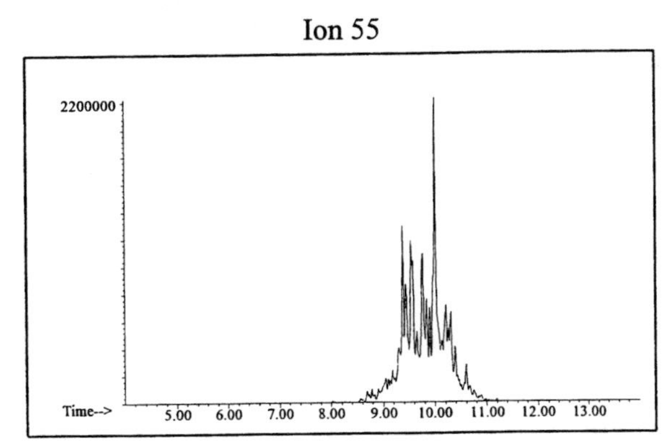

Ion 69

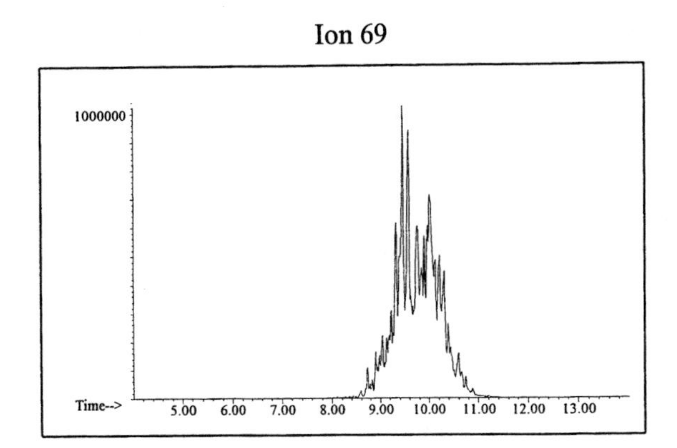

Ion 83

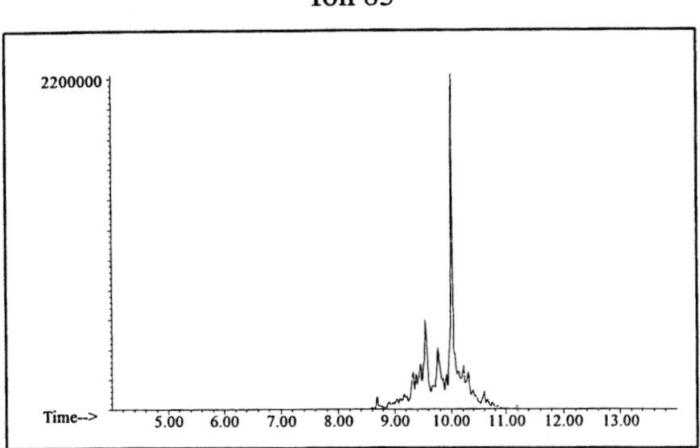

Naphthalene

Ion 128

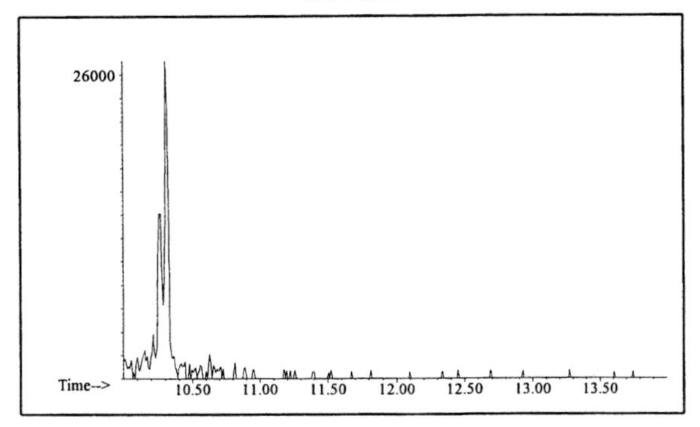

Ion 142

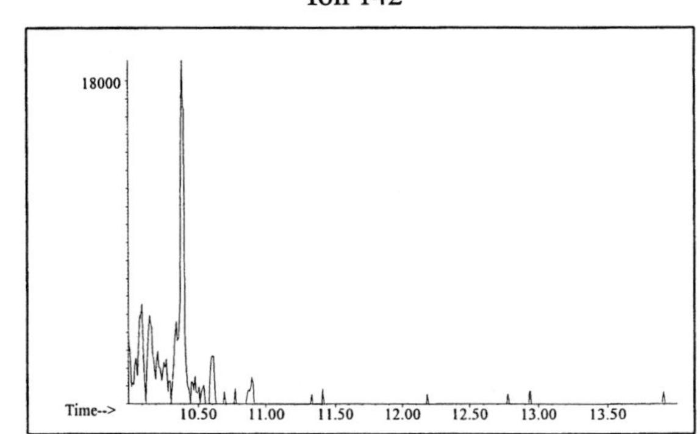

Ion 156

No Useful Data Obtained

Naphthenic-Paraffinic #7

20uL/mL Pentane

Product Displayed: Vista LPA 210

Other Similar Products:

ASTM: Class 0.5 (Naphthenic-paraffinic)

Product Uses: Solvent, feedstock

Macro Code: H

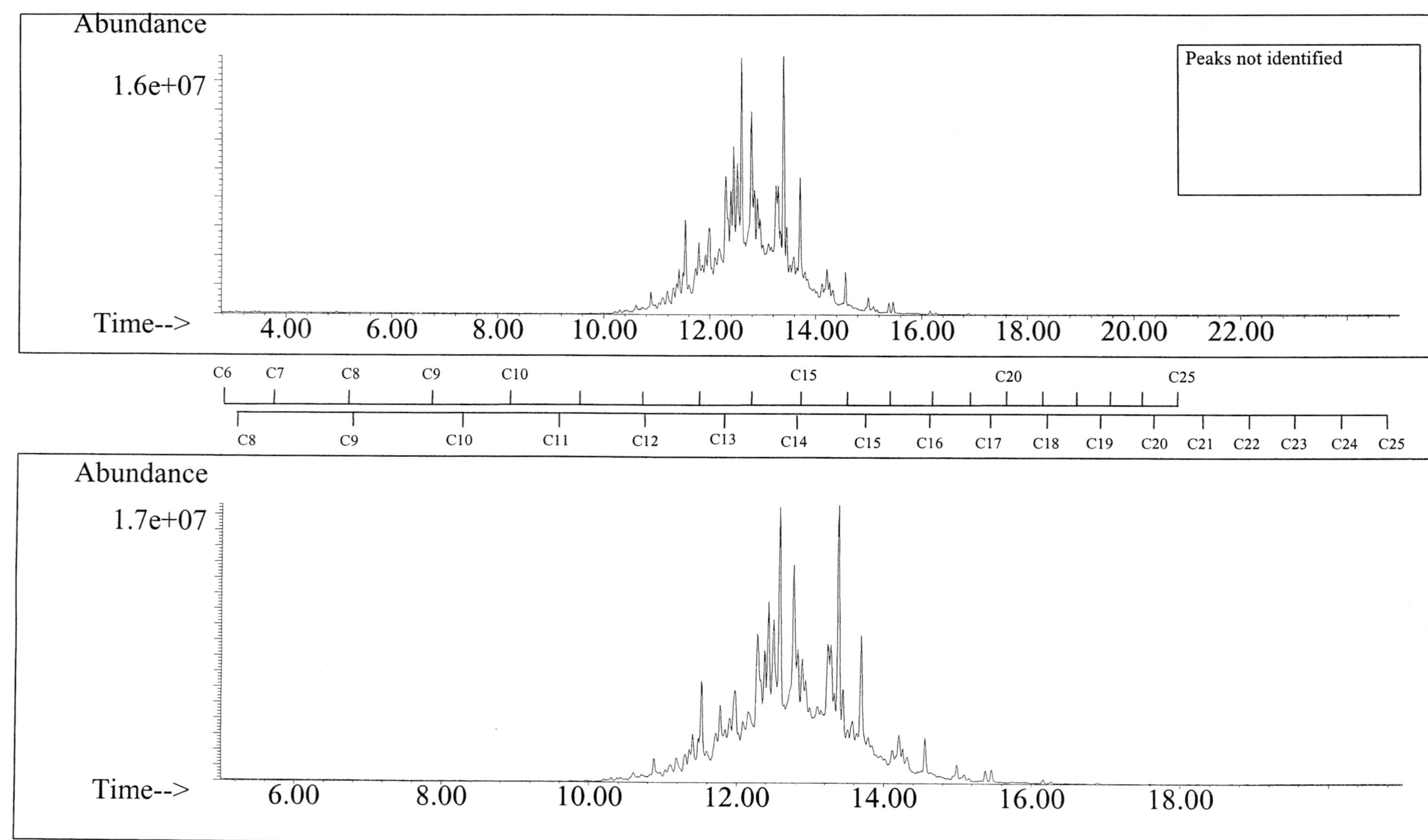

SUMMED PROFILES Naphthenic-Paraffinic #7

ALKANES

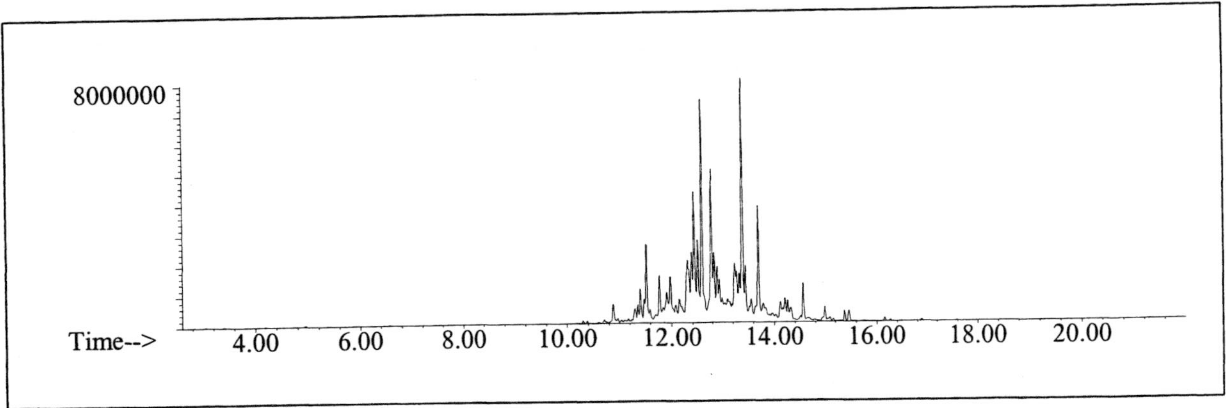

AROMATICS

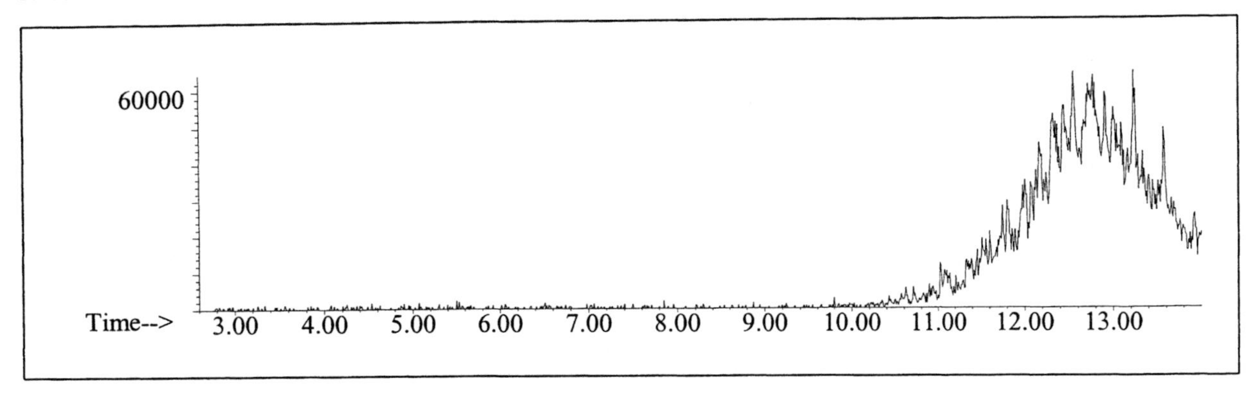

CYCLOPARAFFINS AND ALKENES

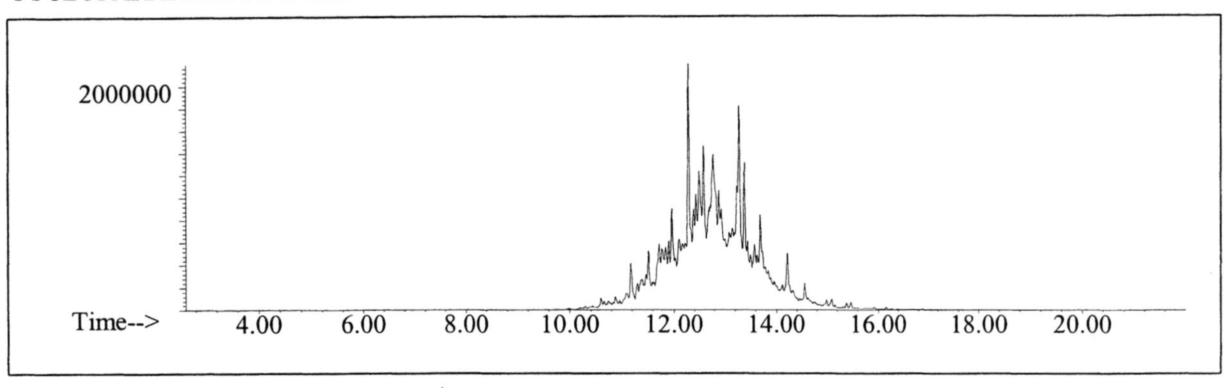

NAPHTHALENES

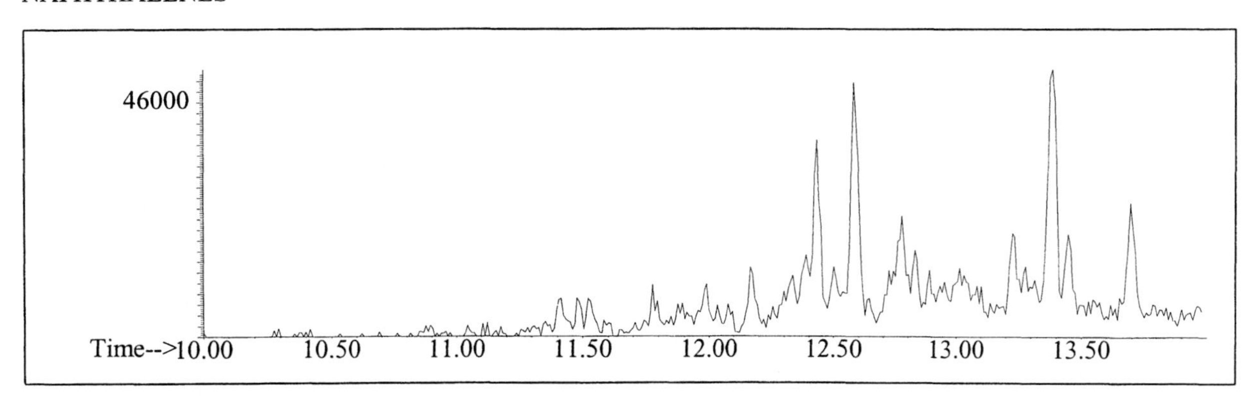

INDIVIDUAL PROFILES Naphthenic-Paraffinic #7

Alkane

Ion 43

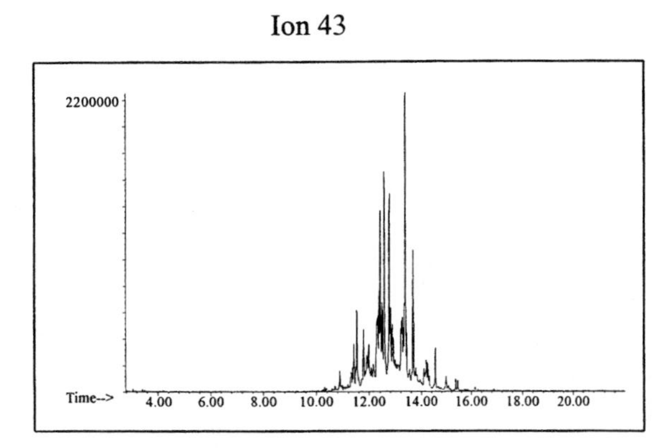

Ion 57

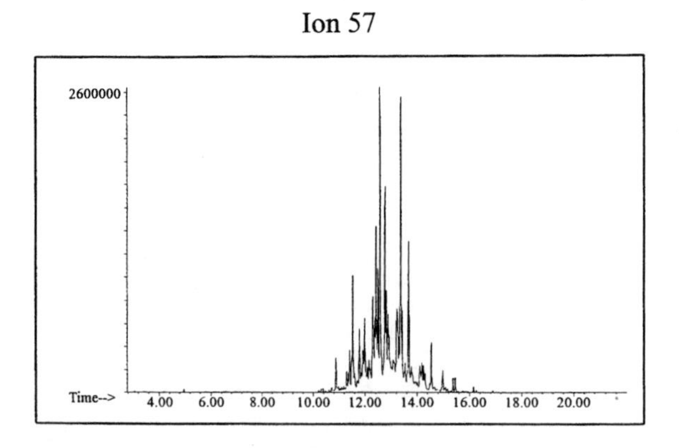

Ion 71

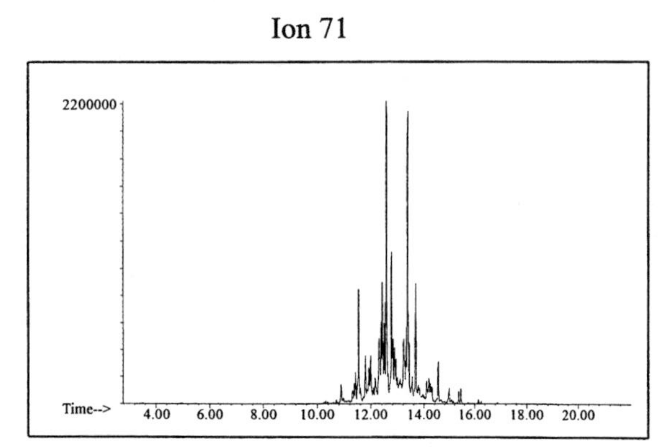

Ion 85

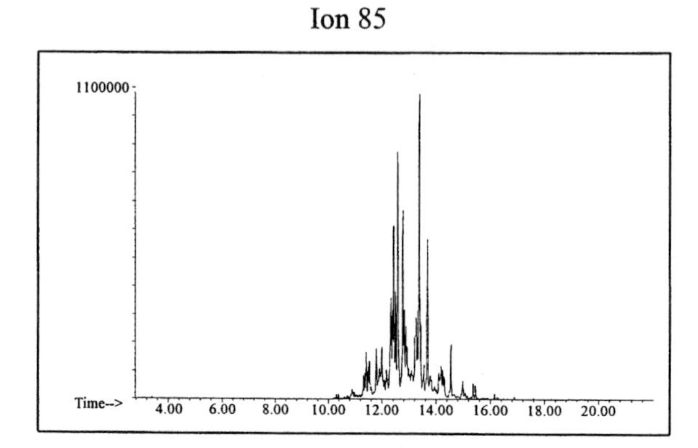

Aromatic

Ion 91

No Useful Data Obtained

Ion 105

No Useful Data Obtained

Ion 119

No Useful Data Obtained

INDIVIDUAL PROFILES Naphthenic-Paraffinic #7

Cycloparaffin

Ion 55

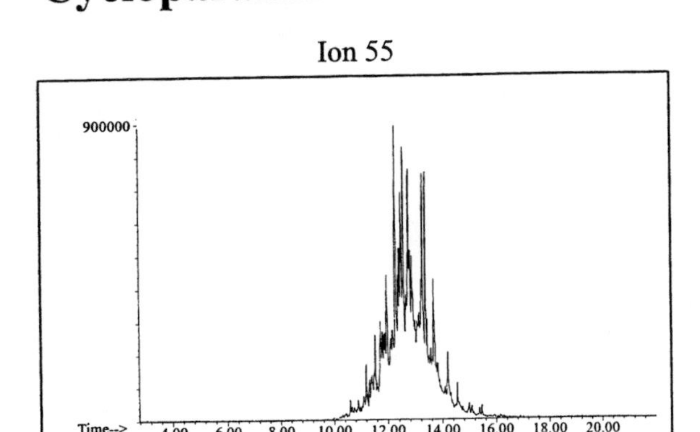

Ion 69

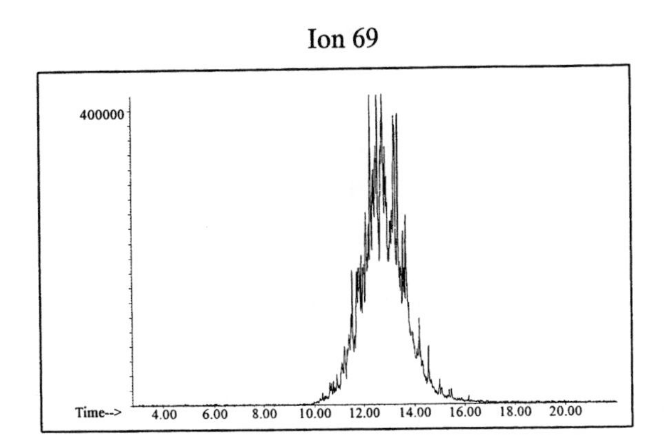

Ion 83

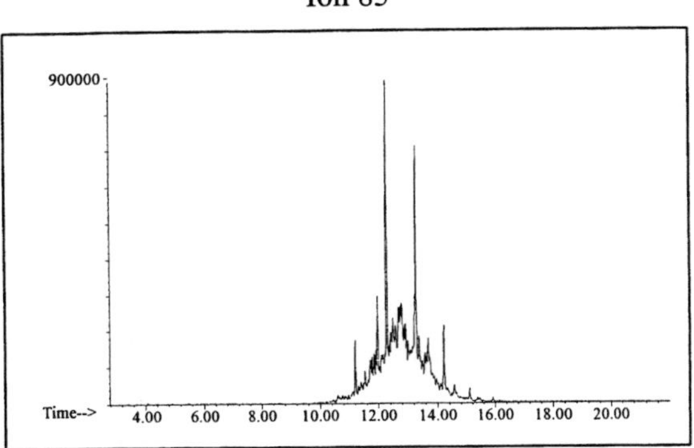

Naphthalene

Ion 128

No Useful Data Obtained

Ion 142

No Useful Data Obtained

Ion 156

No Useful Data Obtained

Miscellaneous and Mixtures

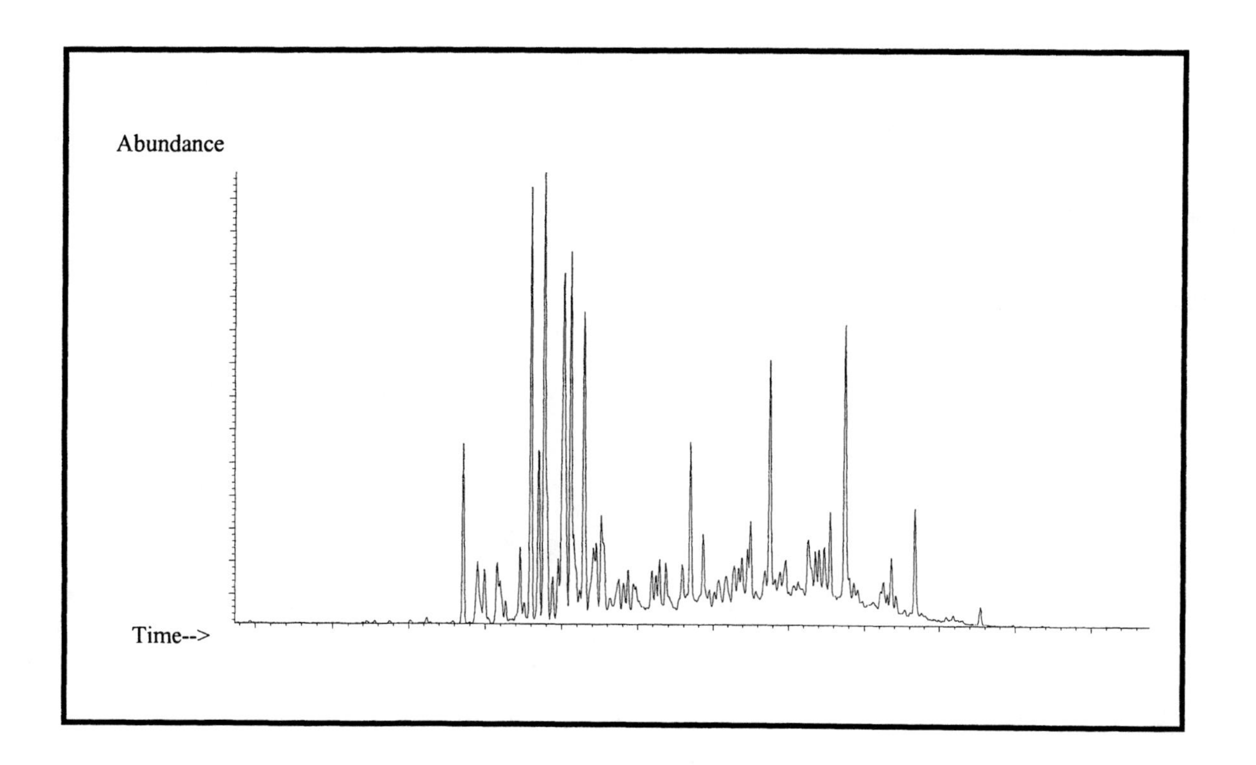

Miscellaneous #1

20uL/mL Pentane

ASTM: Class 0 (Miscellaneous)

Macro Code: L

Product Displayed: B-12 Carburetor and Choke Cleaner

Product Uses: Cleaning solvent

Other Similar Products:

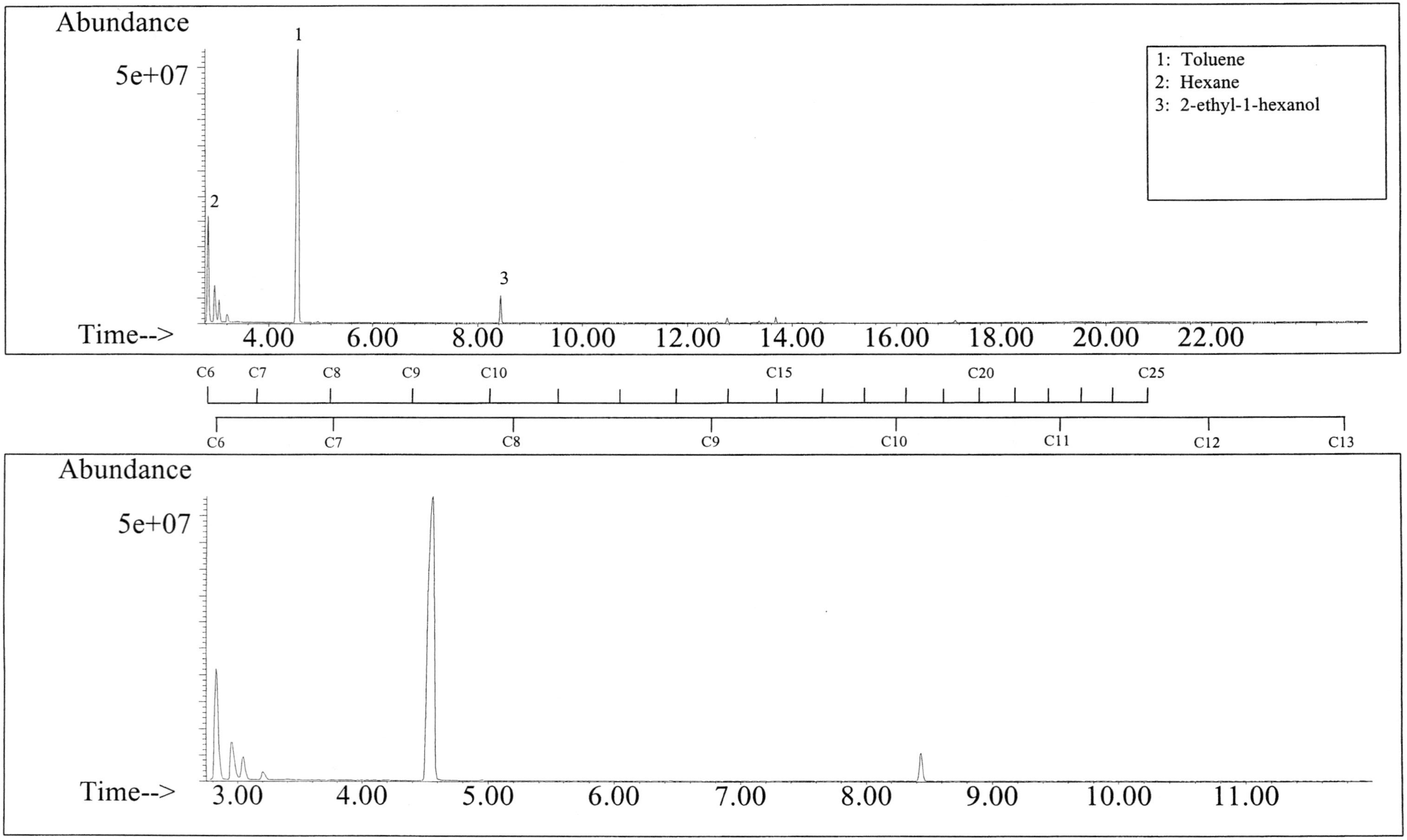

Abundance

5e+07

1: Toluene
2: Hexane
3: 2-ethyl-1-hexanol

Time--> 4.00 6.00 8.00 10.00 12.00 14.00 16.00 18.00 20.00 22.00

C6 C7 C8 C9 C10 C15 C20 C25

C6 C7 C8 C9 C10 C11 C12 C13

Abundance

5e+07

Time--> 3.00 4.00 5.00 6.00 7.00 8.00 9.00 10.00 11.00

ALKANES

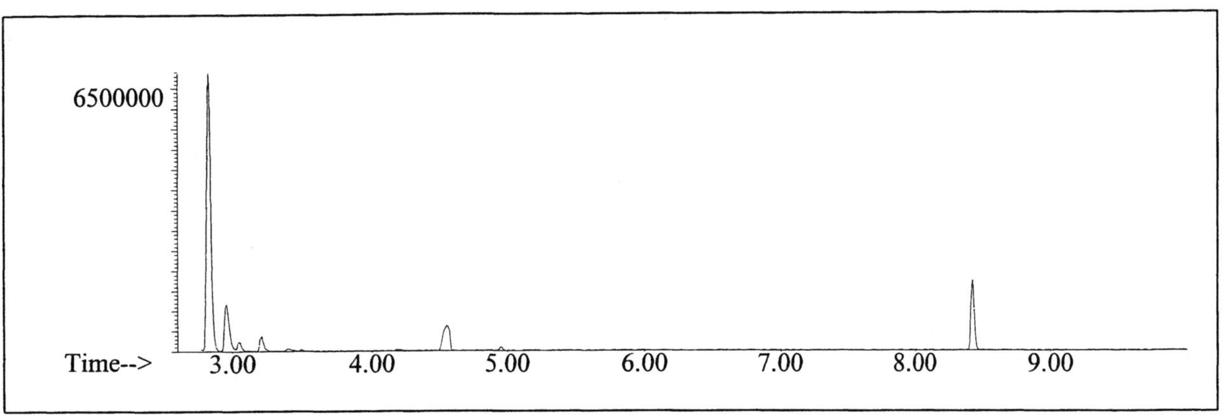

AROMATICS

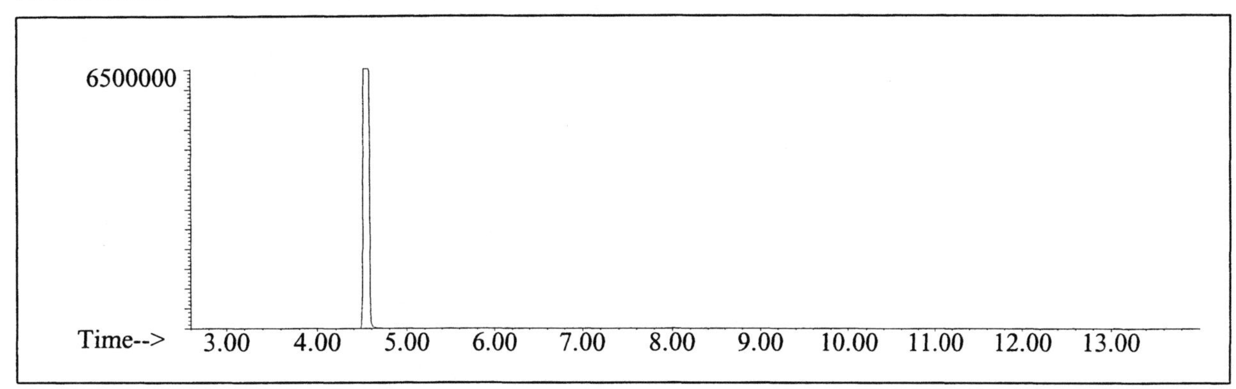

CYCLOPARAFFINS AND ALKENES

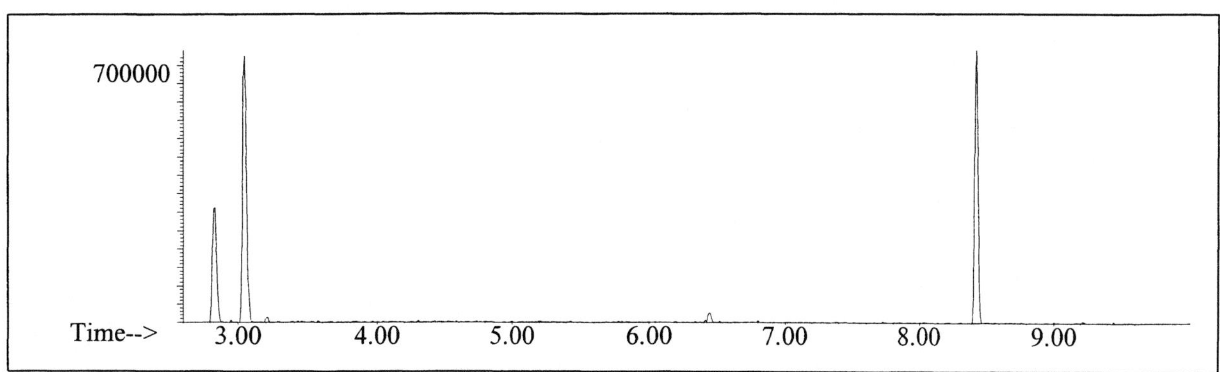

NAPHTHALENES

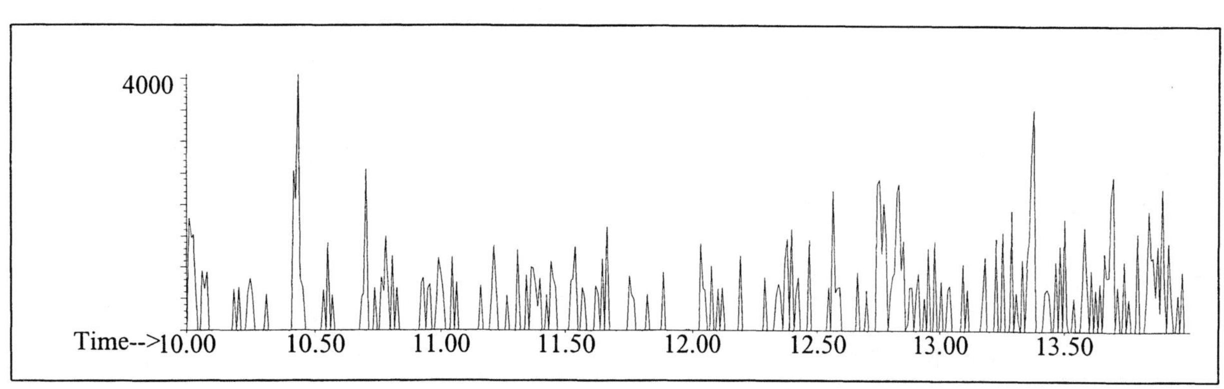

Alkane

Ion 43

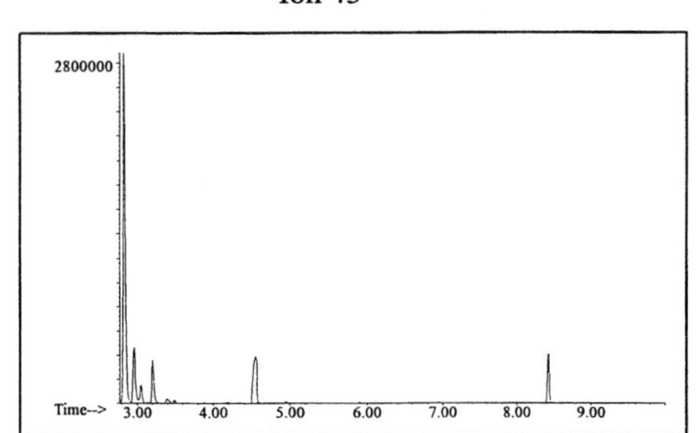

Ion 57

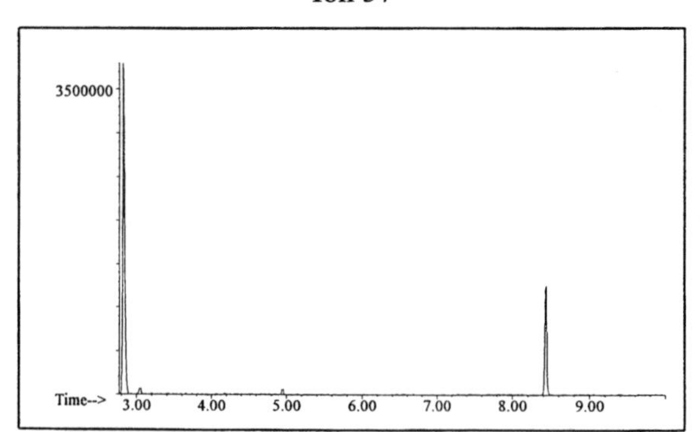

Ion 71

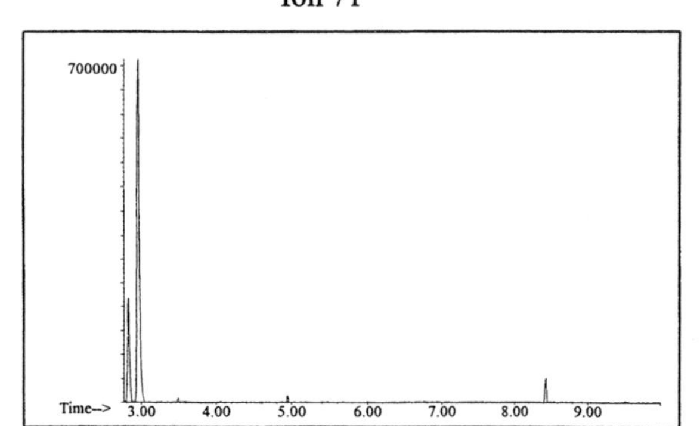

Ion 85

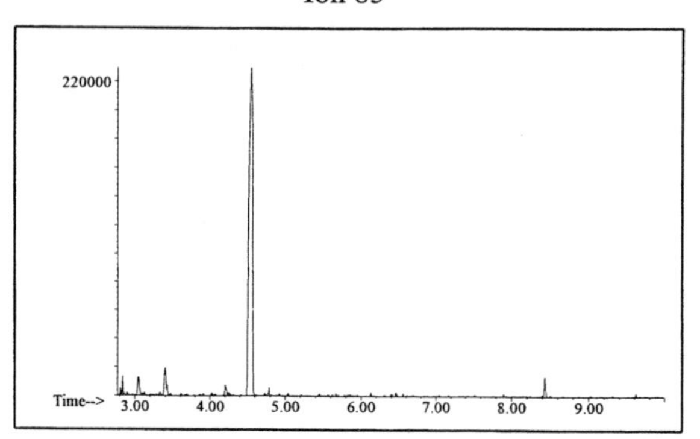

Aromatic

Ion 91

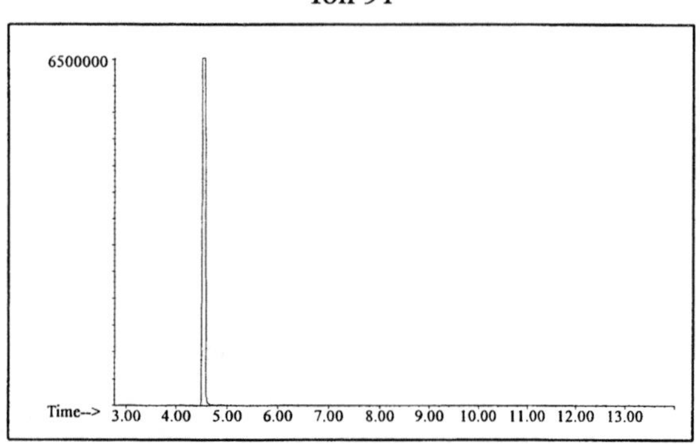

Ion 105

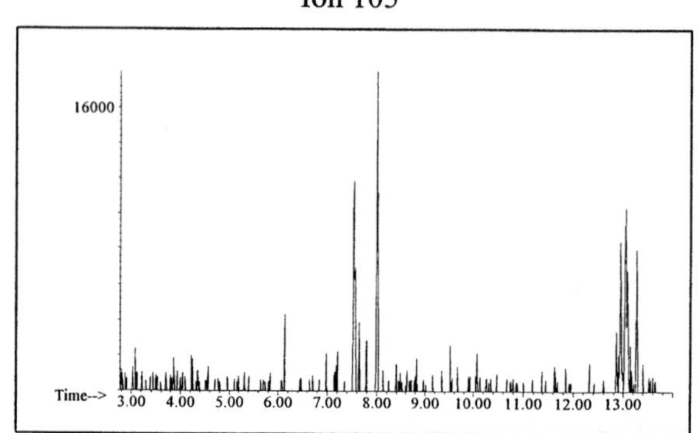

Ion 119

No Useful Data Obtained

Cycloparaffin

Ion 55

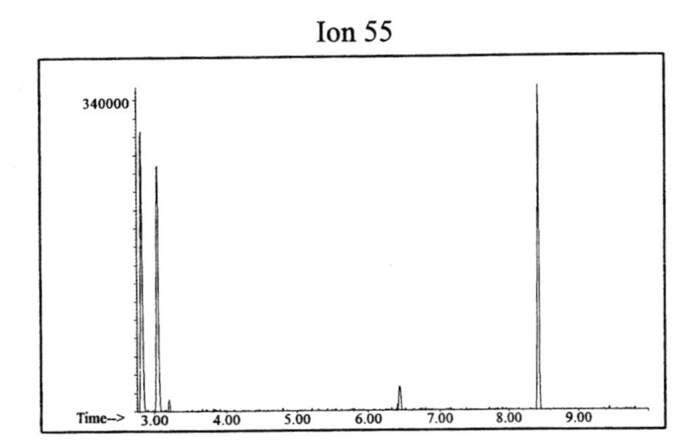

Ion 69

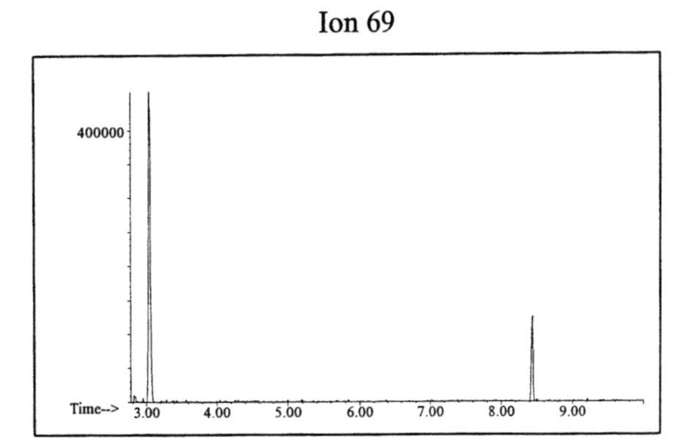

Ion 83

260000
Time--> 3.00 4.00 5.00 6.00 7.00 8.00 9.00

Naphthalene

Ion 128

No Useful Data Obtained

Ion 142

No Useful Data Obtained

Ion 156

No Useful Data Obtained

Miscellaneous #2

20uL/mL Pentane

Product Displayed: DTR 600 Enamel Reducer

Other Similar Products:

ASTM: Class 0 (Miscellaneous)

Product Uses: Enamel reducer

Macro Code: L

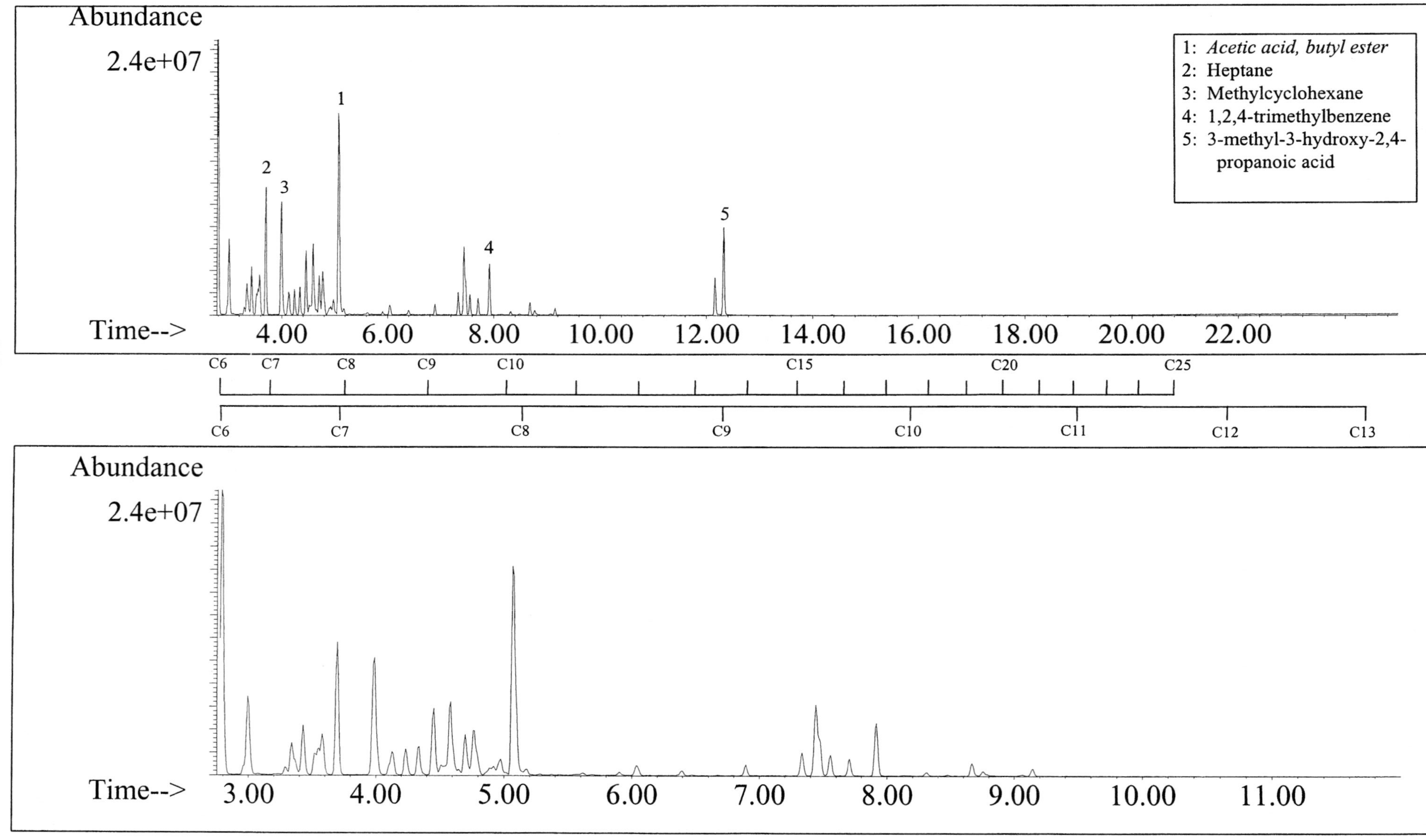

ALKANES

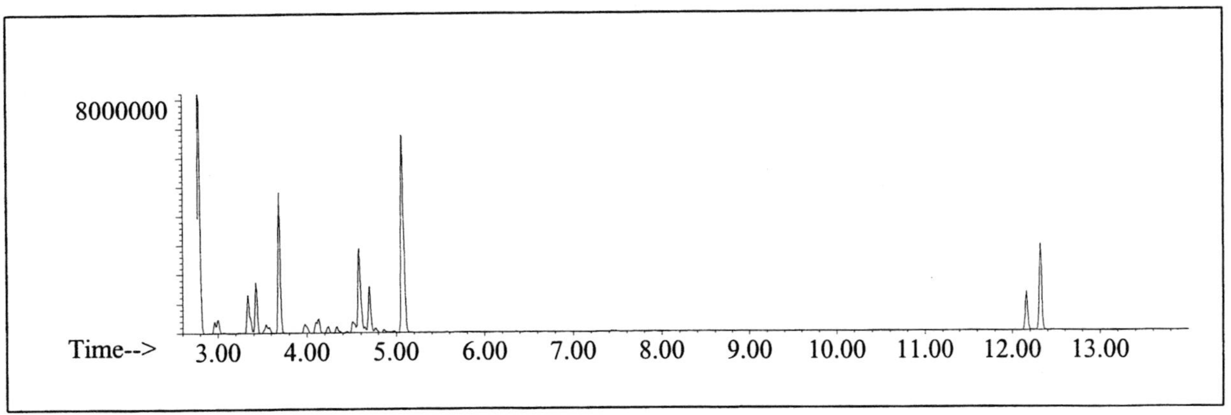

AROMATICS

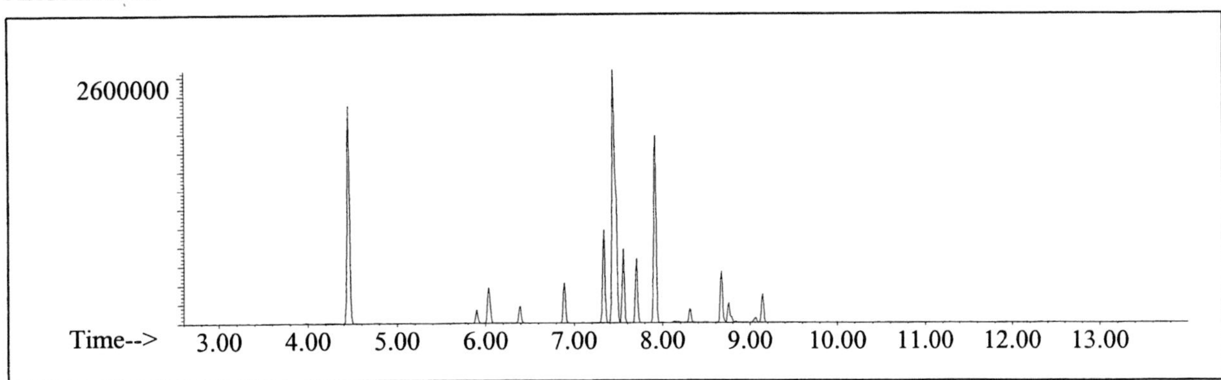

CYCLOPARAFFINS AND ALKENES

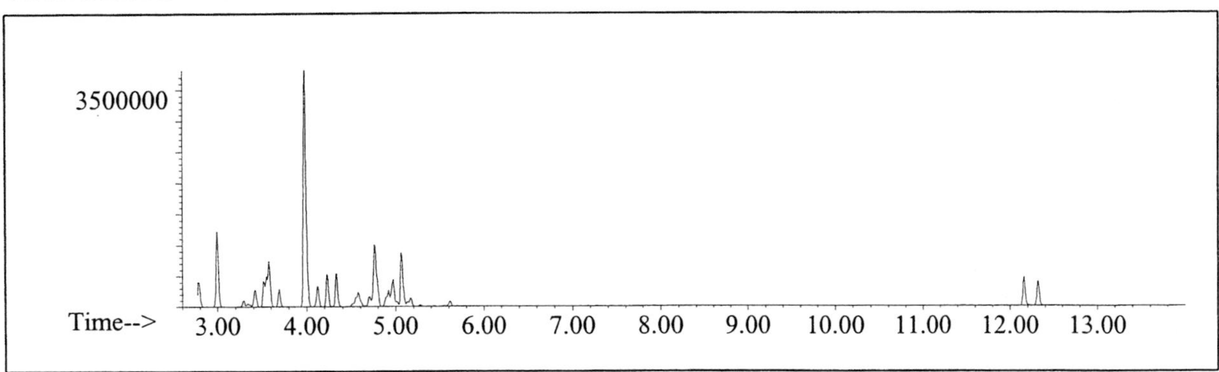

NAPHTHALENES

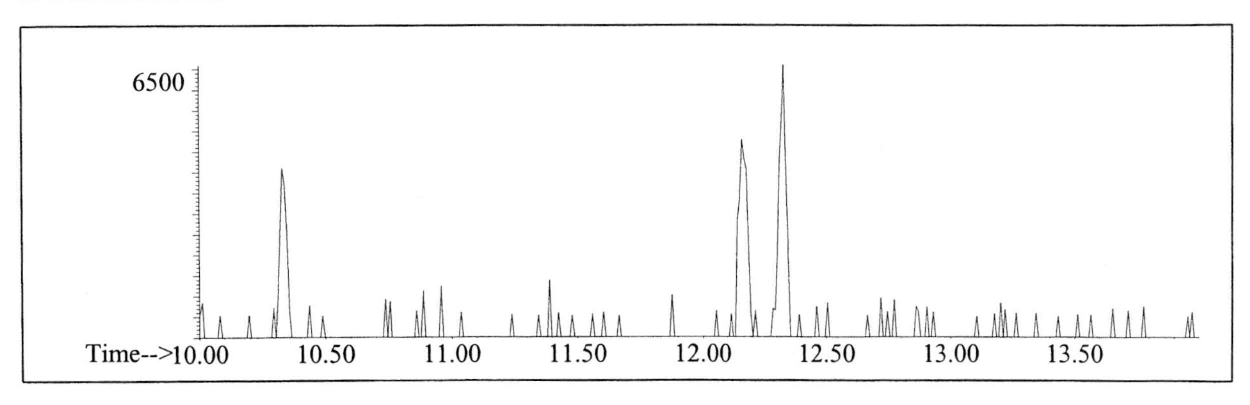

Alkane

Ion 43

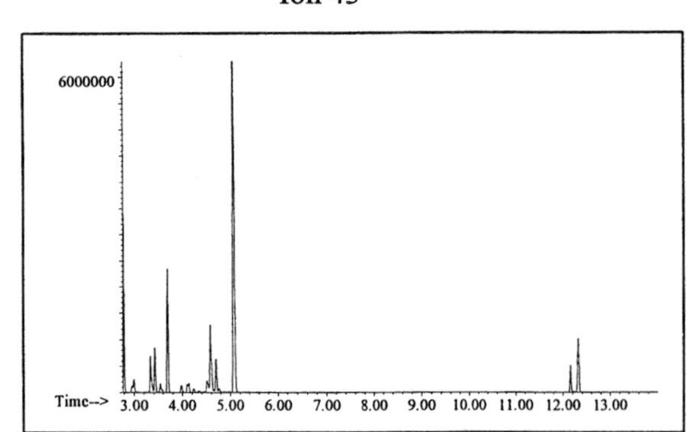

Ion 57

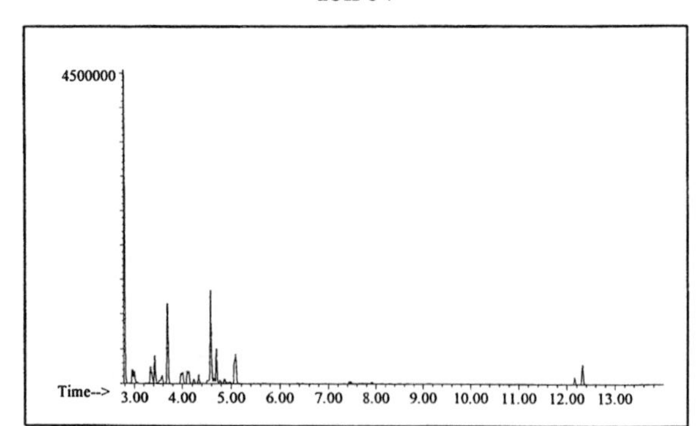

Ion 71

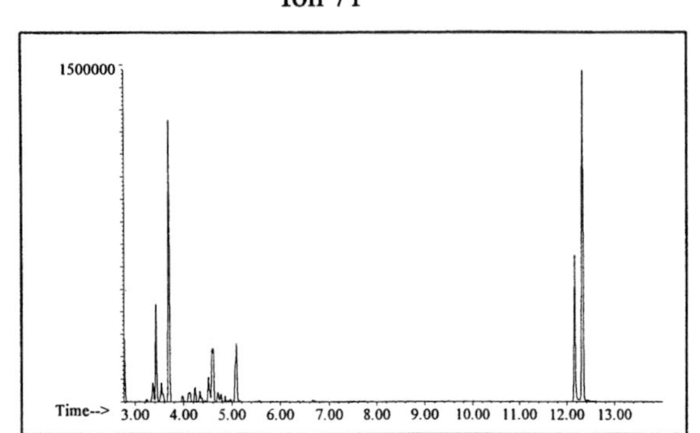

Ion 85

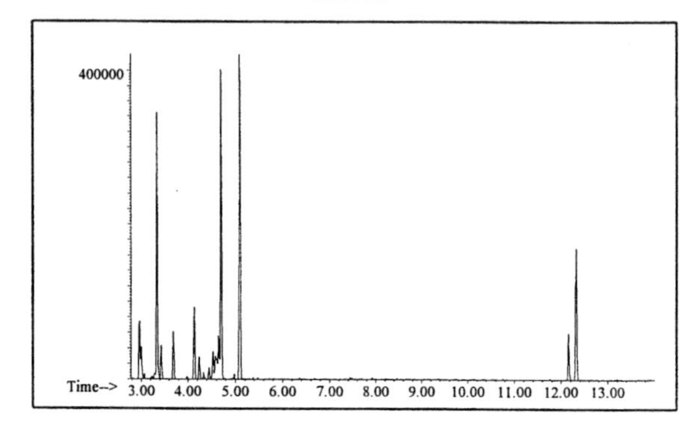

Aromatic

Ion 91

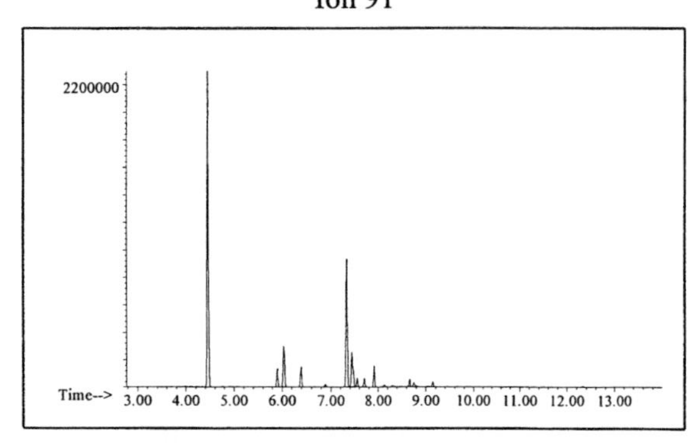

Ion 105

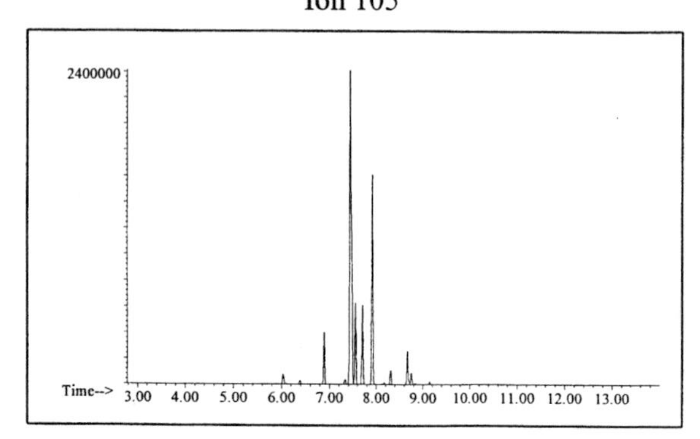

Ion 119

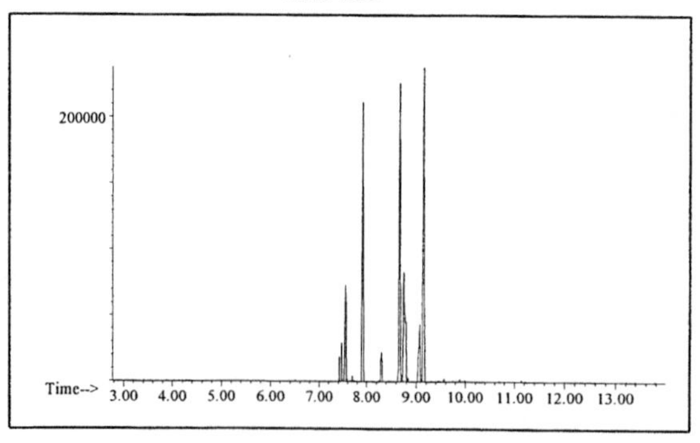

Cycloparaffin

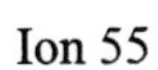

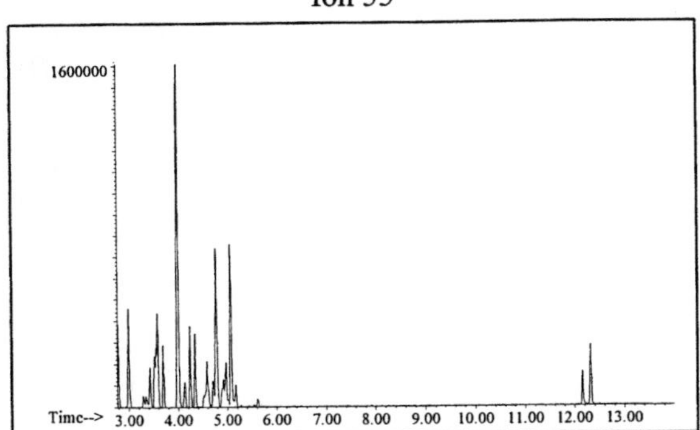

Ion 69

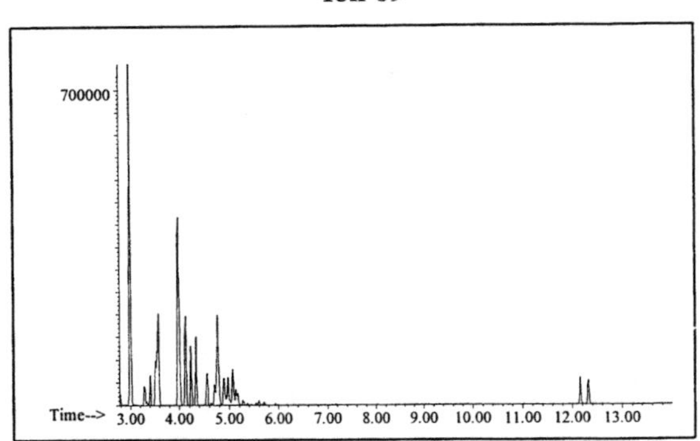

Ion 83

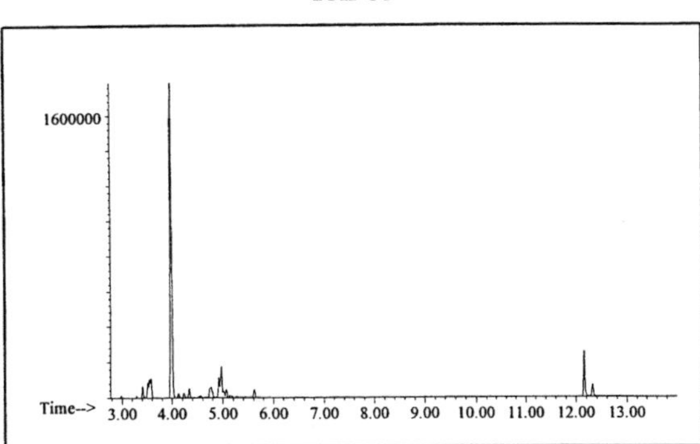

Naphthalene

Ion 128

No Useful Data Obtained

Ion 142

No Useful Data Obtained

Ion 156

No Useful Data Obtained

Miscellaneous #3

20uL/mL Pentane

Product Displayed: Prep-Eez Liquid Sandpaper

Other Similar Products:

ASTM: Class 0 (Miscellaneous)

Product Uses: Surface preparation

Macro Code: L

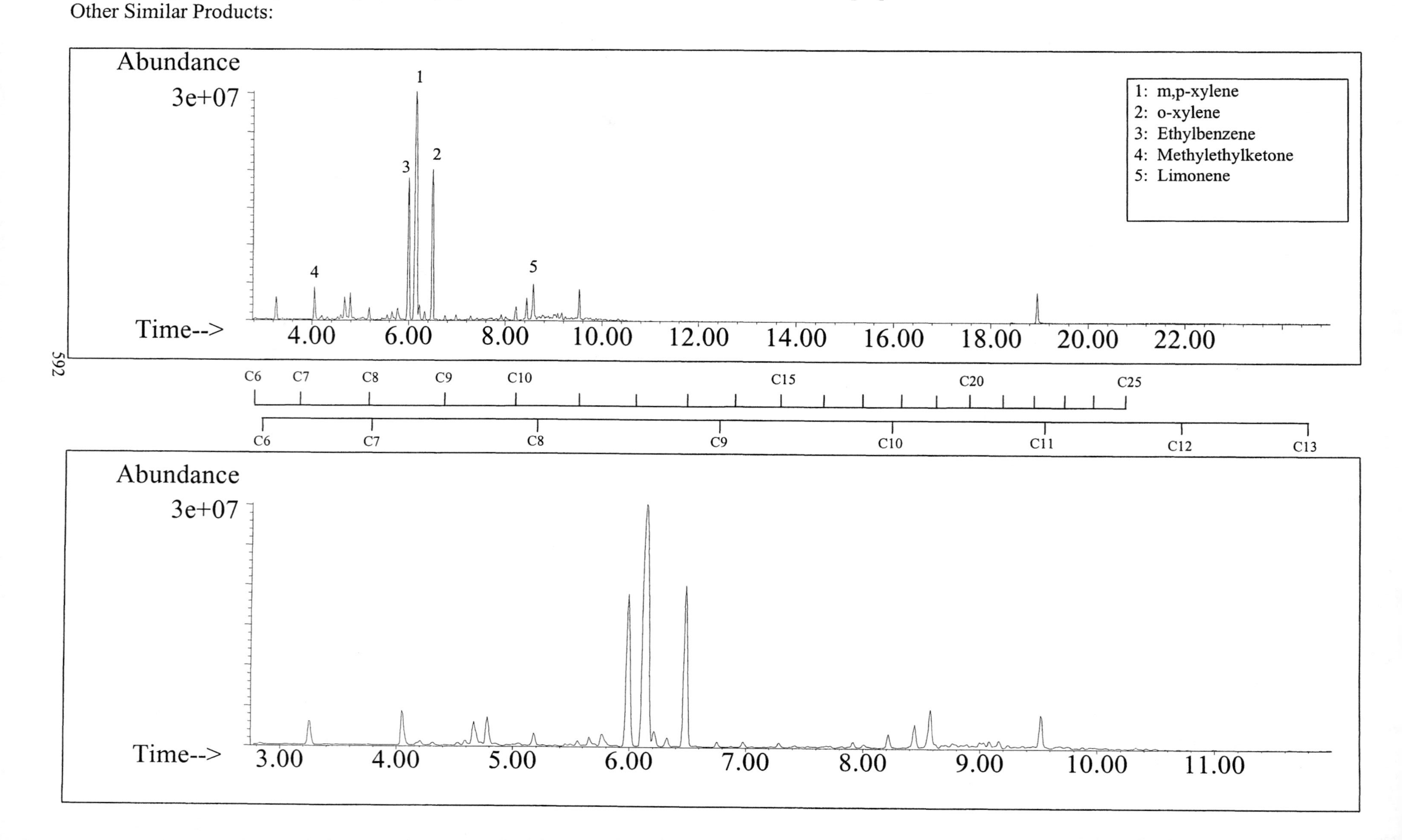

ALKANES

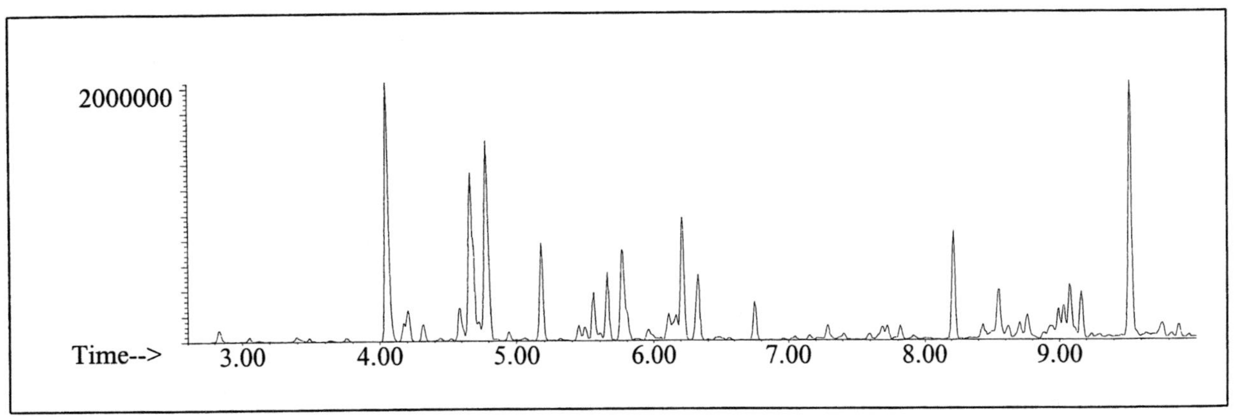

AROMATICS

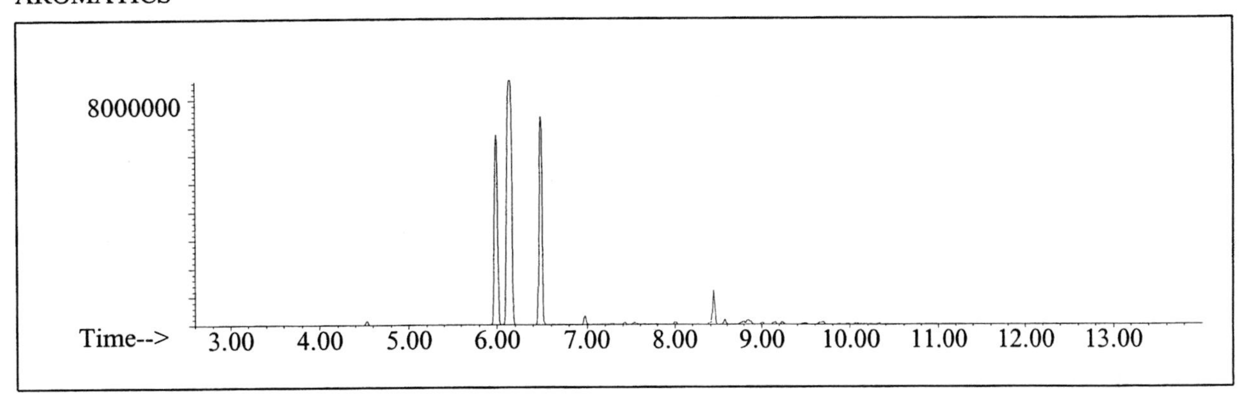

CYCLOPARAFFINS AND ALKENES

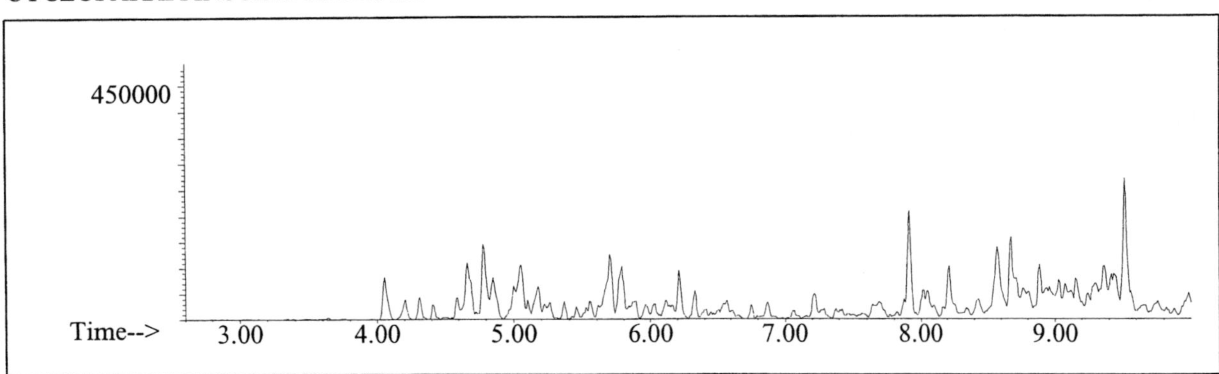

NAPHTHALENES

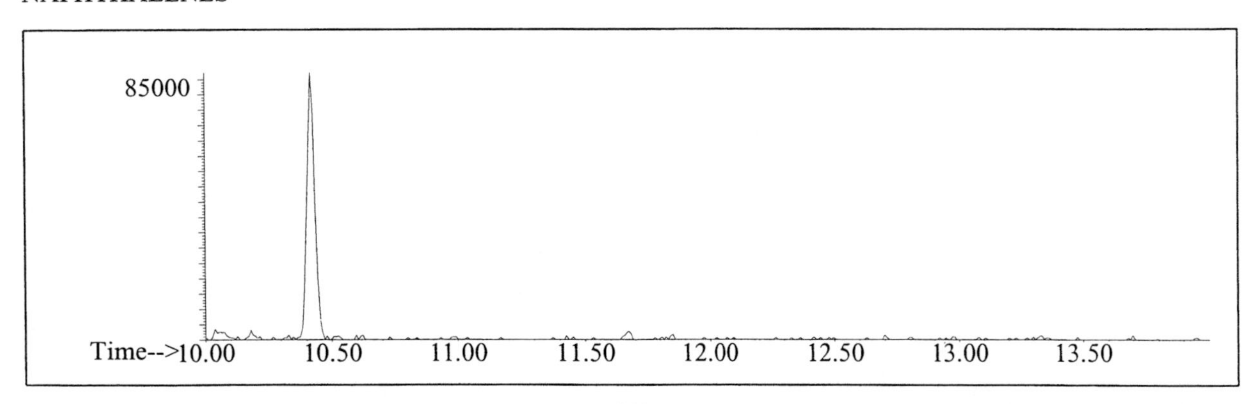

Alkane

Ion 43

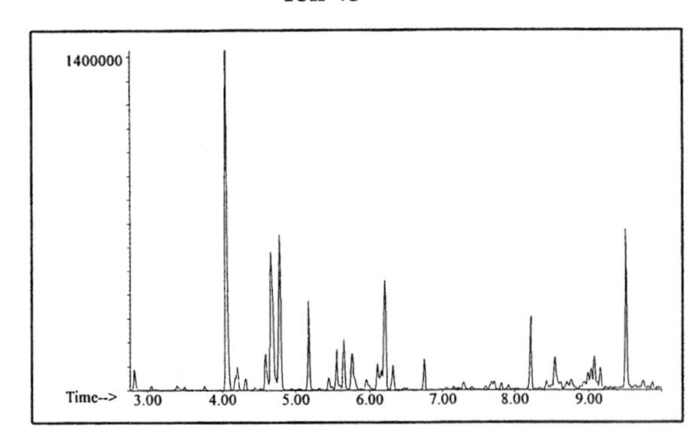

Ion 57

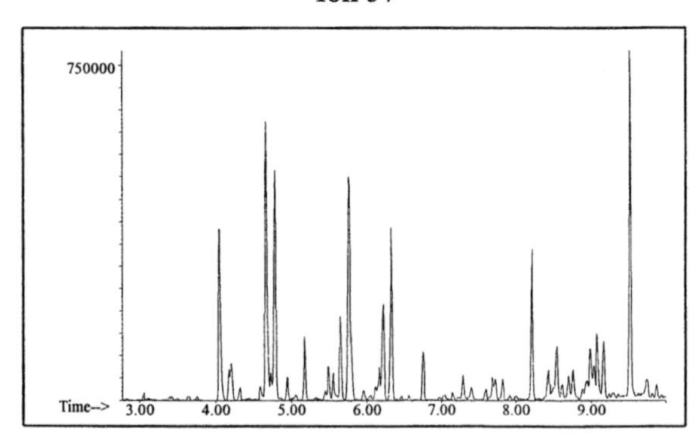

Ion 71

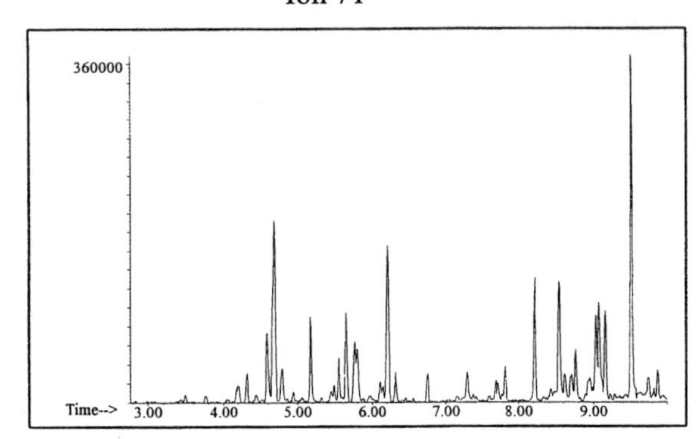

Ion 85

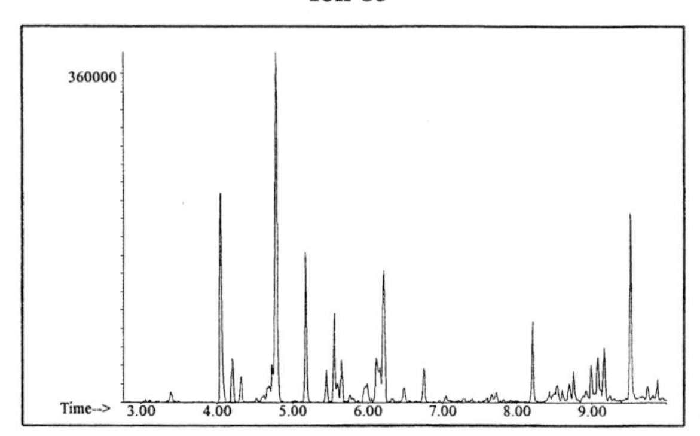

Aromatic

Ion 91

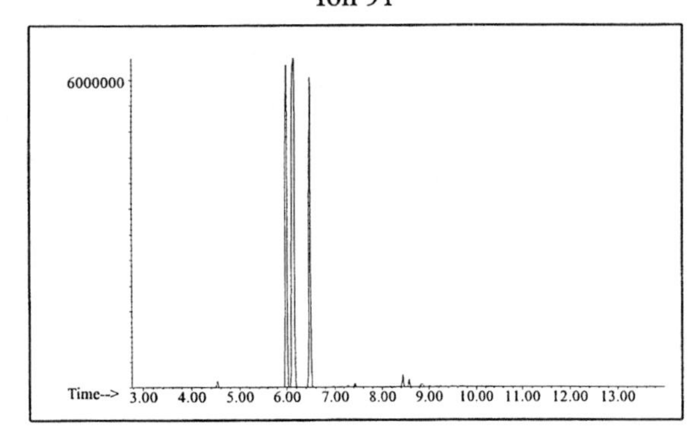

Ion 105

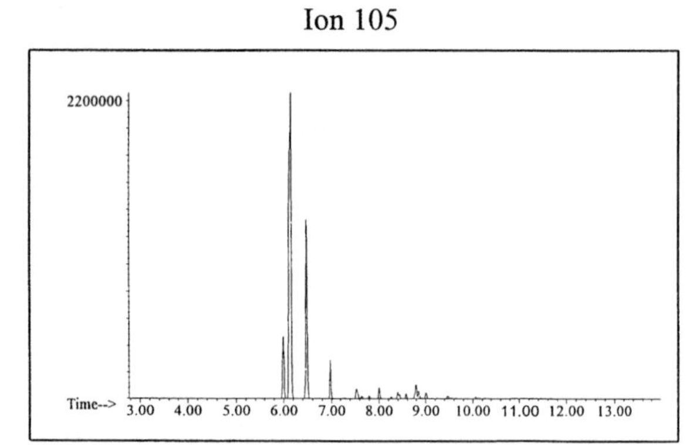

Ion 119

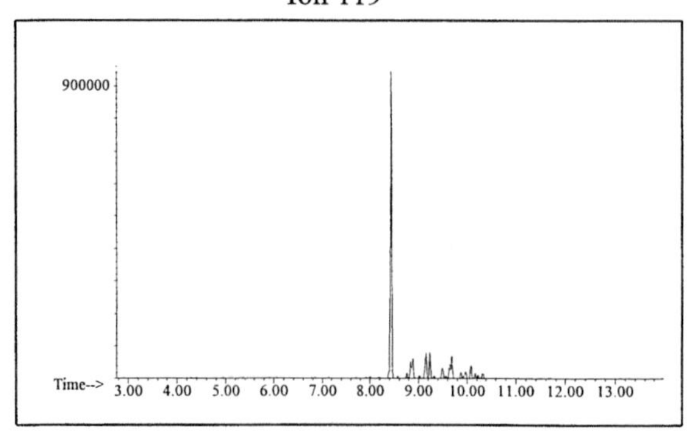

Cycloparaffin

Ion 55

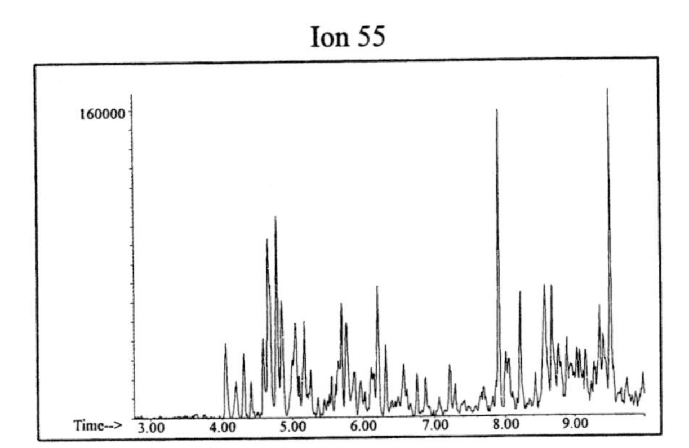

Ion 69

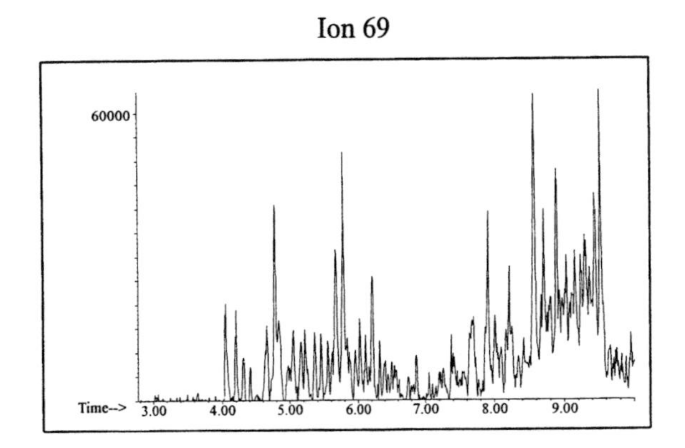

Ion 83

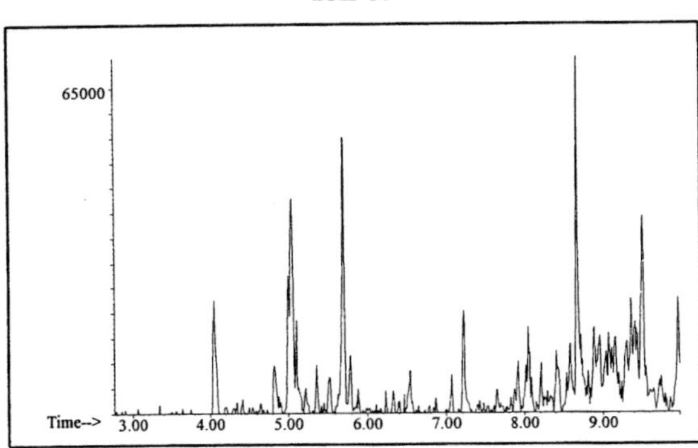

Naphthalene

Ion 128

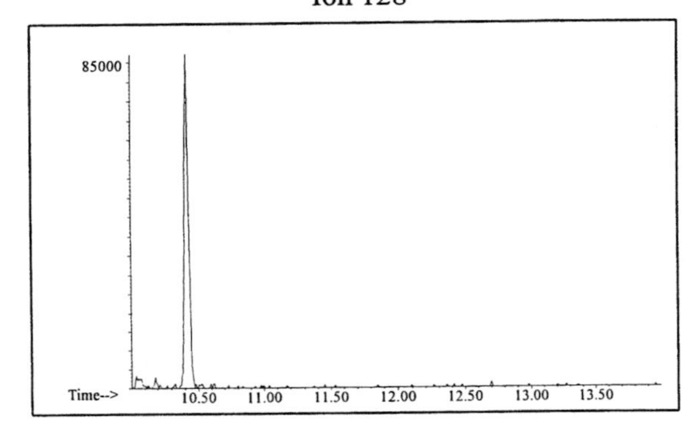

Ion 142

No Useful Data Obtained

Ion 156

No Useful Data Obtained

Miscellaneous #4

20uL/mL Pentane

ASTM: Class 0 (Miscellaneous)

Macro Code: L

Product Displayed: Goof-Off Paint Remover

Product Uses: Paint remover

Other Similar Products:

Abundance

2.6e+07

1: m,p-xylene
2: o-xylene
3: Ethylbenzene
4: 1,1,1-trichloroethane
5: Tetrachloroethylene

Time-->

4.00 6.00 8.00 10.00 12.00 14.00 16.00 18.00 20.00 22.00

C6 C7 C8 C9 C10 C15 C20 C25

C6 C7 C8 C9 C10 C11 C12 C13

Abundance

2.6e+07

Time-->

3.00 4.00 5.00 6.00 7.00 8.00 9.00 10.00 11.00

ALKANES

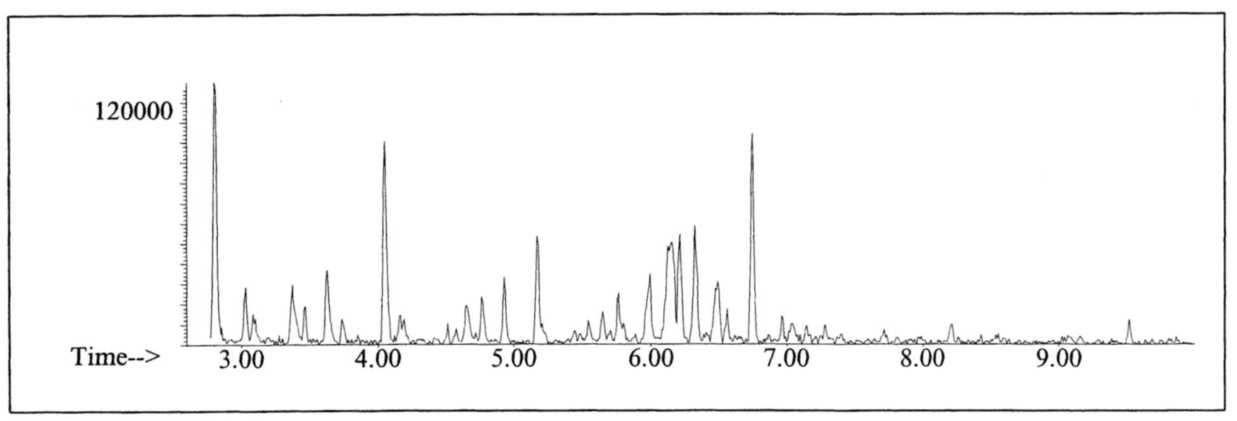

AROMATICS

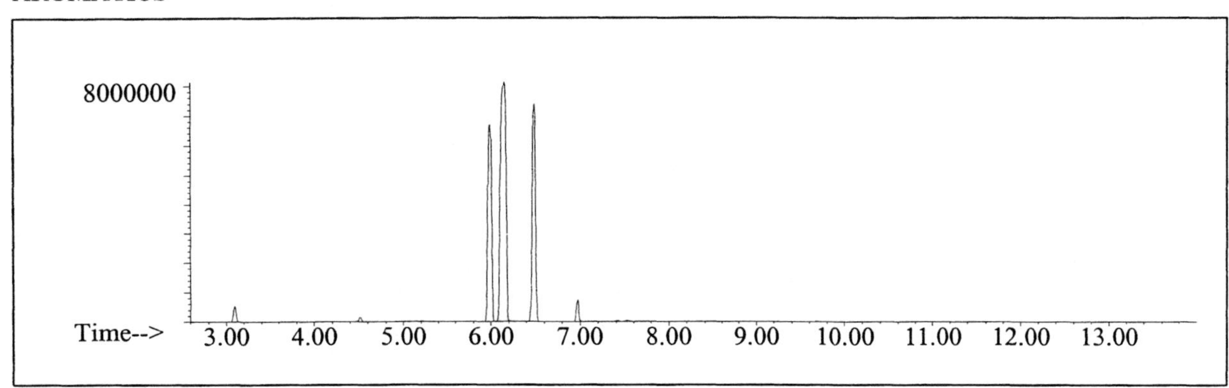

CYCLOPARAFFINS AND ALKENES

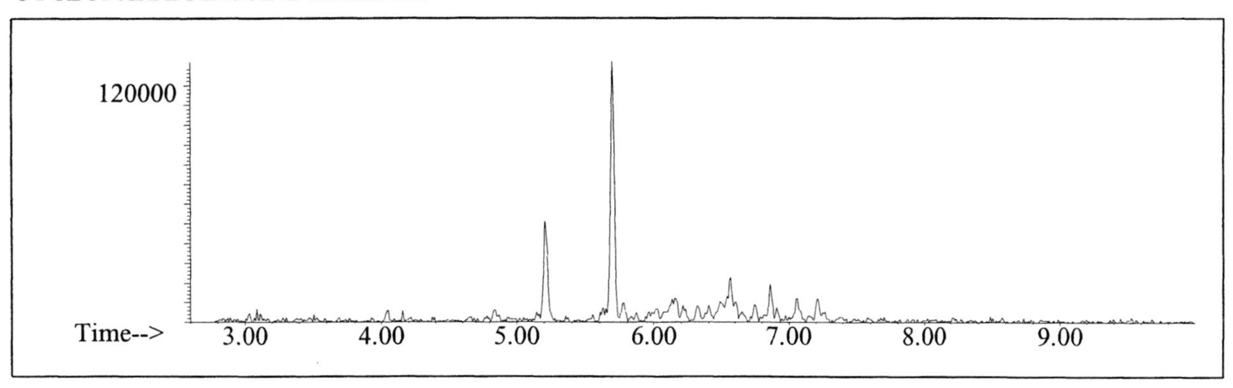

NAPHTHALENES

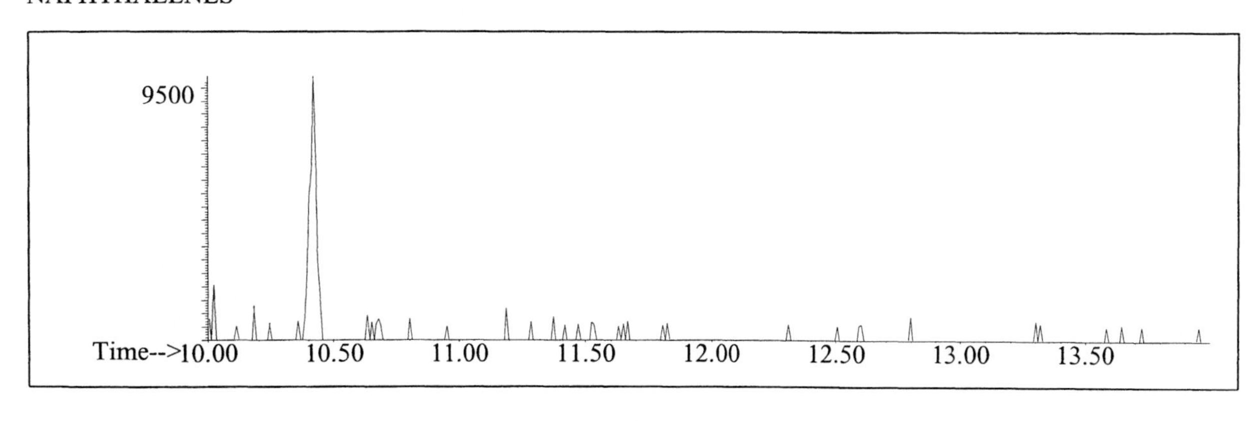

INDIVIDUAL PROFILES Miscellaneous #4

Alkane

Ion 43

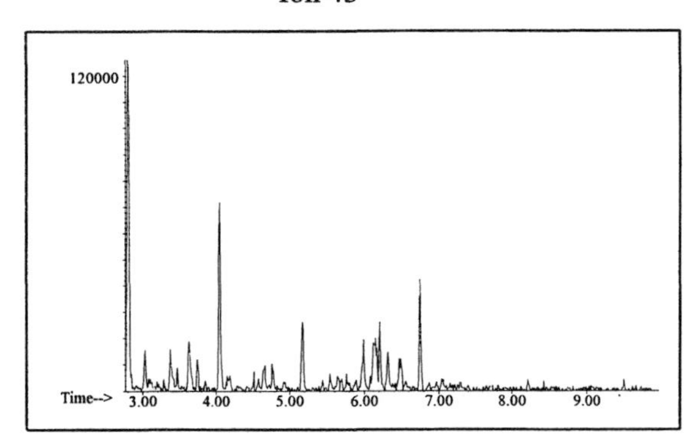

Ion 57

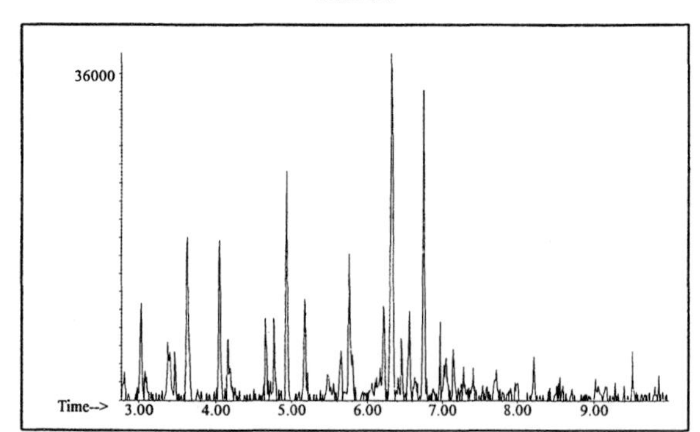

Ion 71

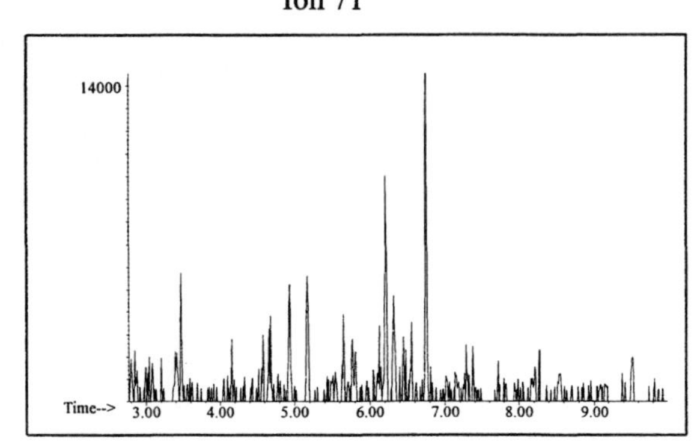

Ion 85

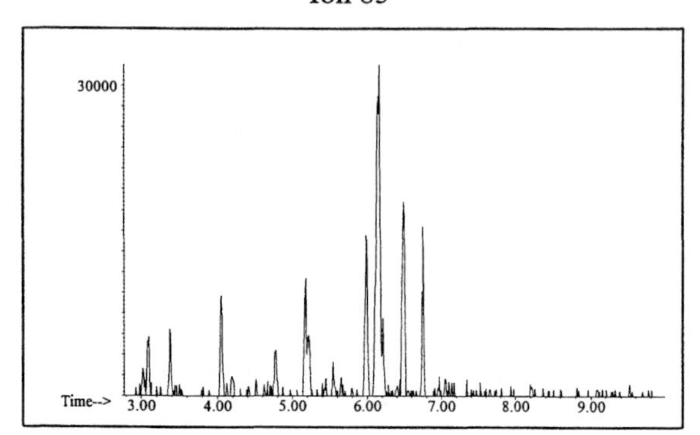

Aromatic

Ion 91

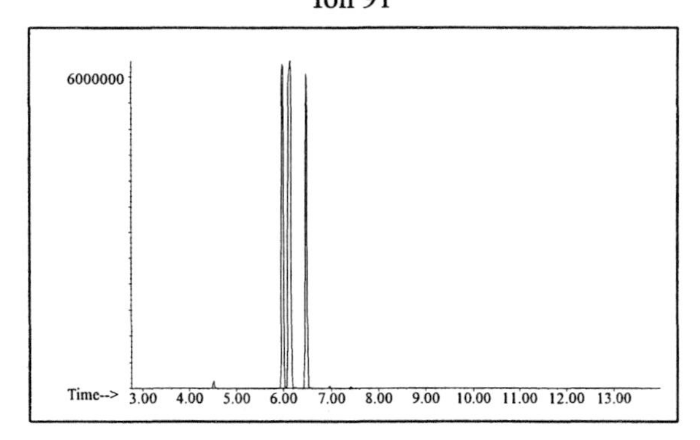

Ion 105

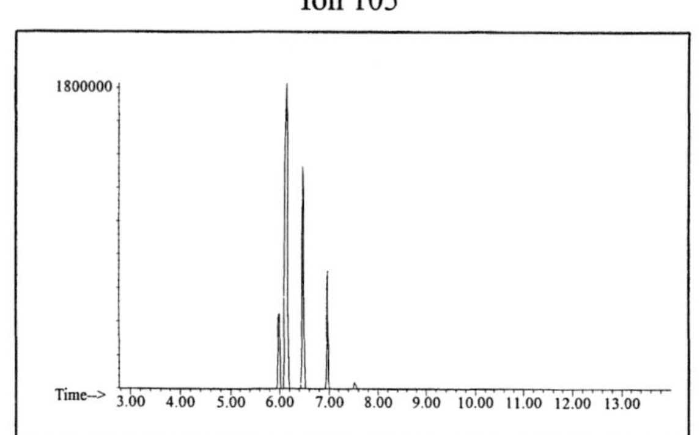

Ion 119

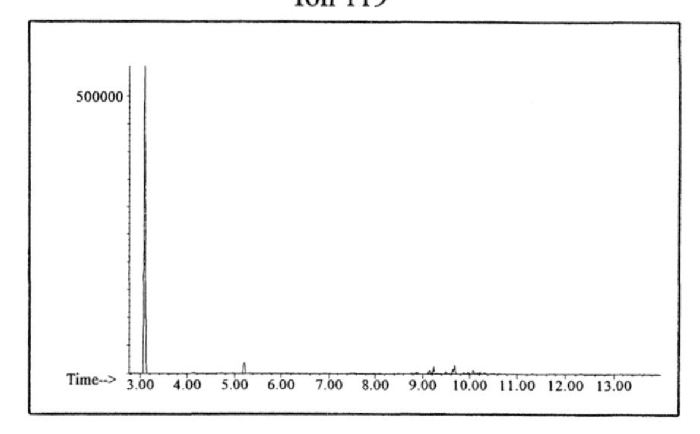

Cycloparaffin

Ion 55

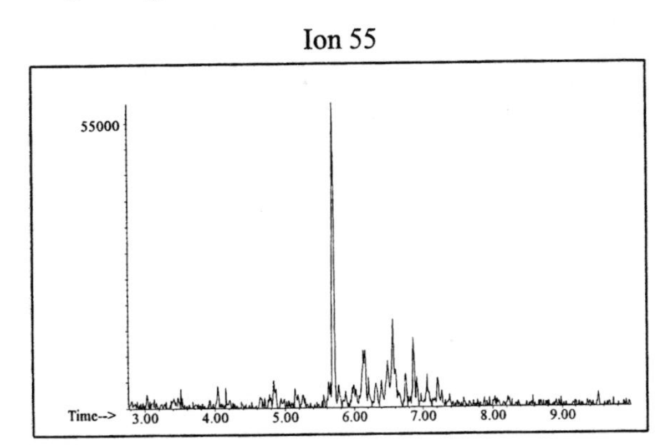

Ion 69

No Useful Data Obtained

Ion 83

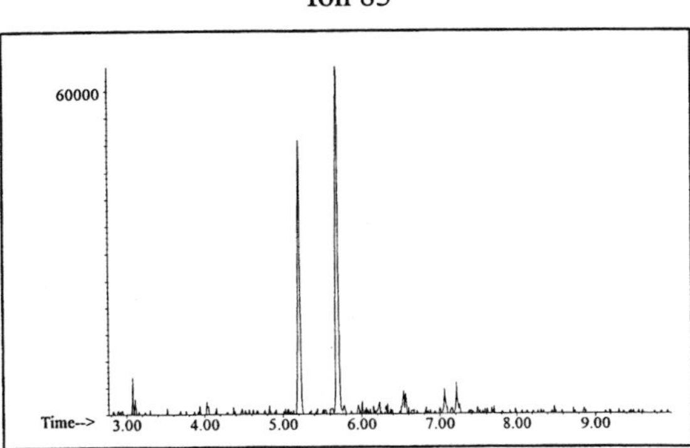

Naphthalene

Ion 128

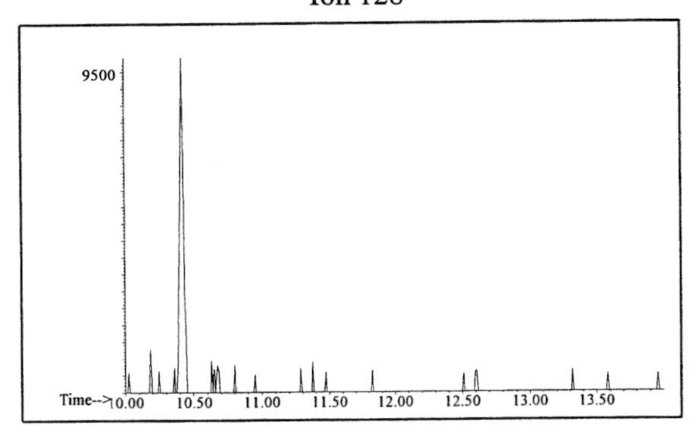

Ion 142

No Useful Data Obtained

Ion 156

No Useful Data Obtained

Miscellaneous #5

20uL/mL Pentane

Product Displayed: DTR 601 Enamel Reducer

Other Similar Products:

ASTM: Class 0 (Miscellaneous)

Product Uses: Enamel reducer

Macro Code: L

1: Toluene
2: Heptane
3: Methylcyclohexane
4: 1,2,4-trimethylbenzene
5: 3-methyl-3-hydroxy-2,4-propanoic acid

ALKANES

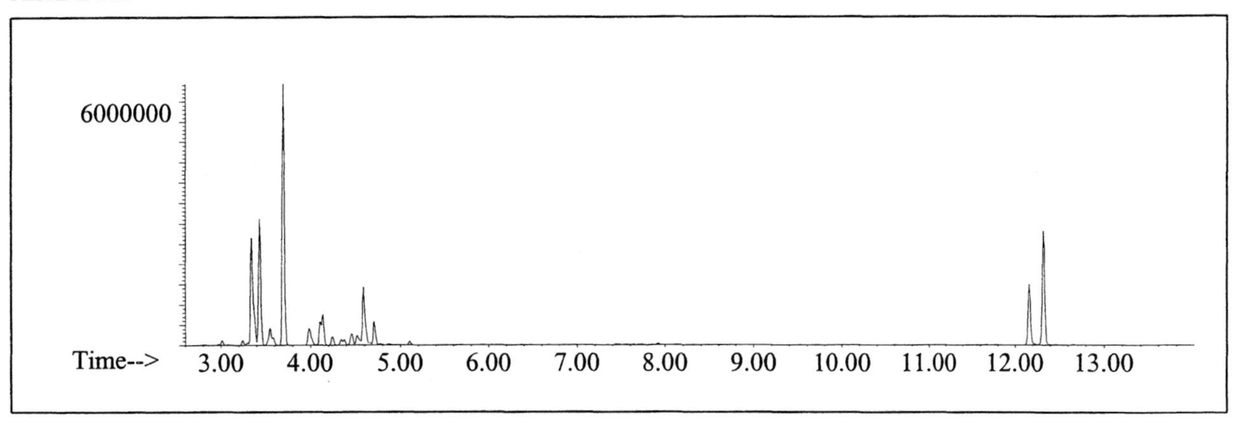

AROMATICS

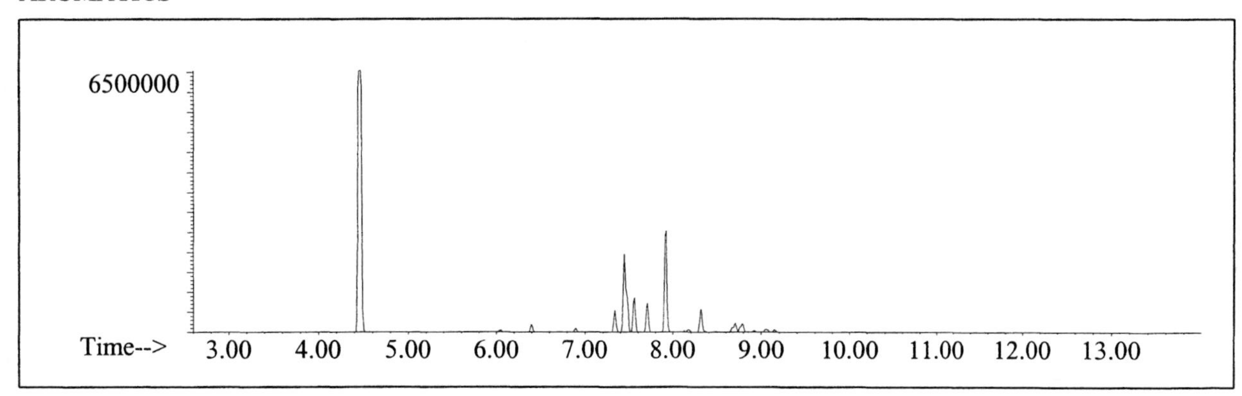

CYCLOPARAFFINS AND ALKENES

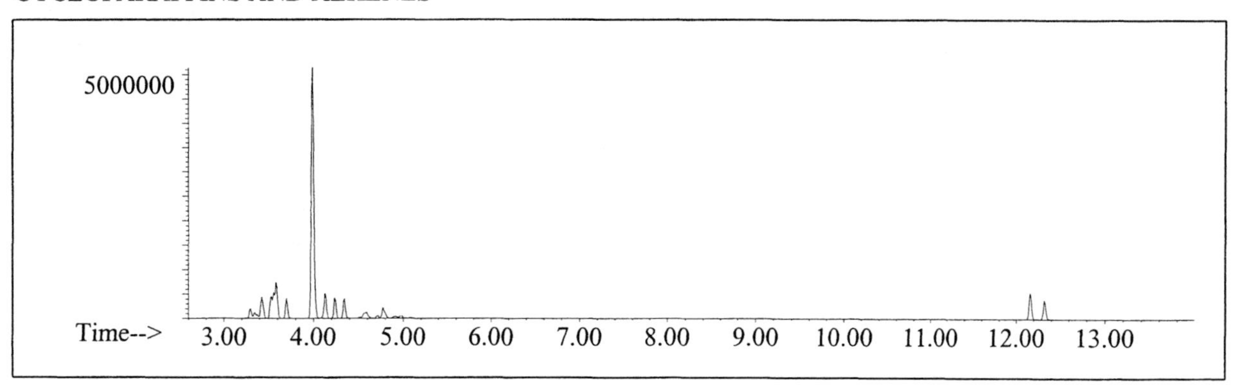

NAPHTHALENES

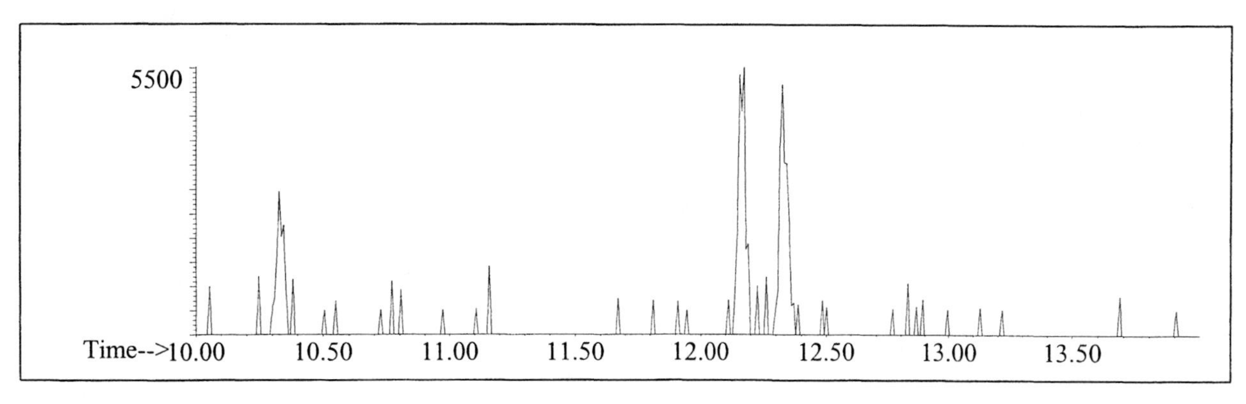

Alkane

Ion 43

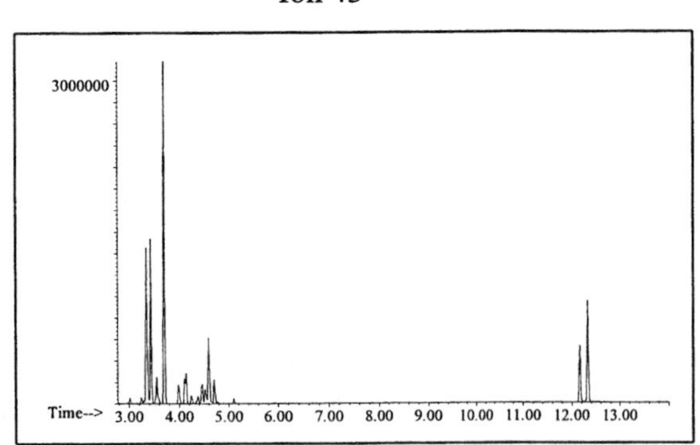

Ion 57

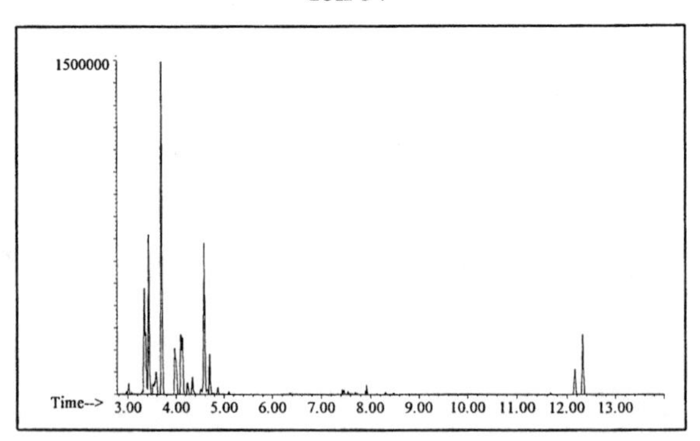

Ion 71

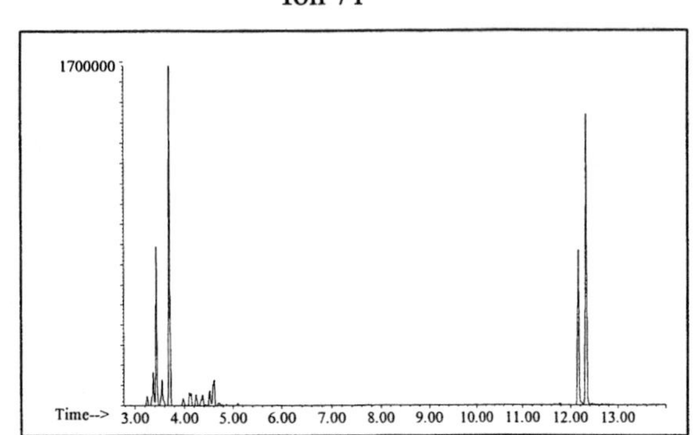

Ion 85

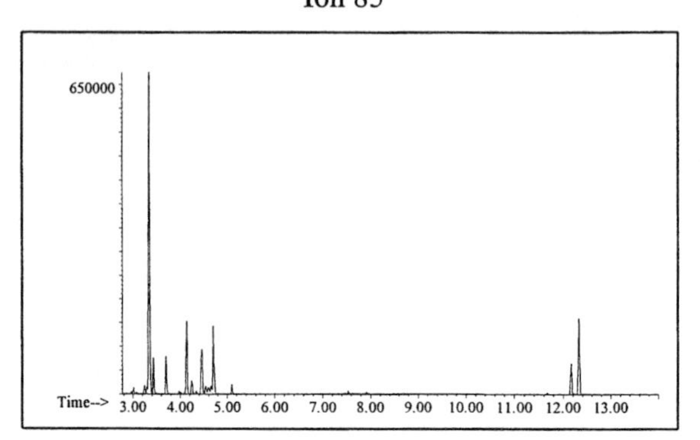

Aromatic

Ion 91

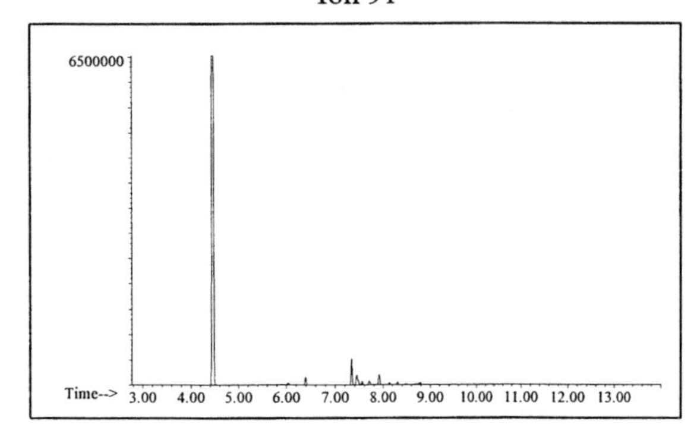

Ion 105

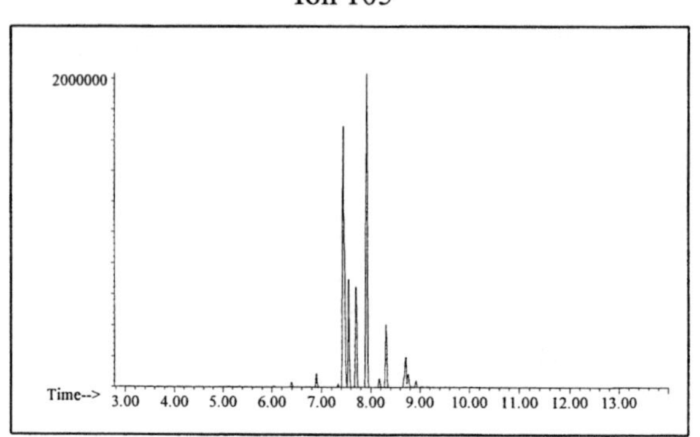

Ion 119

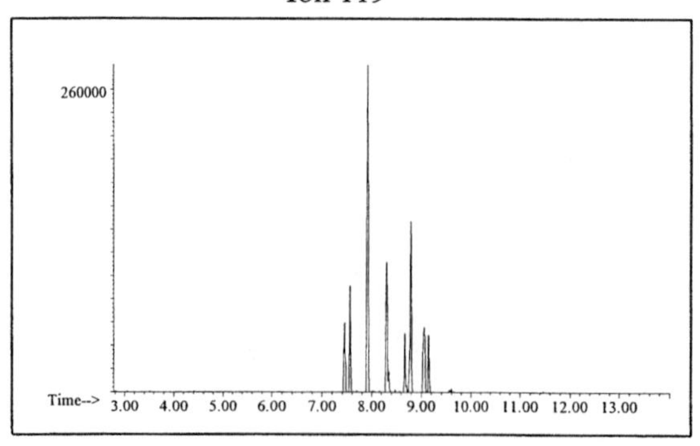

Cycloparaffin

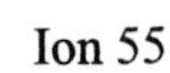

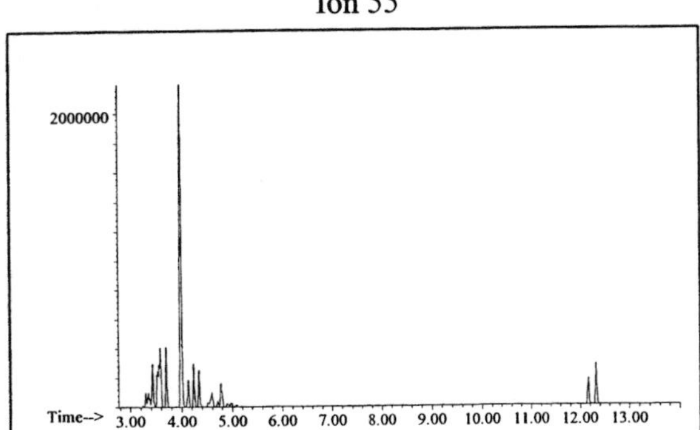

Ion 69

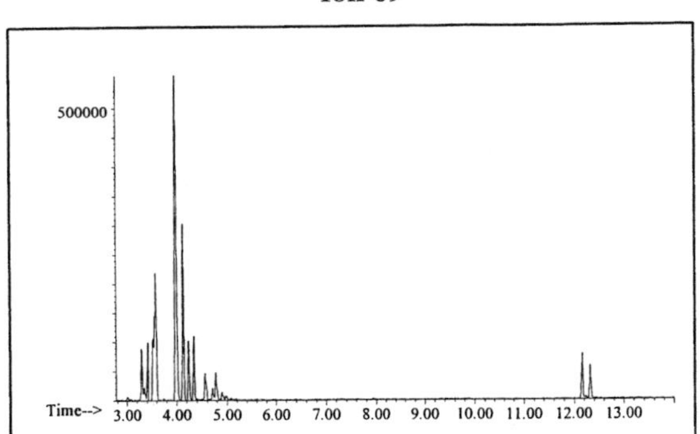

Ion 83

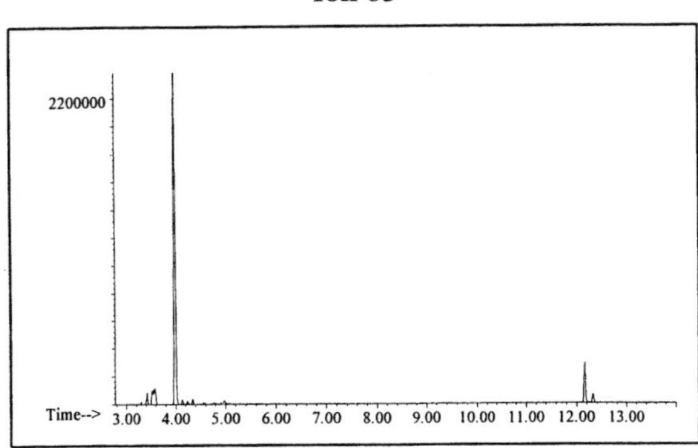

Naphthalene

Ion 128

No Useful Data Obtained

Ion 142

No Useful Data Obtained

Ion 156

No Useful Data Obtained

Miscellaneous #6

20uL/mL Pentane

Product Displayed: DX 330 Enamel Reducer

Other Similar Products:

ASTM: Class 0 (Miscellaneous)

Product Uses: Enamel reducer

Macro Code: L

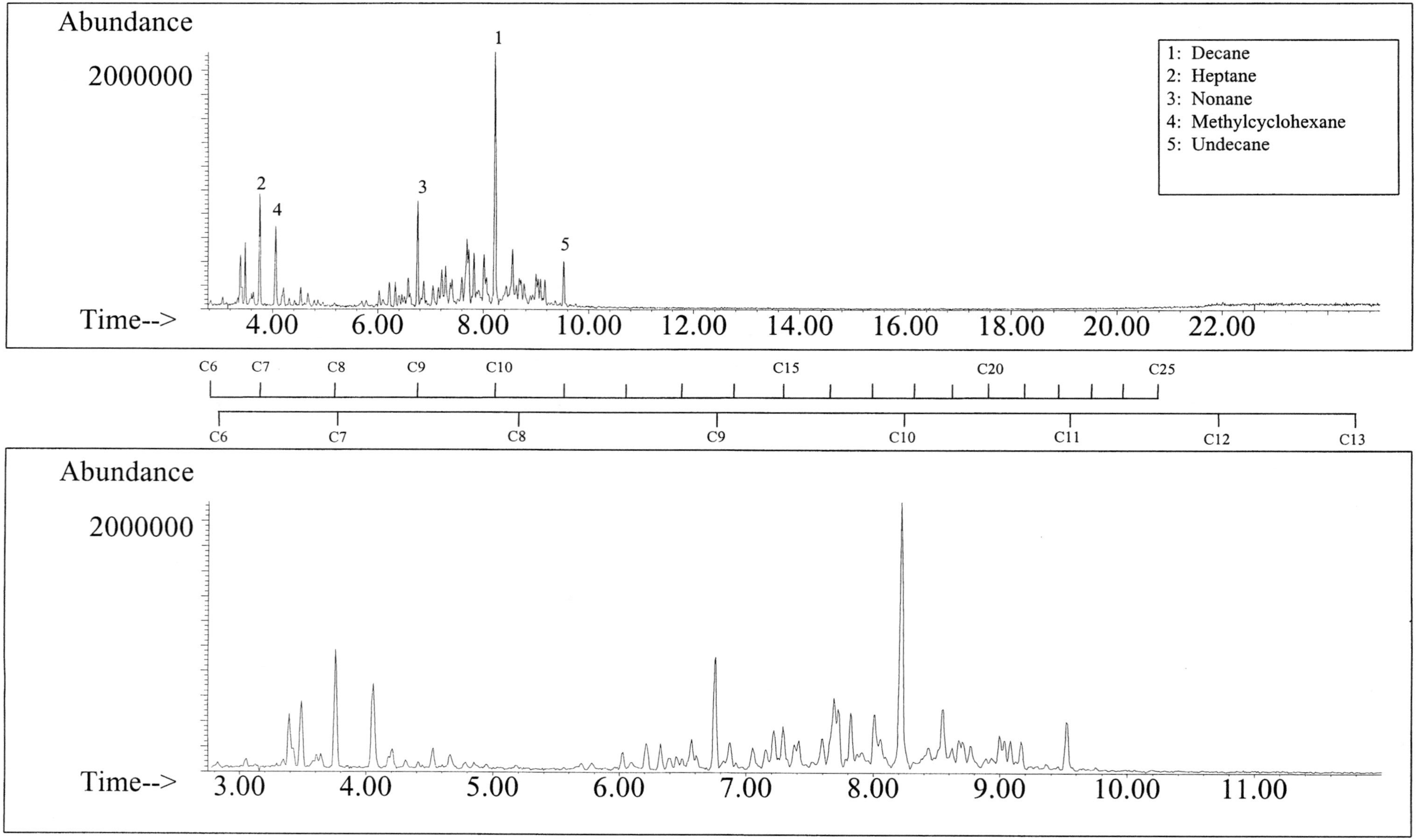

Abundance

2000000

1: Decane
2: Heptane
3: Nonane
4: Methylcyclohexane
5: Undecane

Time--> 4.00 6.00 8.00 10.00 12.00 14.00 16.00 18.00 20.00 22.00

C6 C7 C8 C9 C10 C15 C20 C25

C6 C7 C8 C9 C10 C11 C12 C13

Abundance

2000000

Time--> 3.00 4.00 5.00 6.00 7.00 8.00 9.00 10.00 11.00

ALKANES

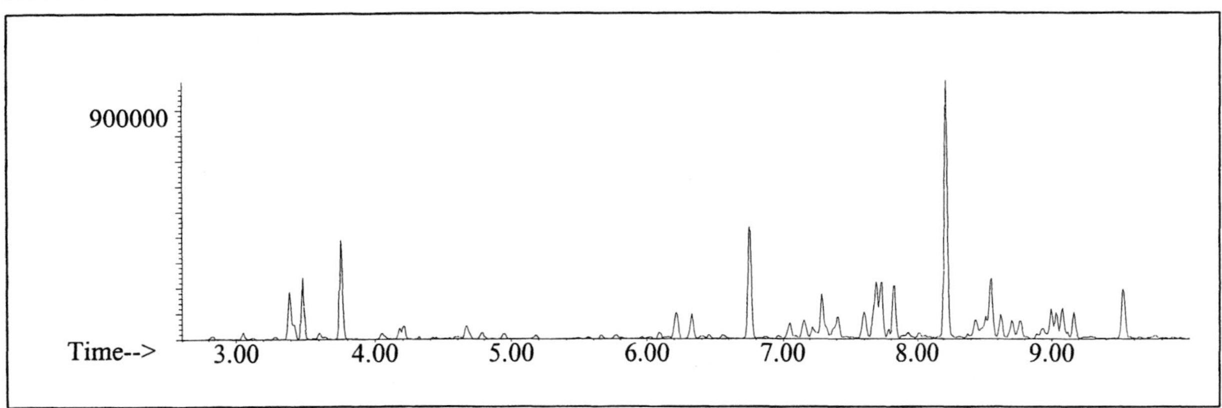

AROMATICS

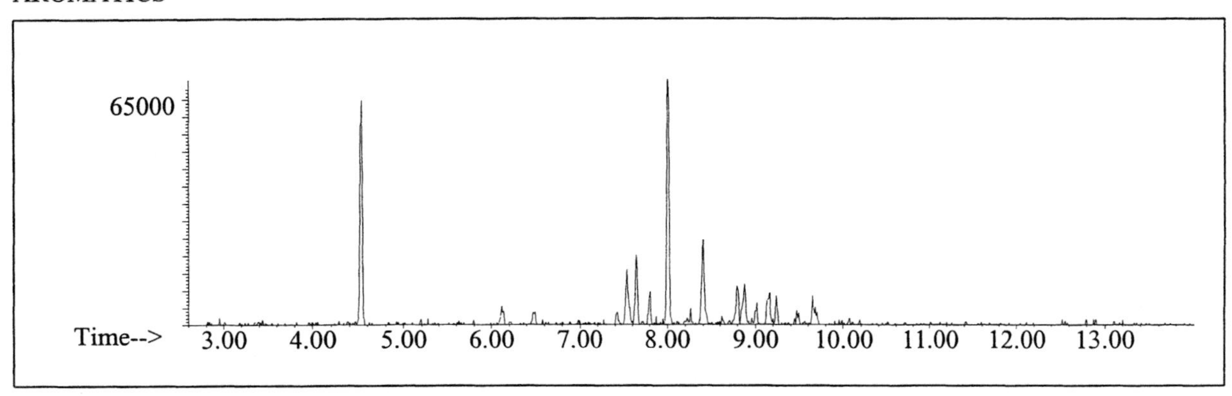

CYCLOPARAFFINS AND ALKENES

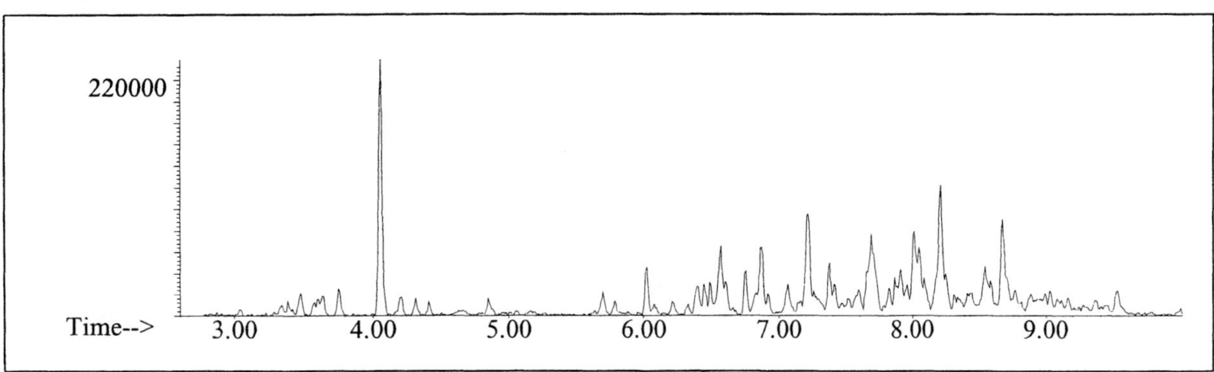

NAPHTHALENES

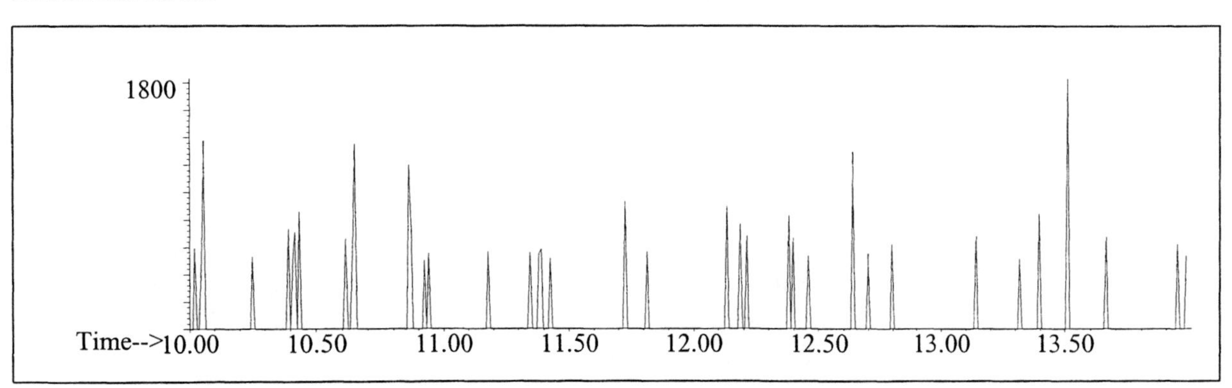

Alkane

Ion 43

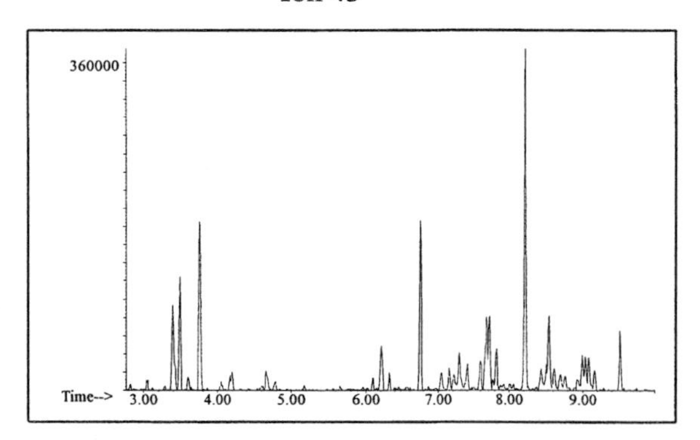

Ion 57

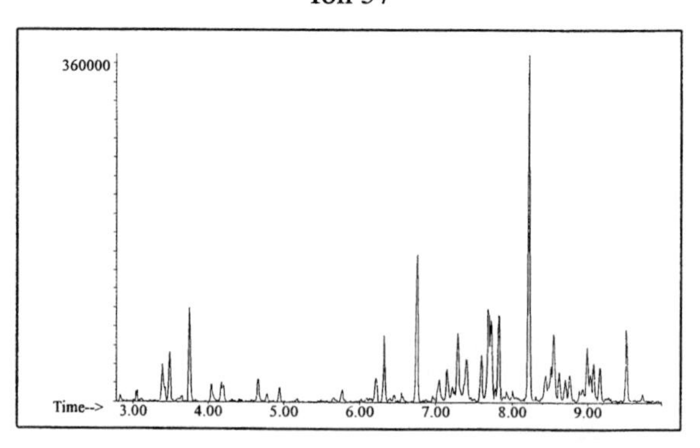

Ion 71

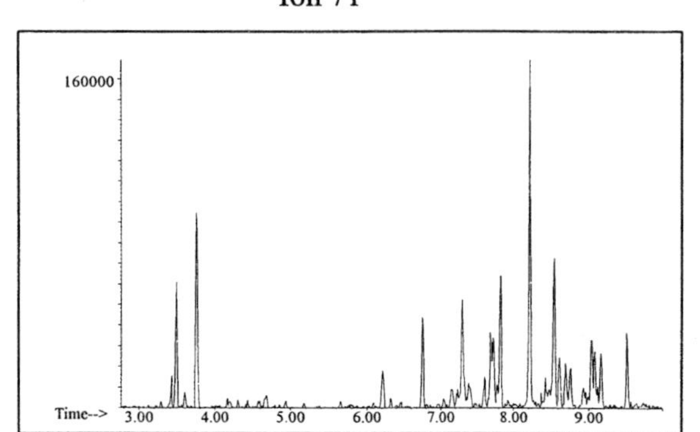

Ion 85

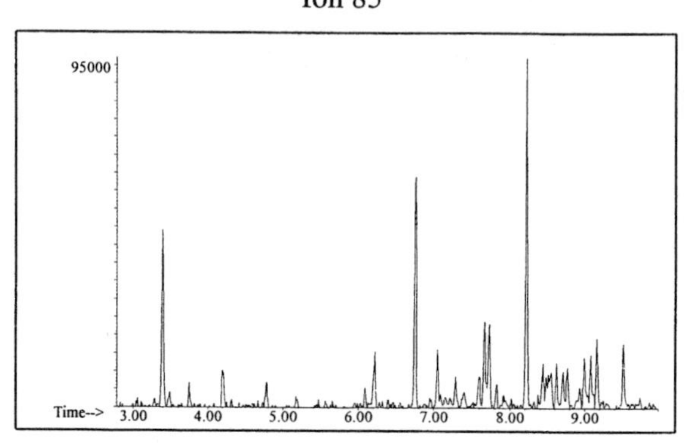

Aromatic

Ion 91

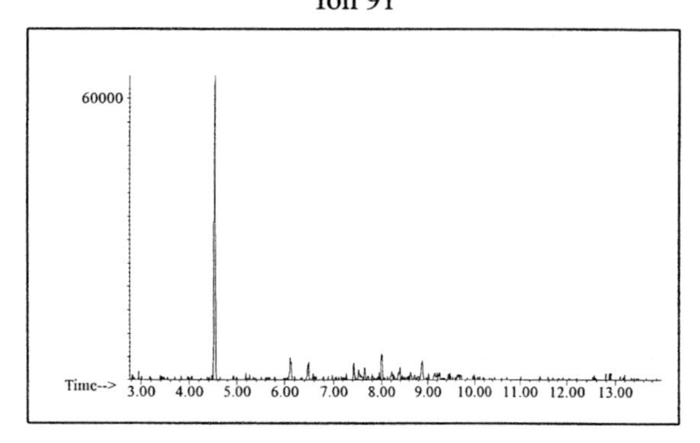

Ion 105

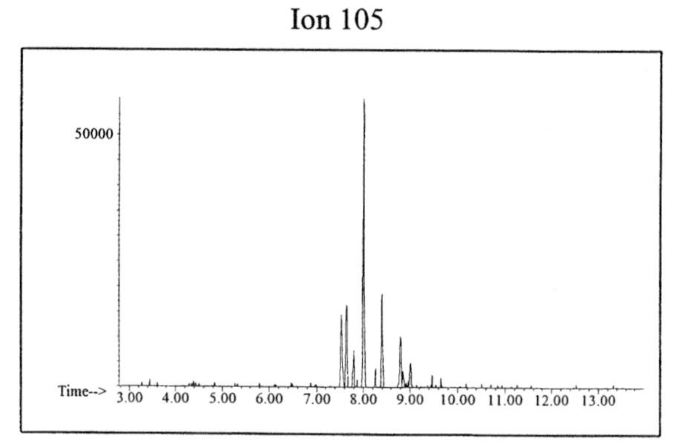

Ion 119

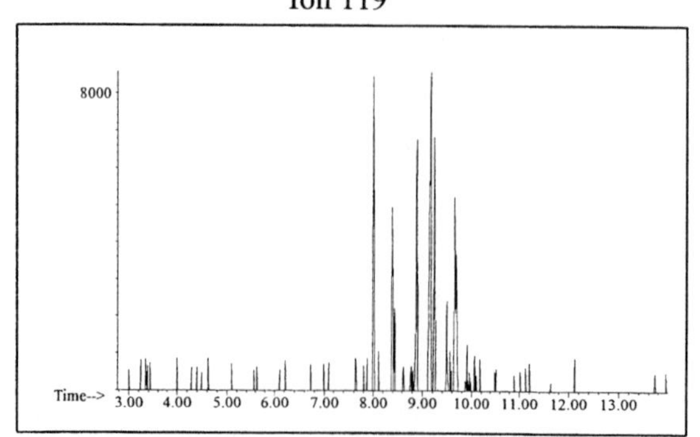

INDIVIDUAL PROFILES

Cycloparaffin

Ion 55

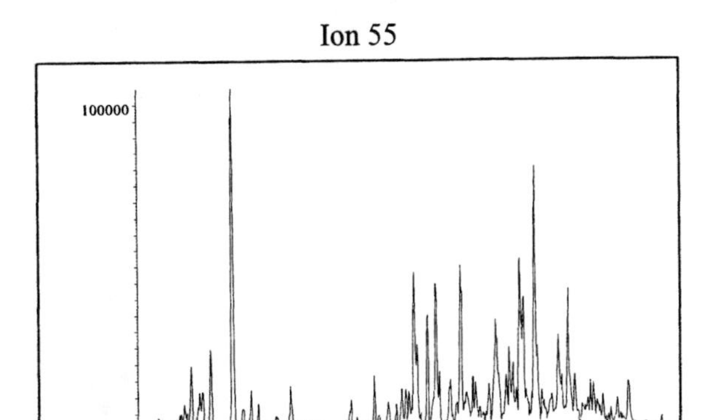

Ion 69

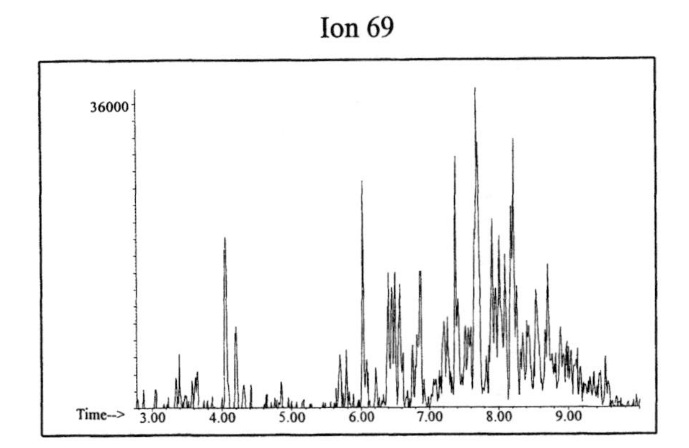

Ion 83

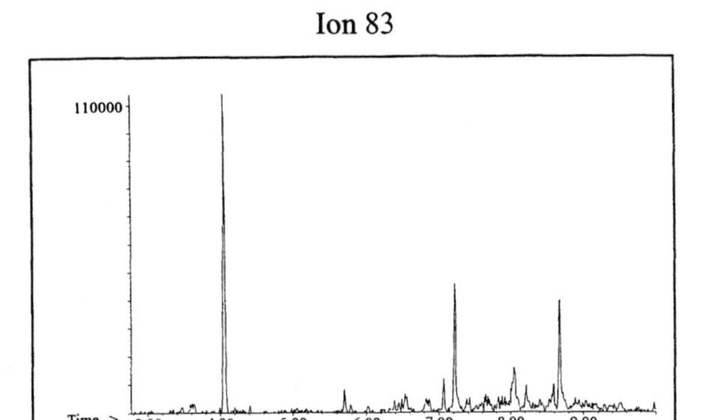

Naphthalene

Ion 128

No Useful Data Obtained

Ion 142

No Useful Data Obtained

Ion 156

No Useful Data Obtained

Miscellaneous #7

20uL/mL Pentane

ASTM: Class 0 (Miscellaneous)

Macro Code: L

Product Displayed: Tru-Test Gasoline Stove Lantern Fuel

Product Uses: Lantern fuel, camping fuel

Other Similar Products:

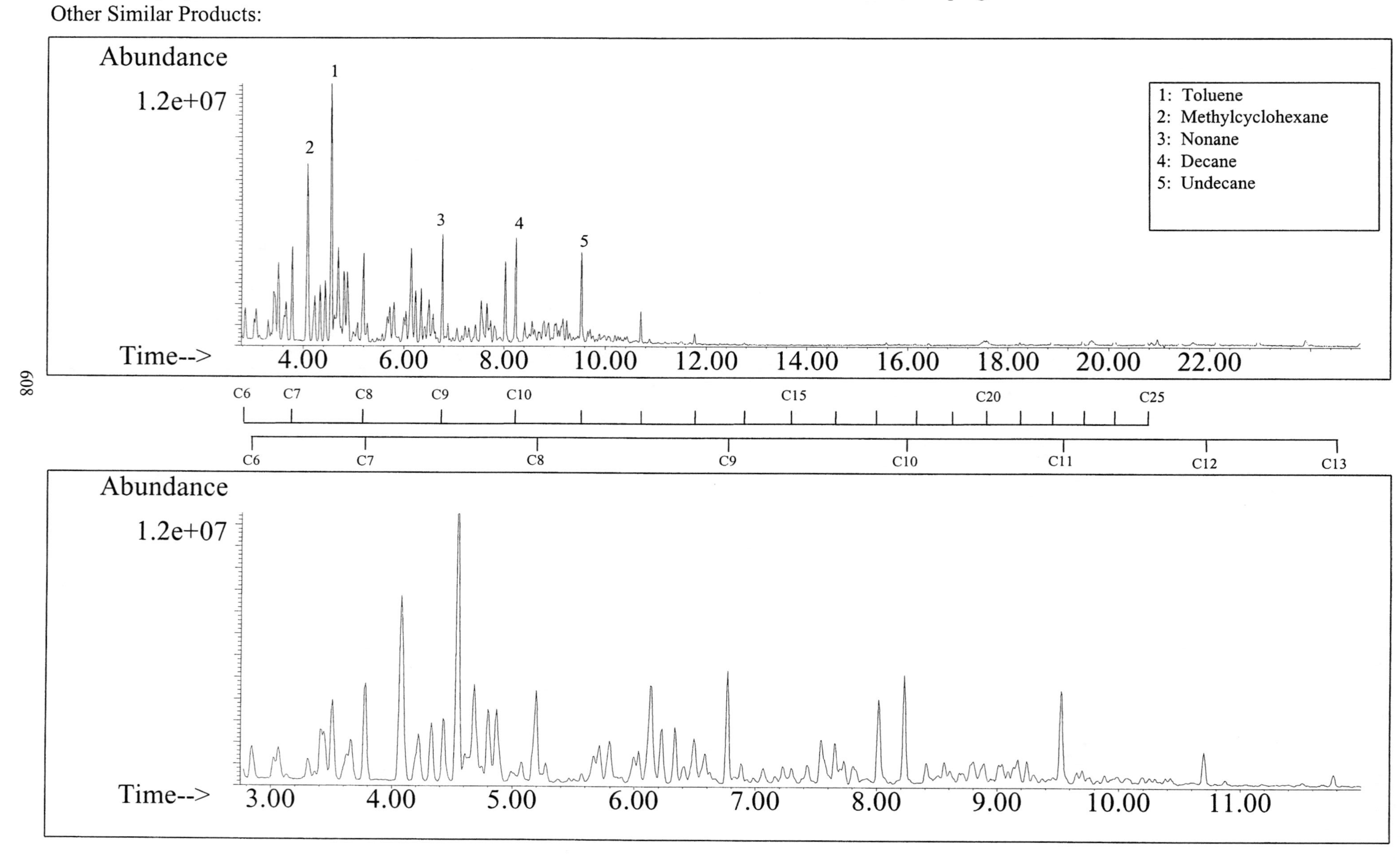

SUMMED PROFILES Miscellaneous #7

ALKANES

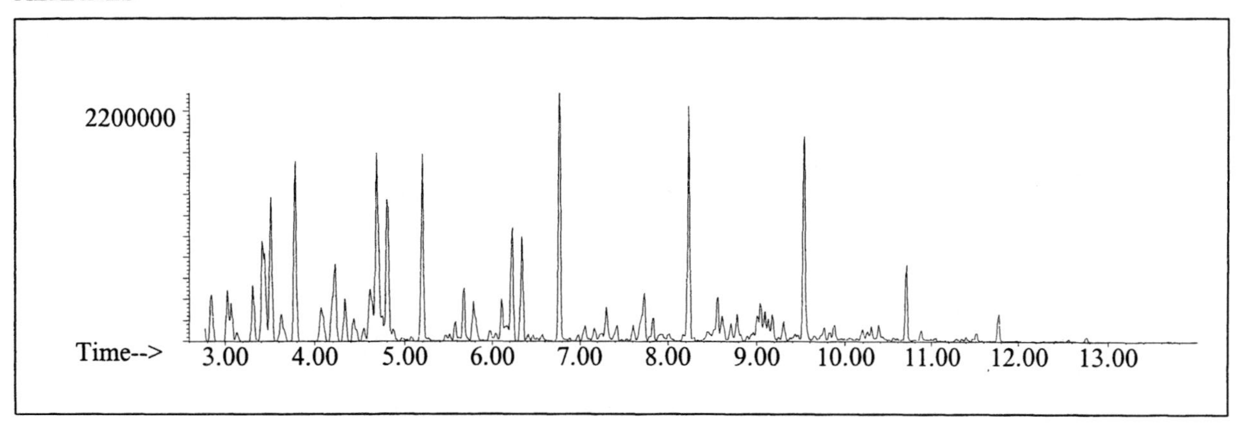

AROMATICS

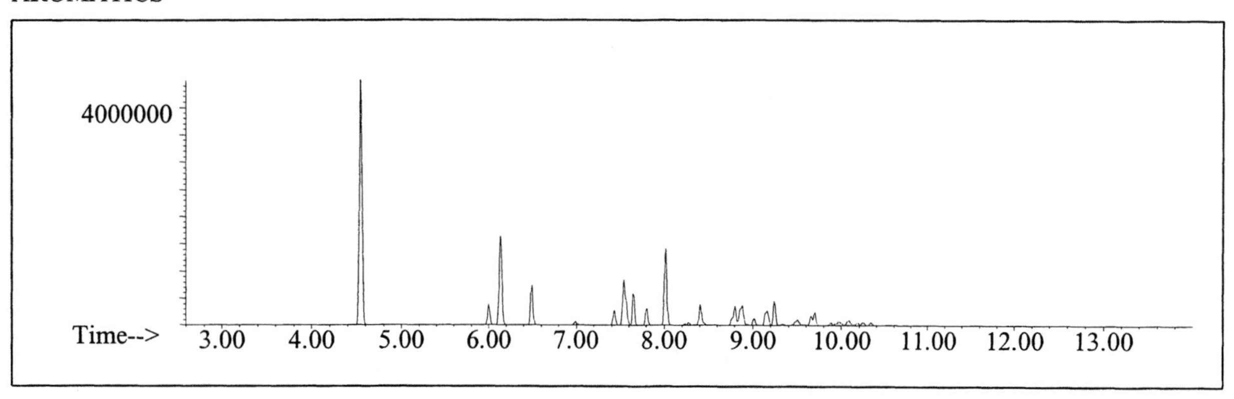

CYCLOPARAFFINS AND ALKENES

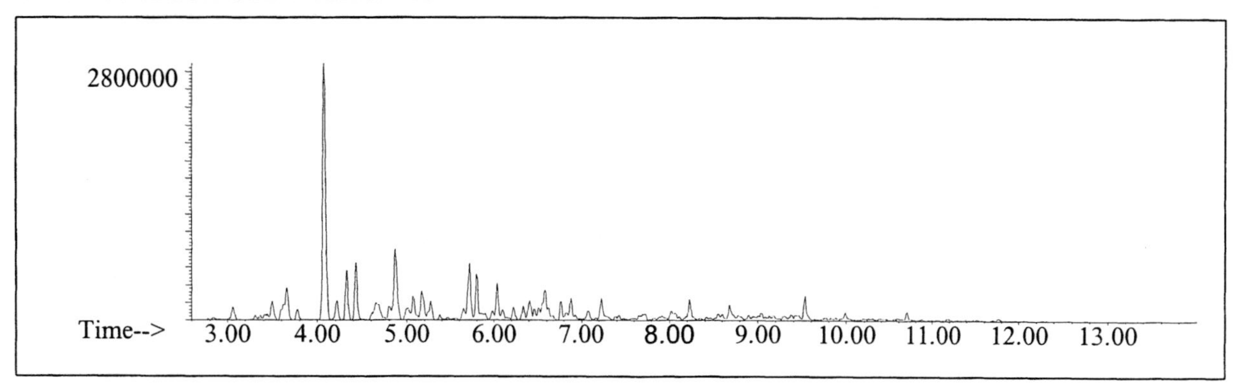

NAPHTHALENES

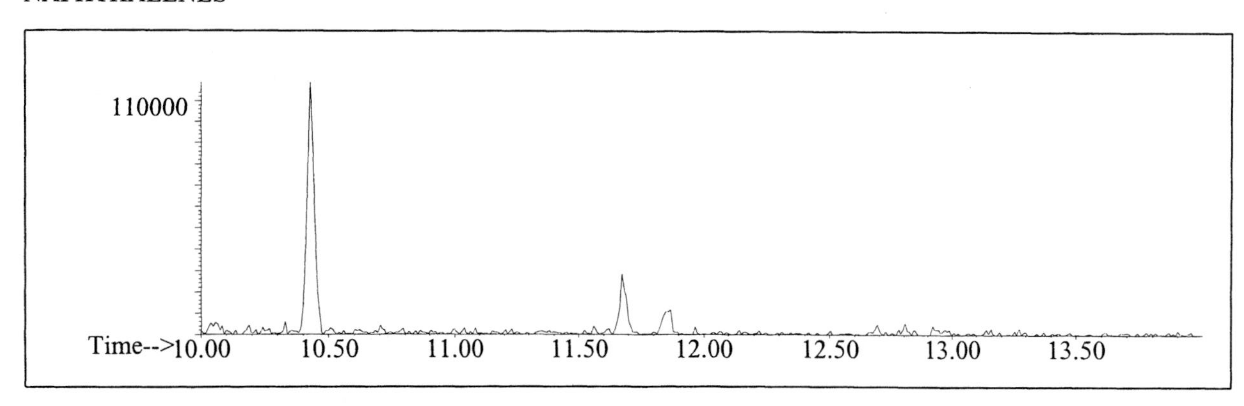

Alkane

Ion 43

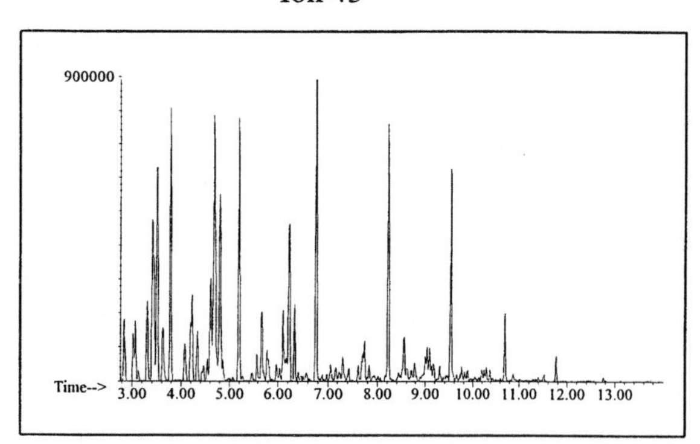

Ion 57

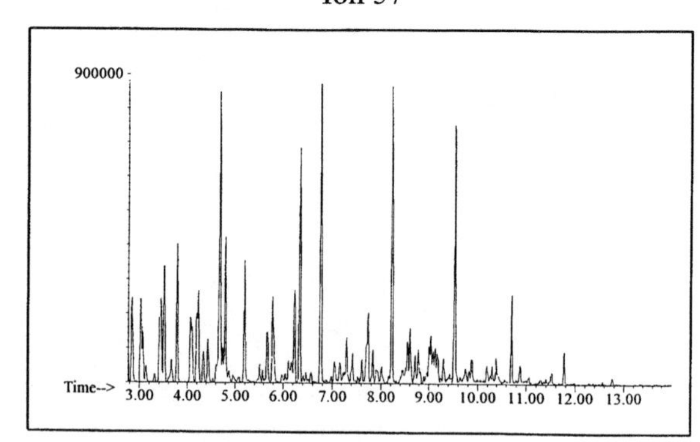

Ion 71

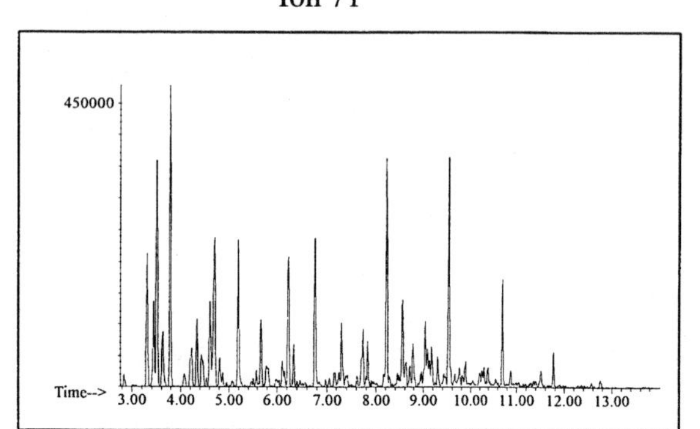

Ion 85

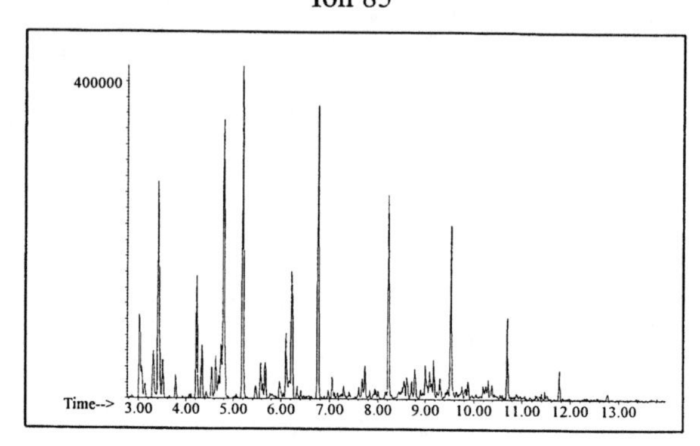

Aromatic

Ion 91

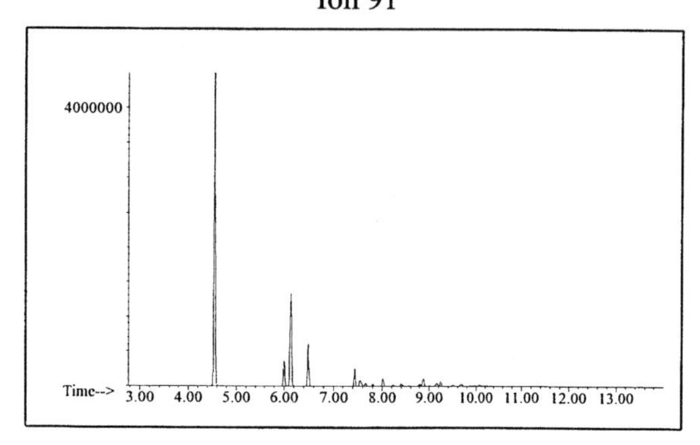

Ion 105

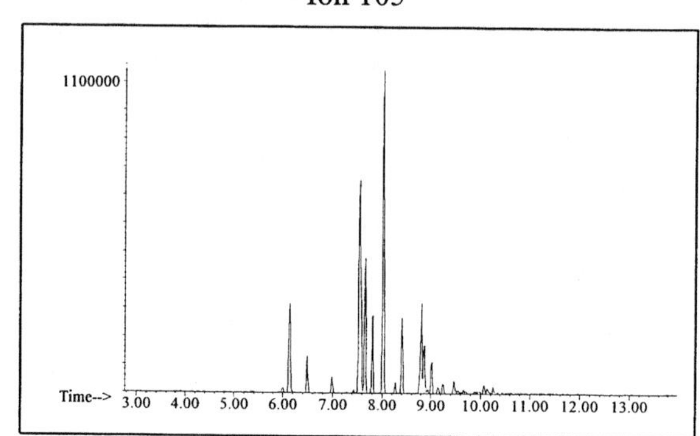

Ion 119

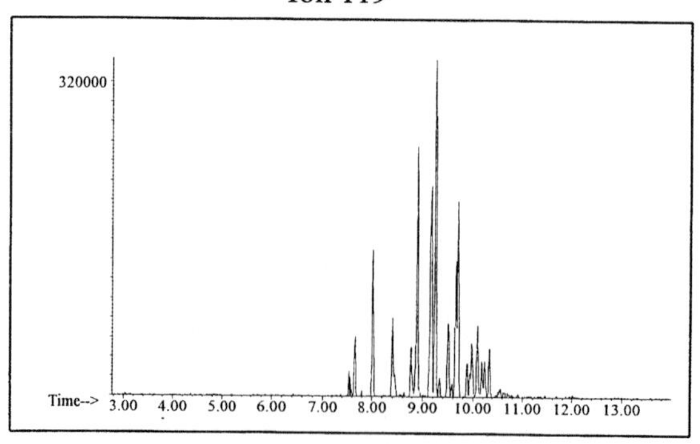

Cycloparaffin

Ion 55

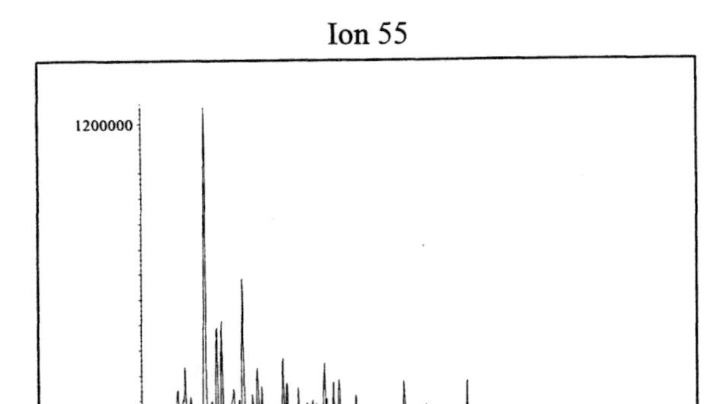

Ion 69

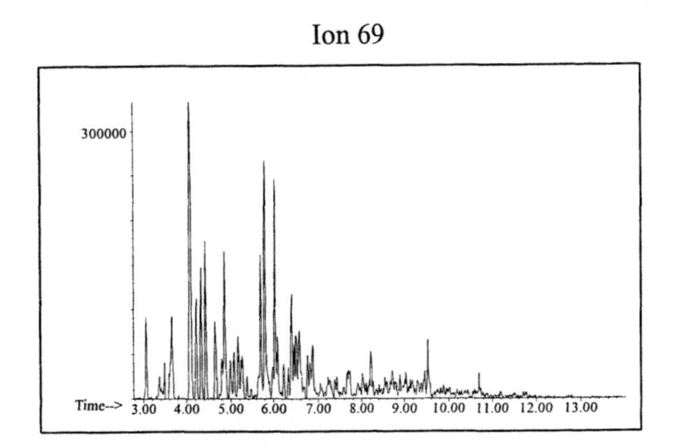

Ion 83

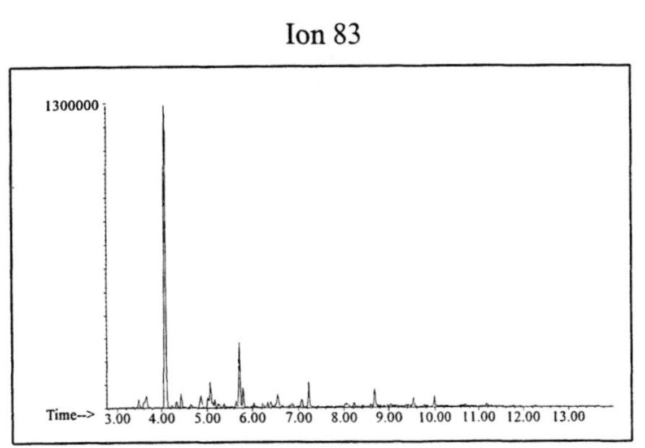

Naphthalene

Ion 128

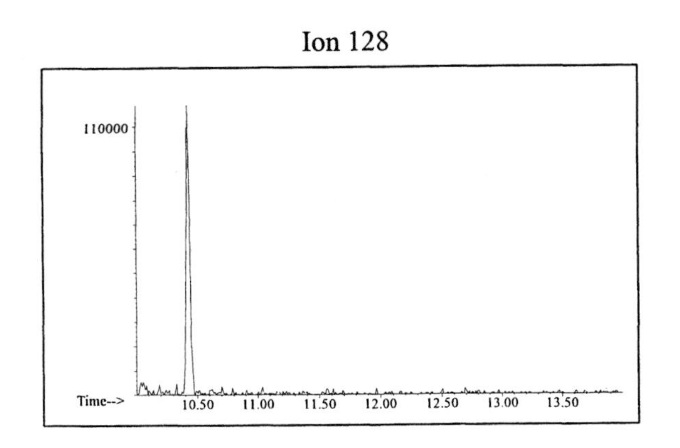

Ion 142

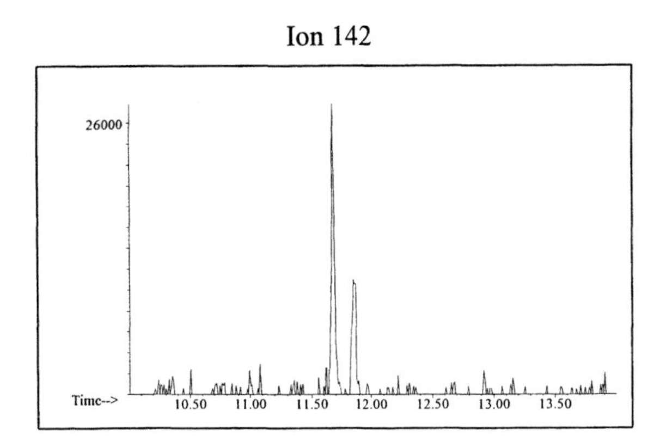

Ion 156

No Useful Data Obtained

Miscellaneous #8

20uL/mL Pentane

ASTM: Class 0 (Miscellaneous)

Macro Code: L

Product Displayed: Glidden Strip Paint Remover

Product Uses: Paint remover

Other Similar Products:

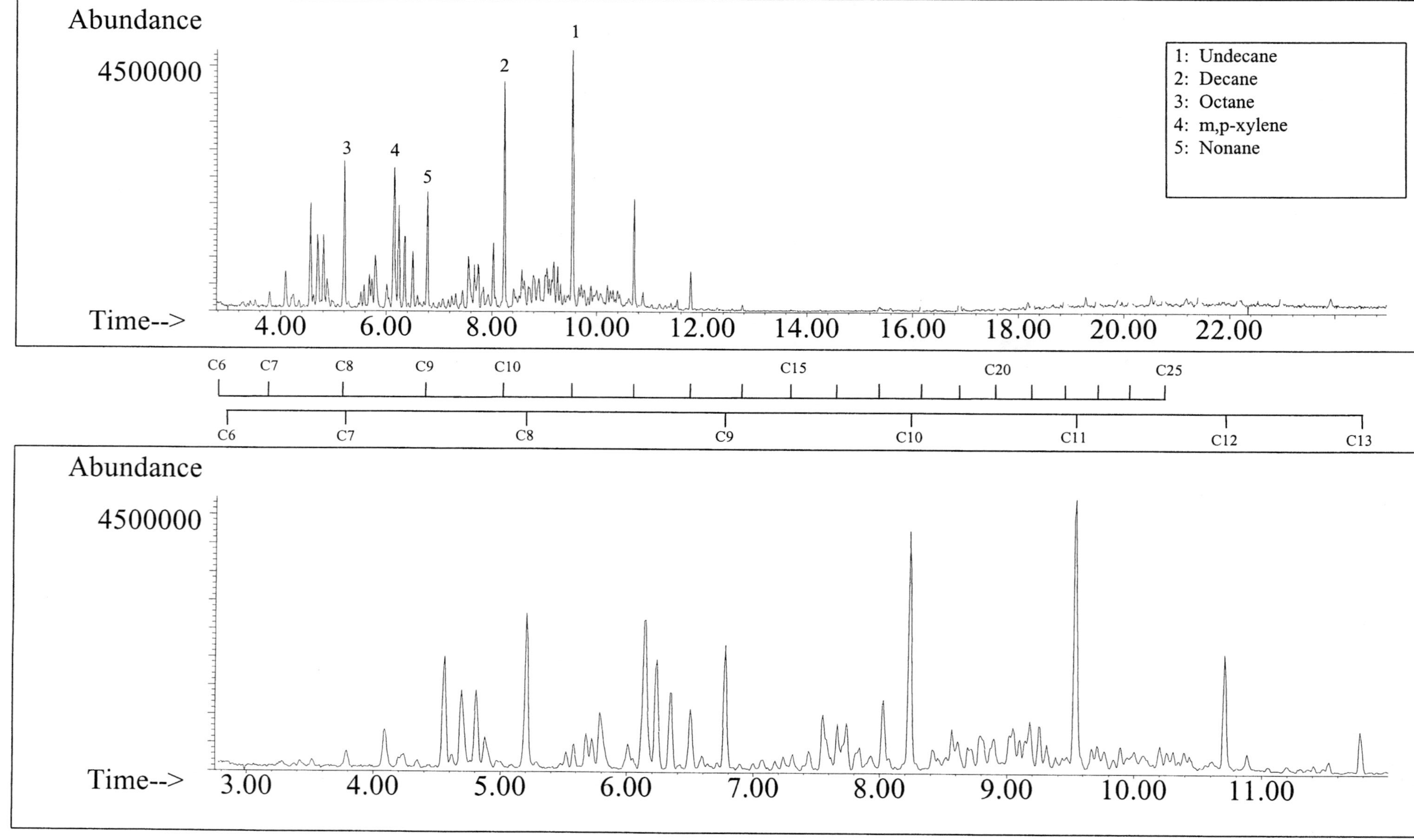

ALKANES

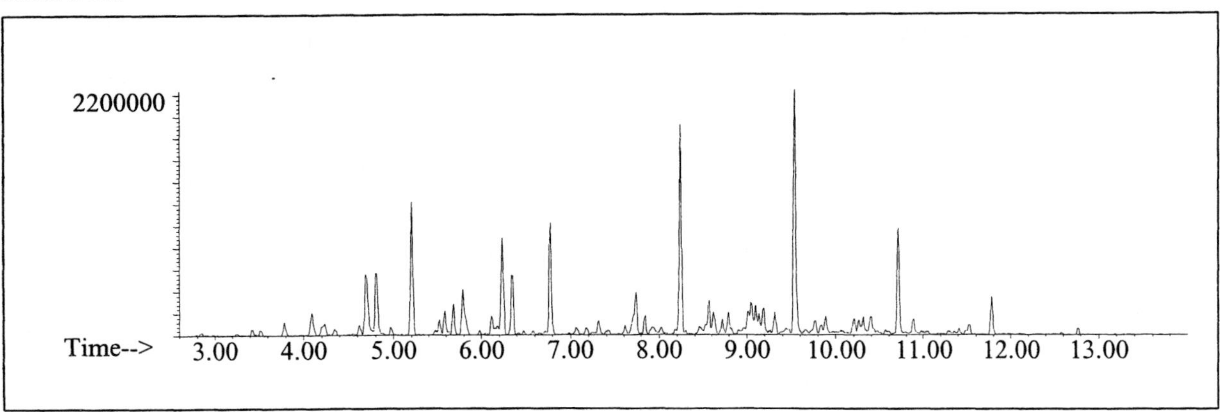

AROMATICS

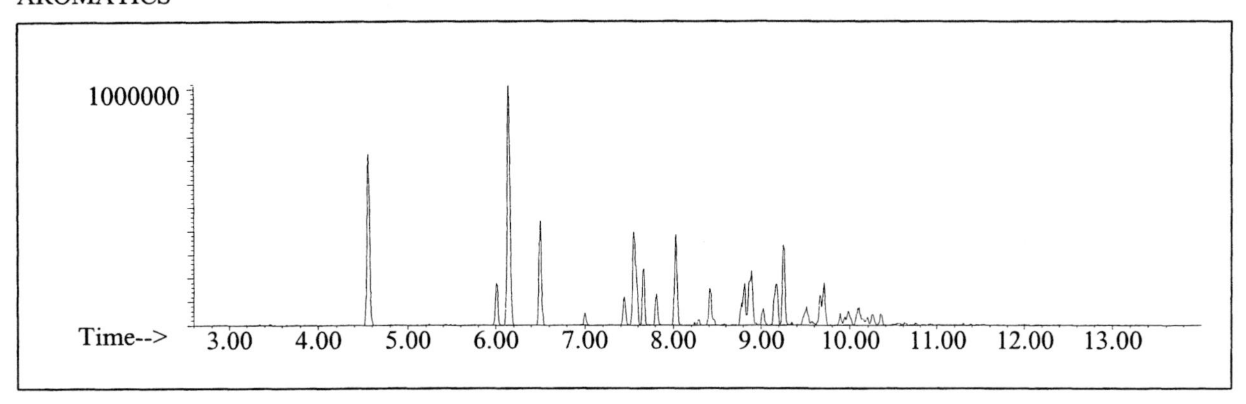

CYCLOPARAFFINS AND ALKENES

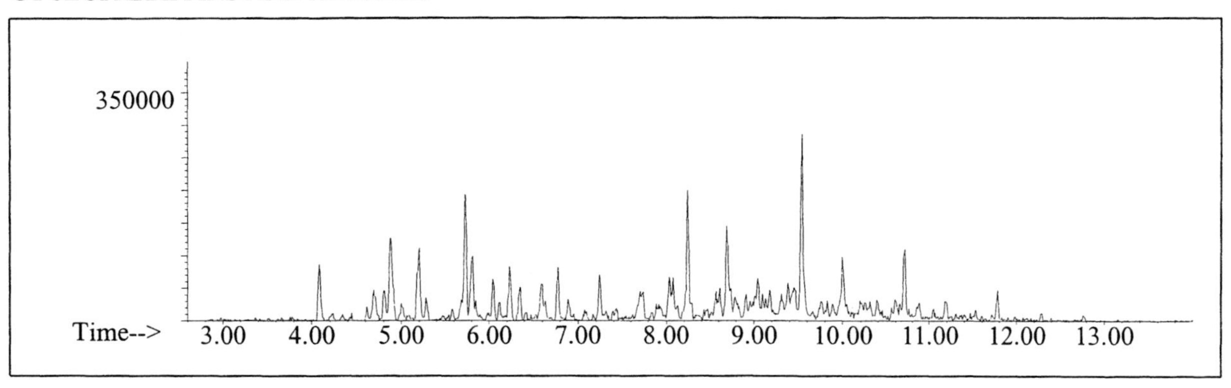

NAPHTHALENES

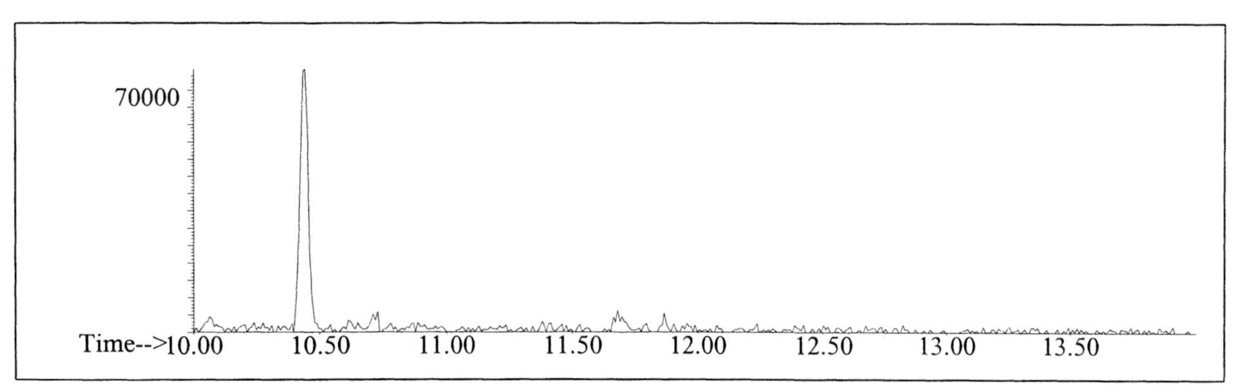

Alkane

Ion 43

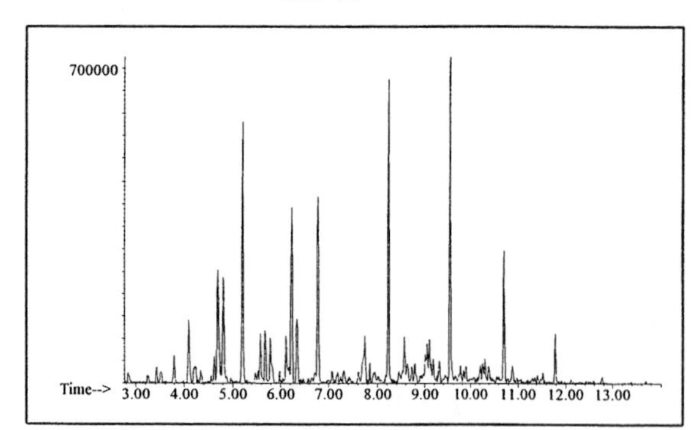

Ion 57

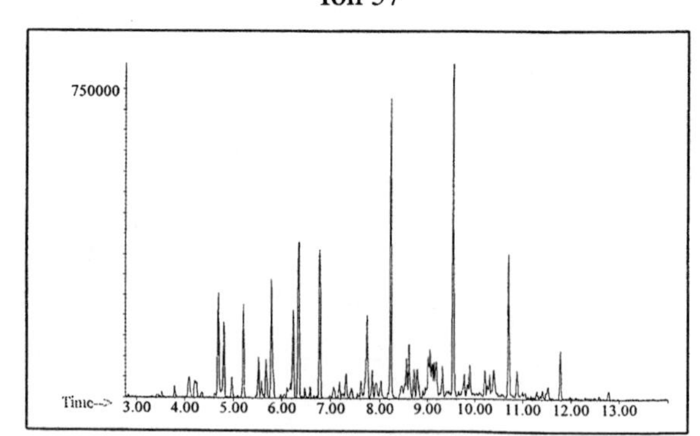

Ion 71

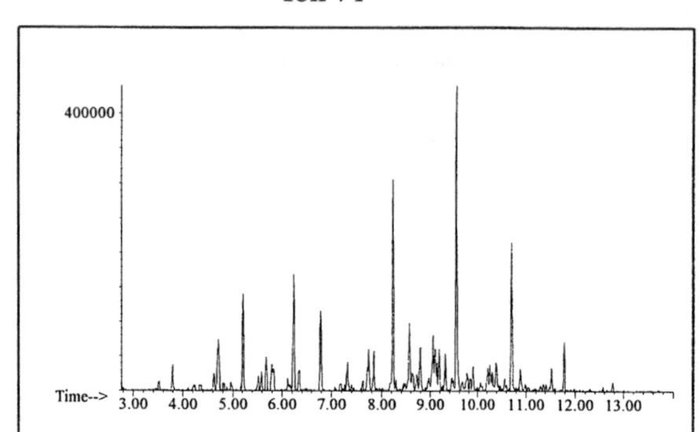

Ion 85

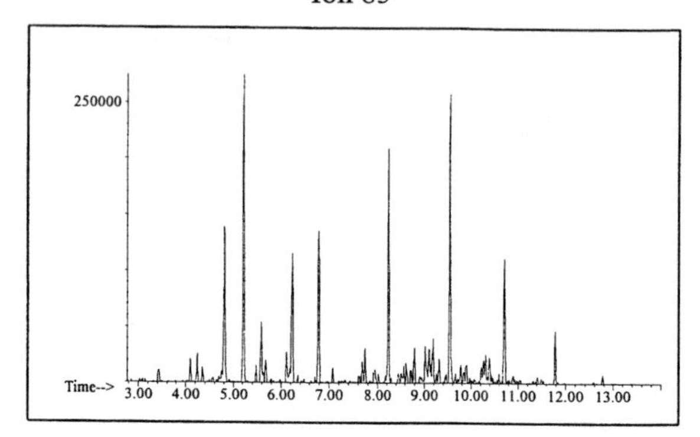

Aromatic

Ion 91

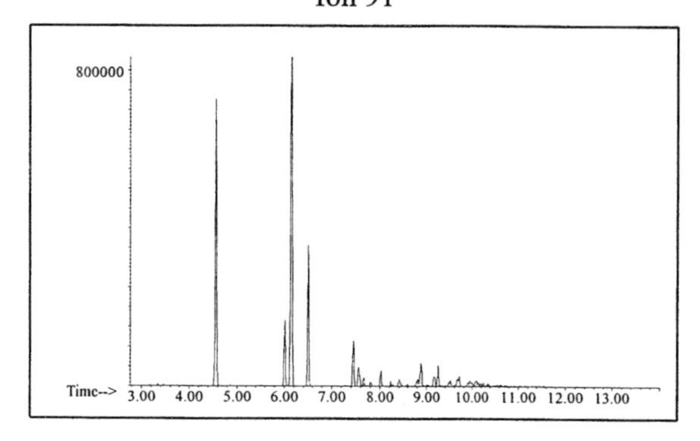

Ion 105

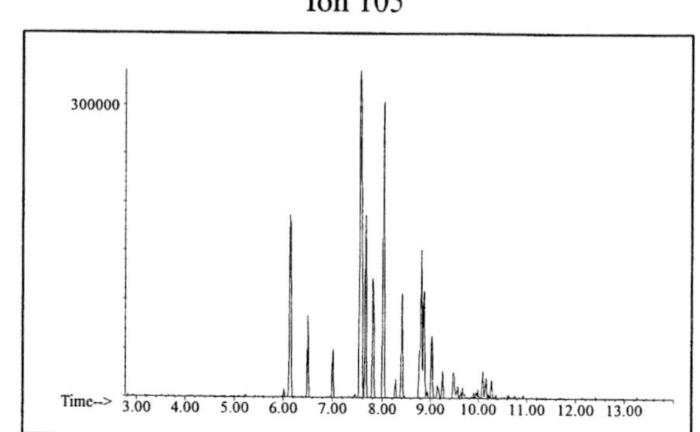

Ion 119

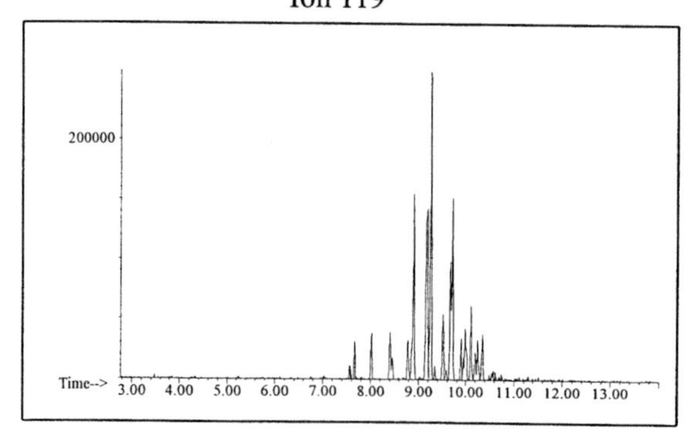

Cycloparaffin

Ion 55

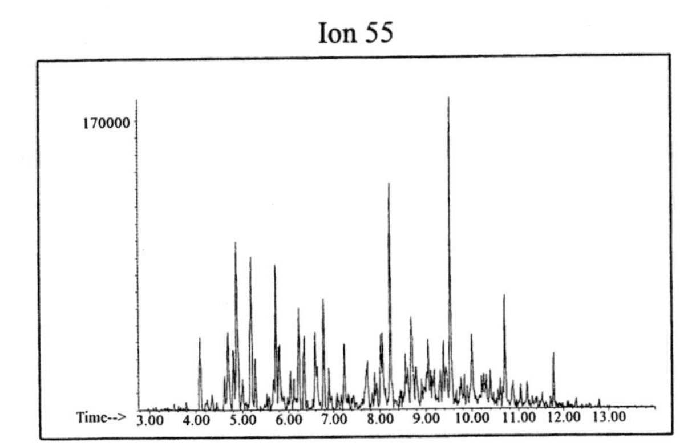

Ion 69

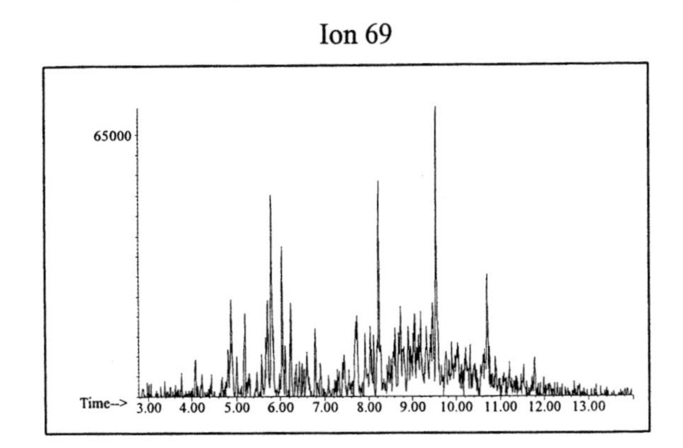

Ion 83

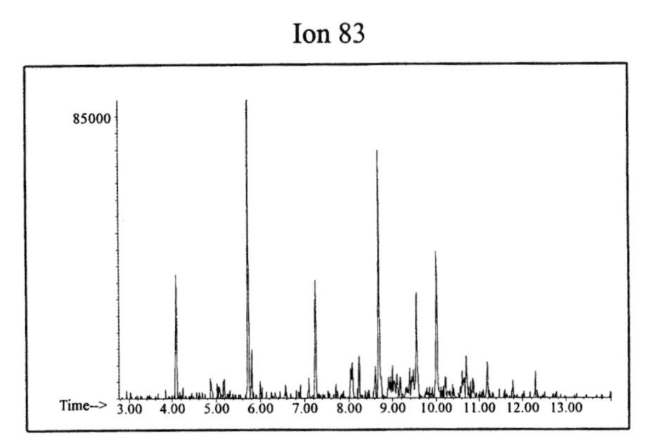

Naphthalene

Ion 128

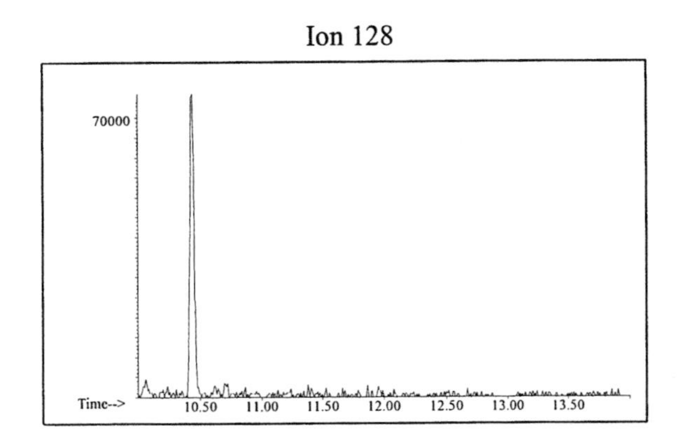

Ion 142

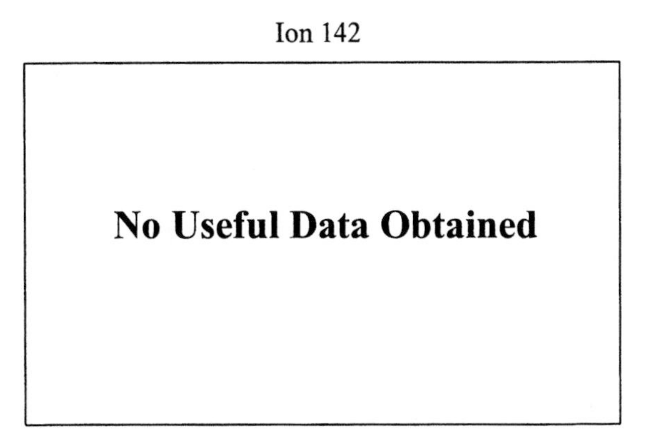

Ion 156

No Useful Data Obtained

Miscellaneous #9

20uL/mL Pentane

Product Displayed: DTR 602 Enamel Reducer

Other Similar Products:

ASTM: Class 0 (Miscellaneous)

Product Uses: Enamel reducer

Macro Code: L

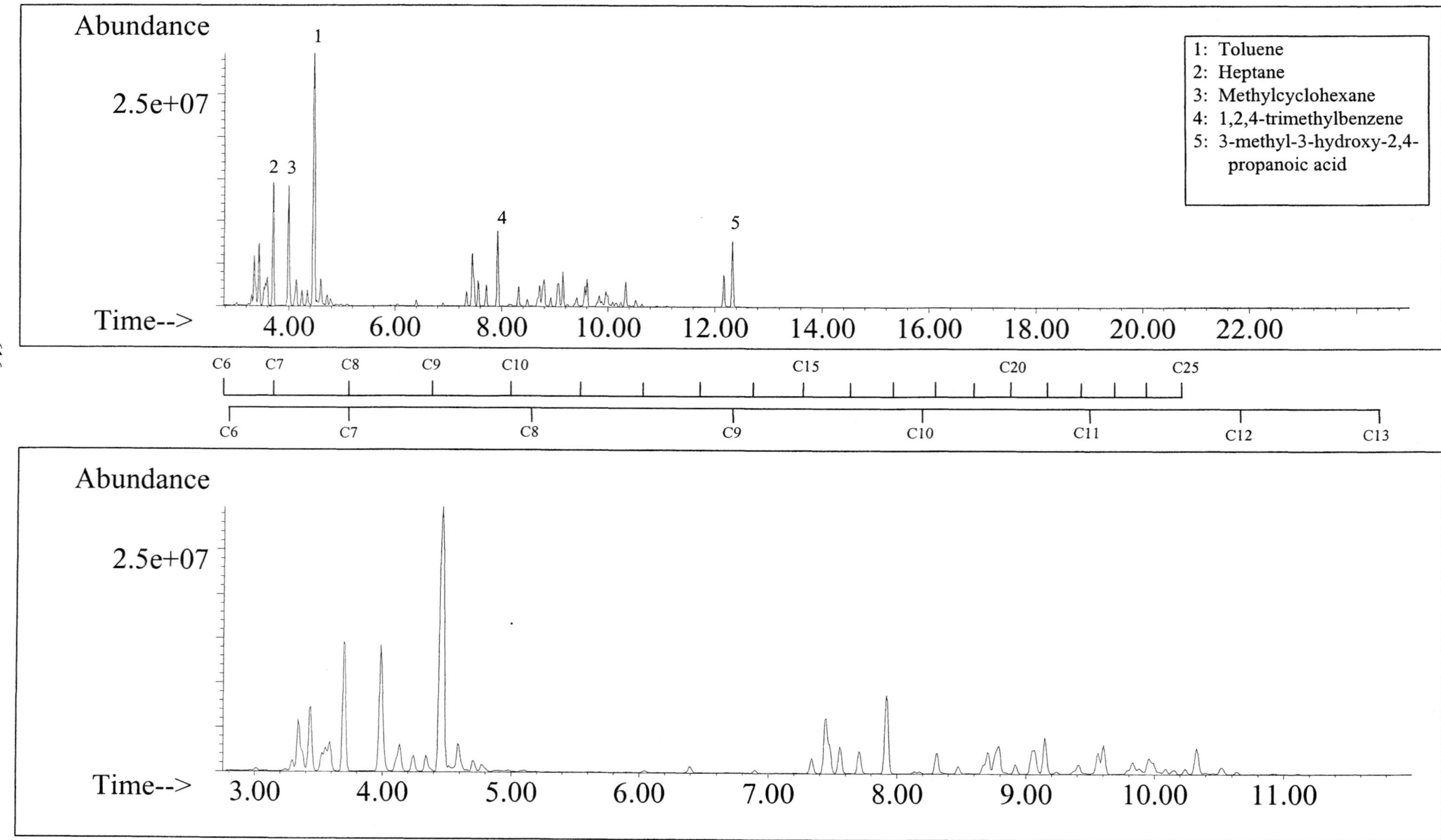

ALKANES

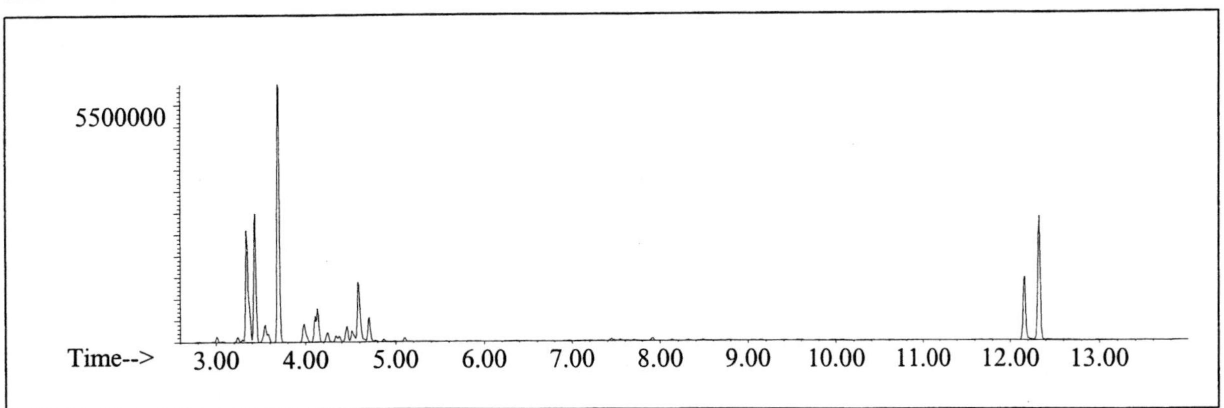

AROMATICS

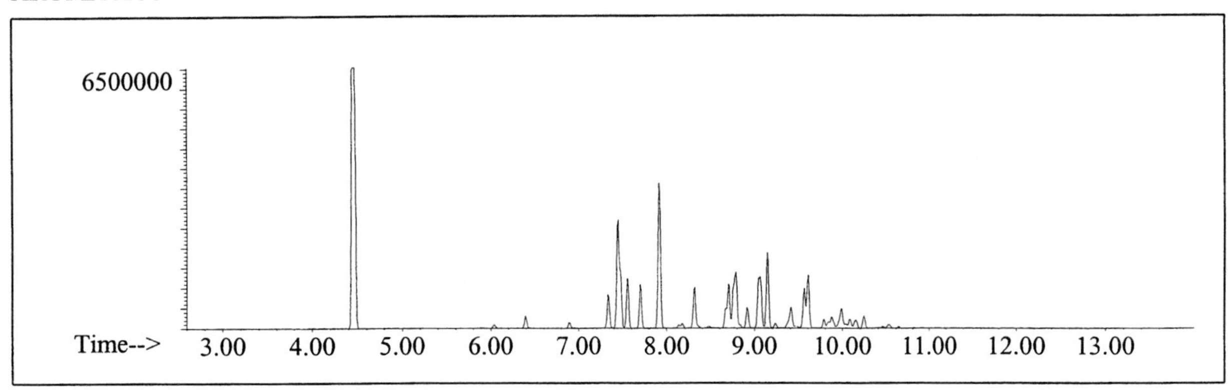

CYCLOPARAFFINS AND ALKENES

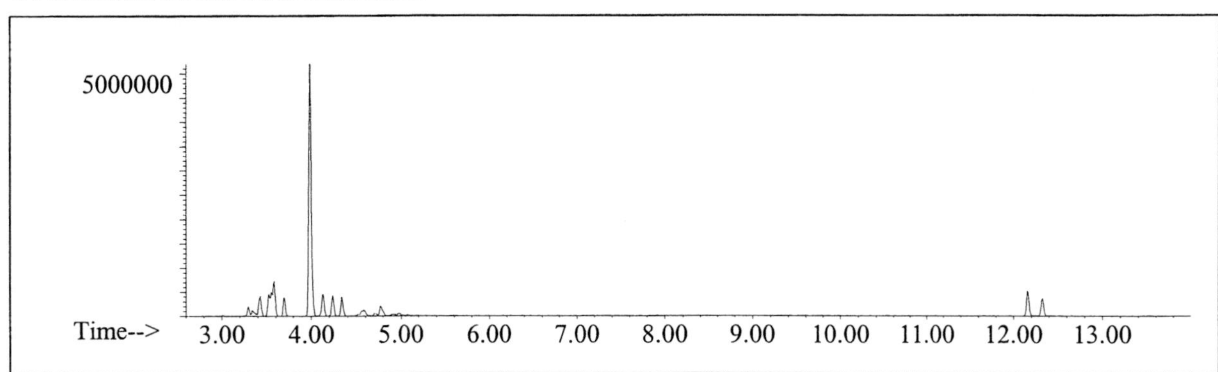

NAPHTHALENES

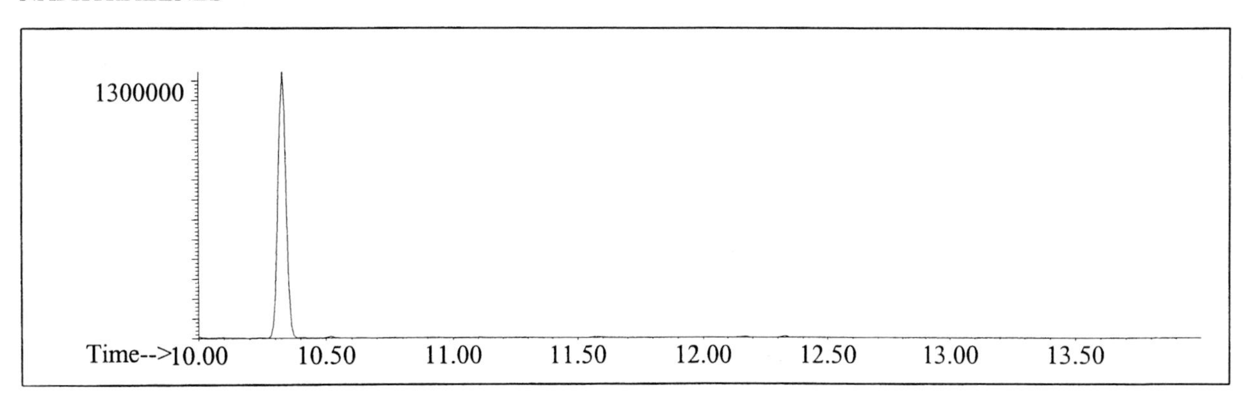

Alkane

Ion 43

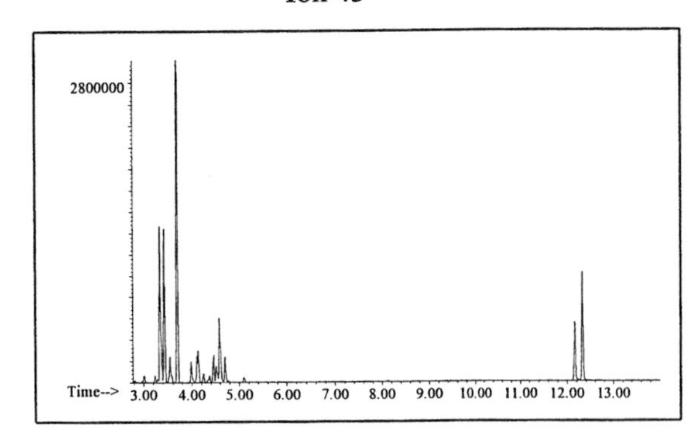

Ion 57

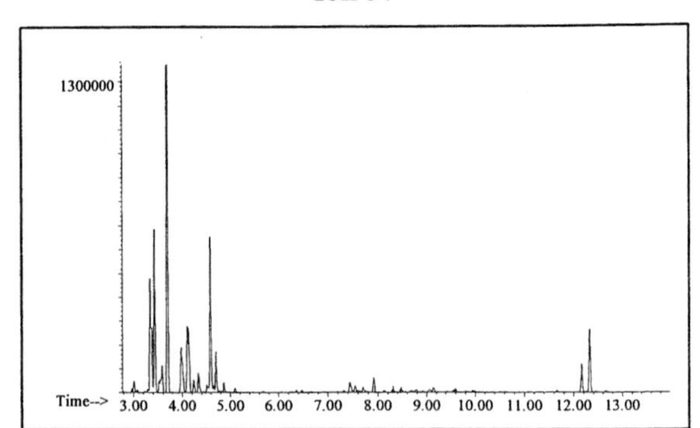

Ion 71

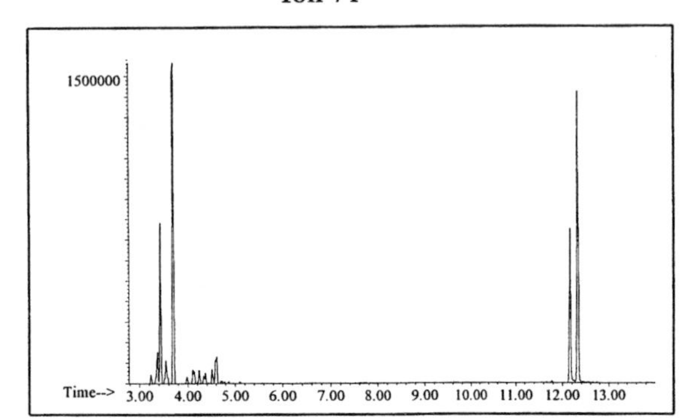

Ion 85

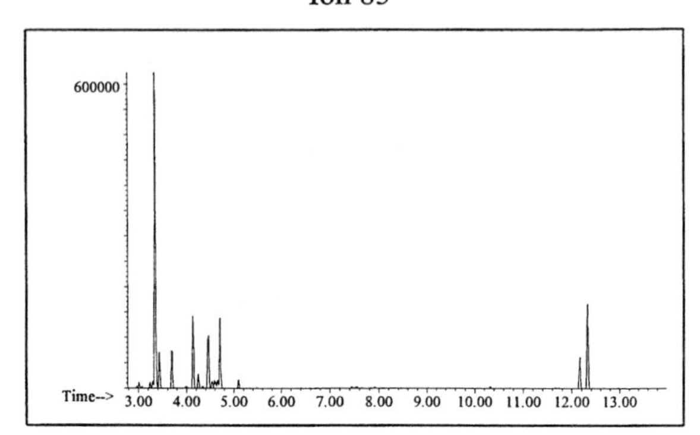

Aromatic

Ion 91

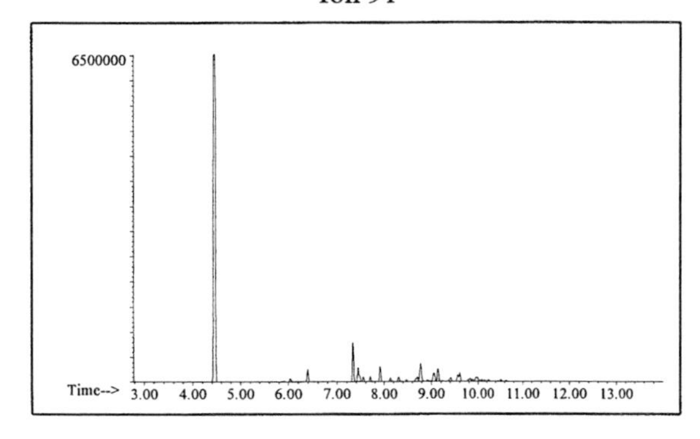

Ion 105

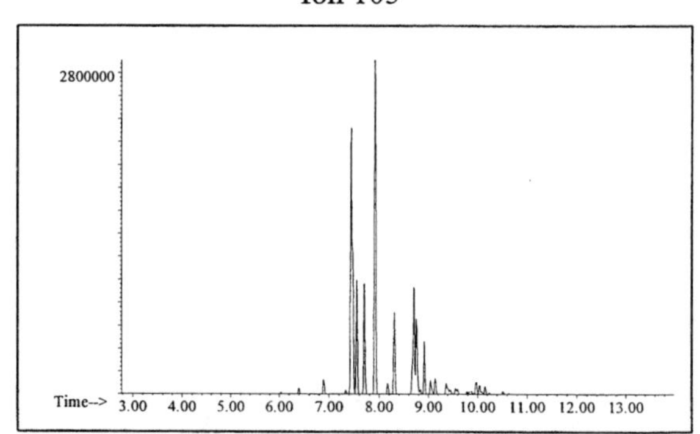

Ion 119

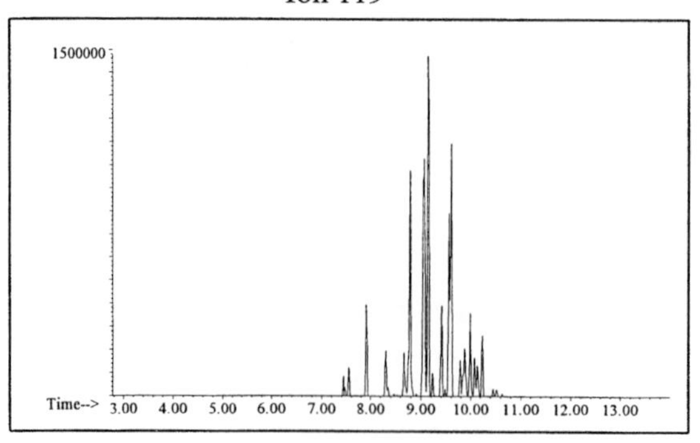

Cycloparaffin

Ion 55

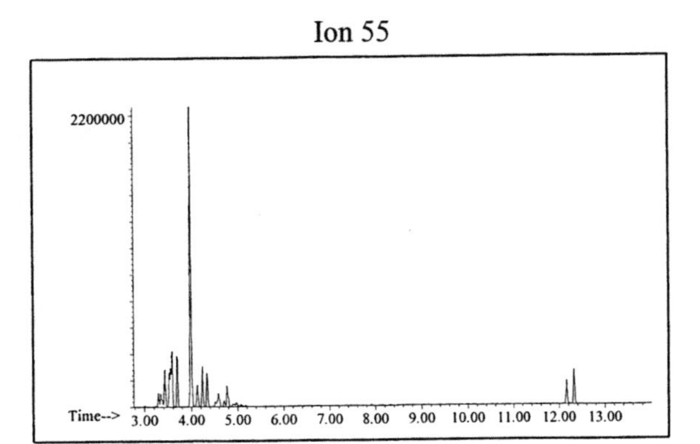

Ion 69

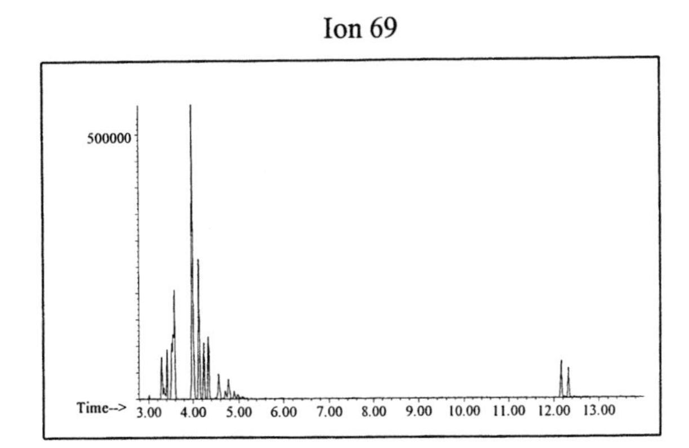

Ion 83

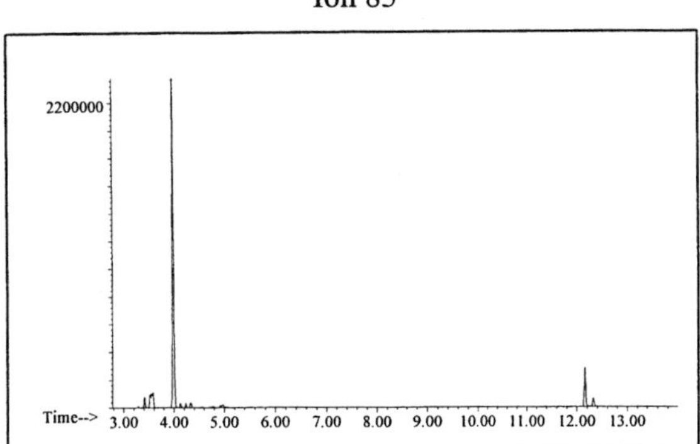

Naphthalene

Ion 128

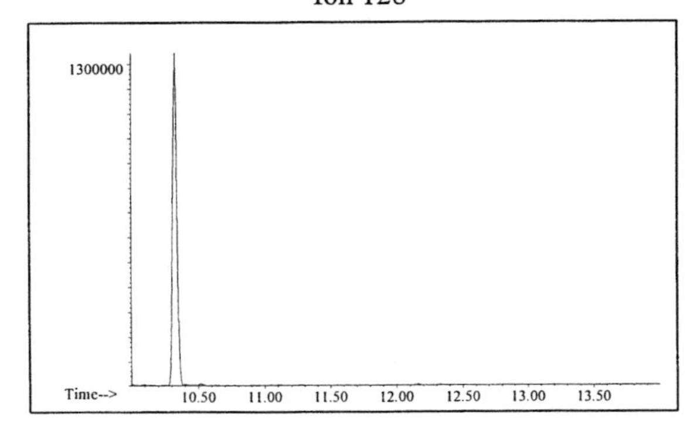

Ion 142

No Useful Data Obtained

Ion 156

No Useful Data Obtained

Miscellaneous #10

20uL/mL Pentane

Product Displayed: DTR 604 Enamel Reducer

Other Similar Products:

ASTM: Class 0 (Miscellaneous)

Product Uses: Enamel reducer

Macro Code: L

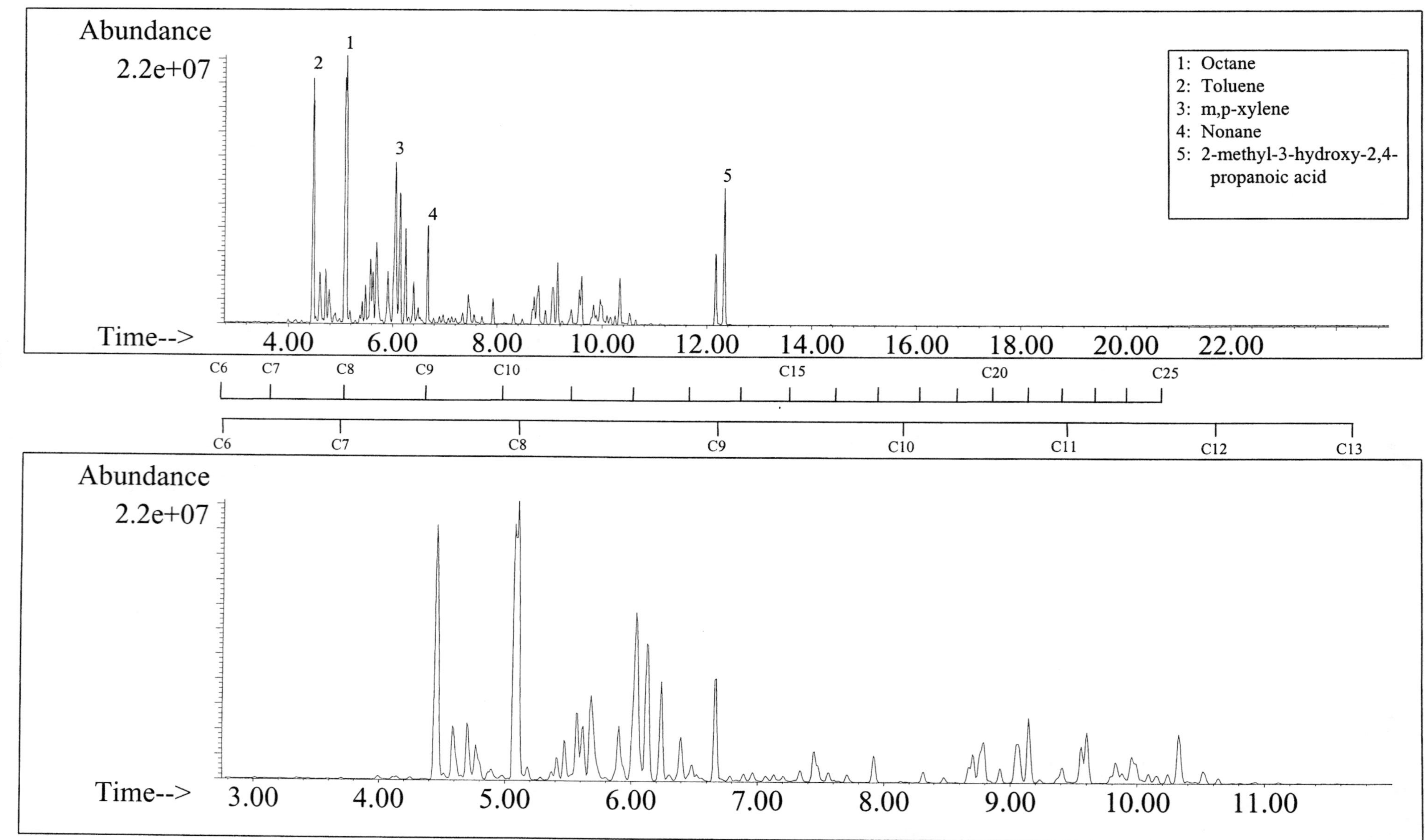

ALKANES

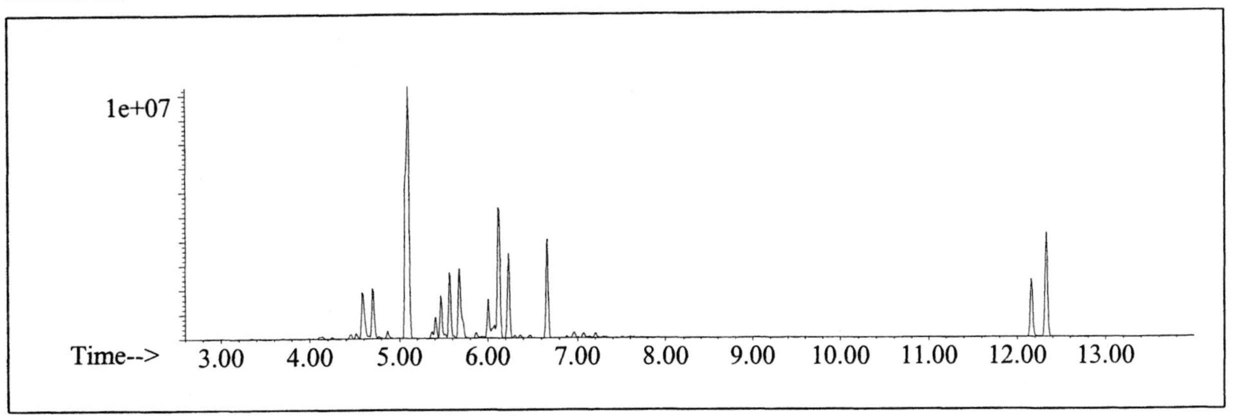

AROMATICS

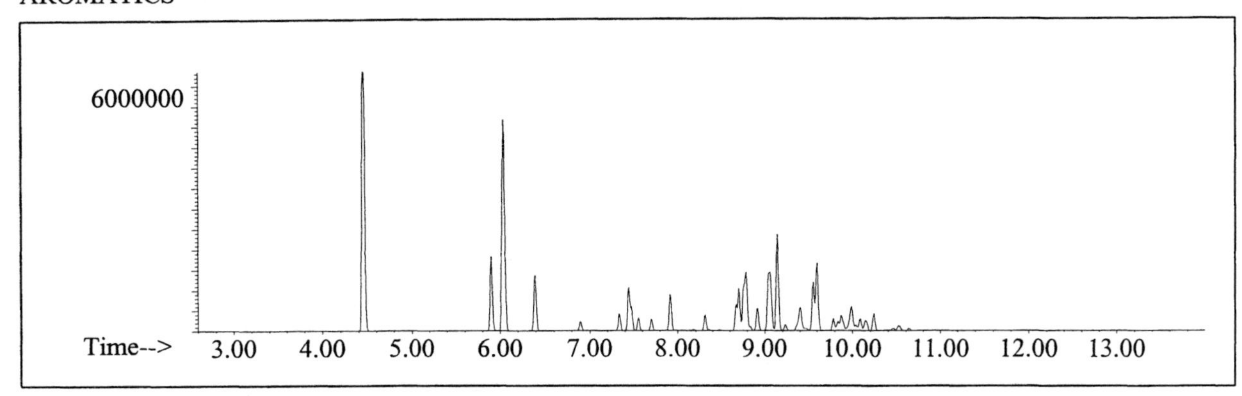

CYCLOPARAFFINS AND ALKENES

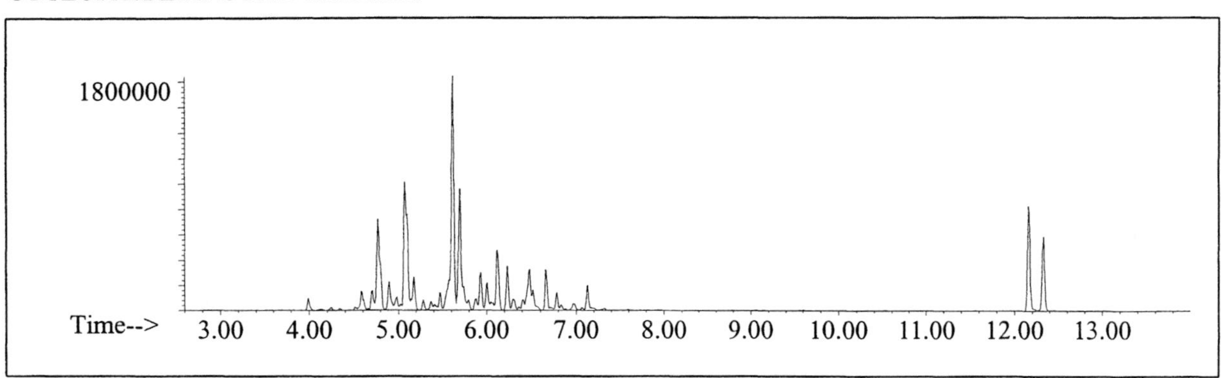

NAPHTHALENES

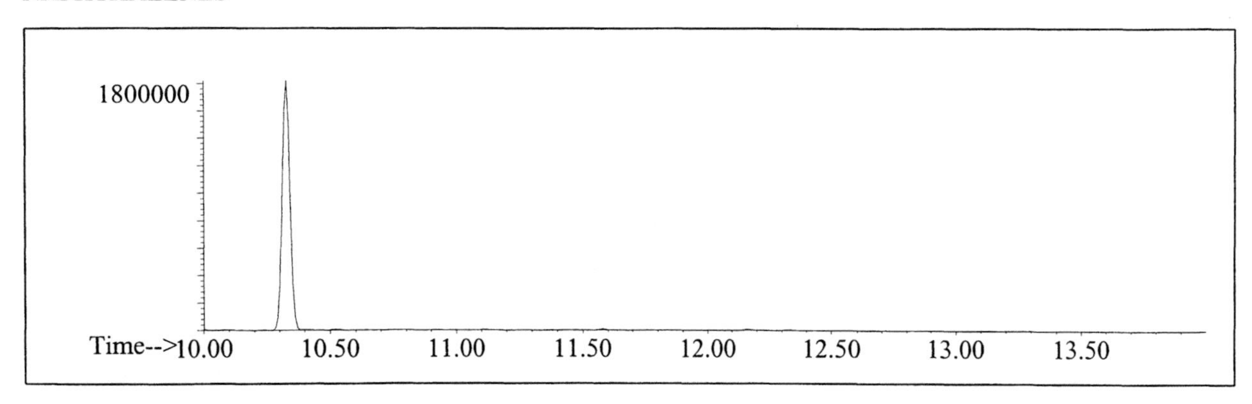

Alkane

Ion 43

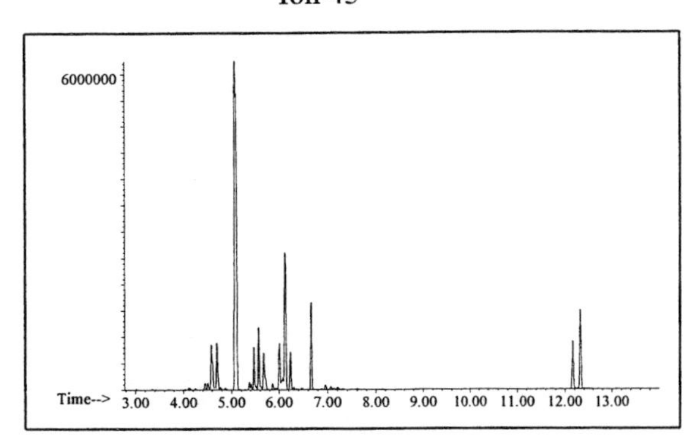

Ion 57

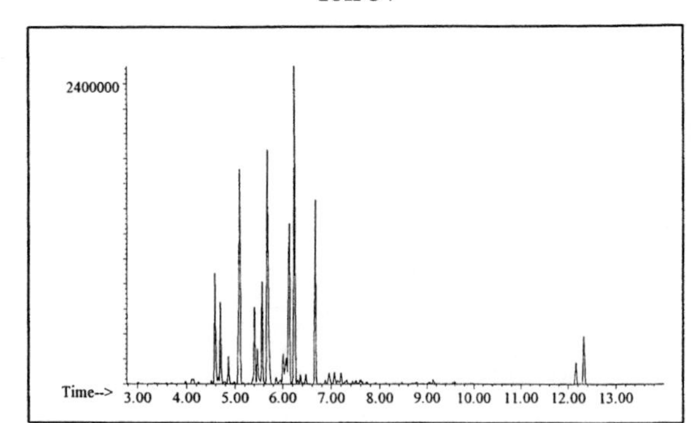

Ion 71

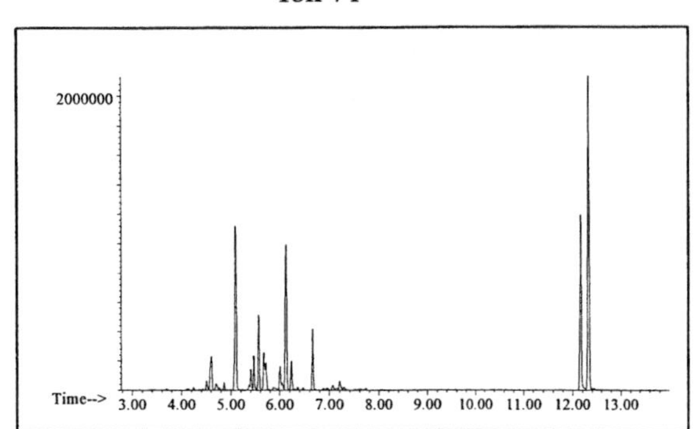

Ion 85

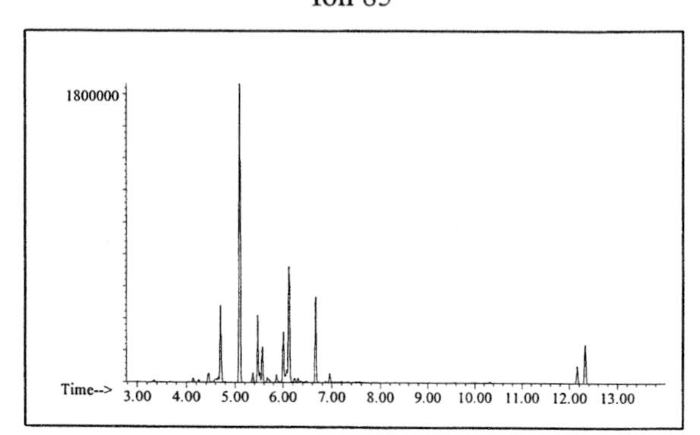

Aromatic

Ion 91

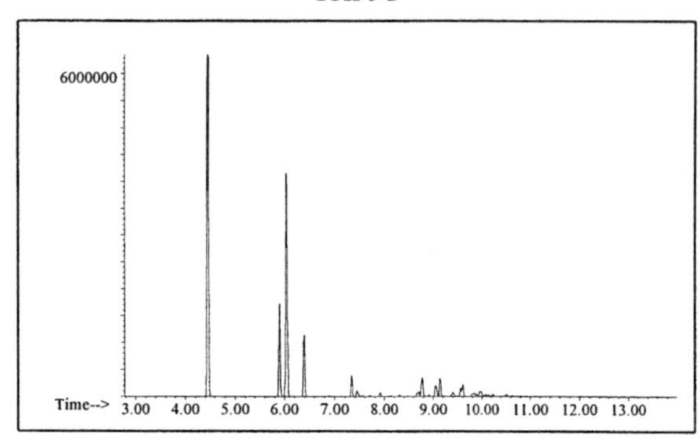

Ion 105

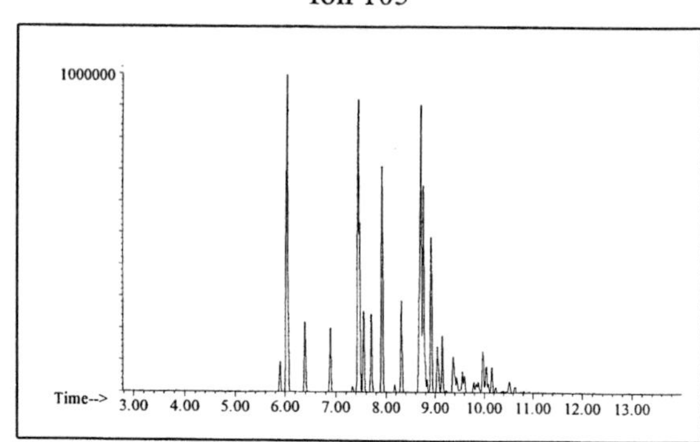

Ion 119

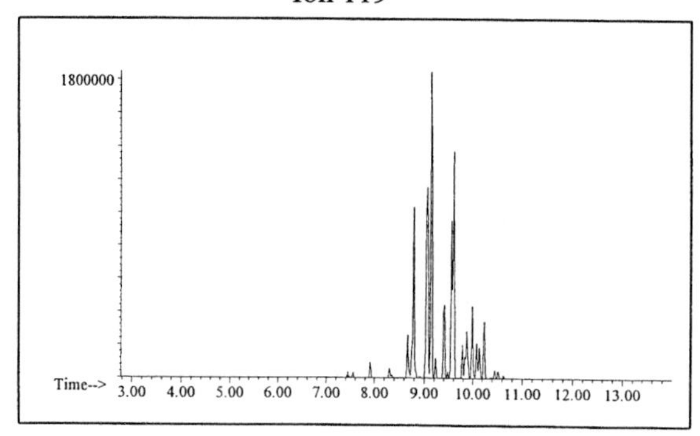

INDIVIDUAL PROFILES

Miscellaneous #10

Cycloparaffin

Ion 55

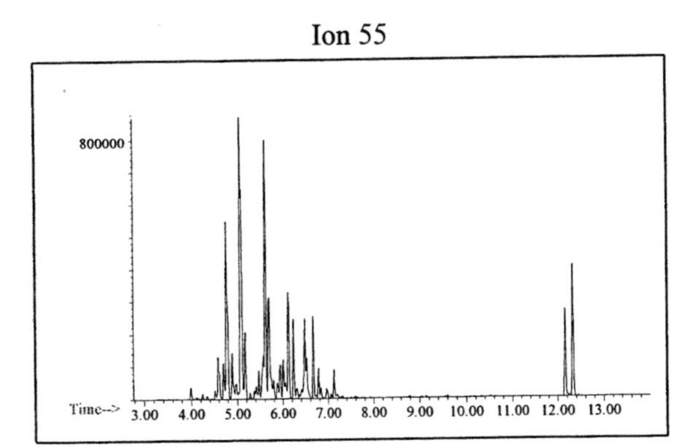

Ion 69

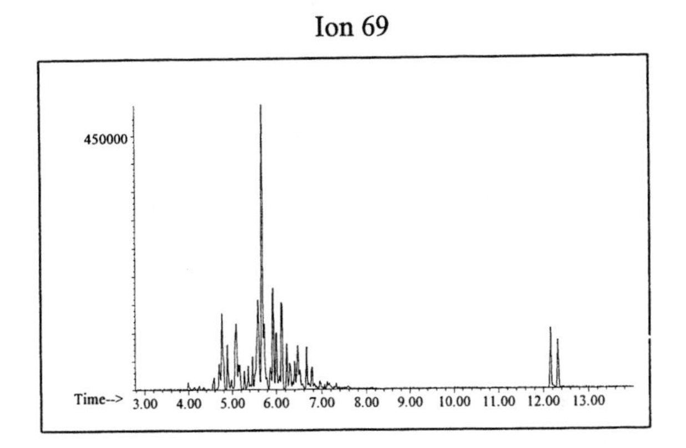

Ion 83

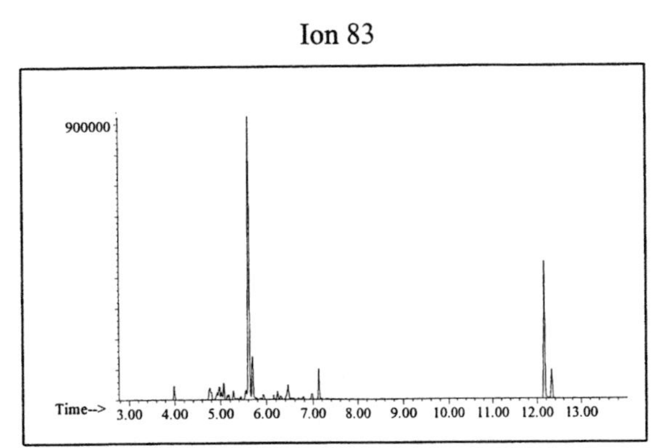

Naphthalene

Ion 128

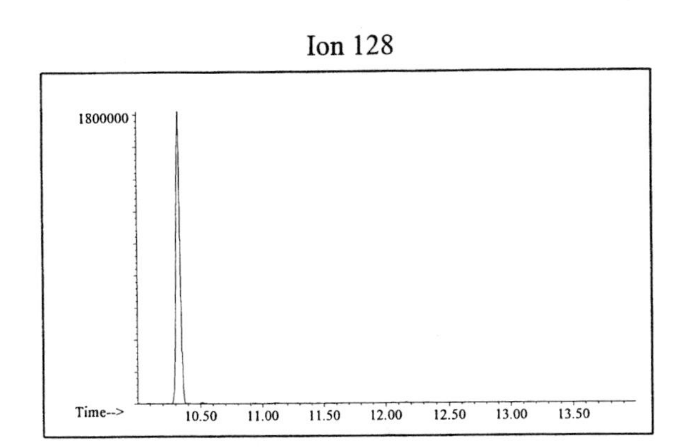

Ion 142

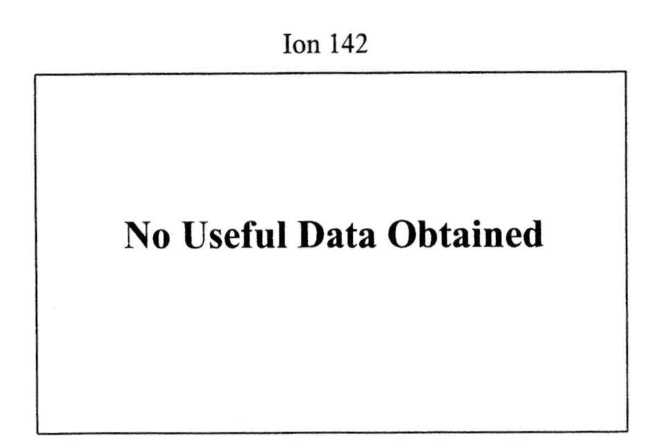

Ion 156

No Useful Data Obtained

Miscellaneous #11

20uL/mL Pentane

ASTM: Class 0 (Miscellaneous)

Macro Code: L

Product Displayed: Parks Quit-N-Time Brush/Roller Cleaner

Product Uses: Cleaning solvent, paint remover

Other Similar Products: Parks Rough and Ready Deglosser, Formby's Paint Remover

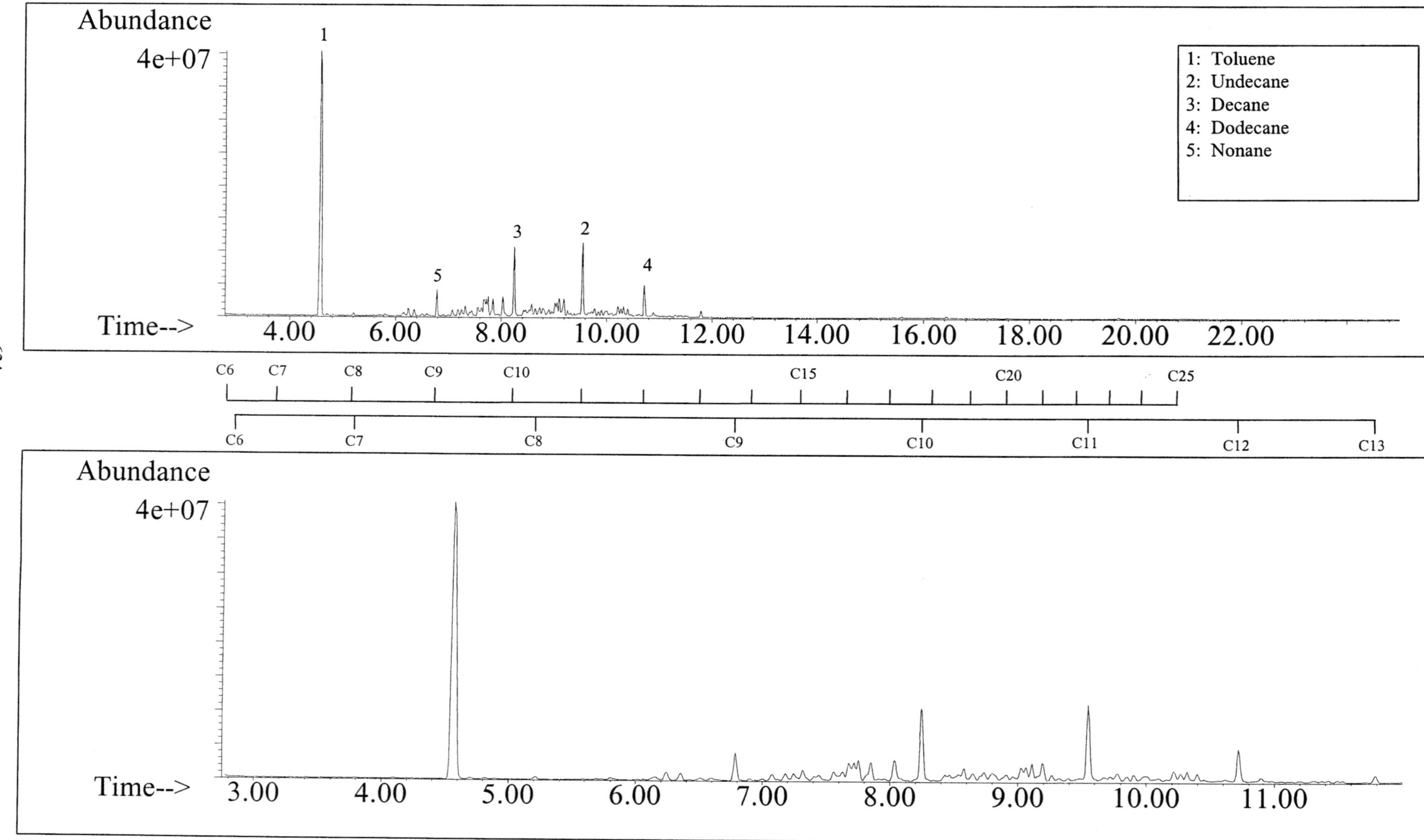

ALKANES

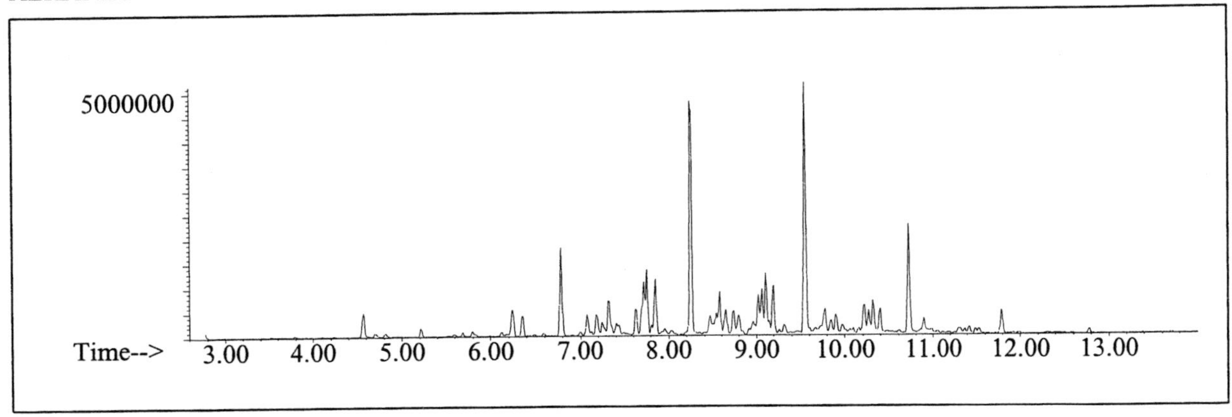

AROMATICS

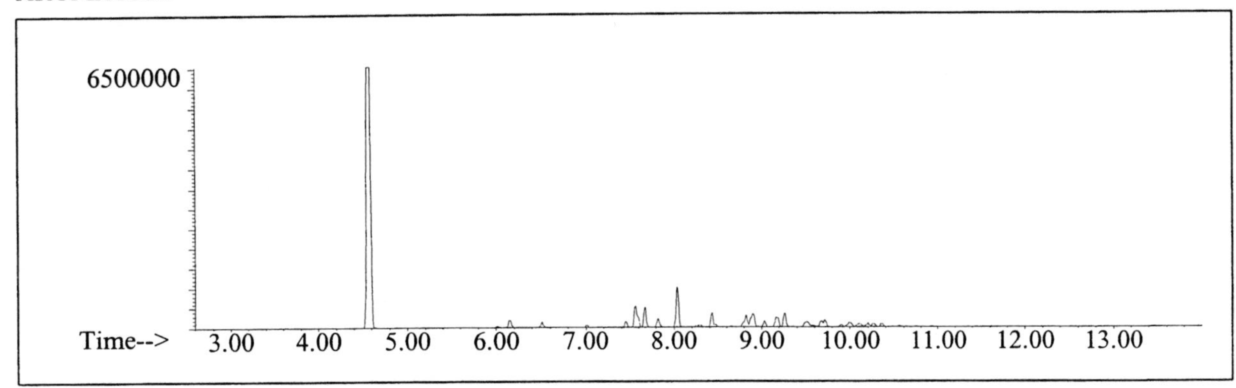

CYCLOPARAFFINS AND ALKENES

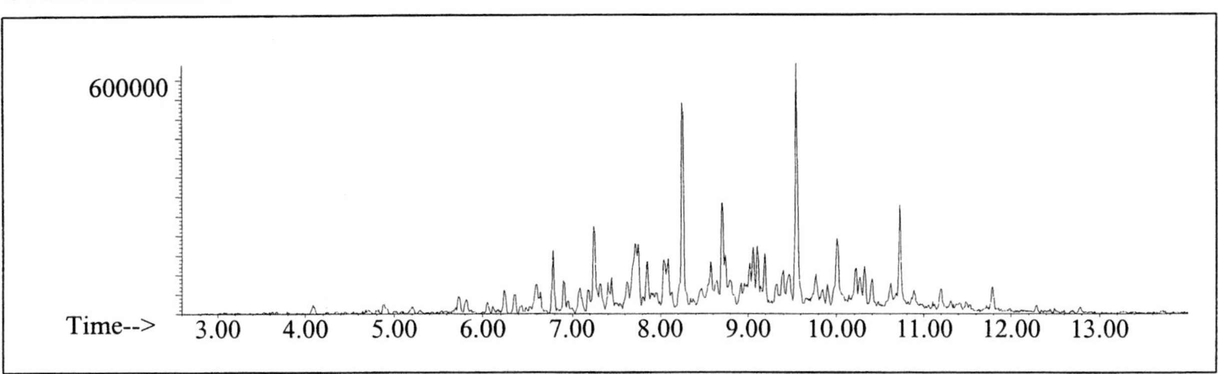

NAPHTHALENES

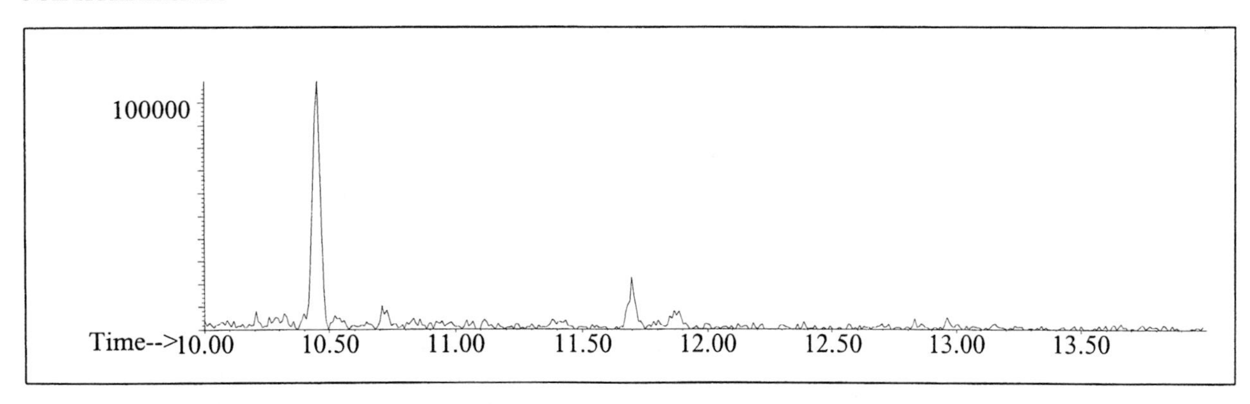

Alkane

Ion 43

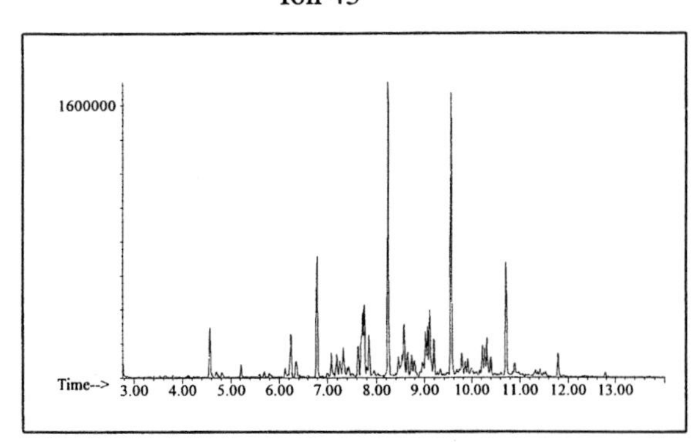

Ion 57

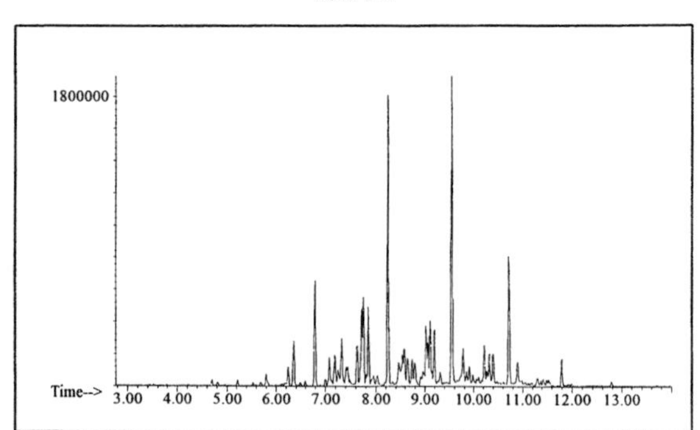

Ion 71

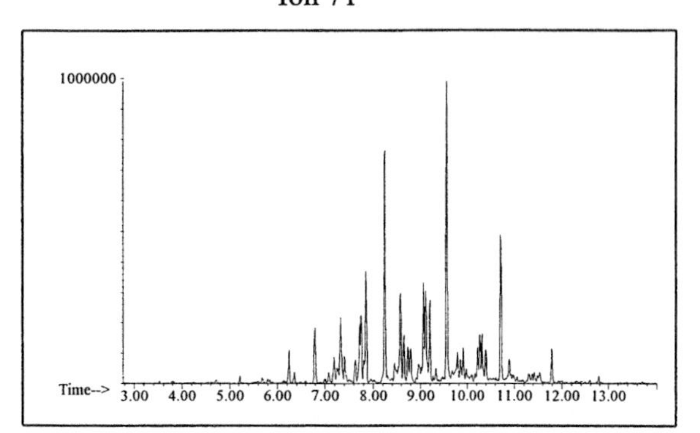

Ion 85

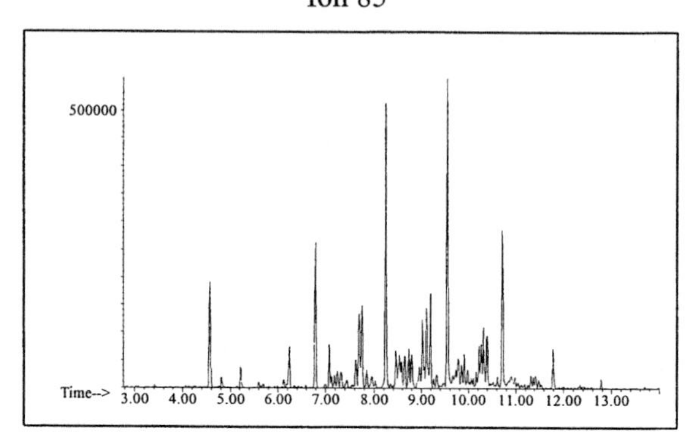

Aromatic

Ion 91

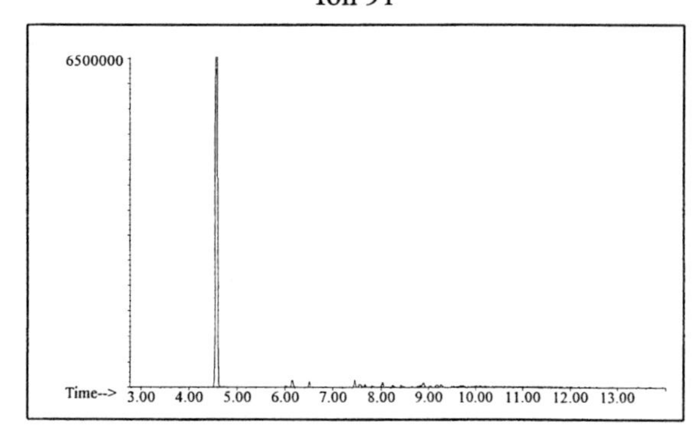

Ion 105

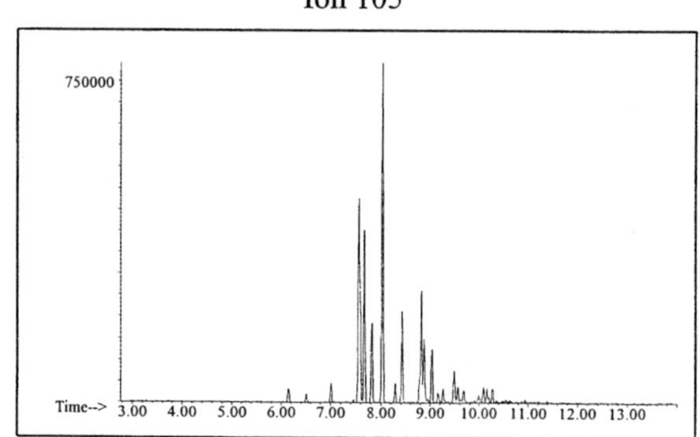

Ion 119

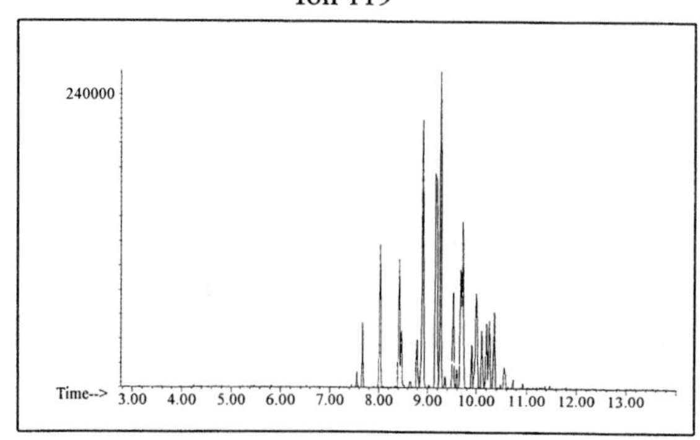

Cycloparaffin

Ion 55

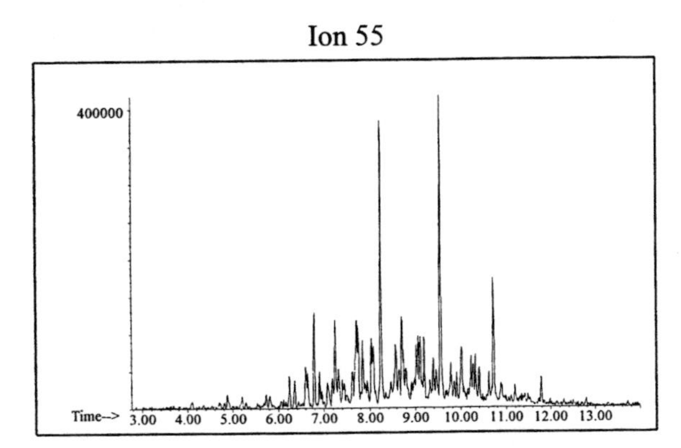

Ion 69

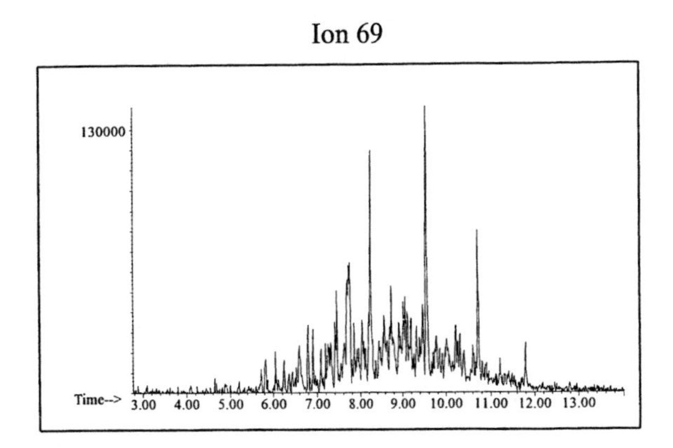

Ion 83

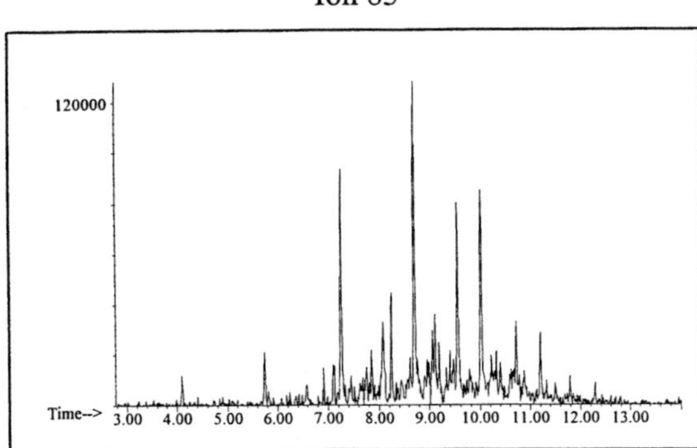

Naphthalene

Ion 128

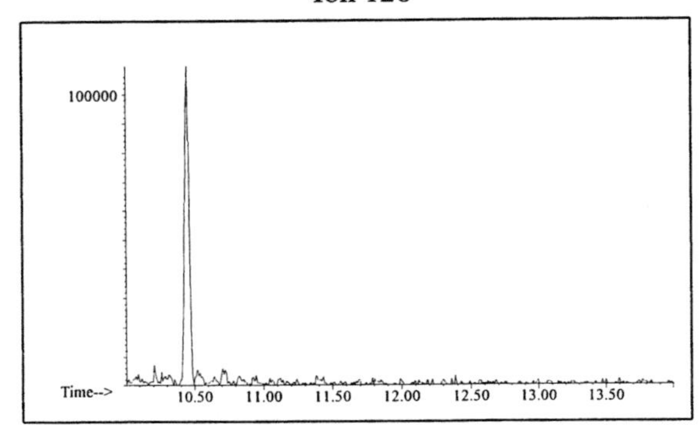

Ion 142

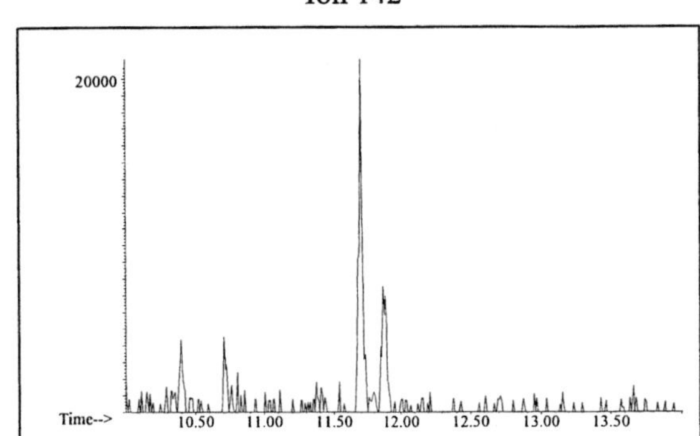

Ion 156

No Useful Data Obtained

Miscellaneous #12

20uL/mL Pentane
Product Displayed: DTR 607 Enamel Reducer
Other Similar Products:

ASTM: Class 0 (Miscellaneous)
Product Uses: Enamel reducer

Macro Code: L

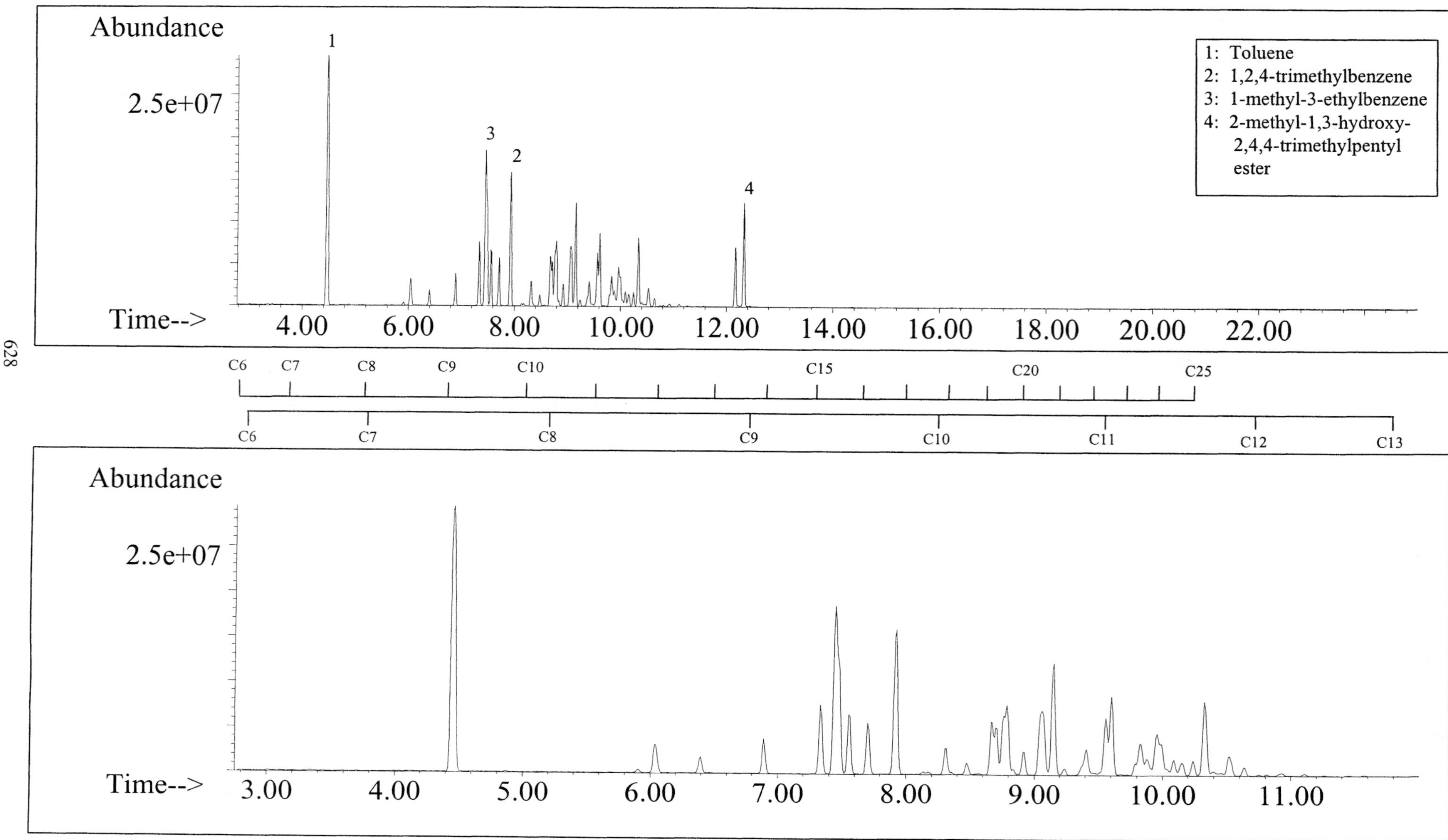

ALKANES

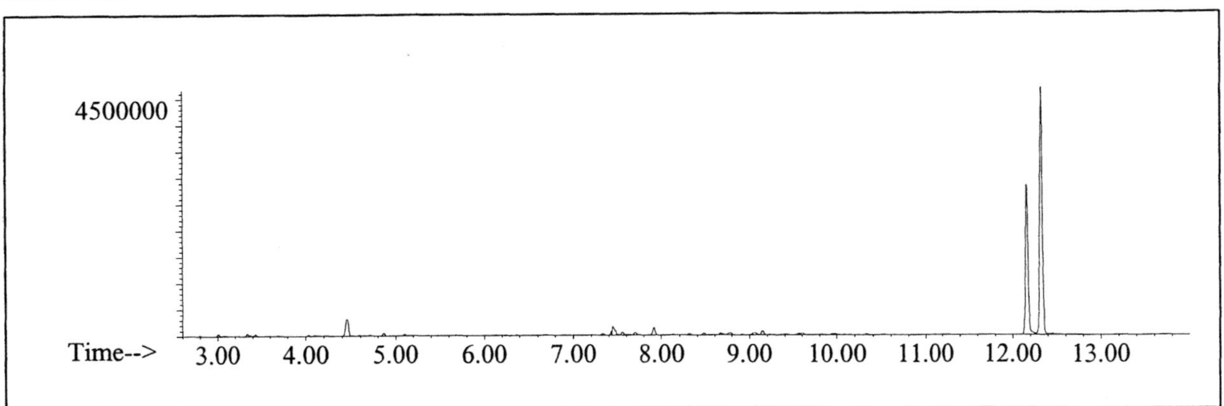

AROMATICS

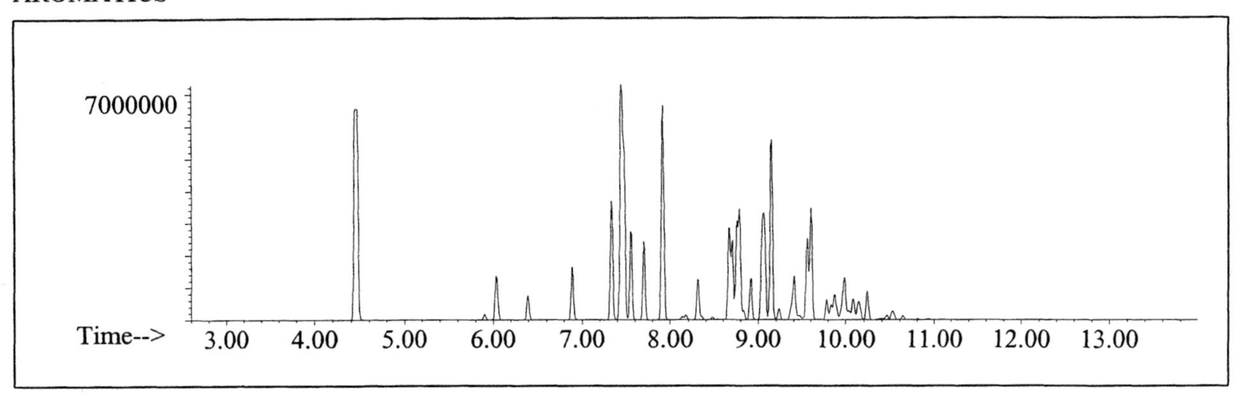

CYCLOPARAFFINS AND ALKENES

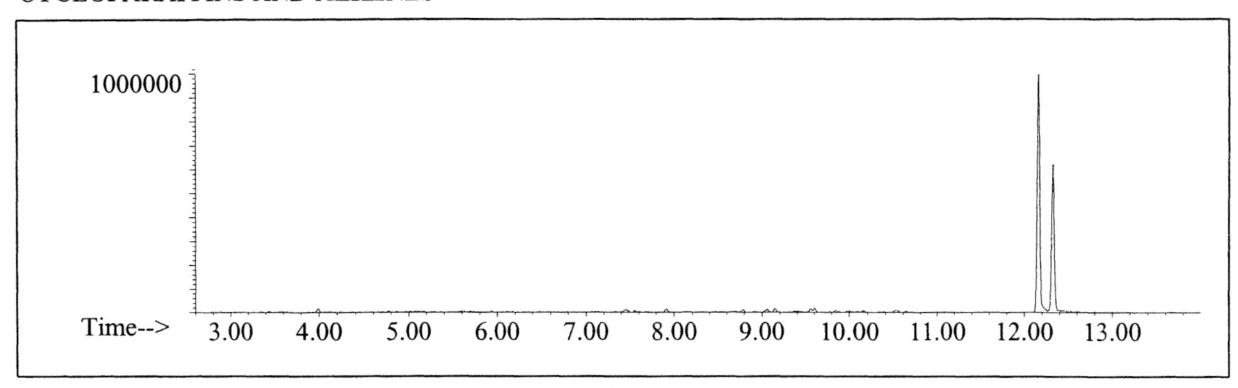

NAPHTHALENES

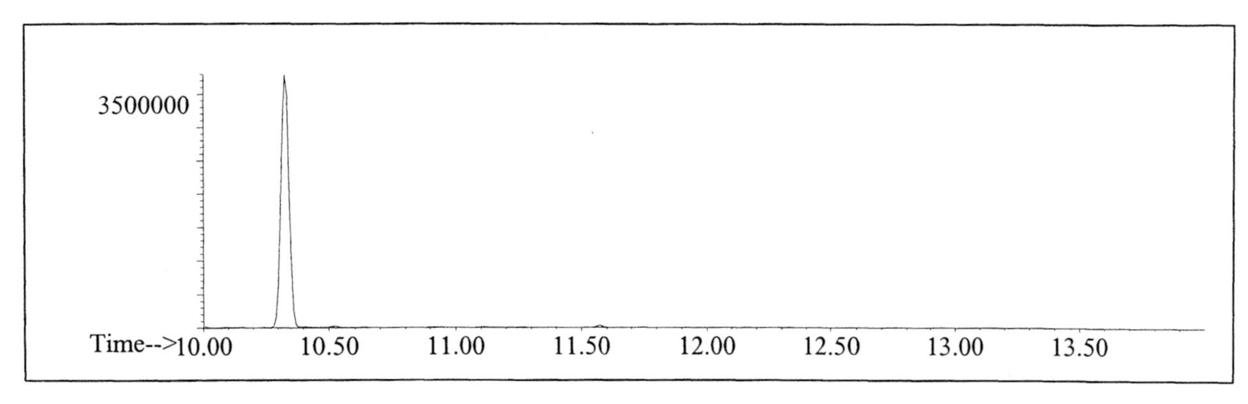

Alkane

Ion 43

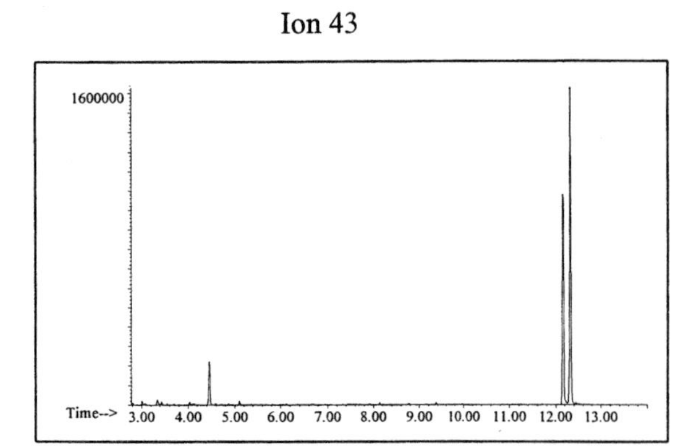

Ion 57

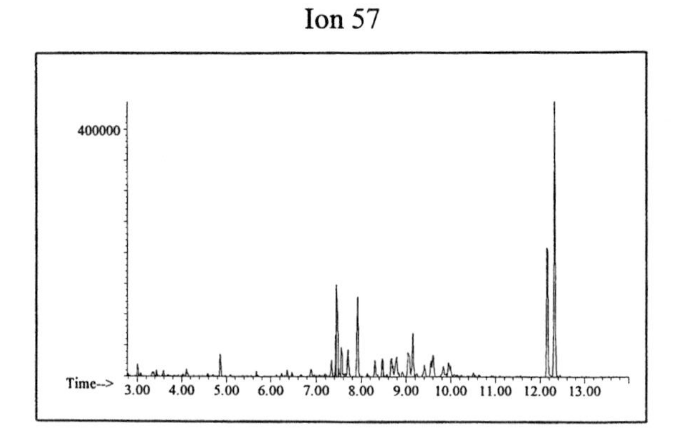

Ion 71

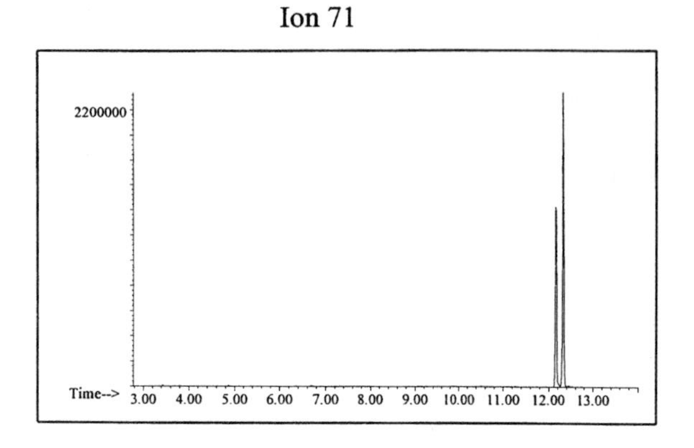

Ion 85

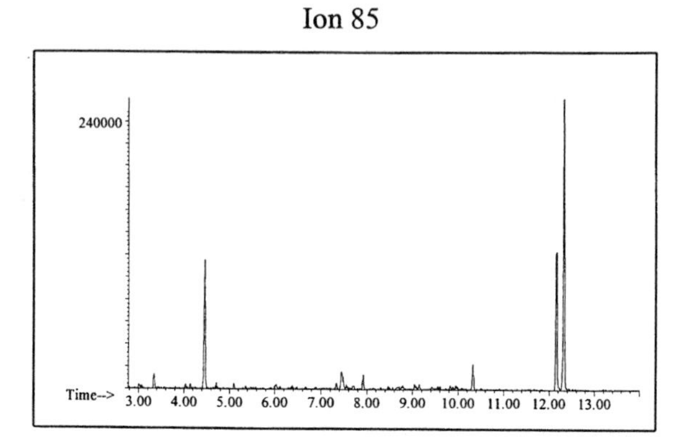

Aromatic

Ion 91

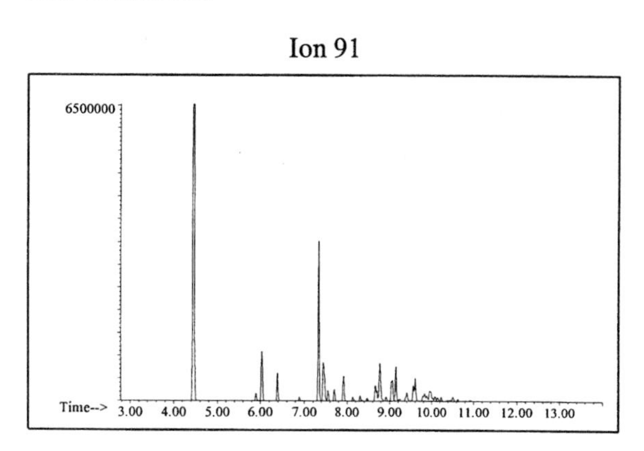

Ion 105

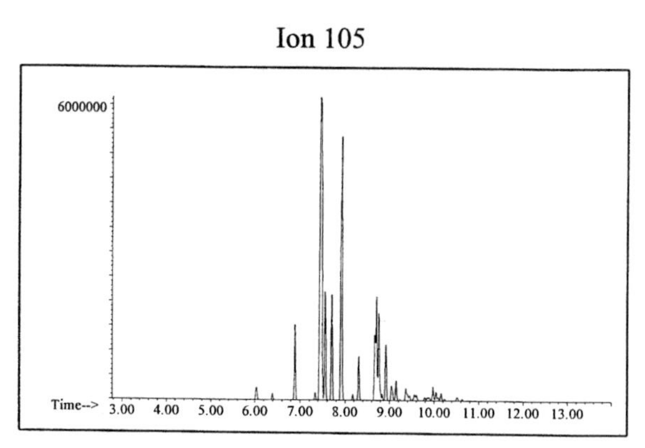

Ion 119

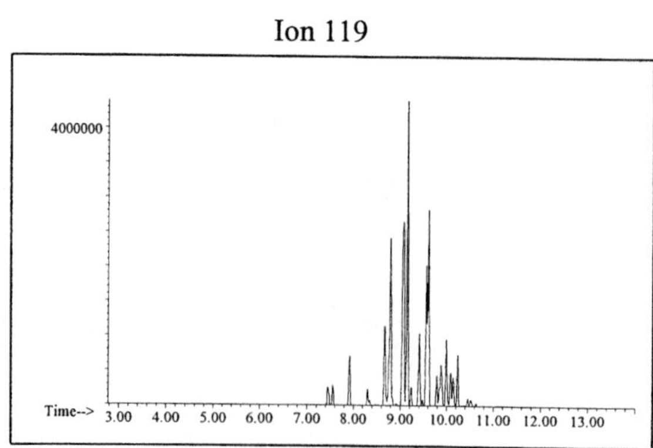

Cycloparaffin

Ion 55

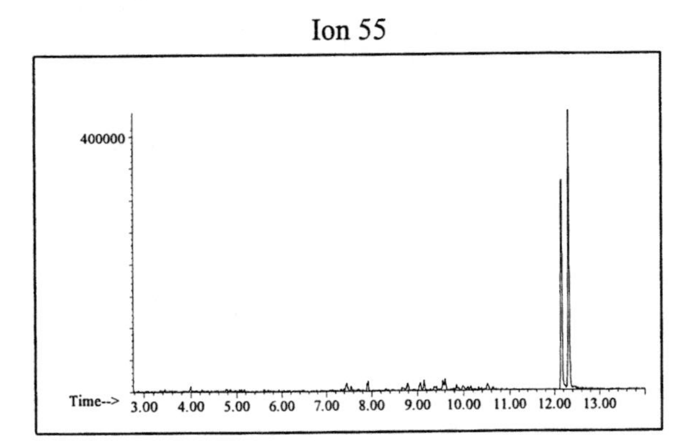

Ion 69

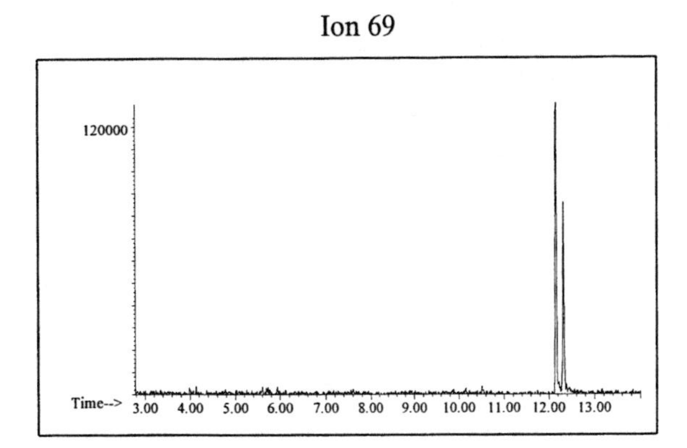

Ion 83

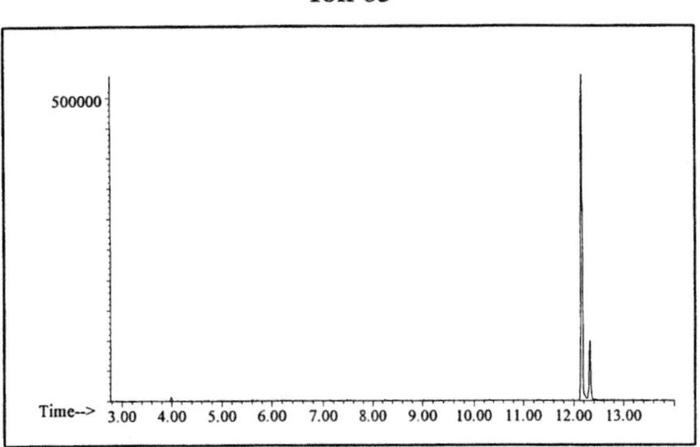

Naphthalene

Ion 128

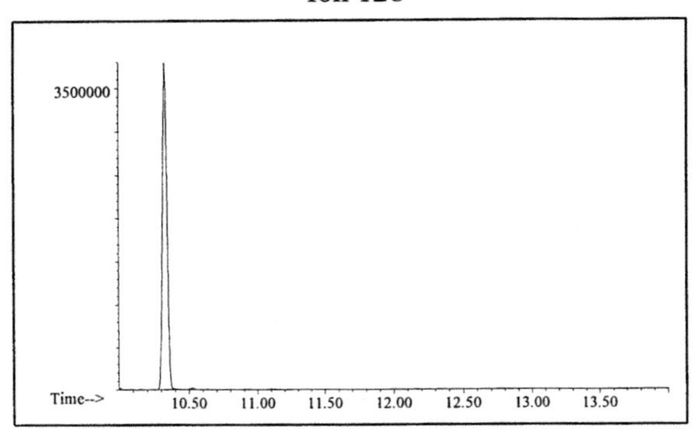

Ion 142

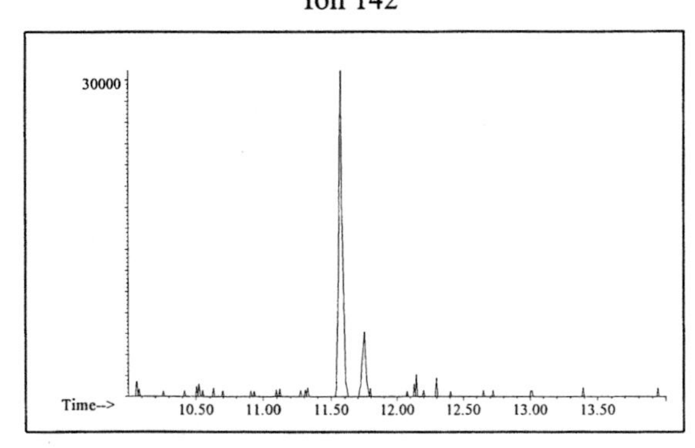

Ion 156

No Useful Data Obtained

Miscellaneous #13

20uL/mL Pentane

ASTM: Class 0 (Miscellaneous)

Macro Code: L

Product Displayed: Crystal Aire Vinyl Top Coat

Product Uses: Top coat

Other Similar Products:

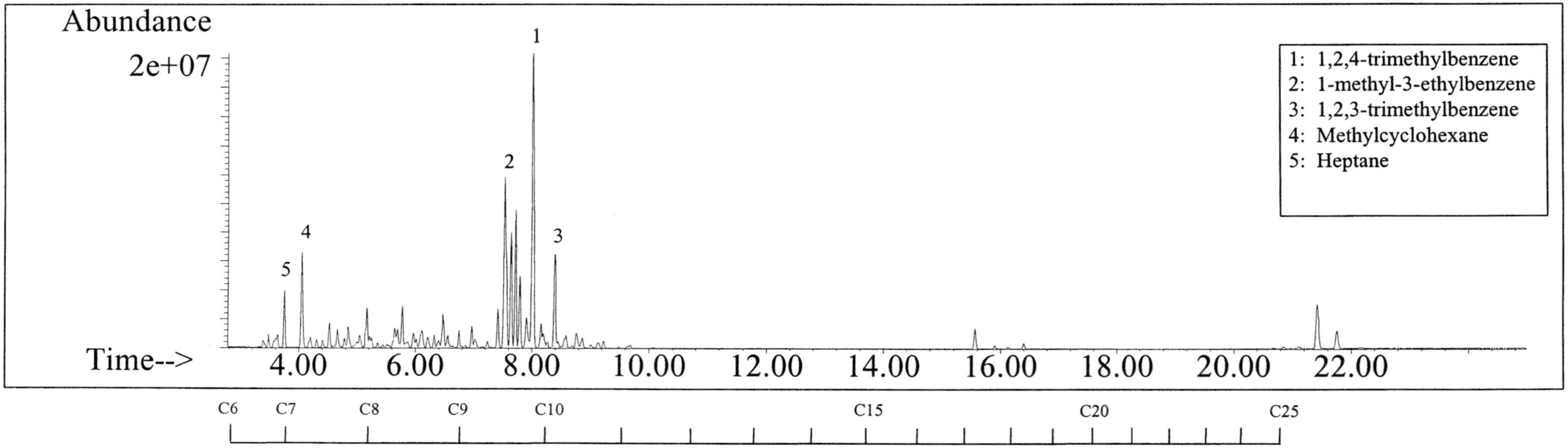

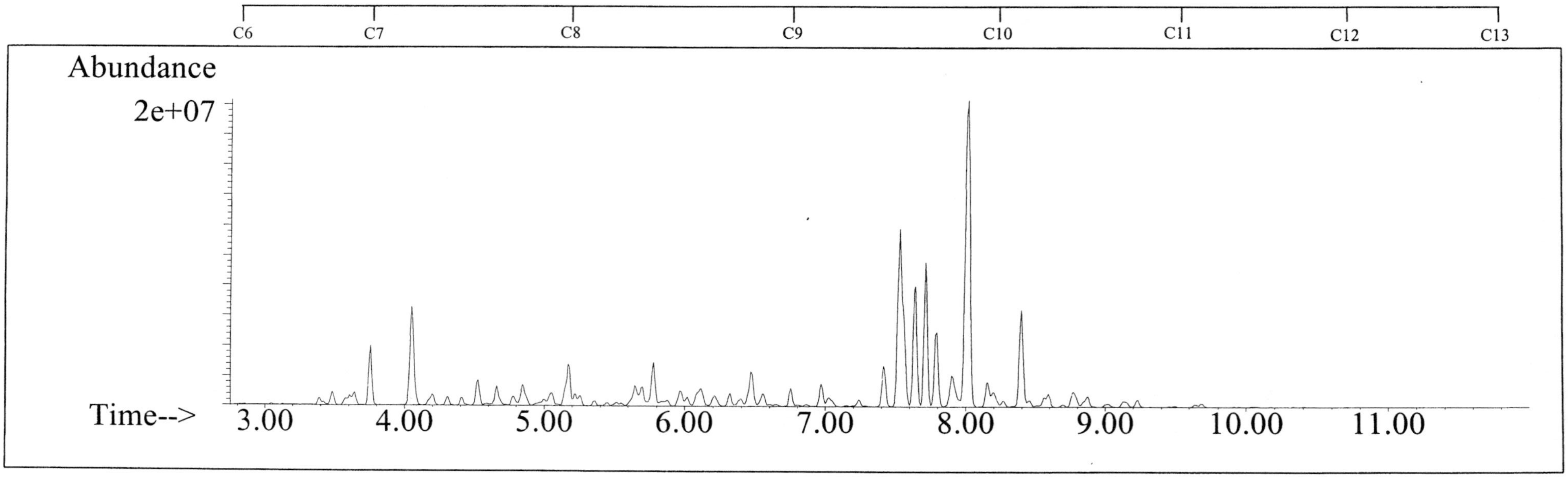

ALKANES

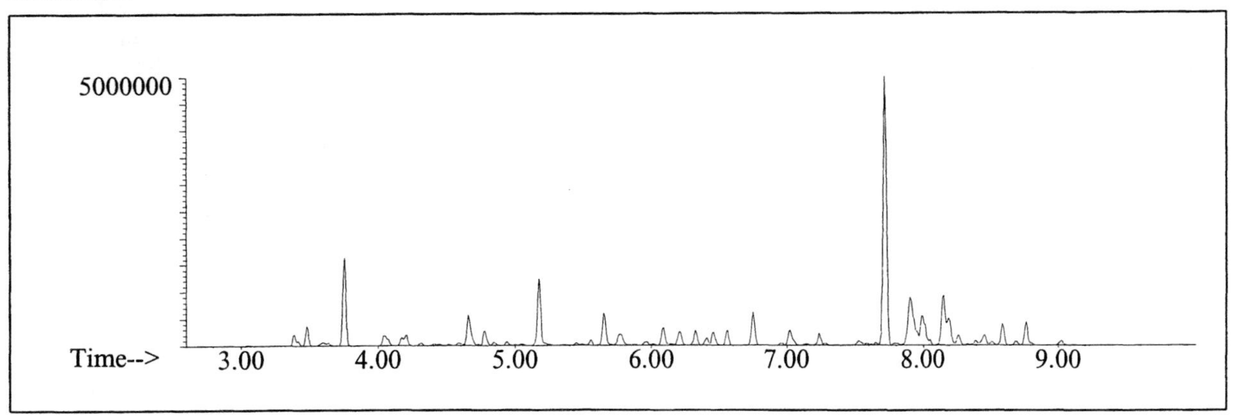

AROMATICS

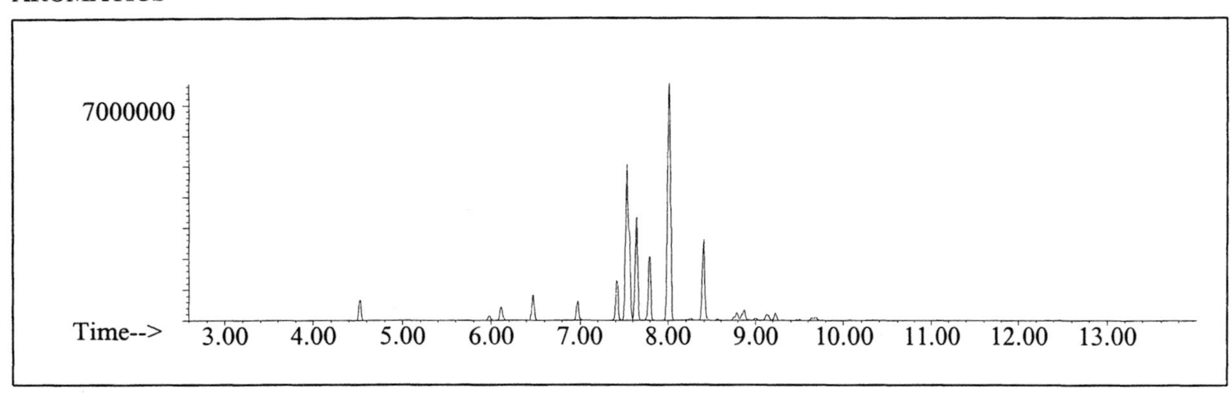

CYCLOPARAFFINS AND ALKENES

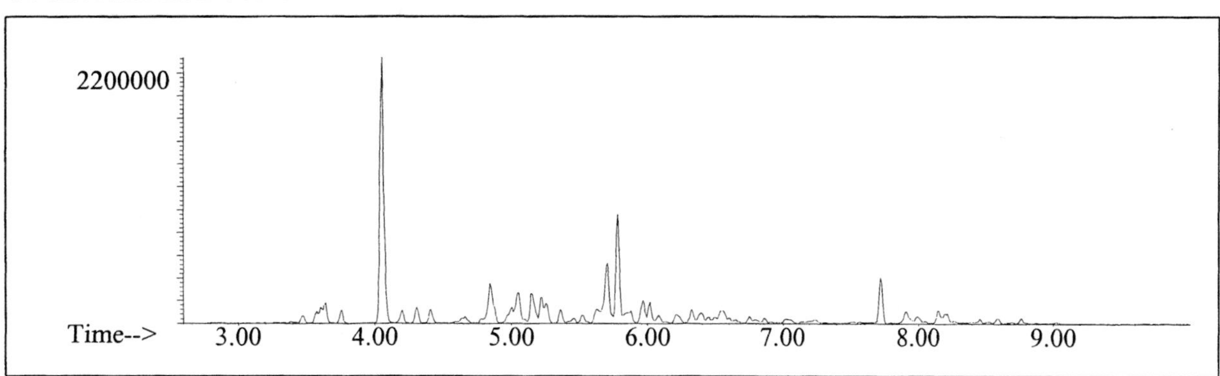

NAPHTHALENES

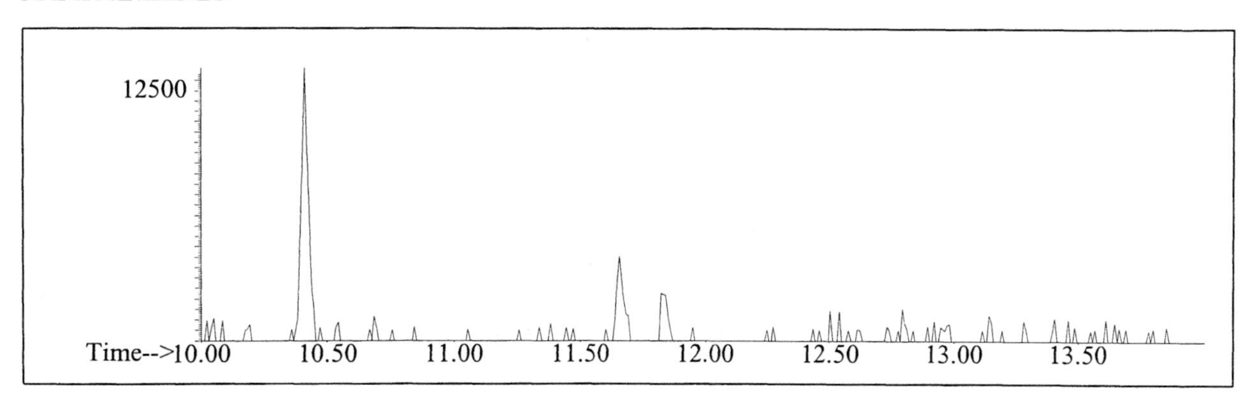

Alkane

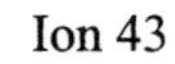

Ion 43

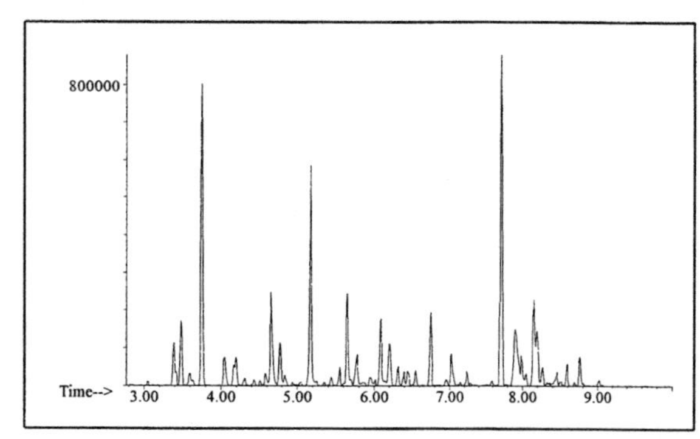

Ion 57

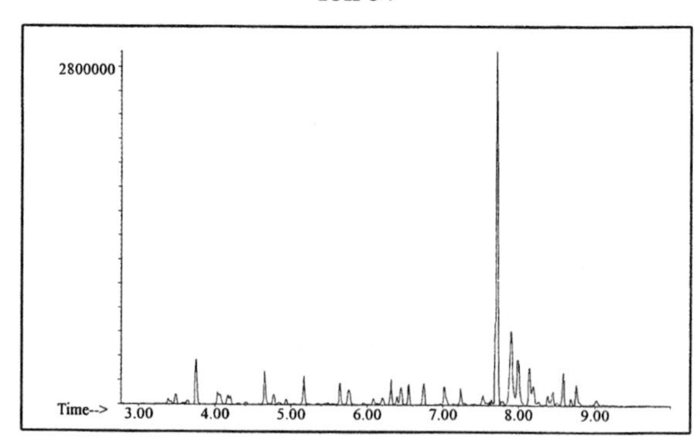

Ion 71

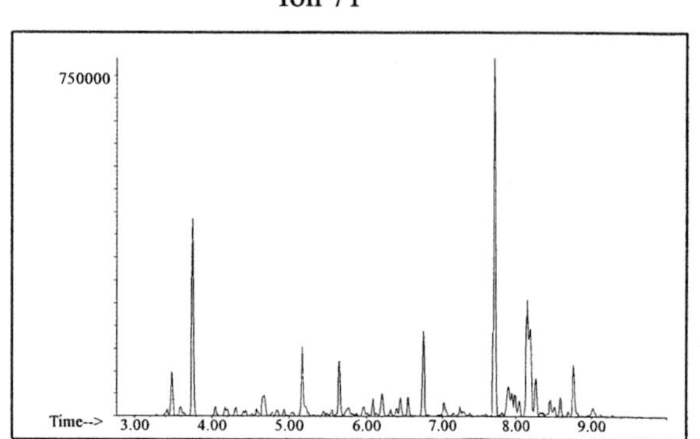

Ion 85

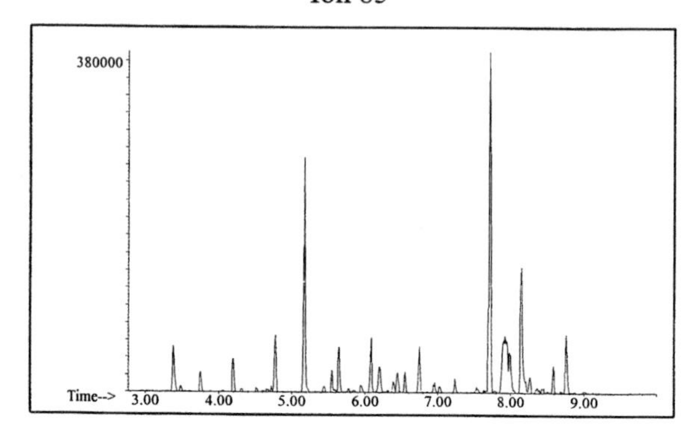

Aromatic

Ion 91

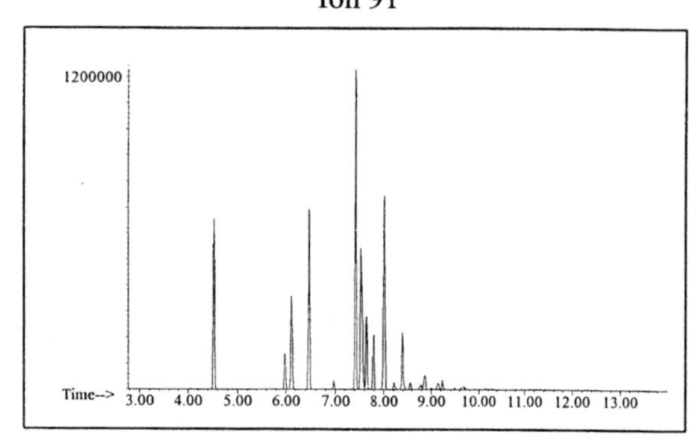

Ion 105

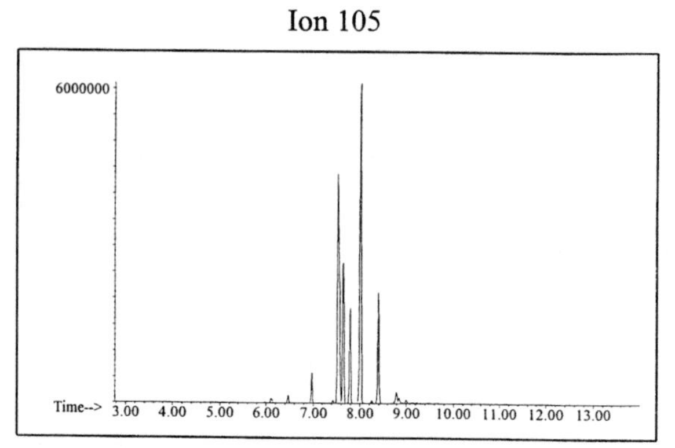

Ion 119

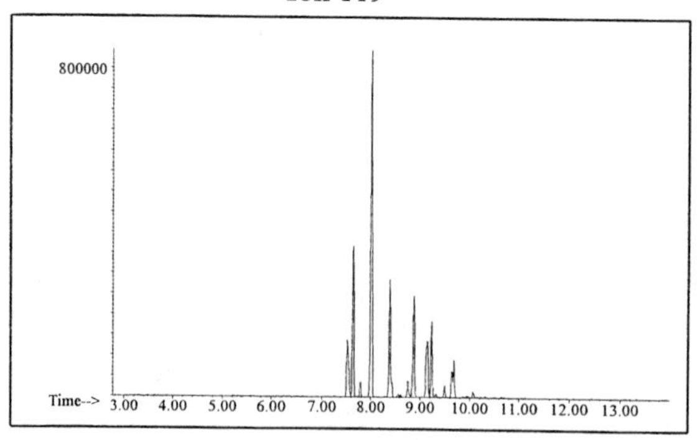

Cycloparaffin

Ion 55

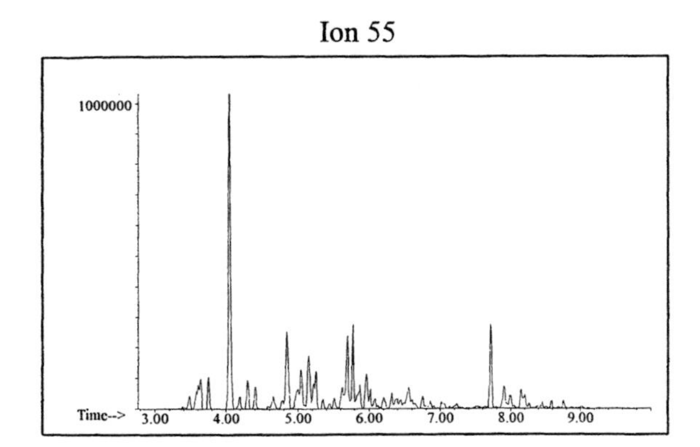

Ion 69

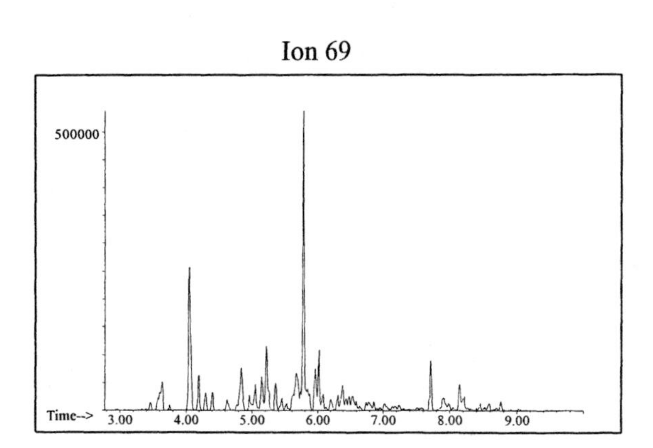

Ion 83

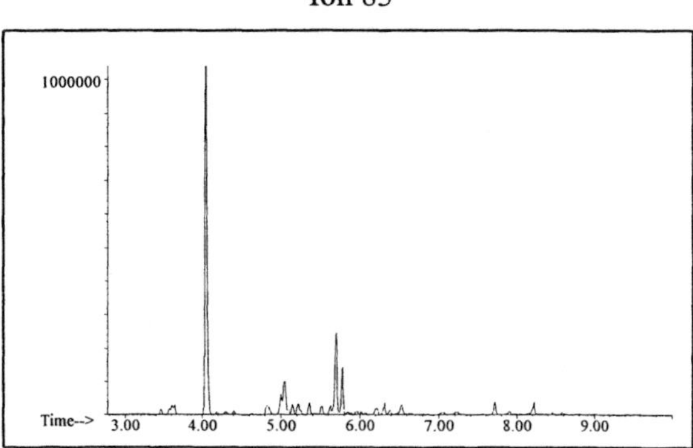

Naphthalene

Ion 128

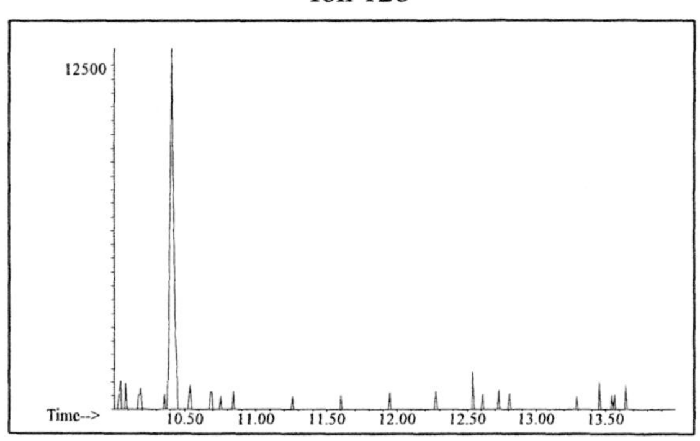

Ion 142

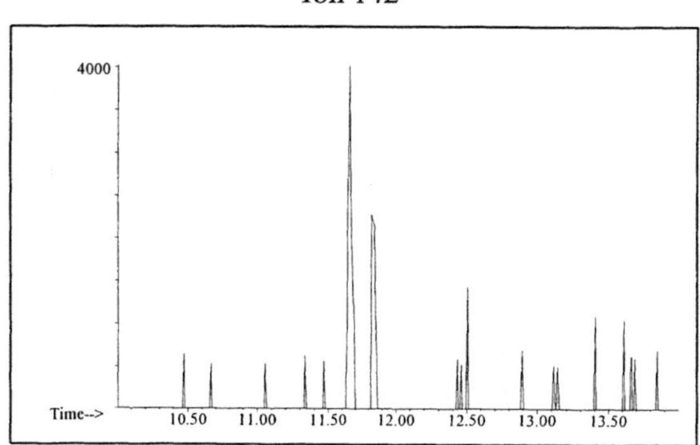

Ion 156

No Useful Data Obtained

Miscellaneous #14

20uL/mL Pentane

ASTM: Class 0 (Miscellaneous)

Macro Code: L

Product Displayed: Brush Top Spot Remover

Product Uses: Cleaning solvent

Other Similar Products:

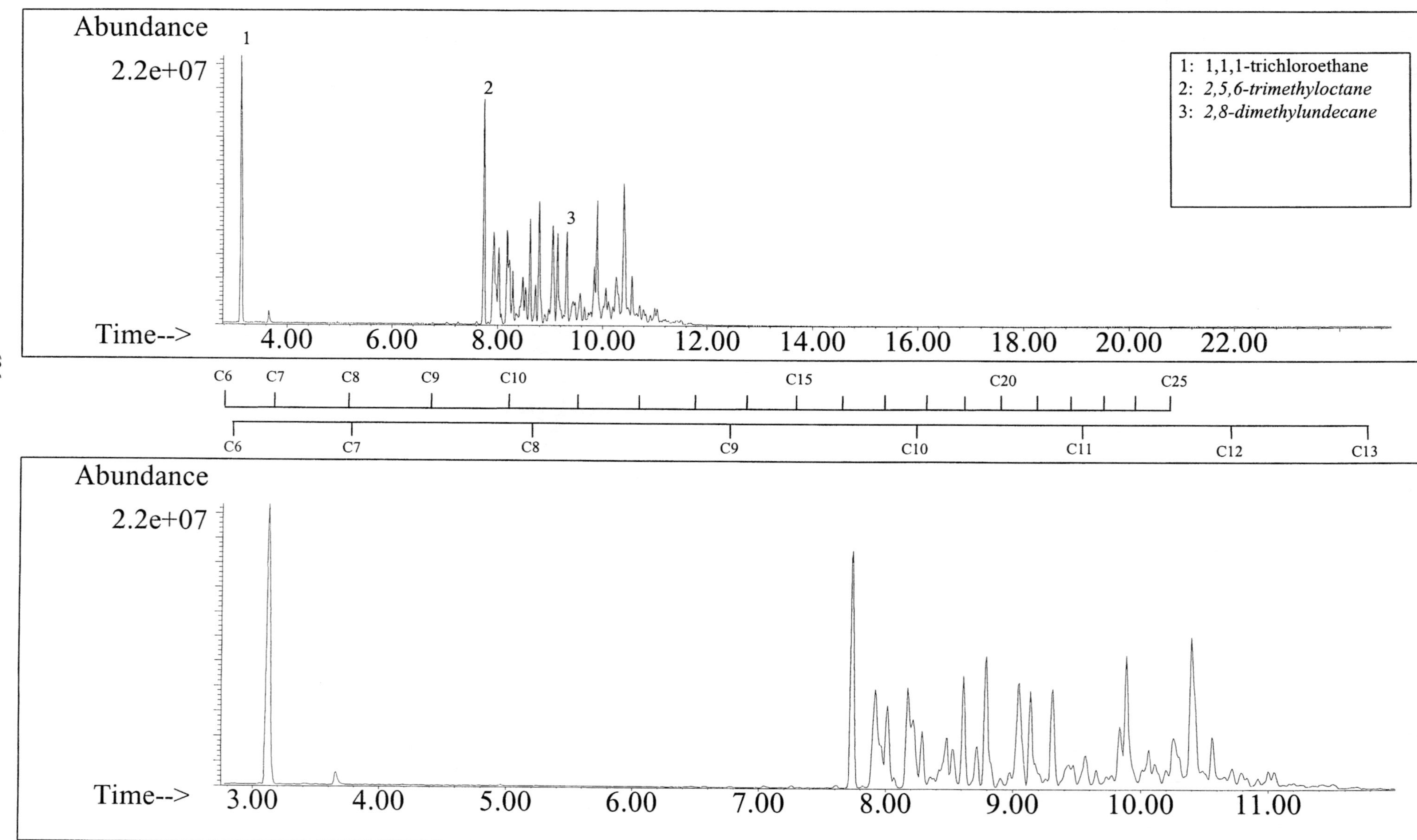

ALKANES

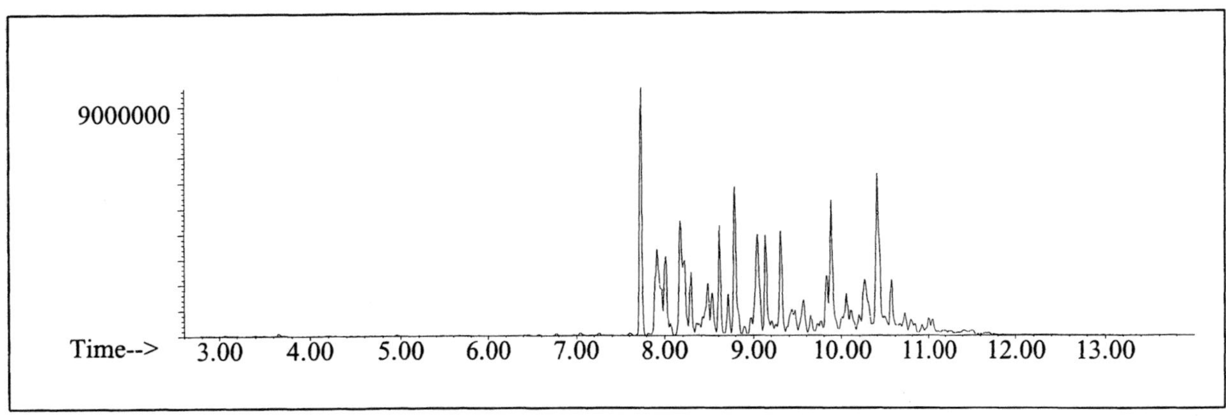

AROMATICS

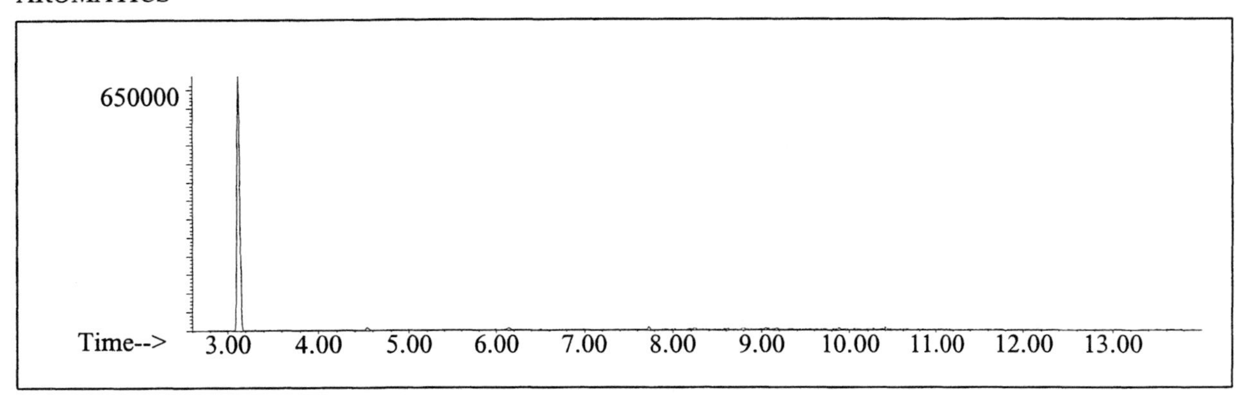

CYCLOPARAFFINS AND ALKENES

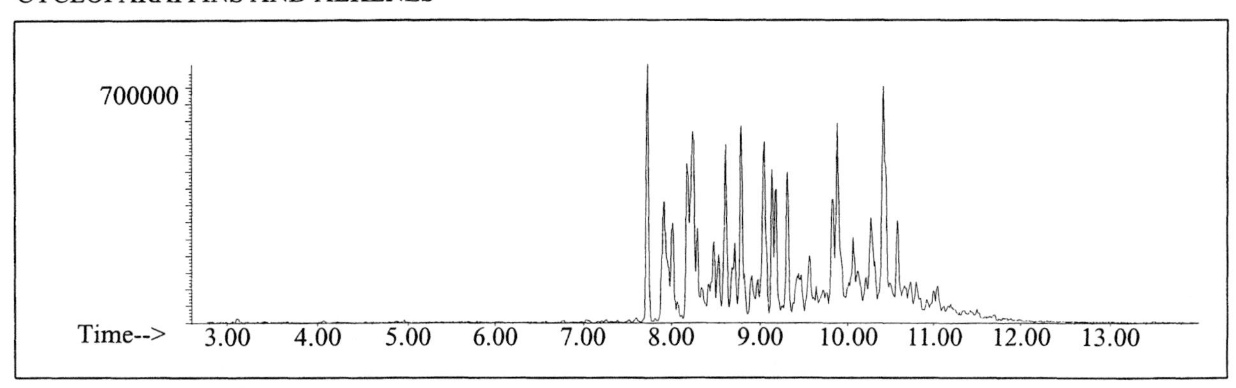

NAPHTHALENES

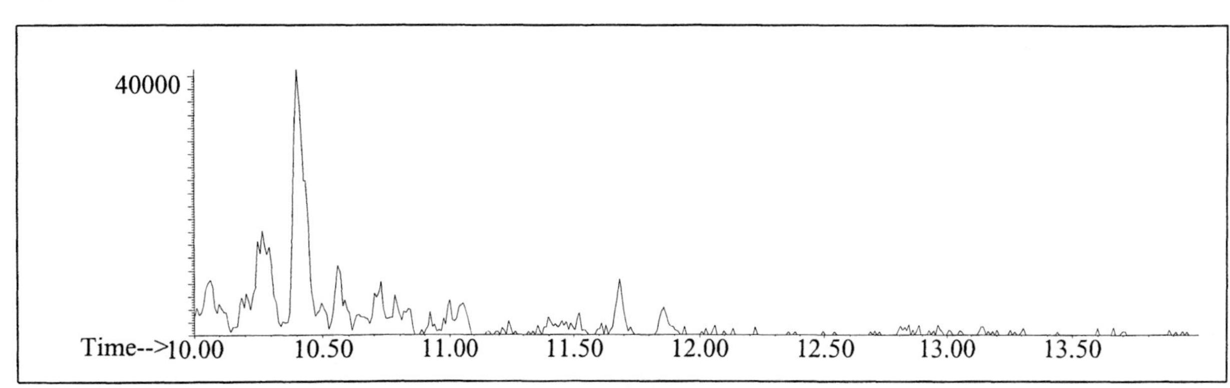

INDIVIDUAL PROFILES

Alkane

Ion 43

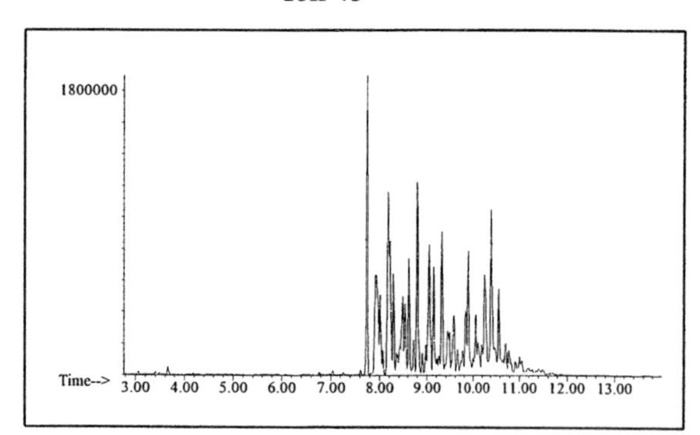

Ion 57

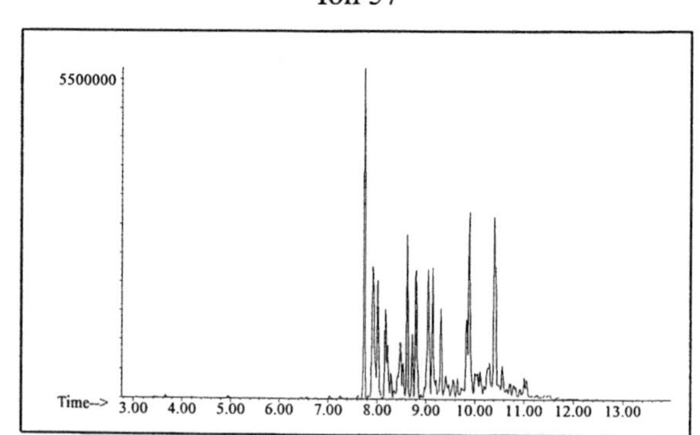

Ion 71

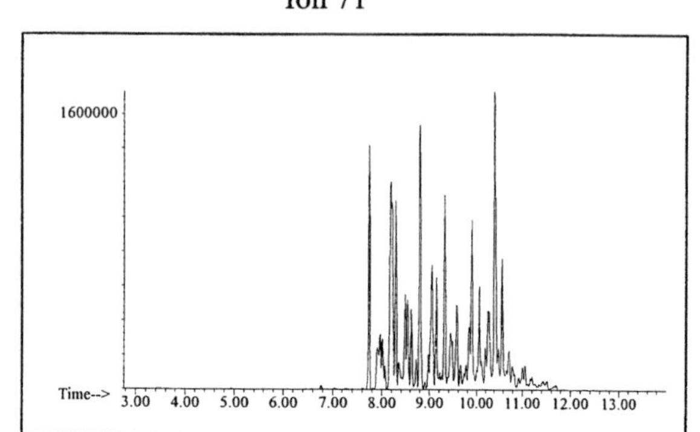

Ion 85

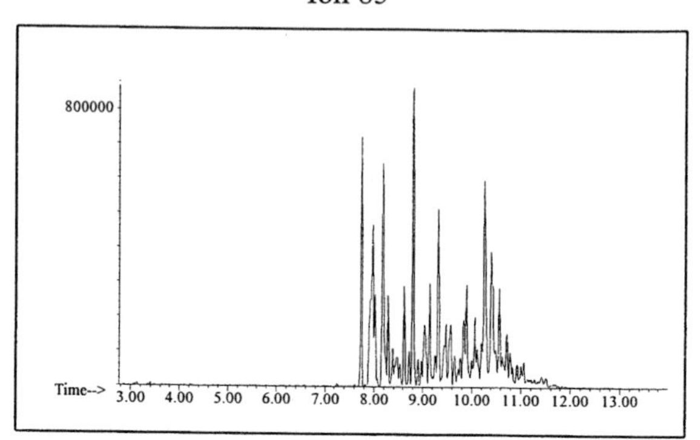

Aromatic

Ion 91

No Useful Data Obtained

Ion 105

No Useful Data Obtained

Ion 119

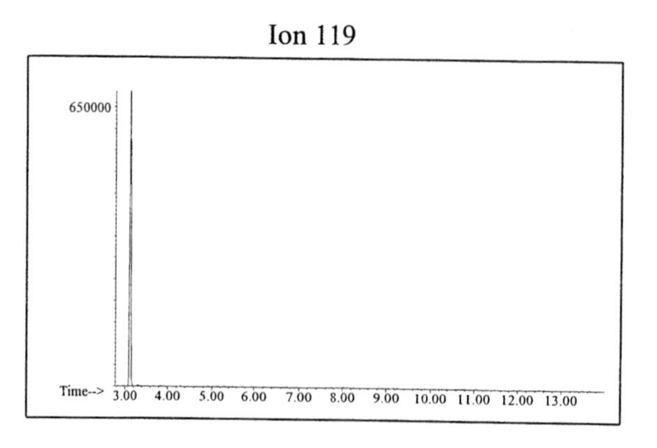

Cycloparaffin

Ion 55

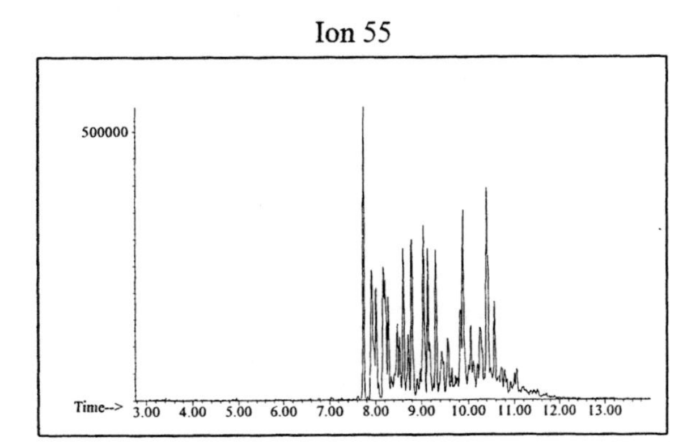

Ion 69

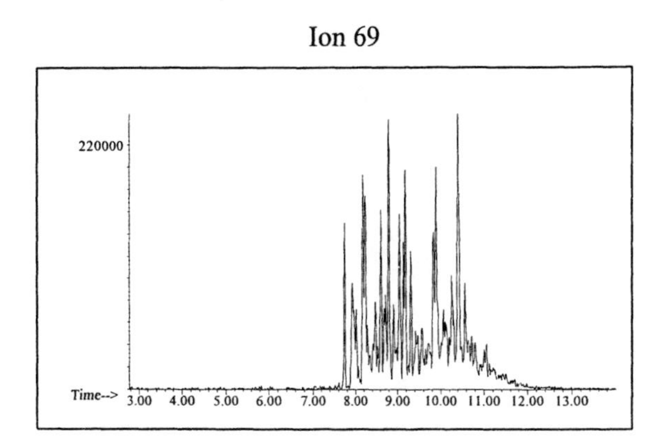

Ion 83

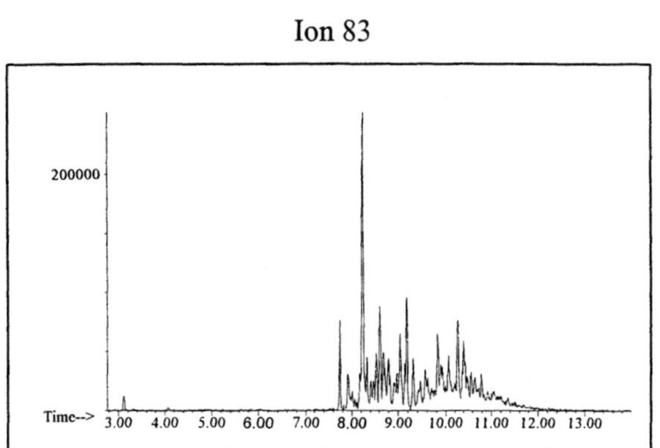

Naphthalene

Ion 128

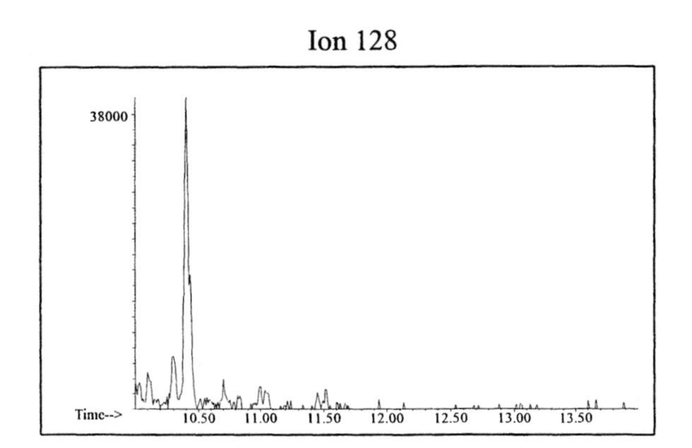

Ion 142

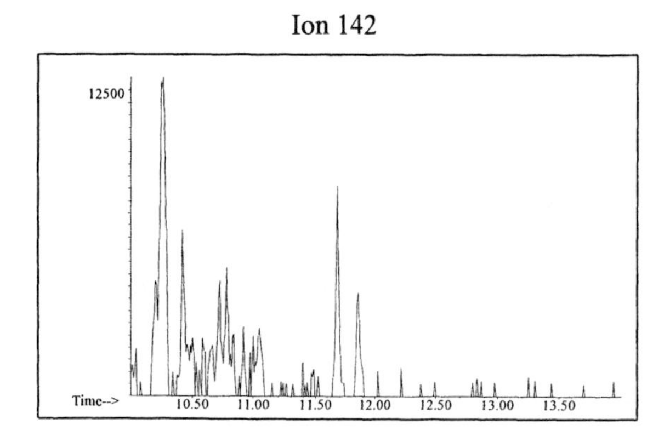

Ion 156

No Useful Data Obtained

Miscellaneous #15

20uL/mL Pentane | ASTM: Class 0 (Miscellaneous) | Macro Code: L

Product Displayed: Penn Champ Impvovco Charcoal Lighter

Product Uses: Charcoal starter

Other Similar Products:

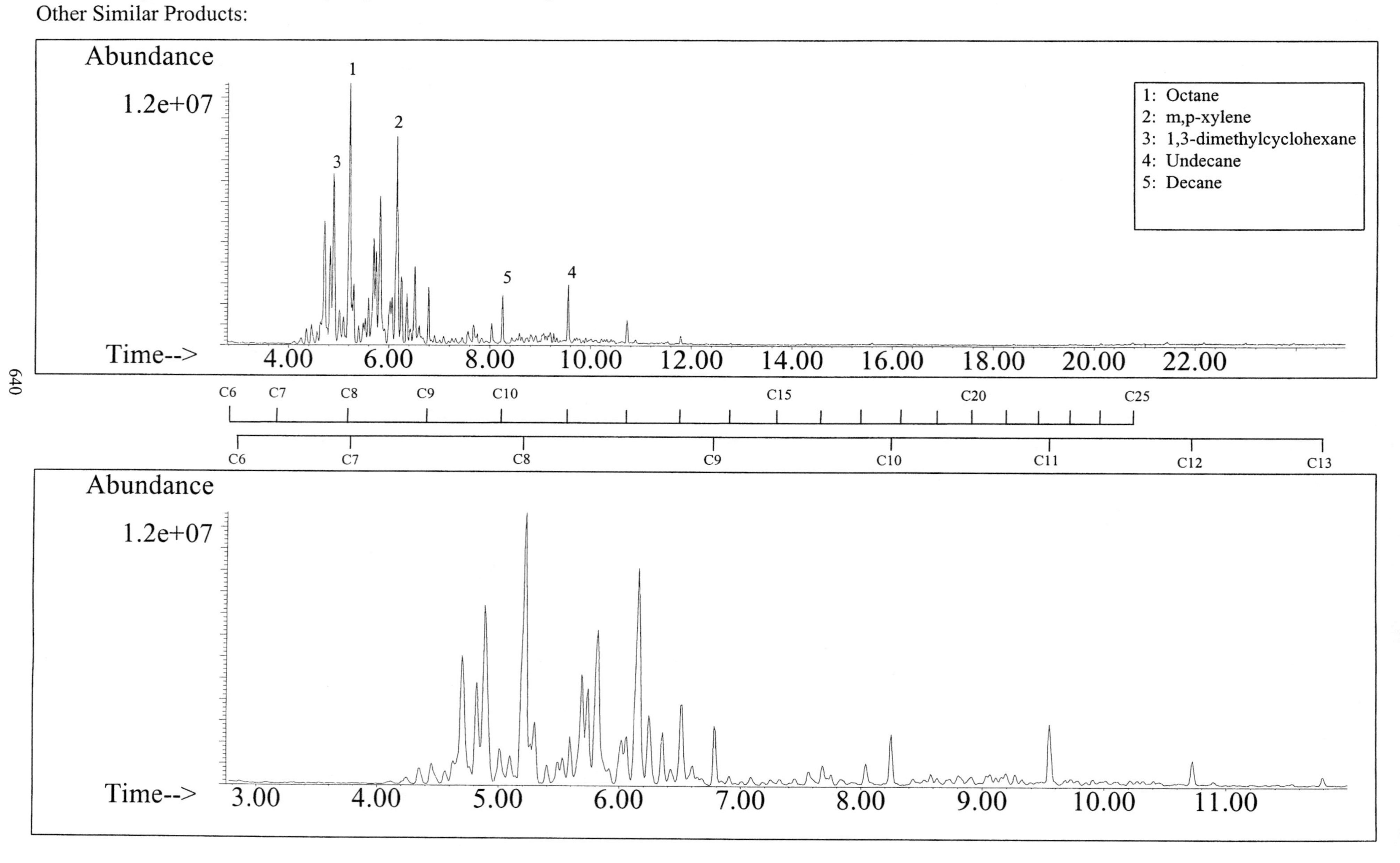

ALKANES

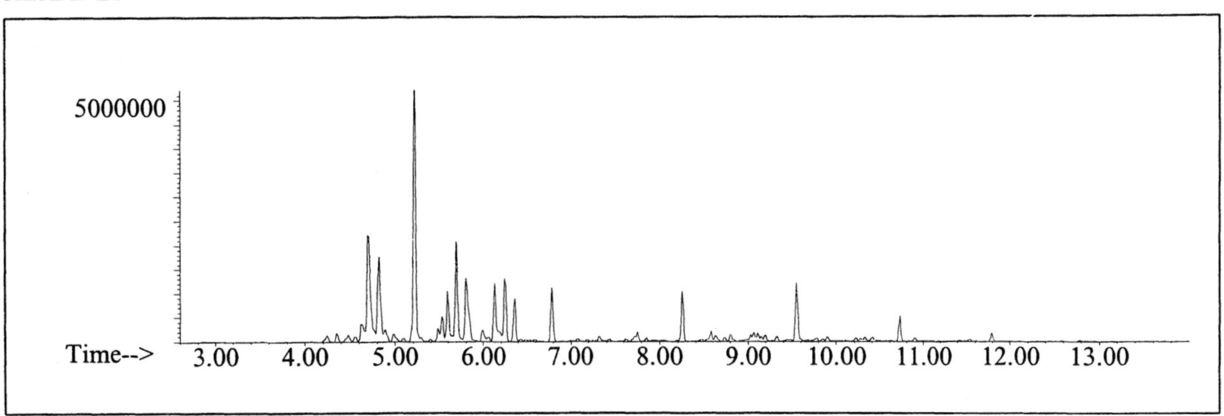

AROMATICS

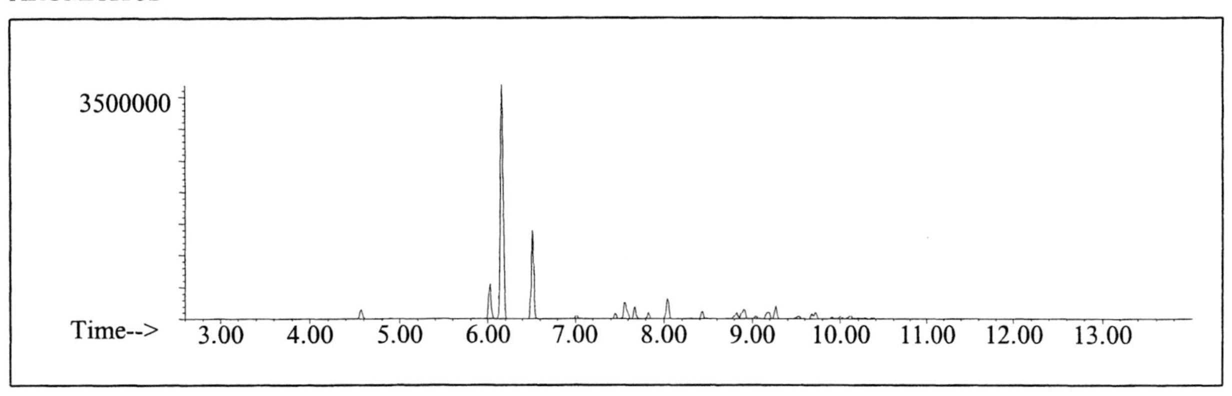

CYCLOPARAFFINS AND ALKENES

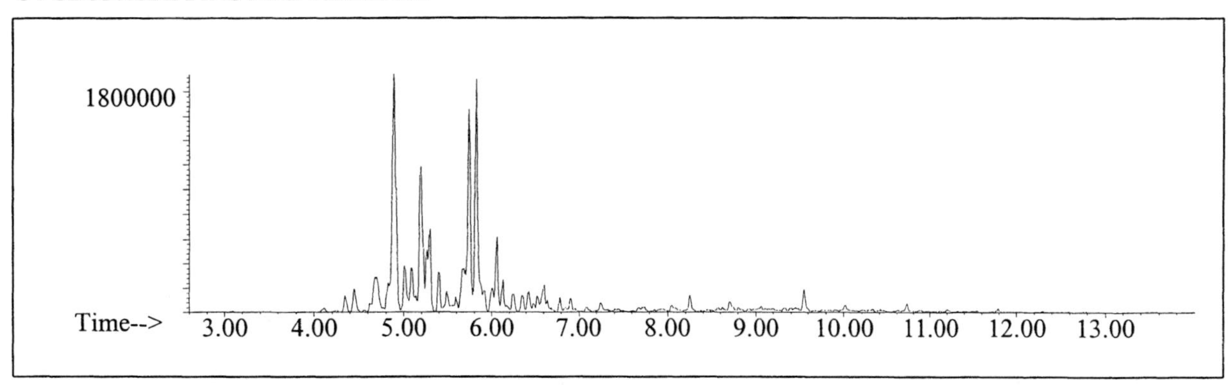

NAPHTHALENES

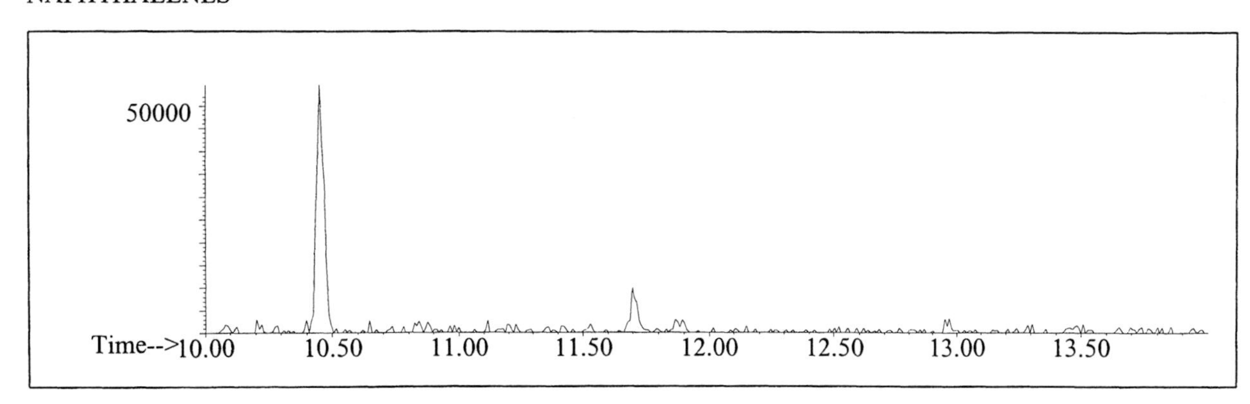

Alkane

Ion 43

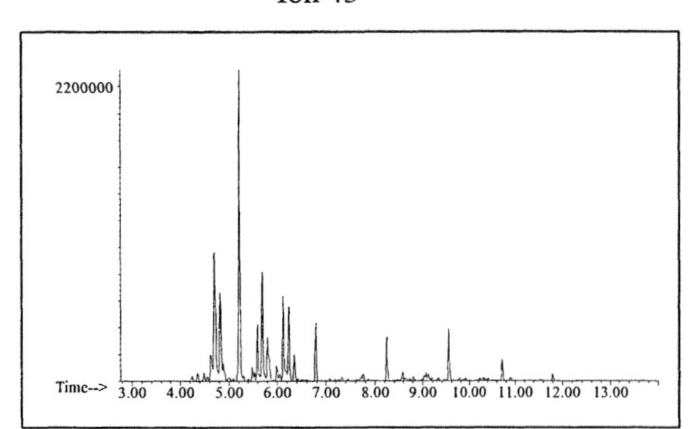

Ion 57

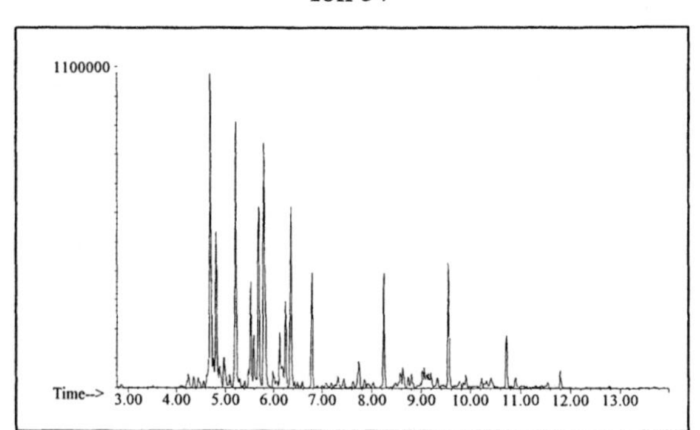

Ion 71

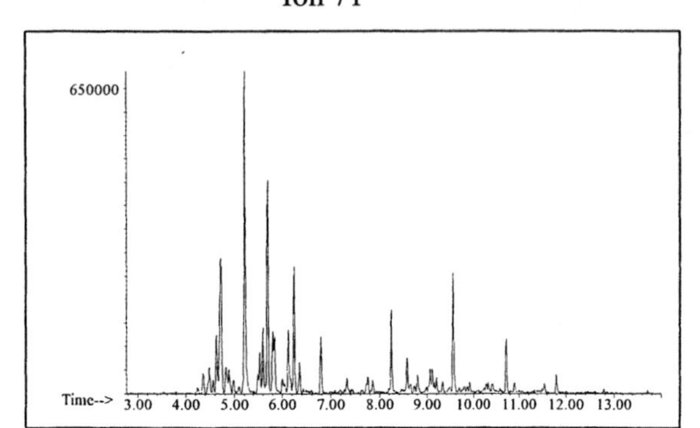

Ion 85

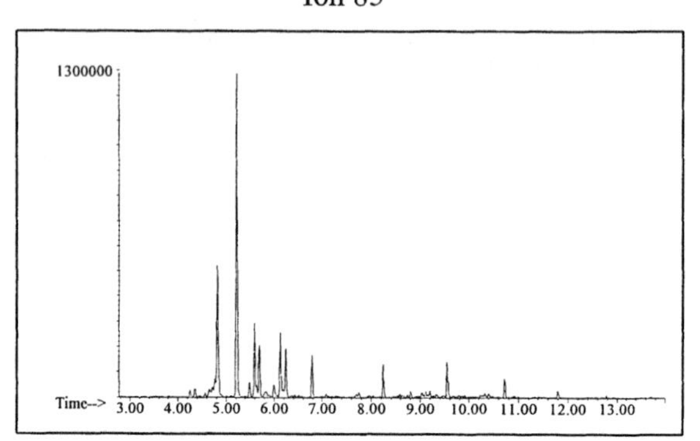

Aromatic

Ion 91

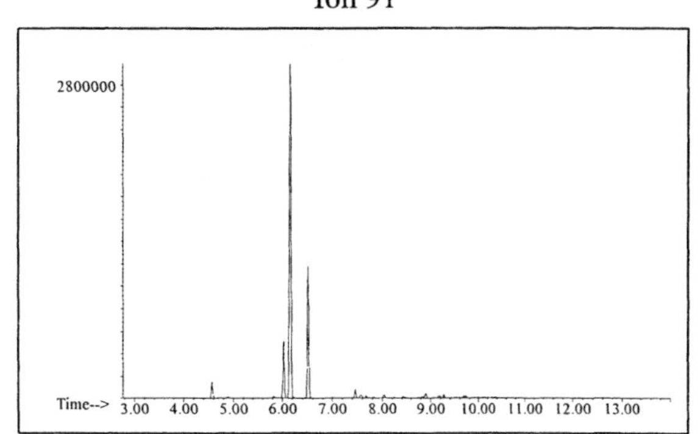

Ion 105

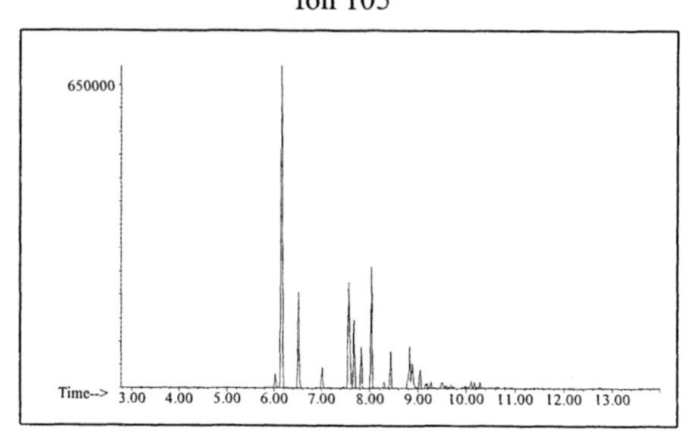

Ion 119

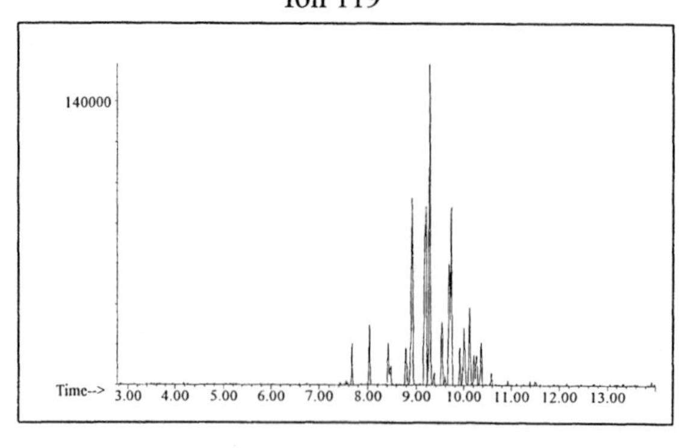

Cycloparaffin

Ion 55

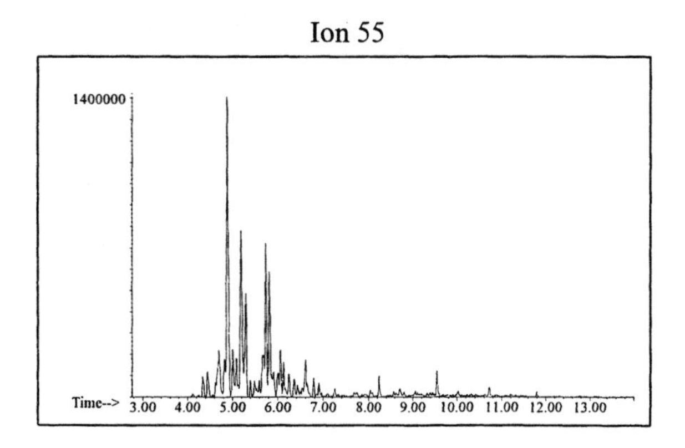

Ion 69

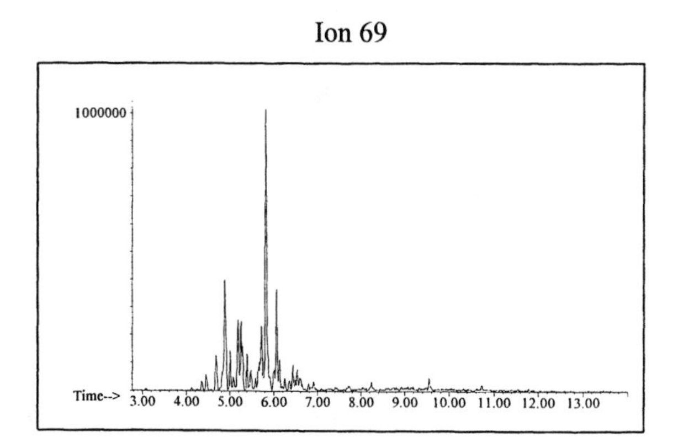

Ion 83

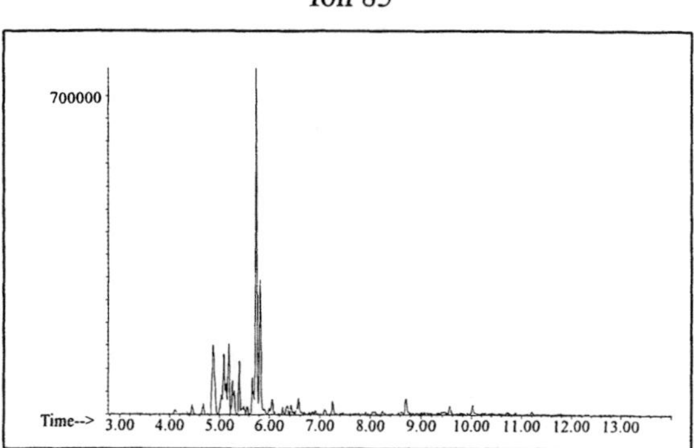

Naphthalene

Ion 128

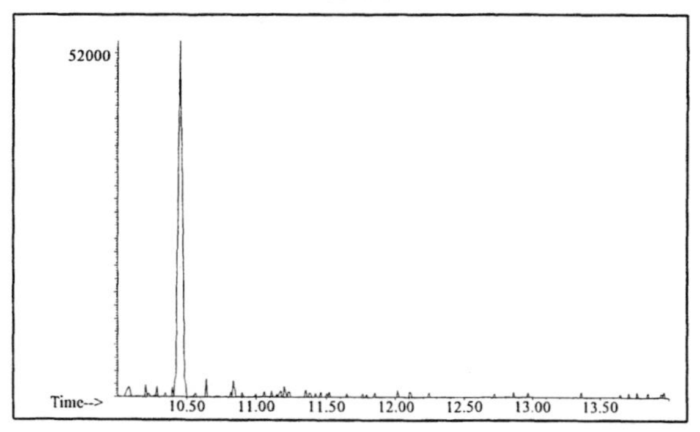

Ion 142

No Useful Data Obtained

Ion 156

No Useful Data Obtained

Miscellaneous #16

20uL/mL Pentane

Product Displayed: RXP Gas Dynamizer

Other Similar Products:

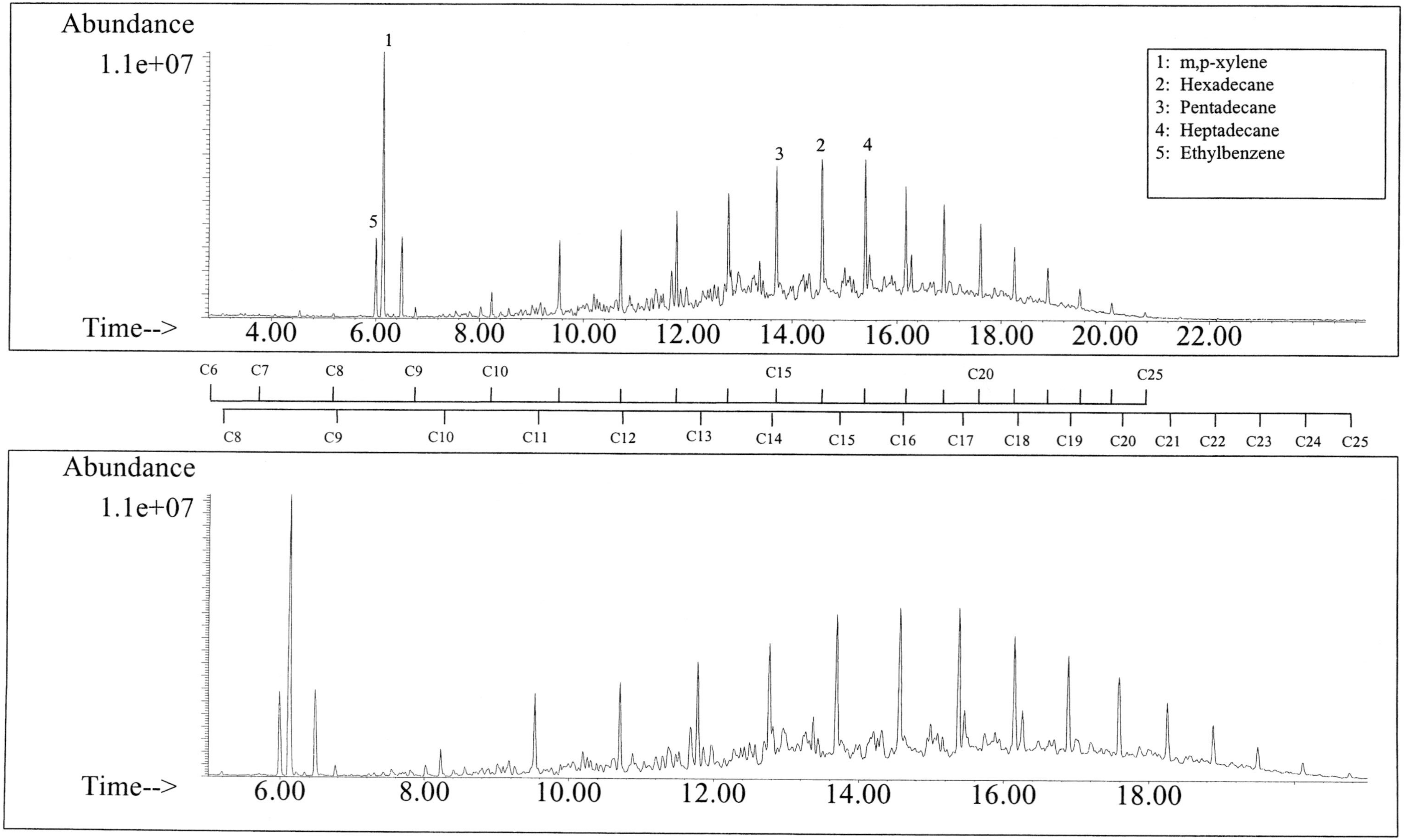

ASTM: Class 0 (Miscellaneous)

Product Uses: Fuel treatment

Macro Code: H

ALKANES

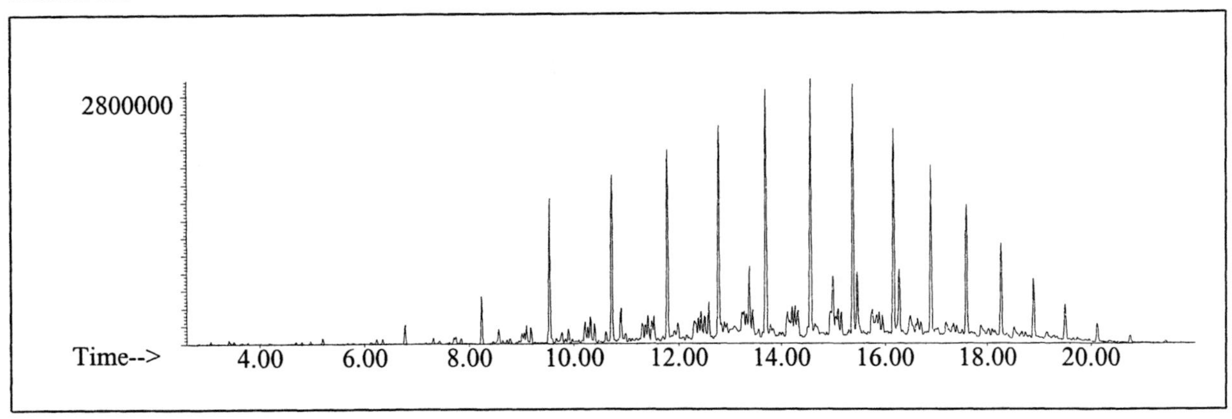

AROMATICS

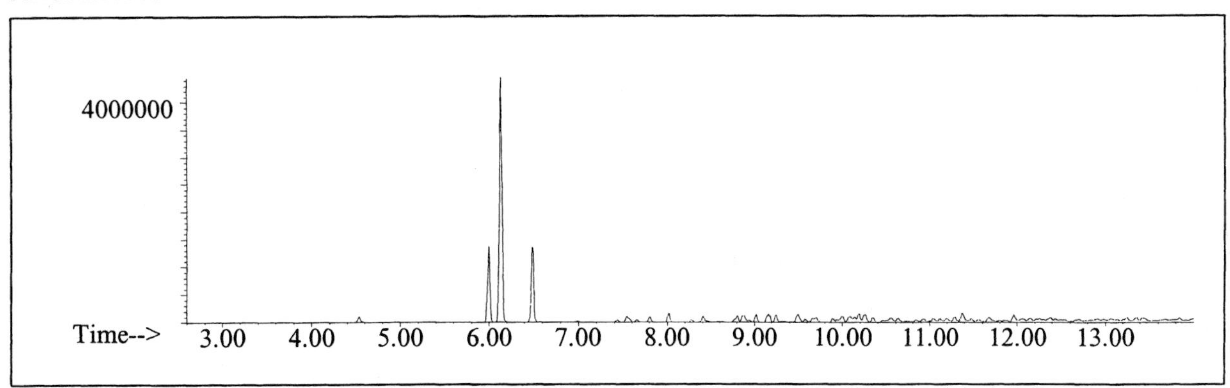

CYCLOPARAFFINS AND ALKENES

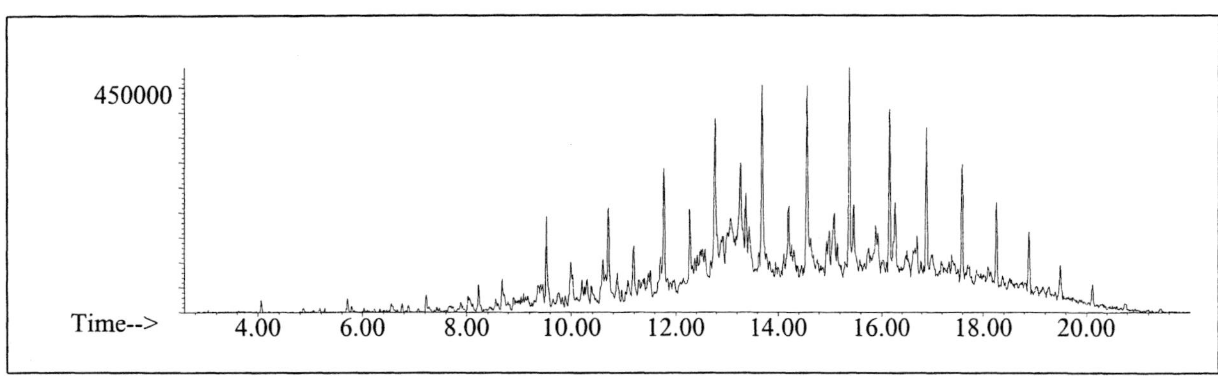

NAPHTHALENES

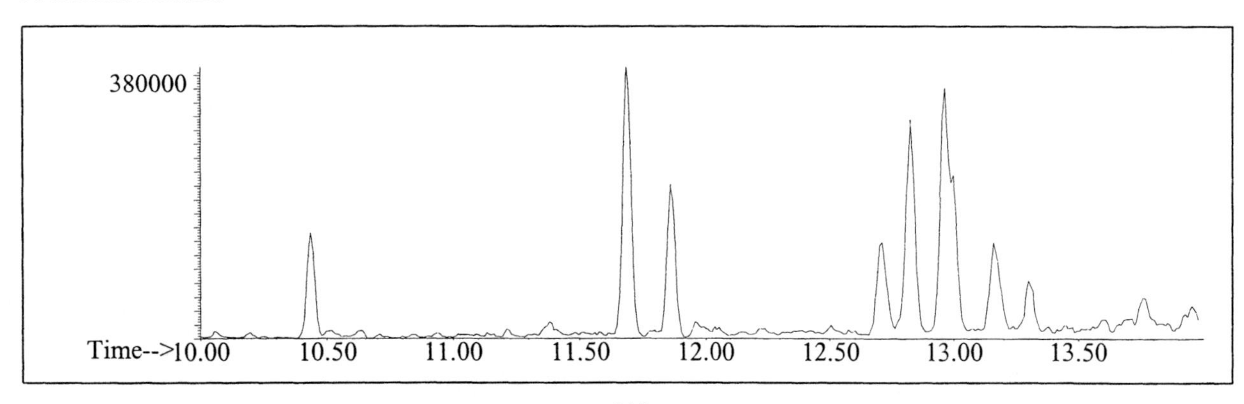

Alkane

Ion 43

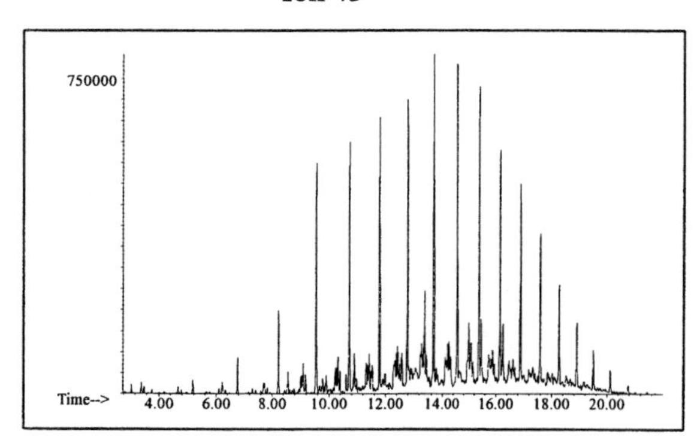

Ion 57

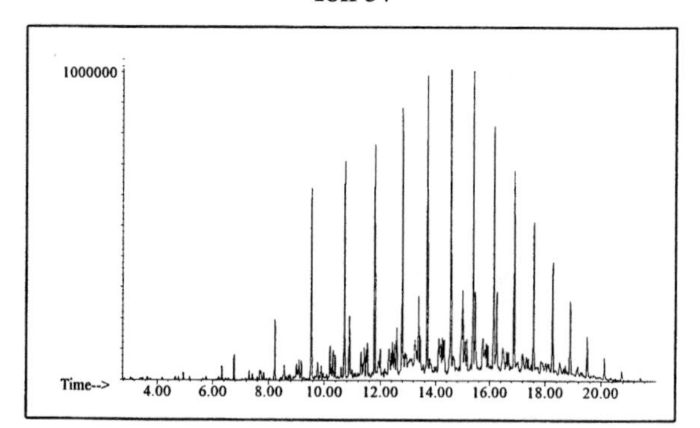

Ion 71

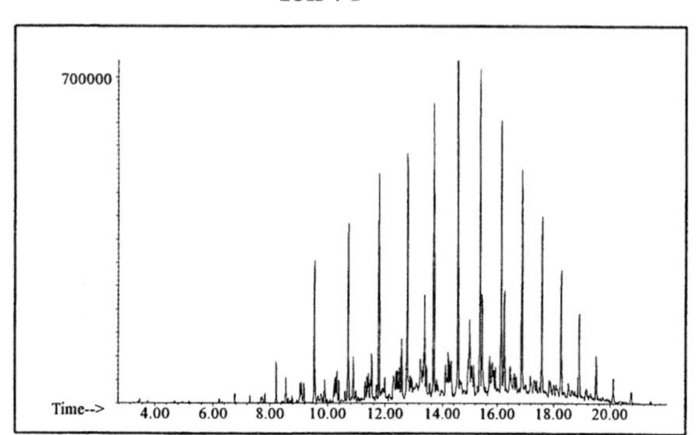

Ion 85

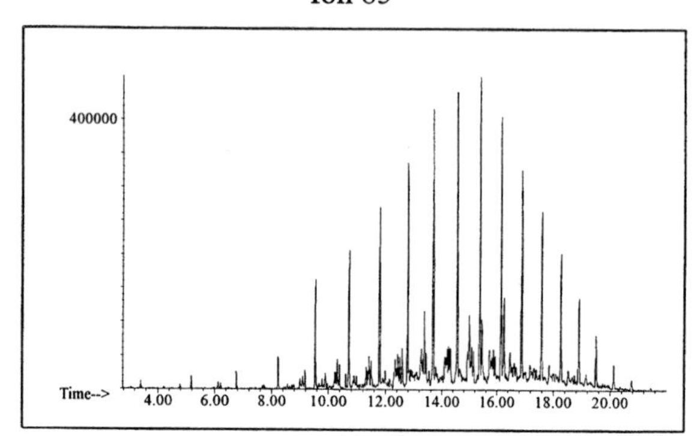

Aromatic

Ion 91

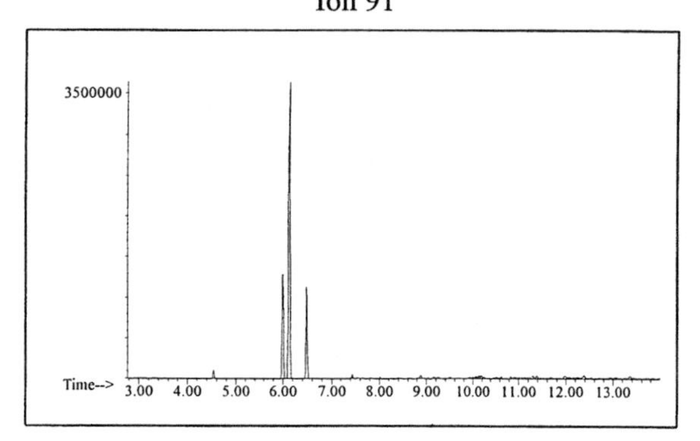

Ion 105

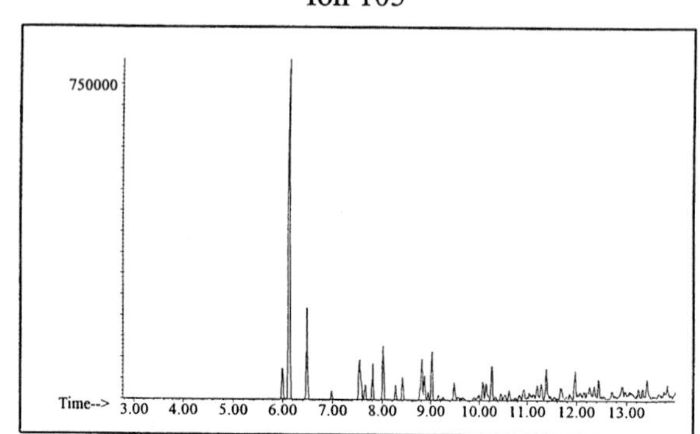

Ion 119

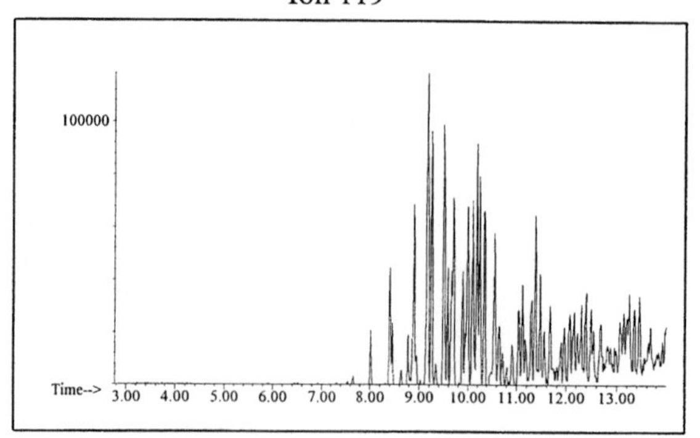

Cycloparaffin

Ion 55

Ion 69

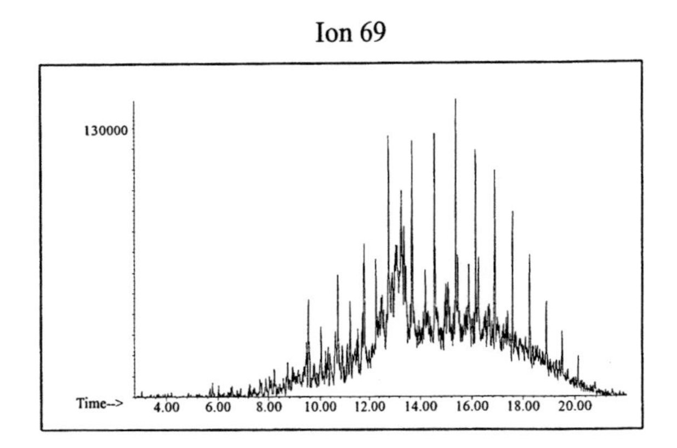

Ion 83

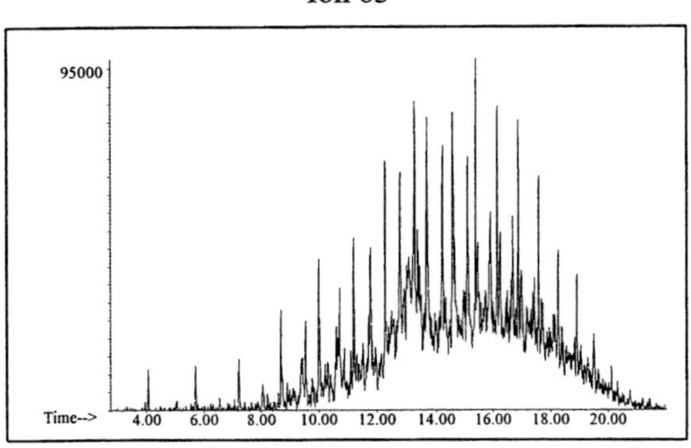

Naphthalene

Ion 128

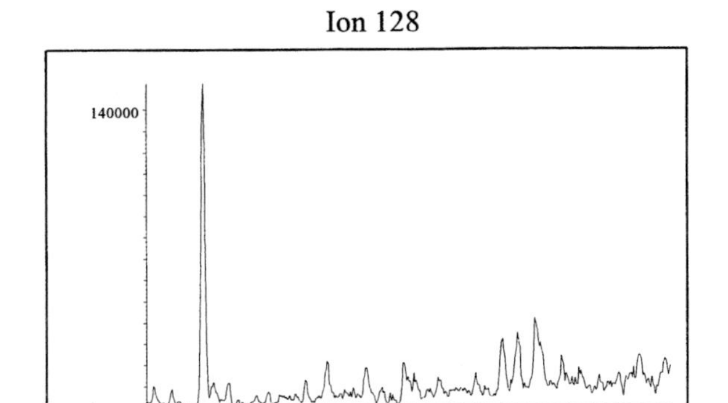

Ion 142

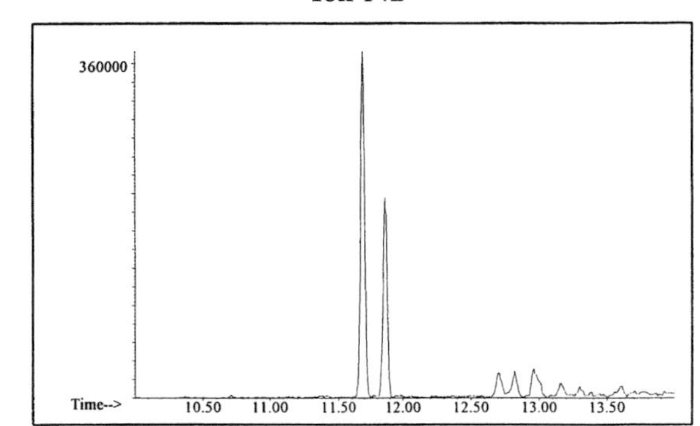

Ion 156

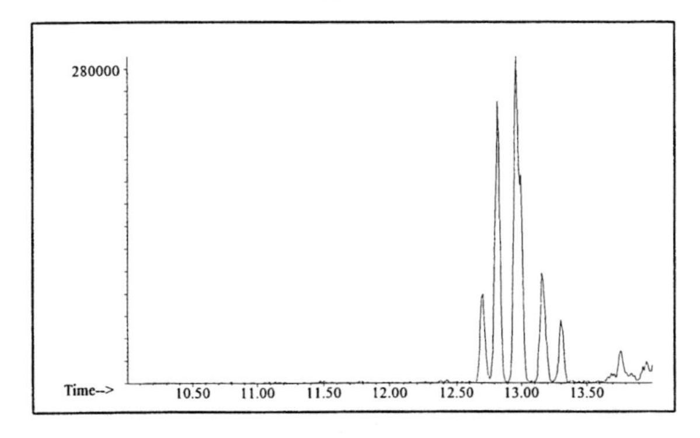

Miscellaneous #17

20uL/mL Pentane

Product Displayed: 3M Adhesive Remover/ Cleaner

Other Similar Products:

ASTM: Class 0 (Miscellaneous)

Product Uses: Cleaning solvent

Macro Code: L

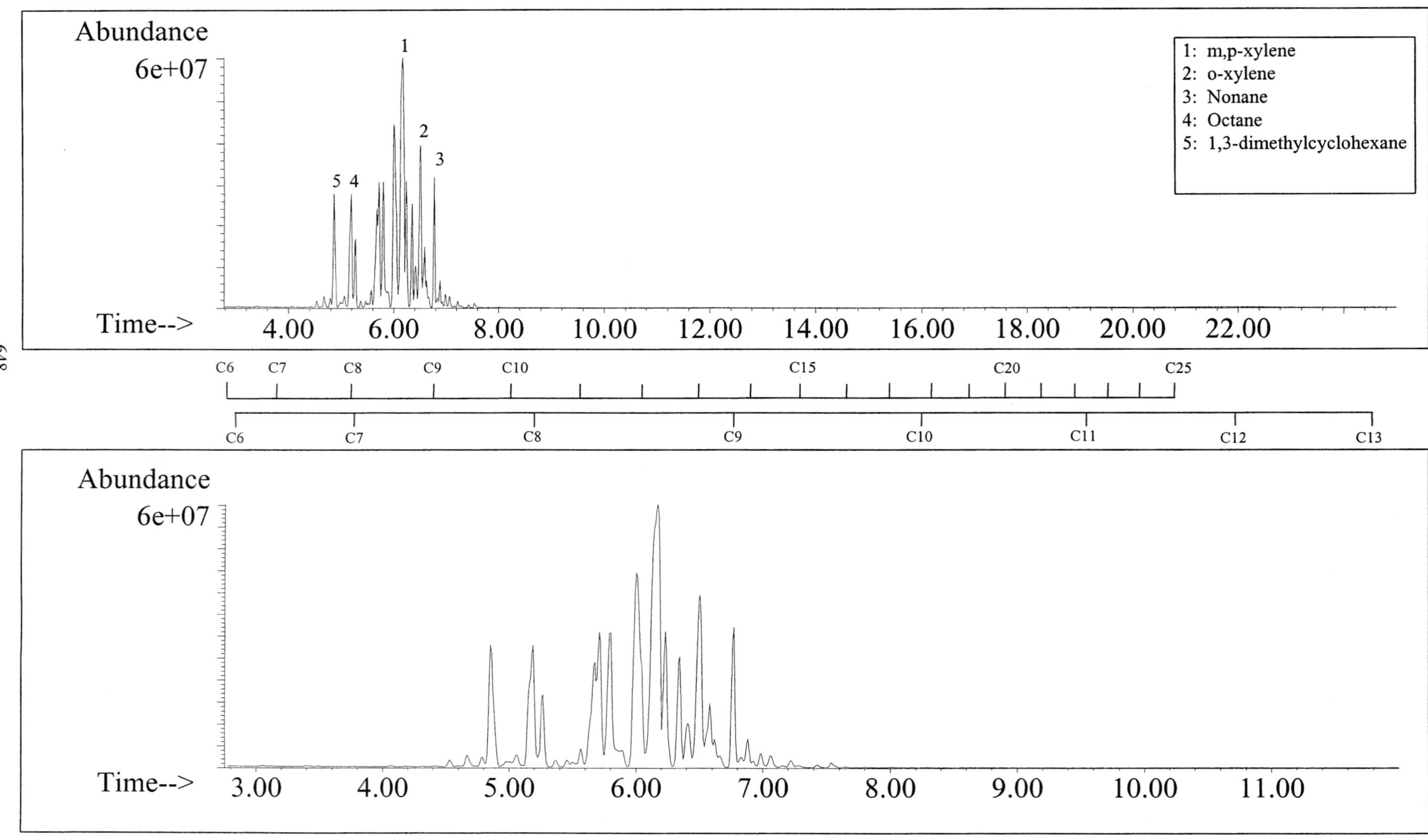

ALKANES

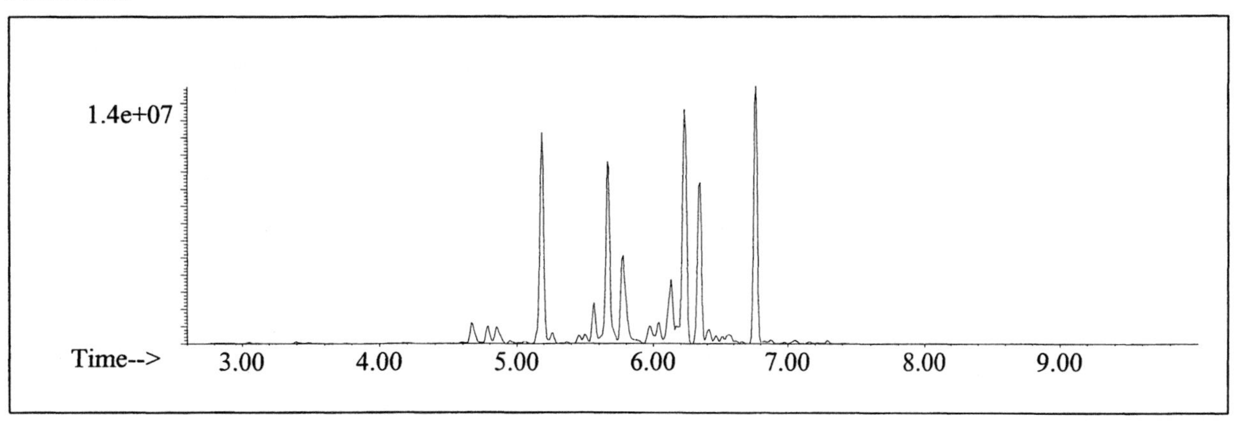

AROMATICS

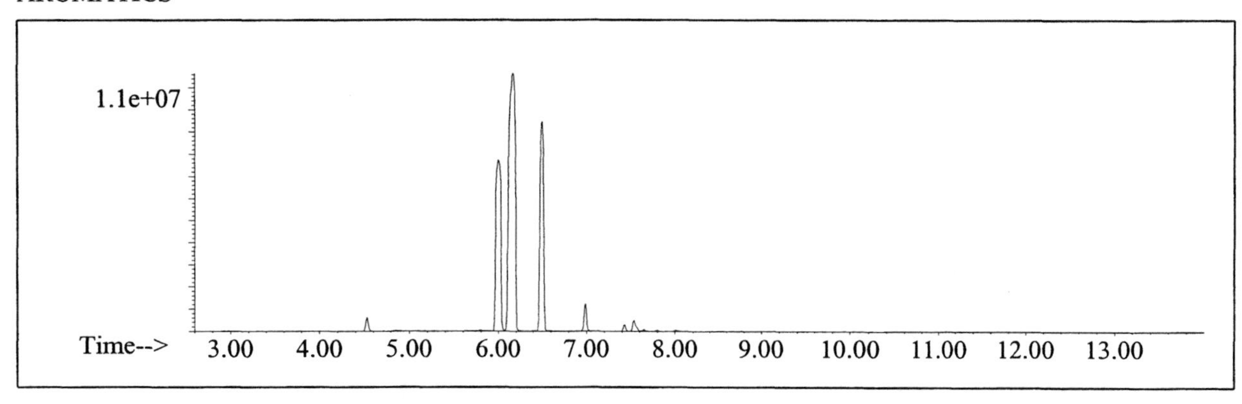

CYCLOPARAFFINS AND ALKENES

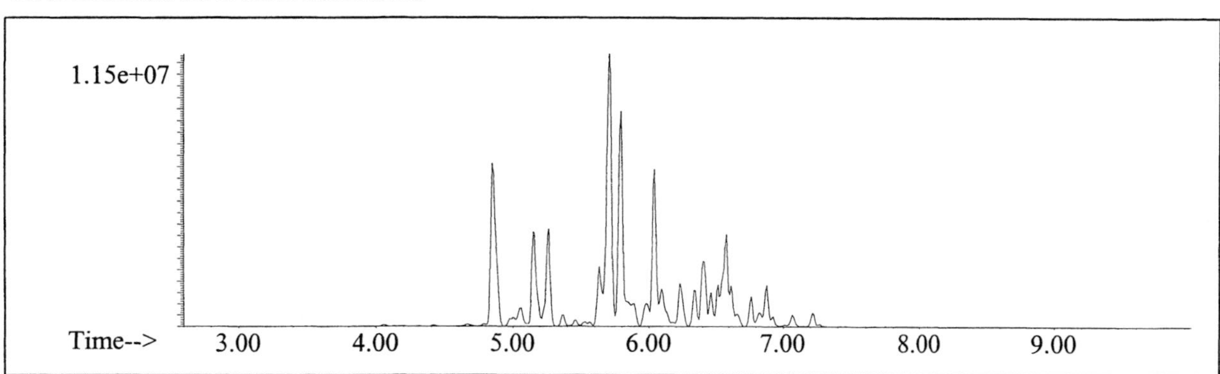

NAPHTHALENES

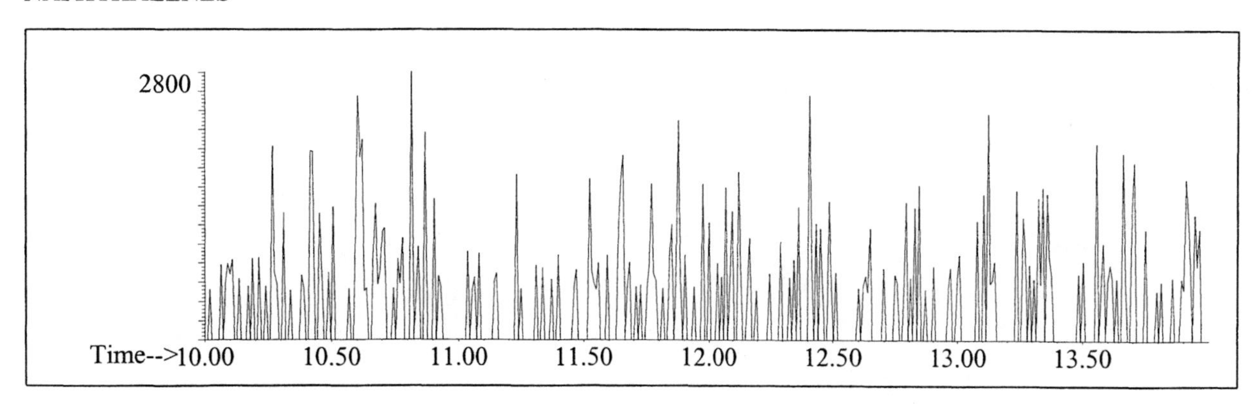

INDIVIDUAL PROFILES

Alkane

Ion 43

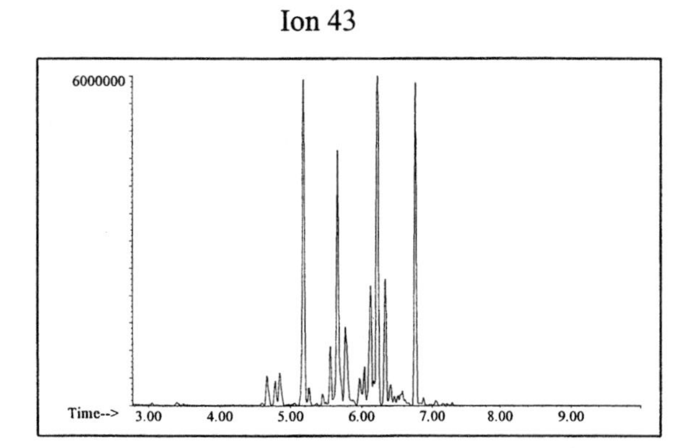

Ion 57

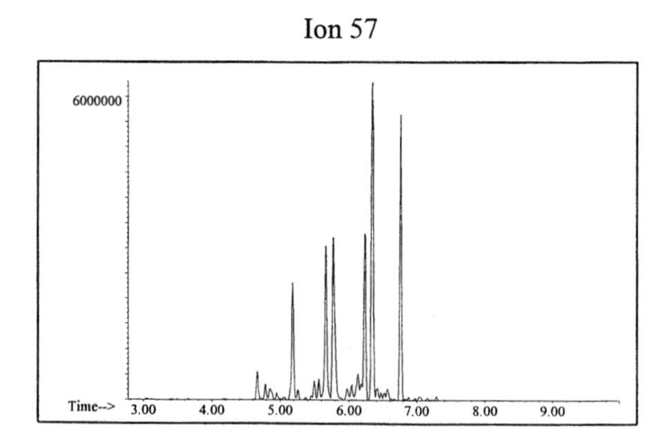

Ion 71

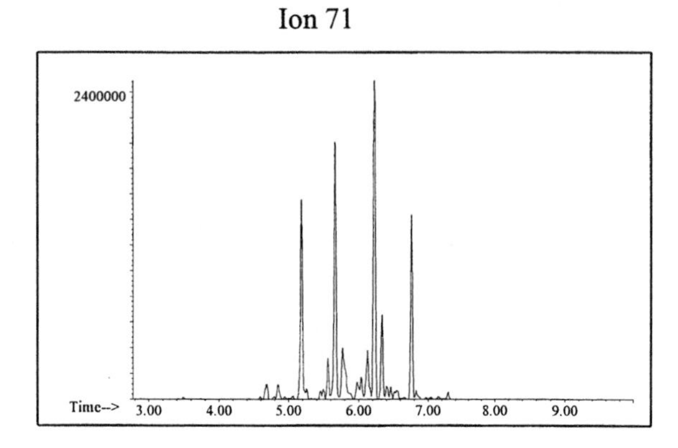

Ion 85

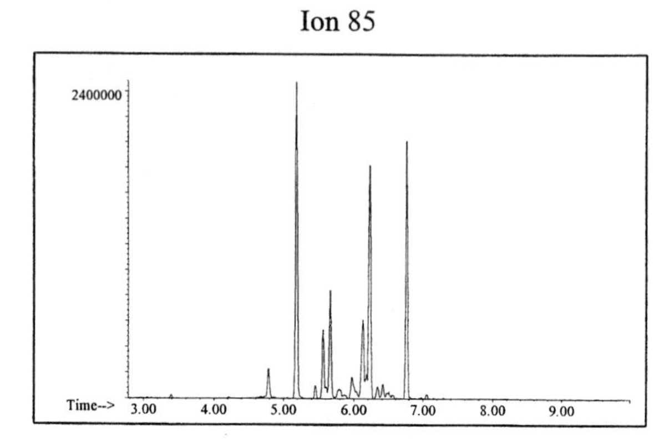

Aromatic

Ion 91

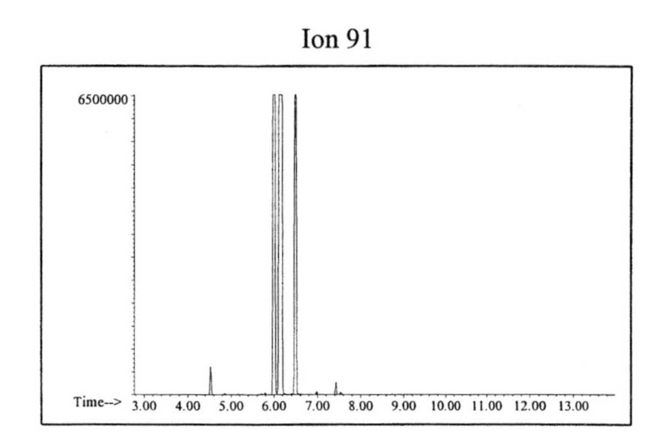

Ion 105

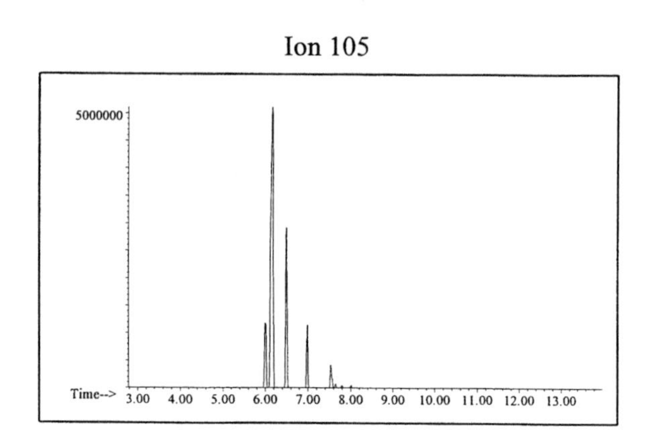

Ion 119

No Useful Data Obtained

Cycloparaffin

Ion 55

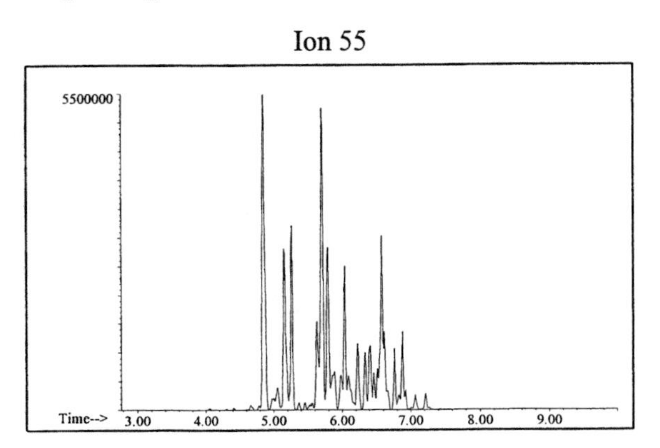

Ion 69

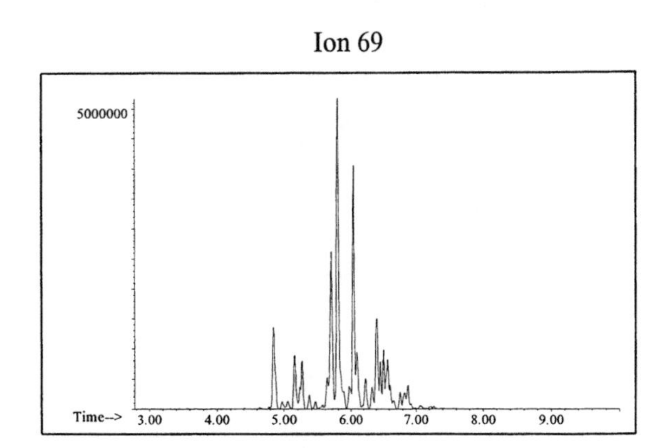

Ion 83

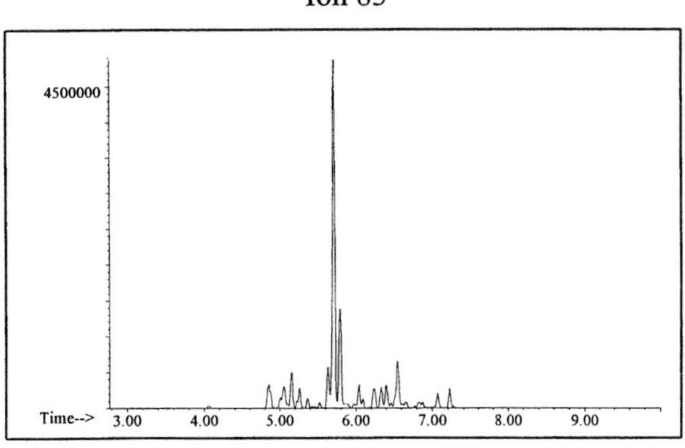

Naphthalene

Ion 128

No Useful Data Obtained

Ion 142

No Useful Data Obtained

Ion 156

No Useful Data Obtained

Miscellaneous #18

20uL/mL Pentane

ASTM: Class 0 (Miscellaneous)

Macro Code: M

Product Displayed: Power Diesel Fuel Substitute

Product Uses: Fuel supplement

Other Similar Products:

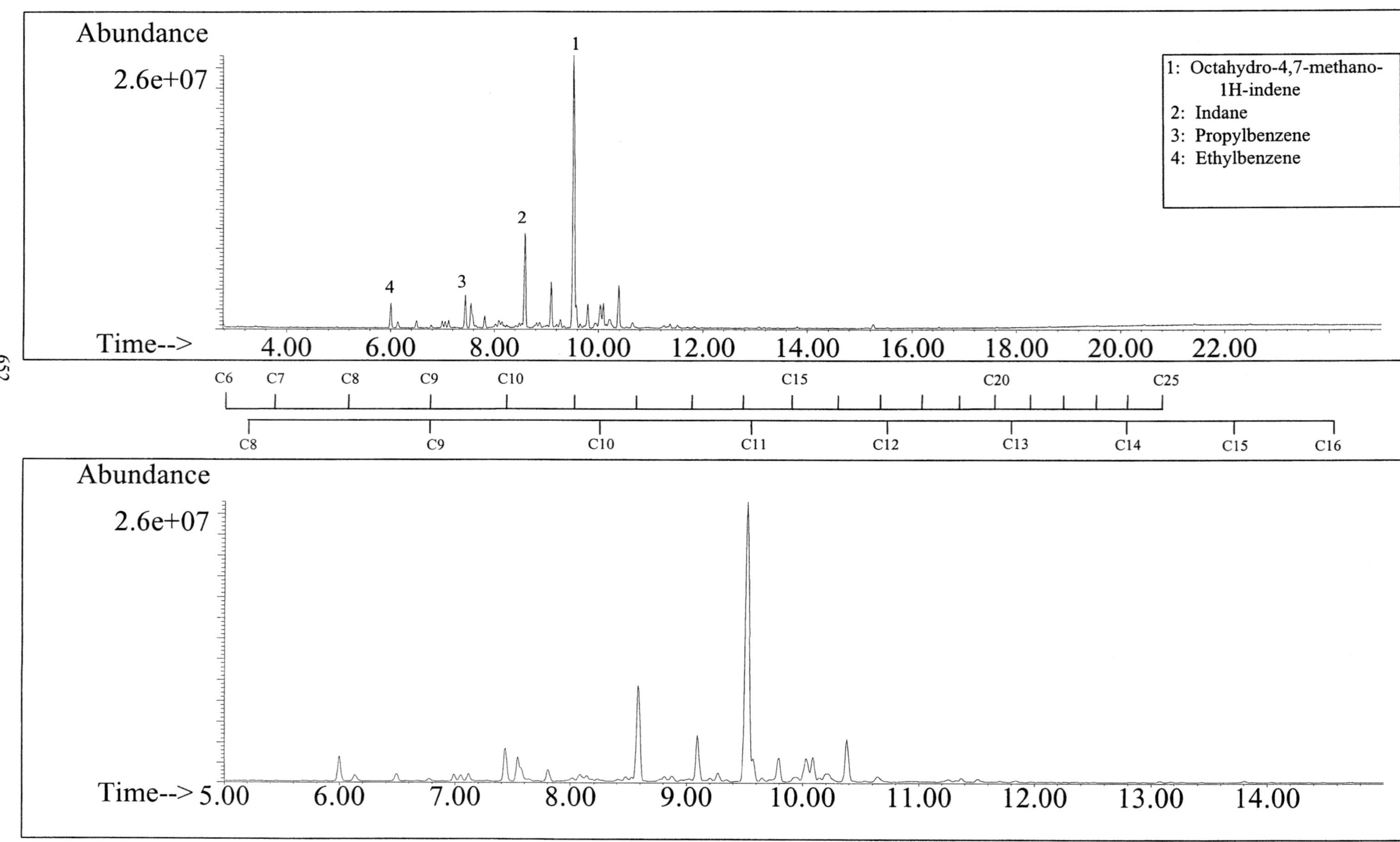

ALKANES

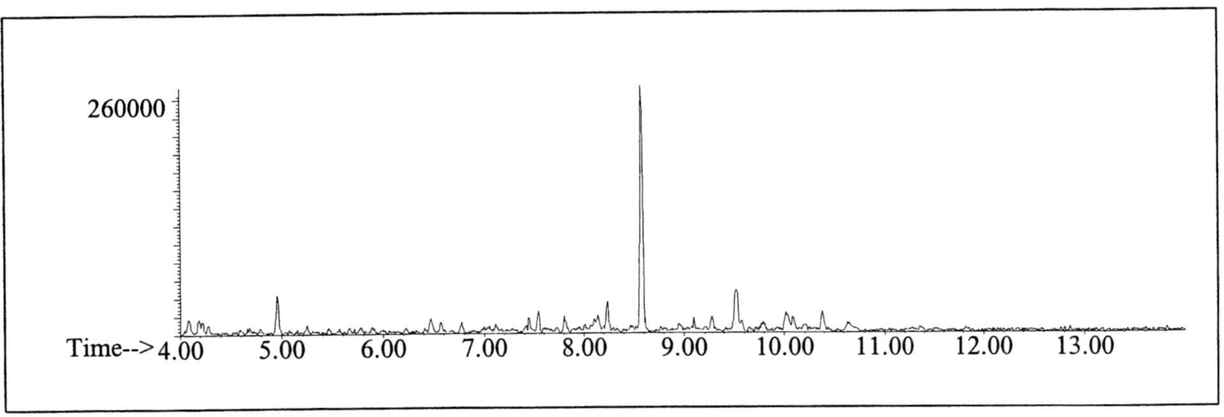

AROMATICS

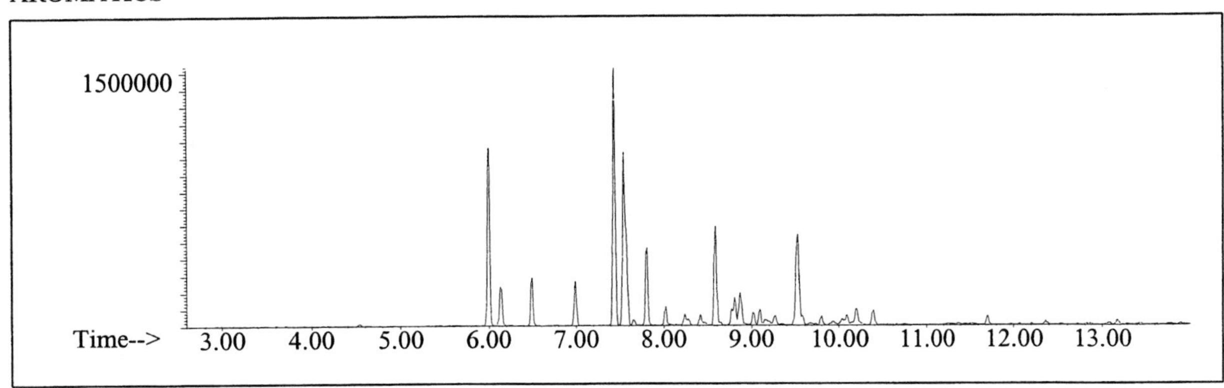

CYCLOPARAFFINS AND ALKENES

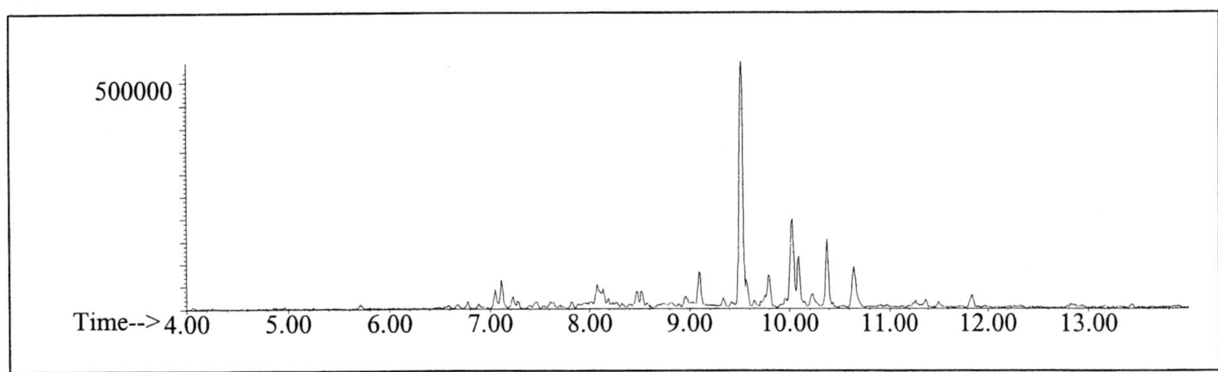

NAPHTHALENES

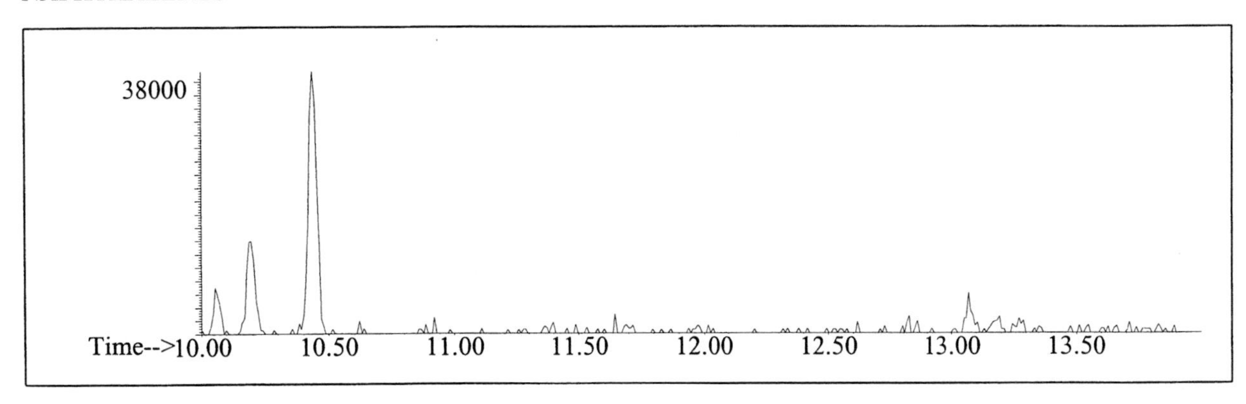

INDIVIDUAL PROFILES

Miscellaneous #18

Alkane

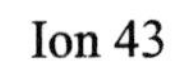

Ion 43

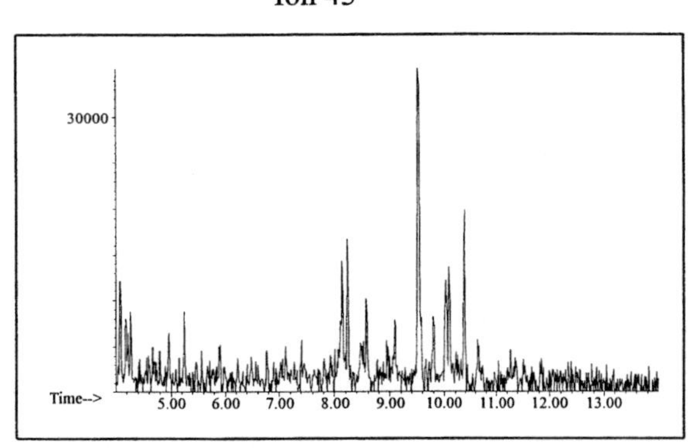

Ion 57

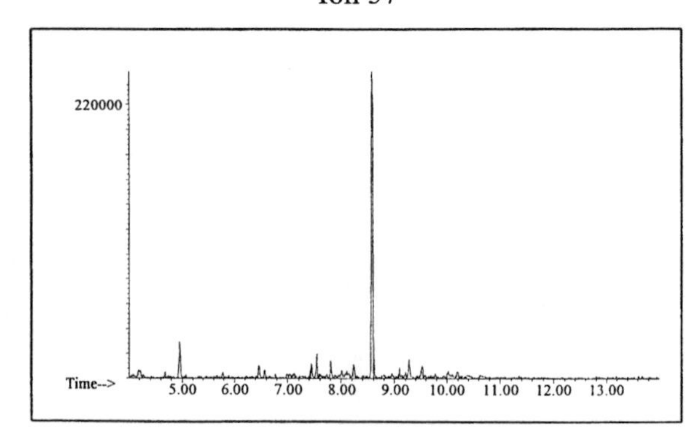

Ion 71

No Useful Data Obtained

Ion 85

No Useful Data Obtained

Aromatic

Ion 91

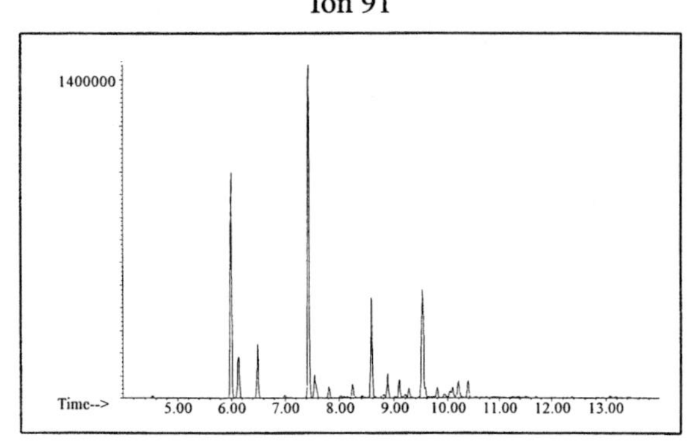

Ion 105

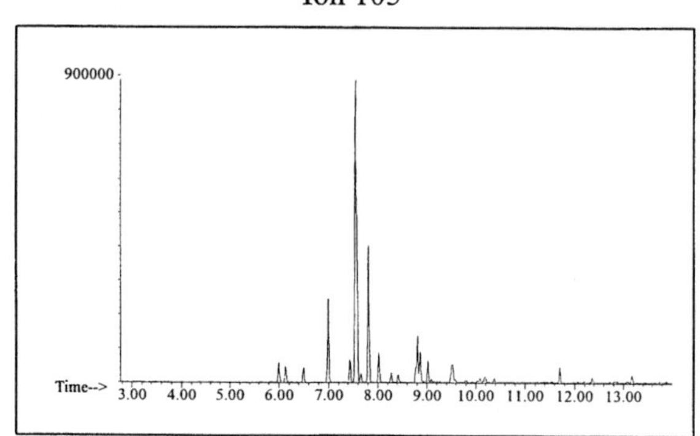

Ion 119

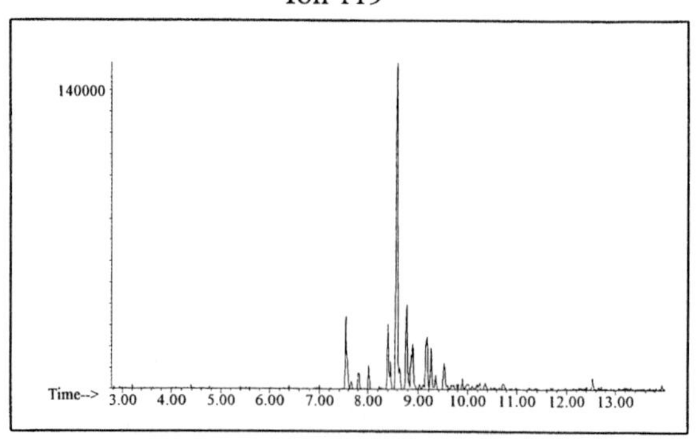

INDIVIDUAL PROFILES

Miscellaneous #18

Cycloparaffin

Ion 55

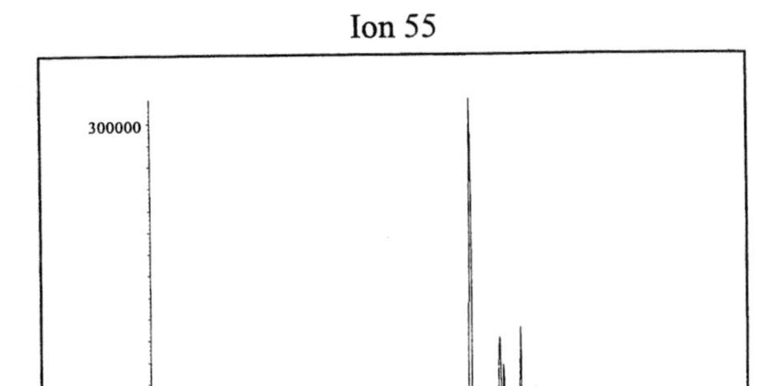

Ion 69

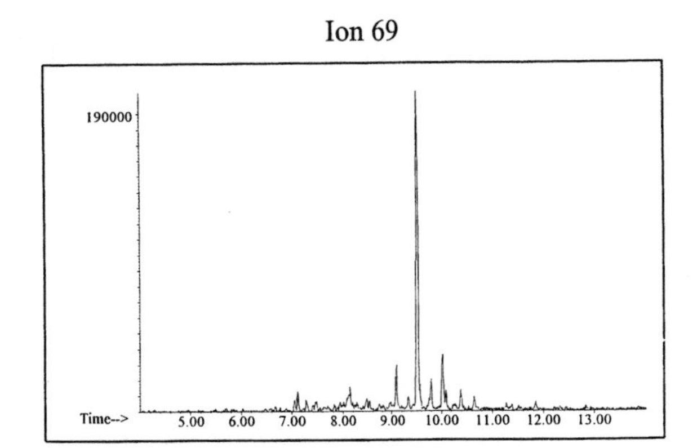

Ion 83

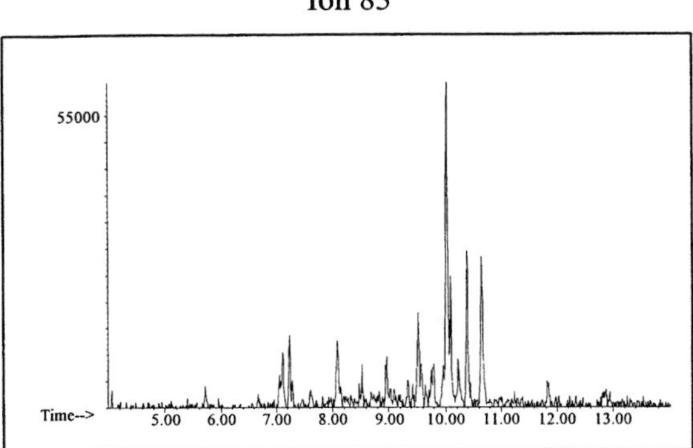

Naphthalene

Ion 128

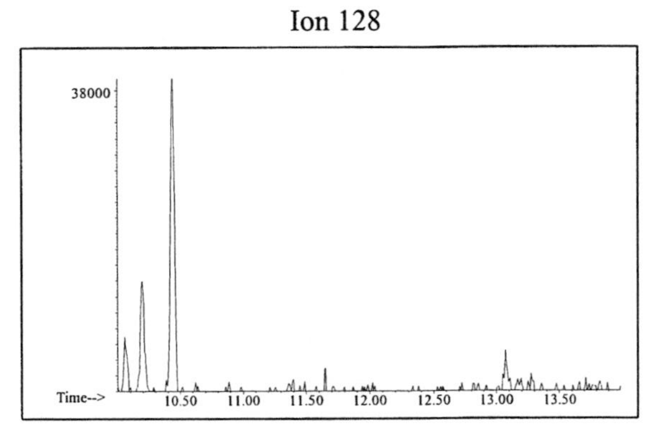

Ion 142

No Useful Data Obtained

Ion 156

No Useful Data Obtained

Miscellaneous #19

20uL/mL Pentane

Product Displayed: Sears Tripolene Thinner

Other Similar Products:

ASTM: Class 0 (Miscellaneous)

Product Uses: Paint thinner

Macro Code: M

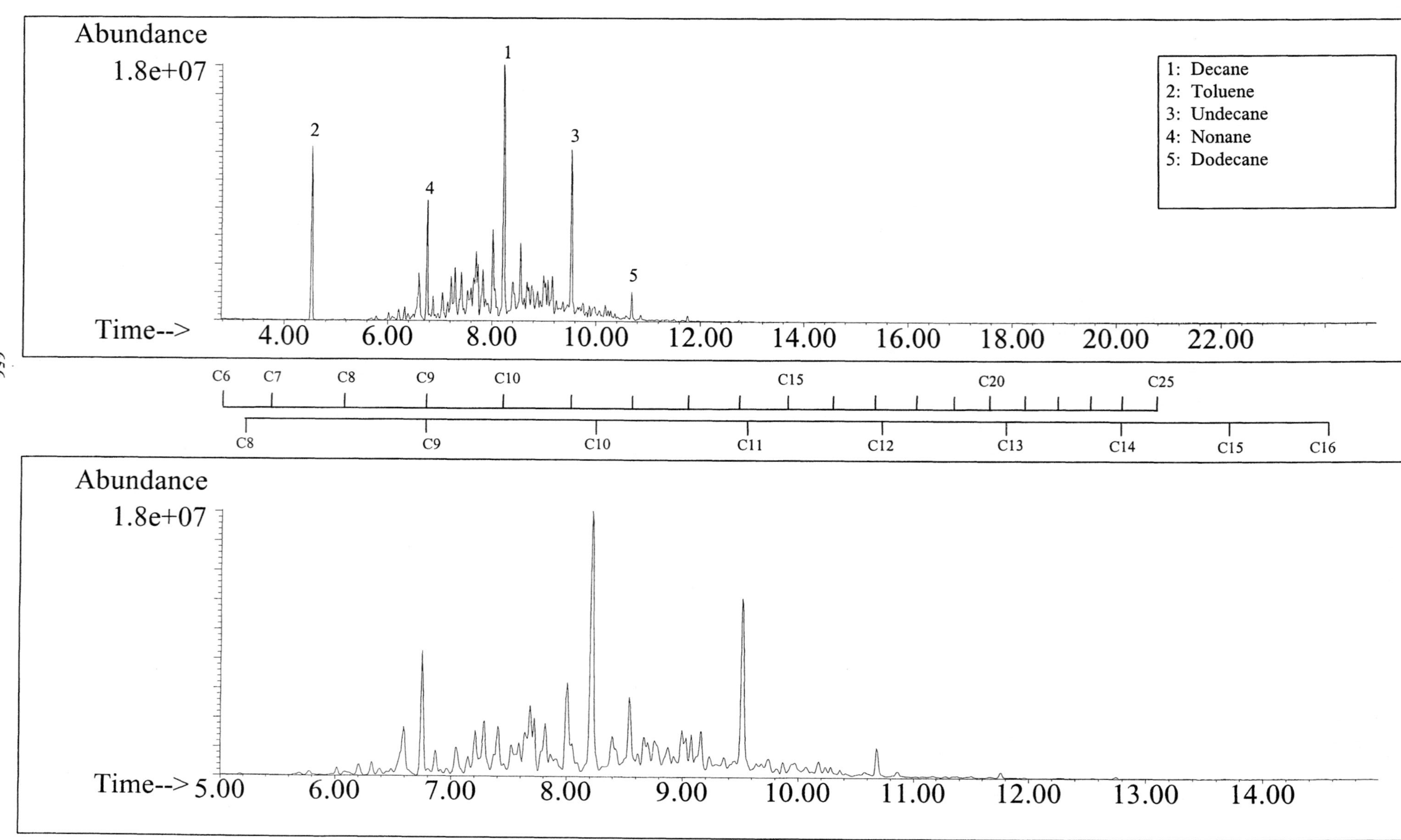

ALKANES

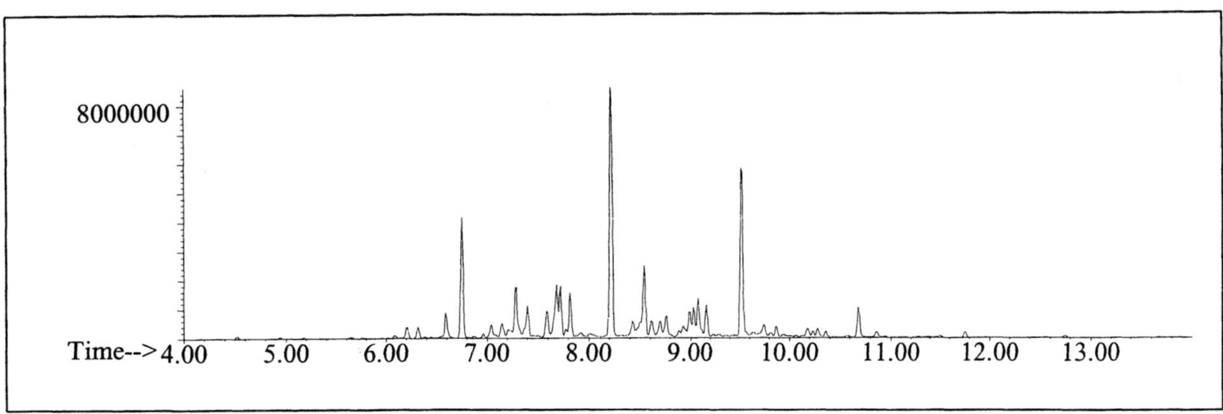

AROMATICS

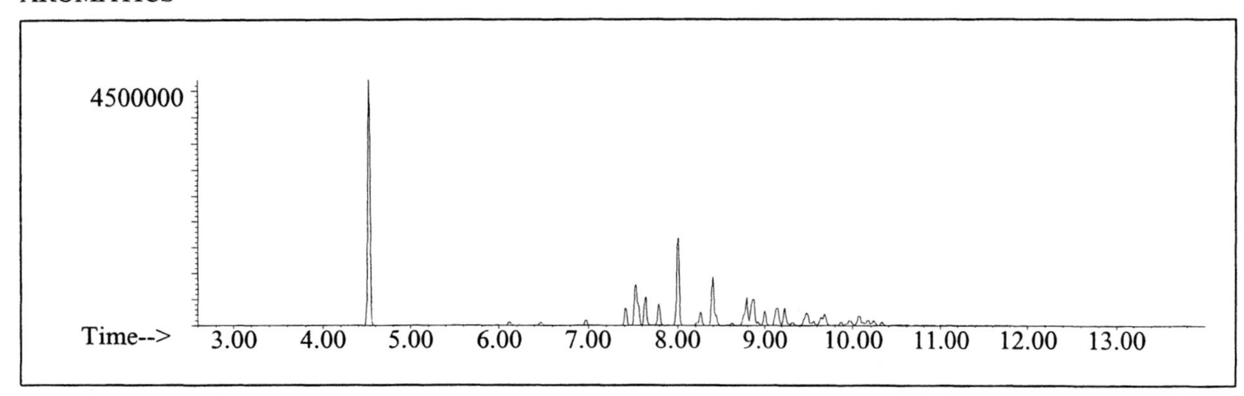

CYCLOPARAFFINS AND ALKENES

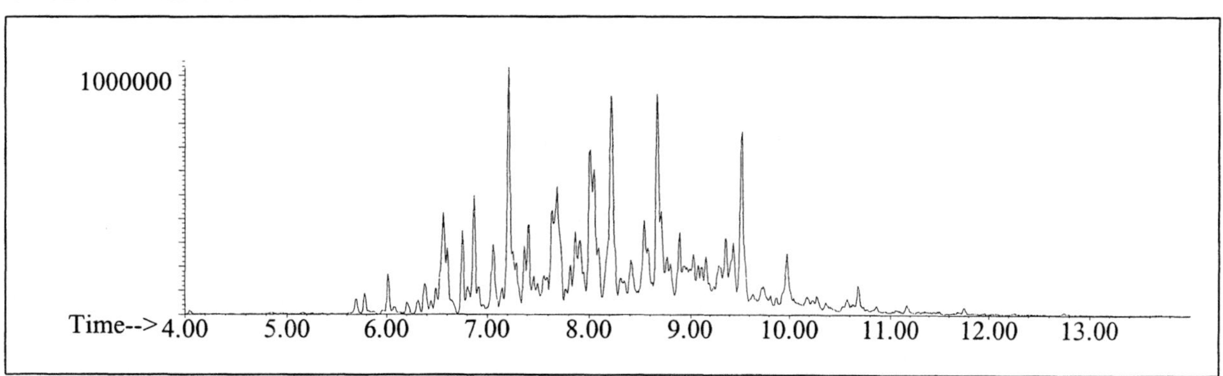

NAPHTHALENES

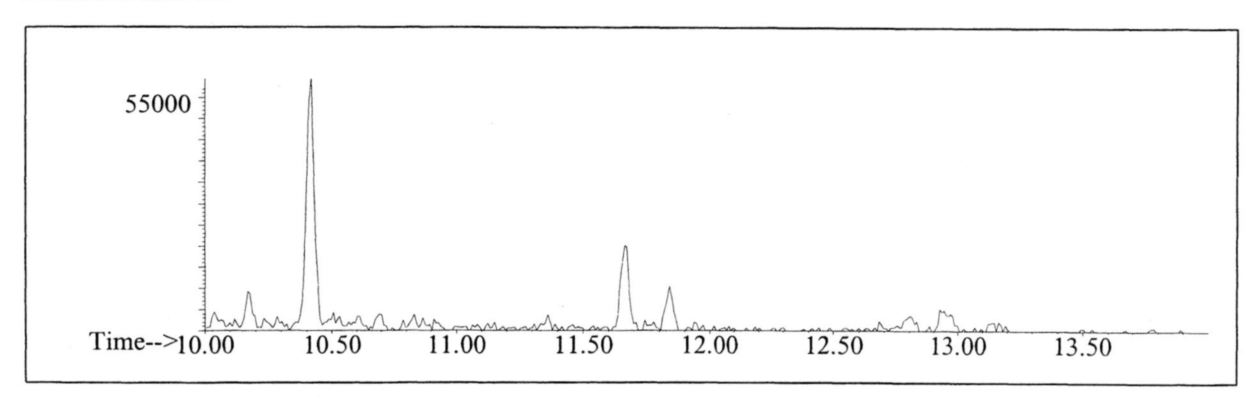

Alkane

Ion 43

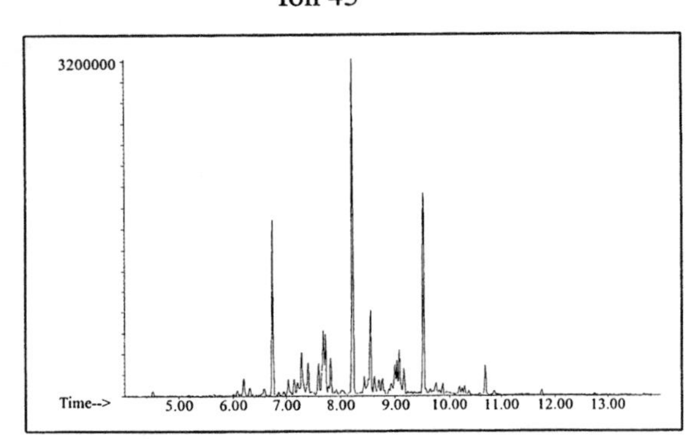

Ion 57

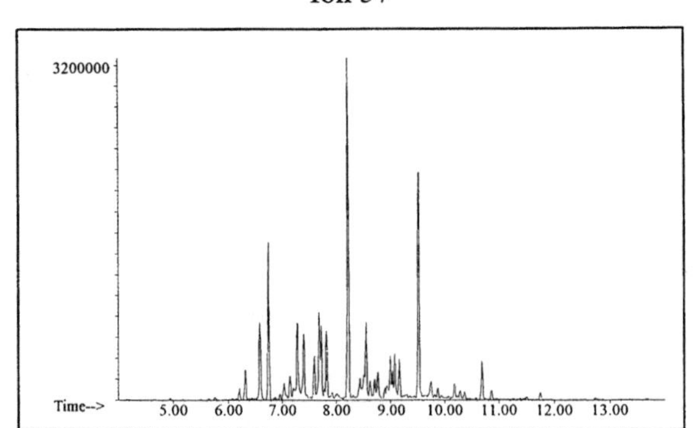

Ion 71

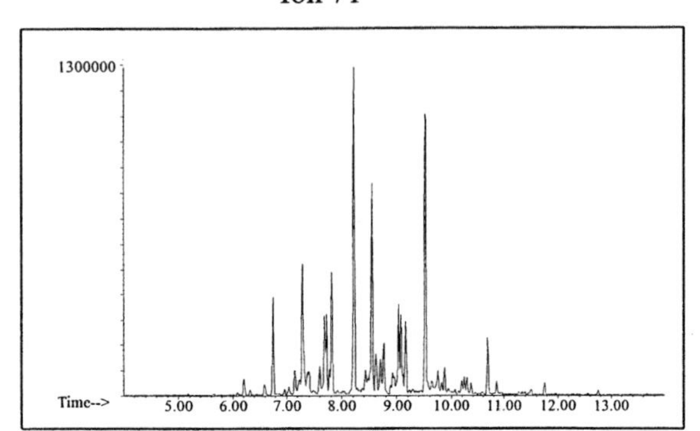

Ion 85

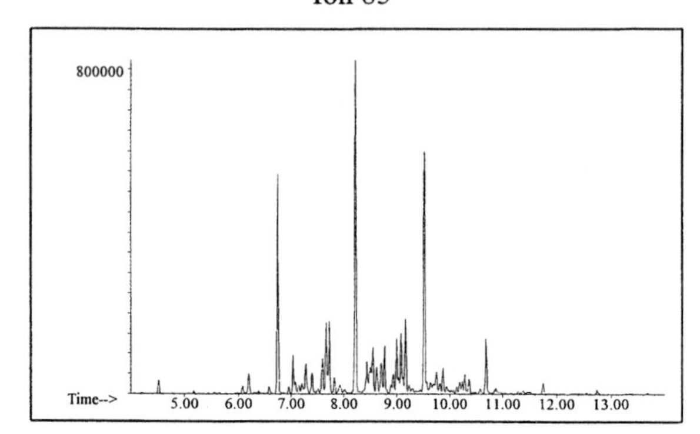

Aromatic

Ion 91

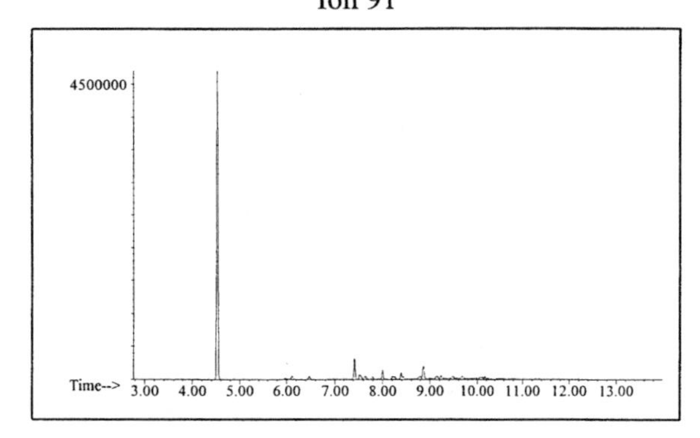

Ion 105

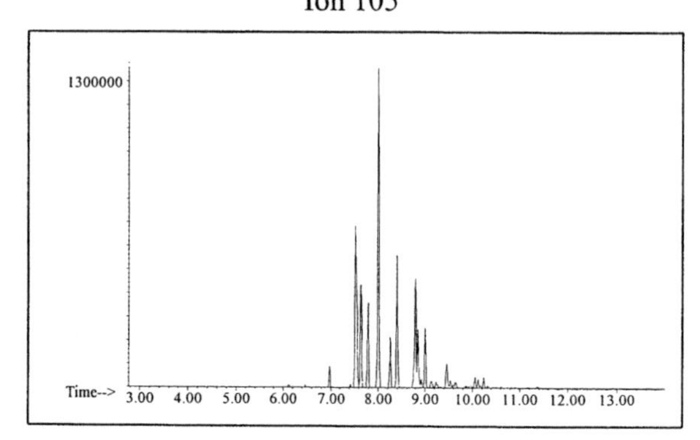

Ion 119

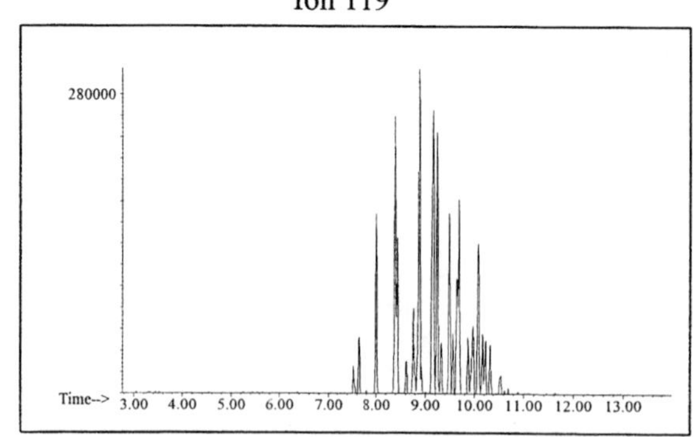

Cycloparaffin

Ion 55

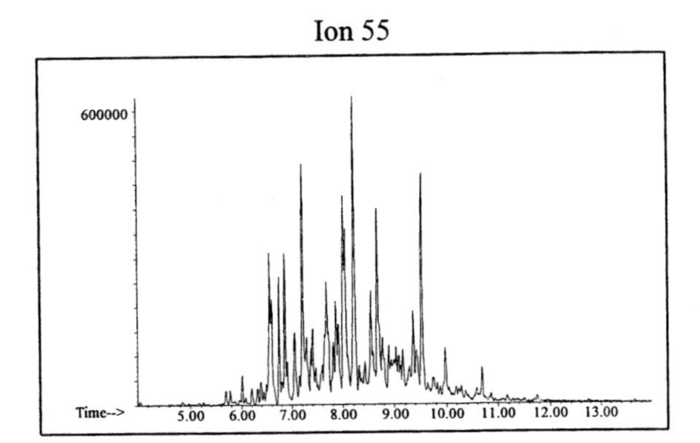

Ion 69

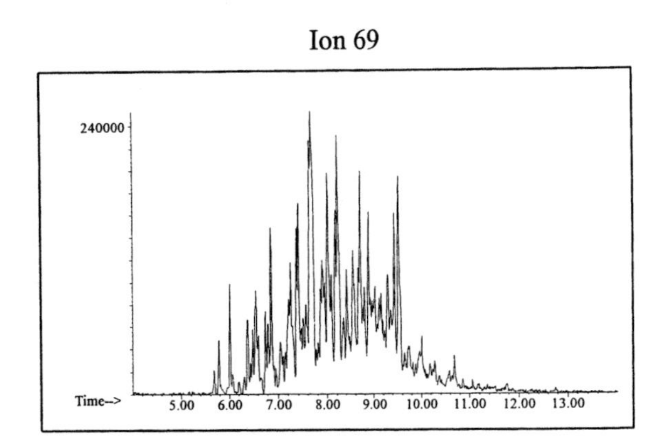

Ion 83

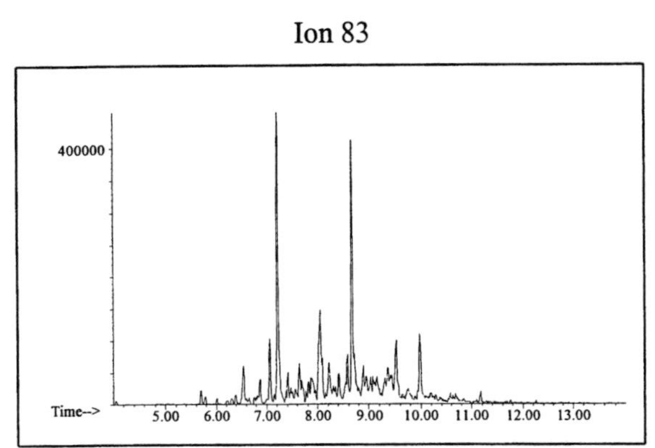

Naphthalene

Ion 128

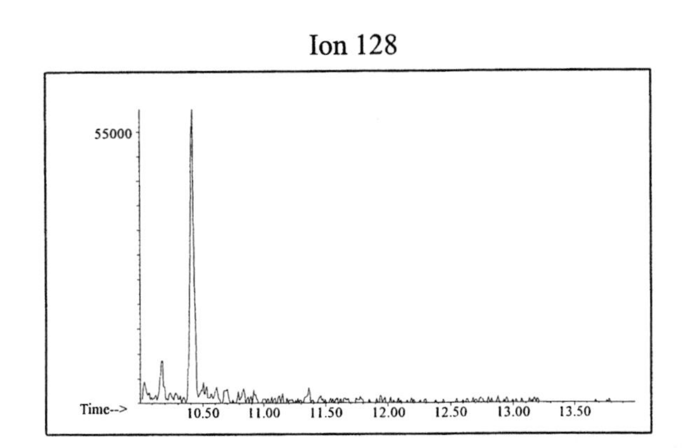

Ion 142

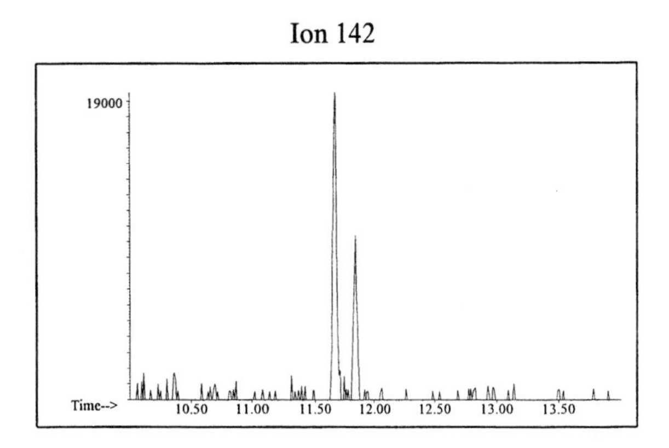

Ion 156

No Useful Data Obtained

Miscellaneous #20

20uL/mL Pentane

ASTM: Class 0 (Miscellaneous)

Macro Code: M

Product Displayed: American Super Glaze Liquid Wax

Product Uses: Wax

Other Similar Products:

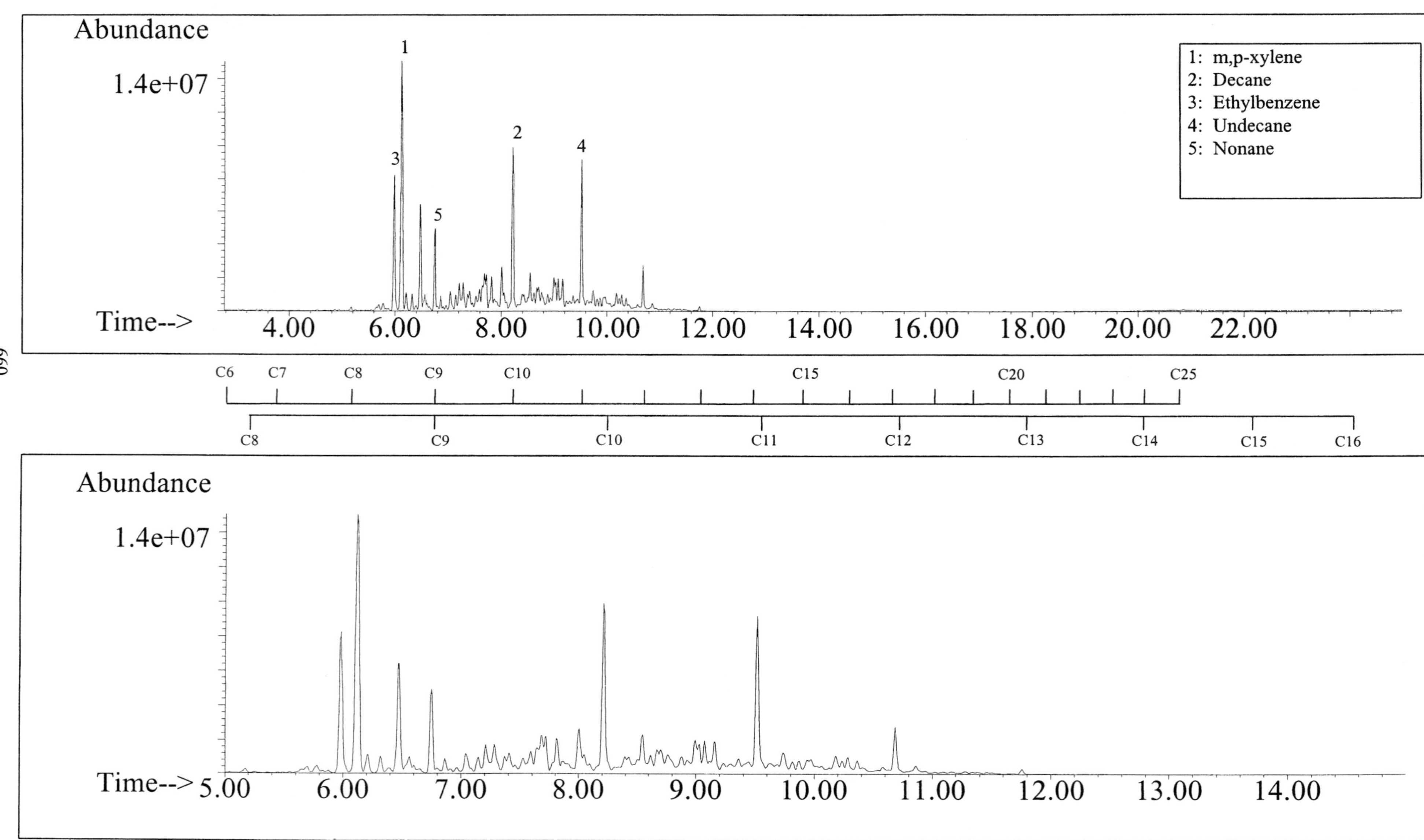

ALKANES

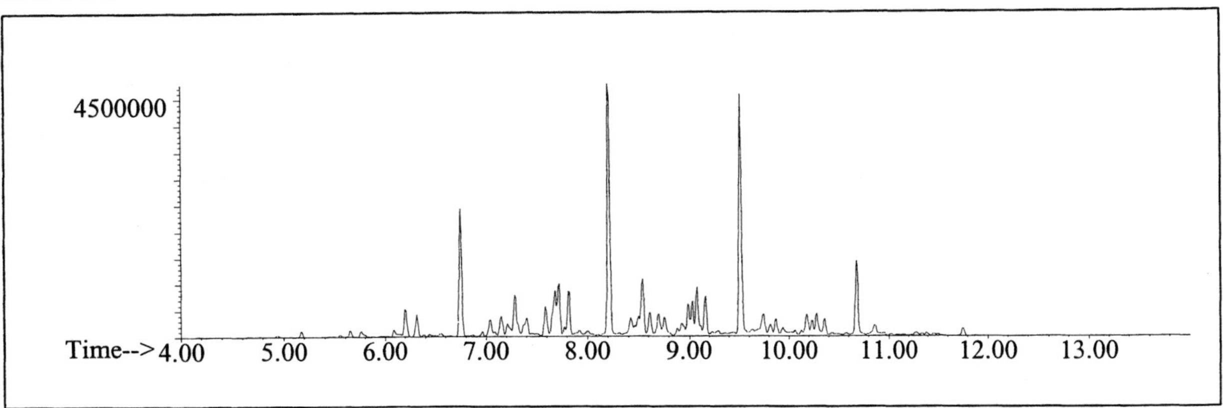

AROMATICS

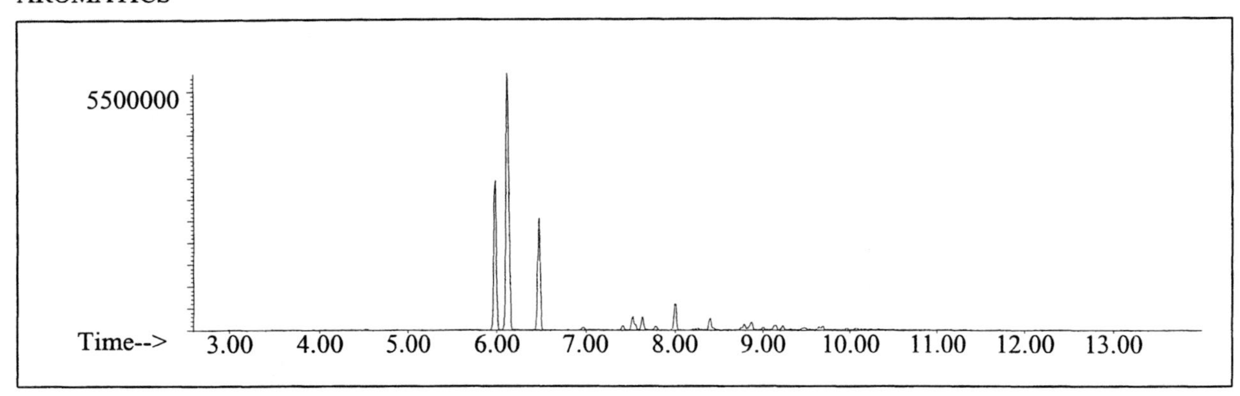

CYCLOPARAFFINS AND ALKENES

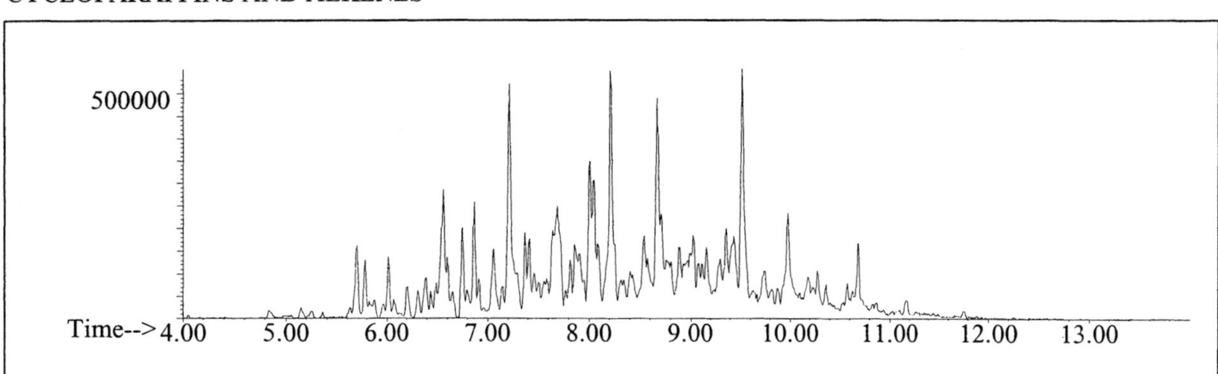

NAPHTHALENES

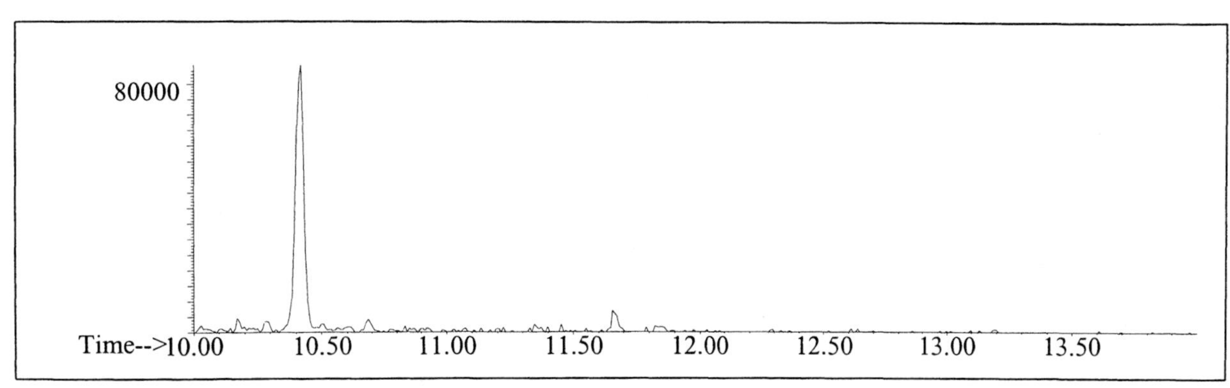

Alkane

Ion 43

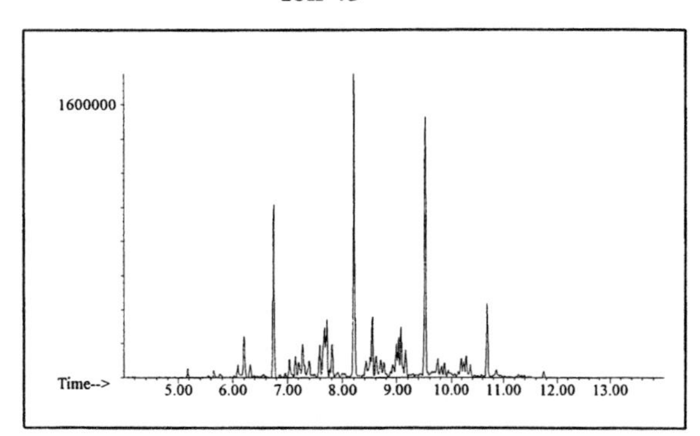

Ion 57

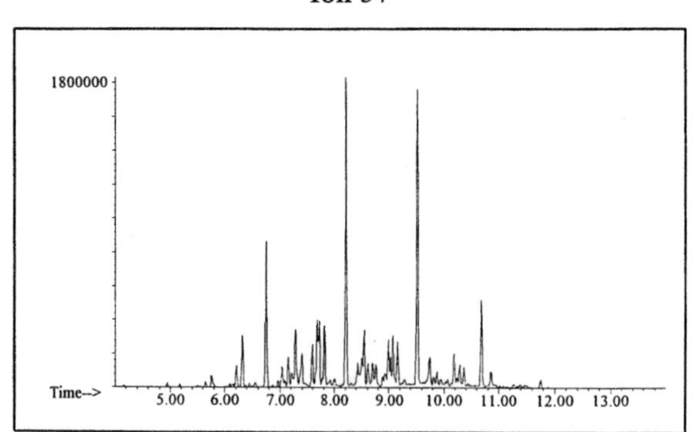

Ion 71

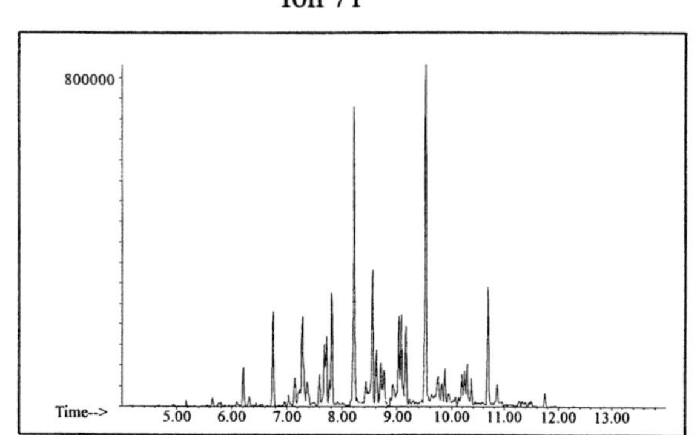

Ion 85

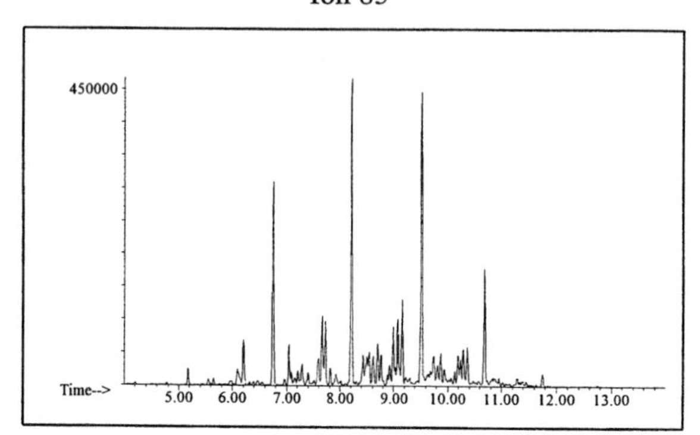

Aromatic

Ion 91

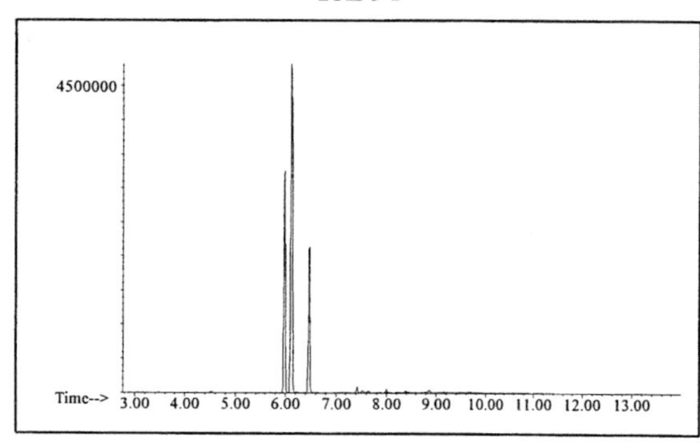

Ion 105

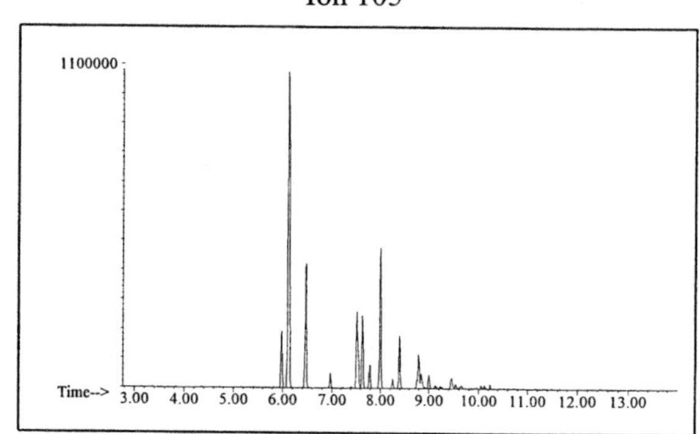

Ion 119

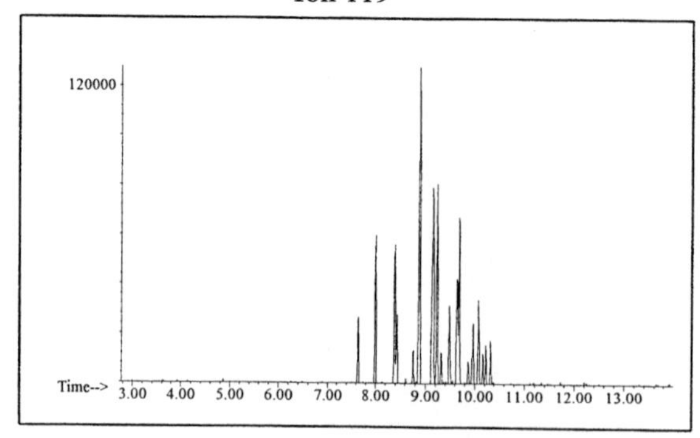

Cycloparaffin

Ion 55

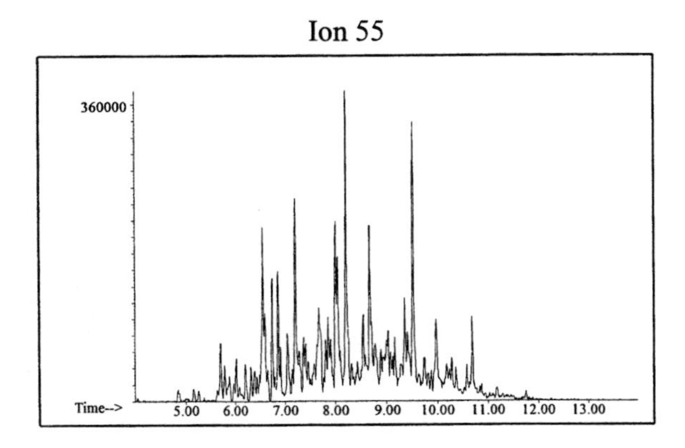

Ion 69

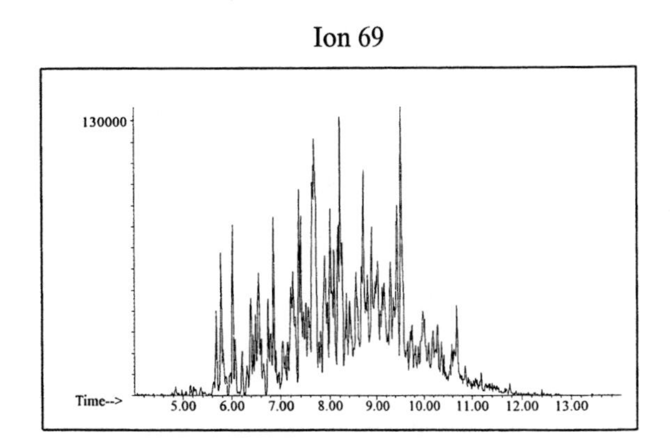

Ion 83

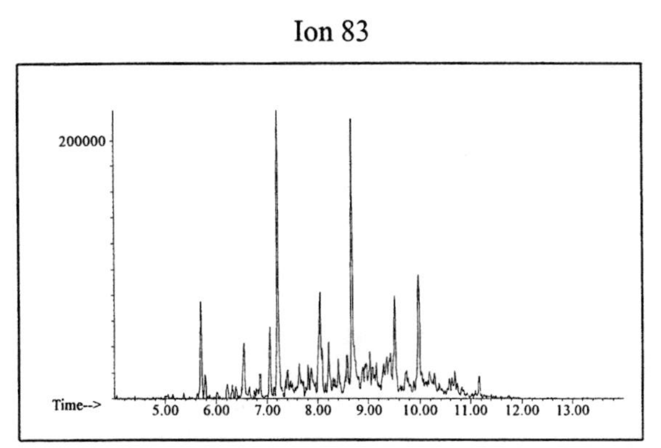

Naphthalene

Ion 128

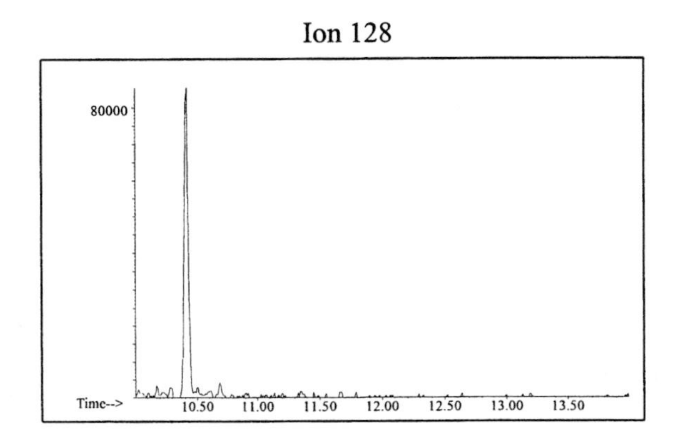

Ion 142

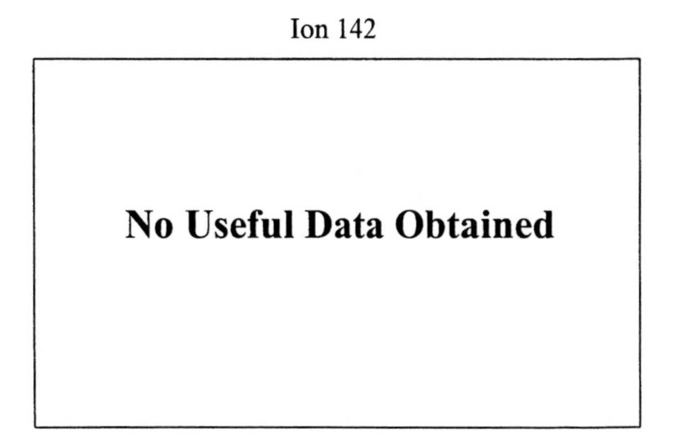

Ion 156

No Useful Data Obtained

Miscellaneous #21

20uL/mL Pentane

ASTM: Class 0 (Miscellaneous)

Macro Code: M

Product Displayed: Reckitt Colman Brand Gulf Lite

Product Uses: Charcoal lighter

Other Similar Products:

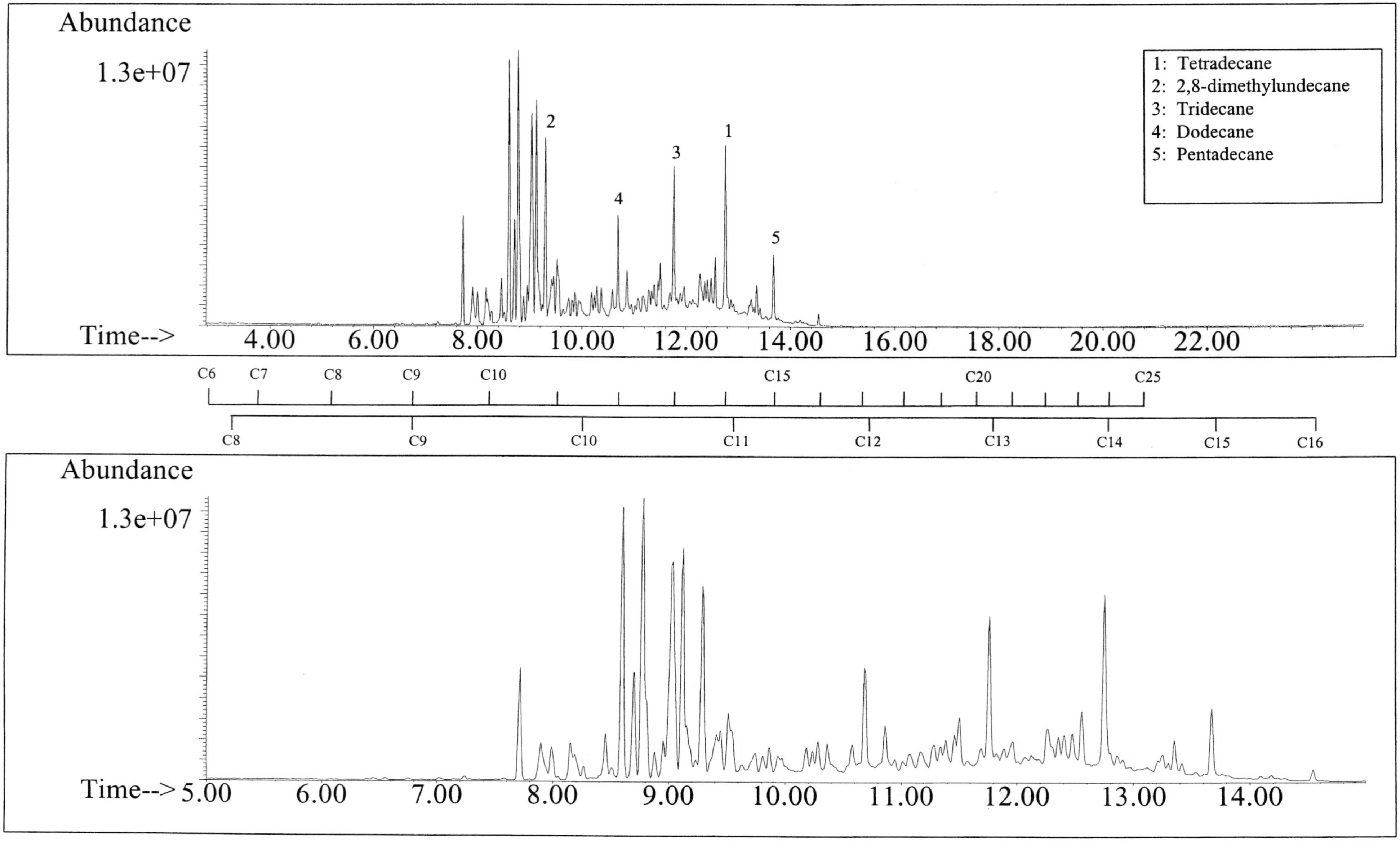

ALKANES

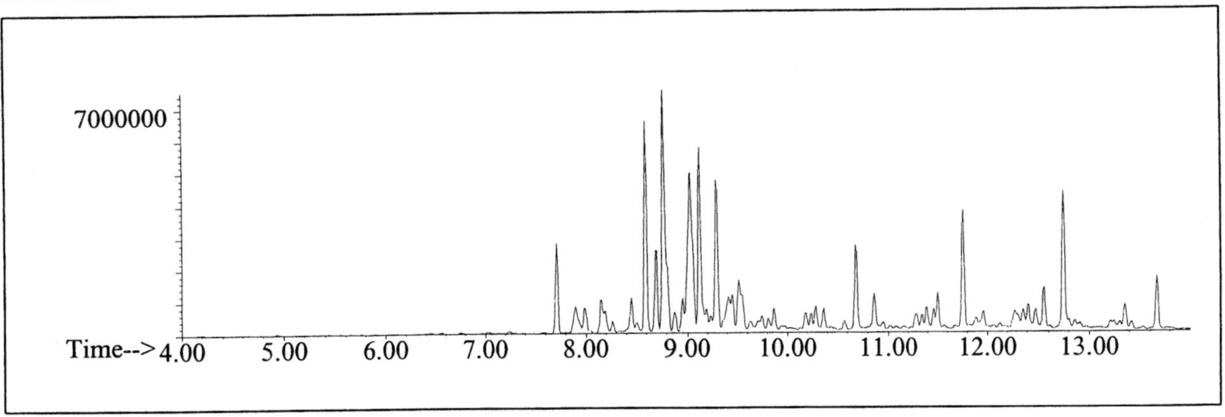

AROMATICS

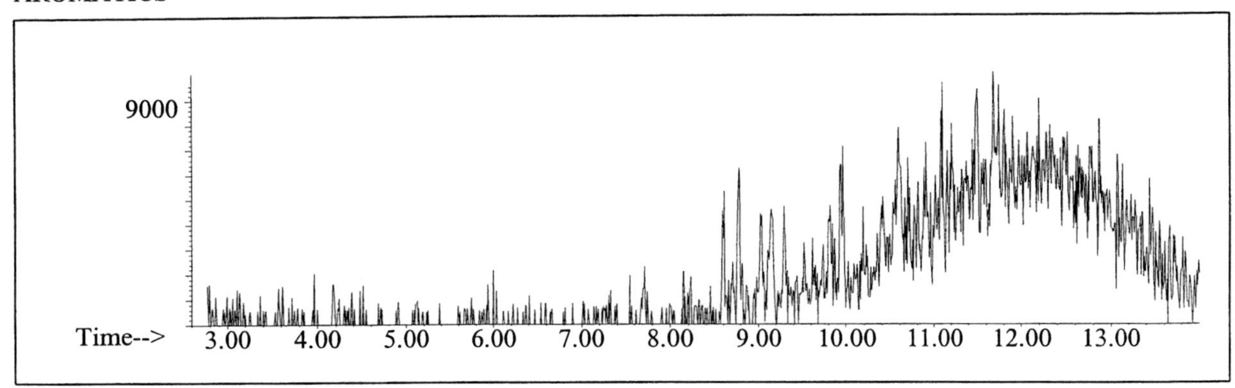

CYCLOPARAFFINS AND ALKENES

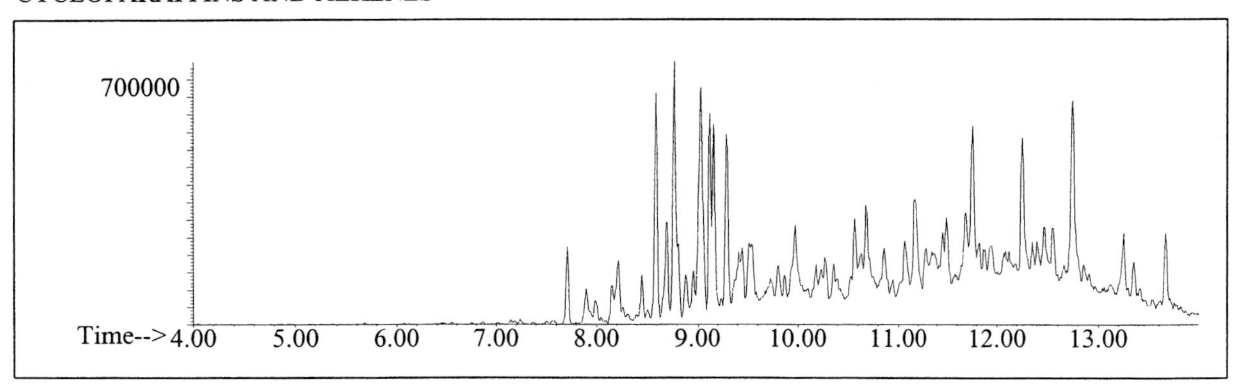

NAPHTHALENES

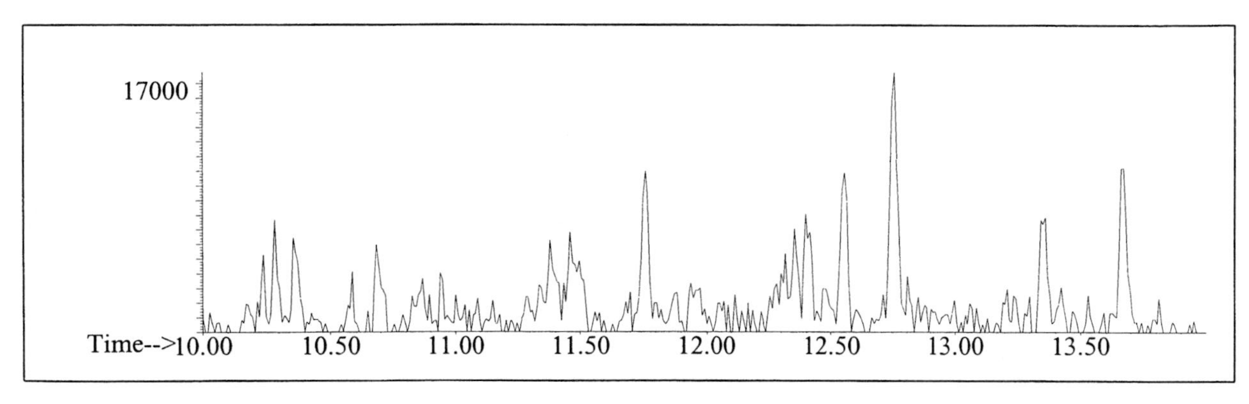

Alkane

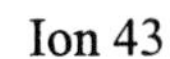
Ion 43

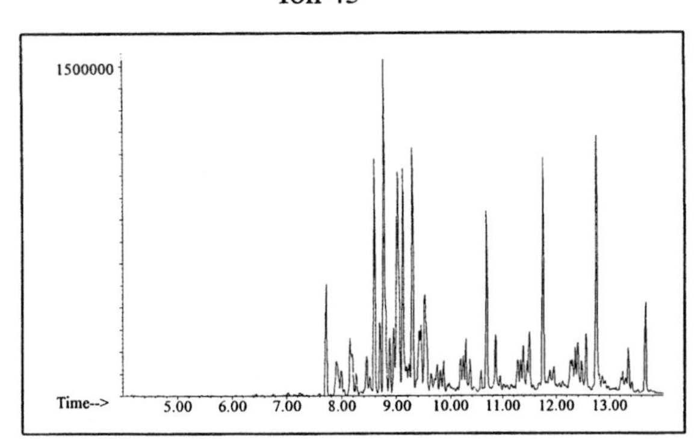

Ion 57

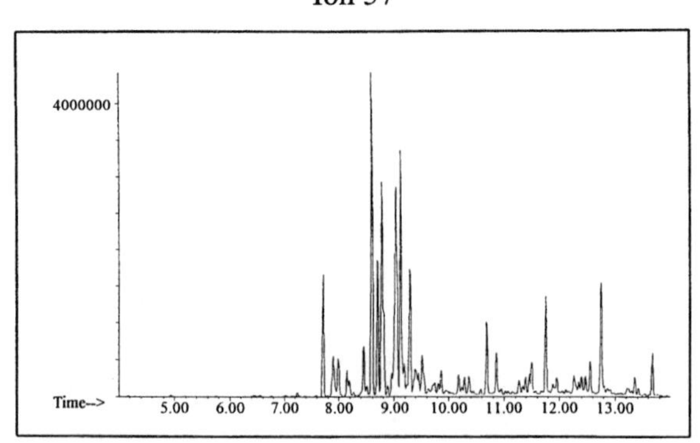

Ion 71

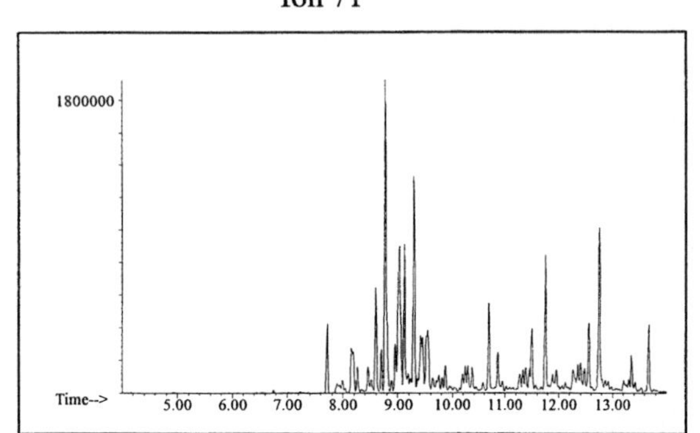

Ion 85

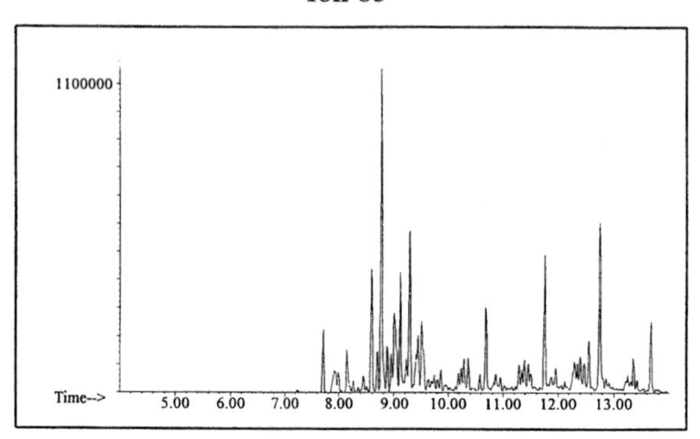

Aromatic

Ion 91

No Useful Data Obtained

Ion 105

No Useful Data Obtained

Ion 119

No Useful Data Obtained

Cycloparaffin

Ion 55

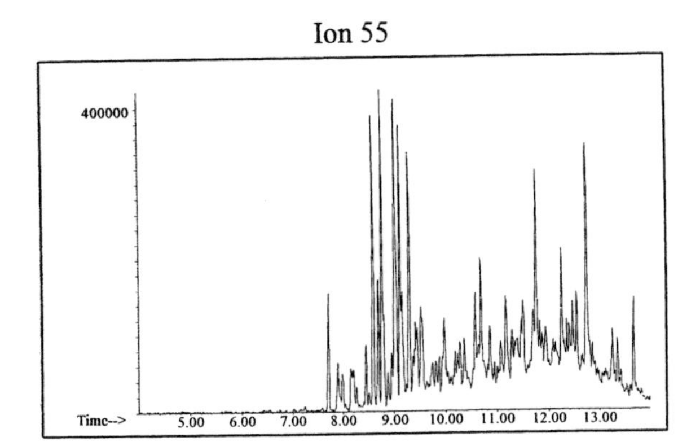

Ion 69

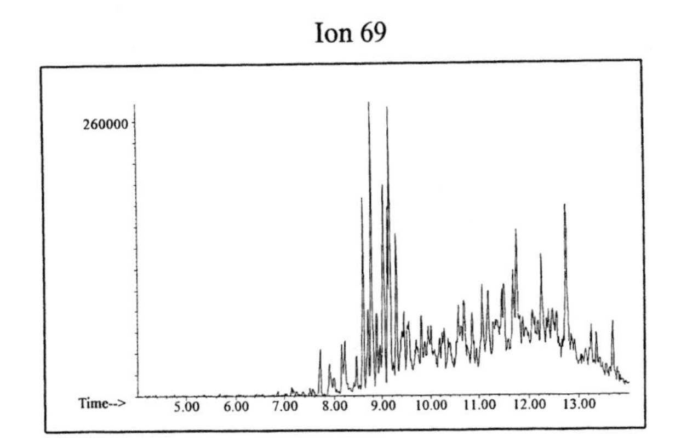

Ion 83

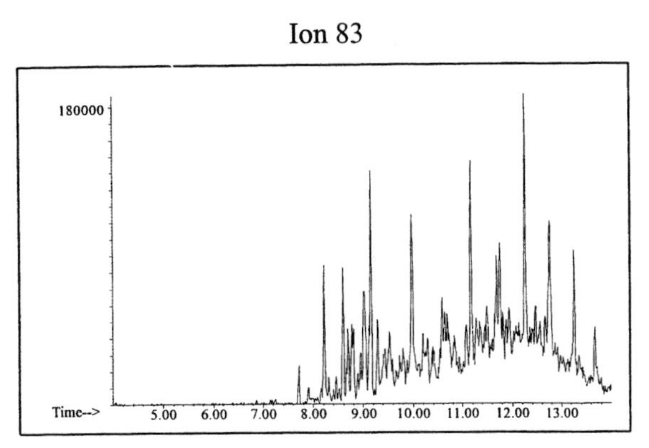

Naphthalene

Ion 128

No Useful Data Obtained

Ion 142

No Useful Data Obtained

Ion 156

No Useful Data Obtained

Miscellaneous #22

20uL/mL Pentane

ASTM: Class 0 (Miscellaneous)

Macro Code: M

Product Displayed: Outer Gun Cleaner

Product Uses: Cleaning solvent

Other Similar Products:

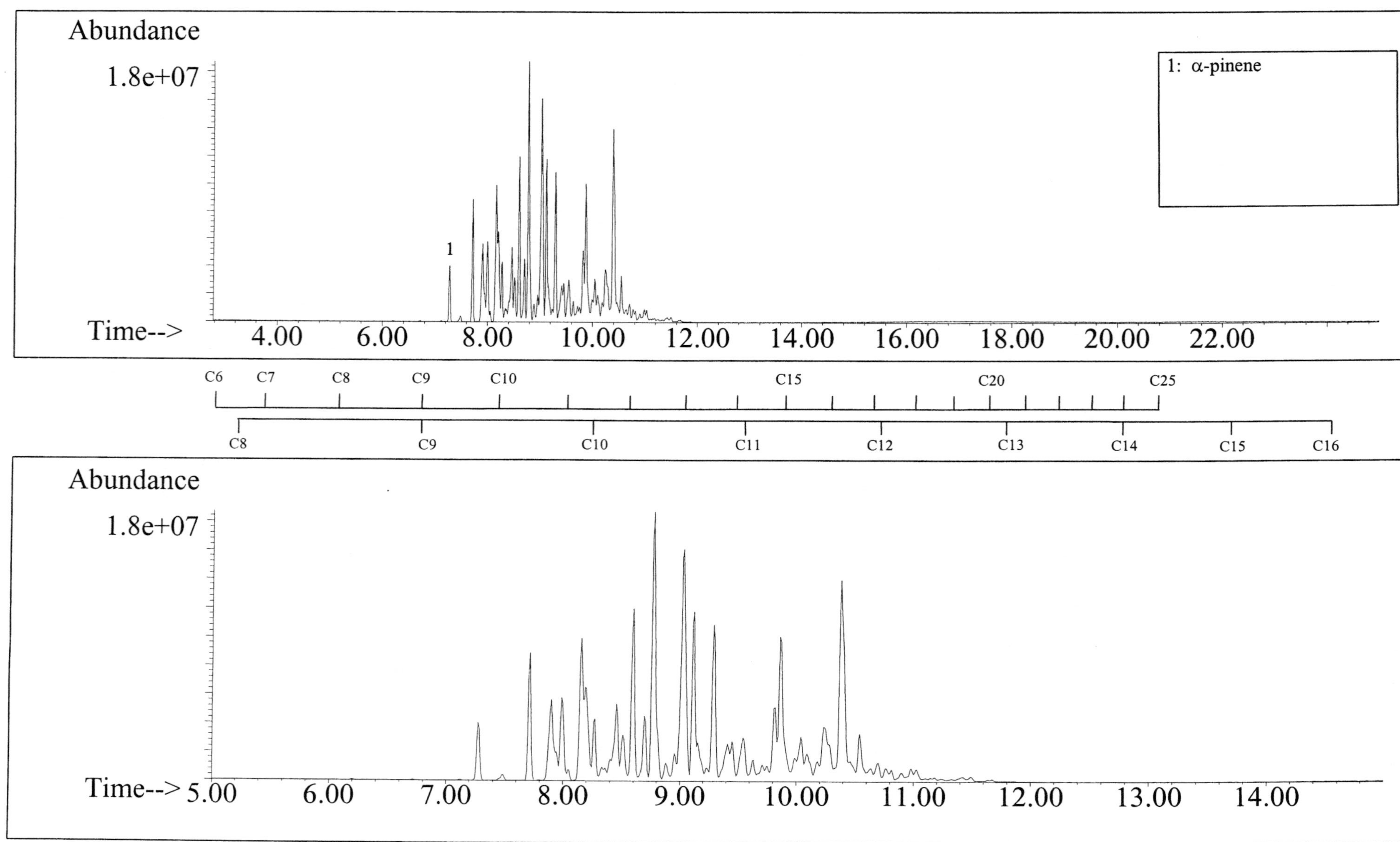

ALKANES

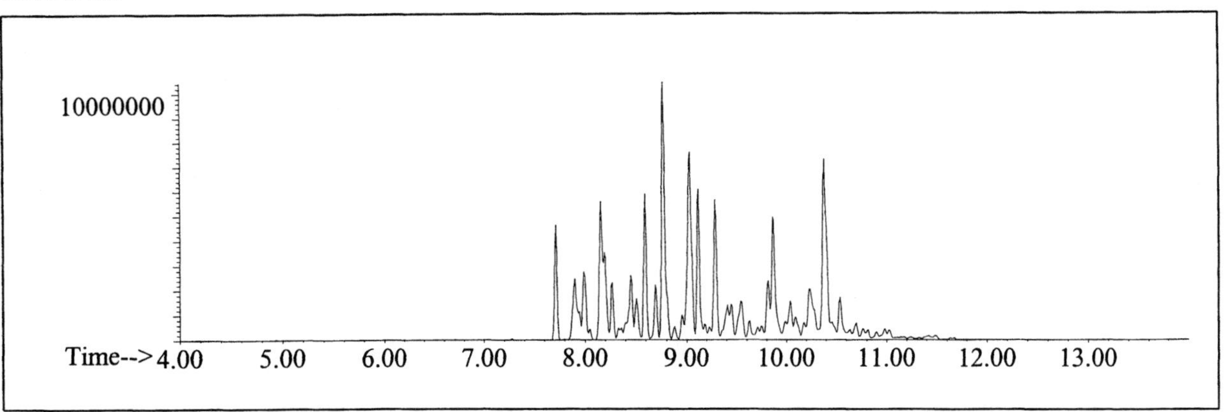

AROMATICS

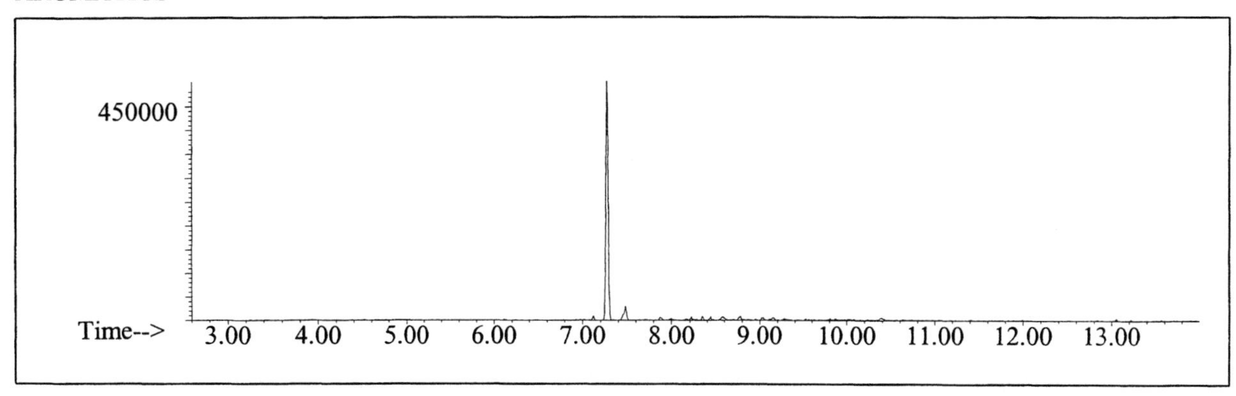

CYCLOPARAFFINS AND ALKENES

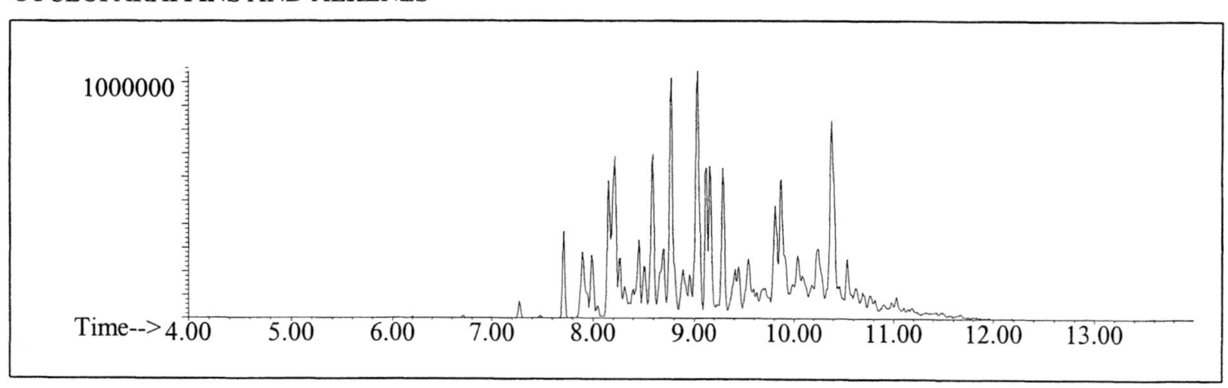

NAPHTHALENES

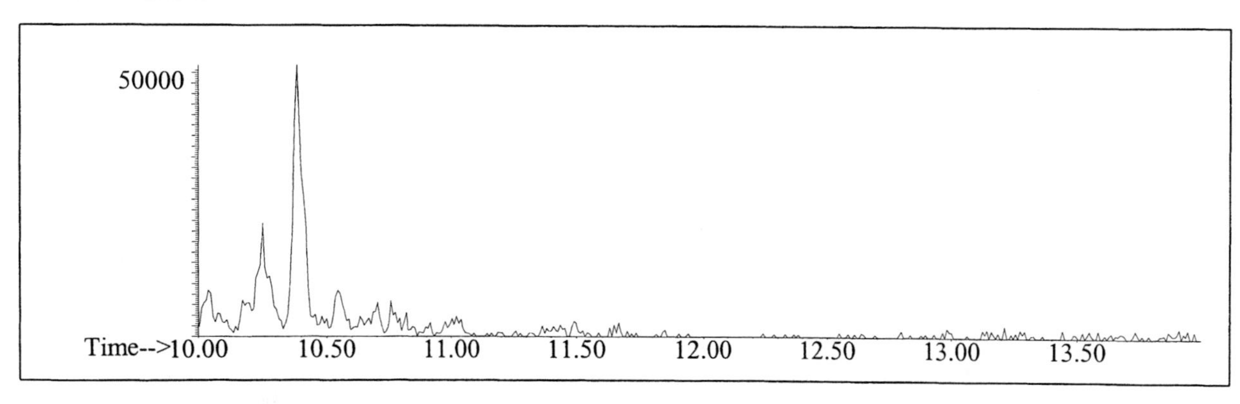

Alkane

Ion 43

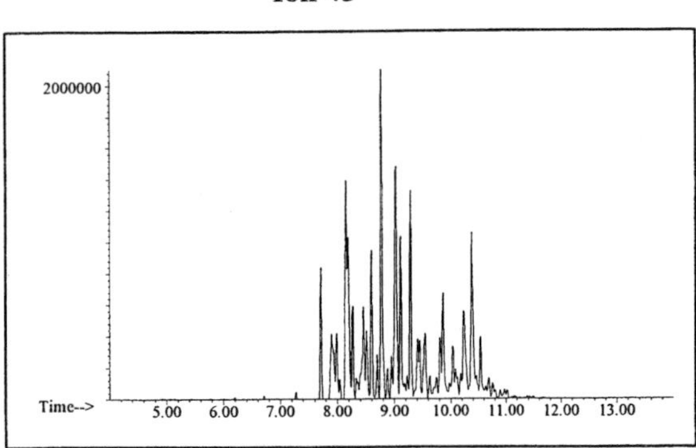

Ion 57

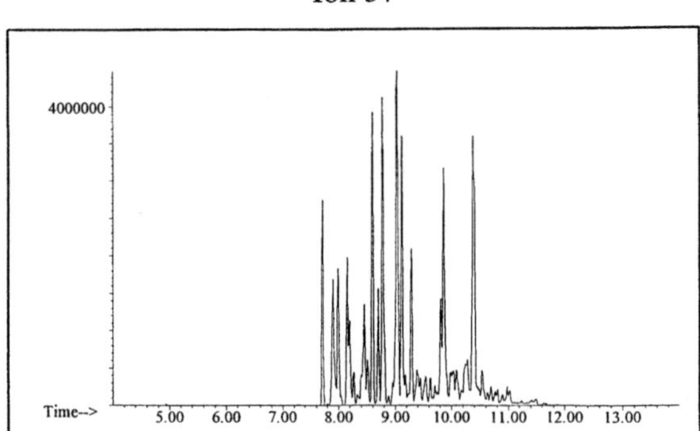

Ion 71

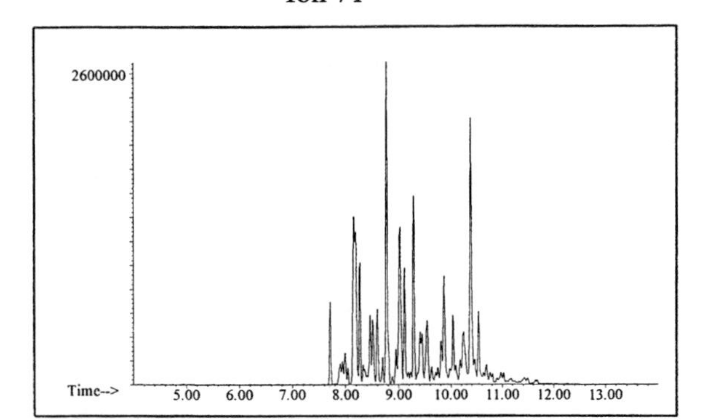

Ion 85

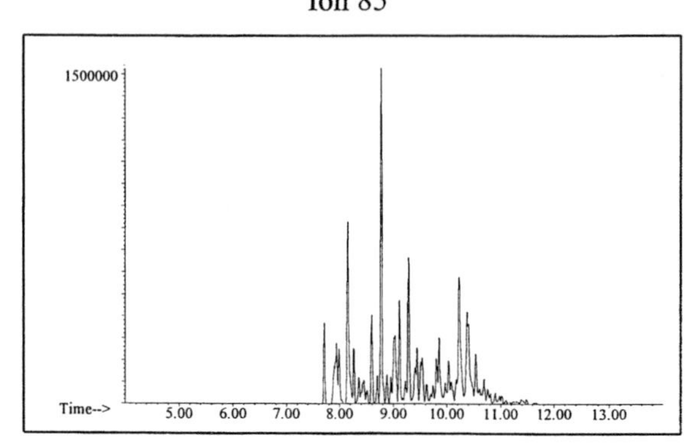

Aromatic

Ion 91

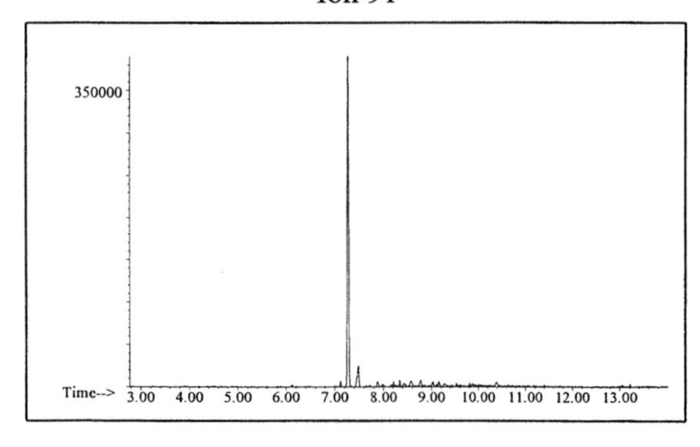

Ion 105

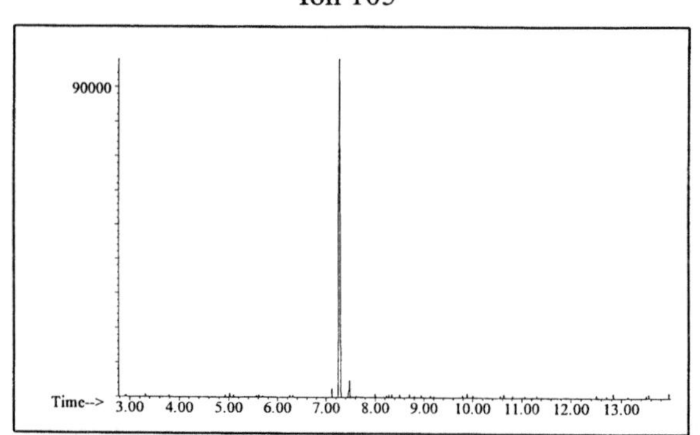

Ion 119

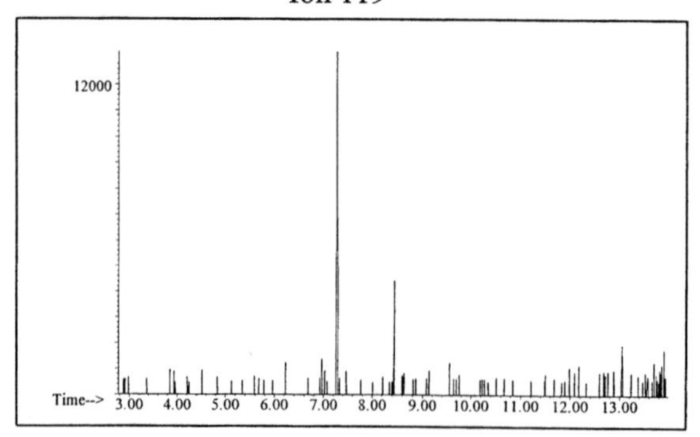

Cycloparaffin

Ion 55

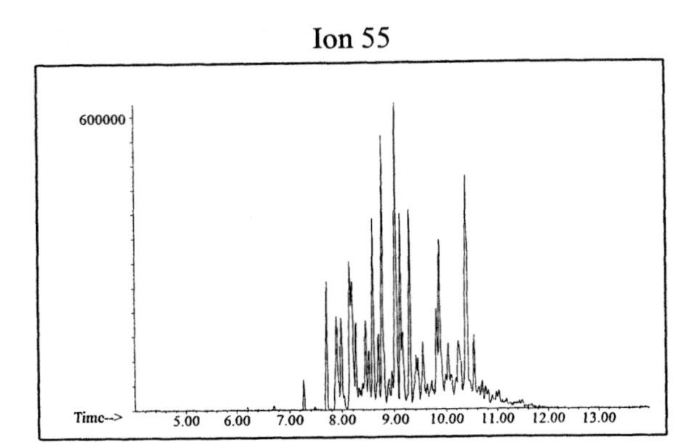

Ion 69

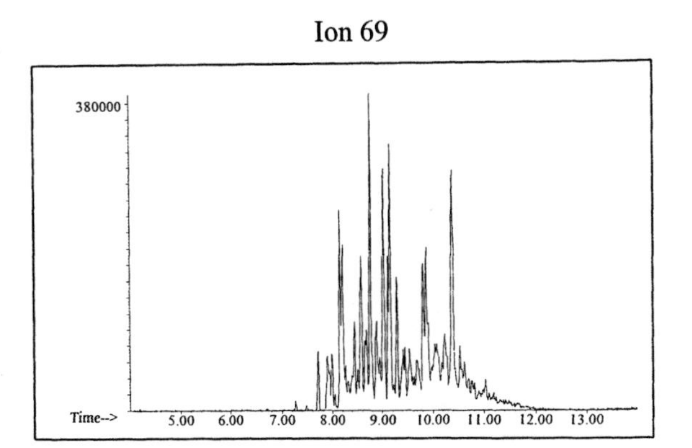

Ion 83

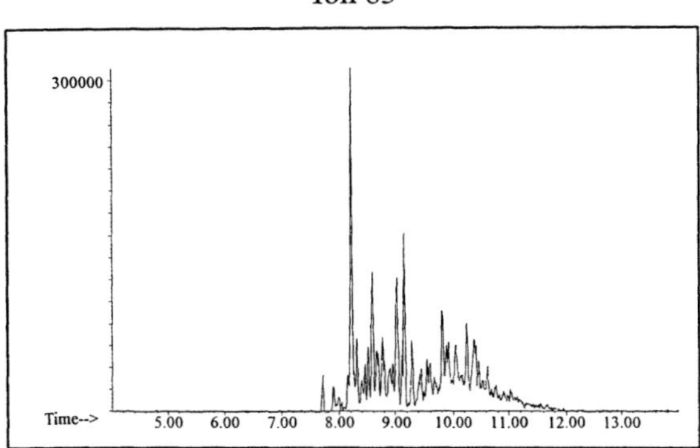

Naphthalene

Ion 128

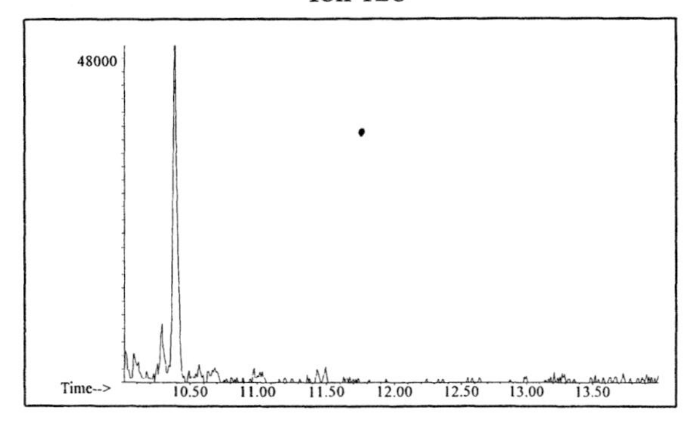

Ion 142

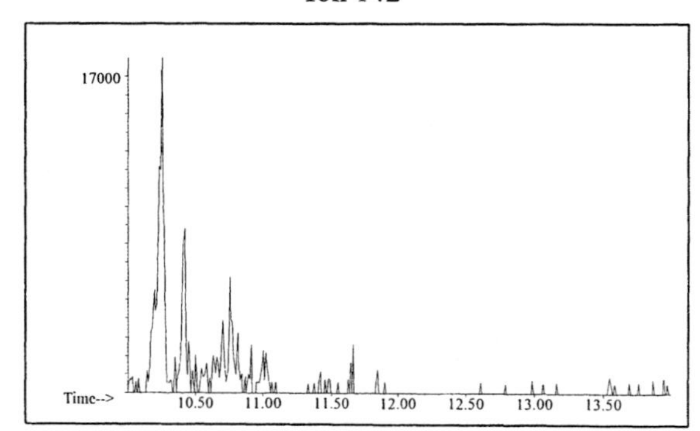

Ion 156

No Useful Data Obtained

Miscellaneous #23

20uL/mL Pentane

Product Displayed: Afta Cleaner/ Degreaser

Other Similar Products:

ASTM: Class 0 (Miscellaneous)

Product Uses: Cleaning solvent

Macro Code: L

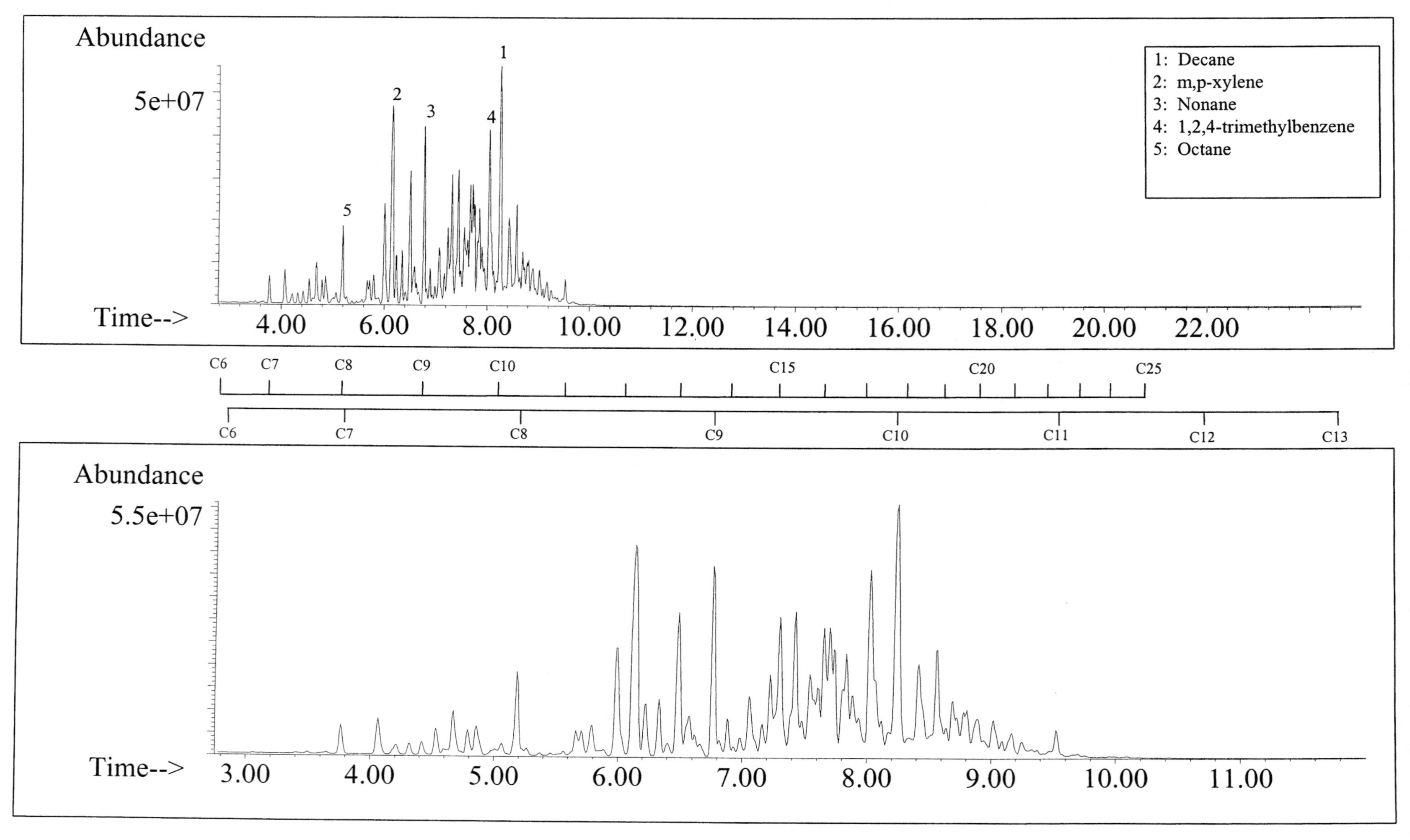

ALKANES

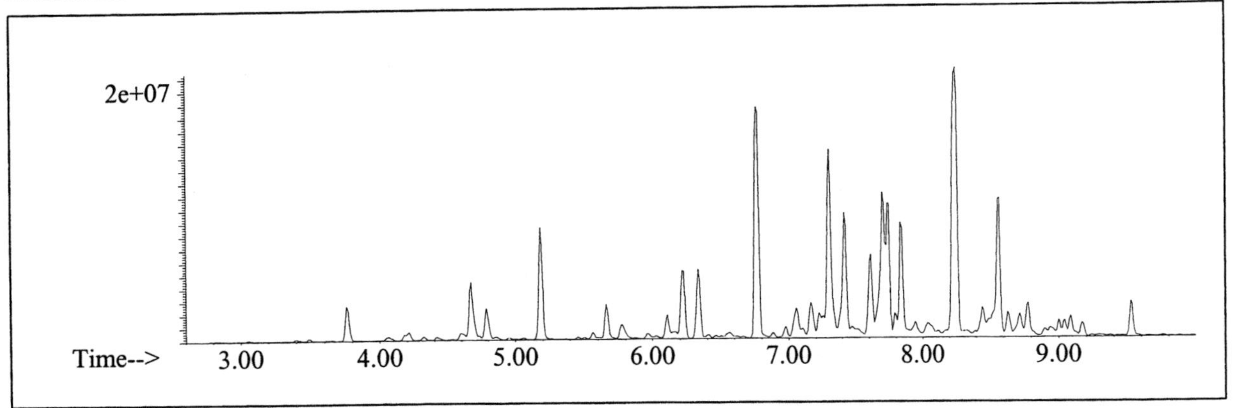

AROMATICS

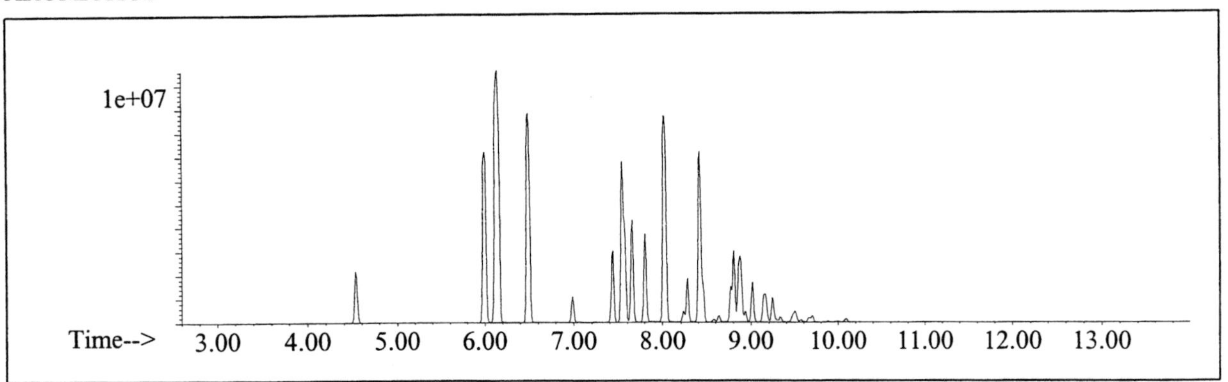

CYCLOPARAFFINS AND ALKENES

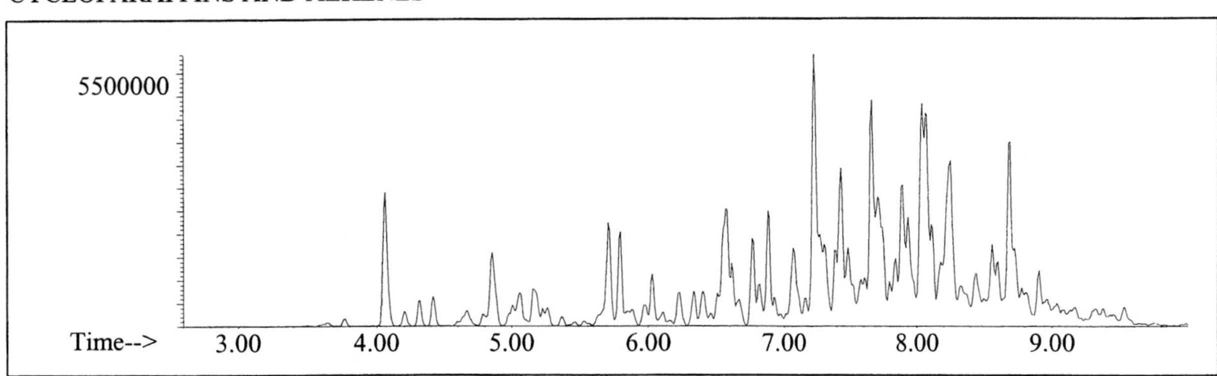

NAPHTHALENES

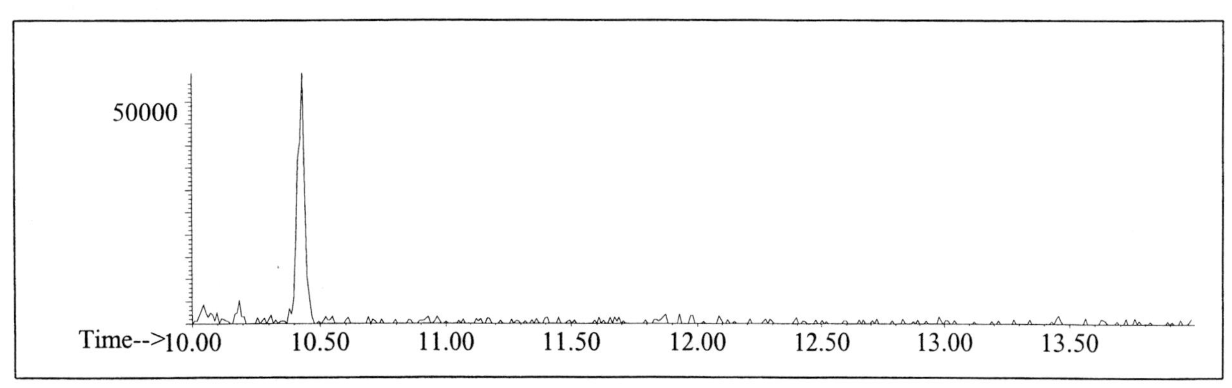

Alkane

Ion 43

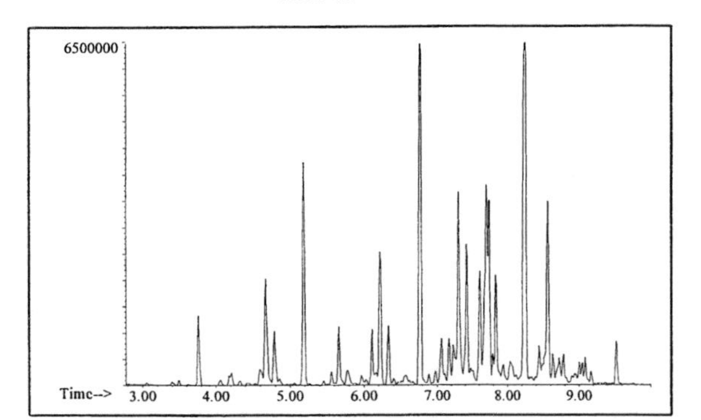

Ion 57

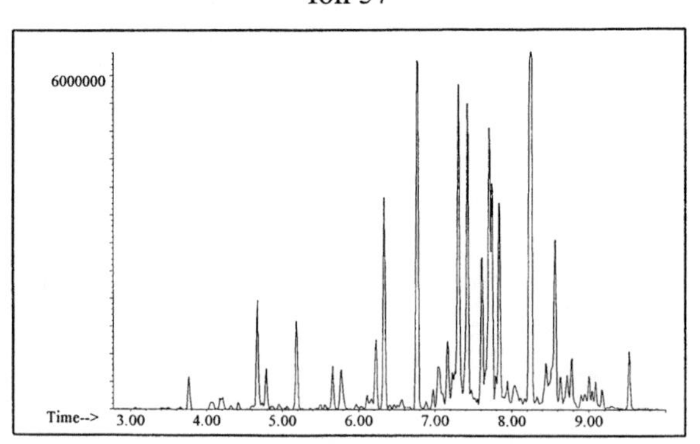

Ion 71

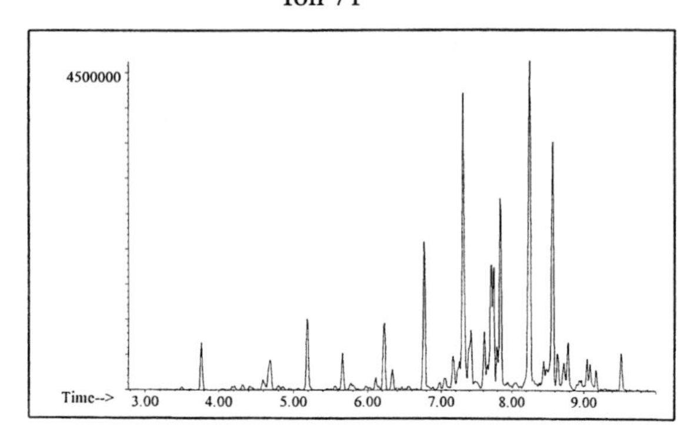

Ion 85

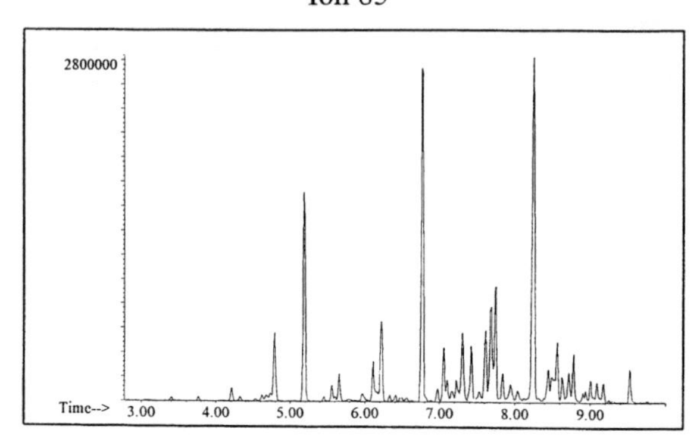

Aromatic

Ion 91

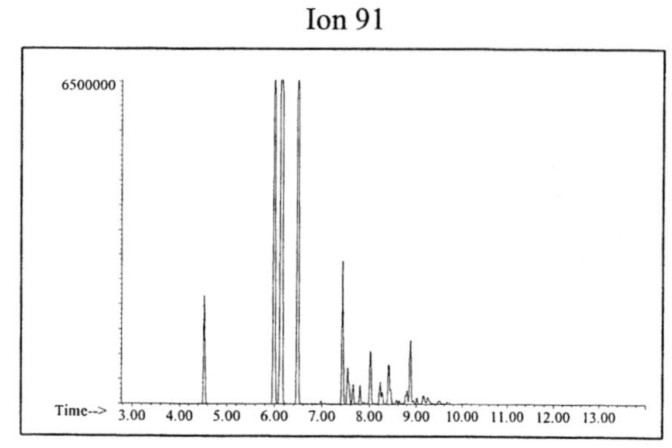

Ion 105

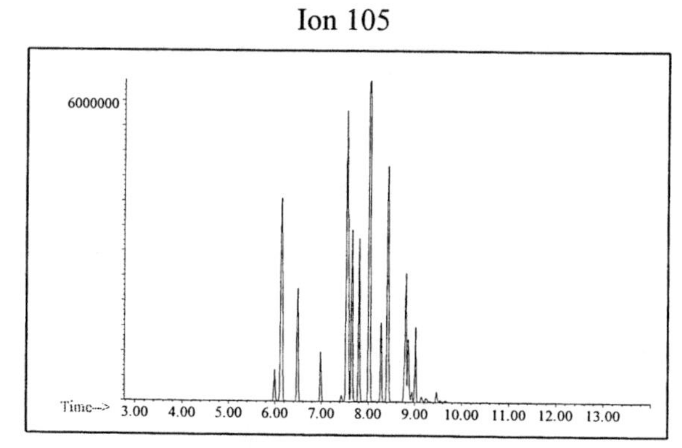

Ion 119

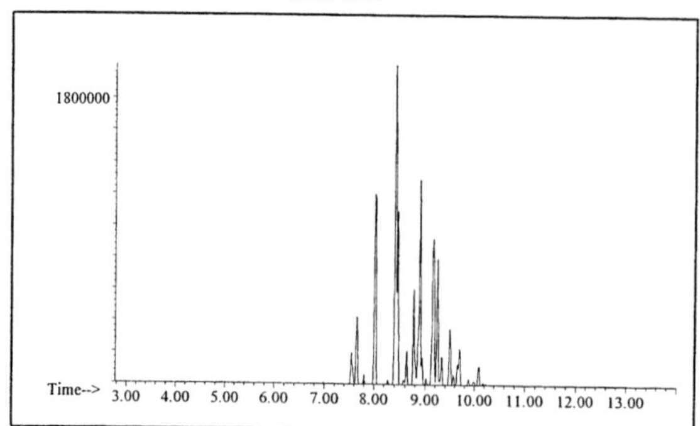

Cycloparaffin

Ion 55

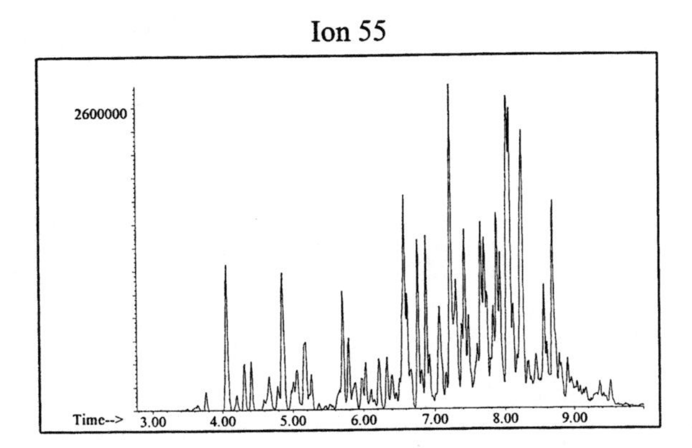

Ion 69

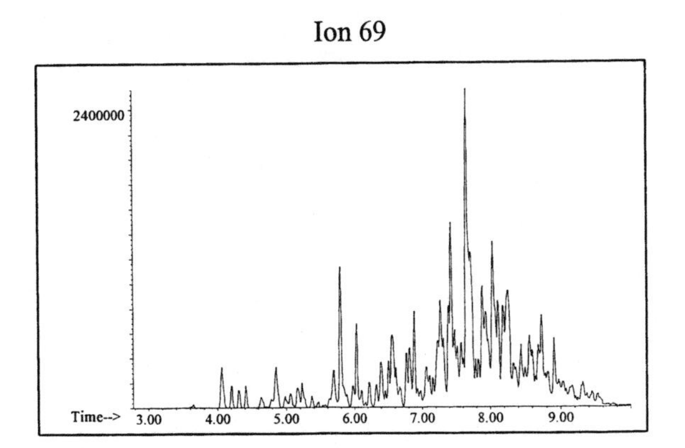

Ion 83

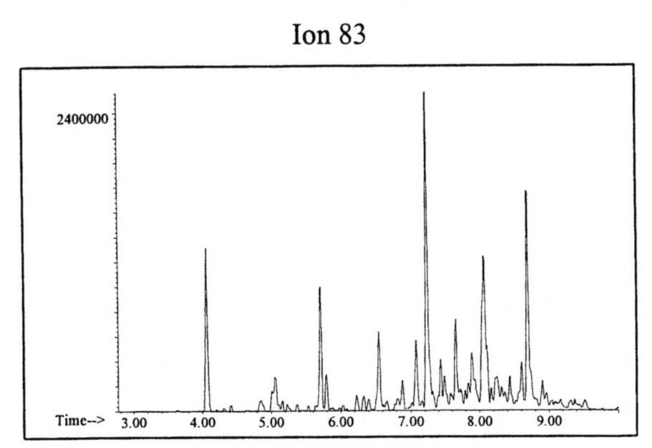

Naphthalene

Ion 128

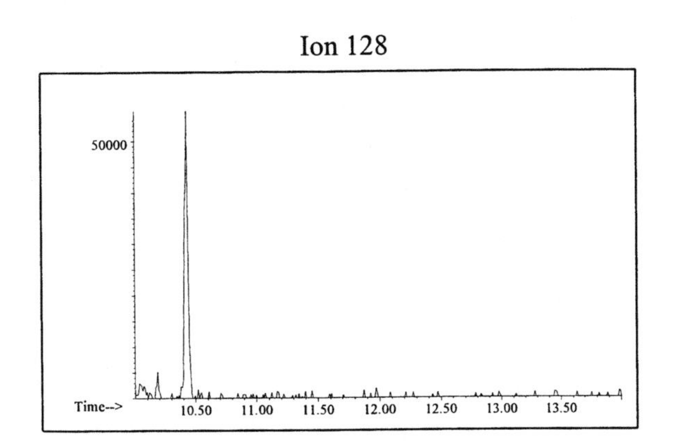

Ion 142

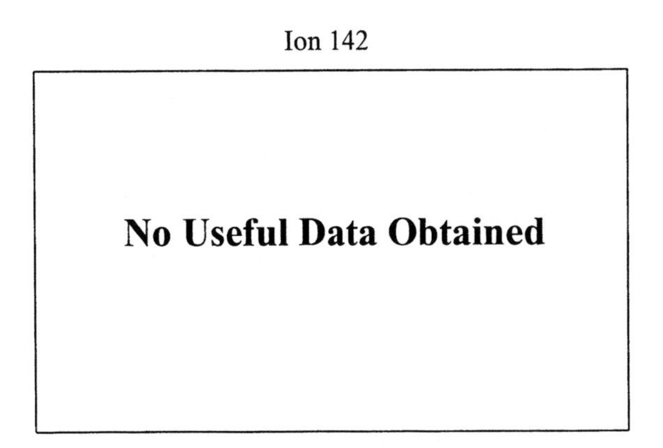

Ion 156

No Useful Data Obtained

Miscellaneous #24

20uL/mL Pentane

Product Displayed: JP-4 Jet Fuel

Other Similar Products:

ASTM: Class 0 (Miscellaneous)

Product Uses: Jet fuel

Macro Code: H

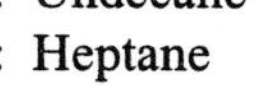
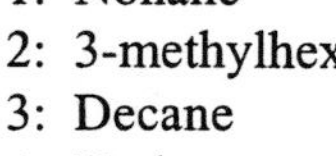
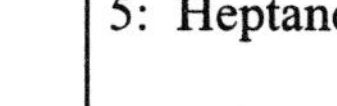
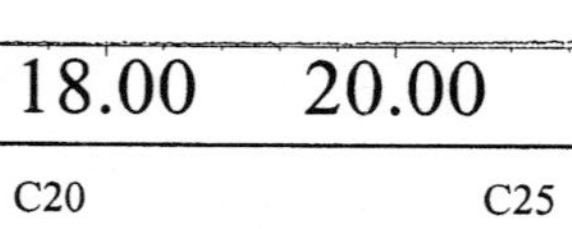
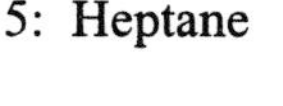
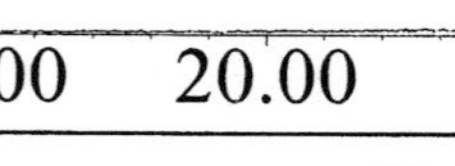
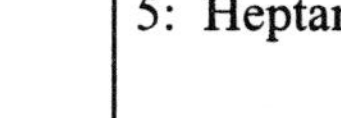
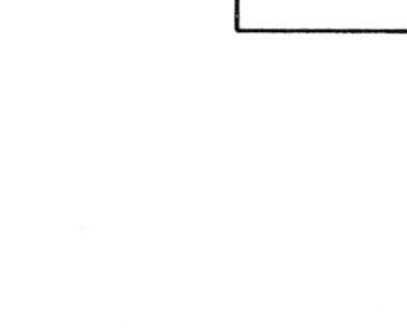
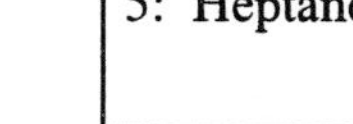

ALKANES

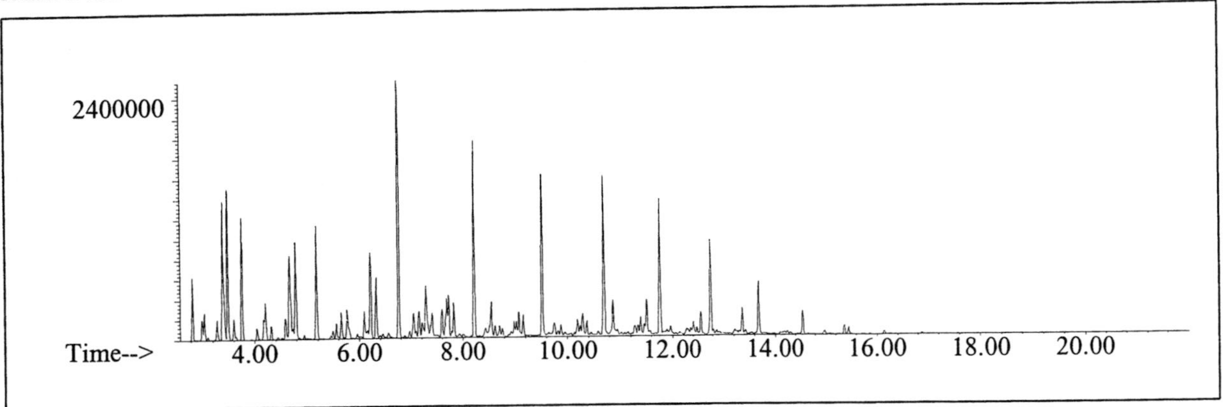

AROMATICS

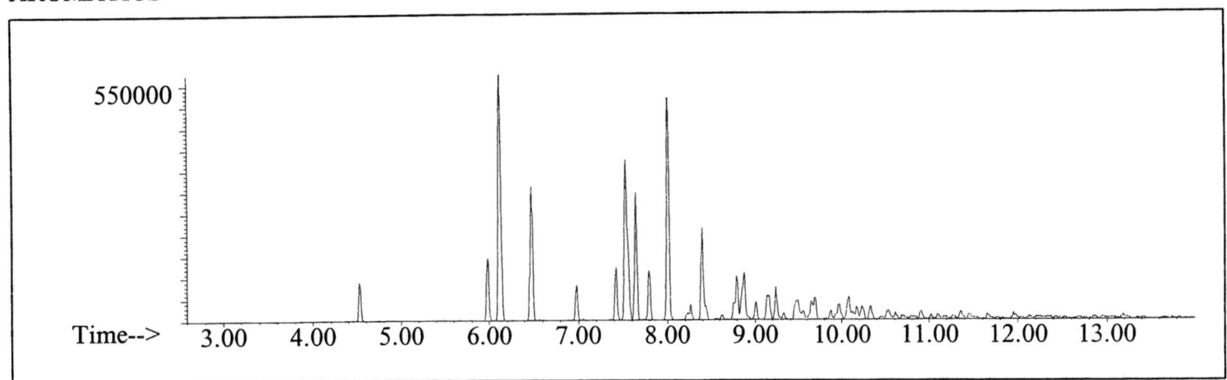

CYCLOPARAFFINS AND ALKENES

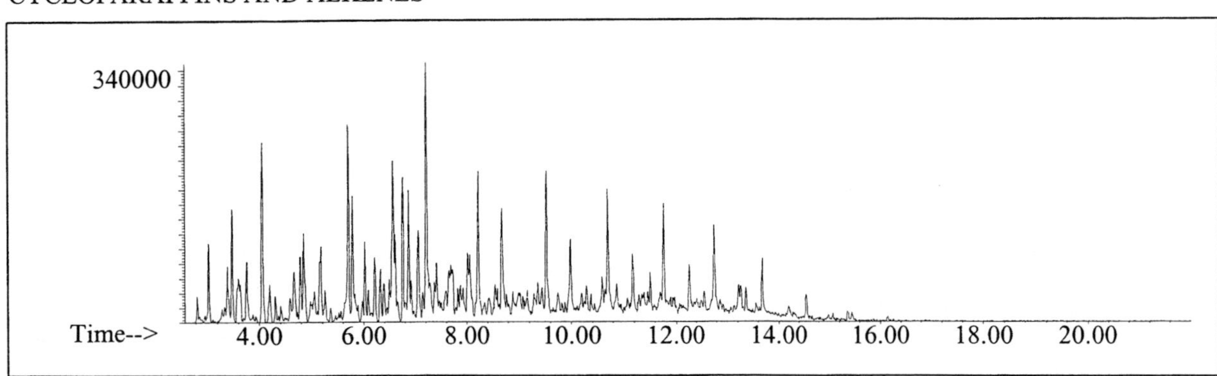

NAPHTHALENES

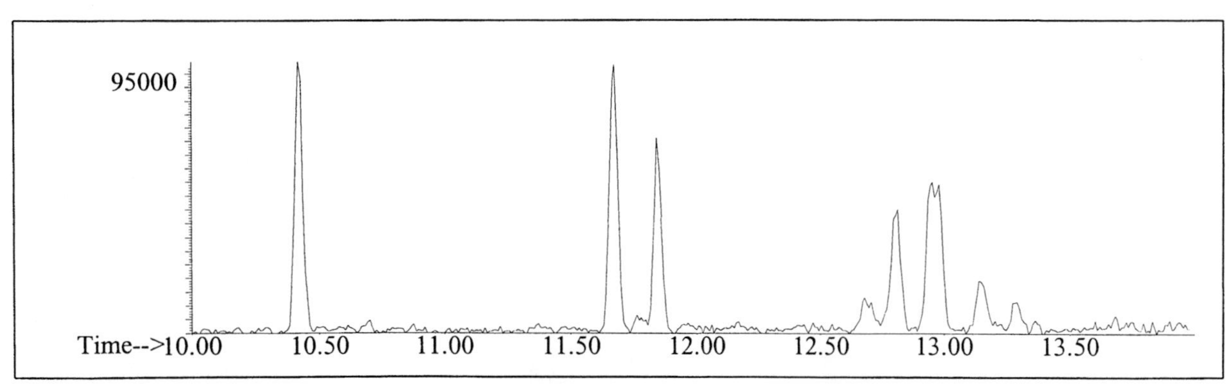

Alkane

Ion 43

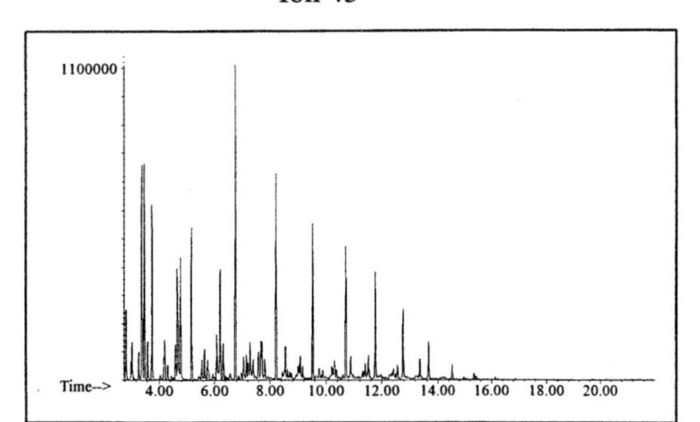

Ion 57

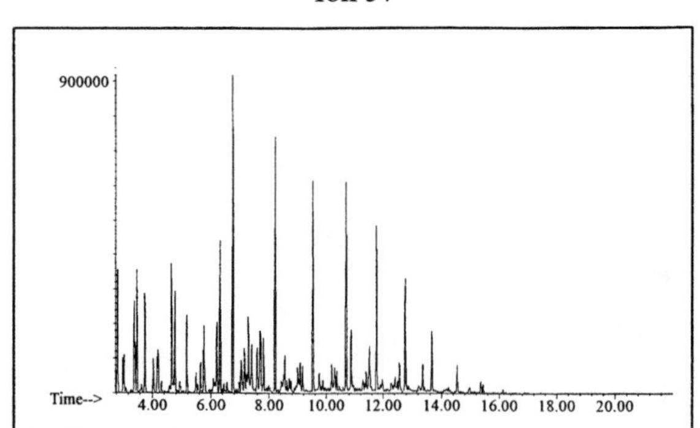

Ion 71

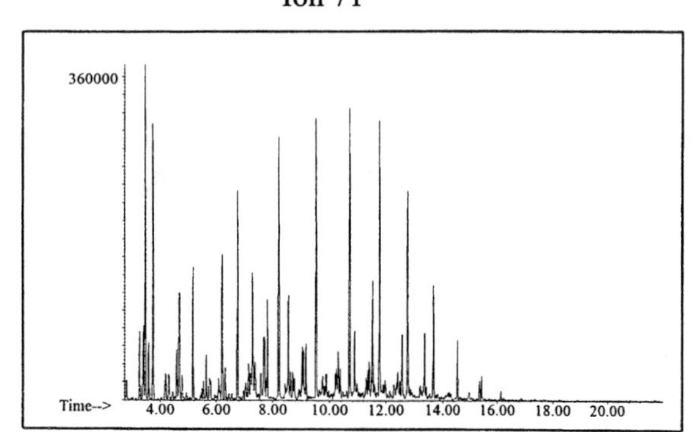

Ion 85

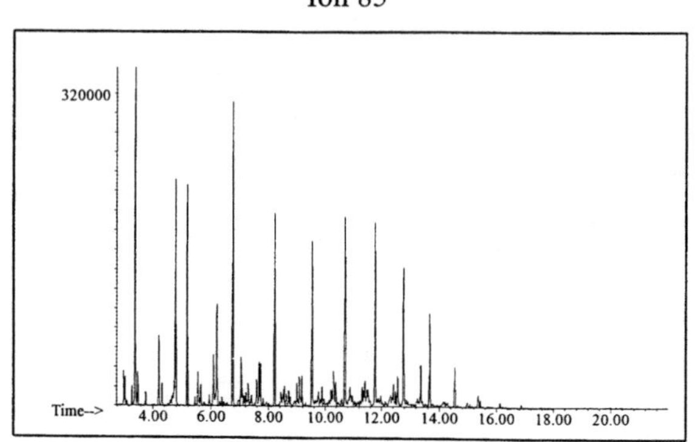

Aromatic

Ion 91

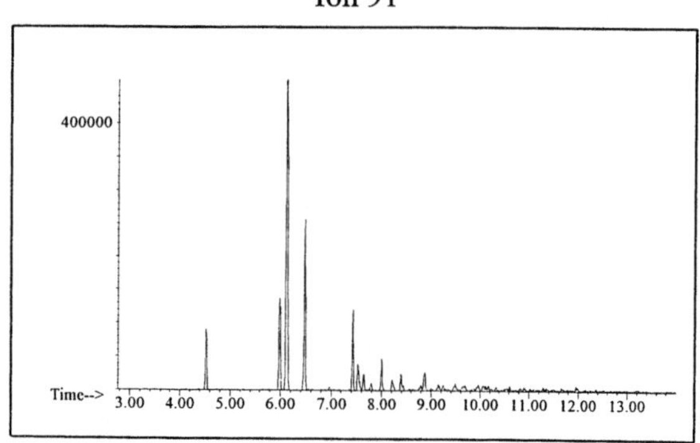

Ion 105

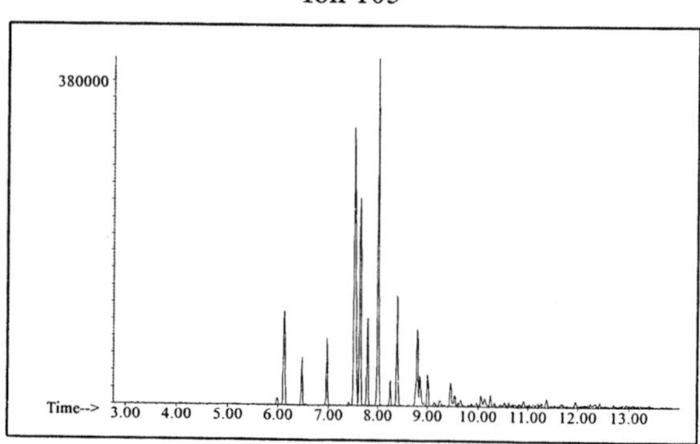

Ion 119

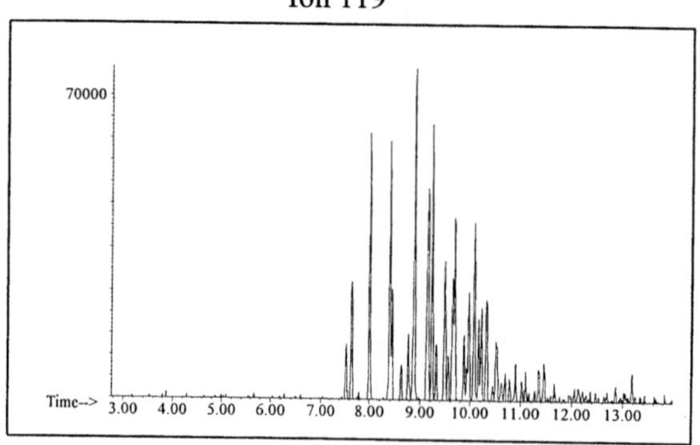

Cycloparaffin

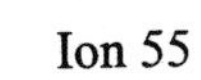

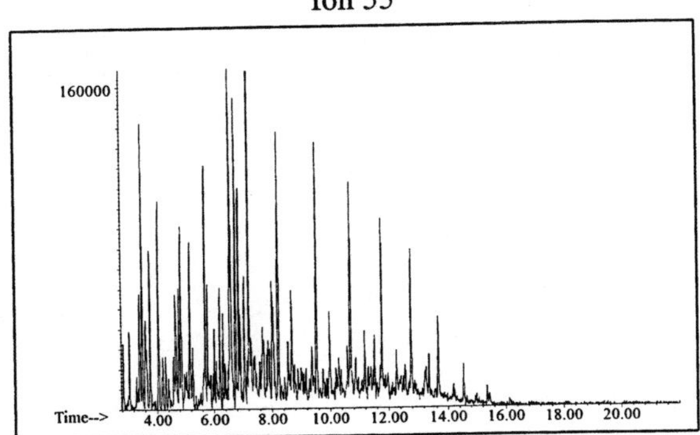

Ion 69

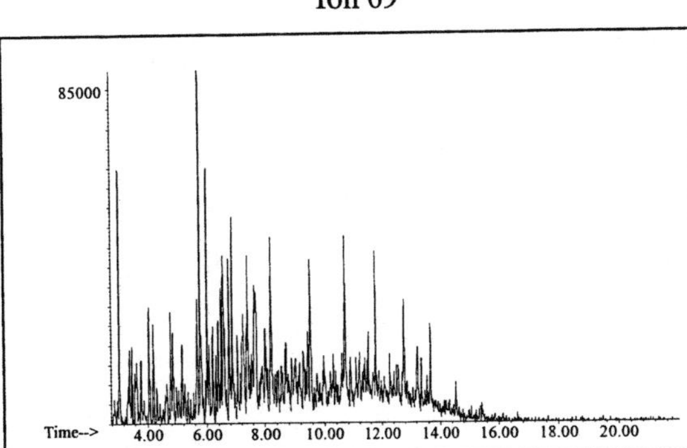

Ion 83

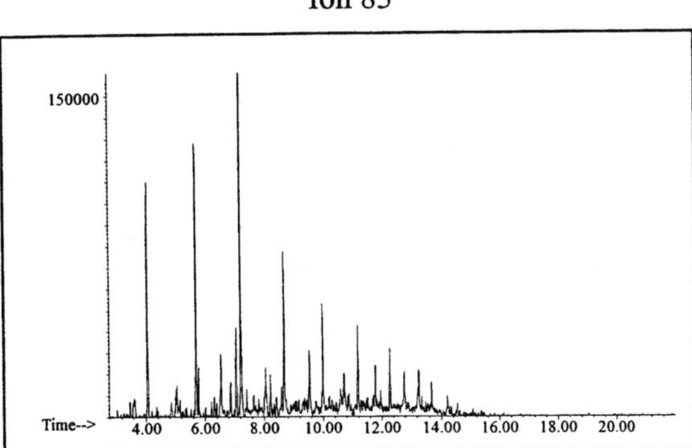

Naphthalene

Ion 128

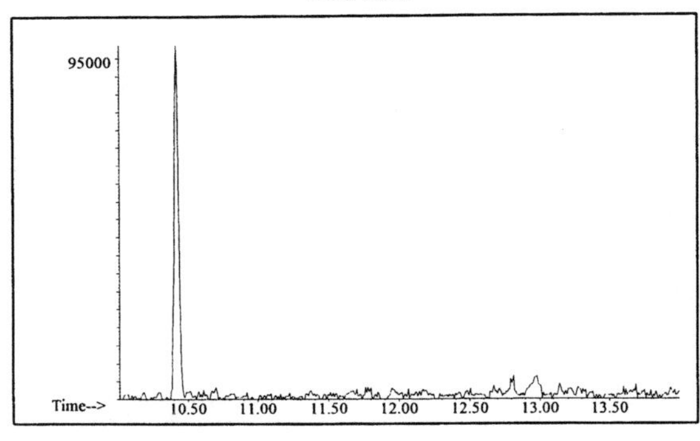

Ion 142

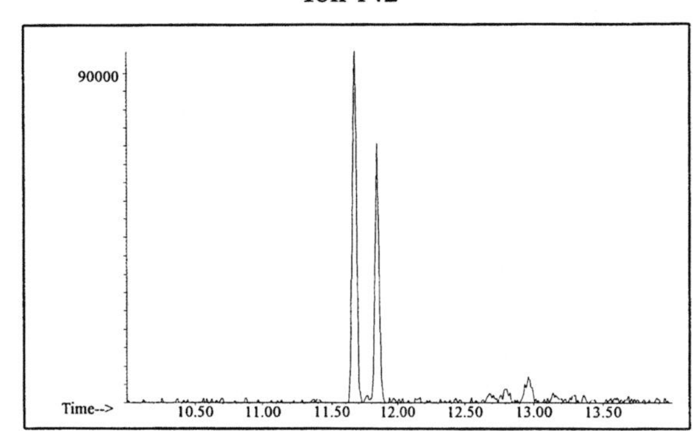

Ion 156

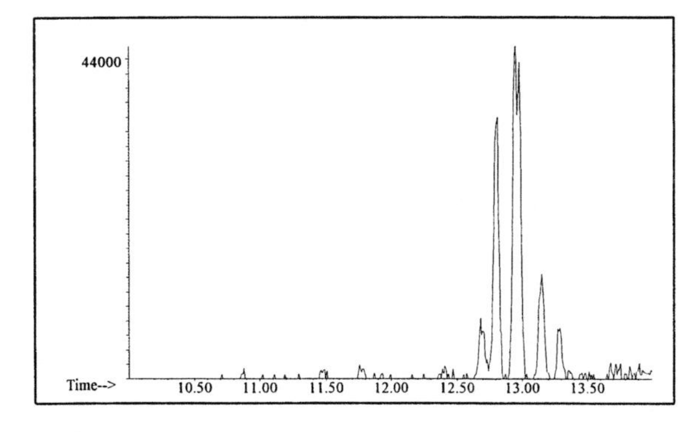

Miscellaneous #25

20uL/mL Pentane

Product Displayed: Montgomery Ward Penetrating Oil

Other Similar Products:

ASTM: Class 0 (Miscellaneous)

Product Uses: Penetrating oil

Macro Code: M

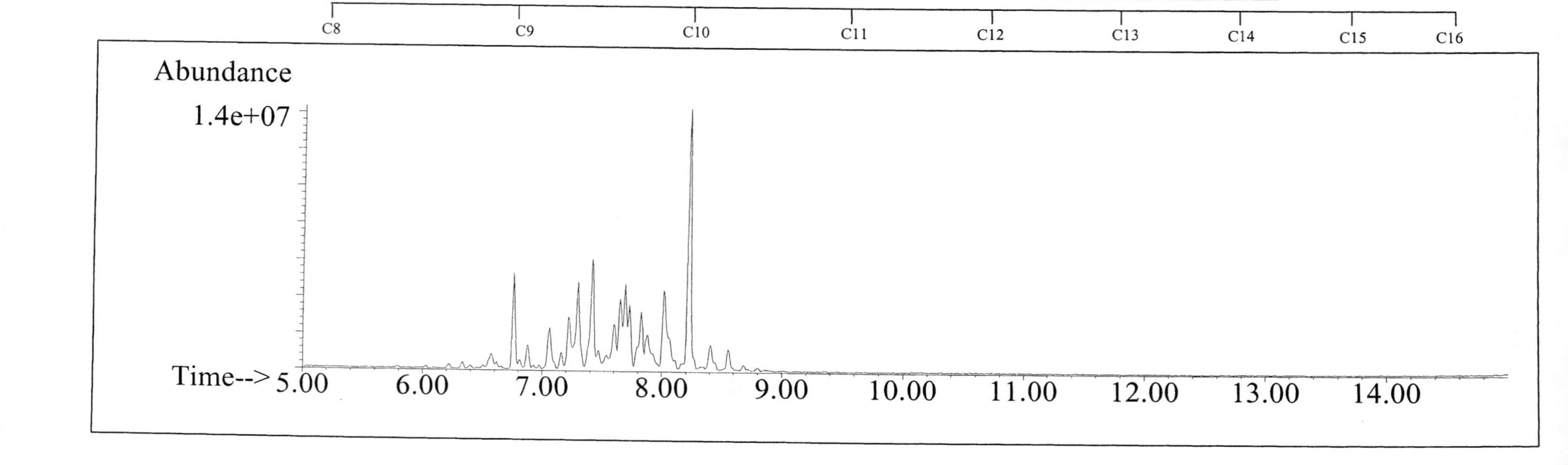

ALKANES

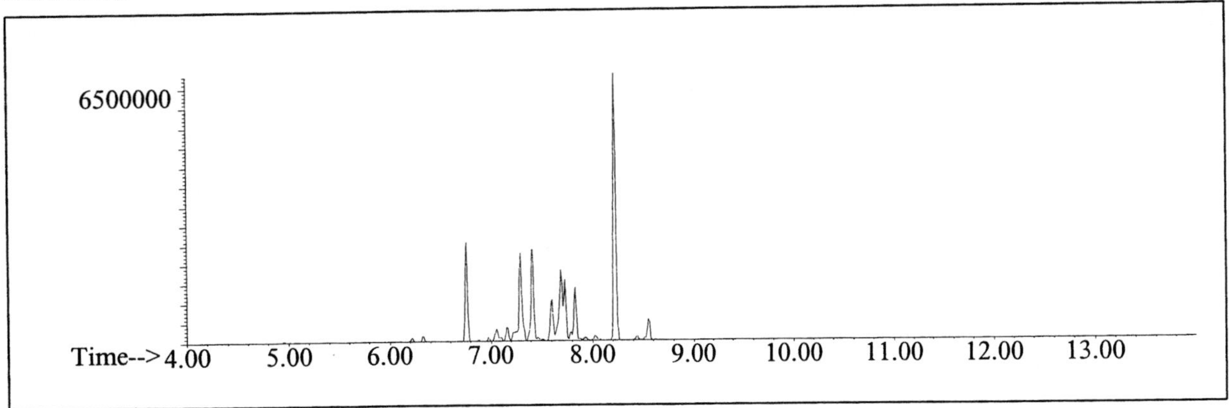

AROMATICS

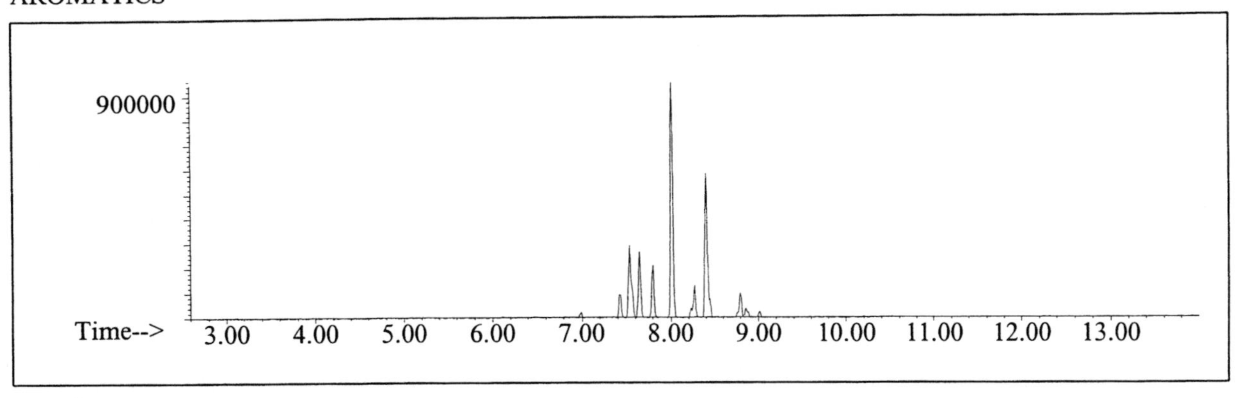

CYCLOPARAFFINS AND ALKENES

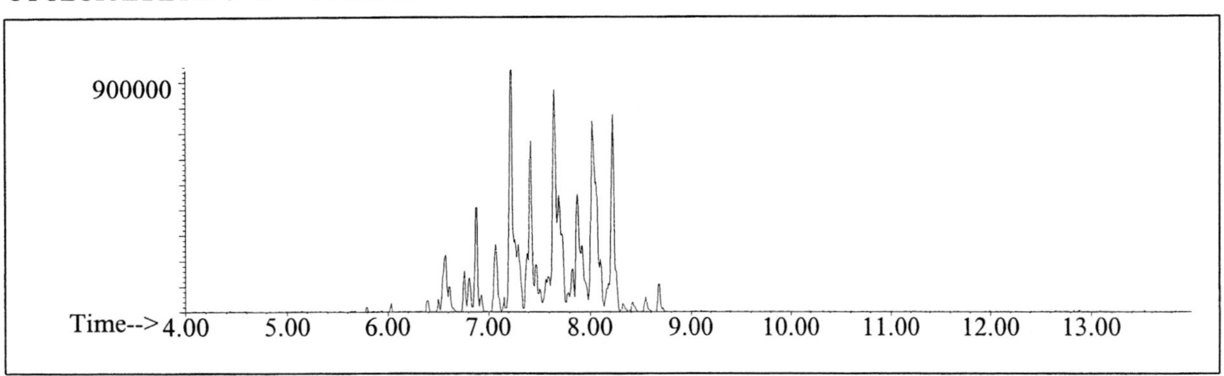

NAPHTHALENES

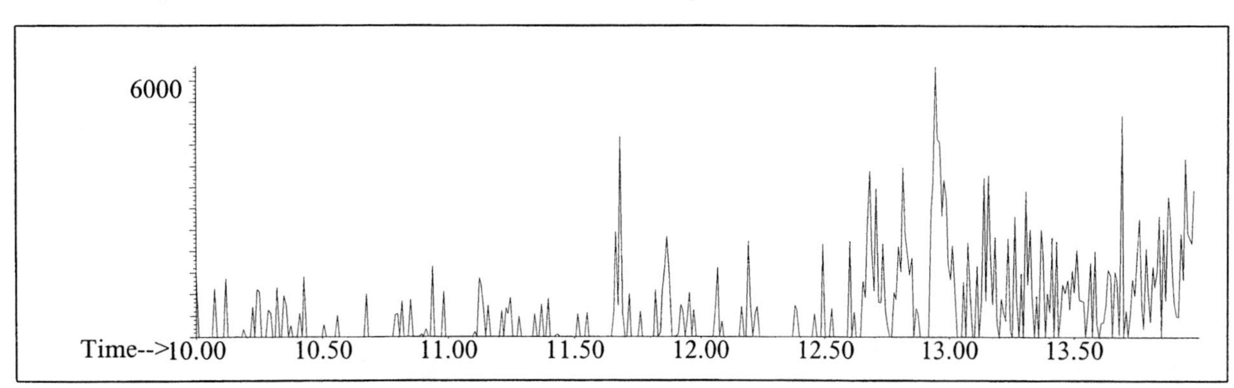

Alkane

Ion 43

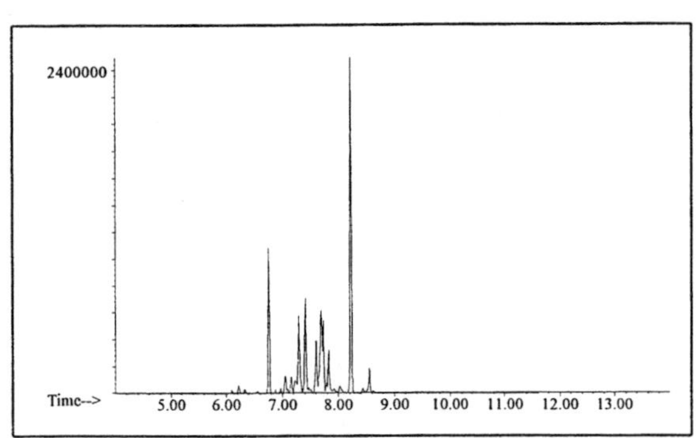

Ion 57

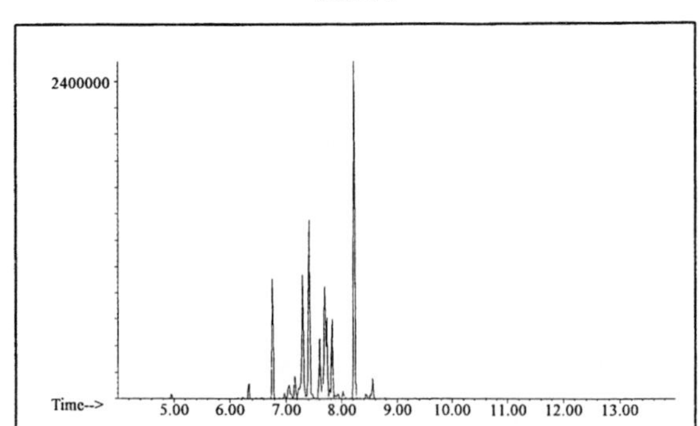

Ion 71

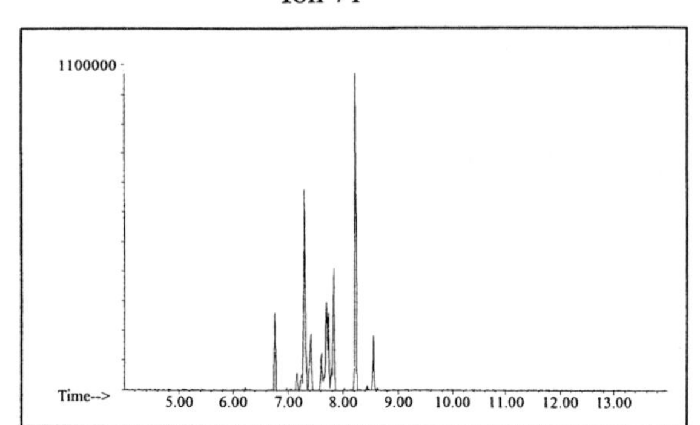

Ion 85

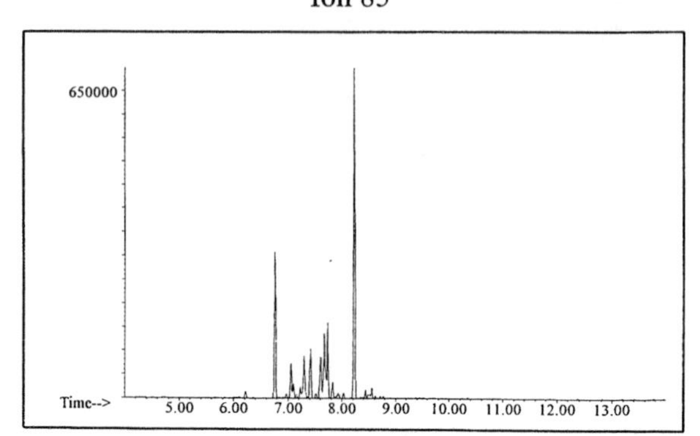

Aromatic

Ion 91

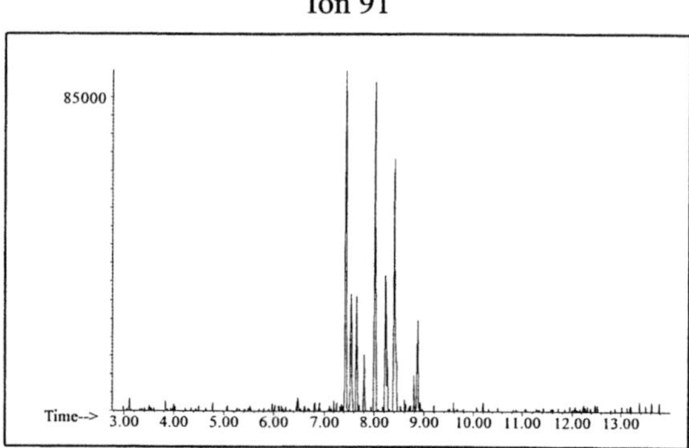

Ion 105

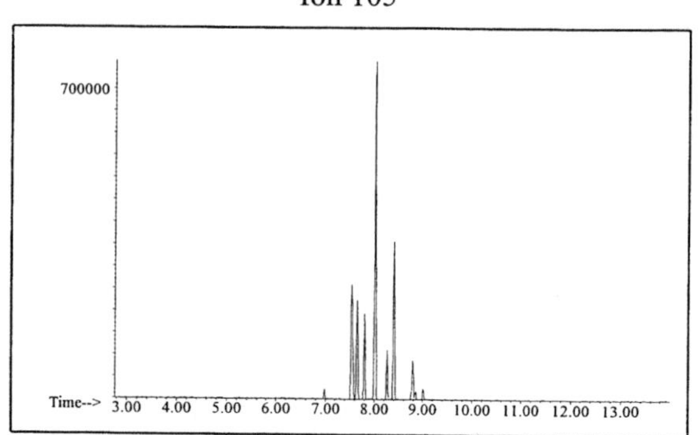

Ion 119

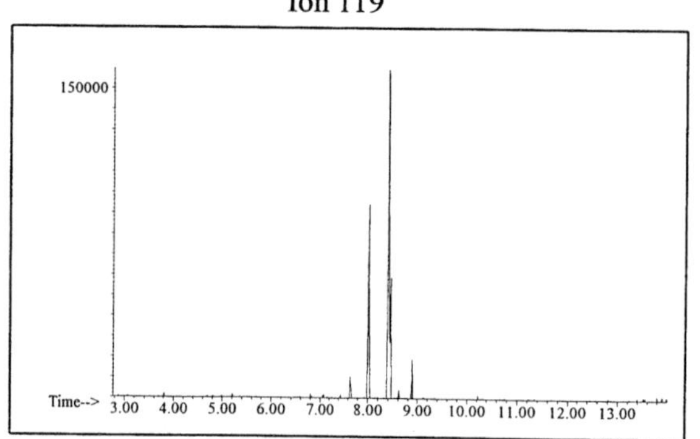

Cycloparaffin

Ion 55

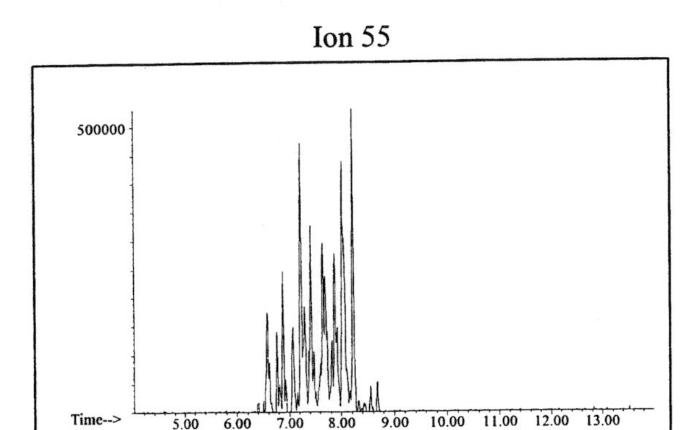

Ion 69

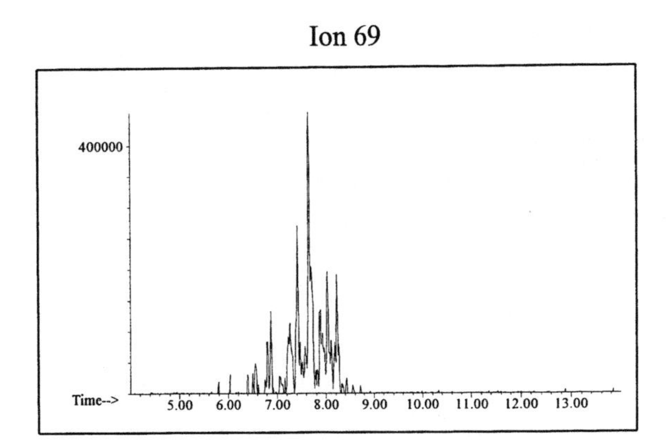

Ion 83

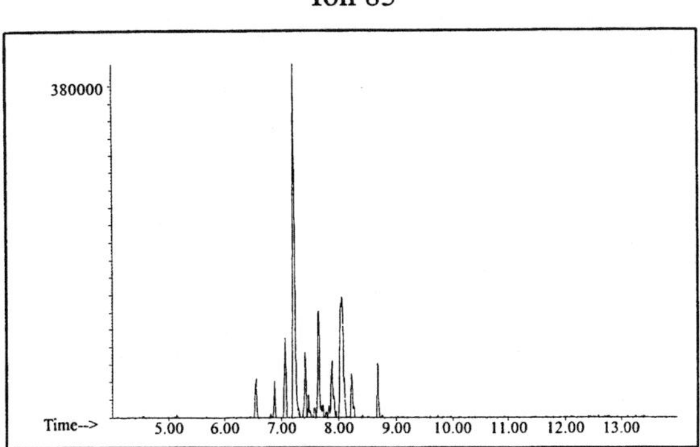

Naphthalene

Ion 128

No Useful Data Obtained

Ion 142

No Useful Data Obtained

Ion 156

No Useful Data Obtained

Miscellaneous #26

20uL/mL Pentane

Product Displayed: WD-40

Other Similar Products:

ASTM: Class 0 (Miscellaneous)

Product Uses: Penetrating oil, cleaning solvent

Macro Code: M

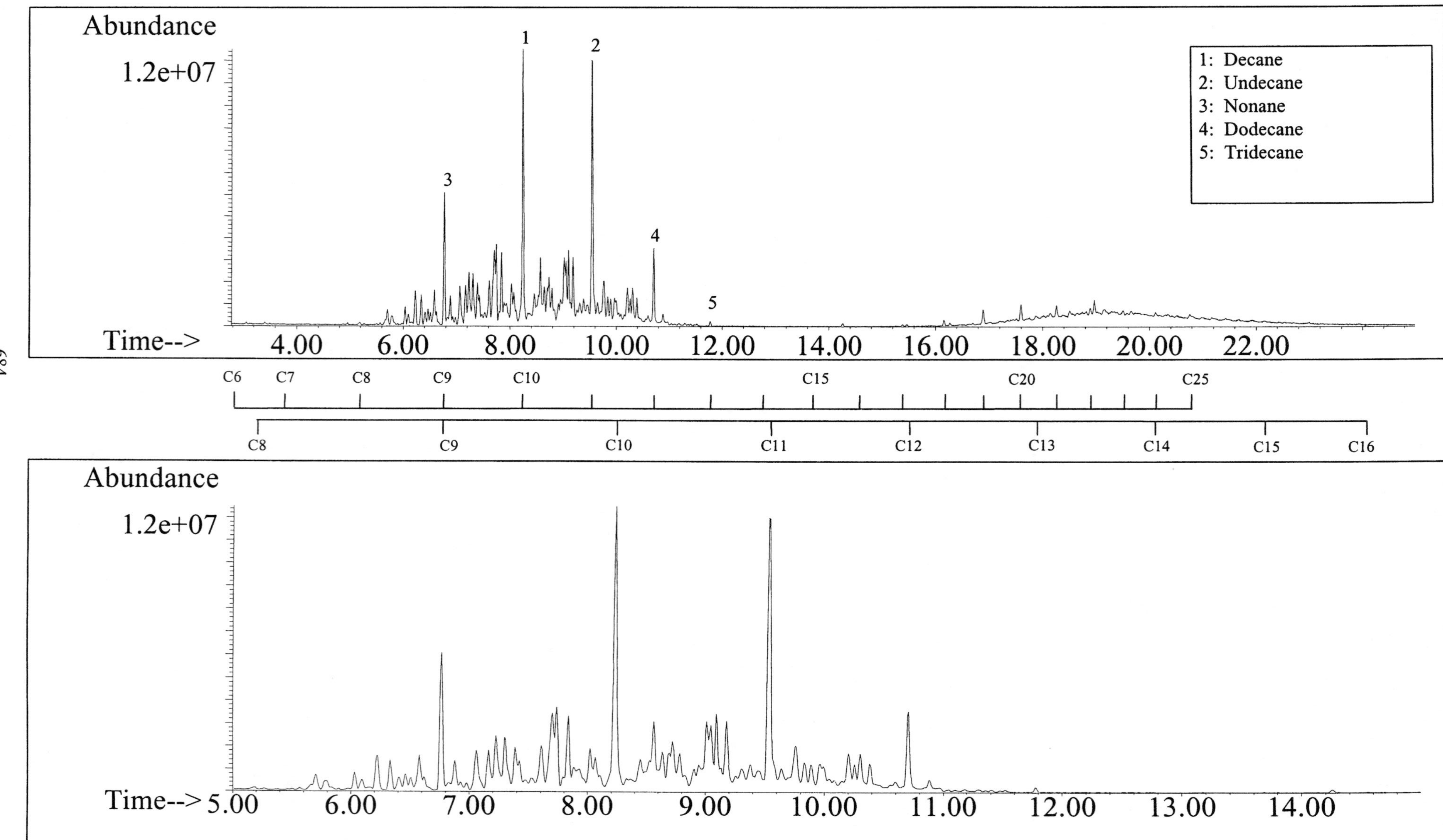

ALKANES

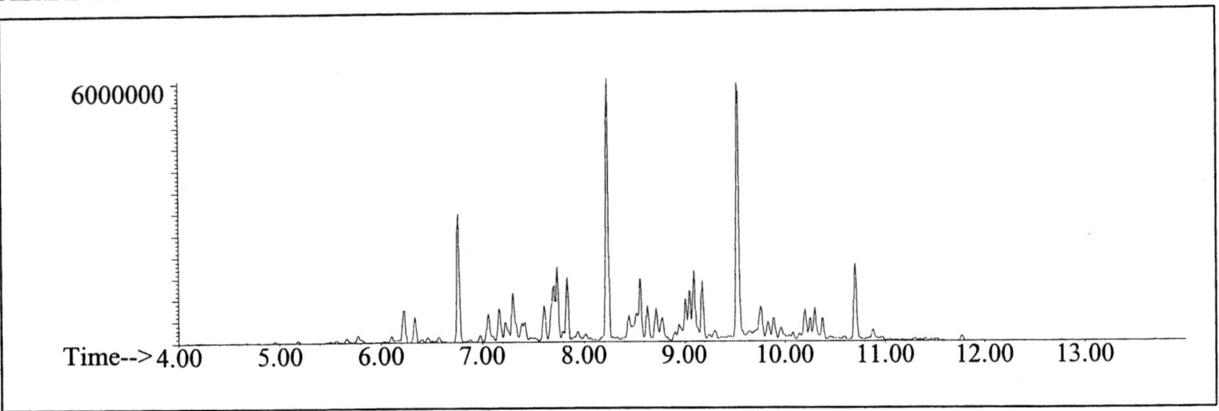

AROMATICS

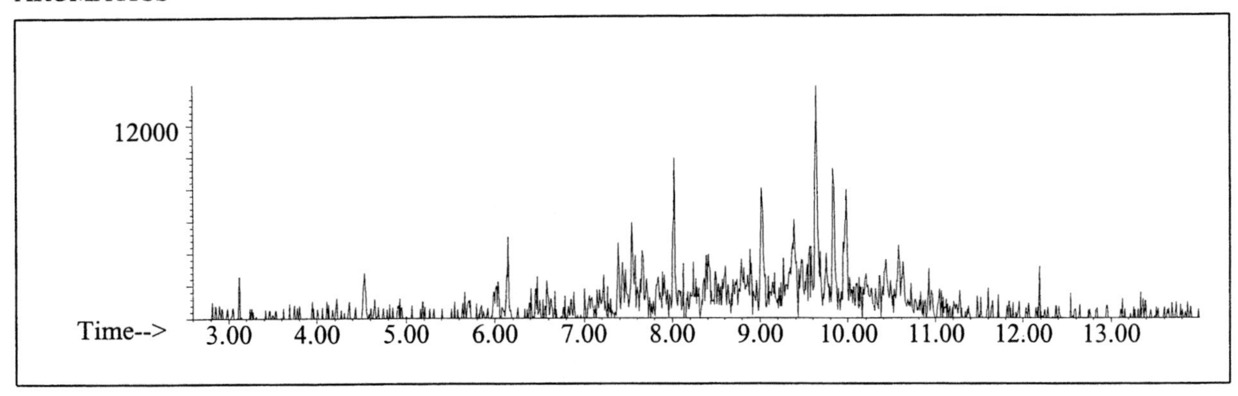

CYCLOPARAFFINS AND ALKENES

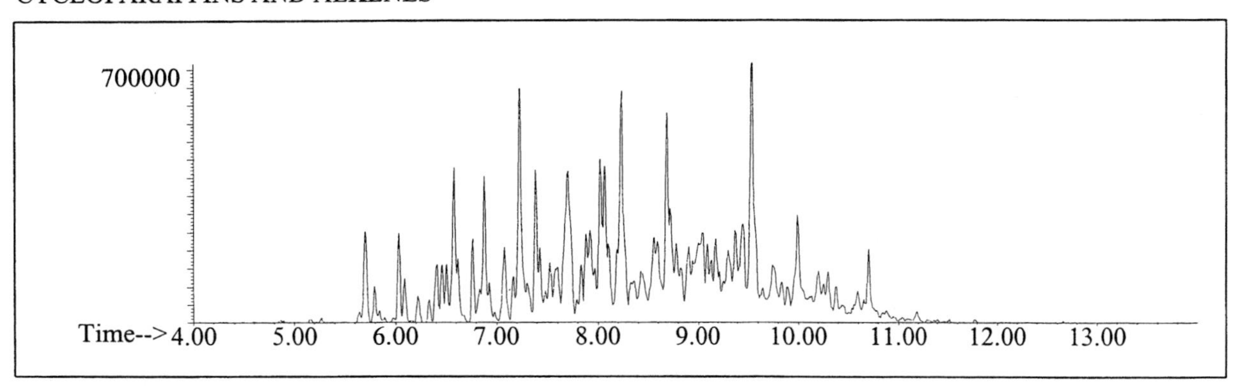

NAPHTHALENES

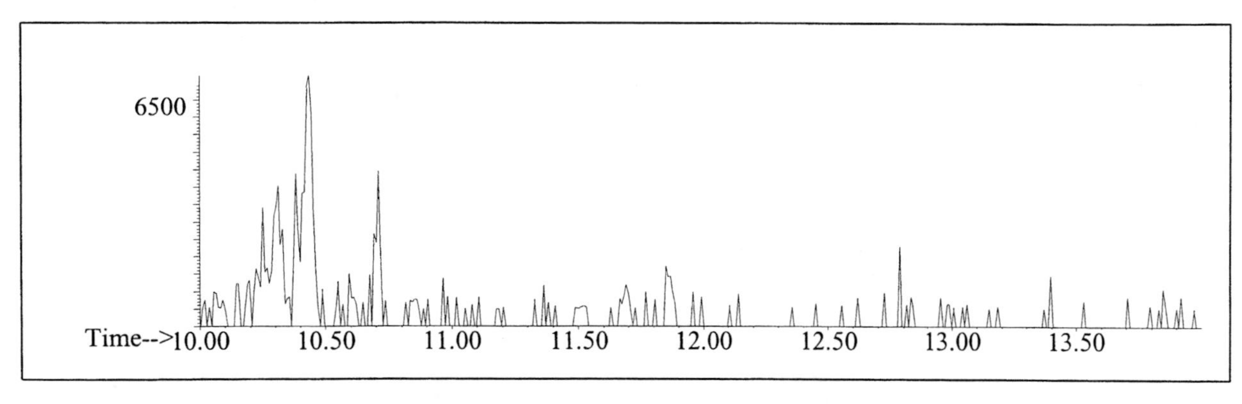

Alkane

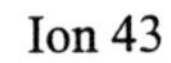

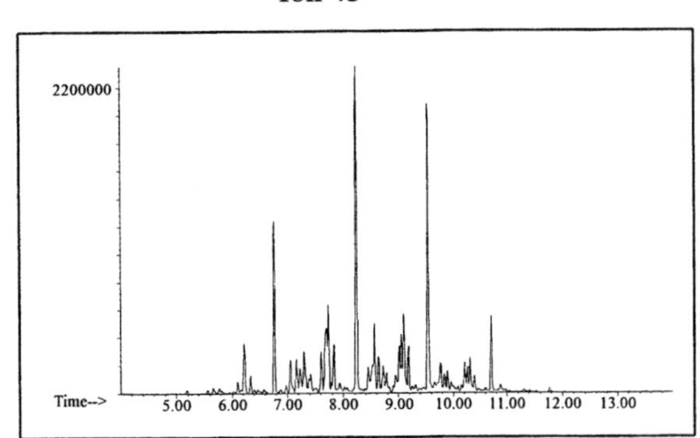

Ion 57

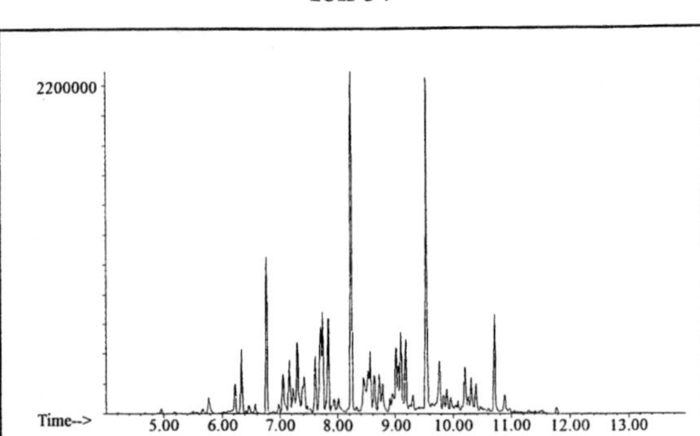

Ion 71

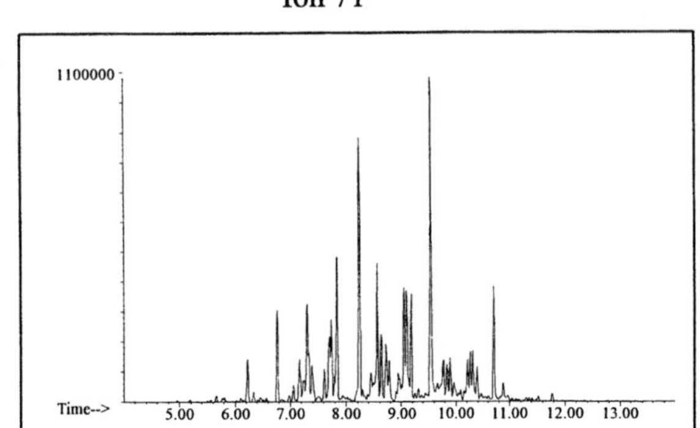

Ion 85

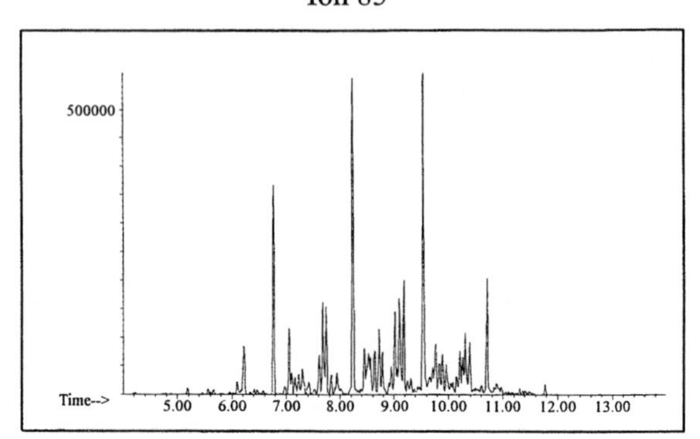

Aromatic

Ion 91

No Useful Data Obtained

Ion 105

No Useful Data Obtained

Ion 119

No Useful Data Obtained

Cycloparaffin

Ion 55

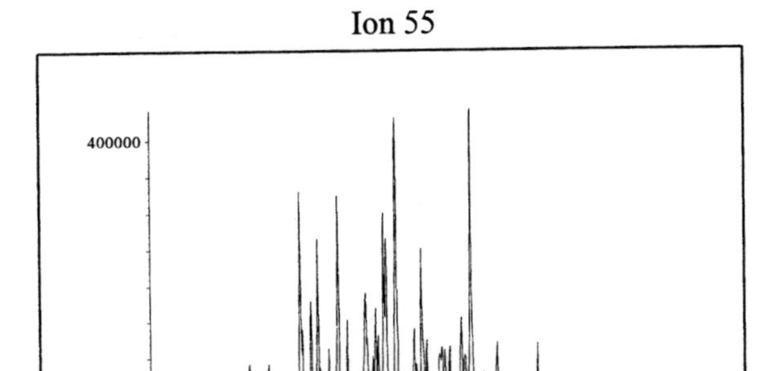

Ion 69

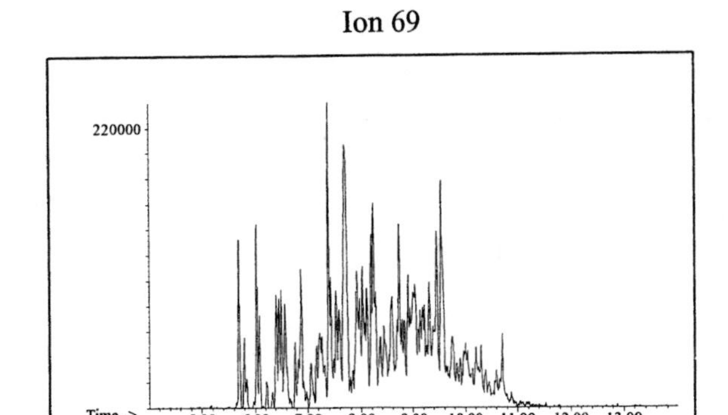

Ion 83

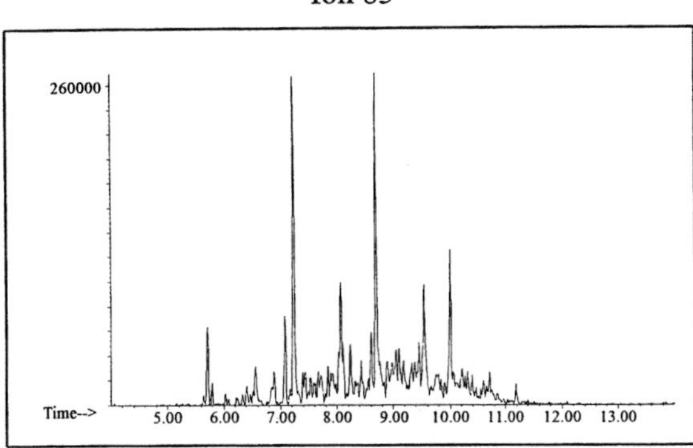

Naphthalene

Ion 128

No Useful Data Obtained

Ion 142

No Useful Data Obtained

Ion 156

No Useful Data Obtained

Miscellaneous #27

20uL/mL Pentane

Product Displayed: Gunk Degreaser

Other Similar Products:

ASTM: Class 5 (Miscellaneous)

Product Uses: Cleaning solvent

Macro Code: H

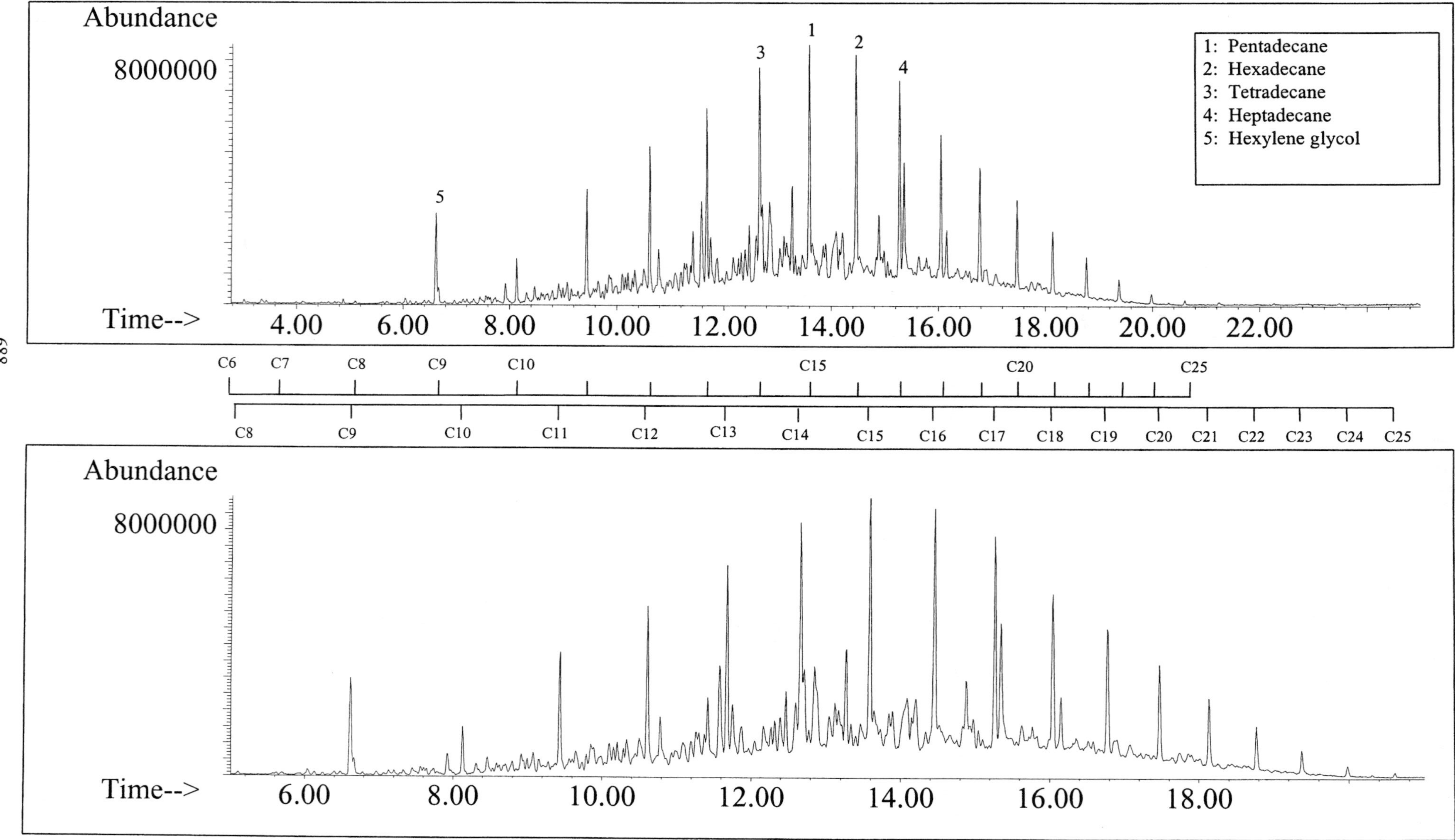

ALKANES

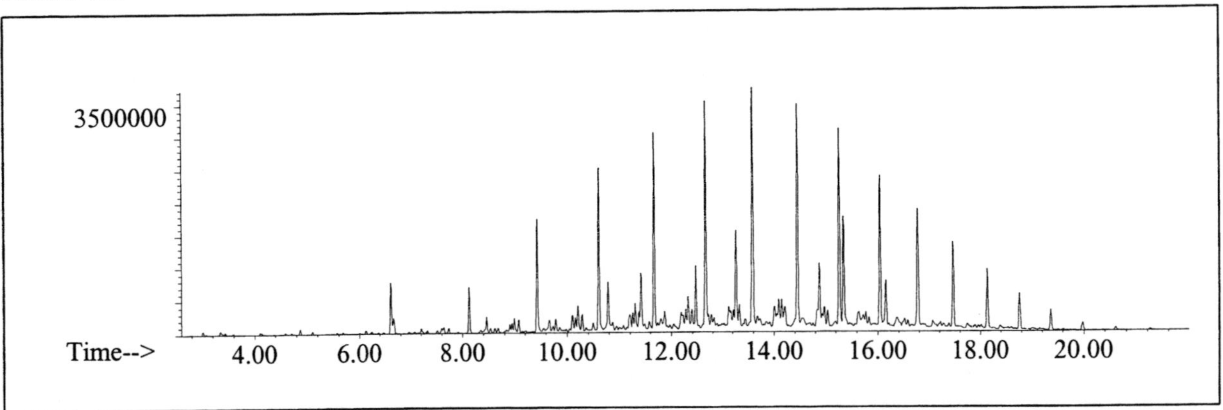

AROMATICS

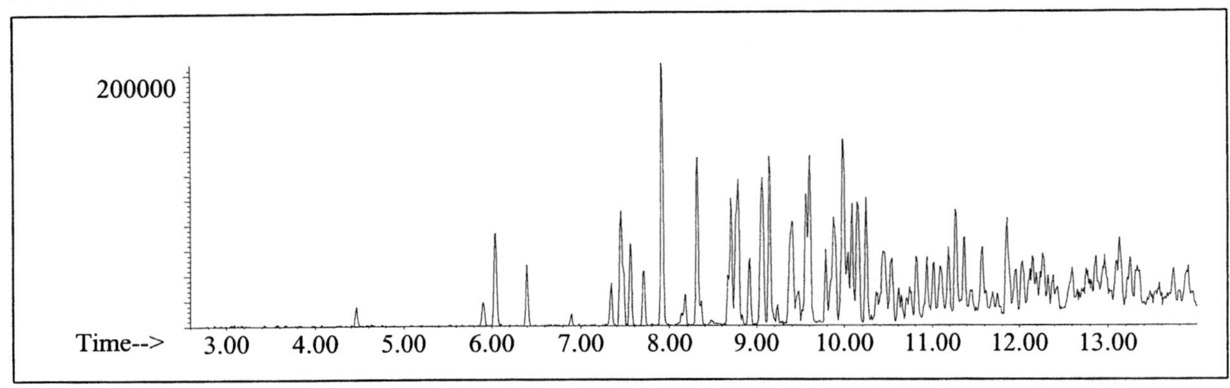

CYCLOPARAFFINS AND ALKENES

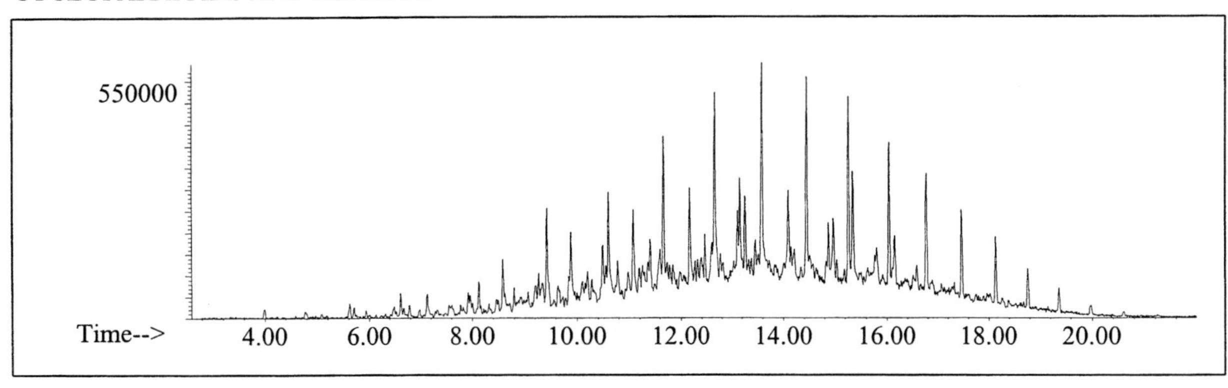

NAPHTHALENES

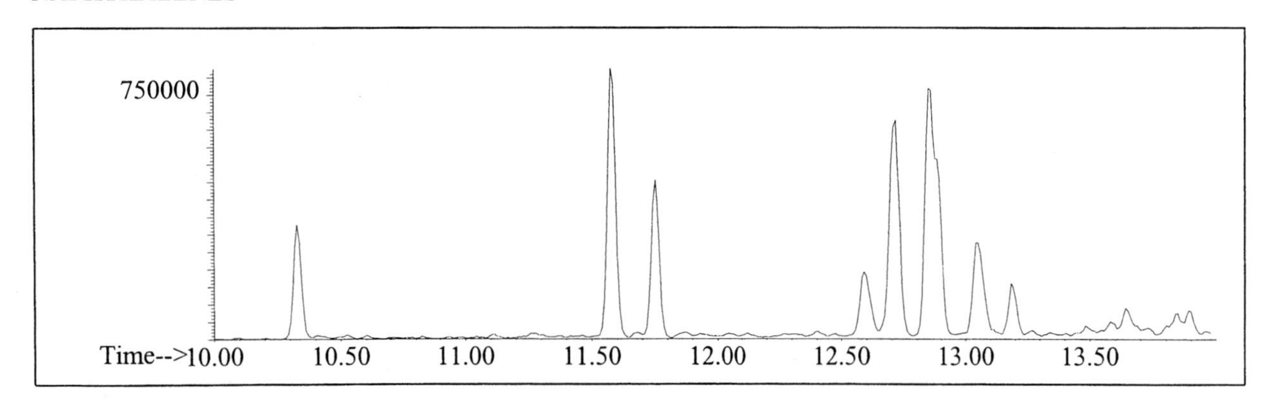

Alkane

Ion 43

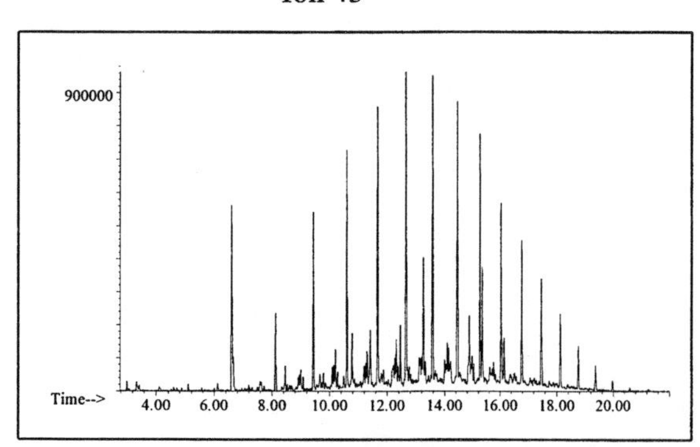

Ion 57

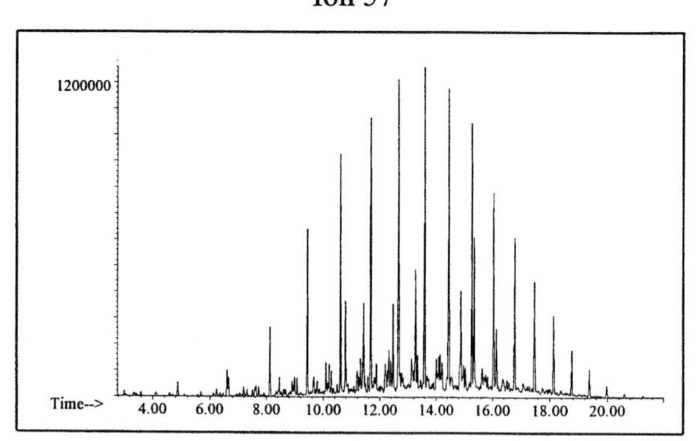

Ion 71

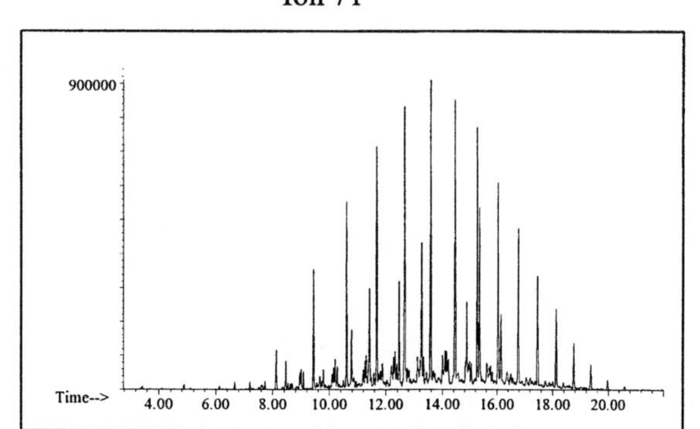

Ion 85

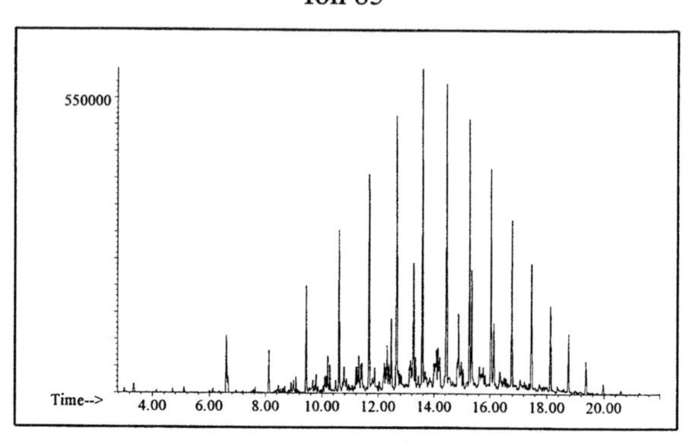

Aromatic

Ion 91

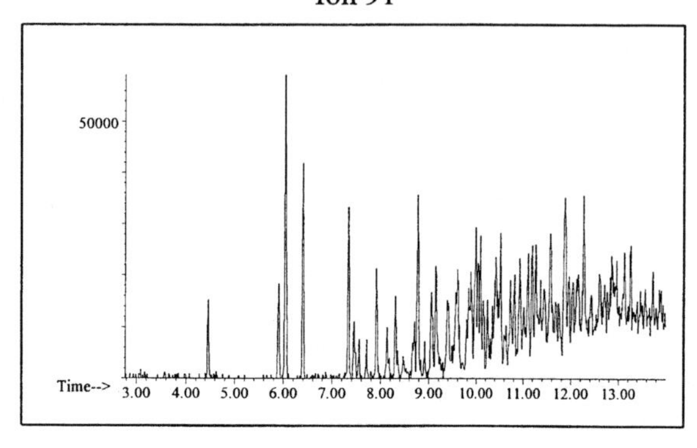

Ion 105

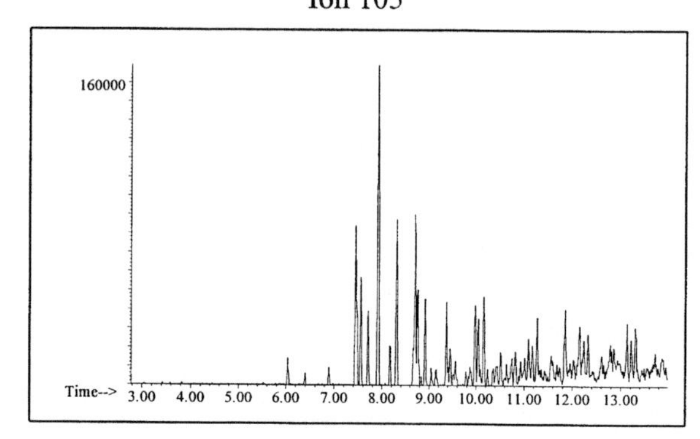

Ion 119

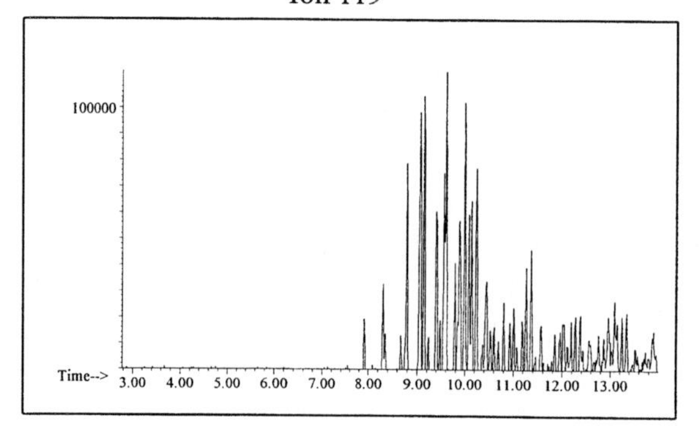

Cycloparaffin

Ion 55

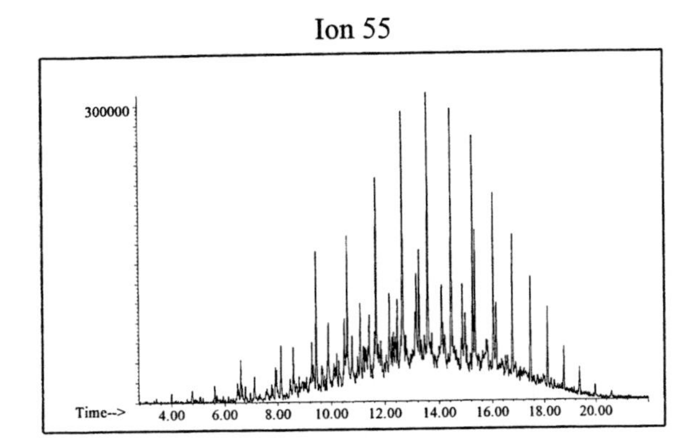

Ion 69

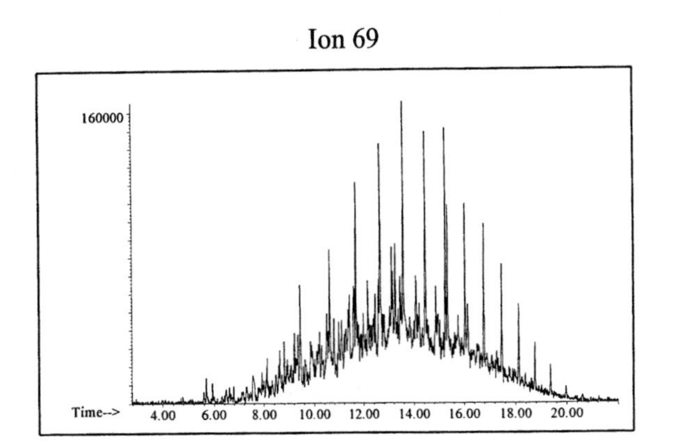

Ion 83

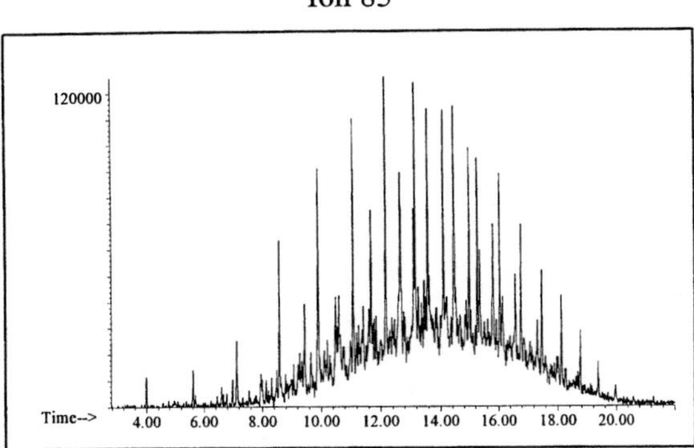

Naphthalene

Ion 128

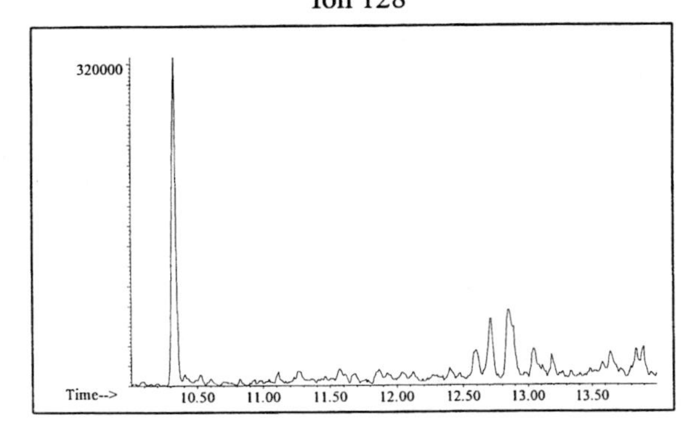

Ion 142

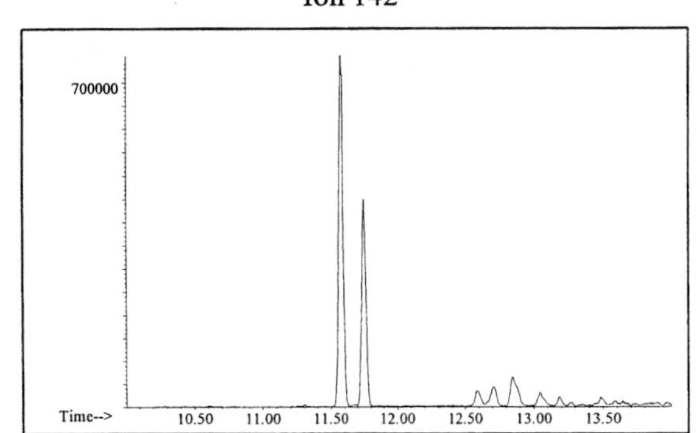

Ion 156

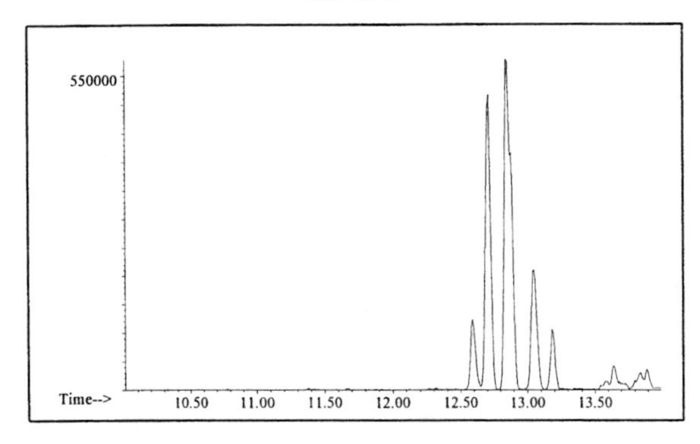

Miscellaneous #28

20uL/mL Pentane

Product Displayed: Prestone Water Remover

Other Similar Products:

ASTM: Class 0 (Miscellaneous)

Product Uses: Fuel treatment

Macro Code: H

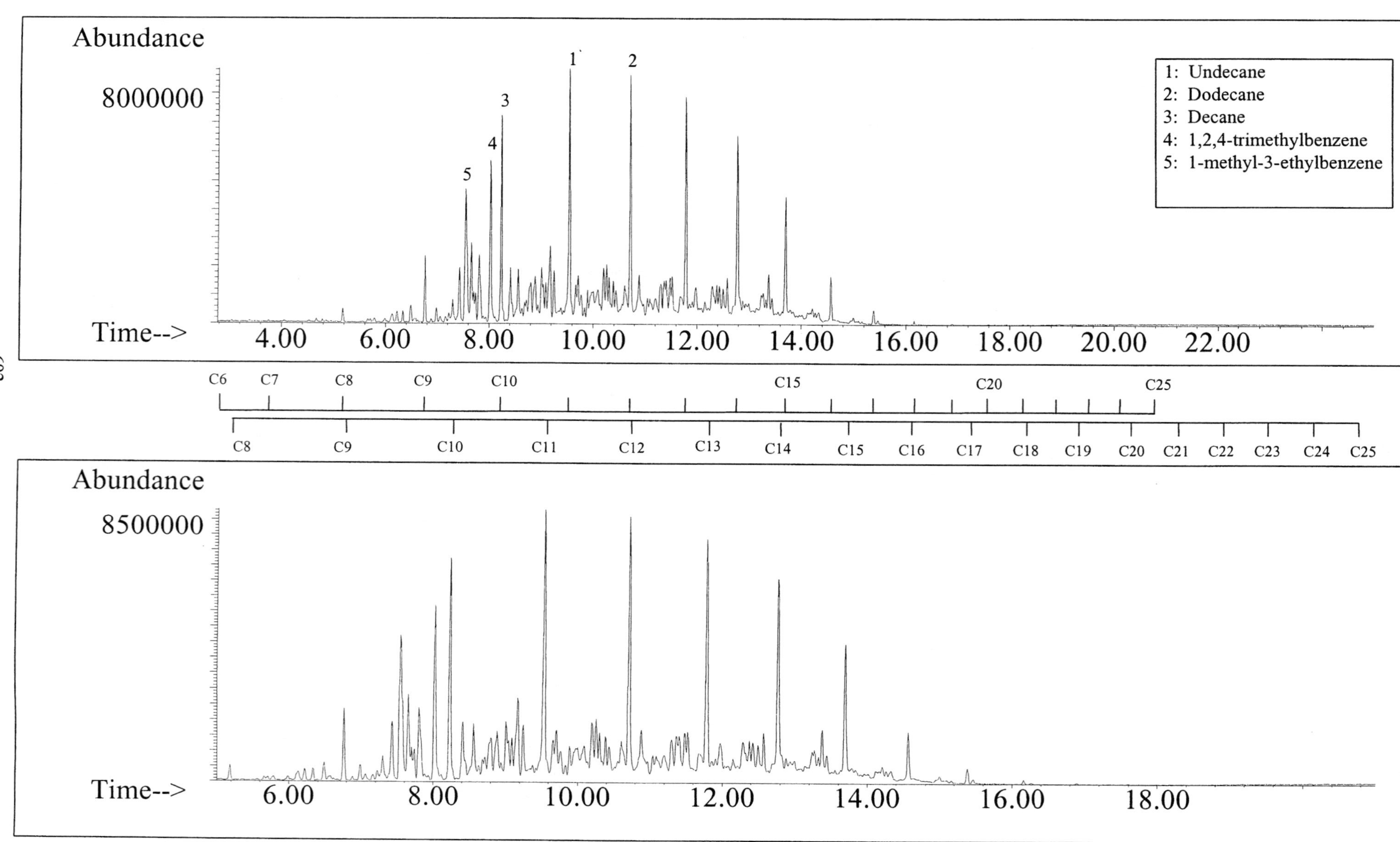

ALKANES

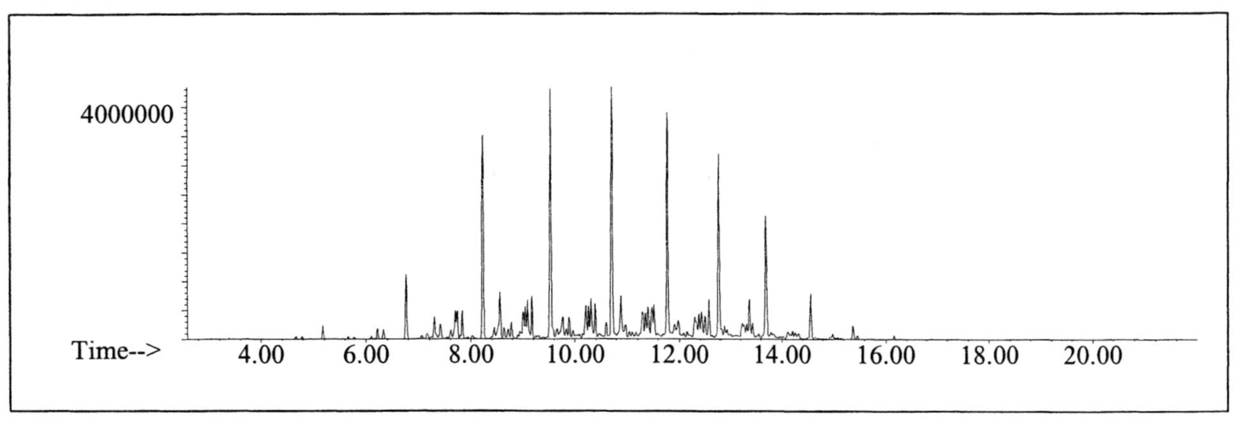

AROMATICS

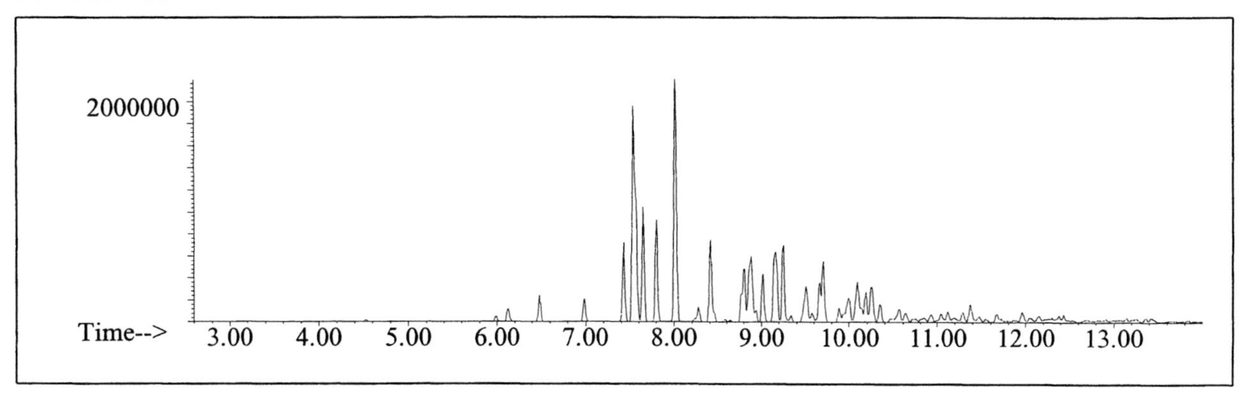

CYCLOPARAFFINS AND ALKENES

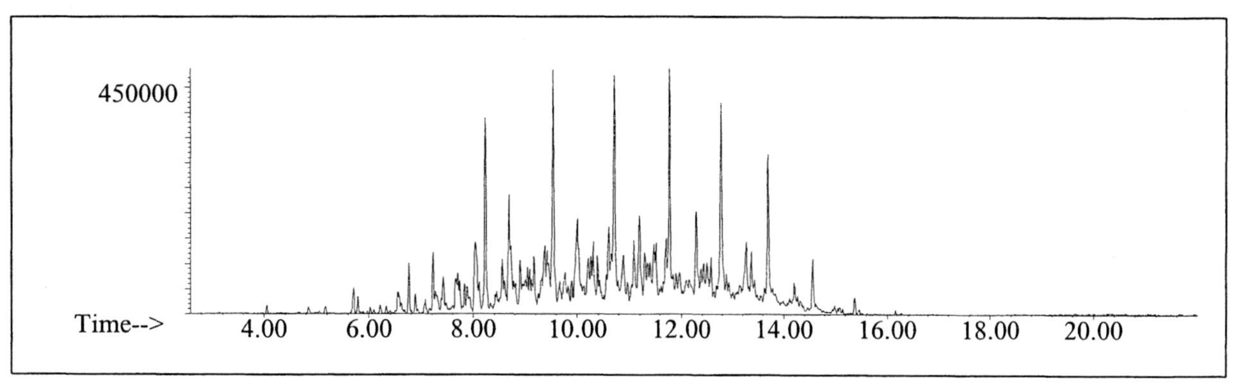

NAPHTHALENES

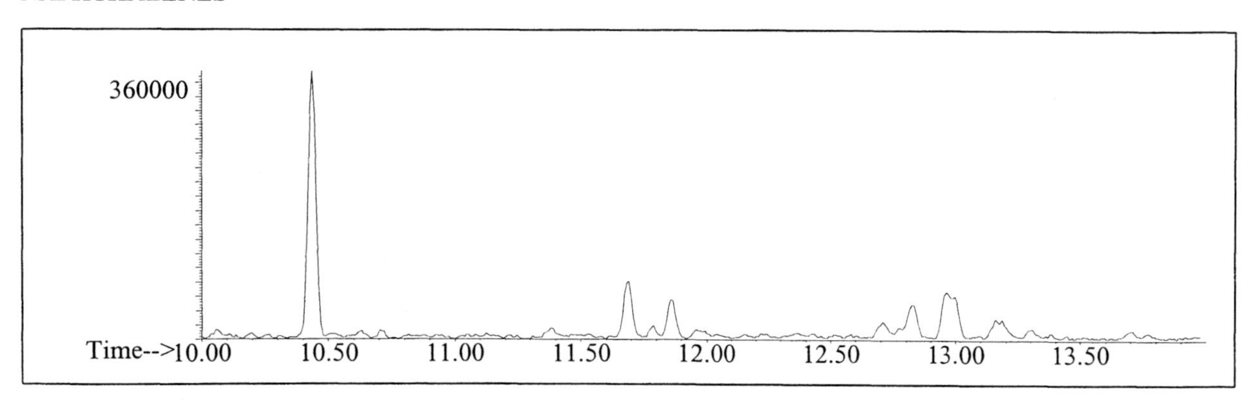

Alkane

Ion 43

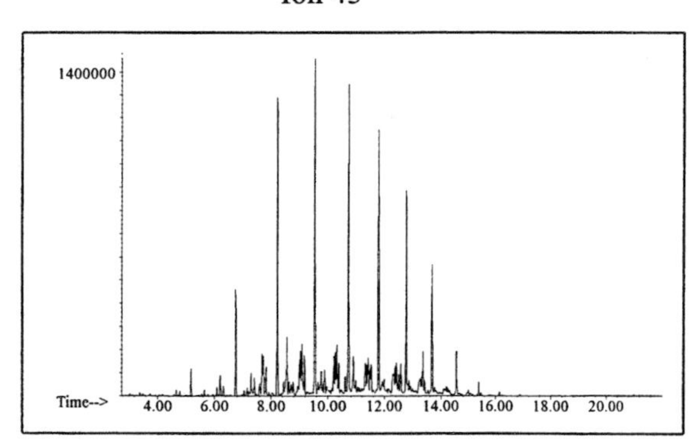

Ion 57

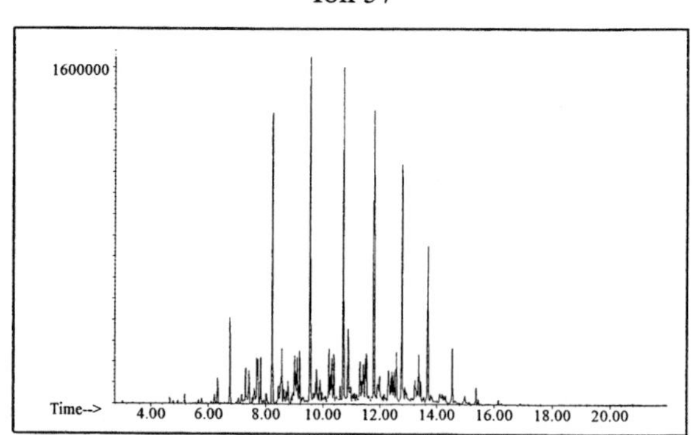

Ion 71

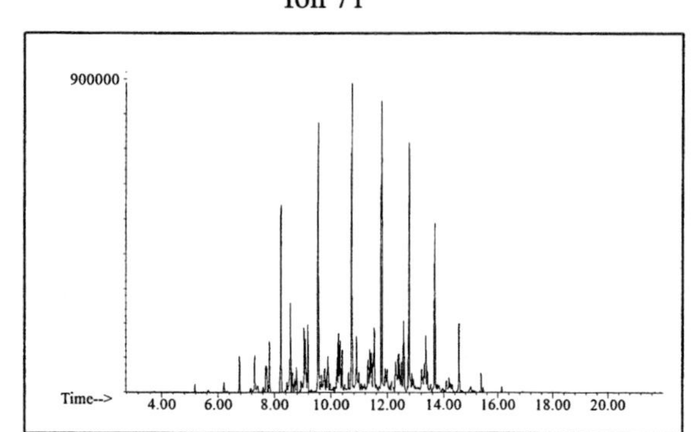

Ion 85

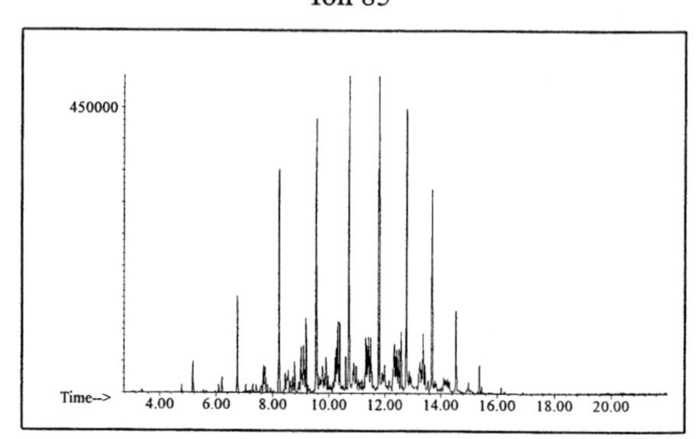

Aromatic

Ion 91

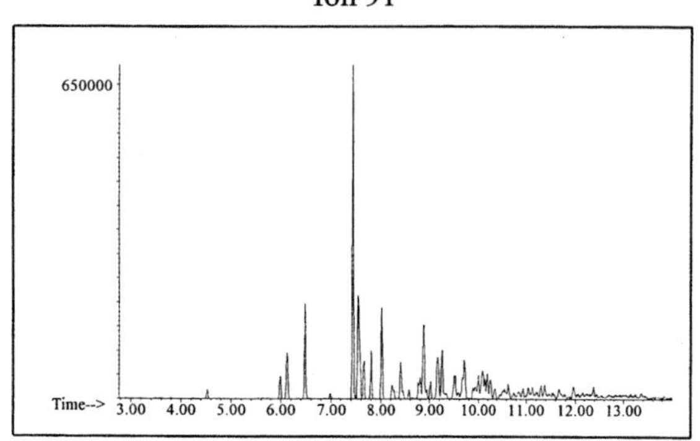

Ion 105

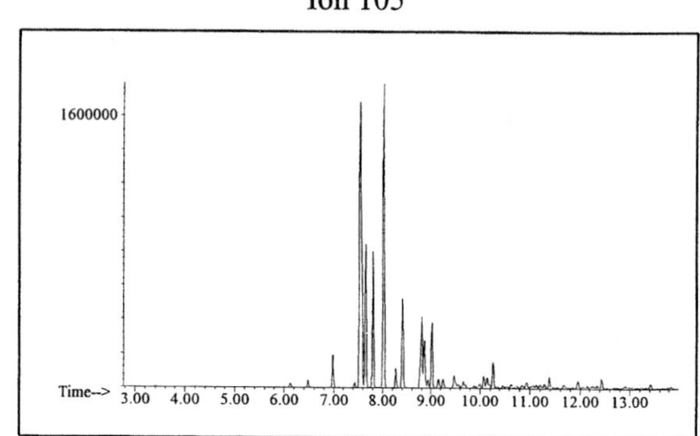

Ion 119

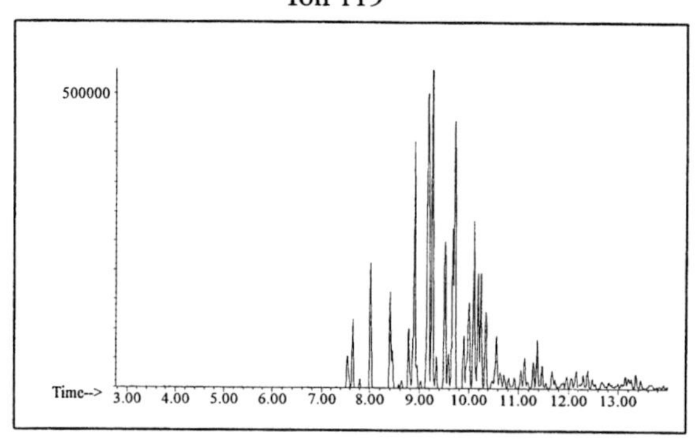

Cycloparaffin

Ion 55

Ion 69

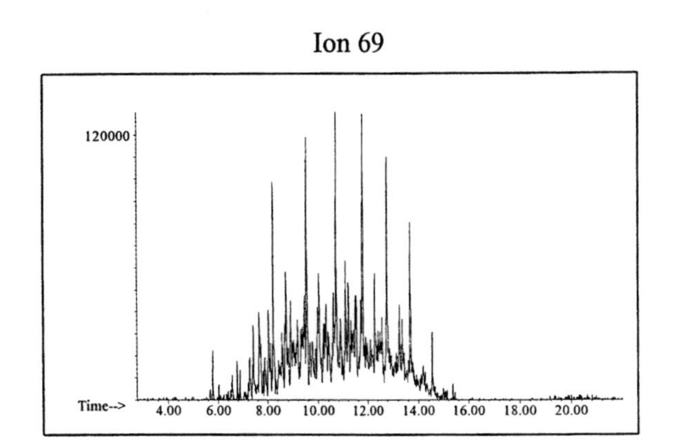

Ion 83

Naphthalene

Ion 128

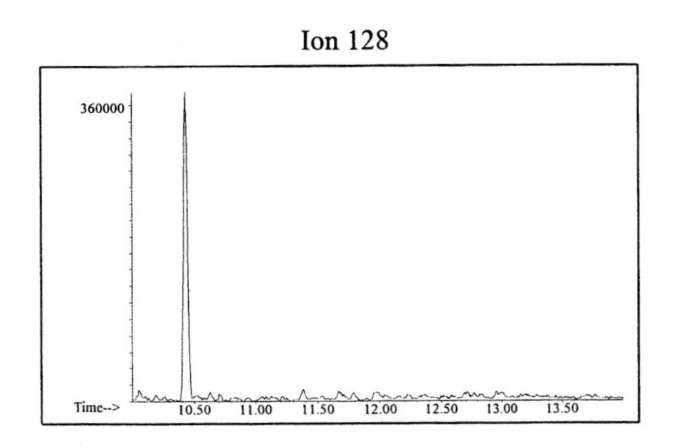

Ion 142

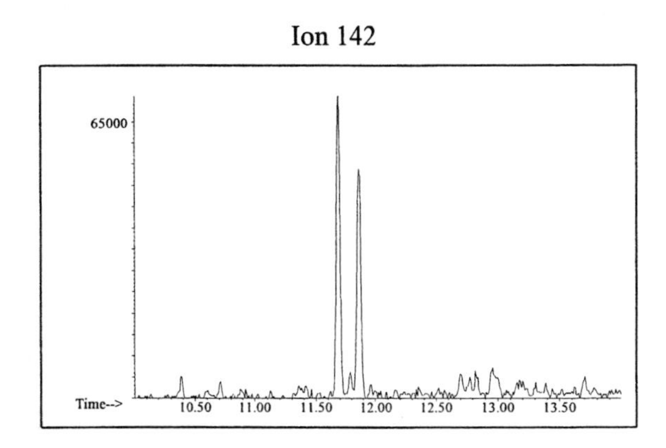

Ion 156

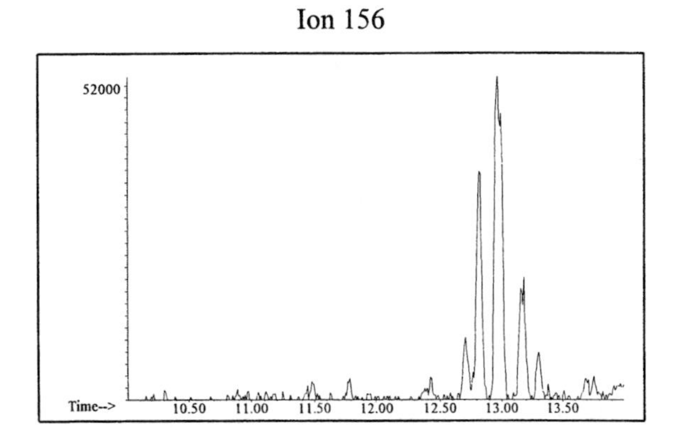

Miscellaneous #29

20uL/mL Pentane

ASTM: Class 0 (Miscellaneous)

Macro Code: L

Product Displayed: Turpentine Gum Spirits

Product Uses: Paint thinner

Other Similar Products:

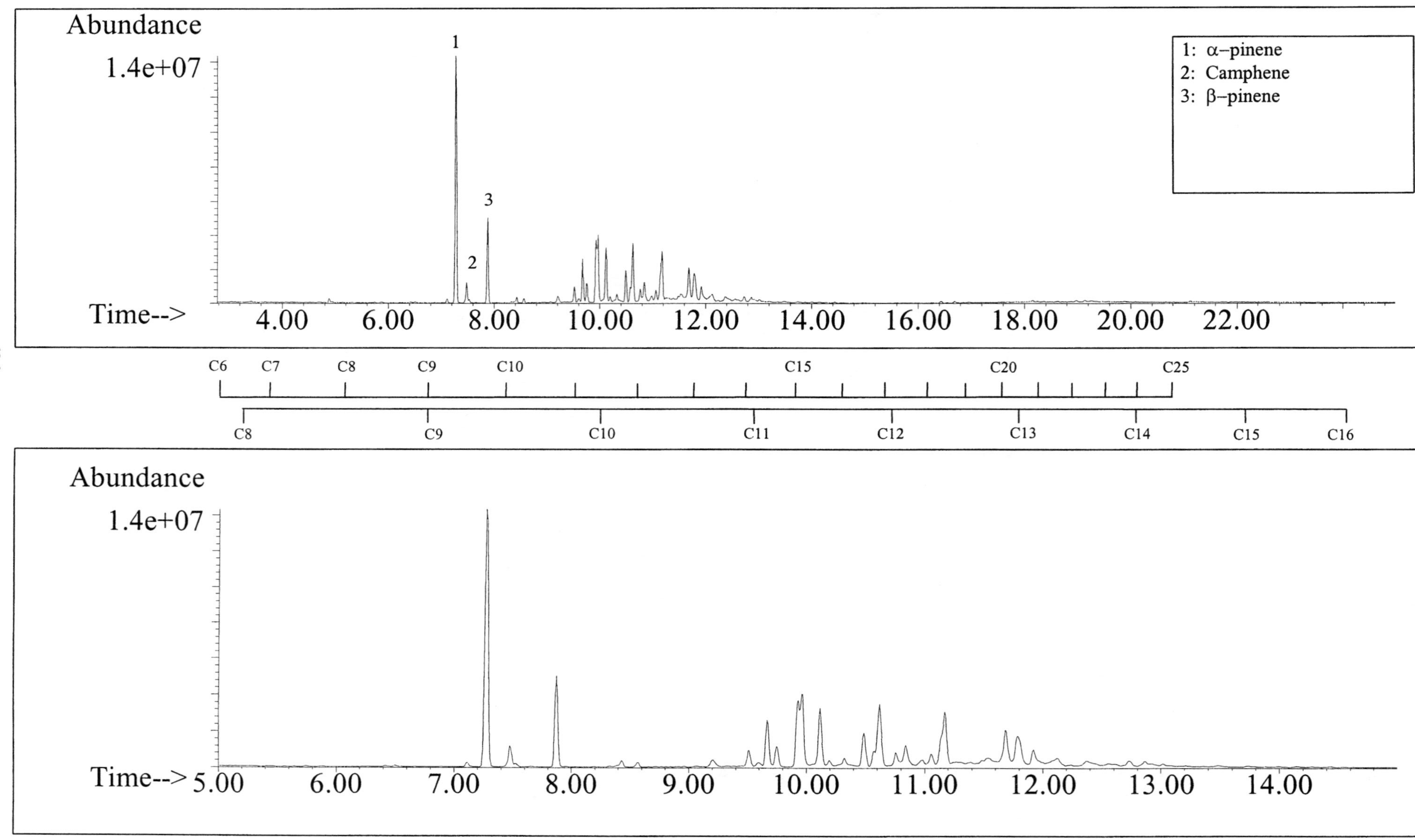

ALKANES

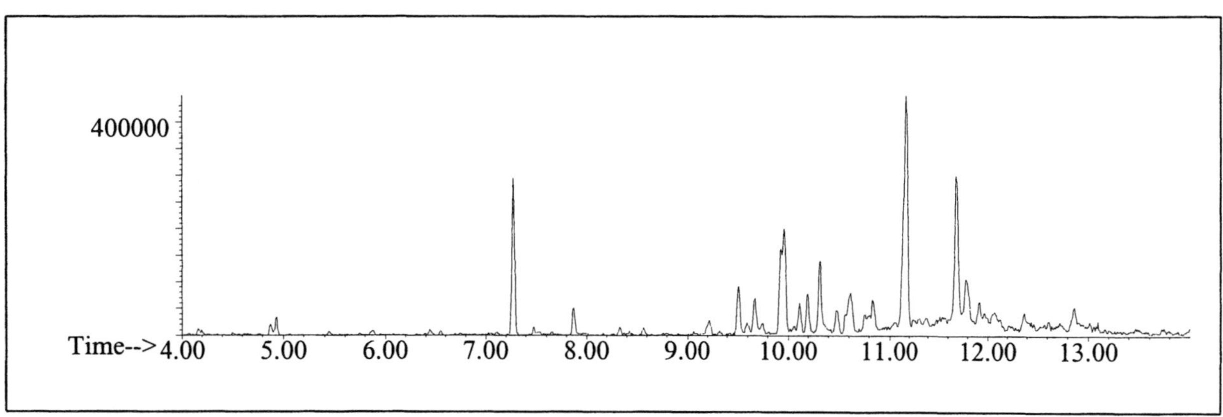

AROMATICS

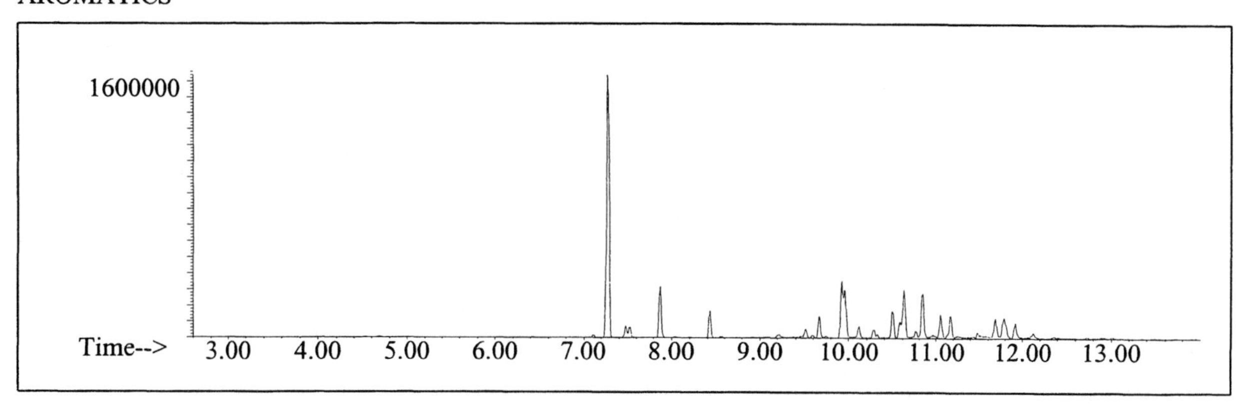

CYCLOPARAFFINS AND ALKENES

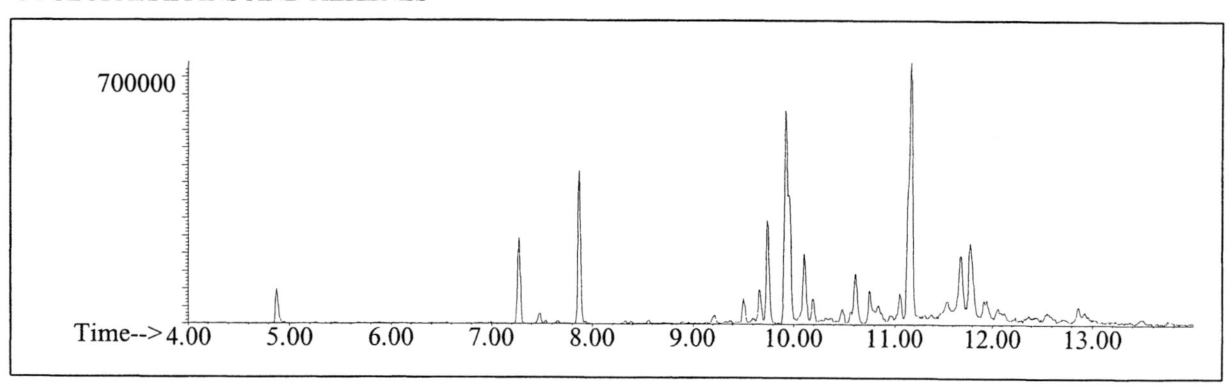

NAPHTHALENES

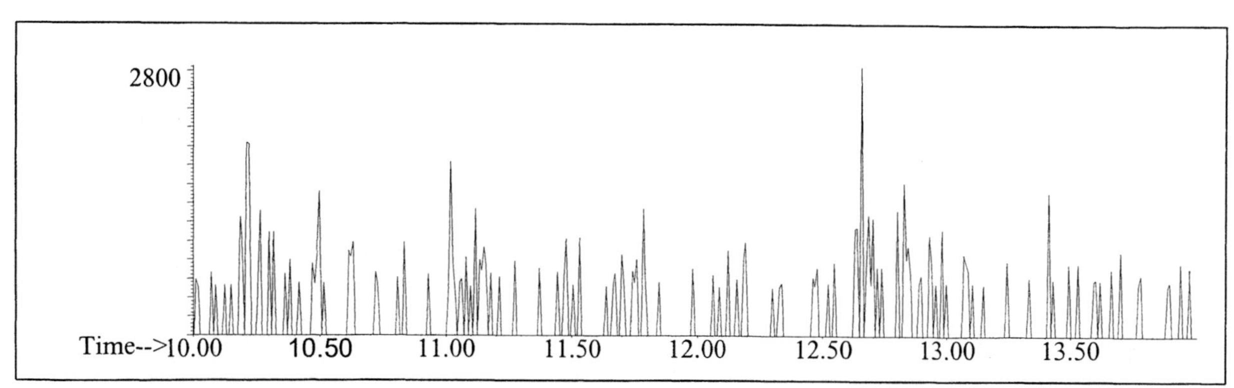

Alkane

Ion 43

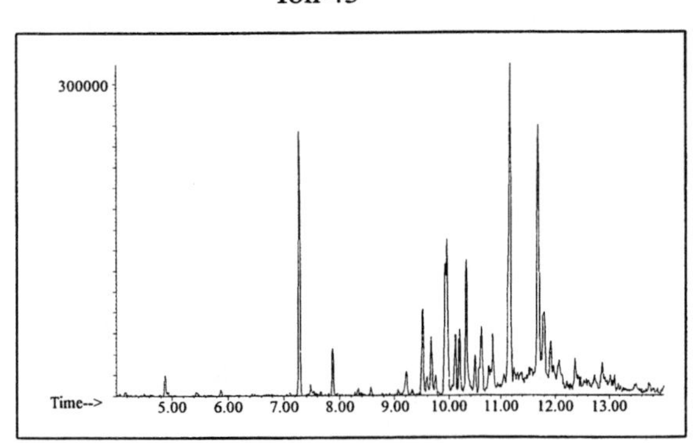

Ion 57

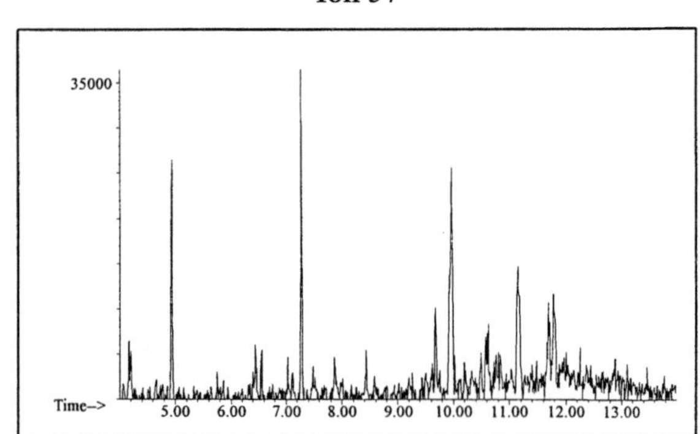

Ion 71

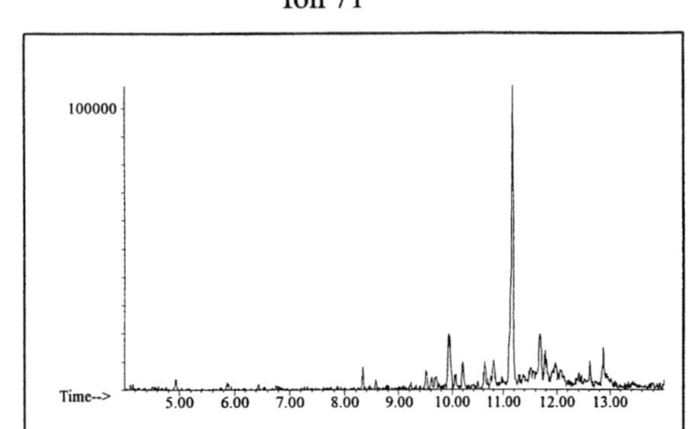

Ion 85

No Useful Data Obtained

Aromatic

Ion 91

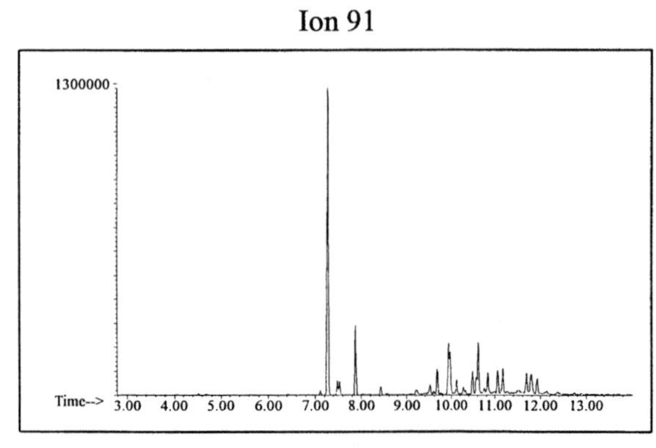

Ion 105

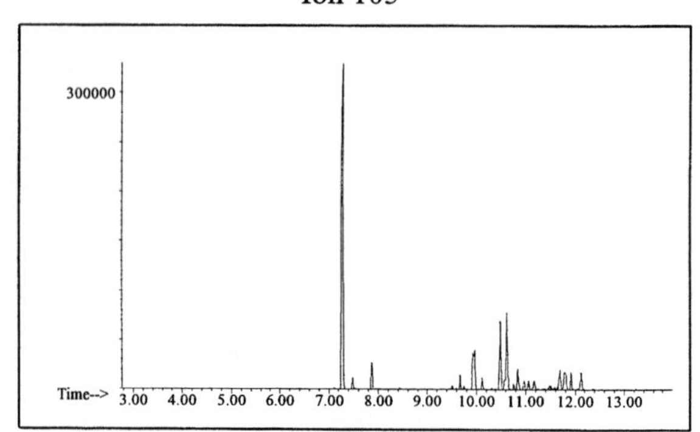

Ion 119

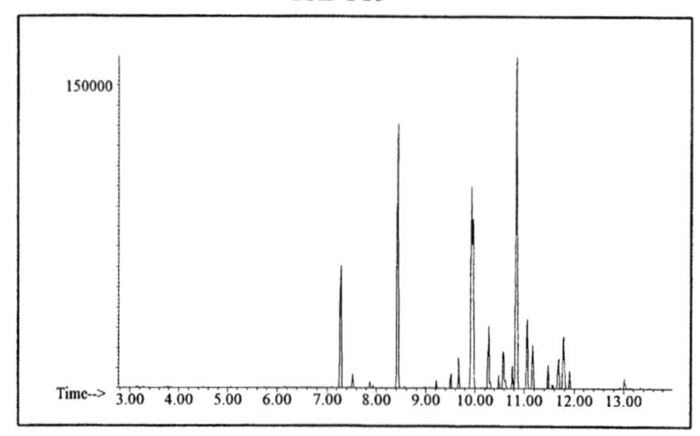

Cycloparaffin

Ion 55

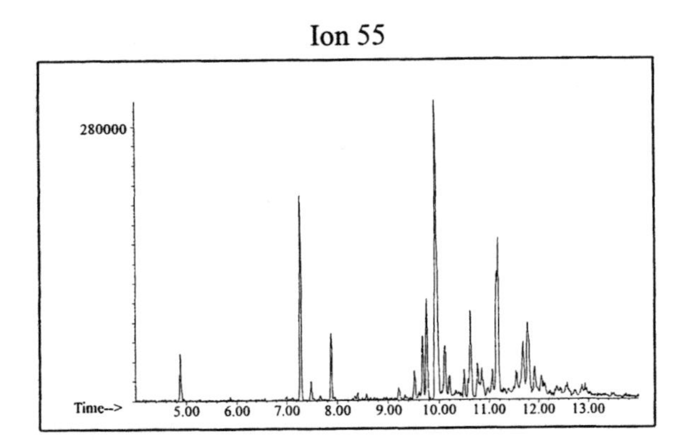

Ion 69

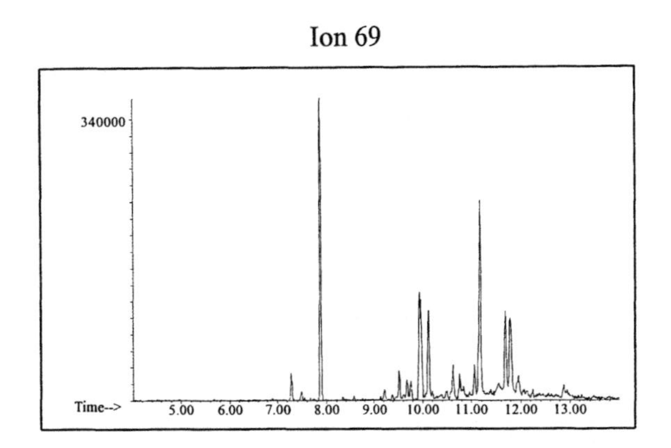

Ion 83

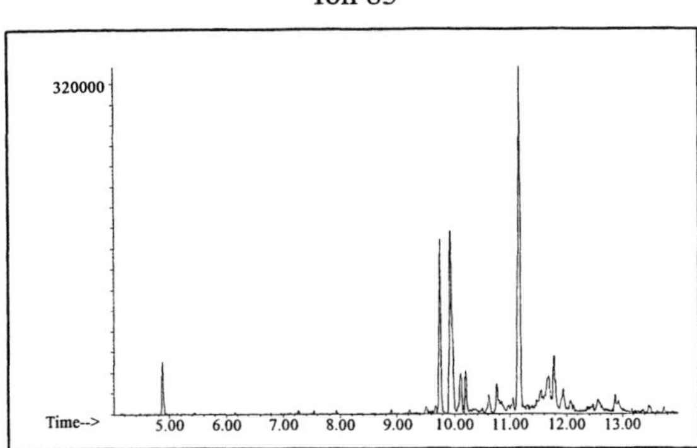

Naphthalene

Ion 128

No Useful Data Obtained

Ion 142

No Useful Data Obtained

Ion 156

No Useful Data Obtained

Miscellaneous #30

20uL/mL Pentane

Product Displayed: Steam Distilled Turpentine

Other Similar Products:

ASTM: Class 0 (Miscellaneous)

Product Uses: Cleaning solvent, paint thinner

Macro Code: L

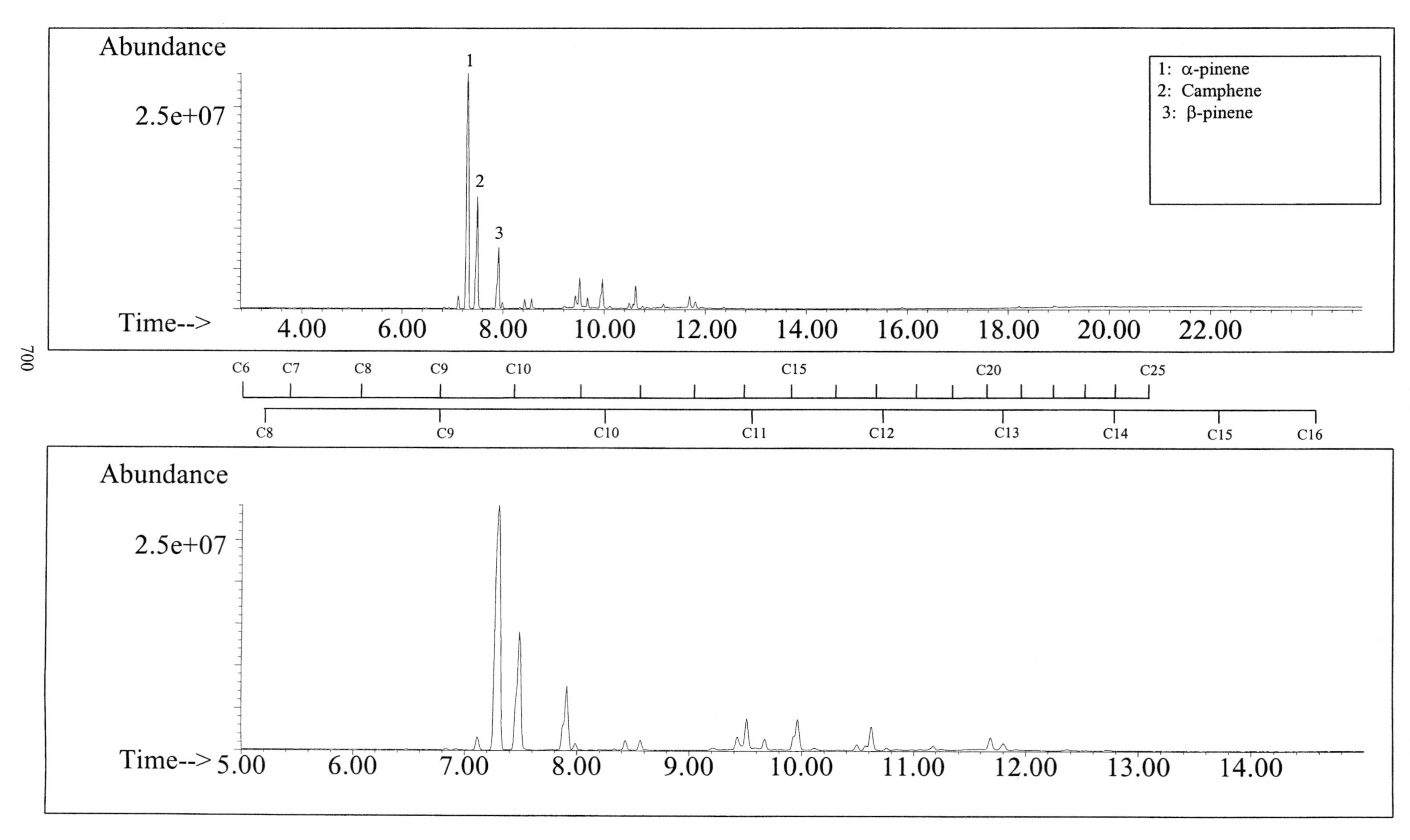

ALKANES

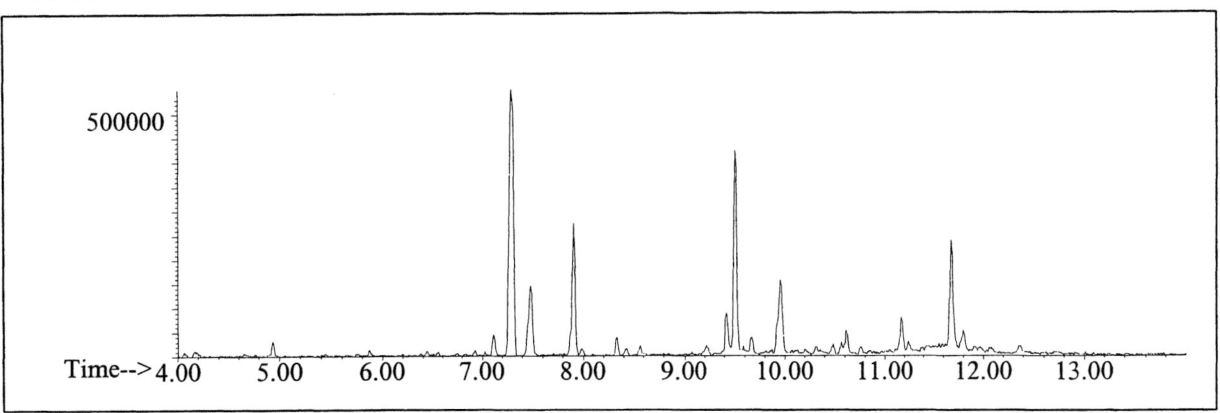

AROMATICS

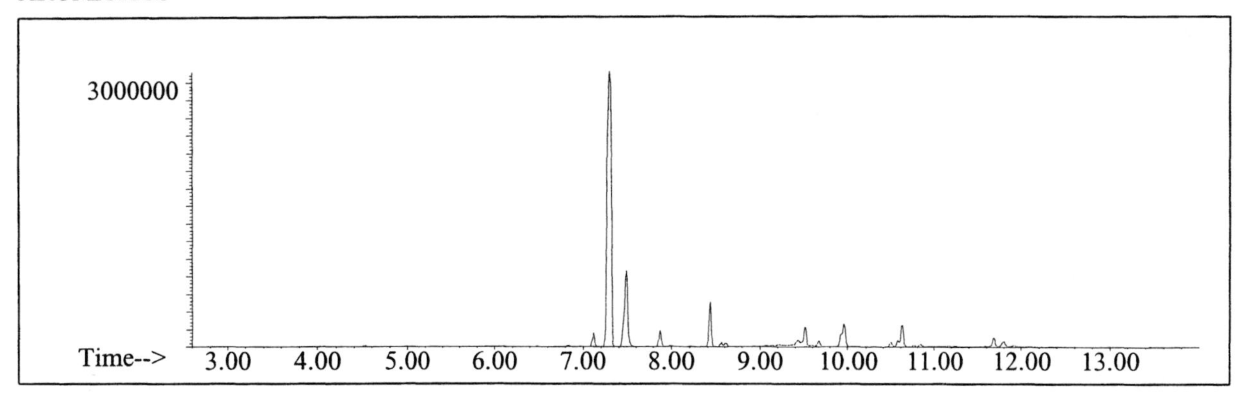

CYCLOPARAFFINS AND ALKENES

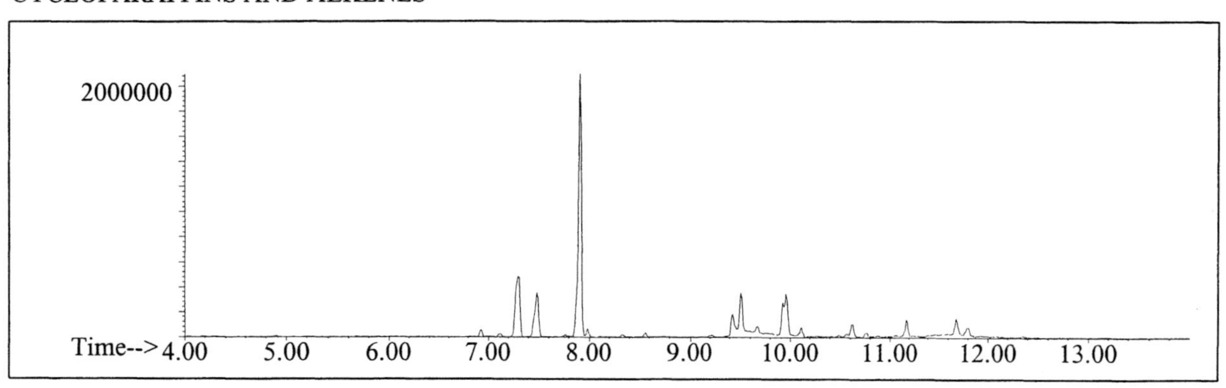

NAPHTHALENES

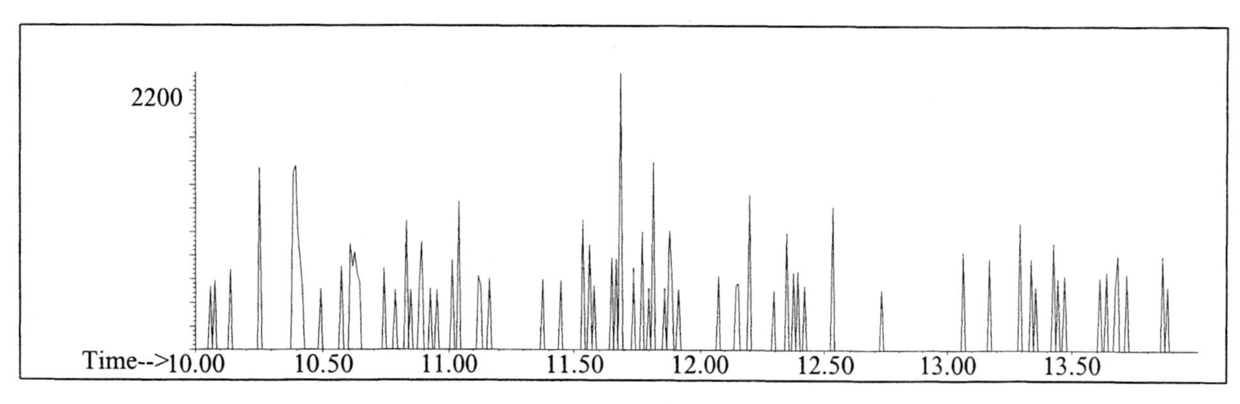

Alkane

Ion 43

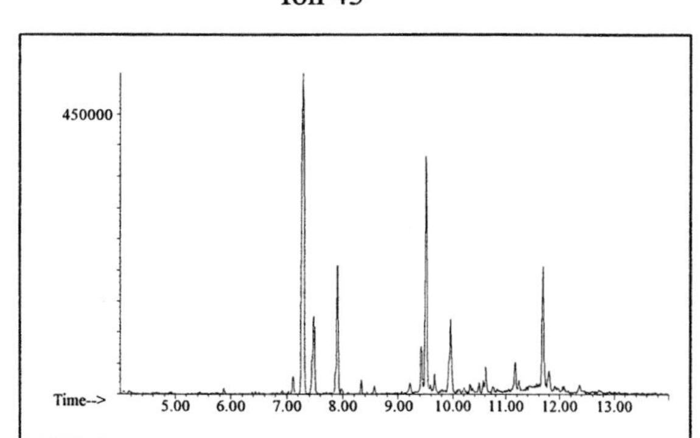

Ion 57

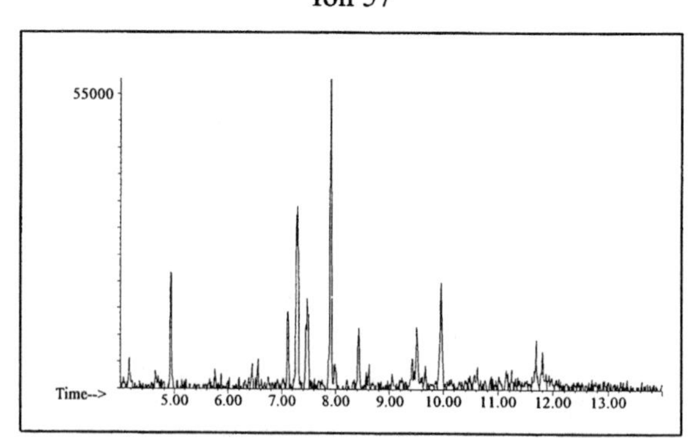

Ion 71

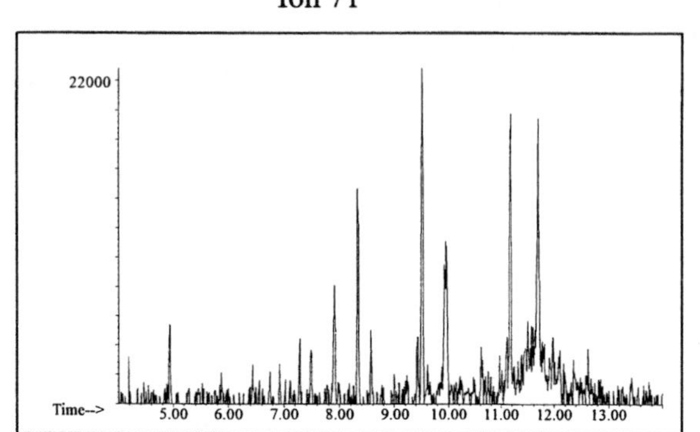

Ion 85

No Useful Data Obtained

Aromatic

Ion 91

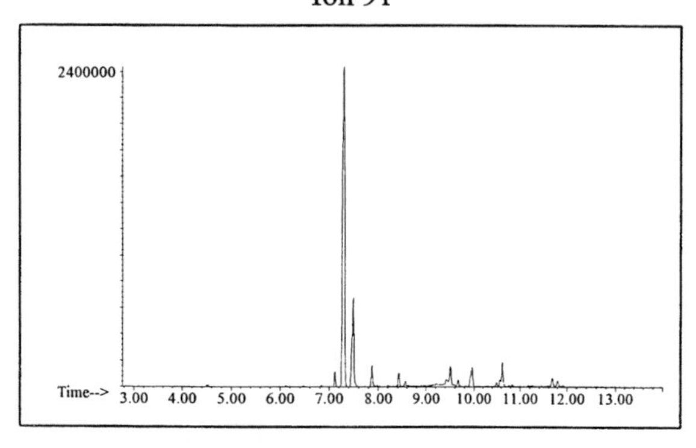

Ion 105

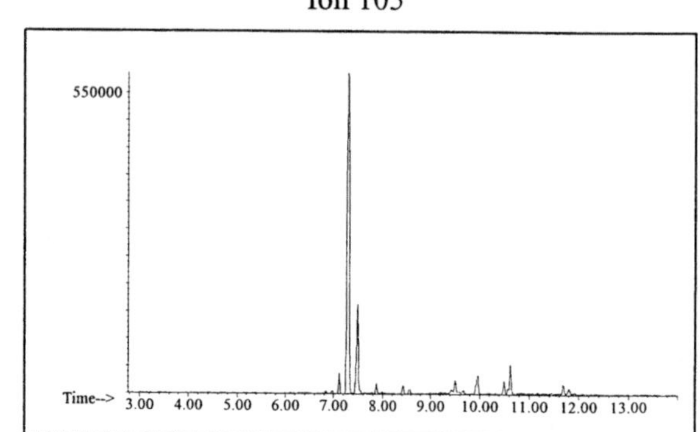

Ion 119

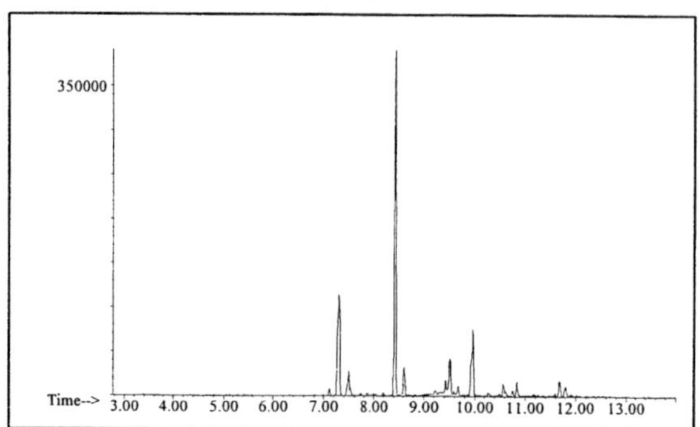

Cycloparaffin

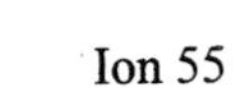

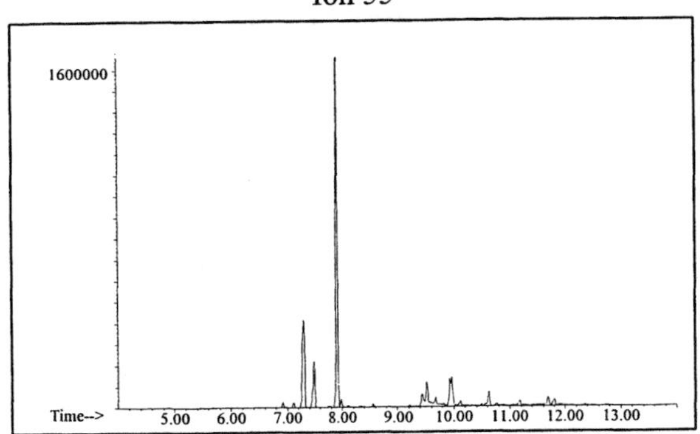

Ion 69

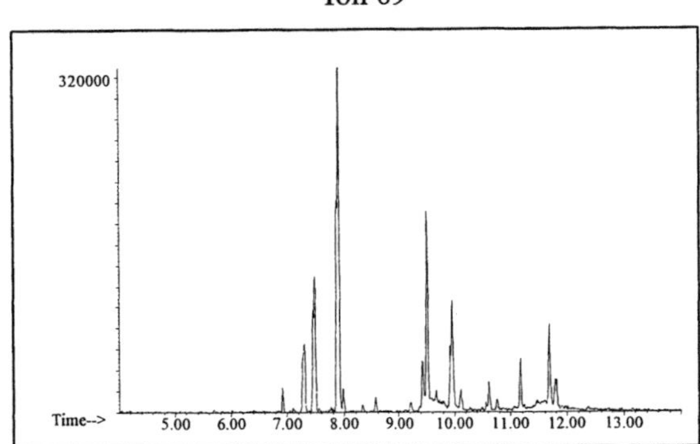

Ion 83

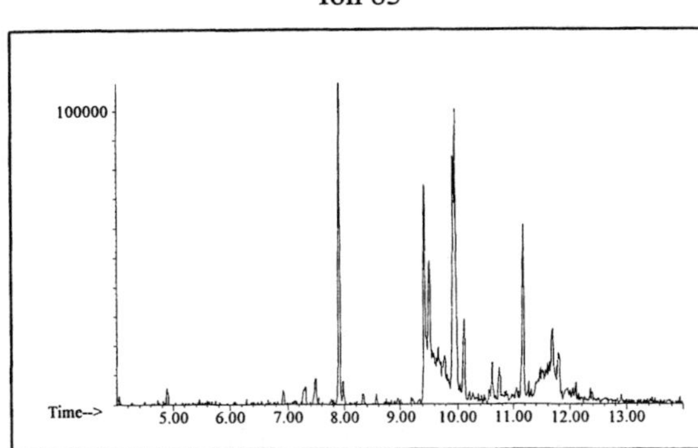

Naphthalene

Ion 128

No Useful Data Obtained

Ion 142

No Useful Data Obtained

Ion 156

No Useful Data Obtained

Miscellaneous #31

20uL/mL Pentane

Product Displayed: Kleen Strip Turpentine Substitute

Other Similar Products:

ASTM: Class 0 (Miscellaneous)

Product Uses: Cleaning solvent, paint thinner

Macro Code: L

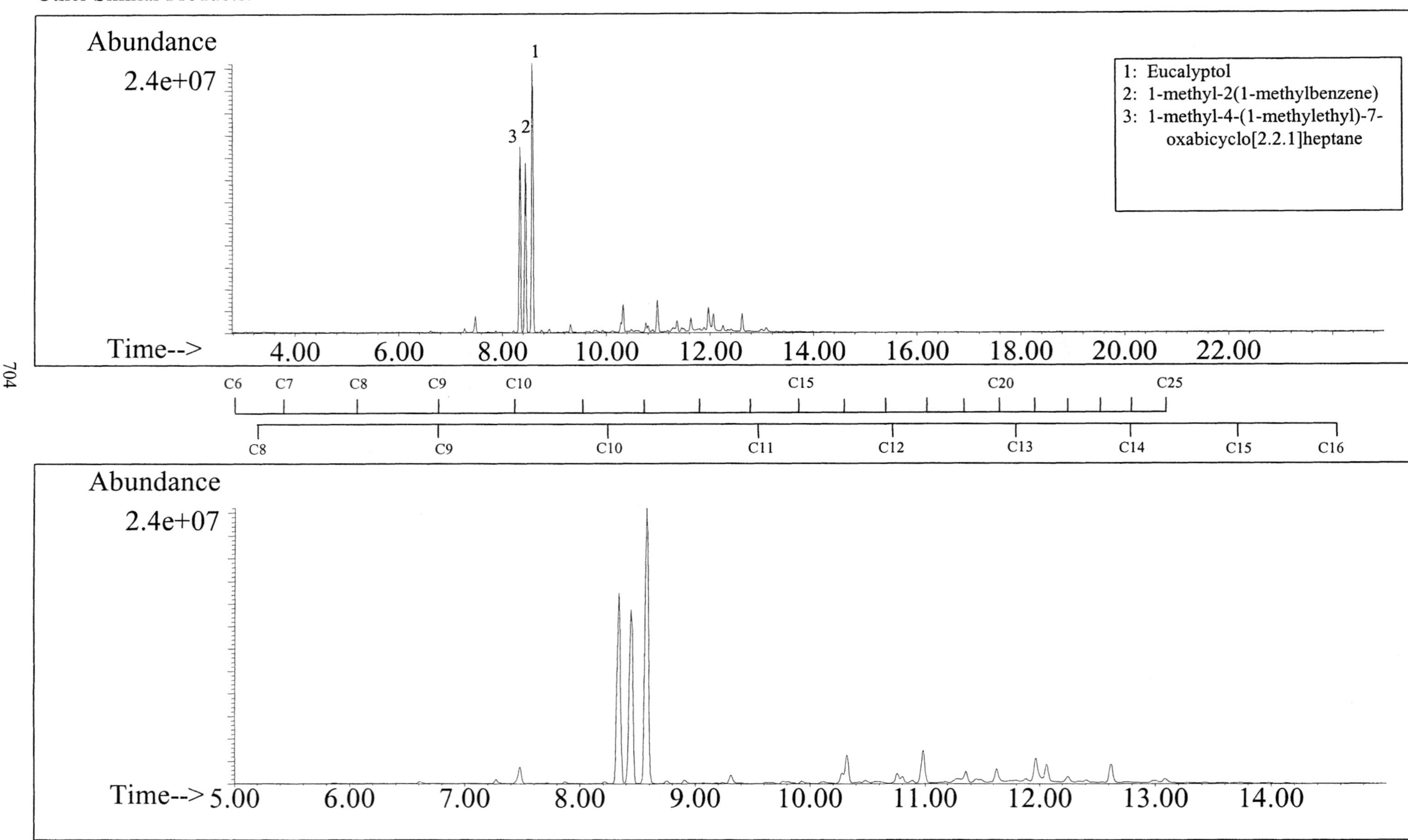

ALKANES

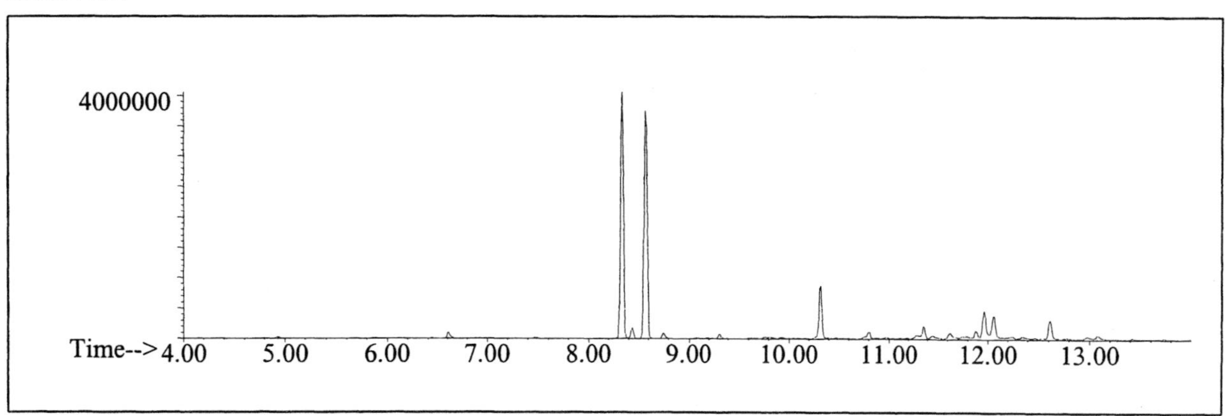

AROMATICS

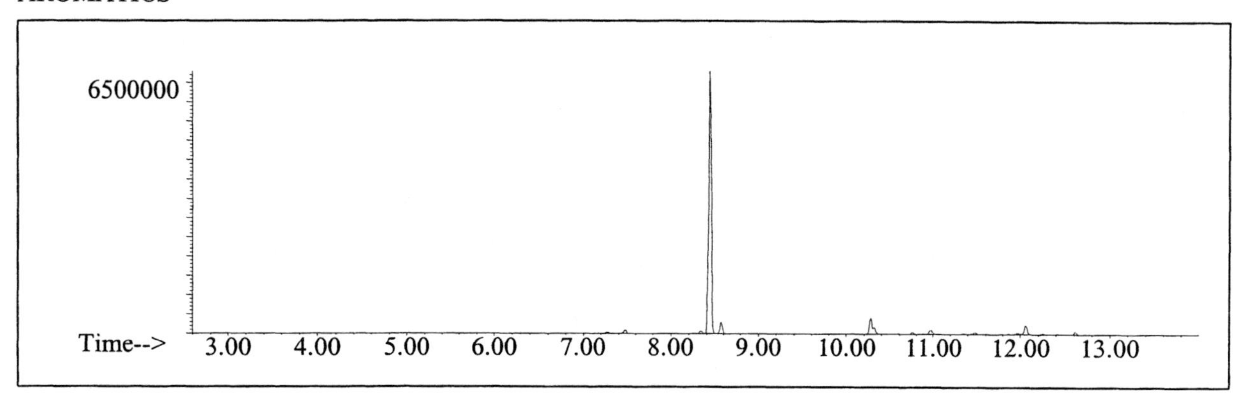

CYCLOPARAFFINS AND ALKENES

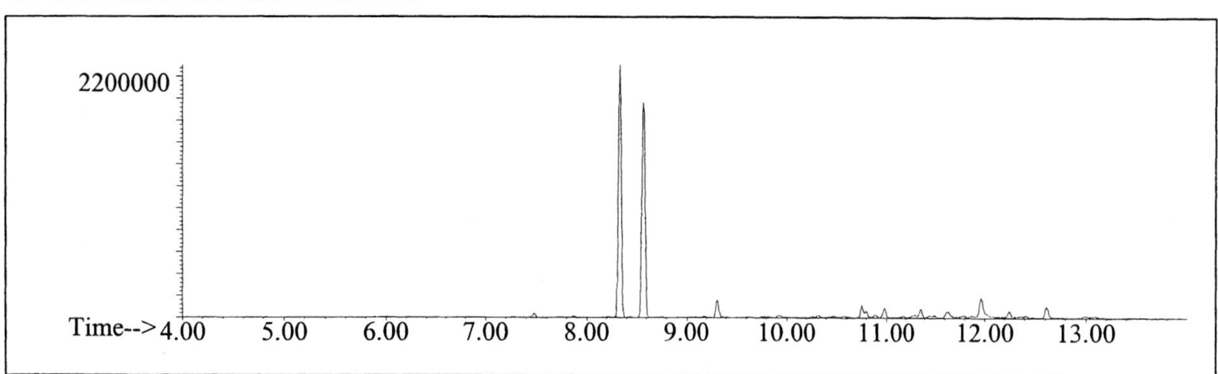

NAPHTHALENES

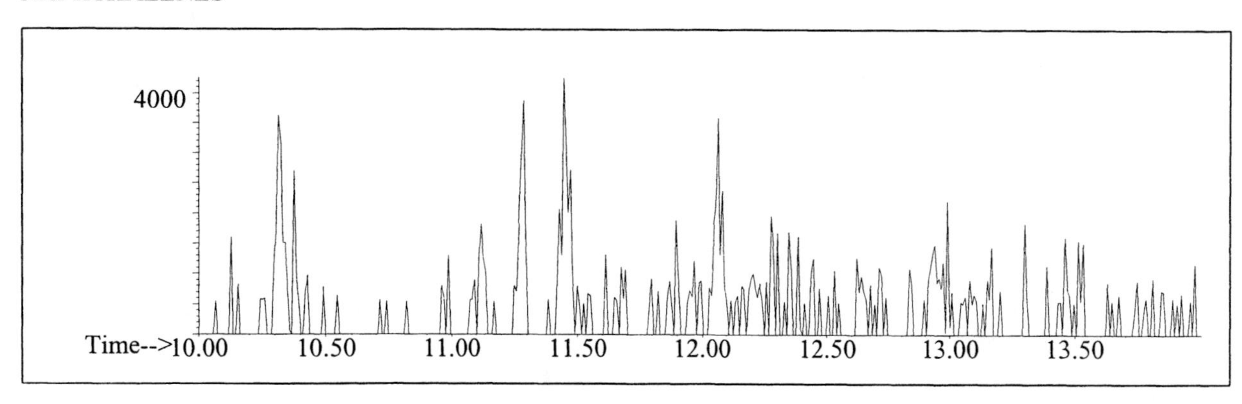

Alkane

Ion 43

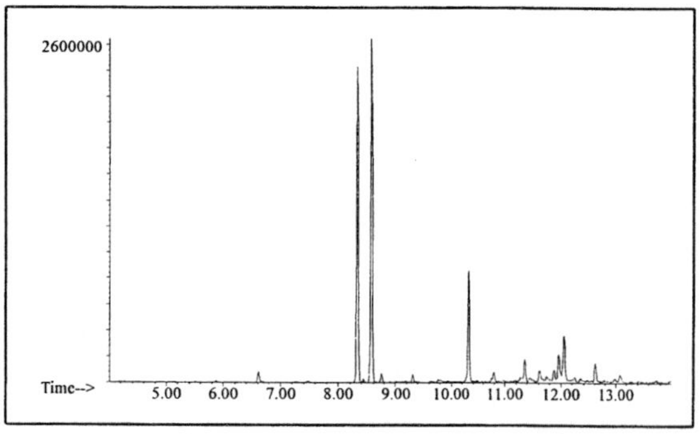

Ion 57

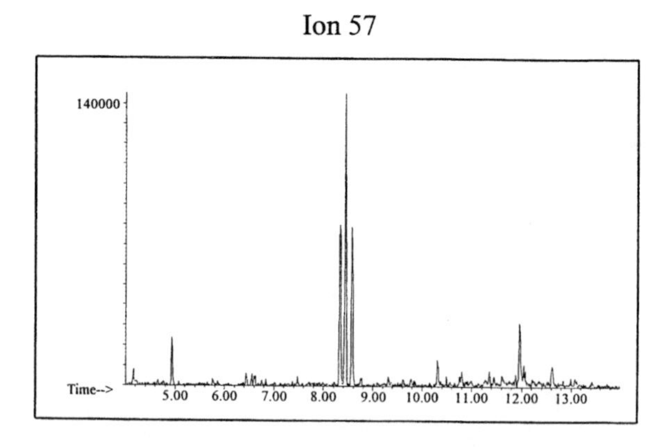

Ion 71

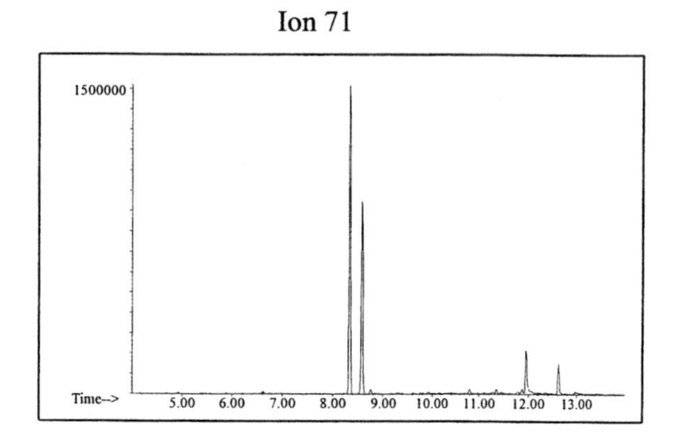

Ion 85

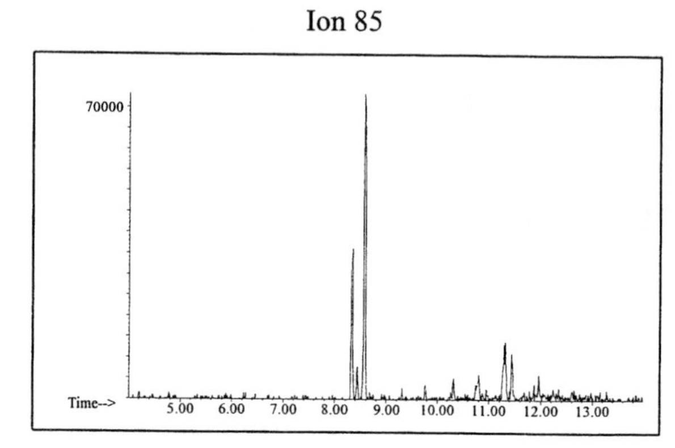

Aromatic

Ion 91

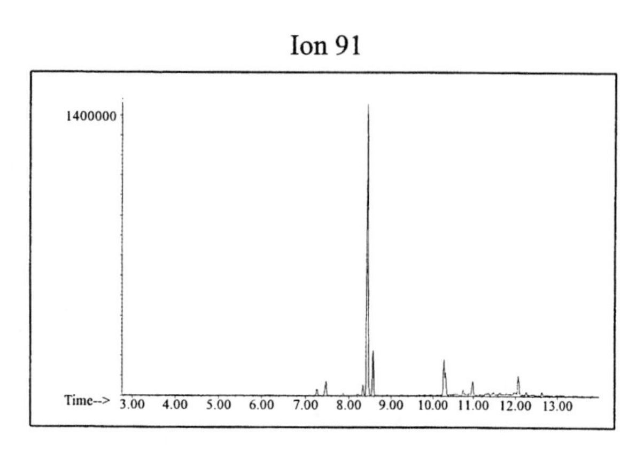

Ion 105

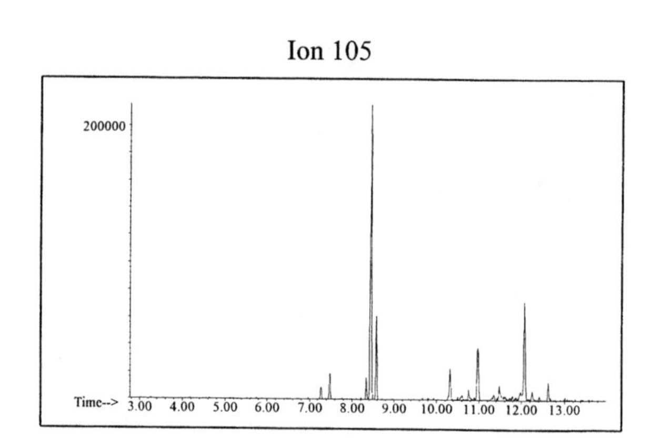

Ion 119

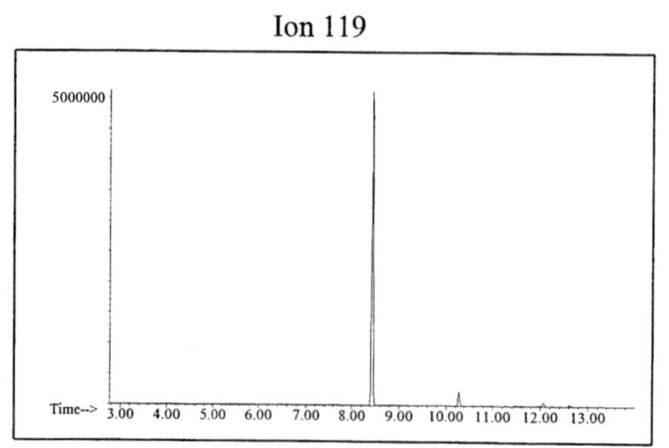

Cycloparaffin

Ion 55

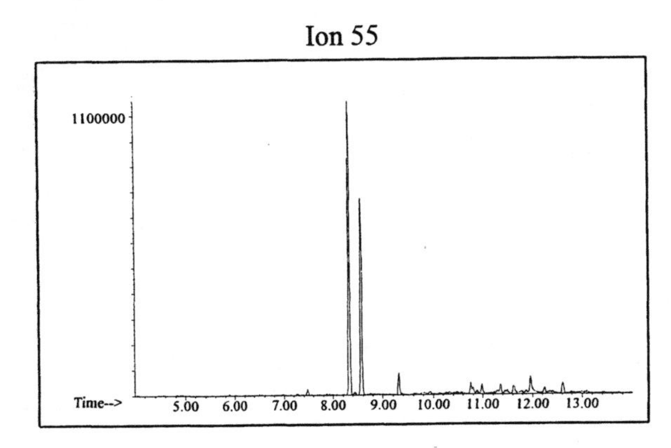

Ion 69

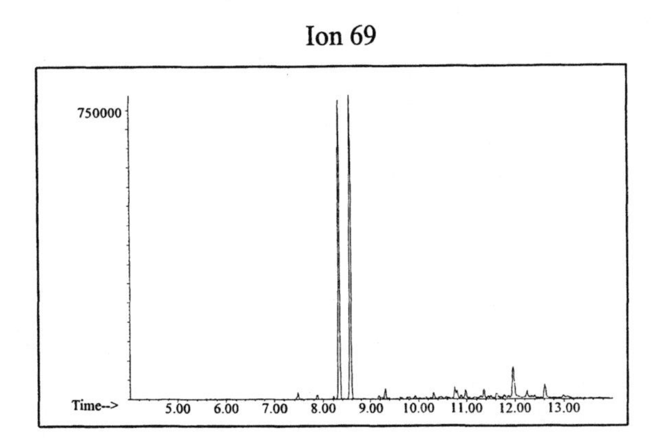

Ion 83

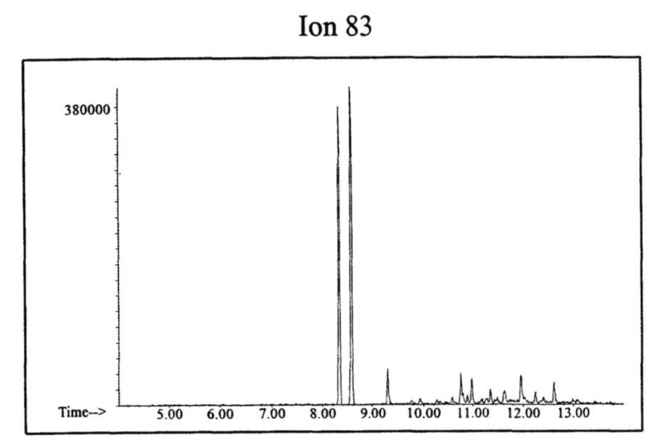

Naphthalene

Ion 128

No Useful Data Obtained

Ion 142

No Useful Data Obtained

Ion 156

No Useful Data Obtained

Miscellaneous #32

20uL/mL Pentane

Product Displayed: Nasco Spirits of Turpentine

Other Similar Products:

ASTM: Class 0 (Miscellaneous)

Product Uses: Cleaning solvent

Macro Code: L

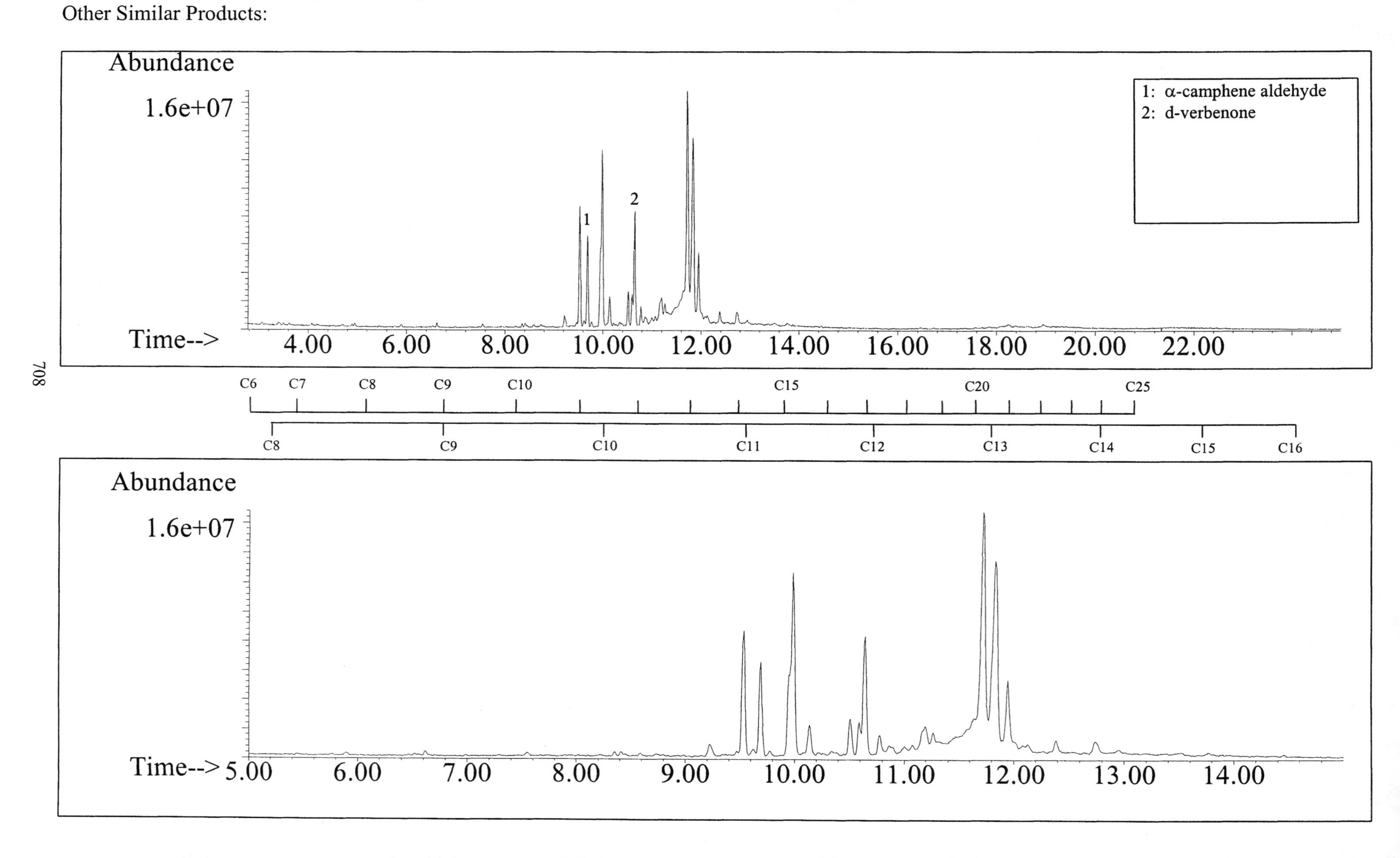

ALKANES

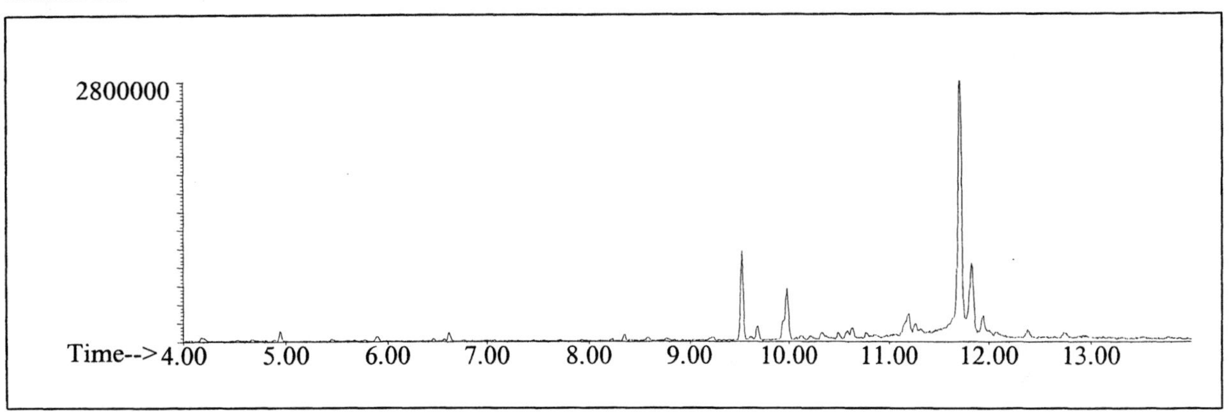

AROMATICS

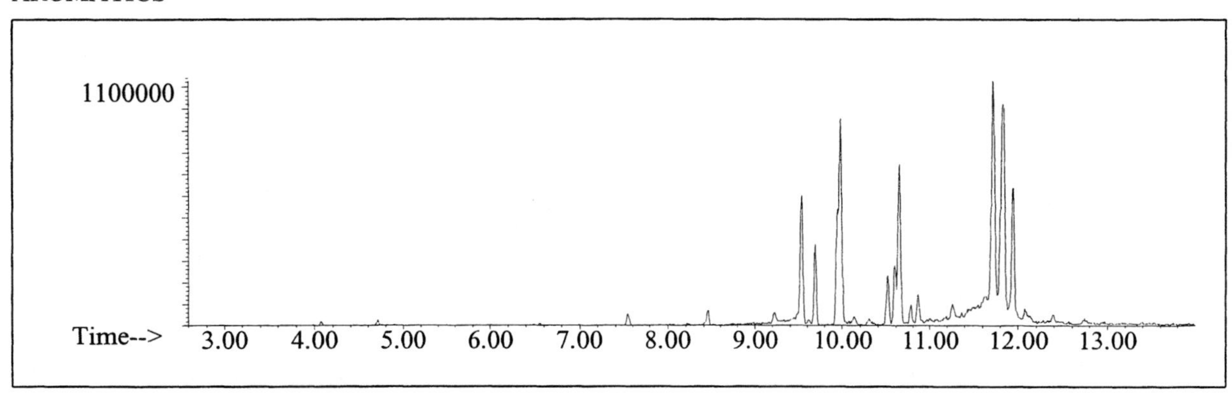

CYCLOPARAFFINS AND ALKENES

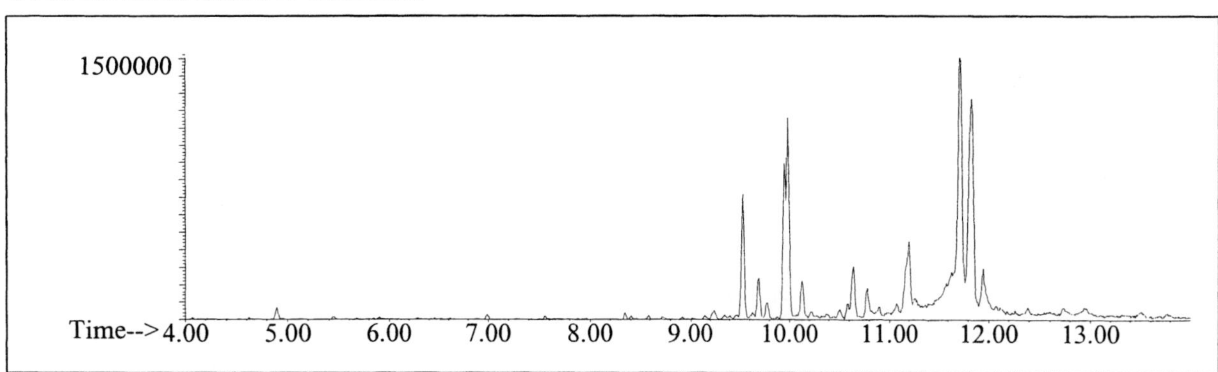

NAPHTHALENES

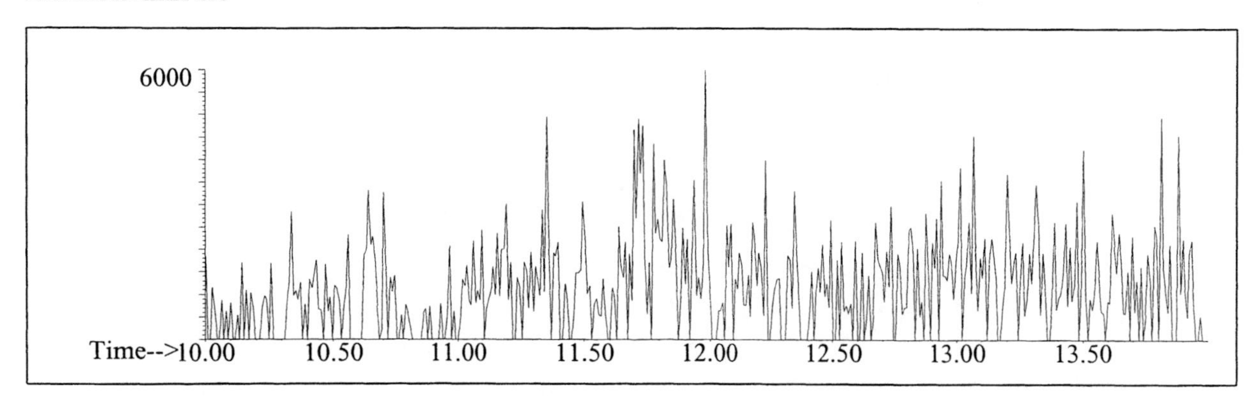

Alkane

Ion 43

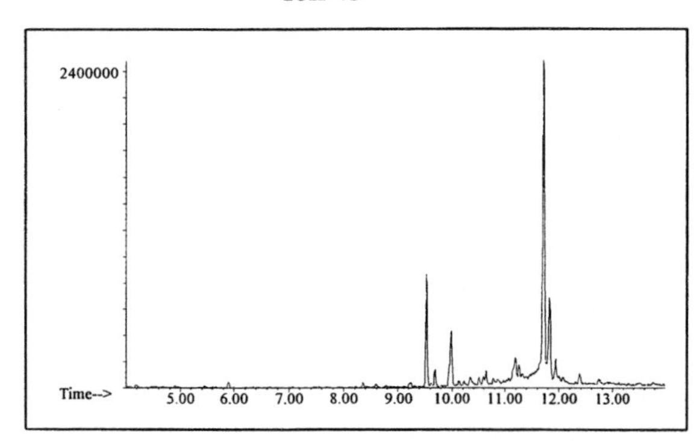

Ion 57

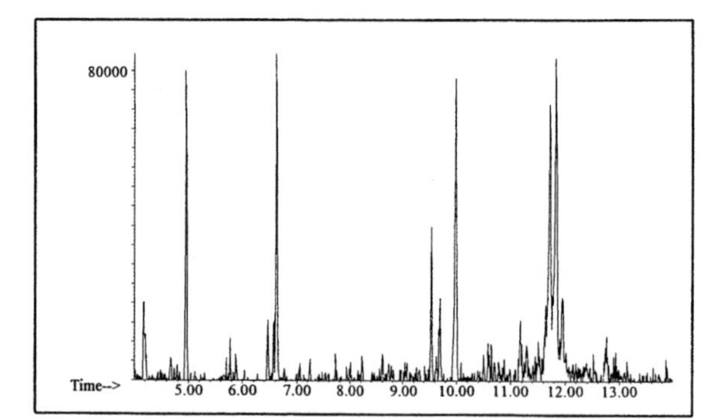

Ion 71

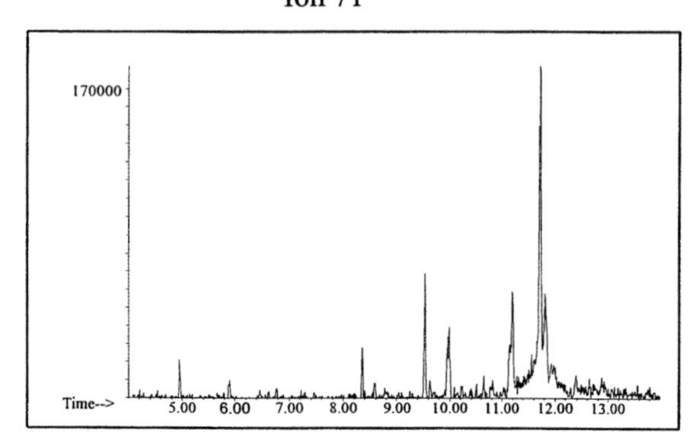

Ion 85

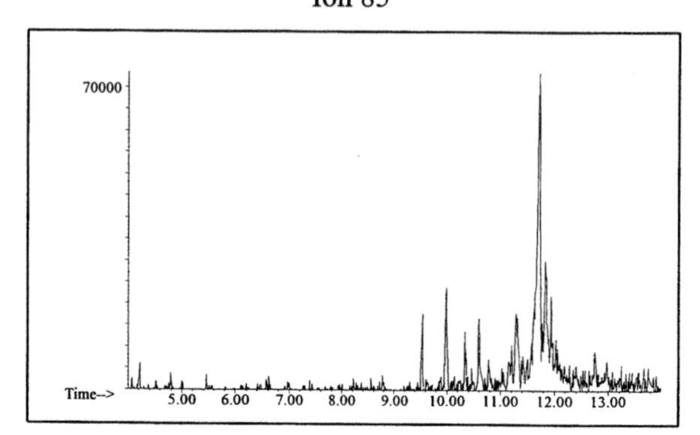

Aromatic

Ion 91

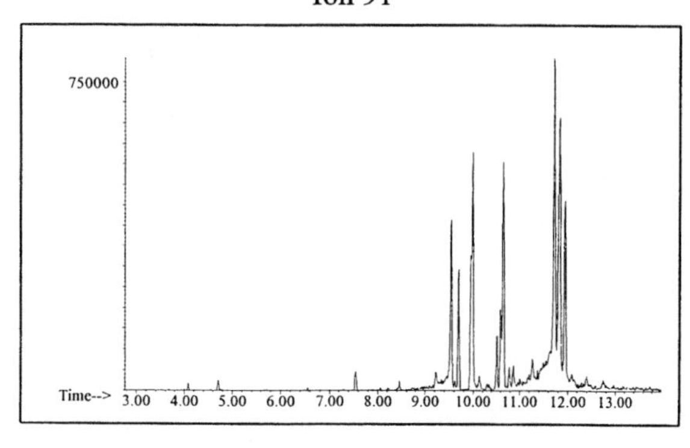

Ion 105

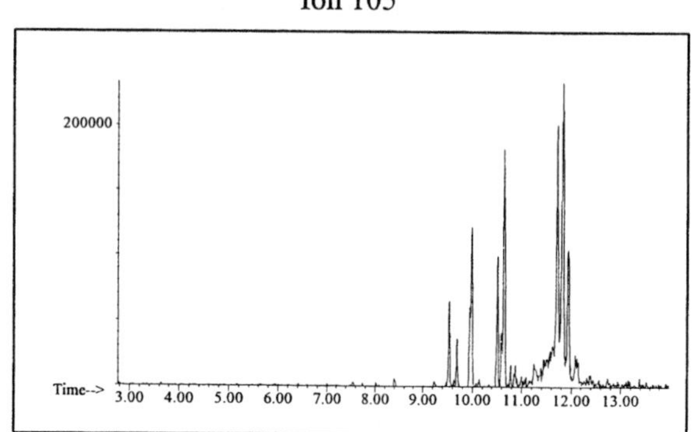

Ion 119

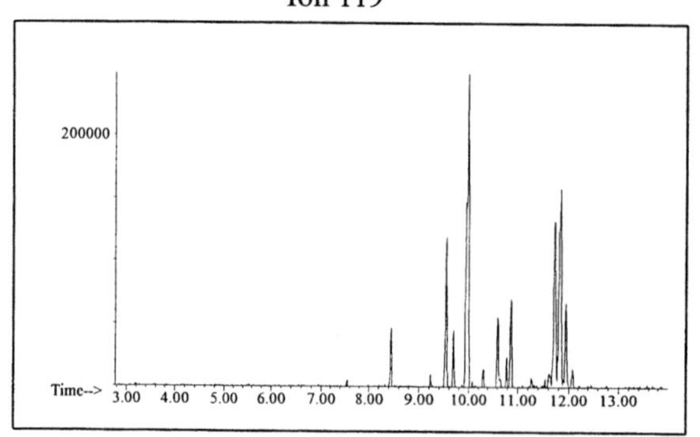

Cycloparaffin

Ion 55

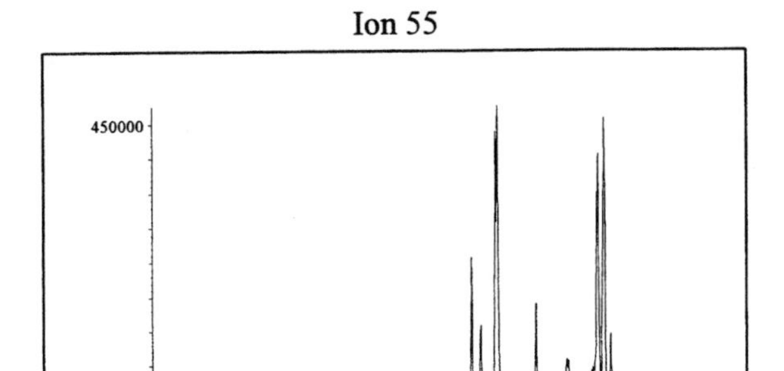

Ion 69

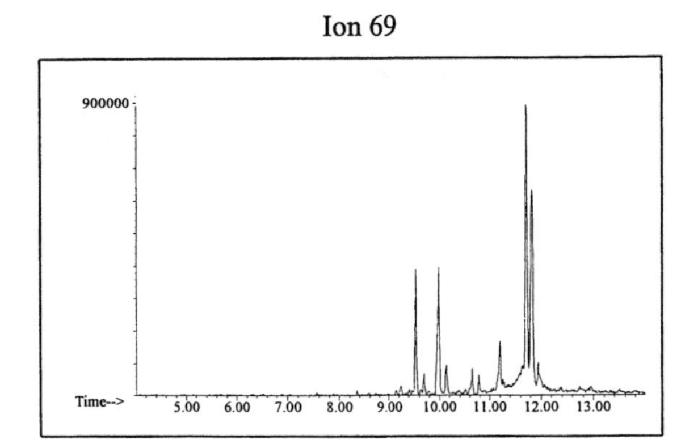

Ion 83

260000
Time-->
5.00 6.00 7.00 8.00 9.00 10.00 11.00 12.00 13.00

Naphthalene

Ion 128

No Useful Data Obtained

Ion 142

No Useful Data Obtained

Ion 156

No Useful Data Obtained

Miscellaneous #33

20uL/mL Pentane

Product Displayed: Wicks and Stix Lamp Oil

Other Similar Products:

ASTM: Class 0 (Miscellaneous)

Product Uses: Lamp oil

Macro Code: M

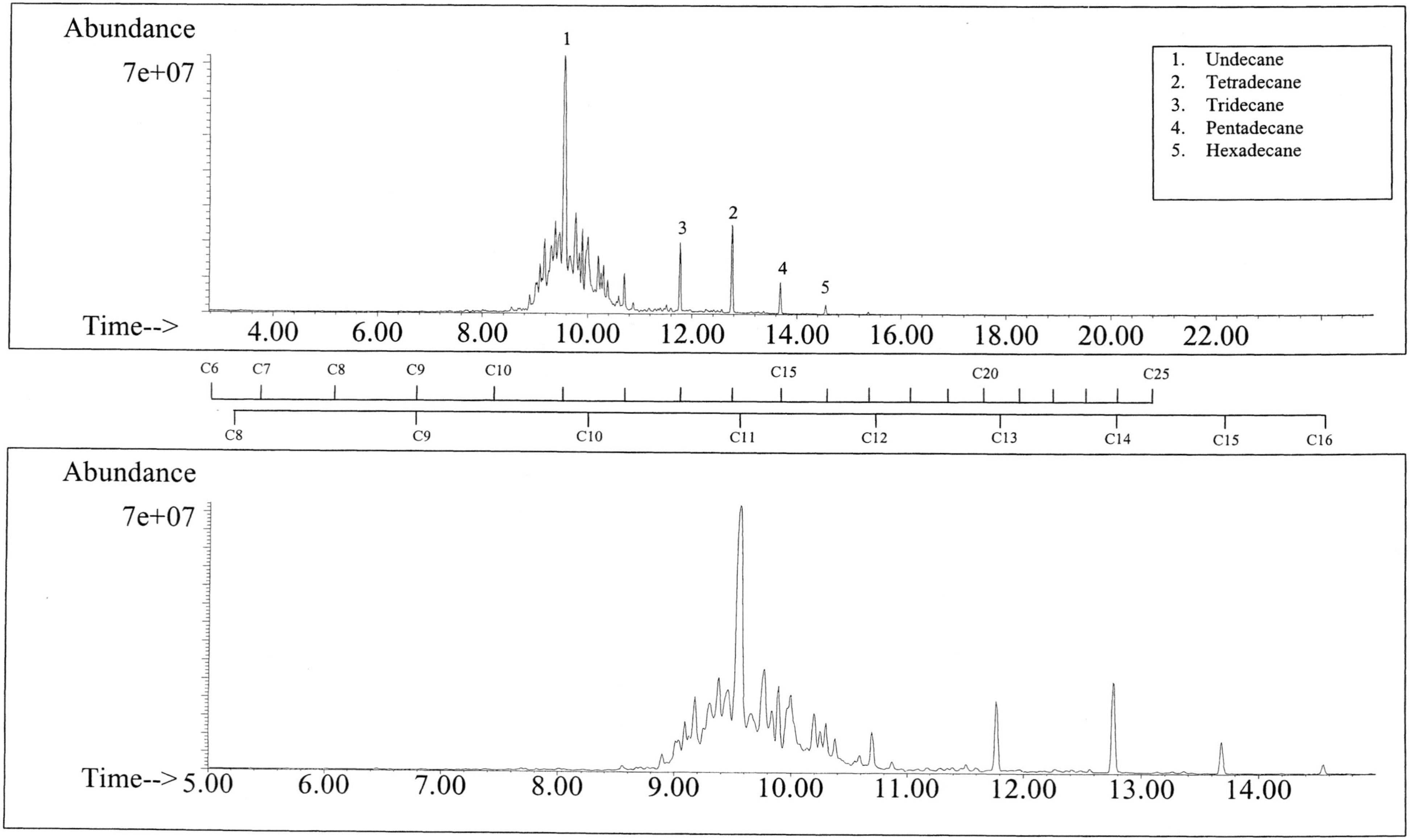

Abundance

7e+07

Time-->

4.00 6.00 8.00 10.00 12.00 14.00 16.00 18.00 20.00 22.00

1. Undecane
2. Tetradecane
3. Tridecane
4. Pentadecane
5. Hexadecane

C6 C7 C8 C9 C10 C15 C20 C25

C8 C9 C10 C11 C12 C13 C14 C15 C16

Abundance

7e+07

Time--> 5.00 6.00 7.00 8.00 9.00 10.00 11.00 12.00 13.00 14.00

ALKANES

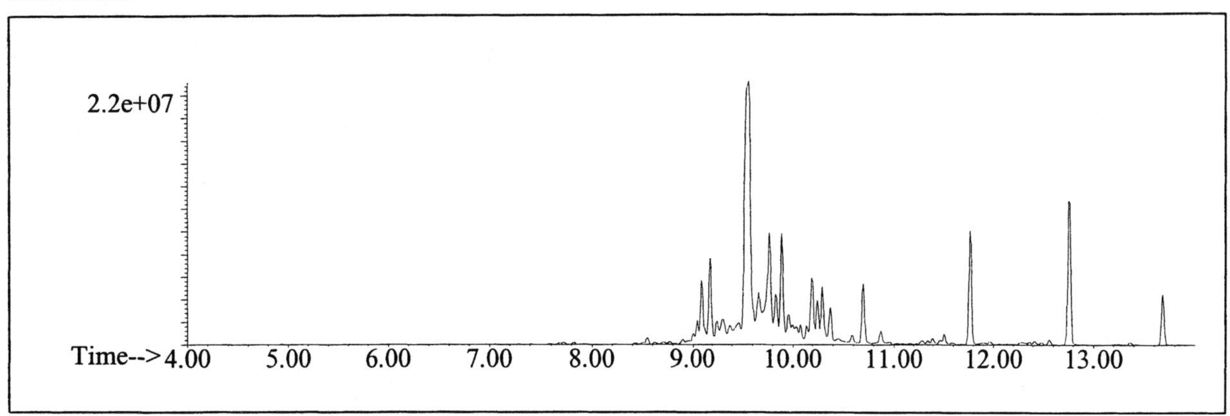

AROMATICS

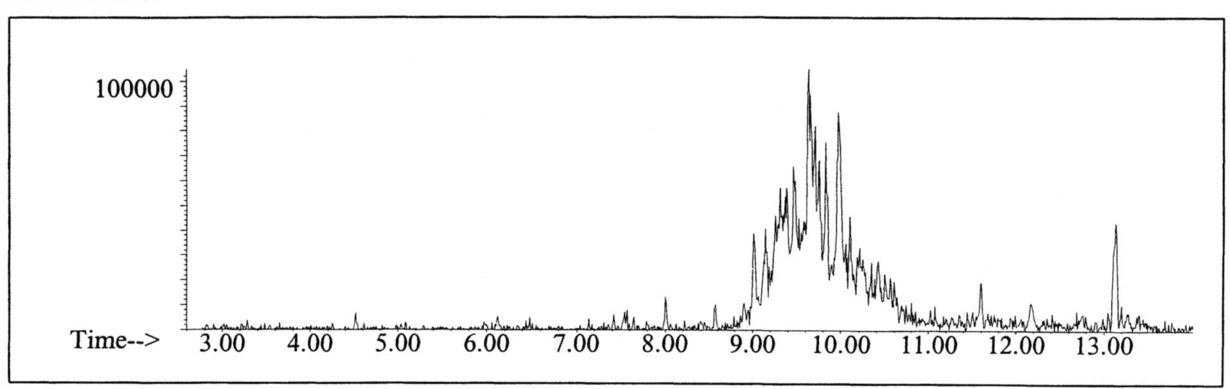

CYCLOPARAFFINS AND ALKENES

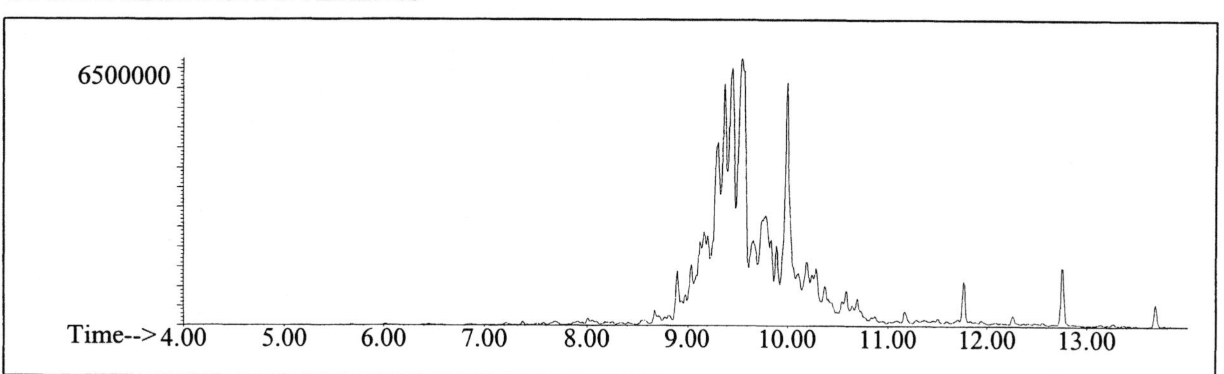

NAPHTHALENES

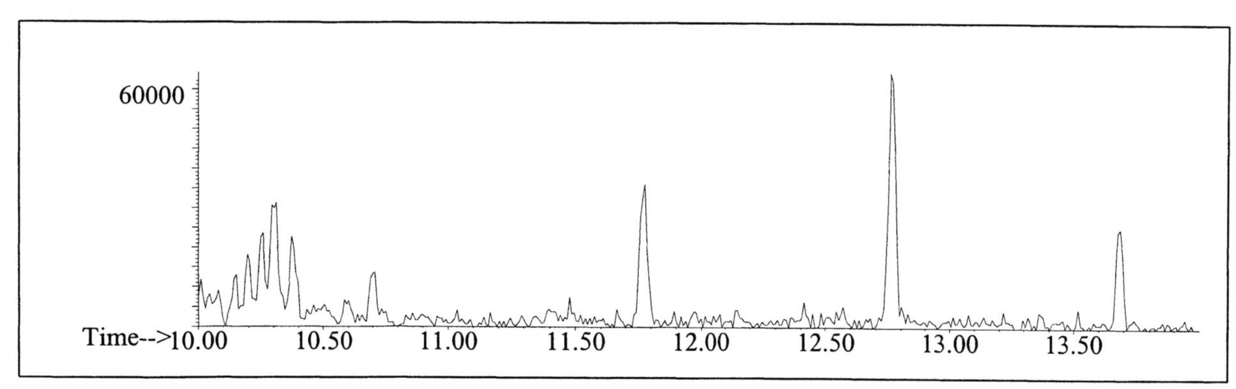

Alkane

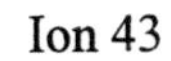

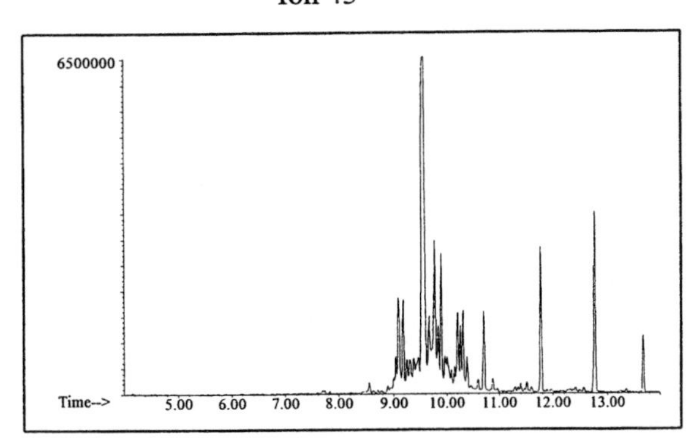

Ion 57

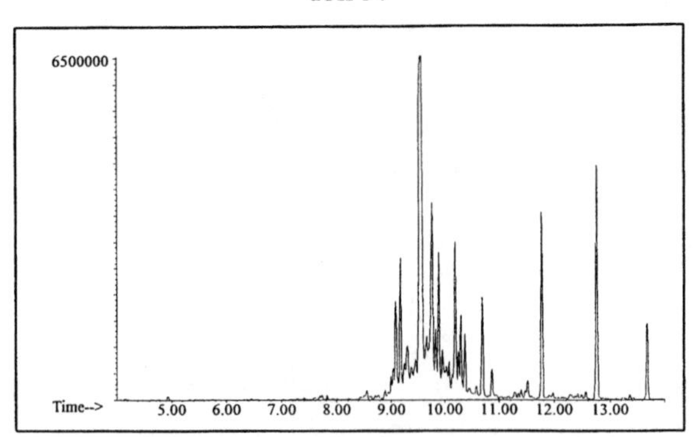

Ion 71

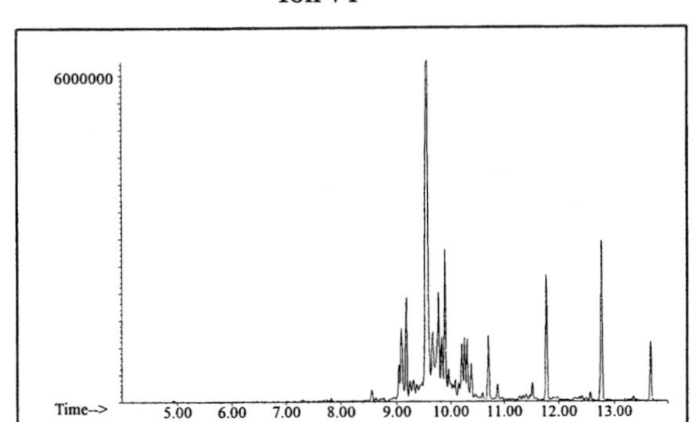

Ion 85

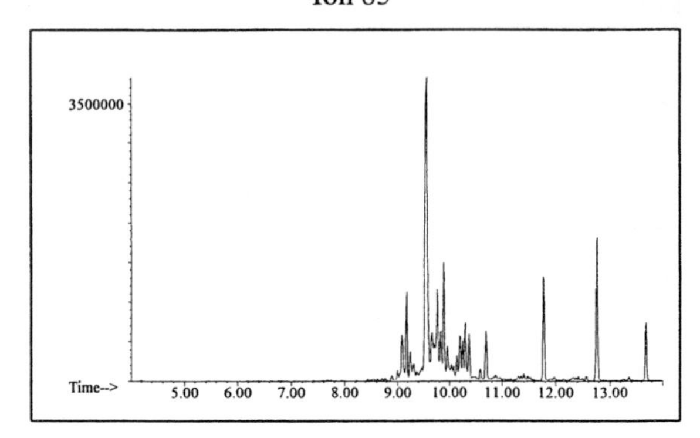

Aromatic

Ion 91

No Useful Data Obtained

Ion 105

No Useful Data Obtained

Ion 119

No Useful Data Obtained

Cycloparaffin

Ion 55

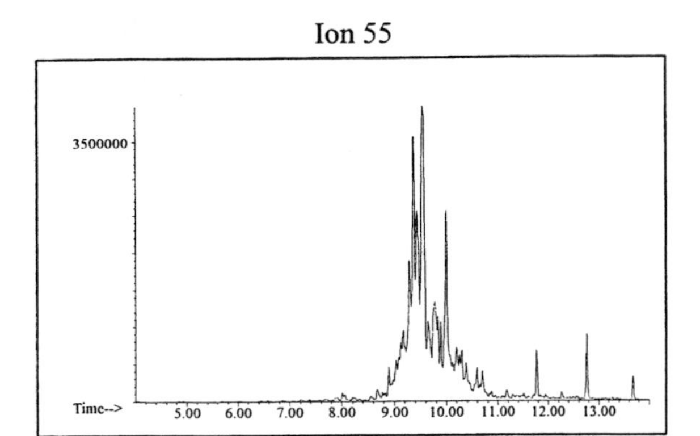

Ion 69

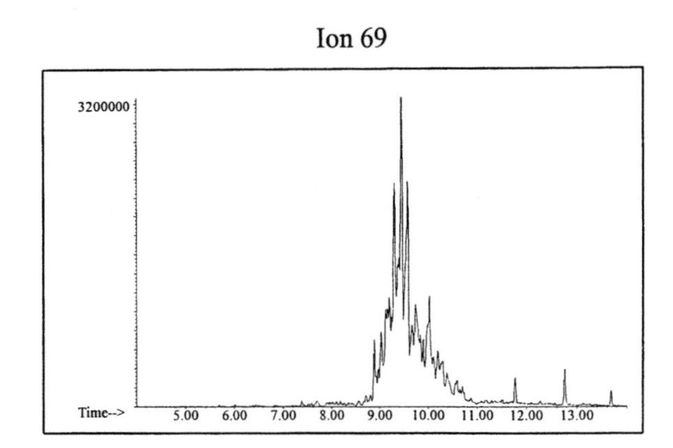

Ion 83

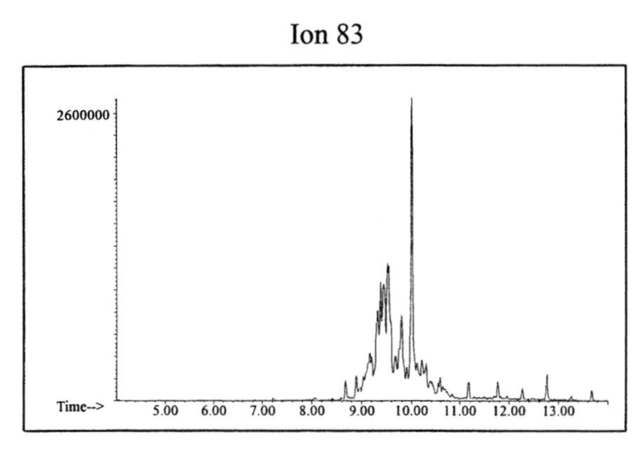

Naphthalene

Ion 128

No Useful Data Obtained

Ion 142

No Useful Data Obtained

Ion 156

No Useful Data Obtained

Miscellaneous #34

20uL/mL Pentane

Product Displayed: Citronella Lamp Oil

Other Similar Products:

ASTM: Class 0 (Miscellaneous)

Product Uses: Lamp oil

Macro Code: H

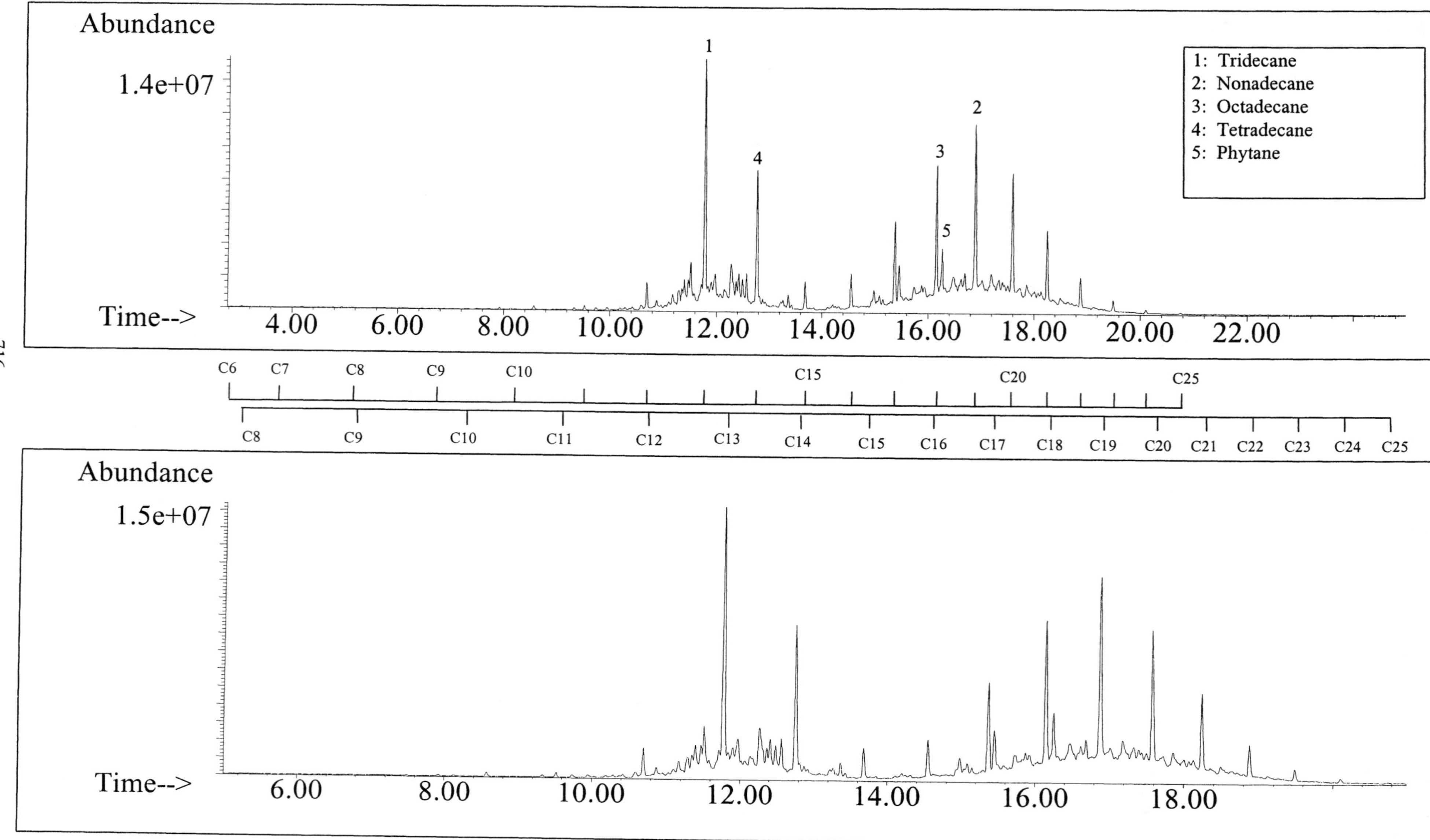

ALKANES

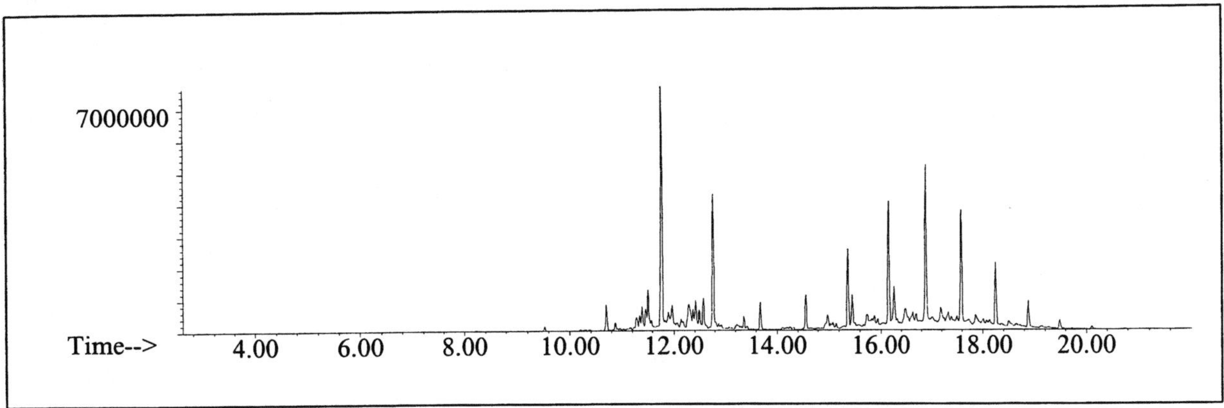

AROMATICS

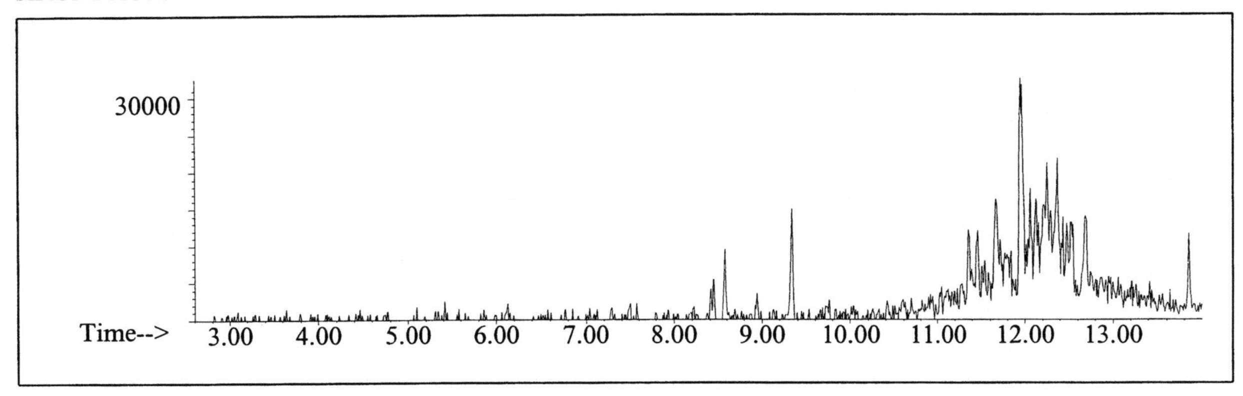

CYCLOPARAFFINS AND ALKENES

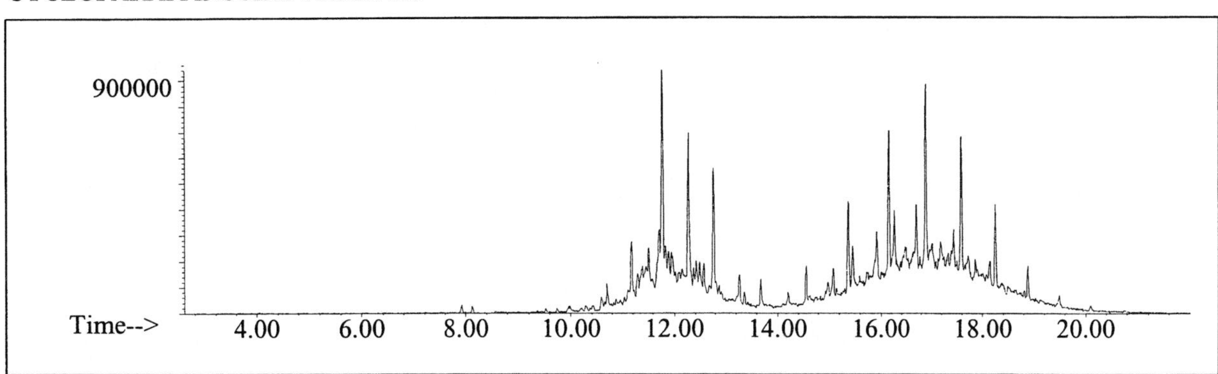

NAPHTHALENES

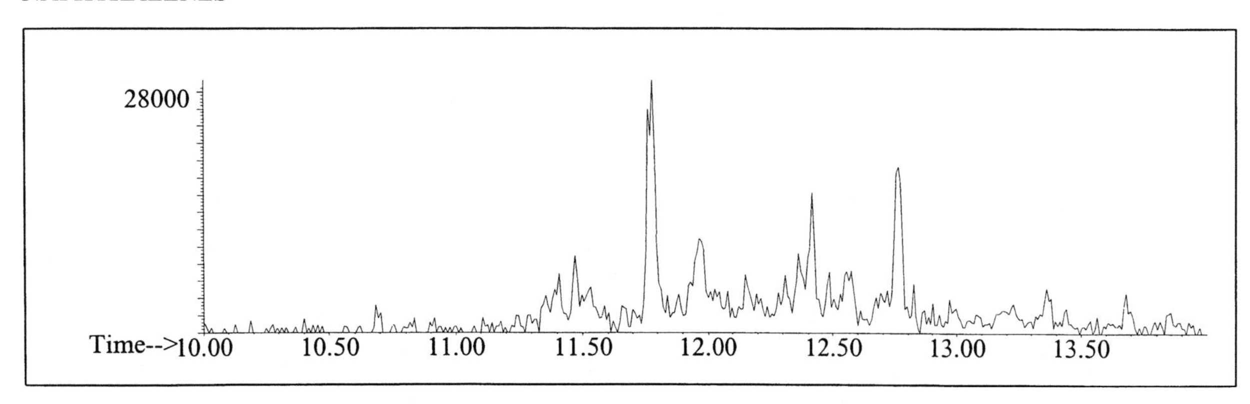

Alkane

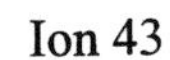
Ion 43

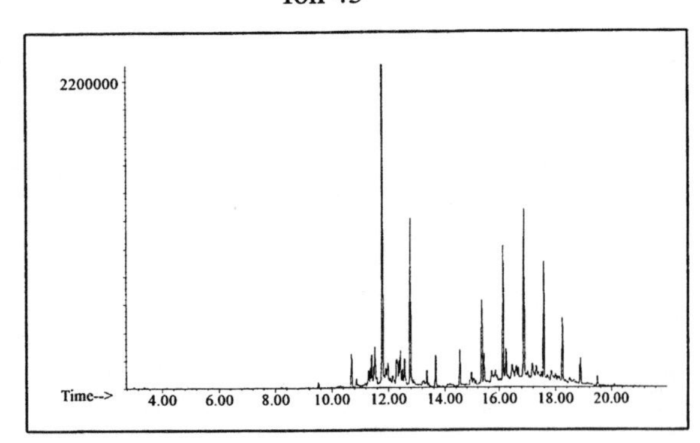

Ion 57

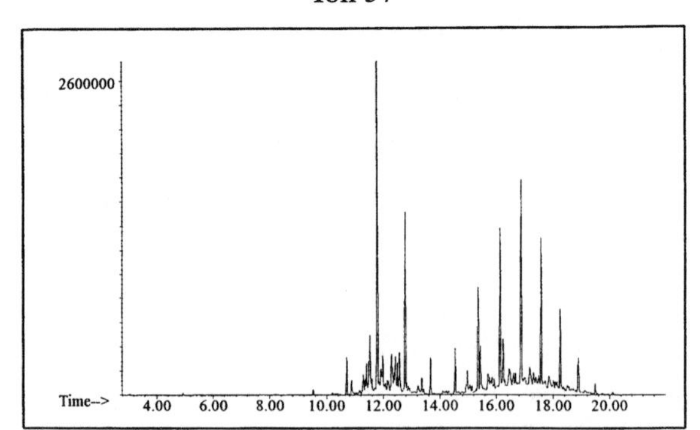

Ion 71

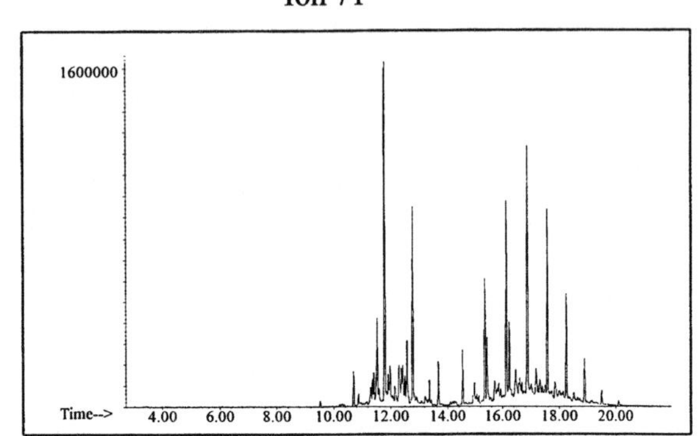

Ion 85

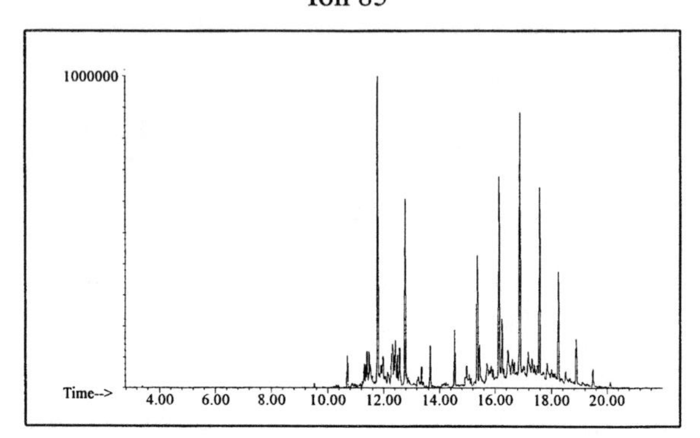

Aromatic

Ion 91

No Useful Data Obtained

Ion 105

No Useful Data Obtained

Ion 119

No Useful Data Obtained

INDIVIDUAL PROFILES — Miscellaneous #34

Cycloparaffin

Ion 55

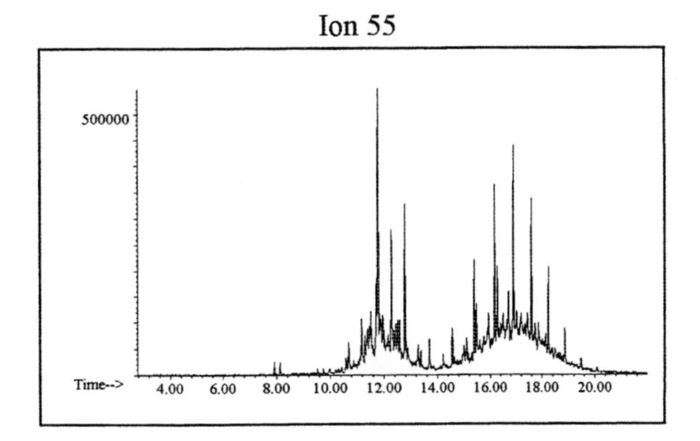

Ion 69

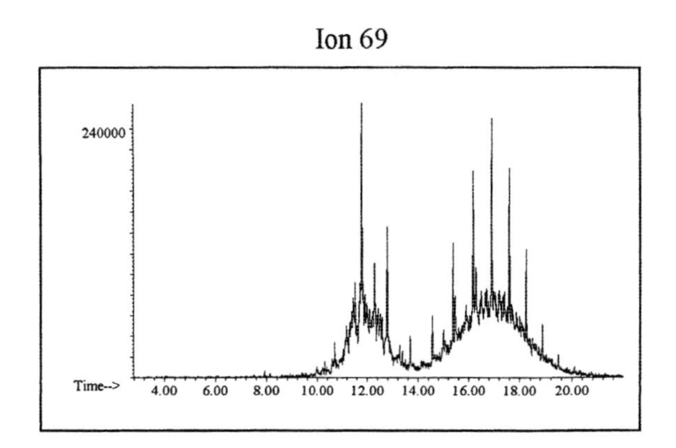

Ion 83

280000
Time-->
4.00 6.00 8.00 10.00 12.00 14.00 16.00 18.00 20.00

Naphthalene

Ion 128

No Useful Data Obtained

Ion 142

No Useful Data Obtained

Ion 156

No Useful Data Obtained

Miscellaneous #35

20uL/mL Pentane
Product Displayed: Liquid Wrench
Other Similar Products:

ASTM: Class 0 (Miscellaneous)
Product Uses: Solvent

Macro Code: H

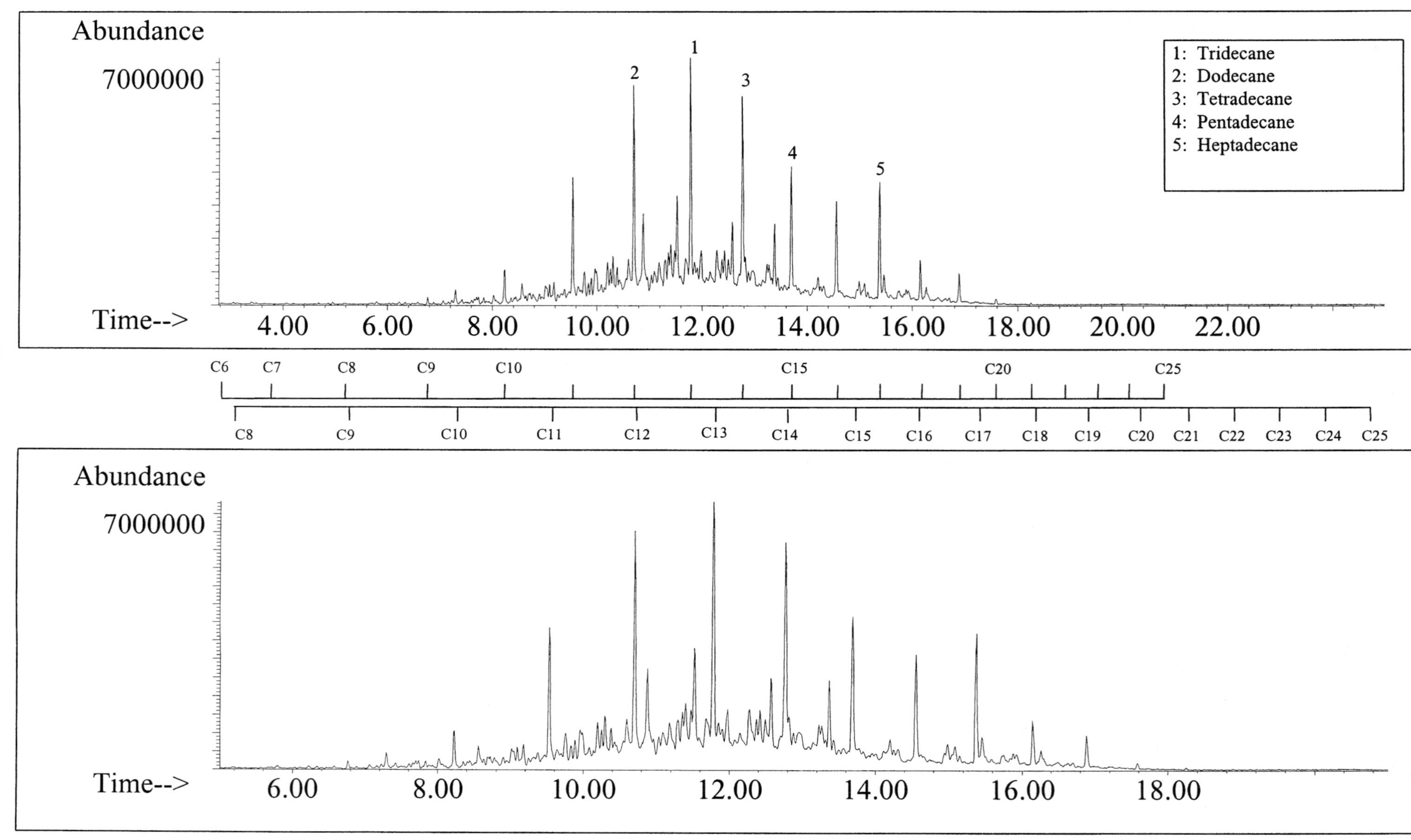

ALKANES

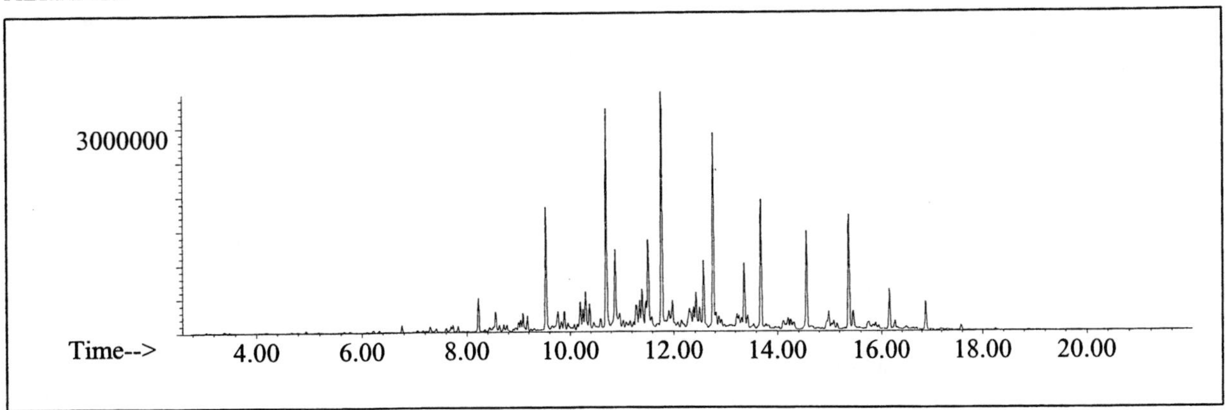

AROMATICS

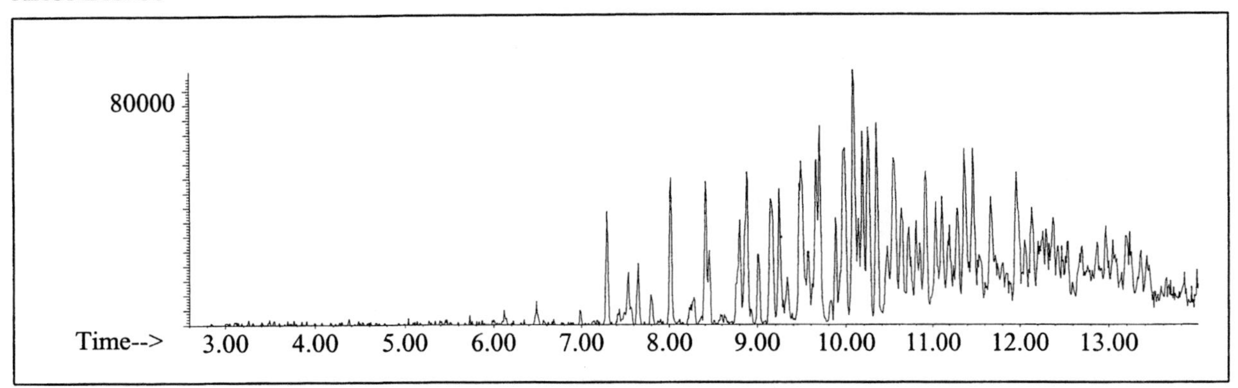

CYCLOPARAFFINS AND ALKENES

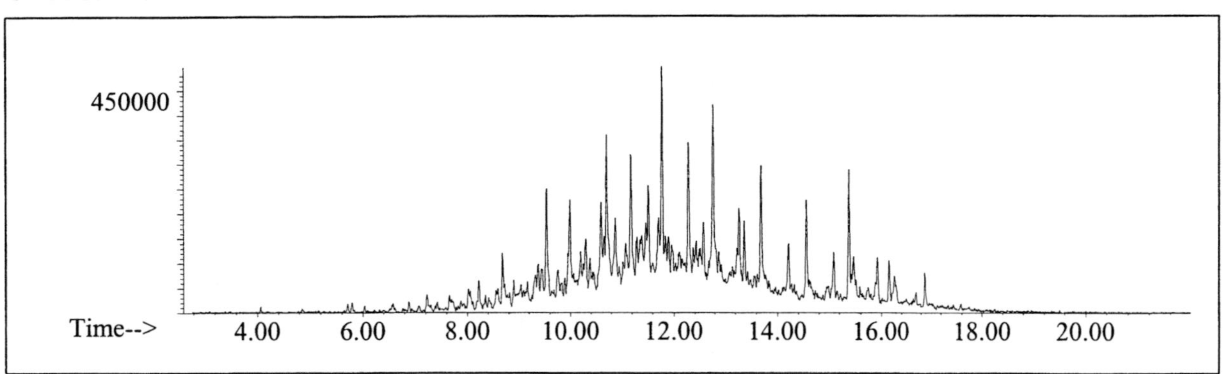

NAPHTHALENES

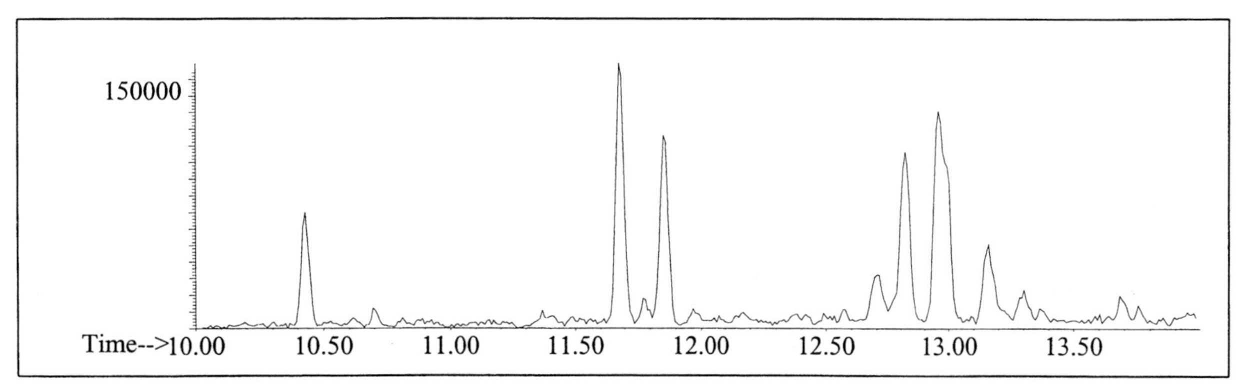

Alkane

Ion 43

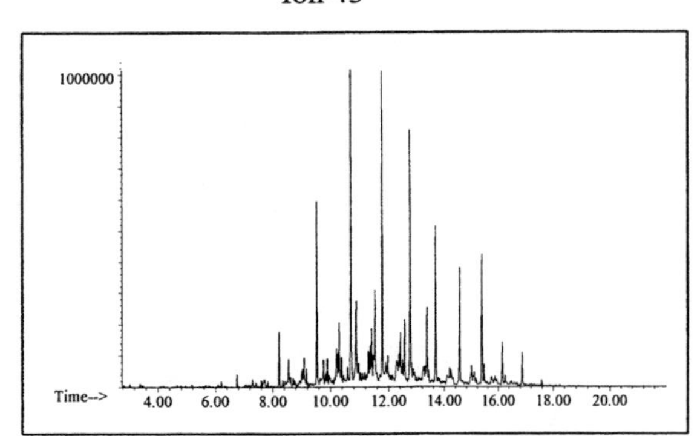

Ion 57

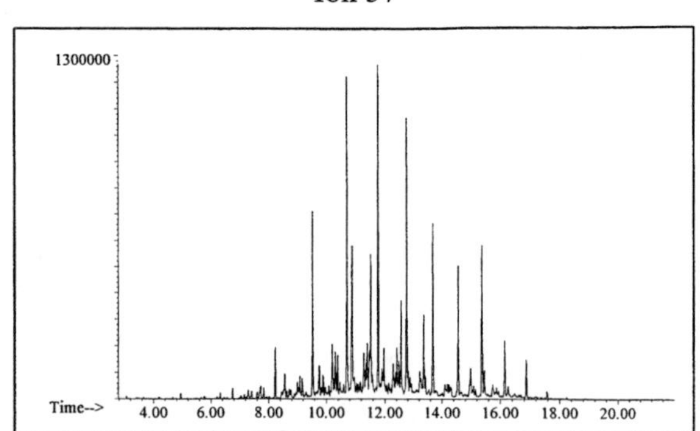

Ion 71

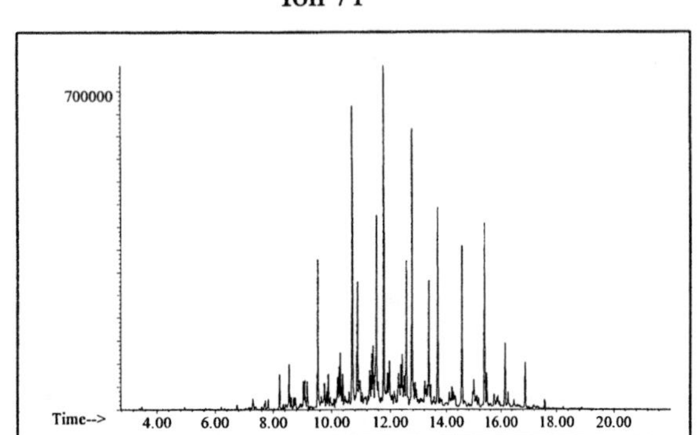

Ion 85

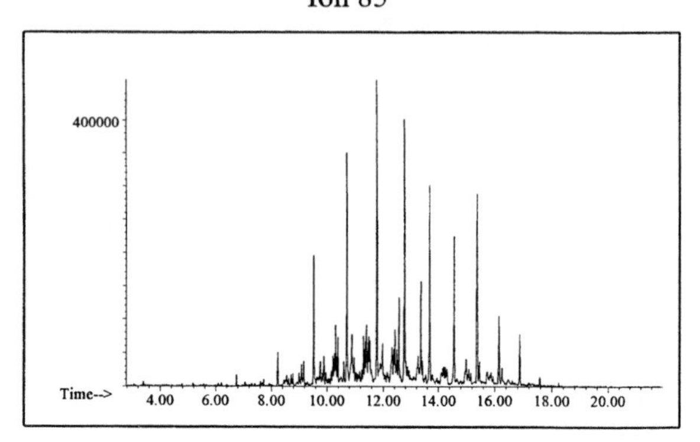

Aromatic

Ion 91

No Useful Data Obtained

Ion 105

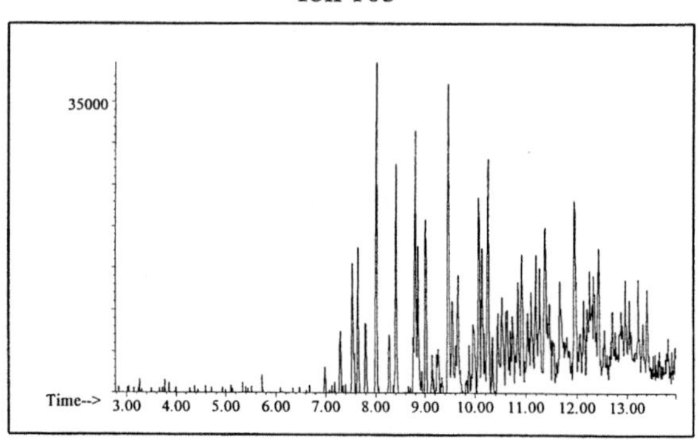

Ion 119

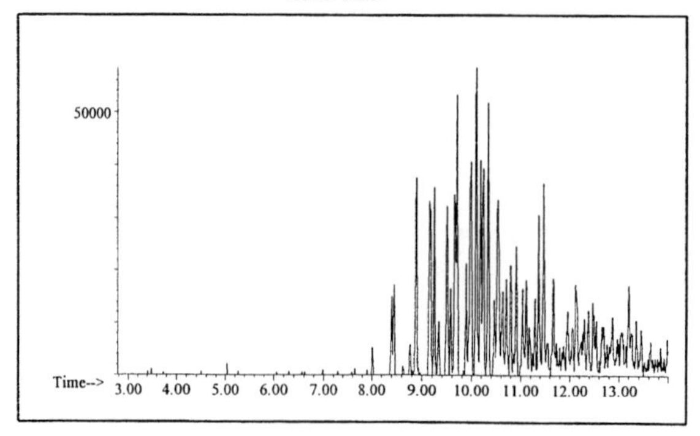

Cycloparaffin

Ion 55

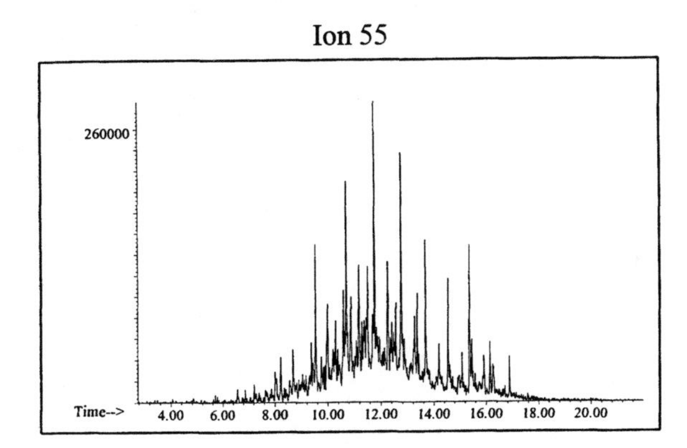

Ion 69

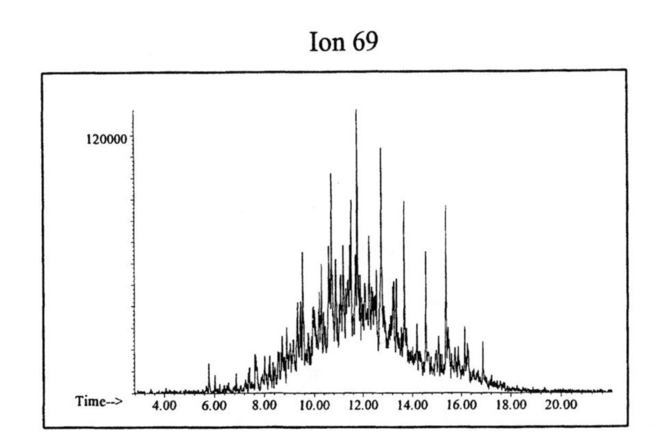

Ion 83

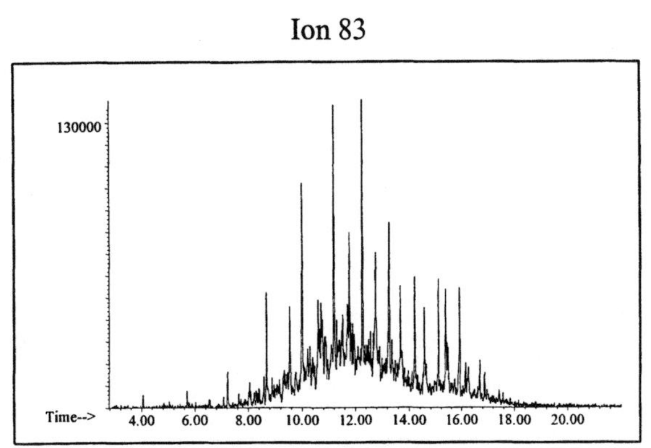

Naphthalene

Ion 128

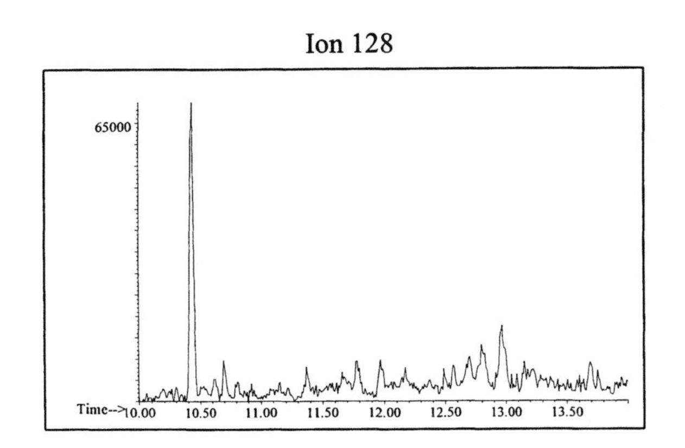

Ion 142

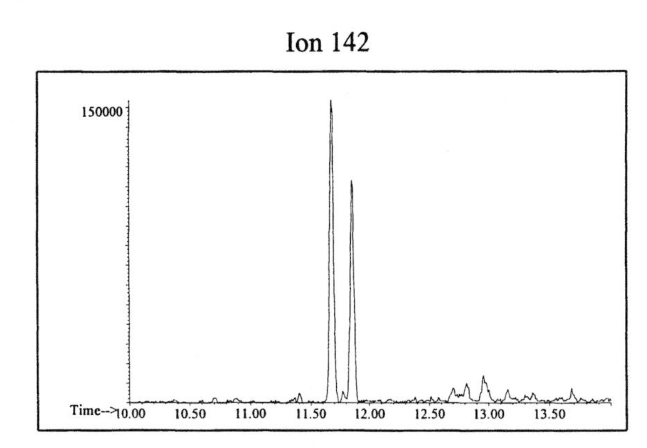

Ion 156

95000

Time-->10.00 10.50 11.00 11.50 12.00 12.50 13.00 13.50

Appendices

APPENDIX A

Computer Program for Total and Extracted Ion Chromatograms

This macro was produced for the Hewlett Packard Chemstation software; however, it can be adapted for other software systems. The program should be written in either "Notepad" or "Write". Advanced word processors are not recommended. The macro must be saved in [drive]:\hpchem\msexe with the extension **.mac;** i.e., c:\hpchemstation\msexe\book.mac. Refer to software manuals for details on the use of macro programs in data analysis.

NOTE: In order to compensate for differences in instrument conditions and parameters, retention time changes must be made in the subprogram Setup_Program where indicated.

```
! This program is designed to produce total ion chromatograms (complete and expanded),
! summed and individual  extracted ion chromatograms.
! Draw parameters correspond to chromatograms displayed in text
! Programs may be adapted to instrumental conditions by making changes in
! SETUP_Program where indicated

NAME DO_HEADER
        STARTPRINT
        PAGESIZE PWID,PLEN
        STRPRINT "FIRE DEBRIS ANALYSIS-ION CHROMATOGRAMS",(PWID-38)/2,4
RETURN

NAME DO_FOOTER
        LOCAL page$
        STRPRINT _Dataname$,5,(PLEN-3)
        PAGE$ = "Page "+val$(pagenum)
        STRPRINT PAGE$,PWID-11,PLEN-3
        PAGENUM = PAGENUM+1
        ENDPRINT
RETURN

NAME DO_TICS_ONLY
        !       Produces and draws total ion and expanded total ion chromatograms
        TIC,X
        DR 1,x
        DR 2,x,dr_xlow:dr_xhigh
        STRING$ = "Total Ion Chromatograms"
        STR$ = "Total Ion Chromatograms"
RETURN
```

```
NAME PRINT_TICS
        !       Outputs TICs to printer  in landscape mode
        ORIENTPR=2
        DO_HEADER
        STRPRINT STRING$,(PWID-23)/2,7
        WINPRINT 1,5,9,94,14
        WINPRINT 2,5,27,94,14
        DO_FOOTER
RETURN

NAME DO_ALKANE_SUMMED
        !       Extracts and sums ions 43  57   71 and 85
        !       Produces summed alkane profile
        EIC ALK_XLOW:ALK_XHIGH, 43
        EIC ALK_XLOW:ALK_XHIGH, 57
        ADD X,Y
        EIC ALK_XLOW:ALK_XHIGH,71
        ADD X,Y
        EIC ALK_XLOW:ALK_XHIGH,85
        ADD X,Y
        EXCHANGE R1,X
        STRING1$= "Alkane profile:  Extracted Ions 43  57  85  99 - Summed"
        STR$ = "Summed Ion Profiles"
RETURN

NAME DO_AROMATICS_SUMMED
        !       Extracts and sums ions 91 105 and 119
        !       Produces summed aromatic profile
        EIC ARO_XLOW:ARO_XHIGH, 91
        EIC ARO_XLOW:ARO_XHIGH, 105
        ADD X,Y
        EIC ARO_XLOW:ARO_XHIGH, 119
        ADD X,Y
        EXCHANGE R2,X
        STRING2$ = "Aromatic Profile: Extracted Ions 91 105 119 - Summed"
RETURN

NAME DO_CYCLOPARAFFINS_SUMMED
        !       Extracts and sums ions 55 69  83
        !       Produces summed cycloparaffin/alkene profile
        EIC ALK_XLOW:ALK_XHIGH, 55
        EIC ALK_XLOW:ALK_XHIGH, 69
        ADD X,Y
        EIC ALK_XLOW:ALK_XHIGH, 83
        ADD X,Y
        EXCHANGE R3,X
        STRING3$ = "Cycloparaffin/Alkene Profile:  Extracted Ions 55 69 83 - Summed"
RETURN
```

```
NAME DO_NAPHTHALENES_SUMMED
        !       Extracts and sums ions 128  142 and 156
        !       Produces summed naphthalene profile
        EIC NAPH_XLOW:NAPH_XHIGH, 128
        EIC NAPH_XLOW:NAPH_XHIGH, 142
        ADD X,Y
        EIC NAPH_XLOW:NAPH_XHIGH, 156
        ADD X,Y
        EXCHANGE R4,X
        STRING4$ = "Naphthalene Profile: Extracted Ions 128 142 156 - Summed"
RETURN

NAME PRINT_SUMMED_PROFILES
        !       PRINTS SUMMED ION PROFILES ON ONE PAGE IN
        !       PORTRAIT CONFIGUATION
        ORIENTPR = 1
        DO_HEADER
        STRPRINT "SUMMED ION PROFILES",(pWID-19)/2,6
        DR 3,R1,ALK_XLOW:ALK_XHIGH
        STRPRINT STRING1$,9,9
        WINPRINT 3,9,10,60,10
        DR 3,R2,ARO_XLOW:ARO_XHIGH
        STRPRINT STRING2$,9,21
        WINPRINT 3,9,22,60,10
        DR 3,R3,ALK_XLOW:ALK_XHIGH
        STRPRINT STRING3$,9,33
        WINPRINT 3,9,34,60,10
        DR 3,R4,NAPH_XLOW:NAPH_XHIGH
        STRPRINT STRING4$,9,45
        WINPRINT 3,9,46,60,10
        DO_FOOTER
RETURN

NAME DO_SINGLE_IONS_ALKANES
        !       EXTRACTS IONS 43  57  71 AND 85 AND PREPARES DATA FOR
        !       INDIVIDUAL DISPLAY/OUTPUT
        HEADSTRING1$="ALKANES"
        EIC ALK_XLOW:ALK_XHIGH,43
        EXCHANGE R1,X
        STRING1$="Ion 43"
        EIC ALK_XLOW:ALK_XHIGH,57
        EXCHANGE R2,X
        STRING2$="Ion 57"
        EIC ALK_XLOW:ALK_XHIGH,71
        EXCHANGE R3,X
        STRING3$="Ion 71"
        EIC ALK_XLOW:ALK_XHIGH,85
        EXCHANGE R4,X
        STRING4$="Ion 85"
```

```
RETURN

NAME DO_SINGLE_AROMATICS
	!	EXTRACTS IONS 91  105  AND 119 AND PREPARES DATA FOR
	!	INDIVIDUAL DISPLAY/OUTPUT
	HEADSTRING2$="AROMATICS"
	EIC ARO_XLOW:ARO_XHIGH, 91
	EXCHANGE R5,X
	STRING5$="Ion 91"
	EIC ARO_XLOW:ARO_XHIGH, 105
	EXCHANGE R6,X
	STRING6$="Ion 105"
	EIC ARO_XLOW:ARO_XHIGH, 119
	EXCHANGE R7,X
	STRING7$="Ion 119"
RETURN

NAME PRINT_ALKANE_AROMATICS
	!	PRINTS INDIVIDUAL ALKANE AND AROMATIC ION
	!	CHROMATOGRAMS ON SINGLE PAGE IN PORTRAIT MODE

	ORIENTPR=1
	DO_HEADER
	STRPRINT HEADSTRING1$,9,6
	STRPRINT STRING1$,9,7
	DR 3,R1,DR_XLOW:DR_XHIGH
	WINPRINT 3,9,8,27,10
	STRPRINT STRING2$,40,7
	DR 3,R2,DR_XLOW:DR_XHIGH
	WINPRINT 3,40,8,27,10
	STRPRINT STRING3$,9,20
	DR 3,R3,DR_XLOW:DR_XHIGH
	WINPRINT 3,9,21,27,10
	STRPRINT STRING4$,40,20
	DR 3,R4,DR_XLOW:DR_XHIGH
	WINPRINT 3,40,21,27,10
	STRPRINT HEADSTRING2$,9,33
	STRPRINT STRING5$,9,34
	DR 3,R5,ARO_XLOW:ARO_XHIGH
	WINPRINT 3,9,35,27,10
	STRPRINT STRING6$,40,34
	DR 3,R6,ARO_XLOW:ARO_XHIGH
	WINPRINT 3,40,35,27,10
	STRPRINT STRING7$,25,46
	DR 3,R7,ARO_XLOW:ARO_XHIGH
	WINPRINT 3,25,47,27,10
	DO_FOOTER

RETURN
```

```
NAME DO_SINGLE_CYCLOPARAFFINS
	!	EXTRACTS IONS 55  69  AND 83 AND PREPARES DATA FOR
	!	INDIVIDUAL DISPLAY/OUTPUT
	HEADSTRING1$="CYCLOPARAFFIN"
	EIC ALK_XLOW:ALK_XHIGH,55
	EXCHANGE R1,X
	STRING1$="Ion 55"
	EIC ALK_XLOW:ALK_XHIGH,69
	EXCHANGE R2,X
	STRING2$="Ion 69"
	EIC ALK_XLOW:ALK_XHIGH,83
	EXCHANGE R3,X
	STRING3$="Ion 83"
RETURN


NAME DO_SINGLE_NAPHTHALENES
	!	EXTRACTS IONS 128 142 AND 156 AND PREPARES DATA FOR
	!	INDIVIDUAL DISPLAY/OUTPUT
	HEADSTRING2$="NAPHTHALENES"
	EIC NAPH_XLOW:NAPH_XHIGH, 128
	EXCHANGE R4,X
	STRING4$="Ion 128"
	EIC NAPH_XLOW:NAPH_XHIGH, 142
	EXCHANGE R5,X
	STRING5$="Ion 142"
	EIC NAPH_XLOW:NAPH_XHIGH, 156
	EXCHANGE R6,X
	STRING6$="Ion 156"
RETURN


NAME PRINT_CYCLOPAR_NAPHTHALENES
	!	PRINTS INDIVIDUAL CYCLOPARAFFIN AND NAPHTHALENES ION
	!	CHROMATOGRAMS ON SINGLE PAGE IN PORTRAIT MODE
	ORIENTPR=1
	DO_HEADER
	STRPRINT HEADSTRING1$,9,6
	STRPRINT STRING1$,9,7
	DR 3,R1,ALK_XLOW:ALK_XHIGH
	WINPRINT 3,9,8,27,10
	STRPRINT STRING2$,40,7
	DR 3,R2,ALK_XLOW:ALK_XHIGH
	WINPRINT 3,40,8,27,10
	STRPRINT STRING3$,25,20
	DR 3,R3,ALK_XLOW:ALK_XHIGH
	WINPRINT 3,25,21,27,10
	STRPRINT HEADSTRING2$,9,33
```

```
        STRPRINT STRING4$,9,34
        DR 3,R4,NAPH_XLOW:NAPH_XHIGH
        WINPRINT 3,9,35,27,10
        STRPRINT STRING5$,40,34
        DR 3,R5,NAPH_XLOW:NAPH_XHIGH
        WINPRINT 3,40,35,27,10
        STRPRINT STRING6$,25,46
        DR 3,R6,NAPH_XLOW:NAPH_XHIGH
        WINPRINT 3,25,47,27,10
        DO_FOOTER
RETURN


NAME SETUP_PROGRAM
        ! SETS UP EXTRACTION PARAMETERS AND DRAW
        !           PARAMETERS BASED ON RANGE
        LOCAL A, LABEL$,BUTTON$,PROFILE,I,ANS$
        SDIM LABEL$,3
        SDIM BUTTON$,2
        DIM PROFILE,3
        LABEL$[1] = "Light Product Range   C6-C13"
        LABEL$[2] = "Medium Product Range  C8-C16"
        LABEL$[3] = "Heavy Product Range   C8-C25"
        BUTTON$[1] = "OK"
        BUTTON$[2] = "CANCEL"
        INPUT6 2,3, "PREDOMINENT ALKANE RANGE",LABEL$,BUTTON$,PROFILE
        IF PROFILE[1] = 1
                DR_XLOW =2.75          approx. retention time C6
                DR_XHIGH =12.00        approx. retention time C13
                ALK_XLOW =2.75         approx. retention time C6
                ALK_XHIGH =10.00       approx. retention time C11
        ENDIF
        IF PROFILE[2] = 1
                DR_XLOW =5.00          approx. retention time C8
                DR_XHIGH=15.00         approx. retention time C16
                ALK_XLOW =4.00         approx. retention time C7
                ALK_XHIGH =13.00       approx. retention time C14
        ENDIF
        IF PROFILE[3] = 1
                DR_XLOW =5.00          approx. retention time C8
                DR_XHIGH=21.00         approx. retention time C25
                ALK_XLOW =2.75         approx. retention time C6
                ALK_XHIGH =14.00       approx. retention time C15
        ENDIF
        ARO_XLOW=2.75                  approx. retention time C6
        ARO_XHIGH=14.00                approx. retention time C15
        NAPH_XLOW=10.00                approx. retention time C11
        NAPH_XHIGH=15.00               approx. retention time C16
        PAGENUM=1
```

```
RETURN

NAME TIC_CONTROL
	!	CONTROL PROGRAM FOR TIC SUBROUTINES
	DO_TICS_ONLY
	PRINT_TICS
RETURN

NAME SUMMED_ION_CONTROL
	!	CONTROL PROGRAM FOR SUMMED ION SUBROUTINES
	DO_ALKANE_SUMMED
	DO_AROMATICS_SUMMED
	DO_CYCLOPARAFFINS_SUMMED
	DO_NAPHTHALENES_SUMMED
	PRINT_SUMMED_PROFILES
RETURN

NAME SINGLE_ION_CONTROL
	!	CONTROL PROGRAM FOR INDIVIDUAL ION SUBROUTINES
	DO_SINGLE_IONS_ALKANES
	DO_SINGLE_AROMATICS
	PRINT_ALKANE_AROMATICS
	DO_SINGLE_CYCLOPARAFFINS
	DO_SINGLE_NAPHTHALENES
	PRINT_CYCLOPAR_NAPHTHALENES
RETURN


NAME CONTROL
	!	PROGRAM CONTOL PROGRAM
	SETUP_PROGRAM
	TIC_CONTROL
	SUMMED_ION_CONTROL
	SINGLE_ION_CONTROL
	PRINT "MACRO COMPLETE"
	REMOVE  DO_HEADER,DO_FOOTER,DO_TICS_ONLY,PRINT_TICS
	REMOVE DO_ALKANE_SUMMED,DO_AROMATICS_SUMMED
	REMOVE DO_CYCLOPARAFFINS_SUMMED, DO_NAPHTHALENES_SUMMED
	REMOVE  PRINT_SUMMED_PROFILES
	REMOVE DO_SINGLE_IONS_ALKANES, DO_SINGLE_AROMATICS
	REMOVE DO_SINGLE_CYCLOPARAFFINS, DO_SINGLE_NAPHTHALENES
	REMOVE PRINT_ALKANE_AROMATICS,PRINT_CYCLOPAR_NAPTHALENES
	REMOVE SETUP_PROGRAM, SUMMED_ION_CONTROL
	REMOVE SINGLE_ION_CONTROL
RETURN
```

APPENDIX B

Hydrocarbon Range and Classification

Product	Hydrocarbon Range	ASTM Class	Description
Oxygenate #1	C6-C7	0.1	oxygenated solvent
Distillate #1	C6-C7	1	distillate
Oxygenate #3	C6-C8	0.1	oxygenated solvent
Oxygenate #4	C6-C8	0.1	oxygenated solvent
Isoparaffin #1	C6-C8	0.2	isoparaffinic product
Isoparaffin #3	C6-C8	0.2	isoparaffinic product
Distillate #2	C6-C8	1	distillate
Distillate #3	C6-C8	1	distillate
Distillate #4	C6-C8	1	distillate
Miscellaneous #4	C6-C9	0	1,1,1-trichloroethane + aromatic solvent
Oxygenate #5	C6-C9	0.1	oxygenated solvent
Oxygenate #6	C6-C9	0.1	oxygenated solvent
Oxygenate #7	C6-C9	0.1	oxygenated solvent
Oxygenate #8	C6-C9	0.1	oxygenated solvent
Distillate #12	C6-C9	1	distillate
Distillate #13	C6-C9	1	distillate
Distillate #6	C6-C9	1	distillate
Distillate #7	C6-C9	1	distillate
Distillate #8	C6-C9	1	distillate
Miscellaneous #1	C6-C10	0	toluene + light distillate
Distillate #5	C6-C10	1	distillate
Miscellaneous #6	C6-C11	0	unidentified compounds + medium petroleum distillate
Oxygenate #11	C6-C11	0.1	oxygenated solvent
Distillate #17	C6-C11	1	distillate
Oxygenate #10	C6-C12	0.1	oxygenated solvent
Miscellaneous #14	C6-C13	0	1,1,1-trichloroethane + isoparaffinic product
Gasolines	C6-C13	2	gasolines
Miscellaneous #2	C6-C14	0	oxygenate + aromatic solvent + light petroleum distillate
Miscellaneous #7	C6-C14	0	aromatic solvent + medium petroleum distillate
Miscellaneous #9	C6-C14	0	toluene + aromatic solvent + light petroleum distillate
Miscellaneous #24	C6-C17	0	light petroleum distillate + medium petroleum distillate
Oxygenate #2	C7-C8	0.1	oxygenated solvent
Aromatic #1	C7-C8	0.4	aromatic solvent
Aromatic #2	C7-C8	0.4	aromatic solvent
Miscellaneous #17	C7-C9	0	aromatic solvent + light petroleum distillate
Oxygenate #12	C7-C9	0.1	oxygenated solvent

APPENDIX B

Hydrocarbon Range and Classification

Product	Hydrocarbon Range	ASTM Class	Description
Oxygenate #13	C7-C9	0.1	oxygenated solvent
Oxygenate #14	C7-C9	0.1	oxygenated solvent
Oxygenate #9	C7-C9	0.1	oxygenated solvent
Isoparaffin #2	C7-C9	0.2	isoparaffinic product
Isoparaffin #4	C7-C9	0.2	isoparaffinic product
Distillate #10	C7-C9	1	distillate
Distillate #11	C7-C9	1	distillate
Distillate #14	C7-C9	1	distillate
Distillate #15	C7-C9	1	distillate
Distillate #9	C7-C9	1	distillate
Miscellaneous #13	C7-C11	0	aromatic solvent + light petroleum distillate
Miscellaneous #23	C7-C11	0	aromatic solvent + medium petroleum distillate
Aromatic #3	C7-C11	0.4	aromatic solvent
Aromatic #5	C7-C11	0.4	aromatic solvent
Aromatic #6	C7-C11	0.4	aromatic solvent
Distillate #19	C7-C11	3	distillate
Miscellaneous #19	C7-C12	0	toluene + medium petroleum distillate
Miscellaneous #11	C7-C13	0	toluene + medium petroleum distillate
Miscellaneous #15	C7-C13	0	light petroleum distillate + medium petroleum distillate
Miscellaneous #3	C7-C13	0	limonene + aromatic solvent
Miscellaneous #8	C7-C13	0	aromatic solvent + medium petroleum distillate
Miscellaneous #10	C7-C14	0	aromatic solvent + light petroleum distillate
Miscellaneous #12	C7-C14	0	toluene + aromatic solvent + unidentified compounds
Miscellaneous #5	C7-C14	0	toluene + aromatic solvent + light petroleum distillate
Aromatic #4	C8-C9	0.4	aromatic solvent
Distillate #16	C8-C9	1	distillate
Aromatic #7	C8-C11	0.4	aromatic solvent
Distillate #23	C8-C12	0	dearomatized distillate
Miscellaneous #18	C8-C12	0	indenes
Aromatic #11	C8-C12	0.4	aromatic solvent
Distillate #20	C8-C12	3	distillate
Miscellaneous #20	C8-C13	0	aromatic solvent + medium petroleum distillate
Distillate #21	C8-C13	3	distillate
Distillate #24	C8-C13	3	distillate
Distillate #25	C8-C13	3	distillate
Distillate #32	C8-C13	3	distillate
Distillate #34	C8-C15	3	distillate

APPENDIX B

Hydrocarbon Range and Classification

Product	Hydrocarbon Range	ASTM Class	Description
Miscellaneous #28	C8-C17	0	aromatic solvent + kerosene
Naphthenic-Paraffinic #5	C8-C18	0.5	naphthenic-paraffinic product
Distillate #43	C8-C18	4	distillate
Distillate #35	C8-C19	4	distillate
Miscellaneous #16	C8-C25	0	aromatic solvent + heavy petroleum distillate
Miscellaneous #26	C8-C25+	0	medium petroleum distillate + oil
Oxygenate #15	C9-C11	0.1	oxygenated solvent
Isoparaffin #5	C9-C11	0.2	isoparaffinic product
Aromatic #8	C9-C11	0.4	aromatic solvent
Miscellaneous #22	C9-C12	0	alpha pinene + isoparaffinic product
Distillate #18	C9-C12	3	distillate
Distillate #22	C9-C12	3	distillate
Distillate #29	C9-C12	3	distillate
Miscellaneous #30	C9-C13	0	terpenes
Distillate #26	C9-C13	3	distillate
Distillate #27	C9-C13	3	distillate
Distillate #30	C9-C13	3	distillate
Distillate #31	C9-C13	3	distillate
Miscellaneous #29	C9-C14	0	terpenes
Miscellaneous #21	C9-C16	0	isoparaffinic product + dearomatized distillate
Naphthenic-Paraffinic #3	C9-C16	0.5	naphthenic-paraffinic product
Naphthenic-Paraffinic #4	C9-C16	0.5	naphthenic-paraffinic product
Distillate #41	C9-C16	4	distillate
Distillate #40	C9-C18	4	distillate
Distillate #44	C9-C18	5	distillate
Miscellaneous #35	C9-C20	0	kerosene + heavy petroleum distillate
Distillate #53	C9-C21	5	distillate
Distillate #45	C9-C22	5	distillate
Miscellaneous #27	C9-C25	0	hexylene glycol + heavy petroleum distillate
Distillate #49	C9-C25	5	distillate
Miscellaneous #25	C9-C25+	0	medium petroleum distillate + oil
Isoparaffin #7	C10-C12	0.2	isoparaffinic product
Isoparaffin #6	C10-C12	0.2	isoparaffinic product
Aromatic #9	C10-C12	0.4	aromatic solvent
Distillate #33	C10-C13	0	dearomatized distillate
Isoparaffin #10	C10-C13	0.2	isoparaffinic product
Isoparaffin #8	C10-C13	0.2	isoparaffinic product

APPENDIX B

Hydrocarbon Range and Classification

Product	Hydrocarbon Range	ASTM Class	Description
Distillate #28	C10-C13	3	distillate
Naphthenic-Paraffinic #1	C10-C14	0.5	naphthenic-paraffinic product
Distillate #42	C10-C15	0	dearomatized distillate
Miscellaneous #31	C10-C15	0	terpenes
Miscellaneous #33	C10-C16	0	normal alkane product + medium petroleum distillate
Naphthenic-Paraffinic #2	C10-C16	0.5	naphthenic-paraffinic product
Distillate #37	C10-C17	4	distillate
Distillate #38	C10-C18	4	distillate
Naphthenic-Paraffinic #6	C11-C13	0.5	naphthenic-paraffinic product
Miscellaneous #32	C11-C14	0	terpenes
Isoparaffin #12	C11-C15	0.2	isoparaffinic product
Normal alkane #1	C11-C15	0.3	normal alkane product
Normal alkane #2	C11-C15	0.3	normal alkane product
Isoparaffin #11	C11-C16	0.2	isoparaffinic product
Aromatic #12	C11-C17	0.4	aromatic solvent
Distillate #46	C11-C17	4	distillate
Isoparaffin #13	C11-C18	0.2	isoparaffinic product
Distillate #48	C11-C20	5	distillate
Distillate #50	C11-C25	5	distillate
Distillate #56	C11-C25	5	distillate
Naphthenic-Paraffinic #7	C12-C17	0.5	naphthenic-paraffinic product
Distillate #39	C12-C18	4	distillate
Miscellaneous #34	C12-C23	0	medium petroleum distillate + heavy petroleum distillate
Distillate #47	C13-C16	0	dearomatized distillate
Normal alkane #3	C13-C17	0.3	normal alkane product
Distillate #54	C13-C21	5	distillate
Distillate #51	C14-C26	5	distillate
Distillate #55	C15-C20	0	dearomatized distillate
Distillate #52	C16-C26	5	distillate

APPENDIX C

DISTILLATE INDEX

Product	Highest n-alkane	Hydrocarbon Range
Distillate #1	C6	C6-C7
Distillate #2	C7	C6-C8
Distillate #3	C7	C6-C8
Distillate #4	C7	C6-C8
Distillate #5	C7	C6-C10
Distillate #6	C8	C6-C9
Distillate #7	C8	C6-C9
Distillate #8	C8	C6-C9
Distillate #12	C8	C6-C9
Distillate #17	C8	C6-C11
Distillate #9	C8	C7-C9
Distillate #10	C8	C7-C9
Distillate #11	C8	C7-C9
Distillate #13	C9	C6-C9
Distillate #14	C9	C7-C9
Distillate #15	C9	C7-C9
Distillate #16	C9	C8-C9
Distillate #19	C10	C7-C11
Distillate #20	C10	C8-C12
Distillate #21	C10	C8-C13
Distillate #23	C10	C8-C12
Distillate #22	C10	C9-C12
Distillate #18	C10	C9-C12
Distillate #24	C10-C11	C8-C13
Distillate #25	C11	C8-C13
Distillate #32	C11	C8-C13
Distillate #34	C11	C8-C15
Distillate #35	C11	C8-C19
Distillate #29	C11	C9-C12
Distillate #30	C11	C9-C13
Distillate #31	C11	C9-C13
Distillate #26	C11	C9-C13
Distillate #27	C11	C9-C13
Distillate #28	C11	C10-C13
Distillate #33	C11	C10-C13
Distillate #36	C12	C8-C17
Distillate #43	C12	C8-C18
Distillate #41	C12	C9-C16
Distillate #40	C12	C9-C18
Distillate #42	C12	C10-C15
Distillate #37	C13	C10-C17

APPENDIX C

DISTILLATE INDEX

Product	Highest n-alkane	Hydrocarbon Range
Distillate #38	C13	C10-C18
Distillate #45	C14	C9-C22
Distillate #46	C14	C11-C17
Distillate #39	C14	C12-C18
Distillate #53	C15	C9-C21
Distillate #48	C15	C11-C20
Distillate #47	C15	C13-C16
Distillate #54	C16	C13-C21
Distillate #49	C17	C9-C25
Distillate #50	C17	C11-C25
Distillate #55	C17	C15-C20
Distillate #56	C18	C11-C25
Distillate #51	C19	C14-C26
Distillate #52	C20	C16-C26

APPENDIX D

MAJOR OXYGENATES

Product	Major Oxygenate	Other Major Component	Other Major Component
Oxygenate #11	butanol	toluene	butoxyethanol
Oxygenate #13	butoxyethanol	toluene	
Oxygenate #5	butylacetate	ethoxyacetate	butanol
Oxygenate #2	butylacetate		
Oxygenate #14	butylacetate	toluene	
Oxygenate #8	heptane	toluene	butoxyethanol
Oxygenate #6	isopropylacetate	propylacetate	toluene
Oxygenate #1	methoxyethanol		
Oxygenate #10	methylethylketone	toluene	butoxyethanol
Oxygenate #9	methylethylketone	toluene	butoxyethanol
Oxygenate #3	methylpropanol	butylacetate	butoxyethanol
Oxygenate #7	methylpropanol	toluene	heptanone
Oxygenate #4	tetrahydrofuran	dimethylformide	cyclohexane
Oxygenate #12	unidentified oxygenate	toluene	xylene
Oxygenate #15	unidentified oxygenate		

APPENDIX E

Miscellaneous and Mixture

Major Components Index

Product	Component 1	Component 2	Component 3
Miscellaneous #4	Aromatic Solvent	1,1,1-trichloroethane	
Miscellaneous #18	Aromatic Solvent	Heavy Petroleum Distillate	
Miscellaneous #9	Aromatic Solvent	Light Petroleum Distillate	Toluene
Miscellaneous #15	Aromatic Solvent	Light Petroleum Distillate	
Miscellaneous #19	Aromatic Solvent	Light Petroleum Distillate	
Miscellaneous #3	Aromatic Solvent	Limonene	
Miscellaneous #7	Aromatic Solvent	Medium Petroleum Distillate	
Miscellaneous #8	Aromatic Solvent	Medium Petroleum Distillate	
Miscellaneous #22	Aromatic Solvent	Medium Petroleum Distillate	
Miscellaneous #14	Aromatic Solvent	Toluene	
Miscellaneous #29	Heavy Petroleum Distillate	Hexylene Glycol	
Miscellaneous #20	Indenes		
Miscellaneous #16	Isoparaffinic Product	1,1,1-trichloroethane	
Miscellaneous #24	Isoparaffinic Product	α-Pinene	
Miscellaneous #23	Isoparaffinic Product	Dearomatized Distillate	
Miscellaneous #30	Kerosene	Aromatic Solvent	
Miscellaneous #37	Kerosene	Heavy Petroleum Distillate	
Miscellaneous #5	Light Petroleum Distillate	Aromatic Solvent	Toluene
Miscellaneous #10	Light Petroleum Distillate	Aromatic Solvent	
Miscellaneous #26	Light Petroleum Distillate	Heavy Petroleum Distillate	
Miscellaneous #1	Light Petroleum Distillate	Toluene	
Miscellaneous #17	Medium Petroleum Distillate	Light Petroleum Distillate	
Miscellaneous #25	Medium Petroleum Distillate	Aromatic Solvent	
Miscellaneous #27	Medium Petroleum Distillate	Heavy Oil	
Miscellaneous #28	Medium Petroleum Distillate	Heavy Oil	
Miscellaneous #36	Medium Petroleum Distillate	Heavy Petroleum Distillate	
Miscellaneous #21	Medium Petroleum Distillate	Toluene	
Miscellaneous #6	Medium Petroleum Distillate	Unknown Compounds	
Miscellaneous #11	Medium Petroleum Distillate	Toluene	
Miscellaneous #12	Medium Petroleum Distillate	Toluene	
Miscellaneous #13	Medium Petroleum Distillate	Toluene	
Miscellaneous #35	Normal Alkane Product	Medium Petroleum Distillate	
Miscellaneous #2	Oxygenated Solvent	Aromatic Solvent	Light Petroleum Distillate
Miscellaneous #31	Pinenes	Terpenes	
Miscellaneous #32	Pinenes	Terpenes	
Miscellaneous #33	Pinenes	Terpenes	
Miscellaneous #34	Pinenes	Terpenes	

Index

E

F

G

H

I

J

K

L

M

N

O